ORGANIC COMPOUNDS

Family						
Ether	**Amine**	**Aldehyde**	**Ketone**	**Carboxylic Acid**	**Ester**	**Amide**
CH_3OCH_3	CH_3NH_2	$\overset{\displaystyle O}{\overset{\displaystyle \|}{CH_3CH}}$	$\overset{\displaystyle O}{\overset{\displaystyle \|}{CH_3CCH_3}}$	$\overset{\displaystyle O}{\overset{\displaystyle \|}{CH_3COH}}$	$\overset{\displaystyle O}{\overset{\displaystyle \|}{CH_3COCH_3}}$	$\overset{\displaystyle O}{\overset{\displaystyle \|}{CH_3CNH_2}}$
Methoxy-methane	Methan-amine	Ethanal	Propanone	Ethanoic Acid	Methyl ethanoate	Ethanamide
Dimethyl ether	Methyl-amine	Acetal-dehyde	Acetone	Acetic acid	Methyl acetate	Acetamide
ROR	RNH_2 R_2NH R_3N	$\overset{\displaystyle O}{\overset{\displaystyle \|}{RCH}}$	$\overset{\displaystyle O}{\overset{\displaystyle \|}{RCR}}$	$\overset{\displaystyle O}{\overset{\displaystyle \|}{RCOH}}$	$\overset{\displaystyle O}{\overset{\displaystyle \|}{RCOR}}$	$\overset{\displaystyle O}{\overset{\displaystyle \|}{RCNH_2}}$ $\overset{\displaystyle O}{\overset{\displaystyle \|}{RCNHR}}$ $\overset{\displaystyle O}{\overset{\displaystyle \|}{RCNR_2}}$
$-\!\overset{\|}{\underset{\|}{C}}\!-\!O\!-\!\overset{\|}{\underset{\|}{C}}\!-$	$-\!\overset{\|}{\underset{\|}{C}}\!-\!N\!-$	$-\!\overset{\displaystyle O}{\overset{\displaystyle \|}{\underset{\|}{C}}}\!-\!H$	$-\!\overset{\|}{\underset{\|}{C}}\!-\!\overset{\displaystyle O}{\overset{\displaystyle \|}{C}}\!-\!\overset{\|}{\underset{\|}{C}}\!-$	$-\!\overset{\displaystyle O}{\overset{\displaystyle \|}{C}}\!-\!OH$	$-\!\overset{\displaystyle O}{\overset{\displaystyle \|}{C}}\!-\!O\!-\!\overset{\|}{\underset{\|}{C}}\!-$	$-\!\overset{\displaystyle O}{\overset{\displaystyle \|}{C}}\!-\!N\!-$

ESHBACH'S
HANDBOOK OF
ENGINEERING
FUNDAMENTALS

ESHBACH'S HANDBOOK OF ENGINEERING FUNDAMENTALS

FOURTH EDITION

Editor

BYRON D. TAPLEY

Department of Aerospace Engineering
and Engineering Mechanics
The University of Texas at Austin
Austin, Texas

Managing Editor

THURMAN R. POSTON

Postscript, Consultants to Publishers
Temenos
Snohomish, Washington

A WILEY-INTERSCIENCE PUBLICATION

John Wiley & Sons, Inc.

NEW YORK / CHICHESTER / BRISBANE / TORONTO / SINGAPORE

Library of Congress Cataloging in Publication Data:

Eshbach, Ovid W. (Ovid Wallace), 1893–1958.
  [Handbook of engineering fundamentals]
  Eshbach's handbook of engineering fundamentals.—4th ed./
editor, Byron D. Tapley; managing editor, Thurman R. Poston.
    p. cm.
  "A Wiley-Interscience publication."
  Bibliography: p.
  1. Engineering—Handbooks, manuals, etc.   I. Tapley, Byron D.
II. Poston, Thurman R.   III. Title.   IV. Title: Handbook of
engineering fundamentals.
TA151.E8 1989
620—dc19   88-20581
                                                         CIP

  ISBN 0-471-89084-7

Printed in the United States of America

10 9 8 7 6 5 4 3 2 1

CONTENTS

CONTENTS

PREFACE

The *Handbook of Engineering Fundamentals* has occupied a prominent place on the desk of practicing engineers for over five decades now. As the fundamentals of engineering practice have evolved, the handbook has been revised to maintain its applicability to contemporary engineering practice. In keeping with the prior practice, the fourth edition of the *Handbook of Engineering Fundamentals* provides revisions and additions to account for the substantial growth in engineering knowledge during the 15 years since the last edition. As in previous editions, the primary objective of the handbook is to bring together under one cover the broad range of fundamentals required to support the breadth of current engineering practices, which range from detailed research and development to large-scale systems design and technological interactions with society.

The topics for the original Eshbach handbook were selected with the objective of covering the fundamentals from the appropriate engineering disciplines and the appropriate foundations from math, physics and chemistry. The subject matter included in the revised edition has been selected in recognition of the explosion in technical information which has occurred since the last edition and the erosion of strict boundaries between the treatment of mechanical and electrical systems in the design process. Integrated design approaches, along with the evolution of the personal computer and/or design workstation, have led to the requirement that both the practicing engineer and the technical manager rely on much broader knowledge. The content of the handbook has been selected in recognition of these facts. The emphasis has been to state the fundamental principles and to describe a selected number of applications. The handbook has been assimilated to serve the needs of

1. the young engineer, who, although a recent graduate, will need to maintain his or her technical competence,
2. the specialist, who needs detailed reference material from other areas,
3. the practicing systems engineer, who of necessity must be concerned with material from a number of areas,
4. the engineering student, who may use the material both as a test for a specific course of study and as a desk reference to support studies in other courses, and
5. the senior engineer or manager, who needs reference material as a basis for technical or management discussion.

In revising the previous edition, recognition has been given to the dramatic change that computers and computer technology have made in the way we generate, receive, and display information. The handbook has been modified to account for this impact in three significant ways: (1) the chapter on mathematical and trigonometric tables has been reduced substantially in recognition of the fact that both small handheld computers and desktop personal computers allow a rapid generation of much of the information contained in this chapter, (2) a specific chapter dealing with the elements of computers and computer science has been added, and (3) specific applications where computers are useful have been included in many of the chapters. This is especially true in the mathematics section, where sections on differential equations and the finite-element method have been added to the traditional topic coverage. In recognition of the role control theory plays in current engineering design and applications, an expanded chapter on control theory has been included. In view of the important role space exploration and application play in our national technology, the previous chapter on aeronautics and astrodynamics has been separated into two distinct chapters. Substantial revisions to the chapters on electromagnetics and circuits, electronics, and radiation, light and acoustics have been made to account for the significant changes in these areas. The chapter on engineering economics, which is a crucial element in the successful application of any design, has been revised extensively. Finally, the adoption of the international standard units throughout and the revision of the references to cite current literature make this edition an important text for the contemporary engineer.

The editor would appreciate receiving comments on the handbook, either on the omissions in coverage or in errors in presentation, which could lead to improvements in future editions.

November 1989   BYRON D. TAPLEY

CONTRIBUTORS

Nikitas A. Alexandritis
Department of Digital Systems
  and Computers
National Technical University of Athens
Athens, Greece

J. Anthony
Department of Astronautics
United States Air Force Academy
Colorado Springs, Colorado 80840

David F. Bantz
Department of Computer Sciences
IBM Thomas J. Watson Research Center
Yorktown Heights, New York 10598

D. N. Barlow
Department of Aeronautics
United States Air Force Academy
Colorado Springs, Colorado 80840

R. R. Bates
Department of Astronautics
United States Air Force Academy
Colorado Springs, Colorado 80840

W. C. Bauman
Department of Aeronautics
United States Air Force Academy
Colorado Springs, Colorado 80840

J. J. Beoddy
Department of Aeronautics
United States Air Force Academy
Colorado Springs, Colorado 80840

Yves H. Berthelot
School of Mechanical Engineering
Georgia Institute of Technology
Atlanta, Georgia 30332

Sheldon S. L. Chang
Department of Mechanical Engineering
State University of New York
Stony Brook, New York 11794

J. T. Clay
Department of Aeronautics
United States Air Force Academy
Colorado Springs, Colorado 80840

S. Czyzak
Department of Aeronautics
United States Air Force Academy
Colorado Springs, Colorado 80840

D. H. Daley
Department of Aeronautics
United States Air Force Academy
Colorado Springs, Colorado 80840

M. L. DeLorenzo
Department of Astronautics
United States Air Force Academy
Colorado Springs, Colorado 80840

George E. Dieter
School of Engineering
University of Maryland
College Park, Maryland 20472

Yusuf Ziga Efe
Department of Electrical Engineering
The Cooper Union College
New York, New York 10003

Neal F. Enke
Department of Engineering Mechanics
University of Wisconsin
Madison, Wisonsin 53706

George C. Feth
IBM Thomas J. Watson Research Center
Yorktown Heights, New York 10598

D. Finkleman
Department of Aeronautics
United States Air Force Academy
Colorado Springs, Colorado 80840

C. A. Forbrich, Jr.
Department of Aeronautics
United States Air Force Academy
Colorado Springs, Colorado 80840

T. J. Forster
Department of Aeronautics
United States Air Force Academy
Colorado Springs, Colorado 80840

Andres Fortino
Department of Electrical Engineering
Temple University
Philadelphia, Pennsylvania 19122

Wallace Fowler
Department of Aerospace Engineering
  and Engineering Mechanics
The University of Texas at Austin
Austin, Texas 78712

R. S. Fraser
Department of Astronautics
United States Air Force Academy
Colorado Springs, Colorado 80840

M. Parker Givens
Institute of Optics
University of Rochester
Rochester, New York 14627

R. B. Griffen
Department of Astronautics
United States Air Force Academy
Colorado Springs, Colorado 80840

W. F. Hallgren
Department of Aeronautics
United States Air Force Academy
Colorado Springs, Colorado 80840

Syed Hamid
Halliburton Services
Duncan, Oklahoma

T. A. Hammond
Department of Aeronautics
United States Air Force Academy
Colorado Springs, Colorado 80840

Emory J. Harry
Tektronix, Inc.
Beaverton, Oregon 97005

J. D. Hines
Department of Astronautics
United States Air Force Academy
Colorado Springs, Colorado 80840

Frank P. Incropera
Department of Mechanical Engineering
Purdue University
West Lafayette, Indiana 47907

Suhada Jayasuriya
Department of Mechanical Engineering
Texas A & M University
College Station, Texas 77843

E. J. Jumper
Department of Aeronautics
United States Air Force Academy
Colorado Springs, Colorado 80840

P. I. King
Department of Aeronautics
United States Air Force Academy
Colorado Springs, Colorado 80840

D. A. Kohl
Department of Chemistry
The University of Texas at Austin
Austin, Texas 78712

Robert L. Kustom
Argonne National Laboratory
Argonne, Illinois 60439

E. L. Larson
Department of Aeronautics
United States Air Force Academy
Colorado Springs, Colorado 80840

R. J. Lisowski
Department of Astronautics
United States Air Force Academy
Colorado Springs, Colorado 80840

R. D. Marker
Department of Aeronautics
United States Air Force Academy
Colorado Springs, Colorado 80840

Velio A. Marsocci
Department of Electrical Engineering
State University of New York
Stony Brook, New York 11794

Dennis J. McBride
Applied Research Department
IBM Thomas J. Watson Research Center
Yorktown Heights, New York 10598

D. E. Mercier
Department of Astronautics
United States Air Force Academy
Colorado Springs, Colorado 80840

Robert E. Metzler
Tektronix, Inc.
Beaverton, Oregon 97005

R. W. Milling
Department of Aeronautics
United States Air Force Academy
Colorado Springs, Colorado 80840

Narenda Mohan
Department of Electrical Engineering
University of Minnesota
Minneapolis, Minnesota 55455

T. Peter Neal
Industrial Development Engineering
  Department
Moog, Inc.
East Aurora, New York 14052

L. M. Nicolai
Department of Aeronautics
United States Air Force Academy
Colorado Springs, Colorado 80840

Harry N. Norton
Jet Propulsion Laboratory
California Institute of Technology
Pasedena, California 91125

R. B. Oliver
Department of Aeronautics
United States Air Force Academy
Colorado Springs, Colorado 80840

B. W. Parkinson
Department of Astronautics
United States Air Force Academy
Colorado Springs, Colorado 80840

Allan D. Pierce
School of Mechanical Engineering
Georgia Institute of Technology
Atlanta, Georgia 30332

A. E. Preyss
Department of Astronautics
United States Air Force Academy
Colorado Springs, Colorado 80840

J. N. Reddy
Department of Engineering Science
  and Mechanics
Virginia Polytechnic Institute
  and State University
Blacksburg, Virginia 24061

Karl N. Reid
College of Engineering
Oklahoma State University
Stillwater, Oklahoma 74078

Albert Rosa
Department of Engineering
University of Denver
Denver, Colorado 80210

Jack L. Rosenfeld
Department of Computer Science
IBM Thomas J. Watson Research Center
Yorktown Heights, New York 10598

J. C. Ruth
Department of Astronautics
United States Air Force Academy
Colorado Springs, Colorado 80840

Bela I. Sandor
Department of Engineering Mechanics
University of Wisconsin
Madison, Wisconsin 53706

M. L. Smith
Department of Aeronautics
United States Air Force Academy
Colorado Springs, Colorado 80840

Kenneth Short
Department of Electrical Engineering
State University of New York
Stony Brook, New York 11794

Kenneth Sohn
Department of Electrical Engineering
New Jersey Institute of Technology
Newark, New Jersey 07102

Krishnaswamy Srinivasan
Department of Mechanical Engineering
The Ohio State University
Columbus, Ohio 43210

R. J. Stiles
Department of Aeronautics
United States Air Force Academy
Colorado Springs, Colorado 80840

Earl E. Swartzlander, Jr.
TRW
Defense Systems Group
Redondo Beach, California 90277

Denny D. Tang
IBM Thomas J. Watson Research Center
Yorktown Heights, New York 10598

R. H. Tate
Department of Astronautics
United States Air Force Academy
Colorado Springs, Colorado 80840

G. E. Thompson
Department of Aeronautics
United States Air Force Academy
Colorado Springs, Colorado 80840

James E. Thorton
Network Systems Corporation
Brooklyn Park, Minnesota 55429

Kenneth J. Thurber
Architecture Technology Corporation
Minneapolis, Minnesota 55406

Kishor S. Trivedi
Department of Computer Science
Duke University
Durham, North Carolina 27706

Gary Z. Watters
School of Engineering
California State University
Chico, California 95929

Jack H. Westbrook
Sci-Tech Knowledge Systems
133 Saratoga Road
Scotia, New York 12302

John A. White
School of Industrial and Systems Engineering
Georgia Institute of Technology
Atlanta, Georgia 30332

R. E. Willes
Department of Aeronautics
United States Air Force Academy
Colorado Springs, Colorado 80840

J. B. Wissler
Department of Aeronautics
United States Air Force Academy
Colorado Springs, Colorado 80840

J. P. Wittry
Department of Astronautics
United States Air Force Academy
Colorado Springs, Colorado 80840

Bernard D. Wood
Department of Mechanical and Aerospace
    Engineering
Syracuse University
Syracuse, New York 13244

CHAPTER 1

MATHEMATICAL AND PHYSICAL UNITS, STANDARDS, AND TABLES*

Jack H. Westbrook

Sci-Tech Knowledge Systems
Scotia, New York

*This chapter is a revision and extension of Sections 1 and 3 of the third edition, which were written by Mott Souders and Ernst Weber, respectively. Section 1.4.4 is derived principally from ASTM's *Standard for Metric Practice*, ASTM E380-82, Philadelphia, 1982 (with permission). Section 1.6.1 is derived from *MIS Newsletter*, General Electric Co., 1980 (with permission).

1.1   SYMBOLS AND ABBREVIATIONS

TABLE 1.1   Greek Alphabet

A	α	Alpha	H	η		Eta	N	ν		Nu	T	τ	Tau
B	β	Beta	Θ	ϑ	θ	Theta	Ξ	ξ		Xi	Υ	υ	Upsilon
Γ	γ	Gamma	I	ι		Iota	O	o		Omicron	Φ	ϕ	Phi
Δ	δ	Delta	K	κ		Kappa	Π	π		Pi	X	χ	Chi
E	ε	Epsilon	Λ	λ		Lambda	P	ρ		Rho	Ψ	ψ	Psi
Z	ζ	Zeta	M	μ		Mu	Σ	σ	s	Sigma	Ω	ω	Omega

TABLE 1.2   Symbols for Mathematical Operations[a]

Addition and Subtraction

$a + b$, a plus b
$a - b$, a minus b
$a \pm b$, a plus or minus b
$a \mp b$, a minus or plus b

Multiplication and Division

$a \times b$, or $a \cdot b$, or ab, a times b

$a \div b$, or $\dfrac{a}{b}$, or a/b, a divided by b

Symbols of Aggregation

( ) parentheses
[ ] brackets
{ } braces
– vinculum

Equalities and Inequalities

$a = b$, a equals b
$a \approx b$, a approximately equals b
$a \neq b$, a is not equal to b
$a > b$, a is greater than b
$a < b$, a is less than b
$a \gg b$, a much larger than b
$a \ll b$, a much smaller than b
$a \geqq b$, a equals or is greater than b
$a \leqq b$, a is less than or equals b
$a \equiv b$, a is identical to b
$a \rightarrow b$, or $a \doteq b$, a approaches b as a limit

Proportion

$a/b = c/d$, or $a : b :: c : d$, a is to b as c is to d
$a \propto b$, $a \sim b$, a varies directly as b
%, per cent

Powers and Roots

a^2, a squared
a^n, a raised to the nth power
$\sqrt{a}$, square root of a
$\sqrt[3]{a}$, cube root of a
$\sqrt[n]{a}$, or $a^{1/n}$, nth root of a
a^{-n}, $1/a^n$
$3.14 \times 10^4 = 31{,}400$
$3.14 \times 10^{-4} = 0.000314$

Miscellaneous

$\bar{a}$, mean value of a
$a!$, $= 1 \cdot 2 \cdot 3 \ldots a$, factorial a
$|a|$ = absolute value of a
$P(n, r) = n(n - 1)(n - 2) \ldots (n - r + 1)$

$C(n, r) = \dfrac{P(n, r)}{r!} = \binom{n}{r}$ = binomial
coefficients

i (or j) $= \sqrt{-1}$, imaginary unit
$\pi = 3.1416$, ratio of the circumference to the
diameter of a circle
∞, infinity

TABLE 1.2 (*Continued*)

Plane Geometry

∠, angle
△, triangle
∥, parallel
⊥ , perpendicular
⊙, circle
▱, parallelogram
∴, therefore
° ′ ″, degree, minute, second
′ ″, feet, inches

Logarithms and Exponentials

$\log a = \log_{10} a$, common logarithm of a or log of
  a to the base 10
$\ln a = \log_e a$, natural logarithm of a or log of a
  to the base e ($e = 2.718$)
$\log^{-1} a$, number whose log is a
$\text{lb } x$ or $\log_2 x$ = binary logarithm of x
exponential of x, $\exp x$, e^x

Trigonometry

$\left.\begin{array}{l} \sin,\ \cos,\ \tan \\ \csc \text{ or csc, sec, cot or ctn} \\ \text{vers, covers} \end{array}\right\}$ trigonometric functions

$\left.\begin{array}{l} \sin^{-1},\ \cos^{-1},\ \text{etc.} \\ \arcsin,\ \arccos, \end{array}\right\}$ inverse of the functions

Analytic Geometry

$x, y, z;\ \xi, \eta, \zeta$, rectangular coordinates
ρ, s, intrinsic coordinates
ρ, radius of curvature
s, length of arc
r, θ, polar coordinates
ψ, angle from radius vector to tangent
r, θ, ϕ, spherical coordinates
θ, co-latitude
ϕ, longitude
r, θ, z, cylindrical coordinates
e, eccentricity in conics
p, semi latus rectum in conics
$l = \cos \alpha,\ m = \cos \beta,\ n = \cos \gamma$, direction
  cosines

Calculus

$y = f(x)$, y is a function of x
$y' = f'(x) = \dfrac{dy}{dx} = D_x y$, derivative of $y = f(x)$
  with respect to x
$y'' = f''(x) = \dfrac{d(y')}{dx} = D_x^2 y = \dfrac{d^2 y}{dx^2}$, second
  derivative of $y = f(x)$ with respect to x

$u = f(x, y)$, u is a function of x and y
$u_x = f_x(x, y) = D_x(u) = \dfrac{\partial u}{\partial x}$, partial derivative
  of $u = f(x, y)$ with respect to x
$u_{xy} = f_{xy}(x, y) = D_y(D_x u) = \dfrac{\partial^2 u}{\partial y \partial x}$, second
  partial derivative of $u = f(x, y)$ with respect to
  x and y
Δy, increment of y
dy, differential of y
δy, variation of y
$\displaystyle\sum_{i=a}^{b}$, summation over i from a to b
$\lim\limits_{x \to a} (y) = b$, $y \to b$ as $x \to a$
$\displaystyle\int$, integral of
$\displaystyle\int_a^b$, definite integral of

Vector Analysis

i, j, k, unit vectors along the axes
  (right-handed system)
$a \cdot b = (ab) = Sab$, scalar product of a and b
$a \times b = [ab] = Vab$, vector product of a and b
Vectors are indicated in print by bold-faced type.
$|A|$, A, absolute value
$\partial/\partial r$, ∇, differential vector operator
grad φ, $\nabla\varphi$, gradient
div A, $\nabla \cdot A$, divergence
curl A, rot A, $\nabla \times A$, curl
$\Delta\varphi$, $\nabla^2\varphi$, Laplacian
$\Box\varphi, \Box^2 = \dfrac{1}{c^2}\dfrac{\partial^2}{\partial t^2} - \nabla^2$, D'Alembertian

Logic and Boolean Algebra

$a \in A$, a is contained in set A
$A \cap B$, $A \cdot B$, logical multiplication. Intersection
  of set A and set B, A AND B
$A \cup B$, $A + B$, logical addition. Union of set A
  and set B. A OR B
$A \oplus B$, exclusive OR
$A \supset B$, logical inclusion. Inclusion of set B
  in set A
$A \ominus B$, complement of set B in set A
$\tilde{A}$, $\overline{A}$, logical complementation. NOT set A.
  Negation
$\emptyset$, 0, logical impossibility. Empty (null) set.
  Zero state
I, 1, logical certainty. Universal set. One state

[a]References: Mathematic signs and symbols for use in the physical sciences and technology, ANSI Y10.20—1975.

TABLE 1.3 Abbreviations[a] for Scientific and Engineering Terms[b]

Name of Term	Abbreviation	Name of Term	Abbreviation
absolute	abs	cubic centimeter	cu cm, cm^3
acre	spell out		(liquid,
acre-foot	acre-ft		meaning
air horsepower	air hp		milliliter, ml)
alternating-current		cubic foot	cu ft
(as adjective)	a-c	cubic feet per minute	cfm
ampere	amp or A	cubic feet per second	cfs
ampere-hour	amp-hr	cubic inch	cu in.
amplitude, an elliptic function	am.	cubic meter	cu m or m^3
Angstrom unit	A	cubic micron	cu μ or μ^3
antilogarithm	antilog		or cu mu
atmosphere	atm	cubic millimeter	cu mm or mm^3
atomic weight	at. wt	cubic yard	cu yd
average	avg	current density	spell out
avoirdupois	avdp	cycles per second	spell out or
azimuth	az or α		cps or Hz
		cylinder	cyl
barometer	bar.		
barrel	bbl	day	spell out
Baumé	Bé	decibel	db
board feet (feet board measure)	fbm	degree[d]	deg or °
boiler pressure	spell out	degree centigrade	C
boiling point	bp	degree Fahrenheit	F
brake horsepower	bhp	degree Kelvin	K
brake horsepower-hour	bhp-hr	degree Rankine	R
Brinell hardness number	Bhn	delta amplitude,	
British thermal unit[c]	Btu or B	an elliptic function	dn
bushel	bu	diameter	diam
		direct-current (as adjective)	d-c
calorie	cal	dollar	$
candle	c	dozen	doz
candle-hour	c-hr	dram	dr
candlepower	cp		
cent	c or ¢	efficiency	eff
center to center	c to c	electric	elec
centigram	cg	electromotive force	emf
centiliter	cl	elevation	el
centimeter	cm	equation	eq
centimeter-gram-second		external	ext
(system)	cgs		
chemical	chem	farad	spell out or F
chemically pure	cp	feet board measure (board feet)	fbm
circular	cir	feet per minute	fpm
circular mils	cir mils	feet per second	fps
coefficient	coef	fluid	fl
cologarithm	colog	foot	ft
concentrate	conc	foot-candle	ft-c
conductivity	cond	foot-Lambert	ft-L
constant	const	foot-pound	ft-lb
continental horsepower	cont hp	foot-pound-second (system)	fps
cord	cd	foot-second (see cubic feet	
cosecant	csc	per second)	
cosine	cos	franc	fr
cosine of the amplitude,		free aboard ship	spell out
an elliptic function	cn	free alongside ship	spell out
cost, insurance, and freight	cif	free on board	fob
cotangent	cot	freezing point	fp
coulomb	spell out or C	frequency	spell out
counter electromotive force	cemf	fusion point	fnp
cubic	cu		

TABLE 1.3   (*Continued*)

Name of Term	Abbreviation	Name of Term	Abbreviation
gallon	gal	longitude	long. or λ
gallons per minute	gpm	low-pressure (as adjective)	l-p
gallons per second	gps	lumen	lm
grain	spell out	lumen-hour	lm-hr
gram	g	lumens per watt	lpw
gram-calorie	g-cal		
greatest common divisor	gcd	mass	spell out
		mathematics (ical)	math
haversine	hav	maximum	max
hectare	ha	mean effective pressure	mep
henry	H	mean horizontal candlepower	mhcp
high-pressure (adjective)	h-p	megacycle	spell out
hogshead	hhd	megohm	spell out
horsepower	hp	melting point	mp
horsepower-hour	hp-hr	meter	m
hour	hr	meter-kilogram	m-kg
hour (in astronomical tables)	h	mho	spell out
hundred	C	microampere	μa or mu a
hundredweight (112 lb)	cwt	microfarad	μf
hyperbolic cosine	cosh	microinch	μin.
hyperbolic sine	sinh	microfarad	$\mu\mu$f
hyperbolic tangent	tanh	micromicron	$\mu\mu$ or mu mu
		micron	μ or mu
inch	in.	microvolt	μv
inch-pound	in-lb	microwatt	μw or mu w
inches per second	ips	mile	spell out
indicated horsepower	ihp	miles per hour	mph
indicated horsepower-hour	ihp-hr	miles per hour per second	mphps
inside diameter	ID	milliampere	ma
intermediate-pressure		milligram	mg
(adjective)	i-p	millihenry	mH
internal	int	millilambert	mL
		milliliter	ml
joule	J	millimeter	mm
		millimicron	mμ or m mu
kilocalorie	kcal	million	spell out
kilocycles per second	kc	million gallons per day	mgd
kilogram	kg	millivolt	mV
kilogram-calorie	kg-cal	minimum	min
kilogram-meter	kg-m	minute	min
kilograms per cubic meter	kg per cu m	minute (angular measure)	
	or kg/m^3	minute (time) (in astronomical	
kilograms per second	kgps	tables)	m
kiloliter	kl	mole	spell out
kilometer	km	molecular weight	mol. wt
kilometers per second	kmps	month	spell out
kilovolt	kv		
kilovolt-ampere	kva	National Electrical Code	NEC
kilowatt	kw		
kilowatthour	kwhr	ohm	spell out or Ω
		ohm-centimeter	ohm-cm
lambert	L	ounce	oz
latitude	lat or ϕ	ounce-foot	oz-ft
least common multiple	lcm	ounce-inch	oz-in.
linear foot	lin ft	outside diameter	OD
liquid	liq		
lira	spell out	parts per million	ppm
liter	L	peck	pk
logarithm (common)	log	penny (pence)	d
logarithm (natural)	$\log_e$ or ln	pennyweight	dwt

TABLE 1.3 (*Continued*)

Name of Term	Abbreviation	Name of Term	Abbreviation
per	(See Fundamental Rules)	sine of the amplitude, an elliptic function	sn
peso	spell out	specific gravity	sp gr
pint	pt	specific heat	sp ht
potential	spell out	spherical candle power	scp
potential difference	spell out	square	sq
pound	lb	square centimeter	sq cm or cm^2
pound-foot	lb-ft	square foot	sq ft
pound-inch	lb-in.	square inch	sq in.
pound sterling	£	square kilometer	sq km or km^2
pounds per brake horsepower-hour	lb per bhp-hr	square meter	sq m or m^2
pounds per cubic foot	lb per cu ft	square micron	sq μ or sq mu or μ^2
pounds per square foot	psf	square millimeter	sq mm or mm^2
pounds per square inch	psi	square root of mean square	rms
pounds per square inch absolute	psia	standard	std
power factor	spell out or pf	steradian	sr
		tangent	tan
quart	qt	temperature	temp
		tensile strength	ts
		thousand	M
radian	spell out	thousand foot-pounds	kip-ft
reactive kilovolt-ampere	kvar	thousand pound	kip
reactive volt-ampere	var	ton	spell out
revolutions per minute	rpm	ton-mile	spell out
revolutions per second	rps		
rod	spell out	versed sine	vers
root mean square	rms	volt	V
		volt-ampere	Va
secant	sec	volt-coulomb	spell out
second	sec		
second (angular measure)	''	watt	W
second-foot (see cubic feet per second)		watthour	Whr
		watts per candle	Wpc
second (time) (in astronomical tables)	s	week	spell out
shaft horsepower	shp	weight	wt
shilling	s	yard	yd
sine	sin	year	yr

[a]These forms are recommended for readers whose familiarity with the terms used makes possible a maximum of abbreviations. For other classes of readers, editors may wish to use less contracted combinations made up from this list. For example, the list gives the abbreviation of the term "feet per second" as "fps." To some readers ft per sec will be more easily understood.

[b]This list of abbreviations is adapted from the recommendations of the American National Standards Institute (See ANSI Y1.1-1972 (R1984)).

[c]Abbreviation recommended by the A.S.M.E. Power Test Codes Committee. B = 1 Btu, kB = 1000 Btu, mB = 1,000,000 Btu. The A.S.H. & V.E. recommends the use of Mb = 1000 Btu and Mbh = 1000 Btu per hr.

[d]There are circumstances under which one or the other of these forms is preferred. In general the sign ° is used where space conditions make it necessary, as in tabular matter, and when abbreviations are cumbersome, as in some angular measurements, i.e., 59° 23' 42''. In the interest of simplicity and clarity the Committee has recommended that the abbreviation for the temperature scale, F, C, K, etc., always be included in expressions for numerical temperatures, but, wherever feasible, the abbreviation for "degree" be omitted; as 69 F.

TABLE 1.4   Symbols for Physical Quantities[a, b]

Name of Quantity	Symbol	Name of Quantity	Symbol
Absorption factor	α	Chézy's coefficient	C
Acceleration		Chord length	c
Angular	α	Circular frequency $(2\pi f)$	ω
Linear, general	a	Circulation, strength of single	
Acceleration due to gravity		vortex	Γ
General	g	Coefficient	
International Adopted Standard	g_0	Absolute	C
Local	g_L	General	C
Gravitational conversion factor	g_c	Of contraction	C_c
Activity	a	Of discharge	C
Activity coefficient, molal basis	γ	Of discharge	C_q
Adiabatic factor	X	Of energy per unit weight in	
Admittance	Y	$\qquad C_e \dfrac{V^2}{29}$	C_e
Advanced ratio of propeller	J		
Altitude	h, z	Of flow (Chézy)	C
Amplitude	A	Of friction (Weisbach-Darcy)	f
Angle	α	Of friction	μ, f
Angle	$\beta\phi$	Of momentum per unit weight	
Blade angle	β	$\qquad \text{in } C_m \dfrac{V}{g}$	C_m
Effective helix	ψ		
Dihedral	Γ	Of roughness (Bazin)	m
Helical angle of advance	ϕ	Of roughness (Kutter and	
Of attack	α	Manning)	n
Of downwash	ε	Of heat transfer overall	V
Of radiation	θ	Of velocity	C_v
Of sideslip	β	Compressibility factor	z
Of sidewash	σ	Concentrated load	F, P, Q
Solid	ω	Concentration	C, c
Angular		Concentration, volumetric	c
Acceleration	α	Concentration factor, stress	K
Displacements	δ	Conductance	
Frequency	ω	Electrical	G
Momentum	H	Thermal	$1/R$
Velocity	ω	Per unit area	$1/RA$
Area	A	Conductivity	
Area[c]	S	Electrical	γ, σ
Aspect ratio	A, AR	Equivalent	Λ
Atomic weight	A	Thermal	k
Attack, angle of	α	Contraction, coefficient of	C_c
Attenuation	a	Correlation coefficient	R
Axes		Coupling coefficient	k
Of aircraft (left handed)		Critical state or indicating critical	
Earth-bound coordinate		value (subscript)	c
system	x, y, z	Current[d]	I
Lateral	Y		
Longitudinal	X	Damping	
Normal	Z	Coefficient	c
		Constant or coefficient	δ
Bazin's coefficient of roughness	m	Factor	λ
Blade width (propellers)	b	Deflection	δ
Boundary layer thickness	δ	Of beam, maximum	δ
Breadth	b	Density	ρ
		Relative to standard air density	σ
Capacitance, capacity	C	Depth	h
Capacitivity	ε	Depth	y
Of evacuated space	ε_v	Of flow, channels	y
Relative	ε_r	Diameter	D
Charge, electric or quantity of		Dielectric constant	ε
electricity	Q	Difference between values	Δ
Charge density		Difference of potential[d, e]	E, e
Line density of charge	λ	Diffusion coefficient	
Surface density of charge	σ	Diffusivity	D_v
Volume density of charge	ρ		

TABLE 1.4 (*Continued*)

Name of Quantity	Symbol	Name of Quantity	Symbol
Diffusivity	α	Fluidity	$1/\mu$
Diffusivity, thermal	α	Flux	
Of vapor	D_v	Density	
Discharge		Magnetic	B
Coefficient of	C_q	Displacement	ψ
Coefficient of	C	Magnetic	Φ
Rate of; or flow	Q	Force	**F**
Per unit width	q	Force	F
Displacement, electric	D	Electromotive[e]	E, e
Distance		Magnetomotive	$M, \mathscr{F}$
From center of gravity to center		Moment of	M
of pressure of horizontal		Normal	N
tail surface	f	Shearing force in beam section	V
Linear	s	Total load	F
Drag, absolute coefficient of	D	Forces or loads, concentrated	P, Q, F
Dynamic (or impact) pressure	q	Fraction	
		By volume	x_v
Eccentricity of application of load	e	By weight	x_w
Efficiency	η	Free energy	
Elastance	S	Gibbs	G
Mutual	S_m, S_{rc}	Helmholtz	A
Self	S, S_{cc}	Frequency	f
Elasticity		Circular ($2\pi f$)	θ
Bulk modulus, of liquids	K	Of radiant energy	ν
Kinematic $\dfrac{K}{\rho}$	e	Reduced (flutter)	k
		Rotational	n
Modulus of	E	Frequency, angular	ω
Elastivity	σ	Friction	
Electric potential[d, e]	E, e	Coefficient of sliding	f, μ
Electricity, quantity of	Q, q	Factor used in expressing pipe-	
Electromotive force[d]	E, e	loss	f
Electronic charge, absolute value	e	In energy balance	F
Electrostatic flux	ψ	Fugacity	f
Elevation			
Above datum	Z	Gas constant	R
Above stream bed	Z_0	Gibbs' function, total potential	
Elongation, total	δ	function[f]	G, g
Emissivity, total	ε	Gyration, radius of	k
Energy	W		
Work total	E	Head	
Energy	E	Atmospheric	h_a
Internal; intrinsic[f]	U, u	Lost[c]	h
Kinetic	E_k, T	Potential	h_{pz}
Per unit time (power)	P	Pressure	h_p
Potential	E_p, V	Velocity	h_v
Enthalpy[f]	H or h	Heat	
Enthalpy	H	Content; enthalpy[f]	H, h
Of dry saturated vapor	h_g	Content of dry saturated vapor;	
Of saturated liquid	h_f	enthalpy of dry saturated	
Per unit weight	h	vapor	h_g
Entropy	S, s	Content of saturated liquid; en-	
Error signal	ε	thalpy of saturated liquid	h_f
Expansion, exponent of polytropic	n	Equivalent of work	$1/J$
Cubical, thermal coefficient	β	Flow rate	q
Linear, thermal coefficient	α	Across a boundary surface	h
		Latent, of evaporation	λ, h_{fg}
Factor of safety	N	Mechanical equivalent of	J
Film thickness, effective	B	Of vaporization at constant	
Flow rate	w	pressure[f]	$H_{fg}, \lambda,$ or h_{fg}
In pounds per unit of time	w		
Volumetric	q	Specific, at constant pressure	c_p
		Specific, at constant volume	c_v

TABLE 1.4 (*Continued*)

Name of Quantity	Symbol	Name of Quantity	Symbol
Ratio of specific heats	γ	Linear expansion, coefficient	α
Ratio of specific	γ, κ, or k	Linear velocity	v
Transfer, overall coefficient of	U	Load	
Transfer, surface coefficient of	h	Concentrated	F, P, Q
Height	h	Eccentricity of application of	e
Crest, weirs	z	Factor	n
Helix, effective angle	ϕ	Per unit distance	w, q
Helmholtz' free energy; internal		Total	W, P
potential function^f	A, a		
Humidity	H	Mach	
Density of water vapor; weight of		Angle	μ
water vapor per unit of		Number	M
volume of space	ρ_H	Magnetic	
Density of water vapor at		Flux	Φ
saturation	ρ_s	Intensity	$\mathbf{H}$
Enthalpy of the mixture minus		Magnetomotive force	$M, \mathscr{F}$
the enthalpy of the liquid		Mass	m
at the temperature of		Flow rate	w
Adiabatic saturation;		Velocity	G
Carrier sigma function	h_Σ	Mean free path	λ
Humid volume, volume of		Mechanical equivalent of heat	J
mixture per unit of		Microscale (turbulence)	λ
weight of dry air	v_H	Modulus	
Partial pressure of water vapor	p_H	Bulk, of elasticity of liquids	K
Percentage humidity by weight	w_H/w_s	Of elasticity	E
Relative humidity; ratio of an		Of elasticity in shear	G
actual partial pressure of		Section	Z
water vapor in air to the		Shear	G
saturation partial pressure	H_R	Molecular weight	M
Saturation pressure of water		Moment	
vapor	p_s	Electric	$\mathbf{p}$
Saturation weight of water vapor		Magnetic	$\mathbf{m}$
per unit of weight of dry air	H_s, w_s	Of any area about a given axis,	
Weight of water vapor per unit		statical	Q
of weight of dry air	H, w_H	Of force, including bending	
Hydraulic radius	R_H	moment	M
Mean in a reach	R_m	Of inertia, polar	J
Of cross-sectional area	R	Rectangular	I
Hydraulic slope	S_w	Mutual inductance	L_m
Impedance	Z	Neutral axis, distance to extreme	
Impulse	I	fiber	c
Inductance	L	Nozzle divergence factor	λ
Magnetic	B	Number in general	N
Mutual	L_m	Of conductors or turns	N
Self	L, L_{cc}	Of moles, pound-moles, kilogram-	
Inertia, moment of	I	moles, etc.	n
Inertia, moment of		Of phases	m
Polar	J	Of poles	p
Rectangular	I	Of revolutions per unit of time	n
Product moment of	I_{xy}	Perimeter, wetted, of a sectional	
Intensity		area	P
Electric	$\mathbf{E, K}$	Period	T
Magnetic	$\mathbf{H}$	Permeability	
Isentropic factor	X	Magnetic	μ
		Of evacuated space	μ_c
Joule-Thomson's coefficient	μ	Relative	μ_r
		Permeance	$\mathscr{P}, \Lambda$
Kutter's coefficient of roughness	n	Permittivity	ε
		Phase	
Length	L	Angle	ϕ
Length	l	Constant	β
Lift	L		

TABLE 1.4 (*Continued*)

Name of Quantity	Symbol	Name of Quantity	Symbol
Displacement	ϕ	Temperature coefficient	α
Pitch, geometric	p	Thermal	R
Planck constant	h	Per unit area	RA
Poisson ratio	μ, ν	Resistivity	
Polarization, magnetic	B_i	Electrical	ρ
Pole strength	m	Thermal	$1/k$
Potential		Revolutions per unit time	n
Electric[d, e]	V	Reynolds' number	R
Function	ϕ	Richness; equivalence ratio	
Function, internal; Helmholtz'		(combustion)	R
free energy	A, a	Rotation	
Function, total; Gibbs' function	G, g	Rate of	n
Magnetic	$\mathbf{M}, \mathscr{F}$	Speed of	n
Magnetic vector	$\mathbf{A}$		
Retarded vector	$\mathbf{A}_r$	Safety factor	N
Power	P	Saturation pressure of water	
Power		vapor	p_s
Active	P	Section modulus	Z
Apparent	S	Self-inductance	L, L_{cc}
Factor	F_p	Set of control surfaces, angle of[c]	δ
Reactive	Q	Shape factor	S
Pressure		Shearing force in beam section	V
Dynamic	q	Slip	s
Intensity; force per unit area	p	Slope	
Relative	δ	Of channel bed	S_0
Saturation of water vapor	p_s	Of cuts and embankments	s
Propagation constant	γ	Of energy grade line	S
Poynting vector	II	Of hydraulic grade line	S_w
		Of lift curve	a
Q factor of a reactor	Q	Solidity, propellers	σ
Quality of vapor	x	Span	b
Quantity		Effectiveness	e
Of electricity	Q	Specific	
Of heat per unit mass or unit		Gravity	G
weight	q	Heat	c
Of heat per unit time	q	Heat at constant pressure	c_p
Of matter	W	Molar	C_p
Total, of a fluid, water, gas, heat		Heat at constant volume	c_v
(by volume)	Q	Molar	C_v
		Heats, ratio	γ
Radiant density	u	Heats, ratio	κ or k
Radiant energy	U	Volume	v
Radiant flux	Φ	Weight	γ
Density	W	Speed	
Radiant intensity	J	Linear	V, v, u
Radiation, intensity of	N	Of rotation	n
Radii	r, R	Spring constant	k
Radius	r	Stefan-Boltzmann constant	σ
Of gyration	k	Strain	
Range	R	Normal	ε
Reactance	X	Shear	γ
Capacitive	X_c	Stream function	ψ
Inductive	X_L	Stress	
Mutual	X_m, X_{rc}	Concentration factor	K
Self	X, X_{cc}	Normal	σ, s
Reactive factor	F_q	Shear	τ, s_s
Recovery factor	η_r	Supercompressibility factor	z
Reduced frequency (flutter)	k	Surface coefficient of heat transfer	h
Reflection factor	ρ	Surface per unit volume	a
Reluctance	$\mathscr{R}$	Surface tension	σ
Reluctivity	ν	Kinematic σ/ρ	ω
Resistance		Susceptance	B
Electrical	R		

TABLE 1.4 (*Continued*)

Name of Quantity	Symbol	Name of Quantity	Symbol
Susceptibility		Mean (Q/A)	V
Dielectric	η	Of light	c
Magnetic	κ	Of sound	a, c
Sweepback angle	Λ	Of uniform flow	V_0
		Of wave celerity	c
Taper ratio	λ	Relative	v
Temperature		Temporal means of components	$\bar{u}, \bar{v}, \bar{w}$
Absolute[g] (°F abs or K)	T or Θ	Vibration constant	p
Ordinary[g] (°F or °C)	t or θ	Viscosity	
Ratio	θ	Absolute; coefficient of	μ
Thermal		Kinematic	ν
Conductance	$1/R$	Relative (to absolute viscosity	
Per unit area;		of water)	μ/μ_w
"unit conductance"	$1/RA$	Relative kinematic	use ν/ν_w
Conductivity	k	Voltage[d]	E, e
Diffusivity	α	Volume	V
Resistance	R	Molar	V, V_m
Resistance of unit area	RA	Specific	v
Resistivity	$1/k$	Total	V, V_L
Transfer factor	j	Volume rate; discharge by volume,	
Transmission	q	fluid rate of flow by volume	q, Q
Thickness	$d, t,$ or h		
Thrust		Wavelength	λ
Stream	F	Constant	β
Propeller	T	Weight	
Time	t	Molecular	M
Time[h]	t or τ	Per unit time per unit area of	
Time constant	τ	cross section; "mass	
Torque	Q	velocity"	G
Torque	T or M	Per unit volume	γ
Transmission		Rate; per unit of power; for unit	
Factor	τ	of time	w
Thermal	q	Rate of flow per unit of breadth	Γ
Turbulence exchange, coefficient	ε	Specific, with g_c	γ
Turbulence scale	L	Total	W
		Weirs	
Vaporization, heat of, at		Crest height	z
constant pressure	H_{fg}, h_{fg}, λ	Crest length	b
Velocity	V	Degree of submergence	N
Velocity	V or v	Wetted perimeter	L_p
Acoustic	V_a	Width (same as breadth)	b
Angular	ω	Of stream bed	b
Average	V	Width, channel surface	b_w
Belanger critical	V_c	Wing setting, angle of (angle	
Components in x, y, z		between the wing chord	
directions, respectively	u, v, w	and the thrust line)	i_w
Linear	v	Work	W
Local	u	External	W_e
Mass, mass-flow, per unit cross-		Heat equivalent of	$1/J$ or A
sectional area, per unit time	G	Per unit weight	w, w_k

[a] The most frequently used American Standard and Tentative Standard Symbols are included in this table. Sources used are publications of the American National Standards Institute shown in bibliography below.

[b] Where possible, capital letters denote total quantities and small letters denote specific quantities, or quantities per unit.

[c] Use with appropriate subscript.

[d] Where distinctions between maximum, instantaneous, effective (root-mean-square), and average values are necessary, E_m, I_m, P_m are recommended for maximum values; e, i, p for instantaneous values; E, I for effective (rms) values; and P for average value.

[e] Where a distinction between electromotive force and difference of electric potential is desirable, the symbols $E, e,$ and V, v, respectively, may be used.

[f] In each instance uppercase italics may be used optionally for values in general or per mole. Molal values may have subscript M. Lowercase italics are to be used for specific values (per pound, gram, liter, etc.). Molecular values may be represented by lowercase italics or by lowercase italics with subscript m.

[g] θ is preferable only when t is used for time in the same discussion. Θ is preferable only when θ is used for ordinary temperature.

[h] τ should be used only when t is used for ordinary temperature in the same discussion.

BIBLIOGRAPHY FOR LETTER SYMBOLS

Acoustics, Letter Symbols and Abbreviations for Quantities Used in, **ANSI / ASME Y10.11-1984.**

Aeronautical Sciences, Letter Symbols for, **ANSI Y10.7-1954.**

Chemical Engineering, Letter Symbols for, **ANSI Y10.12-1955(R1973).**

Glossary of Terms Concerning Letter Symbols, **ANSI Y10.1-1972.**

Heat and Thermodynamics, Letter Symbols for, **ANSI Y10.4-1982.**

Hydraulics, Letter Symbols for, **ANSI Y10.2-1958.**

Illuminating Engineering, Letter Symbols for, **ANSI Y10.18-1967(R1977).**

Letter Symbols for SI Units and Certain Other Units of Measurement, **ANSI / IEEE 260-1978.**

Mathematic Signs and Symbols for Use in Physical Sciences and Technology (includes supplement ANSI Y10.20a-1975), **ANSI Y10.20-1975.**

Mechanics and Time-Related Phenomena, **ANSI / ASME Y10.3M-1984.**

Meteorology, Letter Symbols for, **ANSI Y10.10-1953(R1973).**

Quantities Used in Electrical Science and Electrical Engineering, Letter Symbols for, **ANSI / IEEE 280-1985.**

Selecting Greek Letters Used as Letter Symbols for Engineering Mathematics, Guide for, **ANSI Y10.17-1961(R1973).**

TABLE 1.5   Graphic Symbols (after Dreyfus)

Symbols are a graphical referent to information and have been used for millennia as devices for convenient shorthand notation, to restrict interpretation only to *cognoscenti*, or for compression of data. Examples from several engineering fields are shown here. ISO Recommendations are indicated by ¶ and ISO Draft recommendations by ¶¶.

TABLE 1.5   (*Continued*)

Chemical Engineering

JACKETED REACTOR, Stirred	NUCLEAR REACTOR	PACKED COLUMN	PLATE COLUMN	SECTIONED COLUMN	DISK and DONUT COLUMN
FIXED BED REACTOR	FLUIDIZED BED REACTOR	AUTOCLAVE	CENTRIFUGAL PUMP	RECIPROCATING PUMP	
REBOILER		HEAT EXCHANGER	WATER COOLER	COOLING TOWER	SPRAY DRYER
BLOWER; FAN	BELT CONVEYOR; SHAKER	BUCKET CONVEYOR	SCREW FEEDER	CENTRIFUGE	CYCLONE SEPARATOR
SINGLE–EFFECT EVAPORATOR	BAROMETRIC CONDENSER	ELECTRICAL PRECIPITATOR	PLATE and FRAME FILTER	ROTARY VACUUM FILTER	THICKENER
JET MIXER; EJECTOR	MIXER	SCREENER	BALL MILL	ROLLER CRUSHER	JACKETED VESSEL
ROTARY DRUM DRYER; KILN	ROTARY FILM DRYER; FLAKER	PRESSURE STORAGE TANK	BULK STORAGE TANK	GAS HOLDER STORAGE TANK	GAS FLOW (100 CFM)
TEMPERATURE (200 °F)	PRESSURE (100 psig)	ALL CONTROL VALVES	PRESSURE CONTROLLER (PC)	TEMPERATURE CONTROLLER (TC)	FLOW CONTROLLER (FC)
LEVEL CONTROLLER (LC)		SUMMATION POINT ($R \xrightarrow{+} \ominus \rightarrow R - B$, B)		OPERATIONAL BLOCK ($M \rightarrow \boxed{G} \rightarrow MG$)	

TABLE 1.5 *(Continued)*

Chemical Engineering (continued)

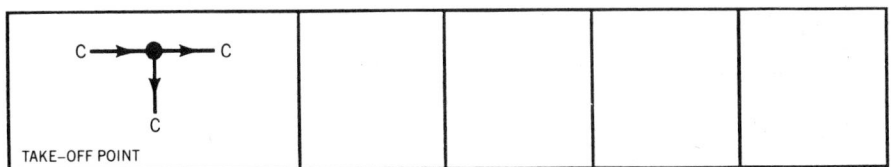

C ——●—— C ↓ C TAKE–OFF POINT				

Electrical Engineering

DIRECT CURRENT (DC)	ALTERNATING CURRENT (AC)	AUDIO FREQUENCY AC	SUPERAUDIO FREQUENCY AC	CROSSED CONDUCTORS	JOINED CONDUCTORS
SINGLE–PHASE	2–PHASE 3–WIRE	2–PHASE 4–WIRE	3–PHASE 3–WIRE (Delta)	3–PHASE 3–WIRE (Star)	3–PHASE 4–WIRE (Star)
2 and 3–PHASE TEE CONNECTED	3–PHASE, 3–WIRE VEE CONNECTED	6–PHASE; FORK with NEUTRAL	START of WINDING	VARIABLE CONTROL	VARIABLE CONTROL by STEPS
PRESET CONTROL	ADJUSTABLE TAPPING	PRESET TAPPING	NON–LINEAR VARIABILITY	SATURABLE PROPERTIES	EARTH (Ground)
CHASSIS of EQUIPMENT	INSULATED COUPLING	UNINSULATED COUPLING	SCREENED CONDUCTOR	RESISTOR OR	
NON–INDUCTIVE RESISTOR (Heater)	ADJUSTABLE CONTACT RESISTOR	INDUCTOR	TRANSFORMER	VACUUM TUBE (Triode)	DIODE
CONTROLLED RECTIFIER	TRANSISTOR (n–––n type)	TRANSISTOR (p–n–p type)	TRANSISTOR, Field–effect (n–channel)	FIXED CAPACITOR	ELECTROLYTIC CAPACITOR

TABLE 1.5 (*Continued*)

ALTERNATING CURRENT SOURCE	BATTERY (Direct Current Source)	PIEZOELECTRIC CRYSTAL UNIT	AMPLIFIER	LOUDSPEAKER ¶¶	MICROPHONE
CATHODE RAY TUBE (TV)	LAMP BULB ¶¶	INDICATOR ¶¶	LIGHTNING ARRESTER ¶¶	SWITCH On / Off OR On / Off ¶¶	
FUSE ¶¶	CIRCUIT BREAKER	RECORDING HEAD	PLAYBACK HEAD	ERASE HEAD	EQUIPMENT OUTLINE ¶¶
General ¶¶ AERIAL (Antenna) OR	Dipole ¶¶ OR	Loop ¶¶ OR	Loop ¶¶	Male Female ¶¶ CONNECTOR OR Male / Female	
PERMANENT MAGNET ¶¶	PHOTO– SENSITIVITY ¶¶	BELL ¶¶	BUZZER ¶¶ OR ¶¶		

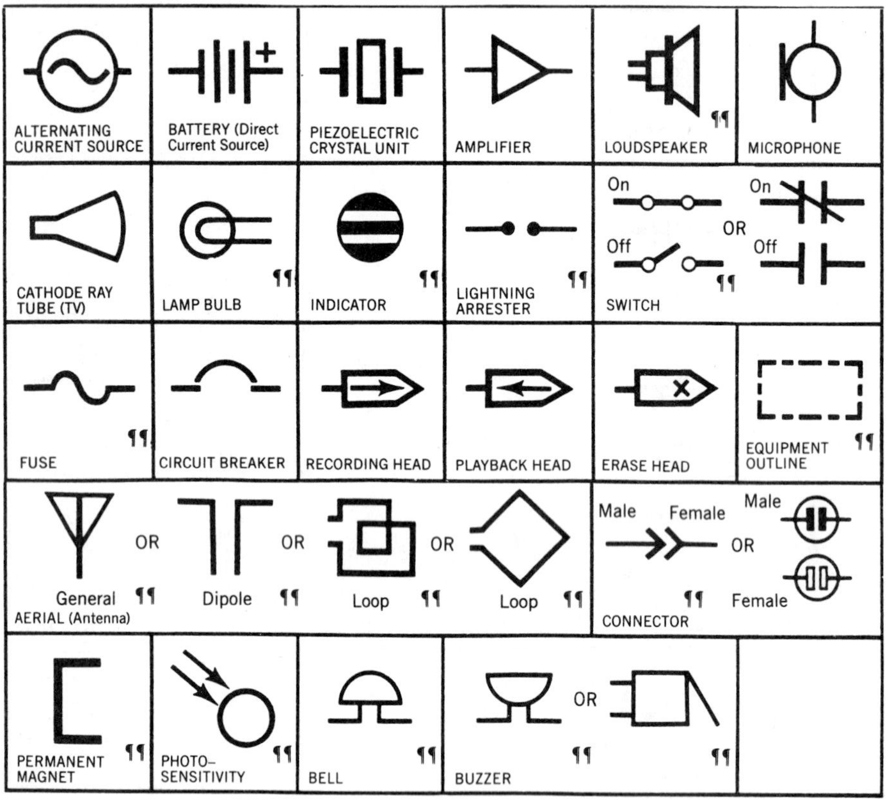

¶¶ Draft ISO Recommendation

COMPRESSOR	PNEUMATIC COMPRESSOR ▲	HYDRAULIC MOTOR ▲	OSCILLATING MOTOR ¶¶ ▲	HYDRAULIC PUMP ▲	PUMP, ROTARY and CENTRIFUGAL ★
ENGINE, Gas	BLOWER, Gas	TURBINE	HEAT EXCHANGER	AIR COOLED CONDENSER	WATER COOLED CONDENSER
PIPE LINE JUNCTION ¶¶	CROSSED ¶¶ PIPE LINES	LOW PRESSURE STEAM SUPPLY	LOW PRESSURE STEAM RETURN	MEDIUM PRESSURE STEAM SUPPLY ■	HIGH PRESSURE STEAM SUPPLY ■

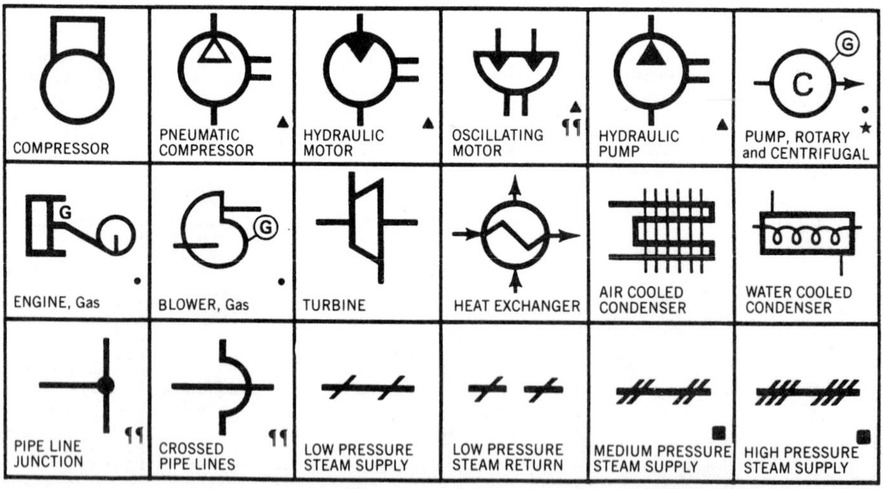

TABLE 1.5 (*Continued*)

Mechanical Engineering (continued)

PNEUMATIC FLOW DIRECTION	HYDRAULIC FLOW DIRECTION	WASTE WATER	COLD WATER	HOT WATER SUPPLY	HOT WATER RETURN
VENT PIPE	CHILLED WATER LINE	FUEL LINE	GAS LINE	VACUUM LINE	THREADED PIPE JOINT
FLANGED PIPE JOINT	WELDED PIPE JOINT	BELL and SPIGOT PIPE JOINT	SOLDERED PIPE JOINT	UNION, Threaded	TEE JOINT, Threaded
CROSS JOINT, Threaded	90° ELBOW, Threaded	LATERAL JOINT, Threaded	ECCENTRIC REDUCER	CONCENTRIC REDUCER	THREADED BUSHING
EXPANSION JOINT FLANGE	CHECK VALVE	SHUT–OFF VALVE; GATE VALVE	GLOBE VALVE	COCK VALVE	DIAPHRAGM VALVE
SAFETY VALVE	STOP COCK	PRESSURE GAUGE	THERMOMETER	Welding	FILLET
PLUG; SLOT	ARC–SPOT; ARC SEAM	BACKING; BACK	MELT–THROUGH	EDGE FLANGE	CORNER FLANGE
SURFACING	SQUARE GROOVE	"V" GROOVE	"U" GROOVE	"J" GROOVE	FLARE "V" GROOVE
FLARE BEVEL GROOVE	BEVEL GROOVE	WELD ALL AROUND	FIELD WELD	FLUSH CONTOUR	CONVEX CONTOUR

TABLE 1.5   (*Continued*)

Mechanical Engineering (continued)

Geometric Tolerances	STRAIGHTNESS ¶	FLATNESS ¶	FLATNESS and STRAIGHTNESS	CIRCULARITY (Roundness) ¶	CYLINDRICITY ¶
PROFILE of any LINE ¶	PROFILE of any SURFACE	PARALLELISM ¶	SQUARENESS (Perpendicularity) ¶	ANGULARITY	POSITION ¶
COAXIALITY; CONCENTRICITY ¶	SYMMETRY ¶	RUN–OUT ¶	SURFACE ROUGHNESS	SURFACE to be FINISHED (Machined)	

▲ Pneumatic machinery is indicated by △ , hydraulic machinery by ▲
• G indicates Gas. Different initial may be substituted to indicate other type of machine; e.g., D (diesel), M (motor), T (turbine), E (steam). Exception: (see Engine, Gas) Steam Engine is indicated by symbol without initial.
★ C indicates Circulating Water. Different initial indicates other type of machine or service; e.g., D (concentrate), F (boiler feed), O (oil), S (service), V (air).
■ "Return" indicated by broken line, as illustrated in Low Pressure Steam Return.
¶¶ Draft ISO Recommendation
▼ Flanged, Welded, Bell and Spigot, or Soldered Union indicated by substituting appropriate markings (see Joints). Example: ✕▏✕ Welded Union.
¶ ISO Recommendation

BIBLIOGRAPHY FOR GRAPHIC SYMBOLS

Arnell, A. *Standard Graphical Symbols—A Comprehensive Guide for Use in Industry, Engineering and Science*, New York, McGraw-Hill 1963.

Dreyfus, H. *Symbol Sourcebook*, New York, McGraw-Hill, 1972.

Electrical and Electronics Diagrams (Including Reference Designation Class Designation Letters), Graphic Symbols for, **ANSI / IEEE 315-1975.**

Electrical and Electronics Diagrams, Graphic Symbols for (supplement to ANSI/IEEE 315-1975), **ANSI / IEEE 315A-1986.**

Electrical and Electronics Parts and Equipments, Reference Designations for, **ANSI / IEEE 200-1975.**

Electrical Wiring and Layout Diagrams Used in Architecture and Building Construction, Graphic Symbols for, **ANSI Y32.9-1972.**

Fire Fighting Operations, Symbols for, **ANSI / NFPA 178-1980.**

Fire Protection Symbols for Risk Analysis Diagrams, **ANSI / NFPA 174-1980.**

Fire-Protection Symbols for Architectural and Engineering Drawings, **ANSI / NFPA 172-1980.**

Fluid Power Diagrams, Graphic Symbols for, **ANSI Y32.10-1967(R1979).**

Grid and Mapping Used in Cable Television Systems, Graphic Symbols for, **ANSI / IEEE 623-1976.**

Heat-Power Apparatus, Graphic Symbols for, **ANSI Y32.2.6M-1984.**

Heating, Ventilating, and Air Conditioning, Graphic Symbols for, **ANSI Y32.2.4M-1984.**

Polon, D. D. (Ed.) *Encyclopedia of Engineering Signs and Symbols*, New York, Odyssey Press, 1965.

Shepard, W. *Shepard's Glossary of Graphic Signs and Symbols*, London, Dent, 1971.

TABLE 1.6 Personal Computer Numeric Codes for Characters and Symbols

IBM PC Character Set (00–7F) Quick Reference

IBM PC Character Set (80–FF) Quick Reference

DECIMAL VALUE ►	HEXADECIMAL VALUE ►	0 / 0	16 / 1	32 / 2	48 / 3	64 / 4	80 / 5	96 / 6	112 / 7	128 / 8	144 / 9	160 / A	176 / B	192 / C	208 / D	224 / E	240 / F
0	0	BLANK (NULL)	►	BLANK (SPACE)	0	@	P	`	p	Ç	É	á	░	└	╨	α	≡
1	1	☺	◄	!	1	A	Q	a	q	ü	æ	í	▒	┴	╤	ß	±
2	2	☻	↕	"	2	B	R	b	r	é	Æ	ó	▓	┬	╥	Γ	≥
3	3	♥	‼	#	3	C	S	c	s	â	ô	ú	│	├	╙	π	≤
4	4	♦	¶	$	4	D	T	d	t	ä	ö	ñ	┤	─	╘	Σ	⌠
5	5	♣	§	%	5	E	U	e	u	à	ò	Ñ	╡	┼	╒	σ	⌡
6	6	♠	▬	&	6	F	V	f	v	å	û	ª	╢	╞	╓	µ	÷
7	7	•	↨	'	7	G	W	g	w	ç	ù	º	╖	╟	╫	τ	≈
8	8	◘	↑	(	8	H	X	h	x	ê	ÿ	¿	╕	╚	╪	Φ	°
9	9	○	↓	)	9	I	Y	i	y	ë	Ö	⌐	╣	╔	┘	Θ	∙
10	A	◙	→	*	:	J	Z	j	z	è	Ü	¬	║	╩	┌	Ω	·
11	B	♂	←	+	;	K	[	k	{	ï	¢	½	╗	╦	█	δ	√
12	C	♀	∟	,	<	L	\	l	|	î	£	¼	╝	╠	▄	∞	ⁿ
13	D	♪	↔	-	=	M	]	m	}	ì	¥	¡	╜	═	▌	φ	²
14	E	♫	▲	.	>	N	^	n	~	Ä	₧	«	╛	╬	▐	ε	■
15	F	☼	▼	/	?	O	_	o	⌂	Å	ƒ	»	┐	╧	▀	∩	BLANK FF

Source: Reprinted by permission from IBM Document 1502243. Copyright © 1985 by International Business Machines Corporation.

TABLE 1.7 Conversions for Number Systems of Different Bases

Radix 16 Hexadecimal	Radix 10 Decimal	Radix 8 Octal	Radix 2 Binary BIT 8765 4321	Radix 16 Hexadecimal	Radix 10 Decimal	Radix 8 Octal	Radix 2 Binary BIT 8765 4321
00	0	00	0000 0000	3A	58	72	0011 1010
01	1	01	0000 0001	3B	59	73	0011 1011
02	2	02	0000 0010	3C	60	74	0011 1100
03	3	03	0000 0011	3D	61	75	0011 1101
04	4	04	0000 0100	3E	62	76	0011 1110
05	5	05	0000 0101	3F	63	77	0011 1111
06	6	06	0000 0110	40	64	100	0100 0000
07	7	07	0000 0111	41	65	101	0100 0001
08	8	10	0000 1000	42	66	102	0100 0010
09	9	11	0000 1001	43	67	103	0100 0011
0A	10	12	0000 1010	44	68	104	0100 0100
0B	11	13	0000 1011	45	69	105	0100 0101
0C	12	14	0000 1100	46	70	106	0100 0110
0D	13	15	0000 1101	47	71	107	0100 0111
0E	14	16	0000 1110	48	72	110	0100 1000
0F	15	17	0000 1111	49	73	111	0100 1001
10	16	20	0001 0000	4A	74	112	0100 1010
11	17	21	0001 0001	4B	75	113	0100 1011
12	18	22	0001 0010	4C	76	114	0100 1100
13	19	23	0001 0011	4D	77	115	0100 1101
14	20	24	0001 0100	4E	78	116	0100 1110
15	21	25	0001 0101	4F	79	117	0100 1111
16	22	26	0001 0110	50	80	120	0101 0000
17	23	27	0001 0111	51	81	121	0101 0001
18	24	30	0001 1000	52	82	122	0101 0010
19	25	31	0001 1001	53	83	123	0101 0011
1A	26	32	0001 1010	54	84	124	0101 0100
1B	27	33	0001 1011	55	85	125	0101 0101
1C	28	34	0001 1100	56	86	126	0101 0110
1D	29	35	0001 1101	57	87	127	0101 0111
1E	30	36	0001 1110	58	88	130	0101 1000
1F	31	37	0001 1111	59	89	131	0101 1001
20	32	40	0010 0000	5A	90	132	0101 1010
21	33	41	0010 0001	5B	91	133	0101 1011
22	34	42	0010 0010	5C	92	134	0101 1100
23	35	43	0010 0011	5D	93	135	0101 1101
24	36	44	0010 0100	5E	94	136	0101 1110
25	37	45	0010 0101	5F	95	137	0101 1111
26	38	46	0010 0110	60	96	140	0110 0000
27	39	47	0010 0111	61	97	141	0110 0001
28	40	50	0010 1000	62	98	142	0110 0010
29	41	51	0010 1001	63	99	143	0110 0011
2A	42	52	0010 1010	64	100	144	0110 0100
2B	43	53	0010 1011	65	101	145	0110 0101
2C	44	54	0010 1100	66	102	146	0110 0110
2D	45	55	0010 1101	67	103	147	0110 0111
2E	46	56	0010 1110	68	104	150	0110 1000
2F	47	57	0010 1111	69	105	151	0110 1001
30	48	60	0011 0000	6A	106	152	0110 1010
31	49	61	0011 0001	6B	107	153	0110 1011
32	50	62	0011 0010	6C	108	154	0110 1100
33	51	63	0011 0011	6D	109	155	0110 1101
34	52	64	0011 0100	6E	110	156	0110 1110
35	53	65	0011 0101	6F	111	157	0110 1111
36	54	66	0011 0110	70	112	160	0111 0000
37	55	67	0011 0111	71	113	161	0111 0001
38	56	70	0011 1000	72	114	162	0111 0010
39	57	71	0011 1001	73	115	163	0111 0011

TABLE 1.7   (*Continued*)

Radix 16 Hexadecimal	Radix 10 Decimal	Radix 8 Octal	Radix 2 Binary	Radix 16 Hexadecimal	Radix 10 Decimal	Radix 8 Octal	Radix 2 Binary
			BIT 6765 4321				BIT 6765 4321
74	116	164	0111 0100	D9	217	331	1101 1001
75	117	165	0111 0101	DA	218	332	1101 1010
76	118	166	0111 0110	DB	219	333	1101 1011
77	119	167	0111 0111	DC	220	334	1101 1100
78	120	170	0111 1000	DD	221	335	1101 1101
79	121	171	0111 1001	DE	222	336	1101 1110
7A	122	172	0111 1010	DF	223	337	1101 1111
7B	123	173	0111 1011	E0	224	340	1110 0000
7C	124	174	0111 1100	E1	225	341	1110 0001
7D	125	175	0111 1101	E2	226	342	1110 0010
7E	126	176	0111 1110	E3	227	343	1110 0011
7F	127	177	0111 1111	E4	228	344	1110 0100
80	128	200	1000 0000	E5	229	345	1110 0101
81	129	201	1000 0001	E6	230	346	1110 0110
82	130	202	1000 0010	E7	231	347	1110 0111
83	131	203	1000 0011	E8	232	350	1110 1000
84	132	204	1000 0100	E9	233	351	1110 1001
85	133	205	1000 0101	EA	234	352	1110 1010
86	134	206	1000 0110	EB	235	353	1110 1011
87	135	207	1000 0111	EC	236	354	1110 1100
88	136	210	1000 1000	ED	237	355	1110 1101
89	137	211	1000 1001	EE	238	356	1110 1110
8A	138	212	1000 1010	EF	239	357	1110 1111
8B	139	213	1000 1011	F0	240	360	1111 0000
8C	140	214	1000 1100	F1	241	361	1111 0001
8D	141	215	1000 1101	F2	242	362	1111 0010
8E	142	216	1000 1110	F3	243	363	1111 0011
8F	143	217	1000 1111	F4	244	364	1111 0100
90	144	220	1001 0000	F5	245	365	1111 0101
91	145	221	1001 0001	F6	246	366	1111 0110
92	146	222	1001 0010	F7	247	367	1111 0111
93	147	223	1001 0011	F8	248	370	1111 1000
94	148	224	1001 0100	F9	249	371	1111 1001
95	149	225	1001 0101	FA	250	372	1111 1010
96	150	226	1001 0110	FB	251	373	1111 1011
97	151	227	1001 0111	FC	252	374	1111 1100
98	152	230	1001 1000	FD	253	375	1111 1101
99	153	231	1001 1001	FE	254	376	1111 1110
9A	154	232	1001 1010	FF	255	377	1111 1111
9B	155	233	1001 1011	AB	171	253	1010 1011
9C	156	234	1001 1100	AC	172	254	1010 1100
9D	157	235	1001 1101	AD	173	255	1010 1101
9E	158	236	1001 1110	AE	174	256	1010 1110
9F	159	237	1001 1111	AF	175	257	1010 1111
A0	160	240	1010 0000	B0	176	260	1011 0000
A1	161	241	1010 0001	B1	177	261	1011 0001
A2	162	242	1010 0010	B2	178	262	1011 0010
A3	163	243	1010 0011	B3	179	263	1011 0011
A4	164	244	1010 0100	B4	180	264	1011 0100
A5	165	245	1010 0101	B5	181	265	1011 0101
A6	166	246	1010 0110	B6	182	286	1011 0110
A7	167	247	1010 0111	B7	183	267	1011 0111
A8	168	250	1010 1000	B8	184	270	1011 1000
A9	169	251	1010 1001	B9	185	271	1011 1001
AA	170	252	1010 1010	BA	186	272	1011 1010
D6	214	326	1101 0110	BB	187	273	1011 1011
D7	215	327	1101 0111	BC	188	274	1011 1100
D8	216	330	1101 1000	BD	189	275	1011 1101

TABLE 1.7   (*Continued*)

Radix 16 Hexadecimal	Radix 10 Decimal	Radix 8 Octal	Radix 2 Binary BIT 8765 4321	Radix 16 Hexadecimal	Radix 10 Decimal	Radix 8 Octal	Radix 2 Binary BIT 8765 4321
BE	190	276	1011 1110	CA	202	312	1100 1010
BF	191	277	1011 1111	CB	203	313	1100 1011
C0	192	300	1100 0000	CC	204	314	1100 1100
C1	193	301	1100 0001	CD	205	315	1100 1101
C2	194	302	1100 0010	CE	206	316	1100 1110
C3	195	303	1100 0011	CF	207	317	1100 1111
C4	196	304	1100 0100	D0	208	320	1101 0000
C5	197	305	1100 0101	D1	209	321	1101 0001
C6	198	306	1100 0110	D2	210	322	1101 0010
C7	199	307	1100 0111	D3	211	323	1101 0011
C8	200	310	1100 1000	D4	212	324	1101 0100
C9	201	311	1100 1001	D5	213	325	1101 0101

TABLE 1.8   Computer Graphics Codes and Standards

Modern computer-aided design (CAD), computer-aided manufacturing (CAM), and computer-aided engineering (CAE) are heavily dependent on computer graphics. A standard computer graphics metafile (CGM) is necessary in order to:

1. Allow picture information to be stored in an organized way on a graphical software system.
2. Facilitate transfer of picture information between different graphical software systems.
3. Enable picture information to be transferred between graphical devices.
4. Enable picture information to be transferred between different computer graphics installations.

More particularly, the CGM should provide these capabilities ion a device-independent manner. To accomplish this, the Standard defines the form (syntax) and functional behavior (semantics) of a set of elements that may occur in the CGM. There are eight classes of elements.

1. Delimiter Elements—delimit significant structures within the metafile.
2. Metafile Descriptor Elements—describe the functional content, default conditions, identification, and characteristics of the CGM.
3. Picture Descriptor Elements—set the interpretation modes of attribute elements for each picture.
4. Control Elements—allow picture boundaries and coordinate representation to be modified.
5. Graphical Primitive Elements—describe the visual components of a picture in the CGM.
6. Attribute Elements—describe the appearance of graphical primitive elements.
7. Escape Elements—describe device- or system-dependent elements used to construct a picture; however, the elements are not otherwise standardized.
8. External Elements—communicate information not directly related to the generation of a graphical image.

A computer graphics metafile is a collection of elements from this standardized set. The BEGIN METAFILE and END METAFILE elements each occur exactly once in a complete metafile; as many or as few of the elements in the other classes may occur as are needed. A metafile needs to be interpreted in order to display its pictorial content on a graphics device. The descriptor elements give the interpreter sufficient data to interpret metafile elements and to make informed decisions concerning the resources needed for display.

A CGM contains delimiter elements; in addition it may include control elements for metafile interpretation, picture descriptor elements for declaring parameter modes of attribute elements, graphical primitive elements for defining graphical entities, attribute elements for defining the appearance of the graphical primitive elements, escape elements for accessing nonstandardized features of particular devices, and external elements for communication of information external to the definition of the pictures in the CGM.

Full description and depiction of all the elements thus far defined in this standardized set is beyond the scope of this handbook. The interested reader is referred to ANSI Standard X3.122-1986 and Smith, B. M. et al. "Initial Graphics Exchange Specification (IGES), Version 2.0," NBS (R82-2631) (AF) Feb. (1983) 26 pp.

1.2 MATHEMATICAL TABLES

TABLE 1.9 Certain Constants Containing e and π [a]

Powers of e			Multiples of π			Fractions of π		
e^n	Value	Logarithm	n_π	Value	Logarithm	π/n	Value	Logarithm
e	2.718282	0.434294	π	3.141593	0.497150	$\pi/2$	1.570780	0.196120
e^{-1}	0.367879	$\bar{1}.565706$	2π	6.283185	0.798180	$\pi/3$	1.047198	0.020029
e^2	7.389057	0.868589	3π	9.424778	0.974271	$\pi/4$	0.785398	$\bar{1}.895090$
e^{-2}	0.135335	$\bar{1}.131411$	4π	12.566371	1.099210	$\pi/180$	0.017453[b]	$\bar{2}.241877$
$e^{1/2}$	1.648721	0.217147	5π	15.707963	1.196120			

Reciprocals of π			Powers of π			Roots of π		
n/π	Value	Logarithm	$\pi^{\pm n}$	Value	Logarithm	$\pi^{\pm 1/n}$	Value	Logarithm
$1/\pi$	0.318310	$\bar{1}.502850$	π^2	9.869604	0.994300	$\sqrt{\pi}$	1.772454	0.248575
$2/\pi$	0.636620	$\bar{1}.803880$	$1/\pi^2$	0.101321	$\bar{1}.005700$	$1/\sqrt{\pi}$	0.564190	$\bar{1}.751425$
$3/\pi$	0.954930	$\bar{1}.979971$	π^3	31.006277	1.491450	$\sqrt[3]{\pi}$	1.464592	0.165717
$180/\pi$	57.295780[c]	1.758123	$1/\pi^3$	0.032252	$\bar{2}.508550$	$1/\sqrt[3]{\pi}$	0.682784	$\bar{1}.834283$

[a] $e = 2.7182818285$; $\pi = 3.1415926536$; $M = \log_{10}e = 0.4342944819$; $M^{-1} = \log_e 10 = 2.3025850930$.
[b] Number of radians per degree.
[c] Number of degrees per radian.

TABLE 1.10 Factorials

n	$n! = 1 \cdot 2 \cdot 3 \ldots n$	$1/n!$	n	$n! = 1 \cdot 2 \cdot 3 \ldots n$	$1/n!$
1	1	1.	11	$399{,}168 \times 10^2$	0.250521×10^{-7}
2	2	0.5	12	$479{,}002 \times 10^3$	0.208768×10^{-8}
3	6	0.166667	13	$622{,}702 \times 10^4$	0.160590×10^{-9}
4	24	0.416667×10^{-1}	14	$871{,}783 \times 10^5$	0.114707×10^{-10}
5	120	0.833333×10^{-2}	15	$130{,}767 \times 10^7$	0.764716×10^{-12}
6	720	0.138889×10^{-2}	16	$209{,}228 \times 10^8$	0.477948×10^{-13}
7	5,040	0.198413×10^{-3}	17	$355{,}687 \times 10^9$	0.281146×10^{-14}
8	40,320	0.248016×10^{-4}	18	$640{,}237 \times 10^{10}$	0.156192×10^{-15}
9	362,880	0.275573×10^{-5}	19	$121{,}645 \times 10^{12}$	0.822064×10^{-17}
10	3,628,800	0.275573×10^{-6}	20	$243{,}290 \times 10^{13}$	0.411032×10^{-19}

TABLE 1.11 Common and Natural Logarithms of Numbers

The common logarithm of a number is the index of the power to which the base 10 must be raised in order to equal the number.

The common logarithm of every positive number not an integral power of 10 consists of an *integral* and a *decimal part*. The integral part or whole number is called the *characteristic* and may be either *positive* or *negative*. The decimal or fractional part is a *positive* number called the *mantissa* and is the same for all numbers which have the same sequential digits.

The characteristic of the logarithm of any positive number greater than one is positive and is one less than the number of digits before the decimal point.

The characteristic of the logarithm of any positive number less than one is negative and is one more than the number of ciphers immediately after the decimal point.

A negative number or number less than zero has no real logarithm.

Examples: $\log_{10} 25400. = 4.404834$ $\log_{10} 0.0254 = \bar{2}.404834$ or $8.404834 - 10$

The two systems of logarithms in general use are the common or Briggsian logarithms, introduced in 1615 by Henry Briggs, a contemporary of John Napier, the inventor of logarithms, and the natural or less appropriately termed Napierian or hyperbolic logarithms, which developed somewhat accidentally from Napier's original work. The latter have a base denoted by e, an irrational number, which is:

$$\lim_{u=\infty}\left(1+\frac{1}{u}\right)^{u} = 1 + 1 + \frac{1}{2!} + \frac{1}{3!} + \frac{1}{4!} + \cdots = 2.7182818$$

To obtain the natural logarithm, the common logarithm is multiplied by $\log_e 10$, which is 2.302585, or $\log_e N = 2.302585 \log_{10} N$.

The natural logarithm of a number is the index of the power to which the base e ($= 2.7182818$) must be raised in order to equal the number.

Example: $\log_e 4.12 = \ln 4.12 = 1.4159$.

Natural logarithms of numbers from 1.00 to 9.99 may be obtained directly; the natural logarithms of numbers outside of that range by the addition or subtraction of the natural logarithms of powers of 10.

Examples: $\log_e 679. = \log_e 6.79 + \log_e 10^2 = 1.9155 + 4.6052 = 6.5207$.
$\log_e 0.0679 = \log_e 6.79 - \log_e 10^2 = 1.9155 - 4.6052 = -2.6897$.

Natural Logarithms of Powers of 10

$\log_e 10 = 2.302\ 585$	$\log_e 10^4 = 9.210\ 340$	$\log_e 10^7 = 16.118\ 096$
$\log_e 10^2 = 4.605\ 170$	$\log_e 10^5 = 11.512\ 925$	$\log_e 10^8 = 18.420\ 681$
$\log_e 10^3 = 6.907\ 755$	$\log_e 10^6 = 13.815\ 511$	$\log_e 10^9 = 20.723\ 266$

To obtain the common logarithm, the natural logarithm is multiplied by $\log_{10} e$, which is 0.434 294, or $\log_{10} N = 0.434\ 294 \log_e N$.

A negative number or number less than zero has no real logarithm.

Tabulations of common and natural logarithms are no longer provided in this handbook because of ready access to them on modern pocket and desk calculators.

Values and Logarithms of Exponentials and Hyperbolic Functions

Many calculators directly give values of e^x, e^{-x}, $\sinh x$, $\cosh x$, and $\tanh x$ for any value of x. These quantities are therefore not tabulated here.

For values of x greater than 6, e^x may be computed from the relationship $e^x = \log^{-1}(x \log_{10} e)$ $= \log^{-1} 0.43429x$; e^{-x} approaches zero; $\sinh x$ and $\cosh x$ are approximately equal and become $0.5e^x$; and $\tanh x$ and $\coth x$ have values approximately equal to unity.

Where more accurate values of the exponentials and functions are required they may be computed from the following relationships.

$$e = 2.7182818285 \qquad \frac{1}{e} = 0.3678794412$$

$$M = \log_{10} e = 0.4342944819 \qquad \frac{1}{M} = \log_e 10 = 2.3025850930$$

$$e^x = \log^{-1} Mx \qquad e^{-x} = \log^{-1} - Mx$$

$$\sinh x = \frac{e^x - e^{-x}}{2} \qquad \cosh x = \frac{e^x + e^{-x}}{2} \qquad \tanh x = \frac{e^x - e^{-x}}{e^x + e^{-x}}$$

$$\operatorname{csch} x = \frac{1}{\sinh x} \qquad \operatorname{sech} x = \frac{1}{\cosh x} \qquad \coth x = \frac{1}{\tanh x}$$

Probability

Let p = probability of an event e in one trial, and q = probability of failure of e. The probability that, in n trials, the event e will occur exactly $(n - t)$ times is $\binom{n}{t} p^{n-t} q^t$. The probability that an

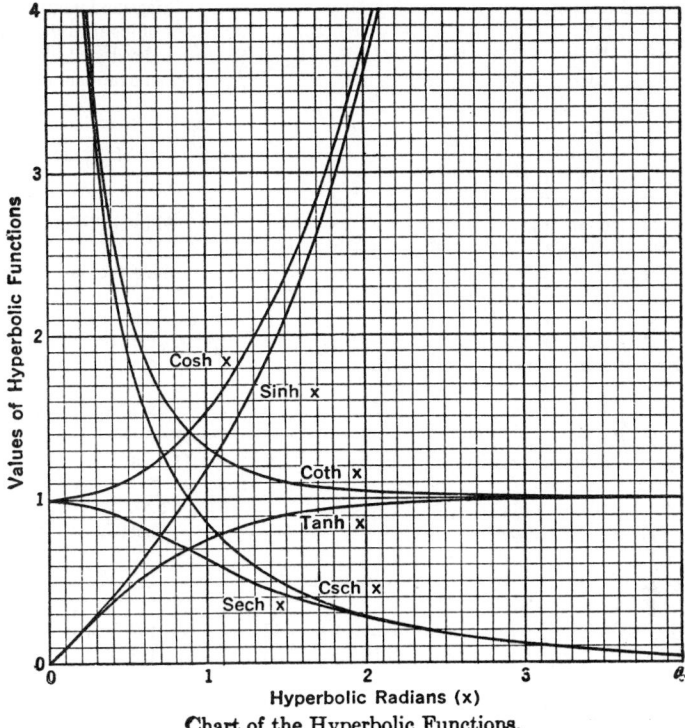

Chart of the Hyperbolic Functions.

Values of the hyperbolic functions are shown in the figure.

TABLE 1.12   Circular Arcs, Chords, and Segments

Central Angle in Degrees	$\dfrac{\text{Arc}}{R}$	$\dfrac{\text{Height}}{R}$	$\dfrac{\text{Chord}}{R}$	$\dfrac{\text{Height}}{\text{Chord}}$	$\dfrac{\text{Area}}{R^2}$	Central Angle in Degrees	$\dfrac{\text{Arc}}{R}$	$\dfrac{\text{Height}}{R}$	$\dfrac{\text{Chord}}{R}$	$\dfrac{\text{Height}}{\text{Chord}}$	$\dfrac{\text{Area}}{R^2}$
1	0.0175	0.0000	0.0175	0.0022	0.00000	21	0.3665	0.0167	0.3645	0.0459	0.00408
2	0.0349	0.0002	0.0349	0.0044	0.00000	22	0.3840	0.0184	0.3816	0.0481	0.00468
3	0.0524	0.0003	0.0524	0.0066	0.00001	23	0.4014	0.0201	0.3987	0.0503	0.00535
4	0.0698	0.0006	0.0698	0.0087	0.00003	24	0.4189	0.0219	0.4158	0.0526	0.00607
5	0.0873	0.0010	0.0872	0.0109	0.00006	25	0.4363	0.0237	0.4329	0.0548	0.00686
6	0.1047	0.0014	0.1047	0.0131	0.00010	26	0.4538	0.0256	0.4499	0.0570	0.00771
7	0.1222	0.0019	0.1221	0.0153	0.00015	27	0.4712	0.0276	0.4669	0.0592	0.00862
8	0.1396	0.0024	0.1395	0.0175	0.00023	28	0.4887	0.0297	0.4838	0.0614	0.00961
9	0.1571	0.0031	0.1569	0.0196	0.00032	29	0.5061	0.0319	0.5008	0.0636	0.01067
10	**0.1745**	**0.0038**	**0.1743**	**0.0218**	**0.00044**	**30**	**0.5236**	**0.0341**	**0.5176**	**0.0658**	**0.1180**
11	0.1920	0.0046	0.1917	0.0240	0.00059	31	0.5411	0.0364	0.5345	0.0680	0.01301
12	0.2094	0.0055	0.2091	0.0262	0.00076	32	0.5585	0.0387	0.5513	0.0703	0.01429
13	0.2296	0.0064	0.2264	0.0284	0.00097	33	0.5760	0.0412	0.5680	0.0725	0.01566
14	0.2443	0.0075	0.2437	0.0306	0.00121	34	0.5934	0.0437	0.5847	0.0747	0.01711
15	0.2618	0.0086	0.2611	0.0328	0.00149	35	0.6109	0.0463	0.6014	0.0770	0.01864
16	0.2793	0.0097	0.2783	0.0350	0.00181	36	0.6283	0.0489	0.6180	0.0792	0.02027
17	0.2967	0.0110	0.2956	0.0372	0.00217	37	0.6458	0.0517	0.6346	0.0814	0.02198
18	0.3142	0.0123	0.3129	0.0394	0.00257	38	0.6632	0.0545	0.6511	0.0837	00.2378
19	0.3316	0.0137	0.3301	0.0415	0.00302	39	0.6807	0.0574	0.6676	0.0859	0.02568
20	**0.3491**	**0.0152**	**0.3473**	**0.0437**	**0.00352**	**40**	**0.6981**	**0.0603**	**0.6840**	**0.0882**	**0.02767**

TABLE 1.12   (*Continued*)

Central Angle in Degrees	Arc / R	Height / R	Chord / R	Height / Chord	Area / R²	Central Angle in Degrees	Arc / R	Height / R	Chord / R	Height / Chord	Area / R²
41	0.7156	0.0633	0.7004	0.0904	0.02976	101	1.7628	0.3639	1.543	0.2358	0.39058
42	0.7330	0.0664	0.7167	0.0927	0.03195	102	1.7802	0.3707	1.554	0.2385	0.40104
43	0.7505	0.0696	0.7330	0.0949	0.03425	103	1.7977	0.3775	1.565	0.2412	0.41166
44	0.7679	0.0728	0.7492	0.0972	0.03664	104	1.8151	0.3843	1.576	0.2439	0.42242
45	0.7854	0.0761	0.7654	0.0995	0.03915	105	1.8326	0.3912	1.587	0.2466	0.43333
46	0.8029	0.0795	0.7815	0.1017	0.04176	106	1.8500	0.3982	1.597	0.2493	0.44439
47	0.8203	0.0829	0.7975	0.1040	0.04448	107	1.8675	0.4052	1.608	0.2520	0.45560
48	0.8378	0.0865	0.8135	0.1063	0.04731	108	1.8850	0.4122	1.618	0.2548	0.46695
49	0.8552	0.0900	0.8294	0.1086	0.05025	109	1.9024	0.4193	1.628	0.2575	0.47844
50	**0.8727**	**0.0937**	**0.8452**	**0.1108**	**0.05331**	**110**	**1.9199**	**0.4264**	**1.638**	**0.2603**	**0.49008**
51	0.8901	0.0974	0.8610	0.1131	0.05649	111	1.9373	0.4336	1.648	0.2631	0.50187
52	0.9076	0.1012	0.8767	0.1154	0.05978	112	1.9548	0.4408	1.658	0.2659	0.51379
53	0.9250	0.1051	0.8924	0.1177	0.06319	113	1.9722	0.4481	1.668	0.2687	0.52586
54	0.9425	0.1090	0.9080	0.1200	0.06673	114	1.9897	0.4554	1.677	0.2715	0.53807
55	0.9599	0.1130	0.9235	0.1223	0.07039	115	2.0071	0.4627	1.687	0.2743	0.55041
56	0.9774	0.1171	0.9389	0.1247	0.07417	116	2.0246	0.4701	1.696	0.2772	0.56289
57	0.9948	0.1212	0.9543	0.1270	0.07808	117	2.0420	0.4775	1.705	0.2800	0.57551
58	1.0123	0.1254	0.9696	0.1293	0.08212	118	2.0595	0.4850	1.714	0.2829	0.58827
59	1.0297	0.1296	0.9848	0.1316	0.08629	119	2.0769	0.4925	1.723	0.2858	0.60116
60	**1.0472**	**0.1340**	**1.0000**	**0.1340**	**0.09059**	**120**	**2.0944**	**0.5000**	**1.732**	**0.2887**	**0.61418**
61	1.0647	0.1384	1.015	0.1363	0.09502	121	2.1118	0.5076	1.741	0.2916	0.62734
62	1.0821	0.1428	1.030	0.1387	0.09958	122	2.1293	0.5152	1.749	0.2945	0.64063
63	1.0996	0.1474	1.045	0.1410	0.10428	123	2.1468	0.5228	1.758	0.2975	0.65404
64	1.1170	0.1520	1.060	0.1434	0.10911	124	2.1642	0.5305	1.766	0.3004	0.66759
65	1.1345	0.1566	1.075	0.1457	0.11408	125	2.1817	0.5383	1.774	0.3034	0.68125
66	1.1519	0.1613	1.089	0.1481	0.11919	126	2.1991	0.5460	1.782	0.3064	0.69505
67	1.1694	0.1661	1.104	0.1505	0.12443	127	2.2166	0.5538	1.790	0.3094	0.70897
68	1.1868	0.1710	1.118	0.1529	0.12982	128	2.2340	0.5616	1.798	0.3124	0.72301
69	1.2043	0.1759	1.133	0.1553	0.13535	129	2.2515	0.5695	1.805	0.3155	0.73716
70	**1.2217**	**0.1808**	**1.147**	**0.1576**	**0.14102**	**130**	**2.2689**	**0.5774**	**1.813**	**0.3185**	**0.75143**
71	1.2392	0.1859	1.161	0.1601	0.14683	131	2.2864	0.5853	1.820	0.3216	0.76584
72	1.2566	0.1910	1.176	0.1625	0.15279	132	2.3038	0.5933	1.827	0.3247	0.78034
73	1.2741	0.1961	1.190	0.1649	0.15889	133	2.3213	0.6013	1.834	0.3278	0.79497
74	1.2915	0.2014	1.204	0.1673	0.16514	134	2.3387	0.6093	1.841	0.3309	0.80970
75	1.3090	0.2066	1.218	0.1697	0.17154	135	2.3562	0.6173	1.848	0.3341	0.82454
76	1.3265	0.2120	1.231	0.1722	0.17808	136	2.3736	0.6254	1.854	0.3373	0.83949
77	1.3439	0.2174	1.245	0.1746	0.18477	137	2.3911	0.6335	1.861	0.3404	0.85455
78	1.3614	0.2229	1.259	0.1771	0.19160	138	2.4086	0.6416	1.867	0.3436	0.86971
79	1.3788	0.2284	1.272	0.1795	0.19859	139	2.4260	0.6498	1.873	0.3469	0.88497
80	**1.3963**	**0.2340**	**1.286**	**0.1820**	**0.20573**	**140**	**2.4435**	**0.6580**	**1.879**	**0.3501**	**0.90034**
81	1.4137	0.2396	1.299	0.1845	0.21301	141	2.4609	0.6662	1.885	0.3534	0.91580
82	1.4312	0.2453	1.312	0.1869	0.22045	142	2.4784	0.6744	1.891	0.3566	0.93135
83	1.4486	0.2510	1.325	0.1894	0.22804	143	2.4958	0.6827	1.897	0.3599	0.94700
84	1.4661	0.2569	1.338	0.1919	0.23578	144	2.5133	0.6910	1.902	0.3633	0.96274
85	1.4835	0.2627	1.351	0.1944	0.24367	145	2.5307	0.6993	1.907	0.3666	0.97858
86	1.5010	0.2686	1.364	0.1970	0.25171	146	2.5482	0.7076	1.913	0.3700	0.99449
87	1.5184	0.2746	1.377	0.1995	0.25990	147	2.5656	0.7160	1.918	0.3734	1.0105
88	1.5359	0.2807	1.389	0.2020	0.26825	148	2.5831	0.7244	1.923	0.3768	1.0266
89	1.5533	0.2867	1.402	0.2046	0.27675	149	2.6005	0.7328	1.927	0.3802	1.0428
90	**1.5708**	**0.2929**	**1.414**	**0.2071**	**0.28540**	**150**	**2.6180**	**0.7412**	**1.932**	**0.3837**	**1.0590**
91	1.5882	0.2991	1.427	0.2097	0.29420	151	2.6354	0.7496	1.936	0.3871	1.0753
92	1.6057	0.3053	1.439	0.2122	0.30316	152	2.6529	0.7581	1.941	0.3906	1.0917
93	1.6232	0.3116	1.451	0.2148	0.31226	153	2.6704	0.7666	1.945	0.3942	1.1082
94	1.6406	0.3180	1.463	0.2174	0.32152	154	2.6878	0.7750	1.949	0.3977	1.1247
95	1.6581	0.3244	1.475	0.2200	0.33093	155	2.7053	0.7836	1.953	0.4013	1.1413
96	1.6755	0.3309	1.486	0.2226	0.34050	156	2.7227	0.7921	1.956	0.4049	1.1580
97	1.6930	0.3374	1.498	0.2252	0.35021	157	2.7402	0.8006	1.960	0.4085	1.1747
98	1.7104	0.3439	1.509	0.2279	0.36008	158	2.7576	0.8092	1.963	0.4122	1.1915
99	1.7279	0.3506	1.521	0.2305	0.37009	159	2.7751	0.8178	1.967	0.4158	1.2084
100	**1.7453**	**0.3572**	**1.532**	**0.2332**	**0.38026**	**160**	**2.7925**	**0.8264**	**1.970**	**0.4195**	**1.2253**

TABLE 1.12   (*Continued*)

Central Angle in Degrees	Arc $\dfrac{}{R}$	Height $\dfrac{}{R}$	Chord $\dfrac{}{R}$	Height $\dfrac{}{Chord}$	Area $\dfrac{}{R^2}$	Central Angle in Degrees	Arc $\dfrac{}{R}$	Height $\dfrac{}{R}$	Chord $\dfrac{}{R}$	Height $\dfrac{}{Chord}$	Area $\dfrac{}{R^2}$
161	2.8100	0.8350	1.973	0.4233	1.2422	171	2.9845	0.9215	1.994	0.4622	1.4140
162	2.8274	0.8436	1.975	0.4270	1.2592	172	3.0020	0.9302	1.995	0.4663	1.4314
163	2.8449	0.8522	1.978	0.4308	1.2763	173	3.0194	0.9390	1.996	0.4704	1.4488
164	2.8623	0.8608	1.981	0.4346	1.2934	174	3.0369	0.9477	1.997	0.4745	1.4662
165	2.8798	0.8695	1.983	0.4385	1.3105	175	3.0543	0.9564	1.998	0.4786	1.4836
166	2.8972	0.8781	1.985	0.4424	1.3277	176	3.0718	0.9651	1.999	0.4828	1.5010
167	2.9147	0.8868	1.987	0.4463	1.3449	177	3.0892	0.9738	1.999	0.4871	1.5184
168	2.9322	0.8955	1.989	0.4502	1.3621	178	3.1067	0.9825	2.000	0.4914	1.5359
169	2.9496	0.9042	1.991	0.4542	1.3794	179	3.1241	0.9913	2.000	0.4957	1.5533
170	**2.9671**	**0.9128**	**1.992**	**0.4582**	**1.3967**	**180**	**3.1416**	**1.0000**	**2.000**	**0.5000**	**1.5708**

TABLE 1.13   Values of Degrees, Minutes, and Seconds in Radians[a]

Degrees	Radians Arc Length $R = 1$	Degrees	Radians Arc Length $R = 1$	Degrees	Radians Arc Length $R = 1$	Radians Arc Length $R = 1$ Minutes	Seconds
0		**40**	**0.69813170**	**80**	**1.39626340**	**0**	
1	0.01745329	41	0.71558499	81	1.41371669	1 0.00029089	0.00000485
2	0.03490659	42	0.73303829	82	1.43116999	2 0.00058178	0.00000970
3	0.05235988	43	0.75049158	83	1.44862328	3 0.00087266	0.00001454
4	0.06981317	44	0.76794487	84	1.46607657	4 0.00116355	0.00001939
5	0.08726646	45	0.78539816	85	1.48352986	5 0.00145444	0.00002424
6	0.10471976	46	0.80285146	86	1.50098316	6 0.00174533	0.00002909
7	0.12217305	47	0.82030475	87	1.51843645	7 0.00203622	0.00003394
8	0.13962634	48	0.83775804	88	1.53588974	8 0.00232711	0.00003879
9	0.15707963	49	0.85521133	89	1.55334303	9 0.00261799	0.00004363
10	**0.17453293**	**50**	**0.87266463**	**90**	**1.57079633**	**10 0.00290888**	**0.00004848**
11	0.19198622	51	0.89011792	91	1.58824962	11 0.00319977	0.00005333
12	0.20943951	52	0.90757121	92	1.60570291	12 0.00349066	0.00005818
13	0.22689280	53	0.92502450	93	1.62315620	13 0.00378155	0.00006303
14	0.24434610	54	0.94247780	94	1.64060950	14 0.00407243	0.000068787
15	0.26179939	55	0.95993109	95	1.65806279	15 0.00436332	0.00007272
16	0.27925268	56	0.97738438	96	1.67551608	16 0.00465421	0.00007757
17	0.29670597	57	0.99483767	97	1.69296937	17 0.00494510	0.00008242
18	0.31415927	58	1.01229097	98	1.71042267	18 0.00523599	0.00008727
19	0.33161256	59	1.02974426	99	1.72787596	19 0.00552688	0.00009211
20	**0.34906585**	**60**	**1.04719755**	**100**	**1.74532925**	**20 0.00581776**	**0.00009696**
21	0.36651914	61	1.06465084	101	1.76278254	21 0.00610865	0.00010181
22	0.38397244	62	1.08210414	102	1.78023584	22 0.00639954	0.00010666
23	0.40142573	63	1.09955743	103	1.79768913	23 0.00669043	0.00011151
24	0.41887902	64	1.11701072	104	1.81514242	24 0.00698132	0.00011636
25	0.43633231	65	1.13446401	105	1.83259571	25 0.00727221	0.00012120
26	0.45378561	66	1.15191731	106	1.85004901	26 0.00756309	0.00012605
27	0.47123890	67	1.16937060	107	1.86750230	27 0.00785398	0.00013090
28	0.48869219	68	1.18682389	108	1.88495559	28 0.00814487	0.00013575
29	0.50614548	69	1.20427718	109	1.90240888	29 0.00843576	0.00014060
30	**0.52359878**	**70**	**1.22173048**	**110**	**1.91986218**	**30 0.00872665**	**0.00014544**
31	0.54105207	71	1.23918377	111	1.93731547	31 0.00901753	0.00015029
32	0.55850536	72	1.25663706	112	1.95476876	32 0.00930842	0.00015514
33	0.57595865	73	1.27409035	113	1.97222205	33 0.00959931	0.00015999
34	0.59341195	74	1.29154365	114	1.98967535	34 0.00989020	0.00016484
35	0.61086524	75	1.30899694	115	2.00712864	35 0.01018109	0.00016968
36	0.62831853	76	1.32645023	116	2.02458193	36 0.01047198	0.00017453
37	0.64577182	77	1.34390352	117	2.04203522	37 0.01076286	0.00017938
38	0.66322512	78	1.36135682	118	2.05948852	38 0.01105375	0.00018423
39	0.68067841	79	1.37881011	119	2.07694181	39 0.01134464	0.00018908

TABLE 1.13 (*Continued*)

Degrees	Radians Arc Length R = 1	Degrees	Radians Arc Length R = 1	Degrees	Radians Arc Length R = 1	Radians Arc Length R = 1 Minutes	Seconds
120	2.09439510	140	2.44346095	160	2.79252680	40 0.01163553	0.00019393
121	2.11184840	141	2.46091424	161	2.80998009	41 0.01192642	0.00019877
122	2.12930169	142	2.47836754	162	2.82743338	42 0.01221730	0.00020362
123	2.14675498	143	2.49582083	163	2.84488668	43 0.01250819	0.00020847
124	2.16420828	144	2.51327413	164	2.86233997	44 0.01279908	0.00021332
125	2.18166157	145	2.53072742	165	2.87979327	45 0.01308997	0.00021817
126	2.19911486	146	2.54818071	166	2.89724655	46 0.01338086	0.00022301
127	2.21656815	147	2.56563401	167	2.91469985	47 0.01367175	0.00022786
128	2.23402145	148	2.58308729	168	2.93215314	48 0.01396263	0.00023271
129	2.25147474	149	2.60054058	169	2.94960643	49 0.01425352	0.00023756
130	2.26892803	150	2.61799388	170	2.96705972	50 0.01454441	0.00024241
131	2.28638133	151	2.63544717	171	2.98451302	51 0.01483530	0.00024725
132	2.30383462	152	2.65290046	172	3.00196631	52 0.01512619	0.00025210
133	2.32128791	153	2.67035375	173	3.01941961	53 0.01541707	0.00025695
134	2.33874121	154	2.68780705	174	3.03687289	54 0.01570796	0.00026180
135	2.35619450	155	2.70526034	175	3.05432619	55 0.01599885	0.00026665
136	2.37364780	156	2.72271363	176	3.07177948	56 0.01628974	0.00027150
137	2.39110107	157	2.74016693	177	3.08923277	57 0.01658063	0.00027634
138	2.40855436	158	2.75762022	178	3.10668607	58 0.01687152	0.00028119
139	2.42600766	159	2.77507351	179	3.12413962	59 0.01716240	0.00028604
				180	3.14159265		

a Lengths of circular arcs, radius unity, for example:

$$\Theta = 30° \, 20' \, 10''$$
$$30° = 0.52359878$$
$$20' = 0.00581776$$
$$10'' = 0.00004848$$
$$\text{Arc length} = \overline{0.52946502} \quad \text{radians}$$

TABLE 1.14   Values of Radians in Degrees

Radian	0.00	0.01	0.02	0.03	0.04	0.05	0.06	0.07	0.08	0.09
	Deg	Deg	Deg	Deg	Deg	Deg	Deg	Deg	Deg	Deg
0.0	0.0000	0.5730	1.1459	1.7189	2.2918	2.8648	3.4377	4.0107	4.5837	5.1566
0.1	5.7296	6.3025	6.8755	7.4485	8.0214	8.5944	9.1673	9.7403	10.3132	10.8862
0.2	11.4591	12.0321	12.6051	13.1780	13.7510	14.3239	14.8969	15.4699	16.0428	16.6158
0.3	17.1887	17.7617	18.3346	18.9076	19.4806	20.0535	20.6265	21.1994	21.7724	22.3454
0.4	22.9183	23.4913	24.0642	24.6372	25.2101	25.7831	26.3561	26.9290	27.5020	28.0749
0.5	28.6479	29.2208	29.7938	30.3668	30.9397	31.1527	32.0856	32.6586	33.2316	33.8045
0.6	34.3775	34.9504	35.5234	36.0963	36.6693	37.2423	37.8152	38.3882	38.9611	39.5341
0.7	40.1070	40.6800	41.2530	41.8259	42.3989	42.9718	43.5448	44.1178	44.6907	45.2637
0.8	45.8366	46.4096	46.9825	47.5555	48.1285	48.7014	49.2744	49.8473	50.4203	50.9932
0.9	51.5662	52.1392	52.7121	53.2851	53.8580	54.4310	55.0039	55.5769	56.1499	56.7228

1 rad. = 57.29578°   2 rad. = 114.59156°   3 rad. = 171.88734°

TABLE 1.15   Decimals of a Degree in Minutes and Seconds

Decimal	0.00		0.01		0.02		0.03		0.04		0.05		0.06		0.07		0.08		0.09	
	Min	Sec	Min	Sec	Min	Sec	Min	Sec	Min	Sec	Min	Sec	Min	Sec	Min	Sec	Min	Sec	Min	Sec
0.0	0	0	0	36	1	12	1	48	2	24	3	0	3	36	4	12	4	48	5	24
0.1	6	0	6	36	7	12	7	48	8	24	9	0	9	36	10	12	10	48	11	24
0.2	12	0	12	36	13	12	13	48	14	24	15	0	15	36	16	12	16	48	17	24
0.3	18	0	18	36	19	12	19	48	20	24	21	0	21	36	22	12	22	48	23	24
0.4	24	0	24	36	25	12	25	48	26	24	27	0	27	36	28	12	28	48	29	24

TABLE 1.15   (*Continued*)

Decimal	0.00		0.01		0.02		0.03		0.04		0.05		0.06		0.07		0.08		0.09	
	Min	Sec	Min	Sec	Min	Sec	Min	Sec	Min	Sec	Min	Sec	Min	Sec	Min	Sec	Min	Sec	Min	Sec
0.5	30	0	30	36	31	12	31	48	32	24	33	0	33	36	34	12	34	48	35	24
0.6	36	0	36	36	37	12	37	48	38	24	39	0	39	36	40	12	40	48	41	24
0.7	42	0	42	36	43	12	43	48	44	24	45	0	45	36	46	12	46	48	47	24
0.8	48	0	48	36	49	12	49	48	50	24	51	0	51	36	52	12	52	48	53	24
0.9	54	0	54	36	55	12	55	48	56	24	57	0	57	36	58	12	58	48	59	24

TABLE 1.16   **Minutes in Decimals of a Degree**

Minutes	0	1	2	3	4	5	6	7	8	9
	Deg	Deg	Deg	Deg	Deg	Deg	Deg	Deg	Deg	Deg
0	0.00000	0.01667	0.03333	0.05000	0.06667	0.08333	0.10000	0.11667	0.13333	0.15000
10	0.16667	0.18333	0.20000	0.21667	0.23333	0.25000	0.26667	0.28333	0.30000	0.31667
20	0.33333	0.35000	0.36667	0.38333	0.40000	0.41667	0.43333	0.45000	0.46667	0.48333
30	0.50000	0.51667	0.53333	0.55000	0.56667	0.58333	0.60000	0.61667	0.63333	0.65000
40	0.66667	0.68333	0.70000	0.71667	0.73333	0.75000	0.76667	0.78333	0.80000	0.81667
50	0.83333	0.85000	0.86667	0.88333	0.90000	0.91667	0.93333	0.95000	0.96667	0.98333

TABLE 1.17   **Seconds in Decimals of a Degree**

Seconds	0	1	2	3	4
	Degrees	Degrees	Degrees	Degrees	Degrees
0	0	0.0002778	0.0005555	0.0008333	0.0011111
10	0.0027778	0.0030555	0.0033333	0.0036111	0.0038888
20	0.0055555	0.0058333	0.0061111	0.0063888	0.0066667
30	0.0083333	0.0086111	0.0088888	0.0091667	0.0094444
40	0.0111111	0.0113888	0.0116667	0.0119444	0.0122222
50	0.0138888	0.0141667	0.0144444	0.0147222	0.0150000

Seconds	5	6	7	8	9
	Degrees	Degrees	Degrees	Degrees	Degrees
0	0.0013888	0.0016667	0.0019444	0.0022222	0.0024999
10	0.0041667	0.0044444	0.0047222	0.0050000	0.0052778
20	0.0069444	0.0072222	0.0075000	0.0077778	0.0080555
30	0.0097222	0.0100000	0.0102778	0.0105555	0.0108333
40	0.0125000	0.0127778	0.0130555	0.0133333	0.0136111
50	0.0152778	0.0155555	0.0158333	0.0161111	0.0163888

TABLE 1.18   **Table of Integrals**

Elementary Indefinite Integrals

1. $\int a\,dx = ax$

2. $\int (u + v + w + \cdots)\,dx = \int u\,dx + \int v\,dx + \int w\,dx + \cdots$

3. $\int u\,dv = uv - \int v\,du$, integration by parts

4. $\int f(x)\,dx = \int f[\phi(y)]\phi'(y)\,dy, \ x = \phi(y)$, change of variable

5. $\int x^n\,dx = \dfrac{x^{n+1}}{n+1}, \quad (n \neq -1)$

6. $\int \dfrac{dx}{x} = \log_e x + c = \log_e c_1 x, \ [\log_e x = \log_e(-x) + (2k+1)\pi i]$

TABLE 1.18   (*Continued*)

Elementary Indefinite Integrals (*Continued*)

7. $\int e^{ax}\, dx = \dfrac{1}{a} e^{ax}$

8. $\int a^x\, dx = \dfrac{a^x}{\log_e a}$

9. $\int a^x \log_e a\, dx = a^x$

10. $\int \sin ax\, dx = -\dfrac{1}{a}\cos ax$

11. $\int \cos ax\, dx = \dfrac{1}{a}\sin ax$

12. $\int \tan ax\, dx = -\dfrac{1}{a}\log_e \cos ax = \dfrac{1}{a}\log_e \sec ax$

13. $\int \cot ax\, dx = \dfrac{1}{a}\log_e \sin ax = -\dfrac{1}{a}\log_e \csc ax$

14. $\int \sec ax\, dx = \dfrac{1}{a}\log_e(\sec ax + \tan ax) = \dfrac{1}{a}\log_e \tan\left(\dfrac{ax}{2} + \dfrac{\pi}{4}\right)$

15. $\int \csc ax\, dx = \dfrac{1}{a}\log_e(\csc ax - \cot ax) = \dfrac{1}{a}\log_e \tan\dfrac{ax}{2}$

16. $\int \dfrac{dx}{\sqrt{a^2 - x^2}} = \sin^{-1}\dfrac{x}{a} = -\cos^{-1}\dfrac{x}{a} \quad (x^2 < a^2)$

17. $\int \dfrac{dx}{a^2 + x^2} = \dfrac{1}{a}\tan^{-1}\dfrac{x}{a} = -\dfrac{1}{a}\cot^{-1}\dfrac{x}{a}$

18. $\int \sinh ax\, dx = \dfrac{1}{a}\cosh ax$

19. $\int \cosh ax\, dx = \dfrac{1}{a}\sinh ax$

20. $\int \tanh ax\, dx = \dfrac{1}{a}\log_e(\cosh ax)$

21. $\int \coth ax\, dx = \dfrac{1}{a}\log_e(\sinh ax)$

22. $\int \mathrm{sech}\, ax\, dx = \dfrac{1}{a}\sin^{-1}(\tanh ax) = \dfrac{1}{a}\tan^{-1}(\sinh ax)$

23. $\int \mathrm{csch}\, ax\, dx = \dfrac{1}{a}\log_e\left(\tanh\dfrac{ax}{2}\right)$

24. $\int \sin^2 ax\, dx = \dfrac{1}{2}x - \dfrac{1}{2a}\sin ax \cos ax = \dfrac{1}{2}x - \dfrac{1}{4a}\sin 2ax$

25. $\int \cos^2 ax\, dx = \dfrac{1}{2}x + \dfrac{1}{2a}\sin ax \cos ax = \dfrac{1}{2}x + \dfrac{1}{4a}\sin 2ax$

26. $\int \tan^2 ax\, dx = \dfrac{1}{a}\tan ax - x$

27. $\int \cot^2 ax\, dx = -\dfrac{1}{a}\cot ax - x$

28. $\int \sec^2 ax\, dx = \dfrac{1}{a}\tan ax$

29. $\int \csc^2 ax\, dx = -\dfrac{1}{a}\cot ax$

30. $\int \sin^{-1} ax\, dx = x \sin^{-1} ax + \dfrac{1}{a}\sqrt{1 - a^2 x^2}$

31. $\int \cos^{-1} ax\, dx = x \cos^{-1} ax - \dfrac{1}{a}\sqrt{1 - a^2 x^2}$

32. $\int \tan^{-1} ax\, dx = x \tan^{-1} ax - \dfrac{1}{2a}\log_e(1 + a^2 x^2)$

TABLE 1.18 (*Continued*)

Elementary Indefinite Integrals (*Continued*)

33. $\int \cot^{-1}ax \, dx = x \cot^{-1}ax + \dfrac{1}{2a}\log_e(1 + a^2x^2)$

34. $\int \sec^{-1}ax \, dx = x \sec^{-1}ax - \dfrac{1}{a}\log_e(ax + \sqrt{a^2x^2 - 1})$

35. $\int \csc^{-1}ax \, dx = x \csc^{-1}ax + \dfrac{1}{a}\log_e(ax + \sqrt{a^2x^2 - 1})$

Integrals Involving $(ax + b)$

36. $\int (ax + b)^n \, dx = \dfrac{1}{a(n+1)}(ax + b)^{n+1} \quad (n \neq -1)$

37. $\int \dfrac{dx}{ax + b} = \dfrac{1}{a}\log_e(ax + b)$

38. $\int x(ax + b)^n \, dx = \dfrac{1}{a^2(n+2)}(ax + b)^{n+2} - \dfrac{b}{a^2(n+1)}(ax + b)^{n+1} \quad (n \neq -1, -2)$

39. $\int \dfrac{x \, dx}{ax + b} = \dfrac{x}{a} - \dfrac{b}{a^2}\log_e(ax + b)$

40. $\int \dfrac{x \, dx}{(ax + b)^2} = \dfrac{b}{a^2(ax + b)} + \dfrac{1}{a^2}\log_e(ax + b)$

41. $\int \dfrac{x^2 \, dx}{ax + b} = \dfrac{1}{a^3}\left[\dfrac{1}{2}(ax + b)^2 - 2b(ax + b) + b^2\log_e(ax + b)\right]$

42. $\int \dfrac{x^2 \, dx}{(ax + b)^2} = \dfrac{1}{a^3}\left[(ax + b) - 2b\log_e(ax + b) - \dfrac{b^2}{ax + b}\right]$

43. $\int \dfrac{x^2 \, dx}{(ax + b)^3} = \dfrac{1}{a^3}\left[\log_e(ax + b) + \dfrac{2b}{ax + b} - \dfrac{b^2}{2(ax + b)^2}\right]$

44. $\int \dfrac{dx}{x(ax + b)} = \dfrac{1}{b}\log_e\dfrac{x}{ax + b}$

45. $\int \dfrac{dx}{x^2(ax + b)} = -\dfrac{1}{bx} + \dfrac{a}{b^2}\log_e\dfrac{ax + b}{x}$

46. $\int \dfrac{dx}{x(ax + b)^2} = \dfrac{1}{b(ax + b)} - \dfrac{1}{b^2}\log_e\dfrac{ax + b}{x}$

47. $\int \dfrac{dx}{x^2(ax + b)^2} = -\dfrac{b + 2ax}{b^2x(ax + b)} + \dfrac{2a}{b^3}\log_e\dfrac{ax + b}{x}$

48. $\int \dfrac{dx}{x\sqrt{ax + b}} = \dfrac{1}{\sqrt{b}}\log_e\dfrac{\sqrt{ax + b} - \sqrt{b}}{\sqrt{ax + b} + \sqrt{b}} \quad (b \text{ positive})$

49. $\int \dfrac{dx}{x\sqrt{ax + b}} = \dfrac{2}{\sqrt{-b}}\tan^{-1}\sqrt{\dfrac{ax + b}{-b}} \quad (b \text{ negative})$

50. $\int \dfrac{\sqrt{ax + b}}{x} \, dx = 2\sqrt{ax + b} + \sqrt{b}\log_e\dfrac{\sqrt{ax + b} - \sqrt{b}}{\sqrt{ax + b} + \sqrt{b}} \quad (b \text{ positive})$

51. $\int \dfrac{\sqrt{ax + b}}{x} \, dx = 2\sqrt{ax + b} - 2\sqrt{-b}\tan^{-1}\sqrt{\dfrac{ax + b}{-b}} \quad (b \text{ negative})$

52. $\int \dfrac{dx}{x^2\sqrt{ax + b}} = -\dfrac{\sqrt{ax + b}}{bx} - \dfrac{a}{2b\sqrt{b}}\log_e\dfrac{\sqrt{ax + b} - \sqrt{b}}{\sqrt{ax + b} + \sqrt{b}} \quad (b \text{ positive})$

53. $\int \dfrac{dx}{x^2\sqrt{ax + b}} = -\dfrac{\sqrt{ax + b}}{bx} - \dfrac{a}{b\sqrt{-b}}\tan^{-1}\sqrt{\dfrac{ax + b}{-b}} \quad (b \text{ negative})$

54. $\int \dfrac{ax + b}{fx + g} \, dx = \dfrac{ax}{f} + \dfrac{bf - cg}{f^2}\log_e(fx + g)$

TABLE 1.18   (*Continued*)

Integrals Involving ($ax + b$)   (*Continued*)

55. $\displaystyle\int \frac{dx}{(ax+b)(fx+g)} = \frac{1}{bf-ag}\log_e\!\left(\frac{fx+g}{ax+b}\right)$ ($ag \neq bf$)

56. $\displaystyle\int \frac{x\,dx}{(ax+b)(fx+g)} = \frac{1}{bf-ag}\left[\frac{b}{a}\log_e(ax+b) - \frac{g}{f}\log_e(fx+g)\right]$ ($ag \neq bf$)

57. $\displaystyle\int \frac{dx}{(ax+b)^2(fx+g)} = \frac{1}{bf-ag}\left(\frac{1}{ax+b} + \frac{f}{bf-ag}\log_e\frac{fx+g}{ax+b}\right)$ ($ag \neq bf$)

Integrals Involving ($ax^n + b$)

58. $\displaystyle\int (ax^2+b)^n x\,dx = \frac{1}{2a}\frac{(ax^2+b)^{n+1}}{n+1}$ ($n \neq -1$)

59. $\displaystyle\int \frac{dx}{ax^2+b} = \frac{1}{\sqrt{ab}}\tan^{-1}\!\left(x\sqrt{\frac{a}{b}}\right)$ (*a* and *b* positive)

60. $\displaystyle\int \frac{dx}{ax^2+b} = \frac{1}{2\sqrt{-ab}}\log_e\frac{x\sqrt{a}-\sqrt{-b}}{x\sqrt{a}+\sqrt{-b}}$ (*a* positive, *b* negative)

$\displaystyle\hphantom{\int \frac{dx}{ax^2+b}} = \frac{1}{2\sqrt{-ab}}\log_e\frac{\sqrt{b}+x\sqrt{-a}}{\sqrt{b}-x\sqrt{-a}}$ (*a* negative, *b* positive)

61. $\displaystyle\int \frac{dx}{x(ax^2+b)} = \frac{1}{2b}\log_e\frac{x^2}{ax^2+b}$

62. $\displaystyle\int \frac{dx}{(ax^2+b)^n} = \frac{1}{2(n-1)\,b}\frac{x}{(ax^2+b)^{n-1}}$
$\displaystyle\quad + \frac{2n-3}{2(n-1)\,b}\int \frac{dx}{(ax^2+b)^{n-1}}$ (*n* integer > 1)

63. $\displaystyle\int \frac{x^2\,dx}{ax^2+b} = \frac{x}{a} - \frac{b}{a}\int \frac{dx}{ax^2+b}$

64. $\displaystyle\int \frac{x^2\,dx}{(ax^2+b)^n} = -\frac{1}{2(n-1)\,a}\frac{x}{(ax^2+b)^{n-1}}$
$\displaystyle\quad + \frac{1}{2(n-1)a}\int \frac{dx}{(ax^2+b)^{n-1}}$ (*n* integer > 1)

65. $\displaystyle\int \frac{dx}{x^2(ax^2+b)^n} = \frac{1}{b}\int \frac{dx}{x^2(ax^2+b)^{n-1}} - \frac{a}{b}\int \frac{dx}{(ax^2+b)^n}$ (*n* = positive integer)

66. $\displaystyle\int \sqrt{ax^2+b}\ dx = \frac{x}{2}\sqrt{ax^2+b} + \frac{b}{2\sqrt{a}}\log_e\frac{x\sqrt{a}+\sqrt{ax^2+b}}{\sqrt{b}}$ (*a* positive)

67. $\displaystyle\int \sqrt{ax^2+b}\ dx = \frac{x}{2}\sqrt{ax^2+b} + \frac{b}{2\sqrt{-a}}\sin^{-1}\!\left(x\sqrt{-\frac{a}{b}}\right)$ (*a* negative)

68. $\displaystyle\int \frac{dx}{\sqrt{ax^2+b}} = \frac{1}{\sqrt{a}}\log_e(x\sqrt{a}+\sqrt{ax^2+b})$ (*a* positive)

69. $\displaystyle\int \frac{dx}{\sqrt{ax^2+b}} = \frac{1}{\sqrt{-a}}\sin^{-1}\!\left(x\sqrt{-\frac{a}{b}}\right)$ (*a* negative)

70. $\displaystyle\int \frac{x\,dx}{\sqrt{ax^2+b}} = \frac{1}{a}\sqrt{ax^2+b}$

71. $\displaystyle\int \frac{\sqrt{ax^2+b}}{x}\,dx = \sqrt{ax^2+b} + \sqrt{b}\log_e\frac{\sqrt{ax^2+b}-\sqrt{b}}{x}$ (*b* positive)

72. $\displaystyle\int \frac{\sqrt{ax^2+b}}{x}\,dx = \sqrt{ax^2+b} - \sqrt{-b}\tan^{-1}\frac{\sqrt{ax^2+b}}{\sqrt{-b}}$ (*b* negative)

TABLE 1.18 (*Continued*)

Integrals Involving $(ax^n + b)$ (*Continued*)

73. $\int x\sqrt{ax^2 + b}\, dx = \dfrac{1}{3a}(ax^2 + b)^{3/2}$

74. $\int x^2\sqrt{ax^2 + b}\, dx = \dfrac{x}{4a}(ax^2 + b)^{3/2} - \dfrac{bx}{8a}\sqrt{ax^2 + b}$

$\qquad - \dfrac{b^2}{8a\sqrt{a}}\log_e(x\sqrt{a} + \sqrt{ax^2 + b})$ (*a* positive)

75. $\int x^2\sqrt{ax^2 + b}\, dx = \dfrac{x}{4a}(ax^2 + b)^{3/2} - \dfrac{bx}{8a}\sqrt{ax^2 + b}$

$\qquad - \dfrac{b^2}{8a\sqrt{-a}}\sin^{-1}\left(x\sqrt{\dfrac{-a}{b}}\right)$ (*a* negative)

76. $\displaystyle\int \dfrac{dx}{x\sqrt{ax^2 + b}} = \dfrac{1}{\sqrt{b}}\log_e\dfrac{\sqrt{ax^2 + b} - \sqrt{b}}{x}$ (*b* positive)

77. $\displaystyle\int \dfrac{dx}{x\sqrt{ax^2 + b}} = \dfrac{1}{\sqrt{-b}}\sec^{-1}\left(x\sqrt{-\dfrac{a}{b}}\right)$ (*b* negative)

78. $\displaystyle\int \dfrac{x^2\, dx}{\sqrt{ax^2 + b}} = \dfrac{x}{2a}\sqrt{ax^2 + b} - \dfrac{b}{2a\sqrt{a}}\log_e(x\sqrt{a} + \sqrt{ax^2 + b})$ (*a* positive)

79. $\displaystyle\int \dfrac{x^2\, dx}{\sqrt{ax^2 + b}} = \dfrac{x}{2a}\sqrt{ax^2 + b} - \dfrac{b}{2a\sqrt{-a}}\sin^{-1}\left(x\sqrt{-\dfrac{a}{b}}\right)$ (*a* negative)

80. $\displaystyle\int \dfrac{\sqrt{ax^2 + b}}{x^2}\, dx = -\dfrac{\sqrt{ax^2 + b}}{x} + \sqrt{a}\log_e(x\sqrt{a} + \sqrt{ax^2 + b})$ (*a* positive)

81. $\displaystyle\int \dfrac{\sqrt{ax^2 + b}}{x^2}\, dx = -\dfrac{\sqrt{ax^2 + b}}{x} - \sqrt{-a}\sin^{-1}\left(x\sqrt{-\dfrac{a}{b}}\right)$ (*a* negative)

82. $\displaystyle\int \dfrac{dx}{x(ax^n + b)} = \dfrac{1}{bn}\log_e\dfrac{x^n}{ax^n + b}$

83. $\displaystyle\int \dfrac{dx}{x\sqrt{ax^n + b}} = \dfrac{1}{n\sqrt{b}}\log_e\dfrac{\sqrt{ax^n + b} - \sqrt{b}}{\sqrt{ax^n + b} + \sqrt{b}}$ (*b* positive)

84. $\displaystyle\int \dfrac{dx}{x\sqrt{ax^n + b}} = \dfrac{2}{n\sqrt{-b}}\sec^{-1}\sqrt{-\dfrac{ax^n}{b}}$ (*b* negative)

Integrals Involving $(ax^2 + bx + d)$

85. $\displaystyle\int \dfrac{dx}{ax^2 + bx + d} = \dfrac{1}{\sqrt{b^2 - 4ad}}\log_e\dfrac{2ax + b - \sqrt{b^2 - 4ad}}{2ax + b + \sqrt{b^2 - 4ad}}$ $(b^2 > 4ad)$

86. $\displaystyle\int \dfrac{dx}{ax^2 + bx + d} = \dfrac{2}{\sqrt{4ad - b^2}}\tan^{-1}\dfrac{2ax + b}{\sqrt{4ad - b^2}}$ $(b^2 < 4ad)$

87. $\displaystyle\int \dfrac{dx}{ax^2 + bx + d} = -\dfrac{2}{2ax + b}$ $(b^2 = 4ad)$

88. $\displaystyle\int \dfrac{dx}{\sqrt{ax^2 + bx + d}} = \dfrac{1}{\sqrt{a}}\log_e\!\left(2ax + b + 2\sqrt{a(ax^2 + bx + d)}\right)$ (*a* positive)

89. $\displaystyle\int \dfrac{dx}{\sqrt{ax^2 + bx + d}} = \dfrac{1}{\sqrt{-a}}\sin^{-1}\dfrac{-2ax - b}{\sqrt{b^2 - 4ad}}$ (*a* negative)

90. $\displaystyle\int \dfrac{x\, dx}{ax^2 + bx + d} = \dfrac{1}{2a}\log_e(ax^2 + bx + d) - \dfrac{b}{2a}\int \dfrac{dx}{ax^2 + bx + d}$

91. $\displaystyle\int \dfrac{x\, dx}{\sqrt{ax^2 + bx + d}} = \dfrac{\sqrt{ax^2 + bx + d}}{a} - \dfrac{b}{2a}\int \dfrac{dx}{\sqrt{ax^2 + bx + d}}$

TABLE 1.18   *(Continued)*

Integrals Involving $(ax^2 + bx + d)$ (Continued)

92. $\displaystyle \int \frac{dx}{x\sqrt{ax^2 + bx + d}} = -\frac{1}{\sqrt{d}} \log_e \left(\frac{\sqrt{ax^2 + bx + d} + \sqrt{d}}{x} + \frac{b}{2\sqrt{d}} \right)$ (d positive)

93. $\displaystyle \int \frac{dx}{x\sqrt{ax^2 + bx + d}} = \frac{1}{\sqrt{-d}} \sin^{-1} \frac{bx + 2d}{x\sqrt{b^2 - 4ad}}$ (d negative)

94. $\displaystyle \int \frac{dx}{x\sqrt{ax^2 + bx}} = -\frac{2}{bx}\sqrt{ax^2 + bx}$

95. $\displaystyle \int \sqrt{ax^2 + bx + d}\; dx = \frac{2ax + b}{4a}\sqrt{ax^2 + bx + d} + \frac{4ad - b^2}{8a}\int \frac{dx}{\sqrt{ax^2 + bx + d}}$

96. $\displaystyle \int x\sqrt{ax^2 + bx + d}\; dx = \frac{(ax^2 + bx + d)^{3/2}}{3a} - \frac{b}{2a}\int \sqrt{ax^2 + bx + d}\; dx$

Integrals Involving $\sin^n ax$

97. $\displaystyle \int \sin^3 ax \; dx = -\frac{1}{a}\cos ax + \frac{1}{3a}\cos^3 ax$

98. $\displaystyle \int \sin^4 ax \; dx = \frac{3}{8}x - \frac{1}{4a}\sin 2ax + \frac{1}{32a}\sin 4ax$

99. $\displaystyle \int \sin^n ax \; dx = -\frac{\sin^{n-1}ax \cos ax}{na} + \frac{n-1}{n}\int \sin^{n-2}ax \; dx$ (n = positive integer)

100. $\displaystyle \int x \sin ax \; dx = \frac{\sin ax}{a^2} - \frac{x \cos ax}{a}$

101. $\displaystyle \int x^2 \sin ax \; dx = \frac{2x}{a^2}\sin ax - \left(\frac{x^2}{a} - \frac{2}{a^3} \right)\cos ax$

102. $\displaystyle \int x^3 \sin ax \; dx = \left(\frac{3x^2}{a^2} - \frac{6}{a^4} \right)\sin ax - \left(\frac{x^3}{a} - \frac{6x}{a^3} \right)\cos ax$

103. $\displaystyle \int x^n \sin ax \; dx = -\frac{x^n}{a}\cos ax + \frac{n}{a}\int x^{n-1}\cos ax \; dx$ ($n > 0$)

104. $\displaystyle \int \frac{\sin ax}{x^n} dx = -\frac{1}{n-1}\frac{\sin ax}{x^{n-1}} + \frac{a}{n-1}\int \frac{\cos ax}{x^n - 1}dx$

105. $\displaystyle \int \frac{dx}{\sin^n ax} = -\frac{1}{a(n-1)}\frac{\cos ax}{\sin^{n-1}ax} + \frac{n-2}{n-1}\int \frac{dx}{\sin^{n-2}ax}$ (n integer > 1)

106. $\displaystyle \int \frac{x \, dx}{\sin^2 ax} = -\frac{x}{a}\cot ax + \frac{1}{a^2}\log_e \sin ax$

107. $\displaystyle \int \frac{dx}{1 + \sin ax} = -\frac{1}{a}\tan\left(\frac{\pi}{4} - \frac{ax}{2} \right)$

108. $\displaystyle \int \frac{dx}{1 - \sin ax} = \frac{1}{a}\cot\left(\frac{\pi}{4} - \frac{ax}{2} \right)$

109. $\displaystyle \int \frac{x \, dx}{1 + \sin ax} = -\frac{x}{a}\tan\left(\frac{\pi}{4} - \frac{ax}{2} \right) + \frac{2}{a^2}\log_e \cos\left(\frac{\pi}{4} - \frac{ax}{2} \right)$

110. $\displaystyle \int \frac{x \, dx}{1 - \sin ax} = \frac{x}{a}\cot\left(\frac{\pi}{4} - \frac{ax}{2} \right) + \frac{2}{a^2}\log_e \sin\left(\frac{\pi}{4} - \frac{ax}{2} \right)$

111. $\displaystyle \int \frac{dx}{b + d \sin ax} = \frac{-2}{a\sqrt{b^2 - d^2}}\tan^{-1}\left[\sqrt{\frac{b - d}{b + d}}\, \tan\left(\frac{\pi}{4} - \frac{ax}{2} \right) \right]$ ($b^2 > d^2$)

112. $\displaystyle \int \frac{dx}{b + d \sin ax} = \frac{-1}{a\sqrt{d^2 - b^2}}\log_e \frac{d + b \sin ax + \sqrt{d^2 - b^2}\cos ax}{b + d \sin ax}$ ($d^2 > b^2$)

113. $\displaystyle \int \sin ax \sin bx \; dx = \frac{\sin(a - b)x}{2(a - b)} - \frac{\sin(a + b)x}{2(a + b)}$ ($a^2 \neq b^2$)

TABLE 1.18   (*Continued*)

Integrals Involving cosⁿax

114. $\int \cos^3 ax \, dx = \dfrac{1}{a}\sin ax - \dfrac{1}{3a}\sin^3 ax$

115. $\int \cos^4 ax \, dx = \dfrac{3}{8}x + \dfrac{1}{4a}\sin 2ax + \dfrac{1}{32a}\sin 4ax$

116. $\int \cos^n ax \, dx = \dfrac{\cos^{n-1}ax \sin ax}{na} + \dfrac{n-1}{n}\int \cos^{n-2}ax \, dx$ (n = positive integer)

117. $\int x \cos ax \, dx = \dfrac{\cos ax}{a^2} + \dfrac{x \sin ax}{a}$

118. $\int x^2 \cos ax \, dx = \dfrac{2x}{a^2}\cos ax + \left(\dfrac{x^2}{a} - \dfrac{2}{a^3}\right)\sin ax$

119. $\int x^3 \cos ax \, dx = \left(\dfrac{3x^2}{a^2} - \dfrac{6}{a^4}\right)\cos ax + \left(\dfrac{x^3}{a} - \dfrac{6x}{a^3}\right)\sin ax$

120. $\int x^n \cos ax \, dx = \dfrac{x^n \sin ax}{a} - \dfrac{n}{a}\int x^{n-1}\sin ax \, dx$ ($n > 0$)

121. $\int \dfrac{\cos ax}{x^n}\,dx = -\dfrac{1}{n-1}\dfrac{\cos ax}{x^{n-1}} - \dfrac{a}{n-1}\int \dfrac{\sin ax}{x^{n-1}}\,dx$

122. $\int \dfrac{dx}{\cos^n ax} = \dfrac{1}{a(n-1)}\dfrac{\sin ax}{\cos^{n-1}ax} + \dfrac{n-2}{n-1}\int \dfrac{dx}{\cos^{n-2}ax}$ (n integer > 1)

123. $\int \dfrac{x\,dx}{\cos^2 ax} = \dfrac{x}{a}\tan ax + \dfrac{1}{a^2}\log_e\cos ax$

124. $\int \dfrac{dx}{1 + \cos ax} = \dfrac{1}{a}\tan\dfrac{ax}{2}$

125. $\int \dfrac{dx}{1 - \cos ax} = -\dfrac{1}{a}\cot\dfrac{ax}{2}$

126. $\int \dfrac{x\,dx}{1 + \cos ax} = \dfrac{x}{a}\tan\dfrac{ax}{2} + \dfrac{2}{a^2}\log_e\cos\dfrac{ax}{2}$

127. $\int \dfrac{x\,dx}{1 - \cos ax} = -\dfrac{x}{a}\cot\dfrac{ax}{2} + \dfrac{2}{a^2}\log_e\sin\dfrac{ax}{2}$

128. $\int \dfrac{dx}{b + d\cos ax} = \dfrac{2}{a\sqrt{b^2 - d^2}}\tan^{-1}\left(\sqrt{\dfrac{b-d}{b+d}}\,\tan\dfrac{ax}{2}\right)$ ($b^2 > d^2$)

129. $\int \dfrac{dx}{b + d\cos ax} = \dfrac{1}{a\sqrt{d^2 - b^2}}\log_e\dfrac{d + b\cos ax + \sqrt{d^2 - b^2}\sin ax}{b + d\cos ax}$ ($d^2 > b^2$)

130. $\int \cos ax \cos bx \, dx = \dfrac{\sin(a-b)x}{2(a-b)} + \dfrac{\sin(a+b)x}{2(a+b)}$ ($a^2 \neq b^2$)

Integrals Involving sinⁿax, cosⁿax

131. $\int \sin ax \cos bx \, dx = -\dfrac{1}{2}\left[\dfrac{\cos(a-b)x}{a-b} + \dfrac{\cos(a+b)x}{a+b}\right]$ ($a^2 \neq b^2$)

132. $\int \sin^2 ax \cos^2 ax \, dx = \dfrac{x}{8} - \dfrac{\sin 4ax}{32a}$

133. $\int \sin^n ax \cos ax \, dx = \dfrac{1}{a(n+1)}\sin^{n+1}ax$ ($n \neq -1$)

134. $\int \sin ax \cos^n ax \, dx = -\dfrac{1}{a(n+1)}\cos^{n+1}ax$ ($n \neq -1$)

135. $\int \sin^n ax \cos^m ax \, dx = -\dfrac{\sin^{n-1}ax \cos^{m+1}ax}{a(n+m)} + \dfrac{n-1}{n+m}\int \sin^{n-2}ax \cos^m ax \, dx$ (m, n pos)

136. $\int \dfrac{\sin^n ax}{\cos^m ax}\,dx = \dfrac{\sin^{n+1}ax}{a(m-1)\cos^{m-1}ax} - \dfrac{n-m+2}{m-1}\int \dfrac{\sin^n ax}{\cos^{m-2}ax}\,dx$ (m, n pos, $m \neq 1$)

TABLE 1.18 *(Continued)*

Integrals Involving sin^{n}ax, cos^{n}ax (Continued)

137. $\int \dfrac{\cos^m ax}{\sin^n ax}\,dx = \dfrac{-\cos^{m+1}ax}{a(n-1)\sin^{n-1}ax} + \dfrac{n-m-2}{(n-1)}\int \dfrac{\cos^m ax}{\sin^{n-2}ax}\,dx$ (m, n pos, n ≠ 1)

138. $\int \dfrac{dx}{\sin ax \cos ax} = \dfrac{1}{a}\log_e \tan ax$

139. $\int \dfrac{dx}{b\sin ax + d\cos ax} = \dfrac{1}{a\sqrt{b^2+d^2}}\log_e \tan \dfrac{1}{2}\left(ax + \tan^{-1}\dfrac{d}{b}\right)$

140. $\int \dfrac{\sin ax}{b + d\cos ax}\,dx = -\dfrac{1}{ad}\log_e(b + d\cos ax)$

141. $\int \dfrac{\cos ax}{b + d\sin ax}\,dx = \dfrac{1}{ad}\log_e(b + d\sin ax)$

Integrals Involving tan^{n}ax, cot^{n}ax, sec^{n}ax, csc^{n}ax

142. $\int \tan^n ax\,dx = \dfrac{1}{a(n-1)}\tan^{n-1}ax - \int \tan^{n-2}ax\,dx$ (n integer > 1)

143. $\int \cot^n ax\,dx = -\dfrac{1}{a(n-1)}\cot^{n-1}ax - \int \cot^{n-2}ax\,dx$ (n integer > 1)

144. $\int \sec^n ax\,dx = \dfrac{1}{a(n-1)}\dfrac{\sin ax}{\cos^{n-1}ax} + \dfrac{n-2}{n-1}\int \sec^{n-2}ax\,dx$ (n integer > 1)

145. $\int \csc^n ax\,dx = -\dfrac{1}{a(n-1)}\dfrac{\cos ax}{\sin^{n-1}ax} + \dfrac{n-2}{n-1}\int \csc^{n-2}ax\,dx$ (n integer > 1)

146. $\int \dfrac{dx}{b + d\tan ax} = \dfrac{1}{b^2+d^2}\left[bx + \dfrac{d}{a}\log_e(b\cos ax + d\sin ax)\right]$

147. $\int \dfrac{dx}{\sqrt{b + d\tan^2 ax}} = \dfrac{1}{a\sqrt{b-d}}\sin^{-1}\left[\sqrt{\dfrac{b-d}{b}}\sin ax\right]$ (b pos, $b^2 > d^2$)

148. $\int \tan ax \sec ax\,dx = \dfrac{1}{a}\sec ax$

149. $\int \tan^n ax \sec^2 ax\,dx = \dfrac{1}{a(n+1)}\tan^{n+1}ax$ (n ≠ −1)

150. $\int \dfrac{\sec^2 ax\,dx}{\tan ax} = \dfrac{1}{a}\log_e \tan ax$

151. $\int \cot ax \csc ax\,dx = -\dfrac{1}{a}\csc ax$

152. $\int \cot^n ax \csc^2 ax\,dx = -\dfrac{1}{a(n+1)}\cot^{n+1}ax$ (n ≠ −1)

153. $\int \dfrac{\csc^2 ax}{\cot ax}\,dx = -\dfrac{1}{a}\log_e \cot ax$

Integrals Involving b^{ax}, e^{ax}, sin bx, cos bx

154. $\int xb^{ax}\,dx = \dfrac{xb^{ax}}{a\log_e b} - \dfrac{b^{ax}}{a^2(\log_e b)^2}$

155. $\int xe^{ax}\,dx = \dfrac{e^{ax}}{a^2}(ax - 1)$

156. $\int x^n b^{ax}\,dx = \dfrac{x^n b^{ax}}{a\log_e b} - \dfrac{n}{a\log_e b}\int x^{n-1}b^{ax}\,dx$ (n positive)

157. $\int x^n e^{ax}\,dx = \dfrac{1}{a}x^n e^{ax} - \dfrac{n}{a}\int x^{n-1}e^{ax}\,dx$ (n positive)

TABLE 1.18 (*Continued*)

Integrals Involving b^{ax}, e^{ax}, sin bx, cos bx (*Continued*)

158. $\int \dfrac{dx}{b + de^{ax}} = \dfrac{1}{ab}[ax - \log_e(b + de^{ax})]$

159. $\int \dfrac{e^{ax}\,dx}{b + de^{ax}} = \dfrac{1}{ad}\log_e(b + de^{ax})$

160. $\int \dfrac{dx}{be^{ax} + de^{-ax}} = \dfrac{1}{a\sqrt{bd}}\tan^{-1}\left(e^{ax}\sqrt{\dfrac{b}{d}}\right)$　(b and d positive)

161. $\int \dfrac{e^{ax}}{x}\,dx = \log_e x + ax + \dfrac{(ax)^2}{2\cdot 2!} + \dfrac{(ax)^3}{3\cdot 3!} + \cdots$

162. $\int \dfrac{e^{ax}}{x^n}\,dx = \dfrac{1}{n-1}\left(-\dfrac{e^{ax}}{x^{n-1}} + a\int \dfrac{e^{ax}}{x^{n-1}}\,dx\right)$　(n integer > 1)

163. $\int e^{ax}\sin bx\,dx = \dfrac{e^{ax}}{a^2 + b^2}(a \sin bx - b \cos bx)$

164. $\int e^{ax}\cos bx\,dx = \dfrac{e^{ax}}{a^2 + b^2}(a \cos bx + b \sin bx)$

165. $\int xe^{ax}\sin bx\,dx = \dfrac{xe^{ax}}{a^2 + b^2}(a \sin bx - b \cos bx)$

$\qquad\qquad - \dfrac{e^{ax}}{(a^2 + b^2)^2}[(a^2 - b^2)\sin bx - 2ab \cos bx]$

166. $\int xe^{ax}\cos bx\,dx = \dfrac{xe^{ax}}{a^2 + b^2}(a \cos bx + b \sin bx)$

$\qquad\qquad - \dfrac{e^{ax}}{(a^2 + b^2)^2}[(a^2 - b^2)\cos bx + 2ab \sin bx]$

Integrals Involving $\log_e ax$

167. $\int \log_e ax\,dx = x\log_e ax - x$

168. $\int (\log_e ax)^n\,dx = x(\log_e ax)^n - n(\log_e ax)^{n-1}\,dx$　(n positive)

169. $\int x^n \log_e ax\,dx = x^{n+1}\left(\dfrac{\log_e ax}{n+1}\right) - \dfrac{1}{(n+1)^2}$,　$n \neq -1$

170. $\int \dfrac{(\log_e ax)^n}{x}\,dx = \dfrac{(\log_e ax)^{n+1}}{n+1}$,　$n \neq -1$

171. $\int \dfrac{dx}{x\log_e ax} = \log_e(\log_e x)$

172. $\int \dfrac{dx}{\log_e ax} = \dfrac{1}{a}\left[\log_e(\log_e ax) + \log_e ax + \dfrac{(\log_e ax)^2}{2\cdot 2!} + \cdots\right]$

173. $\int x^m(\log_e ax)^n\,dx = \dfrac{x^{m+1}(\log_e ax)^n}{m+1} - \dfrac{n}{m+1}\int x^m(\log_e ax)^{n-1}\,dx$,　$m, n \neq 1$

174. $\int \dfrac{x^m\,dx}{(\log_e ax)^n} = -\dfrac{x^{m+1}}{(n-1)(\log_e ax)^{n-1}} + \dfrac{m+1}{n-1}\int \dfrac{x^m\,dx}{(\log_e ax)^{n-1}}$

Some Definite Integrals

1. $\int_0^a \sqrt{a^2 - x^2}\,dx = \dfrac{\pi a^2}{4}$

2. $\int_0^a \sqrt{2ax - x^2}\,dx = \dfrac{\pi a^2}{4}$

TABLE 1.18   (*Continued*)

Some Definite Integrals (*Continued*)

3. $\int_0^\infty \dfrac{dx}{a + bx^2} = \dfrac{\pi}{2\sqrt{ab}}$ (*a* and *b* positive)

4. $\int_0^{\sqrt{a/b}} \dfrac{dx}{a + bx^2} dx = \int_{\sqrt{a/b}}^\infty \dfrac{dx}{a + bx^2} = \dfrac{\pi}{4\sqrt{ab}}$ (*a* and *b* positive)

5. $\int_0^{\sqrt{a/b}} \dfrac{dx}{\sqrt{a - bx^2}} = \dfrac{\pi}{2\sqrt{b}}$ (*a* and *b* positive)

6. $\int_0^\infty \dfrac{\sin bx}{x} dx = \begin{cases} \dfrac{\pi}{2} & (b > 0) \\ 0 & (b = 0) \\ -\dfrac{\pi}{2} & (b < 0) \end{cases}$

7. $\int_0^\infty \dfrac{\tan x}{x} dx = \dfrac{\pi}{2}$

8. $\int_0^{\pi/2} \sin^{2n+1}x \, dx = \int_0^{\pi/2} \cos^{2n+1}x \, dx = \dfrac{2 \cdot 4 \cdot 6 \cdots \cdot 2n}{3 \cdot 5 \cdot 7 \cdots \cdot (2n + 1)}$ (*n* > 0)

9. $\int_0^{\pi/2} \sin^{2n}x \, dx = \int_0^{\pi/2} \cos^{2n}x \, dx = \dfrac{1 \cdot 3 \cdot 5 \cdots \cdot (2n - 1)}{2 \cdot 4 \cdot 6 \cdots \cdot 2n} \cdot \dfrac{\pi}{2}$ (*n* > 0)

10. $\int_0^\pi \sin ax \sin bx \, dx = \int_0^\pi \cos ax \cos bx \, dx = 0$ (*a* ≠ *b*)

11. $\int_0^\pi \sin^2 ax \, dx = \int_0^\pi \cos^2 ax \, dx = \dfrac{\pi}{2}$

12. $\int_0^{\pi/2} \log_e \cos x \, dx = \int_0^{\pi/2} \log_e \sin x \, dx = -\dfrac{\pi}{2} \log_e 2$

13. $\int_0^\infty e^{-ax^2} dx = \dfrac{1}{2} \sqrt{\dfrac{\pi}{a}}$

14. $\int_0^\infty x^n e^{-ax} dx = \dfrac{n!}{a^{n+1}}$ (*a* > 0, *n* = 1, 2, 3, …)

15. $\int_0^1 \dfrac{\log_e x}{1 - x} dx = -\dfrac{\pi^2}{6}$

16. $\int_0^1 \dfrac{\log_e x}{1 + x} dx = -\dfrac{\pi^2}{12}$

17. $\int_0^1 \dfrac{\log_e x}{1 - x^2} dx = \dfrac{\pi^2}{8}$

TABLE 1.19   Haversines[a]

$\theta°$	Value	Log	$\theta°$	Value	Log	$\theta°$	Value	Log	$\theta°$	Value	Log
0	0.00000	—	10	0.00760	0.88059	20	0.03015	0.47934	30	0.06699	0.82599
1	0.00008	0.88168	11	0.00919	0.96315	21	0.03321	0.52127	31	0.07142	0.85380
2	0.00030	0.48371	12	0.01093	0.03847	22	0.03641	0.56120	32	0.07598	0.88068
3	0.00069	0.83584	13	0.01281	0.10772	23	0.03975	0.59931	33	0.08066	0.90668
4	0.00122	0.08564	14	0.01485	0.17179	24	0.04323	0.63576	34	0.08548	0.93187
5	0.00190	0.27936	15	0.01704	0.23140	25	0.04685	0.67067	35	0.09042	0.95628
6	0.00274	0.43760	16	0.01937	0.28711	26	0.05060	0.70418	36	0.09549	0.97996
7	0.00373	0.57135	17	0.02185	0.33940	27	0.05450	0.73637	37	0.10068	0.00295
8	0.00487	0.68717	18	0.02447	0.38867	28	0.05853	0.76735	38	0.10599	0.02528
9	0.00616	0.78929	19	0.02724	0.43522	29	0.06269	0.79720	39	0.11143	0.04699

TABLE 1.19 (*Continued*)

θ°	Value	Log	θ°	Value	Log	θ°	Value	Log	θ°	Value	Log
40	0.11698	0.06810	**80**	0.41318	0.61613	**120**	0.75000	0.87506	**160**	0.96985	0.98670
41	0.12265	0.08865	81	0.42178	0.62509	121	0.75752	0.87939	161	0.97276	0.98801
42	0.12843	0.10866	82	0.43041	0.63389	122	0.76496	0.88364	162	0.97553	0.98924
43	0.13432	0.12815	83	0.43907	0.64253	123	0.77232	0.88780	163	0.97815	0.99041
44	0.14033	0.14715	84	0.44774	0.65102	124	0.77960	0.89187	164	0.98063	0.99151
45	0.14645	0.16568	85	0.45642	0.65937	125	0.78679	0.89586	165	0.98296	0.99254
46	0.15267	0.18376	86	0.46512	0.66757	126	0.79389	0.89976	166	0.98515	0.99350
47	0.15900	0.20140	87	0.47383	0.67562	127	0.80091	0.90358	167	0.98719	0.99440
48	0.16543	0.21863	88	0.48255	0.68354	128	0.80783	0.90732	168	0.98907	0.99523
49	0.17197	0.23545	89	0.49127	0.69132	129	0.81466	0.91098	169	0.99081	0.99599
50	0.17861	0.25190	**90**	0.50000	0.69897	**130**	0.82139	0.91455	**170**	0.99240	0.99669
51	0.18534	0.26797	91	0.50873	0.70648	131	0.82803	0.91805	171	0.99384	0.99732
52	0.19217	0.28368	92	0.51745	0.71387	132	0.83457	0.92146	172	0.99513	0.99788
53	0.19909	0.29905	93	0.52617	0.72112	133	0.84100	0.92480	173	0.99627	0.99838
54	0.20611	0.31409	94	0.53488	0.72825	134	0.84733	0.92805	174	0.99726	0.99881
55	0.21321	0.32281	95	0.54358	0.73526	135	0.85355	0.93123	175	0.99810	0.99917
56	0.22040	0.34322	96	0.55226	0.74215	136	0.85967	0.93433	176	0.99878	0.99947
57	0.22768	0.35733	97	0.56093	0.74891	137	0.86568	0.93736	177	0.99931	0.99970
58	0.23504	0.37114	98	0.56959	0.75556	138	0.87157	0.94030	178	0.99970	0.99987
59	0.24248	0.38468	99	0.57822	0.76209	139	0.87735	0.94318	179	0.99992	0.99997
60	0.25000	0.39794	**100**	0.58682	0.76851	**140**	0.88302	0.94597	**180**	1.00000	0.00000
61	0.25760	0.41094	101	0.59540	0.77481	141	0.88857	0.94869			
62	0.26526	0.42368	102	0.60396	0.78101	142	0.89401	0.95134			
63	0.27300	0.43617	103	0.61248	0.78709	143	0.89932	0.95391			
64	0.28081	0.44842	104	0.62096	0.79306	144	0.90451	0.95641			
65	0.28869	0.46043	105	0.62941	0.79893	145	0.90958	0.95884			
66	0.29663	0.47222	106	0.63782	0.80470	146	0.91452	0.96119			
67	0.30463	0.48378	107	0.64619	0.81036	147	0.91934	0.96347			
68	0.31270	0.49512	108	0.65451	0.81592	148	0.92402	0.96568			
69	0.32082	0.50625	109	0.66278	0.82137	149	0.92858	0.96782			
70	0.32899	0.51718	**110**	0.67101	0.82673	**150**	0.93301	0.96989			
71	0.33722	0.52791	111	0.67918	0.83199	151	0.93731	0.97188			
72	0.34549	0.53844	112	0.68730	0.83715	152	0.94147	0.97381			
73	0.35381	0.54878	113	0.69537	0.84221	153	0.94550	0.97566			
74	0.36218	0.55893	114	0.70337	0.84718	154	0.94940	0.97745			
75	0.37059	0.56889	115	0.71131	0.85206	155	0.95315	0.97016			
76	0.37904	0.57868	116	0.71919	0.85684	156	0.95677	0.98081			
77	0.38752	0.58830	117	0.72700	0.86153	157	0.96025	0.98239			
78	0.39604	0.59774	118	0.73474	0.86613	158	0.96359	0.98389			
79	0.40460	0.60702	119	0.74240	0.87064	159	0.96679	0.98533			

[a]hav $\theta = \frac{1}{2}$ vers $\theta = \frac{1}{2}(1 - \cos\theta) = \sin^2\frac{1}{2}\theta$
hav $(-\theta)$ = hav θ
hav $(180° - \theta)$ = hav $(180° + \theta) = 1 -$ hav θ
Characteristics of the logarithms are omitted.

TABLE 1.20 Complete Elliptic Integrals[a]

$\sin^{-1}k$	K	log K	E	log E	$\sin^{-1}k$	K	log K	E	log E
0°	1.5708	0.196120	1.5708	0.196120	**5°**	1.5738	0.196947	1.5678	0.195293
1	1.5709	0.196153	1.5707	0.196087	6	1.5751	0.197312	1.5665	0.194930
2	1.5713	0.196252	1.5703	0.195988	7	1.5767	0.197743	1.5649	0.194500
3	1.5719	0.196418	1.5697	0.195822	8	1.5785	0.198241	1.5632	0.194004
4	1.5727	0.196649	1.5689	0.195591	9	1.5805	0.198806	1.5611	0.193442

TABLE 1.20　(*Continued*)

$\sin^{-1}k$	K	$\log K$	E	$\log E$	$\sin^{-1}k$	K	$\log K$	E	$\log E$
10	1.5828	0.199438	1.5589	0.192815	**50°**	1.9356	0.286811	1.3055	0.115790
11	1.5854	0.200137	1.5564	0.192121	51	1.9539	0.290895	1.2963	0.112698
12	1.5882	0.200904	1.5537	0.191362	52	1.9729	0.295101	1.2870	0.109563
13	1.5913	0.201740	1.5507	0.190537	53	1.9927	0.299435	1.2776	0.106386
14	1.5946	0.202643	1.5476	0.189646	54	2.0133	0.303901	1.2681	0.103169
15	1.5981	0.203615	1.5442	0.188690	**55**	2.0347	0.308504	1.2587	0.099915
16	1.6020	0.204657	1.5405	0.187668	56	2.0571	0.313247	1.2492	0.096626
17	1.6061	0.205768	1.5367	0.186581	57	2.0804	0.318138	1.2397	0.093303
18	1.6105	0.206948	1.5326	0.185428	58	2.1047	0.323182	1.2301	0.089950
19	1.6151	0.208200	1.5283	0.184210	59	2.1300	0.328384	1.2206	0.086569
20	1.6200	0.209522	1.5238	0.182928	**60**	2.1565	0.333753	1.2111	0.083164
21	1.6252	0.210916	1.5191	0.181580	61	2.1842	0.339295	1.2015	0.079738
22	1.6307	0.212382	1.5141	0.180168	62	2.2132	0.345020	1.1920	0.076293
23	1.6365	0.213921	1.5090	0.178691	63	2.2435	0.350936	1.1826	0.072834
24	1.6426	0.215533	1.5037	0.177150	64	2.2754	0.357053	1.1732	0.069364
25	1.6490	0.217219	1.4981	0.175545	**65**	2.3088	0.363384	1.1638	0.065889
26	1.6557	0.218981	1.4924	0.173876	66	2.3439	0.369940	1.1545	0.062412
27	1.6627	0.220818	1.4864	0.172144	67	2.3809	0.376736	1.1453	0.058937
28	1.6701	0.222732	1.4803	0.170348	68	2.4198	0.383787	1.1362	0.055472
29	1.6777	0.224723	1.4740	0.168489	69	2.4610	0.391112	1.1272	0.052020
30	1.6858	0.226793	1.4675	0.166567	**70**	2.5046	0.398730	1.1184	0.048589
31	1.6941	0.228943	1.4608	0.164583	71	2.5507	0.406665	1.1096	0.045183
32	1.7028	0.231173	1.4539	0.162537	72	2.5998	0.414943	1.1011	0.041812
33	1.7119	0.233485	1.4469	0.160429	73	2.6521	0.423596	1.0927	0.038481
34	1.7214	0.235880	1.4397	0.158261	74	2.7081	0.432660	1.0844	0.035200
35	1.7312	0.238359	1.4323	0.156031	**75**	2.7681	0.442176	1.0764	0.031976
36	1.7415	0.240923	1.4248	0.153742	76	2.8327	0.452196	1.0686	0.028819
37	1.7552	0.243575	1.4171	0.151393	77	2.9026	0.462782	1.0611	0.025740
38	1.7633	0.246315	1.4092	0.148985	78	2.9786	0.474008	1.0538	0.022749
39	1.7748	0.249146	1.4013	0.146519	79	3.0617	0.485967	1.0468	0.019858
40	1.7868	0.252068	1.3931	0.143995	**80**	3.1534	0.498777	1.0401	0.017081
41	1.7992	0.255085	1.3849	0.141414	81	3.2553	0.512591	1.0338	0.014432
42	1.8122	0.258197	1.3765	0.138778	82	3.3699	0.527613	1.0278	0.011927
43	1.8256	0.261406	1.3680	0.136086	83	3.5004	9.544120	1.0223	0.009584
44	1.8396	0.264716	1.3594	0.133340	84	3.6519	0.562514	1.0172	0.007422
45	1.8541	0.268127	1.3506	0.130541	**85**	3.8317	0.583396	1.0127	0.005465
46	1.8691	0.271644	1.3418	0.127690	86	4.0528	0.607751	1.0086	0.003740
47	1.8848	0.275267	1.3329	0.124788	87	4.3387	0.637355	1.0053	0.002278
48	1.9011	0.279001	1.3238	0.121836	88	4.7427	0.676027	1.0026	0.001121
49	1.9180	0.282848	1.3147	0.118836	89	5.4349	0.735192	1.0008	0.000326
					90	∞	∞	1.0000	0.000000

$\sin^{-1}k$		K	$\log K$	$\sin^{-1}k$		K	$\log K$	$\sin^{-1}k$		K	$\log K$
89	**20**	5.840	0.76641	**89**	**40**	6.533	0.81511	**89**	**50**	7.226	0.85890
89	22	5.891	0.77019	89	41	6.584	0.81849	89	51	7.332	0.86522
89	24	5.946	0.77422	89	42	6.639	0.82210	89	52	7.449	0.87210
89	26	6.003	0.77837	89	43	6.696	0.82582	89	53	7.583	0.87984
89	28	6.063	0.78269	89	44	6.756	0.82969	89	54	7.737	0.88857
89	**30**	6.128	0.78732	**89**	**45**	6.821	0.83385	**89**	**55**	7.919	0.89867
89	32	6.197	0.79218	89	46	6.890	0.83822	89	56	8.143	0.91078
89	34	6.271	0.79734	89	47	6.964	0.84286	89	57	8.430	0.92583
89	36	6.351	0.80284	89	48	7.044	0.84782	89	58	8.836	0.94626
89	38	6.438	0.80875	89	49	7.131	0.85315	89	59	9.529	0.97905
								90	**0**	∞	∞

$$^aK = \int_0^{\pi/2} \frac{d\Phi}{\sqrt{1 - k^2 \sin^2\Phi}} = F\left(k, \frac{\pi}{2}\right) \qquad E = \int_0^{\pi/2} \sqrt{1 - k^2 \sin^2\Phi}\; d\Phi = E\left(k, \frac{\pi}{2}\right)$$

TABLE 1.21   Gamma Function[a, b]

n	$\Gamma(n)$	n	$\Gamma(n)$	n	$\Gamma(n)$	n	$\Gamma(n)$
1.00	1.00000	1.25	0.90640	1.50	0.88623	1.75	0.91906
1.01	0.99433	1.26	0.90440	1.51	0.88659	1.76	0.92137
1.02	0.98884	1.27	0.90250	1.52	0.88704	1.77	0.92376
1.03	0.98355	1.28	0.90072	1.53	0.88757	1.78	0.92623
1.04	0.97844	1.29	0.89904	1.54	0.88818	1.79	0.92877
1.05	0.97350	1.30	0.89747	1.55	0.88887	1.80	0.93138
1.06	0.96874	1.31	0.89600	1.56	0.88964	1.81	0.93408
1.07	0.96415	1.32	0.89464	1.57	0.89049	1.82	0.93685
1.08	0.95973	1.33	0.89338	1.58	0.89142	1.83	0.93969
1.09	0.95546	1.34	0.89222	1.59	0.89243	1.84	0.94261
1.10	0.95135	1.35	0.89115	1.60	0.89352	1.85	0.94561
1.11	0.94739	1.36	0.89018	1.61	0.89468	1.86	0.94869
1.12	0.94359	1.37	0.88931	1.62	0.89592	1.87	0.95184
1.13	0.93993	1.38	0.88854	1.63	0.89724	1.88	0.95507
1.14	0.93642	1.39	0.88785	1.64	0.89864	1.89	0.95838
1.15	0.93304	1.40	0.88726	1.65	0.90012	1.90	0.96177
1.16	0.92980	1.41	0.88676	1.66	0.90167	1.91	0.96523
1.17	0.92670	1.42	0.88636	1.67	0.90330	1.92	0.96878
1.18	0.92373	1.43	0.88604	1.68	0.90500	1.93	0.97240
1.19	0.92088	1.44	0.88580	1.69	0.90678	1.94	0.97610
1.20	0.91817	1.45	0.88565	1.70	0.90864	1.95	0.97988
1.21	0.91558	1.46	0.88560	1.71	0.91057	1.96	0.98374
1.22	0.91311	1.47	0.88563	1.72	0.91258	1.97	0.98768
1.23	0.91075	1.48	0.88575	1.73	0.91466	1.98	0.99171
1.24	0.90852	1.49	0.88595	1.74	0.91683	1.99	0.99581
						2.00	1.00000

[a] Values of $\Gamma(n) = \int_0^\infty e^{-x} x^{n-1}\, dx$; $\Gamma(n+1) = n\Gamma(n)$.

For large positive integers, Stirling's formula gives an approximation in which the relative error decreases as n increases:

$$\Gamma(n+1) = (2\pi n)^{1/2}\left(\frac{n}{e}\right)^n$$

[b] From *CRC Standard Mathematical Tables*, Chemical Rubber Publishing Co., 12th ed., 1959. Used by permission.

TABLE 1.22   Bessel Functions

	$J_0(x)$ and $J_1(x)$[a]							
x	$J_0(x)$	$J_1(x)$	x	$J_0(x)$	$J_1(x)$	x	$J_0(x)$	$J_1(x)$
0.0	1.0000	0.0000	1.5	0.5118	0.5579	3.0	−0.2601	0.3391
0.1	0.9975	0.0499	1.6	0.4554	0.5699	3.1	−0.2921	0.3009
0.2	0.9900	0.0995	1.7	0.3980	0.5778	3.2	−0.3202	0.2613
0.3	0.9776	0.1483	1.8	0.3400	0.5815	3.3	−0.3443	0.2207
0.4	0.9604	0.1960	1.9	0.2818	0.5812	3.4	−0.3643	0.1792
0.5	0.9385	0.2423	2.0	0.2239	0.5767	3.5	−0.3801	0.1374
0.6	0.9120	0.2867	2.1	0.1666	0.5683	3.6	−0.3918	0.0955
0.7	0.8812	0.3290	2.2	0.1104	0.5560	3.7	−0.3992	0.0538
0.8	0.8463	0.3668	2.3	0.0555	0.5399	3.8	−0.4026	0.0128
0.9	0.8075	0.4059	2.4	0.0025	0.5202	3.9	−0.4018	−0.0272
1.0	0.7652	0.4401	2.5	−0.0484	0.4971	4.0	−0.3971	−0.0660
1.1	0.7196	0.4709	2.6	−0.0968	0.4708	4.1	−0.3887	−0.1033
1.2	0.6711	0.4983	2.7	−0.1424	0.4416	4.2	−0.3766	−0.1386
1.3	0.6201	0.5220	2.8	−0.1850	0.4097	4.3	−0.3610	−0.1719
1.4	0.5669	0.5419	2.9	−0.2243	0.3754	4.4	−0.3423	−0.2028

TABLE 1.22 (*Continued*)

				$J_0(x)$ and $J_1(x)^a$				
x	$J_0(x)$	$J_1(x)$	x	$J_0(x)$	$J_1(x)$	x	$J_0(x)$	$J_1(x)$
4.5	−0.3205	−0.2311	6.0	0.1506	−0.2767	7.5	0.2663	0.1352
4.6	−0.2961	−0.2566	6.1	0.1773	−0.2559	7.6	0.2516	0.1592
4.7	−0.2693	−0.2791	6.2	0.2017	−0.2329	7.7	0.2346	0.1813
4.8	−0.2404	−0.2985	6.3	0.2238	−0.2081	7.8	0.2154	0.2014
4.9	−0.2097	−0.3147	6.4	0.2433	−0.1816	7.9	0.1944	0.2192
5.0	−0.1776	−0.3276	6.5	0.2601	−0.1538	8.0	0.1717	0.2346
5.1	−0.1443	−0.3371	6.6	0.2740	−0.1250	8.1	0.1475	0.2476
5.2	−0.1103	−0.3432	6.7	0.2851	−0.0953	8.2	0.1222	0.2580
5.3	−0.0758	−0.3460	6.8	0.2931	−0.0652	8.3	0.0960	0.2657
5.4	−0.0412	−0.3453	6.9	0.2981	−0.0349	8.4	0.0692	0.2708
5.5	−0.0068	−0.3414	7.0	0.3001	−0.0047	8.5	0.0419	0.2731
5.6	0.0270	−0.3343	7.1	0.2991	0.0252	8.6	0.0146	0.2728
5.7	0.0599	−0.3241	7.2	0.2951	0.0543	8.7	−0.0125	0.2697
5.8	0.0917	−0.3110	7.3	0.2882	0.0826	8.8	−0.0392	0.2641
5.9	0.1220	−0.2951	7.4	0.2786	0.1096	8.9	−0.0653	0.2559

				$Y_0(x)$ and $Y_1(x)$				
x	$Y_0(x)$	$Y_1(x)$	x	$Y_0(x)$	$Y_1(x)$	x	$Y_0(x)$	$Y_1(x)$
0.0	$(-\infty)$	$(-\infty)$	2.5	0.498	0.146	5.0	−0.309	0.148
0.5	−0.445	−1.471	3.0	0.377	0.325	5.5	−0.340	−0.024
1.0	0.088	−0.781	3.5	0.189	0.410	6.0	−0.288	−0.175
1.5	0.382	−0.412	4.0	−0.017	0.398	6.5	−0.173	−0.274
2.0	0.510	−0.107	4.5	−0.195	0.301	7.0	−0.026	−0.303

$^a J_1(x) = 0$ for $x = 0, 3.832, 7.016, 10.173, 13.324, \ldots$
$J_0(x) = 0$ for $x = 2.405, 5.520, 8.654, 11.792, \ldots$

1.3 STATISTICAL TABLES

TABLE 1.23 Binomial Coefficients

n	$\binom{n}{0}$	$\binom{n}{1}$	$\binom{n}{2}$	$\binom{n}{3}$	$\binom{n}{4}$	$\binom{n}{5}$	$\binom{n}{6}$	$\binom{n}{7}$	$\binom{n}{8}$	$\binom{n}{9}$	$\binom{n}{10}$
0	1										
1	1	1									
2	1	2	1								
3	1	3	3	1							
4	1	4	6	4	1						
5	1	5	10	10	5	1					
6	1	6	15	20	15	6	1				
7	1	7	21	35	35	21	7	1			
8	1	8	28	56	70	56	28	8	1		
9	1	9	36	84	126	126	84	36	9	1	
10	1	10	45	120	210	252	210	120	45	10	1
11	1	11	55	165	330	462	462	330	165	55	11
12	1	12	66	220	495	792	924	792	495	220	66
13	1	13	78	286	715	1287	1716	1716	1287	715	286
14	1	14	91	364	1001	2002	3003	3432	3003	2002	1001
15	1	15	105	455	1365	3003	5005	6435	6435	5005	3003
16	1	16	120	560	1820	4368	8008	11440	12870	11440	8008
17	1	17	136	680	2380	6188	12376	19448	24310	24310	19448
18	1	18	153	816	3060	8568	18564	31824	43758	48620	43758
19	1	19	171	969	3876	11628	27132	50388	75582	92378	92378
20	1	20	190	1140	4845	15504	38760	77520	125970	167960	184756

$$_nC_m = \binom{n}{m} = \frac{n!}{[(n-m)!m!]} = \binom{n}{n-m}, \binom{n}{0} = 1$$

$$(p+q)^n = p^n + \binom{n}{1}p^{n-1}q + \cdots + \binom{n}{s}p^s q^t + \cdots + q^n, \quad s + t = n.$$

event e will happen at least r times in n trials is $\displaystyle\sum_{t=0}^{t=n-r} \binom{n}{t} p^{n-t} q^t$; at most r times in n trials is $\displaystyle\sum_{t=n-r}^{t=n}$ $\binom{n}{t} p^{n-t} q^t$.

In a *point binomial*, $(p + q)^n$, distribution, the mean number of favorable events is np; the mean number of unfavorable events is nq; the *standard deviation* is $\sigma = \sqrt{pqn}$; and $a_3 = (p - q)/\sigma$. The *mean deviation from the mean* = MD is $\sigma\sqrt{2/\pi} = 0.7979\sigma$; the *semiquartile deviation from the mean* = $0.6745\sigma = 0.845$ MD.

The probability that a deviation of an individual measure from the average lies between $y = -a$ and $y = a$ is

$$\frac{1}{\sqrt{\pi}} \int_{y=-a}^{y=a} h e^{-h^2 y^2}\, dy = \frac{1}{\sigma\sqrt{2\pi}} \int_{y=-a}^{y=a} e^{-y^2/2\sigma^2}\, dy = \frac{1}{\sqrt{2\pi}} \int_{x=-b}^{x=b} e^{-x^2/2}\, dx$$

where $x = hy\sqrt{2}$, $b = ha\sqrt{2}$, and $\sigma = 1/h\sqrt{2}$; h is called the *modulus of precision* and σ the *standard (quadratic mean) deviation*.

TABLE 1.24 Probability Functions

Definitions:

$$\tfrac{1}{2}(1 + \alpha) = \int_{-\infty}^{x} \Phi(x)\, dx = \text{area under } \Phi(x) \text{ from } -\infty \text{ to } x$$

$$\alpha = \int_{-x}^{x} \Phi(x)\, dx, \qquad \Phi(x) = \frac{1}{\sqrt{2\pi}} e^{-x^2/2} = \text{normal function}$$

$$\Phi^{(2)}(x) = (x^2 - 1)\Phi(x) \qquad = \text{second derivative of } \Phi(x)$$

$$\Phi^{(3)}(x) = (3x - x^3)\Phi(x) \qquad = \text{third derivative of } \Phi(x)$$

$$\Phi^{(4)}(x) = (x^4 - 6x^2 + 3)\Phi(x) = \text{fourth derivative of } \Phi(x)$$

x	$\tfrac{1}{2}(1+\alpha)$	$\Phi(x)$	$\Phi^{(2)}(x)$	$\Phi^{(3)}(x)$	$\Phi^{(4)}(x)$	x	$\tfrac{1}{2}(1+\alpha)$	$\Phi(x)$	$\Phi^{(2)}(x)$	$\Phi^{(3)}(x)$	$\Phi^{(4)}(x)$
0.00	0.5000	0.3989	−0.3989	0.0000	1.1968	0.25	0.5987	0.3867	−0.3625	0.2840	1.0165
0.01	0.5040	0.3989	−0.3989	0.0120	1.1965	0.26	0.6026	0.3857	−0.3596	0.2941	1.0024
0.02	0.5080	0.3989	−0.3987	0.0239	1.1956	0.27	0.6064	0.3847	−0.3566	0.3040	0.9878
0.03	0.5120	0.3988	−0.3984	0.0359	1.1941	0.28	0.6103	0.3836	−0.3535	0.3138	0.9727
0.04	0.5160	0.3986	−0.3980	0.0478	1.1920	0.29	0.6141	0.3825	−0.3504	0.3235	0.9572
0.05	0.5199	0.3984	−0.3975	0.0597	1.1894	0.30	0.6179	0.3814	−0.3471	0.3330	0.9413
0.06	0.5239	0.3982	−0.3968	0.0716	1.1861	0.31	0.6217	0.3802	−0.3437	0.3423	0.9250
0.07	0.5279	0.3980	−0.3960	0.0834	1.1822	0.32	0.6255	0.3790	−0.3402	0.3515	0.9082
0.08	0.5319	0.3977	−0.3951	0.0952	1.1778	0.33	0.6293	0.3778	−0.3367	0.3605	0.8910
0.09	0.5359	0.3973	−0.3941	0.1070	1.1727	0.34	0.6331	0.3765	−0.3330	0.3693	0.8735
0.10	0.5398	0.3970	−0.3930	0.1187	1.1671	0.35	0.6368	0.3752	−0.3293	0.3779	0.8556
0.11	0.5438	0.3965	−0.3917	0.1303	1.1609	0.36	0.6406	0.3739	−0.3255	0.3864	0.8373
0.12	0.5478	0.3961	−0.3904	0.1419	1.1541	0.37	0.6443	0.3726	−0.3216	0.3947	0.8186
0.13	0.5517	0.3956	−0.3889	0.1534	1.1468	0.38	0.6480	0.3712	−0.3176	0.4028	0.7996
0.14	0.5557	0.3951	−0.3873	0.1648	1.1389	0.39	0.6517	0.3697	−0.3135	0.4107	0.7803
0.15	0.5596	0.3945	−0.3856	0.1762	1.1304	0.40	0.6554	0.3683	−0.3094	0.4184	0.7607
0.16	0.5636	0.3939	−0.3838	0.1874	1.1214	0.41	0.6591	0.3668	−0.3059	0.4259	0.7408
0.17	0.5675	0.3932	−0.3819	0.1986	1.1118	0.42	0.6628	0.3653	−0.3008	0.4332	0.7206
0.18	0.5714	0.3925	−0.3798	0.2097	1.1017	0.43	0.6664	0.3637	−0.2965	0.4403	0.7001
0.19	0.5753	0.3918	−0.3777	0.2206	1.0911	0.44	0.6700	0.3621	−0.2920	0.4472	0.6793
0.20	0.5793	0.3910	−0.3754	0.2315	1.0799	0.45	0.6736	0.3605	−0.2875	0.4539	0.6583
0.21	0.5832	0.3902	−0.3730	0.2422	1.0682	0.46	0.6772	0.3589	−0.2830	0.4603	0.6371
0.22	0.5871	0.3894	−0.3706	0.2529	1.0560	0.47	0.6808	0.3572	−0.2783	0.4666	0.6156
0.23	0.5910	0.3885	−0.3680	0.2634	1.0434	0.48	0.6844	0.3555	−0.2736	0.4727	0.5940
0.24	0.5948	0.3876	−0.3653	0.2737	1.0302	0.49	0.6879	0.3538	−0.2689	0.4785	0.5721

*Tables 1.23–1.25 from Burington, *Handbook of Math Tables and Formulas*, published by McGraw-Hill.

TABLE 1.24 (*Continued*)

x	$\frac{1}{2}(1+\alpha)$	$\Phi(x)$	$\Phi^{(2)}(x)$	$\Phi^{(3)}(x)$	$\Phi^{(4)}(x)$	x	$\frac{1}{2}(1+\alpha)$	$\Phi(x)$	$\Phi^{(2)}(x)$	$\Phi^{(3)}(x)$	$\Phi^{(4)}(x)$
0.50	0.6915	0.3521	−0.2641	0.4841	0.5501	1.05	0.8531	0.2299	0.0236	0.4580	−0.5516
0.51	0.6950	0.3503	−0.2592	0.4895	0.5279	1.06	0.8554	0.2275	0.0281	0.4524	−0.5639
0.52	0.6985	0.3485	−0.2543	0.4947	0.5056	1.07	0.8577	0.2251	0.0326	0.4467	−0.5758
0.53	0.7019	0.3467	−0.2493	0.4996	0.4831	1.08	0.8599	0.2227	0.0371	0.4409	−0.5873
0.54	0.7054	0.3448	−0.2443	0.5043	0.4605	1.09	0.8621	0.2203	0.0414	0.4350	−0.5984
0.55	0.7088	0.3429	−0.2392	0.5088	0.4378	1.10	0.8643	0.2179	0.0458	0.4290	−0.6091
0.56	0.7123	0.3410	−0.2341	0.5131	0.4150	1.11	0.8665	0.2155	0.0500	0.4228	−0.6193
0.57	0.7157	0.3391	−0.2289	0.5171	0.3921	1.12	0.8686	0.2131	0.0542	0.4166	−0.6292
0.58	0.7190	0.3372	−0.2238	0.5209	0.3691	1.13	0.8708	0.2107	0.0583	0.4102	−0.6386
0.59	0.7224	0.3352	−0.2185	0.5245	0.3461	1.14	0.8729	0.2083	0.0624	0.4038	−0.6476
0.60	0.7257	0.3332	−0.2133	0.5278	0.3231	1.15	0.8749	0.2059	0.0664	0.3973	−0.6561
0.61	0.7291	0.3312	−0.2080	0.5309	0.3000	1.16	0.8770	0.2036	0.0704	0.3907	−0.6643
0.62	0.7324	0.3292	−0.2027	0.5338	0.2770	1.17	0.8790	0.2012	0.0742	0.3840	−0.6720
0.63	0.7357	0.3271	−0.1973	0.5365	0.2539	1.18	0.8810	0.1989	0.0780	0.3772	−0.6792
0.64	0.7389	0.3251	−0.1919	0.5389	0.2309	1.19	0.8830	0.1965	0.0818	0.3704	−0.6861
0.65	0.7422	0.3230	−0.1865	0.5411	0.2078	1.20	0.8849	0.1942	0.0854	0.3635	−0.6926
0.66	0.7454	0.3209	−0.1811	0.5431	0.1849	1.21	0.8869	0.1919	0.0890	0.3566	−0.6986
0.67	0.7486	0.3187	−0.1757	0.5448	0.1620	1.22	0.8888	0.1919	0.0890	0.3566	−0.6986
0.68	0.7517	0.3166	−0.1702	0.5463	0.1391	1.23	0.8907	0.1872	0.0960	0.3425	−0.7094
0.69	0.7549	0.3144	−0.1647	0.5476	0.1164	1.24	0.8925	0.1849	0.0994	0.3354	−0.7141
0.70	0.7580	0.3123	−0.1593	0.5486	0.0937	1.25	0.8944	0.1826	0.1027	0.3282	−0.7185
0.71	0.7611	0.3101	−0.1538	0.5495	0.0712	1.26	0.8962	0.1804	0.1060	0.3210	−0.7224
0.72	0.7642	0.3079	−0.1483	0.5501	0.0487	1.27	0.8980	0.1781	0.1092	0.3138	−0.7259
0.73	0.7673	0.3056	−0.1428	0.5504	0.0265	1.28	0.8997	0.1758	0.1123	0.3065	−0.7291
0.74	0.7704	0.3034	−0.1373	0.5506	0.0043	1.29	0.9015	0.1736	0.1153	0.2992	−0.7318
0.75	0.7734	0.3011	−0.1318	0.5505	−0.0176	1.30	0.9032	0.1714	0.1182	0.2918	−0.7341
0.76	0.7764	0.2989	−0.1262	0.5502	−0.0394	1.31	0.9049	0.1691	0.1211	0.2845	−0.7361
0.77	0.7794	0.2966	−0.1207	0.5497	−0.0611	1.32	0.9066	0.1669	0.1239	0.2771	−0.7376
0.78	0.7823	0.2943	−0.1153	0.5490	−0.0825	1.33	0.9082	0.1647	0.1267	0.2697	−0.7388
0.79	0.7852	0.2920	−0.1098	0.5481	−0.1037	1.34	0.9099	0.1626	0.1293	0.2624	−0.7395
0.80	0.7881	0.2897	−0.1043	0.5469	−0.1247	1.35	0.9115	0.1604	0.1319	0.2550	−0.7399
0.81	0.7910	0.2874	−0.0988	0.5456	−0.1455	1.36	0.9131	0.1582	0.1344	0.2476	−0.7400
0.82	0.7939	0.2850	−0.0934	0.5440	−0.1660	1.37	0.9147	0.1561	0.1369	0.2402	−0.7396
0.83	0.7967	0.2827	−0.0880	0.5423	−0.1862	1.38	0.9162	0.1539	0.1392	0.2328	−0.7389
0.84	0.7995	0.2803	−0.0825	0.5403	−0.2063	1.39	0.9177	0.1518	0.1415	0.2254	−0.7378
0.85	0.8023	0.2780	−0.0771	0.5381	−0.2260	1.40	0.9192	0.1497	0.1437	0.2180	−0.7364
0.86	0.8051	0.2756	−0.0718	0.5358	−0.2455	1.41	0.9207	0.1476	0.1459	0.2107	−0.7347
0.87	0.8078	0.2732	−0.0664	0.5332	−0.2646	1.42	0.9222	0.1456	0.1480	0.2033	−0.7326
0.88	0.8106	0.2709	−0.0611	0.5305	−0.2835	1.43	0.9236	0.1435	0.1500	0.1960	−0.7301
0.89	0.8133	0.2685	−0.0558	0.5276	−0.3021	1.44	0.9251	0.1415	0.1519	0.1887	−0.7274
0.90	0.8159	0.2661	−0.0506	0.5245	−0.3203	1.45	0.9265	0.1394	0.1537	0.1815	−0.7243
0.91	0.8186	0.2637	−0.0453	0.5212	−0.3383	1.46	0.9279	0.1374	0.1555	0.1742	−0.7209
0.92	0.8212	0.2613	−0.0401	0.5177	−0.3559	1.47	0.9292	0.1354	0.1572	0.1670	−0.7172
0.93	0.8238	0.2589	−0.0350	0.5140	−0.3731	1.48	0.9306	0.1344	0.1588	0.1599	−0.7132
0.94	0.8264	0.2565	−0.0299	0.5102	−0.3901	1.49	0.9319	0.1315	0.1604	0.1528	−0.7089
0.95	0.8289	0.2541	−0.0248	0.5062	−0.4066	1.50	0.9332	0.1295	0.1619	0.1457	−0.7043
0.96	0.8315	0.2516	−0.0197	−0.521	−0.4228	1.51	0.9345	0.1276	0.1633	0.1387	−0.6994
0.97	0.8340	0.2492	−0.0147	0.4978	−0.4387	1.52	0.9357	0.1257	0.1647	0.1317	−0.6942
0.98	0.8365	0.2468	−0.0098	0.4933	−0.4541	1.53	0.9370	0.1238	0.1660	0.1248	−0.6888
0.99	0.8389	0.2444	−0.0049	0.4887	−0.4692	1.54	0.9382	0.1219	0.1672	0.1180	−0.6831
1.00	0.8413	0.2420	0.0000	0.4839	−0.4839	1.55	0.9394	0.1200	0.1683	0.1111	−0.6772
1.01	0.8438	0.2396	0.0048	0.4790	−0.4983	1.56	0.9406	0.1182	0.1694	0.1044	−0.6710
1.02	0.8461	0.2371	0.0096	0.4740	−0.5122	1.57	0.9418	0.1163	0.1704	0.0977	−0.6646
1.03	0.8485	0.2347	0.0143	0.4688	−0.5257	1.58	0.9429	0.1145	0.1714	0.0911	−0.6580
1.04	0.8508	0.2323	0.0190	0.4635	−0.5389	1.59	0.9441	0.1127	0.1722	0.0846	−0.6511

TABLE 1.24 (*Continued*)

x	$\frac{1}{2}(1+\alpha)$	$\Phi(x)$	$\Phi^{(2)}(x)$	$\Phi^{(3)}(x)$	$\Phi^{(4)}(x)$	x	$\frac{1}{2}(1+\alpha)$	$\Phi(x)$	$\Phi^{(2)}(x)$	$\Phi^{(3)}(x)$	$\Phi^{(4)}(x)$
1.60	0.9452	0.1109	0.1730	0.0781	-0.6441	2.15	0.9842	0.0395	0.1433	-0.1380	-0.1332
1.61	0.9463	0.1092	0.1738	0.0717	-0.6368	2.16	0.9846	0.0387	0.1419	-0.1393	-0.1249
1.62	0.9474	0.1074	0.1745	0.0654	-0.6293	2.17	0.9850	0.0379	0.1405	-0.1405	-0.1167
1.63	0.9484	0.1057	0.1751	0.0591	-0.6216	2.18	0.9854	0.0371	0.1391	-0.1416	-0.1086
1.64	0.9495	0.1040	0.1757	0.0529	-0.6138	2.19	0.9857	0.0363	0.1377	-0.1426	-0.1006
1.65	0.9505	0.1023	0.1762	0.0468	-0.6057	2.20	0.9861	0.0355	0.1362	-0.1436	-0.0927
1.66	0.9515	0.1006	0.1766	0.0408	-0.5975	2.21	0.9864	0.0347	0.1348	-0.1445	-0.0850
1.67	0.9525	0.0989	0.1770	0.0349	-0.5891	2.22	0.9868	0.0339	0.1333	-0.1453	-0.0774
1.68	0.9535	0.0973	0.1773	0.0290	-0.5806	2.23	0.9871	0.0332	0.1319	-0.1460	-0.0700
1.69	0.9545	0.0957	0.1776	0.0233	-0.5720	2.24	0.9875	0.0325	0.1304	-0.1467	-0.0626
1.70	0.9554	0.0940	0.1778	0.0176	-0.5632	2.25	0.9878	0.0317	0.1289	-0.1473	-0.0554
1.71	0.9564	0.0925	0.1779	0.0120	-0.5542	2.26	0.9881	0.0310	0.1275	-0.1478	-0.0484
1.72	0.9573	0.0909	0.1780	0.0065	-0.5452	2.27	0.9884	0.0303	0.1260	-0.1483	-0.0414
1.73	0.9582	0.0893	0.1780	0.0011	-0.5360	2.28	0.9887	0.0297	0.1245	-0.1486	-0.0346
1.74	0.9591	0.0878	0.1780	-0.0042	-0.5267	2.29	0.9890	0.0290	0.1230	-0.1490	-0.0279
1.75	0.9599	0.0863	0.1780	-0.0094	-0.5173	2.30	0.9893	0.0283	0.1215	-0.1492	-0.0214
1.76	0.9608	0.0848	0.1778	-0.0146	-0.5079	2.31	0.9896	0.0277	0.1200	-0.1494	-0.0150
1.77	0.9616	0.0833	0.1777	-0.0196	-0.4983	2.32	0.9898	0.0270	0.1185	-0.1495	-0.0088
1.78	0.9625	0.0818	0.1774	-0.0245	-0.4887	2.33	0.9901	0.0264	0.1170	-0.1496	-0.0027
1.79	0.9633	0.0804	0.1772	-0.0294	-0.4789	2.34	0.9904	0.0258	0.1155	-0.1496	0.0033
1.80	0.9641	0.0790	0.1769	-0.0341	-0.4692	2.35	0.9906	0.0252	0.1141	-0.1495	0.0092
1.81	0.9649	0.0775	0.1765	-0.0388	-0.4593	2.36	0.9909	0.0246	0.1126	-0.1494	0.0149
1.82	0.9656	0.0761	0.1761	-0.0433	-0.4494	2.37	0.9911	0.0241	0.1111	-0.1492	0.0204
1.83	0.9664	0.0748	0.1756	-0.0477	-0.4395	2.38	0.9913	0.0235	0.1096	-0.1490	0.0258
1.84	0.9671	0.0734	0.1751	-0.0521	-0.4295	2.39	0.9916	0.0229	0.1081	-0.1487	0.0311
1.85	0.9678	0.0721	0.1746	-0.0563	-0.4195	2.40	0.9918	0.0224	0.1066	-0.1483	0.0362
1.86	0.9686	0.0707	0.1740	-0.0605	-0.4095	2.41	0.9920	0.0219	0.1051	-0.1480	0.0412
1.87	0.9693	0.0694	0.1734	-0.0645	-0.3995	2.42	0.9922	0.0213	0.1036	-0.1475	0.0461
1.88	0.9699	0.0681	0.1727	-0.0685	-0.3894	2.43	0.9925	0.0208	0.1022	-0.1470	0.0508
1.89	0.9706	0.0669	0.1720	-0.0723	-0.3793	2.44	0.9927	0.0203	0.1007	-0.1465	0.0554
1.90	0.9713	0.0656	0.1713	-0.0761	-0.3693	2.45	0.9929	0.0198	0.0992	-0.1459	0.0598
1.91	0.9719	0.0644	0.1705	-0.0797	-0.3592	2.46	0.9931	0.0194	0.0978	-0.1453	0.0641
1.92	0.9726	0.0632	0.1697	-0.0832	-0.3492	2.47	0.9932	0.0189	0.0963	-0.1446	0.0683
1.93	0.9732	0.0620	0.1688	-0.0867	-0.3392	2.48	0.9934	0.0184	0.0949	-0.1439	0.0723
1.94	0.9738	0.0608	0.1679	-0.0900	-0.3292	2.49	0.9936	0.0180	0.0935	-0.1432	0.0762
1.95	0.9744	0.0596	0.1670	-0.0933	-0.3192	2.50	0.9938	0.0175	0.0920	-0.1424	0.0800
1.96	0.9750	0.0584	0.1661	-0.0964	-0.3093	2.51	0.9940	0.0171	0.0906	-0.1416	0.0836
1.97	0.9756	0.0573	0.1651	-0.0994	-0.2994	2.52	0.9941	0.0167	0.0892	-0.1408	0.0871
1.98	0.9761	0.0562	0.1641	-0.1024	-0.2895	2.53	0.9943	0.0163	0.0878	-0.1399	0.0905
1.99	0.9767	0.0551	0.1630	-0.1052	-0.2797	2.54	0.9945	0.0158	0.0864	-0.1389	0.0937
2.00	0.9772	0.0540	0.1620	-0.1080	-0.2700	2.55	0.9946	0.0154	0.0850	-0.1380	0.0968
2.01	0.9778	0.0529	0.1609	-0.1106	-0.2603	2.56	0.9948	0.0151	0.0836	-0.1370	0.0998
2.02	0.9783	0.0519	0.1598	-0.1132	-0.2506	2.57	0.9949	0.0147	0.0823	-0.1360	0.1027
2.03	0.9788	0.0508	0.1586	-0.1157	-0.2411	2.58	0.9951	0.0143	0.0809	-0.1350	0.1054
2.04	0.9793	0.0498	0.1575	-0.1180	-0.2316	2.59	0.9952	0.0139	0.0796	-0.1339	0.1080
2.05	0.9798	0.0468	0.1563	-0.1203	-0.2222	2.60	0.9953	0.0136	0.0782	-0.1328	0.1105
2.06	0.9803	0.0478	0.1550	-0.1225	-0.2129	2.60	0.9953	0.0136	0.0782	-0.1328	0.1105
2.07	0.9808	0.0468	0.1538	-0.1245	-0.2036	2.62	0.9956	0.0129	0.0756	-0.1305	0.1152
2.08	0.9812	0.0459	0.1526	-0.1265	-0.1945	2.63	0.9957	0.0126	0.0743	-0.1294	0.1173
2.09	0.9817	0.0449	0.1513	-0.1284	-0.1854	2.64	0.9959	0.0122	0.0730	-0.1282	0.1194
2.10	0.9821	0.0440	0.1500	-0.1302	-0.1765	2.65	0.9960	0.0119	0.0717	-0.1270	0.1213
2.11	0.9821	0.0440	0.1500	-0.1302	-0.1765	2.66	0.9961	0.0116	0.0705	-0.1258	0.1231
2.12	0.9830	0.0422	0.1474	-0.1336	-0.1588	2.67	0.9962	0.0113	0.0692	-0.1245	0.1248
2.13	0.9834	0.0413	0.1460	-0.1351	-0.1502	2.68	0.9963	0.0110	0.0680	-0.1233	0.1264
2.14	0.9838	0.0404	0.1446	-0.1366	-0.1416	2.69	0.9964	0.0107	0.0668	-0.1220	0.1279

TABLE 1.24 (*Continued*)

x	$\frac{1}{2}(1+\alpha)$	$\Phi(x)$	$\Phi^{(2)}(x)$	$\Phi^{(3)}(x)$	$\Phi^{(4)}(x)$	x	$\frac{1}{2}(1+\alpha)$	$\Phi(x)$	$\Phi^{(2)}(x)$	$\Phi^{(3)}(x)$	$\Phi^{(4)}(x)$
2.70	0.9965	0.0104	0.0656	−0.1207	0.1293	3.20	0.9993	0.0024	0.0220	−0.0552	0.1107
2.71	0.9966	0.0101	0.0644	−0.1194	0.1306	3.21	0.9993	0.0023	0.0215	−0.0541	0.1093
2.72	0.9967	0.0099	0.0632	−0.1181	0.1317	3.22	0.9994	0.0022	0.0210	−0.0531	0.1080
2.73	0.9968	0.0096	0.0620	−0.1168	0.1328	3.23	0.9994	0.0022	0.0204	−0.0520	0.1066
2.74	0.9969	0.0093	0.0608	−0.1154	0.1338	3.24	0.9994	0.0021	0.0199	−0.0509	0.1053
2.75	0.9970	0.0091	0.0597	−0.1141	0.1347	3.25	0.9994	0.0020	0.0194	−0.0499	0.1039
2.76	0.9971	0.0088	0.0585	−0.1127	0.1356	3.26	0.9994	0.0020	0.0189	−0.0488	0.1025
2.77	0.9972	0.0086	0.0574	−0.1114	0.1363	3.27	0.9995	0.0019	0.0184	−0.0478	0.1011
2.78	0.9973	0.0084	0.0563	−0.1100	0.1369	3.28	0.9995	0.0018	0.0180	−0.0468	0.0997
2.79	0.9974	0.0081	0.0552	−0.1087	0.1375	3.29	0.9995	0.0018	0.0175	−0.0458	0.0983
2.80	0.9974	0.0079	0.0541	−0.1073	0.1379	3.30	0.9995	0.0017	0.0170	−0.0449	0.0969
2.81	0.9975	0.0077	0.0531	−0.1059	0.1383	3.31	0.9995	0.0017	0.0106	−0.0439	0.0955
2.82	0.9976	0.0075	0.0520	−0.1045	0.1386	3.32	0.9996	0.0016	0.0102	−0.0429	0.0941
2.83	0.9977	0.0073	0.0510	−0.1031	0.1389	3.33	0.9996	0.0016	0.0157	−0.0420	0.0927
2.84	0.9977	0.0071	0.0500	−0.1017	0.1390	3.34	0.9996	0.0015	0.0153	−0.0411	0.0913
2.85	0.9978	0.0069	0.0490	−0.1003	0.1391	3.35	0.9996	0.0015	0.0149	−0.0402	0.0899
2.86	0.9979	0.0067	0.0480	−0.0990	0.1391	3.36	0.9996	0.0014	0.0145	−0.0393	0.0885
2.87	0.9979	0.0065	0.0470	−0.0976	0.1391	3.37	0.9996	0.0014	0.0141	−0.0384	0.0871
2.88	0.9980	0.0063	0.0460	−0.0962	0.1389	3.38	0.9996	0.0013	0.0138	−0.0376	0.0857
2.89	0.9981	0.0061	0.0451	−0.0948	0.1388	3.39	0.9997	0.0013	0.0134	−0.0367	0.0843
2.90	0.9981	0.0060	0.0441	−0.0934	0.1385	3.40	0.9997	0.0012	0.0130	−0.0359	0.0829
2.91	0.9982	0.0058	0.0432	−0.0920	0.1382	3.41	0.9997	0.0012	0.0127	−0.0350	0.0815
2.92	0.9982	0.0056	0.0423	−0.0906	0.1378	3.42	0.9997	0.0012	0.0123	−0.0342	0.0801
2.93	0.9983	0.0055	0.0414	−0.0893	0.1374	3.43	0.9997	0.0011	0.0120	−0.0334	0.0788
2.94	0.9984	0.0053	0.0405	−0.0879	0.1369	3.44	0.9997	0.0011	0.0116	−0.0327	0.0774
2.95	0.9984	0.0051	0.0396	−0.0865	0.1364	3.45	0.9997	0.0010	0.0113	−0.0319	0.0761
2.96	0.9985	0.0050	0.0388	−0.0852	0.1358	3.46	0.9997	0.0010	0.0110	−0.0311	0.0747
2.97	0.9985	0.0048	0.0379	−0.0838	0.1352	3.47	0.9997	0.0010	0.0107	−0.0304	0.0734
2.98	0.9986	0.0047	0.0371	−0.0825	0.1345	3.48	0.9998	0.0009	0.0104	−0.0297	0.0721
2.99	0.9986	0.0046	0.0363	−0.0811	0.1337	3.49	0.9998	0.0009	0.0101	−0.0290	0.0707
3.00	0.9987	0.0044	0.0355	−0.0798	0.1330	3.50	0.9998	0.0009	0.0098	−0.0283	0.0694
3.01	0.9987	0.0043	0.0347	−0.0785	0.1321	3.51	0.9998	0.0008	0.0095	−0.0276	0.0681
3.02	0.9987	0.0042	0.0339	−0.0771	0.1313	3.52	0.9998	0.0008	0.0093	−0.0269	0.0669
3.03	0.9988	0.0040	0.0331	−0.0758	0.1304	3.53	0.9998	0.0008	0.0090	−0.0262	0.0656
3.04	0.9988	0.0039	0.0324	−0.0745	0.1294	3.54	0.9998	0.0008	0.0087	−0.0256	0.0643
3.05	0.9989	0.0038	0.0316	−0.0732	0.1285	3.55	0.9998	0.0007	0.0085	−0.0249	0.0631
3.06	0.9989	0.0037	0.0309	−0.0720	0.1275	3.56	0.9998	0.0007	0.0082	−0.0243	0.0618
3.07	0.9989	0.0036	0.0302	−0.0707	0.1264	3.57	0.9998	0.0007	0.0080	−0.0237	0.0606
3.08	0.9990	0.0035	0.0295	−0.0694	0.1254	3.58	0.9998	0.0007	0.0078	−0.0231	0.0594
3.09	0.9990	0.0034	0.0288	−0.0682	0.1243	3.59	0.9998	0.0006	0.0075	−0.0225	0.0582
3.10	0.9990	0.0033	0.0281	−0.0669	0.1231	3.60	0.9998	0.0006	0.0073	−0.0219	0.0570
3.11	0.9991	0.0032	0.0275	−0.0657	0.1220	3.61	0.9999	0.0006	0.0071	−0.0214	0.0559
3.12	0.9991	0.0031	0.0268	−0.0645	0.1208	3.62	0.9999	0.0006	0.0069	−0.0208	0.0547
3.13	0.9991	0.0030	0.0262	−0.0633	0.1196	3.63	0.9999	0.0006	0.0067	−0.0203	0.0536
3.14	0.9992	0.0029	0.0256	−0.0621	0.1184	3.64	0.9999	0.0005	0.0065	−0.0198	0.0524
3.15	0.9992	0.0028	0.0249	−0.0609	0.1171	3.65	0.9999	0.0005	0.0063	−0.0192	0.0513
3.16	0.9992	0.0027	0.0243	−0.0598	0.1159	3.66	0.9999	0.0005	0.0061	−0.0187	0.0502
3.17	0.9992	0.0026	0.0237	−0.0586	0.1146	3.67	0.9999	0.0005	0.0059	−0.0182	0.0492
3.18	0.9993	0.0025	0.0232	−0.0575	0.1133	3.68	.9999	0.0005	0.0057	−0.0177	0.0481
3.19	0.9993	0.0025	0.0226	−0.0564	0.1120	3.69	0.9999	0.0004	0.0056	−0.0173	0.0470

TABLE 1.24   (*Continued*)

x	$\frac{1}{2}(1+\alpha)$	$\Phi(x)$	$\Phi^{(2)}(x)$	$\Phi^{(3)}(x)$	$\Phi^{(4)}(x)$	x	$\frac{1}{2}(1+\alpha)$	$\Phi(x)$	$\Phi^{(2)}(x)$	$\Phi^{(3)}(x)$	$\Phi^{(4)}(x)$
3.70	0.9999	0.0004	0.0054	−0.0168	0.0460	3.95	1.0000	0.0002	0.0024	−0.0081	0.0250
3.71	0.9999	0.0004	0.0052	−0.0164	0.0450	3.96	1.0000	0.0002	0.0023	−0.0079	0.0243
3.72	0.9999	0.0004	0.0051	−0.0159	0.0440	3.97	1.0000	0.0002	0.0022	−0.0076	0.0237
3.73	0.9999	0.0004	0.0049	−0.0155	0.0430	3.98	1.0000	0.0001	0.0022	−0.0074	0.0230
3.74	0.9999	0.0004	0.0048	−0.0150	0.0420	3.99	1.0000	0.0001	0.0021	−0.0072	0.0224
3.75	0.9999	0.0004	0.0046	−0.0146	0.0410	4.00	1.0000	0.0001	0.0020	−0.0070	0.0218
3.76	0.9999	0.0003	0.0045	−0.0142	0.0401	4.05	1.0000	0.0001	0.0017	−0.0059	0.0190
3.77	0.9999	0.0003	0.0043	−0.0138	0.0392	4.10	1.0000	0.0001	0.0014	−0.0051	0.0165
3.78	0.9999	0.0003	0.0042	−0.0134	0.0382	4.15	1.0000	0.0001	0.0012	−0.0043	0.0143
3.79	0.9999	0.0003	0.0041	−0.0131	0.0373	4.20	1.0000	0.0001	0.0010	−0.0036	0.0123
3.80	0.9999	0.0003	0.0039	−0.0127	0.0365	4.25	1.0000	0.0001	0.0008	−0.0031	0.0105
3.81	0.9999	0.0003	0.0038	−0.0123	0.0356	4.30	1.0000	0.0000	0.0007	−0.0026	0.0090
3.82	0.9999	0.0003	0.0037	−0.0120	0.0347	4.35	1.0000	0.0000	0.0006	−0.0022	0.0077
3.83	0.9999	0.0003	0.0036	−0.0116	0.0339	4.40	1.0000	0.0000	0.0005	−0.0018	0.0065
3.84	0.9999	0.0003	0.0034	−0.0113	0.0331	4.45	1.0000	0.0000	0.0004	−0.0015	0.0055
3.85	0.9999	0.0002	0.0033	−0.0110	0.0323	4.50	1.0000	0.0000	0.0003	−0.0012	0.0047
3.86	0.9999	0.0002	0.0032	−0.0107	0.0315	4.55	1.0000	0.0000	0.0003	−0.0010	0.0039
3.87	1.0000	0.0002	0.0031	−0.0104	0.0307	4.60	1.0000	0.0000	0.0002	−0.0009	0.0033
3.88	1.0000	0.0002	0.0030	−0.0100	0.0299	4.65	1.0000	0.0000	0.0002	−0.0007	0.0027
3.89	1.0000	0.0002	0.0029	−0.0098	0.0292	4.70	1.0000	0.0000	0.0001	−0.0006	0.0023
3.90	1.0000	0.0002	0.0028	−0.0095	0.0284	4.75	1.0000	0.0000	0.0001	−0.0005	0.0019
3.91	1.0000	0.0002	0.0027	−0.0092	0.0277	4.80	1.0000	0.0000	0.0001	−0.0004	0.0016
3.92	1.0000	0.0002	0.0026	−0.0089	0.0270	4.85	1.0000	0.0000	0.0001	−0.0003	0.0013
3.93	1.0000	0.0002	0.0026	−0.0086	0.0263	4.90	1.0000	0.0000	0.0001	−0.0003	0.0011
3.94	1.0000	0.0002	0.0025	−0.0084	0.0256	4.95	1.0000	0.0000	0.0000	−0.0002	0.0009

The sum of those terms of

$$(p+q)^n \equiv \sum_{t=0}^{n} \binom{n}{t} p^{n-t} q^t, \; p + q = 1$$

in which t ranges from a to b, inclusive, a and b being integers, ($a \le t \le b$), is (if n is large enough) approximately

$$\int_{x_1}^{x_2} \phi(x) \, dx + \left[\frac{q-p}{6\sigma} \phi^{(2)}(x) + \frac{1}{24} \left(\frac{1}{\sigma^2} - \frac{6}{n} \right) \phi^{(3)}(x) \right]_{x_1}^{x_2}$$

where $x_1 = (a - \frac{1}{2} - qn)/\sigma$, $x_2 = (b + \frac{1}{2} - qn)/\sigma$.

The sum of the first $(t+1)$ terms of

$$(p+q)^n \equiv \sum_{t=0}^{n} \binom{n}{t} p^{n-t} q^t, \; p + q = 1$$

is approximately,

$$\int_{x}^{\infty} \phi(x) \, dx + \frac{q-p}{6\sigma} \phi^{(2)}(x) - \frac{1}{24} \left(\frac{1}{\sigma^2} - \frac{6}{n} \right) \phi^{(3)}(x)$$

where $x = (s - \frac{1}{2} - np)/\sigma$, $s = n - t$. The sum of the last $(s + 1)$ terms is approximately

$$\int_x^\infty \phi(x)\, dx - \frac{q - p}{6\sigma}\phi^{(2)}(x) - \frac{1}{24}\left(\frac{1}{\sigma^2} - \frac{6}{n}\right)\phi^{(3)}(x)$$

where $x = (t - \frac{1}{2} - nq)/\sigma$.

TABLE 1.25 Factors for Computing Probable Errors

n	$\dfrac{1}{\sqrt{n}}$	$\dfrac{1}{\sqrt{n(n-1)}}$	$\dfrac{0.6745}{\sqrt{n-1}}$	$\dfrac{0.6745}{\sqrt{n(n-1)}}$	$\dfrac{0.8453}{n\sqrt{n-1}}$	$\dfrac{0.8453}{\sqrt{n(n-1)}}$
2	0.707 107	0.707 107	0.6745	0.4769	0.4227	0.5978
3	0.577 350	0.408 248	0.4769	0.2754	0.1993	0.3451
4	0.500 000	0.288 675	0.3894	0.1947	0.1220	0.2440
5	0.447 214	0.223 607	0.3372	0.1508	0.0845	0.1890
6	0.408 248	0.182 574	0.3016	0.1231	0.0630	0.1543
7	0.377 964	0.154 303	0.2754	0.1041	0.0493	0.1304
8	0.353 553	0.133 631	0.2549	0.0901	0.0399	0.1130
9	0.333 333	0.117 851	0.2385	0.0795	0.0332	0.0996
10	0.316 228	0.105 409	0.2248	0.0711	0.0282	0.0891
11	0.301 511	0.095 346	0.2133	0.0643	0.0243	0.0806
12	0.288 675	0.087 039	0.2034	0.0587	0.0212	0.0736
13	0.277 350	0.080 064	0.1947	0.0540	0.0188	0.0677
14	0.267 261	0.074 125	0.1871	0.0500	0.0167	0.0627
15	0.258 199	0.069 007	0.1803	0.0465	0.0151	0.0583
16	0.250 000	0.064 550	0.1742	0.0435	0.0136	0.0546
17	0.242 536	0.060 634	0.1686	0.0409	0.0124	0.0513
18	0.235 702	0.057 166	0.1636	0.0386	0.0114	0.0483
19	0.229 416	0.054 074	0.1590	0.0365	0.0105	0.0457
20	0.223 607	0.051 299	0.1547	0.0346	0.0097	0.0434
21	0.218 218	0.048 795	0.1508	0.0329	0.0090	0.0412
22	0.213 201	0.046 524	0.1472	0.0314	0.0084	0.0393
23	0.208 514	0.044 455	0.1438	0.0300	0.0078	0.0376
24	0.204 124	0.042 563	0.1406	0.0287	0.0073	0.0360
25	0.200 000	0.040 825	0.1377	0.0275	0.0069	0.0345
26	0.196 116	0.039 223	0.1349	0.0265	0.0065	0.0332
27	0.192 450	0.037 743	0.1323	0.0255	0.0061	0.0319
28	0.188 982	0.036 370	0.1298	0.0245	0.0058	0.0307
29	0.185 695	0.035 093	0.1275	0.0237	0.0055	0.0297
30	0.182 574	0.033 903	0.1252	0.0229	0.0052	0.0287
31	0.179 605	0.032 791	0.1231	0.0221	0.0050	0.0277
32	0.176 777	0.031 750	0.1211	0.0214	0.0047	0.0268
33	0.174 078	0.030 773	0.1192	0.0208	0.0045	0.0260
34	0.171 499	0.029 854	0.1174	0.0201	0.0043	0.0252
35	0.169 031	0.028 989	0.1157	0.0196	0.0041	0.0245
36	0.166 667	0.028 172	0.1140	0.0190	0.0040	0.0238
37	0.164 399	0.027 400	0.1124	0.0185	0.0038	0.0232
38	0.162 221	0.026 669	0.1109	0.0180	0.0037	0.0225
39	0.160 128	0.025 976	0.1094	0.0175	0.0035	0.0220
40	0.158 114	0.025 318	0.1080	0.0171	0.0034	0.0214
41	0.156 174	0.024 693	0.1066	0.0167	0.0033	0.0209
42	0.154 303	0.024 098	0.1053	0.0163	0.0031	0.0204
43	0.152 499	0.023 531	0.1041	0.0159	0.0030	0.0199
44	0.150 756	0.022 990	0.1029	0.0155	0.0029	0.0194

TABLE 1.25   (*Continued*)

n	$\dfrac{1}{\sqrt{n}}$	$\dfrac{1}{\sqrt{n(n-1)}}$	$\dfrac{0.6745}{\sqrt{n-1}}$	$\dfrac{0.6745}{\sqrt{n(n-1)}}$	$\dfrac{0.8453}{n\sqrt{n-1}}$	$\dfrac{0.8453}{\sqrt{n(n-1)}}$
45	0.149 071	0.022 473	0.1017	0.0152	0.0028	0.0190
46	0.147 442	0.021 979	0.1005	0.0148	0.0027	0.0186
47	0.145 865	0.021 507	0.0994	0.0145	0.0027	0.0182
48	0.144 338	0.021 054	0.0984	0.0142	0.0026	0.0178
49	0.142 857	0.020 620	0.0974	0.0139	0.0025	0.0174
50	0.141 421	0.020 203	0.0964	0.0136	0.0024	0.0171
51	0.140 028	0.019 803	0.0954	0.0134	0.0023	0.0167
52	0.138 675	0.019 418	0.0945	0.0131	0.0023	0.0164
53	0.137 361	0.019 048	0.0935	0.0129	0.0022	0.0161
54	0.136 083	0.018 692	0.0927	0.0126	0.0022	0.0158
55	0.134 840	0.018 349	0.0918	0.0124	0.0021	0.0155
56	0.133 631	0.018 019	0.0910	0.0122	0.0020	0.0152
57	0.132 453	0.017 700	0.0901	0.0119	0.0020	0.0150
58	0.131 306	0.017 392	0.0893	0.0117	0.0019	0.0147
59	0.130 189	0.017 095	0.0886	0.0115	0.0019	0.0145
60	0.129 099	0.016 807	0.0878	0.0113	0.0018	0.0142
61	0.128 037	0.016 529	0.0871	0.0112	0.0018	0.0140
62	0.127 000	0.016 261	0.0864	0.0110	0.0018	0.0138
63	0.125 988	0.016 001	0.0857	0.0108	0.0017	0.0135
64	0.125 000	0.015 749	0.0850	0.0106	0.0017	0.0133
65	0.124 035	0.015 504	0.0843	0.0105	0.0016	0.0131
66	0.123 091	0.015 268	0.0837	0.0103	0.0016	0.0129
67	0.122 169	0.015 038	0.0830	0.0101	0.0016	0.0127
68	0.121 268	0.014 815	0.0824	0.0100	0.0015	0.0125
69	0.120 386	0.014 599	0.0818	0.0099	0.0015	0.0123
70	0.119 523	0.014 389	0.0812	0.0097	0.0015	0.0122
71	0.118 678	0.014 185	0.0806	0.0096	0.0014	0.0120
72	0.117 851	0.013 986	0.0801	0.0094	0.0014	0.0118
73	0.117 041	0.013 793	0.0795	0.0093	0.0014	0.0117
74	0.116 248	0.013 606	0.0789	0.0092	0.0013	0.0115
75	0.115 470	0.013 423	0.0784	0.0091	0.0013	0.0113
76	0.114 708	0.013 245	0.0779	0.0089	0.0013	0.0112
77	0.113 961	0.013 072	0.0773	0.0088	0.0013	0.0111
78	0.113 228	0.012 904	0.0769	0.0087	0.0012	0.0109
79	0.112 509	0.012 739	0.0764	0.0086	0.0012	0.0108
80	0.111 803	0.012 579	0.0759	0.0085	0.0012	0.0106
81	0.111 111	0.012 423	0.0754	0.0084	0.0012	0.0105
82	0.110 432	0.012 270	0.0749	0.0083	0.0012	0.0104
83	0.109 764	0.012 121	0.0745	0.0082	0.0011	0.0103
84	0.109 109	0.011 976	0.0740	0.0081	0.0011	0.0101
85	0.108 465	0.011 835	0.0736	0.0080	0.0011	0.0100
86	0.107 833	0.011 696	0.0732	0.0079	0.0011	0.0099
87	0.107 211	0.011 561	0.0727	0.0078	0.0011	0.0098
88	0.106 600	0.011 429	0.0723	0.0077	0.0010	0.0097
89	0.106 000	0.011 300	0.0719	0.0076	0.0010	0.0096
90	0.105 409	0.011 173	0.0715	0.0075	0.0010	0.0094
91	0.104 828	0.011 050	0.0711	0.0075	0.0010	0.0093
92	0.104 257	0.010 929	0.0707	0.0074	0.0010	0.0092
93	0.103 695	0.010 811	0.0703	0.0073	0.0010	0.0091
94	0.103 142	0.010 695	0.0699	0.0072	0.0009	0.0090

TABLE 1.25   (*Continued*)

n	$\dfrac{1}{\sqrt{n}}$	$\dfrac{1}{\sqrt{n(n-1)}}$	$\dfrac{0.6745}{\sqrt{n-1}}$	$\dfrac{0.6745}{\sqrt{n(n-1)}}$	$\dfrac{0.8453}{n\sqrt{n-1}}$	$\dfrac{0.8453}{\sqrt{n(n-1)}}$
95	0.102 598	0.010 582	0.0696	0.0071	0.0009	0.0089
96	0.102 062	0.010 471	0.0692	0.0071	0.0009	0.0089
97	0.101 535	0.010 363	0.0688	0.0070	0.0009	0.0088
98	0.101 015	0.010 257	0.0685	0.0069	0.0009	0.0087
99	0.100 504	0.010 152	0.0681	0.0069	0.0009	0.0086
100	0.100 000	0.010 050	0.0678	0.0068	0.0008	0.0085

The *probable error* of a single observation in a series of n measures, $t_1, t_2, \ldots, t_n$, the arithmetic mean of which is m, is

$$e = \frac{0.6745}{\sqrt{n-1}}\sqrt{(m - t_1)^2 + (m - t_2)^2 + \cdots + (m - t_n)^2}$$

the probable error of the mean is

$$E = \frac{0.6745}{\sqrt{n(n-1)}}\sqrt{(m - t_1)^2 + (m - t_2)^2 + \cdots + (m - t_n)^2}$$

Approximate values of e and E are

$$e = 0.8453\frac{\sum\limits_{i=1}^{n} d_i}{\sqrt{n(n-1)}}\qquad E = 0.8453\frac{\sum\limits_{i=1}^{n} d_i}{n\sqrt{n-1}}$$

where $\sum_{i=1}^{n} d_i$ is the sum of the deviations $d_i = |t_i - m|$.

TABLE 1.26   **Statistics and Probability Formulas**

$p(x) = dP(x)/dx$	Differential probability (density) function of random variable x. Univariate frequency function
$P(x) = \int_{-\infty}^{x} p(x')\, dx'$	Cumulative probability function of random variable x. Univariate distribution function
$P(A < x < B)$	Cumulative probability that x is between A and B
$P(E \cap F)$	Probability of simultaneous (joint) occurrence of E and F
$P(E \cup F)$	Probability of occurrence of E or F or both
$P(E\vert F) = P(E \cap F)/P(F)$	Conditional probability. Probability of occurrence of E provided F has occurred
$E[f(x)] = \int_{-\infty}^{\infty} f(x)p(x)\, dx$	Expected value of function of a random variable x
$E(x) = \bar{x} = \int_{-\infty}^{\infty} xp(x)\, dx$	Expected (mean) value of random variable x
$\alpha_r = E(x^r)$	rth moment of random variable x. rth moment about the origin
$\mu_r = E(x - \bar{x})^r$	rth moment of random variable x from mean value. rth central moment
$\mathrm{Var}\, x = E[(x - \bar{x})^2] = \overline{(x - \bar{x})^2}$	
$\quad = \overline{x^2} - \bar{x}^2$	Variance value of random variable x
$\sigma = (\mathrm{Var}\, x)^{1/2}$	Standard deviation of random variable x
$M_x(s) = E(e^{sx})$	Moment generating function associated with random variable x
$\psi_x(q) = E(e^{jqx})$	Characteristic function associated with random variable x

TABLE 1.26 (*Continued*)

$\psi_g(q) = E[e^{jqg(x)}]$	Characteristic function of $g(x)$ with random variable x
$p(x, y) = d^2P(x, y)/dx\,dy$	Differential probability (density) function of random variables x and y. Bivariate frequency function
$P(x, y)$ $= \int_{-\infty}^{x}\int_{-\infty}^{y} p(x', y')\,dx'\,dy'$	Cumulative probability function of random variables x and y. Bivariate distribution function
$P(A < x < B, C < y < D)$	Cumulative probability that x is between A and B and that also y is between C and D. Cumulative joint probability
$\text{Cov}(x, y) = E[(x - \bar{x})(y - \bar{y})]$ $= \overline{(x - \bar{x})(y - \bar{y})}$	
$\rho(x, y) = \text{Cov}(x, y)/\sigma_x\sigma_y$	Covariance value of random variables x and y Correlation coefficient of random variables x and y

Source: Giacoletto, *Electronic Designers' Handbook*, Copyright © 1977 by McGraw-Hill, pp. 1–8.

1.4   UNITS AND STANDARDS

1.4.1   Physical Quantities and Their Relations

Mathematics is concerned with relations between numerical quantities, either constant or varying in a specified manner over a specified range of values. The numerical values are unique, absolute, and the same all over the world, being the expression of a fundamental perception of the mind. Any *mathematical equation* defines the values of one numerical quantity, known as the dependent, in terms of constants and one or more other numerical quantities, known as the independent variables, as for example

$$z = r^2 + 3x + 4 \qquad y = c \cdot \int_0^x \frac{x^2}{\cos x}\,dx \qquad (1.1)$$

where z and y in these expressions are dependent variables, r and x, independent variables, and c a constant.

Physics, comprising the knowledge of inanimate nature and its laws, is concerned fundamentally with the measuring of the various quantities founded or created by definition, as for example, *length*, *mass*, *electric charge*. In order to specify a *physical quantity* it is not sufficient to state merely a number. The value of a physical quantity can be determined only by comparison of the sample with a known amount of the same quantity, by the process of *measuring*. The reference amount is called a unit, and the result of any measurement must be a statement of "how many times the sample was found to contain the reference amount." Thus a physical quantity Q naturally appears to be the product of a numerical value N and a unit U,

$$Q = N \cdot U \qquad (1.2)$$

as for example: The length of a particular rod is 3.5 ft, or the rod is $3\frac{1}{2}$ times the length of 1 ft. Obviously, the reproduction of a unit must be possible at any time in order to facilitate correct measurements. This is being done by means of the "standards," which are simply a set of fundamental unit quantities kept under normalized conditions in order to preserve their values as accurately as facilities permit.

Any physical relation must be the result of a more or less obvious measurement, so that equations in physics are not merely numerical relations, but express dependencies between physical quantities. Mathematics does not know "standards"; physics cannot be without "standards." The fact that physics often uses the methods of mathematics must not lead to the identification of the two sciences; it is merely an overlapping in the border regions.

Relations between Units. A unit is a particular amount of the physical quantity to be measured, defined in terms of a standard. The choice of a unit depends on convenience, facility of reproduction, and easy subdivision so as to obtain smaller units if desired. The value of a physical quantity Q must be independent of the units used, so that for two different units of the same type

$$Q = N_1 \cdot U_1 = N_2 \cdot U_2 \qquad (1.3)$$

The size of the unit and the numerical value of the quantity are inversely related: the larger the unit the smaller the number of units.

A unit relation is an equation between two different units of the same type

$$U_1 = N_{12} \cdot U_2 \tag{1.4}$$

and serves to convert from one unit U_1 to a different one U_2. The conversion is achieved by replacing U_1, taken as a factor, by its equivalent according to Eq. (1.4) so that

$$Q = N_1 \cdot U_1 = N_1 \cdot (N_{12} \cdot U_2) = (N_1 \cdot N_{12}) \cdot U_2 \tag{1.5}$$

As an example, express the length 3.5 ft in centimeters. The unit relation is 1 ft = 30.5 cm, and therefore $l = 3.5$ ft = $3.5 \times (30.5$ cm) = 106.75 cm. No error is possible if this rule is followed properly.

Physical Equations. Relations between physical quantities are usually given in the form of equations. It is always possible, by the proper use of unit relations (see previous paragraph), to express each side in the same units. Since units are to be considered as factors, they may be canceled and a numerical identity must result. This fact always can be used to check the proper numerical relations and the consistency of the units used.

There are two fundamental types of physical equations:

1. The **mathematical definition** of a physical quantity determines a new quantity uniquely in terms of known quantities. An example is Newton's definition of mass by $f = m \cdot a$, where f is the force and a the acceleration of a moving body. If f and a are measured, m can be computed as a physical quantity with numerical value $N(f)/N(a)$ and unit $U(f)/U(a) = U(m)$. A definition should be in agreement with all the other known relations in a particular field of science; it can only be of restricted value if it contradicts other relations (see later the "absolute" electric systems).

2. The **statement of proportionality** defines one physical quantity as linearly depending on a combination of other, known quantities. It is always the result of an experimental investigation. An example is Newton's law of the gravitational force $f = k(m_1 m_2/r^2)$ where m_1 and m_2 are the two masses, r their center distance, and k the proportionality factor. In the case of a proportionality it is permissible to choose arbitrary units for all measurable physical quantities involved and to use the equation as a definition of the proportionality constant that, in general, will be a physical constant with numerical value and unit. In the example the value of k would be

$$\frac{N(f) \cdot N(r^2)}{N(m_1) \cdot N(m_2)} \times \frac{U(f) \cdot U(r^2)}{U(m_1) \cdot U(m_2)} = N(k) \cdot U(k)$$

Most of the fundamental laws of physics are statements of proportionalities, leading to universal physical constants, as for instance the gravitational constant k, the Planck constant h, the gas constant R, the absolute permeability of free space μ_v and the absolute dielectric constant of free space ε_v. It may be observed that each branch of physics is represented by at least one fundamental proportionality constant.

Derived physical quantities are, in general, the result of mathematical definitions. The units of derived quantities are expressed from the combinations of the units used in the definition. All proportionality constants are ordinarily considered as derived physical quantities.

Fundamental Physical Quantities. The physical quantities, arbitrarily chosen to define new quantities or derived quantities, are called fundamental physical quantities. Their number may vary according to needs and convenience. There is no possibility to designate any physical quantity as absolutely fundamental, or a priori fundamental. Quantities that appear to be fundamental in some one special field may be derived quantities in some other field.

1.4.2 Dimensions and Dimension Systems

Definition of Dimension. To choose a unit for a physical quantity one has an infinity of possibilities. The numerous units of length that were in use about 100 years ago present a good practical illustration. Yet all these units have in common the quality of being a distinct length and not, for example, a volume. It is convenient to state this fact by representing with the notation $[L]$ any unit of length whatsoever. The measurement of a physical quantity Q, therefore, leads to the statement

$$Q = N \cdot [Q] \tag{1.6}$$

where N is a numeric denoting the number of general units $[Q]$ that constitute the total quantity Q. According to Fourier, who first introduced this concept into the literature, $[Q]$ is called the "dimension" of the quantity Q. Be it clearly understood that dimension is simply the expression of a general unit and therefore a characteristic peculiarity of physical quantities, not occurring in mathematics. Each new physical quantity gives rise to a new "dimension" as for instance time $[T]$, force $[F]$, mass $[M]$, and so on. There are as many dimensions, or general units, as there are kinds of physical quantities.

Derived Dimensions. Many physical quantities have been introduced by mathematical definition. Velocity, for example, is defined as $v = ds/dt$, where s is the length of the path measured from a definite origin and t is the time. A possible expression for the dimension of velocity would be $[V]$. It is customary and convenient, however, to make use of the mathematical definition that is but the rule for the measurement of velocity, and to express the dimension in terms of the more familiar dimensions of length and time as a derived dimension $[V] = [L]/[T] = [L][T]^{-1}$. [Read: velocity is of $(+1)$ dimension in length and (-1) dimension in time.] The use of mathematical definitions, leading to derived dimensions of a composite nature, reduces the number of symbols. Thus the measurement of volume, if scientifically conducted, gives $[Vol] = [L]^3$, or in words, "volume is of $(+3)$ dimensions in length $[L]$."
  Proportionality constants of physics have, in general, *derived dimensions*, as they are defined by the corresponding physical equations.

Fundamental Dimensions. The more familiar dimensions used to express derived dimensions are referred to as fundamental dimensions. It is advantageous to use as few of these fundamental dimensions as possible, not because the physical relations become simpler or clearer, but merely as a matter of economy in symbols. In fact, any dimension can be chosen to be a fundamental dimension in a particular field and a derived dimension in some other field of physics. No fundamental dimension can be made a starting point of natural philosophy.

Dimensional Equations. Since a physical equation constitutes in fact two equations, one for the units and one for the numerics, one can disregard the numerical factors entirely and write the general units or dimensions only, arriving thus at a dimensional equation. For instance, the law of gravitation would read $[F] = [k][M]^2[L]^{-2}$, using $[F]$, $[k]$, $[M]$, $[L]$ as dimensions for force, gravitation constant, mass, and length, respectively. From this dimensional equation a derived dimension can be obtained for any quantity involved. Conversely, dimensional equations are used to check the correctness of physical relations, if all dimensions can be made to cancel. Finally, the validity of dimensional equations leads to the method of dimensional analysis.
  A **set of fundamental dimensions** is any group of fundamental dimensions, convenient and useful to express all the physical quantities of a particular field in terms of derived dimensions. The number of fundamental dimensions to make a set may vary according to the field of application. Whether or not a set of fundamental dimensions can be used beyond the field for which it was originally intended will depend upon its suitability as a dimension system. (See next paragraph.) In no case should it be used where it can lead to confusion.
  A set of fundamental dimensions is *incomplete* when the number of fundamental dimensions composing it is less than the number required for a dimension system. Incomplete sets of fundamental dimensions should not be used outside the very restricted field for which they are defined; they necessarily would lead to confusing relations.
  A **dimension system** is composed of the smallest number of fundamental dimensions that will form a consistent and complete set for a field of science. Since each relation between physical quantities can be split up into one relation of numerics and another one of dimensions (as general units), it is possible to combine all known relations of dimensions. In setting up these relations, all proportionality factors must be taken as physical quantities. If there are m independent relations known, $(m + p)$ dimensions may be involved, of which m dimensions can be expressed by any p "fundamental" dimensions chosen arbitrarily.
  This set of p "fundamental" dimensions is then called a dimension system. From the theory of numbers, therefore, it is known that one generally has a choice of $\binom{m + p}{p}$ possible dimension systems. Thus, if $p = 3$, $m = 3$, then one has $\binom{6}{3} = 20$ different possibilities. A necessary condition, however, is that *each* independent relation involve at least $(p + 1)$ dimensions. If this is not the case, then the number of possible dimension systems is less, so that $\binom{m + p}{p}$ indicates the *upper* limit.
  Any dimension system chosen in the described manner is consistent, as well as correct, and never leads to ambiguity with respect to the expression of physical quantities. Complete dimension systems in mechanics must have three, in thermodynamics four, and in electromagnetism four fundamental

dimensions. It seems, according to present knowledge, that five fundamental dimensions suffice for the entire range of physics, namely, the three fundamental dimensions of mechanics, an additional one for thermodynamics, and another additional one for electromagnetism.

All the known dimension systems use length $[L]$ and time $[T]$ as primary fundamental dimensions, adding various fundamental dimensions from the available physical quantities of the fields of physics. The choice of $[L]$ and $[T]$ reduces at once the maximum number of possible dimension systemss to $\binom{m + p - 2}{p - 2}$.

Why Dimension Systems? Since the proper choice of units is the ultimate goal of any critical analysis of physical quantities, the question may be asked: Why is it necessary to discuss dimension of systems? The answer is that each physical quantity may be measured by an infinite variety of units, but has only one dimension, within a given dimension system. The process of deciding upon the fundamental dimensions before fixing the units within the scope of the fundamental dimensions is, therefore, essentially a matter of economy and logic.

1.4.3   Dimension and Unit Systems

In the past different dimension systems were introduced for various fields of technology (mechanics, heat, electromagnetism) and based on different choices of fundamental dimensions, for example, for mechanics, length and time plus mass or force or energy or gravitational constant gave potentially four different dimension system classes. In turn for a given dimensional system, a unit system could be developed, choosing for each fundamental dimension a unit desirably related to a fundamental standard or standards. In seeking to define units with appropriate size values, relationships, and so on, many different unit systems, for example, cgs, MKS, "absolute," "technical," and so on, have been introduced over the years. (See O. W. Eshbach and M. Souders, *Handbook of Engineering Fundamentals*, 3rd ed., Wiley, New York, 1975, for a detailed exposition of the subject).

In recent years a major step toward simplification and standardization has been taken by the increasing adoption of the International System of Units (SI).

1.4.4   The International System of Units

The SI system of units, composed of six fundamental units has been adopted by the Conference Générale (BIPM Sèvres, Paris, 1954 and 1960) to cover the whole range of physics and one in which all international reports are to be expressed.

Fundamental Units	Name	Symbol	
Length	meter	m	
Mass	kilogram	kg	
Time	second	s	
Intensity of electric current	ampere	A	
Thermodynamic temperature	degree kelvin	K	
Luminous intensity	candela	cd	
Amount of substance	mole	mol	

Derived Units with Special Names			
Area	square meter	m^2	
Volume	cubic meter	m^3	
Frequency	hertz	Hz	
Density (mass density)	kilogram per cubic meter	kg/m^3	
Velocity	meter per second	m/s	
Angular velocity	radian per second	rad/s	
Acceleration	meter per square second	m/s^2	
Angular acceleration	radian per square second	rad/s^2	
Force	newton	N	$kg.m/s^2$
Pressure, stress	newton per square meter	N/m^2	
Kinematic viscosity	square meter per second	m^2/s	
Dynamic viscosity	newton-second per square meter	$N.s/m^2$	
Work, energy, heat (quantity of heat)	joule	J	$N \cdot m$
Power, radiant flux	watt	W	J/s
Plane angle	radian	rad	
Solid angle	steradian	sr	
Electric charge	coulomb	C	$A \cdot s$

Fundamental Units	Name	Symbol	
Derived Units with Special Names			
Electric potential, potential difference, electromotive force	volt	V	W/A
Electric field strength	volt per meter	V/m	
Resistance (to direct current)	ohm	Ω	V/A
Electric conductance	siemens	S	A/V
Capacitance	farad	F	A · s/V
Magnetic flux	weber	Wb	V · s
Inductance	henry	H	V · s/A
Magnetic flux density (magnetic induction)	tesla	T	Wb/m^2
Magnetic field strength	ampere per meter	A/m	
Magnetomotive force	ampere	A	
Luminous flux	lumen	lm	cd · sr
Luminance	candela per square meter	cd/m^2	
Illumination	lux	lx	lm/m^2
Activity (of a radionuclide)	becquerel	Bq	1/S
Absorbed dose	gray	Gy	J/kg
Dose equivalent	sievert	Sv	J/kg
Other Common Derived Units			
Absorbed dose rate	gray per second	Gy/s	
Acceleration	meter per second squared	m/s^2	
Angular acceleration	radian per second squared	rad/s^2	
Angular velocity	radian per second	rad/s	
Area	square meter	m^2	
Concentration (of amount of substance)	mole per cubic meter	mol/m^3	
Current density	ampere per square meter	A/m^2	
Density, mass	kilogram per cubic meter	kg/m^3	
Electric charge density	coulomb per cubic meter	C/m^3	
Electric field strength	volt per meter	V/m	
Electric flux density	coulomb per square meter	C/m^2	
Energy density	joule per cubic meter	J/m^3	
Entropy	joule per kelvin	J/K	
Exposure (X and gamma rays)	coulomb per kilogram	C/kg	
Heat capacity	joule per kelvin	J/K	
Heat flux density } Irradiance	watt per square meter	W/m^2	
Luminance	candela per square meter	cd/m^2	
Magnetic field strength	ampere per meter	A/m	
Molar energy	joule per mole	J/mol	
Molar entropy	joule per mole kelvin	J/(mol · K)	
Molar heat capacity	joule per mole kelvin	J/(mol · K)	
Moment of force	newton meter	N · m	
Permeability (magnetic)	henry per meter	H/m	
Permittivity	farad per meter	F/m	
Power density	watt per square meter	W/m^2	
Radiance	watt per square meter steradian	W/(m^2 · sr)	
Radiant intensity	watt per steradian	W/sr	
Specific heat capacity	joule per kilogram kelvin	J/(kg · K)	
Specific energy	joule per kilogram	J/kg	
Specific entropy	joule per kilogram kelvin	J/(kg · K)	
Specific volume	cubic meter per kilogram	m^3/kg	
Surface tension	newton per meter	N/m	
Thermal conductivity	watt per meter kelvin	W/(m · K)	
Velocity	meter per second	m/s	
Viscosity, dynamic	pascal second	Pa · s	
Viscosity, kinematic	square meter per second	m^2/s	
Volume	cubic meter	m^3	
Wave number	1 per meter	1/m	

Definitions of Derived Units of the International System Having Special Names

Quantity	Unit and Definition
1. Absorbed dose	The *gray* is the absorbed dose when the energy per unit mass imparted to matter by ionizing radiation is one joule per kilogram. Note: The *gray* is also used for the ionizing radiation quantities: specific energy imparted, kerma, and absorbed dose index, which have the SI unit joule per kilogram.
2. Activity	The *becquerel* is the activity of a radionuclide decaying at the rate of one spontaneous nuclear transition per second.
3. Celsius temperature	The *degree Celsius* is equal to the kelvin and is used in place of the kelvin for expressing Celsius temperature (symbol t) defined by the equation $t = T - T_0$ where T is the thermodynamic temperature and $T_0 = 273.15$ K by definition.
4. Dose equivalent	The *sievert* is the dose equivalent when the absorbed dose of ionizing radiation multiplied by the dimensionless factors Q (quality factor) and N (product of any other multiplying factors) stipulated by the International Commission on Radiological Protection is one joule per kilogram.
5. Electric capacitance	The *farad* is the capacitance of a capacitor between the plates of which there appears a difference of potential of one volt when it is charged by a quantity of electricity equal to one coulomb.
6. Electric conductance	The *siemens* is the electric conductance of a conductor in which a current of one ampere is produced by an electric potential difference of one volt.
7. Electric inductance	The *henry* is the inductance of a closed circuit in which an electromotive force of one volt is produced when the electric current in the circuit varies uniformly at a rate of one ampere per second.
8. Electric potential difference, Electromotive force	The *volt* (unit of electric potential difference and electromotive force) is the difference of electric potential between two points of a conductor carrying a constant current of one ampere, when the power dissipated between these points is equal to one watt.
9. Electric resistance	The *ohm* is the electric resistance between two points of a conductor when a constant difference of potential of one volt, applied between these two points, produces in this conductor a current of one ampere, this conductor not being the source of any electromotive force.
10. Energy	The *joule* is the work done when the point of application of a force of one newton is displaced a distance of one meter in the direction of the force.
11. Force	The *newton* is that force that, when applied to a body having a mass of one kilogram gives it an acceleration of one meter per second squared.
12. Frequency	The *hertz* is the frequency of a periodic phenomenon of which the period is one second.
13. Illuminance	The *lux* is the illuminance produced by a luminous flux of one lumen uniformly distributed over a surface of one square meter.
14. Luminous flux	The *lumen* is the luminous flux emitted in a solid angle of one steradian by a point source having a uniform intensity of one candela.
15. Magnetic flux	The *weber* is the magnetic flux that, linking a circuit of one turn, produces in it an electromotive force of one volt as it is reduced to zero at a uniform rate in one second.
16. Magnetic flux density	The *tesla* is the magnetic flux density given by a magnetic flux of one weber per square meter.
17. Power	The *watt* is the power that gives rise to the production of energy at the rate of one joule per second.

18. Pressure or stress	The *pascal* is the pressure or stress of one newton per square meter.
19. Quantity of electricity	The *coulomb* is the quantity of electricity transported in one second by a current of one ampere.

Prefixes. The SI system has adopted the following standard set of prefixes:

Multiplication Factor	Prefix	Symbol
1 000 000 000 000 000 000 = 10^{18}	exa	E
1 000 000 000 000 000 = 10^{15}	peta	P
1 000 000 000 000 = 10^{12}	tera	T
1 000 000 000 = 10^{9}	giga	G
1 000 000 = 10^{6}	mega	M
1 000 = 10^{3}	kilo	k
100 = 10^{2}	hecto^a	h
10 = 10^{1}	deka^a	da
0.1 = 10^{-1}	deci^a	d
0.01 = 10^{-2}	centi^a	c
0.001 = 10^{-3}	milli	m
0.000 001 = 10^{-6}	micro	μ
0.000 000 001 = 10^{-9}	nano	n
0.000 000 000 001 = 10^{-12}	pico	p
0.000 000 000 000 001 = 10^{-15}	femto	f
0.000 000 000 000 000 001 = 10^{-18}	atto	a

^{a}To be avoided where practical; see below.

Application of SI Prefixes

General. In general the SI prefixes should be used to indicate orders of magnitude, thus eliminating nonsignificant digits and leading zeros in decimal fractions, and providing a convenient alternative to the powers-of-ten notation preferred in computation. For example,

$$12\ 300\ \text{mm becomes } 12.3\ \text{m}$$

$$12.3 \times 10^3\ \text{m becomes } 12.3\ \text{km}$$

$$0.00123\ \mu\text{A becomes } 1.23\ \text{nA}$$

Selection. When expressing a quantity by a numerical value and a unit, a prefix should preferably be chosen so that the numerical value lies between 0.1 and 1000. To minimize variety, it is recommended that prefixes representing 1000 raised to an integral power be used. However, three factors may justify deviation:

1. In expressing area and volume, the prefixes hecto-, deka-, deci-, and centi- may be required, for example, square hectometer, cubic centimeter.
2. In tables of values of the same quantity, or in a discussion of such values within a given context, it is generally preferable to use the same unit multiple throughout.
3. For certain quantities in particular applications, one particular multiple is customarily used. For example, the millimeter is used for linear dimensions in mechanical engineering drawings even when the values lie far outside the range 0.1 to 1000 mm; the centimeter is often used for body measurements and clothing sizes.

Prefixes in Compound Units*. It is recommended that only one prefix be used in forming a multiple of a compound unit. Normally the prefix should be attached to a unit in the numerator. One exception to this is when the kilogram occurs in the denominator. For example,

$$\text{V/m, } not \text{ mV/mm, and MJ/kg, } not \text{ kJ/g}$$

---

*A compound unit is a derived unit that is expressed in terms of two or more units rather than by a single special name.

Compound Prefixes. Compound prefixes, formed by the juxtaposition of two or more SI prefixes are not to be used. For example, use

$$1 \text{ nm}, \; not \; 1 \text{ m}\mu\text{m}$$
$$1 \text{ pF}, \; not \; 1\mu\mu\text{F}$$

If values are required outside the range covered by the prefixes, they should be expressed by using powers of ten applied to the base unit.

Powers of Units. An exponent attached to a symbol containing a prefix indicates that the multiple or submultiple of the unit (the unit with its prefix) is raised to the power expressed by the exponent. For example,

$$1 \text{ cm}^3 = \left(10^{-2} \text{ m}\right)^3 \quad = 10^{-6} \text{ m}^3$$
$$1 \text{ ns}^{-1} = \left(10^{-9} \text{ s}\right)^{-1} \quad = 10^9 \text{ s}^{-1}$$
$$1 \text{ mm}^2/\text{s} = \left(10^{-3} \text{ m}\right)^2/\text{s} \quad = 10^{-6} \text{ m}^2/\text{s}$$

Calculations. Errors in calculations can be minimized if the base and the coherent derived SI units are used and the resulting numerical values are expressed in powers-of-ten notation instead of using prefixes.

Other Units

Units from Different Systems. To assist in preserving the advantage of SI as a coherent system, it is advisable to minimize the use with it of units from other systems. Such use should be limited to units listed in this section.

A following section presents conversion factors to and from SI units.

1.4.5 Length, Mass, and Time

The English Units and Standards

Units of Length. The foot (ft) is the *fundamental* unit of length in the foot-pound-second (fps) system. It equals, by definition, one-third of a *yard* (yd), which is the English legalized *standard* unit of length. The *United States yard* was defined by Act of Congress, July 28, 1866, as 3600/3937 the length of the *meter*. (See discussion of metric system for definitions of metric length.)

In Great Britain, the *Imperial yard* is measured by a bronze bar preserved in the Standards Office, Westminster. Its length, in terms of the *international prototype meter*, is 3600/3937.0113 meter. For engineering purposes, the United States and British *yards* may be considered identical.

As subunits, the *inch* (in.) is defined as $\frac{1}{12}$ of one standard foot, and the *mil* as the one-thousandth part of one inch. The *nautical mile* (mi) is defined as one minute of arc on the earth's surface at the equator, whereas the United States mile (U. S. mi statute) is exactly 5280 ft and practically identical with the British mile.

Unit of Capacity (Dry). The bushel (bu) is the *standard* unit of *dry* capacity. The *Winchester bushel* (U. S. standard) has a volume of 2150.42 in.[3]

In Great Britain, the *Imperial bushel* (bu) is defined as the volume of 80 lb of pure water at 62°F, weighed against brass weights in air at the same temperature as the water and with the barometer at 30 in. Its volume is approximately 2219.36 in.[3]

Unit of Capacity (Liquid). The *gallon* (gal) is the *standard* unit of *liquid* capacity. The *United States gallon* has a volume of 231 in.[3]

In Great Britain, the *Imperial gallon* is defined as the volume of 10 lb of pure water at 62°F, weighed against brass weights in air at the same temperature as the water and with the barometer at 30 in. Its volume is approximately 277.420 in.[3] The Imperial gallon (liquid measure) equals exactly one-eighth of the Imperial bushel (dry measure). Subunits are the quart (qt), which is $\frac{1}{4}$ of the standard gallon, and the pint (pt), which is $\frac{1}{2}$ qt.

Units of Mass. The pound (avoirdupois) (lb avdp) is the *fundamental* unit of mass in the fps system.* It is also the English legalized *standard* unit of mass. The *United States pound* (*avoirdupois*)

---

*The slug of mass, which is extensively used by engineers and physicists, is (in the English system) the mass to which an acceleration of one foot per second per second would be given by the application of a one-pound force. Under any gravity conditions, one slug of mass = 32.1739 lb of mass.

was defined by Act of Congress, 1866, as $1/2.2046$ kg, but since 1895 there has been used, for greater accuracy, a value that agrees with that given by law as far as the latter is given; namely, 453.5924277 g. This value is now used by the Bureau of Standards as an exact definition and is the basis of the customary United States weights (Circular 47, Bureau of Standards).

In Great Britain, the *Imperial pound* (*avoirdupois*) is the mass of a *platinum cylinder* preserved in the Standards Office, Westminster. Its legal equivalent is 453.59243 g. For engineering purposes, the United States and British pounds (avoirdupois) may be considered as identical.

Subunits of mass are the grain (gr), defined as $\frac{1}{7000}$ of the standard pound (avoirdupois) and the ounce (avoirdupois) (oz-avdp), which is $\frac{1}{16}$ of the standard pound (avoirdupois). The grain was used as fundamental unit in the so-called foot-gram-second (fgs) system of units prior to 1873.

Weight versus Mass. Unfortunately, the word "weight" is used in two different senses, namely, (1) by the layman (as well as loosely by the scientist) to designate a given *mass* or quantity of matter and (2) by the scientist to designate the *pull* in standard gravitational force units that is exerted by the earth upon a piece of matter. The result of the commercial act of "weighing" a specific quantity is independent of the local gravitational pull of the earth, since both spring scales and balances are calibrated locally by comparison with standard masses.

Auxiliary fundamental units and their principal derived units are defined and discussed under the sections of this handbook pertaining to the topics to which they apply. In general, however, conversion factors are included in the tables of Section 1.5.

For an interesting and rather complete history see *British Weights and Measures*, London, 1910, by Sir C. M. Watson.

The Metric (or SI) Units and Standards
The development of the International System of Units and the operations of the international bodies (BIPM, CIPM and CGPM) having cognizance over weights and measures are described in Appendices to ASTM's *Standard for Metric Practice*, ASTM E380-82, Philadelphia, 1982.

Units of Length. The centimeter (cm) is the *fundamental* unit of length in the cgs system. It equals, by definition, $\frac{1}{100}$ of a *meter* (m). The meter has been standardized by international agreement as 1,650,763.73 times the wavelength in vacuum of the unperturbed transition $(2p_{10} - 5d_5)$ of krypton 86. The *basic* meter for international comparisons is the *international prototype meter*, which is the distance, at zero degrees Centigrade, between two lines on a platinum-iridium bar located at the International Bureau of Weights and Measures, at Sèvres, France. This meter is the nearest to a duplicate, ever constructed, of the *original* meter, which was constructed and deposited in the Archives of the French Republic in 1799. The meter is very nearly equal to one ten-millionth of the distance, measured at sea level, from the equator to either pole.

An interesting history of the development of the *international prototype meter* (as well as the *international prototype kilogram*—see discussion on unit of mass) is given by Wm. Parry, National Bureau of Standards, in *Merriman's Civil Engineers' Handbook* as follows:

The use of the meter as the basis of geodetic surveys had become so general throughout Europe that a conference was called in Paris, France, in 1870, for the purpose of establishing a central bureau where the standards of the different countries could be compared. As a result of this conference an International Bureau of Weights and Measures was established near Paris in 1875, by the concurrent action of the principal nations of the world. One of the first tasks undertaken by the Bureau was the construction of exact copies of the meter and kilogram deposited in the Archives. Thirty-one standard meters of iridio-platinum and forty kilograms of the same alloy were constructed and carefully compared with the standards of the Archives and with one another. This great work was completed in 1889, and the meter and kilogram which agreed most nearly with the original standards were called international prototypes, and were deposited at the International Bureau, where they are maintained today subject to the authority of the International Committee on Weights and Measures. The remaining meters and kilograms were distributed by lot to the different nations which contributed to the support of the Bureau. The United States secured two copies of the meter and two copies of the kilogram, which are in the custody of the Bureau of Standards at Washington. One of the meters, known as No. 27, and one kilogram, No. 20, were selected as the United States standards, while the other meter and kilogram are used as secondary standards. It was the declared intention of the International Committee that the various national prototypes should be returned to the International Bureau at regular intervals for the purpose of recomparing them with the international standards and with one another. In this way all measurements based upon metric standards throughout the world are ultimately referred to the international meter and kilogram.

Unit of Capacity. The liter (L) is the *standard* unit of capacity. It is defined as the volume of one kilogram of pure water at the temperature of maximum density (4°C) under a pressure of 76 cm of mercury. For all practical purposes, the liter may be regarded as the equivalent of the cubic decimeter, although the former is actually slightly greater, in the amount of less than three parts in one hundred thousand.

Unit of Mass. The gram (g) is the *fundamental* unit of mass in the cgs system.* It equals, by definition, $\frac{1}{1000}$ of a *kilogram* (kg), which is the *standard* unit of mass. The *basic* kilogram for international comparisons is the *international prototype kilogram*, which is a cylinder of platinum-iridium located at the International Bureau of Weights and Measures, at Sèvres, France. This mass is the nearest to a duplicate, ever constructed, of the *original* kilogram, which was constructed and deposited in the Archives of the French Republic in 1799. The latter was made as nearly as possible equal to the mass of a cube of pure water at 4°C, the sides of the cube being one-tenth the length of the original meter.

An interesting history of the development of the *international prototype kilogram* was given under the discussion on units of length.

Weight versus Mass. See discussion under this same subheading of the English units and standards.

Auxiliary Fundamental Units and Their Principal Derived Units are defined and discussed under the sections of this handbook pertaining to the topics to which they apply. In general, however, conversion factors are included in Tables 1.27–1.64.

The Standard of Time

Unit of Time. The second has been standardized by international agreement as 1/31,556,925.9747 of the Tropical Year at 12 hr, Ephemeris Time, Jan 0 for the year 1900.0. (This definition has been retained for the time being as an Astronomical Time Standard—the following atomic standard of time interval is 100 times more precise.) The second has been standardized by international agreement as the time taken for 9,192,631,770.0 vibrations of the unperturbed hyperfine transition 4,0—3,0 for the $^2S_{1/2}$ fundamental state of the caesium 133 atom. The Cs^{133} standard has been adopted provisionally (see resolution 5 of the 12th General Conference of Weights and Measures, BIPM, Sèvres, Paris, Oct. 1964). A more accurate hydrogen maser standard may be available in the near future that is 100 times more accurate than the Cs^{133} standard.

Measures of Time. A *solar day* is measured by the rotation of the earth about its axis, with respect to the sun. In *astronomical computations* and in *nautical time* the day commences at noon, and in the former it is counted throughout the 24 h. In *civil computations* the day commences at midnight, and is divided into two parts of 12 h each.

A *solar year* is the time in which the earth makes one revolution around the sun. Its average time, called the *mean solar year*, is 365 days, 5 h, 48 min, and 45.9747 sec, or nearly $365\frac{1}{4}$ days.

1.4.6   Force, Energy, and Power

Dynamical and Gravitational Units. According to the use of two different dimension and unit systems, the dynamical (or physical, or "absolute") system and the gravitational (or technical) system, two different sets of units of force, energy, power, and derived quantities are defined in both the English and the metric systems. *One dynamical unit of force* produces an acceleration of unity on unit standard mass. The *gravitational unit of force* is defined as that force required to give a unit standard mass an acceleration equal to that produced by the gravitational pull of the earth. As the acceleration due to gravity, g, varies with location and altitude,$^\leq$ the gravitational unit of force is not constant, and, therefore, its relation to the dynamical unit of force will vary. By international agreement, the value $g_0 = 980.665$ cm/sec sec = 32.1739 ft/sec sec (British) has been chosen as the standard acceleration of gravity to make invariant the gravitational unit of force.

---

*The slug of mass, which is extensively used by engineers and physicists, is (in the metric system) the mass to which an acceleration of one meter per second per second would be given by the application of a one-kilogram force. Under any gravity conditions, one slug of mass = 9.80665 kg of mass.

†The variation of g with latitude ϕ and altitude H is given approximately by (ϕ in degrees, H in meters): $g = 978.039\ (1 + 0.005295\sin^2\phi) - 0.000307\ H$. See *International Critical Tables*, Vol. 1, p. 395.

The English Units

Units of Force. The dynamical or physical unit of force is the *poundal*, defined as the force required to give a mass of one pound an acceleration of one foot per second per second.

The *pound-force* (or weight of the pound mass) is the gravitational or technical unit of force. It is, by definition, the force required to give a mass of one pound an acceleration of 32.1739 ft/sec sec. If a force is measured by "weighing," the result in pounds weight must be multiplied by g/g_0, the ratio of local to standard acceleration of gravity, in order to obtain the absolute value in pound-force units. For engineering purposes this correction can usually be neglected.

The Unit of Pressure. This is defined as the unit of force acting upon a unit area. The most commonly employed unit is the *pound* (force) *per square inch*.

Standard atmospheric pressure is defined to be the force exerted by a column of mercury 760 mm (29.92 in) high at 0°C. This corresponds to 0.101325 MPa or 14.695 psi. Reference or fixed points for pressure calibration exist and are analogous to the phase changes used for temperature standards. These pressure references are based on phase changes or resistance jumps in selected materials.

Units of Work or Energy. The foot-poundal is the physical unit of work or energy and is defined as the work done by a force of one poundal in moving a body through the distance of one foot in the direction of the force.

The *foot-pound* (*force*) is the technical unit of work or energy and is defined as the work required to raise a mass (or weight) of one pound through a vertical distance of one foot at standard acceleration of gravity g_0. If measurements are made in places where the local value of the acceleration of gravity g is different from g_0, a correct factor g/g_0 must be applied, if the exact value of work or energy is desired.

The *British thermal unit* (Btu) is the quantity of heat required to raise the temperature of a one-pound mass of water either at 39°F (at its maximum density) or at 60°F, and standard pressure, through 1°F. The mean British thermal unit is defined as the $\frac{1}{180}$ part of the heat required to raise the temperature of a one pound mass of water from 32 to 212°F at standard pressure. It is obvious that the reference temperature must be indicated with the unit used.

Units of Power. Power is the time rate at which work is done. Its physical unit is the *foot-poundal per second*, its technical units are the *foot-pound* (*force*) *per second*, or the *British thermal unit per second*. The *horsepower* (hp or Hp) is defined as 33,000 ft-lb (force) per minute or 550 ft-lb (force) per second.

Units of Torque. Torque is the effectiveness of a force to produce rotation. It is defined as the product of the force and the perpendicular distance from its line of action to the instantaneous center of rotation. Its physical unit is the poundal-foot, and its technical unit the pound (force)-foot. (Note the reversal of force and length units in the designation of the units of torque as compared with the units of energy or work.)

The Metric Units

Units of Force. The dynamical, or physical, unit of force is the *dyne*, defined as the force required to give a mass of one gram an acceleration of one centimeter per second per second.

The *newton* is the SI unit of force. It is the force required to give a mass of one kilogram an acceleration of one meter per second per second.

The *kilogram force* (or weight of the kilogram mass) is the gravitational or technical unit of force. It is, by definition, the force required to give a mass of one kilogram an acceleration of 980.665 cm/sec sec. If a force is measured by "weighing," the result in kilograms weight must be multiplied by g/g_0, the ratio of local to standard acceleration of gravity, in order to obtain the absolute value in kilogram-force units. For engineering purposes this correction can usually be neglected.

In the electrotechnical system of units the systematic unit of force is defined as the *joule per meter*, based on the fundamental definition of the joule. (See discussion on metric units of energy.)

Unit of Pressure. This is defined as the unit of force acting upon a unit of area.

The *newton per square meter* is the SI unit of pressure and is called the *pascal*.

The *kilogram force per square meter* is the technical unit of pressure. With respect to correction for local gravity, see discussion on force versus weight.

Pressure is measured also by the height in centimeters of the column of water at 4°C, or of the column of mercury at 0°C, which it supports. (See conversion Table 1.41.)

The *normal atmosphere* (*at*), or the standard atmospheric pressure, is defined as the pressure exerted by a column of 76 cm of mercury at sea level and 0°C at standard acceleration of gravity g_0.

It is equal to 1.01321 bars or 1.0332 kg/cm² force and is used extensively in the engineering literature. Some confusion exists since the unit of 1 kg/cm² is occasionally called 1 practical atmosphere.

Units of Work or Energy. The *joule* is the physical or so-called absolute unit of work or energy. It is defined as the work done by a force of one newton acting through the distance of one meter. A larger unit is the *theoretical* or "absolute" *joule* defined as 10^7 ergs; it is a systematic unit in the practical electrical unit systems that is based on the theoretical unit systems. (See discussion on electrical units.)

The *international joule* is defined as the energy expended during one second by an electric current of one international ampere flowing through a resistance of one international ohm. (See discussion on electrical units.) The latest value of the international joule is equal to 1.000165 theoretical joules.*

The *kilowatt-hour* is the practical unit of energy in electrical metering. It is defined as a theoretical or an international unit (see definition of joule already given) and is equal to 3.6 megajoules.

The *meter-kilogram force* (commonly referred to as the kilogram-meter) is the technical unit of work or energy. It is defined as the work required to raise the mass (or weight) of one kilogram through a vertical distance of one meter at standard acceleration of gravity g_0. If measurements are made in places where the local value of the acceleration of gravity g is different from g_0, a correction factor g/g_0 has to be applied, if the exact value of work or energy is desired. (See discussion on force versus weight.)

The *gram-calorie* or small calorie is the physical unit of heat energy. It is defined as the quantity of heat required to raise the temperature of one gram mass of water either from 14.5 to 15.5°C or from 19.5 to 20.5°C, at standard pressure. The two values are designated as 15°C cal and 20°C cal, respectively. The mean gram-calorie is defined as $\frac{1}{100}$ part of the quantity of heat required to raise the temperature of one gram mass of water from 0 to 100°C at standard pressure. The same definitions apply to kilogram-calorie, or large calorie, if the kilogram mass is used as reference standard mass.

The *Ostwald calorie* is the quantity of heat required to raise the temperature of one gram mass from 0 to 100°C. This unit is frequently used by electrochemists and is equal to 100 mean gram-calories.

The *international kilo-calorie* or international steam-table calorie (I T cal) is defined as the 1/860th part of the international kilowatthour. This new unit avoids any reference to the thermal properties of water and was recommended for international adoption at the first International Steam Table Conference (1929).* Its value is very nearly equal to the mean kilo-calorie, 1 I T cal = 1.00037 kilogram-calories (mean).

Units of Power. Power is the time rate at which work is done. Its physical unit is the watt defined as the power which gives rise to the production of energy at the rate of one joule per second.

The *international watt* is defined as the power expended by an electric current of one international ampere flowing through a resistance of one international ohm. (See discussion on electrical units.) The latest value of the international watt is equal to 1.000165 theoretical watts.†

The *electrical horsepower* is defined as 746 absolute watts and is commonly used in the United States and in England in rating electrical machinery.

The *meter-kilogram force per second* (commonly referred to as the kilogram-meter per second) is the technical unit of power. The *metric horsepower* is defined as 75 kg-m/sec and is the most common mechanical unit of power.

Units of Torque. Torque is the effectiveness of a force to produce rotation. It is defined as the product of the force and the perpendicular distance from its line of action to the instantaneous center of rotation. Its physical unit is the dyne-centimeter, and its technical unit the kilogram force meter. (Note the reversal of force and length units in the designation of the units of torque as compared with the units of energy and work.) *N θ — N-m is correct*

1.4.7 Thermal Units and Standards

Temperature

Definition of Temperature. The *temperature* of a body may be defined as its thermal state considered from the standpoint of its ability to communicate heat to other bodies. When two bodies are placed in thermal communication, the one that loses heat to the other is said to be at the higher temperature.

Mechanical Engineering, Feb., 1930, pp. 122, 139; Nov., 1935, p. 710.
†*Announcement of Changes in Electrical and Photometric Units*, Circular of National Bureau of Standards C459, Washington, D.C., 1947.

Standard Temperatures. Certain thermal states or "temperatures" may be reproduced and recognized by the fact that definite physical phenomena occur at these temperatures. Such thermal states are called "fixed points," and they may, quite apart from any temperature scale, be specified by the physical phenomena characteristic of those temperatures. The two fundamental fixed points are the ice point and the steam point.

The *ice point* is defined as the temperature of melting ice, which is realized experimentally as the temperature at which pure finely divided ice is in equilibrium with pure, air-saturated water, under standard atmospheric pressure. The effect of increased pressure is to lower the freezing point to the extent of 0.007°C per atmosphere.

The *steam point* is defined as the temperature of condensing water vapor at standard atmospheric pressure, and it is realized experimentally by the use of a hypsometer so constructed as to avoid superheat of the vapor around the thermometer, or contamination with air or other impurities. If the desired conditions have been attained, the observed temperature should be independent of the rate of heat supply to the boiler, except as this may affect the pressure within the hypsometer, and of the length of time the hypsometer has been in operation.

Definition of Temperature Scale. The purpose of establishing a temperature scale is to assign a number to every thermal state or temperature, and to provide a means for determining the temperature of any particular body.

A *temperature scale* may be defined by (1) selecting definite numbers for certain fixed points, (2) selecting some physical property of a definite substance that varies with temperature, and (3) selecting a mathematical law expressing temperatures on the scale in question in terms of the selected property of the thermometric substance. For example, on the Centigrade mercury-in-glass scale, the ice and steam points are numbered 0 and 100, respectively, the relative or "apparent" expansion of a volume of mercury enclosed in glass of a definite kind is the property used, and the mathematical relation used to express temperature on this scale is that equal increments of apparent volume of the mercury in this glass correspond to equal increments of temperature. If some other substance is substituted for mercury, or if glass of a different kind is used, another scale is obtained that agrees with it at 0 and 100 but not at other temperatures.

Although, in general, a temperature scale depends on the thermometric substance as well as on the expression for the temperature in terms of some property of this substance, Lord Kelvin has shown that, if the property selected is the availability of energy, the scale so defined is wholly independent of the substance and depends only on the mathematical relation chosen. Any scale so defined is known as a thermodynamic scale.

The Kelvin Temperature Scale. The temperature scale finally chosen by Lord Kelvin is the one on which the temperature interval from the ice point to the steam point is 100° and the ratio of the values of any two temperatures is equal to the ratio of the heat taken in to the heat rejected by a reversible thermodynamic engine working with a source and refrigerator at the higher and lower temperatures, respectively. On this scale, which is also known as the absolute thermodynamic scale, the lowest attainable temperature is 0 and the ice point is found experimentally to be 273.16°. The steam point therefore is 373.16° or 100° higher.

The *degree Kelvin* (°K) or degree of absolute temperature is the absolute unit of temperature and is, for practical purposes, identical with the degree Centigrade (°C) of the international temperature scale.

The Thermodynamic Centigrade Scale. This is derived by subtracting from the Kelvin scale a constant number of the proper magnitude to make the ice point 0°. On this scale, therefore, the ice and steam points are 0 and 100°, respectively, and the so-called absolute zero is $-273.16°$.

The International Centigrade Scale. This is a practical representation of the thermodynamic Centigrade scale to such a degree of accuracy as is possible with present-day apparatus and methods. It was adopted at the General Conference on Weights and Measures at Sèvres, France, in 1927 and is subject to revision and amendment as improved and more accurate methods of measurement are evolved.

The unit of temperature on the international scale is the *degree Centigrade* (°C, or °C int) and is very nearly equal to $\frac{1}{100}$ the difference between the temperature of melting ice and the temperature of condensing water vapor under standard atmospheric pressure. (See discussion on metric units for pressure.)

The standard of the international temperature scale between -190 and $+660°C$ is deduced from the electrical resistance of a standard platinum resistance thermometer by means of a formula connecting the resistance R_t at any temperature $t°C$ within the above range with the resistance R_0 at

0°C. The purity of the platinum of which the thermometer is made should be such that the ratio R_t/R_0 for certain fixed temperatures is within specified limits.*

The degree Centigrade is most widely used in scientific publications and increasingly also in the engineering literature. In many countries in Europe it is the common everyday temperature unit. The subdivision into a hundred degrees of the temperature interval between the ice point and the steam point was first used by Celsius, a German, in 1742; therefore, in the European literature "°C" is read "degree Celsius."

The Fahrenheit Temperature Scale. This scale subdivides the temperature interval between the ice point and the steam point into 180 parts, one part of which is chosen as the unit of temperature and named *degree Fahrenheit* (°F). The ice point is assigned the value 32°F, so that the steam point has a temperature of 212°F.

The Fahrenheit unit of temperature is in common everyday use in the English-speaking countries. It was first introduced in England about 1665 by the physicist Fahrenheit; the choice of 32°F for the ice point has its explanation in the fact that Fahrenheit chose as zero the lowest temperature attainable by means of a salt-ice mixture.

The Rankine Absolute Temperature Scale (°R). This is the thermodynamic Fahrenheit scale where absolute zero is 0°R ($-459.69°$F). The ice point is assigned the value 491.69°R and the steam point 671.69°R.

Relations between the Temperature Scales. The following table shows the interrelations between the various temperature scales in form of equations. X indicates the unknown number of chosen temperature units and t the known number of given temperature units.

$x°F =$		$9/5 \, (t°K - 273.16) + 32$	$9/5 \, (t°C) + 32$
$x°K =$	$5/9 \, (t°F - 32) + 273.16$		$(t°C) + 273.16$
$x°C =$	$5/9 \, (t°F - 32)$	$(t°K) - 273.16$	
$x°R =$	$(t°F) + 459.699$	$9/5 \, (t°K)$	$9/5 \, (t°C) + 491.69$

Quantity of Heat and Some Derived Quantities

Units of Quantity of Heat. Quantity of heat is defined as the energy transferred from one body to another by a thermal process, that is, by radiation or conduction. The units for the quantity of heat are the *British thermal unit* and the *calorie*, as specific thermal units; and the *erg* and *joule*, as general physical units (see discussion on units of energy, metric and English system of units).

Thermal Capacity or Specific Heat of a Substance. This is the quantity of heat required to produce a unit change in temperature in a unit of mass of the substance. The common English unit is the British thermal unit per degree Fahrenheit per pound mass (Btu per °F per lb); the usual metric unit is the gram-calorie per degree Centigrade per gram mass (cal per °C per g); and the general physical unit used in the scientific literature is the erg per degree Centigrade per gram mass (erg per °C per g). In the technical literature thermal capacity of a substance is often expressed in watt-seconds (or joules) per degree Centigrade per kilogram mass (watt-sec per °C per kg) on account of the easy comparison with other technical units.

The Calorimetric or Water Equivalent. This is the quantity of heat required to produce a unit change in temperature of a body or system. It is numerically equivalent to the mass of water (in units as involved in the definition of the unit of quantity of heat used) that could be raised a unit temperature by the same total quantity of heat. The thermal capacity is expressed in British thermal units per degree Fahrenheit (Btu per °F), calories per degree Centigrade (cal per °C), or watt-seconds per degree Centigrade (watt-sec per °C).

Thermal Conductivity. This is the time rate of heat transfer through unit area across unit thickness per unit difference in temperature between the end surfaces. It is measured in British thermal units per second per degree Fahrenheit per inch thickness per square inch cross section (Btu per sec per °F per in. per in.^{2}), in calories per second per degree Centigrade per centimeter thickness per square centimeter cross section (cal per sec per °C per cm per cm^2), or in watts per degree Centigrade per meter thickness per square meter cross section (watts per °C per m per m^2).

Thermal Transmittance. Or surface coefficient of transfer is the time rate of heat emitted by unit area for unit difference in temperature between the surface in question and the surroundings. It is

*See also U.S. Bureau of Standards, *Journal of Research*, Vol. 1, p. 636, 1928.

measured in British thermal units per second per degree Fahrenheit per square inch (Btu per sec per °F per in.^{2}), in calories per second per degree Centigrade per square centimeter (cal per sec per °C per cm^2), or in watts per degree Centigrade per square meter (watts per °C per m^2).

The Joule Equivalent

The **Joule equivalent** is defined as the number of foot-pounds of energy per Btu. The numerical values for the various energy units used in the English and metric systems are as follows:

	Joules "Absolute"	Foot-pounds (force)	Foot-poundals	Meter-kilogram (force)	Kilowatt-hour "International"
1 British thermal unit (Btu) (mean) =	1055.18	778.26	25.040	107.599	2.93019×10^{-4}
1 gram-calorie (cal) (mean) =	4.1873	3.0884	99.366	0.42699	1.16279×10^{-6}
1 International kilocalorie (I T-cal) =	4187.3	3088.4	99.366	426.99	1.16279×10^{-3}
1 Ostwald calorie =	418.73	308.84	9936.6	42.699	1.16279×10^{-4}

1.4.8 Chemical Units and Standards

Atomic Weight. The present definition of atomic weights (1961) is based on ^{12}C, which is the most abundant isotope of carbon and whose atomic weight is defined as exactly 12.

Standard Cell Potential. A very large class of chemical reactions are characterized by the transfer of protons or electrons. Substances losing electrons in a reaction are said to be oxidized, those gaining electrons are said to be reduced. Many such reactions can be carried out in a galvanic cell that forms a natural basis for the concept of the half-cell, that is, the overall cell is conceptually the sum of two half-cells, one corresponding to each electrode. The half-cell potential measures the tendency of one reaction, for example, oxidation, to proceed at its electrode; the other half-cell of the pair measures the corresponding tendency for reduction to proceed at the other electrode. Measurable cell potentials are the sum of the two half-cell potentials. Standard cell potentials refer to the tendency of reactants in their standard state to form products in their standard states. The standard conditions are 1 M concentration for solutions, 101.325 kPa (1 atm) for gases, and for solids, their most stable form at 25°C. Since half-cell potentials cannot be measured directly, numerical values are obtained by assigning the hydrogen gas–hydrogen ion half reaction the half-cell potential of 0 V. Thus, by a series of comparisons referred directly or indirectly to the standard hydrogen electrode, values for the strength of a number of oxidants or reductants can be obtained, and standard reduction potentials can be calculated from established values.

Standard cell potentials are meaningful only when they are calibrated against an electromotive force (emf) scale. To achieve an absolute value of emf, electrical quantities must be referred to the basic metric system of mechanical units. If the current unit A and the resistance unit Ω can be defined, then the volt may be defined by Ohm's law as the voltage drop across a resistor of one standard ohm (Ω) when passing one standard ampere (A) of current. In the ohm measurement, a resistance is compared to the reactance of an inductor or capacitor at a known frequency. This reactance is calculated from the measured dimensions and can be expressed in terms of the meter and second. The ampere determination measures the force between two interacting coils while they carry the test current. The force between the coils is opposed by the force of gravity acting on a known mass; hence, the ampere can be defined in terms of the meter, kilogram, and second. Such a means of establishing a reference voltage is inconvenient for frequent use and reference is made to a previously calibrated standard cell.

Ideally, a standard cell is constructed simply and is characterized by a high constancy of emf, a low temperature coefficient of emf, and an emf close to one volt. The Weston cell, which uses a standard cadmium sulfate electrolyte and electrodes of cadmium amalgam and a paste of mercury and mercurous sulfate, essentially meets these conditions. The voltage of the cell is 1.0183 V at 20°C. The ac Josephson effect, which relates the frequency of a superconducting oscillator to the potential difference between two superconducting components, is used by the National Bureau of Standards to maintain the unit of emf, but the definition of the volt remains the Ω/A derivation described.

Concentration. The basic unit of concentration in chemistry is the mole, which is the amount of substance that contains as many entities, for example, atoms, molecules, ions, electrons, protons, and so on, as there are atoms in 12 g of ^{12}C, that is, Avogadro's number $N_A = 6.022045 \times 10^{23}$. Solution concentrations are expressed on either a weight or volume basis. *Molality* is the concentration of a solution in terms of the number of moles of solute per kilogram of solvent. *Molarity* is the concentration of a solution in terms of the number of moles of solute per liter of solution.

A particular concentration measure of acidity of aqueous solutions is pH, which, usually, is regarded as the common logarithm of the reciprocal of the hydrogen ion concentration (qv). More precisely, the potential difference of the hydrogen electrode in normal acid and in normal alkali solution (-0.828 V at 25°C) is divided into 14 equal parts or pH units; each pH unit is 0.0591 V. Operationally, pH is defined by pH = pH (soln) + E/K, where E is the emf of the cell:

$$H_2 \,|\, \text{solution of unknown pH} \,\|\, \text{saturated KCl} \,\|\, \text{solution of known pH} \,|\, H_2$$

and $K = 2.303\ RT/F$, where R is the gas constant, 8.314 J/(mol/K) [1.987 cal/(mol · K)], T is the absolute temperature, and F is the value of the Faraday, 9.64845×10^4 C/mol. pH usually is equated to the negative logarithm of the hydrogen ion activity, although there are differences between these two quantities outside the pH range 4.0–9.2:

$$-\log q_{H^+} m_{H^+} = \text{pH} + 0.014\ (\text{pH} - 9.2) \text{ for pH} > 9.2$$

$$-\log q_{H^+} m_{H^+} = \text{pH} + 0.009\ (4.0 - \text{pH}) \text{ for pH} < 4.0$$

1.4.9   The Theoretical or "Absolute" Electrical Units

With the general adoption of SI as the form of metric system that is preferred for all applications, further use of cgs units of electricity and magnetism is deprecated. Nonetheless, for historical reasons as well as for comprehensiveness, a brief review is included in 1.4.9 and 1.4.10.

The definitions of the **theoretical, or "absolute," units** are based on a particular choice of the numerical value of either k_e, the constant in Coulomb's electrostatic force law or k_m, the constant in Ampere's electrodynamic force law. The designation "absolute" units is generally used because of historical tradition; an interesting account of the history can be found in Glazebrook's *Handbook for Applied Physics*, Vol. II, "Electricity," pp. 211 ff., 1922. Because of the theoretical background of the unit definitions, they have also been designated as "theoretical" units, which is in good contradistinction to practical units based on physical standards.

The Theoretical Electrostatic Units

The **theoretical electrostatic units** are based on the cgs system of mechanical units and the choice of the numerical value unity for k_{ev} in Coulomb's law. They are frequently referred to as the *cgs electrostatic units*, but no specific unit names are available. In order to avoid the cumbersome writing, for example, one "theoretical electrostatic unit of charge," it had been proposed to use the theoretical "practical" unit names and prefix them with either stat or E.S. as, for example, statcoulomb, or E.S. coulomb. The first alternative will be used here.

The **absolute dielectric constant (permittivity)** of free space is the reciprocal of the Coulomb constant k_{ev} and is chosen as the fourth fundamental quantity in the theoretical electrostatic system of units. Its numerical value is defined as unity, and it is identical with one statfarad per centimeter if use is made of prefixing the corresponding unit of the "practical" series.

The theoretical electrostatic unit of **charge** or the statcoulomb is defined as the quantity of electricity that, when concentrated at a point and placed at one centimeter distance from an equal quantity of electricity similarly concentrated, will experience a mechanical force of one dyne in free space. An alternative definition, based on the concept of field lines, gives the theoretical electrostatic unit of charge as a positive charge from which in free space exactly 4π displacement lines emerge.

The theoretical electrostatic unit of **displacement flux (dielectric flux)** is the "line of displacement flux" or $\frac{1}{4}\pi$ of the theoretical electrostatic unit of charge. This definition provides the basis for graphical field mapping insofar as it gives a definite rule for the selection of displacement lines to represent the distribution of the field quantitatively.

The theoretical electrostatic unit of **displacement**, or **dielectric flux density**, is chosen as one displacement line per square centimeter area perpendicular to the direction of the displacement lines. It can be given also as $\frac{1}{4}\pi$ statcoulomb per square centimeter (according to Gauss's law). In isotropic media the displacement has the same direction as the potential gradient, and the surfaces perpendicular to the field lines become the equipotential surfaces; the theoretical electrostatic unit of displacement can then be defined as one displacement line per square centimeter of equipotential surface.

The theoretical electrostatic unit of **electrostatic potential or the statvolt** is defined as existing at a point in an electrostatic field, if the work done to bring the theoretical electrostatic unit of charge, or the statcoulomb, from infinity to this point equals one erg. This customary definition implies, however, that the potential vanishes at infinite distances and has, therefore, only restricted validity.

As it is fundamentally impossible to give absolute values of potential, the use of potential difference and its unit (see below) should be preferred.

The theoretical electrostatic unit of **electrical potential difference** or **voltage**, or the **statvolt**, is defined as existing between two points in space if the work done to bring the theoretical electrostatic unit of charge, or the statcoulomb, from the one of these points to the other equals one erg. Potential difference is counted positive in the direction in which a negative quantity of electricity would be moved by the electrostatic field.

The theoretical electrostatic unit of **capacitance** or the **statfarad** is defined as the capacitance that maintains an electrical potential difference of one statvolt between two conductors charged with equal and opposite electrical charges of one statcoulomb. In the older literature, the cgs electrostatic unit of capacitance is identified with the "centimeter"; this was replaced by statfarad to avoid confusion.

The theoretical electrostatic unit of **electric potential gradient**, or **field strength** (field intensity), is defined to exist at a point in an electric field, if the mechanical force exerted upon the theoretical electrostatic unit of charge concentrated at this point is equal to one dyne. It is expressed as one statvolt per centimeter.

The theoretical electrostatic unit of **current** or the **statampere** is defined as the time rate of transfer of the theoretical electrostatic unit of charge and is identical with the statcoulomb per second.

The theoretical electrostatic unit of **electrical resistance** or the **statohm** is defined as the resistance of a conductor in which a current of one statampere is produced if a potential difference of one statvolt is applied at its ends.

The theoretical electrostatic unit of **electromotive force (emf)** is defined as equivalent to the theoretical electrostatic unit of potential difference if it produces a current of one statampere in a conductor of one statohm resistance. It is identical with the statvolt but, according to its concept, requires an independent definition.

The theoretical electrostatic unit of **magnetic intensity** is defined as the magnetic intensity at the center of a circle of 4π cm diameter in which a current of one statampere is flowing. This unit is equal to 4π statamperes per centimeter but has no name as the factor 4π excludes the possibility of using the prefixed "practical" unit name.

The theoretical electrostatic unit of **magnetic flux** or the **statweber** is defined as the magnetic flux whose time rate of change through a linear conductor loop (linear conductor is used to designate a conductor of infinitely small cross section) produces in this loop an emf of one statvolt.

The theoretical electrostatic unit of **magnetic flux density**, or **induction**, is defined as the electrostatic unit of magnetic flux per square centimeter area, or the statweber per square centimeter.

The **absolute magnetic permeability of free space** is defined as the ratio of magnetic induction to the magnetic intensity. Its unit is the stathenry per centimeter, as a derived unit.

The theoretical electrostatic unit of **inductance** or the **stathenry** is defined as connected with a conductor loop carrying a steady current of one statampere that produces a magnetic flux of one statweber. A more general definition, applicable to varying fields with nonlinear relation between magnetic flux and current, gives the stathenry as connected with a conductor loop in which a time rate of change in the current of one statcoulomb produces a time rate of change in the magnetic flux of one statweber per second.

The Theoretical Electromagnetic Units

The **theoretical electromagnetic units** are based on the cgs system of mechanical units and Coulomb's law of mechanical force action between two isolated magnetic quantities m_1 and m_2 (approximately true for very long bar magnets) that must be written

$$F_m = \frac{k_m}{2} \frac{m_1 m_2}{r^2} \qquad (1.7)$$

where k_m is the proportionality constant of Ampère's law for force action between parallel currents that is more basic, and amenable to much more accurate measurement, than (1.7). The factor $\frac{1}{2}$ appears here because of the three-dimensional character of the field distribution around point magnets as compared with the two-dimensional field of two parallel currents.

The theoretical electromagnetic units are obtained by defining the numerical value of $k_{mv}/2$ (for vacuum) as unity; they are frequently referred to as the cgs electromagnetic units. Only a few specific unit names are available. In order to avoid cumbersome writing, for example, one "theoretical electromagnetic unit of charge," it had been proposed to use the theoretical "practical" unit names and prefix them with either ab- or E.M. as, for example, abcoulomb, or E.M. coulomb. The first alternative will be used here.

The **absolute magnetic permeability** of free space is the value $k_{mv}/2$ in (1.7) and is chosen as the fourth fundamental quantity in the theoretical electromagnetic system of units. Its numerical value is assumed as unity, and it is identical with one abhenry per centimeter if use is made of prefixing the corresponding unit of the "practical" series.

The theoretical electromagnetic unit of **magnetic quantity** is defined as the magnetic quantity that, when concentrated at a point and placed at one centimeter distance from an equal magnetic quantity similarly concentrated, will experience a mechanical force of one dyne in free space. An alternative definition, based on the concept of magnetic intensity lines, gives the theoretical electromagnetic unit of magnetic quantity as a positive magnetic quantity from which, in free space, exactly 4π magnetic intensity lines emerge.

The theoretical electromagnetic unit of **magnetic moment** is defined as the magnetic moment possessed by a magnet formed by two theoretical electromagnetic units of magnetic quantity of opposite sign, concentrated at two points one centimeter apart. As a vector, its positive direction is defined from the negative to the positive magnetic quantity along the center line.

The theoretical electromagnetic unit of **magnetic induction (magnetic flux density), or the Gauss**, is defined to exist at a point in a magnetic field, if the mechanical torque exerted upon a magnet with theoretical electromagnetic unit of magnetic moment and directed perpendicular to the magnetic field is equal to one dyne-centimeter. The lines to which the vector of magnetic induction is tangent at every point are called induction lines or magnetic flux lines; on the basis of this flux concept, magnetic induction is identical with magnetic flux density.

The theoretical electromagnetic unit of **magnetic flux**, or the **Maxwell**, is the "field line" or line of magnetic induction. In free space, the theoretical electromagnetic unit of magnetic quantity issues 4π induction lines; the unit of magnetic flux, or the maxwell, is then $1/4\pi$ of the theoretical electromagnetic unit of magnetic quantity times the absolute permeability of free space.

The theoretical electromagnetic unit of **magnetic intensity (magnetizing force), or the oersted**, is defined to exist at a point in a magnetic field in free space where one measures a magnetic induction of one gauss.

The theoretical electromagnetic unit of **current**, or the **abampere**, is defined as the current that flows in a circle of one centimeter diameter and produces at the center of this circle a magnetic intensity of one oersted.

The theoretical electromagnetic unit of **inductance**, or the **abhenry**, is defined as connected with a conductor loop in which a time rate of change of one maxwell per second in the magnetic flux produces a time rate of change in the current of one abampere per second. In the older literature, the cgs electromagnetic unit of inductance is identified with the "centimeter"; this should be replaced by a henry to avoid confusion.

The theoretical electromagnetic unit of **magnetomotive force (mmf)** is defined as the magnetic driving force produced by a conductor loop carrying a steady current of $\frac{1}{4}\pi$ abamperes; it has the name one gilbert. The concept of magnetomotive force as the driving force in a "magnetic circuit" permits an alternative definition of the gilbert as the magnetomotive force that produces a uniform magnetic intensity of one oersted over a length of one centimeter in the magnetic circuit. Obviously, one gilbert equals one oersted-centimeter.

The theoretical electromagnetic unit of **magnetostatic potential** is defined as the potential existing at a point in a magnetic field, if the work done to bring the theoretical electromagnetic unit of magnetic quantity from infinity to this point equals one erg. This customary definition implies, however, that the potential vanishes at infinite distances, and the definition has, therefore, only restricted validity. The unit, thus defined, is identical with one gilbert. The difference in magnetostatic potential between any two points is usually called magnetomotive force (mmf).

The theoretical electromagnetic unit of **reluctance** is defined as the reluctance of a magnetic circuit in which a magnetomotive force of one gilbert produces a magnetic flux of one maxwell.

The theoretical electromagnetic unit of **electric charge**, or the **abcoulomb**, is defined as the quantity of electricity that passes through any section of an electric circuit in one second if the current is one abampere.

The theoretical electromagnetic unit of **displacement flux (dielectric flux)** is the "line of displacement flux" or $\frac{1}{4}\pi$ of the theoretical electromagnetic unit of electric charge. This definition provides the basis for graphical field mapping insofar as it gives a definite rule for the selection of displacement lines to represent the character of the field.

The theoretical electromagnetic unit of **displacement**, or **dielectric flux density**, is chosen as one displacement line per square centimeter area perpendicular to the direction of the displacement lines. It can also be given as $\frac{1}{4}\pi$ abcoulombs per square centimeter (according to Gauss's law). In isotropic media the theoretical electromagnetic unit of displacement can be defined as one displacement line per square centimeter of equipotential surface. (See discussion on theoretical electrostatic unit of displacement.)

The theoretical electromagnetic unit of **electrical potential difference** or **voltage**, or the **abvolt**, is defined as the potential difference existing between two points in space if the work done in bringing the theoretical electromagnetic unit of charge, or the abcoulomb, from one of these points to the other equals one erg. Potential difference is counted positive in the direction in which a negative quantity of electricity would be moved by the electrostatic field.

The theoretical electromagnetic unit of **capacitance**, or the **abfarad**, is defined as the capacitance that maintains an electrical potential difference of one abvolt between two conductors charged with equal and opposite electrical quantities of one abcoulomb.

The theoretical electromagnetic unit of **potential gradient, or field strength** (field intensity), is defined to exist at a point in an electric field if the mechanical force exerted upon the theoretical electromagnetic unit of charge concentrated at this point is equal to one dyne. It is expressed as one abvolt per centimeter.

The theoretical electromagnetic unit of **resistance**, or the **abohm**, is defined as the resistance of a conductor in which a current of one abampere is produced if a potential difference of one abvolt is applied at its ends.

The theoretical electromagnetic unit of **electromotive force (emf)** is defined as the electromotive force acting in an electric circuit in which a current of one abampere is flowing and electrical energy is converted into other kinds of energy at the rate of one erg per second. This unit is identical with the abvolt.

The **absolute dielectric constant of free space** is defined as the ratio of displacement to the electric field intensity. Its unit is the abfarad per centimeter, a derived unit.

The Theoretical Electrodynamic Units

The **theoretical electrodynamic units** are based on the cgs system of mechanical units and are therefore frequently referred to as *the cgs electrodynamic units*. In contradistinction to the theoretical electromagnetic units, these units are derived from a significant experimental law, Ampère's experiment on the mechanical force between two parallel currents. The units as proposed by Ampère and used by W. Weber differ from the electromagnetic units by factors of 2 and multiples thereof. They can be made to coincide with the theoretical electromagnetic units by proper definition of the fundamental unit of current. Some of the important definitions will be given for this latter case only.

For the *absolute magnetic permeability* of free space see discussion on theoretical electromagnetic units.

The theoretical electrodynamic unit of **current**, or the **abampere**, is defined as the current flowing in a circuit consisting of two infinitely long parallel wires one centimeter apart when the electrodynamic force of repulsion between the two wires is *two* dynes per centimeter length in free space. If the more natural choice of *one* dyne per centimeter length is made, the original proposal of Ampère is obtained and the unit of current becomes $1/\sqrt{2}$ abampere.

The theoretical electrodynamic unit of **magnetic induction** is defined as the magnetic induction inducing an electromotive force of one abvolt in a conductor of one-centimeter length and moving with a velocity of one centimeter per second, if the conductor, its velocity, and the magnetic induction are mutually perpendicular. The unit thus defined is called one gauss.

The theoretical electrodynamic unit of **magnetic flux**, or the **maxwell**, is defined as the magnetic flux represented by a uniform magnetic induction of one gauss over an area of one square centimeter perpendicular to the direction of the magnetic induction.

The theoretical electrodynamic unit of **magnetic intensity**, or the **oersted**, is defined as the magnetic intensity at the center of a circle of 4π-centimeter diameter in which a current of one abampere is flowing.

All the other unit definitions, which do not pertain to magnetic quantities, are identical with the definitions for the theoretical electromagnetic units.

1.4.10   The Internationally Adopted Electrical Units and Standards

The International Committee on Weights and Measures decided in October, 1946, at Paris to abandon the so-called international practical units based on physical standards (see below) and to adopt, effective January 1, 1948, the so-called absolute practical units for international use.

The Adopted "Absolute" Practical Units

By a series of international actions, the "absolute" practical electrical units are defined as exact powers of 10 of corresponding theoretical electrodynamic and electromagnetic units because they are based on the choice of the proportionality constant in Ampère's law for free space as $k_{mv} = 2 \times 10^{-7}$ *henry/m*.

The **"absolute" practical unit of current, or the "absolute" ampere**, is defined as the current flowing in a circuit consisting of two very long parallel thin wires spaced 1 m apart in free space if the electrodynamic force action between the wires is 2×10^{-7} newton = 0.02 dyne, per meter length. It is 10^{-1} of the theoretical or "absolute" electrodynamic or electromagnetic unit of current and was adopted internationally in 1881.

The "absolute" practical unit of **electric charge**, or the **"absolute" coulomb**, is defined as the quantity of electricity that passes through a cross-sectional surface in one second if the current is one "absolute" ampere. It is 10^{-1} of the theoretical or "absolute" electromagnetic unit of electric charge and was adopted internationally in 1881.

The "absolute" practical unit of **electric potential difference**, or the **"absolute" volt**, is defined as the potential difference existing between two points in space if the work done in bringing an electric charge of one "absolute" coulomb from one of these points to another is equal to one "absolute" joule $= 10^7$ ergs. It is 10^8 of the theoretical or "absolute" electromagnetic unit of potential difference and was adopted internationally in 1881.

The "absolute" practical unit of **resistance**, or the **"absolute" ohm**, is defined as the resistance of a conductor in which a current of one "absolute" ampere is produced if a potential difference of one "absolute" volt is applied at its ends. It is 10^9 of the theoretical or "absolute" electromagnetic unit of resistance and was adopted internationally in 1881.

The "absolute" practical unit of **magnetic flux**, or the **"absolute" weber**, is defined to be linked with a closed loop of thin wire of total resistance one "absolute" ohm if upon removing the wire loop from the magnetic field a total charge of one "absolute" coulomb is passed through any cross section of the wire. It is 10^8 of the theoretical or "absolute" electromagnetic unit of magnetic flux, the maxwell, and was adopted internationally in 1933.

The "absolute" practical unit of **inductance**, or the **"absolute" henry**, is defined as connected with a closed loop of thin wire in which a time rate of change of one "absolute" weber per second in the magnetic flux produces a time rate of change in the current of one "absolute" ampere. It is 10^9 of the theoretical or "absolute" electromagnetic unit of inductance and was adopted internationally in 1893.

The "absolute" practical unit of **capacitance**, or the **"absolute" farad**, is defined as the capacitance that maintains an electric potential difference of one "absolute" volt between two conductors charged with equal and opposite electrical quantities of one "coulomb." It is 10^{-9} of the theoretical or "absolute" electromagnetic unit of capacitance and was adopted internationally in 1881.

The Abandoned "International" Practical Units

The International System of electrical and magnetic units is a system for electrical and magnetic quantities that takes as the four fundamental quantities resistance, current, length, and time. The units of resistance and current are defined by physical standards that were originally aimed to be exact replicas of the "absolute" practical units, namely the "absolute" ampere and the "absolute" ohm. On account of long-range variations in the physical standards, it proved impossible to rely upon them for international use and they recently have been replaced by the "absolute" practical units.

The "international" practical standards are defined as follows:

The **international ohm** is the resistance at 0°C of a column of mercury of uniform cross section, having a length of 106.300 cm and a mass of 14.4521 g.

The **international ampere** is defined as the current that will deposit silver at the rate of 0.00111800 g/sec.

From these fundamental units, all other electrical and magnetic units can be defined in a manner similar to the "absolute" practical units. Because of the inconvenience of the silver voltameter as a standard, the various national laboratories actually used a volt, defining its value in terms of the other two standards.

At its conference in October, 1946, in Paris, the International Committee on Weights and Measures accepted as the best relations between the "international" and the "absolute" practical units the following:

$$1 \text{ mean "international" ohm} = 1.00049 \text{ "absolute" ohms}$$

$$1 \text{ mean "international" volt} = 1.00034 \text{ "absolute" volts}$$

These mean values are the averages of values measured in six different national laboratories. On the basis of these mean values, the specific unit relation for converting "international" units appearing on certificates of the National Bureau of Standards, Washington, D. C., into "absolute" practical units, are as follows:

$$1 \text{ international ampere} = 0.999835 \text{ absolute ampere}$$

$$1 \text{ international coulomb} = 0.999835 \text{ absolute coulomb}$$

$$1 \text{ international henry} = 1.000495 \text{ absolute henries}$$

$$1 \text{ international farad} = 0.999505 \text{ absolute farad}$$

$$1 \text{ international watt} = 1.000165 \text{ absolute watts}$$

$$1 \text{ international joule} = 1.000165 \text{ absolute joules}$$

BIBLIOGRAPHY FOR UNITS AND MEASUREMENTS

Cohen, E. R., Taylor, B. N., "The 1986 Adjustment of the Fundamental Physical Constants," *Report of the CODATA Task Group on Fundamental Constants, November 1986*, CODATA Bulletin No. 63. International Council of Scientific Unions, Committee on Data for Science and Technology, Pergamon Press, 1986.

Hvistendahl, H. S., *Engineering Units and Physical Quantities*, London, Macmillan, 1964.

Jerrard, H. G., and McNeill, D. B., *A Dictionary of Scientific Units*, 2nd ed., London, Chapman Hall, 1964.

Letter Symbols for Units of Measurement, ANSI/IEEE Std 260-1978, Institute of Electrical and Electronic Engineers, New York, 1978.

Quantities, Units, Symbols, Conversion Factors, and Conversion Tables ISO Reference 31, 15 sections, Geneva, 1973–1979.

Standard for Metric Practice, ASTM E 380-82, Philadelphia, 1982.

Young, L., *System of Units in Electricity and Magnetism*, Edinburgh, Oliver and Boyd, 1969.

————, *Research Concerning Metrology and Fundamental Constants*, National Academy Press, Washington, D.C., 1983.

1.5   TABLES OF CONVERSION FACTORS*
By J. G. Brainerd
(revised and extended by J. H. Westbrook)

TABLE 1.27   Temperature Conversion

$$°F = (°C × \tfrac{9}{5}) + 32 = (°C + 40) × \tfrac{9}{5} - 40$$
$$°C = (F - 32) × \tfrac{5}{9} = (°F + 40) × \tfrac{5}{9} - 40$$
$$°R = °F + 459.69$$
$$°K = C + 273.16$$

		Interpolation differences			
°C	Temp	°F	°C	Temp	°F
0.5556	1	1.8	3.3334	6	10.8
1.1111	2	3.6	3.8889	7	12.6
1.6667	3	5.4	4.4445	8	14.4
2.2222	4	7.2	5.0000	9	16.2
2.7778	5	9.0	5.5556	10	18.0

°C	Temp	°F	°C	Temp	°F
− 206.67	− 340		− 123.33	− 190	− 310
− 201.11	− 330		− 117.78	− 180	− 292
− 195.56	− 320		− 112.22	− 170	− 274
− 190.00	− 310		− 106.67	− 160	− 256
− 184.44	− 300		− 101.11	− 150	− 238
− 178.89	− 290		− 95.56	− 140	− 220
− 173.33	− 280		− 90.0	− 130	− 202
− 167.78	− 270	− 454	− 84.44	− 120	− 184
− 162.22	− 260	− 436	− 78.89	− 110	− 166
− 156.67	− 250	− 418	− 73.33	− 100	− 148
− 151.11	− 240	− 400	− 67.78	− 90	− 130
− 145.56	− 230	− 382	− 62.22	− 80	− 121
− 140.00	− 220	− 364	− 56.67	− 70	− 94
− 134.44	− 210	− 346	− 51.11	− 60	− 76
− 128.89	− 200	− 328	− 45.56	− 50	− 58

*Bold face units in Tables 1.28 to 1.63 are SI.

TABLE 1.27   (*Continued*)

°C	Temp	°F	°C	Temp	°F
−40.00	−40	−40	13.33	56	132.8
−34.44	−30	−22	13.89	57	134.6
−28.89	−20	−4	14.44	58	136.4
−23.35	−10	14	15.00	59	138.2
−17.78	0	32	15.56	60	140.0
−17.22	1	33.8	16.11	61	141.8
−16.67	2	35.6	16.67	62	143.6
−16.11	3	37.4	17.22	63	145.4
−15.56	4	39.2	17.78	64	147.2
−15.00	5	41.0	18.33	65	149.0
−14.44	6	42.8	18.89	66	150.8
−13.89	7	44.6	19.44	67	152.6
−13.33	8	46.4	20.00	68	154.4
−12.78	9	48.2	20.56	69	156.2
−12.22	10	50.0	21.11	70	158.0
−11.67	11	51.8	21.67	71	159.8
−11.11	12	53.6	22.22	72	161.6
−10.56	13	55.4	22.78	73	163.4
−10.00	14	57.2	23.33	74	165.2
−9.44	15	59.0	23.89	75	167.0
−8.89	16	60.8	24.44	76	168.8
−8.33	17	62.6	25.00	77	170.6
−7.78	18	64.4	25.56	78	172.4
−7.22	19	66.2	26.11	79	174.2
−6.67	20	68.0	26.67	80	176.0
−6.11	21	69.8	27.22	81	177.8
−5.56	22	71.6	27.78	82	179.6
−5.00	23	73.4	28.33	83	181.4
−4.44	24	75.2	28.89	84	183.2
−3.89	25	77.0	29.44	85	185.0
−3.33	26	78.8	30.00	86	186.8
−2.78	27	80.6	30.56	87	188.6
−2.22	28	82.4	31.11	88	190.4
−1.67	29	84.2	31.67	89	192.2
−1.11	30	86.0	32.22	90	194.0
−0.56	31	87.8	32.78	91	195.8
0	32	89.6	33.33	92	197.6
0.56	33	91.4	33.89	93	199.4
1.11	34	93.2	34.44	94	201.2
1.67	35	95.0	35.00	95	203.0
2.22	36	96.8	35.56	96	204.8
2.78	37	98.6	36.11	97	206.6
3.33	38	100.4	36.67	98	208.4
3.89	39	102.2	37.22	99	210.2
4.44	40	104.0	37.78	100	212.0
5.00	41	105.8	43.33	110	230
5.56	42	107.6	48.89	120	248
6.11	43	109.4	54.44	130	266
6.67	44	111.2	60.00	140	284
7.22	45	113.0	65.56	150	302
7.78	46	114.8	71.11	160	320
8.33	47	116.6	76.67	170	338
8.89	48	118.4	82.22	180	356
9.44	49	120.2	87.78	190	374
10.00	50	122.0	93.33	200	392
10.56	51	123.8	100.00	212	413.6
11.11	52	125.6	104.44	220	428
11.67	53	127.4	110.00	230	446
12.22	54	129.2	115.56	240	464
12.78	55	131.0	121.11	250	482

TABLE 1.27 (*Continued*)

°C	Temp	°F	°C	Temp	°F
126.67	**260**	500	593.3	**1100**	2012
132.22	**270**	518	648.9	**1200**	2192
137.78	**280**	536	704.4	**1300**	2372
143.33	**290**	554	760.0	**1400**	2552
148.89	**300**	572	815.6	**1500**	2732
154.44	**310**	590	871.1	**1600**	2912
160.00	**320**	608	926.7	**1700**	3092
165.56	**330**	626	982.2	**1800**	3272
171.11	**340**	644	1038	**1900**	3452
176.67	**350**	662	1093	**2000**	3632
182.22	**360**	680	1149	**2100**	3812
187.78	**370**	698	1204	**2200**	3992
193.33	**380**	716	1260	**2300**	4172
198.89	**390**	734	1316	**2400**	4352
204.44	**400**	752	1371	**2500**	4532
210.00	**410**	770	1427	**2600**	4712
215.55	**420**	788	1482	**2700**	4892
221.11	**430**	806	1538	**2800**	5072
226.66	**440**	824	1593	**2900**	5252
232.22	**450**	842	1649	**3000**	5432
237.77	**460**	860	1704	**3100**	5612
243.33	**470**	878	1760	**3200**	5792
248.88	**480**	896	1816	**3300**	5972
254.44	**490**	914	1871	**3400**	6152
260.00	**500**	932	1927	**3500**	6332
315.6	**600**	1112	1982	**3600**	6512
371.1	**700**	1292	2038	**3700**	6692
426.7	**800**	1472	2093	**3800**	6872
482.2	**900**	1652	2149	**3900**	7052
537.8	**1000**	1832	2205	**4000**	7232

TABLE 1.28   Length [L]

Multiply Number of → by → to Obtain ↓	Centimeters	Feet	Inches	Kilometers	Nautical Miles	Meters[a]	Mils	Miles	Millimeters	Yards
Centimeters	1	30.48	2.540	10^5	1.853×10^5	100	2.540×10^{-3}	1.609×10^5	0.1	91.44
Feet	3.281×10^{-2}	1	8.333×10^{-2}	3281	6080.27	3.281	8.333×10^{-5}	5280	3.281×10^{-3}	3
Inches	0.3937	12	1	3.937×10^4	7.296×10^4	39.37	0.001	6.336×10^4	3.937×10^{-2}	36
Kilometers	10^{-5}	3.048×10^{-4}	2.540×10^{-5}	1	1.853	0.001	2.540×10^{-8}	1.609	10^{-6}	9.144×10^{-4}
Nautical Miles		1.645×10^{-4}		0.5396	1	5.396×10^{-4}		0.8684		4.934×10^{-4}
Meters[a]	0.01	0.3048	2.540×10^{-2}	1000	1853	1	2.540×10^{-5}	1609	0.001	0.9144
Mils	393.7	1.2×10^4	1000	3.937×10^7		3.937×10^4	1		39.37	3.6×10^4
Miles	6.214×10^{-6}	1.894×10^{-4}	1.578×10^{-5}	0.6214	1.1516	6.214×10^{-4}		1	6.214×10^{-7}	5.682×10^{-4}
Millimeters	10	304.8	25.40	10^6		1000	2.540×10^{-2}		1	914.4
Yards	1.094×10^{-2}	0.3333	2.778×10^{-2}	1094	2027	1.094	2.778×10^{-5}	1760	1.094×10^{-3}	1

[a] Bold face units in all Tables 1.28 to 1.63 are SI.

Length

Land Measure

7.92 inches = 1 link

25 links = 1 rod = 16.5 feet = 5.5 yards (1 rod = 1 pole = 1 perch)

4 rods = 1 chain (Gunther's) = 66 feet = 22 yards = 100 links

10 chains = 1 furlong = 660 feet = 220 yards = 1000 links = 40 rods

8 furlongs = 1 mile = 5280 feet = 1760 yards = 8000 links = 320 rods = 80 chains

Ropes and Cables

2 yards = 1 fathom   120 fathoms = 1 cable's length

Nautical Measure

6080.27 feet = 1 nautical mile = 1.15156 statute miles

3 nautical miles = 1 league (U. S.)   3 statute miles = 1 league (Gr. Britain)

(*Note*: A nautical mile is the length of a minute of longitude of the earth at the equator at sea level. The British Admiralty uses the round figure of 6080 feet. The word "knot" is used to denote "nautical miles per hour.")

Miscellaneous

3 inches = 1 palm   9 inches = 1 span

4 inches = 1 hand $2\frac{1}{2}$ feet = 1 military pace

TABLE 1.29　Area [L^2]

Multiply Number of → by → / to Obtain →	Acres	Circular Mils	Square Centimeters	Square Feet	Square Inches	Square Kilometers	*Square Meters*	Square Miles	Square Millimeters	Square Yards
Acres	1			2.296×10^{-5}		247.1	2.471×10^{-4}	640		2.066×10^{-4}
Circular Mils		1	1.973×10^5	1.833×10^8	1.273×10^6		1.973×10^9		1973	
Square Centimeters		5.067×10^{-6}	1	929.0	6.452	10^{10}	10^4	2.590×10^{10}	0.01	8361
Square Feet	4.356×10^4		1.076×10^{-3}	1	6.944×10^{-3}	1.076×10^7	10.76	2.788×10^7	1.076×10^{-5}	9
Square Inches	6,272,640	7.854×10^{-7}	0.1550	144	1	1.550×10^9	1550	4.015×10^9	1.550×10^{-3}	1296
Square Kilometers	4.047×10^{-3}		10^{-10}	9.290×10^{-8}	6.452×10^{-10}	1	10^{-6}	2.590	10^{-12}	8.361×10^{-7}
Square Meters	4047		0.0001	9.290×10^{-2}	6.452×10^{-4}	10^6	1	2.590×10^6	10^{-6}	0.8361
Square Miles	1.562×10^{-3}		3.861×10^{-11}	3.587×10^{-8}		0.3861	3.861×10^{-7}	1	3.861×10^{-13}	3.228×10^{-7}
Square Millimeters		5.067×10^{-4}	100	9.290×10^4	645.2	10^{12}	10^6		1	8.361×10^5
Square Yards	4840		1.196×10^{-4}	0.1111	7.716×10^{-4}	1.196×10^6	1.196	3.098×10^6	1.196×10^{-6}	1

Area

Land Measure

$30\frac{1}{4}$ square yards = 1 square rod = $272\frac{1}{4}$ square feet

16 square rods = 1 square chain = 484 square yards = 4356 square feet

$2\frac{1}{2}$ square chains = 1 rood = 40 square rods = 1210 square yards

4 roods = 1 acre = 10 square chains = 160 square rods

640 acres = 1 square mile = 2560 roods = 102,400 square rods

1 section of land = 1 square mile; 1 quarter section = 160 acres

Architect's Measure

100 square feet = 1 square

Circular Inch and Circular Mil

A circular inch is the area of a circle 1 inch in diameter = 0.7854 square inch

1 square inch = 1.2732 circular inches

A circular mil is the area of a circle 1 mil (or 0.001 inch) in diameter = 0.7854 square mil

1 square mil = 1.2732 circular mils

1 circular inch = 10^6 circular mils = 0.7854×10^6 square mils

1 square inch = 1.2732×10^6 circular mils = 10^6 square mils

TABLE 1.30 Volume $[L^3]$

Multiply Number of → by → to Obtain →	Bushels (Dry)	Cubic Centimeters	Cubic Feet	Cubic Inches	*Cubic Meters*	Cubic Yards	Gallons (Liquid)	Liters	Pints (Liquid)	Quarts (Liquid)
Bushels (Dry)	1		0.8036	4.651×10^{-4}	28.38			2.838×10^{-2}		
Cubic Centimeters	3.524×10^4	1	2.832×10^4	16.39	10^6	7.646×10^5	3785	1000	473.2	946.4
Cubic Feet	1.2445	3.531×10^{-5}	1	5.787×10^{-4}	35.31	27	0.1337	3.531×10^{-2}	1.671×10^{-2}	3.342×10^{-2}
Cubic Inches	2150.4	6.102×10^{-2}	1728	1	6.102×10^4	46,656	231	61.02	28.87	57.75
Cubic Meters	3.524×10^{-2}	10^{-6}	2.832×10^{-2}	1.639×10^{-5}	1	0.7646	3.785×10^{-3}	0.001	4.732×10^{-4}	9.464×10^{-4}
Cubic Yards		1.308×10^{-6}	3.704×10^{-2}	2.143×10^{-5}	1.308	1	4.951×10^{-3}	1.308×10^{-3}	6.189×10^{-4}	1.238×10^{-3}
Gallons (Liquid)		2.642×10^{-4}	7.481	4.329×10^{-3}	264.2	202.0	1	0.2642	0.125	0.25
Liters	35.24	0.001	28.32	1.639×10^{-2}	1000	764.6	3.785	1	0.4732	0.9464
Pints (Liquid)		2.113×10^{-3}	59.84	3.463×10^{-2}	2113	1616	8	2.113	1	2
Quarts (Liquid)		1.057×10^{-3}	29.92	1.732×10^{-2}	1057	807.9	4	1.057	0.5	1

Volume

Cubic Measure

1 cord of wood = pile cut 4 feet long, piled 4 feet high and 8 feet on the ground
= 128 cubic feet

1 perch of stone = quantity $1\frac{1}{2}$ feet thick, 1 foot high and $16\frac{1}{2}$ feet long
= $24\frac{3}{4}$ cubic feet

(*Note:* A perch of stone is, however, often computed differently in different localities; thus, in most if not all of the states and territories west of the Mississippi, stonemasons figure rubble by the perch of $16\frac{1}{2}$ cubic feet. In Philadelphia, 22 cubic feet are called a perch. In Chicago, stone is measured by the cord of 100 cubic feet. Check should be made against local practice.)

Board Measure. In board measure, boards are assumed to be one inch in thickness. Therefore, feet board measure of a stick of square timber = length in feet × breadth in feet × thickness in inches.

Shipping Measure. For register tonnage or measurement of the entire internal capacity of a vessel, it is arbitrarily assumed, to facilitate computation, that

100 cubic feet = 1 register ton

For the measurement of cargo:

40 cubic feet = 1 U. S. shipping ton = 32.143 U. S. bushels

42 cubic feet = 1 British shipping ton = 32.703 Imperial bushels

Dry Measure. One U. S. Winchester bushel contains 1.2445 cubic feet or 2150.42 cubic inches. It holds 77.601 pounds distilled water at 62°F.

(*Note:* This is a *struck* bushel. A *heaped* bushel in general equals $1\frac{1}{4}$ struck bushels, although for apples and pears it contains 1.2731 struck bushels = 2737.72 cubic inches.)

One U. S. gallon (dry measure) = $\frac{1}{8}$ bushel and contains 268.8 cubic inches.

(*Note:* This is not a legal U. S. *dry measure* and therefore is given for comparison only.)

One British Imperial bushel contains 1.2843 cubic feet or 2219.36 cubic inches. It holds 80 pounds distilled water at 62°F.

1 British Imperial gallon = $\frac{1}{8}$ Imperial bushel and contains 277.42 cubic inches.

1 Winchester bushel = 0.9694 Imperial bushel

1 Imperial bushel = 1.032 Winchester bushels

Same relations as before maintain for gallons (dry measure)
(*Note:* 1 U. S. gallon (dry) = 1.164 U. S. gallons (liquid)).

U. S. Units

2 pints = 1 quart	= 67.2 cubic inches
4 quarts = 1 gallon * = 8 pints	= 268.8 cubic inches
2 gallons * = 1 peck = 16 pints = 8 quarts	= 537.6 cubic inches
4 pecks = 1 bushel = 64 pints = 32 quarts = 8 gallons *	= 2150.42 cubic inches
1 cubic foot contains 6.428 gallons (dry measure)*	

Liquid Measure. One U.S. gallon (liquid measure) contains 231 cubic inches. It holds 8.336 pounds distilled water at 62°F.

*The *gallon* is not a U.S. legal *dry measure*.

One British Imperial gallon contains 277.42 cubic inches. It holds 10 pounds distilled water at 62°F.

$$1 \text{ U.S. gallon (liquid)} = 0.8327 \text{ Imperial gallon}$$

$$1 \text{ Imperial gallon} = 1.201 \text{ U.S. gallons (liquid)}$$

(*Note:* 1 U.S. gallon (liquid) = 0.8594 U.S. gallon (dry)).

U.S. Units

4 gills = 1 pint	= 16 fluid ounces
2 pints = 1 quart = 8 gills	= 32 fluid ounces
4 quarts = 1 gallon = 32 gills = 8 pints	= 128 fluid ounces
1 cubic foot contains 7.4805 gallons (liquid measure)	

Apothecaries' Fluid Measure

60 minims = 1 fluid drachm   8 drachms = 1 fluid ounce

In the United States a fluid ounce is the 128th part of a U.S. gallon, or 1.805 cubic inches or 29.58 cubic centimeter. It contains 455.8 grains of water at 62°F. In Great Britain the fluid ounce is 1.732 cubic inches and contains 1 ounce avoirdupois (or 437.5 grains) of water at 62°F.

TABLE 1.31   Plane Angle [No Dimensions]

Multiply Number of → to Obtain ↓ ⟍ by ↘	Degrees	Minutes	Quadrants	*Radians*[a]	Revolutions[a] (Circumferences)	Seconds
Degrees	1	1.667×10^{-2}	90	57.30	360	2.778×10^{-4}
Minutes	60	1	5400	3438	2.16×10^{4}	1.667×10^{-2}
Quadrants	1.111×10^{-2}	1.852×10^{-4}	1	0.6366	4	3.087×10^{-6}
Radians[a]	1.745×10^{-2}	2.909×10^{-4}	1.571	1	6.283	4.848×10^{-6}
Revolutions[a] (Circumferences)	2.778×10^{-3}	4.630×10^{-5}	0.25	0.1591	1	7.716×10^{-7}
Seconds	3600	60	3.24×10^{5}	2.063×10^{5}	1.296×10^{6}	1

[a] 2π rad = 1 circumference = 360° by definition.

TABLE 1.32 Solid Angle [No Dimensions]

to Obtain ↓ / Multiply Number of → by ↘	Hemispheres	Spheres[a]	Spherical Right Angles	Steradians[b]
Hemispheres	1	2	0.25	0.1592
Spheres[a]	0.5	1	0.125	7.958×10^{-2}
Spherical Right Angles	4	8	1	0.6366
Steradians[b]	6.283	12.57	1.571	1

[a] A sphere is the total solid angle about a point.
[b] 4π steradians = 1 sphere by definition.

TABLE 1.33 Time [T]

to Obtain ↓ / Multiply Number of → by ↘	Days	Hours	Minutes	Months (Average)[a]	Seconds	Weeks
Days	1	4.167×10^{-2}	6.944×10^{-4}	30.42	1.157×10^{-5}	7
Hours	24	1	1.667×10^{-2}	730.0	2.778×10^{-4}	168
Minutes	1440	60	1	4.380×10^{4}	1.667×10^{-2}	1.008×10^{4}
Months (Average)[a]	3.288×10^{-2}	1.370×10^{-3}	2.283×10^{-5}	1	3.806×10^{-7}	0.2302
Seconds	8.64×10^{4}	3600	60	2.628×10^{6}	1	6.048×10^{5}
Weeks	0.1429	5.952×10^{-3}	9.921×10^{-5}	4.344	1.654×10^{-6}	1

[a] One common year = 365 days; one leap year = 366 days; one average month = $\frac{1}{12}$ of a common year.

TABLE 1.34 Linear Velocity [LT^{-1}]

Multiply Number of → by → / to Obtain ↓	Centimeters per Second	Feet per Minute	Feet per Second	Kilometers per Hour	Kilometers per Minute	Knots[a]	Meters per Minute	Meters per Second	Miles per Hour	Miles per Minute
Centimeters per Second	1	0.5080	30.48	27.78	1667	51.48	1.667	100	44.70	2682
Feet per Minute	1.969	1	60	54.68	3281	101.3	3.281	196.8	88	5280
Feet per Second	3.281×10^{-2}	1.667×10^{-2}	1	0.9113	54.68	1.689	5.468×10^{-2}	3.281	1.467	88
Kilometers per Hour	0.036	1.829×10^{-2}	1.097	1	60	1.853	0.06	3.6	1.609	96.54
Kilometers per minute	0.0006	3.048×10^{-4}	1.829×10^{-2}	1.667×10^{-2}	1	3.088×10^{-2}	0.001	0.06	2.682×10^{-2}	1.609
Knots[a]	1.943×10^{-2}	9.868×10^{-3}	0.5921	0.5396	32.38	1	3.238×10^{-2}	1.943	0.8684	52.10
Meters per Minute	0.6	0.3048	18.29	16.67	1000	30.88	1	60	26.82	1609
Meters per Second	0.01	5.080×10^{-3}	0.3048	0.2778	16.67	0.5148	1.667×10^{-2}	1	0.4470	26.82
Miles per Hour	2.237×10^{-2}	1.136×10^{-2}	0.6818	0.6214	37.28	1.152	3.728×10^{-2}	2.237	1	60
Miles per Minute	3.728×10^{-4}	1.892×10^{-4}	1.136×10^{-2}	1.036×10^{-2}	0.6214	1.919×10^{-2}	6.214×10^{-4}	3.728×10^{-2}	1.667×10^{-2}	1

[a] Nautical miles per hour.

Linear Velocity

The Miner's Inch. The miner's inch is used in measuring flow of water. An act of the California legislature, May 23, 1901, makes the standard miner's inch 1.5 ft^3/min, measured through any aperture or orifice.

The term miner's inch is more or less indefinite, for the reason that California water companies do not all use the same head above the center of the aperture, and the inch varies from 1.36 to 1.73 ft^3/min, but the most common measurement is through an aperture 2 in. high and whatever length is required, and through a plank $1\frac{1}{4}$ in. thick. The lower edge of the aperture should be 2 in. above the bottom of the measuring box, and the plank 5 in. high above the aperture, thus making a 6-in. head above the center of the stream. Each square inch of this opening represents a miner's inch, which is equal to a flow of 1.5 ft^3/min.

TABLE 1.35   Angular Velocity [T^{-1}]

Multiply Number of → to Obtain ↓ *by*	Degrees per Second	*Radians per Second*	Revolutions per Minute	Revolutions per Second
Degrees per Second	1	57.30	6	360
Radians per Second	1.745×10^{-2}	1	0.1047	6.283
Revolutions per Minute	0.1667	9.549	1	60
Revolutions per Second	2.778×10^{-3}	0.1592	1.667×10^{-2}	1

TABLE 1.36   Linear Acceleration[a] [LT^{-2}]

Multiply Number of → to Obtain ↓ *by*	Centimeters per Second per Second	Feet per Second per Second	Kilometers per Hour per Second	*Meters per Second per Second*	Miles per Hour per Second
Centimeters per Second per Second	1	30.48	27.78	100	44.70
Feet per Second per Second	3.281×10^{-2}	1	0.9113	3.281	1.467
Kilometers per Hour per Second	0.036	1.097	1	3.6	1.609
Meters per Second per Second	0.01	0.3048	0.2778	1	0.4470
Miles per Hour per Second	2.237×10^{-2}	0.6818	0.6214	2.237	1

[a] The (standard) acceleration due to gravity (g_0) = 980.7 cm/sec sec, = 32.17 ft/sec sec = 35.30 km/h sec = 9.807 m/sec sec = 21.94 mph/sec.

TABLE 1.37   Angular Acceleration $[T^{-2}]$

to Obtain ↓ / Multiply Number of → by →	Radians per Second per Second	Revolutions per Minute per Minute	Revolutions per Minute per Second	Revolutions per Second per Second
Radians per Second per Second	1	1.745×10^{-3}	0.1047	6.283
Revolutions per Minute per Minute	573.0	1	60	3600
Revolutions per Minute per Second	9.549	1.667×10^{-2}	1	60
Revolutions per Second per Second	0.1592	2.778×10^{-4}	1.667×10^{-2}	1

TABLE 1.38   Mass $[M]$ and Weight[a]

to Obtain ↓ / Multiply Number of → by →	Grains	Grams	Kilograms	Milligrams	Ounces[b]	Pounds[b]	Tons (Long)	Tons (Metric)	Tons (Short)
Grains	1	15.43	1.543×10^4	1.543×10^{-2}	437.5	7000			
Grams	6.481×10^{-2}	1	1000	0.001	28.35	453.6	1.016×10^6	$\times 10^6$	9.072×10^5
Kilograms	6.481×10^{-5}	0.001	1	10^{-6}	2.835×10^{-2}	0.4536	1016	1000	907.2
Milligrams	64.81	1000	10^6	1	2.835×10^4	4.536×10^5	1.016×10^9	10^9	9.072×10^8
Ounces[b]	2.286×10^{-3}	3.527×10^{-2}	35.27	3.527×10^{-5}	1	16	3.584×10^4	3.527×10^4	3.2×10^4
Pounds[b]	1.429×10^{-4}	2.205×10^{-3}	2.205	2.205×10^{-6}	6.250×10^{-2}	1	2240	2205	2000
Tons (Long)		9.842×10^{-7}	9.842×10^{-4}	9.842×10^{-10}	2.790×10^{-5}	4.464×10^{-4}	1	0.9842	0.8929
Tons (Metric)		10^{-6}	0.001	10^{-9}	2.835×10^{-5}	4.536×10^{-4}	1.016	1	0.9072
Tons (Short)		1.102×10^{-6}	1.102×10^{-3}	1.102×10^{-9}	3.125×10^{-5}	0.0005	1.120	1.102	1

[a] These same conversion factors apply to the *gravitational* units of force having the corresponding names. The dimensions of these units when used as gravitational units of force are MLT^{-2}; see Table 1.40.
[b] Avoirdupois pounds and ounces.

Avoirdupois Weight. Used Commercially.

27.343 grains	= 1 drachm
16 drachms	= 1 ounce (oz) = 437.5 grains
16 ounces	= 1 pound (lb) = 7000 grains
28 pounds	= 1 quarter (qr)
4 quarters	= 1 hundredweight (cwt) = 112 pounds
20 hundredweight	= 1 gross or long ton*
2000 pounds	= 1 net or short ton
14 pounds	= 1 stone 100 pounds = 1 quintal

*The long ton is used by the U.S. custom houses in collecting duties upon foreign goods. It is also used in freighting coal and selling it wholesale.

Troy Weight. Used in weighing gold or silver.

    24 grains   = 1 pennyweight (dwt)
    20 pennyweights = 1 ounce (oz) = 480 grains
    12 ounces   = 1 pound (lb) = 5760 grains

The grain is the same in Avoirdupois, Troy and Apothecaries' weights. A carat, for weighing diamonds = 3.086 grains = 0.200 gram. (International Standard, 1913.)

    1 pound troy   = .8229 pound avoirdupois
    1 pound avoirdupois = 1.2153 pounds troy

Apothecaries' Weight. Used in compounding medicines.

    20 grains   = 1 scruple (Э)
     3 scruples = 1 drachm (3) = 60 grains
     8 drachms = 1 ounce (3) = 480 grains
    12 ounces   = 1 pound (lb) = 5760 grains

The grain is the same in Avoirdupois, Troy, and Apothecaries' weights.

    1 pound apothecaries = 0.82286 pound avoirdupois
    1 pound avoirdupois = 1.2153 pounds apothecaries

TABLE 1.39   Density or Mass per Unit Volume [ML^{-3}]

to Obtain ↓ \ Multiply Number of → by →	Grams per Cubic Centimeter	*Kilograms per Cubic meter*	Pounds per Cubic Foot	Pounds per Cubic Inch
Grams per Cubic Centimeter	1	0.001	1.602×10^{-2}	27.68
Kilograms per Cubic Meter	1000	1	16.02	2.768×10^{4}
Pounds per Cubic Foot	62.43	6.243×10^{-2}	1	1728
Pounds per Cubic Inch	3.613×10^{-2}	3.613×10^{-5}	5.787×10^{-4}	1
Pounds per Mil Foot[a]	3.405×10^{-7}	3.405×10^{-10}	5.456×10^{-9}	9.425×10^{-6}

[a] Unit of volume is a volume one foot long and one circular mil in cross-section area.

TABLE 1.40   Force[a] [MLT^{-2}] or [F]

to Obtain ↓ \ Multiply Number of → by →	Dynes	Grams	Joules per Centimeter	*Newtons* or Joules per Meter	Kilograms	Pounds	Poundals
Dynes	1	980.7	10^{7}	10^{5}	9.807×10^{5}	4.448×10^{5}	1.383×10^{4}
Grams	1.020×10^{-3}	1	1.020×10^{4}	102.0	1000	453.6	14.10
Joules per Centimeter	10^{-7}	9.807×10^{-5}	1	.01	9.807×10^{-2}	4.448×10^{-2}	1.383×10^{-3}
Newtons, or Joules per Meter	10^{-5}	9.807×10^{-3}	100	1	9.807	4.448	0.1383
Kilograms	1.020×10^{-6}	0.001	10.20	0.1020	1	0.4536	1.410×10^{-2}
Pounds	2.248×10^{-6}	2.205×10^{-3}	22.48	0.2248	2.205	1	3.108×10^{-2}
Poundals	7.233×10^{-5}	7.093×10^{-2}	723.3	7.233	70.93	32.17	1

[a] Conversion factors between absolute and gravitational units apply only under standard acceleration due to gravity conditions. (See Sec. 1.4.)

TABLE 1.41 Pressure or Force per Unit Area $[ML^{-1}T^{-2}]$ or $[FL^{-2}]$

Multiply Number of → by ↓

to Obtain ↓	Atmospheres[a]	Baryes or Dynes per Square Centimeter	Centimeters of Mercury at 0°C[b]	Inches of Mercury at 0°C[b]	Inches of Water at 4°C	Kilograms per Square Meter[c]	Pounds per Square Foot	Pounds per Square Inch	Tons (Short) per Square Foot	Pascal
Atmospheres[a]	1	9.869×10^{-7}	1.316×10^{-2}	3.342×10^{-2}	2.458×10^{-3}	9.678×10^{-5}	4.725×10^{-4}	6.804×10^{-2}	0.9450	9.869×10^{-6}
Baryes or Dynes per Square Centimeter	1.013×10^{6}	1	1.333×10^{4}	3.386×10^{4}	2.491×10^{3}	98.07	478.8	6.895×10^{4}	9.576×10^{5}	10
Centimeters of Mercury at 0°C[b]	76.00	7.501×10^{-5}	1	2.540	0.1868	7.356×10^{-3}	3.591×10^{-2}	5.171	71.83	7.501×10^{-4}
Inches of Mercury at 0°C[b]	29.92	2.953×10^{-5}	0.3937	1	7.355×10^{-2}	2.896×10^{-3}	1.414×10^{-2}	2.036	28.28	2.953×10^{-4}
Inches of Water at 4°C	406.8	4.015×10^{-4}	5.354	13.60	1	3.937×10^{-2}	0.1922	27.68	384.5	4.015×10^{-3}
Kilograms per Square Meter[c]	1.033×10^{4}	1.020×10^{-2}	136.0	345.3	25.40	1	4.882	703.1	9765	0.1020
Pounds per Square Foot	2117	2.089×10^{-3}	27.85	70.73	5.204	0.2048	1	144	2000	2.089×10^{-2}
Pounds per Square Inch	14.70	1.450×10^{-5}	0.1934	0.4912	3.613×10^{-2}	1.422×10^{-3}	6.944×10^{-3}	1	13.89	1.450×10^{-4}
Tons (Short) per Square Foot	1.058	1.044×10^{-6}	1.392×10^{-2}	3.536×10^{-2}	2.601×10^{-3}	1.024×10^{-4}	0.0005	0.072	1	1.044×10^{-5}
Pascal	1.013×10^{5}	10^{-1}	1.333×10^{3}	3.386×10^{3}	2.491×10^{2}	9.807	47.88	6.895×10^{3}	9.576×10^{4}	1

[a] Definition: One atmosphere (standard) = 76 cm of mercury at 0°C.

[b] To convert height h of a column of mercury at t degrees Centigrade to the equivalent height h_0 at 0°C use $h_0 = h\{1 - (m - l)t/(1 + mt)\}$ where $m = 0.0001818$ and $l = 18.4 \times 10^{-6}$ if the scale is engraved on brass; $l = 8.5 \times 10^{-6}$ if on glass. This assumes the scale is correct at 0°C; for other cases (any liquid) see International Critical Tables, Vol. 1, 68.

[c] $1\,\text{g/cm}^2 = 10\,\text{kg/m}^2$.

TABLE 1.42   Torque or Moment of Force $[ML^2T^{-2}]$ or $[FL]^a$

to Obtain ↓ \ Multiply Number of → by ↘	Dyne-Centimeters	Gram-Centimeters	Kilogram-Meters	Pound-Feet	*Newton-Meter*
Dyne-Centimeters	1	980.7	9.807×10^7	1.356×10^7	10^7
Gram-Centimeters	1.020×10^{-3}	1	10^5	1.383×10^4	1.020×10^4
Kilogram-Meters	1.020×10^{-8}	10^{-5}	1	0.1383	0.1020
Pound-Feet	7.376×10^{-8}	7.233×10^{-5}	7.233	1	0.7376
Newton-Meter	10^{-7}	9.807×10^{-4}	9.807	1.356	1

^{a}Same dimensions as energy; more properly torque should be expressed as Newton-meters per radian to avoid this confusion.)

TABLE 1.43   Moment of Inertia $[ML^2]$

to Obtain ↓ \ Multiply Number of → by ↘	Gram-Centimeters Squared	*Kilogram-Meters Squared*	Pound-Inches Squared	Pound-Feet Squared	Slug-Feet Squared
Gram-Centimeters Squared	1	10^7	2.9266×10^3	4.21434×10^5	1.3559×10^7
Kilogram-Meters Squared	10^{-7}	1	2.9266×10^{-4}	4.21434×10^{-2}	1.3559
Pound-Inches Squared	3.4169×10^{-4}	3.4169×10^3	1	144	4.63304×10^3
Pound-Feet Squared	2.37285×10^{-6}	23.7285	6.944×10^{-3}	1	32.1739
Slug-Feet Squared	7.37507×10^{-8}	0.737507	2.15841×10^{-4}	3.10811×10^{-2}	1

TABLE 1.44 Energy, Work and Heat^a [ML^2T^{-2}] or [FL]

Multiply Number of → by → / to Obtain →	British Thermal Units^b	Centimeter-Grams	Ergs or Centimeter-Dynes	Foot-Pounds	Horsepower-Hours	Joules^c or Watt-Seconds	Kilogram-Calories^b	Kilowatt-Hours	Meter-Kilograms	Watt-Hours
British Thermal Units^b	1	9.297×10^{-8}	9.480×10^{-11}	1.285×10^{-3}	2545	9.480×10^{-4}	3.969	3413	9.297×10^{-3}	3.413
Centimeter-Grams	1.076×10^7	1	1.020×10^{-3}	1.383×10^4	2.737×10^{10}	1.020×10^4	4.269×10^7	3.671×10^{10}	10^5	3.671×10^7
Ergs or Centimeter-Dynes	1.055×10^{10}	980.7	1	1.356×10^7	2.684×10^{12}	10^7	4.186×10^{10}	3.6×10^{13}	9.807×10^7	3.6×10^{10}
Foot-Pounds	778.0	7.233×10^{-5}	7.367×10^{-8}	1	1.98×10^6	0.7376	3087	2.655×10^6	7.233	2655
Horsepower-Hours	3.929×10^{-4}	3.654×10^{-11}	3.722×10^{-14}	5.050×10^{-7}	1	3.722×10^{-7}	1.559×10^{-3}	1.341	3.653×10^{-6}	1.341×10^{-3}
Joules^c or Watt-Seconds	1054.8	9.807×10^{-5}	10^{-7}	1.356	2.684×10^6	1	4186	3.6×10^6	9.807	3600
Kilogram-Calories^b	0.2520	2.343×10^{-8}	2.389×10^{-11}	3.239×10^{-4}	641.3	2.389×10^{-4}	1	860.0	2.343×10^{-3}	0.8600
Kilowatt-Hours	2.930×10^{-4}	2.724×10^{-11}	2.778×10^{-14}	3.766×10^{-7}	0.7457	2.778×10^{-7}	1.163×10^{-3}	1	2.724×10^{-6}	0.001
Meter-Kilograms	107.6	10^{-5}	1.020×10^{-8}	0.1383	2.737×10^5	0.1020	426.9	3.671×10^5	1	367.1
Watt-Hours	0.2930	2.724×10^{-8}	2.778×10^{-11}	3.766×10^{-4}	745.7	2.778×10^{-4}	1.163	1000	2.724×10^{-3}	1

^{a}See note at the bottom of Table 1.45.
^{b}Mean calorie and Btu used throughout. One gram-calorie = 0.001 kilogram-calorie. One Ostwald calorie = 0.1 kilogram-calorie. The IT cal, 1000 international steam table calories, has been defined as the 1/860th part of the international kilowatthour (see *Mechanical Engineering*, Nov., 1935, p. 710). Its value is very nearly equal to the mean kilogram-calorie, 1 IT cal-1.00037 kilogram-calories (mean). 1 Btu = 251.996 IT cal.
^{c}Absolute joule, defined as 10^7 ergs. The international joule, based on the international ohm and ampere, equals 1.0003 absolute joules.*

TABLE 1.45 Power or Rate of Doing Work^a $[ML^2T^{-3}]$ or $[FLT^{-1}]$

Multiply Number of → by → to Obtain ↓	British Thermal Units per Minute	Ergs per Second	Foot-Pounds per Minute	Foot-Pounds per Second	Horsepower^a	Kilogram-Calories per Minute	Kilowatts	Metric Horsepower	Watts
British Thermal Units per Minute	1	5.689×10^{-9}	1.285×10^{-3}	7.712×10^{-2}	42.41	3.969	56.89	41.83	5.689×10^{-2}
Ergs per Second	1.758×10^{8}	1	2.259×10^{5}	1.356×10^{7}	7.457×10^{9}	6.977×10^{8}	10^{10}	7.355×10^{9}	10^{7}
Foot-Pounds per Minute	778.0	4.426×10^{-6}	1	60	3.3×10^{4}	3087	4.426×10^{4}	3.255×10^{4}	44.26
Foot-Pounds per Second	12.97	7.376×10^{-8}	1.667×10^{-2}	1	550	51.44	737.6	542.5	0.7376
Horsepower^a	2.357×10^{-2}	1.341×10^{-10}	3.030×10^{-5}	1.818×10^{-3}	1	9.355×10^{-2}	1.341	0.9863	1.341×10^{-3}
Kilogram-Calories per Minute	0.2520	1.433×10^{-9}	3.239×10^{-4}	1.943×10^{-2}	10.69	1	14.33	10.54	1.433×10^{-2}
Kilowatts	1.758×10^{-2}	10^{-10}	2.260×10^{-5}	1.356×10^{-3}	0.7457	6.977×10^{-2}	1	0.7355	10^{-3}
Metric Horsepower	2.390×10^{-2}	1.360×10^{-10}	3.072×10^{-5}	1.843×10^{-3}	1.014	9.485×10^{-2}	1.360	1	1.360×10^{-3}
Watts	17.58	10^{-7}	2.260×10^{-2}	1.356	745.7	69.77	1000	735.5	1

1 Cheval-vapeur = 75 kilogram-meters per second

1 Poncelet = 100 kilogram-meters per second

^{a}The "horsepower" used in these tables is equal to 550 foot-pounds per second by definition. Other definitions are one horsepower equals 746 watts (U.S. and Great Britain) and one horsepower equals 736 watts (continental Europe). Neither of these latter definitions is equivalent to the first; the "horsepowers" defined in these latter definitions are widely used in the rating of electrical machinery.

TABLE 1.46   Quantity of Electricity and Dielectric Flux [Q]

Multiply Number of → to Obtain ↓ by ↘	Abcoulombs	Ampere-Hours	*Coulombs*	Faradays	Stat-Coulombs
Abcoulombs	1	360	0.1	9649	3.335×10^{-11}
Ampere-Hours	2.778×10^{-3}	1	2.778×10^{-4}	26.80	9.259×10^{-14}
Coulombs	10	3600	1	9.649×10^{4}	3.335×10^{-10}
Faradays	1.036×10^{-4}	3.731×10^{-2}	1.036×10^{-5}	1	3.457×10^{-15}
Statcoulombs	2.998×10^{10}	1.080×10^{13}	2.998×10^{9}	2.893×10^{14}	1

TABLE 1.47   Charge per Unit Area and Electric Flux Density [QL^{-2}]

Multiply Number of → to Obtain ↓ by ↘	Abcoulombs per Square Centimeter	Coulombs per Square Centimeter	Coulombs per Square Inch	Statcoulombs per Square Centimeter	*Coulombs per Square Meter*
Abcoulombs per Square Centimeter	1	0.1	1.550×10^{-2}	3.335×10^{-11}	10^{-5}
Coulombs per Square Centimeter	10	1	0.1550	3.335×10^{-10}	10^{-4}
Coulombs per Square Inch	64.52	6.452	1	2.151×10^{-9}	6.452×10^{-4}
Statcoulombs per Square Centimeter	2.998×10^{10}	2.998×10^{9}	4.647×10^{8}	1	2.998×10^{5}
Coulombs per Square Meter	10^{5}	10^{4}	1550	3.335×10^{-6}	1

TABLE 1.48   Electric Current [QT^{-1}]

Multiply Number of → to Obtain ↓ by ↘	Abamperes	*Amperes*	Statamperes
Abamperes	1	0.1	3.335×10^{-11}
Amperes	10	1	3.335×10^{-10}
Statamperes	2.998×10^{10}	2.998×10^{9}	1

TABLE 1.49 Current Density $[QT^{-1}L^{-2}]$

to Obtain ↓ / Multiply Number of → by ↘	Abamperes per Square Centimeter	Amperes per Square Centimeter	Amperes per Square Inch	Statamperes per Square Centimeter	*Amperes per Square Meter*
Abamperes per Square Centimeter	1	0.1	1.550×10^{-2}	3.335×10^{-11}	10^{-5}
Amperes per Square Centimeter	10	1	0.1550	3.335×10^{-10}	10^{-4}
Amperes per Square Inch	64.52	6.452	1	2.151×10^{-9}	6.452×10^{-4}
Statamperes per Square Centimeter	2.998×10^{10}	2.998×10^{9}	4.647×10^{8}	1	2.998×10^{5}
Amperes per Square Meter	10^{5}	10^{4}	1550	3.335×10^{-6}	1

TABLE 1.50 Electric Potential and Electromotive Force $[MQ^{-1}L^2T^{-2}]$ or $[FQ^{-1}L]$

to Obtain ↓ / Multiply Number of → by ↘	Abvolts	Microvolts	Millivolts	Statvolts	*Volts*
Abvolts	1	100	10^{5}	2.998×10^{10}	10^{8}
Microvolts	0.01	1	1000	2.998×10^{8}	10^{6}
Millivolts	10^{-5}	0.001	1	2.998×10^{5}	1000
Statvolts	3.335×10^{-11}	3.335×10^{-9}	3.335×10^{-6}	1	3.335×10^{-3}
Volts	10^{-8}	10^{-6}	0.001	299.8	1

TABLE 1.51   Electric Field Intensity and Potential Gradient $[MQ^{-1}LT^{-2}]$ or $[FQ^{-1}]$

Multiply Number of → / to Obtain ↓	Abvolts per Centimeter	Microvolts per Meter	Millivolts per Meter	Statvolts per Centimeter	Volts per Centimeter	Kilovolts per Centimeter	Volts per Inch	Volts per Mil	Volts per Meter
Abvolts per Centimeter	1	1	1000	2.998×10^{10}	10^{8}	10^{11}	3.937×10^{7}	3.937×10^{10}	10^{6}
Microvolts per Meter	1	1	1000	2.998×10^{10}	10^{8}	10^{11}	3.937×10^{7}	3.937×10^{10}	10^{6}
Millivolts per Meter	0.001	0.001	1	2.998×10^{7}	10^{5}	10^{8}	3.937×10^{4}	3.937×10^{7}	1000
Statvolts per Centimeter	3.335×10^{-11}	3.335×10^{-11}	3.335×10^{-8}	1	3.335×10^{-3}	3.335	1.313×10^{-3}	1.313	3.335×10^{-5}
Volts per Centimeter	10^{-8}	10^{-8}	10^{-5}	299.8	1	1000	0.3937	393.7	10^{-2}
Kilovolts per Centimeter	10^{-11}	10^{-11}	10^{-8}	0.2998	0.001	1	3.937×10^{-4}	0.3937	10^{-5}
Volts per Inch	2.540×10^{-8}	2.540×10^{-8}	2.540×10^{-5}	761.6	2.540	2540	1	1000	2.540×10^{-2}
Volts per Mil	2.540×10^{-11}	2.540×10^{-11}	2.540×10^{-8}	0.7616	2.540×10^{-3}	2.540	0.001	1	2.540×10^{-5}
Volts per Meter	10^{-6}	10^{-6}	10^{-3}	2.998×10^{4}	100	10^{5}	39.37	3.937×10^{4}	1

TABLE 1.52   Electric Resistance $[MQ^{-2}L^2T^{-1}]$ or $[FQ^{-2}LT]$

to Obtain ↓ \ Multiply Number of → by ↘	Abohms	Megohms	Microhms	*Ohms*	Statohms
Abohms	1	10^{15}	1000	10^9	8.988×10^{20}
Megohms	10^{-15}	1	10^{-12}	10^{-6}	8.988×10^5
Microhms	0.001	10^{12}	1	10^6	8.988×10^{17}
Ohms	10^{-9}	10^6	10^{-6}	1	8.988×10^{11}
Statohms	1.112×10^{-21}	1.112×10^{-6}	1.112×10^{-18}	1.112×10^{-12}	1

Electric Conductance $[F^{-1}Q^2L^{-1}T^{-1}]$

1 Siemens = 1 mho = 1 ohm^{-1} = 10^{-6} megmho = 10^6 micromho

TABLE 1.53   Electric Resistivity[a] $[MQ^{-2}L^3T^{-1}]$ or $[FQ^{-2}L^2T]$

to Obtain ↓ \ Multiply Number of → by ↘	Abohm-Centimeters	Microhm-Centimeters	Microhm-Inches	Ohms (Mil, Foot)	Ohms (Meter, Gram)[b]	*Ohm-Meters*
Abohm-Centimeters	1	1000	2540	166.2	$10^5/\delta$	10^{11}
Microhm-Centimeters	0.001	1	2.540	0.1662	$100/\delta$	10^8
Microhm-Inches	3.937×10^{-4}	0.3937	1	6.545×10^{-2}	$39.37/\delta$	3.937×10^7
Ohms (Mil, Foot)	6.015×10^{-3}	6.015	15.28	1	$601.5/\delta$	6.015×10^8
Ohms (Meter, Gram)[b]	$10^{-5}\delta$	0.01δ	$2.540 \times 10^{-2}\delta$	$1.662 \times 10^{-3}\delta$	1	$10^{-6}\delta$
Ohm-Meters	10^{-11}	10^{-8}	2.540×10^{-8}	1.662×10^{-9}	$10^{-6}/\delta$	1

[a] In this table δ is density in grams per cm^3. The following names, corresponding respectively to those at the tops of columns, are sometimes used: abohms per cm cube; microhms per cm cube; microhms per inch cube; ohms per milfoot; ohms per meter-gram. The first four columns are headed by units of *volume* resistivity, the last by a unit of *mass* resistivity. The dimensions of the latter are $Q^{-2}L^6T^{-1}$; not those given in the heading of the table.

[b] One ohm (meter, gram) = 5710 ohms (mile, pound).

TABLE 1.54　Electric Conductivity[a] $[M^{-1}Q^2L^{-3}T]$ or $[F^{-1}Q^2L^{-2}T^{-1}]$

to Obtain ↓ \ Multiply Number of → by	Abmhos per Centimeter	Mhos (Mil, Foot)	Mhos (Meter, Gram)	Micromhos per Centimeter	Micromhos per inch	Siemens per Meter
Abmhos per Centimeter	1	6.015×10^{-3}	$10^{-5}\delta$	0.001	3.937×10^{-4}	10^{-11}
Mhos (Mil, Foot)	166.2	1	$1.662 \times 10^{-3}\delta$	0.1662	6.524×10^{-2}	1.662×10^{-9}
Mhos (Meter, Gram)	$10^5/\delta$	$601.5/\delta$	1	$100/\delta$	$39.37/\delta$	$10^{-6}/\delta$
Micromhos per Centimeter	1000	6.015	0.01δ	1	0.3937	10^{-8}
Micromhos per Inch	2540	15.28	$2.540 \times 10^{-2}\delta$	2.540	1	2.54×10^{-8}
Siemens per Meter	10^{11}	6.015×10^8	$10^6\delta$	10^8	3.937×10^7	1

[a]See footnote of Table 1.53. Names sometimes used are abmho per cm cube, mho per mil-foot, etc. Dimensions of mass conductivity are $Q^2L^{-6}T$.

TABLE 1.55　Capacitance $[M^{-1}Q^2L^{-2}T^2]$ or $[F^{-1}Q^2L^{-1}]$

to Obtain ↓ \ Multiply Number of → by	Abfarads	Farads	Microfarads	Statfarads
Abfarads	1	10^{-9}	10^{-15}	1.112×10^{-21}
Farads	10^9	1	10^{-6}	1.112×10^{-12}
Microfarads	10^{15}	10^6	1	1.112×10^{-6}
Statfarads	8.988×10^{20}	8.988×10^{11}	8.988×10^5	1

TABLE 1.56　Inductance $[MQ^{-2}L^2]$ or $[FQ^{-2}LT^2]$

to Obtain ↓ \ Multiply Number of → by	Abhenries[a]	Henries	Microhenries	Millihenries	Stathenries
Abhenries[a]	1	10^9	1000	10^6	8.988×10^{20}
Henries	10^{-9}	1	10^{-6}	0.001	8.988×10^{11}
Microhenries	0.001	10^6	1	1000	8.988×10^{17}
Millihenries	10^{-6}	1000	0.001	1	8.988×10^{14}
Stathenries	1.112×10^{-21}	1.112×10^{-12}	1.112×10^{-18}	1.112×10^{-15}	1

[a]An abhenry is sometimes called a "centimeter."

TABLE 1.57   Magnetic Flux $[MQ^{-1}L^2T^{-1}]$ or $[FQ^{-1}LT]$

Multiply Number of → / to Obtain ↓ by ↘	Kilolines	Maxwells (or Lines)	Webers
Kilolines	1	0.001	10^5
Maxwells (or Lines)	1000	1	10^8
Webers	10^{-5}	10^{-8}	1

TABLE 1.58   Magnetic Flux Density $[MQ^{-1}T^{-1}]$ or $[FQ^{-1}L^{-1}T]$

Multiply Number of → / to Obtain ↓ by ↘	Gausses (or Lines per Square Centimeter)	Lines per Square Inch	Webers per Square Centimeter	Webers per Square Inch	Tesla (Webers per Square Meter)
Gausses (or Lines per Square Centimeter)	1	0.1550	10^8	1.550×10^7	10^4
Lines per Square Inch	6.452	1	6.452×10^8	10^8	6.452×10^4
Webers per Square Centimeter	10^{-8}	1.550×10^{-9}	1	0.1550	10^{-4}
Webers per Square Inch	6.452×10^{-8}	10^{-8}	6.452	1	6.452×10^{-4}
Tesla (Webers per Square Meter)	10^{-4}	1.550×10^{-5}	10^4	1550	1

TABLE 1.59   Magnetic Potential and Magnetomotive Force $[QT^{-1}]$

Multiply Number of → / to Obtain ↓ by ↘	Abampere-Turns	Ampere-Turns	Gilberts
Abampere-Turns	1	0.1	7.958×10^{-2}
Ampere-Turns	10	1	0.7958
Gilberts	12.57	1.257	1

TABLE 1.60   Magnetic Field Intensity, Potential Gradient, and Magnetizing Force [$QL^{-1}T^{-1}$]

Multiply Number of → to Obtain ↓	Abampere-Turns per Centimeter	Ampere-Turns per Centimeter	Ampere-Turns per Inch	Oersteds (Gilberts per Centimeter)	*Ampere-Turns per Meter*
Abampere-Turns per Centimeter	1	0.1	3.937×10^{-2}	7.958×10^{-2}	10^{-3}
Ampere-Turns per Centimeter	10	1	0.3937	0.7958	10^{-2}
Ampere-Turns per Inch	25.40	2.540	1	2.021	2.54×10^{-2}
Oersteds (Gilberts per Centimeter)	12.57	1.257	0.4950	1	1.257×10^{-2}
Ampere-Turns per Meter	10^3	10^2	39.37	79.58	1

TABLE 1.61   Specific Heat [$L^2T^{-2}t^{-1}$] (t = temperature)

To change specific heat in gram-calories per gram per degree Centigrade to the units given in any line of the following table, multiply by the factor in the last column.

Unit of Heat or Energy	Unit of Mass	Temperature Scale[a]	Factor
Gram-calories	Gram	Centigrade	1
Kilogram-calories	Kilogram	Centigrade	1
British thermal units	Pound	Centigrade	1.800
British thermal units	Pound	Fahrenheit	1.000
Joules	Gram	Centigrade	4.186
Joules	Pound	Fahrenheit	1055
Joules	*Kilogram*	*Kelvin*	4.187×10^3
Kilowatt-hours	Kilogram	Centigrade	1.163×10^{-3}
Kilowatt-hours	Pound	Fahrenheit	2.930×10^{-4}

[a] Temperature conversion formulas:

$$t_c = \text{temperature in Centigrade degrees}$$
$$t_f = \text{temperature in Fahrenheit degrees}$$
$$t_K = \text{temperature in Kelvin degrees}$$
$$1\,F = \tfrac{5}{9}\,°C$$
$$1\,K = 1\,°C$$
$$t_c = \tfrac{5}{9}\left(t_f - 32\right)$$
$$t_f = \tfrac{9}{5}t_c + 32$$
$$t_K = t_c + 273$$

TABLE 1.62 Thermal Conductivity ^{a} $LMT^{-3}t^{-1}$

To ↓ \ From →	Btu·ft per h·ft²·°F	Btu·in per h·ft²·°F	Btu·in per sec·ft²·°F	Joules per m·s·°C	kcal per m·h·°C	erg per cm·s·°C	kcal per m·s·°C	cal per cm·s·°C	W per ft·°C	W per m·K
Btu·ft per h·ft²·°F	1	8.333×10^{-2}	3.0×10^{2}	5.778×10^{-1}	6.720×10^{-1}	5.778×10^{-6}	2.419×10^{3}	2.419×10^{2}	1.895	5.778×10^{-1}
Btu·in per h·ft²·°F	12	1	3.6×10^{3}	6.933	8.064	6.933×10^{-5}	2.903×10^{4}	2.903×10^{3}	2.275×10^{1}	6.933
Btu·in per s·ft²·°F	3.333×10^{-3}	2.778×10^{-4}	1	1.926×10^{-3}	2.240×10^{-3}	1.926×10^{-8}	8.064	8.064×10^{-1}	6.319×10^{-3}	1.926×10^{-3}
Joules per m·s·°C	1.731	1.442×10^{-1}	5.192×10^{2}	1	1.163	1.000×10^{-5}	4.187×10^{3}	4.187×10^{2}	3.281	1.0
kcal per m·h·°C	1.483	1.240×10^{-1}	4.465×10^{2}	8.599×10^{-1}	1	8.599×10^{-6}	3.6×10^{3}	3.6×10^{2}	2.821	8.599×10^{-1}
erg per cm·s·°C	1.731×10^{5}	1.442×10^{4}	5.192×10^{7}	1.0×10^{5}	1.163×10^{5}	1	4.187×10^{8}	4.187×10^{7}	3.281×10^{5}	1.0×10^{5}
kcal per m·s·°C	4.134×10^{-4}	3.445×10^{-5}	1.240×10^{-1}	2.388×10^{-4}	2.778×10^{-4}	2.388×10^{-9}	1	1.0×10^{-1}	7.835×10^{-4}	2.388×10^{-4}
cal per cm·s·°C	4.134×10^{-3}	3.445×10^{-4}	1.240	2.388×10^{-3}	2.778×10^{-3}	2.388×10^{-8}	10	1	7.835×10^{-3}	2.388×10^{-3}
W per ft·°C	5.276×10^{-1}	4.395×10^{-2}	1.582×10^{2}	3.048×10^{-1}	3.545×10^{-1}	3.048×10^{-6}	1.276×10^{3}	1.276×10^{2}	1	3.048×10^{-1}
W per m·K	1.731	1.442×10^{-1}	5.192×10^{2}	1.0	1.163	1.00×10^{-5}	4.187×10^{3}	4.187×10^{2}	3.281	1

^{a}International Table Btu = 1.055056×10^{3} joules; and International Table Cal = 4.1868 joules are used throughout.

TABLE 1.63   Photometric Units

	Common Unit	Multiply by	to Get SI Unit
Luminous intensity	International candle	9.81×10^{-1}	**Candela**
Luminance	Candela/in^2	1.550×10^3	**Candela / m^2**
	Candela/cm^2	1×10^4	**Candela / m^2**
	Foot·lambert	3.4263	**Candela / m^2**
Luminous flux	Candela·steradian	1.0000	**Lumen**
	Candle power (spher.)	12.566	**Lumen**
Quantity of light flux			**Lumen·sec**
Luminous exitance[a]			**Lumens / m^2**
Illuminance[b]	Lambert	3.103×10^3	**Candela / m^2**
	Foot candles	1.0764×10	**Lumens / m^2**
	Lumens per ft^2	1.0764×10	**Lumens / m^2**
	Lux	1.000	**Lumens / m^2**
	Phots	1×10^4	**Lumens / m^2**
Luminous efficacy			**Lumens / watt**

[a] Luminous emittance.
[b] Luminous flux density.

TABLE 1.64   Specific Gravity Conversions

$$^\circ\text{Bé} = 145 - \frac{145}{\text{sp gr}} \text{ (heavier than H}_2\text{O)}^a \qquad ^\circ\text{Bé} = \frac{140}{\text{sp gr}} - 130 \text{ (lighter than H}_2\text{O)}^a$$

$$^\circ\text{Tw} = \frac{\text{sp gr } 60^\circ/60^\circ\text{F} - 1}{0.005}^b \qquad\qquad ^\circ\text{API} = \frac{141.5}{\text{sp gr}} - 131.5$$

Specific Gravity 60°/60°	°Bé	°API	lb/gal at 60°F, wt in air	lb/ft^3 at 60°F, wt in air	Specific Gravity 60°/60°	°Bé	°API	lb/gal at 60°F, wt in air	lb/ft^3 at 60°F wt in air
0.600	103.33	104.33	4.9929	37.350	0.745	57.92	58.43	6.2020	46.394
0.605	101.40	102.38	5.0346	37.662	0.750	56.67	57.17	6.2437	46.706
0.610	99.51	100.47	5.0763	37.973	0.755	55.43	55.92	6.2854	47.018
0.615	97.64	98.58	5.1180	38.285	0.760	54.21	54.68	6.3271	47.330
0.620	95.81	96.73	5.1597	38.597	0.765	53.01	53.47	6.3688	47.642
0.625	94.00	94.90	5.2014	39.910	0.770	51.82	52.27	6.4104	47.953
0.630	92.22	93.10	5.2431	39.222	0.775	50.65	51.08	6.4521	48.265
0.635	90.47	91.33	5.2848	39.534	0.780	49.49	49.91	6.4938	48.577
0.640	88.75	89.59	5.3265	39.845	0.785	48.34	48.75	6.5355	48.889
0.645	87.05	87.88	5.3682	40.157	0.790	47.22	47.61	6.5772	49.201
0.650	85.38	86.19	5.4098	40.468	0.795	46.10	46.49	6.6189	49.513
0.655	83.74	84.53	5.4515	40.780					
0.660	82.12	82.89	5.4932	41.092	0.800	45.00	45.38	6.6606	49.825
0.665	80.53	81.28	5.5349	41.404	0.805	43.91	44.28	6.7023	50.137
0.670	78.96	79.69	5.5766	41.716	0.810	42.84	43.19	6.7440	50.448
0.675	77.41	78.13	5.6183	42.028	0.815	41.78	42.12	6.7857	50.760
0.680	75.88	76.59	5.6600	42.340	0.820	40.73	41.06	6.8274	51.072
0.685	74.38	75.07	5.7017	42.652	0.825	39.70	40.02	6.8691	51.384
0.690	72.90	73.57	5.7434	42.963	0.830	38.67	38.98	6.9108	51.696
0.695	71.44	72.10	5.7851	43.275	0.835	37.66	37.96	6.9525	52.008
					0.840	36.67	36.95	6.9941	52.320
0.700	70.00	70.64	5.8268	43.587	0.845	35.68	35.96	7.0358	52.632
0.705	68.58	69.21	5.8685	43.899	0.850	34.71	34.97	7.0775	52.943
0.710	67.18	67.80	5.9101	44.211	0.855	33.74	34.00	7.1192	53.225
0.715	65.80	66.40	5.9518	44.523	0.860	32.79	33.03	7.1609	53.567
0.720	64.44	65.03	5.9935	44.834	0.865	31.85	32.08	7.2026	53.879
0.725	63.10	63.67	6.0352	45.146	0.870	30.92	31.14	7.2443	54.191
0.730	61.78	62.34	6.0769	45.458	0.875	30.00	30.21	7.2860	54.503
0.735	60.48	61.02	6.1186	45.770	0.880	29.09	29.30	7.3277	54.815
0.740	59.19	59.72	6.1603	46.082	0.885	28.19	28.38	7.3694	55.127

TABLE 1.64 (*Continued*)

Specific Gravity 60°/60°	°Bé	°API	lb/gal at 60°F, wt in air	lb/ft³ at 60°F, wt in air	Specific Gravity 60°/60°	°Bé	°Tw	lb/gal at 60°F, wt in air	lb/ft³ at 60°F, wt in air
0.890	27.30	27.49	7.4111	55.438	1.140	17.81	28	9.4957	71.032
0.895	26.42	26.60	7.4528	55.750	1.145	18.36	29	9.5374	71.344
					1.150	18.91	30	9.5790	71.656
0.900	25.76	25.72	7.4944	56.062	1.155	19.46	31	9.6207	71.968
0.905	24.70	24.85	7.5361	56.374	1.160	20.00	32	9.6624	72.280
0.910	23.85	23.99	7.5777	56.685	1.165	20.54	33	9.7041	72.592
0.915	23.01	23.14	7.6194	56.997	1.170	21.07	34	9.7458	72.904
0.920	22.17	22.30	7.6612	57.410	1.175	21.60	35	9.7875	73.216
0.925	21.35	21.47	7.7029	57.622	1.180	22.12	36	9.8292	73.528
0.930	20.54	20.65	7.7446	57.934	1.185	22.64	37	9.8709	73.840
0.935	19.73	19.84	7.7863	58.246	1.190	23.15	38	9.9126	74.151
0.940	18.94	19.03	7.8280	58.557	1.195	23.66	39	9.9543	74.463
0.945	18.15	18.24	7.8697	58.869					
0.950	17.37	17.45	7.9114	59.181	1.200	24.17	40	9.9960	74.775
0.955	16.60	16.67	7.9531	59.493	1.205	24.67	41	10.0377	75.087
0.960	15.83	15.90	7.9947	59.805	1.210	25.17	42	10.0793	75.399
0.965	15.08	15.13	8.0364	60.117	1.215	25.66	43	10.1210	75.711
0.970	14.33	14.38	8.0780	60.428	1.220	26.15	44	10.1627	76.022
0.975	13.59	13.63	8.1197	60.740	1.225	26.63	45	10.2044	76.334
0.980	12.86	12.89	8.1615	61.052	1.230	27.11	46	10.2461	76.646
0.985	12.13	12.15	8.2032	61.364	1.235	27.59	47	10.2878	76.958
0.990	11.41	11.43	8.2449	61.676	1.240	28.06	48	10.3295	77.270
0.995	10.70	10.71	8.2866	61.988	1.245	28.53	49	10.3712	77.582
					1.250	29.00	50	10.4129	77.894
					1.255	29.46	51	10.4546	78.206
Specific Gravity 60°/60°	°Bé	°Tw	lb/gal at 60°F, wt in air	lb/ft³ at 60°F, wt in air	1.260	29.92	52	10.4963	78.518
					1.265	30.38	53	10.5380	78.830
1.000	10.00	10.00	8.3283	62.300	1.270	30.83	54	10.5797	79.141
1.005	0.72	1	8.3700	62.612	1.275	31.27	55	10.6214	79.453
1.010	1.44	2	8.4117	62.924	1.280	31.72	56	10.6630	79.765
1.015	2.14	3	8.4534	63.236	1.285	32.16	57	10.7047	80.077
1.020	2.84	4	8.4950	63.547	1.290	32.60	58	10.7464	80.389
1.025	3.54	5	8.5367	63.859	1.295	33.03	59	10.7881	80.701
1.030	4.22	6	8.5784	64.171					
1.035	4.90	7	8.6201	64.483	1.300	33.46	60	10.8298	81.013
1.040	5.58	8	8.6618	64.795	1.305	33.89	61	10.8715	81.325
1.045	6.24	9	8.7035	65.107	1.310	34.31	62	10.9132	81.636
1.050	6.91	10	8.7452	65.419	1.315	34.73	63	10.9549	81.948
1.055	7.56	11	8.7869	65.731	1.320	35.15	64	10.9966	82.260
1.060	8.21	12	8.8286	66.042	1.325	35.57	65	11.0383	82.572
1.065	8.85	13	8.8703	66.354	1.330	35.98	66	11.0800	82.884
1.070	9.49	14	8.9120	66.666	1.335	36.39	67	11.1217	83.196
1.075	10.12	15	8.9537	66.978	1.340	36.79	68	11.1634	83.508
1.080	10.74	16	8.9954	67.290	1.345	37.19	69	11.2051	83.820
1.085	11.36	17	9.0371	67.602	1.350	37.59	70	11.2467	84.131
1.090	11.97	18	9.0787	67.914	1.355	37.99	71	11.2884	84.443
1.095	12.58	19	9.1204	68.226	1.360	38.38	72	11.3301	84.755
					1.365	38.77	73	11.3718	85.067
1.100	13.18	20	9.1621	68.537	1.370	39.16	74	11.4135	85.379
1.105	13.78	21	9.2038	68.849	1.375	39.55	75	11.4552	85.691
1.110	14.37	22	9.2455	69.161	1.380	39.93	76	11.4969	86.003
1.115	14.96	23	9.2872	69.473	1.385	40.31	77	11.5386	86.315
1.120	15.54	24	9.3289	69.785	1.390	40.68	78	11.5803	86.626
1.125	16.11	25	9.3706	70.097	1.395	41.06	79	11.6220	86.938
1.130	16.68	26	9.4123	70.409					
1.135	17.25	27	9.4540	70.721	1.400	41.43	80	11.6637	87.250
					1.405	41.80	81	11.7054	87.562
					1.410	42.16	82	11.7471	87.874

TABLE 1.64   (*Continued*)

Specific Gravity 60°/60°	°Bé	°Tw	lb/gal at 60°F, wt in air	lb/ft³ at 60°F, wt in air	Specific Gravity 60°/60°	°Bé	°Tw	lb/gal at 60°F, wt in air	lb/ft³ at 60°F, wt in air
1.415	42.53	83	11.7888	88.186	1.68	58.69	136	13.998	104.72
1.420	42.89	84	11.8304	88.498	1.69	59.20	138	14.082	105.34
1.425	43.25	85	11.8721	88.810					
1.430	43.60	86	11.9138	89.121	1.70	59.71	140	14.165	105.96
1.435	43.95	87	11.9555	89.433	1.71	60.20	142	14.249	106.59
1.440	44.31	88	11.9972	89.745	1.72	60.70	144	14.332	107.21
1.445	44.65	89	12.0389	90.057	1.73	61.18	146	14.415	107.83
1.450	45.00	90	12.0806	90.369	1.74	61.67	148	14.499	108.46
1.455	45.34	91	12.1223	90.681	1.75	62.14	150	14.582	109.08
1.460	45.68	92	12.1640	90.993	1.76	62.61	152	14.665	109.71
1.465	46.02	93	12.2057	91.305	1.77	63.08	154	14.749	110.32
1.470	46.36	94	12.2473	91.616	1.78	63.54	156	14.832	110.95
1.475	46.69	95	12.2890	91.928	1.79	63.99	158	14.916	111.58
1.480	47.03	96	12.3307	92.240					
1.485	47.36	97	12.3724	92.552	1.80	64.44	160	14.999	112.20
1.490	47.68	98	12.4141	92.864	1.81	64.89	162	15.082	112.82
1.495	48.01	99	12.4558	93.176	1.82	65.33	164	15.166	113.45
					1.83	65.77	166	15.249	114.07
1.500	48.33	100	12.4975	93.488	1.84	66.20	168	15.333	114.70
1.51	48.97	102	12.581	94.11	1.85	66.62	170	15.416	115.31
1.52	49.61	104	12.644	94.79	1.86	67.04	172	15.499	115.94
1.53	50.23	106	12.748	95.36	1.87	67.46	174	15.583	116.56
1.54	50.84	108	12.831	95.98	1.88	67.87	176	15.666	117.19
1.55	51.45	110	12.914	96.61	1.89	68.28	178	15.750	117.81
1.56	52.05	112	12.998	97.23					
1.57	52.64	114	13.081	97.85	1.90	68.68	180	15.832	118.43
1.58	53.23	116	13.165	98.48	1.91	69.08	182	15.916	119.06
1.59	53.81	118	13.248	99.10	1.92	69.48	184	16.000	119.68
					1.93	69.87	186	16.083	120.31
1.60	54.38	120	13.331	99.73	1.94	70.26	188	16.166	120.93
1.61	54.94	122	13.415	100.35	1.95	70.64	190	16.250	121.56
1.62	55.49	124	13.498	100.97	1.96	71.02	192	16.333	122.18
1.63	56.04	126	13.582	101.60	1.97	71.40	194	16.417	122.80
1.64	56.59	128	13.665	102.22	1.98	71.77	196	16.500	123.43
1.65	57.12	130	13.748	102.84	1.99	72.14	198	16.583	124.05
1.66	57.65	132	13.832	103.47	2.00	72.50	200	16.667	124.68
1.67	58.17	134	13.915	104.09					

[a] Baumé scale.
[b] Twaddell scale.

1.6 STANDARD SIZES

1.6.1 Preferred Numbers

Selection of standard sizes or ratings of many diverse products can be performed advantageously through the use of a geometrically based progression introduced by C. Renard. He originally adopted as a basis a rule that would yield a 10th multiple of the value a after every 5th step of the series:

$$a \times q^5 = 10a \quad \text{or} \quad q = \sqrt[5]{10}$$

where the numerical series $a, a[\sqrt[5]{10}], a[\sqrt[5]{10}]^2, a[\sqrt[5]{10}]^3, a[\sqrt[5]{10}]^4, 10a$, the values of which, to five significant figures are $a, 1.5849a, 2.5119a, 3.9811a, 6.309a, 10a$.

Renard's idea was to substitute, for these values, more rounded but more practical values. He adopted as a a power of 10, positive, nil, or negative obtaining the series $10, 16, 25, 40, 63, 100$, which may be continued in both directions.

From this series, designated by the symbol R5, the R10, R20, R40 series were formed, each adopted ratio being the square root of the preceding one: $\sqrt[10]{10}, \sqrt[20]{10}, \sqrt[40]{10}$. Thus each series provided Renard with twice as many steps in a decade as the preceding one.

Preferred numbers are immediately applicable to commercial sizes and ratings of products. It is advantageous to minimize the number of initial sizes and also to have adequate provision for logical expansion, if and when additional sizes are required. By making the initial sizes correspond to a coarse series such as R5, unnecessary expense can be avoided if subsequent demand for the product is disappointing. If, on the other hand, the product is accepted, intermediate sizes may be selected in a rational manner by using the next finer series R10, and so on. Such a procedure assures a justifiable relationship between successive sizes and is a decided contrast to haphazard selection.

The application of preferred numbers to raw material sizes and to the dimensions of parts also has enormously important potentialities. Under present conditions, commercial sizes of material are the result of a great many dissimilar gauge systems. The current trend in internationally acceptable metric sizing is to use preferred numbers. Even here, though, in the midst of the greatest opportunity for worldwide standardization through the acceptance of Renard series, we have fallen prey to our individualistic nature. The preferred number 1.6 is used by most nations as a standard 1.6 mm material thickness. German manufacturers, however, like 1.5 mm of ISO 497 for a more rounded preferred number. Similarly in metric screw sizes, 6.3 mm is consistent with the preferred number series; yet, 6.0 mm (more rounded) has been adopted as a standard fastener diameter.

The International Electrochemical Commission, IEC, used preferred numbers to establish standard current ratings in amperes as follows: 1, 1.25, 1.6, 2.5, 3.15, 4.5, 6.3. Notice that R10 series is used except for 4.5, which is a third step R20 series.

The American Wire Gauge size for copper wire is based on a geometric series. However, instead of using 1.1220, the rounded value of $\sqrt[20]{10}$, in $a \times q^{20} = 10a$, the q chosen is 1.123.

A special series of preferred numbers is used for designating the characteristic values of capacitors, resistors, inductors, and other electronic products. Instead of using the Renard series R5, R10, R20, R40, R80 as derived from the geometric series of numbers $10^{N/5}, 10^{N/10}, 10^{N/20}, 10^{N/40}, 10^{N/80}$, the geometric series used is $10^{N/6}, 10^{N/12}, 10^{N/24}, 10^{N/48}, 10^{N/96}, 10^{N/192}$, which are designated respectively E6, E12, E24, E48, E96, E192.

It should be evident that any series of preferred numbers can be generated to serve any specific case. Examples taken from ANSI and ISO standards are reproduced in Tables 1.65–1.68.

TABLE 1.65   Basic Series of Preferred Numbers: R5, R10, R20, and R40 Series

| | | | | Theoretical Values | | Differences between |
| | | | | Mantissas of | Calculated | Basic Series and |
R5	R10	R20	R40	Logarithms	Values	Calculated Values (%)
1.00	1.00	1.00	1.00	000	1.0000	0
			1.06	025	1.0593	+0.07
		1.12	1.12	050	1.1220	−0.18
			1.18	075	1.1885	−0.71
	1.25	1.25	1.25	100	1.2589	−0.71
			1.32	125	1.3335	−1.01
		1.40	1.40	150	1.4125	−0.88
			1.50	175	1.4962	+0.25
1.60	1.60	1.60	1.60	200	1.5849	+0.95
			1.70	225	1.6788	+1.26
		1.80	1.80	250	1.7783	+1.22
			1.90	275	1.8836	+0.87
	2.00	2.00	2.00	300	1.9953	+0.24
			2.12	325	2.1135	+0.31
		2.24	2.24	350	2.2387	+0.06
			2.36	375	2.3714	−0.48
2.50	2.50	2.50	2.50	400	2.5119	−0.47
			2.65	425	2.6607	−0.40
		2.80	2.80	450	2.8184	−0.65
			3.00	475	2.9854	+0.49
	3.15	3.15	3.15	500	3.1623	−0.39
			3.35	525	3.3497	+0.01
		3.55	3.55	550	3.5481	+0.05
			3.75	575	3.7584	−0.22

TABLE 1.65 (*Continued*)

| | | | | Theoretical Values | | Differences between |
| | | | | Mantissas of Logarithms | Calculated Values | Basic Series and Calculated Values (%) |
R5	R10	R20	R40			
4.00	4.00	4.00	4.00	600	3.9811	+0.47
			4.25	625	4.2170	+0.78
		4.50	4.50	650	4.4668	+0.74
			4.75	675	4.7315	+0.39
	5.00	5.00	5.00	700	5.0119	−0.24
			5.30	725	5.3088	−0.17
		5.60	5.60	750	5.6234	−0.42
			6.00	775	5.9566	+0.73
6.30	6.30	6.30	6.30	800	6.3096	−0.15
			6.70	825	6.6834	+0.25
		7.10	7.10	850	7.0795	+0.29
			7.50	875	7.4989	+0.01
	8.00	8.00	8.00	900	7.9433	+0.71
			8.50	925	8.4140	+1.02
		9.00	9.00	950	8.9125	+0.98
			9.50	975	9.4406	+0.63
10.00	10.00	10.00	10.00	000	10.0000	0

TABLE 1.66   Basic Series of Preferred Numbers: R80 Series

1.00	1.80	3.15	5.60
1.03	1.85	3.25	5.80
1.06	1.90	3.35	6.00
1.09	1.95	3.45	6.15
1.12	2.00	3.55	6.30
1.15	2.06	3.65	6.50
1.18	2.12	3.75	6.70
1.22	2.18	3.87	6.90
1.25	2.24	4.00	7.10
1.28	2.30	4.12	7.30
1.32	2.36	4.25	7.50
1.36	2.43	4.37	7.75
1.40	2.50	4.50	8.00
1.45	2.58	4.62	8.25
1.50	2.65	4.75	8.50
1.55	2.72	4.87	8.75
1.60	2.80	5.00	9.00
1.65	2.90	5.15	9.25
1.70	3.00	5.20	9.50
1.75	3.07	5.45	9.75

TABLE 1.67   Expansion of R5 Series

Preferred Number	Divided by 10	Multiplied by 10	Multiplied by 100	Multiplied by 1000
1.0	0.10	10	100	1000
1.6	0.16	16	160	1600
2.5	0.25	25	250	2500
4.0	0.40	40	400	4000
6.3	0.63	63	630	6300

TABLE 1.68   Rounding of Preferred Numbers[a]

Preferred Number	First Rounding	Second Rounding
1.12	1.1	1.1
1.25	1.25	1.2
1.60	1.6	1.5[a]
2.24	2.2	2.2
3.15	3.2	3.0
3.55	3.6	3.5
5.60	5.6	5.5
6.30	6.3	6.0
7.10	7.1	7.0

[a]Rounded only when using the R5 or R10 series.

Applicable Documents

Adoption of Renard's preferred number system by international standardization bodies resulted in a host of national standards being generated for particular applications. The current organization in the United States that is charged with generating American national standards is the American National Standards Institute, ANSI. Accordingly, the following national and international standards are in use in the United States.

ANSI Z17.1-1973	American National Standard for Preferred Numbers.
ANSI C83.2-1971	American National Standard Preferred Values for Components for Electronic Equipment.
EIA Standard RS-385	Preferred Values for Components for Electronic Equipment. (Issued by the Electronics Industries Association. Same as ANSI C83.2-1971.)
ISO 3-1973	Preferred numbers—Series of preferred numbers.
ISO 17-1973	Guide to the use of preferred numbers and of series of preferred numbers.
ISO 497-1973	Guide to the choice of series of preferred numbers and of series containing more rounded values of preferred numbers.

Table 1.67 shows the expansibility of preferred numbers in the positive direction. The same expansibility can be made in the negative direction. Table 1.68 shows a deviation by roundings for cases where adhering to a basic preferred number would be absurd as in 31.5 teeth in a gear when clearly 32 makes sense.

1.6.2   Gages

TABLE 1.69   United States Standard Gage[a] for Sheet and Plate Iron and Steel, and Its Extension[b]

Gage Number	Weight per Square Foot		Weight per Square Meter	Approximate thickness			
				Wrought Iron, 480 lb/ft^3		Steel and Open-hearth Iron, 489.6 lb/ft^3	
	oz.	lb	kg	in.	mm	in.	mm
0000000	320	20.00	97.65	0.500	12.70	0.490	12.45
000000	300	18.75	91.55	0.469	11.91	0.460	11.67
00000	280	17.50	85.44	0.438	11.11	0.429	10.90
0000	260	16.25	79.34	0.406	10.32	0.398	10.12

TABLE 1.69   (*Continued*)

Gage Number	Weight per Square Foot		Weight per Square Meter	Approximate thickness			
				Wrought Iron, 480 lb/ft^3		Steel and Open-hearth Iron, 489.6 lb/ft^3	
	oz.	lb	kg	in.	mm	in.	mm
000	240	15.00	73.24	0.375	9.52	0.368	9.34
00	220	13.75	67.13	0.344	8.73	0.337	8.56
0	200	12.50	61.03	0.312	7.94	0.306	7.78
1	180	11.25	54.93	0.2812	7.14	0.2757	7.00
2	170	10.62	51.88	0.2656	6.75	0.2604	6.62
3	160	10.00	48.82	0.2500	6.35	0.2451	6.23
4	150	9.375	45.77	0.2344	5.95	0.2298	5.84
5	140	8.750	42.72	0.2188	5.56	0.2145	5.45
6	130	8.125	39.67	0.2031	5.16	0.1991	5.06
7	120	7.500	36.62	0.1875	4.76	0.1838	4.67
8	110	6.875	33.57	0.1719	4.37	0.1685	4.28
9	100	6.250	30.52	0.1562	3.97	0.1532	3.89
10	90	5.625	27.46	0.1406	3.57	0.1379	3.50
11	80	5.000	24.41	0.1250	3.18	0.1225	3.11
12	70	4.375	21.36	0.1094	2.778	0.1072	2.724
13	60	3.750	18.31	0.0938	2.381	0.0919	2.335
14	50	3.125	15.26	0.0781	1.984	0.0766	1.946
15	45	2.812	13.73	0.0703	1.786	0.0689	1.751
16	40	2.500	12.21	0.0625	1.588	0.0613	1.557
17	36	2.250	10.99	0.0562	1.429	0.0551	1.400
18	32	2.000	9.765	0.0500	1.270	0.0490	1.245
19	28	1.750	8.544	0.0438	1.111	0.0429	1.090
20	24	1.500	7.324	0.0375	0.952	0.0368	0.934
21	22	1.375	6.713	0.0344	0.873	0.0337	0.856
22	20	1.250	6.103	0.0312	0.794	0.0306	0.778
23	18	1.125	5.493	0.0281	0.714	0.0276	0.700
24	16	1.000	4.882	0.0250	0.635	0.0245	0.623
25	14	0.8750	4.272	0.0219	0.556	0.0214	0.545
26	12	0.7500	3.662	0.0188	0.476	0.0184	0.467
27	11	0.6875	3.357	0.0172	0.437	0.0169	0.428
28	10	0.6250	3.052	0.0156	0.397	0.0153	0.389
29	9	0.5625	2.746	0.0141	0.357	0.0138	0.350
30	8	0.5000	2.441	0.0125	0.318	0.0123	0.311
31	7	0.4375	2.136	0.0109	0.278	0.0107	0.272
32	$6\frac{1}{2}$	0.4062	1.983	0.0102	0.258	0.0100	0.253
33	6	0.3750	1.831	0.0094	0.238	0.0092	0.233
34	$5\frac{1}{2}$	0.3438	1.678	0.0086	0.218	0.0084	0.214
35	5	0.3125	1.526	0.0078	0.198	0.0077	0.195
36	$4\frac{1}{2}$	0.2812	1.373	0.0070	0.179	0.0069	0.175
37	$4\frac{1}{4}$	0.2656	1.297	0.0066	0.169	0.0065	0.165
38	4	0.2500	1.221	0.0062	0.159	0.0061	0.156
39	$3\frac{3}{4}$	0.2344	1.144	0.0059	0.149	0.0057	0.146
40	$3\frac{1}{2}$	0.2188	1.068	0.0055	0.139	0.0054	0.136
41	$3\frac{3}{8}$	0.2109	1.030	0.0053	0.134	0.0052	0.131
42	$3\frac{1}{4}$	0.2031	0.9917	0.0051	0.129	0.0050	0.126
43	$3\frac{1}{8}$	0.1953	0.9536	0.0049	0.124	0.0048	0.122
44	3	0.1875	0.9155	0.0047	0.119	0.0046	0.117

[a] For the Galvanized Sheet Gage, add 2.5 oz to the weight per square foot as given in the table. Gage numbers below 8 and above 34 are not used in the Galvanized Sheet Gage.
[b] Gage numbers greater than 38 were not in the standard as set up by law, but are in general use.

TABLE 1.70   American Wire Gage: Weights of Copper, Aluminum, and Brass Sheets and Plates

Gage Number	Thickness		Approximate Weight,[a] lb/ft²		
	in.	mm	Copper	Aluminum	Commercial (high) brass
0000	0.4600	11.68	21.27	6.49	20.27
000	0.4096	10.40	18.94	5.78	18.05
00	0.3648	9.266	16.87	5.14	16.07
0	0.3249	8.252	15.03	4.58	14.32
1	0.2893	7.348	13.38	4.08	12.75
2	0.2576	6.544	11.91	3.632	11.35
3	0.2294	5.827	10.61	3.234	10.11
4	0.2043	5.189	9.45	2.880	9.00
5	0.1819	4.621	8.41	2.565	8.01
6	0.1620	4.115	7.49	2.284	7.14
7	0.1443	3.665	6.67	2.034	6.36
8	0.1285	3.264	5.94	1.812	5.66
9	0.1144	2.906	5.29	1.613	5.04
10	0.1019	2.588	4.713	1.437	4.490
11	0.0907	2.305	4.195	1.279	3.996
12	0.0808	2.053	3.737	1.139	3.560
13	0.0720	1.828	3.330	1.015	3.172
14	0.0641	1.628	2.965	0.904	2.824
15	0.0571	1.450	2.641	0.805	2.516
16	0.0508	1.291	2.349	0.716	2.238
17	0.0453	1.150	2.095	0.639	1.996
18	0.0403	1.024	1.864	0.568	1.776
19	0.0359	0.9116	1.660	0.506	1.582
20	0.0320	0.8118	1.480	0.451	1.410
21	0.0285	0.7230	1.318	0.402	1.256
22	0.0253	0.6438	1.170	0.3567	1.115
23	0.0226	0.5733	1.045	0.3186	0.996
24	0.0201	0.5106	0.930	0.2834	0.886
25	0.0179	0.4547	0.828	0.2524	0.789
26	0.0159	0.4049	0.735	0.2242	0.701
27	0.0142	0.3606	0.657	0.2002	0.626
28	0.0126	0.3211	0.583	0.1776	0.555
29	0.0113	0.2859	0.523	0.1593	0.498
30	0.0100	0.2546	0.4625	0.1410	0.4406
31	0.00893	0.2268	0.4130	0.1259	0.3935
32	0.00795	0.2019	0.3677	0.1121	0.3503
33	0.00708	0.1798	0.3274	0.0998	0.3119
34	0.00630	0.1601	0.2914	0.0888	0.2776
35	0.00561	0.1426	0.2595	0.0791	0.2472
36	0.00500	0.1270	0.2312	0.0705	0.2203
37	0.00445	0.1131	0.2058	0.0627	0.1961
38	0.00397	0.1007	0.1836	0.0560	0.1749
39	0.00353	0.0897	0.1633	0.0498	0.1555
40	0.00314	0.0799	0.1452	0.0443	0.1383

[a]Assumed specific gravities or densities in grams per cubic centimeter; copper, 8.89; aluminum, 2.71; brass, 8.47.

Wire Gages
The sizes of wires having a diameter less than $\frac{1}{2}$ in. are usually stated in terms of certain arbitrary scales called "gages." The size or gage number of a solid wire refers to the cross section of the wire perpendicular to its length; the size or gage number of a stranded wire refers to the total cross section of the constituent wires, irrespective of the pitch of the spiraling. Larger wires are usually described in terms of their area expressed in circular mils. A circular mil is the area of a circle 1 mil in diameter, and the area of any circle in circular mils is equal to the square of its diameter in mils.

TABLE 1.71 Comparison of Wire Gage Diameters in Mils[a]

Gage No.	American Wire Gage (B. & S.)	Steel Wire Gage	Birming-ham Wire Gage (Stubs')	Old English Wire Gage (London)	Stubs' Steel Wire Gage	(British) Standard Wire Gage	Metric Gage[b]	Gage No.
7-0	—	490.0	—	—	—	500	—	7-0
6-0	—	461.5	—	—	—	464	—	6-0
5-0	—	430.5	—	—	—	432	—	5-0
4-0	460	393.8	454	454	—	400	—	4-0
3-0	410	362.5	425	425	—	372	—	3-0
2-0	365	331.0	380	380	—	348	—	2-0
0	325	306.5	340	340	—	324	—	0
1	289	283.0	300	300	227	300	3.94	1
2	258	262.5	284	284	219	276	7.87	2
3	229	243.7	259	259	212	252	11.8	3
4	204	225.3	238	238	207	232	15.7	4
5	182	207.0	220	220	204	212	19.7	5
6	162	192.0	203	203	201	192	23.6	6
7	144	177.0	180	180	199	176	27.6	7
8	128	162.0	165	165	197	160	31.5	8
9	114	148.3	148	148	194	144	35.4	9
10	102	135.0	134	134	191	128	39.4	10
11	91	120.5	120	120	188	116	—	11
12	81	105.5	109	109	185	104	47.2	12
13	72	91.5	95	95	182	92	—	13
14	64	80.0	83	83	180	80	55.1	14
15	57	72.0	72	72	178	72	—	15
16	51	62.5	65	65	175	64	63.0	16
17	45	54.0	58	58	172	56	—	17
18	40	47.5	49	49	168	48	70.9	18
19	36	41.0	42	42	164	40	—	19
20	32	34.8	35	35	161	36	78.7	20
21	28.5	31.7	32	31.5	157	32	—	21
22	25.3	28.6	28	29.5	155	28	—	22
23	22.6	25.8	25	27.0	153	24	—	23
24	20.1	23.0	22	25.0	151	22	—	24
25	17.9	20.4	20	23.0	148	20	98.4	25
26	15.9	18.1	18	20.5	146	18	—	26
27	14.2	17.3	16	18.75	143	16.4	—	27
28	12.6	16.2	14	16.50	139	14.8	—	28
29	11.3	15.0	13	15.50	134	13.6	—	29
30	10.0	14.0	12	13.75	127	12.4	118	30
31	8.9	13.2	10	12.25	120	11.6	—	31
32	8.0	12.8	9	11.25	115	10.8	—	32
33	7.1	11.8	8	10.25	112	10.0	—	33
34	6.3	10.4	7	9.50	110	9.2	—	34
35	5.6	9.5	5	9.00	108	8.4	138	35
36	5.0	9.0	4	7.50	106	7.6	—	36
37	4.5	8.5	—	6.50	103	6.8	—	37
38	4.0	8.0	—	5.75	101	6.0	—	38
39	3.5	7.5	—	5.00	99	5.2	—	39
40	3.1	7.0	—	4.50	97	4.8	157	40
41	—	6.6	—	—	95	4.4	—	41
42	—	6.2	—	—	92	4.0	—	42
43	—	6.0	—	—	88	3.6	—	43
44	—	5.8	—	—	85	3.2	—	44
45	—	5.5	—	—	81	2.8	177	45
46	—	5.2	—	—	79	2.4	—	46
47	—	5.0	—	—	77	2.0	—	47
48	—	4.8	—	—	75	1.6	—	48
49	—	4.6	—	—	72	1.2	—	49
50	—	4.4	—	—	69	1.0	197	50

[a] Bureau of Standards, Circulars No. 31 and No. 67.
[b] For diameters corresponding to metric gage numbers, 1.2, 1.4, 1.6, 1.8, 2.5, 3.5, and 4.5, divide those of 12, 14, etc., by 10.

1.6.3 Paper Sizes

TABLE 1.72 Standard Engineering Drawing Sizes[a]

Size Designation	Width[c] (Vertical)	Length (Horizontal)	Margin Horizontal	Margin Vertical
		Flat Sizes[b]		
A (horizontal)	8.5	11.0	0.38	0.25
A (vertical)	11.0	8.5	0.25	0.38
B	11.0	17.0	0.38	0.62
C	17.0	22.0	0.75	0.50
D	22.0	34.0	0.50	1.00
E	34.0	44.0	1.00	0.50
F	28.0	40.0	0.50	0.50

Size Designation	Width[b] (Vertical)	Length[c] (Horizontal) Min	Length[c] (Horizontal) Max	Margin[c] Horizontal	Margin[c] Vertical
		Roll Sizes			
G	11.0	22.5	90.0	0.38	0.50
H	28.0	44.0	143.0	0.50	0.50
J	34.0	55.0	176.0	0.50	0.50
K	40.0	55.0	143.0	0.50	0.50

[a]See ANSI Y14.1-1980.
[b]All dimensions are in inches.
[c]Not including added protective margins.

International Paper Sizes

Countries that are committed to the International System of Units (SI) have a standard series of paper sizes for printing, writing, and drafting. These paper sizes are called the "international paper sizes."

The advantages of the international paper sizes are as follows:

1. The ratio of width to length remains constant for every size, namely:

$$\frac{\text{Width}}{\text{Length}} = \frac{1}{\sqrt{2}} \quad \text{or} \quad \frac{1}{1.414} \text{ approximately}$$

Since this is the same ratio as the D aperture in the unitized 35-mm microfilm frame, the advantages are apparent.

2. If a sheet is cut in half, that is, if the $\sqrt{2}$ length is cut in half, the two halves retain the constant width-to-length ratio of $1/\sqrt{2}$. No other ratio could do this.

3. All international sizes are created from the A-0 size by single cuts without waste. In storing or stacking they fit together like parts of a jigsaw puzzle...without waste.

TABLE 1.73 Eleven International Paper Sizes

International Paper Size	Millimeters	Inches, Approximate
A-0	841 × 1189	$33\frac{1}{8} \times 46\frac{3}{4}$
A-1	594 × 841	$23\frac{3}{8} \times 33\frac{1}{8}$
A-2	420 × 594	$16\frac{1}{2} \times 23\frac{3}{8}$
A-3	297 × 420	$11\frac{3}{4} \times 16\frac{1}{2}$
A-4	210 × 297	$8\frac{1}{4} \times 11\frac{3}{4}$
A-5	148 × 210	$5\frac{7}{8} \times 8\frac{1}{4}$
A-6	105 × 148	$4\frac{1}{8} \times 5\frac{7}{8}$
A-7	74 × 105	$2\frac{7}{8} \times 4\frac{1}{8}$
A-8	52 × 74	$2 \times 2\frac{7}{8}$
A-9	37 × 52	$1\frac{1}{2} \times 2$
A-10	26 × 37	$1 \times 1\frac{1}{2}$

1.6.4 Sieve Sizes

TABLE 1.74 The Tyler Standard Screen Scale Sieves

This screen scale has as its base an opening of 0.0029 in., which is the opening in 200-mesh 0.0021-in. wire, the standard sieve, as adopted by the Bureau of Standards of the United States Government, the openings increasing in the ratio of the square root of 2 or 1.414.

Where a closer sizing is required, column 5 shows the Tyler Standard Screen Scale with intermediate sieves. In this series the sieve openings increase in the ratio of the fourth root of 2 or 1.189.

1	2	3	4	5	6	7	8	9
Tyler Standard Screen Scale $\sqrt{2}$ or 1.414 Openings (in.)	Every Other Sieve from 0.0029 in. to 0.742 in. Ratio of 2 to 1	Every Other Sieve from 0.0041 in. to 1.050 in. Ratio of 2 to 1	Every Fourth Sieve from 0.0029 in. to 0.742 in. Ratio of 4 to 1	For Closer Sizing Sieves from 0.0029 in. to 1.050 in. Ratio $\sqrt[4]{2}$ or 1.189	Openings (mm)	Opening in Fractions of Inch (approx.)	Mesh	Diameter of Wire
1.050	—	1.050	—	1.050	26.67	1	—	0.148
—	—	—	—	0.883	22.43	$\frac{7}{8}$	—	0.135
0.742	0.742	—	0.742	0.742	18.85	$\frac{3}{4}$	—	0.135
—	—	—	—	0.624	15.85	$\frac{5}{8}$	—	0.120
0.525	—	0.525	—	0.525	13.33	$\frac{1}{2}$	—	0.105
—	—	—	—	0.441	11.20	$\frac{7}{16}$	—	0.105
0.371	0.371	—	—	0.371	9.423	$\frac{3}{8}$	—	0.092
—	—	—	—	0.312	7.925	$\frac{5}{16}$	$2\frac{1}{2}$	0.088
0.263	—	0.263	—	0.263	6.680	$\frac{1}{4}$	3	0.070
—	—	—	—	0.221	5.613	$\frac{7}{32}$	$3\frac{1}{2}$	0.065
0.185	0.185	—	0.185	0.185	4.699	$\frac{3}{16}$	4	0.065
—	—	—	—	0.156	3.962	$\frac{5}{32}$	5	0.044
0.131	—	0.131	—	0.131	3.327	$\frac{1}{8}$	6	0.036
—	—	—	—	0.110	2.794	$\frac{7}{64}$	7	0.0328
0.093	0.093	—	—	0.093	2.362	$\frac{3}{32}$	8	0.032
—	—	—	—	0.078	1.981	$\frac{5}{84}$	9	0.033
0.065	—	0.065	—	0.065	1.651	$\frac{1}{16}$	10	0.035
—	—	—	—	0.055	1.397	—	12	0.028
0.046	0.046	—	0.046	0.046	1.168	$\frac{3}{64}$	14	0.025
—	—	—	—	0.0390	0.991	—	16	0.0235
0.0328	—	0.0328	—	0.0328	0.833	$\frac{1}{32}$	20	0.0172
—	—	—	—	0.0276	0.701	—	24	0.0141
0.0232	0.0232	—	—	0.0232	0.589	—	28	0.0125
—	—	—	—	0.0195	0.495	—	32	0.0118
0.0164	—	0.0164	—	0.0164	0.417	$\frac{1}{64}$	35	0.0122
—	—	—	—	0.0138	0.351	—	42	0.0100
0.0116	0.0116	—	0.0116	0.0116	0.295	—	48	0.0092
—	—	—	—	0.0097	0.246	—	60	0.0070
0.0082	—	0.0082	—	0.0082	0.208	—	65	0.0072
—	—	—	—	0.0069	0.175	—	80	0.0056
0.0058	0.0058	—	—	0.0058	0.147	—	100	0.0042
—	—	—	—	0.0049	0.124	—	115	0.0038
0.0041	—	0.0041	—	0.0041	0.104	—	150	0.0026
—	—	—	—	0.0035	0.088	—	170	0.0024
0.0029	0.0029	—	0.0029	0.0029	0.074	—	200	0.0021

TABLE 1.75 Nominal Dimensions, Permissible Variations, and Limits for Woven Wire Cloth of Standard Sieves, U.S. Series. ASTM Standard[a]

Size or Sieve Designation		Sieve Opening		Permissible Variations in Av. Opening ($\pm\%$)	Permissible Variations in Max. Opening ($\pm\%$)	Wire Diameter	
Micron	No.	mm	Inch (approx. equivalents)			mm	Inch (approx. equivalents)
5660	3½	5.66	0.233	3	10	1.28 –1.90	0.050 –0.075
4760	4	4.76	0.187	3	10	1.14 –1.68	0.045 –0.066
4000	5	4.00	0.157	3	10	1.00 –1.47	0.039 –0.058
3360	6	3.36	0.132	3	10	0.87 –1.32	0.034 –0.052
2830	7	2.83	0.111	3	10	0.80 –1.20	0.031 –0.047
2380	8	2.38	0.0937	3	10	0.74 –1.10	0.0291–0.0433
2000	10	2.00	0.0787	3	10	0.68 –1.00	0.0268–0.0394
1680	12	1.68	0.0661	3	10	0.62 –0.90	0.0244–0.0354
1410	14	1.41	0.0555	3	10	0.56 –0.80	0.0220–0.0315
1190	16	1.19	0.0469	3	10	0.50 –0.70	0.0197–0.0276
1000	18	1.00	0.0394	5	15	0.43 –0.62	0.0169–0.0244
840	20	0.84	0.0331	5	15	0.38 –0.55	0.0150–0.0217
710	25	0.71	0.0280	5	15	0.33 –0.48	0.0130–0.0189
590	30	0.59	0.0232	5	15	0.29 –0.42	0.0114–0.0165
500	35	0.50	0.0197	5	15	0.26 –0.37	0.0102–0.0146
420	40	0.42	0.0165	5	25	0.23 –0.33	0.0091–0.0130
350	45	0.35	0.0138	5	25	0.20 –0.29	0.0079–0.0114
297	50	0.297	0.0117	5	25	0.170–0.253	0.0067–0.0100
250	60	0.250	0.0098	5	25	0.149–0.220	0.0059–0.0087
210	70	0.210	0.0083	5	25	0.130–0.187	0.0051–0.0074
177	80	0.177	0.0070	6	40	0.114–0.154	0.0045–0.0061
149	100	0.149	0.0059	6	40	0.096–0.125	0.0038–0.0049
125	120	0.125	0.0049	6	40	0.079–0.103	0.0031–0.0041
105	140	0.105	0.0041	6	40	0.063–0.087	0.0025–0.0034
88	170	0.088	0.0035	6	40	0.054–0.073	0.0021–0.0029
74	200	0.074	0.0029	7	60	0.045–0.061	0.0018–0.0024
62	230	0.062	0.0024	7	90	0.039–0.052	0.0015–0.0020
53	270	0.053	0.0021	7	90	0.035–0.046	0.0014–0.0018
44	325	0.044	0.0017	7	90	0.031–0.040	0.0012–0.0016
37	400	0.037	0.0015	7	90	0.023–0.035	0.0009–0.0014

[a]For sieves from the 1000-micron (No. 18) to the 37-micron (No. 400) size, inclusive, not more than 5% of the openings shall exceed the nominal opening by more than one-half of the permissible variation in the maximum opening.

1.6.5 Standard Structural Sizes—Steel

Steel Sections. Tables 1.76 to 1.83 give the dimensions, weights, and properties of *rolled steel* structural sections, including wide-flange sections, American standard beams, channels, angles, tees, and zees. The values for the various structural forms, taken from the eighth edition, 1980, of *Steel Construction*, by the kind permission of the publisher, the American Institute of Steel Construction, give the section specifications required in designing steel structures. The theory of design is covered in Section 5—Mechanics of Deformable Bodies.

Most of the sections can be supplied promptly by the Bethlehem, Carnegie, and Illinois Steel Company mills. Owing to variations in the rolling practice of the different mills, their products are not identical, although their divergence from the values given in the tables is practically negligible. For standardization, only the lesser values are given, and therefore they are on the side of safety.

Further information on sections listed in the tables, together with information on other products and on the requirements for placing orders, may be gathered from mill catalogs.

TABLE 1.76   Properties of Wide-Flange Sections

Nominal Size (in.)	Weight per Foot (lb)	Area (in.^{2})	Depth (in.)	Flange Width (in.)	Flange Thickness (in.)	Web Thickness (in.)	Axis X-X I (in.^{4})	S (in.^{3})	r (in.)	Axis Y-Y I (in.^{4})	S (in.^{3})	r (in.)
$36 \times 16\frac{1}{2}$	300	88.17	36.72	16.655	1.680	0.945	20290.2	1105.1	15.17	1225.2	147.1	3.73
	280	82.32	36.50	16.595	1.570	0.885	18819.3	1031.2	15.12	1127.5	135.9	3.70
	260	76.56	36.24	16.555	1.440	0.845	17233.8	951.1	15.00	1020.6	123.3	3.65
	245	72.03	36.06	16.512	1.350	0.802	16092.2	892.5	14.95	944.7	114.4	3.62
	230	67.73	35.88	16.475	1.260	0.765	14988.4	835.5	14.88	870.9	105.7	3.59
36×12	194	57.11	36.48	12.117	1.260	0.770	12103.4	663.6	14.56	355.4	58.7	2.49
	182	53.54	36.32	12.072	1.180	0.725	11281.5	621.2	14.52	327.7	54.3	2.47
	170	49.98	36.16	12.027	1.100	0.680	10470.0	579.1	14.47	300.6	50.0	2.45
	160	47.09	36.00	12.000	1.020	0.653	9738.8	541.0	14.38	275.4	45.9	2.42
	150	44.16	35.84	11.972	0.940	0.625	9012.1	502.9	14.29	250.4	41.8	2.38
$33 \times 15\frac{3}{4}$	240	70.52	33.50	15.865	1.400	0.830	13585.1	811.1	13.88	874.3	110.2	3.52
	220	64.73	33.25	15.810	1.275	0.775	12312.1	740.6	13.79	782.4	99.0	3.48
	200	58.79	33.00	15.750	1.150	0.715	11048.2	669.6	13.71	691.7	87.8	3.43
$33 \times 11\frac{1}{2}$	152	44.71	33.50	11.565	1.055	0.635	8147.6	486.4	13.50	256.1	44.3	2.39
	141	41.51	33.31	11.535	0.960	0.605	7442.2	446.8	13.39	229.7	39.8	2.35
	130	38.26	33.10	11.510	0.855	0.580	6699.0	404.8	13.23	201.4	35.0	2.29
30×15	210	61.78	30.38	15.105	1.315	0.775	9872.4	649.9	12.64	707.9	93.7	3.38
	190	55.90	30.12	15.040	1.185	0.710	8825.9	586.1	12.57	624.6	83.1	3.34
	172	50.65	29.88	14.985	1.065	0.655	7891.5	528.2	12.48	550.1	73.4	3.30
$30 \times 10\frac{1}{2}$	132	38.83	30.30	10.551	1.000	0.615	5753.1	379.7	12.17	185.0	35.1	2.18
	124	36.45	30.16	10.521	0.930	0.585	5347.1	354.6	12.11	169.7	32.3	2.16
	116	34.13	30.00	10.500	0.850	0.564	4919.1	327.9	12.00	153.2	29.2	2.12
	108	31.77	29.82	10.484	0.760	0.548	4461.0	299.2	11.85	135.1	25.8	2.06
27×14	177	52.10	27.31	14.090	1.190	0.725	6728.6	492.8	11.36	518.9	73.7	3.16
	160	47.04	27.08	14.023	1.075	0.658	6018.6	444.5	11.31	458.0	65.3	3.12
	145	42.68	26.88	13.965	0.975	0.600	5414.3	402.9	11.26	406.9	58.3	3.09
27×10	114	33.53	27.28	10.070	0.932	0.570	4080.5	299.2	11.03	149.6	29.7	2.11
	102	30.01	27.07	10.018	0.827	0.518	3604.1	266.3	10.96	129.5	25.9	2.08
	94	27.65	26.91	9.990	0.747	0.490	3266.7	242.8	10.87	115.1	23.0	2.04
24×14	160	47.04	24.72	14.091	1.135	0.656	5110.3	413.5	10.42	492.6	69.9	3.23
	145	42.62	24.49	14.043	1.020	0.608	4561.0	372.5	10.34	434.3	61.8	3.19
	130	38.21	24.25	14.000	0.900	0.565	4009.5	330.7	10.24	375.2	53.6	3.13
24×12	120	35.29	24.31	12.088	0.930	0.556	3635.3	299.1	10.15	254.0	42.0	2.68
	110	32.36	24.16	12.042	0.855	0.510	3315.0	274.4	10.12	229.1	38.0	2.66
	100	29.43	24.00	12.000	0.775	0.468	2987.3	248.9	10.08	203.5	33.9	2.63
24×9	94	27.63	24.29	9.061	0.872	0.516	2683.0	220.9	9.85	102.2	22.6	1.92
	84	24.71	24.09	9.015	0.772	0.470	2364.3	196.3	9.78	88.3	19.6	1.89
	76	22.37	23.91	8.985	0.682	0.440	2096.4	175.4	9.68	76.5	17.0	1.85
21×13	142	41.76	21.46	13.132	1.095	0.659	3403.1	317.2	9.03	385.9	58.8	3.04
	127	37.34	21.24	13.061	0.985	0.588	3017.2	284.1	8.99	338.6	51.8	3.01
	112	32.93	21.00	13.000	0.865	0.527	2620.6	249.6	8.92	289.7	44.6	2.96
21×9	96	28.21	21.14	9.038	0.935	0.575	2088.9	197.6	8.60	109.3	24.2	1.97
	82	24.10	20.86	8.962	0.795	0.499	1752.4	168.0	8.53	89.6	20.0	1.93
$21 \times 8\frac{1}{4}$	73	21.46	21.24	8.295	0.740	0.455	100.3	150.7	8.64	66.2	16.0	1.76
	68	20.02	21.13	8.270	0.685	0.430	1478.3	139.9	8.59	60.4	14.6	1.74
	62	18.23	20.99	8.240	0.615	0.400	1326.8	126.4	8.53	53.1	12.9	1.71
$18 \times 11\frac{3}{4}$	114	33.51	18.48	11.833	0.991	0.595	2033.8	220.1	7.79	255.6	43.2	2.76

TABLE 1.76 (*Continued*)

Nominal Size (in.)	Weight per Foot (lb)	Area (in.^{2})	Depth (in.)	Flange Width (in.)	Flange Thickness (in.)	Web Thickness (in.)	Axis X-X I (in.^{4})	Axis X-X S (in.^{3})	Axis X-X r (in.)	Axis Y-Y I (in.^{4})	Axis Y-Y S (in.^{3})	Axis Y-Y r (in.)
	105	30.86	18.32	11.792	0.911	0.554	1852.5	202.2	7.75	231.0	39.2	2.73
	96	28.22	18.16	11.750	0.831	0.512	1674.7	184.4	7.70	206.8	35.2	2.71
$18 \times 8\frac{3}{4}$	85	24.97	18.32	8.838	0.911	0.526	1429.9	156.1	7.57	99.4	22.5	2.00
	77	22.63	18.16	8.787	0.831	0.475	1286.8	141.7	7.54	88.6	20.2	1.98
	70	20.56	18.00	8.750	0.751	0.438	1153.9	128.2	7.49	78.5	17.9	1.95
	64	18.80	17.87	8.715	0.686	0.403	1045.8	117.0	7.46	70.3	16.1	1.93
$18 \times 7\frac{1}{2}$	60	17.64	18.25	7.558	0.695	0.416	984.0	107.8	7.47	47.1	12.5	1.63
	55	16.19	18.12	7.532	0.630	0.390	889.9	98.2	7.41	42.0	11.1	1.61
	50	14.71	18.00	7.500	0.570	0.358	800.6	89.0	7.38	37.2	9.9	1.59
$16 \times 11\frac{1}{2}$	96	28.22	16.32	11.533	0.875	0.535	1355.1	166.1	6.93	207.2	35.9	2.71
	88	25.87	16.16	11.502	0.795	0.504	1222.6	151.3	6.87	185.2	32.2	2.67
$16 \times 8\frac{1}{2}$	78	22.92	16.32	8.586	0.875	0.529	1042.6	127.8	6.74	87.5	20.4	1.95
	71	20.86	16.16	8.543	0.795	0.486	936.9	115.9	6.70	77.9	18.2	1.93
	64	18.80	16.00	8.500	0.715	0.443	833.8	104.2	6.66	68.4	16.1	1.91
	58	17.04	15.86	8.464	0.645	0.407	746.4	94.1	6.62	60.5	14.3	1.88
16×7	50	14.70	16.25	7.073	0.628	0.380	655.4	80.7	6.68	34.8	9.8	1.54
	45	13.24	16.12	7.039	0.563	0.346	583.3	72.4	6.64	30.5	8.7	1.52
	40	11.77	16.00	7.000	0.503	0.307	515.5	64.4	6.62	26.5	7.6	1.50
	36	10.59	15.85	6.992	0.428	0.299	446.3	56.3	6.49	22.1	6.3	1.45
14×16	426	125.25	18.69	16.695	3.033	1.875	6610.3	707.4	7.26	2359.5	282.7	4.34
	398	116.98	18.31	16.590	2.843	1.770	6013.7	656.9	7.17	2169.7	261.6	4.31
	370	108.78	17.94	16.475	2.658	1.655	5454.2	608.1	7.08	1986.0	241.1	4.27
	342	100.59	17.56	16.365	2.468	1.545	4911.5	559.4	6.99	1806.9	220.8	4.24
	314	92.30	17.19	16.235	2.283	1.415	4399.4	511.9	6.90	1631.4	201.0	4.20
	287	84.37	16.81	16.130	2.093	1.310	3912.1	465.5	6.81	1466.5	181.8	4.17
	264	77.63	16.50	16.025	1.938	1.205	3526.0	427.4	6.74	1331.2	166.1	4.14
	246	72.33	16.25	15.945	1.813	1.125	3228.9	397.4	6.68	1226.6	153.9	4.12
	237	69.69	16.12	15.910	1.748	1.090	3080.9	382.2	6.65	1174.8	147.7	4.11
	228	67.06	16.00	15.865	1.688	1.045	2942.4	367.8	6.62	1124.8	141.8	4.10
	219	64.36	15.87	15.825	1.623	1.005	2798.2	352.6	6.59	1073.2	135.6	4.08
	211	62.07	15.75	15.800	1.563	0.980	2671.4	339.2	6.56	1028.6	130.2	4.07
	202	59.39	15.63	15.750	1.503	0.930	2538.8	324.9	6.54	979.7	124.4	4.06
	193	56.73	15.50	15.710	1.438	0.890	2402.4	310.0	6.51	930.1	118.4	4.05
	184	54.07	15.38	15.660	1.378	0.840	2274.8	295.8	6.49	882.7	112.7	4.04
	176	51.73	15.25	15.640	1.313	0.820	2149.6	281.9	6.45	837.9	107.1	4.02
	167	49.09	15.12	15.600	1.248	0.780	2020.8	267.3	6.42	790.2	101.3	4.01
	158	46.47	15.00	15.550	1.188	0.730	1900.6	253.4	6.40	745.0	95.8	4.00
	150	44.08	14.88	15.515	1.128	0.695	1786.9	240.2	6.37	702.5	90.6	3.99
	142	41.85	14.75	15.500	1.063	0.680	1672.2	226.7	6.32	660.1	85.2	3.97
	320^a	94.12	16.81	16.710	2.093	1.890	4141.7	492.8	6.63	1635.1	195.7	4.17
$14 \times 14\frac{1}{2}$	136	39.98	14.75	14.740	1.063	0.660	1593.0	216.0	6.31	567.7	77.0	3.77
	127	37.33	14.62	14.690	0.998	0.610	1476.7	202.0	6.29	527.6	71.8	3.76
	119	34.99	14.50	14.650	0.938	0.570	1373.1	189.4	6.26	491.8	67.1	3.75
	111	32.65	14.37	14.620	0.873	0.540	1266.5	176.3	6.23	454.9	62.2	3.73
	103	30.26	14.25	14.575	0.813	0.495	1165.8	163.6	6.21	419.7	57.6	3.72
	95	27.94	14.12	14.545	0.748	0.465	1063.5	150.6	6.17	383.7	52.8	3.71
	87	25.56	14.00	14.500	0.688	0.420	966.9	138.1	6.15	349.7	48.2	3.70
14×12	84	24.71	14.18	12.023	0.778	0.451	928.4	130.9	6.13	225.5	37.5	3.02
	78	22.94	14.06	12.000	0.718	0.428	851.2	121.1	6.09	206.9	34.5	3.00
14×10	74	21.76	14.19	10.072	0.783	0.450	796.8	112.3	6.05	133.5	26.5	2.48
	68	20.00	14.06	10.040	0.718	0.418	724.1	103.0	6.02	121.2	24.1	2.46
	61	17.94	13.91	10.000	0.643	0.378	641.5	92.2	5.98	107.3	21.5	2.45
14×8	53	15.59	13.94	8.062	0.658	0.370	542.1	77.8	5.90	57.5	14.3	1.92
	48	14.11	13.81	8.031	0.593	0.339	484.9	70.2	5.86	51.3	12.8	1.91
	43	12.65	13.68	8.000	0.528	0.308	429.0	62.7	5.82	45.1	11.3	1.89
$14 \times 6\frac{3}{4}$	38	11.17	14.12	6.776	0.513	0.313	385.3	54.6	5.87	24.6	7.3	1.49

TABLE 1.76 (*Continued*)

Nominal Size (in.)	Weight per Foot (lb)	Area (in.²)	Depth (in.)	Flange Width (in.)	Flange Thickness (in.)	Web Thickness (in.)	Axis X-X I (in.⁴)	Axis X-X S (in.³)	Axis X-X r (in.)	Axis Y-Y I (in.⁴)	Axis Y-Y S (in.³)	Axis Y-Y r (in.)
	34	10.00	14.00	6.750	0.453	0.287	339.2	48.5	5.83	21.3	6.3	1.46
	30	8.81	13.86	6.733	0.383	0.270	289.6	41.8	5.73	17.5	5.2	1.41
12 × 12	190	55.86	14.38	12.670	1.736	1.060	1892.5	263.2	5.82	589.7	93.1	3.25
	161	47.38	13.88	12.515	1.486	0.905	1541.8	222.2	5.70	486.2	77.7	3.20
	133	39.11	13.38	12.365	1.236	0.755	1221.2	182.5	5.59	389.9	63.1	3.16
	120	35.31	13.12	12.320	1.106	0.710	1071.7	163.4	5.51	345.1	56.0	3.13
	106	31.19	12.88	12.230	0.986	0.620	930.7	144.5	5.46	300.9	49.2	3.11
	99	29.09	12.75	12.190	0.921	0.580	858.5	134.7	5.43	278.2	45.7	3.09
	92	27.06	12.62	12.155	0.856	0.545	788.9	125.0	5.40	256.4	42.2	3.08
	85	24.98	12.50	12.105	0.796	0.495	723.3	115.7	5.38	235.5	38.9	3.07
	79	23.22	12.38	12.080	0.736	0.470	663.0	107.1	5.34	216.4	35.8	3.05
	72	21.16	12.25	12.040	0.671	0.430	597.4	97.5	5.31	195.3	32.4	3.04
	65	19.11	12.12	12.000	0.606	0.390	533.4	88.0	5.28	174.6	29.1	3.02
12 × 10	58	17.06	12.19	10.014	0.641	0.359	476.1	78.1	5.28	107.4	21.4	2.51
	53	15.59	12.06	10.000	0.576	0.345	426.2	70.7	5.23	96.1	19.2	2.48
12 × 8	50	14.71	12.19	8.077	0.641	0.371	394.5	64.7	5.18	56.4	14.0	1.96
	45	13.24	12.06	8.042	0.576	0.336	350.8	58.2	5.15	50.0	12.4	1.94
	40	11.77	11.94	8.000	0.516	0.294	310.1	51.9	5.13	44.1	11.0	1.94
12 × 6½	36	10.59	12.24	6.565	0.540	0.305	280.8	45.9	5.15	23.7	7.2	1.50
	31	9.12	12.09	6.525	0.465	0.265	238.4	39.4	5.11	19.8	6.1	1.47
	27	7.97	11.95	6.500	0.400	0.240	204.1	34.1	5.06	16.6	5.1	1.44
10 × 10	112	32.92	11.38	10.415	1.248	0.755	718.7	126.3	4.67	235.4	45.2	2.67
	100	29.43	11.12	10.345	1.118	0.685	625.0	112.4	4.61	206.6	39.9	2.65
	89	26.19	10.88	10.275	0.998	0.615	542.4	99.7	4.55	180.6	35.2	2.63
	77	22.67	10.62	10.195	0.868	0.535	457.2	86.1	4.49	153.4	30.1	2.60
	72	21.18	10.50	10.170	0.808	0.510	420.7	80.1	4.46	141.8	27.9	2.59
	66	19.41	10.38	10.117	0.748	0.457	382.5	73.7	4.44	129.2	25.5	2.58
	60	17.66	10.25	10.075	0.683	0.415	343.7	67.1	4.41	116.5	23.1	2.57
	54	15.88	10.12	10.028	0.618	0.368	305.7	60.4	4.39	103.9	20.7	2.56
	49	14.40	10.00	10.000	0.558	0.340	272.9	54.6	4.35	93.0	18.6	2.54
10 × 8	45	13.24	10.12	8.022	0.618	0.350	248.6	49.1	4.33	53.2	13.3	2.00
	39	11.48	9.94	7.990	0.528	0.318	209.7	42.2	4.27	44.9	11.2	1.98
	33	9.71	9.75	7.964	0.433	0.292	170.9	35.0	4.20	36.5	9.2	1.94
10 × 5¾	29	8.53	10.22	5.799	0.500	0.289	157.3	30.8	4.29	15.2	5.2	1.34
	25	7.35	10.08	5.762	0.430	0.252	133.2	26.4	4.26	12.7	4.4	1.31
	21	6.19	9.90	5.750	0.340	0.240	106.3	21.5	4.14	9.7	3.4	1.25
8 × 8	67	19.70	9.00	8.287	0.933	0.575	271.8	60.4	3.71	88.6	21.4	2.12
	58	17.06	8.75	8.222	0.808	0.510	227.3	52.0	3.65	74.9	18.2	2.10
	48	14.11	8.50	8.117	0.683	0.405	183.7	43.2	3.61	60.9	15.0	2.08
	40	11.76	8.25	8.077	0.558	0.365	146.3	35.5	3.53	49.0	12.1	2.04
	35	10.30	8.12	8.027	0.493	0.315	126.5	31.1	3.50	42.5	10.6	2.03
	31	9.12	8.00	8.000	0.433	0.288	109.7	27.4	3.47	37.0	9.2	2.01
8 × 6½	28	8.23	8.06	6.540	0.463	0.285	97.8	24.3	3.45	21.6	6.6	1.62
	24	7.06	7.93	6.500	0.398	0.245	82.5	20.8	3.42	18.2	5.6	1.61
8 × 5¼	20	5.88	8.14	5.268	0.378	0.248	69.2	17.0	3.43	8.5	3.2	1.20
	17	5.00	8.00	5.250	0.308	0.230	56.4	14.1	3.36	6.7	2.6	1.16

a Column core section.

TABLE 1.77   Properties of American Standard Beams

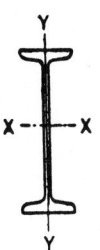

Nominal Size (in.)	Weight per Foot (lb)	Area (in.²)	Depth (in.)	Flange Width (in.)	Flange Thickness (in.)	Web Thickness (in.)	Axis X-X I (in.⁴)	Axis X-X S (in.³)	Axis X-X r (in.)	Axis Y-Y I (in.⁴)	Axis Y-Y S (in.³)	Axis Y-Y r (in.)
$24 \times 7\frac{7}{8}$	120.0	35.13	24.00	8.048	1.102	0.798	3010.8	250.9	9.26	84.9	21.1	1.56
	105.9	30.98	24.00	7.875	1.102	0.625	2811.5	234.3	9.53	78.9	20.0	1.60
24×7	100.0	29.25	24.00	7.247	0.871	0.747	2371.8	197.6	9.05	48.4	13.4	1.29
	90.0	26.30	24.00	7.124	0.871	0.624	2230.1	185.8	9.21	45.5	12.8	1.32
	79.9	23.33	24.00	7.000	0.871	0.500	2087.2	173.9	9.46	42.9	12.2	1.36
20×7	95.0	27.74	20.00	7.200	0.916	0.800	1599.7	160.0	7.59	50.5	14.0	1.35
	85.0	24.80	20.00	7.053	0.916	0.653	1501.7	150.2	7.78	47.0	13.3	1.38
$20 \times 6\frac{1}{4}$	75.0	21.90	20.00	6.391	0.789	0.641	1263.5	126.3	7.60	30.1	9.4	1.17
	65.4	19.08	20.00	6.250	0.789	0.500	1169.5	116.9	7.83	27.9	8.9	1.21
18×6	70.0	20.46	18.00	6.251	0.691	0.711	917.5	101.9	6.70	24.5	7.8	1.09
	54.7	15.94	18.00	6.000	0.691	0.460	795.5	88.4	7.07	21.2	7.1	1.15
$15 \times 5\frac{1}{2}$	50.0	14.59	15.00	5.640	0.622	0.550	481.1	64.2	5.74	16.0	5.7	1.05
	42.9	12.49	15.00	5.500	0.622	0.410	441.8	58.9	5.95	14.6	5.3	1.08
$12 \times 5\frac{1}{4}$	50.0	14.57	12.00	5.477	0.659	0.687	301.6	50.3	4.55	16.0	5.8	1.05
	40.8	11.84	12.00	5.250	0.659	0.460	268.9	44.8	4.77	13.8	5.3	1.08
12×5	35.0	10.20	12.00	5.078	0.544	0.428	227.0	37.8	4.72	10.0	3.9	0.99
	31.8	9.26	12.00	5.000	0.544	0.350	215.8	36.0	4.83	9.5	3.8	1.01
$10 \times 4\frac{5}{8}$	35.0	10.22	10.00	4.944	0.491	0.594	145.8	29.2	3.78	8.5	3.4	0.91
	25.4	7.38	10.00	4.660	0.491	0.310	122.1	24.4	4.07	6.9	3.0	0.97
8×4	23.0	6.71	8.00	4.171	0.425	0.441	64.2	16.0	3.09	4.4	2.1	0.81
	18.4	5.34	8.00	4.000	0.425	0.270	56.9	14.2	3.26	3.8	1.9	0.84
$7 \times 3\frac{5}{8}$	20.0	5.83	7.00	3.860	0.392	0.450	41.9	12.0	2.68	3.1	1.6	0.74
	15.3	4.43	7.00	3.660	0.392	0.250	36.2	10.4	2.86	2.7	1.5	0.78
$6 \times 3\frac{3}{8}$	17.25	5.02	6.00	3.565	0.359	0.465	26.0	8.7	2.28	2.3	1.3	0.68
	12.5	3.61	6.00	3.330	0.359	0.230	21.8	7.3	2.46	1.8	1.1	0.72
5×3	14.75	4.29	5.00	3.284	0.326	0.494	15.0	6.0	1.87	1.7	1.0	0.63
	10.0	2.87	5.00	3.000	0.326	0.210	12.1	4.8	2.05	1.2	0.82	0.65
$4 \times 2\frac{5}{8}$	9.5	2.76	4.00	2.796	0.293	0.326	6.7	3.3	1.56	0.91	0.65	0.58
	7.7	2.21	4.00	2.660	0.293	0.190	6.0	3.0	1.64	0.77	0.58	0.59
$3 \times 2\frac{3}{8}$	7.5	2.17	3.00	2.509	0.260	0.349	2.9	1.9	1.15	0.59	0.47	0.52
	5.7	1.64	3.00	2.330	0.260	0.170	2.5	1.7	1.23	0.46	0.40	0.53

TABLE 1.78 Properties of American Standard Channels

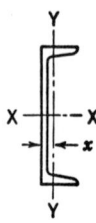

Nominal Size (in.)	Weight per Foot (lb)	Area (in.²)	Depth (in.)	Width (in.)	Flange Average Thickness (in.)	Web Thickness (in.)	Axis X-X I (in.⁴)	S (in.³)	r (in.)	Axis Y-Y I (in.⁴)	S (in.³)	r (in.)	x (in.)
18×4^a	58.0	16.98	18.00	4.200	0.625	0.700	670.7	74.5	6.29	18.5	5.6	1.04	0.88
	51.9	15.18	18.00	4.100	0.625	0.600	622.1	69.1	6.40	17.1	5.3	1.06	0.87
	45.8	13.38	18.00	4.000	0.625	0.500	573.5	63.7	6.55	15.8	5.1	1.09	0.89
	42.7	12.48	18.00	3.950	0.625	0.450	549.2	61.0	6.64	15.0	4.9	1.10	0.90
$15 \times 3\frac{3}{8}$	50.0	14.64	15.00	3.716	0.650	0.716	401.4	53.6	5.24	11.2	3.8	0.87	0.80
	40.0	11.70	15.00	3.520	0.650	0.520	346.3	46.2	5.44	9.3	3.4	0.89	0.78
	33.9	9.90	15.00	3.400	0.650	0.400	312.6	41.7	5.62	8.2	3.2	0.91	0.79
12×3	30.0	8.79	12.00	3.170	0.501	0.510	161.2	26.9	4.28	5.2	2.1	0.77	0.68
	25.0	7.32	12.00	3.047	0.501	0.387	143.5	23.9	4.43	4.5	1.9	0.79	0.68
	20.7	6.03	12.00	2.940	0.501	0.280	128.1	21.4	4.61	3.9	1.7	0.81	0.70
$10 \times 2\frac{5}{8}$	30.0	8.80	10.00	3.033	0.436	0.673	103.0	20.6	3.42	4.0	1.7	0.67	0.65
	25.0	7.33	10.00	2.886	0.436	0.526	90.7	18.1	3.52	3.4	1.5	0.68	0.62
	20.0	5.86	10.00	2.739	0.436	0.379	78.5	15.7	3.66	2.8	1.3	0.70	0.61
	15.3	4.47	10.00	2.600	0.436	0.240	66.9	13.4	3.87	2.3	1.2	0.72	0.64
$9 \times 2\frac{1}{2}$	20.0	5.86	9.00	2.648	0.413	0.448	60.6	13.5	3.22	2.4	1.2	0.65	0.59
	15.0	4.39	9.00	2.485	0.413	0.285	50.7	11.3	3.40	1.9	1.0	0.67	0.59
	13.4	3.89	9.00	2.430	0.413	0.230	47.3	10.5	3.49	1.8	0.97	0.67	0.61
$8 \times 2\frac{1}{4}$	18.75	5.49	8.00	2.527	0.390	0.487	43.7	10.9	2.82	2.0	1.0	0.60	0.57
	13.75	4.02	8.00	2.343	0.390	0.303	35.8	9.0	2.99	1.5	0.86	0.62	0.56
	11.5	3.36	8.00	2.260	0.390	0.220	32.3	8.1	3.10	1.3	0.79	0.63	0.58
$7 \times 2\frac{1}{8}$	14.75	4.32	7.00	2.299	0.366	0.419	27.1	7.7	2.51	1.4	0.79	0.57	0.53
	12.25	3.58	7.00	2.194	0.366	0.314	24.1	6.9	2.59	1.2	0.71	0.58	0.53
	9.8	2.85	7.00	2.090	0.366	0.210	21.1	6.0	2.72	0.98	0.63	0.59	0.55
6×2	13.0	3.81	6.00	2.157	0.343	0.437	17.3	5.8	2.13	1.1	0.65	0.53	0.52
	10.5	3.07	6.00	2.034	0.343	0.314	15.1	5.0	2.22	0.87	0.57	0.53	0.50
	8.2	2.39	6.00	1.920	0.343	0.200	13.0	4.3	2.34	0.70	0.50	0.54	0.52
$5 \times 1\frac{3}{4}$	9.0	2.63	5.00	1.885	0.320	0.325	8.8	3.5	1.83	0.64	0.45	0.49	0.48
	6.7	1.95	5.00	1.750	0.320	0.190	7.4	3.0	1.95	0.48	0.38	0.50	0.49
$4 \times 1\frac{5}{8}$	7.25	2.12	4.00	1.720	0.296	0.320	4.5	2.3	1.47	0.44	0.35	0.46	0.46
	5.4	1.56	4.00	1.580	0.296	0.180	3.8	1.9	1.56	0.32	0.29	0.45	0.46
$3 \times 1\frac{1}{2}$	6.0	1.75	3.00	1.596	0.273	0.356	2.1	1.4	1.08	0.31	0.27	0.42	0.46
	5.0	1.46	3.00	1.498	0.273	0.258	1.8	1.2	1.12	0.25	0.24	0.41	0.44
	4.1	1.19	3.00	1.410	0.273	0.170	1.6	1.1	1.17	0.20	0.21	0.41	0.44

^{a}Car and Shipbuilding Channel; not an American Standard.

TABLE 1.79 Properties of Angles with Equal Legs

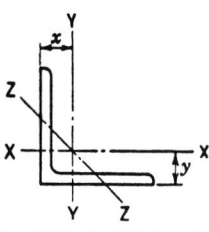

Size (in.)	Thick-ness (in.)	Weight per Foot (lb)	Area (in.²)	Axis X-X and Axis Y-Y				Axis Z-Z
				I (in.⁴)	S (in.³)	r (in.)	x or y (in.)	r (in.)
8 × 8	$1\frac{1}{8}$	56.9	16.73	98.0	17.5	2.42	2.41	1.56
	1	51.0	15.00	89.0	15.8	2.44	2.37	1.56
	$\frac{7}{8}$	45.0	13.23	79.6	14.0	2.45	2.32	1.57
	$\frac{3}{4}$	38.9	11.44	69.7	12.2	2.47	2.28	1.57
	$\frac{5}{8}$	32.7	9.61	59.4	10.3	2.49	2.23	1.58
	$\frac{9}{16}$	29.6	8.68	54.1	9.3	2.50	2.21	1.58
	$\frac{1}{2}$	26.4	7.75	48.6	8.4	2.50	2.19	1.59
6 × 6	1	37.4	11.00	35.5	8.6	1.80	1.86	1.17
	$\frac{7}{8}$	33.1	9.73	31.9	7.6	1.81	1.82	1.17
	$\frac{3}{4}$	28.7	8.44	28.2	6.7	1.83	1.78	1.17
	$\frac{5}{8}$	24.2	7.11	24.2	5.7	1.84	1.73	1.18
	$\frac{9}{16}$	21.9	6.43	22.1	5.1	1.85	1.71	1.18
	$\frac{1}{2}$	19.6	5.75	19.9	4.6	1.86	1.68	1.18
	$\frac{7}{16}$	17.2	5.06	17.7	4.1	1.87	1.66	1.19
	$\frac{3}{8}$	14.9	4.36	15.4	3.5	1.88	1.64	1.19
	$\frac{5}{16}$	12.5	3.66	13.0	3.0	1.89	1.61	1.19
5 × 5	$\frac{7}{8}$	27.2	7.98	17.8	5.2	1.49	1.57	0.97
	$\frac{3}{4}$	23.6	6.94	15.7	4.5	1.51	1.52	0.97
	$\frac{5}{8}$	20.0	5.86	13.6	3.9	1.52	1.48	0.98
	$\frac{1}{2}$	16.2	4.75	11.3	3.2	1.54	1.43	0.98
	$\frac{7}{16}$	14.3	4.18	10.0	2.8	1.55	1.41	0.98
	$\frac{3}{8}$	12.3	3.61	8.7	2.4	1.56	1.39	0.99
	$\frac{5}{16}$	10.3	3.03	7.4	2.0	1.57	1.37	0.99
4 × 4	$\frac{3}{4}$	18.5	5.44	7.7	2.8	1.19	1.27	0.78
	$\frac{5}{8}$	15.7	4.61	6.7	2.4	1.20	1.23	0.78
	$\frac{1}{2}$	12.8	3.75	5.6	2.0	1.22	1.18	0.78
	$\frac{7}{16}$	11.3	3.31	5.0	1.8	1.23	1.16	0.78
	$\frac{3}{8}$	9.8	2.86	4.4	1.5	1.23	1.14	0.79
	$\frac{5}{16}$	8.2	2.40	3.7	1.3	1.24	1.12	0.79
	$\frac{1}{4}$	6.6	1.94	3.0	1.1	1.25	1.09	0.80
$3\frac{1}{2} × 3\frac{1}{2}$	$\frac{1}{2}$	11.1	3.25	3.6	1.5	1.06	1.06	0.68
	$\frac{7}{16}$	9.8	2.87	3.3	1.3	1.07	1.04	0.68
	$\frac{3}{8}$	8.5	2.48	2.9	1.2	1.07	1.01	0.69
	$\frac{5}{16}$	7.2	2.09	2.5	0.98	1.08	0.99	0.69
	$\frac{1}{4}$	5.8	1.69	2.0	0.79	1.09	0.97	0.69
3 × 3	$\frac{1}{2}$	9.4	2.75	2.2	1.1	0.90	0.93	0.58
	$\frac{7}{16}$	8.3	2.43	2.0	0.95	0.91	0.91	0.58
	$\frac{3}{8}$	7.2	2.11	1.8	0.83	0.91	0.89	0.58
	$\frac{5}{16}$	6.1	1.78	1.5	0.71	0.92	0.87	0.59
	$\frac{1}{4}$	4.9	1.44	1.2	0.58	0.93	0.84	0.59
	$\frac{3}{16}$	3.71	1.09	0.96	0.44	0.94	0.82	0.59

TABLE 1.79 *(Continued)*

Size (in.)	Thickness (in.)	Weight per Foot (lb)	Area (in.²)	Axis X-X and Axis Y-Y I (in.⁴)	S (in.³)	r (in.)	x or y (in.)	Axis Z-Z r (in.)
$2\frac{1}{2} \times 2\frac{1}{2}$	$\frac{1}{2}$	7.7	2.25	1.2	0.72	0.74	0.81	0.49
	$\frac{3}{8}$	5.9	1.73	0.98	0.57	0.75	0.76	0.49
	$\frac{5}{16}$	5.0	1.47	0.85	0.48	0.76	0.74	0.49
	$\frac{1}{4}$	4.1	1.19	0.70	0.39	0.77	0.72	0.49
	$\frac{3}{16}$	3.07	0.90	0.55	0.30	0.78	0.69	0.49
2×2	$\frac{3}{8}$	4.7	1.36	0.48	0.35	0.59	0.64	0.39
	$\frac{5}{16}$	3.92	1.15	0.42	0.30	0.60	0.61	0.39
	$\frac{1}{4}$	3.19	0.94	0.35	0.25	0.61	0.59	0.39
	$\frac{3}{16}$	2.44	0.71	0.27	0.19	0.62	0.57	0.39
	$\frac{1}{8}$	1.65	0.48	0.19	0.13	0.63	0.55	0.40
$1\frac{3}{4} \times 1\frac{3}{4}$	$\frac{1}{4}$	2.77	0.81	0.23	0.19	0.53	0.53	0.34
	$\frac{3}{16}$	2.12	0.62	0.18	0.14	0.54	0.51	0.34
	$\frac{1}{8}$	1.44	0.42	0.13	0.10	0.55	0.48	0.35
$1\frac{1}{2} \times 1\frac{1}{2}$	$\frac{1}{4}$	2.34	0.69	0.14	0.13	0.45	0.47	0.29
	$\frac{3}{16}$	1.80	0.53	0.11	0.10	0.46	0.44	0.29
	$\frac{1}{8}$	1.23	0.36	0.08	0.07	0.47	0.42	0.30
$1\frac{1}{4} \times 1\frac{1}{4}$	$\frac{1}{4}$	1.92	0.56	0.08	0.09	0.37	0.40	0.24
	$\frac{3}{16}$	1.48	0.43	0.06	0.07	0.38	0.38	0.24
	$\frac{1}{8}$	1.01	0.30	0.04	0.05	0.38	0.36	0.25
1×1	$\frac{1}{4}$	1.49	0.44	0.04	0.06	0.29	0.34	0.20
	$\frac{3}{16}$	1.16	0.34	0.03	0.04	0.30	0.32	0.19
	$\frac{1}{8}$	0.80	0.23	0.02	0.03	0.30	0.30	0.20

TABLE 1.80 Properties of Angles with Unequal Legs

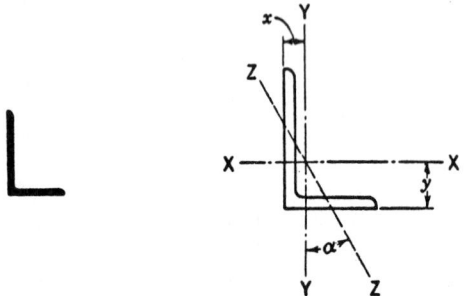

Size (in.)	Thickness (in.)	Weight per Foot (lb)	Area (in.²)	Axis X-X I (in.⁴)	S (in.³)	r (in.)	y (in.)	Axis Y-Y I (in.⁴)	S (in.³)	r (in.)	x (in.)	Axis Z-Z r (in.)	tan α
9×4	1	40.8	12.00	97.0	17.6	2.84	3.50	12.0	4.0	1.00	1.00	0.83	0.203
	$\frac{7}{8}$	36.1	10.61	86.8	15.7	2.86	3.45	10.8	3.6	1.01	0.95	0.84	0.208
	$\frac{3}{4}$	31.3	9.19	76.1	13.6	2.88	3.41	9.6	3.1	1.02	0.91	0.84	0.212
	$\frac{5}{8}$	26.3	7.73	64.9	11.5	2.90	3.36	8.3	2.6	1.04	0.86	0.85	0.216
	$\frac{9}{16}$	23.8	7.00	59.1	10.4	2.91	3.33	7.6	2.4	1.04	0.83	0.85	0.218
	$\frac{1}{2}$	21.3	6.25	53.2	9.3	2.92	3.31	6.9	2.2	1.05	0.81	0.85	0.220
8×6	1	44.2	13.00	80.8	15.1	2.49	2.65	38.8	8.9	1.73	1.65	1.28	0.543
	$\frac{7}{8}$	39.1	11.48	72.3	13.4	2.51	2.61	34.9	7.9	1.74	1.61	1.28	0.547

TABLE 1.80   (*Continued*)

Size (in.)	Thickness (in.)	Weight per Foot (lb)	Area (in.^{2})	Axis X-X I (in.^{4})	S (in.^{3})	r (in.)	y (in.)	Axis Y-Y I (in.^{4})	S (in.^{3})	r (in.)	x (in.)	Axis Z-Z r (in.)	tan α
	$\frac{3}{4}$	33.8	9.94	63.4	11.7	2.53	2.56	30.7	6.9	1.76	1.56	1.29	0.551
	$\frac{5}{8}$	28.5	8.36	54.1	9.9	2.54	2.52	26.3	5.9	1.77	1.52	1.29	0.554
	$\frac{9}{16}$	25.7	7.56	49.3	9.0	2.55	2.50	24.0	5.3	1.78	1.50	1.30	0.556
	$\frac{1}{2}$	23.0	6.75	44.3	8.0	2.56	2.47	21.7	4.8	1.79	1.47	1.30	0.558
	$\frac{7}{16}$	20.2	5.93	39.2	7.1	2.57	2.45	19.3	4.2	1.80	1.45	1.31	0.560
8 × 4	1	37.4	11.00	69.6	14.1	2.52	3.05	11.6	3.9	1.03	1.05	0.85	0.247
	$\frac{7}{8}$	33.1	9.73	62.5	12.5	2.53	3.00	10.5	3.5	1.04	1.00	0.85	0.253
	$\frac{3}{4}$	28.7	8.44	54.9	10.9	2.55	2.95	9.4	3.1	1.05	0.95	0.85	0.258
	$\frac{5}{8}$	24.2	7.11	46.9	9.2	2.57	2.91	8.1	2.6	1.07	0.91	0.86	0.262
	$\frac{9}{16}$	21.9	6.43	42.8	8.4	2.58	2.88	7.4	2.4	1.07	0.88	0.86	0.265
	$\frac{1}{2}$	19.6	5.75	38.5	7.5	2.59	2.86	6.7	2.2	1.08	0.86	0.86	0.267
	$\frac{7}{16}$	17.2	5.06	34.1	6.6	2.60	2.83	6.0	1.9	1.09	0.83	0.87	0.269
7 × 4	$\frac{7}{8}$	30.2	8.86	42.9	9.7	2.20	2.55	10.2	3.5	1.07	1.05	0.86	0.318
	$\frac{3}{4}$	26.2	7.69	37.8	8.4	2.22	2.51	9.1	3.0	1.09	1.01	0.86	0.324
	$\frac{5}{8}$	22.1	6.48	32.4	7.1	2.24	2.46	7.8	2.6	1.10	0.96	0.86	0.329
	$\frac{9}{16}$	20.0	5.87	29.6	6.5	2.24	2.44	7.2	2.4	1.11	0.94	0.87	0.332
	$\frac{1}{2}$	17.9	5.25	26.7	5.8	2.25	2.42	6.5	2.1	1.11	0.92	0.87	0.335
	$\frac{7}{16}$	15.8	4.62	23.7	5.1	2.26	2.39	5.8	1.9	1.12	0.89	0.88	0.337
	$\frac{3}{8}$	13.6	3.98	20.6	4.4	2.27	2.37	5.1	1.6	1.13	0.87	0.88	0.339
6 × 4	$\frac{7}{8}$	27.2	7.98	27.7	7.2	1.86	2.12	9.8	3.4	1.11	1.12	0.86	0.421
	$\frac{3}{4}$	23.6	6.94	24.5	6.3	1.88	2.08	8.7	3.0	1.12	1.08	0.86	0.428
	$\frac{5}{8}$	20.0	5.86	21.1	5.3	1.90	2.03	7.5	2.5	1.13	1.03	0.86	0.435
	$\frac{9}{16}$	18.1	5.31	19.3	4.8	1.90	2.01	6.9	2.3	1.14	1.01	0.87	0.438
	$\frac{1}{2}$	16.2	4.75	17.4	4.3	1.91	1.99	6.3	2.1	1.15	0.99	0.87	0.440
	$\frac{7}{16}$	14.3	4.18	15.5	3.8	1.92	1.96	5.6	1.9	1.16	0.96	0.87	0.443
	$\frac{3}{8}$	12.3	3.61	13.5	3.3	1.93	1.94	4.9	1.6	1.17	0.94	0.88	0.446
	$\frac{5}{16}$	10.3	3.03	11.4	2.8	1.94	1.92	4.2	1.4	1.17	0.92	0.88	0.449
6 × 3$\frac{1}{2}$	$\frac{1}{2}$	15.3	4.50	16.6	4.2	1.92	2.08	4.3	1.6	0.97	0.83	0.76	0.344
	$\frac{3}{8}$	11.7	3.42	12.9	3.2	1.94	2.04	3.3	1.2	0.99	0.79	0.77	0.350
	$\frac{5}{16}$	9.8	2.87	10.9	2.7	1.95	2.01	2.9	1.0	1.00	0.76	0.77	0.352
	$\frac{1}{4}$	7.9	2.31	8.9	2.2	1.96	1.99	2.3	0.85	1.01	0.74	0.78	0.355
5 × 3$\frac{1}{2}$	$\frac{3}{4}$	19.8	5.81	13.9	4.3	1.55	1.75	5.6	2.2	0.98	1.00	0.75	0.464
	$\frac{5}{8}$	16.8	4.92	12.0	3.7	1.56	1.70	4.8	1.9	0.99	0.95	0.75	0.472
	$\frac{1}{2}$	13.6	4.00	10.0	3.0	1.58	1.66	4.1	1.6	1.01	0.91	0.75	0.479
	$\frac{7}{16}$	12.0	3.53	8.9	2.6	1.59	1.63	3.6	1.4	1.01	0.88	0.76	0.482
	$\frac{3}{8}$	10.4	3.05	7.8	2.3	1.60	1.61	3.2	1.2	1.02	0.86	0.76	0.486
	$\frac{5}{16}$	8.7	2.56	6.6	1.9	1.61	1.59	2.7	1.0	1.03	0.84	0.76	0.489
	$\frac{1}{4}$	7.0	2.06	5.4	1.6	1.61	1.56	2.2	0.83	1.04	0.81	0.76	0.492
5 × 3	$\frac{1}{2}$	12.8	3.75	9.5	2.9	1.59	1.75	2.6	1.1	0.83	0.75	0.65	0.357
	$\frac{7}{16}$	11.3	3.31	8.4	2.6	1.60	1.73	2.3	1.0	0.84	0.73	0.65	0.361
	$\frac{3}{8}$	9.8	2.86	7.4	2.2	1.61	1.70	2.0	0.89	0.84	0.70	0.65	0.364
	$\frac{5}{16}$	8.2	2.40	6.3	1.9	1.61	1.68	1.8	0.75	0.85	0.68	0.66	0.368
	$\frac{1}{4}$	6.6	1.94	5.1	1.5	1.62	1.66	1.4	0.61	0.86	0.66	0.66	0.371
4 × 3$\frac{1}{2}$	$\frac{5}{8}$	14.7	4.30	6.4	2.4	1.22	1.29	4.5	1.8	1.03	1.04	0.72	0.745
	$\frac{1}{2}$	11.9	3.50	5.3	1.9	1.23	1.25	3.8	1.5	1.04	1.00	0.72	0.750
	$\frac{7}{16}$	10.6	3.09	4.8	1.7	1.24	1.23	3.4	1.4	1.05	0.98	0.72	0.753
	$\frac{3}{8}$	9.1	2.67	4.2	1.5	1.25	1.21	3.0	1.2	1.06	0.96	0.73	0.755
	$\frac{5}{16}$	7.7	2.25	3.6	1.3	1.26	1.18	2.6	1.0	1.07	0.93	0.73	0.757
	$\frac{1}{4}$	6.2	1.81	2.9	1.0	1.27	1.16	2.1	0.81	1.07	0.91	0.73	0.759

TABLE 1.80 (*Continued*)

Size (in.)	Thickness (in.)	Weight per Foot (lb)	Area (in.^{2})	Axis X-X				Axis Y-Y				Axis Z-Z	
				I (in.^{4})	S (in.^{3})	r (in.)	y (in.)	I (in.^{4})	S (in.^{3})	r (in.)	x (in.)	r (in.)	$\tan \alpha$
4×3	$\frac{5}{8}$	13.6	3.98	6.0	2.3	1.23	1.37	2.9	1.4	0.85	0.87	0.64	0.534
	$\frac{1}{2}$	11.1	3.25	5.1	1.9	1.25	1.33	2.4	1.1	0.86	0.83	0.64	0.543
	$\frac{7}{16}$	9.8	2.87	4.5	1.7	1.25	1.30	2.2	1.0	0.87	0.80	0.64	0.547
	$\frac{3}{8}$	8.5	2.48	4.0	1.5	1.26	1.28	1.9	0.87	0.88	0.78	0.64	0.551
	$\frac{5}{16}$	7.2	2.09	3.4	1.2	1.27	1.26	1.7	0.73	0.89	0.76	0.65	0.554
	$\frac{1}{4}$	5.8	1.69	2.8	1.0	1.28	1.24	1.4	0.60	0.90	0.74	0.65	0.558
$3\frac{1}{2} \times 3$	$\frac{1}{2}$	10.2	3.00	3.5	1.5	1.07	1.13	2.3	1.1	0.88	0.88	0.62	0.714
	$\frac{7}{16}$	9.1	2.65	3.1	1.3	1.08	1.10	2.1	0.98	0.89	0.85	0.62	0.718
	$\frac{3}{8}$	7.9	2.30	2.7	1.1	1.09	1.08	1.9	0.85	0.90	0.83	0.62	0.721
	$\frac{5}{16}$	6.6	1.93	2.3	0.95	1.10	1.06	1.6	0.72	0.90	0.81	0.63	0.724
	$\frac{1}{4}$	5.4	1.56	1.9	0.78	1.11	1.04	1.3	0.59	0.91	0.79	0.63	0.727
$3\frac{1}{2} \times 2\frac{1}{2}$	$\frac{1}{2}$	9.4	2.75	3.2	1.4	1.09	1.20	1.4	0.76	0.70	0.70	0.53	0.486
	$\frac{7}{16}$	8.3	2.43	2.9	1.3	1.09	1.18	1.2	0.68	0.71	0.68	0.54	0.491
	$\frac{3}{8}$	7.2	2.11	2.6	1.1	1.10	1.16	1.1	0.59	0.72	0.66	0.54	0.496
	$\frac{5}{16}$	6.1	1.78	2.2	0.93	1.11	1.14	0.94	0.50	0.73	0.64	0.54	0.501
	$\frac{1}{4}$	4.9	1.44	1.8	0.75	1.12	1.11	0.78	0.41	0.74	0.61	0.54	0.506
$3 \times 2\frac{1}{2}$	$\frac{1}{2}$	8.5	2.50	2.1	1.0	0.91	1.00	1.3	0.74	0.72	0.75	0.52	0.667
	$\frac{7}{16}$	7.6	2.21	1.9	0.93	0.92	0.98	1.2	0.66	0.73	0.73	0.52	0.672
	$\frac{3}{8}$	6.6	1.92	1.7	0.81	0.93	0.96	1.0	0.58	0.74	0.71	0.52	0.676
	$\frac{5}{16}$	5.6	1.62	1.4	0.69	0.94	0.93	0.90	0.49	0.74	0.68	0.53	0.680
	$\frac{1}{4}$	4.5	1.31	1.2	0.56	0.95	0.91	0.74	0.40	0.75	0.66	0.53	0.684
3×2	$\frac{1}{2}$	7.7	2.25	1.9	1.0	0.92	1.08	0.67	0.47	0.55	0.58	0.43	0.414
	$\frac{7}{16}$	6.8	2.00	1.7	0.89	0.93	1.06	0.61	0.42	0.55	0.56	0.43	0.421
	$\frac{3}{8}$	5.9	1.73	1.5	0.78	0.94	1.04	0.54	0.37	0.56	0.54	0.43	0.428
	$\frac{5}{16}$	5.0	1.47	1.3	0.66	0.95	1.02	0.47	0.32	0.57	0.52	0.43	0.435
	$\frac{1}{4}$	4.1	1.19	1.1	0.54	0.95	0.99	0.39	0.26	0.57	0.49	0.43	0.440
	$\frac{3}{16}$	3.07	0.90	0.84	0.41	0.97	0.97	0.31	0.20	0.58	0.47	0.44	0.446
$2\frac{1}{2} \times 2$	$\frac{3}{8}$	5.3	1.55	0.91	0.55	0.77	0.83	0.51	0.36	0.58	0.58	0.42	0.614
	$\frac{5}{16}$	4.5	1.31	0.79	0.47	0.78	0.81	0.45	0.31	0.58	0.56	0.42	0.620
	$\frac{1}{4}$	3.62	1.06	0.65	0.38	0.78	0.79	0.37	0.25	0.59	0.54	0.42	0.626
	$\frac{3}{16}$	2.75	0.81	0.51	0.29	0.79	0.76	0.29	0.20	0.60	0.51	0.43	0.631
$2\frac{1}{2} \times 1\frac{1}{2}$	$\frac{3}{8}$	4.7	1.36	0.82	0.52	0.78	0.92	0.22	0.20	0.40	0.42	0.32	0.340
	$\frac{5}{16}$	3.92	1.15	0.71	0.44	0.79	0.90	0.19	0.17	0.41	0.40	0.32	0.349
	$\frac{1}{4}$	3.19	0.94	0.59	0.36	0.79	0.88	0.16	0.14	0.41	0.38	0.32	0.357
	$\frac{3}{16}$	2.44	0.72	0.46	0.28	0.80	0.85	0.13	0.11	0.42	0.35	0.33	0.364
$2 \times 1\frac{1}{2}$	$\frac{1}{4}$	2.77	0.81	0.32	0.24	0.62	0.66	0.15	0.14	0.43	0.41	0.32	0.543
	$\frac{3}{16}$	2.12	0.62	0.25	0.18	0.63	0.64	0.12	0.11	0.44	0.39	0.32	0.551
	$\frac{1}{8}$	1.44	0.42	0.17	0.13	0.64	0.62	0.09	0.08	0.45	0.37	0.33	0.558
$1\frac{3}{4} \times 1\frac{1}{4}$	$\frac{1}{4}$	2.34	0.69	0.20	0.18	0.54	0.60	0.09	0.10	0.35	0.35	0.27	0.486
	$\frac{3}{16}$	1.80	0.53	0.16	0.14	0.55	0.58	0.07	0.08	0.36	0.33	0.27	0.496
	$\frac{1}{8}$	1.23	0.36	0.11	0.09	0.56	0.56	0.05	0.05	0.37	0.31	0.27	0.506

TABLE 1.81   Properties and Dimensions of Tees

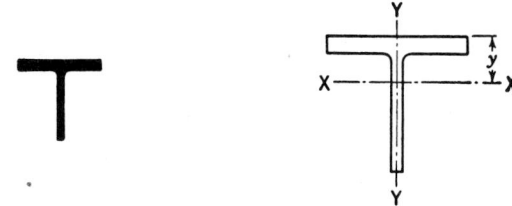

Tees are seldom used as structural framing members. When so used they are generally employed on short spans in flexure. These tables list a few selected sizes, the range of whose section moduli will cover all ordinary conditions. For sizes not listed, the catalogs of the respective rolling mills should be consulted.

Section Number	Weight per Foot (lb)	Area (in.^{2})	Depth of Tee (in.)	Flange Width (in.)	Flange Average Thickness (in.)	Stem Thickness (in.)	Axis X-X I (in.^{4})	S (in.^{3})	r (in.)	y (in.)	Axis Y-Y I (in.^{4})	S (in.^{3})	r (in.)
ST 18 WF[a]	150	44.09	18.36	16.655	1.680	0.945	1222.7	85.9	5.27	4.13	612.6	73.6	3.73
	140	41.16	18.25	16.595	1.570	0.885	1133.3	79.9	5.25	4.07	563.7	67.9	3.70
	130	38.28	18.12	16.555	1.440	0.845	1059.2	75.4	5.26	4.07	510.3	61.6	3.65
	122.5	36.01	18.03	16.512	1.350	0.802	994.3	71.1	5.25	4.04	472.3	57.2	3.62
	115	33.86	17.94	16.475	1.260	0.765	935.8	67.2	5.26	4.02	435.5	52.9	3.59
ST 18 WF	97	28.56	18.24	12.117	1.260	0.770	904.0	67.3	5.63	4.81	177.7	29.3	2.49
	91	26.77	18.16	12.072	1.180	0.725	844.0	63.0	5.61	4.77	163.9	27.1	2.47
	85	24.99	18.08	12.027	1.100	0.680	784.7	58.8	5.60	4.74	150.3	25.0	2.45
	80	23.54	18.00	12.000	1.020	0.653	741.0	56.0	5.61	4.76	137.7	22.9	2.42
	75	22.08	17.92	11.972	0.940	0.625	696.7	53.0	5.62	4.79	125.2	20.9	2.38
ST 16 WF	120	35.26	16.75	15.865	1.400	0.830	822.5	63.2	4.83	3.73	437.2	55.1	3.52
	110	32.36	16.63	15.810	1.275	0.775	754.1	58.4	4.83	3.71	391.2	49.5	3.48
	100	29.40	16.50	15.750	1.150	0.715	683.6	53.3	4.82	3.67	345.8	43.9	3.43
ST 16 WF	76	22.35	16.75	11.565	1.055	0.635	591.9	47.4	5.15	4.26	128.1	22.1	2.39
	70.5	20.76	16.66	11.535	0.960	0.603	551.8	44.7	5.16	4.30	114.9	19.9	2.35
	65	19.13	16.55	11.510	0.855	0.580	513.0	42.1	5.18	4.37	100.7	17.5	2.29
ST 15 WF	105	30.89	15.19	15.105	1.315	0.775	578.0	48.7	4.33	3.31	354.0	46.9	3.38
	95	27.95	15.06	15.040	1.185	0.710	520.4	44.1	4.31	3.26	312.3	41.5	3.34
	86	25.32	14.94	14.985	1.065	0.655	471.0	40.2	4.31	3.23	275.1	36.7	3.30
ST 15 WF	66	19.41	15.15	10.551	1.000	0.615	420.7	37.4	4.66	3.90	92.5	17.5	2.18
	62	18.22	15.08	10.521	0.930	0.585	394.8	35.3	4.65	3.90	84.8	16.1	2.16
	58.0	17.07	15.00	10.500	0.850	0.564	371.8	33.6	4.67	3.94	76.6	14.6	2.12
	54.0	15.88	14.91	10.484	0.760	0.548	349.5	32.1	4.69	4.03	67.6	12.9	2.06
ST 13 WF	88.5	26.05	13.66	14.090	1.190	0.725	391.8	36.7	3.88	2.97	259.4	36.8	3.16
	80	23.72	13.54	14.023	1.075	0.658	351.4	33.1	3.87	2.91	229.0	32.7	3.12
	72.5	21.34	13.44	13.965	0.975	0.600	316.3	29.9	3.85	2.85	203.5	29.1	3.09
ST 13 WF	57	16.77	13.64	10.070	0.932	0.570	288.9	28.3	4.15	3.42	74.8	14.9	2.11
	51	15.01	13.53	10.018	0.827	0.518	257.7	25.4	4.14	3.39	64.8	12.9	2.08
	47	13.83	13.45	9.990	0.747	0.490	238.5	23.7	4.15	3.41	57.5	11.5	2.04
ST 12 WF	80	23.54	12.36	14.091	1.135	0.656	271.6	27.6	3.40	2.51	246.3	35.0	3.23
	72.5	21.31	12.24	14.043	1.020	0.608	246.2	25.2	3.40	2.48	217.1	30.9	3.19
	65	19.11	12.13	14.000	0.900	0.565	222.6	23.1	3.41	2.47	187.6	26.8	3.13
ST 12 WF	60	17.64	12.16	12.088	0.930	0.556	213.6	22.4	3.48	2.62	127.0	21.0	2.68
	55	16.18	12.08	12.042	0.855	0.510	195.2	20.5	3.47	2.57	114.5	19.0	2.66
	50	14.71	12.00	12.000	0.775	0.468	176.7	18.7	3.46	2.54	101.8	17.0	2.63
ST 12 WF	47	13.81	12.15	9.061	0.872	0.516	185.9	20.3	3.67	2.99	51.1	11.3	1.92
	42	12.35	12.04	9.015	0.772	0.470	165.9	18.3	3.66	2.97	44.2	9.8	1.89
	38	11.18	11.95	8.985	0.682	0.440	151.1	16.9	3.68	3.00	38.3	8.5	1.85
ST 10 WF	71	20.88	10.73	13.132	1.095	0.659	177.3	20.8	2.91	2.18	193.0	29.4	3.04
	63.5	18.67	10.62	13.061	0.985	0.588	155.8	18.3	2.89	2.11	169.3	25.9	3.01
	56	16.47	10.50	13.000	0.865	0.527	136.4	16.2	2.88	2.06	144.8	22.3	2.96

TABLE 1.81 (*Continued*)

Section Number	Weight per Foot (lb)	Area (in.²)	Depth of Tee (in.)	Flange Width (in.)	Flange Average Thickness (in.)	Stem Thickness (in.)	Axis X-X I (in.⁴)	S (in.³)	r (in.)	y (in.)	Axis Y-Y I (in.⁴)	S (in.³)	r (in.)
ST 10 WF[a]	48	14.11	10.57	9.038	0.935	0.575	137.1	17.1	3.11	2.55	54.7	12.1	1.97
	41	12.05	10.43	8.962	0.795	0.499	115.4	14.5	3.09	2.48	44.8	10.0	1.93
ST 10 WF	36.5	10.73	10.62	8.295	0.740	0.455	110.2	13.7	3.21	2.60	33.1	7.98	1.76
	34	10.01	10.57	8.270	0.685	0.430	102.8	12.9	3.20	2.59	30.2	7.30	1.74
	31	9.12	10.49	8.240	0.615	0.400	93.7	11.9	3.21	2.59	26.6	6.45	1.71
ST 9 WF	57	16.77	9.24	11.833	0.991	0.595	102.6	13.9	2.47	1.85	127.8	21.6	2.76
	52.5	15.43	9.16	11.792	0.911	0.554	93.9	12.8	2.47	1.82	115.5	19.6	2.73
	48	14.11	9.08	11.750	0.831	0.512	85.3	11.7	2.46	1.78	103.4	17.6	2.71
ST 9 WF	42.5	12.49	9.16	8.838	0.911	0.526	84.4	11.9	2.60	2.05	49.7	11.3	2.00
	38.5	11.32	9.08	8.787	0.831	0.475	75.3	10.6	2.58	1.99	44.3	10.1	1.98
	35	10.28	9.00	8.750	0.751	0.438	68.1	9.67	2.57	1.96	39.2	8.97	1.95
	32	9.40	8.94	8.715	0.686	0.403	61.8	8.82	2.56	1.93	35.2	8.07	1.93
ST 9 WF	30	8.82	9.12	7.558	0.695	0.416	64.8	9.32	2.71	2.17	23.5	6.23	1.63
	27.5	8.09	9.06	7.532	0.630	0.390	59.6	8.63	2.71	2.16	21.0	5.57	1.61
	25	7.35	9.00	7.500	0.570	0.358	53.9	7.85	2.71	2.14	18.6	4.96	1.59
ST 8 WF	48	14.11	8.16	11.533	0.875	0.535	64.7	9.82	2.14	1.57	103.6	18.0	2.71
	44	12.94	8.08	11.502	0.795	0.504	59.5	9.11	2.14	1.55	92.6	16.1	2.67
ST 8 WF	39	11.46	8.16	8.586	0.875	0.529	60.0	9.45	2.28	1.81	43.8	10.2	1.95
	35.5	10.43	8.08	8.543	0.795	0.486	54.0	8.57	2.28	1.77	38.9	9.11	1.93
	32	9.40	8.00	8.500	0.715	0.443	48.3	7.71	2.27	1.73	34.2	8.05	1.91
	29	8.52	7.93	8.464	0.645	0.407	43.6	7.00	2.26	1.70	30.2	7.14	1.88
ST 8 WF	25	7.35	8.13	7.073	0.628	0.380	42.2	6.77	2.40	1.89	17.4	4.92	1.54
	22.5	6.62	8.06	7.039	0.563	0.346	37.8	6.10	2.39	1.87	15.2	4.33	1.52
	20	5.88	8.00	7.000	0.503	0.307	33.2	5.37	2.37	1.82	13.3	3.79	1.50
	18	5.30	7.93	6.992	0.428	0.299	30.7	5.10	2.41	1.90	11.1	3.17	1.45
ST 7 WF	105.5	31.04	7.88	15.800	1.563	0.980	102.2	16.2	1.81	1.57	514.3	65.1	4.07
	101	29.70	7.82	15.750	1.503	0.930	95.7	15.2	1.80	1.53	489.8	62.2	4.06
	96.5	28.36	7.75	15.710	1.438	0.890	90.1	14.4	1.78	1.49	465.1	59.2	4.05
	92	27.04	7.69	15.660	1.378	0.840	83.9	13.4	1.76	1.45	441.4	56.4	4.04
	88	25.87	7.63	15.640	1.313	0.820	80.2	12.9	1.76	1.42	418.9	53.6	4.02
	83.5	24.55	7.56	15.600	1.248	0.780	75.0	12.1	1.75	1.39	395.1	50.7	4.01
	79	23.24	7.50	15.550	1.188	0.730	69.3	11.3	1.73	1.34	372.5	47.9	4.00
	75	22.04	7.44	15.515	1.128	0.695	64.9	10.6	1.72	1.31	351.3	45.3	3.99
	71	20.92	7.38	15.500	1.063	0.680	62.1	10.2	1.72	1.29	330.1	42.6	3.97
ST 7 WF	68	19.99	7.38	14.740	1.063	0.660	60.0	9.89	1.73	1.31	283.9	38.5	3.77
	63.5	18.67	7.31	14.690	0.998	0.610	54.7	9.04	1.71	1.26	263.8	35.9	3.76
	59.5	17.49	7.25	14.650	0.938	0.570	50.4	8.36	1.70	1.22	245.9	33.6	3.75
	55.5	16.33	7.19	14.620	0.873	0.540	46.7	7.80	1.69	1.19	227.4	31.1	3.73
	51.5	15.13	7.13	14.575	0.813	0.495	42.4	7.10	1.67	1.15	209.9	28.8	3.72
	47.5	13.97	7.06	14.545	0.748	0.465	39.1	6.58	1.67	1.12	191.9	26.4	3.71
	43.5	12.78	7.00	14.5	0.688	0.420	34.9	5.88	1.65	1.08	174.8	24.1	3.70
ST 7 WF	42	12.36	7.09	12.023	0.778	0.451	37.4	6.36	1.74	1.21	112.7	18.8	3.02
	39	11.47	7.03	12.000	0.718	0.428	34.8	5.96	1.74	1.19	103.5	17.2	3.00
ST 7 WF	37	10.88	7.10	10.072	0.783	0.450	36.1	6.26	1.82	1.32	66.7	13.3	2.48
	34	10.00	7.03	10.040	0.718	0.418	33.0	5.74	1.81	1.29	60.6	12.1	2.46
	30.5	8.97	6.96	10.000	0.643	0.378	29.2	5.13	1.80	1.25	53.6	10.7	2.45
ST 7 WF	26.5	7.79	6.97	8.062	0.658	0.370	27.7	4.95	1.88	1.38	28.8	7.14	1.92
	24	7.06	6.91	8.031	0.593	0.339	24.9	4.49	1.88	1.35	25.6	6.38	1.91
	21.5	6.32	6.84	8.000	0.528	0.308	22.2	4.02	1.87	1.33	22.6	5.64	1.89
ST 7 WF[a]	19	5.59	7.06	6.776	0.513	0.313	23.5	4.27	2.05	1.56	12.3	3.64	1.49
	17	5.00	7.00	6.750	0.453	0.287	21.1	3.86	2.05	1.55	10.6	3.15	1.46
	15	4.41	6.93	6.733	0.383	0.270	19.0	3.55	2.08	1.59	8.77	2.61	1.41
ST 6 WF	80.5	23.69	6.94	12.515	1.486	0.905	62.6	11.5	1.63	1.47	243.1	38.9	3.20
	66.5	19.56	6.69	12.365	1.236	0.755	48.4	9.03	1.57	1.33	195.0	31.5	3.16

TABLE 1.81 (*Continued*)

Section Number	Weight per Foot (lb)	Area (in.^{2})	Depth of Tee (in.)	Flange Width (in.)	Flange Average Thickness (in.)	Stem Thickness (in.)	Axis X-X I (in.^{4})	S (in.^{3})	r (in.)	y (in.)	Axis Y-Y I (in.^{4})	S (in.^{3})	r (in.)
	60	17.65	6.56	12.320	1.106	0.710	43.4	8.22	1.57	1.28	172.5	28.0	3.13
	53	15.59	6.44	12.230	0.986	0.620	36.7	7.01	1.53	1.20	150.4	24.6	3.11
	49.5	14.54	6.38	12.190	0.921	0.580	33.7	6.46	1.52	1.16	139.1	22.8	3.09
	46	13.53	6.31	12.155	0.856	0.545	31.0	5.98	1.51	1.13	128.2	21.1	3.08
	42.5	12.49	6.25	12.105	0.796	0.495	27.8	5.38	1.49	1.08	117.7	19.5	3.07
	39.5	11.61	6.19	12.080	0.736	0.470	25.8	5.02	1.48	1.06	108.2	17.9	3.05
	36	10.58	6.13	12.040	0.671	0.430	23.1	4.53	1.48	1.02	97.6	16.2	3.04
	32.5	9.55	6.06	12.000	0.606	0.390	20.6	4.06	1.47	0.98	87.3	14.6	3.02
ST 6 WF	29	8.53	6.10	10.014	0.641	0.359	19.0	3.75	1.49	1.03	53.7	10.7	2.51
	26.5	7.80	6.03	10.000	0.576	0.345	17.7	3.54	1.51	1.02	48.0	9.60	2.48
ST 6 WF	25	7.36	6.10	8.077	0.641	0.371	18.7	3.80	1.60	1.17	28.2	6.98	1.96
	22.5	6.62	6.03	8.042	0.576	0.336	16.6	3.40	1.59	1.13	25.0	6.20	1.94
	20	5.89	5.97	8.000	0.516	0.294	14.4	2.94	1.56	1.08	22.0	5.50	1.94
ST 6 WF	18	5.29	6.12	6.565	0.540	0.305	15.3	3.14	1.70	1.26	11.9	3.62	1.50
	15.5	4.56	6.04	6.525	0.465	0.265	13.0	2.69	1.69	1.22	9.9	3.04	1.47
	13.5	3.98	5.98	6.500	0.400	0.240	11.4	2.39	1.69	1.21	8.3	2.55	1.44
ST 6 WF	7	2.07	5.96	3.970	0.224	0.200	7.66	1.83	1.92	1.76	1.13	0.57	0.74
ST 6 I^b	25	7.29	6.00	5.477	0.660	0.687	25.2	6.05	1.85	1.84	7.85	2.87	1.03
	20.4	5.92	6.00	5.250	0.660	0.460	18.8	4.26	1.77	1.57	6.77	2.58	1.06
ST 6 I	17.5	5.10	6.00	5.078	0.544	0.428	17.2	3.95	1.83	1.65	4.93	1.94	0.98
	15.9	4.63	6.00	5.000	0.544	0.350	14.9	3.31	1.78	1.51	4.68	1.87	1.00
ST 5 I	17.5	5.11	5.00	4.944	0.491	0.594	12.5	3.63	1.56	1.56	4.18	1.69	0.90
	12.7	3.69	5.00	4.660	0.491	0.310	7.81	2.05	1.45	1.20	3.39	1.46	0.95
ST 4 I	11.5	3.36	4.00	4.171	0.425	0.441	5.03	1.77	1.22	1.15	2.15	1.03	0.80
	9.2	2.67	4.00	4.000	0.425	0.270	3.50	1.14	1.14	0.94	1.86	0.93	0.83
ST 3.5 I	10	2.92	3.50	3.860	0.392	0.450	3.36	1.36	1.07	1.04	1.58	0.82	0.73
	7.65	2.22	3.50	3.660	0.392	0.250	2.18	0.81	0.99	0.81	1.32	0.72	0.77
ST 3 I	8.625	2.51	3.00	3.565	0.359	0.465	2.13	1.02	0.92	0.91	1.15	0.65	0.67
	6.25	1.81	3.00	3.330	0.359	0.230	1.27	0.55	0.83	0.69	0.93	0.56	0.71
ST 5 WF	56	16.46	5.69	10.415	1.248	0.755	28.8	6.42	1.32	1.21	117.7	22.6	2.67
	50	14.72	5.56	10.345	1.118	0.685	24.8	5.62	1.30	1.14	103.3	20.0	2.65
	44.5	13.09	5.44	10.275	0.998	0.615	21.3	4.88	1.28	1.07	90.3	17.6	2.63
	38.5	11.33	5.31	10.195	0.868	0.535	17.7	4.10	1.25	1.00	76.7	15.1	2.60
	36	10.59	5.25	10.170	0.808	0.510	16.4	3.83	1.24	0.97	70.9	13.9	2.59
	33	9.70	5.19	10.117	0.748	0.457	14.5	3.39	1.22	0.92	64.6	12.8	2.58
	30	8.83	5.13	10.075	0.683	0.415	12.8	3.02	1.21	0.88	58.2	11.6	2.57
	27	7.94	5.06	10.028	0.618	0.368	11.2	2.64	1.18	0.84	51.95	10.4	2.56
	24.5	7.20	5.00	10.000	0.558	0.340	10.1	2.40	1.18	0.81	46.5	9.30	2.54
ST 5 WF	22.5	6.62	5.06	8.022	0.618	0.350	10.3	2.48	1.25	0.91	26.6	6.63	2.00
	19.5	5.74	4.97	7.990	0.528	0.318	8.96	2.19	1.25	0.88	22.5	5.62	1.98
	16.5	4.85	4.88	7.964	0.433	0.292	7.80	1.95	1.27	0.88	18.2	4.58	1.94
ST 5 WF^a	14.5	4.27	5.11	5.799	0.500	0.289	8.38	2.07	1.40	1.05	7.61	2.62	1.34
	12.5	3.67	5.04	5.762	0.430	0.252	7.12	1.77	1.39	1.02	6.34	2.20	1.31
	10.5	3.10	4.95	5.750	0.340	0.240	6.31	1.62	1.43	1.06	4.87	1.69	1.25
ST 4 WF	33.5	9.85	4.50	8.287	0.933	0.575	10.94	3.07	1.05	0.94	44.3	10.7	2.12
	29	8.53	4.38	8.222	0.808	0.510	9.11	2.60	1.03	0.87	37.5	9.10	2.10
	24	7.06	4.25	8.117	0.683	0.405	6.92	2.00	0.99	0.78	30.45	7.50	2.08
	20	5.88	4.13	8.077	0.558	0.365	5.80	1.71	0.99	0.74	24.5	6.05	2.04
	17.5	5.15	4.06	8.027	0.493	0.315	4.88	1.45	0.97	0.69	21.25	5.30	2.03
	15.5	4.56	4.00	8.000	0.433	0.288	4.31	1.30	0.97	0.67	18.5	4.60	2.01
ST 4 WF	14	4.11	4.03	6.540	0.463	0.285	4.22	1.28	1.01	0.73	10.8	3.30	1.62
	12	3.53	3.97	6.500	0.398	0.245	3.53	1.08	1.00	0.70	9.10	2.80	1.61
ST 4 WF	10	2.94	4.07	5.268	0.378	0.248	3.66	1.13	1.12	0.83	4.25	1.61	1.20
	8.5	2.50	4.00	5.250	0.308	0.230	3.21	1.01	1.13	0.84	3.36	1.28	1.16

TABLE 1.81 (*Continued*)

			Dimensions				Axis X-X				Axis Y-Y		
	Weight			Width of	Minimum Thickness								
Nominal Size (in.)	per Foot (lb)	Area (in.²)	Depth (in.)	Flange (in.)	Flange (in.)	Stem (in.)	I (in.⁴)	S (in.³)	r (in.)	y (in.)	I (in.⁴)	S (in.³)	r (in.)
$5 \times 3\frac{1}{8}$	13.6	4.00	$3\frac{1}{8}$	5	$\frac{1}{2}$	$\frac{13}{32}$	2.7	1.1	0.82	0.76	5.2	2.1	1.14
5×3	11.5	3.37	3	5	$\frac{3}{8}$	$\frac{13}{32}$	2.4	1.1	0.84	0.76	3.9	1.6	1.10
$4 \times 4\frac{1}{2}$	11.2	3.29	$4\frac{1}{2}$	4	$\frac{3}{8}$	$\frac{3}{8}$	6.3	2.0	1.39	1.31	2.1	1.1	0.80
4×4	13.5	3.97	4	4	$\frac{1}{2}$	$\frac{1}{2}$	5.7	2.0	1.20	1.18	2.8	1.4	0.84
4×3	9.2	2.68	3	4	$\frac{3}{8}$	$\frac{3}{8}$	2.0	0.90	0.86	0.78	2.1	1.1	0.89
$4 \times 2\frac{1}{2}$	8.5	2.48	$2\frac{1}{2}$	4	$\frac{3}{8}$	$\frac{3}{8}$	1.2	0.62	0.69	0.62	2.1	1.0	0.92
3×3	7.8	2.29	3	3	$\frac{3}{8}$	$\frac{3}{8}$	1.84	0.86	0.89	0.88	0.89	0.60	0.63
3×3	6.7	1.97	3	3	$\frac{5}{16}$	$\frac{5}{16}$	1.61	0.74	0.90	0.85	0.75	0.50	0.62
$3 \times 2\frac{1}{2}$	6.1	1.77	$2\frac{1}{2}$	3	$\frac{5}{16}$	$\frac{5}{16}$	0.94	0.51	0.73	0.68	0.75	0.50	0.65
$2\frac{1}{2} \times 2\frac{1}{2}$	6.4	1.87	$2\frac{1}{2}$	$2\frac{1}{2}$	$\frac{3}{8}$	$\frac{3}{8}$	1.0	0.59	0.74	0.76	0.52	0.42	0.53
$2\frac{1}{2} \times 2\frac{1}{2}$	4.6	1.33	$2\frac{1}{2}$	$2\frac{1}{2}$	$\frac{1}{4}$	$\frac{1}{4}$	0.74	0.42	0.75	0.71	0.34	0.27	0.51
$2\frac{1}{4} \times 2\frac{1}{4}$	4.1	1.19	$2\frac{1}{4}$	$2\frac{1}{4}$	$\frac{1}{4}$	$\frac{1}{4}$	0.52	0.32	0.66	0.65	0.25	0.22	0.46
2×2	4.3	1.26	2	2	$\frac{5}{16}$	$\frac{5}{16}$	0.44	0.31	0.59	0.61	0.23	0.23	0.43
2×2	3.56	1.05	2	2	$\frac{1}{4}$	$\frac{1}{4}$	0.37	0.26	0.59	0.59	0.18	0.18	0.42

[a] WF indicates structural tee cut from wide-flange section.
[b] I indicates structural tee cut from standard beam section.

TABLE 1.82   Properties and Dimensions of Zees

Zees are seldom used as structural framing members. When so used they are generally employed on short spans in flexure. These tables list a few selected sizes, the range of whose section moduli will cover all ordinary conditions. For sizes not listed, the catalogs of the respective rolling mills should be consulted.

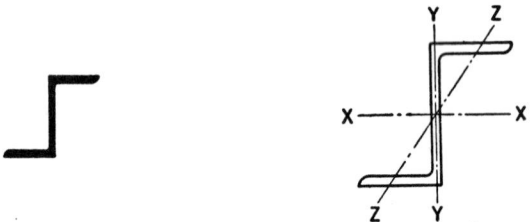

			Dimensions			Axis X-X			Axis Y-Y			Axis Z-Z
	Weight			Width of	Thick-							
Nominal Size (in.)	per Foot (lb)	Area (in.²)	Depth (in.)	Flange (in.)	ness (in.)	I (in.⁴)	S (in.³)	r (in.)	I (in.⁴)	S (in.³)	r (in.)	r (in.)
$6 \times 3\frac{1}{2}$	21.1	6.19	$6\frac{1}{8}$	$3\frac{3}{8}$	$\frac{1}{2}$	34.4	11.2	2.36	12.9	3.8	1.44	0.84
	15.7	4.59	6	$3\frac{1}{2}$	$\frac{3}{8}$	25.3	8.4	2.35	9.1	2.8	1.41	0.83
$5 \times 3\frac{1}{4}$	17.9	5.25	5	$3\frac{1}{4}$	$\frac{1}{2}$	19.2	7.7	1.91	9.1	3.0	1.31	0.74
	16.4	4.81	$5\frac{1}{8}$	$3\frac{3}{8}$	$\frac{7}{16}$	19.1	7.4	1.99	9.2	2.9	1.38	0.77
	14.0	4.10	$5\frac{5}{16}$	$3\frac{3}{16}$	$\frac{3}{8}$	16.2	6.4	1.99	7.7	2.5	1.37	0.76
	11.6	3.40	5	$3\frac{1}{4}$	$\frac{5}{16}$	13.4	5.3	1.98	6.2	2.0	1.35	0.75
$4 \times 3\frac{1}{16}$	15.9	4.66	$4\frac{1}{16}$	$3\frac{1}{8}$	$\frac{1}{2}$	11.2	5.5	1.55	8.0	2.8	1.31	0.67
	12.5	3.66	$4\frac{1}{8}$	$3\frac{3}{16}$	$\frac{3}{8}$	9.6	4.7	1.62	6.8	2.3	1.36	0.69
	10.3	3.03	$4\frac{1}{16}$	$3\frac{1}{8}$	$\frac{5}{16}$	7.9	3.9	1.62	5.5	1.8	1.34	0.68
	8.2	2.41	4	$3\frac{1}{16}$	$\frac{1}{4}$	6.3	3.1	1.62	4.2	1.4	1.33	0.67

TABLE 1.82 *(Continued)*

Nominal Size (in.)	Weight per Foot (lb)	Area (in.²)	Dimensions Depth (in.)	Width of Flange (in.)	Thick-ness (in.)	Axis X-X I (in.⁴)	Axis X-X S (in.³)	Axis X-X r (in.)	Axis Y-Y I (in.⁴)	Axis Y-Y S (in.³)	Axis Y-Y r (in.)	Axis Z-Z r (in.)
$3 \times 2\frac{11}{16}$	12.6	3.69	3	$2\frac{11}{16}$	$\frac{1}{2}$	4.6	3.1	1.12	4.9	2.0	1.15	0.53
	9.8	2.86	3	$2\frac{11}{16}$	$\frac{3}{8}$	3.9	2.6	1.16	3.9	1.6	1.17	0.54
	6.7	1.97	3	$2\frac{11}{16}$	$\frac{1}{4}$	2.9	1.9	1.21	2.8	1.1	1.19	0.55

TABLE 1.83 Properties and Dimensions of H Bearing Piles

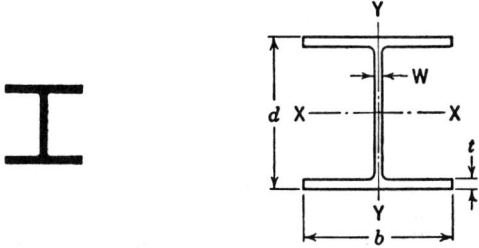

Section Number and Nominal Size	Weight per Foot (lb)	Area A (in.²)	Depth d (in.)	Flange Width b (in.)	Flange Thick-ness t (in.)	Web Thick-ness W (in.)	Axis X-X I (in.⁴)	Axis X-X S (in.³)	Axis X-X r (in.)	Axis Y-Y I' (in.⁴)	Axis Y-Y S' (in.³)	Axis Y-Y r' (in.)
BP 14 14 × 14½	117	34.44	14.234	14.885	0.805	0.805	1228.5	172.6	5.97	443.1	59.5	3.59
	102	30.01	14.032	14.784	0.704	0.704	1055.1	150.4	5.93	379.6	51.3	3.56
	89	26.19	13.856	14.696	0.616	0.616	909.1	131.2	5.89	326.2	44.4	3.53
	73	21.46	13.636	14.586	0.506	0.506	733.1	107.5	5.85	261.9	35.9	3.49
BP 12 12 × 12	74	21.76	12.122	12.217	0.607	0.607	566.5	93.5	5.10	184.7	30.2	2.91
	53	15.58	11.780	12.046	0.436	0.436	394.8	67.0	5.03	127.3	21.2	2.86
BP 10 10 × 10	57	16.76	10.012	10.224	0.564	0.564	294.7	58.9	4.19	100.6	19.7	2.45
	42	12.35	9.720	10.078	0.418	0.418	210.8	43.4	4.13	71.4	14.2	2.40
BP 8 8 × 8	36	10.60	8.026	8.158	0.446	0.446	119.8	29.9	3.36	40.4	9.9	1.95

TABLE 1.84 Square and Round Bars[a]

Size (in.)	Square Weight/ft (lb)	Square Area (in.²)	Round Weight/ft (lb)	Round Area (in.²)	Size (in.)	Square Weight/ft (lb)	Square Area (in.²)	Round Weight/ft (lb)	Round Area (in.²)
0					$\frac{13}{16}$	2.245	0.6602	1.763	0.5185
$\frac{1}{16}$	0.013	0.0039	0.010	0.0031	$\frac{7}{8}$	2.603	0.7656	2.044	0.6013
$\frac{1}{8}$	0.053	0.0156	0.042	0.0123	$\frac{15}{16}$	2.988	0.8789	2.347	0.6903
$\frac{3}{16}$	0.120	0.0352	0.094	0.0276	1	3.400	1.0000	2.670	0.7854
$\frac{1}{4}$	0.213	0.0625	0.167	0.0491	$\frac{1}{16}$	3.838	1.1289	3.015	0.8866
$\frac{5}{16}$	0.332	0.0977	0.261	0.0767	$\frac{1}{8}$	4.303	1.2656	3.380	0.9940
$\frac{3}{8}$	0.478	0.1406	0.376	0.1105	$\frac{3}{16}$	4.795	1.4102	3.766	1.1075
$\frac{7}{16}$	0.651	0.1914	0.511	0.1503	$\frac{1}{4}$	5.313	1.5625	4.172	1.2272
$\frac{1}{2}$	0.850	0.2500	0.668	0.1963	$\frac{5}{16}$	5.857	1.7227	4.600	1.3530
$\frac{9}{16}$	1.076	0.3164	0.845	0.2485	$\frac{3}{8}$	6.428	1.8906	5.049	1.4849
$\frac{5}{8}$	1.328	0.3906	1.043	0.3068	$\frac{7}{16}$	7.026	2.0664	5.518	1.6230
$\frac{11}{16}$	1.607	0.4727	1.262	0.3712	$\frac{1}{2}$	7.650	2.2500	6.008	1.7671
$\frac{3}{4}$	1.913	0.5625	1.502	0.4418	$\frac{9}{16}$	8.301	2.4414	6.519	1.9175

TABLE 1.84 (*Continued*)

Size (in.)	Square Weight/ft (lb)	Area (in.²)	Round Weight/ft (lb)	Area (in.²)	Size (in.)	Square Weight/ft (lb)	Area (in.²)	Round Weight/ft (lb)	Area (in.²)
$\frac{5}{8}$	8.978	2.6406	7.051	2.0739	$\frac{7}{8}$	80.80	23.766	63.46	18.665
$\frac{11}{16}$	9.682	2.8477	7.604	2.2365	$\frac{15}{16}$	82.89	24.379	65.10	19.147
$\frac{3}{4}$	10.413	3.0625	8.178	2.4053	5	85.00	25.000	66.76	19.635
$\frac{13}{16}$	11.170	3.2852	8.773	2.5802	$\frac{1}{16}$	87.14	25.629	68.44	20.129
$\frac{7}{8}$	11.953	3.5156	9.388	2.7612	$\frac{1}{8}$	89.30	26.266	70.14	20.629
$\frac{15}{16}$	12.763	3.7539	10.024	2.9483	$\frac{3}{16}$	91.49	26.910	71.86	21.135
2	13.600	4.0000	10.681	3.1416	$\frac{1}{4}$	93.71	27.563	73.60	21.648
$\frac{1}{16}$	14.463	4.2539	11.359	3.3410	$\frac{5}{16}$	95.96	28.223	75.36	22.166
$\frac{1}{8}$	15.353	4.5156	12.058	3.5466	$\frac{3}{8}$	98.23	28.891	77.15	22.691
$\frac{3}{16}$	16.270	4.7852	12.778	3.7583	$\frac{7}{16}$	100.53	29.566	78.95	23.221
$\frac{1}{4}$	17.213	5.0625	13.519	3.9761	$\frac{1}{2}$	102.85	30.250	80.78	23.758
$\frac{5}{16}$	18.182	5.3477	14.280	4.2000	$\frac{9}{16}$	105.20	30.941	82.62	24.301
$\frac{3}{8}$	19.178	5.6406	15.062	4.4301	$\frac{5}{8}$	107.58	31.641	84.49	24.850
$\frac{7}{16}$	20.201	5.9414	15.866	4.6664	$\frac{11}{16}$	109.98	32.348	86.38	25.406
$\frac{1}{2}$	21.250	6.2500	16.690	4.9087	$\frac{3}{4}$	112.41	33.063	88.29	25.967
$\frac{9}{16}$	22.326	6.5664	17.534	5.1572	$\frac{13}{16}$	114.87	33.785	90.22	26.535
$\frac{5}{8}$	23.428	6.8906	18.400	5.4119	$\frac{7}{8}$	117.35	34.516	92.17	27.109
$\frac{11}{16}$	24.557	7.2227	19.287	5.6727	$\frac{15}{16}$	119.86	35.254	94.14	27.688
$\frac{3}{4}$	25.713	7.5625	20.195	5.9396	6	122.40	36.000	96.13	28.274
$\frac{13}{16}$	26.895	7.9102	21.123	6.2126	$\frac{1}{16}$	124.96	36.754	98.15	28.866
$\frac{7}{8}$	28.103	8.2656	22.072	6.4918	$\frac{1}{8}$	127.55	37.516	100.18	29.465
$\frac{15}{16}$	29.338	8.6289	23.042	6.7771	$\frac{3}{16}$	130.17	38.285	102.23	30.069
3	30.60	9.000	24.03	7.069	$\frac{1}{4}$	132.81	39.063	104.31	30.680
$\frac{1}{16}$	31.89	9.379	25.05	7.366	$\frac{5}{16}$	135.48	39.848	106.41	31.296
$\frac{1}{8}$	33.20	9.766	26.08	7.670	$\frac{3}{8}$	138.18	40.641	108.53	31.919
$\frac{3}{16}$	34.54	10.160	27.13	7.980	$\frac{7}{16}$	140.90	41.441	110.66	32.548
$\frac{1}{4}$	35.91	10.563	28.21	8.296	$\frac{1}{2}$	143.65	42.250	112.82	33.183
$\frac{5}{16}$	37.31	10.973	29.30	8.618	$\frac{9}{16}$	146.43	43.066	115.00	33.824
$\frac{3}{8}$	38.73	11.391	30.42	8.946	$\frac{5}{8}$	149.23	43.891	117.20	34.472
$\frac{7}{16}$	40.18	11.816	31.55	9.281	$\frac{11}{16}$	152.06	44.723	119.43	35.125
$\frac{1}{2}$	41.65	12.250	32.71	9.621	$\frac{3}{4}$	154.91	45.563	121.67	35.785
$\frac{9}{16}$	43.15	12.691	33.89	9.968	$\frac{13}{16}$	157.79	46.410	123.93	36.450
$\frac{5}{8}$	44.68	13.141	35.09	10.321	$\frac{7}{8}$	160.70	47.266	126.22	37.122
$\frac{11}{16}$	46.23	13.598	36.31	10.680	$\frac{15}{16}$	163.64	48.129	128.52	37.800
$\frac{3}{4}$	47.81	14.063	37.55	11.045	7	166.60	49.000	130.85	38.485
$\frac{13}{16}$	49.42	14.535	38.81	11.416	$\frac{1}{16}$	169.59	49.879	133.19	39.175
$\frac{7}{8}$	51.05	15.016	40.10	11.793	$\frac{1}{8}$	172.60	50.766	135.56	39.871
$\frac{15}{16}$	52.71	15.504	41.40	12.177	$\frac{3}{16}$	175.64	51.660	137.95	40.574
4	54.40	16.000	42.73	12.566	$\frac{1}{4}$	178.71	52.563	140.36	41.282
$\frac{1}{16}$	56.11	16.504	44.07	12.962	$\frac{5}{16}$	181.81	53.473	142.79	41.997
$\frac{1}{8}$	57.85	17.016	45.44	13.364	$\frac{3}{8}$	184.93	54.391	145.24	42.718
$\frac{3}{16}$	59.62	17.535	46.83	13.772	$\frac{7}{16}$	188.07	55.316	147.71	43.445
$\frac{1}{4}$	61.41	18.063	48.23	14.186	$\frac{1}{2}$	191.25	56.250	150.21	44.179
$\frac{5}{16}$	63.23	18.598	49.66	14.607	$\frac{9}{16}$	194.45	57.191	152.72	44.918
$\frac{3}{8}$	65.08	19.141	51.11	15.033	$\frac{5}{8}$	197.68	58.141	155.26	45.664
$\frac{7}{16}$	66.95	19.691	52.58	15.466	$\frac{11}{16}$	200.93	59.098	157.81	46.415
$\frac{1}{2}$	68.85	20.250	54.07	15.904	$\frac{3}{4}$	204.21	60.063	160.39	47.173
$\frac{9}{16}$	70.78	20.816	55.59	16.349	$\frac{13}{16}$	207.52	61.035	162.99	47.937
$\frac{5}{8}$	72.73	21.391	57.12	16.800	$\frac{7}{8}$	210.85	62.016	165.60	48.707
$\frac{11}{16}$	74.71	21.973	58.67	17.257	$\frac{15}{16}$	214.21	63.004	168.24	49.483
$\frac{3}{4}$	76.71	22.563	60.25	17.721	8	217.60	64.000	170.90	50.265
$\frac{13}{16}$	78.74	23.160	61.85	18.190					

[a] One cubic inch of rolled steel is assumed to weigh 0.2833 lb.

TABLE 1.85   Dimensions of Ferrous Pipe

Nominal Pipe Size (in.)	Outside Diameter (in.)	Schedule No.	Wall Thickness (in.)	Inside Diameter (in.)	Cross-sectional area Metal (in.²)	Cross-sectional area Flow (ft²)	Circumference, ft, or surface, ft²/ft of length Outside	Circumference, ft, or surface, ft²/ft of length Inside	Capacity at 1 ft/sec Velocity U.S. gal/min	Capacity at 1 ft/sec Velocity lb/hr water	Weight of Plain-end Pipe (lb/ft)
$\frac{1}{8}$	0.405	10S	0.049	0.307	0.055	0.00051	0.106	0.0804	0.231	115.5	0.19
		40ST, 40S	0.068	0.269	0.072	0.00040	0.106	0.0705	0.179	89.5	0.24
		80XS, 80S	0.095	0.215	0.093	0.00025	0.106	0.0563	0.113	56.5	0.31
$\frac{1}{4}$	0.540	10S	0.065	0.410	0.097	0.00092	0.141	0.107	0.412	206.5	0.33
		40ST, 40S	0.088	0.364	0.125	0.00072	0.141	0.095	0.323	161.5	0.42
		80XS, 80S	0.119	0.302	0.157	0.00050	0.141	0.079	0.224	112.0	0.54
$\frac{3}{8}$	0.675	10S	0.065	0.545	0.125	0.00162	0.177	0.143	0.727	363.5	0.42
		40ST, 40S	0.091	0.493	0.167	0.00133	0.177	0.129	0.596	298.0	0.57
		80XS, 80S	0.126	0.423	0.217	0.00098	0.177	0.111	0.440	220.0	0.74
$\frac{1}{2}$	0.840	5S	0.065	0.710	0.158	0.00275	0.220	0.186	1.234	617.0	0.54
		10S	0.083	0.674	0.197	0.00248	0.220	0.176	1.112	556.0	0.67
		40ST, 40S	0.109	0.622	0.250	0.00211	0.220	0.163	0.945	472.0	0.85
		80XS, 80S	0.147	0.546	0.320	0.00163	0.220	0.143	0.730	365.0	1.09
		160	0.188	0.464	0.385	0.00117	0.220	0.122	0.527	263.5	1.31
		XX	0.294	0.252	0.504	0.00035	0.220	0.066	0.155	77.5	1.71
$\frac{3}{4}$	1.050	5S	0.065	0.920	0.201	0.00461	0.275	0.241	2.072	1036.0	0.69
		10S	0.083	0.884	0.252	0.00426	0.275	0.231	1.903	951.5	0.86
		40ST, 40S	0.113	0.824	0.333	0.00371	0.275	0.216	1.665	832.5	1.13
		80XS, 80S	0.154	0.742	0.433	0.00300	0.275	0.194	1.345	672.5	1.47
		160	0.219	0.612	0.572	0.00204	0.275	0.160	0.917	458.5	1.94
		XX	0.308	0.434	0.718	0.00103	0.275	0.114	0.461	230.5	2.44
1	1.315	5S	0.065	1.185	0.255	0.00768	0.344	0.310	3.449	1725	0.87
		10S	0.109	1.097	0.413	0.00656	0.344	0.287	2.946	1473	1.40
		40ST, 40S	0.133	1.049	0.494	0.00600	0.344	0.275	2.690	1345	1.68
		80XS, 80S	0.179	0.957	0.639	0.00499	0.344	0.250	2.240	1120	2.17
		160	0.250	0.815	0.836	0.00362	0.344	0.213	1.625	812.5	2.84
		XX	0.358	0.599	1.076	0.00196	0.344	0.157	0.878	439.0	3.66

TABLE 1.85 (*Continued*)

Nominal Pipe Size (in.)	Outside Diameter (in.)	Schedule No.	Wall Thickness (in.)	Inside Diameter (in.)	Cross-sectional area Metal (in.²)	Cross-sectional area Flow (ft²)	Circumference, ft, or surface, ft²/ft of length Outside	Circumference Inside	Capacity at 1 ft/sec Velocity U.S. gal/min.	Capacity lb/hr water	Weight of Plain-end Pipe (lb/ft)
$1\frac{1}{4}$	1.660	5S	0.065	1.530	0.326	0.01277	0.435	0.401	5.73	2865	1.11
		10S	0.109	1.442	0.531	0.01134	0.435	0.378	5.09	2545	1.81
		40ST, 40S	0.140	1.380	0.668	0.01040	0.435	0.361	4.57	2285	2.27
		80XS, 80S	0.191	1.278	0.881	0.00891	0.435	0.335	3.99	1995	3.00
		160	0.250	1.160	1.107	0.00734	0.435	0.304	3.29	1645	3.76
		XX	0.382	0.896	1.534	0.00438	0.435	0.235	1.97	985	5.21
$1\frac{1}{2}$	1.900	5S	0.065	1.770	0.375	0.01709	0.497	0.463	7.67	3835	1.28
		10S	0.109	1.682	0.614	0.01543	0.497	0.440	6.94	3465	2.09
		40ST, 40S	0.145	1.610	0.800	0.01414	0.497	0.421	6.34	3170	2.72
		80XS, 80S	0.200	1.500	1.069	0.01225	0.497	0.393	5.49	2745	3.63
		160	0.281	1.338	1.429	0.00976	0.497	0.350	4.38	2190	4.86
		XX	0.400	1.100	1.885	0.00660	0.497	0.288	2.96	1480	6.41
2	2.375	5S	0.065	2.245	0.472	0.02749	0.622	0.588	12.34	6170	1.61
		10S	0.109	2.157	0.776	0.02538	0.622	0.565	11.39	5695	2.64
		40ST, 40S	0.154	2.067	1.075	0.02330	0.622	0.541	10.45	5225	3.65
		80ST, 80S	0.218	1.939	1.477	0.02050	0.622	0.508	9.20	4600	5.02
		160	0.344	1.687	2.195	0.01552	0.622	0.436	6.97	3485	7.46
		XX	0.436	1.503	2.656	0.01232	0.622	0.393	5.53	2765	9.03
$2\frac{1}{2}$	2.875	5S	0.083	2.709	0.728	0.04003	0.753	0.709	17.97	8985	2.48
		10S	0.120	2.635	1.039	0.03787	0.753	0.690	17.00	8500	3.53
		40ST, 40S	0.203	2.469	1.704	0.03322	0.753	0.647	14.92	7460	5.79
		80XS, 80S	0.276	2.323	2.254	0.02942	0.753	0.608	13.20	6600	7.66
		160	0.375	2.125	2.945	0.02463	0.753	0.556	11.07	5535	10.01
		XX	0.552	1.771	4.028	0.01711	0.753	0.464	7.68	3840	13.70
3	3.500	5S	0.083	3.334	0.891	0.06063	0.916	0.873	27.21	13,605	3.03
		10S	0.120	3.260	1.274	0.05796	0.916	0.853	26.02	13,010	4.33
		40ST, 40S	0.216	3.068	2.228	0.05130	0.916	0.803	23.00	11,500	7.58
		80XS, 80S	0.300	2.900	3.016	0.04587	0.916	0.759	20.55	10,275	10.25
		160	0.438	2.624	4.213	0.03755	0.916	0.687	16.86	8430	14.31
		XX	0.600	2.300	5.466	0.02885	0.916	0.602	12.95	6475	18.58
$3\frac{1}{2}$	4.0	5S	0.083	3.834	1.021	0.08017	1.047	1.004	35.98	17,990	3.48

4	4.5	10S	0.120	3.760	1.463	0.07711	1.047	0.984	34.61	17,305	4.97
		40ST, 40S	0.226	3.548	2.680	0.06870	1.047	0.929	30.80	15,400	9.11
		80XS, 80S	0.318	3.364	3.678	0.06170	1.047	0.881	27.70	13,850	12.51
5	5.563	5S	0.083	4.334	1.152	0.10245	1.178	1.115	46.0	23,000	3.92
		10S	0.120	4.260	1.651	0.09898	1.178	1.115	44.4	22,200	5.61
		40ST, 40S	0.237	4.026	3.17	0.08840	1.178	1.054	39.6	19,800	10.79
		80XS, 80S	0.337	3.826	4.41	0.07986	1.178	1.002	35.8	17,900	14.98
		120	0.438	3.624	5.58	0.07170	1.178	0.949	32.2	16,100	19.01
		160	0.531	3.438	6.62	0.06647	1.178	0.900	28.9	14,450	22.52
		XX	0.674	3.152	8.10	0.05419	1.178	0.825	24.3	12,150	27.54
5.563		5S	0.109	5.345	1.87	0.1558	1.456	1.399	69.9	34,950	6.36
		10S	0.134	5.295	2.29	0.1529	1.456	1.386	68.6	34,300	7.77
		40ST, 40S	0.258	5.047	4.30	0.1390	1.456	1.321	62.3	31,150	14.62
		80XS, 80S	0.375	4.813	6.11	0.1263	1.456	1.260	57.7	28,850	20.78
		120	0.500	4.563	7.95	0.1136	1.456	1.195	51.0	25,500	27.04
		160	0.625	4.313	9.70	0.1015	1.456	1.129	45.5	22,750	32.96
		XX	0.750	4.063	11.34	0.0900	1.456	1.064	40.4	20,200	38.55
6	6.625	5S	0.109	6.407	2.23	0.2239	1.734	1.677	100.5	50,250	7.60
		10S	0.134	6.357	2.73	0.2204	1.734	1.664	98.9	49,450	9.29
		40ST, 40S	0.280	6.065	5.58	0.2006	1.734	1.588	90.0	45,000	18.97
		80XS, 80S	0.432	5.761	8.40	0.1810	1.734	1.508	81.1	40,550	28.57
		120	0.562	5.501	10.70	0.1650	1.734	1.440	73.9	36,950	36.42
		160	0.719	5.187	13.34	0.1467	1.734	1.358	65.9	32,950	45.34
		XX	0.864	4.897	15.64	0.1308	1.734	1.282	58.7	29,350	53.16
8	8.625	5S	0.109	8.407	2.915	0.3855	2.258	2.201	173.0	86,500	9.93
		10S	0.148	8.329	3.941	0.3784	2.258	2.180	169.8	84,900	13.40
		20	0.250	8.125	6.578	0.3601	2.258	2.127	161.5	80,750	22.36
		30	0.277	8.071	7.260	0.3553	2.258	2.113	159.4	79,700	24.70
		40ST, 40S	0.322	7.981	8.396	0.3474	2.258	2.089	155.7	77,850	28.55
		60	0.406	7.813	10.48	0.3329	2.258	2.045	149.4	74,700	35.66
		80XS, 80S	0.500	7.625	12.76	0.3171	2.258	1.996	142.3	71,150	43.39
		100	0.594	7.437	14.99	0.3017	2.258	1.947	135.4	67,700	50.93
		120	0.719	7.187	17.86	0.2817	2.258	1.882	126.4	63,200	60.69
		140	0.812	7.001	19.93	0.2673	2.258	1.833	120.0	60,000	67.79
		XX	0.875	6.875	21.30	0.2578	2.258	1.800	115.7	57,850	72.42
		160	0.906	6.813	21.97	0.2532	2.258	1.784	113.5	56,750	74.71

TABLE 1.85 (*Continued*)

Nominal Pipe Size (in.)	Outside Diameter (in.)	Schedule No.	Wall Thickness (in.)	Inside Diameter (in.)	Cross-sectional area Metal (in.²)	Cross-sectional area Flow (ft²)	Circumference, ft, or surface, ft²/ft of length Outside	Circumference, ft, or surface, ft²/ft of length Inside	Capacity at 1 ft/sec Velocity U.S. gal/min	Capacity at 1 ft/sec Velocity lb/hr water	Weight of Plain-end Pipe (lb/ft)
10	10.75	5S	0.134	10.842	4.47	0.5993	2.814	2.744	269.0	134,500	15.23
		10S	0.165	10.420	5.49	0.5922	2.814	2.728	265.8	132,900	18.70
		20	0.250	10.250	8.25	0.5731	2.814	2.685	257.0	128,500	28.04
		30	0.307	10.136	10.07	0.5603	2.814	2.655	252.0	126,000	34.24
		40ST, 40S	0.365	10.020	11.91	0.5475	2.814	2.620	246.0	123,000	40.48
		80S, 60XS	0.500	9.750	16.10	0.5185	2.814	2.550	233.0	116,500	54.74
		80	0.594	9.562	18.95	0.4987	2.814	2.503	223.4	111,700	64.40
		100	0.719	9.312	22.66	0.4729	2.814	2.438	212.3	106,150	77.00
		120	0.844	9.062	26.27	0.4479	2.814	2.372	201.0	100,500	89.27
		140, XX	1.000	8.750	30.63	0.4176	2.814	2.291	188.0	94,000	104.13
		160	1.125	8.500	34.02	0.3941	2.814	2.225	177.0	88,500	115.65
12	12.75	5S	0.156	12.438	6.17	0.8438	3.338	3.26	378.7	189,350	22.22
		10S	0.180	12.390	7.11	0.8373	3.338	3.24	375.8	187,900	24.20
		20	0.250	12.250	9.82	0.8185	3.338	3.21	367.0	183,500	33.38
		30	0.330	12.090	12.88	0.7972	3.338	3.17	358.0	179,000	43.77
		ST, 40S	0.375	12.000	14.58	0.7854	3.338	3.14	352.5	176,250	49.56
		40	0.406	11.938	15.74	0.7773	3.338	3.13	349.0	174,500	53.56
		XS, 80S	0.500	11.750	19.24	0.7530	3.338	3.08	338.0	169,000	65.42
		60	0.562	11.626	21.52	0.7372	3.338	3.04	331.0	165,500	73.22
		80	0.688	11.374	26.07	0.7056	3.338	2.98	316.7	158,350	88.57
		100	0.844	11.062	31.57	0.6674	3.338	2.90	299.6	149,800	107.29
		120, XX	1.000	10.750	36.91	0.6303	3.338	2.81	283.0	141,500	125.49
		140	1.125	10.500	41.09	0.6013	3.338	2.75	270.0	135,000	139.68
		160	1.312	10.126	47.14	0.5592	3.338	2.65	251.0	125,500	160.33
14	14	5S	0.156	13.688	6.78	1.0219	3.665	3.58	459	229,500	22.76
		10S	0.188	13.624	8.16	1.0125	3.665	3.57	454	227,000	27.70
		10	0.250	13.500	10.80	0.9940	3.665	3.53	446	223,000	36.71
		20	0.312	13.376	13.42	0.9750	3.665	3.50	438	219,000	45.68
		30, ST	0.375	13.250	16.05	0.9575	3.665	3.47	430	215,000	54.57
		40	0.438	13.124	18.66	0.9397	3.665	3.44	422	211,000	63.37
		XS	0.500	13.000	21.21	0.9218	3.665	3.40	414	207,000	72.09
		60	0.594	12.812	25.02	0.8957	3.665	3.35	402	201,000	85.01
		80	0.750	12.500	31.22	0.8522	3.665	3.27	382	191,000	106.13

Nominal	Schedule	Wall thickness	ID							
	100	0.938	12.124	38.49	0.8017	3.665	3.17	360	180,000	130.79
	120	1.094	11.812	44.36	0.7610	3.665	3.09	342	171,000	150.76
	140	1.250	11.500	50.07	0.7213	3.665	3.01	324	162,000	170.22
	160	1.406	11.188	55.63	0.6827	3.665	2.93	306	153,000	189.12
16	5S	0.165	15.670	8.18	1.3393	4.189	4.10	601	300,500	27.87
	10S	0.188	15.624	9.34	1.3314	4.189	4.09	598	299,000	31.62
	10	0.250	15.500	12.37	1.3104	4.189	4.06	587	293,500	42.05
	20	0.312	15.376	15.38	1.2985	4.189	4.03	578	289,000	52.36
	30, ST	0.375	15.250	18.41	1.2680	4.189	3.99	568	284,000	62.58
	40, XS	0.500	15.000	24.35	1.2272	4.189	3.93	550	275,000	82.77
	60	0.656	14.688	31.62	1.1766	4.189	3.85	528	264,000	107.54
	80	0.844	14.312	40.19	1.1171	4.189	3.75	501	250,500	136.58
	100	1.031	13.938	48.48	1.0596	4.189	3.65	474	237,000	164.86
	120	1.219	13.562	56.61	1.0032	4.189	3.55	450	225,000	192.40
	140	1.438	13.124	65.79	0.9394	4.189	3.44	422	211,000	223.57
	160	1.594	12.812	72.14	0.8953	4.189	3.35	402	201,000	245.22
18	5S	0.165	17.670	9.25	1.7029	4.712	4.63	774	382,000	31.32
	10S	0.188	17.624	10.52	1.6941	4.712	4.61	760	379,400	35.48
	10	0.250	17.500	13.94	1.6703	4.712	4.58	750	375,000	47.39
	20	0.312	17.376	17.34	1.6468	4.712	4.55	739	369,500	59.03
	ST	0.375	17.250	20.76	1.6230	4.712	4.52	728	364,000	70.59
	30	0.438	17.124	24.16	1.5993	4.712	4.48	718	359,000	82.06
	XS	0.500	17.000	27.49	1.5763	4.712	4.45	707	353,500	93.45
	40	0.562	16.876	30.79	1.5533	4.712	4.42	697	348,500	104.76
	60	0.750	16.500	40.64	1.4849	4.712	4.32	666	333,000	138.17
	80	0.938	16.124	50.28	1.4180	4.712	4.22	636	318,000	170.75
	100	1.156	15.688	61.17	1.3423	4.712	4.11	602	301,000	208.00
	120	1.375	15.250	71.82	1.2684	4.712	3.99	569	284,500	244.14
	140	1.562	14.876	80.66	1.2070	4.712	3.89	540	270,000	274.30
	160	1.781	14.438	90.75	1.1370	4.712	3.78	510	255,000	308.55
20	5S	0.188	19.624	11.70	2.1004	5.236	5.14	943	471,500	39.76
	10S	0.218	19.564	13.55	2.0878	5.236	5.12	937	467,500	45.98
	10	0.250	19.500	15.51	2.0740	5.236	5.11	930	465,000	52.73
	20, ST	0.375	19.250	23.12	2.0211	5.236	5.04	902	451,000	78.60
	30, XS	0.500	19.000	30.63	1.9689	5.236	4.97	883	441,500	104.13
	40	0.594	18.812	36.21	1.9302	5.236	4.92	866	433,000	123.06
	60	0.812	18.376	48.95	1.8417	5.236	4.81	826	413,000	166.50

TABLE 1.85 (Continued)

Nominal Pipe Size (in.)	Outside Diameter (in.)	Schedule No.	Wall Thickness (in.)	Inside Diameter (in.)	Cross-sectional area Metal (in.²)	Cross-sectional area Flow (ft²)	Circumference, ft, or surface, ft²/ft of length Outside	Circumference, ft, or surface, ft²/ft of length Inside	Capacity at 1 ft/sec Velocity U.S. gal/min	Capacity at 1 ft/sec Velocity lb/hr water	Weight of Plain-end Pipe (lb/ft)
		80	1.031	17.938	61.44	1.7550	5.236	4.70	787	393,500	208.92
		100	1.281	17.438	75.33	1.6585	5.236	4.57	744	372,000	256.15
		120	1.500	17.000	87.18	1.5763	5.236	4.45	707	353,500	296.37
		140	1.750	16.500	100.3	1.4849	5.236	4.32	665	332,500	341.10
		160	1.969	16.062	111.5	1.4071	5.236	4.21	632	316,000	379.14
24	24	5S	0.218	23.564	16.29	3.0285	6.283	6.17	1350	679,500	55.08
		10, 10S	0.250	23.500	18.65	3.012	6.283	6.15	1350	675,000	63.41
		20, ST	0.375	23.250	27.83	2.948	6.283	6.09	1325	662,500	94.62
		XS	0.500	23.000	36.90	2.885	6.283	6.02	1295	642,500	125.49
		30	0.562	22.876	41.39	2.854	6.283	5.99	1281	640,500	140.80
		40	0.688	22.624	50.39	2.792	6.283	5.92	1253	626,500	171.17
		60	0.969	22.062	70.11	2.655	6.283	5.78	1192	596,000	238.29
		80	1.219	21.562	87.24	2.536	6.283	5.64	1138	569,000	296.53
		100	1.531	20.938	108.1	2.391	6.283	5.48	1073	536,500	367.45
		120	1.812	20.376	126.3	2.264	6.283	5.33	1016	508,000	429.50
		140	2.062	19.876	142.1	2.155	6.283	5.20	965	482,500	483.24
		160	2.344	19.312	159.5	2.034	6.283	5.06	913	456,500	542.09
30	30	5S	0.250	29.500	23.37	4.746	7.854	7.72	2130	1,065,000	79.43
		10, 10S	0.312	29.376	29.10	4.707	7.854	7.69	2110	1,055,000	99.08
		ST	0.375	29.250	34.90	4.666	7.854	7.66	2094	1,048,000	118.65
		20, XS	0.500	29.000	46.34	4.587	7.854	7.59	2055	1,027,500	157.53
		30	0.625	28.750	57.68	4.508	7.854	7.53	2020	1,010,000	196.08

Schedule Nos. 5S, 10S, and 40S ANSI/ASME B.36.19-1985. "Stainless Steel Pipe." ST = standard wall, XS = extra strong wall, XX = double extra strong wall are all taken from ANSI/ASME, B.36.10M-1985, "Welded and Seamless Wrought-steel Pipe." Wrought-iron pipe has slightly thicker walls, approximately 3%, but the same weight per foot, because of lower density. Decimal thicknesses for respective pipe sizes represent their nominal or average wall dimensions. Mill tolerances as high as 12½% are permitted.

Plain-end pipe is produced by a square cut. Pipe is also shipped from the mills threaded, with a threaded coupling on one end, or with the ends beveled for welding, or grooved or sized for patented couplings. Weights per foot for threaded and coupled pipe are slightly greater because of the weight of the coupling, but it is not available larger than 12 in., or lighter than Schedule 30 sizes 8 through 12 in., or Schedule 40 6 in. and smaller.

From *Chemical Engineer's Handbook*, New York, McGraw-Hill, 4th ed., 1963. Used by permission.

TABLE 1.86 Properties and Dimensions of Steel Pipe[a]

	Dimensions						Couplings			Properties		
				Weight per Foot (lb)								
Nom. Diam. (in.)	Outside Diam. (in.)	Inside Diam. (in.)	Thick-ness (in.)	Plain Ends	Thread and Coupling	Threads per Inch	Outside Diam. (in.)	Length (in.)	Weight (lb)	I (in.^{4})	A (in.^{2})	k (in.)
						Schedule 40ST						
$\frac{1}{8}$	0.405	0.269	0.068	0.24	0.25	27	0.562	$\frac{7}{8}$	0.03	0.001	0.072	0.12
$\frac{1}{4}$	0.540	0.364	0.088	0.42	0.43	18	0.685	1-	0.04	0.003	0.125	0.16
$\frac{3}{8}$	0.675	0.493	0.091	0.57	0.57	18	0.848	$1\frac{1}{8}$	0.07	0.007	0.167	0.21
$\frac{1}{2}$	0.840	0.622	0.109	0.85	0.85	14	1.024	$1\frac{3}{8}$	0.12	0.017	0.250	0.26
$\frac{3}{4}$	1.050	0.824	0.113	1.13	1.13	14	1.281	$1\frac{5}{8}$	0.21	0.037	0.333	0.33
1	1.315	1.049	0.133	1.68	1.68	$11\frac{1}{2}$	1.576	$1\frac{7}{8}$	0.35	0.087	0.494	0.42
$1\frac{1}{4}$	1.660	1.380	0.140	2.27	2.28	$11\frac{1}{2}$	1.950	$2\frac{1}{8}$	0.55	0.195	0.669	0.54
$1\frac{1}{2}$	1.900	1.610	0.145	2.72	2.73	$11\frac{1}{2}$	2.218	$2\frac{3}{8}$	0.76	0.310	0.799	0.62
2	2.375	2.067	0.154	3.65	3.68	$11\frac{1}{2}$	2.760	$2\frac{5}{8}$	1.23	0.666	1.075	0.79
$2\frac{1}{2}$	2.875	2.469	0.203	5.79	5.82	8	3.276	$2\frac{7}{8}$	1.76	1.530	1.704	0.95
3	3.500	3.068	0.216	7.58	7.62	8	3.948	$3\frac{1}{8}$	2.55	3.017	2.228	1.16
$3\frac{1}{2}$	4.000	3.548	0.226	9.11	9.20	8	4.591	$3\frac{5}{8}$	4.33	4.788	2.680	1.34
4	4.500	4.026	0.237	10.79	10.89	8	5.091	$3\frac{5}{8}$	5.41	7.233	3.174	1.51
5	5.563	5.047	0.258	14.62	14.81	8	6.296	$4\frac{1}{8}$	9.16	15.16	4.300	1.88
6	6.625	6.065	0.280	18.97	19.19	8	7.358	$4\frac{1}{8}$	10.82	28.14	5.581	2.25
8	8.625	8.071	0.277	24.70	25.00	8	9.420	$4\frac{5}{8}$	15.84	63.35	7.265	2.95
8	8.625	7.981	0.322	28.55	28.81	8	9.420	$4\frac{5}{8}$	15.84	72.49	8.399	2.94
10	10.750	10.192	0.279	31.20	32.00	8	11.721	$6\frac{1}{8}$	33.92	125.4	9.178	3.70
10	10.750	10.136	0.307	34.24	35.00	8	11.721	$6\frac{1}{8}$	33.92	137.4	10.07	3.69
10	10.750	10.020	0.365	40.48	41.13	8	11.721	$6\frac{1}{8}$	33.92	160.7	11.91	3.67
12	12.750	12.090	0.330	43.77	45.00	8	13.958	$6\frac{1}{8}$	48.27	248.5	12.88	4.39
12	12.750	12.000	0.375	49.56	50.71	8	13.958	$6\frac{1}{8}$	48.27	279.3	14.38	4.38
						Schedule 80XS						
$\frac{1}{8}$	0.405	0.215	0.095	0.31	0.32	27	0.582	$1\frac{1}{8}$	0.05	0.001	0.093	0.12
$\frac{1}{4}$	0.540	0.302	0.119	0.54	0.54	18	0.724	$1\frac{3}{8}$	0.07	0.004	0.157	0.16
$\frac{3}{8}$	0.675	0.423	0.126	0.74	0.75	18	0.898	$1\frac{5}{8}$	0.13	0.009	0.217	0.20
$\frac{1}{2}$	0.840	0.546	0.147	1.09	1.10	14	1.085	$1\frac{7}{8}$	0.22	0.020	0.320	0.25
$\frac{3}{4}$	1.050	0.742	0.154	1.47	1.49	14	1.316	$2\frac{1}{8}$	0.33	0.045	0.433	0.32
1	1.315	0.957	0.179	2.17	2.20	$11\frac{1}{2}$	1.575	$2\frac{3}{8}$	0.47	0.106	0.639	0.41
$1\frac{1}{4}$	1.660	1.278	0.191	3.00	3.05	$11\frac{1}{2}$	2.054	$2\frac{7}{8}$	1.04	0.242	0.881	0.52
$1\frac{1}{2}$	1.900	1.500	0.200	3.63	3.69	$11\frac{1}{2}$	2.294	$2\frac{7}{8}$	1.17	0.391	1.068	0.61
2	2.375	1.939	0.218	5.02	5.13	$11\frac{1}{2}$	2.870	$3\frac{5}{8}$	2.17	0.868	1.477	0.77
$2\frac{1}{2}$	2.875	2.323	0.276	7.66	7.83	8	3.389	$4\frac{1}{8}$	3.43	1.924	2.254	0.92
3	3.500	2.900	0.300	10.25	10.46	8	4.014	$4\frac{1}{8}$	4.13	3.894	3.016	1.14
$3\frac{1}{2}$	4.000	3.364	0.318	12.51	12.82	8	4.628	$4\frac{5}{8}$	6.29	6.280	3.678	1.31
4	4.500	3.826	0.337	14.98	15.39	8	5.233	$4\frac{5}{8}$	8.16	9.610	4.407	1.48
5	5.563	4.813	0.375	20.78	21.42	8	6.420	$5\frac{1}{8}$	12.87	20.67	6.112	1.84
6	6.625	5.761	0.432	28.57	29.33	8	7.482	$5\frac{1}{8}$	15.18	40.49	8.405	2.20
8	8.625	7.625	0.500	43.39	44.72	8	9.596	$6\frac{1}{8}$	26.63	105.7	12.76	2.88
10	10.750	9.750	0.500	54.74	56.94	8	11.958	$6\frac{5}{8}$	44.16	211.9	16.10	3.63
12	12.750	11.750	0.500	65.42	68.02	8	13.958	$6\frac{5}{8}$	51.99	361.5	19.24	4.34
						Schedule XX						
$\frac{1}{2}$	0.840	0.252	0.294	1.71	1.73	14	1.085	$1\frac{7}{8}$	0.22	0.024	0.504	0.22
$\frac{3}{4}$	1.050	0.434	0.308	2.44	2.46	14	1.316	$2\frac{1}{8}$	0.33	0.058	0.718	0.28

TABLE 1.86 (*continued*)

	Dimensions						Couplings			Properties		
					Weight per Foot (lb)							
Nom. Diam. (in.)	Outside Diam. (in.)	Inside Diam. (in.)	Thick-ness (in.)	Plain Ends	Thread and Coupling	Threads per Inch	Outside Diam. (in.)	Length (in.)	Weight (lb)	I (in.^{4})	A (in.^{2})	k (in.)
1	1.315	0.599	0.358	3.66	3.68	$11\frac{1}{2}$	1.575	$2\frac{3}{8}$	0.47	0.140	1.076	0.36
$1\frac{1}{4}$	1.660	0.896	0.382	5.21	5.27	$11\frac{1}{2}$	2.054	$2\frac{7}{8}$	1.04	0.341	1.534	0.47
$1\frac{1}{2}$	1.900	1.100	0.400	6.41	6.47	$11\frac{1}{2}$	2.294	$2\frac{7}{8}$	1.17	0.568	1.885	0.55
2	2.375	1.503	0.436	9.03	9.14	$11\frac{1}{2}$	2.870	$3\frac{5}{8}$	2.17	1.311	2.656	0.70
$2\frac{1}{2}$	2.875	1.771	0.552	13.70	13.87	8	3.389	$4\frac{1}{8}$	3.43	2.871	4.028	0.84
3	3.500	2.300	0.600	18.58	18.79	8	4.014	$4\frac{1}{8}$	4.13	5.992	5.466	1.05
$3\frac{1}{2}$	4.000	2.728	0.636	22.85	23.16	8	4.628	$4\frac{5}{8}$	6.29	9.848	6.721	1.21
4	4.500	3.152	0.674	27.54	27.95	8	5.233	$4\frac{5}{8}$	8.16	15.28	8.101	1.37
5	5.563	4.063	0.750	38.55	39.20	8	6.420	$5\frac{1}{8}$	12.87	33.64	11.34	1.72
6	6.625	4.897	0.864	53.16	53.92	8	7.482	$5\frac{1}{8}$	15.18	66.33	15.64	2.06
8	8.625	6.875	0.875	72.42	73.76	8	9.596	$6\frac{1}{8}$	26.63	162.0	21.30	2.76

Large O. D. Pipe

Pipe 14 in. and larger is sold by actual O. S. diameter and thickness.

Sizes 14, 15, and 16 in. are available regularly in thicknesses varying by $\frac{1}{16}$ in. from $\frac{1}{4}$ to 1 in., inclusive.

All pipe is furnished random length unless otherwise ordered, viz: 12 to 22 ft with privilege of furnishing 5% in 6 to 12 ft lengths. Pipe railing is most economically detailed with slip joints and random lengths between couplings.

[a]*Steel Construction*, 1980, A.I.S.C.

1.6.6 Standard Structural Shapes—Aluminum*

TABLE 1.87   Aluminum Association Standard Channels—Dimensions, Areas, Weights, and Section Properties[a]

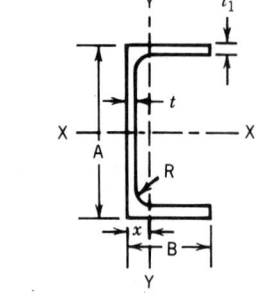

Size				Flange Thickness	Web Thickness	Fillet Radius	Section Properties[d]						
							Axis X–X			Axis Y–Y			
Depth A (in.)	Width B (in.)	Area[b] (in.^{2})	Weight[c] (lb/ft)	t_1 (in.)	t (in.)	R (in.)	I (in.^{4})	S (in.^{3})	r (in.)	I (in.^{4})	S (in.^{3})	r (in.)	x (in.)
2.00	1.00	0.491	0.557	0.13	0.13	0.10	0.288	0.288	0.766	0.045	0.064	0.303	0.298
2.00	1.25	0.911	1.071	0.26	0.17	0.15	0.546	0.546	0.774	0.139	0.178	0.391	0.471
3.00	1.50	0.965	1.135	0.20	0.13	0.25	1.41	0.94	1.21	0.22	0.22	0.47	0.49
3.00	1.75	1.358	1.597	0.26	0.17	0.25	1.97	1.31	1.20	0.42	0.37	0.55	0.62
4.00	2.00	1.478	1.738	0.23	0.15	0.25	3.91	1.95	1.63	0.60	0.45	0.64	0.65
4.00	2.25	1.982	2.331	0.29	0.19	0.25	5.21	2.60	1.62	1.02	0.69	0.72	0.78

*Tables 1.87–1.101 from *Aluminum Standards and Data*. Copyright © 1984 The Aluminum Association.

TABLE 1.87 (*Continued*)

Size				Flange Thickness	Web Thickness	Fillet Radius	Section Properties[d]						
Depth	Width						Axis X–X			Axis Y–Y			
A	B	Area[b]	Weight[c]	t_1	t	R	I	S	r	I	S	r	x
(in.)	(in.)	(in.²)	(lb/ft)	(in.)	(in.)	(in.)	(in.⁴)	(in.³)	(in.)	(in.⁴)	(in.³)	(in.)	(in.)
5.00	2.25	1.881	2.212	0.26	0.15	0.30	7.88	3.15	2.05	0.98	0.64	0.72	0.73
5.00	2.75	2.627	3.089	0.32	0.19	0.30	11.14	4.45	2.06	2.05	1.14	0.88	0.95
6.00	2.50	2.410	2.834	0.29	0.17	0.30	14.35	4.78	2.44	1.53	0.90	0.80	0.79
6.00	3.25	3.427	4.030	0.35	0.21	0.30	21.04	7.01	2.48	3.76	1.76	1.05	1.12
7.00	2.75	2.725	3.205	0.29	0.17	0.30	22.09	6.31	2.85	2.10	1.10	0.88	0.84
7.00	3.50	4.009	4.715	0.38	0.21	0.30	33.79	9.65	2.90	5.13	2.23	1.13	1.20
8.00	3.00	3.526	4.147	0.35	0.19	0.30	37.40	9.35	3.26	3.25	1.57	0.96	0.93
8.00	3.75	4.923	5.789	0.41	0.25	0.35	52.69	13.17	3.27	.7.13	2.82	1.20	1.22
9.00	3.25	4.237	4.983	0.35	0.23	0.35	54.41	12.09	3.58	4.40	1.89	1.02	0.93
9.00	4.00	5.927	6.970	0.44	0.29	0.35	78.31	17.40	3.63	9.61	3.49	1.27	1.25
10.00	3.50	5.218	6.136	0.41	0.25	0.35	83.22	16.64	3.99	6.33	2.56	1.10	1.02
10.00	4.25	7.109	8.360	0.50	0.31	0.40	116.15	23.23	4.04	13.02	4.47	1.35	1.34
12.00	4.00	7.036	8.274	0.47	0.29	0.40	159.76	26.63	4.77	11.03	3.86	1.25	1.14
12.00	5.00	10.053	11.822	0.62	0.35	0.45	239.69	39.95	4.88	25.74	7.60	1.60	1.61

[a]Users are encouraged to ascertain current availability of particular structural shapes through inquiries to their suppliers.
[b]Areas listed are based on nominal dimensions.
[c]Weights per foot are based on nominal dimensions and a density of 0.098 lb/in.³, which is the density of alloy 6061.
[d]I = moment of inertia; S = section modulus; r = radius of gyration.

TABLE 1.88   Aluminum Association Standard I Beams—Dimensions, Areas, Weights, and Section Properties[a]

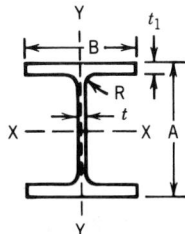

Size				Flange Thickness	Web Thickness	Fillet Radius	Section Properties[d]					
Depth	Width						Axis X–X			Axis Y–Y		
A	B	Area[b]	Weight[c]	t_1	t	R	I	S	r	I	S	r
(in.)	(in.)	(in.²)	(lb/ft)	(in.)	(in.)	(in.)	(in.⁴)	(in.³)	(in.)	(in.⁴)	(in.³)	(in.)
3.00	2.50	1.392	1.637	0.20	0.13	0.25	2.24	1.49	1.27	0.52	0.42	0.61
3.00	2.50	1.726	2.030	0.26	0.15	0.25	2.71	1.81	1.25	0.68	0.54	0.63
4.00	3.00	1.965	2.311	0.23	0.15	0.25	5.62	2.81	1.69	1.04	0.69	0.73
4.00	3.00	2.375	2.793	0.29	0.17	0.25	6.71	3.36	1.68	1.31	0.87	0.74
5.00	3.50	3.146	3.700	0.32	0.19	0.30	13.94	5.58	2.11	2.29	1.31	0.85
6.00	4.00	3.427	4.030	0.29	0.19	0.30	21.99	7.33	2.53	3.10	1.55	0.95
6.00	4.00	3.990	4.692	0.35	0.21	0.30	25.50	8.50	2.53	3.74	1.87	0.97
7.00	4.50	4.932	5.800	0.38	0.23	0.30	42.89	12.25	2.95	5.78	2.57	1.08
8.00	5.00	5.256	6.181	0.35	0.23	0.30	59.69	14.92	3.37	7.30	2.92	1.18
8.00	5.00	5.972	7.023	0.41	0.25	0.30	67.78	16.94	3.37	8.55	3.42	1.20
9.00	5.50	7.110	8.361	0.44	0.27	0.30	102.02	22.67	3.79	12.22	4.44	1.31
10.00	6.00	7.352	8.646	0.41	0.25	0.40	132.09	26.42	4.24	14.78	4.93	1.42
10.00	6.00	8.747	10.286	0.50	0.29	0.40	155.79	31.16	4.22	18.03	6.01	1.44
12.00	7.00	9.925	11.672	0.47	0.29	0.40	255.57	42.60	5.07	26.90	7.69	1.65
12.00	7.00	12.153	14.292	0.62	0.31	0.40	317.33	52.89	5.11	35.48	10.14	1.71

[a]Users are encouraged to ascertain current availability of particular structural shapes through inquiries to their suppliers.
[b]Areas listed are based on nominal dimensions.
[c]Weights per foot are based on nominal dimensions and a density of 0.098 lb/in.³, which is the density of alloy 6061.
[d]I = moment of inertia; S = section modulus; r = radius of gyration.

TABLE 1.89   Standard Structural Shapes–Equal Angles[a]

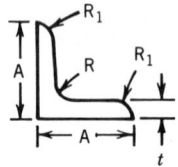

A	t	R	R_1	Area[b] (in.^{2})	Weight per Foot[c] (lb)
$\frac{3}{4}$	$\frac{1}{8}$	$\frac{1}{8}$	$\frac{3}{32}$	0.171	0.201
$\frac{3}{4}$	$\frac{3}{16}$	$\frac{1}{8}$	$\frac{3}{32}$	0.246	0.289
1	$\frac{3}{32}$	$\frac{1}{8}$	$\frac{3}{32}$	0.179	0.211
1	$\frac{1}{8}$	$\frac{1}{8}$	$\frac{3}{32}$	0.234	0.275
1	$\frac{3}{16}$	$\frac{1}{8}$	$\frac{3}{32}$	0.340	0.400
1	$\frac{1}{4}$	$\frac{1}{8}$	$\frac{3}{32}$	0.437	0.514
$1\frac{1}{4}$	$\frac{1}{8}$	$\frac{3}{16}$	$\frac{1}{8}$	0.292	0.343
$1\frac{1}{4}$	$\frac{3}{16}$	$\frac{3}{16}$	$\frac{1}{8}$	0.434	0.510
$1\frac{1}{4}$	$\frac{1}{4}$	$\frac{3}{16}$	$\frac{1}{8}$	0.558	0.656
$1\frac{1}{2}$	$\frac{1}{8}$	$\frac{3}{16}$	$\frac{1}{8}$	0.360	0.423
$1\frac{1}{2}$	$\frac{3}{16}$	$\frac{3}{16}$	$\frac{1}{8}$	0.529	0.619
$1\frac{1}{2}$	$\frac{1}{4}$	$\frac{3}{16}$	$\frac{1}{8}$	0.688	0.809
$1\frac{3}{4}$	$\frac{1}{8}$	$\frac{3}{16}$	$\frac{1}{8}$	0.423	0.497
$1\frac{3}{4}$	$\frac{3}{16}$	$\frac{3}{16}$	$\frac{1}{8}$	0.622	0.731
$1\frac{3}{4}$	$\frac{1}{4}$	$\frac{3}{16}$	$\frac{1}{8}$	0.813	0.956
$1\frac{3}{4}$	$\frac{5}{16}$	$\frac{3}{16}$	$\frac{1}{8}$	0.996	1.171
2	$\frac{1}{8}$	$\frac{1}{4}$	$\frac{1}{8}$	0.491	0.577
2	$\frac{3}{16}$	$\frac{1}{4}$	$\frac{1}{8}$	0.723	0.850
2	$\frac{1}{4}$	$\frac{1}{4}$	$\frac{1}{8}$	0.944	1.110
2	$\frac{5}{16}$	$\frac{1}{4}$	$\frac{1}{8}$	1.160	1.364
2	$\frac{3}{8}$	$\frac{1}{4}$	$\frac{1}{8}$	1.366	1.606
$2\frac{1}{2}$	$\frac{1}{8}$	$\frac{1}{4}$	$\frac{1}{8}$	0.616	0.724
$2\frac{1}{2}$	$\frac{3}{16}$	$\frac{1}{4}$	$\frac{1}{8}$	0.910	1.070
$2\frac{1}{2}$	$\frac{1}{4}$	$\frac{1}{4}$	$\frac{1}{8}$	1.194	1.404
$2\frac{1}{2}$	$\frac{5}{16}$	$\frac{1}{4}$	$\frac{1}{8}$	1.470	1.729
$2\frac{1}{2}$	$\frac{3}{8}$	$\frac{1}{4}$	$\frac{1}{8}$	1.714	2.047
3	$\frac{3}{16}$	$\frac{5}{16}$	$\frac{1}{4}$	1.084	1.275
3	$\frac{1}{4}$	$\frac{5}{16}$	$\frac{1}{4}$	1.432	1.684
3	$\frac{5}{16}$	$\frac{5}{16}$	$\frac{1}{4}$	1.770	2.082
3	$\frac{3}{8}$	$\frac{5}{16}$	$\frac{1}{4}$	2.104	2.474
3	$\frac{7}{16}$	$\frac{5}{16}$	$\frac{1}{4}$	2.428	2.855
3	$\frac{1}{2}$	$\frac{5}{16}$	$\frac{1}{4}$	2.744	3.227
$3\frac{1}{2}$	$\frac{1}{4}$	$\frac{3}{8}$	$\frac{1}{4}$	1.691	1.989
$3\frac{1}{2}$	$\frac{5}{16}$	$\frac{3}{8}$	$\frac{1}{4}$	2.093	2.461
$3\frac{1}{2}$	$\frac{3}{8}$	$\frac{3}{8}$	$\frac{1}{4}$	2.488	2.926
$3\frac{1}{2}$	$\frac{1}{2}$	$\frac{3}{8}$	$\frac{1}{4}$	3.253	3.826
4	$\frac{1}{4}$	$\frac{3}{8}$	$\frac{1}{4}$	1.941	2.283
4	$\frac{5}{16}$	$\frac{3}{8}$	$\frac{1}{4}$	2.406	2.829
4	$\frac{3}{8}$	$\frac{3}{8}$	$\frac{1}{4}$	2.862	3.366
4	$\frac{7}{16}$	$\frac{3}{8}$	$\frac{1}{4}$	3.310	3.893
4	$\frac{1}{2}$	$\frac{3}{8}$	$\frac{1}{4}$	3.753	4.414
4	$\frac{9}{16}$	$\frac{3}{8}$	$\frac{1}{4}$	4.187	4.924
4	$\frac{5}{8}$	$\frac{3}{8}$	$\frac{1}{4}$	4.613	5.425
4	$\frac{11}{16}$	$\frac{3}{8}$	$\frac{1}{4}$	5.032	5.918

TABLE 1.89 (*Continued*)

A	t	R	R_1	Area[b] (in.²)	Weight per Foot[c] (lb)
4	$\frac{3}{4}$	$\frac{3}{8}$	$\frac{1}{4}$	5.441	6.399
5	$\frac{3}{8}$	$\frac{1}{2}$	$\frac{3}{8}$	3.603	4.237
5	$\frac{7}{16}$	$\frac{1}{2}$	$\frac{3}{8}$	4.177	4.912
5	$\frac{1}{2}$	$\frac{1}{2}$	$\frac{3}{8}$	4.743	5.578
5	$\frac{5}{8}$	$\frac{1}{2}$	$\frac{3}{8}$	5.853	6.883
6	$\frac{3}{8}$	$\frac{1}{2}$	$\frac{3}{8}$	4.353	5.119
6	$\frac{7}{16}$	$\frac{1}{2}$	$\frac{3}{8}$	5.052	5.941
6	$\frac{1}{2}$	$\frac{1}{2}$	$\frac{3}{8}$	5.743	6.754
6	$\frac{5}{8}$	$\frac{1}{2}$	$\frac{3}{8}$	7.102	8.352
8	$\frac{1}{2}$	$\frac{5}{8}$	$\frac{3}{8}$	7.773	9.141
8	$\frac{3}{4}$	$\frac{5}{8}$	$\frac{3}{8}$	11.461	13.478
8	1	$\frac{5}{8}$	$\frac{3}{8}$	15.023	17.667

[a] Users are encouraged to ascertain current availability of particular structural shapes through inquiries to their suppliers.
[b] Areas listed are based on nominal dimensions.
[c] Weights per foot are based on nominal dimensions and a density of 0.098 lb/in.³, which is the density of alloy 6061.

TABLE 1.90 Standard Structural Shapes—Unequal Angles[a]

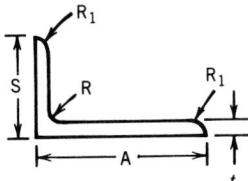

A	B	t	R	R_1	Area[b] (in.²)	Weight per Foot[c] (lb)
$1\frac{1}{4}$	$\frac{3}{4}$	$\frac{3}{32}$	$\frac{3}{32}$	$\frac{3}{64}$	0.180	0.212
$1\frac{1}{4}$	1	$\frac{1}{8}$	$\frac{1}{8}$	$\frac{1}{16}$	0.267	0.314
$1\frac{1}{2}$	$\frac{3}{4}$	$\frac{1}{8}$	$\frac{1}{8}$	$\frac{1}{16}$	0.267	0.314
$1\frac{1}{2}$	$\frac{3}{4}$	$\frac{3}{16}$	$\frac{1}{8}$	$\frac{3}{32}$	0.386	0.454
$1\frac{1}{2}$	1	$\frac{5}{32}$	$\frac{5}{32}$	$\frac{5}{64}$	0.368	0.433
$1\frac{1}{2}$	1	$\frac{1}{4}$	$\frac{3}{16}$	$\frac{1}{8}$	0.563	0.662
$1\frac{1}{2}$	$1\frac{1}{4}$	$\frac{1}{8}$	$\frac{3}{16}$	$\frac{1}{8}$	0.329	0.387
$1\frac{1}{2}$	$1\frac{1}{4}$	$\frac{3}{16}$	$\frac{3}{16}$	$\frac{1}{8}$	0.481	0.566
$1\frac{1}{2}$	$1\frac{1}{4}$	$\frac{1}{4}$	$\frac{3}{16}$	$\frac{1}{8}$	0.624	0.734
$1\frac{3}{4}$	$1\frac{1}{4}$	$\frac{1}{8}$	$\frac{3}{16}$	$\frac{1}{8}$	0.358	0.421
$1\frac{3}{4}$	$1\frac{1}{4}$	$\frac{3}{16}$	$\frac{3}{16}$	$\frac{1}{8}$	0.528	0.621
$1\frac{3}{4}$	$1\frac{1}{4}$	$\frac{1}{4}$	$\frac{3}{16}$	$\frac{1}{8}$	0.688	0.809
2	$1\frac{1}{2}$	$\frac{1}{8}$	$\frac{3}{16}$	$\frac{1}{8}$	0.422	0.496
2	$1\frac{1}{2}$	$\frac{3}{16}$	$\frac{3}{16}$	$\frac{1}{8}$	0.622	0.731
2	$1\frac{1}{2}$	$\frac{1}{4}$	$\frac{3}{16}$	$\frac{1}{8}$	0.813	0.956
2	$1\frac{1}{2}$	$\frac{3}{8}$	$\frac{3}{16}$	$\frac{1}{8}$	1.172	1.378
$2\frac{1}{2}$	$1\frac{1}{2}$	$\frac{3}{16}$	$\frac{1}{4}$	$\frac{1}{8}$	0.723	0.850
$2\frac{1}{2}$	$1\frac{1}{2}$	$\frac{1}{4}$	$\frac{1}{4}$	$\frac{1}{8}$	0.944	1.110
$2\frac{1}{2}$	$1\frac{1}{2}$	$\frac{5}{16}$	$\frac{3}{16}$	$\frac{1}{8}$	1.152	1.355
$2\frac{1}{2}$	2	$\frac{1}{8}$	$\frac{1}{4}$	$\frac{1}{8}$	0.554	0.652
$2\frac{1}{2}$	2	$\frac{3}{16}$	$\frac{1}{4}$	$\frac{1}{8}$	0.817	0.961
$2\frac{1}{2}$	2	$\frac{1}{4}$	$\frac{1}{4}$	$\frac{1}{8}$	1.069	1.257

TABLE 1.90 (*Continued*)

A	B	t	R	R_1	Area[b] (in.^{2})	Weight per Foot[c] (lb)
$2\frac{1}{2}$	2	$\frac{5}{16}$	$\frac{1}{4}$	$\frac{1}{8}$	1.314	1.545
$2\frac{1}{2}$	2	$\frac{3}{8}$	$\frac{1}{4}$	$\frac{1}{8}$	1.554	1.828
3	2	$\frac{3}{16}$	$\frac{5}{16}$	$\frac{3}{16}$	0.911	1.071
3	2	$\frac{1}{4}$	$\frac{5}{16}$	$\frac{3}{16}$	1.193	1.403
3	2	$\frac{5}{16}$	$\frac{5}{16}$	$\frac{3}{16}$	1.471	1.730
3	2	$\frac{3}{8}$	$\frac{5}{16}$	$\frac{3}{16}$	1.740	2.046
3	2	$\frac{7}{16}$	$\frac{5}{16}$	$\frac{3}{16}$	2.001	2.353
3	$2\frac{1}{2}$	$\frac{1}{4}$	$\frac{5}{16}$	$\frac{1}{4}$	1.307	1.537
3	$2\frac{1}{2}$	$\frac{5}{16}$	$\frac{5}{16}$	$\frac{1}{4}$	1.614	1.898
3	$2\frac{1}{2}$	$\frac{3}{8}$	$\frac{5}{16}$	$\frac{1}{4}$	1.916	2.253
$3\frac{1}{2}$	$2\frac{1}{2}$	$\frac{1}{4}$	$\frac{5}{16}$	$\frac{1}{4}$	1.432	1.684
$3\frac{1}{2}$	$2\frac{1}{2}$	$\frac{5}{16}$	$\frac{5}{16}$	$\frac{1}{4}$	1.770	2.082
$3\frac{1}{2}$	$2\frac{1}{2}$	$\frac{3}{8}$	$\frac{5}{16}$	$\frac{1}{4}$	2.104	2.474
$3\frac{1}{2}$	$2\frac{1}{2}$	$\frac{1}{2}$	$\frac{5}{16}$	$\frac{1}{4}$	2.744	3.227
$3\frac{1}{2}$	3	$\frac{1}{4}$	$\frac{3}{8}$	$\frac{1}{4}$	1.566	1.842
$3\frac{1}{2}$	3	$\frac{5}{16}$	$\frac{3}{8}$	$\frac{1}{4}$	1.937	2.278
$3\frac{1}{2}$	3	$\frac{3}{8}$	$\frac{3}{8}$	$\frac{1}{4}$	2.300	2.705
$3\frac{1}{2}$	3	$\frac{1}{2}$	$\frac{3}{8}$	$\frac{1}{4}$	3.003	3.532
4	3	$\frac{1}{4}$	$\frac{3}{8}$	$\frac{1}{4}$	1.691	1.988
4	3	$\frac{5}{16}$	$\frac{3}{8}$	$\frac{1}{4}$	2.091	2.459
4	3	$\frac{3}{8}$	$\frac{3}{8}$	$\frac{1}{4}$	2.488	2.926
4	3	$\frac{7}{16}$	$\frac{3}{8}$	$\frac{1}{4}$	2.874	3.380
4	3	$\frac{1}{2}$	$\frac{3}{8}$	$\frac{1}{4}$	3.253	3.826
4	3	$\frac{5}{8}$	$\frac{3}{8}$	$\frac{1}{4}$	3.988	4.690
4	$3\frac{1}{2}$	$\frac{3}{8}$	$\frac{3}{8}$	$\frac{5}{16}$	2.660	3.128
4	$3\frac{1}{2}$	$\frac{1}{2}$	$\frac{3}{8}$	$\frac{5}{16}$	3.488	4.102
5	3	$\frac{3}{8}$	$\frac{3}{8}$	$\frac{5}{16}$	2.848	3.349
5	3	$\frac{1}{2}$	$\frac{3}{8}$	$\frac{5}{16}$	3.738	4.396
5	$3\frac{1}{2}$	$\frac{5}{16}$	$\frac{7}{16}$	$\frac{5}{16}$	2.558	3.008
5	$3\frac{1}{2}$	$\frac{3}{8}$	$\frac{7}{16}$	$\frac{5}{16}$	3.046	3.582
5	$3\frac{1}{2}$	$\frac{7}{16}$	$\frac{7}{16}$	$\frac{5}{16}$	3.527	4.148
5	$3\frac{1}{2}$	$\frac{1}{2}$	$\frac{7}{16}$	$\frac{5}{16}$	4.000	4.704
5	$3\frac{1}{2}$	$\frac{5}{8}$	$\frac{7}{16}$	$\frac{5}{16}$	4.921	5.787
6	$3\frac{1}{2}$	$\frac{5}{16}$	$\frac{1}{2}$	$\frac{5}{16}$	2.878	3.385
6	$3\frac{1}{2}$	$\frac{3}{8}$	$\frac{1}{2}$	$\frac{5}{16}$	3.433	4.037
6	$3\frac{1}{2}$	$\frac{1}{2}$	$\frac{1}{2}$	$\frac{5}{16}$	4.512	5.306
6	4	$\frac{3}{8}$	$\frac{1}{2}$	$\frac{3}{8}$	3.603	4.237
6	4	$\frac{7}{16}$	$\frac{1}{2}$	$\frac{3}{8}$	4.179	4.915
6	4	$\frac{1}{2}$	$\frac{1}{2}$	$\frac{3}{8}$	4.743	5.578
6	4	$\frac{9}{16}$	$\frac{1}{2}$	$\frac{3}{8}$	5.298	6.230
6	4	$\frac{5}{8}$	$\frac{1}{2}$	$\frac{3}{8}$	5.853	6.883
6	4	$\frac{3}{4}$	$\frac{1}{2}$	$\frac{3}{8}$	6.931	8.151
8	6	$\frac{5}{8}$	$\frac{1}{2}$	$\frac{5}{16}$	8.371	9.844
8	6	$\frac{11}{16}$	$\frac{1}{2}$	$\frac{3}{8}$	9.152	10.763
8	6	$\frac{3}{4}$	$\frac{1}{2}$	$\frac{3}{8}$	9.931	11.679

[a] Users are encouraged to ascertain current availability of particular structural shapes through inquiries to their suppliers.
[b] Areas listed are based on nominal dimensions.
[c] Weights per foot are based on nominal dimensions and a density of 0.098 lb/in.^{3}, which is the density of alloy 6061.

TABLE 1.91 Channels, American Standard[a]

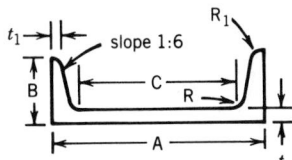

A	B	C	t	t_1	R	R_1	Area[b] (in.²)	Weight per Foot[c] (lb)
3	1.410	$1\frac{3}{4}$	0.170	0.170	0.270	0.100	1.205	1.417
3	1.498	$1\frac{3}{4}$	0.258	0.170	0.270	0.100	1.470	1.729
3	1.596	$1\frac{3}{4}$	0.356	0.170	0.270	0.100	1.764	2.074
4	1.580	$2\frac{3}{4}$	0.180	0.180	0.280	0.110	1.570	1.846
4	1.647	$2\frac{3}{4}$	0.247	0.180	0.280	0.110	1.838	2.161
4	1.720	$2\frac{3}{4}$	0.320	0.180	0.280	0.110	2.129	2.504
5	1.750	$3\frac{3}{4}$	0.190	0.190	0.290	0.110	1.969	2.316
5	1.885	$3\frac{3}{4}$	0.325	0.190	0.290	0.110	2.643	3.108
5	2.032	$3\frac{3}{4}$	0.472	0.190	0.290	0.110	3.380	3.975
6	1.920	$4\frac{1}{2}$	0.200	0.200	0.300	0.120	2.403	2.826
6	1.945	$4\frac{1}{2}$	0.225	0.200	0.300	0.120	2.553	3.002
6	2.034	$4\frac{1}{2}$	0.314	0.200	0.300	0.120	3.088	3.631
6	2.157	$4\frac{1}{2}$	0.437	0.200	0.300	0.120	3.825	4.498
7	2.110	$5\frac{1}{2}$	0.230	0.210	0.310	0.130	3.011	3.541
7	2.194	$5\frac{1}{2}$	0.314	0.210	0.310	0.130	3.599	4.232
7	2.299	$5\frac{1}{2}$	0.419	0.210	0.310	0.130	4.334	5.097
8	2.290	$6\frac{1}{4}$	0.250	0.220	0.320	0.130	3.616	4.252
8	2.343	$6\frac{1}{4}$	0.303	0.220	0.320	0.130	4.040	4.751
8	2.435	$6\frac{1}{4}$	0.395	0.220	0.320	0.130	4.776	5.617
8	2.527	$6\frac{1}{4}$	0.487	0.220	0.320	0.130	5.514	6.484
9	2.430	$7\frac{1}{4}$	0.230	0.230	0.330	0.140	3.915	4.604
9	2.648	$7\frac{1}{4}$	0.448	0.230	0.330	0.140	5.877	6.911
10	2.600	$8\frac{1}{4}$	0.240	0.240	0.340	0.140	4.488	5.278
10	2.886	$8\frac{1}{4}$	0.526	0.240	0.340	0.140	7.348	8.641
12	2.960	10	0.300	0.280	0.380	0.170	6.302	7.411
12	3.047	10	0.387	0.280	0.380	0.170	7.346	8.639
12	3.170	10	0.510	0.280	0.380	0.170	8.822	10.374
15	3.400	$12\frac{3}{8}$	0.400	0.400	0.500	0.240	9.956	11.708
15	3.716	$12\frac{3}{8}$	0.716	0.400	0.500	0.240	14.696	17.282

[a] Users are encouraged to ascertain current availability of particular structural shapes through inquiries to their suppliers.
[b] Areas listed are based on nominal dimensions.
[c] Weights per foot are based on nominal dimensions and a density of 0.098 lb/in.³, which is the density of alloy 6061.

TABLE 1.92   Channels, Shipbuilding and Carbuilding[a]

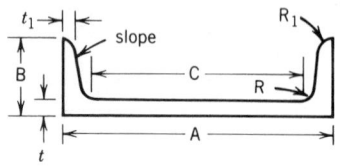

A	B	C	t	t_1	R	R_1	Slope	Area[b] (in.²)	Weight per Foot[c] (lb)
3	2	$1\frac{3}{4}$	0.250	0.250	0.250	0	12 : 12.1	1.900	2.234
3	2	$1\frac{7}{8}$	0.375	0.375	0.188	0.375	0	2.298	2.702
4	$2\frac{1}{2}$	$2\frac{3}{8}$	0.318	0.313	0.375	0.125	1 : 34.9	2.825	3.322
5	$2\frac{7}{8}$	3	0.438	0.438	0.250	0.094	1 : 9.8	4.950	5.821
6	3	$4\frac{1}{2}$	0.500	0.375	0.375	0.250	0	4.909	5.773
6	$3\frac{1}{2}$	4	0.375	0.412	0.480	0.420	1 : 49.6	5.044	5.932
8	3	$5\frac{3}{4}$	0.380	0.380	0.550	0.220	1 : 14.43	5.600	6.586
8	$3\frac{1}{2}$	$5\frac{3}{4}$	0.425	0.471	0.525	0.375	1 : 28.5	6.682	7.858
10	$3\frac{1}{2}$	$7\frac{1}{2}$	0.375	0.375	0.625	0.188	1 : 9	7.298	8.581
10	$3\frac{9}{16}$	$7\frac{1}{2}$	0.438	0.375	0.625	0.188	1 : 9	7.928	9.323
10	$3\frac{5}{8}$	$7\frac{1}{2}$	0.500	0.375	0.625	0.188	1 : 9	8.548	10.052

TABLE 1.93   H Beams[a]

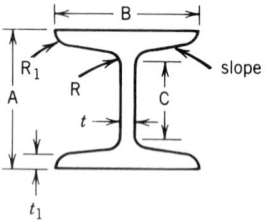

A	B	C	t	t_1	R	R_1	Slope	Area[b] (in.²)	Weight per Foot[c] (lb)
4	4	$2\frac{3}{8}$	0.313	0.290	0.313	0.145	1 : 11.3	4.046	4.758
5	5	$3\frac{3}{8}$	0.313	0.330	0.313	0.165	1 : 13.6	5.522	6.494
6	5.938	$4\frac{3}{8}$	0.250	0.360	0.313	0.180	1 : 15.6	6.678	7.853
8	7.938	$6\frac{1}{4}$	0.313	0.358	0.313	0.179	1 : 18.9	9.554	11.263
8	8.125	$6\frac{1}{4}$	0.500	0.358	0.313	0.179	1 : 18.9	11.050	12.995

[a] Users are encouraged to ascertain current availability of particular structural shapes through inquiries to their suppliers.
[b] Areas listed are based on nominal dimensions.
[c] Weights per foot are based on nominal dimensions and a density of 0.098 lb/in.³, which is the density of alloy 6061.

TABLE 1.94 I Beams[a]

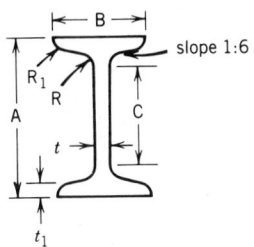

slope 1:6

A	B	C	t	t_1	R	R_1	Area[b] (in.²)	Weight per Foot[c] (lb)
3	2.330	1¾	0.170	0.170	0.270	0.100	1.669	1.963
3	2.509	1¾	0.349	0.170	0.270	0.100	2.203	2.591
4	2.660	2¾	0.190	0.190	0.290	0.110	2.249	2.644
4	2.796	2¾	0.326	0.190	0.290	0.110	2.792	3.283
5	3	3½	0.210	0.210	0.310	0.130	2.917	3.430
5	3.284	3½	0.494	0.210	0.310	0.130	4.337	5.100
6	3.330	4½	0.230	0.230	0.330	0.140	3.658	4.302
6	3.443	4½	0.343	0.230	0.330	0.140	4.336	5.099
7	3.755	5¼	0.345	0.250	0.350	0.150	5.147	6.053
8	4	6¼	0.270	0.270	0.370	0.160	5.398	6.348
8	4.262	6¼	0.532	0.270	0.370	0.160	7.494	8.813
10	4.660	8	0.310	0.310	0.410	0.190	7.452	8.764
12	5	9¾	0.350	0.350	0.450	0.210	9.349	10.994

[a]Users are encouraged to ascertain current availability of particular structural shapes through inquiries to their suppliers.
[b]Areas listed are based on nominal dimensions.
[c]Weights per foot are based on nominal dimensions and a density of 0.098 lb/in.³, which is the density of alloy 6061.

TABLE 1.95 Wide-Flange Beams[a]

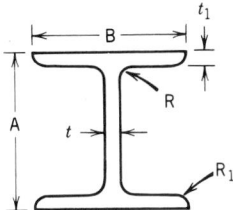

A	B	t	t_1	R	R_1	Area[b] (in.²)	Weight per Foot[c] (lb)
6.000	4.000	0.230	0.279	0.250	—	3.538	4.161
6.000	6.000	0.240	0.269	0.250	—	4.593	5.401
8.000	5.250	0.230	0.308	0.320	—	5.020	5.904
8.000	6.500	0.245	0.398	0.400	—	7.076	8.321
8.000	8.000	0.288	0.433	0.400	—	9.120	10.725
9.750	7.964	0.292	0.433	0.500	—	9.706	11.414
9.900	5.750	0.240	0.340	0.312	0.031	6.205	7.297
11.940	8.000	0.294	0.516	0.600	—	11.772	13.844
12.060	10.000	0.345	0.576	0.600	—	15.593	18.337

[a]Users are encouraged to ascertain current availability of particular structural shapes through inquiries to their suppliers.
[b]Areas listed are based on nominal dimensions.
[c]Weights per foot are based on nominal dimensions and a density of 0.098 lb/in.³, which is the density of alloy 6061.

TABLE 1.96 Tees[a]

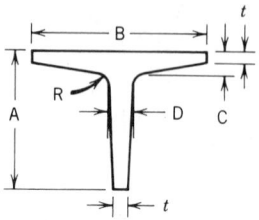

A	B	C	D	t	R	Area[b] (in.²)	Weight per Foot[c] (lb)
2	2	0.312	0.312	0.250	0.250	1.071	1.259
$2\frac{1}{4}$	$2\frac{1}{4}$	0.312	0.312	0.250	0.250	1.208	1.421
$2\frac{1}{2}$	$2\frac{1}{2}$	0.375	0.375	0.312	0.250	1.626	1.912
3	3	0.438	0.438	0.375	0.312	2.310	2.717
4	4	0.438	0.438	0.375	0.500	3.183	3.743

[a] Users are encouraged to ascertain current availability of particular structural shapes through inquiries to their suppliers.
[b] Areas listed are based on nominal dimensions.
[c] Weights per foot are based on nominal dimensions and a density of 0.098 lb/in.³, which is the density of alloy 6061.

TABLE 1.97 Zees[a]

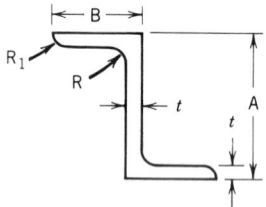

A	B	t	R	R_1	Area[b] (in.²)	Weight per Foot[c] (lb)
3	$2\frac{11}{16}$	0.250	0.312	0.250	1.984	2.333
3	$2\frac{11}{16}$	0.375	0.312	0.250	2.875	3.381
4	$3\frac{1}{16}$	0.250	0.312	0.250	2.422	2.848
$4\frac{1}{16}$	$3\frac{1}{8}$	0.312	0.312	0.250	3.040	3.575
$4\frac{1}{4}$	$3\frac{3}{16}$	0.375	0.312	0.250	3.672	4.318
5	$3\frac{1}{4}$	0.500	0.312	0.250	5.265	6.192
$5\frac{1}{16}$	$3\frac{5}{16}$	0.375	0.312	0.250	4.093	4.813

[a] Users are encouraged to ascertain current availability of particular structural shapes through inquiries to their suppliers.
[b] Areas listed are based on nominal dimensions.
[c] Weights per foot are based on nominal dimensions and a density of 0.098 lb/in.³, which is the density of alloy 6061.

TABLE 1.98 Aluminum Pipe—Diameters, Wall Thicknesses, and Weights

Nominal Pipe Size[a] (in.)	Schedule Number[a]	Outside Diameter (in.)			Inside Diameter (in.)	Wall Thickness (in.)			Weight per Foot (lb)	
		Nom[a]	Min[b, d]	Max[b, d]	Nom	Nom[a]	Min[b]	Max[b]	Nom[c]	Max[b, c]
$\frac{1}{8}$	40	0.405	0.374	0.420	0.269	0.068	0.060	—	0.085	0.091
	80	0.405	0.374	0.420	0.215	0.095	0.083	—	0.109	0.118
$\frac{1}{4}$	40	0.540	0.509	0.555	0.364	0.088	0.077	—	0.147	0.159
	80	0.540	0.509	0.555	0.302	0.119	0.104	—	0.185	0.200
$\frac{3}{8}$	40	0.675	0.644	0.690	0.493	0.091	0.080	—	0.196	0.212
	80	0.675	0.644	0.690	0.493	0.091	0.080	—	0.196	0.212
	5	0.840	0.809	0.855	0.710	0.065	0.053	0.077	0.186	—
	10	0.840	0.809	0.855	0.674	0.083	0.071	0.095	0.232	—
$\frac{1}{2}$	40	0.840	0.809	0.855	0.622	0.109	0.095	—	0.294	0.318
	80	0.840	0.809	0.855	0.546	0.147	0.129	—	0.376	0.406
	160	0.840	0.809	0.855	0.464	0.188	0.164	—	0.453	0.489
	5	1.050	1.019	1.065	0.920	0.065	0.053	0.077	0.237	—
	10	1.050	1.019	1.065	0.884	0.083	0.071	0.095	0.297	—
$\frac{3}{4}$	40	1.050	1.019	1.065	0.824	0.113	0.099	—	0.391	0.422
	80	1.050	1.019	1.065	0.742	0.154	0.135	—	0.510	0.551
	160	1.050	1.019	1.065	0.612	0.219	0.192	—	0.672	0.726
	5	1.315	1.284	1.330	1.185	0.065	0.053	0.077	0.300	—
	10	1.315	1.284	1.330	1.097	0.109	0.095	0.123	0.486	—
1	40	1.315	1.284	1.330	1.049	0.133	0.116	—	0.581	0.627
	80	1.315	1.284	1.330	0.957	0.179	0.157	—	0.751	0.811
	160	1.315	1.284	1.330	0.815	0.250	0.219	—	0.984	1.062
	5	1.660	1.629	1.675	1.530	0.065	0.053	0.077	0.383	—
	10	1.660	1.629	1.675	1.442	0.109	0.095	0.123	0.625	—
$1\frac{1}{4}$	40	1.660	1.629	1.675	1.380	0.140	0.122	—	0.786	0.849
	80	1.660	1.629	1.675	1.278	0.191	0.167	—	1.037	1.120
	160	1.660	1.629	1.675	1.160	0.250	0.219	—	1.302	1.407
	5	1.900	1.869	1.915	1.770	0.065	0.053	0.077	0.441	—
	10	1.900	1.869	1.915	1.682	0.109	0.095	0.123	0.721	—
$1\frac{1}{2}$	40	1.900	1.869	1.915	1.610	0.145	0.127	—	0.940	1.015
	80	1.900	1.869	1.915	1.500	0.200	0.175	—	1.256	1.357
	160	1.900	1.869	1.915	1.338	0.281	0.246	—	1.681	1.815
	5	2.375	2.344	2.406	2.245	0.065	0.053	0.077	0.555	—
	10	2.375	2.344	2.406	2.157	0.109	0.095	0.123	0.913	—
2	40	2.375	2.351	2.399	2.067	0.154	0.135	—	1.264	1.365
	80	2.375	2.351	2.399	1.939	0.218	0.191	—	1.737	1.876
	160	2.375	2.351	2.399	1.687	0.344	0.301	—	2.581	2.788
	5	2.875	2.844	2.906	2.709	0.083	0.071	0.095	0.856	—
	10	2.875	2.844	2.906	2.635	0.120	0.105	0.135	1.221	—
$2\frac{1}{2}$	40	2.875	2.846	2.904	2.469	0.203	0.178	—	2.004	2.164
	80	2.875	2.846	2.904	2.323	0.276	0.242	—	2.650	2.862
	160	2.875	2.846	2.904	2.125	0.375	0.328	—	3.464	3.741
	5	3.500	3.469	3.531	3.334	0.083	0.071	0.095	1.048	—
	10	3.500	3.469	3.531	3.260	0.120	0.105	0.135	1.498	—
3	40	3.500	3.465	3.535	3.068	0.216	0.189	—	2.621	2.830
	80	3.500	3.465	3.535	2.900	0.300	0.262	—	3.547	3.830
	160	3.500	3.465	3.535	2.624	0.438	0.383	—	4.955	5.351
$3\frac{1}{2}$	5	4.000	3.969	4.031	3.834	0.083	0.071	0.095	1.201	—
	10	4.000	3.969	4.031	3.760	0120	0.105	0.135	1.720	—
	40	4.000	3.960	4.040	3.548	0.226	0.198	—	3.151	3.403
	80	4.000	3.960	4.040	3.364	0.318	0.278	—	4.326	4.672
4	5	4.500	4.469	4.531	4.334	0.083	0.071	0.095	1.354	—
	10	4.500	4.469	4.531	4.160	0.120	0.105	0.135	1.942	—
	40	4.500	4.455	4.545	4.026	0.237	0.207	—	3.733	4.031
	80	4.500	4.455	4.545	3.826	0.337	0.295	—	5.183	5.598
	120	4.500	4.455	4.545	3.624	0.438	0.383	—	6.573	7.099
	160	4.500	4.455	4.545	3.438	0.531	0.465	—	7.786	8.409

TABLE 1.98 (*Continued*)

Nominal Pipe Size[a] (in.)	Schedule Number[a]	Outside Diameter (in.)			Inside Diameter (in.)	Wall Thickness (in.)			Weight per Foot (lb)	
		Nom[a]	Min[b, d]	Max[b, d]	Nom	Nom[a]	Min[b]	Max[b]	Nom[c]	Max[b, c]
5	5.563	5.532	5.625	5.345	0.109	0.095	0.123	2.196	—	—
	10	5.563	5.532	5.625	5.295	0.134	0.117	0.151	2.688	—
	40	5.563	5.507	5.619	5.047	0.258	0.226	—	7.057	5.461
	80	5.563	5.507	5.619	4.813	0.375	0.328	—	7.188	7.763
	120	5.563	5.507	5.619	4.563	0.500	0.438	—	9.353	10.10
	160	5.563	5.507	5.619	4.313	0.625	0.547	—	11.40	12.31
6	5	6.625	6.594	6.687	6.407	0.109	0.095	0.123	2.624	—
	10	6.625	6.594	6.687	6.357	0.134	0.117	0.151	3.213	—
	40	6.625	6.559	6.691	6.065	0.280	0.245	—	6.564	7.089
	80	6.625	6.559	6.691	5.761	0.432	0.378	—	9.884	10.67
	120	6.625	6.559	6.691	5.501	0.562	0.492	—	12.59	13.60
	160	6.625	6.559	6.691	5.187	0.719	0.629	—	15.69	16.94
	5	8.625	8.594	8.718	8.407	0.109	0.095	0.123	3.429	—
	10	8.625	8.594	8.718	8.329	0.148	0.130	0.166	4.635	—
	20	8.625	8.539	8.711	8.125	0.250	0.219	—	7.735	8.354
	30	8.625	8.539	8.711	8.071	0.277	0.242	—	8.543	9.227
	40	8.625	8.539	8.711	7.981	0.322	0.282	—	9.878	10.67
8	60	8.625	8.539	8.711	7.813	0.406	0.355	—	12.33	13.31
	80	8.625	8.539	8.711	7.625	0.500	0.438	—	15.01	16.21
	100	8.625	8.539	8.711	7.437	0.594	0.520	—	17.62	19.03
	120	8.625	8.539	8.711	7.187	0.719	0.629	—	21.00	22.68
	140	8.625	8.539	8.711	7.001	0.812	0.710	—	23.44	25.31
	160	8.625	8.539	8.711	6.813	0.906	0.793	—	25.84	27.90
10	5	10.750	10.719	10.843	10.482	0.134	0.117	0.151	5.256	—
	10	10.750	10.719	10.843	10.420	0.165	0.144	0.186	6.453	—
	20	10.750	10.642	10.858	10.250	0.250	0.219	—	9.698	10.47
	30	10.750	10.642	10.858	10.136	0.307	0.269	—	11.84	12.69
	40	10.750	10.642	10.858	10.020	0.365	0.319	—	14.00	15.12
	60	10.750	10.642	10.858	9.750	0.500	0.438	—	18.93	24.07
	80	10.750	10.642	10.858	9.562	0.594	0.520	—	22.29	28.78
	100	10.750	10.642	10.858	9.312	0.719	0.629	—	26.65	28.78
	5	12.750	12.719	12.843	12.438	0.156	0.136	0.176	7.258	—
	10	12.750	12.719	12.843	12.390	0.180	0.158	0.202	8.359	—
	20	12.750	12.622	12.878	12.250	0.250	0.219	—	11.55	12.47
12	30	12.750	12.622	12.878	12.090	0.330	0.289	—	15.14	16.35
	40	12.750	12.622	12.878	11.938	0.406	0.355	—	18.52	20.00
	60	12.750	12.622	12.878	11.626	0.562	0.492	—	25.31	27.33
	80	12.750	12.622	12.878	11.374	0.688	0.602	—	30.66	33.11

[a] In accordance with ANSI Standards B36.10 and B36.19.
[b] Based on standard tolerances for pipe.
[c] Based on nominal dimensions, plain ends, and a density of 0.098 lb/in.^{3}, the density of 6061 alloy. For alloy 6063 multiply by 0.99 and for alloy 3003 multiply by 1.01.
[d] For schedules 5 and 10 these values apply to mean outside diameters.

TABLE 1.99   Aluminum Electrical Conduit—Designed Dimensions and Weights

Nominal or Trade Size of Conduit (in.)	Nominal Inside Diameter (in.)	Outside Diameter (in.)	Nominal Wall Thickness (in.)	Length[a] without Coupling (ft and in.)	Minimum Weight of 10 Unit Lengths with Couplings Attached (lb)
$\frac{1}{4}$	0.364	0.540	0.088	$9\text{–}11\frac{1}{2}$	13.3
$\frac{3}{8}$	0.493	0.675	0.091	$9\text{–}11\frac{1}{2}$	17.8
$\frac{1}{2}$	0.622	0.840	0.109	$9\text{–}11\frac{1}{4}$	27.4
$\frac{3}{4}$	0.824	1.050	0.113	$9\text{–}11\frac{1}{4}$	36.4
1	1.049	1.315	0.133	$9\text{–}11$	53.0
$1\frac{1}{4}$	1.380	1.660	0.140	$9\text{–}11$	69.6
$1\frac{1}{2}$	1.610	1.900	0.145	$9\text{–}11$	86.2
2	2.067	2.375	0.154	$9\text{–}11$	115.7
$2\frac{1}{2}$	2.469	2.875	0.203	$9\text{–}10\frac{1}{2}$	182.5
3	3.068	3.500	0.216	$9\text{–}10\frac{1}{2}$	238.9
$3\frac{1}{2}$	3.548	4.000	0.226	$9\text{–}10\frac{1}{4}$	287.7
4	4.026	4.500	0.237	$9\text{–}10\frac{1}{4}$	340.0
5	5.047	5.563	0.258	$9\text{–}10$	465.4
6	6.065	6.625	0.280	$9\text{–}10$	612.5

TABLE 1.100   Equivalent Resistivity Values

Volume Conductivity percent IACS at 68°F	Equivalent Resistivity at 68°F	
	Volume	
	Ohm—circular mil/ft	Microhm—in.
52.5	19.754	1.2929
53.5	19.385	1.2687
53.8	19.277	1.2617
53.9	19.241	1.2593
54.0	19.206	1.2570
54.3	19.099	1.2501
55.0	18.856	1.2341
56.0	18.520	1.2121
56.5	18.356	1.2014
57.0	18.195	1.1908
59.0	17.578	1.1505
59.5	17.430	1.1408
61.0	17.002	1.1128
61.2	16.946	1.1091
61.3	16.918	1.1073
61.4	16.891	1.1055
61.5	16.863	1.1037
61.8	16.782	1.0983
62.0	16.727	1.0948
62.1	16.700	1.0931
62.2	16.674	1.0913
62.3	16.647	1.0896
62.4	16.620	1.0878

TABLE 1.101 Property Limits—Wire
(Up Thru 0.374 in. Diameter)

Alloy and Temper	Ultimate Strength (ksi)		Electrical Conductivity[a] percent IACS at 68°F min
	Min	Max	
1350			
1350-O	8.5	14.0	61.8
1350-H12 and H22	12.0	17.0	61.0
1350-H14 and H24	15.0	20.0	61.0
1350-H16 and H26	17.0	22.0	61.0
8017			
8017-H212[d]	15.0	21.0	61.0
8030			
8030-H221	15.0	22.0	61.0
8176			
8176-H24	15.0	20.0	61.0
8177			
8177-H221	15.0	22.0	61.0

Alloy and Temper	Specified Diameter (in.)	Ultimate Strength (ksi min)		Elongation Percent min in 10 in.		Electrical Conductivity[a] min percent IACS at 68°F
		Individual[b]	Average[c]	Individual[b]	Average[c]	
1350						
1350-H19	0.0105–0.0500	23.0	25.0	—	—	
	0.0501–0.0600	27.0	29.0	1.2	1.4	
	0.0601–0.0700	27.0	28.5	1.3	1.5	
	0.0701–0.0800	26.5	28.0	1.4	1.6	
	0.0801–0.0900	26.0	27.5	1.5	1.6	
	0.0901–0.1000	25.5	27.0	1.5	1.6	
	0.1001–0.1100	24.5	26.0	1.5	1.6	61.0
	0.1101–0.1200	24.0	25.5	1.6	1.7	
	0.1201–0.1400	23.5	25.0	1.7	1.8	
	0.1401–0.1500	23.5	24.5	1.8	1.9	
	0.1501–0.1800	23.0	24.0	1.9	2.0	
	0.1801–0.2100	23.0	24.0	2.0	2.1	
	0.2101–0.2600	22.5	23.5	2.2	2.3	
5005						
5005-H19	0.0601–0.0700	38.0	40.0	1.3	—	
	0.0701–0.0800	37.5	39.5	1.4	—	
	0.0801–0.0900	37.0	39.0	1.5	—	
	0.0901–0.1000	36.5	38.5	1.5	—	
	0.1001–0.1100	36.0	38.0	1.5	—	
5005-H19	0.1101–0.1200	35.5	37.5	1.6	—	53.5
	0.1201–0.1400	35.0	37.0	1.7	—	
	0.1401–0.1500	35.0	36.5	1.8	—	
	0.1501–0.1600	34.5	36.0	1.9	—	
	0.1601–0.2100	32.5	34.0	2.0	—	
	0.2101–0.2600	31.5	33.0	2.2	—	
6201						
6201-T81	0.0612–0.1327	46.0	48.0	3.0	—	52.5
	0.1328–0.1878	44.0	46.0	3.0	—	
8176						
8176-H24	0.0500–0.2040	15.0	17.0	10.0	—	61.0

[a]To convert conductivity to maximum resistivity use Table 1.100.
[b]Any test in a lot.
[c]Average of all tests in a lot.
[d]Applicable up thru 0.250 in.

1.6.7   Standard Screws*

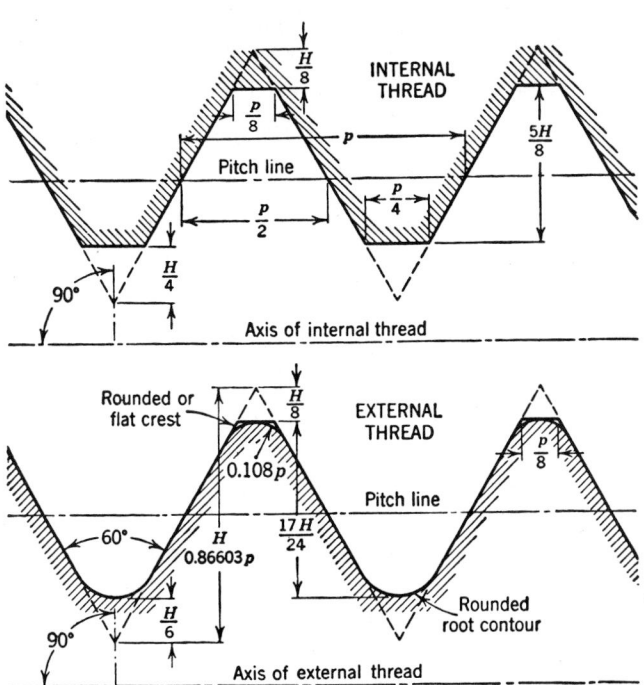

Standard Screw Threads

The Unified and American Screw Threads included in Table 1.102 are taken from the publication of the American Standards Association, ASA B1.1—1949. The *coarse-thread series* is the former United States Standard Series. It is recommended for general use in engineering work where conditions do not require the use of a fine thread. The *fine-thread series* is the former "Regular Screw Thread Series" established by the Society of Automotive Engineers. The *fine-thread series* is recommended for general use in automotive and aircraft work, and where special conditions require a fine thread. The *extra-fine-thread series* is the same as the former SAE fine series and the present SAE extra-fine series. It is used particularly in aircraft and aeronautical equipment where (1) thin-walled material is to be threaded; (2) thread depth of nuts clearing ferrules, coupling flanges, and so on, must be held to a minimum; and (3) a maximum practicable number of threads is required within a given thread length.

The method of designating a screw thread is by the use of the initial letters of the thread series, preceded by the nominal size (diameter in inches or the screw number) and number of threads per inch, all in Arabic numerals, and followed by the classification designation, with or without the pitch diameter tolerances or limits of size. An example of an external thread designation and its meaning is as follows:

Example 1.1

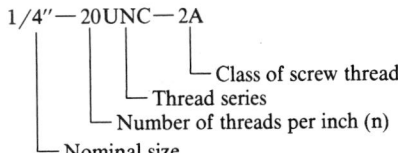

- Class of screw thread
- Thread series
- Number of threads per inch (n)
- Nominal size

A left-hand thread must be identified by the letters "LH" following the class designation. If no such designation is used, the thread is assumed to be right hand.

Classes of thread are distinguished from each other by the amounts of tolerance and allowance specified in ASA B1.1—1949.

*This section extracted, with permission, from EMPIS Materials Selector. Copyright © 1982 General Electric Co.

TABLE 1.102   **Standard Screw Threads**

Sizes	Basic Major Diameter D (in.)	Threads per Inch n	Basic Pitch Diameter[a] E (in.)	Minor Diameter Ext Thds K_s (in.)	Minor Diameter Int Thds K_n (in.)	Section at Minor Diameter at $D - 2h_b$ (in.^{2})	Stress Area[b] (in.^{2})
Coarse-thread Series — UNC and NC (*Basic Dimensions*)							
1 (0.073)	0.0730	64	0.0629	0.0538	0.0561	0.0022	0.0026
2 (0.086)	0.0860	56	0.0744	0.0641	0.0667	0.0031	0.0036
3 (0.099)	0.0990	48	0.0855	0.0734	0.0764	0.0041	0.0048
4 (0.112)	0.1120	40	0.0958	0.0813	0.0849	0.0050	0.0060
5 (0.125)	0.1250	40	0.1088	0.0943	0.0979	0.0067	0.0079
6 (0.138)	0.1380	32	0.1177	0.0997	0.1042	0.0075	0.0090
8 (0.164)	0.1640	32	0.1437	0.1257	0.1302	0.0120	0.0139
10 (0.190)	0.1900	24	0.1629	0.1389	0.1449	0.0145	0.0174
12 (0.216)	0.2160	24	0.1889	0.1649	0.1709	0.0206	0.0240
$\frac{1}{4}$	0.2500	20	0.2175	0.1887	0.1959	0.0269	0.0317
$\frac{5}{16}$	0.3125	18	0.2764	0.2443	0.2524	0.0454	0.0522
$\frac{3}{8}$	0.3750	16	0.3344	0.2983	0.3073	0.0678	0.0773
$\frac{7}{16}$	0.4375	14	0.3911	0.3499	0.3602	0.0933	0.1060
$\frac{1}{2}$	0.5000	13	0.4500	0.4056	0.4167	0.1257	0.1416
$\frac{1}{2}$	0.5000	12	0.4459	0.3978	0.4098	0.1205	0.1374
$\frac{9}{16}$	0.5625	12	0.5084	0.4603	0.4723	0.1620	0.1816
$\frac{5}{8}$	0.6250	11	0.5660	0.5135	0.5266	0.2018	0.2256
$\frac{3}{4}$	0.7500	10	0.6850	0.6273	0.6417	0.3020	0.3340
$\frac{7}{8}$	0.8750	9	0.8028	0.7387	0.7547	0.4193	0.4612
1	1.0000	8	0.9188	0.8466	0.8647	0.5510	0.6051
$1\frac{1}{8}$	1.1250	7	1.0322	0.9497	0.9704	0.6931	0.7627
$1\frac{1}{4}$	1.2500	7	1.1572	1.0747	1.0954	0.8898	0.9684
$1\frac{3}{8}$	1.3750	6	1.2667	1.1705	1.1946	1.0541	1.1538
$1\frac{1}{2}$	1.5000	6	1.3917	1.2955	1.3196	1.2938	1.4041
$1\frac{3}{4}$	1.7500	5	1.6201	1.5046	1.5335	1.7441	1.8983
2	2.0000	$4\frac{1}{2}$	1.8557	1.7274	1.7594	2.3001	2.4971
$2\frac{1}{4}$	2.2500	$4\frac{1}{2}$	2.1057	1.9774	2.0094	3.0212	3.2464
$2\frac{1}{2}$	2.5000	4	2.3376	2.1933	2.2294	3.7161	3.9976
$2\frac{3}{4}$	2.7500	4	2.5876	2.4433	2.4794	4.6194	4.9326
3	3.0000	4	2.8376	2.6933	2.7294	5.6209	5.9659
$3\frac{1}{4}$	3.2500	4	3.0876	2.9433	2.9794	6.7205	7.0992
$3\frac{1}{2}$	3.5000	4	3.3376	3.1933	3.2294	7.9183	8.3268
$3\frac{3}{4}$	3.7500	4	3.5876	3.4433	3.4794	9.2143	9.6546
4	4.0000	4	3.8376	3.6933	3.7294	10.6084	11.0805
Fine-Thread Series — UNF and NF (*Basic Dimensions*)							
0 (0.060)	0.0600	80	0.0519	0.0447	0.0465	0.0015	0.0018
1 (0.073)	0.0730	72	0.0640	0.0560	0.0580	0.0024	0.0027
2 (0.086)	0.0860	64	0.0759	0.0668	0.0691	0.0034	0.0039
3 (0.099)	0.0990	56	0.0874	0.0771	0.0797	0.0045	0.0052
4 (0.112)	0.1120	48	0.0985	0.0864	0.0894	0.0057	0.0065

TABLE 1.102 *(continued)*

Sizes	Basic Major Diameter D (in.)	Threads per Inch n	Basic Pitch Diameter[a] E (in.)	Minor Diameter Ext Thds K_s (in.)	Minor Diameter Int Thds K_n (in.)	Section at Minor Diameter at $D - 2h_b$ (in.²)	Stress Area[b] (in.²)
			Fine-Thread Series—UNF and NF *(Basic Dimensions)*				
5 (0.125)	0.1250	44	0.1102	0.0971	0.1004	0.0072	0.0082
6 (0.138)	0.1380	40	0.1218	0.1073	0.1109	0.0087	0.0101
8 (0.164)	0.1640	36	0.1460	0.1299	0.1339	0.0128	0.0146
10 (0.190)	0.1900	32	0.1697	0.1517	0.1562	0.0175	0.0199
12 (0.216)	0.2160	28	0.1928	0.1722	0.1773	0.0226	0.0257
$\frac{1}{4}$	**0.2500**	**28**	**0.2268**	**0.2062**	**0.2113**	**0.0326**	**0.0362**
$\frac{5}{16}$	**0.3125**	**24**	**0.2854**	**0.2614**	**0.2674**	**0.0524**	**0.0579**
$\frac{3}{8}$	**0.3750**	**24**	**0.3479**	**0.3239**	**0.3299**	**0.0809**	**0.0876**
$\frac{7}{16}$	**0.4375**	**20**	**0.4050**	**0.3762**	**0.3834**	**0.1090**	**0.1185**
$\frac{1}{2}$	**0.5000**	**20**	**0.4675**	**0.4387**	**0.4459**	**0.1486**	**0.1597**
$\frac{9}{16}$	**0.5625**	**18**	**0.5264**	**0.4943**	**0.5024**	**0.1888**	**0.2026**
$\frac{5}{8}$	**0.6250**	**18**	**0.5889**	**0.5568**	**0.5649**	**0.2400**	**0.2555**
$\frac{3}{4}$	**0.7500**	**16**	**0.7094**	**0.6733**	**0.6823**	**0.3513**	**0.3724**
$\frac{7}{8}$	**0.8750**	**14**	**0.8286**	**0.7874**	**0.7977**	**0.4805**	**0.5088**
1	**1.0000**	**12**	**0.9459**	**0.8978**	**0.9098**	**0.6245**	**0.6624**
$1\frac{1}{8}$	**1.1250**	**12**	**1.0709**	**1.0228**	**1.0348**	**0.8118**	**0.8549**
$1\frac{1}{4}$	**1.2500**	**12**	**1.1959**	**1.1478**	**1.1598**	**1.0237**	**1.0721**
$1\frac{3}{8}$	**1.3750**	**12**	**1.3209**	**1.2728**	**1.2848**	**1.2602**	**1.3137**
$1\frac{1}{2}$	**1.5000**	**12**	**1.4459**	**1.3978**	**1.4098**	**1.5212**	**1.5799**
			Extra-Fine-Thread Series—NEF *(Basic Dimensions)*				
12 (0.216)	0.2160	32	0.1957	0.1777	0.1822	0.0242	0.0269
$\frac{1}{4}$	0.2500	32	0.2297	0.2117	0.2162	0.0344	0.0377
$\frac{5}{16}$	0.3125	32	0.2922	0.2742	0.2787	0.0581	0.0622
$\frac{3}{8}$	0.3750	32	0.3547	0.3367	0.3412	0.0878	0.0929
$\frac{7}{16}$	0.4375	28	0.4143	0.3937	0.3988	0.1201	0.1270
$\frac{1}{2}$	0.5000	28	0.4768	0.4562	0.4613	0.1616	0.1695
$\frac{9}{16}$	0.5625	24	0.5354	0.5114	0.5174	0.2030	0.2134
$\frac{5}{8}$	0.6250	24	0.5979	0.5739	0.5799	0.2560	0.2676
$\frac{11}{16}$	0.6875	24	0.6604	0.6364	0.6424	0.3151	0.3280
$\frac{3}{4}$	0.7500	20	0.7175	0.6887	0.6959	0.3685	0.3855
$\frac{13}{16}$	0.8125	20	0.7800	0.7512	0.7584	0.4388	0.4573
$\frac{7}{8}$	0.8750	20	0.8425	0.8137	0.8209	0.5153	0.5352
$\frac{15}{16}$	0.9375	20	0.9050	0.8762	0.8834	0.5979	0.6194
1	1.0000	20	0.9675	0.9387	0.9459	0.6866	0.7095
$1\frac{1}{16}$	1.0625	18	1.0264	0.9943	1.0024	0.7702	0.7973
$1\frac{1}{8}$	1.1250	18	1.0889	1.0568	1.0649	0.8705	0.8993
$1\frac{3}{16}$	1.1875	18	1.1514	1.1193	1.1274	0.9770	1.0074

TABLE 1.102   *(continued)*

Sizes	Basic Major Diameter D (in.)	Threads per Inch n	Basic Pitch Diameter[a] E (in.)	Minor Diameter Ext Thds K_s (in.)	Minor Diameter Int Thds K_n (in.)	Section at Minor Diameter at $D - 2h_b$ (in.^{2})	Stress Area[b] (in.^{2})
			Extra-Fine-Thread Series—NEF (*Basic Dimensions*)				
$1\frac{1}{4}$	1.2500	18	1.2139	1.1818	1.1899	1.0895	1.1216
$1\frac{5}{16}$	1.3125	18	1.2764	1.2443	1.2524	1.2082	1.2420
$1\frac{3}{8}$	1.3750	18	1.3389	1.3068	1.3149	1.3330	1.3684
$1\frac{7}{16}$	1.4375	18	1.4014	1.3693	1.3774	1.4640	1.5010
$1\frac{1}{2}$	1.5000	18	1.4639	1.4318	1.4399	1.6011	1.6397
$1\frac{9}{16}$	1.5625	18	1.5264	1.4943	1.5024	1.7444	1.7846
$1\frac{5}{8}$	1.6250	18	1.5889	1.5568	1.5649	1.8937	1.9357
$1\frac{11}{16}$	1.6875	18	1.6514	1.6193	1.6274	2.0493	2.0929
$1\frac{3}{4}$	1.7500	16	1.7094	1.6733	1.6823	2.1873	2.2382
2	2.0000	16	1.9594	1.9233	1.9323	2.8917	2.9501

[a] British: Effective Diameter.
[b] The stress area is the assumed area of an externally threaded part which is used for the purpose of computing the tensile strength.
   Bold type indicates Unified threads—UNC and UNF.

TABLE 1.103   ASA[a] Standard Bolts and Nuts

Nominal Size	Across Flats (in.)	Across Square Corners (in.)	Across Hex Corners (in.)	Thickness Unfinished (in.)	Thickness Semifinished (in.)
			Regular Bolt Heads		
$\frac{1}{4}$	$\frac{3}{8}$	0.498	0.413	$\frac{11}{64}$	$\frac{5}{32}$
$\frac{5}{16}$	$\frac{1}{2}$	0.665	0.552	$\frac{13}{64}$	$\frac{3}{16}$
$\frac{3}{8}$	$\frac{9}{16}$	0.747	0.620	$\frac{1}{4}$	$\frac{15}{64}$
$\frac{7}{16}$	$\frac{5}{8}$	0.828	0.687	$\frac{19}{64}$	$\frac{9}{32}$
$\frac{1}{2}$	$\frac{3}{4}$	0.995	0.826	$\frac{21}{64}$	$\frac{19}{64}$
$\frac{9}{16}$	$\frac{7}{8}$	1.163	0.966	$\frac{3}{8}$	$\frac{11}{32}$
$\frac{5}{8}$	$\frac{15}{16}$	1.244	1.033	$\frac{27}{64}$	$\frac{25}{64}$
$\frac{3}{4}$	$1\frac{1}{8}$	1.494	1.240	$\frac{1}{2}$	$\frac{15}{32}$
$\frac{7}{8}$	$1\frac{5}{16}$	1.742	1.447	$\frac{19}{32}$	$\frac{9}{16}$
1	$1\frac{1}{2}$	1.991	1.653	$\frac{21}{32}$	$\frac{19}{32}$
$1\frac{1}{8}$	$1\frac{11}{16}$	2.239	1.859	$\frac{3}{4}$	$\frac{11}{16}$
$1\frac{1}{4}$	$1\frac{7}{8}$	2.489	2.066	$\frac{27}{32}$	$\frac{25}{32}$
$1\frac{3}{8}$	$2\frac{1}{16}$	2.738	2.273	$\frac{29}{32}$	$\frac{27}{32}$
$1\frac{1}{2}$	$2\frac{1}{4}$	2.986	2.480	1	$\frac{15}{16}$
$1\frac{5}{8}$	$2\frac{7}{16}$	3.235	2.686	$1\frac{1}{32}$	$1\frac{1}{32}$
$1\frac{3}{4}$	$2\frac{5}{8}$	3.485	2.893	$1\frac{5}{32}$	$1\frac{3}{32}$
$1\frac{7}{8}$	$2\frac{13}{16}$	3.733	3.100	$1\frac{1}{4}$	$1\frac{3}{16}$
2	3	3.982	3.306	$1\frac{11}{32}$	$1\frac{7}{32}$
$2\frac{1}{4}$	$3\frac{3}{8}$	4.479	3.719	$1\frac{1}{2}$	$1\frac{3}{8}$
$2\frac{1}{2}$	$3\frac{3}{4}$	4.977	4.133	$1\frac{21}{32}$	$1\frac{17}{32}$
$2\frac{3}{4}$	$4\frac{1}{8}$	5.476	4.546	$1\frac{53}{64}$	$1\frac{11}{16}$
3	$4\frac{1}{2}$	5.973	4.959	2	$1\frac{7}{8}$

TABLE 1.103   *(continued)*

Heavy Bolt Heads

Nominal Size	Across Flats (in.)	Across Square Corners (in.)	Across Hex Corners (in.)	Thickness Unfinished (in.)	Thickness Semifinished (in.)
$\frac{1}{2}$	$\frac{7}{8}$	1.167	0.969	$\frac{7}{16}$	$\frac{13}{32}$
$\frac{9}{16}$	$\frac{15}{16}$	1.249	1.037	$\frac{15}{32}$	$\frac{7}{16}$
$\frac{5}{8}$	$1\frac{1}{16}$	1.416	1.175	$\frac{17}{32}$	$\frac{1}{2}$
$\frac{3}{4}$	$1\frac{1}{4}$	1.665	1.383	$\frac{5}{8}$	$\frac{19}{32}$
$\frac{7}{8}$	$1\frac{7}{16}$	1.914	1.589	$\frac{23}{32}$	$\frac{11}{16}$
1	$1\frac{5}{8}$	2.162	1.796	$\frac{13}{16}$	$\frac{3}{4}$
$1\frac{1}{8}$	$1\frac{13}{16}$	2.411	2.002	$\frac{29}{32}$	$\frac{27}{32}$
$1\frac{1}{4}$	2	2.661	2.209	1	$\frac{15}{16}$
$1\frac{3}{8}$	$2\frac{3}{16}$	2.909	2.416	$1\frac{3}{32}$	$1\frac{1}{32}$
$1\frac{1}{2}$	$2\frac{3}{8}$	3.158	2.622	$1\frac{3}{16}$	$1\frac{1}{8}$
$1\frac{5}{8}$	$2\frac{9}{16}$	3.406	2.828	$1\frac{9}{32}$	$1\frac{7}{32}$
$1\frac{3}{4}$	$2\frac{3}{4}$	3.655	3.036	$1\frac{3}{8}$	$1\frac{5}{16}$
$1\frac{7}{8}$	$2\frac{15}{16}$	3.905	3.242	$1\frac{15}{32}$	$1\frac{13}{32}$
2	$3\frac{1}{8}$	4.153	3.449	$1\frac{9}{16}$	$1\frac{7}{16}$
$2\frac{1}{4}$	$3\frac{1}{2}$	4.652	3.862	$1\frac{3}{4}$	$1\frac{5}{8}$
$2\frac{1}{2}$	$3\frac{7}{8}$	5.149	4.275	$1\frac{15}{16}$	$1\frac{13}{16}$
$2\frac{3}{4}$	$4\frac{1}{4}$	5.646	4.688	$2\frac{1}{8}$	2
3	$4\frac{5}{8}$	6.144	5.102	$2\frac{5}{16}$	$2\frac{5}{16}$

Regular Nuts and Regular Jam Nuts

Nominal Size	Width Across Flats (in.)	Width Across Corners Square (in.)	Width Across Corners Hex (in.)	Thickness Unfinished Regular Nuts (in.)	Thickness Unfinished Regular Jam Nuts (in.)	Thickness Semifinished Regular Nuts (in.)	Thickness Semifinished Regular Jam Nuts (in.)
$\frac{1}{4}$	$\frac{7}{16}$	0.584	0.484	$\frac{7}{32}$	$\frac{5}{32}$	$\frac{13}{64}$	$\frac{9}{64}$
$\frac{5}{16}$	$\frac{9}{16}$	0.751	0.624	$\frac{17}{64}$	$\frac{3}{16}$	$\frac{1}{4}$	$\frac{11}{64}$
$\frac{3}{8}$	$\frac{5}{8}$	0.832	0.691	$\frac{21}{64}$	$\frac{7}{32}$	$\frac{5}{16}$	$\frac{13}{64}$
$\frac{7}{16}$	$\frac{3}{4}$	1.000	0.830	$\frac{3}{8}$	$\frac{1}{4}$	$\frac{23}{64}$	$\frac{15}{64}$
$\frac{1}{2}$	$\frac{13}{16}$	1.082	0.898	$\frac{7}{16}$	$\frac{5}{16}$	$\frac{27}{64}$	$\frac{19}{64}$
$\frac{9}{16}$	$\frac{7}{8}$	1.163	0.966	$\frac{1}{2}$	$\frac{11}{32}$	$\frac{31}{64}$	$\frac{21}{64}$
$\frac{5}{8}$	1	1.330	1.104	$\frac{35}{64}$	$\frac{3}{8}$	$\frac{17}{32}$	$\frac{23}{64}$
$\frac{3}{4}$	$1\frac{1}{8}$	1.494	1.240	$\frac{21}{32}$	$\frac{7}{16}$	$\frac{41}{64}$	$\frac{27}{64}$
$\frac{7}{8}$	$1\frac{5}{16}$	1.742	1.447	$\frac{49}{64}$	$\frac{1}{2}$	$\frac{3}{4}$	$\frac{31}{64}$
1	$1\frac{1}{2}$	1.991	1.653	$\frac{7}{8}$	$\frac{9}{16}$	$\frac{55}{64}$	$\frac{35}{64}$
$1\frac{1}{8}$	$1\frac{11}{16}$	2.239	1.859	1	$\frac{5}{8}$	$\frac{31}{32}$	$\frac{39}{64}$
$1\frac{1}{4}$	$1\frac{7}{8}$	2.489	2.066	$1\frac{3}{32}$	$\frac{3}{4}$	$1\frac{1}{16}$	$\frac{23}{32}$
$1\frac{3}{8}$	$2\frac{1}{16}$	2.738	2.273	$1\frac{13}{64}$	$\frac{13}{16}$	$1\frac{11}{64}$	$\frac{25}{32}$
$1\frac{1}{2}$	$2\frac{1}{4}$	2.986	2.480	$1\frac{5}{16}$	$\frac{7}{8}$	$1\frac{9}{32}$	$\frac{27}{32}$
$1\frac{5}{8}$	$2\frac{7}{16}$	3.235	2.686	$1\frac{27}{64}$	$\frac{15}{16}$	$1\frac{25}{64}$	$\frac{29}{32}$
$1\frac{3}{4}$	$2\frac{5}{8}$	3.485	2.893	$1\frac{17}{32}$	1	$1\frac{1}{2}$	$\frac{31}{32}$
$1\frac{7}{8}$	$2\frac{13}{16}$	3.733	3.100	$1\frac{41}{64}$	$1\frac{1}{16}$	$1\frac{39}{64}$	$1\frac{1}{32}$
2	3	3.982	3.306	$1\frac{3}{4}$	$1\frac{1}{8}$	$1\frac{23}{32}$	$1\frac{3}{32}$
$2\frac{1}{4}$	$3\frac{3}{8}$	4.479	3.719	$1\frac{31}{32}$	$1\frac{1}{4}$	$1\frac{59}{64}$	$1\frac{13}{64}$
$2\frac{1}{2}$	$3\frac{3}{4}$	4.977	4.133	$2\frac{3}{16}$	$1\frac{1}{2}$	$2\frac{9}{64}$	$1\frac{29}{64}$
$2\frac{3}{4}$	$4\frac{1}{8}$	5.476	4.546	$2\frac{13}{32}$	$1\frac{5}{8}$	$2\frac{23}{64}$	$1\frac{37}{64}$
3	$4\frac{1}{2}$	5.973	4.959	$2\frac{5}{8}$	$1\frac{3}{4}$	$2\frac{37}{64}$	$1\frac{45}{64}$

TABLE 1.103   *(continued)*

Nominal Size	Width Across Flats (in.)	Width Across Corners		Thickness Unfinished Regular		Thickness Semifinished Regular	
		Square (in.)	Hex (in.)	Nuts (in.)	Jam Nuts (in.)	Nuts (in.)	Jam Nuts (in.)
			Heavy Nuts and Heavy Jam Nuts				
$\frac{1}{4}$	$\frac{1}{2}$	0.670	0.556	$\frac{1}{4}$	$\frac{3}{16}$	$\frac{15}{64}$	$\frac{11}{64}$
$\frac{5}{16}$	$\frac{19}{32}$	0.794	0.659	$\frac{5}{16}$	$\frac{7}{32}$	$\frac{19}{64}$	$\frac{13}{64}$
$\frac{3}{8}$	$\frac{11}{16}$	0.919	0.763	$\frac{3}{8}$	$\frac{1}{4}$	$\frac{23}{64}$	$\frac{15}{64}$
$\frac{7}{16}$	$\frac{25}{32}$	1.042	0.865	$\frac{7}{16}$	$\frac{9}{32}$	$\frac{27}{64}$	$\frac{17}{64}$
$\frac{1}{2}$	$\frac{7}{8}$	1.167	0.969	$\frac{1}{2}$	$\frac{5}{16}$	$\frac{31}{64}$	$\frac{19}{64}$
$\frac{9}{16}$	$\frac{15}{16}$	1.249	1.037	$\frac{9}{16}$	$\frac{11}{32}$	$\frac{35}{64}$	$\frac{21}{64}$
$\frac{5}{8}$	$1\frac{1}{16}$	1.416	1.175	$\frac{5}{8}$	$\frac{3}{8}$	$\frac{39}{64}$	$\frac{23}{64}$
$\frac{3}{4}$	$1\frac{1}{4}$	1.665	1.382	$\frac{3}{4}$	$\frac{7}{16}$	$\frac{47}{64}$	$\frac{27}{64}$
$\frac{7}{8}$	$1\frac{7}{16}$	1.914	1.589	$\frac{7}{8}$	$\frac{1}{2}$	$\frac{55}{64}$	$\frac{31}{64}$
1	$1\frac{5}{8}$	2.162	1.796	1	$\frac{9}{16}$	$\frac{63}{64}$	$\frac{35}{64}$
$1\frac{1}{8}$	$1\frac{13}{16}$	2.411	2.002	$1\frac{1}{8}$	$\frac{5}{8}$	$1\frac{7}{64}$	$\frac{39}{64}$
$1\frac{1}{4}$	2	2.661	2.209	$1\frac{1}{4}$	$\frac{3}{4}$	$1\frac{7}{32}$	$\frac{23}{32}$
$1\frac{3}{8}$	$2\frac{3}{16}$	2.909	2.416	$1\frac{3}{8}$	$\frac{13}{16}$	$1\frac{11}{32}$	$\frac{25}{32}$
$1\frac{1}{2}$	$2\frac{3}{8}$	3.158	2.622	$1\frac{1}{2}$	$\frac{7}{8}$	$1\frac{15}{32}$	$\frac{27}{32}$
$1\frac{5}{8}$	$2\frac{9}{16}$	3.406	2.828	$1\frac{5}{8}$	$\frac{15}{16}$	$1\frac{19}{32}$	$\frac{29}{32}$
$1\frac{3}{4}$	$2\frac{3}{4}$	3.656	3.035	$1\frac{3}{4}$	1	$1\frac{23}{32}$	$\frac{31}{32}$
$1\frac{7}{8}$	$2\frac{15}{16}$	3.905	3.242	$1\frac{7}{8}$	$1\frac{1}{16}$	$1\frac{27}{32}$	$1\frac{1}{32}$
2	$3\frac{1}{8}$	4.153	3.449	2	$1\frac{1}{8}$	$1\frac{31}{32}$	$1\frac{3}{32}$
$2\frac{1}{4}$	$3\frac{1}{2}$	4.652	3.862	$2\frac{1}{4}$	$1\frac{1}{4}$	$2\frac{13}{64}$	$1\frac{13}{64}$
$2\frac{1}{2}$	$3\frac{7}{8}$	5.149	4.275	$2\frac{1}{2}$	$1\frac{1}{2}$	$2\frac{29}{64}$	$1\frac{29}{64}$
$2\frac{3}{4}$	$4\frac{1}{4}$	5.646	4.688	$2\frac{3}{4}$	$1\frac{5}{8}$	$2\frac{45}{64}$	$1\frac{37}{64}$
3	$4\frac{5}{8}$	6.144	5.102	3	$1\frac{3}{4}$	$2\frac{61}{64}$	$1\frac{45}{64}$
$3\frac{1}{4}$	5	6.643	5.515	$3\frac{1}{4}$	$1\frac{7}{8}$	$3\frac{3}{16}$	$1\frac{13}{16}$
$3\frac{1}{2}$	$5\frac{3}{8}$	7.140	5.928	$3\frac{1}{2}$	2	$3\frac{7}{16}$	$1\frac{15}{16}$
$3\frac{3}{4}$	$5\frac{3}{4}$	7.637	6.341	$3\frac{3}{4}$	$2\frac{1}{8}$	$3\frac{11}{16}$	$2\frac{1}{16}$
4	$6\frac{1}{8}$	8.135	6.755	4	$2\frac{1}{4}$	$3\frac{15}{16}$	$2\frac{3}{16}$

Nominal Size	Regular Slotted Nuts Semifinished			Heavy Slotted Nuts Semifinished			Slot	
	Width			Width				
	Across Flats (in.)	Across Corners (in.)	Thickness (in.)	Across Flats (in.)	Across Corners (in.)	Thickness (in.)	Width (in.)	Depth (in.)
$\frac{1}{4}$	$\frac{7}{16}$	0.485	$\frac{13}{64}$	$\frac{1}{2}$	0.556	$\frac{15}{64}$	$\frac{5}{64}$	$\frac{3}{32}$
$\frac{5}{16}$	$\frac{9}{16}$	0.624	$\frac{1}{4}$	$\frac{19}{32}$	0.659	$\frac{19}{64}$	$\frac{3}{32}$	$\frac{3}{32}$
$\frac{3}{8}$	$\frac{5}{8}$	0.691	$\frac{5}{16}$	$\frac{11}{16}$	0.763	$\frac{23}{64}$	$\frac{1}{8}$	$\frac{1}{8}$
$\frac{7}{16}$	$\frac{3}{4}$	0.830	$\frac{23}{64}$	$\frac{25}{32}$	0.865	$\frac{27}{64}$	$\frac{1}{8}$	$\frac{5}{32}$
$\frac{1}{2}$	$\frac{13}{16}$	0.898	$\frac{27}{64}$	$\frac{7}{8}$	0.969	$\frac{31}{64}$	$\frac{5}{32}$	$\frac{5}{32}$
$\frac{9}{16}$	$\frac{7}{8}$	0.966	$\frac{31}{64}$	$\frac{15}{16}$	1.037	$\frac{35}{64}$	$\frac{5}{32}$	$\frac{3}{16}$
$\frac{5}{8}$	1	1.104	$\frac{17}{32}$	$1\frac{1}{16}$	1.175	$\frac{39}{64}$	$\frac{3}{16}$	$\frac{7}{32}$
$\frac{3}{4}$	$1\frac{1}{8}$	1.240	$\frac{41}{64}$	$1\frac{1}{4}$	1.382	$\frac{47}{64}$	$\frac{3}{16}$	$\frac{1}{4}$
$\frac{7}{8}$	$1\frac{5}{16}$	1.447	$\frac{3}{4}$	$1\frac{7}{16}$	1.589	$\frac{55}{64}$	$\frac{3}{16}$	$\frac{1}{4}$
1	$1\frac{1}{2}$	1.653	$\frac{55}{64}$	$1\frac{5}{8}$	1.796	$\frac{63}{64}$	$\frac{1}{4}$	$\frac{9}{32}$
$1\frac{1}{8}$	$1\frac{11}{16}$	1.859	$\frac{31}{32}$	$1\frac{13}{16}$	2.002	$1\frac{7}{64}$	$\frac{1}{4}$	$\frac{11}{32}$
$1\frac{1}{4}$	$1\frac{7}{8}$	2.066	$1\frac{1}{16}$	2	2.209	$1\frac{7}{32}$	$\frac{5}{16}$	$\frac{3}{8}$
$1\frac{3}{8}$	$2\frac{1}{16}$	2.273	$1\frac{11}{64}$	$2\frac{3}{16}$	2.416	$1\frac{11}{32}$	$\frac{5}{16}$	$\frac{3}{8}$

TABLE 1.103 (*Continued*)

Nominal Size	Regular Slotted Nuts Semifinished			Heavy Slotted Nuts Semifinished			Slot	
	Width			Width				
	Across Flats (in.)	Across Corners (in.)	Thickness (in.)	Across Flats (in.)	Across Corners (in.)	Thickness (in.)	Width (in.)	Depth (in.)
$1\frac{1}{2}$	$2\frac{1}{4}$	2.480	$1\frac{9}{32}$	$2\frac{3}{8}$	2.622	$1\frac{15}{32}$	$\frac{3}{8}$	$\frac{7}{16}$
$1\frac{5}{8}$	$2\frac{7}{16}$	0.686	$1\frac{25}{64}$	$2\frac{9}{16}$	2.828	$1\frac{19}{32}$	$\frac{3}{8}$	$\frac{7}{16}$
$1\frac{3}{4}$	$2\frac{5}{8}$	2.893	$1\frac{1}{2}$	$2\frac{3}{4}$	3.035	$1\frac{23}{32}$	$\frac{7}{16}$	$\frac{1}{2}$
$1\frac{7}{8}$	$2\frac{13}{16}$	3.100	$1\frac{39}{64}$	$2\frac{15}{16}$	3.242	$1\frac{27}{32}$	$\frac{7}{16}$	$\frac{9}{16}$
2	3	3.306	$1\frac{23}{32}$	$3\frac{1}{8}$	3.449	$1\frac{31}{32}$	$\frac{7}{16}$	$\frac{9}{16}$
$2\frac{1}{4}$	$3\frac{3}{8}$	3.719	$1\frac{59}{64}$	$3\frac{1}{2}$	3.862	$2\frac{13}{64}$	$\frac{7}{16}$	$\frac{9}{16}$
$2\frac{1}{2}$	$3\frac{3}{4}$	4.133	$2\frac{9}{64}$	$3\frac{7}{8}$	4.275	$2\frac{29}{64}$	$\frac{9}{16}$	$\frac{11}{16}$
$2\frac{3}{4}$	$4\frac{1}{8}$	4.546	$2\frac{23}{64}$	$4\frac{1}{4}$	4.688	$2\frac{45}{64}$	$\frac{9}{16}$	$\frac{11}{16}$
3	$4\frac{1}{2}$	4.959	$2\frac{37}{64}$	$4\frac{5}{8}$	5.102	$2\frac{61}{64}$	$\frac{5}{8}$	$\frac{3}{4}$

[a]ANSI standards B18.2.1-1981, B18.2.2-1972 (R1983), B18.6.3-1972 (R1983).

Selection of Screws

By definition, a **screw** is a fastener that is intended to be torqued by the head. Screws are the most widely used method of assembly despite recent technical advances of adhesives, welding, and other joining techniques. Use of screws is essential in those applications that require ease of disassembly for normal maintenance and service. There is no real economy if savings made in factory installation create service problems later. There are many types of screws, and each variety will be treated separately. Material selection is generally common to all types of screws.

Material. Not all materials are suitable for the processes used in the manufacture of fasteners. Large-volume users or those with critical requirements can be very selective in their choice of materials. Low-volume users or those with noncritical applications would be wise to permit a variety of materials in a general category in order to improve availability and lower cost. For example, it is usually desirable to specify low-carbon steel or 18-8 type stainless steel* rather than ask for a specific grade.

Low-carbon steel is widely used in the manufacture of fasteners where lowest cost is desirable and tensile strength requirements are ~ 50,000 psi. If corrosion is a problem, these fasteners can be plated either electrically or mechanically. Zinc or cadmium plating are used in most applications. Other finishes include nickel, chromium, copper, tin, and silver electroplating; electroless nickel and other immersion coatings; hot dip galvanizing; and phosphate coatings.

Medium-carbon steel, quenched, and tempered is widely used in applications requiring tensile strengths from 90,000 to 120,000 psi. Alloy steels are used in applications requiring tensile strengths from 115,000 to 180,000 psi, depending on the grade selected. Where better corrosion resistance is required, 300 series stainless steel can be specified. The 400 series stainless steel is used if it is necessary to have a corrosion resistant material that can be hardened and tempered by heat treatment.

For superior corrosion resistance, materials such as brass, bronze, aluminum, or nickel are sometimes used in the manufacture of fasteners. If strength is no problem, plastics such as nylons are used in severe corrosion applications.

Drivability. When selecting a screw, thought must be given to the means of driving for assembly and disassembly, as well as the head shape. Most screw heads provide either a slot, a recess or a hexagon shape as a means of driving. The slotted screw is the least preferred driving style and serves only when appearance must be combined with ease of disassembly with a common screwdriver. Only a limited amount of torque can be applied with a screwdriver. A slot can become inoperative after

*Manufacturer may use UNS—S30200, S30300, S30400, S30500 (AISI type 302, 303, 304 or 305) depending upon quantity, diameter, and manufacturing process.

repeated disassembly destroys the edge of the wall that the blade of the screwdriver bears against. The hexagon head is preferred for the following reasons:

- Least likely to accidentally spinout (thereby marring the surface of the product).
- Lowest initial cost.
- Adaptable to high-speed power drive.
- Minimum worker fatigue.
- Ease of assembly in difficult places.
- Permits higher driving torque, especially in large sizes where strength is important.
- Contains no recess to become clogged with dirt and interfere with driving.
- Contains no recess to weaken the head.

Unless frequent field disassembly is required, use of the unslotted hex head is preferred.

Appearance is the major disadvantage of the hex head, and this one factor is judged sufficient to eliminate it from consideration for the front or top of products.

The recessed head fastener is widely used and becomes first choice for appearance applications. It usually costs more than a slot or a hexagon shape. There are many kinds of recesses. The Phillips and Phillips POZIDRIV are most widely used. To a lesser extent the Frearson, clutch type, hexagonal, and fluted socket heads are used. For special applications, proprietary types of tamper resistant heads can be selected (Fig. 1.1).

The recessed head has some of the same advantages as the hex head (see preceding list). It also has improved appearance. The Phillips POZIDRIV is slowly replacing the Phillips recess. The POZIDRIV recess can be readily identified by four radial lines centered between each recess slot. These slots are a slight modification of the conventional Phillips recess. This change improves the fit between the driver and the recess thus minimizing the possibility of marring a surface from accidental spinout of the driver as well as increasing the life of the driver. The POZIDRIV design is recommended in high-production applications requiring high driving torques. The POZIDRIV recess usually sells at a slightly higher price than the conventional Phillips recess, but some suppliers will furnish either at the same price. The savings resulting from longer tool life will usually justify the higher initial cost.

A conventional Phillips driver could be used to install or disassemble a POZIDRIV screw. However, a POZIDRIV driver should be used with a POZIDRIV screw in order to take advantage of the many features inherent in the new design. To avoid confusion, it should be clearly understood that a POZIDRIV driver cannot be used to install or remove a conventional Phillips head screw.

A Frearson recess is a somewhat different design than a Phillips recess and has the big advantage that one driving tool can be used for all sizes whereas a Phillips may require four driving tools in the range from #2 (0.086 in.) to 3/8 (0.375 in.) screw size. This must be balanced against the following

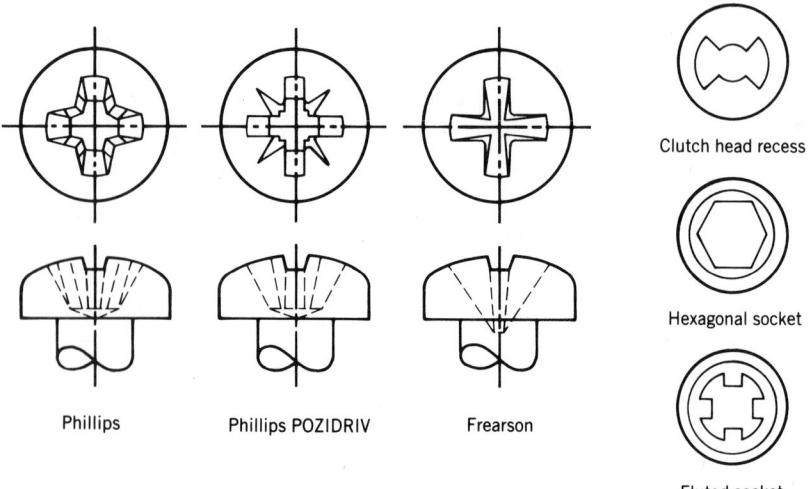

Clutch head recess

Hexagonal socket

Fluted socket

Phillips   Phillips POZIDRIV   Frearson

Fig. 1.1   Recessed head fasteners.

disadvantages as follows:

- Limited availability.
- Greater penetration of the recess means thinner walls between the bottom of the recess and the outer edge of the screw, which tends to weaken the head.
- The sharp point of the driver can easily scratch or otherwise mar the surface of the product if it accidently touches.
- Although one driver can be used for all sizes, for optimum results, different size drivers are recommended for installing various screw sizes, thus minimizing the one real advantage of the Frearson recess.

The hexagon and fluted socket head cap screws are only available in expensive high-strength alloy steel. Its unique small outside diameter or cylindrical head is useful on flanges, counterbored holes, or other locations where clearances are restricted. Such special applications may justify the cost of a socket head cap screw. Appreciable savings can be made in other applications by substitution of a hexagon head screw.

Despite any claims to the contrary, the dimensional accuracy of hexagon socket head cap screws is no better than that of other cold-headed products, and there is no merit in close-thread tolerances, which are advocated by some manufacturers of these products. The high prices, therefore, should be justified solely on the basis of possible space savings in using the cylindrical head.

The fluted socket is not as readily available and should only be considered in the very small sizes where a hexagon key tends to round out the socket. The fluted socket offers spline design so that the key will neither slip nor be subject to excessive wear.

There are many types of special recesses that are tamper resistant. In most of these designs, the recess is an unusual shape requiring a special tool for assembly and disassembly. A readily available driving tool such as a screwdriver or hexagon key would not fit the recess. The purpose of a tamper-resistant fastener is to prevent unauthorized removal of parts and equipment. Their protection is needed on any product located in public places, to discourage vandalism and thievery. They may also be necessary on some consumer products as a safety measure to protect the amateur repairman from injury or to prevent him from causing serious damage to equipment. With product liability mania what it is today, the term "tamperproof" has all but disappeared. Now the fasteners are called "tamper resistant." They are the same as they were under their previous name, but the new term better reflects their true capabilities. Any skilled thief with ample time and proper tools can saw, drill, blast, or otherwise disassemble any tamper-resistant fastener. Therefore, these fasteners are intended only to discourage the casual thief or amateur tinkerer and make it more difficult for a skilled professional. Whatever the choice of fastener design, it is essential that hardened material be specified. No fastener is ever truly tamperproof, but hardened steel helps. Fasteners made of soft material can be disassembled easily by sawing a slot, hammering with a chisel, or drilling a hole and using an extraction bit.

Head Shapes

The following information is equally applicable to all types of recesses as well as a slotted head. For simplification only slotted screws are shown.

The pan head is the most widely used and is intended to replace the round, binding, and truss heads in order to keep varieties to a minimum. It is preferred because it presents the best combination of appearance with adequate head height to minimize weakness due to depth of penetration of the recess (Fig. 1.2).

The round head was widely used in the past (Fig. 1.3). It has since been delisted as an American National Standard. Give preference to pan heads on all new designs. Figure 1.4 shows the superiority of the pan head, the high edge of the pan head at its periphery, where driving action is most effective, provides superior driver-slot engagement and reduces the tendency to chew away the metal at the edge of the slot.

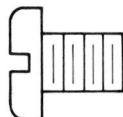

Fig. 1.2 Pan head.

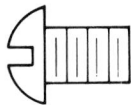

Fig. 1.3 Round head.

Pan head   Round head

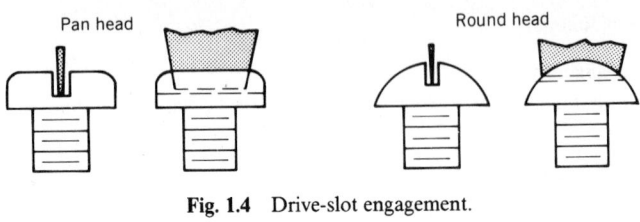

Fig. 1.4 Drive-slot engagement.

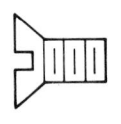   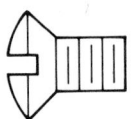   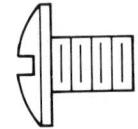

Fig. 1.5 Flat head.   **Fig. 1.6** Oval head.   **Fig. 1.7** Truss head.

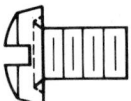   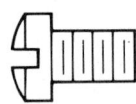

Fig. 1.8 Binding head.   **Fig. 1.9** Fillister head.

The flat head is used where a flush surface is required. The countersunk section aids in centering the screw (Fig. 1.5).

The oval head is similar to a flat head except that instead of a flush surface it presents a low silhouette that improves the appearance (Fig. 1.6).

The truss head is similar to the round head except that the head is shallower and has a larger diameter. It is used where extra bearing surface is required for extra holding power or where the clearance hole is oversized or the material is soft. It also presents a low silhouette that improves the appearance (Fig. 1.7).

The binding head is similar to the pan head and is commonly used for electrical connections where an undercut is usually specified to bind and prevent the fraying of stranded wire (Fig. 1.8).

The fillister head has the smallest diameter for a given shank size. It also has a deep slot that allows a higher torque to be applied during assembly. It is not as readily available or as widely used as some of the other head styles (Fig. 1.9).

The advantages of a hex head are listed in the discussion on drivability. This type head is available in eight variations (Fig. 1.10).

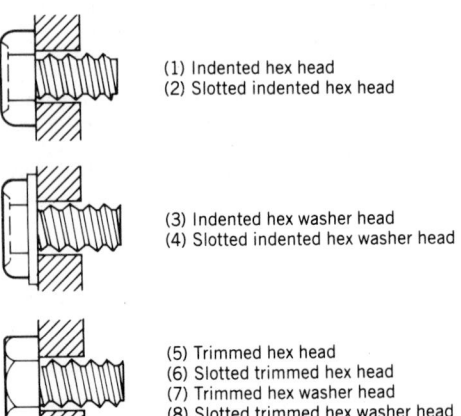

(1) Indented hex head
(2) Slotted indented hex head

(3) Indented hex washer head
(4) Slotted indented hex washer head

(5) Trimmed hex head
(6) Slotted trimmed hex head
(7) Trimmed hex washer head
(8) Slotted trimmed hex washer head

Fig. 1.10 Hex head.

The indented design is lowest cost as the hex is completely cold upset in a counterbore die and possesses an identifying depression in the top surface of the head.

The trimmed design requires an extra operation to produce clean sharp corners with no indentation. Appearance is improved and there is no pocket on top to collect moisture.

The washer design has a larger bearing surface to spread the load over a wider area. The washer is an integral part of the head and also serves to protect the finish of the assembly from wrench disfigurement.

The slot is used to facilitate field service. It adds to the cost, weakens the head, and limits the amount of tightening torque that can be applied. A slot is unnecessary in high-production factory installation.

Any given location should standardize on one or possibly two of the eight variations.

Types of Screws

Machine Screws. Machine screws are meant to be assembled in tapped holes, either into a product or into a nut. The screw threads of a machine screw are readily available in American National Standard Unified Inch Coarse and Fine Thread series. They are generally considered for applications where the material to be joined is too hard, too weak, too brittle, or too thick to take a tapping screw. It is also used in applications where the assembly requires a fastener made of a material that cannot be hardened enough to make its own thread, such as brass or nylon machine screws. Applications requiring freedom from dust or particles of any kind cannot use thread cutting screws and, therefore, must be joined by machine screws or a tapping screw which forms or rolls a thread.

There are many combinations of head styles, shapes, and materials.

Self-Tapping Screws. There are many different types of self-tapping screws commercially available. The following three types are capable of creating an internal thread by being twisted into a smooth hole:

1. Thread-forming screws
2. Thread-cutting screws
3. Thread-rolling screws

The following two types create their own opening before generating the thread:

4. Self-drilling and tapping screws
5. Self-extruding and tapping screws

1. Thread-forming screws. Thread-forming screws create an internal thread by forming or squeezing material. They rely on the pressure of the screw thread to force a mating thread into the work piece. They are applicable in materials where large internal stresses are permissible or desirable to increase resistance to loosening. They are generally used to fasten sheet metal parts. They cannot be used to join brittle materials, such as plastics, because the stresses created in the work piece can cause cracking. The following types of thread-forming screws are commonly used:

Types A and AB. Type AB screws have a spaced thread. This means that each thread is spaced further away from its adjacent thread than the popular machine screw series. They also have a gimlet point for ease in entering a predrilled hole. This type of screw is primarily intended to be used in sheet metal with a thickness from 0.015 (0.38 mm) to 0.05 in. (1.3 mm), resin-impregnated plywood, natural woods, and asbestos compositions.

Type AB screws were introduced several years ago to replace the type A screws. The type A screw is the same as the type AB except for a slightly wider spacing of the threads. Both are still available and can be used interchangeably. The big advantage of the type AB screw is that its threads are spaced exactly as the type B screws to be discussed later. In the interest of standardization it is recommended that Type AB screws be used in place of either the type A or the type B series (Fig. 1.11).

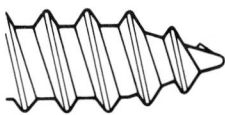

Fig. 1.11   Type AB.

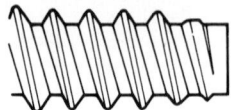

Fig. 1.12   Type B.

Type B. Type B screws have the same spacing as a type AB screw. Instead of a gimlet point, they have a blunt point with incomplete threads at the point. This point makes the type B more suitable for thicker metals and blind holes. The type B screws can be used in any of the applications listed under type AB. In addition the type B screw can be used in sheet metal up to a thickness of 0.200 in. (5 mm) and in nonferrous castings (Fig. 1.12).

Type C. Type C screws look like type B screws except that threads are spaced to be exactly the same as a machine screw thread and may be used to replace a machine screw in the field. They are recommended for general use in metal from 0.030 to 0.100 in. (0.76 to 2.54 mm) thickness. It should be recognized that in specific applications, involving long thread engagement or hard materials, this type of screw requires extreme driving torques.

2. Thread-Cutting Screws. Thread-cutting screws create an internal thread by actual removal of material from the internal hole. The design of the cavity to provide space for the chips and the design of the cutting edge differ with each type. They are used in place of the thread forming type for applications in materials where disruptive internal stresses are undesirable or where excessive driving torques are encountered. The following types of thread cutting screws are commonly used:

Type BT (Formerly Known as Type 25). Type BT screws have a spaced thread and a blunt point similar to the type B screw. In addition they have one cutting edge and a wide chip cavity. These screws are primarily intended for use in very friable plastics such as urea compositions, asbestos, and other similar compositions. In these materials, a larger space between threads is required to produce a satisfactory joint because it reduces the buildup of internal stresses that fracture brittle plastic when a closer spaced thread is used. The wide cutting slot creates a large cutting edge and permits rapid deflection of the chips to produce clean mating threads. For best results all holes should be counterbored to prevent fracturing the plastic. Use of this type screw eliminates the need to use tapped metallic inserts in plastic materials (Fig. 1.13).

Type ABT. Type ABT screws are the same as type BT screws except that they have a gimlet point similar to a type AB screw. This design is not recognized as an American National Standard and should only be selected for large volume applications (over 50,000 pieces of one size and type). It is primarily intended for use in plastic for the same reasons as listed for type BT screws (Fig. 1.14).

Type D (Formerly Known as Type 1). Type D screws have threads of machine screw diameter-pitch combinations approximating unified form with a blunt point and tapered entering threads. In addition a slot is cut off center with one side on the center line. This radial side of the slot creates the sharp serrated cutting edge such as formed on a tap. The slot leaves a thinner section on one side of the screw that collapses and helps concentrate the pressure on the cutting edge. This screw is suitable for use in all thicknesses of metals (Fig. 1.15).

Type F. Type F screws are identical to type D except that instead of one slot there are several slots cut at a slight angle to the axis of the thread. This screw is suitable for use in all thicknesses of metals and can be used interchangeably with a type D screw in many applications. However, the type F screw is superior to the type D screw for tapping into cast iron and permits the use of a smaller pilot hole (Fig. 1.16).

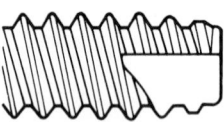

Fig. 1.13   Type BT.

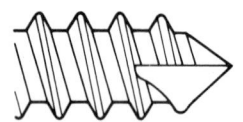

Fig. 1.14   Type ABT.

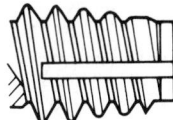

Fig. 1.15 Type D.

Fig. 1.16 Type F.

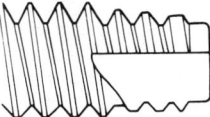

Fig. 1.17 Type T.

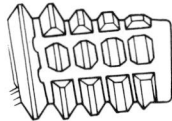

Fig. 1.18 Type BF.

Type D or Type F. Because in many applications these two types can be used interchangeably with the concomitant advantages of simpler inventory and increased availability, a combined specification is often issued permitting the supplier to furnish either type at his option.

Type T (Formerly Known as Type 23). Type T screws are similar to type D and type F except that they have an acute rake angle cutting edge. The cut in the end of the screw is designed to eliminate a pocket that confines the chips. The shape of the slot is such that the chips are forced ahead of the screw as it is driven. This screw is suitable for plastics and other soft materials when a standard machine screw series thread is desired. It is used in place of type D and type F when more chip room is required because of deep penetration (Fig. 1.17).

Type BF. Type BF screws are intended for use in plastics. The wide thread pitch reduces the buildup of internal stresses that fracture brittle plastics when a smaller thread pitch is used. The screw has a blunt point and tapered entering threads with several cutting edges and chip cavity (Fig. 1.18).

3. Thread-Rolling Screws. Thread-rolling screws (see Fig. 1.19) form an internal thread by flowing metal and thus do not cut through nor disrupt the grain flow lines of materials as do thread-cutting screws. The screw compacts and work hardens the material thereby forming strong, smoothly burnished internal threads. The screws have the threads of machine screw diameter-pitch combinations. This type screw is ideal for applications where chips can cause electrical shorting of equipment or jamming of delicate mechanism. Freedom from formation of chips eliminates the costly problem of cleaning the product of chips and burrs as would otherwise be required.

The ratio of driving torque to stripping torque is approximately 1 to 8 for a thread-rolling screw as contrasted to 1 to 3 for a conventional tapping screw. This higher ratio permits the driver torque release to be set well over the required driving torque and yet safely below the stripping torque. This increased ratio minimizes poor fastening due to stripped threads or inadequate seating of the screws.

Plastite is intended for use in filled or unfilled thermoplastics and some of the thermosetting plastics. The other three types are intended for use in metals. At present, there are no data to prove the superiority of one type over another.

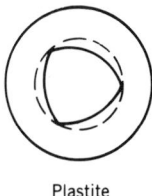

Plastite
Taptite

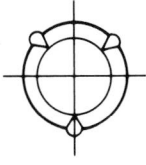

Swageform

Rolox

Fig. 1.19 Thread-rolling screws.

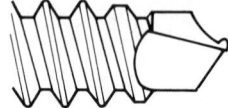

Fig. 1.20   Self-drilling and tapping screws.

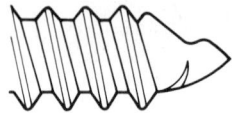

Fig. 1.21   Self-extruding screw.

4. Self-Drilling and Tapping Screws. The self-drilling and tapping screw (Fig. 1.20) drills its own hole and forms a mating thread thus making a complete fastening in a single operation. Assembly labor is reduced by eliminating the need to predrill holes at assembly and by solving the problem of hole alignment. These screws must complete their metal drilling function and fully penetrate the material before the screw thread can engage and begin its advancement. In order to meet this requirement, the unthreaded point length must be equal to or greater than the material thickness to be drilled. Therefore, there is a strict limitation on minimum and maximum material thickness that varies with screw size. There are many different styles and types of self-drilling and tapping screws to meet specific needs.

5. Self-Extruding Screws. Self-extruding screws provide their own extrusion as they are driven into an inexpensively produced punched hole. The resulting extrusion height is several times the base material thickness. This type screw is suitable for material in thicknesses up to 0.048 in. (1.2 mm). By increasing the thread engagement, these screws increase the differential between driving and stripping torque and provide greater pull-out strength. Since they do not produce chips, they are excellent for grounding sheet metal for electrical connections (Fig. 1.21).

There is almost no limit to the variety of head styles, thread forms, and screw materials that are available commercially. The listing only shows representative examples. Users should attempt to keep varieties to a minimum by carefully selecting those variations that best meet the needs of their type of product.

Set Screws. Set screws are available in various combinations of head and point style as well as material and are used as locking, locating, and adjustment devices. The common head styles are slotted headless, square head, hexagonal socket, and fluted socket. The slotted headless has the lowest cost and can be used in a counterbored hole to provide a flush surface. The square head is applicable for location or adjustment of static parts where the projecting head is not objectionable. Its use should be avoided on all rotating parts. The hexagonal socket head can be used in a counterbored hole to provide a flush surface. It permits greater torque to be applied than with a slotted headless design. Fluted sockets are useful in very small diameters, that is, #6 (0.138 in.) and under, where hexagon keys tend to round out the socket in hexagonal socket set screws. Set screws should not be used to transmit large amounts of torque, particularly under shock torsion loads. Increased torsion loads may be carried by two set screws located 120° apart.

The following points are available with the head styles discussed: The cup point (Table 1.104) is the standard stock point for all head shapes and is recommended for all general locking purposes. Flats are recommended on round shafts when close fits are used and it is desirable to avoid interference in disassembling parts because of burrs produced by action of the cup point, or when the flats are desired to increase torque transmission. When flats are not used, it is recommended that the minimum shaft diameter be not less than four times the cup diameter since otherwise the whole cup may not be in contact with the shaft. The self-locking cup point has limited availability. It has counterclockwise knurls to prevent the screw from working loose even in poorly tapped holes (Fig. 1.22).

When oval points are used, the surface it contacts should be grooved or spotted to the same general contour as the point to assure good seating. It is used where frequent adjustment is necessary without excessive deformation of the part against which it bears (Fig. 1.23).

When flat points are used, it is customary to grind a flat on the shaft for better point contact. This point is preferred where wall thickness is thin and on top of plugs made of any soft material (Fig. 1.24).

TABLE 1.104   Holding Power of Flat or Cup Point Set Screws

d (in.)	$\frac{1}{4}$	$\frac{5}{16}$	$\frac{3}{8}$	$\frac{7}{16}$	$\frac{1}{2}$	$\frac{9}{16}$	$\frac{5}{8}$	$\frac{3}{4}$	$\frac{7}{8}$	1	$1\frac{1}{8}$	$1\frac{1}{4}$
P (lb)	100	168	256	366	500	658	840	1,280	1,830	2,500	3,388	4,198

Fig. 1.22 Cup point.

Fig. 1.23 Oval point.

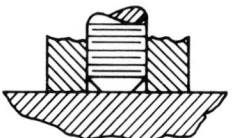

Fig. 1.24 Flat point.

Fig. 1.25 Cone point.

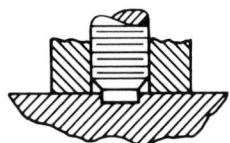

Fig. 1.26 Half-dog point.

TABLE 1.105   Lag Screws

Diameter of screw (in.)	$\frac{1}{4}$	$\frac{5}{16}$	$\frac{3}{8}$	$\frac{7}{16}$	$\frac{1}{2}$	$\frac{5}{8}$	$\frac{3}{4}$	$\frac{7}{8}$	1
No. of threads per inch	10	9	7	7	6	5	$4\frac{1}{2}$	4	$3\frac{1}{2}$
Across flats of hexagon and square heads (in.)	$\frac{3}{8}$	$\frac{15}{32}$	$\frac{9}{16}$	$\frac{21}{32}$	$\frac{3}{4}$	$\frac{15}{16}$	$1\frac{1}{8}$	$1\frac{5}{16}$	$1\frac{1}{2}$
Thickness of hexagon and square heads (in.)	$\frac{3}{16}$	$\frac{1}{4}$	$\frac{5}{16}$	$\frac{3}{8}$	$\frac{7}{16}$	$\frac{17}{32}$	$\frac{5}{8}$	$\frac{3}{4}$	$\frac{7}{8}$

Length of Threads for Screws of All Diameters							
Length of screw (in.)	$1\frac{1}{2}$	2	$2\frac{1}{2}$	3	$3\frac{1}{2}$	4	$4\frac{1}{2}$
Length of thread (in.)	To head	$1\frac{1}{2}$	2	$2\frac{1}{4}$	$2\frac{1}{2}$	3	$3\frac{1}{2}$
Length of screw (in.)	5	$5\frac{1}{2}$	6	7	8	9	10–12
Length of thread (in.)	4	4	$4\frac{1}{2}$	5	6	6	7

When the cone point is used, it is recommended that the angle of countersink be as nearly as possible the angle of screw point for the best efficiency. Cone point set screws have some application as pivot points. It is used where permanent location of parts is required. Because of penetration, it has the highest axial and torsional holding power of any point (Fig. 1.25).

The half-dog point should be considered in lieu of full dog points when the usable length of thread is less than the nominal diameter. It is also more readily obtained than the full-dog point. It can be used in place of dowel pins and where end of thread must be protected (Fig. 1.26).

Lag Screws. Lag screws (Table 1.105) are usually used in wood but also can be used in plastics and with expansion shields in masonry. A 60° gimlet point is the most readily available type. A 60° cone point, not covered in these drawings, is also available. Some suppliers refer to this item as a lag bolt (Fig. 1.27).

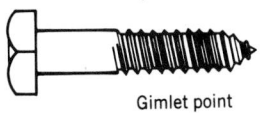

Gimlet point

Fig. 1.27 Lag screws.

TABLE 1.106   Recommended Diameters of Pilot Hole for Types of Wood[a]

Screw Diameter (in.)	White Oak	Southern Yellow Pine Douglas Fir	Redwood Northern White Pine
0.250	0.160	0.150	0.100
0.312	0.210	0.195	0.132
0.375	0.260	0.250	0.180
0.438	0.320	0.290	0.228
0.500	0.375	0.340	0.280
0.625	0.485	0.437	0.375
0.750	0.600	0.540	0.480

[a]Pilot holes should be slightly larger than listed when lag screws of excessive lengths are to be used.

A lag screw is normally used in wood when it is inconvenient or objectionable to use a through bolt and nut. To facilitate the insertion of the screw especially in denser types of wood, it is advisable to use a lubricant on the threads. It is important to have a pilot hole of proper size and following are some recommended hole sizes for commonly used types of wood. Hole sizes for other types of wood should be in proportion to the relative specific gravity of that wood to the ones listed in Table 1.106.

Shoulder Screws. These screws are also referred to as "stripper bolts." They are used mainly as locators or retainers for spring strippers in punch and die operations, and have found some application as fulcrums or pivots in machine designs that involve links, levers, or other oscillating parts. Consideration should be given to the alternative use of a sleeve bearing and a bolt, on the basis of both cost and good design (Fig. 1.28).

Thumb Screws. Thumb screws have a flattened head designed for manual turning without a driver or a wrench. They are useful in applications requiring frequent disassembly or screw adjustment (Fig. 1.29).

Weld Screws. Weld screws come in many different head configurations, all designed to provide one or more projections for welding the screw to a part. Overhead projections are welded directly to the part. Underhead projections go through a pilot hole. The designs in Figs. 1.30 and 1.31 are widely used.

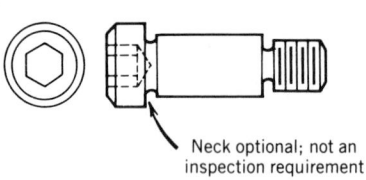

Neck optional; not an inspection requirement

Fig. 1.28   Shoulder screw.

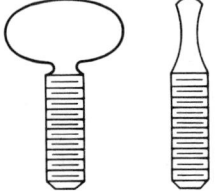

Fig. 1.29   Thumb screws.

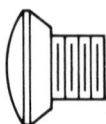

Fig. 1.30   Single projection weld screw.

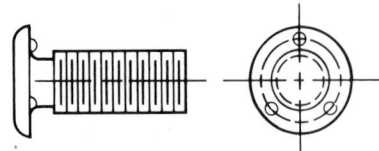

Fig. 1.31   Under head weld screws.

TABLE 1.107   American Standard Wood Screws[a]

Number	0	1	2	3	4	5	6	7	8
Threads per inch	32	28	26	24	22	20	18	16	15
Diameter (in.)	0.060	0.073	0.086	0.099	0.112	0.125	0.138	0.151	0.164
Number	9	10	11	12	14	16	18	20	24
Threads per inch	14	13	12	11	10	9	8	8	7
Diameter (in.)	0.177	0.190	0.203	0.216	0.242	0.268	0.294	0.320	0.372

[a]Included angle of flathead = 82°, see Fig. 1.18.

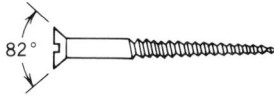

In projection welding of carbon steel screws, care should be observed in applications, since optimum weldability is obtained when the sum, for either parent metal or screw, of $\frac{1}{4}$ the manganese content plus the carbon content does not exceed 0.38. For good weldability with the annular ring type, the height of the weld projection should not exceed $\frac{1}{2}$ the parent metal thickness, as a rule of thumb.

Copper flash plating is provided for applications where cleanliness of screw head is necessary in obtaining good welds.

Wood Screws. Wood screws are (Table 1.107) readily available in lengths from $\frac{1}{4}$ to 5 in. for steel and from $\frac{1}{4}$ to $3\frac{1}{2}$ in. for brass. Consideration should be given to the use of type AB thread-forming screws, which are lower in cost and more efficient than wood screws for use in wood. Wood screws are made with flat, round, or oval heads.

The **resistance of wood screws to withdrawal** from side grain of seasoned wood is given by the formula $P = 2850G^2D$, where P is the allowable load on the screw (lb/ in. penetration of the threaded portion); G is specific gravity of ovendry wood; D is diameter of screw (in.). Wood screws should not be designed to be loaded in withdrawal from end grain.

The **allowable safe lateral resistance** of wood screws embedded 7 diameters in the side grain of seasoned wood is given by the formula $P = KD^2$, where P is the lateral resistance per screw (lb); D is the diameter (in.); and K is 4000 for oak (red and white), 3960 for Douglas fir (coast region) and southern pine, and 3240 for cypress (southern) and Douglas fir (inland region).

The following rules should be observed: (1) the size of the lead hole in soft (hard) woods should be about 70% (90%) of the core or root diameter of the screw; (2) lubricants such as soap may be used without great loss in holding power; (3) long, slender screws are preferable generally, but in hardwood too slender screws may reach the limit of their tensile strength; and (4) in the screws themselves, holding power is favored by thin sharp threads, rough unpolished surface, full diameter under the head, and shallow slots.

SEMS. The machine and tapping screws can be purchased with washers or lock washers as an integral part of the purchased screws. When thus joined together, the part is known as a SEMS unit. The washer is assembled on a headed screw blank before the threads are rolled. The inside diameter of the washer is of a size that will permit free rotation and yet prevent disassembly from the screw after the threads are rolled. If these screws and washers were purchased separately there would be an initial cost savings over the preassembled units. However, these preassembled units reduce installation time because only one hand is needed to position them, leaving the other hand free to hold the driving tool. The time required to assemble a loose washer is eliminated. In addition these assemblies act to minimize installation errors and inspection time because the washer is in place, correctly oriented. Also the use of a single unit, rather than two separate parts simplifies bookkeeping, handling, inventory, and other related operations.

1.6.8   Nominal and Minimum Dressed Sizes of American Standard Lumber

Table 1.108 applies to boards, dimensional lumber, and timbers. The thicknesses apply to all widths and all widths to all thicknesses.

TABLE 1.108   Nominal and Minimum Dressed Sizes of American Standard Lumber

	Thicknesses			Face Widths		
		Minimum Dressed			Minimum Dressed	
Item	Nominal	Dry[a] (in.)	Green (in.)	Nominal	Dry[a] (in.)	Green (in.)
Boards[b]				2	$1\frac{1}{2}$	$1\frac{9}{16}$
				3	$2\frac{1}{2}$	$2\frac{9}{16}$
				4	$3\frac{1}{2}$	$3\frac{9}{16}$
				5	$4\frac{1}{2}$	$4\frac{5}{8}$
				6	$5\frac{1}{2}$	$5\frac{5}{8}$
	1	$\frac{3}{4}$	$\frac{25}{32}$	7	$6\frac{1}{2}$	$6\frac{5}{8}$
	$1\frac{1}{4}$	1	$1\frac{1}{32}$	8	$7\frac{1}{4}$	$7\frac{1}{2}$
	$1\frac{1}{2}$	$1\frac{1}{4}$	$1\frac{9}{32}$	9	$8\frac{1}{4}$	$8\frac{1}{2}$
				10	$9\frac{1}{4}$	$9\frac{1}{2}$
				11	$10\frac{1}{4}$	$10\frac{1}{2}$
				12	$11\frac{1}{4}$	$11\frac{1}{2}$
				14	$13\frac{1}{4}$	$13\frac{1}{2}$
				16	$15\frac{1}{4}$	$15\frac{1}{2}$
Dimension				2	$1\frac{1}{2}$	$1\frac{9}{16}$
				3	$2\frac{1}{2}$	$2\frac{9}{16}$
				4	$3\frac{1}{2}$	$3\frac{9}{16}$
	2	$1\frac{1}{2}$	$1\frac{9}{16}$	5	$4\frac{1}{2}$	$4\frac{5}{8}$
	$2\frac{1}{2}$	2	$2\frac{1}{16}$	6	$5\frac{1}{2}$	$5\frac{5}{8}$
	3	$2\frac{1}{2}$	$2\frac{9}{16}$	8	$7\frac{1}{4}$	$7\frac{1}{2}$
	$3\frac{1}{2}$	3	$3\frac{1}{16}$	10	$9\frac{1}{4}$	$9\frac{1}{2}$
				12	$11\frac{1}{4}$	$11\frac{1}{2}$
				14	$13\frac{1}{4}$	$13\frac{1}{2}$
				16	$15\frac{1}{4}$	$15\frac{1}{2}$
Dimension				2	$1\frac{1}{2}$	$1\frac{9}{16}$
				3	$2\frac{1}{2}$	$2\frac{9}{16}$
				4	$3\frac{1}{2}$	$3\frac{9}{16}$
				5	$4\frac{1}{2}$	$4\frac{5}{8}$
	4	$3\frac{1}{2}$	$3\frac{9}{16}$	6	$5\frac{1}{2}$	$5\frac{5}{8}$
	$4\frac{1}{2}$	4	$4\frac{1}{16}$	8	$7\frac{1}{4}$	$7\frac{1}{2}$
				10	$9\frac{1}{4}$	$9\frac{1}{2}$
				12	$11\frac{1}{4}$	$11\frac{1}{2}$
				14		$13\frac{1}{2}$
				16		$15\frac{1}{2}$
Timbers	5 and thicker	$\frac{1}{2}$ off		5 and wider	$\frac{1}{2}$ off	

[a] Maximum moisture content of 19% or less.

[b] Boards less than the minimum thickness for 1-in. nominal but $\frac{5}{8}$-in. or greater thickness dry ($\frac{11}{16}$ in. green) may be regarded as American Standard Lumber, but such boards shall be marked to show the size and condition of seasoning at the time of dressing. They shall also be distinguished from 1-in. boards on invoices and certificates.

From *American Softwood Lumber Standard*, NBS 20-70, National Bureau of Standards, Washington, D.C., 1970, amended 1986, (available from Superintendent of Documents).

CHAPTER 2
MATHEMATICS*

J. N. Reddy

Department of Engineering Science and Mechanics
Virginia Polytechnic Institute and State University
Blacksburg, Virginia

2.1 ARITHMETIC
  2.1.1 Roman numerals
  2.1.2 Roots of numbers
  2.1.3 Approximate computation
  2.1.4 Interpolation

2.2 ALGEBRA
  2.2.1 Numbers
  2.2.2 Identities
  2.2.3 Binomial theorem
  2.2.4 Approximate formulas
  2.2.5 Inequalities
  2.2.6 Ratio and proportion
  2.2.7 Progressions
  2.2.8 Partial fractions
  2.2.9 Logarithms
  2.2.10 Equations
  2.2.11 Matrices and determinants
  2.2.12 Systems of equations
  2.2.13 Permutations and combinations
  2.2.14 Probability

2.3 SET ALGEBRA
  2.3.1 Sets
  2.3.2 Groups
  2.3.3 Rings, integral domains, and fields

2.4 STATISTICS AND PROBABILITY
  2.4.1 Frequency distributions of one variable
  2.4.2 Correlation
  2.4.3 Statistical estimation by small samples
  2.4.4 Statistical design of experiments
  2.4.5 Precision of measurements

2.5 GEOMETRY
  2.5.1 Geometric concepts
  2.5.2 Mensuration
  2.5.3 Constructions

2.6 TRIGONOMETRY
  2.6.1 Circular functions of plane angles
  2.6.2 Solution of triangles
  2.6.3 Spherical trigonometry
  2.6.4 Hyperbolic trigonometry
  2.6.5 Functions of imaginary and complex angles

2.7 PLANE ANALYTIC GEOMETRY
  2.7.1 Point and line
  2.7.2 Transformation of coordinates
  2.7.3 Conic sections
  2.7.4 Higher plane curves

2.8 SOLID ANALYTIC GEOMETRY
  2.8.1 Coordinate systems
  2.8.2 Point, line, and plane
  2.8.3 Transformation of coordinates
  2.8.4 Quadric surfaces

2.9 DIFFERENTIAL CALCULUS
  2.9.1 Functions and derivatives
  2.9.2 Differentiation formulas
  2.9.3 Partial derivatives
  2.9.4 Infinite series
  2.9.5 Maxima and minima

2.10 INTEGRAL CALCULUS
  2.10.1 Integration
  2.10.2 Definite integrals
  2.10.3 Line, surface, and volume integrals
  2.10.4 Applications of integration

2.11 DIFFERENTIAL EQUATIONS
  2.11.1 Definitions
  2.11.2 First-order equations
  2.11.3 Second-order equations
  2.11.4 Bessel functions
  2.11.5 Linear equations
  2.11.6 Linear algebraic equations
  2.11.7 Partial differential equations

---

*This chapter is a revision and extension of Section 2 of the third edition, which was written by John L. Barnes.

The names of Greek letters are found in Table 1.1; standard mathematical symbols in Table 1.2; and abbreviations for engineering terms in Table 1.3.

2.1 ARITHMETIC

2.1.1 Roman Numerals

Roman Notation. This uses seven letters and a bar; a letter with a bar placed over it represents a thousand times as much as it does without the bar. The letters and rules for combining them to represent numbers are as follows:

I	V	X	L	C	D	M	$\overline{\text{L}}$
1	5	10	50	100	500	1000	50,000

Rule 1. If no letter precedes a letter of greater value, add the numbers represented by the letters.

Example 2.1. XXX represents 30; VI represents 6.

Rule 2. If a letter precedes a letter of greater value, subtract the smaller from the greater; add the remainder or remainders thus obtained to the numbers represented by the other letters.

Example 2.2. IV represents 4; XL represents 40; CXLV represents 145.

Other Illustrations

IX	XIII	XIV	LV	XLII	XCVI	MDCI	$\overline{\text{IV}}$ CCXL
9	13	14	55	42	96	1601	4240

2.1.2 Roots of Numbers

Roots can be found by use of Table 1.7, or logarithms in Section 2.2.9.
To find an nth root by arithmetic, use a method indicated by the binomial theorem expansion of $(a + b)^n$.

$$(a + b)^n = a^n + na^{n-1}b + \frac{n(n-1)}{2}a^{n-2}b^2 + \frac{n(n-1)(n-2)}{3 \cdot 2}a^{n-3}b^3 + \cdots + b^n$$

$$= a^n + bD$$

where $D = na^{n-1} + \frac{1}{2}n(n-1)a^{n-2}b + \cdots + b^{n-1}$.

1. Point off the given number into periods of n figures each, starting at the decimal point and going both ways.
2. Find the largest nth power in the left-hand period and use its root as the first digit of the result. Subtract this nth power from the left-hand period and bring down the next period.
3. Use the quantity D, in which a is 10 times the first digit since the first digit occupies a higher place than the second, as the divisor to obtain the second digit b. As a trial divisor to estimate

b, use the first term in *D*, since it is the largest. Multiply *D* by *b*, subtract, and bring down the next period.

4. To get the next digit use 10 times the first two digits as *a* and proceed as before.

Example 2.3.

1. Square root of 302.980652:

$$3'02.'98'06'52' \quad | \underline{17.406\ +}$$

$$1$$

$$D = 2a + b = 27 \quad \begin{array}{|l} 202 \\ \underline{189} \end{array}$$

$$344 \quad \begin{array}{|l} 1398 \\ \underline{1376} \end{array}$$

$$34806 \quad \begin{array}{|l} 220652 \\ \underline{208836} \end{array}$$

2. Cube root of 1,58,252.632929:

$$158'252'632'929 \quad | \underline{54.09}$$

$$5^3 = \qquad\qquad 125$$

$$\text{Trial divisor} = 3a^2 = 3 \times 50^2 = \quad 7500 \quad | \ 33252$$
$$3ab = 3 \times 50 \times 4 = \quad 600$$
$$b^2 = 4^2 = \qquad\quad 16$$
$$D = 3a^2 + 3ab + b^2 = \quad \underline{8116} \quad | \ 32464$$
$$3 \times 5400^2 = 87480000 \qquad | \ 788632929$$
$$3 \times 5400 \times 9 = \quad 145800$$
$$9^2 = \qquad\qquad 81$$
$$\overline{87625881} \qquad\qquad | \ 788632929$$

2.1.3 Approximate Computation

Standard Notation. $N = a \cdot 10^b$, N is a given number; $1 \leq a < 10$, the figures in a being the *significant figures* in N; b is an integer, positive or negative or zero.

Example 2.4. If $N = 2,953,000$, in which the first five figures are significant, then $N = 2.9530 \times 10^6$.

A number is *rounded* to contain fewer significant figures by dropping figures from the right-hand side. If the figures dropped amount to more than $\frac{1}{2}$ in the last figure kept, this last figure is increased by 1. If the figures dropped amount to $\frac{1}{2}$, the last figure may or may not be increased.

Since the last significant figure used in making a measurement, an estimate, and so on, is not exact, but is usually the nearer of two consecutive figures, an approximate number may represent any value in a range from $\frac{1}{2}$ less in its last significant figure to $\frac{1}{2}$ more. The *absolute error* in an approximate number may be as much as $\frac{1}{2}$ in the last significant figure.

Example 2.5. If $N = 2.9530 \times 10^6$ is an approximate number, then $2.95295 \times 10^6 \leq N \leq 2.95305 \times 10^6$. The absolute error is between -0.00005×10^6 and 0.00005×10^6.

The size of the absolute error depends on the location of the decimal point.

The *relative error* is the ratio of the absolute error to the number. Its size depends on the number of significant figures.

Example 2.6. The relative error in Example 2.5 is at most $0.00005 \times 10^6 / 2.9530 \times 10^6$, or about 1 in 60,000; the percentage error is at most $100 \times (0.00005/2.9530)$, or less than 0.002%.

In the result of a computation with approximate numbers, some figures on the right are doubtful and should be rounded off. In slide-rule computations and to some extent in computations done with tables, rounding is done automatically. It is always possible, by using the bounds of the ranges that approximate numbers represent, to compute exactly the bounds of the range in which the result lies, and then round off the uncertain figures.

Example 2.7. Divide the approximate number 536 by the approximate number 217.4.

<table>
<tr><th></th><th>At least</th><th>At most</th></tr>
<tr><td>$\dfrac{536}{217.4} = 2.47 -$</td><td>$\dfrac{535.5}{217.45} = 246 +$</td><td>$\dfrac{536.5}{217.35} = 2.47 -$</td></tr>
</table>

In the quotient the third figure may be in error. It is useless to carry the division further.

The following rules usually give the largest number of significant figures that it is reasonable to keep.

Addition and Subtraction. Keep as the last significant figure in the result the figure in the last full column. The absolute accuracy of the result is determined by the least absolutely accurate number.

Example 2.8.

$$2.953xx$$
$$0.8942x$$
$$\underline{0.06483}$$
$$3.912xx$$

Multiplication, Division, Powers, and Roots. Keep no more significant figures in the result than the fewest in any number involved. The relative accuracy of the result is determined by that of the least relatively accurate number. Shortcuts as shown in the examples may be used.

Example 2.9

1.
$$2953 \times 413$$
$$2953$$
$$413$$

118	12
3	0
	9

$$122 \quad xxxx = 1.22 \times 10^6$$

2.
$$(1.22 \times 10^6)/2953$$
$$413$$

$$2953 \overline{) 1220000}$$
$$11812$$

| 295 | 388 |
| | 295 |

| 30 | 93 |
| | 90 |

In intermediate results keep one additional figure.

If there is much difference in the relative accuracy, that is, the number of significant figures, of the numbers involved in a computation, round all of them to one more significant figure than the least accurate number has. This procedure may introduce a small error in the last figure kept in the result. A three-digit number beginning with 8 or 9 has about the same relative accuracy as a four-digit number beginning with 1.

Use of Tables. In using a table to find the value of a function corresponding to an approximate value of an argument, it is usually advisable to retain no more significant figures in the function than there are in the argument, although the accuracy of the function varies considerably, depending inversely on the slope of the curve representing the function. However, there is no need for many-place tables if the values of the argument are known only to a few significant figures.

Example 2.10. $\frac{1}{52} = 0.019$; $\cos 61.3° = 0.877$; $\log 3.74 = 0.573$.

To investigate the behavior of the error for any given function, the differential approximation is useful. If $y = f(x)$, then $dy = f'(x)\, dx$ approximates the absolute error, and $dy/y = f'(x)\, dx/f(x)$ the relative error.

For particular approximate values of the arguments, the bounds of the ranges of the functions can be found directly from a table with arguments given to one additional place.

2.1.4 Interpolation

Gregory–Newton Interpolation Formula. Let $f(x)$ be a tabulated function of the argument x, Δx the constant difference between values of x for which the function is tabulated, and p a proper fraction. To find $f(x + p\Delta x)$, use the formula

$$f(x + p\Delta x) = f(x) + p \cdot \Delta f +{_p}C_2 \cdot \Delta_2 f +{_p}C_3 \cdot \Delta_3 f + \cdots$$

in which

$$_pC_r = \frac{p(p-1)\cdots(p-r+1)}{r!}$$

and $\Delta_r f = r$th functional difference.
Binomial coefficients for interpolation:

p	$_pC_2$	$_pC_3$	$_pC_4$	$_pC_5$	p	$_pC_2$	$_pC_3$	$_pC_4$	$_pC_5$
0.1	−0.0450	0.0285	−0.0207	0.0161	0.6	−0.1200	0.0560	−0.0336	0.0228
0.2	−0.0800	0.0480	−0.0336	0.0255	0.7	−0.1050	0.0455	−0.0262	0.0173
0.3	−0.1050	0.0595	−0.0402	0.0297	0.8	−0.0800	0.0320	−0.0176	0.0113
0.4	−0.1200	0.0640	−0.0416	0.0300	0.9	−0.0450	0.0165	−0.0087	0.0054
0.5	−0.1250	0.0625	−0.0391	0.0273					

In ordinary linear interpolation the first two terms of the formula are used.

Example 2.11. Find $\sqrt{15.4}$.

x	$f(x) = \sqrt{x}$	Δf	$\Delta_2 f$	$\Delta_3 f$
15	3.8730			
		0.1270		
16	4.0000		−0.0039	
		0.1231		0.0003
17	4.1231		−0.0036	
		0.1195		
18	4.2426			

$$\Delta x = 1, \quad p = 0.4$$

$$f(15 + 0.4 \times 1) = 3.8730 + 0.4 \times 0.1270 + 0.1200 \times 0.0039 + 0.0640 \times 0.0003$$
$$= 3.9243$$

2.2 ALGEBRA

2.2.1 Numbers

Classification

1. *Real* (*positive* and *negative*).
   a. *Rational*, expressible as the quotient of two integers.
      i. *Integers*, as −1, 2, 53.
      ii. *Fractions*, as $\frac{3}{4}$, $-\frac{5}{2}$.
   b. *Irrational*, not expressible as the quotient of two integers, as $\sqrt{2}$, π.
2. *Imaginary*, a product of a real number and the *imaginary unit* $i(= \sqrt{-1})$. Electrical engineers use j to avoid confusion with i for current. Example: $\sqrt{-2} = \sqrt{2}\,i$.
3. *Complex*, a sum of a real number and an imaginary number, as $a + bi$ (a and b real), $-3 + 0.5i$. A real number may be regarded as a complex number in which $b = 0$, and an imaginary number as one in which $a = 0$.

The absolute value of:

1. A *real number* is the number itself if the number is positive, and the number with its sign changed if it is negative, as, for example, $|3| = |-3| = 3$.
2. A *complex number* $a + bi$ is $\sqrt{a^2 + b^2}$, as, for example, $|-3 + 0.5i| = \sqrt{9 + \frac{1}{4}} = 3.04$.

2.2.2   Identities

Powers

1. $(-a)^n = a^n$, if n is even.

2. $(-a)^n = -a^n$, if n is odd.

3. $a^m \cdot a^n = a^{m+n}$.

4. $\dfrac{a^m}{a^n} = a^{m-n}$.

5. $(ab)^n = a^n b^n$.

6. $\left(\dfrac{a}{b}\right)^n = \dfrac{a^n}{b^n} = \left(\dfrac{b}{a}\right)^{-n} = \dfrac{b^{-n}}{a^{-n}} = a^n b^{-n}$.

7. $a^{-n} = \left(\dfrac{1}{a}\right)^n = \dfrac{1}{a^n}$.

8. $(a^m)^n = a^{mn}$.

9. $a^0 = 1; \; 0^n = 0; \; 0^0$ is meaningless.

Roots

1. $\sqrt[n]{a} = a^{1/n}$.

2. $(\sqrt[n]{a})^n = \sqrt[n]{a^n} = a$.

3. $\sqrt[n]{ab} = \sqrt[n]{a}\,\sqrt[n]{b}$.

4. $\sqrt[n]{\dfrac{a}{b}} = \dfrac{\sqrt[n]{a}}{\sqrt[n]{b}}$.

5. $\sqrt[m]{a}\,\sqrt[n]{a} = a^{(1/m)+(1/n)} = \sqrt[mn]{a^{m+n}}$.

6. $\sqrt[m]{a^n} = (\sqrt[m]{a})^n = a^{n/m}$.

7. $\sqrt[m]{\sqrt[n]{a}} = \sqrt[mn]{a} = \sqrt[n]{\sqrt[m]{a}} = (a^{1/m})^{1/n} = a^{1/mn}$.

8. $\sqrt{a} + \sqrt{b} = \sqrt{a + b + 2\sqrt{ab}}$.

Products

1. $(a \pm b)^2 = a^2 \pm 2ab + b^2$.

2. $(a + b)(a - b) = a^2 - b^2$.

3. $(a + b + c)^2 = a^2 + b^2 + c^2 + 2ab + 2ac + 2bc$.

4. $(a \pm b)^3 = a^3 \pm 3a^2 b + 3ab^2 \pm b^3$.

5. $a^3 \pm b^3 = (a \pm b)(a^2 \mp ab + b^2)$.

Quotients

1. $(a^n - b^n)/(a - b) = a^{n-1} + a^{n-2}b + a^{n-3}b^2 + \cdots + ab^{n-2} + b^{n-1}$, if $a \neq b$.

2. $(a^n + b^n)/(a + b) = a^{n-1} - a^{n-2}b + a^{n-3}b^2 - \cdots - ab^{n-2} + b^{n-1}$, if n is odd.

3. $(a^n - b^n)/(a + b) = a^{n-1} - a^{n-2}b + a^{n-3}b^2 - \cdots + ab^{n-2} - b^{n-1}$, if n is even.

Fractions

Signs. $\quad \dfrac{a}{b} = \dfrac{-a}{-b} = \dfrac{-a}{b} = -\dfrac{a}{-b}$.

Addition and Subtraction. $\quad \dfrac{a}{c} \pm \dfrac{b}{d} = \dfrac{ad \pm bc}{cd}, \quad \dfrac{a}{c} \pm \dfrac{b}{c} = \dfrac{a \pm b}{c}, \quad \dfrac{a}{c} \pm \dfrac{a}{d} = \dfrac{a(d \pm c)}{cd}$,

$$\dfrac{a}{def} + \dfrac{b}{e^3 g} - \dfrac{c}{df^2} = \dfrac{ae^2 fg + bdf^2 - ce^3 g}{de^3 f^2 g}.$$

Multiplication. $\quad \dfrac{a}{b} \times \dfrac{c}{d} = \dfrac{ac}{bd}, \quad \dfrac{a}{b} = \dfrac{ac}{bc}$.

Division. $\quad \dfrac{a}{b} \Big/ \dfrac{c}{d} = \dfrac{a}{b} \times \dfrac{d}{c} = \dfrac{ad}{bc}, \quad \dfrac{a}{b} = \dfrac{a}{c} \Big/ \dfrac{b}{c}$.

Series

1. $1 + 2 + 3 + 4 + \cdots + (n - 1) + n = \dfrac{n(n + 1)}{2}$.

2. $p + (p + 1) + (p + 2) + \cdots + (q - 1) + q = \dfrac{(q + p)(q - p + 1)}{2}$.

3. $2 + 4 + 6 + 8 + \cdots + (2n - 2) + 2n = n(n + 1)$.

4. $1 + 3 + 5 + 7 + \cdots + (2n - 3) + (2n - 1) = n^2$.

5. $1^2 + 2^2 + 3^2 + 4^2 + \cdots + (n - 1)^2 + n^2 = \dfrac{n(n + 1)(2n + 1)}{6}$.

6. $1^3 + 2^3 + 3^3 + 4^3 + \cdots + (n - 1)^3 + n^3 = \dfrac{n^2(n + 1)^2}{4}$.

7. $1^4 + 2^4 + 3^4 + 4^4 + \cdots + (n - 1)^4 + n^4 = \dfrac{n}{30}(n + 1)(2n + 1)(3n^2 + 3n - 1)$.

2.2.3 Binomial Theorem

$$(a \pm b)^n = a^n \pm na^{n-1}b + \frac{n(n - 1)}{1 \cdot 2}a^{n-2}b^2 \pm \frac{n(n - 1)(n - 2)}{1 \cdot 2 \cdot 3}a^{n-3}b^3 + \cdots$$

$$+ (\pm 1)^r \frac{n(n - 1) \cdots (n - r + 1)}{r!}a^{n-r}b^r + \cdots$$

in which the last term shown is the $(r + 1)$th; $r!$, called r *factorial*, equals $1 \cdot 2 \cdot 3 \cdots (r - 1) \cdot r$; and $0! = 1$.

If n is a positive integer, the series is finite; it has $(n + 1)$ terms, the last being b^n; and it holds for all values of a and b. If n is fractional or negative, the series is infinite; it converges only for $|b| < |a|$ (see Section 2.9.4).

The coefficients n, $n(n - 1)/2!$, $n(n - 1)(n - 2)/3!, \ldots$ are called *binomial coefficients*. For brevity the coefficient $n(n - 1) \cdots (n - r + 1)/r!$ of the $(r + 1)$th term is written $\binom{n}{r}$ or $_nC_r$. If n is a positive integer, the coefficients of the rth term from the beginning and the rth from the end are equal.

For any value of n, and $-1 < x < 1$:

$$(1 \pm x)^n = 1 \pm nx + \frac{n(n - 1)}{1 \cdot 2}x^2 \pm \frac{n(n - 1)(n - 2)}{1 \cdot 2 \cdot 3}x^3$$

$$+ \frac{n(n - 1)(n - 2)(n - 3)}{1 \cdot 2 \cdot 3 \cdot 4}x^4 \pm \cdots$$

$$\frac{1}{1 \pm x} = (1 \pm x)^{-1} = 1 \mp x + x^2 \mp x^3 + x^4 \mp x^5 + \cdots$$

$$\sqrt{1 \pm x} = (1 \pm x)^{1/2} = 1 \pm \frac{1}{2}x - \frac{1}{2 \cdot 4}x^2 \pm \frac{1 \cdot 3}{2 \cdot 4 \cdot 6}x^3$$

$$- \frac{1 \cdot 3 \cdot 5}{2 \cdot 4 \cdot 6 \cdot 8}x^4 \pm \frac{1 \cdot 3 \cdot 5 \cdot 7}{2 \cdot 4 \cdot 6 \cdot 8 \cdot 10}x^5 -$$

$$\frac{1}{\sqrt{1 \pm x}} = (1 \pm x)^{-1/2} = 1 \mp \frac{1}{2}x + \frac{1 \cdot 3}{2 \cdot 4}x^2 \mp \frac{1 \cdot 3 \cdot 5}{2 \cdot 4 \cdot 6}x^3 + \cdots$$

2.2.4 Approximate Formulas

(a) If $|x|$ and $|y|$ are small compared with 1:

1. $(1 \pm x)^2 = 1 \pm 2x$.

2. $(1 \pm x)^{1/2} = 1 \pm \dfrac{x}{2}$.

3. $\dfrac{1}{1 \pm x} = 1 \mp x.$

4. $(1 + x)(1 + y) = 1 + x + y.$

5. $(1 + x)(1 - y) = 1 + x - y.$

6. $e^x = 1 + x + \dfrac{x^2}{2}$ (where $e = 2.71828$). ⎫

7. $\log_e(1 \pm x) = \pm x - \dfrac{x^2}{2} \pm \dfrac{x^3}{3}.$ (Last term often may be omitted.)

8. $\log_e\left(\dfrac{1 + x}{1 - x}\right) = 2\left(x + \dfrac{x^3}{3} + \dfrac{x^5}{5}\right).$ ⎭

(b) If $|x|$ is small compared with a and $a > 0$:

9. $a^x = 1 + x\log_e a + \dfrac{x^2}{2}(\log_e a)^2.$ (Last term often may be omitted.)

(c) If a and b are nearly equal and both > 0:

10. $\sqrt{ab} = \dfrac{a + b}{2}.$

(d) If b is small compared with a and both > 0:

11. $\sqrt{a^2 \pm b} = a \pm \dfrac{b}{2a}.$

12. $\sqrt{a^2 \pm b} = a \pm \dfrac{b}{3a^2}.$

13. $\sqrt{a^2 + b^2} = 0.960a + 0.398b.$ This is within 4% of the true value if $a > b$.
A closer approximation is $\sqrt{a^2 + b^2} = 0.9938a + 0.0703b + 0.3567(b^2/a).$

14. $\sqrt{a^2 + b^2 + c^2} = 0.939a + 0.389b + 0.297c.$ This is within 6% of the true value if $a > b > c$. For instance, for the numbers 43, 42, and 41, the error $< 5.2\%.$

(e) If $|x|$ is less than $\pi/18$:

15. $\sin x = x - \dfrac{x^3}{6}$ ⎫

16. $\cos x = 1 - \dfrac{x^2}{2}$ (Last term often may be omitted.)

17. $\tan x = x + \dfrac{x^3}{3}$ ⎭

(*Note:* If $x = 8° = 8\pi/180 = 0.13963$, $\sin x = x - \frac{1}{6}x^3 = 0.13918$, which is one unit in error in the fifth decimal place. If the absolute value of the angle is less than 5°, the values of x and $\sin x$ do not differ more than one unit in the fourth decimal place.)

(f) If $|y|$ is less than $\pi/36$ and small compared with $|x|$:

18. $\sin(x \pm y) = \sin x \pm y\cos x.$

19. $\cos(x \pm y) = \cos x \mp y\sin x.$

20. $\tan(x \pm y) = \tan x \pm \dfrac{y}{\cos^2 x}.$

(g) If $|n| > 1$:

21. $e^{1/n} = 1 + \dfrac{1}{n - 0.5}.$

22. $e^{-1/n} = 1 - \dfrac{1}{n + 0.5}.$

(h) As $n \to \infty$:

23. $\dfrac{1 + 2 + 3 + 4 + 5 \cdots + n}{n^2} \to \dfrac{1}{2}.$

24. $\dfrac{1 + 2^2 + 3^2 + 4^2 + \cdots + n^2}{n^3} \to \dfrac{1}{3}.$

25. $\dfrac{1 + 2^3 + 3^3 + 4^3 + \cdots + n^3}{n^4} \to \dfrac{1}{4}.$

2.2.5 Inequalities

Laws of inequalities for positive quantities:

(a) If $a > b$, then $a + c > b + c$

$$a - c > b - c \qquad b < a$$
$$ac > bc \qquad c - a < c - b$$
$$\frac{a}{c} > \frac{b}{c} \qquad -ca < -cb$$
$$\qquad\qquad \frac{c}{a} < \frac{c}{b}$$

Corollary: If $a - c > b$, then $a > b + c$.

(b) If $a > b$ and $c > d$, then $a + c > b + d$; $ac > bd$; but $a - c$ may be $>$ or $=$ or $< b - d$; a/c may be $>$ or $=$ or $< b/d$.

2.2.6 Ratio and Proportion

Laws of Ratio and Proportion

(a) If $a/b = c/d$, then

$$\frac{a}{c} = \frac{b}{d} \qquad ad = bc$$

$$\frac{ma + nb}{pa + qb} = \frac{mc + nd}{pc + qd} \qquad \left(\frac{a}{b}\right)^n = \left(\frac{c}{d}\right)^n$$

If also $e/f = g/h$, then, $ae/bf = cg/dh$.

(b) If $a/b = c/d = e/f = \cdots$, then

$$\frac{a}{b} = \frac{c}{d} = \frac{e}{f} = \cdots = \frac{pa + qc + re + \cdots}{pb + qd + rf + \cdots}$$

Variation

If $y = kx$, y varies directly as x; that is, y is directly proportional to x.

If $y = k/x$, y varies inversely as x; that is, y is inversely proportional to x.

If $y = kxz$, y varies jointly as x and z.

If $y = k(x/z)$, y varies directly as x and inversely as z.

The constant k is called the proportionality factor.

2.2.7 Progressions

Arithmetic Progression. This is a sequence in which the *difference d* of any two consecutive terms is a constant. If n = number of terms, a = first term, l = last term, s = sum of n terms, then $l = a + (n - 1)d$, and $s = (n/2)(a + l)$. The *arithmetic mean A* of two quantities m, n is the quantity that placed between them makes with them an arithmetic progression; $A = (m + n)/2$.

Example 2.12. Given the series $3 + 5 + 7 + \cdots$ to 10 terms. Here $n = 10$, $a = 3$, $d = 2$; hence $l = 3 + (10 - 1) \times 2 = 21$, and $s = (10/2)(3 + 21) = 120$.

Geometric Progression. This is a sequence in which the *ratio r* of any two consecutive terms is a constant. If n = number of terms, a = first term, l = last term, s = sum of n terms, then $l = ar^{n-1}$, $s = (rl - a)/(r - 1) = a(1 - r^n)/(1 - r)$. The *geometric mean G* of two quantities m, n is the quantity that placed between them makes with them a geometric progression; $G = \sqrt{mn}$.

Example 2.13. Given the series $3 + 6 + 12 + \cdots$ to 6 terms. Here $n = 6$, $a = 3$, $r = 2$; hence $l = 3 \times 2^{6-1} = 96$, and $s = (2 \times 96 - 3)/(2 - 1)$, or $= 3(1 - 2^6)/(1 - 2) = 189$.

If $|r| < 1$, then, as $n \to \infty$, $s \to a/(1 - r)$.

Example 2.14. Given the infinite series $\frac{1}{2} + \frac{1}{4} + \frac{1}{8} + \cdots$. Here $a = \frac{1}{2}$ and $r = \frac{1}{2}$; hence $s \to (\frac{1}{2})/(1 - \frac{1}{2}) = 1$ as $n \to \infty$.

Harmonic Progression. This is a sequence in which the reciprocals of the terms form an arithmetic progression. The *harmonic mean* H of two quantities m, n is the quantity that placed between them makes with them a harmonic progression; $H = 2mn/(m + n)$.

Relation among the arithmetic, geometric, and harmonic means of two quantities; $G^2 = AH$.

2.2.8   Partial Fractions

A *proper* algebraic fraction is one in which the numerator is of lower degree than the denominator. An improper fraction can be changed to the sum of a polynomial and a proper fraction by dividing the numerator by the denominator.

A proper fraction can be resolved into *partial fractions*, the denominators of which are factors, prime to each other, of the denominator of the given fraction.

CASE 1:   The denominator can be factored into real linear factors, $P, Q, R, \ldots$, all different. Let

$$\frac{\text{Num}}{PQR \cdots} = \frac{A}{P} + \frac{B}{Q} + \frac{C}{R} + \cdots$$

Example 2.15

$$\frac{6x^2 - x + 1}{x^3 - x} = \frac{A}{x} + \frac{B}{x - 1} + \frac{C}{x + 1}$$

Clearing fractions,

$$6x^2 - x + 1 = A(x - 1)(x + 1) + Bx(x + 1) + Cx(x - 1) \qquad (2.1)$$

(a) Substitution method. Letting $x = 0$, $A = -1$; $x = 1$, $B = 3$; $x = -1$, $C = 4$. Then

$$\frac{6x^2 - x + 1}{x^3 - x} = -\frac{1}{x} + \frac{3}{x - 1} + \frac{4}{x + 1}$$

(b) Method of undetermined coefficients. Rewriting Eq. (2.1),

$$6x^2 - x + 1 = (A + B + C)x^2 + (B - C)x - A$$

Equating coefficients of like powers of x, $A + B + C = 6$, $B - C = -1$, $-A = 1$. Solving this system of equations, $A = -1$, $B = 3$, $C = 4$.

CASE 2:   The denominator can be factored into real linear factors, $P, Q, \ldots$, one or more repeated. Let

$$\frac{\text{Num}}{P^2 Q^3} = \frac{A}{P} + \frac{B}{P^2} + \frac{C}{Q} + \frac{D}{Q^2} + \frac{E}{Q^3} + \cdots$$

Example 2.16

$$\frac{x + 1}{x(x - 1)^3} = \frac{A}{x} + \frac{B}{x - 1} + \frac{C}{(x - 1)^2} + \frac{D}{(x - 1)^3}$$

Clearing fractions,

$$x + 1 = A(x - 1)^3 + Bx(x - 1)^2 + Cx(x - 1) + Dx$$

A and D can be found by substituting $x = 0$ and $x = 1$. After inserting these numerical values for A and D, B and C can be found by the method of undetermined coefficients.

CASE 3:   The denominator can be factored into quadratic factors, $P, Q, \ldots$, all different, which cannot be factored into real linear factors. Let

$$\frac{\text{Num}}{PQ \cdots} = \frac{Ax + B}{P} + \frac{Cx + D}{Q} + \cdots$$

Example 2.17

$$\frac{3x^2 - 2}{(x^2 + x + 1)(x + 1)} = \frac{Ax + B}{x^2 + x + 1} + \frac{C}{x + 1}$$

Clearing fractions,

$$3x^2 - 2 = (Ax + B)(x + 1) + C(x^2 + x + 1)$$
$$= (A + C)x^2 + (A + B + C)x + (B + C)$$

Use the method of undetermined coefficients to find A, B, C.

CASE 4: The denominator can be factored into quadratic factors, $P, Q, \ldots$, one or more repeated, which cannot be factored into real linear factors. Let

$$\frac{\text{Num}}{P^2 Q^3 \cdots} = \frac{Ax + B}{P} + \frac{Cx + D}{P^2} + \frac{Ex + F}{Q} + \frac{Gx + H}{Q^2} + \frac{Ix + J}{Q^3} + \cdots$$

Example 2.18

$$\frac{5x^2 - 4x + 16}{(x - 3)(x^2 - x + 1)^2} = \frac{A}{x - 3} + \frac{Bx + C}{x^2 - x + 1} + \frac{Dx + E}{(x^2 - x + 1)^2}$$

Clearing fractions,

$$5x^2 - 4x + 16 = A(x^2 - x + 1)^2 + (Bx + C)(x - 3)(x^2 - x + 1) + (Dx + E)(x - 3)$$

Find A by substituting $x = 3$. Then use the method of undetermined coefficients to find B, C, D, E.

2.2.9 Logarithms

If $N = b^x$, then x is the *logarithm* of the number N to the *base* b. For computation, *common*, or Briggs, logarithms to the base 10 (abbreviated $\log_{10}$ or log) are used. For theoretical work involving calculus, *natural*, or Naperian, logarithms to the irrational base $e = 2.71828 \cdots$ (abbreviated ln, $\log_e$, or log) are used. The relation between logarithms of the two systems is:

$$\log_e n = \frac{\log_{10} n}{\log_{10} e} = \frac{\log_{10} n}{0.4343} = 2.303 \log_{10} n$$

The integral part of a common logarithm, called the *characteristic*, may be positive, negative, or zero. The decimal part, called the *mantissa* and given in tables, is always positive.

To find the common logarithm of a number, first find the mantissa from Table 2.10, disregarding the decimal point of the number. Then from the location of the decimal point find the characteristic as follows. If the number is greater than 1, the characteristic is positive or zero. It is one less than the number of figures preceding the decimal point. For a number expressed in standard notation the characteristic is the exponent of 10.

Example 2.19. $\log 6.54 = 0.8156$, $\log 6540 = \log(6.54 \times 10^3) = 3.8156$.

If the number is less than 1, the characteristic is negative and is numerically one greater than the number of zeros immediately following the decimal point. To avoid having a negative integral part and a positive decimal part, the characteristic is written as a difference.

Example 2.20. $\log 0.654 = \log(6.54 \times 10^{-1}) = \bar{1}.8156 = 9.8156 - 10$, $\log 0.000654 = \log(6.54 \times 10^{-4}) = \bar{4}.8156 = 6.8156 - 10$.

To find a number whose logarithm is given, each of the preceding steps is reversed.

The cologarithm of a number is the logarithm of its reciprocal. Hence, $\text{colog } N = \log 1/N = \log 1 - \log N = -\log N$.

Use of Logarithms in Computation.

To multiply a and b	$\log ab = \log a + \log b$
To divide a by b	$\log a/b = \log a - \log b$
To raise a to the nth power	$\log a^n = n \log a$
To find the nth root of a	$\log a^{1/n} = (1/n)\log a$

Example 2.21. (1) $68.31 \times 0.2754 = 18.81$.

$$\log 68.31 = 1.8345$$
$$\log\ 0.2754 = \underline{\ 9.4400 - 10\ }$$
$$11.2745 - 10 = 1.2745 = \log 18.81$$

(2) $0.6841^{1.53} = 0.5582$.

$$\log 0.6831 = 9.8345 - 10$$
$$1.53 \times (9.8345 - 10) = 15.0468 - 15.3$$

To subtract 15.3 from 15.0468, add 10 to 15.0468 and subtract 10 from it.

$$25.0468 - 10$$
$$\underline{15.3\qquad\qquad}$$
$$9.7468 - 10 = \log 0.5582$$

(3) $\sqrt[5]{0.6831} = 0.9266$.

$$\log 0.6831 = 9.8345 - 10$$
$$\frac{49.8345 - 50}{5} = 9.9669 - 10 = \log 0.9266$$

To solve a simple exponential equation of the form $a^x = b$, equate the logarithms of the two sides of the equation:

$$x \log a = \log b$$

from which

$$x = \frac{\log b}{\log a} \quad \text{and} \quad \log x = \log(\log b) - \log(\log a)$$

Example 2.22. $0.6831^x = 27.54$.

$$x = \frac{\log 27.54}{\log 0.6831} = \frac{1.4400}{9.8345 - 10} = \frac{1.4400}{-0.1655} = -8.701$$

2.2.10   Equations

The equation $f(x) = a_0 x^n + a_1 x^{n-1} + a_2 x^{n-2} + \cdots + a_n = 0$, a_i real, is a *polynomial equation* of *degree n* in one variable.

For $n = 1$, the equation $f(x) = ax + b = 0$ is *linear*. It has one root $x_1 = -b/a$.

Quadratic Equation

For $n = 2$, the equation $f(x) = ax^2 + bx + c = 0$ is *quadratic*. It has two roots, both real or both complex, given by the formulas

$$x_1, x_2 = \frac{-b \pm \sqrt{b^2 - 4ac}}{2a} = \frac{2c}{-b \mp \sqrt{b^2 - 4ac}}$$

To avoid loss of precision if $\sqrt{b^2 - 4ac}$ and $|b|$ are nearly equal, use the form that does not involve the difference.

If the quantity $b^2 - 4ac$, called the *discriminant*, is > 0, the roots are real and unequal; if it $= 0$, the roots are real and equal; if it is < 0, the roots are complex.

Cubic Equation

For $n = 3$, the equation $f(x) = a_0 x^3 + a_1 x^2 + a_2 x + a_3 = 0$ is *cubic*. It has three roots, all real or one real and two complex.

Algebraic Solution. Write the equation in the form $ax^3 + 3bx^2 + 3cx + d = 0$. Let

$$q = ac - b^2 \quad \text{and} \quad r = \tfrac{1}{2}(3abc - a^2 d) - b^3$$

Also let

$$s_1 = \left(r + \sqrt{q^3 + r^2}\right)^{1/3} \quad \text{and} \quad s_2 = \left(r - \sqrt{q^3 + r^2}\right)^{1/3}$$

Then the roots are

$$x_1 = \left[(s_1 + s_2) - b\right] \div a$$

$$x_2 = \left[-\frac{1}{2}(s_1 + s_2) + \frac{\sqrt{-3}}{2}(s_1 - s_2) - b\right] \div a$$

$$x_3 = \left[-\frac{1}{2}(s_1 + s_2) - \frac{\sqrt{-3}}{2}(s_1 - s_2) - b\right] \div a$$

If $q^3 + r^2 > 0$, there are one real and two complex roots. If $q^3 + r^2 = 0$, there are three real roots of which at least two are equal. If $q^3 + r^2 < 0$, there are three real roots, but the numerical solution leads to finding the cube roots of complex quantities. In such a case the trigonometric solution is employed.

Example 2.23. Given the equation $x^3 + 12x^2 + 45x + 54 = 0$. Here $a = 1$, $b = 4$, $c = 15$, $d = 54$. $q = 15 - 16 = -1$; $r = \frac{1}{2}(180 - 54) - 64 = 1$; $q^3 + r^2 = -1 + 1 = 0$, $s_1 = s_2 = (-1)^{1/2} = -1$. $s_1 + s_2 = -2$; $s_1 - s_2 = 0$. Hence the roots are $x_1 = (-2 - 4) = -6$; $x_2 = x_3 = [-\frac{1}{2}(-2) - 4] = -3$.

Trigonometric Solution. Write the equation in the form $ax^3 + 3bx^2 + 3cx + d = 0$. Let $q = ac - b^2$ and $r = \frac{1}{2}(3abc - a^2 d) - b^3$ (as in algebraic solution). Then the roots are

$$x_1 = (y_1 - b) \div a$$
$$x_2 = (y_2 - b) \div a$$
$$x_3 = (y_3 - b) \div a$$

where y_1, y_2, and y_3 have the following values (upper of alternative signs being used when r is $+$ and the lower when r is $-$):

CASE 1: If q is $-$ and $q^3 + r^2 \le 0$:

$$y_1 = \pm 2\sqrt{-q} \cos\left[\frac{1}{3}\cos^{-1}\frac{\pm r}{\sqrt{-q^3}}\right]$$

$$y_2 = \pm 2\sqrt{-q} \cos\left[\frac{1}{3}\cos^{-1}\frac{\pm r}{\sqrt{-q^3}} + \frac{2\pi}{3}\right]$$

$$y_3 = \pm 2\sqrt{-q} \cos\left[\frac{1}{3}\cos^{-1}\frac{\pm r}{\sqrt{-q^3}} + \frac{4\pi}{3}\right]$$

CASE 2: If q is $-$ and $q^3 + r^2 \ge 0$:

$$y_1 = \pm 2\sqrt{-q} \cosh\left[\frac{1}{3}\cosh^{-1}\frac{\pm r}{\sqrt{-q^3}}\right]$$

$$y_2 = \mp\sqrt{-q} \cosh\left[\frac{1}{3}\cosh^{-1}\frac{\pm r}{\sqrt{-q^3}}\right] + i\sqrt{-3q}\sinh\left[\frac{1}{3}\cosh^{-1}\frac{\pm r}{\sqrt{-q^3}}\right]$$

$$y_3 = \mp\sqrt{-q} \cosh\left[\frac{1}{3}\cosh^{-1}\frac{\pm r}{\sqrt{-q^3}}\right] - i\sqrt{-3q}\sinh\left[\frac{1}{3}\cosh^{-1}\frac{\pm r}{\sqrt{-q^3}}\right]$$

CASE 3:   If q is $+$:

$$y_1 = \pm 2\sqrt{q}\sinh\left[\frac{1}{3}\sinh^{-1}\frac{\pm r}{\sqrt{q^3}}\right]$$

$$y_2 = \mp\sqrt{q}\sinh\left[\frac{1}{3}\sinh^{-1}\frac{\pm r}{\sqrt{q^3}}\right] + i\sqrt{3q}\cosh\left[\frac{1}{3}\sinh^{-1}\frac{\pm r}{\sqrt{q^3}}\right]$$

$$y_3 = \mp\sqrt{q}\sinh\left[\frac{1}{3}\sinh^{-1}\frac{\pm r}{\sqrt{q^3}}\right] - i\sqrt{3q}\cosh\left[\frac{1}{3}\sinh^{-1}\frac{\pm r}{\sqrt{q^3}}\right]$$

Example 2.24. Given the equation $x^3 + 6x^2 - 9x - 54 = 0$.
Here $a = 1$, $b = 2$, $c = -3$, $d = -54$; $q = -3 - 4 = -7$; $r = \frac{1}{2}(-18 + 54) - 8 = 10$; $q^3 + r^2 = -343 + 100 = -243$. Note that q is $-$; $q^3 + r^2 < 0$; r is $+$. Therefore use Case 1 with upper signs.

$$y_1 = 2\sqrt{7}\cos\left[\frac{1}{3}\cos^{-1}\frac{10}{\sqrt{343}}\right] = 2\sqrt{7}\cos 19.1° = 5$$

Hence, one root is $x_1 = 5 - 2 = 3$. The other roots can be similarly determined.

Quartic Equation
For $n = 4$, the equation $f(x) = a_0x^4 + a_1x^3 + a_2x^2 + a_3x + a_4 = 0$ is *quartic*. It has four roots, all real, all complex, or two real and two complex.
   To solve, first divide the equation by a_0 to put it in the form $x^4 + ax^3 + bx^2 + cx + d = 0$. Find any real root y_1 of the cubic equation:

$$8y^3 - 4by^2 + 2(ac - 4d)y - \left[c^2 + d(a^2 - 4b)\right] = 0$$

Then the four roots of the quartic equation are given by the roots of the two quadratic equations:

$$x^2 + \left[\frac{a}{2} + \sqrt{\frac{a^2}{4} + 2y_1 - b}\right]x + \left(y_1 + \sqrt{y_1^2 - d}\right) = 0$$

$$x^2 + \left[\frac{a}{2} - \sqrt{\frac{a^2}{4} + 2y_1 - b}\right]x + \left(y_1 - \sqrt{y_1^2 - d}\right) = 0$$

nth-Degree Equation

Properties of $f(x) = a_0x^n + a_1x^{n-1} + \cdots + a_n = 0$. a_n's are real.

1. *Remainder Theorem.* If $f(x)$ is divided by $(x - r)$ until a remainder independent of x is obtained, this remainder is equal to $f(r)$, the value of $f(x)$ for $x = r$.
2. *Factor Theorem.* If, and only if, $(x - r)$ is a factor of $f(x)$, then $f(r) = 0$.
3. The equation $f(x) = 0$ has n roots, not necessarily distinct. Complex roots occur in conjugate pairs, $a + bi$ and $a - bi$. If n is odd, there is at least one real root.
4. The sum of the roots is $-a_1/a_0$, the sum of the products of the roots taken two at a time is a_2/a_0, the sum of the products of the roots taken three at a time is $-a_3/a_0$, and so on. The product of all the roots is $(-1)^n a_n/a_0$.
5. If the a_i are integers and p/q is a rational root of $f(x) = 0$, reduced to its lowest terms, then p is a divisor of a_n and q of a_0. If a_0 is 1, the rational roots are integers.
6. If x is replaced by (a) y/m, (b) $-y$, (c) $y + h$, the roots of the resulting equation $\phi(y) = 0$ are (a) m times, (b) the negatives of, (c) less by h than the corresponding roots of $f(x) = 0$.
7. *Descartes' Rule of Signs.* A *variation* of sign occurs in $f(x) = 0$ if two consecutive terms have unlike signs. The number of positive roots is either equal to the number of variations of sign or is less by a positive even integer. For negative roots apply the rule to $f(-x) = 0$.
8. If, for two real numbers a and b, $f(a)$ and $f(b)$ have opposite signs, there is an odd number of roots between a and b.

9. If k is the exponent of the first term with a negative coefficient and G the greatest of the absolute values of the negative coefficients, then an upper bound of the real roots is $1 + \sqrt[n-k]{G/a_0}$.

10. **Sturm's Theorem.** *Let the equation $f(x) = 0$ have no multiple roots. With $f_0 = f(x)$ and $f_1 = f'(x)$, form the sequence $f_0, f_1, f_2, \ldots, f_n$ as follows:*

$$f_0 = q_1 f_1 - f_2 \qquad f_1 = q_2 f_2 - f_3 \qquad f_2 = q_3 f_3 - f_4, \ldots, f_{n-2} = q_{n-1} f_{n-1} - f_n$$

At any step, a function f_i may be multiplied by a positive number to avoid fractions. Let a and b be real numbers, $a < b$, such that $f(a) \neq 0$, $f(b) \neq 0$, and let $V(a)$ be the number of variations of sign in the nonzero members of the sequence $f_0(a), f_i(a), \ldots, f_n$. Then the number of real roots between a and b is $V(a) - V(b)$.

If $f(x) = 0$ has multiple roots, the sequence terminates with the function f_m, $m < n$, when $f_{m-1} = q_m f_m$. For this sequence, $V(a) - V(b)$ is the number of distinct real roots between a and b.

Example 2.25. (1) Locate the real roots of $x^3 - 7x - 7 = 0$.

	$x =$	-2	-1	0	1	2	3	4		$3x - \frac{9}{2}$			x	
$f_0 = x^3 - 7$	$x - 7$	$-$	$-$	$-$	$-$	$-$	$+$	$2x + 3$	$3x^2$		-7	$x^3 -$	$7x -$	7
$f_1 = 3x^2 - 7$		$+$	$-$	$-$	$-$	$+$	$+$	$+$	$6x^2$		-14	$3x^3 - 21x - 21$		
$f_2 = 2x + 3$		$-$	$+$	$+$	$+$	$+$	$+$	$+$	$6x^2 + 9x$			$3x^3 -$	$7x$	
$f_3 = 1$		$+$	$+$	$+$	$+$	$+$	$+$	$+$		$-9x - 14$		$-$	$14x - 21$	
	$V(x) =$	3	1	1	1	1	1	0		$-9x - \frac{27}{2}$			$2x + 3 = f^2$	
										$-\frac{1}{2}$				
										$1 = f_3$				

$$V(-2) - V(-1) = 2 \qquad V(3) - V(4) = 1$$
$$-2 < r_1, r_2 < -1 \qquad 3 < r_3 < 4$$

(2) Locate the real roots of $4x^3 - 3x - 1 = 0$.

	$x =$	-1	0	1	2		$2x$	$-1x$		
$f_0 = 4x^3 - 3x - 1$		$-$	$-$	0	$+$	$2x + 1$	$4x^2 - 1$	$4x^3 - 3x - 1$	
$f_1 = 3(4x^2 - 1)$		$+$	$-$	$-$	$+$		$4x^2 + 2x$	$4x^3 -$	x
$f_2 = 2x + 1$		$-$	$+$	$+$			$-2x - 1$	$-2x - 1$	
	$V(x) =$	2	1	0			$-2x - 1$	$2x + 1 = f_2$	

$$V(-1) - V(0) = 1 \qquad V(0) - V(2) = 1$$
$$-1 < r_1 < 0 \qquad 0 < r_2 < 2$$

r_1 can be found to be a double root.

Synthetic Division. To divide a polynomial $f(x)$ by $(x - a)$, proceed as in Example 2.25. Divide $f(x) = 4x^3 - 7x + 1$ by $x + 2$. Arrange the coefficients in order of descending powers of x, supplying zeros for missing powers. Place a $(= -2)$ to the left. Bring down the first coefficient, multiply it by a, and add the product to the next coefficient. Multiply the sum by a, add the product to the next coefficient, and continue thus.

$$
\begin{array}{r|rrrr}
-2 & 4 + 0 & - 7 & + 1 \\
 & - 8 & + 16 & - 18 \\
\hline
 & 4 - 8 & + 9 & \underline{-17}
\end{array}
$$

The last number is the remainder. It is the value of the polynomial $f(x) = 4x^3 - 7x + 1$ for $x = -2$, or $f(-2) = -17$. The other numbers in the last line are the coefficients of the quotient $4x^2 - 8x + 9$, a polynomial of one degree less than the dividend.

Rational Roots. Possible integral and fractional roots can be found by property 5, and tested by synthetic division. If a rational root r is found, then the remaining roots are roots of $q(x) \equiv f(x)/(x - r) = 0$.

Irrational Roots

Horner's Method. This consists of diminishing a root repeatedly toward zero and adding together the amounts by which it is diminished. This sum approximates the original root. The method is explained by an example.

A root of $x^3 + 4x - 7 = 0$ is located between the successive integers 1 and 2, graphically or by synthetic division, using property 8. First, the roots are diminished by 1 (property 6c) to give an equation $f(y + 1) \equiv \phi(y) = 0$, which has a root between 0 and 1. The method of obtaining the coefficients of $\phi(y)$ by use of successive synthetic divisions is illustrated. The remainders are the required coefficients. The root between 0 and 1 of $\phi(y) = 0$ is then located between successive tenths. Since its value is small, the last two terms set equal to 0 suffice to estimate that it is between 0.2 and 0.3. Next, diminish the roots by 0.2 to obtain an equation with a root between 0 and 0.1. To check that the root was between 0.2 and 0.3, note that the first remainder, which is the value of $\phi(0.2)$, remains negative when $\phi(y)$ is divided by $y - 0.2$, and that the remainder would be found to be positive if $\phi(y)$ were divided by $y - 0.3$. Repeat the process, using the last two terms to estimate that the root of the new equation is between 0.05 and 0.06, and then diminish by 0.05. At the next stage it is frequently possible to estimate two more figures by using the last two terms.

$$
\begin{array}{llll}
1 + 0 & + 4 - & 7 \,|\, 1 & \\
+ 1 & + 1 + & 5 & \\
\hline
1 + 1 & + 5 \,|\!-\! & 2 & \\
+ 1 & + 2 \,| & & \\
\hline
1 + 2 & + 7 & & \\
+ 1 & & & \\
\hline
1 + 3 & + 7 & - \quad 2 & |\, 0.2 \\
+ 0.2 & + 0.64 & + \quad 1.528 & \\
\hline
1 + 3.2 & + 7.64 & - \quad 0.472 & \\
+ 0.2 & + 0.68 & & \\
\hline
1 + 3.4 & + 8.32 & & \\
+ 0.2 & & & \\
\hline
1 + 3.6 & + 8.32 & - \quad 0.472 & |\, 0.05 \\
+ 0.05 & + 0.1825 & + \quad 0.425125 & \\
\hline
1 + 3.65 & + 8.5025 & - \quad 0.046875 & \\
+ 0.05 & + 0.185 & & \\
\hline
1 + 3.70 & + 8.6875 & & \\
+ 0.05 & & & \\
\hline
1 + 3.75 & & &
\end{array}
$$

$$8.6875x - 0.046875 = 0$$
$$x = 0.0054.$$

The root is 1.2554.

To find a *negative* irrational root $-r$ by Horner's method, replace x in $f(x) = 0$ by $-y$, find the positive root r of $\phi(y) = f(-y) = 0$, and change its sign.

Newton's Method. This can be used to find a root of either an algebraic or a transcendental equation. The root is first located graphically between α and β, $f(\alpha)$ and $f(\beta)$ having unlike signs (Fig. 2.1). Assume that there is no maximum, minimum, or inflection point in the interval (α, β), that is, that neither $f'(x)$ nor $f''(x)$ equals zero for any point in (α, β). Take as a first approximation a the end point α or β for which $f(x)$ and $f''(x)$ have the same sign, that is, if the curve is concave up, take the end point at which $f(x)$ is positive, and, if concave down, the end point at which $f(x)$ is negative. The point $a_1 = a - f(a)/f'(a)$, at which the tangent to the curve at $[a, f(a)]$ intersects the x axis, is between a and the root. Then, by using a_1 instead of a, a still better approximation a_2 is obtained, and so forth. If the end point for which $f(x)$ and $f''(x)$ have opposite signs were used, it could happen that the approximation obtained would be better than a_1, but it might be much worse since the tangent would not cross the x axis between the end point used and the root (Fig. 2.1).

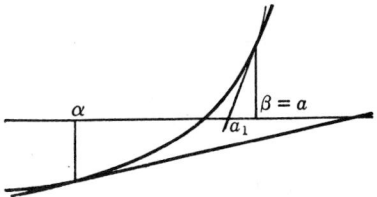

Fig. 2.1

Example 2.26. Find the real root of $x^3 + 4x - 7 = 0$.

$$f(x) = x^3 + 4x - 7$$
$$f'(x) = 3x^2 + 4$$
$$f''(x) = 6x$$

Graphically (Fig. 2.2), $\alpha = 1.2$, $\beta = 1.3$. Since $f(1.2) = -0.472$ and $f(1.3) = 0.397$, and $f''(x)$ is positive in the interval, then $a = 1.3$.

$$a_1 = a - \frac{f(a)}{f'(a)} = 1.3 - \frac{0.397}{9.07} = 1.3 - 0.044 = 1.256$$

$$a_2 = 1.256 - \frac{0.005385}{8.7326} = 1.256 - 0.00062 = 1.25538$$

If Newton's method of using the tangent is not applicable, either because of the presence of a maximum, minimum, or inflection point, or because of difficulty in finding $f'(x)$, the interpolation method using the chord joining $[\alpha, f(\alpha)]$ and $[\beta, f(\beta)]$ can be used. The chord crosses the x axis at $a = \alpha - f(\alpha)(\beta - \alpha)/[f(\beta) - f(\alpha)]$, a better approximation than either α or β. Note that this formula differs from Newton's only in having the difference quotient, which is the slope of the chord, in place of the derivative, which is the slope of the tangent. To get a still better approximation, repeat the procedure, using as one end point a and as the other either α or β, chosen so that $f(x)$ has opposite signs at the end points of the new interval.

Graphical Method of Solution. This can be used to solve any kind of equation if it gives sufficient accuracy. To solve the equation $f(x) = 0$, graph the function $y = f(x)$. The x coordinates of the points at which the graph intersects the x axis are roots of $f(x) = 0$. Another method is to set $f(x)$ equal to any convenient difference $f_1(x) - f_2(x)$, and graph the functions $y = f_1(x)$ and $y = f_2(x)$ on the same axes. The x coordinates of the points of intersection of the two graphs are real roots of $f(x) = 0$. Also, see Sec. 2.2.12.

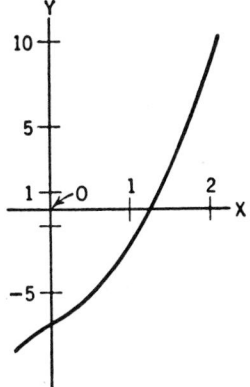

Fig. 2.2

Graeffe's Method for Real and Complex Roots. Let $x_1, x_2, \ldots, x_n$ be the roots of the equation $a_0 x^n + a_1 x^{n-1} + \cdots + a_n = 0$, arranged in descending order of absolute values. Form a sequence of equations such that the roots of each are the negatives of the squares of the roots of the preceding equation. Using the negatives of the squares gives more uniform formulas.

Let A_i be a coefficient of the equation being formed, and a_i a coefficient of the preceding equation.

$$A_0 = a_0 = 1$$
$$A_1 = a_1^2 - 2a_0 a_2 = a_1^2 - 2a_2$$
$$A_2 = a_2^2 - 2a_1 a_3 + 2a_4$$
$$A_3 = a_3^2 - 2a_2 a_4 + 2a_1 a_5 - 2a_6$$
$$\cdots$$
$$A_{n-1} = a_{n-1}^2 - 2a_{n-2} a_n$$
$$A_n = a_n^2$$

Each coefficient is the sum of the square of the preceding and twice the product of all pairs of equidistant coefficients in the preceding equation, taken with alternately minus and plus signs. Missing coefficients are zero. The process is ended when further steps do not affect the nonfluctuating coefficients to the accuracy desired in the roots.

As the successive equations are formed, various cases arise depending on the behavior of the coefficients. Among them are:

CASE 1: Each coefficient approaches the square of the preceding. The roots are real and unequal in absolute value. Let A_i be a coefficient of the equation whose roots are $-x_1^p, -x_2^p, \ldots, -x_n^p$. Then, approximately, $x_1 = \pm \sqrt[p]{A_1}$, $x_2 = \pm \sqrt[p]{A_2/A_1}, \ldots, x_n = \pm \sqrt[p]{A_n/A_{n-1}}$. The signs of the roots are determined by substitution in the original equation. It is usually sufficient to find successive integers between which a root is located.

Example 2.27. $f(x) = x^3 - 2x^2 - 5x + 4 = 0$.

	x^3	x^2	x	x^0
1st	1	-2	-5	4
	1	4	25	16
		10	16	
2nd	1	14	41	16
	1	196	1.681×10^3	256
		-82	-0.448×10^3	
4th	1	1.14×10^2	1.233×10^3	256
	1	1.300×10^4	1.520×10^6	6.554×10^4
		-0.247×10^4	-0.058×10^6	
8th	1	1.053×10^4	1.462×10^6	6.554×10^4
	1	1.109×10^8	2.137×10^{12}	4.295×10^9
		-0.029×10^8	-0.001×10^{12}	
16th	1	1.080×10^8	2.136×10^{12}	4.295×10^9

$$\log x_1 = \tfrac{1}{16} \log 1.080 \times 10^8 = \tfrac{1}{16} \times 8.0334 = 0.5021$$
$$x_1 = \pm 3.177$$

Using synthetic substitution,

$$
\begin{array}{r|rrrr}
3 & 1 & -2 & -5 & +4 \\
  &   & +3 & +3 & -6 \\
\hline
  & 1 & +1 & -2 & -2
\end{array}
\qquad
\begin{array}{r|rrrr}
4 & 1 & -2 & -5 & +4 \\
  &   & +4 & +8 & +12 \\
\hline
  & 1 & +2 & +3 & +16
\end{array}
$$

we have $f(3) = -2$, $f(4) = 16$. Therefore there is a root between 3 and 4, and $x_1 = 3.177$.

$$\log x_2 = \tfrac{1}{16}(\log 2.136 \times 10^{12} - \log 1.080 \times 10^8) = \tfrac{1}{16}(12.3296 - 8.0334)$$
$$= \tfrac{1}{16} \times 4.2962 = 0.2685$$
$$x_2 = \pm 1.856$$

Using synthetic substitution, $f(-2) = -2$, $f(-1) = 6$. Therefore $x_2 = -1.856$.

$$\log x_3 = \tfrac{1}{16}\left(\log 4.295 \times 10^9 - \log 2.136 \times 10^{12}\right) = \tfrac{1}{16}(9.6330 - 12.3296)$$

$$= \tfrac{1}{16}(157.3034 - 160) = 9.8315 - 10$$

$$x_3 = \pm 0.678$$

Since $x_1 + x_2 + x_3 = 2$, $x_3 = 0.678$.

CASE 2: A coefficient fluctuates in sign. There is a pair of complex roots. If the sign of A_i fluctuates, then $x_i = u + iv$ and $x_{i+1} = u - iv$ are complex. Let $r^2 = u^2 + v^2$. Then $r^2 = \sqrt[P]{A_{i+1}/A_{i-1}}$, $2u = -a_1 -$ (sum of real roots), $v = \sqrt{r^2 - u^2}$.

Example 2.28. $f(x) = x^4 - 2x^3 - 4x^2 + 5x - 7 = 0$.

	x^4	x^3	x^2	x	x^0
1st	1	-2	-4	5	-7
	1	4	16	25	49
		8	20	-56	
			-14		
2nd	1	12	22	-31	49
	1	144	484	961	2401
		-44	744	-2156	
			98		
4th	1	100	1326	-1195	2401
	1	1.0000×10^4	1.758×10^6	1.428×10^6	5.765×10^6
		-0.2652×10^4	0.239×10^6	-6.367×10^6	
			0.005×10^6		
8th	1	7.348×10^3	2.002×10^6	-4.939×10^6	5.765×10^6
	1	5.399×10^7	4.008×10^{12}	2.439×10^{13}	3.324×10^{13}
		-0.400×10^7	0.073×10^{12}	-2.308×10^{13}	
			*		
16th	1	4.999×10^7	4.081×10^{12}	1.31×10^{12}	3.324×10^{13}
	1	2.499×10^{15}	1.665×10^{25}	0.017×10^{26}	1.105×10^{27}
		-0.001×10^{15}	*	-2.713×10^{26}	
			*		
32nd	1	2.498×10^{15}	1.665×10^{25}	-2.696×10^{26}	1.105×10^{27}

Since the sign of A_3 fluctuates, x_3 and x_4 are complex.

$$x_1 = \pm \sqrt[32]{2.498 \times 10^{15}} = \pm 3.028 \qquad f(3) = -, \ f(4) = +$$

$$\frac{\log(2.498 \times 10^{15})}{32} = \frac{15.3976}{32} = 0.4812 \qquad x_1 = 3.028$$

$$x_2 = \pm \sqrt[32]{\frac{1.665 \times 10^{25}}{2.498 \times 10^{15}}} = \pm 2.028 \qquad f(-3) = +, \ f(-2) = -$$

25.2214
15.3976
$$\frac{9.8238}{32} = 0.3070 \qquad\qquad x_2 = -2.028$$

$$r^2 = \sqrt[32]{\frac{1.105 \times 10^{27}}{1.665 \times 10^{25}}} = 1.140 \qquad u = \frac{2 - (3.028 - 2.028)}{2} = 0.500$$

27.0434
25.2214
$$\frac{1.8220}{32} = 0.05694 \qquad\qquad v = \sqrt{1.140 - 0.250} = \sqrt{0.890} = 0.943$$

$$x_3, x_4 = 0.5 \pm 0.943i$$

If, for a fourth-degree equation, alternate coefficients, that is, the second and fourth, fluctuate in sign, all four roots are complex. Let the roots be $u_1 \pm iv_1$, $u_2 \pm iv_2$. Then $r_1^2 = \sqrt[p]{A_2}$, $r_2^2 = \sqrt[p]{A_4/A_2}$, $2(u_1 + u_2) = -a_1$, $2(r_2^2 u_1 + r_1^2 u_2) = -a_3$.

Example 2.29. $f(x) = x^4 - 3x^3 - x^2 + 4x + 14 = 0$.

	x^4	x^3	x^2	x	x^0
1st	1	-3	-1	4	14
	1	9	1	16	196
		2	24	28	
			28		
2nd	1	11	53	44	196
	1	121	2809	1936	38416
		-106	-968	-20776	
			392		
4th	1	15	2.233×10^3	-1.884×10^4	3.842×10^4
	1	0.225×10^3	4.986×10^6	3.549×10^8	1.476×10^9
		-4.466×10^3	0.565×10^6	-1.716×10^8	
			0.077×10^6		
8th	1	-4.241×10^3	5.628×10^6	1.833×10^8	1.476×10^9
	1	1.799×10^7	3.1674×10^{13}	3.360×10^{16}	2.178×10^{18}
		-1.126×10^7	0.1555×10^{13}	-1.661×10^{16}	
			0.0003×10^{13}		
16th	1	0.675×10^7	3.323×10^{13}	1.699×10^{16}	2.178×10^{18}

Since A_1 and A_3 fluctuate in sign, there are four complex roots.

$$r_1^2 = \sqrt[16]{3.323 \times 10^{13}} = 7.000 \qquad 2(u_1 + u_2) = 3$$

$$\frac{\log(3.323 \times 10^{13})}{16} = \frac{13.5215}{16} = 0.8451 \qquad 2(2u_1 + 7u_2) = -4$$

$$r_2^2 = \sqrt[16]{\frac{2.178 \times 10^{18}}{3.323 \times 10^{13}}} = 2.000 \qquad u_2 = -1$$

$$\begin{array}{c} 18.3380 \\ 13.5215 \\ \hline 4.8165 \end{array} \qquad u_1 = 2.5$$

$$\frac{4.8165}{16} = 0.3010$$

$$v_2 = \sqrt{r_2^2 - u_2^2} = \sqrt{2 - 1} = 1$$

$$v_1 = \sqrt{r_1^2 - u_1^2} = \sqrt{7 - 6.25} = \sqrt{0.75} = 0.866$$

$$x_1, x_2 = 2.5 \pm 0.866i$$
$$x_3, x_4 = -1 \pm i$$

CASE 3: A coefficient approaches one-half the square of the preceding. There is a double real root, or there are two real roots of equal absolute value. If A_i approaches one-half the square of the preceding coefficient, then $|x_i| = |x_{i+1}| = \sqrt[2p]{A_{i+1}/A_{i-1}}$.

Example 2.30. $f(x) = x^3 + 2.20x^2 - 2.95x + 0.80 = 0$.

	x^3	x^2	x	x^0
1st	1	2.20	-2.95	0.80
	1	4.84	8.703	0.64
		5.90	-3.52	
2nd	1	10.74	5.183	0.64
	1	1.1535×10^2	2.686×10	0.4096
		-0.1037×10^2	-1.375×10	
4th	1	1.050×10^2	1.311×10	0.4096
	1	1.1025×10^4	1.719×10^2	0.1678
		-0.0026×10^4	-0.860×10^2	
8th	1	1.100×10^4	0.859×10^2	0.1678

Since A_2 approaches one-half the square of the preceding coefficient, $|x_2| = |x_3|$.

$$x_1 = \pm\sqrt[8]{1.100 \times 10^4} = \pm 3.20 \qquad\qquad f(-4) = -, \; f(-3) = +$$

$$\frac{\log(1.100 \times 10^4)}{8} = \frac{4.0414}{8} = 0.5052 \qquad\qquad x_1 = -3.20$$

$$|x_2| = |x_3| = \sqrt[16]{\frac{0.1678}{1.100 \times 10^4}} = 0.50 \qquad\qquad f(0.5) = 0, \; f(-0.5) \neq 0$$

$$\frac{9.2248 - 10}{4.0414}$$

$$\frac{155.1834 - 160}{16} = 9.6990 - 10 \qquad\qquad x_2 = x_3 = 0.50$$

For a more complete treatment of Graeffe's method, including other cases, see Doherty and Keller, *Mathematics of Modern Engineering*, John Wiley and Sons, 1936; or Whittaker and Robinson, *The Calculus of Observations*, Blackie and Son, Ltd., 1942.

2.2.11 Matrices and Determinants

Definitions

1. A *matrix* is a system of mn quantities, called *elements*, arranged in a rectangular array of m rows and n columns.

$$A = \begin{pmatrix} a_{11} & a_{12} & \cdots & a_{1n} \\ a_{21} & a_{22} & \cdots & a_{2n} \\ \cdots & \cdots & & \cdots \\ a_{m1} & a_{m2} & & a_{mn} \end{pmatrix} = \begin{Vmatrix} a_{11} & a_{12} & \cdots & a_{1n} \\ a_{21} & a_{22} & \cdots & a_{2n} \\ \cdots & \cdots & & \cdots \\ a_{m1} & a_{m2} & & a_{mn} \end{Vmatrix} = (a_{ij}) = \|a_{ij}\| \quad \begin{matrix} i = 1,\ldots,m \\ j = 1,\ldots,n \end{matrix}$$

2. If $m = n$, then A is a *square* matrix of *order* n.
3. Two matrices are *equal* if and *only if* they have the same number of rows and of columns, and corresponding elements are equal.
4. Two matrices are *transposes* (sometimes called *conjugates*) of each other, if either is obtained from the other by interchanging rows and columns.
5. The *complex conjugate* of a matrix (a_{ij}) with complex elements is the matrix $(\bar{a}_{ij})$. See Section 2.13.1.
6. A matrix is *symmetric* if it is equal to its transpose, that is, if $a_{ij} = a_{ji}$; $i, j = 1,\ldots, n$.
7. A matrix is *skew-symmetric*, or *antisymmetric*, if $a_{ij} = -a_{ji}$; $i, j = 1,\ldots, n$. The diagonal elements $a_{ii} = 0$.
8. A matrix all of whose elements are zero is a *zero matrix*.
9. If the nondiagonal elements a_{ij}, $i \neq j$, of a square matrix A are all zero, then A is a *diagonal matrix*. If, furthermore, the diagonal elements are all equal, the matrix is a *scalar matrix*; if they are all 1, it is an *identity* or *unit matrix*, denoted by I.
10. The *determinant* $|A|$ of a square matrix (a_{ij}); $i, j = 1,\ldots, n$, is the sum of the $n!$ products $a_{1r_1} a_{2r_2} \cdots a_{nr_n}$, in which $r_1, r_2, \ldots, r_n$ is a permutation of $1, 2, \ldots, n$, and the sign of each product is $+$ or $-$ according as the permutation is obtained from $1, 2, \ldots, n$ by an even or an odd number of interchanges of two numbers.
Symbols used are

$$|A| = \begin{vmatrix} a_{11} & a_{12} & \cdots & a_{1n} \\ a_{21} & a_{22} & \cdots & a_{2n} \\ \cdots & \cdots & & \cdots \\ a_{n1} & a_{n2} & & a_{nn} \end{vmatrix} = |a_{ij}| \qquad i, j = 1,\ldots, n$$

11. A square matrix (a_{ij}) is *singular* if its determinant $|a_{ij}|$ is zero.

12. The determinants of the square submatrices of any matrix A, obtained by striking out certain rows or columns, or both, are called the *determinants* or *minors* of A. A matrix is of *rank r* if it has at least one r-rowed determinant that is not zero, while all its determinants of order higher than r are zero. The *nullity d* of a square matrix of order n is $d = n - r$. The zero matrix is of rank 0.

13. The *minor* D_{ij} of the element a_{ij} of a square matrix is the determinant of the submatrix obtained by striking out the row and column in which a_{ij} lies. The *cofactor* A_{ij} of the element a_{ij} is $(-1)^{i+j} D_{ij}$. A *principal minor* is the minor obtained by striking out the same rows as columns.

14. The *inverse* of the square matrix A is

$$A^{-1} = \begin{pmatrix} \dfrac{A_{11}}{|A|} & \cdots & \dfrac{A_{n1}}{|A|} \\ \cdots & \ddots & \cdots \\ \dfrac{A_{1n}}{|A|} & & \dfrac{A_{nn}}{|A|} \end{pmatrix}$$

$$AA^{-1} = A^{-A} = I$$

15. The *adjoint* of A is

$$\text{adj } A = \begin{pmatrix} A_{11} & \cdots & A_{n1} \\ \cdots & \ddots & \cdots \\ A_{1n} & & A_{nn} \end{pmatrix}$$

16. *Elementary transformations* of a matrix are
 a. The interchange of two rows or of two columns.
 b. The addition to the elements of a row (or column) of any constant multiple of the corresponding elements of another row (or column).
 c. The multiplication of each element of a row (or column) by any nonzero constant.

17. Two $m \times n$ matrices A and B are *equivalent* if it is possible to pass from one to the other by a finite number of elementary transformations.
 a. The matrices A and B are equivalent if and only if there exist two nonsingular square matrices E and F, having m and n rows, respectively, such that $EAF = B$.
 b. The matrices A and B are equivalent if and only if they have the same rank.

Matrix Operations

Addition and Subtraction. The sum or difference of two matrices (a_{ij}) and (b_{ij}) is the matrix $(a_{ij} \pm b_{ij})$, $i = 1, \ldots, m$; $j = 1, \ldots, n$.

Scalar Multiplication. The product of the scalar k and the matrix (a_{ij}) is the matrix (ka_{ij}).

Matrix Multiplication. The product (p_{ik}), $i = 1, \ldots, m$; $k = 1, \ldots, q$, of two matrices (a_{ij}), $i = 1, \ldots, m$; $j = 1, \ldots, n$, and (b_{jk}), $j = 1, \ldots, n$; $k = 1, \ldots, q$, is the matrix whose elements are

$$p_{ik} = \sum_{j=1}^{n} a_{ij} b_{jk} = a_{i1} b_{1k} + a_{i2} b_{2k} + \cdots + a_{in} b_{nk}$$

The element in the ith row and kth column of the product is the sum of the n products of the n elements of the ith row of (a_{ij}) by the corresponding n elements of the kth column of (b_{jk}).

Example 2.31

$$\begin{pmatrix} a_{11} & a_{12} \\ a_{21} & a_{22} \end{pmatrix} \begin{pmatrix} b_{11} & b_{12} & b_{13} \\ b_{21} & b_{22} & b_{23} \end{pmatrix} = \begin{pmatrix} a_{11}b_{11} + a_{12}b_{21} & a_{11}b_{12} + a_{12}b_{22} & a_{11}b_{13} + a_{12}b_{23} \\ a_{21}b_{11} + a_{22}b_{21} & a_{21}b_{12} + a_{22}b_{22} & a_{21}b_{13} + a_{22}b_{23} \end{pmatrix}$$

All the laws of ordinary algebra hold for the addition and subtraction of matrices and for scalar multiplication.

Multiplication of matrices is not in general commutative, but it is associative and distributive.

If the product of two or more matrices is zero, it does not follow that one of the factors is zero. The factors are *divisors of zero*.

Example 2.32

$$\begin{pmatrix} a & 0 \\ b & 0 \end{pmatrix}\begin{pmatrix} 0 & 0 \\ c & d \end{pmatrix} = \begin{pmatrix} 0 & 0 \\ 0 & 0 \end{pmatrix}$$

Linear Dependence

1. The quantities $l_1, l_2, \ldots, l_n$ are *linearly dependent* if there exist constants $c_1, c_2, \ldots, c_n$, not all zero, such that

$$c_1 l_1 + c_2 l_2 + \cdots + c_n l_n = 0$$

If no such constants exist, the quantities are *linearly independent*.

2. The linear functions

$$l_i = a_{i1}x_1 + a_{i2}x_2 + \cdots + a_{in}x_n \qquad i = 1, 2, \ldots, m$$

are *linearly dependent* if and only if the matrix of the coefficients is of rank $r < m$. Exactly r of the l_i form a linearly independent set.

3. For $m > n$, any set of m linear functions are linearly dependent.

Consistency of Equations

1. The system of homogeneous linear equations

$$a_{i1}x_1 + a_{i2}x_2 + \cdots + a_{in}x_n = 0 \qquad i = 1, 2, \ldots, m$$

has solutions not all zero if the rank r of the matrix (a_{ij}) is less than n.

If $m < n$, there always exist solutions not all zero. If $m = n$, there exist solutions not all zero if $|a_{ij}| = 0$.

If r of the equations are so selected that their matrix is of rank r, they determine uniquely r of the variables as homogeneous linear functions of the remaining $n - r$ variables. A solution of the system is obtained by assigning arbitrary values to the $n - r$ variables and finding the corresponding values of the r variables.

2. The system of linear equations

$$a_{i1}x_1 + a_{i2}x_2 + \cdots + a_{in}x_n = k_i \qquad i = 1, 2, \ldots, m$$

is consistent if and only if the *augmented* matrix derived from (a_{ij}) by annexing the column $k_1, \ldots, k_m$ has the same rank r as (a_{ij}).

As in the case of a system of homogeneous linear equations, r of the variables can be expressed in terms of the remaining $n - r$ variables.

Linear Transformations

1. If a linear transformation

$$x_i' = a_{i1}x_1 + a_{i2}x_2 + \cdots + a_{in}x_n \qquad i = 1, 2, \ldots, n$$

with matrix (a_{ij}) transforms the variables x_i into the variables x_i', and a linear transformation

$$x_i'' = b_{i1}x_1' + b_{i2}x_2' + \cdots + b_{in}x_n' \qquad i = 1, 2, \ldots, n$$

with matrix (b_{ij}) transforms the variables x_i' into the variables x_i'', then the linear transformation with matrix $(b_{ij})(a_{ij})$ transforms the variables x_i into the variables x_i'' directly.

2. A real *orthogonal* transformation is a linear transformation of the variables x_i into the variables x_i' such that

$$\sum_{i=1}^{n} x_i^2 = \sum_{i=1}^{n} x_i'^2$$

A transformation is orthogonal if and only if the transpose of its matrix is the inverse of its matrix.

3. A *unitary* transformation is a linear transformation of the variables x_i into the variables x_i' such that

$$\sum_{i=1}^{n} x_i \bar{x}_i = \sum_{i=1}^{n} x_i' \bar{x}_i'$$

A transformation is unitary if and only if the transpose of the conjugate of its matrix is the inverse of its matrix.

Quadratic Forms

A *quadratic form* in n variables is

$$\sum_{i,j=1}^{n} a_{ij} x_i x_j = a_{11} x_1^2 + a_{12} x_1 x_2 + \cdots + a_{1n} x_1 x_n$$

$$+ a_{21} x_2 x_1 + a_{22} x_2^2 + \cdots + a_{2n} x_2 x_n$$

$$\cdots$$

$$+ a_{m1} x_n x_1 + a_{n2} x_n x_2 + \cdots + a_{nn} x_n^2$$

in which $a_{ji} = a_{ij}$. The symmetric matrix (a_{ij}) of the coefficients is the *matrix* of the quadratic form and the rank of (a_{ij}) is the *rank* of the quadratic form.

A real quadratic form of rank r can be reduced by a real nonsingular linear transformation to the *normal form*

$$x_1^2 + \cdots + x_p^2 - x_{p+1}^2 - \cdots - x_r^2$$

in which the *index p* is uniquely determined.

If $p = r$, a quadratic form is *positive*, and, if $p = 0$, it is *negative*. If, furthermore, $r = n$, both are *definite*. A quadratic form is positive definite if and only if the determinant and all the principal minors of its matrix are positive.

A method of reducing a quadratic form to its normal form is illustrated.

Example 2.33

$$q = 3x^2 - 4y^2 - z^2 + 4xy - 2xz + 4yz$$

$$q = \left\{ \begin{array}{l} 3x^2 + 2xy - xz \\ +2xy - 4y^2 + 2yz \\ -xz + 2yz - z^2 \end{array} \right\} \begin{array}{l} = \frac{1}{3}(3x + 2y - z)^2 + q_1, \text{ in which the quantity} \\ \text{in parentheses is obtained by factoring } x \text{ out} \\ \text{of the first row} \\ = \frac{1}{3}(9x^2 + 4y^2 + z^2 + 12xy - 6xz - 4yz) + q_1 \end{array}$$

$$q_1 = -\tfrac{4}{3} y^2 - \tfrac{1}{3} z^2 + \tfrac{4}{3} yz - 4y^2 + 4yz - z^2$$

$$= \left\{ \begin{array}{l} -\tfrac{16}{3} y^2 + \tfrac{8}{3} yz \\ +\tfrac{8}{3} yz - \tfrac{4}{3} z^2 \end{array} \right\} = -\tfrac{3}{16}\left(-\tfrac{16}{3} y + \tfrac{8}{3} z\right)^2 + q_2$$

$$q_2 = 0$$

The transformation

$$x' = 3x + 2y - z$$
$$y' = -\tfrac{16}{3} y + \tfrac{8}{3} z$$
$$z' = z$$

reduces q to $\frac{1}{3} x'^2 - \frac{3}{16} y'^2$.

The transformation

$$x'' = \sqrt{3}\, x'$$
$$y'' = \frac{4}{\sqrt{3}} y'$$
$$z'' = z'$$

further reduces q to the normal form $x''^2 - y''^2$ of rank 2 and index 1.

Expressing x, y, z in terms of x'', y'', z'', the real nonsingular linear transformation that reduces q to the normal form is

$$x = \frac{\sqrt{3}}{3}x'' + \frac{1}{2\sqrt{3}}y''$$

$$y = -\frac{\sqrt{3}}{4}y'' + \tfrac{1}{2}z''$$

$$z = z''$$

Hermitian Forms
A *Hermitian form* in n variables is

$$\sum_{i,\,j=1}^{n} a_{ij}x_i\bar{x}_j \qquad a_{ji} = \bar{a}_{ij}$$

The matrix (a_{ij}) is a *Hermitian matrix*. Its transpose is equal to its conjugate. The rank of (a_{ij}) is the *rank* of the Hermitian form.

A Hermitian form of rank r can be reduced by a nonsingular linear transformation to the *normal form*

$$x_1\bar{x}_1 + \cdots + x_p\bar{x}_p - x_{p+1}\bar{x}_{p+1} - \cdots - x_r\bar{x}_r$$

in which the *index p* is uniquely determined.

If $p = r$, the Hermitian form is *positive*, and, if $p = 0$, it is *negative*. If, furthermore, $r = n$, both are *definite*.

Determinants
Second- and third-order determinants are formed from their square symbols by taking diagonal products, down from left to right being positive and up negative.

$$\begin{vmatrix} a_{11} & a_{12} \\ a_{21} & a_{22} \end{vmatrix} = a_{11}a_{22} - a_{21}a_{12}$$

$$\begin{vmatrix} a_{11} & a_{12} & a_{13} \\ a_{21} & a_{22} & a_{23} \\ a_{31} & a_{32} & a_{33} \end{vmatrix} = a_{11}a_{22}a_{33} + a_{12}a_{23}a_{31} + a_{13}a_{32}a_{21}$$
$$- a_{31}a_{22}a_{13} - a_{32}a_{23}a_{11} - a_{33}a_{12}a_{21}$$

Third and higher order determinants are formed by selecting any row or column and taking the sum of the products of each element and its cofactor. This process is continued until second- or third-order cofactors are reached.

$$\begin{vmatrix} a_{11} & a_{12} & a_{13} \\ a_{21} & a_{22} & a_{23} \\ a_{31} & a_{32} & a_{33} \end{vmatrix} = a_{11}\begin{vmatrix} a_{22} & a_{23} \\ a_{32} & a_{33} \end{vmatrix} - a_{21}\begin{vmatrix} a_{12} & a_{13} \\ a_{32} & a_{33} \end{vmatrix} + a_{31}\begin{vmatrix} a_{12} & a_{13} \\ a_{22} & a_{23} \end{vmatrix}$$

The determinant of a matrix A is
1. Zero, if two rows or two columns of A have proportional elements.
2. Unchanged, if
   a. The rows and columns of A are interchanged.
   b. To each element of a row or column of A is added a constant multiple of the corresponding element of another row or column.
3. Changed in sign, if two rows or two columns of A are interchanged.
4. Multiplied by c, if each element of any row or column of A is multiplied by c.
5. The sum of the determinants of two matrices B and C, if A, B, and C have all the same elements, except that in one row or column each element of A is the sum of the corresponding elements of B and C.

Example 2.34

$$\begin{vmatrix} 2 & 9 & 9 & 4 \\ 2 & -3 & 12 & 8 \\ 4 & 8 & 3 & -5 \\ 1 & 2 & 6 & 4 \end{vmatrix} = \begin{vmatrix} 2 & 5 & 9 & 4 \\ 2 & -7 & 12 & 8 \\ 4 & 0 & 3 & -5 \\ 1 & 0 & 6 & 4 \end{vmatrix} = 3\begin{vmatrix} 2 & 5 & 3 & 4 \\ 2 & -7 & 4 & 8 \\ 4 & 0 & 1 & -5 \\ 1 & 0 & 2 & 4 \end{vmatrix}$$

Multiply 1st column   Factor 3 out of
by -2 and add to 2nd.   the 3rd column.

$$= 3 \times (-5)\begin{vmatrix} 2 & 4 & 8 \\ 4 & 1 & -5 \\ 1 & 2 & 4 \end{vmatrix} + 3 \times (-7)\begin{vmatrix} 2 & 3 & 4 \\ 4 & 1 & -5 \\ 1 & 2 & 4 \end{vmatrix} = \qquad 0 \qquad -21\begin{vmatrix} 1 & 1 & 0 \\ 4 & 1 & -5 \\ 1 & 2 & 4 \end{vmatrix}$$

Expand according to 2nd column.   1st and 3rd   Subtract 3rd
     rows are proportional.   row from 1st.

$$= -21\begin{vmatrix} 1 & -5 \\ 2 & 4 \end{vmatrix} - (-21)\begin{vmatrix} 4 & -5 \\ 1 & 4 \end{vmatrix} = -21[(4 + 10) - (16 + 5)] = +147$$

Expand according to 1st row.

2.2.12 Systems of Equations

Linear Systems (also see Section 2.11.6)

Homogeneous. $a_{i1}x_1 + \cdots + a_{in}x_n = 0$, $i = 1, \ldots, m$. Let $r =$ rank of (a_{ij}).

For $m = n$,

$r = n$, $|a_{ij}| \neq 0$; one solution, $x_1 = \cdots = x_n = 0$.
   $r < n$, $|a_{ij}| = 0$; infinite number of solutions.

Nonhomogeneous. $a_{i1}x_1 + \cdots + a_{in}x_n = k_i$, $i = 1, \ldots, m$.

Let $a = (a_{ij})$, an $m \times n$ matrix

$$k = \text{augmented matrix} = \begin{pmatrix} a_{11} & \cdots & a_{1n}k_1 \\ \cdots & \cdot & \cdots \\ a_{m1} & & a_{mn}k_m \end{pmatrix}, \text{ an } m \times (n + 1) \text{ matrix}$$

$r_a =$ rank of a
$r_k =$ rank of k

For $m = n$,

$r_a = r_k$; Consistent.
   (a) $r_a = r_k = n$, $|a_{ij}| \neq 0$; independent. One solution.
   (b) $r_a = r_k < n$, $|a_{ij}| = 0$; dependent. Infinite number of solutions.
$r_a < r_k$; Inconsistent. No solution.

Methods of Solution
Elimination is a practical method of solution for a system of two or three linear equations in as many variables.

Example 2.35

1. By addition and subtraction, solve

$$\begin{cases} 2x + y + 3z = 9 & \text{(2.2)} \\ x - 2y + z = -2 & \text{(2.3)} \\ 3x + 2y + 2z = 7 & \text{(2.4)} \end{cases}$$

$$(2) + (3) \text{ gives: } 4x + 3z = 5 \qquad \text{(2.5)}$$

$$2 \times (1) + (2) \text{ gives: } 5x + 7z = 16 \qquad \text{(2.6)}$$

$$5 \times (4) - 4 \times (5) \text{ gives: } \quad -13z = -39 \quad \text{or} \quad z = 3$$

Putting $z = 3$ in (2.5) or (2.6): $x = -1$
Then from (2.2), (2.3), or (2.4), $y = 2$

2. By substitution, solve

$$\begin{cases} x + 2y - z = 5 & \text{(2.7)} \\ x - y = 2 & \text{(2.8)} \\ 2x + z = 1 & \text{(2.9)} \end{cases}$$

From (2.8), $y = x - 2$, and from (2.9), $z = -2x + 1$. Substituting for y and z in (2.7), $x - 2x - 4 + 2x - 1 = 5$, from which $x = 2$. Then $y = 2 - 2 = 0$, $z = -4 + 1 = -3$.

Determinants can be used to solve a system of n nonhomogeneous linear equations in n variables for which $|a_{i,j}| \neq 0$. To solve for x_j, form a fraction, the denominator of which is the determinant $|a_{i,j}|$ and the numerator the determinant obtained from $|a_{i,j}|$ by replacing its jth column by the constants k_i.

Example 2.36. Solve

$$\begin{cases} 2x + y + 3z = 9 \\ x - 2y + z = -2 \\ 3x + 2y + 2z = 7 \end{cases}$$

$$x = \frac{\begin{vmatrix} 9 & 1 & 3 \\ -2 & -2 & 1 \\ 7 & 2 & 2 \end{vmatrix}}{\begin{vmatrix} 2 & 1 & 3 \\ 1 & -2 & 1 \\ 3 & 2 & 2 \end{vmatrix}} = \frac{\begin{vmatrix} 9 & 1 & 3 \\ -2 & -2 & 1 \\ 5 & 0 & 3 \end{vmatrix}}{\begin{vmatrix} 2 & 1 & 3 \\ 1 & -2 & 1 \\ 4 & 0 & 3 \end{vmatrix}} = \frac{\begin{vmatrix} 9 & 1 & 3 \\ 16 & 0 & 7 \\ 5 & 0 & 3 \end{vmatrix}}{\begin{vmatrix} 2 & 1 & 3 \\ 5 & 0 & 7 \\ 4 & 0 & 3 \end{vmatrix}} = \frac{-(48 - 35)}{-(15 - 28)} = -1$$

Miscellaneous Systems

To be solvable a system of equations must have as many independent equations as variables. A system of two polynomial equations of degrees m and n has mn solutions, real or complex. For systems in general no statement can be made regarding the number of solutions.

Graphical Method of Solution. This is a general method for systems of two equations in two variables. It consists of graphing both equations on the same axes and reading the pairs of coordinates of the points of intersection of the graphs as solutions of the system. This method gives real solutions only.

Example 2.37. Solve

$$\begin{cases} y = \sin x \\ x^2 + y^2 = 2 \end{cases}$$

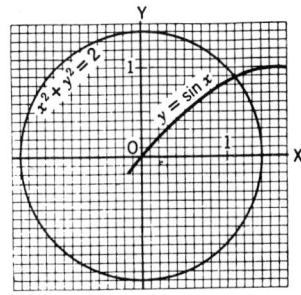

Solution from the graph (Fig. 2.3), $x = 1.1$, $y = 0.9$.
From symmetry, $x = -1.1$, $y = -0.9$, is also a solution.

Fig. 2.3

Method of Elimination of Variables. This is a general method that can be applied to systems composed of any kinds of equations, algebraic or transcendental. However, except in fairly simple cases, practical difficulties are frequently encountered.

Example 2.38. Solve

$$\begin{cases} y = \sin x \\ x^2 + y^2 = 2 \end{cases}$$

Squaring both sides of the first equation and subtracting it from the second to eliminate y, $x^2 = 2 - \sin^2 x$. This equation can be solved by Newton's method. Extraneous solutions introduced by squaring can be eliminated by reference to the graph.

There are numerous devices for eliminating variables in special systems. For example, to solve the system of *two general quadratics*

$$a_1 x^2 + b_1 xy + c_1 y^2 + d_1 x + e_1 y + f_1 = 0 \qquad (2.10)$$

$$a_2 x^2 + b_2 xy + c_2 y^2 + d_2 x + e_2 y + f_2 = 0 \qquad (2.11)$$

eliminate x^2 by multiplying (1) by a_2 and (2) by a_1 and subtracting, solve the resulting equation for x, substitute this expression in either of the given equations, and clear of fractions. The resulting fourth-degree equation in y can be solved by Horner's method. In a similar manner y could have been eliminated instead of x.

2.2.13 Permutations and Combinations

Fundamental Principle. If in a sequence of s events the first event can occur in n_1 ways, the second in $n_2, \ldots$, the sth in n_s, then the number of different ways in which the sequence can occur is $n_1 n_2 \cdots n_s$.

A **permutation** of n objects taken r at a time is an arrangement of any r objects selected from the n objects. The number of permutations of n objects taken r at a time is

$$_nP_r = n(n - 1)(n - 2) \cdots (n - r + 1) = \frac{n!}{(n - r)!}$$

In particular, $_nP_1 = n$, $_nP_n = n!$.
Cyclic permutations are

$$_nP_r^c = \frac{n!}{r(n - r)!} \qquad _nP_n^c = (n - 1)!$$

If the n objects are divided into s sets each containing n_i objects that are alike, the distinguishable permutations are

$$n = n_1 + n_2 + \cdots + n_s, \qquad _nP_n = \frac{n!}{n_1! n_2! \cdots n_s!}$$

A **combination** of n objects taken r at a time is an unarranged selection of any r of the n objects. The number of combinations of n objects taken r at a time is

$$_nC_r = \frac{_nP_r}{r!} = \frac{n!}{r!(n - r)!} = {_nC_{n-r}}$$

In particular, $_nC_1 = n$, $_nC_n = 1$.
Combinations taken any number at a time, $_nC_1 + {_nC_2} + \cdots + {_nC_n} = 2^n - 1$.

2.2.14 Probability

If, in a set M of m events that are mutually exclusive and equally likely, one event will occur, and if in the set M there is a subset N of n events ($n \le m$), then the *a priori probability* p that the event that will occur is one of the subset N is n/m. The probability q that the event that will occur does not belong to N is $1 - n/m$.

Example 2.39. If the probability of drawing one of the 4 aces from a deck of 52 cards is to be found, then $m = 52$, $n = 4$, and $p = \frac{4}{52} = \frac{1}{13}$. The probability of drawing a card that is not an ace is $q = 1 - \frac{1}{13} = \frac{12}{13}$.

If, out of a large number r of observations in which a given event might or might not occur, the event has occurred s times, then a useful approximate value of the *experimental*, or *a posteriori*, *probability* of the occurrence of the event under the same conditions is s/r.

Example 2.40. From the American Experience Mortality Table, out of 100,000 persons living at age 10 years 749 died within a year. Here $r = 100,000$, $s = 749$, and the probability that a person of age 10 will die within a year is $749/100,000$.

If p is the probability of receiving an amount A, then the *expectation* is pA.

Addition Rule (either or). The probability that any one of several mutually exclusive events will occur is the sum of their separate probabilities.

Example 2.41. The probability of drawing an ace from a deck of cards is $\frac{1}{13}$, and the probability of drawing a king is the same. Then the probability of drawing either an ace or a king is $\frac{1}{13} + \frac{1}{13} = \frac{2}{13}$.

Multiplication Rule (both and). (a) The probability that two (or more) independent events will both (or all) occur is the product of their separate probabilities.

(b) If p_1 is the probability that an event will occur, and if, after it has occurred, p_2 is the probability that another event will occur then the probability that both will occur in the given order is $p_1 p_2$. This rule can be extended to more than two events.

Example 2.42. (a) The probability of drawing an ace from a deck of cards is $\frac{1}{13}$, and the probability of drawing a king from another deck is $\frac{1}{13}$. Then the probability that an ace will be drawn from the first deck and a king from the second is $\frac{1}{13} \cdot \frac{1}{13} = \frac{1}{169}$. (b) After an ace has been drawn from a deck of cards, the probability of drawing a king is $\frac{4}{51}$. If two cards are drawn in succession without the first being replaced, the probability that the first is an ace and the second a king is $\frac{1}{13} \cdot \frac{4}{51} = \frac{4}{663}$.

Repeated Trials. If p is the probability that an event will occur in a single trial, then the probability that it will occur exactly s times in r trials is the *binomial* or *Bernoulli* distribution function

$$_rC_s\, p^s (1-p)^{r-s}$$

The probability that it will occur at least s times is

$$p^r +\,_rC_{r-1}\, p^{r-1}(1-p) +\,_rC_{r-2}\, p^{r-2}(1-p)^2 + \cdots +\,_rC_s\, p^s(1-p)^{r-s}$$

Example 2.43. If five cards are drawn, one from each of five decks, the probability that exactly three will be aces is $_5C_3(\frac{1}{13})^3(\frac{12}{13})^2$. The probability that at least three will be aces is $(\frac{1}{13})^5 +\,_5C_4(\frac{1}{13})^4(\frac{12}{13}) +\,_5C_3(\frac{1}{13})^3(\frac{12}{13})^2$.

2.3 SET ALGEBRA

2.3.1 Sets

A **set** is a collection of objects called **elements** that are distinguished by a particular characteristic. Examples are a set of engineers, a set of integers, a set of points. Element e belongs to set $\mathscr{S}$ is written $e \in \mathscr{S}$. If not, $e \notin \mathscr{S}$. A set can be denoted by including the listed elements, or merely by a typical element, in curly brackets: $\{2,4,6\}$; $\{e_1, e_2\}$, $\{e\}$. A set with no elements is called the **null set,** and is denoted by $\varnothing$. A set with one element e_1 is denoted by $\{e_1\}$; and to avoid a paradox in logic, these two ideas must be kept distinct.

Two sets $\mathscr{S}_1$, $\mathscr{S}_2$ may be compared as follows. If every element of set $\mathscr{S}_1$ is also an element of $\mathscr{S}_2$, then $\mathscr{S}_1$ is contained in $\mathscr{S}_2$. This is written $\mathscr{S}_1 \subset \mathscr{S}_2$, and is read "$\mathscr{S}_1$ is **contained in** $\mathscr{S}_2$," or "$\mathscr{S}_1$ is a **subset** of $\mathscr{S}_2$." If, in addition, $\mathscr{S}_2 \subset \mathscr{S}_1$, then their relation is written $\mathscr{S}_1 = \mathscr{S}_2$. On the other hand, if $\mathscr{S}_4$ has at least one element not contained in $\mathscr{S}_3$, but $\mathscr{S}_3 \subset \mathscr{S}_4$, $\mathscr{S}_3$ is a **proper subset** of $\mathscr{S}_4$. If $\mathscr{S}_5$ can contain all the elements of $\mathscr{S}_6$, this can be stressed by writing $\mathscr{S}_5 \subseteq \mathscr{S}_6$. Evidently $\varnothing \subset \mathscr{S}$, for every set $\mathscr{S}$.

If S, called the **space**, is the largest set concerned in a particular discussion, all the other sets are subsets of S. Thus set $\mathscr{A} \subset S$. The **complement** of $\mathscr{A}$, $\mathscr{A}^c$, with respect to space S is the set of elements in S that are not elements of $\mathscr{A}$.

Binary Operations for Sets. The union, $\mathscr{S}_a \cup \mathscr{S}_b$, of sets $\mathscr{S}_a$ and $\mathscr{S}_b$ is the set of elements in $\mathscr{S}_a$ or $\mathscr{S}_b$ or in both. Note that union differs from the idea of sum since in the union the common elements are counted only once. The **intersection**, $\mathscr{S}_a \cap \mathscr{S}_b$, of sets $\mathscr{S}_a$ and $\mathscr{S}_b$ is the set of elements in both $\mathscr{S}_1$ and $\mathscr{S}_2$.

Union   Intersection

Let $\mathscr{S}_a$, $\mathscr{S}_b$, $\mathscr{S}_c$ have their elements in space S.
Boolean algebra has as one representation the following:
UNICITY. Unique union $\mathscr{S}_a \cup \mathscr{S}_b \subset S$. Unique intersection $\mathscr{S}_a \cap \mathscr{S}_b \subset S$.
COMMUTATIVITY. $\mathscr{S}_a \cup \mathscr{S}_b = \mathscr{S}_b \cup \mathscr{S}_a$, $\mathscr{S}_a \cap \mathscr{S}_b = \mathscr{S}_b \cap \mathscr{S}_a$.
ASSOCIATIVITY. $\mathscr{S}_a \cup (\mathscr{S}_b \cup \mathscr{S}_c) = (\mathscr{S}_a \cup \mathscr{S}_b) \cup \mathscr{S}_c$, $\mathscr{S}_a \cap (\mathscr{S}_b \cap \mathscr{S}_c) = (\mathscr{S}_a \cap \mathscr{S}_b) \cap \mathscr{S}_c$.
DISTRIBUTIVITY. $\mathscr{S}_a \cup (\mathscr{S}_b \cap \mathscr{S}_c) = (\mathscr{S}_a \cup \mathscr{S}_b) \cap (\mathscr{S}_a \cup \mathscr{S}_c)$, $\mathscr{S}_a \cap (\mathscr{S}_b \cup \mathscr{S}_c) = (\mathscr{S}_a \cap \mathscr{S}_b) \cup (\mathscr{S}_a \cap \mathscr{S}_c)$.
IDEMPOTENCY. $\mathscr{S}_a \cup \mathscr{S}_a = \mathscr{S}_a$, $\mathscr{S}_a \cap \mathscr{S}_a = \mathscr{S}_a$.
SPACE. $\mathscr{S}_a \cup S = S$, $\mathscr{S}_a \cap S = \mathscr{S}_a$.
NULL SET. $\mathscr{S}_a \cup \varnothing = \mathscr{S}_a$, $\mathscr{S}_a \cap \varnothing = \varnothing$.
SUBSET. $\varnothing \subset \mathscr{S}_a \subset S$, $\mathscr{S}_a \subset (\mathscr{S}_a \cup \mathscr{S}_b)$, $(\mathscr{S}_a \cap \mathscr{S}_b) \subset \mathscr{S}_a$, $\mathscr{S}_a \subset \mathscr{S}_b \Rightarrow \mathscr{S}_a \cup \mathscr{S}_b = \mathscr{S}_b$, and $\mathscr{S}_a \cap \mathscr{S}_b = \mathscr{S}_a$.
COMPLEMENT. To $\mathscr{S}_a \subset S$ there corresponds unique $\mathscr{S}_{a^c} \subset S$. $\mathscr{S}_a \cup \mathscr{S}_{a^c} = S$, $\mathscr{S}_a \cap \mathscr{S}_{a^c} = \varnothing$.
DE MORGAN'S RELATIONS. $(\mathscr{S}_a \cup \mathscr{S}_b)^c = \mathscr{S}_{a^c} \cap \mathscr{S}_{b^c}$, $(\mathscr{S}_a \cap \mathscr{S}_b)^c = \mathscr{S}_{a^c} \cup \mathscr{S}_{b^c}$.
INVARIANT under the **duality transformation**, $\cup \leftrightarrow \cap$, $\subset \leftrightarrow \supset$, $S \leftrightarrow \varnothing$, are all the preceding relations.

2.3.2   Groups

A **group** is a system composed of a *set of elements* $\{a\}$ and a *rule of combination* of any two of them to form a product, such that

1. The *product* of any ordered pair of elements and the square of each element are elements of the set.
2. The *associative* law holds.
3. The set contains an *identity* element I such that $Ia = aI = a$ for any element a of the set.
4. For any element a of the set there is in the set an *inverse* a^{-1} such that $aa^{-1} = a^{-1}a = I$.
5. If, in addition, the *commutative* law holds, the group is *commutative* or *Abelian*.

The *order* of a group is the number n of elements in the group.

2.3.3   Rings, Integral Domains, and Fields

Rings. Space S consists of a set of elements $e_1, e_2, e_3, \ldots$. These elements are compared for **equality** and **order**, and combined by the operations of **addition and multiplication**. These terms in **bold face** are partially defined by the following sets of assumptions.

Equality is a term from logic, and means that if two expressions have this relation, then one may be substituted for the other.

Assumptions of equality E_1. Unicity: either $e_1 = e_2$ or $e_1 \neq e_2$. E_2. Reflexivity: $E_1 = e_1$. E_3. Symmetry: $e_1 = e_2 \Rightarrow e_2 = e_1$. E_4. Transitivity: $e_1 = e_2$, $e_2 = e_3 \Rightarrow e_1 = e_3$.

Assumptions of addition A_1. Closure: $e_1 + e_2 \subset S$. A_2. $e_1 = e_2 \Rightarrow e_1 + e_3 = e_2 + e_3$ and $e_3 + e_1 = e_3 + e_2$. (Invariance under addition.) A_3. Associativity: $e_1 + (e_2 + e_3) = (e_1 + e_2) + e_3$. A_4. Identity Element: There exists an element $z \subset S$, such that $e_1 + z = e_1$, $z + e_1 = e_1$. A_5. Commutativity: $e_1 + e_2 = e_2 + e_1$.

Theorem 1: z is unique.

Negative. To each $e \subset S$, there corresponds an $e' \subset S$ such that $e + e' = z$. e' is called the negative of e and written $-e$.

Theorem 2: e' or $-e$ is unique. Theorem 3: $-(-e) = e$. Theorem 4: $-z = z$. Theorem 5: Equation $x + e_1 = e_2$ has the solution $x = e_2 - e_1$. Theorem 6: $e_1 + e_3 = e_1 \Rightarrow e_3 = z$.

Assumptions of multiplication M_1. Closure: $e_1 \cdot e_2 \subset S$. M_2. $e_1 = e_2 \Rightarrow e_1 \cdot e_3 = e_2 \cdot e_3$ and $e_3 \cdot e_1 = e_3 \cdot e_2$. (Invariance under multiplication.) M_3. Associativity: $e_1(e_2 \cdot e_3) = (e_1 \cdot e_2)e_3$. M_4. Identity element: There exists an element $u \subset S$ such that $e_1 \cdot u = e_1$, $u \cdot e_1 = e_1$. M_5. Commutativity: $e_1 \cdot e_2 = e_2 \cdot e_1$.

Theorem 7: u is unique.

Reciprocal. To each element $e \subset S$ except z, there corresponds an $e'' \subset S$ such that $e \cdot e'' = u$. e'' is called the reciprocal of e and written e^{-1}.

Theorem 8: e'' or e^{-1} is unique.

M_7. Distributivity: $e_1(e_2 + e_3) = e_1 \cdot e_2 + e_1 \cdot e_3$. Theorem 9: $e \cdot z = z$. Theorem 10: $e_1(-e_2) = -(e_1 \cdot e_2) = (-e_1)e_2$. Theorem 11: $(-e_1)(-e_2) = e_1 \cdot e_2$. Theorem 12: If S contains an element besides z then it is $u \neq z$. Theorem 13: $e_1 \cdot e_2 = z \Rightarrow$ either $e_1 = z$ or $e_2 = z$.

A **ring** is a space S having at least two elements for which assumptions E_1 to E_4, A_1 to A_6, M_1 to M_5, and M_7 hold. An example is a residue system modulo 4.

Integral Domain. An **integral domain** is a ring for which, as an assumption, Theorem 13 holds. An example is the set of all integers.

Field. A **field** is an integral domain for which M_6 holds. An example of a field is the set of algebraic numbers.

Assumptions of (*linear*) *order.* O_1. (Contains E_1.) If $e_1, e_2 \subset S$, then either $e_1 < e_2$, $e_1 = e_2$, or $e_2 < e_1$. O_2. $e_1 < e_2 \Rightarrow e_1 + e_3 < e_2 + e_3$. (Invariance under addition.) O_3. Transitivity: $e_1 < e_2$, $e_2 < e_3 \Rightarrow e_1 < e_3$.

Negative. If $e_1 < z$, then e_1 is called **negative**.

Positive. If $z < e_2$, then e_2 is called **positive**.

O_4. $z < e_2 z < e_3 \Rightarrow z < e_2 \cdot e_3$.

An **ordered integral domain** is an integral domain for which O_1 to O_4 hold. An example is the set of all integers.

An **ordered field** is an ordered integral domain for which M_6 holds. An example is the set of all rational numbers.

If an additional order assumption, O_5, known as the Dedekind assumption—see a book on real analysis—be included, then the space S for which assumptions E_1 to E_4, A_1 to A_6, M_1 to M_7, and O_1 to O_5 hold is called the real number space. An example is the set of real numbers. Here z is denoted 0, and u is denoted 1. Another example is the set of points on the real line.

2.4 STATISTICS AND PROBABILITY

2.4.1 Frequency Distributions of One Variable

Definitions

A **frequency distribution** of statistical data consisting of N values of a variable x is a tabulation by intervals, called **classes**, showing the number f_i, called the **frequency** or **weight**, in each class; $N = \Sigma f_i$.

The mid value x_i of a class is the **class mark**.

For equal classes, the **class interval** is $c = x_{i+1} - x_i$.

The **cumulative frequency**, cum f, at any class is the sum of the frequencies of all classes up to and including the given class.

Graphs

Frequency Polygon. Plot the points (x_i, f_i) and draw a broken line through them.

Histogram. Draw a set of rectangles, using as bases intervals representing the classes marked off on a straight line, and using altitudes proportional to the frequencies.

Frequency Curve. Draw a continuous curve approximating a frequency polygon, or such that the region under the curve approximates a histogram. As the class interval c is taken smaller and the total frequency N larger, the approximation becomes better.

Ogive. This is a graph of cumulative frequencies.

Averages

Arithmetic Mean

$$AM = \bar{x} = \frac{1}{N} \sum_{i=1}^{k} f_i x_i$$

in which $N = \sum_{i=1}^{k} f_i$.

Geometric Mean

$$GM = \left(x_1^{f_1} \cdot x_2^{f_2} \cdots x_k^{f_k} \right)^{1/N} \qquad \log GM = \frac{1}{N} \sum_{i=1}^{k} f_i \log x_i$$

Harmonic Mean

$$HM = \frac{N}{\displaystyle\sum_{i=1}^{k} \frac{f_i}{x_i}}$$

Root-Mean-Square

$$rms = \sqrt{\frac{\displaystyle\sum_{i=1}^{k} f_i x_i^2}{N}}$$

Median. (a) For continuously varying data, the value of x for which cum $f = N/2$; (b) for discrete data, the value of x such that there is an equal number of values larger and smaller; for N odd, $N = 2k - 1$, the median is x_k; for N even, $N = 2k$, the median may be taken as $\frac{1}{2}(x_k + x_{k+1})$.

Mode. The value of x that occurs most frequently.

Moments

1. About x_0. In x units

$$\nu_r = \frac{1}{N} \sum_{i=1}^{k} f_i (x_i - x_0)^r \qquad r = 0, 1, \ldots$$

If $x_0 = 0$, $\nu_1 = \bar{x}$, which is the *arithmetic mean*. In u units

$$\nu_r = \frac{1}{N} \sum_{i=1}^{k} f_i u_i^r \qquad r = 0, 1, \ldots \qquad u = \frac{x - x_0}{c} \qquad c = \text{class interval}$$

2. About the *mean*. In x units

$$\mu_r = \frac{1}{N} \sum_{i=1}^{k} f_i (x_i - \bar{x})^r \qquad r = 0, 1, \ldots \qquad \bar{x} = \nu_1 \text{ in } x \text{ units}$$

In u units

$$\mu_r = \frac{1}{N} \sum_{i=1}^{k} f_i (u_i - \bar{u})^r \qquad r = 0, 1, \ldots \qquad \bar{u} = \nu_1 \text{ in } u \text{ units}$$

In either x or u units, the μ's as functions of the ν's:

$$\mu_0 = 1$$
$$\mu_1 = 0$$
$$\mu_2 = \nu_2 - \nu_1^2$$
$$\mu_3 = \nu_3 - 3\nu_1\nu_2 + 2\nu_1^3$$
$$\mu_4 = \nu_4 - 4\nu_1\nu_3 + 6\nu_1^2\nu_2 - 3\nu_1^4$$
$$\mu_r \text{ (in } x \text{ units)} = c^r\mu_r \text{ (in } u \text{ units)} \qquad r = 0, 1, \ldots .$$

In x units, μ_2 is the *variance*; $\sqrt{\mu_2}$ is the *standard deviation* σ. Both are used as measures of dispersion. To compute σ,

$$\sigma = c\sqrt{\dfrac{\sum\limits_{i=1}^{k} f_i u_i^2}{N} - \bar{u}^2}$$

Probable error $= 0.6745\sigma$.

3. In *standard* (deviation) *units*:

$$\alpha_1 = 0$$
$$\alpha_2 = 1$$
$$\alpha_3 = \dfrac{\mu_3}{\sigma^3} \quad \text{(a measure of } skewness \text{)}$$
$$\alpha_4 = \dfrac{\mu_4}{\sigma^4} \quad \text{(a measure of } kurtosis \text{)}$$

The **moment generating function**, or arbitrary-range inverse real Laplace transform, is

$$M(\theta) = \int_a^b e^{\theta x} f(x)\, dx$$

The rth moment is

$$\mu_r = \dfrac{d^r M}{d\theta^r}\bigg|_{\theta=0} \qquad r = 0, 1, 2, \ldots$$

M Tiles
The rth *quartile* Q_r is the value of x for which cum $f/N = r/4$.
$\qquad$ The rth *percentile* P_r is the value of x for which cum $f/N = r/100$.
$\qquad$ For $r = 10s$, $P_r = D_s$, the sth *decile*.

Other Measures of Shape

Dispersion

1. *Range* of x, the difference between the largest and the smallest values of x.

2. *Mean deviation* $= \dfrac{1}{N}\sum\limits_{i=1}^{k} f_i |x_i - \bar{x}|$.

3. *Semiinterquartile range*, or *quartile deviation*,

$$Q = \dfrac{|Q_3 - Q_1|}{2}$$

Skewness. 1. *Quartile coefficient of skewness* $= \dfrac{Q_3 - 2Q_2 + Q_1}{Q}$.

Statistical Hypotheses

A hypothesis concerning one or more statistical distribution parameters is a **statistical hypothesis**. A **test** of such a hypothesis is a procedure leading to a decision to accept or reject the hypothesis. The **significance level** is the probability value below which a hypothesis is rejected.

A **type 1 error** is made if the hypothesis is correct but the test rejects the hypothesis. A **type 2 error** is made if the hypothesis is false but the test accepts the hypothesis.

If the variable x has a distribution function $f(x; \theta)$, with parameter θ, then the **likelihood** function, that is, distribution function of a random sample of size n, is $P(\theta) = f(x_1; \theta)f(x_2; \theta) \cdots f(x_n; \theta)$. The use of $P_{max}(\theta)$ in the estimation of population parameters is the **method of maximum likelihood**. It often consists of solving $dP/d\theta = 0$ for θ.

Random Sampling

A set $x_1, x_2, \ldots, x_n$ of values of x with distribution function $f(x)$ is a *sample of size n* drawn from the population described by $f(x)$. If repeated samples of size n drawn from the population have the x_r's independently distributed in the probability sense and each x_r has the same distribution as the population, then the sampling is *random*.

Normal and Nonnormal Distributions

The normal distribution function in analytic and tabular form is found on pages 2•39 to 2•40. A linear combination of independent normal variables is normally distributed.

The **Poisson distribution**, $P(x) = e^{-m}m^x/x!$, is the limit approached by the binomial distribution (p. 2•29), if the probability p that an event will occur in a single trial approaches zero and the number of trials r becomes infinite in such a way that $rp = m$ remains constant.

If m is the mean of a nonnormal distribution of x, σ the standard deviation, and if the moment generating function exists, then the variable $(\bar{x} - m)n^{1/2}/\sigma$, in which $\bar{x}$ is the mean of a sample of size n, has a distribution that approaches the normal distribution as $n \rightarrow \infty$.

Nonparametric methods are those that do not involve the estimation of parameters of a distribution function. Tchebycheff's inequality (p. 2•42) provides nonparametric tests for the validity of hypotheses. It leads to the *law of large numbers*. Let p be the probability of an event occurring in one trial, and p_n the ratio of the number of occurrences in n trials to the number n. The probability that $|p_n - p| > \varepsilon$ is $\leq pq/n\varepsilon$; this can be made arbitrarily small, however small ε is, by taking n large enough. The ratio p_n converges *stochastically* to the probability p.

Two numbers L_1, L_2 between which a large fraction of a population is expected to lie are **tolerance limits**. If z is the fraction of the population of a variable with a continuous distribution that lies between the extreme values of a random sample of size n from this population, then the distribution of z is $f(z) = n(n - 1)z^{n-2}(1 - z)$.

Statistical Control of Production Processes

A chart on which percentage defective in a sample is graphed as a function of output time can be used for control of an industrial process. Horizontal lines are drawn through the mean m and the controls $m \pm 3\sigma/n^{1/2}$. The behavior of the graph with respect to these control lines is used as an error signal in a feedback system that controls the process. If the graph goes out of the band bounded by the control lines, the process is stopped until the trouble is located and removed.

2.4.2 Correlation

To discover whether there is a simple relation between two variables, corresponding pairs of values are used as coordinates to plot the points of a **scatter diagram**. The simplest relation exists if the scatter diagram can be approximated more or less closely by a straight line.

Least-Square Straight Line. This line, which minimizes the sum of the squares of the y deviations of the points, is

$$\hat{y} - \bar{y} = M(x - \bar{x})$$

in which

$$M = \frac{\Sigma(x - \bar{x})y}{\Sigma(x - \bar{x})^2}$$

(x, y) is a plotted point, and $(x, \hat{y})$ is a point on this *line of regression* of y on x. The **correlation**

coefficient

$$r = \pm \left[1 - \frac{\Sigma(y - \hat{y})^2}{\Sigma(y - \bar{y})^2} \right]^{1/2}$$

is a measure of the usefulness of the regression line. If $r = 0$, the line is useless; if $r = \pm 1$, the line gives a perfect estimate. The percentage of the variance of y that has been accounted for by y's relation to x is equal to r^2.

Polynomial of Degree $n - 1$. This can be passed through n points (x_i, y_i). The method of doing this by *divided differences* is as follows:

Example 2.44. Find the polynomial through $(1, 5)$, $(3, 11)$, $(4, 31)$, $(6, 3)$.
Using the first three values of x, assume the polynomial to be of the form $y = a_1 + a_2(x - 1) + a_3(x - 1)(x - 3) + a_4(x - 1)(x - 3)(x - 4)$. The a_i are the last four numbers in the top diagonal of the following table.

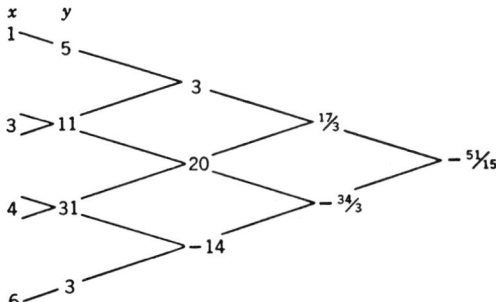

To form the table, put the given (x_i, y_i) in the first two columns. To find a number in any other column, divide the difference of the two numbers just above and below it immediately to the left by the difference of the x's in the two diagonals through it. The polynomial is $y = 5 + 3(x - 1) + \frac{17}{3}(x - 1)(x - 3) - \frac{51}{15}(x - 1)(x - 3)(x - 4)$.

Power Formula. $y = ax^n$ fits well if the points (x_i, y_i) lie approximately on a straight line when plotted on logarithmic (log scales on both horizontal and vertical axes) graph paper. To find a and n use two of the points (x_1, y_1) and (x_2, y_2), preferably far apart.

$$n = \frac{\log y_2 - \log y_1}{\log x_2 - \log x_1}$$

$$\log a = \log y_1 - n \log x_1$$

Exponential Formula. $y = ae^{nx}$ fits well if the points (x_i, y_i) lie approximately on a straight line when plotted on semilogarithmic (log scale on vertical axis) graph paper. To find a and n use two of the points (x_1, y_1) and (x_2, y_2), preferably far apart.

$$n = \frac{\ln y_1 - \ln y_2}{x_1 - x_2} = 2.3026 \frac{\log y_1 - \log y_2}{x_1 - x_2}$$

$$\ln a = \ln y_1 - nx_1 \quad \text{or} \quad \log a = \log y_1 - 0.4343nx_1$$

2.4.3 Statistical Estimation by Small Samples

A statistic is an *unbiased estimate* of a population parameter if its expected value is equal to the population parameter.
In the problem of estimating a population parameter, such as the mean or variance, the interval within which $c\%$ of the sample parameter values lies is the $c\%$ *confidence interval* for the parameter.

The χ^2 distribution function for ν *degrees of freedom* is

$$f(\chi^2) = \frac{1}{2^{\nu/2}\Gamma\left(\dfrac{\nu}{2}\right)}(\chi^2)^{(\nu-2)/2}e^{-\chi2/2}$$

and its moment generating function is

$$M(\theta) = (1 - 2\theta)^{-\nu/2}$$

The sum of the squares of n random sample values of x has a χ^2 distribution with n degrees of freedom if x has a normal distribution with zero mean and unit variance.

The **binomial index of dispersion** is

$$\chi^2 = \sum_{r=1}^{k} \frac{(x_r - \bar{x})^2}{\bar{x}\left(1 - \dfrac{\bar{x}}{n}\right)}$$

For p small and n large, this reduces to the *Poisson index of dispersion*

$$\sum_{r=1}^{k} \frac{(x_r - \bar{x})^2}{\bar{x}}$$

These indices are used to test the hypothesis that k sample frequencies x_r came from the same binomial or Poisson population, respectively.

Student's t distribution for the variable $t = u\nu^{1/2}/v$ is

$$f(t) = c\left(1 + \frac{t^2}{\nu}\right)^{-(\nu+1)/2}$$

ν degrees of freedom, c constant, if u has a normal distribution with zero mean and unit variance and v^2 has a χ^2 distribution with ν degrees of freedom.

The F distribution for the variable $F = (u/\nu_1)/(v/\nu_2)$ is

$$f(F) = \frac{cF^{(\nu_1-2)/2}}{(\nu_2 + \nu_1 F)^{(\nu_1+\nu_2)/2}}$$

ν_1 and ν_2 degrees of freedom, c constant, if u and v have independent χ^2 distributions with ν_1 and ν_2 degrees of freedom, respectively.

Analysis of Variance

Experimental error is the variation in the basic variable remaining after the effects of controlled variables have been removed (p. 2·37). The *analysis of variance* means the resolution of the basic sum of squares into the component that measures the part of the variation being tested and the component that measures the experimental error.

2.4.4   Statistical Design of Experiments

To get *valid* conclusions from an experiment, there is need for proper control of the other variables besides those being investigated, and also for sufficiently large and random samples.

Sampling Inspection. To make an inspection *efficient*, the cost and usually the amount of sampling should be minimized.

It is a common practice in industry for a consumer to accept or reject a lot on the basis of a sample drawn from the lot. There is a maximum fraction of defectives that the consumer will tolerate. This is the **lot tolerance fraction defective** p_t. A random sample of n pieces is selected from a lot of N pieces. The maximum allowable number of defective pieces in an acceptable sample is c. **Single sampling** means: (1) Inspect a sample of n pieces. (2) Accept the lot if the number of defective pieces is c or less; otherwise inspect the remainder of the lot. (3) Replace all defective pieces found by nondefective.

The **consumer's risk**, that is, the probability that a consumer will accept a lot of quality lower than p_t is

$$P_c = \sum_{x=0}^{c} \frac{\binom{Np_t}{x}\binom{N-Np_t}{n-x}}{\binom{N}{n}} \tag{2.12}$$

If a producer has standardized quality at a fractional value $\bar{p}$, the *process average fraction defective*, then the *producer's risk*, that is, the probability that a lot will be erroneously rejected, is

$$P_p = 1 - \sum_{x=0}^{c} \frac{\binom{N\bar{p}}{x}\binom{N-N\bar{p}}{n-x}}{\binom{N}{n}} \tag{2.13}$$

These two risks correspond to errors of type 2 and type 1, respectively.

The average number of pieces inspected per lot for single sampling is $I = n + (N-n)P_p$. The amount of inspection and ordinarily the cost is minimized by finding the pair of values of n and c that satisfy (1) for an assigned value of P_c and minimize I.

Sequential Analysis. An improvement on the fixed-size sampling methods already described results in greater efficiency if the inspection can be conducted on an accumulation-of-information basis. Such sequential methods operate on successive terms of a sequence of observations as they are received. They involve two steps: (1) to accept or reject the hypothesis under test and (2) to continue taking additional observations if the hypothesis is rejected.

For a more extensive treatment of the elementary theory of statistics see *Introduction to Mathematical Statistics* by P. G. Hoel, John Wiley & Sons (1947).

2.4.5 Precision of Measurements

Observations and Errors

The error of an observation is $e_i = m_i - m$, $i = 1, 2, \ldots, n$, where the m_i are the observed values, the e_i the errors, and m the mean value, that is, the arithmetic mean of a very large number (theoretically infinite) of observations.

In a large number of measurements *random errors* are as often negative as positive and have little effect on the arithmetic mean. All other errors are classed as *systematic*. If due to the same cause, they affect the mean in the same sense and give it a definite bias.

Best Estimate and Measured Value. If all systematic errors have been eliminated, it is possible to consider the sample of individual repeated measurements of a quantity with a view to securing the "best" estimate of the mean value m and assessing the degree of reproducibility that has been obtained. The final result will then be expressed in the form $E \pm L$, where E is the best estimate of m, and L the characteristic limit of variation associated with a certain risk. Not merely E, but the entire result $E \pm L$ is the value measured.

The Arithmetic Mean. If a large number of measurements have been made to determine directly the mean m of a certain quantity, all measurements having been made with equal skill and care, the best estimate of m from a sample of n is the arithmetic mean $\bar{m}$ of the measurements in the sample,

$$\bar{m} = \frac{1}{n}\sum_{i=1}^{n} m_i$$

Standard deviation is the *root-mean-square* of the deviations e_i of a set of observations from the mean,

$$\sigma = \left(\frac{1}{n}\sum_{i=1}^{n} e_i^2\right)^{1/2}$$

Since neither the mean m nor the errors of observation e_i are ordinarily known, the deviations from the arithmetic mean, or the *residuals*, $x_i = m_i - \bar{m}$, $i = 1, 2, \ldots, n$, will be referred to as errors.

Likewise, for σ the unbiased value

$$\sigma = (n-1)^{-1/2}\left[\sum_{i=1}^{n}(m_i - \overline{m})^2\right]^{1/2} = (n-1)^{-1/2}\left(\sum_{i=1}^{n}e_i^2\right)^{1/2}$$

will be used, in which n is replaced by $n-1$ since one degree of freedom is lost by using $\overline{m}$ instead of m, $\overline{m}$ being related to the m_i.

The Normal Distribution

Relative Frequency of Errors. The **Gauss–Laplace**, or **normal**, distribution of frequency of errors is (Fig. 2.4)

$$y = \frac{1}{\sigma\sqrt{2\pi}}e^{-x^2/2\sigma^2}$$

or

$$y = \frac{1}{\sqrt{\pi}}he^{-h^2x^2}$$

where $2h^2\sigma^2 = 1$, or $h = 1/(\sqrt{2}\,\sigma)$, and y represents the proportionate number of errors of value x. The area under the curve is unity. The dotted curve is also an error distribution curve with a greater value of the precision index h, which measures the concentration of observations about their mean.

Probability. The fraction of the total number of errors whose values lie between $x = -a$ and $x = a$ is

$$P = \frac{h}{\sqrt{\pi}}\int_{-a}^{+a}e^{-h^2x^2}\,dx = \frac{2}{\sqrt{\pi}}\int_{0}^{ha}e^{-h^2x^2}\,d(hx) \qquad (2.14)$$

that is, P is the probability of an error x having a value between $-a$ and a (see Table 2.1). Similarly, the shaded area represents the probability of errors between b and c.

Probable Error. Results of measurements are sometimes expressed in the form $E \pm r$, where r is the *probable error* of a single observation and is defined as the number that the actual error may with equal probability be greater or less than. From (2.14)

$$\frac{2}{\sqrt{\pi}}\int_{0}^{hr}e^{-h^2x^2}\,d(hx) = 0.50$$

and

$$hr = 0.47694$$

or

$$r = 0.4769 \times \sqrt{2}\,\sigma = 0.6745\sigma$$

Similarly, 5% of the errors x are greater than 2σ, and less than 1% are greater than 3σ. For rapid comparisons the following approximate formula due to Peters is useful:

$$r \approx 0.8453[n(n-1)]^{-1/2}\sum_{i=1}^{n}|x_i|$$

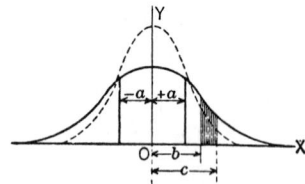

Fig. 2.4

TABLE 2.1 Values of $P = \dfrac{2}{\sqrt{\pi}} \displaystyle\int_0^{ha} e^{-h^2 x^2}\, d(hx)$

ha^a	0	1	2	3	4	5	6	7	8	9
0.0		0.01128	0.02256	0.03384	0.04511	0.05637	0.06762	0.07886	0.09008	0.10128
0.1	0.11246	0.12362	0.13476	0.14587	0.15695	0.16800	0.17901	0.18999	0.20094	0.21184
0.2	0.22270	0.23352	0.24430	0.25502	0.26570	0.27633	0.28690	0.29742	0.30788	0.31828
0.3	0.32863	0.33891	0.34913	0.35928	0.36936	0.37938	0.38933	0.39921	0.40901	0.41874
0.4	0.42839	0.43797	0.44747	0.45689	0.46623	0.47548	0.48466	0.49375	0.50275	0.51167
0.5	0.52050	0.52924	0.53790	0.54646	0.55494	0.56332	0.57162	0.57982	0.58792	0.59594
0.6	0.60386	0.61168	0.61941	0.62705	0.63459	0.64203	0.64938	0.65663	0.66378	0.67084
0.7	0.67780	0.68467	0.69143	0.69810	0.70468	0.71116	0.71754	0.72382	0.73001	0.73610
0.8	0.74210	0.74800	0.75381	0.75952	0.76514	0.77067	0.77610	0.78144	0.78669	0.79184
0.9	0.79691	0.80188	0.80677	0.81156	0.81627	0.82089	0.82542	0.82987	0.83423	0.83851
1.0	0.84270	0.84681	0.85084	0.85478	0.85865	0.86244	0.86614	0.86977	0.87333	0.87680
1.1	0.88021	0.88353	0.88679	0.88997	0.89308	0.89612	0.89910	0.90200	0.90484	0.90761
1.2	0.91031	0.91296	0.91553	0.91805	0.92051	0.92290	0.92524	0.92751	0.92973	0.93190
1.3	0.93401	0.93606	0.93807	0.94002	0.94191	0.94376	0.94556	0.94731	0.94902	0.95067
1.4	0.95229	0.95385	0.95538	0.95686	0.95830	0.95970	0.96105	0.96237	0.96365	0.96490
1.5	0.96611	0.96728	0.96841	0.96952	0.97059	0.97162	0.97263	0.97360	0.97455	0.97546
1.6	0.97635	0.97721	0.97804	0.97884	0.97962	0.98038	0.98110	0.98181	0.98249	0.98315
1.7	0.98379	0.98441	0.98500	0.98558	0.98613	0.98667	0.98719	0.98769	0.98817	0.98864
1.8	0.98909	0.98952	0.98994	0.99035	0.99074	0.99111	0.99147	0.99182	0.99216	0.99248
1.9	0.99279	0.99309	0.99338	0.99366	0.99392	0.99418	0.99443	0.99466	0.99489	0.99511
2.0	0.99532	0.99552	0.99572	0.99591	0.99609	0.99626	0.99642	0.99658	0.99673	0.99688
2.1	0.99702	0.99715	0.99728	0.99741	0.99753	0.99764	0.99775	0.99785	0.99795	0.99805
2.2	0.99814	0.99822	0.99831	0.99839	0.99846	0.99854	0.99861	0.99867	0.99874	0.99880
2.3	0.99886	0.99891	0.99897	0.99902	0.99906	0.99911	0.99915	0.99920	0.99924	0.99928
2.4	0.99931	0.99935	0.99938	0.99941	0.99944	0.99947	0.99950	0.99952	0.99955	0.99957
2.5	0.99959	0.99961	0.99963	0.99965	0.99967	0.99969	0.99971	0.99972	0.99974	0.99975
2.6	0.99976	0.99978	0.99979	0.99980	0.99981	0.99982	0.99983	0.99984	0.99985	0.99986
2.7	0.99987	0.99987	0.99988	0.99989	0.99989	0.99990	0.99991	0.99991	0.99992	0.99992
2.8	0.99992	0.99993	0.99993	0.99994	0.99994	0.99994	0.99995	0.99995	0.99995	0.99996
2.9	0.99996	0.99996	0.99996	0.99997	0.99997	0.99997	0.99997	0.99997	0.99997	0.99998
3.0	0.99998	1.0000	1.0000	1.0000						

$^a ha = 0.47694 \dfrac{a}{r} = \dfrac{1}{\sqrt{2}}\dfrac{a}{\sigma}$

The standard deviation of the arithmetic mean, $\sigma_{\bar{m}}$, as calculated from data, is related to the standard deviation, σ, by the formula

$$\sigma_{\bar{m}} = n^{-1/2}\sigma = [n(n-1)]^{-1/2}\left(\sum_{i=1}^{n} x_i^2\right)^{1/2}$$

From this formula and Tables 2.1 and 2.2 the limits corresponding to given risks can be determined as indicated previously. It is evident that the stability of the mean increases with n, that is, the effect of the erratic behavior of single cases decreases with increase of n.

The probable error of the arithmetic mean as calculated from data, $r_{\bar{m}}$, is then given by

$$r_{\bar{m}} = 0.6745[n(n-1)]^{-1/2}\left(\sum_{i=1}^{n} x_i^2\right)^{1/2}$$

and Peters' formula for the approximate value is

$$r_{\bar{m}} \approx 0.8453[n^2(n-1)]^{-1/2}\sum_{i=1}^{n}|x_i|$$

TABLE 2.2 Values of Functions of n and $(n-1)$ Factors for Computing Actual and Approximate Values of r and $r_{\overline{m}}$

n	$\dfrac{0.6745}{\sqrt{n-1}}$	$\dfrac{0.6745}{\sqrt{n(n-1)}}$	$\dfrac{0.8453}{\sqrt{n(n-1)}}$	$\dfrac{0.8453}{n\sqrt{n-1}}$	n	$\dfrac{0.6745}{\sqrt{n-1}}$	$\dfrac{0.6745}{\sqrt{n(n-1)}}$	$\dfrac{0.8453}{\sqrt{n(n-1)}}$	$\dfrac{0.8453}{n\sqrt{n-1}}$
1					51	0.0954	0.0134	0.0167	0.0023
2	0.6745	0.4769	0.5978	0.4227	52	0.0944	0.0131	0.0164	0.0023
3	0.4769	0.2754	0.3451	0.1993	53	0.0935	0.0128	0.0161	0.0022
4	0.3894	0.1947	0.2440	0.1220	54	0.0926	0.0126	0.0158	0.0022
5	0.3372	0.1508	0.1890	0.0845	55	0.0918	0.0124	0.0155	0.0021
6	0.3016	0.1231	0.1543	0.0630	56	0.0909	0.0122	0.0152	0.0020
7	0.2754	0.1041	0.1304	0.0493	57	0.0901	0.0119	0.0150	0.0020
8	0.2549	0.0901	0.1130	0.0399	58	0.0893	0.0117	0.0147	0.0019
9	0.2385	0.0795	0.0996	0.0332	59	0.0886	0.0115	0.0145	0.0019
10	0.2248	0.0711	0.0891	0.0282	60	0.0878	0.0113	0.0142	0.0018
11	0.2133	0.0643	0.0806	0.0243	61	0.0871	0.0111	0.0140	0.0018
12	0.2034	0.0587	0.0736	0.0212	62	0.0864	0.0110	0.0137	0.0017
13	0.1947	0.0540	0.0677	0.0188	63	0.0857	0.0108	0.0135	0.0017
14	0.1871	0.0500	0.0627	0.0167	64	0.0850	0.0106	0.0133	0.0017
15	0.1803	0.0465	0.0583	0.0151	65	0.0843	0.0105	0.0131	0.0016
16	0.1742	0.0435	0.0546	0.0136	66	0.0837	0.0103	0.0129	0.0016
17	0.1686	0.0409	0.0513	0.0124	67	0.0830	0.0101	0.0127	0.0016
18	0.1636	0.0386	0.0483	0.0114	68	0.0824	0.0100	0.0125	0.0015
19	0.1590	0.0365	0.0457	0.0105	69	0.0818	0.0098	0.0123	0.0015
20	0.1547	0.0346	0.0434	0.0097	70	0.0812	0.0097	0.0122	0.0015
21	0.1508	0.0329	0.0412	0.0090	71	0.0806	0.0096	0.0120	0.0014
22	0.1472	0.0314	0.0393	0.0084	72	0.0800	0.0094	0.0118	0.0014
23	0.1438	0.0300	0.0376	0.0078	73	0.0795	0.0093	0.0117	0.0014
24	0.1406	0.0287	0.0360	0.0073	74	0.0789	0.0092	0.0115	0.0013
25	0.1377	0.0275	0.0345	0.0069	75	0.0784	0.0091	0.0113	0.0013
26	0.1349	0.0265	0.0332	0.0065	76	0.0779	0.0089	0.0112	0.0013
27	0.1323	0.0255	0.0319	0.0061	77	0.0774	0.0088	0.0111	0.0013
28	0.1298	0.0245	0.0307	0.0058	78	0.0769	0.0087	0.0109	0.0012
29	0.1275	0.0237	0.0297	0.0055	79	0.0764	0.0086	0.0108	0.0012
30	0.1252	0.0229	0.0287	0.0052	80	0.0759	0.0085	0.0106	0.0012
31	0.1231	0.0221	0.0277	0.0050	81	0.0754	0.0084	0.0105	0.0012
32	0.1211	0.0214	0.0268	0.0047	82	0.0749	0.0083	0.0104	0.0011
33	0.1192	0.0208	0.0260	0.0045	83	0.0745	0.0082	0.0102	0.0011
34	0.1174	0.0201	0.0252	0.0043	84	0.0740	0.0081	0.0101	0.0011
35	0.1157	0.0196	0.0245	0.0041	85	0.0736	0.0080	0.0100	0.0011
36	0.1140	0.0190	0.0238	0.0040	86	0.0732	0.0079	0.0099	0.0011
37	0.1124	0.0185	0.0232	0.0038	87	0.0727	0.0078	0.0098	0.0010
38	0.1109	0.0180	0.0225	0.0037	88	0.0723	0.0077	0.0097	0.0010
39	0.1094	0.0175	0.0220	0.0035	89	0.0719	0.0076	0.0096	0.0010
40	0.1080	0.0171	0.0214	0.0034	90	0.0715	0.0075	0.0094	0.0010
41	0.1066	0.0167	0.0209	0.0033	91	0.0711	0.0075	0.0093	0.0010
42	0.1053	0.0163	0.0204	0.0031	92	0.0707	0.0074	0.0092	0.0010
43	0.1041	0.0159	0.0199	0.0030	93	0.0703	0.0073	0.0091	0.0009
44	0.1029	0.0155	0.0194	0.0029	94	0.0699	0.0072	0.0090	0.0009
45	0.1017	0.0152	0.0190	0.0028	95	0.0696	0.0071	0.0089	0.0009
46	0.1005	0.0148	0.0186	0.0027	96	0.0692	0.0071	0.0089	0.0009
47	0.0994	0.0145	0.0182	0.0027	97	0.0688	0.0070	0.0088	0.0009
48	0.0984	0.0142	0.0178	0.0026	98	0.0685	0.0069	0.0087	0.0009
49	0.0974	0.0139	0.0174	0.0025	99	0.0681	0.0068	0.0086	0.0009
50	0.0964	0.0136	0.0171	0.0024	100	0.0678	0.0068	0.0085	0.0009

Example 2.45. The following are ten measurements, m_i, of the length of a base line. The values of the residuals, x_i, and their squares are given. m_i: 455.35, 455.35, 455.20, 455.05, 455.75, 455.40, 455.10, 455.30, 455.50, 455.30.

Arithmetic mean, $\overline{m} = 455.330$.

x_i: 0.02, 0.02, -0.13, -0.28, 0.42, 0.07, -0.23, -0.03, -0.17, -0.03.

x_i^2: 0.0004, 0.0004, 0.0169, 0.0784, 0.1764, 0.0049, 0.0529, 0.0009, 0.0289, 0.0009.

Hence

$$\sum_{i=1}^{10} x_i^2 = 0.3610 \quad \text{and} \quad \sum_{i=1}^{10} |x_i| = 1.40$$

So by the standard formulas, $r = 0.6745(9)^{-1/2}(0.3610)^{1/2} = 0.13$, $r_{\overline{m}} = (10)^{-1/2}r = 0.042$. By the approximate formulas, $r \approx 0.8453(90)^{-1/2}(1.40) = 0.12$, $r_{\overline{m}} \approx 0.039$.

For the best estimate of the base line, the result is 455.330 with probable error ± 0.042 (using result given by standard formula), usually written 455.330 ± 0.042. In any considerable number of observations it should be the case, as it is here, that half of the residuals are less than the probable error.

Rounded Numbers. It can be shown that the standard deviation, σ, of a *rounded number* (Section 2.1.3) due to rounding is $\sigma = 0.2887w$, where w is a unit in the last place retained. Consequently, the probable error of a rounded number due to rounding is

$$r = 0.6745 \times 0.2887w = 0.1947w$$

Weighted Observations. Sometimes, notwithstanding the care with which observations are taken, there are reasons for believing that certain observations are better than others. In such cases the observations are given different weights, that is, are counted different numbers of times, the weights or numbers expressing their relative practical worth. If there are n weighted observations, m_i, with weights, p_i, these being made directly on the same quantity, then the best estimate of the mean value m of the quantity is the **weighted arithmetic mean** $\overline{m}$ of the sample,

$$\overline{m} \equiv \frac{\sum\limits_{i=1}^{n} p_i m_i}{\sum\limits_{i=1}^{n} p_i}$$

For the set of weighted observations we have

$$r = 0.6745(n-1)^{-1/2}\left(\sum_{i=1}^{n} p_i x_i^2\right)^{1/2}$$

as the probable error of an observation of unit weight, and

$$r_{\overline{m}} = 0.6745\left[(n-1)\sum_{i=1}^{n} p_i\right]^{-1/2}\left(\sum_{i=1}^{n} p_i x_i^2\right)^{1/2}$$

as the probable error of the arithmetic mean of weighted items, in which

$$x_i \equiv \frac{m_i - \sum\limits_{i=1}^{n} p_i m_i}{\sum\limits_{i=1}^{n} p_i}$$

Example 2.46. Let six observations on the same quantity be made, with weights, p_i, the sum of these weights being 21 (see following tabulation). The sum of the weighted observations, $\sum_{i=1}^{6} p_i m_i$, is 3741.36. The best estimate of the value of m for the observed quantity is $\overline{m} = 3741.36/21 = 178.16$. Subtracting this from each m_i gives the residuals x_i. The sum of the weighted squares of the residuals $\sum_{i=1}^{6} p_i x_i^2$, is 62.95. Then the preceding formulas give the probable error of an observation of weight unity as $r = 2.39$ and the probable error of the weighted mean as $r_{\overline{m}} = 0.52$. The final result then is 178.16 ± 0.52.

p_i	5	4	1	4	3	4
m_i	178.26	176.30	181.06	177.95	176.20	180.85
$p_i m_i$	891.30	705.20	181.06	711.80	528.60	723.40
x_i	0.10	1.86	2.90	0.21	1.96	2.69
x_i^2	0.010	3.460	8.410	0.441	3.842	7.230
$p_i x_i^2$	0.05	13.84	8.41	0.18	11.53	28.94

TABLE 2.3 Combination of Observations

n	20%	50%	n	20%	50%
5	0.64	0.24	30	0.12	0.00014
10	0.40	0.034	40	0.076	8×10^{-6}
15	0.29	0.008	50	0.047	6×10^{-7}
20	0.21	0.0002	100	0.0050	

Probable Error in a Result Calculated from the Means of Several Observed Quantities. Let Z be a sum of n means of observed independent quantities, each taken with a plus or a minus sign. Then, if r_j, $j = 1, 2, \ldots, n$, are the probable errors in these means, the probable error in Z is $(\sum_{j=1}^{n} r_j^2)^{1/2}$.

Let $Z = Az$, where z is the mean of an observed quantity with probable error r, and A an exact number. Then the probable error in Z is Ar.

Let Z be any differentiable function of the means of independently observed quantities z_j with probable errors r_j. Then the probable error in Z is $[\sum_{j=1}^{m}(\partial Z/\partial z_j)^2 r_j^2]^{1/2}$. For example, if $Z = z_1 z_2$, the probable error in Z is $(z_1^2 r_2^2 + z_2^2 r_1^2)^{1/2}$.

Conditions of Applicability. The theory underlying the foregoing development depends on the following assumptions: (1) The sample consists of a large number of observations. (2) The observations have been made with equal care and skill so that (2.1) there are approximately an equal number of readings above and below the mean (except in the case of weighted items), (2.2) the individual deviations from the mean are small in most cases, and (2.3) the number of deviations diminishes rapidly as their size increases.

The extent to which the observed data satisfy these assumptions is a measure of the extent to which we are justified in using the Gauss error distribution curve, which is consistent with the statement that $\overline{m}$ is the best estimate of the mean value m, and which leads to the factor 0.6745 used in computing probable error. Even if we were not justified in assuming the Gaussian distribution of errors, the arithmetic mean still remains the best estimate we have for m. Therefore, there is little difficulty in this regard, especially since "errors" appear to follow the Gaussian distribution as closely as any other we know. Our difficulties enter in connection with the factor 0.6745 and the accuracy of the σ, as estimated from the data.

If the number of observations n in a sample is small the estimate of the standard deviation of the possible infinity of observations with mean m is itself subject to considerable error. For example, for $n = 3$ the standard error of the standard deviation is as large as the standard deviation itself, and hence the probable error calculated from $r = 0.6745\sigma$ would not be very reliable. Table 2.3* will illustrate this. The second and third columns give the probability that the probable error of a single observation should be out 20 and 50%, respectively.

From Table 2.3 it is clear that with 10 observations the odds are only 3 to 2 that the calculated probable error is within 20% of the correct value, and about 30 to 1 that it is within 50% of the correct value. Of course, the probable error of the mean will be correspondingly out.

The use of Table 2.3 is quite legitimate for $100 < n$, and for $30 < n < 100$ the table may be used provided σ is multiplied by $(n - 3)^{-1/2}$. For $n < 30$, a rough estimate can be obtained from the fact that the percentage of cases lying outside the range, $m \pm k\sigma$, is $< 100k^{-2}$ for $1 < k$. A striking property of this inequality due to Tchebycheff is that it is *nonparametric*, which means independent of the nature of the distribution assumed.

2.5 GEOMETRY

2.5.1 Geometric Concepts

Plane Angles

A **degree** (°) is $\frac{1}{360}$ of a revolution (or **perigon**) and is divided into 60 units called *minutes* (') that in turn are divided into 60 units called *seconds* (''').

A **radian** is a *central angle* that intercepts a *circular arc* equal to its *radius*. One radian, therefore, equals $360/2\pi$ degrees or $57.295779513°$, and $1° = 0.017453293$ radian.

An **angle** of 90° is a *right angle*, and the lines that form it are *perpendicular*. An angle less than a right angle is *acute*. An angle greater than a right angle but less than 180° is *obtuse*. If the sum of two angles equals 90°, they are *complementary* to each other, and if their sum is 180°, *supplementary* to each other.

*David Brunt, *The Combination of Observations*, Cambridge University Press (1917).

Polygons

A **polygon**, or **plane rectilinear figure**, is a closed broken line.

A **triangle** is a polygon of three sides. It is *isosceles* if two sides (and their opposite angles) are equal; it is *equilateral* if all three sides (and all three angles) are equal.

A **quadrilateral** is a polygon of four sides. This classification includes the *trapezium* having no two sides parallel; the *trapezoid* having two opposite sides parallel (*isosceles trapezoid* if the nonparallel sides are equal); and the *parallelogram* having both pairs of opposite sides parallel and equal. The parallelogram includes the *rhomboid* having no right angles and, in general, adjacent sides not equal; the *rhombus* having no right angles but all sides equal; the *rectangle* having only right angles and, in general, adjacent sides not equal; and the *square* having only right angles, and all sides equal.

Similar polygons have their respective angles equal and their corresponding sides proportional.

A **regular polygon** has all sides equal and all angles equal. An *equilateral triangle* and a *square* are regular polygons.

Other polygons classified according to number of sides are (5) *pentagon*, (6) *hexagon*, (7) *heptagon*, (8) *octagon*, (9) *enneagon* or *nonagon*, (10) *decagon*, and (12) *dodecagon*. Two regular polygons of the same number of sides are *similar*.

Properties of Triangles

General Triangle. The sum of the angles equals $180°$. $\angle XAB$ (Fig. 2.5) is an *exterior angle* of $\triangle ABC$ and equals the sum of the opposite *interior angles* (i.e., $\angle XAB = \angle B + \angle C$). A *median* of a triangle is a line joining a vertex to the midpoint of the opposite side. The three medians meet at the *center of gravity*, G, and G trisects each median (i.e., $AG = \frac{2}{3}AD$, etc.). *Bisectors of angles* of a triangle (Fig. 2.6) meet in a point M equidistant from all sides. M is the center of the *inscribed circle* (tangent to all sides), or the *incenter* of the triangle. An angle bisector divides the opposite side into segments proportional to the adjacent sides of the angle (i.e., $AK/KC = AB/BC$, etc.). An *altitude* of a triangle is a perpendicular from a vertex to the opposite side. The three altitudes meet in a point called the *orthocenter*. The *perpendicular bisectors* of the sides of a triangle (Fig. 2.7) meet in a point O equidistant from all vertices. O is the center of the *circumscribed circle* (passing through all vertices), or the *circumcenter* of the triangle. The longest side of a triangle is opposite the largest angle, and vice versa. The line joining the midpoints of two sides of a triangle is parallel to the third side and half its length. If two triangles are mutually equiangular, they are *similar*, and their corresponding sides are proportional.

Orthogonal Projection. In Figs. 2.8 and 2.9, AE is the orthogonal projection of AB on AC, BE being perpendicular to AC. The square of the side opposite an acute angle equals the sum of the squares of the other two sides diminished by twice the product of one of those sides by the orthogonal projection of the other side upon it. In Fig. 2.8, $a^2 = b^2 + c^2 - 2b \cdot AE$. The square of the side opposite an obtuse angle equals the sum of the squares of the other two sides increased by twice the product of one of those sides by the orthogonal projection of the other side upon it. In Fig. 2.9, $a^2 = b^2 + c^2 + 2b \cdot AE$.

Right Triangle. In Fig. 2.10, let h be the *altitude* drawn from the vertex of right angle C to the hypotenuse c. Then $\angle A + \angle B = 90°$; $c^2 = a^2 + b^2$; $h^2 = mn$; $b^2 = cm$; $a^2 = cn$; (median from C) $= c/2$.

Isosceles Triangle. Two sides are equal and their opposite angles are equal. If a straight line from the vertex at which the equal sides meet bisects the base, it also bisects the angle at the vertex and is perpendicular to the base.

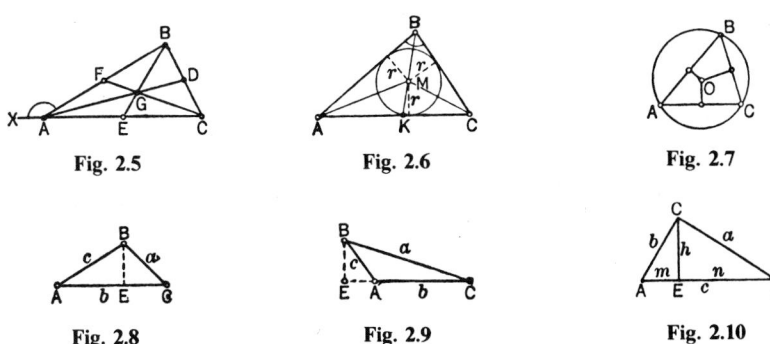

Fig. 2.5 Fig. 2.6 Fig. 2.7

Fig. 2.8 Fig. 2.9 Fig. 2.10

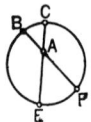 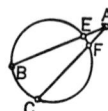 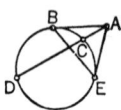 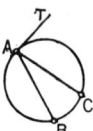

| Fig. 2.11 | Fig. 2.12 | Fig. 2.13 | Fig. 2.14 |

Circles

A **circle** is a closed plane curve, all the points of which are equidistant from a *center* point. A *chord* is a straight line joining two points on a curve, that is, joining the extremities of an *arc*. A *segment* of a circle is the part of its plane included between a concave arc and its chord. An angle *intercepts* an arc cut off by its sides; the arc *subtends* the angle. A *central angle* of a circle is one whose vertex is at the center and whose sides are two radii. A *sector* of a circle is the part of its plane that is included between an arc and two radii drawn to its extremities. A *secant* of a circle is a straight line intersecting it in two points. Parallel secants (or tangents) intercept equal arcs. A tangent line meets a circle in only one point and is perpendicular to the radius to that point. If a radius is perpendicular to a chord, it bisects both the chord and the arc intercepted by the chord. If two circles are tangent to each other, the line of centers passes through the point of contact; if the circles intersect, the line of centers bisects the common chord at right angles. In Fig. 2.11, the product of linear segments AC and AE equals the product of linear segments AB and AF. In Fig. 2.12, the product of the whole secant AB and its external segment AE equals the product of the whole secant AC and its external segment AF. In Fig. 2.13, the product of the whole secant AD and its external segment AC equals the square of tangent AB (or AE). Also $\angle ABE = \angle AEB$.

Angle Measurement. Considering the arc of a circle to be expressed in terms of the central angle that it subtends, the arc may be said to contain a certain number of degrees and hence be used to express the measurement of other angles related to the circle. On this basis, an entire circle equals 360°. The *inscribed angle* formed by two chords intersecting on a circle equals half the arc intercepted by it. Thus, in Fig. 2.14, $\angle BAC = \frac{1}{2}$ arc BC. An angle inscribed in a semicircle is a right angle. The angle formed by a tangent to a circle and a chord having one extremity at the point of contact equals half the arc intercepted by the chord. In Fig. 2.14, $\angle BAT = \frac{1}{2}$ arc BCA. The angle formed by two chords intersecting within a circle equals half the sum of the intercepted arcs. In Fig. 2.11, $\angle BAC$ (or $\angle EAF$) $= \frac{1}{2}$ (arc BC + arc EF). The angle formed by two secants, or two tangents, or a secant and a tangent, intersecting outside a circle, equals half the difference of the intercepted arcs. In Fig. 2.12, $\angle BAC = \frac{1}{2}$(arc BC − arc EF). In Fig. 2.13, $\angle BAE = \frac{1}{2}$(arc BDE − arc BCE), and $\angle BAD = \frac{1}{2}$(arc BD − arc BC).

Coaxal Systems

Types

1. A set of nonintersecting circles having collinear centers and orthogonal to a given circle with center also collinear. The end points of the diameter of the given circle on the line of centers are the limiting points of the system (Fig. 2.15, centers on horizontal line).
2. A set of circles through two given points (Fig. 2.15, centers on vertical line).
3. A set of circles with a common point of tangency.
4. A set of concentric circles.
5. A set of concurrent lines.
6. A set of parallel lines.

Conjugate Systems. Two coaxal systems whose members are mutually orthogonal are *conjugate*. A conjugate pair may consist of (1) a system of type 1 and one of type 2, with the limiting points of one the common points of the other (Fig. 2.15); (2) two systems of type 3; (3) a system of type 4 and one of type 5; (4) two systems of type 6.

Inversion

If the point O is the center of a circle c of radius r, if P and P' are collinear with O, and if $OP \cdot OP' = r^2$, then P and P' are *inverse* to each other with respect to the circle c (Fig. 2.16). The point O is the *center of inversion*.

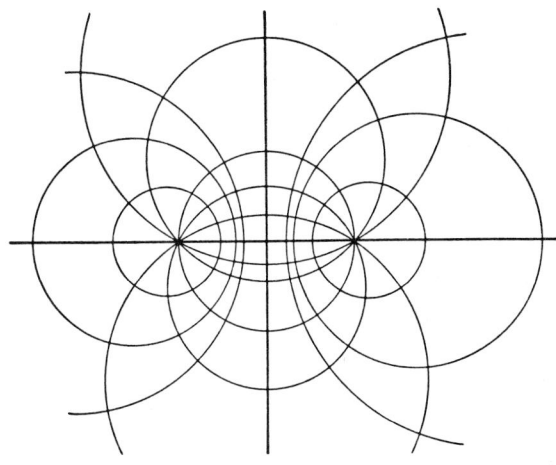

Fig. 2.15

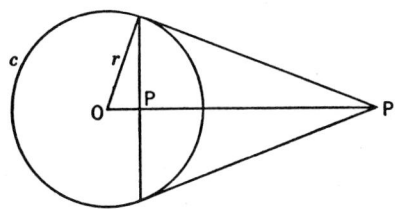

Fig. 2.16

The inverse of a circle not passing through the center of inversion is a circle, the inverse of a circle through the center is a straight line not through the center, and the inverse of a straight line through the center is itself.

Two intersecting curves invert into curves intersecting at the same angle.

Nonplanar Angles

A **dihedral angle** is the opening between two intersecting planes. In Fig. 2.17, $P–BD–Q$ is a dihedral angle of which the two planes are the *faces* and their line of intersection DB is the *edge*. A *plane angle* that measures a dihedral angle is an angle formed by two lines, one in each face, drawn perpendicular to the edge at the same point (as $\angle ABC$). A *right dihedral angle* is one whose plane angle is a right angle. Through a given line oblique or parallel to a given plane, one and only one plane can be passed perpendicular to the given plane. The line of intersection CD (Fig. 2.18) is the *orthogonal projection* of line AB upon plane P. The *angle between a line and a plane* is the angle that the line (produced if necessary) makes with its orthogonal projection on the plane. This angle is the least angle that the line makes with any line in the plane.

A **polyhedral angle** is the opening of three or more planes that meet in a common point. In Fig. 2.19, $O–ABCDE$ is a polyhedral angle of which the intersections of the planes, as OA, OB, and so on, are the *edges*; the portions of the planes lying between the edges are the *faces*; and the common point

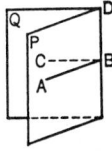

Fig. 2.17

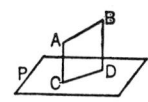

Fig. 2.18

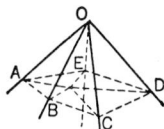

Fig. 2.19

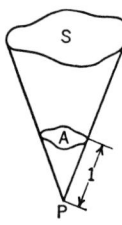

Fig. 2.20

O is the *vertex*. Angles formed by adjacent edges, as angles AOB, BOC, and so on, are *face angles*. A polyhedral angle is called a *trihedral angle* if it has three faces; a *tetrahedral angle* if it has four faces; and so on.

A **solid angle** measures the opening between surfaces, either planar or nonplanar, which meet in a common point. The polyhedral angle is a special case. In Fig. 2.20 the solid angle at any point P, subtended by any surface S, is equal numerically to the portion A of the surface of a sphere of unit radius that is cut out by a conical surface with vertex at P and having the boundary of S for base. The *unit solid angle* is the *steradian* and equals the central solid angle that intercepts a spherical area (of any shape) equal to the (radius)^2. The total solid angle about a point equals 4π steradians.

A **spherical angle** is the opening between two arcs of great circles drawn on a sphere from the same point (*vertex*), and is measured by the plane angle formed by tangents to its sides at its vertex. If the planes of the great circles are perpendicular, the angle is a *right spherical angle*.

Polyhedrons

A **polyhedron** is a convex closed surface consisting of parts of four or more planes, called its *faces*; its faces intersect in straight lines, called its *edges*; its edges at points, called its *vertices*.

A **prism** is a polyhedron of which two faces (the *bases*) are congruent polygons in parallel planes and the other (*lateral*) faces are parallelograms whose planes intersect in the *lateral edges*. Prisms are *triangular, rectangular, quadrangular*, and so on, according as their bases are triangles, rectangles, quadrilaterals, and so on. A *right prism* has its lateral edges perpendicular to its bases. A prism whose bases are parallelograms is a *parallelepiped*; if in addition the edges are perpendicular to the bases, it is a *right parallelepiped*. A *rectangular parallelepiped* is a *right* parallelepiped whose bases are rectangles. A *cube* is a parallelepiped whose six faces are squares. A *truncated prism* is that part of a prism included between a base and a section made by a plane oblique to the base. A *right section* of a prism is a section made by a plane that cuts all the lateral edges perpendicularly.

A **prismatoid** is a polyhedron of which two faces (the *bases*) are polygons in parallel planes and the other (*lateral*) faces are triangles or trapezoids with one side common with one base and the opposite vertex or side common with the other base.

A **pyramid** is a polyhedron of which one face (the *base*) is a polygon and the other (*lateral*) faces are triangles meeting in a common point called the *vertex* of the pyramid and intersecting one another in its *lateral edges*. Pyramids are *triangular, quadrangular*, and so on, according as their bases are triangles, quadrilaterals, and so on. A *regular pyramid* (or *right pyramid*) has for its base a regular polygon whose center coincides with the foot of the perpendicular dropped from the vertex to the base. A *frustum of a pyramid* is the portion of a pyramid included between its base and a section parallel to the base. If the section is not parallel to the base, a *truncated pyramid* results.

A **regular polyhedron** has all faces formed of congruent regular polygons and all polyhedral angles equal. The only regular polyhedrons possible are the five types discussed in the mensuration table (Table 2.4).

A **tetrahedron** is a polyhedron of four faces. It may be described also as a triangular pyramid, and any one of its four triangular faces may be considered as the base. The four perpendiculars erected at circumcenters of the four faces meet in a point equidistant from all vertices, which is the center of the circumscribed sphere. The four *medians*, joining each vertex with the center of gravity of the opposite face, meet in a point, which is the *center of gravity* of the tetrahedron. This point is three-fourths of the distance from each vertex along a median. The four altitudes meet in a point, called the *orthocenter* of the tetrahedron. The six planes bisecting the six dihedral angles meet in a point equidistant from all faces, this being the center of the inscribed sphere.

Solids Having Curved Surfaces

A **cylinder** is a solid bounded by two parallel plane surfaces (the *bases*) and a cylindrical *lateral* surface. A *cylindrical surface* is a surface generated by the movement of a straight line (the *generatrix*) which constantly is parallel to a fixed straight line and touches a fixed curve (the *directrix*) not in the plane of the fixed straight line. The generatrix in any position is an *element* of the cylindrical surface. A *circular cylinder* is one having circular bases. A *right cylinder* is one whose elements are perpendicular to its bases. A *truncated cylinder* is the part of a cylinder included between a base and a section made by a plane oblique to the base. A *right section* of a cylinder is a section made by a plane which cuts all the elements perpendicularly.

A **cone** is a solid bounded by a conic *lateral* surface and a plane (the *base*) that cuts all the elements of the conic surface. A *conic surface* is a surface generated by the movement of a straight line (the *generatrix*) that constantly touches a fixed plane curve (the *directrix*) and passes through a fixed point (the *vertex*) not in the plane of the fixed curve. The generatrix in any position is an *element* of the conic surface. A *circular cone* is one having a circular base. A *right cone* is a circular cone whose center of the base coincides with the foot of the perpendicular dropped from the vertex to the base. A *frustum of a cone* is the portion of a cone included between its base and a section parallel to the base.

A **sphere** is a solid bounded by a surface all points of which are equidistant from a point within called the *center*. Every plane section of a sphere is a circle. This circle is a *great circle* if its plane passes through the center of the sphere; otherwise, it is a *small circle*. *Poles* of such a circle are the extremities of the diameter of the sphere that is perpendicular to the plane of the circle. Through two points on a spherical surface, not extremities of a diameter, one great circle can be passed. The shortest line that can be drawn on the surface of a sphere between two such points is an arc of a great circle less than a semicircumference joining those points. If two spherical surfaces intersect, their line of intersection is a circle whose plane is perpendicular to the line of centers, and whose center lies on this line.

A **spherical sector** is the portion of a sphere generated by the revolution of a circular sector about a diameter of the circle of which the sector is a part. A *hemisphere* is half of a sphere.

A **spherical segment** is the portion of a sphere contained between two parallel plane sections (the *bases*), one of which may be tangent to the sphere (in which case there is only one base). The term "segment" also is applied in an analogous manner to various solids of revolution, the planes in such cases being perpendicular to an axis. A *zone* is the portion of a spherical surface included between two parallel planes.

A **spherical polygon** is a figure on a spherical surface bounded by three or more arcs of great circles. The sum of the angles of a spherical triangle (polygon of three sides) is greater than two right angles and less than six right angles.

Other solids appearing in the mensuration table (Table 2.4), if not sufficiently defined by their figures, may be found discussed in the section on analytic geometry.

2.5.2 Mensuration

Perimeters of similar figures are proportional to their respective linear dimensions, areas to the squares of their linear dimensions, and volumes of similar solids to the cubes of their linear dimensions (see Table 2.4).

TABLE 2.4   Mensuration Formulas

Approximate Decimal Equivalents (for reference)

$\pi = 3.1416$	$\dfrac{1}{\pi} = 0.318$	$\sqrt{2} = 1.414$
$\dfrac{\pi}{2} = 1.5708$	$\dfrac{1}{2\pi} = 0.159$	$\sqrt{3} = 1.732$
$\dfrac{\pi}{4} = 0.7854$	$\dfrac{1}{4\pi} = 0.080$	$\dfrac{1}{\sqrt{2}} = 0.707$
$\dfrac{\pi}{180} = 0.01745$	$\dfrac{180}{\pi} = 57.296$	$\dfrac{1}{\sqrt{3}} = 0.577$
$\dfrac{\pi}{360} = 0.00873$	$\dfrac{360}{\pi} = 114.592$	

TABLE 2.4 (*Continued*)

1a.   Plane Rectilinear Figures

Notation. Lines, $a, b, c, \ldots$; angles, $\alpha, \beta, \gamma, \ldots$; altitude (perpendicular height), h; side, l; diagonals, $d, d_1, \ldots$; perimeter, p; radius of inscribed circle, r; radius of circumscribed circle, R; area, A.

1. Right Triangle	(One angle 90°)

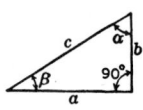

$$p = a + b + c; \; c^2 = a^2 + b^2;$$

$$A = \frac{ab}{2} = \frac{a^2}{2}\tan\beta = \frac{c^2}{4}\sin 2\beta = \frac{c^2}{4}\sin 2\alpha.$$

For additional formulas, see *General Triangle* below, and also trigonometry.

2. General Triangle (and Equilateral Triangle)

For General Triangle:

$$p = a + b + c. \text{ Let } s = \tfrac{1}{2}(a + b + c).$$

$$r = \frac{\sqrt{s(s-a)(s-b)(s-c)}}{s}; \; R = \frac{a}{2\sin\alpha} = \frac{abc}{4rs};$$

$$A = \frac{ah}{2} = \frac{ab}{2}\sin\gamma = \frac{b^2\sin\gamma\sin\alpha}{2\sin\beta} = rs = \frac{abc}{4R}.$$

Length of median to side $c = \tfrac{1}{2}\sqrt{2(a^2 + b^2) - c^2}$.

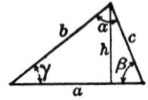

$$\text{Length of bisector of angle } \gamma = \frac{\sqrt{ab\left[(a+b)^2 - c^2\right]}}{a+b}.$$

For Equilateral Triangle ($a = b = c = l$ and $\alpha = \beta = \gamma = 60°$):

(Equal sides and equal angles)

$$p = 3l, \; r = \frac{l}{2\sqrt{3}}; \; R = \frac{l}{\sqrt{3}} = 2r;$$

$$h = \frac{l\sqrt{3}}{2}; \; l = \frac{2h}{\sqrt{3}}; \; A = \frac{l^2\sqrt{3}}{4}.$$

For additional formulas, see trigonometry.

3. Rectangle (and Square)

For Rectangle:

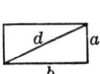

$$p = 2(a + b); \; d = \sqrt{a^2 + b^2}; \; A = ab.$$

For Square ($a = b = l$):

$$p = 4l; \; d = l\sqrt{2}; \; l = \frac{d}{\sqrt{2}}; \; A = l^2 = \frac{d^2}{2}.$$

4. General Parallelogram (and Rhombus)

For General Parallelogram (*Rhomboid*):

(Opposite sides parallel)

$$p = 2(a + b); \; d_1 = \sqrt{a^2 + b^2 - 2ab\cos\gamma};$$

$$d_2 = \sqrt{a^2 + b^2 + 2ab\cos\gamma}; \; d_1^2 + d_2^2 = 2(a^2 + b^2);$$

$$A = ah = ab\sin\gamma.$$

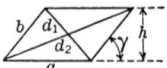

For Rhombus ($a = b = l$):

(Opposite sides parallel and all sides equal)

$$p = 4l; \; d_1 = 2l\sin\frac{\gamma}{2}; \; d_2 = 2l\cos\frac{\gamma}{2}; \; d_1^2 + d_2^2 = 4l^2;$$

$$d_1 d_2 = 2l^2\sin\gamma; \; A = lh = l^2\sin\gamma = \frac{d_1 d_2}{2}.$$

TABLE 2.4   (*Continued*)

1a.   *Plane Rectilinear Figures*

5. General Trapezoid (and Isosceles Trapezoid)

Let mid-line bisecting non-parallel sides $= m$. Then $m = \dfrac{a + b}{2}$.

For General Trapezoid:

(Only one pair of opposite sides parallel)

$$p = a + b + c + d; \quad A = \frac{(a + b)h}{2} = mh.$$

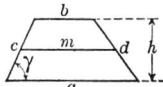

For Isosceles Trapezoid ($d = c$):

(Non-parallel sides equal)

$$A = \frac{(a + b)h}{2} = mh = \frac{(a + b)c \sin \gamma}{2}$$
$$= (a - c \cos \gamma)c \sin \gamma = (b + c \cos \gamma)c \sin \gamma.$$

6. General Quadrilateral (Trapezium)

(No sides parallel)

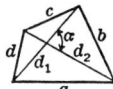

$$p = a + b + c + d$$

$A = \frac{1}{2} d_1 d_2 \sin \alpha = $ sum of areas of the two triangles formed by either diagonal and the four sides.

7. Quadrilateral Inscribed in Circle

(Sum of opposite angles $= 180°$)

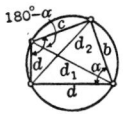

$ac + bd = d_1 d_2$.

Let $s = \frac{1}{2}(a + b + c + d) = \dfrac{p}{2}$ and $\alpha = $ angle between sides a and b.

$$A = \sqrt{(s - a)(s - b)(s - c)(s - d)} = \frac{1}{2}(ab + cd)\sin \alpha.$$

8. Regular Polygon (and General Polygon)

For Regular Polygon:

(Equal sides and equal angles)

Let $n = $ number of sides.

Central angle $= 2\alpha = \dfrac{2\pi}{n}$ radians;

Vertex angle $= \beta = \dfrac{(n - 2)}{n} \pi$ radians.

$p = ns; \quad s = 2r \tan \alpha = 2R \sin \alpha;$

$r = \dfrac{s}{2} \cot \alpha; \quad R = \dfrac{s}{2} \csc \alpha;$

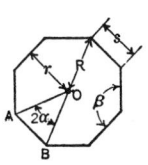

$$A = \frac{nsr}{2} = nr^2 \tan \alpha = \frac{nR^2}{2} \sin 2\alpha = \frac{ns^2}{4} \cot \alpha = \text{sum of areas of the}$$

n equal triangles such as OAB.

For General Polygon:

$A = $ sum of areas of constituent triangles into which it can be divided.

TABLE 2.4 *(Continued)*

1b. Plane Curvilinear Figures

Notation. Lines, $a, b, \ldots$; radius, r; diameter, d; perimeter, p; circumference, c; central angle n radians, θ; arc, s; chord of arc s, l; chord of half arc $s/2$, l'; rise, h; area, A.

9. Circle (and Circular Arc)   *For Circle:*

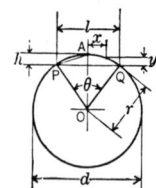

$$d = 2r; \quad c = 2\pi r = \pi d; \quad A = \pi r^2 = \frac{\pi d^2}{4} = \frac{c^2}{4\pi}.$$

For Circular Arc:

Let arc $PAQ = s$; and chord $PA = l'$. Then, $s = r\theta = \dfrac{d\theta}{2}$; $s = \dfrac{8l' - l}{3}$. (The latter equation is Huygen's approximate formula. For θ small; error is very small; for $\theta = 120°$, error is about 0.25%; for $\theta = 180°$, error is less than 1.25%.)

$$l = 2r \sin\frac{\theta}{2}; \quad l = 2\sqrt{2hr - h^2} \text{ (approximate formula)}$$

$$r = \frac{s}{\theta} = \frac{l}{2 \sin\dfrac{\theta}{2}}; \quad r = \frac{4h^2 + l^2}{8h} \text{ (approximate formula)}$$

$$h = r \mp \sqrt{r^2 - \frac{l^2}{4}} \ (- \text{ if } \theta \le 180°; \ + \text{ if } \theta \ge 180°) = r\left(1 - \cos\frac{\theta}{2}\right)$$

$$= r \, \text{versin}\frac{\theta}{2} = 2r \sin^2\frac{\theta}{4} = \frac{l}{2}\tan\frac{\theta}{4} = r + y - \sqrt{r^2 - x^2}.$$

Side ordinate $y = h - r + \sqrt{r^2 - x^2}$.

10. Circular Sector (and Semicircle)   *For Circular Sector:*

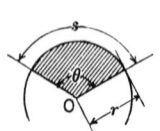

$$A = \frac{\theta r^2}{2} = \frac{sr}{2}.$$

For Semicircle:

$$A = \frac{\pi r^2}{2}.$$

11. Circular Segment

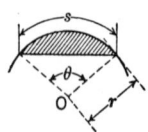

$$A = \frac{r^2}{2}(\theta - \sin\theta)$$

$$= \tfrac{1}{2}[sr \mp l(r - h)] \ (- \text{ if } h \le r; \ + \text{ if } h \ge r).$$

$$A = \frac{2lh}{3} \text{ or } \frac{h}{15}(8l' + 6l). \text{ (Approximate formulas. For } h \text{ small com-}$$

pared with r, error is very small; for $h = \dfrac{r}{4}$, first formula errs about 3.5% and second less than 1.0%.)

12. Annulus   (Region between two concentric circles)

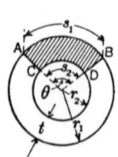

$$A = \pi(r_1^2 - r_2^2) = \pi(r_1 + r_2)(r_1 - r_2);$$

$$A \text{ of sector } ABCD = \frac{\theta}{2}(r_1^2 - r_2^2) = \frac{\theta}{2}(r_1 + r_2)(r_1 - r_2)$$

$$= \frac{t}{2}(s_1 + s_2).$$

TABLE 2.4 (*Continued*)

1b. Plane Curvilinear Figures (Continued)

---

13. Ellipse

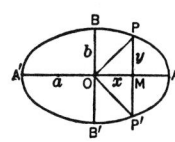

$$p = \pi(a + b)\left(1 + \frac{R^2}{4} + \frac{R^4}{64} + \frac{R^4}{256} + \cdots\right) \text{ where } R = \frac{a - b}{a + b}.$$

$$p = \pi(a + b)\frac{64 - 3R^4}{64 - 16R^2} \text{ (approximate formula).}$$

$$A = \pi ab; \quad A \text{ of quadrant } AOB = \frac{\pi ab}{4};$$

$$A \text{ of sector } AOP = \frac{ab}{2}\cos^{-1}\frac{x}{a}; \quad A \text{ of sector } POB = \frac{ab}{2}\sin^{-1}\frac{x}{a};$$

$$A \text{ of section } BPP'B' = xy + ab\sin^{-1}\frac{x}{a};$$

$$A \text{ of segment } PAP'P = -xy + ab\cos^{-1}\frac{x}{a}.$$

For additional formulas, see analytic geometry.

14. Parabola

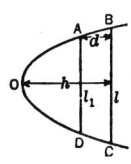

$$\text{Arc } BOC = s = \tfrac{1}{2}\sqrt{l^2 + 16h^2} + \frac{l^2}{8h}\log_e\frac{4h + \sqrt{l^2 + 16h^2}}{l}.$$

Let $R = \dfrac{h}{l}$. Then,

$$s = l\left(1 + \frac{8R^2}{3} - \frac{32R^4}{5} + \cdots\right) \text{ (approximate formula).}$$

$$d = \frac{h}{l^2}(l^2 - l_1^2); \quad l_1 = l\sqrt{\frac{h - d}{h}}; \quad h = \frac{dl^2}{l^2 - l_1^2};$$

$$A \text{ of segment } BOC = \frac{2hl}{3};$$

$$A \text{ of section } ABCD = \frac{2}{3}d\left(\frac{l^3 - l_1^3}{l^2 - l_1^2}\right).$$

For additional formulas, see analytic geometry.

15. Hyperbola

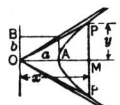

$$A \text{ of figure } OPAP'O = ab\log_e\left(\frac{x}{a} + \frac{y}{b}\right) = ab\cosh^{-1}\frac{x}{a};$$

$$A \text{ of segment } PAP' = xy - ab\log_e\left(\frac{x}{a} + \frac{y}{b}\right) = xy - ab\cosh^{-1}\frac{x}{a}.$$

For additional formulas, see analytic geometry.

16. Cycloid

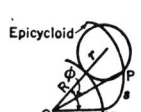

$$\text{Arc } OP = s = 4r\left(1 - \cos\frac{\phi}{2}\right); \quad \text{arc } OMN = 8r;$$

$$A \text{ under curve } OMN = 3\pi r^2.$$

For additional formulas, see analytic geometry.

17. Epicycloid

$$\text{Arc } MP = s = \frac{4r}{R}(R + r)\left(1 - \cos\frac{R\phi}{2r}\right);$$

$$\text{Area } MOP = A = \frac{r}{2R}(R + r)(R + 2r)\left(\frac{R\phi}{r} - \sin\frac{R\phi}{r}\right).$$

For additional formulas, see analytic geometry.

TABLE 2.4   (*Continued*)

18. Hypocycloid

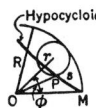

$$\text{Arc } MP = s = \frac{4r}{R}(R - r)\left(1 - \cos\frac{R\phi}{2r}\right);$$

$$\text{Area } MOP = A = \frac{r}{2R}(R - r)(R - 2r)\left(\frac{R\phi}{r} - \sin\frac{R\phi}{r}\right).$$

For additional formulas, see analytic geometry.

19. Catenary

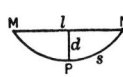

If d is small compared with l:

$$\text{Arc } MPN = s = l\left[1 + \frac{2}{3}\left(\frac{2d}{l}\right)^2\right] \text{ (approximately).}$$

For additional formulas, see analytic geometry.

20. Helix (a skew curve)

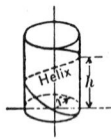

Let length of helix $= s$; radius of coil ($=$ radius of cylinder in figure) $= r$; distance advanced in one revolution $=$ pitch $= h$; and number of revolutions $= n$. Then,

$$s = n\sqrt{(2\pi r)^2 + h^2}\,.$$

21. Spiral of Archimedes

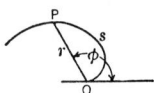

Let $a = \dfrac{r}{\phi}$. Then,

$$\text{Arc } OP = s = \frac{a}{2}\left[\phi\sqrt{1 + \phi^2} + \log_e\left(\phi + \sqrt{1 + \phi^2}\right)\right].$$

For additional formulas, see analytic geometry.

22. Irregular Figure

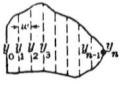

Divide the figure into an *even* number, n, of strips by means of $(n + 1)$ ordinates, y_i, spaced equal distances, w. The area can then be determined approximately by any of the following formulas, which are presented in the order of usual increasing approach to accuracy. In any of the first three cases, the greater the number of strips used, the more nearly accurate will be the result.

(Approximate Formulas)

Trapezoidal Rule

$$A = w\left[\frac{y_0 + y_n}{2} + y_1 + y_2 + \cdots + y_{n-1}\right];$$

Durand's Rule

$$A = w[0.4(y_0 + y_n) + 1.1(y_1 + y_{n-1}) + y_2 + y_3 + \cdots + y_{n-2}];$$

Simpson's Rule
(*n must* be even)

$$A = \frac{w}{3}[(y_0 + y_n) + 4(y_1 + y_3 + \cdots + y_{n-1}) +$$
$$2(y_2 + y_4 + \cdots + y_{n-2})];$$

Weddle's Rule
(for 6 strips only)

$$A = \frac{3w}{10}[5(y_1 + y_5) + 6y_3 + y_0 + y_2 + y_4 + y_6].$$

Areas of irregular regions can often be determined more quickly by such methods as plotting on squared paper and counting the squares; graphical coordinate representation (see analytic geometry); or use of a planimeter.

TABLE 2.4 (*Continued*)

1c. Solids Having Plane Surfaces

Notation. Lines, $a, b, c, \ldots$; altitude (perpendicular height), h; slant height, s; perimeter of base, p_b or p_B; perimeter of a right section, p_r; area of base, A_b or A_B; area of a right section, A_r; total area of lateral surfaces, A_l; total area of all surfaces, A_t; volume, V.

23. Wedge (and Right Triangular Prism) 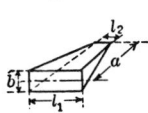	*For Wedge*: (Narrow-side rectangular); $V = \dfrac{ab}{6}(2l_1 + l_2)$. *For Right Triangular Prism* (or wedge having parallel triangular bases perpendicular to sides)$l_2 = l_1 = l$: $$V = \frac{abl}{2}.$$
24. Rectangular Prism (or Rectangular Parallelepiped) (and Cube) 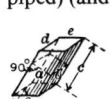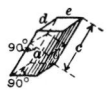	*For Rectangular Prism or Rectangular Parallelepiped*: $A_l = 2c(a + b)$; $A_t = 2(de + ac + bc)$; $V = A_r c = abc$. *For Cube* (letting $b = c = a$): $$A_t = 6a^2; \quad V = a^3; \quad \text{Diagonal} = a\sqrt{3}.$$
25. General Prism 	$A_l = hp_b = sp_r = s(a + b + \cdots + n)$; $V = hA_b = sA_r$.
26. General Truncated Prism (and Truncated Triangular Prism) 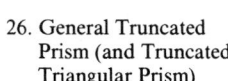	*For General Truncated Prism*: $V = A_r \cdot$ (length of line BC joining centers of gravity of bases). *For Truncated Triangular Prism*: $$V = \frac{A_r}{3}(a + b + c).$$
27. Prismatoid 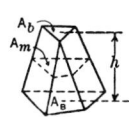	Let area of mid-section $= A_m$. $$V = \frac{h}{6}(A_B + A_b + 4A_m).$$
28. Right Regular Pyramid (and Frustum of Right Regular Pyramid) 	*For Right Regular Pyramid*: $$A_l = \frac{sp_B}{2}; \quad V = \frac{hA_B}{3}.$$ *For Frustum of Right Regular Pyramid*: $$A_l = \frac{s}{2}(p_B + p_b); \quad V = \frac{h}{3}(A_B + A_b + \sqrt{A_B A_b}).$$
29. General Pyramid (and Frustum of Pyramid)	*For General Pyramid*: $$V = \frac{hA_B}{3}.$$ *For Frustum of General Pyramid*: $$V = \frac{h}{3}(A_B + A_b + \sqrt{A_B A_b}).$$

TABLE 2.4   (*Continued*)

30. Regular Polyhedrons

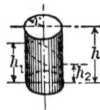

Tetrahedron   Cube   Octahedron

Dodecahedron   Icosahedron

Let edge $= a$, and radius of inscribed sphere $= r$. Then,

$$r = \frac{3V}{A_t}, \text{ and:}$$

Number of Faces	Form of Faces	Total Area A_t	Volume V
4	Equilateral triangle	$1.7321a^2$	$0.1179a^3$
6	Square	$6.0000a^2$	$1.0000a^3$
8	Equilateral triangle	$3.4641a^2$	$0.4714a^3$
12	Regular pentagon	$20.6457a^2$	$7.6631a^3$
20	Equilateral triangle	$8.6603a^2$	$2.1817a^3$

(Factors shown only to four decimal places.)

1d.   Solids Having Curved Surfaces

Notation. Lines, $a, b, c, \ldots$; altitude (perpendicular height), $h, h_1, \ldots$; slant height, s; radius, r; perimeter of base, p_b; perimeter of a right section, p_r; angle in radians, ϕ; arc, s; chord of segment, l; rise, h; area of base, A_b or A_B; area of a right section, A_r; total area of convex surface, A_l; total area of all surfaces, A_t; volume, V.

31. Right Circular Cylinder (and Truncated Right Circular Cylinder)

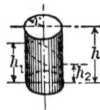

For Right Circular Cylinder:
$A_l = 2\pi rh$; $A_t = 2\pi r(r + h)$;
$V = \pi r^2 h$.

For Truncated Right Circular Cylinder:

$$A_l = \pi r(h_1 + h_2); \quad A_t = \pi r\left[h_1 + h_2 + r + \sqrt{r^2 + \left(\frac{h_1 - h_2}{2}\right)^2}\right];$$

$$V = \frac{\pi r^2}{2}(h_1 + h_2).$$

32. Ungula (Wedge) of Right Circular Cylinder

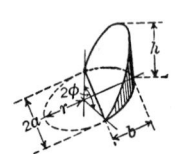

$$A_l = \frac{2rh}{b}[a + (b - r)\phi];$$

$$V = \frac{h}{3b}[a(3r^2 - a^2) + 3r^2(b - r)\phi]$$

$$= \frac{hr^3}{b}\left[\sin\phi - \frac{\sin^3\phi}{3} - \phi\cos\phi\right].$$

For Semicircular Base (letting $a = b = r$):

$$A_l = 2rh; \quad V = \frac{2r^2h}{3}.$$

33. General Cylinder

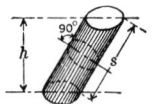

$A_l = p_b h = p_r s$;
$V = A_b h = A_r s$.

34. Right Circular Cone (and Frustum of Right Circular Cone)

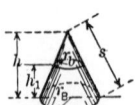

For Right Circular Cone:

$$A_l = \pi r_B s = \pi r_B\sqrt{r_B^2 + h^2}; \quad A_t = \pi r_B(r_B + s);$$

$$V = \frac{\pi r_B^2 h}{3}.$$

TABLE 2.4 *(Continued)*

For Frustum of Right Circular Cone:

$$s = \sqrt{h_1^2 + (r_B - r_b)^2}\,; \quad A_l = \pi s(r_B + r_b);$$

$$V = \frac{\pi h_1}{3}(r_B^2 + r_b^2 + r_B r_b).$$

35. General Cone (and Frustum of General Cone)

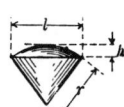

For General Cone:

$$V = \frac{A_B h}{3}.$$

For Frustum of General Cone:

$$V = \frac{h_1}{3}(A_B + A_b + \sqrt{A_B A_b}).$$

36. Sphere

Let diameter = d.

$$A_t = 4\pi r^2 = \pi d^2;$$

$$V = \frac{4\pi r^3}{3} = \frac{\pi d^3}{6}.$$

37. Spherical Sector (and Hemisphere)

For Spherical Sector:

$$A_t = \frac{\pi r}{2}(4h + l); \quad V = \frac{2\pi r^2 h}{3}.$$

For Hemisphere $\left(\text{letting } h = \dfrac{l}{2} = r\right)$:

$$A_t = 3\pi r^2; \quad V = \frac{2\pi r^3}{3}.$$

38. Spherical Zone (and Spherical Segment)

For Spherical Zone Bounded by Two Planes:

$$A_l = 2\pi rh; \quad A_t = \frac{\pi}{4}(8rh + a^2 + b^2).$$

For Spherical Zone Bounded by One Plane $(b = 0)$:

$$A_l = 2\pi rh = \frac{\pi}{4}(4h^2 + a^2);$$

$$A_t = \frac{\pi}{4}(8rh + a^2) = \frac{\pi}{2}(2h^2 + a^2).$$

For Spherical Segment with Two Bases:

$$V = \frac{\pi h}{24}(3a^2 + 3b^2 + 4h^2).$$

For Spherical Segment with One Base $(b = 0)$:

$$V = \frac{\pi h}{24}(3a^2 + 4h^2) = \pi h^2\left(r - \frac{h}{3}\right).$$

39. Spherical Polygon (and Spherical Triangle)

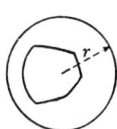

For Spherical Polygon:

Let sum of angles in radians = θ and number of sides = n.

$$A = [\theta - (n - 2)\pi]r^2$$

(The quantity $[\theta - (n - 2)\pi]$ is called "spherical excess.")

For Spherical Triangle $(n = 3)$:

$$A = (\theta - \pi)r^2$$

For additional formuᴠ as, see trigonometry.

TABLE 2.4   (*Continued*)

40. Torus

$A_l = 4\pi^2 Rr$;

$V = 2\pi^2 Rr^2$.

41. Ellipsoid (and Spheroids)

For Ellipsoid:

$$V = \frac{4}{3}\pi abc.$$

For Prolate Spheroid:

Let $c = b$ and $\dfrac{\sqrt{a^2 - b^2}}{a} = e$.

$A_l = 2\pi b^2 + 2\pi ab\dfrac{\sin^{-1}e}{e}$; $V = \dfrac{4}{3}\pi ab^2$.

For Oblate Spheroid:

Let $c = a$ and $\dfrac{\sqrt{a^2 - b^2}}{a} = e$.

$A_l = 2\pi a^2 + \dfrac{\pi b^2}{e}\ln\left(\dfrac{1 + e}{1 - e}\right)$; $V = \dfrac{4}{3}\pi a^2 b$.

42. Paraboloid of Revolution

A_l of segment $DOC = \dfrac{2\pi l}{3h^2}\left[\left(\dfrac{l^2}{16} + h^2\right)^{3/2} - \left(\dfrac{l}{4}\right)^3\right]$.

For Paraboloidal Segment with Two Bases:

$$V \text{ of } ABCD = \frac{\pi d}{8}(l^2 + l_1^2).$$

For Paraboloidal Segment with One Base ($l_1 = 0$ and $d = h$):

$$V \text{ of } DOC = \frac{\pi h l^2}{8}.$$

43. Hyperboloid of Revolution

V of segment $AOB = \dfrac{\pi h}{24}(l^2 + 4l_1^2)$.

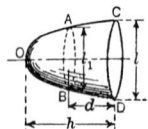

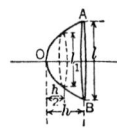

44. Surface and Solid of Revolution

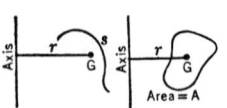

Let perpendicular distance from axis to center of gravity (G) of curve (or surface) = r. Curve (or surface) must not cross axis. Then,

Area of Surface generated by curve revolving about axis:

$A_l = 2\pi rs$.

Volume of Solid generated by surface revolving about axis:

$V = 2\pi rA$.

TABLE 2.4 (*Continued*)

45. Irregular Solid 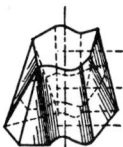	One of the following methods can often be employed to determine the volume of an irregular solid with a reasonable approach to accuracy: (*a*) Divide the solid into prisms, cylinders, etc., and sum their individual volumes. (*b*) Divide one surface into triangles, after replacing curved lines by straight ones and curved surfaces by plane ones. Then multiply the area of each triangle by the mean depth of the section beneath it (which generally approximates the average of the depths at its corners). Sum the volumes thus obtained. (*c*) If two surfaces are parallel, replace any curved lateral surfaces by plane surfaces best suited to the contour and then employ the prismatoidal formula.

2.5.3 Constructions

Lines

1. **To draw a line parallel to a given line.**

CASE 1: At a given distance from the given line (Fig. 2.21). With the given distance as radius and with any centers m and n on the given line AB, describe arcs xy and zw, respectively. Draw CD touching these arcs. CD is the required parallel line.

CASE 2: Through a given point (Fig. 2.22). Let C be the given point and D be any point on the given line AB. Draw CD. With equal radii draw arcs bf and ce with D and C, respectively, as centers. With radius equal to chord bf and with c as center draw an arc cutting arc ce at E. CE is the required parallel line.

2. **To bisect a given line** (Fig. 2.23). Let AB be the given line. With any radius greater than 0.5 AB describe two arcs with A and B as centers. The line CD, through points of intersection of the arcs, is the perpendicular bisector of the given line.

3. **To divide a given line into a given number of equal parts** (Fig. 2.24). Let AB be the given line and let the number of equal parts be five. Draw line AC at any convenient angle with AB, and step

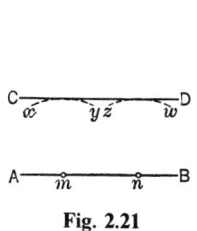

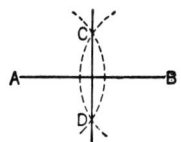

Fig. 2.21

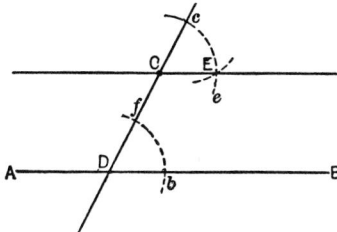

Fig. 2.22

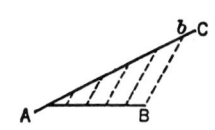

Fig. 2.23

Fig. 2.24

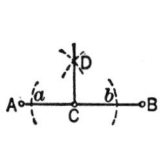

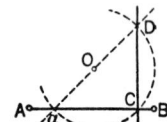

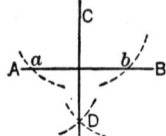

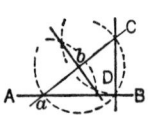

Fig. 2.25   **Fig. 2.26**   **Fig. 2.27**   **Fig. 2.28**

off with dividers five equal lengths from A to b. Connect b with B, and draw parallels to Bb through the other points in AC. The intersections of these parallels with AB determine the required equal parts on the given line.

4. **To divide a given line into segments proportional to a number of given unequal parts.** Follow the same procedure as under item 3 except make the lengths on AC equal to (or proportional to) the lengths of the given unequal parts.

5. **To erect a perpendicular to a given line at a given point in the line.**

CASE 1: Point C is at or near the middle of the line AB (Fig. 2.25). With C as center, describe arcs of equal radii intersecting AB at a and b. With a and b as centers, and any radius greater than Ca, describe arcs intersecting at D. CD is the required perpendicular.

CASE 2: Point C is at or near the extremity of the line AB (Fig. 2.26). With any point O, as center, and radius OC, describe an arc intersecting AB at a. Extend aO to intersect the arc at D. CD is the required perpendicular.

6. **To erect a perpendicular to a given line through a given point outside the line.**

CASE 1: Point C is opposite, or nearly opposite, the middle of the line AB (Fig. 2.27). With C as center, describe an arc intersecting AB at a and b. With a and b as centers, describe arcs of equal radii intersecting at D. CD is the required perpendicular.

CASE 2: Point C is opposite, or nearly opposite, the extremity of the line AB (Fig. 2.28). Through C, draw any line intersecting AB at a. Divide line Ca into two equal parts, ab and bC (method given previously). With b as center, and radius bC, describe an arc intersecting AB at D. CD is the required perpendicular.

Angles

7. **To bisect a given angle.**

CASE 1: Vertex B is accessible (Fig. 2.29). Let ABC be the given angle. With B as center, and a large radius, describe an arc intersecting AB and BC at a and c, respectively. With a and c as centers, describe arcs of equal radii intersecting at D. DB is the required bisector.

CASE 2: The vertex is inaccessible (Fig. 2.30). Let the given angle be that between lines AB and BC. Draw lines ab and bc parallel to the given lines, and at equal distances from them, intersecting at b. Construct Db bisecting angle abc (method given previously). Db is the required bisector.

8. **To construct an angle equal to a given angle if one new side and the new vertex are given** (Fig. 2.31). Let ABC be the given angle; DE the new side; and E the new vertex. With center B and a convenient radius, describe arc ac. With the same radius and center E, draw arc df. With radius equal to chord ac and with center d draw an arc cutting the arc df at F. Draw EF. Then DEF is the required angle.

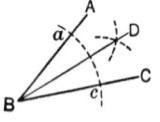

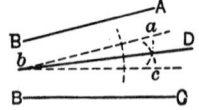

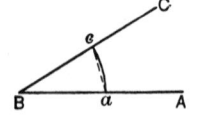

Fig. 2.29   **Fig. 2.30**   **Fig. 2.31**

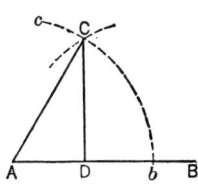

Fig. 2.32

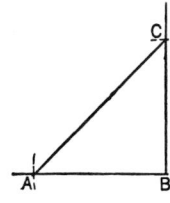

Fig. 2.33

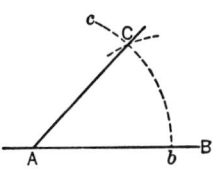

Fig. 2.34

9. **To construct angles of 60° and 30°** (Fig. 2.32). About any point A on a line AB, describe with a convenient radius the arc bc. From b, using an equal radius, describe an arc cutting the former one at C. Draw AC, and drop a perpendicular CD from C to line AB. Then CAD is a 60° angle and ACD is a 30° angle.

10. **To construct an angle of 45°** (Fig. 2.33). Set off any distance AB; draw BC perpendicular and equal to AB; and join CA. Angles CAB and ACB are each 45°.

11. **To draw a line making a given angle with a given line** (Fig. 2.34). Let AB be the given line. With A as the center and with as large a radius as convenient, describe arc bc. Determine from Table 1.12, the length of chord to radius one, corresponding to the given angle. Multiply this chord by the length of Ab, and with the product as a new radius and b as a center, describe an arc cutting bc at C. Draw AC. This line makes the required angle with AB.

Circles

12. **To describe through two given points an arc of a circle having a given radius** (Fig. 2.35). Let A and B be the given points. With the given radius, and these points as centers, describe arcs cutting each other at C. From C, with the same radius, describe arc AB, which is the required arc.

13. **To bisect a given arc of a circle.** Draw the perpendicular bisector of the chord of the arc. The point in which this bisector meets the arc is the required midpoint.

14. **To locate the center of a given circle or circular arc** (Fig. 2.36). Select three points, A, B, C, on the circle (or arc), located well apart. Draw chords AB and BC and erect their perpendicular bisectors. The point O, where the bisectors intersect, is the required center.

15. **To draw a circle through three given points not in the same straight line.**

CASE 1:   Radius small and center accessible (Fig. 2.36). Let A, B, C, be the given points. Draw lines AB and BC and erect their perpendicular bisectors. From point O, where the bisectors intersect, describe a circle of radius OA that is the required circle.

CASE 2:   Radius very long or center inaccessible (Fig. 2.37). Let A, O, A', be the given points (O not necessarily midpoint of AOA'). Draw arcs Aa' and $A'a$ with centers at A' and A, respectively; extend AO to determine a and $A'O$ to determine a'; point off from a on aA' equal parts ab, bc, and so on; lay off $a'b'$, $b'c'$, and so on, equal to ab; join A with any point as b and A' with the corresponding point b'; the intersection P of these joining lines is a point on the required circle.

16. **To lay out a circular arc without locating the center of the circle, having given the chord and the rise** (Fig. 2.37). Let AA' be the chord and QO the rise. (In this case, O is midpoint of AOA'.) The arc can be constructed through the points A, O, A', as under item 15, Case 2.

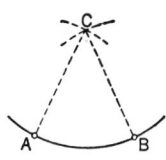

Fig. 2.35

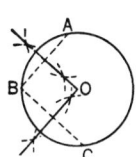

Fig. 2.36

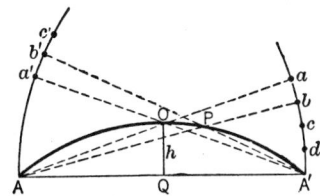

Fig. 2.37

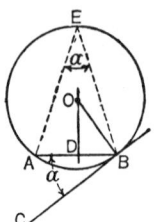

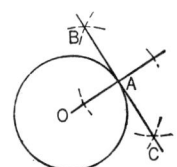
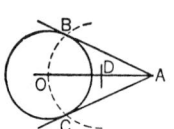
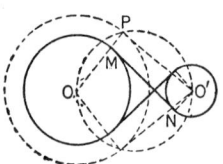

| Fig. 2.38 | Fig. 2.39 | Fig. 2.40 | Fig. 2.41 |

17. **To construct, upon a given chord, a circle in which a given angle can be inscribed** (Fig. 2.38). Let AB be the given chord, and α the given angle. Construct angle ABC equal to angle α. Bisect line AB by the perpendicular at D. Draw a perpendicular to BC from point B. With O, the point of intersection of the perpendiculars, as center, and OB as radius, describe a circle. The angle AEB, with vertex E located anywhere on the arc AEB, equals α, and therefore the circle just drawn is the one required.

18. **To draw a tangent to a given circle through a given point.**

CASE 1:   Point A is on the circle (Fig. 2.39). Draw radius OA. Through A, perpendicular to OA, draw BAC, the required tangent.

CASE 2:   Point A is outside the circle (Fig. 2.40). Two tangents can be drawn. Join O and A. Bisect OA at D, and with D as center and DO as radius, describe an arc intersecting the given circle at B and C. BA and CA are the required tangents.

19. **To draw a common tangent to two given circles.** Let the circles have centers O and O' and corresponding radii r and r' $(r > r')$.

CASE 1:   Common internal tangents (when circles do not intersect) (Fig. 2.41). Construct a circle having the same center O as the larger circle and a radius equal to the sum of the radii of the given circles $(r + r')$. Construct a tangent $O'P$ from center O' of the smaller circle to this circle. Construct $O'N$ perpendicular to this tangent. Draw OP. The line MN joining the extremities of the radii OM and $O'N$ is a common tangent. The figure shows two such common internal tangents.

CASE 2:   Common external tangents (Fig. 2.42). Construct a circle having the same center O as the larger circle and radius equal to the difference of the radii $(r - r')$. Construct a tangent to this circle from the center of the smaller circle. The line joining the extremities M, N, of the radii of the given circles perpendicular to this tangent is a required common tangent. There are two such tangents.

20. **To draw a circle with a given radius that will be tangent to two given circles** (Fig. 2.43). Let r be the given radius and A and B the given circles. About center of circle A with radius equal to r plus radius of A, and about center of B with radius equal to r plus radius of B, draw two arcs cutting each other in C, which is the center of the required circle.

21. **To describe a circular arc touching two given circles, one of them at a given point** (Fig. 2.44). Let AB, FG be the given circles and F the given point. Draw the radius EF, and produce it both ways. Set off FH equal to the radius AC of the other circle; join CH, and bisect it by the perpendicular LT, cutting EF at T. About center T, with radius TF, describe arc FA as required.

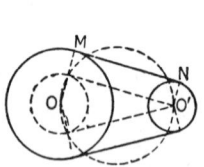
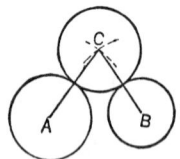
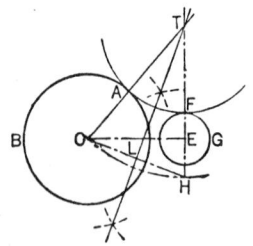

| Fig. 2.42 | Fig. 2.43 | Fig. 2.44 |

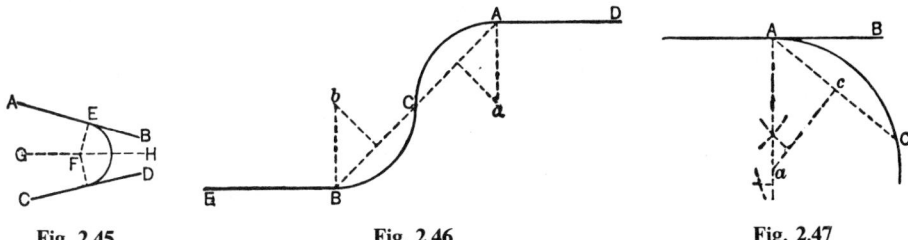

Fig. 2.45   Fig. 2.46   Fig. 2.47

22. **To draw a circular arc that will be tangent to two given lines inclined to one another, one tangential point being given** (Fig. 2.45). Let AB and CD be the given lines and E the given point. Draw the line GH, bisecting the angle formed by AB and CD. From E draw EF at right angles to AB; then F, its intersection with GH, is the center of the required circular arc.

23. **To connect two given parallel lines by a reversed curve composed of two circular arcs of equal radius, the curve being tangent to the lines at given points** (Fig. 2.46). Let AD and BE be the given lines and A and B the given points. Join A and B, and bisect the connecting line at C. Bisect CA and CB by perpendiculars. At A and B erect perpendiculars to the given lines, and the intersections a and b are the centers of the arcs composing the required curve.

24. **To describe a circular arc that will be tangent to a given line at a given point, and pass through another given point outside the line** (Fig. 2.47). Let AB be the given line, A the given point on the line, and C the given point outside it. Draw from A a line perpendicular to the given line. Connect A and C by a straight line, and bisect this line by the perpendicular ca. The point a where these two perpendiculars intersect is the center of the required circular arc.

25. **To draw a circular arc joining two given relatively inclined lines, tangent to the lines, and passing through a given point on the line bisecting their included angle** (Fig. 2.48). Let AB and DE be the given lines and F the given point on the line FC that bisects their included angle. Through F draw DA at right angles to FC; bisect the angles A and D by lines intersecting at C, and about C as a center, with radius CF, draw the arc HFG required.

26. **To draw a series of circles between two given relatively inclined lines, touching the lines, and touching each other** (Fig. 2.49). Let AB and CD be the given lines. Bisect their included angle by the line NO. From a point P in this line draw the perpendicular PB to the line AB, and on P describe the circle BD, touching the given lines and cutting the center line at E. From E draw EF perpendicular to the center line, cutting AB at F; and about F as a center describe an arc EG, cutting AB at G. Draw GH parallel to BP, giving H, the center of the next circle, to be described with the radius HE; and so on for the next circle IN.

27. **To circumscribe a circle about a given triangle** (Fig. 2.50). Construct perpendicular bisectors of two sides. Their point of intersection O is the center (*circumcenter*) of the required circle.

28. **To inscribe a circle in a given triangle** (Fig. 2.51). Draw bisectors of two angles, intersecting in O (*incenter*). From O draw OD perpendicular to BC. Then the circle with center O and radius OD is the required circle.

29. **To circumscribe a circle about a given square** (Fig. 2.52). Let $ACBD$ be the given square. Draw diagonals AB and CD of the square, intersecting at E. On center E, with radius AE, describe the required circle. The same procedure can be used for circumscribing a circle about a given rectangle.

30. **To inscribe a circle in a given square** (Fig. 2.53). Let $ACBD$ be the given square. Draw diagonals AB and CD of the square, intersecting at E. Drop a perpendicular EF from E to one side. On center E, with radius EF, describe the required circle.

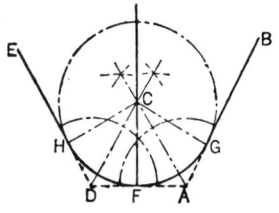

Fig. 2.48

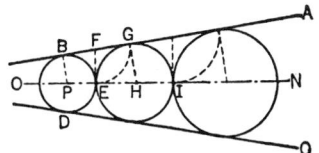

Fig. 2.49

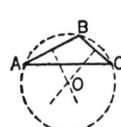

Fig. 2.50

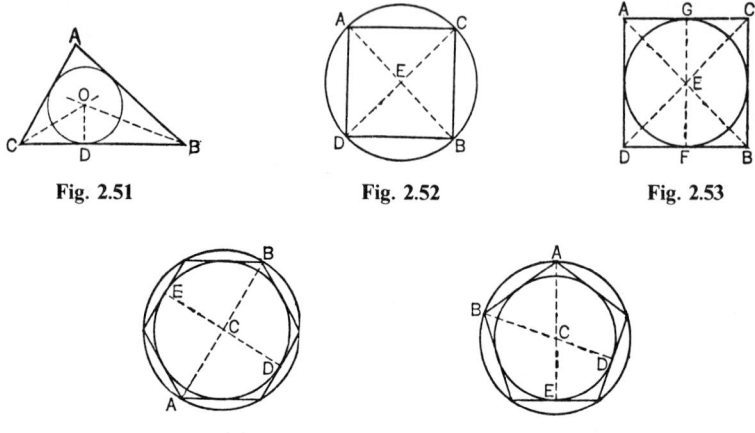

Fig. 2.51   Fig. 2.52   Fig. 2.53

Fig. 2.54   Fig. 2.55

31. **To circumscribe a circle about a given regular polygon.**

CASE 1:   The polygon has an even number of sides (Fig. 2.54). Draw a diagonal AB joining two opposite vertices. Bisect the diagonal by a perpendicular line DE, which is another diagonal or a line bisecting two opposite sides, depending on whether the number of sides is, or is not, divisible by 4. With the midpoint C as the center, and radius CA, describe the required circle.

CASE 2:   The polygon has an odd number of sides (Fig. 2.55). Bisect two of the sides at D and E by the perpendicular lines DB and EA, which pass through the respective opposite vertices and intersect at a point C. With C as the center, and radius CA, describe the required circle.

32. **To inscribe a circle in a given regular polygon** (Figs. 2.54 and 2.55). Locate the center, C, as in item 31. With C as center, and radius CD, describe the required circle.

Polygons

33. **To construct a triangle on a given base, the lengths of the sides being given** (Fig. 2.56). Let AB be the given base and a, b, the given lengths of sides. With A and B as centers, and b and a as respective radii, describe arcs intersecting at C. Draw AC and BC to complete the required triangle.

34. **To construct a rectangle of given base and given height** (Fig. 2.57). Let AB be the base and c the height. Erect the perpendicular AC equal to c. With C and B as centers, and AB and c as respective radii, describe arcs intersecting at D. Draw BD and CD to complete the required rectangle.

35. **To construct a square with a given diagonal** (Fig. 2.58). Let AC be the given diagonal. Draw a circle on AC as diameter and erect the diameter BD perpendicular to AC. Then $ABCD$ is the required square.

36. **To inscribe a square in a given circle** (Fig. 2.58). Draw perpendicular diameters AC and BD. Their extremities are the vertices of an inscribed square.

37. **To circumscribe a square about a given circle** (Fig. 2.59). Draw perpendicular diameters AC and BD. With A, B, C, D, as centers, and the radius of the circle as radius, describe the four semicircular arcs shown. Their outer intersections are the vertices of the required square.

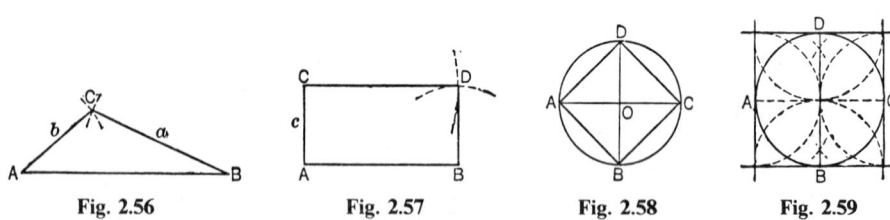

Fig. 2.56   Fig. 2.57   Fig. 2.58   Fig. 2.59

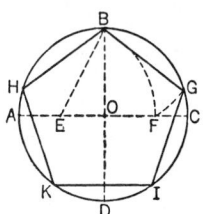

Fig. 2.60

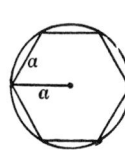

Fig. 2.61

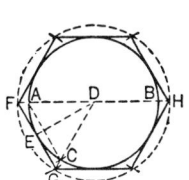

Fig. 2.62

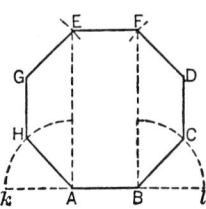

Fig. 2.63

38. **To inscribe a regular pentagon in a given circle** (Fig. 2.60). Draw perpendicular diameters AC and BD intersecting at O. Bisect AO at E and with E as center, and EB as radius, draw an arc cutting AC at F. With B as center and BF as radius, draw an arc cutting the circle at G and H; also with the same radius, step around the circle to I and K. Join the points thus found to form the pentagon.

39. **To inscribe a regular hexagon in a given circle** (Fig. 2.61). Step around the circle with compasses set to the radius and join consecutive divisions thus marked off.

40. **To circumscribe a regular hexagon about a given circle** (Fig. 2.62). Draw a diameter ADB and with center A and radius AD, describe an arc cutting the circle at C. Draw AC and bisect it with the radius DE. Through E, draw FG parallel to AC, cutting diameter AB extended at F. With center D and radius DF, describe the circumscribing circle FH; and within this circle inscribe a regular hexagon as under item 39. This hexagon circumscribes the given circle, as required.

41. **To construct a regular hexagon having a side of given length** (Fig. 2.61). Draw a circle with radius equal to the given length of side and inscribe a regular hexagon (see item 39).

42. **To construct a regular octagon having a side of given length** (Fig. 2.63). Let AB be the given side. Produce AB in both directions, and draw perpendiculars AE and BF. Bisect the external angles at A and B by the lines AH and BC, making them equal to AB. Draw CD and HG parallel to AE, and equal to AB; from the centers G, D, with the radius AB, draw arcs cutting the perpendiculars at E, F, and draw EF to complete the octagon.

43. **To inscribe a regular octagon in a given circle** (Fig. 2.64). Draw perpendicular diameters AC and BD. Bisect arcs AB, BC, etc., and join Ae, eB, and so on, to form the octagon.

44. **To inscribe a regular octagon in a given square** (Fig. 2.65). Draw diagonals of the given square, intersecting at O. With A, B, C, D, as centers, and AO as radius, describe arcs cutting the sides of the square at gn, fk, hm, and ol. Join the points thus found to form the octagon.

45. **To circumscribe a regular octagon about a given circle** (Fig. 2.66). Describe a square about the given circle. Draw perpendiculars ij, kl, and so on, to the diagonals of the squares, touching the circle. Then ij, jk, kl, and so on, form the octagon.

46. **To describe a regular polygon of any given number of sides when one side is given** (Fig. 2.67). Let AB be the given side and let the number of sides be five. Produce the line AB, and with A as center and AB as radius, describe a semicircle. Divide this into as many equal parts as there are to be sides of the polygon—in this case, five. Draw lines from A through the division points a, b, and c (omitting the last). With B and c as centers, and AB as radius, cut Aa at C and Ab at D. Draw cD, DC, and CB, to complete the polygon.

47. **To inscribe a regular polygon of a given number of sides in a given circle.** Determine the central angle subtended by any side by dividing 360° by the number of sides. Lay off this angle successively round the center of the circle by means of a protractor. The radii thus drawn intersect the circle at vertices of the required polygon.

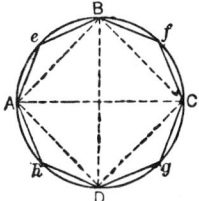

Fig. 2.64

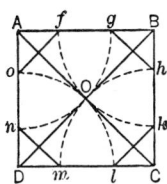

Fig. 2.65

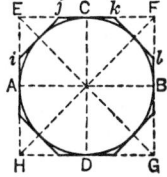

Fig. 2.66

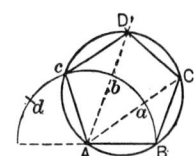

Fig. 2.67

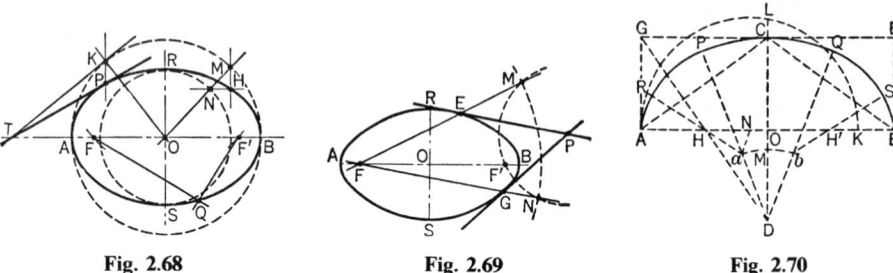

| Fig. 2.68 | Fig. 2.69 | Fig. 2.70 |

Ellipse

An **ellipse** is a curve for which the sum of the distances of any point on it from two fixed points (the *foci*) is constant.

48. **To describe an ellipse for which the axes are given** (Fig. 2.68). Let AB be the *major* and RS the *minor* axis ($AB > RS$). With O as center, and OB and OR as radii, describe circles. From O draw any radial line intersecting the circles at M and N. Through M draw a line parallel to OR, and through N a line parallel to OB. These lines intersect at H, a point on the ellipse. Repeat the construction to obtain other points.

49. **To locate the foci of an ellipse, having given the axes** (Fig. 2.68). With R as center, and radius equal to AO, describe arcs intersecting AB at F and F', the required foci.

50. **To describe an ellipse mechanically, having given an axis and the foci** (Fig. 2.68). A cord of length equal to the major axis is pinned or fixed at its ends to the foci F and F'. With a pencil inside the loop, keeping the cord taut so as to guide the pencil point, trace the outline of the ellipse (Q represents the pencil point and length FQF' the cord). If the minor axis RS is given rather than the major axis AB, the length AB (for the cord) is readily determined as $FR + RF'$.

51. **To draw a tangent to a given ellipse through a given point.**

CASE 1: Point P is on the curve (Fig. 2.68). With O as center, and OB as radius, describe a circle. Through P draw a line parallel to OR, intersecting the circle at K. Through K draw a tangent to the circle, intersecting the major axis at T. PT is the required tangent.

CASE 2: Point P is not on the curve (Fig. 2.69). With P as center, and radius PF', describe an arc. With F as center, and radius AB, describe an arc intersecting the first arc at M and N. Draw FM and FN, intersecting the ellipse at E and G. PE and PG are the required tangents.

52. **To describe an ellipse approximately by means of circular arcs of three radii** (Fig. 2.70). On the major axis AB draw the rectangle BG of altitude equal to half the minor axis, OC; to the diagonal AC draw the perpendicular GHD; set off OK equal to OC, and describe a semicircle on AK; produce OC to L; set off OM equal to CL, and from D describe an arc with radius DM; from A, with radius OL, draw an arc cutting AB at N; from H, with radius HN, draw an arc cutting arc ab at a. Thus the five centers H, a, D, b, H', are found, from which the arcs AR, RP, PQ, QS, SB, are described. The part of the ellipse below axis AB can be constructed in like manner.

Parabola

A **parabola** is a curve for which the distance of any point on it from a fixed line (the *directrix*) is equal to its distance from a fixed point (the *focus*). For a general discussion of its properties, see the section on analytic geometry.

53. **To describe a parabole for which the vertex, the axis, and a point of the curve are given** (Fig. 2.71). Let A be the given vertex, AB the given axis, and M the given point. Construct the rectangle $ABMC$. Divide MC and CA into the same number of equal parts (say four), numbering the divisions consecutively in the manner shown. Connect $A1$, $A2$, and $A3$. Through $1', 2', 3'$, draw parallels to the axis AB. The intersections I, II, and III, of these lines are points on the required curve. A similar construction below the axis will give the other symmetric branch of the curve.

54. **To locate the focus and directrix of a parabola, having given the vertex, the axis, and a point of the curve** (Fig. 2.71). Let A be the given vertex, AB the given axis, and M the given point. Drop the perpendicular MB from M to AB. Bisect it at E and draw AE. Draw ED perpendicular to AE at E and intersecting the axis at D. With A as center and BD as radius, describe arcs cutting the axis at F and J. Then F is the focus, and the line GH, perpendicular to the axis through J, is the directrix.

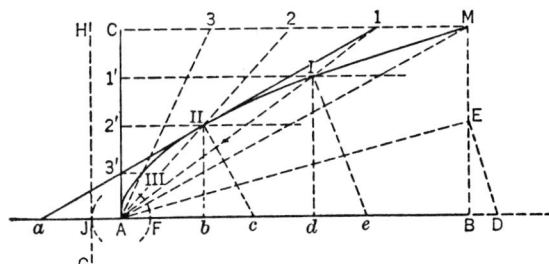

Fig. 2.71   Fig. 2.72

55. **To describe a parabola mechanically, having given the focus and directrix** (Fig. 2.72). Let F be the given focus and EN the given directrix. Place a straight edge to the directrix EN, and apply to it a square, LEG. Fasten to the end G one end of a cord equal in length to the edge EG, and attach the other end to the focus F; slide the square along the straight edge, holding the cord taut against the edge of the square by a pencil D, by which the parabolic curve is described.

56. **To draw a tangent to a given parabola through a given point.**

CASE 1:   The point is on the curve (Fig. 2.71). Let II be the given point. Drop a perpendicular from II to the axis, cutting it at b. Make Aa equal to Ab. Then a line through a and II is the required tangent. The line IIc perpendicular to the tangent at II is the *normal* at that point; bc is the *subnormal*. All subnormals of a given parabola are equal to the distance from the directrix to the focus and hence equal to each other. Thus the subnormal at I is de equal to bc, where d is the foot of the perpendicular dropped from I. The tangent at I can be drawn as a perpendicular to Ie through I.

CASE 2:   The point is off the curve (on the convex side) (Fig. 2.73). Let P be the given point and F the focus of the parabola. With P as center, and PF as radius, draw arcs intersecting the directrix at B and D. Through B and D draw lines parallel to the axis, intersecting the parabola at E and H. PE and PH are the required tangents.

Hyperbola
A **hyperbola** is a curve for which the difference of the distances of any point on it from two fixed points (the *foci*) is constant. It has two distinct branches.

57. **To describe a hyperbola for which the foci and the difference of the focal radii are given** (Fig. 2.74). Let F and F' be the given foci and AOB the given difference of the focal radii. Lay out AOB (the transverse axis) so that $AF = F'B$ and $AO = OB$. A and B are points on the required curve. With centers F and F', and any radius greater than FB or $F'A$, describe arcs aa. With the same centers, and radius equal to the difference between the first radius and the transverse axis AOB, describe arcs bb, intersecting arcs aa at P, Q, R, and S, points on the required curve. Repeat the construction for additional points.

Make $BC = BC' = OF = OF'$, and construct the rectangle $DEFG$; CC' is the conjugate axis. The diagonals DF and EG, produced, are called *asymptotes*. The hyperbola is tangent to its asymptotes at infinity.

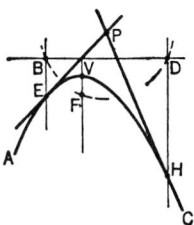

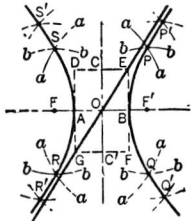

Fig. 2.73   Fig. 2.74

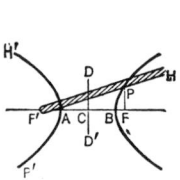

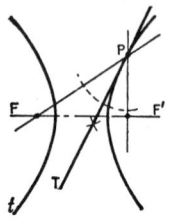

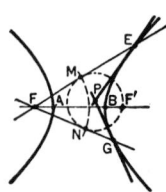

| Fig. 2.75 | Fig. 2.76 | Fig. 2.77 |

58. **To locate the foci of a hyperbola, having given the axes** (Fig. 2.74). With O as center and radius equal to BC, describe arcs intersecting AB extended, at F and F', the required foci.

59. **To describe a hyperbola mechanically, having given the foci and the difference of the focal radii** (Fig. 2.75). Let F and F' be the given foci and AB the given difference of focal radii. Using a ruler longer than the distance $F'F$, fasten one of its extremities at the focus F'. At the other extremity H attach a cord of such a length that the length of the ruler shall exceed the length of the cord by the given distance AB. Attach the other extremity of the cord at the focus F. Press a pencil P against the ruler, and keep the cord constantly taut while the ruler is turned around F' as a center. The point of the pencil will describe one branch of the curve, and the other can be obtained in like manner.

60. **To draw a tangent to a given hyperbola through a given point.**

CASE 1: Point P is on the curve (Fig. 2.76). Draw lines connecting P with the foci. Bisect the angle $F'PF$. The bisecting line TP is the required tangent.

CASE 2: Point P is off the curve on the convex side (Fig. 2.77). With P as center and radius PF', describe an arc. With F as center, and radius AB, describe an arc intersecting the first arc at M and N. Produce lines FM and FN to intersect the curve at E and G. PE and PG are the required tangents.

Cycloid
A **cycloid** is a curve generated by a point on a circle rolling on a straight line.

61. **To describe a cycloid for which the generating circle is given** (Fig. 2.78). Let A be the generating point. Divide the circumference of the generating circle into an even number of equal arcs, as $A1, 1-2, \ldots$, and set off the rectified arcs on the base. Through the points $1, 2, 3, \ldots$, on the circle, draw horizontal lines, and on them set off distances $1a = A1$, $2b = A2$, $3c = A3, \ldots$ The points $A, a, b, c, \ldots$, are points of the cycloid.

An **epicycloid** is a curve generated by a point on one circle rolling on the *outside* of another circle. A **hypocycloid** is a curve generated by the point if the generating circle rolls on the *inside* of the second circle.

Involute of a Circle
An **involute of a circle** is a curve generated by the free end of a taut string as it is unwound from a circle.

62. **To describe an involute of a given circle** (Fig. 2.79). Let AB be the given circle. Through B draw Bb perpendicular to AB. Make Bb equal in length to half the circumference of the circle. Divide

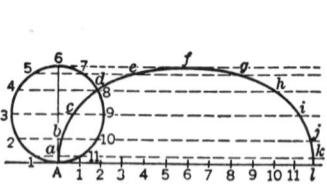

Fig. 2.78

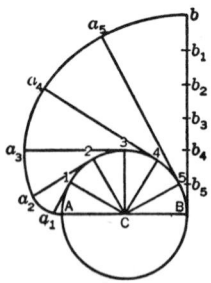

Fig. 2.79

Bb and the semicircumference into the same number of equal parts, say six. From each point of division $1, 2, 3, \ldots$, of the circumference, draw lines to the center C of the circle. Then draw $1a_1$ perpendicular to $C1$; $2a_2$ perpendicular to $C2$; and so on. Make $1a_1$ equal to bb_1; $2a_2$ equal to bb_2; $3a_3$ equal to bb_3; and so on. Join the points A, a_1, a_2, a_3, etc., by a curve; this curve is the required involute.

2.6 TRIGONOMETRY

2.6.1 Circular Functions of Plane Angles

Definitions and Values

Trigonometric Functions. The angle α in Fig. 2.80 is measured in degrees or radians, as defined in Section 2.5.1. The ratio of any two of the quantities x, y, or r determines the extent of the opening between the lines OP and OX. Since these ratios are functions of the angle, they may be used to measure or construct it. The definitions and terms used to designate the functions are as follows:

$$\text{Sine } \alpha = \frac{y}{r} = \sin \alpha$$

$$\text{Cosine } \alpha = \frac{x}{r} = \cos \alpha$$

$$\text{Tangent } \alpha = \frac{y}{x} = \tan \alpha$$

$$\text{Cotangent } \alpha = \frac{x}{y} = \cot \alpha$$

$$\text{Secant } \alpha = \frac{r}{x} = \sec \alpha$$

$$\text{Cosecant } \alpha = \frac{r}{y} = \csc \alpha$$

$$\text{Versine } \alpha = \frac{r - x}{r} = \text{vers } \alpha = 1 - \cos \alpha$$

$$\text{Coversine } \alpha = \frac{r - y}{r} = \text{covers } \alpha = 1 - \sin \alpha$$

$$\text{Haversine } \alpha = \frac{r - x}{2r} = \text{hav } \alpha = \frac{1}{2} \text{vers } \alpha$$

Positive and Negative Values. An angle α (Fig. 2.80), if measured in a *counterclockwise* direction, is said to be *positive*; if measured *clockwise*, *negative*. Following the convention that x is positive if measured along OX to the right of the OY axis and negative if measured to the left, and similarly, y is positive if measured along OY above the OX axis and negative if measured below, the signs of the trigonometric functions are different for angles in the quadrants I, II, III, and IV (Table 2.5).

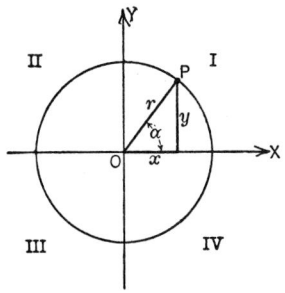

Fig. 2.80

TABLE 2.5　Signs of Trigonometric Functions

Quadrant	sin	cos	tan	cot	sec	csc
I	+	+	+	+	+	+
II	+	−	−	−	−	+
III	−	−	+	+	−	−
IV	−	+	−	−	+	−

TABLE 2.6　Functions of Angles in Any Quadrant in Terms of Angles in the First Quadrant

	$-\alpha$	$90° \pm \alpha$	$180° \pm \alpha$	$270° \pm \alpha$	$360° \pm \alpha$
sin	$-\sin\alpha$	$+\cos\alpha$	$\mp\sin\alpha$	$-\cos\alpha$	$\pm\sin\alpha$
cos	$+\cos\alpha$	$\mp\sin\alpha$	$-\cos\alpha$	$\pm\sin\alpha$	$+\cos\alpha$
tan	$-\tan\alpha$	$\mp\cot\alpha$	$\pm\tan\alpha$	$\mp\cot\alpha$	$\pm\tan\alpha$
cot	$-\cot\alpha$	$\mp\tan\alpha$	$\pm\cot\alpha$	$\mp\tan\alpha$	$\pm\cot\alpha$
sec	$+\sec\alpha$	$\mp\csc\alpha$	$-\sec\alpha$	$\pm\csc\alpha$	$+\sec\alpha$
csc	$-\csc\alpha$	$+\sec\alpha$	$\mp\csc\alpha$	$-\sec\alpha$	$\pm\csc\alpha$

TABLE 2.7　Functions of Certain Angles

	0°	30°	45°	60°	90°	180°	270°	360°
sin	0	$\frac{1}{2}$	$\frac{1}{2}\sqrt{2}$	$\frac{1}{2}\sqrt{3}$	1	0	−1	0
cos	1	$\frac{1}{2}\sqrt{3}$	$\frac{1}{2}\sqrt{2}$	$\frac{1}{2}$	0	−1	0	1
tan	0	$1/3\sqrt{3}$	1	$\sqrt{3}$	∞	0	∞	0
cot	∞	$\sqrt{3}$	1	$1/3\sqrt{3}$	0	∞	0	∞
sec	1	$2/3\sqrt{3}$	$\sqrt{2}$	2	∞	−1	∞	1
csc	∞	2	$\sqrt{2}$	$2/3\sqrt{3}$	1	∞	−1	∞

TABLE 2.8　Functions of an Angle in Terms of Each of the Others[a]

	$\sin\alpha = a$	$\cos\alpha = a$	$\tan\alpha = a$	$\cot\alpha = a$	$\sec\alpha = a$	$\csc\alpha = a$
sin	a	$\sqrt{1-a^2}$	$\dfrac{a}{\sqrt{1+a^2}}$	$\dfrac{1}{\sqrt{1+a^2}}$	$\dfrac{\sqrt{a^2-1}}{a}$	$\dfrac{1}{a}$
cos	$\sqrt{1-a^2}$	a	$\dfrac{1}{\sqrt{1+a^2}}$	$\dfrac{a}{\sqrt{1+a^2}}$	$\dfrac{1}{a}$	$\dfrac{\sqrt{a^2-1}}{a}$
tan	$\dfrac{a}{\sqrt{1-a^2}}$	$\dfrac{\sqrt{1-a^2}}{a}$	a	$\dfrac{1}{a}$	$\sqrt{a^2-1}$	$\dfrac{1}{\sqrt{a^2-1}}$
cot	$\dfrac{\sqrt{1-a^2}}{a}$	$\dfrac{a}{\sqrt{1-a^2}}$	$\dfrac{1}{a}$	a	$\dfrac{1}{\sqrt{a^2-1}}$	$\sqrt{a^2-1}$
sec	$\dfrac{1}{\sqrt{1-a^2}}$	$\dfrac{1}{a}$	$\sqrt{1+a^2}$	$\dfrac{\sqrt{1+a^2}}{a}$	a	$\dfrac{a}{\sqrt{a^2-1}}$
csc	$\dfrac{1}{a}$	$\dfrac{1}{\sqrt{1-a^2}}$	$\dfrac{\sqrt{1+a^2}}{a}$	$\sqrt{1+a^2}$	$\dfrac{a}{\sqrt{a^2-1}}$	a

[a] The sign of the radical is to be determined by the quadrant.

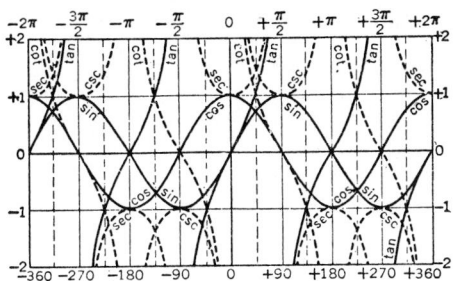

Fig. 2.81

Values of trigonometric functions are periodic, the period of the sin, cos, sec, csc being 2π radians, and that of the tan and cot, π radians (Tables 2.6–2.8). For example, in Fig. 2.81 (n an integer)

$$\sin(\alpha + 2\pi n) = \sin\alpha$$

$$\tan(\alpha + \pi n) = \tan\alpha$$

Inverse, or Antifunctions. The symbol $\sin^{-1}x$ means the angle whose sine is x, and is read inverse sine of x, anti-sine of x, or arc sine x. Similarly for $\cos^{-1}x$, $\tan^{-1}x$, $\cot^{-1}x$, $\sec^{-1}x$, $\csc^{-1}x$, $\mathrm{vers}^{-1}x$, the last meaning an angle α such that $(1 - \cos\alpha) = x$. While the direct functions (sine, etc.) are single valued, the indirect are many valued; thus $\sin 30° = 0.5$, but $\sin^{-1}0.5 = 30°, 150°, \ldots$.

Functional Relations Identities

Functions of the Sum and Difference of Two Angles

$$\sin(\alpha \pm \beta) = \sin\alpha\cos\beta \pm \cos\alpha\sin\beta$$

$$\cos(\alpha \pm \beta) = \cos\alpha\cos\beta \mp \sin\alpha\sin\beta$$

$$\tan(\alpha \pm \beta) = (\tan\alpha \pm \tan\beta)/(1 \mp \tan\alpha\tan\beta)$$

$$\cot(\alpha \pm \beta) = (\cot\beta\cot\alpha \mp 1)/(\cot\beta \pm \cot\alpha)$$

If x is small, say 3° or 4°, then the following are close approximations, in which the quantity x is to be expressed in radians (1° = 0.01745 radian).

$$\sin\alpha \approx \alpha \qquad \cos\alpha \approx 1 \qquad \tan\alpha \approx \alpha$$

$$\sin(\alpha \pm x) \approx \sin\alpha \pm x\cos\alpha \qquad \cos(\alpha \pm x) \approx \cos\alpha \mp x\sin\alpha$$

Functions of Half Angles

$$\sin\tfrac{1}{2}\alpha = \sqrt{\tfrac{1}{2}(1 - \cos\alpha)} = \tfrac{1}{2}\sqrt{1 + \sin\alpha} - \tfrac{1}{2}\sqrt{1 - \sin\alpha}$$

$$\cos\tfrac{1}{2}\alpha = \sqrt{\tfrac{1}{2}(1 + \cos\alpha)} = \tfrac{1}{2}\sqrt{1 + \sin\alpha} + \tfrac{1}{2}\sqrt{1 - \sin\alpha}$$

$$\tan\tfrac{1}{2}\alpha = \sqrt{(1 - \cos\alpha)/(1 + \cos\alpha)} = \frac{1 - \cos\alpha}{\sin\alpha} = \frac{\sin\alpha}{1 + \cos\alpha}$$

$$\cot\tfrac{1}{2}\alpha = \sqrt{(1 + \cos\alpha)/(1 - \cos\alpha)} = \frac{1 + \cos\alpha}{\sin\alpha} = \frac{\sin\alpha}{1 - \cos\alpha}$$

Functions of Multiples of Angles

$$\sin 2\alpha = 2\sin\alpha\cos\alpha$$

$$\tan 2\alpha = \frac{2\tan\alpha}{1-\tan^2\alpha}$$

$$\cos 2\alpha = \cos^2\alpha - \sin^2\alpha = 2\cos^2\alpha - 1 = 1 - 2\sin^2\alpha$$

$$\cot 2\alpha = \frac{\cot^2\alpha - 1}{2\cot\alpha}$$

$$\sin 3\alpha = 3\sin\alpha - 4\sin^3\alpha$$

$$\cos 3\alpha = 4\cos^3\alpha - 3\cos\alpha$$

$$\sin 4\alpha = 8\cos^3\alpha\sin\alpha - 4\cos\alpha\sin\alpha$$

$$\cos 4\alpha = 8\cos^4\alpha - 8\cos^2\alpha + 1$$

$$\sin n\alpha = 2\sin(n-1)\alpha\cos\alpha - \sin(n-2)\alpha$$

$$= n\sin\alpha\cos^{n-1}\alpha - {}_nC_3\sin^3\alpha\cos^{n-3}\alpha + {}_nC_5\sin^5\alpha\cos^{n-5}\alpha - \cdots$$

$$\cos n\alpha = 2\cos(n-1)\alpha\cos\alpha - \cos(n-2)\alpha$$

$$= \cos^n\alpha - {}_nC_2\sin^2\alpha\cos^{n-2}\alpha + {}_nC_4\sin^4\alpha\cos^{n-4}\alpha - \cdots$$

(For ${}_nC_r$, see p. 2•5.)

Products and Powers of Functions

$$\sin\alpha\sin\beta = \tfrac{1}{2}\cos(\alpha-\beta) - \tfrac{1}{2}\cos(\alpha+\beta)$$

$$\cos\alpha\cos\beta = \tfrac{1}{2}\cos(\alpha-\beta) + \tfrac{1}{2}\cos(\alpha+\beta)$$

$$\sin\alpha\cos\beta = \tfrac{1}{2}\sin(\alpha-\beta) + \tfrac{1}{2}\sin(\alpha+\beta)$$

$$\tan\alpha\cot\alpha = \sin\alpha\csc\alpha = \cos\alpha\sec\alpha = 1$$

$$\sin^2\alpha = \tfrac{1}{2}(1-\cos 2\alpha); \quad \cos^2\alpha = \tfrac{1}{2}(1+\cos 2\alpha)$$

$$\sin^3\alpha = \tfrac{1}{4}(3\sin\alpha - \sin 3\alpha); \quad \cos^3\alpha = \tfrac{1}{4}(3\cos\alpha + \cos 3\alpha)$$

$$\sin^4\alpha = \tfrac{1}{8}(3 - 4\cos 2\alpha + \cos 4\alpha); \quad \cos^4\alpha = \tfrac{1}{8}(3 + 4\cos 2\alpha + \cos 4\alpha)$$

$$\sin^5\alpha = \tfrac{1}{16}(10\sin\alpha - 5\sin 3\alpha + \sin 5\alpha)$$

$$\sin^6\alpha = \tfrac{1}{32}(10 - 15\cos 2\alpha + 6\cos 4\alpha - \cos 6\alpha)$$

$$\cos^5\alpha = \tfrac{1}{16}(10\cos\alpha + 5\cos 3\alpha + \cos 5\alpha)$$

$$\cos^6\alpha = \tfrac{1}{32}(10 + 15\cos 2\alpha + 6\cos 4\alpha + \cos 6\alpha)$$

Sums and Differences of Functions

$$\sin\alpha + \sin\beta = 2\sin\tfrac{1}{2}(\alpha+\beta)\cos\tfrac{1}{2}(\alpha-\beta)$$

$$\sin\alpha - \sin\beta = 2\cos\tfrac{1}{2}(\alpha+\beta)\sin\tfrac{1}{2}(\alpha-\beta)$$

$$\cos\alpha + \cos\beta = 2\cos\tfrac{1}{2}(\alpha+\beta)\cos\tfrac{1}{2}(\alpha-\beta)$$

$$\cos\alpha - \cos\beta = -2\sin\tfrac{1}{2}(\alpha+\beta)\sin\tfrac{1}{2}(\alpha-\beta)$$

$$\tan\alpha + \tan\beta = \frac{\sin(\alpha+\beta)}{\cos\alpha\cos\beta}; \quad \cot\alpha + \cot\beta = \frac{\sin(\alpha+\beta)}{\sin\alpha\sin\beta}$$

$$\tan\alpha - \tan\beta = \frac{\sin(\alpha-\beta)}{\cos\alpha\cos\beta}; \quad \cot\alpha - \cot\beta = -\frac{\sin(\alpha-\beta)}{\sin\alpha\sin\beta}$$

$$\sin^2\alpha - \sin^2\beta = \sin(\alpha+\beta)\sin(\alpha-\beta)$$

$$\cos^2\alpha - \cos^2\beta = -\sin(\alpha+\beta)\sin(\alpha-\beta)$$

$$\cos^2\alpha - \sin^2\beta = \cos(\alpha+\beta)\cos(\alpha-\beta)$$

Antitrigonometric or Inverse Functional Relations. In the following formulas the periodic constant is omitted.

$$\sin^{-1}x = -\sin^{-1}(-x) = \frac{\pi}{2} - \cos^{-1}x = \cos^{-1}\sqrt{1-x^2} = \tan^{-1}\frac{x}{\sqrt{1-x^2}}$$

$$= \cot^{-1}\frac{\sqrt{1-x^2}}{x} = \csc^{-1}\frac{1}{x} = \sec^{-1}\frac{1}{\sqrt{1-x^2}}$$

$$\cos^{-1}x = \pi - \cos^{-1}(-x) = \frac{\pi}{2} - \sin^{-1}x = \tfrac{1}{2}\cos^{-1}(2x^2 - 1) = \sin^{-1}\sqrt{1-x^2}$$

$$= \tan^{-1}\frac{\sqrt{1-x^2}}{x} = \cot^{-1}\frac{x}{\sqrt{1-x^2}} = \sec^{-1}\frac{1}{x} = \csc^{-1}\frac{1}{\sqrt{1-x^2}}$$

$$\tan^{-1}x = -\tan^{-1}(-x) = \frac{\pi}{2} - \cot^{-1}x = \sin^{-1}\frac{x}{\sqrt{1+x^2}} = \cos^{-1}\frac{1}{\sqrt{1+x^2}} = \cot^{-1}\frac{1}{x}$$

$$= \sec^{-1}\sqrt{1+x^2} = \csc^{-1}\frac{\sqrt{1+x^2}}{x}$$

$$\cot^{-1}x = \tan^{-1}\frac{1}{x} \qquad \sec^{-1}x = \cos^{-1}\frac{1}{x} \qquad \csc^{-1}x = \sin^{-1}\frac{1}{x}$$

$$\sin^{-1}x \pm \sin^{-1}y = \sin^{-1}\left\{ x\sqrt{1-y^2} \pm y\sqrt{1-x^2} \right\}$$

$$\cos^{-1}x \pm \cos^{-1}y = \cos^{-1}\left\{ xy \mp \sqrt{(1-x^2)(1-y^2)} \right\}$$

$$\sin^{-1}x \pm \cos^{-1}y = \sin^{-1}\left\{ xy \pm \sqrt{(1-x^2)(1-y^2)} \right\} = \cos^{-1}\left\{ y\sqrt{1-x^2} \mp x\sqrt{1-y^2} \right\}$$

$$\tan^{-1}x \pm \tan^{-1}y = \tan^{-1}\frac{x \pm y}{1 \mp xy}$$

$$\tan^{-1}x \pm \cot^{-1}y = \tan^{-1}\frac{xy \pm 1}{y \mp x} = \cot^{-1}\frac{y \mp x}{xy \pm 1}$$

2.6.2 Solution of Triangles

Relations between Angles and Sides of Plane Triangles. Let a, b, c = sides of triangle; α, β, γ = angles opposite, a, b, c, respectively; A = area of triangle; $s = \tfrac{1}{2}(a + b + c)$; r = radius of inscribed circle (Fig. 2.82).

$$\frac{a}{\sin \alpha} = \frac{b}{\sin \beta} = \frac{c}{\sin \gamma} \quad \text{(law of sines)}$$

$$a^2 = b^2 + c^2 - 2bc \cos \alpha \quad \text{(law of cosines)}$$

$$\frac{a - b}{a + b} = \frac{\tan\frac{1}{2}(\alpha - \beta)}{\tan\frac{1}{2}(\alpha + \beta)} \quad \text{(law of tangents)}$$

$$\alpha + \beta + \gamma = 180°$$

$$a = b \cos \gamma + c \cos \beta \qquad b = c \cos \alpha + a \cos \gamma \qquad c = a \cos \beta + b \cos \alpha$$

$$A = \sqrt{s(s - a)(s - b)(s - c)}$$

$$\sin \alpha = \frac{2}{bc}A \qquad \sin \beta = \frac{2}{ca}A \qquad \sin \gamma = \frac{2}{ab}A$$

$$\sin\frac{\alpha}{2} = \sqrt{\frac{(s - b)(s - c)}{bc}} \qquad \sin\frac{\beta}{2} = \sqrt{\frac{(s - c)(s - a)}{ca}} \qquad \sin\frac{\gamma}{2} = \sqrt{\frac{(s - a)(s - b)}{ab}}$$

$$\cos\frac{\alpha}{2} = \sqrt{\frac{s(s - a)}{bc}} \qquad \cos\frac{\beta}{2} = \sqrt{\frac{s(s - b)}{ca}} \qquad \cos\frac{\gamma}{2} = \sqrt{\frac{s(s - c)}{ab}}$$

$$\tan\frac{\alpha}{2} = \sqrt{\frac{(s - b)(s - c)}{s(s - a)}} \qquad \tan\frac{\beta}{2} = \sqrt{\frac{(s - c)(s - a)}{s(s - b)}} \qquad \tan\frac{\gamma}{2} = \sqrt{\frac{(s - a)(s - b)}{s(s - c)}}$$

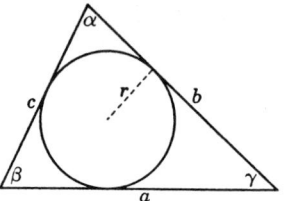

Fig. 2.82

Solution of Plane Oblique Triangles
Given a, b, c: (If logarithms are to be used, use 1.)

1.
$$r = \sqrt{\frac{(s-a)(s-b)(s-c)}{s}} \qquad A = \sqrt{s(s-a)(s-b)(s-c)} = rs$$

$$\tan\frac{\alpha}{2} = \frac{r}{s-a} \qquad \tan\frac{\beta}{2} = \frac{r}{s-b} \qquad \tan\frac{\gamma}{2} = \frac{r}{s-c}$$

2.
$$\cos\alpha = \frac{b^2 + c^2 - a^2}{2bc} \qquad \cos\beta = \frac{a^2 + c^2 - b^2}{2ac}$$

$$\cos\gamma = \frac{a^2 + b^2 - c^2}{2ab} \qquad \text{or} \quad \gamma = 180° - (\alpha + \beta)$$

Given a, b, α:

$$\sin\beta = \frac{b\sin\alpha}{a}$$

(if $a > b$, $\beta < \pi/2$ and has only one value; if $b > a$, β has two values, β_1 and $\beta_2 = 180° - \beta_1$);
$\gamma = 180° - (\alpha + \beta)$; $c = a\sin\gamma/\sin\alpha$; $A = \frac{1}{2}ab\sin\gamma$.

Given a, α, β:

$$b = \frac{a\sin\beta}{\sin\alpha} \qquad \gamma = 180° - (\alpha + \beta) \qquad c = \frac{a\sin\gamma}{\sin\alpha} \qquad A = \frac{1}{2}ab\sin\gamma$$

Given a, b, γ: (If logarithms are to be used, use 1.)

1.
$$\tan\tfrac{1}{2}(\alpha - \beta) = \frac{a-b}{a+b}\cot\tfrac{1}{2}\gamma \qquad \tfrac{1}{2}(\alpha + \beta) = 90° - \tfrac{1}{2}\gamma \qquad c = \frac{a\sin\gamma}{\sin\alpha}$$

$$A = \tfrac{1}{2}ab\sin\gamma$$

2.
$$c = \sqrt{a^2 + b^2 - 2ab\cos\gamma} \qquad \sin\alpha = \frac{a\sin\gamma}{c} \qquad \beta = 180° - (\alpha + \gamma)$$

3.
$$\tan\alpha = \frac{a\sin\gamma}{b - a\cos\gamma} \qquad \beta = 180° - (\alpha + \gamma) \qquad c = \frac{a\sin\gamma}{\sin\alpha}$$

MOLLWEIDE'S CHECK FORMULAS

$$\frac{a-b}{c} = \frac{\sin\tfrac{1}{2}(\alpha - \beta)}{\cos\tfrac{1}{2}\gamma}$$

$$\frac{a+b}{c} = \frac{\cos\tfrac{1}{2}(\alpha - \beta)}{\sin\tfrac{1}{2}\gamma}$$

Solution of Plane Right Triangles. Let $\gamma = 90°$ and c be the hypotenuse. Given any two sides or one side and an acute angle α.

$$a = \sqrt{c^2 - b^2} = \sqrt{(c+b)(c-b)} = b\tan\alpha = c\sin\alpha$$

$$b = \sqrt{c^2 - a^2} = \sqrt{(c+a)(c-a)} = \frac{a}{\tan\alpha} = c\cos\alpha$$

$$c = \sqrt{a^2 + b^2} = \frac{a}{\sin \alpha} = \frac{b}{\cos \alpha}$$

$$\alpha = \sin^{-1}\frac{a}{c} = \cos^{-1}\frac{b}{c} = \tan^{-1}\frac{a}{b}; \ \beta = 90° - \alpha$$

$$A = \frac{ab}{2} = \frac{a^2}{2\tan\alpha} = \frac{b^2\tan\alpha}{2} = \frac{c^2\sin 2\alpha}{4}$$

2.6.3 Spherical Trigonometry

Spherical Trigonometry. Let O be the center of the sphere and a, b, c the sides of a triangle on the surface with opposite angles α, β, γ, respectively, the sides being measured by the angle subtended at the center of the sphere. Let $s = \frac{1}{2}(a + b + c)$, $\sigma = \frac{1}{2}(\alpha + \beta + \gamma)$, $E = \alpha + \beta + \gamma - 180°$, the spherical excess. The following formulas are valid usually only for triangles of which the sides and angles are all between $0°$ and $180°$. To each such triangle there is a polar triangle, whose sides are $180° - \alpha$, $180° - \beta$, $180° - \gamma$, and whose angles are $180° - a$, $180° - b$, $180° - c$.

GENERAL FORMULAS

$$\frac{\sin a}{\sin \alpha} = \frac{\sin b}{\sin \beta} = \frac{\sin c}{\sin \gamma} \quad \text{(Law of sines)}$$

$$\cos a = \cos b \cos c + \sin b \sin c \cos \alpha \quad \text{(Law of cosines)}$$

$$\cos \alpha = -\cos \beta \cos \gamma + \sin \beta \sin \gamma \cos a \quad \text{(Law of cosines)}$$

$$\cos a \sin b = \sin a \cos b \cos \gamma + \sin c \cos \alpha$$

$$\cot a \sin b = \sin \gamma \cot \alpha + \cos \gamma \cos b$$

$$\cos \alpha \sin \beta = \sin \gamma \cos a - \sin \alpha \cos \beta \cos c$$

$$\cot \alpha \sin \beta = \sin c \cot a - \cos c \cos \beta$$

$$\sin\frac{a}{2} = \sqrt{\frac{-\cos\sigma\cos(\sigma - \alpha)}{\sin\beta\sin\gamma}} \qquad\qquad \sin\frac{\alpha}{2} = \sqrt{\frac{\sin(s-b)\sin(s-c)}{\sin b \sin c}}$$

$$\cos\frac{a}{2} = \sqrt{\frac{\cos(\sigma - \beta)\cos(\sigma - \gamma)}{\sin\beta\sin\gamma}} \qquad\qquad \cos\frac{\alpha}{2} = \sqrt{\frac{\sin s \sin(s-a)}{\sin b \sin c}}$$

$$\tan\frac{a}{2} = \sqrt{\frac{-\cos\sigma\cos(\sigma - \alpha)}{\cos(\sigma - \beta)\cos(\sigma - \gamma)}} \qquad\qquad \tan\frac{\alpha}{2} = \sqrt{\frac{\sin(s-b)\sin(s-c)}{\sin s \sin(s-a)}}$$

$$\tan\frac{E}{4} = \sqrt{\tan\frac{s}{2}\tan\frac{(s-a)}{2}\tan\frac{(s-b)}{2}\tan\frac{(s-c)}{2}} \qquad \cot\frac{E}{2} = \frac{\cot\frac{a}{2}\cot\frac{b}{2} + \cos\gamma}{\sin\gamma}$$

$$\tan(\frac{a+b}{2}) = \frac{\cos\left(\frac{\alpha-\beta}{2}\right)}{\cos\left(\frac{\alpha+\beta}{2}\right)}\tan\frac{c}{2} \qquad\qquad \tan(\frac{a-b}{2}) = \frac{\sin\left(\frac{\alpha-\beta}{2}\right)}{\sin\left(\frac{\alpha+\beta}{2}\right)}\tan\frac{c}{2}$$

$$\tan(\frac{\alpha+\beta}{2}) = \frac{\cos\left(\frac{a-b}{2}\right)}{\cos\left(\frac{a+b}{2}\right)}\cot\frac{\gamma}{2} \qquad\qquad \tan(\frac{\alpha-\beta}{2}) = \frac{\sin\left(\frac{a-b}{2}\right)}{\sin\left(\frac{a+b}{2}\right)}\cot\frac{\gamma}{2}$$

$$\cos\left(\frac{\alpha+\beta}{2}\right)\cos\frac{c}{2} = \cos\left(\frac{a+b}{2}\right)\sin\frac{\gamma}{2} \qquad \sin\left(\frac{\alpha+\beta}{2}\right)\cos\frac{c}{2} = \cos\left(\frac{a-b}{2}\right)\cos\frac{\gamma}{2}$$

$$\cos\left(\frac{\alpha-\beta}{2}\right)\sin\frac{c}{2} = \sin\left(\frac{a+b}{2}\right)\sin\frac{\gamma}{2} \qquad \sin\left(\frac{\alpha-\beta}{2}\right)\sin\frac{c}{2} = \sin\left(\frac{a-b}{2}\right)\cos\frac{\gamma}{2}$$

The Right Spherical Triangle. Let $\gamma = 90°$ and c be the hypotenuse.

$$\cos c = \cos a \cos b = \cot \alpha \cot \beta \qquad \cos a = \frac{\cos \alpha}{\sin \beta} \qquad \cos b = \frac{\cos \beta}{\sin \alpha}$$

$$\sin \alpha = \frac{\sin a}{\sin c} \qquad \cos \alpha = \frac{\tan b}{\tan c} \qquad \tan \alpha = \frac{\tan a}{\sin b}$$

2.6.4 Hyperbolic Trigonometry

Hyperbolic Angles. These are defined in a manner similar to circular angles but with reference to an *equilateral hyperbola*. The comparative relations are shown in Figs. 2.83 and 2.84. A *circular angle* is a central angle measured in radians by the ratio s/r or the ratio $2A/r^2$, where A is the area of the sector included by the angle α and the arc s (Fig. 2.83). For the *hyperbola* the radius ρ is not constant and only the value of the *differential hyperbolic angle* $d\theta$ is defined by the ratio ds/ρ. Thus

$$\theta = \int \frac{ds}{\rho} = \frac{2A}{a^2}$$

where A represents the shaded area in Fig. 2.84. If both s and ρ are measured in the same units the angle is expressed in *hyperbolic radians*.

Hyperbolic Functions. These are defined by ratios similar to those defining functions of circular angles and also named similarly. Their names and abbreviations are:

$$\text{Hyperbolic sine } \theta = \frac{y}{a} = \sinh \theta$$

$$\text{Hyperbolic cosine } \theta = \frac{x}{a} = \cosh \theta$$

$$\text{Hyperbolic tangent } \theta = \frac{y}{x} = \tanh \theta$$

$$\text{Hyperbolic cotangent } \theta = \frac{x}{y} = \coth \theta$$

$$\text{Hyperbolic secant } \theta = \frac{a}{x} = \operatorname{sech} \theta$$

$$\text{Hyperbolic cosecant } \theta = \frac{a}{y} = \operatorname{csch} \theta$$

Values and Exponential Equivalents. The values of hyperbolic functions may be computed from their exponential equivalents. The graphs are shown in Fig. 2.85. Values for increments of 0.01 radian are given in Table 1.18.

$$\sinh \theta = \frac{e^\theta - e^{-\theta}}{2} \qquad \cosh \theta = \frac{e^\theta + e^{-\theta}}{2} \qquad \tanh \theta = \frac{e^\theta - e^{-\theta}}{e^\theta + e^{-\theta}}$$

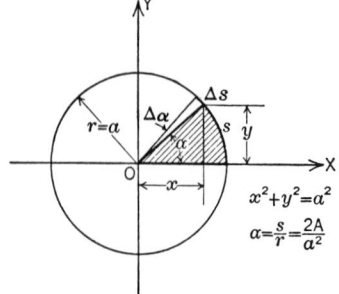

Fig. 2.83   Fig. 2.84

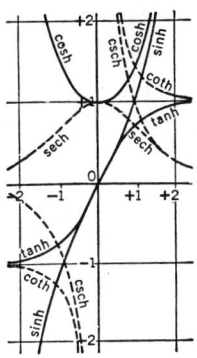

Fig. 2.85

If θ is extremely small, $\sinh\theta \approx \theta$, $\cosh\theta \approx 1$, and $\tanh\theta \approx \theta$. For large values of θ, $\sinh\theta \approx \cosh\theta$, and $\tanh\theta \approx \coth\theta \approx 1$.

Fundamental Identities

$$\operatorname{csch}\theta = \frac{1}{\sinh\theta} \qquad \operatorname{sech}\theta = \frac{1}{\cosh\theta} \qquad \coth\theta = \frac{1}{\tanh\theta}$$

$$\cosh^2\theta - \sinh^2\theta = 1 \qquad \operatorname{sech}^2\theta = 1 - \tanh^2\theta \qquad \operatorname{csch}^2\theta = \coth^2\theta - 1$$

$$\cosh\theta + \sinh\theta = e^\theta \qquad \cosh\theta - \sinh\theta = e^{-\theta}$$

$$\sinh(-\theta) = -\sinh\theta \qquad \cosh(-\theta) = \cosh\theta$$

$$\tanh(-\theta) = -\tanh\theta \qquad \coth(-\theta) = -\coth\theta$$

$$\sinh(\theta_1 \pm \theta_2) = \sinh\theta_1\cosh\theta_2 \pm \cosh\theta_1\sinh\theta_2$$

$$\cosh(\theta_1 \pm \theta_2) = \cosh\theta_1\cosh\theta_2 \pm \sinh\theta_1\sinh\theta_2$$

$$\tanh(\theta_1 \pm \theta_2) = \frac{\tanh\theta_1 \pm \tanh\theta_2}{1 \pm \tanh\theta_1\tanh\theta_2} \qquad \coth(\theta_1 \pm \theta_2) = \frac{1 \pm \coth\theta_1\coth\theta_2}{\coth\theta_1 \pm \coth\theta_2}$$

$$\sinh 2\theta = 2\sinh\theta\cosh\theta = \frac{2\tanh\theta}{1 - \tanh^2\theta}$$

$$\cosh 2\theta = \sinh^2\theta + \cosh^2\theta = 1 + 2\sinh^2\theta = 2\cosh^2\theta - 1 = \frac{1 + \tanh^2\theta}{1 - \tanh^2\theta}$$

$$\tanh 2\theta = \frac{2\tanh\theta}{1 + \tanh^2\theta} \qquad \coth 2\theta = \frac{1 + \coth^2\theta}{2\coth\theta}$$

$$\sinh\frac{\theta}{2} = \sqrt{(\cosh\theta - 1)/2} \qquad \cosh\frac{\theta}{2} = \sqrt{(\cosh\theta + 1)/2}$$

$$\tanh\frac{\theta}{2} = \sqrt{\frac{\cosh\theta - 1}{\cosh\theta + 1}} = \frac{\sinh\theta}{\cosh\theta + 1} = \frac{\cosh\theta - 1}{\sinh\theta}$$

$$\sinh\theta_1 \pm \sinh\theta_2 = 2\sinh\frac{(\theta_1 \pm \theta_2)}{2}\cosh\frac{(\theta_1 \mp \theta_2)}{2}$$

$$\cosh\theta_1 + \cosh\theta_2 = 2\cosh\frac{(\theta_1 + \theta_2)}{2}\cosh\frac{(\theta_1 - \theta_2)}{2}$$

$$\cosh\theta_1 - \cosh\theta_2 = 2\sinh\frac{(\theta_1 + \theta_2)}{2}\sinh\frac{(\theta_1 - \theta_2)}{2}$$

$$\tanh\theta_1 \pm \tanh\theta_2 = \frac{\sinh(\theta_1 \pm \theta_2)}{\cosh\theta_1\cosh\theta_2}$$

$$(\cosh\theta \pm \sinh\theta)^n = \cosh n\theta \pm \sinh n\theta$$

Antihyperbolic or Inverse Functions. The inverse hyperbolic sine of u is written: $\sinh^{-1}u$. Values of the inverse functions may be computed from their logarithmic equivalents.

$$\sinh^{-1}u = \log_e\left(u + \sqrt{u^2 + 1}\,\right) \qquad \cosh^{-1}u = \log_e\left(u + \sqrt{u^2 - 1}\,\right)$$

$$\tanh^{-1}u = \tfrac{1}{2}\log_e\frac{1 + u}{1 - u} \qquad \coth^{-1}u = \tfrac{1}{2}\log_e\frac{u + 1}{u - 1}$$

2.6.5 Functions of Imaginary and Complex Angles

Relations of Hyperbolic to Circular Functions. By comparison of the exponential equivalents of hyperbolic and circular functions the following identities are established ($i = \sqrt{-1}$):

$$\sin\alpha = -i\sinh i\alpha \qquad\qquad \sinh\beta = -i\sin i\beta$$
$$\cos\alpha = \cosh i\alpha \qquad\qquad \cosh\beta = \cos i\beta$$
$$\tan\alpha = -i\tanh i\alpha \qquad\qquad \tanh\beta = -i\tan i\beta$$
$$\cot\alpha = i\coth i\alpha \qquad\qquad \coth\beta = i\cot i\beta$$
$$\sec\alpha = \operatorname{sech} i\alpha \qquad\qquad \operatorname{sech}\beta = \sec i\beta$$
$$\csc\alpha = i\operatorname{csch} i\alpha \qquad\qquad \operatorname{csch}\beta = i\csc i\beta$$

Relations between Inverse Functions

$$\sin^{-1}A = -i\sinh^{-1}iA \qquad\qquad \sinh^{-1}B = -i\sin^{-1}iB$$
$$\cos^{-1}A = -i\cosh^{-1}A \qquad\qquad \cosh^{-1}B = i\cos^{-1}B$$
$$\tan^{-1}A = -i\tanh^{-1}iA \qquad\qquad \tanh^{-1}B = -i\tan^{-1}iB$$
$$\cot^{-1}A = i\coth^{-1}iA \qquad\qquad \coth^{-1}B = i\cot^{-1}iB$$
$$\sec^{-1}A = -i\operatorname{sech}^{-1}A \qquad\qquad \operatorname{sech}^{-1}B = i\sec^{-1}B$$
$$\csc^{-1}A = i\operatorname{csch}^{-1}iA \qquad\qquad \operatorname{csch}^{-1}B = i\csc^{-1}iB$$

Functions of a Complex Angle. In complex notation $c = a + ib = |c|(\cos\theta + i\sin\theta) = |c|e^{i\theta}$, where $|c| = \sqrt{a^2 + b^2}$, $i = \sqrt{-1}$, and $\theta = \tan^{-1}b/a$. $|c|e^{i\theta}$ is frequently written $c\angle\theta$.
$\text{Log}_e|c|e^{i\theta} = \log|c| + i(\theta + 2k\pi)$ and is infinitely many valued. By its principal part will be understood $\log_e|c| + i\theta$. Some convenient identities are

$$\log_e 1 = 0 \qquad \log_e(-1) = i\pi \qquad \log_e i = i\frac{\pi}{2} \qquad \log_e(-i) = i\frac{3\pi}{2}$$

$$(\cos\theta \pm i\sin\theta)^n = \cos n\theta \pm i\sin n\theta \qquad \sqrt[n]{\cos\theta \pm i\sin\theta} = \cos\frac{\theta + 2\pi k}{n} \pm i\sin\frac{\theta + 2\pi k}{n}$$

The use of complex angles occurs frequently in electric circuit problems where it is often necessary to express the functions of them as a complex number.

$$\sin(\alpha \pm i\beta) = \sin\alpha\cosh\beta \pm i\cos\alpha\sinh\beta = \sqrt{\cosh^2\beta - \cos^2\alpha}\,e^{\pm i\theta}$$

where $\theta = \tan^{-1}\cot\alpha\tanh\beta$.

$$\cos(\alpha \pm i\beta) = \cos\alpha\cosh\beta \mp i\sin\alpha\sinh\beta = \sqrt{\cosh^2\beta - \sin^2\alpha}\,e^{\pm i\theta}$$

where $\theta = \tan^{-1}\tan\alpha\tanh\beta$.

$$\sinh(\alpha \pm i\beta) = \sinh\alpha\cos\beta \pm i\cosh\alpha\sin\beta$$
$$= \sqrt{\sinh^2\alpha + \sin^2\beta}\,e^{\pm i\theta} = \sqrt{\cosh^2\alpha - \cos^2\beta}\,e^{\pm i\theta}$$

where $\theta = \tan^{-1}\coth\alpha\tan\beta$.

$$\cosh(\alpha \pm i\beta) = \cosh\alpha\cos\beta \pm i\sinh\alpha\sin\beta$$
$$= \sqrt{\sinh^2\alpha + \cos^2\beta}\,e^{\pm i\theta} = \sqrt{\cosh^2\alpha - \sin^2\beta}\,e^{\pm i\theta}$$

where $\theta = \tan^{-1} \tanh \alpha \tan \beta$.

$$\tan(\alpha \pm i\beta) = \frac{\sin 2\alpha \pm i \sinh 2\beta}{\cos 2\alpha + \cosh 2\beta} \qquad \tanh(\alpha \pm i\beta) = \frac{\sinh 2\alpha \pm i \sin 2\beta}{\cosh 2\alpha + \cos 2\beta}$$

The hyperbolic sine and cosine have the period $2\pi i$; the hyperbolic tangent has the period πi.

$$\sinh(\alpha + 2k\pi i) = \sinh \alpha \qquad \cosh(\alpha + 2k\pi i) = \cosh \alpha$$
$$\tanh(\alpha + k\pi i) = \tanh \alpha \qquad \coth(\alpha + k\pi i) = \coth \alpha$$

Inverse Functions of Complex Numbers

$$\sin^{-1}(A \pm iB) = \sin^{-1}\left[\frac{\sqrt{B^2 + (1 + A)^2} - \sqrt{B^2 + (1 - A)^2}}{2}\right]$$

$$\pm i \cosh^{-1}\left[\frac{\sqrt{B^2 + (1 + A)^2} + \sqrt{B^2 + (1 - A)^2}}{2}\right]$$

$$\cos^{-1}(A \pm iB) = \cos^{-1}\left[\frac{\sqrt{B^2 + (1 + A)^2} - \sqrt{B^2 + (1 - A)^2}}{2}\right]$$

$$\mp i \cosh^{-1}\left[\frac{\sqrt{B^2 + (1 + A)^2} + \sqrt{B^2 + (1 - A)^2}}{2}\right]$$

$$\tan^{-1}(A \pm iB) = \left[\frac{\pi - \tan^{-1}\dfrac{A}{\pm B - 1} + \tan^{-1}\dfrac{A}{\pm B + 1}}{2}\right]$$

$$\pm i\tfrac{1}{4}\log_e \frac{A^2 + (1 \pm B)^2}{A^2 + (1 \mp B)^2}$$

$$\sinh^{-1}(A \pm iB) = \cosh^{-1}\left[\frac{\sqrt{A^2 + (1 + B)^2} + \sqrt{A^2 + (1 - B)^2}}{2}\right]$$

$$\pm i \sin^{-1}\left[\frac{\sqrt{A^2 + (1 + B)^2} - \sqrt{A^2 + (1 - B)^2}}{2}\right]$$

$$\cosh^{-1}(A \pm iB) = \cosh^{-1}\left[\frac{\sqrt{B^2 + (1 + A)^2} + \sqrt{B^2 + (1 - A)^2}}{2}\right]$$

$$\pm i \cos^{-1}\left[\frac{\sqrt{B^2 + (1 + A)^2} - \sqrt{B^2 + (1 - A)^2}}{2}\right]$$

$$\tanh^{-1}(A \pm iB) = \tfrac{1}{2}\tanh^{-1}\frac{2A}{1 + A^2 + B^2} + i\tfrac{1}{2}\tan^{-1}\frac{\pm 2B}{1 - A^2 - B^2}$$

2.7 PLANE ANALYTIC GEOMETRY

2.7.1 Point and Line

Coordinates. The position of a point P_1 in a plane is determined if its distance and direction from each of two lines or axes OX and OY, which are perpendicular to each other, are known. The distances x and y (Fig. 2.86) perpendicular to the axes are called the *Cartesian* or *rectangular coordinates* of the point. The directions to the right of OY and above OX are called *positive*, and opposite directions *negative*. The point O of intersection of OY and OX is called the *origin*.

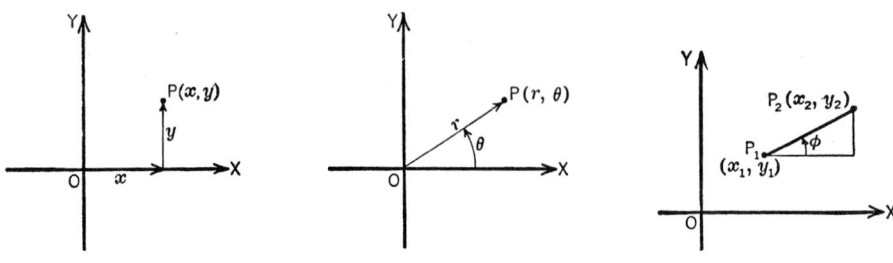

Fig. 2.86   Fig. 2.87   Fig. 2.88

The position of a point P is also given by its radial distance r from the origin and the angle θ between the radius r and the horizontal axis OX (Fig. 2.87). These coordinates r, θ, are called *polar coordinates*.

The distance s between two points $P_1(x_1, y_1)$ and $P_2(x_2, y_2)$ (Fig. 2.88) on a straight line is

$$s = \sqrt{(x_2 - x_1)^2 + (y_2 - y_1)^2} \tag{2.15}$$

In polar coordinates the distance s between $P_1(r_1, \theta_1)$ and $P_2(r_2, \theta_2)$ is

$$s = \sqrt{r_1^2 + r_2^2 - 2r_1 r_2 \cos(\theta_2 - \theta_1)} \tag{2.16}$$

The slope m of the line $P_1 P_2$ is defined as the tangent of the angle ϕ, which the line makes with OX.

$$m = \tan \phi = \frac{y_2 - y_1}{x_2 - x_1} \tag{2.17}$$

To divide the segment $P_1 P_2$ in the ratio c_1/c_2, internally or externally,

$$x = \frac{c_2 x_1 \pm c_1 x_2}{c_2 \pm c_1} \qquad y = \frac{c_2 y_1 \pm c_1 y_2}{c_2 \pm c_1}$$

The midpoint of $P_1 P_2$ is

$$x = \tfrac{1}{2}(x_1 + x_2) \qquad y = \tfrac{1}{2}(y_1 + y_2)$$

Equation of a Straight Line. In Cartesian coordinates the equation of a straight line is of the first degree and is expressed as follows:

$$Ax + By + C = 0 \tag{2.18}$$

where A, B, and C are constants.

Other forms of the equation are

$$y = mx + b \tag{2.19}$$

where m is the slope and b is the y intercept:

$$y - y_1 = m(x - x_1) \tag{2.20}$$

where m is the slope and (x_1, y_1) is a point on the line:

$$\frac{x - x_1}{y - y_1} = \frac{x_1 - x_2}{y_1 - y_2} \tag{2.21}$$

where (x_1, y_1) and (x_2, y_2) are two points on the line:

$$\frac{x}{a} + \frac{y}{b} = 1 \tag{2.22}$$

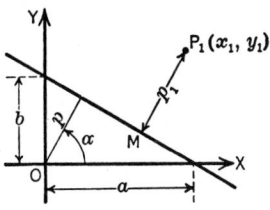

Fig. 2.89

where a and b are the x and y intercepts, respectively:

$$x \cos \alpha + y \sin \alpha - p = 0 \tag{2.23}$$

where α is the angle between OX and the perpendicular from the origin to the line and p is the length of the perpendicular (Fig. 2.89). This is called the *perpendicular* form and is obtained by dividing the general form $Ax + By + C = 0$ by $\pm\sqrt{A^2 + B^2}$. The sign before the radical is taken opposite to that of C if $C \neq 0$ and the same as that of B if $C = 0$.

Equations of lines parallel to the x and y axes, respectively, are

$$y = k \qquad x = k \tag{2.24}$$

The **perpendicular distance of a point** $P_1(x_1, y_1)$ (Fig. 2.89) from the line $Ax + By + C = 0$ is

$$p_1 = \frac{Ax_1 + By_1 + C}{\pm\sqrt{A^2 + B^2}} \tag{2.25}$$

where the sign before the radical is opposite to that of C if $C \neq 0$, and the same as B if $C = 0$.

Parallel Lines. The two lines $y = m_1 x + b_1$, $y = m_2 x + b_2$ are parallel if $m_1 = m_2$. For the form $A_1 x + B_1 y + C_1 = 0$, $A_2 x + B_2 y + C_2 = 0$, the lines are parallel if

$$\frac{A_1}{A_2} = \frac{B_1}{B_2} \tag{2.26}$$

The equation of a line through the point (x_1, y_1) and parallel to the line $Ax + By + C = 0$ is

$$A(x - x_1) + B(y - y_1) = 0 \tag{2.27}$$

Perpendicular Lines. The two lines $y = m_1 x + b_1$ and $y = m_2 x + b_2$ are perpendicular if

$$m_1 = -\frac{1}{m_2} \tag{2.28}$$

For the form $A_1 x + B_1 y + C_1 = 0$, $A_2 x + B_2 y + C_2 = 0$, the lines are perpendicular if

$$A_1 A_2 + B_1 B_2 = 0 \tag{2.29}$$

The equation of a line through the point (x_1, y_1) perpendicular to the line $Ax + By + C = 0$ is

$$B(x - x_1) - A(y - y_1) = 0 \tag{2.30}$$

Intersecting Lines. Let $A_1 x + B_1 y + C_1 = 0$ and $A_2 x + B_2 y + C_2 = 0$ be the equations of two intersecting lines and λ an arbitrary real number. Then

$$(A_1 x + B_1 y + C_1) + \lambda(A_2 x + B_2 y + C_2) = 0 \tag{2.31}$$

represents the system of lines through the point of intersection.

The three lines $A_1x + B_1y + C_1 = 0$, $A_2x + B_2y + C_2 = 0$, $A_3x + B_3y + C_3 = 0$ meet in a point if

$$\begin{vmatrix} A_1 & B_1 & C_1 \\ A_2 & B_2 & C_2 \\ A_3 & B_3 & C_3 \end{vmatrix} = 0 \tag{2.32}$$

The **angle** θ **between two lines** with equations $A_1x + B_1y + C_1 = 0$ and $A_2x + B_2y + C_2 = 0$ can be found from

$$\sin\theta = \frac{A_1B_2 - A_2B_1}{\sqrt{\left(A_1^2 + B_1^2\right)\left(A_2^2 + B_2^2\right)}} \qquad \cos\theta = \frac{A_1A_2 + B_1B_2}{\sqrt{\left(A_1^2 + B_1^2\right)\left(A_2^2 + B_2^2\right)}}$$

$$\tan\theta = \frac{A_1B_2 - A_2B_1}{A_1A_2 + B_1B_2} \tag{2.33}$$

The signs of $\tan\theta$ and $\cos\theta$ determine whether the acute or obtuse angle is meant. If the equations are in the form $y = m_1x + b_1$, $y = m_2x + b_2$, then

$$\sin\theta = \frac{m_2 - m_1}{\sqrt{\left(1 + m_1^2\right)\left(1 + m_2^2\right)}} \qquad \cos\theta = \frac{1 + m_1m_2}{\sqrt{\left(1 + m_1^2\right)\left(1 + m_2^2\right)}} \qquad \tan\theta = \frac{m_2 - m_1}{1 + m_1m_2} \tag{2.34}$$

2.7.2 Transformation of Coordinates

Change of Origin O to O'. Let (x, y) denote the coordinates of a point P with respect to the old axes, and (x', y') the coordinates with respect to the new axes (Fig. 2.90). Then, if the coordinates of the new origin O' with respect to the old axes are $x = h$, $y = k$, the relations between the old and the new coordinates under transformation are

$$\left.\begin{array}{l} x = x' + h \\ y = y' + k \end{array}\right\} \tag{2.35}$$

Rotation of Axes about the Origin. Let θ (Fig. 2.91) be the angle through which the axes are rotated. Then

$$\left.\begin{array}{l} x = x'\cos\theta - y'\sin\theta \\ y = x'\sin\theta + y'\cos\theta \end{array}\right\} \tag{2.36}$$

If the axes are both translated and rotated,

$$\left.\begin{array}{l} x = x'\cos\theta - y'\sin\theta + h \\ y = x'\sin\theta + y'\cos\theta + k \end{array}\right\} \tag{2.37}$$

Coordinate Transformation. The relations between the rectangular coordinates x, y and the polar coordinates r, θ are

$$x = r\cos\theta \qquad y = r\sin\theta \qquad r = \sqrt{x^2 + y^2} \qquad \theta = \tan^{-1}\frac{y}{x} \tag{2.38}$$

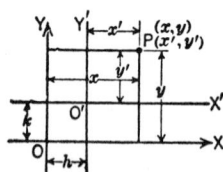

Fig. 2.90

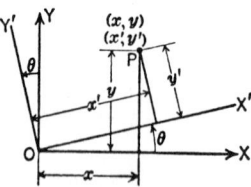

Fig. 2.91

2.7.3 Conic Sections

Conic Section. This is a curve traced by a point P moving in a plane so that the distance PF of the point from a fixed point (*focus*) is in constant ratio to the distance PM of the point from a fixed line (*directrix*) in the plane o. the curve. The ratio, $e = PF/PM$, is called the *eccentricity*. If $e < 1$, the curve is an *ellipse*; $e = 1$, a *parabola*; $e > 1$, a *hyperbola*; and $e = 0$, a *circle*, which is a special case of an ellipse.

The Circle. The equation is

$$(x - x_0)^2 + (y - y_0)^2 = r^2 \tag{2.39}$$

where (x_0, y_0) is the center and r the radius. If the center is at the origin,

$$x^2 + y^2 = r^2 \tag{2.40}$$

Another form is

$$x^2 + y^2 + 2gx + 2fy + c = 0 \tag{2.41}$$

with center $(-g, -f)$ and radius $\sqrt{g^2 + f^2 - c}$.
The equation of the tangent to (2.41) at a point $P_1(x_1, y_1)$ is

$$xx_1 + yy_1 + g(x + x_1) + f(y + y_1) + c = 0 \tag{2.42}$$

The Ellipse (Fig. 2.92). The equation is

$$\frac{(x - x_0)^2}{a^2} + \frac{(y - y_0)^2}{b^2} = 1 \tag{2.43}$$

where (x_0, y_0) is the center, a = semimajor axis, b = semiminor axis. In Fig. 2.92, $(x_0, y_0) = (0, 0)$.
Coordinates of *foci* are $F_1 = (-ae, 0)$, $F_2 = (ae, 0)$; $e^2 = (F_1 P)^2/(MP)^2 = 1 - b^2/a^2 < 1$; and the *directrices* are the lines $x = -a/e$, $x = a/e$.
The chord LL' through F is called the *latus rectum* and has the length $2b^2/a = 2a(1 - e^2)$. If P_1 is any point on the ellipse, $F_1 P_1 = a - ex_1$, $F_2 P_1 = a + ex_1$, and $F_1 P_1 + F_2 P_1 = 2a$ (a constant).
The *area* of the ellipse with semiaxes a and b is

$$A = \pi ab \tag{2.44}$$

The equation of the *tangent* to the ellipse (Fig. 2.92) at the point (x_1, y_1) is

$$\frac{xx_1}{a^2} + \frac{yy_1}{b^2} = 1 \tag{2.45}$$

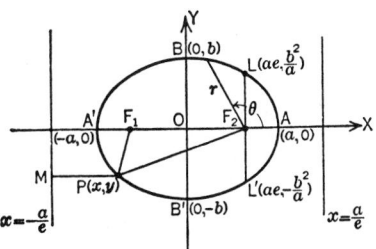

Fig. 2.92

the equation of the tangent with slope m is

$$y = mx \pm \sqrt{a^2m^2 + b^2} \tag{2.46}$$

The equation of the *normal* to the ellipse at the point (x_1, y_1) is

$$a^2y_1(x - x_1) - b^2x_1(y - y_1) = 0 \tag{2.47}$$

Conjugate Diameters. A line through the center of an ellipse is a *diameter*; if the slopes m and m' of the two diameters $y = mx$ and $y = m'x$ are such that $mm' = -b^2/a^2$ each diameter bisects all chords parallel to the other and the diameters are called *conjugate*.

Other Forms of the Equation of the Ellipse

$$\frac{x^2}{a^2} + \frac{y^2}{a^2(1 - e^2)} = 1 \tag{2.48}$$

$$ax^2 + by^2 + 2gx + 2fy + c = 0 \tag{2.49}$$

If a, b, and $g^2/a + f^2/b - c$ have the same sign, (2.49) is an ellipse whose axes are parallel to the coordinate axes.

The parametric form is

$$x = a \cos \phi \qquad y = b \sin \phi \tag{2.50}$$

The Hyperbola (Fig. 2.93). The equation is

$$\frac{(x - x_0)^2}{a^2} - \frac{(y - y_0)^2}{b^2} = 1 \tag{2.51}$$

where (x_0, y_0) is the center, $AA' = 2a$ is the transverse axis, $BB' = 2b$ is the conjugate axis. In Fig. 2.93, $(x_0, y_0) = (0,0)$.

$$e^2 = \frac{(F_1P)^2}{(PM)^2} = 1 + \frac{b^2}{a^2} > 1$$

the coordinates of the *foci*, $F_1 = (-ae, 0)$, $F_2 = (ae, 0)$; and the *directrices* are the lines $x = -a/e$, $x = a/e$.

The chord LL' through F is called the *latus rectum* and has the length $2b^2/a = 2a(e^2 - 1)$. If P_1 is any point on the curve, $F_1P_1 = ex_1 - a$, $F_2P_1 = ex_1 + a$, and $|F_2P_1 - F_1P_1| = 2a$ (a constant).

The equation of the *tangent* to the hyperbola (Fig. 2.93) at the point (x_1, y_1) is

$$\frac{xx_1}{a^2} - \frac{yy_1}{b^2} = 1 \tag{2.52}$$

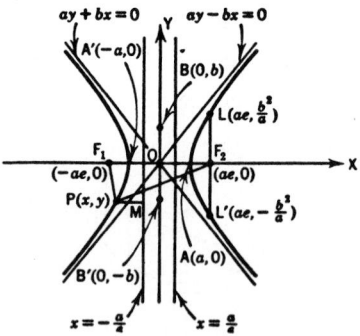

Fig. 2.93

The equation of the tangent whose slope is m is

$$y = mx \pm \sqrt{a^2 m^2 - b^2} \qquad (2.53)$$

The equation of the *normal* to the hyperbola at the point (x_1, y_1) is

$$a^2 y_1 (x - x_1) + b^2 x_1 (y - y_1) = 0 \qquad (2.54)$$

Conjugate Hyperbolas and Diameters. The two hyperbolas

$$\frac{x^2}{a^2} - \frac{y^2}{b^2} = 1 \quad \text{and} \quad \frac{y^2}{b^2} - \frac{x^2}{a^2} = 1$$

are conjugate. The transverse axis of each is the conjugate axis of the other.

If the slopes of the two lines $y = mx$ and $y = m_1 x$ through the center O are connected by the relation $mm_1 = b^2/a^2$, each of these lines bisects all chords of the hyperbola that are parallel to the other line. Two such lines are called *conjugate diameters*. The equation of the hyperbola referred to its conjugate diameters as oblique axes is

$$\frac{x'^2}{a_1^2} - \frac{y'^2}{b_1^2} = 1 \qquad (2.55)$$

where $2a_1$ and $2b_1$ are the conjugate axes.

The Asymptotes. The lines $y = (b/a)x$ and $y = -(b/a)x$ are the *asymptotes* of the hyperbola $x^2/a^2 - y^2/b^2 = 1$. The asymptotes are two tangents whose points of contact with the curve are at an infinite distance from the center. The equation of the hyperbola when referred to its asymptotes as oblique axes is

$$4x'y' = a^2 + b^2 \qquad (2.56)$$

If $a = b$, the asymptotes are the perpendicular lines $y = x$, $y = -x$; the corresponding hyperbola

$$x^2 - y^2 = a^2 \qquad (2.57)$$

is called the *rectangular* or *equilateral hyperbola*.

Other Forms of the Equation of the Hyperbola

$$\frac{x^2}{a^2} - \frac{y^2}{a^2(e^2 - 1)} = 1 \qquad (2.58)$$

$$ax^2 + by^2 + 2gx + 2fy + c = 0 \qquad (2.59)$$

If a and b have unlike signs, (2.59) is a hyperbola with axes parallel to the coordinate axes. The parametric form is

$$x = a \sec \phi \qquad y = a \tan \phi \qquad (2.60)$$

The Parabola. The equation of the parabola is

$$(y - y_0)^2 = 4a(x - x_0) \qquad (2.61)$$

If $(x_0, y_0) = (0,0)$, the *vertex* is at the origin (Fig. 2.94); the *focus* F is on OX, called the *axis of the parabola*, and has the coordinates $(a, 0)$; and the *directrix* is $x = -a$. The chord LL' through F is the *latus rectum* and has the length $4a$. The *eccentricity* $e = FP/PM = 1$.

The *tangent* to the parabola $y^2 = 4ax$ at the point (x_1, y_1) is

$$yy_1 = 2a(x + x_1) \qquad (2.62)$$

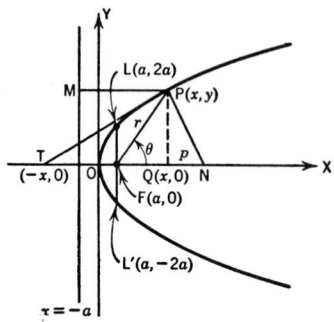

Fig. 2.94

The equation of the tangent whose slope is m is

$$y = mx + \frac{a}{m}$$

(2.63)

The *normal* to the parabola at the point (x_1, y_1) is

$$2a(y - y_1) + y_1(x - x_1) = 0$$

(2.64)

A *diameter* of the curve is a straight line parallel to the axis. It bisects all chords parallel to the tangent at the point where the diameter meets the parabola.

If P_1T is tangent to the curve at (x_1, y_1), then $TQ = 2x_1$ is the *subtangent*, and $QN = 2a$ (a constant) is the *subnormal*, where P_1N is perpendicular to P_1T.

The equation of the form $y^2 + 2gx + 2fy + c = 0$, where $g \neq 0$, is a parabola whose axis is parallel to OX; and the equation $x^2 + 2gx + 2fy + c = 0$, where $f \neq 0$, is a parabola whose axis is parallel to OY.

The parabola referred to the tangents at the extremities of its latus rectum as axes of coordinates is

$$x^{1/2} \pm y^{1/2} = b^{1/2}$$

(2.65)

where b is the distance from the origin to each point of tangency.

Polar Equations of the Conics. If e is the eccentricity, if the directrix is vertical, if the focus is at a distance p to the right or left of it, respectively, and if the polar origin is taken at the focus, the polar equation is

$$r = \frac{ep}{1 \mp e \cos \theta} \quad \text{for ellipse, hyperbola, or parabola}$$

(2.66)

$$r = \frac{a(1 - e^2)}{1 \mp e \cos \theta} \quad \text{for ellipse or circle}$$

(2.67)

$$r = \frac{a(e^2 - 1)}{1 \mp e \cos \theta} \quad \text{for hyperbola}$$

(2.68)

If the directrix is horizontal and the focus is at a distance p above or below it, respectively, the polar equation is

$$r = \frac{ep}{1 \mp e \sin \theta} \quad \text{for ellipse, hyperbola, or parabola}$$

(2.69)

$$r = \frac{a(1 - e^2)}{1 \mp e \sin \theta} \quad \text{for ellipse or circle}$$

(2.70)

$$r = \frac{a(e^2 - 1)}{1 \mp e \sin \theta} \quad \text{for hyperbola}$$

(2.71)

General Equation of a Conic Section. This equation has the form

$$ax^2 + 2hxy + by^2 + 2gx + 2fy + c = 0 \tag{2.72}$$

Let

$$D = \begin{vmatrix} a & h & g \\ h & b & f \\ g & f & c \end{vmatrix} \qquad d = \begin{vmatrix} a & h \\ h & b \end{vmatrix} \qquad \delta = a + b \tag{2.73}$$

Then the following is a classification of conic sections.

1. A parabola for $d = 0$, $D \neq 0$.
2. Two parallel lines (possibly coincident or imaginary) for $d = 0$, $D = 0$.
3. An ellipse for $d > 0$, $\delta D < 0$.
4. No locus (imaginary ellipse) for $d > 0$, $\delta D > 0$.
5. Point ellipse for $d > 0$, $D = 0$.
6. A hyperbola for $d < 0$, $D \neq 0$.
7. Two intersecting lines for $d < 0$, $D = 0$.

Let $A + B = a + b$, $AB = ab - h^2 = d$, and $A - B$ have the same sign as h. Let $c' = D/d$; then the equation of the conic referred to its axes is

$$\frac{x^2}{-c'/A} + \frac{y^2}{-c'/B} = 1 \tag{2.74}$$

To find the center (x_0, y_0) of the conic solve the equations

$$\left. \begin{array}{l} ax_0 + hy_0 + g = 0 \\ hx_0 + by_0 + f = 0 \end{array} \right\} \tag{2.75}$$

To remove the term in xy from (2.64), rotate the axes about the origin through an angle θ such that $\tan 2\theta = 2h/(a - b)$.

2.7.4 Higher Plane Curves

Plane Curves. The point (x, y) describes a plane curve if x and y are continuous functions of a variable t (parameter), as $x = x(t)$, $y = y(t)$. The elimination of t from the two equations gives $F(x, y) = 0$ or in explicit form $y = f(x)$. The angle τ, which a tangent to the curve makes with OX, can be found from

$$\sin \tau = \frac{dy}{ds} \qquad \cos \tau = \frac{dx}{ds} \qquad \tan \tau = \frac{dy}{dx} = y' \tag{2.76}$$

where ds is the element of arc length:

$$ds = \sqrt{dx^2 + dy^2} = \sqrt{1 + y'^2} \, dx \tag{2.77}$$

In polar coordinates,

$$ds = \sqrt{dr^2 + r^2 \, d\theta^2} = \sqrt{\left(\frac{dr}{d\theta}\right)^2 \theta p + r^2} \tag{2.78}$$

From Fig. 2.95, it may be seen that

$$\sin \psi = \frac{r \, d\theta}{ds} \qquad \cos \psi = \frac{dr}{ds} \qquad \tan \psi = \frac{r \, d\theta}{dr} \tag{2.79}$$

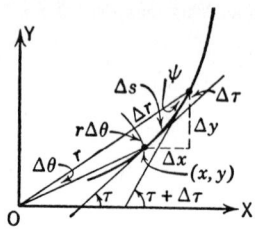

Fig. 2.95

The equation of the *tangent* to the curve $F(x, y) = 0$ at the point (x_1, y_1) is

$$\left(\frac{\partial F}{\partial x}\right)_{x=x_1,\ y=y_1}(x - x_1) + \left(\frac{\partial F}{\partial y}\right)_{x=x_1,\ y=y_1}(y - y_1) = 0 \qquad (2.80)$$

The equation of the *normal* to the curve $F(x, y) = 0$ at the point (x_1, y_1) is

$$\left(\frac{\partial F}{\partial y}\right)_{x=x_1,\ y=y_1}(x - x_1) - \left(\frac{\partial F}{\partial x}\right)_{x=x_1,\ y=y_1}(y - y_1) = 0 \qquad (2.81)$$

The equation of the *tangent* to the curve $y = f(x)$ at the point (x_1, y_1) is

$$y - y_1 = \left(\frac{dy}{dx}\right)_{x=x_1}(x - x_1) \qquad (2.82)$$

The equation of the *normal* to the curve $y = f(x)$ at the point (x_1, y_1) is

$$y - y_1 = -\frac{1}{\left(\dfrac{dy}{dx}\right)_{x=x_1}}(x - x_1) \qquad (2.83)$$

The *radius of curvature* of the curve at the point (x, y) is

$$\rho = \frac{ds}{d\tau} = \frac{\left[1 + \left(\dfrac{dy}{dx}\right)^2\right]^{3/2}}{\dfrac{d^2y}{dx^2}} = \frac{[1 + y'^2]^{3/2}}{y''} \qquad (2.84)$$

The reciprocal $1/\rho$ is called the *curvature of the curve* at (x, y).
The coordinates (x_0, y_0) of the center of curvature for the point (x, y) on the curve (the center of the circle of curvature tangent to the curve at (x, y) and of radius ρ) are

$$\left.\begin{aligned} x_0 &= x - \rho\frac{dy}{ds} = x - y'\frac{[1 + y'^2]}{y''} \\[2mm] y_0 &= y + \rho\frac{dx}{ds} = y + \frac{[1 + y'^2]}{y''} \end{aligned}\right\} \qquad (2.85)$$

A curve has a *singular point* if simultaneously,

$$F(x, y) = 0 \qquad \frac{\partial F}{\partial x} = 0 \qquad \frac{\partial F}{\partial y} = 0 \qquad (2.86)$$

Let

$$D = \left(\frac{\partial^2 F}{\partial x \partial y} \right)^2 - \frac{\partial^2 F}{\partial x^2} \frac{\partial^2 F}{\partial y^2} \tag{2.87}$$

Then for $D > 0$, the curve has a *double point* with two real different tangents.
For $D = 0$, the curve has a *cusp* with two coincident tangents.
For $D < 0$, the curve has an *isolated point* with no real tangent.
See Figs. 2.96–2.100 for special curves.

Semicubic, or Neil's, Parabola	Logarithmic Curve	Exponential Curve

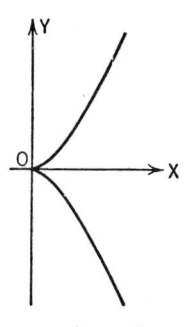

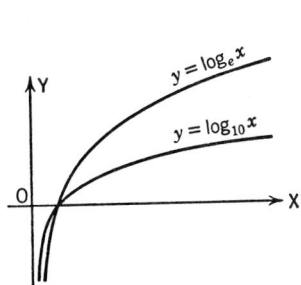

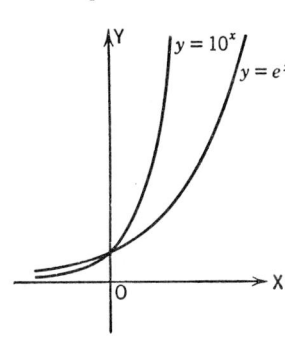

$y^2 = ax^3$

Fig. 2.96

$y = \log_b x$

Fig. 2.97

$y = b^x$

Fig. 2.98

Catenary

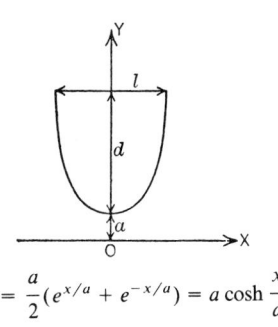

Damped Wave

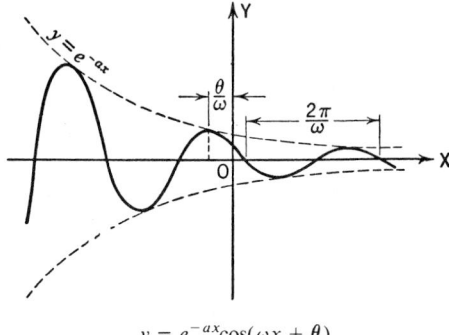

$$y = \frac{a}{2}(e^{x/a} + e^{-x/a}) = a \cosh \frac{x}{a}$$

Fig. 2.99

$y = e^{-ax}\cos(\omega x + \theta)$

Fig. 2.100

For l large compared with d,

$$s \approx l\left[1 + \frac{2}{3}\left(\frac{2d}{l} \right)^2 \right]$$

Trochoid. This is a curve traced by a point at a distance b from the center of a circle of radius a as the circle rolls on a straight line.

$$x = a\phi - b \sin \phi, \quad y = a - b \cos \phi$$

See Figs. 2.101–2.103 for cycloids.

Cycloid

$a = b$

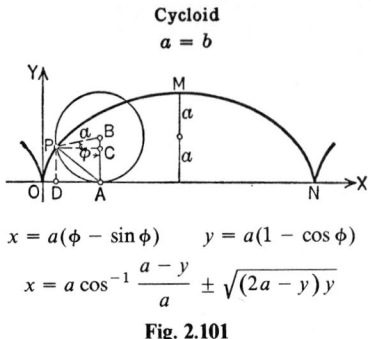

$$x = a(\phi - \sin \phi) \qquad y = a(1 - \cos \phi)$$

$$x = a \cos^{-1} \frac{a - y}{a} \pm \sqrt{(2a - y)y}$$

Fig. 2.101

For one arch, arc length $= 8a$, area $= 3\pi a^2$.

Prolate Cycloid	Curtate Cycloid
$a < b$	$a > b$

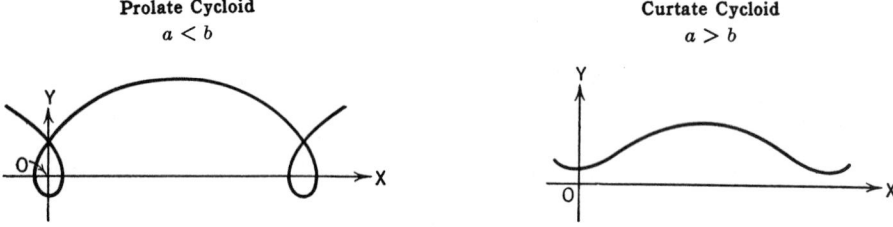

Fig. 2.102   **Fig. 2.103**

Hypotrochoid. This is a curve traced by a point at a distance b from the center of a circle of radius a as the circle rolls on the inside of a fixed circle of radius R.

$$x = (R - a)\cos \phi + b \cos \frac{R - a}{a}\phi \qquad y = (R - a)\sin \phi - b \sin \frac{R - a}{a}\phi$$

Hypocycloid. $b = a$ (Fig. 2.104).

Hypocycloid of Four Cusps, or Astroid

$b = a = {}^1/_4 R$

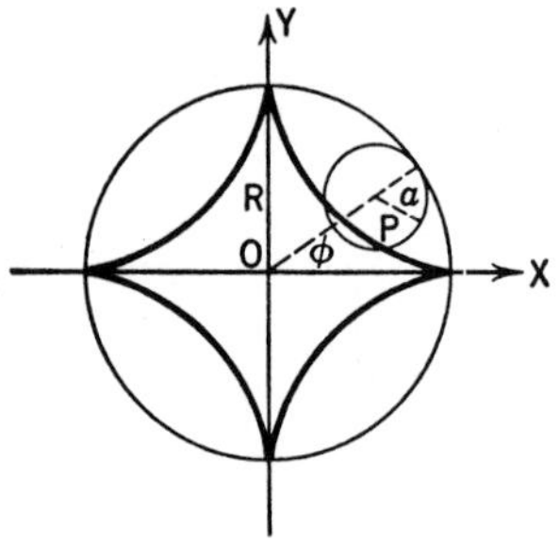

$$x = R \cos^3 \phi, \ y = R \sin^3 \phi$$

$$x^{2/3} + y^{2/3} = R^{2/3}$$

Fig. 2.104

Epitrochoid. This is a curve traced by a point at a distance b from the center of a circle of radius a as the circle rolls on the outside of a fixed circle of radius R. See Figs. 2.105 and 2.106.

Epicycloid
$b = a$

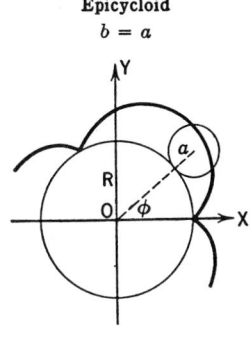

Limaçon of Pascal

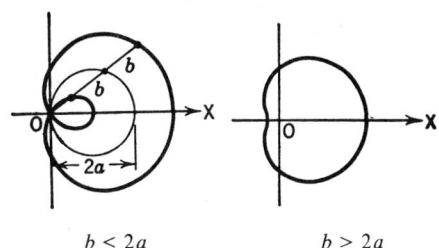

$b < 2a$ $b > 2a$

$r = b + 2a \cos \theta$

Fig. 2.105   Fig. 2.106   Fig. 2.107

Other forms of the right-hand side of the equation, $b + 2a \sin \theta$, $b - 2a \cos \theta$, $b - 2a \sin \theta$, give curves rotated through $1, 2, 3$ right angles, respectively. See Figs. 2.107–2.110.

Involute of a Circle

Cardioid

Limaçon in which $b = 2a$

Epicycloid in which $R = a$

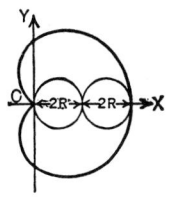

$r = 2a(1 + \cos \theta)$

$(x^2 + y^2 - 2ax)^2 = 4a^2(x^2 + y^2)$

Fig. 2.108

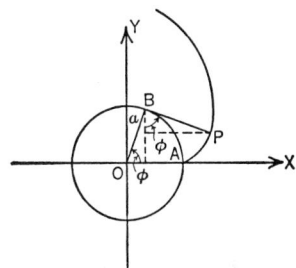

$x = a(\cos \phi + \phi \sin \phi)$

$y = a(\sin \phi - \phi \cos \phi)$

$\theta = \sqrt{r^2/a^2 - 1} - \tan^{-1} \sqrt{r^2/a^2 - 1}$

Spiral traced by the end of a taut string
unwinding from a circle.

Fig. 2.109

Spiral of Archimedes

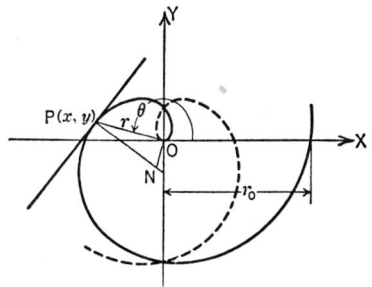

$r = a\theta$

Polar subnormal $ON = a$

Length of arc $OP = s = \frac{1}{2}a(\theta\sqrt{1 + \theta^2}$
$+ \sinh^{-1}\theta)$

For many turns, $s \approx \frac{1}{2}a\theta^2$

Fig. 2.110

Hyperbolic, or Reciprocal, Spiral

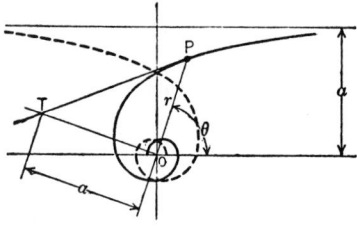

$$r\theta = a$$

Polar subtangent $OT = -a$ or

As $\theta \to \infty$, $r \to 0$. The curve winds an indefinite number of times around the origin.

As $\theta \to 0$, $r \to \infty$. The curve has an asymptote parallel to the polar axis at a distance a.

Fig. 2.111

Logarithmic, or Equiangular, Spiral

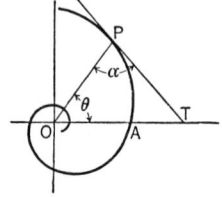

$$r = ae^{m\theta} \qquad m > 0$$

$$\ln \frac{r}{a} = m\theta$$

The tangent to the curve at any point makes a constant angle $\alpha(= \cot^{-1}m)$ with the radius vector.

As $\theta \to -\infty$, $r \to 0$. The curve winds an indefinite number of times around the origin.

Fig. 2.112

Lemniscate of Bernoulli

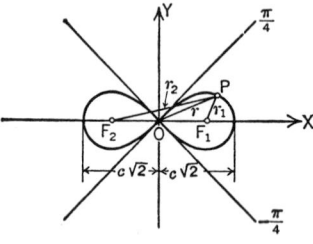

$$(x^2 + y^2)^2 + 2c^2(y^2 - x^2) = 0$$
$$r^2 = 2c^2\cos 2\theta$$

Locus of a point P, the product of whose distances from two fixed points F_1 and F_2 is equal to the square of half the distance between them, $r_1 \cdot r_2 = c^2$.

Fig. 2.113

Three-leaved Roses

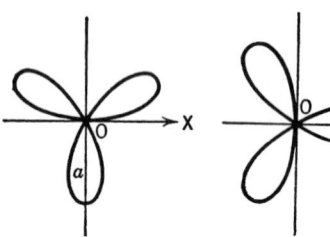

$r = a \sin 3\theta$ $r = a \cos 3\theta$

Fig. 2.114   **Fig. 2.115**

Four-leaved Roses

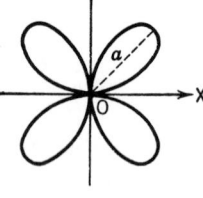   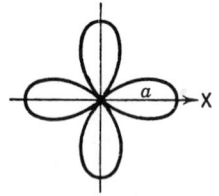

$r = a \sin 2\theta$ $r = a \cos 2\theta$

Fig. 2.116   **Fig. 2.117**

The roses, $r = a \sin n\theta$ and $r = a \cos n\theta$, have, for n even, $2n$ leaves; for n odd, n leaves.

Cissoid of Diocles

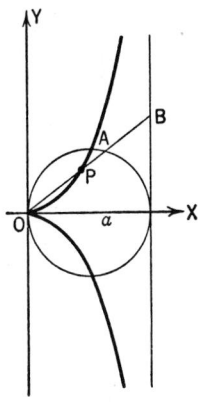

$$y^2 = \frac{x^3}{a - x}$$

$$r = a(\sec \theta - \cos \theta)$$

Locus of point P such that $OP = AB$.

Fig. 2.118

Strophoid

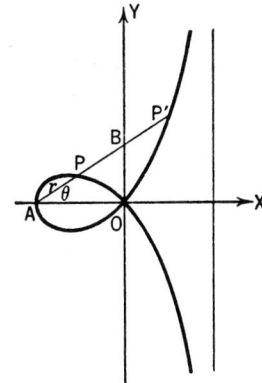

$$y^2 = \frac{x^2(a + x)}{a - x}$$

$$r = a(\sec \theta - \tan \theta)$$

If the line AB rotates about A, intersecting the y axis at B, and if $PB = BP' = OB$, the locus of P and P' is the strophoid.

Fig. 2.119

Conchoid of Nicomedes

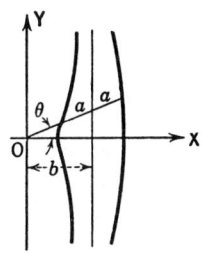

$$(x^2 + y^2)(x - b)^2 = a^2 x^2$$

$$r = b \sec \theta - a$$

Fig. 2.120

Witch of Agnesi

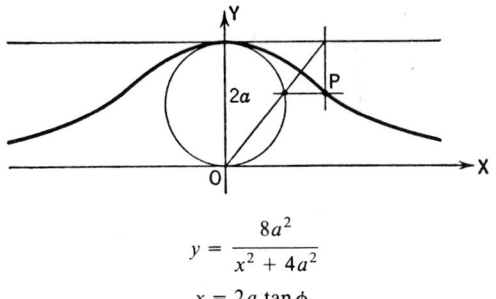

$$y = \frac{8a^2}{x^2 + 4a^2}$$

$$x = 2a \tan \phi$$

$$y = 2a \cos^2 \phi$$

Fig. 2.121

Folium of Descartes

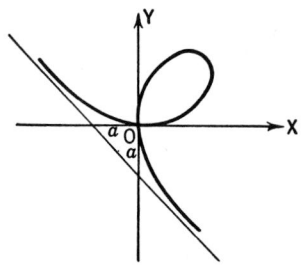

$$x^3 + y^3 - 3axy = 0$$

$$r = \frac{3a \sin\theta \cos\theta}{\sin^3\theta + \cos^3\theta}$$

Fig. 2.122

Tractrix

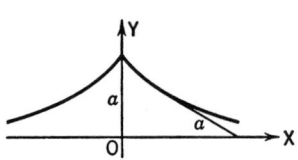

$$x = a \cosh^{-1} \frac{a}{y} - \sqrt{a^2 - y^2}$$

Locus of one end P of tangent line of length a as the other end Q is moved along the x axis.

Fig. 2.123

Circles in Polar Coordinates

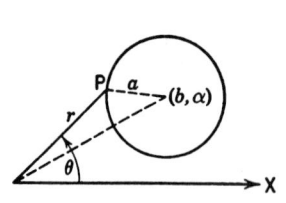

$$r^2 + b^2 - 2rb \cos(\theta - \alpha) = a^2$$

Center at (b, α), radius a

Fig. 2.124

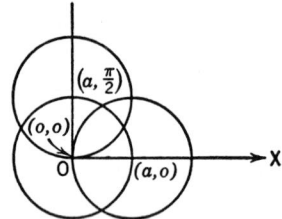

Center $(0,0)$ $r = a$

Center $(a, 0)$ $r = 2a \cos\theta$

Center $\left(a, \dfrac{\pi}{2}\right)$ $r = 2a \sin\theta$

Fig. 2.125

Frequency-modulated Wave

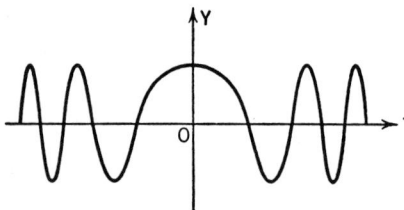

$$y = k \cos[\phi(t)] \qquad \text{instantaneous frequency} = \Omega(t) = \frac{d\phi}{dt}$$

In Fig. 2.126, $y = \cos \pi/2t^2$, $\Omega(t) = \pi t$.

Fig. 2.126

2.8 SOLID ANALYTIC GEOMETRY

2.8.1 Coordinate Systems

Right-Hand Rectangular (Fig. 2.127). The position of a point $P(x, y, z)$ is fixed by its distances x, y, z from the mutually perpendicular planes yz, xz, and xy, respectively.

Spherical, or Polar (Fig. 2.128). The position of a point $P(r, \theta, \phi)$ is fixed by its distance from a given point O, the origin, and its direction from O, determined by the angles θ and ϕ.

Cylindrical (Fig. 2.128). The position of a point $P(\rho, \phi, z)$ is fixed by its distance z from a given plane and the polar coordinates (ρ, ϕ) of the projection Q of P on the given plane.

Relations among coordinates of the three systems:

$$x = r \sin\theta \cos\phi = \rho \cos\phi \tag{2.88}$$

$$y = r \sin\theta \sin\phi = \rho \sin\phi \tag{2.89}$$

$$z = r \cos\theta \tag{2.90}$$

$$\rho = \sqrt{x^2 + y^2} = r \sin\theta \tag{2.91}$$

$$\phi = \tan^{-1}\frac{y}{x} \tag{2.92}$$

$$r = \sqrt{x^2 + y^2 + z^2} = \sqrt{\rho^2 + z^2} \tag{2.93}$$

$$\theta = \tan^{-1}\frac{\sqrt{x^2 + y^2}}{z} = \tan^{-1}\frac{\rho}{z} \tag{2.94}$$

2.8.2 Point, Line, and Plane

Euclidean Distance between Two Points. This distance between $P_1(x_1, y_1, z_1)$ and $P_2(x_2, y_2, z_2)$ is

$$s = \sqrt{(x_2 - x_1)^2 + (y_2 - y_1)^2 + (z_2 - z_1)^2} \tag{2.95}$$

To divide the segment $P_1 P_2$ in the ratio c_1/c_2, internally or externally,

$$x = \frac{c_2 x_1 \pm c_1 x_2}{c_2 \pm c_1} \qquad y = \frac{c_2 y_1 \pm c_1 y_2}{c_2 \pm c_1} \qquad z = \frac{c_2 z_1 \pm c_1 z_2}{c_2 \pm c_1} \tag{2.96}$$

The *midpoint* of $P_1 P_2$ is

$$x = \frac{x_1 + x_2}{2} \qquad y = \frac{y_1 + y_2}{2} \qquad z = \frac{z_1 + z_2}{2} \tag{2.97}$$

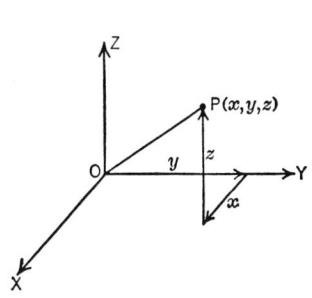

Fig. 2.127

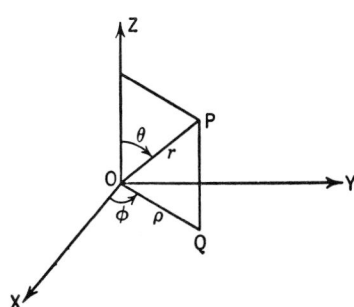

Fig. 2.128

Angles. The angles α, β, γ that the line $P_1 P_2$ makes with the coordinate directions, $x, y, z,$ respectively, are the *direction angles* of $P_1 P_2$. The cosines

$$\cos \alpha = \frac{x_2 - x_1}{s} \qquad \cos \beta = \frac{y_2 - y_1}{s} \qquad \cos \gamma = \frac{z_2 - z_1}{s} \tag{2.98}$$

are the *direction cosines* of $P_1 P_2$, and

$$\cos^2 \alpha + \cos^2 \beta + \cos^2 \gamma = 1 \tag{2.99}$$

If $l : m : n = \cos \alpha : \cos \beta : \cos \gamma$, then

$$\cos \alpha = \frac{l}{\sqrt{l^2 + m^2 + n^2}} \qquad \cos \beta = \frac{m}{\sqrt{l^2 + m^2 + n^2}} \qquad \cos \gamma = \frac{n}{\sqrt{l^2 + m^2 + n^2}} \tag{2.100}$$

The angle θ between two lines in terms of their direction angles $\alpha_1, \beta_1, \gamma_1,$ and $\alpha_2, \beta_2, \gamma_2$ is obtained from

$$\cos \theta = \cos \alpha_1 \cos \alpha_2 + \cos \beta_1 \cos \beta_2 + \cos \gamma_1 \cos \gamma_2 \tag{2.101}$$

If $\cos \theta = 0$, the lines are perpendicular to each other.

Planes. A plane is represented by

$$Ax + By + Cz + D = 0 \tag{2.102}$$

If one of the variables is missing, the plane is parallel to the axis of the missing variable. For example, $Ax + By + D = 0$ represents a plane parallel to the z axis. If two of the variables are missing, the plane is parallel to the plane of the missing variables. For example, $z = k$ represents a plane parallel to the xy plane and k units from it.

A plane through *three points* $P_1(x_1, y_1, z_1)$, $P_2(x_2, y_2, z_2)$, and $P_3(x_3, y_3, z_3)$ has the equation

$$\begin{vmatrix} x & y & z & 1 \\ x_1 & y_1 & z_1 & 1 \\ x_2 & y_2 & z_2 & 1 \\ x_3 & y_3 & z_3 & 1 \end{vmatrix} = 0 \tag{2.103}$$

The equation of a plane whose x, y, z *intercepts* are, respectively, a, b, c (Fig. 2.129) is

$$\frac{x}{a} + \frac{y}{b} + \frac{z}{c} = 1 \tag{2.104}$$

The *perpendicular* form of the equation of a plane, where $OP = p$ is the perpendicular distance of the plane from the origin O and has the direction angles α, β, γ, is

$$x \cos \alpha + y \cos \beta + z \cos \gamma - p = 0 \tag{2.105}$$

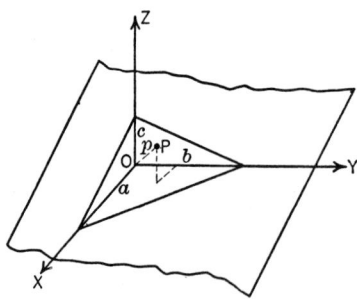

Fig. 2.129

To bring the general form $Ax + By + Cz + D = 0$ into the perpendicular form, divide it by $\pm \sqrt{A^2 + B^2 + C^2}$, where the sign before the radical is opposite to that of D.

The coefficients A, B, C are proportional to the direction cosines λ, μ, ν of a line perpendicular to the plane. Therefore,

$$A(x - x_1) + B(y - y_1) + C(z - z_1) = 0 \tag{2.106}$$

is a plane through $P_1(x_1, y_1, z_1)$ and perpendicular to a line with direction cosines λ, μ, ν proportional to A, B, C.

Perpendicular Distance between Point and Plane. The distance between point P_1 from a plane $Ax + By + Cz + D = 0$ is given by

$$PP_1 = \frac{Ax_1 + By_1 + Cz_1 + D}{\pm \sqrt{A^2 + B^2 + C^2}} \tag{2.107}$$

where the sign before the radical is opposite to that of D.

Parallel Planes. Two planes $A_1 x + B_1 y + C_1 z + D_1 = 0$ and $A_2 x + B_2 y + C_2 z + D_2 = 0$ are parallel if $A_1 : B_1 : C_1 = A_2 : B_2 : C_2$.

$$A(x - x_1) + B(y - y_1) + C(z - z_1) = 0 \tag{2.108}$$

is a plane through the point $P_1(x_1, y_1, z_1)$ and parallel to the plane $Ax + By + Cz + D = 0$.

Angle θ between Two Planes. The angle between $Ax + By + Cz + D = 0$ and $A_1 x + B_1 y + C_1 z + D_1 = 0$ is the angle between two intersecting lines, each perpendicular to one of the planes:

$$\cos \theta = \frac{AA_1 + BB_1 + CC_1}{\pm \sqrt{(A^2 + B^2 + C^2)(A_1^2 + B_1^2 + C_1^2)}} \tag{2.109}$$

The two planes are perpendicular if $AA_1 + BB_1 + CC_1 = 0$.

Points, Planes, and Lines. Four points, $P_k(x_k, y_k, z_k)$ $(k = 1, 2, 3, 4)$, lie in the same plane if

$$\begin{vmatrix} 1 & x_1 & y_1 & z_1 \\ 1 & x_2 & y_2 & z_2 \\ 1 & x_3 & y_3 & z_3 \\ 1 & x_4 & y_4 & z_4 \end{vmatrix} = 0 \tag{2.110}$$

Four planes, $A_k x + B_k y + C_k z + D_k = 0$ $(k = 1, 2, 3, 4)$, pass through the same point if

$$\begin{vmatrix} A_1 & B_1 & C_1 & D_1 \\ A_2 & B_2 & C_2 & D_2 \\ A_3 & B_3 & C_3 & D_3 \\ A_4 & B_4 & C_4 & D_4 \end{vmatrix} = 0 \tag{2.111}$$

A straight line is represented as the intersection of two planes by two first-degree equations

$$\left. \begin{array}{l} A_1 x + B_1 y + C_1 z + D_1 = 0 \\ A_2 x + B_2 y + C_2 z + D_2 = 0 \end{array} \right\} \tag{2.112}$$

The three planes through the line perpendicular to the coordinate planes are its *projecting planes*. The equation of the xy projecting plane is found by eliminating z between the two given equations, and so on. The line can be represented by any two of its projecting planes, for example,

$$\left. \begin{array}{l} y = m_1 x + b_1 \\ z = m_2 x + b_2 \end{array} \right\} \tag{2.113}$$

If the line goes through a *point* $P_1(x_1, y_1, z_1)$ and has the *direction angles* α, β, γ, then

$$\frac{x - x_1}{\cos \alpha} = \frac{y - y_1}{\cos \beta} = \frac{z - z_1}{\cos \gamma} \tag{2.114}$$

and

$$m_1 = \frac{\cos \beta}{\cos \alpha} \qquad m_2 = \frac{\cos \gamma}{\cos \alpha}$$

The equations of a line through *two points* (x_1, y_1, z_1) and (x_2, y_2, z_2) are

$$\frac{x - x_1}{x_2 - x_1} = \frac{y - y_1}{y_2 - y_1} = \frac{z - z_1}{z_2 - z_1} \tag{2.115}$$

A line through a *point* P_1 perpendicular to a plane $Ax + By + Cz + D = 0$ has the equations

$$\frac{x - x_1}{A} = \frac{y - y_1}{B} = \frac{z - z_1}{C} \tag{2.116}$$

Line of Intersection of Two Planes. The direction cosines λ, μ, ν of the line of intersection of two planes $Ax + By + Cz + D = 0$ and $A_1x + B_1y + C_1z + D_1 = 0$ are found from the ratios

$$\lambda : \mu : \nu = \begin{vmatrix} B & C \\ B_1 & C_1 \end{vmatrix} : \begin{vmatrix} C & A \\ C_1 & A_1 \end{vmatrix} : \begin{vmatrix} A & B \\ A_1 & B_1 \end{vmatrix} \tag{2.117}$$

2.8.3   Transformation of Coordinates

Changing the Origin. Let the coordinates of a point P with respect to the original axes be x, y, z and with respect to the new axes x', y', z'. For a parallel displacement of the axes with x_0, y_0, z_0 the coordinates of the new origin

$$x = x_0 + x' \qquad y = y_0 + y' \qquad z = z_0 + z' \tag{2.118}$$

Rotation of the Axes about the Origin. Let the cosines of the angles of the new axes x', y', z', with the x axis be λ_1, μ_1, ν_1, with the y axis be λ_2, μ_2, ν_2, with the z axis be λ_3, μ_3, ν_3. Then

$$\left. \begin{aligned} x &= \lambda_1 x' + \mu_1 y' + \nu_1 z' & x' &= \lambda_1 x + \lambda_2 y + \lambda_3 z \\ y &= \lambda_2 x' + \mu_2 y' + \nu_2 z' & y' &= \mu_1 x + \mu_2 y + \mu_3 z \\ z &= \lambda_3 x' + \mu_3 y' + \nu_3 z' & z' &= \nu_1 x + \nu_2 y + \nu_3 z \end{aligned} \right\} \tag{2.119}$$

The following relations exist:

(1)
$$\lambda_1^2 + \mu_1^2 + \nu_1^2 = 1$$
$$\lambda_2^2 + \mu_2^2 + \nu_2^2 = 1$$
$$\lambda_3^2 + \mu_3^2 + \nu_3^2 = 1$$

(2)
$$\lambda_1^2 + \lambda_2^2 + \lambda_3^2 = 1$$
$$\mu_1^2 + \mu_2^2 + \mu_3^2 = 1$$
$$\nu_1^2 + \nu_2^2 + \nu_3^2 = 1$$

(3)
$$\lambda_1 \lambda_2 + \mu_1 \mu_2 + \nu_1 \nu_2 = 0$$
$$\lambda_2 \lambda_3 + \mu_2 \mu_3 + \nu_2 \nu_3 = 0$$
$$\lambda_3 \lambda_1 + \mu_3 \mu_1 + \nu_3 \nu_1 = 0$$

(4)
$$\lambda_1 \mu_1 + \lambda_2 \mu_2 + \lambda_3 \mu_3 = 0$$
$$\mu_1 \nu_1 + \mu_2 \nu_2 + \mu_3 \nu_3 = 0$$
$$\nu_1 \lambda_1 + \nu_2 \lambda_2 + \nu_3 \lambda_3 = 0$$

(5)
$$\lambda_1 = \mu_2 \nu_3 - \nu_2 \mu_3$$
$$\mu_1 = \nu_2 \lambda_3 - \lambda_2 \nu_3$$
$$\nu_1 = \lambda_2 \mu_3 - \mu_2 \lambda_3$$

(6)
$$\lambda_2 = \nu_1 \mu_3 - \mu_1 \nu_3$$
$$\mu_2 = \lambda_1 \nu_3 - \nu_1 \lambda_3$$
$$\nu_2 = \mu_1 \lambda_3 - \lambda_1 \mu_3$$

(7)
$$\lambda_3 = \mu_1 \nu_2 - \nu_1 \mu_2$$
$$\mu_3 = \nu_1 \lambda_2 - \lambda_1 \nu_2$$
$$\nu_3 = \lambda_1 \mu_2 - \mu_1 \lambda_2$$

(8)
$$\begin{vmatrix} \lambda_1 & \mu_1 & \nu_1 \\ \lambda_2 & \mu_2 & \nu_2 \\ \lambda_3 & \mu_3 & \nu_3 \end{vmatrix} = 1$$

For a combination of displacement and rotation, apply the corresponding equations simultaneously.

2.8.4 Quadric Surfaces

The **general form of the equation of a surface of the second degree** is

$$F(x, y, z) \equiv a_{11}x^2 + 2a_{12}xy + 2a_{13}xz + a_{22}y^2 + 2a_{23}yz + a_{33}z^2 + 2a_{14}x + 2a_{24}y$$
$$+ 2a_{34}z + a_{44} = 0 \tag{2.120}$$

where the a_{ik} are constants, and $a_{ik} = a_{ki}$, that is, $a_{12} = a_{21}$, etc. Let

$$D = \begin{vmatrix} a_{11} & a_{12} & a_{13} & a_{14} \\ a_{21} & a_{22} & a_{23} & a_{24} \\ a_{31} & a_{32} & a_{33} & a_{34} \\ a_{41} & a_{42} & a_{43} & a_{44} \end{vmatrix} \qquad d = \begin{vmatrix} a_{11} & a_{12} & a_{13} \\ a_{21} & a_{22} & a_{23} \\ a_{31} & a_{32} & a_{33} \end{vmatrix}$$

Let $I \equiv a_{11} + a_{22} + a_{33}$ and $J \equiv a_{22}a_{33} + a_{33}a_{11} + a_{11}a_{22} - a_{23}^2 - a_{13}^2 - a_{12}^2$. D, d, I, and J are invariant under coordinate transformation. The following is a classification of the quadratic surfaces, so far as they are real and do not degenerate into curves in one plane:

Ellipsoid, for $D < 0$, $Id > 0$, $J > 0$.
Hyperboloid of two sheets, for $D < 0$, Id and J not both > 0.
Hyperboloid of one sheet, for $D > 0$, Id and J not both > 0.
Cone, for $D = 0$, $d \neq 0$, Id and J not both > 0.
Elliptic paraboloid, for $D < 0$, $d = 0$, $J > 0$.
Hyperbolic paraboloid, for $D > 0$, $d = 0$, $J < 0$.
Cylinder, for $D = 0$, $d = 0$.

Ellipsoid and Hyperboloids. Consider the center of the quadric as the origin and the principal axes of the quadric as the orthogonal coordinate axes. Then

$$\frac{x^2}{a^2} + \frac{y^2}{b^2} + \frac{z^2}{c^2} = 1 \text{ is an } \textit{ellipsoid} \text{ (Fig. 2.130)} \tag{2.121}$$

$$\frac{x^2}{a^2} + \frac{y^2}{b^2} - \frac{z^2}{c^2} = 1 \text{ is a } \textit{hyperboloid of one sheet} \text{ (Fig. 2.131)} \tag{2.122}$$

$$\frac{x^2}{a^2} + \frac{y^2}{b^2} - \frac{z^2}{c^2} = -1 \text{ is a } \textit{hyperboloid of two sheets} \text{ (Fig. 2.132)} \tag{2.123}$$

where a, b, c are the semiaxes.
The length of the semiaxis is found from

$$a^2 = -\frac{D}{\lambda_1 d} \qquad b^2 = -\frac{D}{\lambda_2 d} \qquad c^2 = -\frac{D}{\lambda_3 d} \tag{2.124}$$

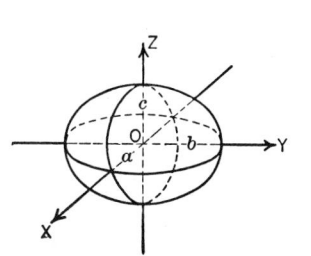

Fig. 2.130

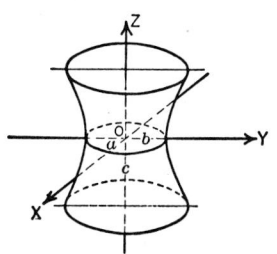

Fig. 2.131

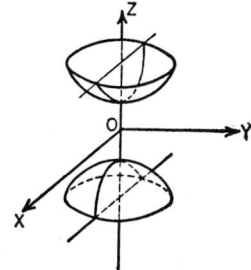

Fig. 2.132

where $\lambda_1, \lambda_2, \lambda_3$ are the real roots of the following cubic equation:

$$\begin{vmatrix} a_{11} - \lambda & a_{12} & a_{13} \\ a_{12} & a_{22} - \lambda & a_{23} \\ a_{13} & a_{23} & a_{33} - \lambda \end{vmatrix} = 0 \qquad (2.125)$$

Cone. The equation

$$ax^2 + by^2 + cz^2 + 2hxy + 2gxz + 2fyz = 0 \qquad (2.126)$$

represents a cone with vertex at the origin. If the cross section of the cone is an ellipse with axes $2a$ and $2b$, whose plane is parallel to the xy plane and at a distance c from the origin, then the equation of the cone with vertex at the origin is

$$\frac{x^2}{a^2} + \frac{y^2}{b^2} - \frac{z^2}{c^2} = 0 \qquad (2.127)$$

If $a = b$, the cross section is circular, and the cone is a cone of revolution.

Sphere. An equation of the form

$$x^2 + y^2 + z^2 + ax + by + cz + d = 0 \qquad (2.128)$$

represents a sphere with radius

$$r = \tfrac{1}{2}\sqrt{a^2 + b^2 + c^2 - 4d} \qquad (2.129)$$

and center

$$x_0 = -\tfrac{1}{2}a \qquad y_0 = -\tfrac{1}{2}b \qquad z_0 = -\tfrac{1}{2}c \qquad (2.130)$$

If (x_0, y_0, z_0) are the coordinates of the center and r is the radius, then the equation of the sphere is

$$(x - x_0)^2 + (y - y_0)^2 + (z - z_0)^2 = r^2 \qquad (2.131)$$

If $x_0 = 0$, $y_0 = 0$, $z_0 = 0$, then the equation is

$$x^2 + y^2 + z^2 = r^2 \qquad (2.132)$$

Paraboloids. The equation

$$\frac{x^2}{a^2} + \frac{y^2}{b^2} = 2cz \qquad (2.133)$$

represents an *elliptic paraboloid* (Fig. 2.133).
    If $a = b$, the equation is of the form

$$x^2 + y^2 = 2cz \quad (\text{a paraboloid of revolution}) \qquad (2.134)$$

The equation

$$\frac{x^2}{a^2} - \frac{y^2}{b^2} = 2cz \text{ represents a } hyperbolic \ paraboloid \ (\text{Fig. 2.134}) \qquad (2.135)$$

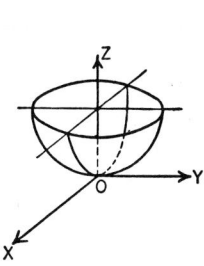

Fig. 2.133

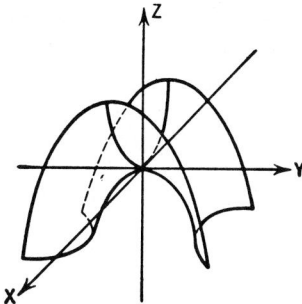

Fig. 2.134

Cylinder. The equation of a cylinder perpendicular to the yz, xz, or xy plane is the same as the equation of a section of the cylinder in the corresponding plane. Thus

$$\frac{x^2}{a^2} + \frac{y^2}{b^2} = 1 \tag{2.136}$$

$$\frac{x^2}{a^2} - \frac{y^2}{b^2} = 1 \tag{2.137}$$

$$y^2 = 4ax \tag{2.138}$$

are *elliptic*, *hyperbolic*, and *parabolic* cylinders, respectively, with elements or generators parallel to OZ.

Tangent Plane. The equation of the tangent plane to any quadric

$$F(x, y, z) \equiv a_{11}x^2 + 2a_{12}xy + 2a_{13}xz + a_{22}y^2 + 2a_{23}yz + a_{33}z^2 + 2a_{14}x$$
$$+ 2a_{24}y + 2a_{34}z + a_{44} = 0 \tag{2.139}$$

at the point (x_1, y_1, z_1) is

$$\left(\frac{\partial F}{\partial x}\right)_{x=x_1, \, y=y_1, \, z=z_1} (x - x_1) + \left(\frac{\partial F}{\partial y}\right)_{x=x_1, \, y=y_1, \, z=z_1} (y - y_1)$$

$$+ \left(\frac{\partial F}{\partial z}\right)_{x=x_1, \, y=y_1, \, z=z_1} (z - z_1) = 0 \tag{2.140}$$

Example 2.47. Find the tangent plane to the hyperboloid of one sheet at point (x_1, y_1, z_1). Given $x^2/a^2 + y^2/b^2 - z^2/c^2 = 1$. Then

$$\left(\frac{\partial F}{\partial x}\right)_{x=x_1, \, y=y_1, \, z=z_1} (x - x_1) + \left(\frac{\partial F}{\partial y}\right)_{x=x_1, \, y=y_1, \, z=z_1} (y - y_1) + \left(\frac{\partial F}{\partial z}\right)_{x=x_1, \, y=y_1, \, z=z_1} (z - z_1)$$

$$= \frac{2x_1(x - x_1)}{a^2} + \frac{2y_1(y - y_1)}{b^2} - \frac{2z_1(z - z_1)}{c^2} = 0$$

$$\frac{xx_1}{a^2} + \frac{yy_1}{b^2} - \frac{zz_1}{c^2} - \frac{x_1^2}{a^2} - \frac{y_1^2}{b^2} + \frac{z_1^2}{c^2} = \frac{xx_1}{a^2} + \frac{yy_1}{b^2} - \frac{zz_1}{c^2} - 1 = 0 \text{ is the tangent plane.}$$

The Normal. The line through a point P_1 on a surface and perpendicular to the tangent plane at P_1 is called the *normal* to the surface at P_1.

The equations of the normal to the surface $F(x, y, z) = 0$ at the point (x_1, y_1, z_1) are

$$\frac{x - x_1}{\left(\dfrac{\partial F}{\partial x}\right)_{x=x_1,\, y=y_1,\, z=z_1}} = \frac{y - y_1}{\left(\dfrac{\partial F}{\partial y}\right)_{x=x_1,\, y=y_1,\, z=z_1}} = \frac{z - z_1}{\left(\dfrac{\partial F}{\partial z}\right)_{x=x_1,\, y=y_1,\, z=z_1}} \qquad (2.141)$$

2.9 DIFFERENTIAL CALCULUS

2.9.1 Functions and Derivatives

Function. If two variables x and y are so related that to each value of x in a given domain there corresponds a value of y, then y is a *function* of x in that domain. The variable x is the *independent* variable and y the *dependent* variable. The symbols $F(x)$, $f(x)$, $\phi(x)$, and so on, are used to represent functions of x; the symbol $f(a)$ represents the value of $f(x)$ for $x = a$.

Limit, Derivative, Differential. The function $f(x)$ approaches the limit 1 as x approaches a if the difference $|f(x) - 1|$ can be made arbitrarily small for all values of x except a within a sufficiently small interval with a as midpoint. In symbols, $\lim_{x \to a} f(x) = 1$.
 The symbols $\lim_{x \to a} f(x) = \infty$ or $\lim_{x \to a} f(x) = -\infty$ mean that, for all values of x except a within a sufficiently small interval with a as midpoint, the values of $f(x)$ can be made arbitrarily large positively or negatively, respectively.
 The symbols $\lim_{x \to \infty} f(x) = 1$ or $\lim_{x \to -\infty} f(x) = 1$ mean that the difference $|f(x) - 1|$ can be made arbitrarily small for all values of x sufficiently large positively or negatively, respectively.
 A change in x is called an *increment* of x and is denoted by Δx. The corresponding change in y is denoted by Δy. If

$$\lim_{\Delta x \to 0} \frac{f(x + \Delta x) - f(x)}{\Delta x}$$

exists, it is called the *derivative* of y with respect to x and is denoted by dy/dx, $f'(x)$, or $D_x y$.
 The geometric interpretation of $f'(x)$ is

$$f'(x) = \frac{dy}{dx} = \tan \theta \qquad (2.142)$$

or $f'(x)$ is equal to the slope of the tangent to the curve $y = f(x)$ at the point $P(x, y)$ (Fig. 2.135).

$$\frac{RS}{PR} = \lim_{PR \to 0} \frac{RQ}{PR} = \lim_{\Delta x \to 0} \frac{\Delta y}{\Delta x} = \lim_{\Delta x \to 0} \frac{f(x + \Delta x) - f(x)}{\Delta x} = \frac{dy}{dx} = f'(x) = \tan \theta \quad (2.143)$$

The differentials of x and y, respectively, are

$$dx = \Delta x$$
$$dy = f'(x)\, dx$$

Continuity. A function is *continuous* at $x = b$ if it has a definite value at b and approaches that value as a limit whenever x approaches b as a limit. The notion of continuity at a point suggests that

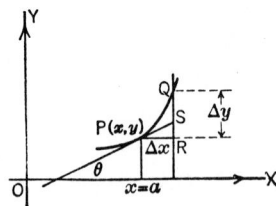

Fig. 2.135

the graph of the function can be drawn without lifting pencil from paper at the point. The analytic conditions that $f(x)$ be continuous at b are that $f(b)$ have a definite value and that for an arbitrarily small positive number ε there exist a $\delta(\varepsilon)$ such that

$$|f(x) - f(b)| < \varepsilon \quad \text{for all values of } x \text{ for which } |x - b| < \delta(\varepsilon) \tag{2.144}$$

A function that is continuous at each point of an interval is said to be continuous in that interval. An example of a continuous function is $f(x) = x^2$. The function $\phi(x) = 1/(x - a)$ is continuous for all values of x except $x = a$, at which point it becomes infinite. Every differentiable function is continuous, although the reverse is not always true.

If, in the preceding definition of continuity, the number δ can be chosen the same for all points in the interval, the function is said to be *uniformly continuous* in that interval.

Derivatives of Higher Order. The *derivative* of the *first derivative* of y with respect to x is called the *second derivative of y with respect to x* and is denoted by

$$\frac{d}{dx}\left(\frac{dy}{dx}\right) = \frac{d^2y}{dx^2} = f''(x) = D_x^2 y \tag{2.145}$$

By successive differentiations the nth derivative

$$\frac{d^n y}{dx^n} = f^{(n)}(x) = D_x^n y \tag{2.146}$$

is obtained.

The nth differential of y is denoted by

$$d^n y = f^{(n)}(x)\, dx^n \tag{2.147}$$

Parametric Differentiation. To find the derivatives of y with respect to x if $y = y(t)$ and $x = x(t)$:

$$y' = \frac{dy}{dx} = \frac{\dfrac{dy}{dt}}{\dfrac{dx}{dt}} \tag{2.148}$$

$$y'' = \frac{d^2y}{dx^2} = \frac{\dfrac{dy'}{dt}}{\dfrac{dx}{dt}} \tag{2.149}$$

$$y^{(n)} = \frac{d^n y}{dx^n} = \frac{\dfrac{dy^{(n-1)}}{dt}}{\dfrac{dx}{dt}} \tag{2.150}$$

Example 2.48. Find the derivatives of y with respect to x for the ellipse $x = a\cos t$, $y = b\sin t$.

$$y' = \frac{dy}{dx} = \frac{b\cos t}{-a\sin t} = -\frac{b}{a}\cot t$$

$$y'' = \frac{dy'}{dx} = \frac{\dfrac{b}{a}\csc^2 t}{-a\sin t} = -\frac{b}{a^2}\csc^3 t$$

$$y''' = \frac{dy''}{dx} = \frac{\dfrac{3b}{a^2}\csc^3 t\cot t}{-a\sin t} = -\frac{3b}{a^3}\csc^4 t\cot t$$

Logarithmic Differentiation for Products and Quotients. If

$$y = \frac{u^l v^m}{w^n} \tag{2.151}$$

take the logarithms of both sides before differentiating.

$$\ln y = l \ln u + m \ln v - n \ln w \tag{2.152}$$

$$\frac{1}{y}\frac{dy}{dx} = \frac{l}{u}\frac{du}{dx} + \frac{m}{v}\frac{dv}{dx} - \frac{n}{w}\frac{dw}{dx} \tag{2.153}$$

$$\frac{dy}{dx} = y\left(\frac{l}{u}\frac{du}{dx} + \frac{m}{v}\frac{dv}{dx} - \frac{n}{w}\frac{dw}{dx}\right) \tag{2.154}$$

Example 2.49. Find dy/dx if

$$y = \frac{\sqrt{x^2 - 25}}{(x - 1)^3 (x + 5)^2}$$

$$\ln y = \tfrac{1}{2}\ln(x^2 - 25) - 3\ln(x - 1) - 2\ln(x + 5)$$

$$\frac{1}{y}\frac{dy}{dx} = \frac{2x}{2(x^2 - 25)} - \frac{3}{x - 1} - \frac{2}{x + 5}$$

$$\frac{dy}{dx} = \frac{y(-4x^2 + 11x + 65)}{(x^2 - 25)(x - 1)}$$

Mean Value Theorem. If $f(x)$ is single valued, continuous in the interval $a \leq x \leq b$, and has a derivative for all values of x between a and b, then there is a value $x = \xi$, $a < \xi < b$, such that

$$f(b) - f(a) = (b - a)f'(\xi) \tag{2.155}$$

Another form is

$$f(x + h) = f(x) + hf'(x + \theta h), \quad 0 < \theta < 1 \tag{2.156}$$

Indeterminate Forms
If a function $f(x)$ for $x = a$ (where a can also be ∞) has no determined value but appears in one of the meaningless forms

$$\frac{0}{0} \qquad \frac{\infty}{\infty} \qquad 0 \cdot \infty \qquad \infty - \infty \qquad 0^0 \qquad \infty^0 \qquad 0^\infty \qquad 1^\infty$$

then it may happen that the $\lim f(x)$ has a definite value. For the determination of this limiting value, if it exists, the following rules can be used:

0/0. If $f(x) = \phi(x)/\psi(x)$, $\phi(a) = 0$, and $\psi(a) = 0$, then

$$\lim_{x \to a} f(x) = \lim_{x \to a} \frac{\phi'(x)}{\psi'(x)} \quad \text{(l'Hospital's rule)} \tag{2.157}$$

If, however, $\phi'(a) = 0$ and $\psi'(a) = 0$, the rule is applied again, with the result

$$\lim_{x \to a} \frac{\phi(x)}{\psi(x)} = \lim_{\xi \to a} \frac{\phi'(\xi)}{\psi'(\xi)} = \frac{\phi''(a)}{\psi''(a)} \tag{2.158}$$

unless $\phi''(a) = 0$ and $\psi''(a) = 0$. In this case, the rule is applied again, and so forth.

Example 2.50. Find the value of $\sin x/x$ for $x = 0$.

$$\lim_{x \to 0} \frac{\sin x}{x} = \lim_{x \to 0} \frac{\cos x}{1} = 1$$

∞ / ∞. If $f(x) = \phi(x)/\psi(x)$, $\phi(a) = \infty$, and $\psi(a) = \infty$, then

$$\lim_{x \to a} \frac{\phi(x)}{\psi(x)} = \lim_{x \to a} \frac{\phi'(x)}{\psi'(x)} \qquad (2.159)$$

as before.

$0 \cdot \infty$. If $f(x) = \phi(x) \cdot \psi(x)$, $\phi(a) = 0$, and $\psi(a) = \infty$, then place $1/\psi(x) = \omega(x)$ and obtain the previous case $0/0$.

$\infty - \infty$. If $f(x) = \phi(x) - \psi(x)$, $\phi(a) = \infty$, and $\psi(a) = \infty$, then place $\phi(x) = 1/u(x)$. $\psi(x) = 1/v(x)$ and obtain

$$f(x) = \frac{v(x) - u(x)}{u(x)v(x)} \qquad (2.160)$$

which takes the form $0/0$.

$0^0, \infty^0, 0^\infty, 1^\infty$. An expression of the type $[\psi(x)]^{\phi(x)}$ may, for $x = a$, give rise to the forms $0^0, \infty^0, 0^\infty, 1^\infty$.

Such an expression may be reduced to a type $0/0$ or ∞/∞ by the use of logarithms. Thus,

$$\left. \begin{array}{l} u = [\psi(x)]^{\phi(x)} \\ \log_e u = \phi(x) \cdot \log_e \psi(x) \end{array} \right\} \qquad (2.161)$$

If $\lim_{x \to a} \phi(x) \cdot \log_e \psi(x)$ can be found by the previous methods, the limit approached by u can be found.

Example 2.51. $u = (1 - x)^{1/x}$ for $x = 0$.

$$\log_e u = \frac{\log_e(1 - x)}{x}$$

$$\lim_{x \to 0} \frac{\log_e(1 - x)}{x} = \lim_{x \to 0} \frac{\dfrac{-1}{1 - x}}{1} = -1$$

Therefore $\lim_{x \to 0} \log_e u = -1$ and $\lim_{x \to 0} u = e^{-1}$

2.9.2 Differentiation Formulas

TABLE 2.9   Differentiation Formulas

Let $u, v, w, \ldots$ be functions of x; a and n be constants; and e be the base of the natural or Naperian logarithms. Then $e = 2.7183$.

$$\frac{d}{dx} a = 0$$

$$\frac{d}{dx}(u + v + w + \cdots) = \frac{du}{dx} + \frac{dv}{dx} + \frac{dw}{dx} + \cdots$$

$$\frac{d}{dx} au = a \frac{du}{dx}$$

$$\frac{d}{dx} uv = u \frac{dv}{dx} + v \frac{du}{dx}$$

$$\frac{d}{dx}(uvw \cdots) =$$

$$\left(\frac{1}{u} \frac{du}{dx} + \frac{1}{v} \frac{dv}{dx} + \frac{1}{w} \frac{dw}{dx} + \cdots \right)(uvw \cdots)$$

$$\frac{d}{dx}\left(\frac{u}{v}\right) = \frac{v \dfrac{du}{dx} - u \dfrac{dv}{dx}}{v^2}$$

$$\frac{d}{dx} u^n = nu^{n-1} \frac{du}{dx}$$

$$\frac{d}{dx} \log_e u = \frac{1}{u} \frac{du}{dx}$$

$$\frac{d}{dx} \log_{10} u = \frac{1}{u} \frac{du}{dx} \log_{10} e = (0.4343)\frac{1}{u} \frac{du}{dx}$$

$$\frac{d}{dx} e^u = e^u \frac{du}{dx}$$

$$\frac{d}{dx} u^v = vu^{v-1} \frac{du}{dx} + u^v \frac{dv}{dx} \log_e u$$

TABLE 2.9 *(Continued)*

$$\frac{d}{dx}f(u) = \frac{df(u)}{du}\cdot\frac{du}{dx}$$

$$\frac{d^2f(u)}{dx^2} = \frac{df(u)}{du}\cdot\frac{d^2u}{dx^2} + \frac{d^2f(u)}{du^2}\left(\frac{du}{dx}\right)^2$$

$$\frac{d}{dx}\sin u = \cos u\frac{du}{dx}$$

$$\frac{d}{dx}\cos u = -\sin u\frac{du}{dx}$$

$$\frac{d}{dx}\tan u = \sec^2 u\frac{du}{dx}$$

$$\frac{d}{dx}\cot u = -\csc^2 u\frac{du}{dx}$$

$$\frac{d}{dx}\sec u = \sec u\tan u\frac{du}{dx}$$

$$\frac{d}{dx}\csc u = -\csc u\cot u\frac{du}{dx}$$

$$\frac{d}{dx}\sin^{-1}u = \frac{1}{\sqrt{1-u^2}}\frac{du}{dx}\left(-\frac{\pi}{2}\le\sin^{-1}u\le\frac{\pi}{2}\right)$$

$$\frac{d}{dx}\cos^{-1}u = -\frac{1}{\sqrt{1-u^2}}\frac{du}{dx}(0\le\cos^{-1}u\le\pi)$$

$$\frac{d}{dx}\tan^{-1}u = \frac{1}{1+u^2}\frac{du}{dx}$$

$$\frac{d}{dx}\cot^{-1}u = -\frac{1}{1+u^2}\frac{du}{dx}$$

$$\frac{d}{dx}\sec^{-1}u = \frac{1}{u\sqrt{u^2-1}}\frac{du}{dx}^a$$

$$\frac{d}{dx}\csc^{-1}u = -\frac{1}{u\sqrt{u^2-1}}\frac{du}{dx}^a$$

$$\frac{d}{dx}\sinh u = \cosh u\frac{du}{dx}$$

$$\frac{d}{dx}\cosh u = \sinh u\frac{du}{dx}$$

$$\frac{d}{dx}\tanh u = \operatorname{sech}^2 u\frac{du}{dx}$$

$$\frac{d}{dx}\coth u = -\operatorname{csch}^2 u\frac{du}{dx}$$

$$\frac{d}{dx}\operatorname{sech} u = -\operatorname{sech} u\tanh u\frac{du}{dx}$$

$$\frac{d}{dx}\operatorname{csch} u = -\operatorname{csch} u\coth u\frac{du}{dx}$$

$$\frac{d}{dx}\sinh^{-1}u = \frac{1}{\sqrt{u^2+1}}\frac{du}{dx}$$

$$\frac{d}{dx}\cosh^{-1}u = \frac{1}{\sqrt{u^2-1}}\frac{du}{dx}$$

$$\frac{d}{dx}\tanh^{-1}u = \frac{1}{1-u^2}\frac{du}{dx}$$

$$\frac{d}{dx}\coth^{-1}u = \frac{1}{1-u^2}\frac{du}{dx}$$

$$\frac{d}{dx}\operatorname{sech}^{-1}u = -\frac{1}{u\sqrt{1-u^2}}\frac{du}{dx}$$

$$\frac{d}{dx}\operatorname{csch}^{-1}u = -\frac{1}{u\sqrt{u^2+1}}\frac{du}{dx}$$

^{a}For angles in the first and third quadrants. Use the opposite sign in the second and fourth quadrants.

2.9.3 Partial Derivatives

Functions of Two Variables. If three variables $f(x, y)$, x, y are so related that to each pair of values of x and y in a given domain there corresponds a value of $f(x, y)$, then $f(x, y)$ is a function of x and y in that domain. If x is considered as the only variable while y is taken as constant, then the derivative of $f(x, y)$ with respect to x is called the *partial derivative* of f with respect to x and is denoted by

$$\frac{\partial f}{\partial x} = f_x = \lim_{\Delta x\to 0}\frac{f(x+\Delta x, y) - f(x, y)}{\Delta x} \qquad (2.162)$$

Likewise, the partial derivative of f with respect to y is obtained by considering x to be constant while y varies:

$$\frac{\partial f}{\partial y} = f_y = \lim_{\Delta y\to 0}\frac{f(x, y+\Delta y) - f(x, y)}{\Delta y} \qquad (2.163)$$

If $\partial f/\partial x$ and $\partial f/\partial y$ are again differentiable, the partial derivatives of the second order may be found.

$$\frac{\partial}{\partial x}\left(\frac{\partial f}{\partial x}\right) = \frac{\partial^2 f}{\partial x^2} = f_{xx} \qquad \frac{\partial}{\partial y}\left(\frac{\partial f}{\partial y}\right) = \frac{\partial^2 f}{\partial y^2} = f_{yy}$$

$$\frac{\partial}{\partial x}\left(\frac{\partial f}{\partial y}\right) = \frac{\partial^2 f}{\partial x\,\partial y} = f_{yx} \qquad \frac{\partial}{\partial y}\left(\frac{\partial f}{\partial x}\right) = \frac{\partial^2 f}{\partial y\,\partial x} = f_{xy} \qquad (2.164)$$

If the derivatives in question are continuous, the order of differentiation is immaterial, that is,

$$\frac{\partial^2 f}{\partial y\,\partial x} = \frac{\partial^2 f}{\partial x\,\partial y} \qquad (2.165)$$

Similarly, the third and higher partial derivatives of $f(x, y)$ may be found. The third partial derivatives, if continuous, are four in number:

$$\frac{\partial}{\partial x}\left(\frac{\partial^2 f}{\partial x^2}\right) = \frac{\partial^3 f}{\partial x^3} \quad \frac{\partial}{\partial x}\left(\frac{\partial^2 f}{\partial y^2}\right) = \frac{\partial}{\partial y}\left(\frac{\partial^2 f}{\partial x\,\partial y}\right) = \frac{\partial^2}{\partial y^2}\left(\frac{\partial f}{\partial x}\right) = \frac{\partial^3 f}{\partial x\,\partial y^2}$$

$$\frac{\partial}{\partial y}\left(\frac{\partial^2 f}{\partial y^2}\right) = \frac{\partial^3 f}{\partial y^3} \quad \frac{\partial}{\partial y}\left(\frac{\partial^2 f}{\partial x^2}\right) = \frac{\partial}{\partial x}\left(\frac{\partial^2 f}{\partial x\,\partial y}\right) = \frac{\partial^2}{\partial x^2}\left(\frac{\partial f}{\partial y}\right) = \frac{\partial^3 f}{\partial x^2\,\partial y} \qquad (2.166)$$

Functions of N Variables. The formulas preceding may be generalized to the case where f is a function of more than two variables, that is, there corresponds a value of $f(x, y, z, \ldots)$ to every set of values of $x, y, z, \ldots$.

If the increments $\Delta x, \Delta y, \Delta z, \ldots$ are assigned to $x, y, z, \ldots$ in $f(x, y, z, \ldots)$, the *total increment* of f is

$$\Delta f = f(x + \Delta x, y + \Delta y, z + \Delta z, \ldots) - f(x, y, z, \ldots) \qquad (2.167)$$

The *total differential* of f is

$$df = \frac{\partial f}{\partial x}\,dx + \frac{\partial f}{\partial y}\,dy + \frac{\partial f}{\partial z}\,dz + \cdots \qquad (2.168)$$

The *second total differential* of f is

$$d^2 f = \frac{\partial^2 f}{\partial x^2}(dx)^2 + \frac{\partial^2 f}{\partial y^2}(dy)^2 + \frac{\partial^2 f}{\partial z^2}(dz)^2 + \cdots + 2\frac{\partial^2 f}{\partial x\,\partial y}\,dx\,dy + \cdots \qquad (2.169)$$

In general,

$$d^n f = \left(\frac{\partial}{\partial x}\,dx + \frac{\partial}{\partial y}\,dy + \frac{\partial}{\partial z}\,dz + \cdots\right)^n f(x, y, z, \ldots) \qquad (2.170)$$

Exact Differential. In order for the expression $P(x, y)\,dx + Q(x, y)\,dy$ to be the *exact* or *complete* differential of a function of two variables, it is necessary and sufficient that

$$\frac{\partial Q}{\partial x} = \frac{\partial P}{\partial y} \quad \text{(integrability condition)} \qquad (2.171)$$

For three variables, $P\,dx + Q\,dy + R\,dz$, the corresponding conditions are

$$\frac{\partial Q}{\partial z} = \frac{\partial R}{\partial y} \qquad \frac{\partial R}{\partial x} = \frac{\partial P}{\partial z} \qquad \frac{\partial P}{\partial y} = \frac{\partial Q}{\partial x} \qquad (2.172)$$

Differentiation of Composite Functions. If $u = f(x, y, z, \ldots, w)$, and $x, y, z, \ldots, w$ are functions of a single variable t, then

$$\frac{du}{dt} = \frac{\partial u}{\partial x}\frac{dx}{dt} + \frac{\partial u}{\partial y}\frac{dy}{dt} + \cdots + \frac{\partial u}{\partial w}\frac{dw}{dt} \qquad (2.173)$$

which is the total derivative of u with respect to t.

Example 2.52. Given: $u = x^2 + y^2 + 3xy$, $x = t^2$, $y = 1/t$.
Then

$$\frac{dx}{dt} = 2t \qquad \frac{dy}{dt} = -\frac{1}{t^2}$$

and

$$\frac{du}{dt} = \left(2t^2 + \frac{3}{t}\right)2t - \left(\frac{2}{t} + 3t^2\right)\frac{1}{t^2}$$

The equation reduces to

$$\frac{du}{dt} = 4t^3 + 3 - \frac{2}{t^3}$$

which expresses the rate of change of u with respect to t as a function of t.

Implicit Functions. The equation $F(x, y) = 0$ defines y as an *implicit* function of x, and x as an implicit function of y. If the equation is solved for y in terms of x, $y = f(x)$, then y is called an *explicit* function of x.

Example 2.53. Implicit function: $F(x, y) = x^2 + y^2 - r^2 = 0$. Explicit function: $y = \pm\sqrt{r^2 - x^2}$.
To find dy/dx, either differentiate $y = f(x)$ or use

$$\frac{dy}{dx} = -\frac{\dfrac{\partial F}{\partial x}}{\dfrac{\partial F}{\partial y}} \qquad \left(\frac{\partial F}{\partial y} \neq 0\right) \qquad (2.174)$$

$$\frac{d^2y}{dx^2} = -\frac{\dfrac{\partial^2 F}{\partial x^2}\left(\dfrac{\partial F}{\partial y}\right)^2 - 2\dfrac{\partial^2 F}{\partial x\,\partial y}\dfrac{\partial F}{\partial x}\dfrac{\partial F}{\partial y} + \dfrac{\partial^2 F}{\partial y^2}\left(\dfrac{\partial F}{\partial x}\right)^2}{\left(\dfrac{\partial F}{\partial y}\right)^3} \qquad \left(\frac{\partial F}{\partial y} \neq 0\right) \qquad (2.175)$$

2.9.4 Infinite Series

Let $a_1, a_2, \ldots, a_n, \ldots$ be a sequence of numbers formed according to some rule. The indicated sum

$$\sum_{n=1}^{\infty} a_n = a_1 + a_2 + \cdots + a_n + \cdots \qquad (2.176)$$

is called an *infinite series*. Let $s_n = a_1 + a_2 + \cdots + a_n$. If the partial sums s_n approach a limit S as $n \to \infty$, then the series is *convergent* and S is the *sum* or *value* of the series. A series that is not convergent is *divergent*.

If the series of absolute values $|a_1| + |a_2| + \cdots + |a_n| + \cdots$ is convergent, then the series (2.176) is *absolutely convergent*. A series that converges, but not absolutely, is *conditionally convergent*. The sum of an absolutely convergent series is not changed by rearrangement of its terms.

Tests for Convergence

Comparison Test. A comparison test for series of positive terms. If there is a convergent series of positive terms $c_1 + c_2 + \cdots + c_n + \cdots$, such that $a_n \leq c_n$ for every n from some term on, then the series (2.176) converges. If there is a divergent series of positive terms $d_1 + d_2 + \cdots + d_n + \cdots$, such that $a_n \geq d_n$ for every n from some term on, then the series (2.176) diverges. Two useful comparison series are

1. The *geometric* series $a + ar + ar^2 + \cdots + ar^{n-1} + \cdots$, which converges for $|r| < 1$ and diverges for $|r| \geq 1$.
2. The p series $1 + 1/2^p + 1/3^p + \cdots + 1/n^p + \cdots$, which converges for $p > 1$ and diverges for $p \leq 1$.

Ratio Test. Let

$$L = \lim_{n \to \infty} \left| \frac{a_{n+1}}{a_n} \right| \tag{2.177}$$

If $L < 1$, the series (2.176) converges absolutely; if L does not exist or if $L > 1$, the series (2.176) diverges; if $L = 1$, the test fails.

Example 2.54

(1)
$$10 + \frac{10^2}{2!} + \frac{10^3}{3!} + \cdots + \frac{10^n}{n!} + \cdots$$

Since

$$L = \lim_{n \to \infty} \frac{\dfrac{10^{n+1}}{(n+1)!}}{\dfrac{10^n}{n!}} = \lim_{n \to \infty} \frac{10}{n+1} = 0,$$

the series converges.

(2)
$$\frac{1+1}{1+3} + \frac{(1+1)(2+1)}{(1+3)(2+3)} + \cdots + \frac{(1+1)(2+1)\cdots(n+1)}{(1+3)(2+3)\cdots(n+3)} + \cdots$$

Since $L = \lim\limits_{n \to \infty} \dfrac{(n+1)+1}{(n+1)+3} = 1$, the test fails. Raabe's test can be used. See Eq. (2.179).

Root Test. Let

$$L = \lim_{n \to \infty} |a_n|^{1/n} \tag{2.178}$$

If $L < 1$, the series (2.176) converges; if $L > 1$, the series (2.176) diverges; if $L = 1$, the test fails.

Example 2.55

$$1 + \frac{1}{(\log 2)^2} + \frac{1}{(\log 3)^3} + \cdots + \frac{1}{(\log n)^n} + \cdots$$

Since

$$L = \lim_{n \to \infty} \frac{1}{\log n} = 0$$

the series converges.

Integral Test. Let $f(n) = a_n$. If $f(x)$ is a positive nonincreasing function of x for $x > k$, then the series (2.176) converges or diverges with the improper integral $\int_k^\infty f(x)\,dx$.

Example 2.56

$$1 + \frac{1}{2(\log 2)^3} + \frac{1}{3(\log 3)^3} + \cdots + \frac{1}{n(\log n)^3} + \cdots$$

Then

$$f(x) = \frac{1}{x(\log x)^3} \text{ for } x \geq 2$$

and

$$\int_2^\infty \frac{dx}{x(\log x)^3} = \lim_{n\to\infty} \frac{1}{2}\left(\frac{1}{(\log 2)^2} - \frac{1}{(\log n)^2}\right) = \frac{1}{2(\log 2)^2}$$

Since the integral is convergent, the series is also.

Raabe's Test. Let

$$L = \lim_{n\to\infty} n\left(\frac{a_n}{a_{n+1}} - 1\right) \tag{2.179}$$

If $L > 1$, the series (2.176) converges; if $L < 1$, the series (2.176) diverges; if $L = 1$, the test fails.

Example 2.57

$$\frac{1+1}{1+3} + \frac{(1+1)(2+1)}{(1+3)(2+3)} + \cdots + \frac{(1+1)(2+1)\cdots(n+1)}{(1+3)(2+3)\cdots(n+3)} + \cdots$$

Since

$$L = \lim_{n\to\infty} n\left(\frac{(n+1)+3}{(n+1)+1} - 1\right) = \lim_{n\to\infty} \frac{2n}{n+2} = 2 > 1$$

the series converges.

Convergence of an Alternating Series. A series

$$a_1 - a_2 + a_3 - + \cdots + (-1)^{n+1}a_n + \cdots \tag{2.180}$$

in which the terms are alternately positive and negative is an *alternating series*. If, from some term on, $|a_{n+1}| \leq |a_n|$ and $a_n \to 0$ as $n \to \infty$, the series converges. The sum of the first n terms differs numerically from the sum of the series by less than $|a_{n+1}|$.

Series of Functions
A *power series* is a series of the form

$$\sum_{n=0}^\infty a_n x^n = a_0 + a_1 x + a_2 x^2 + \cdots + a_n x^n + \cdots \tag{2.181}$$

If $\lim_{n\to\infty} |a_{n-1}/a_n| = r$, the power series converges absolutely for all values of x in the interval $-r < x < r$. For $|x| = r$, it is necessary to use one of the convergence tests for series of numerical terms:

Example 2.58

$$1 - \frac{x}{1\cdot 2} + \frac{x^2}{2\cdot 2^2} - \frac{x^3}{3\cdot 2^3} + \cdots + (-1)^n \frac{x^n}{n\cdot 2^n} + \cdots$$

Since

$$\lim_{n\to\infty} \frac{n\cdot 2^n}{(n-1)2^{n-1}} = 2$$

the interval of convergence is $-2 < x < 2$. For $x = 2$, the series is a convergent alternating series. For $x = -2$, it is a divergent p series.

Taylor's Series. If $f(x)$ has continuous derivatives in the neighborhood of a point $x = a$, then

$$f(x) = f(a) + \frac{f'(a)}{1!}(x - a) + \frac{f''(a)}{2!}(x - a)^2 + \cdots + \frac{f^{(n-1)}(a)}{(n-1)!}(x - a)^{n-1} + \cdots$$

$$(2.182)$$

with the remainder after n terms

$$R_n = \frac{f^{(n)}(\xi)}{n!}(x - a)^n \qquad \xi = a + \theta(x - a) \qquad 0 < \theta < 1 \qquad (2.183)$$

Another form of Taylor's series is

$$f(x + h) = f(x) + \frac{h}{1!}f'(x) + \frac{h^2}{2!}f''(x) + \cdots + \frac{h^{n-1}}{(n-1)!}f^{(n-1)}(x) + \cdots \qquad (2.184)$$

with the remainder after n terms

$$R_n = \frac{h^n}{n!}f^{(n)}(\xi) \qquad \xi = x + \theta h \qquad 0 < \theta < 1 \qquad (2.185)$$

Maclaurin's Series. If $a = 0$ in Eq. (2.182),

$$f(x) = f(0) + \frac{f'(0)}{1!}x + \frac{f''(0)}{2!}x^2 + \cdots + \frac{f^{(n-1)}(0)}{(n-1)!}x^{n-1} + \cdots \qquad (2.186)$$

with the remainder after n terms

$$R_n = \frac{f^{(n)}(\xi)}{n!}x^n \qquad \xi = \theta x \qquad 0 < \theta < 1 \qquad (2.187)$$

A Taylor or Maclaurin series represents a function in an interval if and only if $R_n \to 0$ as $n \to \infty$.

Example 2.59. Expand e^{ax} in powers of x.

$$f(x) = e^{ax} \qquad f'(x) = ae^{ax} \qquad f''(x) = a^2 e^{ax} \qquad f'''(x) = a^3 e^{ax}, \ldots$$
$$f(0) = 1 \qquad f'(0) = a \qquad f''(0) = a^2 \qquad f'''(0) = a^3, \ldots$$
$$f(x) = e^{ax} = 1 + \frac{a}{1!}x + \frac{a^2}{2!}x^2 + \frac{a^3}{3!}x^3 + \cdots$$

Since

$$\lim_{n \to \infty} \frac{\dfrac{a^{n-1}}{(n-1)!}}{\dfrac{a^n}{n!}} = \lim_{n \to \infty} \frac{n}{a} = \infty$$

the series converges for all values of x.

Taylor's Series for Two Variables

$$f(x + h, y + k) = f(x, y) + \frac{1}{1!}\left(h\frac{\partial}{\partial x} + k\frac{\partial}{\partial y} \right) f(x, y)$$

$$+ \frac{1}{2!}\left(h\frac{\partial}{\partial x} + k\frac{\partial}{\partial y} \right)^2 f(x, y) + \cdots$$

$$+ \frac{1}{(n-1)!}\left(h\frac{\partial}{\partial x} + k\frac{\partial}{\partial y} \right)^{n-1} f(x, y) + \cdots \qquad (2.188)$$

with the remainder

$$R_n = \frac{1}{n!}\left(h\frac{\partial}{\partial x} + k\frac{\partial}{\partial y}\right)^n f(x + \theta h, y + \theta k) \qquad 0 < \theta < 1 \qquad (2.189)$$

Fourier Series. If $f(x)$ is of bounded variation over an interval of length $2l$, that is, if it can be expressed as the difference of two nondecreasing or nonincreasing bounded functions, then

$$f(x) = \frac{a_0}{2} + \sum_{n=1}^{\infty}\left(a_n\cos\frac{n\pi x}{l} + b_n\sin\frac{n\pi x}{l}\right)$$

$$= \frac{a_0}{2} + a_1\cos\frac{\pi x}{l} + a_2\cos\frac{2\pi x}{l} + \cdots + b_1\sin\frac{\pi x}{l} + b_2\sin\frac{2\pi x}{l} + \cdots \qquad (2.190)$$

in which

$$a_n = \frac{1}{l}\int_k^{k+2l}f(x)\cos\frac{n\pi x}{l}\,dx \qquad b_n = \frac{1}{l}\int_k^{k+2l}f(x)\sin\frac{n\pi x}{l}\,dx \qquad n = 0,1,2,\dots \quad (2.191)$$

In exponential form

$$f(x) = \sum_{n=-\infty}^{\infty} c_n e^{in\pi x/l} \qquad c_n = \frac{1}{2l}\int_k^{k+2l}f(x)e^{-in\pi x/l}\,dx \qquad n = \dots, -2, -1, 0, 1, 2, \dots$$

$$(2.192)$$

At a point of discontinuity, a Fourier series gives the value at the midpoint of the jump.

Example 2.60. Expand e^x in the interval 0 to 2π.

$$a_0 = \frac{1}{\pi}\int_0^{2\pi}e^x\,dx = \frac{1}{\pi}(e^{2\pi} - 1) \qquad a_n = \frac{1}{\pi}\int_0^{2\pi}e^x\cos nx\,dx = \frac{e^{2\pi} - 1}{\pi(n^2 + 1)}$$

$$b_n = -\frac{n(e^{2\pi} - 1)}{\pi(n^2 + 1)}$$

Hence

$$e^x = \frac{1}{\pi}(e^{2\pi} - 1)\left[\frac{1}{2} + \frac{1}{1^2 + 1}\cos x + \frac{1}{2^2 + 1}\cos 2x + \frac{1}{3^2 + 1}\cos 3x + \cdots\right]$$

$$- \frac{1}{\pi}(e^{2\pi} - 1)\left[\frac{1}{1^2 + 1}\sin x + \frac{2}{2^2 + 1}\sin 2x + \frac{3}{3^2 + 1}\sin 3x + \cdots\right]$$

The expansion is valid only in the interval from 0 to 2π; outside that interval the series repeats itself owing to the periodic property of $\sin nx$ and $\cos nx$.

Fourier Series for Even or Odd Functions. If $f(-x) = f(x)$, it is an *even* function. Then

$$a_n = \frac{2}{l}\int_0^l f(x)\cos\frac{n\pi x}{l}\,dx \qquad n = 0,1,2,\dots$$

$$b_n = 0 \qquad (2.193)$$

If $f(-x) = -f(x)$, it is an *odd* function. Then

$$a_n = 0$$

$$b_n = \frac{2}{l}\int_0^l f(x)\sin\frac{n\pi x}{l}\,dx \qquad n = 0,1,2,\dots \qquad (2.194)$$

Example 2.61. Expand $f(x) = x$ in a cosine series in the interval $(0, \pi)$. Here

$$\frac{1}{2}a_0 = \frac{1}{\pi}\int_0^\pi x\,dx = \frac{\pi}{2}$$

$$a_n = \frac{2}{\pi}\int_0^\pi x\cos nx\,dx = \frac{2}{\pi}\left\{\left[\frac{x\sin nx}{n}\right]_0^\pi - \int_0^\pi \frac{\sin nx}{n}\,dx\right\}$$

$$= \frac{2}{\pi}\left[\frac{1}{n^2}\cos nx\right]_0^\pi = \frac{2}{\pi n^2}(\cos n\pi - 1)$$

Therefore

$$x = \frac{\pi}{2} - \frac{4}{\pi}\left[\cos x + \frac{\cos 3x}{3^2} + \frac{\cos 5x}{5^2} + \cdots\right] \qquad (0 < x < \pi)$$

If $x = 0$, the sum of the series is 0; if $x = \pi$, the sum of the series is π.

Uniform Convergence. Let $R_n(x)$ be the remainder after n terms of the series of functions

$$\sum_{n=1}^{\infty} u_n(x) = u_1(x) + u_2(x) + \cdots + u_n(x) + \cdots \tag{2.195}$$

The series is uniformly convergent in the interval $a \le x \le b$ if, for any $\varepsilon > 0$, there exists an N, dependent on ε but not on x, such that $|R_n(x)| < \varepsilon$ for $n > N$.

If a power series converges in the interval $-r < x < r$, then it converges uniformly in any interval within this interval.

The sum of a uniformly convergent series of continuous functions is also a continuous function.

Weierstrass M Test. If $\sum_{n=1}^{\infty} u_n(x)$ is a series of functions defined in an interval, if $\sum_{n=1}^{\infty} M_n$ is a series of positive constants, and if $|u_n(x)| \le M_n$ for all values of x in the interval, then $\sum_{n=1}^{\infty} u_n(x)$ is absolutely and uniformly convergent in the interval.

Operations

Term by Term Differentiation. If $f(x) = \sum_{n=1}^{\infty} u_n(x)$ is a convergent series of differentiable functions in an interval, and if $\sum_{n=1}^{\infty} u_n'(x)$ is a series of continuous functions that converges uniformly in the interval, then

$$\sum_{n=1}^{\infty} u_n'(x) = f'(x)$$

Term by Term Integration. If

$$f(x) = \sum_{n=1}^{\infty} u_n(x)$$

converges uniformly in an interval, and if a and x are any two values in the interval, then $\sum_{n=1}^{\infty} \int_a^x u_n(t)\,dt$ converges to $\int_a^x f(t)\,dt$. It converges uniformly with respect to x for each fixed value of a.

Two *power series* can be *added*, *subtracted*, or *multiplied* term by term, and the result is a power series that converges when both of the first do and represents the *sum*, *difference*, or *product*, respectively, of the two series. The *product* of two power series $\sum_{n=0}^{\infty} a_n x^n$ and $\sum_{n=0}^{\infty} b_n x^n$ is

$$a_0 b_0 + (a_0 b_1 + a_1 b_0)x + (a_0 b_2 + a_1 b_1 + a_2 b_0)x^2 + \cdots$$
$$+ (a_0 b_n + a_1 b_{n-1} + \cdots + a_n b_0)x^n + \cdots \tag{2.196a}$$

The *quotient* of two convergent power series $\sum_{n=0}^{\infty} a_n x^n$ and $\sum_{n=0}^{\infty} b_n x^x$, $b_0 \ne 0$, is

$$\sum_{n=0}^{\infty} q_n x^n = \frac{a_0}{b_0} + \frac{a_1 b_0 - a_0 b_1}{b_0^2}x + \frac{a_2 b_0^2 - a_1 b_0 b_1 + a_0 b_1^2 - a_0 b_0 b_2}{b_0^3}x^2 + \cdots \tag{2.196b}$$

The interval of convergence of the quotient series must be determined. To obtain $q_3, q_4, \ldots$, solve the equations

$$a_0 = q_0 b_0$$
$$a_1 = q_0 b_1 + q_1 b_0$$
$$a_2 = q_0 b_2 + q_1 b_1 + q_2 b_0$$
$$a_n = q_0 b_n + q_1 b_{n-1} + \cdots + q_n b_0 \qquad\qquad (2.197)$$

See Table 2.10 for series expansions of various functions.

TABLE 2.10   Functions Expanded in Series (log = $\log_e$)

$$(a + x)^n = a^n + na^{n-1}x + \frac{n(n-1)}{2!}a^{n-2}x^2 + \frac{n(n-1)(n-2)}{3!}a^{n-3}x^3 + \cdots \qquad (x^2 < a^2)$$

$$e^x = 1 + x + \frac{x^2}{2!} + \frac{x^3}{3!} + \frac{x^4}{4!} + \cdots \qquad (-\infty < x < \infty)$$

$$a^x = 1 + x \log a + \frac{(x \log a)^2}{2!} + \frac{(x \log a)^3}{3!} + \cdots \qquad (-\infty < x < \infty)$$

$$e^{-x^2} = 1 - x^2 + \frac{x^4}{2!} - \frac{x^6}{3!} + \frac{x^8}{4!} - \cdots \qquad (-\infty < x < \infty)$$

$$e^{\sin x} = 1 + x + \frac{x^2}{2!} - \frac{3x^4}{4!} - \frac{8x^5}{5!} - \frac{3x^6}{6!} + \frac{56x^7}{7!} + \cdots \qquad (-\infty < x < \infty)$$

$$e^{\cos x} = e(1 - \frac{x^2}{2!} + \frac{4x^4}{4!} - \frac{31x^6}{6!} + \cdots) \qquad (-\infty < x < \infty)$$

$$e^{\tan x} = 1 + x + \frac{x^2}{2!} + \frac{3x^3}{3!} + \frac{9x^4}{4!} + \frac{37x^5}{5!} + \cdots \qquad \left(-\frac{\pi}{2} < x < \frac{\pi}{2}\right)$$

$$\log x = \frac{x - 1}{x} + \frac{1}{2}\left(\frac{x - 1}{x}\right)^2 + \frac{1}{3}\left(\frac{x - 1}{x}\right)^3 + \cdots \qquad \left(x > \frac{1}{2}\right)$$

$$\log x = 2\left[\frac{x - 1}{x + 1} + \frac{1}{3}\left(\frac{x - 1}{x + 1}\right)^3 + \frac{1}{5}\left(\frac{x - 1}{x + 1}\right)^5 + \cdots\right] \qquad (x > 0)$$

$$\log(1 + x) = x - \frac{x^2}{2} + \frac{x^3}{3} - \frac{x^4}{4} + \cdots \qquad (-1 < x < 1)$$

$$\log\left(\frac{1 + x}{1 - x}\right) = 2\left[x + \frac{x^3}{3} + \frac{x^5}{5} + \frac{x^7}{7} + \cdots\right] \qquad (-1 < x < 1)$$

$$\log\left(\frac{x + 1}{x - 1}\right) = 2\left[\frac{1}{x} + \frac{1}{3x^3} + \frac{1}{5x^5} + \cdots\right] \qquad (x^2 > 1)$$

$$\log \sin x = \log x - \frac{x^2}{6} - \frac{x^4}{180} - \frac{x^6}{2835} - \cdots \qquad (-\pi < x < \pi)$$

$$\log \cos x = -\frac{x^2}{2} - \frac{x^4}{12} - \frac{x^6}{45} - \frac{17x^8}{2520} - \cdots \qquad \left(-\frac{\pi}{2} < x < \frac{\pi}{2}\right)$$

$$\log \tan x = \log x + \frac{x^2}{3} + \frac{7x^4}{90} + \frac{62x^6}{2835} + \cdots \qquad \left(-\frac{\pi}{2} < x < \frac{\pi}{2}\right)$$

$$\sin x = x - \frac{x^3}{3!} + \frac{x^5}{5!} - \frac{x^7}{7!} + \cdots \qquad (-\infty < x < \infty)$$

$$\cos x = 1 - \frac{x^2}{2!} + \frac{x^4}{4!} - \frac{x^6}{6!} + \cdots \qquad (-\infty < x < \infty)$$

$$\tan x = x + \frac{x^3}{3} + \frac{2x^5}{15} + \frac{17x^7}{315} + \frac{62x^9}{2835} + \cdots \qquad \left(-\frac{\pi}{2} < x < \frac{\pi}{2}\right)$$

$$\cot x = \frac{1}{x} - \frac{x}{3} - \frac{x^3}{45} - \frac{2x^5}{945} - \frac{x^7}{4725} - \cdots \qquad (-\pi < x < \pi)$$

TABLE 2.10 (*Continued*)

$$\sec x = 1 + \frac{x^2}{2!} + \frac{5x^4}{4!} + \frac{61x^6}{6!} + \cdots \qquad \left(-\frac{\pi}{2} < x < \frac{\pi}{2}\right)$$

$$\csc x = \frac{1}{x} + \frac{x}{3!} + \frac{7x^3}{3 \cdot 5!} + \frac{31x^5}{3 \cdot 7!} + \cdots \qquad (-\pi < x < \pi)$$

$$\sin^{-1}x = x + \frac{x^3}{2 \cdot 3} + \frac{3x^5}{2 \cdot 4 \cdot 5} + \frac{3 \cdot 5x^7}{2 \cdot 4 \cdot 6 \cdot 7} + \cdots \qquad (-1 \le x \le 1)$$

$$\cos^{-1}x = \frac{\pi}{2} - \sin^{-1}x$$

$$\tan^{-1}x = \frac{\pi}{2} - \frac{1}{x} + \frac{1}{3x^3} - \frac{1}{5x^5} + \cdots \qquad (x^2 \ge 1)$$

$$= x - \frac{x^3}{3} + \frac{x^5}{5} - \frac{x^7}{7} + \cdots \qquad (-1 \le x \le 1)$$

$$\cot^{-1}x = \frac{\pi}{2} - \tan^{-1}x$$

$$\sec^{-1}x = \frac{\pi}{2} - \frac{1}{x} - \frac{1}{2 \cdot 3x^3} - \frac{3}{2 \cdot 4 \cdot 5x^5} - \frac{3 \cdot 5}{2 \cdot 4 \cdot 6 \cdot 7x^7} - \cdots \qquad (x^2 > 1)$$

$$\csc^{-1}x = \frac{\pi}{2} - \sec^{-1}x$$

$$\sinh x = x + \frac{x^3}{3!} + \frac{x^5}{5!} + \frac{x^7}{7!} + \cdots \qquad (-\infty < x < \infty)$$

$$\cosh x = 1 + \frac{x^2}{2!} + \frac{x^4}{4!} + \frac{x^6}{6!} + \frac{x^8}{8!} + \cdots \qquad (-\infty < x < \infty)$$

$$\tanh x = x - \frac{x^3}{3} + \frac{2x^5}{15} - \frac{17x^7}{315} + \cdots \qquad \left(-\frac{\pi}{2} < x < \frac{\pi}{2}\right)$$

$$\coth x = \frac{1}{x} + \frac{x}{3} - \frac{x^3}{45} + \frac{2x^5}{945} - \frac{x^7}{4725} + \cdots \qquad (-\pi < x < \pi)$$

$$\operatorname{sech} x = 1 - \frac{x^2}{2!} + \frac{5x^4}{4!} - \frac{61x^6}{6!} + \frac{1385x^8}{8!} - \cdots \qquad \left(-\frac{\pi}{2} < x < \frac{\pi}{2}\right)$$

$$\operatorname{csch} x = \frac{1}{x} - \frac{x}{6} + \frac{7x^3}{360} - \frac{31x^5}{15{,}120} + \cdots \qquad (-\pi < x < \pi)$$

$$\sinh^{-1}x = x - \frac{x^3}{2 \cdot 3} + \frac{3x^5}{2 \cdot 4 \cdot 5} - \frac{3 \cdot 5x^7}{2 \cdot 4 \cdot 6 \cdot 7} + \cdots \qquad (-1 < x < 1)$$

$$\sinh^{-1}x = \log 2x + \frac{1}{2 \cdot 2x^2} - \frac{3}{2 \cdot 4 \cdot 4x^4} + \frac{3 \cdot 5}{2 \cdot 4 \cdot 6 \cdot 6 \cdot x^6} + \cdots \qquad (x^2 > 1)$$

$$\cosh^{-1}x = \pm(\log 2x - \frac{1}{2 \cdot 2x^2} - \frac{1 \cdot 3}{2 \cdot 4 \cdot 4x^4} - \frac{1 \cdot 3 \cdot 5}{2 \cdot 4 \cdot 6 \cdot 6x^6} - \cdots) \qquad (x > 1)$$

$$\tanh^{-1}x = x + \frac{x^3}{3} + \frac{x^5}{5} + \frac{x^7}{7} + \cdots \qquad (-1 < x < 1)$$

$$\coth^{-1}x = \frac{1}{x} + \frac{1}{3x^3} + \frac{1}{5x^5} + \frac{1}{7x^7} + \cdots \qquad (x^2 > 1)$$

$$\operatorname{sech}^{-1}x = \pm\left(\log\frac{2}{x} - \frac{1}{2 \cdot 2}x^2 - \frac{1 \cdot 3}{2 \cdot 4 \cdot 4}x - \frac{1 \cdot 3 \cdot 5}{2 \cdot 4 \cdot 6 \cdot 6}x^6 - \cdots\right) \qquad (0 < x < 1)$$

$$\operatorname{csch}^{-1}x = \frac{1}{x} - \frac{1}{2 \cdot 3x^3} + \frac{3}{2 \cdot 4 \cdot 5x^5} - \frac{3 \cdot 5}{2 \cdot 4 \cdot 6 \cdot 7x^7} + \cdots \qquad (x^2 > 1)$$

2.9.5 Maxima and Minima

Function of One Variable. A function $f(x)$ has a relative *maximum* (*minimum*) at a point $x = a$ if at every point in some neighborhood of $x = a$ the values of $f(x)$ are all less (greater) than $f(a)$. Either a maximum or a minimum is an *extreme*. If the derivative exists at a relative extreme, it must be zero, that is, the tangent must be parallel to the x axis. To locate possible extreme points solve the equation $f'(x) = 0$. A solution $x = a$ gives a maximum (minimum) value of $f(x)$ if and only if the

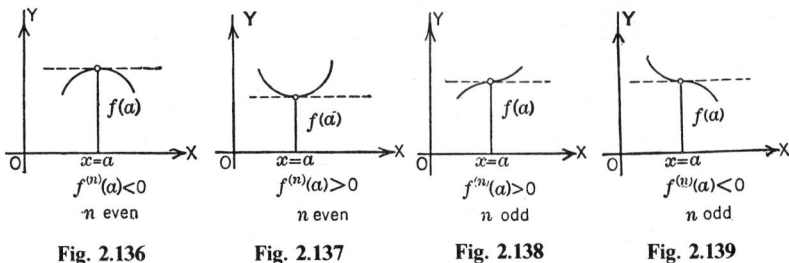

Fig. 2.136   Fig. 2.137   Fig. 2.138   Fig. 2.139

derivative is positive (negative) for $x < a$ and negative (positive) for $x > a$. If the derivative does not change sign, $x = a$ gives a *point of inflection*.

A solution $x = a$ can be tested also by using the higher derivatives of $f(x)$. Let $f^{(n)}(x)$ be the first derivative that does not equal zero for $x = a$, $f^{(n)}(a) \neq 0$. If n is even, there is a maximum (Fig. 2.136) if $f^{(n)}(a) < 0$, and a minimum (Fig. 2.137) if $f^{(n)}(a) > 0$. If n is odd, there is a point of inflection (Figs. 2.138 and 2.139). In many problems physical considerations make testing unnecessary.

Example 2.62. A piece of wire of length 30 in. is bent into a rectangle. Find the maximum area.

Let $x =$ the base, then $\frac{1}{2}(30 - 2x) =$ the altitude.

The area, $A = x(15 - x) = 15x - x^2$.

For a maximum or minimum, $dA/dx = 15 - 2x = 0$, $x = 7.5$.

Then $A = 7.5(15 - 7.5) = 56.25$ in.^{2}

To find whether the area is maximum or minimum, $d^2A/dx^2 = -2$, which is less than 0, and therefore the area 56.25 in.^{2} is a maximum.

Function of Two or More Variables. A function $f(x, y)$ has a relative *maximum* (*minimum*) at a point $(x, y) = (a, b)$ if at every point in some neighborhood of (a, b) the values of $f(x, y)$ are all less (greater) than $f(a, b)$. If the first partial derivatives exist at a relative extreme, it is necessary that

$$\frac{\partial f}{\partial x} = \frac{\partial f}{\partial y} = 0 \tag{2.198}$$

If, furthermore, at the point (a, b)

$$\frac{\partial^2 f}{\partial x^2}\frac{\partial^2 f}{\partial y^2} - \left(\frac{\partial^2 f}{\partial x \, \partial y}\right)^2 > 0 \tag{2.199}$$

then $f(a, b)$ is an extreme value, which is a maximum if

$$\frac{\partial^2 f}{\partial x^2} < 0 \quad \left(\text{and consequently } \frac{\partial^2 f}{\partial y^2} < 0\right) \tag{2.200}$$

and a minimum if

$$\frac{\partial^2 f}{\partial x^2} > 0 \tag{2.201}$$

If

$$\frac{\partial^2 f}{\partial x^2}\frac{\partial^2 f}{\partial y^2} - \left(\frac{\partial^2 f}{\partial x \, \partial y}\right)^2 < 0 \tag{2.202}$$

then $f(x, y)$ does not have an extreme value, but has a saddle point. If

$$\frac{\partial^2 f}{\partial x^2}\frac{\partial^2 f}{\partial y^2} - \left(\frac{\partial^2 f}{\partial x \, \partial y}\right)^2 = 0 \tag{2.203}$$

the test gives no information.

For a function of several variables $f(x, y, z, \ldots)$, necessary conditions for an extreme value are

$$\frac{\partial f}{\partial x} = \frac{\partial f}{\partial y} = \frac{\partial f}{\partial z} = \cdots = 0 \qquad (2.204)$$

2.10 INTEGRAL CALCULUS

2.10.1 Integration

The operation of **integration** is the inverse of differentiation. If $F'(x) = f(x)$, then $F(x)$ is an *indefinite integral* of $f(x)$. Any other indefinite integral of $f(x)$ differs from $F(x)$ at most by an additive constant. In symbols, $F(x) = \int f(x)\,dx + c$; or, more precisely, $F(x) = \int_a^x f(\xi)\,d\xi + c$, where a and c are constants (see Section 2.10.2). The function $f(x)$ is the *integrand*.

The integrals of many functions are given in Table 1.18. Table 2.11 is a convenient short table of fundamental integrals. The constant of integration is usually omitted in tables.

In general, with the exception of the square root of polynomials of the second degree, integrals containing fractional powers of polynomials above the first degree cannot be integrated in terms of the elementary integral forms.

Elliptic Integrals. An elliptic integral has the form

$$\int R\left[x, \sqrt{f(x)}\right] dx \qquad (2.205)$$

where R represents a rational function and $f(x) = a + bx + cx^2 + dx^3 + ex^4$, an algebraic function of the third or fourth degree.

Elliptic Integral of the First Kind

$$F(\phi, k) = \int_0^\phi \frac{d\theta}{\sqrt{1 - k^2 \sin^2\theta}} = \int_0^x \frac{d\xi}{\sqrt{(1 - \xi^2)(1 - k^2\xi^2)}} \qquad x = \sin\phi,\ k^2 < 1 \quad (2.206)$$

Elliptic Integral of the Second Kind

$$E(\phi, k) = \int_0^\phi \sqrt{1 - k^2 \sin^2\theta}\ d\theta = \int_0^x \frac{\sqrt{1 - k^2\xi^2}}{\sqrt{1 - \xi^2}}\ d\xi \qquad x = \sin\phi,\ k^2 < 1 \qquad (2.207)$$

Elliptic Integral of the Third Kind

$$\Pi(\phi, n, k) = \int_0^\phi \frac{d\theta}{(1 + n\sin^2\theta)\sqrt{1 - k^2\sin^2\theta}}$$

$$= \int_0^x \frac{d\xi}{(1 + n\xi^2)\sqrt{(1 - \xi^2)(1 - k^2\xi^2)}} \qquad x = \sin\phi,\ k^2 < 1 \qquad (2.208)$$

The "complete" integrals are

$$K = F\left(\frac{\pi}{2}, k\right) = \frac{\pi}{2}\left[1 + \left(\frac{1}{2}\right)^2 k^2 + \left(\frac{3}{2\cdot 4}\right)^2 k^4 + \left(\frac{3\cdot 5}{2\cdot 4\cdot 6}\right)^3 k^6 + \cdots\right] \qquad (2.209)$$

$$E = E\left(\frac{\pi}{2}, k\right) = \frac{\pi}{2}\left[1 - \left(\frac{1}{2^2}\right) k^2 - \left(\frac{3}{2^2\cdot 4^2}\right) k^4 - \left(\frac{3^2\cdot 5}{2^2\cdot 4^2\cdot 6^2}\right) k^6 - \cdots\right] \qquad (2.210)$$

$$K' = F\left(\frac{\pi}{2}, \sqrt{1 - k^2}\right) \qquad E' = E\left(\frac{\pi}{2}, \sqrt{1 - k^2}\right) \qquad (2.211)$$

They are connected by the relation

$$KE' + EK' - KK' = \frac{\pi}{2} \qquad (2.212)$$

TABLE 2.11 Fundamental Integrals

1. $\int u^n \, du = \dfrac{u^{n+1}}{n+1}$

2. $\int \dfrac{du}{u} = \log u$

3. $\int a^u \, du = \dfrac{a^u}{\log a}$

4. $\int e^u \, du = e^u$

5. $\int \cos u \, du = \sin u$

6. $\int \sin u \, du = -\cos u$

7. $\int \sec^2 u \, du = \tan u$

8. $\int \csc^2 u \, du = -\cot u$

9. $\int \sec u \tan u \, du = \sec u$

10. $\int \csc u \cot u \, du = -\csc u$

11. $\int \tan u \, du = \log \sec u$

      $= -\log \cos u$

12. $\int \cot u \, du = \log \sin u$

      $= -\log \csc u$

13. $\int \sec u \, du = \log(\sec u + \tan u)$

      $= \log \tan\left(\dfrac{\pi}{4} + \dfrac{u}{2}\right)$

14. $\int \csc u \, du = \log(\csc u - \cot u)$

      $= \log \tan \dfrac{u}{2}$

15. $\int \dfrac{du}{u^2 + a^2} = \dfrac{1}{a} \tan^{-1} \dfrac{u}{a}$

      $= -\dfrac{1}{a} \cot^{-1} \dfrac{u}{a}$

16. $\int \dfrac{du}{u^2 - a^2} = \dfrac{1}{2a} \log \dfrac{u-a}{u+a}$ $(u^2 > a^2)$

      $= \dfrac{1}{2a} \log \dfrac{a-u}{a+u}$ $(u^2 < a^2)$

      $= -\dfrac{1}{a} \tanh^{-1} \dfrac{u}{a}$ $(u^2 < a^2)$

      $= -\dfrac{1}{a} \coth^{-1} \dfrac{u}{a}$ $(u^2 > a^2)$

17. $\int \dfrac{du}{\sqrt{a^2 - u^2}} = \sin^{-1} \dfrac{u}{a}$

      $= -\cos^{-1} \dfrac{u}{a}$

18. $\int \dfrac{du}{\sqrt{u^2 \pm a^2}} = \log\left(u + \sqrt{u^2 \pm a^2}\right)$

      $\int \dfrac{du}{\sqrt{u^2 + a^2}} = \sinh^{-1} \dfrac{u}{a}$

      $\int \dfrac{du}{\sqrt{u^2 - a^2}} = \cosh^{-1} \dfrac{u}{a}$

19. $\int \dfrac{du}{u\sqrt{u^2 - a^2}} = \dfrac{1}{a} \sec^{-1} \dfrac{u}{a}$

      $= -\dfrac{1}{a} \csc^{-1} \dfrac{u}{a}$

20. $\int \dfrac{du}{\sqrt{2au - u^2}} = \text{vers}^{-1} \dfrac{u}{a}$

      $= \cos^{-1}\left(1 - \dfrac{u}{a}\right)$

21. $\int \sinh u \, du = \cosh u$

22. $\int \cosh u \, du = \sinh u$

23. $\int \tanh u \, du = \log \cosh u$

24. $\int \coth u \, du = \log \sinh u$

25. $\int \operatorname{sech} u \, du = 2 \tan^{-1} e^u$

26. $\int \operatorname{csch} u \, du = \log \tanh \dfrac{u}{2}$

The inverse function of $u = F(\phi, k)$ is $\phi = \text{am } u$, (am = amplitude)

$$x \equiv \sin \phi \equiv \text{sn } u = u - (1 + k^2)\frac{u^3}{3!} + (1 + 14k^2 + k^4)\frac{u^5}{5!} - \cdots \qquad (2.213)$$

$$\cos \phi \equiv \text{cn } u = 1 - \frac{u^2}{2!} + (1 + 4k^2)\frac{u^4}{4!} - (1 + 44k^2 + 16k^4)\frac{u^6}{6!} + \cdots \qquad (2.214)$$

$$\sqrt{1 - k^2 x^2} \equiv \Delta\phi \equiv \text{dn } u = 1 - k^2\frac{u^2}{2!} + k^2(4 + k^2)\frac{u^4}{4!} - k^2(16 + 44k^2 + k^4)\frac{u^6}{6!} + \cdots$$

$$\qquad (2.215)$$

Methods of Integration

Integration by Parts. The formula

$$\int u \, dv = uv - \int v \, du, \quad u \text{ and } v \text{ functions of } x \tag{2.216}$$

is useful in integrating a product, if factors of the product are a function of x and the derivative of another function of x.

Example 2.63. To find $\int x \sin x \, dx$, let $u = x$, $dv = \sin x \, dx$. Then $du = dx$, $v = -\cos x$, and $\int x \sin x \, dx = -x \cos x + \int \cos x \, dx = -x \cos x + \sin x + c$.

Integration of Rational Fractions. If the quotient of two polynomials

$$R(x) = \frac{P_n(x)}{P_d(x)} \tag{2.217}$$

is not a proper fraction, that is, if the degree of the numerator is not less than that of the denominator, $R(x)$ can be changed, by dividing as indicated, to the sum of a polynomial, which is immediately integrable, and a proper fraction. If the proper fraction cannot be integrated by reference to Table 1.18, use the methods of Section 2.2.8, to resolve it, if possible, into partial fractions. These can be integrated from the table.

Irrational Functions. These can sometimes be put into integrable forms by rationalizing them by a change of variable.

Form	Substitution
$\int [(ax+b)^{p/q}] \, dx$	let $ax + b = y^q$
$\int [(ax+b)^{p/q}(ax+b)^{r/s}] \, dx$	let $ax + b = y^n$, where n is the LCM of q, s
$\int \left[x, \sqrt{x^3 + ax + b} \right] dx$	let $\sqrt{x^2 + ax + b} = y - x$
$\int \left[x, \sqrt{-x^2 + ax + b} \right] dx$	let $\sqrt{-x^2 + ax + b} = \sqrt{(\alpha - x)(\beta + x)}$
	$= (\alpha - x)y \text{ or } = (\beta + x)y$
$\int [\sin x, \cos x] \, dx$	let $\tan \dfrac{x}{2} = y$
$\int \left[x, \sqrt{a^2 - x^2} \right] dx$	let $x = a \sin y$
$\int \left[x, \sqrt{x^2 - a^2} \right] dx$	let $x = a \sec y$ or $x = a \cosh y$
$\int \left[x, \sqrt{x^2 + a^2} \right] dx$	let $x = a \tan y$ or $x = a \sinh y$

2.10.2 Definite Integrals

The **definite integral** of $f(x)$ from a to b is

$$\int_a^b f(x) \, dx = \lim_{\substack{n \to \infty \\ \max \Delta x_\nu \to 0}} \sum_{\nu=1}^n f(\xi_\nu) \, \Delta x_\nu \tag{2.218}$$

in which the interval $a \leq x \leq b$ is divided into n arbitrary parts Δx_ν, $\nu = 1, 2, \ldots, n$, and ξ_ν is an arbitrary point in Δx_ν (Fig. 2.140). A sufficient condition that this integral exists is that $f(x)$ be continuous. However, it is necessary and sufficient only that $f(x)$ be bounded and that its points of discontinuity form a set of Lebesgue measure 0. A set of points is of Lebesgue measure 0 if the points can be enclosed in a set of intervals I_ν, $\nu = 1, 2, 3, \ldots$, finite or infinite in number, such that, for any $\varepsilon > 0$, the sum of the lengths of the I_ν is $< \varepsilon$.

A definite integral of a given function over a given interval is a number. The geometric interpretation of the definite integral of $f(x)$, $f(x) \geq 0$, is the area bounded by the curve $y = f(x)$, the x axis, and the ordinates at $x = a$ and $x = b$.

To evaluate a definite integral, if $F(x) = \int f(x) \, dx$, then $\int_a^b f(x) \, dx = F(b) - F(a)$.

Example 2.64

$$\int_3^5 x^2 \, dx = \left. \frac{x^3}{3} \right|_3^5 = \frac{125}{3} - 9 = \frac{98}{3}$$

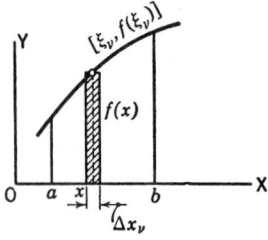

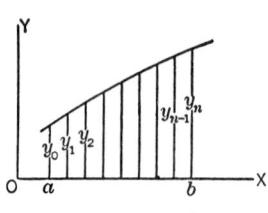

Fig. 2.140   Fig. 2.141

Some Fundamental Theorems.

$$\frac{d}{dx}\int_a^x f(\xi)\,d\xi = f(x)$$

$$\int_a^b cf(x)\,dx = c\int_a^b f(x)\,dx$$

$$\int_a^b [f_1(x) + f_2(x) + \cdots + f_n(x)]\,dx = \int_a^b f_1(x)\,dx + \int_a^b f_2(x)\,dx + \cdots + \int_a^b f_n(x)\,dx$$

$$\int_a^b f(x)\,dx = -\int_b^a f(x)\,dx$$

$$\int_a^b f(x)\,dx = \int_a^c f(x)\,dx + \int_c^b f(x)\,dx \qquad a \le c \le b$$

$$\int_a^b f(x)\,dx = (b-a)f(\xi)$$

for some ξ such that $a \le \xi \le b$ (mean value theorem).

Simpson's Rule for Approximate Integration. To evaluate $\int_a^b f(x)\,dx$ approximately, divide the interval from a to b into an even number n of equal parts with end points $x_0 = a$, $x_1, \ldots, x_n = b$ and let $y_i = f(x_i)$ (Fig. 2.141). Then

$$\int_a^b f(x)\,dx \approx \frac{b-a}{3n}(y_0 + 4y_1 + 2y_2 + 4y_3 + 2y_4 + \cdots + 4y_{n-1} + y_n) \qquad (2.219)$$

Improper Integrals. If *one limit* is *infinite*,

$$\int_a^\infty f(x)\,dx = \lim_{b\to\infty}\int_a^b f(x)\,dx \qquad (2.220)$$

The integral exists, or converges, if there is a number $k > 1$ and a number M independent of x such that $x^k|f(x)| < M$ for arbitrarily large values of x. If $x|f(x)| > m$, an arbitrary positive number, for sufficiently large values of x, the interval diverges.

Example 2.65. The integral $\int_0^\infty x\,dx/(x+x^2)^{3/2}$ exists, since, for $k = 2$ and $M = 1$,

$$x^2\left|\frac{x}{(x+x^2)^{3/2}}\right| = \left(\frac{x^2}{x+x^2}\right)^{3/2} < 1$$

no matter how large the value of x.

If the integrand is *infinite* at the upper limit

$$\int_a^b f(x)\,dx = \lim_{\varepsilon\to 0}\int_a^{b-\varepsilon} f(x)\,dx \qquad 0 < \varepsilon < (b-a) \qquad (2.221)$$

The integral exists if there is a number $k < 1$ and a number M independent of x such that

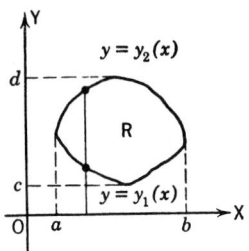

Fig. 2.142

$(b - x)^k |f(x)| < M$ for $a \le x < b$. If there is a number $k \ge 1$ and a number m such that $(b - x)^k |f(x)| > m$ for $a \le x < b$, the integral diverges.

Example 2.66. The integral $\int_0^1 dx/(1 - x)$ diverges, since, for $k = 1$ and $m = \frac{1}{2}$, $(1 - x)/(1 - x) = 1 > \frac{1}{2}$.

If the integrand is infinite at the lower limit, the tests are analogous. If the integrand is infinite at an intermediate point, use the point to divide the interval into two subintervals and apply the preceding tests.

Multiple Integrals. Let $f(x, y)$ be defined in the region R of the xy plane. Divide R into subregions $\Delta R_1, \Delta R_2, \ldots, \Delta R_n$ of areas $\Delta A_1, \Delta A_2, \ldots, \Delta A_n$. Let (ξ_i, η_i) be any point in ΔR_i. If the sum $\sum_{i=1}^{n} f(\xi_i, \eta_i) \Delta A_i$ has a limit as $n \to \infty$ and the maximum diameter of the subregions ΔR_i approaches 0, then

$$\int_R f(x, y) \, dA = \lim_{n \to \infty} \sum_{i=1}^{n} f(\xi_i, \eta_i) \Delta A_i \qquad (2.222)$$

The double integral is evaluated by two successive single integrations, first with respect to y holding x constant, between variable limits of integration, and then with respect to x between constant limits (Fig. 2.142). If $f(x, y)$ is continuous, the order of integration can be reversed.

$$\int_R f(x, y) \, dA = \int_a^b \int_{y_1(x)}^{y_2(x)} f(x, y) \, dy \, dx = \int_c^d \int_{x_1(y)}^{x_2(y)} f(x, y) \, dx \, dy \qquad (2.223a)$$

In polar coordinates,

$$\int_R F(r, \theta) \, dA = \int_\alpha^\beta \int_{r_1(\theta)}^{r_2(\theta)} F(r, \theta) r \, dr \, d\theta = \int_k^l \int_{\theta_1(r)}^{\theta_2(r)} F(r, \theta) r \, d\theta \, dr \qquad (2.223b)$$

The analogous triple integrals are evaluated by three single integrations. In rectangular coordinates,

$$\int_R f(x, y, z) \, dV = \int \int \int f(x, y, z) \, dx \, dy \, dz \qquad (2.224)$$

In spherical coordinates,

$$\int_R F(r, \theta, \phi) \, dV = \int \int \int F(r, \theta, \phi) r^2 \sin \theta \, dr \, d\theta \, d\phi \qquad (2.225)$$

In cylindrical coordinates,

$$\int_R G(\rho, \phi, z) \, dV = \int \int \int G(\rho, \phi, z) \rho \, d\rho \, d\phi \, dz \qquad (2.226)$$

Integrals Containing a Parameter. If $f(x, y)$ is a continuous function of x and y in the closed rectangle $x_0 \le x \le x_1$, $y_0 \le y \le y_1$, and if $f(x, y)$ is integrated with respect to x, with y regarded as

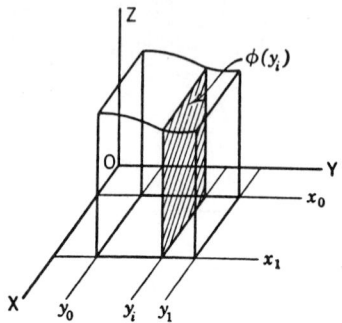

Fig. 2.143

fixed and called a *parameter*, then

$$\int_{x_0}^{x_1} f(x, y)\, dx = \phi(y) \tag{2.227}$$

is a continuous function of y. Geometrically, the function $f(x, y)$ may be plotted as a surface $z = f(x, y)$. Then the value of $\phi(y_i)$ is the area of the section under the surface made by the plane $y = y_i$ (Fig. 2.143). If the limits of integration are continuous functions of y instead of constants, then $\phi(y)$ is continuous.

Differentiation under the Integral Sign. If $\partial f / \partial y$ is a continuous function of x and y in a closed rectangle, then

$$\frac{d\phi}{dy} = \int_{x_0}^{x_1} \frac{\partial f(x, y)}{\partial y}\, dx \tag{2.228}$$

If $\partial f / \partial y$ is continuous and the limits of integration are differentiable functions of y, then

$$\frac{d\phi}{dy} = \frac{d}{dy} \int_{x_0 = g_0(y)}^{x_1 = g_1(y)} f(x, y)\, dx = \int_{g_0(y)}^{g_1(y)} \frac{\partial f(x, y)}{\partial y}\, dx - f(g_0, y)\frac{dg_0}{dy} + f(g_1, y)\frac{dg_1}{dy} \tag{2.229}$$

If $f(x)$ is integrable in the interval $a \le x \le b$ and continuous at a point within the interval, then at that point the function $F(x) = \int_a^x f(\xi)\, d\xi$ has a derivative $F'(x) = f(x)$.

Uniform Convergence and Change of Order of Integration. The improper integral

$$\phi(y) = \int_{x_0}^{\infty} f(x, y)\, dx \tag{2.230}$$

converges uniformly in y in the interval $y_0 \le y \le y_1$, if for any $\varepsilon > 0$ there exists an L, dependent on ε, but not on y, such that

$$\left| \int_l^{\infty} f(x_1 y)\, dx \right| < \varepsilon \quad \text{for } l \ge L \tag{2.231}$$

If $\int_{x_0}^{\infty} f(x, y)\, dx$ is uniformly convergent for $y_0 \le y \le y_1$, then

$$\int_{y_0}^{y_1} \int_{x_0}^{\infty} f(x, y)\, dx\, dy = \int_{x_0}^{\infty} \int_{y_0}^{y_1} f(x, y)\, dy\, dx \tag{2.232}$$

Stieltjes Integral. If $f(x)$ and $\phi(x)$ are defined in the interval (a, b), the Stieltjes integral of $f(x)$ with respect to $\phi(x)$ is

$$\int_a^b f(x)\, d\phi(x) = \lim_{\substack{n \to \infty \\ \max \Delta x_\nu \to 0}} \sum_{\nu=1}^{n} f(\xi_\nu)[\phi(x_\nu) - \phi(x_{\nu-1})] \tag{2.233}$$

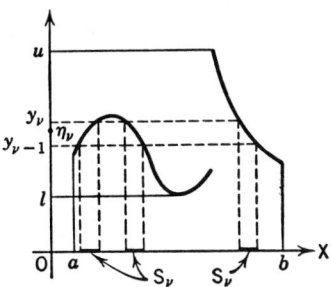

Fig. 2.144

in which the interval (a, b) is divided into n arbitrary parts $\Delta x_\nu = x_\nu - x_{\nu-1}$ by the points $x_0 = a, x_1, \ldots, x_n = b$, and ξ_ν is an arbitrary point in Δx_ν. This limit exists if $f(x)$ is continuous and $\phi(x)$ is of bounded variation, that is, can be expressed as the difference of two nonincreasing or two nondecreasing bounded functions. However, it is not necessary that $f(x)$ be continuous, but only that the variation of $\phi(x)$ over the set of points of discontinuity of $f(x)$ be zero.

Lebesgue Integral. Let S be a set of points in the interval (a, b), and $C(S)$ the complement of S, that is, the set of all the points of (a, b) that do not belong to S. Enclose the points of S in a set of intervals I_ν, $\nu = 1, 2, 3, \ldots$, finite or infinite in number, and let the sum of the lengths of the I_ν be L. The greatest lower bound of all possible values of L is the exterior measure $\overline{m}(S)$ of S. The interior measure of S is $m(S) = (b - a) - \overline{m}[C(S)]$. If $\overline{m}(S) = m(S)$, the set S is measurable and its *measure* is $m(S) = \overline{m}(S)$.

A function $f(x)$ defined in the interval (a, b) is *measurable* if the set of points x for which $y_0 \leq f(x) < y_1$ is measurable for any values of y_0 and y_1.

Let u and l be the upper and lower bounds of a measurable function $f(x)$ defined in the interval (a, b) (Fig. 2.144). Divide the interval (u, l) into n arbitrary subintervals Δy_ν by the points $y_0 = 1, y_1, \ldots, y_n = u$. Let S_ν be the set of points for which $y_{\nu-1} \leq f(x) < y_\nu$, and η_ν any point in the interval Δy_ν. Then the Lebesgue integral of $f(x)$ in the interval (a, b) is

$$\int_a^b f(x)\, dx = \lim_{\substack{n \to \infty \\ \max \Delta y_\nu \to 0}} \sum_{\nu=1}^n \eta_\nu \cdot m(S_\nu) \tag{2.234}$$

If the Riemann integral in the interval (a, b), defined on p. 2·117, exists, the Lebesgue integral does also, and the two are equal, but not conversely.

2.10.3 Line, Surface, and Volume Integrals

Line Integrals. Let $P(x, y)$ and $Q(x, y)$ be functions continuous at all points of a continuous curve C joining the points A and B in the xy plane. Divide the curve C into n arbitrary parts Δs_ν by the points (x_ν, y_ν), let (ξ_ν, η_ν) be an arbitrary point on Δs_ν, and let Δx_ν and Δy_ν be the projections of ΔS_ν on the x and the y axes (Fig. 2.145). The line integral is

$$\int_A^B [P(x, y)\, dx + Q(x, y)\, dy] = \lim_{\substack{n \to \infty \\ \max \Delta x_\nu, \Delta y_\nu \to 0}} \sum_{\nu=1}^n [P(\xi_\nu, \eta_\nu)\, \Delta x_\nu + Q(\xi_\nu, \eta_\nu)\, \Delta y_\nu] \tag{2.235}$$

If the equation of the curve C is $y = f(x)$, $x = \phi(y)$, or the parametric equations $x = x(t)$, $y = y(t)$, the line integral can be evaluated as a definite integral in the one variable x, y, or t, respectively.

Example 2.67. Fine the value of $\int_{0,0}^{1,3} [y^2\, dx + (xy - x^2)\, dy]$ along the paths (a) $y = 3x$, (b) $y^2 = 9x$.

(a) Substitute $y = 3x$, $dy = 3\, dx$ and obtain

$$\int_0^1 [9x^2 + (3x^2 - x^2)3]\, dx = \int_0^1 15x^2\, dx = 5$$

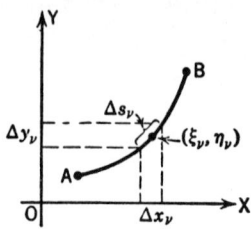

Fig. 2.145

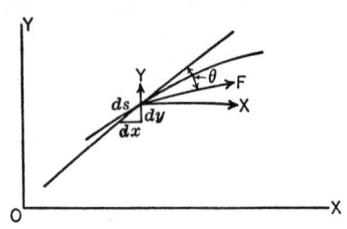

Fig. 2.146

(b) Substitute $y^2 = 9x$, $2y\,dy = 9\,dx$ and obtain

$$\int_0^3 \left[\frac{2}{9}y^3 + \left(\frac{y^3}{9} - \frac{y^4}{81}\right)\right] dy = \left[\frac{1}{12}y^4 - \frac{y^5}{405}\right]_0^3 = 6\tfrac{3}{20}$$

A line integral in the xyz space

$$\int_A^B [P(x, y, z)\, dx + Q(x, y, z)\, dy + R(x, y, z)\, dz]\qquad (2.236)$$

is defined similarly.

Applications

Work. The work done by a constant force F acting on a particle that moves a distance s along a straight line inclined at an angle θ to the force is $W = Fs \cos \theta$. If the path is a curve C and the force variable, the differential of work is $dW = F \cos \theta\, ds$, where ds is the differential of path. Then

$$W = \int dW = \int_C F \cos \theta\, ds = \int_C (X\,dx + Y\,dy)\qquad (2.237)$$

where X and Y are the x and y components of F (Fig. 2.146).

Area. The area of a region bounded by a closed curve C such that a line parallel to the x or y axis meets C in no more than two points is

$$A = \frac{1}{2}\int_C (x\,dy - y\,dx)\qquad (2.238)$$

The formula can be applied to any region that can be divided by a finite number of lines into regions satisfying the preceding condition.

Surface Integrals. Let $P(x, y, z)$ be a function continuous at all points of a region S (bounded by a simple closed curve) of a surface $z = f(x, y)$, which has a continuously turning tangent plane except possibly at isolated points or lines. Let A be the projection of S on the xy plane. Divide S into arbitrary subregions ΔS_ν and let $(\xi_\nu, \eta_\nu, \zeta_\nu)$ be an arbitrary point in ΔS_ν (Fig. 2.147). The surface integral is

$$\lim_{\substack{n \to \infty \\ \text{max diam } \Delta S_\nu \to 0}} \sum_{\nu=1}^n P(\xi_\nu, \eta_\nu, \zeta_\nu)\, \Delta S_\nu = \int_S P(x, y, z)\, dS$$

$$= \int\!\!\int_A P(x, y, z) \sqrt{1 + \left(\frac{\partial z}{\partial x}\right)^2 + \left(\frac{\partial z}{\partial y}\right)^2}\; dx\,dy \quad (2.239)$$

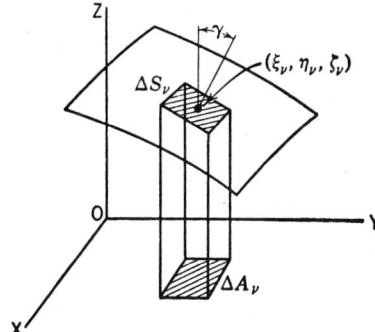

Fig. 2.147

If α, β, γ are the direction angles of the normal to S, the form of the surface integral analogous to the line integral (2.235) is

$$\iint_A (P\,dy\,dz + Q\,dz\,dx + R\,dx\,dy) = \int_S (P\cos\alpha + Q\cos\beta + R\cos\gamma)\,dS \qquad (2.240)$$

Green's Theorem. Let $P(x, y)$ and $Q(x, y)$ be continuous functions with continuous partial derivatives $\partial P/\partial y$ and $\partial Q/\partial x$ in a simply connected region R bounded by a simple closed curve C. Then

$$\iint_R \left(\frac{\partial Q}{\partial x} - \frac{\partial P}{\partial y} \right) dx\,dy = \int_C (P\,dx + Q\,dy) \qquad (2.241)$$

A region is *simply connected* if any closed curve in the region can be shrunk to a point without passing outside the region.

Stokes's Theorem. Let $P(x, y, z), Q(x, y, z), R(x, y, z)$ be continuous functions with continuous first partial derivatives, S a region (bounded by a simple closed curve C) of a surface $z = f(x, y)$, continuous with continuous first partial derivatives. Then

$$\iint_S \left[\left(\frac{\partial R}{\partial y} - \frac{\partial Q}{\partial z} \right) dy\,dz + \left(\frac{\partial P}{\partial z} - \frac{\partial R}{\partial x} \right) dz\,dx + \left(\frac{\partial Q}{\partial x} - \frac{\partial P}{\partial y} \right) dx\,dy \right]$$

$$= \int_C (P\,dx + Q\,dy + R\,dz) \qquad (2.242)$$

The signs are such that an observer standing on the surface with head in the direction of the normal will see the integration around C taken in the positive direction.

Divergence, or Gauss's, Theorem. Let $P(x, y, z), Q(x, y, z), R(x, y, z)$ be continuous functions with continuous first partial derivatives. Let V be a region in the xyz space bounded by a closed surface S with a continuously turning tangent plane except possibly at isolated points or lines. Then

$$\iiint_V \left(\frac{\partial P}{\partial x} + \frac{\partial Q}{\partial y} + \frac{\partial R}{\partial z} \right) dx\,dy\,dz = \iint_S (P\,dy\,dz + Q\,dz\,dx + R\,dx\,dy) \qquad (2.243)$$

Example 2.68. Evaluate $\iint (x\,dy\,dz + y\,dz\,dx + z\,dx\,dy)$ over the cylinder

$$x^2 + y^2 = a^2 \qquad z = \pm b$$

Since

$$P = x \qquad Q = y \qquad R = z$$

$$\frac{\partial P}{\partial x} = 1 \qquad \frac{\partial Q}{\partial y} = 1 \qquad \frac{\partial R}{\partial z} = 1$$

and

$$\int_{-a}^{a} \int_{-\sqrt{a^2-x^2}}^{\sqrt{a^2-x^2}} \int_{-b}^{+b} 3 \, dz \, dy \, dx = 6\pi a^2 b$$

Independence of Path and Exact Differential. Under the conditions of Green's theorem, the following statements are equivalent:

1. $\int_C (P \, dx + Q \, dy) = 0$ for any closed curve C in the region R.
2. The value of $\int_{(a, b)}^{(\xi, \eta)} (P \, dx + Q \, dy)$ is independent of the curve connecting (a, b) and (ξ, η), any points in R.
3. $\partial P/\partial y = \partial Q/\partial x$ at all points of R.
4. There exists a function $F(x, y)$ such that $dF = P \, dx + Q \, dy$.

Under the conditions of Stokes's theorem, the corresponding statements for three dimensions are:

1. $\int_C (P \, dx + Q \, dy + R \, dz) = 0$ for any closed curve C in the region S.
2. The value of $\int_{(a, b, c)}^{(\xi, \eta, \zeta)} (P \, dx + Q \, dy + R \, dz)$ is independent of the curve connecting (a, b, c) and (ξ, η, ζ), any points in S.
3. $\partial P/\partial y = \partial Q/\partial x$, $\partial Q/\partial z = \partial R/\partial y$, $\partial R/\partial x = \partial P/\partial z$ at all points of S.
4. There exists a function $F(x, y, z)$ such that $dF = P \, dx + Q \, dy + R \, dz$.

2.10.4　Applications of Integration

Length of Arc of a Curve. The length s of the arc of a plane curve $y = f(x)$ from the point (a, b) to the point (c, d) is

$$s = \int_a^c \sqrt{1 + \left(\frac{dy}{dx}\right)^2} \, dx = \int_b^d \sqrt{1 + \left(\frac{dx}{dy}\right)^2} \, dy \qquad (2.244)$$

If the equation of the curve is in polar coordinates, $r = f(\theta)$, then the length of the arc from the point (r_1, θ_1) to the point (r_2, θ_2) is

$$s = \int_{\theta_1}^{\theta_2} \sqrt{r^2 + \left(\frac{dr}{d\theta}\right)^2} \, d\theta = \int_{r_1}^{r_2} \sqrt{1 + r^2 \left(\frac{d\theta}{dr}\right)^2} \, dr \qquad (2.245)$$

If the curve is in three dimensions, represented by the equations $y = f_1(x)$, $z = f_2(x)$, the length of arc from $x_1 = a$ to $x_2 = b$ is

$$s = \int_a^b \sqrt{1 + \left(\frac{dy}{dx}\right)^2 + \left(\frac{dz}{dx}\right)^2} \, dx \qquad (2.246)$$

Plane Area. The area bounded by the curve $y = f(x)$, the x axis, and the ordinates at $x = a$, $x = b$ is

$$A = \int_a^b f(x) \, dx \qquad (2.247)$$

where y has the same sign for all values of x between a and b.

In polar coordinates, the area bounded by the curve $r = f(\theta)$, and the two radii $\theta = \alpha$, $\theta = \beta$ (Fig. 2.148) is

$$A = \frac{1}{2} \int_\alpha^\beta r^2 \, d\theta \qquad (2.248)$$

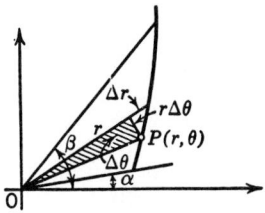

Fig. 2.148

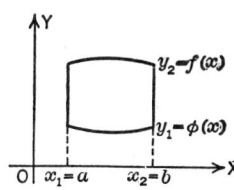

Fig. 2.149

In rectangular coordinates, if the area is bounded by the two curves $y_2 = f(x)$, $y_1 = \phi(x)$, and the lines $x_2 = b$, $x_1 = a$ (Fig. 2.149), then

$$A = \int_a^b dx \int_{\phi(x)}^{f(x)} dy \qquad (2.249)$$

If the area is bounded by the two curves $x_2 = \psi(y)$, $x_1 = \xi(y)$, and the lines $y_2 = d$, $y_1 = c$, then

$$A = \int_c^d dy \int_{\xi(y)}^{\psi(y)} dx \qquad (2.250)$$

If expressed in polar coordinates, the area by double integration is

$$A = \int_{\theta_1}^{\theta_2} d\theta \int_{r_1=f_1(\theta)}^{r_2=f_2(\theta)} r\,dr \quad \text{or} \quad \int_{r_1}^{r_2} r\,dr \int_{\theta_1=\phi_1(r)}^{\theta_2=\phi_2(r)} d\theta \qquad (2.251)$$

Area of a Surface Revolution. The area of the surface of a solid of revolution generated by revolving the curve $y = f(x)$ between $x = a$ and $x = b$:

about the x axis is

$$2\pi \int_a^b y \sqrt{1 + \left(\frac{dy}{dx}\right)^2}\, dx \qquad (2.252)$$

about the y axis is

$$2\pi \int_c^d x \sqrt{1 + \left(\frac{dx}{dy}\right)^2}\, dy \qquad (2.253)$$

where $c = f(a)$ and $d = f(b)$.

Volume. By triple integration:

Rectangular coordinates:

$$V = \int\int\int dx\,dy\,dz \qquad (2.254)$$

Spherical coordinates:

$$V = \int\int\int r^2 \sin\theta\, d\theta\, d\phi\, dr \qquad (2.255)$$

Cylindrical coordinates:

$$V = \int\int\int \rho\, d\rho\, d\phi\, dz \qquad (2.256)$$

(the limits of integration to be supplied).

Volume of a Solid of Revolution. The volume of a solid of revolution generated by revolving the region bounded by the x axis and the curve $y = f(x)$ between $x = a$ and $x = b$: about the x axis is

$$\pi \int_a^b y^2\, dx \qquad (2.257a)$$

about the y axis is

$$\pi \int_c^d x^2 \, dy \tag{2.257b}$$

where $c = f(a)$ and $d = f(b)$.

Surfaces. If the equation of a surface is written in the parametric form $x = f_1(u, v)$, $y = f_2(u, v)$, $z = f_3(u, v)$, the length of arc of a curve $u = u(t)$, $v = v(t)$ on the surface is

$$s = \int \sqrt{E\left(\frac{du}{dt}\right)^2 + 2F\frac{du}{dt}\frac{dv}{dt} + G\left(\frac{dv}{dt}\right)^2} \, dt \tag{2.258}$$

The area S of a region on the surface is

$$S = \int\int \sqrt{EG - F^2} \, du \, dv \tag{2.259}$$

where

$$E = \left(\frac{\partial x}{\partial u}\right)^2 + \left(\frac{\partial y}{\partial u}\right)^2 + \left(\frac{\partial z}{\partial u}\right)^2$$

$$F = \frac{\partial x}{\partial u}\frac{\partial x}{\partial v} + \frac{\partial y}{\partial u}\frac{\partial y}{\partial v} + \frac{\partial z}{\partial u}\frac{\partial z}{\partial v}$$

$$G = \left(\frac{\partial x}{\partial v}\right)^2 + \left(\frac{\partial y}{\partial v}\right)^2 + \left(\frac{\partial z}{\partial v}\right)^2$$

If the equation of the surface is written as $x = u$, $y = v$, $z = f(u, v) = f(x, y)$,

the arc length $\quad s = \int \sqrt{(1 + p^2)\left(\frac{dx}{dt}\right)^2 + 2pq\frac{dx}{dt}\frac{dy}{dt} + (1 + q^2)\left(\frac{dy}{dt}\right)^2} \, dt \tag{2.260}$

the area $\quad S = \int\int \sqrt{1 + p^2 + q^2} \, dx \, dy$, where $p = \dfrac{\partial z}{\partial x}$, $q = \dfrac{\partial z}{\partial y}$ $\tag{2.261}$

(the limits of integration to be supplied).

Moment. The moment of a mass m

$$\left. \begin{aligned} \text{about the } yz \text{ plane,} \quad M_{yz} &= \int x \, dm \\ \text{about the } xz \text{ plane,} \quad M_{xz} &= \int y \, dm \\ \text{about the } xy \text{ plane,} \quad M_{xy} &= \int z \, dm \end{aligned} \right\} \tag{2.262}$$

(the limits of integration to be supplied)

Center of Gravity. The coordinates of the center of gravity of a mass m are

$$x = \frac{\int x \, dm}{\int dm} \qquad y = \frac{\int y \, dm}{\int dm} \qquad z = \frac{\int z \, dm}{\int dm} \tag{2.263}$$

(the limits of integration to be supplied).

Moment of Inertia. The moments of inertia, I, are

$$\left.\begin{array}{l} \text{for a plane curve about the } x \text{ axis,} \quad I_x = \int y^2 \, ds \\[8pt] \text{for a plane curve about the } y \text{ axis,} \quad I_y = \int x^2 \, ds \\[8pt] \text{for a plane curve about the origin,} \quad I_0 = \int (x^2 + y^2) \, ds \end{array}\right\} \quad (2.264)$$

$$\left.\begin{array}{l} \text{for a plane area about the } x \text{ axis,} \quad I_x = \int y^2 \, dA \\[8pt] \text{for a plane area about the } y \text{ axis,} \quad I_y = \int x^2 \, dA \\[8pt] \text{for a plane area about the origin,} \quad I_0 = \int (x^2 + y^2) \, dA \end{array}\right\} \quad (2.265)$$

$$\left.\begin{array}{l} \text{for a solid of mass } m \text{ about the } yz \text{ plane,} \quad I_{yz} = \int x^2 \, dm \\[8pt] \text{for a solid of mass } m \text{ about the } xz \text{ plane,} \quad I_{xz} = \int y^2 \, dm \\[8pt] \text{for a solid of mass } m \text{ about the } xy \text{ plane,} \quad I_{xy} = \int z^2 \, dm \\[8pt] \text{for a solid of mass } m \text{ about the } x \text{ axis,} \quad I_x = I_{xz} + I_{xy}, \text{ etc.} \end{array}\right\} \quad (2.266)$$

(the limits of integration to be supplied).

Fluid Pressure. The total force F against a plane surface perpendicular to the surface of the liquid and between the depths a and b is

$$F = \int_{y=a}^{y=b} \rho y \, dA = \int_a^b \rho yx \, dy \qquad (2.267)$$

where ρ is the weight of the liquid per unit volume and y is the depth beneath the surface of the liquid of a horizontal element of area dA. Usually, $dA = x \, dy$, where x is the width of the vertical surface expressed as a function of y.

Center of Pressure. The depth $\bar{y}$ of the center of pressure against a surface perpendicular to the surface of the liquid and between the depths a and b is

$$\bar{y} = \frac{\displaystyle\int_{y=a}^{y=b} \rho y^2 \, dA}{\displaystyle\int_{y=a}^{y=b} \rho y \, dA} \qquad (2.268)$$

Work. The work W done in moving a particle from $s = a$ to $s = b$ against a force whose component expressed as a function of s in the direction of motion is $F(s)$ is

$$W = \int_{s=a}^{s=b} F(s) \, ds \qquad (2.269)$$

2.11 DIFFERENTIAL EQUATIONS

2.11.1 Definitions

A **differential equation** is an equation containing an unknown function of a set of variables and its derivatives. If the equation has derivatives with respect to one variable only, it is an *ordinary differential equation*, otherwise it is a *partial differential equation*.

Example 2.69

$$\frac{d^2y}{dx^2} + k^2y = 0 \tag{2.270}$$

$$\frac{d^2y}{dx^2} = \sqrt{1 + y^2 + \frac{dy}{dx}} \tag{2.271}$$

$$y\frac{\partial^2 z}{\partial x^2} + zx\frac{\partial^2 z}{\partial x\,\partial y} - \frac{\partial z}{\partial y} = xyz \tag{2.272}$$

$$y - x\frac{dy}{dx} + 3\frac{dx}{dy} = 0 \tag{2.273}$$

Equations (2.270), (2.271), and (2.273) are ordinary differential equations, and (2.272) is a partial differential equation.

The **order** of a differential equation is the order of the highest derivative involved. Thus in Eqs. (2.270)–(2.272), the order is two; in (2.273), the order is one.

The **degree** of a differential equation is the exponent of the highest order appearing in the equation after it is rationalized and cleared of fractions with respect to the derivatives. The degree of (2.270), (2.272), and (2.273) is one; that of (2.271) is two.

A **solution** or **integral** of a differential equation is a relation among the variables that satisfies the equation identically.

A **general solution** of an ordinary differential equation of the nth order is one that contains n independent constants. Thus, $y = \sin x + c$ is a general solution of the equation $dy/dx = \cos x$.

A **particular solution** is one that is derivable from a general solution by assigning fixed values to the arbitrary constants. Thus, $y_1 = \sin x$, $y_2 = \sin x + 4$ are two particular solutions of the preceding equation.

2.11.2   First-Order Equations

Separation of Variables. A differential equation of the first order

$$f\left(x, y, \frac{dy}{dx}\right) = 0 \tag{2.274}$$

can be brought into the form

$$P(x, y)\, dx + Q(x, y)\, dy = 0 \tag{2.275}$$

For the special case where P is a function of x only and Q a function of y only,

$$P(x)\, dx + Q(y)\, dy = 0 \tag{2.276}$$

the variables are separated. The solution is

$$\int P(x)\, dx + \int Q(y)\, dy = c \tag{2.277}$$

Example 2.70. Solve

$$\frac{dy}{dx} = -\frac{x}{y}$$

This can be written as $x\,dx + y\,dy = 0$ and has the solution

$$\int x\,dx + \int y\,dy = \frac{x^2}{2} + \frac{y^2}{2} = c$$

If $c = r^2/2$, then $x^2 + y^2 = r^2$, a set of concentric circles. There are an infinite number of solutions depending on the value of r. Through each point in the plane there passes one circle and only one.

Homogeneous Equations. A function $f(x, y)$ is homogeneous of the nth degree in x and y, if $f(kx, ky) = k^n f(x, y)$. An equation

$$P(x, y)\, dx + Q(x, y)\, dy = 0 \tag{2.278}$$

is homogeneous if the functions $P(x, y)$ and $Q(x, y)$ are homogeneous in x and y. By substituting $y = vx$, the variables can be separated.

Example 2.71. Solve $(x^2 + y^2)\, dx - 2xy\, dy = 0$.

This is of the form $P(x, y)\, dx + Q(x, y)\, dy = 0$, where P and Q are homogeneous functions of the second degree. Making the substitution $y = vx$, the equation becomes $(1 + v^2)\, dx - 2v(x\, dv + v\, dx) = 0$.

Separating variables,

$$\frac{dx}{x} - \frac{2v}{1 - v^2}\, dv = 0$$

Integrating, $\log_e x(1 - v^2) = \log_e c$; replacing $v = y/x, \log(1 - y^2/x^2)x = \log_e c$; and taking exponentials $x^2 - y^2 = cx$.

Linear Differential Equation. The differential equation

$$\frac{dy}{dx} + P(x)y = Q(x) \tag{2.279}$$

in which y and dy/dx appear only in the first degree, and P and Q are functions of x, is a *linear equation of the first order*. This has the general solution

$$y = e^{-\int P(x)\, dx}\left[\int Q(x)e^{\int P(x)\, dx}\, dx + c\right] \tag{2.280}$$

Example 2.72. An equation in the theory of electric networks is

$$L\frac{di}{dt} + Ri = E$$

where i is the current, L the inductance (a constant), R the resistance (a constant), and E the electromotive force, a function of time or constant. If $E = E(t)$

$$i = e^{-(R/L)t}\left[\int \frac{E}{L}e^{(R/L)t}\, dt + c\right]$$

If E is constant and if $i = 0$ at $t = 0$, then

$$i = \frac{E}{R}(1 - e^{-(R/L)t})$$

Bernoulli Equation. This is

$$\frac{dy}{dx} + P(x)y = Q(x)y^n \tag{2.281}$$

in which $n \neq 1$. By making the substitution $z = y^{1-n}$, a linear equation is obtained and the general solution is

$$y = e^{-\int P(x)\, dx}\left[(1 - n)\int e^{(1-n)\int P(x)\, dx}Q(x)\, dx + c\right]^{1/1-n} \tag{2.282}$$

Example 2.73. Solve the equation

$$\frac{dy}{dx} - xy = xy^2$$

Substitute $z = y^{-1}$ and obtain $dz/dx + xz = -x$. The general integral is

$$z = ce^{-x^2/2} - 1 \quad \text{or} \quad y = \frac{1}{ce^{-x^2/2} - 1}$$

Exact Differential Equation. The equation

$$P(x, y)\, dx + Q(x, y)\, dy = 0 \tag{2.283}$$

is an *exact differential equation* if its left side is an exact differential

$$du = P\, dx + Q\, dy \tag{2.284}$$

that is, if $\partial P/\partial y = \partial Q/\partial x$. Then,

$$\int P\, dx + \int \left[Q - \frac{\partial \int P\, dx}{\partial y} \right] dy = c \tag{2.285}$$

is a solution.

Example 2.74. Solve $(x^2 - 4xy - y^2)\, dx + (y^2 - 2xy - 2x^2)\, dy = 0$.
This is an exact equation because $\partial P/\partial y = -4x - 2y = \partial Q/\partial x$.

$$\int (x^2 - 4xy - y^2)\, dx = \frac{x^3}{3} - 2x^2 y - xy^2$$

$$\int \left[(y^2 - 2xy - 2x^2) - (-2x^2 - 2xy) \right] dy = \frac{y^3}{3}$$

The general solution is

$$\frac{x^3}{3} - 2x^2 y - xy^2 + \frac{y^3}{3} = c$$

Integrating Factor. If the left member of the differential equation $P(x, y)\, dx + Q(x, y)\, dy = 0$ is not an exact differential, look for a factor $v(x, y)$ such that $du = v(P\, dx + Q\, dy)$ is an exact differential. Such an *integrating factor* satisfies the equation

$$Q\frac{\partial v}{\partial x} - P\frac{\partial v}{\partial y} + \left(\frac{\partial Q}{\partial x} - \frac{\partial P}{\partial y} \right) v = 0 \tag{2.286}$$

Example 2.75. The equation $(xy^2 - y^3)\, dx + (1 - xy^2)\, dy = 0$ when multiplied by $v = 1/y^2$ becomes $(x - y)\, dx + (1/y^2 - x)\, dy = 0$, of which the left side $du = (x - y)\, dx + (1/y^2 - x)\, dy$ is an exact differential since $\partial P/\partial y = \partial Q/\partial x$. The integration gives $u = x^2/2 - xy - 1/y$. The general solution is $u = c$ or $x^2 y - 2xy^2 - 2cy - 2 = 0$.

Riccati's Equation. This is

$$\frac{dy}{dx} + P(x)y^2 + Q(x)y + R(x) = 0 \tag{2.287}$$

If a particular integral y_1 is known, place $y = y_1 + 1/z$ and obtain a linear equation in z.

2.11.3 Second-Order Equations

The differential equation

$$F\left(x, y, \frac{dy}{dx}, \frac{d^2 y}{dx^2} \right) = 0 \tag{2.288}$$

is of the *second order*. If some of these variables are missing, there is a straightforward method of solution.

CASE 1: With y and dy/dx missing:

$$\frac{d^2y}{dx^2} = f(x) \tag{2.289}$$

This has the solution

$$y = \int dx \int f(x)\, dx + cx + c_1 \tag{2.290}$$

CASE 2: With x and dy/dx missing:

$$\frac{d^2y}{dx^2} = f(y) \tag{2.291}$$

Multiply both sides by $2\, dy/dx$ and obtain

$$x = \int \frac{dy}{\sqrt{c + 2\int f(y)\, dy}} + c_1 \tag{2.292}$$

as a solution.

CASE 3: With x and y missing:

$$\frac{d^2y}{dx^2} = f\left(\frac{dy}{dx}\right) \tag{2.293}$$

Place

$$\frac{dy}{dx} = p \qquad \frac{d^2y}{dx^2} = \frac{dp}{dx} \tag{2.294}$$

Then

$$x = \int \frac{dp}{f(p)} + c$$

Solve for p, replace p by dy/dx, and solve the resulting first order equation.

Example 2.76. The differential equation of the catenary is

$$a\frac{d^2y}{dx^2} = \sqrt{1 + \left(\frac{dy}{dx}\right)^2}$$

Let $p = dy/dx$, then

$$a\frac{dp}{dx} = \sqrt{1 + p^2}$$

By separating variables,

$$\frac{dp}{\sqrt{1 + p^2}} = \frac{dx}{a}$$

which has the solution

$$\sinh^{-1}p = \frac{x + c}{a} \quad \text{or} \quad p = \frac{dy}{dx} = \sinh\frac{x + c}{a}$$

Integrating this latter,

$$y = a \cosh \frac{x + c}{a} + c_1$$

CASE 4: With y missing:

$$\frac{d^2y}{dx^2} = f\left(\frac{dy}{dx}, x\right)$$ (2.295)

Place $dy/dx = p$ and obtain the first-order equation $dp/dx = f(p, x)$. If this can be solved for p, then

$$y = \int p(x)\, dx + c$$ (2.296)

CASE 5: With x missing:

$$\frac{d^2y}{dx^2} = f\left(\frac{dy}{dx}, y\right)$$ (2.297)

Place $dy/dx = p$ and obtain the first-order equation $p\, dp/dy = f(p, y)$. If this can be solved for p, then

$$x = \int \frac{dy}{p(y)} + c$$ (2.298)

2.11.4   Bessel Functions

Wherever the mathematics of problems having circular or cylindrical symmetry appears, it is usually appropriate to consider the solutions of Bessel's differential equation (2.299). Such applications include radiation from a cylindrical antenna, eddy current losses in a cylindrical wire, and sinusoidal angle modulations including phase and frequency modulation.

$$x^2\frac{d^2y}{dx^2} + x\frac{dy}{dx} + (x^2 - n^2)y = 0$$ (2.299)

where n is real, possibly integral or fractional, or complex, and the solution $y(x)$ is said to be of the **first kind** an denoted $J_n(x)$ for $0 \le n$ an integer. Tables of $J_0(x)$ and $J_1(x)$ are available in Table 1.22. Graphs of these are shown in Fig. 2.150.

Bessel functions $J_n(x)$ are **almost periodic functions** that for increasing x have a zero-crossing half "period" approaching π from below. A sequence of these functions can be used to construct an orthogonal series much in the same way that periodic functions, sine, and cosine waves make up a Fourier series.

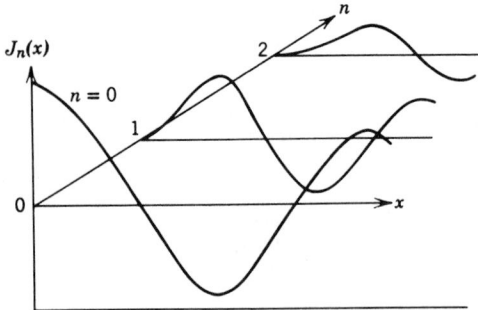

Fig. 2.150   Bessel functions of first kind.

For an extensive set of tables of Bessel functions of many types, see Jahnke, Emde, and Lösch, *Tables of Higher Functions*, Stuttgart, B. G. Teubner Verlag, 1960.

2.11.5 Linear Equations

General Theorem. The differential equation

$$\frac{d^n y}{dx^n} + P_1(x)\frac{d^{n-1}y}{dx^{n-1}} + \cdots + P_{n-1}(x)\frac{dy}{dx} + P_n(x)y = F(x) \tag{2.300}$$

is called the general nth-order linear differential equation. If $F(x) = 0$, the equation is *homogeneous*; otherwise it is *nonhomogeneous*. If $\phi(x)$ is a solution of the nonhomogeneous equation and $y_1, y_2, \ldots, y_n$ are linearly independent solutions of the homogeneous equation, then the *general solution* of (2.300) is

$$y = c_1 y_1 + c_2 y_2 + \cdots + c_n y_n + \phi(x) \tag{2.301}$$

The part $\phi(x)$ is called the *particular integral*, and the part $c_1 y_1 + \cdots + c_n y_n$ is the *complementary function*.

Homogeneous Differential Equation with Constant Coefficients

$$\frac{d^n y}{dx^n} + a_1\frac{d^{n-1}y}{dx^{n-1}} + \cdots + a_{n-1}\frac{dy}{dx} + a_n y = 0 \tag{2.302}$$

A solution of this equation is

$$y_k = ce^{r_k x} \tag{2.303}$$

if r_k is a root of the algebraic equation

$$r^n + a_1 r^{n-1} + \cdots + a_{n-1}r + a_n = 0 \tag{2.304}$$

If all the n roots $r_1, r_2, \ldots, r_n$ of (2.304) are different, then

$$y = c_1 e^{r_1 x} + c_2 e^{r_2 x} + \cdots + c_n e^{r_n x} \tag{2.305}$$

is a general solution of (2.302). If k of the roots are equal, $r_1 = r_2 = \cdots = r_k$ while $r_{k+1}, \ldots, r_n$ are different, then

$$y = \left(c_1 + c_2 x + \cdots + c_k x^{k-1}\right)e^{r_1 x} + c_{k+1}e^{r_{k+1}x} + \cdots + c_n e^{r_n x} \tag{2.306}$$

is a general solution. If $r_1 = p + iq$, $r_2 = p - iq$ are conjugate complex roots of (2.304), then

$$c_1 e^{r_1 x} + c_2 e^{r_2 x} = e^{px}\left(C_1 \cos qx + C_2 \sin qx\right) \tag{2.307}$$

Example 2.77

$$\frac{d^2 y}{dx^2} + 13\frac{dy}{dx} + 40y = 0$$

has the solution $y = c_1 e^{-5x} + c_2 e^{-8x}$.

Example 2.78

$$\frac{d^2 y}{dx^2} + 6\frac{dy}{dx} + 34y = 0$$

has the solution $y = (c_1 \cos 5x + c_2 \sin 5x)e^{-3x}$.

Nonhomogeneous Differential Equation with Constant Coefficients

$$\frac{d^n y}{dx^n} + a_1 \frac{d^{n-1} y}{dx^{n-1}} + \cdots + a_{n-1} \frac{dy}{dx} + a_n y = F(x) \tag{2.308}$$

The complementary function is found as previously. To find the particular integral, replace

$$\frac{dy}{dx} \text{ by } D \qquad \frac{d^2 y}{dx^2} \text{ by } D^2, \ldots, \qquad \frac{d^n y}{dx^n} \text{ by } D^n \tag{2.309}$$

$$P(D)y = \left(D^n + a_1 D^{n-1} + \cdots + a_{n-1} D + a_n \right) y = F(x) \tag{2.310}$$

Particular integrals y_p, in which B_i, A, B are undetermined coefficients, to be determined by substituting y_p in (2.310) and equating coefficients of like terms, are:

(a) If $F(x) = x^n + b_1 x^{n-1} + \cdots + b_{n-1} x + b_n$, then $y_p = x^n + B_1 x^{n-1} + \cdots + B_{n-1} x + B_n$. If D^m is a factor of $P(D)$, then $y_p = (x^n + B_1 x^{n-1} + \cdots + B_{n-1} x + B_n) x^m$.

(b) If $F(x) = b \sin ax$ or $b \cos ax$, then $y_p = A \sin ax + B \cos ax$. If $(D^2 + a^2)^m$ is a factor of $P(D)$, then $y_p = (A \sin ax + B \cos ax) x^m$.

(c) If $F(x) = ce^{ax}$, then $y_p = Ae^{ax}$. If $(D - a)^m$ is a factor of $P(D)$, then $y_p = x^m A e^{ax}$.

(d) If $F(x) = g(x)e^{ax}$, place $y_p = e^{ax} w$ in (2.308), divide out e^{ax}, and solve the equation for w_p as a function of x.

(e) If $F(x)$ is the sum of a number of these functions, then y_p is the sum of the particular integrals corresponding to each of the functions.

(f) If $F(x)$ is not of the type (e), try the method of Laplace transformation, p. 2·168.

Example 2.79. $d^2 y/dx^2 + 4y = x^2 + \cos x$ can be written as $(D^2 + 4)y = (D + 2i)(D - 2i) = x^2 + \cos x$. By (2.307), the complementary function is $y = c_1 \cos 2x + c_2 \sin 2x$. For a particular integral take $y_p = ax^2 + bx + c + f \sin x + g \cos x$ [by (a), (b), (e)].
Then

$$\frac{d^2 y_p}{dx^2} = 2a - f \sin x - g \cos x$$

and substituting in the original equation

$$\frac{d^2 y_p}{dx^2} + 4y_p = 2a - f \sin x - g \cos x + 4ax^2 + 4bx + 4c + 4f \sin x + 4g \cos x$$

$$= x^2 + \cos x$$

Equating coefficients, $a = \frac{1}{4}$, $b = 0$, $c = -\frac{1}{8}$, $f = 0$, $g = \frac{1}{3}$ and the general solution is $y = c_1 \cos 2x + c_2 \sin 2x + x^2/4 - \frac{1}{8} + \frac{1}{3} \cos x$.

Euler's Homogeneous Equation

$$x^n \frac{d^n y}{dx^n} + ax^{n-1} \frac{d^{n-1} y}{dx^{n-1}} + \cdots + a_{n-1} x \frac{dy}{dx} + a_n y = 0 \tag{2.311}$$

Place $x = e^t$, and since

$$x\frac{dy}{dx} = \frac{dy}{dt} \qquad x^2 \frac{d^2 y}{dx^2} = \left[\frac{d}{dt}\left(\frac{d}{dt} - 1 \right) \right] y \qquad x^3 \frac{d^3 y}{dx^3} = \left[\frac{d}{dt}\left(\frac{d}{dt} - 1 \right)\left(\frac{d}{dt} - 2 \right) \right] y, \ldots \tag{2.312}$$

(2.311) is transformed into a linear homogeneous differential equation with constant coefficients.

Depression of Order. If a particular integral of a linear homogeneous differential equation is known, the order of the equation can be lowered. If y_1 is a particular integral of

$$\frac{d^n y}{dx^n} + P_1(x) \frac{d^{n-1} y}{dx^{n-1}} + \cdots + P_{n-1}(x) \frac{dy}{dx} + P_n(x) = 0 \tag{2.313}$$

substitute $y = y_1 z$. The coefficient of z will be zero, and then by placing $dz/dx = u$, the equation is reduced to the $(n - 1)$st order.

Example 2.80. Given

$$\frac{d^2 y}{dx^2} + p(x)\frac{dy}{dx} + q(x)y = 0$$

and y_1, a particular integral of this equation.
Let $y = y_1 z$, then

$$\frac{dy}{dx} = y_1 \frac{dz}{dx} + z\frac{dy_1}{dx} \qquad \frac{d^2 y}{dx^2} = y_1\frac{d^2 z}{dx^2} + 2\frac{dy_1}{dx}\frac{dz}{dx} + z\frac{d^2 y_1}{dx^2}$$

Substituting in the original equation

$$y_1\frac{d^2 z}{dx^2} + 2\frac{dy_1}{dx}\frac{dz}{dx} + z\frac{d^2 y_1}{dx^2} + p\left[y_1\frac{dz}{dx} + z\frac{dy_1}{dx}\right] + qy_1 z = 0$$

and since the coefficient of z is zero, this reduces to

$$y_1\frac{d^2 z}{dx^2} + \left(2\frac{dy_1}{dx} + py_1\right)\frac{dz}{dx} = 0$$

Writing

$$\frac{dz}{dx} = u \qquad \frac{du}{u} + \left(2\frac{dy_1}{dx} + py_1\right)\frac{dx}{y_1} = 0$$

By integrating,

$$\log_e u + \int p\, dx + \log_e y_1^2 = \log_e c \quad \text{or} \quad u = \frac{c}{y_1^2}e^{-\int p\, dx}$$

Another integration gives z. Then

$$y = y_1\int \frac{c}{y_1^2}e^{-\int p\, dx}\, dx + c_1$$

Systems of Linear Differential Equations with Constant Coefficients. For a system of n linear equations with constant coefficients in n dependent variables and one independent variable t, the symbolic algebraic method of solution may be used. If $n = 2$,

$$\left.\begin{array}{l}\left(D^n + a_1 D^{n-1} + \cdots + a_n\right)x + \left(D^m + b_1 D^{m-1} + \cdots + b_m\right)y = R(t)\\ \left(D^p + c_1 D^{p-1} + \cdots + c_p\right)x + \left(D^q + d_1 D^{q-1} + \cdots + d_q\right)y = S(t)\end{array}\right\} \tag{2.314}$$

where $D = d/dt$. The equations may be written as

$$P_1(D)x + Q_1(D)y = R, \qquad P_2(D)x + Q_2(D)y = S \tag{2.315}$$

Treating these as algebraic equations, eliminate either x or y and solve the equation thus obtained.

Example 2.81. Solve the system,

$$(1)\quad \frac{dx}{dt} + \frac{dy}{dt} + 2x + y = 0 \qquad (2)\quad \frac{dy}{dt} + 5x + 3y = 0$$

By using the symbol D these equations can be written

$$(D + 2)x + (D + 1)y = 0, \qquad 5x + (D + 3)y = 0$$

Eliminating x, $(D^2 + 1)y = 0$. From (2.307) (1) this has the solution $y = c_1 \cos t + c_2 \sin t$. Substituting this in (2),

$$x = -\frac{3c_1 + c_2}{5} \cos t + \frac{c_1 - 3c_2}{5} \sin t$$

2.11.6 Linear Algebraic Equations

Consider the set of linear algebraic equations,

$$\sum_{i=1}^{n} a_{ki}\alpha_i = f_k \qquad (k = 1, 2, \ldots, m) \tag{2.316}$$

Equation (2.316) contains m linear algebraic equations in n unknowns, α_i.

Any system of linear equations in which all f_k are zero is called **homogeneous**. Consider the following homogeneous equations associated with the matrix operator A:

Homogeneous Equation: $A\alpha = 0$ (2.317)

Adjoint Homogeneous Equation: $A^*\beta = 0$ (2.318)

where A^* is the adjoint of A. For the linear algebraic equations, $A^* = A^T$, the transpose of A.

The homogeneous adjoint equations can also be written in the form

$$(\mathbf{A}_i, \boldsymbol{\beta}) = 0 \qquad (i = 1, 2, \ldots, n) \tag{2.319}$$

where $(\cdot, \cdot)$ denotes the inner product in Euclidean space, A_i are the column vectors of the matrix $[A]$. From Eqs. (2.316) and (2.319), we deduce the following result, known as the **solvability condition**: The nonhomogeneous equation $A\alpha = \mathbf{f}$ possesses a solution α if and only if the vector $\mathbf{f}$ is orthogonal to all vectors $\boldsymbol{\beta}$ that are the solutions of the homogeneous adjoint equation, $A^*\boldsymbol{\beta} = \mathbf{0}$. In analytical form this statement can be expressed as

$$(\mathbf{f}, \boldsymbol{\beta}) = 0 \tag{2.320}$$

We now consider two cases of linear equations and discuss the existence and uniqueness of solutions of linear equations.

1. If (2.317) has only the trivial (i.e., zero) solution, it follows that $\det A \neq 0$ (otherwise, the trivial solution cannot be determined) and hence $\det A^* \neq 0$. Therefore, the adjoint homogeneous equation (2.318) also has only the trivial solution. Moreover, the solvability conditions are automatically satisfied for any $\mathbf{f}$ [since the only solution of (2.318) is $\boldsymbol{\beta} = \mathbf{0}$], and the nonhomogeneous equation (2.316) has one and only one solution, $\alpha = A^{-1}\mathbf{f}$, where A^{-1} is the inverse of the matrix A.

2. If (2.317) has nontrivial solutions, then $\det A = 0$. This in turn implies that the rows (or columns) of A are linearly dependent. If these linear dependencies are also reflected in the column vector $\mathbf{f}$ (e.g., if the third row of A is the sum of the first and second rows, we must have $f_3 = f_1 + f_2$ in order to have any solutions), then there is a hope of having a solution to the system. If there are $r(\geq n)$ number of independent solutions to (2.316), A is said to have a *r-dimensional null space* (i.e., nullity of A is r). It can be shown that A^* also has a r-dimensional null space, which is in general different from that of A. A necessary and sufficient condition for (2.316) to have solutions is provided by the solvability condition,

$$(f, \beta) \equiv \sum_{i=1}^{n} f_i \beta_i = 0$$

where β is the solution of Eq. (2.318).

Example 2.82. This example has three cases:

1. Consider the following pair of equations in two unknowns α_1 and α_2:

$$3\alpha_1 - 2\alpha_2 = 4$$
$$2\alpha_1 + \alpha_2 = 5$$

or

$$\begin{bmatrix} 3 & -2 \\ 2 & 1 \end{bmatrix} \begin{Bmatrix} \alpha_1 \\ \alpha_2 \end{Bmatrix} = \begin{Bmatrix} 4 \\ 5 \end{Bmatrix} \qquad (A\alpha = f)$$

We note that det $A = 3 + 4 = 7 \neq 0$. The solution is then given by

$$\begin{Bmatrix} \alpha_1 \\ \alpha_2 \end{Bmatrix} = \begin{bmatrix} \frac{1}{7} & \frac{2}{7} \\ -\frac{2}{7} & \frac{3}{7} \end{bmatrix} \begin{Bmatrix} 4 \\ 5 \end{Bmatrix} = \begin{Bmatrix} 2 \\ 1 \end{Bmatrix}$$

The solution of the adjoint equations is trivial, $\beta = 0$, and therefore, the solvability condition is identically satisfied.

2. Next consider the pair of equations,

$$6\alpha_1 + 4\alpha_2 = 4$$
$$3\alpha_1 + 2\alpha_2 = 2$$

or

$$\begin{bmatrix} 6 & 4 \\ 3 & 2 \end{bmatrix} \begin{Bmatrix} \alpha_1 \\ \alpha_2 \end{Bmatrix} = \begin{Bmatrix} 4 \\ 2 \end{Bmatrix} \qquad A\alpha = f$$

We have det $A = 0$, because row 1 (R_1) is equal to two times row 2 (R_2). However, we also have $2f_2 = f_1$. Consequently, we have one linearly independent solution, say $\alpha^{(1)}$, and the other depends on $\alpha^{(1)}$:

$$\alpha^{(1)} = (2, -2)$$

Note that there are many dependent solutions to the pair. For example, $(2, -2)$, $(4, -5)$, $(-2, 4)$, and so on, are solutions of $A\alpha = f$. The solution to the adjoint homogeneous equation

$$\begin{bmatrix} 6 & 3 \\ 4 & 2 \end{bmatrix} \begin{Bmatrix} \beta_1 \\ \beta_2 \end{Bmatrix} = \begin{Bmatrix} 0 \\ 0 \end{Bmatrix}$$

is given by $\beta_2 = -2\beta_1$. Note that $(f, \beta) \equiv f_1\beta_1 + f_2\beta_2 = 4(-\frac{1}{2}\beta_2) + 2\beta_2 = 0$, hence the solvability condition is satisfied.

3. Finally, consider the pair of equations

$$6\alpha_1 + 4\alpha_2 = 3$$
$$3\alpha_1 + 2\alpha_2 = 2$$

or

$$\begin{bmatrix} 6 & 4 \\ 3 & 2 \end{bmatrix} \begin{Bmatrix} \alpha_1 \\ \alpha_2 \end{Bmatrix} = \begin{Bmatrix} 3 \\ 2 \end{Bmatrix}$$

We note that det $A = 0$, because $2R_2 = R_1$. However, $2f_2 \neq f_1$. Hence the pair of equations is inconsistent, and therefore no solutions exist.

Geometrically, we can interpret these three pairs of equations as pairs of straight lines in R^2 with $\alpha_i = x_i$, $i = 1, 2$ (see Fig. 2.151). In part 1, the lines represented by the two equations intersect at the point $(x_1, x_2) = (2, 1)$. In part 2, the lines coincide, or intersect at an infinite number of points, and hence many solutions exist. In part 3, the lines do not intersect at all showing that no solutions exist. From this geometric interpretation, one can see that the lines are nearly parallel (i.e., the angle θ is nearly zero), the determinant of A is nearly zero [because $\tan\theta = (a_{11}a_{22} - a_{12}a_{21})/(a_{11}a_{21} + a_{12}a_{22})$] and therefore it is difficult to obtain an accurate numerical solution. In such cases the system of equations is said to be **ill conditioned**. While these observations can be generalized to a system of n equations, the geometric interpretation becomes complicated.

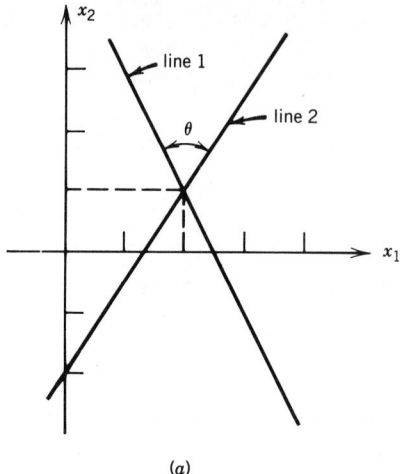

(a)

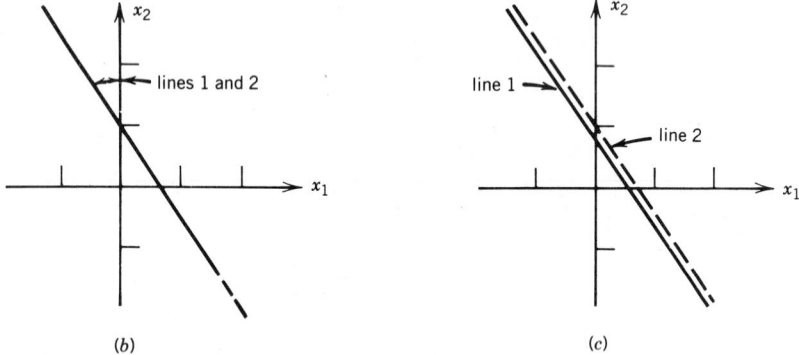

(b)   (c)

Fig. 2.151 Geometric interpretation of the solution of two simultaneous algebraic equations in a plane. (a) Unique solution; (b) many solutions; (c) no solution.

Example 2.83. This example has two cases:
  1. Consider the following set of three equations in three unknowns:

$$\alpha_1 + \alpha_2 + \alpha_3 = 2$$
$$\alpha_1 - \alpha_2 - 3\alpha_3 = 3 \quad \text{or } A\alpha = \mathbf{f}$$
$$3\alpha_1 + \alpha_2 - \alpha_3 = 1$$

The adjoint homogeneous equations become

$$\beta_1 + \beta_2 + 3\beta_3 = 0$$
$$\beta_1 - \beta_2 + \beta_3 = 0 \quad \text{or } A^*\beta = 0$$
$$\beta_1 - 3\beta_2 - \beta_3 = 0$$

Solving for $\boldsymbol{\beta}$, we obtain $\beta_1 = 2\beta_2 = -2\beta_3$. Hence, the null space of A^* is defined by

$$\mathcal{N}(A^*) = \{(2a, a, -a), \quad a \text{ is a real number}\}$$

Clearly $\mathcal{N}(A^*)$ is one dimensional. The null space of A is given by

$$\mathcal{N}(A) = \{(a, -2a, a), \quad a \text{ is a real number}\}$$

Note that $\mathcal{N}(A^*) \neq \mathcal{N}(A)$, but their dimension is the same. Clearly $(2, 1, -1)$ is a solution of $A^*\beta = 0$ while $(1, -2, 1)$ is a solution of $A\alpha = 0$. The solvability condition gives

$$(f, \beta) = 2 \times 2 + 3 \times 1 + 1 \times (-1) \neq 0$$

and therefore $A\alpha = f$ has *no* solution.

2. Reconsider the preceding linear equations with $f = \{-1, 3, 1\}^{\mathsf{T}}$. Then the solvability condition is clearly satisfied. Hence there is one linearly independent solution to $A\alpha = f$ (note that $-2R_1 + R_3 = R_2$ and $-2f_1 + f_3 = f_2$):

$$\alpha \equiv (\alpha_1, \alpha_2, \alpha_3) = (1, -2, 0)$$

Only one of the three α's is arbitrary (not determined), and the remaining two α's are given in terms of the arbitrary α. For example, if α_1 is arbitrary, we have

$$\alpha_1 + \alpha_2 + \alpha_3 = 1$$
$$\alpha_1 + \alpha_2 - 3\alpha_3 = 2 \quad \text{or } A\alpha = f$$
$$3\alpha_1 + \alpha_2 - \alpha_3 = 3$$

We have $\det A \neq 0$. It can be easily verified that $(A) = (A^*) = \{(0, 0, 0)\}$. The unique solution to $A\alpha = f$ is given by

$$\alpha_1 = \tfrac{3}{4} \qquad \alpha_2 = \tfrac{1}{2} \qquad \alpha_3 = -\tfrac{1}{4}.$$

2.11.7 Partial Differential Equations

First Order

Definition. If $x_1, x_2, \ldots, x_n$ are n independent variables, $z = z(x_1, x_2, \ldots, x_n)$ the dependent variable, and if

$$\frac{\partial z}{\partial x_1} = p_1, \ldots, \frac{\partial z}{\partial x_n} = p_n \qquad (2.321)$$

then

$$F(x_1, x_2, \ldots, x_n, z, p_1, p_2, \ldots, p_n) = 0 \qquad (2.322)$$

is a partial differential equation of the first order. An equation

$$f(x_1, x_2, \ldots, x_n, z, c_1, \ldots, c_n) = 0 \qquad (2.323)$$

with n independent constants is a *complete integral* of (2.322) if the elimination of the constants by partial differentiation gives the differential equation (2.322).

Example 2.84

$$F = z^2\left[\left(\frac{\partial z}{\partial x}\right)^2 + \left(\frac{\partial z}{\partial y}\right)^2 + 1\right] - c^2 = 0$$

Then

$$f = (x - h)^2 + (y - k)^2 + z^2 - c^2 = 0$$

is a solution, since by differentiating it with respect to x and y

$$(x - h) + z\frac{\partial z}{\partial x} = 0$$

$$(y - k) + z\frac{\partial z}{\partial y} = 0$$

and substituting the values of $x - h$, $y - k$ from the last two equations in f, expression F is obtained.

If the eliminant obtained by eliminating $c_1, \ldots, c_n$ from the equations $f = 0$, $\partial f/\partial c_1 = 0, \ldots, \partial f/\partial c_n = 0$ satisfies the differential equation, it is a *singular solution*. This differs from a particular integral in that it is usually not obtainable from the complete integral by giving particular values to the constants.

Suppose that the equation

$$F\left(x, y, z, \frac{\partial z}{\partial x}, \frac{\partial z}{\partial y}\right) = 0$$

has the complete integral $f(x, y, z, a, b) = 0$. Let one of the constants $b = \phi(a)$; then $f[x, y, z, a, \phi(a)] = 0$. The *general integral* is the set of solutions found by eliminating a between $f[x, y, z, a, \phi(a)] = 0$ and $d\phi/da = 0$, for all choices of ϕ.

Linear Differential Equations

$$P(x, y, z)p + Q(x, y, z)q = R(x, y, z) \quad \text{where } p = \frac{\partial z}{\partial x} \quad q = \frac{\partial z}{\partial y} \quad (2.324)$$

is a linear partial differential equation. From the system of ordinary equations

$$\frac{dx}{P} = \frac{dy}{Q} = \frac{dz}{R}$$

the two independent solutions $u(x, y, z) = c_1$, $v(x, y, z) = c_2$ are obtained. Then $\Phi(u, v) = 0$, where Φ is an arbitrary function, is the *general solution* of $Pp + Qq = R$.

Example 2.85. Given $xp + yq = z$. The system $dx/x = dy/y = dz/z$ has the solution $u = y/x = c_1$, $v = z/x = c_2$. Then the general solution is

$$\Phi(u, v) = \Phi\left(\frac{y}{x}, \frac{z}{x}\right) = 0$$

Example 2.86. Given $(ny - mz)p + (lz - nx)q = mx - ly$. From

$$\frac{dx}{ny - mz} = \frac{dy}{lz - nx} = \frac{dz}{mx - ly}$$

by using the multipliers l, m, n, and adding the fraction $(l\,dx + m\,dy + n\,dz)/0$ is obtained. Therefore $l\,dx + m\,dy + n\,dz = 0$. This has the solution $lx + my + nz = c_1$. Similarly, $x\,dx + y\,dy + z\,dz = 0$, or $x^2 + y^2 + z^2 = c_2$. Then the general solution is $\Phi(x^2 + y^2 + z^2, lx + my + nz) = 0$.

General Method of Solution. Given $F(x, y, z, p, q) = 0$, the partial differential equation to be solved. Since z is a function of x and y, it follows that $dz = p\,dx + q\,dy$. If another relation can be found among x, y, z, p, q, such as $f(x, y, z, p, q) = 0$, then p and q can be eliminated. The solution of the ordinary differential equation thus formed involving x, y, z, will satisfy the given equation, $F(x, y, z, p, q) = 0$. The unknown function f must satisfy the following linear partial differential equation:

$$\frac{\partial F}{\partial p}\frac{\partial f}{\partial x} + \frac{\partial F}{\partial q}\frac{df}{\partial y} + \left(p\frac{\partial F}{\partial p} + q\frac{\partial F}{\partial q}\right)\frac{\partial f}{\partial z} - \left(\frac{\partial F}{\partial x} + p\frac{\partial F}{\partial z}\right)\frac{\partial f}{\partial p} - \left(\frac{\partial F}{\partial y} + q\frac{\partial F}{\partial z}\right)\frac{\partial f}{\partial q} = 0 \quad (2.325)$$

which is satisfied by any of the solutions of the system

$$\frac{\partial x}{\frac{\partial F}{\partial p}} = \frac{\partial y}{\frac{\partial F}{\partial q}} = \frac{dz}{p\frac{\partial F}{\partial p} + q\frac{\partial F}{\partial q}} = \frac{-dp}{\frac{\partial F}{\partial x} + p\frac{\partial F}{\partial z}} = \frac{-dq}{\frac{\partial F}{\partial y} + q\frac{\partial F}{\partial z}} \qquad (2.326)$$

Example 2.87. Solve $p(q^2 + 1) + (b - z)q = 0$. Here Eqs. (2.326) reduce to

$$\frac{dp}{pq} = \frac{dq}{q^2} = \frac{dz}{3pq^2 + p + (b - z)q} = \frac{dx}{q^2 + 1} = \frac{dy}{-z + b + 2pq}$$

The third fraction, by virtue of the given equation, reduces to $dz/2\,pq^2$. From the first two fractions, by integration, $q = cp$. This and the original equation determine the values of p and q, namely,

$$p = \frac{\sqrt{c_1(z - b)} - 1}{c_1} \qquad q = \sqrt{c_1(z - b)} - 1$$

Substitution of these values in $dz = p\,dx + q\,dy$ gives

$$dz = \left(\frac{dx}{c_1} + dy\right)\sqrt{c_1(z - b)} - 1$$

In this equation the variables are separable; this on integration gives the complete integral $2\sqrt{c_1(z - b)} - 1 = x + c_1 y + c_2$. There is no singular solution. In this work, had another pair of ratios been chosen, say $dq/q^2 = dx/(q^2 + 1)$, another complete integral would have been obtained, namely,

$$(z - b)\left\{\frac{x + k_1}{2} - \sqrt{\left(\frac{x + k_1}{2}\right)^2 + 1}\right\} + y + k_2 = 0$$

Second Order

Definitions. A linear partial differential equation of the second order with two independent variables is of the form

$$L = Ar + 2Bs + Ct + Dp + Eq + Fz = f(x, y) \qquad (2.327)$$

where

$$r = \frac{\partial^2 z}{\partial x^2} \qquad s = \frac{\partial^2 z}{\partial x\,\partial y} \qquad t = \frac{\partial^2 z}{\partial y^2} \qquad p = \frac{\partial z}{\partial x} \qquad q = \frac{\partial z}{\partial y}$$

The coefficients $A, \ldots, F$ are real continuous functions of the real variables x and y. Let $\xi = \xi(x, y)$, $\eta = \eta(x, y)$ be two solutions of the following homogeneous partial differential equation of the first order:

$$Ap^2 + 2Bpq + Cq^2 = 0 \qquad (2.328)$$

If $B^2 - AC = 0$, the homogeneous form of (2.327), $L = 0$, is called the *parabolic* type, and has the normal form

$$\frac{\partial^2 z}{\partial \xi^2} + a\frac{\partial z}{\partial \xi} + b\frac{\partial z}{\partial \eta} + cz = 0 \qquad (2.329)$$

where a, b, c are functions of ξ and η. An example is the equation of heat flow, $\partial u/\partial t = a^2\,\partial^2 u/\partial t^2$ where $u = u(x, t)$ is the temperature, t is the time, a^2 is constant. If $B^2 - AC > 0$ in (2.328), the homogeneous form of (2.327) is the *hyperbolic* type that has as its two normal forms

$$\frac{\partial^2 z}{\partial \xi \, \partial \eta} + a \frac{\partial z}{\partial \xi} + b \frac{\partial z}{\partial \eta} + cz = 0 \tag{2.330}$$

$$\frac{\partial^2 z}{\partial \xi^2} - \frac{\partial^2 z}{\partial \eta^2} + a \frac{\partial z}{\partial \xi} + b \frac{\partial z}{\partial \eta} + cz = 0 \tag{2.331}$$

An example is the equation of a vibrating string

$$\frac{\partial^2 z}{\partial t^2} = a^2 \frac{\partial^2 z}{\partial x^2}$$

where z is the transverse displacement of a point on the string, with abscissa x at time t and a^2 is constant. If $B^2 - AC < 0$, the equation is of the *elliptic* type that has the normal form

$$\frac{\partial^2 z}{\partial \xi^2} + \frac{\partial^2 z}{\partial z^2} + a \frac{\partial z}{\partial \xi} + b \frac{\partial z}{\partial \eta} + cz = 0 \tag{2.332}$$

An example is Laplace's equation

$$\frac{\partial^2 z}{\partial \xi^2} + \frac{\partial^2 z}{\partial \eta^2} = 0$$

usually written $\nabla^2 z = 0$. The two solutions of (2.328) are real in the hyperbolic case and conjugate complex in the elliptic case. That is, in the latter case, $\xi = \frac{1}{2}(\alpha + i\beta)$, $\eta = \frac{1}{2}(\alpha - i\beta)$, where α and β are real, and

$$\frac{\partial^2 z}{\partial \xi \, \partial \eta} = \frac{1}{4} \left(\frac{\partial^2 z}{\partial \alpha^2} + \frac{\partial^2 z}{\partial \beta^2} \right)$$

As in ordinary linear equations, the whole solution consists of the complementary function and the particular integral. Also if $z = z_1$, $z = z_2, \ldots, z = z_n$ are solutions of the homogeneous equation (2.327), $L = 0$, then $z = c_1 z_1 + c_2 z_2 + \cdots + c_n z_n$ is again a solution.

Equations Linear in the Second Derivatives. The general type of second-order equation linear in the second derivatives may be written in the form

$$Ar + Bs + Ct = V \tag{2.333}$$

where A, B, C, V are functions of x, y, z, p, q. From the equations

$$A \, dy^2 - B \, dx \, dy + C \, dx^2 = 0 \tag{2.334}$$
$$A \, dp \, dy + C \, dq \, dx - V \, dx \, dy = 0 \tag{2.335}$$
$$p \, dx + q \, dy = dz \tag{2.336}$$

it may be possible to derive either one or two relations between x, y, z, p, q, called intermediary integrals, and from these to deduce the solution of (2.333). To obtain an intermediary integral, resolve (2.334), supposing the left member not a perfect square, into the two equations $dy - n_1 \, dx = 0$, $dy - n_2 \, dx = 0$. From the first of these and from (2.335) combined, if necessary, with (2.336), obtain the two integrals $u_1(x, y, z, p, q) = a$, $v_1(x, y, z, p, q) = b$; then $u_1 = f_1(v_1)$, where f_1 is an arbitrary function, is now an intermediary integral. In the same way, from $dy - n_2 \, dx = 0$, obtain another pair of integrals $u_2 = a_1$, $v_2 = b_1$; then $u_2 = f_2(v_2)$ is an intermediary integral. For the final integral, if $n_1 = n_2$, the intermediary integral may be integrated. If $n_1 \neq n_2$, solve the two intermediary integrals for p and q, substitute in $p \, dx + q \, dy = dz$, and integrate for the solution.

Example 2.88. Solve

$$r^2 - a^2 t = 0 \tag{2.337}$$

the equation for a vibrating string.

The auxiliary equations are

$$dy - a\,dx = 0 \qquad dy + a\,dx = 0 \qquad dp\,dy - a^2\,dx\,dq = 0 \qquad (2.338)$$

Hence $y + ax = c_1$, $y - ax = c_2$. Combining $y + ax = c_1$ with (2.338), $dp + a\,dq = 0$ is obtained, whereupon $p + aq = c_3 = f_1(y + ax)$. Combining $y - ax = c_1$ with (2.338), $dp - a\,dq = 0$ is obtained, whereupon $p - aq = c_4 = f_2(y - ax)$.

Solving for p and q,

$$p = \frac{1}{2}[f_1(y + ax) + f_2(y - ax)] \qquad q = \frac{1}{2a}[f_1(y + ax) - f_2(y - ax)]$$

Substituting these in $p\,dx + q\,dy = dz$,

$$dz = \frac{1}{2a}[f_1(y + a)(dy + a\,dx) - f_2(y - ax)(dy - a\,dx)]$$

which is an exact differential. Integration gives $z = \phi(y + ax) + \psi(y - ax)$.

Homogeneous Equation with Constant Coefficients

$$\frac{\partial^2 z}{\partial x^2} + A_1\frac{\partial^2 z}{\partial x\,\partial y} + A_2\frac{\partial^2 z}{\partial y^2} = 0 \qquad (2.339)$$

This equation is equivalent to

$$\left(\frac{\partial}{\partial x} - m_1\frac{\partial}{\partial y}\right)\left(\frac{\partial}{\partial x} - m_2\frac{\partial}{\partial y}\right)z = 0 \qquad (2.340)$$

where m_1 and m_2 are roots of the auxiliary equation $X^2 + A_1 X + A_2 = 0$. The general solution of (2.340) is

$$z = f_1(y + m_1 x) + f_2(y + m_2 x) \qquad (2.341)$$

Example 2.89. Solve

$$8\frac{\partial^2 z}{\partial x^2} + 2\frac{\partial^2 z}{\partial x\,\partial y} - 15\frac{\partial^2 z}{\partial y^2} = 0$$

The auxiliary equation is $8X^2 + 2X - 15 = (2X + 3)(4X - 5) = 0$. Hence $m_1 = -\frac{3}{2}$, $m_2 = \frac{5}{4}$. The general solution is $z = f_1(2y - 3x) + f_2(4y + 5x)$.

If the auxiliary equation has multiple factors, the general solution is $z = f_1(y + m_1 x) + xf_2(y + m_1 x)$.

Example 2.90. Solve

$$\frac{\partial^2 z}{\partial x^2} + 6\frac{\partial^2 z}{\partial x\,\partial y} + 9\frac{\partial^2 z}{\partial y^2} = 0$$

The auxiliary equation is $X^2 + 6X + 9 = (X + 3)(X + 3) = 0$. The general solution is $z = f_1(y - 3x) + xf_2(y - 3x)$.

If the coefficients in Eq. (2.339) are real, the complex roots of the auxiliary equation occur in conjugate pairs. Then the general solution will have the form

$$z = f(y + \alpha x + i\beta x) + g(y + \alpha x - i\beta x)$$

Example 2.91. Solve

$$\frac{\partial^2 z}{\partial x^2} - 2\frac{\partial^2 z}{\partial x\,\partial y} + 2\frac{\partial^2 z}{\partial y^2} = 0$$

The auxiliary equation is $X^2 - 2X + 2 = 0$ and $m = 1 \pm i$. The general solution is $z = f(y + x + ix) + g(y + x - ix)$, which can be written as $z = f_1(y + x + ix) + f_1(y + x - ix) + i[g_1(y + x + ix) - g_1(y + x - ix)]$, where f_1 and g_1 are any twice differentiable real functions. If, in particular, $f_1 = \cos u$ and $g_1 = e^u$, it can be shown that $z = 2\cos(x + y)\cosh x - 2e^{x+y}\sin x$.

Method of Separation of Variables. As an example of this method, the solution will be given to Laplace's equation

$$\nabla^2 u = \frac{\partial^2 u}{\partial x^2} + \frac{\partial^2 u}{\partial y^2} = 0 \tag{2.342}$$

Assume that

$$u = X(x) \cdot Y(y) \tag{2.343}$$

where X is a function of x only, and Y a function of y only. By substitution and dividing by $X \cdot Y$, (2.342) becomes

$$\frac{1}{X}\frac{d^2 X}{dx^2} = -\frac{1}{Y}\frac{d^2 Y}{dy^2} \tag{2.344}$$

Since the left side does not contain y, the right side does not contain x, and the two sides are equal, they must equal a constant, say $-k^2$.

$$\frac{1}{X}\frac{d^2 X}{dx^2} = -k^2 \qquad \frac{1}{Y}\frac{d^2 Y}{dy^2} = k^2 \tag{2.345}$$

The solutions of these homogeneous linear differential equations with constant coefficients are

$$X = c_1\cos kx + c_2\sin kx \qquad Y = c_3 e^{ky} + c_4 e^{-ky} \tag{2.346}$$

Hence, from (2.343),

$$u = (c_1\cos kx + c_2\sin kx)(c_3 e^{ky} + c_4 e^{-ky})$$
$$= e^{ky}(k_1\cos kx + k_2\sin kx) + e^{-ky}(k_3\cos kx + k_4\sin kx) \tag{2.347}$$

Since (2.342) is linear, the sum of any number of solutions is again a solution. An infinite number of solutions may be taken provided the series converges and may be differentiated term by term. Then

$$u = \sum_{n=0}^{\infty}\left[e^{ky}(A_n\cos kx + B_n\sin kx) + e^{-ky}(D_n\cos kx + E_n\sin kx)\right] \tag{2.348}$$

is a solution of (2.342). The coefficients of (2.348) are determined by using the series as a Fourier series to fit the boundary conditions.

Functions that satisfy Laplace's equation are *harmonic*. In polar coordinates (2.342) becomes,

$$\nabla^2 u = \frac{\partial^2 u}{\partial r^2} + \frac{1}{r^2}\frac{\partial^2 u}{\partial \theta^2} + \frac{1}{r}\frac{\partial u}{\partial r} = 0 \tag{2.349}$$

In three dimensions, Laplace's equation in rectangular coordinates is

$$\nabla^2 u = \frac{\partial^2 u}{\partial x^2} + \frac{\partial^2 u}{\partial y^2} + \frac{\partial^2 u}{\partial z^2} = 0 \tag{2.350}$$

In cylindrical coordinates,

$$\nabla^2 u = \frac{\partial^2 u}{\partial \rho^2} + \frac{1}{\rho}\frac{\partial u}{\partial \rho} + \frac{1}{\rho^2}\frac{\partial^2 u}{\partial \phi^2} + \frac{\partial^2 u}{\partial z^2} = 0 \tag{2.351}$$

In spherical coordinates,

$$\nabla^2 u = \frac{1}{r^2}\frac{\partial}{\partial r}\left(r^2\frac{\partial u}{\partial r}\right) + \frac{1}{r^2\sin^2\theta}\frac{\partial^2 u}{\partial \phi^2} + \frac{1}{r^2\sin\theta}\frac{\partial}{\partial\theta}\left(\sin\theta\frac{\partial u}{\partial\theta}\right) \qquad (2.352)$$

2.12   THE FINITE-ELEMENT METHOD

2.12.1   Introduction

The finite-element method is a powerful numerical technique that uses variational methods and interpolation theory for solving differential and integral equations of initial and boundary-value problems. The method is so general that it can be applied to a wide variety of engineering problems, including heat transfer, fluid mechanics, solid mechanics, chemical processing, electrical systems, and a host of other fields. The method is also so systematic and modular that it can be implemented on a digital computer and can be utilized to solve a wide range of practical engineering problems by merely changing the data input to the program. The method is naturally suited for the description of complicated geometries and the modeling and simulation of most physical phenomena.

Basic Features

The finite element method is characterised by two distinct features: First, the **domain** of the problem is viewed as a collection of simple subdomains, called **finite elements**. By the word *domain* we refer to a physical structure, system, or region over which the governing equations are to be solved. The collection of the elements is called the *finite-element mesh*. Second, over each element, the solution of the equations being solved is approximated by interpolation polynomials. The first feature, dividing a whole into parts, called **discretization of the domain**, allows the analyst to represent any complex system as one of numerous smaller connected elements, each element being of a simpler shape that permits approximation of the solution by a linear combination of algebraic polynomials. The second feature, *element-wise polynomial approximation*, enables the analyst to represent the solution on an element by polynomials so that the numerical evaluation of integrals becomes easy. The polynomials are typically interpolants of the solution at a preselected number of points, called **nodes**, in the element. The number and location of the nodes in an element depends on the geometry of the element and the degree of the polynomial, which in turn depends on the equation being solved. Since the solution is represented by polynomials on each element, a continuous approximation of the solution of the whole can only be obtained by imposing the continuity of the finite-element solution, and possibly its derivatives, at element interfaces (i.e., at the nodes common to two elements). The procedure of putting the elements together is called the **connectivity** or **assembly**.

Finite-Element Approximation

Beyond the two features already described, the finite-element method is a variational method, like the Ritz, Galerkin, and weighted-residual methods in which the approximate solution is sought in the form

$$u \approx U_N = \sum_{j=1}^{N} c_j \phi_j$$

where ϕ_j are preselected functions and c_j are parameters that are determined using a **variational statement** of the equation governing u. However, the finite-element method typically entails the solution of a very large number of equations for the nodal values of the function being sought. The number of equations is equal to the number of unknown nodal values. In most practical problems the number of unknown nodal values are so large that it is practical only if the calculations are carried on an electronic computer.

2.12.2   One-Dimensional Problems

The finite-element analysis consists of dividing a domain into simple parts (i.e., elements) that are easier to work with. Over each element the method involves representing the solution in terms of its nodal values and the development of a relationship between the nodal values and their counterparts by means of a variational method. Assembly of these relations and solution of the equations after imposing known boundary and initial conditions completes the analysis.

Evaluation of an Integral

Consider the evaluation of the integral

$$I = \int_a^b f(x)\, dx \qquad (2.353)$$

where $f(x)$ is a complicated function whose integration by conventional methods (e.g., exact integration) is not possible. A step-by-step procedure of the numerical evaluation of the integral I by the finite-element method is given later.

Discretization of the Domain. The area can be approximated by representing the interval (domain) $\Omega = (a, b)$ as a finite set of subintervals (see Fig. 2.152). A typical subinterval (element), $\Omega^e = (x_e, x_{e+1})$, is of length $h_e \equiv x_{e+1} - x_e$, with $x_1 = a$, and $x_{N+1} = b$, where N is the number of elements.

Approximation of the Solution. Over each element, the function $f(x)$ is approximated using polynomials of a desired degree. The accuracy increases with increasing N and degree of the approximating polynomial. Over each element Ω^e, the function $f(x)$ can be approximated by a linear polynomial (see Fig. 2.153)

$$f(x) \approx F_e(x) = c_1^e + c_2^e x \qquad (2.354)$$

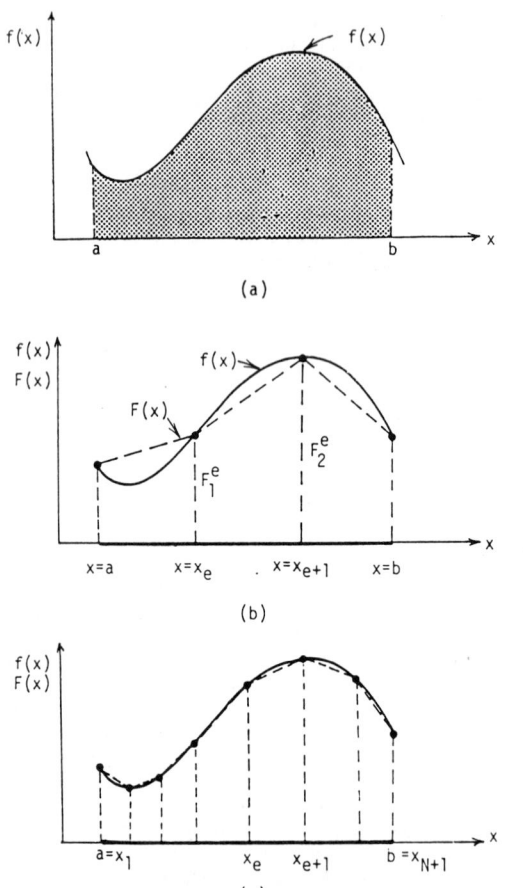

Fig. 2.152 Piecewise approximation of the integral of a function by polynomials.

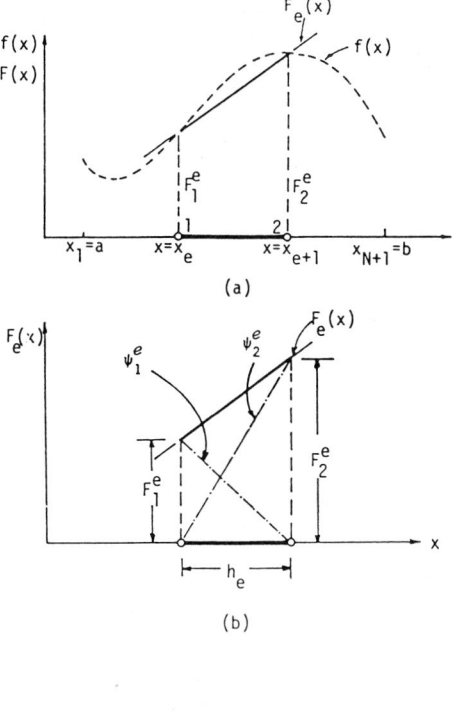

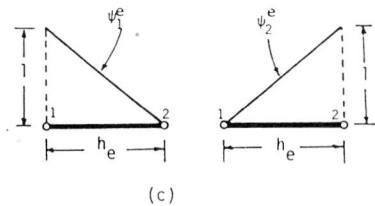

Fig. 2.153 Finite-element approximation of a function $f(x)$ over a typical element.

where c_1^e and c_2^e are constants that can be determined in terms of the values of the function f at the end points, x_e and x_{e+1}, called the *nodes*. Let F_1^e and F_2^e denote the values of $F_e(x)$ at nodes 1 and 2 of element Ω^e:

$$F_1^e = F_e(x_e) \qquad F_2^e = F_e(x_{e+1}) \tag{2.355}$$

Now $F_e(x)$ can be expressed in terms of its values at the nodes as

$$F_e(x) = \frac{x_{e+1} - x}{h_e} F_1^e + \frac{x - x_e}{h_e} F_2^e = \sum_{j=1}^{2} F_j^e \psi_j^e \tag{2.356}$$

where ψ_j^e are called the **element interpolation functions** (see Fig. 2.153),

$$\psi_1^e = \frac{x_{e+1} - x}{h_e} \qquad \psi_2^e = \frac{x - x_e}{h_e} \tag{2.357}$$

Let the approximation of the area I over a typical element Ω^e be denoted by I_e,

$$I_e = \int_{x_e}^{x_{e+1}} F_e(x) \, dx \tag{2.358}$$

Substituting Eq. (2.356) into (2.358) and integrating, one obtains

$$I_e = \sum_{j=1}^{2} F_j^e \int_{x_e}^{x_{e+1}} \psi_j^e \, dx$$

$$= \frac{1}{h_e} \left\{ F_1^e \left[h_e x_{e+1} - \frac{h_e}{2} (x_{e+1} + x_e) \right] + F_2^e \left[\frac{h_e}{2} (x_{e+1} + x_e) - h_e x_e \right] \right\} \qquad (2.359)$$

$$= \frac{h_e}{2} (F_1^e + F_2^e)$$

Thus, the area under the function $F_e(x)$ over the element Ω^e is given by the area of the trapezoid of sides F_1^e and F_2^e, and width h_e (see Fig. 2.153b).

Assembly of Equations. An approximation of the total area I is given by the sum of the areas I_e, $e = 1, 2, \ldots, N$:

$$I = \sum_{e=1}^{N} \int_{x_e}^{x_{e+1}} f(x) \, dx$$

$$\approx \sum_{e=1}^{N} \int_{x_e}^{x_{e+1}} F_e(x) \, dx \qquad (2.360)$$

$$= \sum_{e=1}^{N} I_e = \sum_{e=1}^{N} \frac{h_e}{2} (F_1^e + F_2^e)$$

Incidentally, Eq. (2.360) is known as the **trapezoidal rule**.

The accuracy of the approximation can be improved by increasing the number of elements N (see Fig. 2.152c) or by using higher-order approximation of $f(x)$ over each element. Note that the accuracy can also be improved by using unequal intervals, with smaller elements in areas where function $f(x)$ varies rapidly.

The quadratic interpolation of $f(x)$ over Ω^e is given by

$$f(x) \approx F_e(x) = F_1^e \psi_1^e + F_2^e \psi_2^e + F_3^e \psi_3^e$$

$$= \sum_{j=1}^{3} F_j^e \psi_j^e \qquad (2.361)$$

where ψ_j^e are the quadratic interpolation functions

$$\psi_1^e = \frac{(x - \xi_2)(x - \xi_3)}{(\xi_1 - \xi_2)(\xi_1 - \xi_3)}$$

$$\psi_2^e = \frac{(x - \xi_1)(x - \xi_3)}{(\xi_2 - \xi_1)(\xi_2 - \xi_3)} \qquad (2.362)$$

$$\psi_3^e = \frac{(x - \xi_1)(x - \xi_2)}{(\xi_3 - \xi_1)(\xi_3 - \xi_2)}$$

and ξ_1, ξ_2, and ξ_3 are the coordinates of the three nodes in Ω^e. If nodes are equally spaced within each element (see Fig. 2.154), (ξ_1, ξ_2, ξ_3) take the values

$$\xi_1 = x_{2e-1} \qquad \xi_2 = x_{2e} \qquad \xi_3 = x_{2e+1} \quad (e = 1, 2, \ldots, N)$$

Then Eqs. (2.362) become

$$\psi_1^e = \left(\frac{2\bar{x}}{h_e} - 1 \right) \left(\frac{\bar{x}}{h_e} - 1 \right) \qquad \psi_2^e = -\frac{4\bar{x}}{h_e} \left(\frac{\bar{x}}{h_e} - 1 \right) \qquad \psi_3^e = \frac{\bar{x}}{h_e} \left(\frac{2\bar{x}}{h_e} - 1 \right) \qquad (2.363)$$

where $\bar{x} = x - x_{2e-1}$ and h_e is the length of the element Ω^e.

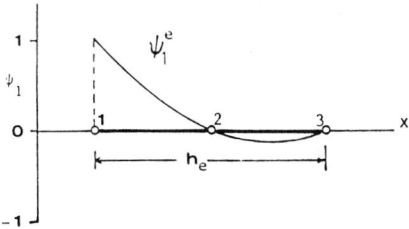

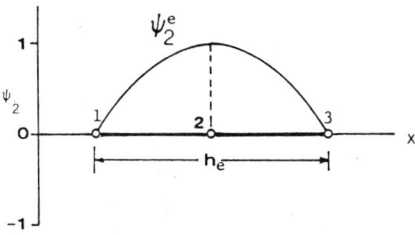

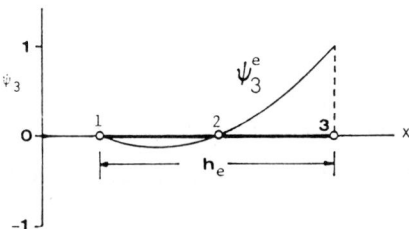

Fig. 2.154 One-dimensional quadratic interpolation functions.

In general, the interpolation functions ψ_j^e satisfy the properties

$$\psi_j^e(\xi_i) = \begin{cases} 0, & \text{if } i \neq j \\ 1, & \text{if } i = j \end{cases} \tag{2.364}$$

Substituting Eq. (2.361) into (2.358) and integrating, one obtains

$$I_e = \sum_{j=1}^{3} F_j^e \int_{x_{2e-1}}^{x_{2e+1}} \psi_j^e(x)\, dx$$

$$= \sum_{j=1}^{3} F_j^e \int_0^{h_e} \psi_j^e(\bar{x})\, d\bar{x}$$

$$= \frac{h_e}{6} \left(F_1^e + 4F_2^e + F_3^e \right)$$

The total area is given by

$$I \approx \sum_{e=1}^{N} I_e = \sum_{e=1}^{N} \frac{h_e}{6} \left(F_1^e + 4F_2^e + F_3^e \right)$$

This equation is known as the **one-third Simpson's rule**.

Example 2.92. Consider the integral of the function

$$f(x) = \sin(2 \cos x) \sin^2 x$$

over the domain $\Omega = (0, \pi/2)$. Table 2.12 contains the finite-element solutions obtained using the linear and quadratic interpolation. It is clear that the accuracy improves as the number of elements or the degree of polynomial is increased.

Solution of a Differential Equation

Model Equation. Consider the differential equation,

$$-\frac{d}{dx}\left[a(x)\frac{du}{dx} \right] - f(x) = 0, \quad 0 < x < L \qquad (2.365)$$

which arises in connection with heat transfer in a heat exchanger fin, where $a(x) = kA$, k is the thermal conductivity, A is the cross-sectional area of the fin, $f(x)$ is the heat source and $u = u(x)$ is temperature to be determined. Equation (2.365) also arises in many fields of engineering. In addition to Eq. (2.365), the function u is required to satisfy certain boundary conditions (i.e., conditions at points $x = 0$ and $x = L$). Equation (2.365), in general, has the following types of boundary conditions:

Specify: either u or $(a\, du/dx)$ at a boundary point

Discretization. The domain $\Omega = (0, L)$ is represented as a collection of line elements, each element having at least two end nodes so that it can be connected to adjacent elements. A two-node element with one unknown per node requires uniquely, a linear polynomial approximation of the variable over the element (see Fig. 2.155).

Approximation. Over a typical element $\Omega^e = (x_e, x_{e+1})$, the function $u(x)$ is approximated by $U_e(x)$, which is assumed to be of the form

$$U_e(x) = \sum_{j=1}^{n} U_j^e \psi_j^e(x) \qquad (2.366)$$

where U_j^e denotes the value of $U_e(x)$ at the jth node, and ψ_j^e are the linear [see Eq. (2.357)], quadratic [see Eq. (2.363)], or higher-order interpolation functions. The values U_j^e are to be determined such that Eq. (2.365), with appropriate boundary conditions, is satisfied in integral sense.

TABLE 2.12   Finite-Element Solutions Using Linear and Quadratic Interpolation

Number of Elements	Linear Interpolation $\bar{I}$	Error[a] (%)	Quadratic Interpolation $\bar{I}$	Error (%)	Exact
2(1)[b]	0.38790	23.6	0.51719	−1.8	0.50797
4(2)	0.48149	5.2	0.51268	−0.9	0.50797
6(3)	0.49640	2.3	0.50865	−0.1	0.50797
8(4)	0.50150	1.3	0.50817	−0.04	0.50797
10(5)	0.50384	0.8	0.50805	−0.02	0.50797

[a]$(1 - \bar{I}/I(100))$.
[b]Numbers in parenthesis indicate number of equivalent quadratic elements.

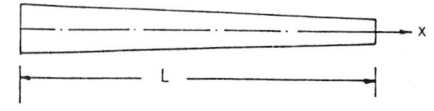

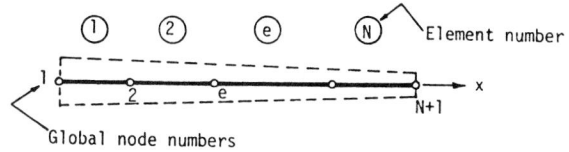

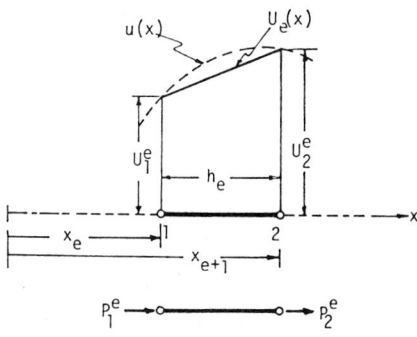

Fig. 2.155 One-dimensional domain, finite-element discretization and finite-element approximation over an element.

Need for a Variational Statement. The difference between the numerical evaluation of an integral and the numerical solution of a differential equation is that in the case of a differential equation one is required to determine a function that satisfies a given differential equation and boundary conditions. It is possible to recast the differential equation as an integral statement, called a **variational statement**. The variational statement of Eq. (2.333), with the aid of a variational method of approximation, gives the same number of algebraic equations as the number of unknowns (n) in the approximation (2.334).

Variational Formulation. The variational statement of Eq. (2.333) over an element $\Omega^e = (x_e, x_{e+1})$ (see Fig. 2.155) is constructed as follows. Multiply Eq. (3.333) with an arbitrary but continuous function W and integrate over the domain of the element to obtain

$$0 = \int_{x_A}^{x_B} W\left[-\frac{d}{dx}\left(a\frac{dU}{dx}\right) - f\right] dx \tag{2.367}$$

The (Ritz) finite-element model uses a **weak form** that can be obtained from Eq. (2.367) by trading differentiation between the weight function W and the variable of approximation U equally:

$$0 = \int_{x_A}^{x_B}\left(a\frac{dW}{dx}\frac{dU}{dx} - Wf\right) dx - \left[W\left(a\frac{dU}{dx}\right)\right]_{x_A}^{x_B} \tag{2.368}$$

which is obtained by integrating the first term in Eq. (2.367) by parts. The phrase *weak form* is appropriate because the solution U of Eq. (2.368) requires weaker continuity conditions on ψ_i than U of Eq. (2.367). Also, the weak formulation allows the incorporation of the boundary conditions of the "flux" type, a dU/dx (the coefficient of the weight function W in the boundary term, called the **natural boundary condition**), into the variational statement (2.368). Boundary conditions on U in the same form as the weight function in the boundary terms are called the **essential boundary conditions**.

Identifying the coefficients of the weight function in the boundary terms (i.e., fluxes) as the **dual variables**,

$$\left(a\frac{dU}{dx}\right)\Bigg|_{x=x_e} = -P_1^e \qquad \left(a\frac{dU}{dx}\right)\Bigg|_{x=x_{e+1}} = P_2^e$$

Eq. (2.368) can be written as

$$0 = \int_{x_e}^{x_{e+1}}\left(a\frac{dW}{dx}\frac{dU}{dx} - Wf\right)dx - W(x_e)P_1^e - W(x_{e+1})P_2^e \qquad (2.369)$$

Equation (2.369) represents the variational statement of Eq. (2.365) for the (Ritz) finite-element model.

As a general rule, the essential boundary conditions of the variational form of a problem indicate what interelement continuity conditions are to be imposed on the function U and its derivatives. This in turn dictates the type and degree of approximation, and hence the element type. For example, Eq. (2.369) indicates that U must be continuous in the interval (x_e, x_{e+1}). A *complete* continuous polynomial in x is a linear polynomial

$$U_e(x) = c_1^e + c_2^e x$$

The constants c_1^e and c_2^e are expressed in terms of the values of U_e at nodes 1 and 2,

$$U_e = \sum_{j=1}^{2} U_j^e \psi_j^e(x)$$

For the $(n-1)$st degree polynomial approximation, U_e is of the form

$$U_e(x) = \sum_{j=1}^{n} U_j^e \psi_j^e(x)$$

The (Ritz) Finite-Element Model. In the Ritz model U_j^e such that Eq. (2.369) is satisfied for each $W = \psi_i^e$ $(i = 1, 2, \ldots, n)$. For each choice of W, an algebraic equation can be obtained:

$$0 = \int_{x_e}^{x_{e+1}}\left[a\frac{d\psi_1^e}{dx}\left(\sum_{j=1}^{n} U_j^e\frac{d\psi_j^e}{dx}\right) - \psi_1^e f\right]dx - \psi_1^e(x_e)P_1^e - \psi_1^e(x_{e+1})P_2^e$$

$$0 = \int_{x_e}^{x_{e+1}}\left[a\frac{d\psi_2^e}{dx}\left(\sum_{j=1}^{n} U_j^e\frac{d\psi_j^e}{dx}\right) - \psi_2^e f\right]dx - \psi_2^e(x_e)P_1^e - \psi_2^e(x_{e+1})P_2^e$$

$$\vdots$$

$$0 = \int_{x_e}^{x_{e+1}}\left[a\frac{d\psi_n^e}{dx}\left(\sum_{j=1}^{n} U_j^e\frac{d\psi_j^e}{dx}\right) - \psi_n^e f\right]dx - \psi_n^e(x_e)P_1^e - \psi_n^e(x_{e+1})P_2^e$$

The ith equation can be written in compact form as

$$0 = \sum_{j=1}^{n} K_{ij}^e U_j^e - F_i^e \qquad (2.370a)$$

where

$$K_{ij}^e = \int_{x_e}^{x_{e+1}} a\frac{d\psi_i^e}{dx}\frac{d\psi_j^e}{dx}dx$$

$$F_i^e = \int_{x_e}^{x_{e+1}} f\psi_i^e dx + \psi_i^e(x_e)P_1^e + \psi_i^e(x_{e+1})P_2^e$$

$$(2.370b)$$

To be more specific, let ψ_i^e be the linear interpolation functions of Eq. (2.357). Because of the interpolation property (2.364) of ψ_j^e, the F_i^e of Eq. (2.371) can be written as

$$F_i^e = \int_{x_e}^{x_{e+1}} f\psi_i^e \, dx + P_i^e \equiv f_i^e + P_i^e$$

For element-wise constant values of a and f, the element coefficient matrix $[K^e]$ and source vector $\{f^e\}$ become

$$[K^e] = \frac{a_e}{h_e}\begin{bmatrix} 1 & -1 \\ -1 & 1 \end{bmatrix} \qquad \{f^e\} = \frac{h_e f_e}{2}\begin{Bmatrix} 1 \\ 1 \end{Bmatrix}$$

Assembly of Elements. The element equations (2.370) must be put together to obtain the equations of the whole domain. Geometrically, the elements are connected together by noting that the second node of element Ω^e is the same as the first node of element Ω^{e+1}. Since the solution and hence it approximation, are single valued throughout the domain, the geometric continuity also implies the continuity of the approximate solution (see Fig. 2.156):

$$U_2^e = U_1^{e+1}, \, e = 1, 2, \ldots, N$$

In addition to the continuity of U_e, the balance of the dual variables P_i at interelement nodes is also

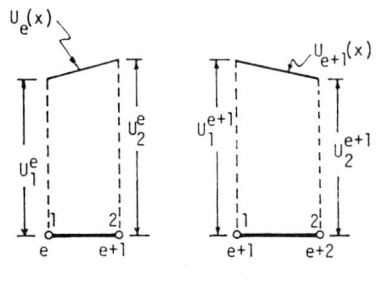

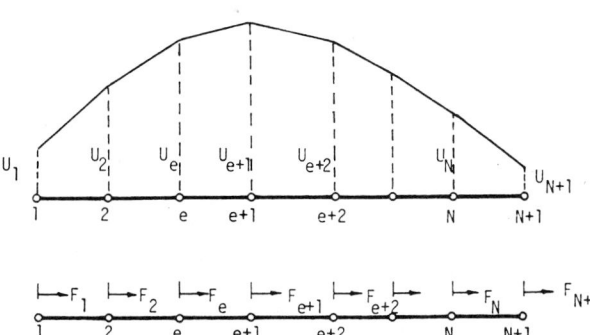

Fig. 2.156 Assembly of finite elements using the continuity of the finite-element approximation between elements.

enforced:

$$P_2^e + P_1^{e+1} = 0, \quad e = 1, 2, \ldots, N$$

Note that this *does not* guarantee the continuity of $a\,dU_e/dx$ at interelement nodes. The finite-element approximation on the entire domain $\Omega = \sum_{e=1}^{N}\Omega^e$ is given by

$$U = \sum_{e=1}^{N} U_e = \sum_{e=1}^{N}\sum_{j=1}^{2} U_j^e \psi_j^e(x)$$

In view of the continuity conditions and the element-wise definition of the interpolation functions ψ_j^e, the finite-element approximation can be written as

$$U = \sum_{J=1}^{N+1} U_J \Phi_J(x) \tag{2.371}$$

where U_J denotes the value of $U(x)$ at the Jth (global) node of the mesh and Φ_J are the **global interpolation functions,** related to the local (or element) interpolation functions by

$$\Phi_1 = \psi_1^1 \qquad 0 = x_1 \leq x \leq x_2$$

$$\Phi_J = \begin{cases} \psi_2^{J-1}, & x_{J-1} \leq x \leq x_J \\ \psi_1^J, & x_J \leq x \leq x_{J+1} \end{cases} \quad (J = 2, 3, \ldots, N)$$

$$\Phi_{N+1} = \psi_2^N \qquad x_N \leq x \leq x_{N+1} = L$$

Note that Φ_J are continuous and defined only on the two elements connected at the global node J.
Analogous to the variational form (2.369) for an element Ω^e, a variational form for the entire domain can be derived as

$$0 = \int_0^L \left(a\frac{dW}{dx}\frac{dU}{dx} - Wf \right) dx - \left[W\left(a\frac{dU}{dx} \right) \right]_{x=0}^{x=L} \tag{2.372}$$

Substitution of Eq. (2.371) for U and $W = \Phi_I$ $(I = 1, 2, \ldots, N + 1)$ into Eq. (2.372) gives

$$0 = \int_0^L \left[a\frac{d\Phi_I}{dx}\left(\sum_{j=1}^{N+1} U_J \frac{d\Phi_J}{dx} \right) - \Phi_I f \right] dx - \left[\Phi_I \sum_{e=1}^{N} a\frac{dU_e}{dx} \right]_{x=0}^{x=L}$$

Since each Φ_I is defined on two neighboring elements, this equation becomes

$$0 = \int_{x_{I-1}}^{x_I} \left[a\frac{d\psi_2^{I-1}}{dx}\left(U_{I-1}\frac{d\psi_1^{I-1}}{dx} + U_I \frac{d\psi_2^{I-1}}{dx} \right) - \psi_2^{I-1} f \right] dx - \psi_2^{I-1}(L) P_2^{I-1}$$

$$+ \int_{x_I}^{x_{I+1}} \left[a\frac{d\psi_1^I}{dx}\left(U_I\frac{d\psi_1^I}{dx} + U_{I+1}\frac{d\psi_2^I}{dx} \right) - \psi_1^I f \right] dx - \psi_1^I(a) P_1^I \tag{2.373}$$

$$= K_{21}^{I-1} U_{I-1} + \left(K_{22}^{I-1} + K_{11}^I \right) U_I + K_{12}^I U_{I+1} - \left(f_2^{I-1} + f_1^I \right)$$

$$- \psi_2^{I-1}(L) P_2^{I-1} - \psi_1^I(0) P_1^I$$

Thus, the equations of the connected elements (i.e., the finite-element equations of the entire domain) are given by setting $I = 1, 2, \ldots, N + 1$ in Eq. (2.373) [set $K_{IJ}^0 = F_I^0 = P_I^0 = 0$]:

$$K_{11}^1 U_1 + K_{12}^1 U_2 = f_1^1 + P_1^1$$

$$K_{21}^1 U_1 + \left(K_{22}^1 + K_{11}^2 \right) U_2 + K_{12}^2 U_3 = f_2^1 + f_1^2 + \left(P_2^1 + P_1^2 \right) \longrightarrow 0$$

$$K_{21}^2 U_2 + \left(K_{22}^2 + K_{11}^3 \right) U_3 + K_{12}^3 U_4 = f_2^2 + f_1^3 + \left(P_2^2 + P_1^3 \right) \longrightarrow 0$$

$$\vdots$$

$$K_{21}^N U_N + K_{22}^N U_{n+1} = f_2^N + P_2^N$$

or in matrix form,

$$
\begin{bmatrix}
K_{11}^1 & K_{12}^1 & 0 & 0 & 0 \\
K_{21}^1 & K_{22}^1 + K_{11}^2 & K_{12}^2 & 0 & 0 \\
0 & K_{21}^2 & K_{22}^2 + K_{11}^3 & K_{12}^3 & 0 \\
\vdots & \vdots & \vdots & \vdots & \vdots \\
0 & 0 & 0 & K_{21}^N & K_{22}^N
\end{bmatrix}
\begin{Bmatrix}
U_1 \\
U_2 \\
U_3 \\
\vdots \\
U_{N+1}
\end{Bmatrix}
=
\begin{Bmatrix}
f_1^1 + P_1^1 \\
f_2^1 + f_1^2 \\
f_2^2 + f_1^3 \\
\vdots \\
f_2^N + P_2^N
\end{Bmatrix}
\qquad (2.374)
$$

One does not repeat the connectivity procedure described in Eqs. (2.371)–(2.374) for every problem, but uses the pattern implied in the final equations (2.374) for all problems described by Eq. (2.365).

Example 2.93. Heat conduction in a long radially symmetric coaxial cylindrical cables can be described by

$$-\frac{d}{dr}\left[a(r)\frac{du}{dr} \right] = 0 \qquad (2.375)$$

where u denotes the temperature and $a = 2\pi rk$, k being the thermal conductivity of the medium.

Equation (2.375) is in the same form as Eq. (2.365). Therefore, Eqs. (2.370) and (2.374) describe the element and global finite-element models of Eq. (2.375). For the choice of linear interpolation functions, we have

$$K_{ij}^e = \int_{r_e}^{r_{e+1}} 2\pi k_e r \frac{d\psi_i^e}{dr} \frac{d\psi_j^e}{dr}\, dr$$

where

$$\psi_i^e = \frac{r_{e+1} - r}{r_{e+1} - r_e} \qquad \psi_2^e = \frac{r - r_e}{r_{e+1} - r_e} \qquad h_e = r_{e+1} - r_e$$

For example, K_{11}^e is given by

$$K_{11}^e = 2\pi k_e \int_{r_e}^{r_{e+1}} r\left(-\frac{1}{h_e} \right)^2 dr$$

$$= \frac{\pi k_e}{h_e}(r_{e+1} + r_e)$$

We have

$$\frac{\pi k_e}{h_e}(r_{e+1} + r_e)\begin{bmatrix} 1 & -1 \\ -1 & 1 \end{bmatrix}\begin{Bmatrix} u_1^e \\ u_2^e \end{Bmatrix} = \begin{Bmatrix} P_1^e \\ P_2^e \end{Bmatrix}$$

where P_i^e denote the internal heats,

$$P_1^e = -2\pi k_e \left(r\frac{dU}{dr} \right)\bigg|_{r=r_e} \qquad P_2^e = 2\pi k_e \left(r\frac{dU}{dr} \right)\bigg|_{r=r_{e+1}}$$

The assembled equations for an N-element case is given by

$$
\begin{bmatrix}
\dfrac{K_1}{h_1} & -\dfrac{K_1}{h_1} & 0 & & & & \\[2ex]
-\dfrac{K_1}{h_1} & \dfrac{K_1}{h_1}+\dfrac{K_2}{h_2} & -\dfrac{K_2}{h_2} & & & & \\[2ex]
0 & -\dfrac{K_2}{h_2} & \dfrac{K_2}{h_2}+\dfrac{K_3}{h_3} & \ddots & & & \\[2ex]
\cdot & \cdot & \ddots & \cdots & -\dfrac{K_N}{h_N} & 0 & \\[2ex]
\cdot & \cdot & & -\dfrac{K_N}{h_N} & \dfrac{K_N}{h_N}+\dfrac{K_{N+1}}{h_{N+1}} & -\dfrac{K_{N+1}}{h_{N+1}} & \\[2ex]
\cdot & \cdot & & 0 & -\dfrac{K_{N+1}}{h_{N+1}} & \dfrac{K_{N+1}}{h_{N+1}} &
\end{bmatrix}
\begin{Bmatrix}
U_1 \\ U_2 \\ U_3 \\ \vdots \\ U_N \\ U_{N+1}
\end{Bmatrix}
$$

$$
=
\begin{Bmatrix}
P_1^1 \\ P_2^1 + P_1^2 \\ P_2^2 + P_1^3 \\ \vdots \\ P_2^{N-1} + P_1^N \\ P_2^N
\end{Bmatrix}
$$

where $K_i = k_i(r_{i+1} + r_i)\pi$.

We now impose the boundary conditions of the problem. Suppose that the domain is the cross section of a coaxial cylinder with two materials (i.e., with different thermal conductivities), as shown in Fig. 2.157. Let the internal and external radii be $r_1 = 20$ mm and $r_{N+1} = 50$ mm, and let the

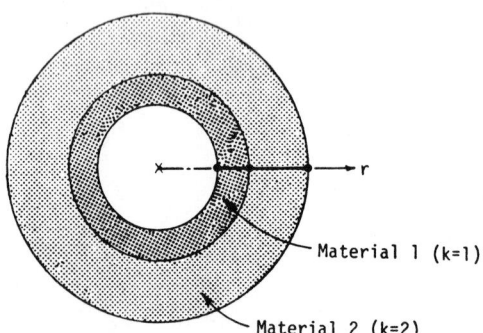

Material 1 (k=1)

Material 2 (k=2)

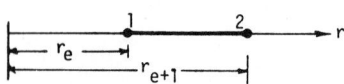

Fig. 2.157 Finite-element representation of a radially symmetric problem with two different materials.

TABLE 2.13 Finite-Element Solutions Obtained by Various Nonuniform Meshes

r	Two Elements	Four Elements	Eight Elements	Analytical Solution
20.0	100.000	100.000	100.00	100.000
22.6	—	—	95.559	95.559
25.1	—	91.745	91.746	91.746
28.4	—	—	87.258	87.257
31.6	83.375	83.377	83.377	83.377
35.7	—	—	61.213	61.210
39.8	—	41.458	41.457	41.457
44.9	—	—	19.551	19.549
50.0	0.000	0.000	0.000	0.000

thickness of the first material be 11.6 mm and that of the second material be 18.4 mm, and the associated material constants (k) be 5 and 1. We assume the boundary conditions to be $u(20) = 100°C$ and $u(50) = 0.0$. These conditions translate to

$$U_1 = 100.0 \qquad U_{N+1} = 0.0$$
$$P_2^1 + P_1^2 = 0, \ldots \qquad P_2^{N-1} + P_1^N = 0$$

For a nonuniform mesh of four elements ($h_1 = 5.1$, $h_2 = 6.5$, $h_3 = 8.2$, $h_4 = 10.2$; equivalently, $r_1 = 20$, $r_2 = 25.1$, $r_3 = 31.6$, $r_4 = 39.8$ and $r_5 = 50.0$), the assembled equations become

$$2\pi \begin{bmatrix} 22.108 & -22.108 & 0 & 0 & 0 \\ -22.108 & 43.916 & -21.808 & 0 & 0 \\ 0 & -21.808 & 26.162 & -4.354 & 0 \\ 0 & 0 & -4.354 & 8.756 & -4.402 \\ 0 & 0 & 0 & -4.402 & 4.402 \end{bmatrix} \begin{Bmatrix} U_1 \\ U_2 \\ U_3 \\ U_4 \\ U_5 \end{Bmatrix} = \begin{Bmatrix} P_1^2 \\ P_2^1 + P_1^2 \\ P_2^2 + P_1^3 \\ P_2^3 + P_1^4 \\ P_2^4 \end{Bmatrix}$$

The boundary and continuity conditions are

$$U_1 = 100.0 \qquad U_5 = 0.0 \qquad P_2^1 + P_1^2 = 0 \qquad P_2^2 + P_1^3 = 0 \qquad P_2^3 + P_1^3 = 0$$

The solution for U_2, U_3, and U_4 is obtained by solving the second, third, and fourth equations of the assembled system:

$$\begin{bmatrix} 43.916 & -21.808 & 0 \\ -21.808 & 26.162 & -4.354 \\ 0 & -4.354 & 8.756 \end{bmatrix} \begin{Bmatrix} U_2 \\ U_3 \\ U_4 \end{Bmatrix} = \begin{Bmatrix} 22.108U_1 \\ 0 \\ 0 \end{Bmatrix}$$

or,

$$U_2 = 91.745°C \qquad U_2 = 83.377°C \qquad U_4 = 41.458°C$$

Table 2.13 contains a comparison of the finite-element solutions obtained by three different nonuniform meshes with the analytical solution. The numerical convergence and accuracy is apparent from the results.

2.12.3 Two-Dimensional Problems

As a model equation, consider the following second-order equation in two dimensions:

$$-\frac{\partial}{\partial x}\left(a_{11}\frac{\partial u}{\partial x}\right) - \frac{\partial}{\partial y}\left(a_{22}\frac{\partial u}{\partial y}\right) + a_0 u = f \text{ in } \Omega \qquad (2.376)$$

The coefficients a_{11}, a_{22}, and a_0, and the source term f are known functions of position (x, y) in the domain Ω. Equation (2.376) arises in the study of a number of engineering problems, including heat transfer, irrotational flow of a fluid, transverse deflection of a membrane, torsion of a cylindrical member, and so on. Also, the Stokes flow and plane elasticity problems are described by a pair of equations of the same form as the model equation. Thus, the finite-element procedure to be described for Eq. (2.376) is applicable to *any* problem that can be formulated as one of solving equations of the form of (2.376).

While the basic ideas are the same as described before, the mathematical complexity for two-dimensional problems increases because of the partial differential equations on two-dimensional domains with possibly curved boundaries. It is necessary to approximate not only the solution of a partial differential equation but also the domain by a suitable finite-element mesh. This latter property is what made the finite-element method a more attractive practical analysis tool over other competing methods.

Discretization of a Domain

Two-dimensional domains can be represented by more than one type of geometric shape. For example, a plane curved domain can be represented by triangular elements or rectangular elements. Without reference to a specific geometric shape, we simply denote a typical element by Ω^e and proceed to discuss the approximation of Eq. (2.376).

The choice of the finite-element mesh depends both on the element characteristics (convergence, computational simplicity, etc.) and the ability to represent the domain accurately. The concept of so-called **isoparametric formulations** allows the representation of the element geometry by the same interpolation as that used in the approximation of the dependent variables. Thus, by identifying nodes on the boundary of the domain one can approximate the domain by suitable collection of elements to a desired accuracy.

Element Equations

Variational Formulation. Consider a typical finite element Ω^e from the finite-element mesh of the domain $\bar{\Omega}$ (see Fig. 2.158). Let ψ_i^e $(i = 1, 2, \ldots, n)$ denote the interpolation functions used to

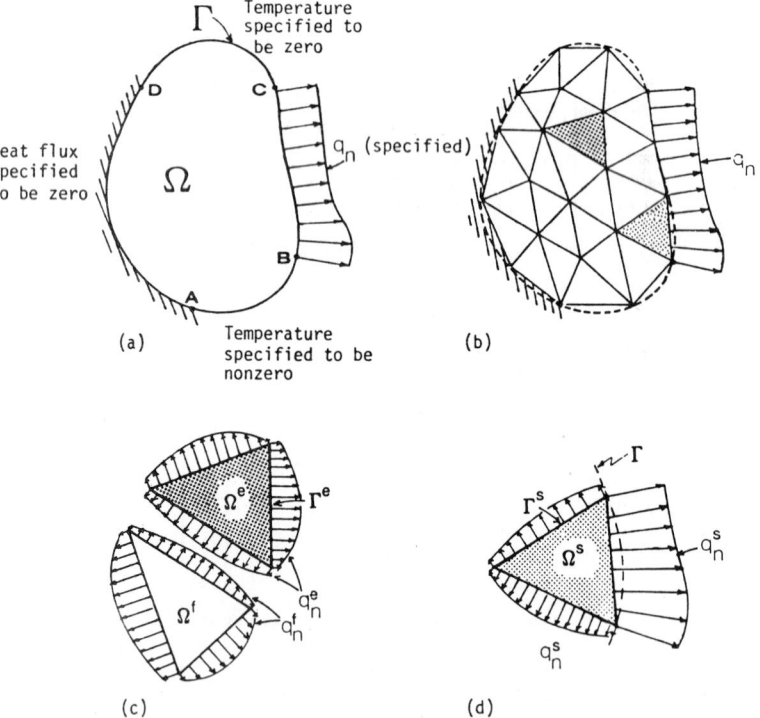

Fig. 2.158 Finite-element representation of a two-dimensional domain with various types of boundary conditions.

approximate u on Ω^e. Multiply Eq. (2.376) with a weight function W, integrate over the element domain Ω^e and use the Green–Gauss theorem to trade differentiation to W to obtain the weak variational form

$$0 = \int_{\Omega^e} \left[\frac{\partial W}{\partial x} \left(a_{11}^e \frac{\partial U}{\partial x} \right) + \frac{\partial W}{\partial y} \left(a_{22}^e \frac{\partial U}{\partial y} \right) + a_0^e W U - W f_e \right] dx\, dy$$

$$- \int_{\Gamma^e} W \left[n_x \left(a_{11}^e \frac{\partial U}{\partial x} \right) + n_y \left(a_{22}^e \frac{\partial U}{\partial y} \right) \right] ds \qquad (2.377)$$

where n_x and n_y are the components (i.e., direction cosines) of the unit normal $\hat{n}$,

$$\hat{n} = n_x \hat{i} + n_y \hat{j} = \cos \alpha \hat{i} + \sin \alpha \hat{j}$$

on the boundary Γ^e, and ds is the elemental arc length along the boundary of the element. From an inspection of the boundary term in Eq. (2.377), it follows that the specification of the coefficient of W,

$$q_n^e \equiv n_x \left(a_{11}^e \frac{\partial U}{\partial x} \right) + n_y \left(a_{22}^e \frac{\partial U}{\partial y} \right) \qquad (2.378)$$

constitutes the natural boundary condition. The variable q_n is of physical interest in most problems. For example, in the case of the heat transfer through an anisotropic medium (where a_{ij} denotes the conductivities of the medium), q_n denotes the heat flux across the boundary of the element (see Fig. 2.158). The variable U is called the **primary variable** and q_n (heat flux) is termed the **secondary variable**.

The variational form in Eq. (2.377) now becomes

$$0 = \int_{\Omega^e} \left[\frac{\partial W}{\partial x} \left(a_{11}^e \frac{\partial U}{\partial x} \right) + \frac{\partial W}{\partial y} \left(a_{22}^e \frac{\partial U}{\partial y} \right) + a_{00}^e W U - W f_e \right] dx\, dy$$

$$- \int_{\Gamma^e} W q_n^e \, ds \qquad (2.379)$$

This variational equation forms the basis of the Ritz finite-element model. The boundary term indicates that W should be continuous at interelement boundaries.

Finite-Element Formulation. The variational form in (2.379) indicates that the approximation chosen for u should be at least bilinear in x and y so that $\partial u / \partial x$ and $\partial u / \partial y$ are nonzero and the interelement continuity of u can be imposed. Suppose that the temperature is approximated by the expression

$$u \approx U_e = \sum_{j=1}^{n} U_j^e \psi_j^e \qquad (2.380)$$

where U_j^e are the values of U_e at the point (x_j, y_j) in Ω^e and ψ_j^e are the interpolation functions with the property

$$\psi_i^e(x_j, y_j) = \delta_{ij}$$

The specific form of ψ_i^e will be derived later for linear triangular and rectangular elements.

Substituting Eq. (2.380) for U_e and ψ_i^e for W into the variational form (2.379), the ith algebraic equation of the model is obtained

$$\sum_{j=1}^{n} K_{ij}^e U_j^e = F_i^e, \ (i = 1, 2, \ldots, n) \qquad (2.381)$$

where

$$K_{ij}^e = \int_{\Omega^e} \left[\frac{\partial \psi_i^e}{\partial x} \left(a_{11}^e \frac{\partial \psi_j^e}{\partial x} \right) + \frac{\partial \psi_i^e}{\partial y} \left(a_{22}^e \frac{\partial \psi_j^e}{\partial y} \right) + a_0^e \psi_i^e \psi_j^e \right] dx\, dy$$

$$F_i^e = \int_{\Omega^e} f_e \psi_i^e \, dx\, dy + \int_{\Gamma^e} q_n^e \psi_i^e \, ds \equiv f_i^e + P_i^e$$

$$(2.382)$$

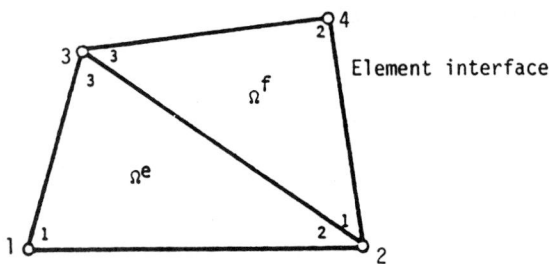

Fig. 2.159 Assembly (or connectivity) of linear triangular elements.

Note that $K_{ij}^e = K_{ji}^e$ (i.e., $[K^e]$ is symmetric). Equation (2.381) is called the finite-element model of Eq. (2.376).

Assembly of Elements

The assembly of finite-element equations is based on the same principle as that employed in one-dimensional problems. We illustrate the procedure by considering a finite-element mesh consisting of two triangular elements (see Fig. 2.159). Let K_{ij}^e and let K_{ij}^f ($i, j = 1, 2, 3$) denote the coefficient matrices and $\{F^e\}$ and $\{F^f\}$ denote the column vectors of three-node triangular elements Ω^e and Ω^f. From the finite-element mesh shown in Fig. 2.159, the following correspondence between the global and element nodal values of the temperature is noted:

$$U_1 = U_1^e \qquad U_2 = U_2^e = U_1^f \qquad U_3 = U_3^e = U_e^f \qquad U_4 = U_2^f$$

The continuity of U at the interelement nodes guarantees its continuity along the *entire* interelement boundary. To see this, consider two linear triangular elements (see Fig. 2.159). The finite-element solution for U is linear along the boundaries of the elements. The interelement boundary is along the line connecting global nodes 2 and 3. Since U_e is linear along side 2-3 of element Ω^e, it is uniquely determined by the two values U_2^e and U_3^e. Similarly, U_f is uniquely determined along side 1-3 of element Ω^f by the two values U_1^f and U_3^f. Since $U_2^e = U_1^f$ and $U_3^e = U_3^f$ it follows that $U_e = U_f$ along the interface. Similar arguments can be presented for higher-order elements.

The coefficient K_{ij}^e is a representation of a physical property of node i with respect to node j of element Ω^e. The assembled coefficient matrix also represents the same property among the global nodes. But the global property comes from the element nodes shared by the global nodes. For example, the coefficient K_{23} of the global coefficient matrix is the sum of the contributions from nodes 2 and 3 of Ω^e and nodes 1 and 3 of Ω^f (see Fig. 2.159):

$$K_{23} = K_{23}^e + K_{13}^f \qquad K_{32} = K_{32}^e + K_{31}^f$$

Similarly,

$$K_{22} = K_{22}^e + K_{11}^f \qquad K_{33} = K_{33}^e + K_{33}^f, \dots$$

If the global nodes I and J do not correspond to nodes in the same element, then $K_{IJ} = 0$. For example, K_{14} is zero because global nodes 1 and 4 do not belong to the same element. The column vectors can be assembled using the same logic:

$$F_2 = F_2^e + F_1^f \qquad F_3 = F_3^e + F_3^f, \dots$$

The complete assembled equations for the two-element mesh is given by

$$\begin{bmatrix} K_{11}^e & K_{12}^e & K_{13}^e & 0 \\ K_{21}^e & K_{22}^e + K_{11}^f & K_{23}^e + K_{13}^f & K_{12}^f \\ K_{31}^e & K_{32}^e + K_{31}^f & K_{33}^e + K_{33}^f & K_{32}^f \\ 0 & K_{21}^f & K_{23}^f & K_{32}^f \end{bmatrix} \begin{Bmatrix} U_1 \\ U_2 \\ U_3 \\ U_4 \end{Bmatrix} = \begin{Bmatrix} F_1^e \\ F_2^e + F_1^f \\ F_3^e + F_3^f \\ F_2^f \end{Bmatrix}$$

Imposition of Boundary Conditions

The boundary conditions on the primary variables (temperatures) and secondary variables (heats) are imposed on the assembled equations in the same way as in the one-dimensional problems. To understand the physical significance of the P's [see Eq. (2.382)], take a closer look at the definition,

$$P_i^e \equiv \int_{\Gamma^e} q_n^e \psi_i^e(s)\, ds \qquad (2.383)$$

where $\psi_i^e(s)$ is the value of $\psi_i^e(x, y)$ on the boundary Γ^e. The heat flux q_n^e [see Eq. (2.378)] is an unknown when Ω^e is an interior element of the mesh (see Fig. 2.158a). However, when the element equations are assembled, the contribution of the heat flux q_n^e to the nodes (namely, P_i^e) of Ω^e get canceled by similar contributions from the adjoining elements (see Fig. 2.158b). If the element Ω^r has any of its sides on the boundary Γ of the domain Ω (see Fig. 2.158c), then on that side the heat flux q_n^r is either specified or unspecified. If q_n^r is specified, then the heat P_i^r at the nodes on that side can be computed using Eq. (2.383). If q_n^r is not specified then the primary variable U_r is known on that portion of the boundary.

The remaining steps of the analysis do not differ from those of one-dimensional problems.

Interpolation Functions

Linear Triangular Element. The simplest finite element in two dimensions is the triangular element. Since a triangle is defined uniquely by three points that form its vertices, the vertex points are chosen as the nodes (see Fig. 2.160a). These nodes will be connected to the nodes of adjoining elements in a finite-element mesh.

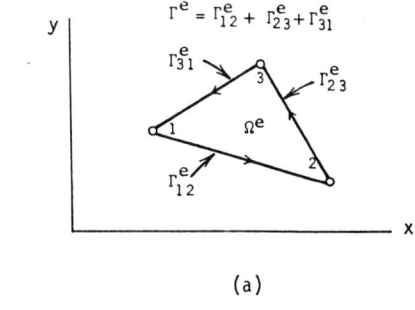

(a)

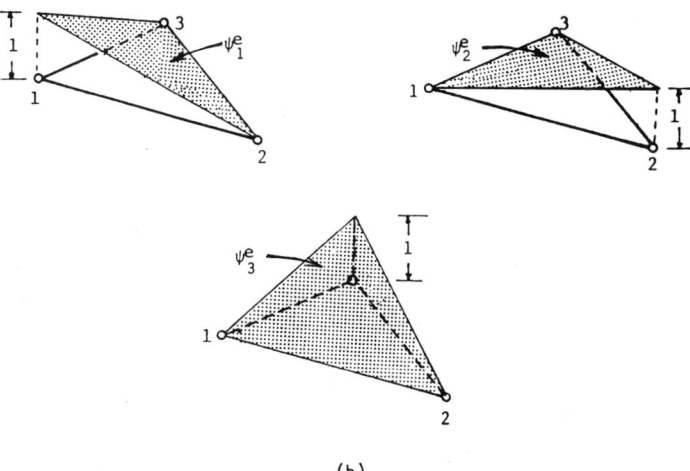

(b)

Fig. 2.160 Typical linear triangular element and associated finite-element interpolations function.

A polynomial in x and y that is uniquely defined by three constants is of the form $p(x, y) = c_0 + c_1 x + c_2 y$. Hence, assume approximation of u_e in the form

$$U_e = c_0^e + c_1^e x + c_2^e y \tag{2.384}$$

Proceeding as in the case of one-dimensional elements, write

$$U_i^e \equiv U_e(x_i, y_i) = c_0^e + c_1^e x_i + c_2^e y_i \qquad i = 1, 2, 3$$

where (x_i, y_i) denote the global coordinates of the element node i in Ω^e. In explicit form this equation becomes

$$\begin{Bmatrix} U_1^e \\ U_2^e \\ U_3^e \end{Bmatrix} = \begin{bmatrix} 1 & x_1 & y_1 \\ 1 & x_2 & y_2 \\ 1 & x_3 & y_3 \end{bmatrix} \begin{Bmatrix} c_0^e \\ c_1^e \\ c_2^e \end{Bmatrix}.$$

Note that the element nodes are numbered counterclockwise. Upon solving for c's and substituting back into Eq. (2.384), one obtains

$$U_e = \sum_{i=1}^{3} U_i^e \psi_i^e(x, y)$$

$$\psi_i^e = \frac{1}{2A_e}(\alpha_i^e + \beta_i^e x + \gamma_i^e y)$$

where A_e represents the area of the triangle, and

$$\alpha_i^e = x_j y_k - x_k y_j \qquad \beta_i^e = y_j - y_k \qquad \gamma_i^e = x_k - x_j \qquad i \neq j \neq k,$$
$$i, j, k = 1, 2, 3$$

and the indices on α_i^e, β_i^e, and γ_i^e permute in a natural order. For example, α_1^e is given by setting $i = 1$, $j = 2$, and $k = 3$:

$$\alpha_1^e = x_2 y_3 - x_3 y_2$$

The sign of the determinant changes if the node numbering is changed to clockwise. The interpolation functions ψ_i^e satisfy the interpolation properties listed in Eq. (2.364). The shape of these functions is shown in Fig. 2.160b.

Note that the derivative of ψ_i^e with respect to x or y is a constant. Hence, the derivatives of the solution evaluated in the postcomputation would be element-wise constant. Also, the coefficient matrix

$$K_{ij}^e = \int_{\Omega^e} \left(a_{11}^e \frac{\partial \psi_i^e}{\partial x} \frac{\partial \psi_j^e}{\partial x} + a_{22}^e \frac{\partial \psi_i^e}{\partial y} \frac{\partial \psi_j^e}{\partial y} \right) dx\, dy \tag{2.385}$$

can be easily evaluated for the linear interpolation functions for a triangle. We have

$$\frac{\partial \psi_i^e}{\partial x} = \frac{\beta_i^e}{2A_e} \qquad \frac{\partial \psi_i^e}{\partial y} = \frac{\gamma_i^e}{2A_e}$$

and, for element-wise constant values of a_{11}^e and a_{22}^e, the coefficients of K_{ij}^e become

$$K_{ij}^e = \frac{1}{4A_e^2}(a_{11}^e \beta_i^e \beta_j^e + a_{22}^e \gamma_i^e \gamma_j^e)\left(\int_{\Omega^e} dx\, dy \right)$$

$$= \frac{1}{4A_e}(a_{11}^e \beta_i^e \beta_j^e + a_{22}^e \gamma_i^e \gamma_j^e)$$

Linear Rectangular Element. A rectangular element is uniquely defined by the four corner points (see Fig. 2.161). Therefore, the four-term polynomial can be used to derive the interpolation functions. Express u_e in the form

$$u_e = c_0^e + c_1^e x + c_2^e y + c_3^e xy \tag{2.386}$$

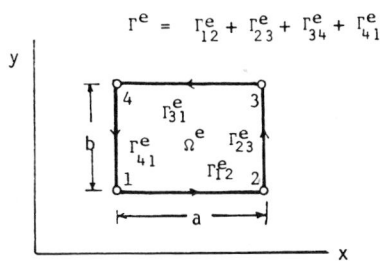

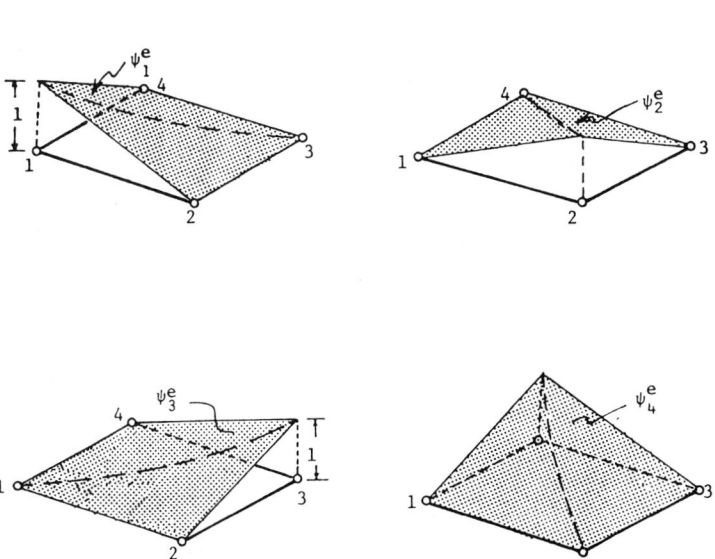

Fig. 2.161 Typical linear rectangular element and associated finite-element interpolation functions.

and obtain

$$
\begin{Bmatrix} U_1^e \\ U_2^e \\ U_3^e \\ U_4 \end{Bmatrix} =
\begin{bmatrix}
1 & x_1 & y_1 & x_1 y_1 \\
1 & x_2 & y_2 & x_2 y_2 \\
1 & x_3 & y_3 & x_3 y_3 \\
1 & x_4 & y_4 & x_4 y_4
\end{bmatrix}
\begin{Bmatrix} c_0^e \\ c_1^e \\ c_2^e \\ c_3^e \end{Bmatrix}
$$

By inverting the equations for c's and substituting into Eq. (2.386), one obtains

$$
\psi_1^e = \left(1 - \frac{\xi}{a}\right)\left(1 - \frac{\eta}{b}\right) \qquad \psi_2^e = \frac{\xi}{a}\left(1 - \frac{\eta}{b}\right)
$$

$$
\psi_3^e = \frac{\xi}{a}\frac{\eta}{b} \qquad \psi_4^e = \left(1 - \frac{\xi}{a}\right)\frac{\eta}{b}
$$

where (ξ, η) are the element coordinates,

$$
\xi = x - x_1 \qquad \eta = y - y_1
$$

The functions are geometrically represented in Fig. 2.161. In calculating element matrices, one finds that the use of the local coordinate system (ξ, η) is more convenient than using the global coordinates (x, y).

For the linear rectangular element, the derivatives of the shape functions are not constant within the element:

$$\frac{\partial \psi_i^e}{\partial x} = \text{linear in } y \qquad \frac{\partial \psi_i^e}{\partial y} = \text{linear in } x$$

The integration of polynomial expressions over a rectangular element is made simple by the fact that $\Omega^e = (0, a) \times (0, b)$:

$$\int_{\Omega^e} f(x, y)\, dx\, dy = \int_b^a \int_0^b f(x, y)\, dx\, dy$$

The coefficients in Eq. (2.385) can be easily evaluated over a linear rectangular element for element-wise constant values of a_{11}^e and a_{22}^e:

$$[K^e] = \frac{b}{6a}\begin{bmatrix} 2 & -2 & -1 & 1 \\ -2 & 2 & 1 & -1 \\ -1 & 1 & 2 & -2 \\ 1 & -1 & -2 & 2 \end{bmatrix} + \frac{a}{6b}\begin{bmatrix} 2 & 1 & -1 & -2 \\ 1 & 2 & -2 & -1 \\ -1 & -2 & 2 & 1 \\ -2 & -1 & 1 & 2 \end{bmatrix}$$

Example 2.94. Consider a computational example of Eq. (2.376) for the case where $a_{11} = a_{22} = 1$, $f = 0$, and Ω is a unit square. Let the boundary conditions be as follows (see Fig. 2.162a)

$$u(0, y) = u(1, y) = 0 \qquad u(x, 0) = 0$$

$$\frac{\partial u}{\partial y}(x, 1) = x$$

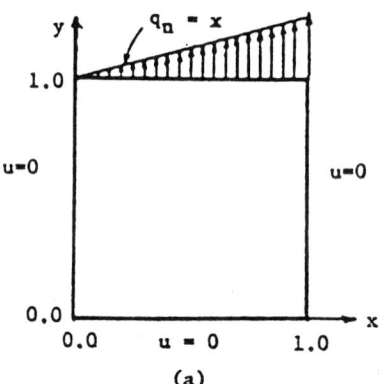

(a)

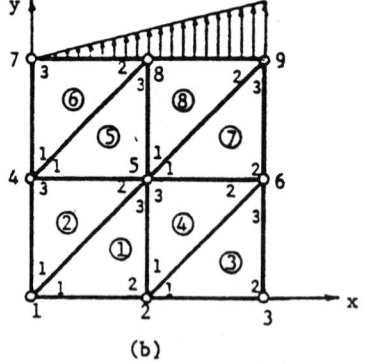

(b)

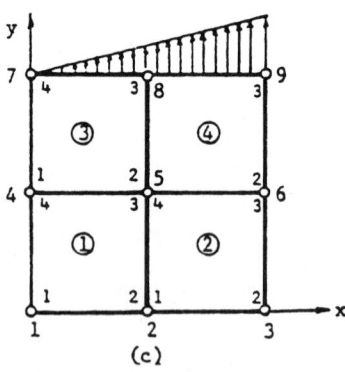

(c)

Fig. 2.162   Domain, boundary conditions, and finite-element meshes.

The finite-element model is given by Eq. (2.381), with

$$K_{ij}^e = \int_{\Omega^e} \left(\frac{\partial \psi_i^e}{\partial x} \frac{\partial \psi_j^e}{\partial x} + \frac{\partial \psi_i^e}{\partial y} \frac{\partial \psi_j^e}{\partial y} \right) dx\,dy \qquad f_i^e = 0$$

Triangular Elements. The 2×2 mesh of triangular elements is shown in Fig. 2.162b. The element coefficient matrices are given by

$$[K^1] = [K^3] = [K^5] = [K^7] = \frac{1}{2}\begin{bmatrix} 1 & -1 & 0 \\ -1 & 2 & -1 \\ 0 & -1 & 1 \end{bmatrix}$$

$$[K^2] = [K^4] = [K^6] = [K^8] = \frac{1}{2}\begin{bmatrix} 1 & 0 & -1 \\ 0 & 1 & -1 \\ -1 & -1 & 2 \end{bmatrix}$$

The assembled equations are given by (refer to Fig. 2.162b)

$$\frac{1}{2}\begin{bmatrix}
1+1 & -1 & \times & -1 & 0 & \times & \times & \times & \times \\
& 2+1+1 & -1 & \times & -1-1 & 0+0 & \times & \times & \times \\
& & 2 & \times & \times & -1 & \times & \times & \times \\
& & & 2+1+1 & -1-1 & \times & -1 & 0+0 & \times \\
& & & & \begin{matrix}2+2+1\\+1+1+1\end{matrix} & -1-1 & \times & -1-1 & 0+0 \\
& & & & & 2+1+1 & \times & \times & -1 \\
& \text{symmetric} & & & & & 2 & -1 & \times \\
& & & & & & & 2+1+1 & -1 \\
& & & & & & & & 1+1
\end{bmatrix}$$

$$\times \begin{Bmatrix} U_1 \\ U_2 \\ U_3 \\ U_4 \\ U_5 \\ U_6 \\ U_7 \\ U_8 \\ U_9 \end{Bmatrix} = \begin{Bmatrix} P_1^1 + P_1^2 \\ P_2^1 + P_1^3 + P_1^4 \\ P_2^3 \\ P_3^2 + P_1^5 + P_1^6 \\ \left(P_3^1 + P_2^2 + P_3^4 + P_2^5 \right. \\ \left. + P_1^7 + P_1^8\right) \\ P_3^3 + P_2^4 + P_2^7 \\ P_3^6 \\ P_3^5 + P_2^6 + P_3^8 \\ P_3^7 + P_2^8 \end{Bmatrix} \begin{matrix} 1 \\ 2 \\ 3 \\ 4 \\ 5 \\ \\ 6 \\ 7 \\ 8 \\ 9 \end{matrix}$$

where $\times$ denotes a zero due to disconnectivity (e.g., $K_{13} = 0$ because global nodes 1 and 3 do not belong to the same element).

The boundary conditions on the primary variables (i.e., U's) are

$$U_1 = U_2 = U_3 = U_4 = U_6 = U_7 = U_9 = 0$$

The known secondary variables are (correspond to nodes 5 and 8)

$$P_3^1 + P_2^2 + P_3^4 + P_2^5 + P_1^7 + P_1^8 = 0 \quad \text{(because no flux is specified at node 5)}$$

$$P_3^5 + P_2^6 + P_3^8 = \int_0^{0.5} q\psi_2^6(x,1)\,dx + \int_{0.5}^{1.0} q\psi_3^8(x,1)\,dx \qquad (2.387)$$

Note that the individual fluxes P_i^e in Eqs. (2.387) are not zero, but their sum is equal to the values indicated. For example, consider P_2^6:

$$P_2^6 = \int_{S^6} q\psi_2^6 \, ds$$

$$= \int_{0.5}^{1.0} q\psi_2^6 (y - 0.5, y)\left(\frac{1}{\sqrt{2}} \, dy\right) + \int_{0.5}^0 q\psi_2^6 (x,1)(-dx) + \int_{1.0}^{0.5} q\psi_2^6 (0, y)(-dy)$$

where $\psi_2^6(x, y) = 2x$. The first integral is nonzero but gets canceled by a similar but negative contribution from P_3^5, the second integral is nonzero and can be evaluated since $q = x$ is known, and the third integral is zero because $\psi_2^6(0, y) = 0$.

Evaluating the integral in Eq. (2.387) [with $\psi_2^6(x, y) = 2x$ and $\psi_3^8 = 2(y - x)$], we obtain

$$P_3^5 + P_2^6 + P_3^8 = \int_0^{0.5} x 2x \, dx + \int_{0.5}^{1.0} 2x(1 - x) \, dx$$

$$= \tfrac{2}{3}(0.5)^3 + (1)^2 - (0.5)^2 - \tfrac{2}{3}\left[(1)^3 - (0.5)^3\right]$$

$$= \tfrac{1}{12} + \tfrac{1}{6} = \tfrac{1}{4}$$

To solve for the unknowns U_5 and U_8, equations 5 and 8 of the assembled equations are used. This choice is dictated by the fact that the remaining equations contain additional unknowns in P's. The solution is given by

$$U_5 = \tfrac{1}{28} \qquad U_8 = \tfrac{2}{14}$$

The internal heat P_i^e can be determined from either the element equations (2.381) or by the definition (2.383). In general, the values computed by the two methods are not the same because P_i^e determined from the element equations is the internal heat in equilibrium with the heat from the neighboring elements, whereas P_i^e computed from the gradient of the approximate temperature field is not.

Rectangular Elements. For the 2×2 mesh of rectangular elements shown in Fig. 2.162c, the element matrices are given by

$$[K^1] = [K^2] = \frac{1}{6}\begin{bmatrix} 4 & -1 & -2 & -1 \\ -1 & 4 & -1 & -2 \\ -2 & -1 & 4 & -1 \\ -1 & -2 & -1 & 4 \end{bmatrix}$$

The assembled equations are

$$\frac{1}{6}\begin{bmatrix}
4 & -1 & \times & -2 & -2 & \times & \times & \times & \times \\
 & 4\times 4 & -1 & -2 & -1-1 & -2 & \times & \times & \times \\
 & & 4 & \times & -2 & -1 & \times & \times & \times \\
 & & & 4+4 & -1-1 & \times & -1 & -2 & \times \\
 & & & & 4+4 & -1-1 & -2 & -1'-1 & -2 \\
 & & & & +4+4 & & & & \\
 & & & & & 4+4 & \times & -2 & -1 \\
\text{symmetric} & & & & & & 4 & -1 & \times \\
 & & & & & & & 4+4 & -1 \\
 & & & & & & & & 4
\end{bmatrix}
\begin{Bmatrix} U_1 \\ U_2 \\ U_3 \\ U_4 \\ U_5 \\ \\ U_6 \\ U_7 \\ U_8 \\ U_9 \end{Bmatrix}
=
\begin{Bmatrix} P_1^1 \\ P_2^1 + P_1^2 \\ P_2^2 \\ P_4^1 + P_1^3 \\ P_3^1 + P_4^2 + P_2^3 + P_1^4 \\ \\ P_3^2 + P_2^4 \\ P_4^3 \\ P_3^3 + P_4^4 \\ P_3^4 \end{Bmatrix}
\begin{matrix} 1 \\ 2 \\ 3 \\ 4 \\ 5 \\ \\ 6 \\ 7 \\ 8 \\ 9 \end{matrix}$$

column labels: 1 2 3 4 5 6 7 8 9

TABLE 2.14   Comparison of Finite-Element Solutions

x	y	Triangles 2 × 2	Triangles 4 × 4	Rectangles 2 × 2	Rectangles 4 × 4	Series Solution
0.25	0.25	—	0.0101	—	0.0095	0.0103
0.50	0.25	—	0.0151	—	0.0136	0.0152
0.75	0.25	—	0.0114	—	0.0097	0.0112
0.25	0.50	—	0.0253	—	0.0254	0.0264
0.50	0.50	0.0357	0.0387	0.0242	0.0370	0.0400
0.75	0.50	—	0.0305	—	0.0270	0.0308
0.25	0.75	—	0.0525	—	0.0552	0.0555
0.50	0.75	—	0.0840	—	0.0882	0.0894
0.75	0.75	—	0.0719	—	0.0675	0.0765
0.25	1.00	—	0.1007	—	0.1059	0.1057
0.50	1.00	0.1429	0.1729	0.1936	0.1851	0.1846
0.75	1.00	—	0.1729	—	0.2027	0.1990

TABLE 2.15   Basic Steps in the Finite-Element Analysis of a Typical Problem

1. *Discretization of a Domain.* Represent the given domain as a collection of a finite number of simple subdomains, called *finite elements.* The number, shape, and type of element depend on the domain and differential equation being solved. The principal parts of this step include:
   a. Number the nodes (see step 2) and elements of the collection, called the *finite-element mesh.*
   b. Generate the coordinates of the nodes in the mesh and the relationship between the element nodes to global nodes (called *the connectivity matrix* that indicates the relative position of each element in the mesh).

2. *Approximation of the Solution*
   a. *Derivation of the Approximating Functions.* For each element in the mesh, derive the approximation functions needed in the variational method. These functions are generally algebraic polynomials generated by interpolating the unknown function in terms of its values at preselected points, the *nodes*, of the element.
   b. *Variational Approximation of the Equation.* Using the functions derived in step 2a and any appropriate variational method, derive the algebraic equations among the nodal values of the primary and secondary variables.

3. *Connectivity (or Assembly) of Elements.* Combine the algebraic equations of all elements in the mesh by imposing the continuity of the primary nodal variables (i.e., the values of the primary variables at a node shared by two or more elements are the same). This can be viewed as putting the elements (which were isolated in steps 2a and 2b from the mesh to derive the algebraic equations) back into their original places. This gives the algebraic equations governing the whole problem.

4. *Imposition of Boundary Conditions.* Impose the boundary conditions, both on primary and secondary variables of the assembled equations.

5. *Solution of Equations.* Solve the equations for the unknown nodal values of the primary variables.

6. *Computation of Additional Quantities.* Using the nodal values of the primary variables, compute the secondary variables (via constitutive equations).

The boundary conditions are given by

$$P_3^1 + P_4^2 + P_2^3 + P_1^4 = 0 \qquad P_3^3 + P_4^4 = 0.25$$

The condensed equations become

$$\frac{1}{6}\begin{bmatrix} 16 & -2 \\ -2 & 8 \end{bmatrix} \begin{Bmatrix} U_5 \\ U_8 \end{Bmatrix} = \begin{Bmatrix} 0 \\ \frac{1}{4} \end{Bmatrix}$$

and the solution is given by

$$U_5 = \tfrac{3}{124} \qquad U_8 = \tfrac{6}{31}$$

The exact solution of the problem is given by

$$u(x, y) = \frac{2}{\pi^2} \sum_{n=1}^{\infty} \frac{(-1)^{n+1}}{n^2 \cosh n\pi} \sin n\pi x \sinh n\pi y$$

A comparison of the finite-element solutions obtained with 2×2 and 4×4 meshes of linear rectangular and triangular elements with the series solution is presented in Table 2.14. The finite-element solution improves as the mesh is refined.

In summary, the finite-element method is a numerical technique of solving field problems of engineering. It is endowed with two unique features: The domain in which the equations are defined is represented by a collection of simple parts (finite elements), and over each element the problem is approximated using any one of the variational methods with polynomials for the approximation functions. The first feature allows approximate representation of geometrically complicated domains by simple geometric shapes, while the second feature enables the approximation of the field variables, evaluation of the coefficient matrices, and solution of the finite-element equations on a computer. A list of basic steps of the finite-element analysis is presented in Table 2.15.

2.13 LAPLACE TRANSFORMATION

2.13.1 Transformation Principles

The Laplace and Fourier transformation methods and the Heaviside operational calculus are in essence different aspects of the same method. This method simplifies the solving of linear constant-coefficient integrodifferential equations and convolution-type integral equations. For brevity the conditions under which the steps of the method may be validly applied will be omitted. Hence the correctness of a final result should be checked in each case by showing that the formal solution satisfies the given equation and conditions.

1. Direct Laplace Transformation. Let t be a real variable, s a complex variable (Section 2.14.2), $f(t)$ a real function of t that equals zero for $t < 0$, $F(s)$ a function of s, and e the base of the natural logarithms. If the Lebesgue integral

$$\int_0^{\infty} e^{-st} f(t)\, dt = F(s) \qquad (2.388)$$

then $F(s)$ is the *direct Laplace transform* of $f(t)$; in simpler notation

$$\mathcal{L}[f(t)] = F(s) \qquad (2.389)$$

2. Inverse Laplace Transformation. Under certain conditions the direct transformation can be inverted, giving as one explicit representation

$$\frac{1}{2\pi i} \int_{c-i\infty}^{c+i\infty} e^{ts} F(s)\, ds (=) f(t) \qquad (2.390)$$

in which c is a real constant chosen so that the path of integration lies to the right of all the singularities of $F(s)$, and $(=)$ means equals except possibly for a set of values of t of measure zero. If this relation holds, then $f(t)$ is the *inverse Laplace transform* of $F(s)$. In simpler notation the transformation is written

$$\mathcal{L}^{-1}[F(s)](=)f(t) \qquad (2.391)$$

3. Transformation of nth Derivative. If $\mathcal{L}[f(t)] = F(s)$, then

$$\mathcal{L}\left[\frac{d^n f(t)}{dt^n}\right] = s^n F(s) - \sum_{k=0}^{n-1} f^{(k)}(0+) \cdot s^{n-1-k} \qquad (2.392)$$

where $f^{(2)}(0+)$ means $d^2 f(t)/dt^2$ evaluated for $t \to 0$, and $f^{(0)}(0+)$ means $f(0+)$, and $n = 1, 2, 3, \dots$.

4. Transformation of nth Integral. If $\mathscr{L}[f(t)] = F(s)$, then

$$\mathscr{L}\left[\overbrace{\int\int\cdots\int}^{n} f(t)\,dt\right] = s^{-n}F(\varepsilon) + \sum_{k=-1}^{-n} f^{(k)}(0+) \cdot s^{-n-1-k} \qquad (2.393)$$

where $n = 1, 2, 3, \ldots$. For example, $f^{(-2)}(0+)$ means $\int\int f(t)\,dt\,dt$ evaluated for $t \to 0$.

5. Inverse Transformation of Product. If

$$\mathscr{L}^{-1}[F_1(s)] = f_1(t) \qquad \mathscr{L}^{-1}[F_2(s)] = f_2(t) \qquad (2.394)$$

then

$$\mathscr{L}^{-1}[F_1(s) \cdot F_2(s)] = \int_0^t f_1(t-\lambda) \cdot f_2(\lambda)\,d\lambda \qquad (2.395)$$

6. Linear Transformations $\mathscr{L}$ and $\mathscr{L}^{-1}$. Let k_1, k_2 be real constants. Then

$$\mathscr{L}[k_1 f_1(t) + k_2 f_2(t)] = k_1\mathscr{L}[f_1(t)] + k_2\mathscr{L}[f_2(t)] \qquad (2.396)$$

and

$$\mathscr{L}^{-1}[k_1 F_2(s) + k_2 F_2(s)] = k_1\mathscr{L}^{-1}[F_1(s)] + k_2\mathscr{L}^{-1}[F_2(s)] \qquad (2.397)$$

2.13.2 Procedure

To illustrate the application of the rules of procedure the following simple initial-value problem will be solved. Given the equation

$$k_1\frac{dy(t)}{dt} + k_2 y(t) + k_3\int y(t)\,dt = u(t)$$

and initial values $y(0)$, $y^{(-1)}(0)$ where $u(t) = 0$ for $t < 0$, and 1 for $0 < t$, and k_1, k_2, k_3 are real constants. Assume that $y(t)$ has a Laplace transform $Y(s)$, that is, $\mathscr{L}[y(t)] = Y(s)$.

Step A. Find the Laplace transform of the equation to be solved and express it in terms of the transform of the unknown function.

Thus,

$$\mathscr{L}\left[k_1\frac{dy(t)}{dt} + k_2 y(t) + k_3\int y(t)\,dt\right] = \mathscr{L}[u(t)]$$

By (2.396) this becomes

$$k_1\mathscr{L}\left[\frac{dy(t)}{dt}\right] + k_2\mathscr{L}[y(t)] + k_3\mathscr{L}\left[\int y(t)\,dt\right] = \mathscr{L}[u(t)]$$

By (2.392) and (2.393) and the given initial conditions of the problem the equation becomes

$$k_1[sY(s) - y(0)] + k_2 Y(s) + k_3[s^{-1}Y(s) + y^{(-1)}(0) \cdot s^{-1}] = \mathscr{L}[u(t)]$$

Step B. Solve the resulting equation for the transform of the unknown function. Thus,

$$Y(s) = \frac{\mathscr{L}[u(t)] + k_1 y(0) - y^{(-1)}(0) \cdot s^{-1}}{k_1 s + k_2 + k_3 s^{-1}}$$

Step C. Evaluate the direct transform of the given function (right member) in the original equation.

Since

$$\mathscr{L}[u(t)] = \frac{1}{s}$$

$$Y(s) = \frac{k_1 y(0) \cdot s - y^{(-1)}(0) + 1}{k_1 s^2 + k_2 s + k_3}$$

Step D. Obtain the solution of the problem by evaluating the inverse Laplace transform of the function obtained by the preceding steps.

One way to carry out step D is to find the inverse transform from the table of Laplace transforms in Section 2.12.3. To use the table, the denominator of the fraction should be factored.

$$y(t) = \mathscr{L}^{-1}[Y(s)] = \mathscr{L}^{-1}\left[\frac{k_1 y(0) \cdot s - y^{(-1)}(0) + 1}{k_1 s^2 + k_2 s + k_3}\right]$$

$$= \mathscr{L}^{-1}\left[\frac{k_1 y(0) \cdot s - y^{(-1)}(0) + 1}{k_1(s + K_1)(s + K_2)}\right]$$

in which

$$K_1 \equiv \frac{k_2}{2k_1} - \frac{1}{2k_1}\left(k_2^2 - 4k_1 k_3\right)^{1/2}$$

$$K_2 \equiv \frac{k_2}{2k_1} + \frac{1}{2k_1}\left(k_2^2 - 4k_1 k_3\right)^{1/2}$$

To find the result it is necessary to distinguish between two cases.

CASE 1: If $K_1 \neq K_2$,

$$y(t) = \left\{\left[k_1 y(0) K_1 + y^{(-1)}(0) - 1\right]e^{-K_1 t} - \left[k_1 y(0) K_2 + y^{(-1)}(0) - 1\right]e^{-K_2 t}\right\}\Big/\left[k_1(K_1 - K_2)\right]$$

for $0 < t$, and $= 0$ for $t < 0$.

CASE 2: If $K_1 = K_2 = K$, then $K = k_2/2k_1$, and

$$y(t) = \mathscr{L}^{-1}\left[\frac{k_1 y(0) \cdot s - y^{(-1)}(0) + 1}{k_1(s + K)^2}\right]$$

From Table 2.16,

$$y(t) = \left\{k_1 y(0) e^{-Kt} - \left[y^{(-1)}(0) - 1 + k_1 y(0) K\right]t e^{-Kt}\right\}\Big/k_1$$

for $0 < t$, and $= 0$ for $t < 0$.

The solutions can be shown to satisfy the original equation and initial conditions.

The use of step C can be avoided by using steps E, F, and G in place of steps C and D in the following way.

Step E. Factor the transform of the unknown function obtained by step B, and evaluate the inverse Laplace transform of each factor.

Note. The inverse transform of a rational fraction can be found only if it is a proper fraction. Thus

$$Y(s) = \frac{k_1 y(0) \cdot s - y^{(-1)}(0)}{k_1 s^2 + k_2 s + k_3} + \frac{s \mathscr{L}[u(t)]}{k_1 s^2 + k_2 s + k_3}$$

Let

$$y_1(t) \equiv \mathscr{L}^{-1}\left[\frac{k_1 y(0) \cdot s - y^{(-1)}(0)}{k_1(s + K_1)(s + K_2)}\right]$$

$$= \left[k_1(y)(0) K_1 + y^{(-1)}(0)\right]e^{-K_1 t} - \frac{\left[k_1 y(0) K_2 + y^{(-1)}(0)\right]e^{-K_2 t}}{k_1(K_1 - K_2)}$$

TABLE 2.16   Laplace Transforms

	Unilateral Laplace Operation-Transform Pairs	
Name	$f(t), 0 \leq t$	$F(s)$
Linearity	$af(t)$ a is constant or variable independent of t $f_1(t) \pm f_2(t)$	$aF(s)$ a is constant or variable independent of s $F_1(s) \pm F_2(s)$
Real differentiation	$\dfrac{df(t)}{dt} \triangleq f'(t)$	$sF(s) - f(0+)$
Multiplication by s	$f'(t)$ if $f(0+) = 0$	$sF(s)$
Real integration	$\int f(t)\, dt \triangleq f^{(-1)}(t)$	$\dfrac{F(s)}{s} + \dfrac{f^{(-1)}(0+)}{s}$
Division by s	$\int_0^t f(t)\, dt \triangleq f^{(-1)}(t) - f^{(-1)}(0+)$	$\dfrac{F(s)}{s}$
Scale change	$f\left(\dfrac{t}{a}\right)$ a is positive constant or positive variable independent of t	$aF(as)$ a is positive constant or positive variable independent of s
Complex multiplication	$\int_0^t f_1(t - \tau)f_2(\tau)\, d\tau \triangleq f_1(t) * f_2(t)$	$F_1(s)F_2(s)$
Real translation	$f(t - a)$ if $f(t - a) = 0, 0 < t < a$ $f(t + a)$ if $f(t + a) = 0, -a < t < 0$ a is a nonnegative real number.	$e^{-as}F(s)$ $e^{as}F(s)$
Complex translation	$e^{-at}f(t)$ $e^{at}f(t)$ a is complex number with nonnegative real part	$F(s + a)$ $F(s - a)$ a is complex number with nonnegative real part.
Second independent variable	$\lim_{a \to a_0} f(t, a)$ a is second variable independent of t	$\lim_{a \to a_0} F(s, a)$ a is second variable independent of s
Differentiation with respect to second independent variable	$\dfrac{\partial}{\partial a} f(t, a)$ a is second variable independent of t	$\dfrac{\partial}{\partial a} F(s, a)$ a is second variable independent of t
Final value	$\lim_{t \to \infty} f(t) = \lim_{s \to 0} s(s)$ if $sF(s)$ is analytic on the axis of imaginaries and in the right half-plane	$\lim_{t \to \infty} f(t) = \lim_{s \to 0} s(s)$ if $sF(s)$ is analytic on the axis of imaginaries and in the right half-plane
Initial value	$\lim_{t \to 0} f(t) = \lim_{s \to \infty} sF(s)$	$\lim_{t \to 0} f(t) = \lim_{s \to \infty} sF(s)$
Complex differentiation	$tf(t)$	$-\dfrac{d}{ds} F(s)$
Complex integration	$\dfrac{1}{t} f(t)$	$\int_s^\infty F(s)\, ds$

TABLE 2.16   (*Continued*)

Unilateral Laplace Operation-Transform Pairs

Name	$f(t), 0 \le t$	$F(s)$
Integration with respect to second independent variable	$\int_{a_0}^{a} f(t, a)\, da$ a is second variable independent of t	$\int_{a_0}^{a} F(s, a)\, da$ a is second variable independent of s
Real multiplication	$f_1(t) f_2(t)$	$\dfrac{1}{2\pi j} \int_{c_1-j\infty}^{c_2+j\infty} F_1(s - w) F_2(w)\, dw \triangleq F_1(s) \otimes F_2(s),$ $\max(\sigma_{a1}, \sigma_{a2}, \sigma_{a1} + \sigma_{a2}) < \sigma, \sigma_{a2} < c_2 < \sigma - \sigma_{a2}$
		$\displaystyle\sum_{k-1}^{q} \frac{A_1(s_k)}{B_1'(s_k)} F_2(s - s_k)$ if $F_1(s) \triangleq \dfrac{A_1(s)}{B_1(s)}$ is rational algebraic fraction having only first-order poles
		$\displaystyle\sum_{k-1}^{n} \sum_{j-1}^{m_k} \frac{(-1)^{m_k-j} K_{kj}}{(m_k - j)!} \left[\frac{d^{m_k-j}}{ds^{m_k-j}} F_2(s) \right]_{s=s-s_k}$ $K_{kj} \triangleq \dfrac{1}{(j - 1)!} \left[\dfrac{d^{j-1}}{ds^{j-1}} (s - s_k)^{m_k} F_1(s) \right]_{s=s_k}$ if $F_1(s)$ is rational algebraic fraction having multiple-order poles

Unilateral Laplace Function-Transform Pairs

$F(s)$	$f(t), 0 \le t$
$\dfrac{A(s)}{B(s)}$ Rational proper fraction; first-order poles only.	$\displaystyle\sum_{k-1}^{q} \frac{A(s_k)}{B'(s_k)} e^{s_k i}$
$\dfrac{A(s)}{B(s)}$ Rational proper fraction; higher-order poles, general case.	$\displaystyle\sum_{k-1}^{n} \sum_{j-1}^{mk} \frac{K_{kj}}{(m_k - j)!} t^{m_k-j} e^{s_k i} \quad m_1 + m_2 + \cdots + m_n = q$ $K_{kj} \triangleq \dfrac{1}{(j - 1)!} \left[\dfrac{d^{j-1}}{ds^{j-1}} \dfrac{(s - s_k)^{m_k} A(s)}{B(s)} \right]_{s=s_k}$ $B(s) \triangleq (s - s_1)^{m_1}(s - s_2)^{m_2} \cdots (s - s_k)^{m_k} \cdots (s - s_n)^{m_n}$
1	$u_1(t) \triangleq \displaystyle\lim_{a \to 0} \frac{u(t) - u(t - a)}{a}$, unit impulse at $t = 0$
s	$u_2(t) \triangleq \displaystyle\lim_{a \to 0} \frac{u(t) - 2u(t - a) + u(t - 2a)}{a^2}$ unit doublet impulse at $t = 0$
$\dfrac{1}{s}$	1, or $u(t)$, unit step at $t = 0$
$\dfrac{1}{s + \alpha}$	$e^{-\alpha t}$
$\dfrac{1}{(s + \alpha)(s + \gamma)}$	$\dfrac{e^{-\alpha t} - e^{-\gamma t}}{\gamma - \alpha}$

$^a \alpha, \beta, \gamma, \delta,$ and λ are real numbers.

TABLE 2.16 (*Continued*)

Unilateral Laplace Function-Transform (Continued)

$F(s)$	$f(t), 0 \le t$
$\dfrac{s + a_0}{(s + \alpha)(s + \gamma)}$	$\dfrac{(a_0 - \alpha)e^{-\alpha t} - (a_0 - \gamma)e^{-\gamma t}}{\gamma - \alpha}$
$\dfrac{1}{s(s + \alpha)(s + \gamma)}$	$\dfrac{1}{\alpha\gamma} + \dfrac{\gamma e^{-\alpha t} - \alpha e^{-\gamma t}}{\alpha\gamma(\alpha - \gamma)}$
$\dfrac{s + a_0}{s(s + \alpha)(s + \gamma)}$	$\dfrac{a_0}{\alpha\gamma} + \dfrac{a_0 - \alpha}{\alpha(\alpha - \gamma)}e^{-\alpha t} + \dfrac{a_0 - \gamma}{\gamma(\gamma - \alpha)}e^{-\gamma t}$
$\dfrac{s^2 + a_1 s + a_0}{s(s + \alpha)(s + \gamma)}$	$\dfrac{a_0}{\alpha\gamma} + \dfrac{\alpha^2 - a_1\alpha + a_0}{\alpha(\alpha - \gamma)}e^{-\alpha t} - \dfrac{\gamma^2 - a_1\gamma + a_0}{\gamma(\alpha - \gamma)}e^{-\gamma t}$
$\dfrac{1}{(s + \alpha)(s + \gamma)(s + \delta)}$	$\dfrac{e^{-\alpha t}}{(\gamma - \alpha)(\delta - \alpha)} + \dfrac{e^{-\gamma t}}{(\alpha - \gamma)(\delta - \gamma)} + \dfrac{e^{-\delta t}}{(\alpha - \delta)(\gamma - \delta)}$
$\dfrac{s + a_0}{(s + \alpha)(s + \gamma)(s + \delta)}$	$\dfrac{a_0 - \alpha}{(\gamma - \alpha)(\delta - \alpha)}e^{-\alpha t} + \dfrac{a_0 - \gamma}{(\alpha - \gamma)(\delta - \gamma)}e^{-\gamma t} + \dfrac{a_0 - \delta}{(\alpha - \delta)(\gamma - \delta)}e^{-\delta t}$
$\dfrac{s^2 + a_1 s + a_0}{(s + \alpha)(s + \gamma)(s + \delta)}$	$\dfrac{\alpha^2 - a_1\alpha + a_0}{(\gamma - \alpha)(\delta - \alpha)}e^{-\alpha t} + \dfrac{\gamma^2 - a_1\gamma + a_0}{(\alpha - \gamma)(\delta - \gamma)}e^{-\gamma t} + \dfrac{\delta^2 - a_1\delta + a_0}{(\alpha - \delta)(\gamma - \delta)}e^{-\delta t}$
$\dfrac{1}{s^2 + \beta^2}$	$\dfrac{1}{\beta}\sin \beta t$
$\dfrac{1}{s^2 - \beta^2}$	$\dfrac{1}{\beta}\sinh \beta t$
$\dfrac{s}{s^2 + \beta^2}$	$\cos \beta t$
$\dfrac{s}{s^2 - \beta^2}$	$\cosh \beta t$
$\dfrac{s + a_0}{s^2 + \beta^2}$	$\dfrac{1}{\beta}(a_0 + \beta^2)^{1/2}\sin(\beta t + \psi)$ $\psi \triangleq \tan^{-1}\dfrac{\beta}{a_0}$
$\dfrac{1}{s(s^2 + \beta^2)}$	$\dfrac{1}{\beta^2}(1 - \cos \beta t)$
$\dfrac{s + a_0}{s(s^2 + \beta^2)}$	$\dfrac{a_0}{\beta^2} - \dfrac{(a_0^2 + \beta^2)^{1/2}}{\beta^2}\cos(\beta t + \psi)$ $\psi \triangleq \tan^{-1}\dfrac{\beta}{a_0}$
$\dfrac{s^2 + a_1 s + a_0}{s(s^2 + \beta^2)}$	$\dfrac{a_0}{\beta^2} - \dfrac{\left[(a_0 - \beta^2)^2 + a_1^2\beta^2\right]^{1/2}}{\beta^2}\cos(\beta t + \psi)$ $\psi \triangleq \tan^{-1}\dfrac{a_1\beta}{a_0 - \beta^2}$
$\dfrac{s + a_0}{(s + \alpha)(s^2 + \beta^2)}$	$\dfrac{a_0 - \alpha}{\alpha^2 + \beta^2}e^{-\alpha t} + \dfrac{1}{\beta}\left[\dfrac{a_0^2 + \beta^2}{\alpha^2 + \beta^2}\right]^{1/2}\sin(\beta t + \psi)$ $\psi \triangleq \tan^{-1}\dfrac{\beta}{a_0} - \tan^{-1}\dfrac{\beta}{\alpha}$

TABLE 2.16 (*Continued*)

Unilateral Laplace Function-Transform (*Continued*)

$F(s)$	$f(t), 0 \leq t$
$\dfrac{s^2 + a_1 s + a_0}{(s+\alpha)(s^2+\beta^2)}$	$\dfrac{\alpha^2 - a_1\alpha + a_0}{\alpha^2 + \beta^2} e^{-\alpha t} + \dfrac{1}{\beta}\left[\dfrac{(a_0-\beta^2)^2 + a_1^2\beta^2}{\alpha^2+\beta^2}\right]^{1/2} \sin(\beta t + \psi)$ $\psi \triangleq \tan^{-1}\dfrac{a_1\beta}{a_0-\beta^2} - \tan^{-1}\dfrac{\beta}{\alpha}$
$\dfrac{s + a_0}{s(s+\alpha)(s^2+\beta^2)}$	$\dfrac{a_0}{\alpha\beta^2} + \dfrac{\alpha - a_0}{\alpha(\alpha^2+\beta^2)} e^{-\alpha t} - \dfrac{1}{\beta^2}\left[\dfrac{a_0^2+\beta^2}{\alpha^2+\beta^2}\right]^{1/2} \cos(\beta t + \psi)$ $\psi \triangleq \tan^{-1}\dfrac{\beta}{a_0} - \tan^{-1}\dfrac{\beta}{\alpha}$
$\dfrac{s^2 + a_1 s + a_0}{s(s+\alpha)(s^2+\beta^2)}$	$\dfrac{a_0}{\alpha\beta^2} - \dfrac{\alpha^2 - a_1\alpha + a_0}{\alpha(\alpha^2+\beta^2)} e^{-\alpha t} - \dfrac{1}{\beta^2}\left[\dfrac{(a_0-\beta^2)^2 + a_1^2\beta^2}{\alpha^2+\beta^2}\right]^{1/2} \cos(\beta t + \psi)$ $\psi \triangleq \tan^{-1}\dfrac{a_1\beta}{a_0-\beta^2} - \tan^{-1}\dfrac{\beta}{\alpha}$
$\dfrac{s^2 + a_1 s + a_0}{(s+\alpha)(s+\gamma)(s^2+\beta^2)}$	$\dfrac{\alpha^2 - a_1\alpha + a_0}{(\gamma-\alpha)(\alpha^2+\beta^2)} e^{-\alpha t} + \dfrac{\gamma^2 + a_1\gamma + a_0}{(\alpha-\gamma)(\gamma^2+\beta^2)} e^{-\gamma t}$ $\qquad + \dfrac{1}{\beta}\left[\dfrac{(a_0-\beta^2)^2 + a_1^2\beta^2}{(\alpha^2+\beta^2)(\gamma^2+\beta^2)}\right]^{1/2} \sin(\beta t + \psi)$ $\psi \triangleq \tan^{-1}\dfrac{a_1\beta}{a_0-\beta^2} - \tan^{-1}\dfrac{\beta}{\alpha} - \tan^{-1}\dfrac{\beta}{\gamma}$
$\dfrac{s^3 + a_2 s^2 + a_1 s + a_0}{(s+\alpha)(s+\gamma)(s^2+\beta^2)}$	$\dfrac{-\alpha^3 + a_2\alpha^2 - a_1\alpha - a_0}{(\gamma-\alpha)(\alpha^2+\beta^2)} e^{-\alpha t} + \dfrac{-\gamma^3 + a_2\gamma^2 - a_1\gamma + a_0}{(\alpha-\gamma)(\gamma^2+\beta^2)} e^{-\gamma t}$ $\qquad + \dfrac{1}{\beta}\left[\dfrac{(a_0-a_2\beta^2)^2 + \beta^2(a_1-\beta^2)^2}{(\alpha^2+\beta^2)(\gamma^2+\beta^2)}\right]^{1/2} \sin(\beta t + \psi)$ $\psi \triangleq \tan^{-1}\dfrac{\beta(a_1-\beta^2)}{a_0-a_2\beta^2} - \tan^{-1}\dfrac{\beta}{\alpha} - \tan^{-1}\dfrac{\beta}{\gamma}$
$\dfrac{s}{(s^2+\beta^2)(s^2+\gamma^2)}$	$\dfrac{\cos\beta t - \cos\lambda t}{\lambda^2 - \beta^2}$
$\dfrac{s}{s^2 + (\beta+\lambda)^2} \times \dfrac{1}{s^2 + (\beta-\lambda)^2}$	$\dfrac{1}{2\lambda\beta} \sin\lambda t \cdot \sin\beta t$

TABLE 2.16 (*Continued*)

<table>
<tr><td colspan="2" align="center">*Unilateral Laplace Function-Transform (Continued)*</td></tr>
<tr><td>$F(s)$</td><td align="center">$f(t),\ 0 \le t$</td></tr>
<tr>
<td>$\dfrac{s^2 + a_1 s + a_0}{(s^2 + \beta^2)(s^2 + \lambda^2)}$</td>
<td>

$$\dfrac{\left[(a_0 - \beta^2)^2 + a_1^2\beta^2\right]^{1/2}}{\beta(\lambda^2 - \beta^2)}\sin(\beta t + \psi_1)$$

$$+ \dfrac{\left[(a_0 - \lambda^2)^2 + a_1^2\lambda^2\right]^{1/2}}{\lambda(\beta^2 - \lambda^2)}\sin(\lambda t + \psi_2)$$

$$\psi_1 \triangleq \tan^{-1}\dfrac{a_1\beta}{a_0 - \beta^2};\ \psi_2 \triangleq \tan^{-1}\dfrac{a_1\lambda}{a_0 - \lambda^2}$$

</td>
</tr>
<tr>
<td>$\dfrac{s^3 + a_2 s^2 + a_1 s + a_0}{(s^2 + \beta^2)(s^2 + \lambda^2)}$</td>
<td>

$$\dfrac{\left[(a_0 - a_2\beta^2)^2 + \beta^2(a_1 - \beta^2)^2\right]^{1/2}}{\beta(\lambda^2 - \beta^2)}\sin(\beta t + \psi_1)$$

$$+ \dfrac{\left[(a_0 - a_2\lambda^2)^2 + \lambda^2(a_1 - \lambda^2)^2\right]^{1/2}}{\lambda(\beta^2 - \lambda^2)}\sin(\lambda t + \psi_2)$$

$$\psi_1 \triangleq \tan^{-1}\dfrac{\beta(a_1 - \beta^2)}{a_0 - a_2\beta^2};\ \psi_2 \triangleq \tan^{-1}\dfrac{\lambda(a_1 - \lambda^2)}{a_0 - a_2\lambda^2}$$

</td>
</tr>
<tr>
<td>$\dfrac{1}{(s + \alpha)^2 + \beta^2}$</td>
<td>$\dfrac{1}{\beta}e^{-\alpha t}\sin\beta t$</td>
</tr>
<tr>
<td>$\dfrac{s + a_0}{(s + \alpha)^2 + \beta^2}$</td>
<td>

$$\dfrac{1}{\beta}[(a_0 - \alpha)^2 + \beta^2]^{1/2}e^{-\alpha t}\sin(\beta t + \psi)$$

$$\psi \triangleq \tan^{-1}\dfrac{\beta}{a_0 - \alpha}$$

</td>
</tr>
<tr>
<td>$\dfrac{s + \alpha}{(s + \alpha)^2 + \beta^2}$</td>
<td>$e^{-\alpha t}\cos\beta t$</td>
</tr>
<tr>
<td>$\dfrac{1}{s\left[(s + \alpha)^2 + \beta^2\right]}$</td>
<td>

$$\dfrac{1}{\beta_0^2} + \dfrac{1}{\beta_0\beta}e^{-\alpha t}\sin(\beta t - \psi)$$

$$\psi \triangleq \tan^{-1}\dfrac{\beta}{-\alpha}$$

$$\beta_0^2 \triangleq \alpha^2 + \beta^2$$

</td>
</tr>
<tr>
<td>$\dfrac{s + a_0}{s\left[(s + \alpha)^2 + (\beta^2)\right]}$</td>
<td>

$$\dfrac{a_0}{\beta_0^2} + \dfrac{1}{\beta\beta_0}[(a_0 - \alpha)^2 + \beta^2]^{1/2}e^{-\alpha t}\sin(\beta t + \psi)$$

$$\psi \triangleq \tan^{-1}\dfrac{\beta}{a_0 - \alpha} - \tan^{-1}\dfrac{\beta}{-\alpha}$$

$$\beta_0^2 \triangleq \alpha^2 + \beta^2$$

</td>
</tr>
<tr>
<td>$\dfrac{s^2 + a_1 s + a_0}{s\left[(s + \alpha)^2 + \beta^2\right]}$</td>
<td>

$$\dfrac{a_0}{\beta_0^2} + \dfrac{1}{\beta\beta_0}[(\alpha^2 - \beta^2 - a_1\alpha + a_0)^2 + \beta^2(a_1 - 2\alpha)^2]^{1/2}e^{-\alpha t}\sin(\beta t + \psi)$$

$$\psi \triangleq \tan^{-1}\dfrac{\beta(a_1 - 2\alpha)}{\alpha^2 - \beta^2 - a_1\alpha + a_0} - \tan^{-1}\dfrac{\beta}{-\alpha}$$

$$\beta_0^2 \triangleq \beta^2 + \alpha^2$$

</td>
</tr>
</table>

TABLE 2.16   (*Continued*)

Unilateral Laplace Function-Transform (Continued)

$F(s)$	$f(t),\ 0 \le t$
$\dfrac{1}{(s + \gamma)\left[(s + \alpha)^2 + \beta^2\right]}$	$\dfrac{e^{-\gamma t}}{(\gamma - \alpha)^2 + \beta^2} + \dfrac{1}{\beta\left[(\gamma - \alpha)^2 + \beta^2\right]^{1/2}}\, e^{-\alpha t}\sin(\beta t - \psi)$ $\psi \triangleq \tan^{-1}\dfrac{\beta}{\gamma - \alpha}$
$\dfrac{s + a_0}{(s + \gamma)\left[(s + \alpha)^2 + \beta^2\right]}$	$\dfrac{a_0 - \gamma}{(\alpha - \gamma)^2 + \beta^2}\, e^{-\gamma t} + \dfrac{1}{\beta}\left[\dfrac{(a_0 - \alpha)^2 + \beta^2}{(\gamma - \alpha)^2 + \beta^2}\right]^{1/2} e^{-\alpha t}\sin(\beta t + \psi)$ $\psi \triangleq \tan^{-1}\dfrac{\beta}{a_0 - \alpha} - \tan^{-1}\dfrac{\beta}{\gamma - \alpha}$
$\dfrac{s^2 + a_1 s + a_0}{(s + \gamma)\left[(s + \alpha)^2 + \beta^2\right]}$	$\dfrac{\gamma^2 - a_1\gamma + a_0}{(\alpha - \gamma)^2 + \beta^2}\, e^{-\gamma t}$ $+ \dfrac{1}{\beta}\left[\dfrac{\left(\alpha^2 - \beta^2 - a_1\alpha + a_0\right)^2 + \beta^2\left(a_1 - 2\alpha\right)^2}{(\gamma - \alpha)^2 + \beta^2}\right]^{1/2} e^{-\alpha t}\sin(\beta t + \psi)$ $\psi \triangleq \tan^{-1}\dfrac{\beta(a_1 - 2\alpha)}{\alpha^2 - \beta^2 - a_1\alpha + a_0} - \tan^{-1}\dfrac{\beta}{\gamma - \alpha}$
$\dfrac{1}{s(s + \gamma)\left[(s + \alpha)^2 + \beta^2\right]}$	$\dfrac{1}{\gamma\beta_0^2} - \dfrac{1}{\gamma\left[(\alpha - \gamma)^2 + \beta^2\right]}\, e^{-\gamma t}$ $+ \dfrac{1}{\beta\beta_0\left[(\gamma - \alpha)^2 + \beta^2\right]^{1/2}}\, e^{-\alpha t}\sin(\beta t - \psi)$ $\gamma \triangleq \tan^{-1}\dfrac{\beta}{-\alpha} + \tan^{-1}\dfrac{\beta}{\gamma - \alpha}$ $\beta_0^2 \triangleq \alpha^2 + \beta^2$
$\dfrac{s + a_0}{s(s + \gamma)\left[(s + \alpha)^2 + \beta^2\right]}$	$\dfrac{a_0}{\gamma\beta_0^2} + \dfrac{\gamma - a_0}{\gamma\left[(\alpha - \gamma)^2 + \beta^2\right]}\, e^{-\gamma t}$ $+ \dfrac{1}{\beta\beta_0}\left[\dfrac{(a_0 - \alpha)^2 + \beta^2}{(\gamma - \alpha)^2 + \beta^2}\right]^{1/2} e^{-\alpha t}\sin(\beta t + \psi)$ $\psi \triangleq \tan^{-1}\dfrac{\beta}{a_0 - \alpha} - \tan^{-1}\dfrac{\beta}{\gamma - \alpha} - \tan^{-1}\dfrac{\beta}{-\alpha}$ $\beta_0^2 \triangleq \alpha^2 + \beta^2$
$\dfrac{s^2 + a_1 s + a_0}{(s + \gamma)(s + \delta)\left[(s + \alpha)^2 + \beta^2\right]}$	$\dfrac{\gamma^2 - a_1\gamma + a_0}{(\delta - \gamma)\left[(\alpha - \gamma)^2 + \beta^2\right]}\, e^{-\gamma t} + \dfrac{\delta^2 - a_1\delta + a_0}{(\gamma - \delta)\left[(\alpha - \delta)^2 + \beta^2\right]}\, e^{-\delta t}$ $+ \dfrac{1}{\beta}\left\{\dfrac{\left(\alpha^2 - \beta^2 - a_1\alpha + a_0\right)^2 + \beta^2\left(a_1 - 2\alpha\right)^2}{\left[(\delta - \alpha)^2 + \beta^2\right]\left[(\gamma - \alpha)^2 + \beta^2\right]}\right\}^{1/2} e^{-\alpha t}\sin(\beta t + \psi)$ $\psi \triangleq \tan^{-1}\dfrac{\beta(a_1 - 2\alpha)}{\alpha^2 - \beta^2 - a_1\alpha + a_0} - \tan^{-1}\dfrac{\beta}{\gamma - \alpha} - \tan^{-1}\dfrac{\beta}{\delta - \alpha}$

TABLE 2.16 (*Continued*)

$F(s)$	$f(t), 0 \le t$
	Unilateral Laplace Function-Transform (Continued)
$\dfrac{1}{(s^2 + \lambda^2)\left[(s + \alpha)^2 + \beta^2\right]}$	$\dfrac{1}{\left[(\beta_0^2 - \lambda^2)^2 + 4\alpha^2\lambda^2\right]^{1/2}}\left[\dfrac{1}{\lambda}\sin(\lambda t - \psi_1) + \dfrac{1}{\beta}e^{-\alpha t}\sin(\beta t - \psi_2)\right]$ $\psi_1 \triangleq \tan^{-1}\dfrac{2\alpha\lambda}{\beta_0^2 - \lambda^2}; \; \psi_2 \triangleq \dfrac{-2\alpha\beta}{\alpha^2 - \beta^2 + \lambda^2}; \; \beta_0^2 \triangleq \alpha^2 + \beta^2$
$\dfrac{s + a_0}{(s^2 + \lambda^2)\left[(s + \alpha)^2 + \beta^2\right]}$	$\dfrac{1}{\lambda}\left[\dfrac{a_0^2 + \lambda^2}{(\beta_0^2 - \lambda^2)^2 + 4\alpha^2\lambda^2}\right]^{1/2}\sin(\lambda t + \psi_1)$ $+\dfrac{1}{\beta}\left[\dfrac{(a_0 - \alpha)^2 + \beta^2}{(\beta_0^2 - \lambda^2)^2 + 4\alpha^2\lambda^2}\right]^{1/2}e^{-\alpha t}\sin(\beta t + \psi_2)$ $\psi_1 \triangleq \tan^{-1}\dfrac{\lambda}{a_0} - \tan^{-1}\dfrac{2\alpha\lambda}{\beta_0^2 - \lambda^2}$ $\psi_2 \triangleq \tan^{-1}\dfrac{\beta}{a_0 - \alpha} - \tan^{-1}\dfrac{-2\alpha\beta}{\alpha^2 - \beta^2 + \lambda^2}$ $\beta^2 \triangleq \alpha^2 + \beta_0^2$
$\dfrac{s^2 + a_1 s + a_0}{(s^2 + \lambda^2)\left[(s + \alpha)^2 + \beta^2\right]}$	$\dfrac{1}{\lambda}\left[\dfrac{(a_0 - \lambda^2)^2 + a_1^2\lambda^2}{(\beta_0^2 - \lambda^2)^2 + 4\alpha^2\lambda^2}\right]^{1/2}\sin(\lambda t + \psi_1)$ $+\dfrac{1}{\beta}\left[\dfrac{(\alpha^2 - \beta^2 - a_1\alpha + a_0)^2 + \beta^2(a_1 - 2\alpha)^2}{(\beta_0^2 - \lambda^2)^2 + 4\alpha^2\lambda^2}\right]^{1/2}e^{-\alpha t}\sin(\beta t + \psi_2)$ $\psi_1 \triangleq \tan^{-1}\dfrac{a_1\lambda}{a_0 - \lambda^2} - \tan^{-1}\dfrac{2\alpha\lambda}{\beta_0^2 - \lambda^2}$ $\psi_2 \triangleq \tan^{-1}\dfrac{\beta(a_1 - 2\alpha)}{\alpha^2 - \beta^2 - a_1\alpha + a_0} - \tan^{-1}\dfrac{-2\alpha\beta}{\alpha^2 - \beta^2 + \lambda^2}$ $\beta_0^2 \triangleq \alpha^2 + \beta^2$
$\dfrac{s + a_0}{(s + \gamma)(s^2 + \lambda^2)}$ $\times \dfrac{1}{(s + \alpha)^2 + \beta^2}$	$\dfrac{a_0 - \gamma}{(\lambda^2 + \gamma^2)\left[(\alpha - \gamma)^2 + \beta^2\right]}e^{-\gamma t}$ $+\dfrac{1}{\lambda}\left\{\dfrac{a_0^2 + \lambda^2}{(\gamma^2 + \lambda^2)\left[(\beta_0^2 - \lambda^2)^2 + 4\alpha^2\lambda^2\right]}\right\}^{1/2}\sin(\lambda t + \psi_1)$ $+\dfrac{1}{\beta}\left\{\dfrac{(a_0 - \alpha)^2 + \beta^2}{\left[(\gamma - \alpha)^2 + \beta^2\right]\left[(\beta_0^2 - \lambda^2)^2 + 4\alpha^2\lambda^2\right]}\right\}^{1/2}e^{-\alpha t}\sin(\beta t + \psi_2)$ $\psi_1 \triangleq \tan^{-1}\dfrac{\lambda}{a_0} - \tan^{-1}\dfrac{\lambda}{\gamma} - \tan^{-1}\dfrac{2\alpha\lambda}{\beta_0^2 - \lambda^2}$ $\gamma^2 \triangleq \tan^{-1}\dfrac{\beta}{a_0 - \alpha} - \tan^{-1}\dfrac{\beta}{\gamma - \alpha} - \tan^{-1}\dfrac{-2\alpha\beta}{\alpha^2 - \beta^2 + \lambda^2}$ $\beta_0^2 \triangleq \alpha^2 + \beta^2$
$\dfrac{1}{s^2}$	t
$\dfrac{1}{s^n}$	$\dfrac{1}{(n - 1)!}t^{n-1}$ n is a positive integer

TABLE 2.16   *(Continued)*

Unilateral Laplace Function-Transform Pairs (Continued)

$F(s)$	$f(t),\, 0 \le t$
$\dfrac{1}{(s+\alpha)s^2}$	$\dfrac{e^{-\alpha t} + \alpha t - 1}{\alpha^2}$
$\dfrac{s + a_0}{(s+\alpha)s^2}$	$\dfrac{a_0 - \alpha}{\alpha^2}e^{-\alpha t} + \dfrac{a_0}{\alpha}t + \dfrac{\alpha - a_0}{\alpha^2}$
$\dfrac{s^2 + a_1 s + a_0}{(s+\alpha)s^2}$	$\dfrac{\alpha^2 - a_1\alpha + a_0}{\alpha^2}e^{-\alpha t} + \dfrac{a_0}{\alpha}t + \dfrac{a_1\alpha - a_0}{\alpha^2}$
$\dfrac{1}{(s+\alpha)^2}$	$te^{-\alpha t}$
$\dfrac{s + a_0}{(s+\alpha)^2}$	$[(a_0 - \alpha)t + 1]e^{-\alpha t}$
$\dfrac{1}{(s+\alpha)^n}$	$\dfrac{1}{(n-1)!}t^{n-1}e^{-\alpha t}\qquad n$ is a positive integer
$\dfrac{s^n}{(s+\alpha)^{n+1}}$	$e^{-\alpha t}\displaystyle\sum_{k=0}^{n}\dfrac{n!(-\alpha)^k}{(n-k)!(k!)^2}t^k \qquad n$ is a non-negative integer
$\dfrac{1}{s(s+\alpha)^2}$	$\dfrac{1 - (1 + \alpha t)e^{-\alpha t}}{\alpha^2}$
$\dfrac{s + a_0}{s(s+\alpha)^2}$	$\dfrac{a_0}{\alpha^2} + \left(\dfrac{\alpha - a_0}{\alpha}t - \dfrac{a_0}{\alpha^2}\right)e^{-\alpha t}$
$\dfrac{s^2 + a_1 s + a_0}{s(s+\alpha)^2}$	$\dfrac{a_0}{\alpha^2} + \left(\dfrac{a_1\alpha - a_0 - \alpha^2}{\alpha}t + \dfrac{\alpha^2 - a_0}{\alpha^2}\right)e^{-\alpha t}$
$\dfrac{1}{(s+\gamma)(s+\alpha)^2}$	$\dfrac{1}{(\gamma - \alpha)^2}e^{-\gamma t} + \dfrac{(\gamma - \alpha)t - 1}{(\gamma - \alpha)^2}e^{-\alpha t}$
$\dfrac{s + a_0}{(s+\gamma)(s+\alpha)^2}$	$\dfrac{a_0 - \gamma}{(\alpha - \gamma)^2}e^{-\gamma t} + \left[\dfrac{a_0 - \alpha}{\gamma - \alpha}t + \dfrac{\gamma - a_0}{(\gamma - \alpha)^2}\right]e^{-\alpha t}$
$\dfrac{s^2 + a_1 s + a_0}{(s+\gamma)(s+\alpha)^2}$	$\dfrac{\gamma^2 - a_1\gamma + a_0}{(\alpha - \gamma)^2}e^{-\gamma t}$ $+ \left[\dfrac{\alpha^2 - a_1\alpha + a_0}{\gamma - \alpha}t + \dfrac{\alpha^2 - 2\alpha\gamma + a_1\gamma - a_0}{(\gamma - \alpha)^2}\right]e^{-\alpha t}$
$\dfrac{s + a_0}{(s+\gamma)(s+\alpha)^3}$	$\dfrac{a_0 - \gamma}{(\alpha - \gamma)^3}e^{-\gamma t} + \left[\dfrac{a_0 - \alpha}{2(\gamma - \alpha)}t^2 + \dfrac{\gamma - a_0}{(\gamma - \alpha)^2}t + \dfrac{a_0 - \gamma}{(\gamma - \alpha)^3}\right]e^{-\alpha t}$
$\dfrac{s + a_0}{s(s+\gamma)(s+\alpha)^2}$	$\dfrac{a_0}{\gamma\alpha^2} + \dfrac{\gamma - a_0}{\gamma(\alpha - \gamma)^2}e^{-\gamma t} + \left[\dfrac{a_0 - \alpha}{\alpha(\alpha - \gamma)}t + \dfrac{2a_0\alpha - \alpha^2 - a_0\gamma}{\alpha^2(\alpha - \gamma)^2}\right]e^{-\alpha t}$
$\dfrac{s^2 + a_1 s + a_0}{s(s+\gamma)(s+\alpha)^2}$	$\dfrac{a_0}{\gamma\alpha^2} - \dfrac{\gamma^2 - a_1\gamma + a_0}{\gamma(\alpha - \gamma)^2}e^{-\gamma t}$ $+ \left[\dfrac{\alpha^2 - a_1\alpha + a_0}{\alpha(\alpha - \gamma)}t + \dfrac{(\gamma - a_1)\alpha^2 + (2\alpha - \gamma)a_0}{\alpha^2(\alpha - \gamma)^2}\right]e^{-\alpha t}$

TABLE 2.16 (*Continued*)

$F(s)$	$f(t),\ 0 \le t$
$\dfrac{s + a_0}{(s + \gamma)(s + \delta)(s + \alpha)^2}$	$\dfrac{a_0 - \gamma}{(\delta - \gamma)(\alpha - \gamma)^2} e^{-\gamma t} + \dfrac{a_0 - \delta}{(\gamma - \delta)(\alpha - \delta)^2} e^{-\delta t}$ $+ \left[\dfrac{a_0 - \alpha}{(\gamma - \alpha)(\delta - \alpha)} t + \dfrac{2a_0\alpha - \alpha^2 - a_0(\gamma + \delta) + \gamma\delta}{(\gamma - \alpha)^2(\delta - \alpha)^2} \right] e^{-\alpha t}$
$\dfrac{s + a_0}{(s + \alpha)(s + \gamma)s^2}$	$\dfrac{a_0 - \alpha}{\alpha^2(\gamma - \alpha)} e^{-\alpha t} + \dfrac{a_0 - \gamma}{\gamma^2(\alpha - \gamma)} e^{\gamma - t} + \dfrac{a_0}{\alpha\gamma} t + \dfrac{\alpha\gamma - a_0(\alpha + \gamma)}{\alpha^2\gamma^2}$
$\dfrac{s^2 + a_1 s + a_0}{(s + \alpha)(s + \gamma)s^2}$	$\dfrac{\alpha^2 - a_1\alpha + a_0}{\alpha^2(\gamma - \alpha)} e^{-\alpha t} + \dfrac{\gamma^2 - a_1\gamma + a_0}{\gamma^2(\alpha - \gamma)} e^{-\gamma t} + \dfrac{a_0}{\alpha\gamma} t$ $+ \dfrac{a_1\alpha\gamma - a_0(\alpha + \gamma)}{\alpha^2\gamma^2}$
$\dfrac{s^2 + a_1 s + a_0}{(s + \alpha)^2 s^2}$	$\left[\dfrac{\alpha^2 - a_1\alpha + a_0}{\alpha^2} t + \dfrac{2a_0 - a_1\alpha}{\alpha^3} \right] e^{-\alpha t} + \dfrac{a_0}{\alpha^2} t + \dfrac{a_1\alpha - 2a_0}{\alpha^3}$
$\dfrac{s + a_0}{(s + \alpha)^2(s + \gamma)^2}$	$\left[\dfrac{a_0 - \alpha}{(\gamma - \alpha)^2} t + \dfrac{\alpha + \gamma - 2a_0}{(\gamma - \alpha)^3} \right] e^{-\alpha t}$ $+ \left[\dfrac{a_0 - \gamma}{(\alpha - \gamma)^2} t + \dfrac{\alpha + \gamma - 2a_0}{(\alpha - \gamma)^3} \right] e^{-\gamma t}$
$\dfrac{s^2 + a_1 s + a_0}{(s + \alpha)^2(s + \gamma)^2}$	$\left[\dfrac{\alpha^2 - a_1\alpha + a_0}{(\gamma - \alpha)^2} t + \dfrac{a_1(\alpha + \gamma) - 2(\alpha\gamma + a_0)}{(\gamma - \alpha)^3} \right] e^{-\alpha t}$ $+ \left[\dfrac{\gamma^2 - a_1\gamma + a_0}{(\gamma - \alpha)^2} t - \dfrac{a_1(\alpha + \gamma) - 2(\alpha\gamma + a_0)}{(\gamma - \alpha)^3} \right] e^{-\gamma t}$
$\dfrac{s^2 + a_1 s + a_0}{(s + \alpha)^3 s^2}$	$\left(\dfrac{\alpha^2 - a_1\alpha + a_0}{2\alpha^2} t^2 + \dfrac{-a_1\alpha + 2a_0}{\alpha^3} t + \dfrac{-a_1\alpha + 3a_0}{\alpha^4} \right) e^{-\alpha t}$ $+ \dfrac{a_0}{\alpha^3} t + \dfrac{a_1\alpha - 3a_0}{\alpha^4}$
$\dfrac{1}{(s^2 + \beta^2)s^2}$	$\dfrac{1}{\beta^2} t - \dfrac{1}{\beta^3} \sin \beta t$
$\dfrac{1}{(s^2 - \beta^2)s^2}$	$\dfrac{1}{\beta^3} \sinh \beta t - \dfrac{1}{\beta^2} t$
$\dfrac{s + a_0}{(s^2 + \beta^2)s^2}$	$\dfrac{a_0}{\beta^2} t + \dfrac{1}{\beta^2} - \dfrac{1}{\beta^3}(a_0^2 + \beta^2)^{1/2} \sin(\beta t + \psi)$ $\psi \triangleq \tan^{-1} \dfrac{\beta}{a_0}$
$\dfrac{s^2 + a_1 s + a_0}{(s^2 + \beta^2)s^2}$	$\dfrac{a_0}{\beta^2} t + \dfrac{a_1}{\beta^2} - \dfrac{1}{\beta^3}[(a_0 - \beta^2)^2 + a_1^2\beta^2]^{1/2} \sin(\beta t + \psi)$ $\psi \triangleq \tan^{-1} \dfrac{a_1\beta}{a_0 - \beta^2}$

Unilateral Laplace Function-Transform Pairs (Continued)

TABLE 2.16   (*Continued*)

Unilateral Laplace Function-Transform Pairs (Continued)

$F(s)$	$f(t),\ 0 \le t$
$\dfrac{1}{(s^2 + \beta^2)s^3}$	$\dfrac{1}{\beta^4}(\cos \beta t - 1) + \dfrac{1}{2\beta^2}t^2$
$\dfrac{1}{(s^2 - \beta^2)s^3}$	$\dfrac{1}{\beta^4}(\cosh \beta t - 1) - \dfrac{1}{2\beta^2}t^2$
$\dfrac{1}{(s^2 + \beta^2)(s + \alpha)^2}$	$\dfrac{1}{\beta(\alpha^2 + \beta^2)}\sin(\beta t - \psi) + \left[\dfrac{1}{\alpha^2 + \beta^2}t + \dfrac{2\alpha}{(\alpha^2 + \beta^2)^2}\right]e^{-\alpha t}$ $\psi \triangleq 2\tan^{-1}\dfrac{\beta}{\alpha}$
$\dfrac{s + a_0}{(s^2 + \beta^2)(s + \alpha)^2}$	$\dfrac{(a_0^2 + \beta^2)^{1/2}}{\beta(\alpha^2 + \beta^2)}\sin(\beta t + \psi) + \left[\dfrac{a_0 - \alpha}{\alpha^2 + \beta^2}t + \dfrac{2a_0\alpha + \beta^2 - \alpha^2}{(\alpha^2 + \beta^2)^2}\right]e^{-\alpha t}$ $\psi \triangleq \tan^{-1}\dfrac{\beta}{a_0} - 2\tan^{-1}\dfrac{\beta}{\alpha}$
$\dfrac{s^2 + a_1 s + a_0}{(s^2 + \beta^2)(s + \alpha)^2}$	$\dfrac{\left[(a_0 - \beta^2)^2 + a_1^2\beta^2\right]^{1/2}}{\beta(\alpha^2 + \beta^2)}\sin(\beta t + \psi)$ $+\left[\dfrac{\alpha^2 - a_1\alpha + a_0}{\alpha^2\beta^2}t + \dfrac{a_1(\beta^2 - \alpha^2) + 2\alpha(a_0 - \beta^2)}{(\alpha^2 + \beta^2)^2}\right]e^{-\alpha t}$ $\psi \triangleq \tan^{-1}\dfrac{a_1\beta}{a_0 - \beta^2} - 2\tan^{-1}\dfrac{\beta}{\alpha}$
$\dfrac{s + a_0}{s(s^2 + \beta^2)(s + \alpha)^2}$	$\dfrac{a_0}{\beta^2\alpha^2} - \dfrac{(a_0^2 + \beta^2)^{1/2}}{\beta^2(\alpha^2 + \beta^2)}\cos(\beta t + \psi)$ $+\left[\dfrac{\alpha - a_0}{\alpha(\alpha^2 + \beta^2)}t + \dfrac{2\alpha^3 - 3a_0\alpha^2 - a_0\beta^2}{\alpha^2(\alpha^2 + \beta^2)^2}\right]e^{-\alpha t}$ $\psi \triangleq \tan^{-1}\dfrac{\beta}{a_0} - 2\tan^{-1}\dfrac{\beta}{\alpha}$
$\dfrac{s^2 + a_1 s + a_0}{(s + \gamma)(s^2 + \beta^2)(s + \alpha)^2}$	$\dfrac{\gamma^2 - a_1\gamma + a_0}{(\gamma^2 + \beta^2)(\alpha - \gamma)^2}e^{-\gamma t} + \dfrac{\left[(a_0 - \beta^2)^2 + a_1^2\beta^2\right]^{1/2}}{\beta(\gamma^2 + \beta^2)^{1/2}(\alpha^2 + \beta^2)}\sin(\beta t + \psi)$ $+\dfrac{\alpha^2 - a_1\alpha + a_0}{(\gamma - \alpha)(\alpha^2 + \beta^2)}t e^{-\alpha t}$ $+\dfrac{(\gamma - \alpha)(\alpha^2 + \beta^2)(a_1 - 2\alpha) - (\alpha^2 - a_1\alpha + a_0)(3\alpha^2 + \beta^2 - 2\alpha\gamma)}{(\gamma - \alpha)^2(\alpha^2 + \beta^2)^2}$ $\times e^{-\alpha t}$ $\psi \triangleq \tan^{-1}\dfrac{a_1\beta}{a_0 - \beta^2} - \tan^{-1}\dfrac{\beta}{\gamma} - 2\tan^{-1}\dfrac{\beta}{\alpha}$
$\dfrac{1}{(s^2 + \beta^2)^2}$	$\dfrac{1}{2\beta^3}(\sin \beta t - \beta t \cos \beta t)$

TABLE 2.16 (*Continued*)

Unilateral Laplace Function-Transform Pairs (*Continued*)	
$F(s)$	$f(t),\ 0 \le t$
$\dfrac{s}{\left(s^2 + \beta^2\right)^2}$	$\dfrac{1}{2\beta} t \sin \beta t$
$\dfrac{s^2}{\left(s^2 + \beta^2\right)^2}$	$\dfrac{1}{2\beta}(\sin \beta t + \beta t \cos \beta t)$
$\dfrac{s^2 - \beta^2}{\left(s^2 + \beta^2\right)^2}$	$t \cos \beta t$
$\dfrac{1}{s\left(s^2 + \beta^2\right)^2}$	$\dfrac{1}{\beta^4}(1 - \cos \beta t) - \dfrac{1}{2\beta^3} t \sin \beta t$
$\dfrac{s^2 + a_1 s + a_0}{s\left(s^2 + \beta^2\right)^2}$	$\dfrac{a_0}{\beta^4} - \dfrac{\left[\left(a_0 - \beta^2\right)^2 + a_1^2\beta^2\right]^{1/2}}{2\beta^3} t \sin(\beta t + \psi_1)$ $- \dfrac{\left(4a_0^2 + a_1^2\beta^2\right)^{1/2}}{2\beta^4}\cos(\beta t + \psi_2)$ $\psi_1 \triangleq \tan^{-1}\dfrac{a_1\beta}{a_0 - \beta^2}; \qquad \psi_2 \triangleq \tan^{-1}\dfrac{a_1\beta}{2a_0}$
$\dfrac{1}{\left[(s + \alpha)^2 + \beta^2\right]s^2}$	$\dfrac{1}{\beta_0^2}\left[t - \dfrac{2\alpha}{\beta_0^2} + \dfrac{1}{\beta}e^{-\alpha t}\sin(\beta t - \psi)\right]$ $\psi \triangleq 2\tan^{-1}\dfrac{\beta}{-\alpha}$ $\beta_0^2 \triangleq \alpha^2 + \beta^2$
$\dfrac{1}{(s + \gamma)^2\left[(s + \alpha)^2 + \beta^2\right]}$	$\dfrac{1}{(\alpha - \gamma)^2 + \beta^2}\left[te^{-\gamma t} + \dfrac{2(\gamma - \alpha)}{(\alpha - \gamma)^2 + \beta^2}e^{-\gamma t} + \dfrac{1}{\beta}e^{-\alpha t}\sin(\beta t - \psi)\right]$ $\psi \triangleq 2\tan^{-1}\dfrac{\beta}{\gamma - \alpha}$
$\dfrac{s^2 + a_1 s + a_0}{(s + \gamma)^2\left[(s + \alpha)^2 + \beta^2\right]}$	$\dfrac{\gamma^2 - a_1\gamma + a_0}{(\alpha - \gamma)^2 + \beta^2}te^{-\gamma t}$ $+ \dfrac{\left[(\alpha - \gamma)^2 + \beta^2\right](a_1 - 2\gamma) - 2(\alpha - \gamma)\left(\gamma^2 - a_1\gamma + a_0\right)}{\left[(\alpha - \gamma)^2 + \beta^2\right]^2}e^{-\gamma t}$ $+ \dfrac{\left[\left(\alpha^2 - \beta^2 - a_1\alpha + a_0\right)^2 + \beta^2\left(a_1 - 2\alpha\right)^2\right]^{1/2}}{\beta\left[(\gamma - \alpha)^2 + \beta^2\right]}e^{-\alpha t}\sin(\beta t + \psi)$ $\psi \triangleq \tan^{-1}\dfrac{\beta(a_1 - 2\alpha)}{\alpha^2 - \beta^2 - a_1\alpha + a_0} - 2\tan^{-1}\dfrac{\beta}{\gamma - \alpha}$
$\dfrac{1}{\left[(s + \alpha)^2 + \beta^2\right]^2}$	$\dfrac{1}{2\beta^3}e^{-\alpha t}(\sin \beta t - \beta t \cos \beta t)$
$\dfrac{s + \alpha}{\left[(s + \alpha)^2 + \beta^2\right]^2}$	$\dfrac{1}{2\beta}te^{-\alpha t}\sin \beta t$

TABLE 2.16   (*Continued*)

Unilateral Laplace Function-Transform Pairs (Continued)

$F(s)$	$f(t), 0 \le t$
$\dfrac{s^2 + a_0}{\left[(s+\alpha)^2 + \beta^2\right]^2}$	$\dfrac{\beta_0^2 + a_0}{2\beta^3} e^{-\alpha t}\sin\beta t - \dfrac{\left[(\alpha^2 - \beta^2 + a_0)^2 + 4\alpha^2\beta^2\right]^{1/2}}{2\beta^2} te^{-\alpha t}\cos(\beta t + \psi)$ $\psi \triangleq \tan^{-1}\dfrac{-2\alpha\beta}{\alpha^2 - \beta^2 + a_0}; \; \beta_0^2 \triangleq \alpha^2 + \beta^2$
$\dfrac{(s+\alpha)^2 - \beta^2}{\left[(s+\alpha)^2 + \beta^2\right]^2}$	$te^{-\alpha t}\cos\beta t$
$\tan^{-1}\dfrac{\beta}{s}$	$\dfrac{\sin\beta t}{t}$
$\ln\dfrac{s+\beta}{s+\alpha}$	$\dfrac{e^{-\alpha t} - e^{-\beta t}}{t}$
$e^{s^2/4a}\text{cerf}\dfrac{s}{2\sqrt{a}}$	$2\sqrt{\dfrac{a}{\pi}}\, e^{-at^2}$ $\text{cerf } y \triangleq 1 - \text{erf } y \triangleq 1 - \dfrac{2}{\sqrt{\pi}}\int_0^y e^{-x^2}\, dx$
$\dfrac{1}{\sqrt{s^2+\alpha^2}}$	$J_0(\alpha t)$
$\dfrac{1}{\sqrt{s^2+\alpha^2}\left(\sqrt{s^2+\alpha^2}+s\right)}$	$\dfrac{1}{\alpha}J_1(\alpha t)$
$\dfrac{1}{\sqrt{s^2+\alpha^2}\left(\sqrt{s^2+\alpha^2}+s\right)^n}$	$\dfrac{1}{\alpha^n}J_n(\alpha t)$ n is a nonnegative integer
$\dfrac{1}{s\sqrt{s^2+\alpha^2}\left(\sqrt{s^2+\alpha^2}+s\right)^n}$	$\dfrac{1}{\alpha^n}\int_0^t J_n(\alpha t)\, dt$ n is a nonnegative integer
$\dfrac{1}{\sqrt{s^2+\alpha^2}+s}$	$\dfrac{1}{\alpha}\dfrac{J_1(\alpha t)}{t}$
$\dfrac{1}{\left(\sqrt{s^2+\alpha^2}+s\right)^n}$	$\dfrac{n}{\alpha^n}\dfrac{J_n(\alpha t)}{t}$ n is a positive integer
$\dfrac{1}{s\left(\sqrt{s^2+\alpha^2}+s\right)^n}$	$\dfrac{n}{\alpha^n}\int_0^t \dfrac{J_n(\alpha t)}{t}\, dt$ n is a positive integer
$\dfrac{1}{s}e^{-as}$	$u(t-a)$
$\dfrac{1}{s^2}e^{-as}$	$(t-a)u(t-a)$
$\left(\dfrac{a}{s} + \dfrac{1}{s^2}\right)e^{-as}$	$tu(t-a)$
$\left(\dfrac{2}{s^3} + \dfrac{2a}{s^2} + \dfrac{a^2}{s}\right)e^{-as}$	$t^2 u(t-a)$

TABLE 2.16   (*Continued*)

Unilateral Laplace Function-Transform Pairs (Continued)

$F(s)$	$f(t), 0 \le t$
$\dfrac{1}{s}(e^{-as} - e^{-bs})$ $a < b$	$u(t - a) - u(t - b)$
$\left(\dfrac{1 - e^{-s}}{s}\right)^2$	$\begin{cases} t & 0 < t < 1 \\ 2 - t & 1 < t < 2 \\ 0 & 2 < t \end{cases}$
$\left(\dfrac{1 - e^{-s}}{s}\right)^3$	$\begin{cases} 0.5t^2 & 0 < t < 1 \\ 0.75 - (t - 1.5)^2 & 1 < t < 2 \\ 0.5(t - 3)^2 & 2 < t < 3 \\ 0 & 3 < t \end{cases}$
$\dfrac{1}{s^2}(1 - e^{-s})$	$\begin{cases} t & 0 < t < 1 \\ 1 & 1 < t \end{cases}$
$\dfrac{1}{s^3}(1 - e^{-s})^2$	$\begin{cases} 0.5t^2 & 0 < t < 0 \\ 1 - 0.5(t - 2)^2 & 1 < t < 2 \\ 1 & 2 < t \end{cases}$
$\dfrac{1}{s(1 + e^{-s})}$	$\displaystyle\sum_{k=0}^{\infty} (-1)^k u(t - k)$
$\dfrac{1}{s \sinh s}$	$\displaystyle 2\sum_{k=0}^{\infty} u(t - 2k - 1)$
$\dfrac{1}{s \cosh s}$	$\displaystyle 2\sum_{k=0}^{\infty} (-1)^k u(t - 2k - 1)$
$\dfrac{1}{s}\tanh s$	$\displaystyle u(t) + 2\sum_{k=1}^{\infty} (-1)^k u(t - 2k)$ or $\displaystyle\sum_{k=0}^{\infty} (-1)^k u(t - 2k)u(2k + 2 - t)$
$\dfrac{e^s - s - 1}{s^2(e^s - 1)}$	$\displaystyle t - \sum_{k=1}^{\infty} u(t - k)$ or $\displaystyle\sum_{k=0}^{\infty} (t - k)u(t - k)u(k + 1 - t)$

for $0 < t$, and $= 0$ for $t < 0$. Also

$$\mathscr{L}^{-1}\left[\frac{s}{k_1(s + K_1)(s + K_2)}\right] = \frac{K_1 e^{-K_1 t} - K_2 e^{-K_2 t}}{k_1(K_1 - K_2)}$$

for $0 < t$, and $= 0$ for $t < 0$. Finally, $\mathscr{L}^{-1}\{\mathscr{L}[u(t)]\} = u(t)$.

Step F. Use condition 5 to find the inverse transform of the product. Thus, by condition 6 and step F,

$$y(t) = y_1(t) + [k_1(K_1 - K_2)]^{-1} \int_0^t \left[K_1 e^{-K_1(t-\tau)} - K_2 e^{-K_2(t-\tau)}\right] u(\tau)\, d\tau$$

Step G. Evaluate the (convolution) integral arising from step F. Thus,

$$y(t) = y_1(t) + [k_1(K_1 - K_2)]^{-1}(e^{-K_2 t} - e^{-K_1 t})$$

for $0 < t$, and $= 0$ for $t < 0$.

For the particular problem treated it is much simpler to use steps C and D than steps E, F, and G. However, for a more complicated right member of the original equation it could happen that step G would be easier to carry out than step C, in which case the second method (A, B, E, F, G) should be used rather than the first (A, B, C, D).

One physical representation of the initial-value problem that we have used for illustration is the problem of finding the current response of a series electric circuit containing constant lumped inductance, resistance, and capacitance to an applied electromotive force $u(t)$, with an initial current in the inductance and an initial charge on the condenser.

The complete method (of which only a part has been given) is not restricted in its field of application to linear equations with constant coefficients, but the solution of this type of equation is most simplified.

2.13.3   Transform Pairs

Table 2.16 of Laplace transforms are applicable in the solution of ordinary integrodifferential and difference equations. These tables are, by permission, from *Transients in Linear Systems*, by M. F. Gardner and J. L. Barnes, John Wiley and Sons, New York, 1942.

2.14   COMPLEX ANALYSIS

2.14.1   Complex Numbers

A *complex number A* is a combination of two real numbers a_1, a_2 in the ordered pair $(a_1, a_2) = A = a_1 + ia_2$, where $i = (-1)^{1/2}$. Real and imaginary numbers are special cases of complex numbers obtained by placing $(a_1, 0) = a_1, (0, a_2) = ia_2$ (see Fig. 2.163).

1. If $a_1 + ia_2 = 0$, then $a_1 = 0$, $a_2 = 0$.
2. If $a_1 + ia_2 = b_1 + ib_2$, then $a_1 = b_1$, $a_2 = b_2$.
3. $a_1 + ia_2$ and $a_1 - ia_2$ are *conjugate* complex numbers. The complex conjugate of A is $\bar{A}$ or A^*.
4. $A + B = (a_1 + ia_2) + (b_1 + ib_2) = (a_1 + b_1) + i(a_2 + b_2)$.
5. $a_1 + ia_2 = |A|(\cos\angle A + i\sin\angle A) = |A|e^{i\angle A}$
   $a_1 - ia_2 = |A|(\cos\angle A - i\sin\angle A) = |A|e^{-i\angle A}$
   where $|A| = \sqrt{a_1^2 + a_2^2}$, $\sin\angle A = a_2/|A|$, $\cos\angle A = a_1/|A|$, $|A|$ is the *absolute value* (*modulus*), and $\angle A$ is the *angle* of A.
6. $AB = (a_1 + ia_2)(b_1 + ib_2) = (a_1 b_1 - a_2 b_2) + i(a_2 b_1 + a_1 b_2) = |A||B|e^{i(\angle A + \angle B)}$.
7. $A\bar{A} = (a_1 + ia_2)(a_1 - ia_2) = a_1^2 + a_2^2 = |A|^2$.
8. $\dfrac{A}{B} = \dfrac{a_1 + ia_2}{b_1 + ib_2} = \dfrac{(a_1 + ia_2)(b_1 - ib_2)}{(b_1 + ib_2)(b_1 - ib_2)} = \dfrac{a_1 b_1 + a_2 b_2}{b_1^2 + b_2^2} + i\dfrac{a_2 b_1 - a_1 b_2}{b_1^2 + b_2^2}$
   $= \dfrac{|A|}{|B|}e^{i(\angle A - \angle B)}$.

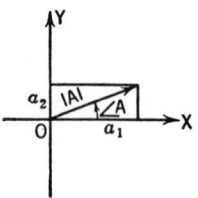

Fig. 2.163

9. $A^n = (a_1 + ia_2)^n = [|A|(\cos \angle A + i \sin \angle A)]^n = |A|^n e^{in\angle A}$
$\qquad = |A|^n(\cos n\angle A + i \sin n\angle A)$.
$\bar{A}^n = (a_1 - ia_2)^n = [|A|(\cos \angle A - i \sin \angle A)]^n = |A|^n e^{-in\angle A}$
$\qquad = |A|^n(\cos n\angle A - i \sin n\angle A)$.

10. $\sqrt[n]{A} = \sqrt[n]{a_1 + ia_2} = \sqrt[n]{|A|}\left(\cos\dfrac{\angle A + 2k\pi}{n} + i \sin\dfrac{\angle A + 2k\pi}{n}\right)$
$\qquad = R|A|n e^{i\angle A + 2k\pi/n}$

where k is an integer. For $k = 0, 1, 2, \ldots, n - 1$, all of the n roots are obtained.

2.14.2 Complex Variables

Analytic Functions of a Complex Variable. A function $w = f(z)$, $z = x + iy$, which has a derivative

$$\frac{df}{dz} = f'(z) = \lim_{h \to 0} \frac{f(z + h) - f(z)}{h}$$

at a point z independent of the manner of approach of $z + h$ to z is *analytic* at z and may be expanded in a convergent power series there. A function that is analytic at every point of a region is *analytic in the region*. If $f(z) = w = u(x, y) + iv(x, y)$ and $f(z)$ is analytic at z, then the Cauchy–Riemann differential equations

$$\frac{\partial u}{\partial x} = \frac{\partial v}{\partial y}, \quad \frac{\partial u}{\partial y} = -\frac{\partial v}{\partial x}$$

hold at z. If in a neighborhood of a point z these four partial derivatives exist and are continuous and if the Cauchy–Riemann equations hold, then $f(z)$ is analytic at z. The functions u and v satisfy Laplace's equation

$$\frac{\partial^2 \phi}{\partial x^2} + \frac{\partial^2 \phi}{\partial y^2} = 0$$

Example 2.95. Examples of analytic functions are: $z, 1/z, e^z$, and $\sin z$. An example of a nonanalytic function is $w = x - iy$.

Conformal Mapping. The function $w = f(z)$, analytic in a region R_z of the z plane, *conformally* maps each point in R_z on a point of the w plane in the region R_w if $f'(z) \neq 0$ at all points of R_z. This mapping is also *isogonal*, that is, the angle between two curves starting at z_0 is equal to the angle between their mapped curves starting at w_0.

Example 2.96. 1. $w = z + b$, b complex, is a *translation* of magnitude $|b|$ in the direction $\angle b$.
2. $w = az$, a complex, is a *rotation* through $\angle a$ and a *magnification* by $|a|$.
3. $w = az + b$, a, b complex, the *integral linear transformation*, is a combination of 1 and 2.
4. $w = 1/z$, the *inversion transformation*, carries the origin of the z plane into the point at infinity in the *enlarged* w plane.
5. $w = (az + b)/(cz + d)$, $ad - bc \neq 0$, the *general linear* or *bilinear transformation*, can be resolved into two linear integral and one inversion transformations.

Integrals of Analytic Functions. If $dF(z)/dz = f(z)$ in a simply connected region R_z, then $F(z) = \int f(z)\, dz$ is analytic throughout R_z. If $f_1(z), f_2(z)$ are analytic in R_z and the path of integration is in R_z, then

1. $\int_{z_0}^{z_0} f_1(z)\, dz = 0$
2. $\int_{z_0}^{z_1}[k_1 f_1(z) + k_2 f_2(z)]\, dz = k_1 \int_{z_0}^{z_1} f_1(z)\, dz + k_2 \int_{z_0}^{z_1} f_2(z)\, dz$
3. $\int_{z_1}^{z_0} f_1(z)\, dz = -\int_{z_0}^{z_1} f_1(z)\, dz$
4. $\int_{z_0}^{z_1} f_1(z)\, dz + \int_{z_1}^{z_2} f_1(z)\, dz = \int_{z_0}^{z_2} f_1(z)\, dz$

Cauchy's Integral Theorem. If $f(z)$ is analytic and single valued on and within a simple closed contour C, then $\int_C f(z)\, dz = 0$.
A *contour* is a continuous curve made up of a finite number of elementary arcs.

If $f(z)$ is continuous on a simple closed contour C and analytic in the region bounded by C, then *Cauchy's integral formula*

$$f(z) = \frac{1}{2\pi i} \int_C \frac{f(\zeta)}{\zeta - z} \, d\zeta$$

holds; also

$$f^{(n)}(z) = \frac{n!}{2\pi i} \int_C \frac{f(\zeta)}{(\zeta - z)^{n+1}} \, d\zeta$$

Laurent Series. A function $f(z)$ has a *zero of order* n at z_1 if it can be put in the form $f(z) = (z - z_1)^n f_1(z)$, n a positive integer, $f_1(z_1) \neq 0$. A function $f(z)$ has a *pole of order* n at z_1 if it can be put in the form $f(z) = f_2(z)/(z - z_1)^n$, n a positive integer, $f_2(z_1) \neq 0$.

If $f(z)$ is analytic in a ring between and on two concentric circles C_1 and C_2 with radii R_1 and R_2, $R_1 < R_2$, and center z_1, then the *Laurent series*

$$f(z) = \sum_{n=-\infty}^{\infty} c_n (z - z_1)^n \qquad R_1 < |z - z_1| < R_2$$

is convergent everywhere in the ring, and

$$c_n = \frac{1}{2\pi i} \int_C \frac{f(z)}{(z - z_1)^{n+1}} \, dz$$

C is circle $|z - z_1| = r$, $R_1 < r < R_2$.

If a single-valued analytic function $f(z)$ is expanded in a Laurent series in the neighborhood of an isolated singularity z_1, then the *residue* of $f(z)$ at z_1 is

$$c_{-1} = \frac{1}{2\pi i} \int_C f(z) \, dz$$

C is any circle with center at z_1 that excludes all other singularities of $f(z)$.

A function is *meromorphic* in a region if it is analytic in the region except for a finite number of poles. If $f(z)$ is analytic on and inside a contour C, except for a finite number of poles, and has no zeros on C, then

$$\frac{1}{2\pi i} \int_C \frac{f'(z)}{f(z)} \, dz = N - P$$

N the total order of the zeros and P the total order of the poles within the contour.

2.15   VECTOR ANALYSIS

2.15.1   Vector Algebra

A *scalar* is a quantity that has magnitude, such as mass, density, and temperature. A *vector* is a quantity that has magnitude and direction, such as force, velocity, and acceleration. A vector may be represented geometrically by an oriented line segment.

Two vectors A and B are equal if they have the same magnitude and direction. A vector may be displaced parallel to itself provided it retains the same magnitude and direction. A vector having the same magnitude but direction opposite to that of A is the negative of A and is written $-A$. If A is a vector of magnitude, or length, a, then $|A| = a$. A vector parallel to A but with magnitude equal to the reciprocal of the magnitude of A is written $A^{-1} = 1/A$. A unit vector $A/|A|$ ($A \neq 0$) has the direction of A and magnitude 1.

The **sum** of two vectors A and B is $A + B$ (Fig. 2.164). Similarly, the sum of three or more vectors can be found by adding them end to end.

The sum of A and $-B$ is $A - B$ (Fig. 2.164), the *difference* of two vectors.

Let A, B, C be vectors and p, q scalars.

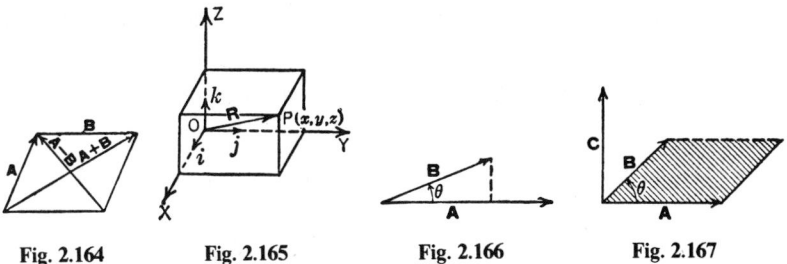

Fig. 2.164   Fig. 2.165   Fig. 2.166   Fig. 2.167

$p\mathbf{A} = \mathbf{A}p$, a vector p times as long as $\mathbf{A}$ with the same direction as $\mathbf{A}$ if p is positive and opposite if p is negative.

$$(p + q)\mathbf{A} = p\mathbf{A} + q\mathbf{A} \qquad p(\mathbf{A} + \mathbf{B}) = p\mathbf{A} + p\mathbf{B}$$
$$\mathbf{A} + \mathbf{B} = \mathbf{B} + \mathbf{A}$$
$$\mathbf{A} + (\mathbf{B} + \mathbf{C}) = (\mathbf{A} + \mathbf{B}) + \mathbf{C}$$

$|\mathbf{A} + \mathbf{B}| \leq |\mathbf{A}| + |\mathbf{B}|$, where the equality sign holds only for $\mathbf{A}$ parallel to $\mathbf{B}$.

Rectangular Coordinates. Figure 2.165 shows a right-hand coordinate system. Let $\mathbf{i}, \mathbf{j}, \mathbf{k}$ be unit vectors with the directions OX, OY, OZ, respectively. The vector $\mathbf{R}$ with initial point O and end point $P(x, y, z)$ can be expressed as the sum of its components

$$\mathbf{R} = \mathbf{i}x + \mathbf{j}y + \mathbf{k}z$$

If $\mathbf{A} = \mathbf{i}a_1 + \mathbf{j}a_2 + \mathbf{k}a_3$ and $\mathbf{B} = \mathbf{i}b_1 + \mathbf{j}b_2 + \mathbf{k}b_3$, then

$$\mathbf{A} + \mathbf{B} = \mathbf{i}(a_1 + b_1) + \mathbf{j}(a_2 + b_2) + \mathbf{k}(a_3 + b_3)$$

The **scalar, inner, or dot product** of two vectors $\mathbf{A}$ and $\mathbf{B}$ is $\mathbf{A} \cdot \mathbf{B} = |\mathbf{A}||\mathbf{B}|\cos \theta$ (Fig. 2.166).

$$\mathbf{A} \cdot \mathbf{B} = \mathbf{B} \cdot \mathbf{A}$$
$$\mathbf{A} \cdot (\mathbf{B} + \mathbf{C}) = \mathbf{A} \cdot \mathbf{B} + \mathbf{A} \cdot \mathbf{C}$$
$$\mathbf{A} \cdot \mathbf{A} = A^2 = |\mathbf{A}|^2$$
$$\mathbf{i} \cdot \mathbf{i} = \mathbf{j} \cdot \mathbf{j} = \mathbf{k} \cdot \mathbf{k} = 1$$
$$\mathbf{i} \cdot \mathbf{j} = \mathbf{j} \cdot \mathbf{k} = \mathbf{k} \cdot \mathbf{i} = 0$$

If $\mathbf{A} \cdot \mathbf{B} = 0$, then either $\mathbf{A} = 0$, $\mathbf{B} = 0$, or $\mathbf{A}$ is perpendicular to $\mathbf{B}$.

If $\mathbf{A} = \mathbf{i}a_1 + \mathbf{j}a_2 + \mathbf{k}a_3$ and $\mathbf{B} = \mathbf{i}b_1 + \mathbf{j}b_2 + \mathbf{k}b_3$, then $\mathbf{A} \cdot \mathbf{B} = a_1b_1 + a_2b_2 + a_3b_3$.

The **vector, restricted outer, or cross product** of two vectors $\mathbf{A}$ and $\mathbf{B}$ is $\mathbf{A} \times \mathbf{B} = \mathbf{C}$, where $\mathbf{C}$ is perpendicular to the plane of $\mathbf{A}$ and $\mathbf{B}$ with the magnitude $|\mathbf{C}| = |\mathbf{A}||\mathbf{B}|\sin \theta$ (the area of the parallelogram made by $\mathbf{A}$ and $\mathbf{B}$, Fig. 2.167) and so directed that a right-hand rotation of less than 180° carries $\mathbf{A}$ into $\mathbf{B}$.

$$\mathbf{A} \times \mathbf{B} = -\mathbf{B} \times \mathbf{A}$$
$$\mathbf{A} \times (\mathbf{B} + \mathbf{C}) = \mathbf{A} \times \mathbf{B} + \mathbf{A} \times \mathbf{C}$$
$$(\mathbf{B} + \mathbf{C}) \times \mathbf{A} = \mathbf{B} \times \mathbf{A} + \mathbf{C} \times \mathbf{A}$$
$$\mathbf{i} \times \mathbf{i} = \mathbf{j} \times \mathbf{j} = \mathbf{k} \times \mathbf{k} = 0$$
$$\mathbf{i} \times \mathbf{j} = \mathbf{k} = -\mathbf{j} \times \mathbf{i}$$
$$\mathbf{j} \times \mathbf{k} = \mathbf{i} = -\mathbf{k} \times \mathbf{j}$$
$$\mathbf{k} \times \mathbf{i} = \mathbf{j} = -\mathbf{i} \times \mathbf{k}$$

If $\mathbf{A} \times \mathbf{B} = 0$, then either $\mathbf{A} = 0$, $\mathbf{B} = 0$, or $\mathbf{A}$ is parallel to $\mathbf{B}$.

If $\mathbf{A} = \mathbf{i}a_1 + \mathbf{j}a_2 + \mathbf{k}a_3$ and $\mathbf{B} = \mathbf{i}b_1 + \mathbf{j}b_2 + \mathbf{k}b_3$, then

$$\mathbf{A} \times \mathbf{B} = \begin{vmatrix} \mathbf{i} & \mathbf{j} & \mathbf{k} \\ a_1 & a_2 & a_3 \\ b_1 & b_2 & b_3 \end{vmatrix}$$

If $A = ia_1 + ja_2 + ka_3$, $B = ib_1 + jb_2 + kb_3$, $C = ic_1 + jc_2 + kc_3$, then the *scalar triple product*

$$A \cdot (B \times C) = (A \times B) \cdot C = B \cdot (C \times A) = (ABC) = \begin{vmatrix} a_1 & a_2 & a_3 \\ b_1 & b_2 & b_3 \\ c_1 & c_2 & c_3 \end{vmatrix}$$

and is equal to the volume of a parallelepiped whose three determining edges are A, B, C.

$$(A \times B) \times C = (A \cdot C)B - (B \cdot C)A = -C \times (A \times B)$$
$$(A \times B) \cdot (C \times D) = (A \cdot C)(B \cdot D) - (A \cdot D)(B \cdot C)$$

2.15.2 Differentiation and Integration of Vectors

Differentiation. A vector function of one or more scalar variables is called a *variable vector* or *field vector*. The derivative is

$$\frac{dF}{dt} = F'(t) = \lim_{\Delta t \to 0} \frac{F(t + \Delta t) - F(t)}{\Delta t} = \lim_{\Delta t \to 0} \frac{\Delta F}{\Delta t} \quad \text{(Fig. 2.168)}$$

If the length of F remains unaltered, then $F \cdot dF = 0$. If the direction of F remains unaltered, then $F \times dF = 0$.

$$d(A + B) = dA + dB$$
$$d(A \cdot B) = A \cdot dB + B \cdot dA$$
$$d(A \times B) = dA \times B + A \times dB = A \times dB - B \times dA$$
$$d(A \cdot B \cdot C) = A \cdot B \cdot dC + B \cdot C \cdot dA + C \cdot A \cdot dB$$

The Derivative Operators

$$\nabla = \text{del} = i\frac{\partial}{\partial x} + j\frac{\partial}{\partial y} + k\frac{\partial}{\partial z}$$

$$\nabla^2 = \text{Laplacian} = \frac{\partial^2}{\partial x^2} + \frac{\partial^2}{\partial y^2} + \frac{\partial^2}{\partial z^2}$$

If V is a scalar function, then

$$\nabla V = \text{grad } V = i\frac{\partial V}{\partial x} + j\frac{\partial V}{\partial y} + k\frac{\partial V}{\partial z}$$

If A is a vector function with components A_x, A_y, A_z, then

$$\nabla \cdot A = \text{div } A = \frac{\partial A_x}{\partial x} + \frac{\partial A_y}{\partial y} + \frac{\partial A_z}{\partial z}$$

$$\nabla \times A = \text{curl } A = \text{rot } A = \begin{vmatrix} i & j & k \\ \dfrac{\partial}{\partial x} & \dfrac{\partial}{\partial y} & \dfrac{\partial}{\partial z} \\ A_x & A_y & A_z \end{vmatrix}$$

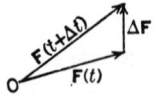

Fig. 2.168

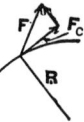

Fig. 2.169

Formulas for Differentiation. Let U and V be scalar functions and $\mathbf{A}$ and $\mathbf{B}$ be vector functions of x, y, z. Then (see: *Advanced Engineering Analysis* by J. N. Reddy and M. L. Rasmussen, Wiley, 1982.):

$$\nabla(U + V) = \nabla U + \nabla V \qquad \nabla \cdot (\mathbf{A} + \mathbf{B}) = \nabla \cdot \mathbf{A} + \nabla \cdot \mathbf{B} \qquad \nabla \times (\mathbf{A} + \mathbf{B}) = \nabla \times \mathbf{A} + \nabla \times \mathbf{B}$$

$$\nabla(UV) = V\nabla U + U\nabla V \qquad \nabla \cdot (U\mathbf{A}) = U\nabla \cdot \mathbf{A} + \mathbf{A} \cdot \nabla U$$

$$\nabla \times (U\mathbf{A}) = \nabla U \times \mathbf{A} + U\nabla \times \mathbf{A}$$

$$\nabla \cdot (\mathbf{A} \times \mathbf{B}) = \mathbf{B} \cdot \nabla \times \mathbf{A} - \mathbf{A} \cdot \nabla \times \mathbf{B}$$

$$\nabla(\mathbf{A} \cdot \mathbf{B}) = \mathbf{A} \cdot \nabla\mathbf{B} + \mathbf{B} \cdot \nabla\mathbf{A} + \mathbf{A} \times (\nabla \times \mathbf{B}) + \mathbf{B} \times (\nabla \times \mathbf{A})$$

$$\nabla \times (\mathbf{A} \times \mathbf{B}) = \mathbf{B} \cdot \nabla\mathbf{A} - \mathbf{A} \cdot \nabla\mathbf{B} + \mathbf{A}(\nabla \cdot \mathbf{B}) - \mathbf{B}(\nabla \cdot \mathbf{A})$$

$$\nabla \times (\nabla \times \mathbf{A}) = \nabla(\nabla \cdot \mathbf{A}) - \nabla^2\mathbf{A}$$

$$\nabla \cdot (\nabla \times \mathbf{A}) = 0$$

$$\nabla \times (\nabla U) = 0$$

If $\mathbf{R} = \mathbf{i}x + \mathbf{j}y + \mathbf{k}z$ (Fig. 2.165), then

$$\nabla \cdot \mathbf{R} = 3 \qquad \nabla \times \mathbf{R} = 0 \qquad \mathbf{A} \cdot \nabla|\mathbf{R}| = |\mathbf{A}| \qquad \nabla \cdot \frac{1}{|\mathbf{R}|} = -\frac{\mathbf{R}}{|\mathbf{R}|^3} \qquad \nabla^2\frac{1}{|\mathbf{R}|} = 0$$

Integration. The line integral of a vector $\mathbf{F}$ along a curve AB denotes the integral of the tangential component of the vector along the curve; thus

$$\int_A^B \mathbf{F} \cdot d\mathbf{R} = \int_A^B |\mathbf{F}_c|\, ds \quad \text{(Fig. 2.169)}$$

where $d\mathbf{R} = \mathbf{i}\, dx + \mathbf{j}\, dy + \mathbf{k}\, dz$.

If $\mathbf{F} = \nabla U$ is the gradient of a single-valued continuous function $U(x, y, z)$ the line integral of $\mathbf{F}$ depends only on the end points. Conversely, if $\mathbf{F}(x, y, z)$ is continuous and $\int_C \mathbf{F} \cdot d\mathbf{R} = 0$ for any closed path C in a three-dimensional region, there is a function $U(x, y, z)$ such that $\mathbf{F} = \nabla U$.

2.15.3   Theorems and Formulas

Let $\mathbf{n}$ be the vector of unit length perpendicular to a surface at a point P and extending on the positive side (the outward normal); dS, the element of surface, and dv, the element of volume.

The Divergence (Gauss) Theorem. If a field vector $\mathbf{F}$ and its first derivatives are continuous at all points in a region of volume v bounded by a closed elementary surface S, then

$$\iint_S \mathbf{F} \cdot \mathbf{n}\, dS = \iiint_v \nabla \cdot \mathbf{F}\, dv$$

Stokes' Theorem. If a field vector $\mathbf{F}$ and its first derivatives are continuous at all points in a region of area S bounded by a closed curve C, then

$$\iint_S \nabla \times \mathbf{F} \cdot \mathbf{n}\, dS = \int_C \mathbf{F} \cdot d\mathbf{R}$$

Green's Theorem. Under the conditions of the divergence theorem,

$$\iint_S \mathbf{n} \cdot U \nabla V \, dS = \iiint_v U \nabla^2 V \, dv + \iiint_v (\nabla U \cdot \nabla V) \, dv$$

$$\iint_S \mathbf{n} \cdot (U \nabla V - V \nabla U) \, dS = \iiint_v (U \nabla^2 V - V \nabla^2 U) \, dv$$

Cylindrical Coordinates

$$x = r \cos \theta \qquad y = r \sin \theta \qquad z = z$$

The element of volume $dv = r \, dr \, d\theta \, dz$. The unit vectors $\mathbf{u}_r, \mathbf{u}_\theta, \mathbf{u}_z$ are perpendicular to each other.

$$\operatorname{grad} V = \nabla V = \frac{\partial V}{\partial r} \mathbf{u}_r + \frac{1}{r} \frac{\partial V}{\partial \theta} \mathbf{u}_\theta + \frac{\partial V}{\partial z} \mathbf{u}_z$$

$$\operatorname{div} \mathbf{F} = \nabla \cdot \mathbf{F} = \frac{1}{r} \frac{\partial}{\partial r} (r \mathbf{F}_r) + \frac{1}{r} \frac{\partial}{\partial \theta} (\mathbf{F}_\theta) + \frac{\partial}{\partial z} (\mathbf{F}_z)$$

$$\operatorname{curl} \mathbf{F} = \nabla \times \mathbf{F} = \begin{vmatrix} \dfrac{\mathbf{u}_r}{r} & \mathbf{u}_\theta & \dfrac{\mathbf{u}_z}{r} \\[2mm] \dfrac{\partial}{\partial r} & \dfrac{\partial}{\partial \theta} & \dfrac{\partial}{\partial z} \\[2mm] \mathbf{F}_r & r\mathbf{F}_\theta & \mathbf{F}_s \end{vmatrix}$$

$$\nabla^2 V = \frac{1}{r} \frac{\partial V}{\partial r} + \frac{\partial^2 V}{\partial r^2} + \frac{1}{r^2} \frac{\partial^2 V}{\partial \theta^2} + \frac{\partial^2 V}{\partial z^2}$$

Spherical Coordinates

$$x = r \cos \phi \sin \theta \qquad y = r \sin \phi \sin \theta \qquad z = r \cos \theta$$

The unit vectors $\mathbf{u}_r, \mathbf{u}_\phi, \mathbf{u}_\theta$ are perpendicular to each other.

$$\operatorname{grad} V = \nabla V = \frac{\partial V}{\partial r} \mathbf{u}_r + \frac{1}{r \sin \theta} \frac{\partial V}{\partial \phi} \mathbf{u}_\phi + \frac{1}{r} \frac{\partial V}{\partial \theta} \mathbf{u}_\theta$$

$$\operatorname{div} \mathbf{F} = \nabla \cdot \mathbf{F} = \frac{1}{r^2} \frac{\partial}{\partial r} (r^2 \mathbf{F}_r) + \frac{1}{r \sin \theta} \frac{\partial \mathbf{F}_\phi}{\partial \phi} + \frac{1}{r \sin \theta} \frac{\partial}{\partial \phi} (\sin \theta \, \mathbf{F}_\theta)$$

$$\operatorname{curl} \mathbf{F} = \nabla \times \mathbf{F} = \begin{vmatrix} \dfrac{\mathbf{u}_r}{r^2 \sin \theta} & \dfrac{\mathbf{u}_\theta}{r \sin \theta} & \dfrac{\mathbf{u}_\phi}{r} \\[2mm] \dfrac{\partial}{\partial r} & \dfrac{\partial}{\partial \theta} & \dfrac{\partial}{\partial \phi} \\[2mm] \mathbf{F}_r & r\mathbf{F}_\theta & r \sin \theta \, \mathbf{F}_\phi \end{vmatrix}$$

$$\nabla^2 V = \frac{1}{r^2} \frac{\partial}{\partial r} \left(r^2 \frac{\partial V}{\partial r} \right) + \frac{1}{r^2 \sin^2 \theta} \frac{\partial^2 V}{\partial \phi^2} + \frac{1}{r^2 \sin \theta} \frac{\partial}{\partial \theta} \left(\sin \theta \frac{\partial V}{\partial \theta} \right)$$

Solid Angle. The lines joining the point P to points of a surface S generate a solid angle. If a is the area intercepted by these lines on a sphere of center P with radius r, then

$$\omega = \frac{a}{r^2}$$

is the measure of the solid angle (Fig. 2.170). If S is a surface that does not pass through P, $\cos \theta$ is nowhere zero, and $\mathbf{n}$ is everywhere continuous on the surface, then

$$\omega = \int_S \frac{\mathbf{R} \cdot \mathbf{n} \, da}{r^3}$$

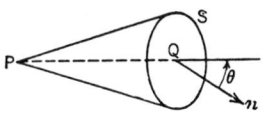

Fig. 2.170

where $\mathbf{R} = PQ$. If S forms the complete boundary of a three-dimensional region, the total solid angle subtended by S at P is zero if P lies outside the region, and 4π if P lies inside the region.

BIBLIOGRAPHY

Beaumont, R., and Pierce, R., *The Algebraic Foundations of Mathematics*, Reading, Mass., Addison-Wesley, 1963.

Beckenbach, E., and Bellman, R., *Introduction to Inequalities*, New York, Random House, 1962.

Becker, E. B., Carey, G. F., and Oden, J. T., *Finite Elements: An Introduction*, Englewood Cliffs, N.J., Prentice-Hall, 1981.

Boole, G., *An Investigation of the Laws of Thoughts*, New York, Dover, 1951.

Brebbia, C. A., and Connor, J. J., *Fundamentals of Finite Element Techniques for Structural Engineers*, London, Butterworth, 1975.

Carnahan, B., Luther, H. A., and Wilkes, J. O., *Applied Numerical Methods*, New York, Wiley, 1969.

Carrier, G. F., and Pearson, C. E., *Ordinary Differential Equations*, Waltham, Mass., Blaisdell, 1968.

Chapra, S. C., and Canale, R. P., *Numerical Methods for Engineers with Personal Computer Applications*, New York, McGraw-Hill, 1985.

Churchill, R. V., *Fourier Series and Boundary Values Problems*, New York, McGraw-Hill, 1941.

Churchill, R. V., *Complex Variables with Applications*, New York, McGraw-Hill, 1960.

Collatz, L., *The Numerical Treatment of Numerical Differential Equations*, Berlin, Springer-Verlag, 1960.

Courant, R., and Hilbert, D., *Methods of Mathematical Physics*, Vols. I and II, New York, Wiley, 1968.

Davis, A. S., *The Finite Element Method, a First Approach*, Oxford, Clarendon, 1980.

Feller, W., *An Introduction to Probability Theory and Its Applications*, New York, Wiley, 1968.

Forsyth, A. R., *Theory of Differential Equations*, New York, Cambridge University Press, 1906, Dover, 1959.

Gantmacher, F. R., *The Theory of Matrices*, New York, Chelsea, 1959.

Garabedian, P., *Partial Differential Equations*, New York, Wiley, 1964.

Hashisaki, J., and Peterson, J., *Theory of Arithmetic*, New York, Wiley, 1971.

Hildebrand, F. B., *Introduction to Numerical Analysis*, New York, McGraw-Hill, 1974.

Hildebrand, F. B., *Methods of Applied Mathematics*, 2nd ed., Englewood Cliffs, N.J., Prentice-Hall, 1965.

Hinton, E., and Owen, D. R. J., *An Introduction to Finite Element Computations*, Swansea, Pineridge Press, 1979.

Householder, A. S., *The Theory of Matrices in Numerical Analysis*, New York, Blaisdell, 1964.

Isaacson, E., and Keller, H. B., *Analysis of Numerical Methods*, New York, Wiley, 1966.

Leeds, H. D., and Weinberg, G. M., *Computer Programming Fundamentals*, New York, McGraw-Hill, 1961.

McCormick, E. M., *Digital Computer Primer*, New York, McGraw-Hill, 1949.

McCracken, D. D., *A Guide to FORTRAN IV Programming*, New York, Wiley, 1965.

McCrea, W., *Analytic Geometry of Three Dimensions*, New York, Interscience, 1948.

Morse, P. M., and Feshbach, H., *Methods of Theoretical Physics*, New York, McGraw-Hill, 1953.

Noble, B., *Applied Linear Algebra*, Englewood-Cliffs, N.J., Prentice-Hall, 1969.

Olmsted, J., *The Real Number System*, New York, Appleton-Century-Crofts, 1962.

Pearson, C. E., *Handbook of Applied Mathematics*, New York, Van Nostrand Reinhold, 1974.

Pipes, L. A., and Harvill, L. R., *Applied Mathematics for Engineers and Physicists*, 3rd ed., New York, McGraw-Hill, 1970.

Ralston, A., and Wilf, H. S., *Mathematical Methods for Digital Computers*, New York, Wiley, 1967.

Reddy, J. N., *An Introduction to the Finite Element Method*, New York, McGraw-Hill, 1984.

Reddy, J. N., *Applied Functional Analysis and Variational Methods in Engineering*, New York, McGraw-Hill, 1986.

Reddy, J. N., and Rasmussen, M. L., *Advanced Engineering Analysis*, New York, Wiley, 1982.

Reddy, J. N., *Energy and Variational Methods in Applied Mechanics*, (with an Introduction to the Finite Element Method), New York, Wiley, 1984.

Rice, J. R., *Numerical Methods, Software and Analysis*, New York, McGraw-Hill, 1983.

Salmon, G., *A Treatise on Conic Sections*, New York, Chelsea, 1954.

Titscmarsh, E. C., *Theory of Fourier Integrals*, 2nd Ed., New York, Oxford University Press, 1948.

Whittaker, E. T., and Watson, G. N., *Modern Analysis*, New York, Cambridge University Press, 1946.

Wilks, S. S., *Mathematical Statistics*, New York, Wiley, 1962.

Zienkiewicz, O. C., *The Finite Element Method in Engineering Science*, London, McGraw-Hill, 1969.

CHAPTER 3

MECHANICS OF RIGID BODIES

Wallace Fowler

Department of Aerospace Engineering and Engineering Mechanics
The University of Texas at Austin
Austin, Texas

3.1 DEFINITIONS

Mechanics is that branch of *science* that treats of forces and motion.

Statics is that branch of *mechanics* that deals with the equilibrium of forces on bodies *at rest* (or moving at a uniform velocity in a straight line).

Kinematics is that branch of *mechanics* that deals with the motion of bodies without consideration of the character of the bodies or of the influence of forces upon their motion. It considers only concepts of *geometry* and *time*.

Kinetics (or **dynamics**) is that branch of *mechanics* that deals with the effect of unbalanced external forces in *modifying the motion* of bodies.

Mass and weight, in the *gravitational system of units* employed by English engineers, are related by the formula $W = Mg$, where W = weight, M = mass, and g = acceleration due to gravity. For a thorough discussion of these terms, see Chapter 1.

Force is that which changes or tends to change the state of rest or motion of a body.

Inertia is that property of a body by virtue of which it tends to continue in the state of rest or motion in which it may be placed, until acted on by some force.

Reaction is that *equal and opposite force* exerted by a body in opposing another force acting upon it.

Newton's Laws of Motion

First Law. If a body is at rest, it will remain at rest, or if in motion, it will move uniformly in a straight line, until acted on by some force.

Second Law. The time rate of change of the linear momentum of a particle (defined as the product of its mass and its acceleration, a vector) is proportional to the force (a vector) acting on the particle.

Third Law. If a force acts to change the state of a body with respect to rest or motion, the body will offer a resistance equal and directly opposed to the force. Or, to every action there is opposed an equal and opposite reaction.
Special terms such as **hydrostatics, aerodynamics,** and so on are used to denote the theory of statics as applied to *liquid bodies*, the theory of dynamics as applied to *gaseous bodies*, and so on. **Mechanics of materials** considers, in addition to *external forces*, the *internal forces* or *stresses* between molecules of a body. Subjects of these types are covered in other sections of this handbook. The present section on *mechanics* is confined, in general, to the discussion of motion of, and *external forces* applied to, *rigid bodies*.

3.2 STATICS

3.2.1 Graphical Representation and Classification of Forces

Graphical Representation of Force
A **force** is completely specified by its *magnitude*, *direction*, and *point of application*. The word **sense** as applied to a force refers to one of the two directions along the line of action of the force. The effect of any force applied to a rigid body at rest is the same, no matter where in its own line of action the force is applied. This is known as the principle of the **transmissibility of force**. A force may be represented graphically in magnitude and direction by a straight line drawn parallel to its line of action, the length being proportional to the magnitude of the force; its sense is indicated by an arrowhead placed on the line. The English engineers' unit of force is the pound, or the earth's pull on a mass of 1 lb. A drawing that indicates the lines of action of the various forces acting on a machine or structure is called a **space diagram**; one in which vectors are drawn to represent the magnitudes and directions of the forces is a **vector diagram**. A force is indicated on a space diagram by two lowercase letters placed on opposite sides of the line of action of the force; the vector, representing its magnitude and direction, by the same capital letters placed at the ends. Thus, in Fig. 3.1, *AB* represents the magnitude and direction of the force *W*, and *ab* its action line. The vector being read as *AB* indicates a downward sense; read as *BA*, an upward sense.

Classification of Systems of Forces
A **system of forces** consists of any number of forces taken collectively.
Classification of systems of forces is made according to the arrangement of their **action lines**. If the action lines lie in the same plane the system is **coplanar**, otherwise **noncoplanar**. If they pass through the same point the system is **concurrent**, otherwise **nonconcurrent**. If two or more forces have the same action line, they are **collinear**. A system of two equal forces, parallel, opposite in sense, and having different action lines is a **couple**. Two or more forces equivalent to a single force are components of the single force. **Resolution** is the operation of replacing a single force by a system of components. The single force is the **resultant** of its components. In general, the resultant of a system of forces is the simplest equivalent system. This may be a *single force*, a *single couple*, or a *noncoplanar force and couple* (or *two skewed forces*). When the resultant is a single force the equilibrant is a force equal in magnitude, having the same line of action but opposite sense. **Composition** is the operation of replacing a system of forces by its resultant.

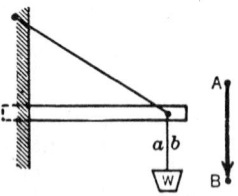

Fig. 3.1 Space diagram.

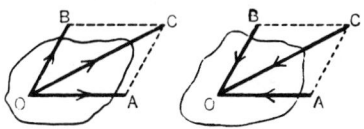

Fig. 3.2 Concurrent forces.

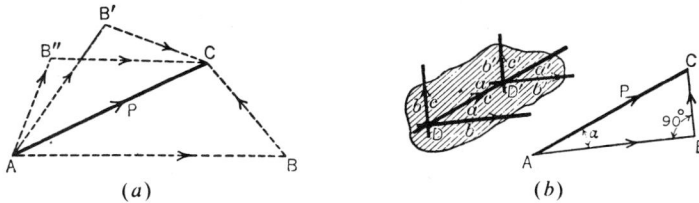

(a)   (b)

Fig. 3.3 Force components.

3.2.2 Addition and Resolution of Concurrent Forces

Addition of Two Concurrent Forces*

Parallelogram Law. If magnitudes, lines of action, and senses of two concurrent forces acting on a rigid body are represented by OA and OB (Fig. 3.2), the magnitude, line of action, and sense of their resultant are represented by the diagonal OC of the parallelogram $OABC$. The points of application of the forces may be anywhere on the body in the lines OA, OB, and OC, or their extensions. The arrowheads on the lines OA, OB, and OC all point toward or all away from the point of concurrence O.

Triangle Law. The triangle law for graphical vector addition of two concurrent forces follows directly from the parallelogram law. In the parallelogram law, vectors are drawn tail to tail and a parallelogram is created. If either half of the parallelogram in Fig. 3.2 (either above or below line OC, with each "half" containing line OC) is examined, it can be seen that each triangle thus considered contains both of the forces being composed (added), laid out in a head-to-tail sequence.

Resolution into Two Concurrent Forces
A force may be resolved into a infinite number of pairs of components by constructing triangles with the given force as one of the sides. In each such triangle, the other two sides of the triangle are the components of the given force in the directions defined by the other two sides of the triangle, as shown in Fig. 3.3a. Of special interest are those cases in which the given force is the hypotenuse of a right triangle, in which case the sides of the triangle are rectangular components of the given force, as shown in Fig. 3.3b. Algebraically, the rectangular components of the force P in Fig. 3.3b in the x and y directions are given by $P_x = P \cos \alpha$ and $P_y = P \sin \alpha$.

Addition of more than Two Coplanar Concurrent Forces

Graphic Method. In Fig. 3.4, consider body G acted on by the four forces shown. Construct a **force polygon** as follows: Plot AB parallel to ab, and scale it to represent 60 lb; from B plot BC parallel to bc, and scale it to represent 80 lb; in like manner plot CD and DE, so that the arrows lead *confluently* from A to E. The resultant of the system is AE in magnitude and sense and equals 114 lb. Its action line is ae. The resultant will be the same regardless of the order in which the forces are plotted. Note particularly that the resultant is not confluent with the component forces.

Algebraic Method. Choose rectangular axes OX and OY. Referring to Fig. 3.4 resolve each force into its x and y components, considering components acting upward or to the right as positive, and those acting downward or to the left as negative. Arrange the results in tabular form, placing the forces in the first column, the x components in the second, and the y components in the third. ΣF_x = algebraic sum of x components, and ΣF_y = algebraic sum of y components.

---

*It is evident that a pair of concurrent forces and their resultant are necessarily coplanar.

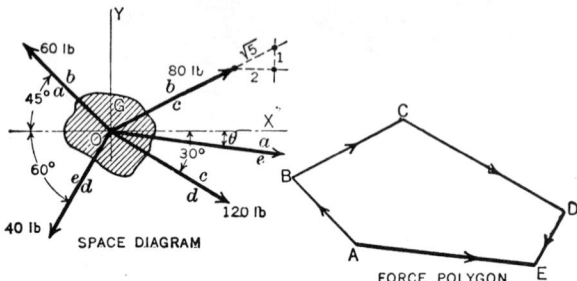

SPACE DIAGRAM

FORCE POLYGON

Fig. 3.4   Force polygon.

F (lb)	F_x (lb)	F_y (lb)
$ab = 60$	$-60 \times 0.707 = -42.4$	$+60 \times 0.707 = +42.4$
$bc = 80$	$+80 \times 2/\sqrt{5} = +71.4$	$+80 \times 1/\sqrt{5} = +35.7$
$cd = 120$	$+120 \times 0.866 = +104$	$-120 \times 0.5 = -60$
$de = 40$	$-40 \times 0.5 = -20$	$-40 \times 0.866 = -34.6$
	$\Sigma F_x = +113$	$\Sigma F_y = -16.5$

Then $R = \sqrt{\Sigma F_x^2 + \Sigma F_y^2} = \sqrt{13{,}041} = 114$ lb. Sense is downward and to the right (Fig. 3.5):

$$\tan \theta = \frac{\Sigma F_y}{\Sigma F_x} = -0.146; \qquad \theta = -8°20'$$

Addition of Noncoplanar Forces

Graphical Methods. Graphical methods for the addition of noncoplanar forces are unnecessarily cumbersome in view of the ease with which the process can be handled algebraically using rectangular components.

Algebraic Method. Consider any number of concurrent forces (e.g., the three forces **F**, **G**, and **H** acting at point O as shown in fig 3.6. Each such force can be resolved into its components with respect to a set of rectangular axes by using the direction angles of the line of action of the force (and their cosines, called direction cosines) as shown in Fig. 3.6. The rectangular components of the force **F** are given by

$$F_x = F \cos \alpha_F \qquad F_y = F \cos \beta_F \qquad F_z = F \cos \gamma_F$$

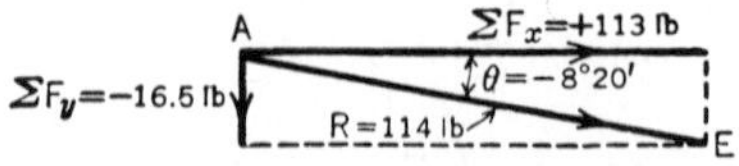

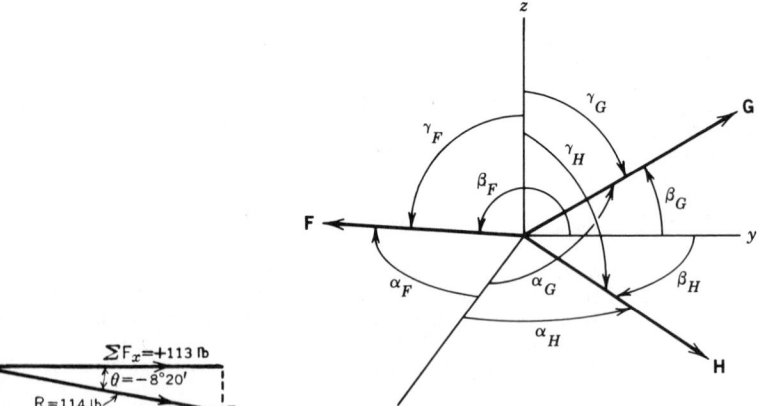

$\Sigma F_x = +113$ lb
$\theta = -8°20'$
$\Sigma F_y = -16.5$ lb
$R = 114$ lb

Fig. 3.5   Rectangular components.

Fig. 3.6   Direction angles.

and the force **F** can be written as

$$\mathbf{F} = F_x\mathbf{i} + F_y\mathbf{j} + F_z\mathbf{k}$$

where **i**, **j**, and **k** are unit vectors in the x, y, and z directions, respectively. Similar forms can be written for the forces **G** and **H**. The resultant or sum of the three forces. **F**, **G**, and **H**, is obtained by adding corresponding components of the forces. Thus, the resultant, **R**, is given by

$$\mathbf{R} = R_x\mathbf{i} + R_y\mathbf{j} + R_z\mathbf{k}$$

where

$$R_x = F_x + G_x + H_x$$
$$R_y = F_y + G_y + H_y$$
$$R_z = F_z + G_z + H_z$$

Nature of Resultant of Concurrent Forces
The **resultant** of any system of *concurrent forces* that are not in equilibrium is a *single force*.

3.2.3   Moments and Couples

Moment (or Torque) of a Force about a Point
Moment of torque of a force about a point is the product of the force magnitude and the distance from the point to its action line. This perpendicular distance is called the *arm* of the force, and the point is the *origin or center of moments*. The product is the measure of the rotational tendency of the force. The name of the unit of moment is a combination of the names of force and distance units, as foot-pound, inch-ton, and so on. (Some writers use lb-ft as a unit of moment of a force to distinguish from ft-lb as the unit of work or energy, similar distinction being made for the other units.)

The computation of moments is most easily carried out by introducing a coordinate system and then taking the vector cross product

$$\mathbf{M} = \mathbf{R} \times \mathbf{F}$$

where **R** is the vector from the reference point to the point of action of the force, and **F** is the vector representation of the force in the same coordinate system. The moment vector, **M**, is perpendicular to the plane of **R** and **F**. This vector operation, in effect, takes into account the angle between **R** and **F** and can be used in two- and three-dimensional situations.

Moment (or Torque) of a Force about a Line
The moment of a force, **F**, about a line is most easily calculated using vectors. Let O be any point on the line and let A be any other point on the line. Let P be the point of application of the force **F**. The moment of the force **F** about the line OA measures the torque of the force about the axis OA. In order to calculate this torque, it is most convenient to break **F** into two components, with one of the components parallel to OA. The component of **F** parallel to OA does not exert a torque around the axis OA. Thus, the torque of the force **F** around the axis OA can be written as

$$\mathbf{M}\ (\text{about } OA) = \mathbf{R} \times (\mathbf{F} - \mathbf{F} \cdot \mathbf{U}_{OA})$$

where **R** is the vector from O to P, **F** is the force vector, and $\mathbf{U}_{OA}$ is a unit vector directed along the axis OA. The quantity $\mathbf{F} \cdot \mathbf{U}_{OA}$ is the component of **F** parallel to the axis OA.

Principle of moments

For a point. The moment of the resultant of any set of forces about a point is the vector sum of the moments of the individual forces about the point.

For an Axis. The moment of the resultant of any set of forces about an axis is the vector sum of the moments of the individual forces about the axis.

Couples

Nature of Couples. Two equal and parallel forces of opposite sense are called a *couple*. The tendency of a couple is to produce rotation only. Since a couple has no single resultant, no single force can balance it. To prevent the rotation of a body acted upon by a couple, the application of two other forces is required, forming a second couple.

The *arm* of a couple is the perpendicular distance between the lines of action of the forces. The *moment* of a couple is constant and independent of the origin of moments; it is equal to one of the forces times the arm of the couple. Its sense is positive or negative according as rotational tendency is counterclockwise or clockwise. Couples of equal moments, in the same or parallel planes, are *equivalent* and may be replaced one by the other. Further, the *center of rotation* for a couple may be anywhere in its plane. Hence, a couple may be turned about in its own plane or moved to a parallel plane or replaced by another couple (having an arm of any given length but the same moment) without altering its effect on a rigid body.

Resultant of Couples. The resultant of any number of coplanar couples is a couple. Its moment and sense are determined by the algebraic sum of the moments of the individual couples.

In vector form, the moment of a couple can be calculated by choosing a point O on the line of action of one of the forces and determining the moment of the other force of the couple about that point. Thus, the moment of a couple is calculated from

$$\mathbf{M} = \mathbf{R} \times \mathbf{F}$$

The resultant of any number of couples is the vector sum of the moments of the individual forces (each calculated as shown), and this resultant is itself a couple.

Composition of Single Force and Couple. A *single force and couple in the same plane (or parallel planes)* may be composed into *another single force* equal and parallel to the original force, at a distance from it equal to the moment of the couple divided by the magnitude of the force and so situated that the moment of the resultant about the point of application of the original force is of the same sign as the moment of the couple. The couple may be brought into the position shown in Fig. 3.7. The resultant of P, $-Q$, and Q is $R(=P)$ acting in a line through point C so that $(P - Q) \times AC = Q \times BC$. From this it follows that

$$AC = \frac{Q(AC + BC)}{P} = \frac{\text{moment of couple}}{P}$$

Resolution into Single Force through Chosen Point and Couple. A *single force* may be resolved into *another single force acting through a chosen point and a couple* (the new force being equal and parallel to the original force). In Fig. 3.8, P_1 is the given force and O the chosen point. Through O apply a pair of forces, opposite in sense, equal and parallel to P_1. As P_2 and P_3 balance, no change is produced in the motion of the body due to the addition. P_1 and P_3 constitute a couple of moment $= P \times a$, which is the same as moment of P_1 about O; and P_2 is a force just like P_1, but acting through the chosen point O.

3.2.4   Addition of Nonconcurrent Forces and Moments

The addition of nonconcurrent forces involves the determination of the force and moment resultants of the forces. The procedure requires that a reference point, O, for the moments be defined and that all position vectors and force vectors be written in terms of vector components. The procedure works for all force and moment sets.

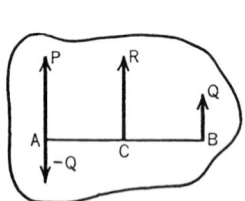

Fig. 3.7   Composition of a force and a couple.

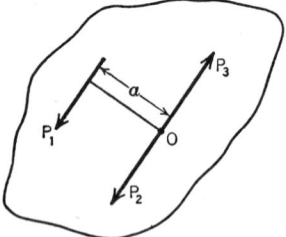

Fig. 3.8   Resolution of a force into a force plus couple.

Force Resultant. The force resultant of a set of nonconcurrent forces is the vector sum of the individual forces.

$$\mathbf{R} = \sum_{i=1}^{n} \mathbf{F}_i = \sum_{i=1}^{n} (\mathbf{F}_{ix})\mathbf{i} + \sum_{i=1}^{n} (\mathbf{F}_{iy})\mathbf{j} + \sum_{i=1}^{n} (\mathbf{F}_{iz})\mathbf{k}$$

Moment Resultant. The moment resultant of a set of nonconcurrent moments about a point O is the vector sum of the moments of the individual forces about O.

$$\mathbf{M} = \sum_{i=1}^{n} \mathbf{M}_i = \sum_{i=1}^{n} (\mathbf{r}_i \times \mathbf{F}_i)$$

3.2.5   Principles of Equilibrium

Forces in Equilibrium. A system of forces is in equilibrium if their combined action produces no change in motion of the body to which they are applied., There is no change in motion if the body remains at rest or moves in a straight line at constant speed. When a force system is in equilibrium, its resultant must be zero. This statement may be called the *general condition of equilibrium*. It implies both zero force and zero couple.

In all cases, the conditions of equilibrium can be written in vector form as

$$\Sigma \mathbf{F} = 0$$
$$\Sigma \mathbf{M} = 0$$

and in scalar form in rectangular Cartesian coordinates as

$$\begin{aligned} \Sigma F_x &= 0 & \Sigma M_x &= 0 \\ \Sigma F_y &= 0 & \Sigma M_y &= 0 \\ \Sigma F_z &= 0 & \Sigma M_z &= 0 \end{aligned}$$

Note: If a problem is planar (e.g., in the xy plane), then the force equation in the z direction and the moment equations in the x and y directions are identically zero and may be ignored.

Body in Equilibrium. A rigid body is in equilibrium if it remains at rest or moves in a straight line at constant speed; that is, if its state of motion does not change. This condition obtains if all the external forces acting upon it (including those due to pull of gravity, friction, etc.) form a system in equilibrium.

Conditions of Equilibrium
In a problem in *statics* a body is known to be in equilibrium; hence the system composed of all the external forces acting upon it must be in equilibrium. In such a case, tests are not needed to ascertain if equilibrium exists, but they are used to set up relations involving unknown forces, distances, or angles, and the unknown elements are then computed provided their number does not exceed the number of independent equations that may be set up by means of the equilibrium conditions. When the number of unknown elements exceeds the number of independent equations, the problem is said to be statically indeterminate.

Special Conditions. If three forces are in equilibrium they must be coplanar and concurrent or parallel; if concurrent, each force is proportional to the sine of the angle between the other two; if parallel, each force is proportional to the distance between the other two. If a force system is in equilibrium, the resultant of any part must balance the resultant of the other part. It follows that if four coplanar nonconcurrent nonparallel forces are in equilibrium, the resultant of any two is concurrent with the other two.

Stability of Equilibrium. When a body (or collection of bodies) is in equilibrium and the state is such that if when displaced slightly in any way the body returns of itself to its original position, the equilibrium is *stable*; if when displaced slightly the body moves further from its original position, the equilibrium is *unstable*; and if when displaced slightly it remains in that displaced position, the equilibrium is *neutral* or *indifferent*. The body or collection is also said to be stable, unstable, or neutral (or indifferent) under these respective conditions. When the body is stable or unstable, the system of forces is changed by the slight displacement and is no longer in equilibrium; hence the further displacement. Only when the stability is neutral is the equilibrium of the force system undisturbed by a slight displacement of the body.

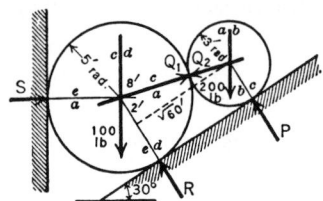

Fig. 3.9   Space diagram of cylinders.

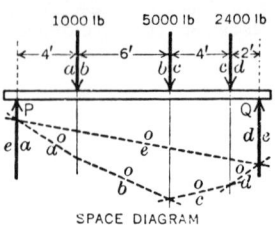

SPACE DIAGRAM

Fig. 3.10   Space diagram of beam.

3.2.6   Equilibrium Problems

General Principles

1. It frequently happens that the external force system acting on a body as a whole cannot be solved directly owing to the presence of more unknown elements (forces, distances, and angles) than there are conditions of equilibrium. In such cases, endeavor to separate the original body into simpler parts that will permit solutions, making use of the unknowns thus determined in solving the force systems acting on other sections until the complete solution, if obtainable, has been found.

2. To facilitate computations, it is desirable if resolution equations are used, to resolve perpendicular to one of the unknown forces; if a moment origin is used, to select it on the action line of an unknown force; if moment axes are used, to select them so as to intersect some of the unknown forces.

3. Assume senses for unknown forces. A plus answer then indicates the sense to have been correctly assumed; a minus answer, incorrect assumption.

4. When force polygons are used, letter action lines of wholly known forces first and those of the remainder last. Draw the force polygon to the end of the last known vector. Vectors required to close it (remembering that the senses must read confluently from the starting point back to the same point) determine unknown magnitudes and senses and/or lines of action.

Typical Problems

I. System of Coplanar Concurrent Forces in Equilibrium. *In this system all forces are known except two whose action lines only are known.* The magnitudes and senses of these two forces are to be determined.*

Example 3.1. Two smooth cylinders rest upon a 30° plane and against a vertical wall as shown in Fig. 3.9. Determine all forces acting on each cylinder. (*a*) The forces involved are 100 lb, 200 lb, *P*, *Q*, *R*, and *S*, the last four being normal to the surfaces of contact (smooth surfaces). (*b*) Consider the two cylinders as a single free body. The external force system is 100 lb, 200 lb, *P*, *R*, and *S* (Q_1 and Q_2 are internal). The system is nonconcurrent, so does not come under typical problem I. Consider the large cylinder as a free body. The external force system is 100 lb, *Q*, *R*, and *S*. While this system is concurrent, it cannot be solved because there are more than two unknown quantities. Next consider the small cylinder as a free body. The force system is 200 lb, *P*, and *Q*, and this is typical problem I.

ALGEBRAIC SOLUTION. Choose *X* and *Y* directions parallel and perpendicular to the plane. $\Sigma F_x = 0 = Q_1(\sqrt{60}/8) - 200 \sin 30°$. Hence $Q_1 = 800/\sqrt{60} = 103.3$ lb. $\Sigma F_y = 0 = P - 200 \cos 30° - 103.3 \times 2/8$. Hence $P = 199$ lb. Consider the large cylinder as a free body. $Q_1 = Q_2 = 103.3$ lb. Use the same *X* and *Y* directions. $\Sigma F_x = 0 = S \cos 30° - 103.3 \times (\sqrt{60}/8) - 100 \sin 30°$. Hence, $S = 173.2$ lb. $\Sigma F_y = 0 = R - 100 \cos 30° - 173.2 \sin 30° + 103.3 \times 2/8$. Hence $R = 147.4$ lb.

II. System of Coplanar Parallel Forces in Equilibrium. *In this system all forces are known except two whose action lines only are known.* The magnitudes and senses of these two forces are to be determined.

Example 3.2. A beam is loaded as shown in Fig. 3.10 and supported at the points *P* and *Q*. Determine the reactions of the supports. Consider the beam as a free body. The external force system

---

*This is a common problem in the determination of the stresses of a roof or bridge truss.

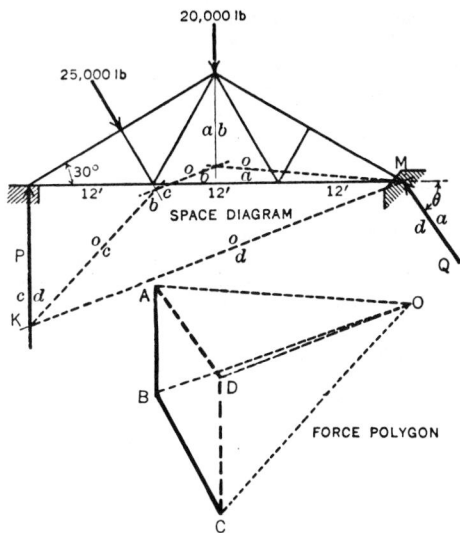

Fig. 3.11   Space diagram of roof truss.

consists of the forces 1000 lb, 5000 lb, 2400 lb, P, and Q. This is a coplanar parallel system and is typical problem II.

ALGEBRAIC SOLUTION. Assume senses for the reactions.

$$\Sigma M_p = 0 = -4 \times 1000 - 10 \times 5000 - 14 \times 2400 + 16Q$$

Hence $Q = 5475$ lb; correct sense was assumed.

$$\Sigma M_q = 0 = 2 \times 2400 + 6 \times 5000 + 12 \times 1000 - 16P$$

Hence $P = 2925$ lb; correct sense assumed. As a check, apply a third equilibrium condition, $\Sigma F = 0$.

$$\Sigma F = 0 = -2925 + 1000 + 5000 + 2400 - 5475$$

III. System of Coplanar Nonparallel Nonconcurrent Forces in Equilibrium. *In this system all forces are known except two, of which the action line of one and a point in the action line of the other are known. The magnitude and sense of the one and the magnitude, sense, and angular direction of the other are to be determined.**

Example 3.3. A roof truss is loaded as in Fig. 3.11. The left end of the truss rests on a smooth horizontal support. The right end is secured to a wall by means of a pin. Determine the reactions. The external forces acting on the truss are the given loads, the left reaction P (vertical, on account of the smooth support), and the right reaction Q (inclined, through point M). The unknown quantities are the reactions P and Q. This is typical problem III.

ALGEBRAIC SOLUTION. Assume P upward.
$\Sigma M_M = 0 = 20{,}000 \times 18 + 25{,}000 \times 24 \cos 30° - 36P$; hence, $P = 24{,}430$ lb; correct sense assumed.
Assume Q upward to the left at angle θ with horizontal.
$\Sigma F_x = 0 = 25{,}000 \sin 30° - Q \cos \theta$; $\Sigma F_y = 0 = -25{,}000 \cos 30° - 20{,}000 + 24{,}430 + Q \sin \theta$.
Solving simultaneously, $Q = 21{,}300$ lb, and $\theta = 54°$. Sense and direction were correctly assumed, hence Q acts upward to the left at 54° to the horizontal. As a check, apply condition $\Sigma M_P = 0$.

$$\Sigma M_P = 0 = -25{,}000 \times 12 \cos 30° - 20{,}000 \times 18 + 21{,}300 \times 36 \sin 54°.$$

*This is a common problem in the determination of the reactions on a roof truss sustaining wind pressures, the truss being fixed at one end and resting on rollers at the other.

GRAPHIC SOLUTION. *ab* and *bc* are the action lines of the given loads, *cd* of the reaction *P* and *da* of the reaction *Q*. Draw the vectors *AB* and *BC*, and a line through *C*, parallel to *cd*. Choose a pole and draw the rays. Construct the funicular polygon, drawing *oa* through *M*, and draw closing string *od* from *K* to *M*. Draw *OD* through *O* parallel to *od* to intersect *CD* at *D*. Draw *DA*. Vectors *CD* and *DA* represent the two unknown forces, *P* = 24,430 lb and *Q* = 21,300 lb. The action line of *Q* is *da*, making angle with horizontal = 54°.

Special Case. A case coming under the preceding classification that requires a variation in treatment when employing the graphic method is one in which the action lines of all three forces are known but their magnitudes and senses are unknown. The procedure is in general similar to methods employed before except that, in the graphic solution, two of the unknown forces that are concurrent must be replaced by their unknown resultant acting through their point of concurrency along an unknown action line. After the magnitude and sense of this resultant have been determined (by the method employed in Example 3.3), it is resolved into its two components along the action lines of the two unknown forces that it had replaced. These components represent the magnitudes and senses of this pair of forces.

IV. System of Noncoplanar Nonparallel Nonconcurrent Forces in Equilibrium. *In this system one force is completely known and action lines (or a point in the action line) of the others are known.* All unknown force magnitudes, action lines and senses are to be determined.

Example 3.4. The crane (Fig. 3.12) is supported by a socket at the foot of the post at *D*; is kept from overturning by the backstays *AB* and *AC*; and carries a load of 600 lb (*E, A, F, G, D,* are in the vertical *XY* plane). Determine the axial components of the reaction on the post at *D* and the tensions in the backstays. The external forces acting on the post are the load, the reaction at *D*, and the tensions in the backstays at *A*. This is typical problem IV. Moment equations are the most convenient to apply for this solution.

$$\Sigma M_{BC} = 0 = 600 \times 40 - 20D_y \qquad D_y = 1200 \text{ lb}$$

$$\Sigma M_{Z_A} = 0 = 600 \times 20 - 16D_x \qquad D_x = 750 \text{ lb}$$

$$\Sigma M_{X_A} = 0 = 600 \times 4 - 16D_z \qquad D_z = 150 \text{ lb}$$

$$\Sigma M_{X_C} = 0 = AB \times \frac{16}{\sqrt{881}} \times 25 - 1200 \times 10 + 600 \times 6 \qquad AB = 622 \text{ lb}$$

$$\Sigma M_{X_B} = 0 = AC \times \frac{16}{\sqrt{756}} \times 25 + 600 \times 19 - 1200 \times 15 \qquad AC = 452 \text{ lb}$$

The senses of all forces are as shown in Fig. 3.12.

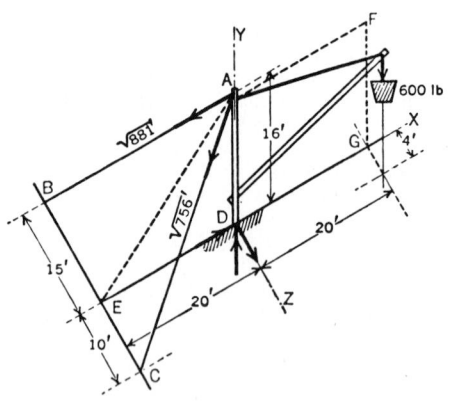

Fig. 3.12   Space diagram of truss.

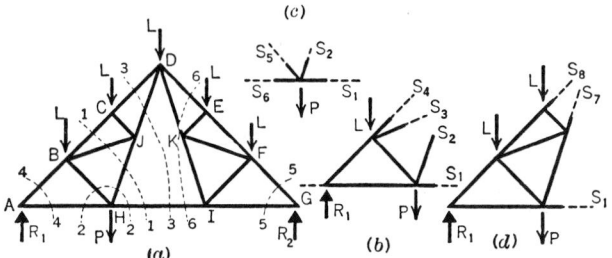

Fig. 3.13 Truss sections.

Truss Analysis

A **truss** is a framework* for carrying loads, each *member* of which is subjected only to tension or compression loads. The members are usually pin jointed with loads applied only at the joints.

The **stress in a member** at any section is the force that either of its two parts exerts internally on the other part as a result of the external forces acting on the member. Longitudinal stresses, like external longitudinal forces, may be either tensile or compressive.

The **analysis of a truss** under a given loading condition refers to the determination of the stresses in its members due to the loads.

Analysis by Method of Sections. First, determine the reactions on the truss due to the loads; second, imagine the truss separated into two distinct parts (i.e., pass a section through the truss) so that the member under consideration is one of the members cut and so that the system of forces, including stresses, acting on either part of the truss is solvable for the desired stress; third, solve the system.

To pass the section, suppose the stress in HI (Fig. 3.13a) is required, the truss being supported at its ends and bearing five loads L and one P, and suppose the reactions determined. Trying section 1–1, the force system on the left part of the truss (Fig. 3.13b) is a nonconcurrent one of seven forces, and includes four unknown stresses S_1, S_2, S_3, and S_4; it is not solvable for the desired stress S_1. Trying section 2–2, the force system on the lower part (Fig. 3.13c) is a concurrent one, and includes four unknown stresses, S_1, S_2, S_5, and S_6; it is not solvable. Trying section 3–3, the force system on the left part (Fig. 3.13d) is nonconcurrent with three unknown stresses, S_1, S_7, and S_8; it is solvable. In some instances different sections may be used, each leading to a solution.

S_1 having been determined, the force system of Fig. 3.13b becomes solvable, and then, with S_2 also determined, the force system of Fig. 3.13c may be solved.

ALGEBRAIC SOLUTION. Following the general method of procedure outlined above, determine the various stresses by employing algebraic conditions of equilibrium in manners similar to those illustrated in Section 3.2.6.

In making the imaginary separations of the truss, care should be taken to cut not more than three members in which the stresses are unknown. It is advantageous to make the separation so that not more than two such members are cut. If this is done, a single force polygon will determine the two unknowns, whereas if three are cut, a force polygon and an equilibrium polygon, or the equivalent, are necessary for determining the three unknowns.

Analysis by Method of Joint Resolution.* Consider the pin at each joint as a body acted upon by forces in equilibrium.

ALGEBRAIC SOLUTION. Determine the various stresses by employing algebraic conditions of equilibrium in manners similar to those illustrated in Section 3.2.6.

GRAPHIC SOLUTION. For each joint, draw the force polygon. In doing so, it will be advantageous to represent the forces in the order in which they occur about the joint. A force polygon so drawn will

---

Redundant frames (i.e., ones having more members than necessary to preserve their shapes under the loading conditions) are not considered in this section, since the stresses in them cannot be determined by elementary static methods.
*This method is not usually so convenient for determining algebraically the stress in a single specified member.

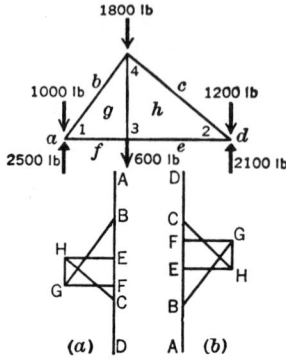

Fig. 3.14   Roof truss/polygon solution.

be called a polygon for the joint; and for brevity, if the order taken is clockwise, the polygon will be called a clockwise polygon, and if counterclockwise, it will be called a counterclockwise polygon. If the polygons for all the joints of a truss are drawn separately, the stress in each member will have been represented twice. It is possible to combine the polygons so that it will not be necessary to represent the stress in any member more than once, thus reducing the number of lines to be drawn. Such a combination of force polygons is called a *stress diagram.* Each triangular space in the truss diagram is marked by a small letter; also the space between consecutive action lines of the loads and reactions. Then the two letters on opposite sides of any line serve to designate that line, and the same large letters are used to designate the magnitude of the corresponding force.

To construct a stress diagram for a truss under given loads: (1) Determine the reactions. (2) Letter the truss diagram as directed. (3) Construct a force polygon for all the external forces applied to the truss (loads and reactions), representing them in the order in which their application points occur about the truss, clockwise or counterclockwise. (4) On the sides of that polygon construct the polygons for the joints. They must be clockwise or counterclockwise according as the polygon for the loads and reactions was drawn clockwise or counterclockwise. (The first polygon drawn must be for a joint at which only two members are fastened; the joints at the supports are usually such. Next, that joint is considered, and its polygon is drawn, at which not more than two stresses are unknown.)

Example 3.5.   Figure 3.14 represents a roof truss sustaining loads of 600, 1000, 1200, and 1800 lb; the right reaction is 2100 lb, and the left 2500 lb. *ABCDEFA* is a polygon for the loads and reactions, these being represented in the order in which their points of application occur about the truss. The polygon for joint 1 is *FABGF;* the force *BG* acts toward the joint, hence *bg* is under compression, and *GF* acts away from the joint, hence *gf* is in tension. The polygon for joint 2 is *CDEHC;* the force *EH* acts away from the joint, hence *eh* is in tension; and *HC* acts toward the joint, hence *hc* is in compression. The polygon for joint 3 is *HEFGH;* the force *GH* acts away from the joint and hence *gh* is in tension. If the work has been done correctly, *GH* is parallel to *gh.* (In Figure 3.14*a* all the polygons are clockwise, and in Fig. 3.14*b*, counterclockwise.)

3.2.7   Center of Gravity

Definitions
The **centroid** of a system of parallel forces having fixed application points is the point through which their resultant will always pass regardless of how the forces may be turned, provided they remain parallel.

The **center of gravity of a body**[†] *or system of bodies* is the *centroid of the forces of gravitation*[‡] acting upon all the particles thereof. Referring the application points of such a force system to a set of coordinate axes, the coordinates of the centroid, or center of gravity (cg), are

$$\bar{x} = \frac{\Sigma F_i \cdot x_i}{\Sigma F_i} = \frac{\int x\, dF}{F} \qquad \bar{y} = \frac{\Sigma F_i \cdot y_i}{\Sigma F_i} = \frac{\int y\, dF}{F}$$

$$\bar{z} = \frac{\Sigma F_i \cdot z_i}{\Sigma F_i} = \frac{\int z\, dF}{F}$$

[†]Sometimes called *center of mass* or *center of inertia.*
[‡]For practical purposes, the forces of gravitation may be considered as parallel.

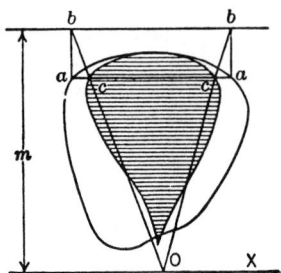

Fig. 3.15 Center of gravity determination.

in which F_i represents the force on (or weight of) one particle and x_i, y_i, z_i are the coordinates of its application point. If a group of bodies is involved, the coordinates of the center of gravity of the group are

$$\bar{x} = \frac{\Sigma W_i \cdot \bar{x}_i}{\Sigma W_i} \qquad \bar{y} = \frac{\Sigma W_i \cdot \bar{y}_i}{\Sigma W_i} \qquad \bar{z} = \frac{\Sigma W_i \cdot \bar{z}_i}{\Sigma W_i}$$

in which W_i represents the weight of one body and $\bar{x}_i$, $\bar{y}_i$, $\bar{z}_i$ are the coordinates of its center of gravity.* A body (or system of bodies), if supported at its center of gravity, will remain at rest in any position.

The **center of gravity of part of a body** may be located by the rule that its moment, with respect to any plane, equals the moment of the whole minus, algebraically, the moment of the remainder.

The **center of gravity of a line, surface, or volume** is that point that would be the center of gravity if the line were replaced by a homogeneous rod of infinitesimal diameter, the surface by a homogeneous plate of infinitesimal thickness, or the volume by a homogeneous body.

Symmetry. Two points are symmetrical with respect to a third point if the line joining the two is bisected by the third. Two points are symmetrical with respect to a line or a plane if the line joining them is perpendicular to the given line or plane and is bisected by it. A body, line, surface, or volume is symmetrical with respect to a point, a line, or a plane if all the points of the body, line, surface, or volume can be paired so that each pair is symmetrical with respect to the point, line, or plane. If a homogeneous body, or a line, surface, or volume is symmetrical with respect to a point, line, or plane, its center of gravity is at the point, in the line or in the plane.

The **static moment** of a body (having weight), a line (having length), a surface (having area), or a solid[†] (having volume) with respect to any plane is the product of the weight, length, area, or volume and the distance of the center of gravity of the body, line, surface, or solid from the plane. The static moment of a plane line or plane surface with respect to a straight line in the plane is the product of the length or area and the distance of the center of gravity of the line or surface from the reference line. A static moment is regarded as positive or negative according as the corresponding center of gravity is on the positive or negative side of the reference plane or line.

Determination of Center of Gravity Location

When practicable, determination of center of gravity location by algebraic or integration methods, based on dividing the sum of the moments by the sum of the forces, is generally the simplest process. For some bodies of nonhomogeneous nature or of very irregular shape, one of the following methods of procedure may be necessary or at least preferable:

Graphic Method. For application to *plane* figures.[‡] Referring to Fig. 3.15, take a point O and a line bb on opposite sides of the figure at any convenient distance m apart; project any width of the figure parallel to bb as aa on bb, connect the projections bb with O, and note the intersections cc;

---

*As F (in the gravitational system of units) equals W, these symbols may be used interchangeably in the two sets of formulas. Also, any one of the expressions may be read as "the sum of the moments divided by the sum of the forces (or weights)."

[†] The word "solid" where used herein denotes "that which has volume." Care should be taken to distinguish this from a "body," which has "mass" (as well as "volume"). Some writers use the word "solid" to denote at various times either volume or mass, which is sometimes confusing. In this section, the word "volume" is frequently used even in preference to "solid" to avoid the possibility of confusion with "mass."

[‡] Including areas or flat homogeneous bodies of uniform thickness.

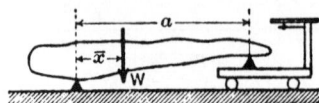

Fig. 3.16   Weighing/knife edge method.

determine other points *cc* and draw a smooth curve through them as shown; measure the area A' within the curve *cc*; then $A'm$ is the static moment of the given figure with respect to OX; if A is the area of the given figure and y the distance of its center of gravity from OX, $y = A'm/A$. In a similar way the distance of the center of gravity from a line perpendicular to OX can be determined and its exact position thus definitely located.

Suspension Method. For application to *plane* figures.[§] Suspend the body (or a model representing it) from a point near its edge and mark on it the direction of a plumb line hung from that point. Repeat this operation, using a second suspension point. The center of gravity is at (or behind) the intersection of the two markings.

Weighing Method. Generally applied where location of the center of gravity in one plane only is required. Determine weight $\overline{W}$ of the body and then support it on a knife edge (Fig. 3.16) and on a point support resting upon a platform scale. Weigh reaction R of the point support and measure horizontal distance a between the point and the knife edge. Then the horizontal distance from the knife edge to the center of gravity is $\bar{x} = Ra/W$.

Balancing Method. For general application. Balance the body (or a model representing it) on a straightedge, marking on the body the vertical plane containing the edge. Repeat for two more balancing positions of the body. The center of gravity is at the point common to the three planes thus determined.

3.2.8   Moment of Inertia

Plane Surfaces—Definitions
The **moment of inertia of a plane surface** (or figure) with respect to (or about) a line (or axis) is the sum of the products obtained by multiplying the area of each element of the surface by the square of its distance from the line.[*] Letting I_x denote moment of inertia about an X axis:

$$I_x = \int y^2 \, dA$$

in which A is the total area and y is the perpendicular distance of any element of area dA from the axis. The moment of inertia of a surface is obviously the sum of the moments of inertia of its parts. The moment of inertia of a plane surface is **rectangular** if the axis used is *in* the plane of the area; it is **polar** if the axis is *perpendicular* to the plane of the area.

The **radius of gyration of a plane surface** with respect to a line is the length whose square multiplied by the area of the surface equals the moment of inertia of the surface with respect to the line. Letting k denote radius of **gyration**:

$$I = k^2 A \quad \text{or} \quad k = \sqrt{I/A}$$

in which I is the moment of inertia and A the area.

The **product of inertia of a plane surface** with respect to a pair of coordinate axes in the plane is the algebraic sum of the products obtained by multiplying the area of each element of the surface by its coordinates.[*] Letting U_{xy} denote product of inertia with respect to X and Y axes:

$$U_{xy} = \int xy \, dA$$

in which A is the total area and x and y are the coordinates of any element of area dA.

[§] Including areas or flat bodies of uniform homogeneous thickness.
[*] Moment of inertia is always positive and never zero. Product of inertia may be positive, zero, or negative, depending on the distribution of the area with respect to the axes. If a surface has an axis of symmetry, its product of inertia with respect to that axis and one perpendicular thereto is zero.
[†] In certain special cases, as for axes through the point in the center of a circular area, the moment of inertia is the same for any axis and therefore there is no principal axis through that point.

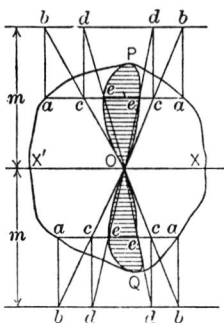

Fig. 3.17   Moments of inertia/graphic method.

The **principal axes of inertia of a plane surface at a particular point** in the plane are the two axes about which the moments of inertia are greater and less than for any other axis, through the point in the plane.[†] The corresponding moments of inertia are called the **principal moments of inertia** of the surface at the point. The principal axes are always at right angles to each other. The product of inertia with respect to them is zero.

The **customary engineer's unit** for both moment and product of inertia of a surface is biquadratic inches (in.^{4}).

Determination of Moment of Inertia of Plane Surfaces
When practicable, determination of moment of inertia with respect to an axis by algebraic or integration methods is generally the simplest process. For some surfaces of very irregular shape, the following graphic method of procedure may be necessary or at least preferable:

Graphic Method. Let *aaaa* (Fig. 3.17) be the outline and *XX'* the axis with respect to which the moment of inertia is desired; at any convenient distance *m* from *XX'* draw two parallels (but if *XX'* does not cut the figure, only one parallel, the one on the opposite side of the figure from *XX'*); draw any line as *aa* parallel to *XX'* and project the points *aa* on the nearer parallel; join the projections *bb* to any point *O* in *XX'*, and note the intersections *cc* on *aa*; project *cc* on the same parallel; join the projections *dd* with *O*, and note the intersections *ee* on *aa*. In a similar manner determine points like *ee* for other widths like *aa*, and connect all points *e* as shown. Then measure the area of the loops *OPO* and *OQO*; denoting this combined area by A'', $I = A''m^2$. (There will be only one loop if only one parallel *bb* is used.)

Transformation Formulas — Plane Surfaces

Parallel Axes Theorems. Let I = *moment of inertia* (either rectangular or polar) of a plane figure with respect to any line or axis, $\bar{I}$ = that with respect to a parallel axis passing through the center of gravity of the figure, d = distance between the axes, k and $\bar{k}$ = the radii of gyration with respect to the same area, respectively, and A = area of the figure; then

$$I = \bar{I} + Ad^2 \quad \text{and} \quad k^2 = \bar{k}^2 + d^2$$

These show that with respect to all parallel axes the moment of inertia and the radius of gyration are least for the one passing through the center of gravity of the figure.

Similarly, let U = *product of inertia* of a plane figure with respect to a pair of coordinate axes in the plane, and $\bar{U}$ = that with respect to a parallel pair whose origin is at the center of gravity; $\bar{x}$, $\bar{y}$ the coordinates of the center of gravity referred to the first pair, and A the area of the figure; then $U = \bar{U} + A\bar{x}\bar{y}$.

Relation of Rectangular and Polar Moments of Inertia. Let I_x, I_y, and J_z = moments of inertia of a plane figure with respect to x, y, and z axes, respectively, the axes being *at right angles* to each other and the x and y axes in the plane; and let k_x, k_y, and k_z = corresponding radii of gyration; then $J_z = I_x + I_y$, $k_z^2 = k_x^2 + k_y^2$.

*If U_{xy} and $(I_y - I_x)$ are both zero, there is no principal axis through the point.

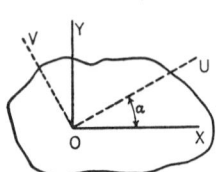

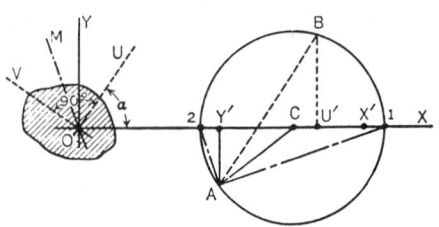

Fig. 3.18   Rotated axes theorem.   Fig. 3.19   Inertia circle.

Rotated Axes Theorem. Let XOY and UOV (Fig. 3.18) be two sets of rectangular coordinate axes with a common origin and in a given plane figure; I_x, I_y, I_u, I_v = moments of inertia of the figure with respect to x, y, u, and v axes, respectively; U_{xy} and U_{uv} = its products of inertia with respect to the sets of axes, respectively; α = angle through which x axis must be rotated to bring it into u axis, regarded as positive or negative according as the turning is counterclockwise or clockwise. Then $I_u + I_v = I_x + I_y$, and

$$I_u = I_x \cos^2\alpha + I_y \sin^2\alpha - U_{xy}\sin 2\alpha$$

$$U_{uv} = \tfrac{1}{2}(I_x - I_y)\sin 2\alpha + U_{xy}\cos 2\alpha$$

If OU and OV are *principal axes* (see definition already given), $U_{uv} = 0$ and therefore $\tan 2\alpha = 2U_{xy}/(I_y - I_x)$. Hence, the principal axes of a figure at a point can be readily found if the moments of inertia and the product of inertia of the figure with respect to two rectangular axes through the point and in the plane are known.* The *principal moments of inertia* are then I_u from the preceding formula and I_v from the same formula after replacing α by $\alpha + \pi/2$. As a check, $I_u + I_v = I_x + I_y$.

Graphic Transformations—Plane Surfaces

The **inertia circle** is a device for determining *graphically* the moment of inertia of a plane figure with respect to any line of the plane through a given point; and the principal axes and principal moments of inertia for the same point. To construct the circle, it is necessary to know the moments of inertia and the product of inertia with respect to two rectangular axes through the point, in the plane figure. Suppose I_x, I_y, and U_{xy} given for the shaded area in Figure 3.19. To convenient scale, plot OX' and OY' to represent I_x and I_y, and $Y'A$ to represent U_{xy} (downward if negative and upward if positive). Center C is midway between X' and Y'. With CA as radius, describe the inertia circle. To find I_u, draw chord AB parallel to axis OU; draw perpendicular BU'. OU' (to scale) = I_u, and BU' (to scale) = U_{uv}. OM, parallel to $A2$, is axis of least I; and a parallel to $A1$, through O, is axis of greatest I. $O2$ (to scale) is the value of least $I = I_2$; and $O1$, value of greatest $I = I_1$. *Least radius of gyration for an axis through point $O = \sqrt{I_2/\text{area}}$*.

Bodies—Definitions

The **moment of inertia of a body** with respect to (or about) a line (or axis) is the sum of the products obtained by multiplying the mass of each elementary part by the square of its distance from the line.* Letting I_x denote moment of inertia about an X axis:

$$I_x = \int y^2 \, dm$$

in which m is the total mass and y is the perpendicular distance of any element of mass dm from the axis.[†] The moment of inertia of a body is obviously the sum of the moments of inertia of its parts.

The **center of gyration of a body** with respect to a line is a point at such a distance from the line that, if the entire mass of the body were concentrated there, its moment of inertia would be the same as that of the body.

---

*Moment of inertia is always positive and never zero. Product of inertia may be positive, zero, or negative, depending on the distribution of the mass with respect to the coordinate planes. If a body has a plane of symmetry, its product of inertia with respect to that plane and one perpendicular thereto is zero.
[†]Strictly speaking, this is the moment of inertia of the *mass* of a body. If m be replaced by w = weight, the result is the moment of inertia of the *weight* of a body.

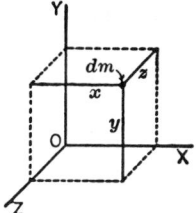

Fig. 3.20   Products of inertia.

The **radius of gyration of a body** with respect to a line is the distance from the center of gyration to the line. Letting k denote radius of gyration:

$$I = k^2 m \quad \text{or} \quad k = \sqrt{I/m}$$

in which I is the moment of inertia and m the mass.

The **product of inertia of a body** with respect to a pair of coordinate planes is the algebraic sum of the products obtained by multiplying the mass of each element of the body by its coordinates with reference to those planes.[†] Thus with respect to YOZ and ZOX (Fig. 3.20), ZOX and XOY, and XOY and YOZ planes, the products of inertia are, respectively:

$$U_{xy} = \int xy\, dm \qquad U_{yz} = \int yz\, dm \qquad U_{zx} = \int zx\, dm$$

The Principal Axes of Inertia of a Body at a Particular Point. The values of moments of inertia of a body for all axes through a given point are in general unequal; for one axis the moment of inertia is greater and for another it is less than for any other axis through the point. These two axes are at right angles, and they together with one at right angles to their plane and passing through the point are **principal axes of inertia of the body at the point**; the corresponding moments of inertia are the **principal moments of inertia of the body at the point**. If the point is the center of gravity of the body, the axes and moments are called *central principal axes* and *central principal moments of inertia*. For a set of principal axes, the three products of inertia, with respect to the **principal planes** determined by them, are zero.

The **customary engineer's unit** for both moment and product of inertia is slug-ft.^{2}

Transformation Formulas—Bodies

Parallel Axes Theorems. Let I = moment of inertia of a body with respect to any line or axis, $\bar{I}$ = that with respect to a parallel axis passing through the center of gravity of the body, d = distance between the axes, k and $\bar{k}$ = the radii of gyration with respect to the same axes, respectively, and m = mass of the body; then

$$I = \bar{I} + md^2 \quad \text{and} \quad k^2 = \bar{k}^2 + d^2$$

Rotated Axes Theorems. Let I_x, I_y, and I_z denote the moments of inertia of a body with respect to rectangular axes x, y, and z, respectively; U_{xy}, U_{yz}, and U_{zx} its products of inertia with respect to yz and zx planes, zx and xy planes, and xy and yz planes, respectively; I the moment of inertia of the body with respect to a line through the origin of coordinates having direction angles α, β, and γ; then

$$I = I_x\cos^2\alpha + I_y\cos^2\beta + I_z\cos^2\gamma - 2U_{yz}\cos\beta\cos\gamma - 2U_{zx}\cos\gamma\cos\alpha$$
$$- 2U_{xy}\cos\alpha\cos\beta$$

If $U_{xy} = U_{yz} = 0$, the y axis is a principal axis at the origin.
If $U_{yz} = U_{zx} = 0$, the z axis is a principal axis at the origin.
If $U_{zx} = U_{xy} = 0$, the x axis is a principal axis at the origin.
If a homogeneous body has a plane of symmetry, any perpendicular to the plane is a principal axis of the body at the point where the line pierces the plane. If it has two planes of symmetry at right

---

[†]Strictly speaking, this is the moment of inertia of the *mass* of a body. If m be replaced by w = weight, the result is the moment of inertia of the *weight* of a body.

angles to each other, their intersection is a principal axis at any point of the intersection, the other two being in the planes of symmetry. If it has three planes of symmetry, their lines of intersection are the central principal axes of the body.

Properties of Various Lines, Surfaces, Volumes, and Bodies

Symbols. I_i = Rectangular moment of inertia; k_i = corresponding radius of gyration; J_o = polar moment of inertia about axis through O perpendicular to plane; k_o = corresponding radius of gyration; m = mass = W/g where W = weight and g = acceleration due to gravity. Moments of inertia of bodies are given in terms of mass. For their values in terms of weight, replace m by W in the formulas.

DECIMAL EQUIVALENTS (*for reference in using Table 3.1*):

$\pi = 3.1416$	$\dfrac{\pi}{128} = 0.0245$	$\sqrt{10} = 3.162$	$\dfrac{1}{\sqrt{6}} = 0.408$
$\dfrac{\pi}{2} = 1.5708$		$\sqrt{12} = 3.464$	
$\dfrac{\pi}{4} = 0.7854$	$\dfrac{1}{\pi} = 0.318$	$\sqrt{18} = 4.242$	$\dfrac{1}{\sqrt{8}} = 0.354$
	$\sqrt{2} = 1.414$	$\dfrac{1}{\sqrt{2}} = 0.707$	$\dfrac{1}{\sqrt{10}} = 0.316$
$\dfrac{\pi}{8} = 0.3927$	$\sqrt{3} = 1.732$		
$\dfrac{\pi}{32} = 0.0982$	$\sqrt{5} = 2.236$	$\dfrac{1}{\sqrt{3}} = 0.577$	$\dfrac{1}{\sqrt{12}} = 0.289$
	$\sqrt{6} = 2.449$		
$\dfrac{\pi}{64} = 0.0491$	$\sqrt{8} = 2.828$	$\dfrac{1}{\sqrt{5}} = 0.447$	$\dfrac{1}{\sqrt{18}} = 0.238$

TABLE 3.1

Lines	
Figure	Centroid Location
1. Any Plane Curve	C.G. is at point having coordinates $\bar{x}$, $\bar{y}$, where $$\bar{x} = \frac{\int x\,ds}{\text{Length}} \text{ where } ds = \sqrt{1 + \left(\frac{dy}{dx}\right)^2}\,dx$$ $$\bar{y} = \frac{\int y\,ds}{\text{Length}} \text{ where } ds = \sqrt{1 + \left(\frac{dx}{dy}\right)^2}\,dy$$
2. Circular Arc	C.G. is on axis of symmetry at $$\bar{x} = \frac{r \sin \alpha}{\alpha} = \frac{rc}{s}. \text{ If } \alpha \text{ is small, distance from C.G. to chord} =$$ approx. $\dfrac{2h}{3}$. (Error is small even for $\alpha = 45°$) For semi-circle: $\bar{x} = \dfrac{2r}{\pi}$ For quadrant: $\bar{x} = \dfrac{2r\sqrt{2}}{\pi}$, and distance from radius drawn to either end of arc $= \dfrac{2r}{\pi}$.

TABLE 3.1 (*Continued*)

Plane Surfaces	
Figure	Centroid Location; Moments of Inertia; Radii of Gyration
3. Any Plane Surface 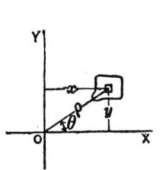	C.G. is at point having coordinates $\bar{x}$, $\bar{y}$, where $$\bar{x} = \frac{\int\int x\,dx\,dy}{\text{Area}} = \frac{\int\int \rho^2 \cos\theta\,d\rho\,d\theta}{\text{Area}};$$ $$\bar{y} = \frac{\int\int y\,dx\,dy}{\text{Area}} = \frac{\int\int \rho^2 \sin\theta\,d\rho\,d\theta}{\text{Area}}$$ $$I_x = \int\int y^2\,dx\,dy; \quad I_y = \int\int x^2\,dx\,dy; \quad J_0 = \int\int \rho^3\,d\rho\,d\theta = I_x + I_y$$ $$k_x = \sqrt{\frac{I_x}{\text{Area}}}; \quad k_y = \sqrt{\frac{I_y}{\text{Area}}}; \quad k_0 = \sqrt{\frac{J_0}{\text{Area}}} = \sqrt{\frac{I_x + I_y}{\text{Area}}}$$
4. Triangle 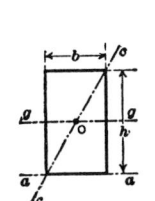	C.G. is at O = intersection of medians. Perpendicular distance from $a - a = \dfrac{h}{3}$. $$I_g = \frac{bh^3}{36}; \quad I_a = \frac{bh^3}{12}; \quad I_c = \frac{bh^3}{4}$$ $$k_g = \frac{h}{3\sqrt{2}}; \quad k_a = \frac{h}{\sqrt{6}}; \quad k_c = \frac{h}{\sqrt{2}}$$
5. Solid Rectangle (or Square) 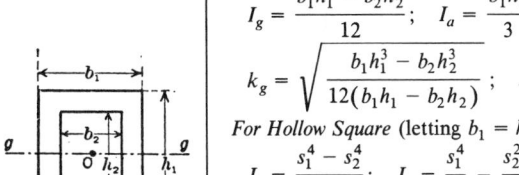	C.G. is at O = intersection of diagonals. *For Rectangle:* $$I_g = \frac{bh^3}{12}; \quad I_a = \frac{bh^3}{3}; \quad I_c = \frac{b^3 h^3}{6(b^2 + h^2)}; \quad J_o = \frac{bh(b^2 + h^2)}{12}$$ $$k_g = \frac{h}{2\sqrt{3}}; \quad k_a = \frac{h}{\sqrt{3}}; \quad k_c = \frac{bh}{\sqrt{6(b^2 + h^2)}}; \quad k_o = \sqrt{\frac{b^2 + h^2}{12}}$$ *For Square* (letting $b = h = s$): $$I_g = \frac{s^4}{12}; \quad I_a = \frac{s^4}{3}; \quad I_c = \frac{s^4}{12}; \quad J_o = \frac{s^4}{6}$$ $$k_g = \frac{s}{2\sqrt{3}}; \quad k_a = \frac{s}{\sqrt{3}}; \quad k_c = \frac{s}{2\sqrt{3}}; \quad k_o = \frac{s}{\sqrt{6}}$$
6. Hollow Rectangle (or Square) 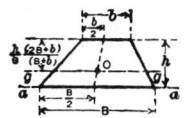	C.G. is at O = intersection of diagonals. *For Hollow Rectangle:* $$I_g = \frac{b_1 h_1^3 - b_2 h_2^3}{12}; \quad I_a = \frac{b_1 h_1^3}{3} - \frac{b_2 h_2(3h_1^2 + h_2^2)}{12}$$ $$k_g = \sqrt{\frac{b_1 h_1^3 - b_2 h_2^3}{12(b_1 h_1 - b_2 h_2)}}; \quad J_o = \frac{b_1 h_1(b_1^2 + h_1^2) - b_2 h_2(b_2^2 + h_2^2)}{12}$$ *For Hollow Square* (letting $b_1 = h_1 = s_1$ and $b_2 = h_2 = s_2$): $$I_g = \frac{s_1^4 - s_2^4}{12}; \quad I_a = \frac{s_1^4}{3} - \frac{s_2^2(3s_1^2 + s_2^2)}{12}$$ $$k_g = \sqrt{\frac{s_1^2 + s_2^2}{12}}; \quad J_o = \frac{s_1^4 - s_2^4}{6}$$ (Note: For a diagonal $c - c$, $I_c = I_g$ and $k_c = k_g$)
7. Trapezoid 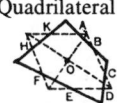	C.G. is at O, located as shown. $$I_g = \frac{h^3(B^2 + 4Bb + b^2)}{36(B + b)}; \quad I_a = \frac{h^3(B + 3b)}{12}$$ $$k_g = \frac{h\sqrt{2(B^2 + 4Bb + b^2)}}{6(B + b)}; \quad k_a = \frac{h}{\sqrt{6}}\sqrt{\frac{B + 3b}{B + b}}$$
8. Quadrilateral	C.G. is at O, located as follows: Divide the sides into thirds and construct the parallelogram with sides passing through the third-points as shown. The intersection of the diagonals of this parallelogram is the desired centroid.

TABLE 3.1 (*Continued*)

Plane Surfaces	
Figure	Centroid Location; Moments of Inertia; Radii of Gyration
9. Regular Polygon 	C.G. is at O = geometrical center. Let $g - g$ be any axis through O and in plane of polygon. Then $$I_g = \frac{\text{Area} \cdot (6R^2 - a^2)}{24} = \frac{\text{Area} \cdot (12r^2 + a^2)}{48};$$ $$J_o = \frac{\text{Area} \cdot (6R^2 - a^2)}{12} = \frac{\text{Area} \cdot (12r^2 + a^2)}{24}$$ $$k_g = \sqrt{\frac{6R^2 - a^2}{24}} = \sqrt{\frac{12r^2 + a^2}{48}};$$ $$k_o = \sqrt{\frac{6R^2 - a^2}{12}} = \sqrt{\frac{12r^2 + a^2}{24}}$$
10. Circle 	C.G. is at O = geometrical center. $$I_g = \frac{\pi r^4}{4} = \frac{\pi d^4}{64}; \quad J_o = \frac{\pi r^4}{2} = \frac{\pi d^4}{32}$$ $$k_g = \frac{r}{2} = \frac{d}{4}; \quad k_o = \frac{r}{\sqrt{2}} = \frac{d}{\sqrt{8}}$$
11. Circular Sector 	C.G. is on axis of symmetry at O. Distance from $$a - a = \frac{2r \sin \alpha}{3\alpha} = \frac{2rc}{3s}.$$ $A = \text{area} = r^2\alpha$ $$I_g = \frac{Ar^2}{4}\left(1 - \frac{\sin\alpha\cos\alpha}{\alpha}\right); \quad I_\alpha = \frac{Ar^2}{4}\left(1 + \frac{\sin\alpha\cos\alpha}{\alpha}\right)$$ $$k_g = \frac{r}{2}\sqrt{1 - \frac{\sin\alpha\cos\alpha}{\alpha}}; \quad k_\alpha = \frac{r}{2}\sqrt{1 + \frac{\sin\alpha\cos\alpha}{\alpha}}$$
12. Semi-circle 	C.G. is on axis of symmetry at O. Distance from $a - a = \dfrac{4r}{3\pi} = 0.424r$ $$I_g = \frac{d^4(9\pi^2 - 64)}{1152\pi} = \frac{r^4(9\pi^2 - 64)}{72\pi} = 0.1098r^4;$$ $$I_a = I_b = \frac{\pi d^4}{128} = \frac{\pi r^4}{4}; \quad J_o = r^4\left(\frac{\pi}{4} - \frac{8}{9\pi}\right) = 0.5025r^4$$ $$k_g = \frac{d\sqrt{9\pi^2 - 64}}{12\pi} = \frac{r\sqrt{9\pi^2 - 64}}{6\pi} = 0.264r; \quad k_a = k_b = \frac{d}{4} = \frac{r}{2};$$ $$k_o = r\sqrt{\frac{1}{2} - \frac{16}{9\pi^2}} = 0.566r$$
13. Circular Segment 	C.G. is on axis of symmetry at O. Distance from $$a - a = \frac{2r^3\sin^3 ga}{3A} = \frac{c^3}{12A} \text{ where } A = \text{area} = \frac{r^2(2\alpha - \sin 2\alpha)}{2}.$$ $$I_g = \frac{Ar^2}{4}\left(1 - \frac{2\sin^3\alpha\cos\alpha}{3(\alpha - \sin\alpha\cos\alpha)}\right); \quad I_a = \frac{Ar^2}{4}\left(1 + \frac{2\sin^3\alpha\cos\alpha}{(\alpha - \sin\alpha\cos\alpha)}\right)$$ $$k_g = \frac{r}{2}\sqrt{1 - \frac{2\sin^3\alpha\cos\alpha}{3(\alpha - \sin\alpha\cos\alpha)}}; \quad k_a = \frac{r}{2}\sqrt{1 + \frac{2\sin^3\alpha\cos\alpha}{(\alpha - \sin\alpha\cos\alpha)}}$$
14. Annulus 	C.G. is at O = geometrical center. $$I_g = \frac{\pi(d_1^4 - d_2^4)}{64} = \frac{\pi(r_1^4 - r_2^4)}{4}; \quad J_o = \frac{\pi(d_1^4 - d_2^4)}{32} = \frac{\pi(r_1^4 - r_2^4)}{2}$$ $$k_g = \frac{\sqrt{d_1^2 + d_2^2}}{4} = \frac{\sqrt{r_1^2 + r_2^2}}{2}; \quad k_o = \sqrt{\frac{d_1^2 + d_2^2}{8}} = \sqrt{\frac{r_1^2 + r_2^2}{2}}$$

TABLE 3.1 (*Continued*)

Plane Surfaces	
Figure	Centroid Location; Moments of Inertia; Radii of Gyration

15. Ellipse

C.G. is at O = geometrical center. *For semi-ellipse* ABB', C.G. is on OA at distance to right of $c - c = \dfrac{4a}{3\pi}$. *For quarter-ellipse* ABO, C.G. is at distance to right of $c - c = \dfrac{4a}{3\pi}$ and at distance above $g - g = \dfrac{4b}{3\pi}$.

$$I_g = \frac{\pi ab^3}{4} = \frac{Ab^2}{4}; \quad I_c = \frac{\pi a^3 b}{4} = \frac{Aa^2}{4}; \quad J_o = \frac{A(a^2 + b^2)}{4}$$

$$k_g = \frac{b}{2}; \quad k_c = \frac{a}{2}; \quad k_o = \frac{\sqrt{a^2 + b^2}}{2}$$

16. Parabolic Segment

C.G. is on axis of symmetry at O. Distance from $c - c = \dfrac{3a}{5}$.

$$I_g = \frac{4ab^3}{15}; \quad I_c = \frac{4a^3 b}{7}$$

$$k_g = \frac{b}{\sqrt{5}} = 0.447b; \quad k_c = a\sqrt{\frac{3}{7}} = 0.654a$$

Homogeneous Bodies (Including Nonplanar Surfaces)	

17. Any Surface or Body of Revolution

Let axis of revolution be X axis. Then generating curve is $y = f(x)$. C.G. is at point having coordinates $\bar{x}, \bar{y}, \bar{z}$.

For Surface:

$$\bar{x} = \frac{\int 2\pi xy \, ds}{\int 2\pi y \, ds} = \frac{\int xy \sqrt{1 + \left(\dfrac{dy}{dx}\right)^2} \, dx}{\int y \sqrt{1 + \left(\dfrac{dy}{dx}\right)^2} \, dx}; \quad \bar{y} = 0; \quad \bar{z} = 0$$

For Body $\left(\text{letting } \delta = \text{density} = \dfrac{m}{\text{volume}}\right)$:

$$\bar{x} = \frac{\int \pi xy^2 \, dx}{\int \pi y^2 \, dx}; \quad \bar{y} = 0; \quad \bar{z} = 0.$$

$$I_x = \frac{\pi \delta}{2} \int y^4 \, dx; \quad I_y = I_z = \pi \delta \int \left(\frac{y^4}{4} + x^2 y^2\right) dx$$

$$k_x = \sqrt{\frac{I_x}{m}}; \quad k_y = k_z = \sqrt{\frac{I_y}{m}} = \sqrt{\frac{I_z}{m}}$$

For Thin Shell having mass:
C.G. coordinates are same as for surface.

$$I_x = 2\pi \delta \int y^3 \, ds = 2\pi \delta \int y^3 \sqrt{1 + \left(\frac{dy}{dx}\right)^2} \, dx; \quad k_x = \sqrt{\frac{I_x}{m}}$$

18. Thin Straight Rod

C.G. is at O = geometrical center. *For Body:*

$$I_g = \frac{ml^2}{12}; \quad I_b = \frac{ml^2}{3}; \quad I_c = \frac{ml^2 \sin^2 a}{12}; \quad I_d = \frac{ml^2 \sin^2 \alpha}{3}$$

$$k_g = \frac{l}{\sqrt{12}}; \quad k_b = \frac{l}{\sqrt{3}}; \quad k_c = \frac{l \sin \alpha}{\sqrt{12}}; \quad k_d = \frac{l \sin \alpha}{\sqrt{3}}$$

19. Thin Rod Bent into Circular Arc

C.G. is on axis of symmetry at $\bar{x} = \dfrac{r \sin \alpha}{\alpha}$. *For Body:*

$$I_x = \frac{mr^2}{2}\left(1 - \frac{\sin \alpha \cos \alpha}{\alpha}\right); \quad I_y = \frac{mr^2}{2}\left(1 + \frac{\sin \alpha \cos \alpha}{\alpha}\right); \quad I_z = mr^2$$

$$k_x = r\sqrt{\frac{1}{2} - \frac{\sin \alpha \cos \alpha}{2\alpha}}; \quad k_y = r\sqrt{\frac{1}{2} + \frac{\sin \alpha \cos \alpha}{2\alpha}}; \quad k_z = r$$

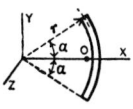

TABLE 3.1   (*Continued*)

Homogeneous Bodies (*Continued*)*	
Figure	Centroid Location; Moments of Inertia; Radii of Gyration

20. Rectangular Parallelepiped (or Cube)

C.G. is at O = geometrical center.
For Parallelepiped:

$$I_g = \frac{m(b^2 + c^2)}{12}; \quad I_d = \frac{m(a^2 + b^2)}{12}; \quad I_e = \frac{m(4a^2 + b^2)}{12}$$

$$k_g = \sqrt{\frac{b^2 + c^2}{12}}; \quad k_d = \sqrt{\frac{a^2 + b^2}{12}}; \quad k_e = \sqrt{\frac{4a^2 + b^2}{12}}$$

For Cube (letting $a = b = c = s$):

$$I_g = I_d = \frac{ms^2}{6}; \quad I_e = \frac{5ms^2}{12}$$

$$k_g = k_d = \frac{s}{\sqrt{6}}; \quad k_e = s\sqrt{\frac{5}{12}}$$

21. Right Rectangular Pyramid

C.G. is on axis of symmetry at O. Distance from base $= \frac{h}{4}$.

Drawing $g - g$ axis through O parallel to side a:

$$I_g = \frac{m}{20}\left(b^2 + \frac{3h^2}{4}\right); \quad I_c = \frac{m}{20}(a^2 + b^2)$$

$$k_g = \sqrt{\frac{4b^2 + 3h^2}{80}}; \quad k_c = \sqrt{\frac{a^2 + b^2}{20}}$$

22. Pyramid (or Frustum of Pyramid)

For Surface of Any Pyramid:
C.G. of surface (base excluded) is on line joining apex with centroid of perimeter of base, at a distance two-thirds its length from the apex.
For Body of Any Pyramid:
C.G. of body is on line joining apex with centroid of base, at a distance three-fourths its length from the apex.
For Surface of Frustum of Pyramid having Rectangular Bases:
Letting R and r be the lengths of sides of the larger and smaller bases respectively, and h the altitude:
C.G. of surface (bases excluded) is at distance from larger base $= \dfrac{h(R + 2r)}{3(R + r)}$.
For Body of Frustum of Any Pyramid:
Letting A and a be the areas of the larger and smaller bases, respectively, and h the altitude:

C.G. of body is at distance from larger base $= \dfrac{h(A + 2\sqrt{Aa} + 3a)}{4(A + \sqrt{Aa} + a)}$.

23. Right Elliptical Cylinder (or Circular Cylinder)

C.G. is at O = geometrical center.
For Right Elliptical Cylinder:

$$I_g = \frac{m}{12}(3b^2 + h^2); \quad I_c = \frac{m}{4}(a^2 + b^2); \quad I_e = \frac{m}{12}(3r^2 + 4h^2)$$

$$k_g = \sqrt{\frac{3b^2 + h^2}{12}}; \quad k_c = \frac{\sqrt{a^2 + b^2}}{2}; \quad k_e = \sqrt{\frac{3r^2 + 4h^2}{12}}$$

For Right Circular Cylinder (letting $a = b = r$):

$$I_g = \frac{m}{12}(3r^2 + h^2); \quad I_c = \frac{mr^2}{2}$$

$$k_g = \sqrt{\frac{3r^2 + h^2}{12}}; \quad k_c = \frac{r}{\sqrt{2}}$$

*"Body" is to be understood unless "Surface" is indicated.

TABLE 3.1 (*Continued*)

Homogeneous Bodies (*Continued*)*	
Figure	Centroid Location; Moments of Inertia; Radii of Gyration
24. Hollow Right Circular Cylinder	C.G. is at O = geometrical center. $$I_g = \frac{m}{12}(3R^2 + 3r^2 + h^2); \quad I_c = \frac{m}{2}(R^2 + r^2); \quad I_e = \frac{m}{12}(3R^2 + 3r^2 + 4h^2)$$ $$k_g = \sqrt{\frac{3R^2 + 3r^2 + h^2}{12}}; \quad k_c = \sqrt{\frac{R^2 + r^2}{2}}; \quad k_e = \sqrt{\frac{3R^2 + 3r^2 + 4h^2}{12}}$$ For Thin Shell (radius R): $$I_g = \frac{m}{12}(6R^2 + h^2); \quad I_c = mR^2; \quad I_e = \frac{m}{6}(3R^2 + 2h^2)$$ $$k_g = \sqrt{\frac{6R^2 + h^2}{12}}; \quad k_c = R; \quad k_e = \sqrt{\frac{3R^2 + 2h^2}{6}}$$
25. Right Circular Cone	C.G. is on axis of symmetry at O. Distance from base = $\frac{h}{4}$. Drawing $g-g$ axis through O and $d-d$ axis through apex, both parallel to base: $$I_g = \frac{3m}{20}\left(r^2 + \frac{h^2}{4}\right); \quad I_c = \frac{3mr^2}{10}; \quad I_d = \frac{3m}{20}(r^2 + 4h^2)$$ $$k_g = \sqrt{\frac{3}{80}(4r^2 + h^2)}; \quad k_c = \frac{3r}{\sqrt{30}}; \quad k_d = \sqrt{\frac{3}{20}(r^2 + 4h^2)}$$
26. Frustum of Right Circular Cone	C.G. is on axis of symmetry at O. $$\text{Distance from base} = \frac{h(R^2 + 2Rr + 3r^2)}{4(R^2 + Rr + r^2)}.$$ $$I_c = \frac{3m(R^5 - r^5)}{10(R^3 - r^3)}; \quad k_c = \sqrt{\frac{3(R^5 - r^5)}{10(R^3 - r^3)}}$$
27. Cone (or Frustum of Cone)	For Surface of Any Cone: C.G. of surface (base excluded) is on line joining apex with centroid of perimeter of base, at a distance two-thirds its length from the apex. For Body of Any Cone: C.G. of body is on line joining apex with centroid of base, at a distance three-fourths its length from the apex. For Surface of Frustum of a Circular Cone: Letting R and r be the radii of the larger and smaller bases, respectively, and h the altitude: C.G. of surface (bases excluded) is at distance from larger base = $\frac{h(R + 2r)}{3(R + r)}$. For Body of Frustum of a Circular Cone: Letting R and r be the radii of the larger and smaller bases, respectively, and h the altitude: C.G. of body is at distance from larger base = $\frac{h(R^2 + 2Rr + 3r^2)}{4(R^2 + Rr + r^2)}$
28. Thin Circular Lamina	C.G. is at O = geometrical center. $$I_g = \frac{mr^2}{4}; \quad I_c = \frac{mr^2}{2} \text{ (where } c-c \text{ axis is perpendicular to the plane).}$$ $$k_g = \frac{r}{2}; \quad k_c = \frac{r}{\sqrt{2}}$$
29. Sphere	C.G. is at O = geometrical center. $$I_g = \frac{2mr^2}{5}$$ $$k_g = \frac{2r}{\sqrt{10}}$$

*"Body" is to be understood unless "Surface" is indicated.

TABLE 3.1 (*Continued*)

Homogeneous Bodies (*Continued*)*	
Figure	Centroid Location; Moments of Inertia; Radii of Gyration
30. Hollow Sphere 	C.G. is at O = geometrical center. $$I_g = \frac{2m}{5}\left(\frac{R^5 - r^5}{R^3 - r^3}\right); \quad k_g = \sqrt{\frac{2}{5}\left(\frac{R^5 - r^5}{R^3 - r^3}\right)}$$ For Thin Shell (radius R): $$I_g = \frac{2mR^2}{3}; \quad k_g = \frac{2R}{\sqrt{6}}$$
31. Spherical Sector 	C.G. is on axis of symmetry at O. Distance from center of sphere = $$\frac{3(2r - h)}{8}$$ $$I_g = \frac{m}{5}(3rh - h^2)$$ $$k_g = \sqrt{\frac{3rh - h^2}{5}}$$
32. Hemisphere 	*For Surface:* C.G. is on axis of symmetry at distance from center of sphere = $\dfrac{r}{2}$. *For Body:* C.G. is on axis of symmetry at distance from center of sphere = $\dfrac{3r}{8}$. $$I_g = \frac{2mr^2}{5}; \quad k_g = \frac{2r}{\sqrt{10}}$$
33. Spherical Segment 	C.G. is on axis of symmetry at distance from center of sphere = $$\frac{3(2r - h)^2}{4(3r - h)}.$$ $$I_g = m\left(r^2 - \frac{3rh}{4} + \frac{3h^2}{20}\right)\frac{2h}{(3r - h)}$$ $$k_g = \sqrt{\left(r^2 - \frac{3rh}{4} + \frac{3h^2}{20}\right)\frac{2h}{3r - h}}$$
34. Torus 	C.G. is at O = geometrical center. $$I_g = \frac{m(4R^2 + 5r^2)}{8}; \quad I_c = \frac{m(4R^2 + 3r^2)}{4}$$ $$k_g = \sqrt{\frac{4R^2 + 5r^2}{8}}; \quad k_c = \frac{\sqrt{4R^2 + 3r^2}}{2}$$
35. Ellipsoid 	C.G. is at O = geometrical center. C.G. of *one octant* is at point having coordinates: $$\bar{x} = \frac{3a}{8}; \quad \bar{y} = \frac{3b}{8}; \quad \bar{z} = \frac{3c}{8}$$ *For Complete Ellipsoid:* $$I_x = \frac{m}{5}(b^2 + c^2); \quad I_y = \frac{m}{5}(a^2 + c^2); \quad I_z = \frac{m}{5}(a^2 + b^2)$$ $$k_x = \sqrt{\frac{b^2 + c^2}{5}}; \quad k_y = \sqrt{\frac{a^2 + c^2}{5}}; \quad k_x = \sqrt{\frac{a^2 + b^2}{5}}$$
36. Paraboloid 	C.G. is on axis of symmetry at O. Distance from base = $\dfrac{h}{3}$. $$I_g = \frac{mr^2}{3}; \quad I_c = \frac{m}{18}(3r^2 + h^2)$$ $$k_g = \frac{r}{\sqrt{3}}; \quad k_c = \sqrt{\frac{3r^2 + h^2}{18}}$$

*"Body" is to be understood unless "Surface" is indicated.

3.3 KINEMATICS

3.3.1 Motions of a Particle

Definitions
Motion of a particle is the continual change of its position with respect to some base point or coordinate set. Particle motion is characterized by a time-dependent position vector, $\mathbf{r}(t)$, that describes the location of the particle with respect to the base point or coordinates.

Rectilinear motion is motion along a straight line.

Curvilinear motion is motion along a curved path that may be planar or in three dimensions.

Displacement of a particle is its change of position and is a vector quantity. The displacement vector associated with the times t_1 and t_2, $\Delta\mathbf{r}$, is given by

$$\Delta\mathbf{r} = \mathbf{r}(t_2) - \mathbf{r}(t_1)$$

Velocity of a particle is the time rate of change of its displacement and is a vector quantity. If the time interval $t_2 - t_1$ is denoted as Δt, then

$$\mathbf{v} = \lim_{\Delta t \to 0} \frac{\Delta\mathbf{r}}{\Delta t} = \frac{d\mathbf{r}}{dt}$$

Acceleration of a particle is the time rate of change of its velocity vector, and is itself a vector quantity.

$$\mathbf{a} = \lim_{\Delta t \to 0} \frac{\Delta\mathbf{v}}{\Delta t} = \frac{d\mathbf{v}}{dt}$$

3.3.2 Rectilinear Motion

Velocity. Let s = distance measured along the path of a particle, s_1 = distance from origin at time t_1, s_2 = distance at a later time t_2, $\Delta s = s_2 - s_1$ = *displacement** in time interval $\Delta t = t_2 - t_1$. Then *average velocity* $= \Delta s/\Delta t$. If the position changes at at uniform rate (which implies no change in sense), actual velocity at any time $= \Delta s/\Delta t$. For every case, *instantaneous velocity* is

$$v = \frac{ds}{dt} = \lim_{\Delta t \to 0} \frac{\Delta s}{\Delta t}$$

Unit of velocity is any distance unit divided by any time unit. *Units* commonly used are feet per second, meters per second, miles per hour, and so on.

Acceleration. Let v_1 = velocity of particle at time t_1, v_2 = velocity at a later time t_2, $\Delta v = v_2 - v_1$ = change in velocity in time interval $\Delta t = t_2 - t_1$. Then *average acceleration* $= \Delta v/\Delta t$. If the velocity changes at a uniform rate, the actual acceleration at any time $= \Delta v/\Delta t$. For every case, *instantaneous acceleration* equals

$$a = \frac{dv}{dt} = \frac{d^2s}{dt^2} = \lim_{\Delta t \to 0} \frac{\Delta v}{\Delta t}$$

Unit of acceleration is distance unit divided by the square of any time unit. Units commonly used are feet per second per second (feet per second squared), meters per second squared, and so on.)

Formulas for Determination of a, v, s, t. If s is given algebraically in terms of t, then v and a may be determined in terms of t by differentiation as indicated previously. If a is given algebraically in terms of t, then v and s may be determined in terms of t by integration. Other relations not involving t may be determined by similar methods. The common formulas are:

$$v = \frac{ds}{dt} \qquad a = \frac{dv}{dt} = \frac{d^2s}{dt^2} \qquad \frac{a}{v} = \frac{dv}{ds}$$

$$s_2 - s_1 = \int_{t_1}^{t_2} v\,dt \qquad v_2 - v_1 = \int_{t_1}^{t_2} a\,dt$$

$$t_2 - t_1 = \int_{s_1}^{s_2} \frac{ds}{v} = \int_{v_1}^{v_2} \frac{dv}{a} \qquad v_2^2 - v_1^2 = 2\int_{s_1}^{s_2} a\,ds$$

*The difference in distances along the path equals the displacement *only* when the path is a straight line. (See definition of displacement.)

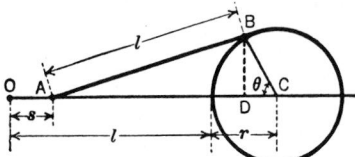

Fig. 3.21 Motion graph.

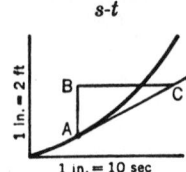

Fig. 3.22 Motion graph.

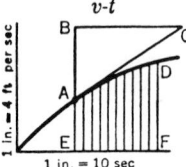

Fig. 3.23 Velocity versus time.

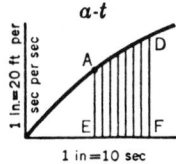

Fig. 3.24 Acceleration versus time.

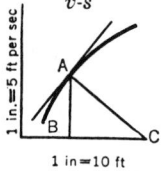

Fig. 3.25 Velocity versus distance.

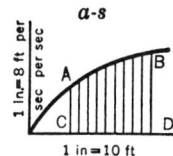

Fig. 3.26 Acceleration versus distance.

For **uniform acceleration**, a = constant; $v = at + v_0$; $s = \frac{1}{2}at^2 + v_0t + s_0$; $v^2 = 2a(s - s_0) + v_0^2$; v_0 being initial velocity and s_0 initial distance.

If algebraic relations between a, v, s, and t are not given but a number of pairs of corresponding values of two of the variables are known, curves may be plotted for the approximate determination of other corresponding pairs of values and of other unknowns, within the range of the data. Such curves are discussed later under "Motion Graphs."

Examples of Rectilinear Motion

Falling Body.[†] If a body *falls from rest* in a vacuum, $v_0 = 0$, $s_0 = 0$, and $a = g = 32.2$ ft/sec^2 (approx.). Hence $v = gt = \sqrt{2gs}$; $s = \frac{1}{2}gt^2$. If a body is *projected upward* at an initial velocity v_0, $a = -g$ and the formulas become $v = -gt + v_0 = \sqrt{-2gs + v_0^2}$, $s = -\frac{1}{2}gt^2 + v_0t$. Total ascent (to highest position) = $v_0^2/2g$, and time required = v_0/g.

Crank and Connecting Rod Mechanism. The problem is to find expressions for the velocity and acceleration of any point in the crosshead, as A in Fig. 3.21. Such a point describes rectilinear motion. Let $c = r/l$, n = revolutions per second (assumed constant), ω = radians of angle described by crank per second, and s = distance of A from its extreme left position, all distances expressed in feet. Then,

$$s = (l + r) - l(1 - c^2\sin^2\theta)^{1/2} - r\cos\theta$$

$$v = r\omega\left(\sin\theta + \frac{c\sin 2\theta}{2(1 - c^2\sin^2\theta)^{1/2}}\right) \qquad a = r\omega^2\left(\cos\theta + \frac{c\cos 2\theta + c^3\sin^4\theta}{(1 - c^2\sin^2\theta)^{3/2}}\right)$$

These formulas are exact; close approximations are:

$$s = r(1 - \cos\theta) + \frac{1}{4}cr(1 - \cos 2\theta)$$

$$v = r\omega\left(\sin\theta + \frac{1}{2}c\sin 2\theta\right)$$

$$a = r\omega^2(\cos\theta + c\cos 2\theta)$$

Motion Graphs

Space-time, velocity-time, acceleration-time, velocity-space, and acceleration-space curves for a particle are graphs showing the relations between magnitudes of s and t, v and t, a and t, v and s, a and s, respectively. Figures 3.22–3.26 illustrate such graphs but do not correspond to the same motion.

[†]Rotation disregarded and body considered as a particle.

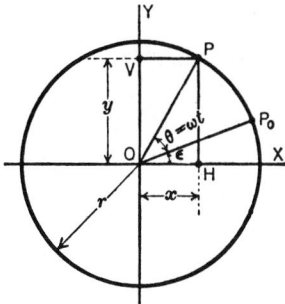

Fig. 3.27   Simple harmonic motion.

Space-Time Diagram. In Fig. 3.22, the *slope* of the curve at any point represents the magnitude of the velocity. If AB and BC are measured by the s and t scales of the drawing, respectively, the slope equals the velocity magnitude; thus if $AB = 0.2$ in. and $BC = 0.4$ in., $v = 0.4/4 = 0.1$ ft/sec.

Velocity-Time Diagram. In Fig. 3.23, the *slope* of the curve at any point represents the magnitude of the acceleration.* If AB and BC are measured by the v and t scales, respectively, the slope equals the acceleration magnitude;* thus if $AB = 0.3$ in. and $BC = 0.5$ in., $a = 1.2/5 = 0.24$ ft/sec sec. The *area* included between any two ordinates (as AE and DF), the curve, and the t axis, represents the displacement[†] of the moving point in the time EF. If area is below the time axis, it is considered minus. If the area is computed by multiplying its average ordinate measured by the velocity scale (this being the average velocity) by EF measured by the time scale, the product equals the displacement; thus if the average ordinate is 0.35 in., and EF is 0.4 in., the displacement $= 1.4 \times 4 = 5.6$ ft.

Acceleration-Time Diagram. In Fig. 3.24, the *slope* represents the rate at which the acceleration is changing.[‡] The *area* (plus above and minus below time axis) included between any two ordinates (as AE and DF), the curve, and the t axis, represents the velocity change in the time EF.[‡] Thus if the average ordinate is 0.3 in. and EF is 0.2 in., the velocity change $= 6 \times 2 = 12$ ft/sec.

Velocity-Space Diagram. In Fig. 3.25, the subnormal represents the acceleration.* If the length of the subnormal is multiplied by the square of the velocity scale number and the product is divided by the space scale number, the result will equal the acceleration;* thus suppose that the subnormal $BC = \frac{1}{3}$ in., then $a = (\frac{1}{3} \times 25)/10 = 0.83$ ft/sec^2.

Acceleration-Space Diagram. In Fig. 3.26, the area (plus above and minus below space axis) included between two ordinates (as AC and BD), the curve, and the s axis, represents the change in the velocity square. If the area is computed by multiplying the mean ordinate measured by the acceleration scale by CD measured by the space scale, the product times two equals the change in the velocity square; thus if the average ordinate $= 0.3$ in., and $CD = 0.4$ in., the change $= 2.4 \times 4 \times 2 = 19.2$ ft^2/sec^2.

Simple Harmonic Motion

Simple Harmonic Motion and Its Motion Graphs. These have wide application in physics and engineering. If a point P moves in a circular path of radius r at uniform speed, its projection on any diameter has *simple harmonic motion*. The radius r is called the *amplitude*. The *period* is the time required for the projection to go from one end of the diameter to the other and back. The *frequency* is the number of periods per unit time, which makes it the reciprocal of the period. Angle XOP (Fig. 3.27) (considered as less than 2π radians) is the *phase angle*. *The displacement* at any time is the distance of the point having simple harmonic motion from the center of its path or range.

When $t = 0$, let P be at P_0. ε is called the *lead angle* (*lag, if negative*). For simple harmonic motion (SHM) of V in the vertical diameter, $y = r \sin(\theta + \varepsilon) = r \sin(\omega t + \varepsilon)$, in which $\omega = d\theta/dt$

---

*For curvilinear motion, this is tangential acceleration only.
[†]For curvilinear motion, this is distance along the path (not displacement).
[‡]For rectilinear motion only.
*For curvilinear motion, this is tangential acceleration only.

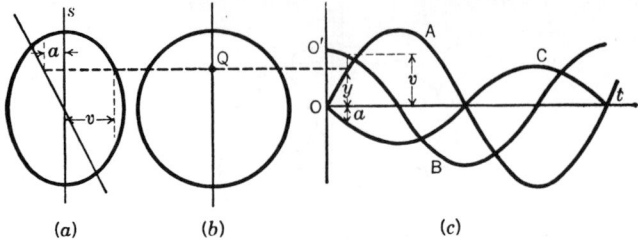

Fig. 3.28   (*a*) Space-time curve; (*b*) velocity-time curve; (*c*) acceleration-time curve.

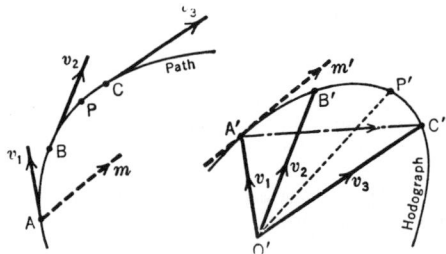

Fig. 3.29   Curvilinear motion.

= radians per unit time (i.e., 2π times the frequency).

$$v_y = r\omega \cos(\omega t + \varepsilon) = \omega x$$

$$a_y = -r\omega^2 \sin(\omega t + \varepsilon) = -\omega^2 y$$

For SHM of H in horizontal diameter:

$$x = r\cos(\theta + \varepsilon) = r\cos(\omega t + \varepsilon) \qquad v_x = -r\omega \sin(\omega t + \varepsilon) = -\omega y$$

$$a_x = -r\omega^2 \cos(\omega t + \varepsilon) = -\omega^2 x$$

If the time is reckoned from the instant when V is in its mid-position, and moving upward, $\varepsilon = 0$. The three curves (Fig. 3.28) OA, $O'B$, and OC are the space-time, velocity-time, and acceleration-time curves, respectively, for one complete period of a simple harmonic motion; $\varepsilon = 0$; Ot represents the period; the values of y, v, and a marked are for position Q, shown. In Fig. 3.28*a* the curve is the velocity-space curve, and the inclined line the acceleration-space curve. They show how v and a vary with the displacement of the moving point; thus for the position Q (Fig. 3.28*b*), v and a have values as marked.

From the preceding equations and curves, it will be noted that simple harmonic motion may be defined also as any rectilinear motion in which the acceleration is always directed toward a fixed point in the path and is proportional to the distance between that point and the moving point.[†]

3.3.3   Curvilinear Motion

Velocity. If s is distance measured along the curved path of a particle, then the magnitude of velocity (speed) at any instant $= ds/dt$; the linear direction of the velocity is tangent to the path at the instantaneous position of the particle; and the sense of the velocity corresponds to the direction of motion of the particle at the instant.

The velocity vector changes in magnitude and direction. In Fig. 3.29, let A, B, C represent positions of particle P in its curved path; s, distance along the path; and v_1, v_2, v_3, velocity vectors at A, B, C. Plot velocity vectors $O'A'$, $O'B'$, $O'C'$, and so on, from any origin O' to represent the

[†]A common example of simple harmonic motion is the motion of a weight suspended from an elastic spring.

velocities at A, B, C, and so on. The curve $A'B'C'$, drawn through the ends of the vectors, is called a *hodograph* for the motion. For every position of P in its path, there is a corresponding position P' in the hodograph; and P' describes distance s' on the hodograph while P describes distance s on the path. Vector $O'P'$ represents the velocity of P. In time Δt, P moves from A to C, its velocity changes from $O'A'$ to $O'C'$, and the velocity change is $A'C'$.

Acceleration. Referring to the hodograph (Fig. 3.29), *average acceleration* for interval Δt, during which particle P moves from A to C, is vector $A'C'/\Delta t$, and it has the direction of the chord $A'C'$. The *instantaneous acceleration* of P at $A = a =$ limit of the average acceleration as Δt approaches zero.

$$a = \lim_{\Delta t \to 0} \left(\frac{\text{vector } A'C'}{\Delta t} \right) = \lim_{\Delta t \to 0} \frac{\text{arc } A'B'C'}{\Delta t} = \frac{ds'}{dt} = \text{speed of } P'$$

on hodograph. The direction of a is along the tangent $A'm'$, and as P' is moving clockwise, the sense is as indicated by arrow at m'. Hence acceleration at A is Am, parallel to $A'M'$ and $= ds'/dt$. Its tangential component is ds/dt' and its normal component is v^2/ρ, ρ being the radius of curvature at A. *Unit of acceleration* is any velocity unit divided by any time unit.

Components of Velocity and Acceleration

Components of Velocity and Acceleration of a Particle for Any Curved Path (*not necessarily planar*). The position of the particle P being defined by its coordinates, x, y, z, the axial components of velocity are $v_x = dx/dt$, $v_y = dy/dt$, and $v_z = dz/dt$. Resultant velocity $v = \sqrt{v_x^2 + v_y^2 + v_z^2}$, and its direction cosines are $\cos\theta_x = v_x/v$, $\cos\theta_y = v_y/v$, and $\cos\theta_z = v_z/v$. Axial components of acceleration are

$$a_x = \frac{dv_x}{dt} = \frac{d^2x}{dt^2} \qquad a_y = \frac{dv_y}{dt} = \frac{d^2y}{dt^2} \qquad a_z = \frac{dv_z}{dt} = \frac{d^2z}{dt^2}$$

Resultant acceleration $a = \sqrt{a_x^2 + a_y^2 + a_z^2}$; and its direction cosines are:

$$\cos\phi_x = \frac{a_x}{a} \qquad \cos\phi_y = \frac{a_y}{a} \qquad \cos\phi_z = \frac{a_z}{a}$$

The tangential and normal components of acceleration are $a_t = dv/dt = d^2s/dt^2$, and $a_n = v^2/\rho$, ρ being the radius of curvature. Resultant acceleration is

$$a = \sqrt{a_t^2 + a_n^2} = \sqrt{a_x^2 + a_y^2 + a_z^2}$$

If the path is a plane curve, $v_z = 0$ and $a_z = 0$.

The preceding discussion shows that velocities and accelerations (like forces) may be composed or resolved according to the parallelogram and parallelepipedon laws.

Motion of a Projectile

Projectile* Describing Plane Curvilinear Motion. In the following formulas air resistance is neglected; $v_0 =$ velocity of projection; $\theta =$ angle of projection (Fig. 3.30); x and $y =$ coordinates of the projectile at any time t after projection; $v =$ velocity; v_x and $v_y = x$ and y components, respectively, of v; $r =$ range on the horizontal plane through O; $\theta_1 =$ value of θ for maximum r; $h =$ greatest height attained; and $T =$ time of flight. The path of the projectile, or the trajectory, is a parabola as represented, and a set of parametric equations for it are

$$x = v_0\cos\theta \cdot t \qquad y = v_0\sin\theta \cdot t - \tfrac{1}{2}gt^2$$

from which

$$y = x\tan\theta - \frac{gx^2}{2v_0^2\cos^2\theta}$$

---

*Rotation disregarded and body considered as a particle.

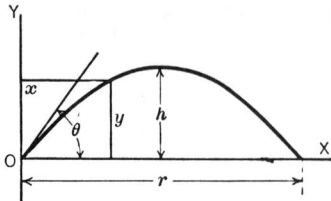

Fig. 3.30   Projectile motion.

Also:

$$v_x = v_0\cos\theta \qquad v_y = v_0\sin\theta - gt \qquad v = \sqrt{v_0^2 - 2gy} \qquad h = \frac{\sin^2\theta \cdot v_0^2}{2g}$$

$$r = \frac{\sin 2\theta \cdot v_0^2}{g} \qquad \theta_1 = 45° \qquad T = \frac{2v_0\sin\theta}{g}$$

If the direction of projection is horizontal, $\theta = 0$; the equation of the path is $y = -gx^2/2v_0^2$; and $x = v_0 t$, and $y = -\frac{1}{2}gt^2$.

The fact that the horizontal component of velocity is constant indicates that the hodograph of the motion of a projectile is a straight vertical line.

Motion Graphs

Motion graphs similar to those previously discussed for rectilinear motion may be constructed for curvilinear motion of a particle. Great care must be exercised, however, in interpreting the significance of slopes, areas, and subnormals when acceleration or distance is involved. In this connection, reference should be made to the footnotes referred to in the previous discussion.

In general, accelerations obtained are tangential components only, while "displacements" must be replaced by "distances along the curve." Thus, in the velocity-time graph (Fig. 3.23), the slope of the curve represents the magnitude of the *tangential component* of the acceleration,[†] while the area under the curve represents the *distance along* the curve.

3.3.4   Motions of a Body

Translation of a rigid body is a motion such that each straight line in it remains fixed in direction. The *paths* of all particles of the body are exactly alike, straight or curved (not necessarily plane curves); the *displacements* of all particles during a given time are the same; the *velocities* of all particles at any instant are the same; and their *accelerations* at any instant are the same. For these reasons, it is customary to use the expressions "velocity of the body" and "acceleration of the body." The motion is described by the same formulas as those previously derived for rectilinear and curvilinear motions of a particle.

Rotation of a rigid body is a motion such that one line of the body, or of its extension, remains fixed. The fixed line is the *axis*. The plane through the mass center perpendicular to the axis is the *plane of rotation*.

Plane motion of a rigid body is a motion such that each particle of the body moves in a plane at a constant distance from a fixed plane through the mass center (called the *plane of motion*), while each line of the body parallel to the plane of motion turns through the same angle in the same time interval.

Three-dimensional motion of a rigid body is a term covering all types of motion in three-dimensional space, including pure translation along a skewed curve as a special case. Even in the most general case, any three-dimensional motion of a rigid body may be regarded as consisting of two components: one, a translation equal to that of the mass center, and the other a rotation about some axis through the mass center.

Angular displacement of a rigid body is the change of angular position of any line in the plane of motion. The angular displacement of a rigid body *cannot* be expressed as a vector quantity.

[†]However, if a velocity-time graph were made for motion along the hodograph of the original motion, the slope of the curve would represent the magnitude of the *total* acceleration of the original particle along the original path.

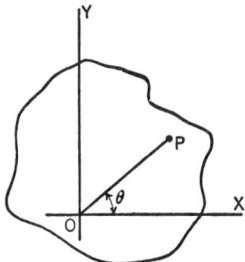

Fig. 3.31   Angular displacement.

Angular velocity of a rigid body is its time rate of angular displacement (i.e., rate of change of angular position). The angular velocity of a rigid body can be expressed as a vector quantity.

Angular acceleration of a rigid body is its time rate of change of angular velocity. The angular acceleration of a rigid body can be expressed as a vector quantity.

3.3.5   Rotation

Angular Velocity. The paths of all particles are circles with centers on the axis. Since all lines of the body parallel to the plane of rotation sweep out equal angles in equal times, it is customary to describe rotation by the behavior of one radial line. In Fig. 3.31, let θ be the angle from the x axis to the radial line OP. $\Delta\theta = \theta_2 - \theta_1$ is the *angular displacement* of the body in the time $\Delta t = t_2 - t_1$, and is expressed in any angular unit.

$$Average\ angular\ velocity = \frac{\theta_2 - \theta_1}{t_2 - t_1} = \frac{\Delta\theta}{\Delta t}$$

If the angle changes at a uniform rate, actual velocity at any time $= \Delta\theta/\Delta t$. For every case, the *instantaneous angular velocity* is

$$\omega = \lim_{\Delta t \to 0}\left(\frac{\Delta\theta}{\Delta t}\right) = \frac{d\theta}{dt}$$

The sign depends on the numerator of the fraction, or the way in which θ is changing. The *unit of angular velocity* is any angular displacement unit divided by any time unit, such as radians per second, revolutions per minute, and so on.

Angular velocity is generally expressed as a vector quantity. The direction of the vector is along the instantaneous axis of rotation, and the magnitude of the vector is equal to the instantaneous angular velocity. The sense along the axis is such that the right-hand rule applies (if the right hand grasps the axis of rotation with the fingers curled in the direction of rotation, the right thumb points in the direction of the angular velocity vector).

Angular Acceleration

$$Average\ angular\ acceleration = \frac{\omega_2 - \omega_1}{t_2 - t_1} = \frac{\Delta\omega}{\Delta t}$$

If the angular velocity changes at a uniform rate, actual angular acceleration at any time $= \Delta\omega/\Delta t$. For every case, the *instantaneous angular acceleration* is

$$\alpha = \lim_{\Delta t \to 0}\left(\frac{\Delta\omega}{\Delta t}\right) = \frac{d\omega}{dt} = \frac{d^2\theta}{dt^2}$$

The sign of α depends on the numerator of the fraction, or on the way in which ω is changing. The *unit of angular acceleration* is any angular velocity unit divided by any time unit, as radians per second per second (i.e., radians per second^2), and so on.

Formulas for Determination of $\alpha, \omega, \theta, t$. The formulas are exactly analogous to those previously derived for rectilinear motion, a, v, and s being replaced by α, ω, and θ, respectively. The formulas are

$$\omega = \frac{d\theta}{dt} \qquad \alpha = \frac{d\omega}{dt} = \frac{d^2\theta}{dt^2} \qquad\qquad \frac{\alpha}{\omega} = \frac{d\omega}{d\theta}$$

$$\theta_2 - \theta_1 = \int_{t_1}^{t_2} \omega \, dt \qquad \omega_2 - \omega_1 = \int_{t_1}^{t_2} \alpha \, dt \qquad t_2 - t_1 = \int_{\theta_1}^{\theta_2} \frac{d\theta}{\omega} = \int_{\omega_1}^{\omega_2} \frac{d\omega}{\alpha} \qquad \omega_2^2 - \omega_1^2 = 2\int_{\theta_1}^{\theta_2} \alpha \, d\theta$$

Angular acceleration is generally expressed as a vector quantity. The magnitude of the vector is equal to the instantaneous angular acceleration. The direction of the angular acceleration vector is perpendicular to the plane formed by successive values of the angular velocity vector. The sense along the axis is such that the right-hand rule applies (if the right hand grasps the axis of rotation with the fingers curled in the direction of increasing angular velocity, the right thumb points in the direction of the angular acceleration vector).

In the case of planar motion, there is no plane formed by successive values of the angular velocity vector. In this case, the angular velocity vector and the angular acceleration vector both lie perpendicular to the plane of motion.

Relations between Rectilinear and Angular Velocities and Accelerations. Let ω and α, respectively, be instantaneous angular velocity and acceleration of a rotating body, and v and a the corresponding instantaneous rectilinear velocity and acceleration of a point P of the body located at distance r from the axis of rotation. Then

$$v = r\omega \qquad a_t = r\alpha \qquad a_n = r\omega^2 \qquad a = r\sqrt{\alpha^2 + \omega^4}$$

Sense of v must agree with sense of ω, and sense of a_t with sense of α. Sense of a_n is always toward axis.

Motion Graphs

Motion graphs analogous to those previously discussed for rectilinear motion may be constructed to show the relations between angular displacement, velocity and acceleration, and time. θ, ω, and α correspond to s, v, and a, respectively.

3.3.6 Plane Motion

Any **displacement** resulting from plane motion may be accomplished by a translation of the body that will bring any one line of it, which is perpendicular to the plane of motion, into final position, followed by a rotation of the body about that line into final position. The necessary amount of translation depends on the line of the body selected as axis of the rotation; the amount of the rotation does not. The **state of motion** of a body at any instant may be regarded as consisting of two components, a translational motion and a rotational motion. Thus a plane motion may be traced by giving the history of the movement of one point of the body (called a base point) in its own curved path, and a description of the rotation of the body about the selected base point.* The point selected as base should be one for which the motion is readily specified. For a wheel rolling along a straight path, the center would be selected as a base point.

Velocity of any point P of the body, at any instant, with respect to a fixed point O, is the vector sum of the velocity of base point A, with respect to O, and of the velocity of P with respect to O due to rotation about A. Thus (Fig. 3.32) O is the fixed point, A the moving base point, and P any other point of the body at distance r from A; v_1 is velocity of A with respect to O, and $v_2 = r\omega$ is velocity of P with respect to A. Resultant velocity of P with respect to $O = v$; or $v_{P \text{ to } O} = v_{P \text{ to } A} + v_{A \text{ to } O}$. In vector form this relation can be written as the vector cross product:

$$\mathbf{V}_P = \mathbf{V}_A + \boldsymbol{\omega} \times \mathbf{r}_P$$

Acceleration of any point P, with respect to a fixed point O, at any instant, has two components; one is that of the base point A with respect to O, and the other that of P with respect to base A. Acceleration of P with respect to A is rotational, and is conveniently replaced by its tangential and

---

*To simplify matters, "points" are referred to throughout this and the following discussion but "lines" through the points perpendicular to the plane of motion should be understood. Thus, in Fig. 3.32 and 3.33, the parallel lines through P, A, and O, perpendicular to the plane of motion, move relative to each other.

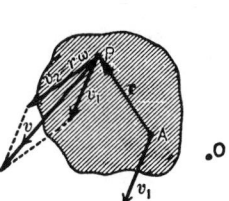

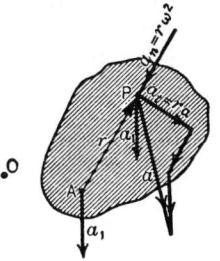

Fig. 3.32 Velocity of a point on a rigid body.   **Fig. 3.33** Acceleration of a point on a rigid body.

normal components, $a_t = r\alpha$ and $a_n = r\omega^2$. Then resultant acceleration of P, with respect to O, is the vector sum of $r\alpha$, $r\omega^2$, and acceleration of A with respect to O. Thus (Fig. 3.33) a_1 is acceleration of A to O; and acceleration P to A is resultant of a_t and a_n. Acceleration of P to O is $a =$ vector sum of a_1, a_t, and a_n. In vector form this relation can be written as three vector cross products:

$$\mathbf{a}_P = \mathbf{a}_A + \alpha \times \mathbf{r} + \omega \times (\omega \times \mathbf{r})$$

Instantaneous Axis. For a body having plane motion, there is always one point in it (or in its extension), at each instant, for which the velocity with respect to A (Fig. 3.32) is equal and opposite to velocity of A with respect to O; that is, its velocity is zero at the instant. This point Q is called the *instantaneous* (*or instant*) *center* of rotation, and a line through Q, perpendicular to the plane of motion, is called the *instantaneous axis*. Since Q is at rest for the instant, the resultant velocities of all points at the instant are purely rotational about the instant axis. The instant center is the intersection of two lines drawn from any two points, C and D, in the plane of the motion, perpendicular to their velocities. If the velocity of the point C is known, ω for the body is determined by dividing v_C by the distance of C from Q, or by r_C. The velocity of any other point E is $\omega \times r_E$, perpendicular to the radius r_E.

The position of Q in the body (or in its extension) is continually changing; its locus is a line (usually curved) fixed in the body and moving with it, called the *body centrode*. The locus of the positions of Q in the fixed plane of motion is a line (usually curved) called the *space centrode*. The plane motion may be considered as produced by the rolling, without slipping, of the body centrode upon the space centrode.

Example 3.6: Rolling Wheel Describing Plane Motion. A wheel of 6-ft radius rolls along a straight horizontal path, and at a certain instant the point P, 2 ft from center of wheel, is in the position shown in Fig. 3.34a. At this instant $\omega = 16.75$ rad/sec and $\alpha = 5.6$ rad/sec^2. Determine the velocity and acceleration of point P with respect to fixed point O at the specified instant.

SOLUTION. Select center A as base point. From relations between v and a of any point of a rotating body, and ω and α of the body:

$v_{A\ \text{to}\ O} = r\omega = 6 \times 16.75 = 100.5$ ft/sec, horizontally toward left.
$v_{P\ \text{to}\ A} = r\omega = 2 \times 16.75 = 33.5$ ft/sec, vertically upward.
Therefore, $v_{P\ \text{to}\ O} = 105.9$ ft/sec, upward to left, at $18°16'$ to horizontal (Fig. 3.34b).

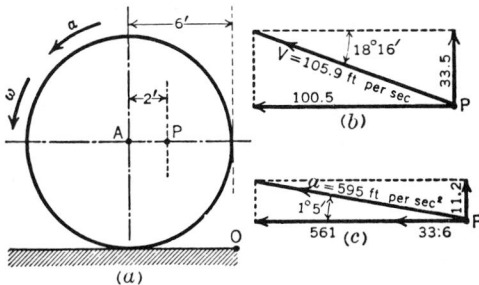

Fig. 3.34 Instantaneous axis of rotation.

$a_{A \text{ to } O} = r\alpha = 6 \times 5.6 = 33.6$ ft/sec^2, horizontally toward left.
$a_{t_{P \text{ to } A}} = r\alpha = 2 \times 5.6 = 11.2$ ft/sec^2, vertically upward.
$a_{n_{P \text{ to } A}} = rw^2 = 2 \times (16.75)^2 = 561$ ft/sec^2, horizontally toward left.
Therefore, $a_{P \text{ to } O} = 595$ ft/sec^2, upward to left, at $1°5'$ to horizontal (Fig. 3.34c).

3.4 KINETICS

3.4.1 Basic Quantities

System of Units*

The **unit of force** is the pound force in the English system and is the newton in the SI system. The pound force is defined as the force required to give one slug of mass an acceleration of one foot per second per second. The newton is defined as the force required to give one kilogram of mass an acceleration of one meter per second per second.

The **unit of acceleration** in the English system is the foot per second per second. In the SI system, the unit is the meter per second per second.

The **unit of mass** in the English system is the slug, which is the mass of 32.1739 standard pounds. The standard pound is a platinum standard kept at the National Bureau of Standards in Washington, D.C. Its mass is 0.45359243 kg. The SI unit of mass is the standard kilogram, a platinum standard kept at the International Bureau of Weights and Measures near Paris, France.

The **relation between force, mass, and acceleration** of a constant mass particle, using either of the preceding systems of units, is expressed by the relation

$$\mathbf{F} = m\mathbf{a} \quad \left(\text{or } \mathbf{F} = \frac{W}{g}\mathbf{a} \right)$$

where $\mathbf{F}$ is the force in pounds or newtons, m is the mass in slugs or kilograms, and $\mathbf{a}$ is acceleration in feet per second per second or meters per second per second.[†]

3.4.2 Derived Quantities and Relations

Work, Power, and Energy

Work. Work of a force, if constant, is the product of the force and the effective displacement of its application point. *Effective displacement* of application point is the component of the displacement parallel to the force. The body exerting the force is also said to do work. In Fig. 3.35, the work of force F as application point describes path $AB = F \times AC$. Since $F \cdot (AB \cos \alpha) = (F \cos \alpha) \cdot AB$, the work is also equal to displacement of application point times component of force parallel to the displacement. *Work of a variable force* in moving a body through distance $\Delta S = (s_2 - s_1)$ is $W = \int_{s_1}^{s_2} F \cos \alpha \, ds = \int_{s_1}^{s_2} F_t \, ds$, in which F is the variable force, ds is the elementary length of path, α is angle between force and element ds, and F_t is tangential component of force. The *sign of work* is positive if force and effective displacement have the same sense; it is negative if they differ in sense. Work done by a body against a force is equal and opposite to work done by the force on the body.

In vector form, work is calculated by integrating the dot product $\mathbf{F} \cdot d\mathbf{r}$ along the path of the application point. In equation form,

$$W = \int_{s_1}^{s_2} \mathbf{F} \cdot d\mathbf{r}$$

where $\mathbf{F}$ is the vector representation of the force and $d\mathbf{r}$ is the differential displacement along the path of the application point. Work is a scalar quantity.

The **unit of work** is any force unit times any distance unit (such as the foot-pound or newton-meter) or any power unit times any time unit (such as the watt-hour).

Work diagram (Fig. 3.36). Plot values of F_t as ordinates; corresponding values of s as abscissas; draw curve AB through ends of ordinates. Area $ABDC$ times mn equals work, in foot-pounds, done by F_t over distance $s_2 - s_1$.

Work of gravity on a body in any motion equals product of weight and change in height of the mass center. *Work of a central force F* (one always directed toward a fixed point), in any displacement

---

[†]With any system of units, $F = Kma$, where K is a constant. In this system $K = 1$.
*The units herein defined are those of the English gravitational system. For a complete discussion of English and metric gravitational and absolute units, see Section 3.2.2.

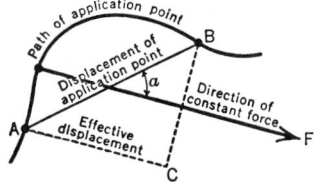
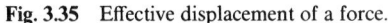

Fig. 3.35   Effective displacement of a force.

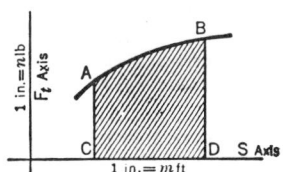

Fig. 3.36   Work of a force.

of its application point, is $\int_{r_1}^{r_2} F\,dr$, in which r_2 and r_1 are the distances of the application point from the center at the beginning and end of the displacement.

Work of a torque T on a rotating body for an angular displacement of $\theta = (\theta_2 - \theta_1)$ radians is $W = \int_{\theta_1}^{\theta_2} T\,d\theta$. If T is constant, $W = T(\theta_2 - \theta_1)$.

Mechanical efficiency of a machine is the ratio of useful output to total input of work. Let W_u = useful work performed, W_f = useless work required to overcome friction or air or any other type of resistance, W_a = work applied to the machine. Then $W_a = W_u + W_f$, and mechanical efficiency = W_u/W_a.

Power. Power of a force is its time rate of doing work. The body exerting the force is also said to have power. Let P = power and W = work. Then instantaneous $P = dW/dt = F_t(ds/dt) = F_t v$, where v is instantaneous velocity of application point of force F.

The unit of power is any work unit divided by any time unit (as foot-pound per second). One horsepower = 550 ft-lb/sec = 33,000 ft-lb/min = 0.7457 W.

Power of a torque at any instant is $P = dW/dt = T(d\theta/dt) = T\omega$ where ω is instantaneous angular velocity of the body.

Energy. Energy* of a *body* (or system of bodies) is the amount of work it can do, by virtue of its motion or position, against forces applied to it, while changing to a standard state.

Potential energy (*PE*) of a body is that possessed by virtue of its configuration. Thus, a body of weight W, located at a height above the earth's surface such that its mass center can descend h feet, has a potential energy PE = Wh.

Kinetic energy (*KE*) of a body is that possessed by virtue of its velocity, and the standard state is zero velocity. KE of a *body in translation* = $\frac{1}{2}mv^2$. KE of a *rotating body* = $\frac{1}{2}I\omega^2 = \frac{1}{2}mk^2\omega^2$, I, k, and ω being moment of inertia, radius of gyration, and angular velocity, respectively, about axis of rotation. KE of a body having plane motion = $\frac{1}{2}I\omega^2 = \frac{1}{2}mk^2\omega^2 = \frac{1}{2}m\bar{v}^2 + \frac{1}{2}\bar{I}\omega^2$, in which I and k are referred to instantaneous axis, $\bar{v}$ = velocity of mass center, and $\bar{I}$ = moment of inertia about axis through mass center perpendicular to plane of motion. *Unit of energy* is same as unit of work. For KE in foot-pounds, use m in slugs, v in feet per second, ω in radians per second, and k in feet.

Principal of Conservation of Energy. If a body or system of bodies is isolated so that it neither receives nor gives out energy, its total store of energy, all forms included, remains constant; there may be a transfer of energy from one part of the system to another, but the total gain or loss in one part is exactly equivalent to the loss or gain in the remainder. This is the *principle of conservation of energy*.

Principle of Work and Kinetic Energy. Total work of the applied forces acting on any body, or on any system of connected bodies, equals the change in the kinetic energy of the body, or bodies. (This assumes no work converted into nonmechanical types of energy.) Work done = ΔKE. ΔKE in translation = $\frac{1}{2}m(v_2^2 - v_1^2)$, v_1 and v_2 being initial and final velocities. ΔKE in rotation = $\frac{1}{2}I(\omega_2^2 - \omega_1^2) = \frac{1}{2}mk^2(\omega_2^2 - \omega_1^2)$, ω_1 and ω_2 being initial and final angular velocities. In plane motion, change in KE is

$$\Delta\text{KE} = \tfrac{1}{2}I\left(\omega_2^2 - \omega_1^2\right) = \tfrac{1}{2}mk^2\left(\omega_2^2 - \omega_1^2\right) = \tfrac{1}{2}m\left(\bar{v}_2^2 - \bar{v}_1^2\right) + \tfrac{1}{2}\bar{I}\left(\omega_2^2 - \omega_1^2\right)$$
$$= \tfrac{1}{2}m\left(\bar{v}_2^2 - \bar{v}_1^2\right) + \tfrac{1}{2}m\bar{k}^2\left(\omega_2^2 - \omega_1^2\right)$$

in which I and k are referred to instantaneous axis, $\bar{I}$ and $\bar{k}$ to a parallel axis through mass center, and $\bar{v}$ is velocity of mass center.

*Mechanical energy (which includes potential and kinetic energy) is referred to in this definition. There are other forms of energy such as thermal, chemical, and electrical.

Example 3.7. Water falling from a height of 120 ft at the rate of 1000 ft^3/min drives a turbine directly connected to an electric generator at 120 rpm. If the total resisting torque due to friction is 250 lb-ft, and the water leaves the turbine blades with a velocity of 15 ft/sec, find the power developed by the generator.

SOLUTION. This is a problem in the conversion of potential energy to work that in turn is converted to useful kinetic energy, wasted kinetic energy, and wasted thermal energy, the total energy of the system of course remaining constant. Assume that 1 ft^3 of water weighs 62.5 lb and $g = 32$ ft/sec^2. In 1 min:

$$\Delta PE = Wh = 1000 \times 62.5 \times 120 = 7,500,000 \text{ ft-lb}$$

$$\text{Wasted } \Delta KE = \tfrac{1}{2}mv^2 = \frac{1000 \times 62.5 \times \overline{15}^2}{2 \times 32} = 219,700 \text{ ft-lb}$$

$$\text{Wasted friction (thermal) } \Delta TE = T\theta = 250 \times 2\pi \times 120 = 188,500 \text{ ft-lb}$$

Therefore

$$\text{Useful } \Delta KE = \Delta PE - \text{wasted } \Delta KE - \text{wasted } \Delta TE$$

$$= 7,500,000 - 219,700 - 188,500 = 7,091,800 \text{ ft-lb}$$

$$P = \frac{7,091,800}{33,000} = 215 \text{ hp or } 215 \times 0.7457 = 160 \text{ kW}$$

Impulse, Momentum, and Impact of a Force

Impulse. Linear impulse of a force is the integral of the force over the time of application of the force. The integral can be separated into components along convenient coordinate directions. Linear impulse is a vector quantity. Thus, the linear impulse of a force **F** can be calculated as

$$\int_{t_1}^{t_2} \mathbf{F} \, dt = \int_{t_1}^{t_2} F_x \, dt \, \mathbf{i} + \int_{t_1}^{t_2} F_y \, dt \, \mathbf{j} + \int_{t_1}^{t_2} F_z \, dt \, \mathbf{k}$$

The direction cosines of the resultant vector are determined in the usual manner. *Unit of impulse* is any unit force times any unit time, as pound (force) seconds.

Angular impulse of a force about a point is the time integral of the moment of that force about the point. If **r** is the vector from the point O to any point on the line of application of the force **F**, then the angular impulse of **F** about O is given by

$$\int_{t_1}^{t_2} (\mathbf{r} \times \mathbf{F}) \, dt$$

The angular impulse of a force about a line is the time integral of the moment of the force about the line. If the line connects points O and A, and if **r** is the vector from O to any point on the line of action of **F**, then the moment of the force **F** about OA, is given by $\mathbf{M}_{OA} = \mathbf{r} \times (\mathbf{F} - \mathbf{F} \cdot \mathbf{U}_{OA})$, and the angular impulse of **F** about OA is

$$\int_{t_1}^{t_2} \mathbf{r} \times (\mathbf{F} - \mathbf{F} \cdot \mathbf{U}_{OA}) \, dt$$

where $\mathbf{U}_{OA}$ is a unit vector along the OA line, selected to be positive when directed from O to A. *Unit of angular impulse* is unit torque times unit time, as pound (force) feet seconds.

Momentum. Linear momentum of a *particle* is the product of its mass and velocity. It is a vector quantity and has the *sense* and *direction* of the velocity. *Unit of momentum* is the same as unit of impulse. Linear momentum of a *body* is the resultant, or vector sum, of the momentums of its particles. In any motion the linear momentum of a body is mv, m being mass of the body and $\bar{v}$ the velocity of its mass center.

Angular momentum of a particle about a point is the moment of its linear momentum about that point. If the linear momentum of the particle is $\mathbf{P} = m\mathbf{v}$, and if O is the reference point and if **r** is the vector from O to the particle, then the angular momentum of the particle about O is

$$\mathbf{h}_O = \mathbf{r} \times \mathbf{P} = \mathbf{r} \times m\mathbf{U}$$

Angular momentum is a vector quantity. Unit of angular momentum is mass times velocity times distance.

Angular momentum of a particle about an axis is the moment of the particle's linear momentum component which is not parallel to the axis about any point on the axis. If OA is the axis, then the angular momentum of the particle about the axis is

$$\mathbf{h}_{OA} = \mathbf{r} \times (\mathbf{P} - \mathbf{P} \cdot \mathbf{U}_{OA})$$

where $\mathbf{U}_{OA}$ is a unit vector among the line OA.

In Fig. 3.37, let mv = momentum of particle P. Resolve the momentum into components parallel and perpendicular to the axis. DE is perpendicular distance from axis to line AP. The angular momentum of $P = mv \cos \alpha \times DE$. The angular momentum of a *body* about an axis is the algebraic sum of the angular momentums of its particles. The angular momentum of a rotating body about the axis of rotation is $I\omega = mk^2\omega$, I and k being moment of inertia and radius of gyration, respectively, about the axis of rotation, and ω the angular velocity. *Unit of angular momentum* is same as unit of angular impulse.

Principle of Conservation of Linear and Angular Momentum. When no external forces are acting upon a body or system of bodies, the component linear momentum along any line and the angular momentum about any line remain constant; this is the *principle of conservation of linear and angular momentum.*

Principle of Impulse and Momentum. For *linear momentum*, the impulse of the resultant force acting for an infinitesimal time upon a body is equal to the change in linear momentum of its mass center during that time parallel to the direction of the force. Referred to coordinate axes, the change in the component of linear momentum parallel to any axis x for any length of time $t_2 - t_1$ equals the algebraic sum of the components of the impulses of the applied forces parallel to the axis in the same time, or, more briefly, $\Delta(m\bar{v}_x) = \Sigma \int_{t_1}^{t_2} F_x \, dt$. Similarly, the change in the *angular momentum* about any axis y in the time $t_2 - t_1$ equals the algebraic sum of the angular impulses of the applied forces about the axis in the same time, or, more briefly, $\Delta(I_y\omega) = \Sigma \int_{t_1}^{t_2} T_y \, dt$.

Example 3.8. A jet of water strikes a concave vessel with a velocity of 80 ft/sec and leaves it with a velocity that has the same magnitude but makes an angle of 120° with the original direction. If the diameter of the jet is 1 in., find the force necessary to hold the vessel in position.

SOLUTION

The sustaining force $\mathbf{F}$ must bisect the acute angle between the lines representing the original and final velocities. Let the line of action of $\mathbf{F}$ (Fig. 3.38) be taken as the X axis. There is no change in the Y component of momentum. The impulse of the force in the X direction in t seconds $= F \times t$ pound-seconds. The weight of water deflected in t seconds is $W = 80\pi \times 62.5t/576$ lb. The component of original momentum in the X direction $= -80W \cos 30°/g$ pound-seconds. The component of final momentum in the X direction $= 80W \cos 30°/g$ pound-seconds. The change in momentum in the X direction $= 160W \cos 30°/g = 5W \cos 30°$ pound-seconds. The fundamental relation gives $F \times t = 5 \times 80\pi \times 62.5 \times \cos 30° \times t/576$, whence $F = 118$ lb. Observe that the sustaining force F does no mechanical work and that the water suffers no loss of kinetic energy.

Impact. Impact occurs when two bodies collide. It is *direct* when the motion is perpendicular to the striking surfaces; otherwise it is *oblique*. It is *central* if the forces that the bodies exert on each other are directed along the line joining the mass centers; otherwise it is *eccentric*. In any collision, the forces that the two bodies exert on each other are equal and opposite at each instant; hence the total

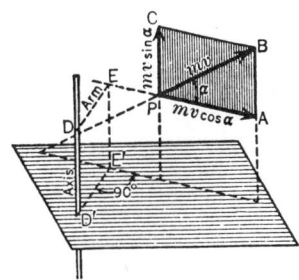

Fig. 3.37 Angular momentum of a particle.

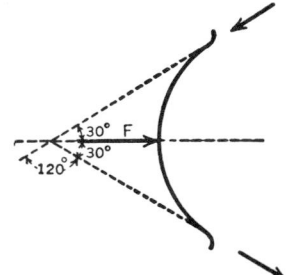

Fig. 3.38 Deflected water jet.

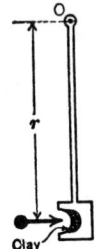

Fig. 3.39 Ballistic pendulum.

impulses of these forces during the collision are equal and opposite, and according to the principle of impulse and momentum the changes in the momentums of the bodies produced by the collision must be equal and opposite; or, otherwise stated, the total momentum of the two bodies is unchanged by the collision. Or, for *direct central impact*:

$$m_1 v_1 + m_2 v_2 = m_1 V_1 + m_2 V_2$$

wherein m_1 and m_2 = masses of the bodies, v_1 and v_2 their velocities before, and V_1 and V_2 their velocities after, the collision; but in numerical substitution, velocities in one direction are given the same sign and those in the other direction the opposite sign.

Experiments on direct central impact of spherical bodies have shown that the relative velocities of spheres after impact are always less than before the impact and that these relative velocities are opposite in direction. The ratio of the relative velocities after impact to that before impact is called *coefficient of restitution*; it seems to depend only on the material of the impinging spheres. For glass the coefficient is $\frac{15}{16}$, for steel and cork $\frac{5}{9}$, ivory $\frac{8}{9}$, wood about $\frac{1}{2}$, clay and putty 0. If e = coefficient, then

$$V_1 - V_2 = -e(v_1 - v_2)$$

This equation and the preceding one solved simultaneously show that

$$V_1 = v_1 - \frac{(1 + e)m_2}{m_1 + m_2}(v_1 - v_2) \qquad V_2 = v_2 - \frac{(1 + e)m_1}{m_1 + m_2}(v_2 - v_1)$$

During impact there is, in general, loss of kinetic energy;* the loss is $\frac{1}{2}(v_1 - v_2)^2(1 - e^2)m_1 m_2/(m_1 + m_2)$. Bodies for which $e = 0$ are said to be *inelastic*; and those for which e is nearly 1 are said to be nearly perfectly *elastic*. When a sphere is dropped on a horizontal surface of a large body from a height h, if H = height of rebound, then $H = e^2 h$. This equation furnishes a means of computing e.

Example 3.9: Ballistic Pendulum. This is a device for determining the velocity of a bullet. The bullet is imbedded in soft material, such a clay, for which $e = 0$. Referring to Fig. 3.39, let m_1 = mass of bullet, m_2 = mass of pendulum, k = radius of gyration about axis of suspension O, v_1 = velocity of bullet (to be determined), $v_2 = r\omega$ = velocity of bullet after impact, ω = angular velocity of pendulum after impact, and assume pendulum stationary before impact. Then angular momentum of system before impact = $m_1 v_1 r$, and angular momentum of system just after impact = $m_1 r^2 \omega + m_2 k^2 \omega$. Since total momentum of system remains constant, $m_1 v_1 r = m_1 r^2 \omega + m_2 k^2 \omega$. If mass center of pendulum rises to height h, $k^2 \omega^2 = 2gh$. Combining last two equations to eliminate ω and solving for v_1, $v_1 = (m_1 r^2 + m_2 k^2)\sqrt{2gh}/m_1 rk$. The quantities on the right-hand side of this equation are easily determined experimentally.

*As energy can be neither created nor destroyed, the total energy remains constant. The "lost" kinetic energy is simply converted into other forms, as into work done in distorting the bodies thermal (heat) energy, etc.

3.4.3 Kinematic and Kinetic Formulas

Symbols. s = distance along path of motion; x, y, z = coordinates of any point; $\bar{x}$, $\bar{y}$, $\bar{z}$ = coordinates of mass center; t = time; a = resultant linear acceleration of any point; $\bar{a}$ = resultant linear acceleration of mass center; $a_{x, y, z}$ = components of resultant acceleration along x, y, z axes; a_t = resultant tangential acceleration of any point; a_n = resultant normal acceleration of any point; v, $\bar{v}$, $v_{x, y, z}$ = linear velocities having corresponding significances; θ = angular displacement; α = angular acceleration; ω = angular velocity; n = revolutions per unit time; r = radius (of curvature); g = acceleration of gravity = 32.2 ft/sec^2 (approx.); m = weight/g = mass; F = resultant force; $F_{x, y, z}$ = components of resultant force along x, y, z axes; F_t = resultant tangential force; F_n = resultant normal force; W = work; Eff = efficiency; P = power; KE = kinetic energy; Imp = linear impulse; Mom = linear momentum; T = resultant torque about axis of rotation; $T_{x, y, z}$ = torques about x, y, z axes; Ang Imp = angular impulse; Ang Mom = angular momentum; I = moment of inertia (for mass) about axis of rotation; $I_{x, y, z}$ = moments of inertia about x, y, z axes; $\bar{I}$ = moment of inertia about axis through mass center; k, $k_{x, y, z}$, $\bar{k}$ = corresponding radii of gyration; d = distance between axes; U = product of inertia; U_{xy} = product of inertia with respect to YOZ and ZOX planes; (U_{yz}, U_{zx} have corresponding significances); Δ indicates "change in."

Gravitation and Inertia Functions. Mass center has coordinates:

$$\bar{x} = \frac{\int x\, dm}{m} \qquad \bar{y} = \frac{\int y\, dm}{m} \qquad \bar{z} = \frac{\int z\, dm}{m}$$

$$\bar{I} = \int r^2\, dm \quad \bar{k} = \sqrt{\bar{I}/m} \qquad I = \bar{I} + md^2 \quad k = \sqrt{I/m} \qquad k^2 = \bar{k}^2 + d^2$$

$$U_{xy} = \int xy\, dm \qquad U_{yz} = \int yz\, dm \qquad U_{zx} = \int zx\, dm$$

*Translation — (Rectilinear Motion)**

$$v = \frac{ds}{dt} \qquad a = \frac{dv}{dt} = \frac{d^2 s}{dt^2} \qquad \frac{a}{v} = \frac{dv}{ds}$$

$$\Delta s = \int_{t_1}^{t_2} v\, dt \qquad \Delta v = \int_{t_1}^{t_2} a\, dt \qquad \Delta t = \int_{s_1}^{s_2} \frac{ds}{v} = \int_{v_1}^{v_2} \frac{dv}{a} \qquad \Delta v^2 = 2\int_{s_1}^{s_2} a\, ds$$

$$F = m\bar{a} \qquad \Delta W = \int_{s_1}^{s_2} F\, ds \qquad KE = \tfrac{1}{2}m\bar{v}^2 \qquad \Delta W = \Delta KE$$

$$P = \frac{dW}{dt} = Fv \qquad \Delta \text{Imp} = \int_{t_1}^{t_2} F\, dt \qquad \text{Mom} = m\bar{v} \qquad \Delta \text{Imp} = \Delta \text{Mom}$$

*Translation — (Curvilinear Motion)**

$$v = \frac{ds}{dt} \qquad v_x = \frac{dx}{dt} \qquad v_y = \frac{dy}{dt} \qquad v_z = \frac{dz}{dt} \qquad v = \sqrt{v_x^2 + v_y^2 + v_z^2}$$

Directional cosines of v are: $\cos\theta_x = \dfrac{v_x}{v} \qquad \cos\theta_y = \dfrac{v_y}{v} \qquad \cos\theta_z = \dfrac{v_z}{v}$

$$a_x = \frac{dv_x}{dt} = \frac{d^2 x}{dt^2} \qquad a_y = \frac{dv_y}{dt} = \frac{d^2 y}{dt^2} \qquad a_z = \frac{dv_z}{dt} = \frac{d^2 z}{dt^2} \qquad a = \sqrt{a_x^2 + a_y^2 + a_z^2}$$

*For a rigid body in translation, accelerations and velocities of all particles are equal. However, $\bar{a}$ and $\bar{v}$ are indicated in certain of the kinetic translation formulas to make them applicable also to nonrigid bodies.

Directional cosines of a are: $\cos \phi_x = \dfrac{a_x}{a}$ $\cos \phi_y = \dfrac{a_y}{a}$ $\cos \phi_z = \dfrac{a_z}{a}$

$$a_t = \frac{dv}{dt} = \frac{d^2 s}{dt^2} \qquad a_n = \frac{v^2}{r} \qquad a = \sqrt{a_t^2 + a_n^2} \qquad \frac{a_t}{v} = \frac{dv}{ds}$$

$$\Delta s = \int_{t_1}^{t_2} v \, dt \qquad \Delta v = \int_{t_1}^{t_2} a_t \, dt \qquad \Delta t = \int_{s_1}^{s_2} \frac{ds}{v} = \int_{v_1}^{v_2} \frac{dv}{a_t} \qquad \Delta v^2 = 2 \int_{s_1}^{s_2} a_t \, ds$$

$$\mathbf{F} = m\bar{a} \qquad F_x = m\bar{a}_x \qquad F_y = m\bar{a}_y \qquad F_z = m\bar{a}_z \qquad F_t = m\bar{a}_t \qquad F_n = m\bar{a}_n$$

$$F = \sqrt{F_x^2 + F_y^2 + F_z^2} \qquad \text{or} \qquad F = \sqrt{F_t^2 + F_n^2}$$

$$\Delta W = \int_{s_1}^{s_2} F_t \, ds \qquad \text{KE} = \tfrac{1}{2} m v^2 \qquad \Delta W = \Delta \text{KE} \qquad P = \frac{dW}{dt} = F_t v$$

$$\Delta \text{Imp}_x = \int_{t_1}^{t_2} F_x \, dt \qquad \Delta \text{Imp}_y = \int_{t_1}^{t_2} F_y \, dt \qquad \Delta \text{Imp}_z = \int_{t_1}^{t_2} F_z \, dt \qquad \text{Imp} = \sqrt{\overline{\text{Imp}_x^2} + \overline{\text{Imp}_y^2} + \overline{\text{Imp}_z^2}}$$

$$\text{Mom}_x = m\bar{v}_x \qquad \text{Mom}_y = m\bar{v}_y \qquad \text{Mom}_z = m\bar{v}_z \qquad \text{Mom} = \sqrt{\text{Mom}_x^2 + \text{Mom}_y^2 + \text{Mom}_z^2}$$

$$\Delta \text{Imp}_x = \Delta \text{Mom}_x \qquad \Delta \text{Imp}_y = \Delta \text{Mom}_y \qquad \Delta \text{Imp}_z = \Delta \text{Mom}_z \qquad \Delta \text{Imp} = \Delta \text{Mom}$$

Directional cosines of $\Delta \text{Imp} = \Delta \text{Mom}$ are: $\cos \psi_x = \dfrac{\Delta \bar{v}_x}{\Delta \bar{v}}$ $\cos \psi = \dfrac{\Delta \bar{v}_y}{\Delta \bar{v}}$ $\cos \psi_z = \dfrac{\Delta \bar{v}_z}{\Delta \bar{v}}$

For kinetic formulas applying to a translated *body* for rotation about an axis not fixed in the body or its extension, use formulas applying to "Rotation of a *Particle*" about its axis, considering entire mass of body as concentrated at the mass center.

Rotation*

$$\omega = \frac{d\theta}{dt} \qquad\qquad \alpha = \frac{d\omega}{dt} = \frac{d^2\theta}{dt^2} \qquad\qquad\qquad \frac{\alpha}{\omega} = \frac{d\omega}{d\theta}$$

$$\Delta\theta = \int_{t_1}^{t_2} \omega \, dt \qquad \Delta\omega = \int_{t_1}^{t_2} \alpha \, dt \qquad \Delta t = \int_{\theta_1}^{\theta_2} \frac{d\theta}{\omega} = \int_{\omega_1}^{\omega_2} \frac{d\omega}{\alpha} \qquad \Delta\omega^2 = 2 \int_{\theta_1}^{\theta_2} \alpha \, d\theta$$

For a "particle" ($\bar{I}$ infinitesimal compared with I for finite body):

$$s = r\theta \qquad\qquad v = r\omega \qquad\qquad a_t = r\alpha \qquad\qquad a_n = r\omega^2$$

$$a = r\sqrt{\alpha^2 + \omega^4}$$

$$T = F_t r = mr^2 \alpha \qquad\qquad\qquad F_t = mr\alpha \qquad\qquad F_n = mr\omega^2$$

$$F = mr\sqrt{\alpha^2 + \omega^4}$$

$$\Delta W = \int_{\theta_1}^{\theta_2} T \, d\theta \qquad\qquad \text{KE} = \tfrac{1}{2} mr^2 \omega^2 \qquad\qquad \Delta W = \Delta \text{KE} \qquad\qquad P = \frac{dW}{dt} = T\omega$$

$$\Delta \text{Ang Imp} = \int_{t_1}^{t_2} T \, dt \qquad\qquad \text{Ang Mom} = mr^2 \omega \qquad\qquad \Delta \text{Ang Imp} = \Delta \text{Ang Mom}$$

For a body:

$$T = I\alpha = mk^2\alpha \qquad \Delta W = \int_{\theta_1}^{\theta_2} T \, d\theta \qquad \Delta W = \Delta \text{KE} \qquad P = \frac{dW}{dt} = T\omega \qquad \text{KE} = \tfrac{1}{2} I\omega^2 = \tfrac{1}{2} mk^2\omega^2$$

$$\Delta \text{Ang Imp} = \int_{t_1}^{t_2} T \, dt \qquad \text{Ang Mom} = I\omega = mk^2\omega \qquad \Delta \text{Ang Imp} = \Delta \text{Ang Mom}$$

*Formulas are for rigid bodies.

Constrained Rotation[†]
Plane of rotation fixed above and parallel to horizontal *XZ* plane; vertical *Y* axis of rotation not passing through mass center (except as special case).
All previous rotation formulas apply if T is replaced by T_y. Additional formulas are:
For a particle ($\bar{I}$ *infinitesimal* compared with I for a finite body):

$$F_x = m\bar{z}\alpha - m\bar{x}\omega^2 \qquad F_y = 0 \qquad F_z = -m\bar{x}\alpha - m\bar{z}\omega^2 \qquad F = \sqrt{F_x^2 + F_y^2 + F_z^2}$$

$$T_x = F_z\bar{y} = -m\overline{xy}\,\alpha - m\overline{yz}\,\omega^2 \qquad T_y = F_x\bar{z} = m\bar{z}^2\alpha - m\overline{xz}\,\omega^2 = mr^2\alpha$$

$$= -F_z\bar{x} = m\bar{x}^2\alpha + m\overline{zx}\,\omega^2 = mr^2\alpha$$

$$T_z = -F_x\bar{y} = -m\overline{yz}\,\alpha + m\overline{xy}\,\omega^2$$

$(\theta, \omega, \alpha$ positive for counterclockwise rotation facing origin from plus point on axis).

For a body:

$$T_x = -U_{xy}\alpha - U_{yz}\omega^2 \qquad T_y = I_y\alpha \qquad T_z = -U_{yz}\alpha + U_{xy}\omega^2 \quad \text{(sign convention as before)}.$$

Center of percussion and center of oscillation of a pendulum are located at distance from center of suspension $= k^2/\bar{z}$, where $\bar{z}$ is distance from center of suspension to mass center.

Plane and Three-Dimensional Motions
For translation of mass center: Consider entire mass as concentrated at mass center. Refer motion to set of fixed axes located outside the body. To determine acceleration of mass center, apply formulas for translation.

For rotation about mass center: Consider mass center as fixed and resultant of forces as a couple. Refer motion to set of central principal axes. To determine components of angular acceleration about these axes, use formulas:

$$T_x = I_x\alpha_x + \left(I_z - I_y\right)\omega_y\omega_z \qquad T_y = I_y\alpha_y + \left(I_x - I_z\right)\omega_z\omega_x \qquad T_z = I_z\alpha_z + \left(I_y - I_x\right)\omega_x\omega_y$$

For complete resultant motion: Combine motion of translation of mass center with motion of rotation about mass center.

Work and kinetic energy changes also are equal to the respective sums of the corresponding changes under the preceding component motions.

3.4.4   Translation

Kinetic formulas for motion of translation follow directly from the kinematic formulas applying to such motion and the previous discussion on kinetic quantities and their relations. The formulas are summarized in section 3.4.3 (p. 3·39) under the heading "Translation." For the solution of a specific problem, careful choice of formulas will often facilitate the computations. As there is no rotation, the resultant force acts through the mass center and there is no couple.

Example 3.10: Motion of Parallel Rod of a Locomotive. The problem is to find the forces acting upon the parallel rod when it is in any position with respect to the wheels. Assume velocity of locomotive constant at 60 mph on a level track; driver diameter 5.5 ft; and crank length 1 ft; and weight of rod 275 lb. The forces acting on the rod are its weight and the pressures of the crank pins at its ends; the latter are represented (Fig. 3.40) by their horizontal and vertical components. Since the resultant of all these forces acts through the mass center, $V_1 = V_2$; also $2V_1 - 275 = (W/g)a_y = 8.55a_y$ and $H_1 - H_2 = (W/g)a_x = 8.55a_x$. To determine a_x and a_y: The velocity of the center of either crank pin relative to the locomotive is $(88 \times 1)/2.75 = 32$ ft/sec (60 mph = 88 ft/sec), and the relative motion of the pin being circular at constant velocity, the relative acceleration is toward the center of the crank pin circle at all times and equals $32^2/1 = 1024$ ft/sec². This is also the absolute acceleration of the crank pin, since the locomotive is assumed to have no acceleration. But the rod has the same acceleration as the crank pin; hence $a_x = 1024 \sin\theta$, and $a_y = 1024 \cos\theta$. Thus $V = \frac{1}{2}(8755\cos\theta + 275)$, and $H_1 - H_2 = 8755\sin\theta$. In the lowest position of the rod, $\theta = 0$, $a_x = 0$, $a_y = 1024$, $H_1 = H_2$, $V = \frac{1}{2}(8755 + 275) = 4515$. In a mid-position when $\theta = 90°$, $a_x = 1024$, $a_y = 0$, $H_1 - H_2 = 8755$, and $V = \frac{1}{2}(275) = 137.5$. In the highest position, $\theta = 180°$, $a_x = 0$, $a_y = -1024$, $H_1 = H_2$, and $V = \frac{1}{2}(275 - 8755) = -4240$, the negative sign meaning that V acts downward on the rod.

[†]In obtaining total forces and torques, effect of weight of body (this effect depending on position of plane of rotation) must not be neglected.

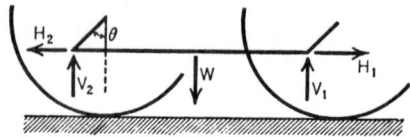
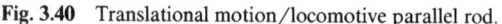

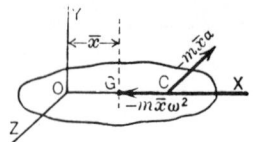

Fig. 3.40 Translational motion/locomotive parallel rod.   **Fig. 3.41** Constrained rotation.

3.4.5 Rotation

Kinetic formulas for motion of rotation follow directly from the kinematic formulas applying to such motion and the previous discussion on kinetic quantities and their relations. The formulas are summarized in section 3.4.3 (p. 3·40) under the heading "Rotation." For the solution of a specific problem, careful choice of formulas will often facilitate the computations. As there is no translation, the resultant force is zero but there is a couple.

Example 3.11. A punch is required to exert a force of 100,000 lb through a distance of $\frac{1}{4}$ in., and the work is to be supplied by a flywheel of radius of gyration $= 1.5$ ft making 120 rpm. Find the weight of the wheel, if the speed is not to be reduced below 100 rpm.

SOLUTION. $\omega_1 = 120$ rpm $= 4\pi$ rad/sec, $\omega_2 = 100$ rpm $= (10/3)\pi$ rad/sec. Work done by punch $= 100,000/48$ ft-lb $=$ reduction in KE of flywheel.
    Change in KE $= \frac{1}{2}mk^2 \Delta\omega^2 = W \times 2.25(\omega_1^2 - \omega_2^2)/64 = W \times 2.25(\omega_1 - \omega_2)(\omega_1 + \omega_2)/64$. Hence $(W \times 2.25)/64(2\pi/3)(22\pi/3) = 100,000/48$, hence $W = 1230$ lb $=$ minimum weight of flywheel.

Constrained Rotation

Constrained rotation refers to rotation of a body about a fixed axis that does not pass through its mass center. Such an axis, since it constrains the motion,* must be held by forces (exerted by bearings) to keep it from shifting position. These bearing reactions depend on the weight of the body, the manner in which the mass of body is distributed about the axis, the applied forces, the angular velocity ω, and the angular acceleration α. Generally, the resultant of the applied forces for such a body is not a single force, but a single force at a selected origin and a couple. Selecting the origin on the axis of rotation, the axial components of the single force and axial components of the couple are given by the following six equations:[†]

$$\Sigma F_x = m\bar{z}\alpha - m\bar{x}\omega^2 \qquad \Sigma T_x = -\alpha\int xy\,dm - \omega^2\int yz\,dm = -U_{xy}\alpha - U_{yz}\omega^2$$

$$\Sigma F_y = 0 \qquad \Sigma T_y = I_y\alpha$$

$$\Sigma F_z = -m\bar{x}\alpha - m\bar{z}\omega^2 \qquad \Sigma T_z = -\alpha\int yz\,dm + \omega^2\int xy\,dm = -U_{yz}\alpha + U_{xy}\omega^2$$

In these equations, the axis of rotation is fundamentally the y axis; $\bar{x}$, $\bar{y}$, and $\bar{z}$ are the instantaneous coordinates of the mass center; $\Sigma F_x, \Sigma F_y, \Sigma F_z$ are the sums of components of all applied forces in the axial directions; $\Sigma T_x, \Sigma T_y, \Sigma T_z$ are the sums of moments of all applied forces about the axes; and the convention of signs for moments of forces, and senses of θ, ω, and α are that counterclockwise rotation, facing the origin from any plus point on an axis, is positive.
    The equations are simultaneous at each instant. They are used more often to determine the forces exerted by the bearings on the axle, than to determine the resultant.

Special Cases (Fig. 3.41). Choose the x axis through an instantaneous location of the mass center and let XZ be a plane of symmetry of a homogeneous body. The resultant is a single force in the plane of symmetry having the Z component $-m\bar{x}\alpha$ and the X component $-m\bar{x}\omega^2$ acting at point C. $OC = k_y^2/\bar{x}$, k_y being radius of gyration about y axis. If $\bar{x} = 0$, the resultant becomes a couple in the XZ plane, of moment $= \Sigma T_y = I_y\alpha$. If $\alpha = 0$ and $\bar{x} \neq 0$, the resultant $= -m\bar{x}\omega^2$, in the sense CO. If $\alpha = 0$ and $\bar{x} = 0$, the resultant vanishes.

*In certain cases, the "physical path" itself constrains the rotation, as the action of the track on a train rounding a curve.
[†]In obtaining total forces and torques, effect of weight of body must not be neglected.

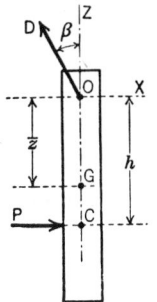

Fig. 3.42 Center of percussion.

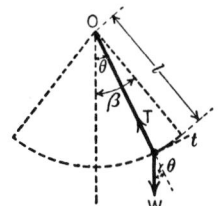

Fig. 3.43 Simple pendulum.

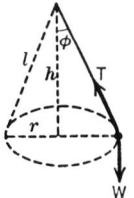

Fig. 3.44 Conical pendulum.

Centrifugal Force. Let any particle of mass m move in a circular path of radius r about a fixed y axis. The resultant of all forces acting on the particle has a normal component $= mr\omega^2$, and a tangential component $= mr\alpha$. The component $mr\alpha$ increases or decreases the speed of the particle; the component $mr\omega^2$ continually changes the direction of the linear velocity. The resultant of such forces for all the particles of the body is equivalent to the resultant specified by the preceding general equations. If ω is constant and $\alpha = 0$, the resultant force acting on the particle to make it rotate in its circular path is $mr\omega^2$ toward the axis, and is called *centripetal force*. *Centrifugal force* for the particle is equal and opposite to centripetal force, and is exerted by the particle upon its neighboring particles, or upon the axis of rotation. *Centrifugal resultant* for a body is the resultant of the centrifugal forces of all its particles. Generally, this resultant is not a single force; it may be computed from the general equations by making $\alpha = 0$ and reversing senses of resultant force and couple.

Center of Percussion. A prismatic bar (Fig. 3.42) is suspended on a horizontal y axis at O. G is the mass center. If a force P, parallel to x axis, is applied to the body, the axle reaction OD will generally be inclined to the z axis at some angle $\pm\beta$, the angle depending on the distance h of P from the axis of rotation. If $h = k_y^2/\bar{z}$, in which k_y is radius of gyration about y axis, P will cause no x component of axle reaction; that is, β will be zero, and the point C, where the action line of P intersects OG, is the *center of percussion*. In impact testing machines, heavy pendulums are used to deliver blows, and proper design requires the striking point to coincide with the center of percussion in order to avoid shock to the axle and detrimental vibration of the pendulum itself.

Examples of Constrained Rotation

Simple Pendulum. This consists of a small heavy bob on a light string (Fig. 3.43).[†] The forces acting on it are the weight, W, and tension, T. The resultant force along the tangent $= -W\sin\theta$; the resultant force along the normal $= T - W\cos\theta$. The force equations are $Wa_t/g = -W\sin\theta$, $Wa_n/g = T - W\cos\theta$. Since $a_n = l\omega^2$, tension $T = W(\cos\theta + l\bar{\omega}^2/g)$.
To determine the motion: $a_t/g = l\alpha/g = -\sin\theta$.
The solution of this equation leads to elliptic functions. An approximate solution for small oscillations can be obtained by putting $\sin\theta = \theta$. (Difference between θ and $\sin\theta$ is less than 1% if θ is less than 14°.) The differential equation becomes $\omega\, d\omega/d\theta = -g\theta/l$. If the pendulum is at the end of its swing when $t = 0$, then $\theta = \beta$, $\omega = 0$. Integrating, $\omega^2 = g(\beta^2 - \theta^2)/l$ and $\omega = d\theta/dt = \pm\sqrt{(g/l)(\beta^2 - \theta^2)}$. Integrating, $\theta = \beta\cos\sqrt{(g/l)}\,t$. Period of oscillation $= 2\pi\sqrt{l/g}$.

A Conical Pendulum.* This consists of a small heavy bob suspended from a fixed point by a light string go that it can be made to rotate about the vertical axis through the fixed point (Fig. 3.44). If the bob rotates with constant angular velocity, ω, the quantities ϕ, r, h are constants. Since there is no vertical acceleration, $T\cos\phi = W$. The force acting inward on the bob is $T\sin\phi$. Hence the force equation gives $T\sin\phi = Wa_n/g = Wv^2/gr$, and $\tan\phi = v^2/gr = r\omega^2/g$. Also $h = g/\omega^2$; $T = Wl\omega^2/g$; period of one revolution $= 2\pi/\omega = 2\pi\sqrt{h/g}$.

[†]Radius of gyration of bob about axis through its mass center parallel to axis of rotation is considered negligible compared with radius of its path.
*The principle of the conical pendulum is employed in the Watt governor for steam engines.

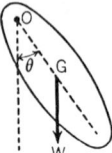

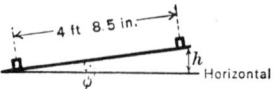

Fig. 3.45 Compound pendulum.   Fig. 3.46 Banking of track on a curve.

Compound (or Physical) Pendulum. This is any rigid body suspended from a horizontal axis about which it may rotate under the action of its own weight. The forces acting on the body are its weight, acting downward at G (Fig. 3.45), and the reaction of the axis at O. Let $\bar{r}$ = distance OG; k = radius of gyration about O. The torque equation gives $Wk^2\alpha/g = -W\bar{r}\sin\theta$, whence $\alpha = -\bar{r}g\sin\theta/k^2$. This is the equation of a simple pendulum (discussed previously) of length $l = k^2/\bar{r}$ called the length of the equivalent simple pendulum. The motion of a compound pendulum is the same as the motion of the equivalent simple pendulum. The point on the compound pendulum located at the distance $k^2/\bar{r}$ from the axis of rotation is called the *center of oscillation*. It coincides with the center of percussion (discussed previously).

Superelevation of Outer Rail of a Railroad Track. This is determined as follows (Fig. 3.46): Let r = radius of curvature in feet and v = speed in feet per second of car. Then horizontal centrifugal force is Wv^2/gr and vertical force is W, acting through mass center.[†] For the resultant to be perpendicular to the track and thus impose no side load on the rails, $\tan\phi = v^2/gr$. For small angles, the sine instead of the tangent may be used and, if h = superelevation of the outer rail in inches, $h = 56.5v^2/gr$.

Skidding and Tipping. Suppose a car (Fig. 3.47) is taking a curve of radius r feet at a speed of v feet per second, G is mass center, N_1 is the vertical and F_1 the horizontal pressure on the outer wheel, and f = coefficient of friction.[†] The problem becomes one of statics by introducing $Wv^2/gr = F_1 + F_2 = F = fW$. If $f < v^2/gr$, the car will skid. Suppose $f > v^2/gr$; then $N_1 = W(\frac{1}{2} + v^2h/dgr)$, $N_2 = W(\frac{1}{2} - v^2h/dgr)$. The critical speed is $v_1 = \sqrt{dgr}/2h$, when the total weight is borne on the outer wheel. If this critical speed is exceeded, the car will tip over.

Note on Use of Rotation Formulas. In practice, nearly all problems of the nature illustrated by the car problems are solved by the use of formulas applying to rotation of a *particle* about an axis. It should be realized, however, that the assumption thus made that the mass is concentrated at the mass center is not strictly correct except in the event that the body has a true motion of translation (as exemplified by the motion of the parallel rod of a locomotive). Seldom is this the case in practice as usually every line of the body lying in the plane of motion makes one complete revolution for each revolution of the body about the center of rotation of its path. Therefore the formulas applying to rotation of a *body* are the only ones giving absolutely correct results.

For example, in the problem above on the motion of a simple pendulum, considering the path as that of the mass center, the actual torque $T = Wk^2\alpha/g$, where k is radius of gyration of bob about horizontal axis of rotation through O. Let $\bar{k}$ = radius of gyration of bob about horizontal axis through mass center parallel to axis through O. Then $k^2 = l^2 + \bar{k}^2$, and actual torque $T = W(l^2 + \bar{k}^2)\alpha/g$. But the approximate assumption made in the problem that $F_t = Wl\alpha/g$ gives $T = Wl^2\alpha/g$, which is too small by the amount $W\bar{k}^2\alpha/g$. However, when $\bar{k}$ is very small compared with l, the results obtained are sufficiently near accurate.

3.4.6 Plane Motion

Kinetic formulas for plane motion are a combination of those for motions of translation and rotation. The procedure for solution of a problem was summarized under the heading "Plane and Three-Dimensional Motions." The general formula given there for determination of angular accelerations reduces to $T_x = I_x\alpha = mk_x^2\alpha$, the x axis being perpendicular to the plane of motion and passing through the mass center. The theory forming the basis for the assumptions regarding mass concentration and arrangement of forces is explained under the general case of "Three-Dimensional Motion."

[†]Radius of gyration of the car about axis through its mass center parallel to axis of rotation is considered negligible compared with radius of its path.

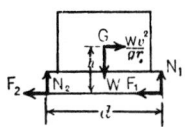

Fig. 3.47 Skidding and tipping on a curve.

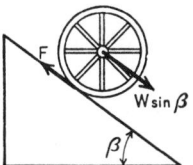

Fig. 3.48 Wheel on an inclined plane.

Example 3.12: Wheel on Inclined Plane. In Fig. 3.48, $\tan \beta = \frac{3}{4}$, the wheel weighs 100 lb, diameter = 4 ft, radius of gyration = 1.6 ft. (a) Find the acceleration of the center, if the wheel rolls without slipping. (b) Find the least coefficient of friction to prevent slipping. (c) If the coefficient of friction = 0.1, find the acceleration of the center and the number of turns made while the center moves 20 ft.

SOLUTION. (a) The forces acting to move the wheel are $W \sin \beta = 60$ lb and friction, F. The equation of motion of the center is $100a/32 = 60 - F$. The force acting to turn the wheel is F. The torque equation is $100 \times 1.6 \times 1.6\alpha/32 = 2F$. Since the wheel does not slip, $a = 2\alpha$. Elimination of F and α gives $a = 11.7$ ft/sec^2.

(b) Friction = min. coeff. of friction × normal pressure, or $F = fW \cos \beta$ = minimum $f \times 80$. From the equation above, $F = 23.4$ lb, thus minimum $f = 0.29$.

(c) The relation between a and α is not known when the wheel slips. $F = 80 \times 0.1 = 8$ lb. The equation of motion of the center is $100a/32 = 60 - 8 = 52$, whence $a = 16.6$ ft/sec^2. Distance moved by center, $x = 8.3t^2$. Time to move 20 ft is given by $t^2 = 20/8.3$. Torque equation is $100 \times 1.6 \times 1.6\alpha/32 = 2 \times 8$, hence $\alpha = 2$ rad/sec^2. The angle turned through, $\theta = t^2 = 20/8.3 = 2.41$ rad = 0.38 rev.

3.4.7 Three-Dimensional Motion

Kinetic formulas for three-dimensional motion are a combination of those for motions of translation and rotation. The procedure for solution of a problem was summarized under the heading "Plane and Three-Dimensional Motions."

Any motion of a body may be regarded as consisting of two components: one, a translation equal to that of the mass center, and the other a rotation about some axis through the mass center. These motions may be said to be produced independently by the forces acting on the body: thus (a) the acceleration of the mass center is the same as if the whole mass were concentrated there and acted upon by forces equal in magnitude to, and the same in direction as, the actual external forces; and (b) the angular acceleration is the same as if the mass center were fixed and the actual external forces applied. The reasonableness of this will be seen from the following: Imagine each force acting on the body replaced by a force acting at the mass center G and a couple (see p.   (see p.   ); the resultant of all the forces acting at G is a single force R, and the resultant of all the couples is a single couple C; R cannot turn the body but gives it a motion of translation only, and C cannot move G but merely turns the body about some line through G. In general, C does not cause turning about a line perpendicular to the plane of C, only so if the plane of C is perpendicular to one of the principal central axes of the body. To determine the acceleration of the mass center, take fixed x, y, and z axes outside the body and resolve all external forces F_1, F_2, and so on, into x, y, and z components; then

$$\Sigma F_x = m\bar{a}_x \qquad \Sigma F_y = m\bar{a}_y \qquad \Sigma F_z = m\bar{a}_z$$

m denoting the mass of the body. To determine the angular acceleration of the body, take moments of all the forces F_1, F_2, and so on, about the three central principal axes; calling the sums of the moments about these axes ΣT_1, ΣT_2, and ΣT_3, the components of the angular acceleration α_1, α_2, and α_3, and the components of the angular velocity ω_1, ω_2, and ω_3, then

$$\Sigma T_1 = I_1\alpha_1 + (I_3 - I_2)\omega_2\omega_3 \qquad \Sigma T_2 = I_2\alpha_2 + (I_1 - I_3)\omega_3\omega_1 \qquad \Sigma T_3 = I_3\alpha_3 + (I_2 - I_1)\omega_1\omega_2$$

wherein I_1, I_2, and I_3 denote the three central principal moments of inertia of the body. In any motion of a body, the kinetic energy may be computed in two parts: (1) the kinetic energy of the whole body moving with a velocity equal to that of the mass center, and (2) the sum of the kinetic energies of the constituent particles of the body due to their velocities relative to an axis through the mass center.

3.4.8 Moving Axes

A **frame of reference** is a set of coordinate axes or coordinate curves with respect to which linear and angular positions, velocities, and accelerations are measured. For brevity, the word *frame* will denote a frame of reference; and a measurement with respect to a frame will be said to be made *in* that frame. A frame may move relative to another frame. In so doing, it has the same freedom of motion as a rigid body.

Forces can be specified and measured in a manner independent of the observer's frame of reference, as for example by spring extensions. It follows that Newton's third law, which is concerned with forces only, holds in all frames: in the mechanical interaction of two bodies, their mutual forces are equal but opposite, and in the same line of action.

An **inertial frame** is a frame in which the first law of Newton holds, that is, that a particle free from forces is unaccelerated. This defines an inertial frame, while experiment shows that inertial frames exist. The second law of Newton, that net force equals mass times acceleration, is experimentally true in all inertial frames and only in inertial frames. There is no particular inertial frame that may be called absolute; but rather there are infinitely many possible inertial frames, and all of them are equivalent. Suppose that a frame R_1 is inertial. Other inertial frames may have any position of their origin in R_1 and their axes may have any orientation in R_1. Another frame R_2 is inertial if and only if its origin has no acceleration in R_1 and in addition its orientation in R_1 is fixed. In other words, R_2 is inertial if and only if its motion in R_1 is at most a pure translation without acceleration. These conditions are necessary and sufficient in order that all accelerations should appear the same in R_2 as in R_1; and they are therefore the conditions that, if R_1 is inertial, R_2 is also. Acceleration in an inertial frame will be called *true acceleration*.

Noninertial frames are of practical interest, for all terrestrial experiments are performed in such frames. Any frame fixed in the earth is noninertial because the earth is both spinning and accelerating relative to an inertial frame. Nevertheless, the observation of phenomena caused by the noninertial character of an earth-fixed frame is a matter of some delicacy, so that to a certain approximation such a frame may be considered inertial. An earth-fixed frame will be referred to simply as *the earth*.

Effects of the noninertial character of a frame are apparent accelerations or forces not accounted for by actual forces. Let a_x, a_y, and a_z be the components of the apparent acceleration of a point mass m observed in a noninertial frame S. Let the true acceleration components in the same directions be a'_x, a'_y, and a'_z. Define components of *acceleration difference*, the difference between apparent and true accelerations:

$$g_x \equiv a_x - a'_x \qquad g_y \equiv a_y - a'_y \qquad g_z \equiv a_z - a'_z$$

The acceleration difference involves the motion of S relative to an inertial frame R, but it also involves the position x, y, z and the velocity v_x, v_y, v_z of m in S. Now, if the mass is free from actual forces, it is observed in S to have the acceleration g_x, g_y, g_z; or, if the actual forces on it are specified, it is observed to have this acceleration in addition to that predicted from Newton's second law by the observer in S. If, on the other hand, the observer in S specifies the motion of the particle relative to his frame, he finds it necessary to exert upon m, in addition to the force that he computes from Newton's second law, the force

$$f_x = -mg_x \qquad f_y = -mg_y \qquad f_z = -mg_z$$

The particle m thus exerts a reaction force having no apparent physical cause:

$$r_x = mg_x \qquad r_y = mg_y \qquad r_z = mg_z$$

The following paragraphs classify the possible acceleration differences.

Translational acceleration of frame S with components a_{0x}, a_{0y}, a_{0z} relative to an inertial frame R produces an acceleration difference

$$g_x = -a_{0x} \qquad g_y = -a_{0y} \qquad g_z = -a_{0z}$$

Thus, in a train that has acceleration a_{0x} along a straight track in the forward direction, a body free to move would have an apparent backward acceleration $g_x = -a_{0x}$. A mass m constrained to remain at rest in the train would exert a backward reaction $-ma_{0x}$.

Centripetal acceleration causes an acceleration difference when the system S rotates relative to R. If the rotation of S is about its z axis with an angular speed ω, then a point mass m at x, y, z has an acceleration difference $\omega^2\sqrt{x^2 + y^2}$ directed perpendicularly away from the z axis. The components of acceleration difference are

$$g_x = \omega^2 x \qquad g_y = \omega^2 y \qquad g_z = 0$$

If m is fixed in S, its reaction force is $r_x = m\omega^2 x$, $r_y = m\omega^2 y$. This is called *centrifugal force*. The rider on a merry-go-round exerts this force on his mount.

Angular acceleration of S in R produces an acceleration difference. Consider first the case when the axis of rotation remains in a fixed direction, and let this coincide with the z axis. Let $\alpha = d\omega/dt$ be the rate of change of the magnitude of the angular velocity ω. The resulting acceleration difference for a mass point m at x, y, z is given by $\alpha\sqrt{x^2 + y^2}$ in a direction tangent to the circle of radius $\sqrt{x^2 + y^2}$ about the z axis. The components of the acceleration difference are

$$g_x = \alpha y \qquad g_y = -\alpha x \qquad g_z = 0$$

If a speck of dust is to cling to a phonograph record as the turntable starts or stops, the frictional force on it must be $-m\alpha y$, $m\alpha x$.

Precessional acceleration occurs when the frame S is rotating in R in such a way that its angular velocity vector changes direction. This change of direction may be described at a given instant as the rotation of the angular velocity vector about some axis perpendicular to itself with an angular velocity of precession Ω. Let the axes of S and R coincide at a particular instant, let the angular velocity ω of S be directed along positive z, and let the precession be along positive y. The right-handed-screw convention may be used to refer both angular velocities to their axis directions. The components of the acceleration difference for a mass point at x, y, z are then

$$g_z = 0 \qquad g_y = \Omega\omega z \qquad g_z = -\Omega\omega y$$

Its magnitude is $\Omega\omega\sqrt{y^2 + z^2}$, and its direction is tangent to a circle about the x axis. The force required to be exerted on mass m is then

$$f_y = -m\Omega\omega z \qquad f_z = m\Omega\omega y$$

If a collection of particles fixed in S forms a flywheel, there must be exerted on the flywheel a torque about the positive x axis in order to supply these forces and produce the precession. This is the well-known gyroscopic torque.

Coriolis acceleration is the only difference acceleration that depends on the velocity v_x, v_y, v_z of a point m in a rotating frame S. It occurs when m has a component of velocity in S that is perpendicular to the axis of rotation. Coriolis acceleration is always perpendicular to the velocity of m in S and is independent of the location of m. Let the rotation of S be about the z axis in the positive sense. The components of the Coriolis acceleration are then

$$g_x = 2\omega v_y \qquad g_y = -2\omega v_x \qquad g_z = 0$$

Coriolis acceleration appears in a body falling freely on the earth, causing it to deviate eastward from the plumb line. If the velocity of fall is v and the latitude of the place is λ, the acceleration difference is $2\omega v \cos \lambda$. As another illustration, let a train run north on a horizontal track with velocity v at the same (north) latitude λ. The train then exerts a Coriolis reaction force eastward, on the track, of magnitude $r = 2\omega v \sin \lambda$.

The general case may involve all the above types of acceleration simultaneously. These accelerations simply add vectorially.

3.4.9 Gyroscopic Motion

A **gyroscope** is essentially a symmetrical rotor that spins rapidly about its axis. The moment of inertia about the axis is made as large as possible within the limitations of weight and size of the instrument. Gyroscopic phenomena are only those that relate changes of direction of the spin axis to applied torque.

Precession is a term for rotation of the axis direction.

The **angular momentum** of the gyroscope about the spin axis is its basic characteristic quantity. If the moment of inertia about the axis is I_1 and the component of the angular velocity parallel to the axis is ω, the component of angular momentum parallel to the axis is

$$\rho = I_1\omega \qquad\qquad (3.1)$$

When precession occurs, the total angular momentum vector is not parallel to the axis; also the value of ω may possibly be affected. In practice, however, the spin is very large compared with precession velocities. It is then a good approximation to consider that the total angular momentum vector is always parallel to the spin axis and has a constant magnitude ρ. This vector will be called $\boldsymbol{\rho}$.

It is necessary to adopt a convention for the sense of ρ. The right-handed-screw convention will be adopted for both angular momentum and torque. Thus ρ has the direction in which a right-handed screw spinning with the rotor would advance. Likewise torque is in the direction of advance of a right-handed screw to which it is applied.

The theory of the gyroscope rests upon the theorem of mechanics that the rate of change of the angular momentum vector equals the applied torque vector. Thus, if $\mathbf{L}$ is the vector torque applied to a gyroscope,

$$\mathbf{L} = \frac{d\rho}{dt} \tag{3.2}$$

Since ρ is assumed to have a constant magnitude in a gyroscope problem, $d\rho/dt$ must always be perpendicular to ρ. Thus the gyroscopic torques that can be applied to a gyroscope, and hence its reaction torques, are always perpendicular to the axis. If the vector ρ is drawn from a fixed origin, its tip moves always in the direction of the torque vector. Thus, suppose a gyroscope to have its ρ vector pointed toward an observer, and to be mounted so as to be free to precess about its center of mass. See Fig. 3.49. If the observer exerts a downward force on the end of the axis projecting toward him, the torque direction is horizontally to the right. Hence, this end of the axis precesses horizontally to the right. If the gyroscope is constrained so that horizontal precession is prevented, no downward force (except that required for ordinary angular acceleration) can be exerted on the end of the axis: the axis yields freely. The constraints are then providing torque to produce this downward precession of the end of the axis.

In **steady precession** about a fixed axis, the tip of the ρ vector drawn from a fixed origin describes a circle about this axis, and the ρ vector sweeps out a half cone about the axis. See Fig. 3.50. In the simple case where the cone degenerates into a plane, that is, when the ρ vector is perpendicular to the precession axis, the magnitude of the torque is given by

$$L = \rho\Omega = I_1\omega\Omega \tag{3.3}$$

where Ω is the angular velocity of precession. See Fig. 3.51. To maintain the precession, the torque axis must rotate with the ρ vector. In the general case of a cone of half angle θ, the torque is given, to the approximation that has been made, by

$$L = (\rho \sin\theta)\Omega \tag{3.4}$$

It should be noticed that the axis of precession is always perpendicular to the torque vector.

Example 3.13. A gyroscope has a rotor that weighs 8 lb, has a radius of gyration of 3 in., and spins at 3000 rpm. A torque of 2 lb-ft is applied, and precession occurs about an axis perpendicular to the spin. Find the velocity of precession.

$$I_1 = mk^2 = \tfrac{8}{32} \text{ slugs} \times \left(\tfrac{1}{4} \text{ ft}\right)^2 = \tfrac{1}{64} \text{ slug-ft}^2$$
$$\omega = 3000 \text{ rpm} = 100\pi \text{ rad/sec}$$

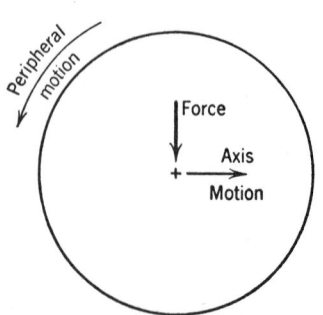

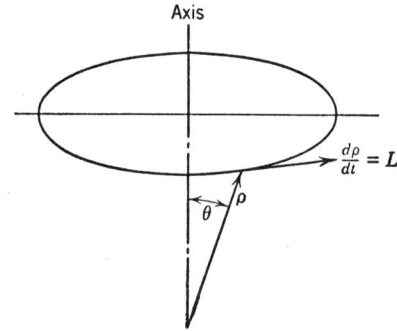

Fig. 3.49 Force/motion relations for gyroscope.   **Fig. 3.50** Steady precession.

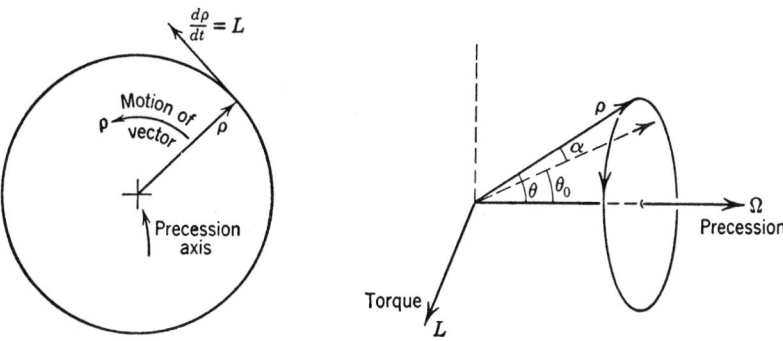

Fig. 3.51 Necessary condition for steady precession. **Fig. 3.52** Gyroscopic "resistance."

Therefore
$$\rho = \frac{100\pi}{64}\text{ slug-ft}^2/\text{sec} = \frac{100\pi}{64}\text{ lb-ft/sec}$$

$$L = 2\text{ lb-ft}$$

Therefore
$$\Omega = \frac{2\text{ lb-ft}}{\dfrac{100\pi}{64}\text{ lb-ft/sec}} = 0.408\text{ rad/sec}$$

The "**resistance**" of a gyroscope to change of its axis direction involves two different types of motion: rotation about the precession axis and rotation about the torque axis. Suppose a torque **L** to be applied to a gyroscope that is free to precess. Let **L** be exerted always about the axis of the angle θ between the spin and precession axes. See Fig. 3.52. A transient rotation about the torque axis occurs when the torque is first applied. After the transient motion dies out, there is a final steady deflection produced by the torque. The gyroscope behaves as if there were a strong spring opposing rotation about the torque axis. The transient motion may be analyzed by means of the general equations for rotation of a body about its mass center. The result is simple if the increment $\alpha \equiv \theta - \theta_0$ is sufficiently small, θ_0 being the value of θ for no applied torque and no precession. The equation is

$$L = I_2\frac{d^2\alpha}{dt^2} + \left(\frac{I_1^2\omega^2}{I_2}\right)\alpha \tag{3.5}$$

Here L is torque related by the right-handed-screw rule to positive sense of α, and I_2 is the moment of inertia of the rotor about a transverse axis. Solutions of (3.7) lead to simple harmonic motion when L is constant. When the friction that has been omitted from this equation has damped out the oscillations, the resulting deflection is

$$\alpha_{\text{final}} = \frac{I_2 L}{I_1^2\omega^2} \tag{3.6}$$

Thus the gyroscope acts like a spring whose torsional stiffness is $I_1^2\omega^2/I_2 = \rho^2/I_2$. The stiffness may be very large in practice.

Example 3.14. For the gyroscope of Example 3.13, find the torsional stiffness, assuming $I_2 = 0.80 I_1$.

$$\rho = \frac{100\pi}{64}\text{ lb-ft/sec} = 4.90\text{ lb-ft/sec}$$

$$\rho^2 = 24\text{ lb}^2\text{-ft}^2/\text{sec}^2 \qquad I_2 = \frac{0.80}{64}\text{ slug-ft}^2 = \frac{1}{80}\text{ slug-ft}^2 = \frac{1}{80}\text{ lb-ft/sec}^2$$

Therefore
$$\frac{\rho^2}{I_2} = \frac{24}{\dfrac{1}{80}}\text{ lb-ft/rad} = 1920\text{ lb-ft/rad}$$

Here the angle unit of the radian has had to be supplied.

The total deflection of the gyroscope about the precession axis depends on the angular impulse received about the torque axis. If the angle α of Fig. 3.52 is small, then in the steady state the precession velocity is

$$\Omega = \frac{\rho}{I_2 \sin \theta_0} \alpha \tag{3.7}$$

If the torque acts for a time Δt, and if $d\alpha/dt$ is the same at the beginning and end of Δt, the resulting deflection about the precession axis is

$$\Delta \psi = \frac{1}{\rho \sin \theta_0} \int_0^{\Delta t} L \, dt$$

$\Delta \psi$ may be made small for a given angular impulse by making ρ large. Thus the gyroscope yields but little in any direction to an applied impulse, provided it is free to precess. In time, however, a steady torque produces an indefinitely large deflection about the precession axis.

Applications of the gyroscope are numerous and important. A few follow:

1. *Maintenance of a fixed orientation in space for a brief time.* The gyroscope's axis has this property when it is mounted in very friction-free gimbals and when it is balanced to remove gravitational torque. Such a free gyroscope is used to control a torpedo or a rocket.

2. *Indication of vertical in an airplane.* This cannot be done by a free gyroscope for any length of time because or rotation of the earth and motion over the earth's surface. Also the irreducible minimum of friction in the mounting would produce precession after long operation. Small correcting torques are applied to the gyroscope to cause it to precess very slowly toward the *apparent* vertical. In this way the time average of the apparent vertical is indicated; and if accelerations are random and relatively short-lived, this average will approximate the true vertical.

3. *Ship stabilization.* A huge gyroscope is mounted so that it can precess about a horizontal axis transverse to the ship. The precession is limited to a small range of angle so that the spin axis remains near the ship's vertical. The torque resulting from this precession is then about a fore-and-aft axis, so that it can combat rolling of the ship. The precession is not free, but is motor driven, and is controlled by a small gyroscope so as to be most effective.

4. *Bicycle operation* depends on gyroscopic action. Turning is a precession about a vertical axis and requires the torque produced by unbalancing the weight toward the inside of the turn. On the other hand, torque applied to the handlebars about a roughly vertical axis produces precession about a horizontal axis, which is tipping of the bicycle. This assists the rider to control his balance.

5. *Indication of the meridian.* The gyrocompass makes use of the rotation of the earth together with the gravitational field in such a way that the gyroscope axis seeks north.

6. *Indication of rate of turn of an airplane.* The gyroscope is mounted with its axis horizontal and is forced to turn with the airplane. This precession produces a reaction torque that works against a spring and actuates an indicator.

3.4.10   Generalized Coordinates

The **configuration** of a mechanical system is that characteristic that is determined by the position and orientation of all its parts. If the system is composed of particles so small relative to the whole that their rotations are irrelevant to the mechanical problem, the configuration of the system may be specified by the Cartesian coordinates of all its particles in some inertial frame of reference. In problems of ordinary mechanics, the atoms of matter of which the system is composed may be treated as particles. The concept of configuration does not involve velocities or accelerations. A mechanical system is solved when its configuration is known as a function of time. From the time dependence of the configuration, all the kinematic properties, such as velocity and acceleration, may be deduced. Generalized coordinates and Lagrange's equations provide a systematic procedure for solving mechanical systems.

A **displacement** of a mechanical system is a change of its configuration.

Constraints are conditions imposed upon a system to limit its possible displacements. For example, a particle may be constrained to remain on a given surface or on a given line. The atoms composing a rigid body may for many purposes be considered to be constrained to remain at fixed distances from one another. A flywheel may be constrained to rotate in fixed bearings. Sometimes constraints change with time, as when a bead is constrained to slide on a moving wire. Such constraints will be called time dependent. Their motion will be supposed to be given; otherwise, the constraining body must be considered a part of the mechanical system to be solved.

Generalized coordinates are quantities describing the configuration of a system. They describe the configuration completely if the constraints are not time dependent; otherwise, the time is needed explicitly, together with the generalized coordinates, to specify the configuration. It is required for what follows that the set of generalized coordinates of a system be so chosen as to contain the least number of quantities necessary to describe the configuration. Thus, for example, the position of three noncollinear points of a rigid body is not a satisfactory set of generalized coordinates for the body: for it requires nine coordinates to locate these points, whereas only six coordinates are required for a rigid body. A true set of generalized coordinates is the location of one point of the body together with a set of three angles, the Euler angles, describing the orientation of the body. A set of generalized coordinates for a mechanical system will be denoted by $q_1, q_2 \cdots, q_j \cdots q_n$.

The **number of degrees of freedom** of a system is equal to the number, n, of generalized coordinates required to specify its configuration, *provided* all n of the increments δq_j are independent. The criterion of independence of the increments δq_j is that any one of them can be specified arbitrarily while all the others are zero. When this is the case, the system is called *holonomic*. This discussion will be limited to holonomic systems. Nonholonomic systems are characterized by the existence of nonintegrable relations between infinitesimal increments δq_j so that they are not independent.

Generalized forces are quantities Q_j such that the work done by the actual forces of the system during an infinitesimal displacement δq_j of only one generalized coordinate is

$$\delta W_j = Q_j \, \delta q_j \tag{3.8}$$

The Q_j are ordinarily functions of the generalized coordinates, but they may also involve the time derivatives of the coordinates.

It is important to notice that any actual forces that do not work during possible displacements of the system may be omitted completely from consideration. This kind of force is often present owing to constraints. For example, a weightless, rigid rod connecting two masses may exert forces along its length, but these forces do no work since the length of the rod is constant. Forces that constrain a body to remain in contact with a surface are workless.

A generalized force does not have the dimensions of force unless its corresponding generalized coordinate has the dimensions of length. For example, the generalized force is a torque when its corresponding coordinate is an angle.

The generalized forces of a system can ordinarily be obtained directly by computing the work done by the applied forces for infinitesimal changes of the generalized coordinates. There is a formula for Q_j in terms of the Cartesian components of force, X_i, Y_i, and Z_i, acting on the particles of the system whose coordinates are x_i, y_i, and z_i:

$$Q_i = \sum_{i=1}^{N} \left(X_i \frac{\partial x_i}{\partial q_j} + Y_i \frac{\partial y_i}{\partial q_j} + Z_i \frac{\partial z_i}{\partial q_j} \right) \tag{3.9}$$

This is summed over all N particles of the system.

In the case of moving constraints, the motion of the constraint is not considered in computing the generalized forces. The displacements δq_j of Eq. (3.8) are assumed to take place in zero time, that is, before the constraint has time to change.

The **kinetic energy** of a mechanical system may be expressed in terms of the generalized coordinates and their time derivatives. The time appears explicitly only if the constraints are time dependent. The usual notation for the time derivatives of the generalized coordinates is

$$\dot{q}_j \equiv \frac{dq_j}{dt} \tag{3.10}$$

The kinetic energy function is then written in general

$$T(q_1, q_2 \cdots q_n; \dot{q}_1, \dot{q}_2 \cdots \dot{q}_n; t) \tag{3.11}$$

If the constraints are not time dependent, this function becomes a homogeneous quadratic form in the $\dot{q}_j$:

$$T = \sum_{j=1}^{n} \sum_{k=1}^{n} A_{jk}(q_1, q_2 \cdots q_n) \dot{q}_j \dot{q}_k \tag{3.12}$$

Equation (3.12) is quite easy to prove, starting with the kinetic energy equation in terms of the ultimate particles of the system

$$T = \tfrac{1}{2} \sum_{l=1}^{N} m_l \left(\dot{x}_l^2 + \dot{y}_l^2 + \dot{z}_l^2 \right) \tag{3.13}$$

and using the functional dependence of the Cartesian coordinates upon the generalized coordinates in the case of fixed constraints:

$$x_l = x_l(q_1 \cdots q_n) \tag{3.14}$$

In an actual problem, the kinetic energy function (3.11) or (3.12) is not found by working from Eq. (3.13) but rather by inspection of the system and addition of the kinetic energies of its large-scale members.

Lagrange Equations

Lagrange equations of motion are second-order differential equations in the generalized coordinates, time being the independent variable. They have the same form for any holonomic system, and there are n such equations of this form for a given system:

$$\frac{d}{dt} \frac{\partial T}{\partial \dot{q}_j} - \frac{\partial T}{\partial q_j} = Q_j \qquad j = 1, 2, \dots, n \tag{3.15}$$

The solution of any holonomic problem is thus systematized. The process involves assigning generalized coordinates; finding the Q_j functions and the T function; use of Equation (3.15) to obtain n differential equations; simultaneous solution and integration of the differential equations; and consideration of boundary conditions.

In the use of Eq. (3.15), the meaning of the partial derivatives must be clearly understood. In the process of finding $\partial T/\partial q_j$, all other q's besides q_j are considered constants; all the $\dot{q}$'s are considered constant, including $\dot{q}_j$; and, if time appears explicitly, it too is considered constant. Similarly, to find $\partial T/\partial \dot{q}_j$, $\dot{q}_j$ is considered the only variable. It is understood, of course, that T has been expressed in the form of Eq. (3.11) to start with.

Two examples of the use of Lagrange's equations will illustrate most of the ideas involved. Consider first the Atwood machine shown in Fig. 3.53. This consists of a frictionless pulley of groove radius R and moment of inertia I, over which is passed a light, inextensible cord tied to two masses m_1 and m_2. The problem is simplest when the masses are constrained by frictionless guides to move only vertically. There is then only one degree of freedom. Let the generalized coordinate q be the downward vertical coordinate to m_1. The work done by gravity for a displacement δq is then

$$\delta W = (m_1 g - m_2 g)\, \delta q$$

whence the generalized force is

$$Q = (m_1 - m_2) g$$

The kinetic energy is the sum of the kinetic energies of the pulley and the weights:

$$T = \frac{1}{2} m_1 \dot{q}^2 + \frac{1}{2} m_2 \dot{q}^2 + \frac{1}{2} I \left(\frac{\dot{q}}{R} \right)^2$$

The Lagrange equation is therefore

$$\frac{d}{dt}\left[\left(m_1 + m_2 + \frac{I}{R^2} \right) \dot{q} \right] = (m_1 - m_2) g$$

Therefore
$$\frac{d^2 q}{dt^2} = \frac{(m_1 - m_2) q}{m_1 + m_2 + I/R^2}$$

As a second example, consider the problem of a point mass m constrained to move in a plane under a central force $F = kr$ toward the origin, where r is the distance from the origin to m and k is constant. Let us choose polar coordinates r, θ as the generalized coordinates, as in Fig. 3.54,

$$q_1 = r$$
$$q_2 = \theta$$

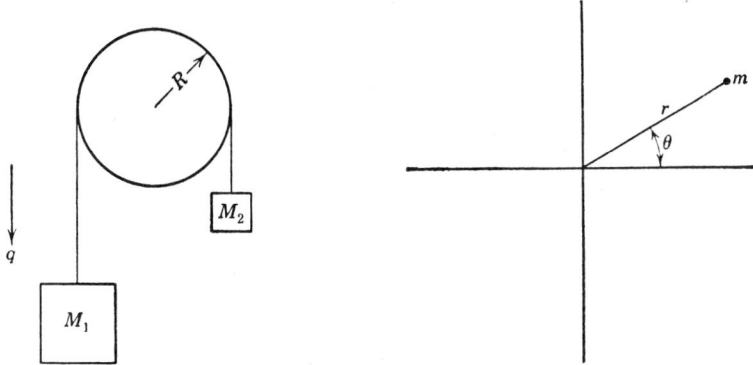

Fig. 3.53   Generalized coordinates/pulley system.   Fig. 3.54   Generalized coordinates/central force motion.

For displacement δq_1, the force F does work

$$W = -F\,\delta r = -kr\,\delta r = -kq_1\,\delta q_1$$

Thus the generalized force Q_1 is $-kq_1$. No work is done during a rotation $\delta\theta$, so $Q_2 = 0$. The kinetic energy $\frac{1}{2}mv^2$ is readily expressed in polar coordinates:

$$T = \tfrac{1}{2}m(\dot{r}^2 + r^2\dot{\theta}^2) = \tfrac{1}{2}m\left(\dot{q}_1^2 + q_1^2\dot{q}_2^2\right)$$

The two Lagrange equations of the motion are therefore

$$\frac{d}{dt}(m\dot{q}_1) - mq_1\dot{q}_2^2 = -kq_1$$

and

$$\frac{d}{dt}\left(mq_1^2\dot{q}_2\right) = 0$$

Potential energy may be used in the case of conservative systems. In such systems, there is a potential energy function V such that, during any displacement of the system, the work done by the force is

$$\delta W = -\delta V \tag{3.16}$$

Since potential energy is energy of position or configuration, it must be expressible as a function $V(q_1, q_2, \ldots q_n)$ of the generalized coordinates. Then we may write for an infinitesimal displacement

$$\delta V = -\delta W = -\left(\frac{\partial V}{\partial q_1}\delta q_1 + \frac{\partial V}{\partial q_2}\delta q_2 + \cdots + \frac{\partial V}{\partial q_n}\delta q_n\right)$$

It follows immediately from definition (3.8) that the generalized forces are

$$Q_1 = -\frac{\partial V}{\partial q_1}\cdots$$

or in general

$$Q_j = -\frac{\partial V}{\partial q_j} \tag{3.17}$$

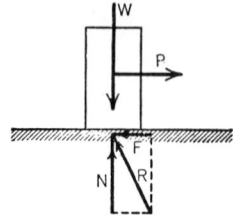

Fig. 3.55   Force diagram/friction.

If the Lagrangian function

$$L(q_1 \cdots q_n; \dot{q}_1 \cdots \dot{q}_n; t) \equiv T(q_1 \cdots q_n; \dot{q}_1 \cdots \dot{q}_n; t) - V(q_1 \cdots q_n) \qquad (3.18)$$

is formed, it may be seen immediately that for a conservative system the Lagrange equation (3.15) may be written in the compact form

$$\frac{d}{dt}\frac{\partial L}{\partial \dot{q}_j} - \frac{\partial L}{\partial \dot{q}_j} = 0 \qquad j = 1, 2, \ldots, n \qquad (3.19)$$

3.5   FRICTION

3.5.1   Static and Kinetic Friction

A smooth surface is one that offers no resistance to the sliding of a body upon it. **A rough surface** does offer resistance to such motion. The **total reaction** (R) (Fig. 3.55) of the surface of one body upon another body is its resultant force. *Friction* (F) is that component of the total reaction (R) that is tangent to the surface. *Normal reaction* (N) is that component that is normal to the surface.

Static frictional force (F) is that friction that opposes motion when there is no slipping. Its value varies as the need for it to prevent motion is developed. *Limiting frictional force* (F') is the value of static frictional force when slipping impends. *Coefficient of static friction* (f) is the ratio F'/N. *Angle of static friction* (ϕ) is defined by $\tan\phi = F'/N = f$. *Angle of repose* is that angle that the surface of one body makes with the horizontal when slipping of another body upon it impends. It applies to the particular rubbing surfaces in contact. It equals the angle of static friction.

Kinetic frictional force (F_k) opposes motion when one body is slipping on the surface of the other. Its value is usually less than that of the limiting friction. *Coefficient of kinetic friction* (f_k) is the ratio F_k/N. *Angle of kinetic friction* (ϕ_k) is defined by $\tan\phi_k = F_k/N = f_k$.

Laws of Friction for Dry Surfaces

1. Friction between two given bodies is directly proportional to the pressure; the coefficient of friction is constant for all pressures.
2. The coefficient and amount of friction for given pressures are independent of the area of contact.
3. The coefficient of friction is independent of the relative velocity, although static friction is greater than kinetic friction.

The preceding laws are only approximately true. The coefficient of friction is slightly greater for small pressures upon large areas than for great pressures upon small areas. The coefficient of friction decreases as the speed increases.*

Coefficients of Static and Kinetic Friction. These are affected by the preceding laws of friction and also by the characters of the surfaces, the kinds of material, and the nature of any lubricant used. Rough averages for a number of materials and conditions are given in Table 3.2.[†]

---

*Recent experiments have proved also that time of contact of the surfaces affects the coefficient of static friction.
[†]Hudson's Manual, p. 102.

TABLE 3.2 Coefficients of Static and Kinetic Friction[a]

Materials	Condition	Sliding Friction		Static Friction	
		ϕ	f	ϕ	f
Cast iron on cast iron or bronze	Wet	$17\frac{1}{4}°$	0.31	—	—
Cast iron on cast iron or bronze	Greased	$4\frac{1}{2}°-5\frac{3}{4}°$	0.08–0.10	$9°$	0.16
Cast iron on oak (fibers parallel)	Dry	$16\frac{3}{4}°-26\frac{1}{2}°$	0.30–0.50	—	—
Cast iron on oak (fibers parallel)	Wet	$12\frac{1}{2}°$	0.22	$33°$	0.65
Cast iron on oak (fibers parallel)	Greased	$10\frac{3}{4}°$	0.19	—	—
Earth on earth	—	—	—	$14°-45°$	0.25–1.0
Earth on earth (clay)	Damp	—	—	$45°$	1.0
Earth on earth (clay)	Wet	—	—	$17\frac{1}{4}°$	0.31
Hemp-rope on rough wood	Dry	$26\frac{1}{2}°$	0.50	$26\frac{1}{2}°-38\frac{3}{4}°$	0.50–0.80
Hemp-rope on polished wood	Dry	—	—	$18\frac{1}{4}°$	0.33
Leather on oak	Dry	$16\frac{3}{4}°-26\frac{1}{2}°$	0.30–0.50	$26\frac{1}{2}°-31°$	0.50–0.60
Leather on cast iron	Dry	$29\frac{1}{4}°$	0.56	$16\frac{3}{4}°-26\frac{1}{2}°$	0.30–0.50
Oak on oak (fibers parallel)	Dry	$25\frac{3}{4}°$	0.48	$31\frac{3}{4}°$	0.62
Oak on oak (fibers crossed)	Dry	$18\frac{3}{4}°$	0.34	$28\frac{1}{4}°$	0.54
Oak on oak (fibers crossed)	Wet	$14°$	0.25	$35\frac{1}{4}°$	0.71
Oak on oak (fibers perpendicular)	Dry	$10\frac{3}{4}°$	0.19	$23\frac{1}{4}°$	0.43
Steel on ice	Dry	—	0.01	$1\frac{1}{2}°$	0.027
Steel on steel	Dry	Vel. 10 ft/sec 0.09 Vel. 100 ft/sec 0.03		$8\frac{1}{2}°$	0.15
Stone masonry on concrete	Dry	—	—	$37\frac{1}{4}°$	0.76
Stone masonry on undisturbed ground	Dry	—	—	$33°$	0.65
Stone masonry on undisturbed ground	Wet	—	—	$16\frac{3}{4}°$	0.30
Wrought iron on wrought iron	Dry	$23\frac{3}{4}°$	0.44	—	—
Wrought iron on wrought iron	Greased	$4\frac{1}{2}°-5\frac{3}{4}°$	0.08–0.10	$6\frac{1}{2}°$	0.11
Wrought iron on cast iron or bronze	Dry	$10\frac{1}{4}°$	0.18	$10\frac{3}{4}°$	0.19
Wrought iron on cast iron or bronze	Greased	—	—	$4°-4\frac{1}{2}°$	0.07–0.08

[a]*Hudson's Manual*, p. 102.

3.5.2 Axle Friction and Lubricated Surfaces

Axle Friction, Nonlubricated

Axle friction is the friction that opposes the turning of an axle in its bearing. For a dry bearing, the axle (Fig. 3.56) will obviously move to the right. It will climb until a point is reached, at some angle ϕ, where the friction force $f_0 N$ on the shaft or journal is balanced by the tangential component F of the vertical load W, that is $f_0 N = F$. Normal force N is equal to $W \cos\phi$, and f_0 is the coefficient of friction for the surfaces in question. Friction force F is equal to $W \sin\phi$, so that the usual relationship, $f_0 = \tan\phi$, for friction prevails. It is customary, however, to define the coefficient of

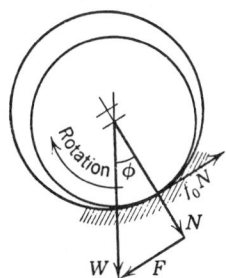

Fig. 3.56 Force diagram/axle friction.

friction f for a journal bearing as the ratio of load W to the tangential friction force F or $f = F/W = \sin\phi$. The relationship between f_0 and f is then represented by $f = f_1\cos\phi$.

Friction of Lubricated Surfaces. The friction of lubricated surfaces is characterized by two types of sliding motion:

1. Hydrodynamic or thick-film lubrication.
2. Boundary of thin-film lubrication.

Hydrodynamic lubrication occurs with a plentiful supply of oil when the thickness of the film is large as compared with the height of the irregularities of the surfaces. Metallic contact or wear does not occur, and the frictional resistance is independent of the kinds of materials composing the bodies.

Although somewhat involved mathematically, the laws for this type of friction are well understood. Figure 3.57 shows a weightless plate of area A being pushed along at velocity U by force F as it rests on an oil film of thickness h. Newton observed that force F was directly proportional to area A and velocity U and inversely proportional to the film thickness h. This can be written in the form of an equation, $F = \mu UA/h$, where the constant of proportionality μ is called the coefficient of viscosity of the lubricant. It has dimensions of lb-sec/in.^{2} or dyne sec/cm^2.

Example 3.15. Find the force necessary to maintain a velocity of 20 ft/sec for a 3 in. × 3 in. plate if the film has a thickness of 0.003 in. A heavy oil of viscosity 0.000005 lb-sec/in.^{2} is used.

SOLUTION. Substitution in Newton's equation gives $F = 0.000005 \times 240 \times 9/0.003 = 3.6$ lb.

Journal Friction, Lubricated

Newton's equation can be adapted to the journal bearing of Fig. 3.58 of radius r and axial length l by substituting $A = 2\pi rl$ and $U = 2\pi rn_s$, where n_s is the revolutions per second. Film thickness h is replaced by radial clearance c. The result,

$$F = 4\pi^2\mu r^2 h_s \frac{l}{c}$$

is known as *Petroff's equation*. It is valid only for very lightly loaded bearings where the shaft is centrally located in the clearance space. The friction is practically unchanged when the bearing is loaded, and an equation for the coefficient of friction, $f = F/W$, is obtained by dividing both sides of Petroff's equation by the load W. The result, $f = 2\pi^2\mu h_s r/pc$, is obtained by the substitution $p = W/2rl$, where p is the load per square inch of projected area of the bearing.

Example 3.16. Find the coefficient of friction for a 2 in. diameter × 3 in. long journal bearing carrying a load of 600 lb at a speed of 1200 rpm. Radial clearance is 0.002 in. Viscosity of the lubricant is 0.000003 lb-sec/in.^{2}

SOLUTION. Substitution gives

$$f = \frac{2\pi^2 0.000003 \times 20 \times 1}{100 \times 0.002} = 0.0059$$

This example shows that the coefficient of friction, under conditions of flooded lubrication, is very much smaller than the value of the coefficient when the surfaces are dry. It should be noted that the coefficient of friction for lubricated surfaces varies directly with the velocity and inversely with the

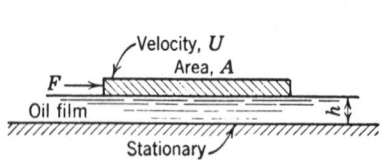

Fig. 3.57 Lubricated surface friction.

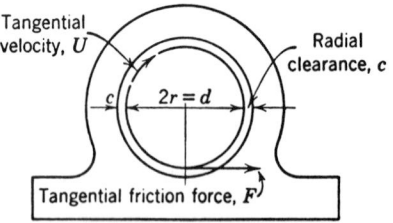

Fig. 3.58 Journal friction.

pressure. This is in contrast with the coefficient of friction for dry surfaces, which is independent of both velocity and pressure.

The plate of Fig. 3.57 could carry a downward load provided that the forward edge would be tipped at a small upward angle. The oil would be drawn into the wedge-shaped opening, and a hydrostatic pressure would be developed in the film sufficient to support the load. When a lubricated journal bearing carries a load, the shaft center will shift sidewise in the direction opposite to that for a dry bearing. The oil film will converge in such a manner as to produce the pressure necessary to carry the load. The mathematical treatment of these phenomena may be found in almost any treatise on lubrication.

Boundary Friction

Boundary friction occurs between surfaces that are slightly oily or greasy but where the oil supply is insufficient to maintain a hydrodynamic film. Although the bodies approach each other closely, the load is carried by the absorbed layers of lubricant attached to the surfaces. The chemical and physical properties of the lubricant and of the materials composing the bodies are now of importance. At the present time, available knowledge for this type of friction is highly specific and cannot be formulated into general laws. It would appear, however, that the friction is relatively independent of the area in contact and the velocity of sliding. The value for the coefficient of friction is usually intermediate between that for dry friction and that for hydrodynamic friction.

In boundary lubrication one type of oil may have less friction than another of the same viscosity. This property has been given the name of *oiliness*. Certain animal and vegetable oils and compounds are superior to mineral oils with respect to oiliness. They are therefore used as additives to mineral oils to decrease the friction.

If the load on a bearing is increased until the absorbed boundary layers of lubricant can no longer be maintained, metal-to-metal contact will occur at the high points of the sliding surfaces and wear will take place. The exact physical and metallurgical process of wear is not clearly understood. The harder metal may gouge off particles of the softer metal. Experimental evidence exists that the surfaces weld together at the points of metallic contact because of the extremely high localized pressures. The welds are broken as soon as made and may account for the major portion of the resistance to the motion. The amount of welding decreases with decrease of solid solubility between the metals of the two bodies. There is thus justification for the long-established rule that the two members of a bearing should be made of dissimilar metals to decrease friction and the possibility of galling and failure. The *running-in* of a bearing is a form of wear process in which an improvement is effected in the surface conditions.

In addition to a scanty oil supply, a bearing operating with boundary friction will be adversely affected by a decrease in the viscosity of the lubricant, a decrease in the speed, or an increase in the pressure. An unfavorable combination of these qualities may cause the coefficient of friction to increase until the bearing overheats and failure becomes imminent.

3.5.3 Rolling Friction

Rolling friction is that friction developed when one body rolls over the surface of another, and depends on the hardness of the surfaces in contact and the radius of the rolling surface. The theory is based on the idea that surfaces are slightly deformed at the place of contact and that the effect of rolling friction is the same as if the surfaces were not deformed and the rolling body passed constantly over a small obstruction. Let P (Fig. 3.59) be the horizontal force required to overcome the small obstruction B. Then $hP = aW$, and, since h is nearly equal to r, $P = aW/r$ (approximately). *Coefficient of rolling friction* is a, and, as an analogy with definitions of static and kinetic friction coefficients, might be defined as the ratio T_r/W, where $T_r = Pr$ is the torque resisting rolling motion and W is the normal force (in this case, the weight of the body). The coefficient is a linear

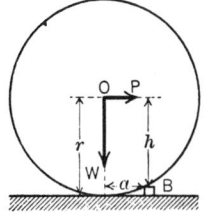

Fig. 3.59 Rolling friction.

distance and is usually given in inches. The values of the coefficient of rolling friction given by various investigators are not in close agreement and should be used with caution.

Coefficients of Rolling Friction. The following are some reported values of coefficients of rolling friction:

Lignum vitae roller on oak track	0.019 in.
Elm roller on oak track	0.032 in.
Cast-iron wheel (20-in. diam.) on cast-iron rail	0.018–0.019 in.
Railroad wheels (39.4-in. diam.)	0.020–0.022 in.
Iron or steel wheels on wood track	0.06–0.10 in.

3.5.4 Pivot Friction

Pivot friction is that friction that opposes the turning of the end of a vertical, or inclined, shaft in its bearing. Some examples of pivot friction, with friction torque and power formulas applying, are shown in Table 3.3.

TABLE 3.3 Pivot Friction[a,b]

Type of Pivot	Torque T in Pound-inches	Power P Lost by Friction in Foot-pounds per Second
Shafts and Journals (180° bearing)	$T = fWr$	$P = \dfrac{2\pi n}{12} fWr.$
Flat Pivot	$T = \tfrac{2}{3} fWr.$	$P = \dfrac{4\pi n}{3 \times 12} fWr$
Collar-bearing	$T = \tfrac{2}{3} fW \dfrac{R^3 - r^3}{R^2 - r^2}$	$P = \dfrac{4\pi n}{3 \times 12} fW \dfrac{R^3 - r^3}{R^2 - r^2}$
Conical Pivot	$T = \tfrac{2}{3} fW \dfrac{r}{\sin \alpha}$	$P = \dfrac{4\pi n f W r}{3 \times 12 \sin \alpha}$
Truncated-Cone Pivot	$T = \tfrac{2}{3} fW \dfrac{R^3 - r^3}{(R^2 - r^2)\sin \alpha}$	$P = \dfrac{4\pi n f W (R^3 - r^3)}{3 \times 12 (R^2 - r^2)\sin \alpha}$

[a] *Hudson's Manual*, p. 105.
[b] f = coefficient of friction, W = load in pounds, T = torque of friction about the axis of the shaft, r = radius in inches, n = revolutions per second.

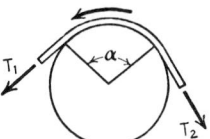

Fig. 3.60   Belt friction.

3.5.5   Belt Friction

Belt or **coil friction** is that friction that opposes the slipping of a belt, rope, brake band, or similar article coiled about a pulley, sheave, post, capstan, or similar device. When power is being transmitted, say by a belt driving a pulley, the tension T_1 on the driving side of the belt is greater than the tension T_2 on the driven side. Neglecting the effect of centrifugal force, which is small at low speeds, the tensions are related by the formula $T_1/T_2 = e^{f\alpha}$ where e = 2.718 + (i.e., base of natural logarithms), f = coefficient of friction between belt and pulley, and α = angle of contact between belt and pulley (Fig. 3.60). Values of T_1/T_2 for various values of f and α are shown in Table 3.4.

TABLE 3.4   Maximum Ratio T_1 / T_2 (Slipping Impending)

α Radians	Values of f (Coefficient of Friction)								
2π	0.10	0.15	0.20	0.25	0.30	0.35	0.40	0.45	0.50
0.1	1.06	1.1	1.13	1.17	1.21	1.25	1.29	1.33	1.37
0.2	1.13	1.21	1.29	1.37	1.46	1.55	1.65	1.76	1.87
0.3	1.21	1.32	1.45	1.60	1.76	1.93	2.13	2.34	2.57
0.4	1.29	1.46	1.65	1.87	2.12	2.41	2.73	3.10	3.51
0.425	1.31	1.49	1.70	1.95	2.23	2.55	2.91	3.33	3.80
0.45	1.33	1.53	1.76	2.03	2.34	2.69	3.10	3.57	4.11
0.475	1.35	1.56	1.82	2.11	2.45	2.84	3.30	3.83	4.45
0.5	1.37	1.60	1.87	2.19	2.57	3.00	3.51	4.11	4.81
0.525	1.39	1.64	1.93	2.28	2.69	3.17	3.74	4.41	5.20
0.55	1.41	1.68	2.00	2.37	2.82	3.35	3.98	4.74	5.63
0.6	1.46	1.76	2.13	2.57	3.10	3.74	4.52	5.45	6.59
0.7	1.52	1.93	2.41	3.00	3.74	4.66	5.81	7.24	9.02
0.8	1.65	2.13	2.73	3.51	4.52	5.81	7.47	9.60	12.35
0.9	1.76	2.34	3.10	4.11	5.45	7.24	9.60	12.74	16.90
1.0	1.87	2.57	3.51	4.81	6.59	9.02	12.35	16.90	23.14
1.5	2.57	4.11	6.59	10.55	16.90	27.08	43.38	69.49	111.32
2.0	3.51	6.59	12.35	23.14	43.38	81.31	152.40	285.68	535.49
2.5	4.81	10.55	23.14	50.75	111.32	244.15	535.49	1,174.5	2,575.9
3.0	6.59	16.90	43.38	111.32	285.68	733.14	1,881.5	4,828.5	12,391.
3.5	9.02	27.08	81.31	244.15	733.14	2199.9	6,610.7	19,851.	59,608.
4.0	12.35	43.38	152.40	535.49	1881.5	6610.7	23,227.	81,610.	286,744.

TABLE 3.5   Coefficients of Friction for Belts and Pulley Materials[a]

Belt Material	Pulley Material					
	Iron–Steel	Wood	Paper	Wet Iron	Greasy Iron	Oily Iron
Oak-tanned leather	0.25	0.30	0.35	0.20	0.15	0.12
Mineral-tanned leather	0.40	0.45	0.50	0.35	0.25	0.20
Canvas stitched	0.20	0.23	0.25	0.15	0.12	0.10
Balata	0.32	0.35	0.40	0.20	—	—
Cotton woven	0.22	0.25	0.28	0.15	0.12	0.10
Camel hair	0.35	0.40	0.45	0.25	0.20	0.15
Rubber friction	0.30	0.32	0.35	0.18	—	—
Rubber covered	0.32	0.35	0.38	0.15	—	—
Rubber on fabric	0.35	0.38	0.40	0.20	—	—

[a]*Machinery*, Vol. 37, 1931, p. 560-A.

Power transmitted is given by $P = (T_1 - T_2)v$, where P is power in foot-pounds per second and v is velocity in feet per second. Coefficients of friction for belts and pulley material are shown in Table 3.5.

ACKNOWLEDGMENT

Chapter 3 was originally a revision of material in previous handbooks published by John Wiley & Sons, most of which was written by Professors C. H. Burnside and E. R. Maurer for *Merriam's Civil Engineers' Handbook* and *Peele's Mining Engineers' Handbook*. Later revisions by Janvier M. Rice, E. E. appeared in the *Handbook of Engineering Fundamentals*, Third Edition, a part of the Wiley Handbook Series. The present section was revised by Professor Fowler, with the primary revisions being focused on the introduction of vector mechanics.

BIBLIOGRAPHY

Beer, F. P., and Russell Johnston, E. Jr., *Vector Mechanics for Engineers, Statics and Dynamics*, 4th ed., McGraw-Hill, New York, 1984.

Ginsberg, J. H., and Genin, J., *Statics and Dynamics*, Combined Version, Wiley, New York, 1984.

Hibbeler, R. C., *Engineering Mechanics, Statics and Dynamics*, 3rd ed., Macmillan, New York, 1983.

Meriam, J. L., *Dynamics*, 2nd ed., SI Version, Wiley, New York, 1975.

Shames, I. H., *Engineering Mechanics, Vol. I, Statics*, 2nd ed., Prentice-Hall, Inglewood Cliffs, NJ, 1966.

Shames, I. H., *Engineering Mechanics, Vol. II, Dynamics*, 2nd ed., Prentice-Hall, Inglewood Cliffs, NJ, 1966.

CHAPTER 4

MECHANICS OF DEFORMABLE BODIES

Neal F. Enke and Bela I. Sandor

Department of Engineering Mechanics
University of Wisconsin
Madison, Wisconsin

4.1 INTRODUCTION TO STRESS AND STRAIN

4.1.1 Definitions of Stress and Strain

Stress at a Point

Stress is the description of the intensity of force. For a general state of loading, such as represented in Fig. 4.1, the stress vector **S** acting on the section D at point Q is given by

$$\mathbf{S} = \lim_{\Delta A \to 0} \frac{\Delta \mathbf{P}}{\Delta A}$$

where $\Delta \mathbf{P}$ is the force acting on the area ΔA. The stress vector **S** can be further divided into a component normal to the section and a component tangential to the section. The normal component is referred to as the normal stress and is usually denoted by σ. The tangential component is referred to as the shear (or shearing) stress and is commonly denoted by τ. It is often useful to further divide τ into two orthogonal components, thus giving an orthogonal triad of stress components. Tensile stresses are normal stresses that tend to increase the length of a member. Compressive stresses are normal stresses that tend to decrease the length of a member. In the U.S. customary system of units, stress is expressed in pounds per square inch (psi). Stress is also commonly expressed in kilopounds per square inch (ksi).

In general, if different sections are considered through the same point in a body under load, different stress vectors will be obtained. In order to completely specify the state of stress at a point, it is sufficient to consider the stresses acting on three mutually orthogonal sections at that point. The common procedure is to envision an infinitesimally small cube at the point of interest (see Fig. 4.2). In the figure, σ_{ij} is the stress acting on the i face in the j direction, where i and j can be any of x, y, or z. The state of stress at a point is often represented in matrix form as

$$[\sigma_{ij}] = \begin{bmatrix} \sigma_{xx} & \sigma_{xy} & \sigma_{xz} \\ \sigma_{yx} & \sigma_{yy} & \sigma_{yz} \\ \sigma_{zx} & \sigma_{zy} & \sigma_{zz} \end{bmatrix} \tag{4.1}$$

The three diagonal terms $(\sigma_{xx}, \sigma_{yy}, \sigma_{zz})$ are the normal stress components in the x, y, and z directions, respectively. The off-diagonal terms represent the shear stress components. The i face is considered positive if its outward normal acts in the same direction as the positive i axis; otherwise, it

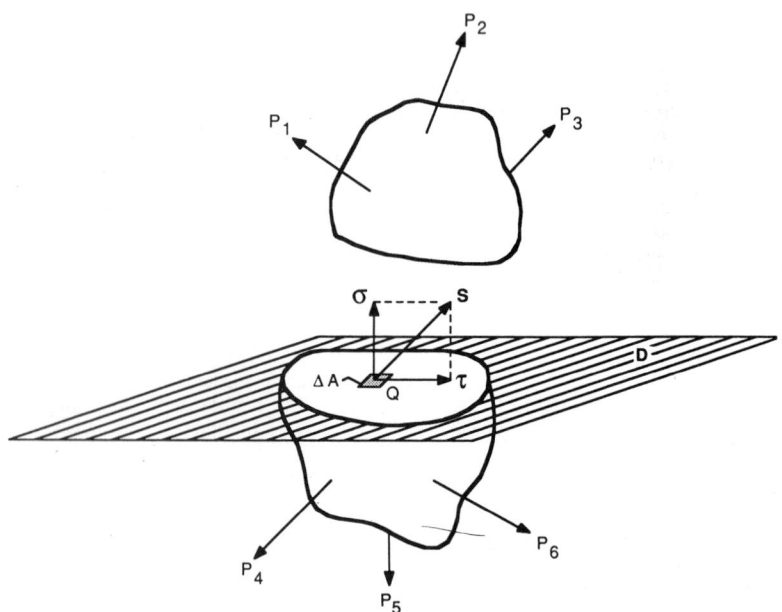

Fig. 4.1 Normal and shear components of stress.

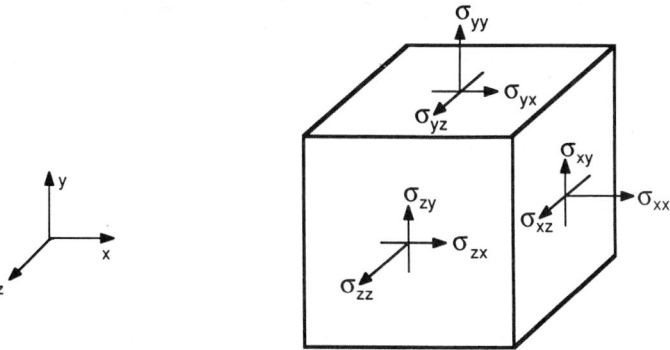

Fig. 4.2   General state of stress.

is considered negative. The stress σ_{ij} is considered positive if it acts on the positive i face in the positive j direction or if it acts on the negative i face in the negative j direction. Thus, the three faces shown in Fig. 4.2 are positive as are the stress components acting on them. A positive normal stress is a tensile stress while a negative normal stress is a compressive stress.

If the point of interest is in a state of static equilibrium, it can be shown (by summing moments about a small element) that the cross-shearing stresses must be equal. That is $\sigma_{xy} = \sigma_{yx}$, $\sigma_{xz} = \sigma_{zx}$, and $\sigma_{yz} = \sigma_{zy}$. The total number of stress components needed to completely define the state of stress at a point is thus reduced to six in this case. This so-called complementary property of shear is not valid for dynamic situations. It is also invalid in situations where surface or body couples are significant, but this condition is rarely encountered in practice.

Strain at a Point
Deformation is the movement of points in a body relative to each other. Strain is the description of the intensity of deformation. The normal strain at a point Q in the x direction, ε_{xx}, is given by

$$\varepsilon_{xx} = \lim_{L \to 0} \frac{\Delta L}{L}$$

Here, L is the original length of a line segment in the x direction centered at Q, and ΔL is the change in length due to the deformation (Fig. 4.3). Normal strains are considered positive if they cause an increase in the length of the line segment. The shear (or shearing) strain associated with the orthogonal x and y directions, γ_{xy}, is defined as the change in angle between infinitesimal line segments originally in the x and y directions before deformation (Fig. 4.4). Shear strains are considered positive if they bring about a reduction in angle. By this definition of shear strain, it is apparent that $\gamma_{xy} = \gamma_{yx}$, provided that no discontinuities (such as void formation) occur at the point of interest.

Tensorial shear strains, ε_{ij}, are defined to be equal to one-half the ordinary shear strains, γ_{ij}. That is, $\varepsilon_{ij} = \gamma_{ij}/2$ for $i \neq j$. Tensorial shear strains are often used in the theory of elasticity and continuum mechanics because they result in algebraically simpler equations than if ordinary shear strains are employed. Utilizing the tensorial shear strains, the state of strain at a point can be

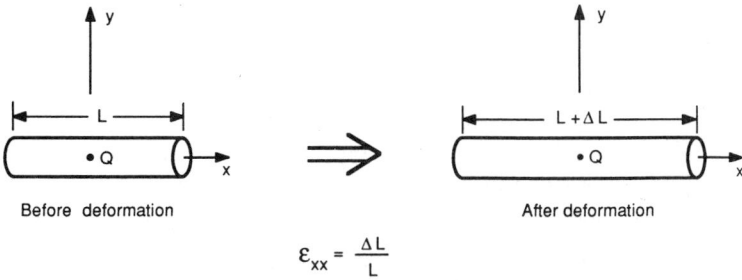

Before deformation   After deformation

$$\varepsilon_{xx} = \frac{\Delta L}{L}$$

Fig. 4.3   Definition of normal strain.

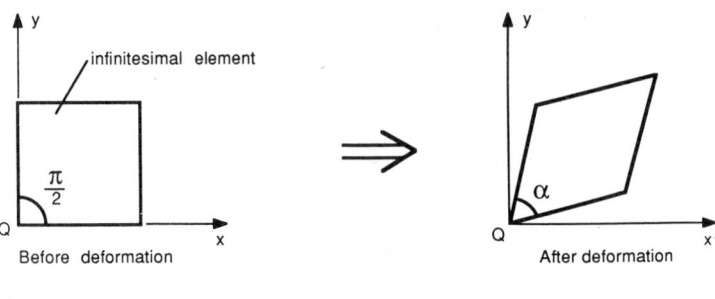

$$\gamma_{xy} = \frac{\pi}{2} - \alpha$$

Fig. 4.4 Definition of shear strain.

represented in matrix form as

$$[\varepsilon_{ij}] = \begin{bmatrix} \varepsilon_{xx} & \varepsilon_{xy} & \varepsilon_{xz} \\ \varepsilon_{yx} & \varepsilon_{yy} & \varepsilon_{yz} \\ \varepsilon_{zx} & \varepsilon_{zy} & \varepsilon_{zz} \end{bmatrix} \qquad (4.2)$$

The three diagonal terms are the normal strain components while the off-diagonal terms are the tensorial shear strain components. Although strain is a dimensionless quantity, it is common to express it in terms of inches per inches or similar units.

Strain-Displacement Relations
Mathematically, strains are defined in terms of derivatives of displacements. Assume a general point Q in a body undergoes a displacement to Q' when the body deforms under application of load. Let u, v, and w represent the displacement components of Q in the x, y, and z directions, respectively. Using the previously given definitions of normal and shear strains, the strain components at the point Q can be shown to be[1]

$$\varepsilon_{xx} = \sqrt{1 + 2\frac{\partial u}{\partial x} + \left(\frac{\partial u}{\partial x}\right)^2 + \left(\frac{\partial v}{\partial x}\right)^2 + \left(\frac{\partial w}{\partial x}\right)^2} - 1$$

$$\varepsilon_{yy} = \sqrt{1 + 2\frac{\partial v}{\partial y} + \left(\frac{\partial u}{\partial y}\right)^2 + \left(\frac{\partial v}{\partial y}\right)^2 + \left(\frac{\partial w}{\partial y}\right)^2} - 1$$

$$\varepsilon_{zz} = \sqrt{1 + 2\frac{\partial w}{\partial z} + \left(\frac{\partial u}{\partial z}\right)^2 + \left(\frac{\partial v}{\partial z}\right)^2 + \left(\frac{\partial w}{\partial z}\right)^2} - 1$$

$$\gamma_{xy} = \gamma_{yx} = \sin^{-1}\left[\frac{\dfrac{\partial u}{\partial y} + \dfrac{\partial v}{\partial x} + \dfrac{\partial u}{\partial x}\dfrac{\partial u}{\partial y} + \dfrac{\partial v}{\partial x}\dfrac{\partial v}{\partial y} + \dfrac{\partial w}{\partial x}\dfrac{\partial w}{\partial y}}{(1 + \varepsilon_{xx})(1 + \varepsilon_{yy})}\right] \qquad (4.3)$$

$$\gamma_{xz} = \gamma_{zx} = \sin^{-1}\left[\frac{\dfrac{\partial w}{\partial x} + \dfrac{\partial u}{\partial z} + \dfrac{\partial w}{\partial z}\dfrac{\partial w}{\partial x} + \dfrac{\partial u}{\partial z}\dfrac{\partial u}{\partial x} + \dfrac{\partial v}{\partial z}\dfrac{\partial v}{\partial x}}{(1 + \varepsilon_{xx})(1 + \varepsilon_{zz})}\right]$$

$$\gamma_{yz} = \gamma_{zy} = \sin^{-1}\left[\frac{\dfrac{\partial v}{\partial z} + \dfrac{\partial w}{\partial y} + \dfrac{\partial v}{\partial y}\dfrac{\partial v}{\partial z} + \dfrac{\partial w}{\partial y}\dfrac{\partial w}{\partial z} + \dfrac{\partial u}{\partial y}\dfrac{\partial u}{\partial z}}{(1 + \varepsilon_{yy})(1 + \varepsilon_{zz})}\right]$$

Equations (4.3) give the strain components for arbitrary displacements. They are known as the strain-displacement relations for large displacements. Slight variations of these equations, due to simplifying assumptions or slightly different definitions of strain, are often found in the literature (see, e.g., Refs. 2, Chapter 33, and 3).

If the displacements and strains are sufficiently small, the higher-order terms in Eqs. (4.3) can be neglected. The equations then reduce to

$$\varepsilon_{xx} = \frac{\partial u}{\partial x} \qquad \gamma_{xy} = \gamma_{yx} = \frac{\partial v}{\partial x} + \frac{\partial u}{\partial y}$$

$$\varepsilon_{yy} = \frac{\partial v}{\partial y} \qquad \gamma_{xz} = \gamma_{zx} = \frac{\partial w}{\partial x} + \frac{\partial u}{\partial z} \qquad (4.4)$$

$$\varepsilon_{zz} = \frac{\partial w}{\partial z} \qquad \gamma_{yz} = \gamma_{zy} = \frac{\partial w}{\partial y} + \frac{\partial v}{\partial z}$$

These equations are known as the strain-displacement relations for small displacements.

Equations of Compatibility
Equations (4.4) allow for the determination of six strain components using only three displacement components. Physically, this implies that the strains must be compatible; that is, the infinitesimal elements of the deformed member must fit together without any voids or overlap being present. Mathematically, this implies that the strains are not arbitrary functions of the coordinates. There must exist three equations relating the strains to one another. These equations of compatibility are as follows:

$$\frac{\partial^2 \gamma_{xy}}{\partial x\,\partial y} = \frac{\partial^2 \varepsilon_{xx}}{\partial y^2} + \frac{\partial^2 \varepsilon_{yy}}{\partial x^2}$$

$$\frac{\partial^2 \gamma_{yz}}{\partial y\,\partial z} = \frac{\partial^2 \varepsilon_{yy}}{\partial z^2} + \frac{\partial^2 \varepsilon_{zz}}{\partial y^2} \qquad (4.5a)$$

$$\frac{\partial^2 \gamma_{zx}}{\partial z\,\partial x} = \frac{\partial^2 \varepsilon_{zz}}{\partial x^2} + \frac{\partial^2 \varepsilon_{xx}}{\partial z^2}$$

Three additional compatibility equations are often used,

$$2\frac{\partial^2 \varepsilon_{xx}}{\partial y\,\partial z} = \frac{\partial}{\partial x}\left(-\frac{\partial \gamma_{yz}}{\partial x} + \frac{\partial \gamma_{zx}}{\partial y} + \frac{\partial \gamma_{xy}}{\partial z}\right)$$

$$2\frac{\partial^2 \varepsilon_{yy}}{\partial z\,\partial x} = \frac{\partial}{\partial y}\left(\frac{\partial \gamma_{yz}}{\partial x} - \frac{\partial \gamma_{zx}}{\partial y} + \frac{\partial \gamma_{xy}}{\partial z}\right) \qquad (4.5b)$$

$$2\frac{\partial^2 \varepsilon_{zz}}{\partial x\,\partial y} = \frac{\partial}{\partial z}\left(\frac{\partial \gamma_{yz}}{\partial x} + \frac{\partial \gamma_{zx}}{\partial y} - \frac{\partial \gamma_{xy}}{\partial z}\right)$$

It can be shown that Eqs. (4.5b) are not independent of Eqs. (4.5a).[2] The equations of compatibility are used in the theory of elasticity. They are also useful in verifying the accuracy of experimentally determined strains such as in the moiré method of strain analysis.

Stress Equations of Equilibrium
By considering the static equilibrium of an infinitesimal element, it can be shown that the following equations must hold:

$$\frac{\partial \sigma_{xx}}{\partial x} + \frac{\partial \sigma_{yx}}{\partial y} + \frac{\partial \sigma_{zx}}{\partial z} + F_x = 0$$

$$\frac{\partial \sigma_{xy}}{\partial x} + \frac{\partial \sigma_{yy}}{\partial y} + \frac{\partial \sigma_{zy}}{\partial z} + F_y = 0 \qquad (4.6)$$

$$\frac{\partial \sigma_{xz}}{\partial x} + \frac{\partial \sigma_{yz}}{\partial y} + \frac{\partial \sigma_{zz}}{\partial z} + F_z = 0$$

where F_x, F_y, and F_z are the body forces per unit volume in the x, y, and z directions, respectively. Equations (4.6) are known as the stress equations of equilibrium.

4.1.2   Linear Elastic Stress-Strain Relationships

Stresses and strains in a solid body are generally not independent but rather can be related to each other. The exact form of these relationships depends on whether the body is behaving in an elastic, plastic, viscoelastic, or some other fashion. The relationships also depend on whether the material is isotropic or anisotropic. An isotropic material displays the same properties in all directions. An anisotropic material displays different properties in different directions. Only isotropic response will be considered here. Anisotropic response is discussed in Section 4.7 on composite materials.

A member is said to experience elastic response if it returns to its original shape upon removal of loads. If, in addition, the relationship between stress and strain can be written in a linear form, the member is said to be experiencing linear elastic response.

Axial Loading
For the case of uniaxial loading of a prismatic member, the stress-strain relationship is

$$\sigma_a = \frac{P}{A} = E\varepsilon_a \tag{4.7}$$

where P is the applied axial force, A is the cross-sectional area of the member, ε_a is the axial strain, and E is a material constant known as the modulus of elasticity or Young's modulus. Equation (4.7) is commonly referred to as Hooke's law. The transverse strain, ε_t, is given by

$$\varepsilon_t = -\nu\varepsilon_a = \frac{-\nu\sigma_a}{E} \tag{4.8}$$

where ν is a material constant known as Poisson's ratio. Its value is always between 0 and $\frac{1}{2}$ for linear elastic, isotropic response.

If the member has a length L, the axial elongation, δ, is given by $\delta = \varepsilon_a L$. In view of Eq. (4.7) this becomes $\delta = PL/AE$. If P, A, or E varies along the length of the member, the deflection must be found by integrating $\varepsilon_a\, dx$ over the length of the member:

$$\delta = \int_0^L \frac{P\, dx}{AE}$$

Multiaxial State of Stress
For the general case of a three-dimensional state of stress, the stress-strain relationships are

$$
\begin{aligned}
\varepsilon_{xx} &= \frac{1}{E}\left[\sigma_{xx} - \nu(\sigma_{yy} + \sigma_{zz})\right] & \gamma_{xy} &= \frac{\sigma_{xy}}{G} \\
\varepsilon_{yy} &= \frac{1}{E}\left[\sigma_{yy} - \nu(\sigma_{xx} + \sigma_{zz})\right] & \gamma_{xz} &= \frac{\sigma_{xz}}{G} \\
\varepsilon_{zz} &= \frac{1}{E}\left[\sigma_{zz} - \nu(\sigma_{xx} + \sigma_{yy})\right] & \gamma_{yz} &= \frac{\sigma_{yz}}{G}
\end{aligned}
\tag{4.9}
$$

where G is a material constant known as the shear modulus or modulus of rigidity. Equations (4.9) are often referred to as the generalized Hooke's law. These equations can be inverted to give the stresses in terms of the strains:

$$
\begin{aligned}
\sigma_{xx} &= \frac{E}{(1+\nu)(1-2\nu)}\left[(1-\nu)\varepsilon_{xx} + \nu(\varepsilon_{yy} + \varepsilon_{zz})\right] & \sigma_{xy} &= G\gamma_{xy} \\
\sigma_{yy} &= \frac{E}{(1+\nu)(1-2\nu)}\left[(1-\nu)\varepsilon_{yy} + \nu(\varepsilon_{xx} + \varepsilon_{zz})\right] & \sigma_{xz} &= G\gamma_{xz} \\
\sigma_{zz} &= \frac{E}{(1+\nu)(1-2\nu)}\left[(1-\nu)\varepsilon_{zz} + \nu(\varepsilon_{xx} + \varepsilon_{yy})\right] & \sigma_{yz} &= G\gamma_{yz}
\end{aligned}
\tag{4.10}
$$

The material constants G, E, and ν are not independent of one another. It is shown in the theory of elasticity that they are related by the equation $G = E/2(1 + \nu)$.

The dilatation, e, represents the change in volume per unit volume. For small strains it is given by $e = \varepsilon_{xx} + \varepsilon_{yy} + \varepsilon_{zz}$. For the case of pure hydrostatic stress, $\sigma_{xx} = \sigma_{yy} = \sigma_{zz} = -p$ and $\sigma_{xy} = \sigma_{xz} = \sigma_{yz} = 0$, where p is the uniform pressure acting on the member. In this case the dilatation is linearly related to the pressure by $e = -p/k$, where k is a material constant known as the bulk modulus. The bulk modulus is related to E and ν by the equation $k = E/3(1 - 2\nu)$.

One other elastic constant that arises in the theory of elasticity is Lamé's constant, λ. Unlike the other constants (E, ν, G, and k), λ has no physical significance. That is, there is no mechanical test that can be used to directly measure λ. Since only two material constants are required to completely describe linear elastic isotropic response, the constants E, ν, G, k, and λ can be interrelated (see Ref. 1, p. 44, for details).

Plane Stress
Plane stress, or two-dimensional state of stress, exists at a point if an orientation can be found such that the stress in one of the three coordinate directions is zero. If this direction of zero stress is arbitrarily taken to be the z direction (i.e., $\sigma_{zz} = \sigma_{xz} = \sigma_{yz} = 0$), then Eqs. (4.9) reduce to

$$\varepsilon_{xx} = \frac{1}{E}(\sigma_{xx} - \nu\sigma_{yy}) \qquad \gamma_{xy} = \frac{\sigma_{xy}}{G}$$

$$\varepsilon_{yy} = \frac{1}{E}(\sigma_{yy} - \nu\sigma_{xx}) \qquad \gamma_{xz} = 0 \qquad (4.11)$$

$$\varepsilon_{zz} = \frac{-\nu}{E}(\sigma_{xx} + \sigma_{yy}) \qquad \gamma_{yz} = 0$$

In addition, ε_{zz} can be written as

$$\varepsilon_{zz} = \frac{-\nu}{1-\nu}(\varepsilon_{xx} + \varepsilon_{yy})$$

Equations (4.10) reduce to

$$\sigma_{xx} = \frac{E}{1-\nu^2}(\varepsilon_{xx} + \nu\varepsilon_{yy}) \qquad \sigma_{xy} = G\gamma_{xy}$$

$$\sigma_{yy} = \frac{E}{1-\nu^2}(\varepsilon_{yy} + \nu\varepsilon_{xx}) \qquad \sigma_{xz} = 0 \qquad (4.12)$$

$$\sigma_{zz} = 0 \qquad \sigma_{yz} = 0$$

Plane Strain
Plane strain occurs at a point if an orientation can be found such that the strain in one of the three coordinate directions is zero. If this direction is arbitrarily taken to be the z direction, then $\varepsilon_{zz} = \gamma_{xz} = \gamma_{yz} = 0$. Equations (4.9) reduce to

$$\varepsilon_{xx} = \frac{1+\nu}{E}[(1-\nu)\sigma_{xx} - \nu\sigma_{yy}] \qquad \gamma_{xy} = \frac{\sigma_{xy}}{G}$$

$$\varepsilon_{xx} = \frac{1+\nu}{E}[(1-\nu)\sigma_{yy} - \nu\sigma_{xx}] \qquad \gamma_{xz} = 0 \qquad (4.13)$$

$$\varepsilon_{zz} = 0 \qquad \gamma_{yz} = 0$$

Equations (4.10) reduce to

$$\sigma_{xx} = \frac{E}{(1+\nu)(1-2\nu)}[(1-\nu)\varepsilon_{xx} + \nu\varepsilon_{yy}] \qquad \sigma_{xy} = G\gamma_{xy}$$

$$\sigma_{yy} = \frac{E}{(1+\nu)(1-2\nu)}[(1-\nu)\varepsilon_{yy} + \nu\varepsilon_{xx}] \qquad \sigma_{xz} = 0 \qquad (4.14)$$

$$\sigma_{zz} = \frac{\nu E}{(1+\nu)(1-2\nu)}[\varepsilon_{xx} + \varepsilon_{yy}] \qquad \sigma_{yz} = 0$$

In addition, σ_{zz} can be written as $\sigma_{zz} = \nu(\sigma_{xx} + \sigma_{yy})$.

Thermal Stresses and Strains
Thermal stresses and strains can occur when a member is subjected to a temperature change. For the case of a uniform, unrestrained slender rod of length L, the total increase in length, δ, due to a temperature change ΔT is given by (Fig. 4.5a):

$$\delta = L\int_{T}^{T+\Delta T}\alpha(T)\, dT$$

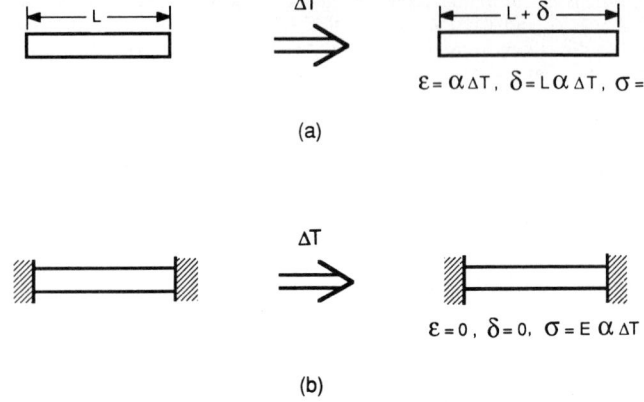

$$\varepsilon = \alpha \Delta T, \quad \delta = L\alpha \Delta T, \quad \sigma = 0$$

(a)

$$\varepsilon = 0, \quad \delta = 0, \quad \sigma = E\alpha \Delta T$$

(b)

Fig. 4.5 Thermally induced stresses and strains: (*a*) Member free to expand; (*b*) member fixed at both ends.

where α is the material's coefficient of thermal expansion. For many materials, α can be considered to be constant over a wide temperature range. In this case δ is given by $\delta = \alpha L \Delta T$, and the thermal strain is given by $\varepsilon = \delta/L = \alpha \Delta T$. If the slender rod is fixed at both ends so that no deformation can occur in the axial direction as the temperature is changed, an internal axial stress will develop (Fig. 4.5*b*). If the magnitude of this stress remains within the linear elastic range, its value is given by

$$|\sigma| = E \int_{T}^{T+\Delta T} \alpha(T)\, dT$$

If the temperature change is small enough so that α can be considered a constant, then $\sigma = E\alpha \Delta T$. Thus, the thermal stress is equal to E times the thermal strain that would develop if the member were free to expand. The determination of thermal stresses and strains for members of more complex geometries can be a formidable task. In such cases, assuming the stresses remain within the elastic range, it is necessary to apply the theory of thermal elasticity (see, e.g., Ref. 4).

4.1.3   Transformations of Stress and Strain

General Equations of Transformation

Given the complete state of stress or strain in a given orientation, it is often desirable to obtain the state of stress or strain in some new orientation. Using the notation that $\cos(i', j)$ represents the direction cosine between the original j coordinate axis and the new i' coordinate axis, the matrix equation for stress transformation from the $Oxyz$ coordinate system to the $Ox'y'z'$ coordinate system is given by

$$[\sigma'] = [\alpha]^{T}[\sigma][\alpha] \tag{4.15}$$

In this equation, $[\sigma]$ is the stress matrix for the original xyz coordinate system [see Eq. (4.1)], $[\sigma']$ is the stress matrix for the new $x'y'z'$ coordinate system, and $[\alpha]$ is the matrix of direction cosines given by

$$[\alpha] = \begin{bmatrix} \cos(x', x) & \cos(y', x) & \cos(z', x) \\ \cos(x', y) & \cos(y', y) & \cos(z', y) \\ \cos(x', z) & \cos(y', z) & \cos(z', z) \end{bmatrix} \tag{4.16}$$

Finally, $[\alpha]^{T}$ is the matrix transpose of $[\alpha]$.

The equation for strain transformation is similarly given by

$$[\varepsilon'] = [\alpha]^{T}[\varepsilon][\alpha] \tag{4.17}$$

Here, $[\varepsilon]$ is the strain matrix for the original xyz coordinate system [see Eq. (4.2)], $[\varepsilon']$ is the strain matrix for the new $x'y'z'$ coordinate system, and $[\alpha]$ is again given by Eq. (4.16).

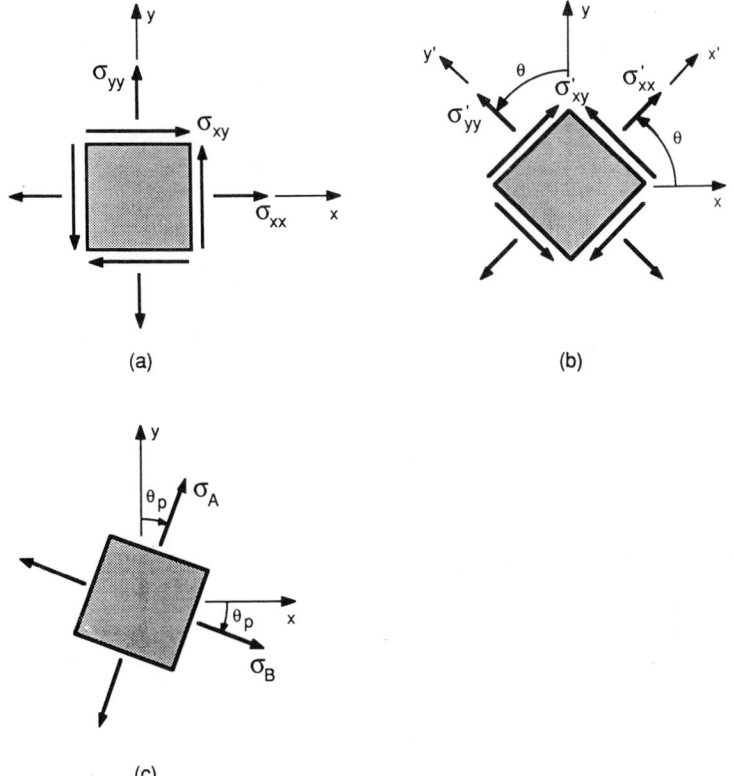

(a)   (b)

(c)

Fig. 4.6 Two-dimensional coordinate rotation: (*a*) Original configuration; (*b*) arbitrary rotation; (*c*) rotation to principal coordinates.

Transformation about a Fixed Axis

If the new orientation is obtained by rotating through an angle θ about the z axis (i.e., $z' = z$), Eqs. (4.15) and (4.16) can be greatly simplified. Using the notation of Fig. 4.6*b*, where θ is measured positive counterclockwise, it follows that

$$\cos(x', x) = \cos(y', y) = \cos\theta$$
$$\cos(y', x) = \cos(90° + \theta)$$
$$\cos(x', y) = \cos(90° - \theta)$$
$$\cos(z', x) = \cos(x', z) = \cos(z', y) = \cos(y', z) = 0$$
$$\cos(z', z) = 1$$

Equation (4.15) reduces to

$$\sigma'_{xx} = \frac{\sigma_{xx} + \sigma_{yy}}{2} + \frac{\sigma_{xx} - \sigma_{yy}}{2}\cos 2\theta + \sigma_{xy}\sin 2\theta$$

$$\sigma'_{yy} = \frac{\sigma_{xx} + \sigma_{yy}}{2} - \frac{\sigma_{xx} - \sigma_{yy}}{2}\cos 2\theta - \sigma_{xy}\sin 2\theta \qquad (4.18)$$

$$\sigma'_{xy} = -\frac{\sigma_{xx} - \sigma_{yy}}{2}\sin 2\theta + \sigma_{xy}\cos 2\theta$$

$$\sigma'_{zz} = \sigma_{zz} \qquad \sigma'_{xz} = \sigma_{xz} \qquad \sigma'_{yz} = \sigma_{yz}$$

Similarly, Eq. (4.17) reduces to

$$\varepsilon'_{xx} = \frac{\varepsilon_{xx} + \varepsilon_{yy}}{2} + \frac{\varepsilon_{xx} - \varepsilon_{yy}}{2}\cos 2\theta + \varepsilon_{xy}\sin 2\theta$$

$$\varepsilon'_{yy} = \frac{\varepsilon_{xx} + \varepsilon_{yy}}{2} - \frac{\varepsilon_{xx} - \varepsilon_{yy}}{2}\cos 2\theta - \varepsilon_{xy}\sin 2\theta \qquad (4.19)$$

$$\varepsilon'_{xy} = -\frac{\varepsilon_{xx} - \varepsilon_{yy}}{2}\sin 2\theta + \varepsilon_{xy}\cos 2\theta$$

$$\varepsilon'_{zz} = \varepsilon_{zz} \qquad \varepsilon'_{xz} = \varepsilon_{xz} \qquad \varepsilon'_{yz} = \varepsilon_{yz}$$

Mohr's Circle for Stress and Strain

Through further mathematical manipulation, Eqs. (4.18) can be written as

$$(\sigma'_{xx} - \sigma_{\text{avg}})^2 + (\sigma'_{xy})^2 = R^2$$

where

$$\sigma_{\text{avg}} = \frac{\sigma_{xx} + \sigma_{yy}}{2} \qquad \text{and} \qquad R = \left[\left(\frac{\sigma_{xx} - \sigma_{yy}}{2}\right)^2 + \sigma_{xy}^2\right]^{1/2}$$

This is the equation of a circle in the $(\sigma_{xx}, \sigma_{xy})$ plane known as Mohr's circle for stress. The horizontal and vertical axes are chosen to represent the applied normal and shear stresses, respectively (Fig. 4.7a). Normal stresses are plotted positive to the right. Shear stresses are plotted positive downward. With this convention, the points $(\sigma_{xx}, \sigma_{xy})$ and $(\sigma_{yy}, -\sigma_{xy})$ should be plotted on the Mohr's circle in order that the positive direction of θ is the same in Figs. 4.6b and 4.7a. Mohr's circle for stress thus consists of the points $(\sigma_{xx}, \sigma_{xy})$ and $(\sigma_{yy}, -\sigma_{xy})$ as 2θ ranges from 0° to 360°. Note that exactly the same Mohr's circle would be obtained if shear stresses were plotted positive upward and the points $(\sigma_{xx}, -\sigma_{xy})$ and $(\sigma_{yy}, \sigma_{xy})$ were plotted. A variety of sign conventions for the shear stress can be found in the literature, but all of them will lead to the same Mohr's circle when used consistently in each system.

It is important to keep in mind that a rotation of the element in Fig. 4.6a through an angle θ to bring about the orientation of Fig. 4.6b corresponds to a rotation of 2θ on the Mohr's circle in Fig. 4.7a. It is also important to realize that the application of Mohr's circle relies only on the fact that the old and new coordinate systems have a common axis of rotation (e.g., the z axis). It is not required that the element also be in a state of plane stress as is often stated in elementary strength of materials texts.

Mohr's circle for strain can be constructed in a manner analogous to Mohr's circle for stress. The appropriate equation is

$$(\varepsilon'_{xx} - \varepsilon_{\text{avg}})^2 + (\varepsilon'_{xy})^2 = R^2$$

where

$$\varepsilon_{\text{avg}} = \frac{\varepsilon_{xx} + \varepsilon_{yy}}{2} \qquad \text{and} \qquad R = \left[\left(\frac{\varepsilon_{xx} - \varepsilon_{yy}}{2}\right)^2 + \varepsilon_{xy}^2\right]^{1/2}$$

Normal strains are plotted horizontally with positive values to the right. Tensorial shear strains are plotted vertically with positive values downward. Mohr's circle for strain thus consists of the points $(\varepsilon_{xx}, \varepsilon_{xy})$ and $(\varepsilon_{yy}, -\varepsilon_{xy})$ as 2θ ranges from 0° to 360° (Fig. 4.7b).

Principal Stresses and Strains

For any state of stress at a point there exists an orientation such that the shear stresses vanish. In this orientation the stress matrix is given by

$$[\sigma] = \begin{bmatrix} \sigma_1 & 0 & 0 \\ 0 & \sigma_2 & 0 \\ 0 & 0 & \sigma_3 \end{bmatrix}$$

where σ_1, σ_2, and σ_3 are known as the principal stresses. It is common to assume that the orientation has been taken so that σ_1 is the largest algebraic stress and σ_3 is the smallest algebraic stress. The

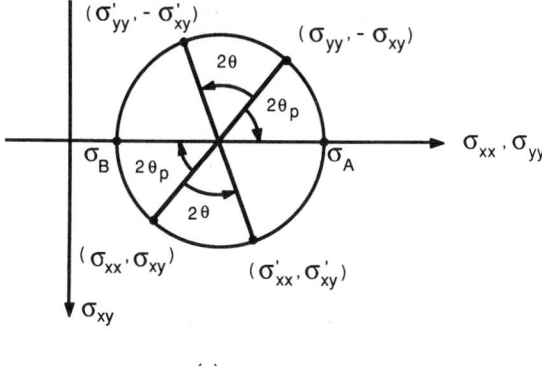

(a)

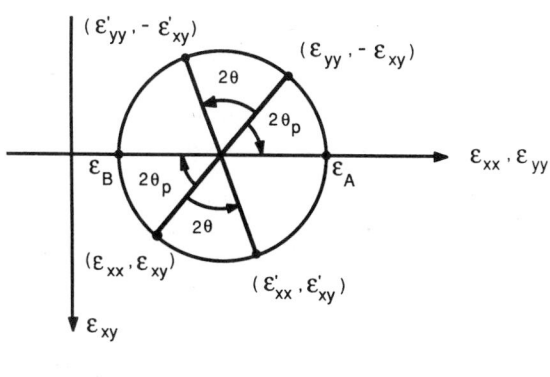

(b)

Fig. 4.7 Mohr's circles for stress and strain: (a) Mohr's circle for stress; (b) Mohr's circle for strain.

principal stresses represent the solutions to the following cubic equation:

$$\sigma_n^3 - I_1\sigma_n^2 + I_2\sigma_n - I_3 = 0 \qquad (4.20)$$

where
$$I_1 = \sigma_{xx} + \sigma_{yy} + \sigma_{zz}$$
$$I_2 = \sigma_{xx}\sigma_{yy} + \sigma_{xx}\sigma_{zz} + \sigma_{yy}\sigma_{zz} - \sigma_{xy}^2 - \sigma_{xz}^2 - \sigma_{yz}^2$$
$$I_3 = \sigma_{xx}\sigma_{yy}\sigma_{zz} - \sigma_{xx}\sigma_{yz}^2 - \sigma_{yy}\sigma_{xz}^2 - \sigma_{zz}\sigma_{xy}^2 + 2\sigma_{xy}\sigma_{xz}\sigma_{yz}$$

I_1, I_2, and I_3 are known as the first, second, and third invariants of stress, respectively. These quantities are independent of the orientation being considered.

Once the principal stresses have been found, their orientations can be determined by substituting each principal stress, σ_n, individually into the following set of equations and solving for the direction cosines associated with that principal stress:

$$(\sigma_{xx} - \sigma_n)\cos(n, x) + \sigma_{yx}\cos(n, y) + \sigma_{zx}\cos(n, z) = 0$$
$$\sigma_{xy}\cos(n, x) + (\sigma_{yy} - \sigma_n)\cos(n, y) + \sigma_{zy}\cos(n, z) = 0$$
$$\sigma_{xz}\cos(n, x) + \sigma_{yz}\cos(n, y) + (\sigma_{zz} - \sigma_n)\cos(n, z) = 0$$
$$\cos^2(n, x) + \cos^2(n, y) + \cos^2(n, z) = 1$$

The last equation is based on the fact that a direction cosine vector has unit magnitude. It is required since the first three equations are not linearly independent and hence are insufficient to explicitly solve for the direction cosines.

Just as for stresses, an orientation can be found such that the shear strains vanish. The principal strains (ε_1, ε_2, and ε_3) represent the solutions to the following cubic equation:

$$\varepsilon_n^3 - J_1\varepsilon_n^2 + J_2\varepsilon_n - J_3 = 0 \tag{4.21}$$

where
$$J_1 = \varepsilon_{xx} + \varepsilon_{yy} + \varepsilon_{zz}$$
$$J_2 = \varepsilon_{xx}\varepsilon_{yy} + \varepsilon_{xx}\varepsilon_{zz} + \varepsilon_{yy}\varepsilon_{zz} - \varepsilon_{xy}^2 - \varepsilon_{xz}^2 - \varepsilon_{yz}^2$$
$$J_3 = \varepsilon_{xx}\varepsilon_{yy}\varepsilon_{zz} - \varepsilon_{xx}\varepsilon_{yz}^2 - \varepsilon_{yy}\varepsilon_{xz}^2 - \varepsilon_{zz}\varepsilon_{xy}^2 + 2\varepsilon_{xy}\varepsilon_{xz}\varepsilon_{yz}$$

J_1, J_2, and J_3 are known as the first, second, and third invariants of strain, respectively. These quantities remain the same regardless of the orientation under consideration.

Once the principal strains have been found, their orientations can be determined by substituting each principal strain, ε_n, individually into the following set of equations and solving for the direction cosines associated with that principal strain:

$$(\varepsilon_{xx} - \varepsilon_n)\cos(n, x) + \varepsilon_{yx}\cos(n, y) + \varepsilon_{zx}\cos(n, z) = 0$$
$$\varepsilon_{xy}\cos(n, x) + (\varepsilon_{yy} - \varepsilon_n)\cos(n, y) + \varepsilon_{zy}\cos(n, z) = 0$$
$$\varepsilon_{xz}\cos(n, x) + \varepsilon_{yz}\cos(n, y) + (\varepsilon_{zz} - \varepsilon_n)\cos(n, z) = 0$$
$$\cos^2(n, x) + \cos^2(n, y) + \cos^2(n, z) = 1$$

Again, the last equation is required since the first three equations are not linearly independent.

For linear elastic isotropic response, $\sigma_{ij} = G\gamma_{ij} = 2G\varepsilon_{ij}$ for $i \neq j$. It thus follows in this case that $\varepsilon_{ij} = 0$ when $\sigma_{ij} = 0$. That is, the principal directions of stress and strain coincide. If the principal stresses are known, the principal strains can be found using Eqs. (4.9) rather than solving Eq. (4.21). Similarly, if the principal strains are known, the principal stresses can be found using Eqs. (4.10) rather than solving Eq. (4.20).

For anisotropic response, such as often occurs in composite materials, the principal strains may have a different orientation than the principal stresses. It is worth noting, however, that Eqs. (4.20) and (4.21) are still valid for anisotropic response, as are Mohr's circles for stress and strain.

For the case of plane stress ($\sigma_{zz} = \sigma_{xz} = \sigma_{yz} = 0$) in a material exhibiting linear elastic isotropic response, it is clear that the z direction is a principal direction. The other two principal directions lie in the xy plane. Since the horizontal axis of Mohr's circle represents a state of zero shear stress, the principal stresses σ_A and σ_B and their directions relative to the x axis (θ_p and $\theta_p + 90°$, respectively) can be found easily from Mohr's circle (Fig. 4.7a). The corresponding rotated element is shown in Fig. 4.6c. The principal strains ε_A and ε_B in this case can also be found easily from the corresponding Mohr's circle for strain (Fig. 4.7b). The principal strain in the z direction can be found from $\varepsilon_{zz} = (-\nu/E)(\sigma_{xx} + \sigma_{yy})$.

Similarly, for the case of plane strain ($\varepsilon_{zz} = \varepsilon_{xz} = \varepsilon_{yz} = 0$), the z direction is again a principal direction. The determination of the principal stresses and strains lying within the xy plane can once more be accomplished using Mohr's circles. The principal stress in the z direction can be found from

$$\sigma_{zz} = \frac{\nu E}{(1 + \nu)(1 - 2\nu)}(\varepsilon_{xx} + \varepsilon_{yy})$$

Maximum Shear Stress
The maximum shear stress at a point is given by $\tau_{max} = (\sigma_1 - \sigma_3)/2$, where σ_1 and σ_3 are the maximum and minimum principal stresses, respectively. The maximum shear stress acts in the 1–3 plane at an angle of $\pm45°$ to the 1 axis.

4.1.4   The Tension Test

The tension test is one of the most fundamental and important of mechanical tests. A wide and diversified set of material properties can be determined from it. Specimens used for tension tests commonly consist of round or rectangular cross sections with a gage region, a gripping region, and an intermediate fillet region. It is important that the gage region has a reduced cross-sectional area; otherwise, the specimen will often fail in or near the region of gripping. The *Annual Book of ASTM Standards*[5] is a multiple-volume reference containing adopted standards for a wide variety of tests. Included within this book are standards for tension tests on a number of different types of materials. These standards describe the required specimen dimensions, gripping arrangement, and testing procedure. In particular, see Volume 3.01, standard E8 for tension testing of metallic materials.

In general, to perform a tension test, the following procedure is used. First, the specimen is appropriately gripped in the testing system. This is usually a screw-driven tension testing machine or the more sophisticated closed-loop, servo-hydraulic mechanical testing system (see Section 4.9.1 for further discussion of these testing systems). Next, a displacement measuring device must be attached to the specimen. When specimen size and rigidity are sufficient, the attachment of an extensometer within the gage region of the specimen is desirable. Finally, the specimen is pulled apart at a constant displacement rate until failure occurs. While the test is proceeding, the load on the specimen and the displacement of the specimen are continuously recorded. The resulting load-displacement plot must be converted to an engineering stress-strain plot or a true stress-strain plot so that mechanical properties can be determined.

Engineering Stress and Strain

Engineering stress S is defined as the instantaneous value of the load P on the specimen divided by the specimen's original cross-sectional area, A_0. That is, $S = P/A_0$. Engineering strain e is defined by $e = (L - L_0)/L_0$, where L is the instantaneous gauge length and L_0 the initial gage length. Engineering stress-strain diagrams for some common materials are shown in Fig. 4.8. For some materials, the engineering stress does not monotonically increase to failure. This is due to the fact that the specimen necks down within the gage area. When necking down occurs, one small region of the specimen begins to deform at a much more rapid rate than the rest of the specimen. This region of the specimen thus has a much smaller cross-sectional area and hence a larger stress. The strain rate within this region is much higher than in the rest of the specimen; as a result, the load applied to the specimen must be lowered in order to maintain a constant displacement rate in the necked-down region. The state of stress within this region is no longer uniaxial, which further complicates matters. When engineering stress and strain are used, this necking-down phenomenon disguises the true nature of the material's mechanical response. In order to obtain a more realistic material response up to the point of failure, the true stress-strain diagram should be determined.

True Stress and Strain

True stress, σ, is defined as $\sigma = P/A_i$, where A_i is the instantaneous minimal cross-sectional area and P is again the instantaneous value of the load acting on the specimen. Before necking down occurs, A_i will be approximately the same along the entire gauge region. Once necking down begins, however, A_i must be measured at the point where the necking down is taking place.

True strain, ε, is defined by $\varepsilon = \log(L_i/L_0)$, where log represents the natural logarithm. For this equation to hold, the strain must be uniform within the region where L_i is being measured. This assumption is obviously invalid when necking down occurs within the gage region. To overcome this problem, it is often assumed that plastic deformation occurs with no change in volume: $A_0 L_0 = A_i L_i$.

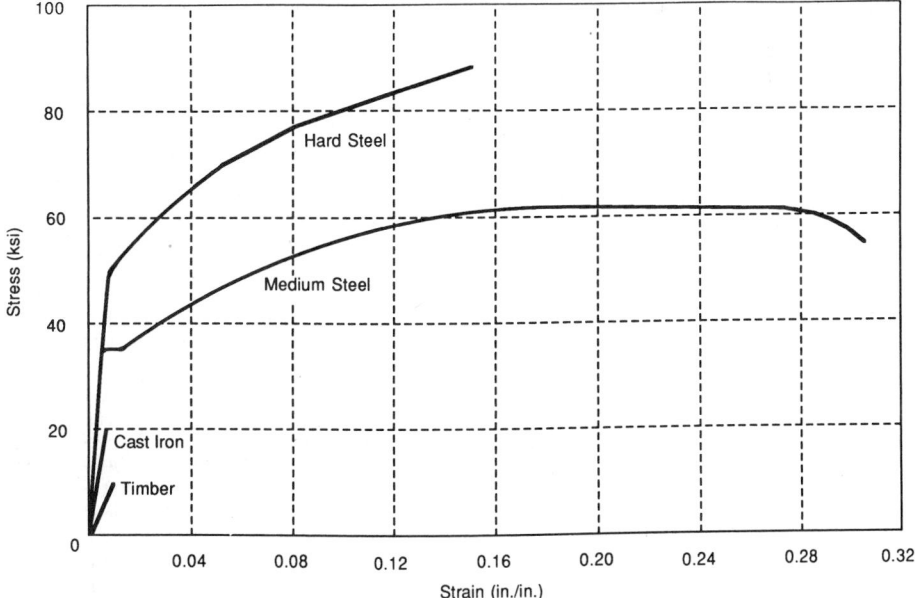

Fig. 4.8 Monotonic engineering stress-strain diagrams for some common materials.

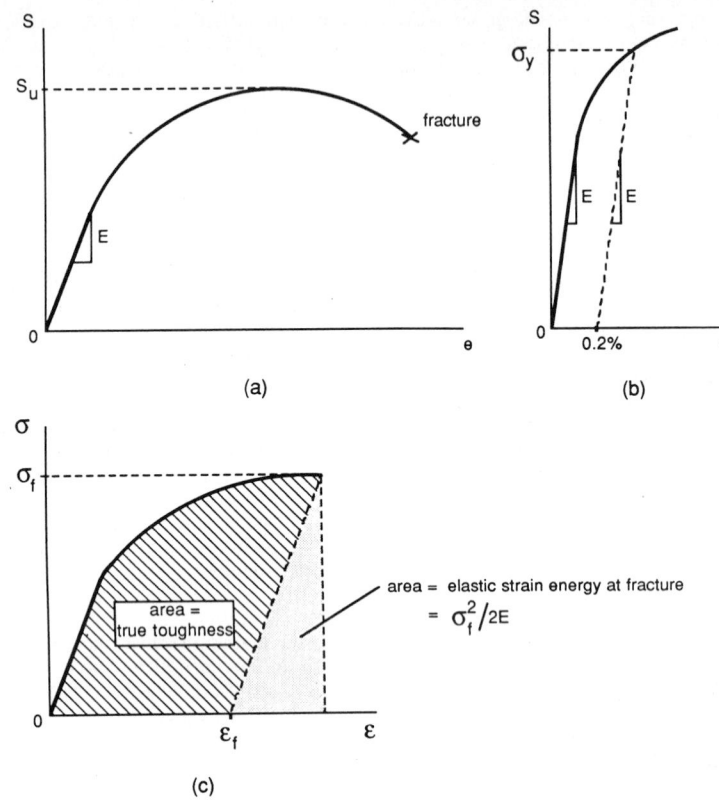

Fig. 4.9 Material properties determined from stress-strain diagrams: (a) Engineering stress-strain diagram; (b) determination of σ_y by the offset method; (c) true stress-strain diagram.

This assumption is fairly reasonable for many ductile metals. By using this assumption, the true strain can be written as $\varepsilon = \log(A_0/A_i)$.

If the instantaneous minimal cross-sectional area can be measured during a test along with P and L_i, and if the constant-volume deformation assumption is valid while plastic deformation is occurring, then a true stress-strain diagram can be constructed. The construction should consist of using the equation $\varepsilon = \log(L_i/L_0)$ for true strain in the elastic and initial plastic regions and the equation $\varepsilon = \log(A_0/A_i)$ once significant plastic deformation has begun. In practice, however, the equation $\varepsilon = \log(A_0/A_i)$ is used for the entire strain range since the amount of error introduced is usually negligible. The equation $\sigma = P/A_i$ for true stress is valid throughout the entire test. Accurate construction of a true stress-strain diagram becomes exceedingly difficult if the plastic deformation cannot be assumed to occur at a constant volume. Finally, it should be mentioned that neither the true stress-strain diagram nor the engineering stress-strain diagram accounts for the fact that the state of stress within the necked-down region is multiaxial. There is no simple way to account for the effect of this multiaxial state of stress on the material's response, so it is customarily ignored. A schematic representation of an engineering stress-strain diagram and the corresponding true stress-strain diagram can be found in Fig. 4.9.

Relationships between Engineering and True Stress and Strain
There are two useful equations for constructing a true stress-strain diagram if the engineering stress-strain diagram is known. The first equation is given by $\varepsilon = \log(1 + e)$. It is based on the assumption of homogeneous strains and is thus valid only up to the point in the tension test when necking down begins. The second equation, $\sigma = S(1 + e)$, assumes a constant-volume process as well as homogeneous strains and is thus valid between the onset of plastic deformation (assuming constant-volume plastic deformation) and the beginning of the necking-down process. In fact, this equation will be slightly inaccurate because of the fact that the elastic deformation that took place prior to the beginning of plastic flow was not a constant-volume process. However, because of the small size of the elastic strains, the error introduced can usually be ignored.

For small strains (such as those that occur within the elastic range of most metals), there is no significant difference between engineering stress-strain and true stress-strain, but for large strains the difference can be considerable. It becomes apparent that the true stress-strain diagram is considerably more difficult to obtain than the engineering stress-strain diagram. As a result, material properties based on the engineering stress-strain diagram are usually reported. However, the importance of the true stress-strain diagram in giving a more representative picture of a material's response should not be overlooked.

Mechanical Properties Determined from the Tension Test

A variety of mechanical properties can be determined from a tension test. Of course, some properties such as the yield strength and the strain-hardening exponent only apply to ductile materials. Table 4.1 lists mechanical properties for wide variety of materials. Many of the mechanical properties that will be discussed are displayed graphically in Fig. 4.9.

The **modulus of elasticity**, E, is given by the slope of the initial straight-line portion of the stress-strain diagram.

The **proportional limit** is the largest stress for which the stress-strain relationship is linear. It is the largest stress for which Hooke's law can be applied for the material. This property depends on the judgment of the observer and is thus of limited value.

The **elastic limit** is the largest stress to which a material can be subjected and still return to its original form upon removal of the load. It can be determined by application and release of a series of increasing loads to the specimen until a permanent deformation is observed upon release of the load. With the advent of equipment that can measure extremely small strains, it has been found that most materials will experience a small amount of permanent deformation at stresses far below what was once thought to be the elastic limit. The determination of this property is thus dependent on the sensitivity of the testing equipment. Its use in design is strongly discouraged.

The **yield strength**, σ_y, is roughly defined as the stress at which the material begins to experience significant plasticity. In order to make this material property of use to the designer, the yield strength is now usually taken to be the stress that will cause a specified amount of permanent set (also called offset). The amount of permanent set specified is arbitrary, and hence the value used should be reported along with the yield strength. The most commonly used offset is 0.2%. The yield strength is determined by drawing a line of slope E on the engineering stress-strain diagram through the point on the strain axis representing the desired amount of offset. The stress corresponding to the intersection of this line with the stress-strain curve represents the yield strength (Fig. 4.9b).

The **ultimate strength**, S_u, is the maximum tensile load a specimen can resist, P_{max}, divided by the specimen's original cross-sectional area. Thus, $S_u = P_{max}/A_0$.

The **percent elongation** equals $100\,(L_f - L_0)/L_0$, where L_f is the final gage length at fracture. The percent elongation varies depending on the initial gage length and is of somewhat limited value.

The **percent reduction of area** is given by $\%RA = 100\,(A_0 - A_f)/A_0$, where A_f is the cross-sectional area of the specimen at the point of fracture. The $\%RA$ is not as strongly dependent on specimen geometry as the percent elongation. It is a good measure of a material's ductility.

The **fracture strength**, σ_f, represents the true stress just prior to fracture. It should not be confused with the ultimate strength.

The **fracture ductility**, ε_f, is the true plastic strain at fracture. If the plastic deformation occurs without volume change, ε_f is related to $\%RA$ by

$$\varepsilon_f = \log\left(\frac{A_0}{A_f}\right) = \log\frac{100}{100 - \%RA}$$

Toughness has several common definitions. In its simplest form, the toughness is the area under the engineering stress-strain diagram.

True toughness represents the irrecoverable work done on a material during plastic deformation. It is given by the area under a plot of true stress versus true plastic strain, ε_p. The true plastic strain is related to σ and ε by $\varepsilon_p = \varepsilon - \sigma/E$. The true toughness is equivalently given by the total area under the true stress-strain diagram minus the elastic strain energy at fracture, $\sigma_f^2/2E$ (Fig. 4.9c).

For ductile materials it is often found that a simple power law relationship exists between the true stress and the true plastic strain. This relationship takes the form

$$\sigma = K\varepsilon_p^n \tag{4.22}$$

where K is known as the **strength coefficient**, and n is the **strain-hardening exponent**. These two quantities are easily determined by plotting $\log \sigma$ versus $\log \varepsilon_p$. The slope of such a plot gives n, while K is given by the value of σ corresponding to $\varepsilon_p = 1$.

TABLE 4.1 Mechanical Properties of Selected Materials

Material	Specific Weight (lb/in.^{3})	E (10^6 psi)	G (10^6 psi)	σ_y (ksi)	σ_u (ksi)	Percentage of Elongation (in 2 in.)	α (10^{-6}/°F)
Aluminum							
2014 T6 (155 BHN)	0.10	10.5	4	67	74	13	12.8
2024 T4	0.1	10.5	4	44	69	20	12.9
6061 T6	0.098	10.0	3.7	37	42	17	13.1
7075 T6 (95 BHN)	0.1	10.3	4	68	84	11	13.1
Cast iron							
Gray #20	0.251	14	—	—	20	—	0.60
Gray #30	0.260	15.2	—	—	30	—	0.60
Gray #40	0.260	18.3	—	—	40	—	0.60
Gray #60	0.270	19	—	—	60	—	0.60
Malleable	0.266	26	8.8	32–45	50–65	—	0.75
Nodular	0.257	23.5	—	45–65	60–100	—	0.66
Concrete (compression)							
Medium strength	0.084	3.6	—	—	4.0	—	5.5
High strength	—	4.5	—	—	6.0	—	5.5
Copper							
High purity, annealed	0.323	17	6.4	10	33	50	9.8
High purity, cold-worked	0.323	17	6.4	45	50	10	9.8
Glass (compression)							
98% silica	0.079	9.6	4.1	—	7	—	44
Magnesium: AZ80A-T5	0.065	6.5	2.4	38	55	7	16.0
Nickel 200	0.321	30	—	25	65	45	7.4
Nylon							
6 (cast)	0.04	0.30	—	—	8	50	45
Molded	0.04	0.55	—	13	13	20	44
Phosphor bronze							
Cold-rolled (510)	0.320	15.9	5.9	75	81	10	9.9
Spring temper (524)	0.317	16	—	—	122	4	10.2
Polyethylene							
Medium density	0.033	—	—	—	2	200	120
Polystyrene (molded)	0.039	—	—	5	5	2–30	40
Rubber							
Natural	0.034	—	—	—	3	800	400
Neoprene	0.045	—	—	—	3	850	350
Steel							
SAE 1005 (90 BHN)	0.283	30	12	38	50	40	6.5
SAE 1005 (125 BHN)	0.283	30	12	60	65	30	6.5
SAE 1020 (90 BHN)	0.283	30	12	35	65	30	6.5
SAE 1045 (225 BHN)	0.283	29	12	92	105	—	6.5
SAE 1045 (450 BHN)	0.283	30	12	220	230	—	6.5
SAE 1045 (650 BHN)	0.283	30	12	270	325	—	6.5
SAE 4142 (670 BHN)	0.283	29	12	240	355	—	6.5
SAE 5160 (430 BHN)	0.283	28	12	222	242	—	6.5
SAE 9262 (260 BHN)	0.283	30	12	200	227	—	6.5
SAE 9262 (410 BHN)	0.281	29	12	300	375	—	6.5
Stainless steel							
AISI 304 (160 BHN)	0.29	27	—	37	108	—	9.6
18 NI Maraging (460 BHN)	0.29	27	—	260	270	—	6.3
17-7 PH (TH-1050)	0.281	29	—	182	193	10	6.0
Titanium							
6%AL, 4% V	0.161	16.5	—	120	130	10	5.3
5%AL, 2.5%Sn	0.161	17	6.2	110	115	—	5.7
8-1-1	0.170	17	—	150	160	15	4.7
Wood (loaded parallel to grain)							
Birch	0.026	2.1	—	8.3	2.0	—	1.1
Douglas Fir	0.019	1.8	—	—	—	—	1.7–2.5
Eastern Spruce	0.016	1.3	—	—	—	—	1.7–2.5
Southern Pine	0.022	1.6	—	—	—	—	1.7–2.5
White Oak	0.028	1.6	—	7.0	1.9	—	2.7

Note: σ_y is the monotonic yield strength. For some materials, the cyclic yield strength (discussed in Section 4.8.4) differs radically from σ_y.

4.2   BEAMS AND BENDING

4.2.1   Shear and Bending Moment in Beams

Classification of Beams
A **beam** is a structural member whose length is large compared to its transverse dimensions and is subjected to forces acting transverse to its longitudinal axis. The following categories of beams are simple abstractions that approximate actual beams used in practice.

A **simple beam**, or simply supported beam (Fig. 4.10a), has a roller support at one end and a pin support at the other. The ends of a simple beam cannot support a bending moment but can support upward and downward vertical loads. Stated differently, the ends are free to rotate but cannot translate in the vertical direction. The end with the roller support is free to translate in the axial direction.

A **cantilever beam** (Fig. 4.10b) has one end rigidly fixed and the other end free. The fixed end can neither translate nor rotate while the free end can do both.

A **continuously supported beam** (Fig. 4.10c) is a beam resting on more than two supports.

A **fixed beam** (Fig. 4.10d) is rigidly fixed at both ends.

A **restrained beam** (Fig. 4.10e) is rigidly fixed at one end and simply supported at the other.

An **overhanging beam** (Fig. 4.10f) projects beyond one or both ends of its supports.

Loads on a beam commonly include concentrated loads (Fig. 4.10a), distributed loads (Fig. 4.10b), concentrated moments (Fig. 4.10c), and combinations of these.

Relations Among Load, Shear, and Bending Moment
At any point along a beam, the shear force and bending moment acting on the beam cross section must be such as to balance the external loads on either side of the cross section. The commonly used

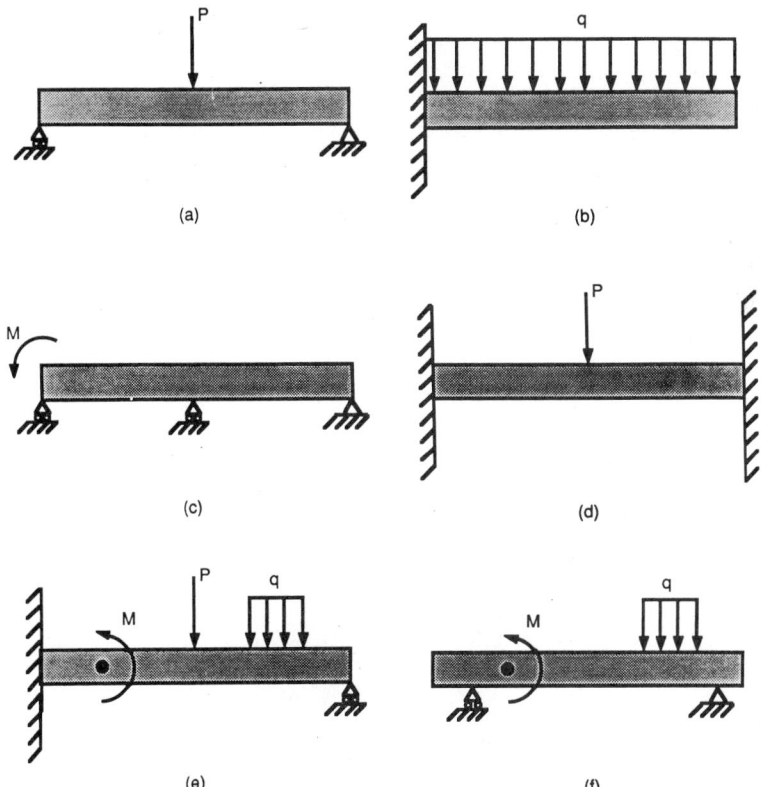

(a)   (b)

(c)   (d)

(e)   (f)

Fig. 4.10   Classification of beams and bending: (a) Simple beam, concentrated load; (b) cantilever beam, distributed load; (c) continuous beam, concentrated moment; (d) fixed beam, concentrated load; (e) restrained beam, combined loading; (f) overhanging beam, combined loading.

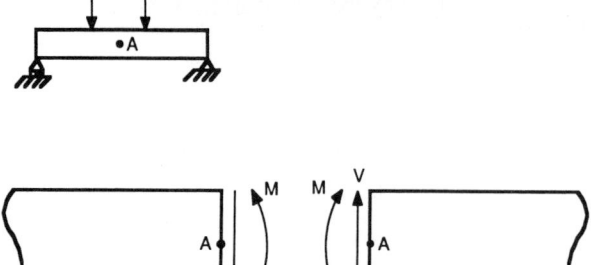

Fig. 4.11   Positive sign convention for internal shear and bending moment.

sign convention for internal shear force and bending moment is shown in Fig. 4.11. It is important to determine the points along a beam where the shear force and bending moment are maximum since it is at these points that the shear stresses and bending stresses, respectively, reach their maximum values. These points are commonly referred to as critical points.

Consider an infinitesimal section of a beam that is under a general state of loading (Fig. 4.12). Equilibrium in the vertical direction requires that

$$\frac{dV}{dx} = -q(x) \tag{4.23a}$$

or

$$V_B - V_A = \Delta V_{A-B} = -\int_{x_A}^{x_B} q(x)\,dx \tag{4.23b}$$

where $q(x)$ is the applied distributed load, A and B are arbitrary points along the beam, V is the internal shear force, and ΔV_{A-B} is the change in shear force between points A and B. There is equilibrium of moments if

$$\frac{dM}{dx} = V(x) \tag{4.24a}$$

or

$$M_B - M_A = \Delta M_{A-B} = \int_{x_A}^{x_B} V(x)\,dx \tag{4.24b}$$

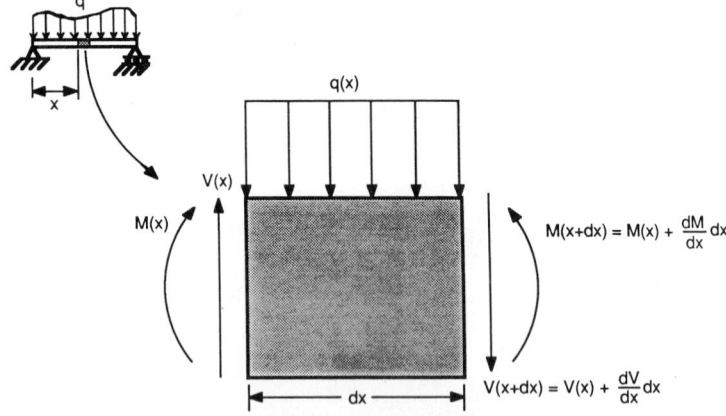

Fig. 4.12   Infinitesimal element of a beam.

where ΔM_{A-B} is the change in bending moment between points A and B. Equation (4.23b) is valid only if no concentrated loads act between points A and B since discontinuities in the shear occur at points of application of concentrated loads. Similarly, Eq. (4.23b) is valid only if no concentrated moments act between points A and B.

Shear and Bending Moment Diagrams

Equations (4.23b) and (4.24b) provide a simple means for determining the shear and bending moment as a function of distance along the beam. In particular, if no distributed or concentrated loads act between two points on a beam, Eq. (4.23b) shows that the shear force is constant in this region. Equation (4.24b) shows that the moment varies linearly in this region, provided that no concentrated moments exist.

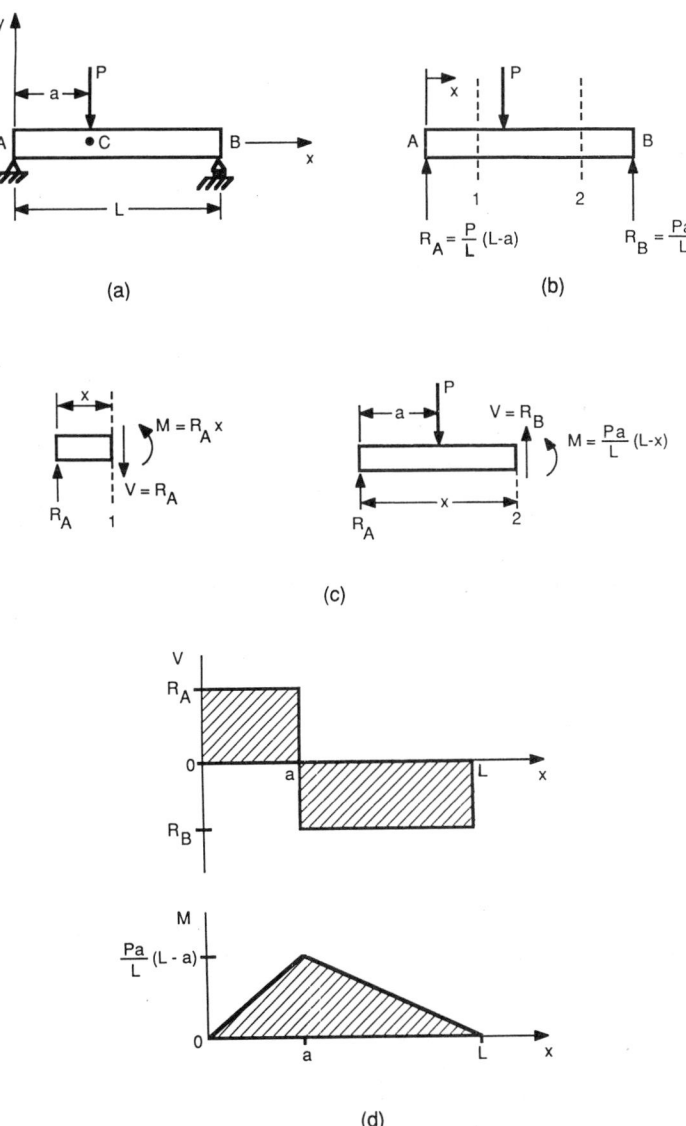

(a)

(b)

(c)

(d)

Fig. 4.13 Construction of shear and bending moment diagrams: (*a*) Beam configuration; (*b*) static equilibrium of forces; (*c*) shear and bending moment in sections AC and CB; (*d*) shear and bending moment diagrams.

A shear diagram is a graphical representation of the vertical shear at every point along a beam. A bending moment diagram is a graphical representation of the bending moment at every point along a beam. It is clear from Eq. (4.24b) that the bending moment at any point along a beam is equal to the area under the shear diagram up to that point, provided that no concentrated moments are present. If concentrated moments do exist, they will cause abrupt changes in the bending moment diagram, but this situation can be accounted for easily.

The general procedure for constructing shear and bending moment diagrams is described with the aid of Fig. 4.13:

1. Using the equations of static equilibrium ($\Sigma F_y = 0$ and $\Sigma M_z = 0$), determine the reactions at the supports (Fig. 4.13b). If the beam is statically indeterminate, the reactions can often be found by considering the deformations of the beam (see Section 4.2.4).

2. Determine the equations for shear and bending moment within each region of the beam in which the loading is continuous (Fig. 4.13c). Every discontinuity in loading, such as the application of a concentrated force, requires that another set of equations for shear and bending moment be determined.

3. Using the equations derived in step 2, draw the shear and bending moment diagrams (Fig. 4.13d).

Shear and bending moment diagrams are useful because they provide a rapid means for determining the critical points in a beam. These are the points at which failure is most likely to occur. With practice, it becomes possible to draw the diagrams directly without the need for first deriving the equations for shear and bending moment (at least for relatively simple loadings).

4.2.2 Theory of Flexure

For a straight beam possessing a plane of symmetry and subjected to moments acting in that plane of symmetry, it is assumed that cross sections perpendicular to the longitudinal axis remain plane. This assumption is very accurate except for short beams subjected to shear forces. It follows from this assumption that the beam deforms into a circular arc subtending an angle θ (Fig. 4.14a). The only stress acting in the beam will be a normal stress component in the longitudinal direction, σ_{xx}. On the concave side, the strain ε_{xx} and stress σ_{xx} are negative (compressive), while on the convex side they are positive (tensile); consequently, there exists a surface in the beam where ε_{xx} and σ_{xx} are zero. The neutral surface is the surface within which the normal stress and strain are zero. The neutral axis is the intersection of the neutral surface with any cross section perpendicular to the longitudinal axis (Fig. 4.14b).

Since the normal strain within the neutral surface is zero, it follows that $L = \rho\theta$. Here, L is the original beam length and ρ is the radius of the arc AB of the neutral surface (Fig. 4.14a). At a radial distance y from the neutral surface (where y is shown as positive on the concave side of the beam), the length of the beam is given by $L' = (\rho - y)\theta$. The strain along this arc is given by $\varepsilon_{xx} = (L' - L)/L$, or equivalently by $\varepsilon_{xx} = -y/\rho$. Thus, the longitudinal strain varies linearly with distance from the neutral axis.

Assuming linearly elastic isotropic response, the longitudinal stress distribution is given by $\sigma_{xx} = E\varepsilon_{xx}$, or

$$\sigma_{xx} = \frac{-Ey}{\rho} \qquad (4.25)$$

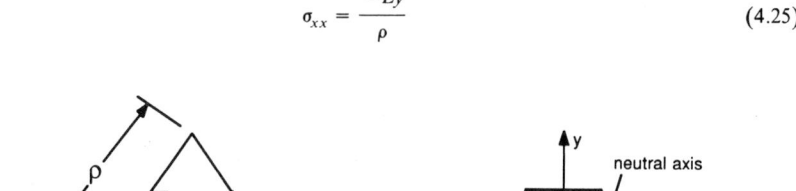

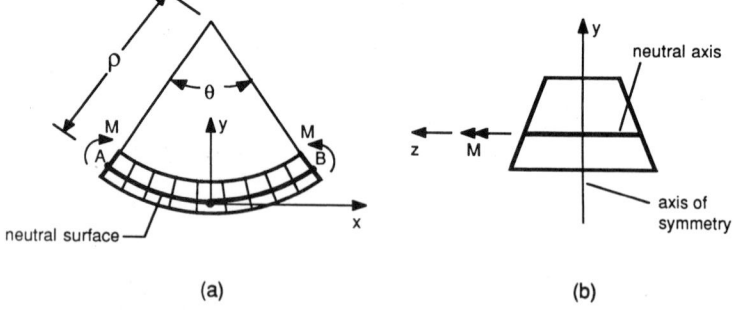

(a)   (b)

Fig. 4.14 Symmetric beam in pure bending: (a) Side view; (b) enlarged cross-sectional view.

From statics, the integral of σ_{xx} over the cross-sectional area of the beam must equal zero since there is no applied longitudinal force. Placing the origin of the coordinate axes at the center of the neutral surface and integrating over the cross-sectional area gives

$$-\frac{E}{\rho}\int y\,dA = 0$$

For finite ρ and nonzero E it follows that

$$\int y\,dA = 0$$

Stated in words, the first moment of the cross-sectional area with respect to the neutral axis must equal zero. The neutral axis therefore passes through the centroid of the cross section.

The applied moment, M, must be equal to the integral of $-y\sigma_{xx}$ over the cross-sectional area:

$$M = -\int y\sigma_{xx}\,dA$$

Substituting Eq. (4.25) into this expression gives

$$M = \frac{E}{\rho}\int y^2\,dA$$

This can be rewritten as

$$\frac{1}{\rho} = \frac{M}{EI_{zz}} \tag{4.26}$$

where I_{zz} is the moment of inertia of the cross section with respect to the neutral axis (the z axis in Fig. 4.14b). Thus, $I_{zz} = \int y^2\,dA$. Substituting for ρ from Eq. (4.25) into (4.26) leads to

$$\sigma_{xx} = \frac{-My}{I_{zz}} \tag{4.27}$$

This is known as the elastic flexure formula. The longitudinal stress is seen to have a maximum value at the farthest distance from the neutral axis. Denoting this distance by c, it follows that $|\sigma_{max}| = Mc/I_{zz}$.

The section modulus, S, is equal to I_{zz}/c. The maximum longitudinal stress in the beam can be written as $|\sigma_{max}| = M/S$. The section modulus represents the ability of a beam section to resist an applied bending moment. When designing beams, S should be made as large as is practical while minimizing the amount and cost of the material used. Table 4.2 gives values for the moment of inertia and section modulus of some common beam cross sections.

In summary, the following conditions must be satisfied if the elastic flexure formula is to be valid:

1. The beam must be straight or nearly straight.
2. The beam must possess a plane of symmetry and be subjected to moments acting in that plane of symmetry.
3. The material must be homogeneous and exhibit linear elastic, isotropic response.
4. The beam should be long enough so that shear forces do not cause warping of the cross section. A length-to-height ratio of 10 or better is normally sufficient to make shearing effects negligible.
5. The beam must have sufficient lateral stiffness so that it does not buckle under the applied loads (see Section 4.5.4 for details).

4.2.3 Shear Stresses in Beams

Symmetric beams that are subjected to a state of pure bending acting in the plane of symmetry experience only a flexural normal stress as given by Eq. (4.27). Thus, $\sigma_{xx} = -My/I_{zz}$, and $\sigma_{yz} = \sigma_{xz} = \sigma_{xy} = \sigma_{yy} = \sigma_{zz} = 0$ (using the coordinate axes given in Fig. 4.14). Beams that are subjected to transverse loads experience shear stresses σ_{xy} in addition to the flexural stress σ_{xx}. The average vertical shear stress throughout a cross section is given by $(\sigma_{xy})_{avg} = V/A$, where V is the resisting vertical shear force on the cross section, and A is the cross-sectional area.

TABLE 4.2 Properties for Common Beam Cross Sections

Beam Section	Moment of Inertia, I_{xx}	Section Modulus, $\dfrac{I_{xx}}{c}$
(rectangle, width b, depth d)	$\dfrac{bd^3}{12}$	$\dfrac{bd^2}{6}$
(hollow rectangle, b_1, d_1, b_2, d_2)	$\dfrac{b_1d_1{}^3 - b_2d_2{}^3}{12}$	$\dfrac{b_1d_1{}^3 - b_2d_2{}^3}{6d_1}$
(I-beam, b, d, t_w, t_f)	$\dfrac{bt_f{}^3}{6} + \dfrac{bt_f}{2}(d-t_f)^2$ $+ \dfrac{t_w}{12}(d-2t_f)^3$	$\dfrac{2I_{xx}}{d}$
(circle, diameter d)	$\dfrac{\pi d^4}{64}$	$\dfrac{\pi d^3}{32}$
(hollow circle, d_1, d_2)	$\dfrac{\pi(d_1{}^4 - d_2{}^4)}{64}$	$\dfrac{\pi(d_1{}^4 - d_2{}^4)}{32d_1}$
(triangle, base b, height d)	$\dfrac{bd^3}{36}$	$\dfrac{bd^2}{24}$

In general, the shear stress is not uniformly distributed over the cross section. To determine the shear stress distribution, it must first be noted that the presence of a shear stress σ_{xy} implies the presence of a shear stress σ_{yx} of equal magnitude. This follows from the complementary property of shearing stresses (see Section 4.1.1).

Consider the beam shown in Fig. 4.15. For simplicity of demonstration, a rectangular cross section has been drawn, but Eq. (4.29), to be derived, is valid for an arbitrary cross-sectional shape provided there is at least one axis of symmetry, and the beam is loaded in that plane of symmetry. A portion CD of the beam is shown enlarged in Fig. 4.15c. Static equilibrium in the horizontal direction leads to

$$\left(\sigma_{yx}\right)_{\text{avg}} Lt = \int_{y=y_0}^{y=h} (\sigma_D - \sigma_C)\, dA \tag{4.28}$$

where L is the length of the beam element CD, t is the thickness of CD, and $\left(\sigma_{yx}\right)_{\text{avg}}$ is the average horizontal shear stress acting on the bottom face of CD. From Eq. (4.24a), $M_D - M_C = VL$, since V is constant between C and D. Note that V is the shear force acting on the cross-sectional area of the whole beam at C, whereas V', shown in Fig. 4.15c, is only that portion of V that acts on the cross-sectional area of element CD. Also, recalling Eq. (4.27), $\sigma = -My/I$, and substituting

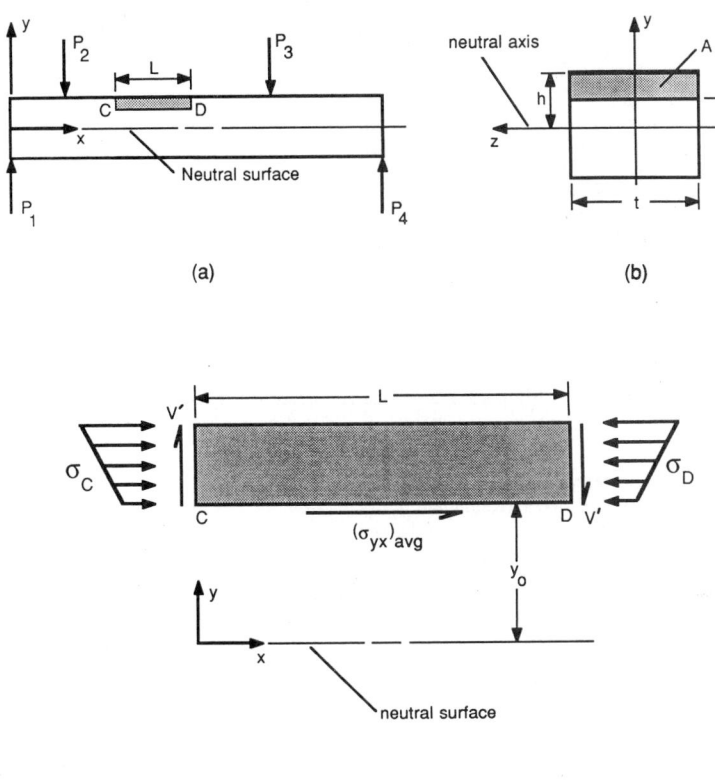

Fig. 4.15 Shear stresses in a beam: (*a*) Side view; (*b*) cross-sectional view; (*c*) enlarged view of segment *CD*.

$M_D - M_C = VL$ gives

$$\sigma_D - \sigma_C = -\frac{(M_D - M_C)y}{I} = -\frac{VLy}{I}$$

Finally, substituting this result into Eq. (4.28) and considering the limit as L goes to zero leads to

$$(\sigma_{yx})_{\text{avg}} = -\frac{VQ}{It} \tag{4.29}$$

where

$$Q = \int_{y=y_0}^{y=h} y \, dA$$

The quantity Q represents the first moment of the shaded area in Fig. 4.15*b* with respect to the neutral axis. The minus sign in Eq. (4.29) implies that σ_{yx} is negative when V is positive (see Sections 4.1.1 and 4.2.1 for discussions of the sign conventions for σ_{yx} and V, respectively).

From the complementary property of shear (Section 4.1.1), it follows that $\sigma_{xy} = \sigma_{yx}$, provided that static equilibrium exists. In particular, along the top and bottom of the beam, $Q = 0$. Thus, $\sigma_{xy} = \sigma_{yx} = 0$ on these surfaces; consequently, the shear stress goes from zero at the top and bottom surfaces of the beam to some maximum value in the interior.

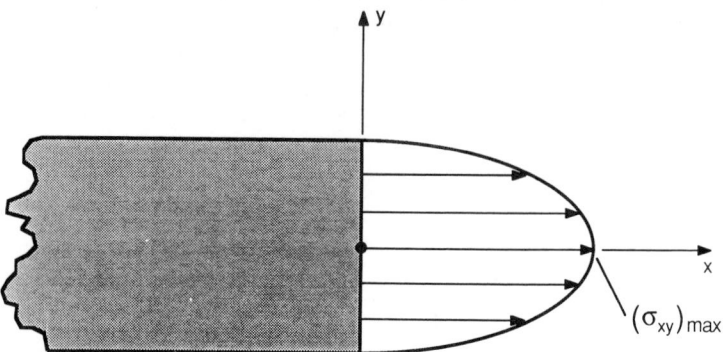

Fig. 4.16 Shear stress distribution in a rectangular beam subjected to transverse loads.

Shear Stresses in Rectangular Beams

For the case of a beam having a rectangular cross section, Eq. (4.29) can be greatly simplified. Denoting the height of the beam by $2h$, it can be shown that the shear stress is given by

$$\sigma_{xy} = \sigma_{yx} = \frac{3}{2}\frac{V}{A}\left(1 - \frac{y^2}{h^2}\right)$$

Clearly, the maximum shear stress is given by $(\sigma_{xy})_{max} = 1.5(V/A)$. This maximum shear stress occurs at $y = 0$ (at the neutral axis, where the normal stress σ_{xx} is zero). The shear stress distribution for a beam of rectangular cross section is shown in Fig. 4.16.

Shear Stresses in Thin-walled Beams

It can be shown that Eq. (4.29) can be used to describe the shear stress on an arbitrary longitudinal cut. For example, consider a section of a T beam as shown in Fig. 4.17a. The shear stress σ_{xy} must equal zero on the top and bottom of the flange since these are free surfaces. Provided the flange thickness is small, σ_{xy} will not become significant through the thickness of the flange. Note that on the bottom of the flange Q is nonzero. Equation (4.29) thus predicts that a shear stress $\sigma_{xy} = \sigma_{yx}$ would exist on the free surface. In reality, this is physically impossible. This demonstrates one of the limitations of Eq. (4.29). This equation also gives incorrect results along the free surface of beams of circular cross section, except along the neutral axis (see Ref. 6 for further details). For a proper analysis that leads to an answer of zero shear stress on the bottom of the flange, the theory of elasticity must be employed. Equation (4.29) should thus be used with care.

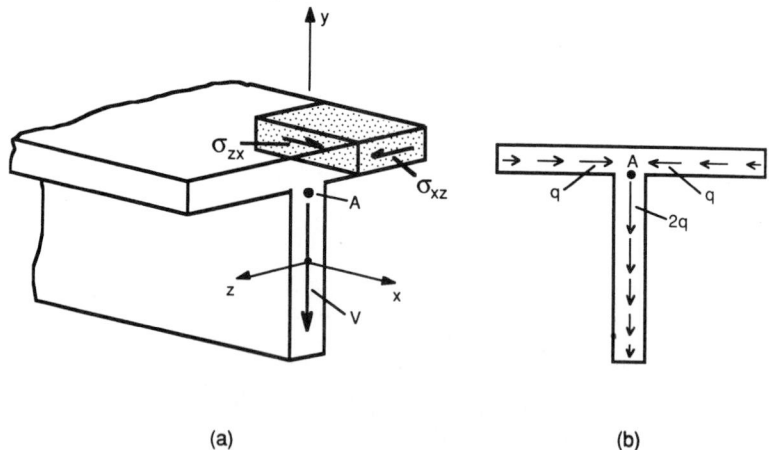

(a)   (b)

Fig. 4.17 Shear flow in a thin-walled beam: (a) Isometric view of a beam; (b) sketch of shear flow.

The shear stresses $\sigma_{xz} = \sigma_{zx}$ in the flange can become large. Their value is again given by Eq. (4.29). As the web is reached (point A in Fig. 4.17a), the horizontal shear stress, σ_{xz}, becomes a vertical shear stress, σ_{xy}. The shear flow, q, is defined as

$$q = \frac{VQ}{I_{zz}} = \tau_{avg} t$$

where V is again the internal resisting shear force, Q is the first moment of area with respect to the neutral axis, I_{zz} is the moment of inertia of the beam cross section, and τ_{avg} is the average horizontal shear stress acting in the web. Shear flow in a thin-walled beam is analogous to the flow of water through a channel. If the beam thickness decreases, τ_{avg} will increase so as to keep q constant. This is analogous to conservation of mass in fluid flow through a channel of variable diameter. Similarly, as the point A is approached in Fig. 4.17b, the shear flows from each flange, q, combine to give a shear flow of $2q$ in the web. This is analogous to water flow through a T section.

The total shear flow acting over the area of the cross section must be such as to equal the resisting shearing force, V. For example, if no external horizontal forces are present, horizontal forces due to shear flow must cancel each other out. This is obvious for the T section of Fig. 4.17b since the horizontal shear flow in one flange exactly cancels the horizontal shear flow in the other flange.

TABLE 4.3   Shear Centers for Common Sections

Beam Cross Section	Location of Q
	$e = \dfrac{4R(\sin\theta - \theta\cos\theta)}{2\theta - \sin2\theta}$
	$e = \dfrac{3t_f d^2}{6 t_f d + t_w h}$
	$e = \dfrac{(t_1 + h^2)(t_2^3 h_2)}{2(t_2^3 h_2 + h_1^3 t_1)}$
	$e = \dfrac{3 t_f (d_2^2 - d_1^2)}{6 t_f (d_1 + d_2) + t_w h}$ provided $d_1 < d_2$

Shear Center for Beams

In general, a beam will undergo both bending and twisting when it is subjected to transverse loads. There does exist a point, however, at which transverse loads can be applied without any twisting occurring. This point is referred to as the shear center of the beam. For thick-walled or solid cross sections, the shear center will usually be located close to the centroid of the cross section. Knowledge of the precise location of the shear center in beams with thick or solid cross sections is generally unimportant since such beams will exhibit considerable resistance to torsional loads. If such a beam is loaded through a point slightly away from the shear center, no significant amount of twisting occurs.

For thin-walled open section beams, the situation is quite different. Such beams exhibit very little torsional resistance; therefore, it is important that the application of transverse loads occurs at the shear center. The shear center is found by balancing out the moments due to the shear flows, q, in the beam and the resisting shear force, V. Details of the procedure can be found in Refs. 7 and 8. Table 4.3 gives the location of the shear center for some common thin-walled cross sections. More extensive tables can be found in Refs. 8 and 9.

4.2.4 Deflection of Beams

A variety of techniques exist for determining the elastic deflection of beams. The techniques to be discussed here are only valid for small deflections. In addition, the assumptions used in deriving the elastic flexure formula must also be satisfied (see Section 4.2.2).

Method of Integration

For small deflections the radius of curvature, ρ, can be written as

$$\frac{1}{\rho} = \frac{d^2y(x)}{dx^2}$$

where $y(x)$ represents the deflection of the beam at a longitudinal distance, x, from the origin of the coordinates. Substituting this result into Eq. (4.26) leads to

$$\frac{d^2y(x)}{dx^2} = \frac{M(x)}{EI_{zz}} \tag{4.30}$$

Equation (4.30) is a second-order linear differential equation for y. Its solution requires a knowledge of the bending moment at every point along the beam as well as two boundary conditions. The equation is only valid provided that no sudden changes in M, E, or I_{zz} occur. When there is a sudden change, such as the application of a concentrated force, it is necessary to divide the beam into several portions and solve Eq. (4.30) within each portion. The following example illustrates the method of integration.

Example 4.1. For the beam shown in Fig. 4.13a, derive the deflection equation by the method of integration. Consider the case where $a = L/2$.

SOLUTION. Since the beam contains a concentrated load at $x = L/2$, it is necessary to write two separate differential equations for the two halves of the beam. Using the moment equations derived in Fig. 4.13c and substituting them into Eq. (4.30), with $a = L/2$, leads to

$$\frac{d^2y_1}{dx^2} = \frac{P}{2EI}x \qquad 0 \le x \le \frac{L}{2}$$

and

$$\frac{d^2y_2}{dx^2} = \frac{-P}{2EI}x + \frac{PL}{2EI} \qquad \frac{L}{2} \le x \le L$$

Integrating these equations gives

$$\frac{dy_1}{dx} = \frac{Px^2}{4EI} + C_1 \qquad\qquad \frac{dy_2}{dx} = \frac{-Px^2}{4EI} + \frac{PL}{2EI}x + C_3$$

$$y_1 = \frac{Px^3}{12EI} + C_1x + C_2 \qquad y_2 = \frac{-Px^3}{12EI} + \frac{PLx^2}{4EI} + C_3x + C_4$$

Finally, the boundary conditions must be applied. Clearly, four boundary conditions are needed to solve for the four constants of integration, C_1 through C_4. These boundary conditions are as follows:

$$\text{at } x = 0 \qquad y_1 = 0 \Rightarrow C_2 = 0$$

$$\text{at } x = L \qquad y_2 = 0 \Rightarrow C_4 = \frac{-PL^3}{6EI} - C_3 L$$

$$\text{at } x = \frac{L}{2} \qquad \frac{dy_1}{dx} = \frac{dy_2}{dx} \Rightarrow \frac{PL^2}{16EI} + C_1 = \frac{3PL^2}{16EI} + C_3$$

$$\text{at } x = \frac{L}{2} \qquad y_1 = y_2 \Rightarrow \frac{PL^3}{96EI} + C_1\frac{L}{2} = \frac{5PL^3}{96EI} + C_3\frac{L}{2} + C_4$$

The last two conditions are due to the fact that the slope and deflection must be continuous at $x = L/2$. Solving these equations for the constants gives

$$C_1 = \frac{-PL^2}{16EI} \qquad C_2 = 0$$

$$C_3 = \frac{-3PL^2}{16EI} \qquad C_4 = \frac{PL^3}{48EI}$$

TABLE 4.4   Deflections and Slopes of Uniform Beams

Loading	Equation Of Elastic Curve	Maximum Deflection	End Slopes
	$y=\frac{P}{48EI}(4x^3-3L^2x)$ for $0\le x\le\frac{L}{2}$	$y_{max}=\frac{PL^3}{48EI}$ at $x=\frac{L}{2}$	$\theta_{x=0}=-\frac{PL^2}{16EI}$ $\theta_{x=L}=\frac{PL^2}{16EI}$
	$y=\frac{Pb}{6EIL}[x^3-(L^2-b^2)x]$ for $0\le x\le a$	$y_{max}=\frac{Pa(L^2-a^2)^{\frac{3}{2}}}{9\sqrt{3}\,EIL}$ at $x=\sqrt{(L^2-a^2)/3}$ provided a<b	$\theta_{x=0}=-\frac{Pb(L^2-b^2)}{6EIL}$ $\theta_{x=L}=\frac{Pa(L^2-a^2)}{6EIL}$
	$y=\frac{P}{6EI}(x^3-3ax(L-a))$ for $0\le x\le a$	$y_{max}=-\frac{Pa}{24EI}(3L^2-4a^2)$ at $x=\frac{L}{2}$	$\theta_{x=0}=-\frac{Pa}{2EI}(L-a)$ $\theta_{x=L}=\frac{Pa}{2EI}(L-a)$
	$y=-\frac{q}{24EI}(x^4-2Lx^3+L^3x)$	$y_{max}=-\frac{5qL^4}{384EI}$ at $x=\frac{L}{2}$	$\theta_{x=0}=-\frac{qL^3}{24EI}$ $\theta_{x=L}=\frac{qL^3}{24EI}$
	$y=-\frac{M}{6EIL}(x^3-L^2x)$	$y_{max}=\frac{ML^2}{9\sqrt{3}\,EI}$ at $x=\frac{L}{\sqrt{3}}$	$\theta_{x=0}=-\frac{ML}{6EI}$ $\theta_{x=L}=\frac{ML}{3EI}$
	$y=\frac{P}{6EI}(x^3-3Lx^2)$	$y_{max}=-\frac{PL^3}{3EI}$ at $x=L$	$\theta_{x=0}=0$ $\theta_{x=L}=-\frac{PL^2}{2EI}$

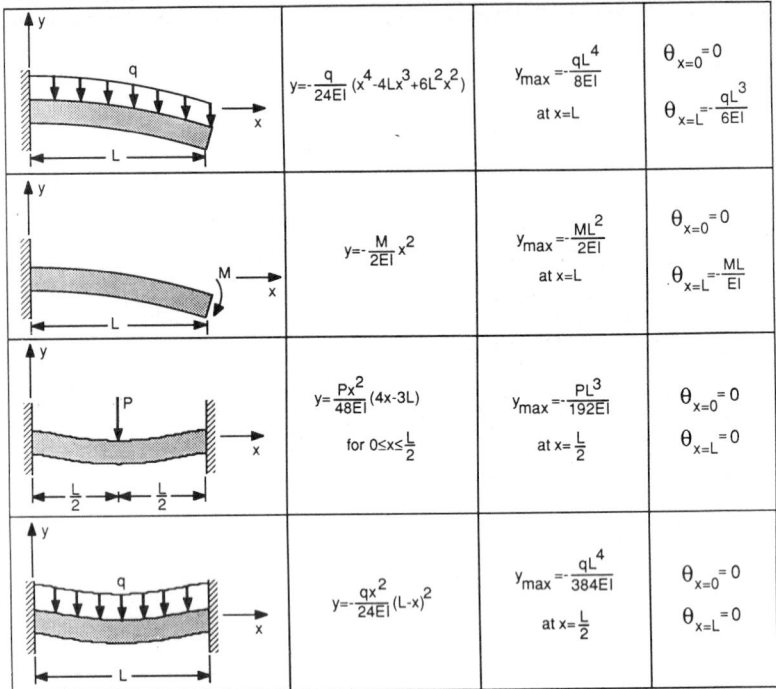

The deflection equations for the two halves of the beam are thus given by

$$y_1 = \frac{P}{48EI}(4x^3 - 3L^2x) \qquad\qquad 0 \le x \le \frac{L}{2}$$

$$y_2 = \frac{P}{48EI}(-4x^3 + 12Lx^2 - 9L^2x + L^3) \qquad \frac{L}{2} \le x \le L$$

In general, if n discontinuities in the loading occur between the end points of the beam, it will be necessary to write and solve $(n + 1)$ differential equations in order to describe the displacement along the entire length of the beam. It is possible to write a single differential equation for even the most general state of loading by using singularity functions.[10] Table 4.4 contains deflection equations for a variety of loading conditions.

Method of Superposition
Since Eq. (4.30), the governing differential equation for beam deflection, is linear, it is possible to solve a beam deflection problem involving many loads by considering the effect of each load as if it acted alone. The total deflection is obtained by adding together the deflections caused by each individual load. The procedure (known as the method of superposition) is illustrated in Example 4.2.

Example 4.2. Consider the case of a simple beam of length L carrying a uniformly distributed load q, and a concentrated load P at its center. Find the center deflection of the beam using the method of superposition.

SOLUTION. From Table 4.4, the center deflection for a simple beam carrying a concentrated load at its center is equal to $-PL^3/48EI$. Similarly, for a simple beam having a uniformly distributed load, the center deflection is equal to $-5qL^4/384EI$. Thus, applying the method of superposition, it follows that the center deflection for a simple beam having both a uniformly distributed load along its length and a concentrated load at its center is given by

$$y_{x=L/2} = \frac{-PL^3}{48EI} - \frac{5qL^4}{384EI} = \frac{-L^3}{384EI}(8P + 5qL)$$

Since the center deflections were the maximum deflections for the individual loadings, it follows that the center deflection is the maximum deflection for the combined state of loading as well.

4.2.5 Unsymmetric Bending

Moments of Inertia of a Plane Area

In order to determine the flexural stress in a beam undergoing unsymmetric bending, it is necessary to compute the moments of inertia of the cross-sectional area of the beam. Assuming the beam cross section lies in the xy plane, the required terms are as follows:

$$I_{xx} = \int y^2 \, dA$$

$$I_{yy} = \int x^2 \, dA \qquad (4.31)$$

$$I_{xy} = \int xy \, dA$$

The terms I_{xx} and I_{yy} are known as the moments of inertia of the cross section, while the term I_{xy} is called the product of inertia of the cross section. Whereas I_{xx} and I_{yy} are always positive, I_{xy} can be positive, negative, or zero. For the cross sections shown in Table 4.2, the products of inertia are zero. This is because each of these cross sections has one or more axes of symmetry, and at least one of the two coordinate axes lies along an axis of symmetry. In fact, every cross section has some orientation for which $I_{xy} = 0$. A Mohr's circle for inertias can even be drawn. Details can be found in most statics and mechanics texts.

Parallel Axis Theorem

The determination of the moments of inertia for a complex cross section using the definitions in Eq. (4.31) is often quite difficult. Instead, it is usually more expeditious to consider the cross section to be made up of several smaller sections of simple shapes and to determine the moments of inertia of these smaller sections separately. After this, the moments of inertia of the smaller sections must somehow be assembled to give the moment of inertia of the whole cross section. This assemblage is accomplished using the parallel axis theorem. This theorem states that the moments of inertia of a cross section with respect to an arbitrary xy coordinate system are related to the moments of inertia of that cross section with respect to a parallel, centroidal $x'y'$ coordinate system by the following:

$$I_{xx} = I'_{xx} + A\bar{y}^2$$

$$I_{yy} = I'_{yy} + A\bar{x}^2$$

$$I_{xy} = I'_{xy} + A\bar{x}\bar{y}$$

where A is the area of the cross section, and $\bar{x}$ and $\bar{y}$ are the distances from the origin of the xy coordinate system to the origin of the $x'y'$ coordinate system.

Example 4.3. For the angle section shown in Fig. 4.18, determine the moments of inertia of the cross section with respect to its centroid.

SOLUTION. First determine the coordinates of the centroid C with respect to the given x^*y^* reference frame. The cross section is considered to be made up of two small rectangles as shown in Fig. 4.18. The centroid of the cross section is given by the coordinates x_C^* and y_C^*:

$$x_C^* = \frac{\Sigma(\bar{x}A)}{\Sigma A} = \frac{1(2)(4) + 3(2)(6)}{(2)(4) + (2)(6)} = 2.2 \text{ in.}$$

$$y_C^* = \frac{\Sigma(\bar{y}A)}{\Sigma A} = \frac{2(2)(4) + 5(2)(6)}{(2)(4) + (2)(6)} = 3.8 \text{ in.}$$

These equations are based on the definition of the centroid of a cross section. Next, place the xy coordinate system at the centroid, C. Using the values for the moments of inertia of a rectangle with

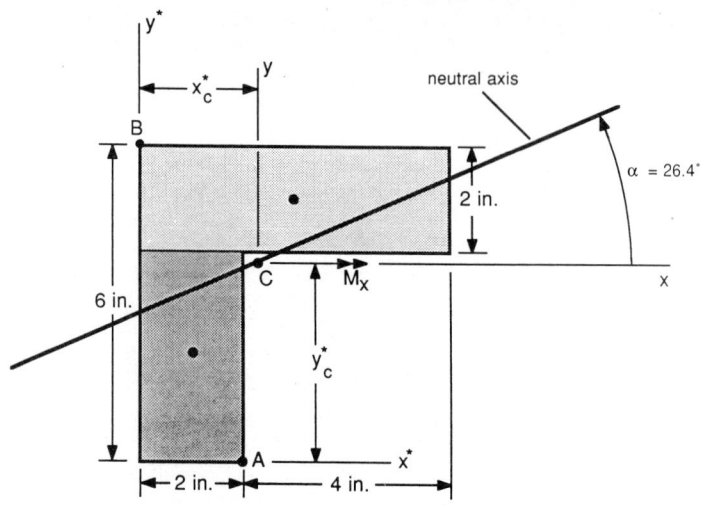

Fig. 4.18 Beam cross section for use with Examples 4.3 and 4.4.

respect to its centroid (from Table 4.2) and applying the parallel axis theorem leads to

$$I_{xx} = \frac{6(2)^3}{12} + 6(2)(1.2)^2 + \frac{2(4)^3}{12} + 2(4)(-1.8)^2 = 57.87 \text{ in.}^4$$

$$I_{yy} = \frac{2(6)^3}{12} + 6(2)(0.8)^2 + \frac{4(2)^3}{12} + 2(4)(-1.2)^2 = 57.87 \text{ in.}^4$$

$$I_{xy} = 0 + 6(2)(1.2)(0.8) + 0 + 2(4)(-1.8)(-1.2) = 28.80 \text{ in.}^4$$

Equation for Flexural Stress

When a beam cross section does not have a plane of symmetry, or if it has a plane of symmetry but the applied moments do not act in this plane, Eq. (4.27) is no longer valid. Provided that the origin of the xy coordinate axes is at the centroid of the cross section, the flexural stress at a point $P(x, y)$ is given by

$$\sigma_{zz} = \frac{(M_x I_{yy} + M_y I_{xy}) y - (M_x I_{xy} + M_y I_{xx}) x}{I_{xx} I_{yy} - I_{xy}^2} \qquad (4.32)$$

where M_x and M_y are the components of the moment acting along the x and y axes, respectively.

The neutral axis can be found by setting $\sigma_{xx} = 0$ in Eq. (4.32). Since $x = y = 0$ satisfies $\sigma_{xx} = 0$, it is clear that the neutral axis passes through the centroid of the cross section. Let α be the angle of the neutral axis with respect to the x axis. Then α satisfies the following:

$$\tan \alpha = \frac{y}{x} = \frac{M_x I_{xy} + M_y I_{xx}}{M_x I_{yy} + M_y I_{xy}} \qquad (4.33)$$

By eliminating M_y between Eqs. (4.32) and (4.33), a simpler expression for the flexural stress is obtained:

$$\sigma_{zz} = \frac{y - x \tan \alpha}{I_{xx} - I_{xy} \tan \alpha} M_x \qquad (4.34)$$

This equation is invalid if $\alpha = \pm 90°$. In such cases, Eq. (4.32) should be used.

Example 4.4. Suppose a beam has the cross-sectional dimensions given in Fig. 4.18. Assume the beam is acted upon by a 90,000 lb-in. moment that acts along the positive x axis. Determine the maximum flexural stress in the beam.

SOLUTION. The moments of inertia for this cross section were previously found in Example 4.3. The values are $I_{xx} = 57.9$ in.^{4}, $I_{yy} = 57.9$ in.^{4}, and $I_{xy} = 28.8$ in.^{4}. The orientation of the neutral axis is found using Eq. (4.33) with $M_y = 0$:

$$\tan \alpha = \frac{28.8}{57.9} \Rightarrow \alpha = 26.4°$$

The neutral axis has been shown in Fig. 4.18. From this figure it is logical to expect that the maximum stress will occur at A or B since these points are furthest from the neutral axis. Using Eq. (4.34), the stresses at these points are given by

$$(\sigma_{zz})_A = \frac{(-3.8) - (-0.2)\tan 26.4°}{57.9 - 28.8 \tan 26.4°}(90,000) = -7640 \text{ psi}$$

$$(\sigma_{zz})_B = \frac{(2.2) - (-2.2)\tan 26.4°}{57.9 - 28.8 \tan 26.4°}(90,000) = 6800 \text{ psi}$$

Thus, the maximum tensile stress in the beam is 6800 psi and the maximum compressive stress is 7640 psi.

4.2.6   Curved Beams

Circumferential Stress
The formula presented in Section 4.2.2 for flexural stress, $\sigma = -My/I$, is applicable for long, initially straight beams. It can still be used with reasonable accuracy for beams with some initial curvature provided the beam depth is small compared to the radius of curvature. When the beam's depth is of the same order of magnitude as its curvature, however, a new equation must be used. The derivation of this equation is too involved to be presented here. The details can be found in almost any modern text on strength of materials; in particular, see Refs. 7 and 8.

The equation for the circumferential stress distribution of a curved member with a plane of symmetry and subjected to moments acting in this plane of symmetry is given by (see Fig. 4.19):

$$\sigma = \frac{-My}{A(R - r_n)r} \tag{4.35}$$

where r = distance from the center of curvature to the point under consideration
r_n = distance from the center of curvature to the neutral axis of the cross section
y = distance from the neutral axis to the point Q under consideration
R = distance from the center of curvature to the centroidal axis of the cross section
A = area of the cross section
M = applied moment

The moment is considered positive if it tends to create compressive stresses on the concave side of the beam cross section. The coordinate y is taken to be positive on the concave side of the neutral axis.

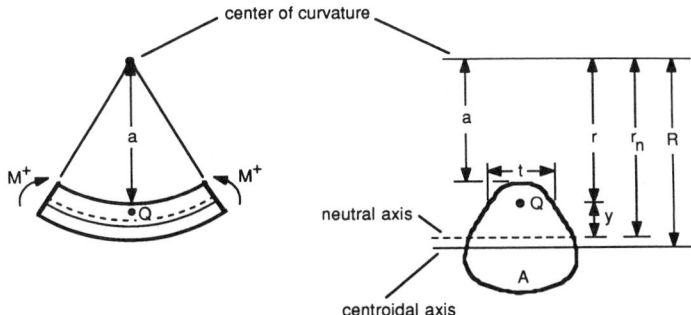

Fig. 4.19   Notation for theory of curved beams.

TABLE 4.5　Cross-Sectional Properties for Curved Beams

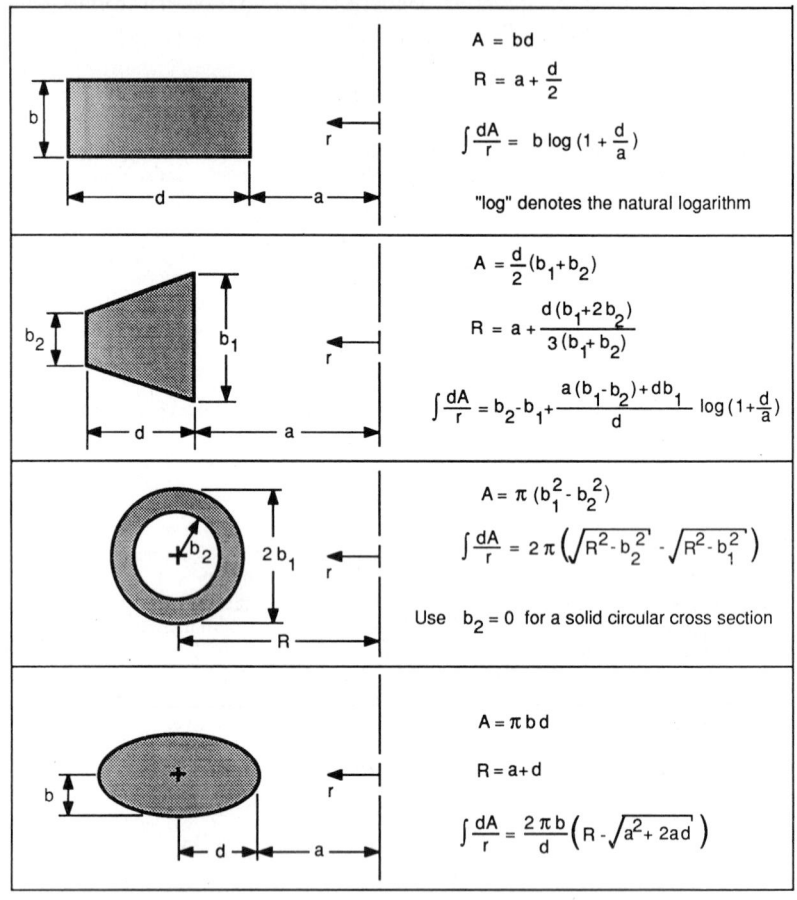

It can be shown that r_n is given by

$$r_n = \frac{A}{\displaystyle\int \frac{dA}{r}}$$

Also, R is given by

$$R = \frac{1}{A}\int r\,dA$$

Thus, the centroidal axis and the neutral axis do not coincide. In general their difference is small; consequently, it is important to determine these distances accurately since their difference occurs in the denominator of Eq. (4.35). Table 4.5 gives required cross-sectional properties for application to curved beams. More complete tables can be found in Refs. 7–9.

Radial Stress

For an initially straight beam, only circumferential stresses develop upon application of a bending moment. For initially curved beams, however, significant radial stresses can develop. An approximate solution for the radial stress is given by[7]:

$$\sigma_r = \frac{-M}{A(R - r_n)tr}\left(r_n\int_a^r \frac{dA}{r} - \int_a^r dA\right) \qquad (4.36)$$

where a is the distance from the center of curvature to the edge of the cross section, and t is the thickness of the cross section at the point under consideration. All other terms were previously defined in conjunction with Eq. (4.35). See Fig. 4.19 for details.

It should be noted that Eq. (4.36) neglects the effect of normal and shear forces on the radial stress. In general, these forces do not have a major effect on the radial stress and can be ignored for most applications.[7]

4.3   TORSION AND SHAFTS

4.3.1   Circular Shafts

Shear Stress Distribution
Circular shafts have the unique property that, when subjected to torsion, every cross section of the shaft remains plane and undistorted. This fact allows for the easy determination of the shear strain distribution. When subjected to torsion, a small square element, A, deforms into a rhombus, A', as shown in Fig. 4.20a. The angle γ shown in Fig. 4.20a is identical to the shear strain that the square element has gone through, Fig. 4.20b.

For small values of γ it is reasonable to assume that $\gamma = r\phi/L$, where L is the shaft length, r is the radius of the shaft at the point under consideration, and ϕ is the angle of twist of the shaft (see Fig. 4.20). Denoting the outer radius of the shaft by c, it follows that

$$\gamma_{max} = \frac{c\phi}{L} \tag{4.37}$$

and hence,

$$\gamma(r) = \frac{r}{c}\gamma_{max} \tag{4.38}$$

The only assumption made in deriving Eq. (4.38) was to assume that the shear strains are small. For linear elastic response, $\tau = G\gamma$. Substituting this relationship into Eq. (4.38) leads to

$$\tau(r) = \frac{r}{c}\tau_{max} \tag{4.39}$$

It follows that, for torsion of an isotropic linear solid elastic circular shaft, the shear stress distribution is linear with respect to radius. This is illustrated in Fig. 4.21a. The shear stress distribution of a hollow circular shaft (Fig. 4.21b) is also given by Eq. (4.39).

From statics, the net resisting torque generated by the internal shear stresses must equal the applied external torque, T:

$$T = \int r\tau \, dA \tag{4.40}$$

Substituting for τ from Eq. (4.39) into (4.40) leads to

$$\tau_{max} = \frac{Tc}{J} \tag{4.41}$$

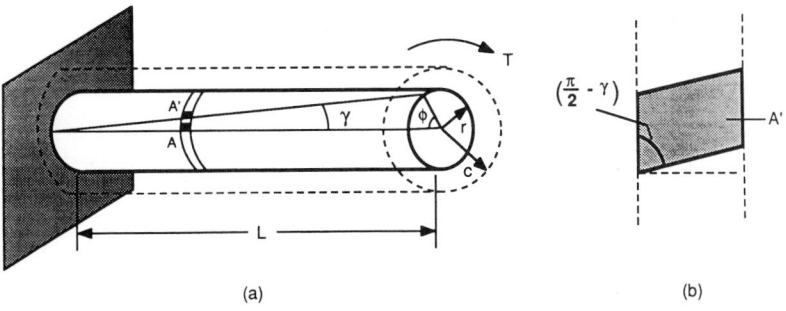

(a)   (b)

Fig. 4.20   Torsion of a circular shaft.

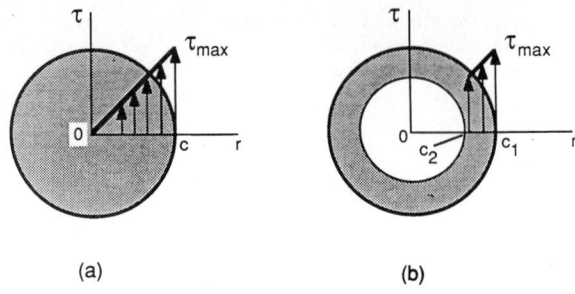

(a)   (b)

Fig. 4.21   Shear stress distribution in a circular shaft: (a) Solid shaft; (b) hollow shaft.

where $J = \int r^2\, dA$. The term J is known as the polar moment of inertia of the cross section with respect to its centroid. For a solid circular cross section, $J = \pi c^4/2$. For a hollow circular tube, $J = \pi(c_1^4 - c_2^4)/2$, where c_1 and c_2 are the outer radius and inner radius of the tube, respectively. Since the shear stress distribution is linear with respect to radius, it follows that the shear stress at a distance r from the centroid of a circular shaft is given by $\tau(r) = Tr/J$.

Angle of Twist

Using Hooke's law, Eq. (4.41) can be rewritten as

$$\gamma_{max} = \frac{\tau_{max}}{G} = \frac{Tc}{JG}$$

Equating this expression with Eq. (4.37) leads to

$$\phi = \frac{TL}{JG} \tag{4.42}$$

Equation (4.42) applies to a uniform shaft subjected to a single torque at its end. If the shaft dimensions vary continuously with axial distance, ϕ must be found by integration:

$$\phi = \int_0^L \frac{T\, dx}{JG}$$

where x is the axial distance coordinate.

If several concentrated torques are applied to the shaft, or if the shaft has sudden changes in cross section, the relative angle of twist must be found within each uniform shaft region. The total angle of twist is found by summing these relative angles of twist:

$$\phi_{Total} = \sum_k \phi_k = \sum_k \frac{T_k L_k}{J_k G_k}$$

Design of Transmission Shafts

Circular shafts are often used for power transmission. The relation between the power transmitted, P, the angular speed of the shaft, ω, and the applied torque, T, is given by $P = T\omega$. If ω is expressed in revolutions per minute and P is expressed in horsepower,

$$T = \frac{63{,}030P}{\omega} \tag{4.43}$$

where T has units of lb-in. For a given power transmission and a given speed of rotation, it is important that the shaft dimensions are large enough so that the torque, as given by Eq. (4.43), does not cause the maximum shear stress in the shaft to exceed allowable limits. Since $\tau_{max} = Tc/J$, it is clear that the quantity c/J should be kept as small as possible.

4.3.2   Thin-walled Closed Hollow Tubes

A tube is considered thin if the shear stress due to an applied torque does not vary significantly across the wall thickness. Put another way, the wall thickness must be small compared to the overall dimensions of the cross section.

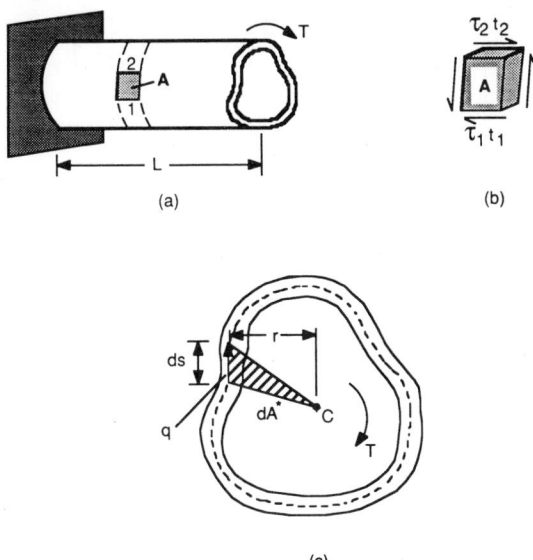

Fig. 4.22   Torsion of a single-celled, thin-walled, closed hollow tube.

Single Cells

Consider the single-celled, thin-walled, closed hollow tube shown in Fig. 4.22 a. By considering equilibrium of horizontal forces for the small element A, as shown in Fig. 4.22 b, it becomes apparent that

$$q = \tau_1 t_1 = \tau_2 t_2 = \text{const.}$$

where q is known as the shear flow and t is the tube thickness. The following equation is obtained when equilibrium of the torsional forces is considered:

$$T = \oint qr\, ds \tag{4.44}$$

where T is the applied torque, and r is the perpendicular distance from the centroid of the cross section to the shear flow q acting on the small section ds (Fig. 4.22 c). The area of the shaded triangle in Fig. 4.22 c is given by

$$dA^* = \tfrac{1}{2} r\, ds \tag{4.45}$$

Substituting for $r\, ds$ from Eq. (4.45) into (4.44) leads to

$$T = 2qA^*$$

where A^* is the area bounded by the centerline of the wall. Recalling that $q = \tau t$ gives

$$\tau = \frac{T}{2tA^*} \tag{4.46}$$

Clearly, the maximum shear stress occurs at the thinnest section of the wall.

Using energy methods it can be shown that the angle of twist is given by (see Ref. 7 for details):

$$\phi = \frac{TL}{4G(A^*)^2} \oint \frac{ds}{t} \tag{4.47}$$

where L is the length of the tube, and G is the shear modulus of the material.

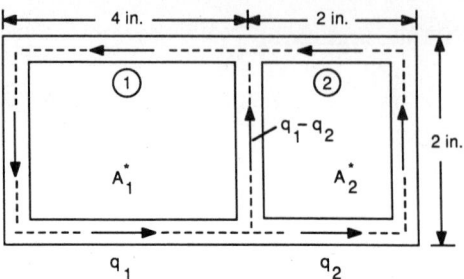

Fig. 4.23   Two-celled rectangular cross section for use with Example 4.5.

Multicells

For a multicell section, the shear flow is no longer constant throughout the cross section. Instead, the shear flow is analogous to the flow of water through pipes or the flow of current in an electric circuit. Consider the rectangular multicell section shown in Fig. 4.23. If the two sections are labeled 1 and 2 as shown, the shear flows around the outer walls of these sections are q_1 and q_2, respectively. The shear flow in the intermediate wall is $q_1 - q_2$.

The torque can be expressed in terms of the shear flows using a procedure similar to that used for single-celled tubes. In fact, it can be shown that for an n-celled tube:

$$T = \sum_{i=1}^{n} A_i^* q_i \tag{4.48}$$

where A_i^* is the area contained within the centerline bounding the ith cell.

For an n-celled section there are $n + 1$ unknowns $(q_1, q_2, \ldots, q_n, \phi)$; consequently, $n + 1$ equations are required in order to solve for these unknowns. Equation (4.48) provides one equation. The other n equations come about by noting that, according to Saint-Venant's torsion theory, cross sections do not warp within their own plane when a torque is applied. In other words, the angle of twist, ϕ, is the same for each cell. It follows that

$$\phi = \frac{L}{2G(A_i^*)} \left(\oint \frac{q\, ds}{t} \right)_i \qquad i = 1, 2, \ldots, n \tag{4.49}$$

Equation (4.49) provides the other n equations needed to solve for all the unknowns.

Example 4.5.   For the cross section shown in Fig. 4.23, assume the wall thickness is uniform at 0.25 in. If the tube length is 12 in. and a torque of 50,000 lb-in. is applied, find the maximum shear stress in the tube. Assume $G = 11.5 \times 10^6$ psi.

SOLUTION.   Writing Eq. (4.49) for the two cells,

$$2G\frac{\phi}{L} = \frac{1}{4(2)} \left[\frac{4 + 2 + 4}{0.25} q_1 + \frac{2}{0.25} (q_1 - q_2) \right]$$

$$2G\frac{\phi}{L} = \frac{1}{2(2)} \left[\frac{2 + 2 + 2}{0.25} q_2 + \frac{2}{0.25} (q_2 - q_1) \right]$$

Using Eq. (4.48),

$$T = 50,000 = 4(2) q_1 + 2(2) q_2$$

Solving these equations for q_1, q_2, and ϕ gives $\phi = 0.0115$ rad, $q_1 = 4310$ lb/in., and $q_2 = 3830$ lb/in. Thus, $\tau_{max} = \tau_1 = 4310/0.25 = 17,250$ psi.

4.3.3   Torsion of Noncircular Cross Sections

Theories for analyzing the torsional response of circular shafts and thin-walled, closed hollow tubes were developed in Sections 4.3.1 and 4.3.2. For solid or thick-walled noncircular cross sections, it is usually necessary to use the theory of elasticity in order to determine torsion formulas. Because of the complexities involved, the most practical means for determining the torsion formula for a noncircular

TABLE 4.6   Torsion Formulas for Solid Sections

Cross Section	Maximum Shear Stress	Angle of Twist
Ellipse (2a, 2b)	$\tau_{max} = \dfrac{2T}{\pi ab^2}$ $(a>b)$	$\theta = \dfrac{T(a^2+b^2)L}{\pi a^3 b^3 G}$
Equilateral Triangle (a)	$\tau_{max} = \dfrac{20T}{a^3}$	$\theta = \dfrac{46.2TL}{a^4 G}$
Rectangle (a, b)	$\tau_{max} = \dfrac{T}{K_1 ab^2}$	$\theta = \dfrac{TL}{K_2 ab^3 G}$
Regular Hexagon (a)	$\tau_{max} = \dfrac{1.09T}{a^3}$	$\theta = \dfrac{0.967TL}{a^4 G}$

$\dfrac{a}{b}$	K_1	K_2
1.0	0.208	0.141
1.5	0.231	0.196
2.0	0.246	0.229
3.0	0.267	0.263
4.0	0.282	0.281
6.0	0.299	0.299
10.0	0.312	0.312
∞	0.333	0.333

T = applied torque, L = length of shaft, G = shear modulus

cross section is to find the appropriate formula in a handbook or other source. Table 4.6 gives formulas for several common cross sections. A much more thorough listing is provided in Ref. 9.

Although quantitative determination of the stress distribution for the torsion of a noncircular cross section is usually quite difficult, an analogy does exist that provides for a qualitative understanding. This is the Prandtl elastic-membrane analogy. In essence, this analogy states that the shear stress at a given point (x, y) in a cross section is proportional to the slope at the point (x, y) of a membrane (such as a soap film) made of the same cross-sectional dimensions, where the membrane is pressurized from one side. Thus, by visualizing the shape of a membrane of the same cross-sectional dimensions, it is possible to obtain an idea of the distribution of shear stresses in the cross section of a shaft undergoing torsional loading. In particular, it becomes obvious that the shear stress is zero at the corners of a shaft made of a rectangular cross section since a membrane of the same geometry and pressurized from one side will have zero slope at its corners. See Refs. 7, 8, and 11 for further details.

4.4   PLATES, SHELLS, AND CONTACT STRESSES

4.4.1   Classical Theory of Elastic, Isotropic Plates

Flat plates are most often used to carry transverse pressure loads. Just as with beams, these pressure loads are carried by out-of-plane shear stresses, which in turn induce a bending moment distribution in the plate. Unlike beams, however, the bending moment distribution in a plate is two dimensional, and a torsional moment distribution is also usually present.

Assumptions Used

In deriving the equations of equilibrium, it is conventional to take the origin of coordinates at the middle surface of the plate, as shown in Fig. 4.24a. The commonly used sign convention for moments and shear forces is shown in Fig. 4.24b. A number of assumptions are used in deriving the classical theory of elastic isotropic plates. These assumptions are listed below:

1. The plate is initially flat.
2. The material is linear elastic, isotropic, and homogeneous.
3. The plate thickness, h, is small compared to its in-plane dimensions, a and b (see Fig. 4.24a). Generally, a and b should be at least 10 times greater than h.
4. Deflections are small and slopes of the deflected middle surface are small compared to unity.
5. The middle surface of the plate is a neutral surface; that is, it remains unstrained during bending.
6. Distortions due to transverse shear are negligible. Thus, initially straight lines normal to the middle surface remain straight and normal after deformation.
7. Stresses normal to the middle surface are negligible compared to other stresses.

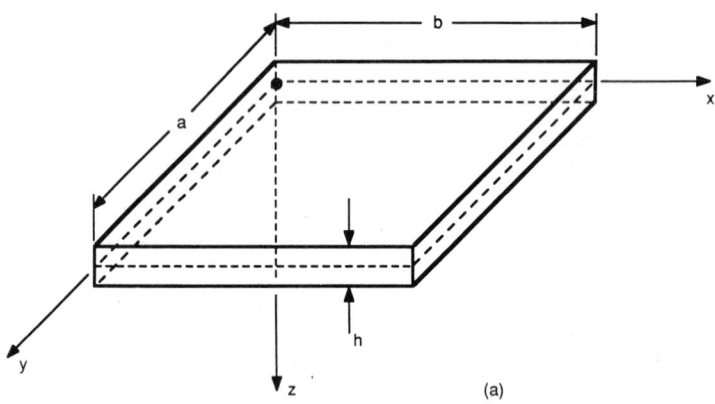

(a)

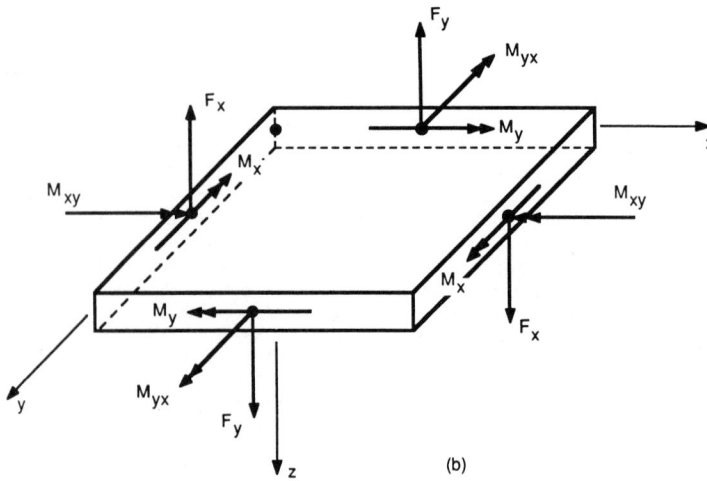

(b)

Fig. 4.24 Positive sign convention for plate deformation: (*a*) Plate dimensions; (*b*) positive sign convention for forces and moments.

Stresses and Moments in Terms of Curvature

The displacements in the x and y directions, u and v, respectively, depend on the z coordinate and the slope components:

$$u = -z\frac{\partial w}{\partial x} \qquad v = -z\frac{\partial w}{\partial y}$$

where w is the displacement component in the z direction. Using the strain-displacement relations, Eqs. (4.4), the strains can be expressed in terms of the curvatures as follows:

$$\varepsilon_{xx} = -z\frac{\partial^2 w}{\partial x^2} \qquad \varepsilon_{yy} = -z\frac{\partial^2 w}{\partial y^2} \qquad \gamma_{xy} = -2z\frac{\partial^2 w}{\partial x \partial y} \tag{4.50}$$

For thin plates, plane stress conditions prevail ($\sigma_{zz} = \sigma_{xz} = \sigma_{yz} = 0$). Substituting from Eqs. (4.50) into (4.11) leads to

$$\sigma_{xx} = \frac{-Ez}{1-v^2}\left(\frac{\partial^2 w}{\partial x^2} + v\frac{\partial^2 w}{\partial y^2}\right)$$

$$\sigma_{yy} = \frac{-Ez}{1-v^2}\left(\frac{\partial^2 w}{\partial y^2} + v\frac{\partial^2 w}{\partial x^2}\right) \tag{4.51}$$

$$\sigma_{xy} = \frac{-Ez}{1+v}\frac{\partial^2 w}{\partial x \partial y}$$

The moments per unit length are given by

$$M_x = \int_{-h/2}^{h/2} \sigma_{xx} z\, dz$$

$$M_y = \int_{-h/2}^{h/2} \sigma_{yy} z\, dz \tag{4.52}$$

$$M_{yx} = M_{xy} = \int_{-h/2}^{h/2} \sigma_{xy} z\, dz$$

Substitution of Eqs. (4.51) into (4.52) gives

$$M_x = -D\left(\frac{\partial^2 w}{\partial x^2} + v\frac{\partial^2 w}{\partial y^2}\right)$$

$$M_y = -D\left(\frac{\partial^2 w}{\partial y^2} + v\frac{\partial^2 w}{\partial x^2}\right) \tag{4.53}$$

$$M_{xy} = -D(1-v)\frac{\partial^2 w}{\partial x \partial y}$$

where

$$D = \frac{Eh^3}{12(1-v^2)}$$

The parameter D is known as the bending stiffness, or flexural rigidity, of the plate.

Equations of Equilibrium

If an infinitesimal section of the plate is considered, equilibrium of moments about the x and y axes leads to

$$\frac{\partial M_x}{\partial x} + \frac{M_{xy}}{\partial y} - F_x = 0 \qquad \frac{\partial M_y}{\partial y} + \frac{M_{xy}}{\partial x} - F_y = 0 \tag{4.54}$$

where F_x and F_y are the shear forces per unit length on the x and y faces, respectively (see Fig. 4.24b). These equations, when combined with Eqs. (4.53), lead to

$$F_x = -D\frac{\partial}{\partial x}\nabla^2 w \qquad F_y = -D\frac{\partial}{\partial y}\nabla^2 w \tag{4.55}$$

where

$$\nabla^2 = \frac{\partial^2}{\partial x^2} + \frac{\partial^2}{\partial y^2}$$

Consideration of equilibrium in the z direction gives

$$\frac{\partial F_x}{\partial x} + \frac{\partial F_y}{\partial x} + p = 0 \tag{4.56}$$

where p is the load per unit area in the z direction acting on the infinitesimal section. Finally, substituting Eqs. (4.55) into (4.56) gives the governing differential equation for equilibrium of a plate:

$$\nabla^2\nabla^2 w = \nabla^4 w = \frac{p}{D} \tag{4.57}$$

where

$$\nabla^4 = \frac{\partial^4}{\partial x^4} + 2\frac{\partial^4}{\partial x^2\,\partial y^2} + \frac{\partial^4}{\partial y^4}$$

Boundary Conditions

When attempting to solve Eq. (4.57) for a given pressure distribution $p(x, y)$, the boundary conditions of the plate edges must be known. The mathematical forms of the three commonly encountered boundary conditions follow for a plate edge parallel to the y axis and at a distance $x = a$.

1. Simply-supported-edge conditions

$$w = 0 \quad \text{at } x = a$$

$$\frac{\partial^2 w}{\partial x^2} + \nu\frac{\partial^2 w}{\partial y^2} = 0 \quad \text{at } x = a$$

2. Fixed-edge conditions

$$w = 0 \quad \text{at } x = a$$

$$\frac{\partial w}{\partial y} = 0 \quad \text{at } x = a$$

3. Free-edge conditions

$$\frac{\partial^2 w}{\partial x^2} + \nu\frac{\partial^2 w}{\partial y^2} = 0 \quad \text{at } x = a$$

$$\frac{\partial^3 w}{\partial x^2} + (2 - \nu)\frac{\partial^3 w}{\partial x\,\partial y^2} = 0 \quad \text{at } x = a$$

Strain Energy

When the elastic energy of a component is written in terms of displacements, it is referred to as the strain energy. For thin plates undergoing small deflections, only the strain energy due to bending is significant. This quantity is given by

$$U = \frac{1}{2}\int_A D\left\{\left(\frac{\partial^2 w}{\partial x^2}\right)^2 + \left(\frac{\partial^2 w}{\partial y^2}\right)^2 + 2\nu\frac{\partial^2 w}{\partial x^2}\frac{\partial^2 w}{\partial y^2} + 2(1 - \nu)\left(\frac{\partial^2 w}{\partial x\,\partial y}\right)^2\right\}dx\,dy$$

This expression can be useful for obtaining approximate solutions to plate problems.

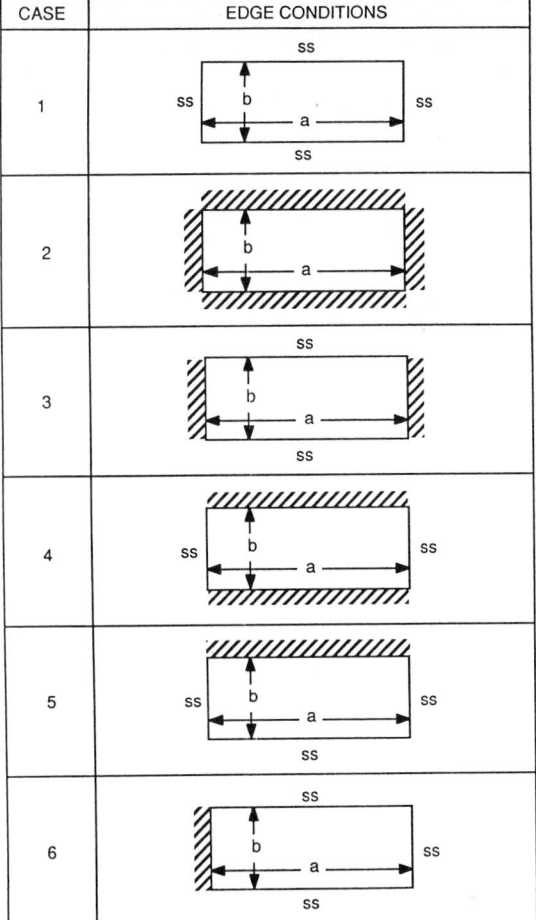

ss = simply supported

////// = fixed

Fig. 4.25a Maximum stress and deflection of linear elastic, isotropic, rectangular plates for $\nu = 0.3$. Plate thickness = h. Elastic modulus = E. Each plate is uniformly loaded over its entire surface with a pressure P:

$$\sigma_{max} = \frac{\alpha P b^2}{h^2} \qquad w_{max} = \frac{\beta P b^4}{E h^3}$$

Values of α and β for the cases shown are plotted in charts A and B, respectively, in Fig. 4.25b.

Solutions to Plate Problems

For relatively simple loadings and edge conditions, solutions for the lateral deflection, w, can be found using classical techniques, such as Fourier series analysis.[12,13] Approximate solutions can be obtained by means of Galerkin's method and energy methods such as the Rayleigh–Ritz method (Ref. 14, Chapter 5). In more recent years, the trend has been to apply numerical methods, most notably the finite-difference and finite-element methods.[15,16]

The most important quantities in plate design are usually the maximum stress and maximum deflection. Figures 4.25 and 4.26 contain this information in graphical form for a variety of loading and edge conditions. More extensive data bases can be found in Refs. 9 (Chapter 10) and 17.

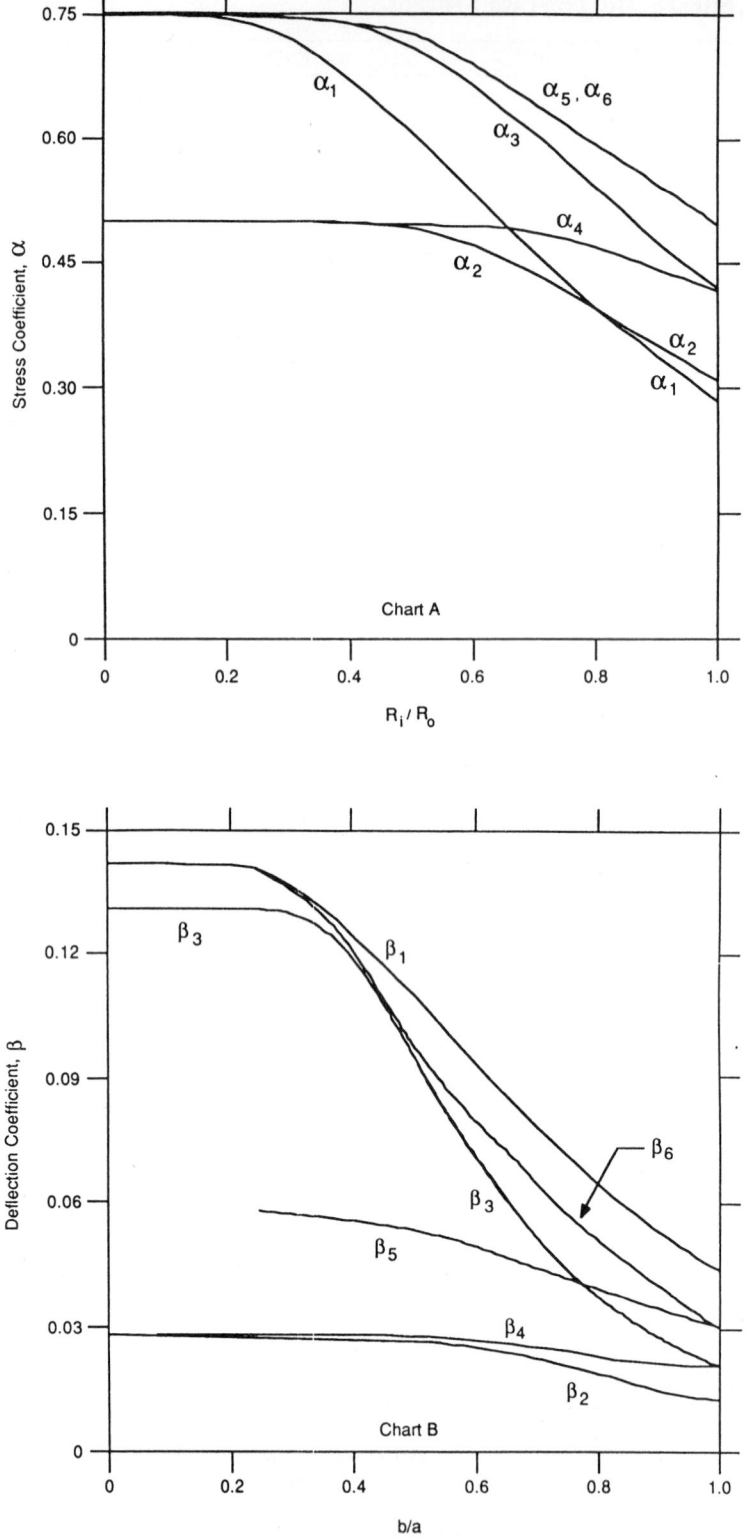

Fig. 4.25b Values of α and β for cases in Fig. 4.25a.

CASE	LOADING AND EDGE CONDITIONS	DESCRIPTION
1	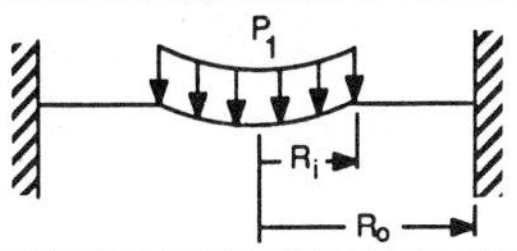	outer edge fixed uniform load over circle of radius R_i
2	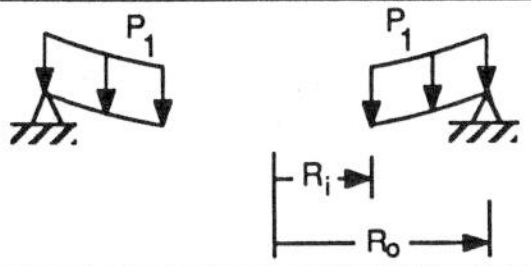	outer edge supported inner edge free uniform load from R_i to R_o
3	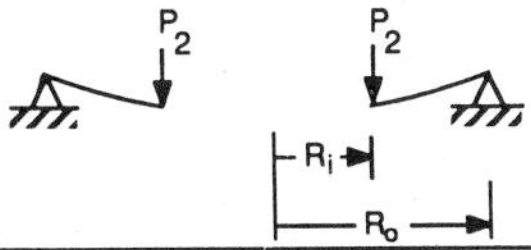	outer edge supported inner edge free line load at R_i
4	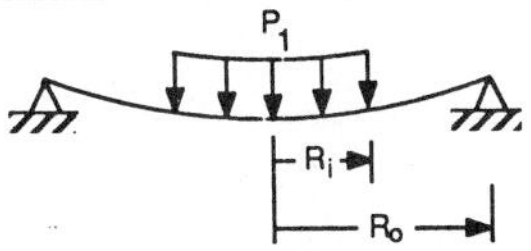	outer edge supported uniform load over R_i
5	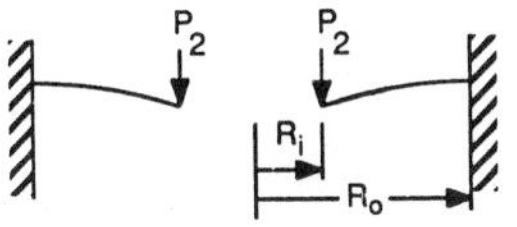	outer edge fixed inner edge free line load at R_i
6	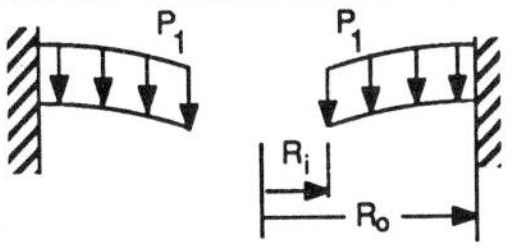	outer edge fixed uniform load from R_i to R_o

R_i = inside radius
R_o = outside radius

Fig. 4.26a Maximum stress and deflection of linear elastic, isotropic, circular plates for $\nu = 0.3$. Plate thickness $= h$. Elastic modulus $= E$:

$$\sigma_{max} = \frac{\alpha P}{h^2} \qquad w_{max} = \frac{\beta P R_0^2}{Eh^3}$$

where P is the applied loading, either a pressure load (P_1) or a line load (P_2). Values of α and β for the cases shown are plotted in charts A and B, respectively, in Fig. 4.26b.

4.4.2 Orthotropic Plates

The discussion of Section 4.4.1 assumed an isotropic plate. That is, the material properties are independent of direction. With the current widespread usage of composite materials, the problem of determining stresses and deflections in anisotropic plates has become quite prevalent. A good treatment of the general theory of anisotropic plates can be found in Ref. 18. Typical examples of anisotropic plates include asymmetrically laminated plates made from multiple plies of fiber-reinforced composite materials, and plates reinforced with stringers and ribs.

While the general theory of anisotropic plates is too complex to present here, one simpler plate configuration, which is easier to analyze, often occurs in practice. This is the case of an orthotropic plate. A material exhibits orthotropic response if there are two orthogonal planes of material property symmetry. These two directions of symmetry normally lie within the horizontal plane of the plate (the xy plane in Fig. 4.24). If these directions of symmetry coincide with the x and y directions, the

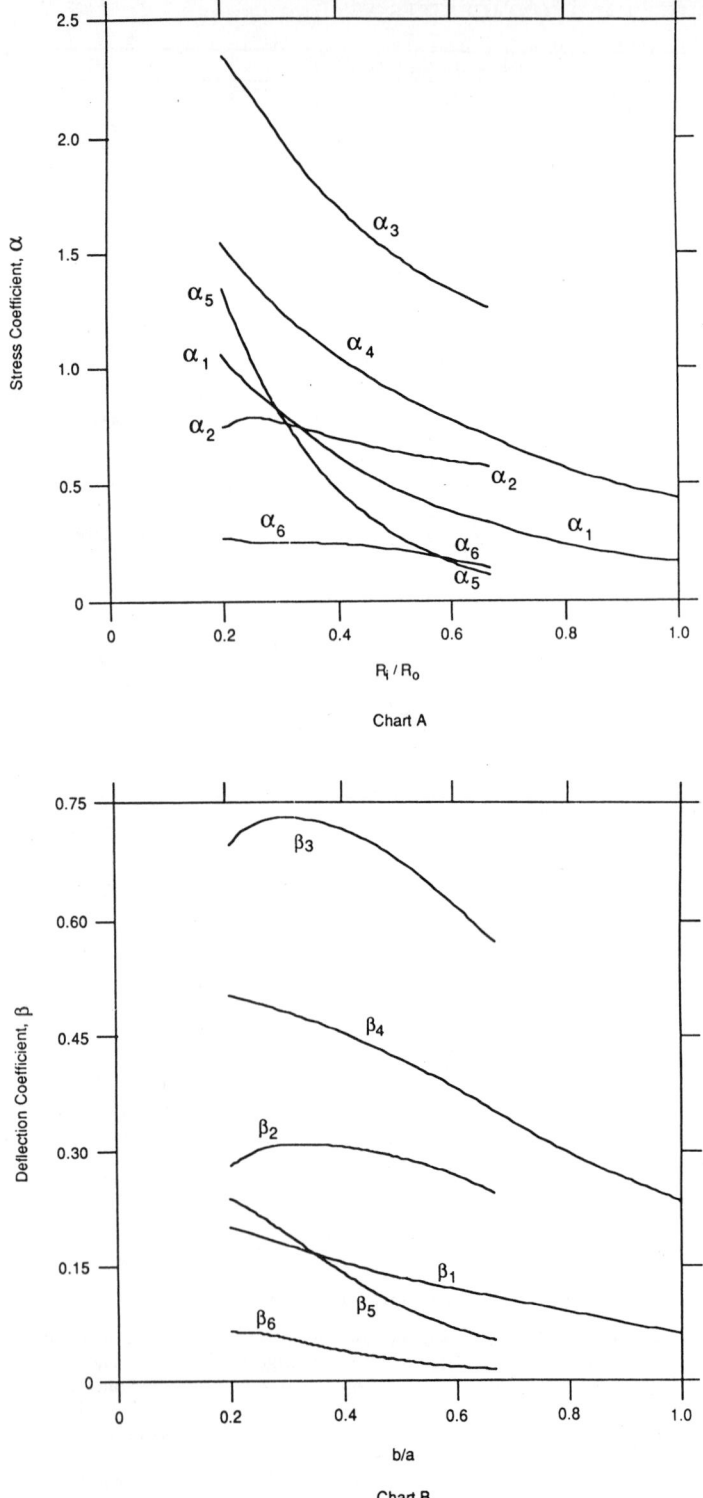

Fig. 4.26b Values of α and β for cases shown in Fig. 4.26a.

governing differential equation for equilibrium of a plate becomes[15]

$$D_{xx}\frac{\partial^4 w}{\partial x^4} + 2H\frac{\partial^4 w}{\partial x^2 y^2} + D_{yy}\frac{\partial^4 w}{\partial y^4} = p \qquad (4.58)$$

where

$$D_{xx} = \frac{h^3 Q_{xx}}{12} \qquad D_{yy} = \frac{h^3 Q_{yy}}{12} \qquad D_{xy} = \frac{h^3 Q_{xy}}{12}$$

$$G_{xy} = \frac{h^3 G}{12} \qquad H = D_{xy} + 2G_{xy}$$

where h is the plate thickness, and G is the shear modulus of the material. The quantities Q_{xx}, Q_{yy}, Q_{xy}, and G characterize the linearly elastic, orthotropic stress-strain response of the material as given by

$$\begin{Bmatrix} \sigma_{xx} \\ \sigma_{yy} \\ \sigma_{xy} \end{Bmatrix} = \begin{bmatrix} Q_{xx} & Q_{xy} & 0 \\ Q_{xy} & Q_{yy} & 0 \\ 0 & 0 & G \end{bmatrix} \begin{Bmatrix} \varepsilon_{xx} \\ \varepsilon_{yy} \\ \gamma_{xy} \end{Bmatrix}$$

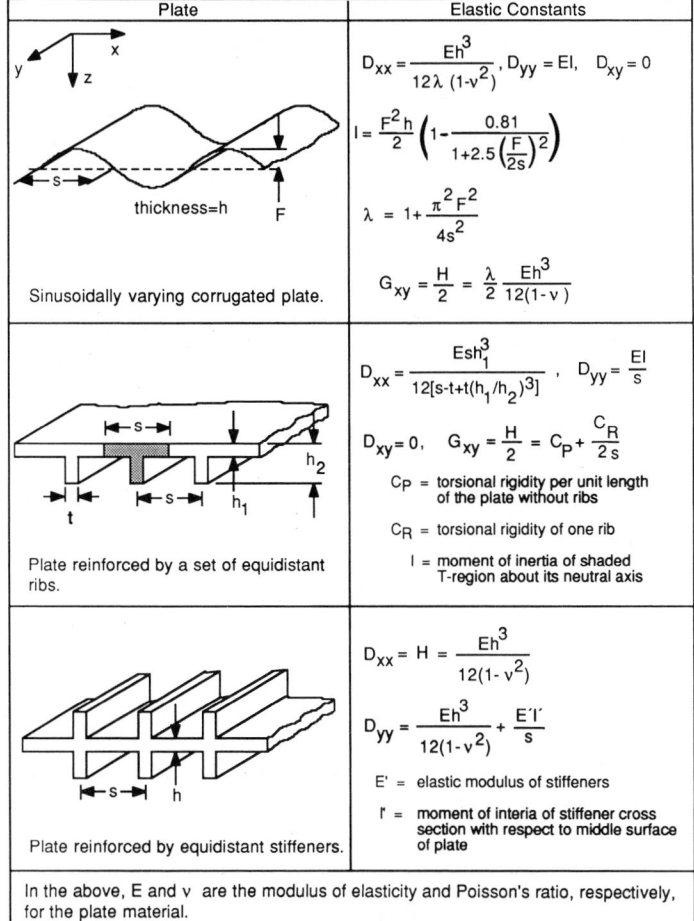

Plate	Elastic Constants
Sinusoidally varying corrugated plate. thickness=h	$D_{xx} = \dfrac{Eh^3}{12\lambda(1-v^2)}$, $D_{yy} = EI$, $D_{xy} = 0$ $I = \dfrac{F^2 h}{2}\left(1 - \dfrac{0.81}{1+2.5\left(\frac{F}{2s}\right)^2}\right)$ $\lambda = 1 + \dfrac{\pi^2 F^2}{4s^2}$ $G_{xy} = \dfrac{H}{2} = \dfrac{\lambda}{2}\dfrac{Eh^3}{12(1-v)}$
Plate reinforced by a set of equidistant ribs.	$D_{xx} = \dfrac{Esh_1^3}{12[s-t+t(h_1/h_2)^3]}$, $D_{yy} = \dfrac{EI}{s}$ $D_{xy} = 0$, $G_{xy} = \dfrac{H}{2} = C_p + \dfrac{C_R}{2s}$ C_p = torsional rigidity per unit length of the plate without ribs C_R = torsional rigidity of one rib I = moment of inertia of shaded T-region about its neutral axis
Plate reinforced by equidistant stiffeners.	$D_{xx} = H = \dfrac{Eh^3}{12(1-v^2)}$ $D_{yy} = \dfrac{Eh^3}{12(1-v^2)} + \dfrac{E'I'}{s}$ E' = elastic modulus of stiffeners I' = moment of interia of stiffener cross section with respect to middle surface of plate
In the above, E and v are the modulus of elasticity and Poisson's ratio, respectively, for the plate material.	

Fig. 4.27 Stiffness coefficients for various orthotropic plates.

(See Section 4.7.2 for further discussion of orthotropic response.) The bending moments in terms of the curvatures have the form

$$M_x = -\left(D_{xx}\frac{\partial^2 w}{\partial x^2} + D_{xy}\frac{\partial^2 w}{\partial y^2} \right)$$

$$M_y = -\left(D_{yy}\frac{\partial^2 w}{\partial y^2} + D_{xy}\frac{\partial^2 w}{\partial x^2} \right)$$

$$M_{xy} = -2G_{xy}\frac{\partial^2 w}{\partial x\,\partial y}$$

It should be mentioned that the preceding equations are valid only for small deflection, linear elastic, orthotropic response. In addition to symmetrically stacked composite laminated plates, corrugated plates, plates reinforced by closely spaced ribs and stiffeners, and reinforced concrete slabs can also be modeled as orthotropic plates.[12] The problem of analyzing orthotropic plates reduces to determining D_{xx}, D_{yy}, D_{xy}, and G_{xy}. Once this has been accomplished, Eq. (4.58) can be solved by classical, energy, or numerical techniques just as is done for isotropic plates. Fig. 4.27 gives approximate values of these constants for corrugated plates and rib and stiffener reinforced plates.

4.4.3   Buckling of Isotropic Plates

Critical Buckling Loads

The classical isotropic plate theory discussed in Section 4.4.1 ignored the effects of in-plane forces. When such forces are large, they can significantly alter the response of the plate. If the in-plane forces are tensile, the plate will exhibit an effective increase in stiffness with respect to lateral loads. If the in-plane forces are compressive, buckling of the plate may occur. Let N_x and N_y represent the in-plane normal forces per unit length in the x and y directions, respectively. Similarly, N_{xy} is the in-plane shear force per unit length. The governing differential equation for plate equilibrium takes the form

$$D\nabla^4 w = N_x\frac{\partial^2 w}{\partial x^2} + N_y\frac{\partial^2 w}{\partial y^2} + 2N_{xy}\frac{\partial^2 w}{\partial x\,\partial y} + p \tag{4.59}$$

By setting $p = N_{xy} = N_y = 0$, the critical buckling load for uniaxial compression can be obtained. Similarly, the critical buckling load for any combination of shear and axial forces can be obtained. See Refs. 12–15 for details on Fourier series, energy, and numerical techniques for solving Eqs. (4.59). Figure 4.28 gives critical buckling stresses for plates with various boundary conditions. More extensive results can be found in Ref. 9.

Postbuckling Strength

The postbuckling behavior of plates is quite different from that of columns. Whereas columns collapse as soon as the critical load is reached, plates can exhibit considerable postbuckling strength. As the deflection of the plate after buckling increases, the plate stiffness also increases. This behavior is due to the development of transverse tensile stresses subsequent to the start of buckling. Thus, while plate buckling is clearly an undesired component response, it does not necessarily result in catastrophic failure of the engineering structure.

4.4.4   Shells

A shell is an object that forms a curved surface. Usually, a shell is thin; that is, the thickness of a shell is small compared to its radius of curvature. It is important to note that this definition of a thin shell is only for classical mechanics analysis of average stresses and deformations. The absolute thickness of a so-called thin shell may be critically important in other areas such as embrittlement of a ductile material caused by thickness and notches or flaws (see Section 4.8.3). Examples of shells include pressure vessels, soap films, and stadium domes. Whereas plates carry load by the development of bending stresses, shells carry load directly. The strength of a shell comes about primarily because of the way it carries load rather than because of the strength of the material of which it is made.

While even very thin shells will have some bending stiffness, the majority of the loads acting on a shell will be carried by stresses acting tangential to the shell's surface. In the classical membrane theory of shells, bending stiffness is neglected completely. Membrane theory is most applicable for shells of revolution such as shells having the shapes of hemispheres, paraboloids, and cylinders. For

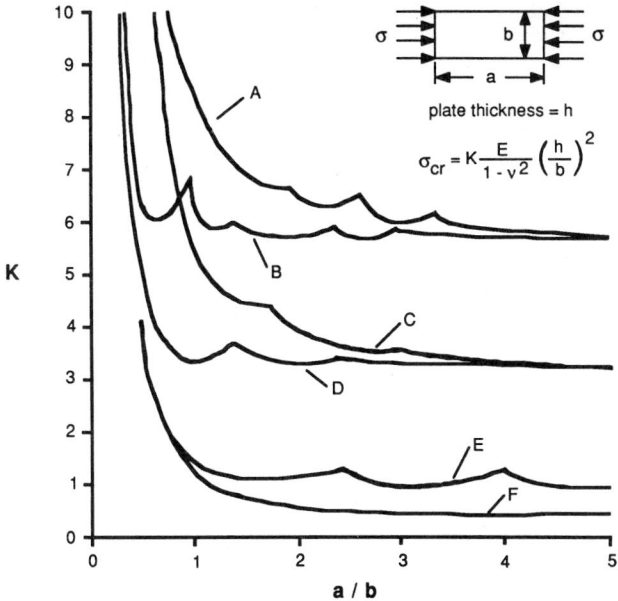

A - All edges clamped
B - Edges a clamped, edges b simply supported
C - Edges a simply supported, edges b clamped
D - All edges simply supported
E - One a edge clamped, one a edge free, b edges simply supported
F - One a edge free, one a edge and b edges simply supported

Fig. 4.28 Critical buckling stress for rectangular plates with various boundary conditions.

shells of these shapes, classical solutions can be obtained.[12, 19] Because of the variety of shell shapes and the loadings imposed on them, it is difficult to compile a useful yet brief table of formulas. One of the more complete listings of results can be found in Ref. 16. Because of the common use of cylindrical shells as pressure vessels, this particular shell type is discussed in detail in Section 4.4.5. The previously mentioned potential problem with absolute thickness is especially important for pressure vessels but cannot be discussed here in detail (see Section 4.8.3 for background on fracture mechanics).

4.4.5   Cylinders

Thin-Walled Cylinders
A cylinder can be considered thin provided its wall thickness is less than one-tenth its radius. Under such conditions, only membrane stresses need to be considered when the cylinder is pressurized internally. In addition, these membrane stresses are essentially constant throughout the wall thickness. If the cylinder is end capped and pressurized with an internal pressure, p, the longitudinal stress is given by

$$\sigma_l = \frac{pr}{2t} \tag{4.60}$$

where r is the inner radius of the cylinder, and t is the wall thickness. Similarly, the transverse stress (hoop stress) is given by

$$\sigma_h = \frac{pr}{t} \tag{4.61}$$

For a thin-walled spherical pressure vessel, the tensile stress in any direction is given by Eq. (4.60).
Equations (4.60) and (4.61) are derived easily by considering the equilibrium forces acting on a cross section of the pressure vessel.

Thick-Walled Cylinders

Thick-walled cylinders are used in many industrial fields. Examples include pressure vessels, gun tubes, and pipes. When the wall thickness is a significant fraction of the cylinder's radius, the tangential stresses can no longer be considered constant through the wall thickness.

For an end-capped cylinder, the stress distribution near the ends can become quite complex due to the effects of welds, joints, localized yielding, and so on. Thus, the stresses near the ends are commonly determined using experimental techniques. For sections far removed from the ends, however, analytical solutions are available. Assuming linear elastic isotropic response, the stresses are given by (Ref. 7, Chapter 3):

$$\sigma_l = \frac{p_1 r_1^2 - p_2 r_2^2}{\left(r_2^2 - r_1^2\right)} + \frac{F}{\pi\left(r_2^2 - r_1^2\right)} \tag{4.62a}$$

$$\sigma_h = \frac{p_1 r_1^2 - p_2 r_2^2}{\left(r_2^2 - r_1^2\right)} + \frac{r_1^2 r_2^2}{r^2\left(r_2^2 - r_1^2\right)}(p_1 - p_2) \tag{4.62b}$$

$$\sigma_r = \frac{p_1 r_1^2 - p_2 r_2^2}{r_2^2 - r_1^2} - \frac{r_1^2 r_2^2}{r^2\left(r_2^2 - r_1^2\right)}(p_1 - p_2) \tag{4.62c}$$

where σ_l = longitudinal stress
  σ_h = hoop stress
  σ_r = radial stress
  p_1 = internal pressure
  p_2 = external pressure
  r_1 = inner radius of cylinder
  r_2 = outer radius of cylinder
  r = radial distance to point under consideration
  F = longitudinal load applied to ends of cylinder

If temperature gradients exist through the thickness of the cylinder, the equations for the stresses are more complex (Ref. 8, Chapter 11).

Failure Criteria

For thin-walled pressure vessels made of a ductile material, the maximum shear stress criterion or maximum distortion energy criterion is applicable (see Section 4.8.2). If a brittle material is used, or if the vessel is thick walled, it is important to consider the possibility of failure due to unstable crack propagation. Fracture-mechanics-based design should be used in such cases (see Section 4.8.3). If the vessel will be under considerable cyclic loading, then fatigue failure may occur (see Section 4.8.4). An extensive set of rules for the design of pressure vessels and piping has been prepared by the American Society of Mechanical Engineers.[20]

Compound Cylinders under Internal Pressure

It is apparent from Eqs. (4.62) that all the stresses cannot be made arbitrarily small by making the outer radius of the cylinder arbitrarily large. In particular, the hoop stress along the inner radius ($r = r_1$) is always greater than the magnitude of the internal pressure, $\sigma_h > p_1$, and the radial stress is always less than $-p_1$, $\sigma_h < -p_1$. Consequently, the maximum shear stress is always larger than p_1. This situation limits the allowable internal pressure. In order to make more efficient use of material, it is common practice to prestress the cylinder. This can be done by shrink-fitting a larger cylinder over the original or prestressing the cylinder to high enough pressures so that plastic flow occurs.[7] Only the former technique will be discussed here.

In order to shrink-fit cylinders, the inner diameter of the larger cylinder must be slightly smaller than the outer diameter of the smaller cylinder. While at the same temperature, the smaller cylinder will not fit inside the larger. If the larger cylinder is heated, however, it will expand in size to the point where the smaller cylinder will fit inside. When the assembly is then brought back to room temperature, a compressive residual hoop stress will exist along the inner radius of the smaller cylinder. This compressive residual stress helps to counteract the tensile hoop stress that develops when internal pressure is applied to the assembly. Details of the procedures to be used for determining the final stress distribution and optimum design of compound cylinders can be found in Refs. 7 and 8.

TABLE 4.7 Stresses and Contact Area for Two Surfaces in Contact

Surface Types	Maximum compressive stress, σ (psi)	Radius, r, or width, b, of contact area (in.)
sphere on a sphere	$\sigma = 0.616 \left[PE^2 \left(\frac{D_1 + D_2}{D_1 D_2} \right)^2 \right]^{\frac{1}{3}}$	$r = 0.881 \left[\frac{P}{E} \left(\frac{D_1 D_2}{D_1 + D_2} \right) \right]^{\frac{1}{3}}$
sphere in a spherical socket	$\sigma = 0.616 \left[PE^2 \left(\frac{D_1 - D_2}{D_1 D_2} \right)^2 \right]^{\frac{1}{3}}$	$r = 0.881 \left[\frac{P}{E} \left(\frac{D_1 D_2}{D_1 - D_2} \right) \right]^{\frac{1}{3}}$
sphere on a plate	$\sigma = 0.616 \left[\frac{PE^2}{D^2} \right]^{\frac{1}{3}}$	$r = 0.881 \left[\frac{PD}{E} \right]^{\frac{1}{3}}$
cylinder on a cylinder	$\sigma = 0.591 \sqrt{pE \left(\frac{D_1 + D_2}{D_1 D_2} \right)}$	$b = 2.15 \sqrt{\frac{p}{E} \left(\frac{D_1 D_2}{D_1 + D_2} \right)}$
cylinder on a plate	$\sigma = 0.591 \sqrt{\frac{pE}{D}}$	$b = 2.15 \sqrt{\frac{pD}{E}}$

Poisson's ratio = 0.3, E = modulus of elasticity (psi), P = load (lb), p = load per unit length (lb/in.)
r, b, D_1, D_2 and D in inches

4.4.6 Contact Stresses

Contact stresses are the stresses that develop when two solid bodies are pressed against each other. Knowledge of the magnitude of these stresses is important in the design of ball bearings, expansion rollers, and so on. The pioneering work in this field was performed by H. Hertz who is better known for his work with radio waves. The contact stress problem is a nonlinear one, and the analysis is difficult. The primary assumptions used in Hertz's theory are as follows:

1. The contacting bodies are linear elastic, homogeneous, and isotropic.
2. The zone of contact is small compared to the dimensions of the bodies.
3. No friction is present.

Hertz's formulas give the maximum compressive stress that occurs in the contact region, but not the maximum shear stress or maximum tensile stress. Table 4.7 gives formulas for the maximum stress and contact radius for some common cases. These formulas are based on the original work of Hertz. References 7–9 give more details on the problem of contact stresses.

4.5   NONLINEAR RESPONSE OF MATERIALS

4.5.1   Introduction to Plasticity

Plastic flow is said to have occurred in a component if the component does not return to its original shape upon removal of the applied loads. Theories of plasticity fall into two major categories: mathematical theories and physical theories. Mathematical theories provide useful, quantitative information on the plastic flow of materials from a macroscopic point of view. Physical theories help one explain why a material plastically flows; this is done by looking at the material's microstructure. From the viewpoint of the design engineer, the phenomenological approach offered by mathematical theories is the more useful of the two approaches.

Idealized Response
The plastic flow of metals is extremely nonlinear as is evident from the monotonic stress-strain curves in Fig. 4.58. Accurate mathematical modeling of such material response is exceedingly difficult. In order to provide useful (but only approximate) predictions, several idealized stress-strain curves are often employed. These are illustrated in Fig. 4.29. The model of rigid, perfectly plastic behavior neglects elastic strains as well as strain-hardening effects. While this model may seem to be overly simplistic, it has proven useful in limit analysis as well as confined metal-forming processes such as extrusion.[3] The model of elastic, perfectly plastic behavior is useful for representing the initial yielding response of mild steels. The model of rigid, linear strain-hardening plastic behavior and the model of elastic, linear strain-hardening plastic behavior are useful for representing large plastic strains in strain-hardening materials.

Bauschinger Effect
Low-carbon steels (mild steels) are among the most commonly used engineering materials. They are in a special class because of their sharp-yielding behavior and related phenomena described in the following.

   If monotonic tension and compression tests are performed on homogeneous specimens of mild steel and the resulting true stress-strain curves are constructed, the two curves look almost identical. In particular, the yield strengths in tension and compression are essentially the same. If a specimen is first loaded in tension into the plastic range and then loaded in compression, the yield strength in compression (σ_{yc}) is generally much lower than the yield strength in tension (σ_{yt}) (Fig. 4.30). This phenomenon is known as the Bauschinger effect. It plays an important role in cyclic plasticity of these metals.

   It is also worthwhile to note that plastic flow is an anisotropic phenomenon. If an initially isotropic material is loaded into the plastic regime and the load is then released, the material will no longer exhibit the same strength properties in all directions. For example, the yield strength of

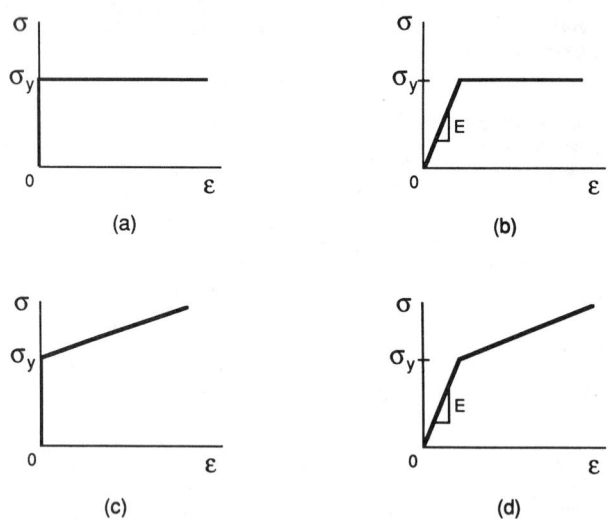

Fig. 4.29   Idealized stress-strain curves: (a) Rigid, perfectly plastic; (b) elastic, perfectly plastic; (c) rigid, linear strain-hardening plastic; (d) elastic, linear strain-hardening plastic.

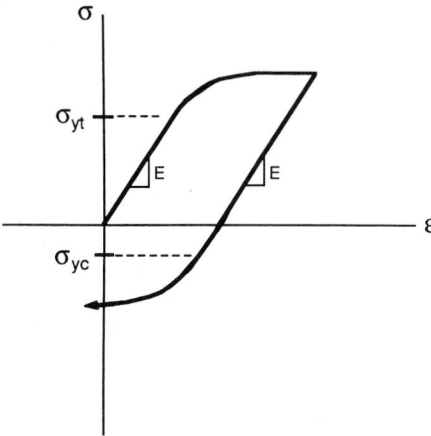

Fig. 4.30 Bauschinger effect.

cold-rolled sheet will be substantially different in the thickness direction from that in the rolling direction.

Effects of Hydrostatic Pressure and Incompressibility
It has been found that, except for extremely large pressures, hydrostatic pressure has a negligible effect on plastic deformations.[21] Many theories of plasticity assume that the hydrostatic component of stress, $\sigma_{xx} + \sigma_{yy} + \sigma_{zz}$, does not influence the yield behavior of the material. Instead of using the stress tensor [Eq. (4.1)], it is often found convenient to work with the deviatoric stress tensor defined by

$$[\sigma_d] = \begin{bmatrix} \sigma_{xx} + p & \sigma_{xy} & \sigma_{xz} \\ \sigma_{yx} & \sigma_{yy} + p & \sigma_{yz} \\ \sigma_{zx} & \sigma_{zy} & \sigma_{zz} + p \end{bmatrix}$$

where $p = -\frac{1}{3}(\sigma_{xx} + \sigma_{yy} + \sigma_{zz})$.

It is also commonly assumed in theories of plasticity that plastic deformation is a constant-volume process. Provided the strains are small enough so that product terms can be neglected, the assumption of incompressibility takes the form $\varepsilon_{xx} + \varepsilon_{yy} + \varepsilon_{zz} = 0$.

Solution procedures for general problems using mathematical theories of plasticity are beyond the scope of this work. Details of such solution procedures can be found in Refs. 2 (Chapter 49) and 22.

Multiaxial Loadings
A variety of theories exist for the yielding of ductile metals under multiaxial loading conditions. The most popular of these are the distortion energy theory and the maximum shear stress theory. Details of these theories are given in Section 4.8.2. Reference 22 provides a comparison of these two theories with experimental data obtained by various researchers.

Limit Analysis
Limit analysis refers to the determination of the magnitude of the load that will cause plastic collapse of a structure. The plastic collapse load is that load at which the increment of deflection per unit increment of load becomes large. This phenomenon is referred to as uncontained plastic flow. In such situations, enough of the engineering component is undergoing plastic flow so that the remaining elastic regions play only a minor role in sustaining the applied loads. Since failure due to plastic collapse is the governing design criterion for many structures, limit analysis has received increasing attention over the years. It represents a more efficient design criterion than, for example, design based on the yield strength. In many structures, of course, other failure modes such as fatigue, brittle fracture, or buckling govern the design. Limit analysis is most useful in the design of beams and rigid-jointed frames made of mild steels.

In order to obtain quantitative results, the following two assumptions are made:

1. The material exhibits elastic, perfectly plastic response.
2. Changes in geometry of the structure for loads less than the limit load are insignificant.

The advantages of limit load analysis, besides making more efficient use of material, is that the limit load can often be obtained through very simple calculations. One of the biggest disadvantages is that the deflections for loads below the plastic collapse load are not determined. For some structures, these deflections may be so large as to govern the design, so that plastic collapse becomes irrelevant. Further details on limit analysis can be found in Refs. 2 and 22.

4.5.2 Plastic Response of Beams and Shafts

When analyzing the plastic response of beams and shafts, it is commonly assumed that the material to be used is elastic-plastic; that is, it behaves in an elastic, perfectly plastic manner (Fig. 4.29b).

Plastic Bending of a Rectangular Beam

To illustrate the analysis procedures for determining plastic response of beams, the case of a beam of rectangular cross section undergoing pure bending is considered. The beam is assumed to have a width b and height $2h$. The moment of inertia is given by $I = b(2h)^3/12 = 2bh^3/3$. The maximum stress, σ_m, in the beam for elastic response occurs at the top and bottom edges ($y = \pm h$) (Fig. 4.31a). Its magnitude is given by Eq. (4.27):

$$|\sigma_m| = \frac{Mh}{I} = \frac{3M}{2bh^2}$$

The bending moment corresponding to the initiation of yield on the top and bottom edges of the beam, M_y, is given by $M_y = (\frac{2}{3})bh^2\sigma_y$, where σ_y is the yield strength of the material.

As the bending moment is further increased, plastic zones develop on the top and bottom of the beam (Fig. 4.31b). The stress is constant in these zones with a value of σ_y. Within the remaining elastic core, the stresses have the form

$$\sigma_x = -\frac{y}{y_e}\sigma_y$$

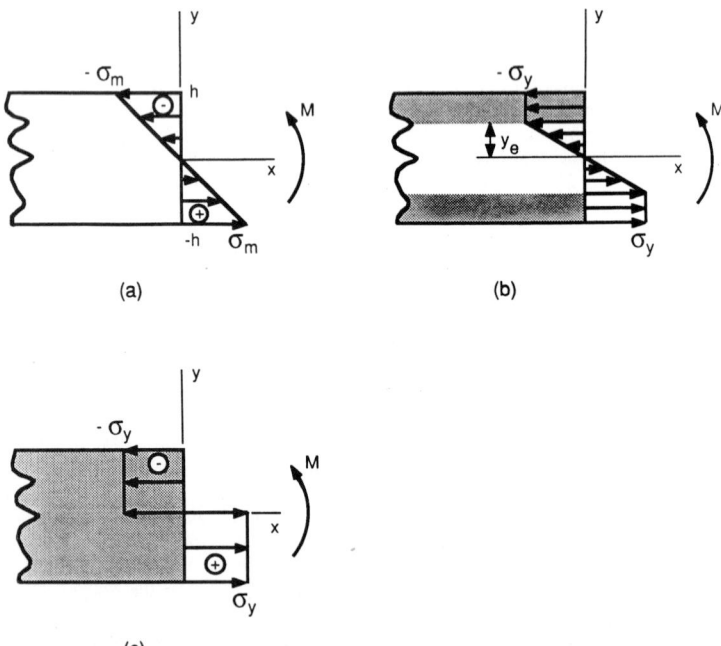

(a)   (b)

(c)

Fig. 4.31 Plastic deformations in a beam undergoing bending: (*a*) Elastic response; (*b*) elastic-plastic response; (*c*) fully plastic response.

where y_e is half the height of the elastic core. The corresponding moment is given by

$$M = -2\int_0^{y_e} y\left(-\frac{y}{y_e}\sigma_y\right)(b\,dy) - 2\int_{y_e}^h y(-\sigma_y)(b\,dy)$$

Integrating this expression and simplifying it leads to

$$M = \frac{3}{2}M_y\left(1 - \frac{1}{3}\frac{y_e^2}{h^2}\right) \tag{4.63}$$

As y_e approaches zero, the beam becomes fully plastic, and plastic collapse is imminent (Fig. 4.31c). The bending moment for fully plastic collapse, M_p, is found by substituting $y_e = 0$ into Eq. (4.63): $M_p = 1.5M_y = bh^2\sigma_y$. By using limit analysis, this result can be found directly:

$$M_p = -2\int_0^h(-y\sigma_y)(b\,dy) = bh^2\sigma_y$$

Similar expressions can be derived for the limit loads of other kinds of beam loadings. It is often convenient to introduce the plastic section modulus, Z, which satisfies the following equation: $M_p = \sigma_y Z$. For the rectangular cross section just considered, $Z = bh^2$. Plastic section moduli for common steel shapes can be found in the *Manual of Steel Construction*.[23]

Plastic Deformations in a Circular Shaft Undergoing Torsion

The circular shaft is assumed to be solid with a radius c. The polar moment of inertia of the cross section is $J = 0.5\pi c^4$. The maximum shear stress for elastic response, τ_m, occurs on the outer surface of the shaft (Fig. 4.32a). Its value is given by Eq. (4.41):

$$\tau_m = \frac{Tc}{J} = \frac{2T}{\pi c^3}$$

Denoting the torque that will initiate yielding by T_y, it follows that

$$T_y = \frac{\pi}{2}c^3\tau_y$$

where τ_y is the yield strength in shear of the material.

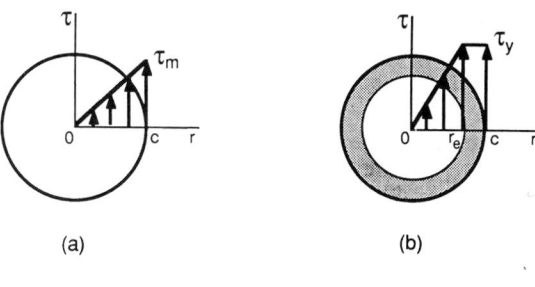

(a)   (b)

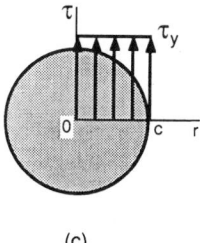

(c)

Fig. 4.32 Plastic deformations in a circular shaft undergoing torsion: (*a*) Elastic response; (*b*) elastic-plastic response; (*c*) fully plastic response.

As the torque is further increased, a plastic zone develops in the outer regions of the shaft (Fig. 4.32b). This plastic zone surrounds an elastic core. Within the plastic zone the shear stress is given by τ_y. In the elastic core, the shear stress has the form $\tau = (r/r_e)\tau_y$, where r is the radial distance to the point under consideration, and r_e is the radial distance to the edge of the elastic core. The corresponding torque can be found with the aid of Eq. (4.40):

$$T = \int_0^{r_e} r\left(\frac{r}{r_e}\tau_y\right)(2\pi r\, dr) + \int_{r_e}^c r(\tau_y)(2\pi r\, dr)$$

Integrating and simplifying this equation leads to

$$T = \frac{4}{3}T_y\left(1 - \frac{1}{4}\frac{r_e^3}{c^3}\right) \tag{4.64}$$

The torque corresponding to fully plastic deformation, T_p, is found by setting r_e equal to zero in Eq. (4.64):

$$T_p = \tfrac{4}{3}T_y = \tfrac{2}{3}\pi c^3\tau_y$$

Using limit analysis, this result can be obtained directly from Eq. (4.40) by setting $\tau = \tau_y$:

$$T_p = \int_0^c r(\tau_y)(2\pi r\, dr) = \frac{2}{3}\pi c^3\tau_y$$

4.5.3   Viscoelasticity

Viscoelasticity is the study of the time-dependent response of solid materials. Time, rate, and temperature effects must be considered when analyzing the viscoelastic response of materials. In particular, it is useful to know the homologous temperature T_h. This is defined as the ratio of the operating temperature, T, to the melting temperature of the material, T_m: $T_h = T/T_m$. Both T and T_m must be expressed as absolute temperatures. For U.S. customary units, absolute temperature is measured in Rankines (°R = °F + 459.67). If T_h is below 0.3, viscoelastic effects can usually be neglected. If T_h is greater than 0.5, viscoelastic phenomena will almost certainly play a role in the mechanical response of the material.

Creep and the Stress Rupture Test
Creep refers to the time-dependent accumulation of strain in a component subjected to a constant stress. In a typical creep test, a slender specimen is subjected to a constant load rather than a constant stress. A schematic diagram of a creep curve that results from a constant-load test is shown in Fig. 4.33. The creep process can usually be divided into three stages. During primary creep, the creep rate is a decreasing function of time. This response is brought about due to strain-hardening effects. During secondary creep, also called steady-state creep, the creep rate is constant. During this stage, strain-hardening effects are balanced by the effects of increasing true stress amplitude due to decreasing cross-sectional area. During tertiary creep, the decreasing cross-sectional area becomes the controlling feature, and the creep rate rises rapidly until failure occurs. This response is somewhat

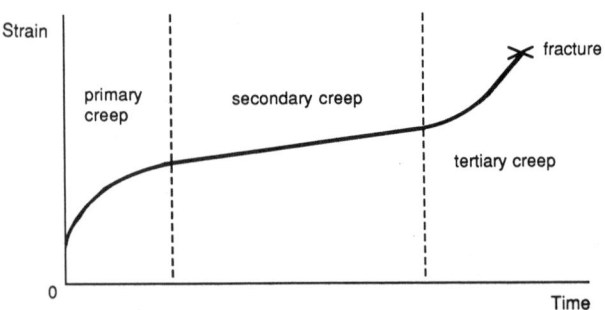

Fig. 4.33   Schematic diagram of a typical creep curve.

analogous to the decrease in load amplitude during a tension test due to the necking down of the specimen (see Section 4.1.4).

It is possible, although difficult, to perform a creep test in which the true stress is maintained constant rather than the load. For such tests, tertiary creep is not observed. This indicates that tertiary creep is indeed due to a mechanical instability rather than representing actual material response. It is often found that the strain at which tertiary creep begins in a constant-load creep test is roughly the same as the strain at which necking down begins in a tensile test. Even secondary creep is probably a misconception. For carefully controlled constant-true-stress creep tests, the creep rate is usually found to continuously decrease with time.[24]

In spite of the aforementioned problems, constant-load creep testing, commonly referred to as stress rupture testing, is still the most common test method used to determine the viscoelastic properties of a material. In a typical testing series, the time to rupture is determined for a range of loads and temperatures. The data are then plotted as the engineering stress versus time to rupture on log-log coordinates.

Stress Relaxation Tests

In a stress relaxation test, the strain is held constant, and the decrease in stress with time is recorded. The advantage of a stress relaxation test over a stress rupture test is that no mechanical instability arises. Thus, the stress relaxation test gives more representative material response than the stress rupture test. The primary disadvantage is that the stress must be continuously adjusted so as to keep the strain constant. This requires the use of sophisticated mechanical testing equipment. A schematic of the typical stress versus time response observed in a stress relaxation test is shown in Fig. 4.34. Unlike the three-stage response of a stress rupture test, a stress relaxation test shows only one general response: a continuing decrease in the stress relaxation rate with time.

Linear Viscoelastic Models

Viscoelastic response is sometimes modeled by using combinations of springs and dashpots. Two of the most commonly used of these models are the Maxwell model and the Kelvin–Voigt model (Fig. 4.35). In this figure, E is the stiffness of the spring, and η is the viscosity coefficient of the dashpot.

The Maxwell model consists of a spring and dashpot in series. This model predicts a linear creep rate (which is representative of secondary creep). The model predicts exponential decay of stress under constant strain conditions. This is the kind of response typically observed in stress relaxation tests of metals.

The Kelvin–Voigt model consists of a spring and dashpot in parallel. The creep response of this model is indicative of primary creep. This model does not exhibit stress relaxation, however.

It is clear that neither the Maxwell model nor the Kelvin–Voigt model is capable of completely describing the creep and stress relaxation response of common materials. Even so, each model is applicable for certain stages of the creep process. These models are not necessarily useful for their ability to accurately predict material response, but rather for the simple visualization of viscoelastic response that they provide. More sophisticated spring-dashpot models are sometimes used. A good discussion of these models can be found in Ref. 25.

In the continuum mechanics approach to viscoelasticity, time-dependent response is modeled by using constitutive equations containing convolution integrals. The stress response, $\sigma(t)$, for an

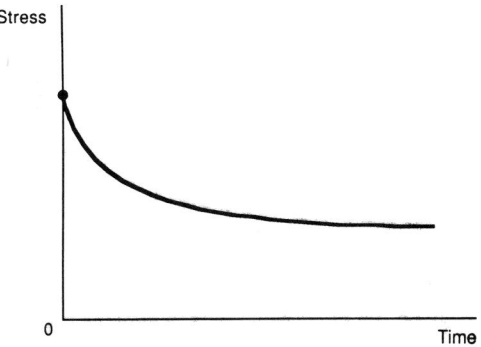

Fig. 4.34 Stress versus time for a stress relaxation test.

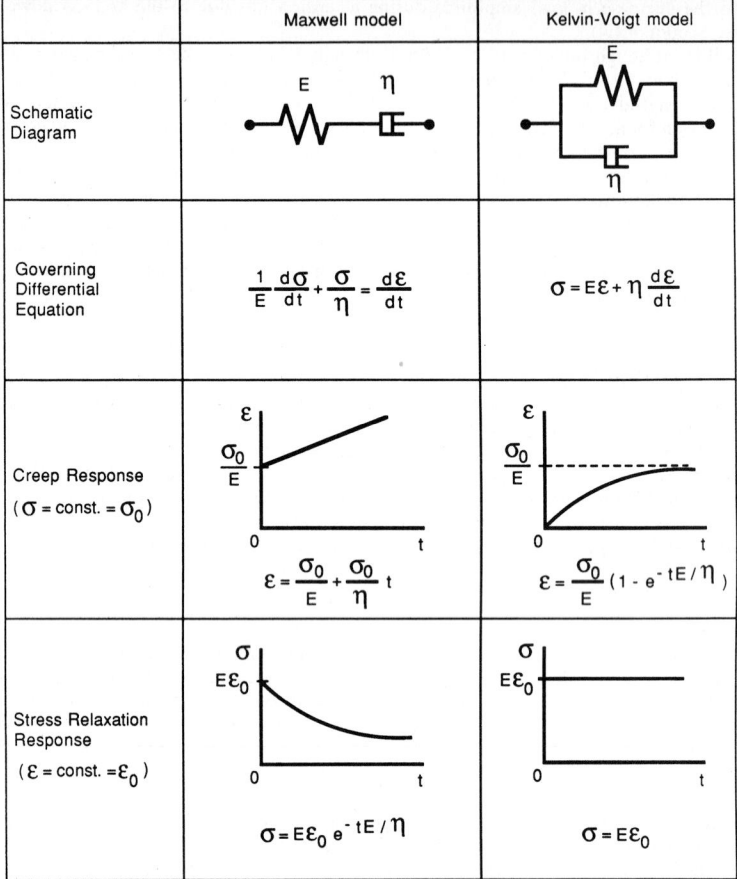

	Maxwell model	Kelvin-Voigt model
Schematic Diagram	E η	E η
Governing Differential Equation	$\dfrac{1}{E}\dfrac{d\sigma}{dt} + \dfrac{\sigma}{\eta} = \dfrac{d\varepsilon}{dt}$	$\sigma = E\varepsilon + \eta\dfrac{d\varepsilon}{dt}$
Creep Response $(\sigma = \text{const.} = \sigma_0)$	$\varepsilon = \dfrac{\sigma_0}{E} + \dfrac{\sigma_0}{\eta}\,t$	$\varepsilon = \dfrac{\sigma_0}{E}(1 - e^{-tE/\eta})$
Stress Relaxation Response $(\varepsilon = \text{const.} = \varepsilon_0)$	$\sigma = E\varepsilon_0\, e^{-tE/\eta}$	$\sigma = E\varepsilon_0$

Fig. 4.35   Models for viscoelastic response.

arbitrary strain input, $\varepsilon(t)$, has the form

$$\sigma(t) = \varepsilon(0)G(t) + \int_0^t G(t - \tau)\frac{d\varepsilon(\tau)}{d\tau}\,d\tau$$

where t represents time, $\varepsilon(0)$ is the strain corresponding to $t = 0$, and $G(t)$ is the relaxation modulus function (which is often a complicated function of time).[26] Similarly, the strain response due to an arbitrary stress input is given by

$$\varepsilon(t) = \sigma(0)J(t) + \int_0^t J(t - \tau)\frac{d\sigma(\tau)}{d\tau}\,d\tau$$

where $\sigma(0)$ is the stress corresponding to $t = 0$, and $J(t)$ is the creep compliance function. It can be shown that $G(t)$ and $J(t)$ satisfy

$$\int_0^t G(t - \tau)\frac{dJ(\tau)}{d\tau}\,d\tau = h(t)$$

where $h(t)$ is the unit step function. Thus, determination of $G(t)$ using a stress relaxation test automatically prescribes $J(t)$. Similarly, determining $J(t)$ using a constant-true-stress creep test automatically prescribes $G(t)$. Therefore, the continuum mechanics approach to viscoelasticity rests on the assumption that creep and stress relaxation are simply different manifestations of the same material response. While the continuum mechanics approach to viscoelasticity can lead to powerful methods of analysis, these methods are generally exceedingly difficult to implement in practice.

4.5.4 Elastic Stability and Column Buckling

When a slender column is subjected to small compressive loads, the column axially shortens according to $\delta = PL/AE$, where δ is the axial shortening, P is the applied load, A is the cross-sectional area, E is the elastic modulus of the material, and L is the column length. If continually larger loads are applied, a load is reached at which the column suddenly bows out sideways. This load is referred to as the critical or buckling load of the column. These sideways deformations are normally too large to be acceptable; consequently, the column is considered to have failed. For slender columns, the axial stress corresponding to the critical load is generally below the yield strength of the material. Since the stresses in the column just prior to buckling are within the elastic range, the failure is referred to as elastic buckling. Of course, once the column has buckled, large sideways deformations may cause some plastic flow to occur. The term elastic stability is commonly used to designate the study of elastic buckling problems. For short columns, failure may be governed by yielding or rupture of the column while it is still axially straight. Failure of short columns may also be caused by inelastic buckling; that is, large sideways deformations that occur when the nominal axial stress is greater than the yield strength.

Besides columns, other structural components that are prone to buckling include plates, shells, and frames. The discussion here will be limited to columns. Buckling of plates is described in Section 4.4.3. Buckling of frames and shells, as well as torsional buckling phenomena, are discussed in Ref. 27.

Theory of Column Buckling

The first studies of the elastic buckling of columns were performed by Euler in the mid-eighteenth century. The governing differential equation for the elastic stability of a pinned-pinned column is given by (see Fig. 4.36):

$$y'' = \frac{d^2y}{dx^2} = \frac{M}{EI} \tag{4.65}$$

where E is the elastic modulus, M is the moment acting on the cross section, and I is the moment of inertia of the cross section. For this equation to govern the buckling of the column, I must be the minimum moment of inertia of the cross section. Also, in using this equation, one assumes linear elastic, isotropic response. From statics, $M = -Py$. Equation (4.65) thus becomes

$$EIy'' + Py = 0 \tag{4.66}$$

This is a second-order, homogeneous, linear differential equation with constant coefficients. Its

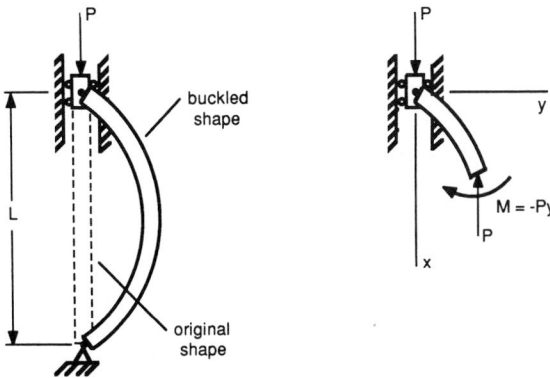

(a) Original and buckled shapes   (b) Static equilibrium of buckled element

Fig. 4.36 Pinned-pinned Euler column.

general solution takes the form

$$y = A \sin \lambda x + B \cos \lambda x \tag{4.67}$$

where $\lambda^2 = P/EI$, and A and B are constants.

Four boundary conditions can be applied to solve for A and B in Eq. (4.67). Two of these conditions are

$$y'' = 0 \quad \text{at } x = 0$$
$$y'' = 0 \quad \text{at } x = L$$

where the primes denote differentiation with respect to x. These conditions are known as natural boundary conditions. They result from the fact that the moments at the pin connections must be zero. Two other boundary conditions are

$$y = 0 \quad \text{at } x = 0$$
$$y = 0 \quad \text{at } x = L$$

These are known as forced boundary conditions. They result from the geometric constraints that exist. Using either the forced or the natural boundary conditions results in the same two possible solutions:

$$A = 0 \quad \text{and} \quad B = 0$$

or

$$\sin \lambda L = 0 \quad \text{and} \quad B = 0$$

The first of these solutions ($A = 0$ and $B = 0$) is known as the trivial solution. It represents the condition that the beam is still straight. The second solution leads to

$$\lambda L = n\pi \quad n = 0, 1, 2, 3, \ldots$$

Using $\lambda^2 = P/EI$,

$$P = \frac{n^2 \pi^2 EI}{L^2} \quad n = 0, 1, 2, 3, \ldots$$

Clearly, $n = 0$ corresponds to no load being applied to the column. The minimum load for which buckling will occur corresponds to $n = 1$. This critical buckling load, P_{cr}, is given by

$$P_{cr} = \frac{\pi^2 EI}{L^2} \tag{4.68}$$

Equation (4.68) is often referred to as Euler's formula. If I is expressed in terms of the radius of gyration and area of the cross section, $I = r^2 A$, Eq. (4.68) can be expressed in terms of the critical buckling stress:

$$\sigma_{cr} = \frac{\pi^2 E}{(L/r)^2}$$

The quantity L/r is referred to as the slenderness ratio of the column. The deflected shape is found from Eq. (4.67): $y = A \sin(\pi x/L)$. The deflected shape is hence sinusoidal in nature. Since the constant A cannot be prescribed, the magnitude of the deflected shape is indeterminate using this approach.

Equation (4.66) is valid only for a pinned-pinned column. For other end conditions, a different second-order differential equation applies. It has been shown by Timoshenko and Gere[28] that a single fourth-order differential equation applicable to all end conditions can be employed. This equation has the form

$$y^{iv} + \lambda^2 y'' = 0 \tag{4.69}$$

This differential equation has the general solution

$$y = A \sin \lambda x + B \cos \lambda x + Cx + D \tag{4.70}$$

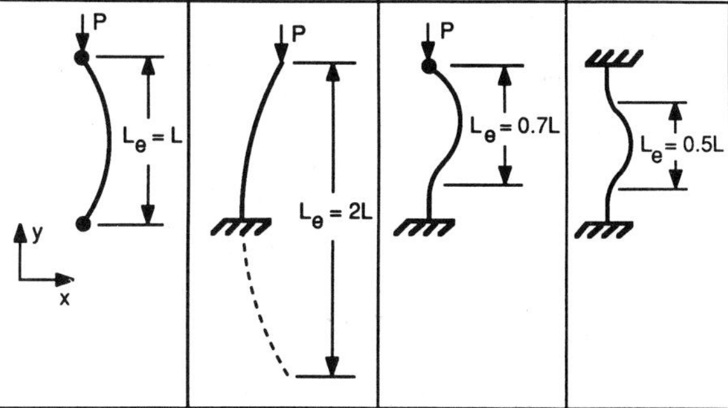

Column type:	Pinned-pinned	Fixed - free	Fixed-pinned	Fixed-fixed
Effective length:	$L_e = L$	$L_e = 2L$	$L_e = 0.7L$	$L_e = 0.5L$
Boundary Conditions:	$x = 0,\ y = 0$ $x = L,\ y = 0$ $x = 0,\ y'' = 0$ $x = L,\ y'' = 0$	$x = 0,\ y = 0$ $x = 0,\ y' = 0$ $x = L,\ y'' = 0$ $x = L,\ y''' + k^2 y' = 0$	$x = 0,\ y = 0$ $x = 0,\ y' = 0$ $x = L,\ y = 0$ $x = L,\ y'' = 0$	$x = 0,\ y = 0$ $x = 0,\ y' = 0$ $x = L,\ y = 0$ $x = L,\ y' = 0$

Fig. 4.37 Effective lengths and boundary conditions for various columns.

where A, B, C, and D are constants. Application of the four available boundary conditions allows for determination of P_{cr} and three of the constants. One of the constants (either A or B) is indeterminate. Thus, the magnitude of the deflected shape cannot be determined using this classical technique either.

Effective Column Length
For end conditions other than pinned-pinned, it is common to express the Euler buckling load in the form

$$P_{cr} = \frac{\pi^2 EI}{L_e^2}$$

where L_e is the effective column length. Figure 4.37 provides effective column lengths and mathematical boundary conditions for a variety of end conditions. The mathematical boundary conditions are used in conjunction with Eq. (4.70) to obtain L_e. See Ref. 27 for further details.

Imperfect Columns
In reality, columns are never perfectly straight. Instead, they have some initial eccentricity, often denoted by e, as shown in exaggerated form in Fig. 4.38. If e is known or can be estimated, the

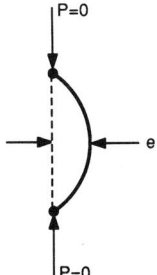

Fig. 4.38 Initially eccentric column.

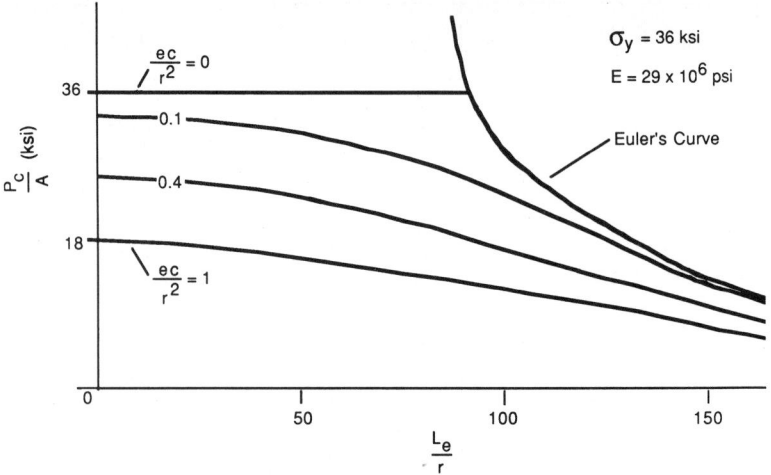

Fig. 4.39 Comparison of Euler and secant formulas.

maximum deflection as a function of load can be obtained as

$$y_{max} = e\left[\sec\left(\frac{L}{2}\sqrt{\frac{P}{EI}}\right) - 1\right]$$

The maximum compressive stress is given by

$$\sigma_{max} = \frac{P}{A}\left[1 + \frac{ec}{r^2}\sec\left(\frac{L}{2}\sqrt{\frac{P}{EI}}\right)\right] \tag{4.71}$$

where c is the distance from the neutral axis to the outermost compressed fiber of the cross section. All other symbols have been previously defined. Equation (4.71) is referred to as the secant formula. It gives the maximum stress that exists in a column for a given applied load (assuming elastic response throughout the column). For design purposes, the maximum stress is set equal to the yield strength (or to the fracture stress for brittle materials), and the corresponding critical load P_{cr} is found by trial and error. The appropriate factor of safety is applied to this load. The factor of safety must not be applied to the stress since the relation between stress and load in Eq. (4.71) is nonlinear. Figure 4.39 gives a comparison of the secant and Euler formulas for a mild steel.

The secant formula can be used for columns that are eccentrically loaded if e is taken as the eccentricity due to the load plus the initial eccentricity of the column.

Inelastic Buckling

For very short columns, the critical stress corresponds to the compressive strength of the material. For very long columns, the Euler critical stress is applicable. For intermediate length columns, failure often occurs at stresses below that predicted by the Euler critical stress or the compressive strength. The effects of residual stresses, especially in rolled carbon steel columns, play an important role in the buckling behavior of intermediate length columns. The tangent modulus theory is sometimes useful in dealing with intermediate length columns. In this theory the elastic modulus in Euler's formula, E, is replaced by the tangent modulus, E_T. The tangent modulus corresponding to a given stress is the slope of the stress-strain curve at that stress level. For elastic response, E and E_T are identical. In equation form, the tangent modulus theory is given by

$$P_{cr} = \frac{\pi^2 E_T I}{L_e^2} \tag{4.72}$$

where E_T is the slope of the stress-strain curve at the critical stress. Equation (4.72) is sometimes referred to as the Engesser equation. It is valid only for a rectangular cross section bent about its weak axis. Because use of the tangent modulus theory requires a detailed knowledge of a material's stress-strain curve, it is difficult to implement in practice. Simplified column formulas are more commonly used.

Simplified Column Formulas
Because of the complexities of the tangent modulus theory, a number of simpler formulas have been proposed:

Parabolic:
$$\frac{P_c}{A} = \sigma_0 - C\left(\frac{L_e}{r}\right)^2 \tag{4.73}$$

Gordon–Rankine:
$$\frac{P_c}{A} = \frac{\sigma_0}{1 + C\left(\dfrac{L_e}{r}\right)^2} \tag{4.74}$$

Straight Line:
$$\frac{P_c}{A} = \sigma_0 - C\left(\frac{L_e}{r}\right) \tag{4.75}$$

where P_c/A is the average buckling stress, σ_0 is the compressive strength of the material, C is a material constant, L_e is the effective column length, and r is the radius of gyration of the cross section. The constants σ_0 and C may not be the same in each formula, even for the same material, since they are adjusted to provide a best fit to experimental data. Equations (4.73) through (4.75) are displayed graphically in Fig. 4.40.

Steel Columns
Because of the variation in cooling rates of different regions of a hot rolled wide-flange member, large compressive residual stresses can develop. Cold rolling of steel members can also result in significant residual stresses. These residual stresses must be allowed for (by appropriate factors of safety) when designing against column buckling. The current AISC specification for buckling of steel columns is as follows.[29]

For slenderness ratios less than C_c, where $C_c = (2\pi^2 E/\sigma_y)^{0.5}$, the column buckling formula is given by

$$\sigma_a = \frac{\sigma_y\left\{1 - \left[(1/2C_c^2)(L_e/r)^2\right]\right\}}{FS} \tag{4.76}$$

where σ_a = allowable compressive stress
L_e/r = slenderness ratio
$FS = \frac{5}{3} + 3(L_e/r)/8C_c - (L_e/r)^3/8C_c^3$

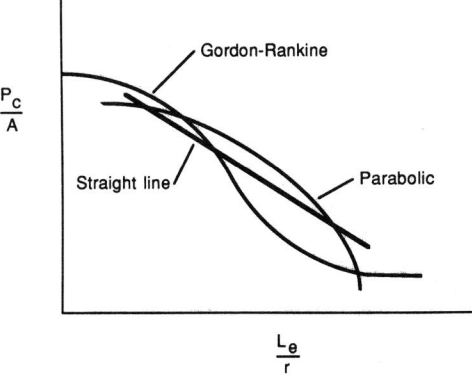

Fig. 4.40 Simplified column formulas.

For slenderness ratios greater than C_c, the Euler equation is used with a built-in factor of safety of 1.92:

$$\sigma_a = \frac{\pi^2 E}{1.92(L_e/r)^2} \tag{4.77}$$

Aluminum Columns
The Aluminum Association provides three formulas for use with each aluminum alloy.[30] For short columns, the allowable compressive stress is constant with respect to slenderness ratio. For intermediate columns, a straight-line relationship similar to Eq. (4.75) is used. For long columns, a Euler formula is used.

Timber Columns
The American Institute of Timber Construction specifies the use of Euler's formula with a factor of safety of 2.727:

$$\sigma_a = \frac{\pi^2 E}{2.727\left(\dfrac{L_e}{r}\right)^2} \tag{4.78}$$

Of course, the allowable stress must not exceed the allowable compressive strength parallel to the grain.[31]

4.6   ENERGY METHODS

4.6.1   Strain Energy and Strain Energy Density

Axial Loading
Consider the case of a uniform rod on which an axial load P is acting. Assume that the axis of the rod is in the x direction. The work done by the load P in extending the rod from its original length L to a length $L + \Delta L$ is given by

$$U = \int_0^{\Delta L} P \, dx$$

where U is known as the strain energy of the rod. It is normally expressed in ft-lb or in-lb.
    The strain energy density, u, is defined as the strain energy per unit volume. For the case of an axially loaded uniform rod,

$$u = \frac{U}{V} = \frac{1}{AL}\int_0^{\Delta L} P \, dx = \int_0^\epsilon \sigma_{xx} \, d\epsilon_{xx} \tag{4.79}$$

where V is the volume of the rod, and A is its cross-sectional area. It is apparent from Eq. (4.79) that the strain energy density for the case of axial loading of a rod represents the area under the stress-strain diagram. Strain energy density has units of in-lb/in^3.
    For linear elastic, isotropic response, $\epsilon_{xx} = \sigma_{xx}/E$, where E is the elastic modulus. Equation (4.79) becomes

$$u = \frac{\sigma_{xx}^2}{2E} \tag{4.80}$$

The total strain energy of the uniform rod is

$$U = uV = \frac{\sigma_{xx}^2}{2E}(AL) = \frac{P^2 L}{2AE}$$

If the cross section of the rod varies along the length of the rod, the total strain energy must be found by integration of Eq. (4.80):

$$U = \int \frac{\sigma_{xx}^2}{2E} \, dV \tag{4.81}$$

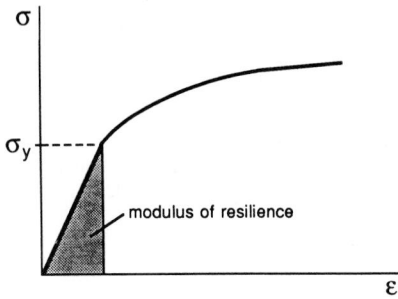

Fig. 4.41 Definition of modulus of resilience.

It should be kept in mind that Eqs. (4.80) and (4.81) apply to a linear elastic, isotropic rod undergoing axial loading.

The modulus of resilience, u_y, represents the area under the stress-strain diagram up to the yield strength (Fig. 4.41). Assuming that the yield strength and proportional limit coincide, u_y can be written as

$$u_y = \frac{\sigma_y^2}{2E} \tag{4.82}$$

where σ_y is the yield strength of the material. The modulus of resilience is important because it represents the energy per unit volume that a material can absorb without yielding.

General State of Loading
Assuming linear elastic, isotropic response, the strain energy density can be expressed in several equivalent forms,

$$u = \frac{1}{2}\left\{\sigma_{xx}\varepsilon_{xx} + \sigma_{yy}\varepsilon_{yy} + \sigma_{zz}\varepsilon_{zz} + \sigma_{xy}\gamma_{xy} + \sigma_{yz}\gamma_{yz} + \sigma_{zx}\gamma_{zx}\right\} \tag{4.83a}$$

$$u = \frac{1}{2E}\left\{\sigma_{xx}^2 + \sigma_{yy}^2 + \sigma_{zz}^2 - 2\nu\left(\sigma_{xx}\sigma_{yy} + \sigma_{yy}\sigma_{zz} + \sigma_{zz}\sigma_{xx}\right)\right.$$
$$\left. + 2(1+\nu)\left(\sigma_{xy}^2 + \sigma_{yz}^2 + \sigma_{zx}^2\right)\right\} \tag{4.83b}$$

$$u = G\left\{\varepsilon_{xx}^2 + \varepsilon_{yy}^2 + \varepsilon_{zz}^2 + \frac{\nu e^2}{1-2\nu} + \frac{1}{2}\left(\gamma_{xy}^2 + \gamma_{yz}^2 + \gamma_{zx}^2\right)\right\} \tag{4.83c}$$

where $e = \varepsilon_{xx} + \varepsilon_{yy} + \varepsilon_{zz}$. The total strain energy of a component is found by integrating any of Eqs. (4.83) over the volume of the component.

Strain Energy in Bending
For the case of a beam in pure bending, the flexural stress is given by $\sigma_x = -My/I$, where M is the applied moment, I is the moment of inertia of the cross section, and y is the distance from the neutral axis to the point under consideration (see Section 4.2.2 for details). All other stresses are zero. Equation (4.83b) becomes

$$u = \frac{M^2 y^2}{2EI^2}$$

and thus

$$U = \int_0^L \frac{M^2}{2EI^2}\left(\int y^2\,dA\right)dx$$

where L is the length of the beam, and dA represents an infinitesimal element of the cross-sectional area. Noting that $I = \int y^2 \, dA$,

$$U = \int_0^L \frac{M^2}{2EI} \, dx \tag{4.84}$$

Strain Energy Due to Transverse Shear

For a beam subjected to transverse loads, the total strain energy consists of strain energy due to bending [as given by Eq. (4.84)] and strain energy due to shear as given by

$$U = \int_0^L \frac{kV(x)^2}{2GA} \, dx \tag{4.85}$$

where A is the cross-sectional area, $V(x)$ is the shear force at a distance x, and K is a constant that depends on the cross section of the beam. Values of K for various cross sections are as follows[7]:

Cross Section	K
Rectangular	1.20
Circular	≈ 1.11
Thin-walled cylinder	2.00
I section	1.20
Closed, thin-walled box section	1.00

For the I section, A should be the area of the flanges. For the closed, thin-walled box section, A should be the area of the webs.

Strain Energy in Torsion

For the case of a circular rod of length L undergoing torsion, the strain energy can be expressed in the form

$$U = \int_0^L \frac{T^2}{2JG} \, dx \tag{4.86}$$

where T is the applied torque, G is the shear modulus, and J is the polar moment of inertia of the cross section. If the cross section is uniform along the length of the rod, Eq. (4.86) reduces to

$$U = \frac{T^2 L}{2JG}$$

Example 4.6. Determine the strain energy of a cantilever beam of length L consisting of a uniform rectangular cross section (width = b, height = h), taking into account both shear and bending effects (see Fig. 4.42).

SOLUTION. The moment of inertia is given by $I = bh^3/12$. The moment at a longitudinal distance x from the applied load $(0 \le x \le L)$ takes the form $M = Px$. The shear force is constant along the length of the beam with a value of P. By Eq. (4.84), the strain energy due to bending is

$$U_b = \int_0^L \frac{P^2 x^2}{2E\left(\frac{1}{12} bh^3\right)} \, dx = \frac{2P^2 L^3}{Ebh^3}$$

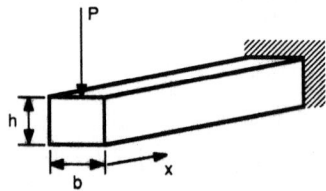

Fig. 4.42   Cantilever beam for use with Examples 4.6–4.9.

From Eq. (4.85), with $K = 1.20$, the strain energy due to shear is

$$U_v = \int_0^L \frac{1.20 P^2}{2Gbh} \, dx = \frac{3P^2 L}{5Gbh}$$

The total strain energy of the beam is thus

$$U = U_b + U_v = U_b \left(1 + \frac{3Eh^2}{10GL^2} \right)$$

For isotropic materials, $E/G < 3$. Thus, provided $h/L < 0.1$, neglecting the strain energy due to shear results in an error of less than 1%. This is why shear effects in long, slender beams are normally ignored.

4.6.2 Castigliano's Theorems

Complementary Energy
In Section 4.6.1, the strain energy density, u, for the case of axial loading was defined to be the area under the stress-strain curve. The complementary energy density, u^*, is defined as the area above the stress-strain diagram (Fig. 4.43). For multiaxial states of stress, the visualization of complementary energy is more difficult. The total complementary energy of a structure, U^*, is given by $U^* = \int u^* \, dV$. See Ref. 14 for details.

Castigliano's First Theorem
Consider linear elastic structures that dissipate no energy and undergo small displacements. Also consider conservative forces, that is, forces that keep their orientations as the structure deforms. Castigliano's first theorem states that the load P_i that corresponds to a displacement D_i is given by the partial derivative of the strain energy with respect to D_i. In equation form,

$$P_i = \frac{\partial U}{\partial D_i} \tag{4.87}$$

Castigliano's Second Theorem
Once again consider linear elastic structures subjected to conservative forces and undergoing small displacements. Castigliano's second theorem states that the displacement D_i due to an applied load P_i equals the partial derivative of the complementary energy with respect to P_i. In equation form,

$$D_i = \frac{\partial U^*}{\partial P_i} \tag{4.88}$$

For linear elastic response, $U = U^*$, so that U and U^* can be interchanged in Eqs. (4.87) and (4.88).

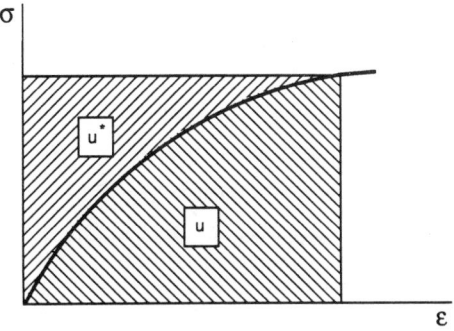

Fig. 4.43 Definitions of strain energy density and complementary energy density.

It can be shown that a similar relationship holds for moments and angular displacements. In equation form,

$$\theta_i = \frac{\partial U^*}{\partial M_i} \tag{4.89}$$

where θ_i is the angular displacement at point i due to the applied moment M_i.

Example 4.7. Using Castigliano's second theorem, determine the end displacement of a cantilever beam loaded at its end. Neglect shear effects.

SOLUTION. From Example 4.6, the strain energy is given by

$$U = U_b = \frac{P^2 L^3}{6EI}$$

For linear elastic response, $U = U^*$, thus

$$\Delta = \frac{\partial U}{\partial P} = \frac{PL^3}{3EI}$$

This answer agrees with the results obtained using beam deflection theory (Section 4.2.4).

Example 4.8. For the beam of Example 4.7, determine the end slope.

SOLUTION. Since no end moment is applied, a fictitious moment, M, is created. Its value will be set equal to zero after differentiating the strain energy expression. The strain energy due to P and M is given by

$$U = \int_0^L \frac{(M + Px)^2}{2EI} \, dx$$

Differentiating this expression with respect to M gives

$$\theta = \frac{\partial U}{\partial M} = \int_0^L \frac{2(M + Px)}{2EI} \, dx$$

Setting $M = 0$ and integrating leads to

$$\theta = \int_0^L \frac{Px}{EI} \, dx = \frac{PL^2}{2EI}$$

Again, this answer agrees with beam deflection theory.

Castigliano's theorems can also be used to solve statically indeterminate problems. Details can be found in most books on advanced mechanics of materials.

4.6.3 Impact Stresses

An interesting application of energy methods is in the determination of stresses due to impact loadings, such as a weight dropped onto a beam. In order to solve impact problems using simple energy methods, the following two assumptions are made:

1. No energy is dissipated during impact.
2. The striking body does not bounce off the structure.

In essence, these two assumptions require that all of the kinetic energy of the striking object is transferred to the structure. To solve an impact problem, it is assumed that the kinetic energy of the striking object equals the strain energy induced in the structure.

Example 4.9. Determine the maximum stress induced in a uniform cantilever beam of rectangular cross section (height $= h$, width $= b$) if a weight W is dropped from a height h_0 above the free end of the beam.

SOLUTION. Assuming the end deflection of the beam is negligible compared to h_0, the energy transferred to the beam by the weight is simply Wh_0. From Example 4.6, the strain energy of the beam (neglecting shear effects) is

$$U = \frac{P_w^2 L^3}{6EI}$$

where P_w is the static load that would bring about the same strain energy in the beam as does the weight. Thus,

$$Wh_0 = \frac{P_w^2 L^3}{6EI} \quad \text{or} \quad P_w = \sqrt{\frac{6EIWh_0}{L^3}}$$

The maximum stress in the beam is given by

$$\sigma_{max} = \frac{Mc}{I} = \frac{P_w L (h/2)}{I} = \frac{h}{2}\sqrt{\frac{6EWh_0}{LI}}$$

In general, for a structure to efficiently resist an impact load, it should have the following properties[10]:

1. Possess a large volume.
2. Have a low modulus of elasticity and high yield strength.
3. Have as uniform a shape as possible so that the stresses are evenly distributed.

4.7 COMPOSITE MATERIALS

4.7.1 Introduction to Composite Materials

Definition of a Composite Material
There is no universally accepted definition of what constitutes a composite material. In everyday usage, a composite is something made up of several parts. In engineering, whether or not a given material should be considered a composite depends on the scale of viewing.

At the atomic level, any material consisting of more than one kind of atom could be considered a composite. With this definition, only pure elements would not be composites.

At the microscopic level, any material composed of more than one phase could be considered a composite. For example, steel is a composite since it is a multiphase alloy of carbon and iron.

At the macroscopic level, only materials consisting of two or more macroscopic constituents are called composites. An example is glass fibers embedded in an epoxy matrix.

The following definition of composite materials is provided by Schwartz[32]:

A composite material is a material system composed of a mixture or combination of two or more macroconstituents differing in form and/or material composition and that are essentially insoluble in each other.

Classification of Composite Materials
Composites can be classified in a variety of ways such as by the matrix constituent (i.e., metal-matrix, ceramic-matrix, and resin-matrix composites). More commonly, classification is based on the structural constituents. This leads to three common types of composite materials:

1. Fiber composites that are composed of fibers embedded in a matrix.
2. Particulate composites that consist of particles embedded in a matrix.
3. Laminated composites that are composed of layers of various materials.

Advantages of Composites
Some of the advantages of composites over conventional materials include the following:

1. Higher specific strength (i.e., strength per unit weight)
2. Higher specific stiffness (i.e., stiffness per unit weight)
3. The ability to custom design the properties of a composite for a specific application.

Many other advantages exist for certain composites as discussed in the technical literature.

4.7.2 Orthotropic Elasticity

While composites offer higher specific strength and stiffness than many conventional materials, they suffer from the fact that their mechanical behavior is more difficult to understand. In the most general case, a material can display anisotropic stress-strain response that requires 21 material constants to describe the elastic behavior. In comparison, only 2 elastic constants are needed to describe isotropic response. For unidirectional, fiber-reinforced materials such as shown in Fig. 4.44, there are 3 mutually orthogonal planes of material symmetry. This reduces the required number of elastic constants from 21 to 9. A material having such symmetry is said to be orthotropic. Assuming the 1, 2, and 3 directions are the principal directions of material symmetry, the stress-strain relationships are

$$
\begin{Bmatrix} \sigma_1 \\ \sigma_2 \\ \sigma_3 \\ \sigma_{23} \\ \sigma_{31} \\ \sigma_{12} \end{Bmatrix}
=
\begin{bmatrix}
C_{11} & C_{12} & C_{13} & 0 & 0 & 0 \\
C_{12} & C_{22} & C_{23} & 0 & 0 & 0 \\
C_{13} & C_{23} & C_{33} & 0 & 0 & 0 \\
0 & 0 & 0 & C_{44} & 0 & 0 \\
0 & 0 & 0 & 0 & C_{55} & 0 \\
0 & 0 & 0 & 0 & 0 & C_{66}
\end{bmatrix}
\begin{Bmatrix} \varepsilon_1 \\ \varepsilon_2 \\ \varepsilon_3 \\ \gamma_{23} \\ \gamma_{31} \\ \gamma_{12} \end{Bmatrix}
\tag{4.90}
$$

where $[C_{ij}]$ is known as the stiffness matrix. The shear strains (γ_{12}, γ_{23}, and γ_{31}) represent engineering shear strains. Equation (4.90) can be inverted to give strains in terms of stresses:

$$
\begin{Bmatrix} \varepsilon_1 \\ \varepsilon_2 \\ \varepsilon_3 \\ \gamma_{23} \\ \gamma_{31} \\ \gamma_{12} \end{Bmatrix}
=
\begin{bmatrix}
S_{11} & S_{12} & S_{13} & 0 & 0 & 0 \\
S_{12} & S_{22} & S_{23} & 0 & 0 & 0 \\
S_{13} & S_{23} & S_{33} & 0 & 0 & 0 \\
0 & 0 & 0 & S_{44} & 0 & 0 \\
0 & 0 & 0 & 0 & S_{55} & 0 \\
0 & 0 & 0 & 0 & 0 & S_{66}
\end{bmatrix}
\begin{Bmatrix} \sigma_1 \\ \sigma_2 \\ \sigma_3 \\ \sigma_{23} \\ \sigma_{31} \\ \sigma_{12} \end{Bmatrix}
\tag{4.91}
$$

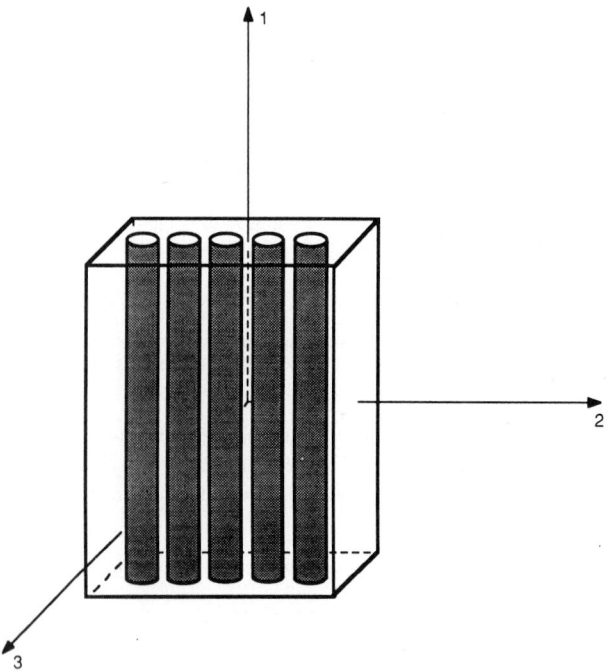

Fig. 4.44 Principal material directions for unidirectionally reinforced lamina.

where $[S_{ij}]$ is known as the compliance matrix. Note that both the stiffness and compliance matrices are symmetric. Also, they are matrix inverses of each other. The components of the stiffness matrix can be related to the components of the compliance matrix by the following expressions:

$$C_{11} = \frac{S_{22}S_{33} - S_{23}^2}{S} \qquad C_{12} = \frac{S_{13}S_{23} - S_{12}S_{33}}{S}$$

$$C_{22} = \frac{S_{33}S_{11} - S_{13}^2}{S} \qquad C_{13} = \frac{S_{12}S_{23} - S_{13}S_{22}}{S} \qquad (4.92)$$

$$C_{33} = \frac{S_{11}S_{22} - S_{12}^2}{S} \qquad C_{23} = \frac{S_{12}S_{13} - S_{23}S_{11}}{S}$$

$$C_{44} = \frac{1}{S_{44}} \qquad C_{55} = \frac{1}{S_{55}} \qquad C_{66} = \frac{1}{S_{66}}$$

where

$$S = S_{11}S_{22}S_{33} - S_{11}S_{23}^2 - S_{22}S_{13}^2 - S_{33}S_{12}^2 + 2S_{12}S_{23}S_{13}$$

The converse relationship (i.e., S_{ij}'s in terms of C_{ij}'s) can be obtained by interchanging the symbols C and S everywhere in Eqs. (4.92).

Engineering Constants

In order to predict material behavior, it is necessary to determine the material constants in either Eq. (4.90) or (4.92). It is convenient to introduce the following engineering constants[33]:

E_1, E_2, E_3 = elastic moduli in the 1, 2, and 3 directions, respectively. Thus, $E_i = \sigma_i/\varepsilon_i$ for $\sigma_i = \sigma$ and all other stresses being zero.

G_{12}, G_{23}, G_{31} = shear moduli in the 1-2, 2-3, and 3-1 planes, respectively.

ν_{ij} = Poisson's ratio for transverse strain in the j direction when the specimen is stressed in the i direction. Hence, $\nu_{ij} = -\varepsilon_j/\varepsilon_i$ for $\sigma_i = \sigma$ and all other stresses being zero.

The compliance matrix can be rewritten as

$$[S_{ij}] = \begin{bmatrix} \dfrac{1}{E_1} & -\dfrac{\nu_{21}}{E_2} & -\dfrac{\nu_{31}}{E_3} & 0 & 0 & 0 \\[2mm] -\dfrac{\nu_{12}}{E_1} & \dfrac{1}{E_2} & -\dfrac{\nu_{32}}{E_3} & 0 & 0 & 0 \\[2mm] -\dfrac{\nu_{13}}{E_1} & -\dfrac{\nu_{23}}{E_2} & \dfrac{1}{E_3} & 0 & 0 & 0 \\[2mm] 0 & 0 & 0 & \dfrac{1}{G_{23}} & 0 & 0 \\[2mm] 0 & 0 & 0 & 0 & \dfrac{1}{G_{31}} & 0 \\[2mm] 0 & 0 & 0 & 0 & 0 & \dfrac{1}{G_{12}} \end{bmatrix} \qquad (4.93)$$

Because of the symmetry of the compliance matrix ($S_{ij} = S_{ji}$), it is apparent that

$$\frac{\nu_{ij}}{E_i} = \frac{\nu_{ji}}{E_j}$$

In terms of the engineering constants, the stiffness matrix components take the form

$$C_{11} = \frac{1 - \nu_{23}\nu_{32}}{E_2 E_3 \Delta}$$

$$C_{12} = \frac{\nu_{21} + \nu_{31}\nu_{23}}{E_2 E_3 \Delta} = \frac{\nu_{12} + \nu_{32}\nu_{13}}{E_1 E_3 \Delta}$$

$$C_{13} = \frac{\nu_{31} + \nu_{21}\nu_{32}}{E_2 E_3 \Delta} = \frac{\nu_{13} + \nu_{12}\nu_{23}}{E_1 E_2 \Delta}$$

$$C_{22} = \frac{1 - \nu_{31}\nu_{13}}{E_1 E_3 \Delta}$$

$$C_{23} = \frac{\nu_{32} + \nu_{12}\nu_{31}}{E_1 E_3 \Delta} = \frac{\nu_{23} + \nu_{21}\nu_{13}}{E_1 E_2 \Delta}$$

$$C_{33} = \frac{1 - \nu_{12}\nu_{21}}{E_1 E_2 \Delta}$$

$$C_{44} = G_{23} \qquad C_{55} = G_{31} \qquad C_{66} = G_{12}$$

where

$$\Delta = \frac{1 - \nu_{12}\nu_{21} - \nu_{23}\nu_{32} - \nu_{31}\nu_{13} - 2\nu_{21}\nu_{32}\nu_{13}}{E_1 E_2 E_3}$$

4.7.3 Plane Stress

For orthotropic response under plane stress conditions, the in-plane stress-strain relationships are given by

$$\begin{Bmatrix} \varepsilon_1 \\ \varepsilon_2 \\ \gamma_{12} \end{Bmatrix} = \begin{bmatrix} S_{11} & S_{12} & 0 \\ S_{12} & S_{22} & 0 \\ 0 & 0 & S_{66} \end{bmatrix} \begin{Bmatrix} \sigma_1 \\ \sigma_2 \\ \sigma_{12} \end{Bmatrix}$$

and

$$\begin{Bmatrix} \sigma_1 \\ \sigma_2 \\ \sigma_{12} \end{Bmatrix} = \begin{bmatrix} Q_{11} & Q_{12} & 0 \\ Q_{12} & Q_{22} & 0 \\ 0 & 0 & Q_{66} \end{bmatrix} \begin{Bmatrix} \varepsilon_1 \\ \varepsilon_2 \\ \gamma_{12} \end{Bmatrix}$$

where the compliance coefficients (S_{ij}) are the same as in Eq. (4.93). The components Q_{ij}, sometimes referred to as the reduced stiffnesses, can be expressed in terms of the compliance coefficients as

$$Q_{11} = \frac{S_{22}}{S_{11}S_{22} - S_{12}^2} \qquad Q_{22} = \frac{S_{11}}{S_{11}S_{22} - S_{12}^2}$$

$$Q_{12} = -\frac{S_{12}}{S_{11}S_{22} - S_{12}^2} \qquad Q_{66} = \frac{1}{S_{66}}$$

4.7.4 Stress-Strain Relationships for Arbitrary Orientations

In Section 4.7.2, the stress-strain relationships with respect to the principal material directions were presented. Unfortunately, the principal material directions often do not coincide with the coordinates that naturally fit the geometry of the structure. For plane stress conditions, Mohr's circle can be used to transform stresses or strains; however, the principal stresses and strains do not necessarily occur at the same orientation. For an xy coordinate system located at an angle θ with respect to the principal material coordinates (Fig. 4.45), the stress-strain relationships take the form

$$\begin{Bmatrix} \sigma_x \\ \sigma_y \\ \sigma_{xy} \end{Bmatrix} = \begin{bmatrix} \bar{Q}_{11} & \bar{Q}_{12} & \bar{Q}_{16} \\ \bar{Q}_{12} & \bar{Q}_{22} & \bar{Q}_{26} \\ \bar{Q}_{16} & \bar{Q}_{26} & \bar{Q}_{66} \end{bmatrix} \begin{Bmatrix} \varepsilon_x \\ \varepsilon_y \\ \gamma_{xy} \end{Bmatrix} \qquad (4.94)$$

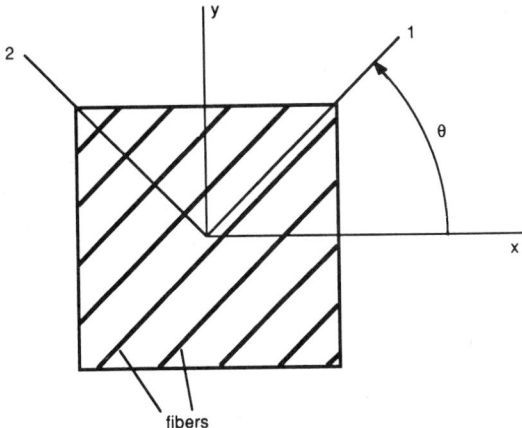

Fig. 4.45 Arbitrary in-plane orientation of xy coordinate axes.

where

$$\bar{Q}_{11} = Q_{11}\cos^4\theta + 2(Q_{12} + 2Q_{66})\sin^2\theta\cos^2\theta + Q_{22}\sin^4\theta$$

$$\bar{Q}_{12} = (Q_{11} + Q_{22} - 4Q_{66})\sin^2\theta\cos^2\theta + Q_{12}(\sin^4\theta + \cos^4\theta)$$

$$\bar{Q}_{22} = Q_{11}\sin^4\theta + 2(Q_{12} + 2Q_{66})\sin^2\theta\cos^2\theta + Q_{22}\cos^4\theta$$

$$\bar{Q}_{16} = (Q_{11} - Q_{12} - 2Q_{66})\sin\theta\cos^3\theta + (Q_{12} - Q_{22} + 2Q_{66})\sin^3\theta\cos\theta$$

$$\bar{Q}_{26} = (Q_{11} - Q_{12} - 2Q_{66})\sin^3\theta\cos\theta + (Q_{12} - Q_{22} + 2Q_{66})\sin\theta\cos^3\theta$$

$$\bar{Q}_{66} = (Q_{11} + Q_{22} - 2Q_{12} - 2Q_{66})\sin^2\theta\cos^2\theta + Q_{66}(\sin^4\theta + \cos^4\theta)$$

or alternatively

$$\begin{Bmatrix} \varepsilon_x \\ \varepsilon_y \\ \gamma_{xy} \end{Bmatrix} = \begin{bmatrix} \bar{S}_{11} & \bar{S}_{12} & \bar{S}_{16} \\ \bar{S}_{12} & \bar{S}_{22} & \bar{S}_{26} \\ \bar{S}_{16} & \bar{S}_{26} & \bar{S}_{66} \end{bmatrix} \begin{Bmatrix} \sigma_x \\ \sigma_y \\ \sigma_{xy} \end{Bmatrix} \qquad (4.95)$$

where

$$\bar{S}_{11} = S_{11}\cos^4\theta + (2S_{12} + S_{66})\sin^2\theta\cos^2\theta + S_{22}\sin^4\theta$$

$$\bar{S}_{12} = S_{12}(\sin^4\theta + \cos^4\theta) + (S_{11} + S_{22} - S_{66})\sin^2\theta\cos^2\theta$$

$$\bar{S}_{22} = S_{11}\sin^4\theta + (2S_{12} + S_{66})\sin^2\theta\cos^2\theta + S_{22}\cos^4\theta$$

$$\bar{S}_{16} = (2S_{11} - 2S_{12} - S_{66})\sin\theta\cos^3\theta - (2S_{22} - 2S_{12} - S_{66})\sin^3\theta\cos\theta$$

$$\bar{S}_{26} = (2S_{11} - 2S_{12} - S_{66})\sin^3\theta\cos\theta - (2S_{22} - 2S_{12} - S_{66})\sin\theta\cos^3\theta$$

$$\bar{S}_{66} = 2(2S_{11} + 2S_{22} - 4S_{12} - S_{66})\sin^2\theta\cos^2\theta + S_{66}(\sin^4\theta + \cos^4\theta)$$

If $\gamma_{xy} = 0$ in Eq. (4.94), the xy orientation corresponds to the principal strain directions. However, since $\sigma_{xy} = Q_{16}\varepsilon_{xx} + Q_{26}\varepsilon_{yy}$, this is seldom the direction of the principal stresses. Similarly, when $\sigma_{xy} = 0$ in Eq. (4.95), it is seldom true that $\gamma_{xy} = 0$. This fact is very important since some theories of lamina strength are based on principal strains, while others are based on principal stresses.

4.7.5 Biaxial Strength Theories

Several theories are available for predicting the strength of lamina under in-plane biaxial loading. Some of the simpler of these theories are now discussed.

Maximum Strength Theory

The maximum strength states that the lamina is safe provided the stresses in the principal material directions are less than the ultimate strengths in these directions. Hence, the lamina is safe provided

$$\sigma_1 < X_t \quad \text{and} \quad \sigma_2 < Y_t \quad \text{and} \quad \sigma_{12} < S$$

for tension, and

$$|\sigma_1| < |X_c| \quad \text{and} \quad |\sigma_2| < |Y_c|$$

for compression. In these inequalities, X_t is the tensile strength in the 1 direction, X_c is the compressive strength in the 1 direction, Y_t is the tensile strength in the 2 direction, Y_c is the compressive strength in the 2 direction, and S is the in-plane shear strength.

Maximum Strain Theory

The maximum strain theory is similar to the maximum strength theory except that the stresses are replaced by strains. Thus, the material is safe provided

$$\varepsilon_1 < \varepsilon_1^{ut} \quad \text{and} \quad \varepsilon_2 < \varepsilon_2^{ut} \quad \text{and} \quad |\gamma_{12}| < S_\varepsilon$$

for tension, and

$$|\varepsilon_1| < |\varepsilon_1^{uc}| \quad \text{and} \quad |\varepsilon_2| < |\varepsilon_2^{uc}|$$

for compression. In these inequalities, ε_1^{ut} is the maximum tensile strain at failure for loading in the 1 direction, ε_1^{uc} is the maximum compressive strain at failure for loading in the 1 direction, ε_2^{ut} is the maximum tensile strain at failure for loading in the 2 direction, ε_2^{uc} is the maximum compressive strain at failure for loading in the 2 direction, and S_ε is the shear strain at failure for an in-plane shear test of the lamina.

Tsai–Hill Theory

It should be noted that the maximum stress and maximum strain theories assume no interaction between modes of failure. As such, these theories can lead to inaccurate predictions of strength. The Tsai–Hill theory states that failure is imminent when the following inequality is satisfied:

$$\left(\frac{\sigma_1}{X} \right)^2 - \frac{\sigma_1 \sigma_2}{X^2} + \left(\frac{\sigma_2}{Y} \right)^2 + \left(\frac{\sigma_{12}}{S} \right)^2 \geq 1$$

where X is the ultimate strength in the 1 direction, Y is the ultimate strength in the 2 direction, and S is the in-plane shear strength of the lamina. The Tsai–Hill theory is easy to use and gives reasonably good results for many kinds of composite lamina. Note that it does not differentiate between tensile and compressive strengths, however.

More sophisticated lamina failure theories are available. A thorough review of these is given in Ref. 34.

Laminates

A lamina is a single layer of composite material with fibers running in one direction. A laminate consists of several individual laminae stacked one on top of another and bonded together to form a single component. Theories for predicting the stress-strain relationships and failure strengths of composite laminates are available, but they are too complex to be presented here (see Ref. 34 for a thorough review of these theories). Further details on the mechanics of composite materials can be found in Refs. 33–38.

4.8   THEORIES OF STRENGTH AND FAILURE

4.8.1   Stress Concentrations

Large increases in stresses occur near abrupt changes in geometry such as holes, notches, fillets, and cracks. With regards to mechanics of materials, these geometrical discontinuities are referred to as stress raisers or stress concentrations. For some simple geometries, the effect of stress raisers on the stress distribution can be determined using the theory of elasticity. For complicated geometries, however, numerical or experimental techniques represent more practical means of determining the stress distribution.

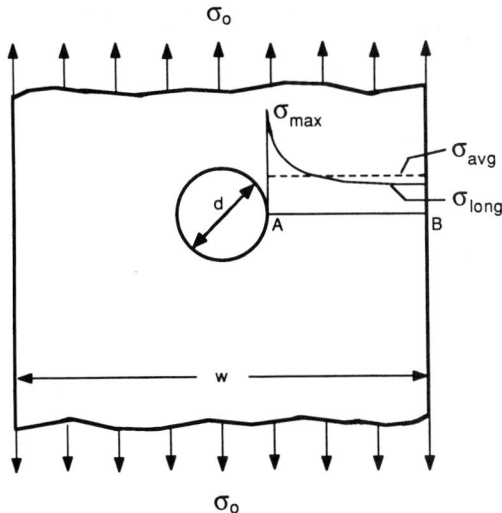

Fig. 4.46 Hole in a semi-infinite plate.

The theoretical stress concentration factor, K_t, is defined by

$$K_t = \frac{\sigma_{max}}{\sigma_{avg}} \tag{4.96}$$

where σ_{max} is the maximum stress that occurs in the vicinity of the stress concentration, and σ_{avg} is the average stress that acts at the minimal cross-sectional area. For example, consider the case of a hole in a semi-infinite plate under uniaxial loading (Fig. 4.46). Far away from the hole, the longitudinal stress is given by $\sigma = P/wt$, where P is the applied load, w is the plate width, and t is the plate thickness. Along the line AB in Fig. 4.46, the average longitudinal stress is higher due to the reduced cross-sectional area caused by the presence of the hole. Thus,

$$\sigma_{avg} = \frac{P}{(w-d)t}$$

where d is the hole diameter. The actual stress distribution along the line AB is not constant at σ_{avg}, but rather it varies as shown in the figure. The maximum longitudinal stress occurs at the edge of the hole.

Stress concentration factors provide a quick method for estimating the severity of a given stress raiser. Extensive tables of these factors for a wide variety of situations are available in the literature.[9, 39] Stress concentration factors should be used with caution in design, however, since several important issues are neglected:

1. Stress concentration factors are normally given for uniaxial states of loading. In many instances, there is a combined state of loading (e.g., biaxial loading of a plate with a hole).
2. Even when the far-field loading is uniaxial, the geometrical discontinuity induces a state of biaxial stresses in its immediate vicinity.
3. If the maximum stress exceeds the yield strength, a plastic zone develops. The maximum stress predicted by Eq. (4.96) is too high in such circumstances.

It is apparent from the last of these statements that the use of the stress concentration factor can lead to overly conservative design in the case of a ductile material. It is possible, however, for normally ductile materials to exhibit brittle behavior. This is especially true if the component thickness is large enough to induce a state of plane strain. In such cases, the use of the stress concentration factor in design can lead to nonconservative predictions. This apparent dilemma can be explained by the application of modern theories of fracture mechanics. Further details are provided in Sections 4.8.2 and 4.8.3.

4.8.2   Classical Theories of Static Failure

Ductile versus Brittle Material Behavior

For materials exhibiting extremely ductile behavior, such as mild steel at room temperature, design against failure is often based on the yield strength. Roughly speaking, a material can be considered to be ductile if it exhibits an elongation of at least 5% in a standard tension test.[40] It is important to recognize, however, that materials are in fact neither ductile nor brittle. Instead, materials are ductile in some circumstances and brittle in others. For example, while most mild steels display considerable ductility at room temperature, the ductility tends to decrease as the temperature is lowered. In fact, a rapid decrease in ductility often occurs over a narrow temperature range (see Section 4.8.3 for further details). In general, several other factors besides temperature can induce brittle behavior in what are normally considered to be ductile materials. These include the presence of notches, corrosive environments, and very rapid rates of loading.[41]

Maximum Shear Stress Theory

The maximum shear stress theory states that a region of material in a component yields when the maximum shear stress at that point equals the maximum shear stress at yield of a tensile specimen of the same material. From Section 4.1.3, $\tau_{max} = (\sigma_1 - \sigma_3)/2$, where σ_1 and σ_3 are the algebraically largest and smallest principal stresses, respectively. For the case of a uniaxial tension specimen at yield, $\sigma_3 = 0$ and $\sigma_1 = \sigma_y$. The maximum shear stress theory takes the form

$$\tau_{max} = \frac{\sigma_1 - \sigma_3}{2} = \frac{\sigma_y}{2} \text{ at yield} \tag{4.97}$$

This theory is relatively simple to use and normally results in conservative predictions. For biaxial states of loading, σ_1 and σ_3 are determined easily using Mohr's circle.

For a state of pure shear, $\tau_{max} = \tau_y$ at yield. The maximum shear stress theory thus predicts a value of 0.5 for the ratio of the yield strength in torsion to the yield strength in tension,

$$\frac{\tau_y}{\sigma_y} = 0.5$$

Distortion Energy Theory

The distortion energy theory, also known as the von Mises–Hencky theory, is based on the assumption that yielding occurs at a point in a component when the total distortional energy per unit volume at that point equals the distortional energy per unit volume in a tensile specimen at yield. The distortion energy per unit volume, u_d, is given by

$$u_d = \frac{1}{12G}\left[(\sigma_{xx} - \sigma_{yy})^2 + (\sigma_{yy} - \sigma_{zz})^2 + (\sigma_{zz} - \sigma_{xx})^2 + 6\left(\sigma_{xy}^2 + \sigma_{zx}^2 + \sigma_{yz}^2\right)\right]$$

or, in terms of the principal stresses,

$$u_d = \frac{1}{6G}\left(\sigma_1^2 + \sigma_2^2 + \sigma_3^2 - \sigma_1\sigma_2 - \sigma_1\sigma_3 - \sigma_2\sigma_3\right)$$

For a tensile specimen at yield, $\sigma_1 = \sigma_y$ and $\sigma_2 = \sigma_3 = 0$. Accordingly, $u_d = \sigma_y^2/6G$ in this case. The distortion energy theory can therefore be rewritten as

$$\sigma_1^2 + \sigma_2^2 + \sigma_3^2 - \sigma_1\sigma_2 - \sigma_1\sigma_3 - \sigma_2\sigma_3 = \sigma_y^2 \text{ at yield} \tag{4.98}$$

For a state of pure shear, $\sigma_1 = \tau$, $\sigma_2 = 0$, and $\sigma_3 = -\tau$. Using the distortion energy theory, the relationship between yield in shear and tension is

$$\tau_y^2 + 0 + \tau_y^2 - 0 - 0 - 0 = \sigma_y^2$$

or

$$\frac{\tau_y}{\sigma_y} = 0.577$$

Comparison of Theories for Yielding

For most ductile materials, the distortion energy theory gives more accurate results than the maximum shear stress theory.[22] For a biaxial state of loading, the out-of-plane principal stress is zero. Denoting the in-plane principal stresses by σ_a and σ_b, it is possible to graphically portray the two theories. This has been done in Fig. 4.47. It is seen that the maximum shear stress theory is

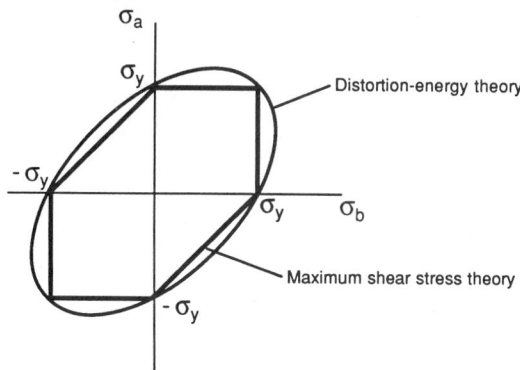

Fig. 4.47 Comparison of theories of yielding for plane stress conditions.

represented by a hexagon, while the distortion energy theory is represented by an ellipse. In general, both theories give reasonably good results for plane stress conditions. While the distortion energy theory is usually more accurate, its use can result in slightly nonconservative predictions. On the other hand, use of the maximum shear stress theory almost always leads to slightly conservative predictions.

Maximum Normal Stress Theory
The maximum normal stress theory is generally used for calculating the failure stress for brittle materials. It states that a component fails when its maximum principal stress, σ_1, exceeds the ultimate strength as determined by a tensile test. This theory should be used with caution. For plane stress situations, this theory gives reasonable results if both in-plane principal stresses are tensile. It can be grossly inaccurate, however, when the in-plane principal stresses are of opposite sign.[42]

Coulomb–Mohr Theory
The Coulomb–Mohr theory is used almost exclusively for predicting failure of brittle materials. It is based on the ultimate strengths of a material in tension and compression, σ_{ut} and σ_{uc}, respectively. In general, brittle materials exhibit a larger ultimate strength in compression than in tension. To apply the Coulomb–Mohr theory, Mohr's circles are drawn corresponding to the states of stress at failure for tension and compression specimens. The tangent lines to these circles are then drawn. The theory states that failure occurs for any state of stress that produces a Mohrs circle as large as the envelope of the σ_{ut} and σ_{uc} circles and their tangent lines. That is, if the Mohrs circle for the state of stress under consideration extends out to or beyond the shaded region in Fig. 4.48, failure is imminent.

Figure 4.48b shows a graphical representation of the Coulomb–Mohr theory for a state of plane stress. It should be noted that this theory is identical to the maximum normal stress theory in the first quadrant. In general, the Coulomb–Mohr theory gives conservative results. It is the best classical theory available for estimating failure of brittle materials. In modern design, it is better to use the theory of fracture mechanics to estimate failure loads rather than the Coulomb–Mohr theory because

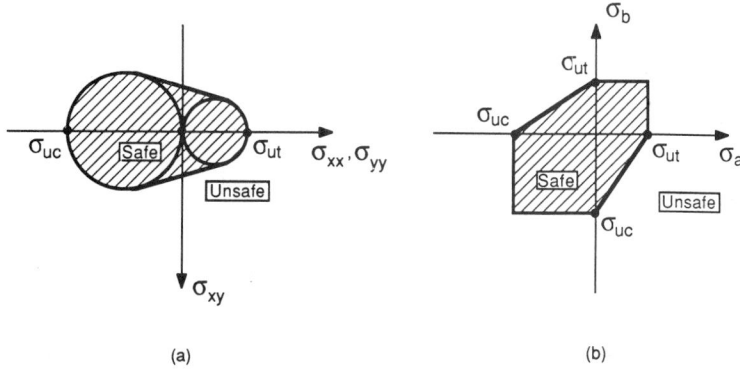

(a) (b)

Fig. 4.48 Coulomb–Mohr theory for brittle failure.

the presence of even a very small flaw in a brittle material can lead to failure at loads much lower than those predicted by the Coulomb–Mohr theory.

4.8.3 Fracture Mechanics

Fracture mechanics is that branch of mechanics that deals with the ability of a material to resist crack propagation under a given set of loading and environmental conditions. Modern fracture mechanics can be divided into two major categories: linear elastic fracture mechanics (LEFM) and elastic-plastic fracture mechanics (EPFM). Before discussing these topics, some background information regarding the nature of brittle fracture must be presented.

Brittle Fracture

Brittle fracture involves the rapid propagation of a crack leading to failure of the component. Little energy is absorbed by the component, and only a small amount of plastic deformation occurs. Crack propagation is conventionally considered under three basic modes of loading as shown in Fig. 4.49. Mode I loading involves an opening or tensile mode. It is the most common loading encountered in practice. Mode II loading represents an in-plane shearing mode. Mode III represents an out-of-plane shearing mode. Because of the prevalence of mode I loading in practice and the relative ease in experimentation and analysis, a majority of investigations have focused on this cracking mode. As mentioned in Section 4.8.2, several factors can cause a normally ductile material to behave in a brittle fashion. These include reduced temperatures, high strain rates, corrosive environments, and the presence of notches.

On a macroscopic level, for uniaxial states of loading, a brittle fracture results in a relatively flat fracture surface with only small amounts of slanted surfaces. The slanted surfaces, commonly referred to as shear lips, are indicative of plastic deformation. In a uniaxial test specimen, these shear lips are at approximately a 45° angle to the applied load axis since the maximum shear stress occurs along planes at this angle. For a ductile fracture, the fracture surface is rougher, and considerable shear lip formations are present.

On a microscopic level, the fracture process can appear quite complicated. For isotropic homogeneous metals, three common microstructural failure modes are microvoid coalescence, cleavage fracture, and intergranular fracture. Cleavage and intergranular fracture are often associated with a brittle failure, while microvoid coalescence is most often associated with ductile failure. This categorization is not always accurate, however. Further information on microscopic fracture analysis can be found in Refs. 43–45.

Linear Elastic Fracture Mechanics

The theory of linear elastic fracture mechanics is based on the assumption that the zone of plastic deformation at the tip of a crack is small. In this case, the stress distributions near the tip of a crack can be determined using the theory of elasticity. Assuming the crack tip is infinitely sharp, the stresses near the crack tip are given by (Fig. 4.50):

$$\sigma_{xx} = \frac{K}{\sqrt{2\pi r}} \cos\frac{\theta}{2}\left(1 - \sin\frac{\theta}{2}\sin\frac{3\theta}{2}\right)$$

$$\sigma_{yy} = \frac{K}{\sqrt{2\pi r}} \cos\frac{\theta}{2}\left(1 + \sin\frac{\theta}{2}\sin\frac{3\theta}{2}\right)$$

$$\sigma_{xy} = \frac{K}{\sqrt{2\pi r}} \left(\sin\frac{\theta}{2}\cos\frac{\theta}{2}\cos\frac{3\theta}{2}\right)$$

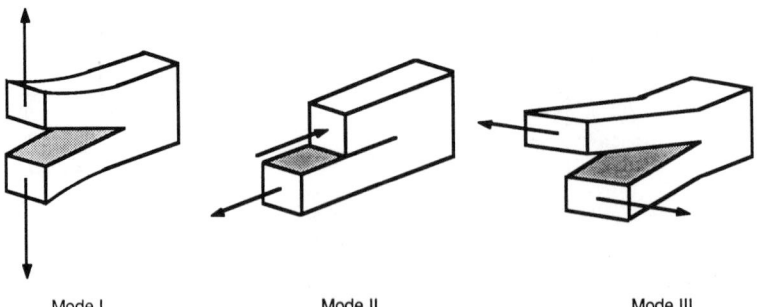

Mode I   Mode II   Mode III

Fig. 4.49 Basic modes of loading for crack propagation.

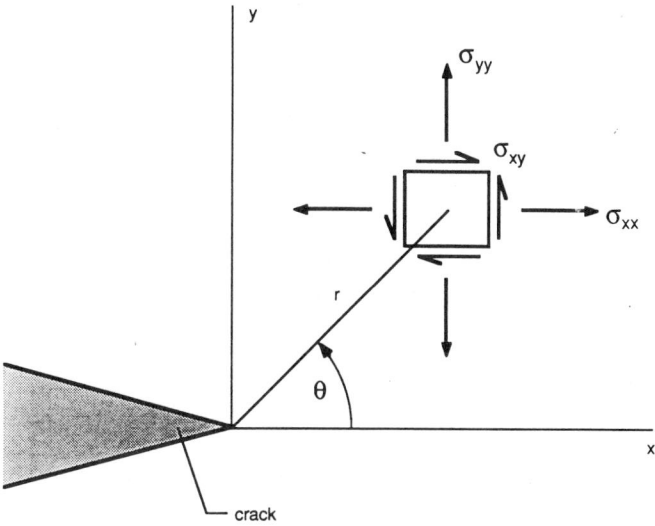

Fig. 4.50 Stresses near a crack tip.

where the parameter K is known as the stress intensity factor. These equations were first derived by Westergaard.[46] It is apparent that these equations predict infinite stresses at the crack tip ($r = 0$). In reality, yielding occurs at the crack tip and limits the maximum stresses attained. The stress intensity factor K is a function of the crack geometry and the applied loads. Extensive tabulations of K for a variety of geometries and loading conditions can be found in the literature.[47, 48] For a center crack of length $2a$ in an infinite sheet, the stress intensity factor is given by $K = \sigma\sqrt{\pi a}$ (Fig. 4.51a). For an edge crack of length a in a semi-infinite sheet, $K \approx 1.1\sigma\sqrt{\pi a}$ (Fig. 4.51b). As the crack length increases, the stress intensity factor rises sharply. In essence, K is a measure of the severity of the crack tip (for a given component, loading, and crack length). If the applied load is increased high

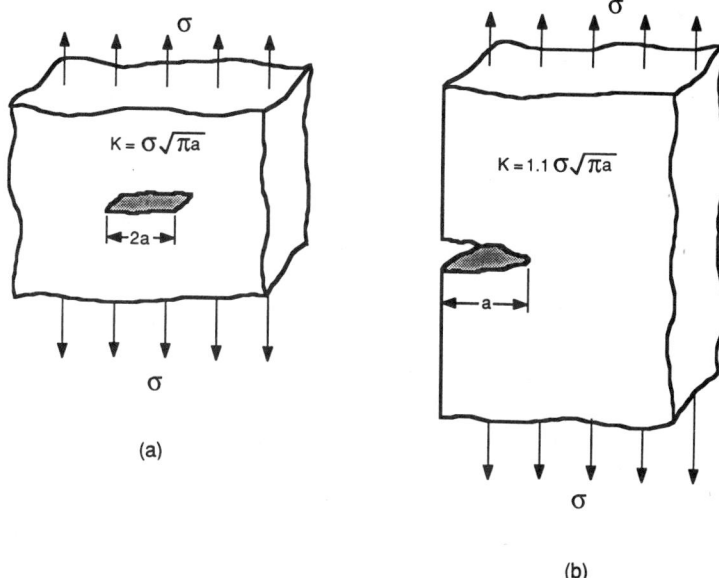

Fig. 4.51 Examples of stress intensity factors: (a) Through-thickness crack in an infinite plate; (b) through-thickness edge crack in a semi-infinite plate.

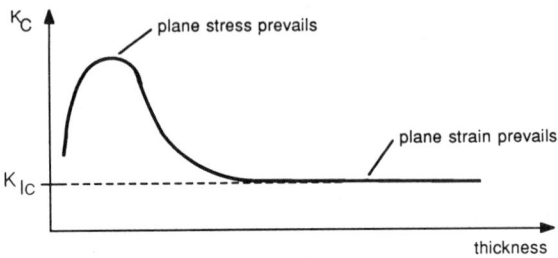

Fig. 4.52   Variation of fracture toughness with thickness.

TABLE 4.8   Plane Strain Fracture Toughness of Selected Alloys

Material	K_{Ic} (ksi $\sqrt{\text{in.}}$)	Yield Strength (ksi)
Aluminum		
2014-T651	22	66
2024-T851	24	66
7075-T651	27	72
Titanium		
Ti-6Al-4V	84	124
Ti-6Al-6V-2Sn; mill annealed (1300°F)	52	145
Ti-6Al-6V-2Sn; solution treated and aged (1050°F)	30	179
Steels		
4340; tempered at 400°F	50	235
4340; tempered at 800°F	76	204
300M; tempered at 600°F	49	260
18 Ni maraging (200); aged 900°F, 6 hr	100	210
Stainless steels		
PH 13-8Mo; H1000	98	215
PH 13-8Mo; H950	54	210

enough, rapid failure due to unstable crack propagation occurs. The value of the stress intensity factor at which failure occurs is known as the fracture toughness of the particular component, K_c. For a component of specified in-plane dimensions, the fracture toughness varies with thickness as shown in Fig. 4.52. For very thin specimens, there is little resistance to cracking, and the fracture toughness is low. For moderate thicknesses where considerable plastic flow can occur, plane stress conditions prevail, and maximum toughness is achieved. As the thickness is further increased, transverse stresses begin to develop near the crack tip. These stresses limit the ability of the material near the crack tip to yield, which in turn reduces the fracture toughness of the component. If the component is made thick enough, a state of plane strain exists in the mid-thickness near the crack tip. The fracture toughness reaches a minimum value at this point. Further increases in thickness do not change its value. This limiting value of toughness is known as the plane strain fracture toughness and is denoted by K_{Ic}. The subscript I indicates that the toughness value is for a mode I loading situation. In general, the fracture toughness, K_c, depends on both component geometry and material, whereas the plane strain fracture toughness is considered a material property when the thickness is sufficiently large. Since K_{Ic} represents a lower bound on the fracture toughness, its use in design should lead to conservative predictions. Table 4.8 lists values of plane strain fracture toughness for a variety of materials along with the associated yield strengths. It must be noted that the values given represent averages. Considerable variation in reported plane strain fracture toughness values is not uncommon. More extensive tables for K_{Ic} can be found elsewhere.[49, 50] Standardized testing procedures for determining K_{Ic} values can be found in Ref. 51.

Transition Temperature Phenomenon
Face-centered cubic metals, such as aluminum, copper, and nickel, show little variation in their mechanical properties with moderate changes in temperature. In contrast, the strength and toughness of body-centered cubic metals, such as ferritic alloys, are often quite sensitive to temperature and strain rate effects. In order to study temperature effects, some means of measuring toughness is

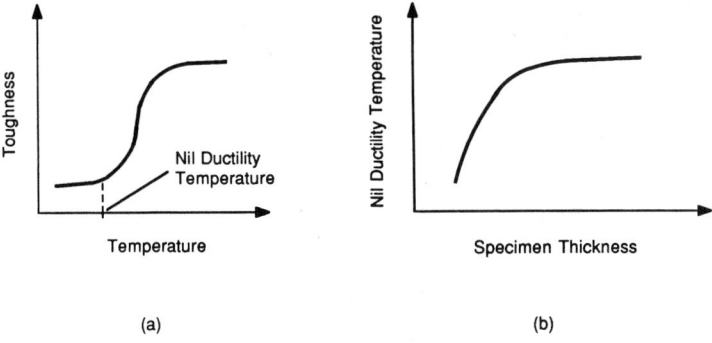

Fig. 4.53   Transition temperature phenomenon.

required. While performing K_{Ic} tests over a range of temperatures would probably provide the most reliable data, the cost of performing these tests often precludes their use. The most common technique employed is the Charpy test. This test consists of dropping an impact hammer pendulum onto a small notched specimen. The energy absorbed by the specimen during fracture is given by $W(h_i - h_f)$ where W is the weight of the hammer, h_i is the initial height of the hammer, and h_f is the maximum height to which the hammer swings after impacting the specimen. The primary disadvantages of the Charpy test are that the small specimen size may allow for considerable plasticity at the crack tip and that it is a rather severe test (a single blow causing fracture). The occurrence of plasticity may result in overly optimistic toughness values. Other means of measuring toughness include the drop weight tear test and the dynamic tear test.[52]

Whatever technique is used for measuring toughness (and the choice is important), one common trend that is observed in low-strength ferritic steels and some other alloys is that a sudden drop in material toughness occurs over a fairly narrow temperature range (Fig. 4.53a). This is known as the transition temperature phenomenon. The temperature corresponding to the lower toughness end of this transition is known as the nil ductility temperature (NDT). In general, the measured value for the NDT depends on the testing technique used, specimen thickness, and material. For thicker specimens, plastic flow is constrained, and the NDT rises (Fig. 4.53b).

Elastic-Plastic Fracture Mechanics
While linear elastic fracture mechanics (LEFM) has proven to be a very powerful design tool, its use is not always appropriate. For high-toughness materials or thin components experiencing plane stress conditions near a crack tip rather than plane strain conditions, LEFM leads to overly conservative design restrictions. Also, large specimen sizes are often required to bring about plane strain conditions for K_{Ic} determination. These large specimens can be expensive to fabricate and difficult to test. For all these reasons and others, considerable research in the field of elastic-plastic fracture mechanics has been done during the past 20 years. The most notable work has been the J integral originally defined by Rice to be a path-independent line integral for two-dimensional crack problems.[53] It was later concluded that the value of the J integral, J, is a measure of the stress and strain singularity near a crack tip. In this regard, J for the plastic case is analogous to the stress intensity factor K for the linear elastic case.

It has been shown that, in the linear elastic range, the critical value of J required to cause unstable crack propagation, J_{Ic}, is related to the plane strain fracture toughness by

$$J_{Ic} = \frac{1 - \nu^2}{E} K_{Ic} \tag{4.99}$$

where E and ν are Young's modulus and Poisson's ratio, respectively. The advantage of determining J_{Ic} instead of K_{Ic} is that much smaller specimen sizes can be used. A standard procedure for determining J_{Ic} has been adopted recently by the American Society for Testing and Materials.[54]

In addition to providing an alternative technique for determining plane strain fracture toughness, the J integral may be used to determine whether a given cracked component is stable with respect to crack initiation and propagation when the material near the crack tip deforms plastically (i.e., when the size of the plastic zone ahead of the crack tip is substantial).[55]

Fracture Mechanics Design

To apply fracture mechanics in design, it is necessary to know the loading conditions, specimen geometry and crack configuration. If no cracks are known to be present, the maximum crack size that could have been overlooked by nondestructive inspection should be used. Once these conditions are known, the stress intensity factor, K, can be determined. This value of K is then compared to the critical stress intensity factor, K_c. If K is larger than K_c, the design is unsafe. In a more general sense, given any two of K_c, σ, and a (fracture toughness, applied stress, and crack length, respectively), the third can be determined since the three are related by the following condition: Unstable crack propagation occurs when the value of K reaches K_c, where $K = K_0 \sigma \sqrt{\pi a}$. Here, K_0 is a constant that depends on the specimen geometry and loading conditions. For example, $K_0 = 1$ for uniaxial loading of an infinite plate with a through-thickness center crack of length $2a$ (Fig. 4.51a). For large component thicknesses, K_c will be equal to K_{1c}. Since K_c values as a function of thickness are not commonly available for most materials, it is common to use K_{1c} values even when component thicknesses are too small to bring about plane strain conditions. This can lead to overly conservative design.

4.8.4   Fatigue

Fatigue investigations concern the effects of cyclic stresses and strains on the mechanical integrity of a structural component. If enough cycles are applied to a component, it is possible for failure to occur even though the maximum stress never exceeded the ultimate strength of the material. It is often useful to classify fatigue problems into two regimes: high-cycle fatigue (cycles to failure $> 10^4$) and low-cycle fatigue (cycles to failure $< 10^4$). This classification needs to be refined in some cases. It is often necessary to further classify the fatigue loading of a given component. If the operating temperature of the material is a sizable fraction of its melting temperature, creep of the material can become substantial. In such cases, theories for creep-fatigue must be employed. Another important distinction is whether the cyclic loads imposed on the component result in uniaxial or multiaxial states of stress.

Looked at simplistically, the fatigue process can be divided into three categories: initiation of a dominant macrocrack, propagation of this crack, and final failure of the component. In rough terms, a macrocrack exists if the crack length is large compared to the microstructure grain size. For high-cycle fatigue, the majority of the life is spent in initiating a macrocrack. In low-cycle fatigue, propagation of a macrocrack can consume over half the life.

Two distinct approaches to fatigue analysis have evolved. In the classical approach, empirical models are developed that relate the imposed loading conditions to the cycles to failure. In the fracture mechanics approach, crack propagation rates are related to the cyclic change in stress intensity factors. The resulting relationship is then integrated to give the cycles to failure. In the application of either approach, two steps need to be performed. First of all, the expected cyclic loading history of the component should be estimated. The obtaining of these data is known as load spectrum analysis. Second, the response of the material to the applied cyclic loading must be known. These data are obtained by laboratory testing under controlled conditions. Finally, the laboratory results must be correlated with the expected service loads so that a life prediction can be made.

High-Cycle Fatigue

In high-cycle fatigue testing, a fully reversed loading is applied to a specimen, and the number of cycles to failure is recorded. When this is done for a variety of load amplitudes, an S-N plot can be constructed, Fig. 4.54. In this diagram, the log of the applied stress amplitude is plotted versus the log of the reversals to failure. For every cycle applied, two reversals in the direction of applied loading must occur (one reversal when the maximum stress is obtained and the other when the minimum stress is obtained). Thus, if the number of cycles to fracture is designated by N_f, the number of reversals to failure is $2N_f$. For many materials, this S-N plot shows a linear relationship over an extensive portion. Within this region, the applied true stress amplitude, σ_a, can be related to the number of reversals to failure by

$$\sigma_a = \sigma_f' \left(2N_f \right)^b \tag{4.100}$$

where σ_f' and b are material constants known as the fatigue strength coefficient and fatigue strength exponent, respectively. The constant b is also commonly referred to as Basquin's exponent. For some materials, this linear relationship is valid from $2N_f = 1$ (which is equivalent to the tension test in the lower limit). In such cases, the fatigue strength coefficient, σ_f', is equal to the true fracture strength, σ_f.

Some materials exhibit a stress amplitude below which no failures occur. This stress amplitude is known as the fatigue limit or endurance limit of the material. This phenomenon is observed in mild steels, for example. Most nonferritic metals behave differently. In these cases the fatigue strength

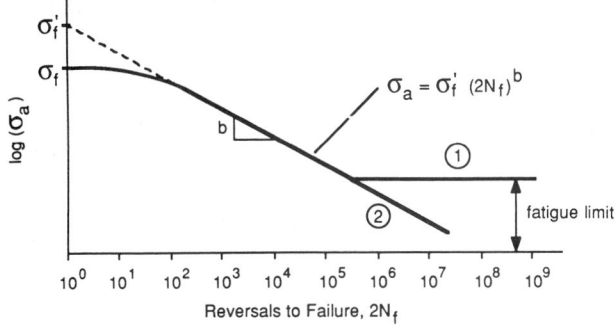

① material exhibiting a fatigue limit
② material not exhibiting a fatigue limit

Fig. 4.54   Schematic representation of an *S-N* diagram.

(instead of a fatigue limit) is defined as the stress amplitude required to bring about failure in a specified number of cycles (normally, $N_f = 10^8$).[56]

When relating the results from laboratory testing, as given by an *S-N* diagram, to structural components, it is necessary to consider many additional factors such as stress concentrations, surface finish quality, residual stresses, and so on. Accurate accounting of the influence of these factors on the fatigue life of the component requires considerable experience. Further details can be found in Ref. 57.

Effect of Mean Stress on Fatigue Life

In the previous discussion, it was assumed that the fatigue loading was fully reversed as shown in Fig. 4.55*a*. It is common in actual components to find a mean stress, σ_m, superimposed on the reversed

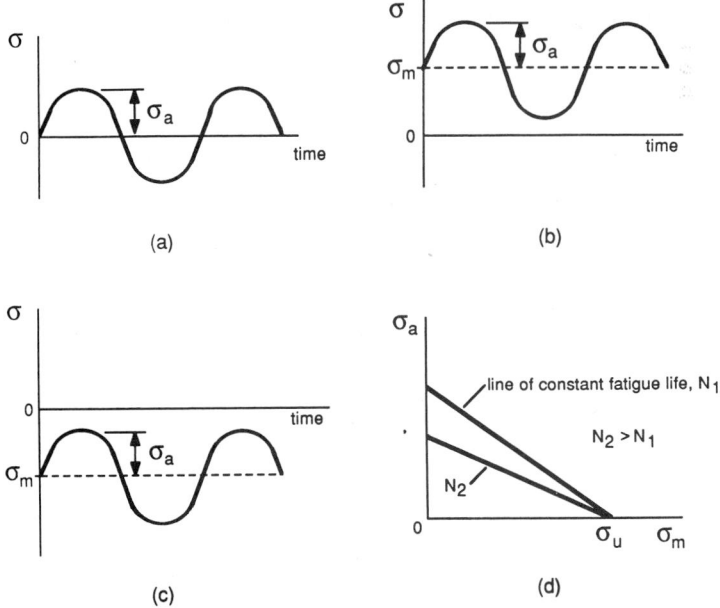

Fig. 4.55   Mean stress effects in fatigue: (*a*) Fully reversed loading; (*b*) tensile mean stress; (*c*) compressive mean stress; (*d*) modified Goodman diagram.

loading. Tensile mean stresses (Fig. 4.55b) tend to reduce the fatigue life while compressive mean stresses (Fig. 4.55c) tend to increase the life. Several empirical relationships have been devised to account for mean stress effects. The most commonly used of these is the modified Goodman diagram (Fig. 4.55d). In this diagram, the alternating stress amplitude is plotted versus the mean stress. The ordinate is chosen to be the stress amplitude for a given cyclic life. The abscissa is taken to be the ultimate strength of the material. The line connecting these two points represents combinations of stress amplitudes and superimposed mean stresses that give the same fatigue life as for the selected fully reversed stress amplitude. Thus, given the desired cyclic life and σ_a, the allowable mean stress can be determined. Similarly, given N_f and σ_m, σ_a can be determined. This empirical technique is easy to use and leads to reasonably accurate results for many materials.

Low-Cycle Fatigue

When performing laboratory tests in the low-cycle fatigue regime ($N_f < 10^4$ cycles), it is common to control the strain amplitude rather than the stress amplitude. Considerable plastic flow can occur in low-cycle fatigue testing. If the stress-strain response for a complete cycle is plotted, the resulting diagram is known as a hysteresis loop (Fig. 4.56). A considerable amount of information is contained in this diagram. The area within the hysteresis loop represents the work per unit volume done on the material in a single cycle. Most of this work is dissipated as heat. The vertical distance between the tips of the hysteresis loop is the stress range, $\Delta\sigma$. The horizontal distance between the tips of the hysteresis loop represents the applied strain range, $\Delta\varepsilon$. The maximum width of the loop is known as the plastic strain range, $\Delta\varepsilon_p$. The elastic strain range, $\Delta\varepsilon_e$, is given by $\Delta\varepsilon_e = \Delta\varepsilon - \Delta\varepsilon_p$. Young's modulus is given by the slope of the initial unloading portion of each half of the hysteresis loop. For many materials, the plastic strain range is not constant for each cycle. Instead, some materials show a cyclic hardening effect (i.e., $\Delta\varepsilon_p$ decreases with increasing cycles), while others show a cyclic softening effect (i.e., $\Delta\varepsilon_p$ increases with increasing cycles). Usually, a material stabilizes by the time the half-life is reached.

If the log of the total strain amplitude, $\log(\Delta\varepsilon/2)$, is plotted versus the log of the reversals to failure, a strain-life curve is obtained (Fig. 4.57). It is often instructive to include the plastic and elastic strain amplitudes, as measured at the half-life, as well. These elastic and plastic strain versus life curves are often straight lines. The plastic strain versus life line is described by the well-known Coffin–Manson equation,[58]

$$\frac{\Delta\varepsilon_p}{2} = \varepsilon_f'\left(2N_f\right)^c \tag{4.101}$$

where ε_f' is the fatigue ductility coefficient and c is the fatigue ductility exponent. For most materials, ε_f' is between $0.5\varepsilon_f$ and $1.5\varepsilon_f$, where ε_f is the true strain at fracture as measured in a tension test (see

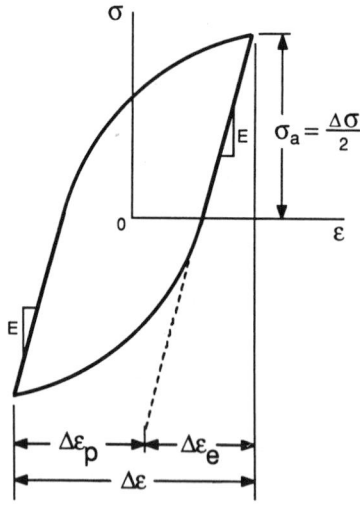

Fig. 4.56　Hysteresis loop.

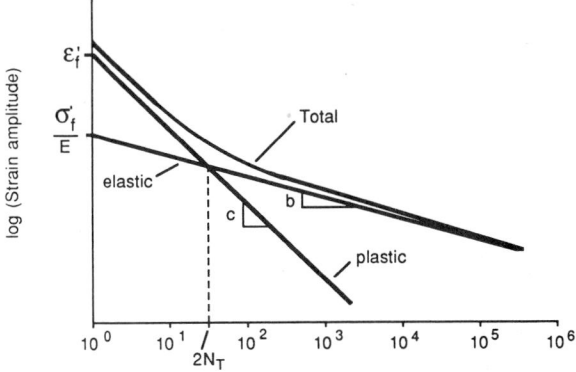

Fig. 4.57 Strain-life curve.

TABLE 4.9 Fatigue Properties of Selected Alloys

Material	E (ksi)	σ_f' (ksi)	ε_f'	b	c
Steels					
SAE 1005-100	30×10^3	77	0.17	-0.066	-0.45
SAE 1015	30×10^3	120	0.96	-0.11	-0.64
SAE 1045-390BHN	30×10^3	230	0.45	-0.074	-0.68
A1SI 4130-365BHN	29×10^3	246	0.89	-0.081	-0.69
SAE 4142-380BHN	30×10^3	265	0.45	-0.08	-0.75
SAE 4340-350BHN	28×10^3	240	0.73	-0.076	-0.62
A1SI 304-327BHN	25×10^3	330	0.89	-0.12	-0.69
Gray cast iron					
Pearlitic-210BHN	16×10^3	59	0.008	-0.084	-0.385
Aluminum alloys					
1100	10×10^3	28	1.80	-0.106	-0.69
2024-T4	10.2×10^3	147	0.21	-0.11	-0.52
7075-T6	10.3×10^3	191	0.19	-0.126	-0.52

Section 4.1.4). The exponent c is usually found to be between -0.5 and -0.7. Thus, if low-cycle fatigue data are lacking, the assumptions $\varepsilon_f' = \varepsilon_f$ and $c = -0.6$ should give reasonable results.

The elastic strain versus life line is described by dividing Eq. (4.100) by Young's modulus, E:

$$\frac{\Delta \varepsilon_e}{2} = \frac{\sigma_a}{E} = \frac{\sigma_f'}{E}(2N_f)^b \qquad (4.102)$$

By combining Eqs. (4.101) and (4.102), a relationship between total strain amplitude and reversals to failure is obtained:

$$\frac{\Delta \varepsilon}{2} = \frac{\Delta \varepsilon_e}{2} + \frac{\Delta \varepsilon_p}{2} = \frac{\sigma_f'}{E}(2N_f)^b + \varepsilon_f'(2N_f)^c$$

The fatigue properties of some selected materials are given in Table 4.9.

Another important parameter obtained from the strain-life curve is the transition fatigue life, $2N_T$. This is the number of reversals corresponding to the intersection of the elastic and plastic strain versus life lines ($\Delta \varepsilon_e/2 = \Delta \varepsilon_p/2$). For lives less than $2N_T$, plastic strains dominate the fatigue process and a material's ductility is of great importance. For lives greater than $2N_T$, elastic strains are the dominant strains, and a material's strength is of greater importance than its ductility. It is not surprising that high-strength, low-ductility materials show a short transition life (sometimes N_T is at a few cycles) while lower-strength, ductile materials tend to have a high transition life. By equating $\Delta \varepsilon_p/2$ to $\Delta \varepsilon_e/2$ in Eqs. (4.100) and (4.101), respectively, the following equation for the transition life is obtained:

$$2N_t = \left(\frac{\varepsilon_f' E}{\sigma_f'}\right)^{1/(b-c)}$$

The transition fatigue life is more satisfactory to distinguish the low- and high-cycle regimes of a particular material than an arbitrary number of cycles such as 10^4.

Cyclic Stress-Strain Diagram

For components that undergo significant dynamic loading, design based on the monotonic stress-strain curve may lead to poor results. This is because many materials exhibit significant changes in strength properties due to cyclic loading. These changes are not necessarily due to the development of a macroscopic crack, but rather represent global changes in the material's microstructure. Some materials exhibit cyclic hardening while others exhibit cyclic softening and still others are relatively stable with respect to cyclic loading. Thus, when designing components that undergo significant dynamic loading, the cyclic stress-strain diagram should be used.

Several techniques exist for construction of a cyclic stress-strain curve. In one commonly used procedure, the tips of stabilized hysteresis loops from a number of strain-controlled fatigue tests are connected together to form a smooth curve. In another single-specimen technique, the specimen is cycled at higher and higher strain amplitudes. At each amplitude, enough cycles are performed to obtain a relatively stable hysteresis loop, but not so many cycles as to significantly damage the specimen. Again, the tips of the hysteresis loops so obtained are connected together to form a smooth curve. Some examples of cyclic stress-strain diagrams and the corresponding monotonic curves are provided in Fig. 4.58.

If true stress and true strain are used to construct the cyclic stress-strain diagram, a relationship similar to Eq. (4.22) is often found to be valid:

$$\sigma_a = K'\left(\frac{\Delta\varepsilon_p}{2}\right)^{n'}$$

where σ_a is the steady-state true stress amplitude, $\Delta\varepsilon_p/2$ is the corresponding plastic strain amplitude, K' is the cyclic strength coefficient, and n' is the cyclic strain-hardening exponent. The value of n' is usually between 0.10 and 0.20. In general, metals with high monotonic strain-hardening exponents ($n > 0.15$) cyclically harden while those with low values of n ($n < 0.15$) cyclically soften. It can be shown, by considering energy quantities, that[59]

$$n' = \frac{b}{c}$$

This equation provides a correlation between fatigue properties and cyclic stress-strain properties. If typical values for c and b of -0.6 and -0.9, respectively, are substituted into this equation, a value for n' of 0.15 results. This is in good agreement with the fact that n' is usually found to be between 0.1 and 0.2 for most metals.

Theories of Cumulative Damage

Constant amplitude cyclic loading does not occur in most real situations. Instead, the loading is most often random (for example, the loadings on a car axle) or consists of distinct blocks of loading at given amplitudes (for example, machinery operated at several discrete speeds). In order to estimate fatigue life under these conditions, a theory of cumulative damage is required.

For simple block-loading situations, the most commonly used theory is the Palmgren–Miner rule (also known as Miner's rule, or as the linear cumulative damage theory). This theory states that the fraction of life exhausted by the ith block loading is given by n_i/N_{fi}, where n_i is the number of cycles applied in the ith block and N_{fi} is the number of cycles to failure of a virgin specimen cycled at the same load amplitude as used in the ith block. Clearly, an S-N or strain-life diagram is required to use this theory since N_f data are needed. Failure is assumed to occur when the total fraction of life exhausted becomes equal to one. That is,

$$\sum_i \frac{n_i}{N_{fi}} = \frac{n_1}{N_{f1}} + \frac{n_2}{N_{f2}} + \frac{n_3}{N_{f3}} + \cdots = 1 \text{ at failure}$$

In using the Palmgren–Miner rule, one assumes that fatigue damage accumulates in a linear fashion. In other words, the damage accumulated per cycle must be the same early and late in the cycling

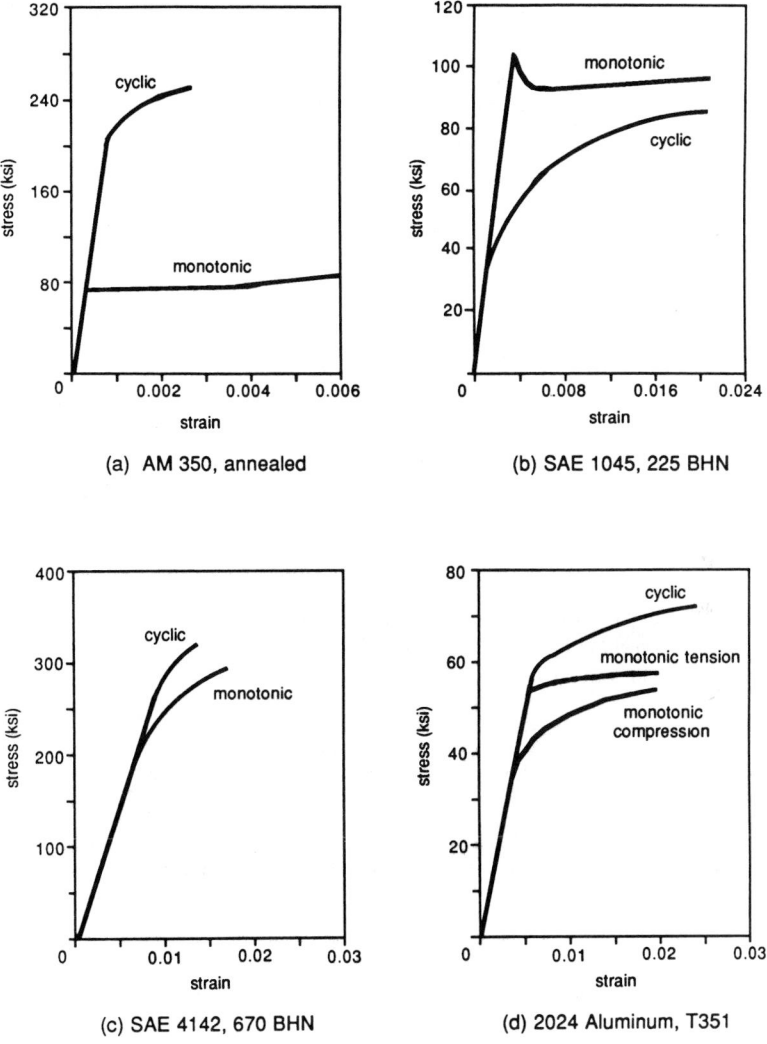

Fig. 4.58   Examples of some monotonic and cyclic stress-strain diagrams.

sequence for a given block loading. In reality, this is often not the case. Even so, this empirical rule or modified versions of it usually give satisfactory results.

For components undergoing random loading, it is necessary to obtain or estimate a typical portion of the loading history of the component. As previously mentioned, this is referred to as load spectrum analysis. Once this is done, more sophisticated models for fatigue damage accumulation, such as rainflow counting, can be applied. An excellent discussion of fatigue life estimation techniques is provided in Ref. 60.

Creep and Fatigue
There are many situations in modern engineering in which a material must be subjected to cyclic loads at temperatures that are relatively near the melting point of the material. Some examples include turbine blades in aircraft jet engines, nuclear pressure vessels, and solder joints used in electronics interconnection technology. Under these circumstances, the standard theories of fatigue presented previously are no longer adequate. Instead, theories for creep-fatigue need to be applied. The two most commonly used theories are the Robinson–Taira theory[61] and the theory of strain range partitioning.[62]

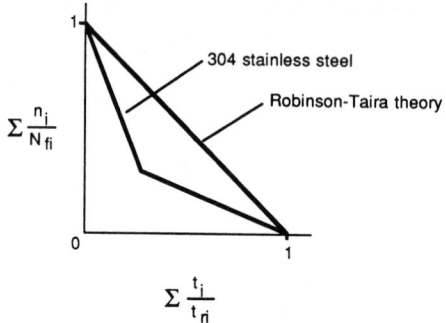

$$\Sigma \frac{n_i}{N_{fi}}$$

$$\Sigma \frac{t_i}{t_{ri}}$$

Fig. 4.59   Robinson–Taira theory for creep-fatigue.

The Robinson–Taira theory has the form

$$\sum_i \frac{n_i}{N_{fi}} + \sum_i \frac{t_i}{t_{ri}} = 1 \text{ at failure}$$

where n_i = number of cycles applied at the ith loading
N_{fi} = number of cycles to failure at the ith loading
t_i = time spent at the ith stress amplitude, σ_i
t_{ri} = time to creep failure at a stress level of σ_i

Note the similarity between the Robinson–Taira equation and the Palmgren–Miner rule. In fact, the Robinson–Taira equation reduces to the Palmgren–Miner rule when creep damage is insignificant. The Robinson–Taira equation is displayed graphically in Fig. 4.59. For some materials (such as 304 stainless steel), this equation can lead to very nonconservative predictions. A modification of the Robinson–Taira equation, which forms the basis for an ASME code, helps alleviate this problem.

The theory of strain range partitioning (SRP) was developed at NASA-Lewis Research Center in the early 1970s. It involves the partitioning of a hysteresis loop obtained from a region of a structural member into four basic strain components: plastic-plastic ($\Delta\varepsilon_{pp}$), creep-creep ($\Delta\varepsilon_{cc}$), plastic-creep ($\Delta\varepsilon_{pc}$) and creep-plastic ($\Delta\varepsilon_{cp}$). The idealized hysteresis loops for these four strain components are shown in Fig. 4.60a. Only one of $\Delta\varepsilon_{pc}$ and $\Delta\varepsilon_{cp}$ can be present in any real hysteresis loop. Strain-life plots for all four types of strain components must be obtained from laboratory testing. Once these strain-life plots are obtained, the predicted life of the structural member, N_{pred}, is obtained using a linear interactive damage rule,

$$\frac{1}{N_{pred}} = \frac{F_{pp}}{N_{pp}} + \frac{F_{cc}}{N_{cc}} + \frac{F_{pc}}{N_{pc}} + \frac{F_{cp}}{N_{cp}}$$

where $F_{pp} = \Delta\varepsilon_{cc}/\Delta\varepsilon_{in}$, $F_{cc} = \Delta\varepsilon_{cc}/\Delta\varepsilon_{in}$, and so on. The quantity $\Delta\varepsilon_{in}$ is the total inelastic strain range of the structural member's hysteresis loop. That is, $\Delta\varepsilon_{in} = \Delta\varepsilon_{pp} + \Delta\varepsilon_{cc} + \Delta\varepsilon_{pc} + \Delta\varepsilon_{cp}$ (where one of $\Delta\varepsilon_{pc}$ and $\Delta\varepsilon_{cp}$ must be zero). N_{pp} is the number of cycles to failure if $\Delta\varepsilon_{in}$ were made up solely of plastic-plastic strains. Similar definitions hold for N_{cc}, N_{pc} and N_{cp}. Figure 4.60b should provide for further understanding of these quantities.

For isothermal fatigue, the modified Robinson–Taira theory and the theory of strain range partitioning both give reasonable results. While the theory of strain range partitioning is more complicated to use and can involve considerable laboratory testing, it usually leads to life predictions that are within a factor of two or three of actual results.

For thermal fatigue situations (i.e., situations in which large temperature fluctuations occur within each cycle), no theories have yet gained broad acceptance. An excellent overview of highly regarded theories is given in Ref. 63.

Multiaxial Fatigue

When attempting to predict fatigue lives under multiaxial loading conditions, many new factors must be considered. First of all, it must be known whether the loading is proportional (i.e., σ_1/σ_2 = constant throughout time) as shown in Fig. 4.61a, or nonproportional. Nonproportional loading can be further

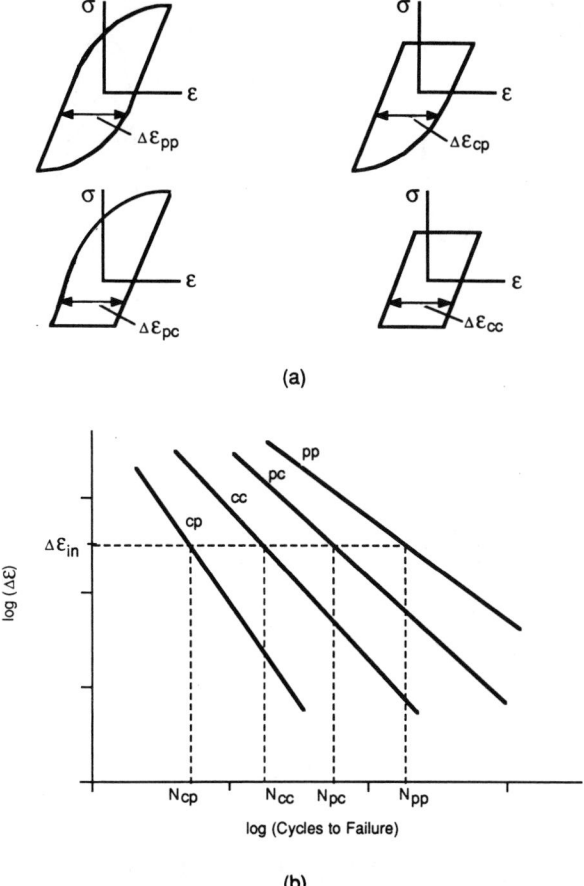

Fig. 4.60 Strain range partitioning components: (*a*) Idealized hysteresis loops for the four strain range components; (*b*) strain-life plots for the four strain range components.

divided into in-phase (Fig. 4.61*b*) and out-of-phase (Fig. 4.61*c*) situations. It must also be known if the cycling is periodic, as shown in Fig. 4.61, or of a block-loading or random-loading nature. If a point on the free surface of a component is being analyzed, it should be noted that the orientation of the principal stresses (or strains) with respect to the free surface can have a major influence on the fatigue life. Only theories for periodic, in-phase, proportional loading are considered here. Excellent reviews of the available theories for multiaxial fatigue can be found in Refs. 64–66.

Some of the earliest work on multiaxial fatigue centered around modifying static yield theories for fatigue situations. The results were the maximum normal stress theory,

$$S_f = \sigma_{a1}$$

the maximum shear stress theory,

$$S_f = \tfrac{1}{2}(\sigma_{a1} - \sigma_{a3})$$

or

$$e_f = \tfrac{1}{2}(\varepsilon_{a1} - \varepsilon_{a3})$$

and the energy-of-distortion theory,

$$S_f = \tfrac{1}{2}\left[(\sigma_{a1} - \sigma_{a2})^2 + (\sigma_{a1} - \sigma_{a3})^2 + (\sigma_{a2} - \sigma_{a3})^2\right]^{1/3}$$
$$e_f = \tfrac{2}{3}\left[(\varepsilon_{a1} - \varepsilon_{a2})^2 + (\varepsilon_{a1} - \varepsilon_{a3})^2 + (\varepsilon_{a2} - \varepsilon_{a3})^2\right]^{1/2}$$

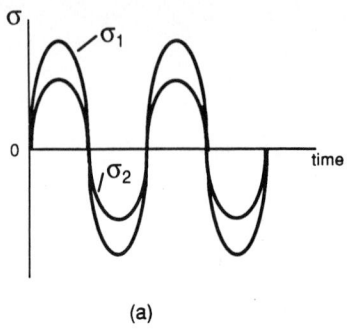

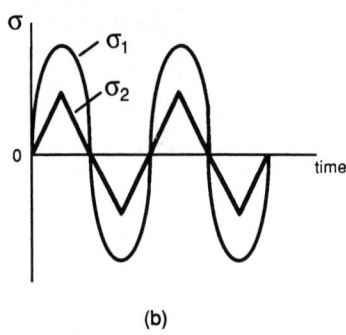

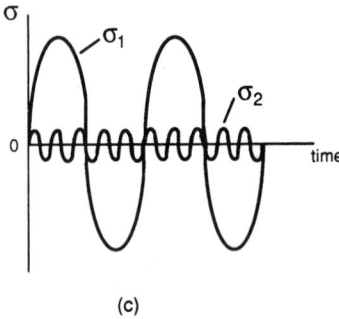

Fig. 4.61 Distinct examples of multiaxial cyclic loading: (*a*) Proportional loading; (*b*) in-phase nonproportional loading; (*c*) out-of-phase nonproportional loading.

In these equations, σ_{ai} is the amplitude of the *i*th principal stress. Similarly, ε_{ai} is the amplitude of the *i*th principal strain and S_f and e_f are stress and strain parameters, respectively, used to predict fatigue life. While each of these theories has been found to be applicable under certain loading conditions, each one can lead to serious errors in life prediction under other conditions.

A more successful approach that has been used by many investigators is to account for the effects of hydrostatic pressure on fatigue life. One of the more useful versions of these theories has the form

$$\left(\frac{\sigma_{a1} - \sigma_{a3}}{2}\right) + K_{N_f}\left(\frac{\sigma_{a1} + \sigma_{a3}}{2}\right) = C_{N_f}$$

where K_{N_f} and C_{N_f} are empirically determined parameters applicable to a certain fatigue life, N_f. The first term in parentheses represents the maximum shear stress amplitude, while the second term represents the normal stress acting on the plane of maximum shear. For low-cycle fatigue analysis, strain amplitudes should be used instead of stress amplitudes. Although this modern approach to multiaxial fatigue tends to give reasonably accurate results, considerable amounts of laboratory data must be acquired for its application.

Damage-Tolerant Design

The damage-tolerant design approach to fatigue analysis is based on linear elastic fracture mechanics (LEFM). In the usual approach, an initial crack size is assumed or known via nondestructive testing. Using LEFM techniques, the number of cycles required to propagate the crack to a size at which component failure occurs is determined. The primary disadvantage of damage-tolerant design is that, in many high-cycle fatigue applications, the initiation of a fatigue crack may entail a majority of the component life. Thus, assuming a certain initial crack size when none is actually present may lead to an overly conservative design. Neglecting this disadvantage, damage-tolerant design can be useful in many applications:

1. Given the desired service life, the smallest initial flaw size that would lead to failure at the end of the service life can be determined. Nondestructive testing that can detect this flaw size must be employed to assure the minimum desired service life.

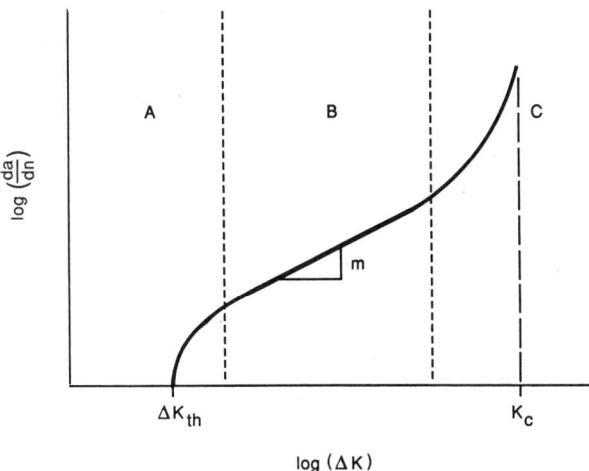

Fig. 4.62 Variation of fatigue-crack growth rate with ΔK.

2. If a flaw is discovered during routine in-service inspection (such as occurs with inspection of aircraft), the remaining safe life of the component before replacement is necessary can be assessed.

The primary relationship used in damage-tolerant design is the Paris equation[67]

$$\frac{da}{dN} = A(\Delta K)^{m} \qquad (4.103)$$

where da/dN is the crack growth rate, ΔK is the change in stress intensity factor in a single cycle, and A and m are material constants. The constant m is usually between 2 and 4 for most metals. Equation (4.103) is not valid for all values of ΔK; instead, it is applicable only within a certain range as shown in Fig. 4.62. Region A in the figure is often referred to as the threshold regime. Equation (4.103) is not valid in this region. In fact, for values of ΔK below the threshold intensity factor range, ΔK_{th}, no fatigue cracks should propagate. There is still some debate over the proper use of ΔK_{th} in design. In region B of Fig. 4.62, material response is relatively stable, and Eq. (4.103) is valid. In region C, crack propagation becomes unstable, and failure is imminent.

In order to apply the Paris equation for life prediction, the initial crack size and stress range must be such that $\Delta K > \Delta K_{th}$. Also, it must be possible to compute ΔK as a function of crack length. If these two things can be done, the Paris equation can be integrated to give the number of cycles until ΔK is so large that unstable crack propagation occurs. A wealth of information on damage-tolerant design can be found in Ref. 68.

4.9   MECHANICAL TESTING OF MATERIALS

4.9.1   Testing Machines

In the past few decades, the development of highly sophisticated mechanical testing equipment has resulted in new techniques for analyzing the mechanical properties of materials. The most notable of these developments was the proliferation of closed-loop, servo-hydraulic test systems in the 1960s. With the advent in recent years of low-cost microcomputers for digital data acquisition and processing, data reduction that used to require many hours to accomplish can now be done in seconds. By combining the closed-loop concept with digital control, powerful and versatile mechanical test systems have been created.

No matter what kind of mechanical test is being performed, or what level of sophistication exists in the testing equipment, all mechanical testing systems have some common properties:

1. A power source must be available to provide for loading of the specimen.
2. Some parameter is controlled, such as load, strain, displacement rate, impact energy, and so on.
3. The controlled parameter, and usually some other parameters, are continuously monitored. The monitoring may be by visual, mechanical, electrical, or digital means.

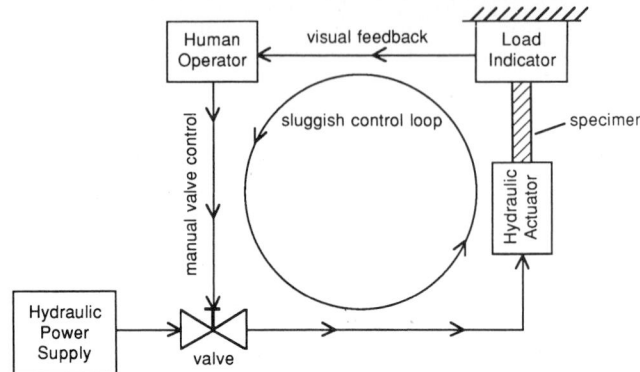

Fig. 4.63   Load control in an open-loop system via a human operator.

Open-Loop Systems

An open-loop system represents the crudest kind of mechanical testing system. There is no feedback to the control mechanism that would allow for continuous adjustment of the controlled parameter. Instead, the parameter is "controlled" by the preset factory adjustments of the control mechanism. For example, a screw-gear tension testing machine is capable of operating at a constant displacement rate. This is accomplished by operating the motor that drives the screws at a constant speed. However, the displacement rate thus obtained is determined based on the factory settings. There is no measurement of the displacement rate as the test proceeds. Also, it would not be possible to control loading rate with such a device since there is no feedback from the load measuring device to the motor driving the screws that would allow for the motor to adjust its speed so as to maintain a specified loading rate.

With some open-loop systems, such as universal hydraulic testing machines, the human operator can provide additional control. For example, the operator might continuously monitor the strain output of a tensile specimen (as measured using an extensometer) and adjust the amount of oil going to the hydraulic actuator that loads the specimen so as to maintain a desired strain rate. Clearly, this is a crude means for obtaining improved equipment response since humans are sluggish and prone to error considering the needs of modern mechanical testing. Figure 4.63 shows a schematic representation of using an open-loop system for controlling load by using a human operator to "close the loop." If the operator is removed, the system becomes completely open loop in nature.

Closed-Loop, Servo-Hydraulic Systems

In a closed-loop, servo-hydraulic testing system (shown schematically in Fig. 4.64), the parameter being controlled is continuously monitored by the testing system via the servo controller. If any error exists between the desired value of the controlled parameter (as measured from the electronic programmer) and the actual value of that parameter (as measured by a transducer or other suitable device), the servo controller applies a command signal to the servovalve. This command signal causes the servovalve to adjust the oil flow to the hydraulic actuator so as to compensate for the error. The speed at which adjustments can be made with this technology is exceedingly fast. In a standard system, approximately 10,000 adjustments per second are made automatically.

In a typical closed-loop system, it is possible to control load, strain, or displacement as functions of time; however, these systems can be relatively easily modified to control other parameters as well. It is possible to perform tension, compression, creep, stress relaxation, fatigue (low and high cycle), fracture mechanics, and many other mechanical tests, all with the same basic testing system. Because of this versatility, these systems are now widely used in university, industrial, and government laboratories. An example of a modern closed-loop, servo-hydraulic testing system is shown in Fig. 4.65.

Screw-Gear Machines

Screw-gear machines apply load to a specimen via a screw and gear mechanism. They are primarily designed for performing simple tension and compression tests.

Torsion Testing Machines

Torsion testing machines are typically used to determine the shear strength of materials. Both solid and hollow specimens are commonly used. In solid specimens, the shear stress varies continuously

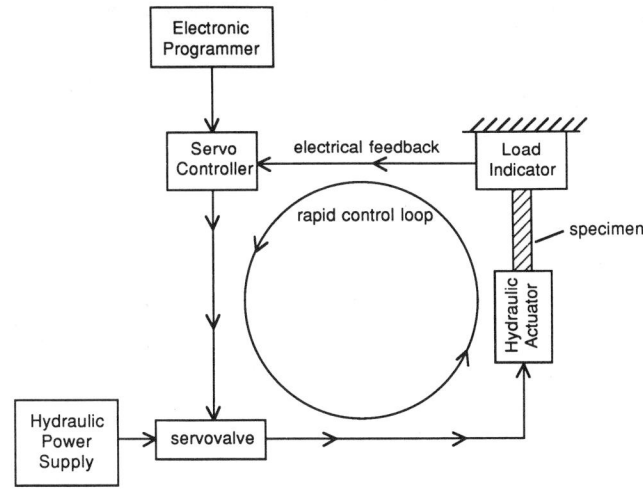

Fig. 4.64 Load control using a closed-loop, servo-hydraulic testing system.

Fig. 4.65 Closed-loop, servo-hydraulic testing system.

with radial distance as discussed in Section 4.3. Thin-walled hollow specimens allow for more accurate determination of material properties than solid specimens since the shear stress is approximately constant throughout the wall thickness.

Both gear-actuated and hydraulic-powered machines are commonly used for performing torsion tests. The advantage of hydraulic-powered machines is that they can be incorporated into a sophisticated, closed-loop testing system. In fact, closed-loop, servo-hydraulic testing systems that allow for simultaneous but independent axial and torsional control have been commercially available for many years.

Impact Testing Machines
Impact testing machines typically consist of a pendulum hammer. The hammer is released from a known height and strikes a small notched specimen, causing it to fracture. The hammer proceeds to

rise to some final maximum height. The difference between the initial and final heights of the hammer is directly proportional to the energy absorbed by the specimen.

In the **Charpy pendulum machine**, the specimen is placed horizontally with the ends supported by an anvil. The hammer strikes the specimen midway between the supports opposite the notch.

In the **Izod machine**, the specimen is mounted vertically and supported from the bottom in a cantilever fashion. The hammer strikes the specimen on its notched side at a specified distance above the notch.

Hardness Testing Machines

The **Brinell hardness testing machine** measures the resistance of a material to the penetration of a hardened steel ball subjected to a standard load. The diameter of the indentation made is measured with the aid of a microscope. A Brinell hardness number is calculated based on this diameter. This number is used to make relative comparisons of the hardness of different materials.

The **Rockwell hardness testing machine** imparts a standard load on a steel ball or diamond indenter. A direct read-out of the Rockwell hardness number is obtained.

The **Vickers hardness testing machine** employs a pyramidal diamond indenter subjected to a standard load. The diagonal length of the indentation is measured with a microscope, and the Vickers hardness number is calculated based on this diagonal length.

The **Shore scleroscope** measures the rebound height of a weight dropped on a specimen. This rebound height provides a relative measure of hardness.

Fatigue Testing Machines

The rotary bending machine and the flat-plate bending machine are among the simplest fatigue testing machines. Most of these utilize cantilever specimens that are bent repeatedly, with constant amplitudes of force or displacement. Although rotary and flat-plate bending machines (and some other simple fatigue testing machines) are still in common use, a large percentage of modern fatigue testing is carried out using closed-loop, servo-hydraulic testing systems. These systems have the advantage of great versatility in terms of allowable specimen designs, operating speeds, and wave shapes available (including realistic random loading).

4.9.2   Transducers

A transducer is an electromechanical device that converts a change in a mechanical quantity, such as displacement or force, into a change in an electrical quantity that can be monitored as a voltage after signal processing.[69]

Differential Transformers

Differential transformers, which are used to measure displacement, are based on a variable-inductance principle. The most popular of these devices is the linear variable differential transformer (LVDT). This device consists of one primary and two secondary coils surrounding a magnetic core (Fig. 4.66). As the core moves, the mutual inductance between the primary and secondary coils changes. This inductance change is detected as a change in the voltage output, E_0. Within the range of operation, the relation between E_0 and core displacement is linear. The sensitivities of LVDTs vary from about 4 to 70 V per inch of displacement. The greater the displacement range capability of an LVDT, the lower its sensitivity is.

Electrical Resistance Strain Gages

Electrical resistance strain gages consist of thin metal-foil grids bonded to a polymer backing. These gages are available in a variety of sizes and configurations. Standard gage resistances are 120 and 350 Ω. In order to use a strain gage to monitor strain, it must first be bonded to the structure under investigation. This is typically accomplished using epoxy adhesives. When the structure is loaded, the strains induced in the structure are transmitted to the strain gage which results in a resistance change in the metal-foil grid. When a current is applied to the metal-foil grid, this resistance change is detected as a change in voltage. The relationship between strain and resistance change is given by

$$\frac{\Delta R_g}{R_g} = S_g \varepsilon$$

where ε is the strain induced in the gage, R_g is the original gage resistance, and S_g is the gage factor for the gage. For Advance alloy, the alloy most commonly used in electrical resistance strain gages, S_g usually has a value of around 2.0.

The output of a strain gage, $\Delta R_g / R_g$, is usually converted to a voltage signal using a Wheatstone bridge (Fig. 4.67). One, two, or four gages are typically employed in the bridge. The advantage of

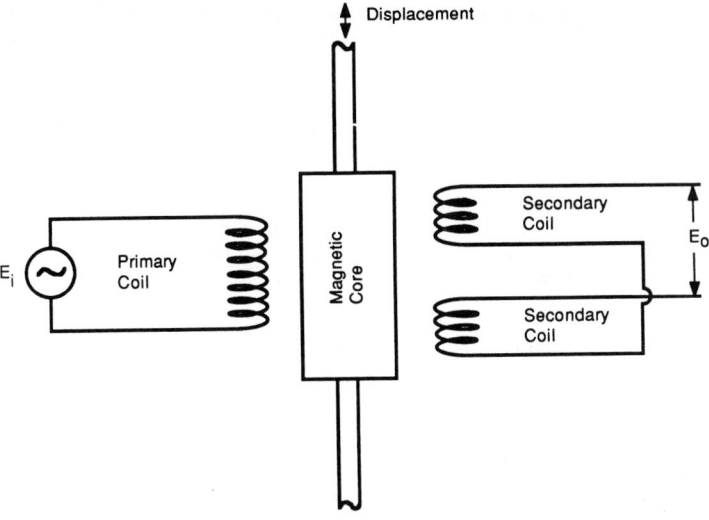

Fig. 4.66 Schematic diagram of the LVDT circuit.

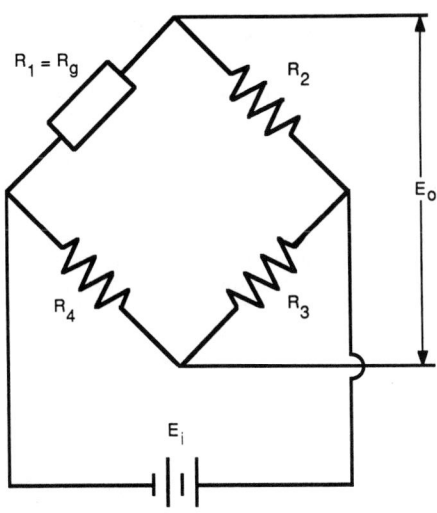

Fig. 4.67 Schematic diagram of the Wheatstone bridge circuit.

employing more than one gage in the bridge is that greater sensitivities can be obtained. For transducers employing electrical resistance strain gages, such as load cells and extensometers, four gages are normally employed in the Wheatstone bridge in order to maximize sensitivity as well to eliminate undesirable effects. For example, a properly designed axial load cell will be insensitive to bending and torsional effects. Further details can be found in Ref. 1.

Capacitance Strain Gages

The capacitance strain gage consists of two parallel plates of equal size separated by a small air gap. If the distance between the plates changes, or one plate moves laterally with respect to the other, a change in the capacitance of the gage occurs. For lateral movements, the relationship between displacement and capacitance change is linear. However, a nonlinear relationship exists between changes in the air gap thickness and the corresponding change in capacitance. These gages are not significantly affected by changes in the pressure or temperature of the surrounding environment.

Load Cells
Load cells are devices that are used for measuring forces. Modern load cells typically employ electrical resistance strain gages that are bonded to an elastic member to which the load to be measured is applied. The elastic member is usually a ring, beam, link, or shear-web. Forces applied to the elastic member cause strains in the member. These strains are transmitted to the strain gages and cause resistance changes in the gages. These resistance changes are converted to a voltage output by means of a Wheatstone bridge. Commercially available load cells employing these designs are available in force capacities ranging from tenths of a pound to several million pounds. These load cells can normally be employed for both static and dynamic measurements.

Extensometers
Extensometers are devices used to measure strains in a test specimen. They commonly consist of a flexure plate to which electrical resistance strain gages are bonded. The flexure plate is connected to the specimen through rigid arms. Strain in the specimen induces a bending strain in the flexure plate that is transmitted to the strain gages. Four strain gages are typically employed so that a full Wheatstone bridge circuit is obtained. Commercially available extensometers have gage lengths ranging from 0.12 in. up to 2 in. The maximum strain that can be measured is usually between 0.15 and 0.50 in./in. The main advantage of using an extensometer rather than mounting strain gages directly on the specimen is that the extensometer can be used for innumerable tests, can be calibrated before and after each test, and requires relatively short preparation. Strain gages are sacrificed in one test. On the other hand, extensometers are relatively expensive and take up more space than strain gages.

4.9.3   Standardized Testing Procedures

In order to obtain reliable and consistent data from one laboratory to another, or from one test to another, it is important to have carefully controlled testing procedures. Variations in test results can come about due to differences in specimen sizes and fabrication procedures, inaccurately calibrated equipment, differences in loading rates, environmental factors such as humidity and temperature, and gripping systems and procedures. When measuring complicated material properties, such as plane strain fracture toughness, seemingly minor variations in specimen quality and dimensions can lead to drastically different results. It is therefore important to adhere to carefully prepared testing standards.

While a variety of engineering societies have adopted mechanical testing standards, the most widely accepted are the standards published by the American Society for Testing and Materials. There is not enough space here for even a brief discussion of the mechanical testing standards that have been adopted by this society. Suffice it to say that standards for most common engineering materials and mechanical tests (tension, compression, fatigue, plane strain fracture toughness, etc.) are available through this society. Complete details of the accepted testing procedures are updated annually and compiled in the *Annual Book of ASTM Standards*.[5]

4.9.4   Experimental Stress Analysis

The expression "experimental stress analysis" is somewhat of a misnomer. In reality, it is usually strains that are measured experimentally and not stresses. Nevertheless, the expression is in general use today. A variety of techniques have been developed to measure strains in structural components and models. These include electrical resistance strain gages (discussed in Section 4.9.2), brittle coatings, photoelasticity, moiré analysis, Thermographic Stress Analysis, laser holography and interferometry, and many more. Some of these techniques are briefly described in the following paragraphs.

Brittle Coatings
In the brittle coating method of stress analysis, a thin layer of resin-based or ceramic-based coating is applied to the surface of a component. When the component is loaded, the strains induced in the component are transmitted to the brittle coating causing it to crack. The technique has the advantages that the brittle coating is applied to the actual component rather than to a model, and full-field data on both the magnitude and direction of principal strains can be obtained in a rapid fashion. The primary disadvantage of resin-based brittle coatings is that accurate determination of stress magnitudes can be difficult due to the sensitivity of the coating to thickness and environmental effects. The primary disadvantage of ceramic-based brittle coatings is that these coatings must be fired onto the component at temperatures ranging from 950 to 1100°F. This severely limits their use with many alloys.[1]

In order to obtain quantitative data with brittle coatings, it is necessary to determine the threshold strain at which the coating will crack. This is accomplished by spraying several calibration bars with the coating at the same time the component is sprayed. At the time the component is to be tested, the calibration bars are loaded in a cantilever fashion, using a special loading fixture, until cracks appear. The threshold strain is then easily determined by means of the flexure formula and Hooke's law.

Both ceramic-based and resin-based brittle coatings suffer from the fact that only tensile strains cause the coatings to crack. It is still possible to detect compressive strains, however, by employing the following technique:

1. Load the component to the desired load level.
2. Spray the component with the brittle coating and allow the coating to dry. At this point, the coating will be stress free.
3. Quickly release some of the load from the component and observe if cracks appear.

When the load is removed, compressively stressed regions of the component will expand. This induces tensile strains in the coating in these regions.

In general, brittle coatings are most often used today for qualitative analysis. They provide a simple means for determining the regions of maximum stress in a component. Once these regions have been isolated, electrical resistance strain gauges are normally applied so that more accurate strain measurements can be made.

Moiré Method of Strain Analysis

The moiré method of strain analysis[70, 71] may be used for determining full-field strains in situations where the magnitudes of the strains are large. This method employs two gratings that consist of closely spaced opaque parallel lines with transparent spacing between the lines. Generally, the width of the transparent spacing is equal to the width of the opaque lines. The number of lines per inch, typically 50 to 1000, is referred to as the pitch of the grating. One grating (the model grating) is applied to the actual component. The second grating is referred to as the reference grating. It is overlayed on top of the model grating. A thin layer of oil or other suitable substance is placed between the two gratings to keep them in contact while at the same time minimizing the transmission of strains from the model to the reference grating.

Initially, the gratings are aligned so that no fringe pattern is apparent. When the component is loaded, the model grating is strained while the reference grating is not. This causes a fringe pattern to develop. The fringes represent curves along which the displacement in a direction perpendicular to the grating lines is constant. The change in length in a 1-in. interval of the component is given by $\Delta L = np$, where ΔL is the change in length in 1 in., p is the pitch grating (lines per inch), and n is the number of fringes in the 1-in. gauge length. By estimating the rate of change of fringe spacing (i.e., $\Delta L/L$), it is possible to obtain displacement gradients ($\partial u/\partial x$ and $\partial u/\partial y$ if the grating lines are perpendicular to the x direction). These can be substituted in Eqs. (4.3) if the strains or displacements are large, or Eqs. (4.4) if the strains and displacements are small, in order to determine the strains in the component. Since standard gratings have lines in one direction only, it is necessary to separately apply two sets of gratings in perpendicular directions in order to determine all four displacement gradients ($\partial u/\partial x$, $\partial u/\partial y$, $\partial v/\partial x$, and $\partial v/\partial y$) needed for obtaining the surface strains, ε_{xx}, ε_{yy}, and γ_{xy}. The displacement gradients $\partial w/\partial x$ and $\partial w/\partial y$ are usually small enough to be neglected. An example of a moiré pattern of a polyethylene plate with a hole is given in Fig. 4.68.

Photoelasticity

The following discussion is intended to serve merely as a brief and elementary overview of the photoelastic method. More thorough discussions of this method can be found in many textbooks and journals. In particular, see Refs. 1, 72–74.

Certain transparent materials, when placed under load, transmit incoming light rays at different speeds depending on the angle of the plane of oscillation of the light ray relative to the directions of the principal stresses. When the loading is removed, all light rays are propagated at equal rates. This phenomenon is known as temporary double refraction, and it forms the basis of the photoelastic method. A polariscope is a device that is used to observe fringe patterns that come about due to the stressing of flat specimens made of a temporarily doubly refracting material.

A plane polariscope consists of a light source, two polarizing elements, and the specimen (which must be made of a temporarily doubly refracting material). The light first passes through one polarizing element (called the polarizer), then through the specimen, and finally through a second polarizing element (called the analyzer) whose axis of polarization is at 90° to the polarizer (Fig. 4.69). No light passes through the analyzer if the specimen is unstressed, and a dark field is obtained. If the specimen is stressed, two sets of fringes are formed. One set of black fringes, referred to as isoclinics, occur at those points along which one of the principal-stress directions coincides with the

Fig. 4.68 Moiré fringes about a hole in a polyethylene bar. The straight lines are for reference. The bar is loaded uniaxially in the vertical direction. (Courtesy T. G. Ebbott, University of Wisconsin-Madison).

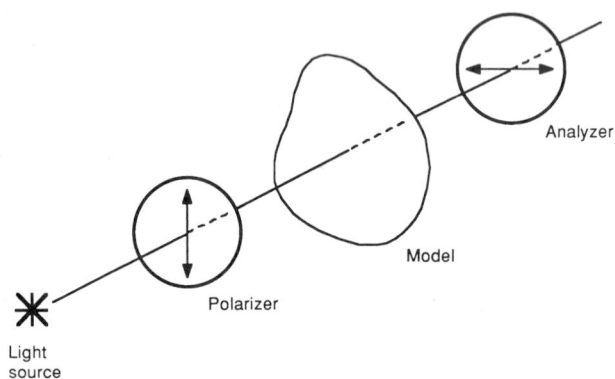

Fig. 4.69 Schematic diagram of a plane polariscope.

axis of the polarizer. A second set of fringes (known as isochromatics) occur at points where the sum of the principal stresses is zero or is sufficient to produce an integral number of wavelengths of retardation:

$$\sigma_1 - \sigma_2 = \frac{N f_\sigma}{h} \qquad N = 0, 1, 2, 3, \ldots$$

where σ_1 and σ_2 are the principal directions in the plane of the specimen ($\sigma_3 = 0$ for a thin specimen), h is the specimen thickness, and f_σ is known as the material fringe value or fringe stress

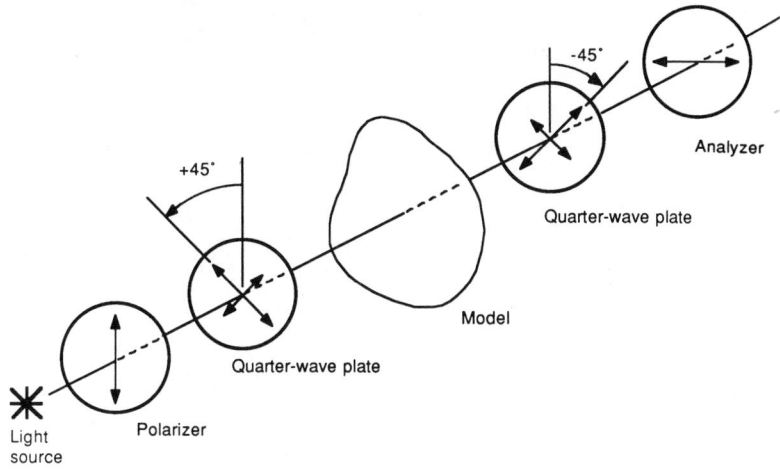

Fig. 4.70 Schematic diagram of a circular polariscope.

coefficient. The material fringe value depends on the wavelength of light being used as well as on the material. If white light is used, the isochromatics appear as colored fringes since different values of $\sigma_1 - \sigma_2$ lead to extinction of different wavelengths of light. Black fringes appear only at points where $\sigma_1 - \sigma_2 = 0$. If a monochromatic light source is used, all the fringes appear black. In addition, monochromatic light produces more sharply defined fringes than white light.

In a **circular polariscope**, quarter-wave plates are inserted between the specimen and the polarizing elements. A quarter-wave plate consists of a permanently doubly refracting material. It transmits a light ray at different speeds in two orthogonal directions. The axis along which the light is transmitted most rapidly is referred to as the fast axis. The axis along which the light is transmitted most slowly is referred to as the slow axis. For the first quarter-wave plate, the fast axis is placed at an angle of $+45°$ to the axis of the polarizer. For the second quarter-wave plate, the fast axis is placed at an angle of $-45°$ to the axis of the polarizer (Fig. 4.70). The thickness of a quarter-wave plate is chosen so that, for a specific wavelength of light, a relative angular phase shift between the fast and slow axes of $90°$ occurs. Clearly, a circular polariscope must employ a monochromatic light source since the quarter-wave plates are designed for a specific wavelength of light.

The effect of adding the quarter-wave plates into a plane polariscope in order to form a circular polariscope is to eliminate the isoclinic fringes. For the circular polariscope shown schematically in Fig. 4.70, a dark field is produced (no light is transmitted through the polariscope) when the specimen is unstressed. If the analyzer is rotated through $90°$, a light field is produced (all the light is transmitted through the polariscope) when the specimen is unstressed. If a light-field configuration is employed, dark fringes are created at points along which

$$\sigma_1 - \sigma_2 = \frac{(N + 0.5)f_\sigma}{h} \qquad N = 0, 1, 2, 3, \ldots$$

Thus, when both light-field and dark-field configurations are employed, the amount of principal-stress difference data available is doubled as compared to using only one of the two configurations. An example of a dark-field isochromatic fringe pattern is shown in Fig. 4.71.

The isochromatic patterns give the principal-stress difference, $\sigma_1 - \sigma_2$, while the isoclinics give the principal-stress directions. A variety of techniques are available that allow for determination of the individual principal-stress values, σ_1 and σ_2. These techniques are referred to as stress separation methods. They include methods based on the stress equilibrium equations, the equations of compatibility, the generalized Hooke's law, and the oblique-incidence method.

The advantages of the method of photoelasticity include the following:

1. Full-field stress analysis is possible.
2. Both the magnitude and the directions of the principal stresses can be obtained.
3. The method can be used for the determination of stress patterns that are three-dimensional in nature by means of a stress-freezing technique (see Ref. 1 for an excellent discussion of this technique).

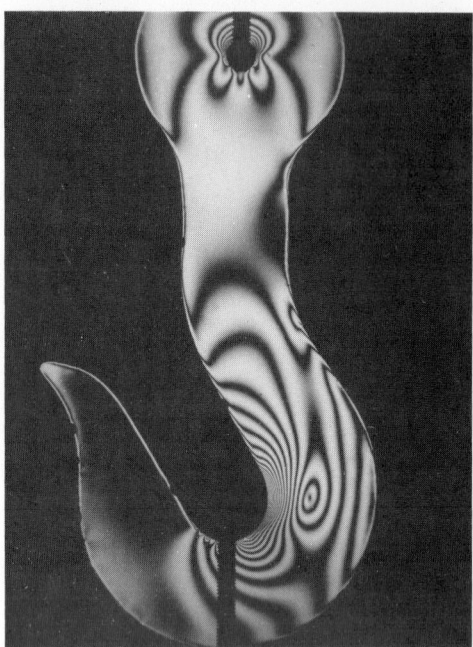

Fig. 4.71 Dark-field isochromatic fringe pattern of a model hook under load. (Courtesy J. Dreger, University of Wisconsin-Madison).

The primary disadvantages of the photoelastic method are the considerable amount of time required to separate the principal stresses and the fact that plastic models must be used rather than actual components.

Thermographic Stress Analysis

When a member is loaded within the elastic range, the member experiences a small change in temperature due to the thermoelastic effect. The thermoelastic equation relates this temperature change to the change in the sum of the principal stresses:

$$(\sigma_1 + \sigma_2 + \sigma_3)_{t_1} - (\sigma_1 + \sigma_2 + \sigma_3)_{t_2} = K(\Delta T)$$

where ΔT is the change in temperature between times t_1 and t_2, and K is a temperature-dependent material parameter. For most solids at room temperature, K can be considered a constant. This equation is valid only for linear elastic, isotropic, adiabatic response. For anistropic, nonadiabatic, or elastic-plastic response, a more elaborate equation is necessary.

In a typical application of Thermographic Stress Analysis (TSA), also referred to as the SPATE method or thermoelastic stress analysis, a member is subjected to sinusoidal cyclic loading. This results in a sinusoidal variation in the member's temperature, as given by the thermoelastic equation. This temperature change causes a change in the infrared flux emitted by the member. This flux change is monitored with an infrared detector which is attached to approiate signal conditioning and post-processing equipment.[75] Using this technique, full-field data can be obtained with strain resolutions approaching those found in strain gages. Another advantage of TSA is that there is no known upper temperature limitation. Data have been obtained at over 1000°C with strain resolutions equal to those obtainable at room temperature. A host of other advantages is emerging. The two primary disadvantages of TSA are that the component must be cyclically loaded, and separation of the principal stresses into individual stress components cannot be done in any straightforward fashion. An example of a full-field TSA scan is shown in Fig. 4.72.

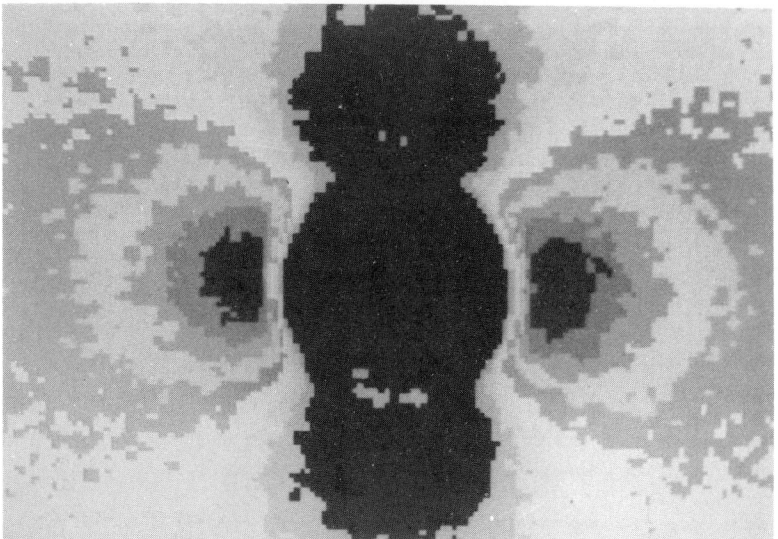

Fig. 4.72 TSA frame scan of an aluminum plate with a hole. Dark circular region at center of photo represents the hole. Dark circular regions on top and bottom of hole represent regions of large compressive stresses. Dark circular regions on right and left of hole represent regions of large tensile stresses. Plate is loaded uniaxially in the vertical direction. More detail is seen in color images.

REFERENCES

1. Dally, J. W., and Riley, W. F., *Experimental Stress Analysis*, 2nd ed., New York, McGraw-Hill, 1978.
2. Flügge, W. (ed.), *Handbook of Engineering Mechanics*, New York, McGraw-Hill, 1962.
3. Malvern, L. E., *Introduction to the Mechanics of a Continuous Medium*, Englewood Cliffs, NJ, Prentice-Hall, 1969.
4. Timoshenko, S. P., and Goodier, J. N., *Theory of Elasticity*, 3rd ed., New York, McGraw-Hill, 1970.
5. American Society for Testing and Materials (ASTM), *Annual Book of ASTM Standards*, Philadelphia, ASTM, 1987.
6. Timoshenko, S., *Strength of Materials*, 2nd ed., New York, Van Nostrand, 1940.
7. Cook, R. D., and Young, W. C., *Advanced Mechanics of Materials*, New York, Macmillan, 1985.
8. Boresi, A. P., Sidebottom, O. M., Seely, F. B., and Smith, J. O., *Advanced Mechanics of Materials*, 3rd ed., New York, Wiley, 1978.
9. Roark, R. J., and Young, W. C., *Formulas for Stress and Strain*, 5th ed., New York, McGraw-Hill, 1982.
10. Beer, F. P., and Johnston, E. R., Jr., *Mechanics of Materials*, New York, McGraw-Hill, 1981.
11. Den Hartog, J. P., *Advanced Strength of Materials*, New York, McGraw-Hill, 1952.
12. Timoshenko, S. P., and Woinowsky-Krieger, S., *Theory of Plates and Shells*, 2nd ed., New York, McGraw-Hill, 1959.
13. Szilard, R., *Theory and Analysis of Plates*, Englewood Cliffs, NJ, Prentice-Hall, 1974.
14. Langhaar, H. L., *Energy Methods in Applied Mechanics*, New York, Wiley, 1962.
15. McFarland, D., Smith, B. L., and Bernhart, W. D., *Analysis of Plates*, New York, Spartan Books, 1972.
16. Cook, R. D., *Concepts and Applications of Finite Element Analysis*, 2nd ed., New York, Wiley, 1981, Chap. 9.
17. Griffel, W., *Plate Formulas*, New York, Frederick Ungar, 1968.
18. Lekhnetskii, S. G., *Anisotropic Plates*, New York, Gordon & Breach, 1968.

19. Vinson, J. R., *Structural Mechanics: The Behavior of Plates and Shells*, New York, Wiley, 1974.

20. *Boiler and Pressure Vessel Code*, Section III, Nuclear Power Plant Components, Division 1, New York, The American Society of Mechanical Engineers, 1980.

21. Bridgman, P. W., *Studies in Large Plastic Flow and Fracture with Special Emphasis on the Effects of Hydrostatic Pressure*, New York, McGraw-Hill, 1952.

22. Mendelson, A., *Plasticity: Theory and Application*, New York, Macmillan, 1968.

23. *Manual of Steel Construction*, (AISC Handbook), 7th ed., New York, American Institute of Steel Construction, 1970.

24. Lubahn, J. D., and Felgar, R. P., *Plasticity and Creep of Metals*, New York, Wiley, 1961.

25. Flügge, W., *Viscoelasticity*, Waltham, MA, Blaisdell, 1967.

26. Christensen, R. M., *Theory of Viscoelasticity*, 2nd ed., New York, Academic Press, 1982.

27. Chajes, A., *Principles of Structural Stability Theory*, Englewood Cliffs, NJ, Prentice-Hall, 1974.

28. Timoshenko, S. P., and Gere, J. M., *Theory of Elastic Stability*, New York, McGraw-Hill, 1952.

29. *AISC Specification for the Design, Fabrication, and Erection of Structural Steel for Buildings*, Chicago, American Institute of Steel Construction, November 1978.

30. *Specifications for Aluminum Structures*, Washington, D.C., Aluminum Association, Inc., 1976.

31. *Timber Construction Manual*, American Institute of Timber Construction, New York, Wiley, 1974.

32. Schwartz, M. M., *Composite Materials Handbook*, New York, McGraw-Hill, 1984.

33. Jones, R. M., *Mechanics of Composite Materials*, New York, McGraw-Hill, 1975.

34. Rowlands, R. E., "Strength (Failure) Theories and Their Experimental Correlation," in *Handbook of Composites*, Vol. 3, Sih, G. C., and Skudra, A. M., (eds.), Amsterdam, North-Holland, 1985, pp. 71–125.

35. Christensen, R. M., *Mechanics of Composite Materials*, New York, Wiley, 1979.

36. Tewary, V. K., *Mechanics of Fibre Composites*, New York, Wiley, 1978.

37. Agarwal, B. D., and Broutman, L. J., *Analysis and Performance of Fiber Composites*, New York, Wiley, 1980.

38. Tsai, S. W., and Hahn, H. T., *Introduction to Composite Materials*, Westport, CT, Technomic Publishing Co., 1980.

39. Peterson, R. E., *Stress Concentration Design Factors*, New York, Wiley, 1953.

40. Blake, A., (ed.), *Handbook of Mechanics, Materials, and Structures*, New York, Wiley, 1985, p. 268.

41. Kanninen, M. F. and Popelar, C. H., *Advanced Fracture Mechanics*, New York, Oxford University Press, 1985.

42. Blake, A., (ed.), *Handbook of Mechanics, Materials, and Structures*, New York, Wiley, 1985, p. 270.

43. Hertzberg, R. W., *Deformation and Fracture Mechanics of Materials*, New York, Wiley, 1976.

44. Hellan, K., *Introduction to Fracture Mechanics*, New York, McGraw-Hill, 1984.

45. *Fatigue and Microstructure*, Metals Park, OH, American Society for Metals, 1979.

46. Westergaard, H. M., *Trans., ASME, J. Appl. Mech.*, Vol. 61, p. 49, 1939.

47. Sih, G. C., *Handbook of Stress Intensity Factors*, Lehigh University, Bethlehem, PA, 1973.

48. Rooke, D. P., and Cartwright, D. J., *Compendium of Stress Intensity Factors*, London, H. M. Stationary Office, 1976.

49. Mathews, W. T., *Plain Strain Fracture Toughness (K_{Ic}) Data Handbook for Metals*, Army Materials and Mechanics Research Center, 1974, AD-773-673.

50. Campbell, J. E., "Plane-Strain Fracture Toughness Data for Selected Metals and Alloys", DMIC Report S-28, 1969.

51. American Society for Testing and Materials (ASTM), *Annual Book of ASTM Standards*, E339-83, Section 3, Vol. 03.01, ASTM, Philadelphia, 1983, pp. 518–553.

52. American Society for Testing and Materials (ASTM), *Annual Book of ASTM Standards*, E436-71T, Part 31, ASTM, Philadelphia, 1971, p. 1005.

53. Rice, J. R., "A Path Independent Integral and the Approximate Analysis of Strain Concentration by Notches and Cracks," *Trans., ASME, J. Appl. Mech.*, Vol. 35, pp. 379–386, 1968.

54. American Society for Testing and Materials (ASTM), *Annual Book of ASTM Standards*, E813-81, Section 3, Vol. 03.01, ASTM, Philadelphia, 1983, pp. 762–780.

55. Kanninen, M. F., and Popelar, C. H., *Advanced Fracture Mechanics*, New York, Oxford University Press, 1985.

56. Sandor, B. I., *Fundamentals of Cyclic Stress and Strain*, Madison, WI, The University of Wisconsin Press, 1972.

57. Shigley, J. E., *Mechanical Engineering Design*, 2nd ed., New York, McGraw-Hill, 1972, pp. 243–259.

58. Coffin, L. F., Jr., "Low Cycle Fatigue: A Review," *Applied Materials Research*, Vol. 1, pp. 129–141, 1962.

59. Morrow, JoDean, "Cyclic Plastic Strain Energy and Fatigue of Metals," ASTM STP 378, American Society for Testing and Materials, 1965, pp. 45–87.

60. Socie, D. F., "Fatigue Life Estimation Techniques," Datamyte Corporation, Minnetonka, MN, 1981.

61. Robinson, E. L., "Effect of Temperature Variation on the Long-Time Strength of Steels," *Trans. ASME*, Vol. 74, pp. 777–781, 1952.

62. Zamrik, S. Y. (ed.), *Design for Elevated Temperature Environment*, New York, ASME, 1971, pp. 12–24.

63. Halford, G. R., "Low-Cycle Thermal Fatigue," NASA TM-87225, 1986.

64. Zamrik, S. Y., and Dietrich, D. (eds.), *ASME Pressure Vessels and Piping: Design Technology—1982. A Decade of Progress*, The American Society of Mechanical Engineers, New York, 1982, pp. 507–518.

65. Brown, M. W., and Miller, K. J., "Two Decades of Progress in the Assessment of Multiaxial Low-Cycle Fatigue Life," ASTM STP 770, American Society for Testing and Materials, 1982, pp. 482–499.

66. Garud, Y. S., "Multiaxial Fatigue: A Survey of the State of the Art," *Journal of Testing and Evaluation*, Vol. 9, pp. 165–178, 1981.

67. Paris, P. C., and Erdogan, F., "A Critical Analysis of Crack Propagation Laws," *Trans., ASME, J. Basic Engineering*, Vol. 85, p. 528, 1963.

68. *USAF Damage Tolerant Design Handbook: Guidelines for the Analysis and Design of Damage Tolerant Aircraft Structures*, Wright-Patterson Air Force Base, OH, 1984.

69. Dally, J. W., Riley, W. F., and McConnell, K. G., *Instrumentation for Engineering Measurements*, New York, Wiley, 1984, Chap. 3.

70. Theocaris, P. S., *Moire Fringes in Strain Analysis*, Oxford, Pergamon Press, 1969.

71. Durelli, A. J., and Parks, V. J., *Moire Analysis of Strain*, Englewood Cliffs, NJ, Prentice-Hall, 1970.

72. Heywood, R. B., *Photoelasticity for Designers*, Oxford, Pergamon Press, 1969.

73. Frocht, M. M., *Photoelasticity*, New York, Wiley, 1941.

74. Durelli, A. J., and Riley, W. F., *Introduction to Photomechanics*, Englewood Cliffs, NJ, Prentice-Hall, 1970.

75. Oliver, D. E., "Stress Pattern Analysis by Thermal Emission," in *Handbook on Experimental Mechanics*, Kobayashi, A. S. (ed.), Prentice-Hall, 1987, Chap. 14.

CHAPTER 5
MECHANICS OF INCOMPRESSIBLE FLUIDS

Gary Z. Watters

College of Engineering, Computer Science, and Technology
California State University
Chico, California

5.1 INTRODUCTION

All substances are compressible to some extent and fluids are no exception. However, in many cases, fluids may be treated as incompressible without introducing unacceptable inaccuracies in either computations or measurements. In the case of liquids, the vast majority of problems may be addressed as incompressible flow problems. Even in situations where pressure changes are significant enough to cause small changes in density (water hammer), incompressible flow techniques are applied to solve problems. For gases where flow velocities are low compared to the local speed of sound (low

TABLE 5.1   Basic Dimensions and Abbreviations in English and SI Units*

Abbreviations

BTU = British Thermal Unit	m = meter (SI) = mile (FSS)
cfs = cubic feet per second	mb = millibar = 10^{-3} bar
fps, ft/sec = feet per second	mm = millimeter = 10^{-3} meter
ft = foot	mm^2 = square millimeter
gpm = gallons per minute	mph = miles per hour
hp = horsepower	mps, m/s = meters per second
h = hour	N = newton
Hz = hertz	Pa = pascal = N/m^2
in. = inch	psi = pounds per square inch
J = joule = N · m	sec, s = second
kg = kilogram = 10^3 gram	W = watt = J/sec
lb = pound force	

Units

Quantity	SI Unit Name (Symbol)	FSS Unit Name (Symbol)
	Basic Units	
Length	Meter (m)	Foot (ft)
Mass	Kilogram (kg)	Slug (slug)
Time	Second (s)	Second (sec)
Temperature	Kelvin (°K)	Rankine (°R)
	[Celsius (°C)]	[Fahrenheit (°F)]
	Derived Units	
Energy	Joule (J)	Foot-pound (ft-lb)
Force	Newton (N)	Pound (lb)
Frequency	Hertz (Hz)	Hertz (Hz)
Power	Watt (W)	Horsepower (hp)
Pressure	Pascal (Pa)	—

Source: Vennard and Street.[1]

Mach number) or where density changes in the system are small, incompressible flow theory may be used to good approximation.

5.1.1   Definition of a Fluid

A **fluid** is defined as a collection of molecules of a substance (or several substances) that cannot support a shearing stress without undergoing permanent and continual angular deformation. Even though the fluid is a collection of molecules, it is generally considered to be a continuous substance without voids, referred to as a continuum. Newtonian fluids are those whose rate of angular deformation is linear with respect to the magnitude of the deforming shear stress. Most common liquids and gases are Newtonian fluids. Non-Newtonian fluids are those whose rate of angular deformation bears a nonlinear relationship to the applied shear stress. Examples of non-Newtonian fluids are blood, paints, and suspensions.

5.1.2   System of Units and Dimensions

The scientific world is in the midst of a conversion from various systems of units to SI (Système International d'Unités) units. However, the English FSS (foot-slug-second) system is still in widespread use throughout the United States. Hence, this work will use the English system as a primary system of units with the SI equivalent provided wherever possible in tables and numerical examples. Table 5.1 provides the units used to identify the basic dimensional quantities in each system, and Table 5.2 gives a conversion table for fluid properties and other commonly used quantities. Table 5.3 provides a list of the symbols used in this chapter along with their definitions.

*Tables 5.1, 5.2, 5.4–5.7 are reproduced, with permission, from J. K. Vennard and R. L. Street, *Elementary Fluid Mechanics*, 5th ed., New York, Wiley, 1975.

TABLE 5.2   Conversion Factors for English FSS and SI Units

Absolute viscosity   1 slug/ft sec = 1 lb-sec/ft^2 = 47.88 N s/m^2 = 47.88 kg/m s = 478.8 poises
Acceleration due to gravity   32.174 ft/sec^2 = 9.80665 m/s^2
Area   1 ft^2 = 0.0929 m^2
$\quad\quad$ 1 in.^{2} = 645.2 mm^2
Density   1 slug/ft^3 = 515.4 kg/m^3
Energy   1 ft-lb = 1.356 J = 1.356 Nm = 3.77 × 10^{-7} kWhr
$\quad\quad$ 1 BTU = 778.2 ft-lb = 1055 J = 2.93 × 10^{-4} kWhr
Flowrate   1 ft^3/sec = 0.02832 m^3/s = 28.32 liter/s
$\quad\quad$ 1 mgd = 1.55 cfs = 0.0438 m^3/s = 43.8 liter/s
Force   1 lb = 4.448 N
Frequency   1 cycle/s = 1 Hz
Kinematic viscosity   1 ft^2/s = 0.0929 m^2/s = 929 Stokes
Length   1 in. = 25.4 mm
$\quad\quad$ 1 ft = 0.3048 m
$\quad\quad$ 1 mile = 1.609 km
Mass   1 slug = 14.59 kg
Power   1 ft-lb/s = 1.356 W = 1.356 J/s
$\quad\quad$ 1 hp = 550 ft-lb/s = 745.7 W
Pressure   1 psi = 6895 N/m^2 = 6895 Pa $\qquad\qquad$ 1 atmosphere = 14.70 psi =
$\quad\quad$ 1 in. Hg = 25.4 mm Hg = 3386 N/m^2 $\qquad$ 29.92 in. Hg = 760 mm Hg =
$\quad\quad$ 1 in. H$_2$O = 249.1 N/m^2 $\qquad\qquad\qquad\qquad$ 101.325 kN/m^2
$\quad\quad$ 1 lb/ft^2 = 47.88 N/m^2 = 47.88 Pa = 0.4788 mb $\quad$ 1 bar = 14.504 psi =
$\qquad\qquad\qquad\qquad\qquad\qquad\qquad\qquad\qquad\qquad\qquad$ 10^5 N/m^2 = 100 kN/m^2
Specific Heat; Engineering Gas Constant   1 ft-lb/slug °R = 0.1672 Nm/kg °K
Specific weight   1 lb/ft^3 = 157.1 N/m^3
Temperature   1°C = 1°K = 1.8°F = 1.8°R
Velocity   1 fps = 0.3048 mps = 0.3048 m/s
$\quad\quad$ 1 mph = 1.609 km/hr = 0.447/m/s
$\quad\quad$ 1 knot = 1.152 mph = 1.689 fps = 0.5155 m/s
Volume   1 ft^3 = 0.02832 m^3
$\quad\quad$ 1 U.S. gallon = 0.1337 ft^3 = 0.003785 m^3 = 3.785 liters

Source: Vennard and Street.[1]

5.2   FLUID PROPERTIES

Properties of fluids generally encountered in engineering practice are presented in tables and figures in this section. Included are the commonly used properties of density, specific weight, elasticity (compressibility), resistance to flow (viscosity), surface tension, and vapor pressure.

5.2.1   Density and Specific Weight

Density is defined as the mass per unit volume ρ. Table 5.4 provides values for a selection of liquids at standard atmospheric pressure, and Table 5.5 provides properties of some common gases. Density varies with temperature and pressure, but for liquids, variation with pressure is generally negligible. The variation of the density of water with temperature is given in Table 5.6. Also included in these tables is the **specific weight** γ, which is the density multiplied by the acceleration of gravity.

For a gas the density depends heavily on the pressure and temperature. Most gases follow closely the ideal gas equation of state

$$\rho = \frac{p}{RT} \tag{5.1}$$

where p is the absolute pressure, T is the absolute temperature, and R is the engineering gas constant. This equation should be used with discretion under conditions of very high temperature or very low pressure or when the gas approaches a liquid. Because all gases at the same temperature and pressure contain the same number of molecules per unit volume (Avogadro's law), the engineering gas constant can be calculated to good accuracy by dividing the "universal" gas constant 49,700 by the molecular weight of the gas. For example, oxygen has the molecular weight of 2 × 16 = 32. The engineering gas constant is calculated as $R = 49,700/32 = 1553$ (compare with Table 5.5).

The standard atmosphere is used in such a variety of engineering situations that it is included as a convenience in Table 5.7.

TABLE 5.3   Symbols and Abbreviations

a	acceleration, wave speed, constant, complex constant	M	metacenter, mass, moment
b	water surface width, constant	O	point designation
c	celerity of wave propagation, constant	P	point designation, force, perimeter
cfs	cubic feet per second	Q	discharge
d	differential coefficient, constant, diameter	R	gas constant, gauge difference, hydraulic radius
e	pipe wall thickness	$\mathbf{R}$	Reynolds number
f	friction factor (Darcy–Weisbach), function	S	surface area, slope
g	acceleration due to gravity	T	absolute temperature, time, torque
gpm	gallons per minute	U	uniform velocity
h	vertical distance, liquid head, head loss	V	uniform velocity, average velocity,
i	$\sqrt{-1}$	$\forall$	volume
k	adiabatic constant, roughness height, permeability coefficient, constant	W	weight
		X	extraneous force component per unit mass, x direction
l	mixing length, length, direction cosine	Y	extraneous force component per unit mass, y direction
log	natural logarithm		
m	mass, constant, direction cosine	Z	extraneous force component per unit mass, z direction
n	integer, Manning roughness factor, direction cosine		
p	pressure intensity	α	angle, constant, kinetic energy correction factor
psi	pounds per square inch		
q	velocity, discharge per unit width	β	momentum correction factor
r	radius, coordinate	γ	specific weight
s	specific gravity	δ	layer thickness
t	temperature, time	ε	kinematic eddy viscosity, diffusion coefficient
u	velocity, velocity component in x direction		
		ζ	complex variable
v	velocity, velocity component in y direction	η	eddy viscosity, real variable, efficiency
		θ	angle
v'	velocity fluctuation	κ	circulation, constant
w	velocity component in x direction, complex potential	μ	viscosity, strength, constant, Poisson's ratio
		ν	kinematic viscosity
x	coordinate axis	$\dot{v}$	normal velocity
y	coordinate axis	ξ	real variable
z	coordinate axis	$\bar{\omega}$	coordinate
		π	constant 3.14159
A	cross-sectional area, point designation	ρ	mass density
B	point designation	σ	surface tension
C	point designation, Hazen–Williams roughness coefficient, constant	τ	shear stress
		ϕ	velocity potential
D	point designation, drag force, diameter	ψ	stream function
E	modulus of elasticity, specific energy	ω	angular velocity
F	Fahrenheit, function, force	Δ	increment
G	center of gravity	Σ	summation
H	head	Ω	extraneous force potential
I	moment of inertia	∇	del, the Laplacian operator
J	point designation		
K	constant, loss coefficient, bulk modulus of elasticity		
L	length, lift		

5.2.2   Elasticity (Compressibility)

The **elasticity** of fluids is confined to their behavior in the mode of compression. The definition of the modulus of elasticity stems from the relative change in volume:

$$K = -\frac{\Delta p}{\dfrac{\Delta\forall}{\forall}} \tag{5.2}$$

TABLE 5.4 Approximate Properties of Some Common Liquids at Standard Atmospheric Pressure

English (FSS) Units

	Temperature T (°F)	Density, ρ (slug/ft^3)	Specific Gravity s.g. —	Modulus of Elasticity K (psi)	Viscosity $\mu \times 10^5$ (lb-sec/ft^2)	Surface Tension[a] σ (lb/ft)	Vapor Pressure ρ_v (psia)
Benzene	68	1.70	0.88	150,000	1.37	0.0020	1.45
Carbon tetrachloride	68	3.08	1.59	160,000	2.035	0.0018	1.90
Crude oil	68	1.66	0.86	—	15.0	0.002	—
Ethyl alcohol	68	1.53	0.79	175,000	2.51	0.0015	0.85
Freon-12	60	2.61	1.35	—	3.10	—	—
	−30	2.91	—	—	3.82	—	—
Gasoline	68	1.32	0.68	—	0.61	—	8.0
Glycerin	68	2.44	1.26	630,000	3,120	0.0043	0.000002
Hydrogen	−431	0.143	—	—	0.0435	0.0002	3.1
Jet fuel (JP-4)	60	1.50	0.77	—	1.82	0.002	1.3
Mercury	60	26.3	13.57	3,800,000	3.26	0.035	0.000025
	600	24.9	12.8	—	1.88	—	6.85
Oxygen	−320	2.34	—	—	0.58	0.001	3.1
Sodium	600	1.70	—	—	0.690	—	—
	1000	1.60	—	—	0.472	—	—
Water	68	1.936	1.00	318,000	2.10	0.0050	0.34

SI Units

	T (°C)	ρ (kg/m^3)	s.g. —	K (kPa)	$\mu \times 10^4$ (Pa · s)	σ (N/m)	ρ_v (kPa)
Benzene	20	876.2	0.88	1,034,250	6.56	0.029	10.0
Carbon tetrachloride	20	1,587.4	1.59	1,103,200	9.74	0.026	13.1
Crude oil	20	855.6	0.86	—	71.8	0.03	—
Ethyl alcohol	20	788.6	0.79	1,206,625	12.0	0.022	5.86
Freon-12	15.6	1,345.2	1.35	—	14.8	—	—
	−34.4	1,499.8	—	—	18.3	—	—
Gasoline	20	680.3	0.68	—	2.9	—	55.2
Glycerin	20	1,257.6	1.26	4,343,850	14,939	0.063	0.000014
Hydrogen	−257.2	73.7	—	—	0.21	0.0029	21.4
Jet fuel (JP-4)	15.6	773.1	0.77	—	8.7	0.029	8.96
Mercury	15.6	13,555	13.57	26,201,000	15.6	0.51	0.00017
	315.6	12,833	12.8	—	9.0	—	47.2
Oxygen	−195.6	1,206.0	—	—	2.78	0.015	21.4
Sodium	315.6	876.2	—	—	3.30	—	—
	537.8	824.6	—	—	2.26	—	—
Water	20	998.2	1.00	2,170,500	10.0	0.073	2.34

Source: Vennard and Street.[1]

[a]In contact with air.

where Δp is the pressure increment which causes a relative decrease in volume $\Delta V/V$. The K value has the dimensions of pressure, and it depends on the pressure and temperature of a liquid. For a gas, the K value depends on the thermodynamic process governing the change in volume resulting from the pressure increment. For example, if the process is isentropic, $K = kp$; if the process is isothermal, $K = p$.

Values of K for a variety of liquids at standard temperature and pressure are given in Table 5.4. For water, K values for a range of temperatures are provided in Table 5.6. Figure 5.1 is provided to demonstrate the variation of K with pressure for pure water. Entrained gas in a liquid can drastically affect the K value. Figure 5.2 illustrates the effect of entrained air on the elasticity of water for relatively small amounts of air.

TABLE 5.5　Approximate Properties of Some Common Gases

	English (FSS) Units				
	Engineering Gas Constant R (ft-lb/slug · °R)	Universal Gas Constant $\mathcal{R} = mR$ (ft-lb/slug · °R)	Adiabatic Exponent k —	Specific Heat at Constant Pressure c_p (ft-lb/slug · °R)	Viscosity at 68°F (20°C) $\mu \times 10^5$ (lb-sec/ft^2)
Carbon dioxide	1,123	49,419	1.28	5,132	0.0307
Oxygen	1,554	49,741	1.40	5,437	0.0419
Air	1,715	49,709	1.40	6,000	0.0377
Nitrogen	1,773	49,644	1.40	6,210	0.0368
Methane	3,098	49,644	1.31	13,095	0.028
Helium	12,419	49,677	1.66	31,235	0.0411
Hydrogen	24,677	49,741	1.40	86,387	0.0189

	SI Units				
	R (J/kg · °K)	$\mathcal{R} = mR$ (J/kg · °K)	k —	c_p (J/kg · °K)	$\mu \times 10^5$ (Pa · s)
Carbon dioxide	187.8	8,264	1.28	858.2	1.47
Oxygen	259.9	8,318	1.40	909.2	2.01
Air	286.8	8,313	1.40	1,003	1.81
Nitrogen	296.5	8,302	1.40	1,038	1.76
Methane	518.1	8,302	1.31	2,190	1.34
Helium	2,076.8	8,307	1.66	5,223	1.97
Hydrogen	4,126.6	8,318	1.40	14,446	0.90

Source: Vennard and Street.[1]

TABLE 5.6　Properties of Water at Various Temperatures

			English (FSS) Units [a]				
Temperature (°F)	Specific Weight[b] γ (lb/ft^3)	Density[b] ρ (slug/ft^3)	Modulus of Elasticity[c, d] $K/10^3$ (psi)	Viscosity[b] $\mu \times 10^5$ (lb-sec/ft^2)	Kinematic Viscosity[b] $\nu \times 10^5$ (ft^2/sec)	Surface Tension[d, e] σ (lb/ft)	Vapor Pressure[f] p_v (psia)
32	62.42	1.940	287	3.746	1.931	0.00518	0.09
40	62.43	1.940	296	3.229	1.664	0.00614	0.12
50	62.41	1.940	305	2.735	1.410	0.00509	0.18
60	62.37	1.938	313	2.359	1.217	0.00504	0.26
70	62.30	1.936	319	2.050	1.059	0.00498	0.36
80	62.22	1.934	324	1.799	0.930	0.00492	0.51
90	62.11	1.931	328	1.595	0.826	0.00486	0.70
100	62.00	1.927	331	1.424	0.739	0.00480	0.95
110	61.86	1.923	332	1.284	0.667	0.00473	1.27
120	61.71	1.918	332	1.168	0.609	0.00467	1.69
130	61.55	1.913	331	1.069	0.558	0.00460	2.22
140	61.38	1.908	330	0.981	0.514	0.00454	2.89
150	61.20	1.902	328	0.905	0.476	0.00447	3.72
160	61.00	1.896	326	0.838	0.442	0.00441	4.74
170	60.80	1.890	322	0.780	0.413	0.00434	5.99
180	60.58	1.883	318	0.726	0.385	0.00427	7.51
190	60.36	1.876	313	0.678	0.362	0.00420	9.34
200	60.12	1.868	308	0.637	0.341	0.00413	11.52
212	59.83	1.860	300	0.593	0.319	0.00404	14.70

TABLE 5.6 *(Continued)*

T (°C)	γ (kN/m³)	ρ (kg/m³)	$K/10^6$ (kPa)	$\mu \times 10^3$ (Pa · s)	$\nu \times 10^6$ (m²/s)	σ (N/m)	p_v (kPa)
			SI Units[b]				
0	9.805	999.8	1.98	1.781	1.785	0.0756	0.61
5	9.807	1,000.0	2.05	1.518	1.518	0.0749	0.87
10	9.804	999.7	2.10	1.307	1.306	0.0742	1.23
15	9.798	999.1	2.15	1.139	1.139	0.0735	1.70
20	9.789	998.2	2.17	1.002	1.003	0.0728	2.34
25	9.777	997.0	2.22	0.890	0.893	0.0720	3.17
30	9.764	995.7	2.25	0.798	0.800	0.0712	4.24
40	9.730	992.2	2.28	0.653	0.658	0.0696	7.38
50	9.689	988.0	2.29	0.547	0.553	0.0679	12.33
60	9.642	983.2	2.28	0.466	0.474	0.0662	19.92
70	9.589	977.8	2.25	0.404	0.413	0.0644	31.16
80	9.530	971.8	2.20	0.354	0.364	0.0626	47.34
90	9.466	965.3	2.14	0.315	0.326	0.0608	70.10
100	9.399	958.4	2.07	0.282	0.294	0.0589	101.33

Source: Vennard and Street.[1]

[a] Compiled from many sources including those indicated, *Handbook of Chemistry and Physics*, 54th Ed., The CRC Press, 1973, and *Handbook of Tables for Applied Engineering Science*, The Chemical Rubber Co., 1970.
[1] Here, if $K/10^3 = 287$, then $K = 287 \times 10^3$ psi: while if $\mu \times 10^5 = 3.746$, then $\mu = 3.746 \times 10^{-5}$ lb-sec/ft², and so on.
[2] Here, if $K/10^6 = 1.98$, then $K = 1.98 \times 10^6$ kPa, while if $\mu \times 10^3 = 1.781$, then $\mu = 1.781 \times 10^{-3}$ Pa · s, and so on.
[b] From "Hydraulic Models," *A.S.C.E. Manual of Engineering Practice*, No. 25, A.S.C.E., 1942. See footnotes 1 and 2.
[c] Approximate values averaged from many sources.
[d] At atmospheric pressure. See footnotes 1 and 2.
[e] In contact with air.
[f] From J. H. Keenan and F. G. Keyes, *Thermodynamic Properties of Steam*, Wiley, New York, 1936.

TABLE 5.7   The U.S. Standard Atmosphere

Altitude (ft)	Temperature (°F)	Absolute Pressure (psia)	Specific Weight (lb/ft³)	Density (slug/ft³)	Viscosity $\times 10^7$ (lb-sec/ft²)
		English (FSS) Units[a]			
0	59.00	14.696	0.07647	0.002377	3.737
5,000	41.17	12.243	0.06587	0.002048	3.637
10,000	23.36	10.108	0.05643	0.001756	3.534
15,000	5.55	8.297	0.04807	0.001496	3.430
20,000	−12.26	6.759	0.04069	0.001267	3.325
25,000	−30.05	5.461	0.03418	0.001066	3.217
30,000	−47.83	4.373	0.02857	0.000891	3.107
35,000	−65.61	3.468	0.02367	0.000738	2.995
40,000	−69.70	2.730	0.01882	0.000587	2.969
45,000	−69.70	2.149	0.01481	0.000462	2.969
50,000	−69.70	1.690	0.01165	0.000364	2.969
55,000	−69.70	1.331	0.00917	0.000287	2.969
60,000	−69.70	1.049	0.00722	0.000226	2.969
65,000	−69.70	0.826	0.00568	0.000178	2.969
70,000	−67.42	0.651	0.00445	0.000139	2.984
75,000	−64.70	0.514	0.00349	0.000109	3.001
80,000	−61.98	0.404	0.00263	0.000086	3.018
85,000	−59.26	0.322	0.00215	0.000067	3.035
90,000	−56.54	0.255	0.00170	0.000053	3.052
95,000	−53.82	0.203	0.00134	0.000042	3.070
100,000	−51.10	0.162	0.00106	0.000033	3.087

TABLE 5.7 *(Continued)*

		SI Units[a]			
Altitude (km)	Temperature (°C)	Absolute Pressure (kPa)	Specific Weight (N/m³)	Density (kg/m³)	Viscosity ×10⁵ (Pa · s)
0	15.00	101.33	12.01	1.225	1.789
2	2.00	79.50	9.86	1.007	1.726
4	−4.49	70.12	8.02	0.909	1.661
6	−23.96	47.22	6.46	0.660	1.595
8	−36.94	35.65	5.14	0.526	1.527
10	−49.90	26.50	4.04	0.414	1.458
12	−56.50	19.40	3.05	0.312	1.422
14	−56.50	14.17	2.22	0.228	1.422
16	−56.50	10.35	1.62	0.166	1.422
18	−56.50	7.57	1.19	0.122	1.422
20	−56.50	5.53	0.87	0.089	1.422
22	−54.58	4.05	0.63	0.065	1.432
24	−52.59	2.97	0.46	0.047	1.443
26	−50.61	2.19	0.33	0.034	1.454
28	−48.62	1.62	0.24	0.025	1.465
30	−46.64	1.20	0.18	0.018	1.475

Source: Vennard and Street.[1]

[a]Data from *U.S. Standard Atmosphere*, 1962, U.S. Government Printing Office, 1962. Data agrees with ICAO standard atmosphere to 20 km and with ICAO proposed extension to 30 km. For atmospheric tables depicting conditions other than mid-latitude mean represented by standard atmosphere, see *U.S. Standard Atmosphere Supplements*, 1966, U.S. Government Printing Office, 1966.

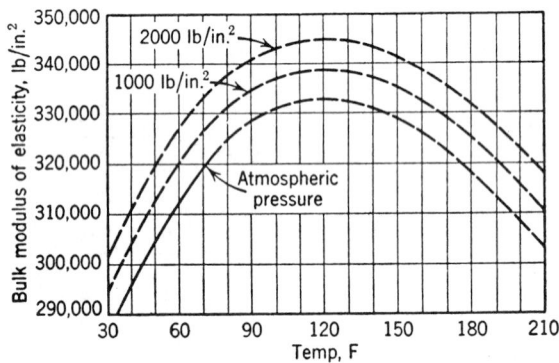

Fig. 5.1 Bulk modulus of elasticity as a function of temperature and pressure intensity. By permission from *Fluid Mechanics for Hydraulic Engineers*, by Hunter Rouse, copyright © 1938, McGraw-Hill, New York, 1938.

One of the common applications of the property of fluid elasticity is in the computation of the acoustic wave speed (speed of sound) in a fluid. The equation applicable to both liquids and gases is

$$\mathbf{a} = \sqrt{\frac{K}{\rho}} \tag{5.3}$$

where **a** is the acoustic wave speed. The value of K is obtained from a table, or in the case of gases, computed using the isentropic thermodynamic process.

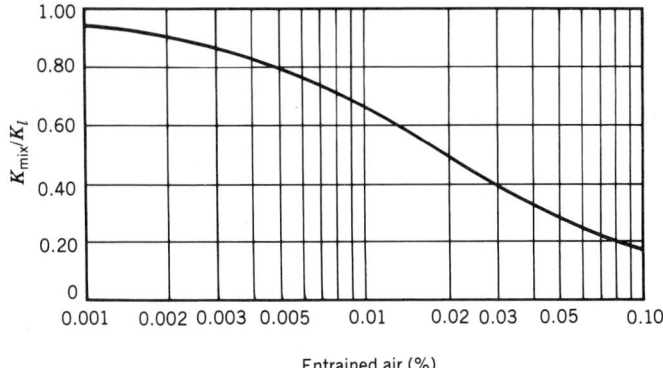

Fig. 5.2   The effect of entrained air on the elasticity of water.

5.2.3   Viscosity

Viscosity is that property of a fluid that measures its resistance to angular deformation under action of a shearing stress. Viscosity is strongly dependent on temperature but is relatively unaffected by pressure.

In liquids, viscosity depends on the strength of cohesive forces between fluid molecules, hence an increase in temperature results in a decrease in viscosity. For gases, viscosity depends on the momentum exchange between layers of gas moving at different velocities. An increase in temperature provides an increase in molecular activity and an increase in viscosity.

A fundamental device for measuring fluid viscosity leads to the definition of viscosity. Figure 5.3 illustrates the flow situation occurring when a moving plate, under the action of a shearing force, slides over a fixed plate separated by a fluid. If the movement is sufficiently slow so that the flow between the plates is laminar, the shear stress is related to the velocity difference by the equation

$$\tau = \mu \frac{dv}{dy} \tag{5.4}$$

where τ is the shear stress, dv/dy is the velocity gradient (rate of angular strain), and μ is the coefficient of viscosity. In a more precise sense, μ is known as the absolute viscosity and has the units lb-sec/ft^2 in the English FSS system. Because the viscosity shows up in many instances divided by the density, the ratio $\nu = \mu/\rho$ is defined as the kinematic viscosity with the English FSS units square feet per second. In the metric system, absolute viscosity is commonly given in poises (dyne-second/square centimeter) where 1 lb-sec/ft^2 = 478.8 poises. The metric equivalent of kinematic viscosity is the stoke (square centimeters/second) where 1 ft^2/sec = 929 stokes.

Figures 5.4–5.6 give absolute and kinematic viscosities for a wide range of liquids and gases. The English FSS values may be converted to the SI values using the conversions

μ: 1 lb-sec/ft^2 = 47.88 N-s/m^2
ν: 1 ft^2/sec = 0.0929 m^2/s

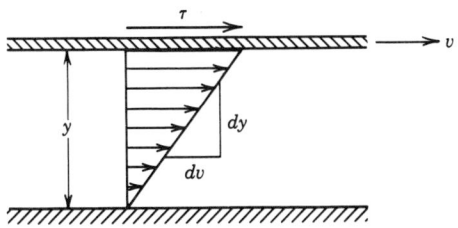

Fig. 5.3   A definition sketch for absolute viscosity.

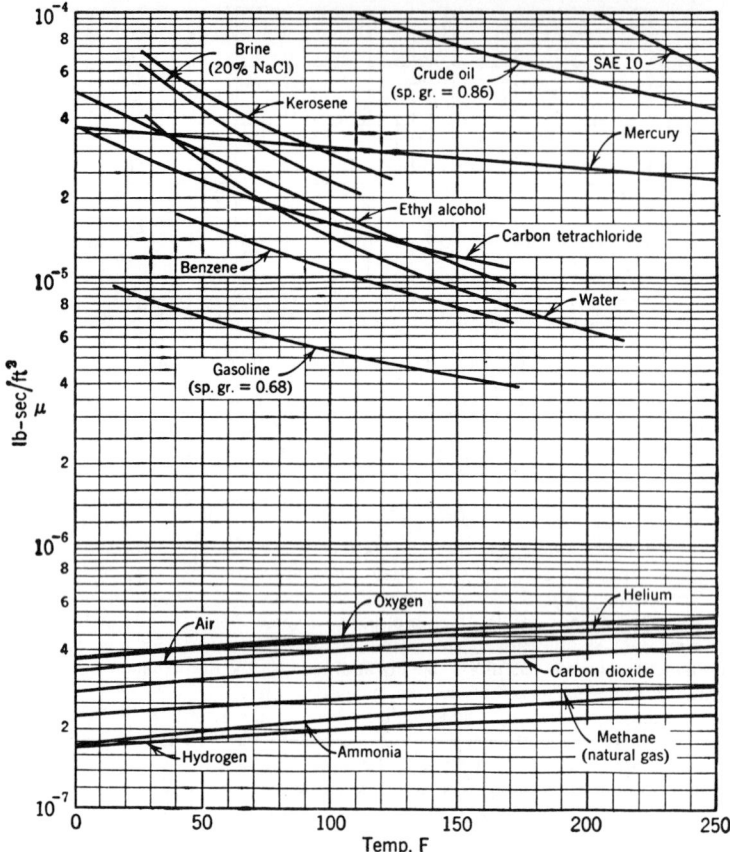

Fig. 5.4 Dynamic viscosity versus temperature for common gases and liquids. (Courtesy of Hunter Rouse, State University of Iowa, Iowa City.)

5.2.4 Surface Tension and Capillarity

Molecules of a liquid exert a mutual cohesive force that causes those molecules in the interior of a liquid to be in a balanced state of cohesive forces, that is, attracted equally in all directions. However, liquid molecules at a surface are attracted by the interior molecules but experience no balancing attraction from above the surface. This imbalance of forces causes the liquid surface to behave as though it were covered with an elastic membrane, hence the term **surface tension**. This phenomenon is also apparent at the interface between the liquids. This is why surface tension values for a liquid must always specify what fluid lies across the interface. As a result of its dependence on molecular cohesion, surface tension σ decreases with increasing temperature. Table 5.4 gives σ values for a variety of liquids and Table 5.6 provides values for water at various temperatures.

Surface tension manifests itself in free liquid jets, bubbles, small waves, and capillary action in small conduits. At the interface between two fluids, surface tension can sustain a pressure discontinuity across the interface. The magnitude of the pressure difference is a function of the surface curvature with the higher pressure on the concave side of the interface.

$$p_i - p_o = \sigma\left(\frac{1}{R_1} + \frac{1}{R_2}\right) \tag{5.5}$$

where R_1, R_2 are the radii of curvatures of the interface in orthogonal directions.

Capillary rise in a circular tube is another important application of surface tension (Fig. 5.7). The equation for capillary rise h is

$$h = \frac{2\sigma \cos \theta}{\gamma r} \tag{5.6}$$

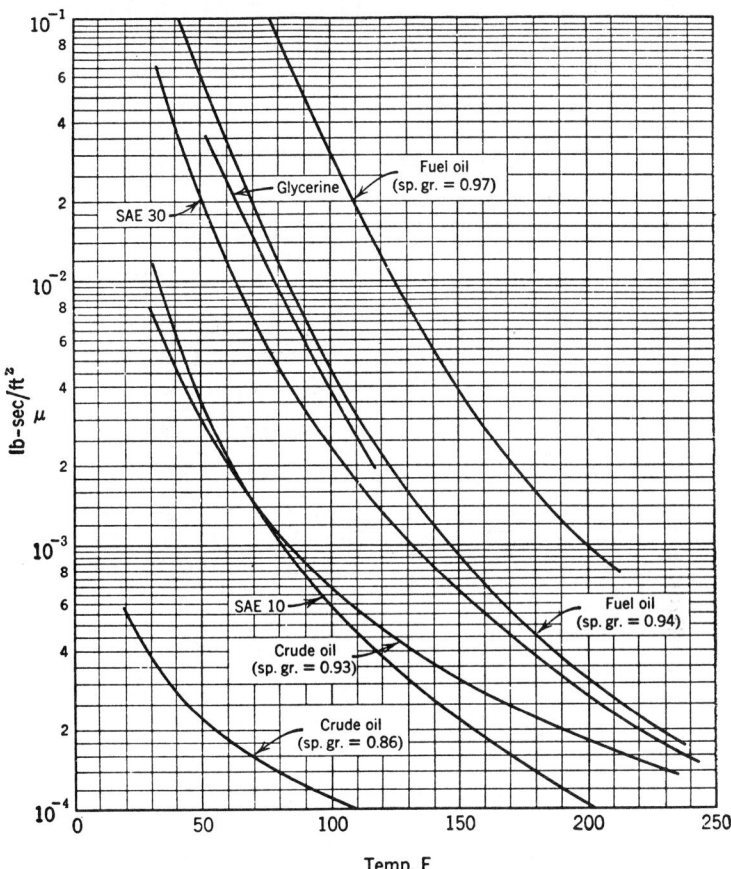

Fig. 5.5 Dynamic viscosity versus temperature for typical grades of oil. (Courtesy of Hunter Rouse, State University of Iowa, Iowa City.)

The value of h depends heavily on the contact angle θ, which in turn depends strongly on the liquid and the tube material as well as the cleanliness of the tube surface. This equation should not be used unless the capillary tube is small enough that the liquid surface is approximately spherical in shape (constant curvature).

5.2.5 Vapor Pressure

A liquid exhibiting a free surface is continually ejecting and absorbing molecules of the liquid across the free surface. If more molecules leave than return, the liquid is said to be evaporating. In an equilibrium situation, the number of molecules expelled equals the number returning. The molecules of the liquid in the overlying gaseous fluid exert a force on the liquid surface, in conjunction with the other gas molecules bombarding the surface, to make up the total surface pressure. That portion of the surface pressure generated by the vapor molecules is called the **partial pressure**, or **vapor pressure**, p_v of the liquid.

The higher the liquid temperature, the more vigorous the molecular activity and the larger the fraction of the total pressure contributed by the liquid vapor. When the liquid temperature is elevated to a level where the vapor pressure is equal to the total pressure, boiling occurs. Boiling is also experienced when the total pressure is reduced to the liquid vapor pressure, for example, in liquid cavitation in pumps, valves, and propellers. Table 5.4 provides values of the vapor pressure p_v for a variety of liquids. Table 5.6 gives values for water over a range of temperatures.

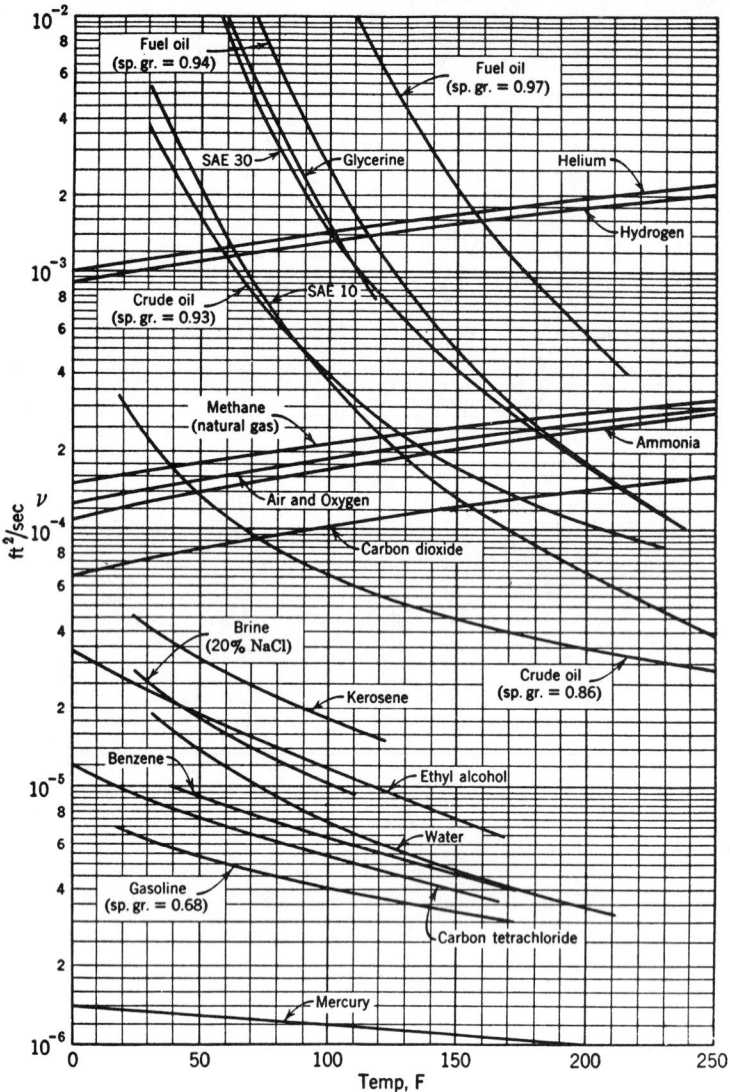

Fig. 5.6 Kinematic viscosity versus temperature for common fluids. (Courtesy of Hunter Rouse, State University of Iowa, Iowa City.)

5.3   FLUID STATICS

5.3.1   Pressure Variation in a Fluid at Rest

By definition, a Newtonian fluid at rest has no internal shear stresses, only normal stresses. As a consequence, pressure does not vary horizontally in a homogeneous fluid, only vertically. The equation describing the vertical variation in pressure is

$$\frac{dp}{dz} = -\gamma \tag{5.7}$$

where z is measured vertically upward. If a fluid has variable density, this expression must be integrated to determine differences in pressure. However, for a constant-density fluid, the commonly

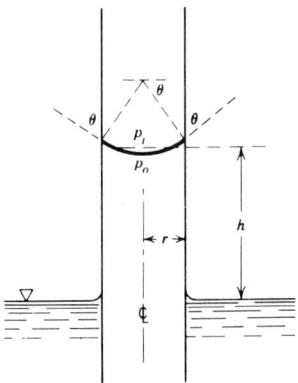

Fig. 5.7 Capillary rise in a small circular tube. By permission from J. K. Vennard and R. L. Street, *Elementary Fluid Mechanics*, 5th ed., Wiley, New York, 1975.

used form of this equation is the hydrostatic pressure variation equation

$$\Delta p = \gamma \Delta h \tag{5.8}$$

where Δh is the difference in elevation between two points and Δp is the corresponding pressure difference.

5.3.2 Basic Pressure Measuring Devices

Pressures are generally measured from one of two datums—absolute zero or the surrounding environment (gauge). Pressures measured above absolute zero, for example, barometric pressure, are known as absolute pressure. Those measured above the local atmospheric pressure are known as gauge pressures. Absolute pressure always equals gauge pressure plus atmospheric pressure. A commonly used mechanical device for measuring both gauge pressure and absolute pressure is the Bourdon-type gauge (Fig. 5.8).

A more fundamental pressure measuring device, which is based on the hydrostatic pressure variation equation, is the manometer (Fig. 5.9). Because one end of this manometer is open to the atmosphere, it is known as an open-end manometer. The gauge pressure at A is calculated as

$$p_A = \gamma_2 y_2 - \gamma_1 y_1 \tag{5.9}$$

Figure 5.10 illustrates a differential manometer. The difference in pressure between A and B can be calculated as

$$p_B - p_A = \gamma_1 y_1 - \gamma_2 y_2 - \gamma_3 (y_3 - y_2) \tag{5.10}$$

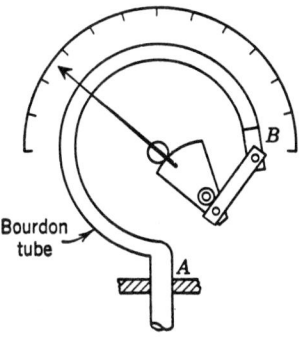

Fig. 5.8 Essential features of a Bourdon gauge.

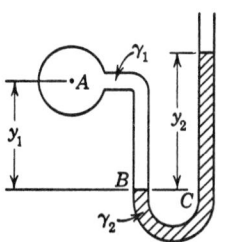

Fig. 5.9 Simple manometer.

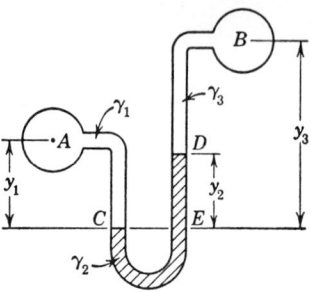

Fig. 5.10 Differential manometer.

5.3.3   Fluid Forces on Plane Surfaces

The magnitude of the total fluid force exerted on a plane (flat) surface can be calculated with the equation

$$F = p_c A \qquad (5.11)$$

where p_c is the pressure at the centroid of the area of the surface and A is the total area of the surface.

If the fluid is a liquid with a free surface,

$$p_c = \gamma h_c \qquad (5.12)$$

where h_c is the distance the centroid lies below the free surface. Locations of centroids of variously shaped areas are shown in Table 5.8.

The location of the resultant of the pressure forces acting on a plane area is known as the center of pressure. For a plane area lying below a liquid surface, the center of pressure location is computed with the equation

$$l_p - l_c = \frac{I_c}{l_c A} \qquad (5.13)$$

where l_p is measured in the plane of the area (slant distance) from the free surface to the center of pressure, l_c is the distance (in the same plane) from the free surface to the centroid of the area, and I_c is the second moment of the area about a horizontal axis through the centroid and in the plane of the area. Values of I_c for common geometric shapes are given in Table 5.8. Note that for deeply submerged plane areas, the center of pressure approaches the centroid of the area.

5.3.4   Fluid Forces on Curved Surfaces

Pressure forces exerted on curved surfaces vary in direction and must usually be divided into horizontal and vertical components to facilitate calculation. The resultants of the horizontal and vertical components of the pressure forces are determined with assistance from the equations developed for plane surfaces. Figure 5.11 illustrates a common situation. The fluid mass ABC is in equilibrium under the action of pressure forces, hence the vertical component of the resultant of the pressure forces is equal to the weight of the fluid mass plus the force F_{AC} exerted on the upper surface of the fluid mass by the overlying fluid. The horizontal component of the resultant of the pressure forces is equal to the force on a vertical projection of the curved surface. The line of action and point of application (center of pressure) for the resultant force is determined by taking moments of the contributing forces about an axis of convenience. The force F_{BC} is located using the formula for plane surfaces. The weight W_{ABC} is concentrated at the centroid of the area ABC (see Table 5.8).

5.3.5 Buoyancy and Stability

An object floating or submerged in a fluid at rest experiences pressure forces from the surrounding fluid. The resultant of these pressure forces is known as the **buoyant force** and is vertical and equal in magnitude to the weight of the displaced fluid. The line of action of the buoyant force is vertical through the centroid of the volume of displaced fluid (center of buoyancy).

TABLE 5.8 Properties of Areas and Volumes

	Sketch	Area or Volume	Location of Centroid	I or I_e
Rectangle		bh	$y_c = \dfrac{h}{2}$	$I_c = \dfrac{bh^3}{12}$
Triangle		$\dfrac{bh}{2}$	$y_c = \dfrac{h}{3}$	$I_c = \dfrac{bh^3}{36}$
Circle		$\dfrac{\pi d^2}{4}$	$y_c = \dfrac{d}{2}$	$I_c = \dfrac{\pi d^4}{64}$
Semicircle [a]		$\dfrac{\pi d^2}{8}$	$y_c = \dfrac{4r}{3\pi}$	$I = \dfrac{\pi d^4}{128}$
Ellipse		$\dfrac{\pi bh}{4}$	$y_c = \dfrac{h}{2}$	$I_c = \dfrac{\pi bh^3}{64}$
Semiellipse		$\dfrac{\pi bh}{4}$	$y_c = \dfrac{4h}{3\pi}$	$I = \dfrac{\pi bh^3}{16}$
Parabola		$\tfrac{2}{3}bh$	$y_c = \dfrac{3h}{5}$ $x_c = \dfrac{3b}{8}$	$I = \dfrac{2bh^3}{7}$
Cylinder		$\dfrac{\pi d^2 h}{4}$	$y_c = \dfrac{h}{2}$	

(continued on next page)

The overturning stability of a floating or submerged body is affected by the relative position of the center of gravity of the body and the center of buoyancy. When the center of gravity lies below the center of buoyancy, the body is always **stable**. If the center of buoyancy is below the center of gravity, the body may or may not be stable in terms of overturning. If the geometric shape of the body is such that a rocking motion results in a shift of the buoyancy center that in turn causes a righting action, the body is stable. For example, a 2 × 4 board floating on its side would be stable. However, the same board, floating on its edge, would be unstable.

5.3.6 Accelerated Fluid Masses without Relative Motion

There are situations when a fluid body is accelerated with no relative motion among fluid particles (zero shear stress). This is true only for constant accelerations. As in a fluid at rest, the pressure will

TABLE 5.8   *(Continued)*

	Sketch	Area or Volume	Location of Centroid	I or I_e
Cone		$\dfrac{1}{3}\left(\dfrac{\pi d^2 h}{4}\right)$	$y_e = \dfrac{h}{4}$	
Paraboloid of revolution		$\dfrac{1}{2}\left(\dfrac{\pi d^2 h}{4}\right)$	$y_c = \dfrac{h}{3}$	
Sphere		$\dfrac{\pi d^3}{6}$	$y_c = \dfrac{d}{2}$	
Hemisphere		$\dfrac{\pi d^3}{12}$	$y_c = \dfrac{3r}{8}$	

[a] For the quarter-circle, the respective values are $\pi d^2/16$, $4r/3\pi$, and $\pi d^4/256$.

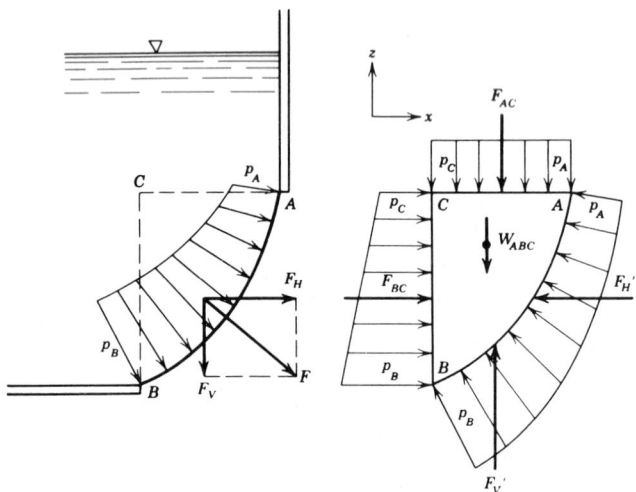

Fig. 5.11 Forces acting on a curved surface. By permission from J. K. Vennard and R. L. Street, *Elementary Fluid Mechanics*, 5th ed., Wiley, New York, 1975.

still vary linearly with elevation, but the pressure gradient will generally not equal the specific weight. Further, there will be a variation in pressure in the horizontal plane in this situation.

For horizontal acceleration of a container of liquid, the free surface will tilt just enough to provide the unbalanced hydrostatic force on the opposite sides of the container required to impart a constant acceleration to the fluid mass. Under these conditions the slope of the free surface is given as

$$\text{Slope} = \frac{a_x}{g} \tag{5.14}$$

where a_x is the constant horizontal acceleration.

For vertical acceleration, the free surface remains horizontal, but the pressure variation with depth is given by the equation

$$\frac{dp}{dz} = -\gamma\left(1 + \frac{a_z}{g}\right) \tag{5.15}$$

Note that if $a_z = -g$ (free fall), the pressure gradient throughout the fluid is zero.

Another situation occurs for a container of liquid undergoing a constant rotational angular velocity. When the relative motion between fluid particles ceases, the free surface forms the shape of a parabola described by the equation

$$z = \frac{\omega^2}{2g}r^2 \tag{5.16}$$

where z is the elevation of the free surface at radius r above that at the center of rotation and ω is the angular velocity of rotation of the container. Pressure in the fluid varies parabolically with radius,

$$p = p_o + \frac{\gamma\omega^2}{2g}r^2 \tag{5.17}$$

where p is the pressure at radius r in a horizontal plane and p_o is the pressure at the center of rotation in the same horizontal plane. Because there is no vertical acceleration, pressure variation in the vertical direction is the same as if the fluid were at rest.

5.4   IDEAL (NONVISCOUS) FLUID DYNAMICS

An **ideal fluid** is a hypothetical substance exhibiting all the characteristics of a real fluid except that it is a continuum (no molecular spaces), it has no viscosity (frictionless), and it can accept pressures down to negative infinity. These characteristics greatly simplify the mathematical treatment of fluid flow and, in many cases, give acceptable engineering results for real fluid situations where viscous effects are not significant.

Although not restricted to ideal fluid flow, some definitions are common to all flow situations. Steady flow occurs when all conditions in the flow at any point do not change with time. In turbulent flows, this definition is extended to include temporal mean values of all conditions. Uniform flow is defined as the condition where the velocity is everywhere the same at a given time. For one-dimensional flow, this definition refers to the average velocity at any cross section of the flow. In nonuniform flow, the velocity varies with position along the conduit at any given time.

Qualitative aspects of a fluid flow situation can be illustrated effectively through the use of streamlines. For steady flow, **streamlines** are tracks of fluid particles that are everywhere tangent to fluid velocity vectors. As a consequence, flow cannot occur across streamlines; so a collection of streamlines defining a closed surface (a streamtube) would act like a conduit within the flow. Further, regions of flow where streamlines are close together represent locally high velocities (and low pressures). Regions where the streamlines are straight and parallel identify situations where there is no acceleration normal to the flow; therefore, the pressure variation normal to the flow is hydrostatic.

5.4.1   One-Dimensional Flow

Conservation of Mass (Continuity)
Conservation of mass (continuity, for constant density flow) requires use of a control volume through which the flow passes. In one-dimensional flow, the control volume typically coincides with the conduit walls and crosses the flow normal to the streamlines at a location where the streamlines are

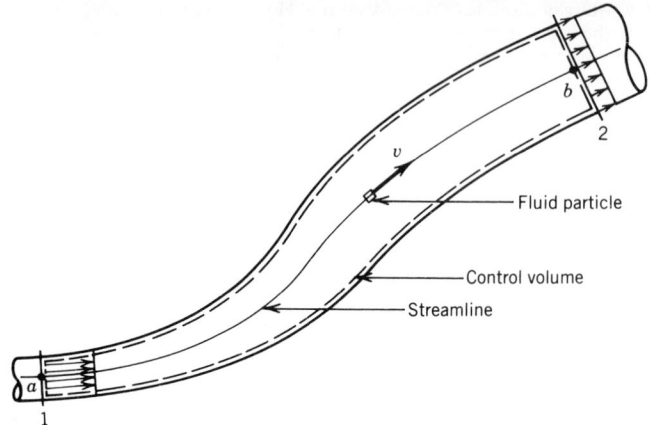

Fig. 5.12   One-dimensional nonviscous flow.

essentially straight and parallel (see Fig. 5.12). For steady flow, the mass flow rate is a constant along the conduit and is written

$$A\rho V = Q\rho = \text{const.} \tag{5.18}$$

where Q is the volume flow rate and V is the average velocity Q/A. In cases where the flow enters the control volume at section 1 and leaves at section 2 (see Fig. 5.12), conservation of mass gives

$$A_1\rho_1V_1 = A_2\rho_2V_2 \tag{5.19}$$

Euler's Equations
Application of Newton's second law to a fluid particle gives Euler's equations of motion. Along a streamline (see Fig. 5.12)

$$\frac{dp}{\gamma} + \frac{v\,dv}{g} + dz = 0 \tag{5.20}$$

where v is the local velocity.
    Normal to the streamline,

$$\frac{dp}{\gamma} - \frac{v^2}{gr}\,dr + dz = 0 \tag{5.21}$$

where r is the local radius of curvature of the streamline along which the fluid particle is moving.

Bernoulli Equation
Integration of the Euler equation along a streamline for constant density yields the Bernoulli equation

$$\frac{p}{\gamma} + \frac{v^2}{2g} + z = \text{const.} \tag{5.22}$$

If the integration is performed along a streamline between points a and b, the Bernoulli equation takes the form (see Fig. 5.12)

$$z_a + \frac{p_a}{\gamma} + \frac{v_a^2}{2g} = z_b + \frac{p_b}{\gamma} + \frac{v_b^2}{2g} \tag{5.23}$$

    The Bernoulli equation can be "expanded" to apply to a full-sized one-dimensional flow (as opposed to a single streamline) if the sections at 1 and 2 (Fig. 5.12) are regions where the streamlines are essentially straight and parallel and the velocity profile uniform (irrotational flow).

$$z_1 + \frac{p_1}{\gamma} + \frac{V_1^2}{2g} = z_2 + \frac{p_2}{\gamma} + \frac{V_2^2}{2g} \tag{5.24}$$

where V is the average velocity. This is true because at a section where the streamlines are straight and parallel, the velocity is constant over the cross section and the pressure variation is hydrostatic $(z + p/\gamma = \text{const.})$ Hence the sum of the three terms on each side of the equation is a constant regardless of the streamline selected. Because 1 and 2 are sections and not points, the quantities z and p/γ are usually evaluated at some arbitrary point in the cross section, for example, at the centerline for a cylindrical pipe.

A further conclusion extends from the uniform-flow section concept. If the sum of the three terms is the same for all streamlines at the special cross section, the sum is everywhere constant throughout the flow.

$$z + \frac{p}{\gamma} + \frac{v^2}{2g} = \text{constant throughout flow field for irrotational flow} \tag{5.25}$$

Work-Energy Equation
An equation similar to the Bernoulli equation can be derived from work-energy principles. However, the work-energy equation has broader application because it can include thermal energy and heat transfer. Utilizing the principle that the work done on a fluid system is equal to the change in energy of the system, and applying this principle to the fluid in the control volume of Fig. 5.12,

$$z_1 + \frac{p_1}{\gamma} + \frac{V_1^2}{2g} = z_2 + \frac{p_2}{\gamma} + \frac{V_2^2}{2g} \tag{5.26}$$

This equation is identical to the Bernoulli equation, however, we have not considered other energies beyond mechanical energies. Further, the terms in the work-energy equation have the following meanings:

$z = $ potential energy per pound of fluid flowing

$\frac{V^2}{2g} = $ kinetic energy per pound of fluid flowing

$\frac{p}{\gamma} = $ flow work done on the fluid in the control volume per pound of fluid flowing (this term is also often referred to as "pressure energy")

Impulse-Momentum Equation

Linear Momentum. Newton's second law was applied to a fluid particle moving along a streamline to produce the Euler equations. If the second law is applied to a finite fluid mass within a control volume (e.g., Fig. 5.12), another set of useful equations are derived. All internal forces cancel so that only external forces due to pressure, shear (for real fluids), and gravity need be considered. The resulting linear momentum equations are vector equations and are written as

$$\sum \mathbf{F}_{\text{ext}} = \sum (Q\rho\mathbf{V})_{\text{out}} - \sum (Q\rho\mathbf{V})_{\text{in}} \tag{5.27}$$

where $\mathbf{F}_{\text{ext}}$ are the external forces acting on the fluid in the control volume. Again, the control volume is generally selected in such a way that the sections where the momentum enters and leaves the control volume are regions where the streamlines are straight and parallel.

In applying the linear impulse momentum equations to a problem, it is customary to use the vector component equations with the orthogonal axes chosen in a convenient orientation. For example, the equation in the x direction would be

$$\sum (F_x)_{\text{ext}} = \sum (Q\rho V_x)_{\text{out}} - \sum (Q\rho V_x)_{\text{in}} \tag{5.28}$$

It should be noted that this equation can be used to determine the resultant force, but not the distribution of that force.

Moment of Momentum. If a derivation similar to that which resulted in the linear impulse-momentum equation is applied to the fluid particles in the control volume but using the moments of the vectors instead, the moment of momentum equation results

$$\sum (\mathbf{r} \times \mathbf{F})_{ext} = \sum [Q\rho(\mathbf{r} \times \mathbf{V})]_{out} - \sum [Q\rho(\mathbf{r} \times \mathbf{V})]_{in} \qquad (5.29)$$

where the terms in parenthesis are vector cross products.

A more directly usable form of the equation is that applicable in two dimensions

$$\sum (M_o)_{ext} = \sum (Q\rho r V_t)_{out} - \sum (Q\rho r V_t)_{in} \qquad (5.30)$$

where M_o are moments of the external forces about an axis through o, V_t is the value of the velocity normal to a vector r extending from the moment center o to the location of the V_t. The most productive area of application is to turbomachinery such as pumps and turbines.

5.4.2   Two- and Three-Dimensional Flow*

The equations describing two- and three-dimensional ideal fluid flow are derived using a small cubic fluid element or control volume of sides dx, dy, and dz.

Conservation of Mass (Continuity)
The continuity equation for an ideal fluid states that the net flow rate into any small volume must be zero, except when the volume contains a singular point. In equation form

$$\frac{\partial u}{\partial x} + \frac{\partial v}{\partial y} + \frac{\partial w}{\partial z} = 0 \qquad (5.31)$$

or in vector notation

$$\nabla \cdot \mathbf{q} = 0 \qquad (5.32)$$

that is, the divergence of the velocity vector $\mathbf{q}$ is everywhere zero except at singular points (see Chapter 2 for vector analysis).

Euler's Equation of Motion
Consideration of forces on a fluid element leads to the Euler equations:

$$X - \frac{1}{\rho}\frac{\partial p}{\partial x} = u\frac{\partial u}{\partial x} + v\frac{\partial u}{\partial y} + w\frac{\partial u}{\partial z} + \frac{\partial u}{\partial t} \qquad (5.33)$$

$$Y - \frac{1}{\rho}\frac{\partial p}{\partial y} = u\frac{\partial v}{\partial x} + v\frac{\partial v}{\partial y} + w\frac{\partial v}{\partial z} + \frac{\partial v}{\partial t} \qquad (5.34)$$

$$Z - \frac{1}{\rho}\frac{\partial p}{\partial z} = u\frac{\partial w}{\partial x} + v\frac{\partial w}{\partial y} + w\frac{\partial w}{\partial z} + \frac{\partial w}{\partial t} \qquad (5.35)$$

where X, Y, Z, are the components of the body (extraneous) forces per unit mass in the xyz directions, respectively; ρ is the mass density; p is the pressure intensity at a point (independent of direction); u, v, w are velocity components in the xyz directions at any point x, y, z. The terms on the right-hand side of the equations are acceleration components, the first three of which are known as the *convective* acceleration and the fourth as the *local* acceleration. The Euler equations must hold for every point in the flow with the exception of singular points.

Boundary Conditions
A kinematic boundary condition must be satisfied at every solid boundary, namely, that the component of velocity normal to the boundary must be equal to the velocity component of the boundary normal to itself. For a stationary boundary

$$lu + mv + nw = 0 \qquad (5.36)$$

---

*Illustrations and material extracted in all or in part from the previous edition by Victor L. Streeter.

where l, m, n, are direction cosines of the normal to the boundary. For a moving boundary

$$lu + mv + nw = \dot{v} \tag{5.37}$$

where $\dot{v}$ is the velocity of the boundary normal to itself. If $F(x, y, z, t)$ is the equation of the boundary surface, the boundary condition may be expressed

$$u\frac{\partial F}{\partial x} + v\frac{\partial F}{\partial y} + w\frac{\partial F}{\partial z} + \frac{\partial F}{\partial t} = 0 \tag{5.38}$$

Dynamic boundary conditions arise when two fluids are in contact. It is necessary that the pressure be continuous across the interface.

Irrotational Flow-Velocity Potential
Rotation of a fluid element may be represented by a vector that has a length proportional to the magnitude of the rotation (radians per second) and a direction parallel to the instantaneous axis of rotation. The right-handed rule is adopted; that is, the positive direction of the vector is the direction a right-handed screw would progress when rotating in the same sense as the element. In vector notation the *curl* of the velocity vector is twice the rotation vector

$$\nabla \times \mathbf{q} = 2\omega \tag{5.39}$$

Scalar components of the rotation vector w_x, w_y, w_z in the directions of the xyz axes may be used in place of the vector itself. Defining a rotation component of an element about an axis as the average angular velocity of two infinitesimal line segments through the point, mutually perpendicular to themselves and the axis,

$$w_x = \frac{1}{2}\left(\frac{\partial w}{\partial y} - \frac{\partial v}{\partial z}\right) \qquad \omega_y = \frac{1}{2}\left(\frac{\partial u}{\partial z} - \frac{\partial w}{\partial x}\right) \qquad \omega_z = \frac{1}{2}\left(\frac{\partial v}{\partial x} - \frac{\partial u}{\partial y}\right) \tag{5.40}$$

Irrotational flow may be defined as that flow where there is an absence of rotation at every point except singular points, hence the following equations must be satisfied

$$\frac{\partial v}{\partial x} = \frac{\partial u}{\partial y} \qquad \frac{\partial w}{\partial y} = \frac{\partial v}{\partial z} \qquad \frac{\partial u}{\partial z} = \frac{\partial w}{\partial x} \tag{5.41}$$

A visual concept of irrotational flow may be obtained by considering as a free body a small element of fluid in the form of a sphere. As the fluid is frictionless, no tangential stresses or forces may be applied to its surface. The pressure forces act normal to its surface, and hence through its center. Extraneous, or body, forces act through its mass center, which is also its geometric center for constant density. Hence, it is evident that no torque may be applied about any diameter of the sphere. The angular acceleration of the sphere must always be zero. If the sphere is initially at rest, it cannot be set in rotation by any means whatsoever; if it is initially in rotation, there is no means of changing its rotation. As this applies to every point in the fluid, one may visualize the fluid elements as being pushed around by boundary movements but not being rotated if initially at rest. Rotation or lack of rotation of the fluid particles is a property of the fluid itself and not its position in space.

The velocity potential ϕ is a scalar function of space such that its negative rate of change with respect to any direction is the velocity component in that direction. In vector notation

$$\mathbf{q} = -\operatorname{grad} \phi \tag{5.42}$$

or in terms of Cartesian coordinates

$$u = -\frac{\partial \phi}{\partial x} \qquad v = -\frac{\partial \phi}{\partial y} \qquad w = -\frac{\partial \phi}{\partial z} \tag{5.43}$$

The assumption of irrotational flow is equivalent to the assumption of a velocity potential.

Bernoulli Equation
Assuming that the flow is irrotational and that the extraneous force components are derivable from a force potential

$$X = -\frac{\partial \Omega}{\partial x} \qquad Y = -\frac{\partial \Omega}{\partial y} \qquad Z = -\frac{\partial \Omega}{\partial z} \tag{5.44}$$

The Euler equations may be integrated, resulting in the Bernoulli equation

$$\frac{1}{2}q^2 - \frac{\partial \phi}{\partial t} + \Omega + \frac{p}{\rho} = F(t) \tag{5.45}$$

where q is the speed at any point, $q^2 = u^2 + v^2 + w^2$, and $F(t)$ is an arbitrary function of the time.

For steady flow, that is, flow such that conditions at any point do not change with the time, $\partial \phi / \partial t = 0$, and $F(t)$ becomes a constant, hence

$$\frac{1}{2}q^2 + \Omega + \frac{p}{\rho} = C \tag{5.46}$$

where C is to be determined from known conditions at some point. With gravity the only extraneous force, and taking the z axis as positive upward,

$$\Omega = gz \tag{5.47}$$

Laplace Equation

The *Laplace equation* results when the continuity equation is written in terms of the velocity potential

$$\frac{\partial^2 \phi}{\partial x^2} + \frac{\partial^2 \phi}{\partial y^2} + \frac{\partial^2 \phi}{\partial z^2} = 0 \tag{5.48}$$

This is usually written $\nabla^2 \phi = 0$, where, in this shortened form, the equation may be in terms of any orthogonal coordinate system. For plane polar coordinates (r, θ)

$$\nabla^2 \phi = \frac{\partial \phi}{\partial r} + r\frac{\partial^2 \phi}{\partial r^2} + \frac{1}{r}\frac{\partial^2 \phi}{\partial \theta^2} = 0 \tag{5.49}$$

For cylindrical coordinates $(x, \bar{\omega}, \omega)$ where $\bar{\omega}$ is the distance from the x axis and ω is the angle a plane through the point and the x axis makes with an initial plane,

$$\nabla^2 \phi = \frac{\partial^2 \phi}{\partial x^2} + \frac{\partial^2 \phi}{\partial \bar{\omega}^2} + \frac{1}{\bar{\omega}}\frac{\partial \phi}{\partial \bar{\omega}} + \frac{1}{\bar{\omega}^2}\frac{\partial^2 \phi}{\partial \omega^2} = 0 \tag{5.50}$$

For spherical polar coordinates (r, θ, ω), where r is the distance from the origin, θ is the polar angle, and ω is the meridian angle,

$$\nabla^2 \phi = \frac{\partial}{\partial r}\left(r^2 \frac{\partial \phi}{\partial r}\right) + \frac{1}{\sin \theta}\frac{\partial}{\partial \theta}\left(\sin \theta \frac{\partial \phi}{\partial \theta}\right) + \frac{1}{\sin^2 \theta}\frac{\partial^2 \phi}{\partial \omega^2} = 0 \tag{5.51}$$

Any ϕ that satisfies the Laplace equation is a possible flow case. In steady flow any streamline may be taken as a solid boundary because it satisfies the boundary condition. The pressure distribution may then be found by use of the Bernoulli equation.

Two-Dimensional Flow

In two-dimensional flow all lines of motion are parallel to a fixed plane, say the xy plane, and the flow patterns (networks of equipotential lines and streamlines) in all planes parallel to this plane are identical. In Cartesian coordinates

$$\nabla^2 \phi = \frac{\partial^2 \phi}{\partial x^2} + \frac{\partial^2 \phi}{\partial y^2} = 0 \tag{5.52}$$

Stream Function. A streamline is a continuous line drawn through the fluid in such a way that at every point it has the direction of the velocity vector. The stream function is a scalar function of space such that the rate of flow across any line connecting two points is given by the difference in values of the stream function at these points.

$$\psi = \psi(x, y, t)$$

is defined as the stream function. Since there is no flow across a streamline, the value of ψ is constant along a streamline. Velocity components, u and v in the x and y directions, may be expressed in terms of ψ

$$u = -\frac{\partial \psi}{\partial y} \qquad v = \frac{\partial \psi}{\partial x} \tag{5.53}$$

From the definition of stream function, it is subject to addition of an arbitrary constant. Substituting these values of u, v into the continuity equation

$$\frac{\partial u}{\partial x} + \frac{\partial v}{\partial y} = 0 \tag{5.54}$$

yields

$$\frac{\partial^2 \psi}{\partial x^2} + \frac{\partial^2 \psi}{\partial y^2} = 0 \tag{5.55}$$

or, since $\nabla^2 \psi = 0$, ψ may also be taken as velocity potential for some case of irrotational flow. Lines in the flow having constant value of the velocity potential are called *equipotential lines*. They are everywhere orthogonal to streamlines, except at singular points, since there is no component of the velocity vector tangent to an equipotential surface. Relations between ϕ and ψ are given by

$$\frac{\partial \phi}{\partial x} = \frac{\partial \psi}{\partial y} \qquad \frac{\partial \phi}{\partial y} = -\frac{\partial \psi}{\partial x} \tag{5.56}$$

Application of Complex Variables to Irrotational Flow. Let $z = x + iy$ be a complex variable where x and y are real and $i = \sqrt{-1}$. Any function of z

$$w = f(z) = f(x + iy) = \phi + i\psi \tag{5.57}$$

that is defined throughout a region and that has a derivative throughout the region gives rise to two possible irrotational flow cases, since it may be shown that

$$\nabla^2 w = \nabla^2 \phi + i \nabla^2 \psi = 0 \tag{5.58}$$

and hence $\nabla^2 \phi = 0$ and $\nabla^2 \psi = 0$. w is called the complex potential, ϕ is the real part of w, and ψ is the pure imaginary part of w. The complex velocity

$$\frac{dw}{dz} = -u + iv \tag{5.59}$$

provides a simple method of finding velocity components from the complex potential. Stagnation points occur at those points in the flow where the velocity is zero, that is,

$$\frac{dw}{dz} = 0 \tag{5.60}$$

Conformal Mapping. Each of the two complex variables w and z introduced in the preceding paragraphs may be represented by plotting on a graph. The w plane is a graph having ϕ as abscissa and ψ as ordinate, showing values of $\phi = \pm nc$, $\psi = \pm nc$, where n is every integer and c is a constant. Figure 5.13 is a flow net of the simplest form showing equipotential and streamlines as series of parallel straight lines. The z plane is a plot of the equipotential and streamlines of the w plane, with x as abscissa and y as ordinate. The particular flow net in the z plane depends entirely on the functional relation between w and z, that is, $w = f(z)$. Since

$$w = f(z) = \phi + i\psi \qquad z = x + iy \tag{5.61}$$

where ϕ, ψ, x, y, are real

$$\phi = \phi(x, y) \qquad \psi = \psi(x, y) \tag{5.62}$$

The values of ϕ = constant and ψ = constant can be plotted on the z plane from the functional relations of ϕ and ψ with x, y.

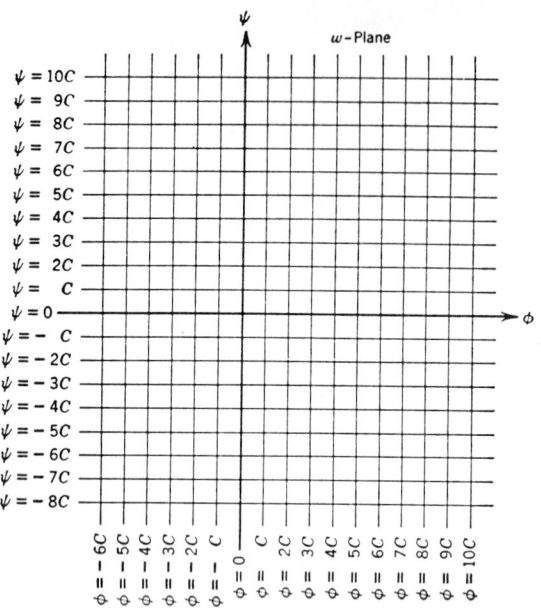

Fig. 5.13   Flow net in the w plane.

Examples of Two-Dimensional Flow. Since the Laplace equation is linear in ϕ, the sum of two solutions is also a solution, and the product of a solution by a constant is a solution. Several of the important solutions are given.

Rectilinear Flow. Uniform straight-line flow is given by

$$w = -Uz + iVz \qquad \phi = -Ux - Vy \qquad \psi = Vx - Uy \tag{5.63}$$

where the character of flow is easily seen from the complex velocity

$$\frac{dw}{dz} = -u + iv = -U + iV \qquad u = U \qquad v = V \tag{5.64}$$

and U, V are the $+x$, y components of the velocity, respectively.

Source or Sink. A source in two-dimensional flow is a straight line (parallel to the z axis) from which fluid flows outward uniformly in all directions normal to the line. A sink is a negative source; that is, the flow is into the line.

$$w = -\mu(z - a) \qquad \phi = -\mu \ln r \qquad \psi = -\theta \qquad \frac{dw}{dz} = -\frac{\mu}{z - a} \tag{5.65}$$

where $2\pi\mu$ is the outward flow from the line per unit length, known as the *strength*, a is the position of the source in the z plane (Fig. 5.14) with r_1, θ_1 shown in the figure. Streamlines are the radial lines of Fig. 5.15, and equipotential lines are the concentric circles. When μ is negative the equations are for a sink.

Vortex. The equations for a source, multiplied by $-i$, yield relations for the vortex

$$w = i\mu \ln(z - a) = i\mu \ln r_1 - \mu\theta_1 \qquad \phi = -\mu\theta_1 \qquad \psi = \mu \ln r_1 \tag{5.66}$$

Figure 5.15 is the flow net for a vortex with the radial lines now equipotential lines and the circles streamlines. The vortex causes circulation about itself. Circulation about any closed curve is defined as the line integral of the velocity around the curve. The circulation about the vortex is $\kappa = 2\pi\mu$, considered positive in the counterclockwise direction around the vortex.

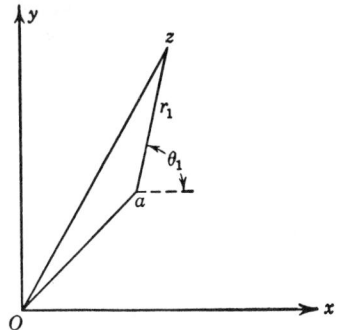

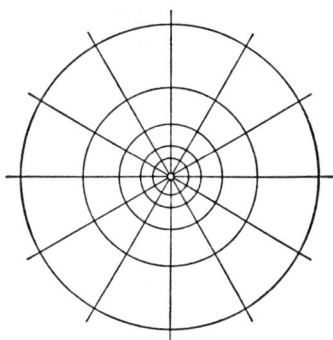

Fig. 5.14   Special coordinates for source at $z = a$.   **Fig. 5.15**   Flow net for source or vortex.

Doublet. The doublet is defined as the limiting case of a source and a sink of equal strength that approach each other such that the product of the strength by the distance between them remains a constant. The constant, μ, is called the strength of the doublet.

$$w = \frac{\mu}{z} \qquad \phi = \frac{\mu x}{x^2 + y^2} \qquad \psi = -\frac{\mu y}{x^2 + y^2} \qquad \frac{dw}{dz} = -\frac{\mu}{z^2} \tag{5.67}$$

Figure 5.16 shows the flow net. The axis of the doublet is in the direction from sink to source and is parallel to the $+x$ axis as given here. The equipotential lines are circles having their centers on the x axis; the streamlines are circles having centers on the y axis.

Flow around a Circular Cylinder. The superposition of a uniform flow in the $-x$ direction on a doublet with axis in the $+x$ direction results in flow around a circular cylinder. Adding the two flows, taking $\mu = Ua^2$,

$$w = \frac{Ua^2}{z} + Uz \qquad \phi = U\left(r + \frac{a^2}{r}\right)\cos\theta \qquad \psi = U\left(r - \frac{a^2}{r}\right)\sin\theta \tag{5.68}$$

where polar coordinates are employed. The streamline $\psi = 0$ is given by $\theta = 0$, π and by $r = a$; hence the circle $r = a$ may be taken as a solid boundary and the flow pattern of Fig. 5.17 is obtained.

Three-Dimensional Flow
Three-dimensional flows in general are more difficult to handle than those in two dimensions, primarily owing to lack of methods comparable to the use of complex variables and conformal mapping. A velocity potential must be found that satisfies the Laplace equation and the boundary

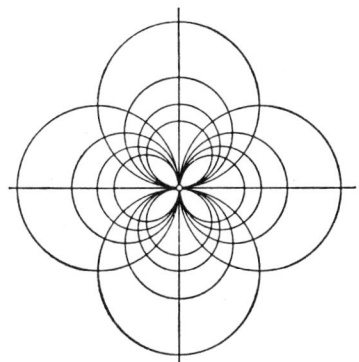

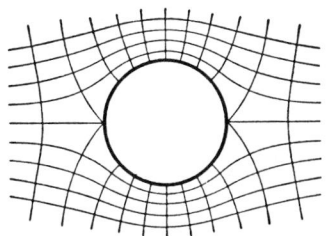

Fig. 5.16   Flow net for doublet, $w = \mu/z$.

Fig. 5.17   Flow pattern for uniform flow around a circular cylinder without circulation.

conditions. When the velocity potential is *single valued*, it may be shown that the solution is unique. For a body moving through an infinite fluid, otherwise at rest, a necessary condition is that the solution be such that the fluid at infinity remain at rest. Flow cases are usually found by investigating solutions of the Laplace equation to determine the particular boundary conditions that they satisfy.

Stokes's Stream Function. The Stokes stream function is defined only for those three-dimensional flow cases that have axial symmetry, that is, where the flow is in a series of planes passing through a given line and where the flow pattern is identical in each of these planes. The intersection of these planes is the axis of symmetry.

In any one of these planes through the axis of symmetry select two points A, P, such that A is fixed and P is variable. Draw a line connecting AP. The flow through the surface generated by rotating AP about the axis of symmetry is a function of the position of P. Let this function be $2\pi\psi$, and let the axis of symmetry be the x axis. Then ψ, which is the Stokes stream function, is a function of x and ω, where

$$\omega = \sqrt{y^2 + z^2} \tag{5.69}$$

is the distance from P to the x axis. The surfaces ψ = constant are stream surfaces. Since A is an arbitrary point, the stream function is always subject to the addition of an arbitrary constant.

Velocity components may be determined in terms of ψ,

$$u = -\frac{1}{\omega}\frac{\partial\psi}{\partial\omega} \qquad v' = \frac{1}{\omega}\frac{\partial\psi}{\partial x} \tag{5.70}$$

where u, v' are in the x, ω directions, respectively. The relations between ϕ and ψ are obtained by equating expressions for velocity components

$$\frac{\partial\phi}{\partial x} = \frac{1}{\omega}\frac{\partial\psi}{\partial\omega} \qquad \frac{\partial\phi}{\partial\omega} = -\frac{1}{\omega}\frac{\partial\psi}{\partial x} \tag{5.71}$$

Stokes's stream function has the dimensions *volume per unit time*. To express ψ in spherical polar coordinates, let r be the distance from the origin and θ be the angle the radius vector makes with the x axis; the meridian angle is not needed because of axial symmetry. Then

$$v_r = -\frac{1}{r^2\sin\theta}\frac{\partial\psi}{\partial\theta} \qquad v\theta = \frac{1}{r\sin\theta}\frac{\partial\psi}{\partial r} \tag{5.72}$$

and

$$\frac{1}{\sin\theta}\frac{\partial\psi}{\partial\theta} = r^2\frac{\partial\phi}{\partial r} \qquad \frac{\partial\psi}{\partial r} = -\sin\theta\frac{\partial\phi}{\partial\theta} \tag{5.73}$$

These equations are useful in dealing with flow about spheres, ellipsoids, and discs, and through apertures.

Examples of Three-Dimensional Flow. Two examples of three-dimensional flow with axial symmetry follow.

Source in a Uniform Stream. A point source is a point from which fluid issues at a uniform rate in all directions. Its strength m is the flow rate from the point, and the velocity potential and stream function for a source at the origin are

$$\phi = \frac{m}{4\pi r} \qquad \psi = \frac{Ur^2}{2}\sin^2\theta \tag{5.74}$$

The flow net is shown in Fig. 5.18. Superposing a uniform flow on a point source results in a *half body*, with the flow equations,

$$\phi = \frac{m}{4\pi r} + Ur\cos\theta \qquad \psi = \frac{m}{4\pi}\cos\theta + \frac{Ur^2}{2}\sin^2\theta \tag{5.75}$$

The resulting flow net is shown in Fig. 5.19. The body extends to infinity in the downstream direction

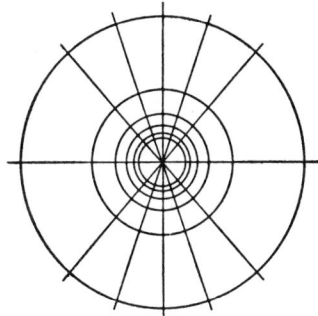

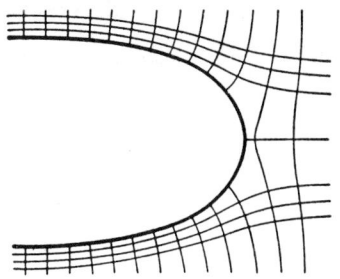

Fig. 5.19 Streamlines and equipotential lines for a half body.

Fig. 5.18 Streamlines and equipotential lines for a source.

and has the asymptotic cylinder $\bar{\omega} = \sqrt{m/(\pi U)}$. The equation of the half body is

$$r = \frac{1}{2}\sqrt{\frac{m}{\pi U}}\, \sec\frac{\theta}{2} \tag{5.76}$$

The pressure intensity on the half body is given by

$$p = \frac{\rho}{2} U^2 \left(\frac{3m^2}{16\pi^2 r^4 U^2} - \frac{m}{2\pi r^2 U} \right) \tag{5.77}$$

showing that the dynamic pressure drops to zero at great distances downstream along the body.

Flow around a Sphere. A doublet is defined as the limiting case as a source and sink approach each other, such that the product of their strength by the distance between them remains a constant. The doublet has directional properties. Its axis is positive from the sink toward the source. For a doublet at the origin, with axis in the $+x$ direction

$$\phi = \frac{\mu}{r^2}\cos\theta \qquad \psi = -\frac{\mu}{r}\sin^2\theta \tag{5.78}$$

where μ is the strength of the doublet.
Superposing a uniform flow, $u = -U$, on the doublet results in flow around a sphere,

$$\phi = \frac{Ua^3}{2r^2}\cos\theta + Ur\cos\theta \qquad \psi = -\frac{Ua^3}{2r}\sin^2\theta + \frac{Ur^2}{2}\sin^2\theta \tag{5.79}$$

where $\mu = Ua^3/2$. This is shown to be the case, since the streamline $\psi = 0$ is satisfied by

$$\theta = 0, \pi \qquad r = a$$

The flow net is shown in Fig. 5.20.

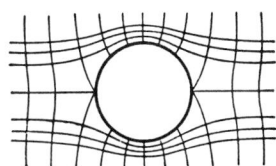

Fig. 5.20 Streamlines and equipotential lines for uniform flow about a sphere at rest.

5.5   VISCOUS FLUID DYNAMICS

The additional consideration of viscosity in the flow of fluids greatly complicates analysis and understanding of flow situations. In the fundamental sense, viscosity only introduces the possibility of shear stress into the flow process. However, the effects are far reaching and for many years have caused researchers and designers to rely heavily on experimental techniques. In recent years the wide availability of supercomputers has led to an increasing use of numerical analysis to predict the behavior of fluid flows.

In general, fluid flows are classified as laminar or turbulent, although in any given flow there may be regions of both laminar and turbulent flow. In laminar flow, fluid particles slide smoothly over one another with no mixing except that normally occurring as the result of molecular activity. Shear stress in laminar flow depends directly on the local velocity gradient and the fluid viscosity

$$\tau = \mu \frac{dv}{dy} \tag{5.80}$$

Shear stress is not affected by boundary roughness in laminar flow so long as the boundary roughness is small in comparison to the flow cross section.

In turbulent flow, a great deal of mixing occurs with random motion of fluid masses of various sizes occurring. Shear stress in turbulent flow depends strongly on the momentum exchange occurring as the result of turbulent mixing. An equivalent expression for shear stress in turbulent flow is

$$\tau = \varepsilon \frac{dv}{dy} \tag{5.81}$$

where ε is the eddy viscosity, which is a function of fluid viscosity and the local structure of the turbulent flow. This expression was unsatisfactory because it was impossible to quantify ε. An analysis of the turbulent momentum exchange process by Reynolds yielded a shear stress equation directly related to the flow turbulence.

$$\tau = -\rho \overline{u'v'} \tag{5.82}$$

where u', v' are the turbulent velocity fluctuations about the mean values and $\overline{u'v'}$ is the time average of the product of the fluctuations in the x and y directions. This expression is also not a very useful practical expression. Research eventually led to the Prandtl–von Karman equation for shear stress:

$$\tau = \rho \kappa^2 \frac{(dv/dy)^4}{(d^2v/dy^2)^2} \tag{5.83}$$

where κ is a dimensionless turbulence constant. This practical form of the shear stress equation can be integrated to generate velocity profiles.

5.5.1   Internal Flows

Flow in pipes and ducts under the action of viscosity requires some changes in basic concepts as well as in the equations for one-dimensional flow. Specifically, the velocity is no longer uniform in the cross section where the streamlines are straight and parallel. The velocity will be zero at the wall and increase to a maximum near the center of the conduit. This nonuniform velocity profile requires correction coefficients in the work-energy and impulse-momentum equations.

In the work-energy equation the kinetic energy term based on the average velocity must be modified by a correction factor

$$\alpha \frac{V^2}{2g} \tag{5.84}$$

where α is the kinetic energy correction factor defined as

$$\alpha = \frac{1}{V^2} \frac{\int^A v^3 \, dA}{\int^A v \, dA} \tag{5.85}$$

This adjustment is necessary because the variation in velocity across the flow cross section means that fluid particles entering the control volume all carry different amounts of kinetic energy.

A similar correction is needed for the impulse-momentum equation. The momentum flux term is modified as follows

$$\beta Q \rho V \tag{5.86}$$

where β is the momentum correction factor defined as

$$\beta = \frac{1}{V} \frac{\int^A v^2 \, dA}{\int^A v \, dA} \tag{5.87}$$

In addition to the need for compensating for the nonuniform velocity profile in one-dimensional flow, it is necessary to account for the energy converted to heat through the frictional processes. If heat transfer is of no interest or concern, then energy conversion to heat is considered a loss in usable energy. The amount of useful energy converted to heat per pound of fluid flowing is designated by h_L, head loss. The one-dimensional work-energy equation now has the form

$$\frac{p_1}{\gamma} + \alpha_1 \frac{V_1^2}{2g} + z_1 = \frac{p_2}{\gamma} + \alpha_2 \frac{V_2^2}{2g} + z_2 + h_{L_{1-2}} \tag{5.88}$$

In actuality, the h_L term reveals itself in an increase in internal energy (temperature) of the fluid or a heat transfer from the fluid.

The impulse-momentum equation has the form

$$\sum \mathbf{F}_{\text{ext}} = \sum (\beta Q \rho \mathbf{V})_{\text{out}} - \sum (\beta Q \rho \mathbf{V})_{\text{in}} \tag{5.89}$$

Another phenomenon stemming directly from the effect of friction is flow separation. Separation occurs when the fluid is forced to decelerate too rapidly or change direction too quickly. The result is the formation of wakes and eddies, poor pressure recovery, and excessive friction losses.

The nonuniform velocity profile occurring in viscous flow also causes vortexes and cross currents to occur at certain locations. Vortexes form in corners and at abrupt changes in cross section or flow direction. Cross currents or secondary flows occur at bends in the conduit. These secondary flows are the consequence of equal pressure gradients in the bend acting on fluid particles moving at different speeds in the flow cross section. The result is a spiral flow(s) superposed on the main flow. Both vortexes and secondary flows generate additional head losses and may cause some additional problems in hydraulic machinery.

5.5.2 External Flows

External flows are generally associated with lift and drag on solid bodies in a fluid of large extent. Although for ideal fluids there is no drag, this is not the case for real fluids. Frictional effects on the surface of the body create skin friction drag. That is, the no-slip fluid condition at the body surface creates a velocity profile and the resulting shear stress at the surface to occur. The frictional effects are confined to the "boundary layer" near the body where the flow may be laminar, turbulent, or both. Theoretical analysis of skin friction drag has been relatively successful. Some specifics on lift and drag are presented in a subsequent section.

Separation in external flows is also caused by "adverse" pressure gradients that are forcing the flow to decelerate faster than it can and still remain "attached" to the boundary. Separation manifests itself as a turbulent wake behind the body, which creates considerable drag on the body largely resulting from the pressure difference between the front and rear. Smooth bodies with gradually changed form tend to generate less separation and lower drag than blunt or sharp-cornered objects.

5.5.3 Navier–Stokes Equations*

Application of Newton's second law to a small fluid particle of dimension $dx\,dy\,dz$ yields a set of equations comparable to the Euler equation previously derived. The equations are referred to as the Navier–Stokes equations, and they take the following form for laminar flow:

$$X - \frac{1}{\rho}\frac{\partial p}{\partial x} + \nu\left(\frac{\partial^2 u}{\partial x^2} + \frac{\partial^2 u}{\partial y^2} + \frac{\partial^2 u}{\partial z^2}\right) = u\frac{\partial u}{\partial x} + v\frac{\partial u}{\partial y} + w\frac{\partial u}{\partial z} + \frac{\partial u}{\partial t} \tag{5.90}$$

$$Y - \frac{1}{\rho}\frac{\partial p}{\partial y} + \nu\left(\frac{\partial^2 v}{\partial x^2} + \frac{\partial^2 v}{\partial y^2} + \frac{\partial^2 v}{\partial z^2}\right) = u\frac{\partial v}{\partial x} + v\frac{\partial v}{\partial y} + w\frac{\partial v}{\partial z} + \frac{\partial v}{\partial t} \tag{5.91}$$

$$Z - \frac{1}{\rho}\frac{\partial p}{\partial z} + \nu\left(\frac{\partial^2 w}{\partial x^2} + \frac{\partial^2 w}{\partial y^2} + \frac{\partial^2 w}{\partial z^2}\right) = u\frac{\partial w}{\partial x} + v\frac{\partial w}{\partial y} + w\frac{\partial w}{\partial z} + \frac{\partial w}{\partial t} \tag{5.92}$$

*Illustrations and material extracted in all or in part from the previous edition by Victor L. Streeter.

X, Y, Z are extraneous force components per unit mass, in the xyz directions, respectively; ρ is the mass density, considered constant; p is the average pressure at a point; ν is the kinematic viscosity; u, v, w are the velocity components in the xyz directions.

These simultaneous, nonlinear, differential equations cannot be integrated except for extremely simple flow cases where many of the terms are neglected. They contain the basic assumption that the stresses on a particle may be expressed as the most general linear function of the velocity gradients.

Boundary Conditions. A real fluid in contact with a solid boundary must have a velocity exactly equal to the velocity of the boundary; this is much more restrictive than for a nonviscous fluid, where no restrictions are placed on tangential velocity components at a boundary.

When two fluids are flowing, a dynamical boundary condition arises at the interface. Applying the equation of motion to a thin layer of fluid enclosing a small portion of the interface shows that the terms containing mass are of higher order of smallness than the surface stress intensities, and hence that the stresses must be continuous through the surface.

In general, the boundary conditions at a solid surface give rise to rotational flow. Although the Navier–Stokes equations are satisfied by a velocity potential, since the viscous terms drop out, the boundary conditions cannot be satisfied.

Continuity. The continuity equation must hold, as in the case of nonviscous flow. It is

$$\frac{\partial u}{\partial x} + \frac{\partial v}{\partial y} + \frac{\partial w}{\partial z} = 0 \tag{5.93}$$

Examples. Several examples of flow at low Reynolds' numbers are given. It is assumed that in each case any turbulent fluctuations are completely damped out by viscous action.

Flow between Parallel Boundaries. For steady flow between fixed parallel boundaries at low Reynolds numbers, the Navier–Stokes equations can be greatly reduced. Taking the coordinates as shown in Fig. 5.21, the differential equations reduce to

$$\frac{\partial}{\partial x}(p + \gamma h) = \mu \frac{\partial^2 u}{\partial z^2} \tag{5.94}$$

$$\frac{\partial}{\partial y}(p + \gamma h) = \mu \frac{\partial^2 v}{\partial z^2} \tag{5.95}$$

$$\frac{\partial}{\partial z}(p + \gamma h) = 0 \tag{5.96}$$

where h is measured vertically upward and γ is the specific weight of the fluid. Integrating and introducing the boundary conditions $u = v = 0$ for $z = \pm b$,

$$u = \frac{z^2 - b^2}{2\mu} \frac{\partial}{\partial x}(p + \gamma h) = \frac{\partial}{\partial x}\left[(p + \gamma h)\left(\frac{z^2 - b^2}{2\mu}\right)\right] = -\frac{\partial \phi}{\partial x} \tag{5.97}$$

$$v = \frac{z^2 - b^2}{2\mu} \frac{\partial}{\partial y}(p + \gamma h) = \frac{\partial}{\partial y}\left[(p + \gamma h)\left(\frac{z^2 - b^2}{2\mu}\right)\right] = -\frac{\partial \phi}{\partial y} \tag{5.98}$$

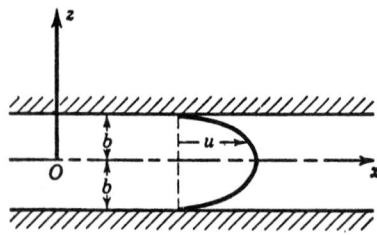

Fig. 5.21   Viscous flow between fixed parallel boundaries.

where $-\phi$ is the quantity in brackets. For this special viscous flow case a velocity potential exists, given by ϕ.

Using this case as an analogy to potential flow, Hele–Shaw[2] constructed an apparatus consisting of two closely spaced glass plates. A transparent fluid is caused to flow between the plates, and dye is continuously injected into the fluid at regular intervals along the upstream edge of the plates. An object placed between the plates causes the fluid to deviate in flowing around it in such a way that the dyed portions of the fluid trace out streamlines for two-dimensional flow. The results are confirmed by potential theory and by other experimental means.

For motion of the upper plate in the x direction with velocity U, the boundary conditions become

$$u = v = 0 \quad \text{for} \quad z = -b \qquad u = U \qquad v = 0 \quad \text{for} \quad z = +b$$

The velocity components are

$$u = \frac{U}{2}\left(1 + \frac{z}{b}\right) + \frac{z^2 - b^2}{2\mu}\frac{\partial}{\partial x}(p + \gamma h) \tag{5.99}$$

$$v = \frac{z^2 - b^2}{2\mu}\frac{\partial}{\partial x}(p + \gamma h) \tag{5.100}$$

The maximum velocity has been displaced from the middle plane. When $p + \gamma h$ is constant, the gradient is zero and flow results due to motion of the upper plate only.

Theory of Lubrication. The equations for two-dimensional viscous flow are applicable to the case of a slider bearing and can be applied to journal bearings. The simple case of a bearing of unit width is developed here, under the assumption that there is no flow out of the sides of the block, that is, normal to the plane of Fig. 5.22, where the clearance b is shown to a greatly exaggerated scale. The motion of a bearing block sliding over a plane surface, inclined slightly so that fluid is crowded between the two surfaces, develops large supporting forces normal to the surfaces. The angle of inclination is very small, therefore the differential equations of the preceding example apply. Since elevation changes also are very small and flow is in the x direction only, the equations reduce to

$$\frac{\partial p}{\partial x} = \mu\frac{\partial^2 u}{\partial z^2} \qquad \frac{\partial p}{\partial z} = 0 \tag{5.101}$$

Considering the inclined block stationary and the plane surface in motion, and taking the pressure at the two ends of the block as zero, the boundary conditions become

$$x = 0, \quad x = L, \quad p = 0; \qquad u = U, \quad z = 0; \qquad u = 0, \quad z = b$$

Integrating the equations, and considering unit width normal to the figure, the discharge Q and pressure distribution are determined,

$$Q = \frac{Ub_1 b_2}{b_1 + b_2} \qquad p = \frac{6\mu Ux(b - b_2)}{b^2(b_1 + b_2)} \tag{5.102}$$

The last relation shows that b must be greater than b_2 for positive pressure buildup in the bearing.

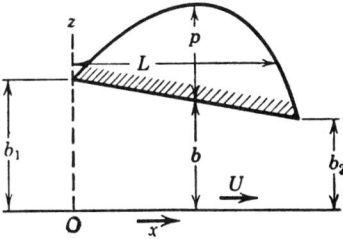

Fig. 5.22 Slider bearing.

The point of maximum pressure intensity and its value are

$$x|_{p_{\max}} = \frac{b_1 L}{b_1 + b_2} \qquad p_{\max} = \frac{3}{2} \frac{\mu UL}{b_1 b_2} \frac{b_1 - b_2}{b_1 + b_2} \tag{5.103}$$

The force P, which the bearing will sustain, is

$$P = \int_0^L p \, dx = \frac{6\mu UL^2}{b_2^2 (k-1)^2} \left[\ln k - \frac{2(k-1)}{k+1} \right] \tag{5.104}$$

The maximum bearing load is obtained for $k = 2.2$, yielding

$$P = 0.16\mu \frac{UL^2}{b_2^2} \qquad D = 0.75\mu \frac{UL}{b_2} \tag{5.105}$$

the ratio

$$\frac{P}{D} = 0.21 \frac{L}{b_2} \tag{5.106}$$

can be made very large, since b_2 is small. The pressure distribution is shown for one case in Fig. 5.22. For $k = 2.2$, the line of action of the bearing load is at $x = 0.58L$. In general the line of action is given by

$$\bar{x} = \frac{L}{2} \left[\frac{2k}{k-1} - \frac{k^2 - 1 - 2k \ln k}{(k^2 - 1)\ln k - 2(k-1)^2} \right] \tag{5.107}$$

Journal bearings are computed in an analogous manner. In general the clearances are so small compared with the radius of curvature of bearing surface that the equations for plane motion can be applied.

Flow through Circular Tubes. For steady flow through circular tubes with values of the Reynolds number $vD\rho/\mu$ less than 2000, the motion is laminar. Equilibrium conditions on a cylindrical element concentric with the tube axis show that the shear stress varies linearly from zero at the pipe axis to a maximum $\Delta p/\Delta L r_0/2$ at the wall, where $\Delta p/\Delta L$ is the drop in pressure intensity per unit length of tube. In one-dimensional laminar flow the relation between shear stress and velocity gradient is given by Newton's law of viscosity,

$$\tau = -\mu \frac{du}{dr} \tag{5.108}$$

Using this relation for value of shear stress, the velocity distribution is found to be

$$u = \frac{\Delta p}{\Delta L} \frac{r_0^2 - r^2}{4\mu} \tag{5.109}$$

where r is the distance from the tube axis. The maximum velocity is at the axis.

$$u_{\max} = \frac{\Delta p \, r_0^2}{\Delta L \, 4\mu} \tag{5.110}$$

The average velocity is half the maximum velocity

$$V = \left(\frac{\Delta p}{\Delta L} \right) \left(\frac{r_0^2}{8\mu} \right) \tag{5.111}$$

The discharge, obtained by multiplying average velocity by cross-sectional area, is

$$Q = \left(\frac{\Delta p}{\Delta L} \right) \left(\frac{\pi D^4}{128\mu} \right) \tag{5.112}$$

which is the *Hagen–Poiseuille law*. The preceding equations are independent of the surface condition in the tube, and they therefore hold for either rough or smooth tubes.

Viscous Flow around a Sphere

Stokes's Law. The flow of an infinite viscous fluid around a sphere at very low Reynolds numbers has been solved by Stokes.[3] The Navier–Stokes equations with the acceleration terms omitted must be satisfied, as well as continuity and the boundary condition that the velocity vanish at the surface of the sphere. Stokes's solution is

$$u = U\left[\frac{3}{4}\frac{ax^2}{r^3}\left(\frac{a^2}{r^2} - 1\right) + 1 - \frac{1}{4}\frac{a}{r}\left(3 + \frac{a^2}{r^2}\right)\right] \tag{5.113}$$

$$v = U\frac{3}{4}\frac{axy}{r^3}\left(\frac{a^2}{r^2} - 1\right) \tag{5.114}$$

$$w = U\frac{3}{4}\frac{axz}{r^3}\left(\frac{a^2}{r^2} - 1\right) \tag{5.115}$$

$$p = -\frac{3}{2}\frac{\mu U ax}{r^3} \tag{5.116}$$

The radius of a sphere is a; the undisturbed velocity is in the x direction, $u = U$; and p is the average dynamic pressure intensity at a point. The drag on a sphere is made up from the pressure difference over the surface of the sphere and from the shear stress. The viscous drag due to shear stress is twice as great as that due to the pressure difference. The total drag is

$$D = 6\pi a\mu U \tag{5.117}$$

This is known as *Stokes's law*.

The settling velocity of small spheres may be obtained by writing the equation for drag, weight of particle, and buoyant force. Solving for settling velocity

$$U = \frac{2}{9}\frac{a^2}{\mu}(\gamma_s - \gamma) \tag{5.118}$$

where γ_s is the specific weight of the solid particle. Stokes's law has been found by experiment to hold for Reynolds numbers below 1, that is,

$$\frac{2a\rho U}{\mu} < 1 \tag{5.119}$$

5.6   SIMILITUDE AND DIMENSIONAL ANALYSIS

Similitude and dimensional analysis are inextricably tied to experimental testing and the presentation of experimental data. Principles of similitude permit the prediction of prototype behavior based on the performance of models. Techniques of dimensional analysis permit the efficient and logical presentation of test data in dimensionless form. Even in recent times when computer simulation is replacing laboratory work at an increasing pace, many flow phenomena are so complex as to warrant fluid modeling.

5.6.1   Similitude

For complete similitude the model and the prototype must be geometrically, kinematically, and dynamically similar. Geometric similarity exists when the model is a photographic reduction (or enlargement) of the prototype. Kinematic similarity occurs when the streamline pattern in the model is a photographic reduction (or enlargement) of that of the prototype. Also ratios of velocity vectors at corresponding points are constant. Dynamic similarity requires that the ratios of similar forces at all corresponding points in the flow be constant.

In general, if geometric similarity and dynamic similarity exist, then kinematic similarity is guaranteed.

Quantitative relationships for dynamic similarity are obtained by considering the contributing forces acting on a fluid particle. Potential contributing forces include viscosity (shear), pressure, gravity, elasticity and surface tension. These forces all contribute to the acceleration of the fluid particle in the prototype flow.

$$\mathbf{F}_v + \mathbf{F}_p + \mathbf{F}_g + \mathbf{F}_e + \mathbf{F}_s = m\mathbf{a} = -\mathbf{F}_I \text{ (inertial force)} \tag{5.120}$$

Dividing by the inertial force,

$$\left(\frac{\mathbf{F}_v}{\mathbf{F}_I}\right)_P + \left(\frac{\mathbf{F}_p}{\mathbf{F}_I}\right)_P + \left(\frac{\mathbf{F}_g}{\mathbf{F}_I}\right)_P + \left(\frac{\mathbf{F}_e}{\mathbf{F}_I}\right)_P + \left(\frac{\mathbf{F}_s}{\mathbf{F}_I}\right)_P = -1 \tag{5.121}$$

If the same process is followed in the model,

$$\left(\frac{\mathbf{F}_v}{\mathbf{F}_I}\right)_m + \left(\frac{\mathbf{F}_p}{\mathbf{F}_I}\right)_m + \left(\frac{\mathbf{F}_g}{\mathbf{F}_I}\right)_m + \left(\frac{\mathbf{F}_e}{\mathbf{F}_I}\right)_m + \left(\frac{\mathbf{F}_s}{\mathbf{F}_I}\right)_m = -1 \tag{5.122}$$

This scaling procedure renders the force polygons at the corresponding points in the model and prototype as congruent provided the force ratios for each source of force are the same in model and prototype. In fact, the scaled polygons are congruent if *all but one* of the forces ratios are equal. This relationship can be summarized as follows:

Viscous Forces:
$$\left(\frac{Vl\rho}{\mu}\right)_m = \left(\frac{Vl\rho}{\mu}\right)_P \tag{5.123}$$

where $Vl\rho/\mu$ = Reynolds number, **R**

Pressure Forces:
$$\left(\frac{\rho V^2}{\Delta p}\right)_m = \left(\frac{\rho V^2}{\Delta p}\right)_P, \tag{5.124}$$

where $V\sqrt{\rho}/\sqrt{2\,\Delta p}$ = Euler number, **E**

Gravity Forces:
$$\left(\frac{V^2}{gl}\right)_m = \left(\frac{V^2}{gl}\right)_P \tag{5.125}$$

where $V/\sqrt{gl}$ = Froude number, **F**

Elastic Forces:
$$\left(\frac{\rho V^2}{K}\right)_m = \left(\frac{\rho V^2}{K}\right)_P \tag{5.126}$$

where $V/\sqrt{K/\rho}$ = Mach number, **M**

Surface Tension Forces:
$$\left(\frac{\rho l V^2}{\sigma}\right)_m = \left(\frac{\rho l V^2}{\sigma}\right)_P \tag{5.127}$$

where $\rho l V^2/\sigma$ = Weber number, **W**

In most fluid problems, some of these forces are negligible and can be ignored in model-prototype relationships. For example, if there is no free surface and flows are well below the sonic velocity, then only viscous and pressure forces are important. In this case, equality of Reynolds numbers would guarantee similarity. Typically, pressure forces are always in existence and the Euler number is the force ratio omitted in specifying similarity (as noted earlier).

There are situations where complete similarity cannot be practically achieved. For example, in the modeling of ship hulls, if water is used as the modeling fluid, the model must be as large as the prototype. This is true because viscous and gravity forces predominate and both Reynolds and Froude numbers must be equal to guarantee similarity. In this case, to get around the problem, the model-prototype relation is determined by the Froude number equality and the viscous effects on hull drag are determined through other means. This situation is known as incomplete similarity.

In other flow situations such as rivers and harbors, the lateral dimensions are so large compared to the vertical dimensions that geometric scaling would make the model so shallow as to be under strong surface tension effects. Also the boundary roughness would be reduced to supersmoothness. In these situations, two different model scales are used, one for horizontal dimensions, another for vertical dimensions. These models generally have to be adjusted for "scale effects" by employing artificial roughness to properly represent prototype behavior.

5.6.2 Dimensional Analysis

Dimensional analysis is based on the principle that meaningful physical relationships between quantities must be dimensionally homogeneous; that is, both sides of an equation must have the same dimensions. The four basic dimensions used in fluid mechanics are force (F), mass (M), length (L), and time (T). In fact, the dimension of force is related to the dimensions of mass, length, and time through Newton's second law

$$F = M\frac{L}{T^2} \qquad (5.128)$$

The units of force are said to be *equivalent* to the units of mass × length over time squared.

The principle of dimensional homogeneity can be used to develop a formal procedure for establishing relations between physical quantities and most importantly, dimensionless groups. Dimensionless groups of quantities are extremely valuable in the presentation of experimental data. Their use permits large amounts of data to be presented on simple graphs, graphs with parametric relationships or in some more complex cases, multicorrelational graphs.

For example, consider the case of the drag force on a sphere falling at constant velocity in a very viscous liquid. Drag (D) is a function of velocity (V), size (d), and viscosity (μ).

$$D = f(V, d, \mu)$$

$$F = f'\left(\frac{L}{T}, L, \frac{F - T}{L^2}\right)$$

The only combination of quantities on the right-hand side that will make the relationship dimensionally correct is

$$D = KV\,d\mu \qquad (5.129)$$

where K is a dimensionless constant. Thus by dimensional reasoning alone, Stokes's law has been deduced.

However, formal dimensional analysis does not always offer a unique approach to generating the desired dimensionless groups. A more modern and intuitive method, called the Π theorem, is more useful. This theorem informs the user as to how many dimensionless groups may be formed and leaves the configuration of each group up to the user. This freedom, coupled with a knowledge of the important similarity principles, permits the user to make a good choice of the proper groups.

For example, when establishing the quantities affecting the drag on a ship's hull, the list includes

$$D \text{ (drag)} = f(l, V, \mu, \rho, g)$$

or

$$f'(D, l, V, \mu, \rho, g) = 0$$

The Π theorem allows the formation of three dimensionless groups. Knowing that viscous and gravitational effects are most significant in affecting drag, the dimensionless groups of Reynolds number and Froude number are selected. Then the drag force D is combined into a third group.

$$g\left(\frac{D}{\rho l^2 V^2}, \frac{Vl\rho}{\mu}, \frac{V^2}{gl}\right) = 0$$

$$g\left(\frac{D}{\rho l^2 V^2}, \mathbf{R}, \mathbf{F}^2\right) = 0$$

Using this knowledge, the experimentor can plot all data on one graph using, for example, $D/\rho l^2 V^2$ as the ordinate, $\mathbf{R}$ as the abscissa, and $\mathbf{F}$ as the parametric variable.

5.7   FLOW IN CLOSED CONDUITS

Flow through a conduit may be steady or unsteady, uniform or nonuniform, and laminar or turbulent. Steady flow refers to flow at constant rate, uniform flow to prismatic sections of conduit, and laminar flow to those cases where viscous forces predominate and the losses are a linear function of the velocity; whereas for turbulent flow the losses vary as the velocity to some power (1.7 to 2.0), depending in part upon the Reynolds number.

The classical methods of hydrodynamics applying to an ideal fluid are of little value in solving flow problems in conduits, although they are extremely useful in connection with flow around immersed bodies. The nature of turbulent flow, on the other hand, is not sufficiently well understood to permit computation of the energy losses for given boundary conditions and rates of flow, and hence recourse must generally be taken to experimentation.

The work-energy equation is of the first importance in solving flow problems. Momentum relationships are of use in certain cases in which the forces acting on the fluid are known, or are desired. In special situations where both the energy and momentum equations are applicable, the energy loss may be computed without recourse to experimentation. The continuity equation in steady flow usually states that the flow past every cross section is the same. When the fluid is compressible, this is a statement of mass or weight flow, but for liquids it is sufficient to deal with volume rates only. The three types of equations (energy, momentum, and continuity), together with the experimentally determined loss relationships, provide the general framework for solving closed-conduit problems.

5.7.1   Velocity Distribution*

The velocity distribution for established laminar flow through round tubes and between parallel plates has been discussed under laminar fluid motion, Section 5.5.

Establishment of Flow. The velocity distribution for fully established uniform flow in a closed conduit is determined by the relationship between the radial velocity gradient and the shear stress. In turbulent flow the velocity distribution cannot be derived exactly, although much has been accomplished in recent years in the analytical approach to rational velocity-distribution equations.

Downstream from any change in cross section or direction, there is a length over which the velocity distribution regains its characteristic form, depending on the shape of cross section, the wall roughness, and the Reynolds number. For example, when the flow passes from a reservoir through a rounded entrance into a conduit, the velocity is practically constant over the section at the upstream end of the conduit. Such flow during the initial stages is therefore practically irrotational, since the boundary layer is very thin. The effect of boundary resistance, however, is to retard the fluid in the wall vicinity, resulting, through lateral transmission of shear, in a continuous growth of the boundary layer with distance from the inlet. Since the mean velocity must nevertheless remain constant, the central portion of the fluid is simultaneously accelerated, until the forces of shear and pressure gradient reach equilibrium as the velocity distribution of uniform flow becomes fully established some distance downstream.

For laminar flow, experiments by Nikuradse give the nominal length L for establishment of flow as

$$\frac{L}{D} = 0.06\mathbf{R}$$

where D is the pipe diameter and the Reynolds number is based on average velocity and diameter.

In turbulent flow, the transition to the established velocity distribution is effected in a much shorter reach, because of the pronounced mixing action that then prevails. Nikuradse's experiments indicate that a distance of 25 to 40 diameters is sufficient, and that the length is not so dependent on the Reynolds number.

Rational Formulas for Established Turbulent Flow. In turbulent flow, the ratio of shear stress to velocity gradient depends not only on physical properties of the fluid but also on characteristics of the flow.

Stanton first stated that the turbulent velocity distribution in the central portions of a conduit has a form that is independent of the wall roughness and viscous effects, provided that the wall shear remains the same. In equation form

$$\frac{v_{\max} - v}{\sqrt{\tau_0/\rho}} = F\left(\frac{r}{r_0}\right)$$

where $v_{\max}$ is the velocity at the pipe axis, v is the velocity at the distance r from the axis, r_0 is the

*Illustrations and material extracted in all or in part from previous edition by Victor L. Streeter.

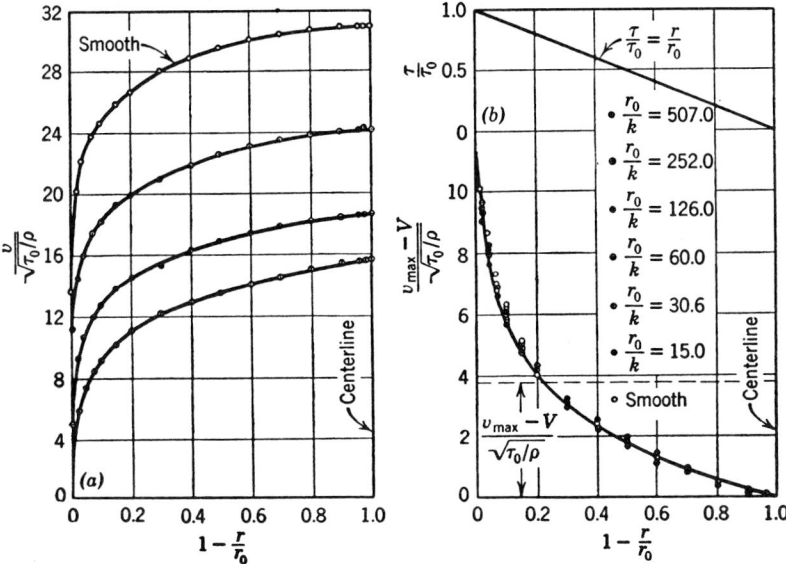

Fig. 5.23   Generalized plot of velocity distribution for smooth and rough pipes.

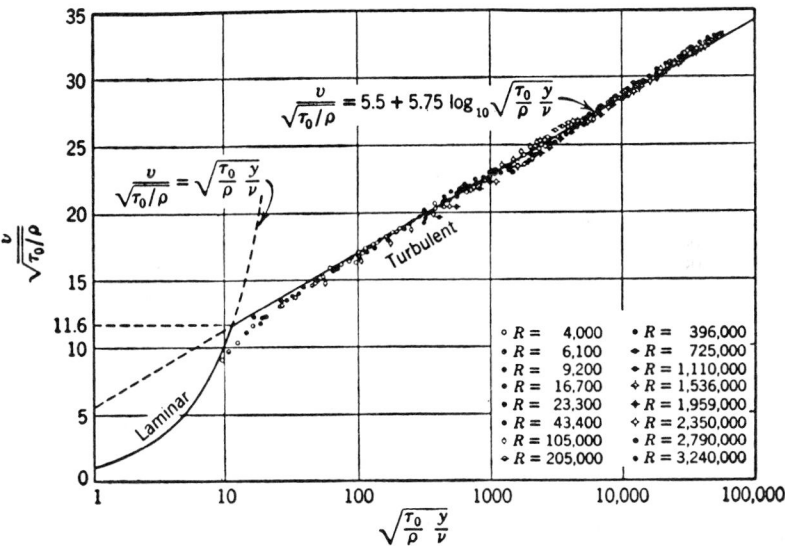

Fig. 5.24   Universal velocity distribution for smooth pipes.

pipe radius, τ_0 is the wall shear, ρ is the mass density of fluid, and F is an unknown function. The proof is evident by an inspection of the Nikuradse data on smooth- and sand-roughened pipes, given in Fig. 5.23; k is the diameter of sand grains cemented to the pipe walls.

Based on the preceding, von Kármán[4] obtained the formula for smooth pipes,

$$\frac{v}{\sqrt{\tau_0/\rho}} = C_1 + \frac{1}{\kappa}\ln\left(\sqrt{\frac{\tau_0}{\rho}}\frac{y}{\nu}\right) \tag{5.130}$$

where κ is a universal constant, having the value 0.40, and ν is the kinematic viscosity. Figure 5.24,

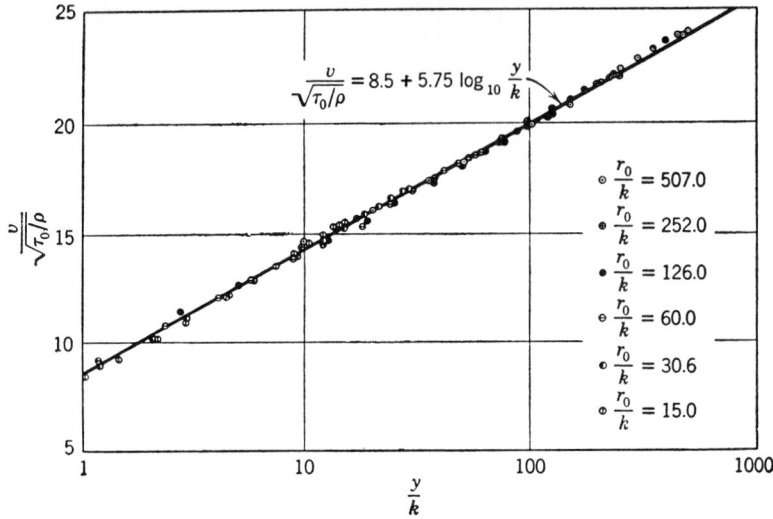

Fig. 5.25   Universal velocity distribution for rough pipes.

based on Nikuradse's tests, shows the value of C_1 to be 5.5 for best agreement with the data. In the immediate vicinity of the pipe wall, through a film called the *laminar sublayer*, the velocity is given closely by $v = y\tau_0/\mu$. This may be written

$$\frac{v}{\sqrt{\tau_0/\rho}} = \sqrt{\frac{\tau_0}{\rho}}\,\frac{y}{\nu} \tag{5.131}$$

plotted in Fig. 5.24. The intersection of the two curves may be taken arbitrarily as the border between the two types of flow, although actually there is a transition from the laminar to the turbulent zone. From the figure, the laminar film has the thickness

$$\delta = \frac{11.6\nu}{\sqrt{\tau_0/\rho}} \tag{5.132}$$

For rough pipes von Kármán obtained the formula

$$\frac{v}{\sqrt{\tau_0/\rho}} = C_2 + \frac{1}{\kappa}\ln\left(\frac{y}{\kappa}\right) \tag{5.133}$$

Figure 5.25 shows the Nikuradse sand-roughened pipe tests. From these data, $C_2 = 8.5$.

The two logarithmic equations do not give a zero slope of the velocity distribution curve at the center line. This is a defect in the formulas that, nevertheless, has little significance from a practical viewpoint. The equations actually portray the true velocity distribution in the central region of the flow very well, although they were derived for the region near the wall.

Energy and Momentum Correction Factors. In writing the work-energy equation between two cross sections of a conduit, it is usually satisfactory to express the mean kinetic energy per unit weight simply as $V^2/2g$, where V is the average velocity at a section. As discussed earlier, however, this is strictly true only when the velocity is constant over both cross sections. For laminar flow, the correction α by which $V^2/2g$ must be multiplied to give the true mean value is 2.0, and for sections where there is back flow the factor may be even larger. Values of α and β based on Prandtl–Kármán turbulent velocity distribution equations are given in Fig. 5.26.

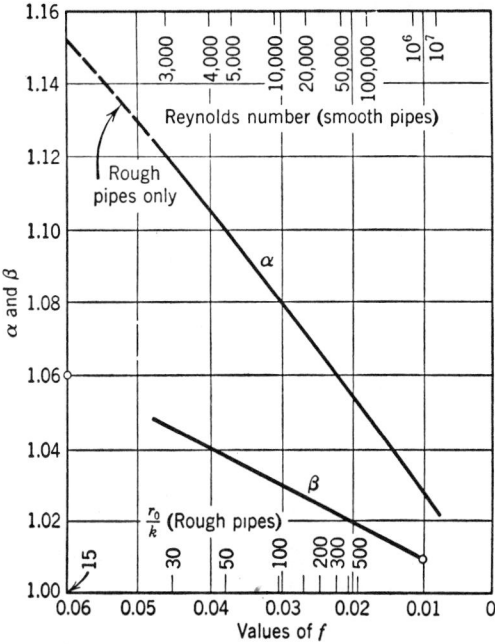

Fig. 5.26 Energy and momentum correction factors for smooth and rough pipes.

5.7.2 Pipe Friction*

A general equation for solving one-dimensional pipe flow problems is the work-energy equation

$$\frac{p_1}{\gamma} + \alpha_1 \frac{V_1^2}{2g} + z_1 = \frac{p_2}{\gamma} + \alpha_2 \frac{V_2^2}{2g} + z_2 + h_{L_{1-2}} \tag{5.134}$$

where $\Sigma h_{L_{1-2}}$ is the total of all the friction losses between the cross sections 1 and 2. These losses can generally be divided into two general categories—pipe friction and minor losses. Pipe friction losses are those caused by the continuing viscous action along the conduit. Minor losses are the result of the additional friction losses over and above the pipe friction. These losses are caused by flow separation, eddies, and wakes that are generated by changes in flow direction or cross section and pipeline appurtenances. This section deals with pipe friction.

Pipe friction losses for circular cylindrical pipes depend on the flow velocity, pipe size, fluid viscosity, and the wall roughness. In regard to wall roughness, friction loss specifically depends on *relative* roughness, that is, the *absolute* roughness compared with the pipe diameter. The various formulas for pipe friction loss differ in how they incorporate wall roughness into the head loss equation. The formulas in most common use today are the following:

Darcy–Weisbach: $\quad\quad\quad\quad\quad\quad h_f = f\dfrac{L}{D}\dfrac{V^2}{2g}$ $\quad\quad\quad\quad\quad\quad\quad\quad\quad\quad\quad\quad$ (5.135)

Hazen–Williams: $\quad\quad\quad\quad\quad V = 0.55CD^{0.63}S^{0.54}$ $\quad\quad\quad\quad\quad\quad\quad\quad\quad\quad\quad$ (5.136)

Manning: $\quad\quad\quad\quad\quad\quad\quad\quad V = \dfrac{0.59}{n}D^{2/3}S^{1/2}$ $\quad\quad\quad\quad\quad\quad\quad\quad\quad\quad$ (5.137)

where f, C, n are friction coefficients; S is head loss per unit length of pipe, D is inside pipe diameter and L is pipe length. The Darcy–Weisbach formula, commonly referred to as the Darcy

*Material and illustrations used by permission from *Analysis and Control of Unsteady Flow in Pipelines* by G. Z. Watters, Butterworth, Stoneham, Mass., 1984.

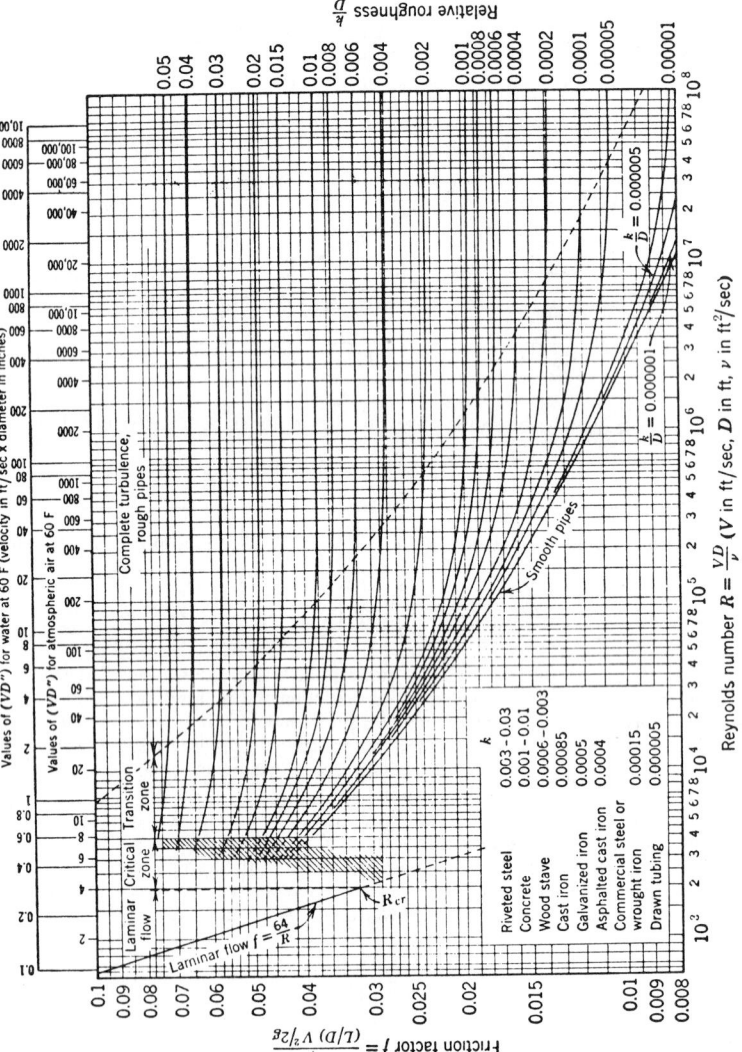

Fig. 5.27 Resistance diagram. Reproduced with permission from *Transactions ASME*, November, 1944, "Friction Factor for Pipe Flow," by Lewis F. Moody, Princeton, N.J.

TABLE 5.9   Roughness Values for Commercial Pipes

Pipe Material	k (in.)	C (Hazen–Williams)	n (Manning)
Riveted steel	0.036–0.36	110	0.013–0.017
Concrete	0.012–0.12	120–140	0.011–0.014
Cast iron (new)	0.010	130	0.013
Cast iron (old)	—	100	0.015–0.035
Galvanized iron	0.0060	—	0.016
Asphalted iron	0.0048	—	0.013
Welded steel	0.0018	120	0.012
Asbestos cement	—	140	0.011
Copper, aluminum tube	Smooth	150	0.010
PVC, plastic	Smooth	150	0.009

formula, is the most general in application. It can be used for a variety of liquids and gases, for laminar and turbulent flow, and for rough or smooth pipes. Its main disadvantage is the fact that the friction factor f is often dependent on one of the design unknowns (pipe diameter or discharge) and a trial solution results. However, engineers are increasingly using this formula because of its breadth of application.

The Hazen–Williams formula was developed for the computation of friction losses for water flowing in distribution system pipes. It works well for moderately smooth pipes (such as cast iron), but it is not accurate for rough pipes, small pipes, or laminar low.

The main task is to find the friction factor f. The f value depends on two parameters—Reynolds number and relative roughness. The Reynolds number measures the effect of viscosity on f and is defined as

$$R = \frac{VD}{\nu} \qquad (5.138)$$

where ν = kinematic viscosity of the fluid. The relative roughness measures the roughness of the pipe wall relative to the pipe diameter and is expressed as k/D, where k is a measure of pipe wall roughness.

The Reynolds number and the relative roughness are used in conjunction with the Moody diagram (Fig. 5.27) to find the f value. For convenience, Table 5.9 was compiled from several sources to provide assistance in selecting roughness values for various pipe materials. It should be noted that, typically, there are several inconsistencies between the roughness values given in Table 5.9 for the different formulas and those generated using the following equations. Table 5.6 lists kinematic viscosities for water over a normal range of temperatures.

The use of the Moody diagram can best be shown by example.

Example 5.1. Compute the friction factor f for the flow of 1500 gpm of water at normal temperature in a 10-in. cast-iron pipe.

$$V = \frac{1500}{449\frac{\pi}{4}\left(\frac{10}{12}\right)^2} = 6.13 \text{ fps}$$

From Table 5.6,

$$\nu(60°F) = 1.22 \times 10^{-5} \text{ ft}^2/\text{sec}$$

From Table 5.9

$$\frac{k}{D} = \frac{0.01}{10} = 0.001$$

From Eq. (5.138)

$$R = \frac{6.13 \times \dfrac{10}{12}}{1.22 \times 10^{-5}} = 418{,}000$$

From Figure 5.27 $f = 0.020$.

Both the Hazen–Williams and Manning formulas can be manipulated into the same form as the Darcy formula. The resulting f expressions may be compared with the Darcy f value to deduce the situations where these formulas may be confidently applied.

Hazen–Williams:
$$h_f = \left(\frac{1090}{C^{1.85}R^{0.15}}\right)\frac{L}{D}\frac{V^2}{2g} \tag{5.139}$$

Manning:
$$h_f = \left(\frac{185n^2}{D^{1/3}}\right)\frac{L}{D}\frac{V^2}{2g} \tag{5.140}$$

Both C and n are roughness values that can be obtained from tables such as Table 5.9. For additional information on C values, see Davis and Sorensen[5] and for n values, see Chow[6].

The areas of applicability of the three head loss formulas can be deduced. It is clear that the Manning formula is indeed a rough pipe formula because the Reynolds number does not appear. On the other hand, the Hazen–Williams formula is a relatively smooth pipe formula because the head loss varies with discharge in the same manner as smooth pipes. Their use should be strictly limited to these specific hydraulic conditions. However, in actual design use of the formulas, the results are not quite so dramatic. Consider Example 5.2.

Example 5.2. Select a pipe to convey 20 cfs of water between two reservoirs 5 miles apart and 200 ft different in elevation. Use welded steel pipe.

   Hazen–Williams:
From Table 5.9 $C = 120$.
From Eq. 5.136,

$$Q = AV = \frac{\pi}{4}D^2 \times 0.55CD^{0.63}S^{0.54}$$

$$20 = \frac{\pi}{4}D^2 \times 0.55 \times 120 \times D^{0.63}\left(\frac{200}{5 \times 5280}\right)^{0.54}$$

$$D = 1.90 \text{ ft or } 22.8 \text{ in.}$$
$$\text{Design } D = 24 \text{ in.}$$
$$\textit{Equivalent } f \textit{ value} = 0.019$$

   Manning:
From Table 5.9 $n = 0.012$.
From Eq. 5.137,

$$Q = AV = \frac{\pi}{4}D^2 \times \frac{0.59}{n}D^{2/3}S^{1/2}$$

$$20 = \frac{\pi}{4}D^2\frac{0.59}{0.012}D^{2/3}\left(\frac{200}{5 \times 5280}\right)^{1/2}$$

$$D = 1.95 \text{ ft or } 23.4 \text{ in.}$$
$$\text{Design } D = 24 \text{ in.}$$
$$\textit{Equivalent } f \textit{ value} = 0.021$$

   Darcy–Weisbach:
This is a trial solution because Reynolds number and relative roughness cannot be found unless diameter is known. Because we know roughly what the diameter is from the previous solution, we will use that as an estimate.

From an estimated $D \approx 24$ in., from Table 5.9, $k/D = 0.000075$.

$$V = \frac{20}{\frac{\pi}{4}2^2} = 6.4 \text{ fps} \qquad R = \frac{6.4 \times 2}{1.2 \times 10^{-5}} = 1.07 \times 10^6$$

From Fig. 5.27, $f = 0.0135$.

$$D = 1.85 \text{ ft or 22 in.}$$
$$\text{Design } D = 22 \text{ in.}$$

Even though there is a rather dramatic variance in f value (variation of almost 50%), the resulting effect on the design pipe diameter is quite a lot less. This is a consequence of the fact that diameter varies at about the $\frac{1}{5}$ power of the f value, so dramatic differences in f value have greatly reduced impact.

For purposes of computer application, it is advantageous to express the information on the Moody diagram in algebraic form. This is most commonly done with the equations of Colebrook[7].

$$\frac{1}{\sqrt{f}} = -2\log_{10}\left(\frac{2.51}{R\sqrt{f}}\right) \quad \text{(smooth)} \tag{5.141}$$

$$\frac{1}{\sqrt{f}} = 1.74 - 2\log_{10}\left(2\frac{k}{D} + \frac{18.7}{R\sqrt{f}}\right) \quad \text{(transition)} \tag{5.142}$$

$$\frac{1}{\sqrt{f}} = -2\log_{10}\left(0.266\frac{k}{D}\right) \quad \text{(rough)} \tag{5.143}$$

These equations can then be solved iteratively by computer to determine f for a given R value and k/D value.

Effects of Aging*. The values of k given for the various pipe materials of Fig. 5.27 and Table 5.9 are for new, clean pipes. In general, pipes become increasingly rough with age, owing to deposition or corrosion. Colebrook and White have determined an approximately linear increase in absolute roughness with time, which may be expressed as

$$k = k_0 + \alpha t \tag{5.144}$$

where k_0 is the absolute roughness of the new material, α is a constant, and k is the absolute roughness at time t.

Example 5.3. After 10 yrs of service, a 10-in. cast-iron pipe line in water service has a drop of 3.13 psi per 1000 ft for a flow of 1000 gpm. What is the estimated pressure drop for 1200 gpm after 20 yrs of service?

$$V = \frac{1000}{7.48 \times 60\frac{\pi}{4}\left(\frac{5}{6}\right)^2} = 4.1 \text{ ft/sec} \qquad \frac{V^2}{2g} = 0.262 \text{ ft}$$

$$h_f = \frac{3.13 \times 144}{62.4} = f\frac{1000}{5/6}0.262 \qquad f = 0.023$$

$$R = \frac{4.1 \times 5}{6 \times 1.2 \times 10^{-5}} = 285{,}000 \left(\text{taking } \nu = 1.2 \times 10^{-5} \text{ ft}^2/\text{sec}\right)$$

From Fig. 5.27 for the values of f and R, $k/D = 0.0017$ and $k = 0.00142$ ft. For new cast iron take $k_0 = 0.001$ ft, hence for 10 yr

$$0.00142 = 0.001 + 10\alpha$$
$$\alpha = 0.000042 \text{ ft/yr}$$

*Material extracted from previous edition by Victor L. Streeter.

and for 20 yr

$$k = 0.001 + 20 \times 0.000042 = 0.00184 \text{ ft}$$

$$\frac{k}{D} = 0.0022 \quad V = \frac{1200}{1000} \times 4.1 = 4.92 \text{ fps} \quad R = \frac{4.92}{4.1} \times 285{,}000 = 342{,}000$$

From Fig. 5.27, $f = 0.025$; then

$$h_f = 0.025 \times \frac{1000}{\frac{5}{6}} \frac{4.92^2}{64.4} = 11.3 \text{ ft}$$

$$\Delta p = \frac{11.3 \times 62.4}{144} = 4.9 \text{ psi}$$

Conduits of Noncircular Cross Section. The Darcy–Weisbach equation may also be applied to noncircular conduits if the diameter D is replaced by some equivalent linear measure of the cross section. The hydraulic radius R, widely used in open channel equations, can be related to D for the circular cross section; this relationship is usually assumed to be a valid replacement of D in the pipe formula. The hydraulic radius is defined as the ratio of the cross-sectional area to the wetted perimeter. For a circular cross section

$$R = \frac{\pi D^2/4}{\pi D} = \frac{D}{4} \tag{5.145}$$

Hence the diameter may be replaced by four times the hydraulic radius in the Reynolds number, the relative roughness, and the resistance equation; the resistance equation becomes

$$h_f = f \frac{L}{4R} \frac{V^2}{2g} \tag{5.146}$$

Although satisfactory for conduits which are reasonably comparable to pipes in cross-sectional form, this equation cannot be expected to give accurate results for cross sections that are at great variance therefrom.

Example 5.4. Find the head loss per 1000 ft for a flow of 200 gpm of water at 150°F through a clear cast-iron conduit of rectangular cross section 3 in. by 6 in.
From Table 5.6, $\nu = 4.8 \times 10^{-6} \text{ ft}^2/\text{sec}$

$$R = \frac{3 \times 6}{12 + 6} = 1 \text{ in.} = \frac{1}{12} \text{ ft} \quad Q = \frac{200}{7.48 \times 60} = 0.446 \text{ cfs}$$

$$V = \frac{0.446}{18} \times 144 = 3.57 \text{ ft/sec} \quad R = \frac{V4R}{\nu} = 248{,}000$$

$$\frac{k}{D} = \frac{k}{4R} = \frac{0.00085}{4/12} = 0.00255$$

From Fig. 5.27, $f = 0.025$; hence

$$h_f = f \frac{L}{4R} \frac{V^2}{2g} = \frac{0.025 \times 1000}{4 \times \frac{1}{12}} \frac{3.57^2}{64.4} = 14.88 \text{ ft}$$

Minor losses are the second type of friction losses and are caused by excessive turbulence beyond that occurring as a result of normal pipe friction. While the losses occur over a finite length of pipe, they are usually assumed to be concentrated at the location of the causative valve, fitting, and so on. This definition is depicted in Fig. 5.28.
The formula for minor losses of all kinds is generally of the form

$$h_m = K_L \frac{V^2}{2g} \tag{5.147}$$

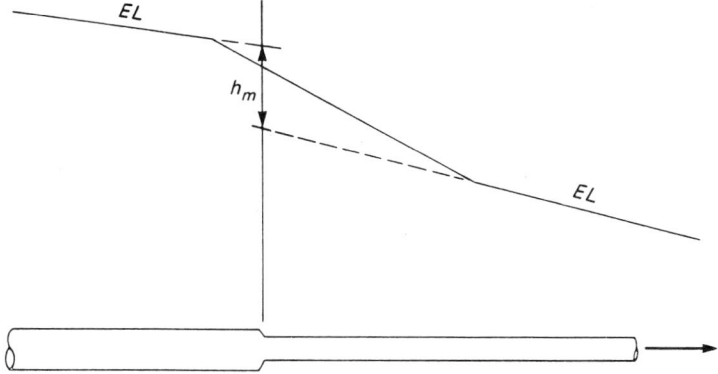

Fig. 5.28 Definition sketch for minor loss by permission from G. Z. Watters, *Analysis and Control of Unsteady Flow in Pipelines*, Butterworth, Stoneham, Mass., 1984.

TABLE 5.10   Loss Coefficients for Enlargements and Contractions Based on Velocity in Small Pipe[a]

Diameter ratio	0	0.1	0.2	0.3	0.4	0.5	0.6	0.7	0.8	0.9	1.0
Contraction K_L	0.5	0.48	0.45	0.42	0.38	0.34	0.29	0.22	0.12	0.04	0.0
Enlargement K_L	1.0	0.98	0.92	0.83	0.70	0.57	0.40	0.26	0.12	0.04	0.0

[a] Courtesy of Crane Company.[8]

TABLE 5.11   Selected Minor Loss Coefficients[a]

Minor Loss Device	K_L	Minor Loss Device	K_L
Pipe entrances		Valves	
Inward projecting	0.78	Globe	$340f$ [b]
Sharp edged	0.50	Angle	$145f$
Slightly rounded	0.23	Ball or plug	$3f$
Well rounded	0.04	Butterfly	$40f$
Pipe exits (all types)	1.00	Gate (fully open)	$13f$
Bends		(75% open)	$35f$
90° miter bends	$58f$ [b]	(50% open)	$160f$
45° miter bends	$15f$	(25% open)	$900f$
Fittings			
90° standard elbow	$30f$ [b]		
45° standard elbow	$16f$		

[a] Courtesy of Crane Company.[8]
[b] f is the friction factor for the pipe.

where K_L = loss coefficient whose value depends on the device causing the loss and V = velocity of flow in the smaller pipe. Virtually all hydraulics or fluid mechanics textbooks have tables of loss coefficients for various types of devices. In addition, Davis and Sorensen[5] list several tables of K_L values. The reference book published by the Crane Company[8] is also a good source for loss coefficients, and a selection of K_L values for common situations has been presented in Tables 5.10 and 5.11. For an exhaustive collection of loss coefficients, see Idelchik[9].

To demonstrate the use of the tables and a complete friction loss situation in pipeline analysis, the following example is presented.

Example 5.5. A series pipe conveys water at 50°F between two reservoirs. The pipe is asphalted cast iron and the entrance to the pipe is sharp edged.

Compute the discharge in the pipeline and draw the energy line (EL) and hydraulic grade line (HGL), labeling the elevation breaks.

When the discharge is unknown, a flow velocity must be estimated to permit calculating **R** and finding an f value.

Assume the velocity in the 12-in. pipe is 7 fps. From Table 5.6

$$\nu = 1.41 \times 10^{-5} \text{ ft}^2/\text{sec}$$

$$R_{12} = \frac{VD}{\nu} = \frac{7 \times 1.0}{1.41 \times 10^{-5}} = 496{,}000$$

$$R_6 = \frac{28 \times 0.5}{1.41 \times 10^{-5}} = 993{,}000$$

From Table 5.9

$$k = 0.0048 \text{ in.}$$

$$\left(\frac{k}{D}\right)_{12} = \frac{0.0048}{12} = 0.0004$$

$$\left(\frac{k}{D}\right)_6 = \frac{0.0048}{6} = 0.0008$$

From Fig. 5.27

$$f_{12} = 0.017$$
$$f_6 = 0.019$$

From the work-energy equation,

$$z_1 + \frac{p_1}{\gamma} + \frac{V_1^2}{2g} = z_2 + \frac{p_2}{\gamma} + \frac{V_2^2}{2g} + \sum h_{L_{1-2}}$$

$$4230 + 0 + 0 = 4200 + 0 + 0 + \sum h_{L_{1-2}}$$

$$\sum h_{L_{1-2}} = 30$$

Summarizing frictional head losses,

$$\sum h_{L_{1-2}} = h_{m_{en}} + h_{f_{12}} + h_{m_c} + h_{f_6} + h_{m_{ex}}$$

From Table 5.10

$$h_{m_{en}} = 0.5\frac{V_{12}^2}{2g} \qquad h_{m_c} = 0.34\frac{V_6^2}{2g} \qquad h_{m_{ex}} = 1.0\frac{V_6^2}{2g}$$

$$\sum h_{L_{1-2}} = 0.5\frac{V_{12}^2}{2g} + 0.017 \times \frac{150}{1.0}\frac{V_{12}^2}{2g} + 0.34\frac{V_6^2}{2g} + 0.019 \times \frac{8}{\frac{6}{12}}\frac{V_6^2}{2g} + 1.0\frac{V_6^2}{2g}$$

From continuity considerations,

$$V_6 = 4V_{12}$$

$$\sum h_{L_{1-2}} = (0.5 + 2.55 + 5.44 + 4.86 + 16.0)\frac{V_{12}^2}{2g}$$

Solving for V_{12},

$$29.35\frac{V_{12}^2}{2g} = 30$$

$$V_{12} = 8.11 \text{ fps} \qquad V_6 = 32.45 \text{ fps} \qquad Q = 6.37 \text{ cfs}$$

Now a check must be made on the accuracy of the initial assumption used to find the f values.

$$R_{12} = 575{,}000 \qquad f_{12} = 0.0169$$
$$R_6 = 1{,}150{,}000 \qquad f_6 = 0.0190$$

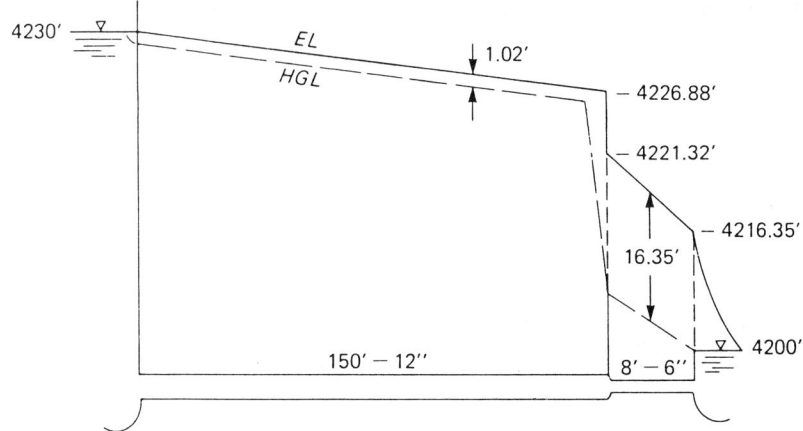

Fig. 5.29 By permission from G. Z. Watters, *Analysis and Control of Unsteady Flow in Pipelines,* Butterworth, Stoneham, Mass., 1984.

Because these values are as close to the initial estimates as one could reasonably expect, we will consider the first estimates as the final values.

The EL and HGL are shown on Fig. 5.29. The "break" elevations are labeled to the hundredth of a foot only to demonstrate how the losses add up to the difference in reservoir elevations.

Another form of expressing minor losses is often used by the manufacturers of valves. This formula is

$$Q = C_v\sqrt{\Delta p} \tag{5.148}$$

where C_v = flow coefficient for the valve in question and Δp and Q are expressed in psi (pounds per square inch) and gpm (gallons per minute), respectively. Of course C_v and K_L measure the same thing. They are approximately related by the equation

$$K_L = \frac{890}{C_v^2}D^4 \tag{5.149}$$

for the units already described with diameter expressed in inches.

Transitions are sections of conduit that connect one prismatic portion to another by a gradual change in cross section. Since, owing to the inherent stability of accelerated flow, losses are small in

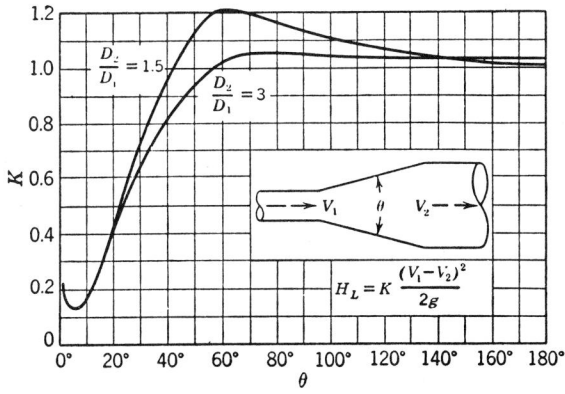

Fig. 5.30 Loss coefficients for conical diffusers.

gradual contractions, transition design is usually determined by factors other than energy loss. For example, it is often important that the pressure decrease continuously to that of the reduced section, so that the sections will be both *separation-proof* and *cavitation-proof*.

In expanding transitions or diffusers, wherein kinetic energy is converted into potential energy, it is even more essential that separation be avoided. The slowly moving fluid near the wall, which is retarded by surface resistance, is also retarded by the adverse pressure gradient due to the flow expansion. If the adverse pressure gradient acts over a sufficient length, it is certain to result in boundary layer separation. Once separation occurs, with the backflow and eddies that accompany it, the losses become high. A series of experiments was conducted by Gibson[10] on conical diffusers, the results of which are shown in Fig. 5.30.

5.7.4 Steady-State Pipeline Analysis

The design situation in pipelines is generally that of selecting a pipe diameter that will convey a prescribed amount of discharge between two locations of known elevation with specified reservoir surface elevations or pressure requirements. The attractiveness of the Hazen–Williams and Manning formulas is immediately clear because, if minor losses are neglected, both formulas lead to direct solutions for the required diameter. Use of the Darcy formula requires several iterations beginning with an assumed diameter and ending with a final check on f, $\mathbf{R}$ and k/D to ensure all are consistent in the final solution. Techniques and skills can be developed to streamline this Darcy equation to determine head loss in a variety of systems with the purpose of refamiliarizing the reader with the techniques.

Single Pipelines. Single-pipeline problems are solved by applying the work-energy equation between two points of known energy levels. An example best illustrates the procedure.

Example 5.6. An 18-in. welded steel pipeline 1200 ft long connects two reservoirs differing in elevation by 20 ft (see Fig. 5.31). The pipe has a sharp-edged entrance and a wide-open globe valve at the downstream end near the location where it enters the lower reservoir. Find the flow rate in the pipe. From the work-energy equation,

$$z_1 + \frac{p_1}{\gamma} + \frac{V_1^2}{2g} = z_2 + \frac{p_2}{\gamma} + \frac{V_2^2}{2g} + \sum h_{L_{1-2}}$$

Using the lower reservoir surface as a datum,

$$20 + 0 + 0 = 0 + 0 + 0 + \sum h_{L_{1-2}}$$

where $\sum h_{L_{1-2}}$ = (entrance + friction + valve + exit) losses. From Tables 5.10 and 5.11,

$$\sum h_{L_{1-2}} = 0.5\frac{V^2}{2g} + f\frac{L}{D}\frac{V^2}{2g} + 340f\frac{V^2}{2g} + \frac{V^2}{2g}$$

$$= \left(0.5 + f\frac{L}{D} + 340f + 1.0\right)\frac{V^2}{2g}$$

From Table 5.9,

$$k = 0.0018 \qquad \frac{k}{D} = 0.0001$$

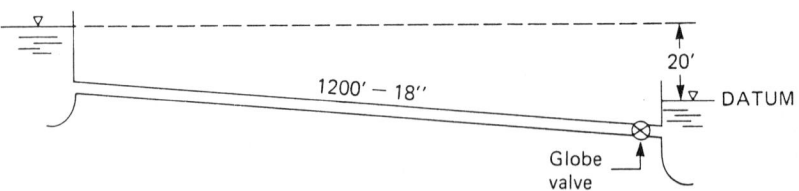

Fig. 5.31 By permission from G. Z. Watters, *Analysis and Control of Unsteady Flow in Pipelines*, Butterworth, Stoneham, Mass., 1984.

From Table 5.6, assuming

$$V \approx 8 \text{ fps} \qquad R = \frac{8 \times 1.5}{1.2 \times 10^{-5}} = 1 \times 10^6$$

From Fig. 5.27, $f = 0.0135$. Substituting into the preceding work-energy equation,

$$20 = 0 + \left(0.5 + 0.0135 \times \frac{1200}{1.5} + 340 \times 0.0135 + 1.0\right)\frac{V^2}{2g}$$

$$= (0.5 + 10.8 + 4.6 + 1.0)\frac{V^2}{2g}$$

$$= 16.9\frac{V^2}{2g}$$

$$V = 8.73 \text{ fps}$$

Checking on initial assumptions,

$$R = \frac{8.73 \times 1.5}{1.2 \times 10^{-5}} = 1.1 \times 10^6$$

$$f = 0.0135, \text{ so solution is acceptable}$$

$$Q = AV = 0.7854 \times 1.5^2 \times 8.73$$

$$Q = 15.4 \text{ cfs or } 6927 \text{ gpm}$$

Series pipelines, that is, pipelines of different diameters connected end to end, are handled in a manner similar to the previous example. It is only required that the single pipe friction term be replaced by a summation of the pipe friction losses in each of the pipes including the additional minor losses between pipes.

Single Pipelines with Pumps. The pumped pipeline is a common design situation with which engineers are confronted. Pumps occur at the upstream end of pipelines (source pumps) and at intermediate locations in the pipeline (booster pumps). In each case, the discharge through the pump and the head increase across the pump are affected by the pipe system in which the pump is installed.

To analyze pumped pipelines, the work-energy equation must be modified to include the energy added by the pump.

$$z_1 + \frac{p_1}{\gamma} + \frac{V_1^2}{2g} + H_p = z_2 + \frac{p_2}{\gamma} + \frac{V_2^2}{2g} + \sum h_{L_{1-2}} \qquad (5.150)$$

where H_p = energy added to each pound of liquid passing through the pump. In the case of series or multistaged pumps, H_p is the sum of the head increases across each pump or stage.

The head increase across a pump is a function of discharge through the pump and is determined experimentally by the manufacturer. The information is presented graphically on a diagram known as the characteristic diagram. Information on the power requirements and pump efficiencies at varying discharges is also included. An example of a typical pump characteristic diagram is shown on Fig. 5.32.

Because power requirements for a pumping situation are of interest, it is important to establish the relation between energy added (H_p) and brake horsepower required. The power added to the liquid by the pump can be expressed as

$$\text{WHP} = \frac{Q\gamma H_p}{550} \qquad (5.151)$$

where WHP = horsepower added to the water. Of course a greater amount of power must be added to the pump shaft because of friction and other losses in the pumping process. The power that must be supplied to a pump shaft (brake horsepower, BHP) in order to provide a given water horsepower (WHP), is related to the hydraulic parameters by the equation

$$\text{BHP} = \frac{Q\gamma H_p}{550\eta} \qquad (5.152)$$

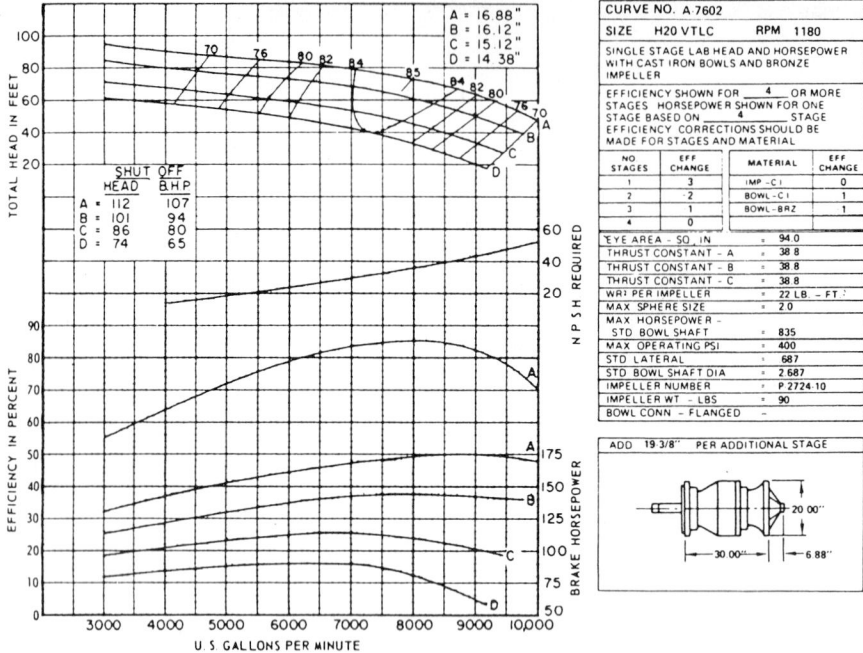

Fig. 5.32 Typical pump characteristics diagram for a vertical turbine pump. (Courtesy of Allis-Chalmers.) By permission from G. Z. Watters, *Analysis and Control of Unsteady Flow in Pipelines*, Butterworth, Stoneham, Mass., 1984.

where η = overall pump efficiency. Both overall pump efficiency and brake horsepower are displayed on the pump characteristic diagram on Fig. 5.32.

Two examples follow that demonstrate the use of the characteristic diagrams in pipeline analysis.

Example 5.7. A single-stage pump with the characteristics shown in Fig. 5.32 (curve A) is used to pump water from a reservoir of elevation 1350 ft to another reservoir at elevation 1400 ft. The line is 6000 ft long and 24 in. in diameter with an f value of 0.021.

Neglecting minor losses, compute the discharge in the pipeline. From the work-energy equation

$$z_1 + \frac{p_1}{\gamma} + \frac{V_1^2}{2g} + H_p = z_2 + \frac{p_2}{\gamma} + \frac{V_2^2}{2g} + \sum h_{L_{1-2}}$$

$$1350 + 0 + 0 + H_p = 1400 + 0 + 0 + f\frac{L}{D}\frac{V^2}{2g}$$

$$H_p = 50 + 0.021 \times \frac{6000}{2}\frac{V^2}{2g} = 50 + 63.0\frac{V^2}{2g}$$

This equation must be solved by trial in conjunction with the head-versus-discharge characteristic for the pump given in Fig. 5.32. In the solution process we neglect losses in the pump discharge column and head, which would normally be included in an analysis.

The solution is best approached using a trial solution table.

Assume Q (gpm)	V (fps)	$\dfrac{V^2}{2g}$	$63\dfrac{V^2}{2g}$	RHS	H_p (ft)
8000	5.67	0.50	31.5	81.5	73
7000	4.96	0.38	24.1	74.1	79
7500	5.32	0.44	27.7	77.7	76
7400	5.25	0.43	26.9	76.9	77

The solution is Q = 7400 gpm.

Example 5.8. Solve the problem of Example 5.7 if two three-stage parallel pumps with curve C characteristics were employed.

The pipeline analysis would remain the same.

$$H_p = 50 + 63.0 \frac{V^2}{2g}$$

However, H_p is now the total head put out by the three stages in each pump. Also the pipeline discharge must be halved to obtain the amount passing through each pump.

The resulting trial solution table is as follows:

Assume Pump Q (gpm)	Q_{pipe}	V (fps)	$\dfrac{V^2}{2g}$	RHS	H_p/stage	H_p (ft)
7000	14,000	9.93	1.53	146	54	159
7500	15,000	10.63	1.76	161	50	150
7250	14,500	10.28	1.64	153	51	153

The solution is $Q = 7250$ gpm.

5.7.5 Pipe Network Analysis

The steady-state analysis of flows in pipe networks can be a very complex problem. Devices such as pressure-reducing valves, minor losses, booster pumps, and supply pumps, as well as reservoirs, all serve to complicate the analysis. The subject is covered comprehensively by Jeppson[11] and the reader is referred to that text for detailed information. The presentation here will be introductory and will apply only to relatively simple systems. However, we shall discuss all three of the most popular analysis methods—the Hardy Cross method, the linear theory method, and the Newton–Raphson method. In addition, a good summary of the application of these three methods is given by Wood and Rayes.[12]

Hardy Cross Method. Because of its simplicity of application, its easily understood theory, and its amenability to hand calculation, the Hardy Cross method has enjoyed (and still enjoys) considerable popularity among practicing engineers.

The first step is to estimate flow rates in all the pipes in a network so that continuity is satisfied at each junction (node). Of course it is unlikely that the EL-HGL is continuous throughout the network because the original estimates of the flow rates are always erroneous to some degree. This method assumes that there can be found a unique flow rate adjustment that can be applied to each loop in the network that will cause the EL-HGL to be continuous around each loop. In hydraulic terms this continuity is expressed as

$$\sum_{i=1}^{N} h_{L_i} = 0 \qquad (5.153)$$

around each loop where i is the pipe number and N is the number of pipes in the loop. Assuming that the head loss can be written in the form

$$h_{L_i} = K_i Q_i^n \qquad (5.154)$$

and assuming a correction ΔQ is being added to each pipe flow in the loop to satisfy the requirement that the sum of the head loss equals zero, this equation becomes

$$\sum_{i=1}^{N} K_i (Q_i + \Delta Q)^n = 0 \qquad (5.155)$$

where Q_i is the most recent estimate for the discharge in each pipe in the loop. It remains only to solve for ΔQ.

Because n is generally a noninteger, the preceding is generally expanded by the binomial theorem to yield an equation for ΔQ. Retaining only the first two terms of the binomial expansion, the following equation for ΔQ is derived.

$$\Delta Q = \frac{\sum\limits_{i=1}^{N} K_i Q_i^n}{-n \sum\limits_{i=1}^{N} K_i Q_i^{n-1}} \tag{5.156}$$

To produce the proper sign on ΔQ, the denominator is kept negative and the terms in the summation in the numerator are positive or negative, depending on whether one moves with or against the flow while proceeding clockwise around the loop.

Once the ΔQ is computed for each loop, it is added (or subtracted) from the flow rates in each member of the loop to get a better estimate of the true flow rate. Because the decomposition of $(Q_i + \Delta Q)^n$ with the binomial theorem was not exact and because pipes that are common to more than one loop have multiple ΔQ corrections, the calculated ΔQ's will not be correct. Therefore, the process is iterative and must be continued until the error is acceptably small (or no convergence to a solution occurs).

Althogh this numerical method is not so sophisticated as the other methods, the results are just as valid, provided convergence is obtained. Actually, a more careful investigation would reveal that Hardy Cross analysis is a decoupled Newton–Raphson analysis.

Linear Theory Method. The linear theory method is a technique for solving a set of network equations, some of which are nonlinear, for the unknown flow rates in the pipes. The equations are generated by writing continuity equations for flow into and out of each junction and by specifying that the algebraic sum of the head losses around each loop is zero. Solving a set of nonlinear equations is an iterative process and there are many techniques for doing this. In the linear theory approach the nonlinear equations for the sum of the head losses around each loop are linearized. Then the complete set of linear equations (the continuity equations are already linear) is solved.

To understand how the procedure works, look at the equations involved. For each loop in the network the following equation is valid:

$$\sum_{i=1}^{N} h_{L_i} = \sum_{i=1}^{N} K_i Q_i^n = 0 \tag{5.157}$$

where N is the number of pipes in the loop. To linearize this equation, Q_i^n is decomposed into two parts so that this equation becomes

$$\sum_{i=1}^{N} \left(K_i Q_i^{n-1} \right) Q_i = \sum_{i=1}^{N} K_i' Q_i = 0 \tag{5.158}$$

Of course K_i' is now a function of Q_i so the process is still iterative.

As the set of linear-plus-linearized equations is successively solved, the estimate of K_i' is revised after each solution. After several iterations the values of Q_i and K_i' should converge to their final values. The mathematical form of the iteration equation is

$$\sum_{i=1}^{N} K_i \left[Q_i^{(j-1)} \right]^{n-1} Q_i^j = 0 \tag{5.159}$$

where j is the iteration number. For example, if we are making the 8th iteration ($j = 8$), then we would calculate K_i''s from the results of the 7th iteration.

Experience with the linear theory has shown that the numerical solution tends to oscillate around the final values. To damp out this numerical oscillation, the iteration equation is altered to include the last two iterations for Q_i in computing K_i'.

$$\sum_{i=1}^{N} K_i \left[\frac{Q_i^{(j-1)} + Q_i^{(j-2)}}{2} \right]^{n-1} Q_i^{(j)} = 0 \tag{5.160}$$

When starting an analysis, only the direction of flow (not the quantity) has to be specified. This is a substantial savings in effort over the Hardy Cross approach. For the first iteration $K_i'^{(1)}$ is assumed to be equal to K_i. For the second iteration $K_i'^{(2)}$ will equal $K_i \left[Q_i^{(1)} \right]^{n-1}$. Thereafter, Eq. (5.160) will be used for each loop.

Newton–Raphson Method. The Newton–Raphson technique has the same conceptual basis as the Hardy Cross method. Flow rates in each pipe are assumed that satisfy continuity, and these flow rates are corrected so that the sum of the head losses around each loop approaches zero. In the Hardy Cross method the flow rates in each pipe are corrected after each ΔQ computation. In the Newton–Raphson method the equations containing ΔQ are written for each loop, then this nonlinear set of equations is solved successively for the final value of ΔQ in each loop. When the solution is complete, only then are the initial flow rates in each pipe adjusted to their final value.

The method gets its name from the technique used to solve the nonlinear set of equations. The Newton–Raphson technique is a frequently used, powerful method of numerical analysis. In operation, it adjusts successive approximations to the solution by computing the way the solution is moving with respect to each variable and then, based on that computation, calculates new trial values for the unknowns.

The Newton–Raphson technique in two or more dimensions (two or more equations with two or more unknowns) is most conveniently expressed in matrix form

$$\{ F^{(j-1)} \} + [J^{(j-1)}]\{ x^{(j)} - x^{(j-1)} \} = \{0\} \tag{5.161}$$

where $\mathbf{J}$ is a $K \times K$ matrix of $\left(\partial F^{(j-1)} / \partial x_k \right)$ known as the Jacobian. Converting this to another form,

$$\{ x^{(j)} \} = \{ x^{(j-1)} \} - [J^{(j-1)}]^{-1}\{ F^{(j-1)} \} \tag{5.162}$$

Because all the F_k can be differentiated, the Jacobian can be evaluated at each new approximation for x_i and the inverse computed. However, this is a very large computational task for large systems of equations, hence, a slightly different approach is employed when working with hydraulic networks.

In hydraulic networks,

$$F_k = \sum_{i=1}^{N} K_i Q_i^n = 0 \tag{5.163}$$

However, because the Q_i's are unknown, a value in each pipe must be estimated and a search for ΔQ's, which will correct the Q_0's to the proper value, must be made. The loop equations now are of the form

$$F_k = \sum_{i=1}^{N} K_{L_i} \left[Q_{0_i} + \sum_{l=1}^{K} \Delta Q_l \right]^n = 0, \ k = 1, \ K \tag{5.164}$$

because any pipe in a given loop may be a member of other loops and their ΔQ's must be included. Then, in general,

$$F_k (\Delta Q_1, \Delta Q_2, \ldots, \Delta Q_k) = 0, \ k = 1, \ K \tag{5.165}$$

If $\Delta Q^{(j)} - \Delta Q^{(j-1)}$ is now represented as $\delta Q^{(j)}$ in each loop, then

$$[J^{(j-1)}]\{ \delta Q^{(j)} \} = -\{ F^{(j-1)} \} \tag{5.166}$$

This equation is now solved for $\{ \delta Q^{(j)} \}$ and

$$\{ \Delta Q^{(j)} \} = \{ \Delta Q^{(j-1)} \} + \{ \delta Q^{(j)} \} \tag{5.167}$$

after each iteration. When the δQ's become small enough, an acceptable solution has been obtained.

It should be noted here that this method only requires the solution to a set of equations equal in number to the number of loops. The linear theory must solve a set of equations equal in number to the number of unknown flow rates. Consequently, the Newton–Raphson technique may require substantially less storage space for solution.

5.7.6 Unsteady Flow in Pipe Systems

Unsteady flow in pipe systems is important because it can result in serious problems. Some of these are

1. Pipe rupture
2. Pipe collapse

3. Vibration
4. Excessive pipe displacements
5. Pipe fitting and support deformation or failure
6. Vapor cavity formation (cavitation, column separation)

Some of the primary causes of unsteady flow are

1. Valve closure (or opening)
2. Flow demand changes
3. Pump shutdown (or power failure to the pump)
4. Pump startup
5. Air venting from lines
6. Failure of flow or pressure regulators
7. Pipe rupture

Unsteady flow analysis in pipe systems is generally divided into two categories.

1. Rigid water column theory (surge theory) where the fluid and pipe are inelastic, pressure changes propagate instantaneously, and the differential equation of motion is "ordinary."
2. Elastic theory (water hammer) where the elasticity of fluid and pipe affect pressure changes, pressure changes propagate with wave speed **a** (1000–4700 fps) and the differential equations of motion are partial and nonlinear.

A simple problem is used to demonstrate phenomena and introduce concepts. A steady flow situation is shown in Fig. 5.33 where velocity V is caused by head H in reservoir. Friction is neglected and the energy line (EL) and hydraulic grade line (HGL) are coincident because water hammer pressures are large compared to velocity head.

The valve is closed suddenly causing a pressure head to propagate upstream at speed **a**. The sequence of events shown in Fig. 5.33 are as follows:

1. Pressure head increase ΔH reaches reservoir at $L/\mathbf{a}$ seconds. Velocity = 0 and pressure head = $H + \Delta H$ throughout pipe. Pipe is stretched. Water is compressed.
2. High pressure in pipe ejects water into reservoir. At 2 $L/\mathbf{a}$ seconds, velocity = $-V$, pressure head = H throughout pipe.
3. Upstream flow suddenly stopped at valve. Negative wave propagates upstream. At 3 $L/\mathbf{a}$ seconds, velocity = 0, pressure head = $H - \Delta H$ throughout pipe.
4. High pressure in reservoir forces water into pipe giving downstream flow. At 4 $L/\mathbf{a}$, velocity = V, pressure = H and wave period is complete.
5. As long as valve remains closed, these cycles repeat. Any friction in system would cause damping of ΔH until it eventually disappears.

The following are important ideas:

1. $L/\mathbf{a}$ is important time parameter in water hammer situations.
2. Pressure head at the valve reaches its maximum if the valve is closed in any time less than 2 $L/\mathbf{a}$ seconds. Valve need not be suddenly closed to create maximum water hammer pressures.

Column Separation. When pressure change is severe enough to drop pressure in the pipe to the vapor pressure of water, "column separation" occurs. Dissolved gases come out of solution, water vapor cavities occur, and liquid columns "separate." Eventually cavity closure causes water hammer pressure "shocks" of a magnitude difficult to calculate and potentially destructive.

Column separation could be caused by simple valve closure. In Fig. 5.33, if ΔH were large enough, column separation would occur on both sides of the valve, although at different times.

Rigid Water Column Theory. This analysis uses Newton's second law $F = ma$. For unsteady flow the equation is

$$\frac{p_1}{\gamma} - \frac{p_2}{\gamma} - \frac{fL}{2gD}V^2 = \frac{L}{g}\frac{dV}{dt} \tag{5.168}$$

where p = pressure, f = Darcy–Weisbach friction factor, L = pipe length, γ = specific weight of

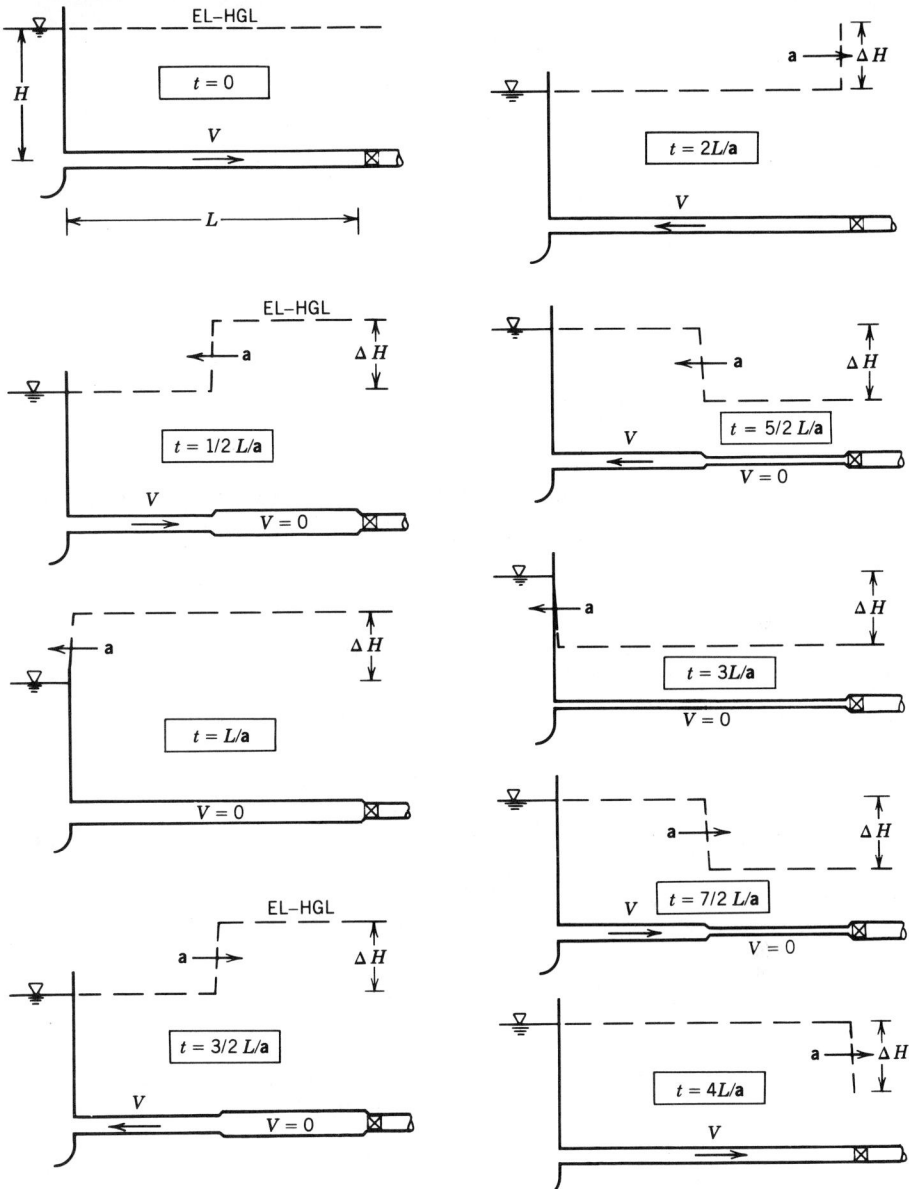

Fig. 5.33 Pressure wave propagation in a simple pipe system. By permission from G. Z. Watters, *Analysis and Control of Unsteady Flow in Pipelines*, Butterworth, Stoneham, Mass., 1984.

water, D = pipe diameter, V = flow velocity, g = acceleration of gravity, and dV/dt = liquid acceleration.

Example 5.9. Flow establishment in a pipe. The physical situation is shown in Fig. 5.34 with free discharge at the valve.

The equation applied to this problem is

$$H_0 - \frac{fL}{2gD} V^2 = \frac{L}{g} \frac{dV}{dt}$$

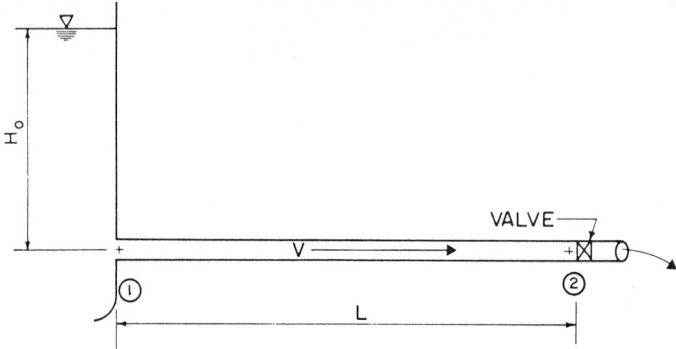

Fig. 5.34 Simple system for applying rigid water column theory. By permission from G. Z. Watters, *Analysis and Control of Unsteady Flow in Pipelines*, Butterworth, Stoneham, Mass., 1984.

This expression can be integrated in closed form. The time to reach steady flow is infinite. To reach 99% of V is

$$t_{99} = 2.65 \frac{LV_0}{gH_0}$$

Pipe systems may not be single, constant-diameter pipes. The equivalent pipe technique may be used to reduce complex pipe systems to single pipes. Criteria for equivalence is friction loss equality and similar dynamic behavior.

Series pipes:
$$\left[\frac{fL}{D^5} \right]_{eq} = \sum_{i=1}^{N} \left[\frac{f_i L_i}{D_i^5} \right] \tag{5.169}$$

$$\left[\frac{L}{D^2} \right]_{eq} = \sum_{i=1}^{N} \left[\frac{L_i}{D_i^2} \right] \tag{5.170}$$

Parallel pipes:
$$\left[\frac{D_{eq}^5}{F_{eq} L_{eq}} \right]^{1/2} = \sum_{i=1}^{N} \left[\frac{D_i^5}{f_i L_i} \right]^{1/2} \tag{5.171}$$

$$\left[\frac{D_{eq}^2}{L_{eq}} \right] = \sum_{i=1}^{N} \left[\frac{D_i^2}{L_i} \right] \tag{5.172}$$

Example 5.10. A three-unit pumped storage facility is shown in Fig. 5.35. Flow through the turbines is shut down so that penstock velocities decrease from 60 fps to zero linearly in 30 sec. Compute p_{max} if the f value is the same for all pipes.

Using the parallel pipe equations,

$$\frac{D_{eq}^2}{L_{eq}} = 3\left(\frac{8^2}{800} \right) = 0.240 \qquad \left[\frac{D_{eq}^5}{L_{eq}} \right]^{1/2} = 3\left[\frac{8^5}{800} \right]^{1/2} = 19.20$$

The parallel pipes can be replaced by a single pipe $D = 11.54$ ft and $L = 555$ ft.

$$L = 2000' \qquad\qquad L = 555'$$

$$D = 24' \qquad\qquad D = 11.54'$$

Using the series pipe equation,

$$\frac{L_{eq}}{D_{eq}^2} = \frac{2000}{24^2} + \frac{555}{11.54^2} = 7.64 \qquad \frac{L_{eq}}{D_{eq}^5} = \frac{2000}{24^5} + \frac{555}{11.54^5} = 0.00296$$

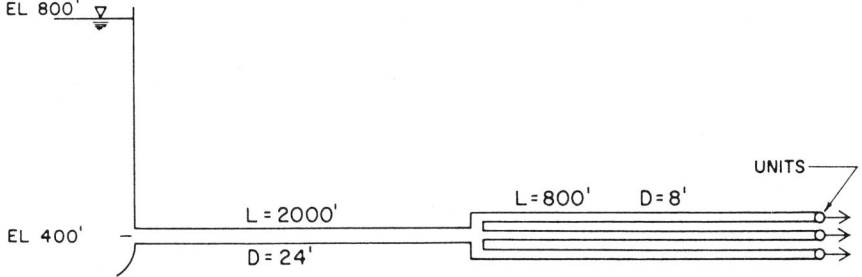

Fig. 5.35 Schematic sketch of a pumped storage facility. By permission from G. Z. Watters, *Analysis and Control of Unsteady Flow in Pipelines*, Butterworth, Stoneham, Mass., 1984.

This series pipe is replaced by single pipe $D = 13.71$ ft and $L = 1437$ ft. Steady flow velocity in the equivalent pipe is

$$V_{eq} = \frac{Q}{A_{eq}} = 61.3 \text{ fps}$$

From the unsteady flow equation,

$$400 - \frac{p_2}{\gamma} - \frac{fLV^2}{2gD} = \left(\frac{1437}{32.2}\right)\frac{0 - 61.3}{30}$$

$$\frac{p_2}{\gamma} = 400 + 91.2 - \frac{fLV^2}{2gD}$$

A maximum pressure head of 491 ft occurs at the instant the valve is completely closed.

Elastic Theory. Elastic theory includes the effect of water and pipe elasticity on pressures and velocities. The impulse-momentum equation and the conservation of mass equation are used.

The impulse-momentum equation is used to develop the equation for ΔH. The x component of the equation is

$$\left(\sum F_{ext}\right)_x = Q\rho\left(V_{out} - V_{in}\right) \tag{5.173}$$

where x direction is along pipe, F_x is forces in x direction, and Q is discharge.

To make the unsteady case steady, the coordinate system is moved along pipe at the wave speed so the wave appears to be standing still. The resulting analysis gives

$$\Delta H = -\frac{a}{g}\Delta V \tag{5.174}$$

It is clear that ΔH depends on the wave speed a.
Conservation of mass is used to find an equation for wave speed.
The result for thin-walled pipes is

$$a = \frac{[K/\rho]^{1/2}}{\left[1 + \frac{K}{E}\frac{D}{e}(C)\right]^{1/2}} \tag{5.175}$$

where $K =$ bulk modulus of elasticity of liquid, $E =$ modulus of elasticity of pipe, $D =$ pipe diameter, $e =$ pipe wall thickness, and $C =$ restraint coefficient.

$C = \frac{5}{4} - \mu$ if pipe is free to stretch in longitudinal direction as a pressure vessel.
$C = 1 - \mu^2$ if no longitudinal stretching occurs.
$C = 1.0$ if functioning expansion joints occur, where $\mu =$ Poisson's ratio.

TABLE 5.12   E and μ Values for Common Pipe Materials

Steel	$E = 30 \times 10^6$ psi	$\mu \approx 0.30$
Ductile cast iron	$E = 24 \times 10^6$ psi	$\mu \approx 0.28$
Copper	$E = 16 \times 10^6$ psi	$\mu \approx 0.36$
Brass	$E = 15 \times 10^6$ psi	$\mu \approx 0.34$
Aluminum	$E = 10.5 \times 10^6$ psi	$\mu \approx 0.33$
PVC	$E \approx 4 \times 10^5$ psi	$\mu \approx 0.45$
Fiberglass-reinforced plastic (FRP)	$E_2 = 4.0 \times 10^6$ psi	$\mu_2 = 0.27{-}0.30$
	$E_1 = 1.3 \times 10^6$ psi	$\mu_1 = 0.20{-}0.24$
Asbestos cement	$E \approx 3.4 \times 10^6$ psi	$\mu \approx 0.30$
Concrete	$E = 57,000 \sqrt{f_c'}\,^a$	$\mu \approx 0.24$ (dynamically)

^{a}Where $f_c' = $ 28-day strength.

Thin-wall pipes have D/e greater than about 40. Suggested values of E and μ are shown in Table 5.12.

For water $[K/\rho]^{1/2} = 4720$ fps. Most buried pipe situations are close to $C = 1 - \mu^2$, however, for $\mu \approx 0.3$, the result is about the same regardless of restraint.

Example 5.11. A 10,000-ft pipe has $V = 10$ fps and a wave speed of 3220 fps. Compute head increase at valve for sudden valve closure.

$$\Delta H = \frac{a}{g}\,\Delta V = \frac{3220}{32.2}\,10 = 1000' \quad \text{or} \quad 433 \text{ psi}$$

Note that $L/a = 3.1$ sec. If the valve is closed in less than 6 sec, full water hammer pressure is developed.

The previous C values for thin-walled pipes must be modified when D/e is less than 40. For homogeneous pipes,

Case (a): $C = \dfrac{1}{1 + \dfrac{e}{D}}\left[\left(\dfrac{5}{4} - \mu\right) + 2\dfrac{e}{D}(1 + \mu)\left(1 + \dfrac{e}{D}\right)\right]$ (5.176)

Case (b): $C = \dfrac{1}{1 + \dfrac{e}{D}}\left[(1 - \mu^2) + 2\dfrac{e}{D}(1 + \mu)\left(1 + \dfrac{e}{D}\right)\right]$ (5.177)

Case (c): $C = \dfrac{1}{1 + \dfrac{e}{D}}\left[1 + 2\dfrac{e}{D}(1 + \mu)\left(1 + \dfrac{e}{D}\right)\right]$ (5.178)

These C values are used in the wave speed equation to compute wave speed.

When air or other dissolved gases come out of solution and form small bubbles, wave speed is affected dramatically. This occurs because:

1. Low pressure at pipeline summit allows air release.
2. Pump sump is aerated by improper inflow design.

If the fraction of the air volume is known, the wave speed can be estimated from the following equation:

$$a = \frac{\sqrt{K_l/\rho_{\text{ave}}}}{\sqrt{1 + \dfrac{K_l}{E}\dfrac{D}{e}C + (\text{void fraction})\dfrac{K_l}{K_a}}}.$$ (5.179)

Amounts of air as low as 0.5% can reduce wave velocity to 25% of its unaerated value.

The previous approach permits calculation of pressure head increase at a point where velocity changes suddenly. Generally, it is necessary to find head H and velocity V at any section of pipe system at any time t.

To accomplish this Newton's second law and conservation of mass is applied to a differential length of pipe through which the water hammer wave is passing. The result is two partial differential equations

$$\frac{dV}{dt} + g\frac{\partial H}{\partial s} + \frac{1}{2}\frac{F}{D}V|V| = 0 \qquad (5.180)$$

$$\frac{a^2}{g}\frac{\partial V}{\partial s} + V\left[\frac{\partial H}{\partial s} - \frac{\partial z}{\partial s}\right] + \frac{\partial H}{\partial t} = 0 \qquad (5.181)$$

where s = location along pipe and z = elevation above datum.

This type of equation can be solved using the "method of characteristics." First, the equations are simplified by replacing dV/dt with $\partial V/\partial t$ and deleting $\partial H/\partial s$ in Eqs. (5.180) and (5.181). This approximation had been shown to have negligible effects on accuracy. If independent variables s and t follow certain relationships in Eqs. (5.180) and (5.181), namely

$$\frac{ds}{dt} = \pm a \qquad (5.182)$$

then *partial* differential equations can be written as *total* differential equations.

$$C^+: \quad \frac{dV}{dt} + \frac{g}{a}\frac{dH}{dt} + \frac{f}{2D}V|V| = 0 \qquad (5.183)$$

if

$$\frac{ds}{dt} = a \qquad (5.184)$$

and

$$C^-: \quad \frac{dV}{dt} - \frac{g}{a}\frac{dH}{dt} + \frac{f}{2D}V|V| = 0 \qquad (5.185)$$

if

$$\frac{ds}{dt} = -a \qquad (5.186)$$

Equations (5.184) and (5.186) called the characteristics of the C^+ and C^- equations, respectively.

Physical meaning of C^+, C^- and characteristic equations is that changes in pressure caused by disturbances (valve closing) propagate at wave speeds upstream and downstream in the pipe ($ds/dt = \pm a$). If this rule is followed, the partial differential equations become total differential equations. The equations are solved numerically as described in Watters,[13] Wylie and Streeter,[14] and Chaudhry.[15]

5.8   FLOW IN OPEN CHANNELS*

Flow in open channels is similar to that in pipes in that flow can be laminar or turbulent, have smooth or rough boundaries, be uniform or nonuniform. Open channel flow has the one unique characteristic that the pressure is zero on the free surface. Laminar flow is quite rare and will not be addressed here. This work will consider primarily uniform and nonuniform flow in the turbulent rough boundary mode.

5.8.1   Uniform Flow

In steady uniform flow, the slope of channel bottom, free surface (hydraulic grade line), and energy grade line are the same (tan θ, Fig. 5.36). For very wide channels, the shear stress varies linearly with distance from the free surface y, given by $\tau = \gamma y \sin \theta$. For other channels, the average shear stress τ_0 at the solid boundary is $\tau_0 = \gamma R \sin \theta$; where R is the hydraulic radius, which is defined as the ratio of area of cross section A to wetted perimeter P. The liquid velocity at the solid boundary is zero; it increases generally with distance from a boundary. The maximum velocity is usually below the free surface.

The *Manning formula* is the most commonly used open-channel formula,

$$V = \frac{1.49}{n}R^{2/3}S^{1/2} \qquad (5.187)$$

*Material and illustrations extracted in whole or in part from previous edition by Victor L. Streeter.

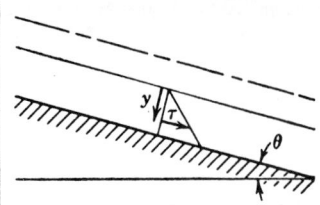

Fig. 5.36   Uniform flow.

TABLE 5.13   Average Manning n Values for Selected Boundaries

Closed Conduits Flowing Partially Full	
Welded steel	0.012
Coated cast iron	0.013
Uncoated cast iron	0.014
Corrugated metal storm drain	0.024
Cement mortar	0.013
Concrete culvert	0.011
Finished concrete	0.012
Unfinished concrete (smooth wood form)	0.014
Vitrified sewer pipe	0.014
Lined Open Channels	
Painted steel	0.013
Cement mortar	0.013
Planed, untreated wood	0.012
Unfinished concrete	0.017
Gunite concrete (good)	0.019
Glazed brick	0.013
Cemented rubble	0.025
Smooth asphalt	0.013
Excavated Channels	
Earth, straight, uniform and clean	0.018
Gravel, straight, uniform and clean	0.025
Earth with short grass and a few weeds	0.027
Dredged channel	0.028
Smooth rock cuts	0.035
Jagged rock cuts	0.040

Source: V. T. Chow, *Open Channel Hydraulics*, McGraw-Hill, New York, 1959. Copyright 1959 McGraw-Hill. Reprinted by permission.

where V is the average velocity; R is the hydraulic radius; S is the $\sin \theta$ (Fig. 5.36); and n is an absolute roughness factor, having the dimensions $L^{1/6}$, whose values for different surfaces are determined experimentally. Table 5.13 lists many of these values. Since the constant in the Manning formula is not dimensionless, it is necessary to use the foot-pound-second system of units.
   Multiplying the formula by A,

$$Q = \frac{1.49}{n} AR^{2/3}S^{1/2} \qquad (5.188)$$

When the cross section is known, the equation may be solved directly for any one of the other quantities that is unknown. For determination of depth of flow in a given section, with Q, n, S given, the solution is effected by trial.

Example 5.12. Find the depth of flow in trapezoidal channel of roughness 0.012, bottom width 10 ft and side slopes 1 : 1 for 650 cfs. The channel slope is 0.0009.
 Writing

$$AR^{2/3} = \frac{A^{5/3}}{P^{2/3}}$$

from the Manning formula

$$\frac{Qn}{1.49S^{1/2}} = \frac{A^{5/3}}{P^{2/3}} = \frac{650 \times 0.012}{1.49 \times 0.03} = 174.7$$

$A = 10D + D^2$, $P = 10 + 2\sqrt{2}\,D$, hence

$$f(D) = \frac{(100 + D^2)^{5/3}}{(10 + 2\sqrt{2}\,D)^{2/3}} = 174.7$$

Trying $D = 5$, $f(D) = 160$; hence D must be larger. Trying $D = 5.5$, $f(D) = 191$. By straight-line interpolation $D = 5.24$, $f(D) = 174$, which is a satisfactory check. Hence $D = 5.24$ ft is the answer sought.
 The cross section having the least perimeter for given conditions is called the *most efficient* cross section. The semicircular section is most efficient of all cross sections since it has the least perimeter for a given area. The most efficient *rectangular* channel has a bottom width twice the depth. The most efficient *trapezoidal* channel is half of a hexagon.

Specific Energy — Critical Depth
The mechanical energy per unit weight, with elevation datum taken as the bottom of the channel, is called *specific energy*. It is simply the sum of depth of flow and velocity head. In steady uniform flow, when all cross sections are identical, the specific energy is constant along the channel.
 Referring to Fig. 5.37, the specific energy is

$$E = y + \frac{V^2}{2g} \tag{5.189}$$

assuming uniform distribution of velocity over the cross section. For a given discharge Q, the specific energy varies with the depth of flow. Substituting $V = Q/A$, where A is the cross-sectional area and a function of y,

$$E = y + \frac{Q^2}{2gA^2} \tag{5.190}$$

For a unit width of rectangular channel, with q the discharge per unit width,

$$E = y + \frac{q^2}{2gy^2} \tag{5.191}$$

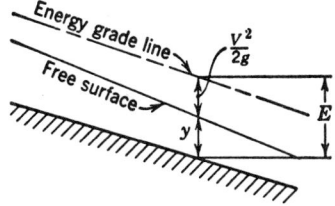

Fig. 5.37 Specific energy.

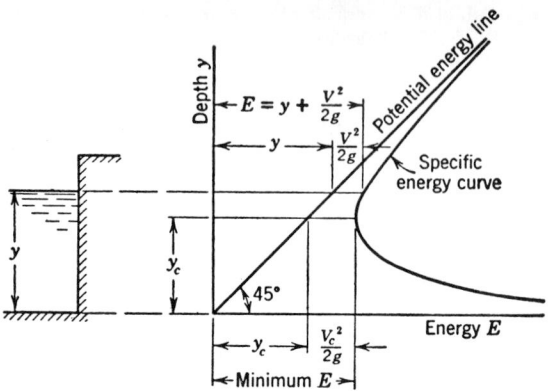

Fig. 5.38 Specific energy diagram. (Courtesy of R. A. Dodge, University of Michigan, Ann Arbor.)

A plot of specific energy against depth, Fig. 5.38, for a constant q, reveals that a certain minimum specific energy is required for the flow, found by setting $dE/dy = 0$. Calling this depth y_c the *critical depth*, we have

$$y_c = \left(\frac{q^2}{g}\right)^{1/3} \tag{5.192}$$

In terms of the velocity, $V_c = \sqrt{gy_c}$. Hence critical depth is the depth at which the velocity of flow V_c is just equal to the velocity of an elementary wave $\sqrt{gy}$ in still liquid. Greater specific energy is required for both greater and lesser depths of flow. It is obvious from Fig. 5.38 that there are two depths at which the flow has the same specific energy.

For nonrectangular channels the critical depth occurs when

$$\frac{Q^2}{g} = \frac{A^3}{b} \tag{5.193}$$

where b is the top width of the cross section at the liquid surface.

5.8.2 Steady, Nonuniform Flow

Gradually varied channel flow is steady flow in which changes in depth, section, slope, and roughness with respect to length along the channel are small. By assuming that the energy loss at any section is the same as in uniform flow at the same discharge and the same depth, a differential equation for change in depth as a function of distance along the channel can be developed:

$$\frac{dy}{dl} = \frac{S_0 - \dfrac{n^2 Q^2}{(1.49)^2 A^2 R^{4/3}}}{1 - \dfrac{Q^2 b}{g A^3}} \tag{5.194}$$

where y is the depth, l the distance along the channel, S_0 the sine of the angle the bottom makes with the horizontal, n the Manning roughness factor, Q the discharge, A the cross-sectional area, R the hydraulic radius, and b the width of cross section at the liquid surface. Solving for l

$$l = \int \frac{1 - \dfrac{Q^2 b}{g A^3}}{S_0 - \dfrac{n^2 Q^2}{(1.49)^2 A^2 R^{4/3}}}\, dy \tag{5.195}$$

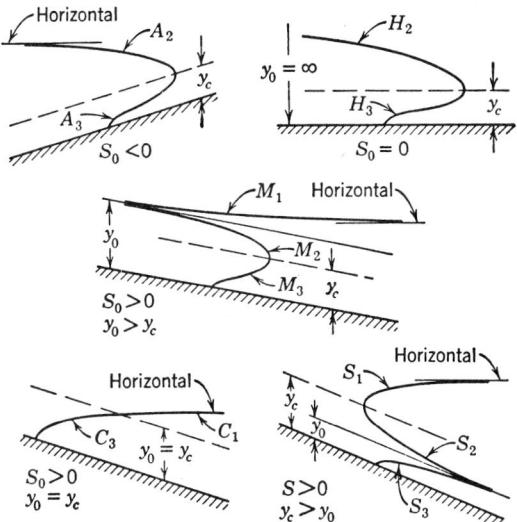

Fig. 5.39 Surface profiles on adverse, horizontal, mild, critical, and steep slopes. (Courtesy of Hunter Rouse, State University of Iowa, Iowa City.)

For constant S_0 and n, the integrand is a function of y only, and l may be determined as a function of y, usually by numerical integration. When the integrand is zero, $Q^2 b/gA^3 = 1$, which is the condition for critical depth. Hence for a change in depth there is no change in l; that is, neglecting the effects of the curvature of streamlines and the nonhydrostatic pressure distribution, the liquid surface is vertical as the flow goes through critical. When the denominator is zero, uniform flow occurs, and there is no change in depth along the channel.

The various possible free-surface profiles given by the preceding equation are shown in Fig. 5.39. In each case the flow is from left to right. y_0 is the normal depth, that is, the depth given by the Manning uniform flow equation. y_c is the critical depth. When the normal depth is greater than the critical depth, the slope of channel is *mild*; when normal depth equals *critical* depth, the slope is critical; when normal depth is less than critical depth the slope is *steep*. The two other cases are *horizontal* and *adverse*.

The determination of surface profiles from the equation are effected by starting the numerical integration at a *control* section. When the flow is above critical depth, the control is always downstream, and the depth is evaluated first for the control section, and then use is made of the gradually varied flow equation. Writing the equation in the form

$$l = \int F(y)\, dy$$

a plot of $F(y)$ as ordinate against y as abscissa is made, starting with the control depth and varying y in the direction indicated by the characteristic curves. This plot (Fig. 5.40) gives the value of l from the control section to the new depth y as the area under the curve between the values of y. In this manner the whole profile may be worked out.

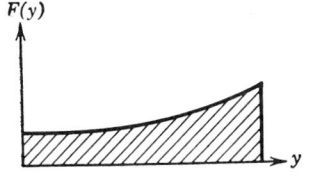

Fig. 5.40 Plot for determination of liquid surface profile.

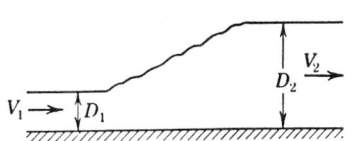

Fig. 5.41 Hydraulic jump.

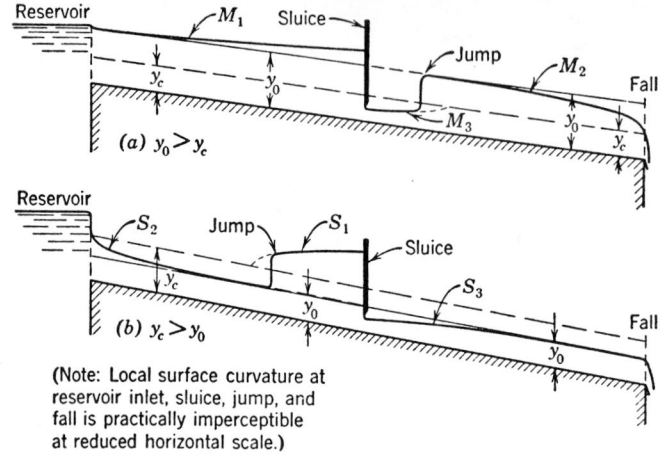

Fig. 5.42 Examples of surface profiles. (Courtesy of Hunter Rouse, State University of Iowa, Iowa City.)

When the depth of flow is less than critical, the control section is upstream (i.e., flow is out from under a gate), and the integration is handled in a similar fashion for determination of profile downstream from the control.

A phenomenon known as the *hydraulic jump* occurs under certain conditions in channel flow. The flow prior to the jump must always be below critical depth, and when the downstream depth is such that the momentum equation is satisfied for the liquid contained in the jump, the hydraulic jump will occur. The momentum equation applied to the liquid between sections 1 and 2 of Fig. 5.41 for a rectangular channel, yields the relation between depths

$$D_2 = -\frac{D_1}{2} + \sqrt{\frac{2D_1 V_1^2}{g} + \frac{D_1^2}{4}} \qquad (5.196)$$

Examples of occurrence of the surface profiles, including situations where the jump results are given in Fig. 5.42.

5.8.3   Unsteady Nonuniform Flow

Any change in discharge in an open channel results in an unsteady nonuniform flow. The changes in flow result in gravity waves moving through the system. In certain cases, waves of fixed form propagate along the channel. The most common is the surge wave depicted in Fig. 5.43. In a rectangular channel, the celerity of the surge can be computed by the equation

$$c = \sqrt{gy_1} \left[\frac{1}{2} \frac{y_2}{y_1} \left(\frac{y_2}{y_1} + 1 \right) \right]^{1/2} \qquad (5.197)$$

where c is the surge celerity relative to the undisturbed fluid velocity V_1.

There are other waves of fixed form, including the monoclinal rising flood wave, the solitary wave, and roll waves. All of these are the result of special flow or channel conditions and can be found in the literature.

The most general approach to open-channel unsteady flow is through a procedure similar to that used in pipe flow. The principles of continuity and momentum are employed to develop a pair of nonlinear partial differential equations

$$\frac{\partial A}{\partial t} + V\frac{\partial A}{\partial x} + A\frac{\partial V}{\partial x} = q \qquad (5.198)$$

$$g\frac{\partial y}{\partial x} + \frac{\partial V}{\partial t} + V\frac{\partial V}{\partial x} = g(S_0 - S_f) - \frac{V}{A}q \qquad (5.199)$$

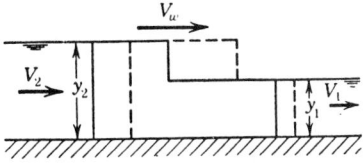

Fig. 5.43 Surge wave.

where A = cross-sectional area, V = velocity of flow, y = depth of flow, S = slope of energy gradient, and q = lateral inflow per unit length.

In a manner similar to that in pipe flow these two equations can be transformed into ordinary differential equations so that

$$\frac{dV}{dt} + \frac{g}{c}\frac{dy}{dt} + g(S - S_0) + \frac{q}{A}(V + c) = 0 \qquad (5.200)$$

if

$$\frac{dx}{dt} = V + c \qquad (5.201)$$

and

$$\frac{dV}{dt} - \frac{g}{c}\frac{dy}{dt} + g(S - S_0) + \frac{q}{A}(V - c) = 0 \qquad (5.202)$$

if

$$\frac{dx}{dt} = V - c \qquad (5.203)$$

where $c = \sqrt{gA/b}$ and b = surface width of the channel.

The equations can now be solved numerically by finite difference methods. This application of the method of characteristics requires that considerable care be exercised to guarantee that the Courant condition relating time step, length step, and wave celerity

$$\Delta t \le \frac{\Delta x}{|V| + c} \qquad (5.204)$$

be satisfied. The work of Wylie and Streeter[14] detail this general approach to the analysis of unsteady flow in open channels.

5.9 FLOW ABOUT IMMERSED OBJECTS

Flow about immersed objects has been discussed in Section 5.4 for the case of an ideal (frictionless) fluid. This section considers the effects of viscosity.

When a viscous fluid flows past an object, the fluid exerts a shear stress on the surface of the object as well as a normal pressure force. If the components of the surface shear and pressure in the direction of the flow are summed, the resulting force is known as the **drag**. The drag consists of a contribution from shear (skin friction drag) and pressure (form drag). In the case of well-formed bodies, the skin friction drag is the most significant. For blunt bodies, form drag dominates.

If the components of shear and pressure forces normal to the oncoming flow are summed, the **lift** on the body results. Typically, shear forces play a minor *direct* role on lift. Pressure is the dominant contributor. However, viscous forces can have a considerable indirect effect on lift and drag by causing boundary layer separation. It is common to represent lift L and drag D in terms of lift and drag coefficients

$$D = C_D \tfrac{1}{2}\rho A V_o^2 \qquad (5.205)$$

$$L = C_L \tfrac{1}{2}\rho A V_o^2 \qquad (5.206)$$

where C_D and C_L = drag and lift coefficients, respectively, ρ = fluid density, A = frontal area of the object, and V_o = free stream velocity.

In 1904 Prandtl developed the concept of the boundary layer and thus forged the link between ideal fluid mechanics and viscous flow. For fluids of relatively small viscosity, the effects of fluid friction are confined to a thin layer of fluid adjacent to the boundary known as the boundary layer. The flow outside the boundary layer can be determined with the tools of ideal fluid flow analysis. It is important to note that the boundary layer is thin and there is little normal acceleration; hence, the pressure variation along the body is determined, for all practical purposes, by the ideal fluid flow. This revelation led to the first analytical approach to the calculation of drag.

Flat Plate Boundary Layer. Flow across a flat plate parallel to the flow direction is subject to boundary layer growth. The forward portion of the plate develops a laminar boundary layer. The laminar boundary layer then "breaks down" forming a turbulent boundary layer that continues downstream indefinitely (see Fig. 5.44). The laminar boundary layer exists until a Reynolds number of 3900 occurs.

$$\mathbf{R} = \frac{V\delta}{\nu} \tag{5.207}$$

where δ = boundary layer thickness. Note that if the approaching flow is turbulent or the leading edge of the plate is rough, the laminar boundary layer may be considerably shorter.

Analysis of the flow on a flat plate provides the following values of shear stress and drag for the laminar flow portion:

$$\tau_o = c_f \tfrac{1}{2}\rho V_o^2 \tag{5.208}$$

where

$$c_f = \sqrt{\frac{8}{15\mathbf{R}_x}} \quad \text{and} \quad \mathbf{R}_x = \frac{V_o x}{\nu} \tag{5.209}$$

where x = distance from forward edge of the plate. The total drag force D is given by

$$D = C_f \tfrac{1}{2}A\rho V_o^2 \tag{5.210}$$

where

$$C_f = \sqrt{\frac{32}{15\mathbf{R}_x}} \tag{5.211}$$

These formulas are valid for Reynolds numbers based on $\mathbf{R}_x$ up to 500,000 where

$$\mathbf{R}_x = \frac{Vx}{\nu} \tag{5.212}$$

If the flow over the flat plate is largely turbulent, then the drag coefficient C_f can be expressed as

$$\frac{1}{\sqrt{C_f}} = 4.13 \log C_f \mathbf{R}_x \tag{5.213}$$

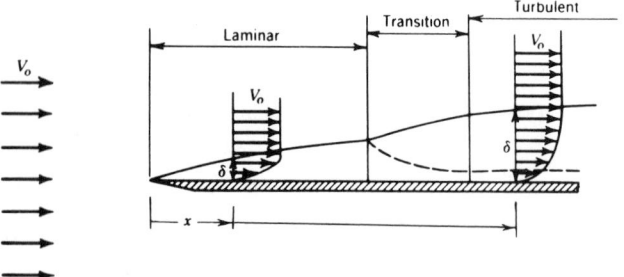

Fig. 5.44 Boundary layers on a flat plate. By permission from J. K. Vennard and R. L. Street, *Elementary Fluid Mechanics*, 5th ed., Wiley, 1975.

In general the C_f values in Fig. 5.45 can be used in the drag force equation to compute total drag on a smooth flat plate.

Drag on Immersed Objects. The total drag on an immersed object may be dominated by skin friction or form drag, but in any case, is generally obtained by experiment. The data is presented as drag coefficients and plotted as C_D versus Reynolds number. The total drag force can then be calculated from Eq. (5.205). Data on drag coefficients for common objects is presented in Figs. 5.46 and 5.47.

Lift. Lift is also presented through graphs of C_L. However, Chapter 6 adequately covers lift, and the reader should refer to that section for more detailed information.

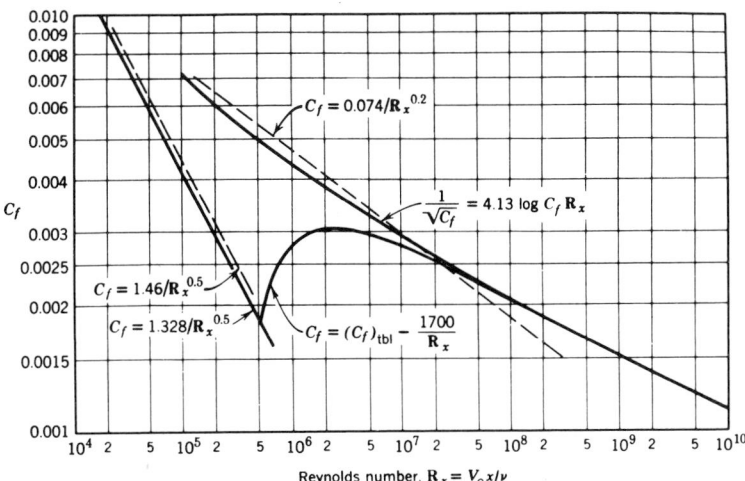

Fig. 5.45 Drag coefficients for smooth, flat plates. By permission from J. K. Vennard and R. L. Street, *Elementary Fluid Mechanics*, 5th ed., Wiley, 1975.

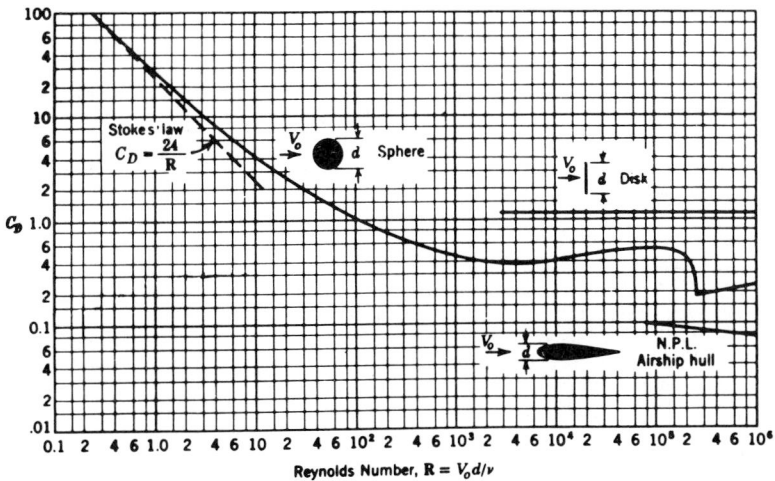

Fig. 5.46 Drag coefficients for sphere, disk, and streamlined body. By permission from J. K. Vennard and R. L. Street, *Elementary Fluid Mechanics*, 5th ed., Wiley, 1975.

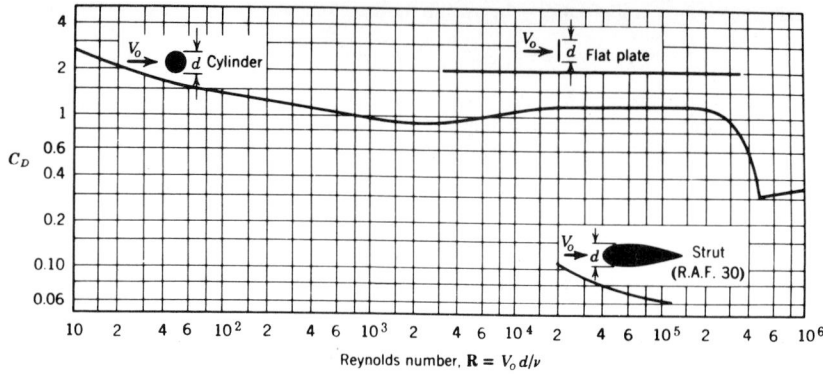

Fig. 5.47 Drag coefficients for circular cylinders, flat plates, and streamlined struts of infinite length. By permission from J. K. Vennard and R. L. Street, *Elementary Fluid Mechanics*, 5th ed., Wiley, 1975.

5.10 FLUID MEASUREMENTS*

In spite of the advances in numerical analysis, analytical techniques, and computer power in recent years, the complex phenomena of fluid flow must still be addressed empirically. The measurement of pressure, shear, discharge, velocity, and so forth remain necessary skills to the serious fluid mechanician. Many techniques and devices remain relatively unchanged over the years. Others, particularly in velocity measurement, reflect the recent advances in technology. This section presents an abbreviated review of the range of devices available.

5.10.1 Fluid Property Measurement

Specific Weight or Density. Measurement of the specific weight of a liquid relies on fundamental concepts and devices that generally need no calibration. Figure 5.48 illustrates three methods that depend on buoyancy calculations and one [Fig. 5.48(d)] that utilizes hydrostatics. Probably the simplest method is not illustrated—simply weighing a known volume of liquid. The device or method selected depends on the availability of the required equipment.

Figure 5.48(a) utilizes the submerged weight of a known volume of liquid. Figure 5.48(b), in a somewhat reversed approach, submerges a known weight and volume into the unknown liquid. Figure 5.48(c) uses calibrated hydrometers. Figure 5.48(d) illustrates the U-tube method, which requires no calibration.

Viscosity Measurement. Viscosity measuring devices (viscometers) generally employ one of three approaches to measuring viscosity—the falling sphere, the flowing tube, or the rotating cylinder. All the devices require that laminar flow be maintained throughout the measurement period.

The falling-sphere device is illustrated in Fig. 5.49. Stokes's law for a sphere falling in a viscous liquid is employed. The time required for a sphere of known size and weight to fall a specified distance is measured. The absolute viscosity can be calculated from the equation

$$\mu = \frac{d^2(\gamma_s - \gamma_l)}{18V} \tag{5.214}$$

where d = sphere diameter, V = sphere velocity and γ_s, γ_l = specific weight of the sphere and liquid, respectively.

The flowing tube devices (Ostwald, Saybolt, Bingham, Redwood, and Engler viscometers) are typified by the Ostwald and Saybolt devices in Fig. 5.50. All these devices depend on the laminar unsteady flow of a liquid. The time required for a given volume of liquid to flow through a small tube is measured, and an empirical formula based on laminar flow principles is used to calculate kinematic viscosity. For example, in the Saybolt device, the time required for the liquid level to drop from level

---

*This section follows closely the treatment in Vennard and Street.[1]

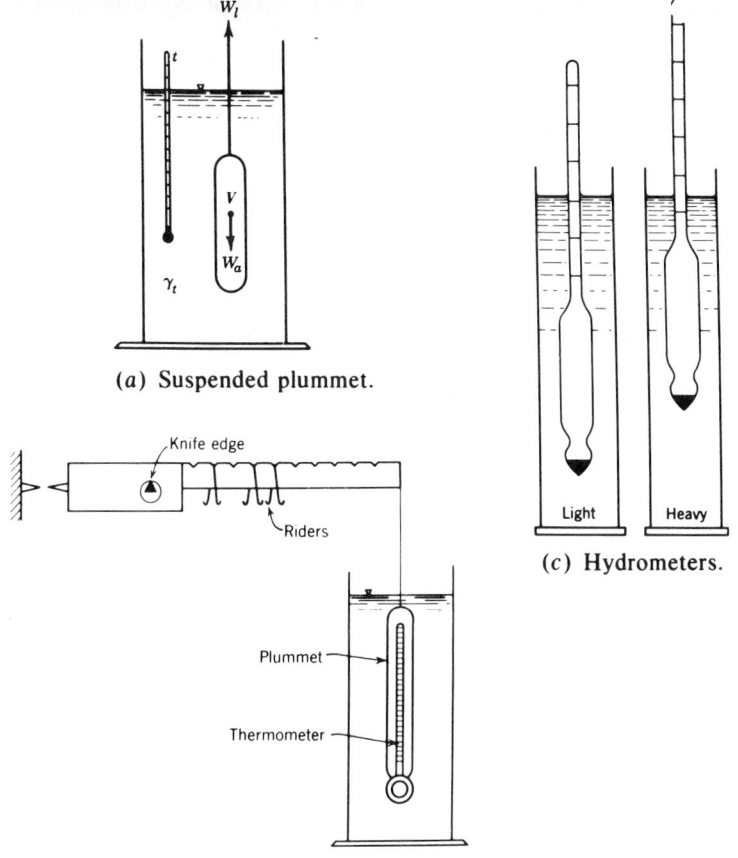

(a) Suspended plummet.

(b) Westphal balance.

(c) Hydrometers.

Fig. 5.48 Devices for density measurement. By permission from J. K. Vennard and R. L. Street, *Elementary Fluid Mechanics*, 5th ed., Wiley, 1975.

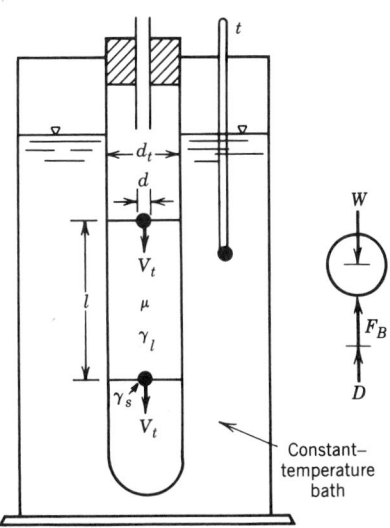

Fig. 5.49 Falling-sphere viscometer. By permission from J. K. Vennard and R. L. Street, *Elementary Fluid Mechanics*, 5th ed., Wiley, 1975.

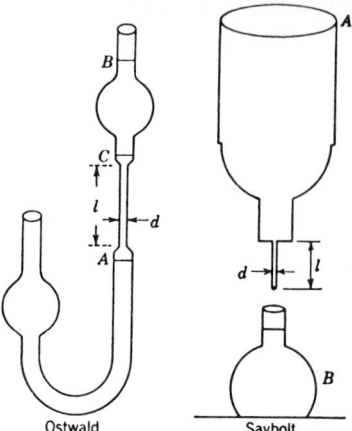

Fig. 5.50 Tube viscometers. By permission from J. K. Vennard and R. L. Street, *Elementary Fluid Mechanics*, 5th ed., Wiley, 1975.

B to C is recorded. The kinematic viscosity is calculated from the following equation:

$$\nu\left(\text{ft}^2/\text{sec}\right) = 0.0000023651t - \frac{0.001935}{t} \tag{5.215}$$

where t = the time in seconds required for the liquid level to drop.

The other devices will not be discussed here as the principles are similar and instructions and calibrated equations are supplied with the devices.

The rotating cylinder devices (Stormer, MacMichael, Brookfield) generally employ a fixed cylindrical container and a rotating inner cylinder (see Fig. 5.51). The space between the two cylinders and the speed of rotation is purposely kept small enough to maintain laminar flow. Measurement of the time required to complete a given number of revolutions under a constant torque leads to a calculation of absolute viscosity. The fundamental equation used is

$$T = \frac{2\pi R^2 h\mu V}{\Delta R} + \frac{\pi R^3 \mu V}{2\,\Delta h} \tag{5.216}$$

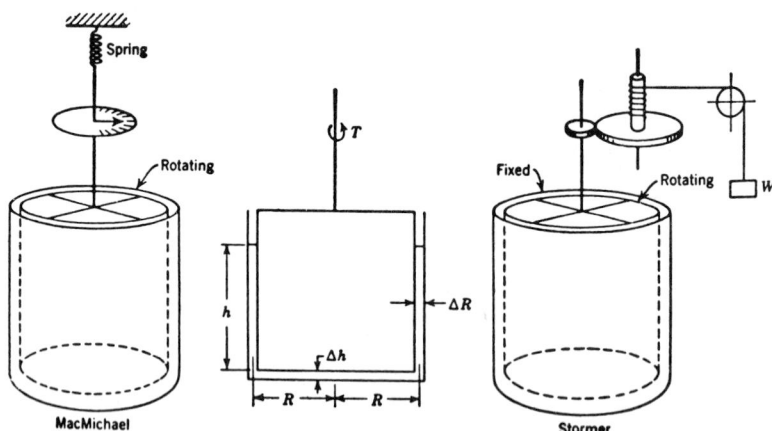

Fig. 5.51 Rotational viscometers (schematic). By permission from J. K. Vennard and R. L. Street, *Elementary Fluid Mechanics*, 5th ed., Wiley, 1975.

5.10.2   Pressure Measurement

Pressure measurement in a fluid at rest is relatively easy to accomplish. The manometers and Bourdon gauges discussed in Section 5.3 are common devices used for this purpose. If the fluid is moving, pressure measurement is more difficult.

Of primary concern is the pressure sensing connection between the fluid and the pressure measuring system. In measuring *static* pressure (pressure unaffected by the velocity of flow), an opening is made in the conduit wall or the pressure probe, such as a pitot-static tube (see Fig. 5.52), so that the pressure in the flow at the surface may be conducted to the pressure measuring device. The surface opening must be small (less than 1 mm) and well finished (square edged, no burr) so that the flow is not disturbed. In the case of a probe, the device must be small enough to not disturb the flow and alter the pressure situation and properly oriented to produce the true static pressure.

The pressure measuring devices used in fluid flows may be manometers or gauges. However, for electronic recording as well as measurement of fluctuating pressures, a pressure transducer is commonly used. A typical pressure transducer is a small diaphragm to which is attached a strain gauge to measure deflection of the diaphragm. The gauge is part of a Wheatstone bridge circuit and is calibrated to an electrical output that can be processed to produce a plot, directed to a computer for further analysis, or simply stored on some electronic device.

Piezoelectric transducers are also used wherein the sensing device is a piezoelectric crystal that produces an electric field when deformed. Care must be taken to install the sensors in a shock-free environment as they are very sensitive to any vibration of the facility hardware.

5.10.3   Velocity Measurement

One of the commonest and simplest devices for the measurement of velocity is a pitot-static tube (see Fig. 5.53). The device senses the difference in pressure between the tip of the pitot-static tube and the side of the tube. This pressure difference can be used to calculate the velocity from the following equation:

$$V_o = \sqrt{\frac{2(p_s - p_o)}{\rho}} = \sqrt{\frac{2g(p_s - p_o)}{\gamma}} \qquad (5.217)$$

where $p_s - p_o$ is the aforementioned difference in pressure.

In very low velocity flows, the anemometer or current meter is often used (see Fig. 5.54). These are calibrated devices that relate flow velocity to the number of revolutions per minute of the rotating element in the meter. Anemometers are typically used to measure wind speeds, and current meters are employed to measure water velocities in rivers.

Measurement of rapidly fluctuating velocities in air and water require devices that can respond quickly to changes in velocity. The hot-wire anemometer is commonly used to measure velocity fluctuations in air. A thin wire connected between two supports passes an electric current that heats the wire [see Fig. 5.55(a)]. Air moving perpendicular to the wire cools the wire in proportion to the flow velocity. The electronics of the system measures the increased voltage necessary to keep the wire temperature constant and relates that voltage to the flow velocity. In some devices the current flowing through the probe is kept constant and the voltage change required for this to occur is related to the flow velocity. However, the constant-temperature device is by far the most popular.

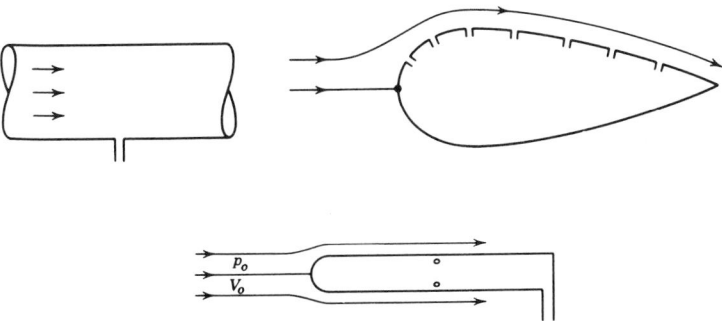

Fig. 5.52 Static tube. By permission from J. K. Vennard and R. L. Street, *Elementary Fluid Mechanics*, 5th ed., Wiley, 1975.

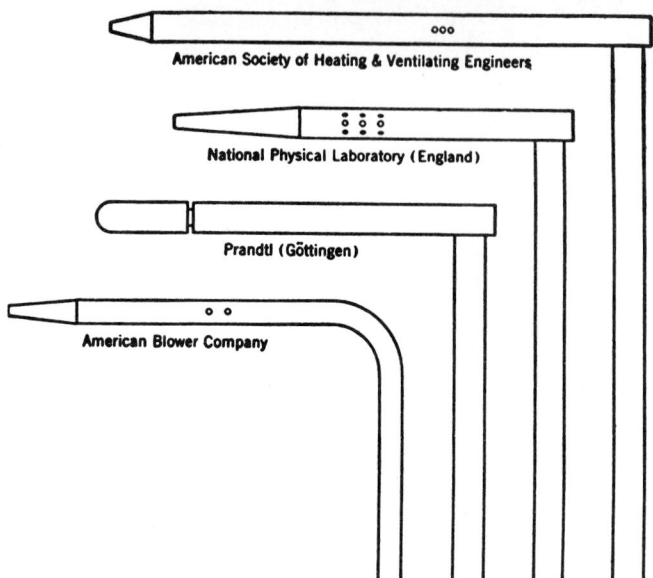

Fig. 5.53 Pitot-static tubes (to scale). By permission from J. K. Vennard and R. L. Street, *Elementary Fluid Mechanics*, 5th ed., Wiley, 1975.

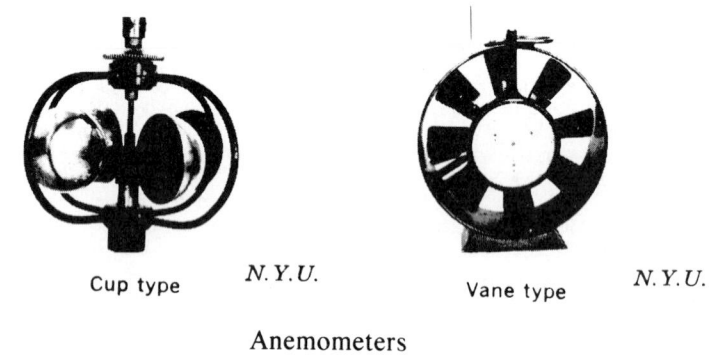

Cup type   *N.Y.U.*   Vane type   *N.Y.U.*

Anemometers

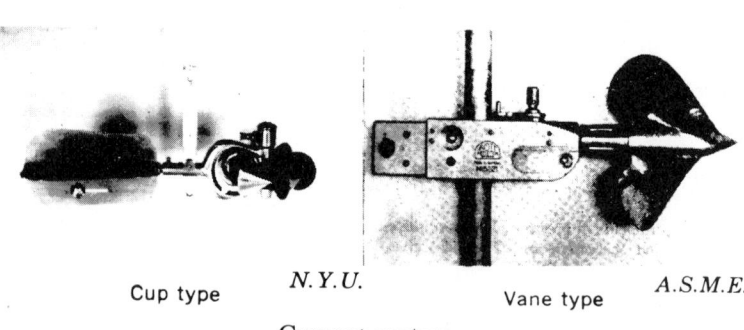

Cup type   *N.Y.U.*   Vane type   *A.S.M.E.*

Current meters

Fig. 5.54 By permission from J. K. Vennard and R. L. Street, *Elementary Fluid Mechanics*, 5th ed., Wiley, 1975.

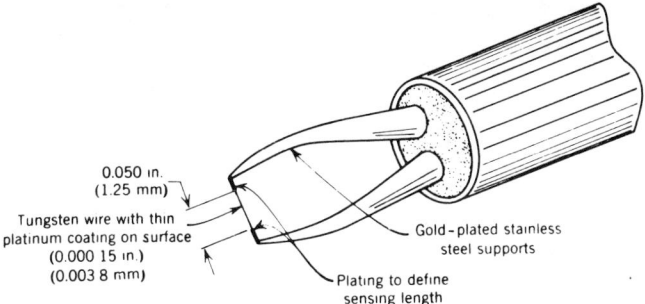

0.050 in.
(1.25 mm)

Tungsten wire with thin
platinum coating on surface
(0.000 15 in.)
(0.003 8 mm)

Gold-plated stainless
steel supports

Plating to define
sensing length

(*a*) Hot-wire sensor and support needles.

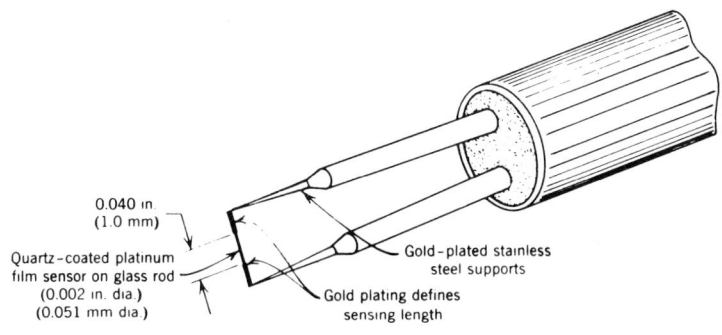

0.040 in.
(1.0 mm)

Quartz-coated platinum
film sensor on glass rod
(0.002 in. dia.)
(0.051 mm dia.)

Gold-plated stainless
steel supports

Gold plating defines
sensing length

(*b*) Hot-film sensor and support needles.

Fig. 5.55 Anemometer sensors. Reproduced from TB5, Thermo-Systems, Inc., 2500 Cleveland Ave. North, St. Paul, Minnesota, 55113. By permission from J. K. Vennard and R. L. Street, *Elementary Fluid Mechanics*, 5th ed., Wiley, New York, 1975.

A variation on the hot-wire anemometer is the hot-film anemometer. The hot-film device is similar to the hot-wire except the wire is coated to protect it from contaminated environments. In addition to coated wires, hot-film anemometry employs probes of other shapes that are sturdier and less likely to trap impurities in the flow (lint, small pieces of organic material, etc.). Figure 5.56 illustrates a few of these different types of probes. Hot-film devices are commonly used in liquid flows. Both these devices are capable of measuring velocity fluctuation frequencies of better than 300 KHz.

One of the most recent techniques for measuring velocities is laser-Doppler velocimetry. The fundamental basis for the technique is the Doppler shift of light that is scattered from extremely small particles in a moving fluid. One valuable attribute of laser-Doppler velocimetry is that it is nonobtrusive, that is, the sensing device is not in the flow. It only needs visual access to the flow for the required light beams. Refer to Goldstein[16] for a thorough treatment of the subject.

5.10.4   Flow Rate Measurement

Methods of measuring flow rate or discharge can generally be categorized into total quantity measurements, pressure drop or pressure difference measuring devices, tracer transport techniques, and devices that induce critical flow conditions or simply changes in water surface depth in open channels. Another indirect approach is the measurement of velocity at several points and a numerical integration of velocity times area to calculate discharge.

Total Quantity Methods. The success of these methods depends on the availability of a means to collect and measure (or weigh) the amount of liquid captured in the container in a given time period. Advantages are the lack of any need to calibrate a device. This approach is often used to calibrate other flow rate measuring devices.

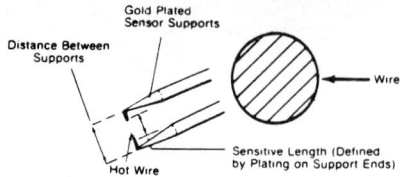

(*a*) **CROSS SECTION OF HOT-WIRE SENSOR**

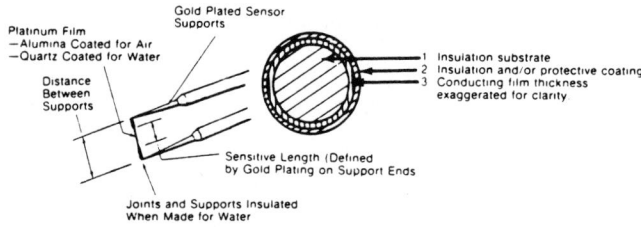

(*b*) **CROSS SECTION OF HOT-FILM SENSOR**

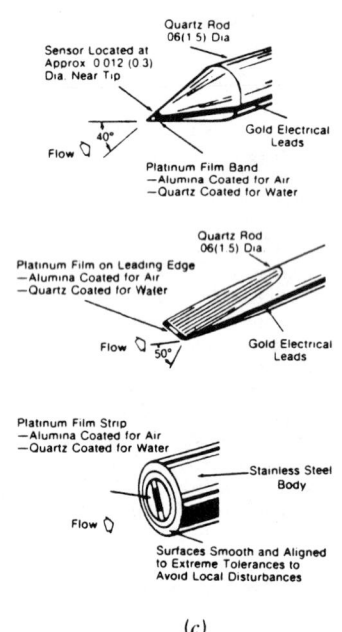

(*c*)

Fig. 5.56 Thermal sensor configuration. By permission from R. J. Goldstein, *Fluid Mechanics Measurements*, Hemisphere, New York, 1983.

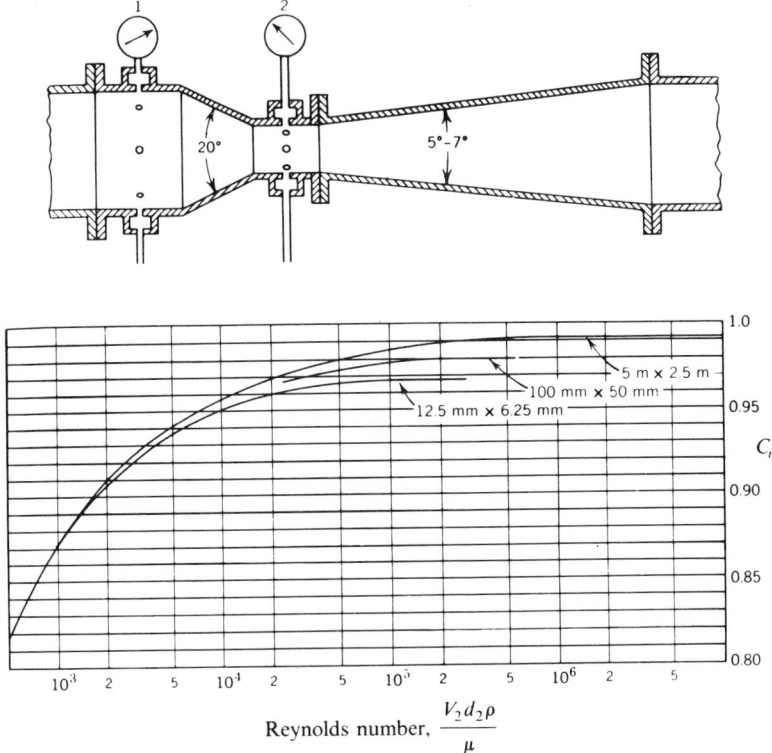

Fig. 5.57 Venturi meter and coefficients. By permission from J. K. Vennard and R. L. Street, *Elementary Fluid Mechanics*, 5th ed., Wiley, New York, 1975.

Pressure Difference Methods. All the devices that create pressure differences in a fluid flow to permit calculation of flow rate also create friction losses in the flow. One of the devices that has the lowest friction loss is the venturi meter (see Fig. 5.57). In the venturi meter the Bernoulli effect is used to generate a pressure difference between the entrance and the throat of the device. This pressure difference is related to the flow rate through the equation

$$Q = \frac{C_v A_2}{\sqrt{1 - (A_2/A_1)^2}} \sqrt{2g\left(\frac{p_1}{\gamma} + z_1 - \frac{p_2}{\gamma} - z_2\right)} \qquad (5.218)$$

where A = cross-section area of the meter and C_v is a calibration coefficient shown on Fig. 5.57. Venturi meters are relatively expensive but produce very little friction loss and are quite accurate.

A device based on a similar technique to the venturi meter is the flow nozzle. This device resembles the upstream portion of a venturi meter (see Fig. 5.58). Pressure recovery experienced in the downstream section of the venturi meter is not realized here so the flow nozzle creates a larger friction loss than the venturi meter. The calculation for discharge is made with the same equation used for the venturi meter with the C_v value taken from Fig. 5.58.

One further step in the direction of simplicity (and lower cost) is the orifice meter (see Fig. 5.59). This device is simply a circular plate with a hole cut out of its center and installed in a pipe, generally in a flanged connection. The orifice meter creates considerably more friction loss than the venturi meter or flow nozzle but its low cost makes it attractive in many instances. Flow rates through an orifice meter may be calculated with the equation

$$Q = CA\sqrt{2g\left(\frac{p_1}{\gamma} + z_1 - \frac{p_2}{\gamma} - z_2\right)} \qquad (5.219)$$

The value of C for a given flow rate is found from Fig. 5.60.

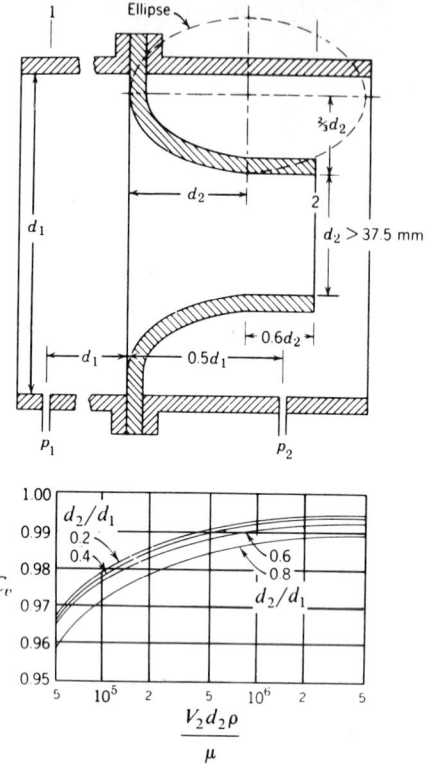

Fig. 5.58 ASME flow nozzle and coefficients. By permission from J. K. Vennard and R. L. Street, *Elementary Fluid Mechanics*, 5th ed., Wiley, New York, 1975.

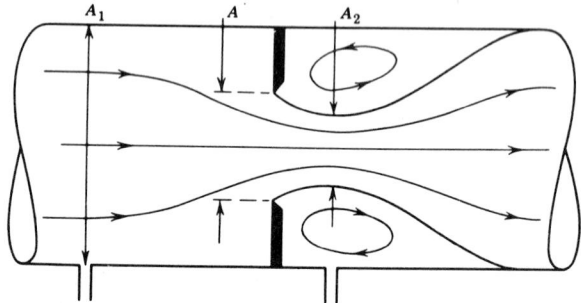

Fig. 5.59 Definition sketch for orifice meter. By permission from J. K. Vennard and R. L. Street, *Elementary Fluid Mechanics*, 5th ed., Wiley, New York, 1975.

There are certain special cases of "orifice meters" shown in Fig. 5.61 where discharges through the orifice can be related to liquid levels on one (or two) side(s) of the opening. The equations used to compute the discharge are

$$Q = A_2 V_2 = C_c C_v A \sqrt{2g(h_1 - h_2)} = CA \sqrt{2g(h_1 - h_2)} \qquad (5.220)$$

$$Q = C_c C_v A \sqrt{2gh} = CA \sqrt{2gh} \qquad (5.221)$$

All the C values for various types of orifices can be found in Fig. 5.62. More precise data for C values of sharp-edged orifices over a large range of heads are given in Fig. 5.63.

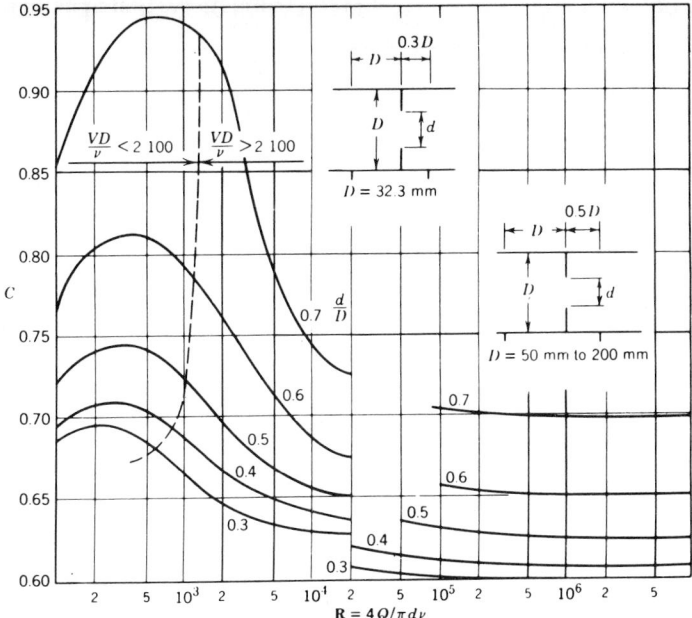

Fig. 5.60 Orifice meter coefficients. By permission from J. K. Vennard and R. L. Street, *Elementary Fluid Mechanics*, 5th ed., Wiley, New York, 1975.

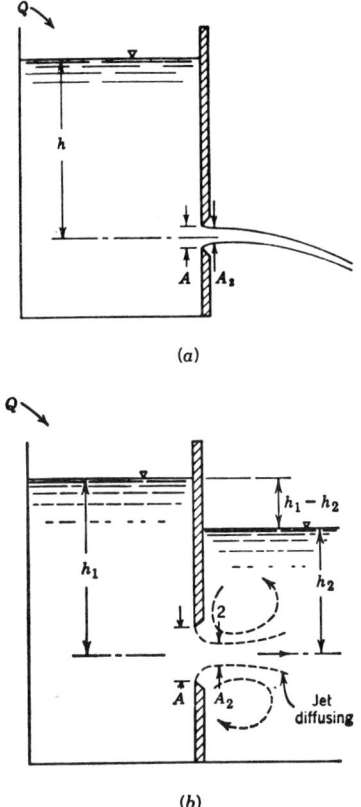

(a)

(b)

Fig. 5.61 (a) Orifice discharging freely; (b) submerged orifice. By permission from J. K. Vennard and R. L. Street, *Elementary Fluid Mechanics*, 5th ed., Wiley, New York, 1975.

Orifices and their Nominal Coefficients				
	Sharp edged	Rounded	Short tube	Borda
	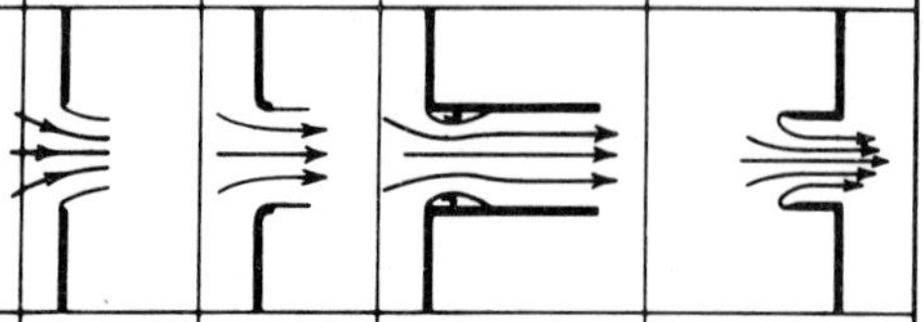			
C	0.61	0.98	0.80	0.51
C_c	0.62	1.00	1.00	0.52
C_v	0.98	0.98	0.80	0.98

Fig. 5.62 By permission from J. K. Vennard and R. L. Street, *Elementary Fluid Mechanics*, 5th ed., Wiley, New York, 1975.

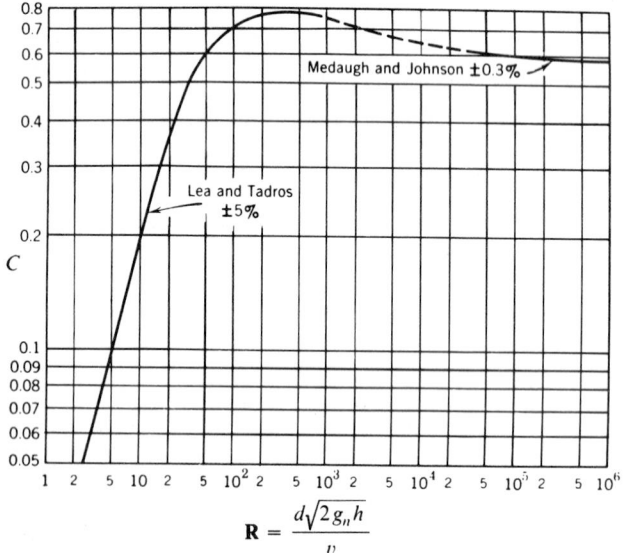

Fig. 5.63 Coefficient for sharp-edged orifices under static head ($h/d > 5$). By permission from J. K. Vennard and R. L. Street, *Elementary Fluid Mechanics*, 5th ed., Wiley, New York, 1975.

Probably the most inexpensive technique for calculating flow rate by the pressure difference method is the elbow meter (see Fig. 5.64). The only requirements are two pressure taps on the inner and outer portions of the elbow. The flow rate is calculated from the equation

$$Q = CA \sqrt{2g \left(\frac{p_o}{\gamma} + z_o - \frac{p_i}{\gamma} - z_i \right)} \qquad (5.222)$$

Because the value of C depends strongly on the elbow geometry, it must generally be determined through a calibration procedure.

Tracer Transport Techniques. This technique is based on the premise that the discharge in a conduit can be measured by the time of travel of the average concentration of a tracer between two points. The tracer could be as simple as salt where concentration is measured via conductivity instruments (see Fig. 5.65), although any other tracer may be employed. Fluorescent dyes are

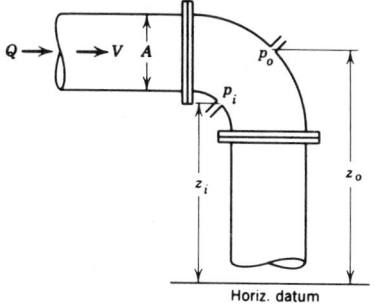

Fig. 5.64 Elbow meter. By permission from J. K. Vennard and R. L. Street, *Elementary Fluid Mechanics*, 5th ed., Wiley, New York, 1975.

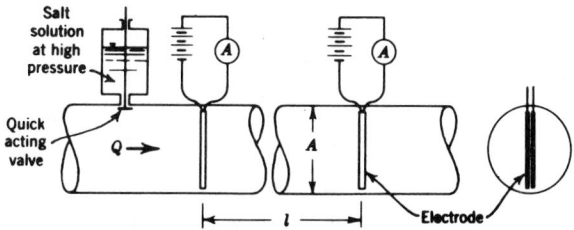

Fig. 5.65 Salt dilution discharge measuring. By permission from J. K. Vennard and R. L. Street, *Elementary Fluid Mechanics*, 5th ed., Wiley, New York, 1975.

commonly used because they have little effect on water quality and are detectable in minute quantities. From Fig. 5.65, the average velocity is computed from dividing the length l between sensors by the time t required for the centroid of the concentration curve to travel over the distance. This approach is particularly useful in open channels of relatively constant cross section and conduits where, for one reason or another, other techniques cannot be used.

Open-Channel Flow Measuring Devices. In free surface flows, the common techniques for flow measurement center on either (1) constricting the flow to create differences in water surface elevations that can be related to flow rate or (2) creating critical flow conditions (see Sec. 5.8) that provide a strong analytical connection to discharge calculation.

The simplest devices are venturi flumes, which are the open-channel version of venturi meters. The drop in water surface in the flume throat is measured and related to the flow rate. Although not extremely accurate, this approach does not create large friction losses.

A simpler version similar to the flow nozzle in closed conduits is the cutthroat flume where the diverging recovery section is not present.

A somewhat more complicated version of the venturi flume, which causes critical or near-critical flow to occur, is the Parshall flume (see Fig. 5.66). This device has been commonly used in irrigation systems for well over 70 years. The discharge is calculated from the equation

$$Q = 4Bh_a^{1.522B^{0.26}} \tag{5.223}$$

Another category of open-channel flow measuring devices is weirs. These devices generally generate more friction loss than the previously described devices, but they are relatively simple and accurate. A typical sharp-crested rectangular weir is shown in Fig. 5.67. The other types of weirs are broad-crested (Fig. 5.68) and triangular (Fig. 5.69). The general form of the discharge equation per foot of width for weirs with two-dimensional flow is

$$q = C_w \frac{2}{3}\sqrt{2g}\, H^{3/2}. \tag{5.224}$$

where H is defined in Figs. 5.67 and 5.68 as the head on the weir. The value of C_w for sharp-crested

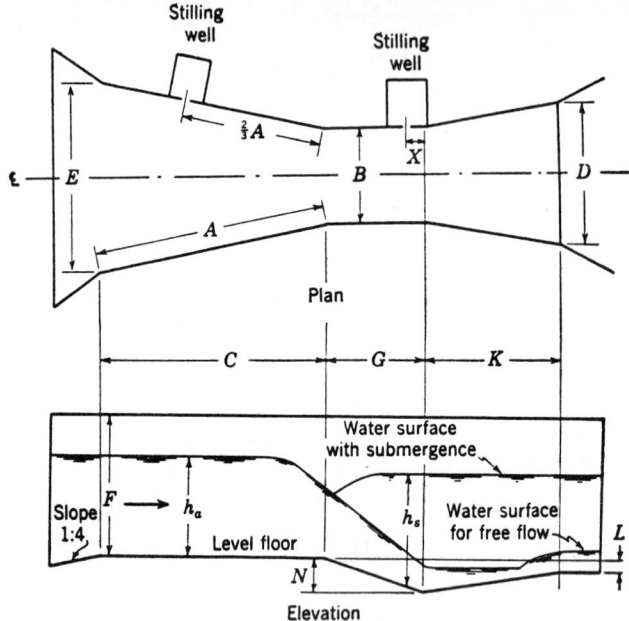

Fig. 5.66 Parshall measuring flume. From R. K. Linsley and J. B. Franzini, *Elements of Hydraulic Engineering*, McGraw-Hill, New York, 1955. Copyright 1955 McGraw-Hill. Reprinted by permission.

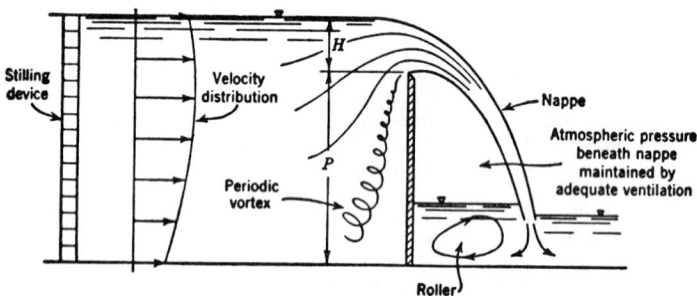

Fig. 5.67 Weir flow (actual). By permission from J. K. Vennard and R. L. Street, *Elementary Fluid Mechanics*, 5th ed., Wiley, New York, 1975.

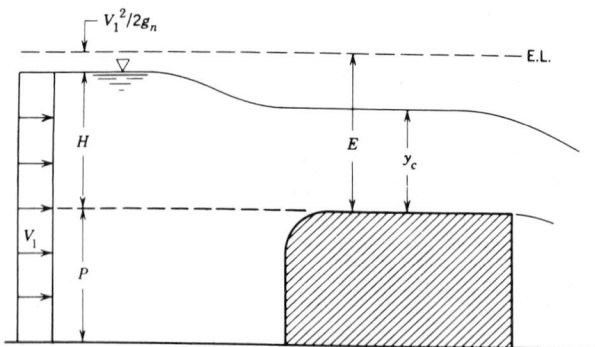

Fig. 5.68 Broad-crested weir. By permission from J. K. Vennard and R. L. Street, *Elementary Fluid Mechanics*, 5th ed., Wiley, New York, 1975.

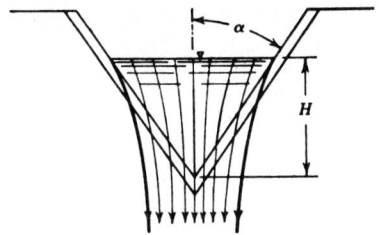

Fig. 5.69 Triangular weir. By permission from J. K. Vennard and R. L. Street, *Elementary Fluid Mechanics*, 5th ed., Wiley, New York, 1975.

well-ventilated weirs (Fig. 5.67) is given by the equation

$$C_w = 0.605 + 0.08\frac{H}{P} + \frac{1}{305H} \tag{5.225}$$

For broad-crested weirs (Fig. 5.68), the flow rate per foot of width is given by

$$q = \sqrt{g\left(\frac{2E}{3}\right)^3} = \left(\frac{2}{3}\right)^{3/2}\sqrt{g}\,E^{3/2} \tag{5.226}$$

or Eq. (5.224) where

$$C_w = \frac{1}{\sqrt{3}}\left(\frac{E}{H}\right)^{3/2} \tag{5.227}$$

where E is defined on Fig. 5.68.

For triangular notch weirs (Fig. 5.69) the equation for computing discharge is

$$Q = C_w \tfrac{8}{15}\tan\alpha\sqrt{2g}\,H^{5/2} \tag{5.228}$$

where C_w depends on the notch angle α. For $\alpha = 90°$,

$$C_w = 0.56 + \frac{0.70}{\mathbf{R}^{0.165}\mathbf{W}^{0.170}} \tag{5.229}$$

REFERENCES

1. Vennard, J. K., and Street, R. L., *Elementary Fluid Mechanics*, 5th ed., New York, Wiley, 1975.
2. Hele-Shaw, H. J. S., Investigation of the Nature of the Surface Resistance of Water and of Streamline Motion under Certain Experimental Conditions, *Trans. Inst. Naval Architects* **40** (1898).
3. Stokes, G., *Trans. Cambridge Phil. Soc.* **8** (1845) and **9** (1851).
4. Kármán, Th. von, Turbulence and Skin Friction, *J. Aeronautical Sci.* **1**(1), 1 (1934).
5. Davis, C. V., and Sorenson, K. E., *Handbook of Applied Hydraulics*, 3rd ed., New York, McGraw-Hill, 1969.
6. Chow, V. T., *Open Channel Hydraulics*, New York, McGraw-Hill, 1959.
7. Colebrook, C. F., Turbulent Flow in Pipes, with Particular Reference to the Transition Region between the Smooth and Rough Pipe Laws, *J. Inst. Civil Engineers, London* **11** (1938–39).
8. Crane Company, "Flow of Fluids Through Valves, Fittings and Pipe," Tech. Paper No. 410, Crane Co., 300 Park Ave., New York, 1969.
9. Idelchik, J. E., *Handbook of Hydraulic Resistance*, 2nd ed., New York, Hemisphere Publishing, 1986.
10. Gibson, A. H., *Hydraulics and Its Applications*, London, Constable, 1912.

11. Jeppson, R. W., *Analysis of Flow in Pipe Networks*, Ann Arbor, Mich., Ann Arbor Science, 1976.
12. Wood, D. J. and Rayes, A. G., Reliability of Algorithms for Pipe Network Analysis, *J. Hydraulics Div., ASCE*, **107**(10) (1981).
13. Watters, G. Z., *Analysis and Control of Unsteady Flow in Pipelines*, Stoneham, Mass. Butterworth Publishers, 1984.
14. Wylie, E. B., and Streeter, V. L., *Fluid Transients*, New York, McGraw-Hill, 1978.
15. Chaudhry, M. H., *Applied Hydraulic Transients*, New York, Van Nostrand Reinhold, 1979.
16. Goldstein, R. J., *Fluid Mechanics Measurements*, New York, Hemisphere Publishing, 1983.

BIBLIOGRAPHY

Moody, L. F., Friction Factors for Pipe Flow, *Trans. ASME*, **66** (1944).

Linsley, R. K. and Franzini, J. B., *Elements of Hydraulic Engineering*, New York, McGraw-Hill, 1955.

Rouse, H., *Fluid Mechanics for Hydraulic Engineers*, New York, McGraw-Hill, 1938.

Streeter, V. L., *Fluid Dynamics*, 3rd ed., McGraw-Hill, New York, 1962.

Vennard, J. K., *Elementary Fluid Mechanics*, 4th ed., New York, Wiley, 1961.

CHAPTER 6
AERONAUTICS

E. J. Jumper, C. A. Forbrich, Jr., L. M. Nicolai, W. C. Bauman,
D. H. Daley, J. B. Wissler, R. B. Oliver, E. L. Larson, D. N.
Barlow, R. D. Marker, R. W. Milling, R. J. Stiles, W. F. Hallgren,
D. Finkleman, J. T. Clay, M. L. Smith, J. J. Beoddy, P. I. King,
S. Czyzak, R. E. Willes, T. J. Forster, G. E. Thompson,
T. A. Hammond

**Department of Aeronautics, United States Air Force Academy,
Colorado Springs, Colorado**

6.1   ATOMIC AND MOLECULAR THEORY OF GASES

E. J. Jumper, C. A. Forbrich, Jr., and L. M. Nicolai

Aeronautics might be thought of as the study of the consequences of fluid dynamic interactions with aerodynamic shapes. As such it is intimately related to the study of fluids, in particular, gases. The normal approach to fluid dynamics is to treat the fluid as a continuum; but even then, the continuum treatment requires the use of state and transport properties that are best understood, if not obtained, from analysis of the fluid as being made up of individual particles (atoms and molecules) in constant motion. The understanding of still other gas properties rely on an analysis that treats the internal structure of the particles. The study of rarefied gases, once important only in other fields, is becoming ever more important to the fields of aeronautics and astronautics with the evolution of true aerospace vehicles designed to operate in environments that range in gas pressures from atmospheres to vacuum. This means that in some portion of the envelope they are subject to flow regimes classified as free-molecular flow. Again, this flow regime requires the treatment of the gas as made up of individual particles.

The ability to treat gases as made up of individual particles has its origin in the nineteenth century with the work of Clausius, Maxwell,[1] and Boltzmann.[2] The two main analytic tools for such treatment are kinetic theory of gases and statistical thermodynamics (c.f. Section 6.12.1). The first of these treats the gas particles using the notions of equilibrium and classical mechanics to derive such things as the kinetic definition of temperature and pressure, as well as extremely good formulas for the various transport properties such as viscosity, thermal conductivity, and mass diffusivity.[3-5] Statistical thermodynamics, on the other hand, makes extensive use of quantum mechanics and has led to important formulas for state properties such as internal energy, specific heats, and ionized-electron density.[2,6] Use of these and hybrid theories combining kinetic theory and statistical mechanics continue to provide understanding to many important areas of interest to aeronautics and astronautics (see, for example, Ref. 7).

6.1.1   Kinetic Theory of Gases (Preliminary Discussion)

Most treatments of kinetic theory begin with the development of a kinetic formula for pressure based on a simplistic model of a gas-surface interaction.[2,8] While such a treatment is far too simplistic for all but the determination of pressure, it gives some appreciation for how powerful a tool kinetic theory is for understanding the role that individual particle motions (microscopic view) play in the thermodynamic properties used in fluid mechanics (macroscopic view). While it is not the purpose of this section to do more than acquaint the reader with some of the more important results of kinetic theory (the reader is directed to the references for in-depth treatments), the simplest treatment of pressure gives some feel for the methods of kinetic theory and is thus worth reviewing.

The most simplistic model of a gas[8] assumes that all the gas particles are not only identical in mass m, but all have the same speed v, taken to be the root-mean-square (rms) kinetic speed of the real gas being modeled, and is given in terms of velocity components by

$$v^2 = v_x^2 + v_y^2 + v_z^2 \tag{6.1}$$

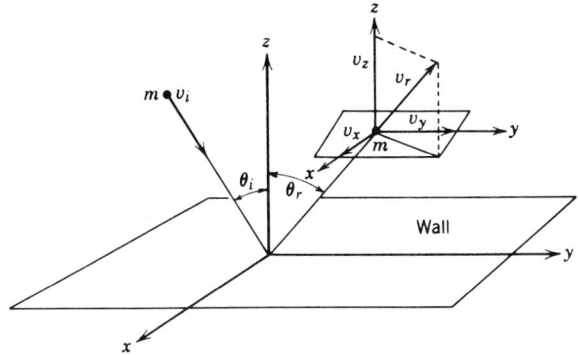

Fig. 6.1 Coordinates for molecular wall collision.

where v_x, v_y, and v_z are the x, y, and z components of velocity, respectively. For the purpose of deriving a formula for the pressure, the gas made up of these uniform particles is assumed to be confined in a cubic enclosure (box) of side *l*. In order to simplify the analysis further, particle-particle collisions are ignored, and collisions of particles with the walls of the box are treated as specular and elastic. Using a coordinate system oriented with respect to one of the walls, as shown in Fig. 6.1, classical mechanics yields an impulse *I* to the wall for a single colliding particle of

$$I = 2mv_z$$

directed normal to the wall. If the particle is assumed to collide only with the walls of the box, it will collide with this same wall $v_z/(2l)$ times per second. The impulse per second delivered to the wall by this single particle might be thought of as exerting an "average force" to the wall, F_i, of

$$F_i = \frac{mv_z^2}{l}$$

A similar analysis for each of the particles in the box will yield an identical result so that the total force delivered to the wall of the box by all of the particles in the box would just be $N \cdot F_i$, where N is the total number of particles in the box. The force per unit wall area or equivalent pressure would be

$$P = \frac{Nmv_z^2}{l^3}$$

But N/l^3 is just n, the number density, and if the average speed in each direction is taken to be equal, the pressure may be written in terms of the rms speed as

$$P = \tfrac{1}{3}nmv^2 \qquad (6.2)$$

Using a principle akin to the correspondence principle in quantum mechanics, the pressure predicted in Eq. (6.2) via kinetic theory must match that of classical thermodynamics given by

$$P = nmRT \qquad (6.3)$$

where R is the gas constant and T is the absolute temperature, so that

$$T = \frac{\tfrac{1}{3}v^2}{R}$$

or in terms of the Boltzmann constant k,

$$T = \frac{\tfrac{1}{3}v^2}{k/m}$$

Since the average kinetic energy of a particle, e_{tr}, is $\frac{1}{2}mv^2$,

$$e_{tr} = \tfrac{3}{2}kT \tag{6.4}$$

Equation (6.4) is sometimes referred to as the kinetic definition of temperature.
The value of Boltzmann's constant depends on the system of units of the problem, for example

$$1.3803 \times 10^{-23} \frac{\text{joules}}{\text{molecule K}} \quad \text{(MKS system)}$$

$$1.3803 \times 10^{-16} \frac{\text{ergs}}{\text{molecule K}} \quad \text{(cgs system)}$$

$$7.23 \times 10^{-27} \frac{\text{Btu}}{\text{molecule °R}} \quad \text{(Engineering)}$$

6.1.2 Collision Theory

The treatment in Section 6.1.1 assumed that each particle interacted only with the wall, and that interaction produced an elastic and specular collision. A more general analysis allows for particle-particle collisions as well as a removal of the requirement for wall collisions to be elastic and specular. It is, however, interesting to note that the more general treatment yields the same results in the case of pressure as the simplistic treatment of Section 6.1.1 (although results other than those for pressure yield different results). The reason that the pressure results remain the same can be attributed to a consequence of collision theory known as detailed balancing[2] that states that for every first collision that removes a particle of a particular velocity distribution, some other collision in the "immediate vicinity" restores a different particle of the same type to the velocity lost by the particle of the first collision. Thus, the analysis of Section 6.1.1 yields the correct pressure results because on average the system of particles acts as if all had the rms speed, specular collisions and, indeed, a random direction distribution.

In collision theory, a distribution function is assumed to exist but of unknown form. The notion of equilibrium is invoked that, when applied to the collision process, requires that, whatever the distribution function is, it must remain unchanging in time (assuming the pressure and temperature of the gas remain constant). The particles are treated as spherical of diameter d, which may be a function of the impact velocity (i.e., the gas temperature). Under the spherical assumption a collision integral is constructed that accounts for the geometry of the collision and is integrated over all possible geometries. The result yields the principle of detailed balancing. When the principle of detailed balancing is coupled with classical mechanics, which requires that mass, momentum, and energy (assumed to be kinetic energy only) are conserved, the outcome is the Maxwell–Boltzmann velocity distribution function, $f(v_i)$, given by[2]

$$f(v_i) = \left(\frac{m}{2\pi kT}\right)^{3/2} \exp\left[-\frac{m}{2kT}\left(v_x^2 + v_y^2 + v_z^2\right)\right] \tag{6.5}$$

The distribution function of Eq. (6.5) is termed the Maxwell–Boltzmann distribution because both kinetic theory (Maxwell) and statistical thermodynamics (Boltzmann) yield the same velocity distribution in the so-called Boltzmann limit. This point is mentioned here to point out that while real atoms/molecules do contain internal structure (ergo, they can have "internal" as well as kinetic energy), are not spherical, and interact quantum mechanically rather than classically, the consequences are such that they appear to preserve the results of classical mechanics.

Two of the more important results of collision theory are the functional forms for the bimolecular collision rate, Z_{AB}, the mean free path, λ, given by[2]

$$Z_{AB} = \frac{n_A n_B}{s} d_{AB}^2 \left(\frac{8\pi kT}{m_{AB}^*}\right)^{1/2} \tag{6.6}$$

where s (nondimensional) is 1 for unlike molecules ($A \neq B$) and 2 for like molecules ($A = B$), $d_{AB} = (d_A + d_B)/2$, and m_{AB}^* is the reduced mass, $m_{AB}^* = (m_A m_B)/(m_A + m_B)$; and

$$\lambda = \frac{1}{\sqrt{2}\,\pi d^2 n} \tag{6.7}$$

respectively. The dimensions of Z_{AB} are numbers per second per unit volume.

6.1.3 Distribution Function Theory

The velocity distribution function of Eq. (6.5) may be used to give the probability of a molecule being in the range of velocities between v_i and $v_i + dv_i$; $f(v_i)$ has the properties of a probability distribution function, as a consequence, the number of molecules in the velocity range dv_i, where integration on dv_i is understood to be equal to $dv_x\, dv_y\, dv_z$, is given by

$$dN = N_0 f(v_i)\, dv_i \qquad (6.8)$$

where N_0 is the total number of molecules in the system under study. Integrating Eq. (6.8) over all molecular velocities (where the integral from $-\infty$ to ∞ is understood to be a triple integral)

$$N_0 = \int_0^{N_0} dN = \int_{-\infty}^{\infty} N_0 f(v_i)\, dv_i$$

or

$$1 = \int_{-\infty}^{\infty} f(v_i)\, dv_i \qquad (6.9)$$

This equation describes the mathematical characteristics of the distribution function. With this definition, any property of the system that is a function of the molecular velocities can be calculated; for example, the mean or average velocity $\bar{v}_i$ is

$$\bar{v}_i = \frac{\int_{-\infty}^{\infty} v_i N_0 f(v_i)\, dv_i}{\int_{-\infty}^{\infty} N_0 f(v_i)\, dv_i} = \int_{-\infty}^{\infty} v_i f(v_i)\, dv_i = 0 \qquad (6.10)$$

and the mean square of the molecular velocity v_i is

$$\overline{v_i^2} = \int_{-\infty}^{\infty} v_i^2 f(v_i)\, dv_i$$

and the mean molecular energy in the ith direction is

$$\frac{1}{2} \overline{mv_i^2} = \int_{-\infty}^{\infty} \frac{1}{2} mv_i^2 f(v_i)\, dv_i$$

Other molecular distribution functions can be obtained from the velocity distribution function of Eq. (6.5) by mathematical transformations; for example, the distribution function for the molecular speed v is

$$f(v) = \frac{4v^2}{\sqrt{\pi}} \left(\frac{m}{2kT}\right)^{3/2} \exp\left(-\frac{mv^2}{2kT}\right) = \begin{array}{l}\text{Maxwell–Boltzmann}\\ \text{speed distribution function}\end{array} \qquad (6.11)$$

Also, the molecular energies are distributed as

$$f(E) = 2\left(\frac{E}{\pi}\right)^{1/2} \left(\frac{1}{kT}\right)^{3/2} \exp\left(-\frac{E}{kT}\right) = \begin{array}{l}\text{Maxwell–Boltzmann}\\ \text{energy distribution}\\ \text{function}\end{array}$$

Using Eq. (6.11), the mean or average speed of a particle is

$$\bar{v} = \int_0^{\infty} v f(v)\, dv = \sqrt{\frac{8kT}{\pi m}} \qquad (6.12)$$

and the mean square speed is

$$\overline{v^2} = \int_0^{\infty} v^2 f(v)\, dv = \frac{3kT}{m}$$

6.1.4 Transport Phenomena

One of the most significant contributions of kinetic theory of gases is the ability to theoretically predict transport properties (viscosity, thermal conductivity, and mass diffusivity). The reader is directed to the beautiful treatment of the development of the form of these properties in Ref. 8 (pp. 156–224), where the theory is developed first at the simplest level and then at each increasing level of complexity. Reference 2 (pp. 15–23) also gives a good simple treatment of the development with a concise discussion of what the complications do to the form of the equations. Reference 3 (pp. 3–30, pp. 243–264, pp. 495–518) gives concise parallel developments of viscosity, thermal conductivity, and mass diffusivity. And finally, Ref. 4 is the single best reference for the development and use of the transport properties.

The following equations for transport properties are taken from Ref. 3.

Momentum Transport (Viscosity). The simplistic theory gives a viscosity, μ, of

$$\mu = \tfrac{1}{3} nm\bar{v}\lambda \tag{6.13}$$

where $\bar{v}$ is given by Eq. (6.12) and λ by Eq. (6.7). Accounting for temperature-dependent collision cross sections gives

$$\mu = 2.6693 \times 10^{-5} \frac{\sqrt{MT}}{\sigma^2 \Omega\mu} \tag{6.14}$$

where the constant has units such that μ has units of $\text{gcm}^{-1}\ \text{sec}^{-1}$; T, the temperature, is in K; M, the molecular weight, is in g-moles; σ, the collision cross section, is in Angstroms; and $\Omega\mu$, the collision integral for viscosity, is dimensionless.

Energy Transport (Thermal Conductivity). The simplistic theory gives a thermal conductivity, κ

$$\kappa = \tfrac{1}{6} n\bar{v}fk\lambda \tag{6.15}$$

where f is 3 for a monatomic gas and 7 for a diatomic gas, and λ is given by Eq. (6.7). Accounting for temperature-dependent collision cross sections and the Eucken approximation[2,3] gives

$$\kappa = \begin{cases} \tfrac{5}{2} c_v \mu & \text{(monatomic)} \\[2mm] \left(c_p + \tfrac{5}{4}\dfrac{R}{M} \right)\mu & \text{(polyatomic)} \end{cases} \tag{6.16}$$

where c_v and c_p are the specific heats at constant volume and pressure, respectively, and the viscosity is computed using Eq. (6.14).

Mass Transport (Mass Diffusivity). The simplistic theory gives (self-diffusion[8]) a diffusivity, $\mathcal{D}$, of

$$\mathcal{D} = \tfrac{1}{3}\bar{v}\lambda \tag{6.17}$$

where $\bar{v}$ and λ are given by Eqs. (6.12) and (6.7), respectively. Using the more accurate Chapman–Enskog kinetic theory[3] gives for a binary mixture

$$\mathcal{D}_{AB} = 0.0018583 \frac{\sqrt{T^3(1/M_A + 1/M_B)}}{P\sigma_{AB}^2 \Omega_{\mathcal{D},\,AB}} \tag{6.18}$$

where the constant has units such that $\mathcal{D}_{AB}$ has units of $\text{cm}^2\ \text{sec}^{-1}$; M is in g-moles; T is K; p is in atm; σ_{AB} is Angstroms; and $\Omega_{\mathcal{D}}$, the collision integral for diffusion, is dimensionless.

6.1.5 The Boltzmann Equation

Using the concept of a distribution function described earlier, it is possible to determine the velocity and position of each molecule in a system as a function of time. Consider the distribution function

$f(x_i, c_i, t)$ where

x_i = position vector

c_i = total velocity of particle

$\quad = u_i + v_i$

u_i = macroscopic velocity of system

v_i = thermal or molecular velocity of particle

The quantity $f(x_i, c_i, t) \, dc_i \, dx_i$ is the probability of a particle having velocity in the range c_i to $c_i + dc_i$ and position in the range x_i to $x_i + dx_i$ at time t. Thus, if $f(x_i, c_i, t)$ is known, the system of particles is completely described.

The rate of change of $f(x_i, c_i, t)$ with time is given by the Boltzmann equation[2, 9, 10]

$$\frac{\partial (nf)}{\partial t} + c_i \frac{\partial (nf)}{\partial x_i} + a_i \frac{\partial (nf)}{\partial c_i} = \left[\frac{\partial (nf)}{\partial t} \right]_{\text{collision}} \qquad (6.19)$$

where a_i = acceleration vector due to an external force and $[\partial(nf)/\partial t]_{\text{collision}}$ is the change in the distribution function due to collisions. Equation (6.19) is a nonlinear integrodifferential equation (due to the collision term) and is very difficult to solve. The Boltzmann equation has been solved for a few restricted cases such as equilibrium flow or by approximating the collision term with simple expressions. It must be emphasized that $f(x_i, c_i, t)$ is the quantity being sought and the Boltzmann equation is the governing equation for f.

6.1.6 The Knudsen Number

It is useful to nondimensionalize the Boltzmann equation and examine its character. The appropriate characteristic quantities are a macroscopic length R, a microscopic length λ, and the macroscopic free-stream velocity. The nondimensional Boltzmann equation (starred quantities are nondimensional) is

$$\frac{\partial f^*}{\partial t^*} + c_i^* \frac{\partial f^*}{\partial x_i^*} + a_i^* \frac{\partial f^*}{\partial c_i^*} = \frac{1}{\sqrt{2}\,\pi} \frac{1}{K_n} \left[\frac{\partial f^*}{\partial t^*} \right]_{\text{collision}} \qquad (6.20)$$

where $K_n = \lambda/R$, the Knudsen number, is the ratio of mean free path to a macroscopic characteristic length.

The character of the Boltzmann equation can be categorized according to the magnitude of K_n.[10] These categories are shown in Table 6.1. For small K_n (less than 0.01) the fluid is collision dominated and is the continuum regime. For large K_n (greater than about 100) the collisions between particles are insignificant compared to the incident particle impact on a surface and the flow regime is called free molecular or rarefied gas (see Section 6.3.1).

In the continuum flow regime the governing similarity parameters are Reynolds number R_e and Mach number M. These parameters are related to the Knudsen number by

$$K_n = 1.88\sqrt{\gamma}\,\frac{M}{R_e}$$

6.1.7 Continuum Flow Conservation Equations

The Boltzmann equation is multiplied by a quantity $Q(c_i)$ and then integrated over all c_i (i.e., integrated over all velocity space). The quantity $Q(c_i)$ is chosen to be equal to the mass, momentum, and energy that are conserved during collisions. When the collision term [right-hand side of Eq.

TABLE 6.1   Categories of Flow

K_n	Flow Regime	Influence of Collisions
< 1	Continuum	Very significant, no slip condition valid
< 1	Slip flow	Significant, slip at the surface
1	Transition	Same order of magnitude as convection
> 1	Free molecular	Not significant

(6.19)] is multiplied by $Q(c_i)$ and integrated over $\underline{c_i}$, the result is zero. The left-hand side is simplified by using the fact that $\int_{-\infty}^{\infty} Q(c_i) f(x_i, c_i, t) \, dc_i = \bar{Q}(x_i, t)$ where the bar indicates a quantity averaged over velocity space. Thus $\int_{-\infty}^{\infty} c_i f(x_i, c_i, t) \, dc_i = u_i \int_{-\infty}^{\infty} f(x_i, c_i, t) \, dc_i + \int_{-\infty}^{\infty} v_i f(x_i, c_i, t) = u_i + 0 = u_i$ and the general result is the following:[1,5]

1. For $Q(c_i) = m \ldots$, the continuity equation

$$\frac{\partial \rho}{\partial t} + \frac{\partial}{\partial x_i}(\rho u_i) = 0 \tag{6.21}$$

2. For $Q(c_j) = mc_j = m(u_j + v_j) \ldots$, the momentum equation

$$\frac{\partial(\rho u_j)}{\partial t} + \frac{\partial}{\partial x_i}(\rho u_j u_i) = \rho a_j - \frac{\partial}{\partial x_i}\left(\rho \overline{v_i v_j}\right) \tag{6.22}$$

The term $\rho \overline{v_i v_j}$ is a tensor quantity representing the transport of momentum by molecular motion. The diagonal terms are identified, using Eq. (6.2), as the pressure p. The off-diagonal terms are identified by the Chapman–Enskog solution[2] to be the shear stress tensor τ_{ij} (see Section 6.3.2).

3. For $Q(c_i) = (\frac{1}{2})mc_i c_i \ldots$, the energy equation

$$\frac{\partial}{\partial t}\left[\frac{1}{2}\rho u^2 + \frac{1}{2}\rho \overline{v^2}\right] + \frac{\partial}{\partial x_i}\left[u_i\left(\frac{1}{2}\rho u^2 + \frac{1}{2}\rho \overline{v^2}\right)\right]$$

$$= \rho a_i u_i - \frac{\partial}{\partial x_i}\left(\frac{1}{2}\rho \overline{v_i v^2}\right) - \frac{\partial}{\partial x_i}\left(\rho \overline{v_i v_k} u_k\right) \tag{6.23}$$

The term $\frac{1}{2}\rho u^2$ is the macroscopic kinetic energy of the system and the term $\frac{1}{2}\rho \overline{v^2} = \frac{3}{2}nkT$ is the thermal or mean molecular energy of the particles in the system. The term $\frac{1}{2}\rho \overline{v_i v^2}$ is the transport by molecular motion of the mean molecular energy. This term is identified as heat transfer by conduction at the macroscopic level. The last term, $\rho \overline{v_i v_k} u_k$, is the rate of work done by the shear stress and the pressure.[2]

Equations (6.21)–(6.23) are called the continuum conservation equations and form a most useful set of equations. However, it must be remembered that this set is an approximation to a more accurate system;[11] that is, the Boltzmann equation. The process of integrating over all velocity space is an averaging process and removes the velocity as an independent variable.

6.2 CONSERVATION EQUATIONS OF CONTINUUM FLOW
L. M. Nicolai

6.2.1 The Continuum Approach

In this section the conservation equations in both integral and differential form that describe the continuum fluid flow phenomena are presented. We consider the fluid as a continuum since in many engineering applications our primary interest lies not in motions of molecules but rather in the gross behavior of the fluid thought of as a continuous material. This means that the fluid region under study must contain enough molecules so that statistically averaged properties are meaningful.

The system of equations can be derived from two general approaches.[12] One is by the method of kinetic theory (Section 6.1.1) where the behavior of the molecules is of interest, and the velocity and position of the molecules are given by a distribution function. By assuming a simplified model for the molecular interactions an equation can be formulated,[1-3] that describes the rate of change, with respect to position and time, of the distribution function. This equation is called the Boltzmann equation[13-15] [see Eq. (6.19)]. If velocity moments of the Boltzmann equation are integrated over the velocity space, the resulting equations are the general form of the continuum conservation equation[1,2] [see Eqs. (6.21)–(6.23)].

The second approach is to replace the detailed molecular structure that is actually present in any gas or liquid by a continuous model of matter with statistically averaged properties at every point.

6.2.2 Integral Equations

The second, or continuum, approach is the one from which the conservation equations will be derived. A control region of volume V, surface area S, and surface normal n_i is considered fixed with respect to the coordinate system. We then examine the fluid property fluxes through the control region, the influences of body and surface forces, and the heat and work interactions at the surface.

The conservation of mass equation follows then from the statement that the time rate of change of mass within V plus the flux of mass across S is equal to zero:

$$\int_V \frac{\partial \rho}{\partial t}\, dV + \int_S \rho(u_i n_i)\, dS = 0 \tag{6.24}$$

where u_i = mean macroscopic motion of fluid.

The conservation of momentum equation follows from Newton's laws of motion for a system of fixed mass. The system boundaries are assumed coincident with the control region boundaries[16] so that the time rate of change of the momentum within the system boundaries is equal to the external forces on the system, which is equal to the local time rate of change of momentum and the momentum flux through the control region V. This is expressed as

$$\int_V \frac{\partial (\rho u_i)}{\partial t}\, dV + \int_S \rho u_i(u_j n_j)\, dS = \int_V \rho f_i\, dV + \int_S \sigma_{ij} n_j\, dS \tag{6.25}$$

where f_i represents the body forces per unit mass (gravity, Lorentz forces, etc.) and σ_{ij} is a symmetric tensor representing the surface stresses.

The conservation of energy equation follows from the first law of thermodynamics for a system. Here again the system boundaries coincide with the boundaries of the control region[16] so that the rate of heat and work interaction at the surface S is equal to the local rate of change of total energy and the flux of total energy through the volume V. The total energy is the sum of the internal, kinetic, and potential energies. This conservation of energy is expressed in the integral form as

$$\int_V \frac{\partial}{\partial t}\left[\rho\left(e + \frac{u_i u_i}{2} + gz\right)\right] dV + \int_S \rho\left(e + \frac{u_i u_i}{2} + gz\right)u_j n_j\, dS$$

$$= -\int_S q_i n_i\, dS + \int_V \rho w_x\, dV + \int_V \dot{Q}\, dV + \int_V \rho u_i f_i\, dV + \int_S (\sigma_{ij} u_i n_j)\, dS \tag{6.26}$$

where q_i represents the rate of energy transfer into the system by conduction, w_x represents the rate of work done per unit mass on the system by a work reservoir of some other system, $\dot{Q}$ is the rate at which energy is generated internal to the system (such as by Joule heating), and the last term is the rate of work done on the system by pressure and shearing forces at the surface. Equation (6.26) neglects the energy transfer by radiation.

6.2.3 Differential Equations

The differential form of the conservation equations follows directly from the integral form by assuming that the parameters have sufficiently many derivatives and using the divergence theorem of calculus to transform the surface integrals into volume integrals.

The conservation of mass equation becomes

$$\frac{\partial \rho}{\partial t} + \frac{\partial}{\partial x_i}(\rho u_i) = 0 \tag{6.27}$$

Before writing the momentum equation in the form usually called the Navier–Stokes equation, we must express the surface stress term σ_{ij} in a more useful form. We assume:

1. Newtonian fluid—the rate of strain of a fluid element is relatively small such that a linear relationship exists between the rate of strain and the rate of stress.
2. The normal stress in the absence of deformation is equal to the negative of the thermodynamic pressure P.
3. The properties of the fluid are isotropic.

The stress tensor σ_{ij} is expressed as[15]

$$\sigma_{ij} = -P\delta_{ij} + \tau_{ij} = -P\delta_{ij} + \mu\left(\frac{\partial u_i}{\partial x_j} + \frac{\partial u_j}{\partial x_i}\right) + \left(\mu' - \frac{2}{3}\mu\right)\frac{\partial u_k}{\partial x_k}\delta_{ij} \tag{6.28}$$

where μ is the ordinary viscosity coefficient, μ' is the dilational coefficient,[12,15] and δ_{ij} is the kronecker delta function.

The momentum equation or Navier–Stokes equation is finally expressed as

$$\rho \frac{\partial u_i}{\partial t} + \rho u_j \frac{\partial u_i}{\partial x_j} = -\frac{\partial P}{\partial x_i} + \rho f_i + \frac{\partial \tau_{ij}}{\partial x_j} \tag{6.29}$$

The differential form for the conservation of energy equation is (omitting the potential energy term):

$$\frac{\partial}{\partial t}\left[\rho\left(e + \frac{1}{2}u_i u_i\right)\right] + \frac{\partial}{\partial x_j}\left[\rho u_j\left(e + \frac{1}{2}u_i u_i\right)\right]$$

$$= -\frac{\partial q_i}{\partial x_i} + \rho u_i f_i - \frac{\partial (Pu_i)}{\partial x_i} + \frac{\partial (u_j \tau_{ij})}{\partial x_i} + \rho w_x + \dot{Q} \tag{6.30}$$

The heat conduction term is usually expressed as $q_i = -k(\partial T/\partial x_i)$ according to Fourier's heat conduction laws. Also, the pressure work term can be expressed as $-(D/Dt)(P/\rho) + (1/\rho)(\partial P/\partial t)$ using the continuity equation.

Finally, the energy equation is rewritten as

$$\rho \frac{D}{Dt}\left(h + \frac{1}{2}u_i u_i\right) = \frac{\partial}{\partial x_i}\left(k \frac{\partial T}{\partial x_i}\right) + \rho u_i f_i + \frac{\partial P}{\partial t} + \frac{\partial}{\partial x_i}(u_j \tau_{ij}) + \rho w_x + \dot{Q} \tag{6.31}$$

where $h = e + (P/\rho)$, the static enthalpy per unit mass.

6.3 ATMOSPHERIC STRUCTURE
W. C. Bauman

6.3.1 Properties of the Atmosphere

The first modern standard atmosphere was developed in the United States and Europe in the 1920s to satisfy a need for standardizing aircraft instruments and performance. Subsequently, several groups have met to obtain international adoption of past work. With the advent of reliable instrumentation in rockets and satellites it has been possible to extend the standard atmosphere to altitudes near 1000 km. The material presented here has been drawn from the results of the United States Committee on Extension of the Standard Atmosphere (COESA), which represented 29 U.S. scientific and engineering organizations.

The atmosphere enclosing the earth is composed of at least 17 different gases. The 2 principal gases are nitrogen and oxygen, 78.084 and 20.947% by volume, respectively. In describing the atmosphere it is convenient to classify the various layers by their thermal structure. The names of the regions and boundaries, up to 100 km, are shown in Fig. 6.2 and are those adopted by the World Meterological Organization.

The *U.S. Standard Atmosphere, 1962*, is defined in geopotential height up to 90 km (300,000 ft) and in geometric height above 90 km.

When using atmospheric properties, two basic equations are used to relate pressure, temperature, and density. The hydrostatic equation is

$$dp = -\rho g\, dz = -\rho g_0\, dh$$

where p is pressure, ρ is density, g is the acceleration of gravity at height z, g_0 is standard acceleration of gravity at sea level, z is the geometric height, and h is the geopotential height, defined by the equation $dh = g/g_0\, dz$.

The equation of state for air is

$$p = \frac{\rho R T_m}{M_0} = \rho R_a T$$

where ρ is the density, R is the universal gas constant, R_a is the gas constant for air, T is the temperature (kinetic temperature), and M_0 is the average molecular weight of air at sea level. The average molecular weight from sea level up to approximately 90 km $(30 \times 10^5 \text{ ft})$ is constant and equal to 82.964. It then decreases to 22.52 at 10^6 ft and 16.77 at 2×10^6 ft. Above 90 km the

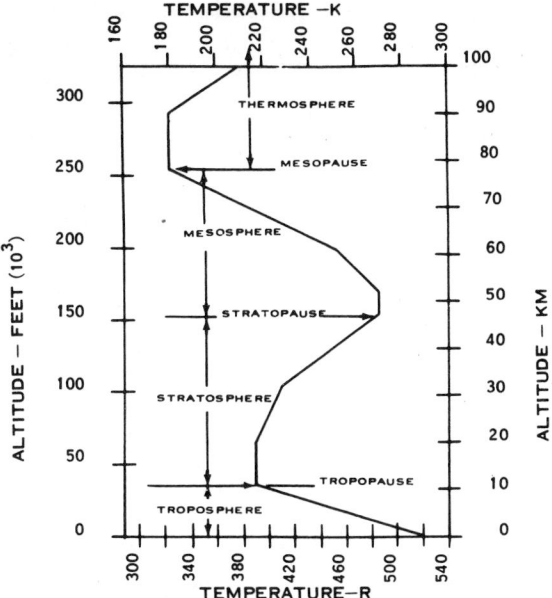

Fig. 6.2   Variation of temperature in 1962 standard atmosphere.

molecular weight begins to decrease because of molecular dissociation and this decrease is approximately linear with altitude. The molecular temperature T_m is defined as

$$T_m = \frac{M_0}{MT}$$

where M is the average molecular weight of air at the specified altitude, M_0 is already defined.

The composition of the atmosphere is virtually constant up to 90 km, above which molecular dissociation increases with altitude (see Chapter 6, Ref. 2). The pressure and density decrease approximately exponentially with altitude (see Fig. 6.3). Water vapor decreases irregularly with altitude except for a slight increase between 35,000 and 115,000 ft.

The coefficient of viscosity is defined as the coefficient of internal friction developed where gas regions move adjacent to one another at different velocities. Two of the more widely used formulas for determining the coefficient of viscosity (μ) are Sutherland's law and the power law. Sutherland's equation gives reasonably good results for the temperature range 180 to 3400°R (at atmospheric pressure) and was used for the *U.S. Standard Atmosphere, 1962.* Sutherland's equation is

$$\mu = \frac{\beta T^{3/2}}{T + S}$$

where T is absolute temperature, S is Sutherland's constant, and β is a constant; both values are listed in Table 6.2 for the English and metric system of units. The power law is somewhat easier to use and yields good results in the temperature range of 300 to 900°R.

$$\mu = \mu_0 \left(\frac{T}{T_0} \right)^{0.76}$$

where μ_0 and T_0 are reference values for viscosity and absolute temperature, $\mu_0 = 3.02 \times 10^{-7}$ lb-sec/ft^2 and $T_0 = 392$°R.

The properties of air at standard sea level pressure (14.696 psia) are given in Table 6.3. The generally accepted method of computing viscosity (μ) is by use of Sutherland's equation. Beyond the upper temperature limit of 3400°R the properties are based on molecular dissociation having taken place.

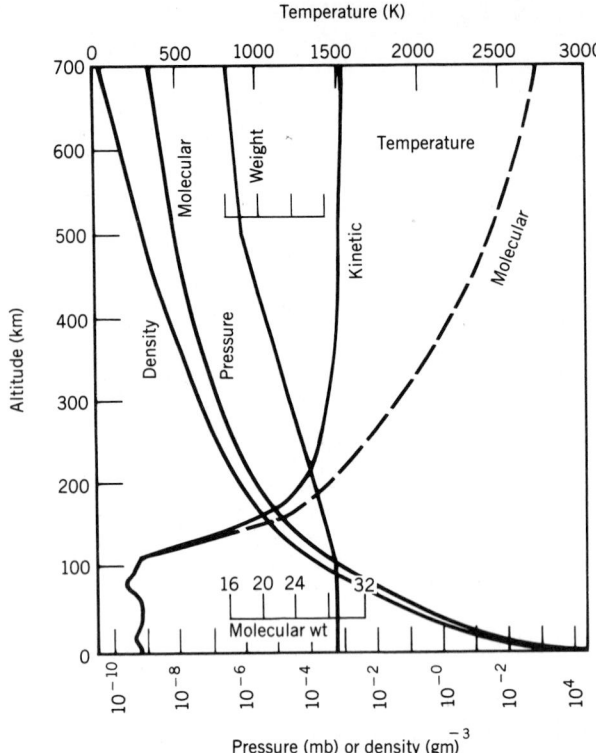

Fig. 6.3   Variation in properties in 1962 standard atmosphere.

An abbreviated listing of the various properties of the atmosphere (*U.S. Standard Atmosphere, 1962*[17]) up to 2,300,000 ft is given in Table 6.4. Above 300,000 ft (90 km) the speed of sound (c), viscosity (μ), and thermal conductivity (k) are omitted as a result of the increased mean free path of the air molecules and the lack of meaning of these conventional properties.

The *U.S. Standard Atmosphere, 1962*, represents an average of the (spring and fall) properties at approximately 45° north latitude. Atmospheric properties for various north latitudes for the different seasons of the year are contained in Ref. 19.

Those individuals and agencies requiring computer-oriented functions for computation of a standard atmospheric pressure and density are referred to Ref. 19, Part 4.

TABLE 6.2[a]   Properties of Air at Standard Atmospheric Pressure

Symbol	English Units (ft-lb-sec)	Metric Units (MKS)
P_0	2116.22 lbf-ft^{-2}	1.013250×10^5 Newtons-M^{-2}
ρ_0	0.076474 lbm-ft^{-3}	1.2250 kg-m^{-3}
t_0	59.0°F	15°C
g_0	32.1741 ft-sec^{-2}	9.80665 m-s^{-2}
S	198.72°R	110.4 K
β	7.3025×10^{-7} lbf-ft^{-1}-sec^{-1}-R$^{-1/2}$	1.458×10^{-6} kg-s^{-1}-m^{-1}-k$^{-1/2}$
T_i	491.67°R	273.15 K
N	2.73179×10^{26} (lbm-mole)$^{-1}$	6.02257×10^{26} (kg-mole)$^{-1}$
R	1545.31 lbf-ft-(lbm-mole)$^{-1}$R^{-1}	8.31432 joules-K^{-1}-mole^{-1}

[a]Subscript 0 refers to standard sea level conditions. p = pressure, ρ = density, t = temperature, g = acceleration due to gravity, S and β = constants for Sutherland's equation, T_i = ice point temperature, N = Avogadro's number, R = universal gas constant.

TABLE 6.3 Properties of Air at Standard Atmospheric Pressure[a]

Temp (°F), t	Density (lbm/ ft^3), $\rho \times 10^2$	Specific Heat (Btu/ lbm-°F), $C_p \times 10$	Viscosity (lbm/ sec-ft), $\mu \times 10^5$	Kinematic Viscosity (ft^2/sec), $\nu \times 10^3$	Thermal Conductivity (Btu/ hr-ft-°F), $K \times 10^2$	Thermal Diffusivity (ft^2/hr), a	Prandtl Number, Pr
−280	22.48	2.452	0.4653	0.020700	0.5342	0.09691	0.770
−200	15.64	2.416	0.6659	0.043856	0.7648	0.20863	0.755
−100	11.04	2.403	0.8930	0.080620	1.0450	0.39390	0.739
0	8.66	2.401	1.0926	0.10960	1.3124	0.54874	0.720
100	7.10	2.404	1.2750	0.18102	1.5647	0.92477	0.706
200	5.99	2.414	1.4413	0.24213	1.8047	1.25633	0.694
300	5.23	2.429	1.5951	0.29293	2.0320	1.53656	0.686
400	4.62	2.450	1.7390	0.36471	2.2481	1.92489	0.681
500	4.14	2.474	1.8743	0.45420	2.4570	2.40600	0.680
600	3.75	2.512	2.0027	0.53587	2.6536	2.82656	0.680
700	3.42	2.538	2.1231	0.62122	2.8431	3.27811	0.682
800	3.14	2.568	2.2390	0.71310	3.0220	3.74800	0.684
900	2.92	2.596	2.3498	0.80562	3.2003	4.22567	0.686
1000	2.71	2.628	2.4569	0.90602	3.3710	4.72967	0.690
1100	2.54	2.659	2.5600	1.0075	3.5313	5.22667	0.693
1200	2.39	2.690	2.6569	1.1117	3.6911	5.74333	0.697
1300	2.25	2.717	2.7513	1.2222	3.8428	6.28333	0.701
1400	2.13	2.742	2.8450	1.3380	3.9933	6.84667	0.703
1500	2.02	2.767	2.9367	1.4547	4.1472	7.42111	0.706
1600	1.93	2.791	3.0234	1.5722	4.2811	7.97067	0.710
1700	1.84	2.815	3.1090	1.6900	4.4100	8.51400	0.714
1800	1.76	2.840	3.1918	1.8222	4.5383	9.11844	0.718
1900	1.68	2.864	3.2736	1.9542	4.6676	9.72622	0.722
2000	1.61	2.887	3.3513	2.0853	4.8003	10.34633	0.725
2100	1.55	2.909	3.4280	2.2171	4.9284	10.9600	0.728
2200	1.49	2.930	3.5030	2.3499	5.0496	11.5600	0.732
2300	1.44	2.953	3.5780	2.4880	5.1813	12.1600	0.736
2400	1.39	2.977	3.6530	2.6297	5.3202	12.7600	0.740
2500	1.34	3.002	3.7231	2.7731	5.4378	13.3778	0.745
2600	1.30	3.028	3.7920	2.9170	5.5500	14.0000	0.749
2700	1.26	3.054	3.8603	3.0648	5.6611	14.6056	0.755
2800	1.22	3.081	3.9277	3.2150	5.7678	15.2356	0.760
2900	1.18	3.110	3.9910	3.3750	5.8567	15.9633	0.764
3000	1.15	3.143	4.0599	3.5414	5.9567	16.6244	0.771
3100	1.11	3.181	4.1371	3.7176	6.0733	17.1856	0.779
3200	1.08	3.223	4.2123	3.8917	6.1867	17.7267	0.790
3300	1.06	3.269	4.2862	4.0644	6.2978	18.2544	0.801
3400	1.03	3.328	4.3441	4.2319	6.3822	18.6667	0.815
3500	1.00	3.390	4.3980	4.3980	6.4600	19.0500	0.831
3600	0.98	3.474	4.4619	4.5663	6.5544	19.3611	0.849
3700	0.95	3.565	4.5239	4.7419	6.6500	19.6444	0.869
3800	0.93	3.686	4.5783	4.9463	6.7500	19.8167	0.898
3900	0.90	3.795	4.6295	5.1211	6.8473	19.9587	0.923
4000	0.89	3.886	4.6759	5.2515	6.9407	20.0553	0.942
4100	0.88	3.977	4.7222	5.3818	7.0340	20.1520	0.961

[a]Symbols: t = temperature; ρ = density; C_p = specific heat at constant pressure; μ = dynamic viscosity; ν = kinematic viscosity; K = thermal conductivity; a = thermal diffusivity ($a = K/\rho C_p$); Pr = Prandtl number ($Pr = \nu/a$). Note: To determine property value, use equation at top of column and set equal to tabulated value. Example: Viscosity at 1000°F, $\mu \times 10^5$ = 2.4569. Therefore μ = 2.4569 × 10^{-5} lbm/sec-ft. See Ref. 18.

TABLE 6.4 Properties of the U.S. Standard Atmosphere, 1962[a]

Geometric Altitude, z (ft)	Temperature, t (°F)	Temperature, t (°C)	Speed of Sound (ft/sec)	Pressure, P (lbf/ft²)	Density, ρ (lbm/ft³)	Coefficient of Viscosity, μ (lbm/ft·sec)	Kinematic Viscosity, ν (ft²/sec)	Thermal Conductivity, K (Btu/sec·ft·°F)	Acceleration Due to Gravity, g (ft/sec²)
0	59.0	15.0	1116.5	2.112 + 3	7.647 − 2	1.202 − 5	1.572 − 4	4.067 − 6	32.174
1,000	55.4	13.0	1112.6	2.041 + 3	7.426 − 2	1.196 − 5	1.611 − 4	4.042 − 6	32.171
2,000	51.9	11.0	1108.8	1.968 + 3	7.209 − 2	1.190 − 5	1.650 − 4	4.017 − 6	32.168
3,000	48.3	9.1	1104.9	1.897 + 3	6.998 − 2	1.183 − 5	1.691 − 4	3.992 − 6	32.165
4,000	44.7	7.1	1101.0	1.828 + 3	6.792 − 2	1.177 − 5	1.732 − 4	3.967 − 6	32.162
5,000	41.2	5.1	1097.1	1.761 + 3	6.590 − 2	1.170 − 5	1.776 − 4	3.942 − 6	32.159
6,000	37.6	3.1	1093.2	1.696 + 3	6.393 − 2	1.164 − 5	1.820 − 4	3.917 − 6	32.156
7,000	34.0	1.1	1089.3	1.633 + 3	6.200 − 2	1.157 − 5	1.866 − 4	3.891 − 6	32.152
8,000	30.5	−0.8	1085.3	1.572 + 3	6.012 − 2	1.150 − 5	1.914 − 4	3.866 − 6	32.149
9,000	26.9	−2.8	1081.4	1.513 + 3	5.828 − 2	1.144 − 5	1.963 − 4	3.840 − 6	32.146
10,000	23.4	−4.8	1077.4	1.456 + 3	5.648 − 2	1.137 − 5	2.013 − 4	3.815 − 6	32.143
11,000	19.8	−6.8	1073.4	1.400 + 3	5.473 − 2	1.131 − 5	2.066 − 4	3.789 − 6	32.140
12,000	16.2	−8.8	1069.4	1.346 + 3	5.302 − 2	1.124 − 5	2.120 − 4	3.764 − 6	32.137
13,000	12.7	−10.7	1065.4	1.294 + 3	5.135 − 2	1.117 − 5	2.175 − 4	3.738 − 6	32.134
14,000	9.1	−12.7	1061.4	1.244 + 3	4.973 − 2	1.110 − 5	2.233 − 4	3.713 − 6	32.131
15,000	5.5	−14.7	1057.4	1.195 + 3	4.814 − 2	1.104 − 5	2.293 − 4	3.687 − 6	32.128
16,000	2.0	−16.7	1053.3	1.148 + 3	4.659 − 2	1.097 − 5	2.354 − 4	3.661 − 6	32.125
17,000	−1.6	−18.7	1049.2	1.102 + 3	4.508 − 2	1.090 − 5	2.418 − 4	3.636 − 6	32.122
18,000	−5.1	−20.6	1045.2	1.058 + 3	4.361 − 2	1.083 − 5	2.484 − 4	3.610 − 6	32.119
19,000	−8.7	−22.6	1041.1	1.015 + 3	4.217 − 2	1.076 − 5	2.553 − 4	3.584 − 6	32.115
20,000	−12.3	−24.6	1036.9	9.733 + 2	4.077 − 2	1.070 − 5	2.623 − 4	3.558 − 6	32.112
21,000	−15.8	−26.6	1032.8	9.333 + 2	3.941 − 2	1.063 − 5	2.697 − 4	3.532 − 6	32.109
22,000	−19.4	−28.5	1028.7	8.946 + 2	3.808 − 2	1.056 − 5	2.772 − 4	3.506 − 6	32.106
23,000	−22.9	−30.5	1024.5	8.573 + 2	3.679 − 2	1.049 − 5	2.851 − 4	3.480 − 6	32.103
24,000	−26.5	−32.5	1020.3	8.212 + 2	3.553 − 2	1.042 − 5	2.932 − 4	3.454 − 6	32.100
25,000	−30.0	−34.5	1016.1	7.863	3.431	1.035	3.017	3.428	32.097
26,000	−33.6	−36.5	1011.9	7.527	3.311	1.028	3.104	3.401	32.094
27,000	−37.2	−38.4	1007.7	7.203	3.195	1.021	3.195	3.375	32.091
28,000	−40.7	−40.4	1003.4	6.890	3.082	1.014 − 5	3.289	3.349	32.088
29,000	−44.3	−42.4	999.1	6.588	2.974	1.007 − 5	3.387	3.322	32.085
30,000	−47.8	−44.4	994.9	6.297	2.866	9.996 − 6	3.488	3.296	32.082
31,000	−51.4	−46.3	990.6	6.016	2.762	9.925	3.594	3.270	32.079
32,000	−54.9	−48.3	986.2	5.746	2.661	9.853	3.703	3.243	32.076
33,000	−58.5	−50.3	981.9	5.486	2.563	9.781	3.817	3.217	32.072
34,000	−62.1	−52.3	977.5	5.235	2.468	9.709	3.935	3.190	32.069

Altitude (ft)									
35,000	−65.6	−54.2	973.1	4.994	2.375	9.637	4.057	3.163	32.066
36,000	−69.2	−56.2	968.7	4.761	2.285	9.564	4.185	3.137	32.063
37,000	−69.7	−56.5	968.1	4.539	2.181	9.553 −6	4.379	3.133	32.060
38,000	−69.7	−56.5	968.1	4.326	2.079	9.553	4.550	3.133	32.057
39,000	−69.7	−56.5	968.1	4.124	1.982	9.553	4.819	3.133	32.054
40,000	−69.7	−56.5	868.1	3.931	1.889	9.553	5.056	3.133	32.051
41,000	−69.7	−56.5	968.1	3.748	1.801	9.553	5.304	3.133	32.048
42,000	−69.7	−56.5	968.1	3.572 +2	1.717 −2	9.553	5.564 −4	3.133	32.045
43,000	−69.7	−56.5	968.1	3.405	1.637	9.553	5.837	3.133	32.042
44,000	−69.7	−56.5	968.1	3.246	1.560	9.553	6.123	3.133	32.039
45,000	−69.7	−56.5	968.1	3.095	1.487	9.553	6.423	3.133	32.036
46,000	−69.7	−56.5	968.1	2.950	1.418	9.553	6.738	3.133	32.033
47,000	−69.7	−56.5	968.1	2.812	1.352	9.553	7.068	3.133	32.030
48,000	−69.7	−56.5	968.1	2.681	1.288	9.553	7.414	3.133	32.026
49,000	−69.7	−56.5	968.1	2.555	1.228	9.553	7.778	3.133	32.023
50,000	−69.7	−56.5	968.1	2.436	1.171 −2	9.553	8.159 −4	3.133	32.020
55,000	−69.7	−56.5	968.1	1.918	9.219 −3	9.553	1.036 −3	3.133	32.005
60,000	−69.7	−56.5	968.1	1.510	7.259	9.553	1.316	3.133	31.990
65,000	−67.4	−55.2	970.9	1.189 +2	5.716 −3	9.553 −6	1.671 −3	3.133	31.974
70,000	−64.7	−53.7	974.2	9.373 +1	4.479 −3	9.599 −6	2.143 −3	3.150 −6	31.959
75,000	−62.0	−52.2	977.6	7.399	3.511	9.655	2.750	3.170	31.944
80,000	−59.3	−50.7	980.9	5.851	2.758	9.711	3.521	3.191	31.929
85,000	−56.5	−49.2	984.3	4.635	2.170	9.766	4.501	3.211	31.913
90,000	−53.8	−47.7	987.6	3.678	1.710	9.821	5.743	3.231	31.898
95,000	−51.1	−46.2	990.9	2.923	1.350	9.876	7.316	3.251	31.883
100,000	−46.2	−44.2	995.9	2.327	1.068 −3	9.931 −6	9.302 −3	3.272	31.868
110,000	−41.3	−40.7	1002.7	1.484 +1	6.647 −4	1.013 −5	1.524 −2	3.345	31.837
120,000	−26.1	−32.3	1020.8	9.601 +0	4.151	1.043	2.512	3.457	31.807
130,000	−10.9	−23.8	1038.5	6.310	2.635	1.072	4.069	3.568	31.777
140,000	4.3	−15.4	1055.9	4.206	1.699	1.101	6.480 −2	4.678	31.746
150,000	19.4	−7.0	1073.0	2.842	1.112	1.130	1.016 −1	3.787	31.716
160,000	27.5	−2.5	1082.0	1.942	7.471 −4	1.145	1.532	3.845	31.686
170,000	27.5	−2.5	1082.0	1.330 +0	5.116 −5	1.145	2.238	3.845	31.656
180,000	18.9	−7.3	1072.4	9.105 −1	3.558	1.129	3.173	3.783	31.626
190,000	8.1	−13.3	1060.3	6.155	2.466	1.109	4.495	3.705	31.596
200,000	−2.7	−19.3	1048.0	4.135	1.696	1.088	6.416	3.628	31.566
210,000	−22.0	−30.0	1025.6	2.744	1.175 −5	1.051	8.944 −1	3.487	31.536
220,000	−43.5	−41.9	1000.1	1.784	8.036 −6	1.008 −5	1.255 0	3.328	31.506
230,000	−64.9	−53.9	974.0	1.135 −1	5.389	9.650 −6	1.791	3.168	31.476
240,000	−86.4	−65.8	947.1	7.041 −2	3.535	9.207	2.604	3.007	31.446
250,000	−107.8	−77.7	919.5	4.248	2.263	8.753	3.868	2.843	31.42
260,000	−129.3	−89.6	891.1	2.484	1.409 −6	8.289	5.881 0	2.678	31.39

TABLE 6.4 *(Continued)*

Geometric Altitude, z (ft)	Temperature, t (°F)	Temperature, t (°C)	Speed of Sound (ft/sec)	Pressure, P (lbf/ft²)	Density, ρ (lbm/ft³)	Coefficient of Viscosity, μ (lbm/ft-sec)	Kinematic Viscosity, ν (ft²/sec)	Thermal Conductivity, K (Btu/sec-ft-°F)	Acceleration Due to Gravity, g (ft/sec²)
270,000	−134.5	−92.5	884.0	1.418 − 2	8.172 − 2	8.173 − 7	1.000 + 1	2.638	31.36
280,000	−134.5	−92.5	884.0	8.086 − 3	4.661	8.173	1.754	2.638	31.33
290,000	−134.5	−92.5	884.0	4.615	2.660	8.173	3.073	2.638	31.30
300,000	−126.8	−88.2	894.5	2.643 − 3	1.488 − 7	8.343 − 6	5.608 − 1	2.698 − 6	31.27
350,000	−24.5	−31.4		2.372 − 4	1.012 − 8				31.12
400,000	233.9	112.2		4.459 − 5	1.164 − 9				30.97
450,000	734.1	390.1		1.757 − 5	2.600 − 10				30.83
500,000	1203.8	651.0		9.760 − 6	1.020 − 10				30.68
550,000	1491.7	810.9		6.217	5.453 − 10				30.54
600,000	1647.2	897.3		4.185	3.350				30.40
650,000	1754.9	957.2		2.909	2.177				30.26
700,000	1835.7	1002.1		2.067	1.467				30.12
750,000	1912.5	1044.8		1.496	1.009 − 11				29.98
800,000	1964.3	1073.5		1.100 − 6	7.135 − 12				29.84
850,000	2010.6	1099.2		8.197 − 7	5.121				29.70
900,000	2053.4	1123.0		6.180	3.724				29.57
950,000	2092.9	1144.9		4.710	2.742				29.43
1,000,000	2124.6	1162.5		3.627	2.046				29.30
1,100,000	2160.3	1182.4		2.205	1.179 − 12				29.03
1,200,000	2189.3	1198.5		1.380 − 7	7.020 − 13				28.77
1,300,000	2214.6	1212.5		8.874 − 8	4.301				28.51
1,400,000	2217.2	1214.0		5.834	2.724				28.25
1,500,000	2221.2	1212.2		3.907	1.761				28.00
1,600,000	2232.1	1222.3		2.661	1.160 − 13				27.75
1,700,000	2233.7	1223.1		1.840	7.814 − 14				27.50
1,800,000	2232.9	1222.7		1.286 − 8	5.351				27.26
1,900,000	2241.4	1227.4		9.089 − 9	3.704 − 14				27.02
2,000,000	2250.8	1232.7		6.486	2.596				26.79
2,100,000	2251.7	1233.2		4.664	1.843				26.55
2,200,000	2253.9	1234.4		3.378	1.318 − 14				26.32
2,300,000	2254.0	1234.4		2.463 − 9	9.491 − 15				26.10

[a]The digit preceded by plus or minus indicates the power of 10 by which the tabulated values should be multiplied. See also Ref. 17.

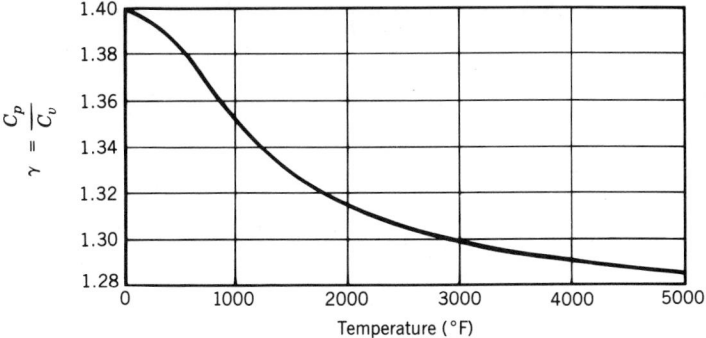

Fig. 6.4 Specific heat ratio of pure dry air.

Figure 6.4 presents the ratio γ of the specific heat at constant pressure to that at constant volume for air, obtained from measurements by Heck.[20]

At lower altitudes the air is dense enough so that it may be considered a continuum under practically all flight conditions. However, at higher altitudes the air may be so rare that it may no longer be considered a continuum as in ordinary gas dynamics but rather an aggregate of rapidly moving molecules. When the mean free path λ of the molecules is small but not negligible compared to the dimensions of a body moving through the air, the phenomenon of slip flow appears; that is, the gas no longer adheres to the body surface but actually slips over the surface with a finite velocity. Furthermore, if the mean free path is much larger than the body dimensions, that is, $K_N > 10$, the flow becomes free-molecule, in which event the body is sprayed, as it were, with pellets that do not interfere with one another after they strike the body. Figures 6.5–6.7 present the mean particle speed $\bar{v}$, collision frequency, and mean free path λ for air.

6.4   STEADY ONE-DIMENSIONAL GAS DYNAMICS
D. H. Daley with contributions by J. B. Wissler

6.4.1   Generalized One-Dimensional Gas Dynamics

The steady one-dimensional flow of a chemically inert perfect gas with constant specific heats is conveniently described and governed by the following definitions and physical laws.[6]

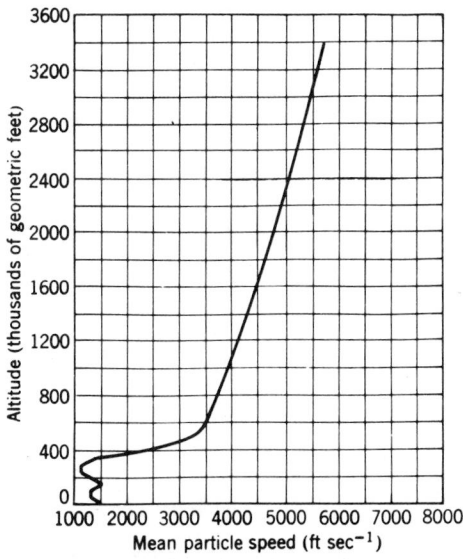

Fig. 6.5 Particle speed versus geopotential altitude.

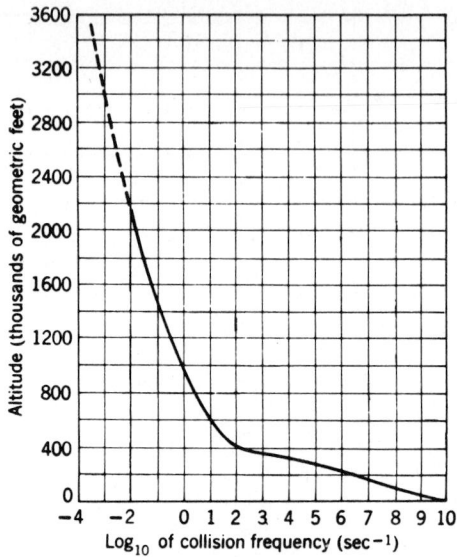

Fig. 6.6   Collision frequency versus geopotential altitude.

Definitions

Perfect Gas:	$p = \rho R T$	(6.32)
Mach Number:	$M = \dfrac{u}{a}$	(6.33)
Stagnation Temperature:	$T_0 = T\left(1 + \dfrac{\gamma - 1}{2} M^2\right)$	(6.34)
Stagnation Pressure:	$p_0 = p\left(1 + \dfrac{\gamma - 1}{2} M^2\right)^{(\gamma/\gamma - 1)}$	(6.35)

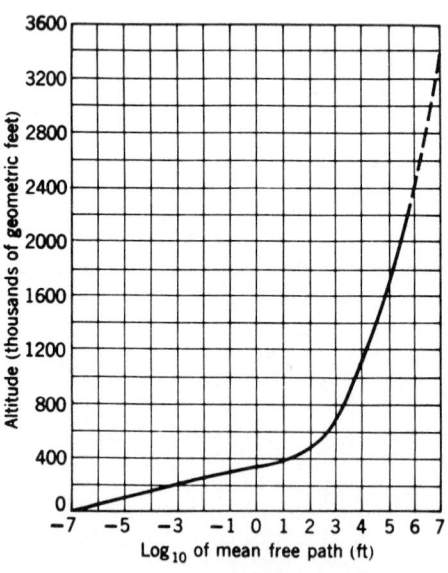

Fig. 6.7   Mean free path versus geopotential altitude.

where p = pressure, ρ = density, R = gas constant, T = temperature, M = Mach number, u = velocity, a = speed of sound, T_0 = stagnation temperature, γ = ratio of specific heat at constant pressure to specific heat at constant volume, and p_0 = stagnation pressure. It is conventional to denote those stream properties at the point in the flow where $M = 1$ by p^*, u^*, and so on.

Physical Laws

For one-dimensional flow through a control volume having the single inlet and exit flow Sections 1 and 2, respectively, we have

Continuity Equation:	$\rho_1 A_1 u_1 = \rho_2 A_2 u_2$	(6.36)
Momentum Equation:	$F_{\text{frict}} = \left(pA + \rho Au^2 \right)_1 - \left(pA + \rho Au^2 \right)_2$	(6.37)
Energy Equation:	$q = c_p \left(T_{02} - T_{01} \right)$ (shaft work = 0)	(6.38)
Entropy Equation:	$s_2 \geq s_1$ (adiabatic flow)	(6.39)

where F_{frict} = frictional force of a solid control surface boundary on the flowing gas, A = the flow cross-sectional area normal to u, q = the heat flow per unit mass flow, c_p = specific heat at constant pressure, and s = entropy per unit mass flow.

The application of (6.32)–(6.37) to flow in the presence of the simultaneous effects of area-change, heating, and friction (Fig. 6.8) results in the following set of equations:

Perfect Gas:	$\dfrac{dp}{p} - \dfrac{d\rho}{\rho} - \dfrac{dT}{T} = 0$	(6.40)
Stagnation Temperature:	$\dfrac{dT}{T} + \dfrac{\dfrac{\gamma - 1}{2} M^2}{1 + \dfrac{\gamma - 1}{2} M^2} \dfrac{dM^2}{M^2} = \dfrac{dT_0}{T_0}$	(6.41)
Continuity:	$\dfrac{d\rho}{\rho} + \dfrac{dA}{A} + \dfrac{du}{u} = 0$	(6.42)
Stagnation Pressure:	$\dfrac{dp}{p} + \dfrac{\dfrac{\gamma M^2}{2}}{1 + \dfrac{\gamma - 1}{2} M^2} \dfrac{dM^2}{M^2} = \dfrac{dp_0}{p_0}$	(6.43)
Momentum:	$\dfrac{dp}{p} + \gamma M^2 \dfrac{du}{u} + \dfrac{\gamma M^2}{2} \dfrac{4f\,dx}{D} = 0$	(6.44)
Mach Number:	$2\dfrac{du}{u} - \dfrac{dT}{T} = \dfrac{dM^2}{M^2}$	(6.45)

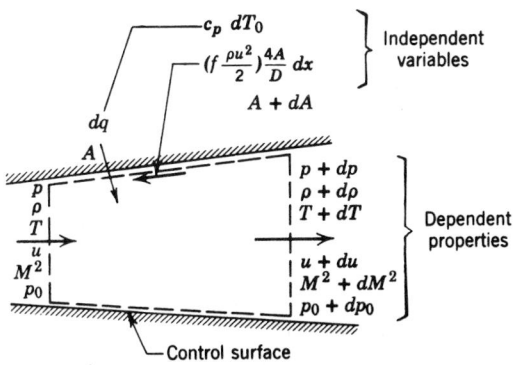

Fig. 6.8 Independent and dependent variables for generalized one-dimensional flow, where f = friction coefficient, D = hydraulic diameter.

In these equations heat effects are measured in terms of the stagnation temperature change according to Eq. (6.38). The entropy condition of Eq. (6.39) is also applicable if $dT_0 = 0$. If $dT_0 \neq 0$, then the entropy requirement is $ds \geq (dq/T)$.

The six dependent variables M^2, u, p, ρ, T, and p_0 in this set of six linear algebraic equations may be expressed in terms of the three independent variables, dA/A, dT_0/T_0, and $4f\,dx/D$. The solution of this set of equations is given in Table 6.5.

General conclusions can be made relative to the variation of the stream properties of the flow with each of the independent variables by the relations of Table 6.5.[6,9] An example of the relation given for (du/u) at the bottom of the table indicates that, in a constant area adiabatic flow, friction will increase the stream velocity in subsonic flow and will decrease the velocity in supersonic flow. Similar reasoning may be applied to determine the manner in which any dependent property varies with a single independent variable.

6.4.2 Simple Flows

A simple flow is defined as one in which all but one of the independent variables in Table 6.5 are zero. Three types of simple flows are summarized in Table 6.6 by presenting for each simple flow (i) the independent effects present, (ii) a schematic of the flow situation, (iii) the locus on a temperature-entropy diagram of the possible states attained for each flow, and (iv) useful functions obtained by integration of the relations of Table 6.5 or the basic Eqs. (6.32)–(6.37). These useful functions are tabulated in Ref. 21.

In the temperature-entropy diagrams of Table 6.6 the path lines of states corresponding to simple area flow, simple heating flow, and simple friction flow, respectively, are shown. These path lines are called the isentrope, Rayleigh, and Fanno lines, respectively.

TABLE 6.5   Influence Coefficients for Steady One-Dimensional Flow

	Independent		
Dependent	$\dfrac{dA}{A}$	$\dfrac{dT_0}{T_0}$	$\dfrac{4f\,dx}{D}$
$\dfrac{dM^2}{M^2}$	$-\dfrac{2\left(1 + \dfrac{\gamma-1}{2}M^2\right)}{1 - M^2}$	$\dfrac{(1+\gamma M^2)\left(1 + \dfrac{\gamma-1}{2}M^2\right)}{1 - M^2}$	$\dfrac{\gamma M^2\left(1 + \dfrac{\gamma-1}{2}M^2\right)}{1 - M^2}$
$\dfrac{du}{u}$	$-\dfrac{1}{1 - M^2}$	$\dfrac{1 + \dfrac{\gamma-1}{2}M^2}{1 - M^2}$	$\dfrac{\gamma M^2}{2(1 - M^2)}$
$\dfrac{dp}{p}$	$\dfrac{\gamma M^2}{1 - M^2}$	$\dfrac{-\gamma M^2\left(1 + \dfrac{\gamma-1}{2}M^2\right)}{1 - M^2}$	$\dfrac{-\gamma M^2\left[1 + (\gamma-1)M^2\right]}{2(1 - M^2)}$
$\dfrac{d\rho}{\rho}$	$\dfrac{M^2}{1 - M^2}$	$-\dfrac{\left(1 + \dfrac{\gamma-1}{2}M^2\right)}{1 - M^2}$	$\dfrac{-\gamma M^2}{2(1 - M^2)}$
$\dfrac{dT}{T}$	$\dfrac{(\gamma-1)M^2}{1 - M^2}$	$\dfrac{(1 - \gamma M^2)\left(1 + \dfrac{\gamma-1}{2}M^2\right)}{1 - M^2}$	$\dfrac{-\gamma(\gamma-1)M^4}{2(1 - M^2)}$
$\dfrac{dp_0}{p_0}$	0	$-\dfrac{\gamma M^2}{2}$	$-\dfrac{\gamma M^2}{2}$

Table is read:

$$\frac{du}{u} = \left[-\frac{1}{1 - M^2}\right]\frac{dA}{A} + \left[\frac{1 + \dfrac{\gamma-1}{2}M^2}{1 - M^2}\right]\frac{dT_0}{T_0} + \left[\frac{\gamma M^2}{2(1 - M^2)}\right]\frac{4f\,dx}{D}$$

TABLE 6.6 Simple Flows

TYPE OF FLOW		ISENTROPIC	RAYLEIGH	FANNO
		SIMPLE AREA	SIMPLE HEATING	SIMPLE FRICTION
EFFECTS	AREA	PRESENT	0	0
	HEATING	0	PRESENT	0
	FRICTION	0	0	PRESENT
SCHEMATIC OF FLOW SITUATION				

TABLE 6.6   (Continued)

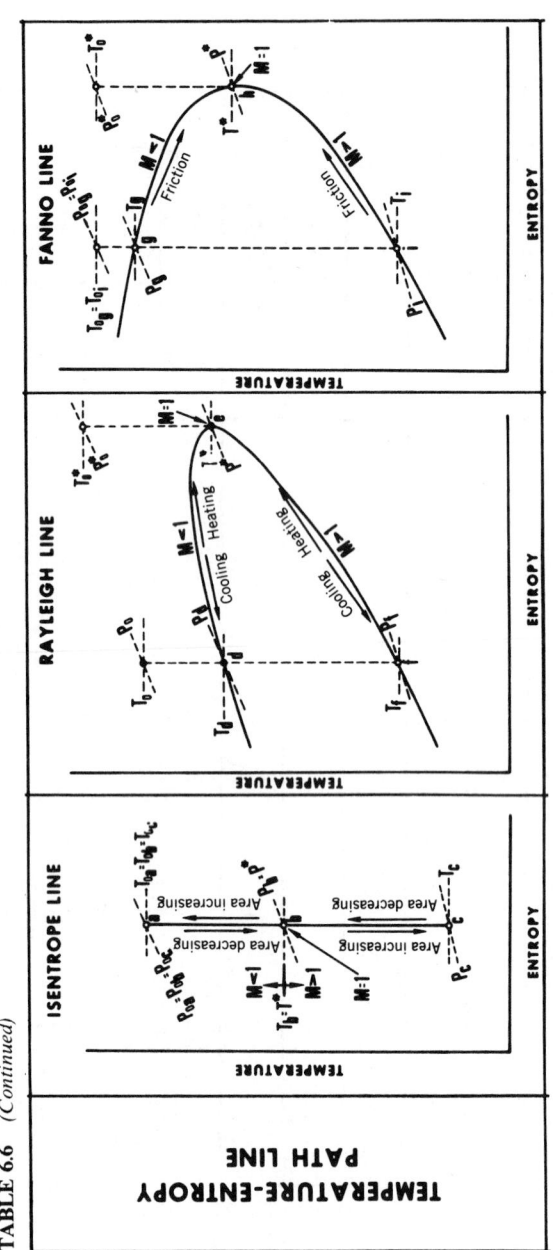

TABLE 6.6 (Continued)

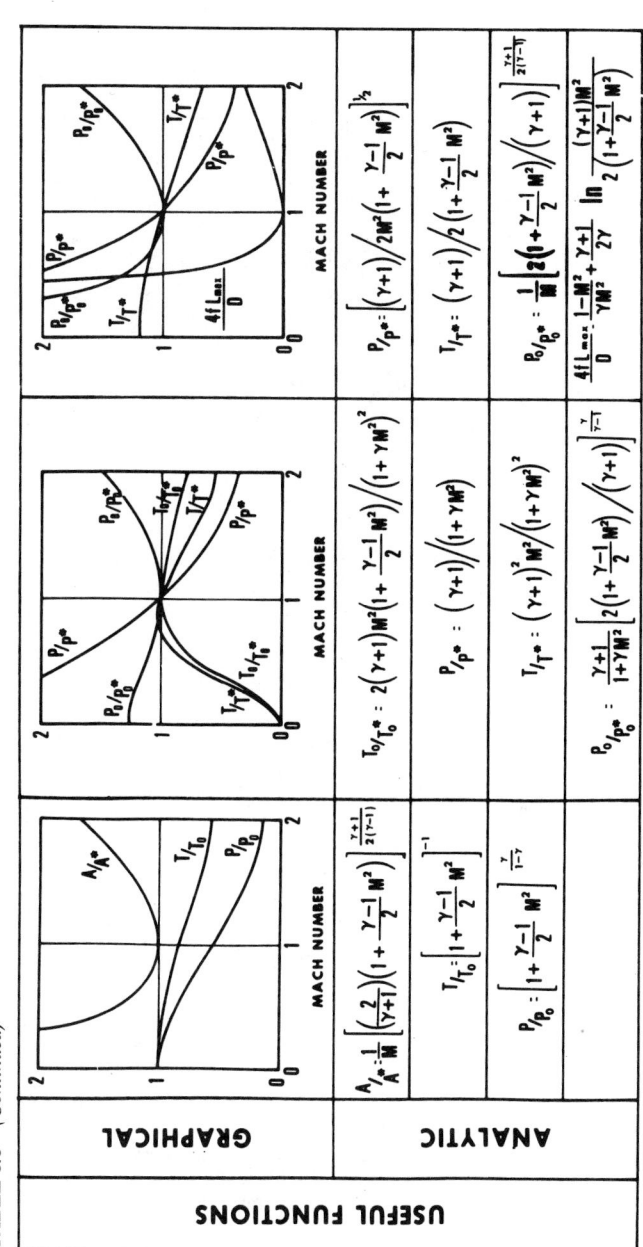

USEFUL FUNCTIONS			
GRAPHICAL	(graph: A/A^*, T/T_0, p/p_0 vs MACH NUMBER)	(graph: p_0/p_0^*, T_0/T_0^*, T/T^*, p/p^*, ρ_0/ρ_0^*, ρ/ρ^* vs MACH NUMBER)	(graph: p_0/p_0^*, T/T^*, p/p^*, ρ_0/ρ_0^*, ρ/ρ^*, $\dfrac{4f L_{max}}{D}$ vs MACH NUMBER)
ANALYTIC	$\dfrac{A}{A^*}=\dfrac{1}{M}\left[\left(\dfrac{2}{\gamma+1}\right)\left(1+\dfrac{\gamma-1}{2}M^2\right)\right]^{\frac{\gamma+1}{2(\gamma-1)}}$	$\dfrac{T_0}{T_0^*}=2(\gamma+1)M^2\left(1+\dfrac{\gamma-1}{2}M^2\right)\Big/(1+\gamma M^2)^2$	$\dfrac{p}{p^*}=\left[(\gamma+1)\Big/2M^2\left(1+\dfrac{\gamma-1}{2}M^2\right)\right]^{\frac{1}{2}}$
	$\dfrac{T}{T_0}=\left[1+\dfrac{\gamma-1}{2}M^2\right]^{-1}$	$\dfrac{p}{p^*}=(\gamma+1)\big/(1+\gamma M^2)$	$\dfrac{T}{T^*}=(\gamma+1)\Big/2\left(1+\dfrac{\gamma-1}{2}M^2\right)$
	$\dfrac{p}{p_0}=\left[1+\dfrac{\gamma-1}{2}M^2\right]^{\frac{\gamma}{1-\gamma}}$	$\dfrac{T}{T^*}=(\gamma+1)^2M^2\big/(1+\gamma M^2)^2$	$\dfrac{p_0}{p_0^*}=\dfrac{1}{M}\left[2\left(1+\dfrac{\gamma-1}{2}M^2\right)\Big/(\gamma+1)\right]^{\frac{\gamma+1}{2(\gamma-1)}}$
		$\dfrac{p_0}{p_0^*}=\dfrac{\gamma+1}{1+\gamma M^2}\left[2\left(1+\dfrac{\gamma-1}{2}M^2\right)\Big/(\gamma+1)\right]^{\frac{\gamma}{\gamma-1}}$	$\dfrac{4f L_{max}}{D}=\dfrac{1-M^2}{\gamma M^2}+\dfrac{\gamma+1}{2\gamma}\ln\dfrac{(\gamma+1)M^2}{2\left(1+\dfrac{\gamma-1}{2}M^2\right)}$

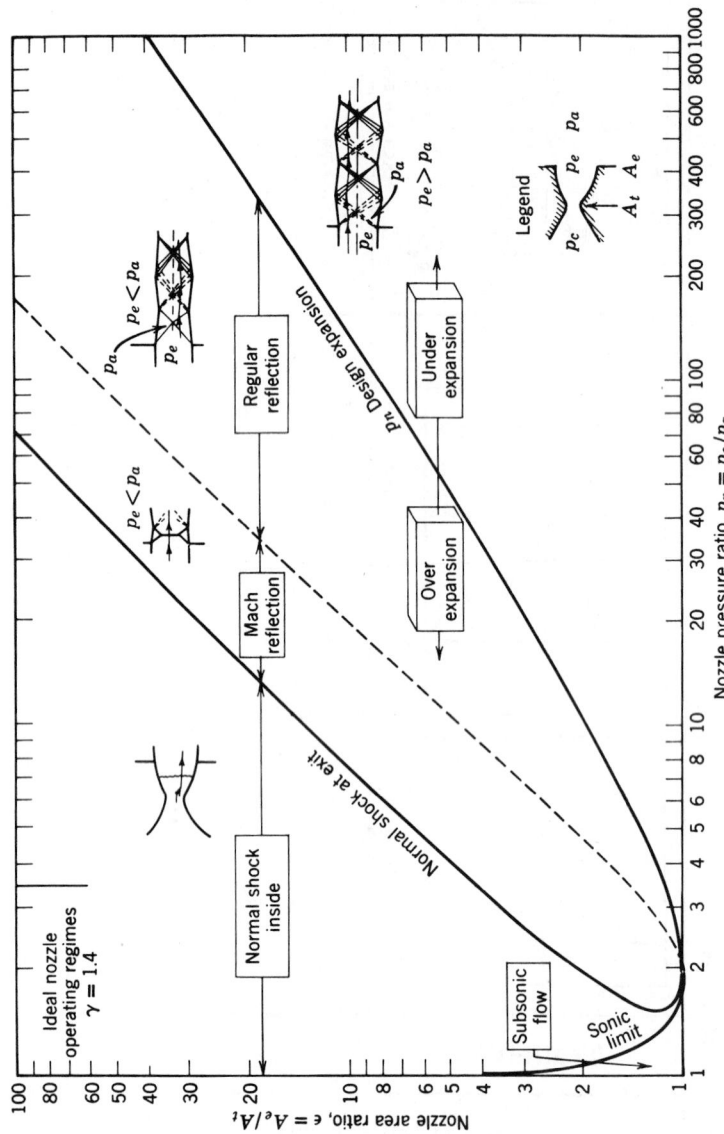

Fig. 6.9 Nozzle operating diagram (from Ref. 22).

To proceed downward along the *isentrope line* from point a of the diagram, the flow area is decreased until the sonic point at b is reached. The area must be increased after point b in order to continue down to point c. By proper adjustments in the flow area and boundary pressure the flow may be made to proceed through point b in either direction along the isentrope line. Point a represents the isentropic stagnation condition for all points on the isentrope ac.

The *Rayleigh line* shows the series of possible states in a steady, frictionless, constant area flow. Motion along the Rayleigh line is caused by changes in the stagnation temperature produced by heating effects that, in turn, produce entropy changes in the manner indicated on the line. Heating in an initially subsonic flow (point d) causes the flow Mach number to approach one (point e). Neither heating nor cooling alone can continuously alter the flow from subsonic to supersonic, or from supersonic to subsonic speeds.

The *Fanno line* represents the possible series of states in a steady, constant area, constant stagnation temperature flow. Frictional effects alone produce motion along the Fanno line. Consequently the flow progression along the line must always be one of increasing entropy toward the sonic point h. The flow is subsonic on the Fanno line above h and supersonic below. Since the entropy decreases along the Fanno line from point h, it is impossible in simple friction flow to proceed by continuous changes through sonic conditions at point h.

6.4.3   Nozzle Operating Characteristics

The operating characteristics of a nozzle are governed by the ratio of its exit area to throat area (A_e/A_t) and by the ratio of its reservoir chamber pressure to exhaust region ambient pressure (p_c/p_a). The operating regimes of a nozzle for isentropic (except for shock waves), steady one-dimensional flow of the perfect gas air are depicted in Fig. 6.9 on a diagram with nozzle pressure ratio and nozzle area ratio as the coordinate axes.

The curve in Fig. 6.9 with the two branches labeled "design expansion" and "sonic limit" is obtained from the simple area isentropic functions of Table 6.6 by plotting A/A^* versus the reciprocal of p/p_0. The design expansion line corresponds to sets of values of nozzle pressure ratio and nozzle area ratio for which the flow is shock free with the nozzle exit section Mach number supersonic and with the exit pressure p_e equal to the exhaust region ambient pressure p_a. For any given area ratio the sonic limit line locates the nozzle pressure ratio, which will produce sonic conditions at the throat and subsonic flow elsewhere. Alternatively, the sonic limit branch corresponds to the minimum nozzle pressure ratio that will give the maximum flow through a given area ratio nozzle with specified throat area and reservoir conditions. The flow is subsonic in a nozzle that has its operating point located below the sonic limit line. In the region to the right of the design expansion line $p_e > p_a$, the nozzle is underexpanded, and the flow expands to ambient conditions in the exhaust jet.

The area between the sonic limit and the design expansion curves is divided into two regions wherein shock waves occur inside the nozzle or in the jet exhaust. The common boundary of these two regions, labeled "normal shock at exit," is the locus of pressure ratios, which will produce a normal shock at the nozzle exit. Between the shock exit line and the sonic limit line, normal shock waves occur inside the nozzle. Between the shock exit and the design expansion lines oblique shocks, as shown, occur in the jet exhaust.

6.4.4   Normal Shock Waves

A discontinuous change in stream properties can be sustained in a one-dimensional flow. When this phenomenon occurs normal to the flow direction, it is called a normal shock wave (Fig. 6.10). The process in this case is irreversible adiabatic and, therefore, with increasing entropy. The velocity of the flow entering the shock is supersonic, and that leaving is subsonic.

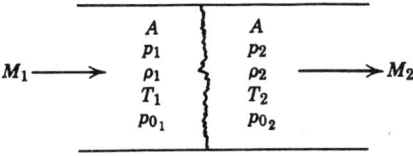

Fig. 6.10   Normal shock.

Various important relations are

$$\frac{p_2}{p_1} = \left(\frac{2\gamma}{\gamma+1}\right)M_1^2 - \left(\frac{\gamma-1}{\gamma+1}\right) \tag{6.46}$$

$$\frac{T_2}{T_1} = \frac{\left(\dfrac{2\gamma}{\gamma-1}M_1^2 - 1\right)\left(1 + \dfrac{\gamma-1}{2}M_1^2\right)}{\dfrac{(\gamma+1)^2}{2(\gamma-1)}M_1^2} \tag{6.47}$$

$$\frac{\rho_2}{\rho_1} = \frac{u_1}{u_2} = \left[\frac{\gamma-1}{\gamma+1} + \frac{2}{(\gamma+1)M_1^2}\right]^{-1} \tag{6.48}$$

$$\frac{p_{0_2}}{p_{0_1}} = \frac{A_1^*}{A_2^*} = \left[\frac{\dfrac{\gamma+1}{2}M_1^2}{1 + \dfrac{\gamma-1}{2}M_1^2}\right]^{(\gamma/\gamma-1)} \left[\frac{2\gamma}{\gamma+1}M_1^2 - \frac{\gamma-1}{\gamma+1}\right]^{(1/1-\gamma)} \tag{6.49}$$

$$M_2 = \left[\frac{M_1^2 + \dfrac{2}{\gamma-1}}{\dfrac{2\gamma}{\gamma-1}M_1^2 - 1}\right]^{1/2} \tag{6.50}$$

and the Rayleigh pitot formula,

$$\frac{p_{0_2}}{p_1} = \left[\frac{\gamma+1}{2}M_1^2\right]^{(\gamma/\gamma-1)} \left[\frac{2\gamma}{\gamma+1}M_1^2 - \frac{\gamma-1}{\gamma+1}\right]^{(1/1-\gamma)} \tag{6.51}$$

Numerical values for Eqs. (6.48)–(6.51) are given in Refs. 21 and 23–25. Equations (6.48)–(6.50) are plotted in Fig. 6.11 and Eq. (6.51) is shown on Fig. 6.12.

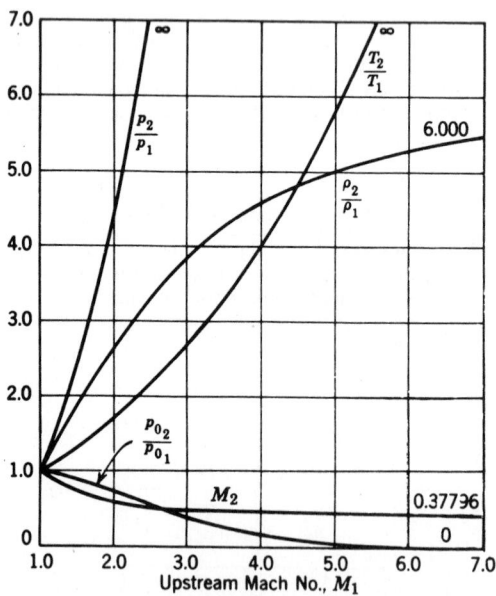

Fig. 6.11   Property variations across a normal shock.

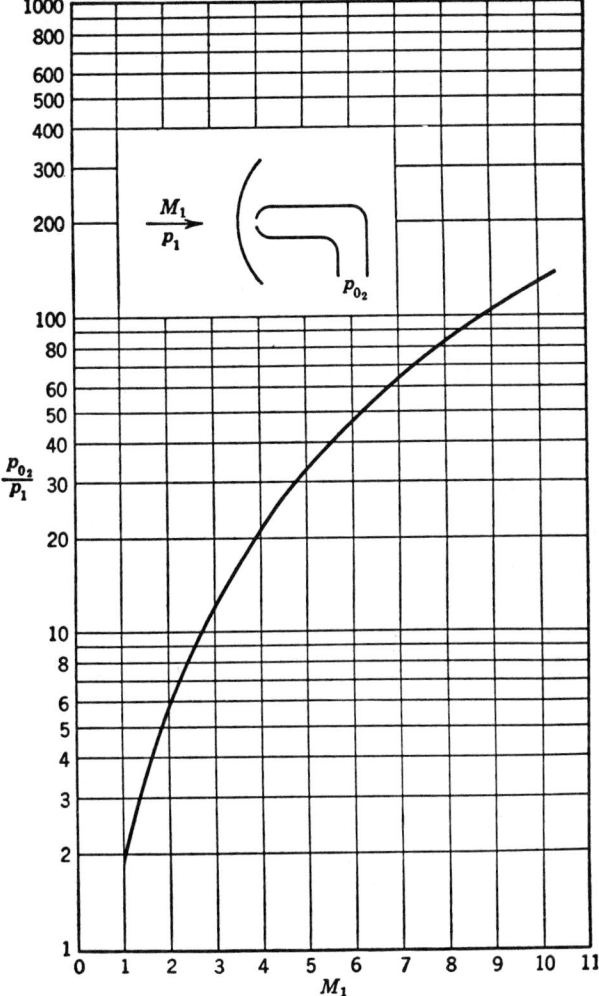

Fig. 6.12 Pitot-static pressure ratio for supersonic flow of air ($\gamma = 1.4$).

6.4.5 Plane Oblique Shock Waves

A plane shock wave oblique to the oncoming flow will turn the flow sharply into the oncoming flow. Figure 6.13 shows an oblique shock wave turning the flow at a sudden change in wall direction. As with a normal shock, the flow process is nonisentropic.

The stream deflection angle θ_s, the wave angle θ_w, and the approach Mach number M_1 of an oblique shock wave are related as follows

$$\tan(\theta_w - \theta_s) = \frac{\dfrac{2}{\gamma + 1} + \dfrac{\gamma - 1}{\gamma + 1} M_1^2 \sin^2\theta_w}{M_1^2 \sin\theta_w \cos\theta_w} \tag{6.52}$$

This relation is plotted in Fig. 6.14. Observe from the figure that two waves are possible. However, when the shock is attached to the corner (Fig. 6.13), only the wave with the smaller wave angle, the so-called weak wave, can exist. In detached shocks, such as the curved shock in front of a blunt wedge, both solutions, strong and weak, occur. The line of minimum M_1 for constant θ_s is the locus of shock-wave detachment.

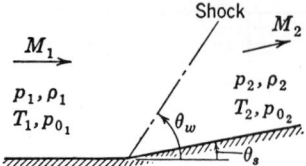

Fig. 6.13 Plane oblique shock wave.

Relations that relate the pressures, temperatures, densities, stagnation pressures, and Mach numbers before and after plane oblique shocks can be obtained in terms of M_1 and θ_w. In fact, when the products $M_1 \sin \theta_w$ and $M_2 \sin(\theta_w - \theta_s)$ are substituted for M_1 and M_2, respectively, in the normal shock equations [(6.46)–(6.50)], they become directly applicable to flow through plane oblique shocks. Thus by entering the graph of Fig. 6.11 with $M_1 \sin \theta_w$ (where θ_w may be obtained from Fig. 6.14 for known values of M_1 and θ_s), the values of the property ratios plotted correspond to those for a plane oblique shock wave. Similarly, the curve labeled M_2 in Fig. 6.11 gives $M_2 \sin(\theta_w - \theta_s)$ for a plane oblique shock wave. The variation of pressure ratio, downstream Mach number, and total pressure ratio with flow deflection and upstream Mach number is plotted in Figs. 6.15 and 6.16. References 21, 23, 25, and 26 present the oblique shock functions in tabular or graphical form. Reference 26 also contains numerous examples.

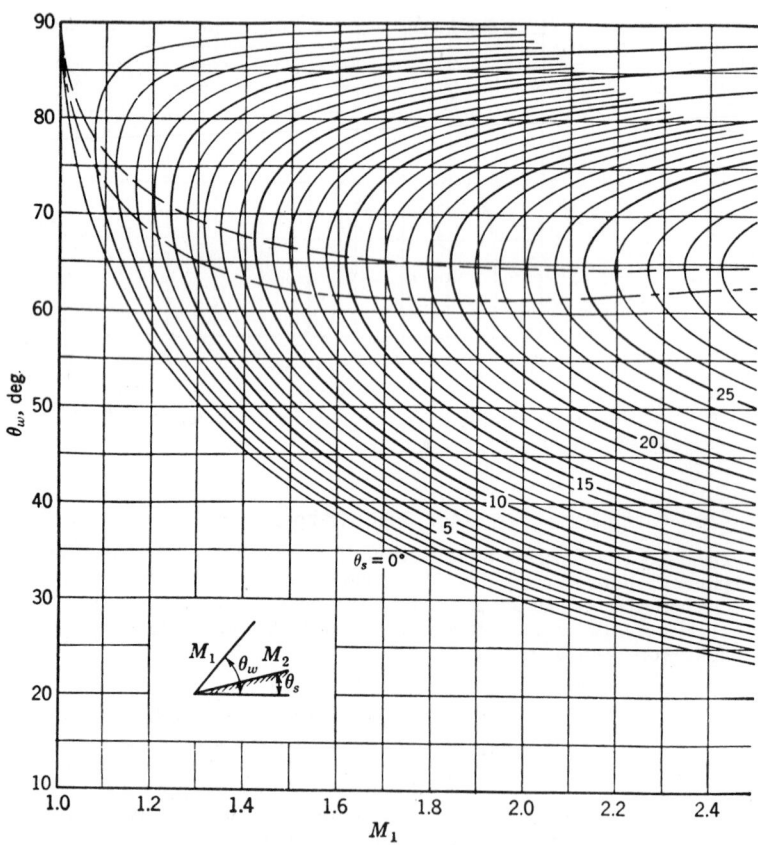

Fig. 6.14 Wave angle for a plane oblique shock wave ($\gamma = 1.4$).

6.4.6 Conical Shock Waves

Conditions on the surface of a cone, Fig. 6.17, with attached shock wave have been calculated on the basis of the Taylor–Maccoll theory.[28] All fluid properties are constant along radial lines through the nose of the cone. Data for the surface stream properties on a cone with an attached conical shock wave are tabulated in Refs. 23 and 27. The shock-wave angles and surface pressure coefficients are plotted in Figs. 6.18 and 6.19.

6.4.7 Prandtl–Meyer Expansion

Supersonic flow turning away from the oncoming flow results in expansion of a gas to a higher Mach number. Figure 6.20 shows such an expansion around a corner starting with an initial Mach number of unity. The flow is here assumed two-dimensional and the process is downward from $M = 1$ along the isentrope line of Table 6.6.

The angle through which a stream must turn in order to expand from $M_1 = 1$ to M_2 is

$$\theta_s = \sqrt{\frac{\gamma + 1}{\gamma - 1}} \tan^{-1} \sqrt{\frac{\gamma - 1}{\gamma + 1} (M_2^2 - 1)} - \tan^{-1} \sqrt{M_2^2 - 1} \qquad (6.53)$$

Numerical values of Eq. (6.53) are given in Refs. 21 and 23 and Table 6.7 for air ($\gamma = 1.4$).

Values given in Table 6.7 can be applied to isentropic flow along any convex curved surface. The downstream Mach number M_2 after expanding M_1 through an angle $\Delta\theta$ can be found by adding $\Delta\theta$ to the θ_s corresponding to M_1 and then finding the M_2 corresponding to $\theta_s + \Delta\theta$ in Table 6.7.

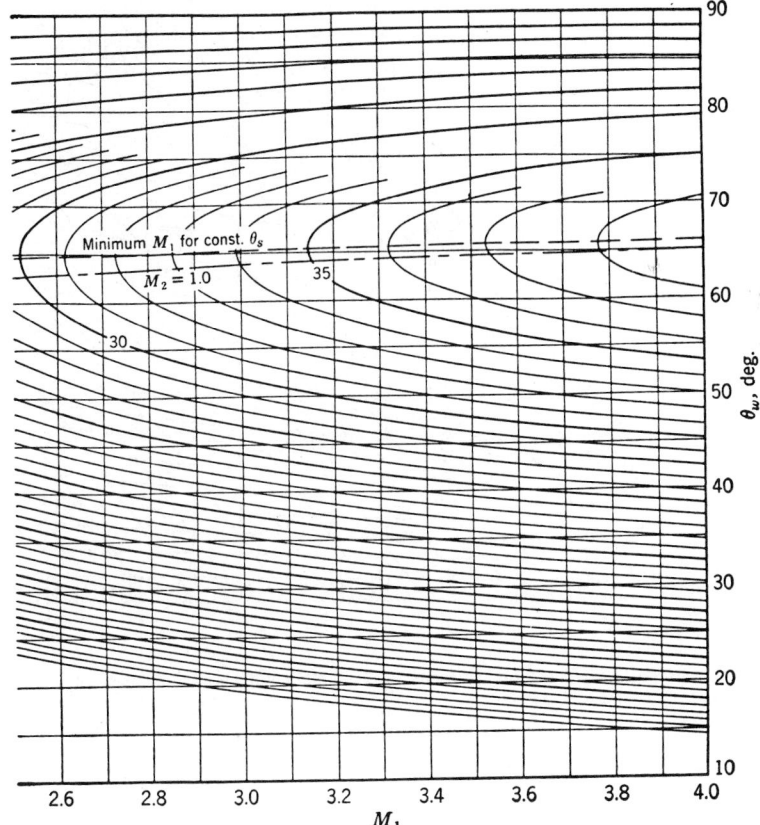

Fig. 6.14 (Continued)

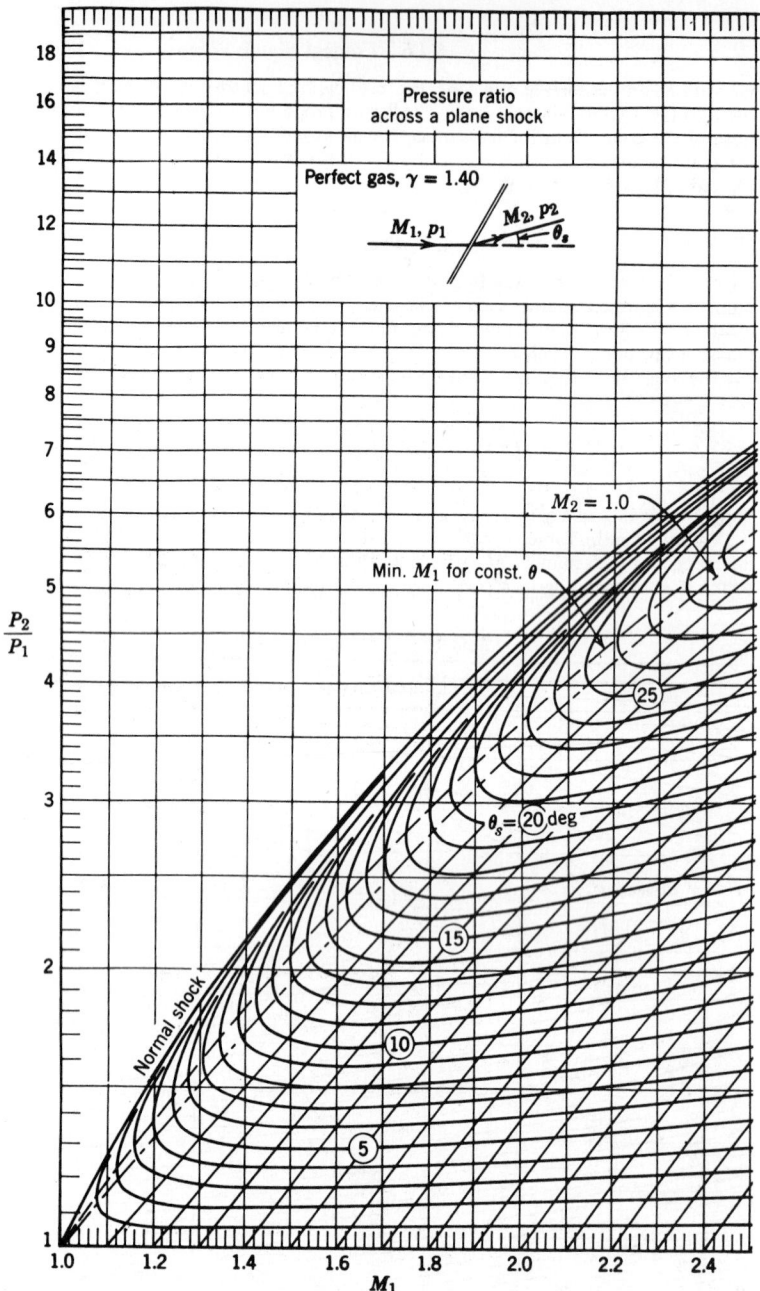

Fig. 6.15 Variation of pressure ratio and downstream Mach number with flow deflection angle and upstream Mach number M (from Ref. 24).

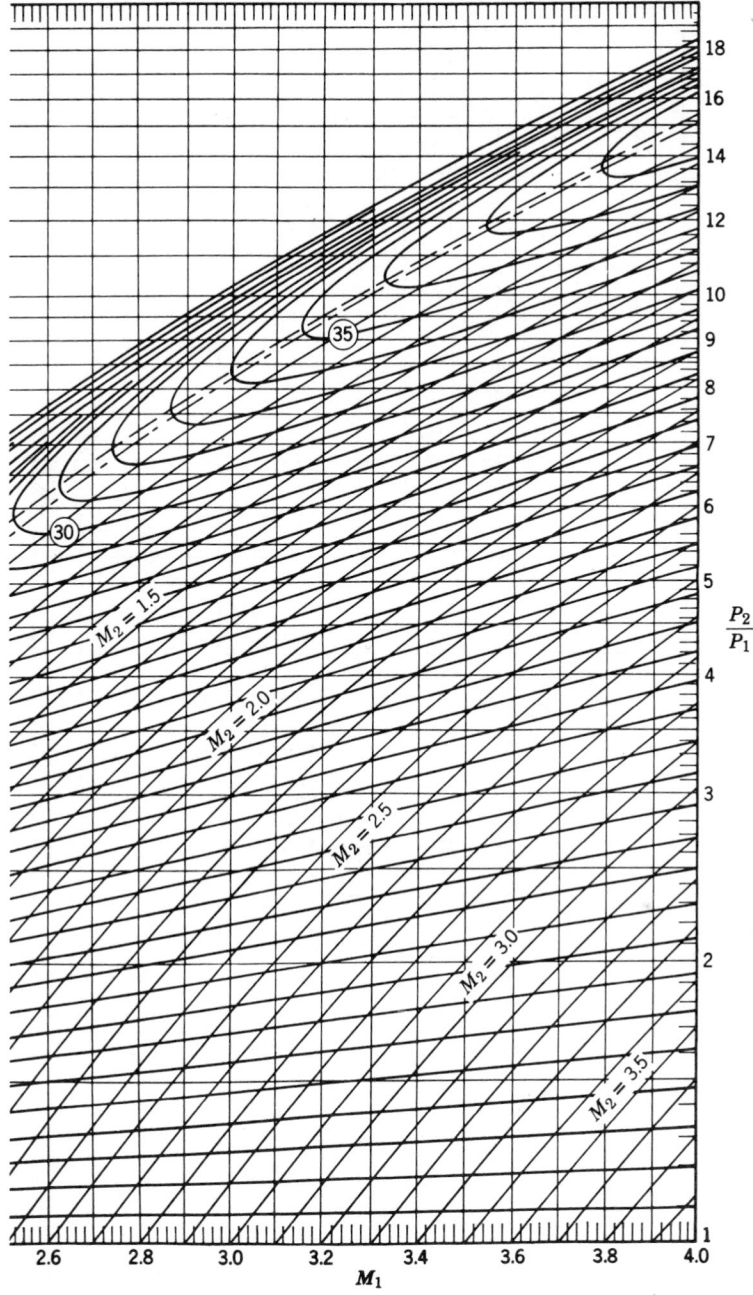

Fig. 6.15 (Continued)

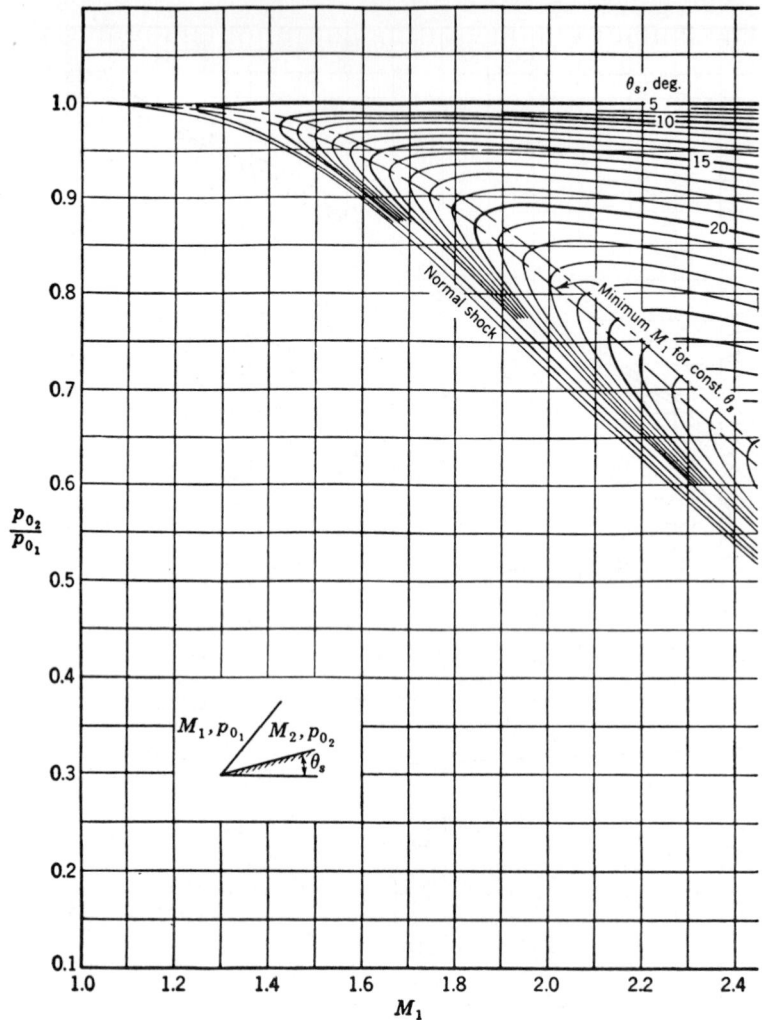

Fig. 6.16   Stagnation pressure ratio across a plane oblique shock wave ($\gamma = 1.4$).

6.5   PROPULSION

R. B. Oliver, E. L. Larson, and D. N. Barlow

6.5.1   General

Several types of thermal air propulsion systems have been developed in addition to the conventional reciprocating aircraft engine. Included in this category are the ramjet, scramjet, pulsejet, turbojet, turboprop, and turbofan. Compounding of reciprocating engines, which consists of gearing the torque of an exhaust gas turbine back into the drive shaft, is also being developed. The rocket, although a self-contained system not of the thermal air type, is included herein for comparative purposes. These engines, or variations of them, singly or in combination, provide power for practically all present-day aircraft, either piloted or pilotless.

The several types are compared briefly[29] in Fig. 6.21. The individual parameters are described in detail later in the text.

The rocket is a self-contained power plant that does not depend on atmospheric air for operation. Fuel for a rocket is either liquid or solid, depending on the type. An oxidizer is carried, along with the fuel, and an ignition system is provided for starting combustion if such is necessary. Most rockets

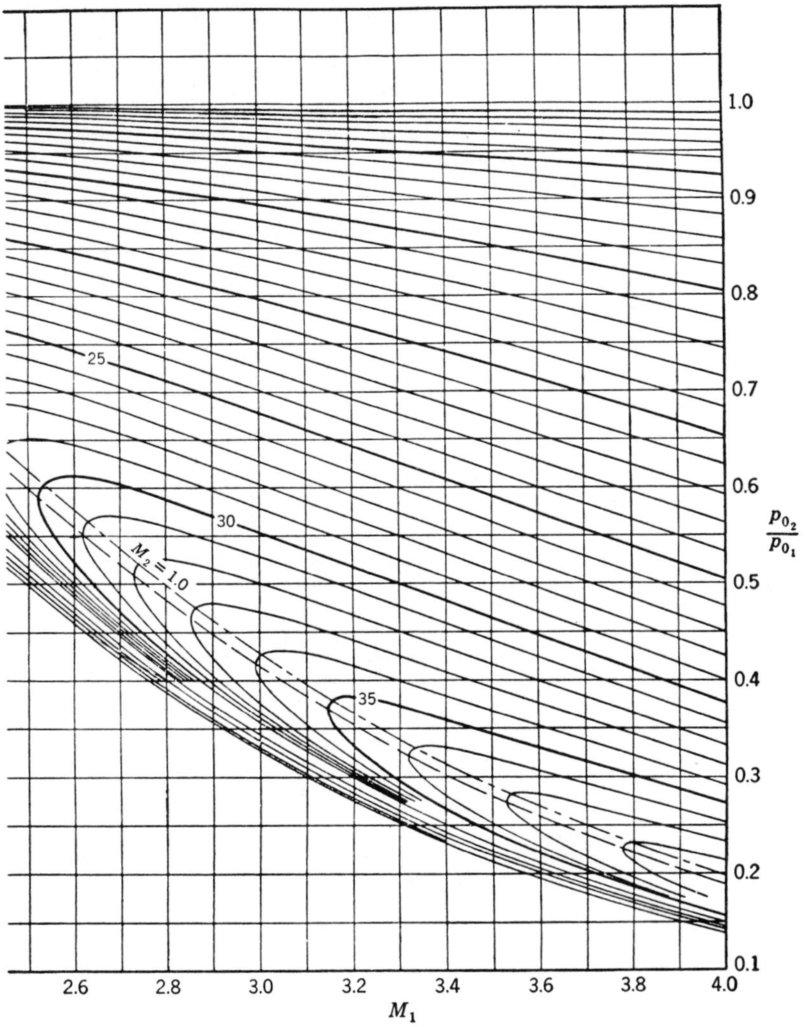

1.0

0.9

0.8

25

0.7

0.6

$\dfrac{p_{0_2}}{p_{0_1}}$

30

0.5

$M_2 = 1.0$

0.4

35

0.3

0.2

0.1

2.6   2.8   3.0   3.2   3.4   3.6   3.8   4.0

M_1

Fig. 6.16   *(Continued)*

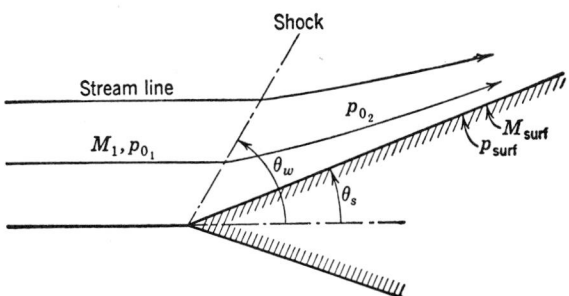

Shock

Stream line

p_{0_2}

M_1, p_{0_1}

M_{surf}

p_{surf}

θ_w

θ_s

Fig. 6.17   Conical shock wave.

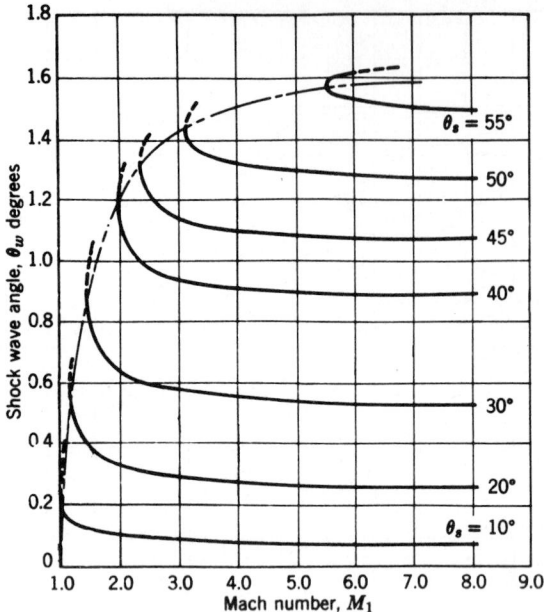

Fig. 6.18 Conical shock-wave angle versus Mach number for various cone angles.

burn fuel and oxidizer at a constant rate and hence produce nearly a constant thrust. The burning gases expand out of a nozzle and produce mechanical energy. The static pressure at the nozzle exit influences the thrust; since expansion to low pressure aids in the conversion of thermal energy into mechanical energy, thrust is increased with altitude.

The ramjet is perhaps the simplest of engines in principle. Air is compressed through a diffuser into a combustion chamber by relative motion of the engine and the air. Fuel is introduced and

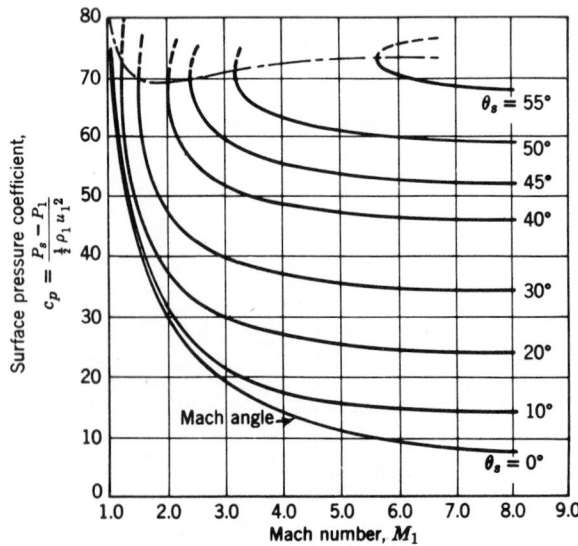

Fig. 6.19 Surface pressure coefficient versus Mach number for various cone angles.

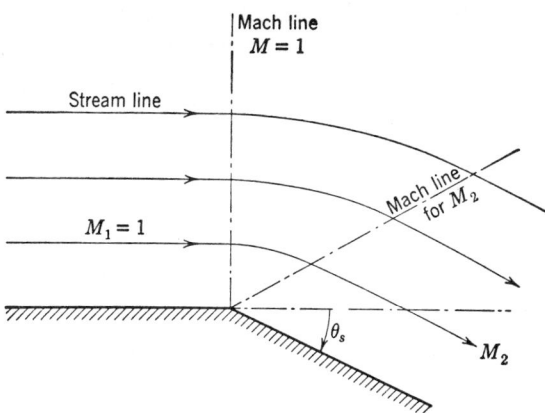

Fig. 6.20 Prandtl–Meyer expansion at a corner.

burned in the combustion chamber. The hot products of combustion expand out of the nozzle, producing an exit velocity in excess of the free-stream velocity of the engine. The resulting increase in momentum of the combustion products determines the thrust produced by the unit. Since the combustion chamber pressure must be greater than ambient, the ramjet cannot produce thrust without relative velocity between it and the air. However, a modification of the ramjet, the pulsejet, is capable of producing static thrust, because intake valves are provided that alternately open and close as a function of the intake pressure and back pressure from combustion, thus making possible a working cycle. Since the pulsejet vibrates badly and operates at a rather high noise level, its use will probably be limited to pilotless aircraft. The specific fuel consumption of the pulsejet is somewhat higher than that of the turbojet for subsonic flight speeds. This tends to limit the usefulness of the

TABLE 6.7   Mach Number Variation in Expanding Flow

θ_s	M_2	θ_s	M_2	θ_s	M_2	θ_s	M_2
0	1.0000	27	2.0222	54	3.2293	81	5.470
1	1.0808	28	2.0585	55	3.2865	82	5.595
2	1.1328	29	2.0957	56	3.3451	83	5.724
3	1.1770	30	2.1336	57	3.4055	84	5.867
4	1.2170	31	2.1723	58	3.4675	85	6.008
5	1.2554	32	2.2105	59	3.5295	86	6.155
6	1.2935	33	2.2492	60	3.5937	87	6.311
7	1.3300	34	2.2885	61	3.6610	88	6.472
8	1.3649	35	2.3288	62	3.7288	89	6.643
9	1.4005	36	2.3688	63	3.7980	90	6.820
10	1.4350	37	2.4108	64	3.8690	91	7.008
11	1.4688	38	2.4525	65	3.9417	92	7.202
12	1.5028	39	2.4942	66	4.0164	93	7.407
13	1.5365	40	2.5372	67	4.0940	94	7.623
14	1.5710	41	2.5810	68	4.1738	95	7.852
15	1.6045	42	2.6254	69	4.2543	96	8.093
16	1.6380	43	2.6716	70	4.3385	97	8.350
17	1.6723	44	2.7179	71	4.4257	98	8.622
18	1.7061	45	2.7643	72	4.5158	99	8.907
19	1.7401	46	2.8120	73	4.6086	100	9.210
20	1.7743	47	2.8610	74	4.7031	101	9.539
21	1.8090	48	2.9105	75	4.7979	102	9.887
22	1.8445	49	2.9616	76	4.9032	103	10.260
23	1.8795	50	3.0131	77	5.009	104	10.658
24	1.9150	51	3.0660	78	5.119	105	11.081
25	1.9502	52	3.1193	79	5.232		
26	1.9861	53	3.1737	80	5.349		

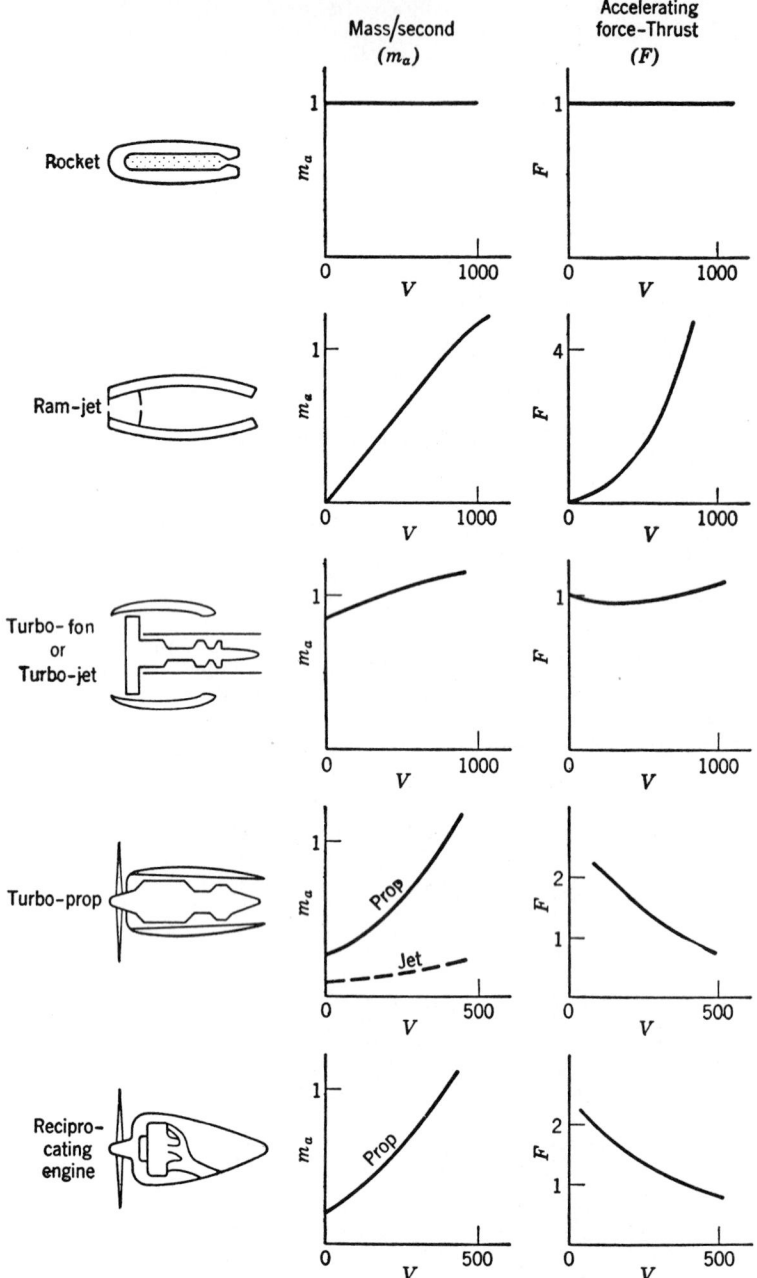

Fig. 6.21 Power plant characteristics. V = remote velocity relative to the engine. (Method of presentation suggested by Westinghouse "Jet Propulsion" brochure, B3834, 5M-2-4, see Perkins and Hage[29]).

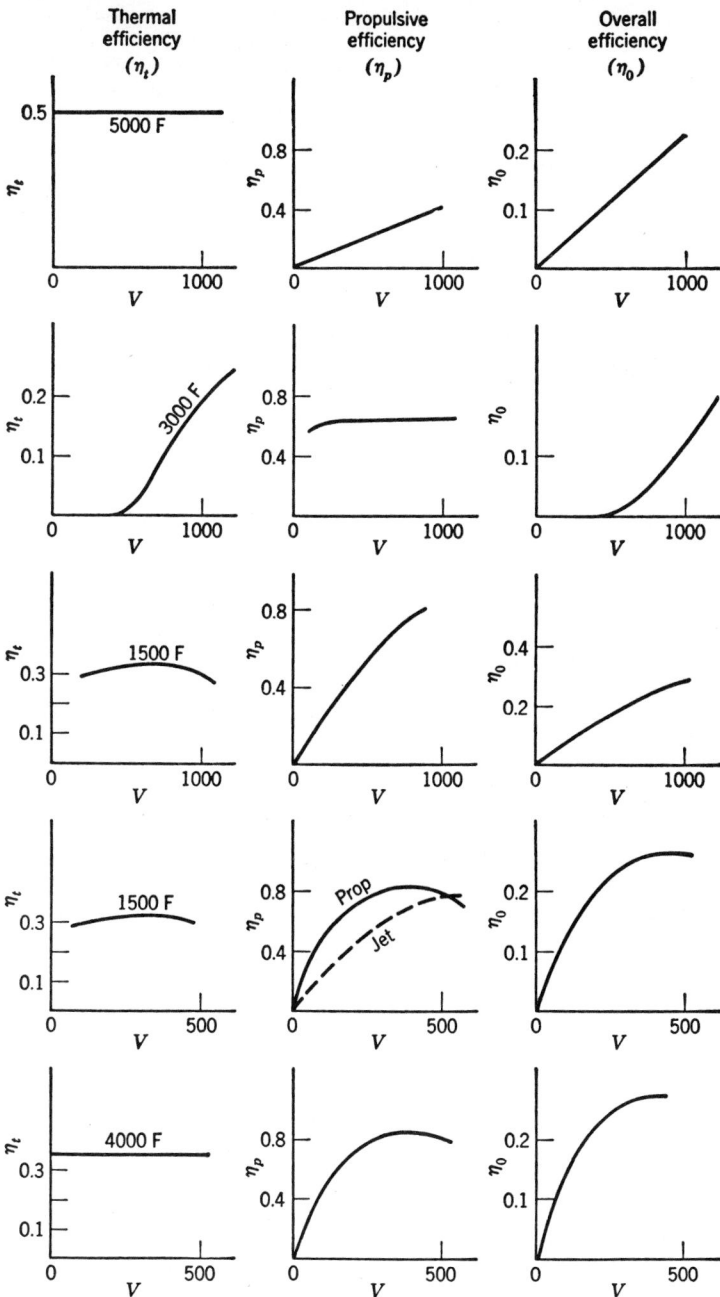

Fig. 6.21   *(Continued)*

pulsejet, inasmuch as at supersonic flight speeds the ramjet appears more promising than the pulsejet. This usefulness is limited however to Mach numbers less than about 6. Above Mach 6, shocks result in large total pressure losses when the ram air is slowed to subsonic velocities for burning in the combustion bomber. Also, this slowing results in high static temperatures before and after the combustion process.

The scramjet, supersonic combustion ramjet, is under development to provide a propulsion system for hypersonic applications. This engine is similar to a ramjet, except that the ram air is not slowed to subsonic speeds for the combustion process. As a result normal shocks are not encountered nor are the associated total pressure losses. In addition, at the higher internal Mach numbers more energy can be added to the flow before reaching limiting metal (static) temperatures. One challenge of this system is maintaining constant and efficient combustion in a supersonic flow.

The turbojet engine differs from the ramjet principally in only one major aspect: the incoming ram air is further compressed by a compressor before fuel is introduced and burned in the combustion chamber. A turbine placed behind the combustion chamber, and integrally connected with the compressor, is driven by the expanding products of combustion. This in turn drives the compressor. Inasmuch as the turbojet can produce pressure above ambient in the combustion chamber even when the engine is at rest, a static thrust can be produced. This desirable characteristic is obtained without the vibration associated with the pulsejet, mentioned previously. Because the combustion pressure is higher than for a ramjet at the same speed, the efficiency is greater, up to Mach numbers between 2 and 3, where allowable temperatures for turbine and compressor blades limit the present-day turbojet.

The turboprop is essentially a turbojet with a propeller attached to the same shaft as the compressor and turbine. Of course, reduction gearing is necessary between the actual shaft and the propeller, which produces the majority of the thrust. The rest of the thrust is obtained from jet action. The high efficiency of this engine at relatively low speeds, along with its small frontal area, makes it desirable for long-range patrol planes and the like.

The turbofan is essentially a turbojet with an additional compressor or fan attached to a second turbine. The fan is located in front of the core compressor. Air passes through the fan and then is split. A portion of the air is bypassed around the core and through a nozzle to produce thrust. The remaining mass flow proceeds through the core and is also expanded through a nozzle to produce thrust. Like the turboprop this engine provides higher efficiencies than the turbojet at medium and low Mach numbers. Because its frontal area and blade tip speeds are lower than a turboprops, it is efficient to higher speeds than the turboprop.

The reciprocating engine is discussed in detail elsewhere. Little need be said here of its characteristics, except that for aircraft application it is doubtful that this engine will be used for air speeds much in excess of 500 mph, frontal area, weight, cooling drag, and propeller tip losses being limiting factors. However, reciprocating aircraft engines will undoubtedly be widely used for many years to come for a large number of airplanes.

6.5.2 Development of Efficiency

The efficiencies referred to in Fig. 6.21 are based on the following definitions (applicable to all previously mentioned engines):

$$\eta_o = \text{overall efficiency} = \frac{\text{useful work done on vehicle}}{\text{thermal energy released by fuel}} = \eta_p \cdot \eta_t$$

$$\eta_t = \text{thermal efficiency} = \frac{\text{power in the exhaust}}{\text{thermal energy released by fuel}}$$

$$\eta_p = \text{propulsive efficiency} = \frac{\text{useful work done on vehicle}}{\text{power in the exhaust}}$$

If $\dot{m}_0$ is the mass of air entering and leaving the ramjet, turbojet, or propeller disk of a conventional engine in unit time, and $\dot{m}_f$ is the mass of fuel leaving the ramjet, turbojet, or rocket (including oxidizer mass) in unit time, the thrust F produced on the supporting structure for ambient entrance (except for the rocket) and exit pressure is

$$F = \left(\dot{m}_0 + \dot{m}_f \right) V_e - \dot{m}_0 V_0 \tag{6.54}$$

where V_e is the exit or wake velocity relative to the engine, and V_0 is the remote velocity relative to the engine. Of course, $\dot{m}_f$ is zero for air passing through a propeller disk, and $\dot{m}_0$ is zero for the rocket. The power output is

$$P = F V_0 = \left[\left(\dot{m}_0 + \dot{m}_f \right) V_e - \dot{m}_0 V_0 \right] V_0 \tag{6.55}$$

The overall efficiency is therefore

$$\eta_o = \frac{\left[\left(\dot{m}_0 + \dot{m}_f\right)V_e - \dot{m}_0 V_0\right]V_0}{h\dot{m}_f} \tag{6.56}$$

where h is heating value of the fuel (for the rocket, the oxidizer is included with the fuel).

The power produced by the system is the time rate of change of kinetic energy produced by the system.

$$\tfrac{1}{2}\left[\left(\dot{m}_f + \dot{m}_0\right)V_e^2 - \dot{m}_0 V_0^2\right]$$

The thermal efficiency is then

$$\eta_t = \frac{\tfrac{1}{2}\left[\left(\dot{m}_f + \dot{m}_0\right)V_e^2 - \dot{m} V_0^2\right]}{h\dot{m}_f} \tag{6.57}$$

and the propulsive efficiency is

$$\eta_p = \frac{\left[\left(\dot{m}_0 + \dot{m}_f\right)V_e - \dot{m}_0 V_0\right]V_0}{\tfrac{1}{2}\left[\left(\dot{m}_f - \dot{m}_0\right)V_e^2 - \dot{m}_0 V_0^2\right]} \tag{6.58}$$

For the ramjet and turbojet, $\dot{m}_f$ is usually quite small compared to $\dot{m}_0$. If $\dot{m}_f$ is neglected in the mechanical energy expression, simple expressions result that apply to the ramjet, turbojet, or propeller. The thermal efficiency reduces to

$$\eta_t = \frac{\tfrac{1}{2}\dot{m}_0\left(V_e^2 - V_0^2\right)}{h\dot{m}_f} \tag{6.59}$$

and the propulsive efficiency reduces to

$$\eta_p = \frac{2}{1 + V_e/V_0} \tag{6.60}$$

For the rocket there is no incoming air since the oxidizer is carried along with the fuel. Hence $\dot{m}_0$ will be zero in the equations. The thermal efficiency reduces to*

$$\eta_t = \frac{1}{2}\frac{V_e^2 + V_0^2}{h} \tag{6.61}$$

and the propulsive efficiency to

$$\eta_p = \frac{2(V_e/V_0)}{1 + (V_e/V_0)^2} \tag{6.62}$$

6.5.3   Rocket Motor

The rocket (Fig. 6.22) consists of a fuel and oxidizer injection system (if liquid fuel is used), a combustion chamber, and a convergent-divergent exit nozzle. A convergent-divergent nozzle is a nozzle with a throat and expanding exit so that if a sufficient ratio of internal pressure to external pressure exists the gases may expand beyond the throat, thus yielding supersonic speeds.

Calculation of the thrust produced by a rocket can be made if the combustion pressure P_{t_c}, combustion temperature T_{t_c}, and the mass flow of fuel and oxidizer $\dot{m}_f$ are known. Denoting the exit pressure, temperature, and Mach number by P_e, T_e, and M_e, respectively, the following relations hold where full expansion of the gases to ambient pressure, $P_e = P_0$, is assumed (subscript t = total

---

*The velocity V_0 in Eq. (6.61), representing the energy of motion of the fuel before combustion, is omitted by some authors. (See Fig. 6.21, in which η_t is sketched independent of V_0.)

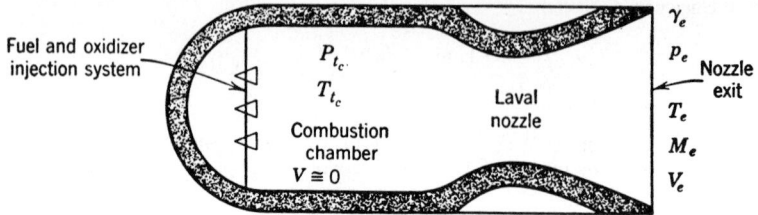

Fig. 6.22   Schematic diagram of rocket motor.

conditions):

$$P_{t_e} = \rho_e P_{t_c} = \rho_e P_0 \left(1 + \frac{\gamma - 1}{2} M_e^2 \right)^{\gamma/(\gamma-1)}$$

$$T_{t_c} = T_{t_e} = T_e \left(1 + \frac{\gamma - 1}{2} M_e^2 \right)$$

For proper use of these expressions, γ should be a mean value. The value of the mean γ depends on the static and total temperatures at the exit. If a value of γ is assumed, and P_{t_e}, P_0, and ρ_e are known, M_e can be calculated. Use of the known T_{t_c}, the assumed mean γ, and the calculated M_e will allow determination of T_e. The mean γ associated with any T_e and T_{t_e} can be calculated for each combination of rocket propellants under consideration. A check of the mean γ for the T_e and T_{t_e} obtained will indicate how closely the assumed γ was matched. If too much discrepancy exists, an iteration should be carried out. A mean γ near 1.2 is often obtained.

When T_e and the value of γ associated with it are known, along with the gas constant R_e, for the particular chemical combination used, the local exit acoustic velocity, a_e, can be obtained,

$$a_e = \sqrt{\gamma_e R_e T_e}$$

It follows that

$$V_e = M_e a_e$$

and

$$F = \dot{m}_f V_e$$

where V_e and F are the exit velocity and thrust, respectively. The specific impulse, I (sometimes called specific thrust), defined as the thrust F divided by the weight flow of fuel w_f, is

$$I = \frac{F}{w_f} = \frac{\dot{m}_f V_e}{\dot{m}_f g} = \frac{V_e}{g}$$

where g is the acceleration due to gravity.

6.5.4   Ramjet Engine

A ramjet engine (Fig. 6.23) consists essentially of an inlet diffuser (1–2), a flame holder and fuel injector (2), a combustion chamber (2–3), and a convergent-divergent exit nozzle (3–5). Performance of a ramjet is usually expressed in terms of specific thrust and specific fuel consumption. Specific thrust is defined as the thrust (F) divided by the incoming mass flow ($\dot{m}_0$). Thrust-specific fuel consumption (S) is defined as the mass flow of fuel per hour divided by the thrust. These two parameters allow easy comparison of different engines by comparing thrust per inlet size and fuel used per pound of thrust produced.

If the free-stream capture area A_0 is equal to the diffuser entrance area, the airflow is known. At other conditions the airflow may be more difficult to obtain (see Ref. 30).

Following is a calculation method that allows easy calculation of specific thrust and specific fuel consumption based on known technology levels. Losses occur in the ramjet due to nonisentropic

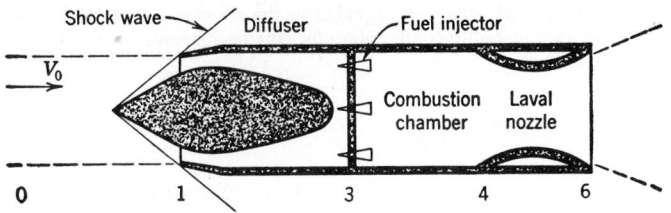

Fig. 6.23   Schematic sketch of ramjet.

processes in each section of the engine. In the diffuser viscous and shock losses occur. In the combustion chamber viscous losses and total pressure losses due to heat addition at finite Mach numbers occur. In the nozzle viscous losses are present. These losses result in total pressure losses in each section of the engine. These total pressure losses result in a lower pressure ratio across the nozzle resulting in lower exit velocities and thrust.

NOTATION

P = pressure
T = temperature
t = total (stagnation) conditions
0 = far upstream conditions
b = burner
d = diffuser
n = nozzle
M = Mach number
R = gas constant
C_{P_c} = specific heat at constant pressure upstream of the combustor
C_{P_t} = specific heat at constant pressure downstream of the combustor
γ = ratio of specific heats
f = fuel-to-air ratio by mass, $\dot{m}_f/\dot{m}_0$

$$\pi = \frac{\text{stagnation pressure leaving component}}{\text{stagnation pressure entering component}}$$

$$\tau = \frac{\text{stagnation temperature leaving component}}{\text{stagnation temperature entering component}}$$

Examples:

$$\pi_b = \text{burner stagnation pressure ratio, } P_{t_3}/P_{t_2}$$
$$\tau_b = \text{burner stagnation temperature ratio, } T_{t_3}/T_{t_2}$$

Exceptions:

$$\tau_r = 1 + \frac{\gamma_c - 1}{2} M_0^2 = T_{t_0}/T_0$$

$$\pi_r = \left(1 + \frac{\gamma_c - 1}{2} M_0^2\right)^{\gamma_c/(\gamma_c - 1)} = P_{t_0}/P_0$$

π_r and τ_r represent the effects of the flight Mach

$$\tau_\lambda = \frac{C_{P_t} T_{t_4}}{C_{P_c} T_0}$$

τ_λ introduces the effect of a design limitation, the maximum allowable total temperature in the combustion chamber.

In addition the burner efficiency, η_b is included in the set of equations to follow. The burner efficiency takes into account that all fuel released in the combustion chamber will not be burned. The burner efficiency is defined as

$$\eta_b = \frac{1}{\dot{m}_f h}\left[(\dot{m}_0 + \dot{m}_f)h_{t_4} - \dot{m}_0 h_{t_3}\right]$$

where h is the heating value of the fuel, h_t is the stagnation enthalpy ($C_P T_t$), and $\dot{m}_f$ is the fuel mass flow rate.

A first law analysis of the physics involved results in the following set of equations.

Inputs: T_0 (°R), γ_c, γ_t, C_{P_c} (Btu/lbm − °R), C_{P_t}, h(Btu/lbm), π_d, π_b, π_n, η_b, $P_6/P)_0$, T_{t_4} (°R), M_0.

Outputs: $F/\dot{m}_0$ (lbf/lbm sec), S (lbm fuel/hr/lbf thrust), f.

$$R_c = \frac{\gamma_c - 1}{\gamma_c} C_{P_c}(25{,}050)$$

$$a_0 = \sqrt{\gamma_c R_c T_0}$$

$$T_\lambda = \frac{T_{t_4} C_{P_t}}{T_0 C_{P_c}}$$

$$\tau_r = 1 + \frac{\gamma_c - 1}{2} M_0^2$$

$$\pi_r = \tau_r^{\gamma_c/\gamma_c - 1}$$

$$f = \frac{T_\lambda - \tau_r}{\left(h\eta_b/C_{P_c} T_0\right) - T_\lambda}$$

$$\frac{P_{t_6}}{P_0} = \frac{P_0}{P_6}\pi_r \pi_d \pi_b \pi_n$$

$$\frac{T_6}{T_0} = \frac{C_{P_c}/C_{P_t}}{\left(\dfrac{P_{T_6}}{P_6}\right)^{\gamma_t - 1/\gamma_t}}$$

$$\frac{V_6}{V_0} = \left\{\frac{T_\lambda}{\tau_r - 1}\left[1 - \left(\frac{P_{t_6}}{P_6}\right)^{-(\gamma_t - 1)/\gamma_t}\right]\right\}^{1/2}$$

$$\frac{F}{\dot{m}_0} = a_0 M_0\left\{\left[(1+f)\frac{V_6}{V_0} - 1\right] + (1+f)\frac{1}{V_6/V_0}\frac{1}{M_0^2}\frac{1}{\gamma_t}\frac{\gamma_t - 1}{\gamma_c - 1}\frac{C_{P_t}}{C_{P_c}}\frac{T_6}{T_0}\left(1 - \frac{P_6}{P_0}\right)\right\}$$

$$S = \frac{f}{F/\dot{m}_0}(10^6)$$

These equations are easily programmed and make parametric investigations of input parameters practical.

To help determine gas property inputs for a cycle analysis refer to Figs. 6.24 and 6.25.

Example 6.1. Using military specification 5008B

$$\pi_d = \left[1 - 0.075(M_0 - 1)\right] \cdot \pi_{d_{max}} \qquad M_0 > 1$$

where $\pi_{d_{max}}$ is the value of total pressure loss achieved in the inlet at subsonic conditions. This is the maximum pressure loss due to viscous effects, which will occur in the inlet at supersonic conditions. Below $M_0 = 1$, $\pi_{d_{max}} = \pi_d$ should be used. Like the other total pressure ratios, it is a measure of technology. Military Spec 5008B must be met by inlet designers of military aircraft.

An engine designer attempts to match P_6 and P_0. If this is achieved, the engine nozzle is considered to be on-design and the maximum thrust is achieved. Notice, the thrust equation greatly

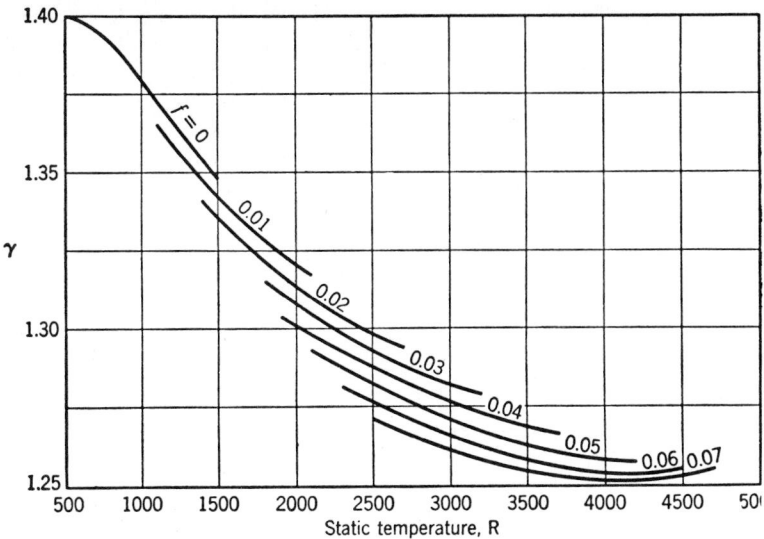

Fig. 6.24 Effect of static temperature on the ratio of specific heats of the combustion products.

simplifies. For example, with the following typical values:

$$\gamma_c = 1.4 \qquad\qquad \pi_{d_{max}} = 0.98$$

$$\gamma_t = 1.3 \qquad\qquad \pi_b = 0.97$$

$$C_{p_c} = 0.238 \text{ Btu/lbm} - \text{°R} \qquad \pi_n = 0.98$$

$$C_{p_t} = 0.295 \text{ Btu/lbm} - \text{°R} \qquad \frac{P_6}{P_0} = 1$$

$$h = 18{,}000 \text{ Btu/lbm}$$

$$T_{t_4} = 3500\text{°R}$$

Figs. 6.26 and 6.27 result.

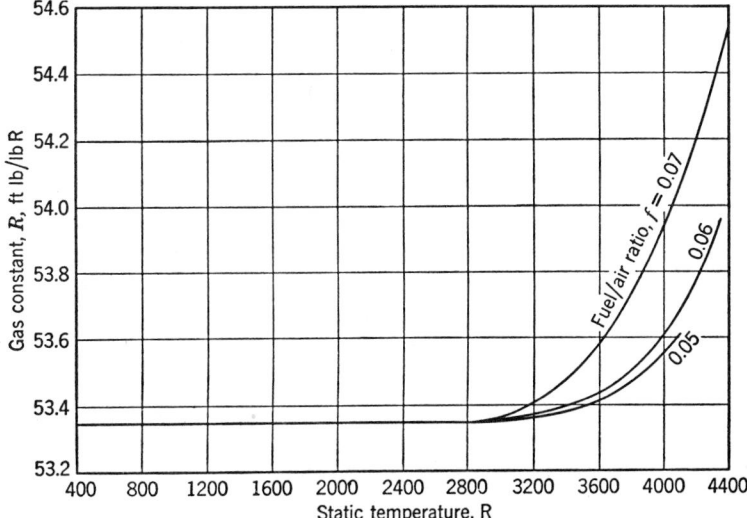

Fig. 6.25 Effect of static temperature on the gas constant of the combustion products.

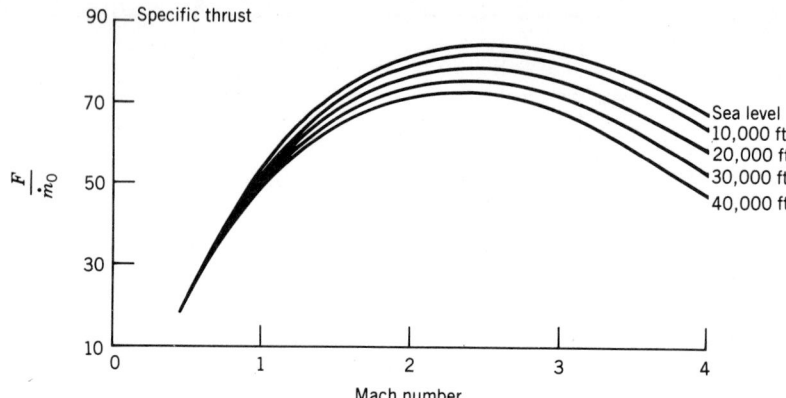

Fig. 6.26 Specific thrust as a function of altitude and Mach number.

6.5.5 Turbojet Engine

A turbojet engine (Fig. 6.28) consists of an air inlet (1), a compressor (2–3), a fuel injector system (3), a combustion chamber (3–4), a turbine on the same shaft as the compressor (4–5), and an exit nozzle (5–6). A turbojet adapted for supersonic flight would probably be fitted with some type of supersonic diffuser in order to recover the maximum possible percentage of free-stream total pressure. Two types of compressors are used in turbojets: the centrifugal and the axial flow. The centrifugal flow compressor was developed first, but the smaller frontal area of the axial flow type for a given airflow has prompted its development. The pressure ratio developed by the compressor of the axial flow type is a function of the blade design, the shaft speed, and the number of stages. The thrust output and the efficiency of the turbojet depend on the relative effects of the turbine and compressor losses caused by driving the compressor and the increased thermodynamic efficiency brought about by having a large pressure differential in the working cycle. The allowable operating temperatures for turbine material limit the compression ratios that can be used, especially for the higher flight speeds.

Analysis of the turbojet engine is similar to the ramjet with the addition of efficiencies for the compressor and turbine.

The compressor efficiency (η_c) is defined by

$$\eta_c = \frac{\text{ideal work interaction for a given pressure ratio}}{\text{actual work interaction for a given pressure ratio}}$$

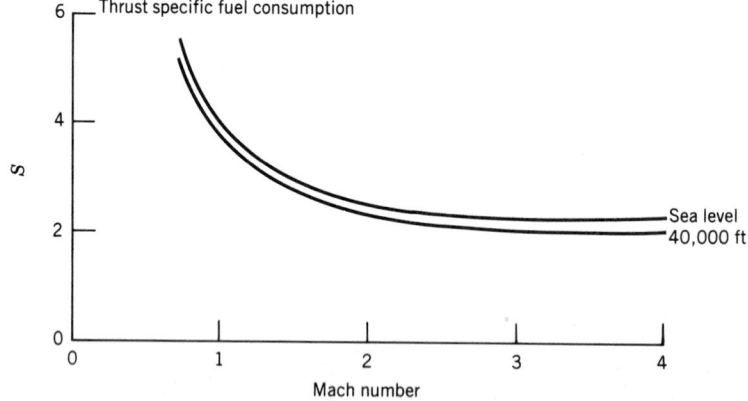

Fig. 6.27 Specific fuel consumption as a function of altitude and Mach number.

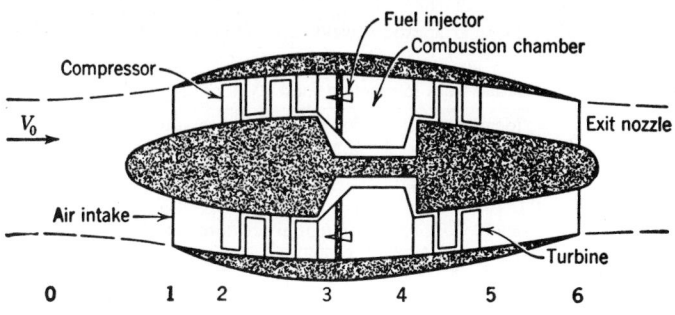

Fig. 6.28 Schematic sketch of a turbojet.

The ideal process is one in which no losses (viscous or thermal) exist. Such a process is referred to as an isentropic process and the following relationship between temperature and pressure applies

$$\frac{T_{t_{3_i}}}{T_{t_2}} = \left(\frac{P_{t_3}}{P_{t_2}}\right)^{(\gamma_c - 1)/\gamma_c} = \pi_c^{(\gamma_c - 1)/\gamma_c}$$

For an actual (nonideal) process the temperature increase will always exceed the ideal temperature rise for a given energy addition to achieve the desired pressure rise. Or conversely, the pressure rise is less for a given energy addition in the compressor than the ideal process would predict.

Using the definition of compressor efficiency,

$$\eta_c = \frac{C_p\left(T_{t_{3_i}} - T_{t_2}\right)}{C_p\left(T_{t_3} - T_{t_2}\right)} = \frac{T_{t_{3_i}}/T_{t_2} - 1}{T_{t_3}/T_{t_2} - 1}$$

Combining this with the isentropic relationship gives

$$\eta_c = \frac{\pi_c^{(\gamma_c - 1)/\gamma_c} - 1}{\tau_c - 1}$$

Although this relationship is very useful to the engineer, it has a practical limit since η_c is a function of the presure ratio of the compressor. As a result it is difficult to relate the efficiency of one compressor to another without considering its size. In order to describe technology levels regardless of compressor size, the compressor polytropic efficiency (e_c) is introduced:

$$e_c = \frac{\text{ideal work of compressor for a differential pressure change}}{\text{actual work of compressor for a differential pressure change}}$$

This concept permits easy comparison of compressors regardless of size. The polytropic efficiency depends on the level of aerodynamic design technology and not how many stages are stacked in the compressor. This concept permits current technology levels to easily be introduced into our calculations. Using the isentropic relation and the definition of e_c, the following relationship can easily be derived (see Ref. 30):

$$\tau_c = \pi_c^{(\gamma_c - 1)/\gamma_c e_c}$$

For a state-of-the-art design, the polytropic efficiency is essentially constant through the compressor. Combining this with the compressor efficiency equation results in

$$\eta_c = \frac{\pi_c^{(\gamma_c - 1)/\gamma_c} - 1}{\pi_c^{(\gamma_c - 1)/\gamma_c e_c} - 1}$$

By choosing a compressor pressure ratio and compressor technology level, the stagnation temperature ratio (τ_c) and overall compressor efficiency (η_c) may be calculated.

Likewise for the turbine, a similar approach yields the following relationships (see Ref. 30):

$$\eta_t = \frac{\text{actual work interaction for a given pressure ratio}}{\text{ideal work interaction for a given pressure ratio}}$$

$$\eta_t = \frac{1 - \tau_t}{1 - \pi_t^{(\gamma_t - 1)/\gamma_t}}$$

$$e_t = \frac{\text{actual work for a differential pressure change}}{\text{ideal work for a differential pressure change}}$$

$$\pi_t = \tau_t^{\gamma_t/(\gamma_t - 1)e_t}$$

$$\eta_t = \frac{1 - \pi_t^{(\gamma_t - 1)/\gamma_t e_t}}{1 - \pi_t^{(\gamma_t - 1)/\gamma_t}}$$

Once τ_c is determined from the preceding, τ_t can be calculated knowing that the work done by the turbine is used to power the compressor ($\dot{W}_c = \dot{W}_t$). Then by assuming an appropriate value for turbine polytropic efficiency π_t may be calculated.

It is interesting to note that with increased engine size (higher π_c) that η_c and η_t change inversely with constant e_c and e_t (see Fig. 6.29).

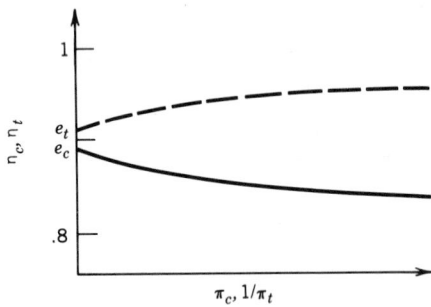

Fig. 6.29   Polytropic efficiency behavior with varying π_c and π_t.

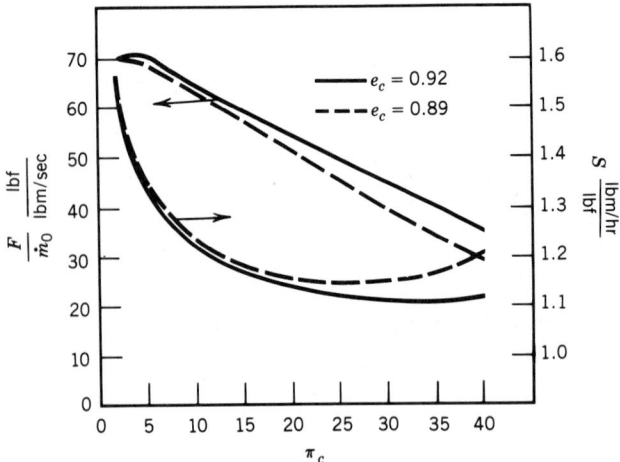

Fig. 6.30   Turbojet performance with varying π_c.

Notation

The notation is the same as for the ramjet except for the following:

$$c = \text{compressor}$$
$$t = \text{turbine}$$
$$\eta = \text{isentropic efficiency}$$
$$e = \text{polytropic efficiency}$$

Example:

$$\pi_c = \text{compressor stagnation pressure ratio, } P_{t_3}/P_{t_2}$$

This information and a first law analysis of the physics involved results in the following set of equations.

Inputs: T_0, γ_c, C_{P_c}, γ_t, C_{P_t}, π_d, π_b, π_n, η_b, e_c, e_t, P_6/P_0, h, T_{t_4}, π_c, M_0.

Outputs: $F/\dot{m}_0$ (lbf/lbm/sec), S (lbm fuel/hr/lbf thrust), f (lbm fuel/sec/lbm air/sec), η_c, η_t.

$$R_c = \frac{\gamma_c - 1}{\gamma_c} C_{P_c} (2.505 \times 10^4)$$

$$a_0 = \sqrt{\gamma_c R_c T_0}$$

$$\tau_r = 1 + \frac{\gamma_c - 1}{2} M_0^2$$

$$\tau_\lambda = \frac{C_{P_t} T_{t_4}}{C_{P_t} T_0}$$

$$\tau_c = \pi_c^{(\gamma_c - 1)/\gamma_c e_c}$$

$$\eta_c = \frac{\pi_c^{(\gamma_c - 1)/\gamma_c} - 1}{\tau_c - 1}$$

$$f = \frac{\tau_\lambda - \tau_r \tau_c}{\left(\dfrac{\eta_b h}{C_{P_c} T_0} \right) - \tau_\lambda}$$

$$\tau_t = 1 - \frac{1}{1 + f} \frac{\tau_r}{\tau_\lambda} (\tau_c - 1)$$

$$\pi_t = \tau_t^{\gamma_t/(\gamma_t - 1)e_t}$$

$$\eta_t = \frac{1 - \tau_t}{1 - \tau_t^{1/e_t}}$$

$$\frac{P_{t_6}}{P_0} = \frac{P_0}{P_6} \pi_r \pi_d \pi_c \pi_b \pi_t \pi_n$$

$$\frac{C_{P_t} T_6}{C_{P_c} T_0} = \frac{\tau_\lambda \tau_t}{\left(\dfrac{P_{t_6}}{P_6} \right)^{(\gamma_t - 1)/\gamma_t}}$$

$$\frac{V_6}{V_0} = \left\{ \frac{\tau_\lambda \tau_t}{\tau_r - 1} \left[1 - \left(\frac{P_{t_6}}{P_6} \right)^{-(\gamma_t - 1)/\gamma_t} \right] \right\}^{1/2}$$

$$\frac{F}{\dot{m}_0} = \frac{a_0 M_0}{32.17} \left\{ \left[(1 + f) \frac{V_6}{V_0} - 1 \right] + (1 + f) \frac{1}{V_6/V_0} \frac{1}{M_0^2} \frac{1}{\gamma_t} \frac{\gamma_t - 1}{\gamma_c - 1} \frac{C_{P_t}}{C_{P_c}} \frac{T_6}{T_0} \left(1 - \frac{P_0}{P_6} \right) \right\}$$

$$S = \frac{3600 f}{F/\dot{m}_0}$$

Notice as with the ramjet that if $P_6 = P_0$, nozzle on design, the specific thrust equation simplifies significantly.

Example 6.2. For the following typical design inputs

$$
\begin{array}{lll}
T_0 = 420°R & \pi_d = 0.9425 & e_c = 0.92, 0.89 \\
\gamma_c = 1.4 & \pi_b = 0.98 & e_t = 0.91 \\
C_{p_c} = 0.238 & \pi_n = 0.99 & P_6/P_0 = 1 \\
\gamma_t = 1.35 & \eta_b = 0.97 & h = 19{,}500 \text{ BTU/lbm} \\
C_{p_t} = 0.262 & & T_{t_4} = 2940°R
\end{array}
$$

The results of an investigation of a turbojet engine flying at Mach 2 are in Fig. 6.30.

It can be seen from Fig. 6.30 that a prospective designer would be immediately confronted with a design choice, because the compressor pressure ratio leading to maximum thrust per unit mass flow is far from that leading to minimum fuel consumption. Clearly, a short-range interceptor would better suit a low compressor pressure ratio with resultant high thrust per unit mass flow and lightweight (small compressor) engine. Conversely, the designer of a long-range transport would favor a high compressor pressure ratio, low specific fuel consumption engine. Thus, we see what should be obvious before an engine can be correctly designed—the mission for which it is being designed must be precisely understood.

Consider a second example in which the Mach number is varied. Figures 6.31–6.33, result. Conclusions are left to the reader.

6.5.6 Propellers

Thrust can be produced by rotating propellers in the same way as lift can be developed by the motion of a wing through the air. The airfoil at a local station on the propeller must be at an angle of attack with respect to the resultant relative wind at that point. The resultant velocity (see Fig. 6.34) is a function of the forward speed of translation and the tangential speed of rotation. The local angle of attack α_b is determined by the local geometric blade angle β and the angle ϕ that the resultant velocity vector V_r makes with a plane perpendicular to the axis of rotation of the propeller. ϕ is obtained from

$$
\phi = \tan^{-1} \frac{V}{\pi n D}
$$

where V is the speed of translation, n the speed of rotation, and D the diameter of the propeller. The advance ratio J, which is generally used to account for the effects of the angle ϕ, is defined as V/nD.

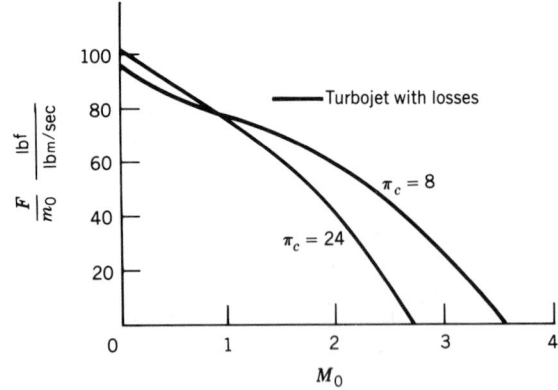

Fig. 6.31 Specific thrust as a function of Mach number.

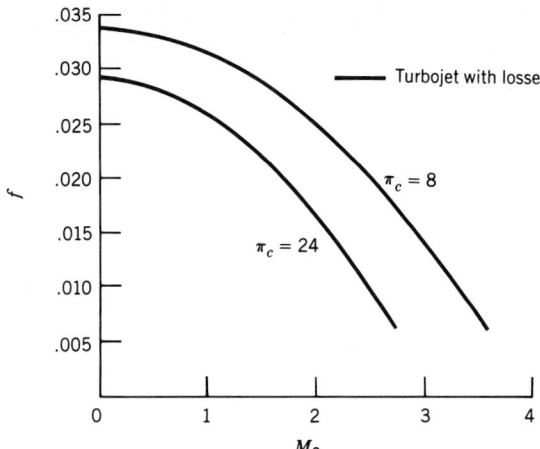

Fig. 6.32 Fuel to air ratio as a function of Mach number.

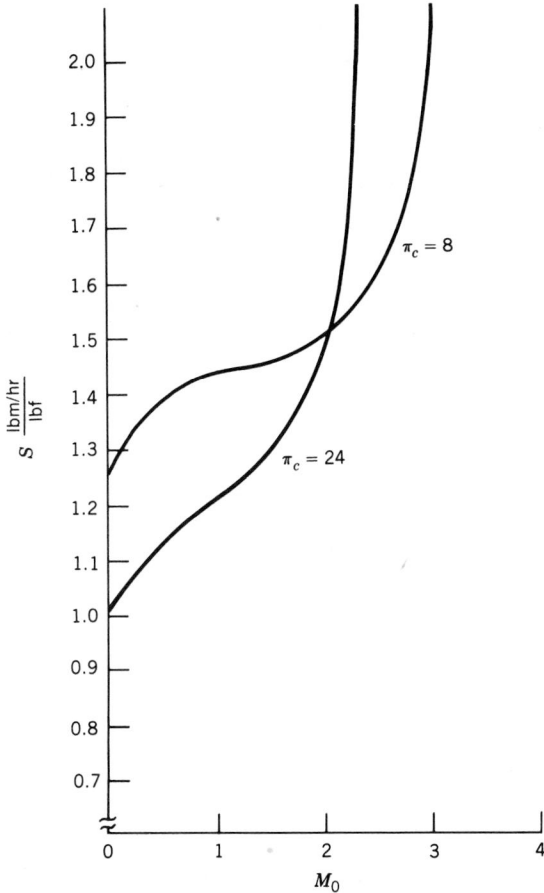

Fig. 6.33 Specific fuel consumption as a function of Mach number.

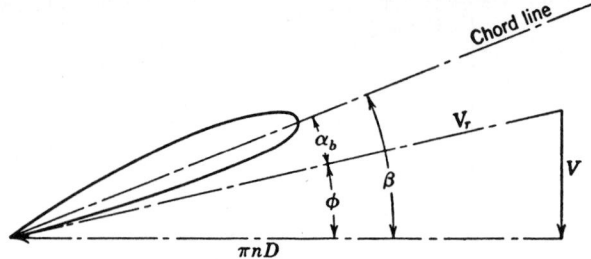

Fig. 6.34   Cross section of a rotating propeller.

More commonly used is

$$J = \frac{88V}{ND}$$

where V is in miles per hour, N is in revolutions per minute, and D is in feet.

Dimensional analysis shows that forward force F and torque Q can be expressed as follows:

$$F = K_{F_p} D^2 V^2$$
$$Q = K_{Q_p} D^3 V^2$$

where K_F and K_Q are functions of $\rho VD/\mu$ and V/nD, ρ and μ being the density and viscosity of the fluid, respectively. Alternative expressions are

$$F = C_{F_p} n^2 D^4$$
$$Q = C_{Q_p} n^2 D^5$$

The power output is

$$P = 2\pi n Q = C_{P_p} n^3 D^5$$

Propeller propulsive efficiency can be defined as

$$\eta = \frac{FV}{P} = \frac{C_{F_p} n^2 D^4 V}{C_{P_p} n^3 D^5} = \frac{C_F}{C_P} \frac{V}{nD} = \frac{C_F J}{C_P}$$

If engineering units are used,

$$C_F = 0.1518 \frac{F/1000}{\sigma (N/1000)^2 (D/10)^4}$$

$$C_P = 0.5 \frac{BHP/1000}{\sigma (N/1000)^3 (D/10)^5}$$

where σ is the ratio of the air density at altitude to the air density at sea level, F is in pounds, N in revolutions per minute, and D in feet.

Propeller data[31] obtained from tests are generally plotted as in Fig. 6.35.

The capacity of a propeller blade to absorb power is indicated by the "activity factor," which is defined as

$$\text{Activity factor} = \text{No. blades} \times \frac{100{,}000}{16} \int_{0.2}^{1.0} \left(\frac{r}{R}\right) \left(\frac{b}{D}\right) \left(d\frac{r}{R}\right)$$

where r = distance measured radially outward from axis of rotation
b = blade width at a given r
D = propeller diameter
R = propeller radius

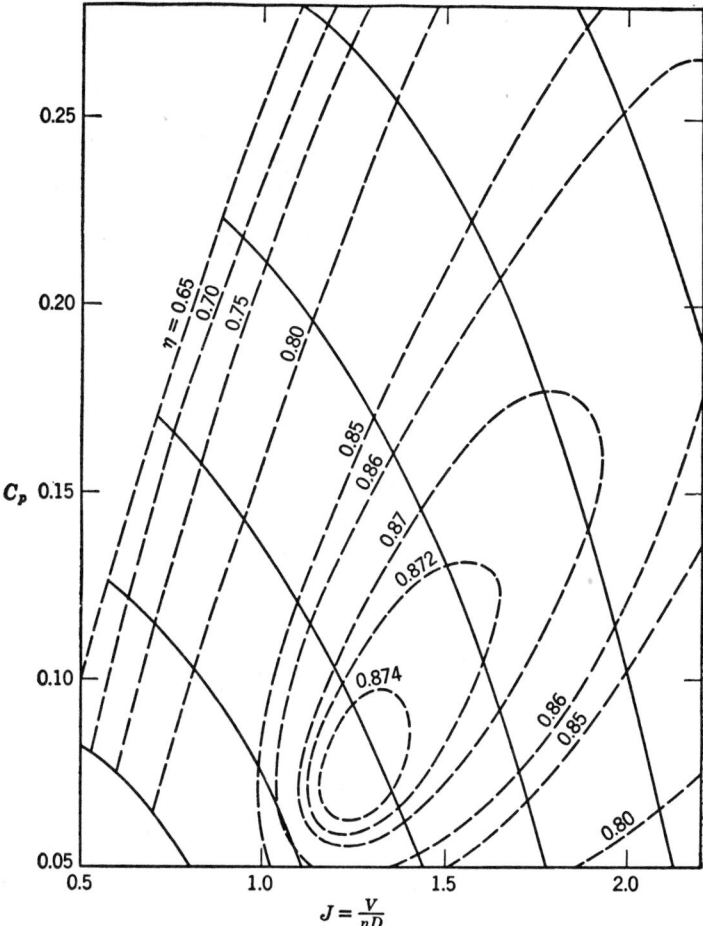

Fig. 6.35   Typical propeller characterization chart. Ref. NACA ARR 3125.

A convenient method of incorporating the effect of the activity factor into the propulsive efficiency is given in Figs. 6.36 and 6.37.

6.5.7   Electric Propulsion

Electrical rocket propulsion has application in space vehicles that can use low thrust devices where the massive power conversion and conditioning equipment currently necessary can be tolerated. The thrust levels range from micropounds to pounds, but have operation times of long duration. Electrical rocket propulsion systems break down into three broad classifications[32]—electrostatic, electrothermal, and electromagnetic.

In electrostatic propulsion systems, ions or colloid particles are accelerated to high velocity in an electrostatic field. The typical ion rocket uses an ion production chamber into which an uncharged propellant is pumped. The propellant is ionized and ions and electrons are withdrawn in separate streams. The ions are then accelerated by electrostatic field to produce thrust. The field is normally established by maintaining the ionization chamber at a positive potential and the final electrode at "ground." After acceleration the beam is neutralized by electrons expelled from the rocket. Ion engines can be further classified by ion source; either bombardment ionization or contact ionization. The principle of operation of colloidal rockets is similar. However, attempts to form uniform charged colloidal particles (1 to 0.01 μ in diameter) have been unsuccessful. In addition to the thrust producing mechanism, electrical rockets require a propellant storage and supply system and an electrical power source.

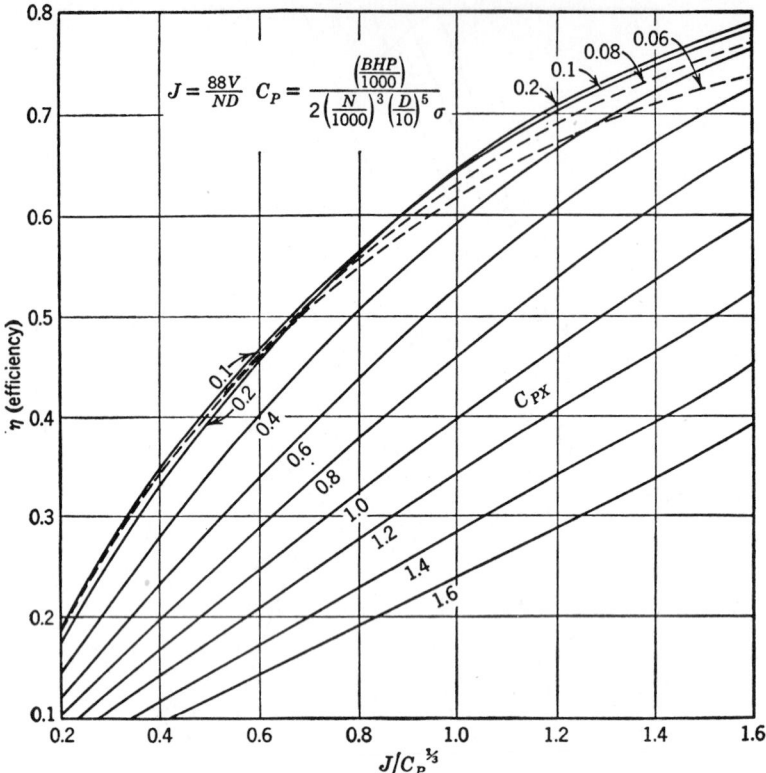

$$J = \frac{88V}{ND} \quad C_P = \frac{\left(\frac{BHP}{1000}\right)}{2\left(\frac{N}{1000}\right)^3 \left(\frac{D}{10}\right)^5 \sigma}$$

Fig. 6.36 General propeller chart 1 (from Perkins and Hage[29]).

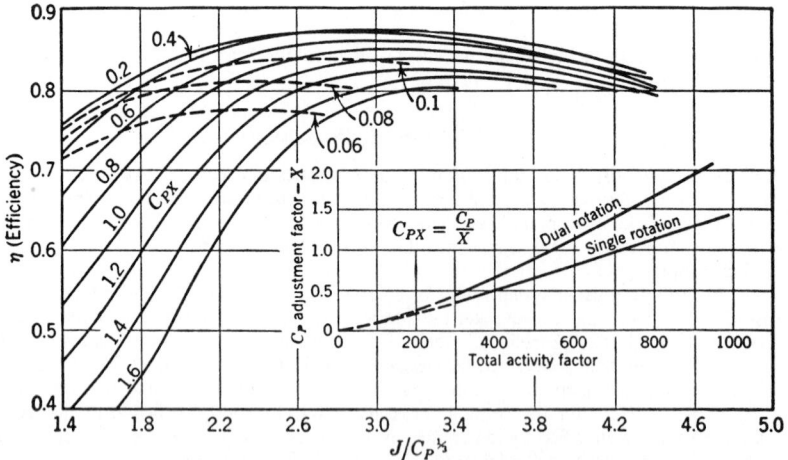

$$C_{PX} = \frac{C_P}{X}$$

Fig. 6.37 General propeller chart 2 (from Perkins and Hage[29]).

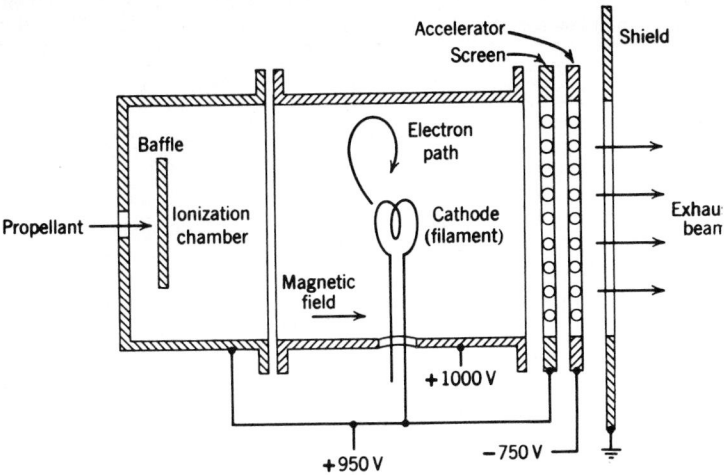

Fig. 6.38 Schematic diagram of a low-density ion engine (from Ref. 33).

Figure 6.38 is a schematic of a typical ion rocket showing the ion engine, power source, and propellant system. Some important relations for ion engines are

$$\text{Specific impulse} = I_{sp} = \frac{1}{g_e}\sqrt{2\frac{q}{m}V_a}$$

where q = charge of particle
$\quad m$ = particle mass
$\quad V_a$ = potential difference particle passes through
$\quad g_e$ = acceleration due to gravity at surface of earth

$$\text{Beam power} = P_b = IV_a = \frac{\dot{m}u_e^2}{2}$$

where I = total beam current $= (q/m)\dot{m}$
$\quad \dot{m}$ = particle mass flow rate
$\quad u_e$ = exhaust velocity

$$\text{Thrust} = \tau = \frac{2P_b}{g_e I_{sp}} = \dot{m}u_e$$

$$\text{Power efficiency } \eta = \frac{\text{beam power}}{\text{input electrical power}}$$

$$\eta \approx \frac{\frac{1}{2}mg_e^2 I_{sp}^2}{\frac{1}{2}mg_e^2 I_{sp}^2 + e_l}$$

where $\frac{1}{2}mg_e^2 I_{sp}^2$ = exhaust kinetic energy per particle
$\quad e_l$ = energy loss/ion, charging energy

The ability to ionize an atom depends on the electronic structure of the atom. The alkali metal elements lithium, sodium, potassium, rubidium, and cesium have single electrons in unfilled outer shells and are easiest to ionize. Table 6.8 indicates the first ionization potentials for these elements and others for reference.

TABLE 6.8   Ionization Potentials

	Element	Atomic No.	First Ionization Potential, c_v
Alkali Metals	Li	3	5.363
	Na	11	5.12
	K	19	4.318
	Rb	37	4.159
	Cs	55	3.87
Inert Elements	He	2	24.46
	Ne	10	21.47
	A	18	15.68
	H	1	13.527
	C	6	11.217
	Hg	80	10.39

Bombardment ionization is accomplished by an electron collision with an atom. Ionization occurs when the electron's kinetic energy equals or is greater than the atom's ionization potential.

$$\text{Rate of ion formation} = \frac{dn_i}{dt} = Q_i N_e n_a \text{ ions/sec unit volume}$$

where Q_i = ionization cross section

N_e = electron flux per second per unit volume

n_a = atom density

The ion engine shown schematically in Fig. 6.21 is a low current density device using an electron bombardment ion source. The propellant flows axially through the ionization chamber where it is bombarded by electrons generated from the cathode. As the electrons flow from the cathode to the anode (case), they are diverted into curved paths by the magnetic field and held in the chamber until at least one collision occurs. The approximate strength of the magnetic field is given by[32, 33]

$$B = \frac{\sqrt{8\left(\frac{m_e}{e}\right)\Delta V_a}}{r_a\left(1 - \frac{r_c^2}{r_a^2}\right)}$$

where B = magnetic field strength

m_e = electron mass

e = electron charge

r_a = anode radius

r_c = cathode radius

ΔV_a = anode potential relative to cathode

Engines similar to the one shown in Fig. 6.21 have produced ions with an energy consumption of the order of 450 V per ion and propellant efficiencies (ratio of ion mass flow to total propellant flow) greater than 90%. Beam interception by the accelerating electrode has been less than 5% and current densities of the order of 1.5 to 2.0 mA/cm^2 have been achieved.

Contact ionization will occur if the work function of the contact surface is greater than the ionization potential of the propellant. Therefore combinations of low work function alkali metals with high work function surfaces are utilized. The most frequently used combination is cesium and tungsten. The cesium contact ion engine has produced current densities of 20–80 mA/cm^2 at reasonable power efficiency and with very low beam interception by the acceleration system.

Electrothermal propulsion uses an arcjet or resistance jet. Figure 6.39 is a schematic of an arcjet. The propellant is heated to a high temperature in an electric arc and then expanded in a conventional nozzle.

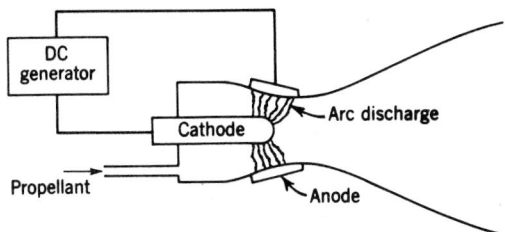

Fig. 6.39 Schematic diagram of an arcjet (from Ref. 33).

The high-current-density arc discharge is maintained by a large voltage difference between cathode and anode. Problem areas include heat transfer to the electrodes, which is especially severe at the cathode, anode erosion due to electron impact, and cathode erosion due to ion impact.

The efficiency of an arcjet is the ratio of kinetic energy flux in the exhaust to electrical power supplied. Efficiencies in the region of 50% for specific impulses of 1200 sec have been attained.

The resistance jet, or resistojet, operates in much the same way as the arcjet, except the propellant is heated by an electrical heating element instead of an arc. Resistojets have been used functionally on several military satellites.

The fraction f of the total energy supplied to the propellant that appears as thermal energy is

$$f = \frac{e_k}{e_k + \alpha_d e_d + \alpha_I e_I + \alpha_{II} e_{II}}$$

where e_k = internal energy added to the propellant per unit mass

e_d = energy of dissociation

e_I = energy of first ionization

e_{II} = energy of second ionization

α_d = dissociation fraction

α_I = first ionization fraction

α_{II} = second ionization fraction

Helium, hydrogen, and ammonia are possible propellants in an arcjet. Helium is superior to hydrogen for specific impulses up to 2000 sec. It has no dissociation losses, is chemically inert, has a high ionization potential, produces a high arc column resistance at high temperature giving good efficiency, and the electrode potential drops are low. Ammonia is heavy, storable, and is a liquid at normal operating temperatures. It dissociates when heated but is the preferred propellant for use in resistojets.

An example of an electromagnetic propulsion device is the steady crossed-field accelerator shown in Fig. 6.40.

The accelerator is bounded on diverging walls by continuous electrodes. A plasma, from a source such as an arcjet, enters the acceleration region. If Hall effects* are negligible, the current j_y, which flows through the plasma as a result of the applied and induced electrostatic field, is

$$j_y = \sigma(E - uB) \quad (J_y \text{ is conventional current} + \text{to} -)$$

where σ = plasma conductivity

B = magnetic field intensity

E = electrostatic field intensity

u = plasma velocity

---

*A resultant current or electron drift component to the right in the schematic is due to the Lorentz force and is called the Hall current.[32] This current produces a desired momentum exchange, but reduces the apparent conductivity, changes the direction of the net force on the gas, and can adversely affect the thrust. It can be significantly reduced by appropriate arrangement of the electrodes. In some thrusters the Hall effect is used to advantage in accelerating the plasma.

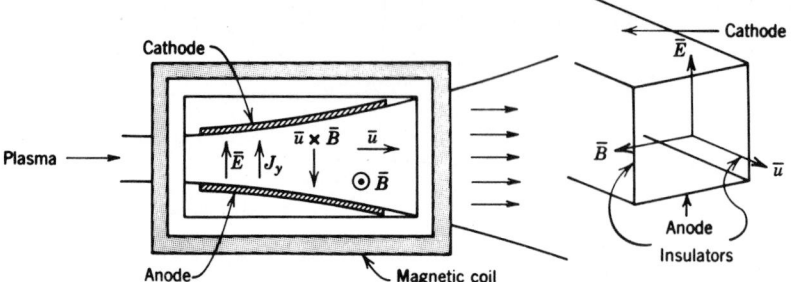

Fig. 6.40   Schematic of a typical steady-flow magnetogasdynamic accelerator.

In order to achieve useful conductivities (1 mho/cm or more) it may be necessary to seed the propellant gas with an alkali metal. One combination that can be used is argon seeded with potassium. Acceleration of the plasma occurs primarily as a result of the Lorentz force per unit volume, which acts on the moving electrons, in the current flow j_y where

$$\bar{F} = \bar{j} \times \bar{B}$$

The electrons collide with molecules in the plasma, thus transmitting kinetic energy to the plasma. Pressure forces may contribute some momentum increase.

The thrust developed by the accelerator can be determined from

$$\tau = BhI_t$$

where B = magnetic field intensity
h = channel height parallel to j
I_t = total current

The efficiency of the accelerator is

$$\eta = \frac{\tau^2/2m + \tau u_1}{p}$$

where p = total electrical power
u_1 = entrance velocity of plasma
m = plasma mass flow rate

The pulsed-plasma accelerator is another form of electromagnetic propulsion device. This accelerator does not require an electromagnet since the plasma current is used to generate the magnetic field that yields the Lorentz force. These devices produce high exhaust velocities and relatively low electrode erosion.

The traveling-wave accelerator is a third type of plasma accelerator that requires neither external magnets nor electrodes. It relies on currents being induced in the plasma by a traveling magnetic wave. The magnetic wave is established by a number of sequentially energized external conductor loops along the duct.

6.6   EXPERIMENTAL METHODS
By R. D. Marker, R. W. Milling and R. J. Stiles

6.6.1   Subsonic and Supersonic Wind Tunnels

Wind tunnels are devices that produce a controlled flow of gas for the study of aerodynamic phenomena. They are generally classified according to their speed range and test section size. Most subsonic tunnels are similar to the closed-circuit design shown in Fig. 6.41. Heat exchangers are seldom employed in low-speed subsonic wind tunnels to remove the heat of compression from the drive system; however, for large high-speed facilities they must be employed.

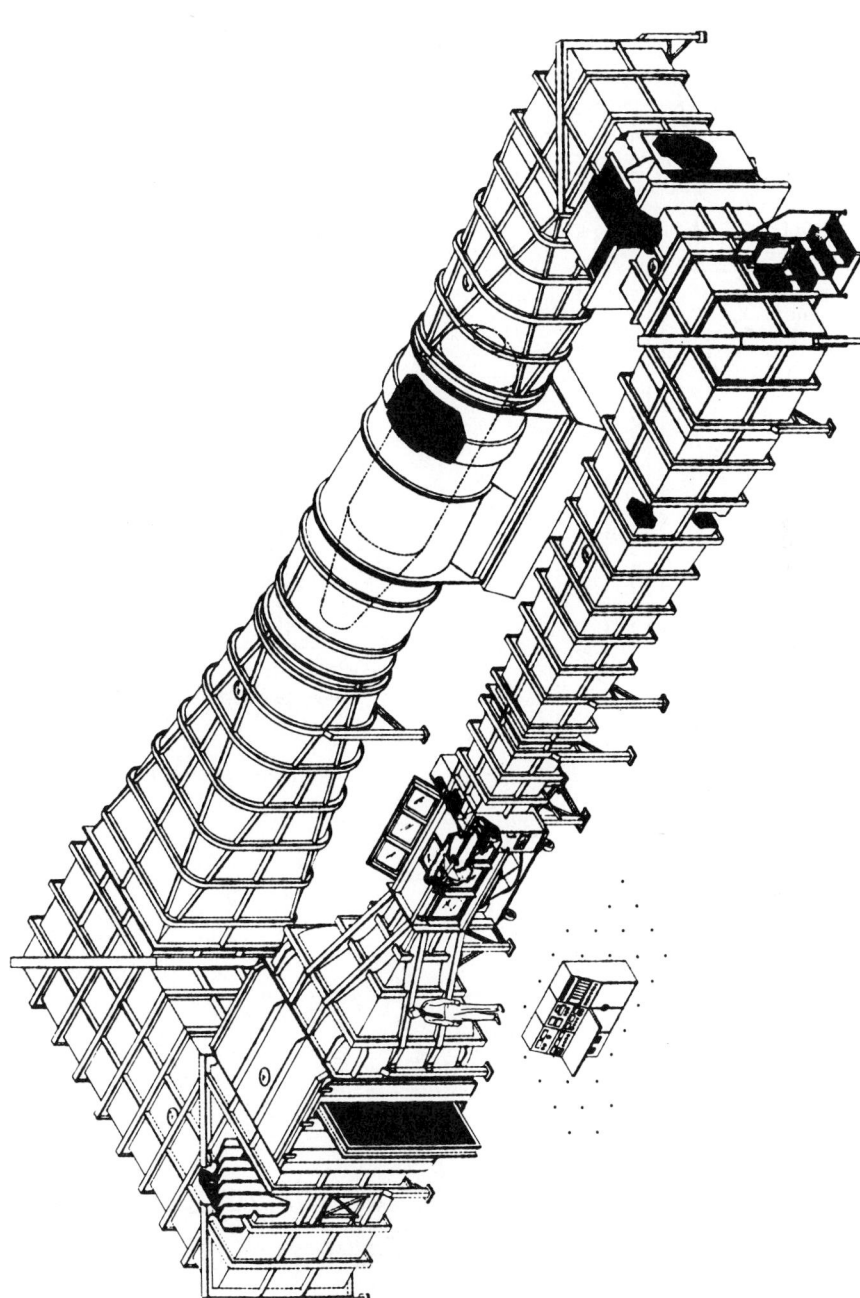

Fig. 6.41   USAF Academy 3 × 3 ft subsonic wind tunnel.

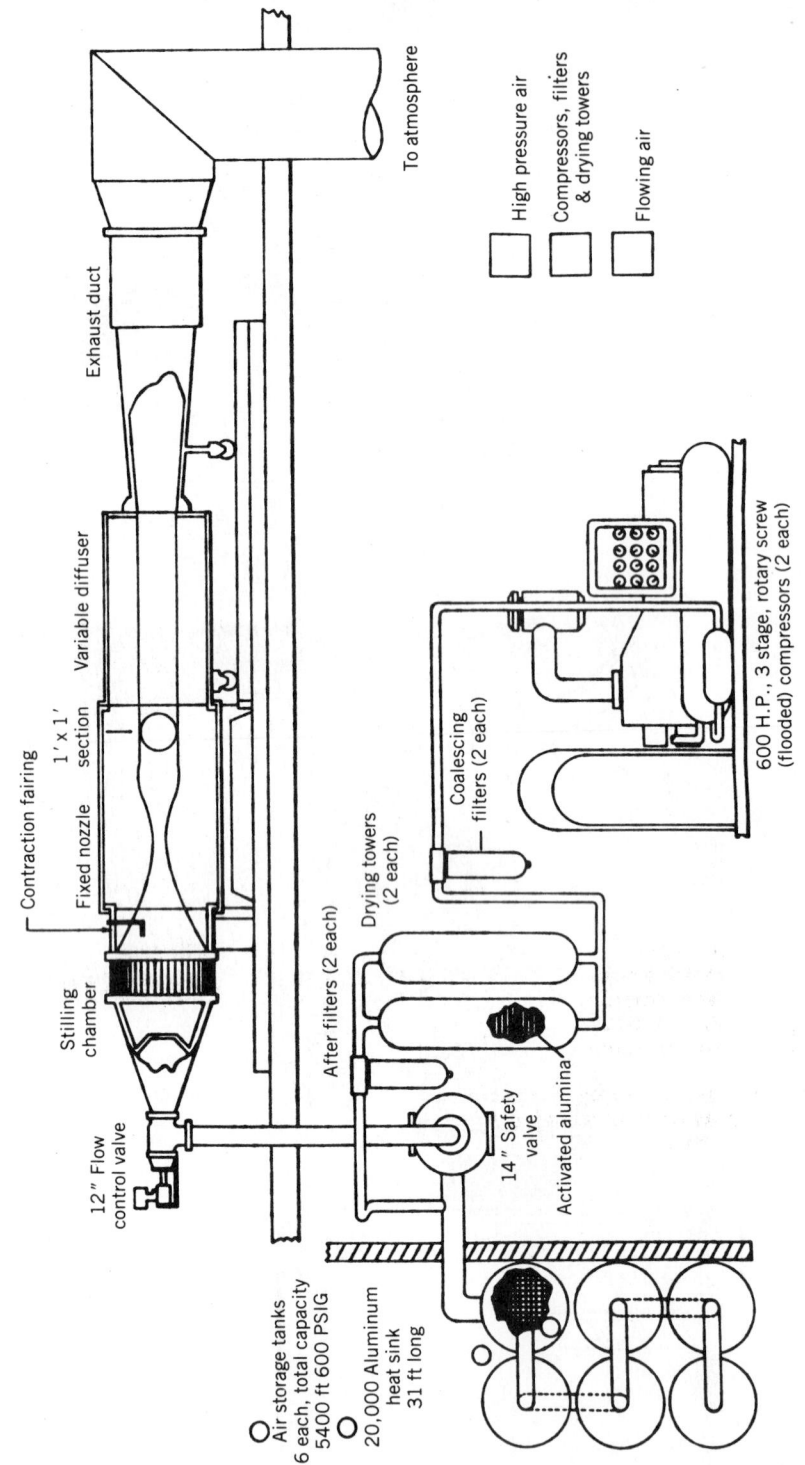

Fig. 6.42   Schematic of USAF Academy trisonic wind tunnel.

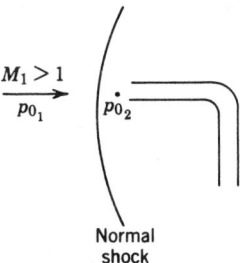

$M_1 > 1$

p_{0_1}

$\dot{p_{0_2}}$

Normal
shock

Fig. 6.43 Schematic of a total pressure probe in a supersonic flow.

One of the largest supersonic wind tunnels in the world is the 16 × 16 ft Propulsion Wind Tunnel Facility at the USAF Arnold Engineering and Development Center, Arnold Air Force Station, Tennessee. This facility consists of two adjacent continuous-flow circuits that share a 200-million-watt drive system to operate separate axial-flow compressors. The transonic circuit develops flows with Mach numbers as high as 1.6 while the supersonic circuit can produce Mach 4.75 flows. Both circuits have 16 × 16 × 40 ft long test sections. Reference 34 provides a summary description of this as well as many other wind tunnel facilities in use in the West.

The cost of building and operating large, continuous-flow transonic and supersonic wind tunnels is prohibitive for most users. Therefore, a blowdown wind tunnel design is commonly employed to achieve subsonic, transonic, and supersonic (trisonic) Mach numbers. Figure 6.42 is a schematic of a 1 × 1 ft trisonic wind tunnel. Compressed air is stored in a reservoir and then allowed to expand through a two-dimensional, fixed-contour nozzle into the test section. Interchangeable nozzle blocks and a transonic test section are used to achieve a broad range of Mach numbers. The test model is mounted on a model support system that provides remote control of the model attitude in the tunnel. Force balance systems and electrical transducers are used to obtain aerodynamic data during the relatively short test runs.

Precision instruments are used to calibrate the test section flow and to measure the various forces, pressures, and temperatures associated with a test. In a subsonic or transonic tunnel, the Mach number is determined from Eq. (6.63) where the total and static pressures, p_0 and p, are measured with a pitot-static probe:

$$\frac{p}{p_0} = \left(1 + \frac{\gamma - 1}{2} M^2\right)^{-\gamma/(\gamma - 1)} \tag{6.63}$$

At supersonic speeds, a single total pressure probe as shown in Fig. 6.43 is used. Since a detached normal shock wave forms in front of the probe, the probe actually senses the total pressure downstream of a normal shock (p_{0_2}), which can be used in conjunction with the free-stream total pressure in Eq. (6.49) to obtain the Mach number. Static pressures on the surface of a model are measured through small orifices in the surface, connected to multiple-tube manometers or pressure transducers.

Aerodynamic forces and moments on the model are generally measured with a multiple-component internal strain-gage balance such as the one shown in Fig. 6.44. The balance is installed inside the model and supported from the rear by a movable sting support system as in Fig. 6.45.

6.6.2 Hypervelocity Tunnels

As the velocity of vehicles traveling within the atmosphere has increased, the need for higher velocity test facilities has become evident. These facilities are variously called hypersonic or hypervelocity wind tunnels, depending on whether the Mach numbers or the velocity is extreme. A difficult problem in the design of such facilities is achieving the extremely high stagnation pressures and temperatures to properly simulate the desired flow field, as illustrated in Fig. 6.46. Many methods of heating air at high pressure have been devised. These include the use of pebble bed storage heaters, electrical resistance heaters, and combustion heaters. Of the more common methods currently in use, three will be discussed here: (1) the shock tunnel, (2) the "hot shot" tunnel, and (3) the arc-heated plasma tunnel.

The shock tunnel is an extension of the shock tube concept; it consists of a shock tube, a nozzle attached to the end of the driven section, and a diaphragm between the driven tube and the nozzle. The test gas, which has been compressed by both the incident and reflected shock waves, provides a reservoir of high-pressure, high-temperature gas of constant properties of a short duration. Stagnation

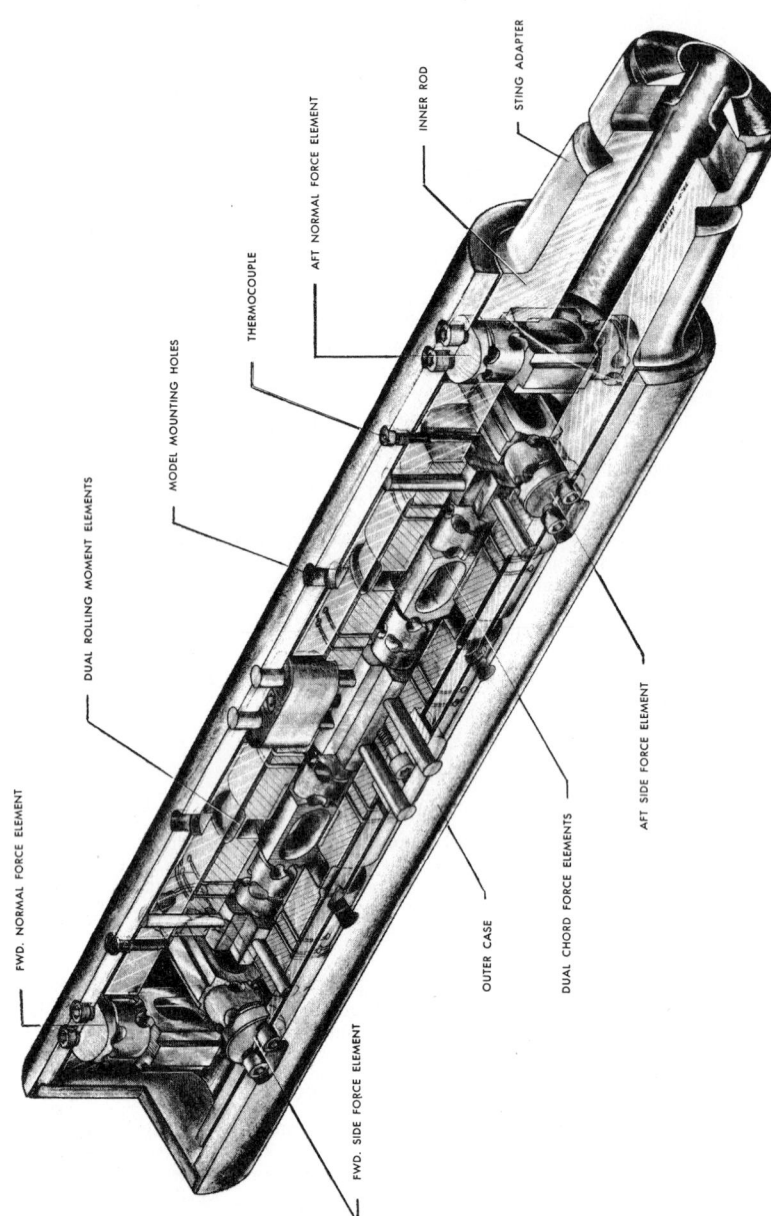

Fig. 6.44 Six-component internal strain-gage balance. (Courtesy of the Able Corp., Anaheim, California.)

Fig. 6.45   Model on sting in the USAF Academy transonic wind tunnel.

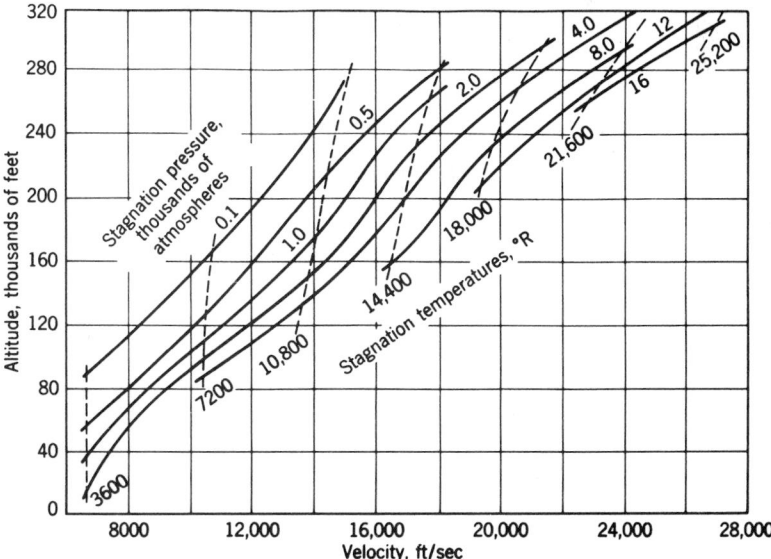

Fig. 6.46   Stagnation pressure and temperature required to duplicate flight conditions over a wide range of speeds and altitudes (from Ref. 35).

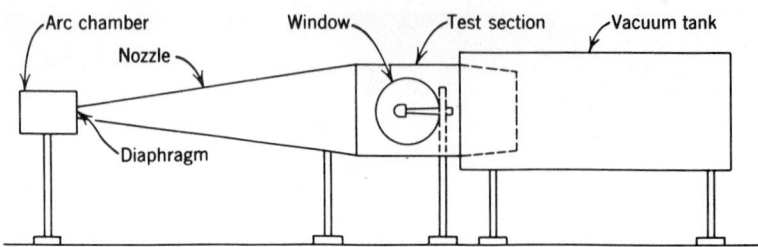

Fig. 6.47   Hot shot hypervelocity tunnel.

conditions of 2000 atm pressure at 7800 K for test periods of 6 or 7 msec have been obtained in this type facility.

The "hot shot" tunnel is illustrated in Fig. 6.47. The arc chamber is initially filled with air (or nitrogen) at pressures up to 680 atm while the remainder of the circuit is exhausted to a very low pressure (several microns). To initiate the flow, electrical energy from a capacitance or inductance storage system is discharged into the arc chamber. This addition of energy causes the pressure and temperature within the arc chamber to increase until the diaphragm ruptures, thus starting the flow. The high velocity flow in the test section may last for as long as 100 msec, but the flow properties vary continuously due to the decay of the pressure and temperature in the arc chamber with time. Relatively constant flow properties can be obtained for periods of 10 to 20 msec. Even this short duration is useful with high-response instrumentation. Operating conditions of 2000 atm stagnation pressure and 4000 K stagnation temperature are obtained, thus giving flow Mach numbers in excess of 20.

Arc-heated plasma tunnels use an electric current to heat the working fluid. There are many designs of arc heaters, both direct and alternating current; the high-voltage design shown schematically in Fig. 6.48 is included to illustrate the basic principles. Air is introduced tangentially into the arc chamber, producing a vortex flow that offers some heat insulation to the electrodes. The arc is drawn down the center of the two cylindrical water-cooled electrodes in a long column, thus giving efficient transfer of energy to the test gas. The arc is rotated by the magnetic field of the coil around the anode, thus reducing contamination of the flow and increasing the life of the electrodes. After passing through the heater, the air is expanded through an axisymmetrical nozzle that is usually conical.

Arc-heated plasma tunnels may be operated essentially continuously and are rated in thousands or even millions of watts. One of the largest facilities of this type uses a 50-million-watt arc heater and produces a stagnation enthalpy of up to 23,000 KJ/kg, stagnation pressures up to 136 atm, and core flow diameters of up to 1.5 m. The stream velocity in the plasma tunnel is generally very high (6100 m/s); however, the Mach number is usually less than 3.0 since the speed of sound is also high. The primary advantages of the plasma tunnel are continuous operation and the ability to develop very high velocities and high heating rates (3.4 MW/m^2). The disadvantages are the existence of nozzle throat erosion at high power levels and contamination of the flow by electrode material.

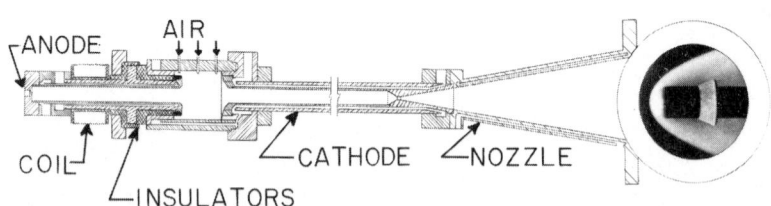

Fig. 6.48   Schematic of typical high-voltage arc air heater.

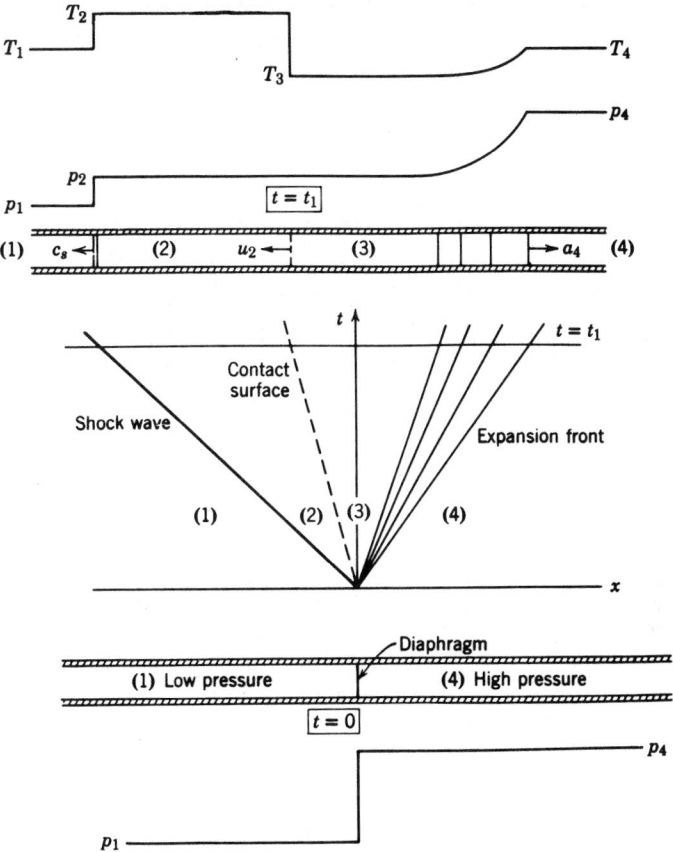

Fig. 6.49 Motion in a shock tube. (From H. W. Liepman and A. Roshko, *Elements of Gasdynamics*, New York, Wiley, 1957.)

6.6.3 Shock Tube

The shock tube provides a simple means of producing rapid changes in the state of a gas so that one may investigate problems in chemical physics including dissociation and ionization. In its simplest form the shock tube is a straight tube divided by a diaphragm into two sections, the driver and the driven tubes. The driver side is pressurized until the diaphragm ruptures and produces a shock wave that propagates through the test gas in the driven tube, thus raising its temperature and pressure. The nomenclature and flow regions are shown in Fig. 6.49. Region (1) is the undisturbed test gas; region (2), the shock heated test gas; region (3), the expanding driver gas; and region (4) is the undisturbed driver gas. The test time is generally determined by the time between the shock-wave passage and the arrival of the contact surface; however, near the end of the tube the testing time may be terminated by the arrival of the reflected shock wave.

The shock velocity is given by

$$c_s = M_1 a_1 = a_1 \left(\frac{\gamma - 1}{2\gamma} + \frac{\gamma + 1}{2\gamma} \frac{p_2}{p_1} \right)^{1/2} \tag{6.64}$$

The shock strength, p_2/p_1, is implicitly related to the diaphragm pressure ration, p_4/p_1, by

$$\frac{p_4}{p_1} = \frac{p_2}{p_1} \left[1 - \frac{(\gamma_4 - 1)(a_1/a_4)(p_2/p_1 - 1)}{\sqrt{2\gamma_1}\sqrt{2\gamma_1 + (\gamma_1 + 1)(p_2/p_1 - 1)}} \right]^{-2\gamma_4/(\gamma_4 - 1)} \tag{6.65}$$

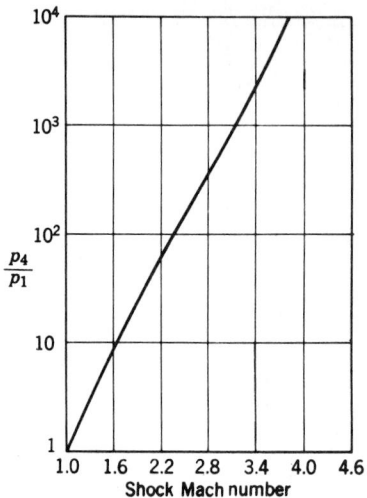

Fig. 6.50 Diaphragm pressure ratio, air driving air. [Courtesy I. I. Glase and G. N. Patterson, *JAS* **22** (1955).]

Other conditions across the shock may be found from the Rankine–Hugoniot relations of Eqs. (6.46)–(6.48).

The maximum shock velocity obtainable is limited by the physical requirements dictated by the need for very high diaphragm pressure ratios, as illustrated in Fig. 6.50 for the case of air driving air. Since the required diaphragm pressure ratio depends on the ratio of the speeds of sound of the driver and driven gases, the use of light driver gases (such as helium or hydrogen) and driver preheating are common.

6.6.4   Flow Visualization

The visualization of the flow about a model in subsonic flow can be achieved with short strings (tufts) attached to the model or held in the vicinity of the model on a grid network. Oil films can also be painted on the model to detect flow patterns and separation. Smoke tunnels are also used to visualize low speed flow.

In transonic and supersonic wind tunnels shadowgraph and Schlieren systems are used extensively. Both systems make use of the fact that the index of refraction of a gas is directly related to the gas density by $n = 1 + k\rho$, where n is the index of refraction, k is the Gladstone–Dale constant, and ρ the density.

Details of the Schlieren system are shown in Fig. 6.51. Light from the source is collimated by lens 1 and then passes through the test section. Lens 2 forms an image of the source at the knife edge and an image of the test section at the screen. As the knife edge is moved into the light beam the image of the test section on the screen is uniformly darkened since the knife edge is located at the image of the source. With air flowing over the model in the test section the light rays from the source will be

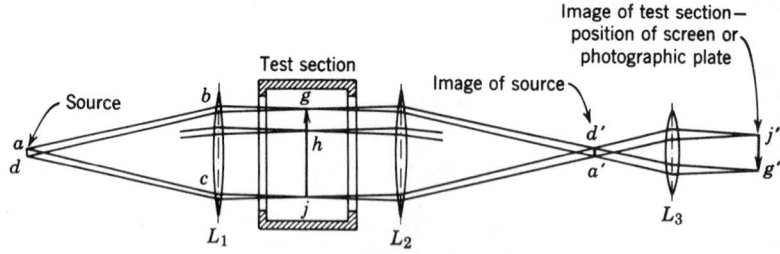

Fig. 6.51 Schlieren system.

refracted by density gradients. Those rays that are refracted onto the knife edge will cause dark areas on the screen while those being refracted off the knife edge will cause light areas on the screen. Schlieren systems are sensitive enough to show the density gradients of the boundary layer as well as those due to shock waves and expansion regions in supersonic flow.

6.6.5 Instrumentation

Instrumentation systems for aerodynamic testing have changed dramatically with the development of computer-based data acquisition and major advances in electronic instrumentation. Modern systems offer speed, versatility, and accuracy that was simply not available in the past. Our treatment in this section is very limited in scope and restricted to a brief discussion of instruments commonly employed for the measurement of velocity and pressure. References 36–38 are recommended for a complete treatment of instrumentation systems.

The hot-wire anemometer is frequently used in wind tunnel testing for measurements of flow velocity, flow direction, and turbulence. The sensing element and probe are very small; wire diameters of several microns are common. The wire, normally tungsten, is attached to two needlelike supports and is heated by the presence of an electrical current. As flow passes by the wire, it is cooled by convective heat transfer. Since the resistance of the wire is a function of wire temperature and the convective heat transfer is a function of the flow velocity, either the voltage or current supplied to the wire may be used to determine the velocity.

Hot-wire anemometers are built using one of two different designs: constant temperature or constant current. In the constant-temperature design, the wire is maintained at a constant temperature (resistance) by controlling the current passing through the wire. In this case, velocity is determined from a calibration of current versus velocity. For a constant-current anemometer, the wire attains an equilibrium temperature when the i^2R energy in the wire is balanced by the convective heat loss. Therefore, as the flow velocity changes, so does the wire temperature, and one can relate wire resistance to flow velocity for this design. Constant current anemometers have a serious drawback in that they are easily burned out if the flow velocity drops suddenly.

Perhaps the most important development in flow measurement in recent years has been the laser Doppler velocimeter (LDV). With this instrument, laser beams are focused at a point of interest in the flow and the incident light is scattered by particles passing through the point. The velocity of the particles causes a Doppler shift in the frequency of the scattered light. The scattered light is sensed by a photodetector that produces a signal that can be directly related to particle velocity. LDVs provide a much smaller sensing volume and significantly better frequency response than hot wires. Also, the LDV is a nonintrusive instrument; it does not disturb the flow. One must weigh these advantages against the increased cost and complexity of LDVs with respect to conventional instrumentation. LDVs also require test section windows that are optically flat and, very often, the flow must be seeded by inserting small particles to scatter the laser light.

Pressure sensors must be carefully selected, based on the range, accuracy, time response, and the physical size of the sensor required. (See Table 6.9.)

One of the most common pressure measuring devices is a liquid manometer in which the height of a column of liquid and the density of the liquid is used to determine an unknown pressure by

$$p = (\Delta h)9.802 \times 10^3 \, \mathrm{N/m^3} \, (\mathrm{sp\ gr})$$

where p = pressure, $\mathrm{N/m^2}$ gauge

Δh = column height, m

sp gr = specific gravity of the liquid compared to water

TABLE 6.9 Pressure and Vacuum Measuring Devices

Type	Usable Pressure Range	Accuracy	Time Response
Manometer	4 atm–0.01 Torr	Excellent	Long
Bourdon tube	Down to 1 mm	Good	Long
Null balance transducer	± 35 atm down to ± 0.1 atm	0.5% of full scale	Long
Strain gage transducer	350 atm down to 0.1 atm	1.0% of full scale	Short
Piezoelectric transducer	Above 350 atm to less than 0.1 atm	1.0% of full scale	Very short (1 kHz)
McLeod gage	5 to 10^{-5} Torr	See text	Long
Pirani gage	2 to 10^{-3} Torr	See text	Short
Thermocouple gage	1 to 10^{-3} Torr	See text	Short
Ionization gage	10^{-3} to 10^{-10} Torr	See text	Short

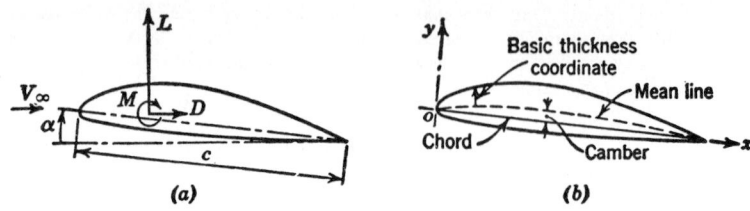

Fig. 6.52   Airfoil notation and geometry.

6.7   AERODYNAMICS OF WINGS
L. M. Nicolai, with contributions by W. F. Hallgren

6.7.1   Introduction

The fundamental problem of aerodynamics is to determine the aerodynamic force and moment on a body (i.e., airfoil, wing, body, or combination) immersed in a moving fluid.* The aerodynamic force is resolved into a force normal to the free-stream velocity (V_∞) called the lift and a force parallel to V_∞ called the drag (see Fig. 6.52a). The integration of the pressure and shear stress distribution over the body will yield the lift, drag, and moment. This stress distribution is the solution of the governing fluid equations and the boundary conditions describing the body surface and far-field property values. These governing equations, discussed in Section 6.2.3, are a complicated set of nonlinear partial differential equations.

The governing equation can be simplified by ignoring the viscous and heat conduction terms and considering what is called an inviscid flow situation. This seemingly drastic action is justified as follows: It can be shown[39] that the lift and moment for a thin body at a small to moderate angle of attack α (up to about 10°) depends primarily on the pressure distribution over the body; the tangential shearing stress contributes a negligible amount. In 1904 Prandtl hypothesized that for large Reynolds number (R_e see Sec. 6.1.6) the effects of viscosity are confined to a thin region next to the surface of the body called a boundary layer. It can be shown[39] that for slender bodies at moderate α and large R_e the static pressure at the surface is approximately the same as the pressure at the edge of the boundary layer; that is, the pressure is constant across the thin boundary layer. Thus, if we seek only the pressure distribution, we can consider the flow at the edge of the boundary layer, which is the inviscid flow region. In general, conventional aircraft operate at large (greater than 10^6) Reynolds number.

These arguments justify the basic approach for estimating the aerodynamic forces and moments over wings and bodies. The flow is broken into an inviscid region (outside the boundary layer) and a viscous region (boundary layer and wake). The inviscid flow theory will give us the pressure distribution from which we can determine the lift, moment, and inviscid drag. The viscous flow theory will give us the skin friction drag, drag due to separation, stall phenomena, and aerodynamic heating.

Sections 6.7.2–6.8.1 will discuss inviscid flow theory. Sections 6.10.1–6.11.1 will present viscous flow theory.

6.7.2   General Inviscid Flow Theory

In this section, we examine the flow field outside the boundary layer—the inviscid region. Additionally, the flow is assumed to be steady, adiabatic, and irrotational[†] (see Sec. 6.1.6) with a free-stream velocity V_∞. A slender solid body is placed in this uniform stream and disturbs the basic motion as shown in Fig. 6.53.

The total velocity $\bar{V}$ in the perturbed flow has components $V_\infty + u$, v, and w in the x, y, and z directions, respectively, where u, v, and w are called "perturbation velocity components."

---

*It is assumed a reference frame transformation may always be applied such that the body of interest is stationary while the fluid flows past it.

[†]Since the field of flow about an airfoil is uncontaminated by viscosity (except in the thin boundary layer), the fluid motion can be considered irrotational. This follows from the fact that irrotational motion in the approaching stream cannot become rotational owing to pressure forces in the field alone.

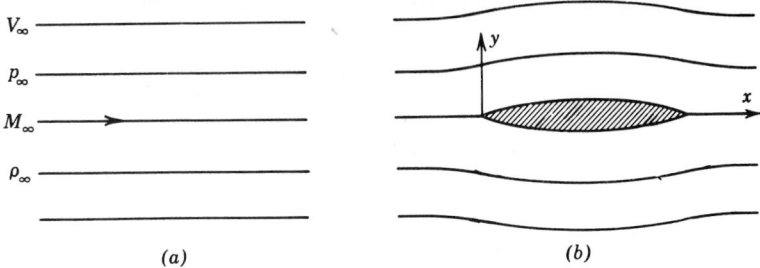

Fig. 6.53 Perturbation of a uniform flow by a slender body: (*a*) uniform flow and (*b*) perturbed flow.

The governing equations for steady, adiabatic, and irrotational flow are (from Section 6.2.3)

$$\frac{\partial}{\partial x_i}(\rho u_i) = 0 \quad \text{(Continuity)} \tag{6.66}$$

$$\rho u_j \frac{\partial u_i}{\partial x_j} = -\frac{\partial p}{\partial x_i} \quad \text{(Momentum)} \tag{6.67}$$

$$\frac{u_i u_i}{2} + h = h_0 \quad \text{(Energy)} \tag{6.68}$$

Since an adiabatic and reversible flow is also isentropic (no dissipative phenomena occur: flow is assumed inviscid and irrotational), we can use the useful isentropic relation:

$$\frac{p}{p_0} = \left(\frac{\rho}{\rho_0}\right)^\gamma \tag{6.69}$$

and the definition of the acoustic velocity (speed of sound)

$$a^2 = \left(\frac{\partial p}{\partial \rho}\right)_s \tag{6.70}$$

where s denotes isentropic.

Equations (6.66)–(6.70) are combined to give the single governing differential equation[40] (SGDE)

$$\left(1 - M_\infty^2\right)\frac{\partial u}{\partial x} + \frac{\partial v}{\partial y} + \frac{\partial w}{\partial z} = M_\infty^2\left[(\gamma + 1)\frac{u}{V_\infty} + \frac{\gamma + 1}{2}\frac{u^2}{V_\infty^2} + \frac{\gamma - 1}{2}\frac{v^2 + w^2}{V_\infty^2}\right]\frac{\partial u}{\partial x}$$

$$+ M_\infty^2\left[(\gamma - 1)\frac{u}{V_\infty} + \frac{\gamma + 1}{2}\frac{v^2}{V_\infty^2} + \frac{\gamma - 1}{2}\frac{w^2 + u^2}{V_\infty^2}\right]\frac{\partial v}{\partial y}$$

$$+ M_\infty^2\left[(\gamma - 1)\frac{u}{V_\infty} + \frac{\gamma + 1}{2}\frac{w^2}{V_\infty^2} + \frac{\gamma - 1}{2}\frac{u^2 + v^2}{V_\infty^2}\right]\frac{\partial w}{\partial z}$$

$$+ M_\infty^2\left[\frac{v}{V_\infty}\left(1 + \frac{u}{V_\infty}\right)\left(\frac{\partial u}{\partial y} + \frac{\partial v}{\partial x}\right) + \frac{w}{V_\infty}\left(1 + \frac{u}{V_\infty}\right)\left(\frac{\partial u}{\partial z} + \frac{\partial w}{\partial x}\right)\right.$$

$$\left. + \frac{vw}{V_\infty^2}\left(\frac{\partial w}{\partial y} + \frac{\partial v}{\partial z}\right)\right] \tag{6.71}$$

where $M_\infty = V_\infty/a_\infty$ is the free-stream Mach number.

The pressure coefficient is defined by

$$C_p = \frac{p - p_\infty}{\frac{1}{2}\rho_\infty V_\infty^2} = \frac{2}{\gamma M_\infty^2}\left(\frac{P}{P_\infty} - 1\right)$$

Using isentropic results, the expression for C_p is

$$C_p = \frac{2}{\gamma M_\infty^2} \left\{ \left[1 + \frac{\gamma - 1}{2} M_\infty^2 \left(1 - \frac{(V_\infty + u)^2 + v^2 + w^2}{V_\infty^2} \right) \right]^{\gamma/\gamma - 1} - 1 \right\} \qquad (6.72)$$

which may be approximated for slender bodies[40] at small angles of attack as

$$C_p = - \left[\frac{2u}{V_\infty} + (1 - M_\infty^2) \frac{u^2}{V_\infty^2} + \frac{v^2 + w^2}{V_\infty^2} \right] \qquad (6.73)$$

Since the flow is irrotational, there exists a velocity potential ϕ (see Section 6.1.6) such that $u = \partial\phi/\partial x$, $v = \partial\phi/\partial y$, and $w = \partial\phi/\partial z$. Thus Eq. (6.71) is a single differential equation for ϕ. The solution of Eq. (6.71) for ϕ, with appropriate boundary conditions, will yield the distribution of perturbation velocities. Equation (6.73) can then be used to determine the pressure coefficient distribution over the body. This, in turn leads to the lift coefficient.

Equation (6.71) cannot be solved in practice because of its nonlinearity. Thus the flow field is broken down into regimes, each having its own set of conditions and assumptions. The SGDE is then solved according to simplifications appropriate to the particular flow regime. These approximate flow regimes are:[41]

1. Incompressible: $M \leq 0.3$ (everywhere in flow)
2. Subsonic: $M < 1.0$ (everywhere in flow)
3. Transonic: $M < 1$ and $M > 1$ (mixed sub/supersonic flow)
4. Supersonic: $M > 1.0$ (everywhere in flow)
5. Hypersonic: $M_\infty \geq 5$

M (without subscript) denotes a "local" Mach number.

The Mach number for the hypersonic flow regime is not clearly defined. When the free-stream Mach number becomes large enough such that viscous interaction or chemically reacting effects begin to dominate the behavior of the flow field, the flow is called "hypersonic." $M_\infty \geq 5$ is frequently used as a guideline.[41]

6.7.3 Incompressible Flow Theory

When $M_\infty \approx 0$, the right-hand side of Eq. (6.71) vanishes and the governing equation becomes Laplace's equation

$$\nabla^2 \phi = 0$$

where

$$u = \frac{\partial\phi}{\partial x} \qquad w = \frac{\partial\phi}{\partial z}$$

$$v = \frac{\partial\phi}{\partial y} \qquad (6.74)$$

Thus the irrotational, incompressible flow is governed by a second-order linear partial differential equation. Physically, the flow velocity is so low that the changes in pressure are small, which results in infinitesimal changes in density, that is, $\rho \approx$ constant. Also, the kinetic energy of the fluid is very small compared to the internal energy so that the fluid temperature is constant in spite of the velocity perturbations.

Two-Dimensional Flow

The continuity equation for the two-dimensional flow of an incompressible fluid is

$$\frac{\partial u}{\partial x} + \frac{\partial v}{\partial y} = 0 \qquad (6.75)$$

As a consequence of continuity, in steady flow, a stream function ψ exists (see Section 6.1.6) such that

$$u = \frac{\partial \psi}{\partial y} \quad \text{and} \quad v = -\frac{\partial \psi}{\partial x} \tag{6.76}$$

The condition for irrotationality is $\nabla \times V = 0$. For two-dimensional flow, this reduces to

$$\frac{\partial v}{\partial x} - \frac{\partial u}{\partial y} = 0 \tag{6.77}$$

Substituting Eq. (6.76) into (6.77) yields

$$\frac{\partial^2 \psi}{\partial x^2} + \frac{\partial^2 \psi}{\partial y^2} = 0 \tag{6.78}$$

This, too, is a second-order linear partial differential equation. The potential lines (constant ϕ) and streamlines (constant ψ) are everywhere perpendicular to each other. This forms the basis of the conformal mapping discussed in Section 6.1.6.

Circulation
A concept called *circulation* is very useful in incompressible flow theory. It is defined as the line integral of tangential velocity components along a closed path, thus (Fig. 6.54)

$$\Gamma = \oint_C \mathbf{V} \cdot ds = \oint_C V \cos \theta \, ds \tag{6.79}$$

taken positive in the counterclockwise direction. It is seen from Fig. 6.54 that

$$\Gamma = \Sigma \, \Delta\Gamma \tag{6.80}$$

since internal line integrals cancel one another.
Integrating around the fluid element in Fig. 6.55 one obtains

$$d\Gamma = \left(\frac{\partial v}{\partial x} - \frac{\partial u}{\partial y} \right) dx \, dy \tag{6.81}$$

whence

$$\frac{d\Gamma}{dx \, dy} = \frac{\partial v}{\partial x} - \frac{\partial u}{\partial y} = 2\omega \tag{6.82}$$

which states that the circulation per unit area is equal to twice the angular velocity (ω) at that point.

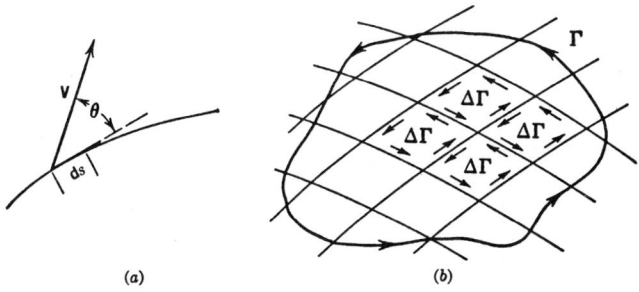

(a)   (b)

Fig. 6.54   Circulation.

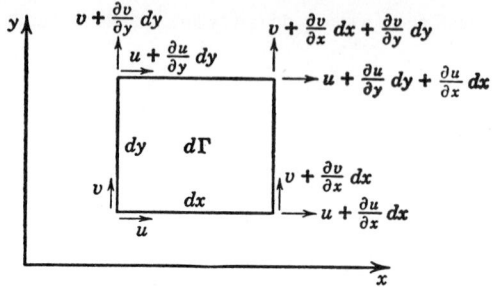

Fig. 6.55   Velocity components on a fluid element.

From Eqs. (6.80) and (6.81)

$$\Gamma = \oiint_s \left(\frac{\partial v}{\partial x} - \frac{\partial u}{\partial y} \right) dx\, dy \qquad (6.83)$$

Therefore, if the flow field is irrotational

$$\left(\frac{\partial v}{\partial x} - \frac{\partial u}{\partial y} = 0 \right)$$

the circulation will equal zero.

Kelvin's Theorem

Kelvin's circulation theorem states the circulation around a closed curve composed of fluid particles and lying wholly in a frictionless region remains constant with time, that is, $d\Gamma/dt = 0$. This is to be expected. Pressure and body forces are not tangential forces and therefore cannot produce a net change in the angular momentum of the enclosed system.[42]

Forces on a Body in a Frictionless (Inviscid) Fluid

Any irrotational physically possible* two-dimensional flow has a stream function and velocity potential that satisfy Laplace's equation (see Section 6.1.6). Conversely, any solution of Laplace's equation in two dimensions represents the stream function or velocity potential of an irrotational physically possible flow. Because Laplace's equation is linear, the sum of any number of solutions is also a solution. Thus, the solution to complicated flow patterns can be obtained by adding together the flow patterns of simple potential (irrotational) flows. For example, Section 6.1.6 shows how the solution for the flow about a rotating cylinder can be obtained by adding together the solution for a uniform stream, a doublet, and a free vortex flow.

The pressure distribution over a body of arbitrary shape can be found by using Bernoulli's equation (see Section 6.1.6):

$$p + \tfrac{1}{2}\rho V^2 = p_\infty + \tfrac{1}{2}\rho V_\infty^2 = \text{const} \quad \text{(total pressure, } P_0 \text{)} \qquad (6.84)$$

and the velocity potential or stream function obtained by superposing appropriate potential solutions. This pressure distribution, when integrated over the surface of the body, yields[42]

$$L' = \rho V_\infty \Gamma \quad \text{(where } L' \text{ is the lift per unit span)}$$
$$D' = 0 \quad \text{(where } D' \text{ is the drag per unit span)} \qquad (6.85)$$

Γ is the circulation about the body. Note, inviscid theory predicts zero drag on a body; this is called "D'Alembert's paradox."[43] Therefore, lift is directly proportional to circulation. This very important result is called the Kutta–Joukowski theorem.

Joukowski Airfoils

The Joukowski airfoils represent the simplest class of theoretically determinable airfoils. They are obtained by means of the mathematical method called conformal transformation (see Section 6.1.6).

*Any flow that satisfies continuity is termed physically possible.

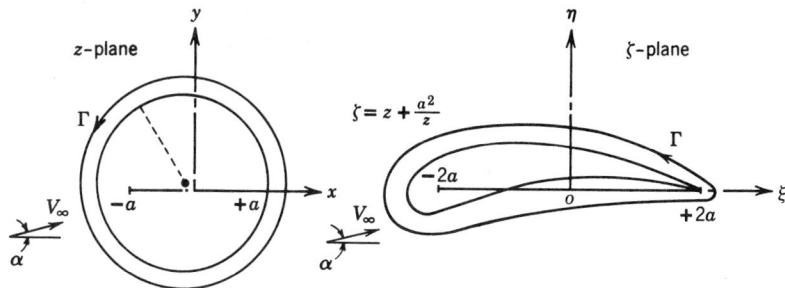

Fig. 6.56   Joukowski transformation.

By this method, the flow about a cylinder is transformed into the flow about a body having the shape of an airfoil (Fig. 6.56). The transformation is $\zeta = \xi + i\eta = z + a^2/z$, where a is a constant and $z = x + iy$.

The Joukowski profiles, however, are not the best for practical purposes. Structurally, the nose is too massive and the tail cusped. Aerodynamically, the pressure distribution is too asymmetric, which precludes high-speed or low-drag airfoils. Furthermore, the Joukowski family of airfoils is restricted in shape to the cambered type and hence does not include such shapes as the reflexed type generally necessary for the flying wing. Although Joukowski airfoils are not used in practice, their theoretical analysis predicts such general characteristics as (a) an aerodynamic center exists (see Properties of Airfoils); (b) the aerodynamic center is near the quarter-chord point for thin airfoils at small angles of attack (α); (c) the moment about the aerodynamic center is proportional to camber for thin airfoils; (d) lift coefficient $= 2\pi\alpha$ for thin airfoils at small angles of attack (α).

Generation of Lift
The theoretical calculation of the lift on a Joukowski airfoil, or any airfoil with sharp trailing edge, requires the adjustment of the circulation about the airfoil such that the rear stagnation point coincides with the trailing edge. This will ensure the fluid will flow smoothly off the trailing edge of the airfoil. This requirement is called the Kutta condition. The Kutta condition is satisfied physically through the action of viscosity.

Consider the course of events upon sudden starting of an airfoil. (See Fig. 6.57 and note that in airfoil practice the circulation Γ is positive clockwise.) At the first instant of starting, the flow past the airfoil is such that the circulation immediately around the airfoil is zero; no lift occurs on the airfoil and the rear stagnation point is found on the upper surface (Fig. 6.57a); also, the circulation around a closed path far from the airfoil is zero. However, as time progresses, the viscosity of the fluid takes effect. At the trailing edge, the fluid separates (Fig. 6.57b) and curls up, forming a vortex that is swept downstream with the flow (Fig. 6.57c). Yet the circulation along the fluid line, which is far from the body and unaffected by viscous action, remains zero by Kelvin's theorem. Hence, circulation must exist along a closed path immediately around but outside of the boundary layer of the airfoil; its value must be equal and opposite to that around the shed vortex, the so-called starting vortex. The lift (per unit span) in Fig. 6.57c, is equal to $\rho V_\infty \Gamma$.

Properties of Airfoils
Lift and drag (see Fig. 6.52) are defined as the components of the aerodynamic force normal to and parallel to the free-stream velocity, respectively. A "nose-up" rotation is defined as a positive

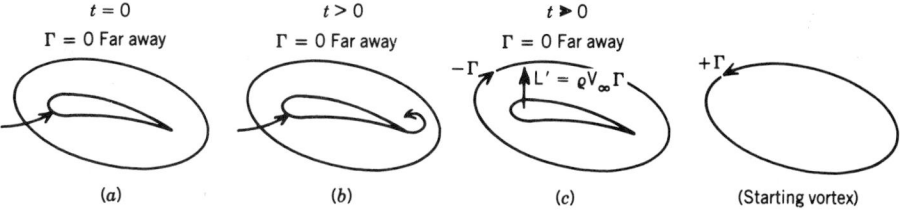

Fig. 6.57   Development of circulation lift.

moment, as shown in Fig. 6.52. The coefficients of lift, drag, and moment on an airfoil are defined by

$$C_l = \frac{L'}{\frac{1}{2}\rho_\infty V_\infty^2 c} \qquad C_d = \frac{D'}{\frac{1}{2}\rho_\infty V_\infty^2 c} \qquad C_m = \frac{M'}{\frac{1}{2}\rho_\infty V_\infty^2 c^2} \qquad (6.86)$$

respectively. L', D', and M' are the lift, drag, and moment per unit span. For three-dimensional bodies such as wings, these coefficients are

$$C_L = \frac{L}{\frac{1}{2}\rho_\infty V_\infty^2 S} \qquad C_D = \frac{D}{\frac{1}{2}\rho_\infty V_\infty^2 S} \qquad C_M = \frac{M}{\frac{1}{2}\rho_\infty V_\infty^2 S c} \qquad (6.87)$$

Notice the use of the lowercase subscripts for two-dimensional coefficients and uppercase subscripts for the three dimensional. These definitions are used in compressible flow as well as incompressible. For three-dimensional wings, S is the planform area (mean chord times span). The chord is the straight line connecting the leading and trailing edges of the airfoil. The angle of attack (α) is the angle between the free-stream velocity vector and the chord line (see Fig. 6.52a).

In the description of the geometry of an airfoil the basic thickness distribution and mean camber line coordinates are usually given. The mean camber line is a line that is equidistant from the upper and lower surfaces of the airfoil as measured perpendicular to the mean camber line itself. The camber is the distance of the mean camber line from the chord and is a function of x along the chord (camber is positive as shown in Fig. 6.52a).

The aerodynamic center (AC) of an airfoil is the point about which the moment coefficient is independent of angle of attack. Such a point exists for any two-dimensional airfoil,[39] as long as the lift is proportional to the angle of attack. (Experiment shows that an aerodynamic center exists also for finite wings below stall.) For thin airfoils, this point is located approximately one-quarter chord length behind the leading edge. For practical airfoils, the aerodynamic center usually lies in the neighborhood of the quarter-chord point; hence, the moment data are usually given about the quarter-chord point and the aerodynamic center.

Thin Airfoil Theory

In thin airfoil theory use is made of the superposition of potential solutions. The airfoil section is replaced by a distribution of vortices along its camber line. The camber line is given as some function $z = f(x)$. The theory gives the following useful results for cambered airfoils:[39]

$$C_l = 2\pi(\alpha - \alpha_{0L}) \qquad (6.88)$$

where α_{0L} is the angle of attack for $C_l = 0$ and the slope of the lift curve is 2π per radian.

$$\alpha_{0L} = -\frac{1}{\pi} \int_0^\pi \left(\frac{dz}{dx}\right)(\cos\theta - 1)\, d\theta \qquad (6.89)$$

$$C_{m_{AC}} = \frac{1}{2} \int_0^\pi \left(\frac{dz}{dx}\right)(\cos 2\theta - \cos\theta)\, d\theta \qquad (6.90)$$

where the dz/dx is the local slope of the camber line line. The θ appears through the change of variable $x = c/2(1 - \cos\theta)$ where x is the distance from leading edge. The aerodynamic center is located at $0.25c$ behind the leading edge.

From Eqs. (6.89) and (6.90) it can be seen that α_{0l} and $C_{m_{AC}}$ will always be a negative number for an airfoil whose mean camber line lies above the chord line ($z > 0$, positive camber). The shape of the camber in the vicinity of the trailing edge (TE) has a very powerful effect on α_{0L} and $C_{m_{AC}}$ whereas the camber line near the leading edge (LE) has very little influence on these properties. It is for this reason that flaps are put on the TE.

The results of thin airfoil theory agree very well with experimental results on two-dimensional airfoils. Some experimental results are shown in Figure 6.58 and Table 6.10.

Most of the NACA (now NASA) airfoils are classified among the three types, the four-digit, the five-digit, and the series 6 series. The meanings of these designations are illustrated by the following examples:

NACA 4415

4—The maximum camber of the mean line is $0.04c$.
4—The position of the maximum camber is at $0.4c$.
15—The maximum thickness is $0.15c$.

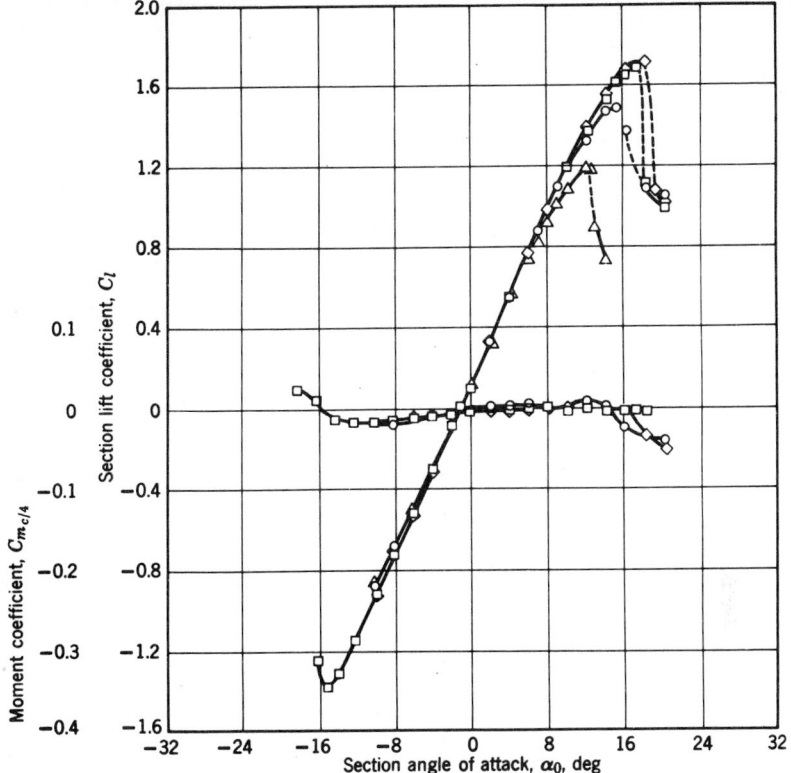

Fig. 6.58   Aerodynamic characteristics of the NACA 23015 airfoil section, 24-in. chord.

NACA 23012

2—The maximum camber of the mean line is approximately $0.02c$. The design lift coefficient is 0.15 times the first digit for this series.
30—The position of the maximum camber is at $0.30/2 = 0.15c$.
12—The maximum thickness is $0.12c$.

NACA 65_3-421

6—Series designation.
5—The minimum pressure is at $0.5c$.
3—The drag coefficient is near its minimum value over a range of lift coefficients of 0.3 above and below the design lift coefficient.
4—The design lift coefficient is 0.4.
21—The maximum thickness is $0.21c$.

Finite Wing Theory
In Prandtl's lifting line theory, the wing is replaced by a bundle of vortex segments at its aerodynamic center. Since the circulation varies along the span, the lengths of the vortex segments must be such that the sum of the strengths (circulations) at each spanwise point of the wing is equal to the circulation around the wing at that section. Furthermore, since a vortex filament cannot end in a fluid (according to Helmholz), it is assumed that "trailing" vortices extend downstream indefinitely from the ends of the vortices "bound" to the wing line. The strengths of the trailing vortices are equal to the strengths of the bound vortices. The pattern of each vortex filament is that of a horseshoe extending to infinity (see Fig. 6.59)

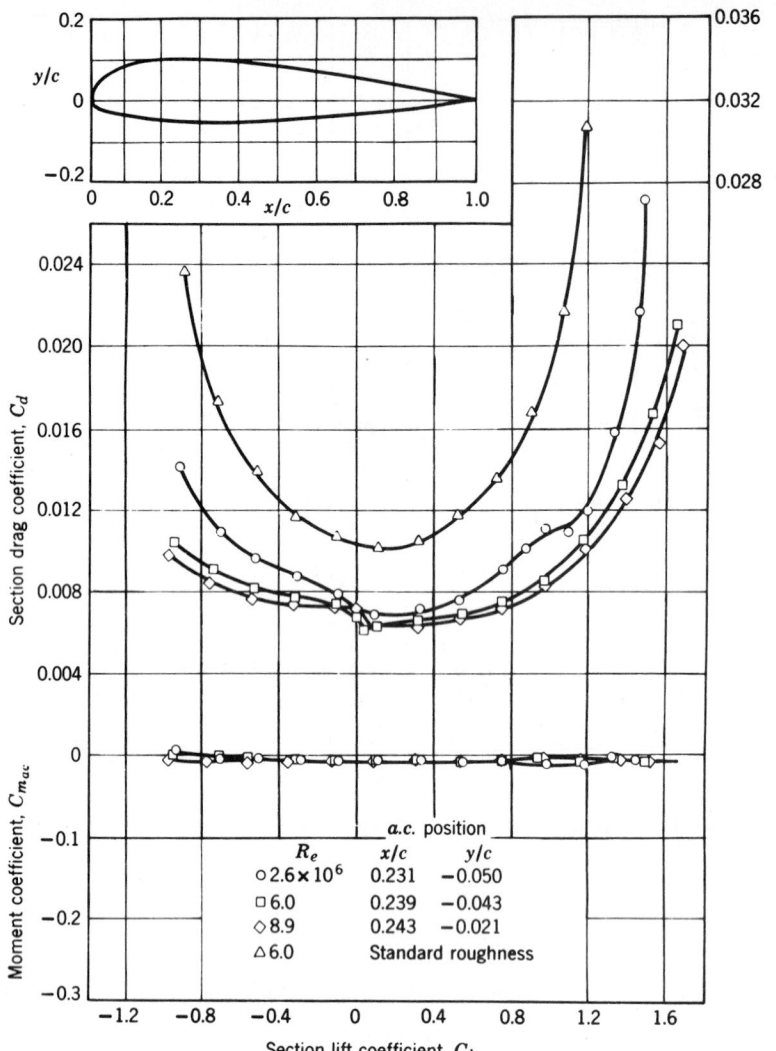

Fig. 6.58 *(Continued)*

TABLE 6.10 Representative Experimental Values of the Section Characteristics

Section Designation	$\dfrac{m_0}{2\pi}$	α_{0L} (degrees)	ac x/c aft of LE	$c_{m_{ac}}$
0009	0.995	0	0.25	0
2412	0.985	−1.9	0.243	−0.05
2415	0.97	−1.9	0.246	−0.05
2418	0.935	−1.85	0.242	−0.05
2421	0.925	−1.85	0.239	−0.045
2424	0.895	−1.8	0.228	−0.04
4412	0.985	−3.9	0.246	−0.095
23012	0.985	−1.2	0.241	−0.015
64_3–418	1.06	−2.9	0.271	−0.07
65_3–418	1.03	−2.5	0.266	−0.06
66_3–418	1.00	−2.5	0.264	−0.065

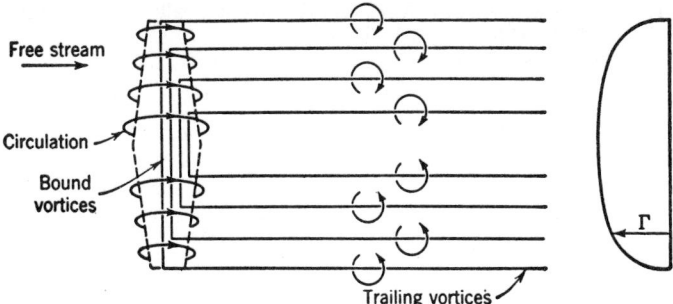

Fig. 6.59 Vortex system for lifting wing.

Induced Drag

The trailing vortices induce downward velocities (downwash) along the lifting line and behind the wing. At each wing section the downward velocity (w), combined with the free-stream velocity (V_∞) yields a resultant velocity V_{res}, inclined downward at an induced angle of attack, α_i. Therefore, the local resultant lift force vector L_0 (per unit length of span), being perpendicular to the local resultant velocity vector V_{res}, is inclined downstream, thus giving rise to a local induced drag, D_i (Fig. 6.60). Since α_i is small in practice, the local lifting force L is approximately equal to the local resultant lifting force L_0; also $D_i \approx \alpha_i L$. Summation of the local conditions (strip theory) over the span then yields an effective angle of attack (α_0) and a resultant induced drag for the entire wing. Symbols for the aerodynamics of the whole wing remain the same as for the local sections.

Prandtl found the planform yielding minimum induced drag (D_i) is elliptic and that the corresponding spanwise lift (circulation) distribution is also elliptic. This results in

$$C_{Di} = \frac{C_L^2}{\pi AR} \tag{6.91}$$

where AR is the aspect ratio defined as the ratio of the span squared to the wing area, and C_L is the resultant lift coefficient. The total drag coefficient is then given by

$$C_D = C_{Dp} + \frac{C_L^2}{\pi AR} \tag{6.92}$$

in which C_{Dp} is the airfoil section drag coefficient due to separation; that is, C_{Dp} is the drag coefficient of a two-dimensional (infinite aspect ratio) wing with the same airfoil section. For low and moderate values of C_l the airfoil section drag coefficient can be approximated very well by

$$C_d = C_{dp\,min} + K''C_l^2 \tag{6.93}$$

where K'' is the section drag due to lift factor and $C_{dp\,min}$ is the minimum separation drag coefficient

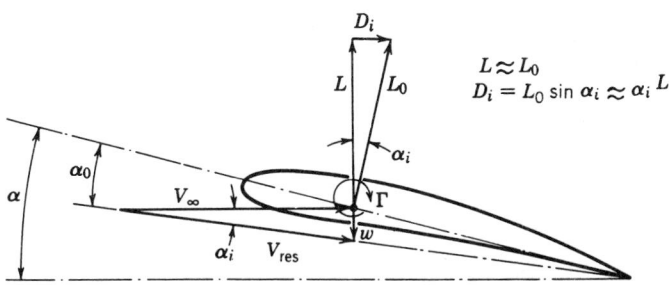

Fig. 6.60 Induced drag vector diagram.

(usually the value at $C_l = 0$). The value of K'' is found experimentally and has a value of 0.003 at $R_e = 9 \times 10^6$ for the airfoil shown in Fig. 6.58.

Equation 6.92 is often expressed as

$$C_D = C_{D_0} + \left[K'' \left(\frac{m_0}{m} \right)^2 + \frac{1}{\pi AR} \right] C_L^2 \qquad (6.94)$$

where the first term, C_{D0}, is the zero drag coefficient and the second term is the drag coefficient due to lift. The drag coefficient due to lift consists of a viscous separation term plus the inviscid (induced) term. In subsonic flow, C_{D0} is made up of $C_{dp\,min}$ and skin friction.

The formula for angle of attack is

$$\alpha = \alpha_0 + \alpha_i = \alpha_0 + \frac{C_L}{\pi AR} \qquad (6.95)$$

where α_0 is the effective angle of attack of the entire wing, that is, the angle of attack for infinite aspect ratio at the same lift coefficient.

Equation (6.95) can be rewritten in terms of a three-dimensional lift curve slope, m:

$$m = \frac{dC_L}{d\alpha} = \frac{m_0}{1 + m_0/\pi AR} \qquad (6.96)$$

m_0 is the two-dimensional lift curve slope value. The value of m_0 is close to the theoretical value of 2π for most airfoils (see Table 6.10).

Thus far it has been assumed that the wings have an elliptical lift distribution, that is, a constant downwash along the aerodynamic center. A wing having an elliptical planform has this elliptical lift distribution. For wings not having such a distribution the downwash and induced angle of attack will vary along the span as will the section lift coefficient C_l. The effect of nonelliptical distribution may be accounted for by the insertion of appropriate correction factors into the equations for the elliptical wing. Letting τ and δ represent the correction factors for induced angle of attack and induced drag, respectively, we get

$$\alpha = \alpha_0 + \frac{C_L(1 + \tau)}{\pi AR} \qquad (6.97)$$

$$C_{Di} = \frac{C_L^2(1 + \delta)}{\pi AR} \qquad (6.98)$$

$$m = \frac{dC_L}{d\alpha} = \frac{m_0}{1 + \dfrac{m_0(1 + \tau)}{\pi AR}} \qquad (6.99)$$

Values of τ and δ are given in Fig. 6.61 as functions of taper ratio and aspect ratio. The fact that τ and δ are always greater than or equal to zero illustrates that minimum induced drag is obtained for

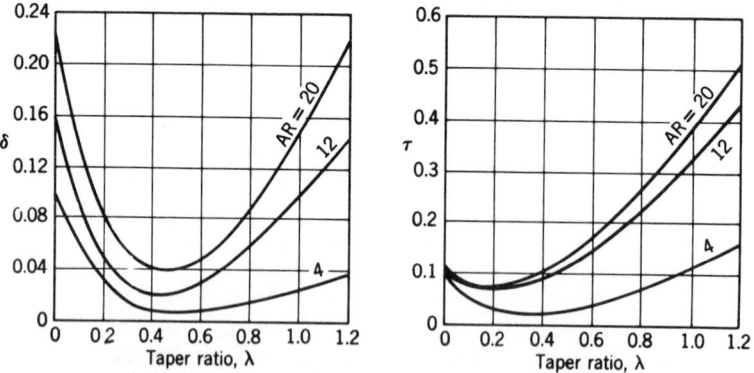

Fig. 6.61   Values of t and d for various aspect and taper ratios.

an elliptical wing since, then, $\tau = \delta = 0$. Note that taper ratios in the range of 0.2 to 0.5 give results very close to the elliptic wing.

Helmbold[44] suggested a first-order correction to Eq. (6.96) to give better agreement with experimental results on wings of moderate thickness ratio. The modified expression for the finite wing lift curve slope is

$$\frac{dC_L}{d\alpha} = \frac{m_0}{\dfrac{m_0}{\pi AR} + \sqrt{1 + \left(\dfrac{m_0}{\pi AR}\right)^2}} \tag{6.100}$$

Equation (6.100) predicts the lift curve slope for moderate and high-aspect ratio wings very well up to the region of stall. For low-aspect ratio wings, (aspect ratio of 2 or less), the normal force due to the separated flow around the wing tip is significant compared to the wing lift. This normal force is called the nonlinear lift. At large α (greater than about 10°) it causes the lift curve slope to be nonlinear. In this case the expression for wing C_L is given by

$$C_L = \left(\frac{dC_L}{d\alpha}\right)\alpha + (\text{nonlinear term})$$

This nonlinear term for low-aspect ratio wings is discussed in Ref. 45.

6.7.4 Subsonic Flow Theory

Here the flow is considered compressible but $M < 1.0$ everywhere in the flow field and bodies are slender in shape such that the perturbation velocities, u, v, and w are all very small compared to V_∞. The assumption of small perturbations, that is, $u/V_\infty \ll 1$, $v/V_\infty \ll 1$, and $w/V_\infty \ll 1$, permits neglecting the entire right-hand side of Eq. (6.71) as long as M_∞ is not close to 1 (i.e., the flow is not transonic). The SGDE reduces to the linear partial differential governing equation:

$$\left(1 - M_\infty^2\right)\frac{\partial^2\phi}{\partial x^2} + \frac{\partial^2\phi}{\partial y^2} + \frac{\partial^2\phi}{\partial z^2} = 0 \tag{6.101}$$

Later it will be shown Eq. (6.101) also holds for supersonic ($M > 1$ everywhere) flow.

Two-Dimensional Subsonic Flow
Equation (6.101) in two dimensions is an elliptic partial differential equation. It can be solved (e.g., separation of variables) for given boundary conditions. Ackeret's solution for a wavy wall is a classical and very useful example.[46]

By a simple change of variables, Eq. (6.101) (in two dimensions) can be transformed into Laplace's equation. Thus an incompressible flow past an airfoil can be related to the compressible flow past a similar shape and subsonic Mach number by the use of similarity rules. Three useful rules are:[47]

Rule 1

$$\text{If } \tau_c = \beta\tau_i \text{ and } \alpha)_c = \beta\alpha)_i, \text{ then } C_{pc} = C_{pi}$$

Rule 2

$$\text{If } \tau_c = \tau_i \text{ and } \alpha)_c = \alpha)_i, \text{ then } C_{pc} = \frac{C_{pi}}{\beta}$$

Rule 3

$$\text{If } \tau_c = \frac{\tau_i}{\beta} \text{ and } \alpha)_c = \frac{\alpha)_i}{\beta}, \text{ then } C_{pc} = \frac{C_{pi}}{\beta^2}$$

where C_p is the pressure coefficient, τ is the thickness ratio, α is the angle of attack, and β is the quantity $\sqrt{1 - M_\infty^2}$. The subscript c denotes the compressible case at free-stream Mach number M_∞ and the subscript i denotes the incompressible ($M \to 0$) case. The bodies can have different τ's and angles of attack but must be of similar shape (i.e., affinely related). The rules are valid only for thin airfoils at small angles of attack. They are not valid near stagnation points. Furthermore, they apply only to subcritical speeds, that is, only to those flight speeds that produce local velocities less than those of sound. The same rules apply to the lift coefficient and lift curve slope.

Rule 2 (thickness ratios and angles of attack the same) is called the Prandtl–Glauret rule. It expresses directly the effect of compressibility on the airfoil section characteristics.

$$C_{pc} = \frac{C_{pi}}{\sqrt{1 - M_\infty^2}} \tag{6.102}$$

$$C_{lc} = \frac{C_{li}}{\sqrt{1 - M_\infty^2}} \tag{6.103}$$

$$\left(\frac{dC_l}{d\alpha}\right)_c = \frac{(dC_l/d\alpha)_i}{\sqrt{1 - M_\infty^2}} \tag{6.104}$$

The comparison of Eqs. (6.103) and (6.104) with experimental data is shown in Figs. 6.62 and 6.63.

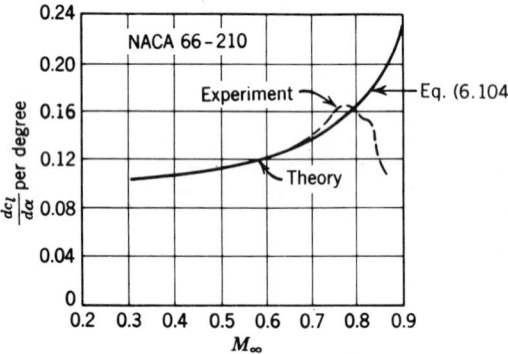

Fig. 6.62 Comparison of theoretical and experimental values of dC_i/da for a NACA series 6 airfoil of 10% thickness ratio.

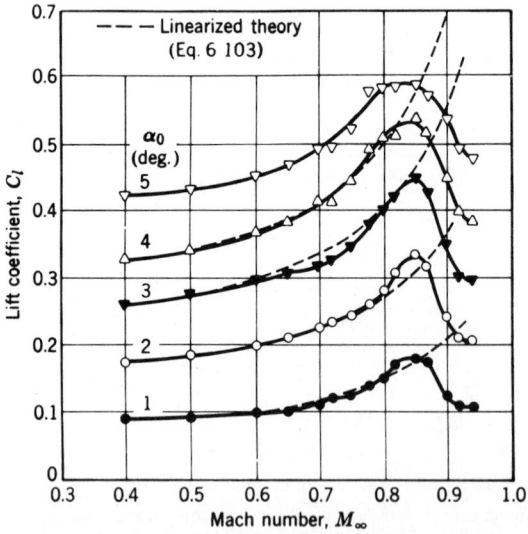

Fig. 6.63 Effect of compressibility on the lift of the NACA 0006-34 airfoil.

Laitone's Modification of Prandtl–Glauret Rule

The Laitone compressibility relation for two-dimensional flow is[41]

$$C_{pc} = \frac{C_{pi}}{\sqrt{1 - M_\infty^2} + \dfrac{M_\infty^2 \left(1 + \dfrac{\gamma - 1}{2} M_\infty^2\right)}{2\sqrt{1 - M_\infty^2}} C_{pi}} \tag{6.105}$$

This relation reduces to the Prandtl–Glauret rule for low Mach numbers but gives higher values than the Prandtl–Glauret rule at higher Mach numbers. Equation (6.105) agrees with experimental C_p data very well.

Three-Dimensional Subsonic Flow (Goethert Rule)

In the calculation of the subsonic compressibility effect on the characteristics of three-dimensional wings (straight or swept wings with finite span), it is necessary to use the more general correction formula developed by Goethert.[48] The rule, based upon a similarity solution of Eq. (6.101), is as follows:

The pressure coefficient at a given point for the compressible flow at Mach number M_∞ past a body of thickness τ is $1/\beta^2$ times as large as the pressure coefficient at the corresponding point for incompressible flow past a thinner affine body of thickness $\beta\tau$.

Consider a wing of thickness ratio τ_c, aspect ratio AR_c, and angle of attack $\alpha)_c$ in a stream of Mach number M_∞. Goethert's rule may be written as (subscripts i and c denote incompressible and compressible, respectively):

If

$$\tau_c = \frac{\tau_i}{\beta} \qquad AR_c = \frac{AR_i}{\beta} \qquad \alpha)_c = \frac{\alpha)_i}{\beta}$$

then

$$C_{pc} = \frac{C_{pi}}{\beta^2}$$

and

$$C_{Lc} = \frac{C_{Li}}{\beta^2}$$

Application of this rule gives the following expression for the compressible three-dimensional lift curve slope:[47]

$$\left(\frac{dC_L}{d\alpha}\right)_c = \frac{m_0}{\sqrt{1 - M_\infty^2} + \dfrac{m_0}{\pi AR}} \tag{6.106}$$

where m_0 is the incompressible two-dimensional lift curve slope (generally, $m_0 = 2\pi$ per radian).

6.7.5 Transonic Flow

A body is considered to be in the transonic flow regime when locally sonic flow occurs on the body surface. The lower limit of transonic flow is some M_∞ less than unity; the specific value depends on the thickness of the body. The upper limit is generally considered to be about $M_\infty = 1.2$, or when $M > 1.0$ everywhere in the flow field (M is the "local" Mach number).

Consider a conventional subsonic airfoil shape as shown in Fig. 6.64. If this airfoil is at a flight Mach number of 0.50 and a slight positive angle of attack, the maximum local velocity on the surface will be greater than the flight speed but most likely less than sonic speed. Assume an increase in flight Mach to 0.72 would produce the first evidence of local sonic flow. This condition would be the highest flight speed possible without regions of supersonic flow; the free-stream Mach number (0.72) is termed the "critical Mach number." Thus the critical Mach number is the boundary between subsonic and transonic flow and is an important point of reference for all compressible effects encountered in transonic flight.

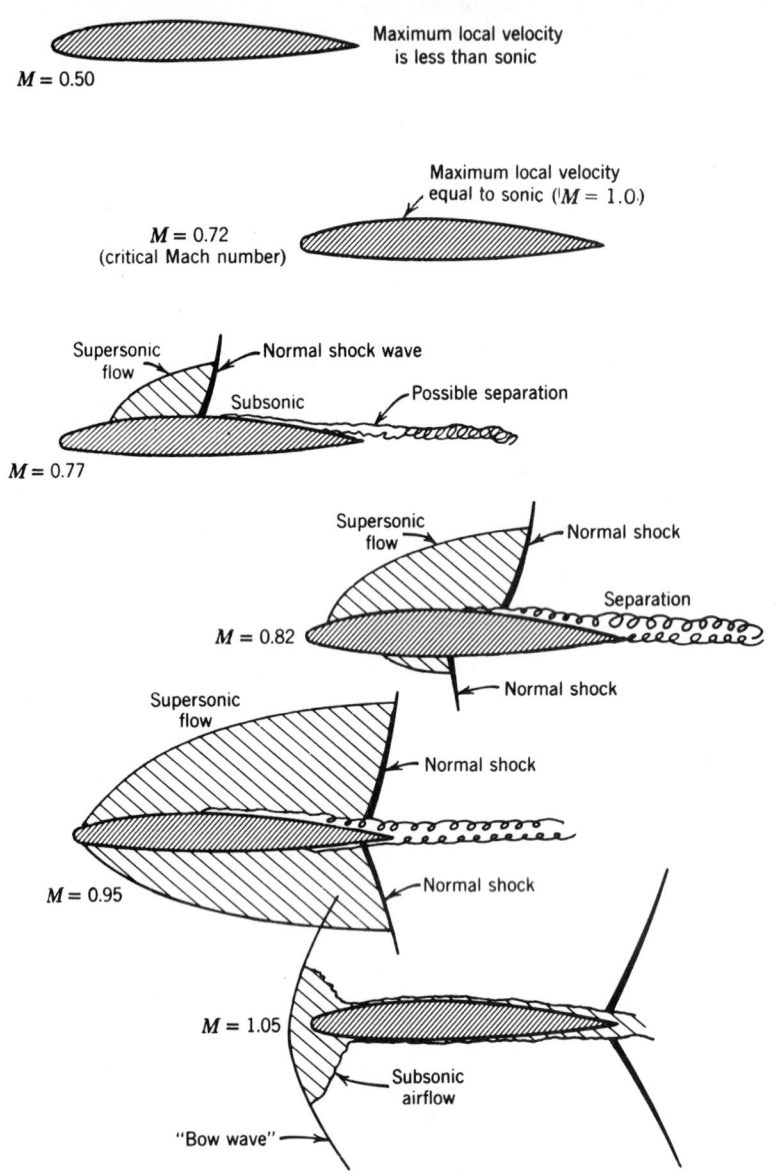

Fig. 6.64   Transonic flow patterns. (Courtesy NASA.)

As critical Mach number is exceeded, an area of supersonic flow is created on the wing surface. The acceleration of the airflow from subsonic to supersonic is smooth and unaccompanied by any shock waves. However, the transition from supersonic to subsonic occurs through a shock wave. Since there is no dramatic change in direction of the flow, the wave formed is typically a normal shock wave.

One of the principal effects of the normal shock wave is to produce a large increase in the static pressure of the airstream behind the wave. At speeds only slightly beyond the critical Mach number the shock wave formed is not strong enough to cause separation or any noticeable change in the aerodynamic force coefficients. If the shock wave is strong, the boundary layer may not have sufficient kinetic energy to withstand the large adverse pressure gradient and separation will occur. Such a flow condition is shown in Fig. 6.64 by the flow pattern for $M = 0.77$. Notice that a further increase in

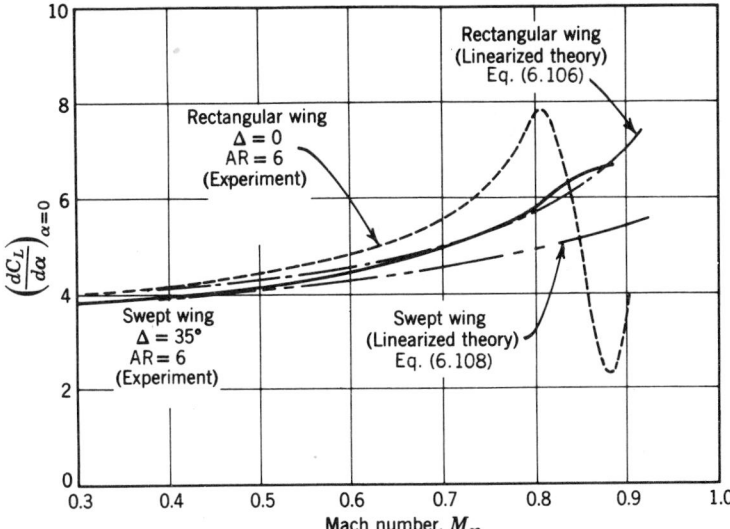

Fig. 6.65 Effect of compressibility and sweepback on lift of NACA 00012 airfoil with aspect ratio 6.

Mach number to 0.82 can enlarge the supersonic area on the upper surface and form an additional area of supersonic flow and a normal shock wave on the lower surface.

As the flight speed approaches the speed of sound, the areas of supersonic flow enlarge and the shock waves become stronger and move nearer the trailing edge. When the flight speed exceeds the speed of sound a "bow" wave forms at the leading edge as illustrated in Fig. 6.64 at $M = 1.05$. If the speed is increased to some higher supersonic value, all oblique portions of the wave incline more greatly and the detached normal shock portion of the bow shock wave moves closer to the leading edge.

The airflow separation induced by the shock-wave formation can create significant variations in the aerodynamic force coefficients. Some typical effects are an increase in the section drag coefficient and a decrease in the section lift coefficient for a given angle of attack (see Figs. 6.65 and 6.66). Accompanying the variations in C_l and C_d is a change in the pitching moment coefficient.

The Mach number that first produces a large increase in the drag coefficient is termed the "drag divergence Mach number." For most airfoils it exceeds the critical Mach number by 5–10%. This condition is also referred to as "drag divergence" or "drag rise."

Associated with the transonic drag rise are buffet, trim, and stability changes, and in general a decrease in the effectiveness of control surfaces. Conventional aileron, rudder, and elevator surfaces subjected to this high-frequency buffet may "buzz" and changes in moments may produce undesirable control forces. Also, when airflow separation occurs on the wing due to shock-wave formation, there will be a loss of lift and subsequent loss of downwash aft of the affected area. If the wings shock unevenly due to physical shape differences or sideslip, a rolling moment may be created and can contribute to control difficulty. If the shock-induced separation occurs symmetrically near the wing root, the resulting decrease in downwash on the horizontal tail will create a diving moment and the aircraft will "tuck under."

Since most of the difficulties of transonic flight are associated with shock-wave-induced flow separation, any means of delaying or alleviating this separation will improve the aerodynamic characteristics of an aircraft.

Wing Sweep
One of the most effective means of delaying and reducing the effects of shock-wave-induced flow separation is the use of sweep. Generally the effect of wing sweep will apply either to sweep back or sweep forward. While the swept forward wing has been used (e.g., X-29), sweepback has been found to be more practical (for structural reasons) for ordinary application.

A method of visualizing the effect of sweepback is shown in Fig. 6.67. The swept wing shown has the streamwise velocity vector resolved into components perpendicular and parallel to the leading edge. The component parallel to the leading edge could be visualized as moving across constant sections and, in doing so, does not contribute to the pressure distribution on the wing. The

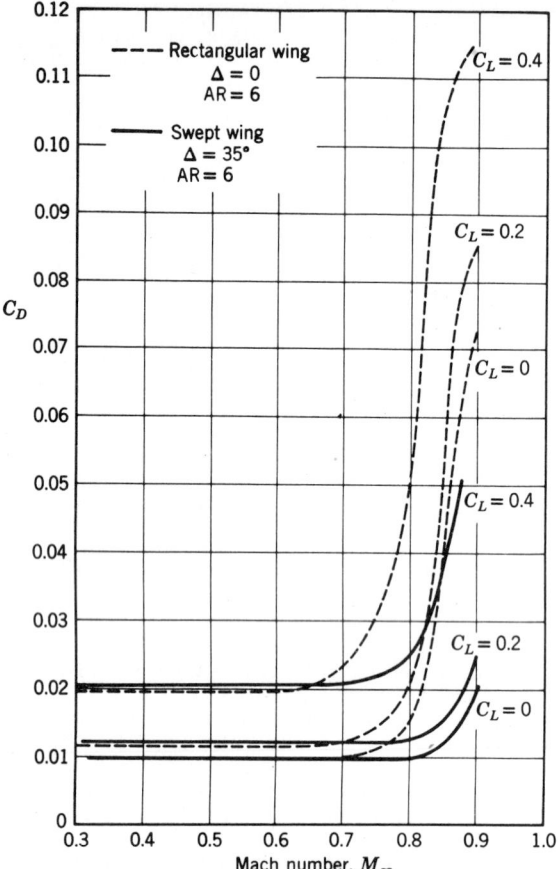

Fig. 6.66 Effect of compressibility and sweepback on drag of NACA 00012 airfoil with aspect ratio 6.

component perpendicular to the leading edge ($V_\infty \cos \Delta_{LE}$) is less than free-stream velocity; it is this component that determines the magnitude of the pressure distribution and the aerodynamic force coefficients.

Hence, sweep of a surface in high-speed flight produces a beneficial effect, since higher flight speeds may be obtained before components of velocity perpendicular to the leading edge produce critical conditions on the wing. Thus sweepback will increase the critical Mach number, drag divergence Mach number, and increase the Mach number at which the drag rise will peak. In other words, sweep will delay the onset of compressibility effects. The critical Mach number M_{cr} is increased by $(M_{cr})_{\Delta=0}/\cos \Delta_{LE}$.

In addition to the delay of the onset of compressibility effects, sweepback will reduce the magnitude of the changes in force coefficients due to compressibility. Since the component of velocity perpendicular to the leading edge is less than free-stream velocity, the magnitude of all pressure forces on the wing will be reduced (approximately by the square of the sweep angle). Since compressibility force divergence occurs due to change in pressure distribution, the use of sweepback will "soften" the force divergence. This effect is illustrated by the graph of Fig. 6.67, which shows the typical variation of drag coefficient with Mach number for various sweepback angles. The straight wing shown begins drag rise at about $M_\infty = 0.70$, reaches a peak near $M_\infty = 1.0$, and begins a continual drop past $M_\infty = 1.0$. Note that use of sweepback then delays the drag rise to some higher Mach number and reduces the magnitude of the rise in drag coefficient. This does not imply drag is decreasing. Recall that drag is equal to $D = C_D \frac{1}{2} \rho_\infty V_\infty^2$. Drag is proportional to the velocity squared. It is evident from the figure that small angles of sweep provide very little benefit. If sweep is to be used at all, at least 35° to 45° should be used.

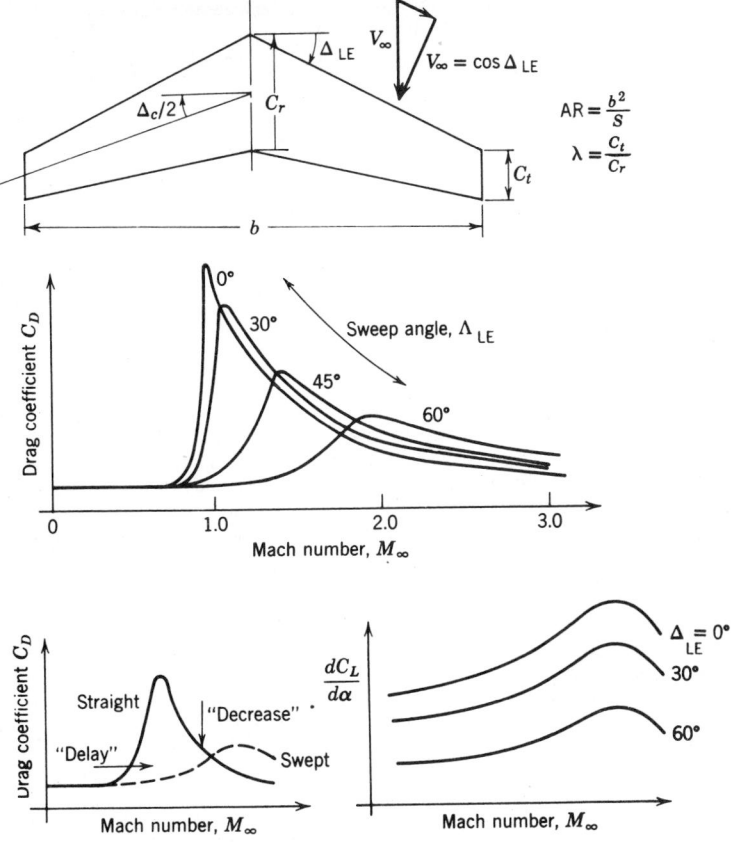

Fig. 6.67   General effects of sweepback.

A disadvantage of wing sweep is the decrease in wing lift curve slope. Reference 49 shows this effect to be approximately

$$\left(\frac{dC_L}{d\alpha}\right)_{\Lambda_{LE}} = \left(\frac{dC_L}{d\alpha}\right)_{\Lambda=0} \cos \Lambda_{LE} \tag{6.107}$$

This means a swept wing aircraft will have to land and takeoff at higher angles of attack than a straight wing aircraft.

If we combine the corrections for sweep [Eq. (6.107)] and Mach number [Eq. (6.104)] with Eq. (6.100), we get the following very useful expression for the finite wing lift curve slope:

$$\frac{dC_L}{d\alpha} = \frac{2\pi AR}{2 + \sqrt{4 + AR^2\beta^2\left(1 + \dfrac{\tan^2 \Delta c/2}{\beta^2}\right)}} \tag{6.108}$$

where $\beta = \sqrt{1 - M_\infty^2}$ and $\Delta_{c/2}$ is the sweep of the half chord line. Equation (6.108) predicts the linear portion of the lift curve slope quite well for a wide range of planform shapes and $M_\infty < 1$. (See Fig. 6.65.)

Other disadvantages to wing sweep are a reduction in $C_{L \max}$ and tip stall. The early flow separation at the tip is due to the spanwise flow (from root toward wing tip) causing a thickening of the boundary layer near the tips and hastening flow separation.

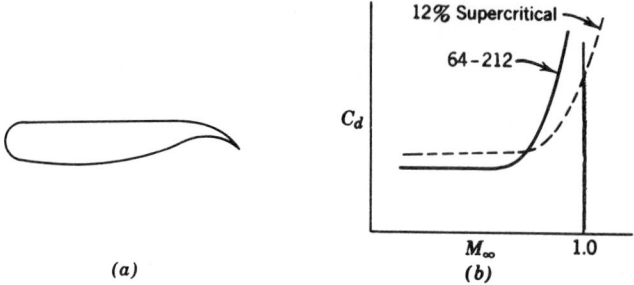

$$C_d$$

12% Supercritical

64-212

M_∞ 1.0

(a)   (b)

Fig. 6.68 Supercritical wing shape and effect of divergence Mach number: (a) 12% thick supercritical airfoil and (b) comparison of two 12% airfoils.

Supercritical Wing

Another way of delaying the drag rise due to shock-wave-induced separation is by using an airfoil shape called a supercritical section. A typical supercritical airfoil shape is shown in Fig. 6.68a. The section is shaped such that the normal shocks occur at locations where the surface pressure is decreasing[43] (favorable pressure gradient) or zero. Thus the boundary layer is better able to cope with the pressure jump across the normal shock and resist the tendency to separate. For a given thickness ratio, the critical Mach number stays the same but the divergence Mach number can be delayed by a supercritical shape as shown in Fig. 6.68b.

Transonic Similarity Solutions

Thus far the discussion of transonic flow effects has largely been based on experimental observations. This is because the transonic flow theory is very meager. Equation (6.71) cannot be linearized by assumptions of small perturbation for flow near $M_\infty = 1$. The governing flow equation is highly nonlinear and general solutions or approximations are not possible.

Equation (6.71) can be simplified somewhat by assuming small perturbations and discarding terms of small orders of magnitude. The resulting approximate equation for two-dimensional flow is

$$\left(1 - M_\infty^2\right)\frac{\partial^2\phi}{\partial x^2} + \frac{\partial^2\phi}{\partial y^2} = \frac{(\gamma + 1)M_\infty^2}{V_\infty}\frac{\partial\phi}{\partial x}\frac{\partial^2\phi}{\partial x^2} \qquad (6.109)$$

A similarity solution of Eq. (6.109) yields the following similarity rule for similar shaped bodies in transonic flow.

Consider two different transonic flow situations denoted by subscripts 1 and 2. If two similar shaped bodies of different thickness ratios τ_1 and τ_2 have their Mach numbers and gases (denoted by ratio of specific heats γ) related such that the following condition is satisfied:

$$\frac{\tau_1 M_1^2(\gamma_1 + 1)}{\left(1 - M_1^2\right)^{3/2}} = \frac{\tau_2 M_2^2(\gamma_2 + 1)}{\left(1 - M_2^2\right)^{3/2}} \qquad (6.110)$$

then the pressure coefficients on the two bodies are related as

$$C_{p2} = C_{p1}\left(\frac{\gamma_1 + 1}{\gamma_2 + 1}\right)\left(\frac{M_1}{M_2}\right)^2\left(\frac{1 - M_2^2}{1 - M_1^2}\right)$$

Here, as before, a known flow solution is extended to another flow solution for affinely related shapes. However, notice that the transonic similarity rule is much more restrictive than the subsonic rules. For example, the same body ($\tau_1 = \tau_2$) cannot be compared at different Mach numbers unless the gas is changed (i.e., γ_1 and γ_2 are different values).

6.7.6 Supersonic Flow Theory

Supersonic Thin Airfoil Theory (Two Dimensional)

In supersonic flow Eq. (6.71) can again be linearized if the airfoil is thin and at small angles of attack. Not only must the flow be greater than unity everywhere but also it must be less than some $M_{\text{hypersonic}}$. Clearly, if M_∞ were large ($M_\infty \geq 5$), there are some terms on the right-hand side of Eq.

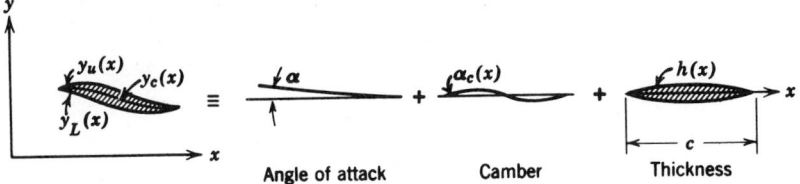

Fig. 6.69 Linear resolution of arbitrary airfoil into angle of attack, camber, and thickness.

(6.71) that could not be considered negligible compared to particular terms on the left-hand side. $M_{\text{hypersonic}}$ will be better defined in Section 6.9.1 but generally $M_{\text{hypersonic}} \approx 5$.

Thus, for thin airfoils,* small angles of attack, and $1 < M_\infty < M_{\text{hypersonic}}$, Eq. (6.71) can be approximated in two dimensions by

$$\left(1 - M_\infty^2\right)\frac{\partial^2 \phi}{\partial x^2} + \frac{\partial^2 \phi}{\partial y^2} = 0 \qquad (6.111)$$

This equation is a hyperbolic partial differential equation and its solution is of the wave type. Ackeret's solution for a wavy wall is a classical and very useful example.[40]

The solution of Eq. (6.111) for the pressure coefficient is

$$C_p = \frac{2\theta}{\sqrt{M_\infty^2 - 1}} \qquad (6.112)$$

where θ is the local flow deflection angle (in radians).

Since the governing equation is linear, the pressure distributions from different shapes can be added to give the pressure distribution over a particular shape. Thus the pressure distribution over an arbitrary airfoil shape can be resolved into contributions due to angle of attack, camber, and thickness. This is shown in Fig. 6.69.

The lift coefficient is given by

$$C_l = \frac{L'}{q_\infty c} = \int_0^c \frac{P_L - P_u}{q_\infty c}\, dx = \frac{1}{c}\int_0^c \left(C_{PL} - C_{Pu}\right) dx$$

where the subscripts L and u denote lower and upper surface, respectively. $q_\infty = \frac{1}{2}\rho_\infty V_\infty^2$.

For the angle of attack (in radians) contribution the integral is

$$C_l = \frac{4\alpha}{\sqrt{M_\infty^2 - 1}} \qquad (6.113)$$

There is no contribution from the mean camber line because the integral of the slope is zero. Also, the thickness envelope is nonlifting. Therefore Eq. (6.113) represents the entire lift coefficient for any thin airfoil.

The drag coefficient is given by (using small angle approximations)

$$C_d = \frac{D'}{q_\infty c} = \frac{1}{c}\int_0^c \left[C_{PL}\theta_L - C_{Pu}\theta_u\right] dx$$

All three components shown in Fig. 6.69 contribute to the drag and the result is[46]

$$C_d = \frac{4}{\sqrt{M_\infty^2 - 1}}\left[\alpha^2 + \overline{\alpha_c^2(x)} + \overline{\left(\frac{dh}{dx}\right)^2}\right] \qquad (6.114)$$

*The airfoils also have sharp leading edges such that the shock waves are attached oblique shocks. Round leading edges would result in detached shock (normal) waves and the perturbation velocities would not be small in the vicinity of the stagnation point.

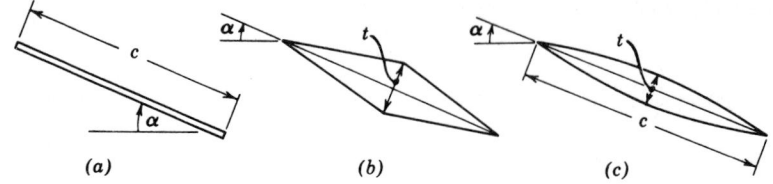

Fig. 6.70   Typical supersonic airfoil sections: (a) flat plate, (b) double wedge and (c) biconvex.

where

$$\overline{\alpha_c^2(x)} = \frac{1}{c}\int_0^c \alpha_c^2(x)\,dx \tag{6.115}$$

and

$$\overline{\left(\frac{dh}{dx}\right)^2} = \frac{1}{c}\int_0^c \left(\frac{dh}{dx}\right)^2 dx \tag{6.116}$$

The drag expressed by Eq. (6.114) is an inviscid drag and is called the "wave drag." It is caused by the shock-wave formation over a body immersed in a supersonic flow. It consists of drag due to lift, a drag due to camber, and a drag due to thickness. For the supersonic shapes shown in Fig. 6.70, we have:

Shape	$\overline{\left(\dfrac{dh}{dx}\right)^2}$
Flat plate	0
Double wedge	$\dfrac{t}{c}$
Biconvex	$\dfrac{4}{3}\dfrac{t}{c}$

If the airfoil shape is symmetric (no camber) such as those in Fig. 6.70, $\overline{\alpha_c^2(x)} = 0$.

Equation (6.114) is often rewritten in terms of a zero lift part and a drag due to lift part [see Eq. (6.93)] as

$$C_d = C_{d0} + C_{dL} = C_{d0} + KC_l^2 \tag{6.117}$$

where

$$C_{d0} = C_f + C_{dw}$$

$$C_f = \text{skin friction (see Section 6.10.4)}$$

$$C_{dw} = \text{wave drag} = \frac{4}{\sqrt{M_\infty^2 - 1}}\left[\overline{\alpha_c^2(x)} + \overline{\left(\frac{dh}{dx}\right)^2}\right]$$

$$C_{dL} = \text{drag due to lift} = \frac{4\alpha^2}{\sqrt{M_\infty^2 - 1}} = \frac{\sqrt{M_\infty^2 - 1}}{4}C_l^2 \tag{6.118}$$

The moment about the leading edge is

$$C_{mLE} = \frac{M'_{LE}}{q_\infty c^2} = -\frac{1}{c^2}\int_0^c (C_{PL} - C_{Pu})x\,dx$$

The moment contribution due to thickness is zero. The result is

$$C_{m\text{LE}} = -\frac{2\alpha}{\sqrt{M_\infty^2 - 1}} + \frac{4}{\sqrt{M_\infty^2 - 1}}\overline{(\alpha_c x)} \qquad (6.119)$$

where

$$\overline{\alpha_c x} = \frac{1}{c^2}\int_0^c \alpha_c(x)x\,dx \qquad (6.120)$$

and $\overline{\alpha_c x}$ is zero for symmetric airfoils. Since the mean camber line contributes no lift (in this linear theory), its moment contribution is a pure couple. Thus Eq. (6.119) indicates that the midchord is the point about which C_m is independent of angle of attack. Therefore the aerodynamic center is located at the midchord for all thin airfoils in supersonic flow. Recall, in subsonic flow the aerodynamic center is in the vicinity of the quarter chord.

The thin airfoil theory is a solution to the linearized equation and as such is an approximate method. The results agree very well with exact solutions for local flow deflections up to about 10°; up to θ's of 15° the thin airfoil theory results are still satisfactory. An exact solution is the useful shock expansion method discussed next.

Shock Expansion

The shock expansion method makes use of the exact oblique shock solutions and Prandtl–Meyer expansion discussed in Sections 6.4.5 and 6.4.7, respectively. The supersonic flow over a two-dimensional body is broken into regions of compression and expansion as shown in Fig. 6.71.

For the airfoil in Fig. 6.71, regions 1 and 6 are compression regions and Fig. 6.15 can be used to find the pressures and Mach numbers in these regions. Regions 2, 3, 4, and 5 are expansion regions, and Table 6.7 can be used to find the Mach numbers in these regions. The pressures in the expansion regions can then be determined using the isentropic results of Section 6.4.2. A slip surface (or slip line) is a surface across which the pressures and flow directions are the same but the temperatures, densities, and entropy levels may be different.[50] Such a surface is shown in Fig. 6.71 and exists behind supersonic shapes developing lift.

Finite Wings

Rectangular Wings. The flow field about a thin rectangular wing moving at supersonic speed in a direction normal to its leading edge is made up of two parts: (1) the regions within the Mach cones (infinitesimal wave fronts) emanating from the wing tips; and (2) the region between the Mach cones where the flow is two dimensional (see Fig. 6.72).

The effect of the wing tips on the wave drag of a rectangular wing depends[51, 52] on the product $AR\sqrt{M_\infty^2 - 1}$. If $AR\sqrt{M_\infty^2 - 1} \geq 1$, that is, the Mach cones from the leading edge corners fall within the trailing edge corners, there is no effect of wing tip regardless of wing section. When $AR\sqrt{M_\infty^2 - 1} < 1$, that is, the Mach cones fall outside of the trailing edge corners, the wave drag is less

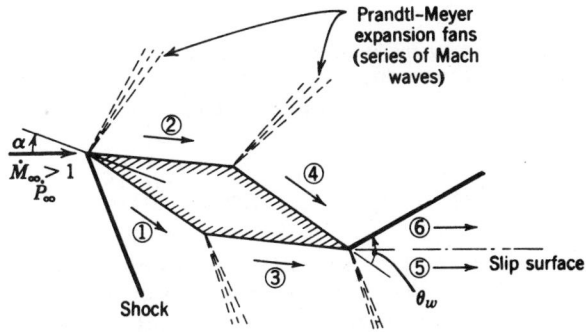

Fig. 6.71 Double-wedge airfoil in supersonic flow.

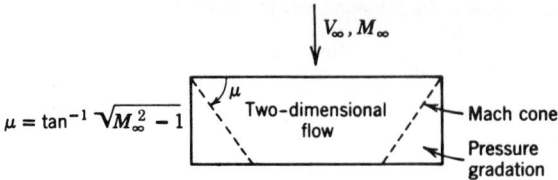

Fig. 6.72 Rectangular wing in supersonic flow.

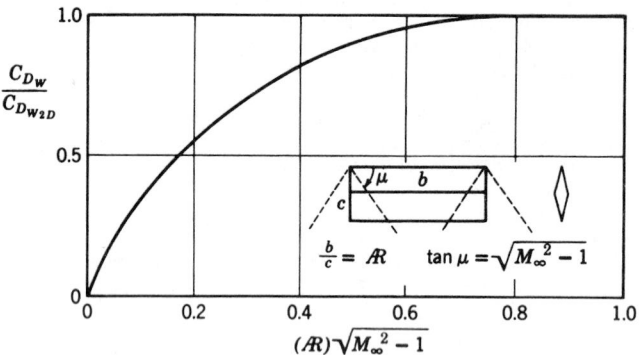

Fig. 6.73 Wave drag of double-wedge rectangular airfoil.

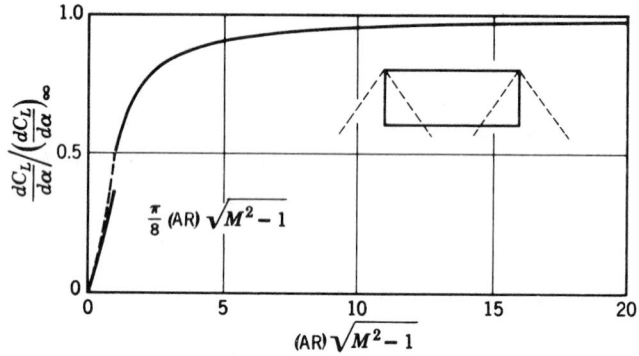

Fig. 6.74 Lift of a rectangular wing.

than that of a two-dimensional wing, the variation depending on the wing section. For a double-wedge wing (symmetric fore and aft) the wave drag is shown in Fig. 6.73.

In the case of lift, if $AR\sqrt{M_\infty^2 - 1} \geq 1$, the slope of the lift curve is given by[53]

$$\frac{dC_l}{d\alpha} = \frac{4}{\sqrt{M_\infty^2 - 1}} \left(1 - \frac{1}{2AR\sqrt{M_\infty^2 - 1}} \right) \tag{6.121}$$

and plotted in Figs. 6.74 and 6.75. The lift is independent of the wing section. As $AR\sqrt{M_\infty^2 - 1}$ decreases below unity, the theory becomes increasingly more complicated[54,55] and difficult to evaluate numerically. As $AR\sqrt{M_\infty^2 - 1} \to 0$, presumably[56] $dC_L/d\alpha \to (\pi/2)AR$, which is indicated in Figs. 6.74 and 6.75.

For rectangular wings, the drag due to lift is obtained by multiplying the lift by the angle of attack, thus $C_{DL} = \alpha C_L$.

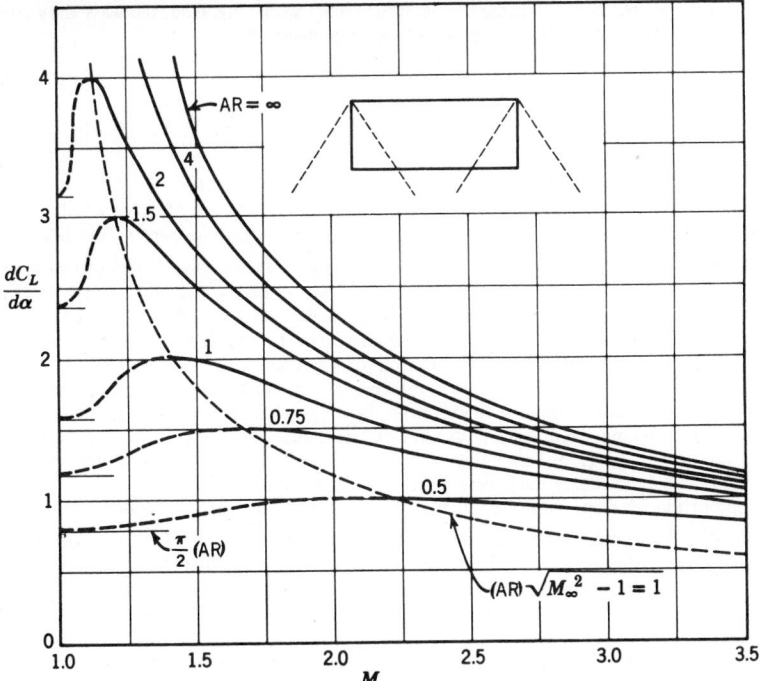

Fig. 6.75   Lift of a rectangular wing.

The center of pressure (CP) of the rectangular wing is located at

$$\frac{CP}{c} = \frac{AR\sqrt{M_\infty^2 - 1} - \frac{2}{3}}{2AR\sqrt{M_\infty^2 - 1} - 1} \tag{6.122}$$

from the leading edge. (Since in the linearized theory the CP does not change with angle of attack, it has the same position as the aerodynamic center.) According to Eq. (6.122), the CP travels forward from the midchord point at $AR = \infty$ to the third-chord point at $AR\sqrt{M_\infty^2 - 1} = 1$. On the other hand, for subsonic flight, the ac moves forward very slowly from the quarter-chord point as aspect ratio decreases. Hence, a transonic stability problem arises that can be alleviated by use of low-aspect-ratio wings.

Delta Wings. Low-aspect-ratio wings are important in supersonic missile design for both aerodynamic and structural reasons. The delta wing offers some definite advantages over the low-aspect-ratio straight wing, and therefore its properties[57] are presented here.

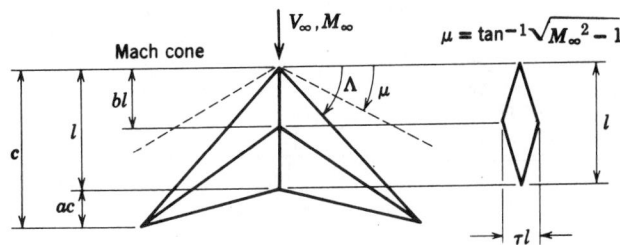

Fig. 6.76   Delta wing.

Linearized theory shows that for the double-wedge delta wing (Fig. 6.76) the wave drag coefficient C_{DW} is generally related to the Mach number and wing geometry by

$$\frac{C_{Dw}\sqrt{M_\infty^2 - 1}}{\tau^2} = f\left(\frac{\sqrt{M_\infty^2 - 1}}{\tan \Lambda}, a, b\right) \qquad (6.123)$$

in which the symbols are defined in Fig. 6.76. The ratio $\sqrt{M_\infty^2 - 1}/\tan \Lambda$, which is equal to $\tan \mu/\tan \Lambda$, indicates the relative position of the leading edge with respect to the Mach cone. If $\sqrt{M_\infty^2 - 1}/\tan \lambda$ is greater, equal to, or less than unity, the leading edge is designated supersonic, sonic, or subsonic, respectively. The function [Eq. (6.121)] is plotted in Figs. 6.77 and 6.78. The peaks of the curves indicate when the Mach cone coincides successively with the trailing edge, maximum thickness line, and leading edge as Mach number increases. In Fig. 6.79 wave drag coefficients are plotted for wings with 60° and 75° sweepback, each for $a = 0$ and 0.5. (The value $b = 0.2$ was chosen for Fig. 6.79 because, at about this value of b, the drag is found to be the minimum as long as the line of maximum thickness is subsonic.) It is seen from these figures that the drag is appreciably less than the two-dimensional-wing value only if the line of maximum thickness is subsonic. However, the maximum wave drag peak is far below that of the two-dimensional wing (infinite when $M_\infty = 1$) and considerably delayed, this being the general effect of sweepback.

The slope of the lift curve is plotted in Figs. 6.80 and 6.81. Because of the difficulty of obtaining solutions to the lift problem when the trailing edge is subsonic, that is, $\sqrt{M_\infty^2 - 1}/\tan \Lambda \geq |a|$, the curves begin at the point where the trailing edge is sonic. It is noted that, for a wide range of Mach number, the delta wing with a sweptback trailing edge has a higher lift curve slope than the two-dimensional wing. In the limit when $a = 0$ and $\tan \Lambda \to \pi/2$, it will be found[58] that $dC_L/d\alpha \to (\pi/2)R$.

The travel of the CP (measured from the nose) with Mach number and wing geometry is shown in Fig. 6.82. The graph shows that the effect of Mach number is generally not great and the CP (or the AC) is located at the center of gravity of the planform.

When the leading edge of the wing is supersonic, the drag due to lift is obtained by merely multiplying the lift by the angle of attack as is done for rectangular wings in supersonic flight. However, when the leading edge is subsonic, a suction occurs at the leading edge (as for subsonic flight of a two-dimensional airfoil), resulting in a leading edge thrust or decrease in drag below that

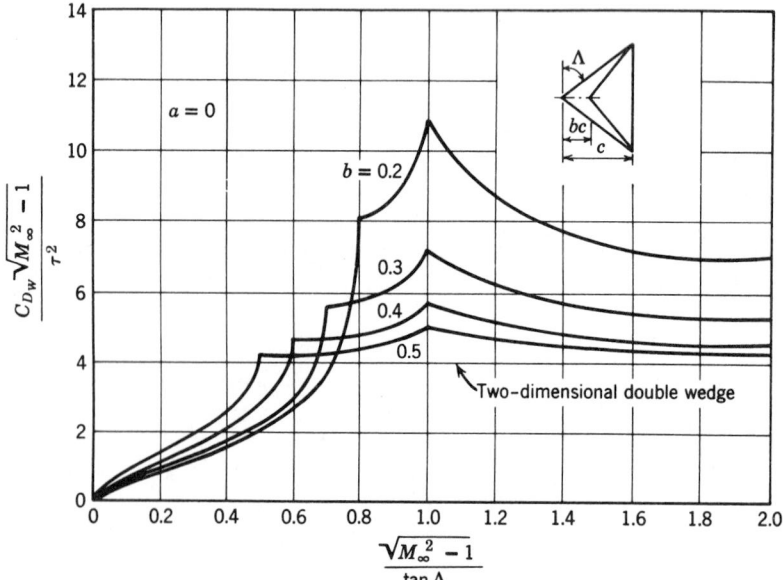

Fig. 6.77 Wave drag of delta wing with straight trailing edge.

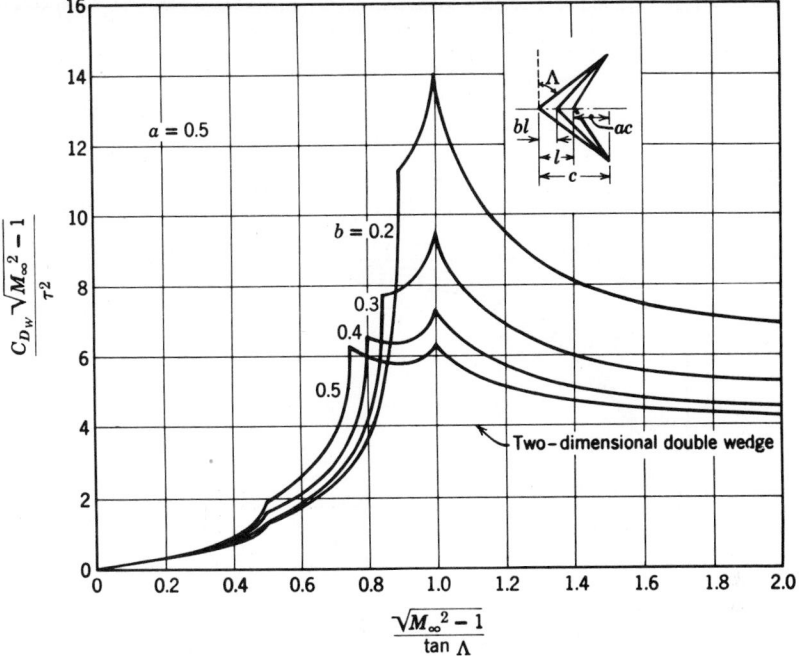

Fig. 6.78 Wave drag of delta wing with swept trailing edge.

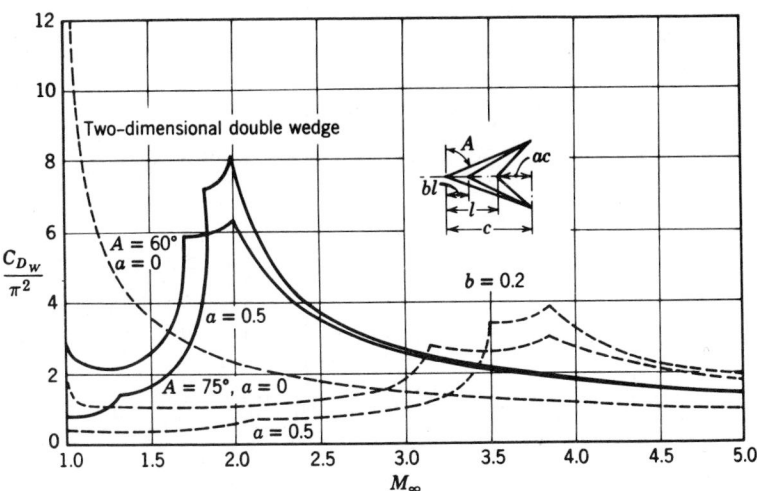

Fig. 6.79 Wave drag of delta wing.

given by $C_{DL} = \alpha C_L$. The corrected drag is plotted in Fig. 6.83 for the range $|a| \leq \sqrt{M_\infty^2 - 1}\,/\tan \Lambda \leq 1$.

Comparison of low-aspect-ratio delta and straight wings will show that: (1) delta wings have considerably less drag than straight wings in transonic and moderate supersonic speeds; (2) the lift-drag ratio of delta wings is better than that of straight wings at moderate supersonic speeds but worse at subsonic and high supersonic speeds; and (3) the stability and control changes of delta wings are smaller than those of straight wings.

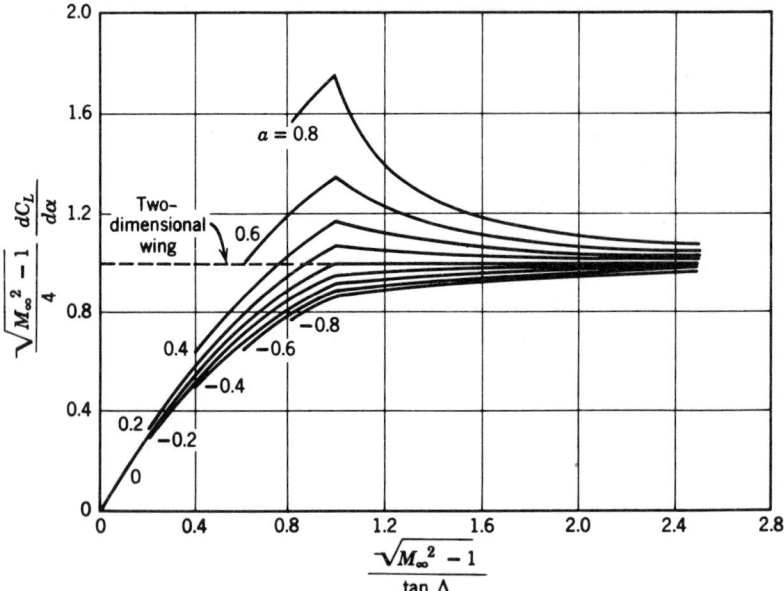

Fig. 6.80 Lift of delta wing.

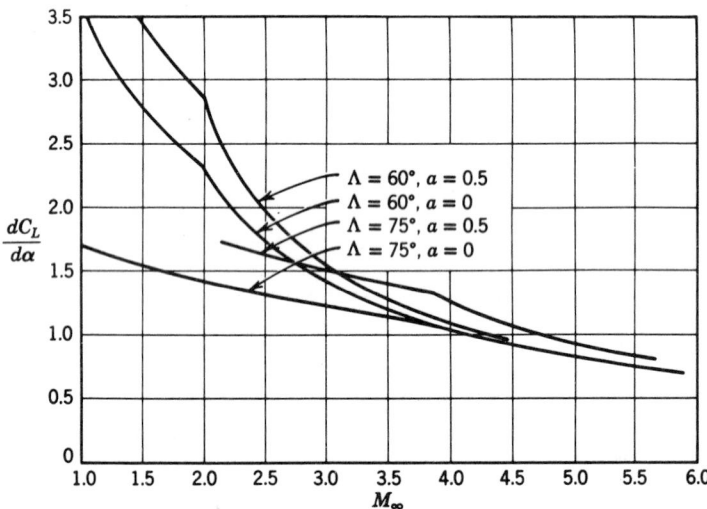

Fig. 6.81 Lift of delta wing.

6.8   AERODYNAMICS OF BODIES
D. Finkleman

6.8.1   Force Components

Steady flow past a body of revolution inclined at angle of attack α with free-stream velocity V_∞ may be resolved into an *axial flow* with velocity $V\cos\alpha$ and a *crossflow* with velocity $V_\infty\sin\alpha$ (Fig. 6.84). The axial flow produces an *axial force* X, the crossflow a *normal force* N and *pitching moment* M_0

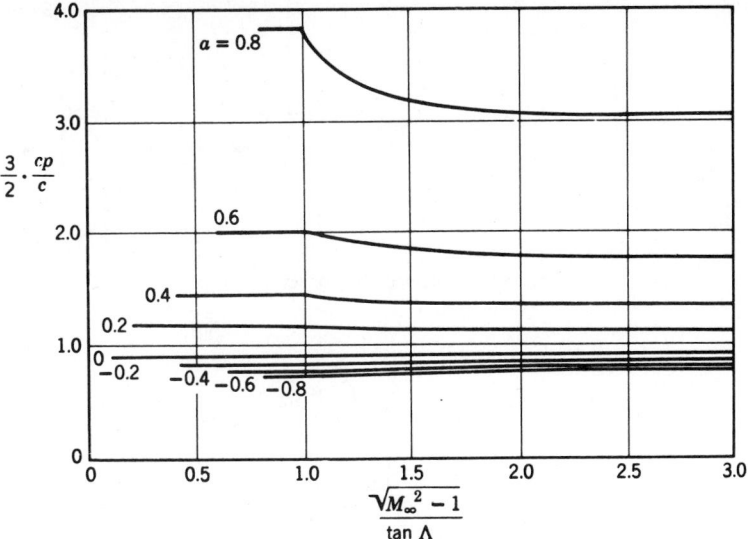

Fig. 6.82 Center of pressure location for delta wing.

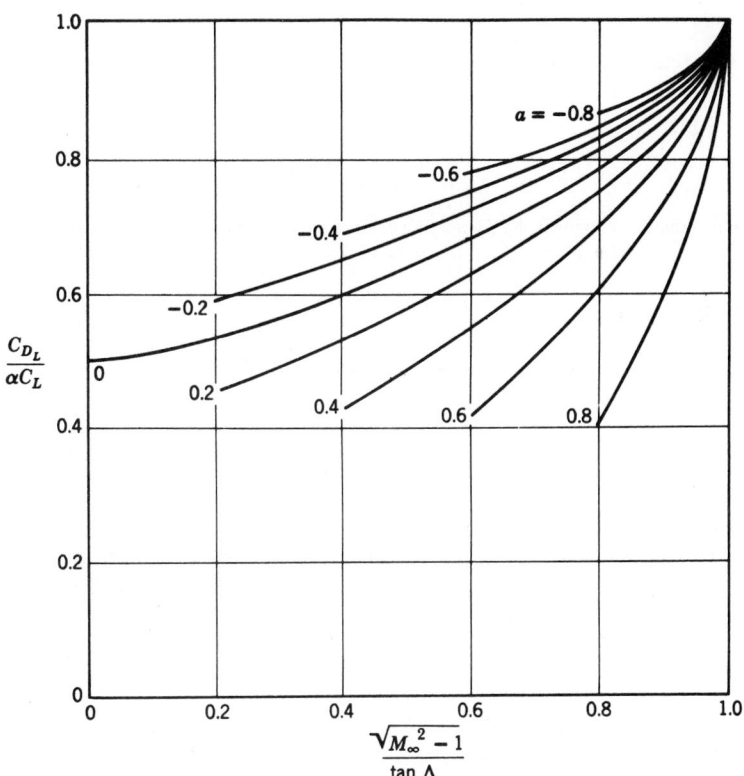

Fig. 6.83 Drag due to lift of delta wing.

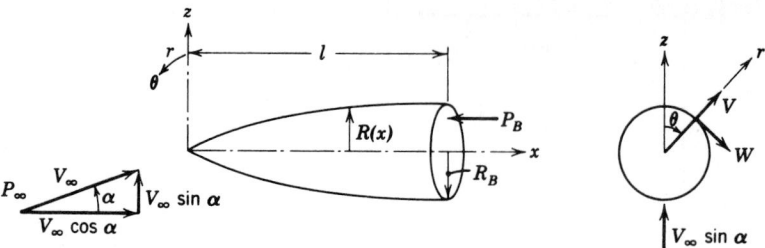

Fig. 6.84   Inclined body of revolution.

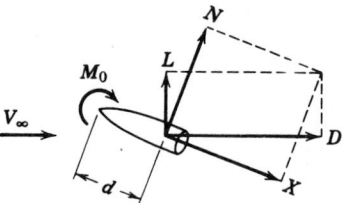

Fig. 6.85   Force system for body of revolution.

about the nose (Fig. 6.85). Often the force is resolved into *drag D* and *lift L*, which are related by

$$D = X \cos \alpha + N \sin \alpha$$
$$L = N \cos \alpha - X \sin \alpha$$
$$X = D \cos \alpha - L \sin \alpha$$
$$N = L \cos \alpha + D \sin \alpha$$

Corresponding relations connect the *force coefficients*, which are usually referred to base area S_B:

$$C_x = \frac{X}{\frac{1}{2}\rho V_\infty^2 S_B} \qquad C_N = \frac{N}{\frac{1}{2}\rho V_\infty^2 S_B} \qquad C_D = \frac{D}{\frac{1}{2}\rho V_\infty^2 S_B} \qquad C_L = \frac{L}{\frac{1}{2}\rho V_\infty^2 S_B}$$

The pitching moment coefficient is referred to the body length l:

$$C_{M_0} = \frac{M_0}{\frac{1}{2}\rho V_\infty^2 S_B l}$$

were $M_0 = -dN$.

Since the governing equation for flow past a slender body at a small angle of attack can be approximated by a linear equation, the solution to the problem can be expressed as the sum of an axial flow solution and a two-dimensional crossflow solution. In terms of perturbation velocity potentials, the solution is

$$\phi(x, r, \theta) = \phi_a(x, r) + \phi_c(x, r, \theta) \tag{6.124}$$

where

$$u = \frac{\partial \phi}{\partial x} \qquad v = \frac{\partial \phi}{\partial r} \qquad w = \frac{1}{r}\frac{\partial \phi}{\partial \theta} \tag{6.125}$$

The local resultant velocity $\bar{V}$ is expressed as (see Fig. 6.84),

$$\bar{V}^2 = (V_\infty \cos \alpha + u)^2 + (V \sin \alpha \cos \theta + v)^2 + (V_\infty \sin \alpha \sin \theta - w)^2 \tag{6.126}$$

and the local pressure coefficient is

$$C_p = \frac{2}{\gamma M_\infty^2} \left[\left(1 + \frac{\gamma - 1}{2} \right) M_\infty^2 \left(1 - \frac{\bar{V}^2}{V_\infty^2} \right)^{\gamma/\gamma-1} \right] \tag{6.127}$$

Pressure coefficients on bodies of revolution can be obtained theoretically for inviscid fluids by using the method of sources and sinks[59, 60] for the axial flow component of the free-stream velocity, and doublets[61, 62] for the crossflow component. The first and most difficult part of the procedure is the determination of the source-sink and doublet distributions necessary to form separately the given body. The axial flow solutions will be discussed in Sections 6.8.2 and 6.8.3, and the crossflow case in Section 6.8.4.

6.8.2 Slender Body Theory, Axial Flow Case

Here the body is considered to be very slender, that is, the ratio of its maximum thickness R_B to length l is very much less than unity. The governing equation [see Eq. (6.71)] is again linearized by assuming small perturbations. For the axisymmetric case the linearized equation[63] is

$$\left(1 - M_\infty^2 \right) \frac{\partial^2 \phi}{\partial x^2} + \frac{\partial^2 \phi}{\partial r^2} + \frac{1}{r} \frac{\partial \phi}{\partial r} = 0 \tag{6.128}$$

which is valid as long as M_∞ is not close to unity or too large. The pressure coefficient, correct to first order, is

$$C_p = -\frac{2 \partial\phi/\partial x}{V_\infty} + \left(\frac{\partial\phi/\partial y}{V_\infty} \right)^2 \tag{6.129}$$

Incompressible Flow

For incompressible flow, $M_\infty \to 0$, Eq. (6.128) is Laplace's equation, which has the basic solution

$$\phi_i = \frac{-A}{\sqrt{x^2 + r^2}} \tag{6.130}$$

Equation (6.130) represents a source of strength A (see Section 6.1.6). If $f(\xi)$ is the source strength per unit length, the effect of such sources distributed along the x axis is

$$\phi_i(x, r) = -\frac{1}{4\pi} \int_0^l \frac{f(\xi)\, d\xi}{\sqrt{(x - \xi)^2 + r^2}} \tag{6.131}$$

Equation (6.131) is an integral equation for the "airship problem" of incompressible flow theory.[59, 61] In a given problem, $f(\xi)$ is determined by satisfying the boundary conditions. The solution is usually a numerical one where the integral is approximated by a finite sum of sources and sinks such as Eq. (6.130).

Using the boundary condition that the flow at the surface must be tangent to the surface, it can be shown that the solution for the source strength is

$$f(x) = V_\infty R \frac{dR}{dx} = \frac{V_\infty}{2\pi} \frac{dS}{dx} \tag{6.132}$$

where $R(x)$ is the r distance to the body surface and $S(x) = \pi R^2$ is the cross-sectional area of the body at x. Equation (6.132) follows also from the reasoning that the volumetric outflow per unit length should be proportional to the streamwise rate of change of cross-sectional area.

Laitone[64] derived the following approximate formula for the pressure coefficient at the surface of a body of revolution:

$$C_p = \frac{1}{\pi} \left[\frac{1 - 2x/l}{1 - x/l} \frac{S'}{2x} - \left(1 + \ln\frac{R}{2l\sqrt{x(1 - x/l)/l}} S'' \right) \right.$$
$$\left. + \frac{l}{4} \left(1 - \frac{2x}{l} \right) S''' + \frac{l^2}{24} \left(1 - \frac{2x}{l} + \frac{2x^2}{l^2} \right) S'''' + \cdots \right] \tag{6.133}$$

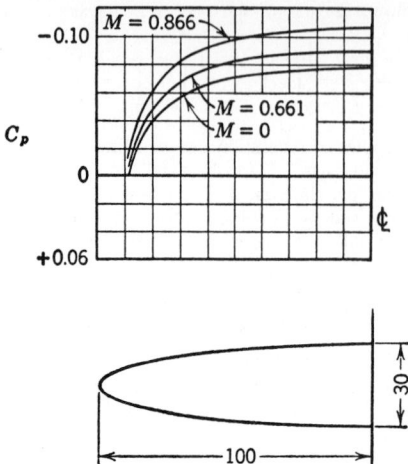

Fig. 6.86 Pressure coefficient along the meridian of an ellipsoid of revolution of fineness ratio 3.3 according to linearized theory.

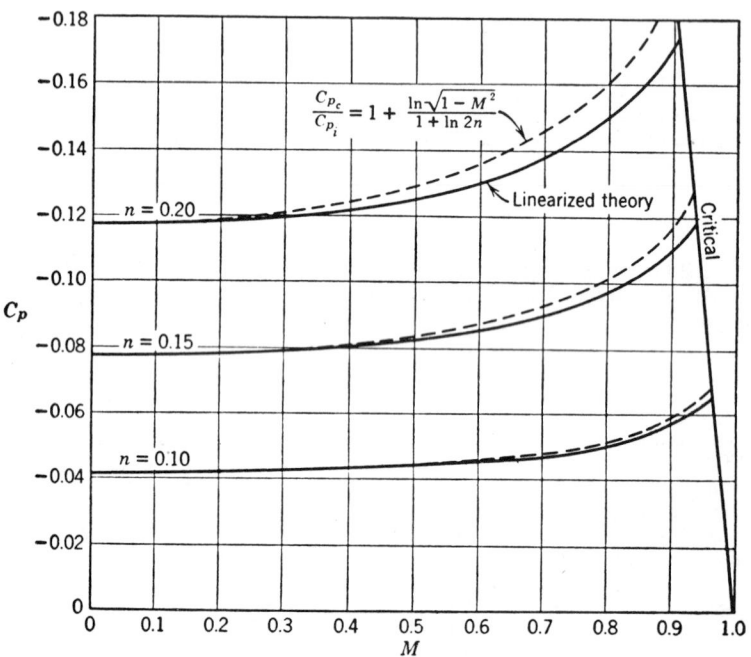

Fig. 6.87 Pressure coefficient at midsection of ellipsoids of revolution.

where l in the length of the body, x is the distance aft of the nose, $S' = dS/dx$, $S'' = d^2S/dx^2$, and so on.

Figures 6.86 and 6.87 show axial flow results for ellipsoids of revolution.

High Subsonic Flow
For compressible subsonic flow, the coordinate transformation $r' = \beta r$, where $\beta^2 = 1 - M_\infty^2 < 0$, transforms Eq. (6.128) into Laplace's equation. Thus the general solution for the subsonic potential is

$$\phi_{sub}(x, r) = -\frac{1}{4\pi} \int_0^l \frac{f(\xi)\, d\xi}{\sqrt{(x - \xi)^2 + \beta^2 r^2}} \qquad (6.134)$$

The method of solving this integral equation for a given body shape is the same as discussed for Eq. (6.131).

Another method for determining the subsonic compressible pressure coefficient is to extend an affinely related incompressible result by means of the Goethert rule[65] (generalized Prandtl–Glauert rule). The rule states that the linearized subsonic compressible flow pressure coefficient on a given slender three-dimensional body at small angle of attack is equal to $1/(1 - M_\infty^2)$ times the linearized incompressible flow pressure coefficient on another body whose thickness ratio and angle of attack are $(1 - M_\infty^2)^{1/2}$ times as great as those of the given body. M_∞ is the free-stream Mach number. For further discussion and experimental investigation, see Ref. 66. The following formula[67, 68] expresses the effect of compressibility on the maximum pressure coefficient for ellipsoids of revolution at zero angle of attack:

$$\frac{C_{pc}}{C_{pi}} = 1 + \frac{\ln\left(1 - M_\infty^2\right)^{1/2}}{1 + \ln 2n} \tag{6.135}$$

where n is the thickness ratio (maximum diameter to length) and the subscripts c and i refer to compressible and incompressible flow, respectively. This formula, which is plotted in Fig. 6.8, can be used to give an indication of the first-order effect of compressibility on pressure at the central portion of very slender bodies of revolution in subsonic motion.

Supersonic
For supersonic flow $\beta^2 = M_\infty^2 - 1$ and Eq. (6.128) becomes the wave equation

$$\frac{\partial^2 \phi}{\partial r^2} + \frac{1}{r}\frac{\partial \phi}{\partial r} - \beta^2 \frac{\partial^2 \phi}{\partial x^2} = 0 \tag{6.136}$$

By a similar analogy with the subsonic solution, it can be shown[63, 69] that the general solution of Eq. (6.136) is

$$\phi_{\text{sup}}(x, r) = -\frac{1}{4\pi}\int_0^{x-\beta r}\frac{f(\xi)\,d\xi}{(x - \xi)^2 - \beta^2 r^2} \tag{6.137}$$

Combining Eq. (6.132) with Eqs. (6.134) and (6.137) yields the axial flow solution for subsonic and supersonic flow,

$$\phi_{\text{sub}} = \frac{V_\infty S'(x)}{2\pi}\ln r + \frac{V_\infty S'(x)}{4\pi}\ln\left[\frac{\beta^2}{4x(l - x)}\right] - \frac{1}{4\pi}\int_0^l \frac{S'(\xi) - S'(x)}{|x - \xi|}\,d\xi \tag{6.138}$$

$$\phi_{\text{sup}} = -\frac{V_\infty S'(x)}{2\pi}\ln\frac{2}{\beta r} - \frac{V_\infty}{2\pi}\int_0^x S''(\xi)\ln(x - \xi)\,d\xi \tag{6.139}$$

where $\beta = \sqrt{|M_\infty^2 - 1|}$.

6.8.3   Supersonic Flow

Cones
Exact solutions for supersonic flow past inclined cones have been calculated numerically. Kopal[70-72] gives detailed tabulations of the flow variables expanded in powers of α up to α^2. Typical results are plotted in Figs. 6.88 and 6.89.

Method of Characteristics
Supersonic flow past slender bodies can be calculated numerically by the method of characteristics. The reader is referred to Refs. 73–75 for a detailed discussion of this very useful method.

Slender Body Theory
Using Eq. (6.139), the pressure coefficient over an axisymmetric body is[63, 69]

$$C_p = \frac{S''(x)}{\pi}\log\frac{2}{\beta R} + \frac{1}{\pi}\frac{d}{dx}\int_0^x S''(x)\log(x - \xi)\,d\xi - \left(\frac{dR}{dx}\right)^2 \tag{6.140}$$

The zero lift drag coefficient, referenced to the maximum cross-sectional area S_{max}, is given by

$$C_{D0} = \frac{D}{q_\infty S_{\text{max}}} = \frac{1}{S_{\text{max}}}\int_0^l C_p\frac{dS}{dx}\,dx + C_{DB} + C_F \tag{6.141}$$

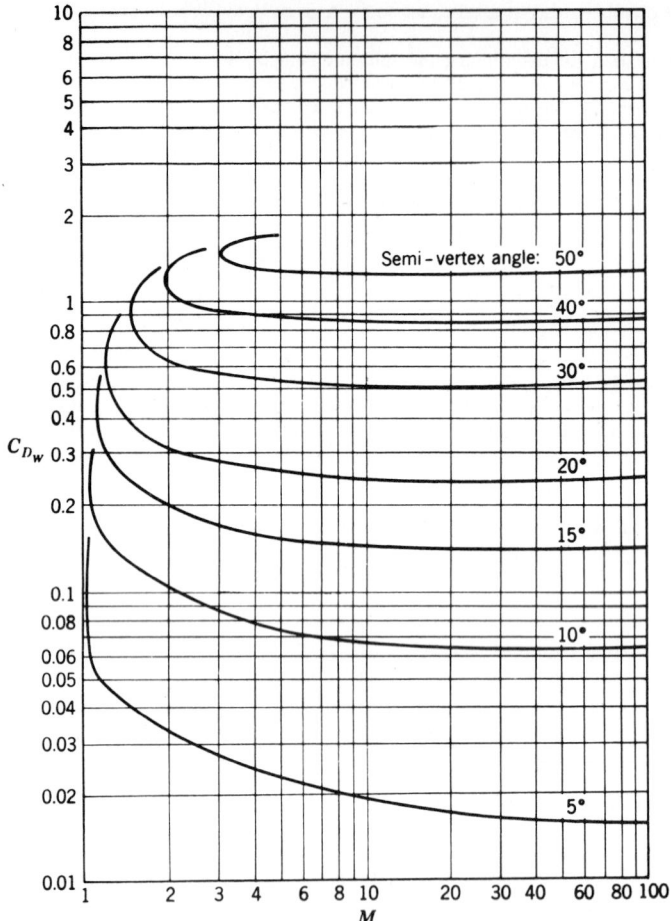

Fig. 6.88 Variation of drag coefficient with Mach number for various cones at zero angle of attack.

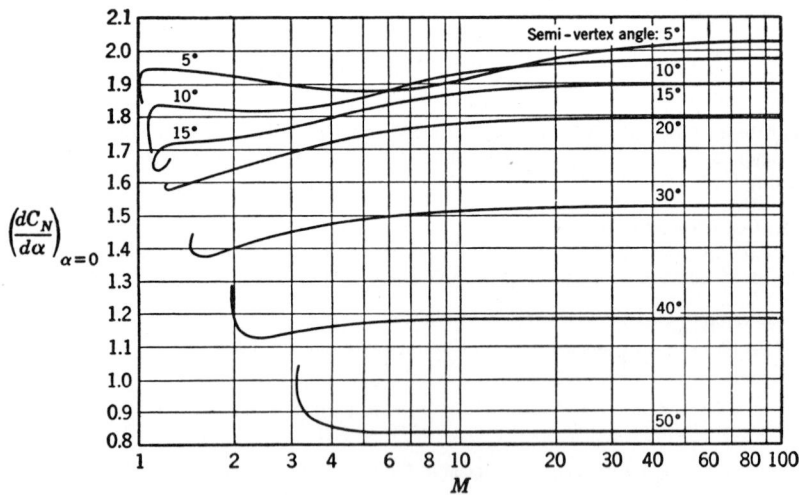

Fig. 6.89 Variation of initial normal-force slope with Mach number for various cones.

where the first term is called the wave drag C_{DW}, C_{DB} is the drag due to the base pressure P_b, and C_F is the skin friction drag. The C_{DB} is determined by the mechanics of the viscous wake and is discussed further in Section 6.8.5. If the slender body is closed at the base, C_{DB} is approximately zero. The C_F is discussed in Section 6.10.3.

Combining Eqs. (6.140) and (6.141) and performing the integration for a body with a pointed nose and either a pointed base or zero slope at the base yields

$$C_{DW} = -\frac{1}{2\pi S_{max}} \int_0^l \int_0^l S''(\xi) S''(x) \log |x - \xi| \, d\xi \, dx \tag{6.142}$$

Equation (6.142) indicates that the wave drag coefficient of a slender body in supersonic flow is independent of Mach number.

Minimum Wave Drag Bodies

It is of interest to find area distributions of slender bodies that, for given length, volume, or some other constraint, give the lowest possible wave drag. Standard variational methods may be applied to Eq. (6.142) with appropriate constraint equations. Some results are:[69, 76]

1. Half body of given length and diameter (von Kármán ogive):

$$\left(\frac{R}{R_{max}}\right)^2 = \frac{S}{S_{max}} = \frac{1}{\pi}\left[\frac{2x}{l}\sqrt{1 - \left(\frac{2x}{l}\right)^2} + \cos^{-1}\left(\frac{-2x}{l}\right)\right]$$

for

$$-\frac{l}{2} \leq x < \frac{l}{2} \qquad \text{volume} = \frac{1}{2}lS_{max} \qquad C_{DW} = \frac{D_W}{\frac{1}{2}\rho V_\infty^2 S_{max}} = \frac{4S_{max}}{\pi l^2}$$

2. Complete body of given length and volume (Sears-Haack body):

$$\left(\frac{R}{R_{max}}\right)^2 = \frac{S}{S_{max}} = \left[1 - \left(\frac{2x}{l}\right)^2\right]^{3/2}$$

for

$$-\frac{l}{2} \leq x \leq \frac{l}{2} \qquad \text{volume} = \frac{3}{16}\pi lS_{max} \qquad C_{DW} = \frac{9}{2}\frac{\pi}{l^2}S_{max}$$

6.8.4 Supersonic Lift of Slender Bodies

For slender bodies at small angle of attack, it can be shown[69] that the crossflow case is independent of β near the body and independent of x. In other words the subsonic and supersonic crossflow can be considered incompressible and two dimensional. For a circular cross section the solution is the potential flow about a cylinder (see Section 6.1.6), i.e.,

$$\phi_c(r, \theta) = V_\infty \alpha R^2 \frac{\cos \theta}{r} \tag{6.143}$$

The crossflow pressure coefficient is[63]

$$C_{PC} = -4\alpha \frac{dR}{dx} \cos \theta + (1 - 4\sin^2 \theta)\alpha^2 \tag{6.144}$$

and the formal force coefficient, reference to S_{max}, is

$$C_N = 2\alpha \tag{6.145}$$

The crossflow contributes a term to the axial force that is similar to the induced drag of a subsonic finite wing. The total supersonic drag coefficient, referenced to S_{max}, is

$$C_D = C_{D_0} + \alpha^2 \tag{6.146}$$

6.8.5 Drag of Various Bodies

The drag on a body immersed in a moving fluid is expressed as

$$D = C_D \left(\text{shape, orientation, } R_e, M_\infty\right)\tfrac{1}{2}\rho V_\infty^2 S$$

where C_D is the drag coefficient, R_e the Reynolds number based on a characteristic dimension of the body, M_∞ the free-stream Mach number, and S the reference area of the body. Figures 6.90 and 6.91 show the variation of drag coefficient with Reynolds number for spheres and cylinders, respectively, when the compressibility effect is negligible. The sudden drop in C_D between R_e of 10^5 to 10^6 is due to the boundary layer transitioning from laminar to turbulent and the associated delay in the flow separation on the surface of the bodies.

Figure 6.92 shows the effect of compressibility on the drag coefficient for various bodies.

At the rear of a body with a blunt base in supersonic flow, the flow tries to expand 90°. Inviscid flow theory would predict that the base pressure P_B would be zero. However, in a viscous fluid the base pressure is not zero but is some value less than ambient pressure P_∞. This is due to the boundary layer bleeding into the separated flow region at the base giving a turbulent or "dead water" region and a $0 < P_B < P_\infty$. Experimental values of base pressure coefficients for two- and three-dimensional bodies are shown on Fig. 6.93. Notice that the C_{DB} of Eq. (6.141) is equal to the negative of C_{PB}.

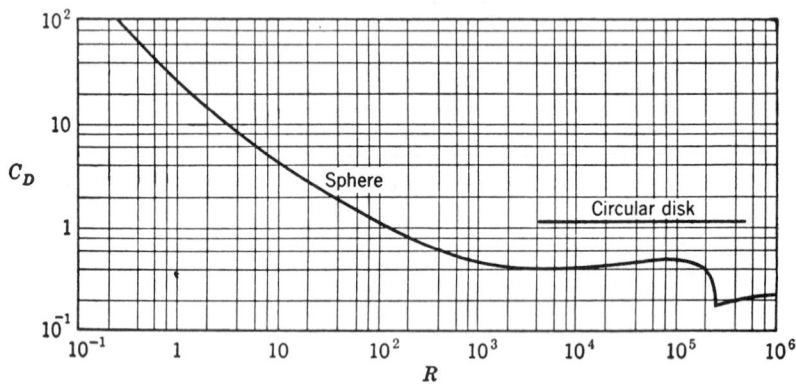

Fig. 6.90 Drag of a sphere in incompressible flow.

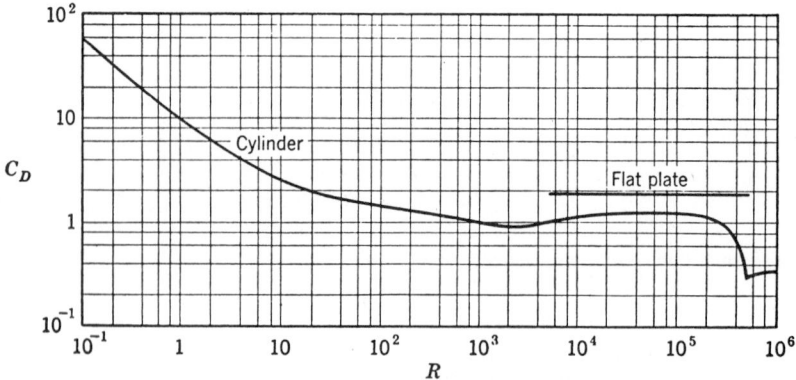

Fig. 6.91 Drag of a cylinder in incompressible flow.

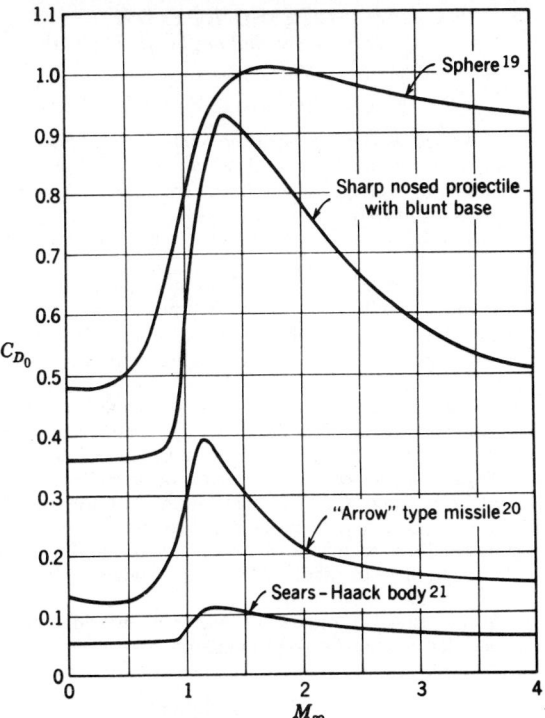

Fig. 6.92 Zero lift drag coefficient for various bodies showing effect of compressibility (C_{D_0} referenced to S_{max}).

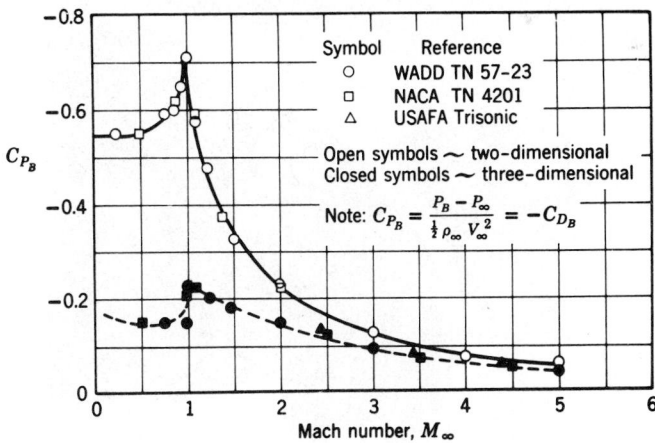

Fig. 6.93 Experimental values of base pressure coefficient for two- and three-dimensional bodies.

6.9 HYPERSONIC FLOW
J. T. Clay with Contributions by M. L. Smith

6.9.1 General Features of Hypersonic Flow

Hypersonic flow can be defined as existing when the nonlinearity of the equation [i.e., Eq. (6.71)] describing the supersonic flow over an object becomes an essential feature due to the largeness of M_∞.[77,78] Thus the breakdown of supersonic linear theory (Section 6.7.6) serves almost as a definition for hypersonic flow. Supersonic linear theory breaks down when the hypersonic similarity parameter $M_\infty \tau \geq 1$, where M_∞ is the free-stream Mach number and τ is proportional to the thickness ratio or flow deflection angle of the body.

A certain Mach number, M_∞, might be hypersonic for one object immersed in the flow while it is supersonic for another object. For very slender bodies, hypersonic flow is usually present for $M_\infty \geq 5$. On the other hand, a blunt object such as a sphere is in a hypersonic flow situation at $M_\infty = 3^4$. Thus we admit to a certain arbitrariness in the term hypersonic; however, hypersonic flow is generally assumed to be $M_\infty > 5$.

The manner of analyzing the flow situation in the hypersonic flow regime is very much different than for the other flow regimes. Potential flow theory used in treating slender bodies and thin airfoils in supersonic flow is not applicable at hypersonic speed. In hypersonic flow the shock waves lie close to the body (see Fig. 6.94) and are curved, thus there are large lateral entropy gradients and the flow field is highly rotational. In addition, the high temperatures that are generated behind the strong shock waves cause real gas effects such as dissociation and ionization and must be accounted for.[79-82]

The effect of compressibility is to thicken the boundary layer. Thus at hypersonic speeds the boundary layer may thicken to the extent that it cannot be considered thin and there may be a shock wave-boundary layer interaction.

The remainder of the discussion will consider the fluid as an inviscid perfect gas; however, the real gas features mentioned previously should be kept in mind.

6.9.2 Exact Solutions

For shapes composed of wedges, cones, and straight sections the pressure coefficient on the body surface can be determined using the plane oblique shock, conical shock, and Prandtl–Meyer expansion solutions discussed earlier in Sections 6.4.5–6.4.7. These solutions are exact[83,84] and the method is the same as for supersonic flow.

Fig. 6.94 Schlieren photograph of flow past a 2-in. hemisphere at $M_\infty = 4.38$ (USAFA trisonic tunnel).

The method of characteristics, combined with a method for defining the flow field between the shock wave and the body at some station downstream of the nose, can be used to solve the problem of flow around more complex bodies in hypersonic flight. Reference 80 discusses the manner in which gas property deviations from a perfect gas can be incorporated.

6.9.3 Approximate Solutions

Newtonian Impact Theory
In this theory it is assumed that the momentum normal to the surface is entirely converted into pressure so that[85]

$$C_P = \frac{P_S - P_\infty}{\frac{1}{2}\rho_\infty V_\infty^2} = 2\sin^2\theta \qquad (6.147)$$

where θ is the local flow deflection angle. The theory also assumes that the pressure coefficient in any expansion region is zero.

A more accurate pressure distribution may be found by replacing the 2 in Eq. (6.147) by the value of the stagnation pressure coefficient on the body. The modified expression is called the modified Newtonian and is expressed as

$$C_P = C_{P_{stag}}\sin^2\theta \qquad (6.148)$$

where $C_{P_{stag}}$ is given by the Rayleigh pitot formula (see Section 6.4.4)

$$C_{P_{stag}} = \frac{2}{\gamma M_\infty^2}\left[\left(\frac{\gamma+1}{2}M_\infty^2\right)^{\gamma/\gamma-1}\left(\frac{\gamma+1}{2\gamma M_\infty^2 - \gamma + 1}\right)^{1/1-\gamma} - 1\right] \qquad (6.149)$$

The modified Newtonian approximation gives quite good results when the shock wave is close to the body, which is the condition necessary for the assumption that the momentum of the gas normal to the surface is lost. Normally the Newtonian approximation is applied to predicting the pressure distribution on the nose of blunt bodies and gives reliable estimates down to $M_\infty = 3.0$.[80] Figure 6.95 shows the measure and calculated [using Eq. (6.148)] values of pressure coefficient around a hemisphere at $M_\infty = 4.38$.

Newtonian flow theory can be applied to flat surfaces to provide an approximation for the aerodynamic coefficients. For a flat plate at an angle of attack α, the coefficients are

$$C_L = 2\sin^2\alpha\cos\alpha \qquad (6.150)$$

$$C_{D_L} = \text{drag coefficient due to lift} = 2\sin^3\alpha \qquad (6.151)$$

$$C_D = C_{D_0} + 2\sin^3\alpha \qquad (6.152)$$

Tangent-Wedge and Tangent-Cone Approximations
The tangent-wedge and tangent-cone approximations assume that the pressure coefficient at any point on the surface of a body is the same as that on a wedge or cone that has the same surface inclination

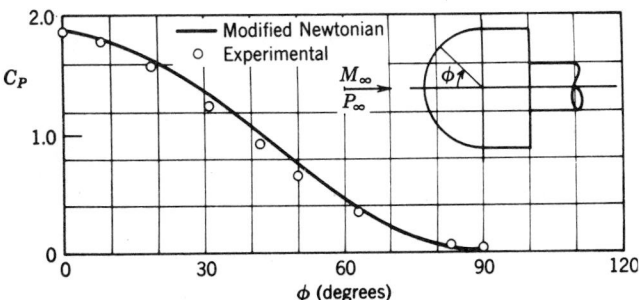

Fig. 6.95 Measured and calculated (modified Newtonian) pressure coefficient distribution for hemisphere in Fig. 6.94 (data from USAFA trisonic tunnel).

to the free stream as the point of interest. The equation for the surface pressure coefficient for a wedge in hypersonic flow can be approximated by[78,85]

$$C_P \approx \theta^2{}_W \left[\frac{\gamma + 1}{2} + \sqrt{\left(\frac{\gamma + 1}{2} \right)^2 + \frac{4}{(M_\infty \theta_W)^2}} \right] \qquad (6.153)$$

This equation is applicable to a two-dimensional body that has small surface inclination θ_W.

Lees[86] developed an approximate equation for the surface pressure on a cone in hypersonic flow that is applicable to determining the surface pressure on bodies of revolution. This expression is

$$C_P = \theta_c^2 \left[\frac{4}{\gamma + 1} \left(\frac{K_s^2 - 1}{K_c^2} \right) + \frac{2(K_s - K_c)^2}{K_c^2} \left(\frac{\gamma + 1}{\gamma - 1 + 2/K_c^2} \right) \right] \qquad (6.154)$$

where θ_c = semivertex angle

$K_c = M_\infty \theta_c$, hypersonic similarity parameter

$$K_s = \frac{\gamma + 1}{\gamma + 3} K_c + \sqrt{\left(\frac{\gamma + 1}{\gamma + 3} \right)^2 K_c^2 + \frac{2}{\gamma + 3}}$$

This formula applies when the shock wave lies close to the body and when $M_\infty \theta_c \geq 1$.

Shock Expansion Method

The shock expansion method is applicable to slender, pointed, two-dimensional, or axisymmetric bodies with attached shock waves. In this method the expansion waves generated by a curved body in a supersonic flow are assumed to be absorbed by the shock wave and will not reflect to interact with the flow field or body downstream of the original wave. Thus, if the body shape is known, then the flow conditions at the nose immediately behind the shock wave can be determined from known solutions for a wedge[83] or cone[83,84] in supersonic flow. The Prandtl–Meyer expansion is then used to determine the flow characteristics downstream of the nose by expanding through the appropriate expansion angle. Although the Prandtl–Meyer expansion is for two-dimensional flows, Eggers, Savin, and Syverson[87] showed the shock expansion method could be used for bodies of revolution as long as the flow could be considered locally two dimensional. It is also possible to apply this method to bodies at angle of attack as long as the shock wave remains attached and the flow can be considered locally two dimensional.

Figure 6.96 shows results of the shock expansion method compared with experimental values for ogives at $\alpha = 15°$. It is noted that this method gives good results when the value of $M_\infty \tau$ is greater than 1.

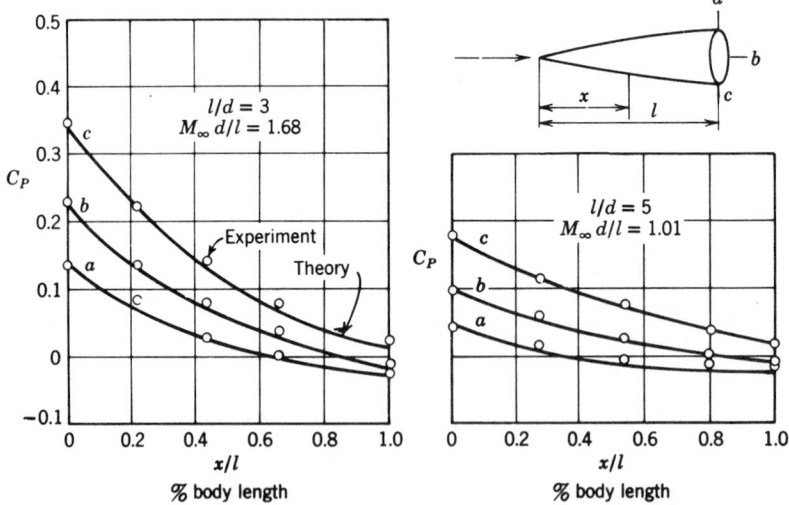

Fig. 6.96   Pressure coefficient distributions over two ogives at $M_\infty = 5.05$ and $\alpha = 15$.

Small Disturbance Theory

Small disturbance theory encompasses a variety of hypersonic flow theories appropriate for slender bodies. Even though the bodies are slender and the perturbation velocities are small compared to the free-stream velocity, they are not small compared to the free-stream speed of sound. Thus the small disturbance theory is essentially a nonlinear one.

If the substitution $x = V_\infty t$ is made in the small perturbation equations and boundary conditions,[85] then they become identical with the equations for unsteady flow in one less space dimension. This relationship is the "equivalence principle."[88] The equivalence principle is important because it means that a number of existing methods and solutions for unsteady flow problems become available for steady hypersonic flows, and an important extension of the principle has resulted in methods of treating the effects of nose blunting on the flow past a slender body.[80,85]

The small disturbance theory leads to the important conclusion that in the hypersonic small perturbation flow equations the Mach number and the body-shape parameter τ appears only in the combination $M_\infty \tau$, the "hypersonic similarity parameter." This means that for flow past affinely related slender bodies that have the same value of $M_\infty \tau$, the values of the nondimensional variables should be similar. This hypersonic similarity rule is discussed next.

Hypersonic Similarity

The hypersonic similarity parameter $M_\infty \tau$ makes it possible to extend the experimental data of a slender body at a particular Mach number to other bodies of similar shape but at different Mach numbers. Two bodies are said to be similar (geometrically) if their nondimensional thickness

TABLE 6.11   Correlation Parameters for Nose Bluntness effects*

Geometrical Shape	Aerodynamic Quantity	Correlation
Blunted cylinder	Surface pressure	$\dfrac{p}{p_\infty}$ vs. $\dfrac{1}{M_\infty^2\sqrt{C_{d_n}}}\left(\dfrac{x}{d_n}\right)$
	Shock-wave shape	$\dfrac{1}{M_\infty\sqrt{C_{d_n}}}\left(\dfrac{r_s}{d_n}\right)$ vs. $\dfrac{1}{M_\infty\sqrt{C_{d_n}}}\left(\dfrac{x}{d_n}\right)$
Blunted flat plate	Surface pressure	$\dfrac{p}{p_\infty}$ vs. $\left\{\dfrac{1}{M_\infty^3 C_{d_n}}\left(\dfrac{x}{d_n}\right)\right\}^{2/3}$
	Shock-wave shape	$\dfrac{1}{M_\infty^2 C_{d_n}}\left(\dfrac{y_s}{d_n}\right)$ vs. $\dfrac{1}{M_\infty^3 C_{d_n}}\left(\dfrac{x}{d_n}\right)$
Blunted cone	Surface pressure	$\dfrac{C_p}{2\theta_c^2}$ vs. $\dfrac{\theta_c^2}{\sqrt{\varepsilon C_{d_n}}}\dfrac{x}{d_n}$
	Heat transfer	$\dfrac{(\varepsilon C_{d_n})^{1/4} C_H\sqrt{Re_\infty, d}}{M_\infty\theta_c^2\sqrt{C^*}}$ vs. $\dfrac{\theta_c^2}{\sqrt{\varepsilon C_{d_n}}}\dfrac{x}{d_n}$
Blunted wedge	Surface pressure	$\dfrac{C_p}{2\theta_w^2}$ vs. $\theta_w^2\left(\dfrac{x}{C_{d_n}\varepsilon d_n}\right)^{2/3}$

C_{d_n} = Nose drag coefficient.

C_H = Coefficient of heat transfer.

C_p = Pressure coefficient.

d_n = Nose diameter.

M_∞ = Free-stream Mach number.

p = Surface pressure.

p_∞ = Free stream static pressure.

r_s = Shock-wave radius.

Re_∞, d = Reynolds number based on nose diameter.

x = Axial distance.

y_s = Distance of shock wave from the centerline.

$\varepsilon = (\gamma - 1)/(\gamma + 1)$.

γ = Ratio of specific heats.

θ_c = Cone half-angle.

θ_w = Wedge half-angle.

ψ = Nose bluntness ratio.

distributions $h(\xi)$ are the same. The surface pressure coefficient is given as[78, 80, 85]

$$C_P = \tau^2 f\left(\gamma, M_\infty\tau, h, \frac{\alpha}{\tau}\right)$$

where γ is the ratio of specific heats and α is the angle of attack. As before, τ is the thickness ratio of the slender body.

If two similar bodies A and B (i.e., $h_A = h_B$) have their M_∞ and α adjusted such that $(M_\infty\tau)_A = (M_\infty\tau)_B$ and $(\alpha/\tau)_A = (\alpha/\tau)_B$, then $C_P/\tau^2)_A = C_P/\tau^2)_B$ and the lift and drag coefficients are related as follows:[85]

1. Two dimensional bodies (coefficients based upon planform area)

$$\left(\frac{C_L}{\tau^2}\right)_A = \left(\frac{C_L}{\tau^2}\right)_B \qquad \left(\frac{C_D}{\tau^3}\right)_A = \left(\frac{C_D}{\tau^3}\right)_B$$

2. Three-dimensional bodies (coefficients based on cross-sectional area)

$$\left(\frac{C_L}{\tau}\right)_A = \left(\frac{C_L}{\tau}\right)_B \qquad \left(\frac{C_D}{\tau^2}\right)_A = \left(\frac{C_D}{\tau^2}\right)_B$$

Nose Bluntness Effects

Nose blunting has a significant effect on the flow field about a body in hypersonic flow because of the strong pressure gradients generated in the nose region. This effect can be significant for hundreds of nose diameters downstream. The blast wave analogy of Lees[89] for treating flow past blunt-nosed cylinders and flat plates and the methods of Chernyi[78] and Cheng[90] for treating flow about blunt-nosed cones and wedges are significant because of the data correlation parameters that result from these theories. Important aerodynamic quantities such as surface pressures, heat transfer coefficient, and shock-wave shape can be correlated against parameters involving the geometry of the aerodynamic configuration. Table 6.11 shows a partial listing of the correlation parameters resulting from these theories.

Experimental evidence of the validity of these correlation parameters is shown in Fig. 6.97 in which $c_p/2\theta_c^2$ is plotted versus $\theta_c^2/\sqrt{\epsilon C_{d_n}}(x/d_n)$ for blunt cones with several bluntness ratios, cone angles, and free-stream Mach numbers. Note that the pressure data correlate quite well and are relatively insensitive to Mach number. These data may be used to obtain estimates of the pressure distributions over cones that have cone angles and bluntness ratios different from those tested. Similar agreement has been found for the other dynamic parameters and for other geometries.

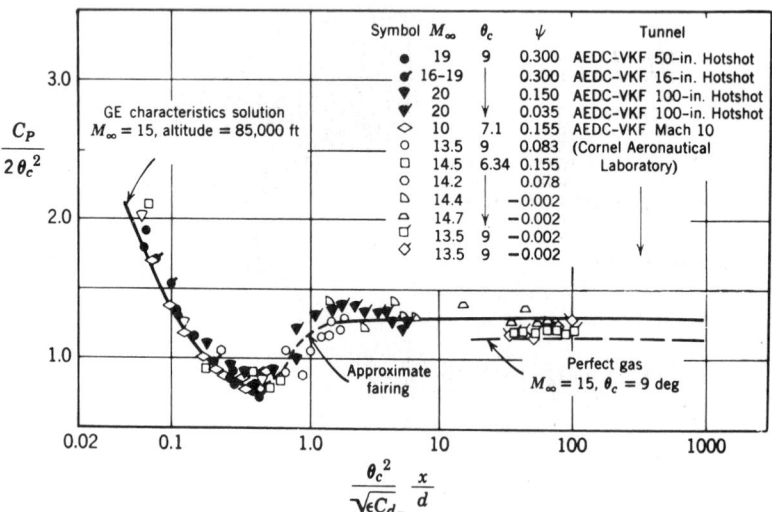

Fig. 6.97 Correlation of zero-lift cone pressures, $M_\infty = 10$ to 20 (Griffith and Lewis[91]); GE characteristics solution from Ref. 92; approximate fairing from Ref. 93.

6.10 VISCOUS FLOWS
L. M. Nicolai and J. J. Beoddy with Contributions by P. I. King

6.10.1 Introduction

The purpose of this section is to present the fundamental ideas of viscous continuum fluid flow over aircraft and hypervelocity vehicles. The discussions will consider the theory of viscous flows and methods for predicting skin friction and aerodynamic heating. For the discussions the flow may be assumed to be laminar if the local Reynolds number is

$$R_e = \frac{\rho_e u_e x}{\mu_e} < 5 \times 10^5$$

where x is the distance from the leading edge to the point of interest and the subscript e denotes conditions at the outer edge of the boundary layer.

Separation and transition to turbulence will not be discussed in this section. These items are extremely elusive and not amenable to brief discussions of fundamentals. References 94 and 95 provide excellent discussions of these topics.

6.10.2 Viscous Flow Theory

Laminar Flow
For a body immersed in a moving fluid of large $R_{eL} = \rho_\infty V_\infty L/\mu_\infty$, the influence of the viscosity is confined to a very thin layer near the surface of the body. Prandtl introduced this idea at the turn of the twentieth century and hypothesized that the boundary layer thickness δ was very small compared to a characteristic body dimension L and that the velocity at the surface was zero. If the continuum two-dimensional equations governing the fluid motion are nondimensionalized with respect to δ, L, and flow properties evaluated at the edge of boundary layer (subscript e) and an order of magnitude analysis performed,[94, 95] the result is the following set of simplified equations valid for the boundary layer region at large R_e:

$$\frac{\partial \rho}{\partial t} + \frac{\partial}{\partial x}(\rho u) + \frac{\partial}{\partial y}(\rho v) = 0 \tag{6.155}$$

$$\rho \frac{\partial u}{\partial t} + \rho u \frac{\partial u}{\partial x} + \rho v \frac{\partial u}{\partial y} = -\frac{\partial p}{\partial x} + \frac{\partial}{\partial y}\left(\mu \frac{\partial u}{\partial y}\right) \tag{6.156}$$

$$\frac{\partial p}{\partial y} = 0 \tag{6.157}$$

$$\rho \frac{\partial H}{\partial t} + \rho u \frac{\partial H}{\partial x} + \rho v \frac{\partial H}{\partial y} = -\frac{\partial}{\partial y}\left(k \frac{\partial T}{\partial y}\right) + \frac{\partial}{\partial y}\left(\mu u \frac{\partial u}{\partial y}\right) \tag{6.158}$$

where $H = h + (u^2 + \tau^2)/2$, total enthalpy, and $h = \int_0^T C_p \, dT$, static enthalpy.
   The boundary conditions are

$$u = 0 \text{ (no slip condition) and } v = 0 \text{ at } y = 0$$
$$u = u_e \quad \text{at } y \rightarrow \infty \tag{6.159}$$

Equation (6.157) is the y-momentum equation and expresses the important result that the pressure along the body surface is equal to the pressure p_e at the edge of the boundary layer. Since the flow outside the boundary layer can be considered inviscid, the pressure gradient term is given by

$$\rho \frac{\partial u_e}{\partial t} + \rho u_e \frac{\partial u_e}{\partial x} = -\frac{\partial p}{\partial x} \tag{6.160}$$

Incompressible.

Zero Pressure Gradient, dp / dx = 0. The solution of the velocity field for the incompressible flat plate at zero angle of attack was determined by *H*. Blasius in 1908. The system of equations under consideration has no preferred length, thus it is reasonable to suppose that the velocity profiles at varying distances from the leading edge are similar to each other. An appropriate choice for the reference velocity and distance are $u_e(x)$ and $\delta(x)$, respectively. Thus we seek a function g such that

$$\frac{u}{u_e} = g(\eta) \tag{6.161}$$

where $\eta = y/\delta$. It can be shown[94] that $\delta \sim \sqrt{\mu x/\rho u_e}$ so that we define $\eta = y\sqrt{\rho u_e/\mu x}$.
A nondimensional stream function $f(\eta)$ is defined as

$$\psi = \sqrt{\frac{\mu x u_e}{\rho}}\, f(\eta) \tag{6.162}$$

The velocity components become

$$u = \frac{\partial \psi}{\partial y} = \frac{\partial \psi}{\partial \eta}\frac{\partial \eta}{\partial y} = u_e f'(\eta) \tag{6.163}$$

$$v = \frac{1}{2}\sqrt{\frac{\mu u_e}{\rho x}}\,(\eta f' - f) \tag{6.164}$$

and our function g in Eq. (6.161) is $f'(\eta)$.

Inserting our new variables into the steady incompressible flat plate continuity and momentum equations gives us

$$2f''' + f''f = 0 \tag{6.165}$$

with $f = f' = 0$ at $\eta = 0$ and $f' = 1$ at $\eta \to \infty$. The details of the solution of Eq. (6.165) are in Ref. 94 and are presented in tabular form in Table 6.12.

If we define our boundary layer thickness δ as that distance from the wall where $u = -0.99u_e$, we see from Table 6.12 that $f'(\eta) = 0.99$ at $\eta = 5.0$. Thus

$$\delta = 5\sqrt{\frac{\mu x}{\rho_e \mu_e}} = \frac{5x}{\sqrt{R_{ex}}} \tag{6.166}$$

The shearing stress at the wall is

$$\tau_w = \left(\mu\frac{\partial u}{\partial y}\right)_w = \mu u_e\sqrt{\frac{\rho u_e}{\mu x}}\, f''(0)$$

TABLE 6.12 The Function f, f' and f'' for Laminar Boundary Layer with $dp\,/\,dx = 0$ (Blasius Solution)

η	$f(\eta)$	$f'(\eta)$	$f''(\eta)$
0	0	0	0.3321
0.2	0.0066	0.0664	0.3320
0.4	0.0266	0.1328	0.3315
0.6	0.0597	0.1989	0.3301
0.8	0.1061	0.2647	0.3274
1.0	0.1656	0.3298	0.3230
1.5	0.3716	0.4867	0.3023
2.0	0.6500	0.6298	0.2668
2.5	0.9953	0.7518	0.2173
3.0	1.3968	0.8460	0.1614
3.5	1.8282	0.9128	0.1080
4.0	2.3058	0.9555	0.0642
4.5	2.7852	0.9794	0.0343
5.0	3.2833	0.9916	0.0159
5.5	3.7806	0.9969	0.0068
6.0	4.2796	0.9990	0.0024
6.5	4.7793	0.9997	0.0008
7.0	5.2793	0.9999	0.0002

and from Table 6.12, $f''(0) = 0.332$, such that the local skin friction drag coefficient for one side of a flat plate is

$$C_f = \frac{\tau_w}{\frac{1}{2}\rho_e u_e^2} = \frac{0.664}{\sqrt{R_{ex}}} \qquad (6.167)$$

and the average skin friction coefficient for one side of a flat plate is

$$C_F = \frac{1}{L}\int_0^L C_f\, dx = \frac{1.328}{\sqrt{R_{eL}}} \qquad (6.168)$$

Nonzero Pressure Gradient. Let us consider that the velocity u_e at the outer edge of the boundary layer is given by[96]

$$u_e(x) = cx^m \qquad (6.169)$$

where c is a constant and x is measured from the stagnation point. This $u_e(x)$ describes the velocity along a wedge of included angle $\pi\beta (0 < \beta < 2, m > 0)$ where $\beta = 2m/(m + 1)$. Following the idea of the last section we transform the independent variable y into

$$\eta = y\sqrt{\frac{m+1}{2}\frac{\rho u_e}{\mu x}} = y\sqrt{\frac{m+1}{2}\frac{c\rho x^{m-1}}{\mu}} \qquad (6.170)$$

The continuity equation is satisfied by the introduction of a stream function

$$\psi(x,y) = \sqrt{\frac{2}{m+1}\frac{\mu c}{\rho}}\, x\frac{m+1}{2} f(\eta)$$

The momentum equation is transformed into

$$f''' + ff'' + \beta(1 - f'^2) = 0 \qquad (6.171)$$

with boundary condition $f = f' = 0$ at $\eta = 0$ and $f' = 1$ at $\eta \to \infty$. Equation (6.171) was first introduced by Falkner and Skan.[96] We observe that when $\beta = 0$ the equation reduces to the Blasius expression for a flat plate [when Eq. (6.170) is put in the Blasius form for η]. The $\beta = 1$, $m = 1$ case is the situation for two-dimensional stagnation flow.

The numerical solution of Eq. (6.171) is presented in Ref. 97. The value of the local skin friction on a wedge can be determined from

$$C_f = \frac{\tau_w}{\frac{1}{2}\rho u_e^2} = 2R_{ex}^{-1/2}\sqrt{\frac{m+1}{2}}\, f''(0)$$

where $f''(0)$ is the slope of the velocity distribution $u/u_e = f'(\eta)$ at $\eta = 0$.

Compressible. Prandtl number $P_r = \mu C_p/k = 1$ and $dp/dx = 0$. The right-hand side of Eq. (6.158) can be expressed in terms of total enthalpy H giving

$$\rho\left(u\frac{\partial H}{\partial x} + v\frac{\partial H}{\partial y}\right) = \frac{\partial}{\partial y}\left(\frac{\mu}{P_r}\cdot\frac{\partial H}{\partial y}\right) + \frac{\partial}{\partial y}\left[\mu u\left(1 - \frac{1}{P_r}\right)\frac{\partial u}{\partial y}\right] \qquad (6.172)$$

For $P_r = \mu C_p/k = 1$ and $dp/dx = 0$ we compare the momentum and energy equations:

$$\rho\left(u\frac{\partial u}{\partial x} + v\frac{\partial u}{\partial y}\right) = \frac{\partial}{\partial y}\left(\mu\frac{\partial u}{\partial y}\right)$$

$$\rho\left(u\frac{\partial H}{\partial x} + v\frac{\partial H}{\partial y}\right) = \frac{\partial}{\partial y}\left(\mu\frac{\partial H}{\partial y}\right) \qquad (6.173)$$

with boundary conditions

$$u = 0 \quad \text{and} \quad H = h_w \quad \text{at } y = 0$$
$$u = u_e \quad \text{and} \quad H = H_e \quad \text{at } y = \delta$$

The similarity of the two equations (both diffusion-type equations) indicates a solution for the energy equation of

$$\frac{H - h_w}{H_e - h_w} = \frac{u}{u_e} \tag{6.174}$$

The solution in terms of static enthalpy is

$$h = h_w + (H_e - h_w)\frac{u}{u_e} - \frac{u^2}{2} \tag{6.175}$$

and we note that since $P_r = 1$, H_e is equal to the recovery enthalpy or adiabatic wall enthalpy h_{aw}. The heat transfer per unit area is

$$-q = k\frac{\partial T}{\partial y} = \frac{k}{c_p}\frac{\partial h}{\partial y} = \frac{k}{c_p}\left[\frac{H_e - h_w}{u_e}\frac{\partial u}{\partial y} - u\frac{\partial u}{\partial y}\right]$$

and at the wall $y = 0$, $u = 0$ such that

$$-q_w = \frac{k}{c_p}\frac{(H_e - h_w)}{u_e}\left(\frac{\partial u}{\partial y}\right)_w = \frac{k}{c_p}\frac{(h_{aw} - h_w)}{u_e}\left(\frac{\partial u}{\partial y}\right)_w \tag{6.176}$$

In terms of the Stanton number

$$St = \frac{-q}{\rho_e u_e (h_{aw} - h_w)} \quad \text{and} \quad C_f = \frac{\tau_w}{\frac{1}{2}\rho_e u_e^2}$$

we have Reynolds analogy

$$St = \tfrac{1}{2}C_f \tag{6.177}$$

which postulates that energy and momentum are transferred by the same mechanism.

As a result of $P_r = 1$ and $dp/dx = 0$ we determined the interesting results (due to Crocco) $St = \tfrac{1}{2}C_f$ and that $H(y)$ is a linear function of $u(y)$, but we have not solved for the velocity field yet. This will come in the next section.

$P_r \neq 1$ and $dp/dx = 0$. The steady momentum and energy Eqs. (6.156) and (6.172) are transformed using a linear viscosity law $\mu = BT$ and the transformation (credit to Dorodnitzen or Howarth[95])

$$S = x \qquad n = \int_0^y \frac{\rho}{\rho_e} dy \tag{6.178}$$

which gives an equivalent incompressible set of equations[95] (i.e., momentum equation is independent of the temperature field). Then we introduce the Blasius nondimensional variables

$$\eta = \frac{n}{\delta} = n\sqrt{\frac{\rho_e u_e}{\mu_e S}} \tag{6.179}$$

and

$$f(\eta) = \frac{\psi}{\sqrt{\mu_e u_e S/\rho_e}} \tag{6.180}$$

The resulting set of equations is

$$f''' + 2ff'' = 0 \tag{6.181}$$

$$H'' + \tfrac{1}{2}P_r fH' = u_e^2(P_r - 1)(f''^2 + f'f''') \tag{6.182}$$

with boundary conditions

$$f = f' = 0 \quad \text{at } \eta = 0 \qquad f' = 1 \quad \text{at } \eta \to \infty$$
$$H = h_w \quad \text{at } \quad \eta = 0 \qquad H = H_e \quad \text{at } \eta \to \infty$$

The Dorodnitzen–Howarth transformation and the linear viscosity law have uncoupled the momentum and energy equation. Equations (6.181) and (6.182) put the solution for the velocity field in the Blasius form. Thus the solution for u/u_e is tabulated in Table 6.12 in terms of the transformed variable n. Equation (6.182) is solved numerically using the Blasius solution for $f(\eta)$. The details of the solution are discussed in Ref. 95. The solution for the local skin friction at the wall is

$$C_f = 0.664 \frac{1}{\sqrt{\rho_e u_e x}} \sqrt{\frac{\mu_w T_e}{\mu_e T_w}} = \frac{0.664}{\sqrt{R_{ex}}} \sqrt{\frac{\mu_w T_e}{\mu_e T_w}} \tag{6.183}$$

Clearly the linear viscosity law is an approximation to the more correct Sutherland's equation (see Section 6.3.1). The law $\mu = BT = C(\mu_e/T_e)T$, where C is a constant, can satisfactorily describe the variation in viscosity over reasonable temperature intervals if $C < 1$. From Eq. (6.183)

$$C_{fe} = \frac{0.664}{\sqrt{R_{ex}}} \sqrt{C} = C_{fi}\sqrt{C}$$

where the subscripts c and i denote compressible and incompressible, respectively. Thus the effect of compressibility, that is, dissipation and heat transfer, is a decrease in the local skin friction. This conclusion agrees with experimental results and is indicated on Fig. 6.98.

Another conclusion imbedded in the transformed y coordinate is that compressibility thickens the boundary layer. This result is shown on Fig. 6.99.

The solution for the heat transfer at the wall is

$$St = \frac{0.332}{P_r^{2/3}} \frac{1}{\sqrt{\rho_e u_e x/\mu_e}} \sqrt{\frac{\mu_w T_e}{\mu_e T_w}} \tag{6.184}$$

Combining C_f and St we obtain an extended Reynolds analogy for general P_r,

$$\frac{St}{C_f} = \frac{P_r^{-2/3}}{2} \tag{6.185}$$

Cohen and Reshotko[98] considered a compressible laminar boundary layer with a nonzero pressure gradient. Their analysis parallels the preceding discussion except that their x and y coordinates are transformed according to

$$X = \int_0^x B \frac{p_e a_e}{p_0 a_0} dx \qquad Y = \frac{a_e}{a_0} \int_0^y \frac{\rho}{\rho_0} dy$$

where a is the speed of sound, B is the constant in the linear viscosity law, and the subscript 0 denotes free-stream stagnation values. The power law velocity distribution of Falkner and Skan, Eq. (6.168) is used so that $p_e(x)$ is known. The details of their analysis and solution are presented in Ref. 98. One important result from Cohen and Reshotko's work is the variation of Reynolds' analogy with the pressure gradient parameter β. For $\beta \neq 0$ (i.e., a nonzero pressure gradient) Eq. (6.185) is not valid.

Stagnation Point Heating. The steady continuity, momentum, and energy (in the form of Eq. (6.172) equations are transformed as before using a linear viscosity law, the Dorodnitzen–Howarth transformation and Eqs. (6.179) and (6.180). The pressure gradient term is determined by the steady form of Eq. (6.160). Near the forward stagnation point the velocity u_e just outside the boundary layer may be expressed as

$$u_e = \frac{du_e}{dx} x$$

where du_e/dx is considered constant and x is the distance along the body from the stagnation point. The momentum and energy equations are[95]

$$f''' + ff' + f'^2 + \frac{\rho_e}{\rho} = 0 \tag{6.186}$$

$$\frac{H''}{P_r} + fH' - \left(\frac{1}{P_r} - 1\right) u_e^2 (f''^2 + f'f''') = 0 \tag{6.187}$$

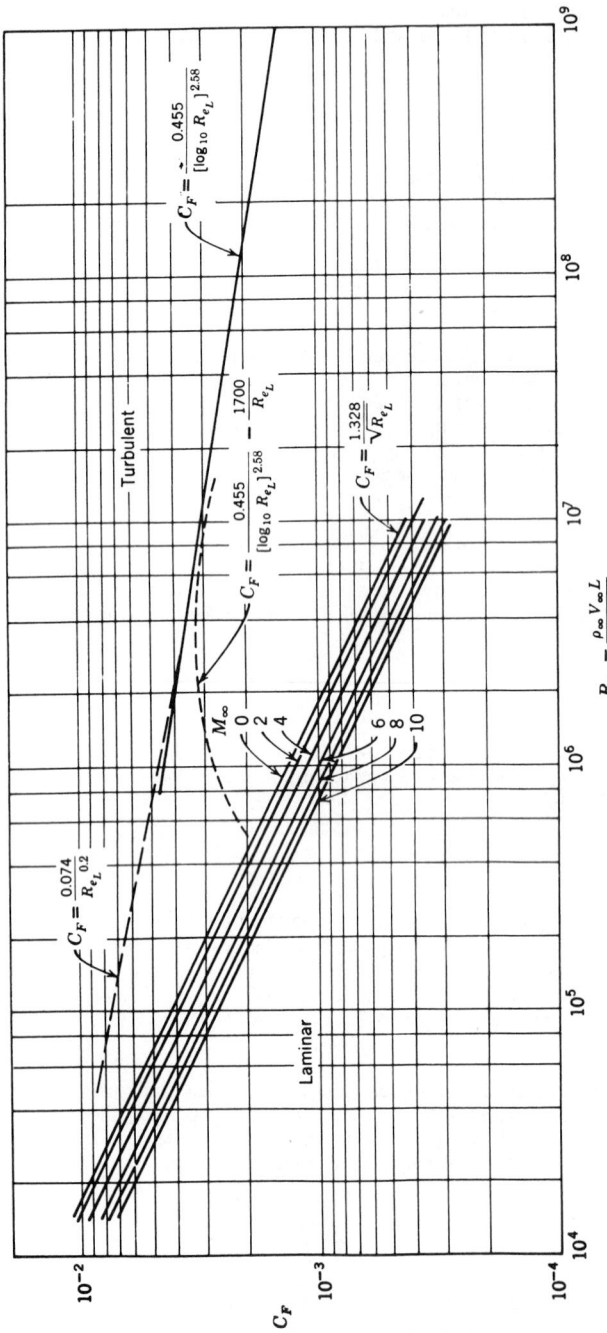

Fig. 6.98 Averaged skin friction coefficient versus Reynolds number for flow over an insulated flat plate.

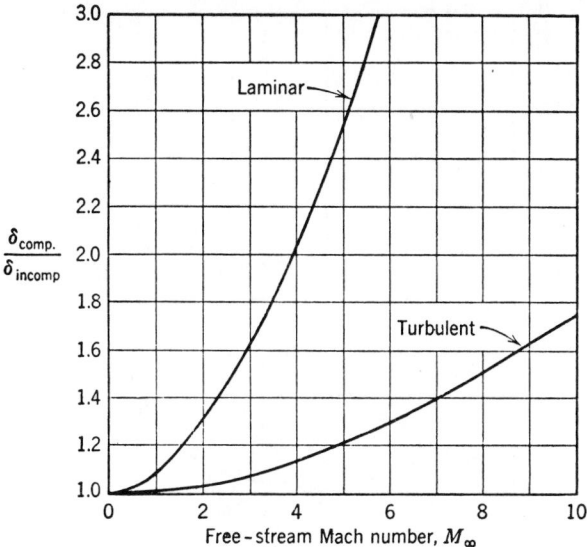

Fig. 6.99   Effect of compressibility on the thickness of the boundary layer on an integrated flat plate.

Since we are interested in solutions near the stagnation point, we let u_e become small such that the energy equation simplifies to

$$H'' + P_r fH' = 0 \tag{6.188}$$

The numerical solution[95] of Eqs. (6.186) and (6.188) provides the following stagnation point heat transfer relation:

$$q_{sp} = \frac{0.5}{P_r^{2/3}} \frac{\mu_w \rho_w}{\sqrt{\mu_e \rho_e}} \sqrt{\frac{du_e}{dx}} (H_e - h_w) \tag{6.189}$$

Hypersonic Flow. At hypersonic speeds the temperature variations in the boundary layer are very large, and chemical reactions such as dissociation and ionization may be present. In these processes more than one chemical species are present and concentration gradients occur in the boundary layer. Energy is then transported by the diffusion of the chemical species as well as by heat conduction.[95, 99]
  The heating of a laminar stagnation point with dissociation effects was considered by Fay and Riddel.[98] They carried out a numerical solution of the governing equations for a stagnation point boundary layer using tabulated values of the properties of high-temperature air in dissociation equilibrium. Their results are given by the expression

$$q_{sp} = \frac{0.54\sqrt{k+1}}{P_r^{0.6}} (H_e - H_w)(\rho_w \mu_w)^{0.1}(\rho_e \mu_e)^{0.4} \sqrt{\frac{du_e}{dx}} \left[1 + (L_e^m - 1)\frac{h_D}{H_e} \right] \tag{6.190}$$

where h_D is the enthalpy of dissociation, L_e is the Lewis number (relative rate of mass diffusion to heat conduction), $k = 0$ for a two-dimensional body, $k = 1$ for an axisymmetrical body, $m = 0.52$ for equilibrium flow, and $m = 0.63$ for frozen flow.

Turbulent Flow

Exact solutions to the governing differential equations of the boundary layer are not readily obtained for the turbulent flow case. With turbulence the particles of fluid experience highly irregular fluctuations that lead to high-frequency variations in both pressure and velocity at a fixed point in the flow. Since the flow is a fluctuating or random process, we work in statistical mechanics, seeking not an explicit account of the motion of every particle of fluid, but only a statistical description in terms of various averaged functions.[100]
  The fluid particles in the turbulent boundary layer undergo large-scale eddy motion. The transfer mechanisms of energy and momentum are microscopic molecular transport (as in the laminar

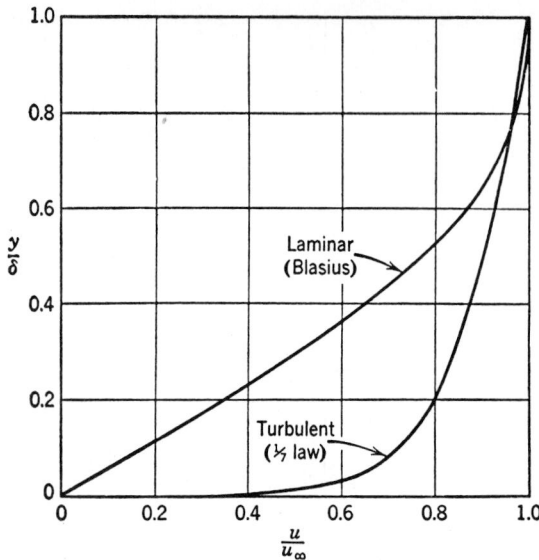

Fig. 6.100   Velocity profiles for laminar and turbulent layers (incompressible flow).

boundary layer) and macroscopic eddy motion; the macroscopic transfer being much larger than the molecular transport. Since the macroscopic transfer mechanism of momentum is the same as that for energy, that is, the large-scale motion of lumps of fluid,[100] we would expect a turbulent Reynolds' analogy to be valid in the fully turbulent region away from the wall.

A very thin region near the body surface has a laminar character and is called the laminar sublayer. The presence of this sublayer is essential since the large-scale eddy motion cannot extend all the way to the wall. In this sublayer the viscous action and heat transfer take place under circumstances like those in a laminar boundary layer.

Experimentally it is observed that the analogy between heat transfer and skin friction expressed by Eq. (6.185) retains its validity everywhere in the turbulent boundary layer. Also, the analogy remains approximately true in cases where the pressure gradient is different from zero.

Because of the macroscopic fluid motion in the fully turbulent region, the velocity distribution is fairly uniform (compared to the laminar boundary layer distribution) away from the wall. This results in a steep velocity gradient at the wall (through the laminar sublayer) and large skin friction. The velocity distribution in the turbulent boundary layer can be approximated by

$$\frac{u}{u_e} = \left(\frac{y}{\delta}\right)^{1/n}$$

where $n = 7$ at $R_e = 5 \times 10^5$ and increases slightly with increasing R_e.[94] The turbulent velocity profile is compared with the laminar profile in Fig. 6.100.

Using the empirical 1/7 power law, Blasius developed the following very useful expressions for turbulent boundary layer thickness δ_T and local skin friction coefficient:[94]

$$\delta_T = \frac{0.37x}{R_{ex}^{0.2}} \tag{6.191}$$

$$C_f = \frac{0.059}{R_{ex}^{0.2}} \tag{6.192}$$

6.10.3   Skin Friction Methods

Laminar

Incompressible. For smooth surfaces with zero pressure gradient the average, integrated skin friction drag coefficient for *one* side of a flat plate is given by the Blasius solution:

$$C_F = \frac{1.328}{\sqrt{R_{eL}}} = 2C_{fx-L} \tag{6.193}$$

where $R_{eL} = \rho_e u_e L / \mu_e$ is evaluated for a characteristic dimension L and the reference area is the area wetted by the flow (on one side).

Compressible. The most useful method for estimating the compressible laminar skin friction is the reference temperature method.[101,102] Essentially this method relies on the empirical observation that the incompressible formulas for heat transfer and skin friction (for laminar and turbulent flows) may be used provided that all physical properties are evaluated at a temperature (or enthalpy for large variations in temperature) intermediate between the values at the surface (T_w) and outer edge of the boundary layer (T_e). Equation (6.168) is used with all of the properties evaluated at a reference temperature T^* given by

$$T^* = 0.5(T_e + T_w) + 0.22r\left(\frac{\gamma - 1}{2}\right)M_e^2 T_e$$

where γ is the ratio of specific heats, M is the Mach number, and $r = P_r^{1/2}$ the recovery factor for laminar flow. The laminar curves on Fig. 6.98 can also be used to obtain a first-order estimate of the effect of compressibility.

Hypersonic. Several methods have been presented in the literature for the prediction of laminar skin friction on flat plates in supersonic and hypersonic flow. These methods are usually complex and laborious to apply; they often require evaluation of the flow field behind the compression shock wave. One of the more successful methods is the reference enthalpy method discussed in Refs. 101 and 102. Schmidt[103] used this method to predict skin friction over a flat plate for a wide range of flight conditions. To simplify further, the prediction of laminar skin friction an empirical curve was fitted to the data presented by Schmidt.[104] This curve is represented by

$$C_f \sqrt{R_e} = 0.45 \cos \alpha + \frac{4.65 u_\infty}{10,000} \sin \alpha \cos^{2.2} \alpha \qquad (6.194)$$

where α is the surface angle of attack and u_∞ is the free-stream velocity in feet per second.

This equation deviates by no more than 20% from the data given by Schmidt for low altitude, high α flight and is more accurate (approximate by $\pm 10\%$) for the rest of the altitude and α range in the hypersonic operating regime.

Extension to Cones and Cylinders. The relations developed so far have been for flat plates with constant pressure in the flow direction. Such constant pressures also arise along the surfaces of wedges, cones, and cylinders in supersonic flow. The flat plate results can be extended to cones having identical flow conditions outside the boundary layer using the Mangler transformation

$$C_{f\,cone} = \sqrt{3}\,C_{f\,flat\ plate} \quad \text{for local skin friction}$$

$$C_{F\,cone} = \frac{2}{\sqrt{3}}\,C_{F\,flat\ plate} \quad \text{for total skin friction}$$

Skin friction coefficients on the surface of a cylinder whose axis is aligned with the free stream are identical with the ones for a flat plate, providing the radius of the cylinder is large compared with the boundary layer thickness.

Turbulent

Incompressible. For turbulent flow the local and averaged skin friction coefficients can be estimated from the Blasius and Prandtl–Schlichting empirical relationships.[94] These expressions are

$$C_f = \frac{0.059}{R_{ex}^{0.2}} \qquad (6.195)$$

$$C_F = \frac{0.455}{\left(\log_{10} R_{eL}\right)^{2.58}} \qquad (6.196)$$

The Prandtl–Schlichting expression for C_F agrees with measured results very well up to $R_e = 10^9$ and is shown on Fig. 6.98.

The R_{eL} at which transition occurs is variable, depending on the turbulence level in the free stream and the roughness of the surface. Assuming transition at $R_e = 5 \times 10^5$, the following expression can be used to estimate the averaged skin friction coefficient in the transition region (see Fig. 6.98).

$$C_F = \frac{0.455}{(\log_{10} R_{eL})^{2.58}} - \frac{1700}{R_e}$$

Compressible. Similar to the laminar compressible flow, the turbulent compressible skin friction can be determined using the incompressible expression (Prandtl–Schlichting) but evaluating the fluid properties at a reference temperature T^*:

$$T^* = 0.5(T_e + T_w) + 0.22r\left(\frac{\gamma - 1}{2}\right)M_e^2 T_e$$

where $r = P_r^{1/3}$, the recovery factor for turbulent flow. Experimental results show that the effect of compressibility is to decrease the turbulent skin friction. Hayes and Probstein[105] found that an excellent fit for the experimental results is given by

$$\frac{C_{Fc}}{C_{Fi}} = \left(1 + 0.144M_\infty^2\right)^{-0.65}$$

Hypersonic. Schmidt[106] used the reference enthalpy method to construct a comprehensive series of graphs for the prediction of turbulent skin friction on a flat plate for a wide range of flight parameters. Hankey and Neumann[104] determined a curve fit to Schmidt's analysis that results in the very useful expression for the local turbulent skin friction coefficient

$$C_f\left(R_e\right)^{0.2} = 0.48 \sin(4.5\alpha) + 0.7\frac{u_\infty}{10,000}\cos^{2.25}\alpha \sin^{1.5}\alpha \qquad (6.197)$$

where

$$R_e = \frac{\rho_e u_\infty L}{\mu_e}$$

and α = surface angle of attack.

The average integrated skin friction coefficient is then

$$C_F = 1.39\left(\frac{S_w}{S}\right)S_\alpha \qquad (6.198)$$

where the characteristic length is the distance to the trailing edge. The factor 1.39 is used for swept flat plates or delta wings. Also

$$S = \text{reference wing area}$$

$$S_w = \text{compression surface wetted area}$$

$$S_\alpha = 1 + \frac{1}{|\alpha| + 1.15^{|\alpha|}} \quad \text{where } \alpha \text{ is in degrees}$$

Extension to Cones. Van Driest[107] developed what is referred to as the "cone rule" for turbulent flow by extending the momentum integral method to the case of the cone. This rule can be expressed as

$$C_{F \text{cone}} = 1.15 C_{F \text{ flat plate}}$$

This rule can be used for other axisymmetric bodies near $\alpha = 0°$.

6.10.4 High-Speed Aerodynamic Heating

Flat Plate—Laminar and Turbulent

The method for estimating the heat transfer at the surface of a body (with zero pressure gradient) is the reference temperature (or enthalpy) method and the extended Reynolds analogy. The discussion here is equally valid for both laminar and turbulent boundary layers in high speed ($M_\infty > 2$) flow.

The local Stanton number is again defined as

$$St = \frac{-q}{\rho_e u_e (h_{aw} - h_w)}$$

and the adiabatic wall enthalpy or recovery enthalpy is

$$h_{aw} = h_e + r\frac{u_e^2}{2} \qquad (6.199)$$

where

$$r = \begin{cases} P_r^{1/2} & \text{for laminar flow} \\ P_r^{1/3} & \text{for turbulent flow} \end{cases}$$

The heat transfer per unit area per unit time at the wall is

$$q_w = -k\left(\frac{\partial T}{\partial y}\right)_w = h_c'(h_{aw} - h_w) \qquad (6.200)$$

where h_c' is the forced convection heat transfer coefficient. If the temperature variation in the boundary layer is not large such that C_p is fairly constant, then Eq. (6.200) is expressed as

$$q_w = h_c(T_{aw} - T_w) \qquad (6.201)$$

where

$$h_c = C_p h_c'$$

The laminar and turbulent extended Reynolds analogy is[94, 102]

$$\frac{St}{C_f} = \frac{P_r^{-2/3}}{2} \qquad (6.202)$$

which is valid for $dp/dx \approx 0$.

The reference temperature or enthalpy is determined from

$$T^* = 0.5(T_w + T_e) + 0.22r\left(\frac{\gamma - 1}{2}\right)M_e^2 T_e$$

$$h^* = 0.5(h_w + h_e) + 0.22r\left(\frac{\gamma - 1}{2}\right)M_e^2 h_e$$

Next the local skin friction coefficient C_f is determined using the methods presented in the last section. Then, using the extended Reynolds analogy, Eq. (6.202), the local Stanton number is obtained and finally the local heat transfer q_w.

This procedure is in widespread use today and is referred to as the reference enthalpy method. Its accuracy has been checked by comparison with the results obtained by boundary layer solutions.[103, 104, 106] Agreement within $\pm 4\%$ has been found for $400 < T_\infty < 800°R$, $400 < T_w < 3000°R$, and Mach numbers up to 16.

For surfaces having laminar and turbulent regions, the average heat transfer for the surface is obtained by integrating the local values over the surface. We consider an abrupt transition to turbulence at a local Reynolds number of 5×10^5. The expression for the average Stanton number over a surface is

$$St\, P_r^{2/3} = \frac{\int_0^L \frac{1}{2} C_f\, dx}{\int_0^L dx}$$

$$= \frac{1}{L}\left[\int_0^{x_c} \frac{0.332}{\sqrt{R_e}}\, dx + \int_{xc}^L \frac{0.228}{(\log_{10} R_e)^{2.58}}\, dx\right]$$

where $R_e = \rho u_\infty x / \mu$ and x_c corresponds to $R_e = 5 \times 10^5$.

If $dp/dx \neq 0$ for the laminar case, the results of Cohen and Reshotko[98] should be used to determine the proper expression for Reynolds analogy.

Stagnation Point

The heat transfer rate per unit area at the stagnation point q_{sp} can be determined using Eq. (6.189). An estimate of the velocity gradient du_e/dx can be obtained using the modified Newtonian (see Section 6.11.1)

$$\frac{p_e(x)}{p_{sp}} = \cos^2\theta$$

where θ is measured from the stagnation point to the point x on the blunt surface. Using $u_e^2 = (2/\rho_e)(p_{sp} - p_e)$ we obtain

$$\frac{du_e}{dx} \approx \frac{u_\infty}{R}\sqrt{\frac{\gamma-1}{\gamma}}$$

where R is the radius of the blunt surface.

Detra and Hidalgo[108] used empirical data for the air properties and obtained the following approximate expression for Eq. (6.189):

$$q_{sp} = \frac{865}{\sqrt{R}}\left(\frac{u_\infty}{10^4}\right)^{3.15}\sqrt{\frac{\rho_\infty}{\rho_s L}} \tag{6.203}$$

where $\rho_s L$ is the density at sea level, R is the nose radius in feet, u_∞ is the free-stream velocity in feet per seconds, and q_{sp} is the heating rate in Btu/ft^2 sec. Equation (6.203) agrees $\pm 10\%$ with experimental results for $6{,}000 \leq u_\infty \leq 26{,}000$ ft/sec and sea level to 400,000 ft.

The distribution of heating rate about a blunt surface behaves similar to the pressure distribution. Thus, using modified Newtonian

$$\frac{q}{q_{sp}} \approx \cos\theta$$

The importance of a blunt surface with a large radius R in order to decrease the stagnation point heating is readily apparent from Eq. (6.203).

The expression developed by Fay and Riddel [Eq. (6.190)] should be used for estimating the stagnation point heating rate when dissociation effects are present.

6.11 AERODYNAMICS OF WING-BODY COMBINATIONS
L. M. Nicolai

6.11.1 Wing-Body Lift

The combined wing-body lift characteristics are expressed as

$$C_L = (C_{L\alpha})_{WB}\alpha + C_1\alpha^2 \tag{6.204}$$

where $(C_{L\alpha})_{WB}$ is the wing-body linear lift curve slope and C_1 is the nonlinear lift factor. For wings with sharp leading edges and aspect ratios less than 2, the nonlinear lift is quite pronounced. Values of C_1 are given in Refs. 109 and 110.

The lift characteristics of a wing and a body do not add directly to give the wing-body lift. Rather, there are interference effects of one component on the other.[111] A method that gives good results for the wing-body linear lift curve slope is

$$(C_{L\alpha})_{WB} = F(C_{L\alpha})_W \tag{6.205}$$

where $(C_{L\alpha})_W$ is the linear lift curve slope (based upon the exposed wing area) of the wing alone and F is a wing-body lift interference factor given on Fig. 6.101. The $(C_{L\alpha})_W$ is determined using Eq. (6.108) for subsonic flow and the methods of Section 6.7.6 for supersonic flow. The curve for F on Fig. 6.101 was determined using the method of Ref. 112. The $(C_{L\alpha})_{WB}$ is referenced to the exposed wing area S_e.

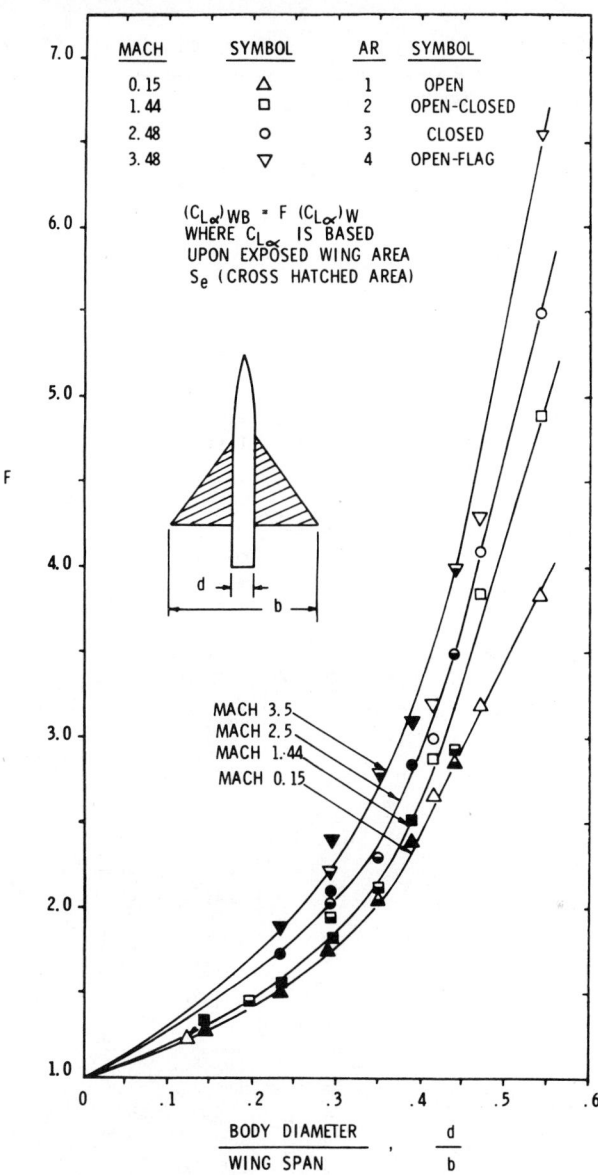

Fig. 6.101 Wing-body lift interference factor.

6.11.2 Wing-Body Drag

The combined wing-body drag characteristics can be expressed as [see Eq. (6.94)]

$$C_D = C_{D_0} + KC_L^2 \tag{6.206}$$

where C_{D_0} = zero lift drag coefficient = $C_{D_p} + C_F + C_{DW}$

C_{D_p} = pressure drag coefficient due to flow separation at $C_L = 0$. Base drag C_{DB} is included here

C_F = skin friction drag coefficient

C_{DW} = wave drag coefficient, due to thickness and camber

K = drag due to lift factor

Subsonic Drag

The subsonic C_{D_0} for slender wing-body combinations is primarily due to skin friction ($C_{DW} = 0$ for subsonic flow). The C_F's for the wing alone and body alone are determined using the methods of Section 6.12.2 and then added together as

$$C_F = 1.05 \left[(C_F)_{\text{body}} \frac{S_{BW}}{S_{\text{ref}}} + (C_F)_{\text{wing}} \frac{S_{WW}}{S_{\text{ref}}} \right] \qquad (6.207)$$

where S_{BW} is the body wetted area, S_{WW} is the wing wetted area (usually $2S_e$), and S_{ref} is the reference area in the expression

$$\text{Drag} = C_D \tfrac{1}{2}\rho_\infty V_\infty^2 S_{\text{ref}} \qquad (6.208)$$

The coefficient of 1.05 in Eq. (6.207) accounts for a drag interference of 5%.

The drag due to lift factor is discussed in Section 6.7.3 and shown in Fig. 6.102 for wing-body combinations having delta planforms.[113,114] A very useful empirical expression for the subsonic drag due to lift factor K for $\mathbb{R} > 2$ is given by[115]

$$K_{WB} = \frac{0.95}{(C_{L\alpha})_{WB}} \qquad (6.209)$$

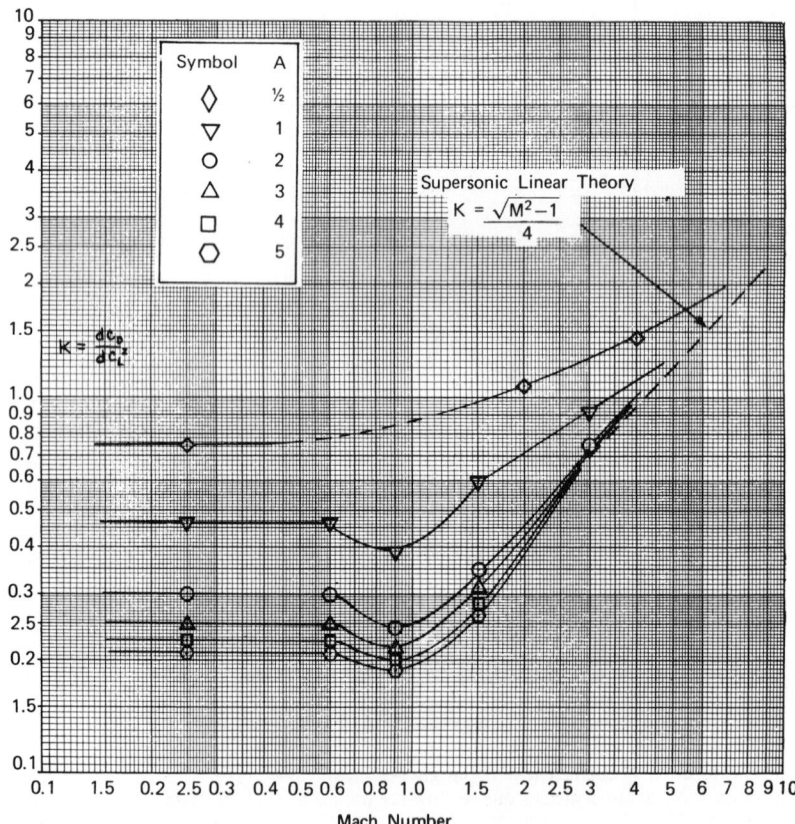

Fig. 6.102   Drag due to lift factor for wing-body combinations delta planforms. (Data from USAFA trisonic tunnel Refs. 113 and 114.)

Transonic and Supersonic Drag

Wave drag interference effects in the transonic and supersonic range are greater than those in the subsonic region because of the higher local velocities of the individual components and the greater propagation of these perturbations from this source. The most successful and by far the most systematic method for predicting the transonic and supersonic drag is the area-rule concept.

The area-rule method is based on the supersonic slender body theory discussed in Section 6.8.3. Reference 116 shows that this supersonic theory can be extended down to a Mach number of 1 as a limiting case. It can be assumed that at large distances from the body the disturbances are independent of the arrangement of the components and only a function of the cross-section area distribution.[117] This means that the drag of a wing-body combination can be calculated as though the combination were a body of revolution with equivalent-area cross sections. This is shown in Fig. 6.103 for a Mach number of 1.

At Mach numbers greater than 1, the cross-section areas $S(x)$ are along planes inclined at the angle $\mu = \arcsin 1/M_\infty$ to the x axis. There is a different $S(x)$ for each roll angle Φ. This is shown in Fig. 6.104.

Once the area distribution $S(x)$ is determined [one $S(x)$ for each Φ angle for $M_\infty > 1$], the wave drag is calculated using Eq. (6.142). For $M_\infty > 1$ the C_{DW} is determined for each roll angle Φ and

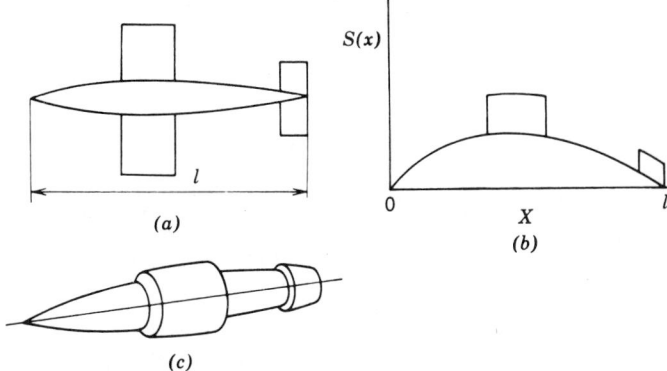

Fig. 6.103 Equivalent body for a wing-body-tain combination at $M_\infty = 1$: (a) wing-body, (b) cross-section area distributions, and (c) equivalent body.

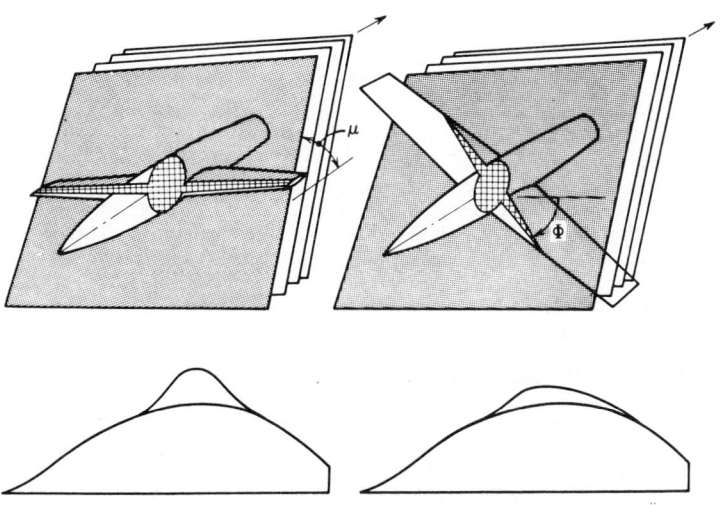

Fig. 6.104 Area distribution given by intersection of Mach$_\infty > 1$.

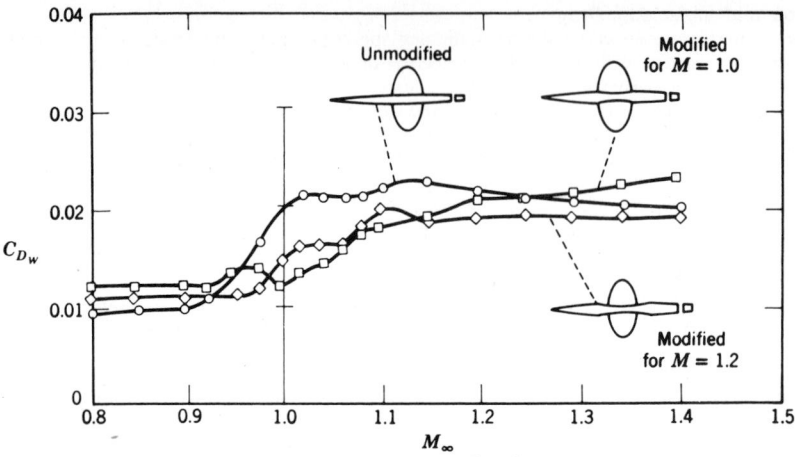

Fig. 6.105   Wave drag of bodies with elliptic wings.

then averaged. Application of the area-rule method usually requires automatic computing equipment.[118]

The area-rule method indicates the most desirable way to arrange the vehicle components for minimum wave drag at a particular M_∞. The most common example of this is to indent or "coke bottle" the fuselage enough to permit the wing to be added without a sharp discontinuity appearing on the $S(x)$ distribution. If the cross-section area distribution of a wing-body combination at a particular M_∞ is the same as a Sears-Haack distribution (see Section 6.8.3), the configuration produces minimum wave drag at that M_∞. Thus a wing-body can be configured to give minimum wave drag at one Mach number but will usually aggravate the wave drag at other Mach numbers. Figure 6.105 demonstrates this.

The wing-body skin friction coefficient is determined in the same manner as for subsonic flow except that compressibility effects are considered. If the wing-body has a blunt base, the C_{DB} is determined from Section 6.7.6.

The drag due to lift factor is determined using the charts in Section 6.7.6 or estimated from Fig. 6.102. Notice that K approaches the value $1/C_{L\alpha}$ at $M_\infty > 1$.

6.12   REAL GASES
D. Finkleman

6.12.1   Thermodynamics of Real Gases

Most classical analyses in aerodynamics assume the medium to behave ideally. Although the results of such investigations have been quite accurate in the past, no fluid behaves as an inviscid gas that is calorically and thermally perfect. Deviation from ideal behavior is due to the presence of molecular transport phenomena that manifest themselves in the macroscopic viscosity, thermal conductivities, and diffusion coefficients (see Section 6.1.4). In this section these effects are considered only indirectly, and nonideal thermodynamics is of major importance. For instance, all gases encountered in practice are mixtures, their chemical composition may change with temperature and pressure, and internal degrees of freedom of their individual atoms and molecules may become excited, thus altering the distribution of energy. The thermodynamics of real gases will first be discussed and then application to practical problems will be indicated.

Equilibrium Statistical Thermodynamics
The thermodynamics and transport properties of gases may be determined only if one considers the detailed motion of their constituent particles. Two approaches (kinetic theory and statistical mechanics), both statistical in nature, are possible. Kinetic theory requires detailed consideration of the mechanics of interparticle encounters. Such analyses rapidly grow cumbersome, thus only relatively simple microscopic situations may be analyzed. Mainly because of a lack of knowledge about the details of both elastic and inelastic encounters of polyatomic molecules, the results one may obtain from kinetic theory are presently limited. Statistical mechanics is far more powerful in the thermody-

namic information that it yields, even though its application should strictly be limited to equilibrium states of completely uniform volumes of material. Although theories of nonequilibrium statistical thermodynamics exist, most rely upon the postulate that the thermodynamic properties of a system subject to nonequilibrium processes are, at each point the same functions of local state variables that they would be if the gas were in equilibrium at those conditions.[119]

At the heart of all thermodynamics is the concept of entropy. For most purposes entropy may be regarded as a means of keeping track of the processes that may occur within a system. Statistically, this is reflected by enumerating each possible energy state in which each of the particles may reside under given macroscopic conditions with the Boltzmann relation

$$S = k \ln \Omega$$

The evaluation of Ω depends on the manner in which the particles behave. If no two particles of the same type may reside in the same energy level, so-called Fermi–Dirac statistics apply. Thus if the number of particles of energy E is N_j and there are g_j possible states with that energy, the total number of ways of assigning the particles to the states is

$$W_j = \prod_j \frac{(g_j)!}{(g_j - N_j)! N_j!}$$

On the other hand, if any number of particles may be crowded into a single energy level, Bose–Einstein statistics apply and

$$W_j = \prod_j \frac{(N_j + g_j - 1)!}{(g_j - 1)! N_j!}$$

A particular choice of the N_j's for a total number of particles, N, is called a distribution. When all possible distributions are considered, the total number of microstates of a system is

$$\Omega = \sum_{\substack{\text{all possible} \\ \text{sets of } N_j \\ \text{such that}}} W_j \quad \left(\begin{array}{l} N = \sum_j N_j \\ E = \sum_j N_j \varepsilon_j \end{array} \right.$$

It can be shown that only the largest term in the summation contributes significantly.[120] For large numbers of particles and an even larger number of possible states (so that the energy levels are sparsely populated) it follows that if all microstates are equally probable in equilibrium (when S is a maximum), the most probable distribution of particles among the energy states is

$$\frac{N_j^*}{N} = \frac{g_j e^{-(\varepsilon_j/kT)}}{\sum_j g_j e^{-(\varepsilon_j/kT)}} \tag{6.210}$$

independent of the type of statistics applied. The denominator in the exponent, kT, and the identification of k as Boltzmann's constant are a consequence of comparison of the statistical entropy expression and resulting reciprocity relations with the first law of thermodynamics. The quantity in the denominator of Eq. (6.210) is the sum-over-states (Zustandsumme) or partition function and is often denoted by Q. We see that the various terms in the summation are proportional to the probabilities of the states to which they refer. The partition functions may be found only if the energies ε_j and degeneracies g_j are known, and they are determined using quantum mechanics.

Partition Functions
The partition function is defined as

$$Q = \sum_j g_j e^{-(\varepsilon_j/kT)}$$

It is assumed that the energies partitioned in the various degrees of freedom are all independent. Therefore the total partition function can be represented as

$$Q = Q_{\text{trans}} Q_{\text{vib}} Q_{\text{rot}} Q_{\text{elect}} \cdots$$

Thus we can look at the individual energy degrees of freedom and determine the Q contribution of each energy mode independently.

In a gas of structureless particles (particles with neither internal degrees of freedom nor interparticle potentials) contained in a rectangular "potential box" of dimensions a_1, a_2, a_3, the allowed energies are[120]

$$\varepsilon_{n_1, n_2, n_3} = \frac{h^2}{8m} \left(\frac{n_1^2}{a_1^2} + \frac{n_2^2}{a_2^2} + \frac{n_3^2}{a_3^2} \right)$$

where all integral n's are allowed and the levels are nondegenerate. The translational partition function is

$$Q_{\text{trans}} = \sum_{n_1=1}^{\infty} \sum_{n_2=1}^{\infty} \sum_{n_3=1}^{\infty} e^{-\varepsilon_{n_1, n_2, n_3}/kT} = V \left(\frac{2\pi mkT}{h^2} \right)^{3/2} \qquad (6.211)$$

where h is Planck's constant. In monatomic gas the only other mode of excitation is electronic.

Electronic excitation requires large amounts of energy, and electrons that deviate large distances from their nuclei may fall under the influence of other particles. This depends on the density of the gas, which provides a density-dependent cutoff to the electronic partition function. This is important only at high temperatures since at moderate temperatures the exponential terms in the electronic partition function

$$Q_{\text{elect}} = \sum_{l=j}^{\infty} g_j e^{-(\varepsilon_j/kT)} \qquad (6.212)$$

rapidly become small with increasing j. Values for g_j and ε_j are given in Ref. 121.

Diatomic gases may possess rotational and vibrational energy as well as translational and electronic energy. The Schroedinger equation appropriate to a rigid rotator requires that the molecule exist in energy levels

$$\varepsilon_l = \frac{h^2}{8\pi^2 I} l(l+1)$$

whose degeneracy is $g_l = (2l + 1)$. The degeneracy is the number of orientations of a quantized angular momentum vector of length l. Thus the expression for the rotational partition function is

$$Q_{\text{rot}} = \sum_{l=0}^{\infty} (2l + 1)\exp\left[\frac{-l(l+1)h^2}{8\pi^2 IkT} \right] \qquad (6.213)$$

The energies of the nondegenerate vibrational levels are

$$\varepsilon_{\text{vib}} = \left(n + \tfrac{1}{2} \right) h\nu$$

For a gas with a finite number of equally spaced vibrational levels below the dissociation limit, it can be shown that

$$Q_{\text{vib}} = \frac{e^{-h\nu/kT}(1 - e^{-D/kT})}{1 - \exp(-h\nu/kT)} \qquad (6.214)$$

In the previous expressions I is the moment of inertia about a centroidal axis, ν is the vibrational frequency, m is the mass of the particle, D the dissociation energy, and all of the electronic energies are properties of the particle in question.[121] They may be obtained from spectroscopic data or from observation of other quantities that depend on the internuclear or interparticle potentials. Table 6.13 presents some of these quantities.

Thermodynamics in Terms of Partition Functions

Once partition functions are available, all of the thermodynamic properties of a substance may be found. For a gas all of whose modes of excitation are characterized by a single temperature, the appropriate form of the first law of Thermodynamics is[120]

$$T\,dS = dE + P\,dV - \mu\,dN \qquad (6.215)$$

TABLE 6.13   Molecular Vibrational and Rotational Data[a]

Molecule	$1/\lambda$, cm^{-1}	B, cm^{-1}
O_2	1560	1.45
H_2	4400	60.8
CO	2170	1.93
NO	1900	1.70
N_2	2360	1.99
CN	2070	1.90

[a] $1/\lambda$ = wave number = ν/c. B = band spectral constant = $h/8\pi^2 Ic$.

where E and μ are the internal energy and chemical potential (Gibbs free energy) per particle. The statistical entropy relation leads to

$$S = Nk\left\{ \ln\frac{Q}{N} + 1 + T\frac{\partial \ln Q}{\partial T} \right\} \tag{6.216a}$$

$$E = NkT^2 \frac{\partial \ln Q}{\partial T} \tag{6.216b}$$

$$P = NkT\frac{\partial \ln Q}{\partial V} \tag{6.216c}$$

$$\mu = -kT \ln\frac{Q}{N} \tag{6.216d}$$

Note that since only the translational partition function depends on the volume, V, only translational motion can contribute to hydrostatic pressure. Use of Eq. (6.211) in Eq. (6.216c) yields the thermally perfect equation of state. Again it is stressed that the treatment outlined here applies only to weakly interacting particles. If the particles interact strongly, the quantum states of the system cannot be described in terms of those of the individual particles.

The principles applied to a system with only one constituent may be used in the study of gas mixtures as well (see Ref. 122).

Nonequilibrium Thermodynamics

Thus far only the thermodynamics of gases in equilibrium has been discussed. Certainly equilibrium exists in many instances, but it must be realized that when physical conditions change abruptly, equilibrium is approached at a finite rate. On the microscopic scale there is a certain number of collisions required among the individual particles for the energies of the various modes of excitation to be equibrated. From previous comments on the activation energies of the various degrees of freedom, one may deduce that for particles whose masses do not differ greatly, translational degrees of freedom are equilibrated first and electronic ones last. Flow with translational nonequilibrium, that is, with nonuniformity in the distribution of translational energy, is most appropriately dealt with when one considers the Boltzmann equation and approximations thereto. These are discussed in Section 6.1.5. A discussion of rate processes would draw upon these concepts.

Surveys of relaxation processes in gases, that is, their approach to equilibrium, may be found in Ref. 120 on an elementary level and in Ref. 122 in a more advanced form.

6.12.2   Real Gas Dynamics

Governing Equations

The control volume derivation of the phenomenological relations that govern all gas dynamics may be found in Section 6.2.1. The extended thermodynamic and state relations previously mentioned assure the applicability of these equations to completely general situations. The conservation of mass, momentum, and energy are, in the absence of transport phenomena,

$$\frac{D\rho}{Dt} + \rho \, \text{div } \mathbf{V} = 0 \tag{6.217}$$

$$\frac{D\mathbf{V}}{Dt} + \frac{1}{\rho}\text{grad } p = 0 \tag{6.218}$$

$$\rho\frac{Dh}{Dt} - \frac{Dp}{Dt} = -\text{div } \mathbf{q} \tag{6.219}$$

where, in the absence of molecular transport phenomena, the heat flux is due only to radiation. For a general system one may allow for N individual species, each with a number of nonequilibrium internal processes denoted by α_{si} for the ith nonequilibrium variable associated with species s. By examining the net rate of flow into a control volume and the rate of progress of each reaction therein, all but translational nonequilibrium phenomena are governed by an equation of the form

$$\frac{D\alpha_{sj}}{Dt} = \frac{\chi_{sj}}{\tau_{sj}} \tag{6.220}$$

where χ and τ may depend on all properties of the fluid. The rate of production of species s, is governed by an equation of this form also. However, since the overall density is

$$\rho = \sum_{s=1}^{N} \rho_s$$

only $(N-1)$ of the species "rate" equations are independent if Eq. (6.217) is employed. If the molecular masses of the various species differ widely, it must be expected that their translational energies will not be equilibrated with each other. If one assigns the translational temperature T_s to each species, the appropriate equation must be derived from consideration of the work done on the boundaries of the control volume due to normal momentum transfer. The relaxation of translational temperatures is governed by

$$\rho_s C_{s \text{ trans}} \frac{DT_s}{Dt} + p_s \text{div } \mathbf{V} = \sum_i \frac{\chi_{si}}{\tau_{si}} + \left(\frac{V^2}{2} - e_s \right) \frac{D\rho_s}{Dt} + Q_s \tag{6.221}$$

where Q_s is the energy lost from species s due to collisional and radiative processes. Note the additional rate-of-work term in Eq. (6.221) that does not appear in Eq. (6.220). Since the enthalpy depends on all of the internal and translational contributions to each of the e_s, one of the equations of the type will not be independent of the others if the overall energy equation is employed. Translational nonequilibrium of this type is important mostly in ionized gases.

Equations (6.217)–(6.221), in addition to $(N-1)$ chemical "rate" equations, complete the relations that govern the motion of real gases. If radiation is included in the formulation, then the radiative transfer equation (see Section 6.13.1) is required as well. The boundary conditions for the fluid mechanics are those normally employed. If the nonequilibrium parameters are specified in the undisturbed medium, it can be proved that data cannot be prescribed for them if diffusion is not allowed. The fluid mechanics and nonequilibrium chemistry are, in most cases, intimately bound together.

Throughout this discussion, when a choice of two variables to specify the state of a gas is to be made, we shall choose pressure and temperature. Experience dictates that density is an inconvenient quantity to deal with. One reason is that the density hardly varies behind a strong shock wave, hence if density and temperature are employed it is difficult to obtain accurate pressure distributions. Furthermore, temperature must be retained if radiative processes are included.

At this point the stage is set for the study of interesting real gas problems such as the structure of shock waves of the nonequilibrium flow in nozzles. These and other real gas problems are discussed in detail in Ref. 120.

6.13   RADIATION GAS DYNAMICS
C. A. Forbrich, Jr., with contributions by S. Czyzak

6.13.1   General Features of Radiation Gas Dynamics

High-temperature gases emit energy in the form of electromagnetic radiation resulting from rotational, vibrational, and electronic energy level transitions in the gas atoms. When radiative energy transfer occurs, the motion of the radiating gas may be affected, that is, if the gas is at rest initially, it may have some velocity after the energy transfer. Radiation gas dynamics is the study of this interaction between the radiative transfer occurring and the motion of the gas.

In the analysis of radiative gas dynamics, it is assumed that the radiation field is composed of photon particles having energy h_ν moving in various directions all at a speed c, the speed of light. The radiation field interacts with the molecular field (the gas) through the emission and absorption of radiation, and the motion of the gas. When the energy absorbed by a gas is just equal to the energy emitted at every frequency, then the gas is in radiative equilibrium. When the energy absorbed is not equal to that emitted, the gas is in radiative nonequilibrium. Nonequilibrium effects can be significant and must be considered in many cases.

As a first step in studying the interaction between the gas motion and the radiative energy transfer, it is necessary to write a conservation equation for the photons so that the rate of energy transfer due to photon fluxes can be determined. Such an equation is given by[123]

$$\frac{1}{c}\frac{\partial I_\nu}{\partial t} + l_j \frac{\partial I_\nu}{\partial x_j} = \rho(j_\nu - K_\nu I_\nu) \tag{6.222}$$

which is the equation of radiative transfer, with I_ν the specific radiation intensity, l_j the direction cosine, ρ the mass density, j_ν the mass emission coefficient, and K_ν the mass absorption coefficient. Thus the specific radiation intensity I_ν is governed by a first-order partial differential equation, which in principle can be solved to give $I_\nu(x_j, l_j, t)$ explicitly.

Emission and Absorption of Radiation

Radiative transfer is essentially a microscopic phenomena. It is possible to consider microscopic processes alone to write an explicit expression for the right-hand side of Eq. (6.222) in terms of the actual atomic transitions that occur.

There are three types of transitions to be considered related to atomic or molecular structure of the gas. The more complicated molecular transitions are excluded here for simplicity. The first of these possible transitions is the *bound-bound* transition that occurs between two bound energy states of the atom. These transitions are characterized by discrete spectral line emission or absorption. A *bound-free* transition occurs when an atom having an electron in a bound state is so highly excited that the atom is ionized and the electron is free. A transition of this type is characterized by continuous spectral radiation emission (or absorption). *Free-free* transitions occur as a result of interactions between two free electrons in the gas or by the interaction of a free electron and an ionized particle and are characterized by continuous radiation.

To describe the term on the right-hand side of Eq. (6.222) in terms of microscopic processes, only *bound-bound* transition will be considered. In this case *Einstein transition probability coefficients* can be introduced to characterize the radiation field. There are three Einstein transition probability coefficients: the Einstein coefficient for spontaneous radiation emission A_{nm}, defined to be the probability per unit time for a spontaneous transition from state n to m (where n is the upper energy state and m is the lower energy state) to occur for a single atom; B_{mn} the probability per unit time for an absorption to occur for a single atom resulting in transition from state m to n; and B_{nm} defined to be the probability per unit time for induced emission to occur for a single atom resulting in a transition from state n to m. Induced emission occurs in the presence of radiation field with a given direction and frequency that corresponds to the transition frequency.

The Einstein coefficients are not independent and can be shown to be related at radiative equilibrium in the following manner:

$$\frac{B_{mn}}{B_{nm}} = 1 \qquad \frac{A_{nm}}{B_{nm}} = \frac{2h\nu^3}{c^2} \tag{6.223}$$

where h is Planck's constant and ν is the radiation frequency. Letting n equal the number of atoms in a particular state, the mass emission term ρj_ν on the right-hand side of Eq. (6.222) is equivalent to $n_n(A_{nm} + B_{nm}I_\nu)h\nu$, which gives the rate at which radiative energy is emitted, and the mass absorption term $\rho K_\nu I_\nu$ is equivalent to $n_m N_{mn}I_\nu h\nu$, which gives the rate at which energy is absorbed. The assumption of local thermodynamic equilibrium (LTE) can be used to further simplify the equation of radiative transfer. LTE is characterized by the gas atoms having local rotational equilibrium (thermodynamic temperature) not necessarily in equilibrium with the radiation field. This can occur because the population of the atomic states locally is governed by atomic collisions resulting in an equilibrium distribution of states corresponding to a local temperature that may vary from point to point in the gas. With the assumption of LTE there is a Boltzmann distribution of equilibrium states corresponding to the local temperature and the equation of radiative transfer becomes

$$\frac{1}{c}\frac{\partial I_\nu}{\partial t} + l_j \frac{\partial I_\nu}{\partial x_j} = \rho K_\nu \left[1 - \exp\left(-\frac{h\nu}{kT}\right)\right][B_\nu - I_\nu]$$

where B_ν is defined as the Planck function representing the equilibrium specific radiation intensity given by

$$B_\nu \equiv \text{Planck function} \equiv \frac{2h\nu^3/c^2}{\exp\{h\nu/kT\} - 1}$$

Formal Solution of the Equation of Radiative Transfer

The equation of radiative transfer can be rewritten in terms of the volume absorption coefficient α_ν as

$$\frac{1}{c}\frac{\partial I_\nu}{\partial t} + l_j\frac{\partial I_\nu}{\partial x_j} = -\alpha_\nu(I_\nu - B_\nu) \tag{6.224}$$

where

$$\alpha_\nu \equiv \rho k_\nu\left[1 - \exp\left(-\frac{h\nu}{kT}\right)\right]$$

To solve Eq. (6.224) the time derivative term is neglected. This can be done because characteristic velocities (not necessarily the flow velocity) in radiative gasdynamics problems are small compared to the speed of light c.

If a coordinate r is chosen such that it is measured in a direction opposite to the direction of radiation propagation as shown in Fig. 6.106, then

$$l_j\frac{\partial I_\nu}{\partial x_j} = -\frac{\partial I_\nu}{\partial r}$$

Under these conditions the equation of radiative transfer becomes

$$\frac{\partial I_\nu}{\partial r} = \alpha_\nu(I_\nu - B_\nu)$$

This equation can be formally integrated to obtain

$$I_\nu(r) = I_\nu(r_s)\exp\{-\tau_\nu(r_s)\} + \int_0^{r_s}\alpha_\nu B_\nu(r)\exp\{-\tau_\nu(r)\}\,dr \tag{6.225}$$

where $\tau_\nu(r) \equiv \int_0^r \alpha_\nu\,dr$ is defined as the optical depth and $r = 0$ is the point in question. In this formal solution the first term on the right-hand side is due to the contribution of the boundary attenuated by the net absorption from the boundary to the point in question, while the integral term accounts for self-emission, absorption, and attenuation within the gas.

It remains to evaluate the radiation heat addition term that will appear in the energy equation of a radiative gas dynamics problem. The heat addition is given by the negative divergence of the radiative heat flux vector q_j^R. The heat addition term is given by

$$-\frac{\partial q_j^R}{\partial x_j} = \int_0^\infty \alpha_\nu\left(\int_0^{4\pi}I_\nu\,d\Omega - 4\pi B_\nu\right)d\nu \tag{6.226}$$

In certain cases approximations to this formal solution are possible to obtain simplified results. Some of these approximations and simplifications will now be discussed with the radiation heat addition terms evaluated.

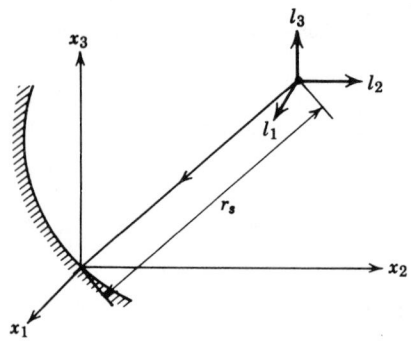

Fig. 6.106   Radiation field coordinate geometry.

Emission-Dominated Approximation. Many physical situations exist in which the specific intensity I_ν is much less than its equilibrium value B_ν. This is the case when the energy being emitted from the gas is much larger than that being absorbed, so the gas is far from radiative equilibrium. When $I_\nu \ll B_\nu$, the equation of radiative transfer is

$$\frac{\partial I_\nu}{\partial r} = -\alpha_\nu B_\nu$$

which can be readily integrated to give

$$I_\nu(r) = \int_0^{r_s} \alpha_\nu B_\nu \, dr$$

in which the radiation contribution resulting from radiating boundaries $I_\nu(r_s)$ is neglected. The radiation heat addition term $\partial q_j^R/\partial x_j$ is given by

$$-\frac{\partial q_j^R}{\partial x_j} = -4\pi \int_0^\infty \alpha_\nu B_\nu \, d\nu = -4\alpha_{P_0} T^4$$

where α_P is the Planck mean absorption coefficient defined by

$$\alpha_P(p, T) \equiv \frac{\int_0^\infty \alpha_\nu B_\nu \, d\nu}{\int_0^\infty B_\nu \, d\nu} = \frac{\pi}{\sigma T^4} \int_0^\infty \alpha_\nu B_\nu \, d\nu$$

Optically Thin Gas Approximation. The optically thin gas is characterized by the condition that the optical depth τ_ν is small compared to unity, that is, $\tau_\nu \ll 1$. In this case the general solution to the radiation equation becomes

$$I_\nu(r) = I_\nu(r_s) + \int_0^{r_s} \alpha_\nu B_\nu(r) \, d\nu$$

This equation states that the specific intensity at the point r is given by the specific intensity $I_\nu(r_s)$ at the radiating boundary plus the unattenuated sum of all emission along the direction of propagation. The relation between the optically thin gas and the emission-dominated gas can be given by the further condition that

$$I_\nu(r) = I_\nu(r_s) + \int_0^{r_s} \alpha_\nu B_\nu(r) \, dr \ll B_\nu$$

If $B_\nu(r) \cong B_\nu$ at $r = 0$, then this relationship reduces to a condition on $I_\nu(r_s)$ compared to B_ν since the optically thin assumption requires the integral term to be small, leading again to the condition $I_\nu(r_s) \ll B_\nu$. So it is seen that the assumption of a gas being optically thin is a necessary but not sufficient condition for the radiation to be emission dominated. In the case of an optically thin, emission-dominated gas, the radiation heat addition term is the same as that for the emission-dominated gas.

Optically Thick Gas Approximation. An optically thick gas is one in which the optical depth τ_ν is large compared to unity, that is, $\tau_\nu \gg 1$, for all frequencies. When the point in the gas under consideration is far enough from any radiating boundaries, the radiation contribution due to the boundaries $I_\nu(r_s)$ is neglected and the formal solution to the equation of radiative transfer at $r = 0$ is

$$I_\nu(0) = \int_0^{r_s} \alpha_\nu(r) B_\nu(r) \exp\{-\tau_\nu(r)\} \, dr$$

Expanding $B_\nu(r)$ in terms of a Taylor series expansion about $r = 0$, the integral can be evaluated yielding for the solution

$$I_\nu(0) = B_\nu(0)\left[1 + \frac{1}{\alpha_\nu(0) B_\nu(0)} \left(\frac{dB_\nu}{dt}\right)_{r=0} \left(\frac{\partial T}{\partial r}\right)_{r=0} + \cdots\right]$$

This result indicates that $I_\nu(0)$ is made up of the equilibrium radiation at the point of interest $B_\nu(0)$ plus a small additional amount due to emission from the gas in the immediate neighborhood of the point of interest. The optically thick condition exists when I_ν differs only a slight amount from B_ν, thus the departure from radiative equilibrium is small as opposed to the cases of emission-dominated and optically thin gases. It should be mentioned that the optically thick gas is somewhat unrealistic because a gas cannot be thick at all frequencies if it radiates primarily in lines or bands.

The radiation heat addition term for the case of the optically thick gas can be shown (see Ref. 123, page 469) to be given by

$$-\frac{\partial q_j^R}{\partial x_j} = \frac{16\sigma T^3}{3\alpha_R} \frac{\partial^2 T}{\partial x_j \partial x_j}$$

where α_R is the Rosseland mean absorption coefficient defined as

$$\alpha_R \equiv \frac{\int_0^\infty \frac{\partial B_\nu}{\partial T} d\nu}{\int_0^\infty \frac{1}{\alpha_\nu} \frac{\partial B_\nu}{\partial T} d\nu} = \frac{4\sigma T^3}{\pi \int_0^\infty \frac{1}{\alpha_\nu} \frac{\partial B_\nu}{\partial T} d\nu}$$

Gray-Gas Approximation. The gray-gas approximation is based on the ad hoc assumption that there exists some unspecified, frequency-averaged absorption coefficient $\alpha(p, T)$ that remains a function of the gas pressure p and temperature T. This assumption does not come from a logical theoretical limit as do the previous assumptions; however, it provides a means of studying radiative gasdynamics from the pedagogical viewpoint and eliminates in the most part extensive use of computers for solutions. In this case the formal solution to the equation of radiative transfer is

$$I(r) = I(r_s)\exp\{-\tau_\nu(r_s)\} + \frac{\sigma}{\pi} \int_0^{r_s} \alpha T^4 \exp\{-\tau_\nu(r)\} dr$$

where $I \equiv \int_0^\infty I_\nu \, d\nu$. The radiation heat addition term is given by

$$-\frac{\partial q_j^R}{\partial x_j} = \int_0^{4\pi} \frac{\partial I}{\partial r} d\Omega = \alpha \left[\int_0^{4\pi} I \, d\Omega - 4\sigma T^4 \right].$$

where σ is the Stephan–Boltzmann constant and $d\Omega$ is the differential solid angle subtended.

Other Approximations. There exist other approximate solutions to the equation of radiative transfer. They will be only mentioned here; however, in many cases they are the most important approximations in the field since they can more accurately describe a realistic physical situation. Among these approximations are:

1. The *one-dimensional approximation*, in which the specific intensity I_ν is a function of one space coordinate only. In many cases the integral expression in the radiation heat addition term can be evaluated and the full problem solved explicitly.

2. The *exponential approximation*, which arises in the case of a gray gas in the one-dimensional approximation. In this case the expression for the radiation heat addition term reduces to exponential integrals that can be evaluated by approximate exponential functions. This approximation has been used to advantage many times in one-dimensional radiative gasdynamics and astrophysics.[124]

3. The *differential approximation*, which is a generally applicable method used in three-dimensional problems including the case of nonisotropic radiation. This approximation is concerned with satisfying certain moments of the equation of radiative transfer and appropriately truncating the assumed series solution usually in terms of spherical harmonics.[125] This method produces a general result that yields the other approximations, that is, optically thin, optically thick, and so on, with an appropriate choice of parameters arising in the solution as discussed in Vincenti and Kruger.[123]

6.13.2 Basic Nonlinear Equations of Radiative Gas Dynamics

The conservation equations of radiative gas dynamics will now be presented. In doing this the radiation energy density and pressure will be neglected, being insignificant except in cases of astrophysical interest and nuclear explosions. In this case the governing equations in a radiative gas dynamics problem are

$$\frac{D\rho}{Dt} + \rho\frac{\partial u_j}{\partial x_j} = 0 \tag{6.227a}$$

$$\rho\frac{Du_i}{Dt} + \frac{\partial p}{\partial x_i} = 0 \tag{6.227b}$$

$$\rho\frac{Dh}{Dt} - \frac{Dp}{Dt} = -\frac{\partial q_j^R}{\partial x_j} \tag{6.227c}$$

where h is the enthalpy and p is the pressure and $D(\ )/Dt$ is the substantial derivative defined by $D(\ )/Dt = \partial(\ )/\partial t + u_j[\partial(\ )/\partial x_j]$. Assuming that the gas is perfect, an equation of state can be written in the form

$$h = \frac{\gamma}{\gamma - 1}\frac{p}{\rho} \tag{6.227d}$$

where p is given by the ideal gas law $p = \rho RT$, with R the ordinary gas constant and γ the ratio of specific heats assumed constant. The radiation heat addition term on the right-hand side of the energy equation is given by Eq. (6.226), which has been evaluated in several asymptotic limits and for certain simplified cases earlier.

These equations constitute a set of 11 scalar equations with 11 unknowns ρ, u_i, p, h, q_j^R, T, and I_ν. They are a set of coupled nonlinear integrodifferential equations to which no closed form solution exists.

6.13.3 Linearized Equations of Radiative Gas Dynamics

Linearization of the complete set of conservation equations governing a problem in radiation gas dynamics is made relative to a uniform gas at rest. Thus it is possible to write $u = u_i'$, $p = p_0 + p'$, $T = T_0 + T'$, $q_i^R = q_{i0}^R + q_i^{R'}$ and so on, where the subscript 0 represents uniform conditions in an undisturbed gas and primed quantities denote small perturbations of the particular quantity of interest. Substituting into the conversion equations, a set of linearized equations is obtained of the form

$$\frac{\partial\rho'}{\partial t} + \rho_0\frac{\partial u_j'}{\partial x_j} = 0 \tag{6.228a}$$

$$\rho_0\frac{\partial u_i'}{\partial t} + \frac{\partial p'}{\partial x_i} = 0 \tag{6.228b}$$

$$\rho_0\frac{\partial h'}{\partial t} - \frac{\partial p'}{\partial t} = -\frac{\partial q_j^{R'}}{\partial x_j} \tag{6.228c}$$

$$-\frac{\partial q_i^{R'}}{\partial x_j} = \alpha_0\left[I_A' - 16\sigma T_0^3 T'\right] \tag{6.228d}$$

$$\frac{\partial I_A'}{\partial x_i} = -3\alpha_0 q_i^{R'} \tag{6.228e}$$

Introducing the velocity potential ϕ, the set of equations can be combined into one fifth-order partial differential equation

$$\frac{\partial^3 A_s}{\partial t\partial x_k\partial x_k} + \frac{16}{B_0}\alpha_0 a_{s0}\frac{\partial^2 A_T}{\partial x_k\partial x_k} - 3\alpha_0\frac{\partial A_s}{\partial t} = 0 \tag{6.229}$$

where

$$A_s = \frac{1}{a_{s0}} \frac{\partial^2 \phi}{\partial t^2} - \frac{\partial^2 \phi}{\partial x_j \, dx_j}$$

$$A_T = \frac{1}{a_{T0}} \frac{\partial^2 \phi}{\partial t^2} - \frac{\partial^2 \phi}{\partial x_j \, \partial x_j}$$

$$a_{s0} = \sqrt{\gamma R T} = \text{isentropic speed of sound}$$

$$a_{T0} = \sqrt{R T} = \text{isothermal speed of sound}$$

$$B_0 = \frac{\gamma R_{\rho 0} a_{s0}}{(\gamma - 1)\sigma T_0^3} = \text{Boltzmann number}$$

The Boltzmann number B_0 is the most important dimensionless parameter in radiation gas dynamics, providing a measure of the relative importance of convective and radiative process in the energy flux.

The two sets of Eqs. (6.227) and (6.228), which govern, respectively, a generalized gas dynamic problem and a linearized radiation gas dynamics problem, can now be solved with the appropriate boundary conditions for a large range of problems. Particular solutions to these equations will not be presented here because any such treatment would by its brevity do an injustice to the solutions readily available in the literature and textbooks. A comprehensive list of references with solutions to some of the classical problems of fluid mechanics, including radiation effects, can be found in the references and bibliography.

6.14   MAGNETOGASDYNAMICS
C. A. Forbrich, Jr.

6.14.1   Charged Particle Motion in a Plasma

Magnetogasdynamics deals with a unique type of matter called a plasma. A plasma is a fluid in which there are charge carriers but that as a whole is electrically neutral because the numbers of positive and negative charges are equal. The presence of the charge carriers renders the plasma electrically conducting. Generally a gaseous plasma is created by the ionization of some or all of the gaseous molecules by collisional processes.

To study the interaction of a plasma with electric or magnetic fields it is necessary to consider certain aspects of the microscopic motion of the particles composing the plasma in order to understand the macroscopic phenomena observed. On a microscopic scale, certain regions of a plasma may have large deviations from electrical neutrality as a result of applied electrical or magnetic fields. If the mean kinetic energy of a particle due to its thermal motion is compared to the potential energy of the field produced by the ionized particles, a characteristic length is revealed, giving the relationship[126, 127]

$$h \equiv \sqrt{\frac{kT}{4\pi n e^2}} \equiv \text{Debye length}$$

where k is the Boltzmann constant, n the number density of electrons, e the electron charge and T the absolute temperature. The Debye length h is a measure of the extent of deviation from electrical neutrality. Assuming the microscopic region of gas being studied can be enclosed in a sphere of radius L, then if $h \ll L$, the gas can be considered as electrically neutral and the region of gas can be studied using a macroscopic description. The condition $h \ll L$ defines an ionized gas as a plasma. If $h \geq L$, then the gas cannot be considered electrically neutral and can be studied only from a microscopic point of view. For gas dynamic considerations it is most convenient to study the coupling action between a magnetic field and a plasma using the macroscopic approach (the condition $h \ll L$ being implied) so the conventional equations of macroscopic gasdynamics can be used when appropriately modified.

If a magnetic field **B** is impressed on a plasma, the charged particles will spiral around the magnetic lines of force. The radius of gyration r of the charged particles about the magnetic force lines is given by

$$r = \frac{3kTm}{e|\mathbf{B}|}$$

where m is the charged mass, ion, or electron, all other terms having been previously defined. The electron obviously has a smaller radius of gyration r_e than the more massive ion radius of gyration r_i. The ion and electron rotate about the magnetic force lines in opposite directions, which can be deduced from the preceding relation for the radius of gyration, since the electron charge is $-e$ and that of an ion is $+e$ for singly ionized particles.

If, in addition to the external magnetic field $\mathbf{B}$, an external electric field $\mathbf{E}$ is applied to a plasma, the charged particles drift sideways in the direction $\mathbf{E} \times \mathbf{B}$ rather than following the lines of force. This effect is called $\mathbf{E} \times \mathbf{B}$ *drift*. The electrical conductivity of the plasma will vary, with and without $\mathbf{E} \times \mathbf{B}$ drift, having in general a tensor character as a result of the various drifts that occur.[126,128]

6.14.2 Magnetogasdynamic Equations of Motion

Having established the condition that the dimensions of the sphere of radius L under study are large compared to the Debye length h, a macroscopic description of a plasma is justified. The electromagnetic character of a plasma is governed by Maxwell's equations although displacement currents are usually neglected in laboratory plasma physics studies. The fluid motion is governed by the usual conservation equations of gasdynamics, however, with additional terms in the momentum and energy equations. The equation for the conservation of mass remains unchanged. In the momentum equation, the external forces must include the Lorentz force density $\mathbf{j} \times \mathbf{B}$, where $\mathbf{j}$ is the electrical current in the field. In place of this term the Maxwell stress tensor may also be used.[128] The Maxwell stress tensor used in the momentum equation leads directly to a Bernoulli equation for magnetogasdynamics contributing a new term known as the magnetic pressure,[127] given by $B^2/8\pi$. The energy equation must include additional energy terms accounting for the electromagnetic work and Joule heating occurring in the system. These considerations lead to a complex set of coupled differential equations given by:

MAXWELLS EQUATIONS   PHENOMENOLOGICAL EQUATIONS

$$\operatorname{curl}\mathbf{H} = \mathbf{j} + \varepsilon\frac{\partial\mathbf{D}}{\partial t} \qquad \mathbf{D} = \varepsilon\mathbf{E}$$

$$\operatorname{div}\mathbf{B} = 0 \qquad \mathbf{B} = \mu\mathbf{H}$$

$$\operatorname{curl}\mathbf{E} = -\frac{\partial\mathbf{B}}{\partial t} \qquad \mathbf{j} = \sigma(\mathbf{E} + \mathbf{V} \times \mathbf{B})$$

$$\operatorname{div}\mathbf{D} = \omega \qquad \sigma = \sigma(T, p)$$

FLUIDMECHANICAL EQUATIONS

Mass: $\qquad \dfrac{\partial\rho}{\partial t} + \operatorname{div}(\rho\mathbf{V}) = 0$

Momentum: $\qquad \rho\dfrac{D\mathbf{V}}{Dt} = -\operatorname{grad} p + \operatorname{div}(\tau_{ij} + T_{ij})$

Energy: $\qquad \rho\dfrac{D[h + (V^2/2)]}{Dt} = \dfrac{\partial p}{\partial t} + \mathbf{j} \cdot \mathbf{E} + \operatorname{div}\mathbf{V}\tau_{ij} - \operatorname{div}\mathbf{q}$

State: $\qquad p = (1 + \alpha)NkT = \rho(1 + \alpha)RT; \ \alpha = \alpha(p, T)$

Heat: $\qquad \mathbf{q} = -K\nabla T$

where α is the degree of ionization, K the heat conductivity, T_{ij} the Maxwell stress tensor, and τ_{ij} the viscous stress tensor, with all other terms in the fluidmechanical equations having their usual meaning. In the electromagnetic and phenomenological equations, $\mathbf{H}$ is the magnetic field strength, $\mathbf{j}$ the electric current, ε the dielectric constant, $\mathbf{D}$ the displacement current, $\mathbf{B}$ the magnetic induction, $\mathbf{E}$ the electric field strength, ω the charge density, μ the magnetic permeability, and σ the electrical conductivity. Transforming these equations by eliminating $\mathbf{j}$ and $\mathbf{E}$ in favor of ρ, $\mathbf{V}$, p, and $\mathbf{B}$ as variables, the governing equations of magnetogasdynamics become

Mass: $\qquad \dfrac{D\rho}{Dt} + \rho\operatorname{div}\mathbf{V} = 0$

Momentum: $\qquad \rho\dfrac{D\mathbf{V}}{Dt} = -\operatorname{grad} p - \operatorname{grad}\left(\dfrac{B^2}{8\pi\mu}\right) + \dfrac{1}{4\pi\mu}(\mathbf{B} \cdot \operatorname{grad})\mathbf{B}$

Induction:
$$\frac{D\mathbf{B}}{Dt} + \mathbf{B}\,\mathrm{div}\,\mathbf{V} - (\mathbf{B}\cdot\mathrm{grad})\mathbf{V} = \frac{c^2}{4\pi\mu\sigma}\,\mathrm{grad}\,\mathrm{div}\,\mathbf{B}$$

State:
$$\frac{D}{Dt}\left(\frac{p}{\rho^{\gamma}}\right) = 0$$

with the condition that $\mathrm{div}\,\mathbf{B} = 0$ and where $D(\ )/Dt = \partial(\ )/\partial t + \mathbf{V}\cdot\mathrm{grad}(\ )$. Thus this set of differential equations describes the plasma gas motion when appropriate boundary conditions are applied.

Applications

A classic example of a magnetofluidynamic flow situation is the Hartmann problem,[126] where an incompressible conducting fluid in a channel is subjected to external magnetic and electric fields. This problem points out many of the fundamental interaction phenomena between a plasma and external fields. The results show that the velocity profile becomes flatter and the velocity gradient near the channel wall becomes steeper as the magnetic field strength is increased.

A plasma can be accelerated to very high velocities by the Lorentz force resulting from crossed electric and magnetic fields.[126] Such devices, sometimes called Hall accelerators, are used as propulsion devices (see Section 6.5.7) and high-speed wind tunnels (see Section 6.6.2).

It is possible to "push" the ionized bow shock away from the nose of a reentry vehicle[129] and to balance an electric arc in a supersonic flow[130] by the proper orientation and magnitude of external magnetic fields. These and other interesting examples are found in the references.

6.14.3   Alfvén Waves

The wave phenomena in magnetogasdynamics is a very complicated study except in the simplest of cases. The classical wave analysis is that of Alfvén who studied plasma waves in an incompressible plasma assuming an applied magnetic field $\mathbf{B}$ perturbing the plasma slightly in one direction, say z. The assumption of incompressibility reduces the continuity equation to $\mathrm{div}\,\mathbf{V} = 0$. Further assuming the fluid is inviscid having infinite electrical conductivity σ, the governing equations of magnetogasdynamics can be linearized to obtain

Mass:
$$\frac{\partial v_i}{\partial x_i} = 0$$

Momentum:
$$\frac{\partial^2}{\partial x_i\,\partial x_i}\left[\frac{p}{\rho} + \frac{B^2}{8\pi\rho}\right] = 0$$

Energy:
$$\frac{\partial v_i}{\partial t} = \frac{B_0}{4\pi\rho}\frac{\partial b_i}{\partial z} - \frac{\partial}{\partial x_i}\left[\frac{p}{\rho} + \frac{B^2}{8\pi\rho}\right]$$

Induction:
$$\frac{\partial b_i}{\partial t} = B_0\frac{\partial v_i}{\partial z}$$

Taking the gradient of the energy equation and combining it with the induction equation results in an equation in terms of one unknown function b_i or v_i:

$$\frac{\partial^2 v_i}{\partial t^2} = V_a^2\frac{\partial^2 v_i}{\partial z^2}$$

$$\frac{\partial^2 b_i}{\partial t^2} = V_a^2\frac{\partial^2 b_i}{\partial t^2}$$

where $V_a^2 = B_0^2/4\pi\rho = $ (Alfvén speed).[131] These are typical wave equations describing transverse waves traveling at a speed $+V_a$ in the direction of the applied magnetic field $\mathbf{B}_0$. These waves are called Alfvén waves and V_a is the Alfvén wave speed. It is emphasized that this is the simplest type of wave motion possible under rather restrictive assumptions. It can be shown by a slightly more complicated analysis that Alfvén waves are damped by considering a viscous fluid having a finite electrical conductivity.

If the fluid is compressible but nonviscous having an infinite electrical conductivity σ, the magnetogasdynamic equations reveal that there is a combination of transverse Alfvén waves and the standard longitudinal sound waves of gasdynamics. If the analysis is carried out in full, the wave

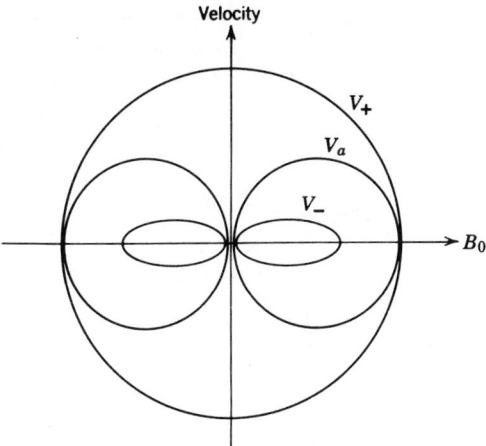

Fig. 6.107   Magnetofluid mechanical waves.

speed at some angle θ with respect to the applied magnetic field $\mathbf{B}_0$ is given by Thompson[127] as

$$V_{\pm}^2 = \tfrac{1}{2}\left[\left(a_s^2 + V_a^2\right) \pm \sqrt{\left(a_s^2 + V_a^2\right)^2 - 4V_a^2 a_s^2 \cos^2\theta}\right]$$

where V_+ are fast (+) or slow (−) waves, a_s is the wave speed of sound, and V_a is the Alfvén wave speed. This equation can be plotted to show the magnitude of the fast and slow waves as given in Fig. 6.107.

6.15   FLIGHT DYNAMICS
R. E. Willes, T. J. Forster, and G. E. Thompson, with contributions by T. A. Hammond

The motion of most flight vehicles can be divided into two specific types; motion of rapid nature about the vehicle's center of mass and a somewhat slower motion of the vehicle's center of mass itself. The motion of the center of mass is the subject of performance analysis[132–136] while the motion about the center of mass is the topic of stability and control analysis.[136–140]

6.15.1   Aircraft Performance
The central problem of aircraft performance analysis is to determine the values of particular performance parameters (rate of climb, speed, range, endurance, turn rate, turn radius, etc.) as a function of other parameters that control the motion of the aircraft's center of mass (aircraft aerodynamics, engine performance, Mach number, altitude, and load factor). Because of the largely numerical format of data concerning both the propulsive and the aerodynamic forces on an aircraft, it is convenient to perform these calculations numerically. The usual method of displaying these calculations is to plot the performance indices as a function of the aircraft's Mach number M, its altitude h, and its load factor n for some engine speed N. A number of typical performance parameter plots are shown in the accompanying figures.

An aircraft can exchange kinetic for potential energy much more rapidly than it can change the combined total energy by excess thrust. The combined kinetic and potential energy per unit weight can be expressed in units of energy altitude h_e where

$$\frac{PE + KE}{W} = h_e = \frac{h + V^2}{2g}$$

is a convenient parameter indicating the range of altitudes/velocities available to the aircraft with the h_e. The excess power per unit weight, which is equal to the rate at which energy altitude can be

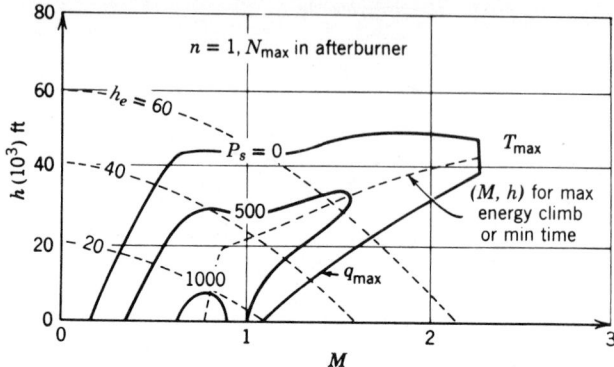

Fig. 6.108 Specific excess power (ft/sec) and energy altitude (ft).

changed (the specific excess power), P_s, is

$$P_s = \frac{F - D}{W} V = \frac{dh_e}{dt}$$

where F is the engine thrust, D is the aircraft total drag, W is the aircraft weight, and V is the aircraft velocity. In Fig. 6.108 contours of constant energy altitude and specific excess power are plotted as functions of Mach number and altitude, at a load factor of one and at maximum rotational engine speed (rpm) in full afterburner. The maximum Mach number normally is limited at high altitudes by temperature and at low altitudes by dynamic pressure. The maximum energy climb is obtained by selecting the altitude–Mach number combination that results in the largest P_s value for the existing energy level, h_e.

The energy gain per unit fuel expended, specific energy efficiency E_s, is

$$E_s \equiv \frac{dh_e}{dW_f} = \frac{F - D}{W} \frac{V}{Fc}$$

where W_f is the weight of fuel and c is the thrust specific fuel consumption. Plots of energy efficiency with the same constraints of load factor and engine speed are shown in Fig. 6.109. A minimum fuel climb (maximum energy efficiency) is determined by selecting the altitude–Mach number pair that results in the highest E_s value for the existing energy level h_s.

The range per unit fuel expended, the specific range R_s is

$$R_s \equiv \frac{dR}{dW_f} = \frac{V}{cF}$$

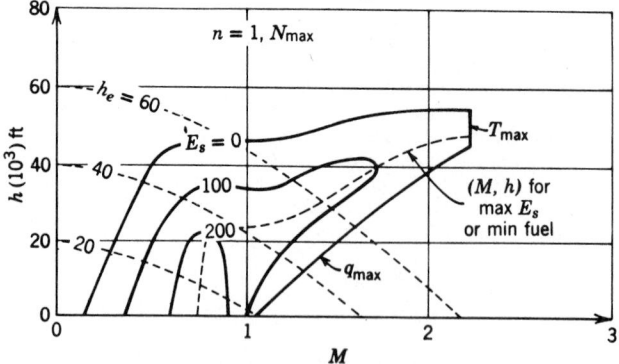

Fig. 6.109 Specific energy efficiency (ft/lb fuel).

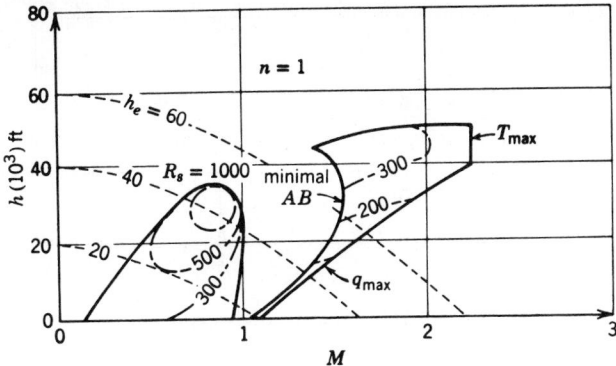

Fig. 6.110   Specific range (nautical miles/10,000 lb fuel).

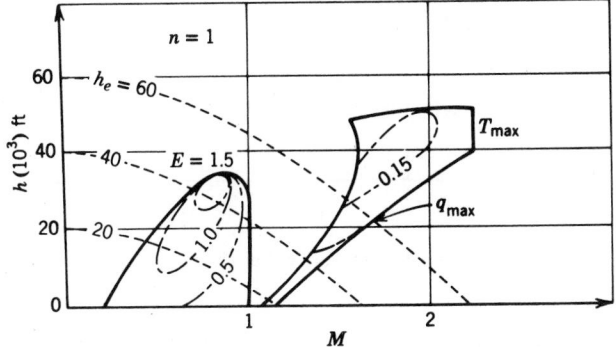

Fig. 6.111   Specific endurance (hr/10,000 lb fuel).

Figure 6.110 shows contours of specific range plotted as a function of Mach number and altitude for a load factor of one and the engine thrust set equal to drag. The maximum range occurs near the maximum altitude and just below Mach 1 for subsonic flight; minimum afterburner and near maximum altitude will result in maximizing R_s in supersonic flight.

Specific endurance is defined as a time per unit fuel used and is given by

$$E \equiv \frac{dt}{dW_f} = \frac{1}{cF}$$

Endurance is plotted as a function of Mach number and altitude (with load factor and thrust as before) in Fig. 6.111. Maximum endurance occurs near the tropopause and near Mach 1 subsonically, and at minimum operating afterburner limits supersonically.

Turn rate ω and radius r are approximately given by

$$\omega \cong \frac{ng}{v} \qquad r \cong \frac{v^2}{ng}$$

Contours of constant turn rate and constant radius are shown in Figs. 6.112 and 6.113 along with typical lift coefficient and load factor limits. Observe that the maximum turn rate and the minimum turn radius occur at the maximum lift coefficient–load factor intersection. The data for these plots is obtained in the following manner. Numerical data for engine thrust F,

$$F = \frac{p}{p_0} F_0(M, N_c)$$

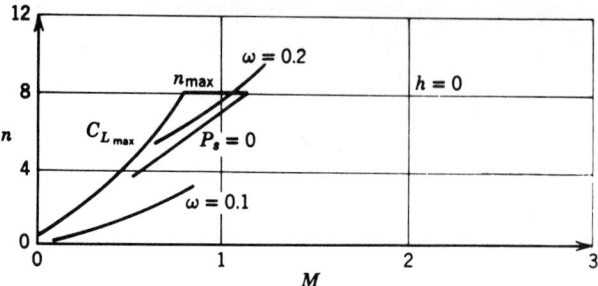

Fig. 6.112   Turn rate (rad/sec).

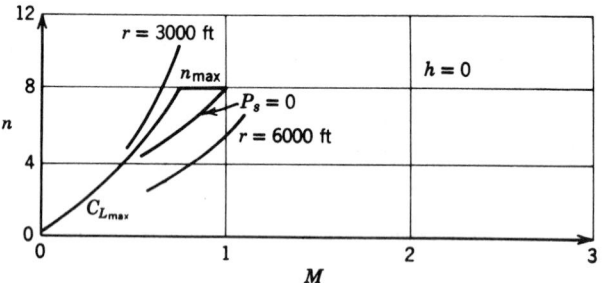

Fig. 6.113   Turn radius (ft).

where

$$N_c = N\sqrt{\frac{T_0}{T}}$$

and specific consumption c,

$$c = \sqrt{\frac{T_0}{T}}\, c_0(M, N_c)$$

are tabulated as a function of Mach number M, corrected engine speed N_c, pressure p, and temperature T. The subscript 0 indicates the value of the quantity at the planetary surface. Typical dependences may be seen in Fig. 6.114.

Atmospheric pressure p and temperature T are functions of altitude. As a first approximation the temperature is constant. The pressure then approximately satisfies the relation

$$p = p_0 e^{-\beta h}$$

where h is the altitude, and β is the inverse atmospheric scale height.

$$\beta = \frac{g}{RT}$$

The drag coefficient, C_D, is a function of Mach number M and lift coefficient C_L. See Figs. 6.115 and 6.116. The lift coefficient C_L is specified by requiring that the lift L equal the normal load nW and is computed from

$$C_L = \frac{nW}{(\gamma/2)\,pM^2 S} = \frac{2nW}{\rho V^2 S}$$

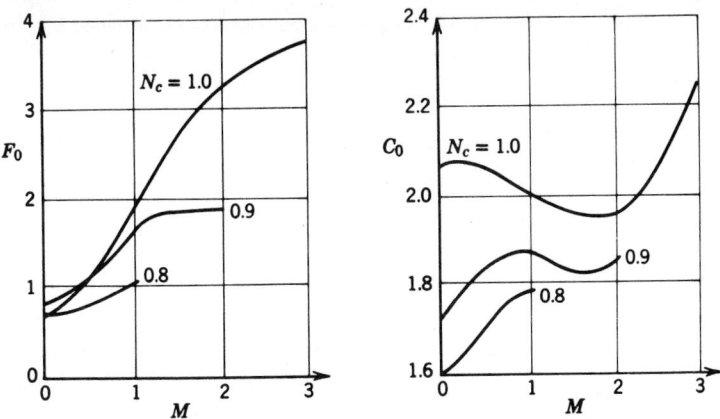

Fig. 6.114   Thrust (10,000 lb) and thrust specific fuel consumption (hr^{-1}) for a typical turbojet in full afterburner.

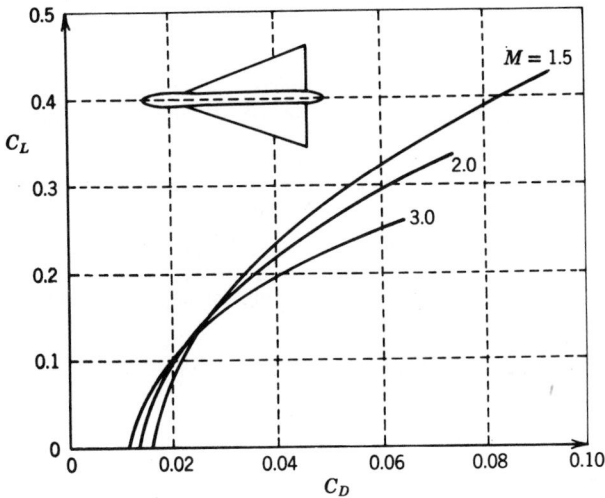

Fig. 6.115   Lift coefficient versus drag coefficient and Mach number of a typical supersonic aircraft.

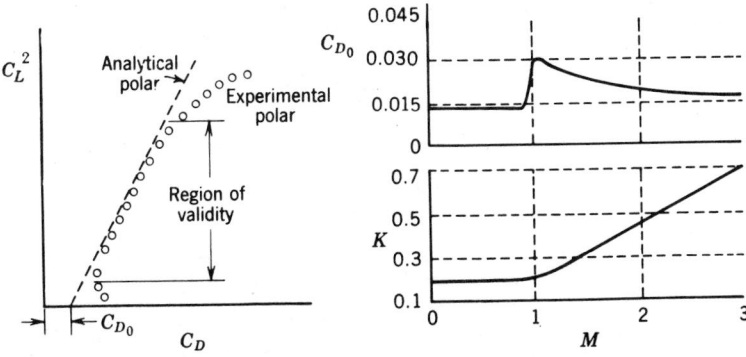

Fig. 6.116   Approximate quadratic drag polar for a typical supersonic aircraft.

where γ is the ratio of specific heats and S is the wing area. The drag by definition is

$$D = C_D \frac{\gamma}{2} pM^2 S = C_D \tfrac{1}{2} \rho V^2 S$$

Aircraft velocity V is related to the speed of sound a and Mach number M by

$$V = Ma$$

where a is the speed of sound

$$a = \sqrt{\gamma RT}$$

and R is the gas constant.

It is desirable to have analytical relations that express performance indices as a function of aircraft configuration. These analytical relations are based on approximating the drag polar with the quadratic form

$$C_D = C_{D0} + KC_L^2$$

where C_{D0} and K are judiciously picked to conform to the actual drag polar over a range of M and C_L of interest. See Fig. 6.116. K is sensitive to angle of attack subsonically after separation occurs and may double before maximum C_L is reached. For high-aspect ratio wings, it is approximately

$$K \cong \frac{1}{\pi \, \text{AR}}$$

where aspect ratio AR is defined as

$$\text{AR} = \frac{b^2}{S} = \frac{b}{\bar{c}}$$

where b is the span, and $\bar{c}$ is the mean aerodynamic chord. See Fig. 6.117. The value of K for low-aspect ratio wings is double that value found for high-aspect ratios.

Using these relations and discarding small terms, the following performance estimates can be obtained:

1. The maximum Mach number attainable by a subsonic turbojet aircraft is given by

$$M \cong \left(\frac{2}{\gamma C_{D0}} \frac{F_0}{p_0 S} \right)^{1/2}$$

while that for a supersonic aircraft is given by

$$M \cong \frac{2}{\gamma C_{D0}} \frac{dF_0/dM}{p_0 S}$$

At lower altitudes this maximum Mach number is usually constrained by the dynamic pressure such

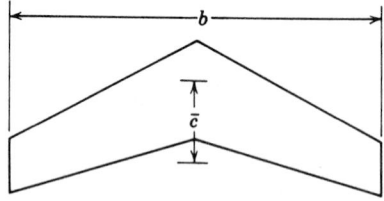

Fig. 6.117 Mean aerodynamic chord and span.

that

$$M_{\text{limit}} = \sqrt{\frac{2}{\gamma p} q_{\max}}$$

and at high altitudes by the maximum allowable stagnation temperature, $T_{\max}$, such that

$$M_{\text{limit}} = \sqrt{\frac{2}{\gamma - 1}\left(\frac{T_{\max}}{T_{\text{atm}}} - 1\right)}$$

2. The stall velocity at a load factor, n,

$$V_s = \left(\frac{2nW}{\rho C_{L\,\max} S}\right)^{1/2}$$

is determined by the largest lift coefficient $C_{L\,\max}$, that can be developed by the aircraft. This stall velocity is usually less than the minimum sustainable speed (due to excessive drag at high values of C_L), so that

$$M_{\min} = \left(\frac{2 K n^2 W^2 p_0}{\gamma p^2 F_0 S}\right)^{1/2}$$

3. The subsonic maximum rate of climb, or equivalently the maximum energy rate of climb for high-performance aircraft occurs at the transonic drag rise while the maximum angle of climb occurs at the velocity for L/D)maximum, V_R. At L/D)maximum the parasite drag equals the induced drag $C_{D_0} = KC_L^2$ and the corresponding velocity V_R is

$$V_R = \sqrt{\frac{2W}{\rho S}} \sqrt[4]{\frac{K}{C_{D_0}}}$$

The rate of climb for the maximum angle of climb is

$$\frac{dh}{dt} = \frac{dh_e}{dt} = \left(\frac{p}{p_0}\frac{F_0}{W} - 2\sqrt{KC_{D_0}}\right)V_R$$

The maximum rate of climb is given by

$$\frac{dh}{dt} = \frac{dh_e}{dt} = \frac{pMa}{W}\left(\frac{F_0}{p_0} - \frac{1}{2}C_{D_0\gamma}M^2 S\right)$$

4. The ceiling (maximum altitude) occurs at the altitude

$$h = -\frac{1}{\beta}\ln\left[2\sqrt{KC_{D_0}}\frac{W}{F_0}\right]$$

which corresponds to the ambient pressure

$$p = 2\sqrt{KC_{D_0}}\frac{W}{F_0}p_0$$

5. Fuel consumption per unit time for the maximum rate of climb trajectory is

$$\frac{dW_f}{dt} = \frac{cp}{p_0}F_0$$

6. In level flight, maximum range R is

$$R = \frac{1.14V_{Ri}}{c\sqrt{KC_{D_0}}}\left(1 - \sqrt{\frac{W}{W_i}}\right)$$

when flown at the following velocity

$$V = 1.316V_R$$

7. In level flight, maximum endurance E is flown at L/D a maximum and is

$$E = \frac{1}{2c}\frac{\ln[W_i/(W_i - W)]}{\sqrt{KC_{D_0}}}$$

and is flown at

$$V = V_R$$

8. Somewhat longer range and endurance are usually obtained if the aircraft is slowly "cruise climbed" at constant engine speed. The profile

$$V = 1.189V_R$$

results in maximum range

$$R = \frac{0.384V_{Ri}}{c}(KC_{D_0})^{-3/4}\left(\frac{F}{W_i}\right)^{1/2}\ln\left(\frac{W_i}{W_i - W}\right)$$

The altitude increase from start to end of the cruise climb is

$$h_f - h_i = \frac{a^2}{Kg}\ln\left(\frac{W_i}{W_i - W}\right)$$

9. The rate of turn ω is asymptotically a constant for high-performance jet aircraft turning at large load factors with fixed thrust and no loss of energy. The rate of turn ω for this case is approximately

$$\omega \cong \frac{ng}{V} = \frac{g}{W}\left(\frac{F\rho S}{2K}\right)^{1/2}$$

The turn radius decreases for increasing C_L and reaches a minimum at the maximum lift coefficient boundary. The minimum turn radius r is given by

$$r \cong \frac{V^2}{ng} = \frac{2}{g\rho S}\left(\frac{KnW^3}{F}\right)^{1/2}$$

It can be seen from these relations that turn rate and turn radius improve dramatically with increased wing area, increased thrust, and lower altitude.

6.15.2 Atmospheric Exit and Reentry

To obtain orbital velocity from the surface of a planet or to land on the surface of a planet from orbital velocity, the planetary atmosphere must be traversed. This transition is the topic of atmospheric exit and reentry.

Launch Trajectories
Aerodynamic forces may be ignored when exiting from planetary atmospheres if the vehicle is kept at zero angle of attack. This is usually necessary when passing through maximum dynamic pressure to assure that the vehicle is not destroyed by lateral loads. Once free of the planetary atmosphere, it can be shown that a desired velocity is obtained with minimum fuel if the thrust vector remains at a

constant orientation angle. The launch trajectory is, therefore, usually divided into three parts, a vertical ascent, a gravity turn at zero angle of attack, and a constant thrust angle phase.

Vertical Ascent. During the vertical ascent, the rocket's nondimensional mass m, velocity v, and altitude h, are described by the following integrals of the dynamic equations

$$m = 1 - \dot{m}t$$

$$v = -\ln m - t$$

$$h = +\frac{m}{\dot{m}}\ln m + t - \frac{t^2}{2}$$

where the nondimensional mass m, mass flow $\dot{m}$ (a positive constant), time t, velocity v, and altitude h are defined as

$$m = \frac{m'}{m_0'} \qquad \dot{m} = \frac{\dot{m}'v_e}{mg} \qquad t = \frac{t'g}{v_e} \qquad v = \frac{v'}{v_e} \qquad h = \frac{h'g}{v_e^2}$$

where m' is the dimensional mass, t' is dimensional time, m_0' is the initial rocket mass, v_a is the rocket exhaust velocity, and g is the acceleration of gravity. Notice that the magnitude of m should be as large as possible to gain maximum v or h with fixed expenditure of m.

Gravity Turn. Aerodynamic forces reach their largest value during the gravity turn, but both aerodynamic and gravity accelerations along the flight path are small in comparison to thrust accelerations. Approximate integrals of the dynamic equations are obtainable, but not as simple functions, though they may be straightforwardly expressed as a series. They are

$$m = 1 - \dot{m}t$$

$$v = -\ln m - t$$

$$\sin \gamma = \frac{z - 1}{z + 1} \qquad z = -\int_1^m \frac{d\tau}{\tau(-\ln \tau + 1)}$$

$$h = -\int_1^m -\frac{\ln \tau + 1}{\tau}\left[1 - \left(\frac{z - 1}{z + 1}\right)^2\right]^{1/2} d\tau$$

where γ is the angle between the velocity vector and the horizontal, the nondimensional down range distance x is

$$x = \frac{x'g}{v_c^2}$$

where v_c is the circular orbit velocity and all other variables are as previously defined.

Maximum dynamic pressure, "max q," occurs where $\rho v^2/2$ is a maximum or where

$$m(1 - \ln m)^2 = \frac{2\dot{m}}{\beta}$$

The nondimensional atmospheric scale height, $1/\beta$, is defined as

$$\beta = \frac{\beta'v_c^2}{g}$$

where β' is dimensional inverse atmospheric scale height.

Constant Thrust Inclination. The minimum fuel path to a given velocity is flown with the thrust at some constant inclination to the horizontal. If such a path is flown after the vehicle escapes the sensible atmosphere in a gravity turn, the integrals of the equation of motion that describe this path are

$$m = 1 - \dot{m}t$$
$$v_x = \cos \gamma_T \ln m + v_{x0}$$
$$v_y = -t - \sin \gamma_T \ln m + v_{y0}$$
$$x = v_{x0}t - \cos \gamma_T \left(\frac{m \ln m}{\dot{m}} - t \right)$$
$$h = v_{y0}t - \tfrac{1}{2}t^2 - \sin \gamma_T \left(\frac{m \ln m}{\dot{m}} - t \right)$$

where v_x and v_y are the components of the velocity vector in the horizontal and vertical direction, respectively, γ_T is the inclination of the thrust vector, and all other variables are as previously defined.

After the rocket has obtained the required velocity for orbit or escape, the engine is cut off, and from this point forward the principles of orbital mechanics apply.

Multistage Rockets
Single-stage chemical rockets have a fundamental limit to the maximum velocity they can obtain that is associated with the rocket's exhaust velocity and the portion of the rocket that must be allocated to structure. If velocities greater than this velocity are desired, the rocket must be multistaged.

Neglecting the usually small drag and gravity acceleration along the flight path, the equation describing the final velocity obtained by the k th stage of a chemical rocket is

$$v_{fk} = v_{0k} + v_{ek'} \ln \frac{m_{fk}}{m_{0k}}$$

The total velocity acquired by n stages is

$$v = \sum_{k=1}^{n} v_{ek} \ln \frac{m_{0k}}{m_{fk}}$$

Defining the structural efficiency factor, ε, and the π payload ratio for each stage as

$$\varepsilon = \frac{m_s}{m_s + m_p}$$

$$\pi = \frac{m_l}{m_0}$$

where m_s is the mass of the structure, m_p is the mass of the propellant, m_l is the mass of the payload, m_0 is the initial mass, and m_f is the final mass; it is then possible to rewrite the equation for v as

$$v = \sum_{k=1}^{n} v_{ek} \ln \frac{1}{\varepsilon_k + (1 - \varepsilon_k)\pi k}$$

It is easily seen that, for maximum v, v_e should be as large as possible and ε_k as small as possible for each stage. For the case where v_e and ε are equal for all stages, it can be shown that v is a maximum if all π are equal so that

$$\frac{m_f}{m_{01}} = \pi^n$$

$$\frac{m_f}{m_{01}} = \left[\frac{\exp[-(v/nv_e)] - \varepsilon}{1 - \varepsilon} \right]^n$$

which is the ratio of payload to initial mass that can be taken to a velocity v by n stages with a structural efficiency of ε. This has an upper level as $n \to \infty$ of

$$\frac{m_f}{m_{01}} = \exp\left(-\frac{v/v_s}{1 - \varepsilon} \right)$$

which gives the upper limit of the payload ratio that can be gained by staging.

In the limit of no payload,

$$v = nv_e \ln \frac{1}{\varepsilon}$$

which gives the maximum velocity that can be obtained with n stages of a rocket with structural efficiency ε.

Atmospheric Entry

The objectives of atmospheric entry are to decelerate a space vehicle from orbital or superorbital velocities with acceptable aerodynamic and heating loads and land on some specified portion of the planetary surface or to maneuver inside the planetary atmosphere to obtain a change of orbital elements. This type of flight poses some unusual problems.

All planetary atmospheres are thin in comparison to the planetary radius. A measure of this thinness is ε, the ratio of the atmosphere scale height $1/\beta$ to the planetary radius r_0:

$$\varepsilon = \frac{1}{\beta r_0}$$

See Table 6.14.

The time required for a vehicle at orbital velocity to pass through one atmospheric scale height at moderate flight path angles is on the order of ε times the orbital period. A vehicle can quickly pass through many scale heights if the flight path angle is not kept extremely small, and thus encounter severe heating and deceleration problems.

Aerodynamic accelerations are large for vehicles having velocities of the order of orbital velocity when deep within any planetary atmosphere. Thus entry trajectories must be kept shallow at entry altitudes where the atmosphere is sparse; however, the danger of skipping out of the sensible atmosphere also exists at these shallow entry trajectories.

At orbital velocities the entry Mach number M_0 is approximately

$$M_0 \cong \sqrt{\frac{1}{\varepsilon}}$$

and is therefore large. This implies that aerodynamic heating, convective and/or radiative, is severe during entry.

Equilibrium Glide. The earliest analytical solutions for flight at near orbital velocity were produced by Sänger.[141] They describe deceleration at small flight path angles where the difference between gravity and centrifugal forces is balanced by lift. The appropriate analytical solutions are

$$v^2 = \frac{1}{(C_L/2)\varepsilon p + 1}$$

$$\gamma = -2\varepsilon \frac{C_D}{C_L}$$

$$\theta = \theta_0 + \frac{1}{2}\frac{C_L}{C_D}\ln\left(\frac{1 - v^2}{1 - v_0^2}\right)$$

TABLE 6.14   Planetary Flight Data

Planet	Equatorial Radius, r_0 (ft)	Equatorial Gravitational Acceleration, g_0 (ft/sec^2)	Inverse Atmospheric Scale Height, β_0 (ft^{-1})	Equatorial Radius/ Atmospheric Scale Height, $1/\varepsilon = \beta_0 r_0$	Orbital Velocity Mach No., M_0	Deceleration loads when $p = 0(\varepsilon)$ $g_0/g_{0\oplus}$	Deceleration loads when $p = 0(1)$ $g_0/g_{0\oplus}$
Venus	2.03×10^7	28.3	4.9×10^{-5}	1006	26	0.88	1000
Earth	2.09×10^7	32.2	4.3×10^{-5}	900	25	1.0	900
Mars	1.11×10^8	12.2	1.0×10^{-5}	132	9.5	0.39	5
Jupiter	2.27×10^8	83.7	1.7×10^{-5}	3600	49	2.61	9100
Saturn	1.87×10^8	36.6	1.6×10^{-5}	3000	45	1.12	3360

where velocity V and pressure p have been nondimensionalized as

$$v = \frac{v'}{\sqrt{r_0 g_0}}$$

$$p = \frac{p'}{mg_0/A}$$

$$p' = p_0^{\exp(-\beta h')}$$

where g_0 is the acceleration of gravity at the planet's surface, p_0 is the value of pressure at the planet's surface, h is height, m is the vehicle's mass, and A is the vehicle's reference area. Other symbols are defined in Fig. 6.109.

Load factor n, due to total aerodynamic force, measured in multiples of earth gravity acceleration, is

$$n = \frac{1}{2\varepsilon}\left(C_L^2 + C_D^2\right)^{1/2} pv^2 \frac{g_0}{g_{0\,\text{earth}}}$$

The load factor stays low for equilibrium glide but slowly increases as the trajectory descends.

Heating loads per unit time are relatively small with small radiative heating loads dominating at high altitudes and velocities, and convective loads dominating at lower altitudes. The total energy absorbed can be quite large if attempts are not made to reradiate the energy to free space. The heating loads can be evaluated using empirical relations of the form

$$\dot{q} = \frac{C_{Q} \rho^i v^{2j}}{r^k}$$

where for stagnation point heating (laminar flow)

$$C_Q = 1.55 \times 10^{-5} \qquad i = 0.5, \; j = 1.5, \; k = 0.5$$

and for sonic point heating (turbulent flow)

$$C_Q = 7.45 \times 10^{-4} \qquad i = 0.8, \; j = 1.5, \; k = 0.2$$

and for stagnation point radiant heating

$$C_Q = 1.23 \times 10^{-25} \qquad i = 1.6, \; j = 4.2, \; k = -1$$

where C_Q is a heat transfer coefficient and r is the vehicle nose radius. For equilibrium glide the maximum heating rate occurs deep in the atmosphere.

Ballistic and Skip Entry. The ballistic and skip solution, associated with the names of Allen and Eggers,[142] is valid when the aerodynamic drag dominates the flight path component of gravity and when the difference between centrifugal and gravity accelerations are small, or when all normal forces are dominated by lift.

For ballistic entry where there is no lift, the solution is

$$v^2 = v_0^2 \exp\left[\frac{C_D(p - p_0)}{\sin \gamma_0}\right]$$

$$\cos \gamma = \cos \gamma_0$$

$$\theta = \theta_0 + \cot \gamma_0 (h - h_0)$$

where all symbols are as previously defined. The maximum load factor due to deceleration occurs at the altitude where

$$p = \frac{\sin \gamma_0}{C_D}$$

and has a value

$$n = \frac{\sin \gamma_0}{\varepsilon} \frac{v_0^2}{e}$$

The maximum heating rate occurs at the altitude where

$$p = -\frac{i}{j} \frac{\sin \gamma_0}{C_D}$$

and has a value

$$\dot{q} = C_Q p v_0^{2/j} \exp\left(-j C_D \frac{p}{\sin \gamma_0} \right)$$

When lift is present (skip entry) the solution is

$$v^2 = v_0^2 \exp\left[\frac{2C_D}{C_L} (\gamma - \gamma_0) \right]$$

$$\cos \gamma = \cos \gamma_0 + \frac{C_L}{2} (p - p_0)$$

Maximum heating rate and deceleration occur near the bottom of the skip where $\gamma = 0$ at the altitude corresponding to

$$p = p_0 + \frac{2}{C_L} (1 - \cos \gamma_0)$$

The load factor due to total aerodynamic force on the vehicle at this point is given by

$$n = \frac{1}{2\varepsilon} (C_L^2 + C_D^2)^{1/2} p v_0^2 \exp\left\{ -2\gamma_0 \frac{C_D}{C_L} \right\} \frac{g_0}{g_{0\,\text{earth}}}$$

and the heating rate is obtained from the expression given for ballistic entry.

Conclusion
A number of other relations are available for both two- and three-dimensional flight. One important use of the relations presented is in the calculations of "entry corridors" from superorbital speed. The corridor is defined by the class of Keplerian orbits that are bounded on the bottom by a skip at maximum lift-up that will not exceed heating or acceleration limits and on the top by an equilibrium glide trajectory with maximum lift-down.

REFERENCES

1. Everitt, C. W. F., *James Clerk Maxwell: Physicist and Natural Philosopher*, New York, Charles Scribner's, 1975.
2. Vincenti, W. G., and Kruger, C. H. Jr., *Introduction to Physical Gas Dynamics*, New York, Wiley, 1965.
3. Bird, R. B., Stewart, W. E., and Lightfoot, E. N., *Transport Phenomena*, New York, Wiley, 1960.
4. Hirschfelder, J. O., Curtiss, C. F., and Bird, R. B., *Molecular Theory of Gases and Liquids*, New York, Wiley, 1967.
5. Svehla, R. A., "Estimated Viscosities and Thermal Conductivities of Gases at High Temperatures," National Aeronautics and Space Administration Report NASA TR-R-132, Washington, D.C., Government Printing Office, 1963.
6. Ames Research Staff, "Equations, Tables, and Charts for Compressible Flow," National Advisory Committee for Aeronautics report NACA Report 1135, Ames Aeronautical Laboratory, Moffett Field, Calif., 1953.
7. Jumper, E. J., Ultee, C. J., and Dorko, E. A., A Model for Fluofrine Atom Recombination on a Nickel Surface, *Journal of Physical Chemistry*, **84**, 41–50 (1980).

8. Jeans, J., *An Introduction to the Kinetic Theory of Gases*, Cambridge, Cambridge University Press, 1940, reissued 1982.

9. Chapman, S., and Cowling, T. G., *The Mathematical Theory of Non-Uniform Gases*, London, Cambridge University Press, 1952.

10. Emmons, H. W. (ed.), *High Speed Aerodynamics and Jet Propulsion, Volume III, Fundamentals of Gas Dynamics*, Princeton, New Jersey, Princeton University Press, 1958.

11. Moore, F. K., *High Speed Aerodynamics and Jet Propulsion, Volume IV, Theory of Laminar Flows*, Princeton, New Jersey, Princeton University Press, 1964.

12. Moore, F. K., *High Speed Aerodynamics and Jet Propulsion, Volume IV, Theory of Laminar Flows*, Princeton, New Jersey, Princeton University Press, 1964.

13. Vincenti, W. G., and Kruger, C. H., Jr., *Introduction to Physical Gas Dynamics*, New York, Wiley, 1965.

14. Uman, M. A., *Introduction to Plasma Physics*, New York, McGraw-Hill, 1964.

15. Emmons, H. W., *High Speed Aerodynamics and Jet Propulsion, Volume III, Fundamentals of Gas Dynamics*, Princeton, New Jersey, Princeton University Press, 1958.

16. Pao, R. H. F., *Fluid Dynamics*, Columbus, Ohio, Merrill, 1967.

17. *U.S. Standard Atmosphere, 1962*, Washington 25, D.C., Bureau of Documents, Washington, D.C., Government Printing Office,

18. *Tables of Thermodynamic and Transport Properties of Air, Argon, Carbon Dioxide, Carbon Monoxide, Hydrogen, Nitrogen, Oxygen and Steam*, originally published as National Bureau of Standard's Circular 564.

19. *U.S. Standard Atmosphere Supplements, 1966*, Washington 25, D.C., Bureau of Documents, Washington, D.C., Government Printing Office,

20. Heck, R. C. H., "The New Specific Heats," *Mech. Eng.*, **63**, 126–135 (1941).

21. Keenan, J. H., and Kaye, J., *Gas Tables*, New York, Wiley, 1957.

22. Daley, D. H., A Nozzle Operating Diagram, *Bull. Mech. Eng. Educ.*, **6**, 293–300 (1967).

23. *Equations, Tables and Charts for Compressible Flow*, NACA Report 1135, Washington, D.C., Government Printing Office, 1953.

24. Dailey, and Wood, *Computation Curves for Compressible Fluid Problems*, New York, Wiley, 1949.

25. Zucrow, M. J., and Hoffman, J. D., *Gas Dynamics, Vol I*, New York, Wiley, 1976.

26. Dennard, J. S., and Spencer, P. B., "Ideal-Gas Tables for Oblique Shock Flow Parameters in Air Mach Numbers From 1.05 to 12.0," NASA TN D-2221, March 1964.

27. Kopal, Z., *Tables and Supersonic Flow Around Cones*, Cambridge, Massachusetts, MIT Press, 1947.

28. Taylor, G. I., and Maccoll, J. W., The Air Pressure on a Cone Moving at High Speeds, *Proc. Roy. Soc.* (London), Ser. A, **139**(838) (February 1933).

29. Perkins, C. D., and Hage, R. E., *Airplane Performance, Stability, and Control*, Chapter 3, New York, Wiley, 1949.

30. Oates, G. C., "Aerothermodynamics of Gas Turbine and Rocket Propulsion," American Institute of Aeronautics and Astronautics, Inc., 1984.

31. Gray, W. H., and Mastrocola, N., "Representative Operating Charts of Propellers Tested in the NACA 20 ft. Propeller Research Tunnel," NACA, ARR 3I25, September, 1943.

32. Jahn, R. G., *Physics of Electric Propulsion*, New York, McGraw-Hill, 1968.

33. Hill, P. G., and Peterson, C. R., *Mechanics and Thermodynamics of Propulsion*, Reading, Mass., Addison-Wesley, 1965.

34. Penaranda, F. E., and Freda, M. S., "Wind Tunnels," NASA RP-1132, Washington, D.C., 1985.

35. Pope, A., and Goin, K. L., *High Speed Wind Tunnel Testing*, New York, Wiley, 1965.

36. Holman, J. P., and Gajda, W. J., *Experimental Methods for Engineers*, 4th ed., New York, McGraw-Hill, 1984.

37. Dally, J. W., Riley, W. F., and McConnell, K. G., *Instrumentation for Engineering Measurements*, New York, Wiley, 1984.

38. Doeblin, E. O., *Measurement Systems, Application and Design*, 2nd ed., New York, McGraw-Hill, 1983.

39. Kuethe A. M., and Chow, C. Y., *Foundations of Aerodynamics: Bases of Aerodynamic Design*, 4th ed., New York, Wiley, 1986.

40. Liepmann, H. W., and Roshko, A., *Elements of Gasdynamics*, New York, Wiley, 1957.

41. Anderson, J. D., Jr., *Fundamentals of Aerodynamics*, New York, McGraw-Hill, 1984.

42. Bertin, J. J., and Smith, M. L., *Aerodynamics for Engineers*, Englewood Cliffs, N.J., Prentice-Hall, 1979.

43. Anderson, J. D., Jr., *Introduction to Flight*, 2nd. ed., New York, McGraw-Hill, 1985.

44. Helmbold, H. B., "Der unverwundene Ellipsenflügelals tragende Flache," in *Jahrbuch 1942 ser Deutshcen Luftfahrtforschung*, Munich, R. Oldenbuourg, pp. 111–113.

45. Gersten, K., A Non-linear Lifting Surface Theory Especially for Low Aspect Ratio Wings, *J. Aeronaut. Sci.*, **30** (April 1963).

46. John, James E. A., *Gas Dynamics*, 2nd ed., Newton, Mass., Allyn and Bacon, 1984.

47. Shapiro, A. H., *The Dynamics and Thermodynamics of Compressible Fluid Flow*, Vol. I, New York, Ronald, 1954.

48. Goethert, B., "Plane and Three-Dimensional Flow at High Subsonic Speeds," NACA TM No. 1105, 1946. (Translated from *Lilienthal Gesellshaft*, Vol. 127, 1940.)

49. Lowry, J. G., and Polhamus, E., NACA TN 3911, 1957.

50. Anderson, J. D., Jr., *Modern Compressible Flow with Historical Perspective*, New York, McGraw-Hill, 1982.

51. Anon "Handbook of Supersonic Aerodynamics for Three-Dimensional Airfoils," NAVORD Report 1488, Vol. 3, Section 7, Bureau of Ordnance Publication, 1958.

52. Von Kármán, Th., Supersonic Aerodynamics-Principles and Applications, *J. Aeronautics Sciences*, **14**(7) (July 1947).

53. Bonney, E. A., *Engineering Supersonic Aerodynamics*, New York, McGraw-Hill, 1950.

54. Coleman, T. F., "Supersonic Lift Solutions Obtained by Extending the Simple Linearized Conical Flow Theory," Applied Physics Laboratory, Johns Hopkins Univ. NAA/CM 440, February 1948.

55. Goodman, T. R., The Lift Distribution on Conical and Non-Conical Flow Regions of Thin Finite Wings in a Supersonic Stream, *J. Aeronaut. Sci.*, **16**(6) (June 1949).

56. De Young, J., "Spanwise Loading for Wings and Control Surfaces of Low Aspect Ratio," NACA TN No. 2011, 1950.

57. Puckett, A. E., and Stewart, H. J., Aerodynamic Performance of Delta Wings at Supersonic Speeds, *J. Aeronaut. Sci.*, **14**(10) (October 1947).

58. Jones, R. T., "Properties of Low-Aspect-Ratio Pointed Wings at Speeds Below and Above the Speed of Sound," NACA Rept. 835, 1946.

59. Munk, M. M., "The Aerodynamic Forces on Airship Hulls," NACA TR 184, 1924.

60. Lamb, H., *Hydrodynamics*, New York, Dover, 6th ed., 1945.

61. Von Kármán, Th., "Calculation of Pressure Distributions on Airship Hulls," NACA TM 574, 1930.

62. Lotz, I., "Calculation of Potential Flow Past Arbitrary Bodies in Yaw," NACA TM 675, 1932.

63. Liepmann, H. W., and Roshko, A., *Elements of Gasdynamics*, New York, Wiley, 1957.

64. Laitone, E. V., The Subsonic Axial Flow About a Body of Revolution, *Quart. Appl. Math.* **5**(2), 227 (1947).

65. Goethert, B., "Plane and Three-Dimensional Flow at High Subsonic Speeds," NACA TM 1105, 1946. (Translated from *Lilienthal Gesselschaft*, Vol. 127, 1940.)

66. Van Driest, E. R., "Die linearisierte Theorie der dreidimensionalen komptessiblen Unterschall-stroemung und die experimentelle Untersuchung von Rotationskoerpern in einem geschlossenen Windkanel," *Mitteil. Inst. Aerodynamik*, No. 16, Zurich, Verlag Leemann, 1949.

67. Lees, L., "A Discussion of the Application of the Prandtl-Glauert Method to Subsonic Compressible Flow over a Slender Body of Revolution," NACA TN 1127, 1946.

68. Schmieden, C., and Kawalki, K. H., "Einfluss der Kompressibilitaet bei rotationssymmetrischer Umstroemung eines Ellipsoids," Forschungsbericht No. 1633, Duesche Luftfahrtforschung, 1942. (See also NACA TM 1233, 1949.)

69. Ashley, H., and Landahl, M., *Aerodynamics of Wings and Bodies*, Reading, Mass., Addison-Wesley, 1965.

70. Kopal, Z., "Supersonic Flow of Air Around Cones," MIT Tech. Rept. 1, 1947.

71. Kopal, Z., "Supersonic Flow Around Yawing Cones," MIT Tech. Rept. 3, 1947.

72. Kopal, Z., "Supersonic Flow Around Cones of Large Yaw," MIT Tech. Rept. 5, 1949.

73. Ferri, A., *Elements of Aerodynamics of Supersonic Flows*, New York, Macmillan, 1949.

74. Shapiro, A. H., *The Dynamics and Thermodynamics of Compressible Fluid Flow*, New York, Ronald, 1953.

75. Courant, R., and Friedrichs, K. O., *Supersonic Flow and Shock Waves*, New York, Interscience, 1948.

76. Sears, W. R., On Projectiles of Minimum Wave Drug, *Quart. Appl. Math.* **6**, 361–366 (1947).

77. Hayes, W. D., and Probstein, R. F., *Hypersonic Flow Theory*, New York, Academic, 1959.

78. Cherny, G. G., *Introduction to Hypersonic Flow*, New York, Academic, 1961.

79. Dorrance, W. H., *Viscous Hypersonic Flow*, New York, McGraw-Hill, 1962.

80. Hayes, W. D., and Probstein, R. F., *Hypersonic Flow Theory, Volume I, Inviscid Flow*, New York, Academic, 1966.

81. Anderson, J. D., Jr., "A Survey of Modern Research in Hypersonic Aerodynamics," AIAA-84-1578, a paper presented at the AIAA 17th Fluid Dynamics Plasma Dynamics, and Lasers Conference, June 25–27, 1984, Snowmass, Colorado.

82. Anderson, J. D., Jr., Hypersonic Aerodynamics, unpublished notes, 1986.

83. Ames Research Staff, "Equations Tables and Charts for Compressible Flow," NACA Report 1135, 1953.

84. Kopal, Z., "Tables of Supersonic Flow Around Coves," MIT Tech. Rept. No. 1, 1947.

85. Cox, R. N., and Crabtree, L. F., *Elements of Hypersonic Aerodynamics*, London, English Univ. Press, 1965.

86. Lees, L., Note on the Hypersonic Similarity Law for an Unyawed Cone, *J. Aeronaut. Sci.*, **18**, 700–702 (1951).

87. Eggers, A. J., Savin, R. C., and Syvertson, C. A., The Generalised Shock Expansion Method and its Application to Bodies Traveling at High Supersonic Airspeeds, *J. Aeronaut. Sci.*, **22**, 231–238, 248 (1955).

88. Hayes, W. D., On Hypersonic Similitude, *Quar. Appl. Math.*, **5**, 105–106 (1947).

89. Lees, L., "Inviscid hypersonic flow over blunt noses slender bodies," GALCIT Hypersonic Research Project, Memo No. 31, 1956.

90. Cheng, H. K., "Hypersonic Flow With Combined Leading-Edge Bluntness and Boundary-Layer Displacement Effects," Cornell Aeronautical Laboratory Report No. AF-1285-A-4, 1960.

91. Griffith, B. J., and Lewis, C. H., "A Study of Laminar Heat Transfer to Spherically Blunted Cones and Hemisphere-Cylinders at Hypersonic Conditions," Arnold Engineering Development Center Technical Documentary Report No. AEDC-TDR-63-102, 1963.

92. Lewis, C. H., "Pressure Distribution and Shock Shape over Blunt Slender Cones at Mach Numbers from 16 to 19," Arnold Engineering Development Center Technical Note No. AEDC-TN-61-81, 1961.

93. Whitfield, J. D., and Norfleet, G. D., "Source Flow Effects in Conical Hypervelocity Nozzles," Arnold Engineering Development Center Technical Documentary Report No. AEDC-TDR-62-116, 1962.

94. Schlichting, H., *Boundary Layer Theory*, 7th ed., New York, McGraw-Hill, 1979.

95. Moore, F. K. (ed.), *Theory of Laminar Flows, Vol. IV, High Speed Aerodynamics and Jet Propulsion*, Princeton, New Jersey, Princeton Univ. Press, 1964.

96. Falkner, V. M., and Skan, S. W., Some Approximate Solutions of the Boundary Layer Equations, *Phil. Mag.*, **12**, 865 (1931).

97. White, F. M., *Viscous Fluid Flow*, New York, McGraw-Hill, 1974.

98. Hartnett, J. P. (ed.), *Recent Advances in Heat and Mass Transfer*, New York, McGraw-Hill, 1961.

99. Cox, R. N., and Crabtree, L. F., *Elements of Hypersonic Aerodynamics*, New York, Academic, 1965.

100. Hinze, J. O., *Turbulence*, New York, McGraw-Hill, 1959.

101. Eckert, E. R. G., "Survey of Heat Transfer at High Speeds," WADC TR 54-70, 1964.

102. Eckert, E. R. G., "Survey of Boundary Layer Heat Transfer at High Velocities and High Temperatures," WADC TR 59-624, 1960.

103. Schmidt, H., "Laminar Skin Friction and Heat Transfer Parameters for a Flat Plate at Hypersonic Speeds in Terms of Free-Stream Flow Properties," NASA TN D-8, September, 1959.

104. Hankey, W. L., and Neumann, R. D., "Design Procedures for Computing Aerodynamic Heating at Hypersonic Speeds," WADC TR 59-610, 1960.

105. Hayes, W. D., and Probstein, R. F., *Hypersonic Flow Theory*, New York, Academic, 1959.

106. Schmidt, H., "Turbulent Skin Friction and Heat Transfer Coefficients for an Inclined Plate at Hypersonic Speeds in Terms of Free-Stream Flow Properties," NASA TN D-869, May, 1961.

107. Van Driest, E. R., Turbulent Boundary Layer on a Cone in Supersonic Flow at Zero Angle of Attack, *J. Aeronaut. Sci.* **19**(1), 55 (1952).

108. Detra, R. W., and Hidalgo, H., *ARS J.* (March, 1961).

109. Gersten, K., A Non-Linear Lifting Surface Theory Especially for Low Aspect Ratio Wing, *J. Aeronaut. Sci.* (April 1963).

110. Gersten, K., "Calculation of Non-Linear Aerodynamic Stability Derivatives of Aeroplanes," AGARD Rept. 342, 1961.

111. Pitts, W. C., Nielsen, J. N., and Kaattari, G. E., "Lift and Center of Pressure of Wing-Body-Tail Combinations at Subsonic, Transonic and Supersonic Speeds," NACA Report 1307, 1959.

112. Nicolai, L. M., and Sanchez, F., Correlation of Wing-Body Combination Lift Data, *AIAA J. Aircraft* (February 1973).

113. Hall, C. F., "Lift, Drag and Pitching Moment of Low Aspect Ratio Wings at Subsonic and Supersonic Speeds," NACA RM A53A30, 1958.

114. Osborne, R. S., and Kelly, T. C., "A Note on the Drag Due to Lift of Delta Wings at Mach Numbers Up to 2.0," NASA TN D-545, November 1960.

115. Benepe, D. B., Kouri, B. G., and Webb, J. B., "Aerodynamic Characteristics of Non-Straight-Taper Wings," AFFDL-TR-66-73, Air Force Flight Dynamics Laboratory, Wright-Patterson AFB, Ohio, 1966.

116. Ashley, H., and Landahl, M., *Aerodynamics of Wings and Bodies*, Reading, Mass., Addison-Wesley, 1965.

117. Nelson, R. L., and Welsh, C. J., "Some Examples of the Application of the Transonic and Supersonic Area Rules to the Prediction of Wave Drag," NASA TN D-446, September 1960.

118. Cahn, M. S., and Olstad, W. B., "A Numerical Method for Evaluating Wave Drag," NACA TN 4258, June 1958.

119. Prigogine, I., *Introduction to Thermodynamics of Irreversible Processes*, 2nd ed., New York, Wiley, 1962.

120. Vincenti, W. G., and Kruger, C. H., *Introduction to Physical Gas Dynamics*, New York, Wiley, 1965.

121. Lee, J. F., Sears, F. W., and Turcotte, D. L., *Statistical Thermodynamics*, Reading, Mass., Addison-Wesley, 1963.

122. Clarke, J. F., and McChesney, M., *The Dynamics of Real Gases*, London, Butterworths, 1964.

123. Vincenti, W. G., and Kruger, C. H., Jr., *Introduction to Physical Gas Dynamics*, New York, Wiley, 1965.

124. Vincenti, W. G., and Baldwin, B. S., Effect of Thermal Radiation On the Propogation of Plane Acoustic Waves, *J. Fluid, Mech.*, **12**(Part 3), 449 (1962).

125. Cheng, P., "Study of the Flow of a Radiating Gas by a Differential Approximation," Ph.D. Dissertation, Stanford University, 1965.

126. Sutton, G. W., and Sherman, A., *Engineering Magnetohydrodynamics*, New York, McGraw-Hill, 1965.

127. Thompson, W. B., *An Introduction to Plasma Physics*, Reading, Mass., Addison-Wesley, 1962.

128. Longmire, C., *Elementary Plasma Physics*, New York, Wiley, 1963.

129. Parker, E. N., *Interplanetary Dynamical Processes*, New York, Wiley, 1963.

130. Nicolai, L. M., "An Experimental and Theoretical Analysis of the Convected Balanced Arc," NASA CR-1267, February 1969.

131. Cambel, A., *Plasma Physics and Magnetofluidmechanics*, New York, McGraw-Hill, 1963.

132. Hale, F. J., *Aircraft Performance, Selection and Design*, New York, Wiley, 1984.

133. Lan, E. C., and Roskam, J., "Airplane Aerodynamics and Performance," Ottawa, US, Roskan Aviation and Engineering, 1981.

134. McCormick, B. W., *Aerodynamics, Aeronautics, and Flight Mechanics*, New York, Wiley, 1979.

135. Miele, A., Flight Mechanics, Reading, Mass., Addison-Wesley, 1962.

136. Smetana, F. O., *Computer Assisted Analysis of Aircraft Performance, Stability and Control*, New York, McGraw-Hill, 1984.

137. Etkin, B., *Dynamics of Flight, Stability and Control*, London, Wiley, 1982.

138. Perkins, C. D., and Hage, R. E., *Airplane Performance Stability and Control*, London, Wiley, 1949.

139. Roskam, J., "Airplane Flight Dynamics and Automatic Flight Controls, Part I and II," Ottawa, US, Roskam Aviation and Engineering, 1979.
140. Seckel, E., *Stability and Control of Airplanes and Helicopters*, New York, Academic, 1964.
141. Loh., W. H., *Dynamics and Thermodynamics of Planetary Entry*, Englewood Cliffs, N.J., Prentice-Hall, 1963.
142. Willes, R. E., Francisco, M. C., Reid, J. G., and Lun, W. K., "An Application of Matched Asymptotic Expansion to Hypervelocity Flight," AIAA Paper No. 67-598, August 67.

BIBLIOGRAPHY FOR ATMOSPHERIC STRUCTURE

Handbook of Geophysics and Space Environment, 1965, Washington 25, D.C., Bureau of Documents, Washington, D.C., Government Printing Office,

BIBLIOGRAPHY FOR STEADY ONE-DIMENSIONAL GAS DYNAMICS

John, J. E. A., *Gas Dynamics*, 2nd ed., Boston, Allyn and Bacon, 1984.
Shapiro, A. H., and Hawthorne, W. R., The Mechanics and Thermodynamics of Steady One-Dimensional Flow, *J. Appl. Math.*, **14**(4), A317 (1947).

BIBLIOGRAPHY FOR PROPULSION

Sutton, G. P., *Rocket Propulsion Elements*, New York, Wiley, 3rd ed., 1963.
Topps, J. E. C., "The Thermal Properties of the Combustion Products of Kerosene Air Mixtures," National Gas Turbine Establishment, Report No. R14, April, Z947.

BIBLIOGRAPHY FOR EXPERIMENTAL METHODS

Liepman, H. W., and Roshko, A., *Elements of Gasdynamics*, New York, Wiley, 1957.
Rae, W. H., Jr., and Pope, A., *Low Speed Wind Tunnel Testing*, 2nd ed., New York, Wiley, 1984.

BIBLIOGRAPHY FOR AERODYNAMICS OF WINGS

Abbott, L. H., von Doenhoff, A. E., and Stivers, L. S., Jr., "Summary of Airfoil Data," NACA Report 824, 1945.
Laitone, E. V., New Compressibility Correction for Two-Dimensional Subsonic Flow, *J. Aeronaut. Sci.*, **18**(5), 350 (1951).

BIBLIOGRAPHY FOR AERODYNAMICS OF BODIES

Charters, A. C., and Thomas, R. N., The Aerodynamic Performance of Small Spheres From Subsonic to High Supersonic Velocities, *J. Aeronaut. Sci.*, **12**(10) (October 1945).
Hall, C. F., "Lift, Drag and Pitching Moment of Low Aspect Ratio Wings at Subsonic and Supersonic Speeds," NACA RM A53A30, January 1958.
Seifert, H. S., Mills, M. W., and Summerfield, M., Physics of Rockets—Dynamics of Long Range Rockets, *Amer. J. Phys.*, May–June 1947.

BIBLIOGRAPHY FOR HYPERSONIC FLOW

Wagner, R. D., and Watson, R., "Induced pressures and Shock Shapes on Blunt Cones in Hypersonic Flow," NASA Technical Note, TN-D-2182, 1964.

BIBLIOGRAPHY FOR VISCOUS FLOW

Francis, W. L., Malvestuto, F. S., and Stuart, J. W., "Study to Determine Skin-Friction Drag in Hypersonic Low Density Flow," ASD-TR 61-433, Vol. I, 1962.

BIBLIOGRAPHY FOR AERODYNAMICS OF WING-BODY COMBINATIONS

Nielsen, J., Katzen, E., and Tang, K., "Lift and Pitching Moment Interference Between a Painted Cylindrical Body and Triangular Wings of Various Aspect Ratios at Mach Numbers of 1.50 and 2.02," NACA TN 3795, 1956.

BIBLIOGRAPHY FOR REAL GASES

Bond, J. W., Watson, K. M., and Welch, J. A., *Atomic Theory of Gas Dynamics*, Reading, Mass., Addison-Wesley, 1965.

BIBLIOGRAPHY FOR GAS DYNAMICS

Benedict, R. P., *Fundamentals of Gas Dynamics*, New York, Wiley, 1983.

Biberman, L. M., Iakubou, I. T., Norman, G. E., and Vorobyov, V. S., Radiation Heating under Hypersonic Flow, *Astronaut. Acta*, **10**, 238 (1964).

Cess, R. D., and Sparrow, E. M., *Radiation Heat Transfer*, New York, McGraw-Hill, 1978.

Clarke, J. F., and McChesney, M., *Dynamics of Relaxing Gases*, Boston, Butterworths, 1976.

Gray, W. A., and Miller, R., *Engineering Calculations in Radiative Heat Transfer*, New York, Pergamon Press, 1974.

Holman, J. P., *Heat Transfer*, New York, McGraw-Hill, 1986.

Zel'dovich, Ya., B., and Raizer, Yu. P., *Physics of Shock Waves and High Temperature Hydrodynamic Phenomena*, New York, Academic, 1966.

BIBLIOGRAPHY FOR MAGNETOGASDYNAMICS

Alfvén, H., and Falthammar, G. G., *Cosmical Electrodynamics*, New Jersey, Oxford University Press, 1963.

Stix, T., *The Theory of Plasma Waves*, New York, McGraw-Hill, 1962.

BIBLIOGRAPHY FOR FLIGHT DYNAMICS

Halfman, R. L., *Dynamics: Particles, Rigid Bodies, and Systems*, Reading, Mass., Addison-Wesley, 1962.

CHAPTER 7
ASTRONAUTICS

**R. B. Giffen, A. E. Preyss, R. R. Bate, J. D. Hines,
D. E. Mercier, R. H. Tate, J. Anthony, J. P. Wittry,
B. W. Parkinson, J. C. Ruth, R. J. Lisowski, R. S. Fraser, and
M. L. DeLorenzo**

**Department of Astronautics
United States Air Force Academy
Colorado Springs, Colorado**

7.1 FOUNDATION OF ORBITAL MECHANICS
A. E. Preyss, R. R. Bate, J. D. Hines, D. E. Mercier, R. H. Tate, and J. Anthony

Newton's laws of motion and his law of universal gravitation form the foundation of orbital mechanics. With these laws the motion of a body orbiting in a gravitational field can be described by

TABLE 7.1 Basic Characteristics of the Solar System[3]

Body	Symbol	Semimajor Axis to Sun (AU)[a]	Period Earth-Years ($\oplus = 1$)	Mean Diameter ($\oplus = 1$)	Mass ($\oplus = 1$)	No. of Natural Satellites	Surface Escape Velocity ($\oplus = 1$)
Sun	☉	—	—	109.0	3×10^5	—	55.0
Mercury	☿	0.387	0.241	0.38	0.054	0	0.371
Venus	♀	0.723	0.616	0.97	0.815	0	0.915
Earth	⊕	1.00	1.00	1.00	1.00	1	1.00
Mars	♂	1.52	1.88	0.52	0.108	2	0.449
Jupiter	♃	5.20	11.9	11.0	318.0	12	5.38
Saturn	♄	9.54	29.5	9.03	95.1	9	3.26
Uranus	♅	19.2	84.0	3.72	14.5	5	1.97
Neptune	♆	30.1	165.0	3.38	17.0	2	2.24
Pluto	♇	39.5	248.0	1.02?	0.8	0	0.85?
Earth's Moon	☾	—	0.075	0.27	0.012	0	0.212

[a] 1 AU = 92,959,670 mi.

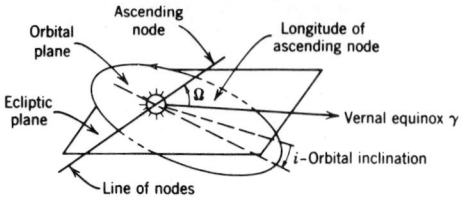

TABLE 7.2   Orbital Characteristics of the Planets[3]

Symbol Planet	☿ Mercury	♀ Venus	⊕ Earth	♂ Mars
Semimajor axis a, AU	0.387099	0.723332	1.000000	1.523691
Perihelion distance, AU $= a(1-e)$	0.307501	0.718422	0.983277	1.381416
Aphelion distance, AU $= a(1+e)$	0.466697	0.728242	1.016723	1.665966
Orbital eccentricity e	0.205628	0.006788	0.016723	0.093375
Mean orbital velocity ($\oplus = 1$)	1.607271	1.175794	1.00	0.806855
km/sec	47.90	35.05	29.77	24.02
NM/sec	25.87	18.92	16.80	12.97
ft/sec	157,186.0	114,958.0	97,702.1	78,805.7
Sidereal mean daily motion, sec	14,732.4	5767.670	3548.193	1886.519
Period of revolution ($\oplus = 1$)	0.2411	0.6156	1.00	1.8822
Orbital inclination i to ecliptic, deg	7.00412	3.39431	0	1.84989
Inclination of equatorial plane to orbit, deg		< 10	23.443597	23.99
Mean longitude of ascending node Ω, deg	47.94364	76.38541		49.30530
True longitude of perihelion $\bar{\omega}$, deg (Epoch 1967 April 20.0)	76.94664	131.11097	102.3781	335.45705
Mean anomaly M, deg	253.906	353.764	105.02313	240.222
Axial rotational period	$59^d \pm 2^d$	$242.6^d \pm 0.6^d$ (retrograde)	$23^h 56.07^m$	$24^h 37.38^m$
Escape velocity $(2\mu/R)^{1/2}$, ft/sec	13,600	33.500	36,675	16,500

means of a vector differential equation of second order. One way of stating Newton's law of universal gravitation is as follows:

One particle attracts another with a force that is directed along a line connecting them and is proportional to the product of their masses and inversely proportional to the square of the distance separating them.

Thus, for a system comprised of n particles wherein only gravitational forces are acting, the motion of the ith particle with respect to an inertial frame of reference is governed by the equation

$$m_i \frac{d^2 \mathbf{r}_i}{dt^2} = \sum_{\substack{j=1 \\ j \neq i}}^{n} G \frac{m_i m_j}{r_{ij}^3} (\mathbf{r}_j - \mathbf{r}_i) \qquad (7.1)$$

where the proportionality factor G is the universal gravitation constant, m_i is the mass of the ith particle, $\mathbf{r}_i$ is the position vector of the ith particle, and r_{ij} is the distance between the ith and jth particles, $r_{ij} = |\mathbf{r}_j - \mathbf{r}_i|$. Since the motion of the ith particle depends on the relative positions of all n particles, it is necessary, in general, to write down a set of n equations, each having the same form as Eq. (7.1) and to solve them simultaneously in order to describe the behavior of the system of particles or of any one particle. Only in the case of two particles is it possible to obtain analytically a complete solution to this problem. For the case of three particles a useful first integral of the motion, the Jacobi integral, has been derived.[1,2] For the general case of n particles certain first integrals are also available and are found in Refs. 2 and 4. Some physical and orbital parameters for bodies in the solar system are listed in Tables 7.1–7.3.

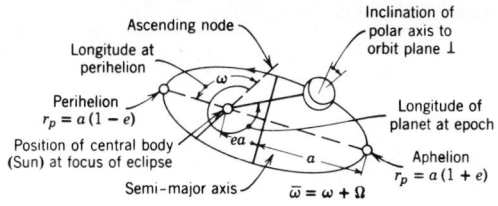

$2\!\!\!\!\downarrow$ Jupiter	$\hbar$ Saturn	$\overset{\star}{\delta}$ Uranus	Ψ Neptune	P Pluto
5.202694	9.538836	19.25290	30.04427	39.64092
4.951857	9.012979	18.285114	29.765219	29.719670
5.453531	10.064693	20.220686	30.323321	49.562170
0.048213	0.055128	0.050267	0.009288	0.250278
0.438411	0.323782	0.2283249	0.1823988	0.1590757
13.05	9.64	6.797	5.43	4.73
7.047	5.20	3.6705	2.93	2.55
42,817.6	31,595.2	22,302.0	17,802.7	15,560
299.128	120.455	42.235	21.532	14.283
11.86	29.46	84.0	164.8	247.7
1.30601	2.48875	0.77250	1.77310	17.12631
3.067	26.733	97.884	28.8	
100.1176	113.4342	73.9208	131.3902	109.7729
13.5550	92.3756	169.1552	56.2011	222.7818
108.0054	277.3342	3.3525	176.5732	328.0351
$9^h 50^m$	$10^h 14^m$	$10^h 49^m$	$15^h 48^m$	6^d
197,500	119,500	72,500	82,400	31,300

7.1.1 Gravitational Potentials

Although the bodies of the solar system can usually be treated as mass particles, there are important situations, such as a satellite orbiting in close proximity to a planet, in which the finite size of one body has a significant influence on the motion of another. When the finite dimensions of one body must be considered, it is useful to introduce the concept of a potential function, the gradient of which gives the attractive force. For example, the gravitational potential produced by the distribution of the earth's mass can be expressed in terms of Legendre polynomials as follows:

$$V(r, \phi) = \frac{Gm_E}{r}\left[1 - \sum_{k=2}^{\infty} J_k \left(\frac{r_E}{r}\right)^k P_k (\cos \phi)\right] \tag{7.2}$$

where the spherical coordinates r and ϕ are the radius and latitude, respectively, of the point at which the potential is to be computed, r_E is the earth's radius, m_E is the mass of the earth, and the J_k are coefficients of the series expansion that must be determined experimentally. Typical values are tabulated in Table 7.4.

Rodrigues's formula can be used for the explicit determination of the Legendre polynomials,

$$P_k (x) = \frac{1}{2^k k!} \frac{d^k (x^2 - 1)^k}{dx^k} \tag{7.3}$$

Equation (7.2) is based on the assumption of symmetry about the polar axis. Contributions to the potential from terms in the summation are small and decay rapidly with increasing distance from the earth. Therefore, when $r \gg r_E$ the sum makes a negligible contribution to the potential and the earth then attracts essentially like a point mass.

TABLE 7.3 Physical Characteristics of the Planets[3] [a]

Symbol		☿	♀	⊕	♂
Planet		Mercury	Venus	Earth	Mars
No. of Natural Satellites		0	0	1	2
Apparent Equatorial Angular Diameter, sec		4.6–12.7	9.9–64.5		3.5–25.1
Equatorial radius (☉ = 1)		0.00346	0.00873	0.00915	0.00488
	(⊕ = 1)	0.379	0.956	1.00	0.535
	km	2.42^3	6.05^3	6.37816^3	3.41^3
	miles	1.50^3	3.78^3	3.96320^3	2.12^3
	NM	1.31^3	3.26^3	3.44393^3	1.84^3
Oblateness f		0	0	1/298.3	1/192
J_2		0	0	1.082^{-3}	1.92^{-3}
Volume (⊕ = 1)		0.054	0.87	1.00	0.153
Mean density (☉ = 1)		4.06	3.68	3.94	2.78
	(⊕ = 1)	1.03	0.952	1.00	0.705
	g/cm^3	5.7	5.25	5.52	3.89
	lb/ft^3	355	329	344.6	243
Mass (☉ = 1)		1.64^{-7}	2.4477^{-6}	3.04039^{-6}	3.236^{-7}
(including) (⊕ = 1)		0.0546	0.81485	1.01230	0.1077
satellites					
$GM = \mu = g_0 R^2$					
$\mu = \dfrac{aV_e^2}{2} = $ $\begin{cases} km^3/sec^2 \\ mi^3/sec^2 \\ ft^3/sec^2 \\ AU^3/day^2 \end{cases}$		2.18^4 5.23^3 7.70^{14} 4.85^{-11}	3.2485^5 7.7936^4 1.1472^{16} 7.2430^{-10}	3.98604^5 9.56302^4 1.40766^{16} 8.88757^{-10}	4.293^4 1.0299^4 1.516^{15} 9.576^{-11}
Equatorial Surface (⊕ = 1)		0.380	0.893	1.00	0.377
Gravity g_e (cm/sec²)		372	873	978.031	369
Albedo		0.06	0.76	0.36	0.15
Maximum surface temperature, °F		750	210	140	90

[a] Superscripts denote exponents of 10; e.g., $4.6^{-8} = 4.6 \times 10^{-8}$. $G = 6.6695 \times 10^{-8}$ (cgs units) = universal gravitational constant.

If a completely general representation of the earth's gravitational potential is required, then the assumption of axial symmetry must be removed and Eq. (7.2) replaced by

$$V(r, \phi, \theta) = \frac{Gm_E}{r} + \sum_{k=1}^{\infty} \frac{A_k}{r^{k+1}} P_k (\cos \phi) + \sum_{k=1}^{\infty} \sum_{j=1}^{k} \frac{B_k^j}{r^{k+1}} P_k^j (\cos \phi) \cos j\theta$$

$$+ \sum_{k=1}^{\infty} \sum_{j=1}^{k} \frac{C_k^j}{r^{k+1}} P_k^j (\cos \phi) \sin j\theta \tag{7.4}$$

where the constant coefficients can be evaluated from the expressions

$$A_k = G \iiint \rho^{k+2} D(\rho, \beta, \lambda) P_k (\cos \phi) \sin \beta \, d\rho \, d\beta \, d\lambda \tag{7.5}$$

$$B_k^j = 2G \frac{(k-j)!}{(k+j)!} \iiint \rho^{k+2} (\rho, \beta, \lambda) P_k^j (\cos \beta) \cos j\lambda \sin \beta \, d\rho \, d\beta \, d\lambda \tag{7.6}$$

$$C_k^j = 2G \frac{(k-j)!}{(k+j)!} \iiint \rho^{k+2} D(\rho, \beta, \lambda) P_k^j (\cos \beta) \sin j\lambda \sin \beta \, d\rho \, d\beta \, d\lambda \tag{7.7}$$

$$m_E = \iiint D(\rho, \beta, \lambda) \rho^2 \sin \beta \, d\rho \, d\beta \, d\lambda \tag{7.8}$$

♃ Jupiter	♄ Saturn	⛢ Uranus	Ψ Neptune	P Pluto
1	9	5	2	
30.8–50.0	14.9–20.6	3.4–4.2	2.2–2.4	0.4–0.6
0.102	0.0865	0.0337	0.0320	0.010
11.14	9.47	3.69	3.50	1.1
7.14^4	6.04^4	2.35^4	2.23^4	7.3
4.43^4	3.75^4	1.46^4	1.39^4	4.3
3.85^4	3.26^4	1.27^4	1.21^4	4.2
1/16.1	1/10.4	1/16	1/50	0
1.47^{-2}	0.67^{-2}	1.5^{-2}	$5.^{-3}$	0
1400	850	50.	43	1.3
0.89	0.44	1.14	1.6	2.4
0.227	0.112	0.290	0.40	0.6
1.25	0.62	1.60	2.2	3.3
78	39	100	138	200
9.5475^{-4}	2.857^{-4}	4.360^{-5}	$5.^{-5}$	$(2.5 \pm 0.3)^{-6}$
317.89	95.12	14.52	17	0.8 ± 0.1
1.2671^8	3.792^7	5.788^6	6.8^6	3.2^5
3.0399	9.098^6	1.388^6	1.6^6	7.7^4
4.4747^{18}	1.339^{18}	2.044^{17}	2.4^{17}	1.1^{16}
2.8252^{-7}	8.454^{-8}	1.290^{-8}	1.5^{-8}	7.4^{-10}
2.54	1.06	1.07	1.4	0.7
2490	1040	1050	1400	700
0.51	0.50	0.66	0.62	0.16
-200	-240	-270	-330	-370

TABLE 7.4　J_k, Coefficients of the Potential Function ($\times 10^6$)

$J_2 = 1{,}082.28 \pm 0.03$	$J_5 = -0.2 \pm 0.1$
$J_3 = -2.3 \pm 0.2$	$J_6 = 1.0 \pm 0.8$
$J_4 = -2.12 \pm 0.05$	

where $D(\rho, \beta, \lambda)$ is the density distribution of matter, and r, ϕ, and θ are spherical coordinates. Due to current emphasis on near earth satellites for communications, weather reconnaissance, and so on, generalized representations such as Eq. (7.4) have received much attention.[2,4,5] Considerable effort has been expended on the experimental determination of the coefficients in the leading terms of the expansion and on the determination of suitable approximate models of gravitational potentials for use in analytic studies.

7.1.2　Two-Body Problem

When, as is often the case, only the relative motion of a two-particle system is of interest, a general analytical solution can be derived. Since such a solution requires six integrations, having ten known integrals makes this possible. By writing down Eq. (7.9) for each particle and letting $\mathbf{r} = \mathbf{r}_2 - \mathbf{r}_1$, an equation of relative motion is obtained:

$$\frac{d^2\mathbf{r}}{dt^2} + \frac{\mu}{r^3}\mathbf{r} = 0 \tag{7.9}$$

where $\mu = G(m_1 + m_2)$. For the two-body problem to adequately represent a system, it must satisfy two assumptions. The bodies are assumed to be spherically symmetric, thereby becoming point masses, and no system forces are acting other than the gravitational forces. A first integral of the motion results from crossing $\mathbf{r}$ with Eq. (7.9) and integrating to find that

$$\mathbf{h} = \mathbf{r} \times \frac{d\mathbf{r}}{dt} = \text{const.} \tag{7.10}$$

where $\mathbf{h}$ is angular momentum per unit mass and, since it is constant, the motion is planar. Another integral is obtained by taking the cross product of $\mathbf{h}$ with Eq. (7.9) and integrating,

$$\frac{d\mathbf{r}}{dt} \times \mathbf{h} = \frac{\mu}{r}(\mathbf{r} + r\mathbf{e}) \tag{7.11}$$

where $\mathbf{e}$ is the eccentricity vector and is a constant. When the scalar product between $\mathbf{r}$ and Eq. (7.11) is taken, the result is

$$r = \frac{h^2/\mu}{1 + e \cos \nu} \tag{7.12}$$

where ν is the angle between $\mathbf{e}$ and $\mathbf{r}$ and is called the true anomaly. This is the equation of a conic whose semilatus rectum or parameter p is given by h^2/μ and whose classification depends on the magnitude of $\mathbf{e}$:

Circle: $e = 0$

Ellipse: $0 < e < 1$

From Eq. (7.12), it is apparent that r is a minimum when $\nu = 0$ and, therefore, the eccentricity vector points in the direction of the periapsis, that is, where the two bodies come closest to each other during their orbits. It also can be shown that each of the two bodies describes a conical path with respect to a fixed origin as well as describing a conical path with respect to each other.

Since $\mathbf{h} \cdot \mathbf{e} = 0$, $\mathbf{h}$ and $\mathbf{e}$ provide only five independent constants of integration, and these specify the size and shape of the conic, its orientation in the plane of motion, and the orientation in the plane. A sixth constant of integration is obtained when determining the time dependency of the relative motion. Time and position can be related by recognizing that the angular momentum per unit mass can be expressed in the form

$$h = r^2 \frac{d\nu}{dt} \tag{7.13}$$

and thus that r can be eliminated between Eq. (7.12) and (7.13) to give

$$\sqrt{\frac{\mu}{p^3}}\, dt = \frac{d\nu}{(1 + e \cos \nu)^2} \tag{7.14}$$

Integrating this expression yields the final constant, which is often taken to be the time when $\nu = 0$ and the bodies are nearest each other.

7.1.3   Orbital Geometry

As discussed in Section 7.1.2, the solution to the two-body problem results in the equation of a conic section [Eq. (7.12)]. This means that all conic sections (hyperbola, circle, parabola, or ellipse) make physical sense as orbital paths. Indeed, for an object to escape the influence of the earth, it must follow a parabolic or hyperbolic path.

Characteristics of an Ellipse, Parabola, and Hyperbola
Since satellite orbital geometry has its roots in planetary astronomy, the terms and definitions that were in use hundreds of years ago are still common (See Figs. 7.1–7.3).

r = magnitude of the position vector
F, F' = primary and vacant focal points
r_p = radius at the periapsis point
r_a = radius at the apoapsis point
a = semimajor axis
$2c$ = distance between the foci
ν = true anomaly (measured in the direction of satellite motion)
ϕ = flight path angle (measured from local horizontal to velocity)
p = semilatus rectum

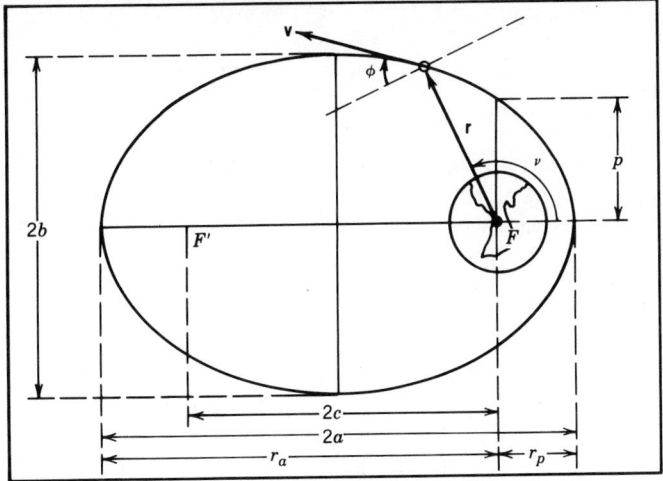

Fig. 7.1 Geometry of an ellipse.

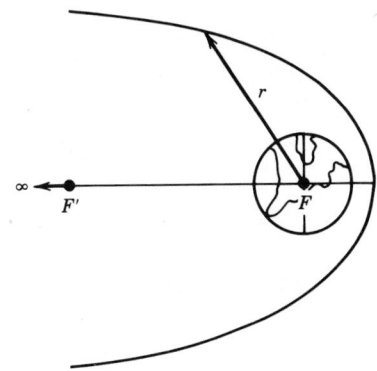

Fig. 7.2 Geometry of a parabola.

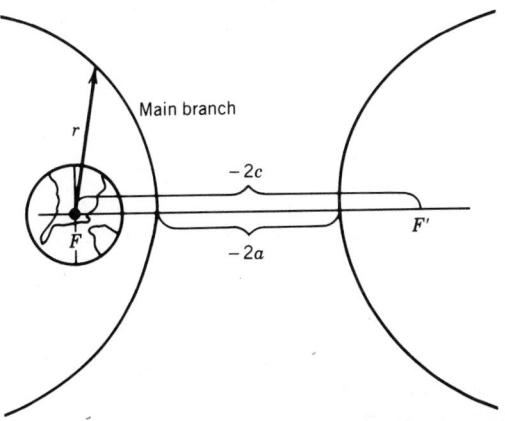

Fig. 7.3 Geometry of a hyperbola.

For a parabola

$$2a = \infty$$
$$2c = \infty$$
$$r_a = \infty$$
$$e = 1$$

For a hyperbola, a and c are negative numbers by convention and $e > 1$.

Useful Two-Body Orbit Equations

Several useful two-body equations can be developed by rearranging Eqs. (7.10)–(7.14). These equations are

$$r_a = a(1 + e)$$
$$r_p = a(1 - e)$$
$$p = a(1 - e^2) = \frac{h^2}{\mu}$$
$$\mathbf{h} = \mathbf{r} \times \mathbf{v} \qquad h = rv \cos \phi$$
$$\mathbf{n} = \mathbf{k} \times \mathbf{h}$$
$$\text{Orbit period} = 2\pi \sqrt{\frac{a^3}{\mu}}$$
$$v = \sqrt{2\left(\varepsilon + \frac{\mu}{r} \right)} \qquad \varepsilon = \frac{-\mu}{2a}$$
$$v_{\text{cir}} = \sqrt{\frac{\mu}{r}} \qquad v_{\text{esc}} = \sqrt{\frac{2\mu}{r}}$$
$$\mathbf{e} = \left(\frac{v^2}{\mu} - \frac{1}{r} \right)\mathbf{r} - \left(\frac{\mathbf{r} \cdot \mathbf{v}}{\mu} \right)\mathbf{v}$$
$$e = \frac{2c}{2a} = \frac{r_a - r_p}{r_a + r_p}$$

7.1.4　Orbital Elements

It is common practice to refer to the six constants of integration of two-body motion as orbital elements. Since two-body motion is governed by Eq. (7.9) these six constants are $\mathbf{r}(t_0)$ and $d\mathbf{r}(t_0)/dt$, where t_0 is some reference time or epoch. Historically, however, different sets of constants related to $\mathbf{r}_0$ and $\mathbf{v}_0$ and frequently having some simple geometric interpretation are selected as orbital elements.

A classical choice for such a set of orbital elements is illustrated in Fig. 7.4. Three Euler angles, the inclination i, the longitude of the ascending node Ω, and the argument of periapsis ω, are used to specify the spatial orientation of the orbit. Size and shape of the orbit are fixed by the semimajor axis a and either the semilatus rectum p or the eccentricity e. Position of the body in the orbit at epoch is established by ν_0, the true anomaly at the reference time t_0. The relationship between this set of elements and the constants $\mathbf{r}_0$ and $\mathbf{v}_0$ is readily obtained from the geometry of Fig. 7.4. The four angles i, Ω, ω, and ν_0 may be defined by taking the scalar product between the appropriate pair of vectors.

Inclination, i, is given by

$$\cos i = \frac{\mathbf{h} \cdot \mathbf{k}}{h} \qquad (7.15)$$

where $\mathbf{h} = \mathbf{r}_0 \times \mathbf{v}_0$ and the inclination is always less than $180°$.

Longitude of the ascending node Ω is given by

$$\cos \Omega = \frac{\mathbf{n} \cdot \mathbf{i}}{n} \qquad (7.16)$$

where $\mathbf{n} = \mathbf{k} \times \mathbf{h}$ is a vector in the direction of the ascending node and defines the so-called line of apsides. A quadrant rule for Ω is: if $n_j > 0$, then $\Omega < 180°$.

Argument of periapsis ω is given by

$$\cos \omega = \frac{\mathbf{n} \cdot \mathbf{e}}{ne} \qquad (7.17)$$

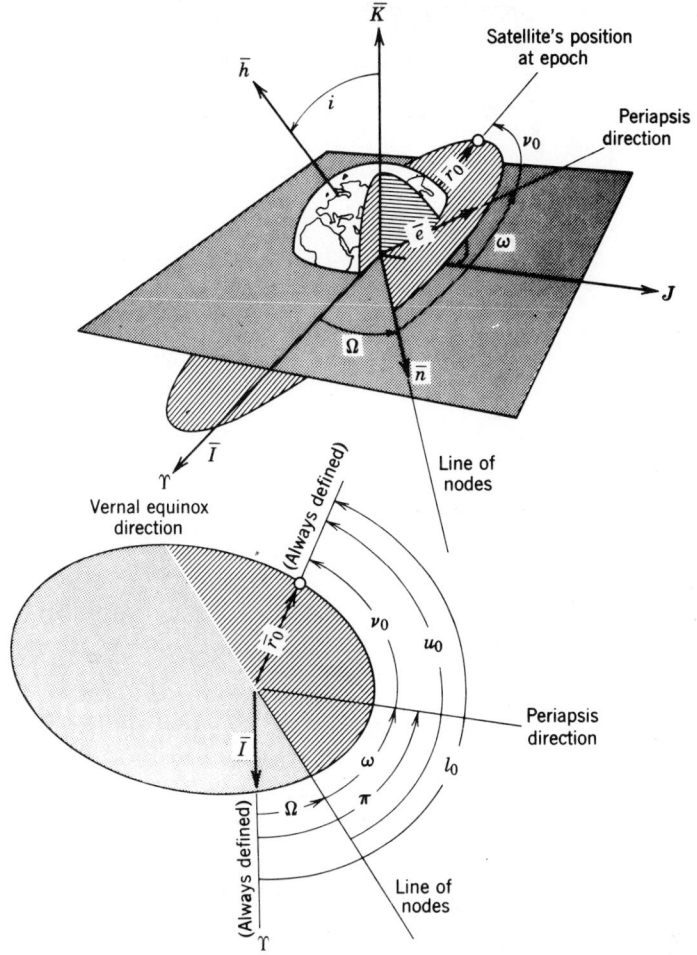

Fig. 7.4 Classical orbital elements.

where the eccentricity vector is given by $\mathbf{e} = (1/\mu)\{[v^2 - (\mu/r)]\mathbf{r} - (\mathbf{r} \cdot \mathbf{v})\mathbf{v}\}$ and the quadrant rule for ω is: if $e_k > 0$, then $\omega < 180°$.

True anomaly at epoch, ν_0, is given by

$$\cos \nu_0 = \frac{\mathbf{r}_0 \cdot \mathbf{e}}{r_0 e} \tag{7.18}$$

where the quadrant rule for ν_0 is: if $\mathbf{r} \cdot \mathbf{v} > 0$, then $\nu < 180°$.

This classical set of orbital elements suffers from the fact that one or more of the elements may not be defined by these equations for certain orbits. In these cases it is customary to introduce other elements. When Ω and ω are defined, the longitude of periapsis π is equal to their sum, $\pi = \Omega + \omega$, but when Ω is not defined (i.e., $i = 0$) the angle π is given by

$$\cos \pi = \frac{\mathbf{i} \cdot \mathbf{e}}{e} \tag{7.19}$$

and the quadrant rule is: if $e_j > 0$, then $\pi < 180°$. When ω and ν are defined, the argument of latitude u_0 is equal to their sum, $u_0 = \omega + \nu_0$, but when ω is not defined (i.e., circular orbits for which $e = 0$), the angle u_0 is given by

$$\cos u_0 = \frac{\mathbf{n} \cdot \mathbf{r}_0}{n r_0} \tag{7.20}$$

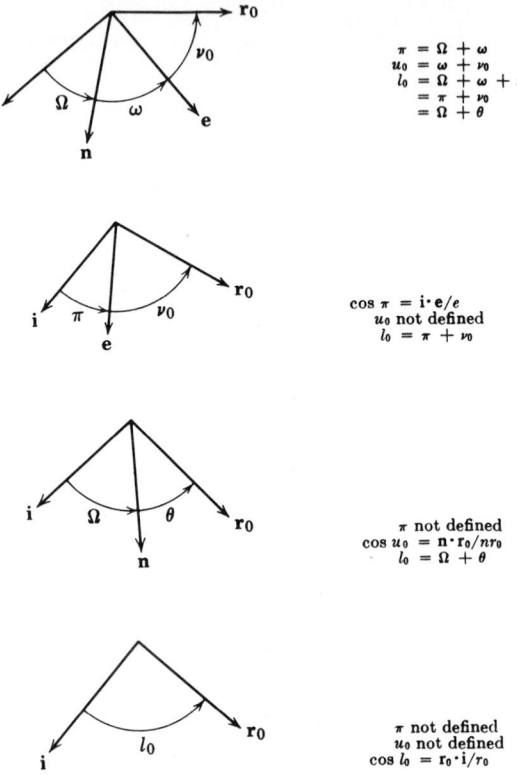

$$\pi = \Omega + \omega$$
$$u_0 = \omega + \nu_0$$
$$l_0 = \Omega + \omega + \nu_0$$
$$= \pi + \nu_0$$
$$= \Omega + \theta$$

$$\cos \pi = \mathbf{i} \cdot \mathbf{e}/e$$
u_0 not defined
$$l_0 = \pi + \nu_0$$

π not defined
$$\cos u_0 = \mathbf{n} \cdot \mathbf{r}_0/nr_0$$
$$l_0 = \Omega + \theta$$

π not defined
u_0 not defined
$$\cos l_0 = \mathbf{r}_0 \cdot \mathbf{i}/r_0$$

Fig. 7.5 Classical orbital elements—special cases.

and the quadrant rule is: if $r_k > 0$, then $u_0 < 180°$. When Ω, ω, and ν_0 are defined the true longitude at epoch, l_0 is equal to their sum, $l_0 = \Omega + \omega + \nu_0$, but when Ω is not defined, $l_0 = \pi + \nu_0$. If $e = 0$, ω is not defined, nor is π. In this case, $l_0 = \Omega + u_0$. When $\mathbf{n}$ and $\mathbf{e}$ are both zero, the true longitude at epoch is given by

$$\cos l_0 = \frac{\mathbf{r}_0 \cdot \mathbf{i}}{r_0} \tag{7.21}$$

where the quadrant rule is: if $r_j > 0$, then $l_0 < 180°$. A summary of the four possible cases and the definitions that apply in each situation is given in Fig. 7.5.

From Eq. (7.12) it is already known that the parameter p is given by

$$p = \frac{h^2}{\mu} \tag{7.22}$$

and therefore, since $\mathbf{h} = \mathbf{r}_0 \times \mathbf{v}_0$, the connection between this orbital element and the constants of integration $\mathbf{r}_0$ and $\mathbf{v}_0$ is clear. Equation (7.22) may be interpreted as saying that the angular momentum determines the shape of the orbit.

7.1.5 Kepler's Laws

Since the time of Aristotle, who taught that circular motion was the only perfect and natural motion and that the heavenly bodies, therefore, necessarily moved in circles, the planets were assumed to revolve in circular paths or combinations of smaller circles moving on larger ones (epicycles). However, early in the seventeenth century Johann Kepler, working with the accurate observations of Tycho Brahe, found immense difficulty in reconciling any such theory with the actual data on celestial motion. From 1601 until 1609 he tried fitting various geometrical curves to Brahe's measurements on the motion of Mars. Finally, after struggling for almost a year to remove a discrepancy of only 8 minutes of arc, Kepler hit upon the ellipse as a possible solution. It fit. The orbit was found and, in 1609, Kepler published his first laws of planetary motion. The third law followed in 1619. These laws, which mark an epoch in the history of mathematical science, are as follows:

Kepler's Laws

First Law. The orbit of each planet is an ellipse, with the sun as a focus.

Second Law. The line joining the planet to the sun sweeps out equal areas in equal times.

Third Law. The square of the period of a planet is proportional to the cube of its mean distance from the sun.

Kepler's laws provide only a description, not an explanation, of planetary motion. Each of these laws, however, may be derived analytically from the equation of motion for the two-body problem. Therefore, Kepler's laws, based on the observational data of Brahe, represent an experimental verification of Newton's laws of motion and his universal law of gravitation, which did not appear until 1687. This, then, is their historical significance in the development of orbital mechanics. A final connection between the geometry of the orbit and the constants of integration can be established through the vis viva integral

$$\varepsilon = \frac{v_0^2}{2} - \frac{\mu}{r_0} \qquad (7.23)$$

which is simply a statement of the conservation of energy. Since it can be shown that

$$\varepsilon = -\frac{\mu}{2a} \qquad (7.24)$$

it is clear that the size of the orbit as specified by a is determined by the energy as specified by r_0 and v_0. A useful relation between size and shape parameters, easily derived from Eq. (7.12), is given by

$$p = a(1 - e^2) \qquad (7.25)$$

7.1.6 Classical Orbit Position Prediction

Kepler's first two laws of planetary motion were originally published in *Astronomia Nova* in 1609. His results showed that the planets did not move with uniform speed, but faster or slower according to their distance from the sun. Later Kepler was able to write an empirical expression for the time of flight of a planet from one point in its orbit to another—although he still did not know the true reason why it should move in an orbit at all.

Time of Flight on an Orbit
Using the geometry of a circle circumscribed about an elliptical orbit, as shown in Figure 7.6, where a is the semimajor axis, v is the true anomaly, e is the eccentricity, and E is the eccentric anomaly, Kepler geometrically solved for the time of flight to point P on the ellipse as

$$(t - T) = \sqrt{\frac{a^3}{\mu}} (E - e \sin E) \qquad (7.26)$$

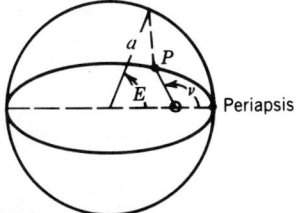

Fig. 7.6 Geometry of a circle circumscribed about an elliptical orbit.

where T is the time of periapsis passage. He also introduced the definition

$$M = E - e \sin E \tag{7.27}$$

where M is called the "mean anomaly." If you also use the definition

$$n = \sqrt{\frac{\mu}{a^3}} \tag{7.28}$$

where n is called the "mean motion," then the mean anomaly may be written as

$$M = n(t - T) = E - e \sin E \tag{7.29}$$

which is often referred to as Kepler's equation. A useful equation relating ν to E is

$$\cos E = \frac{e + \cos \nu}{1 + e \cos \nu} \tag{7.30}$$

or alternately

$$\cos \nu = \frac{e - \cos E}{e \cos E - 1} \tag{7.31}$$

In general, time of flight between any two arbitrary points on an ellipse can be expressed as

$$(t - t_0) = \sqrt{\frac{a^3}{\mu}} \left[2\pi K + (E - e \sin E) - (E_0 - e \sin E_0) \right] \tag{7.32}$$

where K is the number of times the object passes through periapsis en route from t_0 to t. Similarly, the time of flight along a parabola can be expressed as

$$(t - t_0) = \frac{1}{2\sqrt{\mu}} \left[\left(pD + \frac{D^3}{3} \right) - \left(pD_0 + \frac{D_0^3}{3} \right) \right] \tag{7.33}$$

where $D = \sqrt{p} \tan(\nu/2)$ called the "parabolic eccentric anomaly" and the time of flight along a hyperbola can be expressed as

$$(t - t_0) = \sqrt{\frac{-a^3}{\mu}} \left[(e \sinh F - F) - (e \sinh F_0 - F_0) \right] \tag{7.34}$$

where $\cosh F = (e + \cos \nu)/(1 + e \cos \nu)$. Whenever ν is between 0 and π, F should be taken as positive and whenever ν is between π and 2π, F should be taken as negative.

Position Prediction

The problem of predicting a new position (**r**) and velocity (**v**), or a new set of classical orbital elements at a future time given some initial conditions, becomes a solution of Kepler's equation

$$M = E - e \sin E$$

for the eccentric anomaly (E). M can be propagated forward in time using

$$M = n(t - t_0) - 2K\pi + M_0 \tag{7.35}$$

E must now be found using some kind of iteration. A Newton iteration scheme works well in this case

$$E_{n+1} = E_n + \frac{M - (E_n - e \sin E_n)}{1 - e \cos E_n} \tag{7.36}$$

when the difference $(E_{n+1} - E_n)$ becomes acceptably small, iteration can cease. However, since the slope of the M versus E curve approaches zero at $E = 0$ or $E = 2\pi$ when e is close to 1, this method has convergence difficulties for near parabolic orbits. See Fig. 7.7.

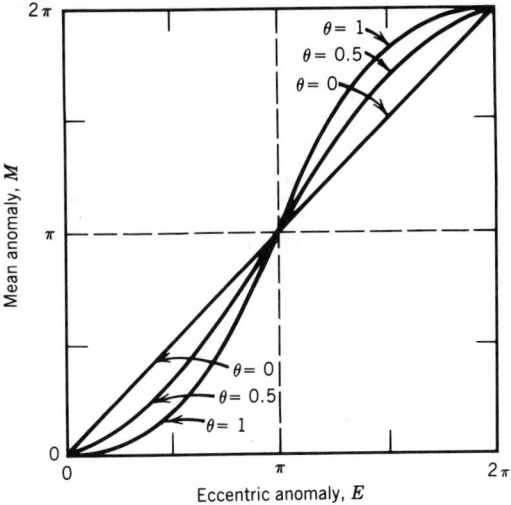

Fig. 7.7 Eccentric versus mean anomaly plot.

7.1.7 Orbital Maneuvering

To achieve mission goals and objectives, it is often necessary to change spacecraft's orbits with maneuvers. The spacecraft orbit size, shape, or orientation relative to the earth can be adjusted by changing velocity. Each time an orbital maneuver is performed, some orbital elements are changed. This section examines fundamental orbital maneuvers and discusses the following:

Velocity calculation
Hohmann transfer
Bielliptical transfer
Fast transfer
Plane changes
Rendezvous
Phasing

The velocity of a satellite is a function of E, the specific mechanical energy of the orbit, and r, the satellite's distance from the center of the earth.

The velocity change, when performing a maneuver, is the difference between the current velocity and the desired velocity:

$$\Delta V = V_{des} - V_{cur} \tag{7.37}$$

To identify orbits during maneuvers (see Fig. 7.8), the initial orbit is labeled 1, the transfer orbit is

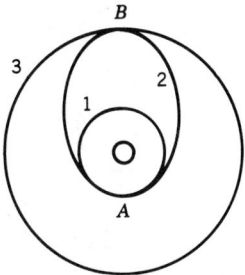

Fig. 7.8 Orbit nomenclature.

labeled 2, and the final orbit is labeled 3. The places where orbital maneuvers occur are lettered alphabetically, starting with A.

Velocity Calculation

The basic requirement in orbital maneuvering is to calculate velocity, both present and required. The following equations are used to determine velocity:

$$\varepsilon = \frac{-\mu}{2a} \tag{7.38}$$

$$r = \frac{a(1 - e^2)}{1 + e \cos \nu} \tag{7.39}$$

$$v = \sqrt{2\left(\varepsilon + \frac{\mu}{r}\right)} \tag{7.40}$$

If an orbit is circular, the velocity is constant:

$$v_{\text{circ}} = \sqrt{\frac{\mu}{r}} \tag{7.41}$$

The Hohmann Transfer

For typical low-earth missions, the most energy-efficient, but longest, direct method of transferring between two coplanar orbits is the Hohmann transfer (Fig. 7.9). The Hohmann transfer requires two ΔV's applied at perigee and apogee of the transfer ellipse.

The following algorithm describes the Hohmann transfer problem.

1. Calculate orbit energies:

$$\varepsilon_1 = \frac{-\mu}{2a_1} \qquad \varepsilon_2 = \frac{-\mu}{r_1 + r_3} \qquad \varepsilon_3 = \frac{-\mu}{2a_3}$$

2. Calculate parking orbit velocity at first maneuver point:

$$V_{1A} = \sqrt{2\left(\varepsilon_1 + \frac{\mu}{r_A}\right)}$$

3. Calculate transfer orbit velocity at perigee:

$$V_{2A} = \sqrt{2\left(\varepsilon_2 + \frac{\mu}{r_A}\right)}$$

4. Calculate change in velocity at first burn:

$$\Delta V_A = V_{2A} - V_{1A}$$

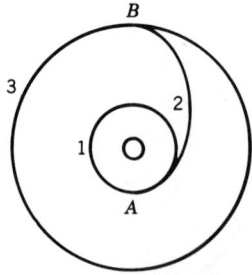

Fig. 7.9 Hohmann transfer.

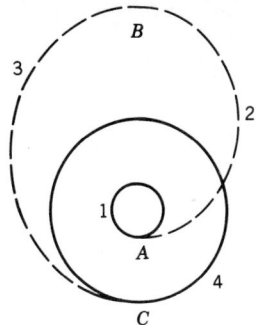

Fig. 7.10 Bielliptic transfer.

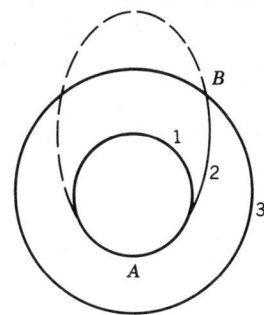

Fig. 7.11 Fast orbit transfer.

5. Calculate transfer orbit velocity at apogee:

$$V_{2B} = \sqrt{2\left(\varepsilon_2 + \frac{\mu}{r_B}\right)}$$

6. Calculate final orbit velocity at second maneuver point:

$$V_{3B} = \sqrt{2\left(\varepsilon_3 + \frac{\mu}{r_B}\right)}$$

7. Calculate change in velocity at second burn:

$$\Delta V_B = V_{3B} - V_{2B}$$

8. Calculate time of flight (TOF) of transfer:

$$\text{TOF} = \pi\sqrt{\frac{a_2^3}{\mu}}$$

The Bielliptic Transfer

The Hohmann transfer is the minimum energy transfer if the ratio of the final orbit radius to the parking orbit radius is less than 11.94 ($r_3/r_1 < 11.94$). For cases where the ratio of final to parking orbit radius is greater than 15.58, the two ellipse, or bielliptic transfer (see Fig. 7.10), is the most fuel-efficient transfer, yet requires three burns.

The bielliptic transfer algorithm is developed using the same velocity equations previously presented. However, various candidate radii must be examined for ΔV_B to find the optimum, most fuel-efficient, transfer orbit. When $r_{\text{final}}/r_{\text{park}} > 15.58$, the bielliptic ΔV is always less than the Hohmann ΔV, no matter what r_B is selected. For the case of $11.94 < r < 15.58$, this is not true. For some r_B values, the Hohmann transfer is cheaper while for other, greater values of r_B, the bielliptic transfer is cheaper. The time of flight of the bielliptic transfer is the sum of half the periods of transfer orbits 2 and 3.

Fast Transfer

For a transfer between two coplanar circular orbits accomplished in a shorter time than the Hohmann transfer, the fast transfer can be used (see Fig. 7.11). The first burn is a tangent ΔV with a magnitude greater than the ΔV to initiate the Hohmann transfer.

With the greater orbital velocity, the intercept occurs in less than 180° of travel. The angle between the position vector at transfer initiation and interception is ν_{2B}. For the fact transfer at final orbit intercept, the magnitude and direction of the velocity are changed.

1. Calculate parking orbit velocity:

$$V_{1A} = \sqrt{\frac{\mu}{r_A}}$$

2. Calculate a and e of fast transfer orbit:

$$e_2 = \frac{r_1/r_3}{\cos \nu_{2B} - r_1/r_3} \qquad a_2 = \frac{r_1}{1 - e_2}$$

3. Calculate transfer orbit energy:

$$\varepsilon_2 = \frac{-\mu}{2a_2}$$

4. Calculate velocity to initiate transfer:

$$V_{2A} = \sqrt{2\left(\varepsilon_2 + \frac{\mu}{r_A}\right)}$$

5. Determine change in velocity at first burn:

$$\Delta V_A = V_{2A} - V_{1A}$$

6. Calculate velocity to terminate transfer: The transfer orbit velocity magnitude and direction must be changed to match the final orbit at the intercept point. To calculate ΔV_B, the true anomaly at intercept, ν_{2B}, must be found and then ϕ_{2B} and V_{2B} are calculated:

$$\phi_{2B} = \tan^{-1}\left[\frac{e_2 \sin \nu_{2B}}{1 + e_2 \cos \nu_{2B}}\right]$$

$$\nu_{2B} = \frac{a_2(1 - e_2)}{1 + e \cos \nu_{2B}}$$

$$V_{2B} = \sqrt{2\left(\varepsilon_2 + \frac{\mu}{r_B}\right)}$$

7. Calculate change in velocity at second burn: Notice V_{3B} is the velocity of the final circular orbit. Therefore, ϕ_{3B} is $0°$.

$$\Delta V_B = \sqrt{V_{3B}^2 + V_{2B}^2 - 2V_{3B}V_{2B}\cos(\phi_{3B} - \phi_{2B})}$$

8. Calculate transfer flight time:

$$\text{TOF} = \sqrt{\frac{a_2^3}{\mu}}\left[(E_B - e_2 \sin E_B) - (E_A - e_2 \sin E_A)\right]$$

where E_B is the eccentric anomaly at interception and E_A is the eccentric anomaly at transfer initiation:

$$E_B = \cos^{-1}\left[\frac{e_2 + \cos \nu_{2B}}{1 + e_2 \cos \nu_{2B}}\right]$$

Orbital Plane Changes
A simple plane change (Fig. 7.12) is a maneuver performed at a node to change only orbit inclination. For circular orbits, to change the inclination a given amount, Δi, use the following equation:

$$\Delta V_s = 2V \sin \frac{\Delta i}{2}$$

The s denotes simple plane change and V is the current velocity in the circular orbit. For noncircular orbits, the flight path angle, ϕ, must also be considered:

$$\Delta V_s = 2V \cos \phi \sin \frac{\Delta i}{2}$$

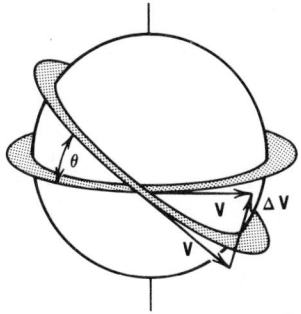

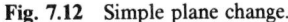

Fig. 7.12 Simple plane change.

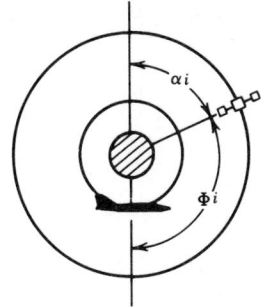

Fig. 7.13 Initial chaser/target configuration.

A combined plane change changes orbit plane and size simultaneously. This maneuver must be made at a node. Considering circular orbits, with subscript c denoting current and subscript d denoting desired velocities,

$$\Delta V_c = \sqrt{V_c^2 + V_d^2 - 2V_c V_d \cos \Delta i}$$

To determine a combined plane change from a noncircular orbit,

$$\Delta V_c = \sqrt{V_c^2 + V_d^2 - 2V_c V_d \left(\sin \phi_c \sin \phi_d + \cos \phi_c \cos \phi_d \cos \Delta i\right)}$$

Rendezvous
Rendezvous is the problem of a chaser vehicle joining up with a target, achieving the same orbit and arriving at the same place at the same time as the target spacecraft. Using a Hohmann transfer, the fundamental problem is to determine how much time the chaser must wait before initiating the transfer.

In order to calculate wait time, the initial chaser/target positions must be examined. Then, knowing the proper chaser/target configuration to begin rendezvous, the wait time can be calculated by determining relative angular rates and positions.

The initial chaser/target configuration is found by analyzing the relative positions of the chaser and target (Fig. 7.13). The initial angle between the target and chaser is ϕ_i. The angular rate of each spacecraft in circular orbits is its mean motion, n:

$$n = \sqrt{\frac{\mu}{r^3}}$$

The proper chaser/target configuration is found by determining the position of the target relative to the chaser so that if the chaser initiated a Hohmann transfer the two would link up after the chaser's 180° of travel in the transfer orbit. The angular distance the target moves during the transfer time of flight is the lead angle, α.

$$\alpha = (n_{\text{targ}}) \times \text{TOF}$$

The final phase angle, α_f, is when the chaser and target are in proper configuration for rendezvous:

$$\phi_f = \pi - \alpha$$

Figure 7.14 depicts the chaser/target configuration for rendezvous initiation.

The wait time is determined by dividing the relative angular positions by the relative angular closure rates:

$$\text{Wait time} = \frac{\phi_i - \phi_f + 2\pi(x)}{n_{\text{chas}} - n_{\text{targ}}}$$

The parameter, x, is a rendezvous opportunity counter. The chaser can wait an additional whole number of rendezvous opportunities.

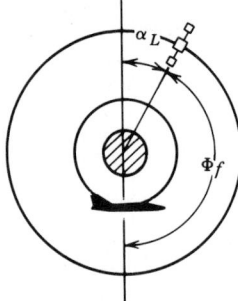

Fig. 7.14 Chaser/target configuration at rendezvous initiation.

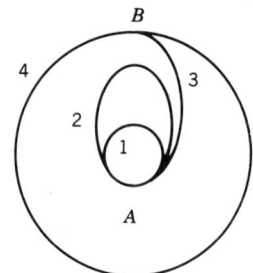

Fig. 7.15 Phasing orbit.

Phasing

Phasing is a similar problem to rendezvous, except it is for noncoplanar orbits. Therefore, the chaser is restricted to initiating a transfer from a node. If at each node the phase angle is evaluated and compared with the phase angle necessary to commence the rendezvous, the wait time may be extremely long or does not exist. One solution is to change the chaser orbit by a simple plane change to be coplanar with the target orbit. This is not the most fuel-efficient method. Another way would be to use an intermediate phasing orbit (Fig. 7.15). Orbit 2 is the phasing orbit. The period of the phasing orbit is such that when the chaser arrives back at the node, the phase angle is then proper to initiate the Hohmann transfer and begin rendezvous.

The period of the phasing orbit is found by

$$\mathbb{P}_2 = \frac{\alpha_i - \alpha_1 + 2\pi x}{n_{\text{targ}}}$$

The desired lead angle, α_1, is found by multiplying the time of flight of the transfer orbit by the mean motion of the target orbit. Since $\phi_i + \alpha_1 = \pi$ radians, α_1 can be found by evaluating the initial chaser/target configuration. When α_i is less than α_1, increment x until the phasing period is acceptable.

The ΔV necessary to enter the phasing orbit is found by determining the phasing orbit energy. The semimajor axis of the phasing orbit is found by

$$a_2 = \left| \frac{\mathbb{P}_2^2 \mu}{4\pi^2} \right|^{1/3}$$

The velocity to enter the phasing orbit, V_{2A}, is calculated using the fundamental velocity equation.

After a_2 is calculated, check to verify $a_1 < a_2 < a_3$ to ensure efficient use of the ΔV on the phasing orbit. Since the most efficient position to perform a plane change is at apogee, it is best to have the phasing orbit in the plane of the parking orbit.

After one phasing orbit the transfer is initiated and the rendezvous is completed at the next nodal crossing of the transfer orbit. A combined plane change maneuver is required to complete the rendezvous.

7.1.8 Orbit Determination

One fundamental problem in celestial mechanics is that of determining the character of an orbit from observational data on the trajectory. Depending on the nature of the observational data, the solution to this problem may follow one or more approaches, some of which have their origin back in the times of Galileo. With the advent of radar and the ability to measure both position and velocity of celestial bodies, the orbit determination problem has been considerably simplified. Since a measurement of r and v at a given time provides the six constants of integration for the motion in the two-body case, the orbit is completely determined once these quantities are known. Without this type of information, orbit determination from such observational data as angular displacements of one body relative to some "fixed" direction is more difficult.[2,4,5]

Following are four methods of orbit determination.

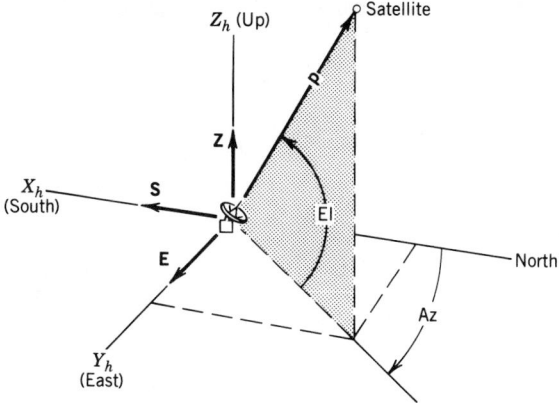

Fig. 7.16 Topocentric—horizon coordinate frame.

Complete Fix Method.[6] This method is used with a single radar sighting of slant range (ρ), azimuth (Az), and elevation (El) and the rates at which these three change. To convert this data to **r** and **v**, it is necessary to define a local coordinate frame centered at the observation site, with the orthogonal axes oriented in the south, east and, zenith directions (Fig. 7.16).

With this frame, the range vector from the site to the object is

$$\boldsymbol{\rho} = \rho_S \hat{S} + \rho_E \hat{E} + \rho_Z \hat{Z} \tag{7.42}$$

where
$$\rho_S = -\rho \cos \text{El} \cos \text{Az}$$
$$\rho_E = \rho \cos \text{El} \sin \text{Az} \tag{7.43}$$
$$\rho_Z = \rho \sin \text{El}$$

The position vector, $\mathbf{r} = \mathbf{R}_s + \boldsymbol{\rho}$, where $\mathbf{R}_s$ is the vector from the center of the earth to the origin of the observation frame. The velocity vector, **v**, is a combination of the rate at which the slant range is changing and the rotation rate of the observation frame. Therefore,

$$\mathbf{v} = \dot{\boldsymbol{\rho}} + \boldsymbol{\omega}_0 \times \mathbf{r} \tag{7.44}$$

The slant range rate in the observation frame is the rate of change of each component. Therefore,

$$\dot{\boldsymbol{\rho}} = \dot{\rho}_S \hat{S} + \dot{\rho}_E \hat{E} + \dot{\rho}_Z \hat{Z}$$

where

$$\dot{\rho}_S = -\dot{\rho} \cos \text{El} \cos \text{Az} + \rho \sin \text{El}(\dot{\text{El}}) \cos \text{Az} + \rho \cos \text{El} \sin \text{Az}(\dot{\text{Az}})$$
$$\dot{\rho}_E = \dot{\rho} \cos \text{El} \sin \text{Az} - \rho \sin \text{El}(\dot{\text{El}}) \sin \text{Az} + \rho \cos \text{El} \cos \text{Az}(\dot{\text{Az}})$$
$$\dot{\rho}_Z = \dot{\rho} \sin \text{El} + \rho \cos \text{El}(\dot{\text{El}})$$

Gibbs Method.[6] This method requires three position sightings with the angle between each sighting greater than 10°. This method can therefore be used with data obtained from several radar sites. The position vectors must be coplanar. Using the same approach as discussed in the complete fix method, radar data can be converted to position vectors. After that is accomplished, the following algorithm provides the velocity at the time of the second observation.

1. Generate test vectors:

$$\mathbf{D} = \mathbf{r}_1 \times \mathbf{r}_2 + \mathbf{r}_2 \times \mathbf{r}_3 + \mathbf{r}_3 \times \mathbf{r}_1$$
$$\mathbf{N} = r_3(\mathbf{r}_1 \times \mathbf{r}_2) + r_1(\mathbf{r}_2 \times \mathbf{r}_3) + r_2(\mathbf{r}_3 \times \mathbf{r}_1)$$
$$\mathbf{S} = (r_2 - r_3)\mathbf{r}_1 + (r_3 - r_1)\mathbf{r}_2 + (r_1 - r_2)\mathbf{r}_3$$

2. Ensure $\mathbf{D} \neq 0$, $\mathbf{N} \neq 0$, and $\mathbf{D} \cdot \mathbf{N} > 0$ to demonstrate two-body motion.

3. Check $r_1 \cdot (r_2 \times r_3) = 0$ to ensure position vectors are coplanar.
4. Calculate v_2:

$$B = D \times r_2$$

$$L = \sqrt{\frac{\mu}{DN}}$$

$$v_2 = \frac{L}{r_2} B + LS$$

Herrick–Gibbs Method.[6] This method requires three position sightings and their respective times, with the angle between each sighting less than 10°. Using a Taylor series expansion, where $\Delta t_{ij} = t_i - t_j$, the velocity at the second sighting is approximated by

$$v_2 = -\Delta t_{32} \left(\frac{1}{\Delta t_{21} \Delta t_{31}} + \frac{\mu}{12 r_1^3} \right) r_1 + (\Delta t_{32} - \Delta t_{21}) \left(\frac{1}{\Delta t_{21} \Delta t_{32}} + \frac{\mu}{12 r_2^3} \right) r_2$$

$$+ \Delta t_{21} \left(\frac{1}{\Delta t_{32} \Delta t_{31}} + \frac{\mu}{12 r_3^3} \right) r_3$$

Gauss Method.[6] This method uses two positions: the time of flight between the positions and the direction of travel. The process then calculates velocities at both positions.

The first step is to define two scalar relations, f and g where

$$f = 1 - \frac{r_2}{p} (1 - \cos \Delta \nu) \tag{7.45}$$

$$g = \frac{r_1 r_2 \sin \Delta \nu}{\sqrt{\mu p}} \tag{7.46}$$

where $\cos \Delta \nu$ is the dot product of r_1 and r_2 divided by the magnitudes. Then,

$$v_1 = \frac{r_2 - f r_1}{g} \tag{7.47}$$

Since these equations are transcendental, one solution approach is the p-iteration method.

1. Guess a value of p.
2. Determine f and g.
3. Calculate v_1.
4. Knowing r_1 and v_1, solve for the orbital elements and p.
5. Repeat steps 1–4 with the new value of p until p converges.

Gauss's Method Using Universal Variables.[6] Another way to solve this problem is to use the universal variable method, which avoids much of the awkwardness of special orbits like parabolas and hyperbolas. Similar to the previous method, this one uses two position vectors (r_1, r_2) and the time of flight between them. The first step is to define

$$A = \frac{\sqrt{r_1 r_2} \sin \Delta \nu}{\sqrt{1 - \cos \Delta \nu}}$$

$$S = \frac{1}{3!} - \frac{z}{5!} + \frac{z^2}{7!} - \frac{z^3}{9!} + \cdots$$

$$C = \frac{1}{2!} - \frac{z}{4!} + \frac{z^2}{4!} - \frac{z^3}{8!} + \cdots$$

$$y = r_1 r_2 A \frac{(1 - zS)}{\sqrt{C}}$$

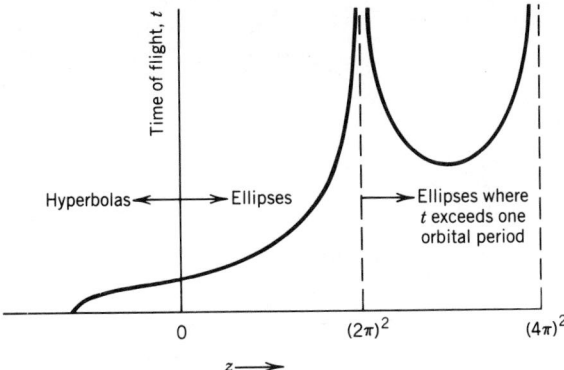

Fig. 7.17 Typical t versus z plot.

$$x = \sqrt{\frac{y}{C}}$$

$$\sqrt{\mu}\, t = x^3 S + A\sqrt{y}$$

$$f = 1 - \frac{y}{r_1}$$

$$g = A\sqrt{\frac{y}{\mu}}$$

$$\dot{g} = 1 - \frac{y}{r_2}$$

$$\mathbf{v}_1 = \frac{\mathbf{r}_2 - f\mathbf{r}_1}{g}$$

$$\mathbf{v}_2 = \frac{\dot{g}\mathbf{r}_2 - \mathbf{r}_1}{g}$$

These equations are transcendental in z, but a simple algorithm for solving the problem is as follows:

1. Evaluate A.
2. Pick a trial value for z. Since $z = \Delta E^2$ and $-z = \Delta F^2$, this amounts to guessing the change in eccentric anomaly. Referring to Fig. 7.17, normal values of z range from minus values to $(2\pi)^2$ with $z = 0$ for a parabola and $z < 0$ for a hyperbola.
3. Evaluate the functions S and C for the selected value of z.
4. Determine the variables y and x.
5. Check the value of z by computing the time of flight (t) and comparing it with the desired time of flight. If it is not nearly the same, pick a new value of z using a bisection or other interpolation scheme and repeat until the desired value of t is obtained. (*Note:* for large negative values of z, the value of y will also become negative resulting in an imaginary value for x and a negative time of flight. Because the z function is continuous, there is always a solution for a positive time of flight.)
6. When the method has converged to a solution, evaluate f, g, and $\dot{g}$, then compute $\mathbf{v}_1$ and $\mathbf{v}_2$.

7.1.9 Perturbations

The dominant force acting on a satellite is the earth's gravity. It is this central inverse-square force that results in two-body or Keplerian motion. All other forces are much smaller; these other outside forces are called perturbations.

There are many perturbation sources. This section discusses the effects of three most significant sources of perturbations: the oblateness, or polar flattening, of the earth, the atmosphere extending high above the earth's surface, and the effect of the sun and moon on an orbiting satellite.

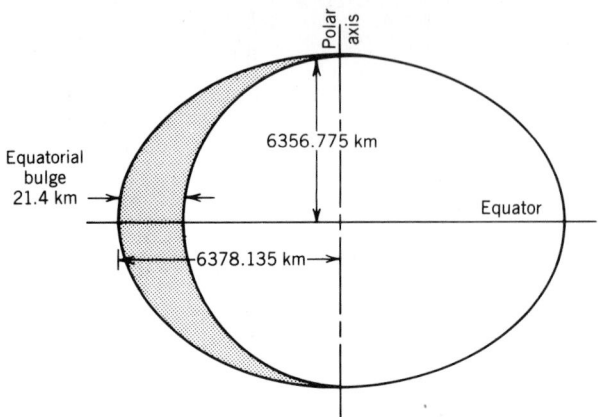

Fig. 7.18 Earth oblateness.

The earth is not a spherically homogeneous mass but has a bulge around the equatorial region as shown in Fig. 7.18. This additional mass causes the gravitational pull on the satellite not to be directed toward the center of the earth. Even though the atmosphere at orbital altitudes is not very dense, it still has a accumulating impact on satellites orbiting at altitudes less than 300 nautical miles (nmi). The earth is not the only source of gravitational attraction since the gravitational fields of the sun and the moon act on the satellite.

Other minor perturbation sources, not to be addressed in this section, include solar radiation pressure, electromagnetic interaction of a charged satellite with the earth's magnetic field, and radiation belts, tidal influences, venting, and thrust.

This section examines the effects of the three primary perturbation sources: geopotential, atmospheric drag, and lunar-solar gravity attraction. The effects and equations presented are not all inclusive, but do represent parameters of greatest concern to a person analyzing orbital motion or planning a space mission.

Geopotential

The two principal effects due to the earth's polar flattening, or equatorial bulge, are regression of the line of nodes and rotation of the line of apsides (major axis). Nodal regression is a rotation of the plane of the orbit about the earth's axis of rotation at a rate that depends on orbital inclination, semimajor axis, and eccentricity. Equation (7.48) is used to determine the nodal rate due to the equatorial bulge. Summary information accompanies this nodal rate expression:

$$\dot{\Omega} = \frac{-3}{2} J_2 \frac{a_e^2}{a^2} \frac{n \cos(i)}{(1 - e^2)^2} \tag{7.48}$$

where

$n = \sqrt{\dfrac{\mu}{a^3}}$, mean motion

a_e = 6378.135 km = 3443.93 nmi, mean equatorial; radius of the earth

J_2 = 1082.63 × 10^{-6}, this is the zonal harmonic coefficient for the earth oblateness model

Note the following scenarios concerning nodal movement:

$\dot{\Omega} < 0$ when $0° < i < 90°$, nodal regression
$\dot{\Omega} > 0$ when $90° < i < 180°$, nodal advance
$\dot{\Omega} = 0$ when $i = 90°$

Figure 7.19 provides a general summary of nodal regression rate. The rotation of the line of apsides is only applicable to noncircular orbits. The rate and direction at which the line of apsides or perigee position moves depends on orbital inclination, semimajor axis, and eccentricity. Equation (7.49) is used to determine the movement of perigee. Summary information accompanies this perigee move-

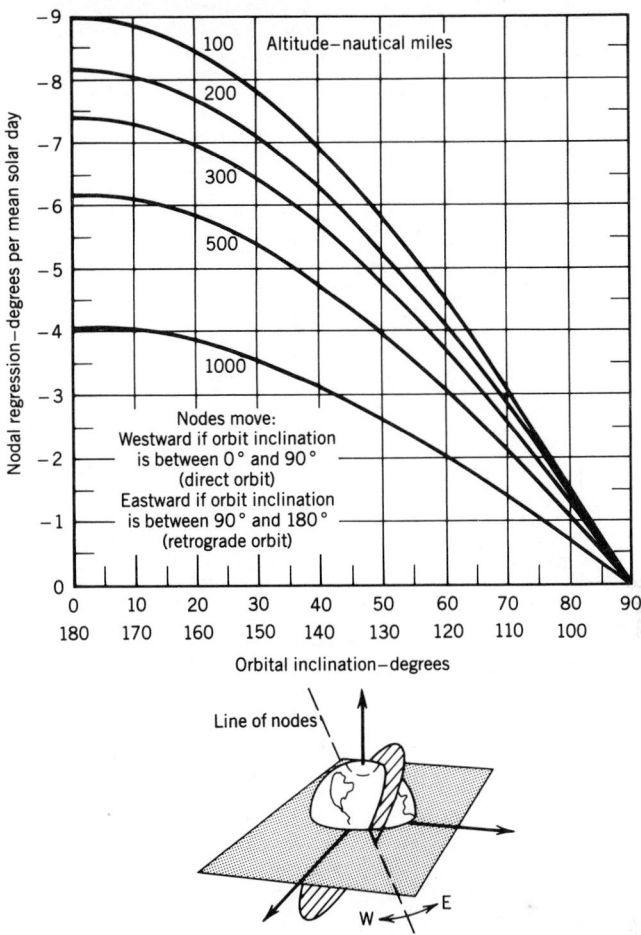

Fig. 7.19 Nodal movement due to earth oblateness.

ment expression:

$$\dot{\omega} = \frac{3}{4} J_2 \left(\frac{a_e}{a} \right)^2 \frac{n \left[5 \cos^2 (i) - 1 \right]}{\left(1 - e^2 \right)^2} \tag{7.49}$$

Note the following scenarios concerning perigee movement:

$\dot{\omega} < 0$ when $63.4° < i < 116.6°$, perigee regression

$\dot{\omega} > 0$ when $\left\{ \begin{array}{l} 0° < i < 63.4° \\ 116.6° < i < 180° \end{array} \right\}$, perigee advance

The inclinations of 63.4° and 116.6° are called the critical inclinations and result in no perigee movement. Figure 7.20 provides a general summary of perigee movement rate.

Atmospheric Drag
Next to earth oblateness, atmospheric drag has the greatest influence on the motion of a near-earth satellite; and during the last few revolutions of a satellite's life it is extremely important.

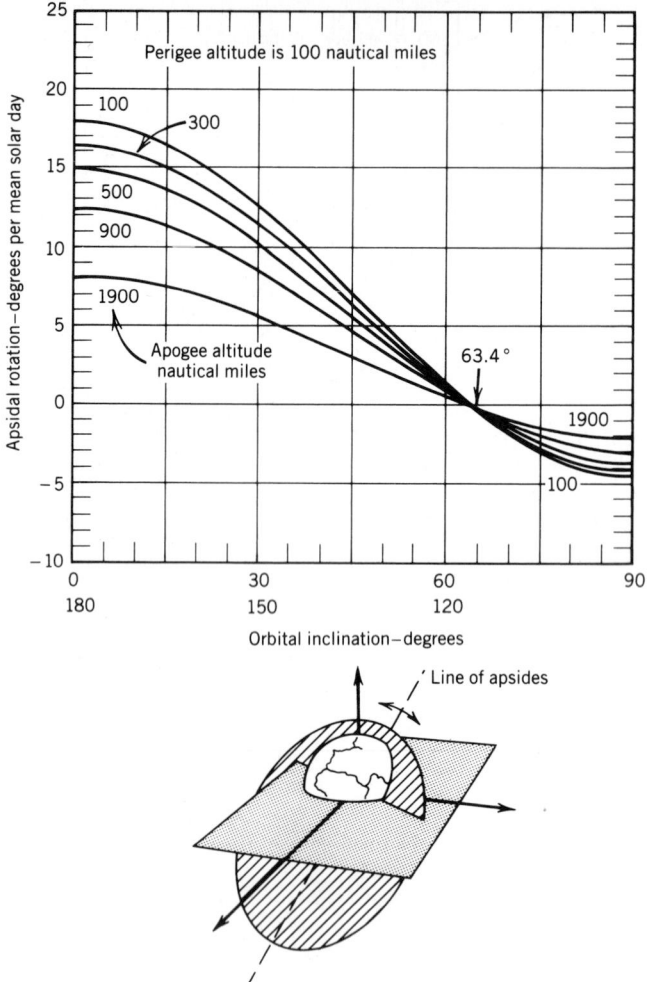

Fig. 7.20   Perigee movement due to earth oblateness.

The acceleration due to atmospheric drag is modeled using Eq. (7.50):

$$a_{\text{drag}} = \frac{1}{2} \frac{C_D A}{m} \rho V^2 \tag{7.50}$$

where

  C_D = drag coefficient, $C_D = 2$ for satellites in upper atmosphere
  A = effective cross-sectional areas of the satellite
  ρ = atmospheric density
  V = speed of the satellite relative to the local atmosphere

The simplest model of the atmosphere assumes a spherically symmetric distribution of atmospheric molecules in which the density ρ, varies exponentially using Eq. (7.51) with Table 7.5 reference data.

$$\rho = \rho_0 e^{[-(h-h_0)/H]} \tag{7.51}$$

TABLE 7.5   Reference Data

h	ρ (g/cm^3)	H
150	2.141×10^{-12}	24.1 km
200	2.706×10^{-13}	40.8
250	7.316×10^{-14}	52.6
300	2.653×10^{-14}	61.2
400	5.192×10^{-15}	73.2
500	1.316×10^{-15}	81.9
600	3.818×10^{-16}	91.0
700	1.216×10^{-16}	105.3
800	4.262×10^{-17}	130.8

where

h = height above earth surface
ρ_0 = density at height h_0
H = scale height

The primary effect the atmosphere has on a satellite is a decrease in eccentricity or circularization. As this happens, the semimajor axis of the orbit decreases. After the orbit is circularized, spiral decay commences and as contact with the atmospheric particles increases with altitude loss entry into denser layer of the atmosphere is imminent. Equation (7.52) is useful for analyzing the effect of semimajor axis decay on the orbital period.

$$\Delta \mathbb{P} = \frac{3\pi}{an} \Delta a \tag{7.52}$$

If particular importance is estimation of satellite lifetime due to atmospheric drag. For an initially circular orbit the lifetime estimate can be found using Eq. (7.53):

$$L = -\frac{H}{\dot{a}} \tag{7.53}$$

Where $\dot{a}$ is the rate of change of the semimajor axis. For noncircular orbits ($e > 0.02$), Eq. (7.54) should be used to estimate satellite lifetime:

$$L = -\frac{e\mathbb{P}}{\dot{\mathbb{P}}} F(e) \tag{7.54}$$

where

$$F(e) = \frac{3}{4} \left[1 + \frac{7e}{6} + \frac{5e^2}{16} + \frac{H}{2ae} \left(1 + \frac{11e}{12} + \frac{3H}{4ae} + \frac{eH^2}{4a^2e^2} \right) \right]$$

Estimating satellite lifetime is very difficult and subject to many uncertainties. These expressions are useful for estimating lifetimes; however, due to the uncertainties involved in modeling drag and accounting for the many variables contributing to satellite motion in the atmosphere error should be anticipated.

Lunar-Solar Gravitational Effects

Just as earth oblateness placed a torque on the satellite orbit so does the gravitational attraction of the sun and the moon have on an orbit. Where the earth's equatorial bulge caused nodal and perigee movement about the earth's polar axis, the sun causes this movement about the pole of the ecliptic, and the moon causes movement about an axis normal to the moon's orbit plane. For close satellites the dominance of the oblateness effect means that the orbit will precess essentially about the polar axis, and for higher altitude orbits the regression will be about some mean pole lying between the earth's pole and the ecliptic pole. The equations describing nodal and perigee movement due to lunar and solar gravity are based on the equator as the reference plane.

Nodal movement due to the sun or moon can be calculated using Eq. (7.55):

$$\dot{\Omega} = -\frac{3K(2 + 3e^2)(2 - 3\sin^2 i_d)}{16n(1 - e^2)^{1/2}}\cos i \qquad (7.55)$$

where if calculating $\dot{\Omega}$ due to solar gravity the following constants are used:

$$i_d = 23.5° = \text{obliquity of the ecliptic}$$

$$K = 0.97 \ (\text{deg/day})^2$$

and n, e, and i are orbital elements associated with the given orbit.
  For calculating $\dot{\Omega}$ due to lunar gravity the following constants are used:

$$i_d = 28.6°$$

$$K = 2.13(\text{deg/day})^3$$

Perigee movement is calculated using Eq. (7.56):

$$\dot{\omega} = \frac{3K(2 - 3\sin^2 i_D)}{16n(1 - e^2)^{1/2}}(e^2 + 4 - 5\sin^2 i) \qquad (7.56)$$

The same constant information holds true for evaluating perigee movement due to solar or lunar gravity.
  An interesting note is the coupling of lunar-solar perturbations with drag perturbations for satellites that have perigee within the appreciable atmosphere.

Summary
The preceding discussion of orbital perturbations due to earth oblateness, atmospheric drag, and lunar-solar gravity is not at all intended to be all inclusive. The primary effects of each perturbation and methods to examine the effect on the orbital elements have been presented. References cited will provide the reader with a much more detailed examination of the subject of perturbation theory and effects.

7.1.10   Statistical Orbit Determination
A preceding section discussed the problem of determining an orbit with a minimum number of observations. Statistical orbit determination allows further observations to be used to refine the initial orbit. This is done using differential correction, based on the concept of residuals. A residual is the difference between an actual observation and what is predicted based on previous observation data.
  The differential correction process is just a six-dimensional Newton iteration where we are trying, by trial and error, to find the value of the orbital elements at time t_0. Assuming the residuals are small, the six first-order equations are

$$\Delta\dot{\rho}_1 = \frac{\delta\dot{\rho}_1}{\delta r_I}\Delta r_I + \frac{\delta\dot{\rho}_1}{\rho r_J}\Delta r_J + \frac{\delta\dot{\rho}_1}{\delta r_K}\Delta r_K + \frac{\delta\dot{\rho}_1}{\delta v_I}\Delta v_I + \frac{\delta\dot{\rho}_1}{\delta v_J}\Delta v_J + \frac{\delta\dot{\rho}_1}{\delta v_K}\Delta v_K$$

through to

$$\Delta\dot{\rho}_6 = \frac{\delta\dot{\rho}_6}{\delta r_I}\Delta r_I + \frac{\delta\dot{\rho}_6}{\delta r_J}\Delta r_J + \frac{\delta\dot{\rho}_6}{\delta r_K}\Delta r_K + \frac{\delta\dot{\rho}_6}{\delta v_I}\Delta v_I + \frac{\delta\dot{\rho}_6}{\delta v_J}\Delta v_J + \frac{\delta\dot{\rho}_6}{\delta v_K}\Delta v_K$$

The partial derivatives can be numerically evaluated by adding a small variation (of 1 or 2%), such as Δr_1, to each of the original orbital elements, in turn, and computing the resulting variation in each of the predicted $\dot{\rho}$'s. Then, for example,

$$\frac{\delta\dot{\rho}_1}{\delta r_I} \approx \frac{\dot{\rho}_1(r_I + \Delta r_I, r_J, r_K, \dots, v_K) - \dot{\rho}_1(r_I, \dots, v_K)}{\Delta r_I}$$

The only requirement for this method is that at least six independent observations are necessary.

Given n number of observations and p number of elements (at least six for orbit problems),

If $p > n$, there is not enough data to solve the problem.
If $p = n$, the exact solution can be solved for explicitly.
If $p < n$, there is no unique solution. In this case, the solution that gives the minimum sum of the squares of the residuals is sought.

The solution to the last case, weighted least-squares iterative differential correction, is

$$\Delta \tilde{z} = (\tilde{A}^T \tilde{W} \tilde{A})^{-1} A^T \tilde{W} \tilde{b} \qquad (7.57)$$

$\Delta \tilde{z} = p \times 1$ matrix of computed corrections to our estimates of the elements.
$\tilde{A} = n \times p$ matrix of partial derivatives of each quantity with respect to each of the elements measured.
$\tilde{W} = n \times n$ diagonal matrix whose elements are the inverses of the variances of the measuring devices.
$\tilde{b} = n \times 1$ matrix of residuals based on the previous estimate of the elements.

7.2 INERTIAL NAVIGATION
J. P. Wittry, B. W. Parkinson, J. C. Ruth, R. J. Lisowski

7.2.1 Concepts of Inertial Navigation

Navigation is the process of continually determining one's position and velocity vectors relative to some chosen reference frame. In inertial navigation this process is accomplished by continuous on-board determination of vehicle acceleration, from which inertial acceleration is computed and twice integrated, with appropriate initial conditions, to yield velocity and position. The problem is three dimensional in most of its applications; namely aircraft, missiles, spacecraft, and submarines.

To illustrate the basic concept of inertial navigation, Fig. 7.21 pictures a single-degree-of-freedom inertial navigation system operating in a vehicle moving along the earth's surface. A linear accelerometer continuously measures specific force (nongravitational force/unit mass) along the instrument's sensitive axis. However, this measurement of acceleration is useless unless the direction of the measured acceleration is known. In the example shown, the gyro is used to provide a stable accelerometer platform on the vehicle. The gyro senses any platform rotation due to irregular vehicle rotation by sensing relative displacement between the gyro spin axis and the gyro case. This relative displacement is sensed as an electrical error signal, amplified, and used to drive a gimbal torque motor, which rotates the platform relative to the vehicle so as to null the error signal from the gyro. This continuous feedback action from the gyro to the platform maintains the platform, and hence the sensitive axis of the accelerometer, in its initial angular orientation with respect to an inertial

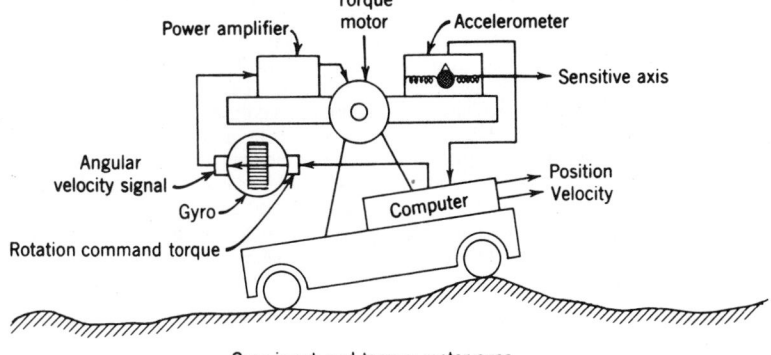

Gyro input and torque motor axes

Fig. 7.21 Simple single-degree-of-freedom inertial navigational system (small scale).

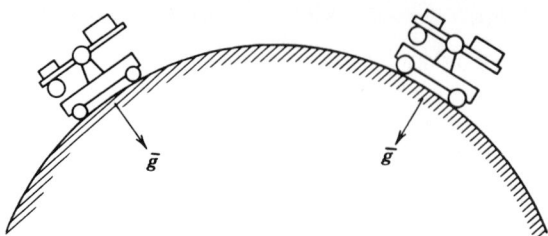

Fig. 7.22 Simple single-degree-of-freedom inertial navigation system (large scale).

reference frame. Thus the direction of the measured acceleration is always known, and the vehicle velocity and position can be computed.

Figure 7.22 shows the same problem in larger perspective to further illustrate the concept. As the vehicle moves along the surface of the curved earth, it is desired that the sensitive axis of the accelerometer be kept oriented in a direction not fixed with respect to an inertial frame, but rather fixed along the local horizontal so that velocity and position along the earth's surface can be computed. To accomplish this, the platform is caused to rotate as a function of changing position as determined by the navigation computer. The desired platform orientation change is effected by the same gyro-platform servo loop previously described. The appropriate platform alignment signal generated by the computer causes a current to be fed to a torque generator in the gyro (Fig. 7.21), causing the gyro spin axis to precess and thus generate an "artificial" error signal that causes the desired platform rotation. This same system provides for initial alignment of the platform to the desired orientation.

The concept outlined for a single-degree-of-freedom inertial navigation system can readily be implemented in three degrees of freedom by providing a multigimballed platform carrying three accelerometers with their sensitive axes arranged in an orthogonal triad, and with three gyros mounted on the platform to sense platform rotation about three axes. In practice, two two-degree-of-freedom gyros are sometimes used in place of three single-degree-of-freedom gyros; in systems that must navigate for more than 10 or 15 min, only horizontal accelerometers may be used, with navigation in the vertical accomplished by another means. It should be noted that the process of platform stabilization described here is continuous and functions on the basis of nearly infinitesimal error signals from the gyro, so that precise platform orientation is continuously maintained even in the presence of rapid and large changes of vehicle attitude.

The concept described above is based on the mechanization of a stabilized accelerometer platform. An alternative method is the "strapdown" system in which the accelerometers are fastened to the vehicle so that the measured acceleration components are in a vehicle fixed reference frame. Gyros are also fastened to the vehicle to measure the attitude changes of the vehicle, hence of the vehicle fixed reference frame. The gyro-derived attitude information is input to the navigation computer along with the measured accelerations. The computer calculates coordinate transformations so that the measured accelerometer readings can be properly interpreted.

Strapdown systems offer some advantages over stabilized platform systems. Because elaborate gimballing is not necessary, the strapdown system can be smaller in both mass and volume, requires less power, and is more reliable and maintainable. The computer algorithm for navigation can be integrated with algorithms for vehicle attitude control and even flight and fire control to form a versatile multifunction software package. Such versatility, however, results in the disadvantage of increased computational demands on the flight computer. Additionally, instrument error must be compensated within the software. Requirements for alignment and calibration of the gyros and accelerometers are stringent. Because the gyros must be operated in the torque-to-balance mode, either very high speed analog-to-digital conversion or sophisticated digital rebalance circuitry is needed to achieve even modest accuracy in position and velocity determination.

In either the stabilized platform or strapdown mechanizations, the inertial navigation system is completely self-contained and independent of external signals or information after the initial alignment.

Accelerometers

The external forces acting on the vehicle are of two basic types: (1) that due to gravitational attraction, and (2) nongravitational forces such as thrust and aerodynamic lift and drag. That is,

$$\bar{f} + \bar{g} = \bar{a} \tag{7.58}$$

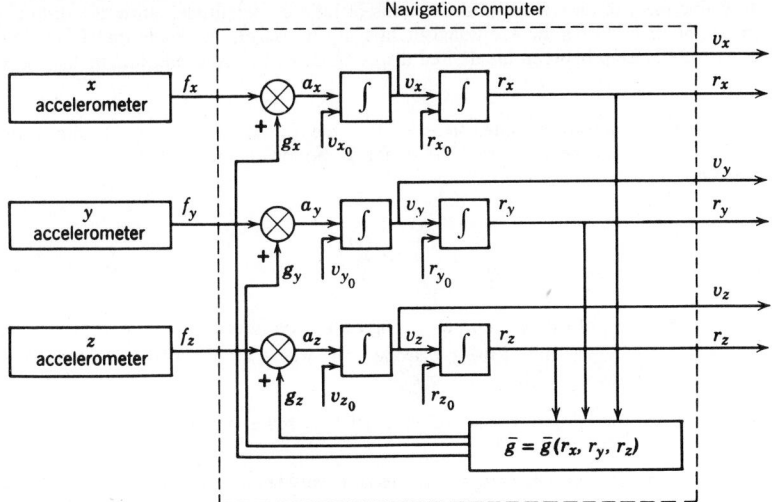

Fig. 7.23 Typical inertial navigational problem.

where $\bar{f}$ = summation of nongravitational specific forces acting
$\bar{g}$ = gravitational specific force
$\bar{a}$ = inertial acceleration

Unfortunately, an accelerometer measures the nongravitational specific force, output $= \bar{f} = \bar{a} - \bar{g}$, and cannot distinguish between the inertial acceleration, $\bar{a}$, and gravitational acceleration, $\bar{g}$. Hence, the accelerometer cannot by itself provide the determination of inertial acceleration required for navigational purposes. Therefore to get the total acceleration vector, accelerometer triad outputs must be added to the gravitational acceleration vector, which is calculated from the known navigator position and attitude:

$$\bar{a} = \overline{\text{output}} + \bar{g}(\bar{r}) \tag{7.59}$$

Figure 7.23 shows how this problem is solved in a typical inertial navigation system. The three components of measured acceleration (nongravitational specific force), f_x, f_y, f_z, are combined with computed components of gravitational specific force, g_x, g_y, g_z, to provide the three components of inertial acceleration, a_x, a_y, a_z. The navigation computer doubly integrates each of these components of inertial acceleration to determine velocity and position and also continuously computes new values of the components of the gravitational specific force according to a stored algebraic function relating this specific force to the position.

The position and velocity outputs shown in Fig. 7.23 relate to some defined internal coordinate frame. The actual choice of a navigation frame involves a trade-off between complexity of the equations and convenience of the output. For example, since accelerometers and gyroscopes measure quantities that are relative to inertial space, an earth-centered inertial coordinate frame would require little or no transformation of coordinates as demonstrated by Fig. 7.23. Such a frame lends itself nicely to an ICBM or space shuttle navigation problem. However, inertial coordinates mean nothing to the pilot of an airliner flying over a rotating earth. An earth-fixed coordinate frame would be better suited for such a problem. In this case, the outputs are meaningful while the equations become more complex because of the need to represent velocities and accelerations with respect to a rotating coordinate frame. Finally, since most navigation on or near the surface of the earth is done at relatively constant altitudes, a local level frame is the most relevant. Such a frame involves commanding a gyro-stabilized platform in order to maintain a local level attitude, offers the most convenient position and velocity reference, requires a barometric or radar altimeter for altitude determination and involves integration of the most complex set of equations of all. As indicated in Fig. 7.23, regardless of the coordinate frame used, a model for gravitational acceleration as a function of position is necessary, which adds to the complexity of the equations.

The position and velocity outputs shown in Fig. 7.23 relate to some defined xyz inertial coordinate frame. If it is desired to present this navigation information in terms of some other coordinate frame (e.g., to an aircraft pilot in terms of latitude, longitude, altitude coordinates), the navigation computer performs the additional mathematical task of coordinate transformation based on the known relationship between the desired display coordinate frame and the inertial frame.

Gyroscopes

A gyroscope is a spinning rotor mounted so as to allow the spin axis to rotate freely about either one or two axes orthogonal to the spin axis. The dynamic motion of the rotor is described by Euler's equation, namely,

$$\mathbf{T} = \frac{d}{dt}(\mathbf{H}) \tag{7.60}$$

where $\mathbf{T}$ = summation of external torques acting on the gyroscope

$\mathbf{H}$ = angular momentum of the gyroscope = $\mathbf{I\Omega}$ where $\mathbf{I}$ is the inertia tensor, and $\mathbf{\Omega}$ is the angular velocity of the gyroscope

$\frac{d}{dt}(\ )$ = time derivative with respect to an inertial reference frame

In a practical gyro the angular momentum of the rotor is very high, accounting for virtually all of the total gyro angular momentum, and $\mathbf{H}$ can be taken to be the angular momentum of the rotor itself.

In accordance with basic vector kinematics (the "law" of Coriolis):

$$\frac{d}{dt}(\mathbf{H})_{\substack{\text{inertial} \\ \text{frame}}} = \frac{d}{dt}(\mathbf{H})_{\substack{\text{Gimbal} \\ \text{frame}}} + \boldsymbol{\omega} \times \mathbf{H} \tag{7.61}$$

where $d/dt(\mathbf{H})_{\substack{\text{Gimbal} \\ \text{frame}}}$ is the time derivative of the rotor angular momentum with respect to the gimbal containing the rotor, and $\boldsymbol{\omega}$ is the angular velocity (precessional velocity) of this gimbal with respect to an inertial reference frame. In practice, $d/dt(\mathbf{H})_{\substack{\text{Gimbal} \\ \text{frame}}} \cong 0$ by careful control of the rotor spin velocity so that the practical equation describing the gyroscope's behavior is

$$\mathbf{T} = \boldsymbol{\omega} \times \mathbf{H} \tag{7.62}$$

Figure 7.24 schematically represents a single-degree-of-freedom gyro mounted on a platform and showing the relationship between the rotor angular momentum vector, applied torque vector (due to platform rotation, for example), and the precession vector. Since the precession is uniquely related to the applied torque, measurement of the precession by the signal generator provides a self-contained measurement of platform rotation about one axis.

In practice, this precession angle remains very small, either by nulling with platform torquers (gimballed system) or by nulling using a gyro torquer (strapdown system). Other torques due to

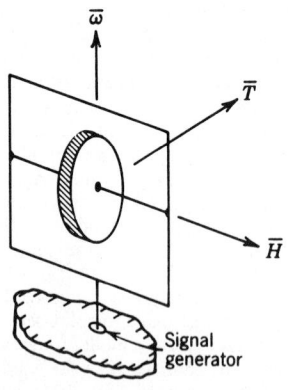

Fig. 7.24   Schematic of a single-degree-of-freedom gyro.

various mechanical imperfections in a real gyro also cause precession, referred to as gyro drift because these unwanted torques cause an undesired drift of the spin axis away from its assumed orientation in space. In practice, other intentional torques are introduced by spring restraint and viscous damping on the precession axis to modify the gyro's dynamic behavior.

Incorporating an additional gimbal allows the gyro to receive inputs and experience precession about two axes. Of course, this two-degree-of-freedom gyro accurately senses rotations along two orthogonal axes.

Vertical Reference

In many mechanizations of inertial navigation systems it is convenient to provide a vertical reference, vertical being the direction of the local gravity vector. A pendulum is a good vertical indicator when mounted on a stationary base, but on an accelerating base the pendulum moves as a result of the gravitational and inertial forces. If it were possible to construct a pendulum that would rotate (when accelerated) at just the rotation rate of local gravity vector as it moves over the earth, it would always indicate the vertical in spite of the base acceleration. This problem is present in the process of flying over a spherical earth and was first presented, along with a theoretical solution, by Dr. M. Schuler in 1923. (See Appendix A of Ref. 7 for an English translation of Dr. Schuler's paper.) For flight over the earth's surface the pendulum must be constructed so as to have an undamped period of 84.4 min. It is impossible to achieve a physical free pendulum with this characteristic, but it is practical to construct a feedback control system using gyroscopes and accelerometers mounted on the stable platform of the inertial navigation system so as to achieve an equivalent to the physical Schuler pendulum. Such a system is referred to as "Schuler tuned" and will provide a true vertical reference in spite of the base acceleration.

Aided Navigation

Pure inertial navigation presents a unique problem. Regardless of the coordinate frame chosen, Fig. 7.23 indicates that the navigation system solves the vector equation of Eq. (7.58)

$$\mathbf{a} = \ddot{\mathbf{r}} = \mathbf{f} + \mathbf{g}(\mathbf{r}) \tag{7.63}$$

If the platform is perfectly aligned, the equations are perfectly initialized, and the accelerometers and gyros are perfect, then Eq. (7.63) will yield stable results for velocity, $\mathbf{v}$, and position, $\mathbf{r}$. However, instruments are not perfect and initial conditions can not be determined precisely, and the following error equation (to first order) can be derived,

$$\delta\ddot{\mathbf{r}} = \delta\mathbf{f} + \frac{\partial\mathbf{g}(\mathbf{r})}{\partial\mathbf{r}}\delta\mathbf{r} \tag{7.64}$$

where $\delta\mathbf{r}$ = error in position

$\delta\mathbf{f}$ = error in accelerometer measurement

$\dfrac{\delta\mathbf{g}}{\partial F}$ = error in gravitational acceleration due to error in position

Choosing the simple gravitational model

$$\mathbf{g} = \frac{-\mu}{r^3}\mathbf{r} \tag{7.65}$$

where μ is the earth's gravitational parameter yields the error equation

$$\delta\ddot{\mathbf{r}} = \delta\mathbf{f} - \frac{\mu}{r^3}\delta\mathbf{r} + \frac{3\mu}{r^5}(\mathbf{r}^T\delta\mathbf{r})\mathbf{r} \tag{7.66}$$

Considering only horizontal and local vertical errors, respectively, yields two forms for Eq. (7.66):

$$\delta\ddot{\mathbf{r}}_h + \frac{\mu}{r^3}\delta\mathbf{r}_h = \delta\mathbf{f} \quad \text{(horizontal)} \tag{7.67}$$

$$\delta\ddot{\mathbf{r}}_v - \frac{2\mu}{r^3}\delta\mathbf{r}_v = \delta\mathbf{f} \quad \text{(vertical)} \tag{7.68}$$

Equation (7.67) is the equation for an oscillator forced by either $\delta\mathbf{f}_h$ or initial conditions on $\delta\mathbf{r}_h(0)$. Its solution indicates that $\delta\mathbf{r}_h$ oscillates at precisely the Schuler frequency. However, Eq. (7.68) is an unstable equation whose solution grows without bound for any nonzero δf or $\delta r_v(0)$. This instability in the vertical direction comes from the navigation system's dependency on r for the calculation of g. The way around this problem is to use measurements from navigation aids, such as radar altimeters or barometric altimeters either to calculate altitude entirely or to correct the navigation computer's calculations periodically.

7.3   SPACECRAFT ATTITUDE DYNAMICS AND CONTROL
R. S. Fraser and M. L. DeLorenzo

7.3.1   Introduction

Spacecraft attitude dynamics and control deals with the rotational motion of a space vehicle. This involves the examination of the behavior of the orientation of the spacecraft as well as what needs to be done to modify or control the orientation. For the most part, the rotational motion of a satellite, or motion about the center of mass, can be treated as being independent of the translational motion, or motion of the center of mass in space. The former is the concern of this section, while the latter is studied in orbital mechanics or astrodynamics. However, for very low altitude orbits, where signifi-cant aerodynamic forces (and torques) are present, the rotational and translational motion cannot be considered to be decoupled. This case will not be examined here, as those orbits represent a special case that tend to decay fairly rapidly. Before continuing, it is useful to define some of the terms that will be used., Various references have slightly different definitions of these interrelated terms. Wertz[8] uses the following:

Attitude Control: process of achieving and maintaining an orientation in space.

Attitude Maneuver: process of reorienting the spacecraft from one attitude to another.

Attitude Acquisition: attitude maneuver in which the initial attitude is known.

Attitude Stabilization: process of maintaining an existing attitude relative to some external reference frame.

The following sections deal with the dynamics of spacecraft orientation and approaches to controlling the orientation.

7.3.2   Spacecraft Attitude Dynamics

Todays spacecraft missions require that the spacecraft's orientation with respect to one or more reference frames be precisely known and controlled. Figure 7.25 depicts a set of reference frames in which precise spacecraft orientation could be required. The fundamental attitude control problem is to measure and control the orientation of the spacecraft's various body frames (i.e., F_B, F_S, etc.) with respect to the earth-fixed or orbit-fixed frames (i.e., F_e, F_o, etc.). This attitude control process requires a knowledge of how internal and external torques affect the spacecraft orientation (i.e., attitude dynamics) and how these dynamics affect the relationship between spacecraft and external reference frames (i.e., attitude parameters). This section includes basic information on the common sets of attitude parameters used in spacecraft attitude control problems and the fundamental equations used to describe spacecraft rigid body and flexible body dynamics.

Attitude Parameters
Before examining the dynamics of rotational motion, it is useful to examine some of the methods for relating the various coordinate frames.

    Direction Cosine Matrix. One of the most common attitude parameter sets for spacecraft attitude control problems is the direction cosine matrix (DCM) or coordinate transformation matrix. Using the DCM, a vector quantity expressed in one frame may be expressed in another frame by the following matrix expression:

$$x^A = C_B^A x^B \tag{7.69}$$

where x^A is the vector quantity x expressed in the A frame and x^B is the vector quantity x expressed in the B frame. The matrix C_B^A represents the DCM. Additionally, if the two frames are orthogonal, which is generally the case, the DCM has the following useful property:

$$x^B = \left[C_B^A \right]^T x^A \tag{7.70}$$

where T denotes matrix transposition. For illustration, let the A frame be represented by an ordered

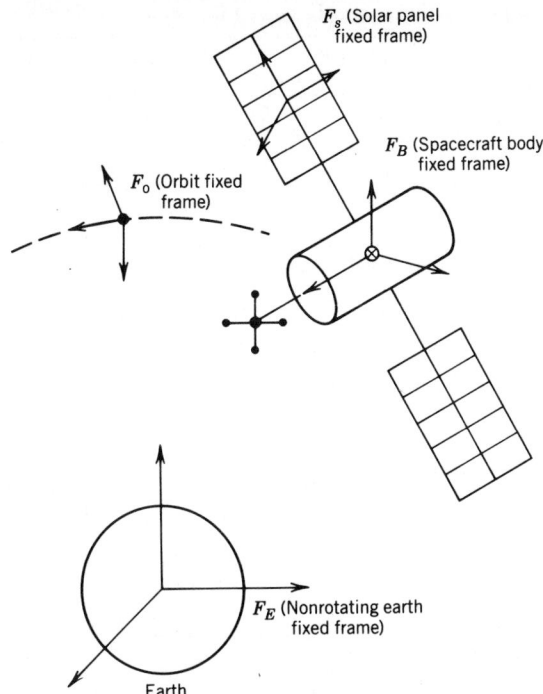

Fig. 7.25 Common spacecraft reference frames.

set of axes $(\hat{1}, \hat{2}, \hat{3})$ where $\hat{1} \times \hat{2} = \hat{3}$, $\hat{2} \times \hat{3} = \hat{1}$ and $\hat{3} \times \hat{1} = \hat{2}$ and let the B frame be represented by a similar set of ordered axis $(\hat{1}', \hat{2}', \hat{3}')$. Table 7.6 shows the DCM for frames A and B separated by a single rotation. (*Note:* $C\theta \Rightarrow \cos\theta$, $S\theta \Rightarrow \sin\theta$ and $+\theta$ implies a positive rotation as defined by the right-hand rule.)

A DCM for frames separated by several rotations can be readily constructed by multiplying the matrices in the *appropriate* order. Table 7.7 demonstrates the process.

It should be noted that the coordinate axes of the B frame change direction under each rotation. In other words, the $+\phi$ rotation about the $\hat{2}'$ axis occurs after the $\hat{2}'$ axis has been rotated from its original direction by an angle θ around the $\hat{1}'$ axis.

For most spacecraft applications, the DCM, which relates a particular body frame to an external frame, changes with time. An expression for the DCM as a function of time becomes essential and is

TABLE 7.6   DCM Frames Separated by Single Rotation

Rotation Sequence to Get from B to A	C_B^A
$+\theta$ about $\hat{1}'$	$\begin{bmatrix} 1 & 0 & 0 \\ 0 & C\theta & S\theta \\ 0 & -S\theta & C\theta \end{bmatrix}$
$+\theta$ about $\hat{2}'$	$\begin{bmatrix} C\theta & 0 & -S\theta \\ 0 & 1 & 0 \\ S\theta & 0 & C\theta \end{bmatrix}$
$+\theta$ about $\hat{3}'$	$\begin{bmatrix} C\theta & S\theta & 0 \\ -S\theta & C\theta & 0 \\ 0 & 0 & 1 \end{bmatrix}$

TABLE 7.7   DCM Frames Separated by Several Rotations

Rotation Sequence to Get from B to A	C_B^A
$+\theta$ about $\hat{1}'$; $+\phi$ about $\hat{2}'$ $+\gamma$ about $\hat{3}'$	$\begin{bmatrix} C\gamma & S\gamma & 0 \\ -S\gamma & C\gamma & 0 \\ 0 & 0 & 1 \end{bmatrix}\begin{bmatrix} C\phi & 0 & -S\phi \\ 0 & 1 & 0 \\ S\phi & 0 & C\phi \end{bmatrix}\begin{bmatrix} 1 & 0 & 0 \\ 0 & C\theta & S\theta \\ 0 & -S\theta & C\theta \end{bmatrix}$

obtained through the following differential equation:

$$\dot{C}_B^A = C_B^A \begin{bmatrix} 0 & -\omega_z & \omega_y \\ \omega_z & 0 & -\omega_x \\ -\omega_y & \omega_x & 0 \end{bmatrix} \tag{7.71}$$

where $[\omega_x \quad \omega_y \quad \omega_z]^T$ is the angular velocity vector (ω_{AB}^B) of the B frame with respect to the A frame expressed in the B frame. For spacecraft applications (ω_{AB}^B) is usually determined from rate gyros aligned to the B frame (i.e., a body frame). For small Δt, the following first-order integration scheme proves useful for updating C_B^A as a function of time:

$$C_B^A(t + \Delta t) = C_B^A(t)\begin{bmatrix} 1 & -\omega_z \Delta t & -\omega_y \Delta t \\ +\omega_z \Delta t & 1 & -\omega_x \Delta t \\ -\omega_y \Delta t & +\omega_x \Delta t & 1 \end{bmatrix} \tag{7.72}$$

The Quaternion. Updating the DCM as a function of time requires tracking nine parameters (i.e., nine differential equations). The quaternion or Euler parameters as it is sometimes called is a four-parameter set that provides the same attitude information as the DCM. The quaternion is defined as follows:

$$q = \begin{bmatrix} a \\ b \\ c \\ d \end{bmatrix} = \begin{bmatrix} a \\ \mathbf{p} \end{bmatrix}$$

where

$$\mathbf{p} = \sin\left(\frac{\phi}{2}\right)\hat{1}_\phi \qquad a = \cos\left(\frac{\phi}{2}\right)$$

The unit vector $\hat{1}_\phi$ represents the axis about which the B frame could be rotated by an angle ϕ and align with the A frame. The quaternion can be expressed as a function of time by using the following differential equation:

$$\dot{q} = \frac{1}{2}\begin{bmatrix} 0 & -\omega_x & -\omega_y & -\omega_z \\ \omega_x & 0 & \omega_z & -\omega_y \\ \omega_y & -\omega_z & 0 & \omega_x \\ \omega_z & \omega_y & -\omega_x & 0 \end{bmatrix} q \tag{7.73}$$

where, as in the DCM expression, $\omega_{AB}^B = [\omega_x \quad \omega_y \quad \omega_z]^T$. Again, for reasonable Δt the following first-order integration scheme can be used to update the quaternion as a function of time:

$$q(t + \Delta t) = \frac{1}{2}\begin{bmatrix} 2 & -\omega_x \Delta t & -\omega_y \Delta t & -\omega_z \Delta t \\ \omega_x \Delta t & 2 & \omega_z \Delta t & -\omega_y \Delta t \\ \omega_y \Delta t & -\omega_z \Delta t & 2 & -\omega_x \Delta t \\ \omega_z \Delta t & \omega_y \Delta t & -\omega_x \Delta t & 2 \end{bmatrix} q(t) \tag{7.74}$$

With the quaternion as a function of time, the DCM can be constructed using this identity:

$$C_B^A = \begin{bmatrix} a^2 + b^2 - c^2 - d^2 & 2(ad + bc) & 2(db - ac) \\ 2(bc - ad) & a^2 - b^2 + c^2 - d^2 & 2(ab + cd) \\ 2(ac + bd) & 2(cd - ab) & a^2 - b^2 - c^2 + d^2 \end{bmatrix} \qquad (7.75)$$

Rigid-Body Dynamics
The rigid-body rotation of a spacecraft is accurately described by Newton's rotational law:

$$\sum \mathbf{T}_{ext} = \frac{d\mathbf{H}}{dt}\bigg|_{Inert} \qquad (7.76)$$

that is, the sum of external torques ($\sum \mathbf{T}_{ext}$) acting on a rigid body is directly proportional to the time rate of change of the angular momentum (**H**) of the body as viewed from inertial space. For spacecraft applications it is typically necessary to describe motion in the spacecraft body frame that can be noninertial. Therefore applying the Coriolis theorem we have the following form for Newton's law expressed in noninertial body frame coordinates:

$$\sum \mathbf{T}_{ext}^B = \frac{d\mathbf{H}^B}{dt}\bigg|_B + \omega_{AB}^B \times \mathbf{H}^B \qquad (7.77)$$

where B represents the spacecraft body frame and A represents some inertial frame. For a rigid spacecraft, $\mathbf{H}^B$ can be defined as

$$\mathbf{H}^B = I\omega_{AB}^B \qquad \omega_{AB}^B = \begin{bmatrix} \omega_1 & \omega_2 & \omega_3 \end{bmatrix}^T$$

I is the inertia tensor (matrix) for the spacecraft:

$$I = \begin{bmatrix} I_{11} & -I_{12} & -I_{13} \\ -I_{12} & I_{22} & -I_{23} \\ -I_{13} & -I_{23} & I_{33} \end{bmatrix}$$

If the body frame is aligned to principal axes of the spacecraft, the inertia tensor has this form:

$$I = \begin{bmatrix} I_1 & 0 & 0 \\ 0 & I_2 & 0 \\ 0 & 0 & I_3 \end{bmatrix}$$

Assuming the body frame is aligned to a set of spacecraft principal axes and expanding the Coriolis expression yields a form of Euler's moment equations that describes spacecraft rigid-body motion:

$$T_1 = I_1\dot{\omega}_1 - (I_2 - I_3)\omega_2\omega_3$$
$$T_2 = I_2\dot{\omega}_2 - (I_3 - I_1)\omega_3\omega_1$$
$$T_3 = I_3\dot{\omega}_3 - (I_1 - I_2)\omega_1\omega_2$$

This first-order, coupled, nonlinear set of differential equations can be used to predict spacecraft rigid-body motion under the influence of environmental or control torques. These equations do not normally have a closed-form or analytic solution, and must be solved numerically. However, there are several special cases that can be solved analytically.

CASE 1:   Torque-free (**T = 0**) axis symmetric ($I_1 = I_2$) motion with $I_1 < I_3$. For this case, Euler's moment equations become

$$\dot{\omega}_3 = 0 \qquad \omega_3(t) = C \quad \text{(i.e., a constant)}$$
$$\dot{\omega}_1 + \lambda\omega_2 = 0 \qquad \lambda = \frac{(I_3 - I_1)}{I_1}C$$
$$\dot{\omega}_2 - \lambda\omega_1 = 0$$

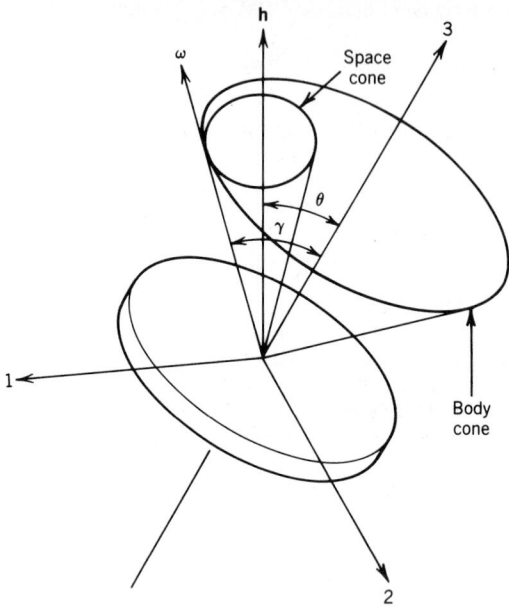

Fig. 7.26   Torque-free axis symmetric motion for $I_1 < I_3$.

and these equations have the following closed-form solution:

$$\omega_3(t) = \omega_3(0) = C$$
$$\omega_1(t) = \omega_1(0) \cos \lambda t - \omega_2(0) \sin \lambda t$$
$$\omega_2(t) = \omega_1(0) \sin \lambda t + \omega_2(0) \cos \lambda t$$

A geometrical representation of the motion of $\omega_{AB}^B(t)$ for a rigid body satisfying case 1 conditions is shown in Fig. 7.26. The body cone attached to the principal axes coordinate system and centered around the 3 axis rotates without slipping on the space cone, which is centered on **H** and fixed in inertial space. The $\omega_{AB}^B(t)$ motion is described by the motion of the tangency point of the two cones. The angles θ and γ are also of interest and are defined as follows:

$$\tan \theta = \frac{I_1 \sqrt{\omega_1^2(0) + \omega_2^2(0)}}{I_3 \omega_3(0)}$$

$$\tan \gamma = \frac{\sqrt{\omega_1^2(0) + \omega_2^2(0)}}{\omega_3(0)}$$

$$\tan \theta = \frac{I_1}{I_3} \tan \gamma \qquad (\theta < \gamma)$$

The angle θ is referred to as the nutation angle and provides the angular relationship between the 3 axis and the inertially fixed **H**.

CASE 2:   Torque-free $(T = 0)$ axis symmetric motion $(I_1 = I_2)$ with $I_1 > I_3$. Euler's moment equations become:

$$\dot{\omega}_3 = 0 \qquad \omega_3(t) = C$$

$$\dot{\omega}_1 - \lambda \omega_2 = 0 \qquad \lambda = \frac{I_1 - I_3}{I_1} C$$

$$\dot{\omega}_2 + \lambda \omega_1 = 0$$

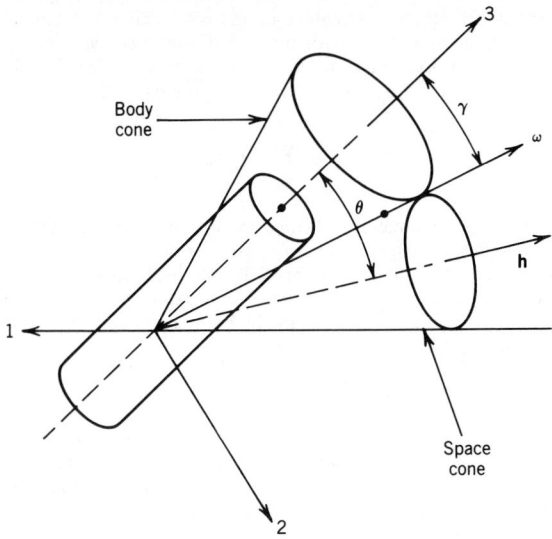

Fig. 7.27   Torque-free axis symmetric motion for $I_1 > I_3$.

with the following closed-form solution:

$$\omega_3(t) = \omega_3(0) = C$$
$$\omega_1(t) = \omega_1(0) \cos \lambda t + \omega_2(0) \sin \lambda t$$
$$\omega_2(t) = \omega_1(0) \sin \lambda t - \omega_2(0) \cos \lambda t$$

Figure 7.27 is the geometrical representation of the motion of $\omega_{AB}^B(t)$ for a rigid body satisfying case 2 conditions. As in the case 1 example, the body cone rotates without slipping on the space cone. The angles θ and γ are defined identically to case 1. Therefore γ must be less than θ as indicated by Fig. 7.27.

Summary. Cases 1 and 2 cover the two possible situations for torque-free axis symmetric motion. The following general properties exist.

1. The angular momentum vector **H** is constant in magnitude and inertial direction.
2. The magnitude of ω_{AB}^B is constant.
3. The angles θ and γ are constant.
4. The vectors **H**, ω_{AB}^B, and the 3 axis are *always* coplanar.

It should also be noted that the solution for $\omega_{AB}^B(t)$ in these two cases can be readily related to inertial space by a DCM and its derivative, which is a function of ω_{AB}^B. (See the previous section.)

As a final note two important properties of spinning rigid bodies will be presented. The proofs of these properties can be found in Refs. 9 and 10.

Property 1. A rigid body has spin stability only about the major and minor axis of inertia. That is, spin about an intermediate axis of inertia will degenerate into spin about a major or minor axis of inertia under the slighest perturbation. However, perturbed spin around a major or minor axis will always be bounded by the amplitude of the perturbation.

Property 2. If a spinning body experiences energy dissipation (i.e., from booms, fuel sloshing, etc.) the spinning motion will *always* degenerate to spin around the *major* axis of inertia.

Flexible Spacecraft Dynamics
Large space structures (LSS) are now an imminent reality. These structures will be measured in kilometers and, of necessity, will be lightweight and highly flexible (i.e., light damping and nonrigid). The problem of modeling the dynamics of an LSS is well discussed in the literature. A representative

example is Ref. 11. The modeling process centers around discretizing a system of partial differential equations that describes the motion of the structure in a six-dimensional time-space continuum. The most common discretization process involves a separation of variables that uses a finite sum of approximate mode shapes multiplied by a time function to represent the time-space continuum:

$$O(t, s) \approx \sum_{i=1}^{N} \hat{\phi}_i(s) q_i(t) \tag{7.79}$$

where $O(t, s)$ represents the six-dimensional time-space continuum defined by the LSS. The t represents time, while s represents the spatial coordinate. The function $\hat{\phi}_i(s)$ is the approximation to the ith mode shape of the structure and is typically obtained through numerical finite-element techniques. The integer N represents the number of approximate mode shapes used to estimate $O(t, s)$, and $q_i(t)$ is the time function for the ith mode. Putting this approximation into the system of partial differential equations gives the following result:

$$M\ddot{q}(t) + D\dot{q}(t) + Kq(t) = Fu(t) \tag{7.80}$$

The quantity $q(t)$ is a vector of the N time functions. The expressions $\dot{q}(t)$ and $\ddot{q}(t)$ represent the first and second time derivatives of $q(t)$, respectively. The matrix M is a diagonal mass/inertia matrix (i.e., $M = \text{diag}(m_1, m_2, \ldots, m_n)$). This matrix is often referred to as the lumped-mass matrix. The matrices D and K are also diagonal with the following components:

$$D = \text{diag}\left[2\zeta_1 \lambda_1^{1/2}, \ldots, 2\zeta_N \lambda_N^{1/2}\right]$$
$$K = \text{diag}[\lambda_1, \ldots, \lambda_N]$$

where ζ_i is the approximate damping ratio of the ith mode, $\lambda_i = m_i \omega_{n_i}^2$ and ω_{n_j} is the natural frequency of the ith mode. The vector $Fu(t)$ represents the forcing or control function for the LSS. The N rows of the matrix F are the approximate mode shape functions of the LSS evaluated at the spatial coordinate s, which specifies the location of the forcing function.

For most LSS control problems it is convenient to transform the matrix differential equation of (7.80) to the following:

$$\ddot{\eta} + \mathscr{D}\dot{\eta} + \Lambda\eta = \mathscr{F}u(t) \tag{7.81}$$

where

$$\mathscr{D} = T^{T}DT = \text{diag}\left[2\zeta_1 \omega_{n_1}, \ldots, 2\zeta_N \omega_{n_N}\right]$$
$$\Lambda = T^{T}KT = \text{diag}\left[\omega_{n_1}^2, \ldots, \omega_{n_N}^2\right]$$
$$\mathscr{F} = T^{T}F$$

and

$$q(t) = Tn(t)$$
$$T^{T}MT = I_N \quad \text{(an } N \times N \text{ identity matrix)}$$

It then is a straightforward matter to convert (7.81) to the following state space form:

$$\dot{x}(t) = Ax(t) + Bu(t)$$

where

$$x(t) = \begin{bmatrix} \eta \\ \dot{\eta} \end{bmatrix} \quad A = \begin{bmatrix} 0 & I_N \\ -\Lambda & -\mathscr{D} \end{bmatrix} \quad B = \begin{bmatrix} 0 \\ \mathscr{F} \end{bmatrix}$$

The LSS dynamics are now expressed as $2N$ first-order ordinary differential equations and a great wealth of control theory techniques are available for attacking the satellite attitude control problem.

7.3.3 Spacecraft Attitude Control

As noted in the last section for spacecraft, the need to control attitude is often directly related to the vehicle's mission. Some of the primary factors that affect the design of the attitude control system are the mission duration, payload, pointing accuracy, stability requirements, power requirements, and maneuvering requirements. Earth observation satellites typically have their sensors pointed in the nadir direction (directly down) with an accuracy of a fraction of a degree. Geosynchronous communications satellites have a pointing accuracy on the order of one degree, while astronomical satellites must be able to point their sensors to within arcseconds. Pointing accuracy will be one of the keys to the effectiveness of space-based weapons such as lasers and particle beams. Some secondary factors that affect attitude control are determined by the other systems on board the spacecraft. The size of the antenna and the amount of power broadcast by the communications system can be reduced if the antenna can be very accurately aimed at the receiving station. Satellites that use solar power must have their solar panels or arrays facing the sun. Finally, in order to fire thrusters for station keeping or orbital maneuvering, the thrusters must be aimed in the proper direction.

Disturbance Torques Affecting the Spacecraft

As was developed in the previous section, the rotational dynamics of the spacecraft are governed by Euler's equations. The left-hand side of (7.78) represents the sum of the torques acting on the spacecraft. These torques fall into two main categories, disturbance torques and actuator torques. The actuator torques will be generated by the spacecraft attitude control system and will subsequently be examined in greater detail. Disturbance torques can be one of two types, internal or environmental. Internal torques are the result of motion inside the satellite such as fuel movement, crew movement, or mechanical motion (e.g., a tape recorder). Environmental torques result from the space environment. Magnetic torques arise from interaction of any magnetic field within the spacecraft with the earth's magnetic field. The spacecraft will act like the needle of a magnetic compass and align with the earth's magnetic field. Aerodynamic torques are present in low-altitude orbits when the center of mass and center of pressure of the spacecraft are not coincident. Solar radiation (both particulate and electromagnetic) can exert a torque, similar to the aerodynamic torque, that causes the spacecraft to "weather-vane" toward the sun. Gravity gradient torques result from the differential attraction of gravity along the vertical direction on the spacecraft.

Control Systems

Using control theory terminology, spacecraft attitude control systems (ACS) can be divided into two major classifications, open loop and closed loop. An open-loop, or passive, system is simply one where the input, or control signal, to the system is predetermined. Accuracy for this type of system will be degraded by errors in the system model and any disturbance torques that act on the system. A closed-loop, or active control, system uses a sensor to measure some system state (angular orientation/velocity in this application). This measurement is compared to the desired state and an appropriate control signal is generated.

It is appropriate to further divide active control systems into two subcategories. The first is "automatic," which implies that the closed-loop system has all of the required equipment on board and will function without outside intervention. The second can be called "man-in-the-loop" and is common to many satellites. One such system has sensors on board to measure the satellite's orientation. This information is relayed to the ground. When the orientation exceeds the desired limits, the ground station will transmit a signal to the satellite that will activate an actuator. This type of system works where the disturbance torques are relatively low and corrections can be made infrequently. Although the man-in-the-loop system requires a communication link, this link is already available due to the requirements of other systems, and the on-board processing requirements are reduced.

Passive Control Systems. Passive control systems tend to be very simple, reliable, and as a result economical. While there have been some very limited applications of aerodynamic and magnetic passive attitude control, the two major techniques are spin stabilization and gravity gradient stabilization. With spin stabilization, the satellite is rotated with a constant angular velocity about an axis, giving it some significant angular momentum. If any disturbance torque, such as a meteor strike, acts on the spacecraft, it will precess as predicted by the law of precession:

$$\mathbf{T}_d = \boldsymbol{\omega}_e \times \mathbf{H}$$

where $\mathbf{T}_d$ = disturbance torque

$\boldsymbol{\omega}_e$ = error angular velocity (precession rate)

$\mathbf{H}$ = satellite angular momentum

If the torque is perpendicular to the angular momentum vector (the worst case possible) the magnitude of the precession rate will be

$$\omega_e = \frac{T_d}{H}$$

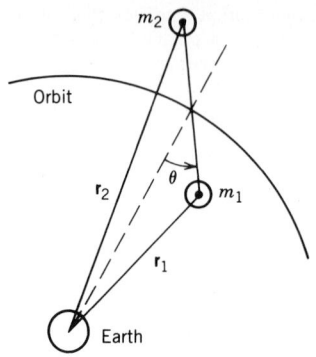

Fig. 7.28   Gravity gradient stabilization.

It should be clear that the greater the angular momentum, the smaller the effect of the disturbance torque. The satellite spin axis will start out pointing in a constant direction with respect to inertial space. But because this is a passive system, disturbance torques will cause the spin axis to deviate from this original direction, and there will be no means available to correct this. However, if the direction of the disturbance torques are randomly distributed, the net effect after a long period of time will be to cancel each other out. The satellite will approximately maintain its original orientation.

Any sensors aboard a spinning satellite will rotate with the vehicle, which imposes serious mission restrictions. A solution to this problem is the dual-spinner, which has a spinning portion for angular momentum and a "despun" sensor or antenna platform. This platform typically rotates at one revolution per orbit for a satellite with its angular momentum vector perpendicular to the orbital plane. This enables the sensors on the despun platform to maintain a local vertical orientation.

Gravity gradient stabilization can best be understood by considering a dumbbell-shaped satellite shown in Fig. 7.28. The only forces acting on the satellite are gravitational. The forces on m_1 and m_2 will be

$$F_1 = \frac{-\mu m_1}{r_1^2}$$

$$F_2 = \frac{-\mu m_2}{r_2^2}$$

If $m_1 = m_2$, F_1 will be greater than F_2 because r_1 is less than r_2. If the satellite is deflected from the vertical by an angle θ, a torque will act on the satellite that will tend to restore it to the local vertical orientation. Because there is no damping (there are no aerodynamic forces or torques) in space, the satellite will librate, or oscillate, about the local vertical. It is possible to incorporate an internal damping device, such as a pendulum with friction or a mass that moves in fluid-filled container. This will reduce the effect of libration. This type of attitude control system is often used when the mission requires the satellite to constantly face the earth. No matter what orientation the satellite is deployed in, it will always align with the local vertical. However, provisions must be made to ensure that initially the correct end points toward the earth, as the satellite is also stable "upside down." Obviously, a surveillance device pointing away from the earth has little utility. For missions that require higher pointing accuracy or more mission flexibility, an active attitude control system must be considered.

Active Control Systems. The three major elements of an active satellite attitude control system (ACS) are shown in Fig. 7.29. As in the case of the stabilized platform in an inertial navigation system, a torque is applied to the satellite by the actuator to cancel out any disturbance torques. There are many different types of sensors that are used to detect, or measure, the attitude of the satellite for comparison to the desired attitude. Some of these include sun sensors, earth sensors, star sensors, magnetometers, and gyroscopes. Most sensors (except gyros) will make measurements to find the relationship between a measured direction and some known direction in the spacecraft. The type of sensor employed will be determined by such factors as accuracy requirements, mission profile, cost, and contractor preference. An earth sensor will vary considerably from a star sensor. For example, in a low earth orbit the earth might occupy up to 40% of the field of view, whereas a star could be treated as a point source of light. Earth sensors must be able to scan the earth and discriminate between where the earth surface actually stops and where the atmosphere starts (horizon). They can

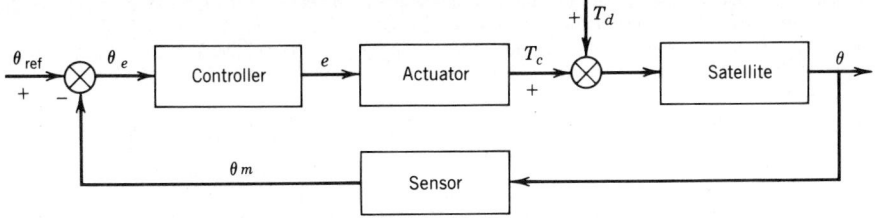

Fig. 7.29 Satellite ACS block diagram (single axis).

θ_{ref} = reference or desired attitude of satellite $\quad e$ = control signal

θ_m = measured attitude of satellite $\qquad\qquad T_c$ = commanded torque

θ_e = error signal $\qquad\qquad\qquad\qquad\qquad\quad T_d$ = disturbance torque

θ = actual attitude of satellite

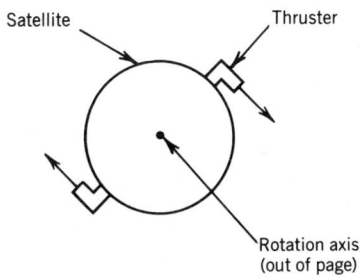

Fig. 7.30 Thruster pairs for satellite attitude control.

be optical or infrared and are subject to atmospheric disturbances. A magnetometer can provide only crude north-south information. Gyroscopes provide good short-term information, but due to long-term drift, they must be periodically updated using other means.

Active satellite attitude control systems are usually classified by actuator type or method used to apply the torque. The most common methods are (a) mass expulsion, (b) (angular) momentum exchange, and (c) magnetic torquing.

The most common mass expulsion technique involves using thrusters to apply a force about some moment arm to create a torque. The thrusters must be used in pairs to create a couple as shown in Fig. 7.30, so that there is no net force acting on the vehicle that would change the orbital dynamics.

As shown, the thruster would affect rotation about a single axis in a single direction. To be able to rotate in both directions about all three axes, a minimum of six pairs of thrusters are required. The thruster typically expels a gas that is either cold or hot. The cold gas is stored under pressure and released. Hot gases usually are the result of combining a fuel and an oxidizer (effectively a small rocket engine). The disadvantage of thrusters is the requirement for a large propellant supply for long-duration missions. It is also possible to accelerate mass electrically to provide the thrust. Electrical thrusters are primarily in the experimental stage, though.

Thruster control systems can be designed using phase plane analysis. To achieve high accuracy, thrusters with very small thrust are required. However, for rapid attitude maneuvers, large thrust is required, which may result in increased complexity due to multiple thruster sizes. Also, due to low structural damping, systems that use thrusters tend to limit cycle about the desired attitude.

One of the more common ways to torque a satellite involves exchanging momentum between the vehicle and an internal device that has angular momentum, such as a large gyroscope wheel. If no torques are acting on the satellite, the total angular momentum will be constant. Thus the satellite can be reoriented by exchanging the angular momentum of the vehicle with the angular momentum of the internal device. If a disturbance torque causes a change in the total angular momentum of the satellite, the vehicle can pass that momentum onto the momentum exchange device. Control moment gyroscopes, reaction wheels, and (bias) momentum wheels are common momentum exchange devices. The control moment gyroscope spins with a constant angular velocity and is mounted on a single or double gimbal, allowing one or two degrees of freedom of momentum exchange. Momentum is exchanged by changing the direction of the angular momentum vector of the gyro with respect to the vehicle. Reaction wheels and (bias) momentum wheels have a fixed orientation in the spacecraft, but change wheel speed to exchange angular momentum. The reaction wheel has a nominal spin angular velocity of zero, while the momentum wheel operates about some nonzero wheel speed, giving it a bias momentum.

Magnetic torquing is also used for attitude control. If the satellite contains electromagnets, they will interact with the earth's magnetic field, resulting in a torque. By using an orthogonal set of electromagnets and varying the strength of the magnetic field, the proper torques can be applied to the satellite. This approach is very limited in terms of the directions of the applied torque, so it is rarely used as a primary attitude control actuator. However, many satellites use magnetic torquing to get rid of excess momentum built up in the reaction wheels.

An important element of the active attitude control system is the control law, which consists of the computer software that determines how much actuator torque should be applied as a function of the difference between the measured and the desired state. The controller will be designed to allow the satellite to meet certain performance specifications such as speed of response, accuracy, and system stability. Control law design is straightforward for rigid bodies, although it is complicated by coupling between axes. Most modern space vehicles are large and flexible, characterized by low-frequency bending modes. The former is adequately handled using classical control theory, while the latter must resort to modern control techniques, including optimal filter theory and adaptive control. The actual controller design is beyond the scope of this handbook.

Redundancy

Due to the extremely high initial cost of satellite systems, the need for flight safety on manned missions, and the limited applications and high cost of on-orbit repair or recovery of failed satellites, there is a requirement for high reliability of satellite systems. This is done for the ACS using redundancy, both of individual components within an ACS and of the ACS itself. This requires a redundancy management system, which can be on board or controlled from the ground, that allows the satellite to be reconfigured when a failure occurs. For example, many satellites that use reaction wheels as actuators will have four of them in a conical array. If any one of the wheels fails, the three remaining wheels are sufficient to control the spacecraft. The space shuttle has 44 thrusters for actuator redundancy. A common design for a redundant ACS uses momentum exchange with magnetic torquers for momentum dumping as the primary ACS, with a thruster system for a backup ACS. If the satellite has to go to the secondary system, it will probably undergo degradation in system performance. Yet, it will not be inoperative, which would be the case if it experienced a failure and had no redundancy.

Conclusion

This section examined the need for and several techniques of controlling the attitude of spacecraft. The satellite attitude control system must be integrated with all of the other systems (e.g., mission sensors, power, thermal management, propulsion, communication, structures, etc.) and is usually complex, costly, and has a large impact on mission success.

REFERENCES

1. Battin, R. H., *Astronautical Guidance*, New York, McGraw-Hill, 1964.

2. Danby, J. M. A., *Fundamentals of Celestial Mechanics*, New York, Macmillan, 1962.

3. Kendrick, J. B. (ed.), *TRW Space Data*, Redondo Beach, California, TRW, 1967.

4. Brouwer, D., and Clemence, G. M., *Methods of Celestial Mechanics*, New York, Academic, 1961.

5. Escobal, P. R., *Methods of Orbit Determination*, New York, Wiley, 1965.

6. Bate, R. R., Mueller, D. D., and White, J. E., *Fundamentals of Astrodynamics*, Dover Publications, New York, 1971.

7. Pitman, G. R., *Inertial Guidance*, New York, Wiley, 1962.

8. Wertz, J. R. (ed.), *Spacecraft Attitude Determination and Control*, Dordrecht, Holland, Reidel Publishing, 1978.

9. Hughes, P. C., *Spacecraft Attitude Dynamics*, New York, Wiley, 1986.

10. Kaplan, M. H., *Modern Spacecraft Dynamics and Control*, New York, Wiley, 1976.

11. Likins, P. W., "Dynamics and Control of Flexible Space Vehicles" Jet Propulsion Laboratory, TR32-1329, Rev. 1, Jan. 1970.

BIBLIOGRAPHY FOR FOUNDATION OF ORBITAL MECHANICS

Baker, R. M. L., Jr., *Astrodynamics*, New York, Academic, 1967.

Baker, R. M. L., Jr., and Makemson, M. W., *An Introduction to Astrodynamics*, New York, Academic, 1960.

Battin, R. H., "Universal Formulae for Conic Trajectory Calculations," MIT Instrumental Lab. Rept. R-382, September 1962.

Dubiago, A., *The Determination of Orbits*, New York, Macmillan, 1961.

Gelb, A. (ed.), *Applied Optimal Estimation*, Cambridge, Mass., The M.I.T. Press, 1979.

Wintner, A., *Analytical Foundations of Celestial Mechanics*, Princeton, N.J., Princeton Univ. Press, 1941.

BIBLIOGRAPHY FOR INERTIAL NAVIGATION

Britting, K., *Inertial Navigation Systems Analysis*, New York, Wiley-Interscience, 1971.

Broxmeyer, C., *Inertial Navigation Systems*, New York, McGraw-Hill, 1964.

Draper, C., Wrigley, W., and Hovorka, J., *Inertial Guidance*, New York, Pergamon Press, 1960.

Farrell, J., *Integrated Aircraft Navigation*, New York, Academic Press, 1976.

Fernandez, M., and Macomber, G., *Inertial Guidance Engineering*, Englewood Cliffs, N.J., Prentice-Hall, 1962.

Leondes, C. T., *Guidance and Control of Aerospace Vehicles* (University of California Engineering and Science Extension Series), New York, McGraw-Hill, 1963.

Markey, W., and Hovorka, J., *The Mechanics of Inertial Position and Heading Indication*, New York, Wiley, 1961.

O'Donnell, C. G., *Inertial Navigation Analysis and Design*, New York, McGraw-Hill, 1964.

Parvin, R. H., *Inertial Navigation*, New York, Van Nostrand Co., 1962.

Savant, C., Jr., Howard, R., Solloway, C., and Savant, C. A., *Principles of Inertial Navigation*, New York, McGraw-Hill, 1961.

Slater, J. M., *Inertial Guidance Sensors*, New York, Reinhold, 1964.

Wrigley, W., Hollister, W. M., and Denhard, W. G., *Principles of Instrumentation*, Cambridge, Mass., MIT Press, 1969.

BIBLIOGRAPHY FOR SPACECRAFT ATTITUDE DYNAMICS AND CONTROL

Kane, T. R., Likins, P. W., and Levinson, D. A., *Spacecraft Dynamics*, New York, McGraw-Hill, 1983.

CHAPTER 8*
AUTOMATIC CONTROL

Part I
Suhada Jayasuriya

Department of Mechanical Engineering
Texas A & M University
College Station, Texas

Part II
Karl N. Reid

College of Engineering
Oklahoma State University
Stillwater, Oklahoma

Syed Hamid

Halliburton Services
Duncan, Oklahoma

Part III
T. Peter Neal

Aircraft Controls Division
Moog, Inc.
East Aurora, New York

Part IV
Krishnaswamy Srinivasan

Department of Mechanical Engineering
The Ohio State University
Columbus, Ohio

PART I   FEEDBACK CONTROL FUNDAMENTALS

This chapter was originally developed for *Instrumentation and Control*, edited by Chester Nachtigal, to be published by John Wiley & Sons, Inc. in 1990. The editors wish to give special recognition to Chester Nachtigal for his valuable help in the development and completion of this chapter.

PART II   ELECTROMECHANICAL AND ELECTROHYDRAULIC SERVOACTUATOR MODELING

PART I

FEEDBACK CONTROL FUNDAMENTALS
Suhada Jayasuriya

8.1 INTRODUCTION

The field of control has a rich heritage of intellectual depth and practical achievement. From the water clock of Ctesibius in ancient Alexandria, where feedback control was used to regulate the flow of water, to the space exploration and the automated manufacturing plants of today, control systems have played a very significant role in technological and scientific development. James Watt's flyball governor (1769) was essential for the operation of the steam engine, which was, in turn, a technology fueling the Industrial Revolution. The fundamental study of feedback begins with James Clerk Maxwell's analysis of system stability of steam engines with governors (1868). Giant strides in our understanding of feedback and its use in design resulted from the pioneering work of Black, Nyquist, and Bode at Bell Labs in the 1920s. Minorsky's work on ship steering was of exceptional practical and theoretical importance. Tremendous advances occurred during World War II in response to the pressing problems of that period. The technology developed during the war years lead, over the next 20 years, to practical applications in many fields.

Since the 1960s, there have been many challenges and spectacular achievements in space. The guidance of the Apollo spacecraft on an optimized trajectory from the earth to the moon and the soft landing on the moon depended heavily on control engineering. Today, the shuttle relies on automatic control in all phases of its flight. In aeronautics, the aircraft autopilot, the control of high-performance jet engines, and ascent/descent trajectory optimization to conserve fuel are typical examples of control applications. Currently, feedback control makes it possible to design aircraft that are aerodynamically unstable (such as the X-29) so as to achieve high performance. The National Aerospace Plane will rely on advanced control algorithms to fly its demanding missions.

Control systems are providing dramatic new opportunities in the automotive industry. Feedback controls for engines permit federal emission levels to be met, while antiskid braking control systems provide enhanced levels of passenger safety. In consumer products, control systems are often a critical factor in performance and thus economic success. From simple thermostats that regulate temperature in buildings to the control of the optics for compact disk systems, from garage door openers to the head servos for computer hard disk drives, and from artificial hearts to remote manipulators, control applications have permeated every aspect of life in industrialized societies.

In process control, where systems may contain hundreds of control loops, adaptive controllers have been available commercially since 1983. Typically, even a small improvement in yield can be quite significant economically. Multivariable control algorithms are now being implemented by several large companies. Moreover, improved control algorithms also permit inventories to be reduced, a particularly important consideration in processing dangerous material. In nuclear reactor control, improved control algorithms can have significant safety and economic consequences. In power systems, coordinated computer control of a large number of variables is becoming common. Over 30,000 computer control systems have been installed in the United States alone. Again, the economic impact of control is vast.

Accomplishments in the defense area are legion. The accomplishments range from the antiaircraft gunsights and the bombsights of World War II to the missile autopilots of today and to the identification and estimation techniques used to track and designate multiple targets.

A large body of knowledge has come into existence as a result of these developments and continues to grow at a very rapid rate. A cross section of this body of knowledge is collected in this

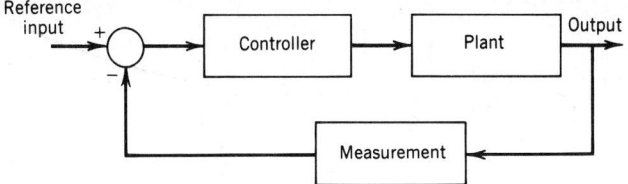

Fig. 8.1   Closed-loop system configuration.

chapter. It includes analysis tools and design methodologies based on classical and state-space techniques. In presenting this material, some familiarity with general control system design principles is assumed. As a result, detailed derivations have been kept to a minimum with appropriate references. The developments up to the 1970s are included since they are by now well understood. The developments in the 1980s are still at a research stage and are not included.

This chapter is organized into four broad sections. Parts I–III deal with classical techniques of control system design based on the notion of the transfer function. Only single-input, single-output systems are treated here. In Part IV multivariable design methods based on state-space techniques are given.

8.1.1   Closed-Loop versus Open-Loop Control

In a closed-loop control system the output is measured and used to alter the control inputs applied to the plant under control. Figure 8.1 shows such a closed-loop system.

If output measurements are not utilized in determining the plant input, then the plant is said to be under open-loop control. An open-loop control system is shown in Fig. 8.2.

An open-loop system is very sensitive to any plant parameter perturbations and any disturbances that enter the plant. So an open-loop control system is effective only when the plant and disturbances are exactly known. In real applications, however, neither plants nor disturbances are exactly known a priori. In such situations open-loop control does not provide satisfactory performance. When the loop is closed as in Fig. 8.1 any plant perturbations or disturbances can be sensed by measuring the outputs, thus allowing plant input to be altered appropriately.

The main reason for closed-loop control is the need for systems to perform well in the presence of uncertainties. It can reduce the sensitivity of the system to plant parameter variations and help reject or mitigate external disturbances. Among other attributes of closed-loop control is the ability to alter the overall dynamics to provide adequate stability and good tracking characteristics. Consequently, a closed-loop system is more complex than an open-loop system due to the required sensing of the output. Sensor noise, however, tends to degrade the performance of a feedback system thus making it necessary to have accurate sensing. Therefore sensors are usually the most expensive devices in a feedback control system.

8.1.2   Supervisory Control Computers[1]

With the advent of digital computer technology computers have found their way into control applications. The earliest applications of digital computers in industrial process control were in a so-called supervisory control mode, and the computers were generally large-scale mainframes (minis and micros had not yet been developed). In supervisory control the individual feedback loops are locally controlled by typical analog devices used prior to the installation of the computer. The main

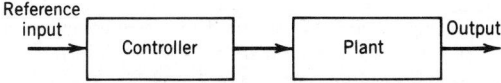

Fig. 8.2   Open-loop system configuration.

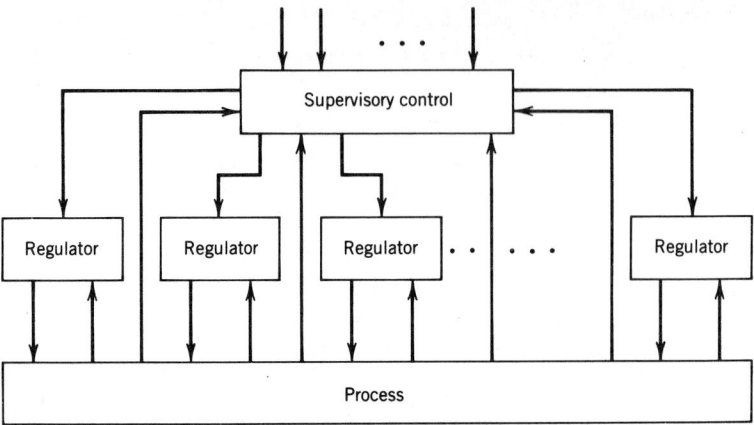

Fig. 8.3   Supervisory control configuration.

function of the computer is to gather information on how the entire process is operating, then feed this into an overall process model located in the computer memory, and then periodically send signals to the set points of individual local analog controllers so that the overall performance of the system is enhanced. A conceptual block diagram of such supervisory control is shown in Fig. 8.3.

8.1.3   Hierarchical Control Computers[1]

In order to continuously achieve optimum overall performance the function of the supervisory control computer can be broken up into several levels and operated in a hierarchical manner. Such a multilevel approach becomes useful in controlling very complex processes, with large numbers of control loops with many inputs and controlled variables. In the hierarchical approach the system is subdivided into a hierarchy of simpler control design problems rather than attempting direct synthesis of a single comprehensive controller for the entire process. Thus the controller on a given "level of control" can be less complex due to the existence of lower-level controllers that remove frequently occurring disturbances. At each higher level in the hierarchy, the complexity of the control algorithm increases, but the required frequency of execution decreases. Such a system is shown in Fig. 8.4.

8.1.4   Direct Digital Control (DDC)

In direct digital control all the analog controllers at the lowest level or intermediate levels are replaced by a central computer serving as a single time-shared controller for all the individual feedback loops. Conventional control modes such as PI or PID (proportional integral differential) are still used for each loop, but the digital versions of the control laws for each loop reside in software in the central computer. In DDC the computer input and output are multiplexed sequentially through the list of control loops, updating each loop's control action in turn and holding this value constant until the next cycle. A typical DDC configuration is shown in Fig. 8.5.

8.1.5   Hybrid Control

Combinations of analog and digital methods based on individual loop controllers is known as hybrid control. It is clear that any real control system would have both analog and digital features. For example, in Fig. 8.5, the plant outputs and inputs are continuous (or analog) quantities. The inputs and outputs of the digital computer are nevertheless digital. The computer processes only numbers. All control algorithms, however, need not be implemented digitally. As an example, for most feedback control loops, the PID algorithm implemented with analog circuitry is superior to the PID algorithm implemented in digital software. The derivative mode for instance requires very high resolution to approach the quality of analog performance. Thus it is sometimes advantageous to use hybrid systems for controller implementation.

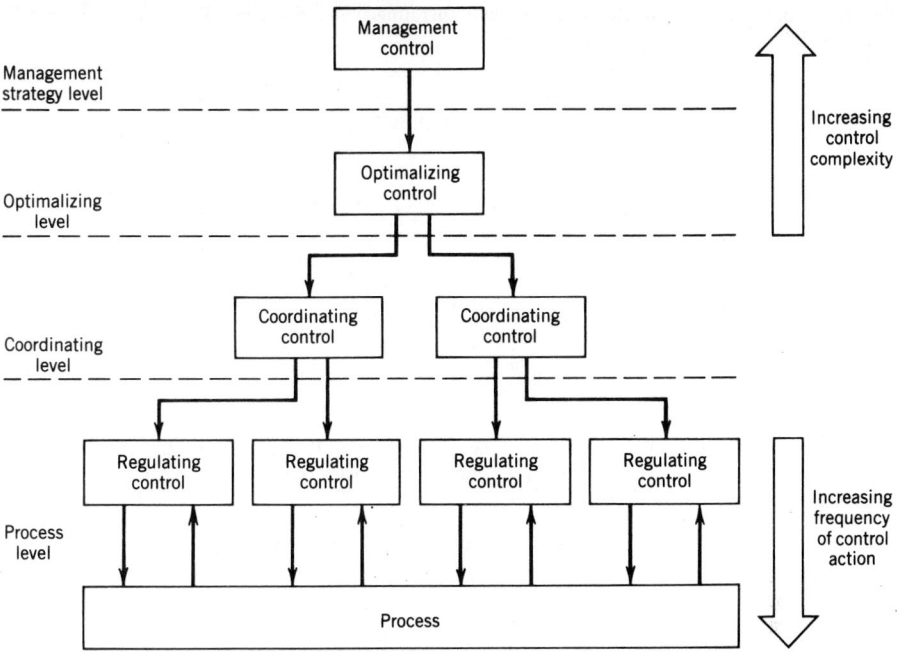

Fig. 8.4 Multilevel hierarchical control structure.

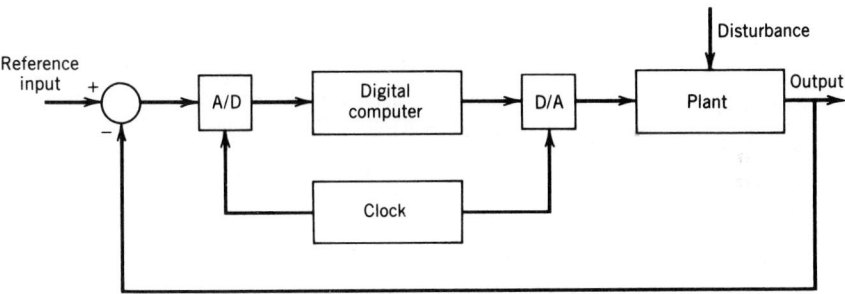

Fig. 8.5 Direct digital control (DDC) configuration.

8.1.6   Real Systems with Digital Control

Real systems are typically mixtures of analog and digital signals. Typical plant inputs and outputs are analog signals, that is, actuators and sensors are analog systems. If a digital computer is included in the control loop, then digitized quantities are needed for their processing. These are usually accomplished by using analog-to-digital (A/D) converters. In order to use real-world actuators, computations done as numbers need to be converted to analog signals by employing digital to analog (D/A) converters. One of the main advantages of digital control is the ease with which an algorithm can be changed on line by altering the software rather than hardware.

8.2   LAPLACE TRANSFORMS

Often in designing control systems, linear time-invariant (LTI) differential equation representations of physical systems to be controlled are sought. These are typically arrived at by appropriately linearizing the nonlinear equations about some operating point. The general form of such an LTI ordinary differential equation representation is

$$a_n \frac{d^n y(t)}{dt^n} + a_{n-1} \frac{d^{n-1} y(t)}{dt^{n-1}} + \cdots + a_1 \frac{dy(t)}{dt} + a_0 y(t)$$

$$= b_m \frac{d^m}{dt^m} u(t) + b_{m-1} \frac{d^{m-1}}{dt^{m-1}} u(t) + \cdots + b_1 \frac{du(t)}{dt} + b_0 u(t) \qquad (8.1)$$

where $y(t)$ = output of the system
  $u(t)$ = input to the system
   t = time
  a_j, b_j = physical parameters of the system

and $n \geq m$ for physical systems.

The ability to transform systems of the form given by Eq. (8.1) to algebraic equations relating the input to the output is the primary reason for employing Laplace transform techniques.

8.2.1   Single-Sided Laplace Transform

The Laplace transform $\mathcal{L}[f(t)]$ of the time function $f(t)$ defined as

$$f(t) = \begin{cases} 0 & t < 0 \\ f(t) & t \geq 0 \end{cases}$$

is given by

$$\mathcal{L}[f(t)] = F(s) = \int_0^\infty e^{-st} f(t)\, dt \qquad t > 0 \qquad (8.2)$$

where s is a complex variable = $\sigma + j\omega$. The integral of Eq. (8.2) cannot be evaluated in a closed form for all $f(t)$ but when it can, it establishes a unique pair of functions, $f(t)$, in the time domain and its companion $F(s)$ in the s domain. It is conventional to use uppercase letters for s functions and lowercase for t functions.

Example 8.1. Determine the Laplace transform of the unit step function $u_s(t)$:

$$u_s(t) = \begin{cases} 0 & t < 0 \\ 1 & t \geq 0 \end{cases}$$

By definition

$$\mathcal{L}[u_s(t)] = U_s = \int_0^\infty e^{-ts} 1\, dt = \frac{1}{s}$$

Example 8.2. Determine the Laplace transform of $f(t)$:

$$f(t) = \begin{cases} 0 & t < 0 \\ e^{-\alpha t} & t \geq 0 \end{cases}$$

$$\mathcal{L}[f(t)] = F(s) = \int_0^\infty e^{-ts} e^{-\alpha t}\, dt = \frac{1}{s + \alpha}$$

Example 8.3. Determine the Laplace transform of the function $f(t)$ given by

$$f(t) = \begin{cases} 0 & t < 0 \\ t & 0 \leq t \leq T \\ T & T \leq t \end{cases}$$

By definition

$$F(s) = \int_0^\infty e^{-ts} f(t)\, dt$$

$$= \int_0^T e^{-ts} t\, dt + \int_T^\infty e^{-ts} T\, dt$$

$$= \frac{1}{s^2} - \frac{e^{-Ts}}{s^2} = \frac{1}{s^2}(1 - e^{-Ts})$$

In transforming differential equations, entire equations need to be transformed. Several theorems useful in such transformations are given next without proof.

T1. Linearity Theorem

$$\mathscr{L}[\alpha f(t) + \beta g(t)] = \alpha \mathscr{L}[f(t)] + \beta \mathscr{L}[g(t)] \tag{8.3}$$

T2. Differentiation Theorem

$$\mathscr{L}\left[\frac{df}{dt}\right] = sF(s) - f(0) \tag{8.4}$$

$$\mathscr{L}\left[\frac{d^2f}{dt^2}\right] = s^2 F(s) - sf(0) - \frac{df}{dt}(0) \tag{8.5}$$

$$\mathscr{L}\left[\frac{d^n f}{dt^n}\right] = s^n F(s) - s^{n-1} f(0) - s^{n-2}\frac{df}{dt}(0) - \cdots - \frac{d^{n-1}}{dt^{n-1}}(0) \tag{8.6}$$

T3. Translated Function (Fig. 8.6)

$$\mathscr{L}[f(t - \alpha)u_s(t - \alpha)] = e^{-\alpha s} F(s + \alpha) \tag{8.7}$$

T4. Multiplication of $f(t)$ by $e^{-\alpha t}$

$$\mathscr{L}[e^{-\alpha t} f(t)] = F(s + \alpha) \tag{8.8}$$

T5. Integration Theorem

$$\mathscr{L}[\smallint f(t)\, dt] = \frac{F(s)}{s} + \frac{f^{-1}(0)}{s} \tag{8.9}$$

where $f^{-1}(0) = \smallint f(t)\, dt$ evaluated at $t = 0$.

T6. Final-Value Theorem. If $f(t)$ and $df(t)/dt$ are Laplace transformable, if $\lim_{t\to\infty} f(t)$ exists, and if $F(s)$ is analytic in the right-half s plane including the $j\omega$ axis, except for a single pole at the origin,

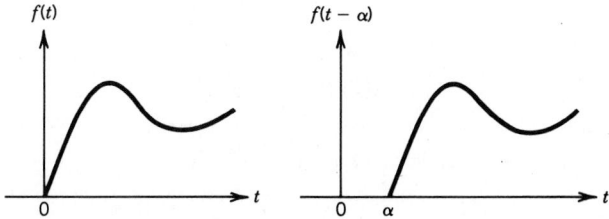

Fig. 8.6 Plots of $f(t)$ and $f(t - \alpha)u_s(t - \alpha)$.

then

$$\lim_{t \to \infty} f(t) = \lim_{s \to 0} sF(s) \tag{8.10}$$

T7. Initial Value Theorem. If $f(t)$ and $df(t)/dt$ are both Laplace transformable, and if $\lim_{s \to \infty} sF(s)$ exists, then

$$f(0) = \lim_{s \to \infty} sF(s) \tag{8.11}$$

Example 8.4. The time function of Example 8.3 can be written as

$$f(t) = tu_s(t) - (t - T)u_s(t - T) \tag{8.12}$$

and

$$\mathscr{L}[f(t)] = \mathscr{L}[tu_s(t)] - \mathscr{L}[(t - T)u_s(t - T)] \tag{8.13}$$

But

$$\mathscr{L}[tu_s(t)] = \frac{1}{s^2} \tag{8.14}$$

By using Eqs. (8.14) and (8.7) in Eq. (8.13), we get

$$F(s) = \frac{1}{s^2} - e^{-Ts}\frac{1}{s^2} = \frac{1}{s^2}(1 - e^{-Ts})$$

8.2.2 Transforming LTI Ordinary Differential Equations

The Laplace transform method yields the complete solution (the particular solution plus the complementary solution) of linear differential equations. Classical methods of finding the complete solution of a differential equation require the evaluation of the integration constants by use of the initial conditions. In the case of the Laplace transform method, initial conditions are automatically included in the Laplace transform of the differential equation. If all initial conditions are zero, then the Laplace transform of the differential equation is obtained simply by replacing d/dt with s, d^2/dt^2 with s^2, and so on.

Consider the differential equation

$$5\frac{d^2y}{dt^2} + 3y = f(t) \tag{8.15}$$

with $\dot{y}(0) = 1$, $y(0) = 0$. By taking the Laplace transform of Eq. (8.15), we get

$$\mathscr{L}\left[5\frac{d^2y}{dt^2} + 3y\right] = \mathscr{L}[f(t)]$$

Now using T1 and T2 yield

$$5[s^2Y(s) - sy(0) - y(0)] + 3Y(s) = F(s)$$

Thus

$$Y(s) = \frac{F(s) + 5}{5s^2 + 3} \tag{8.16}$$

For a given $f(t)$, say $f(t) = u_s(t)$, the unit step function

$$F(s) = \frac{1}{s} \tag{8.17}$$

Substituting Eq. (8.17) in Eq. (8.16) gives

$$Y(s) = \frac{5s + 1}{s(5s^2 + 3)}$$

$y(t)$ can then be found by computing the inverse Laplace transform of $Y(s)$. That is

$$y(t) = \mathscr{L}^{-1}[Y(s)] = \mathscr{L}^{-1}\left\{\frac{5s + 1}{s(5s^2 + 3)}\right\}$$

This will be discussed in Section 8.2.4.

8.2.3 Transfer Function

The transfer function of a linear time-invariant system is defined to be the ratio of the Laplace transform of the output to the Laplace transform of the input under the assumption that all initial conditions are zero.

For the system described by the LTI differential equation (8.1),

$$\frac{\mathscr{L}[\text{output}]}{\mathscr{L}[\text{input}]} = \frac{\mathscr{L}[y(t)]}{\mathscr{L}[u(t)]} = \frac{Y(s)}{U(s)}$$

$$= \frac{b_m s^m + b_{m-1} s^{m-1} + \cdots + b_1 s + b_0}{a_n s^n + a_{n-1} s^{n-1} + \cdots + a_1 s + a_0} \qquad (8.18)$$

It should be noted that the transfer function depends only on the system parameters characterized by a_i's and b_i's in Eq. (8.18). The highest power of s in the denominator of the transfer function defines the order n of the system. The order n of a system is typically greater than or equal to the order of the numerator polynomial m.

Equation (8.18) can be further written in the form

$$\frac{Y(s)}{U(s)} = G(s) = \frac{b_m}{a_n} \frac{\prod_{j=1}^{m}(s + z_i)}{\prod_{j=1}^{n}(s + p_j)} \qquad (8.19)$$

The values of s making Eq. (8.19) equal to zero are called the system zeros, and the values of s making Eq. (8.19) go to ∞ are called poles of the system. Hence $s = -z_i$, $i = 1, \ldots, m$ are the system zeros and $s = -p_j$, $j = 1, \ldots, n$ are the system poles.

8.2.4 Partial Fraction Expansion and Inverse Transform

The mathematical process of passing from the complex variable expression $F(s)$ to the time domain expression $f(t)$ is called an inverse transformation. The notation for the inverse Laplace transformation is $\mathscr{L}^{-1}$, so that

$$\mathscr{L}^{-1}[F(s)] = f(t)$$

Example 8.5. The inverse Laplace transform of $1/s$ is the unit step function:

$$\mathscr{L}[u_s(t)] = \frac{1}{s}$$

Hence

$$\mathscr{L}^{-1}\left(\frac{1}{s}\right) = u_s(t)$$

Time functions for which Laplace transforms are found in a closed form can be readily inverse Laplace transformed by writing them in pairs. Such pairs are listed in Table 8.1. When transform

TABLE 8.1   Laplace Transform Pairs

	$f(t)$	$F(s)$
1	Unit impulse $\delta(t)$	1
2	Unit step $u_s(t)$	$\dfrac{1}{s}$
3	t	$\dfrac{1}{s^2}$
4	e^{-at}	$\dfrac{1}{s+a}$
5	te^{-at}	$\dfrac{1}{(s+a)^2}$
6	$\sin \omega t$	$\dfrac{\omega}{s^2+\omega^2}$
7	$\cos \omega t$	$\dfrac{s}{s^2+\omega^2}$
8	$t^n \quad (n=1,2,3,\ldots)$	$\dfrac{n!}{s^{n+1}}$
9	$t^n e^{-at} \quad (n=1,2,3,\ldots)$	$\dfrac{n!}{(s+a)^{n+1}}$
10	$\dfrac{1}{b-a}(e^{-at}-e^{-bt})$	$\dfrac{1}{(s+a)(s+b)}$
11	$\dfrac{1}{b-a}(be^{-bt}-ae^{-at})$	$\dfrac{s}{(s+a)(s+b)}$
12	$\dfrac{1}{ab}\left[1+\dfrac{1}{a-b}(be^{-at}-ae^{-bt})\right]$	$\dfrac{1}{s(s+a)(s+b)}$
13	$e^{-at}\sin \omega t$	$\dfrac{\omega}{(s+a)^2+\omega^2}$
14	$e^{-at}\cos \omega t$	$\dfrac{s+a}{(s+a)^2+\omega^2}$
15	$\dfrac{1}{a^2}(at-1+e^{-at})$	$\dfrac{1}{s^2(s+a)}$
16	$\dfrac{\omega_n}{\sqrt{1-\zeta^2}}e^{-\zeta\omega_n t}\sin \omega_n\sqrt{1-\zeta^2}\,t$	$\dfrac{\omega_n^2}{s^2+2\zeta\omega_n s+\omega_n^2}$
17	$\dfrac{-1}{\sqrt{1-\zeta^2}}e^{-\zeta\omega_n t}\sin(\omega_n\sqrt{1-\zeta^2}\,t-\phi)$ $\phi=\tan^{-1}\dfrac{\sqrt{1-\zeta^2}}{\zeta}$	$\dfrac{s}{s^2+2\zeta\omega_n s+\omega_n^2}$
18	$1-\dfrac{1}{\sqrt{1-\zeta^2}}e^{-\zeta\omega_n t}\sin(\omega_n\sqrt{1-\zeta^2}\,t+\phi)$ $\phi=\tan^{-1}\dfrac{\sqrt{1-\zeta^2}}{\zeta}$	$\dfrac{\omega_n^2}{s(s^2+2\zeta\omega_n s+\omega_n^2)}$

pairs are not found in tables, other methods have to be used to find the inverse Laplace transform. One such method is the partial fraction expansion of a given Laplace transform.

If $F(s)$, the Laplace transform of $f(t)$, is broken up into components

$$F(s) = F_1(s) + F_2(s) + \cdots + F_n(s)$$

and if the inverse Laplace transforms of $F_1(s)$, $F_2(s)$,..., $F_n(s)$ are readily available, then

$$\mathcal{L}^{-1}[F(s)] = \mathcal{L}^{-1}[F_1(s)] + \mathcal{L}^{-1}[F_2(s)] + \cdots + \mathcal{L}^{-1}[f_n(s)]$$
$$= f_1(t) + f_2(t) + \cdots + f_n(t) \tag{8.20}$$

where $f_1(t)$, $f_2(t)$,..., $f_n(t)$ are the inverse Laplace transforms of $F_1(s)$, $F_2(s)$,..., $F_n(s)$, respectively.

For problems in control systems, $F(s)$ is frequently in the following form:

$$F(s) = \frac{B(s)}{A(s)}$$

where $A(s)$ and $B(s)$ are polynomials in s, and the degree of $B(s)$ is equal to or higher than that of $A(s)$.

If $F(s)$ is written as in Eq. (8.19) and if the poles of $F(s)$ are distinct, then $F(s)$ can always be expressed in terms of simple partial fractions as follows:

$$F(s) = \frac{B(s)}{A(s)} = \frac{\alpha_1}{s + p_1} + \frac{\alpha_2}{s + p_2} + \cdots + \frac{\alpha_n}{s + p_n} \tag{8.21}$$

where the α_j's are constant. Here α_j is called the residue at the pole $s = -p_j$. The value of α_j is found by multiplying both sides of Eq. (8.21) by $(s + p_j)$ and setting $s = -p_j$, giving

$$\alpha_j = \left[(s + p_j) \frac{B(s)}{A(s)} \right]_{s = -p_j} \tag{8.22}$$

Noting that

$$\mathcal{L}^{-1}\left[\frac{1}{s + p_j} \right] = e^{-p_j t}$$

the inverse Laplace transform of Eq. (8.21) can be written as

$$f(t) = \mathcal{L}^{-1}[F(s)] = \sum_{j=1}^{n} \alpha_j e^{-p_j t} \tag{8.23}$$

Example 8.6. Find the inverse Laplace transform of

$$F(s) = \frac{s + 1}{(s + 2)(s + 3)}$$

The partial fraction expansion of $F(s)$ is

$$F(s) = \frac{s + 1}{(s + 2)(s + 3)} = \frac{\alpha_1}{s + 2} + \frac{\alpha_2}{s + 3}$$

where α_1 and α_2 are found by using Eq. (8.22) as follows:

$$\alpha_1 = \left[(s + 2) \frac{s + 1}{(s + 2)(s + 3)} \right]_{s = -2} = -1$$

$$\alpha_2 = \left[(s + 3) \frac{s + 1}{(s + 2)(s + 3)} \right]_{s = -3} = 2$$

Thus

$$f(t) = \mathcal{L}^{-1}\left[\frac{-1}{s+2}\right] + \mathcal{L}^{-1}\left[\frac{2}{s+3}\right]$$

$$= -e^{-2t} + 2e^{-3t} \qquad t \geq 0$$

Partial fraction expansion when $F(s)$ involves multiple poles: Consider

$$F(s) = \frac{K\prod_{i=1}^{m}(s+z_i)}{(s+p_1)^r \prod_{j=r+1}^{n}(s+p_j)} = \frac{B(s)}{A(s)}$$

where the pole at $s = -p_1$ has multiplicity r. The partial fraction expansion of $F(s)$ may then be written as

$$F(s) = \frac{B(s)}{A(s)} = \frac{\beta_1}{s+p_1} + \frac{\beta_2}{(s+p_1)^2} + \cdots + \frac{\beta_r}{(s+p_1)^r}$$

$$+ \sum_{j=r+1}^{n}\frac{\alpha_j}{s+p_j} \qquad (8.24)$$

The α_j's can be evaluated as before using Eq. (8.22). The β_i's are given by

$$\beta_r = \left[\frac{B(s)}{A(s)}(s+p_1)^r\right]_{s=-p_1}$$

$$\beta_{r-1} = \left\{\frac{d}{ds}\left[\frac{B(s)}{A(s)}(s+p_1)^r\right]\right\}_{s=-p_1}$$

$$\vdots$$

$$\beta_{r-j} = \frac{1}{j!}\left\{\frac{d^j}{ds^j}\left[\frac{B(s)}{A(s)}(s+p_1)^r\right]\right\}_{s=-p_1}$$

$$\vdots$$

$$\beta_1 = \frac{1}{(r-1)!}\left\{\frac{d^{r-1}}{ds^{r-1}}\left[\frac{B(s)}{A(s)}(s+p_1)^r\right]\right\}_{s=-p_1}$$

The inverse Laplace transform of Eq. (8.24) can then be obtained as follows:

$$f(t) = \left[\frac{\beta_r}{(r-1)!}t^{r-1} + \frac{\beta_{r-1}}{(r-2)!}t^{r-2} + \cdots + \beta_2 t + \beta_1\right]e^{-p_1 t}$$

$$+ \sum_{j=r+1}^{n}\alpha_j e^{-p_j t} \qquad t \geq 0 \qquad (8.25)$$

Example 8.7. Find the inverse Laplace transform of the function

$$F(s) = \frac{s^2 + 2s + 3}{(s+1)^3}$$

Expanding $F(s)$ into partial fractions, we obtain

$$F(s) = \frac{B(s)}{A(s)} = \frac{\beta_1}{s+1} = \frac{\beta_2}{(s+1)^2} + \frac{\beta_3}{(s+1)^3}$$

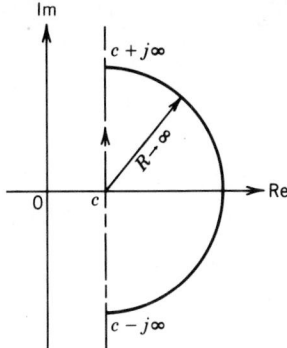

Im

$c + j\infty$

$R \to \infty$

0 c → Re

$c - j\infty$

Fig. 8.7 Path of integration.

where β_1, β_2, and β_3 are determined as already shown.

$$\beta_3 = \left[\frac{B(s)}{A(s)} (s + 1)^3 \right]_{s = -1} = 2$$

similarly $\beta_2 = 0$ and $\beta_1 = 1$. Thus we get

$$f(t) = \mathscr{L}^{-1} \left[\frac{1}{s + 1} \right] + \mathscr{L}^{-1} \left[\frac{2}{(s + 1)^3} \right]$$

$$= e^{-t} + t^2 e^{-t} \qquad t \geq 0$$

8.2.5 Inverse Transform by a General Formula

Given the Laplace transform $F(s)$ of a time function $f(t)$, the following expression holds true:

$$f(t) = \frac{1}{2\pi j} \int_{c - j\infty}^{c + j\infty} F(s) e^{ts} \, ds \qquad (8.26)$$

where c, the abscissa of convergence, is a real constant and is chosen larger than the real parts of all singular points of $F(s)$. Thus the path of integration is parallel to the $j\omega$ axis and is displaced by the amount c from it (see Fig. 8.7). This path of integration is to the right of all singular points. Equation (8.26) need be used only when other simpler methods cannot provide the inverse Laplace transformation.

8.3 BLOCK DIAGRAMS

A control system consists of a number of components. In order to show the functions performed by each component, we commonly use a diagram called a **block diagram**.

In a block diagram, all system variables are linked to each other through blocks. The block is a symbol for the mathematical operation on the input signal to the block that produces the output. The transfer functions of the components are usually entered in the blocks, with blocks connected by arrows to indicate the direction of the flow of signals. A basic assumption in block diagrams is that there is no loading between blocks. Figure 8.8 shows a simple block diagram with two blocks in cascade. The arrowheads pointing toward blocks indicate inputs and those pointing away indicate outputs. These arrows are referred to as signals. The following basic components allow the generation of many complex block diagrams.

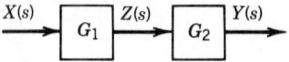

Fig. 8.8 Two blocks in cascade.

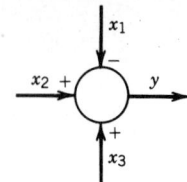

Fig. 8.9 Summing point.

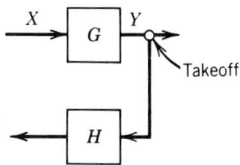

Fig. 8.11 Takeoff point.

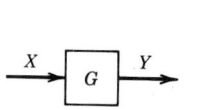

Fig. 8.10 Multiplication $Y = GX$.

Addition. Signal addition is represented by a summing point as shown in Fig. 8.9.

It should be noted that there can be only one signal leaving a summing point though any number of signals can enter a summing point. For the summing point shown in Fig. 8.9, we have

$$-x_1 + x_2 + x_3 = y$$

The sign placed near an arrow indicates whether the signal is to be added or subtracted.

Multiplication. Multiplication is denoted by a symbol as shown in Fig. 8.10. Here X the input and Y the output are related by the expression

$$Y = GX$$

Takeoff Point. If a signal becomes an input to more than one element, then a takeoff point as shown in Fig. 8.11 is employed.

A typical block diagram using these elements is shown in Fig. 8.12.

In the block diagram of Fig. 8.12 it is assumed that G_2 will have no back reaction (or loading) on G_1; G_1 and G_2 usually represent two physical devices. If there are any loading effects between the devices, it is necessary to combine these components into a single block. Such a situation is given in Section 8.3.2.

8.3.1 Block Diagram Reduction

Any number of cascaded blocks representing nonloading components can be replaced by a single block, the transfer function of which is simply the product of the individual transfer functions. For example, the two elements in cascade shown in Fig. 8.12 can be replaced by a single element $G_0 = G_1 G_2$. Some fundamental block diagram reduction rules are shown in Fig. 8.13.

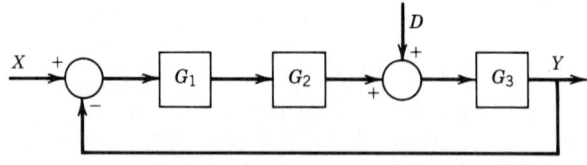

Fig. 8.12 Typical block diagram.

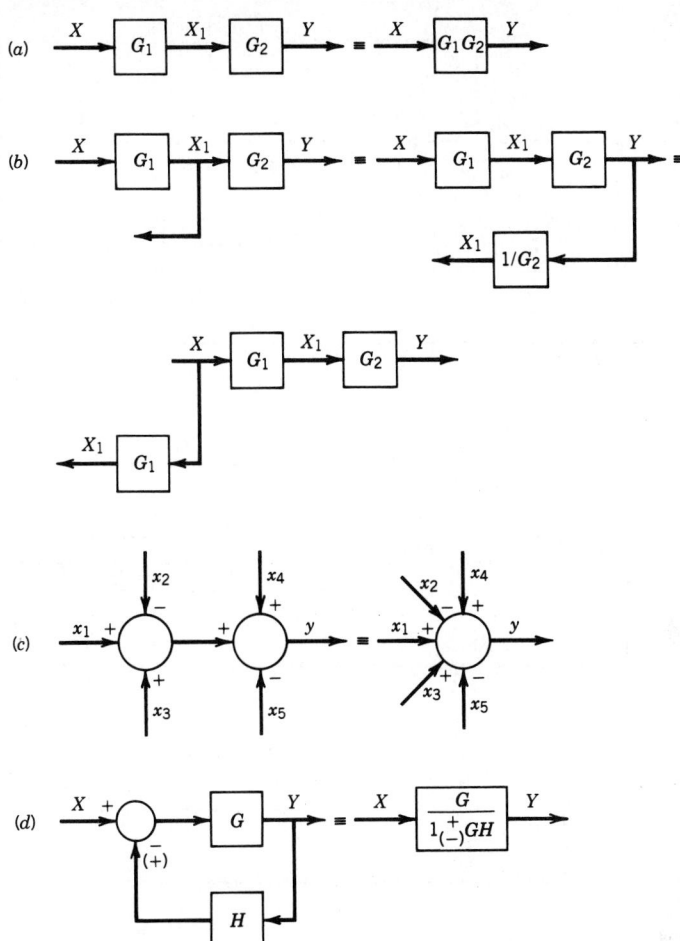

Fig. 8.13 Basic block diagram reduction rules.

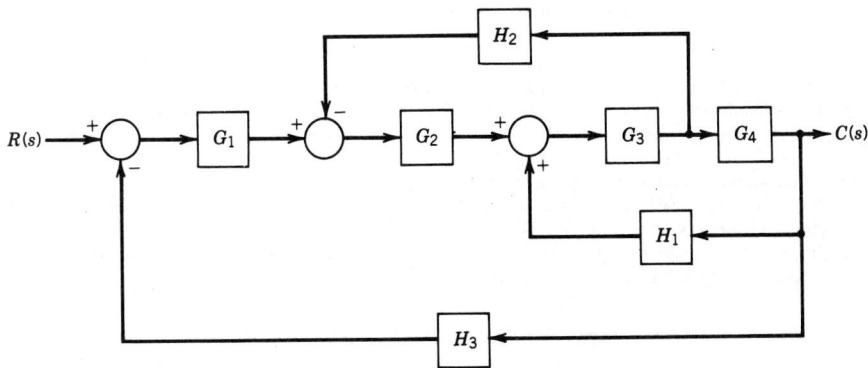

Fig. 8.14 Multiple-loop feedback control system.

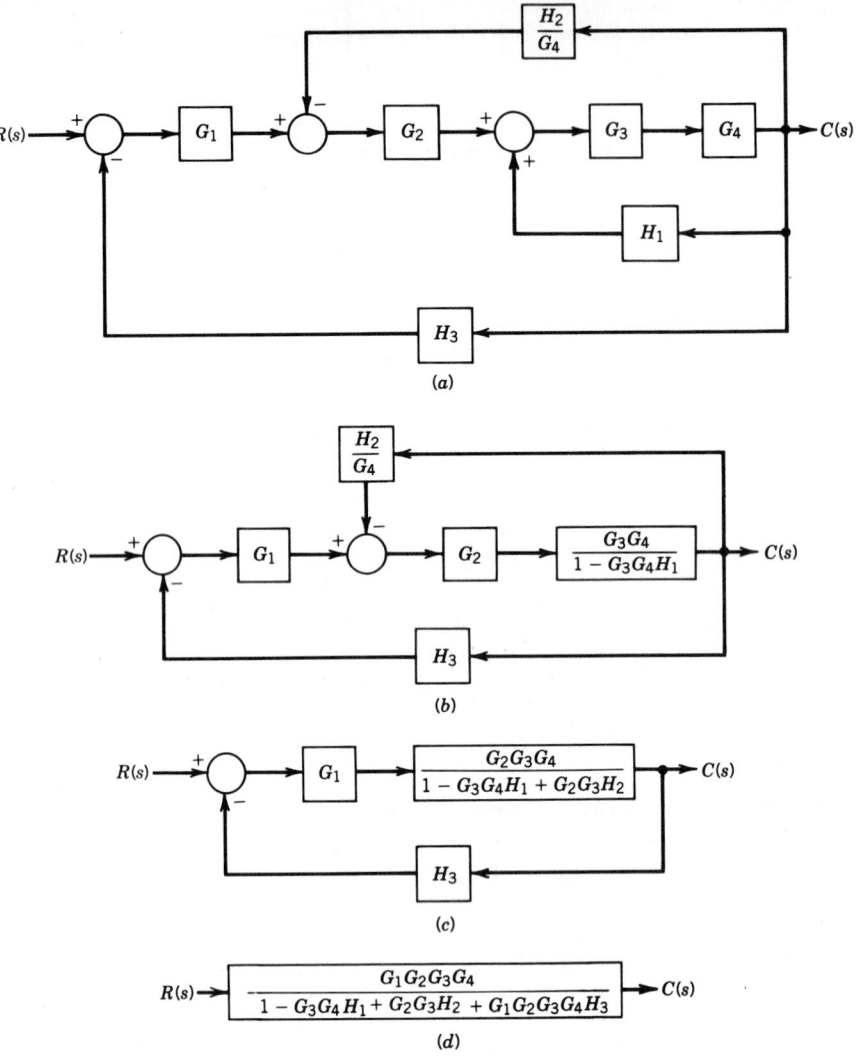

Fig. 8.15 Block diagram reduction of the system of Fig. 8.14.

In simplifying a block diagram a general rule is to first move takeoff points and summing points to form internal feedback loops as shown in Fig. 8.13(d). Then remove the internal feedback loops and proceed in the same manner.

Example 8.8. Reduce the block diagram[2] shown in Fig. 8.14.

First, in order to eliminate the loop $G_3 G_4 H_1$, we move H_2 behind block G_4 and therefore obtain Fig. 8.15(a). Eliminating the loop $G_3 G_4 H_1$, we obtain Fig. 8.15(b). Then eliminating the inner loop containing H_2/G_4, we obtain Fig. 8.15(c). Finally, by reducing the loop containing H_3, we obtain the closed-loop system transfer function as shown in Fig. 8.15(d).

8.3.2 Transfer Functions of Cascaded Elements[3]

Many feedback systems have components that load each other. Consider the system of Fig. 8.16. Assume that e_i is the input and e_o is the output. In this system the second stage of the circuit ($R_2 C_2$ portion) produces a loading effect on the first stage ($R_1 C_1$ portion). The equations for this system are

$$\frac{1}{C_1}\int (i_1 - i_2)\, dt + R_1 i_1 = e_i \qquad (8.27)$$

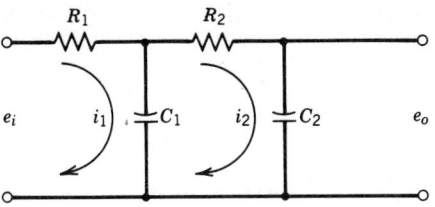

Fig. 8.16 Electrical system.

and

$$\frac{1}{C_1} \int (i_1 - i_2)\, dt + R_2 i_2 = -\frac{1}{C_2} \int i_2 \, dt = -e_0 \tag{8.28}$$

Taking the Laplace transform of Eqs. (8.27) and (8.28), assuming zero initial conditions, and simplifying yield

$$\frac{E_o(s)}{E_i(s)} = \frac{1}{R_1 C_1 R_2 C_2 s^2 + (R_1 C_1 + R_2 C_2 + R_1 C_2)s + 1} \tag{8.29}$$

The term $R_1 C_2 s$ in the denominator of the transfer function represents the interaction of two simple RC circuits.

This analysis shows that if two RC circuits are connected in cascade so that the output from the first circuit is the input to the second, the overall transfer function is not the product of $1/(R_1 C_1 s + 1)$ and $1/(R_2 C_2 s + 1)$. The reason for this is that when we derive the transfer function for an isolated circuit, we implicitly assume that the output is unloaded. In other words, the load impedance is assumed to be infinite, which means that no power is being withdrawn at the output. When the second circuit is connected to the output of the first, however, a certain amount of power is withdrawn and then the assumption of no loading is violated. Therefore if the transfer function of this system is obtained under the assumption of no loading, then it is not valid.

8.4   z TRANSFORMS

One of the mathematical tools commonly used to deal with discrete-time systems is the z transform. The role of the z transform in discrete-time systems is similar to that of the Laplace transform in continuous-time systems. Laplace transforms allow the conversion of linear ordinary differential equations with constant coefficients into algebraic equations in s. The z transformation transforms linear difference equations with constant coefficients into algebraic equations in z.

8.4.1   Single-Sided z Transform

If a signal has discrete values $f_0, f_1, \ldots, f_k, \ldots$, we define the z transform of the signal as the function

$$
\begin{aligned}
F(z) &= \mathscr{Z}\{f(k)\} \\
&= \sum_{k=0}^{\infty} f(k) z^{-k} \qquad r_0 \le |z| \le R_0
\end{aligned} \tag{8.30}
$$

and is assumed that one can find a range of values of the magnitude of the complex variable z for which the series of Eq. (8.30) converges. This z transform is referred to as the one-sided z transform. The symbol $\mathscr{Z}$ denotes "the z transform of." In the one-sided z transform it is assumed $f(k) = 0$ for $k < 0$, which in the continuous-time case corresponds to $f(t) = 0$ for $t < 0$.

Expansion of the right-hand side of Eq. (8.30) gives

$$F(z) = f(0) + f(1)z^{-1} + f(2)z^{-2} + \cdots + f(k)z^{-k} + \cdots \tag{8.31}$$

The last equation implies that the z transform of any continuous-time function $f(t)$ may be written in the series form by inspection. The z^{-k} in this series indicates the instant in time at which the amplitude $f(k)$ occurs. Conversely, if $F(z)$ is given in the series form of Eq. (8.31), the inverse z transform can be obtained by inspection as a sequence of the function $f(k)$ that corresponds to the values of $f(t)$ at the respective instants of time. If the signal is sampled at a fixed sampling period T, then the sampled version of the signal $f(t)$ given by $f(0), f(1), \ldots, f(k)$ correspond to the signal values at the instants $0, T, 2T, \ldots, kT$.

8.4.2 Poles and Zeroes in the z Plane

If the z transform of a function takes the form

$$F(z) = \frac{b_0 z^m + b_1 z^{m-1} + \cdots + b_m}{z^n + a_1 z^{n-1} + \cdots + a_n} \tag{8.32}$$

or

$$F(z) = \frac{b_0 (z - z_1)(z - z_2) \cdots (z - z_m)}{(z - p_1)(z - p_2) \cdots (z - p_n)}$$

then p_i's are the poles of $F(z)$ and z_i's are the zeroes of $F(z)$.

In control studies, $F(z)$ is frequently expressed as a ratio of polynomials in z^{-1} as follows:

$$F(z) = \frac{b_0 z^{-(n-m)} + b_1 z^{-(n-m+1)} + \cdots + b_m z^{-n}}{1 + a_1 z^{-1} + a_2 z^{-2} + \cdots + a_n z^{-n}} \tag{8.33}$$

where z^{-1} is interpreted as the unit delay operator.

8.4.3 z Transforms of Some Elementary Functions

Unit Step Function. Consider the unit step function

$$u_s(t) = \begin{cases} 1 & t \geq 0 \\ 0 & t < 0 \end{cases}$$

whose discrete representation is

$$u_s(k) = \begin{cases} 1 & k \geq 0 \\ 0 & k < 0 \end{cases}$$

From Eq. (8.30)

$$\mathscr{Z}\{u_s(k)\} = \sum_{k=0}^{\infty} 1 \cdot z^{-k} = \sum_{k=0}^{\infty} z^{-k} = 1 + z^{-1} + z^{-2} + \cdots$$

$$= \frac{1}{1 - z^{-1}} = \frac{z}{z - 1} \tag{8.34}$$

Note that the series converges for $|z| > 1$. In finding the z transform, the variable z acts as a dummy operator. It is not necessary to specify the region of z over which $\mathscr{Z}\{u_s(k)\}$ is convergent. The z transform of a function obtained in this manner is valid throughout the z plane except at the poles of the transformed function. The $u_s(k)$ is usually referred to as the unit step sequence.

Exponential Function. Let

$$f(t) = \begin{cases} e^{-at} & t \geq 0 \\ 0 & t < 0 \end{cases}$$

The sampled form of the function with sampling period T is

$$f(kT) = e^{-akT} \qquad k = 0, 1, 2, \ldots$$

By definition

$$F(z) = \mathscr{Z}\{f(k)\}$$

$$= \sum_{k=0}^{\infty} f(k)z^{-k}$$

$$= \sum_{k=0}^{\infty} e^{-akt}z^{-k}$$

$$= 1 + e^{-at}z^{-1} + e^{-2at}z^{-2} + \cdots$$

$$= \frac{1}{1 - e^{-at}z^{-1}}$$

$$= \frac{z}{z - e^{-at}}$$

8.4.4   Some Important Properties and Theorems of the z Transform

In this section some useful properties and theorems are stated without proof. It is assumed that the time function $f(t)$ is z transformable and that $f(t)$ is zero for $t < 0$.

P1. If

$$F(z) = \mathscr{Z}\{f(k)\}$$

then

$$\mathscr{Z}\{af(k)\} = aF(z) \tag{8.35}$$

P2. If $f_1(k)$ and $g_1(k)$ are z transformable and α and β are scalars, then $f(k) = \alpha f_1(k) + \beta g_1(k)$ has the z transform:

$$F(z) = \alpha F_1(z) + \beta G_1(z) \tag{8.36}$$

where $F_1(z)$ and $G_1(z)$ are the z transforms of $f_1(k)$ and $g_1(k)$, respectively.

P3. If

$$F(z) = \mathscr{Z}\{f(k)\}$$

then

$$\mathscr{Z}\{a^k f(k)\} = F\left(\frac{z}{a}\right) \tag{8.37}$$

T1. Shifting Theorem. If $f(t) \equiv 0$ for $t < 0$ and $f(t)$ has the z transform $F(z)$, then

$$\mathscr{Z}\{f(t - nT)\} = z^{-n}F(z)$$

and

$$\mathscr{Z}\{f(t + nT)\} = z^n\left[F(z) - \sum_{k=0}^{n-1} f(kT)z^{-k}\right] \tag{8.38}$$

T2. Complex Translation Theorem. If

$$F(z) = \mathscr{Z}\{f(t)\}$$

then

$$\mathscr{Z}\{e^{-at}f(t)\} = F(ze^{at}) \tag{8.39}$$

T3. Initial-Value Theorem. If $F(z) = \mathcal{Z}\{f(t)\}$ and if $\lim_{z \to \infty} F(z)$ exists, then the initial value $f(0)$ of $f(t)$ or $f(k)$ is given by

$$f(0) = \lim_{z \to \infty} F(z) \tag{8.40}$$

T4. Final-Value Theorem. Suppose that $f(k)$, where $f(k) \equiv 0$ for $k < 0$, has the z transform $F(z)$ and that all the poles of $F(z)$ lie inside the unit circle, with the possible exception of a simple pole at $z = 1$. Then the final value of $f(k)$ is given by

$$\lim_{k \to \infty} f(k) = \lim_{z \to 1} \left[(1 - z^{-1}) F(z) \right] \tag{8.41}$$

8.4.5 Pulse Transfer Function

Consider the linear time-invariant discrete-time system characterized by the following linear difference equation:

$$y(k) + a_1 y(k-1) + \cdots + a_n y(k-n) = b_0 u(k) + b_1 u(k-1) + \cdots + b_m u(k-m) \tag{8.42}$$

where $u(k)$ and $y(k)$ are the system's input and output, respectively, at the kth sampling or at the real time kT; T is the sampling period. To convert the difference Eq. (8.42) to an algebraic equation, take the z transform of both sides of Eq. (8.42) by definition:

$$\mathcal{Z}\{y(k)\} = Y(z) \tag{8.43a}$$
$$\mathcal{Z}\{u(k)\} = U(z) \tag{8.43b}$$

By referring to Table 8.2, the z transform of Eq. (8.42) becomes

$$Y(z) + a_1 z^{-1} Y(z) + \cdots + a_n z^{-n} Y(z) = b_0 U(z) + b_1 z^{-1} U(z) + \cdots + b_m z^{-m} U(z)$$

or

$$\left[1 + a_1 z^{-1} + \cdots + a_n z^{-n} \right] Y(z) = \left[b_0 + b_1 z^{-1} + \cdots + b_m z^{-m} \right] U(z)$$

TABLE 8.2   Table of z Transforms [a]

No.	$\mathcal{F}(s)$	$f(nT)$	$F(z)$
1	—	$1, n = 0; 0, n \neq 0$	1
2	—	$1, n = k; 0, n \neq k$	z^{-k}
3	$\dfrac{1}{s}$	$1(nT)$	$\dfrac{z}{z-1}$
4	$\dfrac{1}{s^2}$	nT	$\dfrac{Tz}{(z-1)^2}$
5	$\dfrac{1}{s^3}$	$\dfrac{1}{2!}(nT)^2$	$\dfrac{T^2}{2}\dfrac{z(z+1)}{(z-1)^3}$
6	$\dfrac{1}{s^4}$	$\dfrac{1}{3!}(nT)^3$	$\dfrac{T^3}{6}\dfrac{z(z^2+4z+1)}{(z-1)^4}$
7	$\dfrac{1}{s^m}$	$\lim_{a \to 0}\dfrac{(-1)^{m-1}}{(m-1)!}\dfrac{\partial^{m-1}}{\partial a^{m-1}}e^{-anT}$	$\lim_{a \to 0}\dfrac{(-1)^{m-1}}{(m-1)!}\dfrac{\partial^{m-1}}{\partial a^{m-1}}\dfrac{z}{z-e^{-aT}}$
8	$\dfrac{1}{s+a}$	e^{-anT}	$\dfrac{z}{z-e^{-aT}}$
9	$\dfrac{1}{(s+a)^2}$	nTe^{-anT}	$\dfrac{Tze^{-aT}}{(z-e^{-aT})^2}$

TABLE 8.2 (*Continued*)

No.	$\mathscr{F}(s)$	$f(nT)$	$F(z)$
10	$\dfrac{1}{(s+a)^3}$	$\dfrac{1}{2}(nT)^2 e^{-anT}$	$\dfrac{T^2}{2}e^{-aT}\dfrac{z(z+e^{-aT})}{(z-e^{-aT})^3}$
11	$\dfrac{1}{(s+a)^m}$	$\dfrac{(-1)^{m-1}}{(m-1)!}\dfrac{\partial^{m-1}}{\partial a^{m-1}}(e^{-anT})$	$\dfrac{(-1)^{m-1}}{(m-1)!}\dfrac{\partial^{m-1}}{\partial a^{m-1}}\dfrac{z}{z-e^{-aT}}$
12	$\dfrac{a}{s(s+a)}$	$1-e^{-anT}$	$\dfrac{z(1-e^{-aT})}{(z-1)(z-e^{-aT})}$
13	$\dfrac{a}{s^2(s+a)}$	$\dfrac{1}{a}(anT-1+e^{-anT})$	$\dfrac{z[(aT-1+e^{-aT})z+(1-e^{-aT}-aTe^{-aT})]}{a(z-1)^2(z-e^{-aT})}$
14	$\dfrac{b-a}{(s+a)(s+b)}$	$(e^{-anT}-e^{-bnT})$	$\dfrac{(e^{-aT}-e^{-bT})z}{(z-e^{-aT})(z-e^{-bT})}$
15	$\dfrac{s}{(s+a)^2}$	$(1-anT)e^{-anT}$	$\dfrac{z[z-e^{-aT}(1+aT)]}{(z-e^{-aT})^2}$
16	$\dfrac{a^2}{s(s+a)^2}$	$1-e^{-anT}(1+anT)$	$\dfrac{z[z(1-e^{-aT}-aTe^{-aT})+e^{-2aT}-e^{-aT}+aTe^{-aT}]}{(z-1)(z-e^{-aT})^2}$
17	$\dfrac{(b-a)s}{(s+a)(s+b)}$	$be^{-bnT}-ae^{-anT}$	$\dfrac{z[z(b-a)-(be^{-aT}-ae^{-bT})]}{(z-e^{-aT})(z-e^{-bT})}$
18	$\dfrac{a}{s^2+a^2}$	$\sin anT$	$\dfrac{z\sin aT}{z^2-(2\cos aT)z+1}$
19	$\dfrac{s}{s^2+a^2}$	$\cos anT$	$\dfrac{z(z-\cos aT)}{z^2-(2\cos aT)z+1}$
20	$\dfrac{s+a}{(s+a)^2+b^2}$	$e^{-anT}\cos bnT$	$\dfrac{z(z-e^{-aT}\cos bT)}{z^2-2e^{-aT}(\cos bT)z+e^{-2aT}}$
21	$\dfrac{b}{(s+a)^2+b^2}$	$e^{-anT}\sin bnT$	$\dfrac{ze^{-aT}\sin bT}{z^2-2e^{-aT}(\cos bT)z+e^{-2aT}}$
22	$\dfrac{a^2+b^2}{s((s+a)^2+b^2)}$	$1-e^{-anT}\left(\cos bnT+\dfrac{a}{b}\sin bnT\right)$	$\dfrac{z(Az+B)}{(z-1)(z^2-2e^{-aT}(\cos bT)z+e^{-2aT})}$
			$A=1-e^{-aT}\cos bT-\dfrac{a}{b}e^{-aT}\sin bT$
			$B=e^{-2aT}+\dfrac{a}{b}e^{-aT}\sin bT-e^{-aT}\cos bT$

[a] $\mathscr{F}(s)$ is the Laplace transform of $f(t)$ and $F(z)$ is the z transform of $f(nT)$. Unless otherwise noted, $f(t)=0$, $t<0$, and the region of convergence of $F(z)$ is outside a circle $r<|z|$ such that all poles of $F(z)$ are inside r.

which can be written as

$$\frac{Y(z)}{U(z)} = \frac{b_0 + b_1 z^{-1} + \cdots + b_m z^{-m}}{1 + a_1 z^{-1} + \cdots + a_n z^{-n}} \tag{8.44}$$

Consider the response of the linear discrete-time system given by Eq. (8.44), initially at rest when the input $u(t)$ is the delta "function" $\delta(kT)$

$$\delta(kT) = \begin{cases} 1 & k = 0 \\ 0 & k \neq 0 \end{cases}$$

since

$$\mathcal{Z}\{\delta(kT)\} = \sum_{k=0}^{\infty} \delta(kT)z^{-k} = 1$$

$$U(z) = \mathcal{L}\{\delta(kT)\} = 1$$

and

$$Y(z) = \frac{b_0 + b_1 z^{-1} + \cdots + b_m z^{-m}}{1 + a_1 z^{-1} + \cdots + a_n z^{-n}} = G(z) \qquad (8.45)$$

Thus $G(z)$ is the response of the system to the delta input (or unit impulse) and plays the same role as the transfer function in linear continuous-time systems. The function $G(z)$ is called the pulse transfer function.

8.4.6 Zero- and First-Order Hold

Discrete-time control systems may operate partly in discrete time and partly in continuous time. Replacing a continuous-time controller with a digital controller necessitates the conversion of numbers to continuous-time signals to be used as true actuating signals. The process by which a discrete-time sequence is converted to a continuous-time signal is called data hold.

In a conventional sampler, a switch closes to admit an input signal every sample period T. In practice, the sampling duration is very small compared with the most significant time constant of the plant. Suppose the discrete-time sequence is $f(kT)$, then the function of the data hold is to specify the values for a continuous equivalent $h(t)$ where $kT \leq t < (k+1)T$. In general, the signal $h(t)$ during the time interval $kT < t < (k+1)T$ may be approximated by a polynomial in τ as follows:

$$h(kT + \tau) = a_n \tau^n + a_{n-1} \tau^{n-1} + \cdots + a_1 \tau + a_0 \qquad (8.46)$$

where $0 \leq \tau < T$. Since the value of the continuous equivalent must match at the sampling instants, one requires

$$h(kT) = f(kT)$$

Hence Eq. (8.46) can be written as

$$h(kT + \tau) = a_n \tau^n + a_{n-1} \tau^{n-1} + \cdots + a_1 \tau + f(kT) \qquad (8.47)$$

If an nth-order polynomial as in Eq. (8.47) is used to extrapolate the data, then the hold circuit is called an nth-order hold. If $n = 1$, it is called a first-order hold (the nth-order hold uses the past $n + 1$ discrete data $f[(k - n)T]$). The simplest data hold is obtained when $n = 0$ in Eq. (8.47), that is, when

$$h(kT + \tau) = f(kT) \qquad (8.48)$$

where $0 \leq \tau < T$ and $k = 0, 1, 2, \ldots$. Equation (8.48) implies that the circuit holds the amplitude of the sample from one sampling instant to the next. Such a data hold is called a zero-order hold. The output of the zero-order hold is shown in Fig. 8.17.

Zero-Order Hold

Assuming that the sampled signal is 0 for $k < 0$, the output $h(t)$ can be related to $f(t)$ as follows:

$$\begin{aligned} h(t) &= f(0)[u_s(t) - u_s(t - T)] + f(T)[u_s(t - T) - u_s(t - 2T)] \\ &\quad + f(2T)[u_s(t - 2T)] - u_s(t - 3T)] + \cdots \\ &= \sum_{k=0}^{\infty} f(kT)\{u_s(t - kT) - u_s[t - (k+1)T]\} \end{aligned} \qquad (8.49)$$

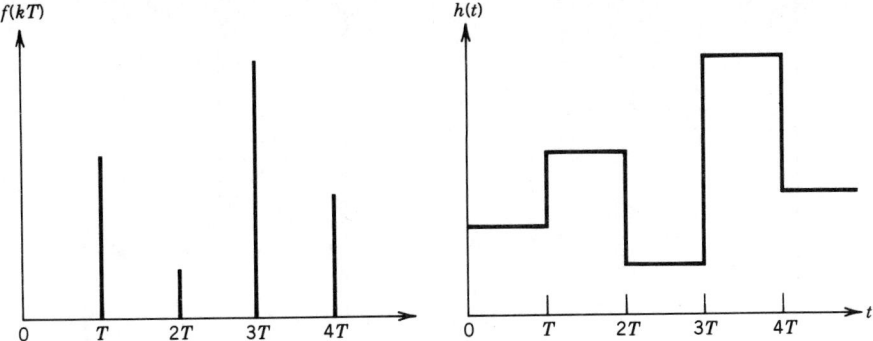

Fig. 8.17 Input $f(kt)$ and output $h(t)$ of the zero-order hold.

Since the $\mathcal{L}[u_s(t - kT)] = e^{-kTs}/s$, the Laplace transform of Eq. (8.49) becomes

$$\mathcal{L}[h(t)] = H(s) = \sum_{k=0}^{\infty} f(kT) \frac{e^{-kTs} - e^{-(k+1)Ts}}{s}$$

$$= \frac{1 - e^{-Ts}}{s} \sum_{k=0}^{\infty} f(kT)e^{-kTs} \qquad (8.50)$$

The right-hand side of Eq. (8.50) may be written as the product of two terms:

$$H(s) = G_{h0}(s)F^*(s) \qquad (8.51)$$

where

$$G_{h0}(s) = \frac{1 - e^{-Ts}}{s}$$

$$F^*(s) = \sum_{k=0}^{\infty} f(kT)e^{-kTs} \qquad (8.52)$$

In Eq. (8.51), $G_{h0}(s)$ may be considered the transfer function between the output $H(s)$ and the input $F^*(s)$ (see Fig. 8.18). Thus the transfer function of the zero-order hold device is

$$G_{h0}(s) = \frac{1 - e^{-Ts}}{s} \qquad (8.53)$$

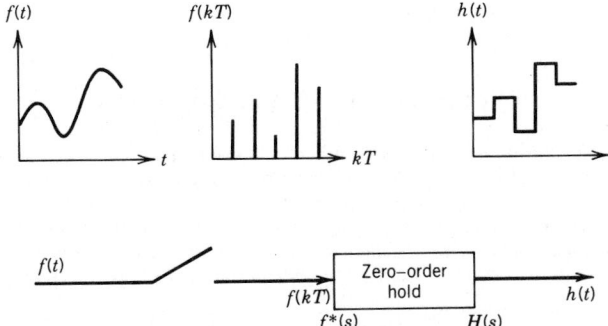

Fig. 8.18 Sampler and zero-order hold.

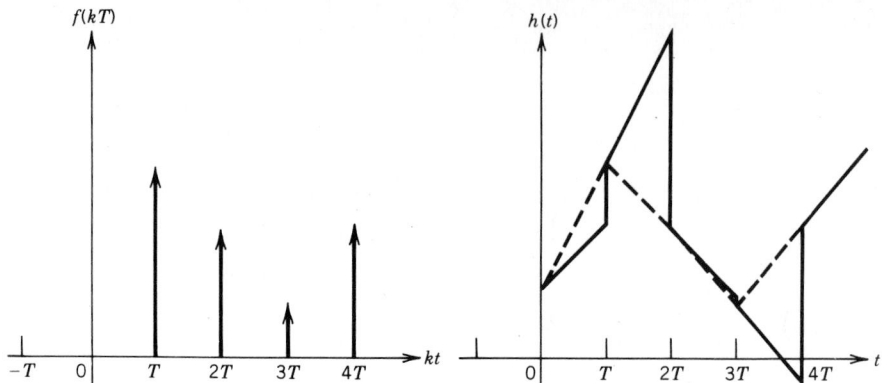

Fig. 8.19 Input $f(kt)$ and output $h(t)$ of a first-order hold.

First-Order Hold

The transfer function of a first-order hold is given by

$$G_{h1}(s) = \frac{1 - e^{-Ts}}{s} \frac{Ts + 1}{T} \tag{8.54}$$

From Eq. (8.47) for $n = 1$,

$$h(kT + \tau) = a_1\tau + f(kT) \tag{8.55}$$

where $0 \leq \tau < T$ and $k = 0, 1, 2, \dots$. By using the condition $h[(k - 1)T] = f[(k - 1)T]$ the constant a_1 can be determined as follows:

$$h[(k - 1)T] = -a_1T + f(kT) = f[(k - 1)T]$$

or

$$a_1 = \frac{f(kT) - f[(k - 1)T]}{T}$$

Hence Eq. (8.55) becomes

$$h(kT + \tau) = f(kT) + \frac{f(kT) - f[(k - 1)T]}{T}\tau \tag{8.56}$$

where $0 \leq \tau < T$. The extrapolation process of the first-order hold is based on Eq. (8.56) and is a piecewise-linear signal as shown in Fig. 8.19.

8.5   CLOSED-LOOP REPRESENTATION

The typical feedback control system has the feature that some output quantity is measured and then compared with a desired value, and the resulting error is used to correct the system output. A block diagram representation of a closed-loop or feedback system is shown in Fig. 8.20.

In this figure, r is the reference input, w is a disturbance, and y is the output. Transfer functions G_p, H, and G_c denote, respectively, the plant dynamics, sensor dynamics, and the controller. The influence of r and w on the output y can be determined using elementary block diagram algebra as

$$Y(s) = \frac{G_cG_p}{1 + G_cG_pH}R(s) + \frac{G_p}{1 + G_cG_pH}W(s) \tag{8.57}$$

where $R(s) = \mathscr{L}[r(t)]$, $W(s) = \mathscr{L}[w(t)]$, and $Y(s) = \mathscr{L}[y(t)]$.

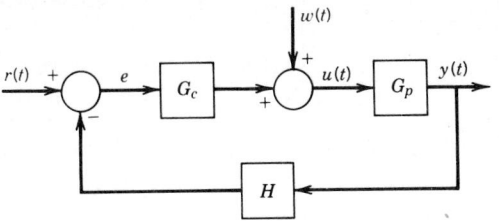

Fig. 8.20   Closed-loop system with disturbance.

8.5.1   Closed-Loop Transfer Function

In Eq. (8.57) if $W(s) = 0$, then

$$Y(s) = \frac{G_c G_p}{1 + G_c G_p H} R(s)$$

or alternatively a transfer function called the closed-loop transfer function between the reference input and the output is defined:

$$\frac{Y(s)}{R(s)} = G_{cl}(s) = \frac{G_c G_p}{1 + G_c G_p H} \qquad (8.58)$$

8.5.2   Open-Loop Transfer Function

The product of transfer functions within the loop, namely $G_c G_p H$, is referred to as the open-loop transfer function or simply the loop transfer function:

$$G_{ol} = G_c G_p H \qquad (8.59)$$

8.5.3   Characteristic Equation

The overall system dynamics given by Eq. (8.57) is primarily governed by the poles of the closed-loop system or the roots of the closed-loop characteristic equation (CLCE):

$$1 + G_c G_p H = 0 \qquad (8.60)$$

It is important to note that the closed-loop characteristic equation is simply

$$1 + G_{ol} = 0 \qquad (8.61)$$

This latter form is the basis of root locus and frequency domain design techniques discussed later. The roots of the characteristic equation are referred to as poles. Specifically, the roots of the open-loop characteristic equation are referred to as open-loop poles and those of the closed loop are called closed-loop poles.

Example 8.9. Consider the block diagram shown in Fig. 8.21. The open-loop transfer function

$$G_{ol} = \frac{K_1(s + 1)(s + 4)}{s(s + 2)(s + 3)}$$

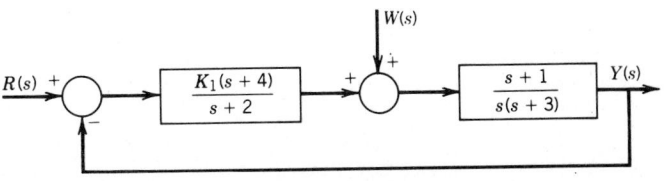

Fig. 8.21   Closed-loop system.

The closed-loop transfer function of the system is

$$G_{cl} = \frac{K_1(s+1)(s+4)}{s(s+2)(s+3) + K_1(s+1)(s+4)}$$

The open-loop characteristic equation (OLCE) is the denominator polynomial of G_{ol} set equal to zero. Hence

$$OLCE \equiv s(s+2)(s+3) = 0$$

and the open-loop poles are 0, -2, and -3.
The closed-loop characteristic equation is

$$s(s+2)(s+3) + K_1(s+1)(s+4) = 0$$

and its roots are the closed-loop poles appropriate for the specific gain value K_1.

If the transfer functions are rational (i.e., are ratios of polynomials), then they can be written as

$$G(s) = \frac{K\prod_{i=1}^{m}(s+z_i)}{\prod_{j=1}^{n}(s+p_j)} \tag{8.62}$$

When the poles of the transfer function $G(s)$ are distinct, $G(s)$ may be written in partial fraction form as

$$G(s) = \sum_{j=1}^{n}\frac{A_j}{s+p_j} \tag{8.63}$$

Hence

$$g(t) = \mathcal{L}^{-1}[G(s)] = \mathcal{L}^{-1}\left[\sum_{j=1}^{n}\frac{A_j}{s+p_j}\right]$$

$$= \sum_{j=1}^{n}A_j e^{-p_j t} \tag{8.64}$$

Since a transfer function $G(s) = \mathcal{L}[\text{output}]/\mathcal{L}[\text{input}]$, $g(t)$ of Eq. (8.64) is the response of the system depicted by $G(s)$ for a unit impulse $\delta(t)$, since $\mathcal{L}[\delta(t)] = 1$.
    The impulse response of a given system is key to its internal stability. The term *system* here is applicable to any part of (or the whole) closed-loop system.
    It should be noted that the zeros of the transfer function only affect the residues A_j. In other words, the contribution from the corresponding transient term $e^{-p_j t}$ may or may not be significant depending on the relative size of A_j. If, for instance, a zero $-z_k$ is very close to a pole $-p_l$ then the transient term $Ae^{-p_l t}$ would have a value close to zero for its residue A_l. As an example, consider the unit impulse response of the two systems:

$$G_1(s) = \frac{1}{(s+1)(s+2)} = \frac{1}{s+1} - \frac{1}{s+2} \tag{8.65}$$

$$G_2(s) = \frac{(s+1.05)}{(s+1)(s+2)} = \frac{0.05}{s+1} + \frac{0.95}{s+2} \tag{8.66}$$

From Eq. (8.65), $g_1(t) = e^{-t} - e^{-2t}$, and from Eq. (8.66), $g_2(t) = 0.05e^{-t} + 0.95e^{-2t}$.
    Note that the effect of the term e^{-t} has been modified from a residue of 1 in G_1 to a residue of 0.05 in G_2. This observation helps to reduce the order of a system when there are poles and zeros close together. In $G_2(s)$, for example, little error would be introduced if the zero at -1.05 and the pole at -1 are neglected and the transfer function approximated by

$$\tilde{G}_2(s) = \frac{1}{s+2}$$

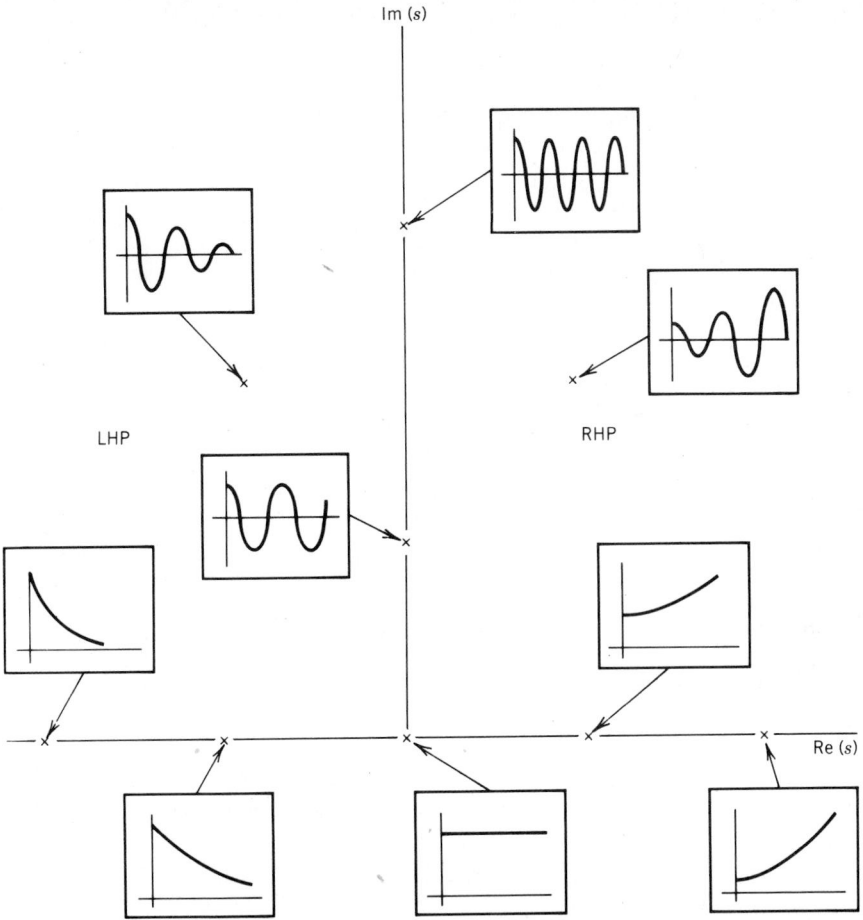

Fig. 8.22 Impulse responses associated with pole locations.

From Eq. (8.64) it can be observed that the shape of the impulse response is determined primarily by the pole locations. A sketch of several pole locations and corresponding impulse responses is given in Fig. 8.22.

The fundamental responses that can be obtained are of the form $e^{-\alpha t}$ and $e^{-\gamma t}\sin \beta t$ with $\beta > 0$. It should be noted that for a real pole its location completely characterizes the resulting impulse response. When a transfer function has complex conjugate poles the impulse response is more complicated.

8.5.4 Standard Second-Order Transfer Function

A standard second-order transfer function takes the form

$$G(s) = \frac{\omega_n^2}{s^2 + 2\zeta\omega_n s + \omega_n^2} = \frac{1}{\dfrac{s^2}{\omega_n^2} + 2\zeta\dfrac{s}{\omega_n} + 1} \tag{8.67}$$

Parameter ζ is called the damping ratio, and ω_n is called the undamped natural frequency. The poles

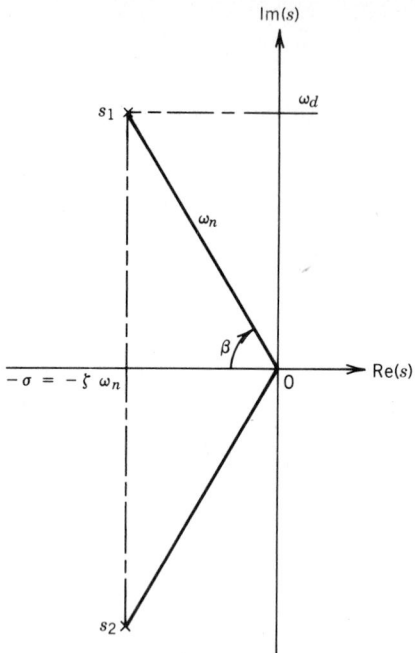

Fig. 8.23 Complex conjugate poles in the s plane.

of the transfer function given by Eq. (8.67) can be determined by solving its characteristic equation:

$$s^2 + 2\zeta\omega_n s + \omega_n^2 = 0 \tag{8.68}$$

giving the poles

$$s_{1,2} = -\zeta\omega_n \pm j\sqrt{1 - \zeta^2}\,\omega_n = -\sigma \pm j\omega_d \tag{8.69}$$

where $\sigma = \zeta\omega_n$ and $\omega_d = \omega_n\sqrt{1 - \zeta^2}$, $\zeta < 1$.

The two complex conjugate poles given by Eq. (8.69) are located in the complex s plane as shown in Fig. 8.23.

It can be easily seen that

$$|s_{1,2}| = \sqrt{\zeta^2\omega_n^2 + \omega_d^2} = \omega_n \tag{8.70}$$

and

$$\cos\beta = \zeta \qquad 0 \le \beta \le \pi/2 \tag{8.71}$$

When the system has no damping, $\zeta = 0$, the impulse response is a pure sinusoidal oscillation. In this case the undamped natural frequency ω_n is equal to the damped natural frequency ω_d.

8.5.5 Step Input Response of a Standard Second-Order System

When the transfer function has complex conjugate poles, the step input response rather than the impulse response is used to characterize its transients. Moreover, these transients are almost always used as time domain design specifications. The unit step input response of the second-order system

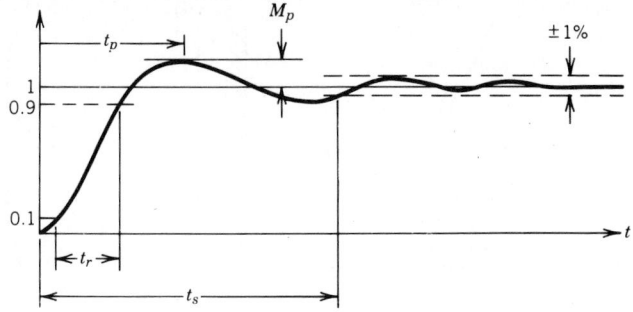

Fig. 8.24 Definition of the rise time t_r, settling time t_s, and overshoot time M_p.

given by Eq. (8.67) can be easily shown to be

$$y(t) = 1 - e^{-\sigma t}\left(\cos \omega_d t + \frac{\sigma}{\omega_d}\sin \omega_d t\right) \tag{8.72}$$

where $y(t)$ is the output.

A typical unit step input response is shown in Fig. 8.24.

From the time response of Eq. (8.72) several key parameters characterizing the shape of the curve are usually used as time domain design specifications. They are the rise time t_r, the settling time t_s, the peak overshoot M_p, and the peak time t_p.

Rise Time, t_r. The time taken for the output to change from 10% of the steady-state value to 90% of the steady-state value is usually called the rise time. There are other definitions of rise time.[3] The basic idea, however, is that t_r is a characterization of how rapid the system responds to an input.

A rule of thumb for t_r is

$$t_r \simeq 1.8/\omega_n \tag{8.73}$$

Settling Time, t_s. This is the time required for the transients to decay to a small value so that $y(t)$ is almost at the steady-state level. Various measures of "smallness" are possible: 1, 2, and 5% are typical.

$$\text{For 1\% settling,} \qquad t_s \simeq \frac{4.6}{\zeta\omega_n}$$

$$\text{For 2\% settling,} \qquad t_s \simeq \frac{4}{\zeta\omega_n} \tag{8.74}$$

$$\text{For 5\% settling,} \qquad t_s \simeq \frac{3}{\zeta\omega_n}$$

Peak Overshoot, M_p. The peak overshoot is the maximum amount by which the output exceeds its steady-state value during the transients. It can be easily shown that

$$M_p = e^{-\pi\zeta/\sqrt{1-\zeta^2}} \tag{8.75}$$

Peak Time, t_p. This is the time at which the peak occurs. It can be readily shown that

$$t_p = \frac{\pi}{\omega_d} \tag{8.76}$$

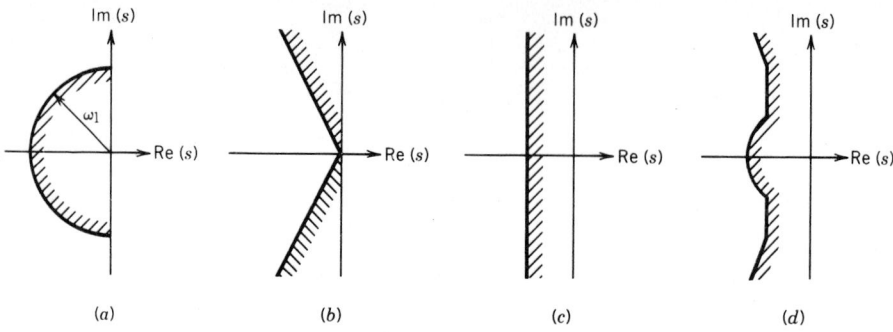

Fig. 8.25 Graphs of regions in the s plane for certain transient requirements to be met: (a) rise-time requirements, (ω_n); (b) overshoot requirements, (ζ); (c) settling time requirements, (ζ); (d) composite of (a), (b), and (c).

It is important to note that given a set of time domain design specifications, they can be converted to an equivalent location of complex conjugate poles. Figure 8.25 shows the allowable regions for complex conjugate poles for different time domain specifications.

For design purposes the following synthesis forms are useful:

$$\omega_n \geq \frac{1.8}{t_r} \tag{8.77a}$$

$$\zeta \geq 0.6(1 - M_p) \quad \text{for} \quad 0 \leq \zeta \leq 0.6 \tag{8.77b}$$

$$\sigma \geq \frac{4.6}{t_s} \tag{8.77c}$$

8.5.6 Effects of an Additional Zero and an Additional Pole[3]

If the standard second-order transfer function is modified due to a zero as

$$G_1(s) = \frac{(s/\alpha\zeta\omega_n + 1)\omega_n^2}{s^2 + 2\zeta\omega_n + \omega_n^2} \tag{8.78}$$

or due to an additional pole as

$$G_2(s) = \frac{\omega_n^2}{(s/\alpha\zeta\omega_n + 1)(s^2 + 2\zeta\omega_n s + \omega_n^2)} \tag{8.79}$$

it is important to know how close the resulting step input response is to the standard second-order step response. Following are several features of these additions:

1. For a second-order system with no finite zeros, the transient parameters can be approximated by Eq. (8.77).
2. An additional zero as in Eq. (8.78) in the left-half-plane will increase the overshoot if the zero is within a factor of 4 of the real part of the complex poles. A plot is given in Fig. 8.26.
3. An additional zero in the right half plane will depress the overshoot (and may cause the step response to undershoot). This is referred to as a nonminimum phase system.
4. An additional pole in the left-half-plane will increase the rise time significantly if the extra pole is within a factor of 4 of the real part of the complex poles. A plot is given in Fig. 8.27.

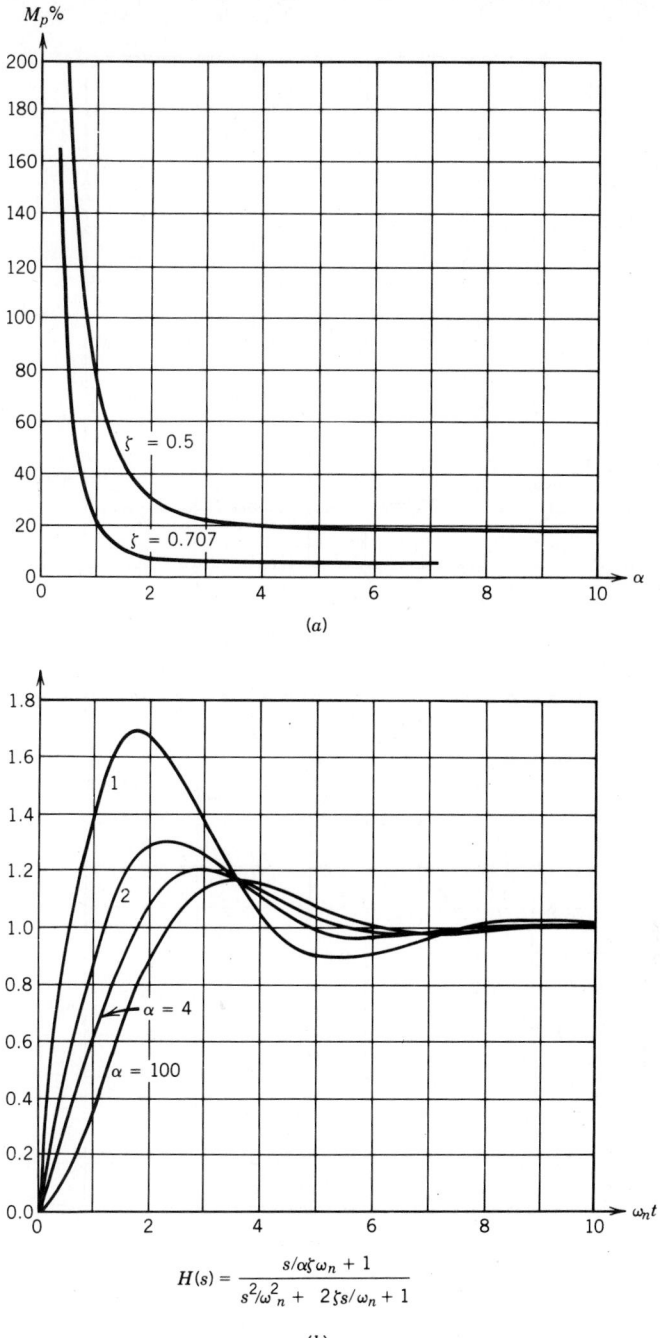

$$H(s) = \frac{s/\alpha\zeta\omega_n + 1}{s^2/\omega^2_n + 2\zeta s/\omega_n + 1}$$

(b)

Fig. 8.26 Effect of an extra zero on a standard second-order system: (a) overshoot M_p versus α; (b) step response versus α for $\zeta = 0.5$.

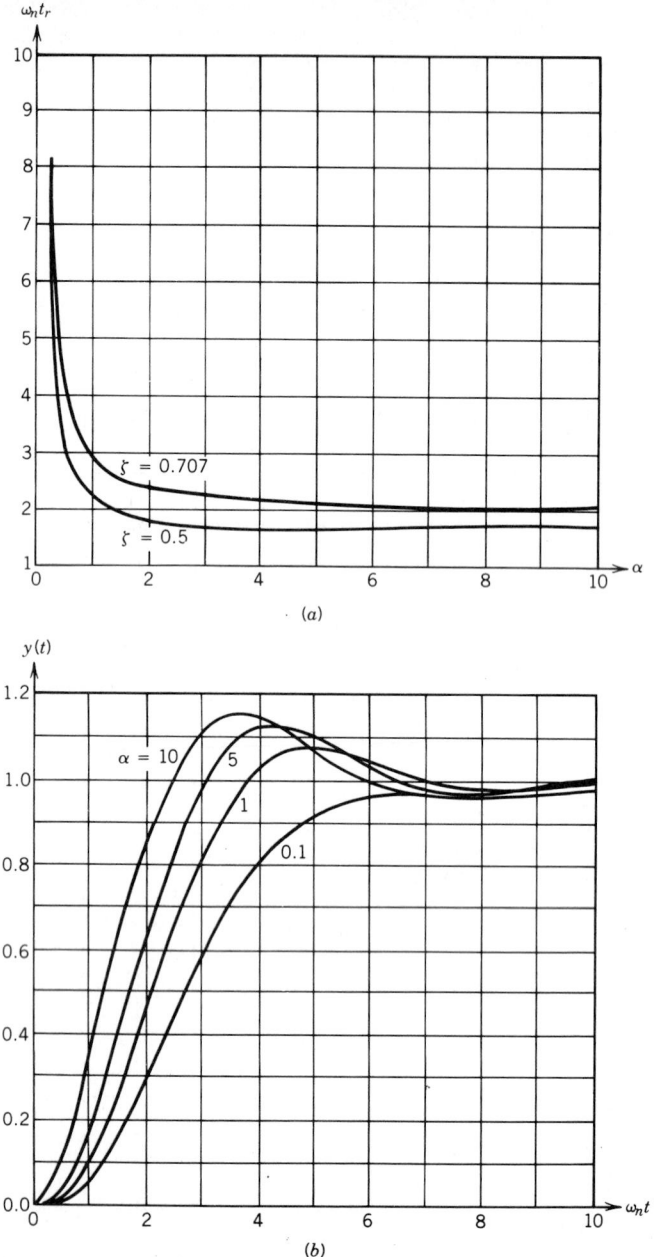

Fig. 8.27 Effect of an additional pole on a standard second-order system: (a) normalized rise time versus α; (b) step response versus α, for $\zeta = 0.5$.

8.6 STABILITY

We shall distinguish between two types of stabilities: external and internal stability. The notion of external stability is concerned with whether or not a bounded input gives a bounded output. In this type of stability we notice that no reference is made to the internal variables of the system. The implication here is that it is possible for an internal variable to grow without bound while the output remains bounded. Whether or not the internal variables are well behaved is typically addressed by the notion of internal stability. Internal stability requires that in the absence of an external input the internal variables stay bounded for any perturbations of these variables. In other words internal stability is concerned with the response of the system due to nonzero initial conditions. It is reasonable to expect that a well-designed system should be both externally and internally stable.

The notion of asymptotic stability is usually discussed within the context of internal stability. Specifically, if the response due to nonzero initial conditions decays to zero asymptotically, then the system is said to be asymptotically stable. A linear time-invariant (LTI) system is asymptotically stable if and only if all the system poles lie in the open left-half-plane (i.e., the left half s plane excluding the imaginary axis). This condition also guarantees external stability for LTI systems. So in the case of LTI the notions of internal and external stability may be considered equivalent.

For LTI systems, knowing the locations of the poles or the roots of the characteristic equation would suffice to predict stability. The Routh–Hurwitz stability criterion is frequently used to obtain stability information without explicitly computing the poles for LTI. This criterion will be discussed in Section 8.6.1.

For nonlinear systems, stability cannot be characterized that easily. As a matter of fact there are many definitions and theorems for assessing stability of such systems. A discussion of these topics is beyond the scope of this handbook. Interested reader however may refer to Ref. 9.

8.6.1   Routh–Hurwitz Stability Criterion

This criterion allows one to predict the status of stability of a system by knowing the coefficients of its characteristic polynomial. Consider the characteristic polynomial of an nth-order system:

$$P(s) = a_n s^n + a_{n-1} s^{n-1} + a_{n-2} s^{n-2} + \cdots + a_1 s + a_0$$

A necessary condition for asymptotic stability is that all the coefficients $\{a_i\}$'s be positive. If any of the coefficients are missing (i.e., are zero) or negative, then the system will have poles located in the closed right half plane. If the necessary condition is satisfied, then the so-called Routh array needs to be formed to make conclusions about the stability. A necessary and sufficient condition for stability is that all the elements in the first column of the Routh array be positive.

To determine the Routh array, the coefficients of the characteristic polynomial are arranged in two rows, each beginning with the first and second coefficients and followed by even-numbered and odd-numbered coefficients as follows:

s^n	a_n	a_{n-2}	a_{n-4} $\cdots$
s^{n-1}	a_{n-1}	a_{n-3}	a_{n-5} $\cdots$

The following rows are subsequently added to complete the Routh array:

s^n	a_n	a_{n-2}	a_{n-4} $\cdots$
s^{n-1}	a_{n-1}	a_{n-3}	a_{n-5} $\cdots$
s^{n-2}	b_1	b_2	b_3 $\cdots$
s^{n-3}	c_1	c_2	c_3 $\cdots$
.	.	.	. $\cdots$
s^2	*	*	
s	*		
s^0	*		

where the elements from the third row on are computed as follows:

$$b_1 = \frac{a_{n-1}a_{n-2} - a_n a_{n-3}}{a_{n-1}} \qquad c_1 = \frac{b_1 a_{n-3} - a_{n-1} b_2}{b_1}$$

$$b_2 = \frac{a_{n-1}a_{n-4} - a_n a_{n-5}}{a_{n-1}} \qquad c_2 = \frac{b_1 a_{n-5} - a_{n-1} b_3}{b_1}$$

$$b_3 = \frac{a_{n-1}a_{n-6} - a_n a_{n-7}}{a_{n-1}} \qquad c_3 = \frac{b_1 a_{n-7} - a_{n-1} b_4}{b_1}$$

Note that the elements of the third row and of rows thereafter are formed from the two previous rows using the two elements in the first column and other elements for successive columns. Normally there will be $n + 1$ elements in the first column when the array is completed.

The Routh–Hurwitz criterion states that the number of roots of $P(s)$ with positive real parts is equal to the number of changes of sign in the first column of the array. This criterion requires that there be no changes of sign in the first column for a stable system. A pattern of $+, -, +$ is counted as two sign changes, one going from $+$ to $-$ and another from $-$ to $+$.

If the first term in one of the rows is zero or if an entire row is zero, then the standard Routh array cannot be formed and the use of special techniques, described in the following discussion become necessary.

Special Cases

1. Zero in the first column, while some other elements of the row containing a zero in the first column are nonzero.
2. Zero in the first column, and the other elements of the row containing the zero are also zero.

CASE 1: In this case the zero is replaced with a small positive constant $\varepsilon > 0$, and the array is completed as before. The stability criterion is then applied by taking the limits of entries of the first column as $\varepsilon \to 0$. For example, consider the following characteristic equation:

$$s^5 + 2s^4 + 2s^3 + 4s^2 + 11s + 10 = 0$$

The Routh array is then

$$
\begin{array}{ccc}
1 & 2 & 11 \\
2 & 4 & 10 \\
\text{-}0\text{-----}& \text{-}6\text{-----} & \text{-}0\text{-} \\
\varepsilon & 6 & 0 \qquad \text{first column zero replaced by } \varepsilon \\
c_1 & 10 & \\
d_1 & 0 &
\end{array}
$$

where

$$c_1 = \frac{4\varepsilon - 12}{\varepsilon} \quad \text{and} \quad d_1 = \frac{6c_1 - 10\varepsilon}{c_1}$$

As $\varepsilon \to 0$, we get $c_1 \approx -12/\varepsilon$, and $d_1 \approx 6$. There are two sign changes due to the large negative number in the first column. Therefore the system is unstable, and two roots lie in the right half plane. As a final example consider the characteristic polynomial

$$P(S) = s^4 + 5s^3 + 7s^2 + 5s + 6$$

The Routh array is

$$
\begin{array}{ccc}
1 & 7 & 6 \\
5 & 5 & \\
6 & 6 & \\
\varepsilon & \leftarrow \text{zero replaced by } \varepsilon > 0 \\
6 &
\end{array}
$$

If $\varepsilon > 0$, there are no sign changes. If $\varepsilon < 0$, there are two sign changes. Thus if $\varepsilon = 0$, it indicates that there are two roots on the imaginary axis, and a slight perturbation would drive the roots into the right half plane or the left half plane. An alternative procedure is to define the auxiliary variable,

$$z = \frac{1}{s}$$

and convert the characteristic polynomial so that it is in terms of z. This usually produces a Routh array with nonzero elements in the first column. The stability properties can then be deduced from this array.

CASE 2: This case corresponds to a situation where the characteristic equation has equal and opposite roots. In this case if the ith row is the vanishing row, an auxiliary equation is formed from the previous $i - 1$ row as follows:

$$P_1(s) = \beta_1 s^{i+1} + \beta_2 s^{i-1} + \beta_3 s^{i-3} + \cdots$$

where $\{\beta_i\}$'s are the coefficients of the $(i - 1)$th row of the array. The ith row is then replaced by the coefficients of the derivative of the auxiliary polynomial, and the array is completed. Moreover, the roots of the auxiliary polynomial are also roots of the characteristic equation. As an example, consider

$$s^5 + 2s^4 + 5s^3 + 10s^2 + 4s + 8 = 0$$

for which the Routh array is

$$
\begin{array}{llll}
s^5 & 1 & 5 & 4 \\
s^4 & 2 & 10 & 8 \\
\cancel{s^3} & \cancel{0} & \cancel{0} & \quad \text{Auxiliary equation: } 2s^4 + 10s^2 + 8 = 0 \\
s^3 & 8 & 20 \\
s^2 & 5 & 8 \\
s^1 & 7.2 \\
s^0 & 8 \\
\end{array}
$$

There are no sign changes in the first column. Hence all the roots have nonpositive real parts with two pairs of roots on the imaginary axis, which are the roots of

$$2s^4 + 10s^2 + 8 = 0$$
$$= 2(s^2 + 4)(s^2 + 1)$$

Thus equal and opposite roots indicated by the vanishing row are $\pm j$ and $\pm 2j$.

The Routh–Hurwitz criterion may also be used in determining the range of parameters for which a feedback system remains stable. Consider, for example, the system described by the closed-loop characteristic equation

$$s^4 + 3s^3 + 3s^2 + 2s + K = 0$$

The corresponding Routh array is

$$
\begin{array}{lll}
s^4 & 1 & 3 \quad K \\
s^3 & 3 & 2 \\
s^2 & \frac{7}{3} & K \\
s^1 & 2 - (9/7)K \\
s^0 & K \\
\end{array}
$$

If the system is to remain asymptotically stable, we must have

$$0 < K < 14/9$$

The Routh–Hurwitz stability criterion can also be used to obtain additional insights. For instance information regarding the speed of response may be obtained by a coordinate transformation of the form $s + a$ where $-a$ characterizes rate. For this additional detail the reader is referred to Ref. 10.

8.6.2 Polar Plots

The frequency response of a system described by a transfer function $G(s)$ is given by

$$y_{ss}(t) = X|G(j\omega)| \sin\left[\omega t + \underline{/G(j\omega)}\,\right]$$

where $X \sin \omega t$ is the forcing input and $y_{ss}(t)$ is the steady-state output. A great deal of information about the dynamic response of the system can be obtained from a knowledge of $G(j\omega)$. The complex quantity $G(j\omega)$ is typically characterized by its frequency-dependent magnitude $|G(j\omega)|$ and phase $\underline{/G(j\omega)}$. There are many different ways in which this information is handled. The manner in which the phasor $G(j\omega)$ traverses the $G(j\omega)$ plane as a function of ω is represented by a polar plot.

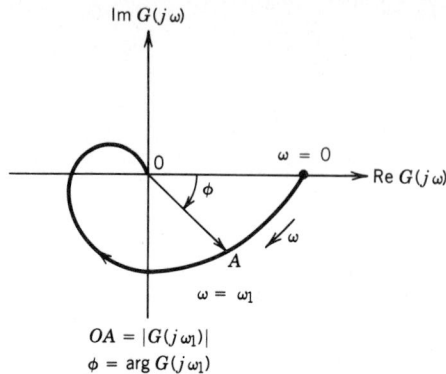

$$OA = |G(j\omega_1)|$$
$$\phi = \arg G(j\omega_1)$$

Fig. 8.28   Polar plot.

This is represented in Fig. 8.28. The phasor OA corresponds to a magnitude $|G(j\omega_1)|$ and the phase $\angle G(j\omega_1)$. The locus of point A as a function of the frequency is the polar plot. This representation is useful in stating the Nyquist stability criterion.

8.6.3   Nyquist Stability Criterion

This is a method by which the closed-loop stability of a linear time-invariant system can be predicted by knowing the frequency response of the loop transfer function. A typical closed-loop characteristic equation may be written as

$$1 + GH = 0$$

where GH is referred to as the loop transfer function.

The Nyquist stability criterion[3] is based on a mapping of the so-called Nyquist contour by the loop transfer function. Figure 8.29(a) shows the Nyquist contour, and Fig. 8.29(b) shows its typical mapped form in the GH plane. Let C_1 be the contour in the s plane to be mapped. Utilizing this information and the poles of GH, the Nyquist stability criterion can be stated as follows:

A closed-loop system is stable if and only if the mapping of C_1 in the GH plane encircles the $(-1, 0)$ point N number of times in the counterclockwise direction, where N is the number of poles of GH with positive real parts.

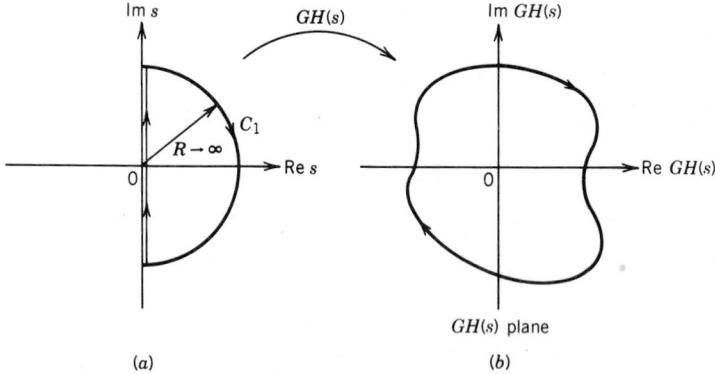

(a)   (b)

Fig. 8.29   Nyquist contour and its mapping by $GH(s)$.

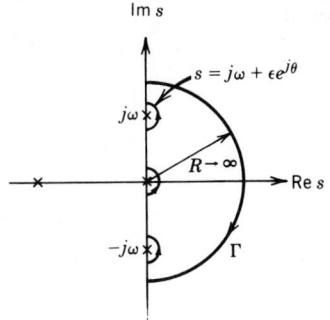

Fig. 8.30   Nyquist contour with indentations.

Clearly if GH does not have any poles in the right half-plane, then C_1 should not encircle the $(-1,0)$ point if the closed-loop system is to be stable.

If there are poles of GH on the imaginary axis, then the Nyquist contour should be modified with appropriate indentations as shown in Fig. 8.30. An example is given to map such an indented contour with the corresponding loop transfer function. Consider

$$GH(s) = \frac{1}{s(s+1)}$$

The contour Γ in the s plane is shown in Fig. 8.31(a), where an indentation is affected by a small semicircle of radius ε, where $\varepsilon > 0$. When mapped by $GH(s)$, the contour of Fig. 8.31(b) is obtained. In order to effect the mapping, the points on the semicircles are represented as $s = \varepsilon e^{j\phi}$ on the small indentation with $\phi \varepsilon [-\pi/2, \pi/2]$ and $\varepsilon \to 0$ and $s = Re^{j\theta}$ with $\theta \varepsilon [-\pi/2, \pi/2]$ and $R \to \infty$ on the infinite semicircle. We observe that the $(-1,0)$ point is not encircled. Since $GH(s)$ does not have any poles in the right half s plane from the Nyquist stability criterion, it follows that the closed-loop system is stable. In Table 8.3 several loop transfer functions, with the appropriate Nyquist contour and mapped contours are given.

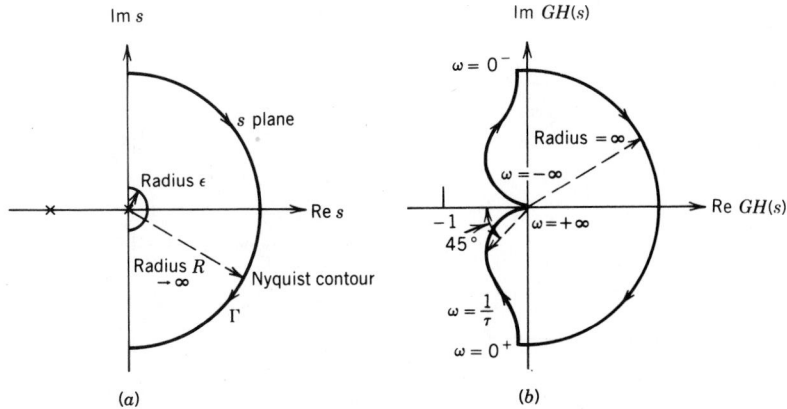

Fig. 8.31   Nyquist contour and mapping for $GH(s) = K/s(\tau s + 1)$.

TABLE 8.3   Loop Transfer Functions

	$G(s)$	Nyquist Contour	Polar plot
1.	$\dfrac{K}{s\tau_1 + 1}$		
2.	$\dfrac{K}{(s\tau_1 + 1)(s\tau_2 + 1)}$		
3.	$\dfrac{K}{(s\tau_1 + 1)(s\tau_2 + 1)(s\tau_3 + 1)}$		
4.	$\dfrac{K}{s}$		
5.	$\dfrac{K}{s(s\tau_1 + 1)}$		

TABLE 8.3 (*Continued*)

	$G(s)$	Nyquist Contour	Polar plot
6.	$\dfrac{K}{s(s\tau_1 + 1)(s\tau_2 + 1)}$		
7.	$\dfrac{K(s\tau_a + 1)}{s(s\tau_1 + 1)(s\tau_2 + 1)}$ $\tau_a < \dfrac{\tau_1 \tau_2}{\tau_1 + \tau_2}$		
8.	$\dfrac{K}{s^2}$		
9.	$\dfrac{K}{s^2(s\tau_1 + 1)}$		
10.	$\dfrac{K(s\tau_1 + 1)}{s^2(s\tau_1 + 1)}$ $\tau_a > \tau_1$		

TABLE 8.3 (*Continued*)

	$G(s)$	Nyquist Contour	Polar plot
11.	$\dfrac{K}{s^3}$		
12.	$\dfrac{K(s\tau_a + 1)}{s^3}$		
13.	$\dfrac{K(s\tau_a + 1)(s\tau_b + 1)}{s^3}$		
14.	$\dfrac{K(s\tau_a + 1)(s\tau_b + 1)}{s(s\tau_1 + 1)(s\tau_2 + 1)(s\tau_3 + 1)(s\tau_4 + 1)}$		
15.	$\dfrac{K(s\tau_a + 1)}{s^2(s\tau_1 + 1)(s\tau_2 + 1)}$		

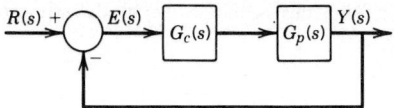

Fig. 8.32   Closed-loop configuration.

8.7   STEADY-STATE PERFORMANCE AND SYSTEM TYPE

In the design of feedback control systems the steady-state performance is also of importance in many instances. This is in addition to the stability and transient performance requirements. Consider the closed-loop system shown in Fig. 8.32.

Here $G_p(s)$ is the plant and $G_c(s)$ is the controller. When steady-state performance is important, it is advantageous to consider the system error due to an input. From the block diagram

$$E(s) = R(s) - Y(s) \tag{8.80}$$

$$Y(s) = G_c(s)G_p(s)E(s) \tag{8.81}$$

Substituting Eq. (8.81) in (8.80) yields

$$\frac{E(s)}{R(s)} = \frac{1}{1 + G_c(s)G_p(s)} \tag{8.82}$$

called the error transfer function. It is important to note that the error dynamics are described by the same poles as those of the closed-loop transfer function. Namely, the roots of the characteristic equation $1 + G_c(s)G_p(s) = 0$.

Given any input $r(t)$ with $\mathscr{L}[r(t)] = R(s)$, the error can be analyzed by considering the inverse Laplace transform of $E(s)$ given by

$$e(t) = \mathscr{L}^{-1}\left[\frac{1}{1 + G_c(s)G_p(s)} R(s)\right] \tag{8.83}$$

For the system's error to be bounded, it is important to first assure that the closed-loop system is asymptotically stable. Once the closed-loop stability is assured, the steady-state error can be computed by using the final-value theorem. Hence

$$\lim_{t \to \infty} e(t) = e_{ss} = \lim_{s \to 0} sE(s) \tag{8.84}$$

By substituting for $E(s)$ in Eq. (8.84)

$$e_{ss} = \lim_{s \to 0} s \frac{1}{1 + G_c(s)G_p(s)} R(s) \tag{8.85}$$

8.7.1   Step Input

If the reference input is a step of magnitude c, then from Eq. (8.85)

$$e_{ss} = \lim_{s \to 0} s \frac{1}{1 + G_c(s)G_p(s)} \frac{c}{s} = \frac{c}{1 + G_c G_p(0)} \tag{8.86}$$

Equation (8.86) suggests that in order to have small steady-state error, the low-frequency gain of the open-loop transfer function $G_c G_p(0)$ must be very large. It is typical to define

$$K_p = G_c G_p(0) \tag{8.87}$$

as the position error constant. With this definition the steady-state error due to a step input of magnitude c can be written as

$$e_{ss} = \frac{c}{1 + K_p} \tag{8.88}$$

Thus a high value of K_p corresponds to a low steady-state error. If the steady-state error is to be zero, then $K_p = \infty$. The only way that $K_p = \infty$ is when the open-loop transfer function has at least one pole at the origin, that is, $G_c G_p(s)$ must be of the form

$$G_c G_p(s) = \frac{1}{s^N} \frac{\prod\limits_{i=1}^{m}(s + z_i)}{\prod\limits_{j=1}^{n}(s + p_j)} \tag{8.89}$$

where $N \geq 1$. When $N \geq 1$,

$$G_c G_p(0) = \frac{1}{0} \frac{\prod\limits_{i=1}^{m} z_i}{\prod\limits_{j=1}^{n} p_j} \to \infty$$

Hence, it can be concluded that for the steady-state error due to a step input to be zero, the open-loop transfer function must have at least one free integrator. The value of N specifies the type of the system. If $N = 1$ it is called a type I, when $N = 2$ it is called a type II system, and so on. So to get zero steady-state error for a step input, the system loop transfer function must be at least type I.

8.7.2 Ramp Input

If the reference input is a ramp $ctu_s(t)$ where $u_s(t)$ is the unit step, then from Eq. (8.85)

$$e_{ss} = \lim_{s \to 0} s \frac{1}{1 + G_c G_p(s)} \frac{c}{s^2} = \lim_{s \to 0} \frac{c}{s G_c G_p(s)} \tag{8.90}$$

From Eq. (8.90) for small steady-state errors $\lim\limits_{s \to 0} s G_c G_p(s) = K_v$ must be large. K_v is called the velocity error constant and

$$e_{ss} = \frac{c}{K_v} \tag{8.91}$$

As in the case with the step input for e_{ss} to be small, K_v must be very large. For zero steady-state error with a ramp input, $K_v = \infty$. From Eq. (8.90) it is clear that for $K_v = \infty$, $G_c G_p(s)$ must be at least type II. Thus

$$K_v = \lim_{s \to 0} s G_c G_p(s) = \lim_{s \to 0} s \frac{1}{s^2} \frac{\prod\limits_{i=1}^{m}(s + z_i)}{\prod\limits_{j=1}^{n}(s + p_j)} = \infty$$

8.7.3 Parabolic Input

If the reference input is a parabolic input of the form

$$r(t) = c \frac{t^2}{2} u_s(t)$$

then the steady-state error becomes

$$e_{ss} = \lim_{s \to 0} \frac{c}{s^2 G_c G_p(s)} \tag{8.92}$$

TABLE 8.4   Steady-State Error in Terms of Error Constants

Type	$c\,u_s(t)$	$ct\,u_s(t)$	$\dfrac{t^2}{2}u_s(t)$
0	$\dfrac{c}{1+K_p}$	∞	∞
1	0	$\dfrac{c}{K_v}$	∞
2	0	0	$\dfrac{c}{K_a}$
3	0	0	0

From Eq. (8.92) for small steady-state errors

$$K_a = \lim_{s\to 0} s^2 G_c G_p(s) \tag{8.93}$$

must be very large. K_a is called the parabolic error constant or the acceleration error constant. For zero steady-state error due to a parabolic input, $K_a = \infty$. Therefore the system open-loop transfer function must be at least type III.

Table 8.4 shows the steady-state errors in terms of the error constants K_p, K_v, and K_a.

In steady-state performance considerations it is important to guarantee both closed-loop stability and the system type. It should also be noticed that having a very high loop gain as given by Eq. (8.87) is very similar to having a free integrator in the open-loop transfer function. This notion is useful in regulation type problems. In tracking systems the problem is compounded by the fact that the system type must be increased. Increasing the system type is usually accompanied by instabilities. To illustrate this consider the following example.

Example 8.10. Synthesize a controller for the system shown in Fig. 8.33 so that the steady-state error due to a ramp input is zero.

The open-loop system is unstable. For tracking purposes transform the system to a closed-loop one as shown in Fig. 8.33(b). To have zero steady-state error for a ramp, the system's open-loop transfer function must be at least type II and must be of a form that guarantees closed-loop stability. Therefore let

$$G_c(s) = \frac{G_1(s)}{s^2} \tag{8.94}$$

The closed-loop characteristic equation (CLCE) is

$$1 + \frac{G_1(s)}{s^2}\,\frac{1}{s-2} = 0 \tag{8.95}$$

That is, $s^3 - 2s^2 + G_1(s) = 0$.

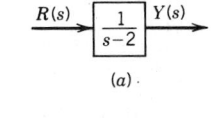

(a)

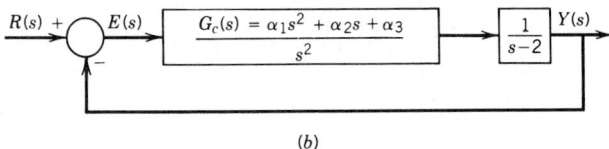

(b)

Fig. 8.33 Tracking system: (a) open-loop system; (b) closed-loop system.

Now for closed-loop stability, $G_1(s)$ must be of the form

$$G_1(s) = \alpha_1 s^2 + \alpha_2 s + \alpha_3$$

yielding a CLCE:

$$s^3 + (\alpha_1 - 2)s^2 + \alpha_2 s + \alpha_3 = 0 \tag{8.96}$$

From the Routh–Hurwitz stability criterion it can be shown that for asymptotic stability $\alpha_1 > 2$, $\alpha_3 > 0$, and $(\alpha_1 - 2)\alpha_2 - \alpha_3 > 0$. So pick $\alpha_1 = 4$, $\alpha_2 = 1$, $\alpha_3 = 1$ yielding the controller

$$G_c(s) = \frac{4s^2 + s + 1}{s^2} \tag{8.97}$$

In an actual design situation the choice of α_1, α_2, and α_3 will be dictated by the transient design specifications.

8.7.4 Indices of Performance

Sometimes the performance of control systems is given in terms of a performance index. A performance index is a number that indicates the "goodness" of system performance. Such performance indices are usually employed to get an optimal design in the sense of minimizing or maximizing the index of performance. The optimal parameter values depend directly upon the performance index chosen. Typical performance indices are $\int_0^\infty e^2(t)\, dt$, $\int_0^\infty te^2(t)\, dt$, $\int_0^\infty |e(t)|\, dt$, $\int_0^\infty t|e(t)|\, dt$. In control system design the task is to select the controller parameters to minimize the chosen index.

8.7.5 Integral-Square-Error (ISE) Criterion

According to the ISE, the quality of system performance is evaluated by minimizing the integral

$$J = \int_0^\infty e^2(t)\, dt \tag{8.98}$$

A system designed by this criterion tends to show a rapid decrease in a large initial error. Hence the response is fast and oscillatory. Thus the system has poor relative stability. ISE is of practical significance because $\int e^2(t)\, dt$ resembles power consumption for some systems.

8.7.6 Integral of Time-Multiplied Absolute-Error (ITAE) Criterion

According to this criterion, the optimum system is the one that minimizes the performance index:

$$J = \int_0^\infty t|e(t)|\, dt \tag{8.99}$$

This criterion weighs large initial errors lightly, and errors occurring late in the transient response are penalized heavily. A system designed by use of ITAE has a characteristic that the overshoot in the transient response is small and oscillations are well damped.

8.7.7 Comparison of Various Error Criteria[3]

Figure 8.34 shows several error performance curves. The system considered is

$$\frac{C(s)}{R(s)} = \frac{1}{s^2 + 2\zeta s + 1} \tag{8.100}$$

The curves of Fig. 8.34 indicate that $\zeta = 0.7$ corresponds to near optimal value with respect to each of the performance indices used. At $\zeta = 0.7$ the system given by Eq. (8.100) results in rapid response to a step input with approximately 5% overshoot.

Table 8.5 summarizes the coefficients that will minimize the ITAE performance criterion for a step input to the closed-loop transfer function[5]

$$\frac{C(s)}{R(s)} = \frac{a_0}{s^n + a_{n-1}s^{n-1} + \cdots + a_1 s + a_0} \tag{8.101}$$

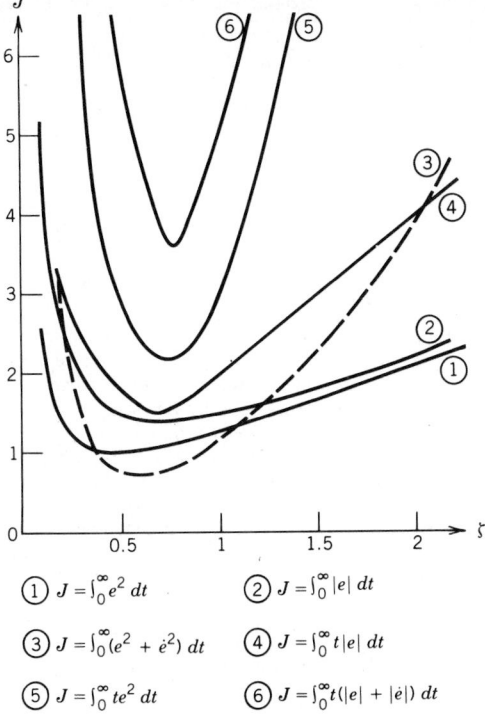

$$\text{①} \quad J = \int_0^\infty e^2 \, dt \qquad \text{②} \quad J = \int_0^\infty |e| \, dt$$

$$\text{③} \quad J = \int_0^\infty (e^2 + \dot{e}^2) \, dt \qquad \text{④} \quad J = \int_0^\infty t|e| \, dt$$

$$\text{⑤} \quad J = \int_0^\infty te^2 \, dt \qquad \text{⑥} \quad J = \int_0^\infty t(|e| + |\dot{e}|) \, dt$$

Fig. 8.34 Error performance curves.

Table 8.6[2] summarizes the coefficients that will minimize the ITAE performance criterion for a ramp input applied to the closed-loop transfer function:

$$\frac{C(s)}{R(s)} = \frac{b_1 s + b_0}{s^n + b_{n-1}s^{n-1} + \cdots + b_1 s + b_0} \tag{8.102}$$

Figure 8.35 shows the response resulting from optimum coefficients for a step input applied to the normalized closed-loop transfer function given in Eq. (8.101), for ISE, IAE (integral absolute-error), and ITAE.

TABLE 8.5 Optimal Form of the Closed-Loop Transfer Function Based on the ITAE Criterion for Step Inputs

$$\frac{C(s)}{R(s)} = \frac{a_0}{s^n + a_{n-1}s^{n-1} + \cdots + a_1 s + a_0}, \qquad a_0 = \omega_n^n$$

$$s + \omega_n$$

$$s^2 + 1.4\omega_n s + \omega_n^2$$

$$s^3 + 1.75\omega_n s^2 + 2.15\omega_n^2 s + \omega_n^3$$

$$s^4 + 2.1\omega_n s^3 + 3.4\omega_n^2 s^2 + 2.7\omega_n^3 s + \omega_n^4$$

$$s^5 + 2.8\omega_n s^4 + 5.0\omega_n^2 s^3 + 5.5\omega_n^3 s^2 + 3.4\omega_n^4 s + \omega_n^5$$

$$s^6 + 3.25\omega_n s^5 + 6.60\omega_n^2 s^4 + 8.60\omega_n^3 s^3 + 7.45\omega_n^4 s^2 + 3.95\omega_n^5 s + \omega_n^6$$

TABLE 8.6   Optimum Coefficients of $T(s)$ Based on the ITAE Criterion for a Ramp Input

$$s^2 + 3.2\omega_n s + \omega_n^2$$

$$s^3 + 1.75\omega_n s^2 + 3.25\omega_n^2 s + \omega_n^3$$

$$s^4 + 2.41\omega_n s^3 + 4.93\omega_n^2 s^2 + 5.14\omega_n^3 s + \omega_n^4$$

$$s^5 + 2.19\omega_n s^4 + 6.50\omega_n^2 s^3 + 6.30\omega_n^3 s^2 + 5.24\omega_n^4 s + \omega_n^5$$

8.8   SIMULATION FOR CONTROL SYSTEM ANALYSIS

Often in control system design the governing differential equations are reduced to a form that facilitates the shaping of a controller. The model order reductions involve things such as linearization, neglecting fast dynamics, parasitics, time delays, and so on. Once a controller is synthesized for a system, after making simplifying assumptions, it becomes necessary to try out the controller on the actual system. This involves experimentation. Computer simulation may also be viewed as experimentation where the synthesized controller is inserted into a very sophisticated model of the actual physical system and the system response computed. Extensive simulation studies can reduce the cost of experimentation and may even at times serve as the final controller to be implemented on the actual physical plant. So in simulation the burden of experiment is transformed into generating accurate models for all the components in a system without over simplifying assumptions. Simulations can be performed on analog, digital, or hybrid computers.[6]

8.8.1   Analog Computation[7]

An analog computer is a machine in which several physical components can be selected and interconnected in such a way that the equations describing the operation of the computer are analogous to that of the actual physical system to be studied. It is a continuous-time device operating in a real-time parallel mode, making it particularly suitable for the solution of differential equations and hence for the simulation of dynamic systems. The most commonly used analog computer is the electronic analog computer in which voltages at various places within the computer are proportional to the variable terms in the actual system. The ease of use, and the direct interactive control over the running of such a computer, allows full scope for engineering intuition and makes it a valuable tool for the analysis of dynamic systems and the synthesis of any associated controllers. A facility frequently useful is the ability to slow down or speed up the problem solution. The accuracy of solution, since it is dependent on analog devices, is generally of the order of a few percent but, for the purposes of system analysis and design, higher accuracy is seldom necessary; also, this accuracy often matches the quality of the available input data.

The basic building block of the analog computer is the high-gain dc amplifier, represented schematically by Fig. 8.36. When the input voltage is $e_i(t)$, the output voltage is given by

$$e_0(t) = -Ae_i(t) \tag{8.103}$$

where A, the amplifier voltage gain, is a large constant value.

If the voltage to the input of an amplifier, commonly referred to as the summing junction, exceeds a few microvolts, then the amplifier becomes saturated or overloaded because the power supply cannot force the output voltage high enough to give the correct gain. Therefore, if an amplifier is to be operated correctly, its summing junction must be very close to ground potential, and is usually treated as such in programming.

When this is used in conjunction with a resistance network as shown in Fig. 8.37, then the resulting circuit can be used to add a number of voltages.

If

$$R_1 = R_2 = R_3$$

then

$$V_0 = -(V_1 + V_2 + V_3) \tag{8.104}$$

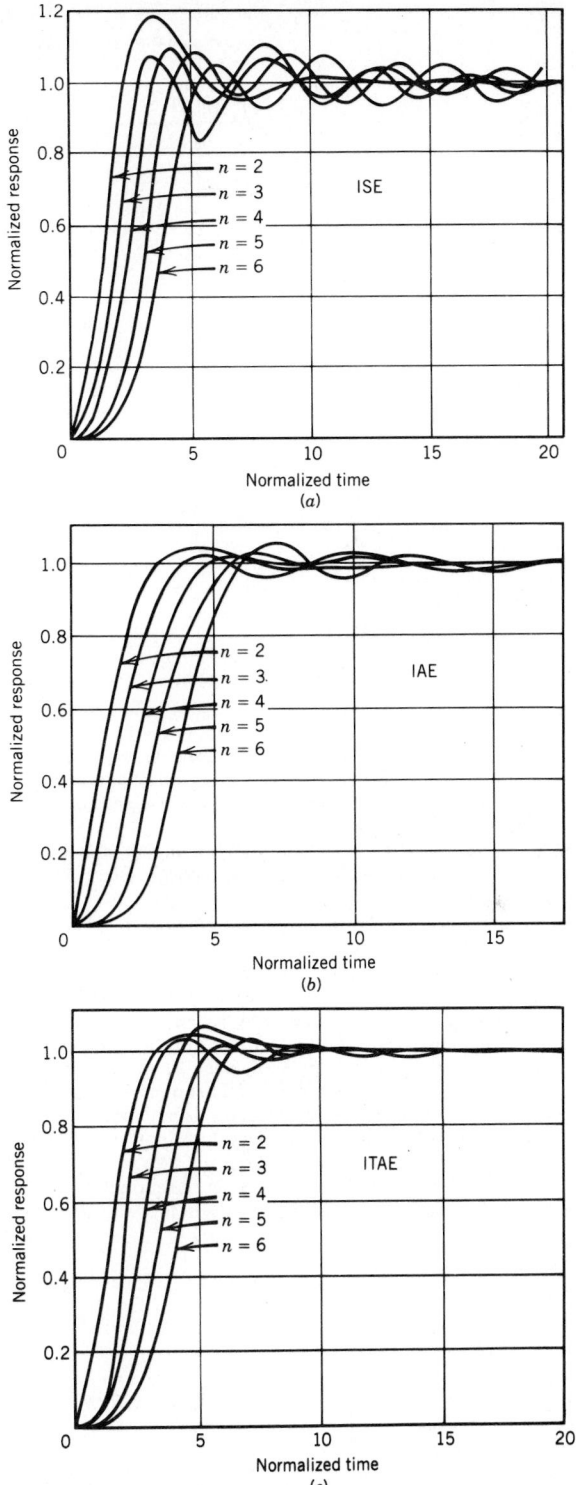

Fig. 8.35 (*a*) Step response of a normalized transfer function using optimum coefficients for ISE, (*b*) IAE, and (*c*) ITAE.

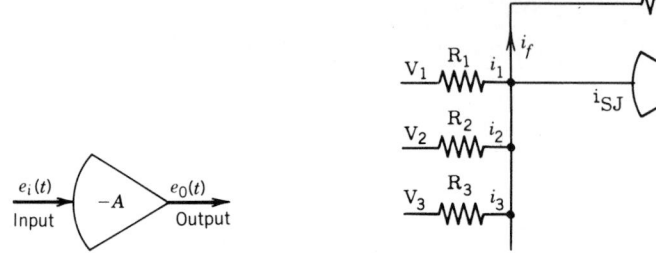

Fig. 8.36 High-gain dc amplifier.   Fig. 8.37 Current balance.

$$\sum \frac{V_n}{R_n} = i_{SJ} \approx 0$$

If R_1, R_2, R_3 are arbitrary, then

$$V_0 = -\left(\frac{R_f}{R_1} V_1 + \frac{R_f}{R_2} V_2 + \frac{R_f}{R_3} V_3 \right) \tag{8.105}$$

If there is only one voltage input, then

$$V_0 = -\frac{R_f}{R_1} V_1 \ldots \quad \text{(multiplication by a constant)} \tag{8.106}$$

It should be noted that in all cases there is a sign inversion. Usually the available ratios R_f/R_1 and so on are standardized to 1 and 10, the appropriate gain being selectable as required. The complete circuit comprising high-gain amplifier, input resistors, and feedback element is termed an **operational amplifier**. It is given symbolically as shown in Fig. 8.38.

   In order to multiply a voltage by a constant other than 10, use is made of a grounded potentiometer (usually a 10-turn helical potentiometer), as shown in Fig. 8.39. This permits multiplication by a constant in the range 0 to 1.

   Since electrical circuits are very nearly linear, the generation of even the simplest nonlinearity, like a product, requires special consideration. One way this can be done is to use the feedback control concept as shown in Fig. 8.40. Two or more potentiometers are mounted rigidly with a common shaft turning the sliders exactly together. A negative unit voltage, which commonly is 10–100 V, is applied across one of the potentiometers. The slider voltage is added to the voltage representing one of the factors. If they differ, then the motor turns the shaft in the direction that eliminates the error. The second factor voltage is applied across the other potentiometer. Its slider then reads the product. This device is called a servomultiplier. It is quite slow in operation because of the motor. Faster multiplication requires the use of special nonlinear networks using diodes. A diode allows current to flow in one direction only if polarities are in the indicated directions in the symbolic representations of Fig. 8.41. Combinations of diodes and resistors can be made such that the current-voltage curve is a series of straight-line segments. The circuit of Fig. 8.42 gives the current indicated there. Almost any shape of curve can be approximated by straight-line segments in this way, in particular a network can

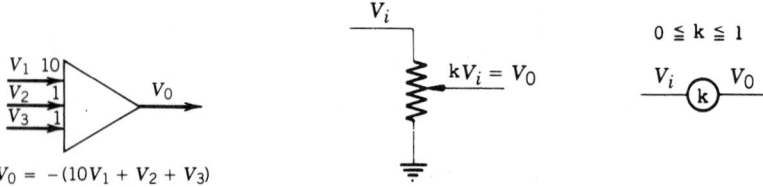

$V_0 = -(10V_1 + V_2 + V_3)$

Fig. 8.38 Symbol for summing amplifier.   Fig. 8.39 Potentiometer.

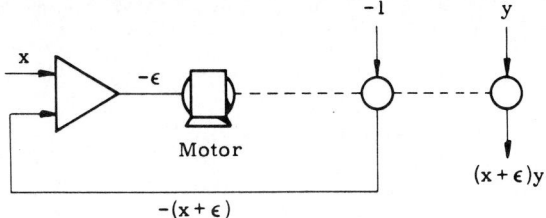

Fig. 8.40 Servomultiplier.

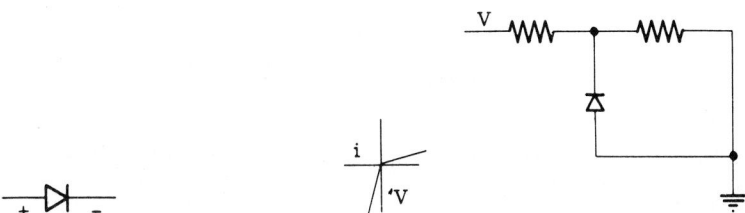

Fig. 8.41 Diode.   **Fig. 8.42** Diode network.

be made in which the current into the summing junction of an amplifier is approximately proportional to the square of the voltage applied.

The quarter-square multiplier uses two of these circuits, based on the identity

$$xy = \tfrac{1}{4}\left[(x + y)^2 - (x - y)^2\right] \qquad (8.107)$$

Its symbol is shown in Fig. 8.43. In most modern computers both polarities of input voltage must be provided. This has been indicated by two amplifiers on the inputs.

Division of voltages is accomplished by using a multiplier in the feedback path of an amplifier, as indicated in Fig. 8.44.

With the quarter-square multiplier, multiplication and division may be performed accurately at high frequencies. These complete the description of how the ordinary algebraic operations are performed on an analog computer.

Next, we turn to the operations of the calculus, which enable differential equations to be solved. Consider the circuit of Fig. 8.45. The charge on a capacitor is the integral of the current, therefore the derivative of the charge is the current. The charge is the potential times the capacitance, which is fixed, thus current balance around the summing junction gives

$$\frac{V_1}{R} + C\dot{V}_2 = 0 \qquad (8.108)$$

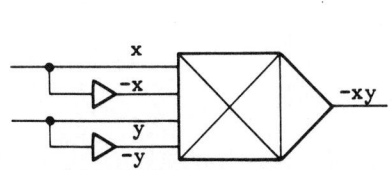

Fig. 8.43 Quarter-square multiplier.

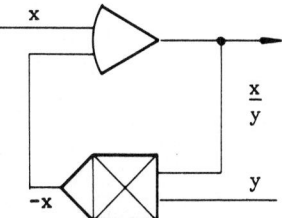

Fig. 8.44 Division.

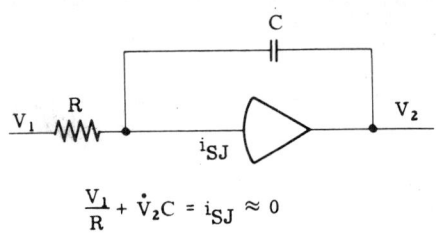

$$\frac{V_1}{R} + \dot{V}_2 C = i_{SJ} \approx 0$$

Fig. 8.45 Current balance.

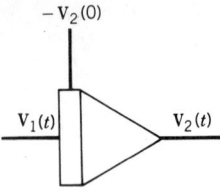

$$V_2(t) = V_2(0) - \int_0^t V_1(\tau)\,d\tau$$

Fig. 8.46 Integration.

or

$$V_2(t) = V_2(0) - \frac{1}{RC}\int_0^t V_1(\tau)\,d\tau \qquad (8.109)$$

Thus this circuit is an accurate integrator with a proportionality factor or time constant of RC. A common symbol is shown in Fig. 8.46. The initial condition can be set in with a special network furnished with most computers.

Differentiation can be performed with an input capacitor and a feedback resistor; however, it is not generally recommended because input signals are often contaminated by high-frequency noise, and the differentiator circuit amplifies this noise in direct proportion to its frequency. Since differential equations can always be written in terms of integrals, this causes no inconvenience.

The circuit of Fig. 8.47 generates a simple time lag, as can be derived from its current balance:

$$\frac{V_1}{R_1} + \dot{V}_2 C + \frac{V_2}{R_2} = 0 \qquad (8.110)$$

or

$$\dot{V}_2 + \frac{1}{R_2 C} V_2 = -\frac{V_1}{R_1 C} \qquad (8.111)$$

The transfer function of this circuit is

$$G(s) = \frac{V_2}{V_1}(s) = \frac{-R_2/R_1}{R_2 Cs + 1} = \frac{-R_2/R_1}{\tau s + 1}$$

Its time response to a pulse input is a simple exponential decay, with its time constant equal to $R_2 C$. The circuit corresponding to a damped sinusoid is shown in Fig. 8.48. Its differential equation is

$$\ddot{y} + 2\zeta\omega_n \dot{y} + \omega_n^2 y = \omega_n^2 f(t) \qquad (8.112)$$

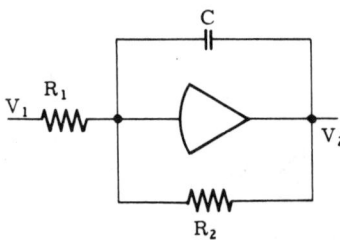

Fig. 8.47 Time lag.

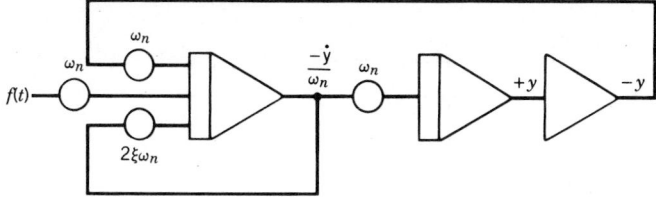

Fig. 8.48 Sinusoid.

The transfer function is

$$G(s) = \frac{\omega_n^2}{s^2 + 2\omega_n \zeta s + \omega_n^2} = \frac{1}{\dfrac{s}{\omega_n^2} + \dfrac{2\zeta}{\omega_n}s + 1} \qquad (8.113)$$

A particular precaution that must be taken with analog computers is in scaling voltages to represent other physical variables. This must be done consistently for all variables and must be done in such a way that no amplifier is ever overloaded. Since amplifier outputs can never be fully predicted, some problems may require rescaling because of a bad initial guess about the range of a dependent variable. This is particularly troublesome if nonlinear equipment is involved in the simulation. The necessity for proper prior scaling is one of the major disadvantages of analog computation.

With these basic circuit elements, the control engineer can proceed directly from a block diagram of a physical system to a wired program board without writing down differential equations because the circuits corresponding to the common transfer functions are combinations of those shown in Figs. 8.46–8.48. Special nonlinearities can be simulated by special networks. Simple ones such as multiplication, squaring, exponentiation, and generation of logarithms are often prewired directly to the program board.

Therefore, by suitable interconnection of components, even the most complicated system can be simulated in full dynamic character. Frequency response, stability limits, sensitivity to parameter changes, and effects of noise can all be tested and studied. Adjustments of the simulated controls for optimum rejection of noise can be done. Different types of control can be tried experimentally. If care has been taken in formulating the system dynamics, then the analog computer can be made to behave precisely like the device it simulates.

Figures 8.49–8.53 show some special diode networks for simulating some of the special discontinuous nonlinearities that have been discussed previously.

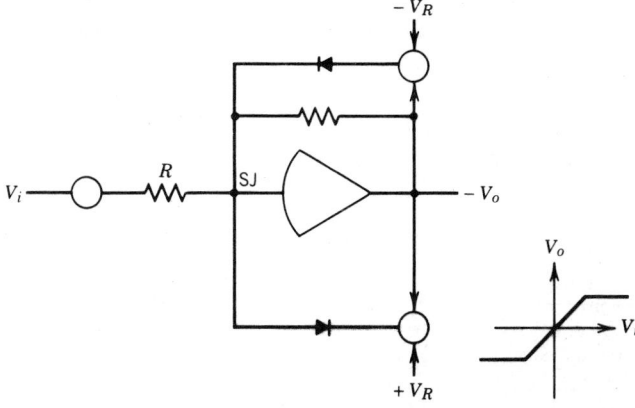

Fig. 8.49 Saturation.

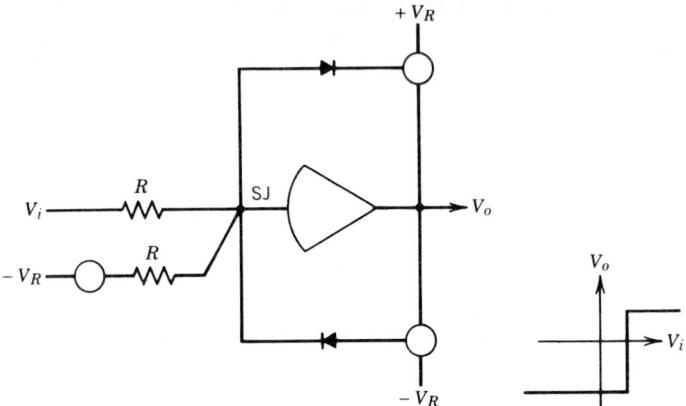

Fig. 8.50   On-off controller.

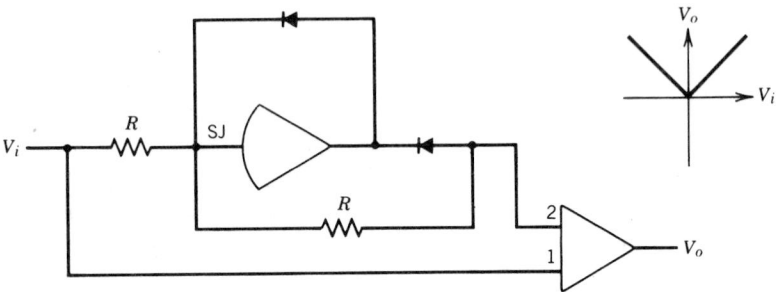

Fig. 8.51   Rectifier.

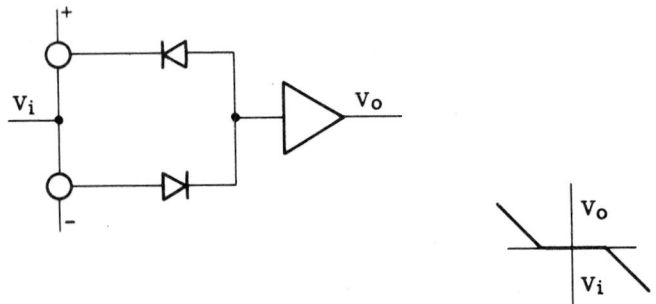

Fig. 8.52   Deadzone.

Modern analog computers are generally equipped with additional types of special equipment, some of which are now listed and described below:

1. Comparator: equivalent to the on-off control of Fig. 8.50.
2. Resolver: a circuit for rapidly and continuously converting from polar to Cartesian coordinates or the reverse.
3. Relays and switches: operated by voltages generated in the simulation.

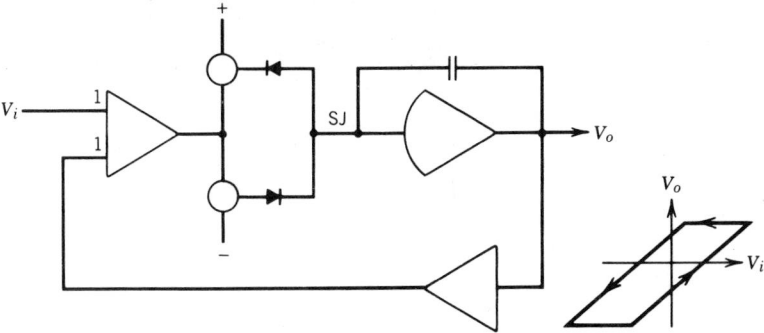

Fig. 8.53 Hysteresis or backlash.

4. Memory amplifiers: An integrator into which an initial condition is set, but to which no integrating input is patched, will "remember" the initial condition.
5. Repetitive operation at high speed: solutions can be generated to repeat hundreds of times per second.
6. Outputs can be displayed on an oscilloscope, plotted on coordinate paper, or on continuous recording equipment.

All these features make an analog computer a highly useful tool for design of control systems.

8.8.2 Digital Computation

An analog computer simultaneously solves all of the differential equations that model a physical system, and continuous voltage signals represent the variables of interest. This enables the machine to operate in real time. Significant disadvantages of analog simulation are the high cost of the computer due to the multiplicity of elements with demanding performance specifications, difficulties with scaling to avoid overloading of amplifiers, and relatively limited accuracy and repeatability due in part to amplifier drift. As a consequence of the very rapid development of digital computer hardware and software giving ever greater capability and flexibility at reducing cost, system simulation is inevitably being carried out more and more on the digital computer. There is effectively no problem of overloading, enabling wide ranges of parameter variation to be accommodated.

The solution of a differential equation involves the process of integration, and for the digital computer analytical integration must be replaced by some numerical method that yields an approximation to the true solution. A number of special programming languages referred to as continuous system simulation languages (CSSL) or simply as simulation languages are available as analytical tools to study the dynamic behavior of a wide range of systems without the need for a detailed knowledge of computing procedures. The languages are designed to be simple to understand and use, and they minimize programming difficulty by allowing the program to be written as a sequence of relatively self-descriptive statements. Numerous different languages with acronyms such as ACSL, CSMP, CSSL, DYNAMO, DARE, MIMIC, TELSIM, ENPORT[8], SCEPTRE, and SIMNON have been developed by computer manufacturers, software companies, universities, and others, some for specific families of machines and others for wider applications. Due to standardization a number of the languages tend to be broadly similar. Symbolic names are used for the system variables, and the main body of the program is written as a series of simple statements based on the system state equations, block diagrams, or bond graphs. To these are added statements specifying initial parameter values and values of system constants, and simple command statements controlling the running of the program and specifying the form in which the output is required. The short user-written program is then automatically translated into a FORTRAN (or other high-level language) program that is then compiled, loaded, and executed to produce a time history of the variables of interest in print or plot form. System constants and initial conditions can be altered and the program rerun without the need to retranslate and recompile. Many of the languages are designed to be run interactively from a graphics terminal and have the facility of displaying on the screen whichever output is of interest and, if the solution is not developing as desired, the capability of interrupting the program, changing parameters, and rerunning immediately.

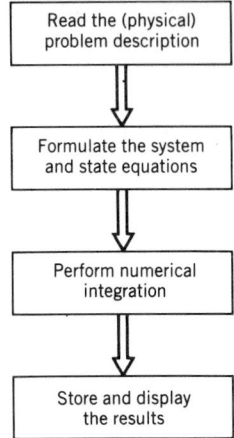

Fig. 8.54 Typical simulation flowchart of the CSSL.

The typical steps to be taken by a CSSL are shown in Fig. 8.54. The major process blocks include:

1. Reading the initial problem description, in which the circuit, bond graph, schematic, or block diagram information is communicated to the computer.
2. Formulating the system and state equations.
3. Performing the numerical integration of the state equations by a suitable method.
4. Storing and displaying results.

Most simulation languages differ with regard to use by the engineer in steps 1 and 2; for example, SCEPTRE for circuit descriptions, ENPORT for bond graph descriptions, CSMP for block diagrams, and DARE for state equations. The best guide to a program's characteristics is its user's manual.

In summary, the advantages of digital simulation are (i) simple program writing, (ii) accuracy and reproducibility, (iii) cost-effectiveness, (iv) where interactive facilities exist, keyboard entry of program and running, (v) inspection of plots on graphics display screen, and (vi) on-line modification and rerunning.

8.8.3 Hybrid Computation

Hybrid computation is a combination of analog and digital computations. In hybrid computations analog and digital machines are used in such a way that a problem can be programmed by exploiting the most efficient features of each. Much of the cost of such a machine arises from the rather complicated interface equipment required to make the necessary conversions between analog and digital signals and vice versa. The cost of an analog or hybrid computer is now justified only for certain large simulations where fast computing speed attained by parallel operation is important. Even this advantage may disappear as parallel computation is introduced into digital systems.

8.9   GAIN AND PHASE MARGIN

The Nyquist stability criterion may be conveniently used to define certain measures of relative stability or robustness. We note that $(-1, 0)$ point in the GH plane plays a crucial role in determining the closed-loop stability of a system. If a system's stability status is known, one might be interested in knowing how stable the system is due to changes in parameters. For example, if the system remains stable despite large changes in parameters, then it is said to possess a high degree of relative stability or robustness. Gain margin and phase margin are two measures typically employed to characterize this robustness. They characterize how close the Nyquist plot is to encircling the $(-1, 0)$ point in the GH plane.

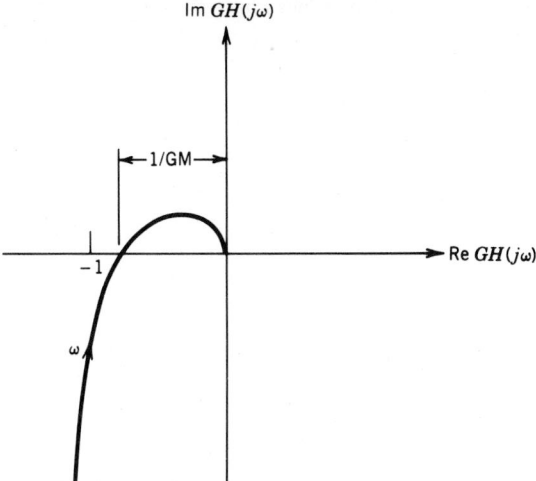

Fig. 8.55   Gain margin.

8.9.1   Gain Margin[4]

This is a measure of how much the loop gain can be raised before closed-loop instability results. The basic definition of gain margin (GM) is apparent from Fig. 8.55.

8.9.2   Phase Margin[4]

The phase margin (PM) is the difference between the phase of $GH(j\omega)$ and 180° when the $GH(j\omega)$ crosses the circle with unit magnitude. A positive phase margin corresponds to a case where the Nyquist locus does not encircle the $(-1, 0)$ point. This is shown in Fig. 8.56.

A stable system corresponds to gain and phase margins that are positive. In some cases, however, the PM and GM notions break down. For first- and second-order systems, the phase never crosses the 180° line; hence the gain margin is always ∞. For higher-order systems, it is possible to have more than one crossing of the unit amplitude circle and more than one crossing of the 180° line. In such situations the GM and PM are somewhat misleading. Furthermore, nonminimum phase systems exhibit stability criteria that are opposite to those previously defined.

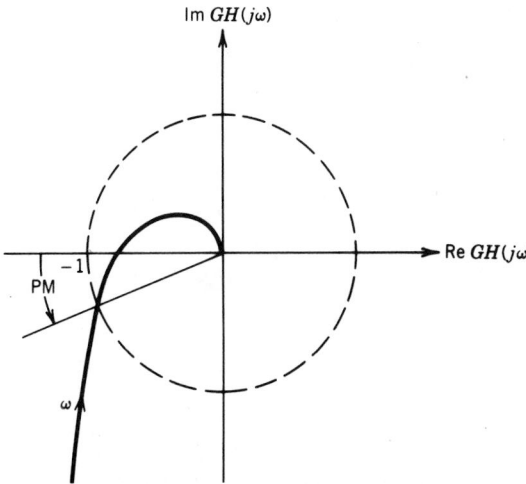

Fig. 8.56   Phase margin.

8.9.3   Gain-Phase Plots

The graphical representation of the frequency response of the system $G(s)$ using either

$$G(j\omega) = G(s)\big|_{s=j\omega} = \operatorname{Re} G(j\omega) + j\operatorname{Im} G(j\omega)$$

or

$$G(j\omega) = |G(j\omega)| e^{j\phi(\omega)}$$

where

$$\phi(\omega) = \underline{/G(j\omega)}$$

is known as the polar plot. The coordinates of the polar plot are the real and imaginary parts of $G(j\omega)$ as shown in Fig. 8.57.

Example 8.11. Obtain the polar plot of the transfer function

$$G(s) = \frac{K}{s(\tau s + 1)}$$

The frequency response is given by

$$G(j\omega) = \frac{K}{j\omega(j\tau\omega + 1)}$$

Then the magnitude and the phase can be written as

$$|G(j\omega)| = \frac{K}{\omega\sqrt{\omega^2\tau^2 + 1}}$$

and

$$\underline{/G(j\omega)} = -\frac{\pi}{2} - \tan^{-1}\omega\tau$$

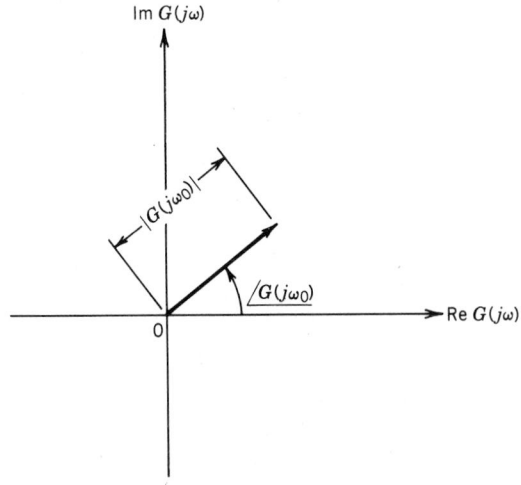

Fig. 8.57   Polar representation.

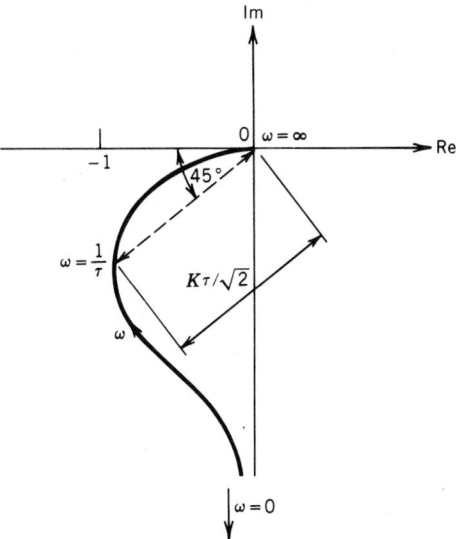

Fig. 8.58 Polar plot for $G(s) = K/s(\tau s + 1)$.

If $|G(j\omega)|$ and $\underline{/G(j\omega)}$ are computed for different frequencies an accurate plot can be obtained. A quick idea can, however, be gained by simply doing a limiting analysis at $\omega = 0$, $\omega = \infty$, and the corner frequency $\omega = 1/\tau$.

We note that for

$$\omega = 0 \qquad |G(j\omega)| \to \infty \qquad \underline{/G(j\omega)} = \frac{-\pi}{2}$$

$$\omega = \infty \qquad |G(j\omega)| \to 0 \qquad \underline{/G(j\omega)} = -\pi$$

$$\omega = \frac{1}{\tau} \qquad |G(j\omega)| = \frac{K\tau}{\sqrt{2}} \qquad \underline{/G(j\omega)} = \frac{-3\pi}{4}$$

The polar plot is shown in Fig. 8.58.

A gain-phase plot is where the frequency response information is given with respect to a Cartesian frame with vertical axis for gain and the horizontal for phase.

8.9.4 Polar Plot as a Design Tool in the Frequency Domain

As a design tool its best use is in determining relative stability with respect to gain and phase margin. If the uncertainty in the transfer function can be characterized by bounds on the gain-phase plot, then it allows one to determine what type of compensation needs to be provided for the system to perform in the presence of such uncertainties. As an example consider the closed-loop system shown in Fig. 8.59.

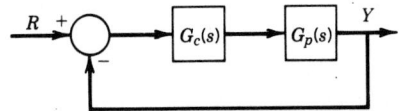

Fig. 8.59 Closed-loop system.

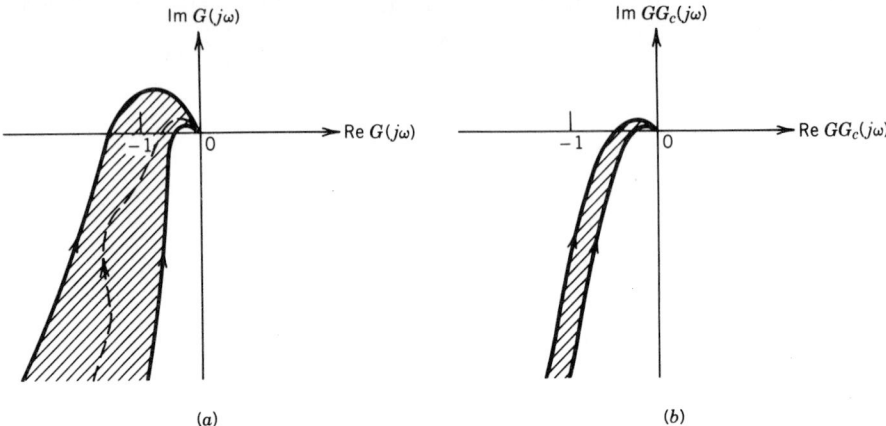

Fig. 8.60 Polar plots for an uncertain system.

Suppose the gain-phase plot is known to lie in the shaded region in the $G(j\omega)$ plane as shown in Fig. 8.60(a).

Since the shaded region includes the $(-1,0)$ point, and the true gain-phase plot for the plant can lie anywhere in the shaded region, the system can potentially be unstable. If stabilization in the presence of uncertainty is the primary design issue, then we would require a $G_c(s)$ so that it would reshape the high-frequency part of the polar plot with a reduced band of uncertainty. The reduction in the band of uncertainty is a required feature of any sound feedback system design. Qualitatively one would expect to reshape the polar plot to something that looks like what is shown in Fig. 8.60(b). Moreover, knowing the important frequency ranges will allow one to be more concerned with relevant portions of the gain-phase plot for reshaping.

To further illustrate the basic philosophy of a typical design in the frequency domain, consider the plant transfer function

$$G_p(s) = \frac{K}{s(1 + s)(1 + 0.0125s)}$$

in the feedback configuration shown in Fig. 8.61.

It is required that when a ramp input is applied to the closed-loop system, the steady-state error of the system does not exceed 1% of the amplitude of the input ramp. Using steady-state error computations we find that the minimum K should be such that

$$\text{Steady-state error} = e_{ss} = \lim_{s \to 0} \frac{1}{sG_p(s)} = \frac{1}{K} \le 0.01$$

that is, $K \ge 100$.

It can be easily verified that with $G_c(s) = 1$ the system is unstable for $K > 81$, implying that a controller $G_c(s)$ must be designed to satisfy the steady-state performance and relative stability

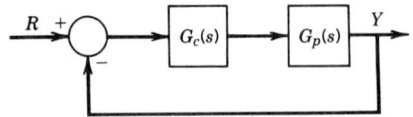

Fig. 8.61 Closed-loop system with $G(s) = K/s(s + 1)(0.0125s + 1)$.

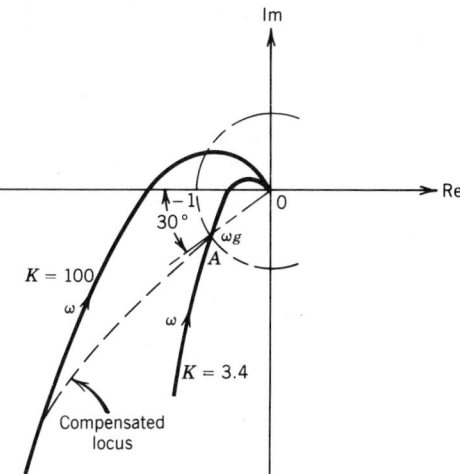

Fig. 8.62   Polar plot for open-loop system transfer function of Fig. 8.61.

requirements. Putting it another way, the controller must be able to keep the zero-frequency gain of $sG_pG_c(s)$ effectively at 100 while maintaining a prescribed degree of relative ability. The principal of the design in the frequency domain is best illustrated by the polar plot of $G_p(s)$ shown in Fig. 8.62. In practice, the Bode diagram is preferred for design purposes because it is simpler to construct. The polar plot is used mainly for analysis and added insight.

As shown in Fig. 8.62, when $K = 100$, the polar plot of $G_p(s)$ encloses the $(-1, 0)$ point, and the closed-loop system is unstable. Let us assume that we wish to realize a PM $= 30°$. This means that the polar plot must pass through point A (with magnitude 1 and phase $-150°$). If K is the only adjustable parameter to achieve this PM, the desired value of $K \simeq 3.4$, as shown in Fig. 8.62. But, K cannot be set to 3.4 since the ramp error constant would only be 3.4 sec^{-1}, and the steady-state error requirement will not be satisfied.

Since the steady-state performance of the system is governed by the characteristics of the transfer function at low frequency, and the damping or the transient behavior of the system is governed by the relatively high-frequency characteristics, as Fig. 8.62 shows, to simultaneously satisfy the transient and the steady-state requirements, the frequency locus of $G_p(s)$ has to be reshaped so that the high-frequency portion of the plot follows the $K = 3.4$ trajectory and the low-frequency portion follows the $K = 100$ trajectory. The significance of this reshaping of the frequency locus is that the compensated locus shown in Fig. 8.62 will be coincident with the high-frequency portion yielding PM $= 30°$, while the zero-frequency gain is maintained at 100 to satisfy the steady-state requirement.

When we inspect the loci of Fig. 8.62, we see that there are at least two alternatives in arriving at the compensated locus:

1. Starting from the $K = 100$ locus and reshaping the locus in the region near the gain cross-over frequency ω_g, while keeping the low-frequency region of $G_p(s)$ relatively unaltered.
2. Starting from the $K = 3.4$ locus and reshaping the low-frequency portion of $G_p(s)$ to obtain an error constant $= 100$ while keeping the locus near $\omega = \omega_g$ relatively unchanged.

In the first approach, the high-frequency portion of $G_p(s)$ is pushed in the counterclockwise (ccw) direction, which means that more phase is added to the system in the positive direction in the proper frequency range. This scheme is basically phase-lead compensation and controllers used for this purpose are often of the high-pass filter type. The second approach apparently involves the shifting of the low-frequency part of the $K = 3.4$ trajectory in the clockwise (cw) direction, or alternatively, reducing the magnitude of $G_p(s)$ with $K = 100$ at the high-frequency range. This scheme is often referred to as phase-lag compensation since more phase lag is introduced to the system in the low-frequency range. The controllers used for this purpose are often referred to as low-pass filters.

8.10 HALL CHART

In typical frequency response design only the open-loop transfer function is plotted. Therefore it is useful to know how the closed-loop performance is related to the open loop. Hall charts provide a convenient way of carrying out a frequency response design with closed-loop performance specifications. One important consideration is the maximum closed-loop gain. Another is the closed-loop phase. A Hall chart primarily consists of constant closed-loop gain loci and constant closed-loop phase loci. A design would then proceed by drawing the open-loop polar plot on the Hall chart.

For a unity negative feedback system as shown in Fig. 8.63 the closed-loop transfer function is

$$\frac{C(s)}{R(s)} = \frac{G(s)}{1 + G(s)} \tag{8.114}$$

In the following discussion we assume that the polar plot of $G(j\omega)$ is known.

8.10.1 Constant-Magnitude Circles

The loci on which the closed-loop magnitude

$$\left|\frac{C(s)}{R(s)}\right| = \left|\frac{G(s)}{1 + G(s)}\right| = M = \text{const.}$$

are referred to as constant-magnitude loci. In fact these loci are circles in the $G(j\omega)$ plane. This can be established by noting a typical point on the $G(j\omega)$ plot as $X + jY$. Then

$$M = \frac{|X + jY|}{|1 + X + jY|}$$

and

$$M^2 = \frac{X^2 + Y^2}{(1 + X)^2 + Y^2}$$

Hence

$$X^2 + \frac{2M^2}{M^2 - 1}X + \frac{M^2}{M^2 - 1} + Y^2 = 0$$

which can be written as

$$\left(X + \frac{M^2}{M^2 - 1}\right)^2 + Y^2 = \frac{M^2}{(M^2 - 1)^2} \tag{8.115}$$

Equation (8.115) is the equation of a circle with center at $X = -M^2/(M^2 - 1)$, $Y = 0$ and with radius $|M/(M^2 - 1)|$. A family of constant M circles is shown in Fig. 8.64. Given a point $P \equiv (X_1, Y_1)$ on an open-loop polar plot $G(j\omega)$, the corresponding closed-loop magnitude can be determined by locating the M circle passing through that point.

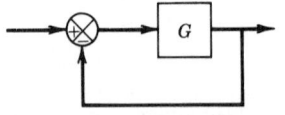

Fig. 8.63   Unity negative feedback system.

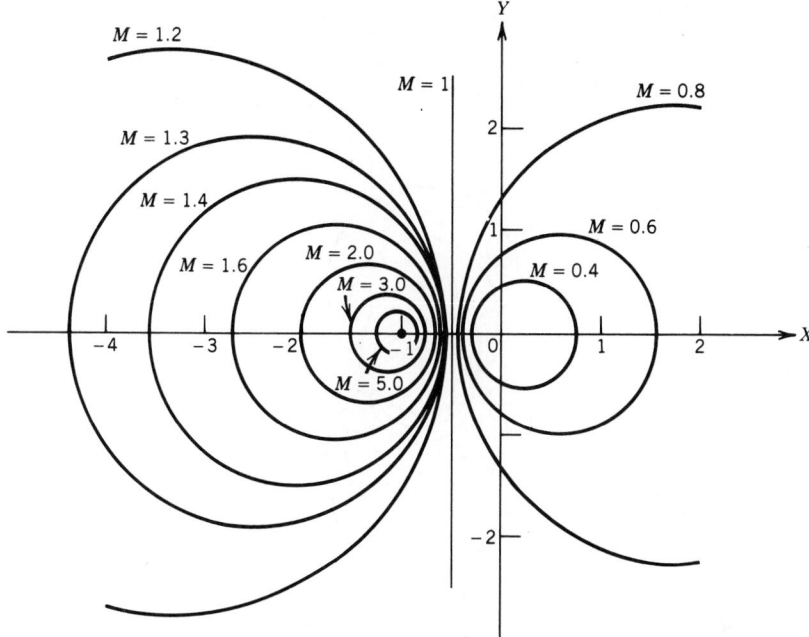

Fig. 8.64   Family of constant *M* circles.

Graphically the intersection of the $G(j\omega)$ plot and the constant *M* locus gives the value of *M* at the frequency denoted on the $G(j\omega)$ curve. If it is desired to keep the value of the maximum closed-loop gain M_r less than a certain value, the $G(j\omega)$ curve must not intersect the corresponding *M* circle at any point, and at the same time must not enclose the $(-1, j0)$ point. The constant *M* circle with the smallest radius that is tangent to the $G(j\omega)$ curve gives the value of M_r, and the resonant frequency ω_r is read off at the tangent point on the $G(j\omega)$ curve.

8.10.2   Constant-Phase Circles

The loci of constant phase of the closed-loop system can also be determined in the $G(j\omega)$ plane by a method similar to that used for constant *M* loci. With reference to Eq. (8.114) the phase of the closed-loop system corresponding to the point $P = X + jY$ is written as

$$\phi = \tan^{-1}\left(\frac{Y}{X}\right) - \tan^{-1}\left(\frac{Y}{1+X}\right) \tag{8.116}$$

Taking the tangent on both sides of Eq. (8.116) and rearranging yields

$$\left(X + \frac{1}{2}\right)^2 + \left(Y - \frac{1}{2N}\right)^2 = \frac{1}{4} + \left(\frac{1}{2N}\right)^2 \tag{8.117}$$

where $N = \tan\phi$.

Equation (8.117) represents a family of circles with center at $(-1/2, 1/2N)$ and with radius $\sqrt{1/4 + 1/(2N)^2}$. The constant-phase loci are shown in Fig. 8.65.

The use of constant magnitude and phase circles enables one to find the entire closed-loop frequency response from the open-loop frequency response $G(j\omega)$ without calculating the magnitude and phase of the closed-loop transfer function at each frequency. The intersections of the $G(j\omega)$ locus and the *M* circles and *N* circles give the values of *M* and *N* at frequency points on the $G(j\omega)$ locus.

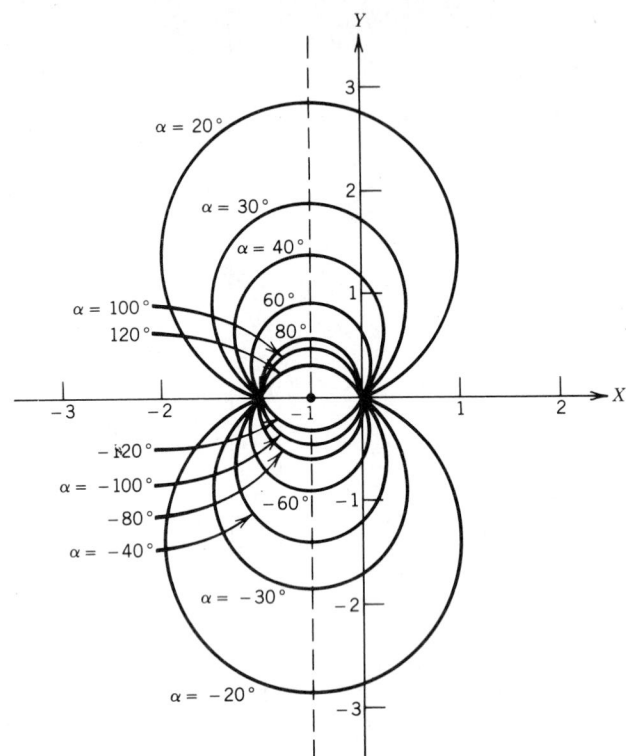

Fig. 8.65   Family of constant N circles.

8.10.3   Closed-Loop Frequency Response for Nonunity Feedback Systems

The constant M and N circles are limited to closed-loop systems with unity negative feedback, whose transfer function is given by Eq. (8.114). When a system has nonunity feedback, the closed-loop transfer function is

$$\frac{C(s)}{R(s)} = \frac{G(s)}{1 + G(s)H(s)} \tag{8.118}$$

and constant M loci derived earlier cannot be directly applied. However, with a slight modification constant M and N loci can still be applied to systems with nonunity feedback. We modify Eq. (8.118) as

$$\frac{C(s)}{R(s)} = \frac{1}{H(s)} \frac{G(s)H(s)}{1 + G(s)H(s)}$$

The magnitude and phase angle of $G_1(s)/[1 + G_1(s)]$ where $G_1(s) = G(s)H(s)$, may be obtained easily by plotting the $G_1(j\omega)$ locus and reading the values of M and N at various frequency points. The closed-loop frequency response $C(j\omega)/R(j\omega)$ may then be obtained by multiplying $G_1(j\omega)/[1 + G_1(j\omega)]$ by $1/H(j\omega)$.

8.10.4   Closed-Loop Amplitude Ratio

In obtaining suitable performance, the adjustment of gain is usually the first consideration. The adjustment of gain is usually based on the maximum closed-loop gain or the resonant peak. That is the gain K which must be chosen so that over the entire frequency range the closed-loop amplitude ratio M_r is not exceeded.

Consider first isolating the circle corresponding to M_r as shown in Fig. 8.66. Then a tangent line to the M_r circle is drawn from the origin, which makes an angle ψ with the real line.

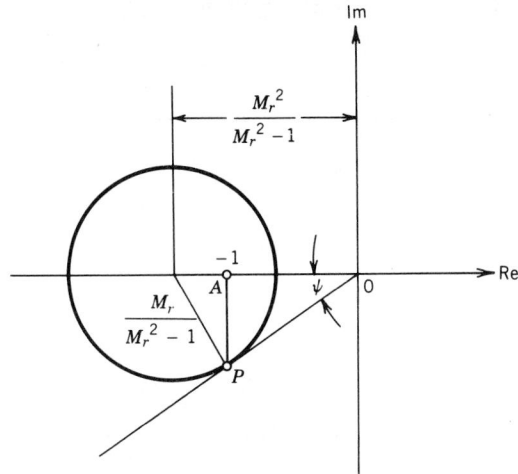

Fig. 8.66 *M* circle.

If $M_r > 1$, then

$$\sin \psi = \left| \frac{M_r/(M_r^2 - 1)}{M_r^2/(M_r^2 - 1)} \right| = \frac{1}{M_r}$$

It can be shown that the line drawn from P, perpendicular to the negative real axis, intersects this axis at the $(-1,0)$ point. These two facts, namely $\sin \psi = 1/M_r$ and that the normal from P passes through $(-1,0)$, can be used to determine the appropriate gain K.

Example 8.12. Consider the system shown in Fig. 8.67(a): determine K so that $M_r = 1.4$. First sketch the polar plot of

$$\frac{G(j\omega)}{K} = \frac{1}{j\omega(1 + j\omega)}$$

as shown in Fig. 8.67(b). The value of ψ corresponding to $M_r = 1.4$ is obtained from

$$\psi = \sin^{-1} \frac{1}{M_r} = \sin^{-1} \frac{1}{1.4} = 45.6°$$

The next step is to draw a line OP that makes an angle $\psi = 45.6°$ with the negative real axis. Then draw the circle that is tangent to both the $G(j\omega)/K$ locus and the line OP. The perpendicular line drawn from the point P intersects the negative real axis at $(-0.63, 0)$. Then the gain K of the system is determined as follows:

$$K = \frac{1}{0.63} = 1.58$$

8.11 NICHOLS CHART

Both the gain and phase plots are generally required to analyze the performance of a closed-loop system. A major disadvantage in working with polar plots is that the curve no longer retains its original shape when a simple modification such as the change of the loop gain is made to the system. In design, however, not only the loop gain must be altered but often series or feedback controllers are

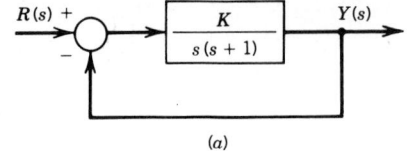

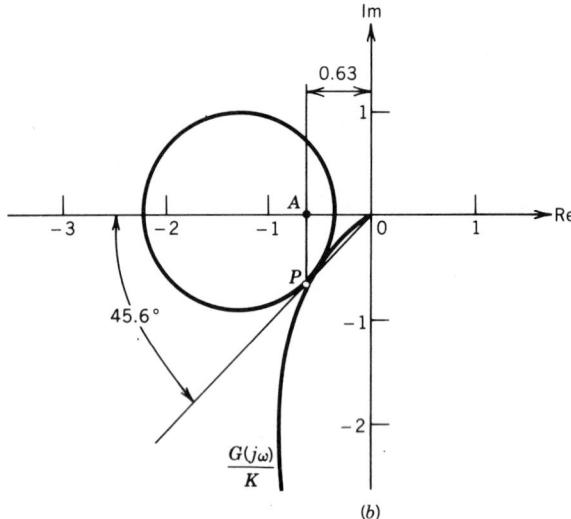

Fig. 8.67 (*a*) Closed-loop system; (*b*) determination of the gain K using an M circle.

to be added to the original system that require the complete reconstruction of the resulting open-loop transfer function. For design purposes it is more convenient to work with Bode diagrams or gain-versus-phase plots. The latter representation with corresponding M and N circles superimposed on it is referred to as the **Nichols chart**. In a gain-versus-phase plot the entire $G(j\omega)$ is shifted up or down vertically when the gain is altered. A Nichols chart is shown in Fig. 8.68. This chart is symmetric about the $-180°$ axis. The M and N loci repeat for every 360°, and there is symmetry at every 180° interval. The M loci are centered about the critical point $(0 \text{ db}, -180°)$.

8.11.1 Closed-Loop Frequency Response from that of Open Loop

It is quite easy to determine the closed-loop frequency response from that of the open loop by using the Nichols chart. If the open-loop frequency response curve is superimposed on the Nichols chart, the intersections of the open-loop frequency response curve $G(j\omega)$ and the M and N loci give the magnitude M and phase angle ϕ of the closed-loop frequency response at each frequency point. If the $G(j\omega)$ locus does not intersect the $M = M_r$ locus but is tangent to it, then the resonant peak value of the closed-loop frequency response is given by M_r. The resonant frequency is given by the frequency at the point of tangency.

As an example consider the unity negative feedback system with the following open-loop transfer function:

$$G(s) = \frac{K}{s(s+1)(0.5s+1)}; \qquad K = 1$$

In order to find the closed-loop frequency response by use of the Nichols chart, the $G(j\omega)$ locus is first constructed. (It is easy to first construct the Bode diagram and then to transfer values to the Nichols chart.) The closed-loop frequency response curves (gain and phase) may be constructed by reading the magnitude and phase angles at various frequency points on the $G(j\omega)$ locus from the M and N loci as shown in Fig. 8.69. Since the $G(j\omega)$ locus is tangent to $M = 5$ db locus, the peak value of the closed-loop frequency response is $M_r = 5$ db, and the resonant frequency is 0.8 rad/sec.

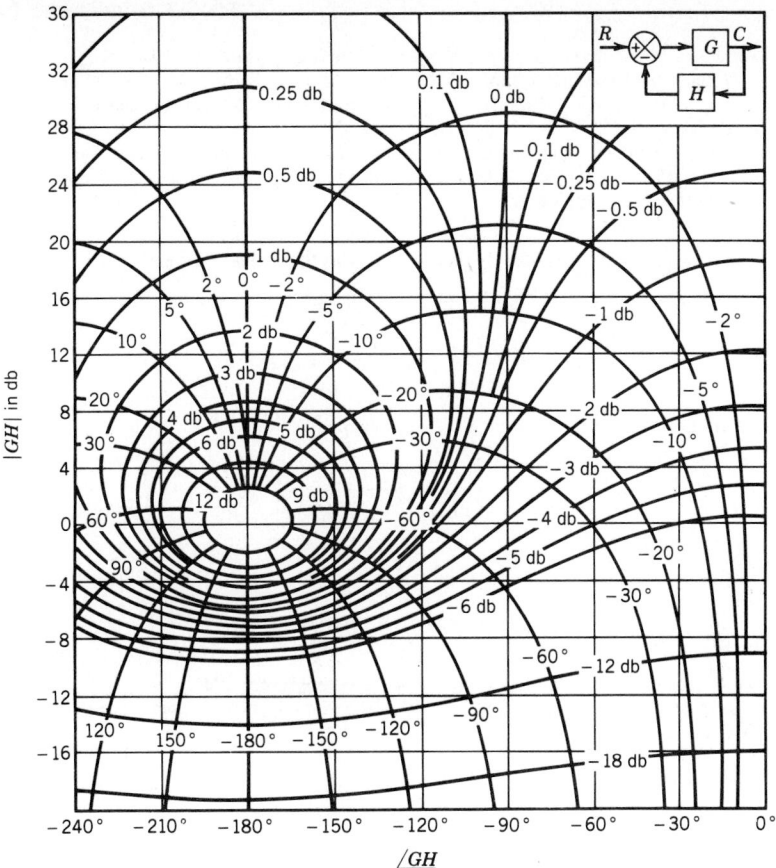

Fig. 8.68 Nichols chart.

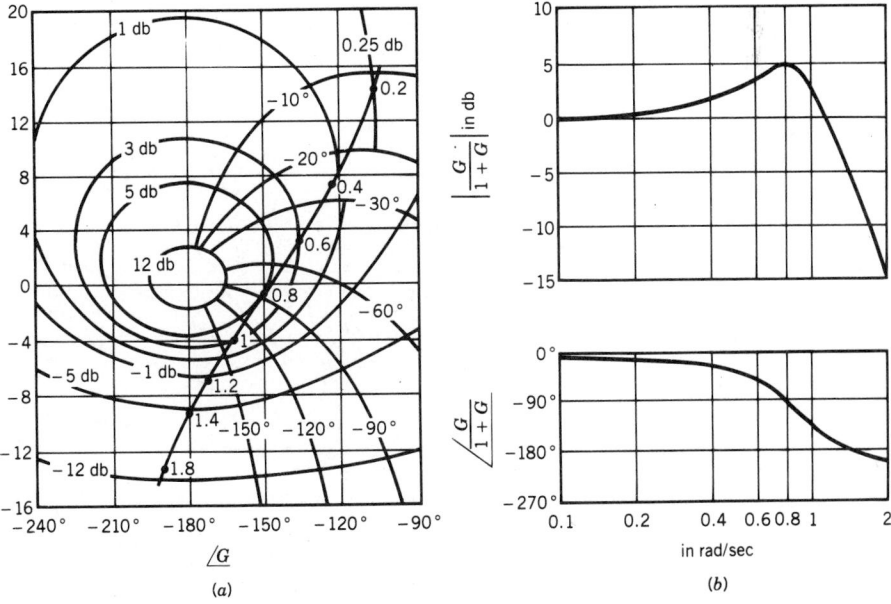

Fig. 8.69 (a) Plot of $G(j\omega)$ superimposed on Nichols chart; (b) closed-loop frequency response curves.

The bandwidth of the closed-loop system can easily be found from the $G(j\omega)$ locus in the Nichols chart. The frequency at the intersection of the $G(j\omega)$ locus and the $M = -3$ db locus gives the bandwidth. The gain and phase margins can be read directly from the Nichols chart.

If the open-loop gain K is varied, the shape of the $G(j\omega)$ locus in the Nichols chart remains the same, but is shifted up (for increasing K) or down (for decreasing K) along the vertical axis. Therefore the modified $G(j\omega)$ locus intersects the M and N loci differently, resulting in a different closed-loop frequency response curve.

8.11.2 Sensitivity Analysis Using the Nichols Chart[12]

Consider a unity feedback system with the transfer function

$$\frac{C(s)}{R(s)} = \frac{G(s)}{1 + G(s)} = G_{cl}(s)$$

The sensitivity of $G_{cl}(s)$ with respect to $G(s)$ is defined as

$$S_G^{G_{cl}}(s) = \frac{dG_{cl}(s)/G_{cl}(s)}{dG(s)/G(s)}$$

which yields

$$S_G^{G_{cl}}(s) = \frac{1}{1 + G(s)} \qquad (8.119)$$

Clearly the sensitivity function is a function of the complex variable s.

To design a system with a prescribed sensitivity the Nichols chart is quite convenient. Equation (8.119) is written as

$$S_G^{G_{cl}}(j\omega) = \frac{G^{-1}(j\omega)}{1 + G^{-1}(j\omega)}$$

which clearly indicates that the magnitude and phase of $S_G^{G_{cl}}(j\omega)$ can be obtained by plotting $G^{-1}(j\omega)$ on the Nichols chart and making use of the constant M loci for constant sensitivity

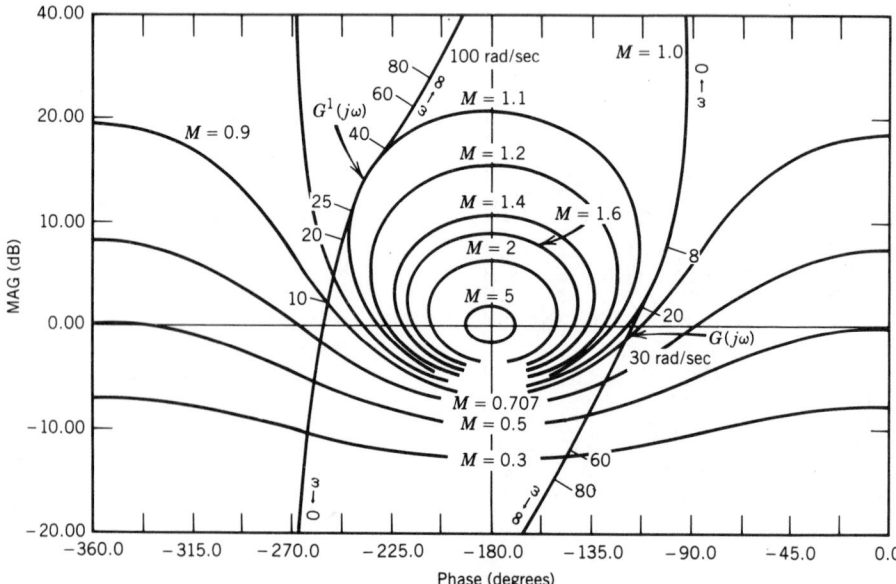

Fig. 8.70 Determination of the sensitivity function S_G^M in the Nichols chart.

function. Since the vertical coordinate of the Nichols chart is in decibels, the $G^{-1}(j\omega)$ curve on the Nichols chart can be easily obtained if $G(j\omega)$ is already available since

$$\left|G^{-1}(j\omega)\right|_{db} = -\left|G(j\omega)\right|_{db}$$

$$\angle G^{-1}(j\omega) = -\angle G(j\omega)$$

As an example consider the unity feedback system with the open-loop transfer function

$$G(s) = \frac{400,000K}{s(s+49)(s+991)}$$

the function $G^{-1}(j\omega)$ is plotted on the Nichols chart as shown in Fig. 8.70, for $K = 2.94$. The intersections of $G^{-1}(j\omega)$ curve with M loci give the magnitude of $S_G^{Gcl}(j\omega)$ at the corresponding frequencies. Figure 8.70 indicates several interesting points with regard to the sensitivity function of the feedback system. The sensitivity function approaches 0 db or unity as $\omega \to \infty$: $S_G^{Gcl} \to 0$ as $\omega \to 0$. A peak value of 1.1 db is reached at $\omega = 25$ rad/sec. This means that the closed-loop system is most sensitive to a change of $G(j\omega)$ at this frequency and more generally in this frequency range.

8.12   ROOT LOCUS

Poles and zero locations of a dynamic system characterize the system performance in a significant way. The root locus method allows one to investigate the closed-loop pole patterns of a dynamic system with respect to a single parameter.

A typical closed-loop characteristic equation of a feedback system can be written as

$$1 + G(s)H(s) = 0 \tag{8.120}$$

where $G(s)H(s)$ is the open-loop transfer function.

If $G(s)H(s)$ has a single parameter K as a variable, then by rewriting as

$$1 + KG(s)H(s) = 0 \tag{8.121}$$

a standard procedure for obtaining the closed-loop poles corresponding to any K is the Evans root locus method.

8.12.1   Angle and Magnitude Conditions

The closed-loop characteristic equation (8.121) can be written as

$$KG(s)H(s) = -1 = e^{j(2\pi l \pm \pi)} \tag{8.122}$$

Thus, any point s_0 satisfying the condition

$$\angle KG(s_0)H(s_0) = (2l \pm 1)\pi \tag{8.123}$$

satisfies Eq. (8.121). If $K > 0$, then Eq. (8.123) reduces to

$$\angle G(s_0)H(s_0) = (2l \pm 1)\pi \tag{8.124}$$

and is commonly called the angle condition. All points s_0 in the complex plane satisfying this angle condition satisfy the closed-loop characteristic equation and hence are said to lie on the root locus.

If s_0 is a point on the root loci, then the corresponding value of K may be computed by noting that

$$|K||G(s_0)||H(s_0)| = 1 \tag{8.125}$$

which is called the amplitude condition.

By studying the angle condition in detail of the closed-loop characteristic equation,

$$\underline{/1 + KG(s)H(s)} = 1 + K\frac{\prod_{i=1}^{m}(s + z_i)}{\prod_{j=1}^{n}(s + p_j)}$$

a set of rules can be developed for constructing the root locus easily. These rules are given next without proof.[3]

Rule 1. The system root loci have n branches originating at the n open-loop poles $-p_j$, $j = 1,\ldots, n$ with the value of $K = 0$.

Rule 2. Out of the n branches m number of branches will terminate on m finite zeros $-z_i$, $i = 1, 2,\ldots, m$ of the open-loop transfer function at $K = \infty$.

Rule 3. The remaining $n - m$ branches will go to ∞ along asymptotes as $K \to \infty$. The asymptotes are straight lines meeting at a point on the real line called the hub with specific orientation as given in Rule 4.

Rule 4. a. The asymptotes meet at the hub

$$\sigma = \frac{\sum_{j=1}^{n}\text{poles} - \sum_{i=1}^{m}\text{zeros}}{n - m}$$

$$= \frac{\sum_{j=1}^{n}(-p_j) - \sum_{i=1}^{m}(-z_i)}{n - m}$$

b. The $n - m$ asymptote angles are given by

$$\theta_N = \pm\frac{180° \, N}{n - m}$$

where N takes on values $1, 3, 5, 7,\ldots$. For each N, two angles are computed and the procedure repeated until $n - m$ angles are obtained.

Rule 5. If to the right of a point on the real axis there lies an odd number of open-loop poles and zeros, then it is a point on the root loci.

Rule 6. If two open-loop poles or two open-loop zeros are connected, then there must be a break point between the two (Fig. 8.71).

If an open-loop pole $-p_l$ and an open-loop zero $-z_q$ are connected, in most cases it may be considered as a full branch of the root loci, that is, that the closed-loop pole corresponding to the

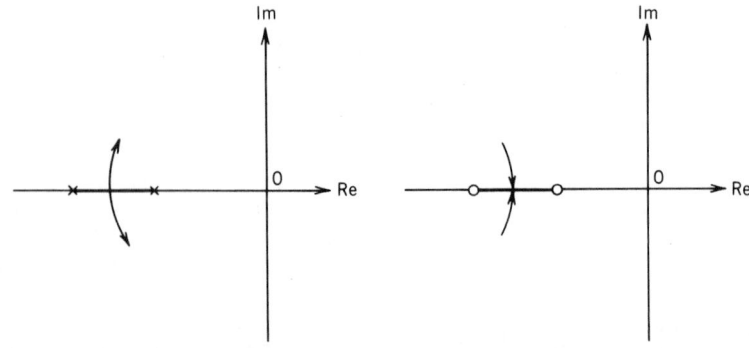

Fig. 8.71 Breakaway and break-in points.

open-loop pole $-p_l$ starts at $-p_l$ for $K = 0$ and reaches the closed-loop pole signified by the open-loop zero $-z_q$ as $K = \infty$.

Note. Exceptions to this rule exist. Some typical situations are depicted in Fig. 8.72(a).

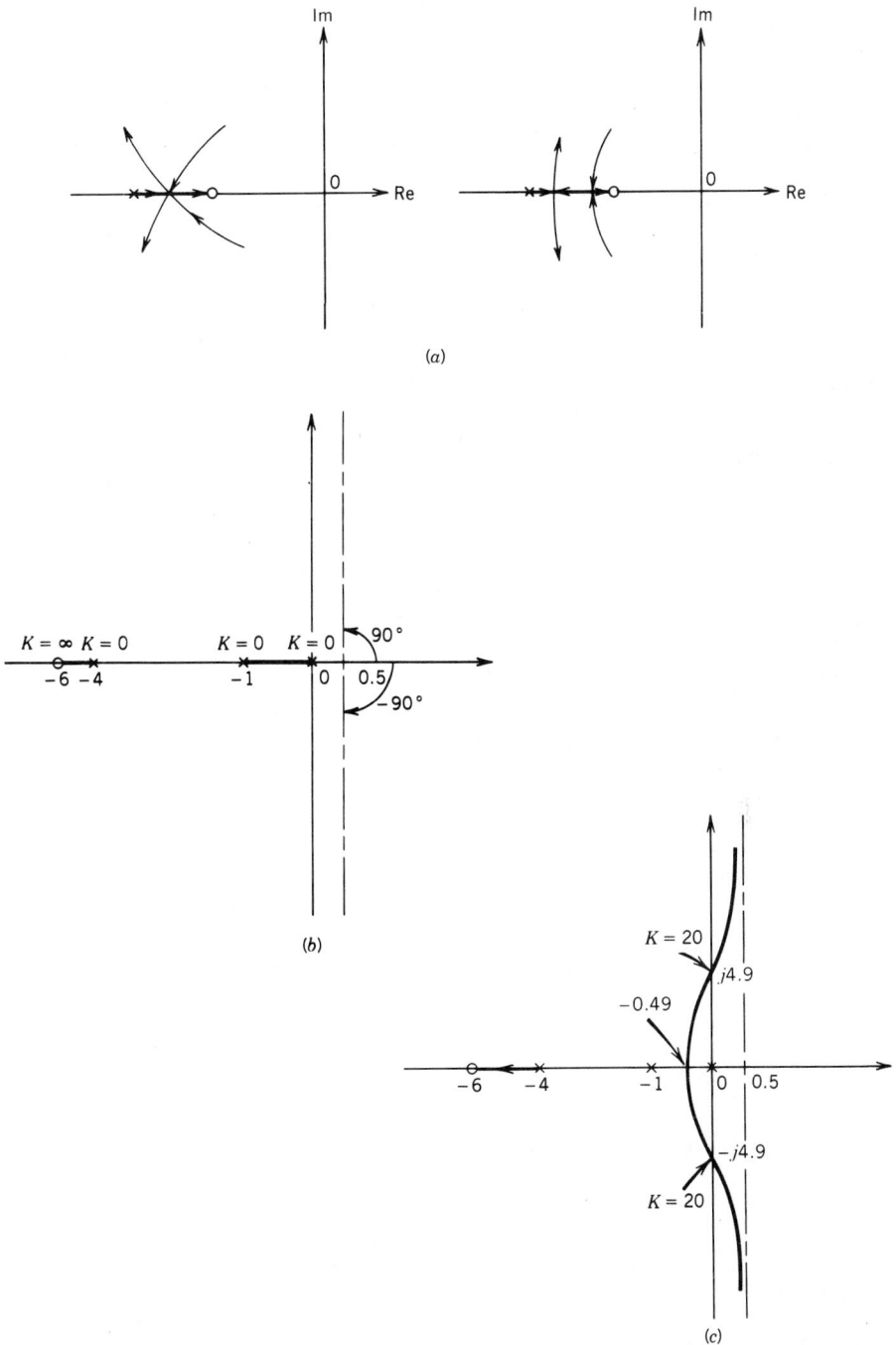

(a)

(b)

(c)

Fig. 8.72 (a) Breakaway and break-in possibilities; (b) sketch of root loci of Example 8.13 resulting from Rules R1–R4; (c) root loci for Example 8.13 where $G(s) = K(s + 6)/s(s + 1)(s + 4)$.

In order to determine the occurrence of such multiple break points, the next rule may be used.

Rule 7. The break points may be computed by determining points for which $dK/ds = 0$. Since

$$1 + KGH = 0$$

$$-K = \frac{1}{GH} = \frac{B(s)}{A(s)}$$

$$\frac{dK}{ds} = B(s)\frac{dA(s)}{ds} - A(s)\frac{dB(s)}{ds} = 0$$

Break points coupled with information from rule 6 makes it rather easy to pin down the branches.

Rule 8. The points at which the branches cross the imaginary axis can be determined by letting $s = j\omega$ in the characteristic equation.

Rule 9. The angle condition is made use of to determine the angle by which a branch would depart from a pole or would arrive at a zero as $K \to \infty$.

A point s_0 is considered very near the pole (zero) and the angle $G(s_0)H(s_0)$ is computed. The fact that s_0 is very near the pole (zero) makes all but one angle fixed. Thus by employing the angle condition, the unknown angle of departure (arrival) can be computed.

An example is given next to illustrate the various rules for constructing a root locus.

Example 8.13. Consider the closed-loop characteristic equation (CLCE)

$$1 + KG(s)H(s) = 1 + \frac{K(s + 6)}{s(s + 1)(s + 4)}$$

R1: $n = 3 \Rightarrow 3$ branches originating at $0, -1, -4$ at $K = 0$.
R2: $m = 1 \Rightarrow 1$ branch terminates at -6, at $K = \infty$.
R3: $n - m = 2$ branches approach ∞ along asymptotes.
R4: Hub $\sigma = (0 - 1 - 4 - (-6))/2 = 0.5$

$$\text{Asymptote angles} \quad \pm \frac{180° \ N}{n - m} = \pm \frac{180° \ N}{2}$$

$$\text{Set } N = 1 \Rightarrow \pm 90°$$

R1–R4: Yield the sketch of Fig. 8.72(b).
R5: Sections on the real line are 0 to -1 and -4 to -6.
R6: There must be a breakaway point between 0 and -1.
R7: Break points $dK/ds = 0$.

$$\Rightarrow (s + 6)(3s^2 + 10s + 4) - (s^3 + 5s^2 + 4s) = 0$$

$$s^3 + 11.5s^2 + 30s + 12 = 0$$

$$(s + 0.49)(s + 7.89)(s + 3.12) = 0$$

$s = -7.89$ and -3.12 are unacceptable from R5. Therefore the only breakaway point is at -0.49.

A sketch of the root loci is given in Fig. 8.72(c).

R8: Imaginary axis crossings:

$$\text{CLCE} \quad s^3 + 5s^2 + (4 + K)s + 6K = 0$$

Now let $s = j\omega$:

$$(j\omega)^3 + 5(j\omega)^2 + (4 + K)j\omega + 6K = 0$$

$$(6k - 5\omega^2) + j\omega[(4 + K) - \omega^2] = 0$$

Therefore

$$6K = 5\omega^2$$
$$\omega[4 + K - \omega^2] = 0$$

yielding

$$K = 0 \qquad \omega = 0$$

and

$$K = 20 \qquad \omega = \pm 4.9$$

Example 8.14. Consider the unity negative feedback system shown in Fig. 8.73(*a*). Obtain the loci of the closed-loop poles as α is varied from 0 to ∞, that is, obtain the root loci for $0 \le \alpha \le \infty$.

SOLUTION. The root loci is the points s satisfying the closed-loop characteristic equation:

$$1 + \frac{750}{(s + 5)(s + 10)(s + \alpha)} = 0$$

In order to utilize the rules for constructing the root loci the CLCE is rearranged in the form of Eq. (8.121).
By expanding and rearranging we get

$$\text{CLCE} = s(s + 5)(s + 10) + \alpha(s + 5)(s + 10) + 750 = 0$$

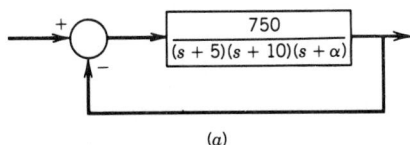

(*a*)

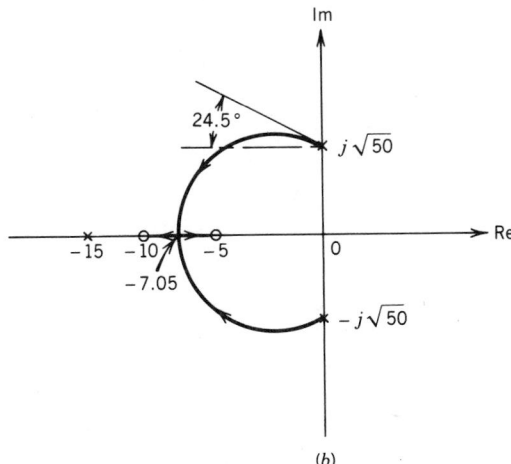

(*b*)

Fig. 8.73 (*a*) Closed-loop system of Example 8.14; (*b*) root loci of system of Example 8.14 where $G(s) = \alpha(s + 5)(s + 10)/(s + 15)(s + 50)$.

or

$$1 + \frac{\alpha(s + 5)(s + 10)}{(s + 15)(s^2 + 50)} = 0 \qquad (8.126)$$

Equation (8.126) has 3 poles at -15, $+j\sqrt{50}$, $-j\sqrt{50}$ and two finite zeros at -5, -10.

R1: $n = 3$ implies there are three branches originating at -15, $-j\sqrt{50}$, $+j\sqrt{50}$ at $K = 0$.
R2: $m = 2$ implies two of the three branches terminate on -5, -10 at $K = \infty$.
R3: $n - m = 1$ implies that there is one asymptote.
R4: For a single asymptote a hub does not exist

$$\text{asymptote angle} = \pm \frac{180° \, N}{n - m} = \pm \frac{180° \, N}{1}$$

that is, $\theta_1 = 180°$

R5: There is a section of the root loci between $-\infty$ and -15 and between -10 and -5.
R6: Since the two zeros -5 and -10 are connected, there must be a break-in point between -5 and -10. The section -15 to $-\infty$ forms a full branch.
R7: Break points: Since

$$G(s)H(s) = \frac{A(s)}{B(s)} = \frac{s^2 + 15s + 50}{s^3 + 15s^2 + 50s + 750}$$

$$\frac{d\alpha}{ds} = (s^2 + 15s + 50)(3s^2 + 30s + 50) - (s^3 + 15s^2 + 50s + 750)(2s + 15) = 0$$

that is,

$$s^4 + 30s^3 - 85s^2 - 8750 = 0 \qquad (8.127)$$

$s = -7.05$ is a break point.

Remark. To obtain the break points the fourth-order polynomial in s given in (8.127) must be factored. However, knowing that there must be a break point in the range -5 and -10 (Rule 6), it is quite easy to find the breakaway point. If all roots of (8.127) are found anyway, then only those points yielding $\alpha \leq 0$ are admissible as break points.

For this example R1–R6 give all the essential information to sketch the root loci of Fig. 8.73(b). If the angles of departure are needed we can employ rule 9.

Consider the point s_0 closer to $+j\sqrt{50}$ and write down the angle condition:

$$\angle(s_0 + 5) + \angle(s_0 + 10) - \angle(s_0 + 15) - \angle(s_0 + j\sqrt{50})$$
$$- \angle(s_0 - j\sqrt{50}) = \pi$$

Now let $s_0 \to j\sqrt{50}$ to yield

$$\angle(5 + j\sqrt{50}) + \angle(10 + j\sqrt{50}) - \angle(15 + j\sqrt{50} - j2\sqrt{50}) - \theta = \pi$$

$$\tan^{-1}\frac{\sqrt{50}}{5} + \tan^{-1}\frac{\sqrt{50}}{10} - \tan^{-1}\frac{\sqrt{50}}{15} - \frac{\pi}{2} - \theta = \pi$$

$$\theta = -3.57 \text{ rad} = -204.5°$$

Some typical root loci plots are shown in Table 8.7.

TABLE 8.7 Typical Root Loci Plots

No.	$G_0(s)$	Root loci	No.	$G_0(s)$	Root loci
1	$\dfrac{1}{s-p_1}$	*(plot)*	6	Same as 5 $\quad p_2 < z_1 < p_1$	*(plot)*
2	$\dfrac{s-z_1}{s-p_1}$	*(plot)*	7	Same as 5 $\quad p_1, p_2 < z_1$	*(plot)*
3	$\dfrac{1}{(s-p_1)(s-p_2)}$	*(plot)*	8	Same as 5	*(plot)*
4	$\dfrac{1}{(s-p_1)(s-p_2)}$ $\quad p_1, p_2$ complex	*(plot)*	9	$\dfrac{1}{(s-p_1)(s-p_2)(s-p_3)}$ $\quad p_1, p_2$ complex	*(plot)*
5	$\dfrac{(s-z_1)}{(s-p_1)(s-p_2)}$ $\quad z_1 < p_1, p_2$	*(plot)*	10	$\dfrac{1}{s(s-p_2)(s-p_3)}$	*(plot)*

TABLE 8.7 (Continued)

No.	$G_0(s)$	Root loci
11	$\dfrac{1}{(s-p)^3}$	
12	$\dfrac{1}{(s-p_1)(s-p_2)(s-p_3)}$ p_2, p_3 complex	
13	Same as 12	
14	Same as 12	
15	$\dfrac{(s-z_1)}{(s-p_1)(s-p_2)(s-p_3)}$	
16	Same as 15	
17	Same as 15	
18	Same as 15	
19	$\dfrac{(s-z_1)(s-z_2)}{(s-p_1)(s-p_2)(s-p_3)}$	
20	Same as 19	

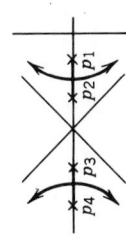

25 $\dfrac{1}{(s-p_1)(s-p_2)(s-p_3)(s-p_4)}$

All poles real

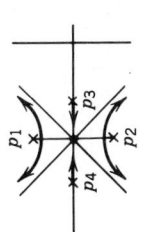

26 Same as 25

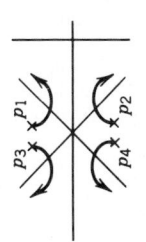

27 2 poles real
2 poles complex
Same as 26

28 Same as 25
All poles complex

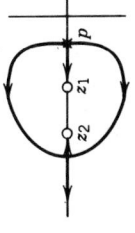

21 $\dfrac{(s-z_1)(s-z_2)}{(s-p)^3}$

$z_2 < z_1 < p$

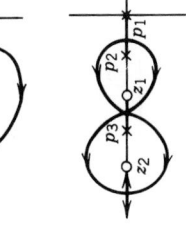

22 $\dfrac{(s-z_1)(s-z_2)}{s(s-p_2)(s-p_3)}$

$z_2 < p_3 < z_1 < p_2 < 0$

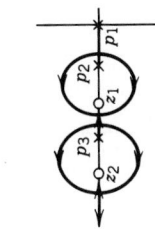

23 Same as 22

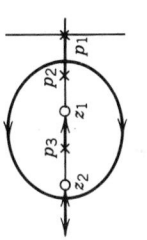

24 Same as 22

8.12.2 Time-Domain Design Using the Root Locus

Time domain performance specifications can often be related in an approximate sense to closed-loop pole locations. If suitable pole locations for a certain time domain performance can be effectively identified, then the root loci can be used to locate the closed-loop poles at those locations by appropriate compensation. Compensation can be provided by introducing additional dynamics into the feedback system in the form of additional poles and zeros (PID control, lead, lag, lead-lag, etc.). We shall now consider some examples to illustrate this time domain design philosophy.

Example 8.15. Obtain the root loci for a system with an open-loop transfer function

$$G(s) = \frac{K}{s(s+2)}$$

a. Indicate the location of closed-loop poles when $K = 4$ and determine the damping ratio ζ and the natural frequency ω_n corresponding to $K = 4$.

b. It is now required to double the natural frequency while keeping the same damping ratio. Design a compensator for satisfying the new design specifications.

SOLUTION. (a) Suppose the closed-loop system is as shown in Fig. 8.74(*a*). Then its root loci are as shown in Fig. 8.74(*b*). To find the poles at $K = 4$, solve the closed-loop characteristic equation (CLCE)

$$1 + \frac{4}{s(s+2)} = 0 \quad \text{or} \quad s^2 + 2s + 4 = 0$$

the roots are $s_{1,2} = -1 \pm j\sqrt{3}$:

$$\zeta = \cos \beta = 0.5$$

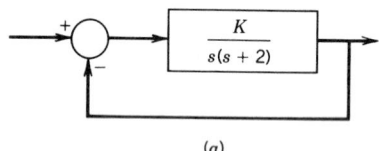

(a)

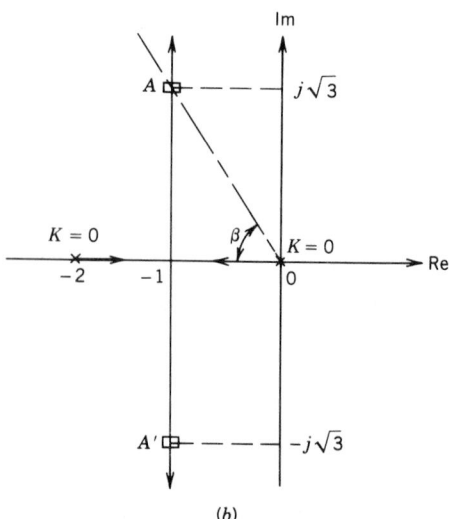

(b)

Fig. 8.74 (*a*) Time-domain design example; (*b*) sketch of root loci.

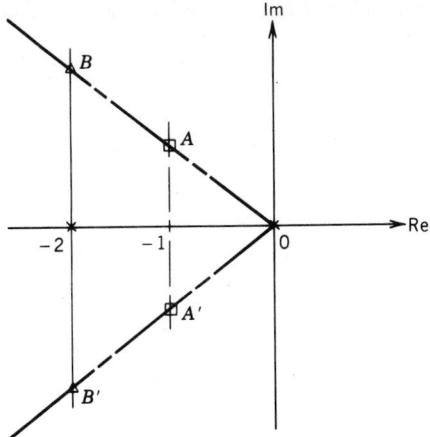

Fig. 8.75 Desired pole locations.

(b) Since the natural frequency is to be doubled keeping the same damping ratio, we need to move the closed-loop poles at A and A' in Fig. 8.75 so that $OB = 2OA$.

In order to satisfy the design specifications, the modified root loci must be made to pass through B and B'. To reshape the original root loci, additional poles and zeros are required. We give below a simple way to appropriately modify the root loci. Conceptually, the modification takes place as shown in Fig. 8.76.

If $G_c(s)$ is chosen to cancel the pole at -2 by selecting

$$G_c(s) = \frac{s+2}{s+p}$$

then we only need to find p so that the modified root locus passes through B. This is quite easy to do

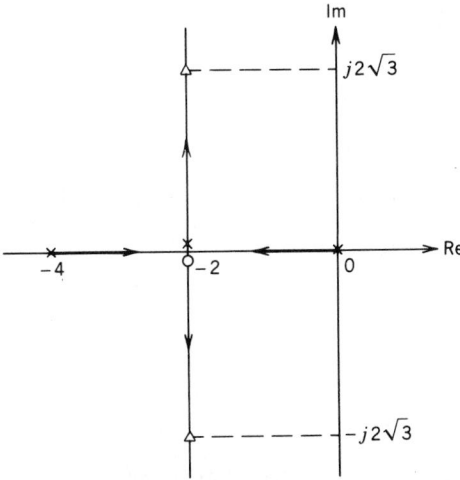

Fig. 8.76 Root loci with pole-zero cancellation.

by noting that the pole-zero cancellation at -2 leaves us with a second-order system with the two open-loop poles at 0 and $-p$.

By selecting $p = +4$ it is easy to verify that the modified root loci are as shown in Fig. 8.76. So the compensator

$$G_c(s) = \frac{s + 2}{s + 4}$$

will work.

Remark. Pole-zero cancellation as was done here must be avoided if it lies in the right half plane. Since any real system model has parameter uncertainty, exact cancellation is almost impossible to achieve. When this is the case such an attempted cancellation will leave an uncompensated unstable mode in the closed-loop system. Even in the case of a stable approximate cancellation the dynamics can change. In order to see this consider in Example 8.15 the pole at -2 to be uncertain (say $-2 \pm \varepsilon$, $\varepsilon > 0$) and that a zero is exactly located at -2. Let us consider the two cases with the pole at $-2 + \varepsilon$ and $-2 - \varepsilon$ (Fig. 8.77).

We note that the modified root loci do not pass through B, B' when $\varepsilon \ne 0$, implying that the time domain performance will be affected.

Example 8.16. Consider the system shown in Fig. 8.78(a) where $K \geq 0$ and α and β are unknown constants. In order to identify K, α, and β, the following information about the system is provided:

1. All the poles of the open-loop transfer function are in the closed left half s plane.
2. When the closed-loop system is excited by the input $r(t) = tu_s(t)$, the trace of Fig. 8.78(b) is obtained ($\infty > e_1 > 0$).
3. When the gain K is doubled, the impulse response of Fig. 8.78(c) is observed.

Determine K, α, and β.

SOLUTION. Since the closed-loop system has a finite steady-state error $e_1 < \infty$ for a ramp input, the system should be type I. Thus we require either α or β to be zero. Let $\alpha = 0$. So it only remains to determine β and K.

Now the root loci for the system can be sketched as shown in Fig. 8.78(d).

From the root locus it is clear that at a certain gain value the system goes unstable. From Fig. 8.78(c) we know that when the gain is doubled, the system has two closed-loop eigenvalues at A and A'. From the impulse response trace the frequency of oscillation is

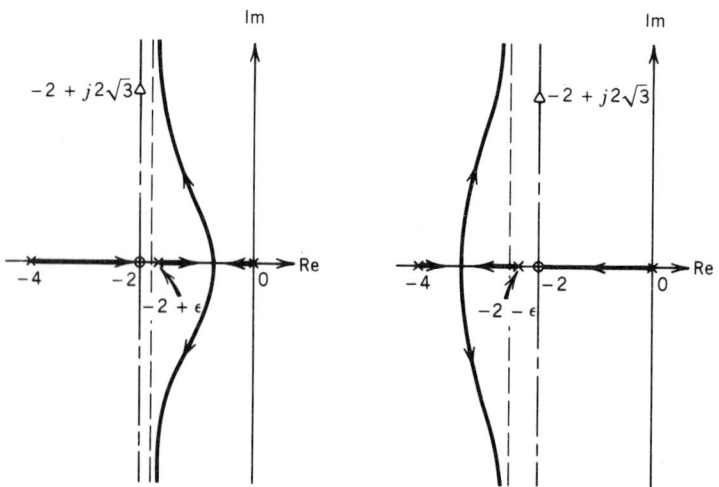

Fig. 8.77 Effect of nonexact pole-zero cancellation on the root loci.

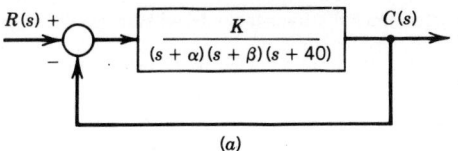

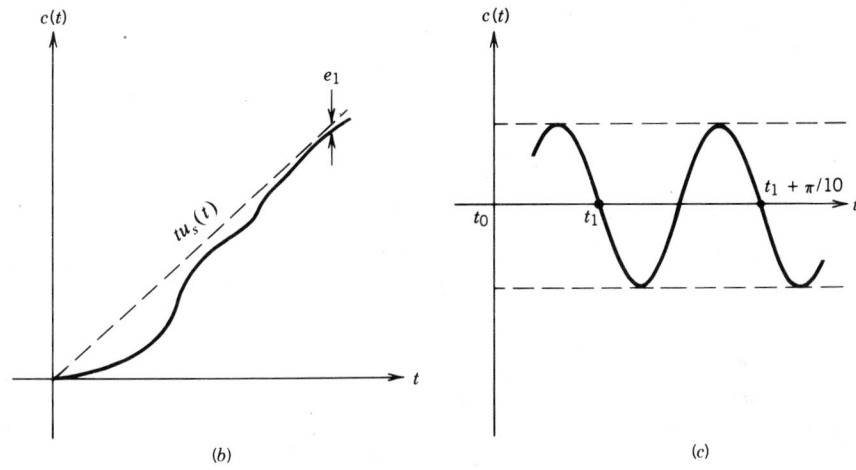

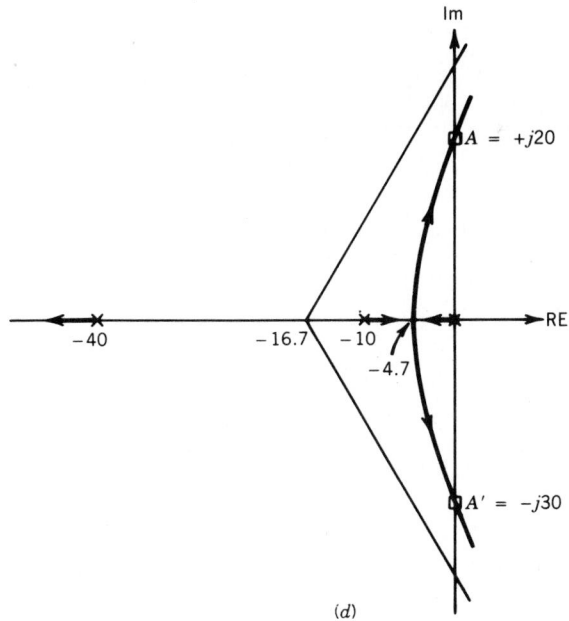

Fig. 8.78 (*a*) System of Example 8.16; (*b*) response due to a ramp; (*c*) response due to an impulse; (*d*) sketch of root loci for (*a*).

A'. From the impulse response trace the frequency of oscillation is

$$\omega = \frac{2\pi}{\pi/10}$$

or

$$\omega = 20 \text{ rad/s}$$

Thus we know that when the gain is doubled, there are two closed-loop poles at $\pm j20$. The corresponding CLCE therefore is

$$P(s) = (s + a)(s + j20)(s - j20) = 0$$

or

$$s^3 + as^2 + 400s + 400a = 0$$

We also know that

$$P(s) = 1 + \frac{2K}{s(s + \beta)(s + 40)} = 0$$

or

$$s^3 + (40 + \beta)s^2 + 40\beta s + 2K = 0$$

By matching coefficients

$$40 + \beta = a$$
$$40\beta = 400$$
$$2K = 400a$$

Therefore $\beta = 10$, $a = 50$, and $K = 10,000$. Hence the open-loop transfer function is

$$G(s) = \frac{10,000}{s(s + 10)(s + 40)}$$

8.12.3   Time-Domain Response versus s Domain Pole Locations

Given a transfer function $G(s)$ the pole locations can be found. These pole locations essentially describe the type of time response to be expected. The basic response can be effectively characterized by the impulse response $g(t)$ given by

$$g(t) = \mathcal{L}^{-1}[G(s)]$$

If

$$G(s) = \frac{K\prod_{i=1}^{m}(s + z_i)}{\prod_{j=1}^{n}(s + p_j)} \qquad n \geq m$$

then

$$g(t) = \sum_{j=1}^{n} a_j e^{-p_j t}$$

We note that any real pole contributes an exponential behavior into the time response and a complex conjugate pair contributes an exponential oscillation. A pure imaginary pair of poles leads to a sustained oscillation. Various components to be expected are shown in Fig. 8.79.

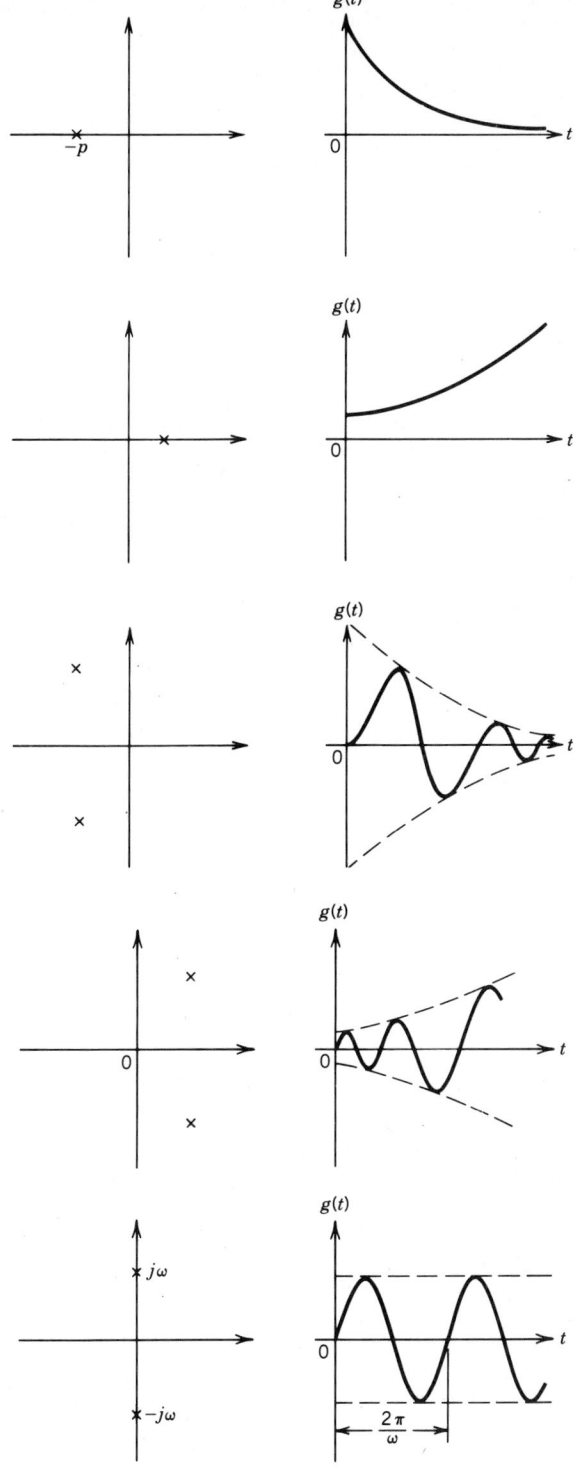

Fig. 8.79   Pole locations and corresponding impulse responses.

The role of zeros of the transfer function is to affect the relative weights a_j in the impulse response. For example, if a pole and a zero are close together, the net contribution to the overall response from such a pair will be negligible. If they cancel each other (say $-p_k$ by $-z_l$), then the coefficient associated with the term $e^{-p_k t}$, is zero. This idea can often be used to reduce the order of a dynamic system, that is, remove all pole-zero pairs close to one another. However, care should be exercised not to remove right half plane poles and zeros. [See Example 8.15 of Section 8.12.2.)

To note the effect of zero locations on the time response consider a second-order oscillatory system with a single zero, that is, consider the transfer function written in the normalized form

$$G(s) = \frac{(s/\alpha\zeta\omega_n) + 1}{(s/\omega_n)^2 + 2\zeta s/\omega_n + 1} = (s/\alpha\zeta\omega_n + 1)G_0(s)$$

The zero is located at $s = -\alpha\zeta\omega_n$, so if α is large, the zero is far removed from the poles and will have little effect on the response of $G_0(s)$. If $\alpha = 1$, the zero is at the value of the real part of the poles and could be expected to have a substantial influence on the response of $\dot{G}_0(s)$. The step response curves for $\zeta = 0.5$ and for several values of α are plotted in Fig. 8.80. We see that the major effect of the zeros is to increase the overshoot M_p with very little influence on the settling time. A plot of M_p versus α is given in Fig. 8.81. If α is negative, then the zero is in the right half s plane. In this case an undershooting phenomenon as shown in Fig. 8.82 occurs.

In addition, it is useful to know the effect of an extra pole on the standard second-order response $G_0(s)$. In this case consider the transfer function

$$G(s) = \frac{1}{(s/\alpha\zeta\omega_n + 1)}G_0(s)$$

Plots of the step response for this case are shown in Fig. 8.83 for $\zeta = 0.5$ and for several values of α. In this case the major effect is to increase the rise time, shown in Fig. 8.84. For a detailed discussion of the effect of a zero and a pole location on a standard second-order response the reader may refer to Ref. 4.

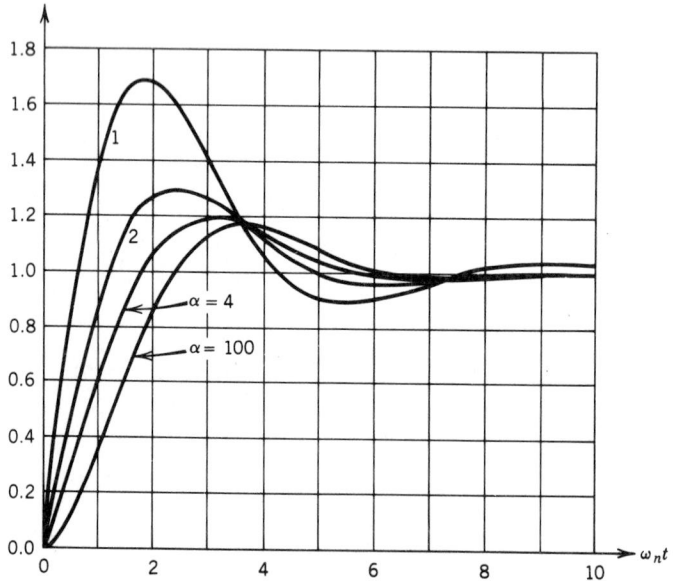

Fig. 8.80 Plots of the step response of a second-order system with an extra zero ($\zeta = 0.5$).

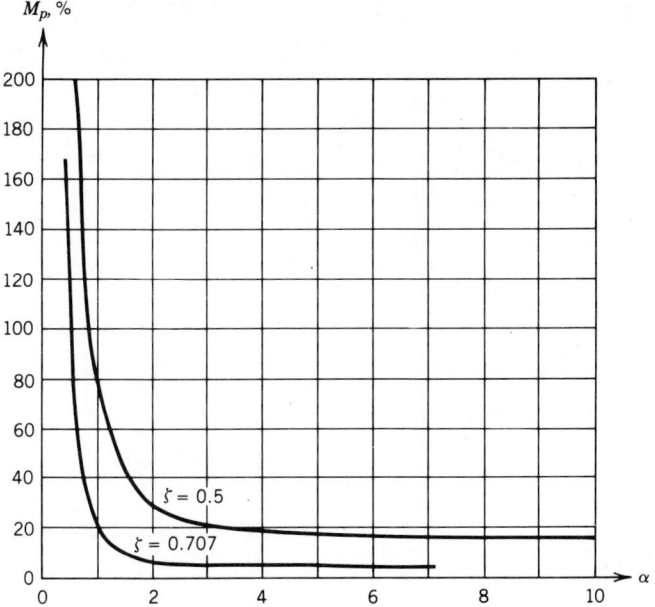

Fig. 8.81 Plot of overshoot M_p as a function of normalized zero location α. At $\alpha = 1$, the real part of the zero equals the real part of the pole.

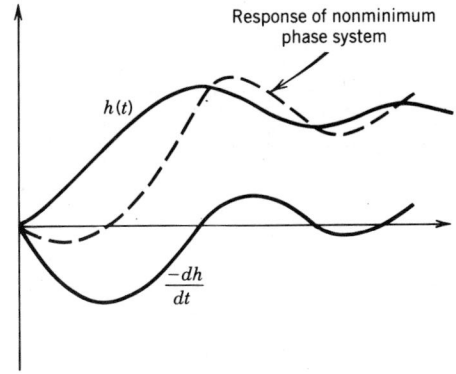

Fig. 8.82 Plot of the response of a second-order system with a right half-plane zero: a nonminimum phase system.

8.13   POLE LOCATIONS IN THE z DOMAIN

For discrete-time systems the input-output relation is given by the pulse transfer function. A typical pulse transfer function $G(z)$ is of the form

$$G(z) = \frac{K \prod_{i=1}^{m} (z - \beta_i)}{\prod_{j=1}^{n} (z - p_j)} \qquad n \geq m$$

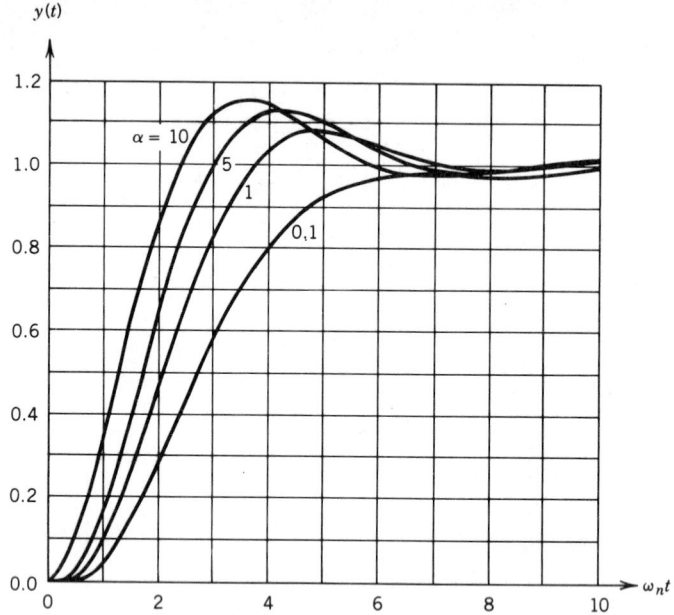

Fig. 8.83   Plot of step response for several third-order systems with $\zeta = 0.5$.

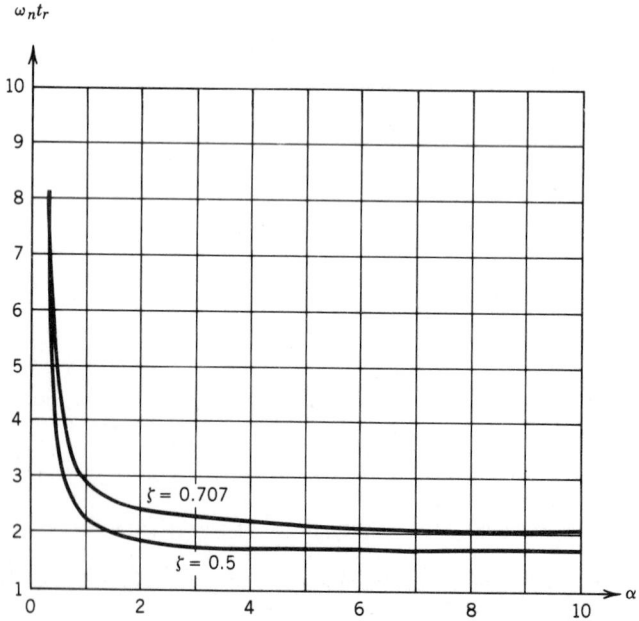

Fig. 8.84   Plot of normalized rise time for several locations of an additional pole.

The poles of the pulse transfer function are p_j, $j = 1, \ldots, n$. As in the case of continuous time, the pole locations determine the stability properties of the system represented by its pulse transfer function. In the z domain poles have to lie inside the unit circle $|z| = 1$ for asymptotic stability. Thus the open left half s plane is equivalent to the interior of the unit circle in the z domain. The exterior of the unit circle (i.e., $|z| > 1$) represents the unstable region in the z domain.

8.13.1 Stability Analysis of Closed-Loop Systems in the z Domain

Consider a unity negative feedback system with the closed-loop pulse transfer function

$$\frac{C(z)}{R(z)} = \frac{G(z)}{1 + G(z)} \tag{8.128}$$

The stability of the system defined by Eq. (8.128), as well as of other types of discrete-time control system, may be determined from the locations of the closed-loop poles in the z plane or the roots of the closed-loop characteristic equation

$$P(z) = 1 + G(z) = 0$$

as follows:

1. For the system to be stable, the closed-loop poles or the roots of the characteristic equation must lie within the unit circle in the z domain. Any closed-loop pole outside the unit circle makes the system unstable.
2. If a simple pole lies at $z = 1$ or $z = -1$, then the system becomes marginally stable. Also, the system becomes marginally stable if a single pair of complex conjugate poles lie on the unit circle in the z domain. Any multiple closed-loop pole on the unit circle makes the system unstable.
3. Closed-loop zeros do not affect the absolute stability and therefore may be located anywhere in the z plane.

Thus, a linear time-invariant single-input–single-output discrete-time closed-loop system becomes unstable if any of the closed-loop poles lies outside the unit circle or any multiple closed-loop pole lies on the unit circle in the z domain.

8.13.2 Performance Related to Proximity of Closed-Loop Poles to the Unit Circle

In the continuous-time case or in the s domain the transient performance of a system can be characterized by the s plane pole locations. Recall that the overshoot is related to the damping ratio ζ.

Damping Ratio ζ. In the s plane a constant damping ratio may be represented by a radial line from the origin. A constant damping ratio locus (for $0 < \zeta < 1$) in the z plane is a logarithmic spiral. Figure 8.85 shows constant ζ loci in both the s plane and the z plane.

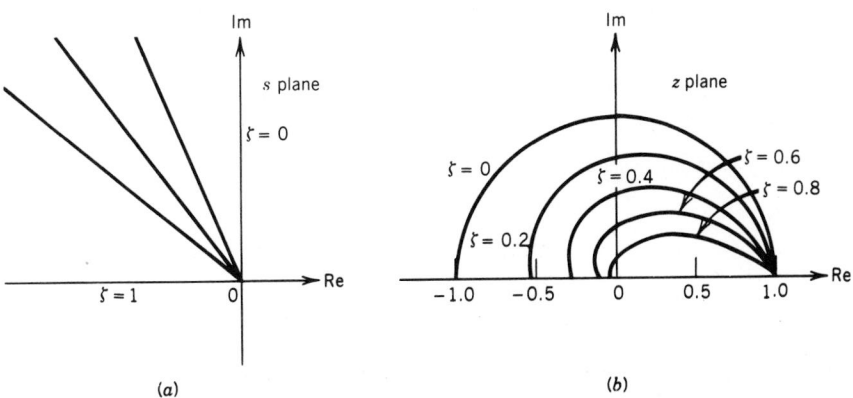

Fig. 8.85 (*a*) Constant ζ loci in the s plane; (*b*) constant ζ loci in the z plane.

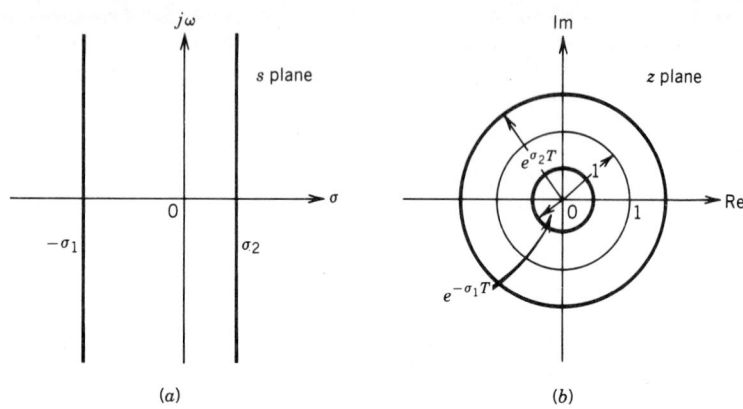

Fig. 8.86 (*a*) Constant attenuation lines in the *s* plane; (*b*) the corresponding loci in the *z* plane.

If all the poles in the *s* plane are specified as having a damping ratio not less than a specified value ζ_1, then the poles must lie to the left of the constant damping ratio line in the *s* plane (shaded region). In the *z* plane, the poles must lie in the region bounded by logarithmic spirals corresponding to $\zeta = \zeta_1$, (shaded region).

Damped Natural Frequency ω_d. The rise time or the speed of response depends on the damped natural frequency ω_d and the damping ratio ζ of the dominant complex conjugate closed-loop poles. In the *s* plane the constant ω_d loci are horizontal lines, while in the *z* plane they are radial lines emanating from the origin.

Settling Time t_s. The settling time is determined by the value of attenuation σ of the dominant closed-loop poles $-\sigma \pm j\omega_d$. If the settling time is specified, it is possible to draw a line, $\sigma = -\sigma_1$, in the *s* plane corresponding to a given settling time. The region to the left of the line, $\sigma = -\sigma_1$, in the *s* plane corresponds to the interior of a circle with radius $e^{-\sigma_1 T}$ in the *z* plane as shown in Fig. 8.86.

Remark. To transform *s*-plane pole locations to the *z* domain, the transformation $z = e^{sT}$ where *T* is the sampling time is employed.

8.13.3 Root Locus in the *z* Domain

The root locus method for continuous-time systems can be extended to discrete-time systems without modifications, except that the stability boundary is changed from the $j\omega$ axis in the *s* plane to the unit circle in the *z* plane. The reason for being able to extend the root locus method is that the characteristic equation for the discrete-time system is of the same form as that for the root loci in the *s* plane. For the discrete-time case the closed-loop characteristic equation is

$$1 + G(z) = 0$$

Exactly the same rules as used for the continuous-time case in the *s* plane can be used for the discrete case too. (See Section 8.12 for root loci construction in the *s* plane.)

Example 8.17. Consider the closed-loop characteristic equation

$$1 + K\frac{(z+1)(z-0.5)}{(z-1)(z-0.9)(z+0.6)} = 0$$

R1: The system has $n = 3$ poles indicating three branches. They start at 1, 0.9, and -0.6 with $K = 0$.

R2: There are two finite zeros ($m = 2$). Hence two of the branches terminate on the zeros -0.9 and 0.5 at $K = \infty$.

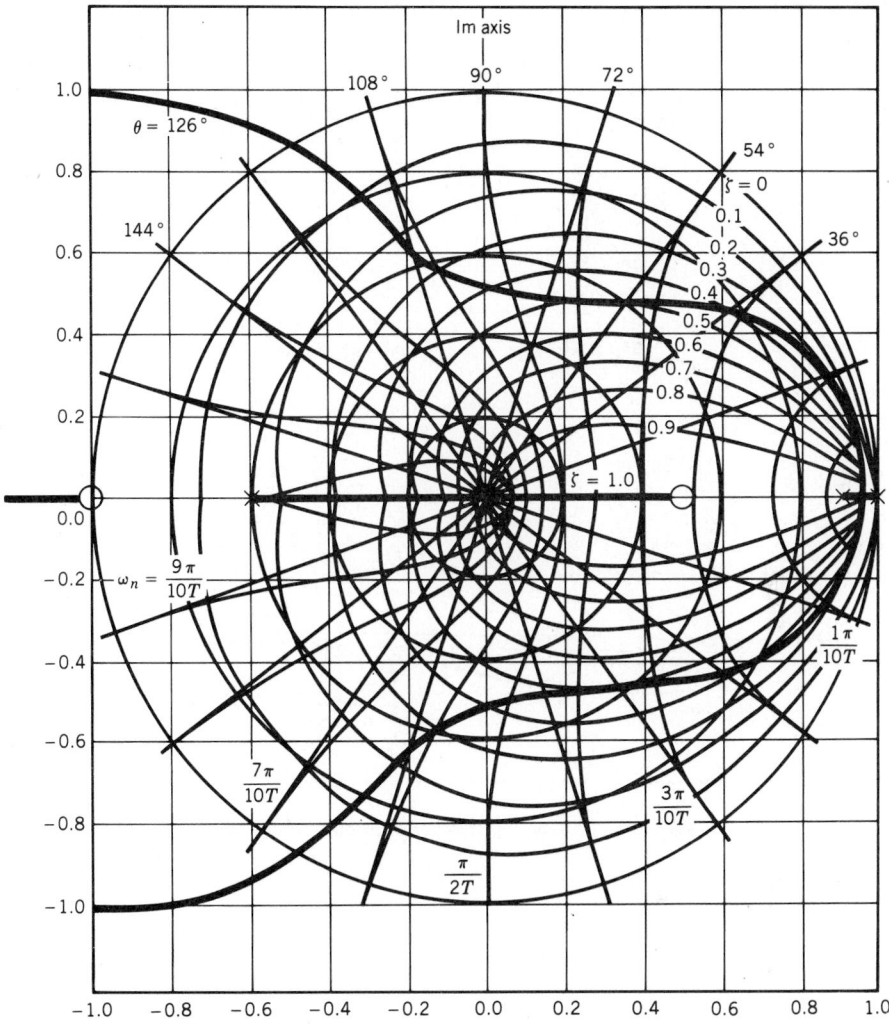

Fig. 8.87 Root loci for Example 8.17.

R3: One branch ($n - m = 1$) will go to ∞ along an asymptote.

R4: Sections of the root loci on the real line are between 0.9 and 1.0, -1.0 and $-\infty$, and -0.6 and 0.5; with this information the sketch shown in Fig. 8.87 can be easily obtained. If additional features are needed, the rest of the root loci construction rules can be applied without change.

8.14   SUMMARY

In the previous Sections (8.1–8.13) some useful tools for designing single-input–single-output systems were given. Once the open-loop system is described, either by its set of poles and zeros or by its frequency response, the root locus method or frequency response method will indicate whether the

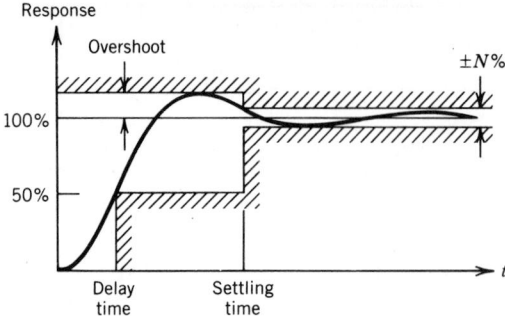

Fig. 8.88   Unit step response specifications.

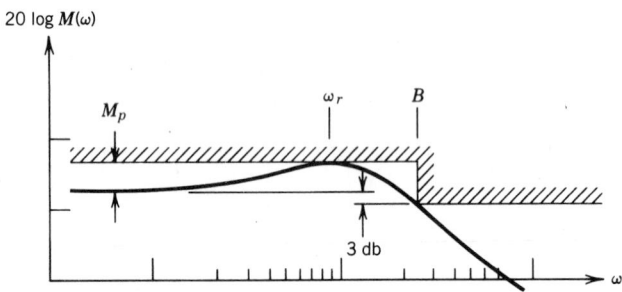

Fig. 8.89   Closed-loop frequency response specifications.

feedback system can be given an acceptable transient response by adjustment of the loop gain. The steady-state accuracy can then be determined.

Often good transient performance and good steady-state performance cannot both be achieved simply by adjusting a single parameter such as a loop gain. When this is the case, it is necessary to modify the system dynamics. Either the dynamic properties of some components in the loop need to be altered or additional components need to be inserted into the loop. The process of modifying the system dynamics so as to allow the performance specifications to be met by subsequent loop gain adjustment is known as compensation.

Transient specifications are typically based on a step input response. By specifying the rise time, overshoot, and settling time, the response is confined to within the shaded region of Fig. 8.88. It is then assumed that a system whose step response satisfies these constraints will have an acceptable transient response to any kind of input.

In the frequency domain the bandwidth and the resonant peak of the closed-loop frequency response are measures roughly corresponding to rise time and overshoot, respectively. Specification of these parameters constrains the magnitude of the closed-loop frequency response to the region shown in Fig. 8.89. An alternative way of constraining the transient response by frequency domain criteria is to stipulate the smallest acceptable gain and phase margins.

Often used compensators are the so-called three-term controllers (PID), lag, and lead compensators. These controller or compensator designs are discussed later. They are often done on a trial-and-error basis and can be designed in the s domain or the z domain depending on the type of application. Continuous-time or s-domain compensators can often be converted to equivalent z-domain compensators by techniques such as pole-zero maps, hold equivalence, and Butterworth pole configurations.[11]

REFERENCES FOR PART I

1. Doebelin, E. O., *Control System Principles and Design*, New York, Wiley, 1985.
2. Dorf, R. C., *Modern Control Systems*, Reading, Mass., Addison-Wesley, 1986.
3. Ogata, K., *Modern Control Engineering*, Englewood Cliffs, N.J., Prentice-Hall, 1970.
4. Franklin, G. F., Powell, J. D., and Emami-Naeini, A., *Feedback Control of Dynamic Systems*, Reading, Mass., Addison-Wesley, 1986.
5. Graham, D., and Lathrop, R. C., "The Synthesis of Optimum Response: Criteria and Standard Forms," *AIEE Transactions*, Part II, 72, pp. 273–288, 1953.
6. Korn, G. A., and Wait, J. V., *Digital Continuous System Simulation*, Englewood Cliffs, N.J., Prentice-Hall, 1978.
7. Eshbach, O. W., and Souders, M. (Eds.)., *Handbook of Engineering Fundamentals*, 3rd ed., New York, Wiley, 1975.
8. Rosenberg, R. C., and Karnopp, D. C., *Introduction to Physical System Dynamics*, New York, McGraw-Hill, 1983.
9. Vidyasagar, M., *Nonlinear Systems Analysis*, Englewood Cliffs, N.J., Prentice Hall, 1978.
10. Takahashi, Y., Rabins, M. J., and Auslander, D. M., *Control and Dynamic Systems*, Reading, Mass., Addison-Wesley, 1972.
11. Franklin, G. F., and Powell, J. D., *Digital Control of Dynamic Systems*, Reading, Mass., Addison-Wesley, 1980.
12. Kuo, B. C., *Automatic Control Systems*, Englewood Cliffs, N.J., Prentice-Hall, 1982.

BIBLIOGRAPHY FOR PART I

Bode, H. W., *Network Analysis and Feedback Amplifier Design*, Van Nostrand, 1945.

Chestnut, H., and Mayer, R. W., *Servomechanisms and Regulating Systems Design*, 2nd ed., Vol. 1, New York, Wiley, 1959.

D'Azzo, J. J., and Houpis, C. H., *Linear Control System Analysis and Design*, New York, McGraw-Hill, 1988.

Distefano, J. J. III, Stubberud, A. R., and Williams, I. J., *Feedback and Control Systems* (Schaum's Outline Series), New York, Shaum Publishing, 1967.

Dransfield, P., *Engineering Systems and Automatic Control*, Englewood Cliffs, N.J., Prentice-Hall, 1968.

Elgerd, O. I., *Control Systems Theory*, New York, McGraw-Hill, 1967.

Evans, W. R., *Control-System Dynamics*, New York, McGraw-Hill, 1954.

Eveleigh, V. W., *Introduction to Control Systems Design*, New York, McGraw-Hill, 1972.

Horowitz, I. M., *Synthesis of Feedback Systems*, New York, Academic Press, 1963.

Houpis, C. H., and Lamont, G. B., *Digital Control Systems: Theory, Hardware, Software*, New York, McGraw-Hill, 1985.

Kuo, B. C., *Digital Control Systems*, New York, Holt, Rinehart and Winston, 1980.

Melsa, J. L., and Schultz, D. G., *Linear Control Systems*, New York, McGraw-Hill, 1969.

Nyquist, H., "Regeneration Theory," *Bell System Tech. J.*, Vol. II, pp. 126–147, 1932.

Palm, N. J. III, *Modeling, Analysis and Control of Dynamic Systems*, New York, Wiley, 1983.

Phillips, C. L., and Nagle, H. T., Jr., *Digital Control System Analysis and Design*, Englewood Cliffs, N.J., Prentice-Hall, 1984.

Ragazzini, J. R., and Franklin, G. F., *Sampled Data Control Systems*, New York, McGraw-Hill, 1958.

Raven, F. H., *Automatic Control Engineering*, 4th ed., New York, McGraw-Hill, 1987.

Shinners, S. M., *Modern Control Systems Theory and Application*, Reading, Mass., Addison Wesley, 1972.

Truxal, J. G., *Automatic Feedback Control Synthesis*, New York, McGraw-Hill, 1955.

PART II

ELECTROMECHANICAL AND ELECTROHYDRAULIC SERVOACTUATOR MODELING
Karl N. Reid and Syed Hamid

8.15 INTRODUCTION

8.15.1 Definitions

A servoactuator is an open-loop system that controls the linear or rotary motion of a load in response to an input command [Fig. 8.90(a)]. Feedback may be used with a servoactuator to produce a closed-loop system referred to as a servosystem [Fig. 8.90(b)]. Servoactuators are normally "rate-type" systems in that an input command results in an output velocity for steady-state operation. Position feedback must be used with the rate-type system to produce a servosystem for position control. If high-accuracy velocity control is required, velocity feedback may be used with the servoactuator.

The modulator provides a conversion of the low power input command (for the open-loop servo) or the error signal (for the closed-loop servo) to a high power output to operate the servomotor. The servomotor is a general term employed here to designate the various types of higher-level energy convertors such as electrical and hydraulic motors. The servomotor provides the muscle function of the servoactuator. The "transducer" provides the feedback in the case of the servosystem. The input to the servo can be electronic, mechanical, hydraulic, or pneumatic. And depending on the energy conversion medium, servoactuators can be of the electromechanical, electrohydraulic, electropneu-matic, or hydromechanical types. The early development of servoactuators was predominantly in electropneumatics (in the process control industry). With the advent of microprocessors and the development of high coercive strength magnetic materials (such as Samarium Cobalt and Neodium), electromechanical servoactuators find the largest applications in modern industry.[12] Table 8.8 describes the servo components (modulator, servomotor, and transducer) for the various implementations.

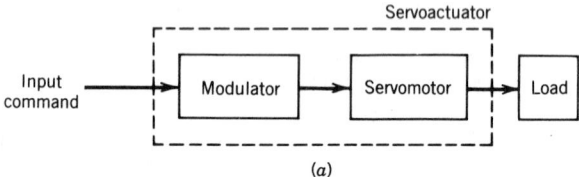

(a)

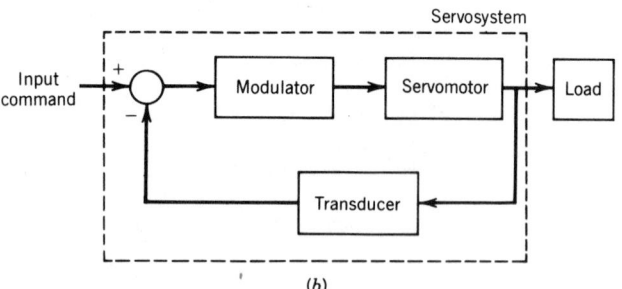

(b)

Fig. 8.90 Servoactuator and servosystem: (a) servoactuator (open loop), (b) servosystem (closed loop).

TABLE 8.8 Servosystem Components

System Type	Modulator	Servomotor	Transducer
Electromechanical	Amplifier	Servomotor AC, DC Linear/Rotary	
	Driver	Brushless servomotor	Position, velocity, torque
	Translator/driver	Stepper motors	
Electrohydraulic	Servovalve	Hydraulic motors cylinders	Position, velocity, force, pressure, torque
Electropneumatic	Servovalve	Airmotors, cylinders	Position, velocity, torque, force

describes the servo components (modulator, servomotor, and transducer) for the various implementations.

8.15.2 Applications

Servoactuators range from fairly simple open-loop actuators such as the hydraulic controls on a backhoe, to complex feedback servoactuators (i.e., servosystems) in robotics and aerospace applications. Figure 8.91 shows a typical linear output servosystem. A three-phase brushless motor is modulated by a pulse-width-modulated (PWM) controller (not shown in Fig. 8.91). A ballscrew is used to convert rotary motion to linear displacement.[11]

8.16 ELECTRICAL SERVOMOTORS

Electrical servomotors may be classified as follows:

1. Type of power (DC or AC)
2. Type of motion executed (continuous or discrete, rotary or translatory)
3. Type of commutation (mechanical or electronic)
4. Method of magnetic field generation (permanent magnet or electromagnetic)

Accordingly there are DC and AC servomotors of both the permanent magnet and field-wound types. Stepper motors belong to the discrete motion type. The rather uncommon linear motor executes translatory motion. Brushless DC motors are of the electronic commutation type. For the sake of simplicity, electrical servomotors are broadly classified here into four categories: DC and AC servomotors, stepper motors, and linear servomotors.

Electrical servomotors offer several advantages over their hydraulic and pneumatic counterparts. These include (a) compactness (facilitated by availability of high coercive strength magnetic materials such as Samarium Cobalt or Neodium), (b) low cost, (c) high reliability, (d) cleanliness, (e) ease of control function implementation, (f) portability due to operation at low DC voltage levels, and (g) large bandwidth due to high torque/inertia ratios.

8.17 DIRECT-CURRENT SERVOMOTORS

The DC servomotors offer certain advantages over alternating-current (AC) servomotors. These are higher reliability, smaller size, portability, and lower cost. Use of epoxy resins and improved brush designs combined with superior magnetic materials contribute to these advantages. Direct-current servomotors are also compatible with thyristors (SCR) and transistor amplifiers, which facilitate control implementations. Typical DC servomotors range in power from fractional horsepower to several thousand horsepower.

Conventional brushed DC motors theoretically can be used as servomotors. However, in lower horsepower levels (10 hp or less) they are not preferred.

(a)

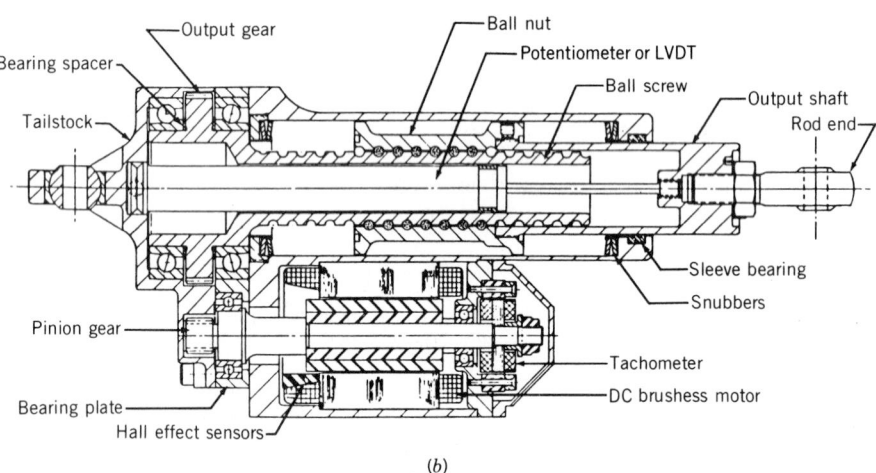

(b)

Fig. 8.91 Electromechanical servoactuator with linear displacement output (Courtesy of Moog, Inc., East Aurora, New York): (a) electromechanical servosystem, (b) cross section of an electromechanical servoactuator.

8.17.1 Brushed DC Servomotors

The principle of operation of the DC servomotor is straightforward. It consists of the interaction of two magnetic fields (either one or both generated electrically) that results in mechanical motion. A typical permanent-magnet DC motor is illustrated in Fig. 8.92. The permanent magnet is sometimes replaced by a field winding to generate the magnetic field. The field winding may be connected in three different ways to the armature winding: series, shunt, or compound. Table 8.9 summarizes the

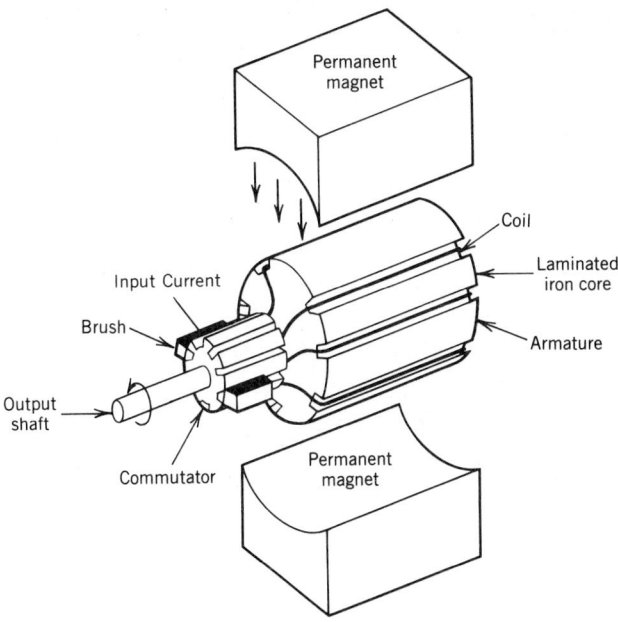

Fig. 8.92 Conventional permanent-magnet motor.

TABLE 8.9 DC Servomotor Classification

Motor Type	Configuration	Typical Steady-State Characteristics	Salient Features
Permanent magnet	$i \rightarrow$ N [ ] S	*Current* / *Speed* vs *Torque*	No power required for field generation Runs cooler Torque-speed characteristic is linear Compactness
Straight series	$i \rightarrow$	*Current* / *Speed* vs *Torque*	Large starting torque
Split series	$i \rightarrow$	*Current* / *Speed* vs *Torque*	Allows quick reversing

TABLE 8.9 (*Continued*)

Motor Type	Configuration	Typical Steady-State Characteristics	Salient Features

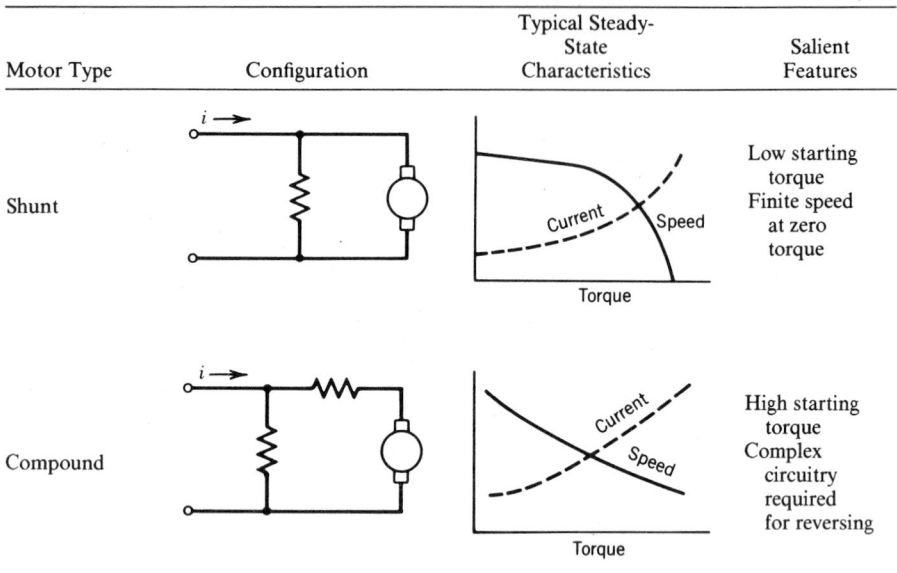

Shunt — Low starting torque; Finite speed at zero torque

Compound — High starting torque; Complex circuitry required for reversing

TABLE 8.10 **Upper Limits of DC Servomotor Performance**

Motor Type	Maximum Power (hp)	Maximum Speed (rpm)	Torque/Inertia Ratio ft-lb./lb-ft^2	Maximum Bandwidth (rads/s)
Moving coil	0.5–1	4500–5500	200–250	1500
Printed circuit	7	3000–4000	130–220	1000
Permanent magnet	10–15	850–3000	15–30	100

Reproduced from Ref. 4.

basic features of the various configurations along with the resultant performance characteristics. Table 8.10 shows typical upper limits of DC servomotor performance specifications of these servomotors.[4]

The split-series wound-field motor (No. 3 in Table 8.9) has two windings, one for each direction of rotation. A manual switch is usually employed to activate the appropriate winding. The two windings of the compound motor are always excited and result in a high starting torque with good linearity. All of the field-wound motors are self-excited with the residual magnetism.

Permanent-Magnet (PM) Motors

Permanent-magnet (PM) motors are the most extensively used for servomotors because they generate less heat and have higher efficiency and more compactness than field-excited motors. Furthermore, the PM motor lends itself to easy servo design calculations.

There are three types of PM motors with mechanical commutation: (1) iron core, (2) surface wound, and (3) moving coil. Figure 8.93 shows the construction of the three types. Details of the advantages and disadvantages of each type may be found in Ref. 18.

Mathematical Model of a PM Motor

Mathematical models of the various components of a servoactuator system are needed for component selection to meet a given set of performance specifications. These specifications may consist of moving a given load through a given displacement or velocity profile in a specified time, or equivalently, following displacement or velocity commands generated by other subsystems or by an operator. The mathematical model of a servomotor describes the steady-state and dynamic performance characteris-

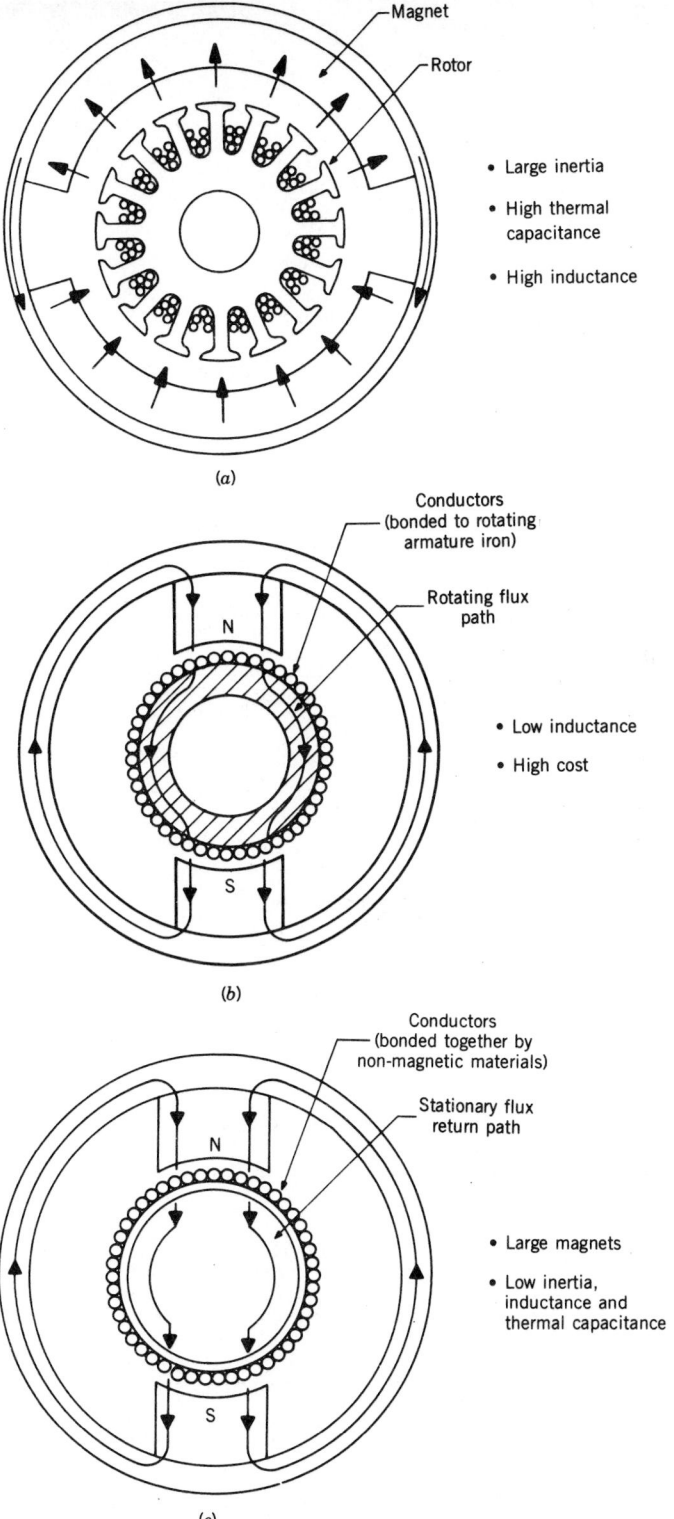

Magnet

Rotor

- Large inertia

- High thermal capacitance

- High inductance

(a)

Conductors
(bonded to rotating
armature iron)

Rotating flux
path

N

S

- Low inductance

- High cost

(b)

Conductors
(bonded together by
non-magnetic materials)

Stationary flux
return path

N

S

- Large magnets

- Low inertia,
inductance and
thermal capacitance

(c)

Fig. 8.93 Types of brushed permanent-magnet DC motors (Taken from Benjamin J. Kuo and Jacob Tal, *DC Motors and Control Systems*, SRL Publishing, pp. 86–87): (*a*) iron core, (*b*) surface wound, (*c*) moving coil.

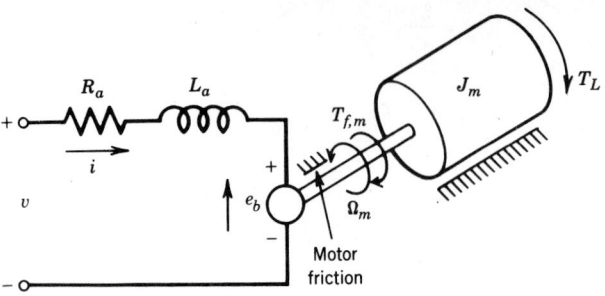

Fig. 8.94   Lumped-parameter model of a permanent-magnet motor.

tics.[13,14,15] An illustrative example given at the end of this section describes the use of the mathematical model.

The mathematical model of a permanent-magnet DC motor is obtained by lumping the inductance and resistance of the armature winding as shown in Fig. 8.94. The resulting equations are:

$$\textit{Electrical Equations:} \qquad v_a = L_a \frac{di}{dt} + R_a i + e_b \tag{8.129}$$

$$e_b = K_E \Omega_m \tag{8.130}$$

$$\textit{Motor Shaft Equation:} \qquad K_T i = J_m \frac{d\Omega_m}{dt} + B_m \Omega_m + T_{f,m} + T_m \tag{8.131}$$

Taking Laplace transforms of Eqs. (8.129)–(8.131) gives, after algebraic manipulation,

$$\Omega_m(s) = G_1(s)V_a(s) - G_2(s)\big[T_{f,m}(s) + T_m(s)\big] \tag{8.132}$$

where the transfer functions G_1 and G_2 are given by

$$G_1(s) = \frac{K_T}{R_a B_m(\tau_e s + 1)(\tau_m s + 1) + K_T K_E} \tag{8.133}$$

and

$$G_2(s) = \frac{R_a(\tau_e s + 1)}{R_a B_m(\tau_e s + 1)(\tau_m s + 1) + K_T K_E} \tag{8.134}$$

The parameters are defined as follows:

v_a = Applied voltage (V)
K_E = $z'\phi P/60$, motor voltage constant or back EMF constant (V/rpm)
e_b = back EMF (V)
z' = Number of conductors per parallel path in armature
ϕ = Magnetic flux per pole (Wb)
P = Number of poles
K_T = $z'\phi P/2\pi$, motor torque constant (N · m/A)
Ω_m = Motor speed (rpm)
L_a = Armature inductance (H)
i = Current through armature (A)
R_a = Armature resistance (ohms)
J_m = Polar moment of inertia of armature (N · m/rpm)
t = Time (s)

B_m = Viscous damping in motor (N · m · s/rpm)
$T_{f,m}$ = Coulomb friction torque in motor (N · m)
T_a = Total load torque reflected to motor shaft (N · m)
$\Omega_m(s)$ = Laplace transform of speed
s = Laplace operator
θ_m = Angular position of motor shaft (radians)
$\tau_e = L_a/R_a$, electrical time constant (s)
$\tau_m = J_m/B_m$, mechanical time constant (s)*
$V_a(s)$ = Laplace transform of armature voltage $v_a(t)$

This second-order transfer function further reduces to a first-order equation [Eq. (8.135)] if the armature inductance is small (making the electrical time constant τ_e negligible) and the Coulomb friction and load torque are assumed zero. Then

$$\frac{\Omega_m(s)}{V_a(s)} = \frac{K_m}{\tau s + 1} \tag{8.135}$$

where

$$K_m = \frac{K_T}{R_a B_m + K_T K_E} \quad \text{(motor constant)} \tag{8.136}$$

and

$$\tau = \frac{R_a J_m}{R_a B_m + K_T K_E} \tag{8.137}$$

Reference 33 discusses cases where the electrical time constant cannot be neglected.
    The above transfer functions assume a voltage input. For applications where a constant-current amplifier is used, the approximate model given by Eq. (8.138) should be used.

$$\frac{\Omega_m(s)}{I(s)} = \frac{K_T}{B_m(\tau_m s + 1)} = \frac{K_T/B_m}{\tau_m s + 1} \tag{8.138}$$

    In principle, the models developed above can be applied to all of the DC motors of the various types with the appropriate input conditions. The above transfer functions describe the open-loop response. For closed-loop systems with velocity or position feedback, an appropriate closed-loop transfer function can be derived easily by making use of the motor transfer function. Examples of closed-loop systems are given in Sections 16.10 and 16.11 of Ref. 33.

Numerical Example
For a Motomatic PM servomotor model number E350-MG, the following specifications are given:

K_T = 3.4 in · oz/A
K_E = 2.5 V/krpm
R_a = 12.4 ohms
$J_m = 2.5 \times 10^{-4}$ in · oz · s^2/rad
B_m = 0.015 in · oz/krpm
$T_{f,m}$ = 0.5 in · oz
T_{max} = 2.5 in · oz
I_{max} = 0.75 A
Ω_{max} = 10,500 rpm at no load
L_a = 3.1 mH

*Some servomotor manufacturers define τ_m differently. For example, Electrocraft defines the mechanical time constant as $(R_a J_m)/(K_T K_E)^1$.

$\tau_e = 0.25$ ms

R_{th} = Thermal resistance = 13°C/Watt

The mechanical time constant can be computed as

$$\tau_m = \frac{J_m}{B_m}$$

$$= 1.75 \text{ s} \tag{8.139}$$

Assuming $\tau_e \ll \tau_m$ and neglecting Coulomb friction and load torque, Eq. (8.135) can be used to determine the dynamic response. Thus,

$$K_m = \frac{K_T}{R_a B_m + K_T K_E}$$

$$= 0.39 \text{ krpm/V} \tag{8.140}$$

The time constant is

$$\tau = \frac{R_a J_m}{R_a B_m + K_T K_E}$$

$$= 0.037 \text{ s} \tag{8.141}$$

The transfer function of Eq. (8.135) then gives

$$\Omega_m(s) = \frac{\Omega_m(s)}{V_a(s)} = \frac{0.39}{0.037s + 1} \tag{8.142}$$

For a 1-volt step input, the motor speed is given by

$$\Omega_m(s) = \frac{1}{s} \frac{0.39}{0.037s + 1} \tag{8.143}$$

The inverse Laplace Transform gives

$$\Omega_m(t) = 0.39(1 - e^{-t/0.037}) \tag{8.144}$$

8.17.2 Brushless DC Servomotors

The development of brushless DC servomotors was an outgrowth of semiconductor devices even though the first patent was obtained with vacuum tube technology.[20] The basic construction of a brushless DC motor involves elimination of the mechanical commutation. Instead, the commutation process is accomplished electronically with no moving contacts. Hence, the problems associated with mechanical commutation such as brush wear particles, EMI, or arcing are eliminated. Elimination of arcing makes DC servomotors excellent candidates for applications requiring explosion-proof safety classification.

Construction
Typically, brushless motors have an inner rotor and outer stator configuration such as the one shown in Fig. 8.95(a). However, the other configuration, i.e., inner stator and outer rotor, is also possible (see Fig. 8.95(b)). The former configuration with the outer stator carrying electrical windings provides excellent thermal dissipation characteristics, since both the iron and copper losses now occur in the stator, which is better exposed to the ambient for convective heat transfer. This allows brushless motors to be operated at higher speeds and hence provides higher power-to-weight ratio.

Brushless DC motors range from 1 to 40 in. (25 to 1016 mm) in diameter with 6 in · oz to 1650 ft-lb of torque capability (see Table 8.11). Typical applications include memory disk drives, videotape recorders, and as position servos in cryogenic compressors and fuel pumps.

The rotors are permanent magnets made from one of three primary materials: ceramics, ALNICO, and rare earths such as Samarium Cobalt. Ceramic rotors are used in applications where cost consideration is important. Rare-earth magnets are the most expensive but provide exceptional performance. ALNICO magnets are of medium cost and provide medium magnetic strengths.

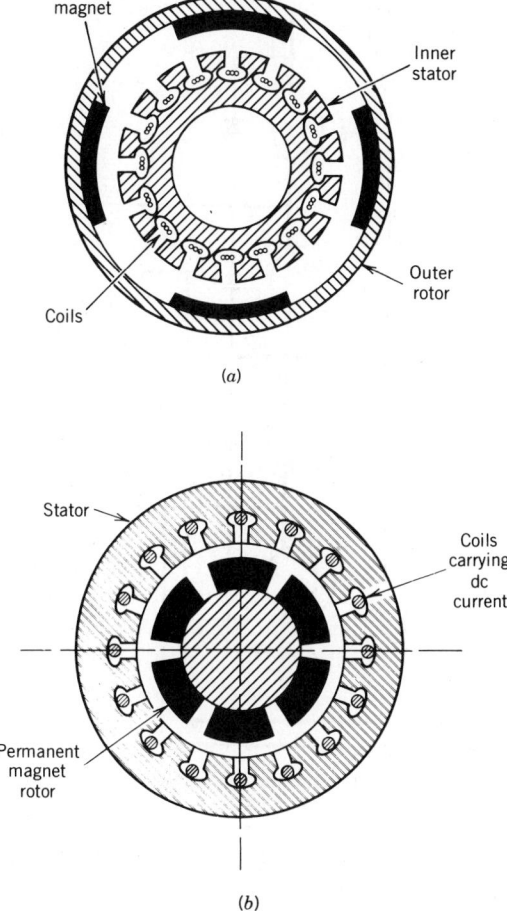

Fig. 8.95   Cross section of typical brushless DC motors: (*a*) inner rotor–outer stator type, (*b*) inner stator–outer rotor type.

TABLE 8.11   Brushless DC Motor Performance Data

Magnet Type	Power (W)	Peak Torque (in · oz)	Electrical Time Constant (s)	Mechanical Time Constant (s)	Torque/Inertia (s^{-2})
Ceramic	25–900	10–600	0.0002–0.0016	0.0221–0.7400	413–11,400
Alnico	20–280	6–5,000	0.0001–0.0030	0.0065–0.1330	465–57,500
Rare earth	25–6,000	10–316,000	0.0001–0.0140	0.0024–0.0291	137–100,000

Operation

The brushless motor is operated by generating a rotating magnetic field that is 90° (electrical) out of phase with the rotor. Position sensors are used to determine the rotor position. These position sensors are of three types: photo transistor–LED, electromagnetic, and Hall effect generators.

Commutation

An electronic module consisting of logic circuits and power amplification circuits is used to drive the motor.[4,11,21,22] This module receives the rotor position information from the position sensors. The

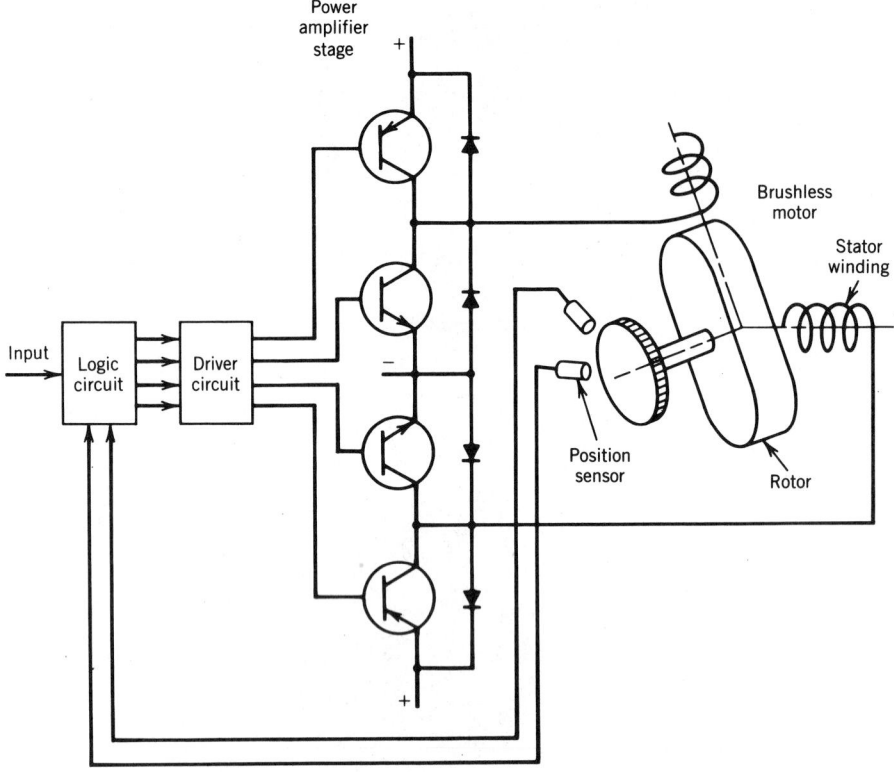

Fig. 8.96 Two-phase brushless motor with driver electronics.

angle through which the rotor turns during the firing of a winding is called the "conduction angle."
Figure 8.96 shows schematically a two-phase brushless motor with the driver electronics.

Figure 8.97 shows the controller circuit for a three-phase brushless motor. Each phase requires a pair of switches for commutation. Since the cost of the motor is dependent on the number of switches, there is a tendency to keep the number of phases to a minimum. Typically three-phase motors with six switches are used. The current through the windings may be varied in a sinusoidal or a square-wave manner. The latter excitation results in a small torque ripple (17% average to peak for a two-phase motor, and 7% for a three-phase motor).

Ideally, a sinusoidal torque function results in a constant torque. But sinusoidal torque function generation is technically difficult and uneconomical. The alternate approach is to design the spatial variation of the magnetic field (possible by means of high coercive strength magnets) to obtain a trapezoidal torque function while the input current has a square waveform (easily generated by simple transistor control circuitry such as shown in Fig. 8.97). The motor torque is then approximately constant and is proportional to the maximum value of current during each cycle. The trapezoidal torque generation scheme also results in higher efficiency.

The locations of the position sensors relative to the rotor are aligned to result in appropriate timing for proper commutation. When properly commutated, a brushless motor duplicates the torque-speed characteristics of a brush-type DC motor.

The power output of the brushless motor is effectively controlled by pulse-width-modulation (PWM) or pulse-frequency modulation (PFM) methods. A linear (i.e., class A) power amplifier can also be used for power control. However, it is susceptible to causing viscous damping resulting from back emf conduction during the zero-voltage portions of the voltage modulation. This can be eliminated by operating in a constant-current mode rather than voltage (see Ref. 1).

Figure 8.98 shows a cross-section of a brushless DC motor developed for use as a fin actuator. Hall effect sensors are used for position measurement.

Mathematical Models
The mathematical model required to represent a brushless DC motor is identical to that of a brush-type DC motor. Therefore, equations given in Section 8.17.1 are applicable.

Three-phase, three-step, half-wave motor controller

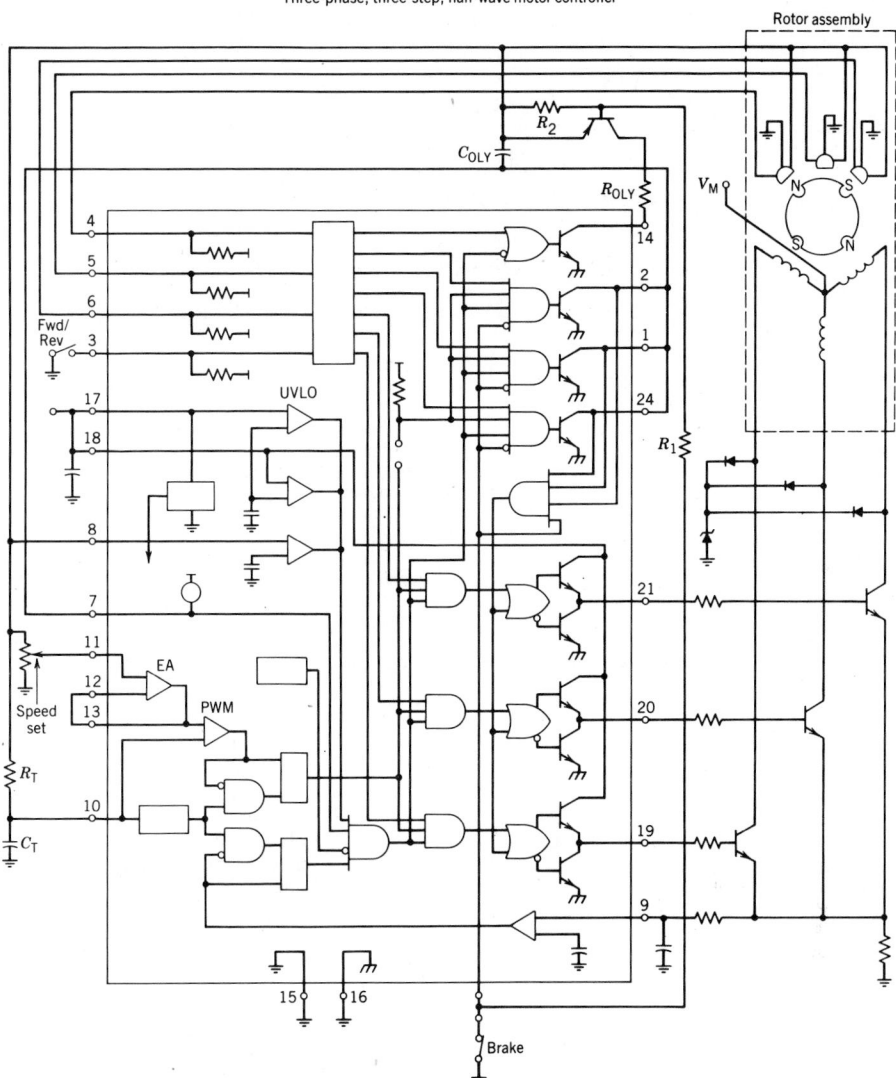

Fig. 8.97 Three-phase brushless motor controller circuit (Taken from *Machine Design*, June 9, 1988, p. 140).

Numerical Example

Table 8.12 shows the specifications for a ceramic magnet, inside rotation-type DC brushless motor manufactured by Magnetic Technology.[23, 24]

If this motor is operating at peak torque, the steady-state motor speed is given by [from Eq. (8.135), dropping the dynamic terms]

$$\Omega_m = K_m v_a$$

$$= \frac{K_T v_a}{R_a B_m + K_T K_E}$$

$$= 309 \text{ rad/s} = 2{,}953 \text{ rpm} \tag{8.145}$$

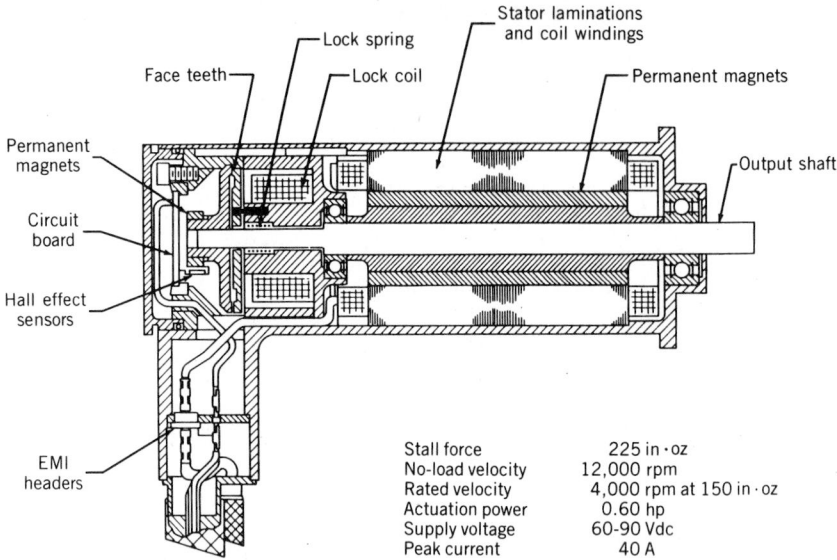

Stall force	225 in·oz
No-load velocity	12,000 rpm
Rated velocity	4,000 rpm at 150 in·oz
Actuation power	0.60 hp
Supply voltage	60-90 Vdc
Peak current	40 A

Fig. 8.98   Cross section of a brushless DC motor (Courtesy of Moog, Inc., East Aurora, New York).

TABLE 8.12   Performance Data for Magnetic Technology Model 2800-153-084 Brushless Motor DC Servomotor

Peak torque = 40 in · oz.	Weight = 19 oz
Power at peak torque = 175 W	Number of poles = 8
Electrical time constant = 0.0005 s	Number of phases = 2
Mechanical time constant = 0.0544 s	Resistance = 8.4 Ω
Damping factor = 0.064 in · oz · s/rad	Inductance = 4.2 mH
Moment of inertia = 0.0035 in · oz · s²	Voltage at peak torque = 38.2 V
Total breakaway torque = 1.5 in · oz	Current at peak torque = 4.57 A
Temperature rise = 2°C/W	Torque constant = 8.7 in · oz/A
Maximum allowable winding temperature = 155°C	Voltage constant = 0.0617 V · s/rad

and the temperature rise is

$$\Delta T = 2°C/W \times 175 \text{ W} = 350°C \tag{8.146}$$

This temperature rise is greater than the 155°C maximum allowable winding temperature. Thus the motor cannot be operated at peak torque indefinitely. The ambient temperature should be added to the temperature rise to arrive at the operating temperature of the winding.

8.18   ALTERNATING-CURRENT SERVOMOTORS

Alternating-current servomotors are used in applications requiring smooth speed control. These motors find widespread use in stationary industrial applications due to the ready availability of AC power.[2,3] In a majority of these applications, two-phase AC servomotors are used because of simplicity of the associated controls. Figure 8.99 shows a schematic of a two-phase AC servomotor. The operation of the two-phase AC servomotor is similar to an induction motor except the voltages applied to the two windings (fixed phase and control phase) are generally unequal and out of phase. The AC voltage applied to the fixed phase is held constant, and the one applied to the control phase is varied to control the motor speed. The two phases are generally 90° out of phase. Changing the phase angle from +90° to −90° reverses the direction of rotation of the motor. The appropriate phase angle is achieved through capacitors or other phase-shift circuits.

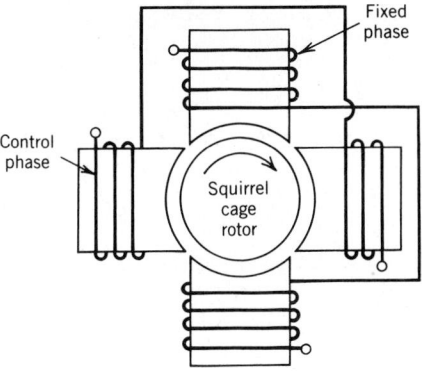

Fig. 8.99 Two-phase AC servomotor.

8.18.1   Types of AC Servomotors

Depending on the rotor construction, AC servomotors are classified into three types: (1) squirrel cage, (2) solid iron, and (3) drag cup (see Fig. 8.100). The squirrel cage construction of the rotor is exactly the same as that of a standard induction motor. The inherent disadvantage of the squirrel-cage construction is cogging, which is minimized by skewing the bars of the cage relative to the rotor axis. The solid-iron rotor eliminates cogging. However, it has a lower torque-to-inertia ratio. The torque-to-inertia ratio is improved by the use of drag-cup construction (see Fig. 8.100). The efficiency of AC servomotors is fairly low (5–20%), which necessitates external cooling.

8.18.2   Mathematical Model

Steady-State Model
Assuming linearity (i.e., operation without magnetic saturation) and using the method of symmetrical components,[9] mathematical models can be developed. Figure 8.101 shows an equivalent circuit of one phase of the AC servomotor. The torque developed by a two-phase servomotor in which the control phase lags the reference phase by an angle θ is given by

$$T_m = \tfrac{1}{4}\left[F_1(1 + 2k\sin\theta + k^2) - F_2(1 - 2k\sin\theta + k^2) \right]|v_m|^2 \qquad (8.147)$$

where F_1 and F_2 are functions of the rotor slip ratio S_R (defined below),

$$F_1 = \frac{2}{\Omega_s}\left| \frac{Z_m}{Z_1(Z_2 + Z_m) + Z_2 Z_m} \right|^2 \frac{R_2'}{S_R} \qquad (8.148)$$

$$F_2 = \frac{2}{\Omega_s}\left| \frac{Z_m}{Z_1(Z_2' + Z_m) + Z_2' Z_m} \right|^2 \frac{R_2'}{2 - S_R} \qquad (8.149)$$

and the parameters are defined as follows

$\quad S_R$ = Slip ratio = $(\Omega_s - \Omega_m)/\Omega_s$
$\quad \Omega_s$ = Synchronous speed (rpm)
$\quad \Omega_m$ = Motor speed (rpm)
$\quad\ f$ = Frequency of AC voltages (Hz)
$\quad P$ = Number of poles
$\quad k = v_c/v_m$
$\quad v_c$ = Voltage applied to the control phase (V)
$\quad v_m$ = Voltage applied to the fixed phase (V)

$$Z_m = \left[\frac{1}{R_m} + \frac{1}{jX_m}\right]^{-1} = \text{Impedance due to magnetic field generation in the stator}$$

$$Z_1 = R_1 + jX_1 \qquad\ \ = \text{Stator impedance}$$

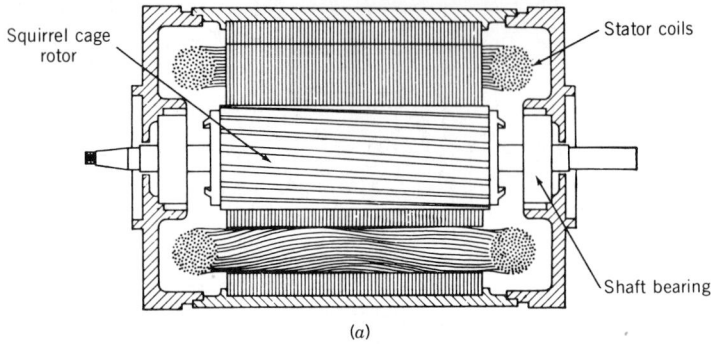

(a)

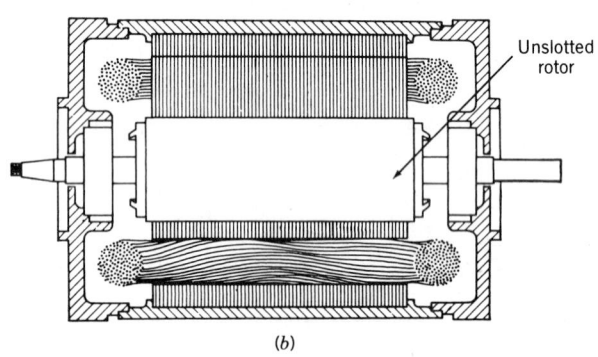

(b)

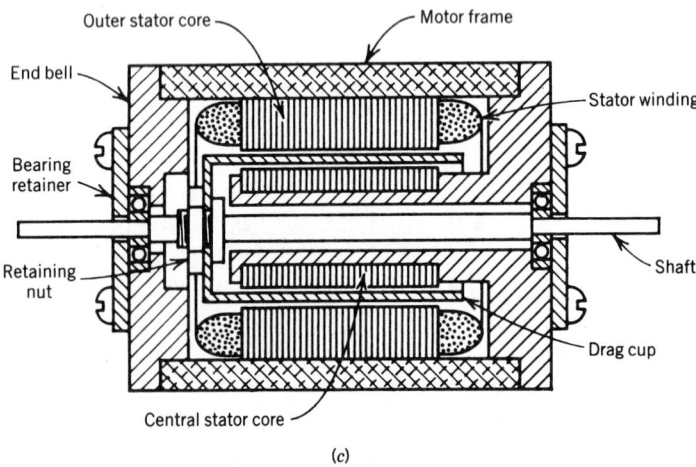

(c)

Fig. 8.100 Types of AC servomotors (Taken from John E. Gibson and Franz B. Tuteur, *Control System Components*, McGraw-Hill, pp. 279–280): (*a*) squirrel-cage rotor, (*b*) solid-iron rotor, (*c*) drag-cup rotor.

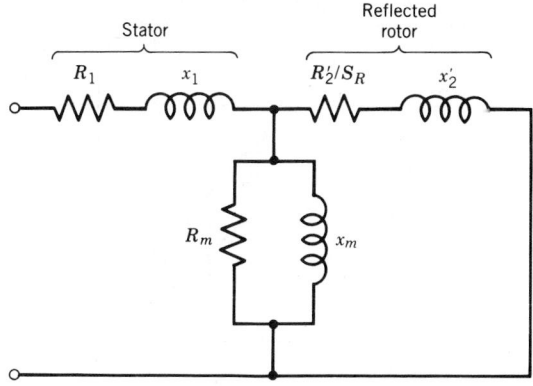

Fig. 8.101 Equivalent circuit diagram of one phase of an AC servomotor.

$$Z_2 = \frac{R'_2}{S_R} + jX'_2 \qquad = \text{Reflected impedance of rotor at slip ratio of } S_R$$

$$Z'_2 = \frac{R'_2}{2 - S_R} + jX'_2 \qquad = \text{Reflected impedance at slip ratio of } (2 - S_R)$$

$\theta =$ Phase angle between fixed and control voltages

$R'_2 = R_2(n_1/n_2)^2$; $R_m =$ Resistance due to magnetic field (ohms)

$x'_2 = x_2(n_1/n_2)^2$; $x_m =$ Inductive reactance due to magnetic field (ohms)

$R_1 =$ Resistance of stator (ohms); $R_2 =$ Resistance of rotor

$X_1 =$ Inductive reactance of stator (ohms); $X_2 =$ Inductive reactance of rotor

$n_1 =$ Number of turns in winding of one pole
of the stator

$n_2 =$ Number of turns of rotor
winding ($n_2 = 1$ for squirrel-cage-type rotor)

Figure 8.102 shows typical torque-speed characteristics for a two-phase AC motor. The characteristics are clearly nonlinear. However, about the origin they may be treated as linear.

Dynamic Model
Combining the equation of motion, the electrical lag (due to stator inductance and resistance), and the torque-speed characteristic of Eq. (8.147) will yield a nonlinear dynamic model. For an approximate static analysis, Fig. 8.102 and Eq. (8.147) may be linearized about an operating point. For servo applications the most interesting and useful operating point is $k = 0$ and $S_R = 1.0$. This linearization yields

$$\Delta\Omega_m = A\,\Delta v_c - \frac{1}{B}\,\Delta T_m \tag{8.150}$$

where

$$A = \frac{\partial\Omega_m}{\partial v_c} \tag{8.151}$$

$$\frac{1}{B} = -\frac{\partial\Omega_m}{\partial T_m} \tag{8.152}$$

and Δ indicates small variations from the operating point. Including the electrical and mechanical dynamics gives the transfer function

$$\Delta\Omega_m(s) = \frac{A\,\Delta V_c(s) - (1/B)\,\Delta T_m(s)}{(\tau_e s + 1)(\tau_m s + 1)} \tag{8.153}$$

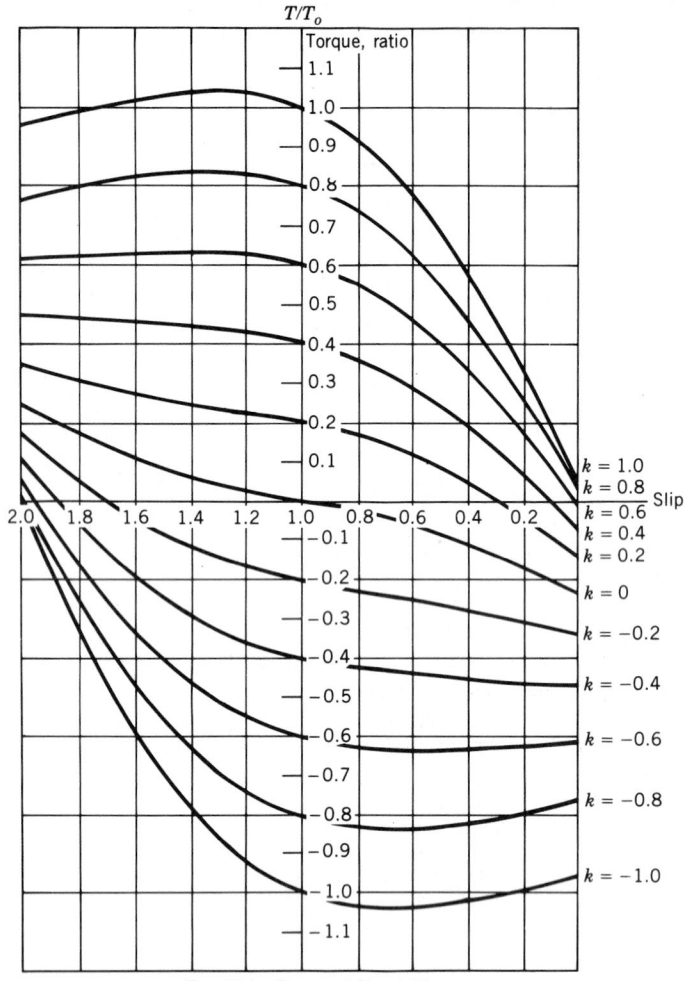

T_o = Motor Torque at Standstill

Fig. 8.102 Torque-speed characteristics of a two-phase AC servomotor (Taken from John E. Gibson and Franz B. Tueter, *Control System Components*, McGraw-Hill, p. 288).

where $\Delta\Omega_m(s)$ = Laplace transform of speed variation $\Delta\Omega_m$
 s = Laplace variable
 $\Delta T_m(s)$ = Laplace transform of change in load torque reflected to motor shaft
 τ_e = Electrical time constant (s) = L_1/R_1
 $\Delta T(s)$ = Laplace transform of the developed torque change
 L_1 = Inductance of stator (H)
 R_1 = Resistance of stator (ohms)
 τ_m = Mechanical time constant (s) = J_m/B_m
 J_m = Polar moment of inertia of rotor (N · m · s/rpm)
 B_m = Damping coefficient of motor (N · m · s)

Computation of the constants A and B of Eq. (8.150) from the torque expression of (Eq. 8.147) is rather tedious and requires measurements of the rotor impedances over a speed range of $-\Omega$ to $+\Omega_s$. As an alternative, an approximate expression for the torque has been developed (see Ref. 9) as follows:

$$T_m = \frac{S_R}{C_1 + C_2 S_R^2} \qquad (8.154)$$

where C_1 and C_2 are constants determined from two points on an experimentally measured torque-speed characteristic. The constants A and B are given by

$$A = \frac{2\Omega_s}{v_m} \frac{C_1 + C_2}{C_1 - C_2} \sin\theta \qquad (8.155)$$

$$B = \frac{1}{2\Omega_s} \frac{C_1 - C_2}{(C_1 + C_2)^2} \qquad (8.156)$$

Numerical Example

Specifications for a typical two-phase AC motor are as follows:

Number of poles $= P = 4$
Stator resistance $= R_1 = 10$ ohms
Stator inductance $= L = 3$ mH
Moment of inertia of rotor $= J_m = 5.4 \times 10^{-5}$ in · oz · s/rpm
Locked torque $= 9.5$ in · oz
Torque at 1000 rpm $= 5.4$ in · oz
Voltages: $v_c = 110$ V at $90°$; $v_m = 110$ V at $0°$
Frequency $= 60$ Hz
Synchronous speed $\Omega_s = 120f/P = 1800$ rpm
Slip ratio at standstill $= S_{R1} = 1$
Slip ratio at 1000 rpm $= S_{R2} = \dfrac{\Omega_s - \Omega}{\Omega_s} = 0.44 \qquad (8.157)$

Substituting $T = 9.5$ in · oz at $S_R = 1$ in Eq. (8.154) gives

$$\frac{1}{C_1 + C_2} = 9.5 \qquad (8.158)$$

Similarly for $T = 5.4$ in · oz at $S_R = 0.44$ gives

$$\frac{0.44}{C_1 + C_2(0.44^2)} = 5.4 \qquad (8.159)$$

Solving Eqs. (8.158) and (8.159) gives

$$C_1 = 0.0758 \ 1/\text{in} \cdot \text{oz}$$
$$C_2 = 0.0295 \ 1/\text{in} \cdot \text{oz}$$

Substituting numerical values in Eqs. (8.155) and (8.156) gives

$$A = 74.4 \ \text{rpm/V}$$
$$B = 0.00166 \ \text{in} \cdot \text{oz/rpm}$$
$$\tau_e = 3 \times 10^{-4} \ \text{s}$$
$$\tau_m = 4.6 \times 10^{-2} \ \text{s}$$

From Eq. (8.153) the transfer function is given as

$$\Delta\Omega_m(s) = \frac{74.4 \, \Delta V_c(s) - 862 \, \Delta T_m(s)}{(3 \times 10^{-4}s + 1)(4.6 \times 10^{-2}s + 1)} \qquad (8.160)$$

Generally the mechanical time constant is much greater than the electrical time constant, as is the case for this example. The term $(\tau_e s + 1)$ in the transfer function often may be neglected without introducing significant error.

8.19   STEPPER MOTORS

Stepper Motors (also called step motors or steppers) represent a significant breakthrough in the area of electromechanical actuation. These are incremental motors that by their very nature are compatible with digital systems. A stepper motor converts an electrical pulse into an equivalent rotary displacement. Since their introduction in the early 1930s, stepper motors have evolved into sophisticated designs.[5,7,8,19]

The stepper motor possesses some inherent advantages over a conventional servomotor: (1) it is compatible with digital processors; (2) open-loop control is possible, which eliminates stability problems associated with closed-loop servos; (3) the step error is noncumulative; (4) the brushless design provides easy maintenance and ruggedness. As a result of these advantages, stepper motors find widespread usage in industrial applications such as drives for TV antennas, NC machines, computer drives, hydraulic valve positioning, and other high performance feedback control systems. The primary disadvantage of stepper motors is their low efficiency, which restricts their use to fractional horsepower applications.

8.19.1   Operation

The principle of operation of stepper motors is illustrated in Fig. 8.103, which shows a single-stack, variable-reluctance, three-phase motor. The stator has six poles that are wound in a three-phase configuration. The soft-iron rotor has four poles. When phase 1 is powered by a DC voltage, one pair of rotor poles will line up with the phase 1 stator poles. When phase 1 is switched off and phase 2 is turned on, the rotor will turn clockwise through 30° until one pair of teeth align with the phase 2 poles at C. Similarly if phase 2 is switched off and phase 3 is turned on, the rotor will rotate another 30° in the clockwise direction. Thus, with each switching the rotor advances through one step of 30°. So the "step angle" is 30° (the most commonly used step angle is 1.8°). The direction of rotation may be reversed by switching in a 1.3.2.1.3.2 sequence.

In these schemes only one phase winding is switched on at a time. This is termed "one-phase-on" switching. An alternative switching scheme is called "two-phase-on" switching. In two-phase-on switching, two windings are turned on simultaneously. Referring to Fig. 8.103, if the switching sequence is 12.23.31.12, the rotor will rotate in a clockwise direction one step of 30° at a time. So, the two-phase-on scheme does not alter the step size for a three-phase stepper motor such as shown in Fig. 8.103. However, for a four-phase variable-reluctance (VR) motor, a switching sequence such as 1.12.2.23.3.34.4.41.1 results in reducing the step size by half. This is called "half stepping." Half stepping results in about 41% more torque (for a four-phase motor) compared to the single-stepping scheme.

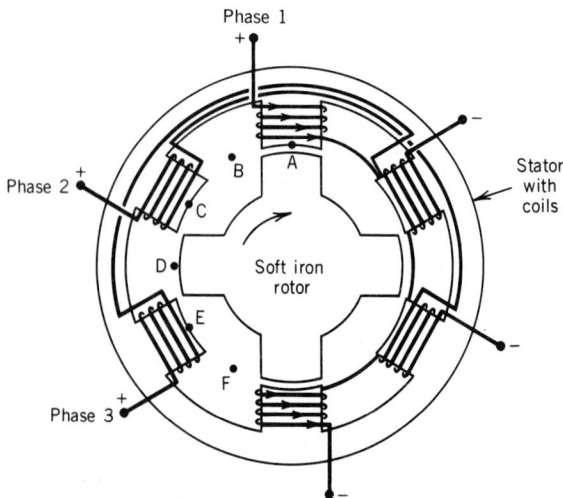

Fig. 8.103   Single-stack variable-reluctance three-phase stepper motor.

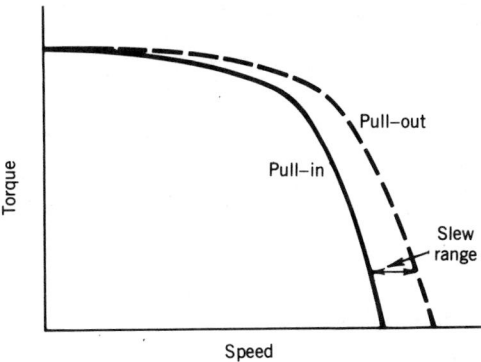

Fig. 8.104 Torque variation of a stepper motor.

By varying the relative magnitude of the voltages applied to the two windings, the rotor can be made to rotate in fractional increments of a step. This is termed "microstepping".

When the motor is energized, the holding torque of a stepper motor varies with the rotor position theoretically as a sinusoidal. Figure 8.104 shows a torque-speed characteristic for a typical stepper motor. The torque above which the stepper motor definitely loses steps is termed the "pull-out torque." The torque below which a stepper does not lose any steps (in an open-loop mode) is termed the "pull-in torque." The speed range between the pull-in and pull-out torques is called the "slew range."

Types of Stepper Motors

In the early years of stepper motor development, the stepper design included mechanical detenting and solenoid controls.[5] However, these designs have been replaced by more rugged and efficient designs. The latter designs may be broadly classified as (1) permanent-magnet (PM) steppers, (2) variable-reluctance (VA) steppers, (3) electromechanical steppers, (4) electrohydraulic steppers, and (5) hybrid steppers. Figure 8.105 shows configurations of each of these various types of stepper motors. The most commonly used are the permanent-magnet and the 3- and 4-phase variable-reluctance types of stepper motors.

The stator windings of variable-reluctance stepper motors sometimes are wound with two windings of opposite polarity per pole. This approach is termed a "bifilar winding." Figure 8.106 shows a bifilar-wound stepper motor. This arrangement provides more torque and improves damping. However, the disadvantage is more complex circuitry.

8.19.2   Mathematical Model of Permanent-Magnet Stepper Motor

The mathematical model of a stepper is generally much more complex than a conventional DC motor since the voltages applied to the various phases change in a discontinuous fashion. These discontinuities in the applied voltages result directly in corresponding discontinuities in the phase currents. This is further complicated by the spatial variation of the magnetic reluctance. Reference 19 gives detailed mathematical models for stepper motors of PM and VR types. Computer codes (in FORTRAN IV) are available in reference 19 for the PM and VR stepper motors.

The mathematical models of stepper motors are inherently nonlinear in nature due to discontinuities in input voltages and also due to the transcendental spatial variation of the self and mutual inductances. Hence these models do not lend themselves to a frequency-domain analysis.

8.19.3   Numerical Example

Table 8.13 gives the specifications of a Crouzet Stepper Motor model no. 82930.0. The motor is to be used to drive a rotary viscometer that has a rotary inertia of 1.5 oz in.^{2}, a constant frictional torque of 1.3 in · oz and a viscous damping coefficient of 0.1 in · oz/rpm. The motor is required to accelerate the viscometer from 50 to 125 rpm in a maximum of 100 ms.

The maximum torque imposed on the motor may be estimated approximately as follows:

$$T_m = (J_m + J_L)\dot{\Omega}_m + B_m\Omega_m + T_f \qquad (8.161)$$

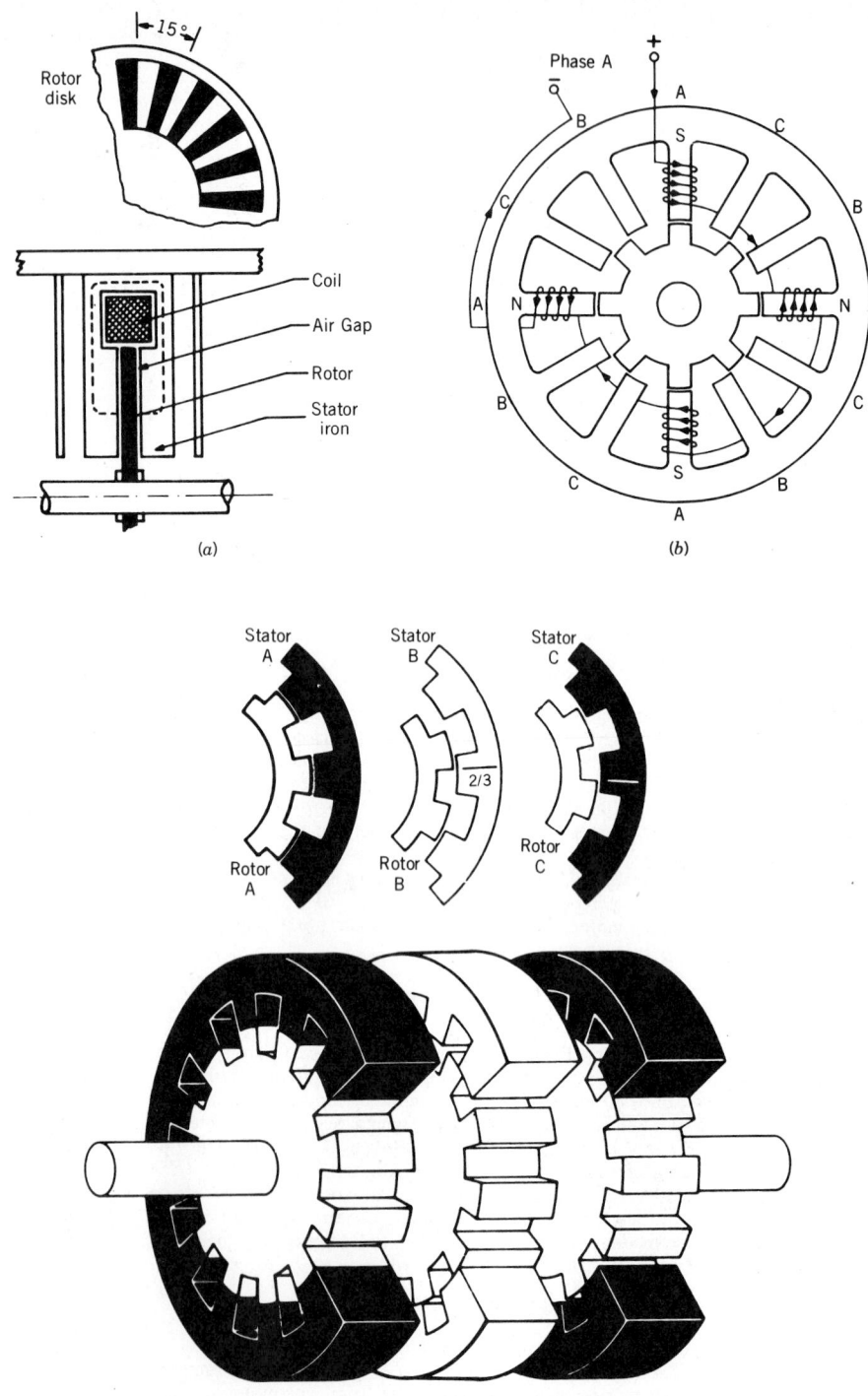

Fig. 8.105 Stepper motor configurations. N_s = number of teeth in stator, N_r = number of teeth in rotor, P = number of stator teeth per phase, n = number of phases, ψ = step angle in degrees. (Taken from Benjamin C. Kuo, *Incremental Motion Control*, Volume II: *Step Motors and Control Systems*, SRL Publishing Co., Champaign, IL, pp. 12–45): (*a*) Single-stack, variable-reluctance, axial-gap stepper motor. Features: (1) compact (2) low inertia. (*b*) Single-stack, variable-reluctance,

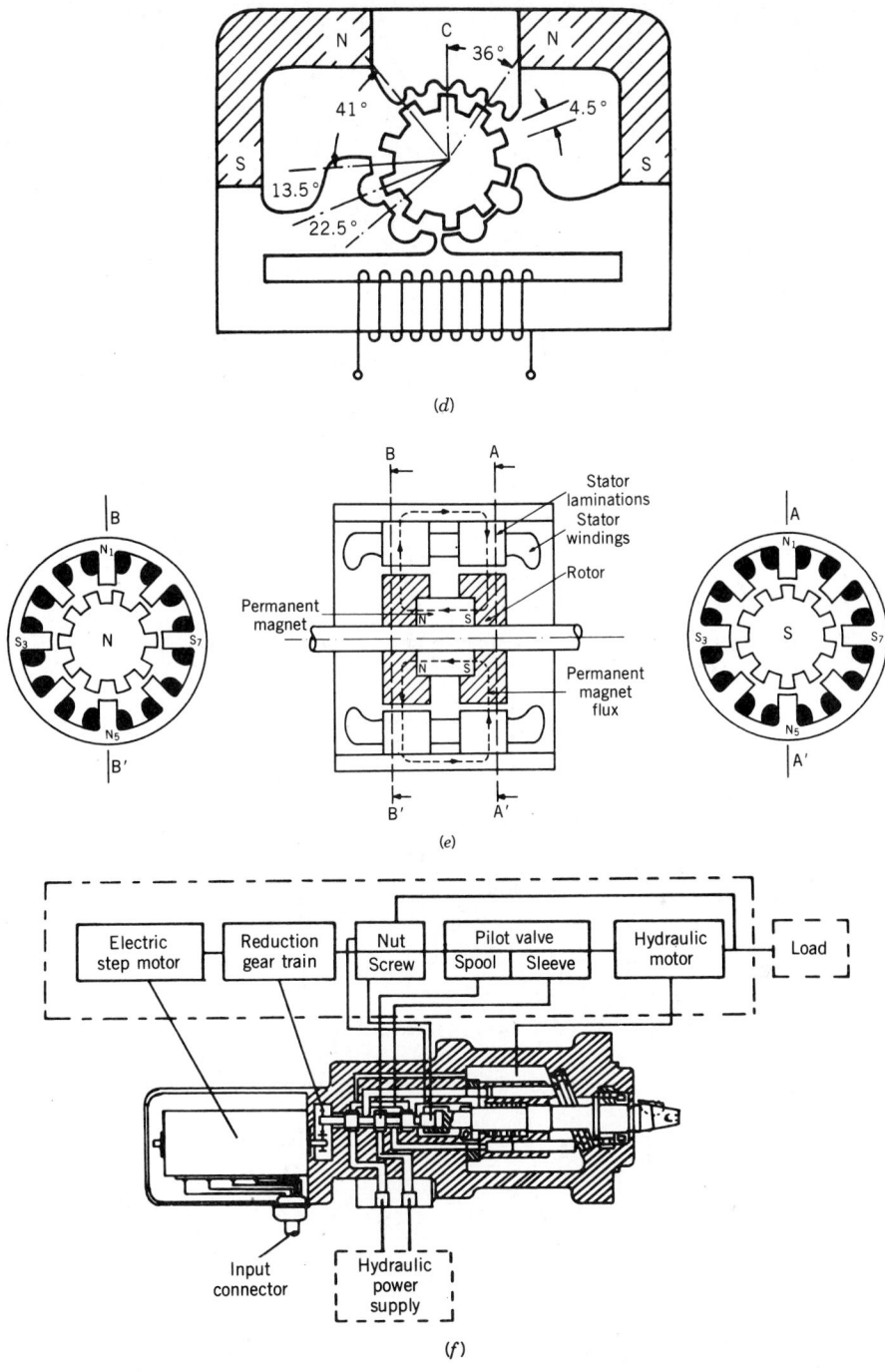

(d)

(e)

(f)

radial-gap stepper motor. Features: (1) less expensive than multi-stack, (2) step angle $\psi = 360|(N_s - N_r)/N_s N_r|$, number of stator teeth per phase $P = N_s N_r/|N_s - N_r| n[n - (N_r - N_s)/|N_r - N_s|]$. (c) Multi-stack, variable-reluctance, radial-gap stepper motor. Features: (1) step angle $\psi = 360/np$, (2) provides high-speed operation (30,000 rpm possible). (d) Permanent-magnet stator stepper motor. Features: (1) provides holding torque at no voltage, (2) simple construction, (3) more efficient, (4) only unidirectional, (5) not suited for high speed. (e) Hybrid stepper motor. (f) Electrohydraulic stepper motor. (g) Electromechanical stepper motor.

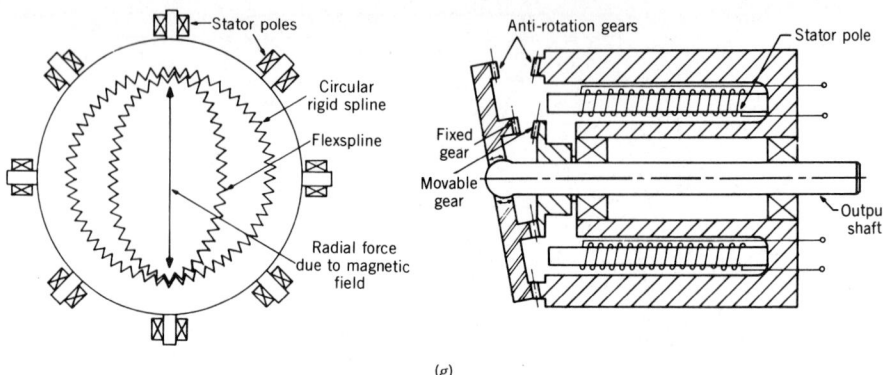

Fig. 8.105   (Continued)

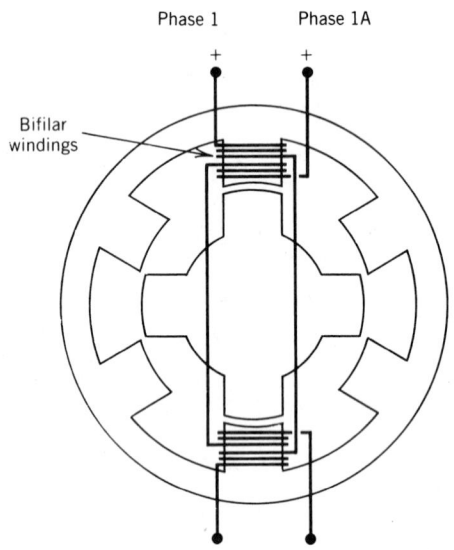

Fig. 8.106   Bipolar-wound stepper motor.

From the specifications and Eq. (8.161):

$$J_m + J_L = 5.07 \times 10^{-3} \text{ in} \cdot \text{oz} \cdot \text{s}^2/\text{rad}$$
$$\hat{\Omega} = 78.5 \text{ rad}/\text{s}^2$$
$$T_m = 14.2 \text{ in} \cdot \text{oz}$$

From the torque-speed characteristics of Table 8.13, it can be seen that the characteristic labeled b can be used. So a series resistance of 9 ohms should be used to meet the torque requirements.

The electrical and mechanical time constants may be estimated as follows:

$$\tau_e \approx \frac{L}{R} = 2.7 \text{ ms}$$

$$\tau_m \approx \frac{J_m + J_L}{B_m} = 5.3 \text{ ms}$$

TABLE 8.13 Stepper Motor Specifications (Crouzet Model 82930.0)

Step Angle = 7.5°
Number of phases = 2 (bipolar)
Resistance per phase = 9 Ω
Inductance per phase = 24 mH
Maximum input power = 10 W
Maximum voltage = 6.7 V (continuous duty)
Current per phase = 0.55 A
Holding torque at 6 V = 21.9 in · oz
Detent torque = 1.67 in · oz
Rotor inertia = 0.46 in · oz · s²/rad
Maximum coil temperature = 248°F

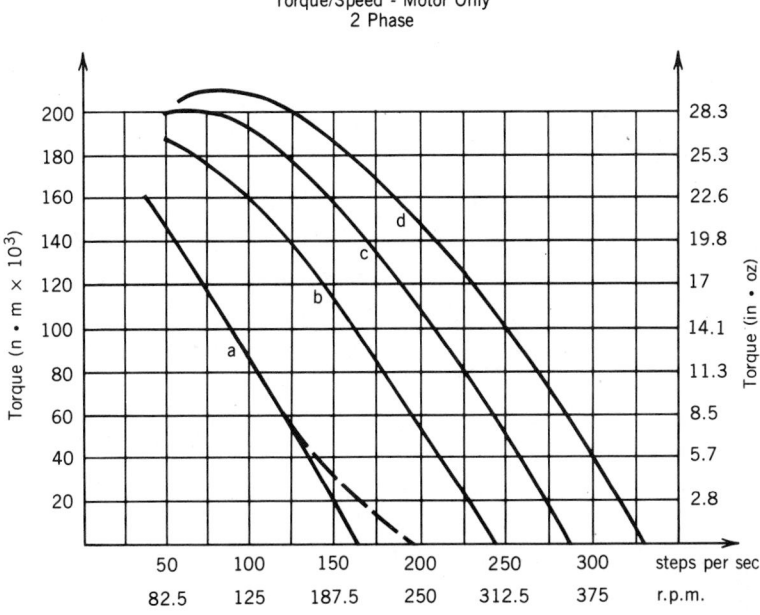

Torque/Speed - Motor Only
2 Phase

a = Using constant voltage drive with R_S (series resistance) = 0 (L/R)
b = Using constant voltage drive with R_S (series resistance) = R Motor $(L/2R)$
c = Using constant voltage drive with R_S (series resistance) = $2R$ Motor $(L/3R)$
d = Using constant voltage drive with R_S (series resistance) = $3R$ Motor $(L/4R)$

Since the time allowed for the acceleration of the load is 100 ms, which is at least an order of magnitude greater than the two time constants, the motor should have adequate dynamic response.

For a more exact transient response determination, the motor parameters may be determined experimentally (as described in Chapter 6 of Ref. 19) and used in the dynamic models.

8.20 ELECTRICAL MODULATORS

This section describes electrical modulators used for the various types of servomotors described in Sections 8.17 through 8.19. The term "modulator," as defined earlier, designates components employed for conversing the command signal into appropriate means for modulating the power flow to the servomotor (see Fig. 8.90). Modulators for the various types of electrical servomotors differ significantly.

8.20.1 Direct-Current Motor Modulators

Modulators used in servoactuator applications with DC motors usually contain two stages of amplification: a first-stage voltage amplifier followed by a second-stage power amplifier. Voltage amplifiers are generally quite linear in performance. The power amplifiers are used in two different configurations, namely type T and type H as shown in Fig. 8.107. The type T configuration employs two power sources and only two power transistors. The type H configuration uses only one power source but four transistors. The type T configuration lends itself readily to current feedback schemes. However, the type H is more commonly used because of a single power source requirement. Type H configurations can be operated in two different modes, resulting in bipolar and unipolar drives.

Linear Amplifiers
Linear amplifiers that are DC amplifiers used in the output stage (as H or T configuration) provide gains typically in the range of 2 to 10. They suffer from excessive heat generation, thus requiring external cooling such as a fan above 200 W. These are particularly severe in high-performance servo applications that typically have low-impedance rotors, operating at low speed and high torque conditions. Sometimes these problems are overcome by incorporating a dual-mode current-limiting device that permits high currents for short durations (for overcoming inertia) and then imposes a lower limit for longer durations.

Linear amplifiers are generally configured as voltage amplifiers, but in some applications a current source configuration is employed. Reference 18 discusses the details of voltage and current source

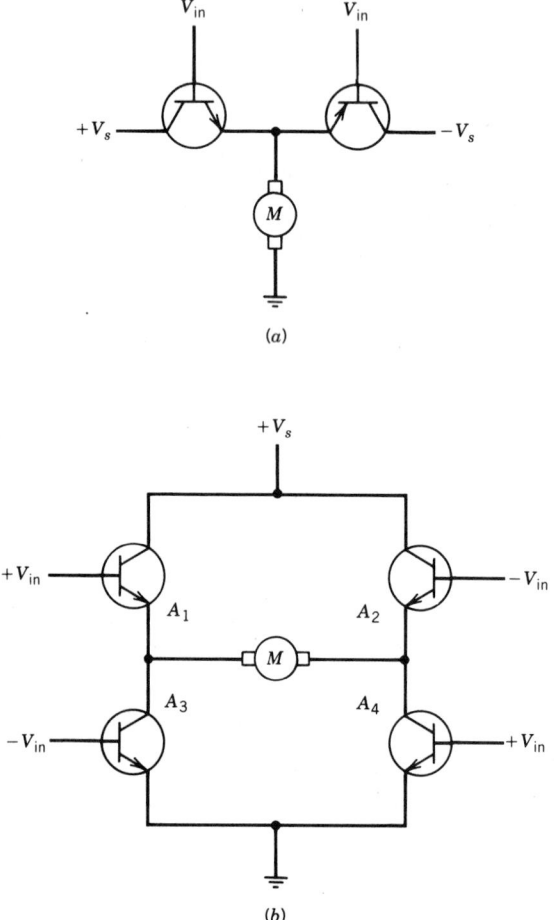

(a)

(b)

Fig. 8.107 Electrical modulator configurations: (*a*) type H, (*b*) type T.

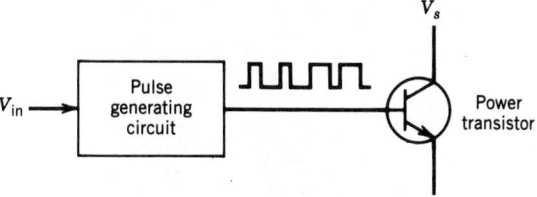

Fig. 8.108 PWM amplifier configuration.

configurations, the associated mathematical models, and the influence of the two configurations on the dynamic response of the motor.

Switching amplifiers overcome heat generation problems by switching the voltage on and off at high frequencies. This switching limits the maximum allowable inductance and results in shorter time constants and increased bandwidth. Also, the RFI noise level generated is much higher than with linear amplifiers. There are predominantly two types of switching used in servoactuator applications: (1) pulse-width modulation (PWM) and (2) frequency modulation (FM); PWM amplifiers are most commonly used.

PWM Amplifiers

In pulse-width-modulation amplifiers, the voltage applied to the servomotor is varied by changing the pulse width of a high constant frequency (typically 10 kHz) train of pulses. Figure 8.108 shows a schematic of a PWM amplifier and Fig. 8.109 shows a typical PWM servoamplifier package. Figure 8.110 shows the firing sequence for unipolar and bipolar configurations.

Mathematical Model—Bipolar Drive PWM Amplifier

The average voltage V_{ma} applied over one cycle of the PWM signal is given by (see Ref. 1, Section 4.5.3)

$$V_{ma} = \frac{2V_s T_a}{T_p} - V_s \qquad (8.162)$$

and the change in current over one cycle is given by

$$\Delta I = \frac{2V_s T_a}{L}\left(1 - \frac{T_a}{T_p}\right) \qquad (8.163)$$

where I = Current

L = Inductance of armature winding

T_a = Actuation time (see Fig. 8.110)

  = $(T_p/2) - T_0 - (T_p V_{in}/V_c)$

T_0 = Time delay (see Fig. 8.110)

T_p = Time period of one cycle of PWM signal

V_c = Peak-to-peak amplitude of a triangular wave voltage signal of time period T_p generated within the amplifier

V_{in} = Input voltage signal applied to the amplifier

V_s = Supply voltage

Equations (8.162) and (8.163) are valid for operating conditions where the instantaneous motor current I remains positive throughout a cycle. The actuation time T_a is determined from the pulse generation circuit characteristics in conjunction with the input signal (see Ref. 1).

Mathematical Model—Unipolar Drive PWM Amplifier

The average voltage and current amplitude for a PWM amplifier in a unipolar drive mode are given by

$$V_{ma} = \frac{T_a V_s}{T_p} \qquad (8.164)$$

(a)

R =	75K	4K	1K	OK
A	8A	4A	2A	0A

Fig. 8.109   PWM Servo Amplifier IC (Courtesy of Advanced Motion Controls, Van Nuys, CA).

and

$$\Delta I = \frac{V_s T_a}{L}\left(1 - \frac{T_a}{T_p}\right) \tag{8.165}$$

For operating conditions where the instantaneous motor current becomes zero for a part of the cycle, the mathematical models can be found in Refs. 16–19.

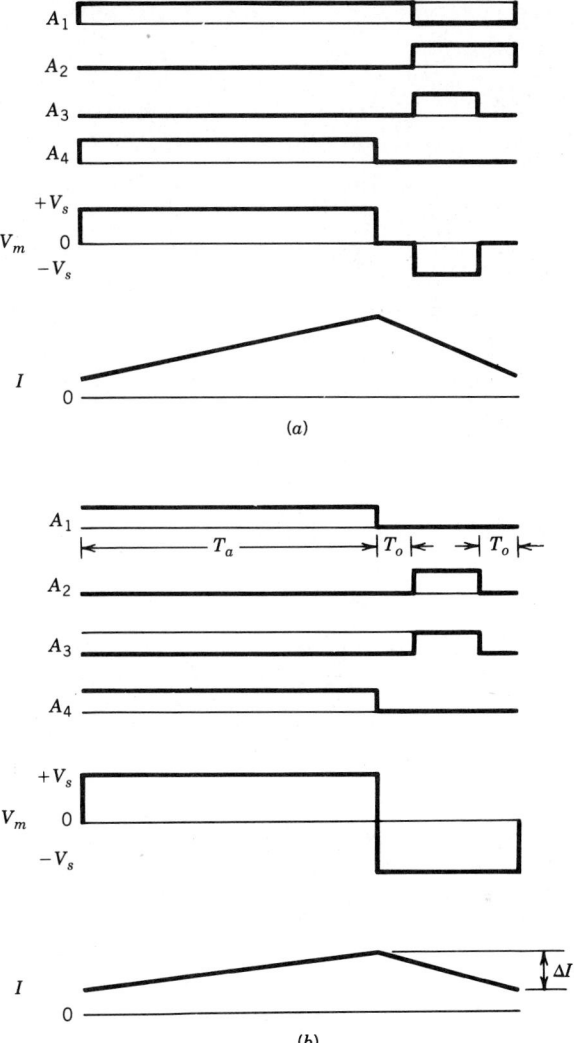

Fig. 8.110 Phases, motor voltage, and current in PWM amplifier: (*a*) unipolar drive, (*b*) bipolar drive.

8.20.2 Stepper Motor Modulators

The modulators for stepper motors are also called "drives." There are three types: oscillator-transistors, indexers, and microstepping controls. Oscillators provide a variable frequency pulse train to the translator. The function of the translator is to convert the pulse train into appropriate signals to operate the power transistors, thereby directing power to the stepper motor. The variable-frequency (3,000–15,000 pulses per second) capability of the oscillators provides for accurate manual control of speed. Similarly, a pulse generator in conjunction with a translator provides manual position control. Figure 8.111 shows a typical oscillator-translator package. Most commercial units also provide for half-stepping and electronic damping.

The indexers are generally programmable microprocessors that perform the functions of the oscillator-translator package, in addition to providing features such as numerical processing, programming, and communications (RS232 and RS274) with computers. The software controls provide such features as setting upper and lower limits of stepper motion and controlling slewing rate. Figure 8.112 shows a diagram of a typical indexer. These generally operate on 20–100 VDC power.

Fig. 8.111 Oscillator-translator modulator for stepper motor control (Courtesy of Superior Electric Co., Bristol, CT).

Stepper motors suffer from mechanical resonances (generally in the range of 50–250 steps per minute) due to the excitation resulting from the square shape of the current pulses. Microstepping eliminates this problem by energizing multiple windings simultaneously and by controlling the current flows. Hence, modulators for microstepping essentially control the simultaneous current flows to the various phases, achieving as high as 50,000 microsteps per revolution. Switching frequencies range from 20 to 200 kHz.

8.21 HYDRAULIC SERVOMOTORS

Hydraulic servomotors or "actuators" convert hydraulic power into mechanical motion. This mechanical motion can be either linear or rotary. Because hydraulic servomotors provide large actuating force or torque capabilities, they are commonly used in heavy-equipment industries. Also, due to their high mechanical stiffness, fast dynamic response, and high power-to-weight ratios, hydraulic servomotors find widespread use in aircraft, missiles, and space vehicles, as well as critical industrial systems where high performance and high power control are needed.[25]

8.21.1 Linear Motion Servomotors

Linear motion servomotors* provide a translatory motion of a load along a straight line. The motion of the load can be controlled by modulating the flow of hydraulic fluid into or out of the actuator. There are two basic types of linear motion servomotor designs: unbalanced and balanced design as shown schematically in Fig. 8.113. The unbalanced design is shorter, while the balanced design has equal extension and return rates when the servomotor is driven by a symmetrical modulator (control valve). High-performance hydraulic and electrohydraulic servosystems require servomotors with low leakage and low friction. Correspondingly, seals are critical elements in servomotor design. Construction details and mounting arrangements are discussed in Refs. 32, 34, and 35.

Since servomotors must move heavy loads quickly and accurately, they should act as very stiff structural members. In a well-designed servomotor, the fluid columns on either side of the piston are the most compliant portions of the structure. The spring rate of one fluid column is given by $K = A\beta/L$, where A is the net column area, L is the column length, and β is the fluid bulk modulus of elasticity. Servomotors used for high dynamic performance systems are designed to have the minimum stroke and the shortest permissible connecting passages in order to minimize the volume of fluid under compression. Modulators often are integrated into the servomotor body to minimize lengths of connecting passages.

*Also called actuators.

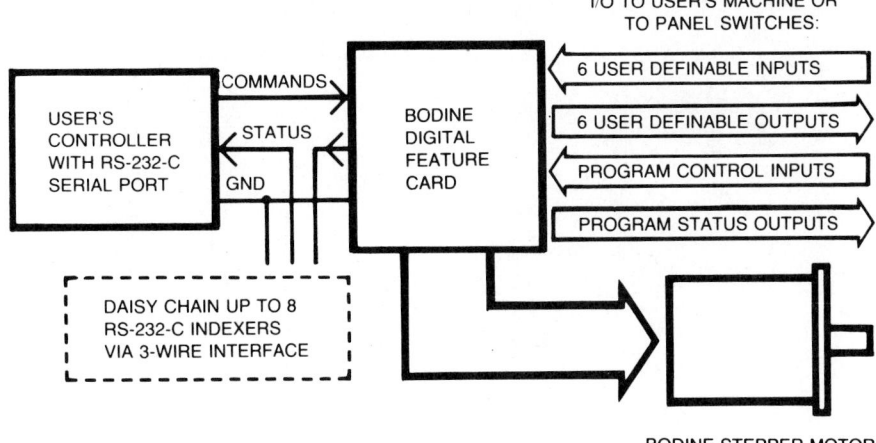

Fig. 8.112 Indexer-type modulator for stepper motor control (Courtesy of Bodine Electric Co., Chicago, IL).

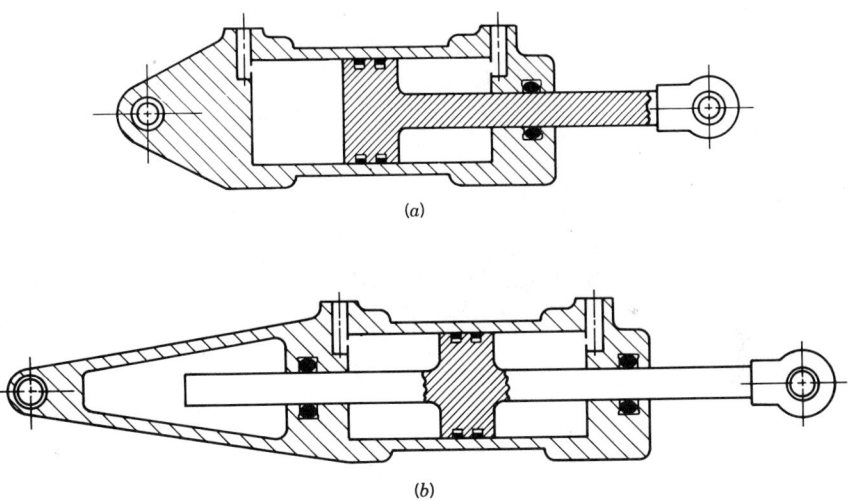

Fig. 8.113 Linear-motion hydraulic servomotors (actuators) (*a*) double-acting unbalanced actuator, (*b*) double-acting balanced actuator.

8.21.2 Rotary Motion Servomotors

Rotary motion servomotors are functionally more versatile than their linear motion counterparts. They are commonly employed even in linear motion applications through a rack-and-pinion type kinematic conversion. However, linear motion servomotors also are employed for rotary applications through corresponding kinematic conversion. Rotary actuators can be either simple reversible hydraulic motors of continuous rotation type or high-performance types with limited angular displacement. Most rotary servomotors used in medium- and high-performance hydraulic and electrohydraulic servosystems are vane- (continuous or limited rotation) or piston-type (continuous rotation) servomotors.

Rotary vane-type servomotors are commonly used in industrial applications where medium to high performance is required and size and weight are not at a high premium. Such applications include earth-moving equipment, agricultural machinery, and materials processing and handling

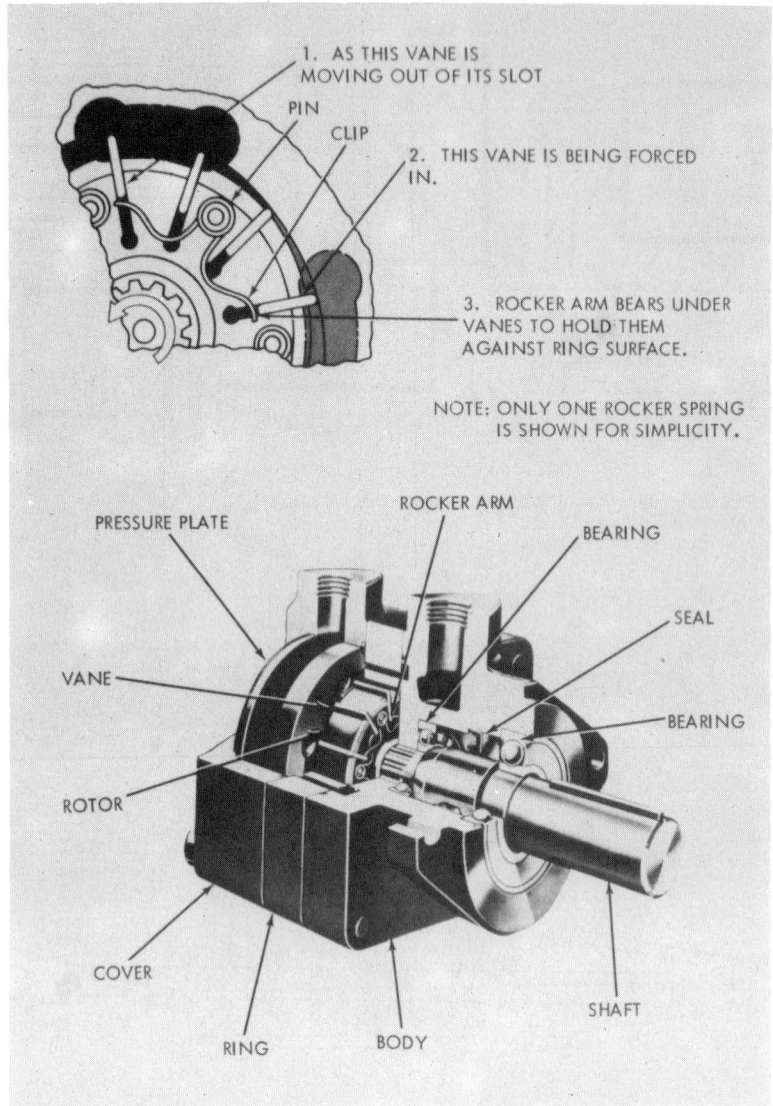

Fig. 8.114   Balanced vane-type hydraulic servomotor (Courtesy of Vickers, Inc., Troy, MI).

equipment. Figure 8.114 shows a typical continuous-rotation rotary-vane servomotor capable of operating at speeds up to 4000 rpm and pressures up to 2500 psi.

In aircraft, missile, and spacecraft applications where high performance and small size and weight are required, piston-type servomotors are commonly used. Figure 8.115 shows a typical "in-line" piston-type servomotor and Figure 8.116 shows a typical "bent-axis" piston-type servomotor. Such servomotors are capable of operating at speeds up to 8,000 rpm and pressures up to 5,000 psi.

8.21.3   Mathematical Models

An hydraulic servomotor (either linear or rotary motion type) is a rate-type device. A given flow rate results in a certain velocity (or speed for a rotary servomotor). Consequently, feedback is required for position control.

Figure 8.117 shows schematically a linear and a rotary actuator with definitions of the variables.

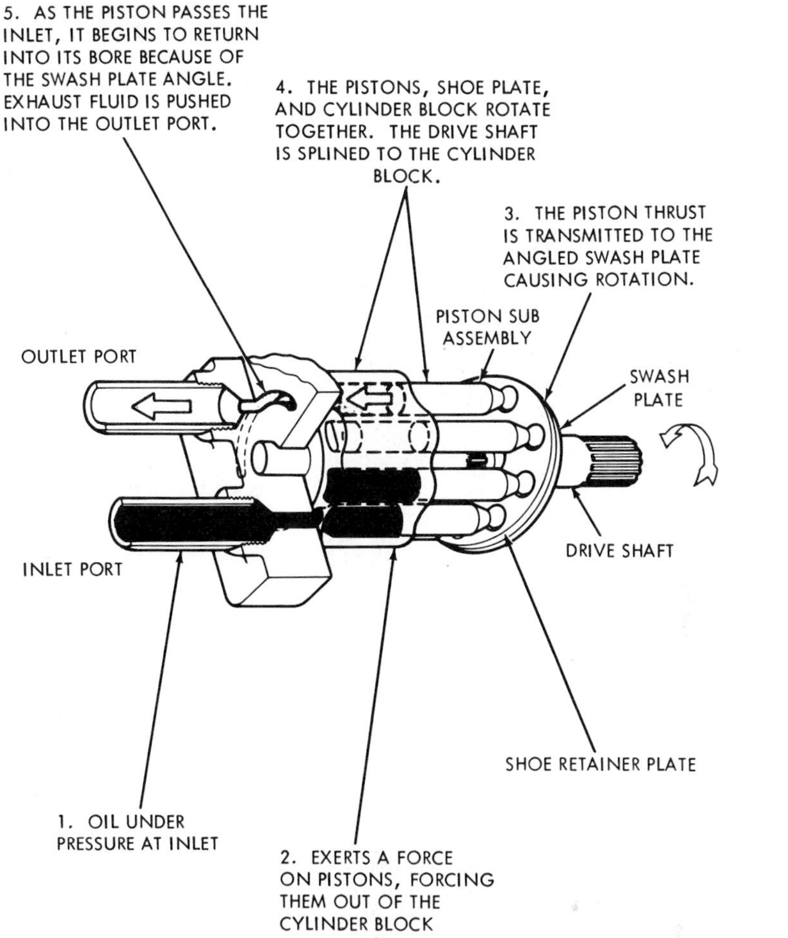

5. AS THE PISTON PASSES THE
INLET, IT BEGINS TO RETURN
INTO ITS BORE BECAUSE OF
THE SWASH PLATE ANGLE.
EXHAUST FLUID IS PUSHED
INTO THE OUTLET PORT.

4. THE PISTONS, SHOE PLATE,
AND CYLINDER BLOCK ROTATE
TOGETHER. THE DRIVE SHAFT
IS SPLINED TO THE CYLINDER
BLOCK.

3. THE PISTON THRUST
IS TRANSMITTED TO THE
ANGLED SWASH PLATE
CAUSING ROTATION.

PISTON SUB
ASSEMBLY

SWASH
PLATE

OUTLET PORT

DRIVE SHAFT

INLET PORT

SHOE RETAINER PLATE

1. OIL UNDER
PRESSURE AT INLET

2. EXERTS A FORCE
ON PISTONS, FORCING
THEM OUT OF THE
CYLINDER BLOCK

(a)

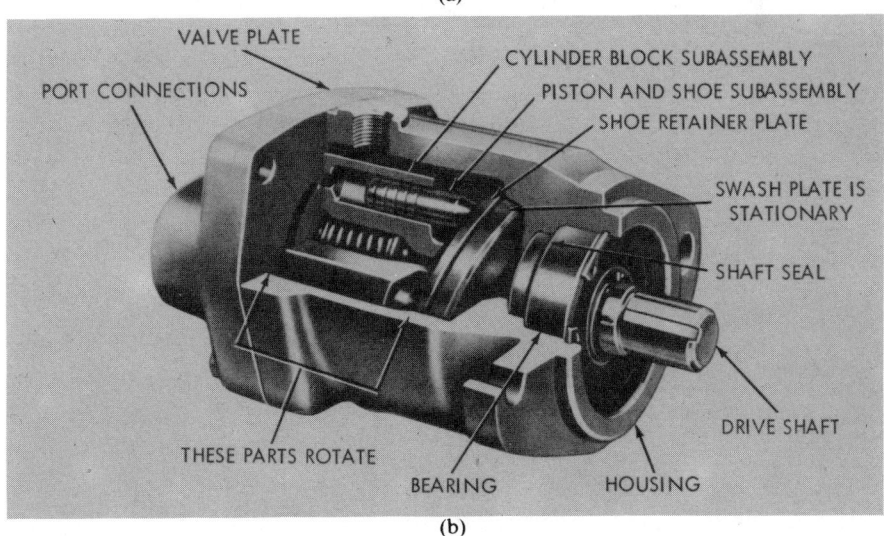

VALVE PLATE

CYLINDER BLOCK SUBASSEMBLY
PISTON AND SHOE SUBASSEMBLY
SHOE RETAINER PLATE

PORT CONNECTIONS

SWASH PLATE IS
STATIONARY

SHAFT SEAL

DRIVE SHAFT

THESE PARTS ROTATE

BEARING   HOUSING

(b)

Fig. 8.115 Inline Piston Servomotor (Courtesy of Vickers Inc., Troy, MI): (*a*) operation of servomotor, (*b*) cutaway of servomotor.

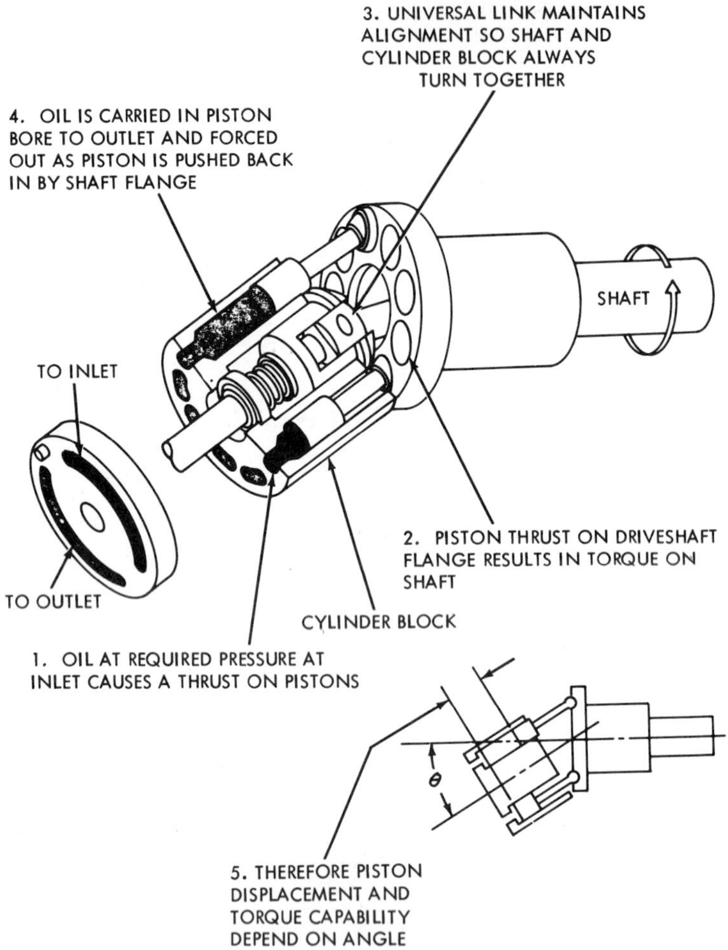

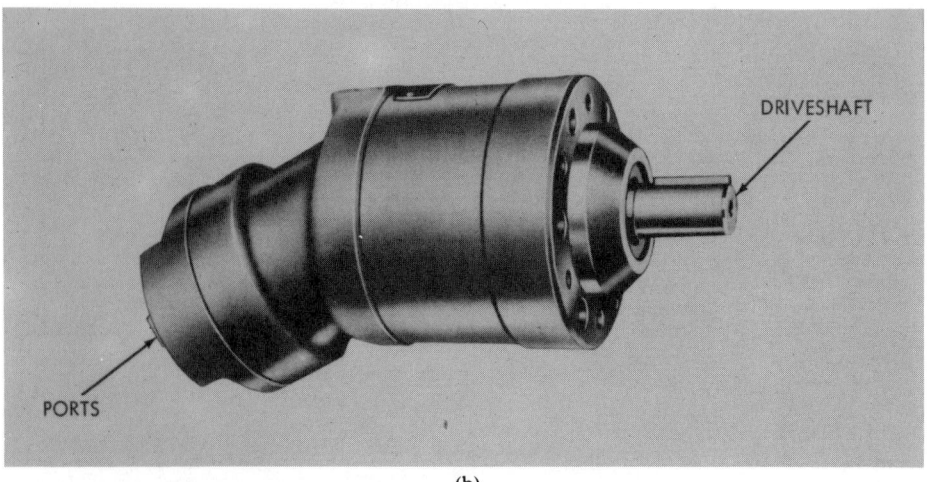

3. UNIVERSAL LINK MAINTAINS
ALIGNMENT SO SHAFT AND
CYLINDER BLOCK ALWAYS
TURN TOGETHER

4. OIL IS CARRIED IN PISTON
BORE TO OUTLET AND FORCED
OUT AS PISTON IS PUSHED BACK
IN BY SHAFT FLANGE

SHAFT

TO INLET

TO OUTLET

2. PISTON THRUST ON DRIVESHAFT
FLANGE RESULTS IN TORQUE ON
SHAFT

CYLINDER BLOCK

1. OIL AT REQUIRED PRESSURE AT
INLET CAUSES A THRUST ON PISTONS

θ

5. THEREFORE PISTON
DISPLACEMENT AND
TORQUE CAPABILITY
DEPEND ON ANGLE

(a)

DRIVESHAFT

PORTS

(b)

Fig. 8.116 Bent-axis piston servomotor (Courtesy of Vickers Inc., Troy, MI): (*a*) operation of servomotor, (*b*) photograph of servomotor.

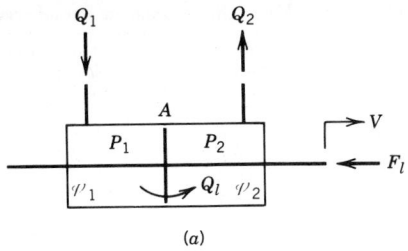

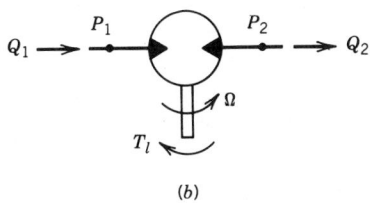

Fig. 8.117   Servomotor nomenclature.

General Dynamic Equations:
Linear Actuator

$$\left.
\begin{aligned}
\frac{dV}{dt} &= \frac{1}{m_a}\left(P_m A - F_v - F_f - F_c - F_e \right) \\[6pt]
\rho_1(Q_1 - AV) - \frac{(\rho_1 + \rho_2)}{2} Q_L &= \frac{\mathcal{V}_1 \rho_1}{\beta}\frac{dP_1}{dt} \\[6pt]
\frac{(\rho_1 + \rho_2)}{2} Q_L + \rho_2 AV - \rho_2 Q_2 &= \frac{\mathcal{V}_2 \rho_2}{\beta}\frac{dP_2}{dt}
\end{aligned}
\right\}
\qquad (8.166)$$

Rotary Actuator

$$\left.
\begin{aligned}
\frac{d\Omega}{dt} &= \frac{1}{J_a}\left(P_m D - T_v - T_f - T_c - T_e \right) \\[6pt]
\rho_1(Q_1 - D\Omega) - \frac{(\rho_1 + \rho_2)}{2} Q_L &= \frac{\mathcal{V}_1 \rho_1}{\beta}\frac{dP_1}{dt} \\[6pt]
\frac{(\rho_1 + \rho_2)}{2} Q_L + \rho_2 D\Omega - \rho_2 Q_2 &= \frac{\mathcal{V}_2 \rho_2}{\beta}\frac{dP_2}{dt}
\end{aligned}
\right\}
\qquad (8.167)$$

where V = Velocity of linear servomotor
m_a = Mass of servomotor piston, piston rods, and other moving parts
$P_m = P_1 - P_2$ = Pressure drop across servomotor ports
P_1, P_2 = Inlet and exhaust pressures
F_v, T_v = Force and torque due to viscous damping in servomotor
F_f, T_f = Force and torque due to mechanical friction (such as seals, etc.) in servomotor
F_c, T_c = Force and torque due to Coulomb friction in servomotor
ρ_1, ρ_2 = Fluid mass densities corresponding to inlet and exhaust pressures
$\mathcal{V}_1, \mathcal{V}_2$ = Fluid volume under compression in the inlet and exhaust sides of the servomotor
Ω = Angular velocity of rotary servomotor
J_a = Polar moment of inertia of rotating parts of the rotary servomotor
D = Displacement of the rotary servomotor
A = Effective area of the linear servomotor piston
Q_L = Internal leakage flow rate in actuator

TABLE 8.14 Hydraulic Servomotor Parameters

Parameter or Variable	Linear Motion Servomotor	Rotary Motion Servomotor
Y	V	Ω
K	A/B_m	D/B_r
τ	m_a/B_m	J_a/B_r

$B_l = F_v/V =$ Viscous damping coefficient
$B_r = T_v/\Omega =$ Viscous damping coefficient
$Q_1, Q_2 =$ Flow rates at inlet and exhaust
$F_e, T_e =$ Force and torque due to external load
$\beta =$ Fluid bulk modulus of elasticity

Equations (8.166) and (8.167) assume the following: (1) The linear actuator is symmetrical, i.e., equal area piston and (2) structural and metal components are rigid.

Simplified Model

A simplified mathematical model can be obtained by assuming that (1) Coulomb friction is negligible, (2) zero leakage, and (3) the load force (or torque) is zero, and (4) the fluid is incompressible.

$$\frac{Y}{P_m}(s) = \frac{K}{\tau s + 1} \qquad (8.168)$$

where the parameters are as defined in Table 8.14.

8.22 HYDRAULIC MODULATORS

Two basic approaches are used to modulate the flow of a high-pressure fluid to a work-producing device (servomotor). First, modulation may be accomplished by varying the displacement of a rotary pump which supplies fluid directly to a rotary servomotor. A variation on this approach is to use a variable-displacement servomotor and a fixed-displacement pump. Such systems, referred to as pump-displacement controlled (or motor-displacement controlled) servoactuators, are not as commonly used in high-performance applications as are valve-controlled servoactuators. The reader is referred to Refs. 29, 36, 37, 38, and 39 for more detailed discussion of pump-displacement controlled servoactuators. The discussion here is limited to the second modulation approach, i.e., servoactuators that employ servovalves as modulators.

8.22.1 Servovalve Design and Operation

Modern servovalves employ one or more of several types of metering devices: flapper-nozzle, poppet, spool, sliding plate, rotary "plate," and jet pipe. Servovalves are typically made in one-, two-, or three-stage configurations. Single- and two-stage configurations are the most common. Generally, the flapper-nozzle valve or the jet-pipe valve is used in the first stage of a two-stage servovalve. The spool-type valve is the most commonly used for single-stage servovalves and for the second stage of two-stage servovalves. Servovalves may be of the two-way, three-way, or four-way type.[29, 36, 37, 38, 39] The four-way type is used in most servosystems where bidirectional motion is required. Three-way servovalves can be used where only unidirectional motion is required. Depending on the internal design and application, servovalves may provide flow-rate control or pressure control. Special designs are available which employ flow, pressure, or dynamic pressure feedback within the valve.[40, 41]

Servovalves may have a mechanical, hydraulic, pneumatic, or electrical input. Most servosystems used in high-performance applications today use electrohydraulic servovalves where an electrical input signal is converted to a mechanical motion through a torque motor. In single-stage valves, the torque motor actuates the control valve, which in turn modulates the flow of hydraulic fluid under pressure from a high-pressure source to a linear or rotary motion servomotor. In two-stage valves, the torque motor actuates the first-stage (or pilot) valve which is typically a flapper-nozzle or jet-pipe valve. The hydraulic output from the first stage drives the second stage (typically a spool-type valve), which in turn modulates the flow from the source to the servomotor. Reference 42 is a detailed history of electrohydraulic servomechanisms with special emphasis on electrohydraulic servovalves.

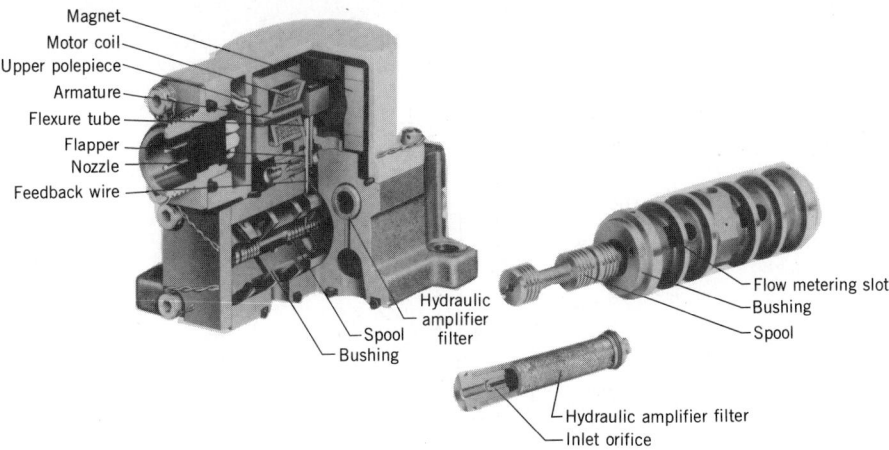

Magnet
Motor coil
Upper polepiece
Armature
Flexure tube
Flapper
Nozzle
Feedback wire

Hydraulic
amplifier
filter
Spool
Bushing

Flow metering slot
Bushing
Spool

Hydraulic amplifier filter
Inlet orifice

Fig. 8.118   Design features of a two-stage electrohydraulic servovalve (Courtesy of Moog, Inc., East Aurora, NY).

Figures 8.118 and 8.119 show the design features of a modern two-stage electrohydraulic servovalve. The double-coil, double-air-gap torque motor is "dry"; i.e., it is in an environmentally sealed compartment isolated from hydraulic fluid by the flexure tube. The first stage or pilot valve is a symmetrical, double-nozzle (four-way) flapper-nozzle valve. The flapper is attached to the upper (free) end of the flexure tube. The second stage is a spool valve that slides in a bushing with mating rectangular slots formed by electric discharge machining. The spool-to-bushing tolerance is held to 0.5 μm. Mechanical force feedback from the second-stage spool to the torque motor is provided by a cantilever spring attached to the flapper at the upper end and to the spool through a ball joint.

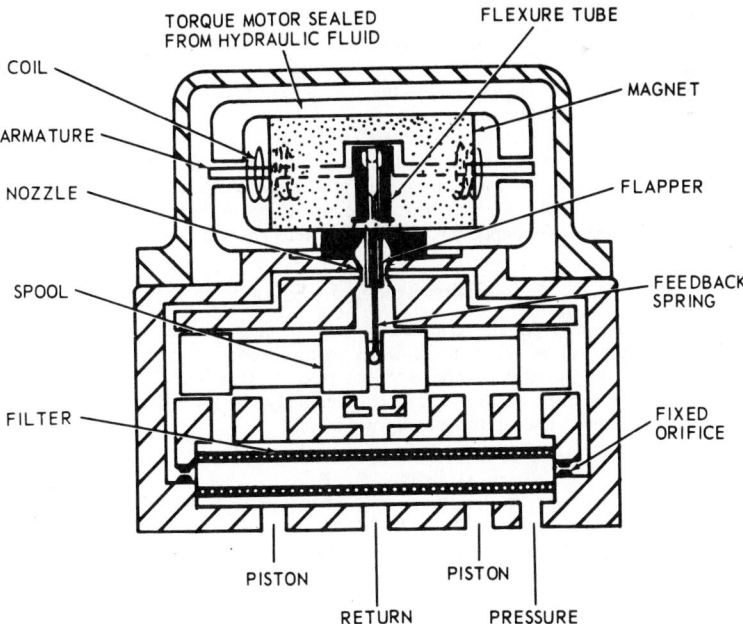

TORQUE MOTOR SEALED
FROM HYDRAULIC FLUID

FLEXURE TUBE

COIL

MAGNET

ARMATURE

FLAPPER

NOZZLE

SPOOL

FEEDBACK
SPRING

FILTER

FIXED
ORIFICE

PISTON

PISTON

RETURN

PRESSURE

Fig. 8.119   Cross section of a two-stage electrohydraulic servovalve with a flapper-nozzle first stage and spool second stage (Courtesy of Moog Inc., East Aurora, NY).

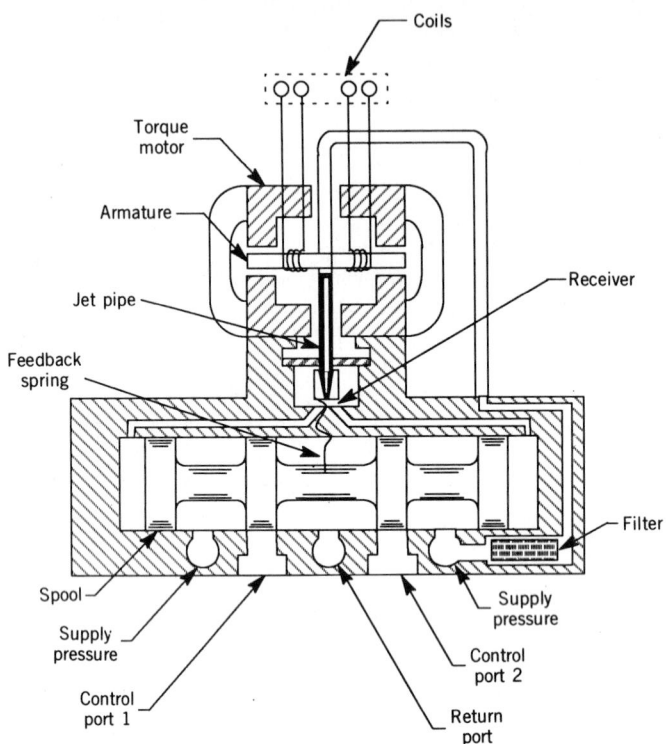

Fig. 8.120   Cross section of a two-stage electrohydraulic servovalve with a jet-pipe first stage and a spool second stage (Courtesy of Abex Corporation, Aerospace Division, Oxnard, CA).

Figure 8.120 is a cross section of a two-stage electrohydraulic servovalve that employs a jet-pipe valve as the first stage. Otherwise the valve is virtually the same in design and operation as the valve shown in Figs. 8.118 and 8.119. The jet-pipe first stage can pass contamination particles as large as 200 μm, whereas the flapper-nozzle valve can only pass 50-μm particles. Good fluid filtration can negate the importance of these differences.

A cutaway diagram of a single-stage "swing plate" servovalve is shown in Figure 8.121(a); a cross section of the valve is shown in Fig. 8.121(b). This is an industrial valve that has a high dynamic response (natural frequency about four times that of the two-stage servovalves above). It is a considerably heavier valve than the two-stage aerospace-type valves in Figs. 8.119 and 8.120.

Table 8.15 shows typical performance capabilities of various types of servovalves.

8.22.2   Mathematical Model of a Spool-Type Valve

Spool-type valves are the most popular due to their ease of construction. They are also easier to analyze than other types of valves. Figure 8.122 shows a typical spool-valve configuration and defines the important variables and parameters. The valve has three "energy ports" where energy or power flows from or to the environment of the valve. Correspondingly, the valve can be modeled using the three mathematical equations shown in functional form below:

$$Q_m = f(x, P_m, P_s) \tag{8.169}$$

$$Q_s = f(x, P_m, P_s) \tag{8.170}$$

$$F_v = f(x, P_m, P_s) \tag{8.171}$$

Equation (8.169) represents the pressure-flow-displacement characteristics of the valve. These characteristics are needed in the dynamic analysis of a servoactuator which employs the valve. Equation (8.170) is used to compute the flow rate required from the source and will not be treated further here.

(a)

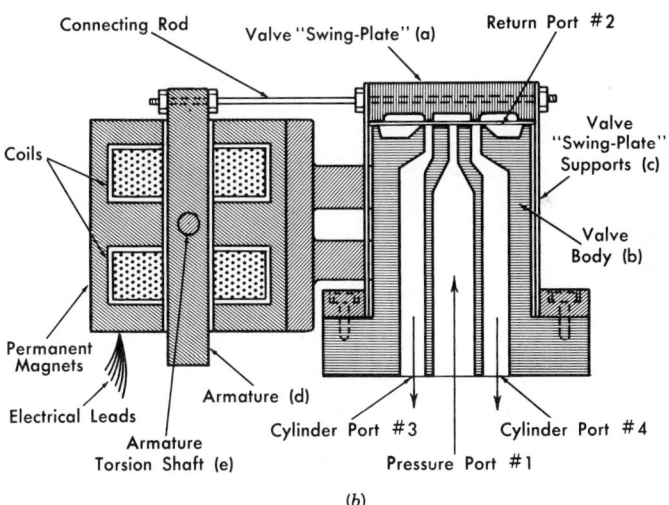

(b)

Fig. 8.121 Cross section of a single-stage, "swing-plate" electrohydraulic servovalve (Courtesy of Oilgear).

Equation (8.171) is used to calculate the force required to move the spool (e.g., the size of the torque motor in the case of an electrohydraulic servovalve).

The steady-state pressure-flow-displacement characteristics of the spool valve are characterized by the nonlinear orifice equation[27, 29]:

$$Q_m = C_d wx \sqrt{\frac{P_s - P_m}{\rho}} \qquad (8.172)$$

TABLE 8.15   Performance Specifications of Servovalves

Valve Type	Maximum Working Pressure (psi)	Maximum Flow at 1000 psi Pressure Drop (gpm)	Frequency at 90° Phase Lag (Hz)	Hysteresis (%)	Resolution (%)
Spool					
One stage	5000	3500	200	0.1	0.01
Two stage	7000	1000	200	1	0.01
Three stage	4500	300	200	1	0.01
Flapper-nozzle					
One and two stage	5000	1000	500	0.2	0.01
Three stage	5000	1000	500	1	0.01
Jet pipe					
One and two stage	4500	300	500	2	0.1
Three stage	4500	300	200	2	0.1
Sliding plate					
One and two stage	3000	40	150	3	0.1

Source: From Ref. 26.

where Q_m = flow rate to the servomotor (see Fig. 8.122),

w = port width of the valve,

x = displacement of the spool from its neutral position,

C_d = effective coefficient of discharge of the valve orifice,

P_s = supply pressure,

P_m = pressure drop across the servomotor (see Fig. 8.122),

ρ = fluid mass density.

This model assumes that the flow through the metering orifices is steady, the fluid is incompressible, and the valve exhaust pressure $P_e = 0$. A linearized form of this model facilitates the dynamic analysis of a servoactuator containing the valve. The nonlinear model may be linearized by considering small changes of all variables about an initial steady-state operating point.

Using a Taylor-series expansion with higher order terms set to zero results in:

$$\Delta Q_m = K_1 \, \Delta x - C_1 \, \Delta P_m \tag{8.173}$$

where

$$K_1 = \left.\frac{\partial Q_m}{\partial x}\right|_{P_{mo}} = C_d w \sqrt{\frac{P_s - P_{mo}}{\rho}}$$

$$\tag{8.174}$$

$$C_1 = -\left.\frac{\partial Q_m}{\partial P_m}\right|_{x_o} = \frac{C_d w x_o}{2\sqrt{\rho(P_s - P_{mo})}}$$

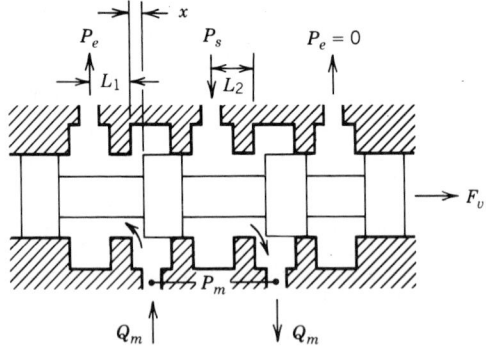

Fig. 8.122 Spool valve nomenclature.

The terms ΔQ_m, Δx, ΔP_m represent small changes of the corresponding variables about the steady-state operating point x_o, P_{mo}. The constants K_1 and C_1 are evaluated at the operating point. These expressions assume that the valve-port shape does not vary with displacement.

The static and dynamic behavior of the valve spool [Eq. (8.171)] can be modeled by considering the forces which act on the spool. These forces include the externally imposed force (input) as well as steady and unsteady flow forces resulting from flow through the orifices. Additional forces that may be present include the viscous damping between the spool and the sleeve and any mechanical spring forces acting on the spool (not shown in Fig. 8.122). The force balance equation for the spool shown in Fig. 8.122 is

$$F_{tm} = m_s \ddot{x} + \frac{\mu A_s}{h} \dot{x} + \left[C_d w \sqrt{2\rho \left(\frac{P_s - P_m}{2} \right)} \, (L_1 - L_2) \right] \dot{x}$$

$$+ \left[2 C_d C_v w \left(\frac{P_s - P_m}{2} \right) \cos \theta_j \right] x \qquad (8.175)$$

where F_{tm} = external force on spool (e.g., imposed by torque motor),
$\quad P_s$ = supply pressure,
$\quad P_m$ = pressure drop across servomotor,
$\quad x$ = displacement of spool,
$\quad \dot{x}$ = velocity of spool,
$\quad \ddot{x}$ = acceleration of spool,
$\quad A_s$ = net shear area of spool lands,
$\quad h$ = radial clearance between spool and bore
$\quad C_v$ = velocity coefficient,[29]
L_1, L_2 = length of fluid column to be accelerated (see Fig. 8.122),
$\quad m_s$ = mass of spool,
$\quad \mu$ = absolute viscosity of fluid,
$\quad \theta_j$ = effective angle of fluid jet.[29]

The fourth term on the right-hand side of Eq. (8.175) is the steady flow-induced force and the third term on the right-hand side is an unsteady flow-induced force. The steady flow force is a "spring-like" force that always opposes the motion of the spool and hence is a stabilizing force. The unsteady flow force is a "damping-like" force that changes its direction of action depending on the flow direction, and hence it can be a stabilizing or destabilizing force. Specifically, the valve is dynamically stable if $L_1 - L_2 > 0$. A more complete discussion of the dynamic modeling of the valve spool is given in Ref. 29.

8.22.3 Mathematical Models for an Electrohydraulic Servovalve

The steady-state pressure-flow characteristics for an electrohydraulic servovalve of the type shown in Fig. 8.119 are identical to those of the "idealized" spool-type valve in the previous section except the input x is replaced by the current I. That is, in the steady state, the motion of the spool in the electrohydraulic servovalve is directly proportional to the current input to the valve. The steady-state pressure-flow-current characteristics for the electrohydraulic servovalve (Fig. 8.119) are given by the equation

$$Q_m = K_v I \sqrt{\frac{P_s - P_m}{\rho}} \qquad (8.176)$$

where Q_m = flow rate to the servomotor ("control flow"),
$\quad K_v$ = a size factor,
$\quad P_s$ = supply pressure,
$\quad P_m$ = pressure drop across the servomotor,
$\quad \rho$ = fluid mass density.

It is assumed that the exhaust pressure $P_e = 0$. Equation (8.176) may be linearized for operation about an initial steady-state operating point with the result

$$\Delta Q_m = K \Delta I - C \Delta P_m \qquad (8.177)$$

where

$$K = \frac{\partial Q_m}{\partial I}\bigg|_{P_{mo}, I_o}$$

$$(8.178)$$

$$C = -\frac{\partial Q_m}{\partial P_m}\bigg|_{I_o, P_{mo}}$$

The terms ΔQ_m, ΔI, and ΔP_m represent small changes of the corresponding variables about the initial steady-state operating point I_o, P_{mo}. The constants K and C are evaluated at the operating point.

Typical steady-state pressure-flow-current characteristics for the "idealized" electrohydraulic servovalve that employs a spool-valve second stage [governed by Eq. (8.176)] are shown in Fig. 8.123. Characteristics for other types of electrohydraulic servovalves are given in Refs. 29, 40, and 41.

Another important characteristic of an electrohydraulic servovalve is its hysteresis due to the characteristics of the permanent magnets of the torque motor. The hysteresis characteristic is determined from a measurement of the output flow rate as a function of the input current for a constant (usually zero) pressure drop across the valve (load pressure drop). A typical hysteresis characteristic is shown in Fig. 8.124. The slope of the flow-current curve is the "flow sensitivity" of the valve, i.e., K in Eqs. (8.178).

It is often convenient in the dynamic analysis of servoactuators to have an approximate dynamic model for the servovalve. Experience has shown that linearized transfer functions based on empirical approximations from measured servovalve responses are adequate for most systems design. Reference 43 outlines considerations underlying the determination of approximate transfer function models for electrohydraulic servovalves. Figure 8.125 shows typical frequency-response plots for electrohydraulic flow-control servovalves, along with approximate transfer functions. For a frequency range of 0–50 Hz, the following first-order expression has been found to be adequate for flow-control servovalves:

$$\frac{\Delta Q_m}{\Delta I}(s) = \frac{K}{\tau s + 1} \qquad (8.179)$$

where Q_m = flow rate to the servomotor (in^3/s),
$\quad I$ = current input to servovalve (mA),
$\quad K$ = servovalve static flow sensitivity at zero load pressure drop (in^3/s · mA),
$\quad \tau$ = apparent servovalve time constant (s),
$\quad s$ = Laplace operator.

Typical time constants for electrohydraulic servovalves are given in Table 8.16.

If a good approximation is desired over a wider frequency range, the following second-order model may be preferred:

$$\frac{\Delta Q_m}{\Delta I}(s) = \frac{K}{\left(\dfrac{s}{\omega_n}\right)^2 + \left(\dfrac{2\zeta}{\omega_n}\right)s + 1} \qquad (8.180)$$

where $\omega_n = 2\pi f_n$ = apparent natural frequency (rad/s),
$\quad \zeta$ = apparent damping ratio (dimensionless)

Typical values of f_n and ζ for electrohydraulic servovalves are given in Table 8.16.

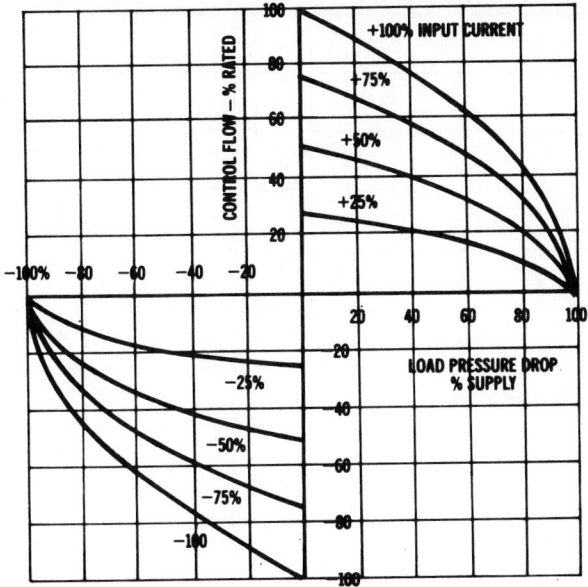

Fig. 8.123 Typical steady-state pressure-flow characteristics of an electrohydraulic servovalve.

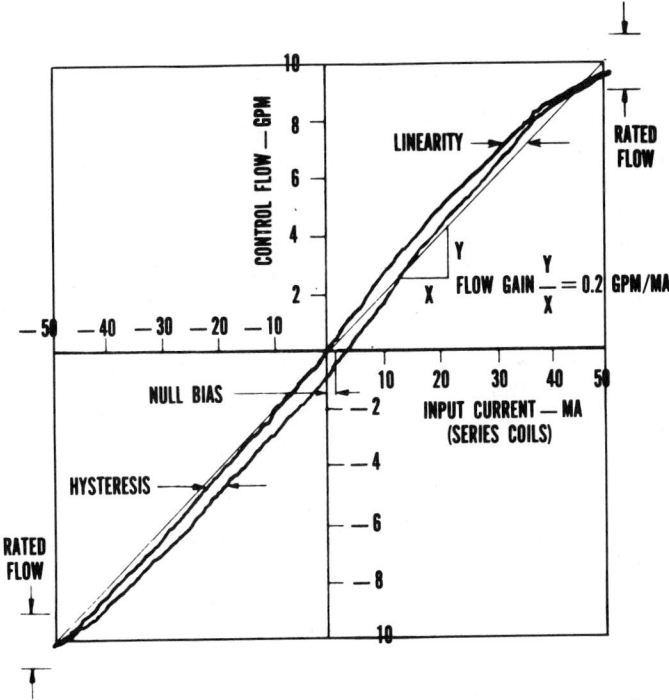

Fig. 8.124 Typical steady-state flow-current characteristics of an electrohydraulic servovalve.

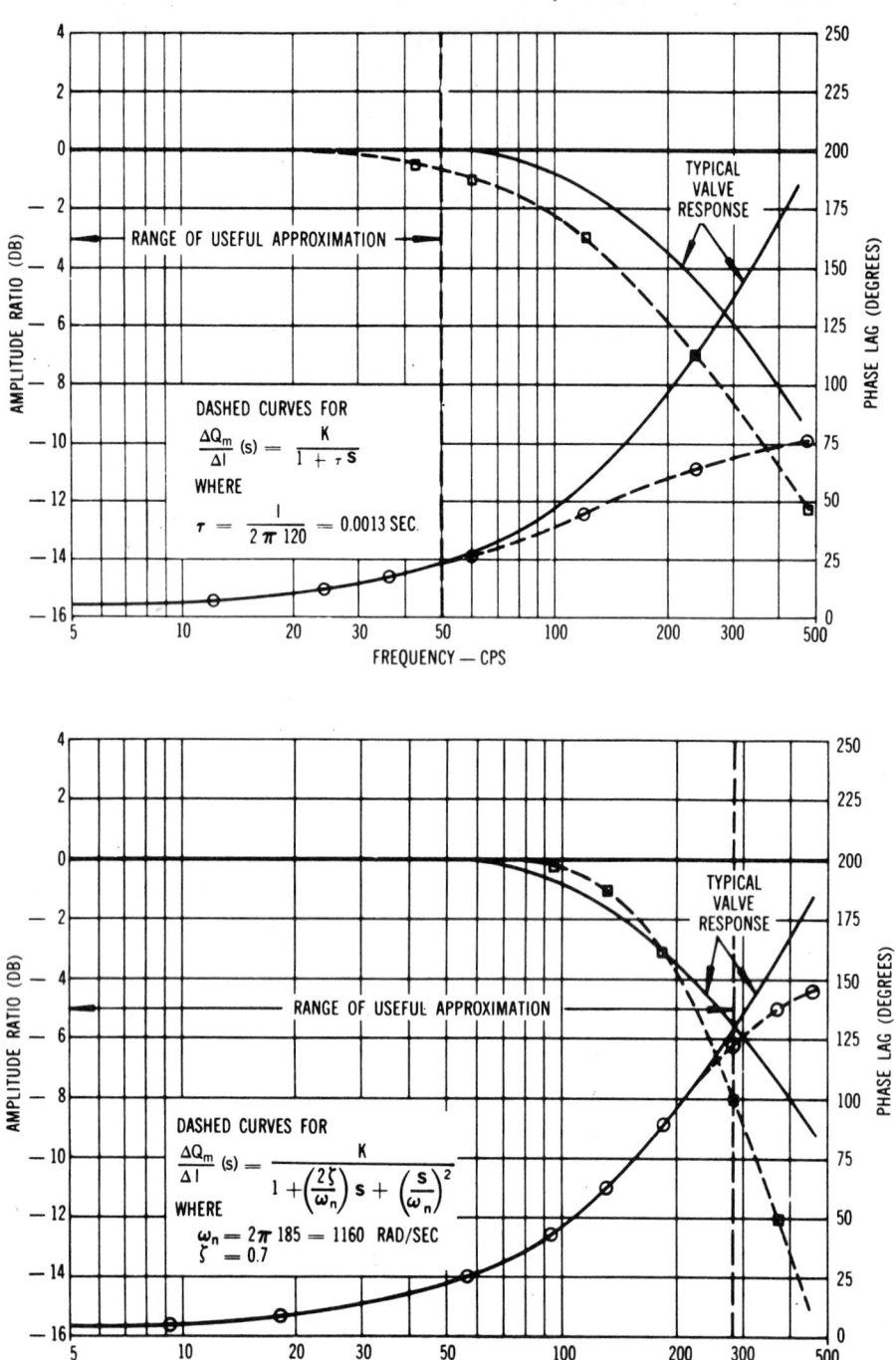

Fig. 8.125 Typical dynamic behavior of electrohydraulic servovalves.

TABLE 8.16 Dynamic Characteristics of Two-Stage Electrohydraulic Servovalves

Flow-Control Servovalve	Max. Flow Capacity at 3000 psi (gpm)	1st Order τ (s)	Approximate Dynamics 3000 psi 100°F P-P Input at 50% Rated Current — 2nd Order f_n (Hz)	ζ
A	2	0.0013	240	0.5
B	6	0.0015	200	0.5
C	12	0.0020	160	0.55
D	18	0.0023	140	0.6
E	30	0.0029	110	0.65

8.23 ELECTROHYDRAULIC SERVOSYSTEMS

A servoactuator, comprising an electrohydraulic servovalve and a servomotor, may be combined with an electronic servoamplifier and an appropriate feedback transducer to form a high-performance servosystem. Schematic diagrams of three typical electrohydraulic servosystems are shown in Figure 8.126.

Figure 8.127 shows a photograph and a cross section of a high-performance servosystem which incorporates an electrohydraulic servovalve, an axial piston servomotor, and a tachometer. The servoamplifier is not shown. Direct manifold mounting of the servovalve results in a small compressed oil volume, and therefore high torsional stiffness and fast dynamic response.

Digital control is becoming an important technique for producing optimum performance from electrohydraulic servosystems. Figure 8.128 shows a cutaway and a block diagram of a fully integrated digital electrohydraulic position control system.[31] A two-stage electrohydraulic servovalve drives a linear-motion servomotor. A microcomputer and other electronics integrated within the servovalve housing drive the valve and provide signal conditioning for the position feedback transducer. A ferromagnetic digital position-measurement transducer is mounted within the servomotor housing. The feedback management technique uses the microcomputer to model the system and provide digital velocity and acceleration signals without the use of separate sensors (i.e., a digital observer is used). Eight gains are required for accurate and smooth control, and these are calculated within the digital controller to generate the control signal to the torque motor.

8.24 COMPARISON OF ELECTROMECHANICAL AND ELECTROHYDRAULIC SERVOSYSTEMS

Precision motion and force control can be achieved using either electromechanical or electrohydraulic servosystems. The models presented in this section have been demonstrated to predict actuation performance with acceptable accuracy.[36, 44, 45] The actual choice between electromechanical and electrohydraulic must be based on a number of factors and tradeoffs. Studies by Moog, Inc., have produced the following conclusions[44]:

1. Brushless motors with high-energy SmCo magnets make possible lightweight electromechanical actuators having good response and high efficiency.
2. Electromechanical actuation is currently an alternative to electrohydraulic actuation for applications requiring up to approximately 3–4 hp. Higher-power electromechanical actuation systems are limited at the present time by the lack of reliable, compact, lightweight electronics.
3. Electrohydraulic actuation has a proven record in a variety of aerospace and industrial applications requiring high power levels.

Table 8.17 compares advantages and disadvantages of electromechanical and electrohydraulic actuation systems.

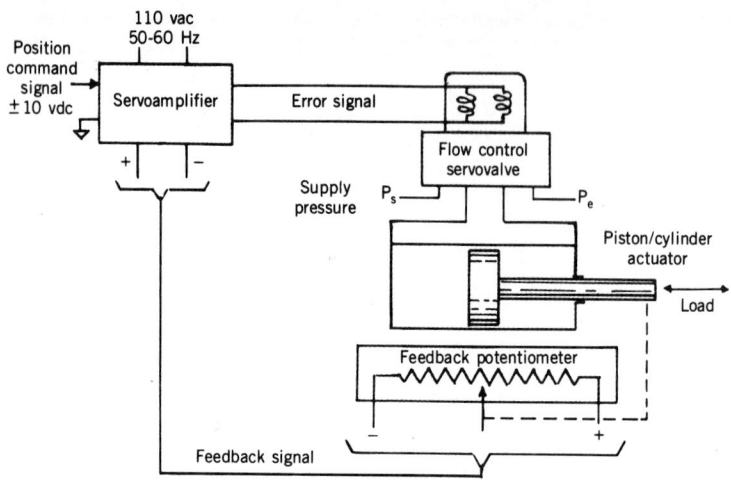

(a) Linear position servosystem

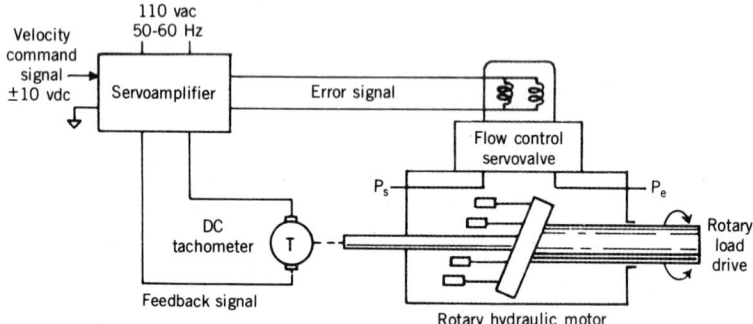

(b) Rotary velocity servosystem

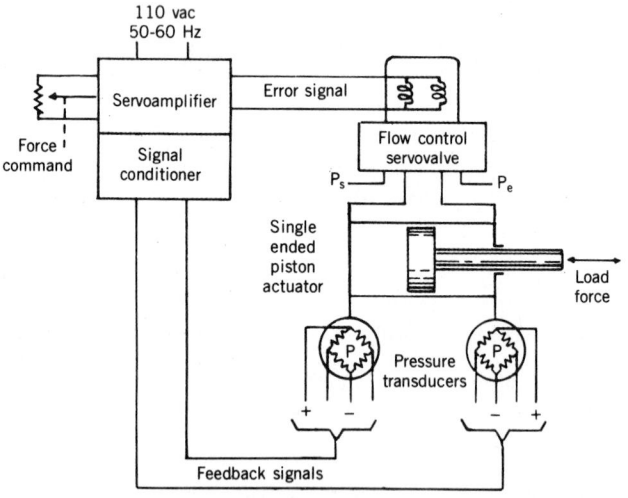

(c) Force servosystem

Fig. 8.126 Typical electrohydraulic servosystems (Courtesy of Moog Inc., East Aurora, NY): (a) linear position servosystem, (b) rotary velocity servosystem, (c) force servosystem.

(a)

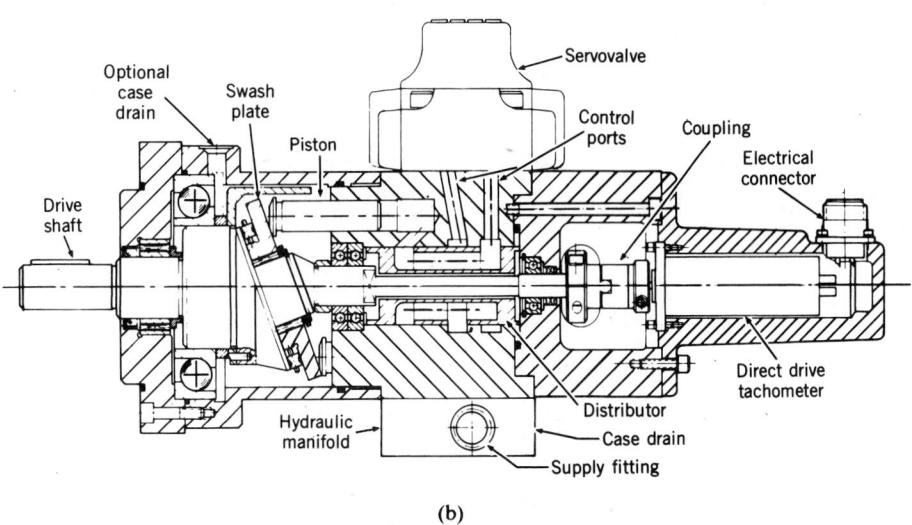

(b)

Fig. 8.127 Servosystem with rotary actuator, tachometer, and electrohydraulic servovalve (Courtesy of Moog Inc., East Aurora, NY): (*a*) photograph of Moog-Donzelli servosystem, (*b*) cross section of Moog-Donzelli servosystem.

(a)

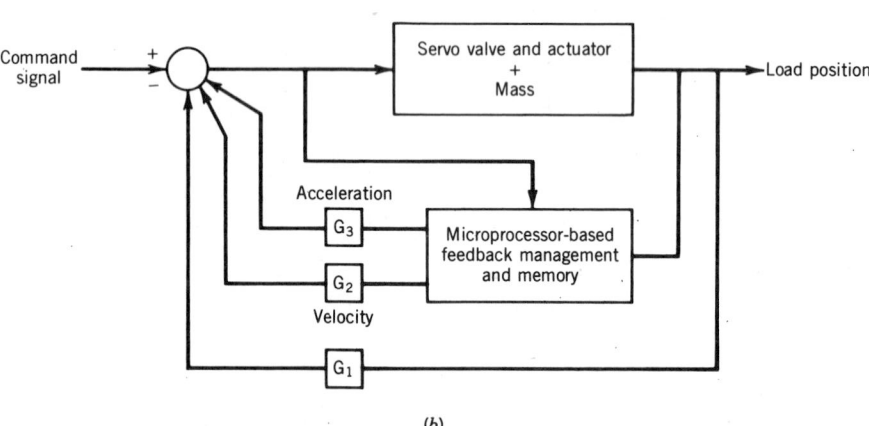

(b)

Fig. 8.128 Servosystem with linear actuator, position transducer, and microprocess-based electrohyraulic servovalve (Courstesy of Vickers Inc., Troy, MI): (a) cutaway of servosystem, (b) block diagram of servosystem.

TABLE 8.17   Comparison of EM and EH Servosystems

Electromechanical	Electrohydraulic
Advantages	*Advantages*
lower cost than electrohydraulic	mature technology
momentary overdrive capability	very high reliability
low quiescent power	highest actuation performance
low system weight in low HP range	smaller and lighter weight in high HP range
packaging flexibility	continuous power output capability
(conventional or pancake motors)	continuous stall torque capability
(different types of gear reduction)	wide temperature capability
easy check-out	high vibration and acceleration capability
single responsibility for servoelectronics	proven long term storability
and actuator	nuclear hardenable
	no EMI generation
	simple, low power servoelectronics
Disadvantages	*Disadvantages*
more complex electronics	usually higher cost
(commutation logic for brushless motors)	generally requires more complex power
(high power drive with current limiting)	conversion equipment
motor inertia-into-stops problem	requires clean hydraulic fluid
overheating with high static loads	quiescent power loss
requires motion reduction/conversion	
generates EMI	
more difficult nuclear hardening	
high power electromechanical actuation	
not yet proven	

REFERENCES FOR PART II

1. Electro-craft Corp., *DC Motors, Speed Controls, Servo Systems*, 5th ed., Hopkins, MN, 1980.

2. Bose, B. K., "Adjustable Speed AC Drives—A Technology Status Review," *Proceedings of the IEEE*, Vol. 70, No. 2, February 1982.

3. Kusko, A., and Galler, D. G., "Survey of Microprocessors in Industrial Motor Drive Systems," *IEEE, IAS*, pp. 435–438, 1982.

4. "Electrical and Electronics 1986," *Machine Design*, Vol. 58, No. 12, May 15, 1986, pp. 6–48.

5. Kuo, B. C., *Theory and Applications of Step Motors*, New York, West Publishing, 1974.

6. Kusko, A., *Solid-State DC Motor Drives*, Cambridge, Mass., MIT Press, 1969.

7. "1.8 deg. PM Hybrid Stepper Motors and Controls," Catalogs ST-1. Bodine Electric Co., 2500 W. Bradley Pl., Chicago, 1981.

8. Carlisle, B. H., "Stepping Motors: Edging into Servomotor Territory," *Machine Design*, Vol. 58, No. 26, November 6, 1986, pp. 88–100.

9. Gibson, J. E., and Tuteur, F. B., *Control System Components*, New York, McGraw-Hill, 1958, pp. 276–304.

10. Canfield, E. B., *Electromechanical Control Systems and Devices*, New York, Wiley, 1965, pp. 143–197.

11. Davis, M. A., "High Performance Electromechanical Servoactuation Using Brushless DC Motors," Technical Bulletin 150, Moog Inc., East Aurora, NY, April 1984.

12. Marx, M. F., and Lewis, T. D. "Electromagnetic Force Motor Design Using Rare Earth-Cobalt Permanent Magnets," NAECON 1977, RECORD 1119–1126.

13. Leonard, J. B., "Electromechanical Primary Flight Control Activation Systems for Fighter/Attack Aircraft," Paper No. 821435, Society of Automotive Engineers, 1982.

14. Sawyer, B., and Edge, J. T., "Design of a Somarium Cobalt Brushless DC Motor for Electromechanical Automator Applications," NAECON 1977, RECORD 1108–1112.

15. Cronin, M. J., "Design Aspects of Systems in All-Electric Aircraft," Paper No. 821436, Society of Automotive Engineers, 1982.

16. Lewis, M. A., "Design Strategies for High-Performance Incremental Servos," Proceedings, Sixth Annual Symposium on Incremental Motion Control Systems and Devices, edited by B. C. Kuo, University of Illinois, pp. 141–151, 1977.

17. "Motor Selector," DC Motor Selection and Servo System—Design Software for the IBM PC and Compatibles by Par Tech Engineering, 51 Ward Road, Windham, ME 04062.

18. Kuo, B. C., and Tal, J., *Incremental Motion Control—DC Motors and Control Systems*, Vol. I, Champaign, Ill., SRL Publishing, 1978.

19. Kuo, B. C., *Incremental Motion Control—Step Motors and Control Systems*, Vol. II, Champaign, Ill., SRL Publishing, 1979.

20. Aha, E., "Brushless DC Motors—A Tutorial Study," *Motion*, March/April 1987, pp. 20–26.

21. Meshkat, S., "Servo System Design—A Tutorial Study. Vectorial Control of Brushless DC Motor and AC Induction Motors," *Motion*, Sept./Oct. 1986, pp. 19–23.

22. Benzer, R., "Single-Chip Brushless Motor Controller," *Machine Design*, June 9, 1988, pp. 140–144.

23. *Motion Control Engineering Handbook—DC Servo/Tachometers/Brushless DC*, Magnetic Technology, Inc., Canoga Park, CA, 1985.

24. *Direct Drive Engineering Handbook—DC Torque Motors/Tachometers/Brushless DC*, Magnetic Technology, Inc., Canoga Park, CA, 1985.

25. *Machine Design*, Fluid Power Reference Issue, September 17, 1987.

26. *Fluid Power Handbook & Directory*, edited by Hydraulics & Pneumatics, Penton/IPC, Cleveland, OH, 1983.

27. Reid, K. N., "Fluid Power Control I," Course Notes, School of Mechanical and Aerospace Engineering, Oklahoma State University, Stillwater, OK, 1987.

28. *Machine Design*, Fluid Power, Reference Issue, 4th ed., September 19, 1968.

29. Blackburn, J. F., Reethof, G., and Shearer, J. L., *Fluid Power Control*, Cambridge, Mass., MIT Press, 1960.

30. Yeaple, F., *Fluid Power Design Handbook*, Basel, NY, Marcel Dekker, 1984.

31. Blickley, G. J., "Servo Valve Becomes Digital Actuator," *Control Engineering*, June 1986, pp. 76–77.

32. Lambeck, R. P., *Hydraulic Pumps and Motors—Selection and Application for Hydraulic Power Control Systems*, Marcel Dekker, 1983.

33. Nachtigal, C. L., ed., *Handbook of Instrumentation and Control*, New York, Wiley, 1990, Chapter 16.

34. Parker-Hannifin Corp., *Fluid Power Design Engineers Handbook*, Cleveland, OH, 1973.

35. Keller, G. B., *Hydraulic System Analysis*, Cleveland, OH, Industrial Publishing Co., 1969.

36. Merritt, H. E., *Hydraulic Control Systems*, New York, Wiley, 1967.

37. Lewis, E., and H. Stern, *Design of Hydraulic Control Systems*, New York, McGraw-Hill, 1962.

38. Guillon, M., *Hydraulic Servo Systems*, New York, Plenum, 1969.

39. Watton, J., *Fluid Power Systems*, Prentice-Hall International, Hertfordshire, 1989.

40. Moog, Inc., *Electrohydraulic Servomechanisms in Industry*, East Aurora, NY (undated).

41. Geyer, L. H., "Controlled Damping through Dynamic Pressure Feedback," Technical Bulletin 101, Moog, Inc., East Aurora, NY, April 1972.

42. Maskrey, R. H. and Thayer, W. J., "A Brief History of Electrohydraulic Servomechanisms," *ASME Transactions, Journal of Dynamic Systems, Measurement and Control*, Vol. 100, June 1978.

43. Thayer, W. J., "Transfer Functions for Moog Servovalves," Technical Bulletin 103, Moog, Inc., East Aurora, NY, January 1965.

44. *Speed and Position Control Systems Distributor Catalog*, Robbins Myers, Inc., Hopkins, MN, Form no. MN-7400-00, January 1988.

45. Neal, T. P., "Performance Estimation for Electrohydraulic Control Systems," Technical Bulletin 126, Moog, Inc., East Aurora, NY, November 1974.

BIBLIOGRAPHY FOR PART II

Conference Proceedings of *Conference on Small Electrical Machines*, 30–31 March 1976, Institution of Electrical Engineers, Savoy Place, London.

Small and Special Electrical Machines, Conference Publication no. 202 22–24 Sept. 1981, Institution of Electrical Engineers, London.

Fitzgerald, A., and Kingley, C., Jr., *Electric Machinery*, New York, McGraw-Hill, 1961.

Puchstein, Lloyd and Conrad, *AC Machines*, 3rd ed., New York, Wiley, 1954.

Koopman, "Operating Characteristics of 2-phase Servomotors," *Trans. AIEE*, Vol. 68, pp. 319–329, 1949.

Persson, E. K., *Brushless DC Motors in High-Performance Servo Systems*, Proceedings, Fourth Annual Symposium on Incremental Motion Control Systems and Devices, edited by B. C. Kuo, University of Illinois, pp. T1–T16, 1975.

Schept, Bob, "Servo System Design—A Tutorial Study," Part 3 of 8, *Motion*, 4th Quarter, pp. 22–30, 1985.

Chitayat, Anwar, "Brushless DC Linear Motors," *Motion*, Sept./Oct., pp. 22–23, 1987.

Mazurkiewicz, John, "Brushless Motors and Brushless Motor Controllers," *Motion*, Nov./Dec., pp. 14–19, 1986.

Tal, Jacob, "Quantization Errors in Digital Servo Systems," *Motion*, Sept./Oct., pp. 10–13, 1986.

Meshkat, Saied, "Vectonal Control of Brushless DC Motors and AC Induction Motors," *Motion*, Sept./Oct., pp. 19–23, 1986.

Marinko, Joseph A., "PWM Servo-Amplifiers: A Tutorial Study," *Motion*, July/Aug., pp. 3–8, 1987.

Humphrey, William M., *Introduction to Servomechanical System Design*, Englewood Cliffs, N.J., Prentice-Hall, 1973.

PART III

CONTROLLER DESIGN
T. Peter Neal

8.25 INTRODUCTION

The purpose of this section is to provide a basis for the specification and functional design of electronic servocontrollers. No attempt is made to treat the subject of electronic circuit design. Instead, the goal is to aid the engineer in selecting and applying a suitable off-the-shelf controller or in specifying the controller requirements to a circuit designer. The emphasis is on position, velocity, or force control of mechanical loads, although many of the techniques are applicable to controller design in general. Specialized subjects, such as multiaxis control and adaptive control are beyond the scope of this section.

As a starting point, it is presumed that a servoactuator has been selected and mounted, together with a suitable power supply, drive amplifier, and mechanical drive mechanism. In addition, the primary feedback transducer has been chosen, and a simple loop closure has been analyzed to determine whether the specified closed-loop performance can be obtained. The process of accomplishing these tasks is treated in Part II. If a simple loop closure provides adequate performance, the controller design problem primarily consists of making some basic decisions concerning electronic implementation.

In many applications, a simple loop closure is inadequate; and more elaborate controller functions are required. These latter cases are the primary subject of this section. The thrust of the discussion is synthesis of the controller function, rather than analysis of an existing design. For this reason, classical frequency domain techniques will be used extensively, all starting from block diagrams and transfer functions based upon the Laplace operator.[1,2,3] These techniques, many of which are graphical, are particularly useful in the early design stages. Since most people intuitively relate to time domain responses, relationships between the frequency domain results and time histories will be discussed as appropriate.

If the control system is to be implemented in digital hardware or software, it is certainly possible to handle the entire design task using the mathematics of digital control systems.[4,5] This approach has not been used here because the so-called classical techniques based on the Laplace transform are more illustrative and generally more familiar. For the control system applications being considered here, it is generally necessary to keep the resolution high and the sampling interval small. In this case, controller characteristics described as transfer functions in Laplace form can be accurately transformed into various mathematical forms appropriate for digital implementation, e.g., the z transform.

8.26   FUNDAMENTALS OF CLOSED-LOOP PERFORMANCE

To properly design a servocontroller, it is necessary to maintain a clear picture of the desired end result, namely, the achievement of some predetermined performance goals. These performance specifications should be established early in the design. As a minimum, they should define the desired static and dynamic accuracy, bandwidth (response time), and stability. The following sections offer a brief review of these important factors and how they relate to basic loop parameters.

8.26.1   Accuracy and Loop Gain

The most basic requirement of a servomechanism is probably static accuracy; that is, the controlled variable must accurately hold the command set point. Referring to the generalized block diagram of Fig. 8.129, several sources of inaccuracy can be described. An external disturbance can cause the load to move without any change in the command signal. The load will continue to move until the resulting error signal causes the actuator to balance the disturbance. Anomalies in the actuator and load must also be offset by a finite error signal. Examples are temperature-induced null shifts, hysteresis, threshold, friction, and lost motion. The magnitude of these error signals is minimized if the amplifier gain is high.

Ideally, the amplifier gain would be set high enough that the accuracy of the servo becomes dependent only upon the accuracy of the transducer itself. In practice, however, the amplifier gain is limited by stability considerations. Therefore, it is desirable to provide high gains at low frequencies for accuracy and low gains at high frequencies to minimize stability problems. Since the rate of amplitude roll-off with frequency is directly related to phase lag, excessive roll-off can create more stability problems than it solves. A good compromise is to make the entire forward path look like an integrator over the frequency range of interest (a type I system). This technique is very commonly used to give nearly infinite static gain and a linear gain roll-off with frequency, at the cost of 90° phase lag.

It is important to note that some servoloops contain an inherent integrator, which complicates the accuracy-versus-stability problem. For example, many actuators are inherently rate devices when operated open loop, so that a steady input results in a proportional velocity output. A velocity servo using such an actuator will inherently have a proportional forward loop, and an integrating servoamplifier can be used (Fig. 8.130). However, the corresponding position servo will inherently have an integration in the forward loop, as shown in Fig. 8.131. In this latter case, the use of an integrating servoamplifier can cause severe stability problems. From these figures, transfer functions can be written for the closed-loop responses to command and disturbance inputs. Note that the dynamic response characteristics of all elements have been neglected, and it is assumed that the load includes no spring to ground:

$$\frac{V}{e_c} = \frac{1}{K_v} \frac{1}{s/K_{vv} + 1} \tag{8.181}$$

$$\frac{V}{F_d} = K_4 \frac{s}{s + K_{vv}} \tag{8.182}$$

$$\frac{X}{e_c} = \frac{1}{K_x} \frac{1}{s/K_{vx} + 1} \tag{8.183}$$

$$\frac{X}{F_d} = \frac{K_4}{K_{vx}} \frac{1}{s/K_{vx} + 1} \tag{8.184}$$

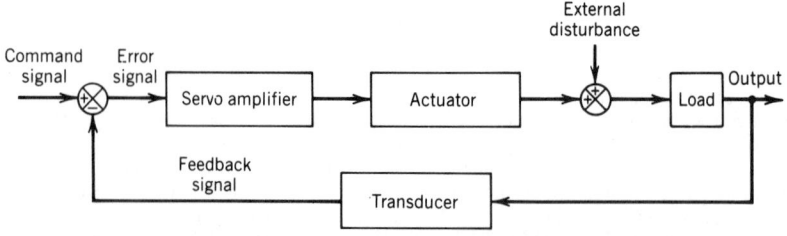

Fig. 8.129   Generalized servomechanism.

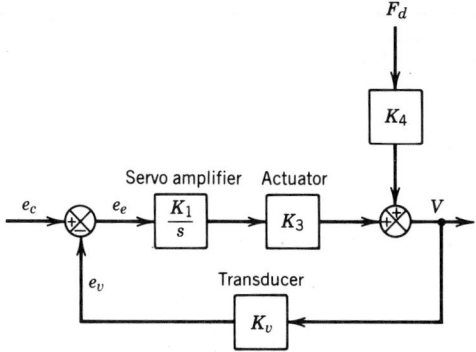

Fig. 8.130 Simplified velocity servo.

where e_c = command signal, volts
e_e = error signal, volts
e_v = velocity feedback signal, volts
e_x = position feedback signal, volts
F_d = disturbance force applied to the load, N
K_1 = integrating servoamplifier gain, (volts/sec)/volt
K_2 = proportional servoamplifier gain, volts/volt
K_3 = actuator gain, (mm/sec)/volt
K_4 = actuator velocity droop due to force disturbance, (mm/sec)/N
K_v = velocity transducer gain, volts/(mm/sec)
K_x = position transducer gain, volts/mm
K_{vv} = open-loop gain of velocity servo = $K_1 K_3 K_v$, sec^{-1}
K_{vx} = open-loop gain of position servo = $K_2 K_3 K_x$, sec^{-1}
X = load position, mm
$V = \dot{X}$, mm/sec

As shown in Fig. 8.132, the velocity and position responses to commands are both characterized by a first-order lag having a break frequency equal to the open-loop gain. However, the responses to disturbance forces are quite different in the two cases (Fig. 8.133). When the disturbance is downstream of the integrator (velocity servo), the servo error is $(K_4 F_d)$ at high frequencies, but rolls off at frequencies (in radians per second) below K_{vv}, and is zero statically. When the disturbance is upstream of the integrator (position servo), there is a static error inversely proportional to the open-loop gain, which rolls off at frequencies above K_{vx}. Note that Eqs. (8.181)–(8.184) remain

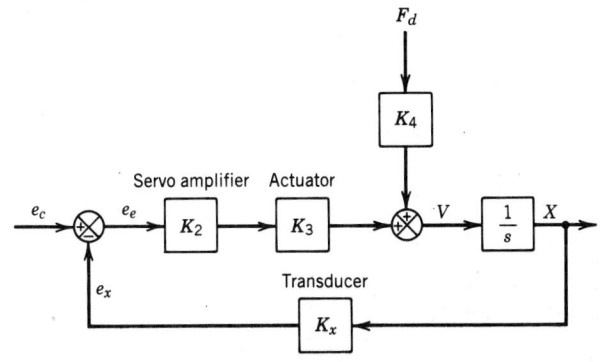

Fig. 8.131 Simplified position servo.

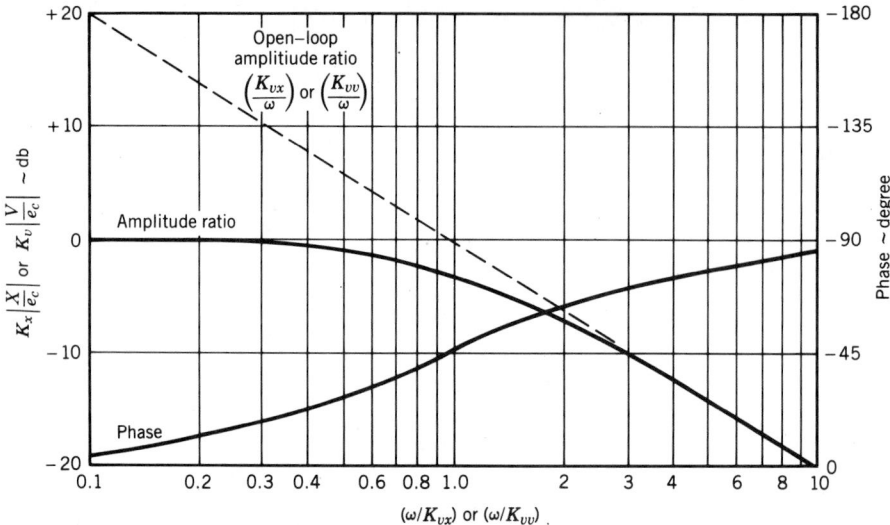

Fig. 8.132 Simplified response to commands.

reasonably valid when the dynamic response characteristics of the various open-loop elements are considered, for those cases in which K_{vv} (or K_{vx}) is well below the lowest break frequencies of those elements. At higher loop gains, the closed-loop dynamics can change considerably, as shown in the following discussion.

The conclusions regarding the effects of disturbance forces on servo accuracy can be generalized to any forward-loop offset or uncertainty. Referring again to Fig. 8.130, it can be seen that the integrating amplifier will compensate for any forward-loop offset downstream of the integrator, so that the static errors are zero. From Fig. 8.131, it is apparent that static errors due to offsets upstream of the integration can be quantified as

$$\frac{X_e}{V_0} = \frac{1}{K_{vx}} \qquad (8.185)$$

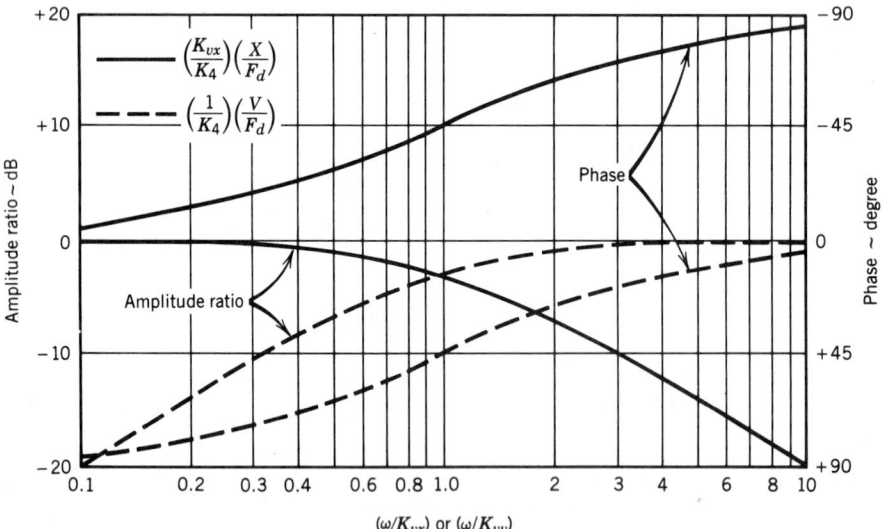

Fig. 8.133 Simplified response to disturbances.

where X_e = output error = $X_c - X$, mm

$\quad\quad V_0$ = forward-loop offset, converted to an equivalent open-loop offset in V, mm/sec

$\quad\quad X_c$ = position command = e_c/K_x, mm

Even when no forward-loop offsets or disturbances are present, servos exhibit following errors, sometimes called tracking errors. In servos having forward-loop integrations, these quasi-static errors result whenever the command signal changes at a constant rate, as in a position-tracking servo. For the position servo of Fig. 8.131, the following error is

$$\frac{e_e}{\dot{e}_c} = \frac{X_e}{\dot{X}_c} = \frac{1}{K_{vx}} \tag{8.186}$$

For the velocity servo of Fig. 8.130, the following error is

$$\frac{e_e}{\dot{e}_c} = \frac{V_e}{\dot{V}_c} = \frac{1}{K_{vv}} \tag{8.187}$$

where V_e = output error = $V_c - V$, mm/sec

$\quad\quad V_c$ = velocity command = e_c/K_v, mm/sec

Note that the following error for the position servo is the position error resulting from a steady rate of change of position command. For the velocity servo, the following error is the velocity error resulting from a steady rate of change of velocity command.

The servo errors discussed thus far have been those that can be minimized by a tight servoloop (high loop gain). To these must be added errors in the transducer mechanism. Even if infinite loop gain were achievable, the servo can be no more accurate than the transducer itself. The most important types of transducer inaccuracies are repeatability, resolution, and linearity. Errors due to transducer location and mounting geometry must also be taken into account.

Many of the foregoing concepts can be applied to servos in general. For example, a force or pressure servo working against a spring load is similar to a position servo in the sense that output force is proportional to actuator position. Also, temperature control servos tend to behave like position servos since the controlling device tends to provide heat flow proportional to temperature error, and thermal loads tend to produce temperature rate of change proportional to heat flow.

8.26.2 Dynamic Response and Stability

As discussed in Section 8.26.1, open-loop gain has a strong influence on servo accuracy. High loop gains also provide fast dynamic response in most cases. However, stability considerations will limit the maximum useful loop gain. The dynamic response and stability of a servo are determined by the dynamic characteristics of the various loop components. In many situations, the forward-loop dynamics are dominated by a relatively small number of low-frequency lag elements, and the transducer dynamics are negligible. In these cases, it is often possible to obtain an adequate estimate of servo performance by approximating the combined forward-loop characteristics with an integrator plus a first-order or second-order lag. The adequacy of this approximation can be determined by the match of the frequency response gain and phase for the frequency range in which the phase lag is less than 180°. Using these two rather basic dynamic forms, the relationships among loop gain, stability, and dynamic response are easily seen.

A block diagram using the basic dynamic forms is shown in Fig. 8.134,

where U = generalized controlled variable

$\quad\quad U_c$ = generalized command input

$\quad\quad D$ = generalized disturbance input

$\quad\quad G_d$ = open-loop response of U to D

The first-order lag is typical of simple temperature control systems and dc servos having short electrical time constants. The second-order lag is often representative of electrohydraulic servos and dc servos having long electrical time constants. The closed-loop responses to command inputs are

$$\frac{U_1}{U_c} = \frac{1}{(\tau_1/K_{u1})s^2 + (1/K_{u1})s + 1} \tag{8.188}$$

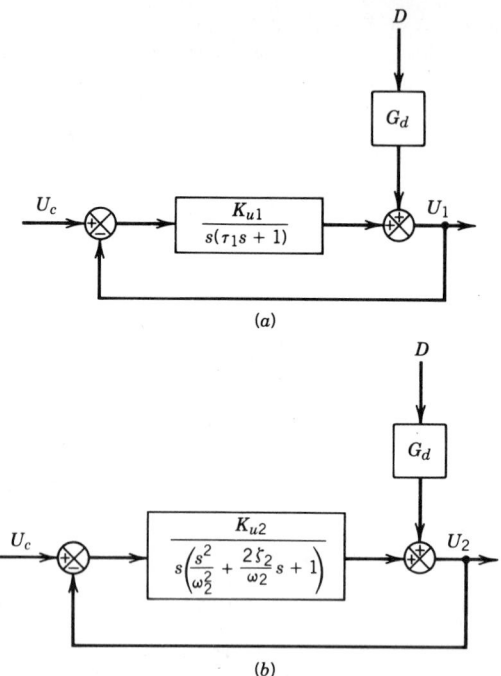

Fig. 8.134 Basic dynamic configurations: (*a*) first-order lag, (*b*) second-order lag.

and

$$\frac{U_2}{U_c} = \frac{1}{s^3/K_{u2}\omega_2^2 + (2\zeta_2/K_{u2}\omega_2)s^2 + (1/K_{u2})s + 1} \tag{8.189}$$

Representative root loci are shown in Fig. 8.135 for both forms. In Fig. 8.135(*a*), the lag combines with the integrator to produce second-order closed-loop poles. The closed-loop natural frequency increases with loop gain, while the damping ratio decreases. In the case of Fig. 8.135(*b*), the closed-loop transfer function consists of a first-order and a second-order lag. The break frequency of the first-order lag increases with loop gain, while the second-order damping ratio rapidly decreases. In both loop closures, there are clearly trade-offs between closed-loop bandwidth and stability.

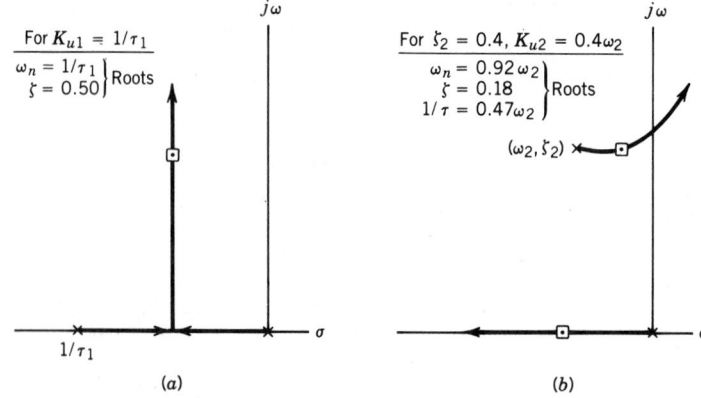

Fig. 8.135 Root loci for basic dynamic configurations: (*a*) first-order lag, (*b*) second-order lag.

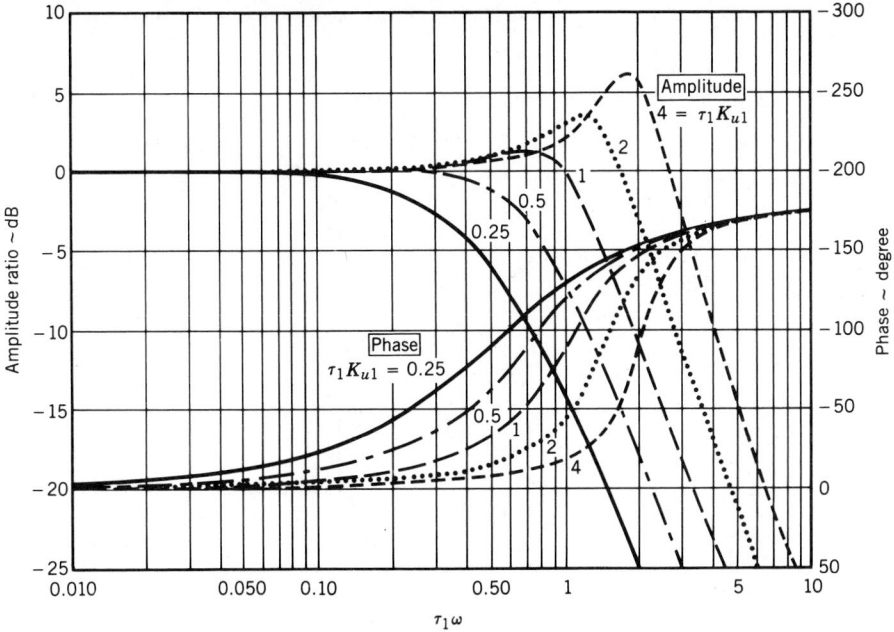

Fig. 8.136   Closed-loop frequency responses for U_1/U_c.

There are numerous methods for quantifying the relationships between bandwidth and stability. Closed-loop frequency responses to command inputs are shown for both basic forms in Figs. 8.136 and 8.137, while Figs. 8.138 and 8.139 present the corresponding step responses. Useful numerical measures of stability are phase margin, gain margin, and damping ratio of the closed-loop complex pair. These are given in Figs. 8.140 and 8.141. Note that the gain margin for Fig. 8.140 is infinite.

Referring to Fig. 8.134, closed-loop responses to disturbance inputs can be written as

$$\frac{U_1}{D} = \frac{G_d}{K_{u1}} \frac{s(\tau_1 s + 1)}{(\tau_1/K_{u1})s^2 + (1/K_{u1})s + 1} \tag{8.190}$$

and

$$\frac{U_2}{D} = \frac{G_d}{K_{u2}} \frac{s[s^2/\omega_2^2 + (2\zeta_2/\omega_2)s + 1]}{s^3/K_{u2}\omega_2^2 + (2\zeta_2/K_{u2}\omega_2)s^2 + (1/K_{u2})s + 1} \tag{8.191}$$

To determine the final dynamic form of these responses, it is necessary to have a transfer function for G_d. This transfer function can be obtained from the physical model of the system, by deriving the response of the controlled variable to the disturbance input with the controller output equal to zero. As an example, consider a dc motor driving an inertial load and having a short electrical time constant. For an integrating velocity loop, the disturbance transfer function has the form

$$G_d = \frac{K_d}{\tau_1 s + 1} \tag{8.192}$$

and

$$\frac{U_1}{D} = \frac{K_d}{K_{u1}} \frac{s}{(\tau_1/K_{u1})s^2 + (1/K_{u1})s + 1} \tag{8.193}$$

For a position loop

$$G_d = \frac{K_d}{s(\tau_1 s + 1)} \tag{8.194}$$

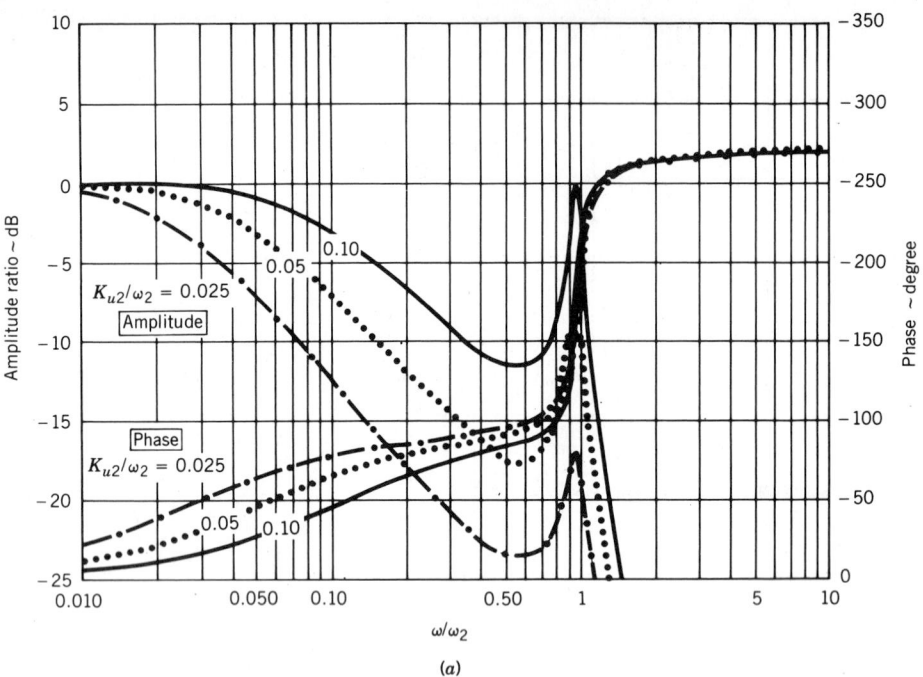

(a)

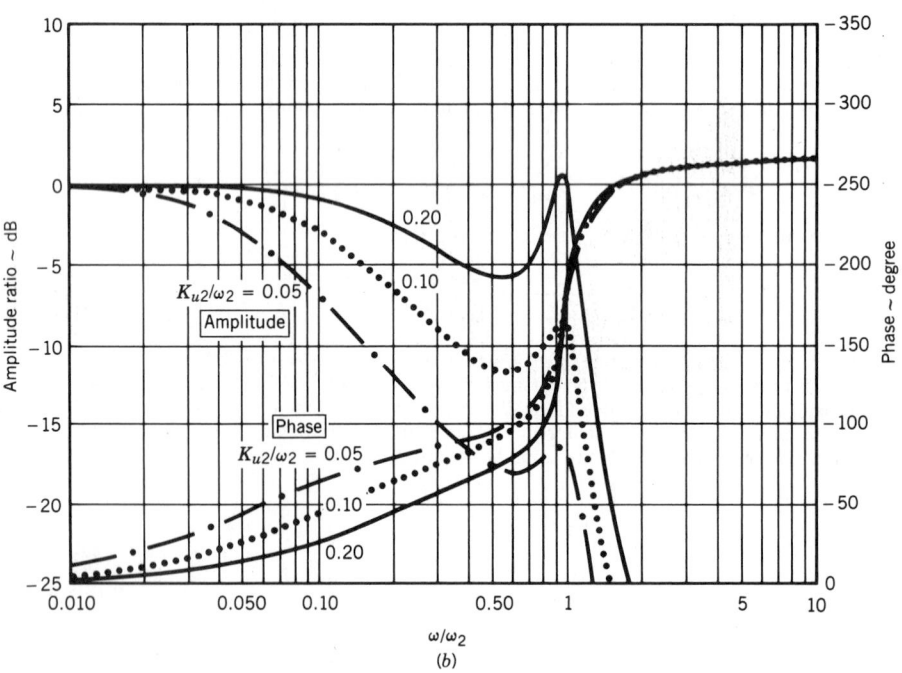

(b)

Fig. 8.137 Closed-loop frequency responses for U_2/U_c: (a) $\zeta_2 = 0.1$, (b) $\zeta_2 = 0.2$, (c) $\zeta_2 = 0.4$, (d) $\zeta_2 = 0.8$.

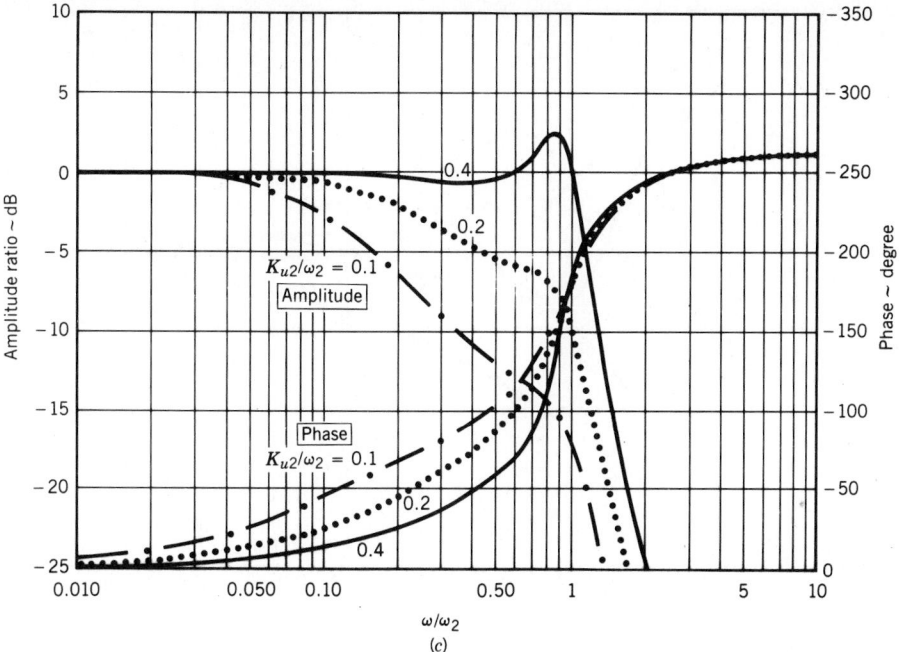

(c)

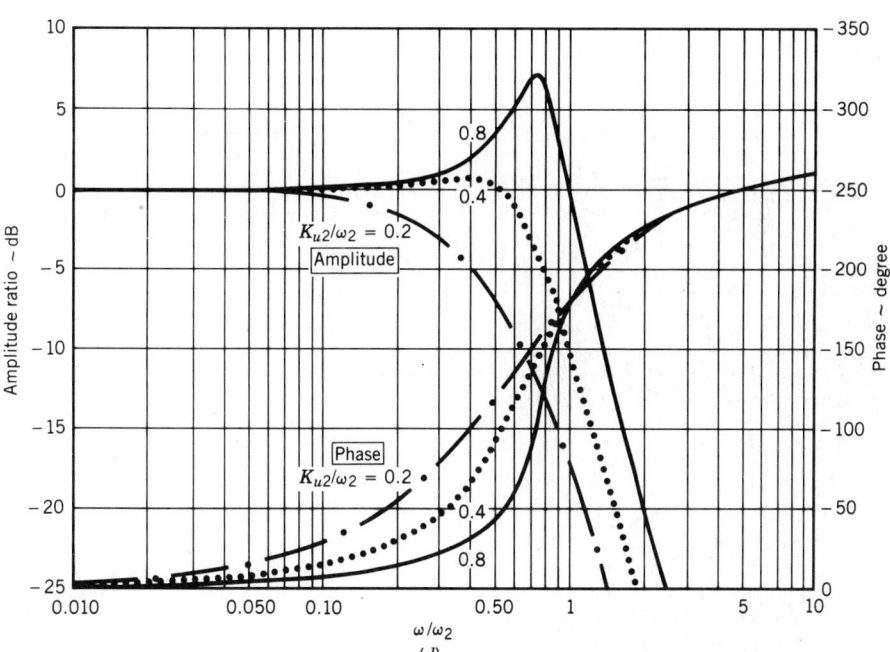

(d)

Fig. 8.137 (*Continued*)

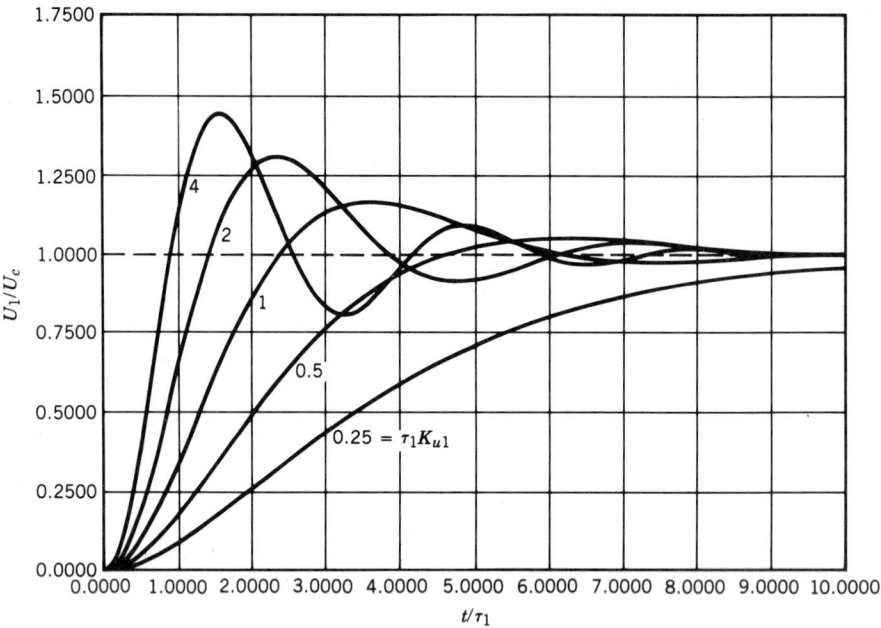

Fig. 8.138   Closed-loop step responses for U_1/U_c.

and

$$\frac{U_1}{D} = \frac{K_d}{K_{u1}} \frac{1}{(\tau_1/K_{u1})s^2 + (1/K_{u1})s + 1} \tag{8.195}$$

It is instructive to compare Eqs. (8.193) and (8.195) with Eqs. (8.182) and (8.184), respectively.

8.27   FREQUENCY COMPENSATION TO IMPROVE OVERALL PERFORMANCE

In Section 8.26, it is clear that open-loop dynamic characteristics impose profound limitations upon closed-loop performance. However, it is often possible to extend these limitations by modifying the inherent open-loop dynamics with frequency compensation (shaping). There are many techniques for designing compensators. The best technique to use is a function of the particular open-loop dynamics under consideration, as well as closed-loop performance goals. The following sections describe some techniques that are useful in various commonly encountered situations.

8.27.1   Well-Damped Systems

As mentioned in Section 8.26.1, an ideal form for the combined forward-loop transfer function is an integrator. This ensures very high gains at low frequencies, a linear gain roll-off with frequency, and only 90° of phase lag. For systems in which the dominant open-loop poles are reasonably well damped, loop gain is usually limited by phase lag. This is clearly illustrated by Figs. 8.140 and 8.141, in which phase margins deteriorate faster than gain margins (except when ζ_2 is low).

An obvious way to improve phase margins is to make the open-loop transfer function look like an integrator out to higher frequencies. This can be accomplished by using a lead compensator, whose zeros are identical to the dominant forward-loop poles. To make the compensator physically realizable, it must have at least as many poles as zeros, but these poles can be placed at higher frequencies. The net effect of such a lead compensator is to move the break frequencies of the forward-loop poles to higher frequencies. For the example of Fig. 8.134(b), the form of the

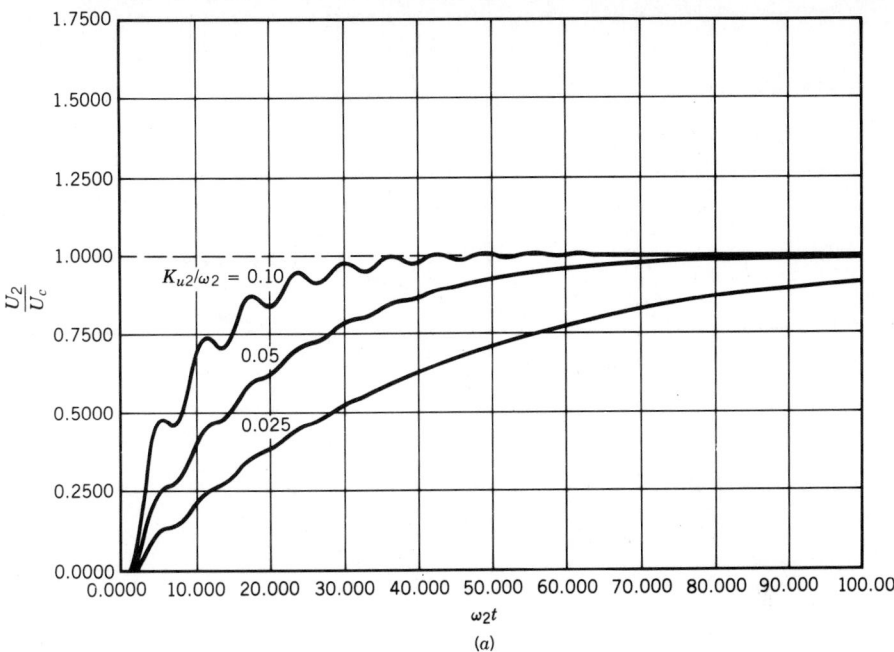

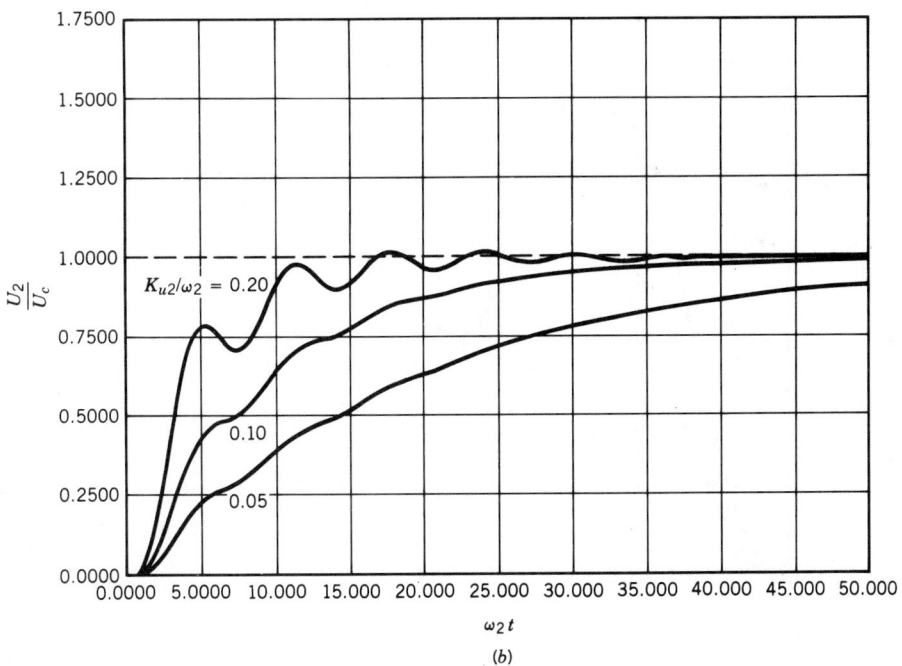

Fig. 8.139   Closed-loop step responses for U_2/U_c: (a) $\zeta_2 = 0.1$, (b) $\zeta_2 = 0.2$.

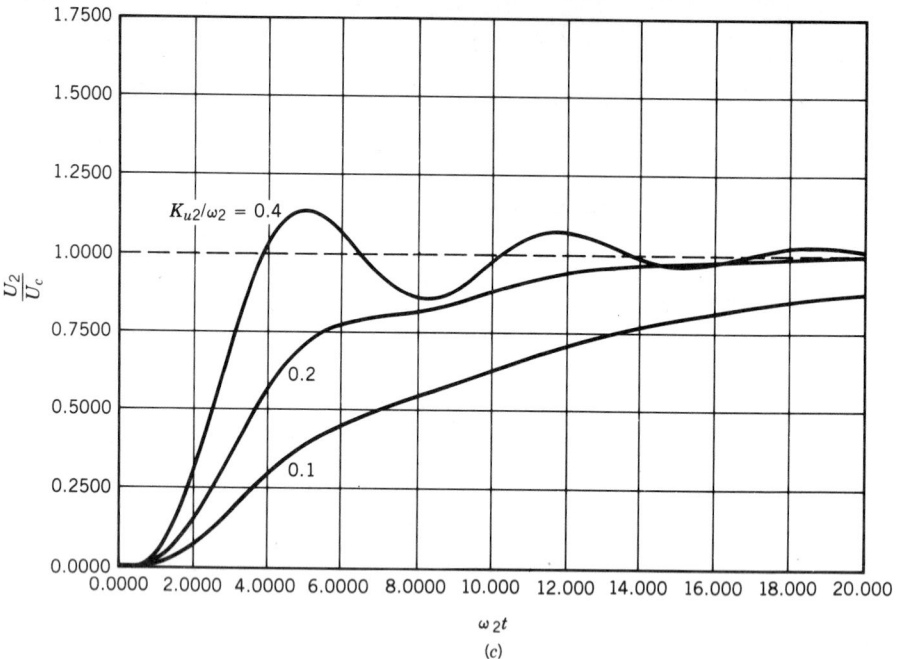

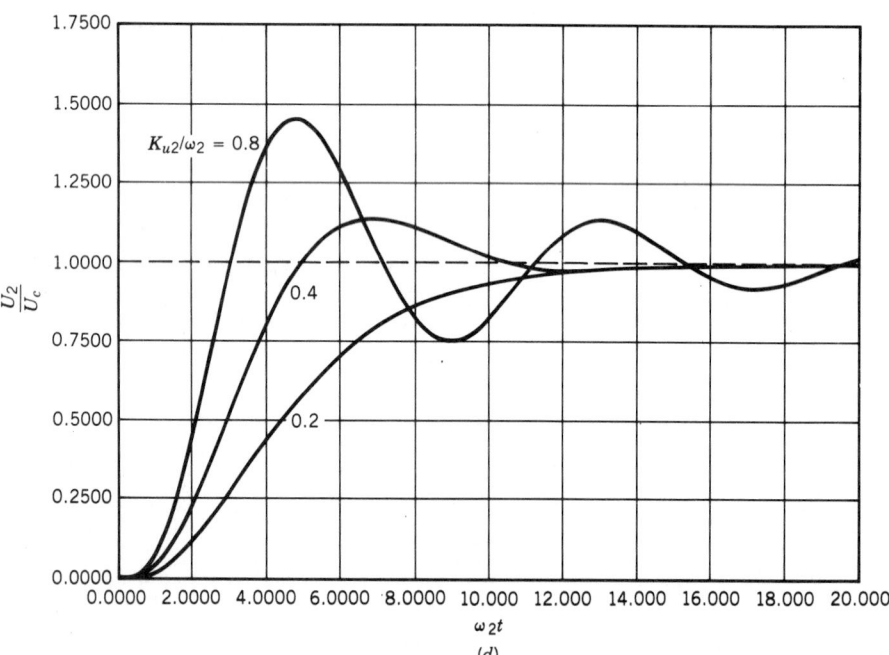

Fig. 8.139 (*Continued*) (*c*) $\zeta_2 = 0.4$, (*d*) $\zeta_2 = 0.8$.

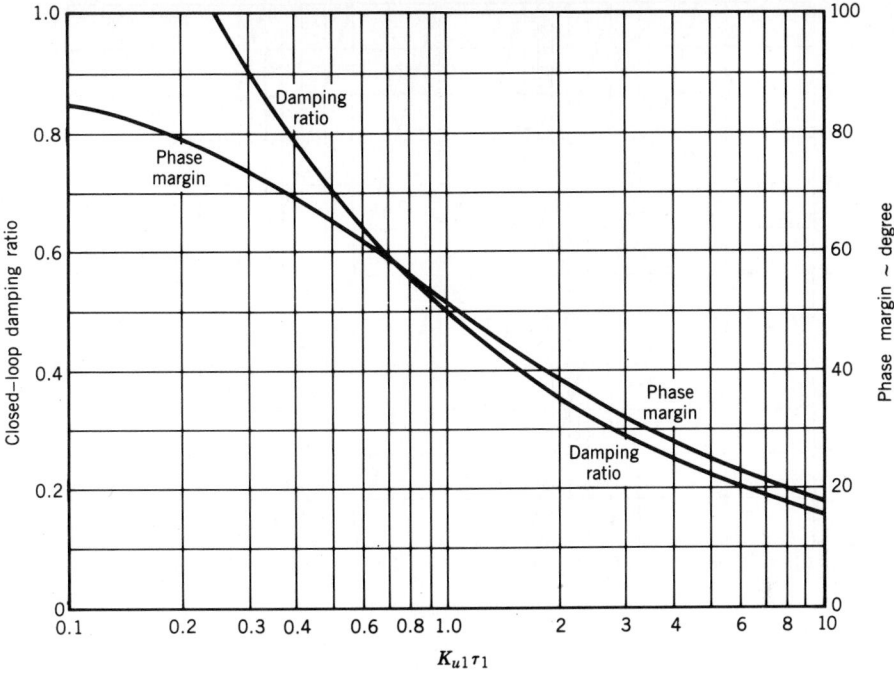

Fig. 8.140 Stability parameters for U_1/U_c.

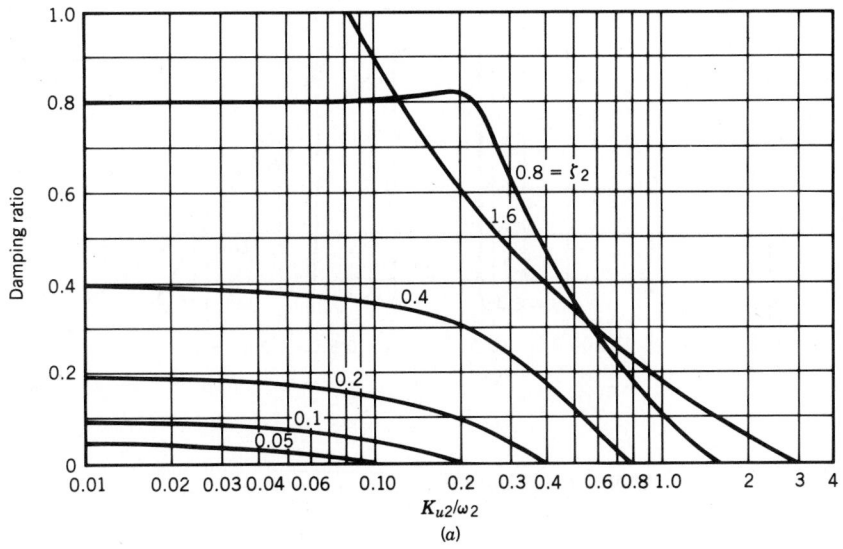

(a)

Fig. 8.141 Stability parameters for U_2/U_c: (a) closed-loop damping ratio.

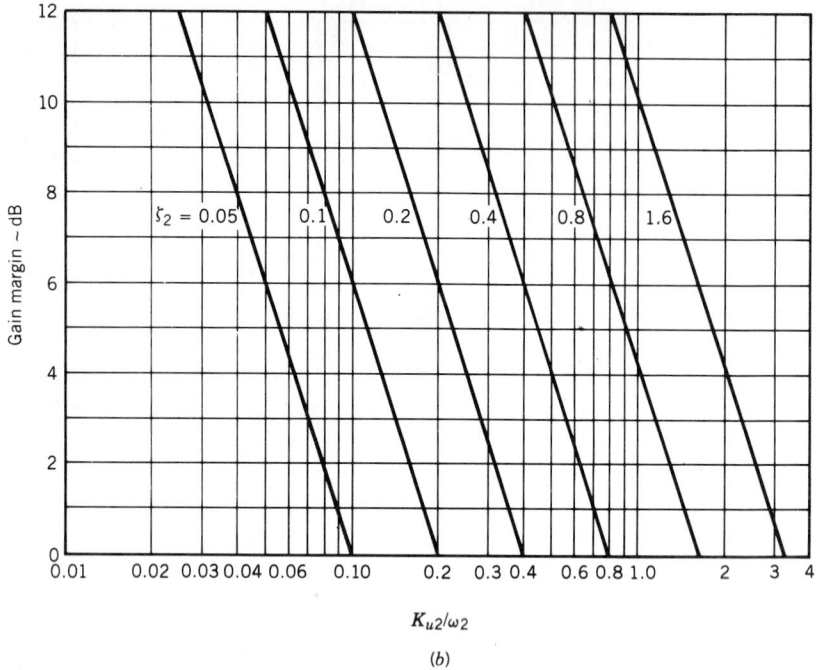

(b)

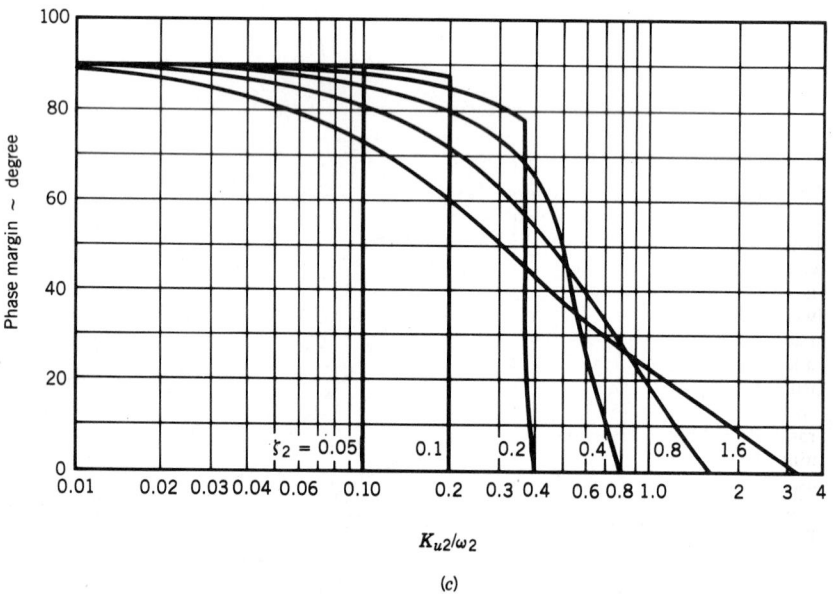

(c)

Fig. 8.141 (*Continued*) (*b*) gain margin, (*c*) phase margin.

compensator would be

$$G_{c1} = \frac{\tau_{cz}s + 1}{\tau_{cp}s + 1} \qquad (8.196)$$

where $\tau_{cz} = \tau_1$. For the example of Fig. 8.134(b), the form would be

$$G_{c2} = \frac{s^2/\omega_{cz}^2 + (2\zeta_{cz}/\omega_{cz})s + 1}{s^2/\omega_{cp}^2 + (2\zeta_{cp}/\omega_{cp})s + 1} \qquad (8.197)$$

where $\omega_{cz} = \omega_2$ and $\zeta_{cz} = \zeta_2$. In both cases, the closed-loop performance is now determined by the compensator poles since the original poles are canceled by the compensator zeros. This augmented performance can be quantified by using Figs. 8.136–8.141, with τ_{cp}, ω_{cp}, ζ_{cp} substituted for τ_1, ω_2, ζ_2, respectively.

There are a number of practical limitations on the use of lead compensation, primarily related to the large high-frequency gain of the compensator itself. For the examples of Eqs. (8.196) and (8.197), the high-frequency gains are τ_{cp}/τ_{cz} and $(\omega_{cp}/\omega_{cz})^2$, respectively. As a minimum, this characteristic will amplify any high-frequency electrical noise in the system. With reasonable care in the electrical design, high-frequency compensator gains of 10 or more are often practical. For a first-order compensator, this means that the forward-loop break frequencies can be boosted by a factor of 10, while the boost is only the square root of this factor for a second-order compensator. Another problem associated with the gain boost of a lead compensator is that poorly damped high-frequency modes can be excited or even destabilized. This latter effect will be further discussed in Section 8.27.3. The practicality of lead compensation in any given application can be best determined experimentally (additional high-frequency lags are sometimes needed).

The high-frequency noise situation is improved considerably for systems in which the integrator is electronic (e.g., the velocity servo described in Section 8.26.1). In this case, the integrator can replace one of the compensator poles so that Eqs. (8.196) and (8.197) are replaced by

$$G_{c3} = \frac{\tau_{z3}s + 1}{s} \qquad (8.198)$$

and

$$G_{c4} = \frac{s^2/\omega_{z4}^2 + (2\zeta_{z4}/\omega_{z4})s + 1}{s(\tau_{p4}s + 1)} \qquad (8.199)$$

Because of the noise-attenuating effect of the electronic integrator, $1/\tau_{p4}$ can often be set a factor of 10 greater than ω_{z4}. As previously discussed, the ratio of ω_{cp}/ω_{cz} in Eq. (8.197) is often limited to $\sqrt{10}$. As shown in the frequency responses of Figs. 8.142 and 8.143, the open-loop characteristics of a lead-compensated system will exhibit substantially improved phase characteristics when the integration is electronic rather than inherent. This makes it possible to greatly improve the closed-loop bandwidth of the system for a given level of stability.

In some cases, the open-loop dynamics may be dominated by a low-frequency lead term rather than a lag. When this occurs, a canceling lag compensator can often prevent instabilities due to poorly damped high-frequency modes. Lag compensation is normally well-behaved and suffers none of the noise problems that limit the use of lead compensation.

Transfer functions describing the system open-loop characteristics are not always available. Sometimes the only available system description is an experimental frequency response. When this is the case, a compensator can often be designed by graphically subtracting the experimental amplitude ratio (in decibels) and phase from those of an integrator having the same low-frequency gain. This "ideal" compensator can then be approximated with an appropriate transfer function.

8.27.2 Poorly Damped Systems

Section 8.27.1 discussed the benefits of lead compensation whose zeros are identical to the dominant forward-loop poles of the system. Theoretically, this technique can be used for any forward-loop transfer function. However, when the dominant forward-loop poles are second order and poorly damaged, practical considerations render the technique highly risky in many cases. The basic problem is that the amplitude and phase of a poorly damped pair of poles change very rapidly with frequency in the vicinity of the resonant peak. If the compensator zeros are not precisely matched to the poles,

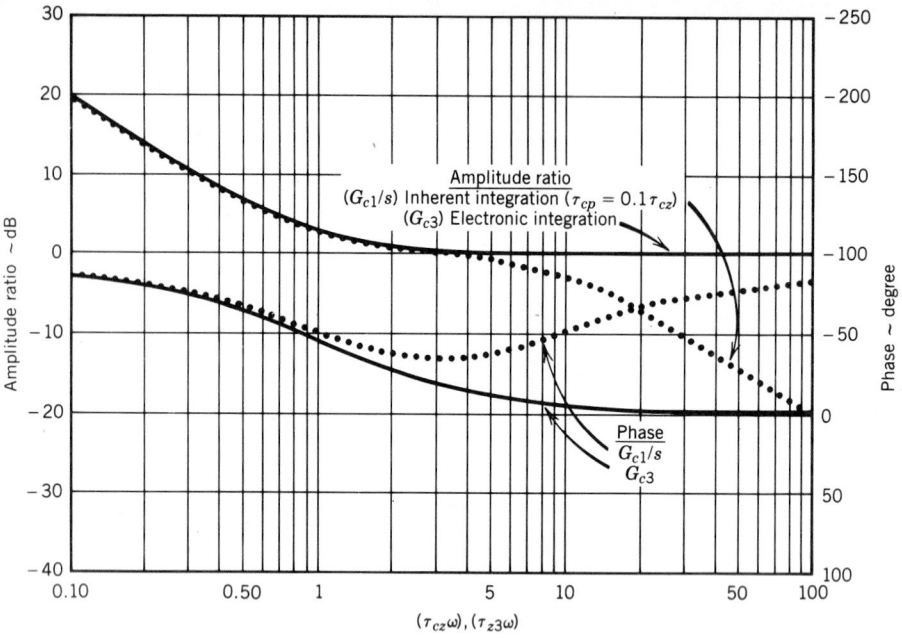

Fig. 8.142   First-order lead compensation.

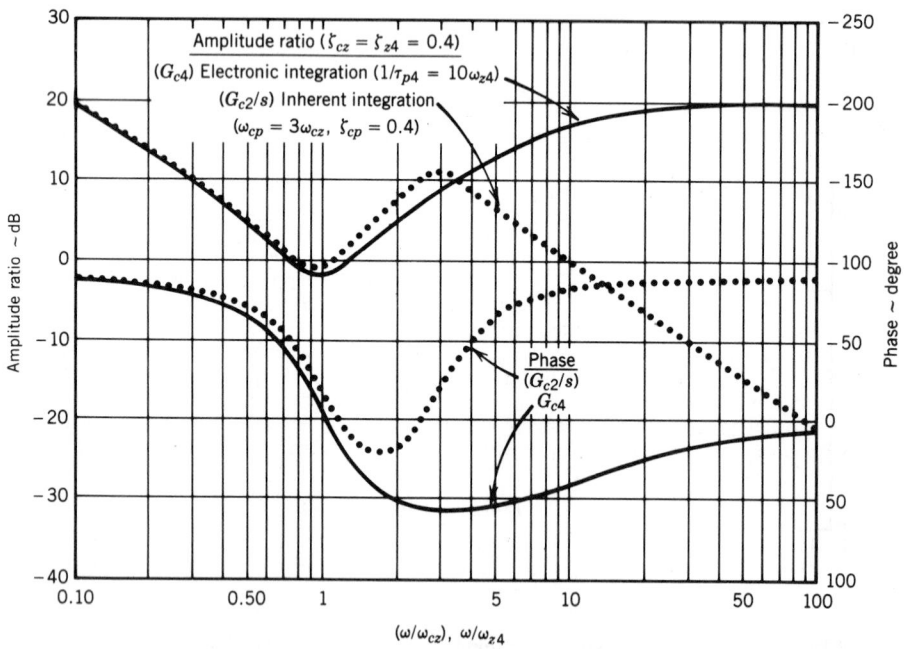

Fig. 8.143   Second-order lead compensation.

the combined forward-loop transfer function can easily exhibit 180° of phase lag in the vicinity of substantial local peaking.

To illustrate the potential stability problem, consider the system of Fig. 8.134(b), with $\zeta_2 = 0.10$. Suppose that the lead compensator of Eq. (8.197) is added with $\omega_{cz} = \omega_2$, $\zeta_{cz} = 0.10$, $\omega_{cp} = 3\omega_{cz}$, and $\zeta_{cp} = 0.80$. Theoretically, the forward loop is now dominated by the integrator and the compensator poles. Referring to Fig. 8.137(d), it can be seen that a well-behaved response can be obtained with a loop gain $K_{u2} = 1.2\omega_{cz}$. However, suppose that ω_2 shifts to a lower value. For example, with an electrohydraulic servoactuator driving an inertial load, the "hydraulic resonance" can change 50% or more over the stroke range of the cylinder. Figure 8.144 shows how the closed-loop roots and the open-loop frequency response change with variations in ω_2.

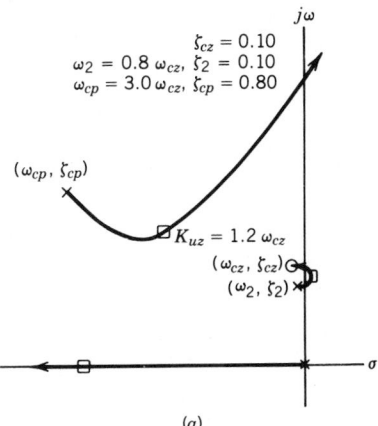

(a)

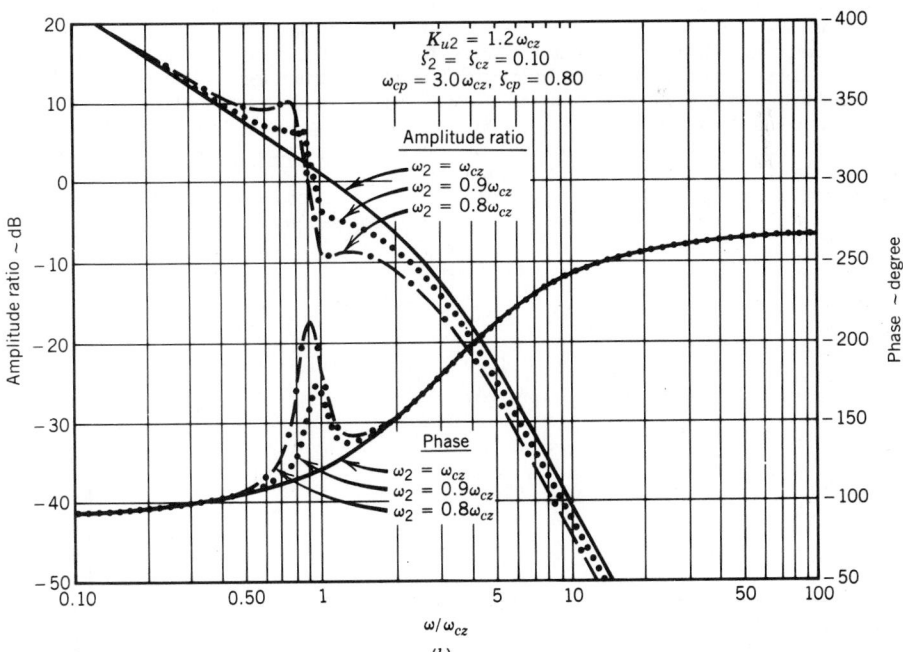

(b)

Fig. 8.144 Effects of variations in forward-loop poles on stability of a lead compensated loop: (a) root locus, (b) open-loop frequency response.

Note that a reduction of ω_2 to $0.89\omega_{cz}$ will cause the closed-loop system to become unstable (0 dB at $-180°$ phase). Even if the natural frequency of the forward-loop poles does not change at all, the poles remain poorly damped in the closed-loop transfer function. Although they are masked by the compensator zeros with regard to command inputs, they may be excited by disturbance inputs to the system. The same general comments apply to the use of notch filters. The notch is intended to attenuate the resonant peak, without the bandwidth boost of the lead compensator described in the present example ($\omega_{cp} = \omega_{cz}$, instead of $3\omega_{cz}$).

However, it is possible to design a compensator that will improve the damping of these poles. To accomplish this, it is useful to recall that for any system in which the number of poles exceeds the number of zeros by two or more, the sum of the real parts of all the poles is not changed when the loop is closed.[1] In this case, closing the loop will cause some roots to become more stable and others to become less stable (usually those that were poorly damped to begin with). On the other hand, if the number of poles exceeds the number of zeros by one or less, it is possible for the loop closure to improve the stability of all the roots. Therefore, to be effective in damping a poorly damped dynamic mode, lead compensation of sufficiently high order is required. To illustrate this concept, again consider the system of Fig. 8.134(b), with $\zeta_2 = 0.1$. Even if we consider ideal lead compensators having no poles, Fig. 8.145 makes it clear that damping cannot be improved unless the order of the lead is 2 or more.

As mentioned in Section 8.27.1, there are important practical constraints on the use of second-order lead compensation. If the forward-loop integrator is electronic in nature, making it practical to achieve a ratio of compensator pole-zero break frequencies on the order of 10, the improvements in damping indicated by Fig. 8.145(c) are indeed possible. Such a compensator is defined by Eq. (8.199), with $\omega_{z4} = \omega_2$, $\zeta_{z4} = 0.4$, and $1/\tau_{p4} = 10\omega_{z4}$. The resulting root locus, shown in Fig. 8.146(a), indicates that open-loop gain on the order of $8\omega_2$, is possible with good stability (gain margin = 6 dB and phase margin = 50°). To achieve the same gain margin without compensation, the open-loop gain would be limited to $0.1\omega_2$. Closed-loop frequency responses to commands for both cases are given in Fig. 8.146(b).

If the forward-loop integration is inherent rather than electronic, the ratio of compensator pole-zero break frequencies may be limited to values as low as 3 (Section 8.27.1). Using the compensator of Eq. (8.197), a good compromise is $\omega_{cz} = \omega_2$, $\zeta_{cz} = 0.4$, $\omega_{cp} = 3\omega_{cz}$, and $\zeta_{cp} = 0.4$. The resulting impact on performance is shown in Fig. 8.147. To maintain a 6-dB gain margin, the open-loop gain must be reduced to $1.0\omega_2$. Even with this reduced gain, the phase margin is only 30°. If the compensator is located in the forward loop, the closed-loop response to commands exhibits considerable peaking as shown in Fig. 8.147(b). This figure also illustrates that if the compensator is located in the feedback loop, the compensator zeros do not appear in the closed-loop response to commands, and the response exhibits less peaking with more phase lag.

A more straightforward method for improving the performance of poorly damped systems is by the use of lag compensation. A first-order low-pass filter in the forward loop will slow the degradation of closed-loop damping ratio as loop gain is increased. In addition, the lag compensator will attenuate abrupt command or disturbance inputs to the system, thereby reducing their ability to excite the poorly damped mode. Figure 8.148 shows the effects of a lag compensator optimized for the system of Fig. 8.134 with $\zeta_2 = 0.10$. The effects of an optimized lag compensator on various stability parameters are illustrated in Fig. 8.149 for several values of ζ_2. Comparisons with Fig. 8.141 show that lag compensation can provide a substantial improvement in loop gain for low ζ_2 but is of little use for

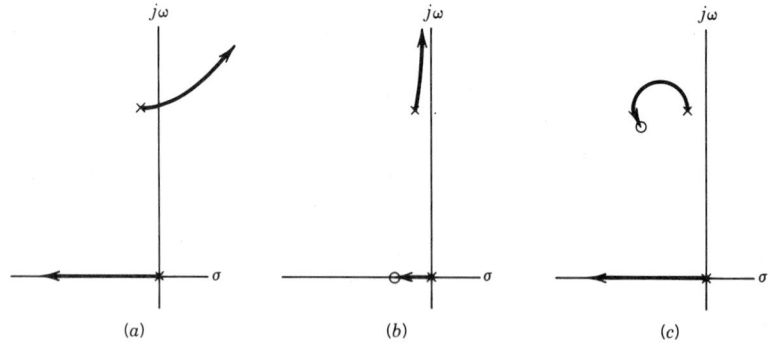

(a) (b) (c)

Fig. 8.145 Effect of lead compensation order on closed-loop roots: (a) no compensation, (b) first-order lead, (c) second-order lead.

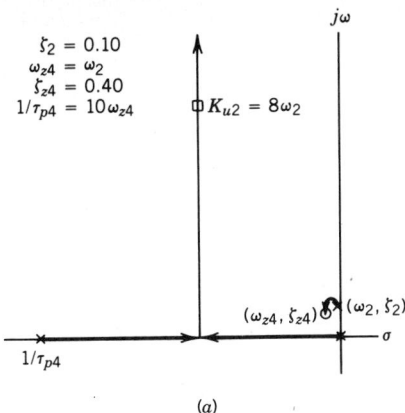

(a)

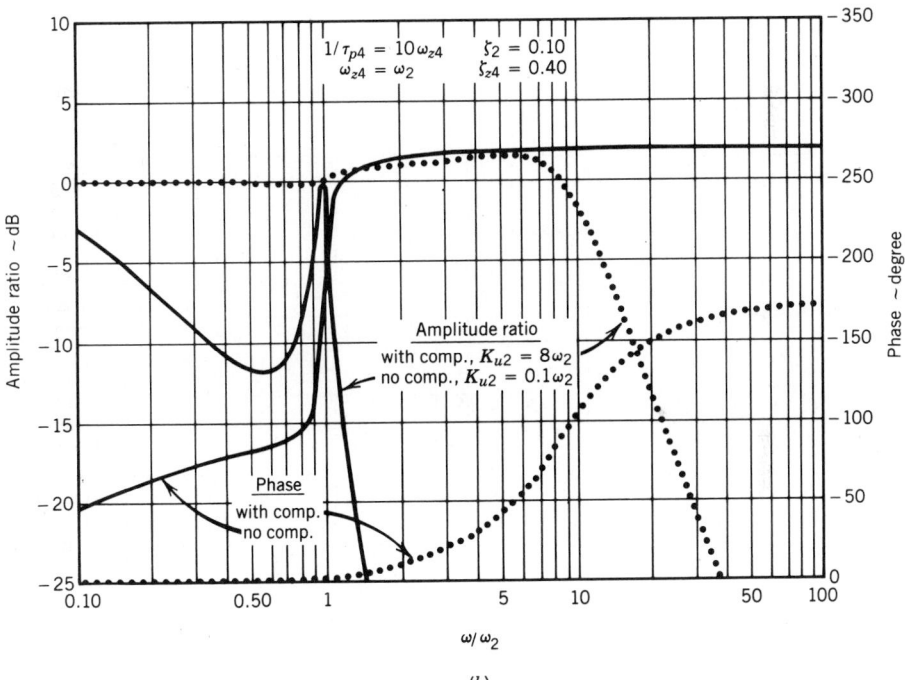

(b)

Fig. 8.146 Effects of second-order lead compensation on a poorly damped system (electronic integration): (*a*) root locus, (*b*) closed-loop frequency response.

$\zeta_2 > 0.3$. For $\zeta_2 = 0.10$, lag compensation allows K_{u2} to be increased from 0.10 to 0.27 sec^{-1} with comparable levels of stability (gain margin = 6 dB). It should be noted that very little advantage is provided by the use of a higher-order low-pass filter instead of the first-order type.

In summary, the use of lead compensation to improve the damping of a poorly damped system is not straightforward, and the end result may be a marginal improvement in closed-loop performance. Lag compensation is more straightforward but offers only a modest improvement in performance. If substantial improvement in system damping and performance is required, the use of inner feedback loops is generally more effective, as explained in Section 8.28.

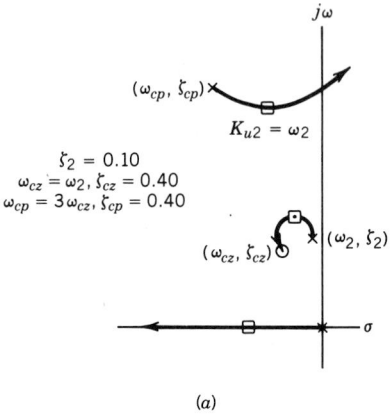

(a)

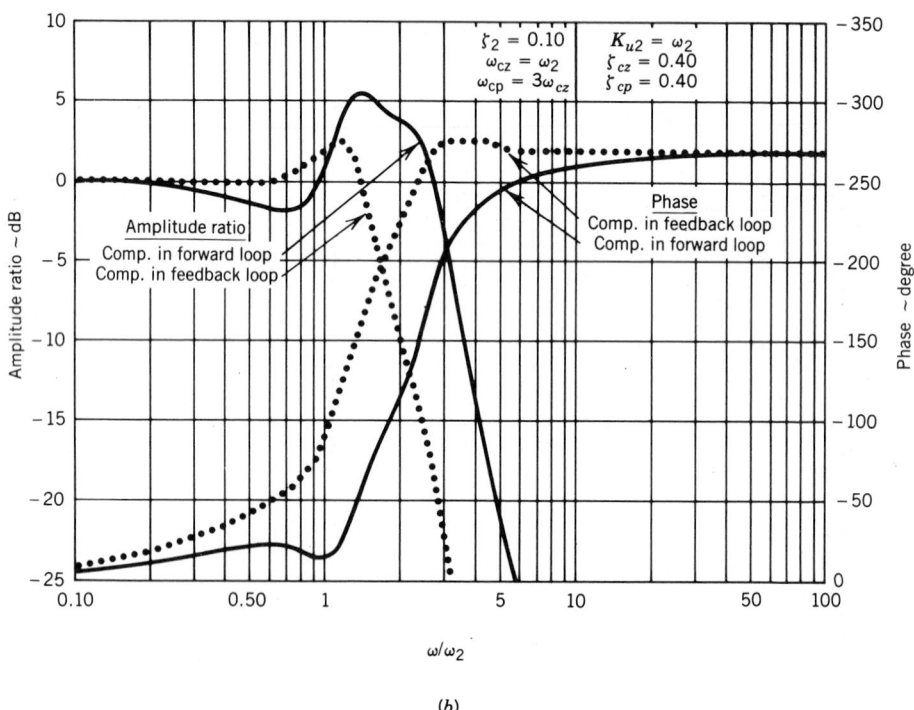

(b)

Fig. 8.147 Effects of second-order lead compensation on a poorly damped system (inherent integration): (a) root locus, (b) closed-loop frequency response.

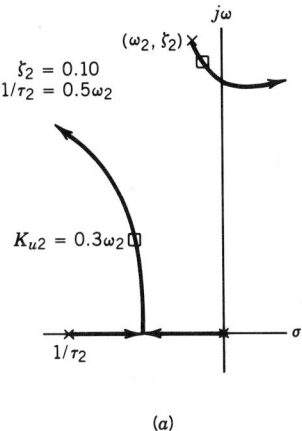

(a)

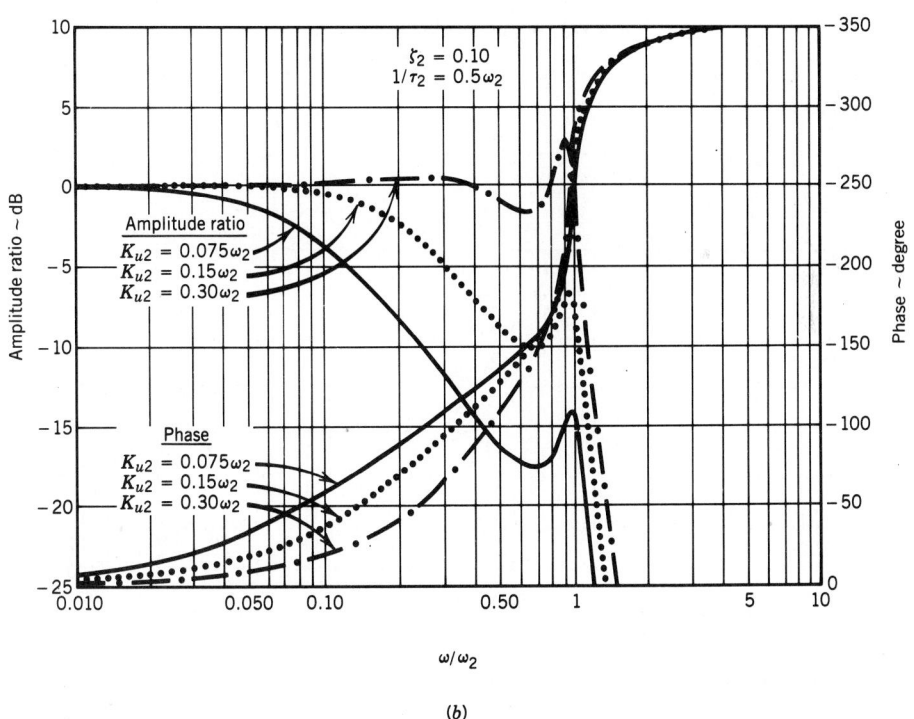

(b)

Fig. 8.148 Effects of a simple lag compensator on the closed-loop response of a poorly damped system: (a) root locus, (b) frequency response.

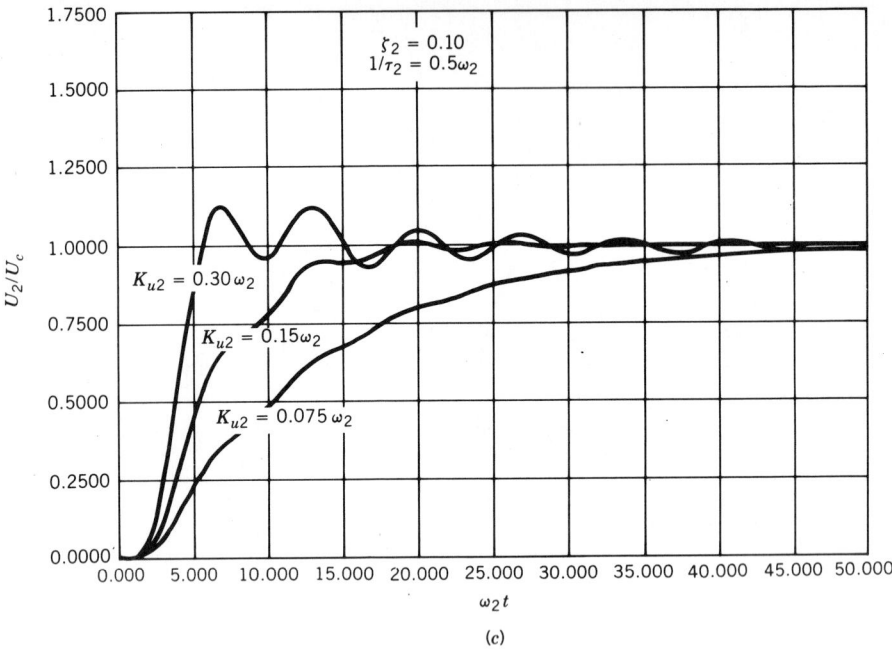

Fig. 8.148   (*Continued*) (*c*) step response.

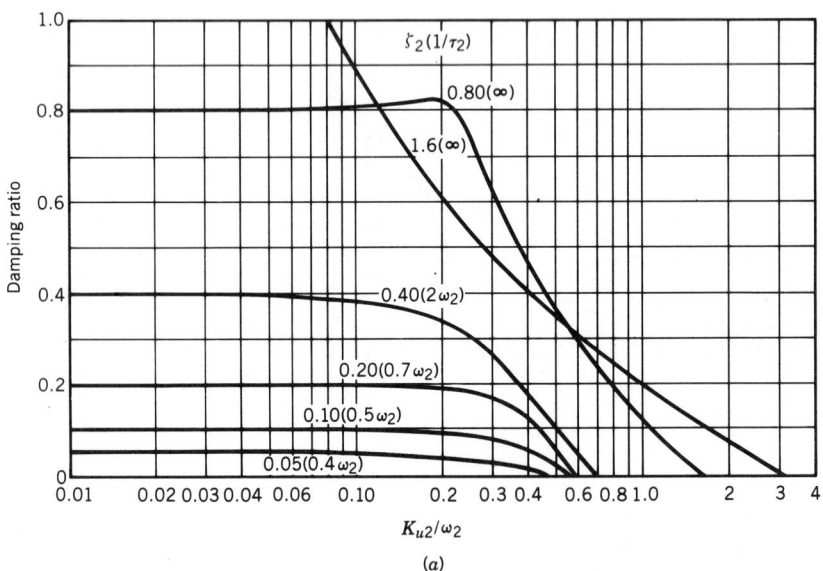

Fig. 8.149   Effects of a simple lag compensator on the stability parameters of a poorly damped system: (*a*) closed-loop damping ratio, (*b*) gain margin, (*c*) phase margin.

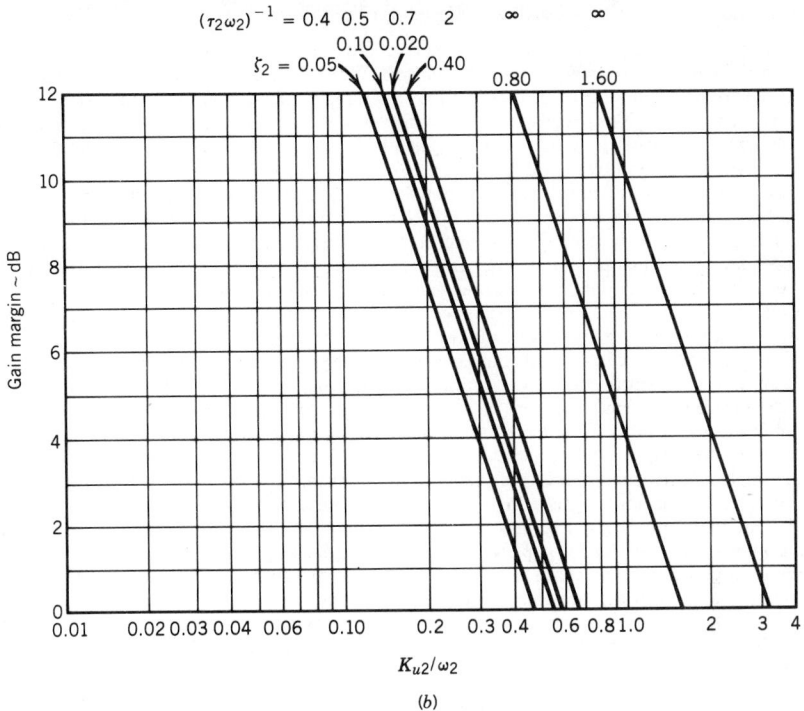

(b)

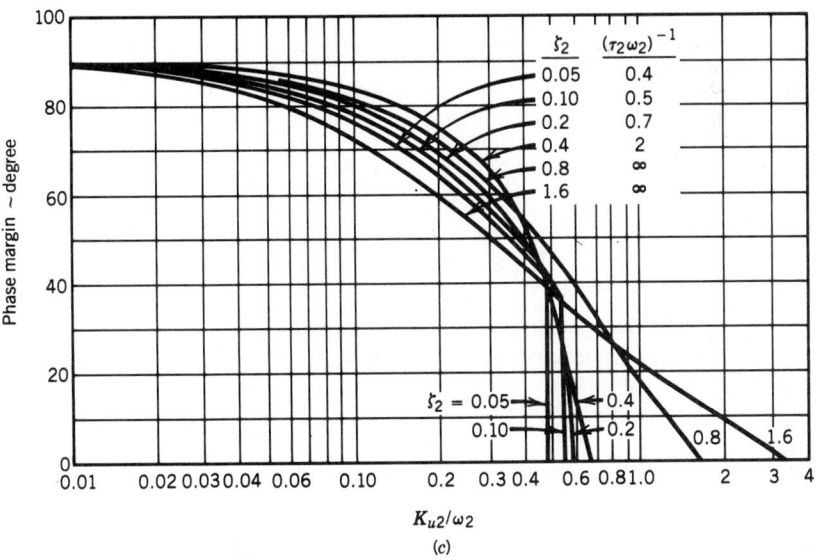

(c)

Fig. 8.149   (*Continued*)

8.27.3 Higher-Order Effects

In the foregoing sections, the discussion has centered around systems whose forward-loop dynamics can be approximated by relatively simple lag elements. While this approach is often entirely adequate for controller design and performance estimation, the designer should be aware of several potential limitations. These limitations usually involve the higher-order, higher-frequency dynamic modes that were unknown or neglected in the early stages of the design. Examples are structural modes, transducer dynamics, and drive amplifier dynamics.

Higher-order modes associated with actuator mounting structure, actuator-load mechanical connections, and transducer mounting are usually second order and poorly damped. Often these modes are characterized as poorly damped pole-zero combinations. In either case, loop gains selected to obtain adequate performance and stability from the dominant low-frequency modes could be high enough to drive the poorly damped high-frequency modes unstable. This is particularly likely if lead compensation is used to improve the dynamics of the dominant modes. High-frequency stability is probably best assessed by a root locus plot or Bode plot that includes estimates of all the system dynamic modes. Often, accurate description of the higher-order modes is difficult at the design stage. For this reason, it is worthwhile investigating the effects of various high-frequency compensation techniques that can be added when the system is first evaluated experimentally. For example, notch filters are often useful in reducing the effects of high-frequency structural modes at the cost of some low-frequency phase lag.

Many higher-order modes are reasonably well damped. The dynamics of drive amplifiers, transducer signal conditioners, and notch filters are typical examples. The primary influence of such modes is that they introduce phase lags that can destabilize the low-frequency dominant modes or limit the effectiveness of lead compensation used to improve low-frequency behavior. A useful approximation of these effects can be made if there is reasonable frequency separation of the higher-order modes from the dominant lower-frequency modes (a factor of 5 or more). In this case, the system's low-frequency dynamic behavior can usually be assessed by replacing the high-frequency modes with a single first-order lag. The time constant of this first-order approximation, τ_3, can be determined as follows:

$$\tau_3 = \frac{\phi_3}{57.3\omega_3} \tag{8.200}$$

where ϕ_3 is the net phase lag (degrees) of all the combined high-frequency modes, measured at ω_3 (radians per second). The frequency ω_3 should be approximately equal to the highest natural frequency of the lower-frequency dominant modes. It is probably best not to use this simplification if ϕ_3 approaches $30°$.

If frequency-response data are available, they can be used to estimate ϕ_3. However, at the design stage, it is likely that rough estimates of higher-order natural frequencies and damping ratios are the only information available. In this case, ϕ_3 can be determined by adding the phase lag contributions of the individual high-frequency modes:

$$\phi_3 = \sum \phi_4 + \sum \phi_5 \tag{8.201}$$

$$\phi_4 = 57.3\tau_4\omega_3 \tag{8.202}$$

$$\phi_5 = 115\zeta_5\frac{\omega_3}{\omega_5} \tag{8.203}$$

where τ_4 is the time constant of a first-order mode, while ω_5 and ζ_5 are the natural frequency and damping ratio of a second-order mode (ϕ_4 and ϕ_5 due to lag terms add to ϕ_3, while lead terms subtract from ϕ_3). The approximations of Eqs. (8.202) and (8.203) are accurate to one degree or better for $1/\tau_4 > 3\omega_3$ and $\omega_5 > 3\omega_3$, respectively.

8.28 INNER FEEDBACK LOOPS

If the desired servo performance cannot be achieved using frequency compensation, as described in Section 8.27, the addition of inner feedback loops can often provide the needed improvement. This is particularly true when the open-loop damping ratio is poor and the forward loop has an inherent integration. Although inner feedback loops require additional transducers, they offer more flexibility in modifying the servo dynamics than does frequency compensation alone. The following sections discuss the merits of feeding back derivatives of the controlled variable, feeding back variables dynamically different than the controlled variable, and nonelectronic mechanizations of inner loops.

8.28.1 Derivatives of the Controlled Variable

Section 8.27 illustrates the benefits of lead compensation but also shows that its effectiveness is limited by the high-frequency gain amplification of the compensator itself. This problem can be alleviated by the use of transducers that directly measure derivatives of the controlled variable. To illustrate the potential benefits, consider again the example of Fig. 8.134(b) with additional feedback of the first and second derivatives of the controlled variable, as illustrated in Fig. 8.150(a). As shown in Fig. 8.150(b), the three feedbacks can be mathematically combined into a single loop having a pure second-order pair of zeros, without the added lag normally associated with lead compensators. The result is that much higher forward-loop gains can be achieved for a given level of closed-loop stability. Of course, each transducer will introduce higher-frequency dynamic effects that will eventually limit the maximum forward-loop gain, but these effects are usually less restrictive than those imposed by electronic lead compensators. Note that the natural frequency and damping ratio of the feedback

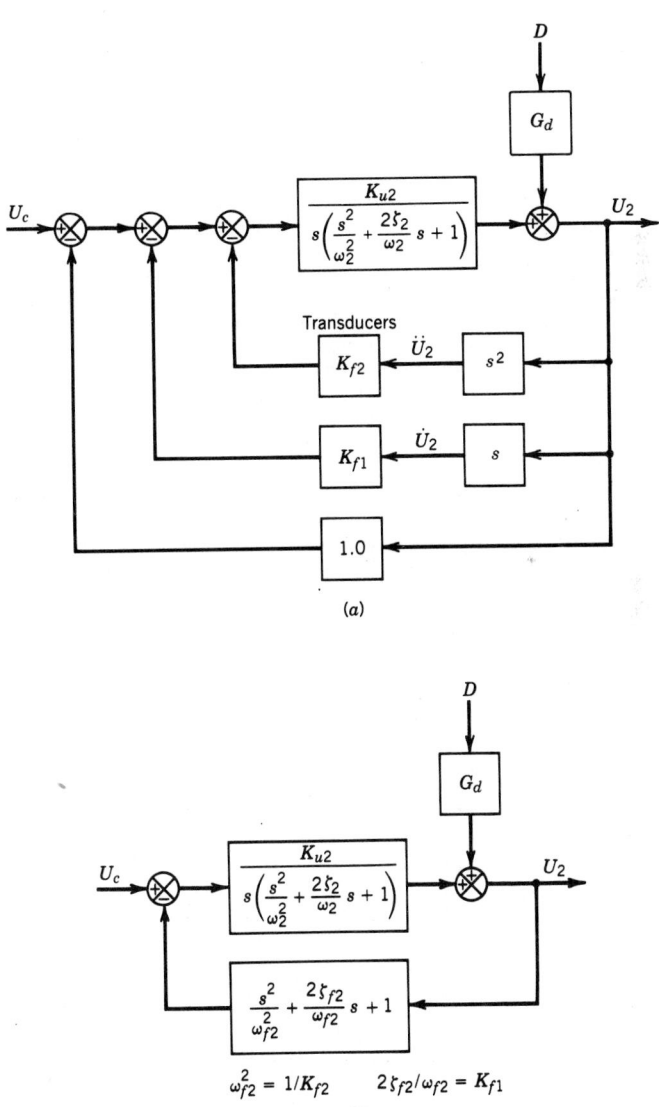

Fig. 8.150 Feedback of controlled-variable derivatives: (a) inner loops to improve damping, (b) combined feedback loops.

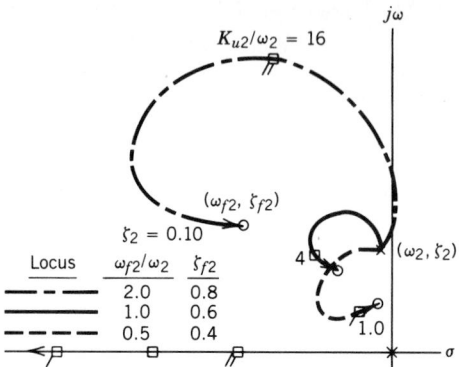

Fig. 8.151 Effects of derivative feedback on closed-loop roots.

zeros in Fig. 8.150 are determined by the magnitude of the derivative feedback gains relative to the primary feedback gain, which is 1.0 in this case.

There are a variety of uses for derivative feedback loops, including improved closed-loop accuracy, bandwidth, and stability, as well as reduced sensitivity to changes in system parameters. These uses can be illustrated with the aid of Figs. 8.151 and 8.152, which show the effects of various sets of feedback zeros on the closed-loop dynamics of Fig. 8.150(b). Figure 8.151 gives root loci for the different zero locations. Because electrical noise and excitation of high-frequency dynamic modes usually limit the gain in the highest derivative loop, a forward-loop gain is selected for each locus that holds ($K_{u2}K_{f2}$) constant. Closed-loop frequency responses are then given in Fig. 8.152. It can be seen that closed-loop damping is improved in all cases. Placement of the feedback zeros at high frequency (low derivative feedback gains) yields high closed-loop bandwidth, but the second-order poles can easily be destabilized by a reduction in forward-loop gain or the presence of the inevitable

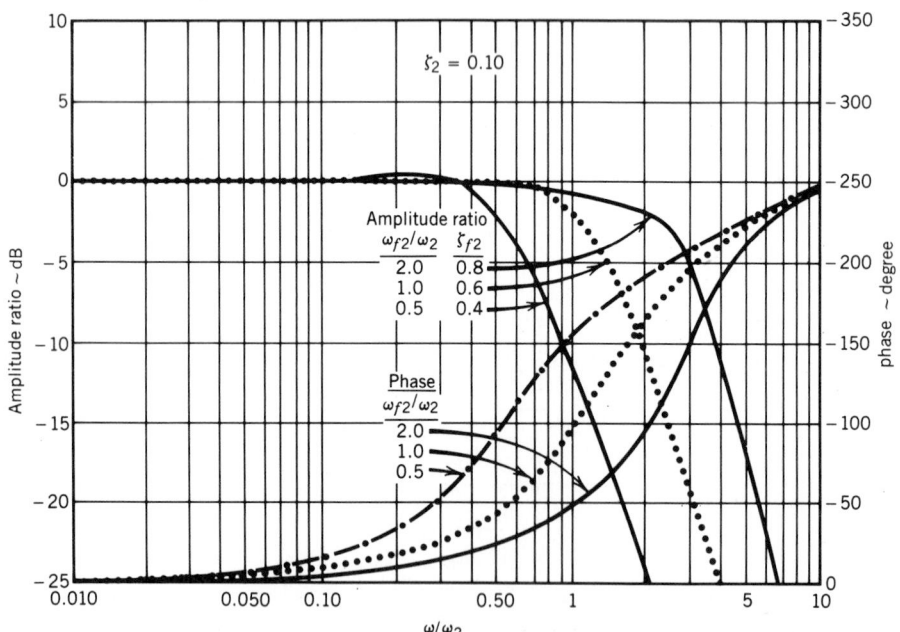

Fig. 8.152 Effects of derivative feedback on closed-loop frequency response of U_2/U_c.

higher-order modes described in Section 8.27.3. On the other hand, low-frequency feedback zeros (high derivative feedback gains) result in lower closed-loop bandwidth but offer greatly reduced gain sensitivity and are little affected by high-frequency modes. Also in the latter case, the closed-loop response is dominated by the low-frequency second-order poles, which are nearly the same as the feedback zeros. Therefore, the closed-loop characteristics do not vary significantly with large changes in the plant parameters $(K_{u2}, \omega_2, \zeta_2)$. A good compromise between bandwidth and parameter sensitivity is to set ω_{f2} approximately equal to ω_2.

Comparison of Figs. 8.151 and 8.152 with Figs. 8.146 and 8.147 show that derivative feedback offers better stability and more flexibility than lead compensation, particularly when the forward loop has an inherent integration. In addition, higher forward-loop gains can generally be used with derivative feedback, which improves the static accuracy of the system. However, forward-loop lead compensation can often produce higher closed-loop bandwidth if the forward-loop integration is electronic in nature, as illustrated in Fig. 8.146.

A useful way to optimize the system of Fig. 8.150(a) is to first establish rough estimates of the feedback and forward-loop gains using the combined feedback approach of Figs. 8.150(b) and 8.151. It is often helpful to include a first-order approximation of the phase lags caused by the higher-order modes [Eq. (8.200)]. Once the approximate gains are established, the closed-loop response and stability can be checked by analysis of a complete multiloop model, with the higher-order modes described more completely and placed in the appropriate loops.

If the forward-loop integrator is electronic in nature, the benefits of derivative feedback can be achieved with one less derivative than if the integrator is inherent. This is accomplished by feeding back the inner loops downstream of the integrator, as illustrated in Fig. 8.153. The static and dynamic characteristics of Fig. 8.153 are entirely equivalent to those of Fig. 8.150(a) and can also be mathematically reduced to the single-loop configuration of Fig. 8.150(b). Note, however, that the practical implementation of Fig. 8.153 requires one less transducer than Fig. 8.150(a).

Another way in which derivative feedback can be useful is to provide very smooth and repeatable dynamic response and high static accuracy when the primary control loop has an inherent integration. This is particularly useful when closed-loop bandwidth is not a major concern. The technique involves closing a tight integrating loop around the first derivative of the controlled variable, as illustrated in Fig. 8.154. This loop submerges the effects of forward-loop gain variations, static offsets, and external disturbances. The primary control loop gain can then be set at relatively low levels to ensure smooth, repeatable dynamic response without the usual concerns about reduced static accuracy. This technique is also useful if the mounting arrangement of the primary transducer results in gain-limiting higher-order dynamic characteristics in the outer feedback loop.

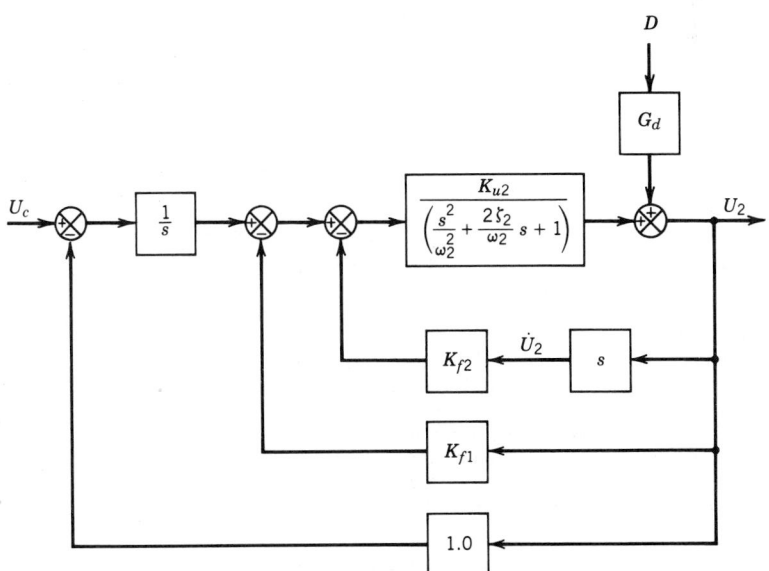

Fig. 8.153 Rearrangement of Fig. 8.150(a) inner loops (electronic integration).

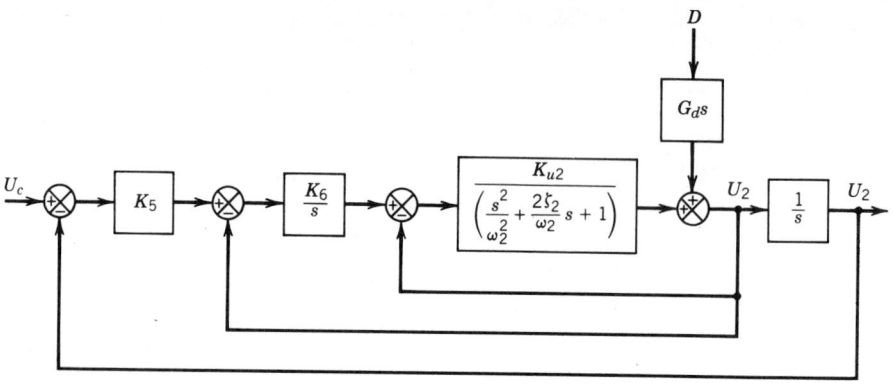

Fig. 8.154   Rearrangement of Fig. 8.134 to give smooth, repeatable, and accurate response (inherent integration).

8.28.2   Alternative Inner-Loop Variables

Sometimes it is advantageous to close inner feedback loops around variables that are dynamically different than derivatives of the primary controlled variable. This might be done because of practical problems related to accurately transducing derivatives of the controlled variable or because it offers an inherent advantage related to feedback dynamic characteristics. Usually, there are advantages and disadvantages of using alternative feedback loops, as illustrated in the following example.

Consider an electrohydraulic position servo that has rather demanding requirements for dynamic response. Suppose that envelope or environmental constraints make it very difficult to mount velocity and acceleration transducers or that the mounting arrangement itself introduces undesirable higher-order dynamics. In this case, consideration might be given to transducing cylinder differential pressure in place of the velocity or acceleration feedbacks. If the load can be represented as primarily a mass, cylinder pressure will be proportional to load acceleration. If cylinder or load friction is large, it may be necessary to use a load cell rather than a pressure transducer. However, the load is often more complex than a simple mass. For example, the load might also have substantial stiffness to ground and viscous damping. In this case, cylinder pressure would have components proportional to load acceleration, velocity, and position. If these components were of the proper relative size and not highly variable, a pressure transducer alone could replace the acceleration and velocity transducers. However, if the load dynamics were complex or highly variable, the use of cylinder pressure as an inner loop might do more harm than good.

Another potential problem associated with the use of alternative inner feedback loops is the influence of external disturbances. Consider again the example of the electrohydraulic position servo. Feedback of either velocity or acceleration will have no affect on closed-loop static stiffness because, by definition, all derivatives of position are zero in the steady state. However, an external force applied to the load will change the cylinder pressure, even in the steady state. Therefore, pressure feedback will reduce closed-loop stiffness unless the pressure feedback signal is high passed, which introduces its own set of dynamic characteristics. Of course, there are applications in which closed-loop stiffness is not critical in the first place.

It should also be mentioned that the mounting of the primary transducer can introduce its own dynamic peculiarities. For example, the controlled variable might be load position relative to ground, and the transducer might be integrally mounted within the servoactuator assembly. If there were substantial compliance in the structure to which the actuator is mounted, the position feedback loop would contain structural zeros. An integral velocity transducer would have the same problem, but a load-mounted accelerometer would not. In this case and in others where the derivative feedback loops are dynamically different from one another, the simplified techniques of Section 8.27.1 are of limited use, and the complete multiloop model must be analyzed directly. Again it should be noted that these more complex derivative feedback loops may result in better or poorer closed-loop performance than comparable inner loops that feed back pure derivatives of the controlled variable. Proper assessment of these trade-offs requires a good physical model of the system, showing the proper relationships of all the feedback variables being considered.

8.28.3   Nonelectronic Inner Loops

Occasionally, it is useful to implement inner-loop feedbacks by mechanical design rather than electronic means. For example, it may be possible to mount the servoactuator or primary transducer so that structural deflections under load produce favorable feedback zeros that improve closed-loop damping. In an electrohydraulic servo, improved damping can often be obtained by using hydraulic pressure feedback, implemented with a cross-port orifice or laminar leakage path. Both of the schemes will suffer loss of closed-loop stiffness. Sometimes it is possible to mount an electrohydraulic servoactuator so that its rod attaches to the mounting structure and its body attaches to the load. If a mass is then attached to the control valve spool and the spool is aligned with the actuator centerline, a form of acceleration feedback can be achieved (valve porting must be arranged to give proper feedback polarity). Mechanical feedback schemes offer the potential advantages of reduced costs and complexity, as well as alleviating the need for high open-loop bandwidth in the actuator and controller. However, the design of servos using these techniques often requires manipulation of rather complex physical models, which is beyond the scope of this section.

8.29   PID CONTROLLERS

A very popular form of controller is called PID (proportional integral differential). It is very simple in concept and is relatively easy to mechanize. Many essays have been written that describe rules of thumb for "tuning" the controller. Unfortunately these tuning procedures can be rather tedious and are usually applicable for only very simple actuator-load dynamics. The purpose of this section is to offer a unified rationale for applying PID controllers that is useful in synthesizing a control system. This rationale also provides insight for systematically adjusting PID parameters on actual hardware.

8.29.1   Equivalence to Frequency Compensation

The basis for the ensuing discussion is that a PID controller is simply a particular form of forward-loop frequency compensation. This can be seen from the generalized controller shown in Fig. 8.155. Note that the differential path is filtered to limit the high-frequency amplitude ratio. Combining the parallel paths of Fig. 8.155, a single transfer function for the controller can be written:

$$
\begin{aligned}
G_{\text{pid}} &= \left(\frac{K_d s}{\tau_d s + 1} \right) + K_p + \left(\frac{K_i}{s} \right) \\
&= \left(\frac{K_i}{s} \right) \left\{ \frac{\left[K_d/K_i + (K_p/K_i)\tau_d \right] s^2 + \left[K_p/K_i + \tau_d \right] s + 1}{\tau_d s + 1} \right\} \\
&= \left(\frac{K_i}{s} \right) \left[\frac{s^2/\omega_{\text{pid}}^2 + (2\zeta_{\text{pid}}/\omega_{\text{pid}})s + 1}{\tau_d s + 1} \right]
\end{aligned}
\tag{8.204}
$$

Note that Eq. (8.204) is simply the transfer function of a lead compensator combined with an integrator, as given by Eq. (8.199). As discussed in Section 8.27.1, it is usually possible to place the lag

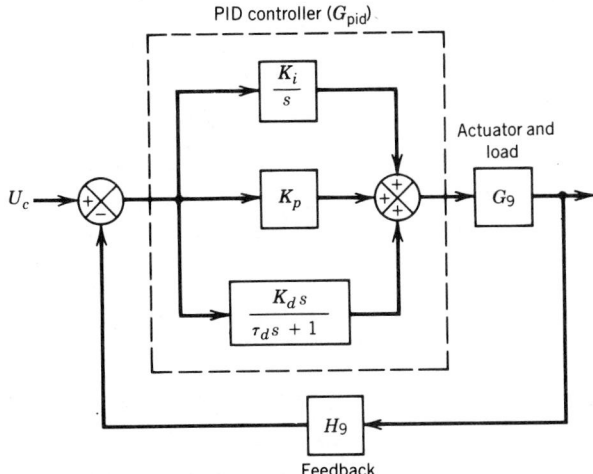

Fig. 8.155   Generalized PID controller.

AUTOMATIC CONTROL: PART III

break frequency a factor of 10 above the lead natural frequency. In this case, the τ_d terms in the numerator are usually small:

$$\omega_{\text{pid}} = \sqrt{\frac{K_i}{K_d}} \quad \text{for } \tau_d \ll \frac{K_d}{K_p} \tag{8.205}$$

$$2\zeta_{\text{pid}}\omega_{\text{pid}} = \left(\frac{K_p}{K_d}\right) \quad \text{for } \tau_d \ll \frac{K_p}{K_i} \tag{8.206}$$

In some applications, second-order lead compensation is not required. In such cases, a simplified version of the PID controller can often be useful. This so-called proportional-integral (PI) controller is formed by setting K_d to zero. The resulting transfer function can then be derived from Eq. (8.204):

$$G_{\text{pi}} = \left(K_p + \frac{K_i}{s}\right) = \left(\frac{K_i}{s}\right)\left(\frac{K_p}{K_i}s + 1\right)$$

$$= \left(\frac{K_i}{s}\right)(\tau_{\text{pi}}s + 1) \tag{8.207}$$

This result is similar to Eq. (8.198).

Another simplified version of the PID controller is obtained by setting $K_i = 0$. The transfer function of this so-called proportional-differential (PD) controller can also be derived from Eq. (8.204):

$$G_{\text{pd}} = \left(\frac{K_d s}{\tau_d s + 1}\right) + K_p$$

$$= K_p\left[\frac{(K_d/K_p + \tau_d)s + 1}{\tau_d s + 1}\right]$$

$$= K_p\left[\frac{\tau_{\text{pd}}s + 1}{\tau_d s + 1}\right] \tag{8.208}$$

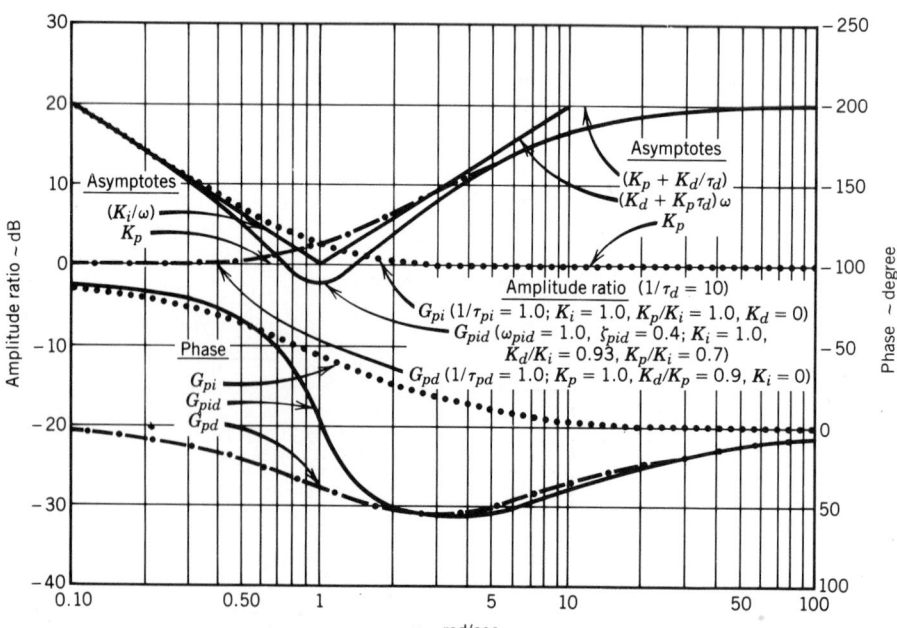

Fig. 8.156 Frequency response comparisons of PID, PI, and PD controllers.

This transfer function is the same as the first-order lead compensator of Eq. (8.196) and is normally used in systems having an inherent integration. Note that the lead break frequency can be a factor of 10 lower than the lag break frequency. In this case the τ_d term in the numerator of Eq. (8.208) is small:

$$\tau_{pd} = \left(\frac{K_d}{K_p} \right) \quad \text{for } \tau_d \ll \frac{K_d}{K_p} \tag{8.209}$$

Frequency responses of representative PID, PI, and PD controllers are given in Fig. 8.156, which also shows the effects of the various controller parameters.

8.29.2 Systems Having No Inherent Integration

Electronic integrators are normally used in the forward loops of systems having no inherent integrations, as explained in Section 8.26.1. Section 8.27 explains the various ways in which lead compensation can be usefully applied. A PID controller offers a convenient method for combining the electronic integration with lead compensation. As discussed in Section 8.29.1, the PI scheme provides first-order lead, while the complete PID scheme offers second-order lead.

8.29.3 Systems Having an Inherent Integration

Section 8.26.1 explains that an electronic integrator in the forward loop minimizes static servo errors. However, the addition of an electronic integrator to a system that already has an inherent integration will usually cause dynamic instability. To prevent this type of instability, lead compensation must be combined with the electronic integrator. This can be accomplished with a PI controller.

As shown in Fig. 8.156, a PI controller contributes nearly 90° of phase lag at low frequencies. If this is added to the 90° of lag already contributed by the integrator inherent in the system, the total low-frequency lag approaches 180°. Since the other system dynamics will add even more phase lag at high frequencies, the PI break frequency must be set low enough to ensure an intermediate frequency range over which the phase lag is reduced. The open-loop frequency responses of Fig. 8.157(a)

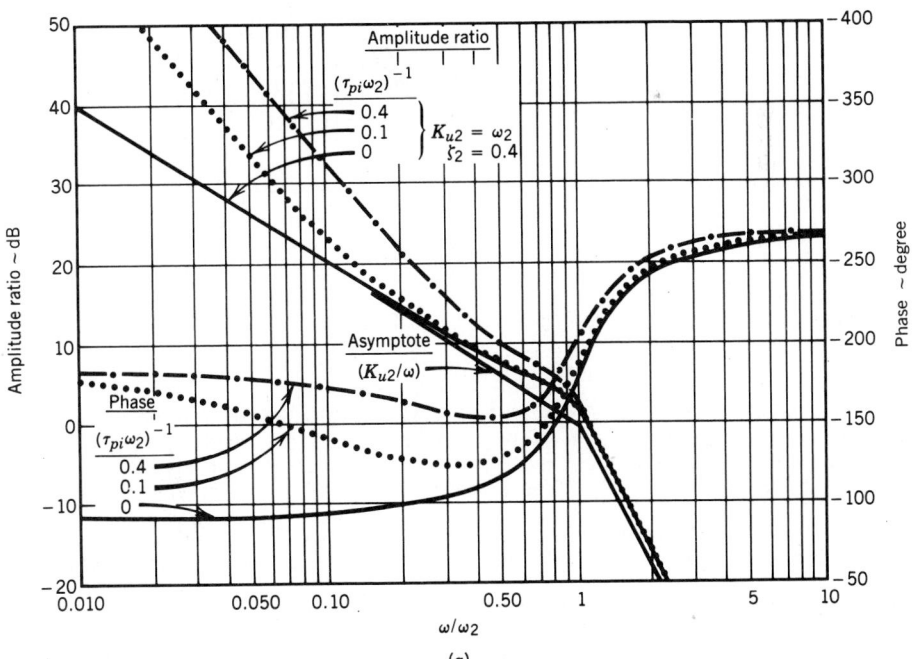

Fig. 8.157 PI controller added to the system of Fig. 8.134(b)—effects of lead break frequency: (a) open-loop frequency response.

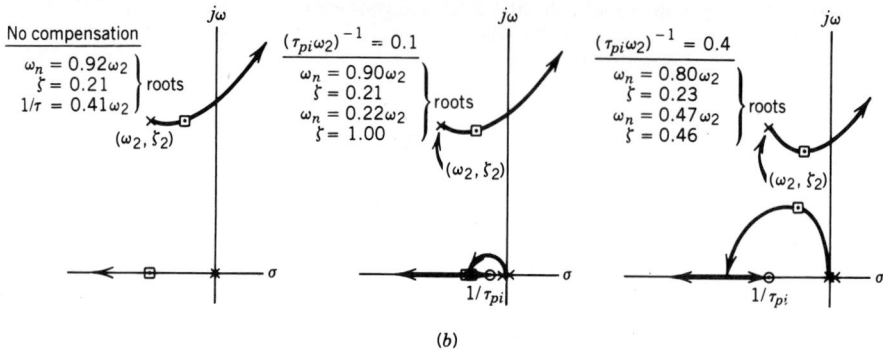

(b)

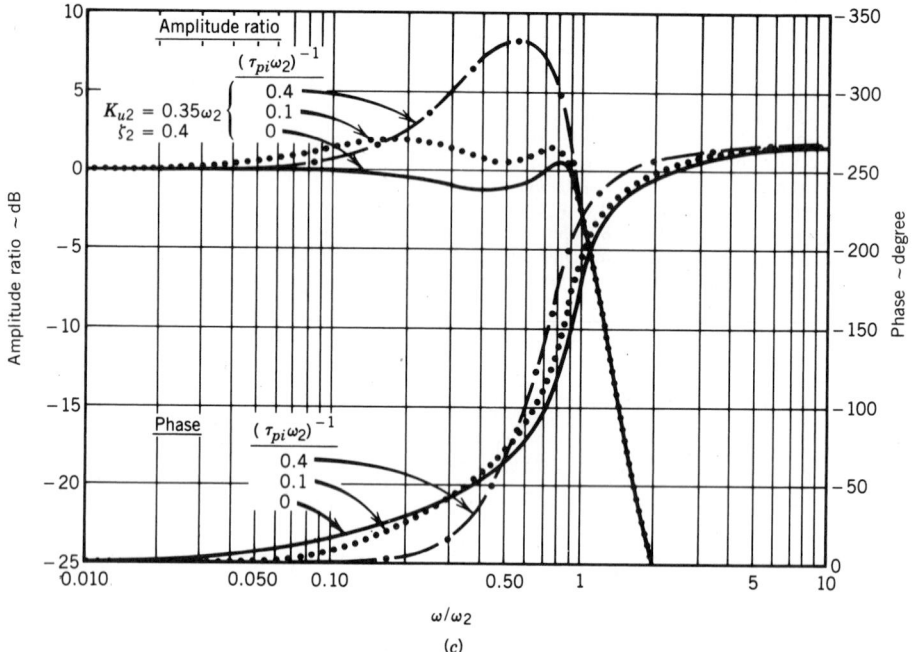

ω/ω_2

(c)

Fig. 8.157 (*Continued*) (*b*) root loci ($\zeta_2 = 0.4$, $K_{u2} = 0.35\omega_2$), (*c*) closed-loop frequency response.

illustrates this effect for a system whose inherent characteristics consist of an integrator and a second-order lag. Generally, a PI break frequency greater than 10% of the system's lowest lag frequency will substantially reduce closed-loop stability. This is shown by the phase plots of Fig. 8.157(*a*), by the root loci of Fig. 8.157(*b*), and by the closed-loop frequency responses of Fig. 8.157(*c*).

Using the 10% rule of thumb for the PI controller, it is interesting to examine some time histories of the closed-loop system. Figure 8.158 shows time responses to step and ramp commands for the system of Fig. 8.157. As illustrated in Fig. 8.158(*a*), the PI controller degrades the step response. Figure 8.158(*b*) shows that the PI's double integration at low frequencies eliminates the following error but causes larger overshoot of the steady state.

The 10% rule also applies to systems whose inherent characteristics consist of an integrator and a first-order lag. Frequency responses, root loci, and time histories for such a system are given in Fig. 8.159. It should also be noted that the effective lag break frequency of the system can be increased by using lead compensation techniques, as described in Section 8.27. For the system of Fig. 8.159, this

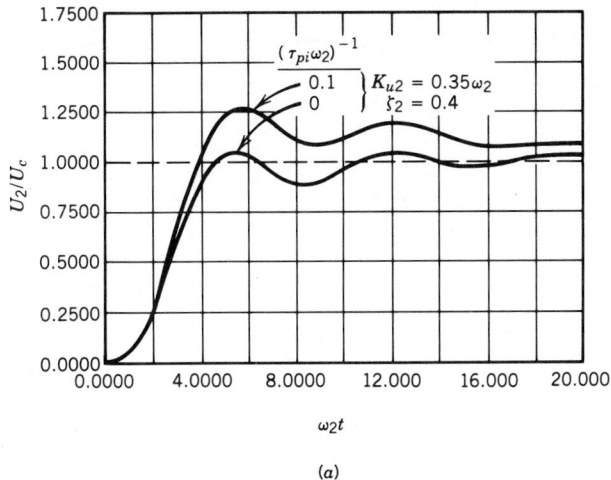

(a)

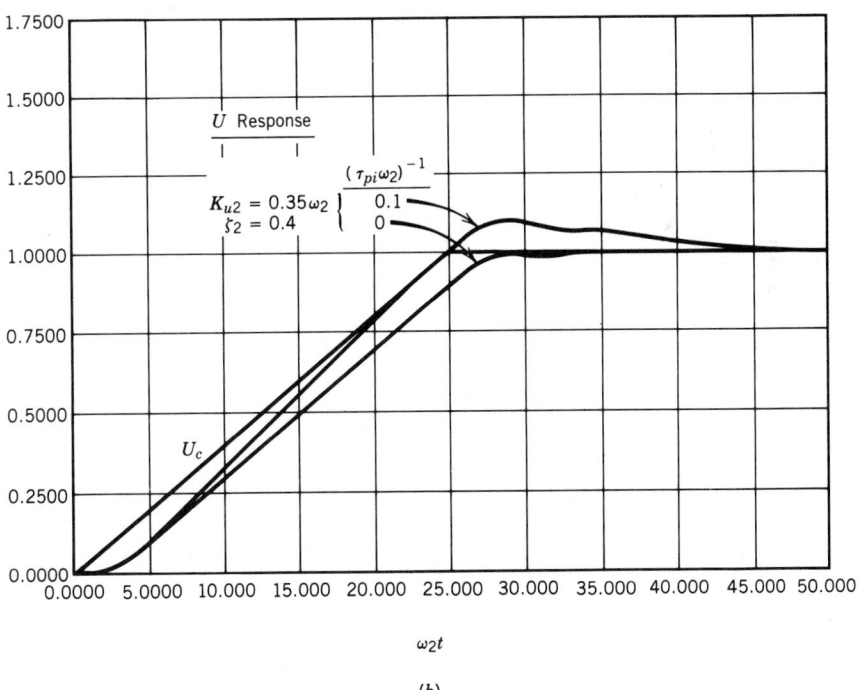

(b)

Fig. 8.158 PI controller added to the system of Fig. 8.134(b)—time histories: (a) step response, (b) ramp response.

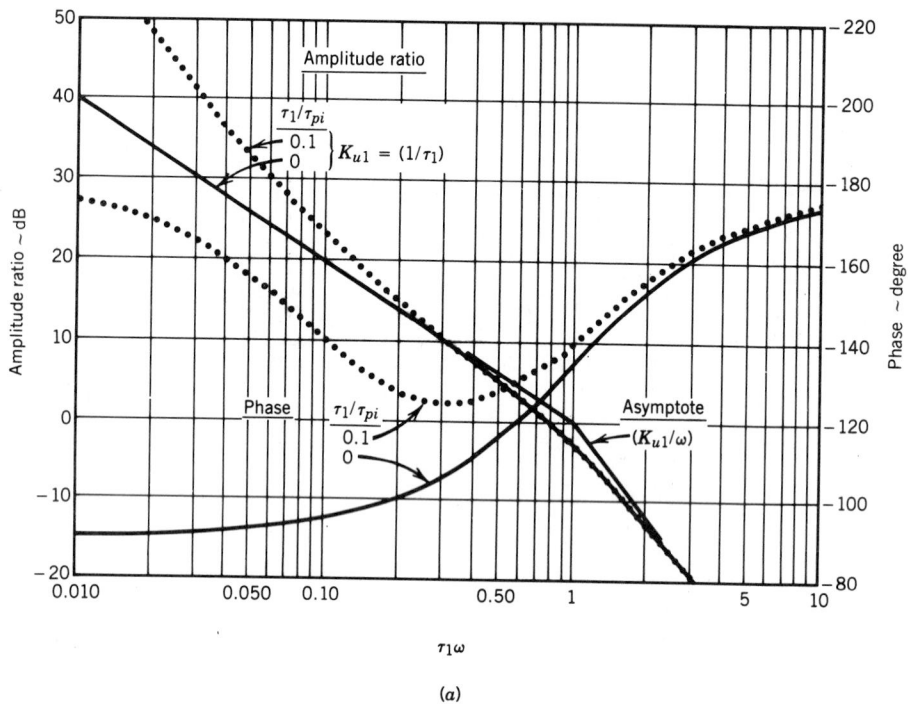

$$\tau_1 \omega$$

(a)

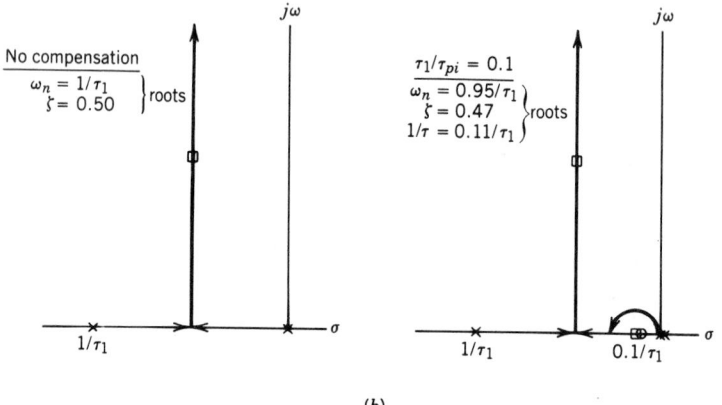

(b)

Fig. 8.159 PI controller added to the system of Fig. 8.134(*a*): (*a*) open-loop frequency response, (*b*) root loci ($K_{u1} = 1/\tau_1$), (*c*) closed-loop frequency response, (*d*) ramp response.

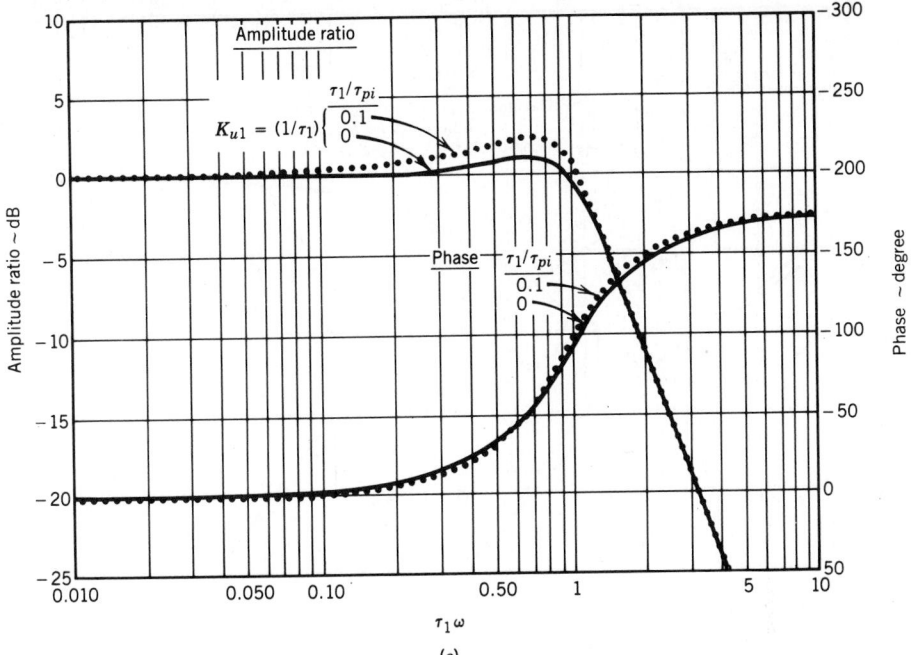

(c)

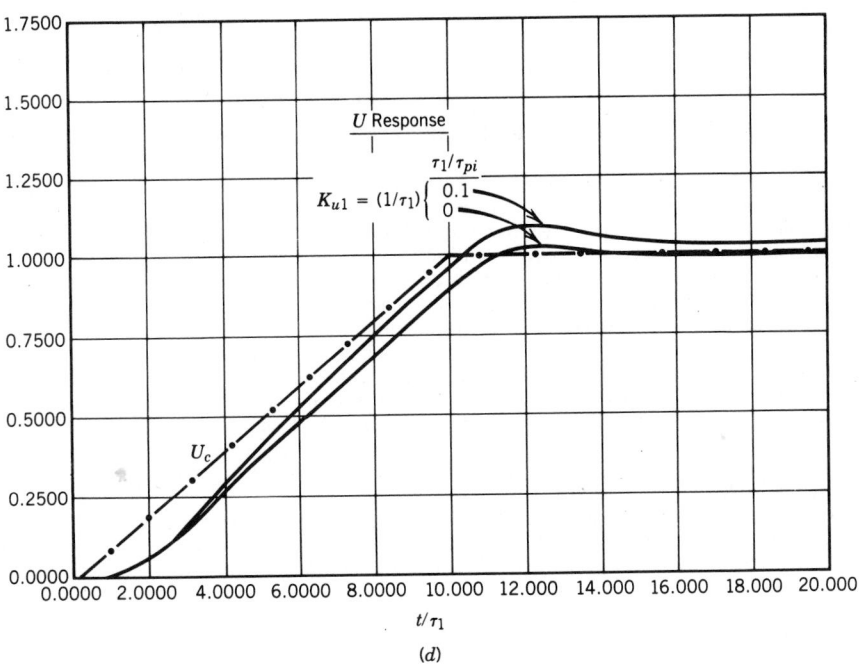

(d)

Fig. 8.159 (*Continued*)

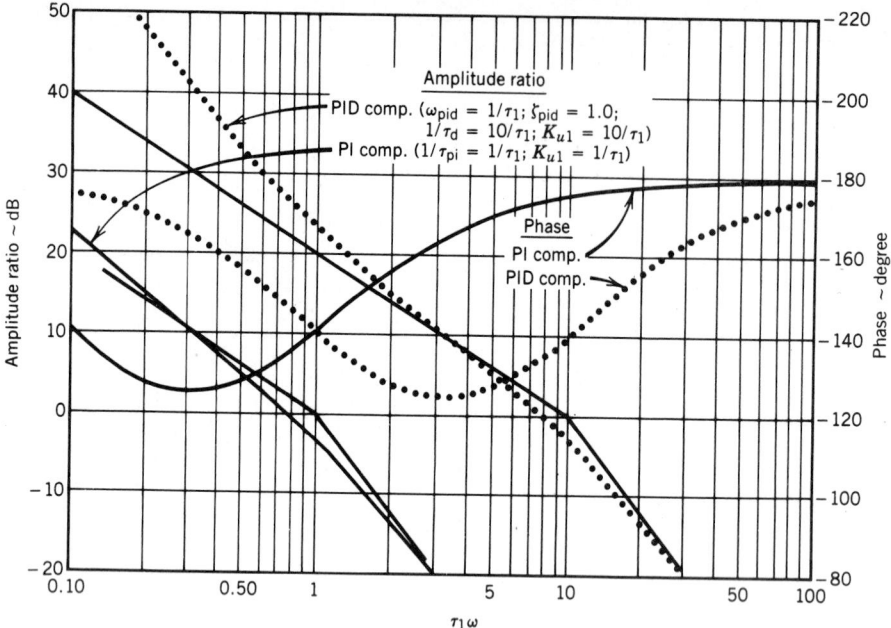

Fig. 8.160 Effects of lead compensation added to the system of Fig. 8.159.

can be accomplished by using a PID, rather than PI, controller. Reexamining the example of Fig. 8.159 using the PID characteristics of Fig. 8.156, a 10-fold increase in effective system bandwidth can be achieved by setting $\omega_{pid} = 1/\tau_1$ and $\zeta_{pid} = 1.0$. In this case, one of the two PID lead terms cancels the system lag at $1/\tau_1$. The resulting open-loop frequency response is shown in Fig. 8.160.

8.30   EFFECTS OF NONLINEARITIES

The previous sections have concentrated on the design of controllers for linear systems. In practice, physical systems are never truly linear. If the nonlinearities are not large, design of the controller using linear techniques is very useful. However, if they are substantial, computer simulation may be necessary to properly predict system response and stability (see Section 8.8).

Figure 8.161 illustrates idealized forms of several nonlinearities that are commonly encountered in systems controlling mechanical loads. Saturating nonlinearities can occur in transducers, electronics, and the servoactuator itself. Deadzone is the lack of output for small changes in input and is generally most significant in servoactuators and transducers. Resolution is the availability of a limited number of output values and is typical of digital electronics and many types of transducers, including encoders and wire-wound potentiometers. Most servoactuators provide output velocities that are force dependent or output forces that are velocity dependent. The load-velocity curves shown in Fig. 8.161(d) are typical of an electrohydraulic servoactuator and are highly nonlinear. Coulomb friction is a constant force that always opposes motion. Static friction (stiction) is often larger than Coulomb friction but is very difficult to model.

Mechanical backlash is motion lost when the direction of motion is reversed, as in gear trains and bearings. Since most servoactuators make use of electromagnetic elements, magnetic hysteresis effects can cause some system performance anomalies. The width of the hysteresis band is dependent upon the amplitude of the input signal (the output is a function of the input's prior history as well as its present value).

8.31   CONTROLLER IMPLEMENTATION

As mentioned, it is not the intent of this section to address the electronic design of a servocontroller. Nevertheless, some basic understanding of controller implementation is required to properly specify and select a controller. The following discussion describes several basic implementation approaches, together with their relative advantages and disadvantages.

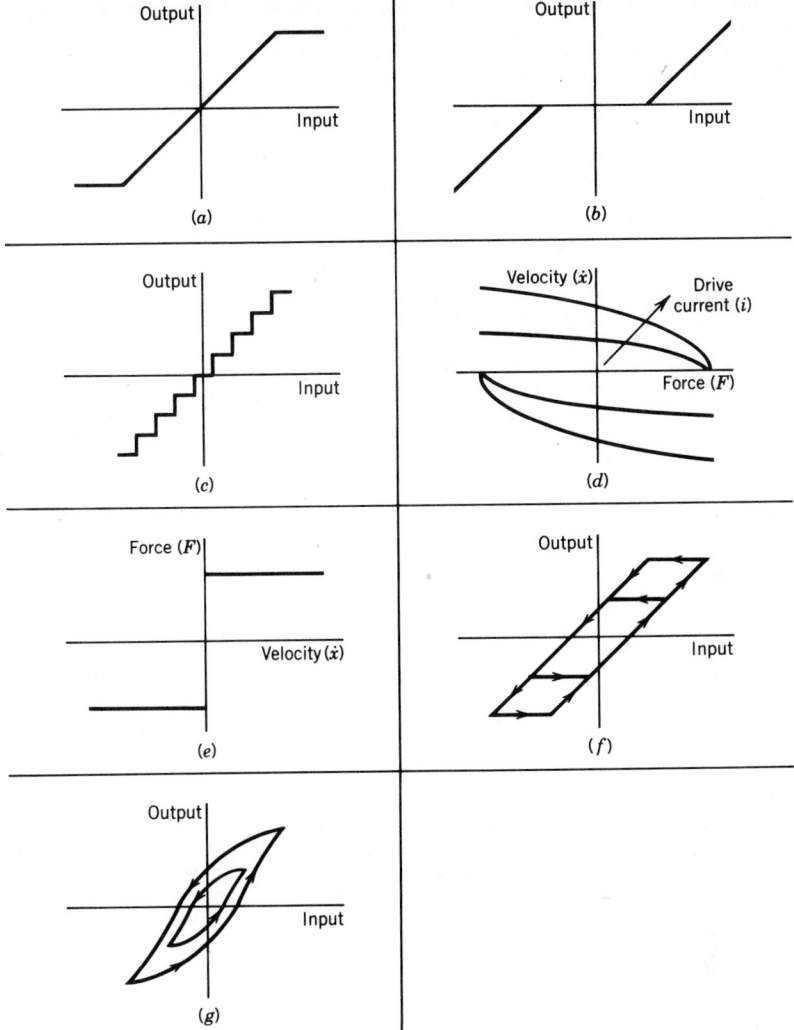

Fig. 8.161 Common nonlinearities: (*a*) saturation, (*b*) deadzone or threshold, (*c*) resolution, (*d*) load-velocity curves, (*e*) Coulomb friction, (*f*) mechanical backlash, (*g*) magnetic hysteresis.

Since control of a mechanical load is inherently a continuous process, the use of a dc analog controller typically provides the highest servo bandwidth and smoothest operation. Furthermore, basic servoloops can be implemented with very simple circuits. Hard-wired digital controllers offer the potential for increased overall accuracy at the cost of degraded resolution and more complex electronic hardware. Microprocessor-based digital controllers offer increased flexibility, versatility, accuracy, and computer-interface capability but typically suffer reduced bandwidth as well as degraded resolution. The electronic hardware associated with processor-based controllers is typically more complex than a simple analog controller. However, in complex control applications, large amounts of analog circuitry can be replaced with software.

8.31.1 Analog Controllers

Analog controllers are typically used in servoloops for which high closed-loop bandwidth and smooth operation are required. Modern operational amplifiers have bandwidths of several hundred kilohertz and virtually infinite resolution. Because of this, they are free of the sampling delays, phase lags, and resolution problems associated with microprocessor-based controllers. Furthermore, they are relatively tolerant of electrical noise, and troubleshooting can be accomplished with simple equipment. On the other hand, analog implementation of complex controller functions such as nonlinearities, automatic gain changing, elaborate command processing, and complex failure detection can cause the electronic circuitry to become extremely complicated. In addition, analog controllers typically require periodic adjustment and calibration.

The basic functional elements of a typical analog controller are shown in Fig. 8.162. The functions of compensation networks have been discussed previously in this chapter. Some form of signal conditioning is usually required for transducers. In the case of a simple dc transducer such as a potentiometer, the conditioning may consist of dc excitation together with an output buffer amplifier. Buffer amplifiers can be designed to protect against large voltages erroneously connected to the electronics, to provide consistent loading of the transducer output, to reject electrical cabling noise (electromagnetic interference), and to filter transducer ripple. Low-output transducers, such as those that employ strain gage elements, require high-gain, low-drift amplifiers. Linear variable differential transformers (LVDTs), resolvers, and other ac transducers require ac excitation, demodulation, and filtering to remove ripple. Greatly increased servo accuracy can be obtained with a combination of ac command generation and ac feedback, such as when synchros are utilized, but this approach has largely been replaced by the use of digital controllers and transducers. In any case, transducer specifications should be carefully studied to determine the proper signal conditioner characteristics. The dynamic characteristics of the signal conditioners, as well as the transducers themselves, can have significant impact on the stability and performance of the servoloops, as explained previously in this chapter.

There are many sources of information concerning the design of analog controllers. Reference 6 is an excellent source for the design of operational amplifier circuits for a wide variety of purposes, including compensation and signal conditioning. Furthermore, most manufacturers of operational amplifiers publish useful application handbooks. Also, application literature from transducer manufacturers often discusses signal conditioning techniques in some detail.

The design of power amplifiers varies widely with the type of servoactuator being driven. In the case of conventional electrohydraulic servoactuators and other types requiring low-power electrical inputs, the power amplifier can be a very simple linear (proportional) circuit. Typically, it utilizes an operational amplifier and a power boost stage consisting of a complimentary pair of transistors. Some operational amplifiers have enough output capability to provide the required electrical input directly. In the case of electromechanical devices that must provide a direct electrical-to-mechanical energy conversion, such as dc servomotors, the power amplifier can become very complex. In this case, its design should be left to an experienced electronics engineer. Linear amplifiers are still the most straightforward and offer the best servo performance but are severely limited in the size of the motor they can control because of the large amount of heat they must dissipate. Since switching transistors typically generate little heat in their full-on or full-off states, various time-modulated on/off power drivers have been developed. The most popular type for servo control applications seems to be pulsewidth modulation (PWM). In this approach, the power devices are switched on and off at a very high fixed frequency. The percent on-time during each cycle is proportional to the dc input voltage from the upstream analog controller circuitry. If the PWM frequency is high enough, only the average cycle voltage will affect the servoactuator output, thereby resulting in nearly proportional control. To accomplish this, the PWM frequency must usually be at least one order of magnitude higher than the bandwidth of the innermost feedback loop (typically a current feedback loop around the power amplifier) or 2 to 3 orders of magnitude higher than the bandwidth of the primary control loop. If PWM frequencies of this magnitude are impractical, it may be necessary to reduce bandwidths of the various control loops to prevent unacceptable servo output at the PWM frequency. Design of a PWM amplifier can be difficult, particularly with regard to polarity switching around zero.

It should be noted that servoactuators requiring high-power electrical inputs can create problems related to voltage saturation in the drive electronics. The reason for this is that the input inductive characteristics of many actuators tend to cause long L/R time constants. Current feedback is often used to improve the current response of the actuator, and this leads to large transient voltages for abrupt inputs to the power amplifier. Practical limits on amplifier voltage capability can result in voltage saturation during transients. It is often necessary to design an active voltage-limiting circuit to prevent stability problems during saturation. Careful attention must be devoted to tailoring the current feedback loop and voltage limiter circuits to the servoactuator's electrical dynamics if amplifier stability and performance problems are to be avoided.

Multiturn potentiometers are usually included in analog controller circuitry to allow proper calibration of transducer scale factors, compensate for electrical and mechanical offsets, and to adjust loop gains. The desire for system accuracy often drives the designer to providing many adjustments, but this practice can greatly complicate maintenance procedures. If adequate performance cannot be achieved with a limited number of well-placed adjustments, then serious consideration should be given to the use of a digital controller.

The use of integrating amplifiers in control loops requires some special consideration. First, pure integrators often cause low-frequency oscillations or "hunting" when backlash, deadzone, or friction

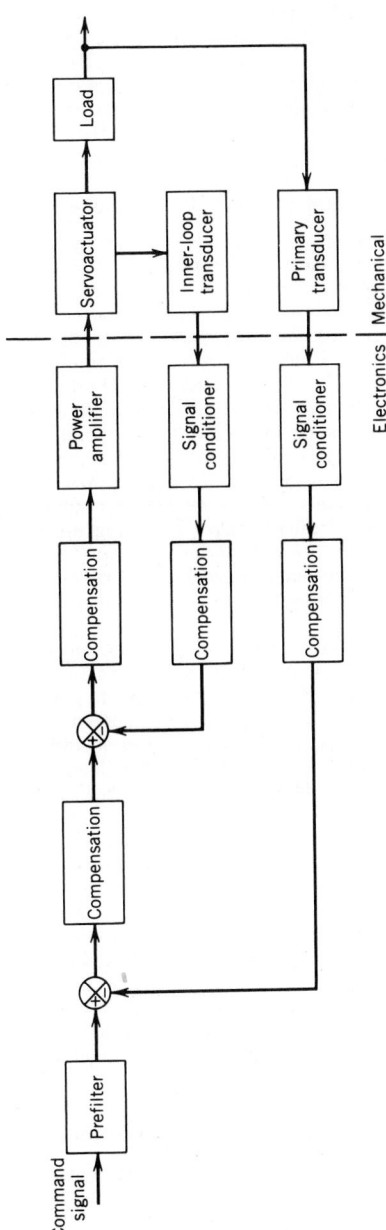

Fig. 8.162 Typical analog controller.

exists in the system. This behavior can often be controlled by adding a large resistance across the integrator's feedback capacitor. This limits the amplifier's gain and makes it look proportional at low frequencies, while preserving an integrating characteristic in the crossover frequency range. If it is possible for an integrating amplifier to saturate during abrupt commands, it may "latch up" and exhibit a long recovery period, which can result in large servo overshoots. This behavior can be prevented by proper gain distribution in the servoloops or by providing the amplifier with a diode limiter. Of course, an integrating amplifier can drift into saturation if it is powered up before the servoactuator is allowed to move. For example, electronics are often powered up prior to releasing a mechanical brake or applying hydraulic power. In this case, integrator saturation can cause a large engagement transient. This can be prevented by shorting the integrator's feedback capacitor with a relay contact or electronic switch. The short is then opened when the actuator is mechanically or hydraulically engaged.

8.31.2   Hard-Wired Digital Controllers

The overall accuracy of a servomechanism can be greatly enhanced by the use of a digital transducer and a digital controller. Furthermore, the need for periodic calibration and adjustment can be virtually eliminated. Hard-wired digital electronics can provide this improved accuracy with bandwidths comparable to analog electronics. However, these digital circuits are considerably more complex than comparable analog circuits, are more susceptible to electrical noise, and have finite resolution. For these reasons, hard-wired digital electronics are usually used only in the primary control loop (accuracy is typically not critical in the inner loops). Furthermore, frequency compensation is difficult to implement in digital hardware and is usually left to analog circuitry.

Hard-wired digital electronics are commonly used with high-resolution incremental encoders, as illustrated in Fig. 8.163(a). Two pulse trains are generated by the encoder, 90° out of phase with one another. The pulse-conditioning circuitry squares up the incoming pulses, determines the transducer's direction of motion, and often increases the resolution by a factor of four. The asynchronous counter is incremented up or down by each feedback pulse, depending upon the direction of motion. Similarly, command pulses also increment the counter up or down. The net count at any particular time represents the difference between the number of command and feedback pulses since the counter was initialized. After digital-to-analog conversion, this count becomes the error signal transmitted to the analog electronics.

The command pulse train can be generated by additional digital hardware or by a computer. However, if the transducer resolution is high and the desired maximum command rate is high, a computer may be hard-pressed to provide the required pulse rates. In this case, hardware comparators and rate multipliers may prove more satisfactory. It should be noted that an incremental system has no inherent knowledge of its absolute position. Therefore, power shutdowns and electrical noise can cause such a system to lose track of where it is. For this reason a "marker pulse" is often provided at some known position to reinitialize the counter periodically. Alternatively, the servo occasionally can be commanded to a mechanical "home" position. As with most transducers having finite resolution, encoders will usually cause limit cycling with an amplitude equal to the least significant bit.

Absolute digital systems can also be implemented in electronic hardware, as shown in Fig. 8.163(b). The encoding transducer outputs a digital word that represents its absolute position at all times. For an optical encoder, the resolution is typically between 12 and 24 bits. After buffering, the feedback word is digitally subtracted from a digital command, and the resulting error is converted to an analog signal that is transmitted to the analog electronics. The digital summing junction can be implemented in a number of ways, including the use of an arithmetic logic unit (ALU), which operates at very high speeds. The digital command and feedback information can be transmitted to the summing junction as parallel digital words or as serial data that must be multiplexed and then decoded. The command information can be generated from a computer or digital thumbwheels.

8.31.3   Computer-Based Digital Controllers

The hard-wired digital controllers of Fig. 8.163 have limited flexibility and functional capability. The use of a microprocessor may reduce electronic hardware complexity when elaborate system functions are required. There are many such functions that are well suited to microprocessor implementation:

- Command processing (nonlinear functions, limiting, switching, and communication with other computers)
- Redundancy management (fault detection, isolation, and reconfiguration)
- Adaptive control (self-adjustment of control loop parameters as operating conditions or environmental factors change)
- Built-in test (BIT) features

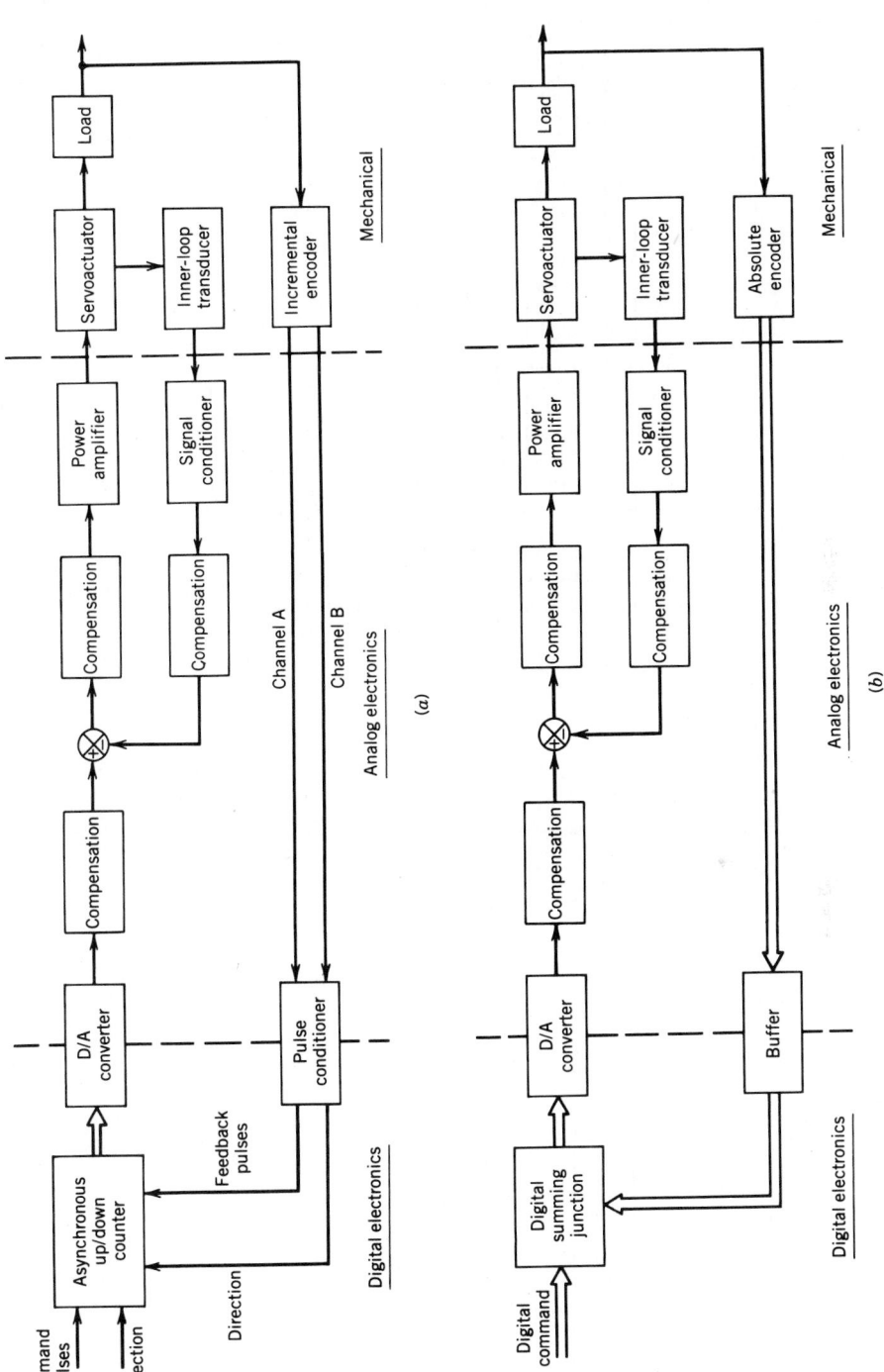

Fig. 8.163 Typical hard-wired digital controller: (*a*) incremental system, (*b*) absolute system.

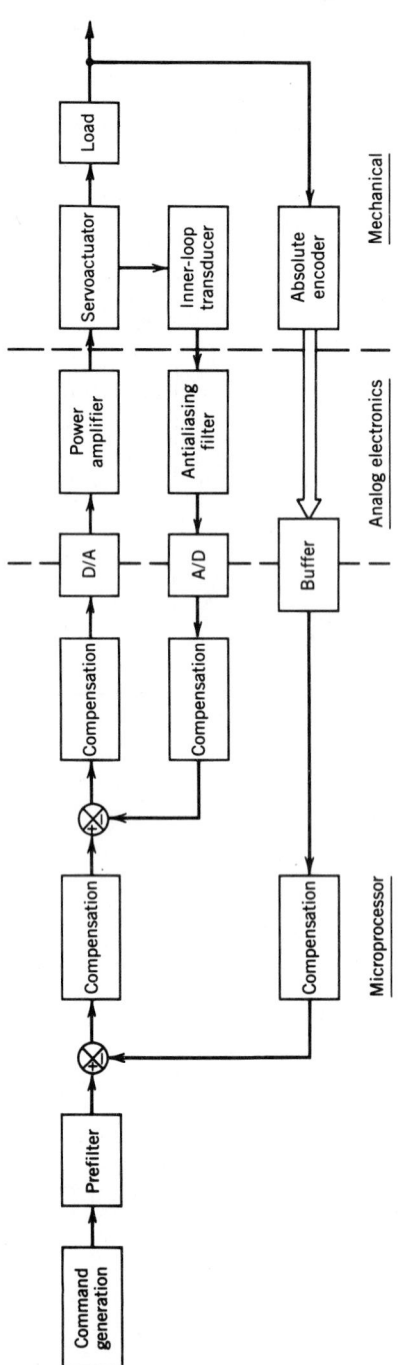

Fig. 8.164   Typical computer-based controller.

Once the need for a microprocessor is established, it may also become reasonable to implement the servoloops in software. Note that if the application requires only the closure of simple servoloops without the need for elaborate additional functions, the use of a microprocessor will usually result in more complex hardware than on all-analog system and may be more complex than a hard-wired digital system.

The loop closure architecture of a microprocessor-based controller is illustrated in Fig. 8.164. This block diagram implements virtually all the loop functions in software, including frequency compensation. Of course, it is not necessary to use a digital outer-loop transducer, but use of an analog transducer compromises the potential accuracy advantages of the digital controller. Several methods can be used to generate the software compensator designs. Perhaps the most straightforward technique is to generate Laplace transfer functions using the continuous frequency domain techniques outlined in Sections 8.25–8.29. These transfer functions can then be converted to equivalent z transforms, from which difference equations can be generated for implementation in software.[5]

It should be noted that a microprocessor-based controller is a sampled-data system, and the digital-to-analog converter usually operates as a zero-order hold (ZOH). The sampling nature of the system together with computation times introduce time delay into the control loops, which can have a profound influence on system performance and stability. Figure 8.165 shows a frequency response of a ZOH operated in a sampled-data system. If the system has been designed using frequency domain techniques, this figure represents the additional phase lag and amplitude that will result from computer implementation of the design. The phase lag is linear with frequency, and an approximate transfer function is a first-order lag with a break frequency equal to (f_s/π), where f_s is the sampling frequency of the computer in hertz. This approximation is very accurate out to a frequency of $f_s/2\pi$. Note that $10°$ of phase lag exists at a frequency of $f_s/18$. This suggests that the sampling frequency should be at least 20 times the crossover frequency (in Hertz) of the loop being implemented if the impact on phase margin is to be minimized. Furthermore, smooth operation of the servo may require heavy filtering at the output of the D/A converter to reduce sampling-induced ripple. This will add additional phase lag in the servoloops.

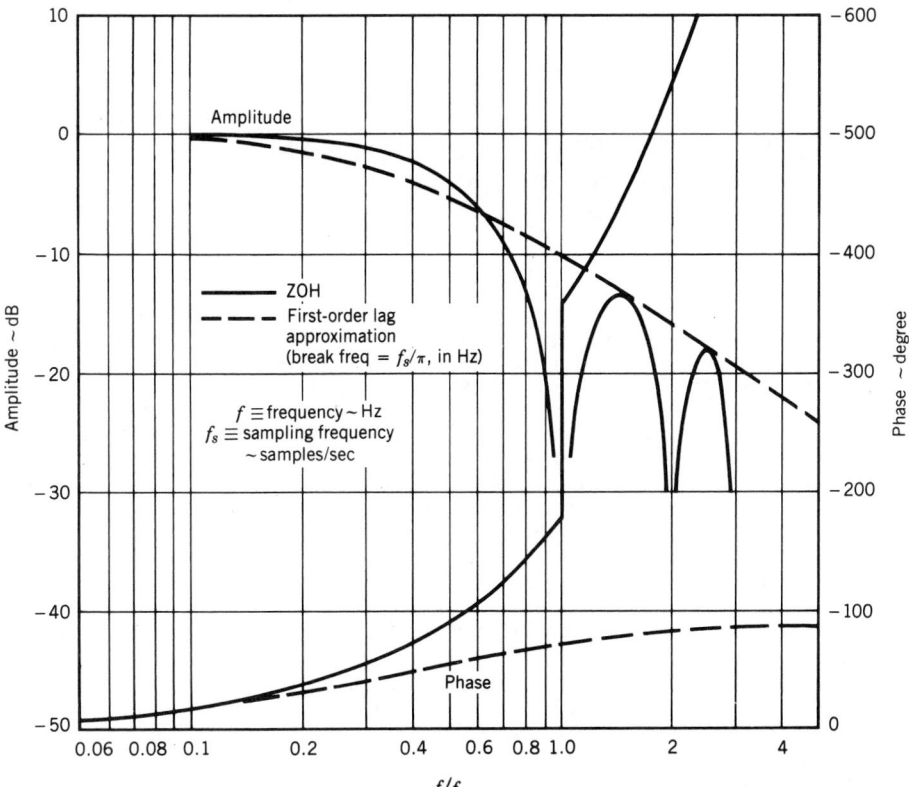

Fig. 8.165 Frequency response characteristics of a zero-order hold.

The need to minimize phase lags can place severe restrictions on the complexity of computations that the microprocessor can handle in one sampling interval. This problem can be partially overcome by using a separate processor to perform loop closure computations or by implementing compensators in the analog circuitry. Alternatively, the need for high sampling rates can be reduced by implementing high-gain inner feedback loops in analog circuitry, as shown in Fig. 8.163(b). If the inner loops utilize analog transducers, the problem of aliasing[7] adds another reason for using analog electronics. To properly utilize the output of an analog transducer in the computer, an antialiasing filter is required at the input to the A/D converter. These filters are often first order with a break frequency equal to f_s/π. This doubles the effective phase lag of the computer.

REFERENCES FOR PART III

1. D'Azzo, J. J., and Houpis, C. H. *Feedback Control System Analysis and Synthesis*, New York, McGraw-Hill, 1966.
2. Kuo, B. C., *Automatic Control Systems*, Englewood Cliffs, N.J., Prentice-Hall, 1982.
3. Doebelin, E. O., *Dynamics Analysis and Feedback Control*, New York, McGraw-Hill, 1962.
4. Kuo, B. C., *Digital Control Systems*, New York, Holt, Rinehart and Winston, 1980.
5. Cadzow, J. A., and Martens, H. R., *Discrete-Time and Computer Control Systems*, Englewood Cliffs, N.J., Prentice-Hall, 1970.
6. Graeme, J. G., Tobey, G. E., and Huelsman, L. P., *Operational Amplifiers, Design and Applications*, New York, McGraw-Hill, 1971.
7. Doebelin, E. O., *System Modeling and Response*, New York, Wiley, 1980.

PART IV

STATE-SPACE METHODS FOR DYNAMIC SYSTEMS ANALYSIS AND CONTROL
Krishnaswamy Srinivasan

8.32 INTRODUCTION

The use of the state-space approach for the analysis and control of dynamic systems results in analysis and design techniques based in the time domain, as opposed to frequency-domain-based transform techniques. The state-space approach has the following characteristics:

1. It employs a more complete internal representation of dynamic systems as compared to transform methods that use input-output representations. The state of a system represents complete information about the current dynamic condition of the system. It incorporates the effect of all past inputs on the system. When combined with a complete description of the system dynamics in the form of state-space equations and knowledge of all future inputs, the future behavior of the system can be determined. More precise definitions of the notion of state are given in standard textbooks.[1-7]

2. It offers a unified approach to the analysis and synthesis of linear and nonlinear, time-invariant and time-varying, continuous-time and discrete-time, single-input and single-output, and multiple-input and multiple-output systems. Available techniques, however, are more plentiful for some categories of systems.

3. State-space-based methods rely more heavily on digital computers than classical transform-based techniques for dynamic systems analysis and control. In fact, the availability of digital computers both for analysis and control synthesis and for implementation of the controllers has been an important factor underlying the growing use of state-space-based methods.

4. State-space-based methods have the potential to improve the performance of controlled systems if such systems can be modeled accurately. They have been less successful in cases where system models are characterized by significant uncertainty. Classical transform-based techniques have been

and continue to be widely used in such cases. In fact, one of the more encouraging trends in control systems development has been the establishment of links between state-space-based methods and transform-based methods.[3]

Though the concept of state has been invoked by a number of methods of classical mechanics and is implicit in the phase-plane concept used for nonlinear system stability analysis, the effective application of state-space-based methods for analysis and control of dynamic system behavior has occurred only over the last three decades. Pioneering theoretical work by Kalman[8-11] and others and the availability of digital computers for performing analysis and design computations have been important underlying factors. State-space methods have been most successful in aerospace control applications and less so in a variety of industrial control applications. Among the factors favoring increased emphasis in the future on state-space methods are:

1. The emphasis on controlled system performance improvement resulting from imperatives such as improved efficiency of energy utilization and improved productivity
2. The increasing availability of inexpensive but powerful digital computers for off-line analysis and design computations and on-line control computations

In Sections 8.33–8.38, methods for analysis and control of dynamic systems using state-space methods are described using the continuous-time formulation. Extension of these results to discrete-time problem formulations used for sampled data control systems is straightforward in most cases. The fact that digital computers are being used more frequently for controller implementation makes such discrete-time formulations important. Sources for such formulations are therefore cited at appropriate locations in the following text.

8.33   STATE-SPACE EQUATIONS FOR CONTINUOUS-TIME SYSTEMS

The differential equations describing the input-output behavior of an nth-order, continuous-time, nonlinear, time-varying, lumped-parameter system can be written in the form of a first-order vector ordinary differential equation and a vector output equation:

$$\mathbf{x}(t) = \mathbf{f}[\mathbf{x}(t), \mathbf{u}(t), t] \qquad t \geq t_0 \tag{8.210}$$

$$\mathbf{y}(t) = \mathbf{g}[\mathbf{x}(t), \mathbf{u}(t), t] \qquad t \geq t_0 \tag{8.211}$$

where $\mathbf{x}(t) = n$-dimensional state vector

$\mathbf{y}(t) = p$-dimensional output vector

$\mathbf{u}(t) = r$-dimensional input vector

$\mathbf{f}, \mathbf{g}$ = vectors of appropriate dimension whose elements are single-valued nonlinear functions of the arguments noted

Equation (8.210) is the state equation and Eq. (8.211) is the output equation. The state, output, and input vectors are

$$\mathbf{x}(t) = \begin{bmatrix} x_1(t) \\ x_2(t) \\ \vdots \\ x_n(t) \end{bmatrix} \qquad \mathbf{y}(t) = \begin{bmatrix} y_1(t) \\ y_2(t) \\ \vdots \\ y_p(t) \end{bmatrix} \qquad \mathbf{u}(t) = \begin{bmatrix} u_1(t) \\ u_2(t) \\ \vdots \\ u_r(t) \end{bmatrix} \tag{8.212}$$

The elements $x_1(t), x_2(t), \ldots, x_n(t)$ of the state vector are the state variables of the system.

Formulation of the higher-order system differential equations as a set of first-order differential equations has the advantage that the latter are easier to solve by numerical methods than the former. If the functions $\mathbf{f}$ and $\mathbf{g}$ are linear functions of $\mathbf{x}(t)$ and $\mathbf{u}(t)$, the system can be described by linear ordinary differential equations. Matrix notation can then be employed to simplify their representation:

$$\dot{\mathbf{x}}(t) = \mathbf{A}(t)\mathbf{x}(t) + \mathbf{B}(t)\mathbf{u}(t) \qquad t \geq t_0 \tag{8.213}$$

$$\mathbf{y}(t) = \mathbf{C}(t)\mathbf{x}(t) + \mathbf{D}(t)\mathbf{u}(t) \qquad t \geq t_0 \tag{8.214}$$

where $\mathbf{A}(t) = n \times n$ system matrix

$\mathbf{B}(t) = n \times r$ input-state coupling matrix

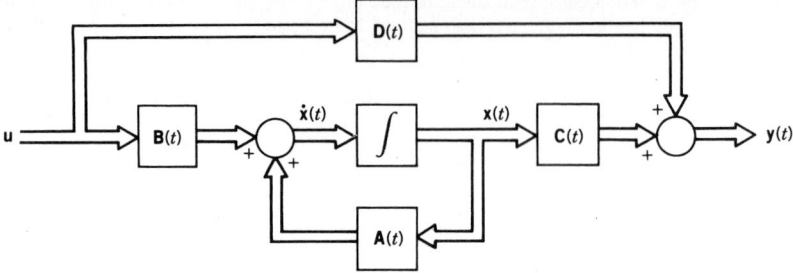

Fig. 8.166   Linear continuous-time system.

$C(t) = p \times n$ state-output coupling matrix
$D(t) = p \times r$ input-output transmission matrix

A block diagram representation of Eqs. (8.213) and (8.214) is given in Fig. 8.166, using standard symbols appropriate for simulation diagrams. If the system is linear and time invariant (LTI), the matrices noted become constant matrices, as indicated by the following equations:

$$\dot{x}(t) = Ax(t) + Bu(t) \qquad t \geq t_0 \qquad (8.215)$$

$$y(t) = Cx(t) + Du(t) \qquad t \geq t_0 \qquad (8.216)$$

Parallel formulations for discrete-time systems are given in standard texts.[2, 4, 7, 12]

8.34   STATE-VARIABLE SELECTION AND CANONICAL FORMS

The state vector of a system is comprised of the minimum set of variables necessary to describe the system behavior in the form of the state-space equations already given. It can be shown that the selection of the state vector for a system is not unique.

For linear continuous-time systems, the following development shows that any vector $q(t)$ related to a valid state vector selection $x(t)$ by a constant nonsingular transformation matrix T is also a valid state vector:

$$q(t) = Tx(t) \qquad (8.217)$$

where T is a nonsingular $n \times n$ matrix. Equations (8.213) and (8.214) may be rewritten in terms of the vector $q(t)$ as

$$\dot{q}(t) = \left[TA(t)T^{-1} \right] q(t) + \left[TB(t) \right] u(t) \qquad t \geq t_0 \qquad (8.218)$$

$$y(t) = \left[C(t)T^{-1} \right] q(t) + \left[D(t) \right] u(t) \qquad t \geq t_0 \qquad (8.219)$$

$q(t)$ satisfies the definition of a state vector since it has the same dimension as $x(t)$. Equations (8.218) and (8.219) are the state-space equations in terms of $q(t)$. The matrices within parentheses in these equations are the modified system and coupling matrices.

Since the state vector of a system is not unique, the selection of state variables for a given application is governed by considerations such as ease of measurement of state variables or simplification of the resulting state-space equations. If the independent energy storage elements in the system of interest are readily identified, selection of state variables directly related to energy storage in the system is appropriate. An nth-order system has n independent energy storage elements that would enable the selection of n state variables. Examples of energy storage elements are springs and masses in mechanical systems, capacitors and inductors in electrical systems, and capacitance and inertance elements in fluid (hydraulic and pneumatic) systems.

Consider the RLC circuit shown in Fig. 8.167. Let $e_{in}(t)$ be the input and $e_{out}(t)$ be the output. The current i_{out} is assumed to be negligible. Kirchhoff's voltage law for the loop yields

$$e_{in}(t) - Ri_{in}(t) - L\frac{di_{in}(t)}{dt} - \frac{1}{C_1}\int i_{in}(t)\, dt = 0 \qquad (8.220)$$

The current $i_{in}(t)$ through the inductor and the voltage $e_{out}(t)$ across the capacitor are directly

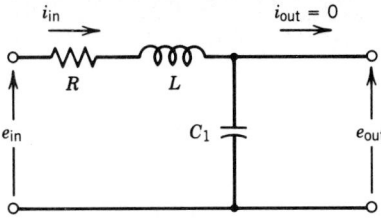

Fig. 8.167 *RLC* circuit.

related to energy storage in the system and are chosen as state variables:

$$x_1(t) = i_{in}(t) \tag{8.221}$$

$$x_2(t) = e_{out}(t) = \frac{1}{C_1} \int i_{in}(t) \, dt \tag{8.222}$$

The state equations can be determined from Eqs. (8.220)–(8.222) as

$$\dot{x}_1(t) = -\frac{R}{L}x_1(t) - \frac{1}{L}x_2(t) + \frac{e_{in}(t)}{L} \tag{8.223}$$

$$\dot{x}_2(t) = \frac{x_1(t)}{C_1} \tag{8.224}$$

The output equation is

$$y(t) = e_{out}(t) = x_2(t) \tag{8.225}$$

The system and coupling matrices for the electrical circuit are

$$\mathbf{A} = \begin{bmatrix} -\dfrac{R}{L} & -\dfrac{1}{L} \\ \dfrac{1}{C_1} & 0 \end{bmatrix} \qquad \mathbf{B} = \begin{bmatrix} \dfrac{1}{L} \\ 0 \end{bmatrix} \tag{8.226}$$

$$\mathbf{C} = [0 \ \ 1] \qquad \mathbf{D} = 0$$

For high-order, single-input–single-output (SISO) systems or multiple-input–multiple-output (MIMO) systems, the number of elements in the system matrices is large. Selection of state variables that simplify the state-space representation is thus desirable. Such representations of the state-space equations also. exhibit significant properties of the system more clearly and are referred to as canonical forms. However, the names and corresponding structures of the canonical forms are not completely standardized.

The controllable canonical form is useful in control system design applications. The controllable canonical form for an nth-order, SISO, LTV, continuous-time system is described in Table 8.18. The selection of state variables $x_1, x_2, \ldots, x_n$ that results in the controllable canonical form of the state equations is indicated on the simulation diagram in the table. For the case where all the β_i, $i = 1, n$ are zero and $\beta_0 \neq 0$, the transfer function $Y(s)/U(s)$ has no numerator dynamics. The n state variables are then simply the output and $n - 1$ successive derivatives of the output. These are referred to as the phase variables. The phase-variable canonical form is thus a special case of the controllable canonical form. The system and coupling matrices for the controllable canonical form are listed in Table 8.18. The special form of the $\mathbf{A}$ matrix is referred to as the companion form. $\mathbf{I}_{n-1}$ is the $(n-1) \times (n-1)$ identity matrix.

For a SISO, LTI system described by Eqs. (8.215) and (8.216), the state transformation matrix $\mathbf{T}$ in Eq. (8.217), which transforms the state-space equations into the controllable canonical form exists if the controllability matrix $\mathbf{P}_c$ in Eq. (8.227) is nonsingular[10]:

$$\mathbf{P}_c = [\mathbf{B} \ \mathbf{AB} \ \ \cdots \ \ \mathbf{A}^{n-1}\mathbf{B}] \tag{8.227}$$

The transformation matrix $\mathbf{T}$ is defined in Table 8.19.

TABLE 8.18 State-Space Canonical Forms for SISO Continuous-Time Systems

$$H(s) = \frac{Y(s)}{U(s)} = \frac{\beta_n s^n + \beta_{n-1} s^{n-1} + \cdots + \beta_1 s + \beta_0}{s^n + \alpha_{n-1} s^{n-1} + \cdots + \alpha_1 s + \alpha_0}$$

I. Controllable Canonical Form
    Simulation diagram

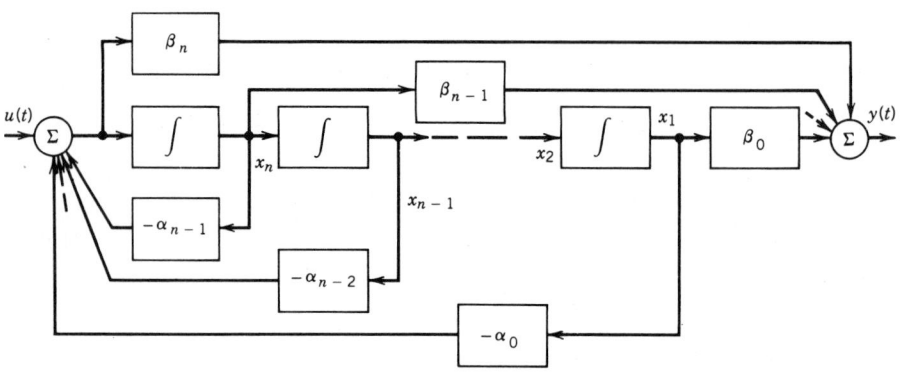

System matrices

$$A = \begin{bmatrix} 0 & & & \\ 0 & & & \\ \vdots & & I_{n-1} & \\ 0 & & & \\ -\alpha_0 & -\alpha & \cdots & -\alpha_{n-1} \end{bmatrix} \qquad B = \begin{bmatrix} 0 \\ 0 \\ \vdots \\ 0 \\ 1 \end{bmatrix} \qquad D = \beta_n$$

$$C = [\beta_0 - \beta_n \alpha_0 \quad \beta_1 - \beta_n \alpha_1 \quad \cdots \quad \beta_{n-1} - \beta_n \alpha_{n-1}]$$

II. Observable Canonical Form
    Simulation diagram

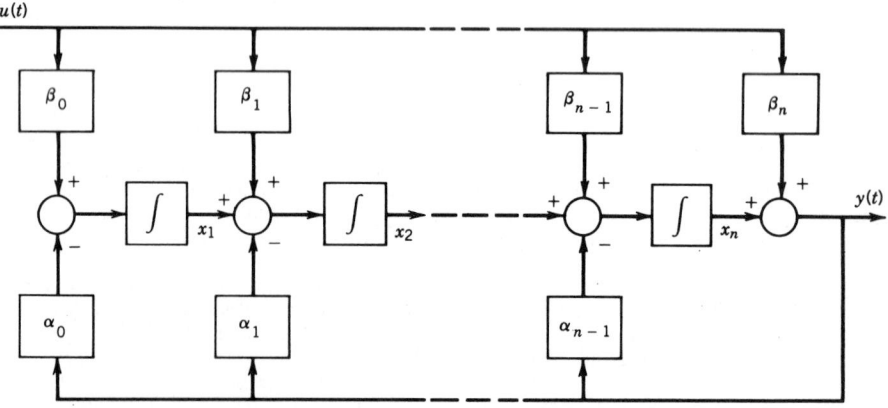

System matrices

$$A = \begin{bmatrix} 0 & 0 & \cdots & 0 & -\alpha_0 \\ & & & & -\alpha_1 \\ & & I_{n-1} & & \vdots \\ & & & & -\alpha_{n-1} \end{bmatrix} \qquad B = \begin{bmatrix} \beta_0 - \beta_n \alpha_0 \\ \beta_1 - \beta_n \alpha_1 \\ \vdots \\ \beta_{n-1} - \beta_n \alpha_{n-1} \end{bmatrix}$$

$$C = [0 \quad 0 \quad \cdots \quad 0 \quad 1] \qquad D = \beta_n$$

TABLE 8.18   (*Continued*)

III. Normal or Diagonal Jordan Canonical Form
    Related conditions
        (i) Characteristic equation roots s_i, $i = 1, \ldots, n$ are real and distinct
    Simulation diagram

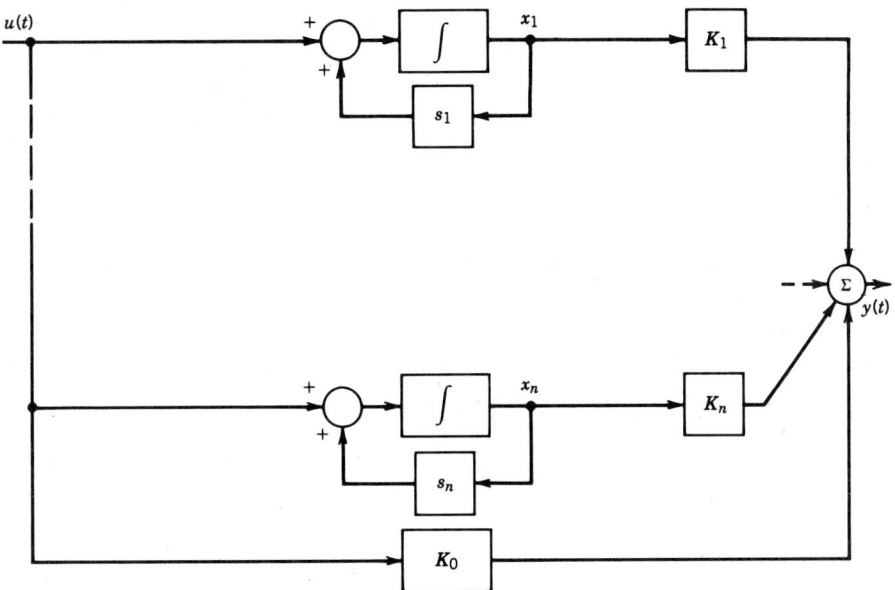

    System matrices

$$A = \begin{bmatrix} s_1 & & & \\ & s_2 & & 0 \\ & & \ddots & \\ 0 & & & s_n \end{bmatrix} \qquad B = \begin{bmatrix} 1 \\ 1 \\ \vdots \\ 1 \end{bmatrix}$$

$$C = [\, K_1 \quad K_2 \quad \cdots \quad K_n \,] \qquad\qquad D = K_0$$

    where $K_0 = \lim_{s \to \infty} [H(s)]$ and $K_i = \lim_{s \to s_i} [(s - s_i)H(s)] \qquad i = 1, \ldots, n.$

IV. Near-Normal Canonical Form
    Related conditions
        (i) One pair of complex conjugate characteristic equation roots

$$s_k = s_{kr} + js_{ki}$$

$$s_{k+1} = s_{kr} - js_{ki}$$

        (ii) All other roots are real and distinct.

TABLE 8.18 (*Continued*)

Simulation diagram

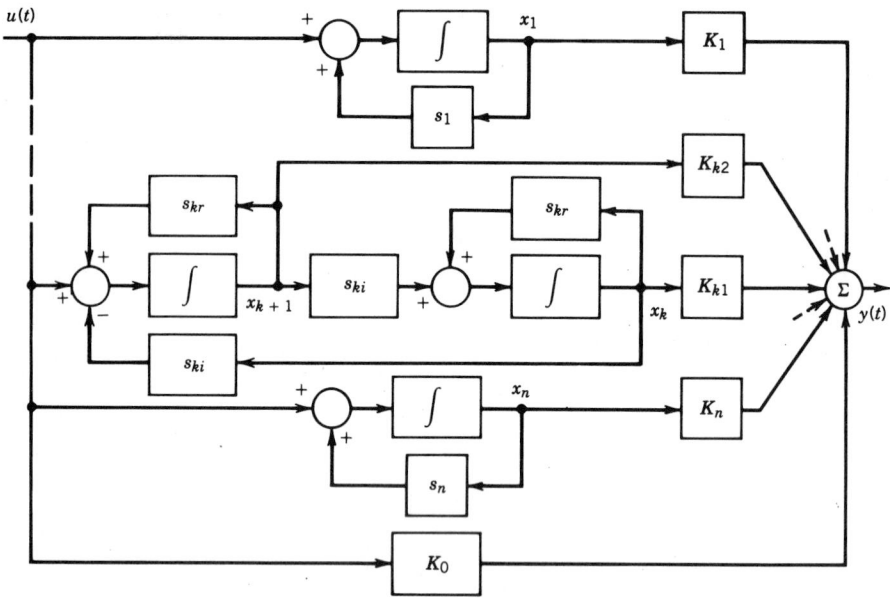

System matrices

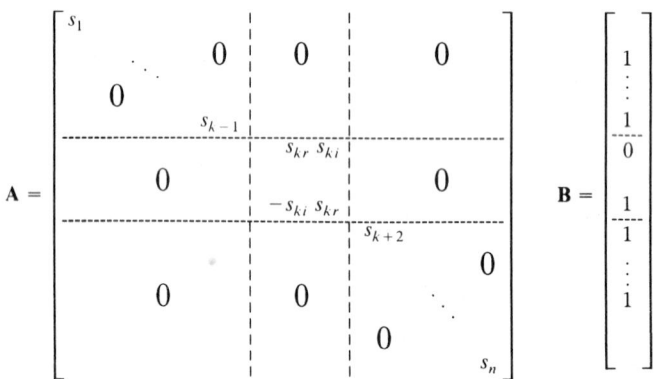

$$\mathbf{C} = [K_1 \quad \cdots \quad K_{k-1} \quad K_{k1} \ K_{k2} \ K_{k+2} \quad \cdots \quad K_n] \qquad \mathbf{D} = K_0$$

where $K_0 = \lim_{s \to \infty} [H(s)]$ $K_i = \lim_{s \to s_i} [(s - s_i)H(s)]$

$$i = 1, \ k - 1 \text{ and } i = k + 2, \ldots, n$$

$$K_{k1} = -\tfrac{1}{2}\operatorname{Im}[(s - s_k)H(s)]_{s=s_k} \quad K_{k2} = \tfrac{1}{2}\operatorname{Re}[(s - s_k)H(s)]_{s=s_k}$$

V. Nondiagonal Jordan Canonical Form
   Related conditions
     (i) One real characteristic equation root s_k repeated m times
        (i.e., $s_k = s_{k+1} = \cdots = s_{k+m-1}$).
     (ii) All other roots real and distinct.

TABLE 8.18   (*Continued*)

Simulation diagram

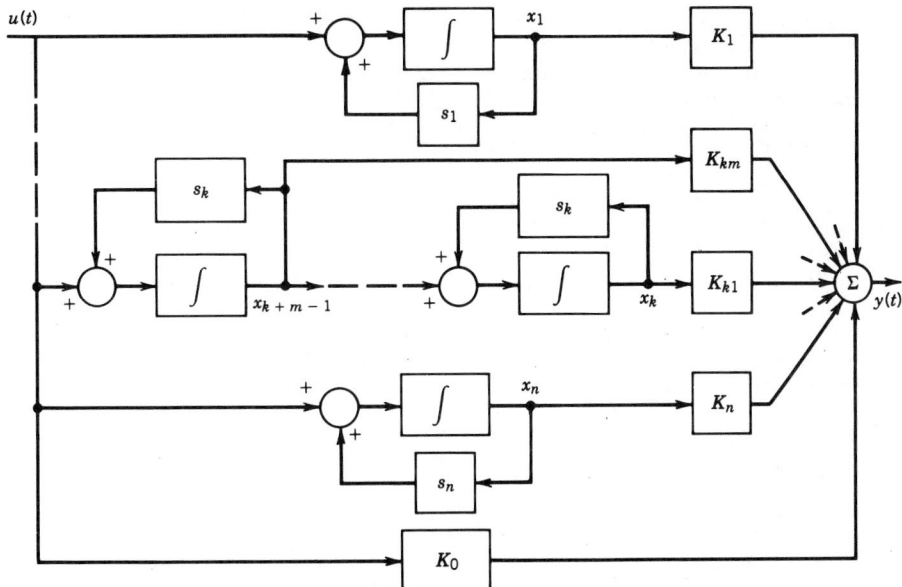

System matrices

$$
\mathbf{A} = \begin{bmatrix}
s_1 & & & & & & \\
& \ddots & 0 & & 0 & & 0 \\
0 & & s_{k-1} & & & & \\
\hline
& & & s_k & 1 & 0 & \\
0 & & & & \ddots & 1 & 0 \\
& & & 0 & & s_k & \\
\hline
& & & & & s_{k+m} & 0 \\
& 0 & & & 0 & 0 & \ddots \\
& & & & & & s_n
\end{bmatrix}
\qquad
\mathbf{B} = \begin{bmatrix}
1 \\
\vdots \\
1 \\
\hline
0 \\
\vdots \\
1 \\
\hline
1 \\
\vdots \\
1
\end{bmatrix}
$$

$\mathbf{C} = [K_1 \quad \cdots \quad K_{k-1} \; K_{k1} \; \cdots \; K_{km} \; K_{k+m} \quad \cdots \quad K_n]$ $\mathbf{D} = K_0$

where $K_0 = \lim\limits_{s \to \infty} [H(s)]$ $K_i = \lim\limits_{s \to s_i} [(s - s_i) H(s)]$

$$i = 1, \, k - 1 \text{ and } i = k + m, \dots, n$$

and $K_{kj} = \dfrac{1}{(j-1)!} \left\{ \dfrac{d^{j-1}}{ds^{j-1}} \left[(s - s_k)^m H(s) \right] \right\}_{s = s_k}$ $j = 1, \dots, m$

TABLE 8.19   Transformation Matrices for Continuous-Time State-Space Canonical Form

State-space equations (SISO, LTI system)	Characteristic equation
$\dot{x}(t) = Ax(t) + Bu(t)$ $y(t) = Cx(t) + Du(t)$	$\det(sI - A) =$ $s^n + \alpha_{n-1}s^{n-1} + \cdots + \alpha_1 s + \alpha_0 = 0$

I.   Controllable Canonical Form
   Transformation conditions
      (i) $P_c = [B \quad AB \quad \ldots \quad A^{n-1}B]$ must be nonsingular
   Transformation matrices
      (i) $q = Tx, T = R^{-1}P_c^{-1}$ (ii) New state matrix $= TAT^{-1}$
      where

$$
R = \begin{bmatrix}
\alpha_1 & \alpha_2 & \cdots & \alpha_{n-1} & 1 \\
\alpha_2 & \alpha_3 & \cdots & 1 & 0 \\
\vdots & \vdots & & \vdots & \vdots \\
\alpha_{n-1} & 1 & \cdots & 0 & 0 \\
1 & 0 & \cdots & 0 & 0
\end{bmatrix}
\qquad = \left[\begin{array}{c|c}
0 & \\
0 & \\
\vdots & I_{n-1} \\
0 & \\
\hline
-\alpha_0 & -\alpha_1 \quad \cdots \quad -\alpha_{n-1}
\end{array}\right]
$$

II.   Observable Canonical Form
   Transformation conditions

      (i) $P_0 = \begin{bmatrix} C \\ CA \\ \vdots \\ CA^{n-1} \end{bmatrix}$ must be nonsingular

   Transformation matrices
      (i) $q = Tx, T = RP_0$ (ii) New state matrix $= TAT^{-1}$
      where

$$
R = \begin{bmatrix}
\alpha_1 & \alpha_2 & \cdots & \alpha_{n-1} & 1 \\
\alpha_2 & \alpha_3 & \cdots & 1 & 0 \\
\vdots & \vdots & & \vdots & \vdots \\
\alpha_{n-1} & 1 & \cdots & 0 & 0 \\
1 & 0 & \cdots & 0 & 0
\end{bmatrix}
\qquad = \left[\begin{array}{c|c}
0 \quad 0 \quad \cdots \quad 0 & -\alpha_0 \\
\hline
 & -\alpha_1 \\
I_{n-1} & \vdots \\
 & -\alpha_{n-1}
\end{array}\right]
$$

III.   Normal or Diagonal Jordan Canonical Form
   Transformation conditions
      (i) A matrix has only distinct, real eigenvalues s_i, $i = 1, \ldots, n$.

   Transformation matrices
      (i) $q = Tx, T^{-1} = M = [v_1 \quad v_2 \quad \cdots \quad v_n]$
         where (a) v_i are the n linearly independent eigenvectors corresponding to s_i and
            (b) v_i are taken to be equal or proportional to any nonzero column of $\text{Adj}(s_i I - A)$
      (ii) New state matrix $= TAT^{-1}$

$$
= \begin{bmatrix}
s_1 & & & \\
& s_2 & & \text{\huge 0} \\
& & \ddots & \\
\text{\huge 0} & & & s_n
\end{bmatrix}
$$

IV.   Normal or Diagonal Jordan Canonical Form
   Transformation conditions
      (i) A matrix has one repeated, real eigenvalue s_k of multiplicity m (i.e., $s_k = s_{k+1} = \cdots = s_{k+m-1}$). All other eigenvalues are real and distinct.
      (ii) Degeneracy $d = n - \text{rank}(s_k I - A) = m$. Full degeneracy.

TABLE 8.19   (Continued)

Transformation matrices
(i) $\mathbf{q} = \mathbf{Tx}, \mathbf{T}^{-1} = \mathbf{M} = [\mathbf{v}_1 \quad \mathbf{v}_2 \quad \cdots \quad \mathbf{v}_n]$
   where (a) $\mathbf{v}_i$, $i = 1, k - 1$ and $i = k + m, n$ are the linearly independent eigenvectors corresponding to the real, distinct eigenvalues,
   (b) $\mathbf{v}_i$, $i = 1, k - 1$ and $i = k + m, n$ are taken to be equal or proportional to any nonzero column of Adj $(s_i \mathbf{I} - \mathbf{A})$, and
   (c) $\mathbf{v}_i$, $i = k, k + m - 1$ are the m linearly independent eigenvectors corresponding to the repeated eigenvalue. They are equal or proportional to the nonzero linearly independent columns of

$$\left\{ \frac{d^{m-1}}{ds^{m-1}} [\text{Adj}\,(s\mathbf{I} - \mathbf{A})] \right\}_{s=s_k}$$

(ii) New state matrix $= \mathbf{TAT}^{-1}$

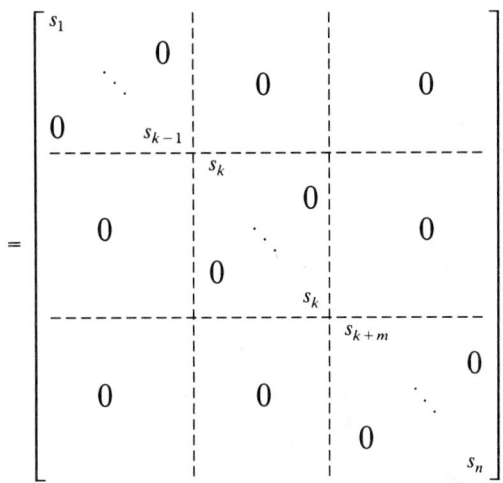

V.   Near-Normal Canonical Form
     Transformation conditions
         (i) **A** matrix has one pair of complex, conjugate eigenvalues, s_k, s_{k+1}

$$s_k = s_{kr} + js_{ki}$$
$$s_{k+1} = s_{kr} - js_{ki}$$

     (ii) All other eigenvalues are real and distinct.

Transformation matrices
(i) $\mathbf{q} = \mathbf{Tx}, \mathbf{T}^{-1} = [\mathbf{v}_1 \quad \cdots \quad \mathbf{v}_{k-1} \quad \mathbf{v}_{kr} \quad \mathbf{v}_{ki} \quad \mathbf{v}_{k+2} \quad \cdots \quad \mathbf{v}_n]$
   where (a) $\mathbf{v}_i$, $i = 1, k - 1$ and $i = k + 2, n$ are the linearly independent eigenvectors corresponding to the real, distinct eigenvalues,
   (b) $\mathbf{v}_i$, for $i = 1,\ldots, n$ are taken to be equal or proportional to any nonzero column of Adj $(s_i \mathbf{I} - \mathbf{A})$, and
   (c) $\mathbf{v}_k = \mathbf{v}_{kr} + j\mathbf{v}_{ki}$
       $\mathbf{v}_{k+1} = \mathbf{v}_{kr} - j\mathbf{v}_{ki}$
       are the complex conjugate eigenvectors corresponding to s_k and s_{k+1}, respectively.

TABLE 8.19 (*Continued*)

(ii) New state matrix = $\mathbf{TAT}^{-1}$

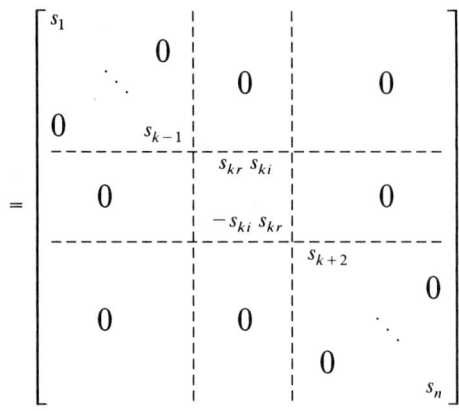

VI. Nondiagonal Jordan Canonical Form
   Transformation conditions
   (i) **A** matrix has one repeated, real eigenvalue s_k of multiplicity m (i.e., $s_k = s_{k+1} = \cdots = s_{k+m-1}$). All other eigenvalues are real and distinct.
   (ii) Degeneracy $d = n - \text{rank}(s_k\mathbf{I} - \mathbf{A}) = 1$. Simple degeneracy.

   Transformation matrices
   (i) $\mathbf{q} = \mathbf{Tx}, \mathbf{T}^{-1} = [\mathbf{t}_1 \quad \mathbf{t}_2 \quad \cdots \quad \mathbf{t}_n]$
      where (a) $\mathbf{t}_i$, $i = 1, k-1$ and $i = k+m, n$ are the linearly independent eigenvectors corresponding to the real, distinct eigenvalues,
         (b) $\mathbf{t}_i$, $i = 1, k-1$ and $i = k+m, n$ are taken to be equal or proportional to any nonzero column of Adj$(s_i\mathbf{I} - \mathbf{A})$, and
         (c) $\mathbf{t}_i$, $i = k, k+m-1$ are obtained by solution of the equation $\mathbf{AT}^{-1} = \mathbf{T}^{-1}\mathbf{J}$, where **J** is the Jordan canonical matrix given. Each $\mathbf{t}_i$ is determined to within a constant of proportionality.
   (ii) New state matrix = $\mathbf{J} = \mathbf{TAT}^{-1}$

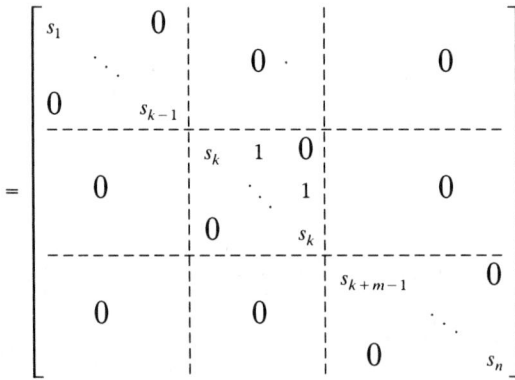

VII. Nondiagonal Jordan Canonical Form
   Transformation conditions
   (i) **A** matrix has one repeated, real eigenvalue of multiplicity m (i.e., $s_k = s_{k+1} = \cdots = s_{k+m-1}$). All other eigenvalues are real and distinct.
   (ii) Degeneracy $d = n - \text{rank}(s_k\mathbf{I} - \mathbf{A})$ is between 1 and m. General degeneracy.

TABLE 8.19   (*Continued*)

Transformation matrices
(i) $\mathbf{q} = \mathbf{Tx}, \mathbf{T}^{-1} = [\mathbf{t}_1 \quad \mathbf{t}_2 \quad \cdots \quad \mathbf{t}_n]$
   where (a) $\mathbf{t}_i$, $i = 1$, $k - 1$ and $i = k + m$, n are the linearly independent eigenvectors corresponding to the real distinct eigenvalues,
   (b) $\mathbf{t}_i$, $i = 1$, $k - 1$ and $i = k + m$, n are taken to be equal or proportional to any nonzero column of Adj$(s_i \mathbf{I} - \mathbf{A})$,
   (c) $\mathbf{t}_i$, $i = k$, $k + m - 1$ are m linearly independent vectors, corresponding to the repeated eigenvalue of multiplicity m. Only d of these vectors are eigenvectors, and
   (d) $\mathbf{t}_i$, $i = k$, $k + m - 1$ obtained by solution of the equation $\mathbf{AT}^{-1} = \mathbf{T}^{-1}\mathbf{J}$, where $\mathbf{J}$ is a Jordan canonical matrix for the problem, with d Jordan blocks. There are d possible choices for $\mathbf{J}$. Each of these choices needs to be tried and the $\mathbf{t}_i$ vectors solved for. Only the correct $\mathbf{J}$ will give the m linearly independent vectors $\mathbf{t}_i$, $i = k$, $k + m - 1$. Each $\mathbf{t}_i$ is determined to within a constant of proportionality.
(ii) New state matrix $\mathbf{J} = \mathbf{TAT}^{-1}$. Correct $\mathbf{J}$ determined by trial-and-error, as previously described.

The observable canonical form is useful in state-estimator or observer design applications. The observable canonical form for an nth-order, SISO, LTI system is described in Table 8.18 in a manner similar to the controllable canonical form. The corresponding $\mathbf{A}$ matrix is the transpose of the $\mathbf{A}$ matrix for the controllable canonical form and is also referred to as a companion matrix. The state transformation matrix $\mathbf{T}$ in Eq. (8.217), which transforms given state-space Eqs. (8.215) and (8.216) into the observable canonical form, exists if the observability matrix $\mathbf{P}_0$ is nonsingular[10]:

$$\mathbf{P}_0 = \begin{bmatrix} \mathbf{C} \\ \mathbf{CA} \\ \vdots \\ \mathbf{CA}^{n-1} \end{bmatrix} \tag{8.228}$$

The transformation matrix $\mathbf{T}$ is defined in Table 8.19.

State variables can also be chosen to diagonalize or nearly diagonalize the state matrix $\mathbf{A}$. The resulting state-space equations are completely or almost completely decoupled from one another and hence show very clearly the effect of initial conditions or forcing inputs on the different characteristic modes of the system response. The resulting physical insight into the system behavior makes the corresponding form of the system equations, called the normal form or diagonal Jordan form, valuable in vibration analysis applications and in control applications involving modal control. The normal form of the state-space equations for a SISO, LTI system with real, distinct characteristic roots is given in Table 8.18. The diagonal elements of the $\mathbf{A}$ matrix in the table are the system characteristic roots or eigenvalues. The state variables $x_1, x_2, \ldots, x_n$ lie along the eigenvectors of the $\mathbf{A}$ matrix, in state space. As the corresponding simulation diagram indicates, the behavior of each state variable is governed solely by one eigenvalue, the initial condition on that state variable, and the forcing input.

If some of the distinct characteristic roots of a SISO, LTI system are complex, the matrices $\mathbf{A}$ and $\mathbf{C}$ in Eqs. (8.215) and (8.216) have complex elements when represented in the normal form just described. Since this could be inconvenient in subsequent matrix manipulations, a nearly diagonal $\mathbf{A}$ matrix can be obtained for cases where the complex characteristic roots occur in complex conjugate pairs. This would be the case for system differential equations with only real coefficients. The near-normal form of the system equations for a system with one pair of complex conjugate characteristic roots is given in Table 8.19. Extension to the case of multiple complex root pairs is straightforward. The complex characteristic roots result in a few nonzero off-diagonal elements in the $\mathbf{A}$ matrix; otherwise, the decoupled nature of the system equations is retained.

If one characteristic root of a SISO, LTI system is real and repeated m times, the state equations can only be partially decoupled by appropriate state-variable selection, as shown in Table 8.19. The resulting state-space equations are said to be in the Jordan canonical form. The corresponding $\mathbf{A}$ matrix has one submatrix with the repeated eigenvalue at the diagonal positions, ones immediately to the right of the repeated diagonal elements within the submatrix and zero elements at all other nondiagonal positions.[2] The $\mathbf{A}$ matrix is then said to have one Jordan block. The extension of the result in Table 8.19 to the case of many different repeated characteristic roots is straightforward.

For SISO or MIMO, LTI systems described by state-space Eqs. (8.215) and (8.216), the transformation matrix $\mathbf{T}$ in Eq. (8.217), which transforms the state-space equations into the diagonal or nondiagonal Jordan form, can be determined. As in Table 8.19, there are a number of different cases to be considered.[2] If the $\mathbf{A}$ matrix has real, distinct eigenvalues s_i, $i = 1, \ldots, n$, the eigenvectors are linearly independent and can be used to form the modal matrix $\mathbf{M}$ as indicated in Table 8.19. The transformation matrix $\mathbf{T}$ is then taken to be $\mathbf{M}^{-1}$. If the $\mathbf{A}$ matrix has one pair of complex, conjugate eigenvalues and if system matrices with real elements only are desired, the transformation matrix $\mathbf{T}$ is defined in a slightly different form as indicated in Table 8.19. The resulting transformed state matrix will have two nonzero off-diagonal elements as indicated.

The transformed state-space equations may be in the nondiagonal Jordan canonical form if the $\mathbf{A}$ matrix has repeated eigenvalues. The procedure for determining the transformation matrix depends on the degeneracy of the matrix $s_k \mathbf{I} - \mathbf{A}$ corresponding to the repeated eigenvalue s_k. If the degeneracy d of $s_k \mathbf{I} - \mathbf{A}$, defined in Table 8.19, is equal to m where m is the multiplicity of the repeated eigenvalue s_k, m linearly independent eigenvectors can be found for the repeated eigenvalue. The procedure for doing so is indicated in Table 8.19. The transformed state matrix is then diagonal. If the degeneracy of $s_k \mathbf{I} - \mathbf{A}$ is one, only one eigenvector can be determined. Since it can be shown that the degeneracy is equal to the number of Jordan blocks associated with the eigenvector, the transformed state matrix $\mathbf{J}$ has only one Jordan block and is uniquely defined.[2] The nonsingular transformation matrix $\mathbf{T}$ is then determined as indicated in Table 8.19. If the degeneracy d of $s_k \mathbf{I} - \mathbf{A}$ is greater than one but less than m, there are d linearly independent eigenvectors and d Jordan blocks associated with the eigenvalue s_k. In this case, the transformed state matrix $\mathbf{J}$ cannot be uniquely defined but can be one of a finite number of possibilities. A trial-and-error formulation of $\mathbf{J}$ and solution for $\mathbf{T}$, as indicated in Table 8.19, is necessary until a nonsingular transformation matrix $\mathbf{T}$ is obtained.[2]

With the exception of the procedures for selecting transformation matrices to convert state-space equations for MIMO, LTI systems to the Jordan canonical form, the procedures and results presented in Tables 8.18 and 8.19 are restricted to SISO, LTI systems. A procedure for representing a SISO, linear time-varying (LTV) system, described by a differential equation, in controllable canonical form has been described by DeRusso, Roy, and Close.[2]

Canonical forms for MIMO, LTI systems cannot, in general, be specified uniquely as is the case for SISO systems. Kailath,[3] Fortmann and Hitz,[1] and Kalman[9] have specified some canonical forms for MIMO systems and have described procedures for representing such systems in these forms, given their transfer function matrix descriptions or state-space equations. The problem of selection of state variables given the transfer function matrix description of MIMO systems is the problem of realization and is considered in Section 8.38.

8.35 SOLUTION OF SYSTEM EQUATIONS

The state-space Eqs. (8.210) and (8.211) for time-varying, nonlinear systems described by ordinary differential equations can be solved by numerical integration techniques. Such numerical integration would, however, have to be repeated if the initial conditions $\mathbf{x}(t_0)$ or the forcing function $\mathbf{u}(t)$ were to be changed. The computational burden can be reduced for linear systems by using the concept of the state transition matrix.

For LTI systems described by the state Eqs. (8.215) and (8.216), the solution is given by

$$\mathbf{x}(t) = \boldsymbol{\phi}(t - t_0)\mathbf{x}(t_0) + \int_{t_0}^{t} \boldsymbol{\phi}(t - \tau)\mathbf{B}\mathbf{u}(\tau)\,d\tau \qquad (8.229)$$

$$\mathbf{y}(t) = \mathbf{C}\boldsymbol{\phi}(t - t_0)\mathbf{x}(t_0) + \int_{t_0}^{t} \mathbf{C}\boldsymbol{\phi}(t - \tau)\mathbf{B}\mathbf{u}(\tau)\,d\tau + \mathbf{D}\mathbf{u}(t) \qquad (8.230)$$

where the $n \times n$ matrix $\boldsymbol{\phi}(t)$ is defined as the state transition matrix of the system. Derivation of this result is available in many standard textbooks on state-space methods.[2,4] The first terms on the right-hand sides of the preceding equations represent the response of the homogeneous system to the initial condition $\mathbf{x}(t_0)$ whereas the second terms represent the forced response of the system. Comparison of the second term in Eq. (8.230), for the case $\mathbf{D} = 0$, with Eq. (8.231) for the forced response of a SISO, LTI system indicates that the matrix $\mathbf{C}\boldsymbol{\phi}(t)\mathbf{B}$ is a matrix of impulse responses:

$$y(t_0) = y(t_0) + \int_{t_0}^{t} h(t - \tau)u(\tau)\,d\tau \qquad (8.231)$$

The variable $h(t)$, in Eq. (8.231), is the impulse response of the system. The interpretation of $\mathbf{C}\boldsymbol{\phi}(t)\mathbf{B}$ as a matrix of impulse responses forms the basis of one of the methods for determining the elements of the transition matrix.[2,4]

The state transition matrix $\phi(t - t_0)$ is the solution of the matrix differential equation:

$$\dot{\phi}(t - t_0) = A\phi(t - t_0) \qquad t \geq t_0 \qquad (8.232)$$

with the initial condition

$$\phi(t_0 - t_0) = \phi(0) = I \qquad (8.233)$$

It has the following properties:

$$\phi(t + \tau) = \phi(t)\phi(\tau) = \phi(\tau)\phi(t) \qquad (8.234)$$

$$\phi^{-1}(t) = \phi(-t) \qquad (8.235)$$

The following expressions for $\phi(t)$ can be verified and are useful in its evaluation:

$$\phi(t) = e^{At} = I + At + \frac{A^2 t^2}{2!} + \frac{A^3 t^3}{3!} + \cdots \qquad (8.236)$$

and

$$\phi(t) = \mathscr{L}^{-1}\left[(sI - A)^{-1}\right] \qquad (8.237)$$

where $\mathscr{L}^{-1}$ denotes the inverse Laplace transform. Details related to Eqs. (8.232)–(8.237) have been described by DeRusso et al.[2] and Brogan.[4]

Knowledge of the state transition matrix for a given system simplifies the task of determining the response of the system to a variety of initial conditions $x(t_0)$ and forcing functions $u(t)$. A number of analytical and numerical techniques for its evaluation are available.

Equation (8.236) forms the basis for a numerical method of determining $\phi(t)$. Closed-form evaluation of e^{At} is possible only for special forms of the A matrix. For example, if A is a diagonal matrix with diagonal elements equal to the eigenvalues s_i, it can be shown that $\phi(t)$ is also diagonal[4] and is given by Eq. (8.238):

$$\phi(t) = e^{At} = \begin{bmatrix} e^{s_1 t} & & & \\ & e^{s_2 t} & & 0 \\ & 0 & \ddots & \\ & & & e^{s_n t} \end{bmatrix} \qquad (8.238)$$

Closed-form evaluation of e^{At} is only slightly more complex if A is in the nondiagonal Jordan canonical form.[4] If the transformation matrix T [Eq. (8.217)] was used to obtain the diagonal or nondiagonal Jordan matrix A, the transition matrix, for the original state vector $T^{-1}x$, is $T^{-1}e^{At}T$.

Equation (8.237) provides the basis for an analytical evaluation of $\phi(t)$ that is suitable for low-order dynamic systems. This method requires the inversion of the $n \times n$ matrix $sI - A$, followed by the inverse Laplace transformation of the n^2 elements. The matrix inversion is especially cumbersome since the elements of the matrix are functions of s. The matrix inversion can be avoided altogether by using simulation diagrams of the system, in conjunction with block diagram reduction techniques,[2] to determine elements of the matrix $(sI - A)^{-1}$. Alternative analytical techniques for the evaluation of $\phi(t)$ based on Sylvester's theorem and the Cayley–Hamilton theorem have been described by DeRusso et al.[2] and Brogan.[4]

Numerical evaluation of $\phi(t)$, for a specified value of t, can be performed using Eq. (8.236) and retaining a finite number of terms from the series expansion. The number of terms retained increases with the desired degree of accuracy. An iterative procedure for determining the number of terms to be retained for a specified degree of accuracy has been described by Shinners.[5]

For linear, time-varying systems described by state-space Eqs. (8.213) and (8.214), the solution is given by[2,4]

$$x(t) = \phi(t, t_0)x(t_0) + \int_{t_0}^{t} \phi(t, \tau)B(\tau)u(\tau)\,d\tau \qquad t \geq t_0 \qquad (8.239)$$

$$y(t) = C(t)\phi(t, t_0)x(t_0) + \int_{t_0}^{t} C(t)\phi(t, \tau)B(\tau)u(\tau)\,d\tau + D(t)u(t) \qquad (8.240)$$

where the $n \times n$ state transition matrix $\phi(t, t_0)$ for the time-varying system depends on both arguments t and t_0 and not merely on the difference between these two time instants as in the time-invariant system.

The state transition matrix $\phi(t, t_0)$ is the solution of the partial differential equation[2,4]

$$\frac{\partial \phi(t, t_0)}{\partial t} = A(t)\phi(t, t_0) \tag{8.241}$$

with the initial condition

$$\phi(t_0, t_0) = I \tag{8.242}$$

It has the following properties:

$$\phi(t_2, t_0) = \phi(t_2, t_1)\phi(t_1, t_0) \tag{8.243}$$

$$\phi(t_1, t_0) = \phi^{-1}(t_0, t_1) \tag{8.244}$$

Techniques for evaluating the state transition matrix for time-varying systems are considerably more involved than for time-invariant systems and are less widely applicable. A number of analytical methods for determining $\phi(t, t_0)$ for special cases of linear time-varying systems have been described by DeRusso et al.[2]

A simple numerical procedure has been suggested by Palm[6] for computing the transition matrix when analytical determination is not possible. Let the ith column of $\phi(t, t_0)$ be denoted by $\psi_i(t)$, for a specified value of t_0. The matrix partial differential Eq. (8.241) becomes n vector ordinary differential equations:

$$\dot{\psi}_i(t) = A(t)\psi_i(t) \qquad i = 1, \ldots, n \qquad t \ge t_0 \tag{8.245}$$

with the initial conditions

$$[\psi_1(t_0)\psi_2(t_0)\ldots\psi_n(t_0)] = I \tag{8.246}$$

Numerical solution of the ordinary differential equations gives $\psi_i(t)$ and hence $\phi(t, t_0)$. Note that the computed $\phi(t, t_0)$ would be different for different values of t_0 for time-varying systems.

Procedures for solution of discrete-time system equations are similar to those described above. DeRusso et al.[2] and Brogan[4] give details.

8.36 STABILITY

Since state-space formulation is applicable to a large class of dynamic systems, the question of stability for systems represented in state space is quite a complex one. A more general consideration of stability than that used for SISO, LTI systems would indicate that stability of dynamic systems is not really a property of the systems but is more properly associated with isolated equilibrium points of dynamic systems.[4] A particular point x_e in state space is an equilibrium point of a dynamic system if, in the absence of inputs, the system state x is equal to x_e for time $t \ge t_0$ for continuous-time systems. For linear systems described by the state-space equations given in Section 8.33, the only isolated equilibrium point is at the origin in state space. For nonlinear systems, there may be a number of isolated equilibrium points. Any isolated equilibrium point can be shifted to the origin in state space by a simple change of state variables.[4] The stability definitions to be given assume therefore that the equilibrium point is at the origin in state space and that the system is unforced. Only the more commonly used types of stability will be defined.

The origin is a stable equilibrium point if, for any given value $\varepsilon > 0$, there exists a number $\kappa(\varepsilon, t_0) > 0$ such that if the norm $\|x(t_0)\| < \kappa$, then the norm $\|x(t)\| < \varepsilon$ for all $t > t_0$. The norm of a vector x is defined as

$$\|x(t)\| = \sqrt{\sum_{i=1}^{n} x_i^2(t)} \tag{8.247}$$

The origin is asymptotically stable if, in addition to being stable, there exists a number $\gamma(t_0) > 0$ such that whenever $\|x(t_0)\| < \gamma(t_0)$, the following condition is satisfied:

$$\lim_{t \to \infty} \|x(t)\| = 0 \tag{8.248}$$

TABLE 8.20   Stability Criteria for Linear, Time-Invariant Continuous-Time Systems

	$\dot{x}(t) = Ax(t) + Bu$ Eigenvalues of A are $s_i = \alpha_{ic} \pm j\omega_{ic}$
Asymptotically stable	$\alpha_{ic} < 0$ for all roots
Stable	$\alpha_{ic} < 0$ for all repeated roots and $\alpha_{ic} \le 0$ for all simple roots
Unstable	$\alpha_{ic} > 0$ for any simple root or $\alpha_{ic} \ge 0$ for any repeated root

If κ and γ are not functions of t_0 in the previous definitions, the origin is said to be uniformly stable or uniformly asymptotically stable, respectively. If $\gamma(t_0)$ can be arbitrarily large, the origin is said to be globally asymptotically stable. Additional types of stability that depend on the inputs to the system have been defined by Brogan[4] and Kuo.[7]

For LTI systems, the conditions for stability reduce to conditions on the eigenvalues of the system matrix A and are summarized in Table 8.20. These eigenvalues are the roots of the system characteristic equation as well, as shown in Section 8.38. They may be computed explicitly by numerical methods. Alternatively, stability criteria such as the Routh–Hurwitz criterion[5] for continuous-time systems or the Jury test[7] for discrete-time systems may be applied. The conditions for asymptotic stability of such systems can also be shown to be sufficient for other types of stability depending on the input, such as bounded-input, bounded-output stability.[4]

For continuous-time, LTV systems, the necessary and sufficient condition for the origin to be a stable equilibrium point is that there exists a number $N(t_0)$ such that the norm of the transition matrix satisfies the following condition:

$$\|\phi(t, t_0)\| \le N(t_0) \quad \text{for } t \ge t_0 \tag{8.249}$$

If, in addition, $\|\phi(t, t_0)\| \to 0$ as $t \to \infty$, the system is globally asymptotically stable.[4] The norm of the matrix ϕ is defined as

$$\|\phi(t, t_0)\| = \sqrt{\max_{\|x\|=1} (x^T \phi^T \phi x)} \tag{8.250}$$

Time-varying systems that satisfy the property that the state converges exponentially with time to the zero state are said to be exponentially stable.[12] For LTI systems, of course, asymptotic stability is the same as exponential stability.

Stability considerations for nonlinear systems are more complex. For unforced second-order nonlinear systems, the phase-plane method is useful for examining the stability of equilibrium points of the system. The phase plane has the state variables as the coordinates. The state-space equations are used to derive analytical expressions for the trajectories or to draw the trajectories by graphical means. The phase portraits can then be examined to determine the equilibrium points and their stability. Application of the phase-plane method is described by DeRusso et al.[2] for continuous-time systems.

Stability analysis of high-order nonlinear systems represented in state space can be done using the second method of Lyapunov. This is a technique requiring considerable ingenuity for effective use and provides sufficient conditions for stability rather than necessary and sufficient conditions.[8]

Lyapunov's method for nonlinear, unforced, time-invariant systems requires the definition of a scalar function of state $V(x)$ called the Lyapunov function. The latter may be thought of as a generalized energy function. The requirement on the Lyapunov function is that it be positive definite in some region about the origin in state space, the origin having been assumed to be an isolated equilibrium point here. A function $V(x)$, which is continuous and has continuous partial derivatives, is said to be positive (negative) definite in some region about the origin if it is zero at the origin and greater than (less than) zero everywhere else in the specified region. If the function is greater than (less than) or equal to zero everywhere in the specified region, it is said to be positive (negative) semidefinite.[4]

Consider the unforced continuous-time system represented by the state equation

$$\dot{x}(t) = f[x(t)] \tag{8.251}$$

where

$$\mathbf{f}(0) = 0 \tag{8.252}$$

If a positive definite function $V(\mathbf{x})$ can be determined in some region Γ about the origin such that its derivative with respect to time is negative semidefinite in Γ, then the origin is a stable equilibrium point. If dV/dt is negative definite, the origin is asymptotically stable. If the region Γ can be arbitrarily large and the conditions for asymptotic stability hold and if, in addition, $V(\mathbf{x}) \to \infty$ as $\|\mathbf{x}\| \to \infty$, the origin is a globally asymptotically stable equilibrium point. Extensions of the stability conditions for time-varying systems have been described by Kalman and Bertram[8] and DeRusso et al.[2]

As an example of the application of the second method of Lyapunov, consider the following nonlinear system[5]:

$$\dot{x}_1 = x_2$$
$$\dot{x}_2 = -a_0 x_2 - b_0 x_2^3 - x_1 \tag{8.253}$$

where $a_0, b_0 \geq 0$ and both are not zero. The origin is an equilibrium point for this system since, if both x_1 and x_2 are zero,

$$\dot{x}_1 = \dot{x}_2 = 0 \tag{8.254}$$

Consider the following Lyapunov function:

$$V(x_1, x_2) = x_1^2 + x_2^2 \tag{8.255}$$

It satisfies the conditions for positive definiteness in an arbitrarily large region about the origin:

$$\frac{dV(x_1, x_2)}{dt} = 2x_1 \dot{x}_1 + 2x_2 \dot{x}_2$$
$$= -2\left(a_0 x_2^2 + b_0 x_2^4 \right) \tag{8.256}$$

after using the state equations to substitute for $\dot{x}_1$ and $\dot{x}_2$. If a_0, b_0 satisfy the inequalities stated, dV/dt is negative definite in an arbitrarily large region about the origin. Since $V(x_1, x_2)$ becomes infinitely large for infinitely large x_1 or x_2, the origin is a global asymptotically stable equilibrium point.

The limitations of Lyapunov's method are that the Lyapunov function is not unique for a system and there are no systematic procedures for finding a suitable Lyapunov function. Since only sufficient conditions for stability are determined, some choices of Lyapunov functions are better in that they provide more information about system stability then others. Also, appropriate choice of the Lyapunov function can lead to an estimate of the system speed of response.[8] In practice, therefore, the second method of Lyapunov is used primarily to analyze the stability of systems such as high-order, nonlinear systems for which other methods of stability analysis are not available.

Parallel results on the stability analysis of discrete-time systems are given by Brogan[4] and Kuo.[7]

8.37 CONTROLLABILITY AND OBSERVABILITY

The controllability of a linear system is a measure of the coupling between the inputs to the system and the system state. The concept of state controllability was introduced by Kalman[11] in order to clarify conditions for the existence of solutions to specific control problems.

A linear, continuous-time system is said to be state controllable at time t_0 if there exists a finite time $t_1 > t_0$ and a control function $\mathbf{u}(t)$, $t_0 < t < t_1$ that drives the system state $\mathbf{x}(t_0)$ to the origin at $t = t_1$. If the system is controllable for all times t_0 and all initial states $\mathbf{x}(t_0)$, the system is completely state controllable.[4] An additional form of controllability for continuous-time LTV systems is that of uniformly complete state controllability. The mathematical definition of this form of controllability may be found in Kalman.[11] This property implies that the control effort and time interval required to derive the system state to the origin is relatively independent of the initial time. For LTI systems, of course, complete state controllability is the same as uniformly complete state controllability.

Though the control problems formulated above are open-loop control problems, the property of controllability has very significant implications for closed-loop control problems. Section 8.40 indicates that the closed-loop poles of a completely state-controllable time-invariant system can be specified and placed arbitrarily in the complex s plane by proportional state-variable feedback.

Moreover, satisfaction of the controllability conditions to be defined in this section for time-invariant systems ensures that the optimal control law for a quadratic performance index is a proportional state-variable feedback law and yields an asymptotically stable closed-loop system.[10]

Direct application of the definition of state controllability to LTI systems yields controllability conditions involving the transition matrices. Simple algebraic conditions are usually available for such systems and are used more often in practice to evaluate controllability.

The controllability condition for LTI systems with distinct eigenvalues may be stated very simply if the state equations are transformed to the diagonal Jordan canonical form. Such systems are completely controllable if there are no zero rows in the transformed $\mathbf{B}$ matrix for continuous-time systems.[4] The presence of a zero row in either of these matrices would indicate that the inputs are not coupled to and cannot control the corresponding mode. Algebraic controllability conditions for systems with repeated eigenvalues are given by Palm.[6]

The controllability conditions for LTI systems in general are stated in terms of the matrix $\mathbf{P}_c$ referred to as the controllability matrix in Section 8.34 and are summarized in Table 8.21. The $n \times nr$ controllability matrix for a MIMO system is defined by

$$\mathbf{P}_c = [\mathbf{B} \ \mathbf{AB} \ \dots \ \mathbf{A}^{n-1}\mathbf{B}] \qquad (8.257)$$

for continuous-time systems. The condition for complete state controllability is simply that the matrix $\mathbf{P}_c$ has rank n. The controllable canonical form for SISO systems, described in Section 8.34, derives its name from the fact that transformation to that form is possible if and only if the system is completely state controllable. The transformation matrix $\mathbf{T}$ in Table 8.19 required to transform the state-space equations for a SISO system to the controllable canonical form exists if and only if the $n \times n$ matrix $\mathbf{P}_c$ is nonsingular; that is, the system is controllable. Equivalent controllability conditions that are simpler to evaluate than the one previously stated are also listed in Table 8.21 along with a necessary (but not sufficient) condition for complete controllability.

TABLE 8.21 Controllability Conditions for Continuous Time Linear Dynamic Systems

Time-Invariant System	
Necessary and sufficient condition for state controllability	(i) rank $(\mathbf{B} \ \mathbf{AB} \ \ \dots \ \ \mathbf{A}^{n-1}\mathbf{B})$ $= \text{rank}\,(\mathbf{B} \ \mathbf{AB} \ \ \dots \ \ \mathbf{A}^{n-r}\mathbf{B})$ $= n$ or (ii) $\det(\mathbf{P}_c\mathbf{P}_c^{\mathrm{T}}) \neq 0$
Necessary condition for state controllability	rank $(\mathbf{B} \ \mathbf{A}) = n$
Necessary and sufficient condition for output controllability	(i) rank $(\mathbf{C}\mathbf{P}_c) = p$ or (ii) $\det(\mathbf{C}\mathbf{P}_c\mathbf{P}_c^{\mathrm{T}}\mathbf{C}^{\mathrm{T}}) \neq 0$
Time-Varying System: Time Interval of Interest $[t_0, t_1]$	
Necessary and sufficient condition for state controllability	(i) $\mathbf{W}_c(t_1, t_0)$ is positive definite or (ii) Zero is not an eigenvalue of $\mathbf{W}_c(t_1, t_0)$ or (iii) $\lvert \mathbf{W}_c(t_1, t_0) \rvert \neq 0$ where $\mathbf{W}_c(t_1, t_0)$ $\triangleq \int_{t_0}^{t_1} \boldsymbol{\phi}(t_1, \tau)\mathbf{B}(\tau)\mathbf{B}^{\mathrm{T}}(\tau)$ $\times \boldsymbol{\phi}^{\mathrm{T}}(t_1, \tau)\,d\tau$
Necessary and sufficient condition for output controllability	$\det \mathbf{W}_y(t_1, t_0) \neq 0$ where $\mathbf{W}_y(t_1, t_0)$ $\triangleq \int_{t_0}^{t_1} \mathbf{C}(\tau)\boldsymbol{\phi}(t_1, \tau)\mathbf{B}(\tau)\mathbf{B}^{\mathrm{T}}(\tau)$ $\times \boldsymbol{\phi}^{\mathrm{T}}(t_1, \tau)\mathbf{C}^{\mathrm{T}}(\tau)\,d\tau$

The controllability conditions for LTV systems over a specified time interval are more cumbersome to evaluate in practice as they involve the system transition matrix.[4] These conditions are listed in Table 8.21. In contrast to time-invariant systems, the controllability of time-varying systems depends on the time interval under consideration.

The concept of output controllability, as opposed to state controllability described earlier, was introduced by Kreindler and Sarachik.[13] A linear, continuous-time system is said to be output controllable at time t_0 if there exists a finite time $t_1 > t_0$ and a control function $u(t)$, $t_0 < t < t_1$, that drives the system output from $y(t_0)$ to $y(t) = 0$ at $t = t_1$. If this condition holds true for all times t_0 and all initial outputs $y(t_0)$, the system is completely output controllable. Extension of the concept to linear, discrete-time systems is straightforward as before.

Output controllability conditions[4] for linear systems that are purely dynamic [i.e., the matrix $D = 0$ in Eqs. (8.214) and (8.216)] are summarized in Table 8.21. These conditions are weaker than the corresponding conditions for state controllability if the number of outputs p is less than the number of state variables n. Since this is true in practice, state controllability implies output controllability. On the other hand, output controllability does not imply state controllability in general. It can be shown, however, that for time-invariant systems if the matrix (CC^T) is nonsingular, output controllability is equivalent to state controllability.

The observability of a linear system is a measure of the coupling between the system state and its outputs. The concept of observability was introduced by Kalman[11] and is relevant to the problem of estimation of system state based on the output vector. The output vector is usually chosen to correspond to measurable variables.

A linear, continuous-time system is said to be observable[4] at time t_0 if there exists a finite time $t_1 > t_0$ such that $x(t_0)$ can be determined from the history of inputs $u(t)$ and outputs $y(t)$ over the time interval $t_0 \leq t \leq t_1$. If the system is observable for all times t_0 and all initial states $x(t_0)$, the system is completely observable. A stronger form of observability for LTV systems is that of uniformly complete observability. The mathematical definition of this form of observability is given by Kalman.[11] This property guarantees that the time interval required to estimate the state is relatively independent of the initial time. For LTI systems, of course, complete observability is the same as uniformly complete observability. A property complementary to observability for LTV systems, is that of reconstructibility,[12, 14] which concerns the estimation of the state of the system from past measurements of the state. In contrast to this, observability concerns the estimation of the state from future measurements of the output. For time-invariant systems, the two properties of reconstructibility and observability are identical to one another.

As was the case for controllability, direct application of the definition of observability already stated yields conditions involving the transition matrix.[4] Simpler algebraic conditions are available for time-invariant systems. The observability condition for LTI systems with distinct eigenvalues can be stated very simply if the state equations are transformed to the Jordan canonical form. Such systems are completely observable if each column in the transformed C matrix has at least one nonzero element.[4] The presence of a column of zeros in this matrix would indicate that the corresponding state variable cannot be estimated from the measured output and input vectors.

More general observability conditions for LTI systems are stated in terms of a matrix P_0 referred to as the observability matrix and are summarized in Table 8.22. The $np \times n$ observability matrix is defined by

$$P_0 = \begin{bmatrix} C \\ CA \\ \vdots \\ CA^{n-1} \end{bmatrix} \tag{8.258}$$

for continuous-time systems. The condition for complete observability is simply that the matrix P_0 have rank n. The observable canonical form for SISO systems, described in Section 8.34, derives its name from the fact that transformation to that form is possible if and only if the system is observable. The transformation matrix T in Table 8.19, required to transform the state-space equations for a SISO system to the observable canonical form, exists if and only if the $n \times n$ matrix P_0 is nonsingular, that is, the system is observable. Equivalent observability conditions, which are simpler to evaluate than the one stated previously are also listed in Table 8.22 along with a necessary (but not sufficient) condition for complete observability. It should be noted that the observability conditions are independent of time for time-invariant systems. In contrast, the observability conditions for time-varying systems over a specified time interval involve the system transition matrix and hence depend on the time interval.[4] They are also listed in Table 8.22.

Though the definition of observability above involves an open-loop state estimation problem, the property of observability has important implications for closed-loop realizations of the state estimation problem. It will be shown in Section 8.43 that, if a time-invariant system is completely observable, a closed-loop state estimator can be constructed such that the estimation error transients can be made to decay to zero as rapidly as possible.

TABLE 8.22   Observability Conditions for Continuous Time Linear Dynamic Systems

Time-Invariant System

Necessary and sufficient condition for observability	(i) $\operatorname{rank}[\mathbf{C}^T \quad \mathbf{A}^T\mathbf{C}^T \quad \cdots \quad (\mathbf{A}^T)^{n-1}\mathbf{C}^T]$ $= \operatorname{rank}[\mathbf{C}^T \quad \mathbf{A}^T\mathbf{C}^T \quad \cdots \quad (\mathbf{A}^T)^{n-p}\mathbf{C}^T]$ $= n$ or (ii) $\det(\mathbf{P}_0^T\mathbf{P}_0) \neq 0$
Necessary condition	$\operatorname{rank}(\mathbf{C}^T \mathbf{A}^T) = n$

Time-Varying System:
Time Interval of Interest $[t_0, t_1]$

Necessary and sufficient condition for observability	Observable at t_0 if and only if there exists a finite time t_1, $t_1 > t_0$ such that (i) $\mathbf{W}_0(t_1, t_0)$ is positive definite or (ii) Zero is not an eigenvalue of $\mathbf{W}_0(t_1, t_0)$ or (iii) $\lvert \mathbf{W}_0(t_1, t_0)\rvert \neq 0$ where $\mathbf{W}_0(t_1, t_0)$ $\triangleq \int_{t_0}^{t_1} \boldsymbol{\phi}^T(\tau, t_0)\mathbf{C}^T(\tau)\mathbf{C}(\tau)$ $\times \boldsymbol{\phi}(\tau, t_0)\, d\tau$

The conditions for controllability and observability noted in Tables 8.21 and 8.22 have obvious similarities. The two properties can be shown to be duals of each other by formulating the concept of the dual of a dynamic system. Interested readers are referred to Kalman et al.[11, 14]

A linear system can, in general, be divided into four subsystems as indicated by Fig. 8.168. The state vector $\mathbf{x}$ can be written as

$$\mathbf{x} = \mathbf{x}_C + \mathbf{x}_{CO} + \mathbf{x}_N + \mathbf{x}_O \qquad (8.259)$$

where the subscripts have the meaning assigned in the figure. The corresponding state-space equations for a time-invariant, continuous-time system are

$$\begin{bmatrix} \dot{\mathbf{x}}_C \\ \dot{\mathbf{x}}_{CO} \\ \dot{\mathbf{x}}_N \\ \dot{\mathbf{x}}_O \end{bmatrix} = \begin{bmatrix} \mathbf{A}_{11} & \mathbf{A}_{12} & \mathbf{A}_{13} & \mathbf{A}_{14} \\ 0 & \mathbf{A}_{22} & 0 & \mathbf{A}_{24} \\ 0 & 0 & \mathbf{A}_{33} & \mathbf{A}_{34} \\ 0 & 0 & 0 & \mathbf{A}_{44} \end{bmatrix} \begin{bmatrix} \mathbf{x}_C \\ \mathbf{x}_{CO} \\ \mathbf{x}_N \\ \mathbf{x}_O \end{bmatrix} + \begin{bmatrix} \mathbf{B}_{11} \\ \mathbf{B}_{21} \\ 0 \\ 0 \end{bmatrix} \mathbf{u} \qquad (8.260)$$

$$\mathbf{y} = \begin{bmatrix} 0 & \mathbf{C}_{12} & 0 & \mathbf{C}_{14} \end{bmatrix} \begin{bmatrix} \mathbf{x}_C \\ \mathbf{x}_{CO} \\ \mathbf{x}_N \\ \mathbf{x}_O \end{bmatrix} + \mathbf{Du}$$

The zero matrices in the $\mathbf{B}$ matrix correspond to the fact that $\mathbf{x}_N$ and $\mathbf{x}_O$ are not controllable. The zero matrices in the $\mathbf{C}$ matrix correspond to the fact that $\mathbf{x}_C$ and $\mathbf{x}_N$ are not observable. If the eigenvalues of the $\mathbf{A}$ matrix are distinct, all off-diagonal elements in the $\mathbf{A}$ matrix would be zero. The procedure for determining the transformation matrix to convert the state-space equations into the canonical form [Eq. (8.260)] has been described by Kalman.[9] For time-varying systems, the state-space decomposition is a function of time but is similar in structure to that already described.

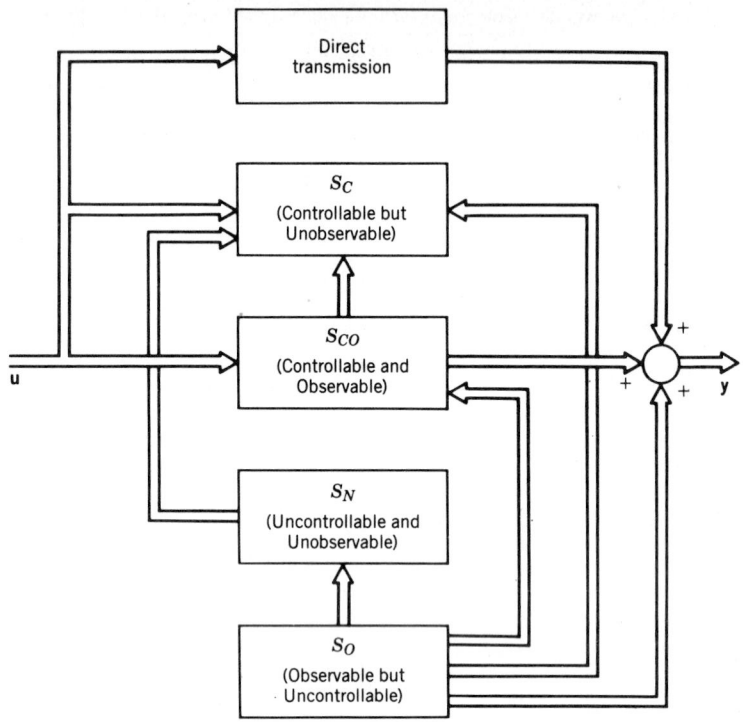

Fig. 8.168 Decomposition of linear system based on controllability and observability.

The significance of the system decomposition as shown in Fig. 8.168 is that it helps relate the state-space description of linear dynamic systems to transfer function or transfer function matrix descriptions of such systems. The transfer function matrix relating **y** to **u** is a description only of the controllable and observable part of the system and masks other modes that are either not observable or not controllable or neither controllable nor observable. The relationship of the state-space description of dynamic systems to the transfer function matrix description of such systems is discussed in greater detail in Section 8.38.

Loss of controllability or observability could occur when controllable and observable subsystems are connected together to form composite systems. Gilbert[15] has formulated rules relating the composite system properties to those of the individual open-loop systems. These rules provide greater insight into the conditions leading to loss of controllability or observability than the simple application of the conditions noted in Tables 8.21 and 8.22.

The concepts of controllability and observability are obviously very important for MIMO systems since the complexity of such systems frequently masks the nature of the coupling of the system state to the inputs and outputs. For SISO systems, lack of complete controllability or observability is a less common occurrence. Conclusions concerning controllability and observability are obvious in many of these cases but less so in others.

Consider the electrical circuit in Fig. 8.169. The state-space equation for the system is given by:

$$
\begin{bmatrix} \dot{x}_1 \\ \dot{x}_2 \end{bmatrix} =
\begin{bmatrix} -\dfrac{R_1 + R_3}{L_1} & \dfrac{R_3}{L_1} \\[2ex] \dfrac{R_3}{L_2} & -\dfrac{R_3 + R_2}{L_2} \end{bmatrix}
\begin{bmatrix} x_1 \\ x_2 \end{bmatrix} +
\begin{bmatrix} \dfrac{1}{L_1} \\[2ex] 0 \end{bmatrix} u(t)
\qquad (8.261)
$$

The system is completely controllable except for the trivial case where R_3 is zero. Similarly, if x_1 or x_2 is chosen as the only output, the system is completely observable as long as R_3 is nonzero. If the

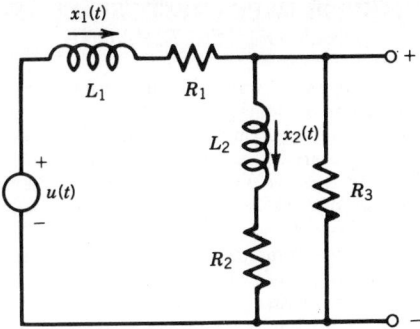

Fig. 8.169   State variables for RL circuit.

voltage across the resistor R_3 is chosen as the output, the corresponding output equation is

$$y(t) = [R_3 \; - R_3] \begin{bmatrix} x_1 \\ x_2 \end{bmatrix} \tag{8.262}$$

The observability matrix $\mathbf{P}_0$ in Eq. (8.258) is then nonsingular and the system is observable if and only if

$$\frac{R_1}{L_1} \neq \frac{R_2}{L_2} \tag{8.263}$$

This observability condition is not an obvious one and is equivalent to the requirement that the time constants associated with the two $R - L$ pairs not be identical. However, it should be noted that, given component tolerances in practice, the inequality (8.263) will be satisfied almost always and the corresponding system will be observable.

Despite the fact that lack of complete controllability or observability is infrequent for SISO systems, these concepts have practical significance for SISO as well as MIMO systems because of the relationship of these concepts to closed-loop control and state estimation problems. Measures of the degree of controllability and observability can be defined for time-invariant and time-varying systems. The controllability and observability conditions in Tables 8.21 and 8.22 relate these properties to the nonsingularity of square matrices for SISO systems. Measures of the degree of controllability and observability are related to the closeness of these matrices to the singularity condition. Such measures have been defined for time-invariant systems by Johnson[16] and Friedland[17] and are significant for SISO as well as MIMO systems. A system with a better degree of controllability can in general be controlled more effectively. Similarly, a better degree of observability implies that state estimation can be performed more accurately. The proposed measures of the degree of controllability and observability are not in common use but have the potential to quantitatively evaluate proposed control strategies and measurement schemes.[18]

Additional concepts of degrees of controllability and observability for time-varying systems have been described by Silverman and Meadows[19] and for MIMO systems by Kreindler and Sarachik.[13] Properties weaker than state controllability and observability have also been defined[12] and are useful in ensuring that closed-loop control and state estimation problems are well-posed. A linear system is said to be stabilizable if the uncontrollable subsystems S_N and S_O in the decomposition of Fig. 8.168 are stable. Similarly, if the unobservable subsystems S_C and S_N are stable, the system is said to be detectable.[12]

The preceding discussion is limited to continuous-time systems, but the concepts apply to discrete-time formulations also. In fact, extension of the concepts to discrete-time systems involves replacing time instant t by the sequence number k and the matrices $\mathbf{A}$, $\mathbf{B}$, $\mathbf{C}$, and $\mathbf{D}$ by corresponding matrices in the discrete-time state-space equations. Readers are referred to DeRusso et al.,[2] Brogan,[4] and Kuo[7] for details.

8.38   RELATIONSHIP BETWEEN STATE-SPACE AND TRANSFER FUNCTION DESCRIPTIONS

The state-space representation of dynamic systems is an accurate representation of the internal structure of a system and its coupling to the system inputs and outputs. For LTI systems, transfer functions (for SISO systems) or transfer function matrices (for MIMO systems) are useful in practice since the dimensions of these matrices are invariably smaller than the dimensions of the corresponding system matrix $\mathbf{A}$ in Eq. (8.215). Analysis and design procedures based on the transfer function matrix descriptions are therefore simpler. The relationship between these two alternative descriptions of LTI systems is described in this section.

Determination of the transfer function matrix from the state-space equations is straightforward. For continuous-time systems, Laplace transformation of Eqs. (8.215) and (8.216) with zero initial conditions x(0) and elimination of $\mathbf{X}(s)$ yields

$$\mathbf{H}(s) = \frac{\mathbf{Y}(s)}{\mathbf{U}(s)} = \mathbf{C}(s\mathbf{I}_n - \mathbf{A})^{-1}\mathbf{B} + \mathbf{D} \qquad (8.264)$$

where $\mathbf{I}_n$ is the $n \times n$ identity matrix. The transfer function matrix corresponding to a given state-space description is therefore unique. However, as indicated in the previous section, the former represents only the controllable and observable part of a system. Unless the entire system is completely controllable and observable, a transfer function matrix description is not a complete characterization of the system dynamic behavior. It can be shown that, for SISO systems, a necessary and sufficient condition for controllability and observability of the system is that there are no pole-zero cancellations between the numerator and denominator of the transfer function matrix in Eq. (8.264). For MIMO systems, this is only a sufficient condition and not a necessary one.[4]

The determination of the state-space description corresponding to a given transfer function matrix description is more complex and is referred to as the problem of realization. Since the transfer function matrix represents only the controllable and observable part of a system, the problem of realization does not have a unique solution. In fact, the transfer function matrix description does not even determine the dimension of the corresponding state vector uniquely. The minimal dimension of the state vector corresponding to a given transfer function matrix is, however, uniquely determined. The associated state-space equations are said to constitute the minimal or irreducible realization of the transfer function matrix. It can be shown that a realization is minimal if and only if it is both controllable and observable.[4] Minimal realizations are not unique. However, any two different minimal realizations of a given transfer function matrix are equivalent in that the corresponding state vectors are related by a nonsingular transformation matrix.

The canonical forms of the state-space equations for SISO systems in Table 8.18 represent minimal realizations. Techniques for obtaining minimal realizations for MIMO systems are more involved. Brogan[4] has described a procedure for obtaining a Jordan form realization for a given transfer function matrix. When applied to cases where elements of the transfer function matrix have only simple poles, the resulting realization is controllable and observable as shown in the following example. If one or more elements of the transfer function matrix have repeated poles, the realization that results is controllable but may or may not be observable.

Consider a continuous-time system with two inputs and two outputs and the following transfer function matrix:

$$\mathbf{H}(s) = \begin{bmatrix} \dfrac{1}{s+1} & \dfrac{s}{(s+1)(s+3)} \\[2ex] \dfrac{1}{s+3} & \dfrac{1}{s+1} \end{bmatrix} \qquad (8.265)$$

Expand $\mathbf{H}(s)$ using a matrix version of partial fraction expansion as

$$\mathbf{H}(s) = \frac{\begin{bmatrix} 1 & -\frac{1}{2} \\ 0 & 1 \end{bmatrix}}{s+1} + \frac{\begin{bmatrix} 0 & \frac{3}{2} \\ 1 & 0 \end{bmatrix}}{s+3}$$

$$= \frac{\begin{bmatrix} 1 \\ 0 \end{bmatrix}\begin{bmatrix} 1 & -\frac{1}{2} \end{bmatrix} + \begin{bmatrix} 0 \\ 1 \end{bmatrix}\begin{bmatrix} 0 & 1 \end{bmatrix}}{s+1} + \frac{\begin{bmatrix} 1 \\ 0 \end{bmatrix}\begin{bmatrix} 0 & \frac{3}{2} \end{bmatrix} + \begin{bmatrix} 0 \\ 1 \end{bmatrix}\begin{bmatrix} 1 & 0 \end{bmatrix}}{s+3} \qquad (8.266)$$

It should be noted that the number of vector products each coefficient matrix is factored into is equal to the rank of the matrix. $\mathbf{H}(s)$ is then written in a form that indicates the matrices $\mathbf{A}, \mathbf{B}, \mathbf{C}$ clearly, by comparison with $\mathbf{C}(s\mathbf{I} - \mathbf{A})^{-1}\mathbf{B}$:

$$\mathbf{H}(s) = \begin{bmatrix} 1 & 0 & 1 & 0 \\ 0 & 1 & 0 & 1 \end{bmatrix} \begin{bmatrix} \dfrac{1}{s+1} & & & \\ & \dfrac{1}{s+1} & & 0 \\ & & \dfrac{1}{s+3} & \\ 0 & & & \dfrac{1}{s+3} \end{bmatrix} \begin{bmatrix} 1 & -\frac{1}{2} \\ 0 & 1 \\ 0 & \frac{3}{2} \\ 1 & 0 \end{bmatrix}$$

$$= \begin{bmatrix} 1 & 0 & 1 & 0 \\ 0 & 1 & 0 & 1 \end{bmatrix} \begin{bmatrix} s+1 & & & \\ & s+1 & & 0 \\ 0 & & s+3 & \\ & & & s+3 \end{bmatrix}^{-1} \begin{bmatrix} 1 & -\frac{1}{2} \\ 0 & 1 \\ 0 & \frac{3}{2} \\ 1 & 0 \end{bmatrix} \qquad (8.267)$$

Thus, the corresponding realization is

$$\mathbf{A} = \begin{bmatrix} -1 & & & \\ & -1 & & 0 \\ 0 & & -3 & \\ & & & -3 \end{bmatrix} \qquad \mathbf{B} = \begin{bmatrix} 1 & -\frac{1}{2} \\ 0 & 1 \\ 0 & \frac{3}{2} \\ 1 & 0 \end{bmatrix} \qquad (8.268)$$

$$\mathbf{C} = \begin{bmatrix} 1 & 0 & 1 & 0 \\ 0 & 1 & 0 & 1 \end{bmatrix}$$

The realization is controllable and observable and hence is minimal. Modifications of this procedure for cases where $\mathbf{H}(s)$ has elements with repeated poles are described by Brogan.[4]

An alternative two-step procedure for determining a minimal realization for a transfer function matrix involves obtaining a nonminimal realization by any one method, as the first step. For example, one of the many realizations in Table 8.18 can be chosen to represent each of the elements of the transfer function matrix. The state-space descriptions of the elements can then be combined to get the state-space equations for the MIMO system. The resulting realization would, in general, be nonminimal. The second step requires transformation of the state-space equations to the form given by Eq. (8.260). Techniques for selecting the transformation matrix are described by Kalman[9] and Fortmann and Hitz.[1] The minimal realization is then given by the controllable and observable subsystem in Fig. 8.168. The resulting equations for a continuous-time system are

$$\dot{\mathbf{x}}_{CO} = \mathbf{A}_{22}\mathbf{x}_{CO} + \mathbf{B}_{21}\mathbf{u} \qquad (8.269)$$

$$\mathbf{y} = \mathbf{C}_{12}\mathbf{x}_{CO} + \mathbf{D}\mathbf{u} \qquad (8.270)$$

Similar results for discrete-time systems are given by Brogan,[4] Kuo,[7] and Kalman.[9]

8.39 SYNTHESIS OF CONTROLLERS USING STATE-SPACE METHODS

The advantages of feedback control in achieving desired input/output relationships are well known. Control system theory based on a frequency domain approach[20] illustrates clearly that the following aspects of SISO system performance can be improved by feedback: (1) the ability to follow reference inputs accurately in the steady state or under transient conditions and (2) the ability to reject disturbance inputs and reduce sensitivity of the overall controlled system behavior to plant parameter variations and modeling errors. For MIMO systems, the coupling between individual inputs and outputs can be modified in a desired manner, in addition to the performance features already mentioned, by appropriate control system design.[21]

State-space methods for control system design result in solutions that utilize the state of the system most effectively for feedback. The resulting state-variable feedback control systems improve the same aspects of system performance as previously mentioned. However, the available state-space design procedures accommodate some performance specifications more readily than others. For instance, performance specifications in the form of desired closed-loop pole locations are readily accommodated. Similarly, performance specifications in the form of an index of performance to be optimized can be accommodated by optimal control theory if the index of performance belongs to a restricted class of performance measures. In fact, recent efforts in control system design using state-space methods have been directed at enhancing the problem formulation to accommodate a greater variety of performance specifications. In spite of these enhancements, performance specifications such as sensitivity of the controlled system performance to plant parameter variations and modeling errors are accommodated more readily by frequency-domain-based design procedures than by state-space or time-domain-based design procedures. Thus, control system design techniques based on frequency domain and time domain approaches should be viewed as being complementary to each other in some ways.

8.40 THE POLE PLACEMENT DESIGN METHOD
FOR THE REGULATOR PROBLEM

It can be shown that, if a linear time-invariant system is completely state controllable and if linear instantaneous state-variable feedback is used, the associated feedback gains can be chosen to place the closed-loop poles of the controlled system at any arbitrarily specified locations in the s plane.[12] Thus, if the continuous-time system described by Eq. (8.215) is completely state controllable and the control law is given by (Fig. 8.170)

$$\mathbf{u} = -\mathbf{Kx} \qquad (8.271)$$

then the eigenvalues of the matrix $\mathbf{A} - \mathbf{BK}$ are the closed-loop pole locations and can be assigned any specified locations in the complex plane by appropriate selection of the gain matrix $\mathbf{K}$. If $\mathbf{K}$ is constrained to be a real matrix, the desired eigenvalues should be specified either as real or as complex conjugate pairs. The resulting design procedure is referred to as the pole placement method and is useful for regulation problems where the objective of the controller is to return the system to equilibrium conditions following an initial disturbance. Specification of the closed-loop poles is equivalent to specification of the damping and speed of response of the closed-loop system transients as the system returns to equilibrium.

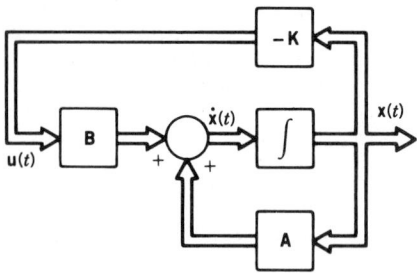

Fig. 8.170 Linear state-variable feedback for continuous-time system.

For single-input systems, specification of the desired closed-loop pole locations uniquely specifies the gain vector $\mathbf{K}$. A formula for the gain vector $\mathbf{K}$ is

$$\mathbf{K} = (0 \ldots 01)(\mathbf{B} \, \mathbf{AB} \ldots \mathbf{A}^{n-1} \mathbf{B})^{-1} \alpha_c(\mathbf{A}) \tag{8.272}$$

for continuous-time systems. In these equations,

$$\alpha_c(\mathbf{A}) = \mathbf{A}^n + \sum_{i=0}^{n-1} \alpha_i \mathbf{A}^i \tag{8.273}$$

The α_i's are the coefficients of the desired characteristic equations of the closed-loop systems. For continuous-time systems we have

$$\alpha_c(s) = \det(s\mathbf{I} - \mathbf{A} + \mathbf{BK})$$
$$= s^n + \sum_{i=0}^{n-1} \alpha_i s^i = 0 \tag{8.274}$$

Computer-aided control system design (CACSD) packages supporting state-space methods usually support pole placement designs[22-24] and require only that the designer input information about the system matrices and the desired closed-loop pole locations. The gain vector $\mathbf{K}$ is then computed and output to the designer.

For multi-input systems, specification of the closed-loop poles does not specify the gain matrix $\mathbf{K}$ uniquely. The additional freedom in the gain matrix selection can be used to assign eigenvectors (or generalized eigenvectors) or individual transfer function zeros to improve the transient response to nonzero reference inputs.[3] Alternative criteria for gain matrix selection are optimization of feedback gain magnitudes and stability of the closed-loop system in the absence or failure of some of the inputs. Brogan[4] has outlined a procedure for gain matrix selection for multi-input systems, based on closed-loop eigenvector specification in addition to eigenvalue specification. For continuous-time systems described by Eqs. (8.215) and (8.271), the feedback gain matrix is given by

$$\mathbf{K} = -(\mathbf{e}_{j_1} \, \mathbf{e}_{j_2} \ldots \mathbf{e}_{j_3}) \big[\boldsymbol{\psi}_{j_1}(s_1) \, \boldsymbol{\psi}_{j_2}(s_2) \ldots \boldsymbol{\psi}_{j_n}(s_n) \big]^{-1} \tag{8.275}$$

where the desired closed-loop eigenvalues and the corresponding eigenvectors are s_i and $\boldsymbol{\psi}_{j_i}(s_i)$, $i = 1, \ldots, n$, respectively. The eigenvectors are chosen to be n linearly independent columns from the $n \times nr$ matrix $[\boldsymbol{\psi}(s_1) \, \boldsymbol{\psi}(s_2) \ldots \boldsymbol{\psi}(s_n)]$ where

$$\boldsymbol{\psi}(s) = (s\mathbf{I}_n - \mathbf{A})^{-1} \mathbf{B} \tag{8.276}$$

If the desired s_i are distinct, it will always be possible to find n linearly independent columns as already described. Here $\mathbf{e}_{j_i}$ is defined as the j_ith column of the $r \times r$ identity matrix $\mathbf{I}_r$ and is uniquely determined once j_i is determined. When repeated eigenvalues are desired, the procedure for specifying n linearly independent generalized eigenvectors is different and has been described by Brogan.[4] Alternative methods for gain matrix selection for multi-input systems have been described by Kailath.[3]

If a single-input, LTV system is completely state controllable, linear state-variable feedback can be used to ensure that the closed-loop transition matrix corresponds to that of any desired nth-order linear differential equation with time-varying coefficients. The state-variable feedback gains are time varying in general and can be computed using a procedure described by Wiberg.[25]

If the complete state is not available for feedback, linear instantaneous feedback of the measured output can be used to place some of the closed-loop poles at specified locations in the complex plane. If the continuous-time system described by Eqs. (8.215) and (8.216) satisfies the output controllability conditions listed in Table 8.21, then p of the n eigenvalues of the closed-loop system can approach arbitrarily specified values to within any degree of accuracy but not always exactly. The control law is

$$\mathbf{u} = -\mathbf{Ky} \tag{8.277}$$

where $\mathbf{K}$ is a $r \times p$ gain matrix. Brogan[4] has described an algorithm for computing $\mathbf{K}$, given the desired values of p closed-loop eigenvalues. The corresponding characteristic equation is

$$\det\big[s\mathbf{I}_n - \mathbf{A} + \mathbf{BK}(\mathbf{I}_p + \mathbf{DK})^{-1}\mathbf{C} \big] = 0 \tag{8.278}$$

for continuous-time systems.

An alternative approach to control system design in the case of incomplete state measurement is to use an observer or a Kalman filter for state estimation. The estimated state is then used for feedback. This procedure is discussed in Section 8.43.

The advantages of the pole placement design method already described are that the controller achieves desired closed-loop pole locations without using pole-zero cancellation and without increasing the order of the system. The desired pole locations can be chosen to ensure a desired degree of stability or damping and speed of response of the closed-loop system. However, there is no convenient way to ensure a priori that the closed-loop system satisfies other important performance specifications such as a desired level of insensitivity to plant parameter variations, acceptable disturbance rejection, and compatibility of control effort with actuator limitations. In addition, for single-input LTI systems, instantaneous state feedback of the form given by Eq. (8.271) does not affect the locations of zeros of the transfer functions between the system input and system outputs.[12] Thus, the pole placement design method does not afford complete control over the system response to the reference input or disturbance inputs. For multi-input systems, the available freedom in the gain matrix selection can be used to assign individual transfer function zeros, in addition to achieving desired closed-loop pole locations. However, systematic procedures to do this are not available. The consequence of these limitations of the pole placement method is that the design process involves considerable trial and error.

Parallel results on pole placement design for discrete-time systems are given by Franklin and Powell[26] and Brogan.[4]

8.41   THE STANDARD LINEAR QUADRATIC REGULATOR PROBLEM

Controller design in regulation applications using pole placement specifications emphasizes only the transient behavior of the state variables as the system returns to equilibrium. There is no explicit consideration of the required control effort. Control effort can be considered if the controller design problem is formulated as an optimal control problem with weighting of both control effort and state-variable transients. For regulation applications, the index of performance to be optimized, J, is usually chosen to be a quadratic function of the control inputs and the state variables. The resulting optimal control law for the control input, when expressed in feedback form, is a linear function of the system state. Hence, this approach to control system design is referred to as the linear quadratic regulator (LQR) problem.

Consider the continuous-time, LTV system described by Eq. (8.213) and the initial condition, for regulation problems, of

$$\mathbf{x}(t_0) = \mathbf{x}_0 \qquad (8.279)$$

The controller design problem is to choose the control input $\mathbf{u}(t)$ to minimize the quadratic index of performance[12]:

$$J = \int_{t_0}^{t_1} \left[\mathbf{x}^\mathrm{T}(t)\mathbf{R}_1(t)\mathbf{x}(t) + \mathbf{u}^\mathrm{T}(t)\mathbf{R}_2(t)\mathbf{u}(t) \right] dt + \mathbf{x}^\mathrm{T}(t_1)\mathbf{P}_f\mathbf{x}(t_1) \qquad (8.280)$$

$\mathbf{R}_1(t)$ is a positive semidefinite symmetric weighting matrix on the state variables, $\mathbf{R}_2(t)$ is a positive definite symmetric weighting matrix on the control inputs, and $\mathbf{P}_f$ is a positive semidefinite symmetric weighting matrix on the terminal state. Times t_0, t_1 are initial and terminal time instants. The problem, as previously formulated, is a finite time, deterministic, LQR problem. Kwakernaak and Sivan[12] have also considered a more general J that includes an additional term of the form $\mathbf{x}^\mathrm{T}(t)\mathbf{R}_{12}(t)\mathbf{u}(t)$ within the integral. They have shown that this J can be reduced to the form of Eq. (8.280) by appropriate redefinition of the weighting matrices and control vector.

The solution of this control problem, using methods from calculus of variations, can be obtained from standard textbooks on optimal control,[12,27] along with conditions for its existence and uniqueness. The optimal control law, given in feedback form, is a linear, time-varying function of the system state:

$$\mathbf{u}(t) = -\mathbf{R}_2^{-1}(t)\mathbf{B}^\mathrm{T}(t)\mathbf{P}(t)\mathbf{x}(t) \qquad (8.281)$$

where $\mathbf{P}(t)$ is a $n \times n$ symmetric positive semidefinite matrix satisfying the matrix Riccati equation:

$$-\dot{\mathbf{P}}(t) = \mathbf{R}_1(t) - \mathbf{P}(t)\mathbf{B}(t)\mathbf{R}_2^{-1}(t)\mathbf{B}^\mathrm{T}(t)\mathbf{P}(t) + \mathbf{P}(t)\mathbf{A}(t) + \mathbf{A}^\mathrm{T}(t)\mathbf{P}(t) \qquad (8.282)$$

and the terminal condition

$$\mathbf{P}(t_1) = \mathbf{P}_f \qquad (8.283)$$

Numerical solution of the matrix Riccati equation is a subject of great importance and of considerable research. Some useful techniques have been briefly described by Kwakernaak and Sivan.[12]

Solution of the LQR problem simplifies as the terminal time t_1 approaches infinity. It can be shown then that the solution of the matrix Riccati equation approaches a steady-state solution $P_s(t)$ that is independent of P_f. The resulting steady-state control law

$$\mathbf{u}(t) = -\mathbf{R}_2^{-1}\mathbf{B}^{\mathrm{T}}(t)\mathbf{P}_s(t)\mathbf{x}(t) \tag{8.284}$$

results in an exponentially stable closed-loop system if:

1. The linear system of Eq. (8.213) is uniformly completely state controllable.
2. The pair $\mathbf{A}(t), \mathbf{H}_r^{\mathrm{T}}(t)$ is uniformly completely reconstructible where $\mathbf{H}_r(t)$ is any matrix such that $\mathbf{H}_r(t)\mathbf{H}_r^{\mathrm{T}}(t)$ equals $\mathbf{R}_1(t)$.

The matrix Riccati equation for the steady-state LQR problem simplifies to an algebraic equation and $\mathbf{P}_s$ is a constant if the system matrices and the weighting matrices in the index of performance are constant. The resulting algebraic Riccati equation is

$$\mathbf{R}_1 - \mathbf{P}_s\mathbf{B}\mathbf{R}_2^{-1}\mathbf{B}^{\mathrm{T}}\mathbf{P}_s + \mathbf{A}^{\mathrm{T}}\mathbf{P}_s + \mathbf{P}_s\mathbf{A} = 0 \tag{8.285}$$

$\mathbf{P}_s$ is a unique positive definite solution of Eq. (8.285) and the resulting time-invariant closed-loop system is asymptotically stable if:

1. The linear system of Eq. (8.215) is completely state controllable.
2. The pair $\mathbf{A}, \mathbf{H}_r^{\mathrm{T}}$ is completely observable (reconstructible), where $\mathbf{H}_r$ is any matrix such that $\mathbf{H}_r\mathbf{H}_r^{\mathrm{T}}$ equals $\mathbf{R}_1$.

Another version of the LQR problem involves minimization of the quadratic index of performance for an LTI system over a finite time interval. If the weighting matrices are also time invariant, in many cases the optimal feedback gains are constant over most of the time interval of interest and vary with time only near the terminal time. Since constant feedback gains are easier to implement in practice, implementation of constant gains over the entire time interval would represent a nearly optimal solution that is practically more convenient.[12]

The results of the LQR problem for discrete-time systems parallel those for continuous-time systems already stated. They are described in detail by Kwakernaak and Sivan.[12]

As in the case of continuous-time systems, the optimal feedback gains are nearly constant even for finite-time LQR problems, if the system matrices and weighting matrices in J are constant.[26] Also, a number of techniques for solving the difference or algebraic equations for the optimal feedback gains are described by Kuo.[7]

An important consideration in the practical usefulness of the optimal control laws for the LQR problems described is the implication of these laws for performance features of the controlled systems not included in J, such as relative stability and sensitivity of the controlled system to unmodeled dynamics or plant parameter variations. Reference has already been made to the fact that the optimal control laws for the continuous-time, infinite-time LQR problems described result in asymptotically stable closed-loop systems provided that specified controllability and reconstructibility or observability conditions are satisfied. Closed-loop systems with a prescribed degree of stability can be obtained by modifying the J, as will be shown for linear, time-invariant, continuous-time systems.[27]

$$J = \int_0^\infty e^{2\alpha t}\left(\mathbf{x}^{\mathrm{T}}\mathbf{R}_1\mathbf{x} + \mathbf{u}^{\mathrm{T}}\mathbf{R}_2\mathbf{u}\right) dt \tag{8.286}$$

where α is a positive scalar constant. If the pair $\mathbf{A}, \mathbf{B}$ is completely state controllable and the pair $\mathbf{A}, \mathbf{H}_r^{\mathrm{T}}$ is completely observable where $\mathbf{H}_r\mathbf{H}_r^{\mathrm{T}}$ is equal to $\mathbf{R}_1$, the solution to this LQR problem results in a finite value of J. Hence, the transients decay at least as rapidly as $e^{-\alpha t}$. Larger values of α would therefore ensure a more rapid return of the system to equilibrium. The corresponding algebraic Riccati equation is

$$\mathbf{R}_1 - \mathbf{P}_s\mathbf{B}\mathbf{R}_2^{-1}\mathbf{B}^{\mathrm{T}}\mathbf{P}_s + \mathbf{A}^{\mathrm{T}}\mathbf{P}_s + \mathbf{P}_s\mathbf{A} + 2\alpha\mathbf{P}_s = 0 \tag{8.287}$$

and the optimal feedback control law is given by Eq. (8.284). A similar procedure for discrete-time LTI systems is described by Franklin and Powell.[26]

Additional results concerning the stability properties of the optimal control law for continuous-time, LTI systems described by Eq. (8.215) and employing only constant weighting matrices in the index of performance, Eq. (8.280), are available and will be summarized. Anderson and Moore[27] have shown that for single-input systems the optimal control law for the infinite-time LQR problem has $\pm 60°$ phase margin, an infinite gain margin, and 50% gain reduction tolerance before the closed-loop system becomes unstable. These results are best explained with the aid of Fig. 8.171(a), where $G_p(s)$ is normalized to be unity at $s = 0$. The transfer function $K_p G_p(s)$ characterizes the modeling accuracy and is unity for an exact model. The result stated previously indicates that modeling errors that result either in phase shifts $\diagup G_p(j\omega)$ of less than 60° in magnitude for all frequencies or in values of the magnitude ratio K_p greater than one-half would not destabilize the closed-loop system. The result on gain margins extends also to static nonlinear gain relationships between u_f and $u.$[27]

Safonov[28] has extended these results on gain and phase margins to multi-input infinite-time LQR problems. As shown in Fig. 8.171(b), the quantities $\mathbf{u}_f(s)$ and $\mathbf{u}(s)$ are vectors and the modeling accuracy, if linear and time invariant, is represented by a transfer matrix $\mathbf{K}_p\mathbf{G}_p(s)$. For an exact model, $\mathbf{K}_p\mathbf{G}_p(s)$ is the identity matrix. The results stated here are special cases of the results derived by Safonov[28] and are valid for the case where $\mathbf{K}_p$ and $\mathbf{G}_p(s)$ are diagonal matrices and $\mathbf{G}_p(s)$ consists of normalized transfer functions $\mathbf{G}_p(s)$ [i.e., $\mathbf{G}_p(0)$ is the identity matrix]. For such a case, as long as all of the phase shifts $\diagup G_{pj}(j\omega)$ are less than 60° in magnitude for all frequencies or as long as all of the elements of the $\mathbf{K}_p$ matrix are greater than one-half, the closed-loop system using the optimal control law is asymptotically stable. More general robustness results, which are more difficult to apply, are also available.[28]

Perkins and Cruz[29] have examined frequency domain characterizations of the infinite-time LQR problem for continuous-time LTI systems and have shown that feedback realization of the optimal control law results in lower sensitivity to plant parameter variations than an equivalent open-loop realization. Kwakernaak and Sivan[12] have provided other results that are somewhat more useful in relating the sensitivity properties of optimal control laws to the weighting matrices in J. As the elements of the control effort weighting matrix $\mathbf{R}_2$ are decreased, the control law sensitivity decreases or improves since the optimal feedback gains are higher. However, higher feedback gains naturally imply greater likelihood of actuator saturation. The relative sensitivities of the different state variables depend on the elements of the state weighting matrix $\mathbf{R}_1$. State variables that are weighted more heavily would have lower sensitivity. Also, the sensitivity characteristics of the optimal control law for non-minimum-phase systems are shown to be inferior to that of minimum-phase systems. Finally, Kwakernaak and Sivan[12] have illustrated that the sensitivity results described do not necessarily extend to discrete-time systems.

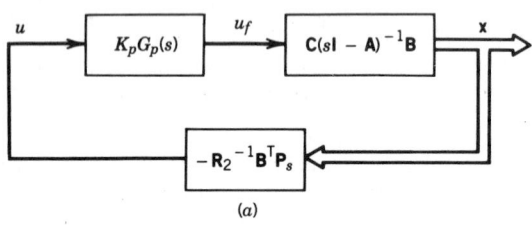

(a)

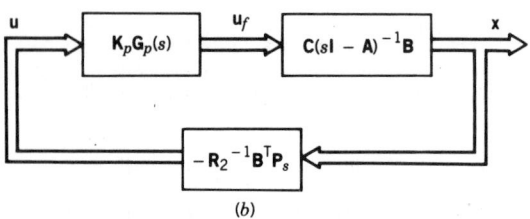

(b)

Fig. 8.171 Robustness of optimal LQR control (a) single input system and (b) multi-input system.

8.42 EXTENSIONS OF THE LINEAR QUADRATIC REGULATOR PROBLEM

The optimal control law for the LQR problem and the pole placement design method described in the preceding sections have a number of limitations. First, as Horowitz[30] has pointed out, control system design by pole placement or quadratic performance index minimization obscures some practically important aspects and objectives. Among these are sensor noise, loop bandwidths, and sensitivity of system performance to significant plant parameter variations. Rosenbrock and McMorran[31] have pointed out that unconditional stability (i.e., stability for all values of the control gains between zero and design values) is essential for industrial control systems but is not guaranteed by the optimal control law. Hence, for multivariable optimal control systems, the failure of a single feedback measuring instrument could destabilize the closed-loop system. Moreover, the achievement of more modest sensitivity requirements is complicated by the lack of clear guidelines for weighting matrix selection. Available procedures for weighting matrix selection enable the achievement of desired transient response characteristics either by specification of a few dominant closed-loop system poles[27] or by implicit model reference following methods.[32] In the latter case, the reference model is chosen to have desired transient response characteristics. However, there is no available method to ensure a priori that the optimal control law has other desirable performance characteristics such as low sensitivity. The consequence of the lack of satisfactory guidelines for weighting matrix selection is that practical control system design using the LQR formulation involves considerable trial and error.

Second, the formulation of the standard LQR problem needs to be extended to be able to effectively handle control problems other than regulation. Examples of such problems include regulation in the presence of persistent disturbances, tracking problems, and vibration control problems. Even though some of these problems can be handled by simple extensions of the LQR problem, effective solutions to these problems require significant extensions of the LQR problem formulation.

Extensions of the standard LQR problem formulation addressing some of its limitations will be described here. The extensions involve alternative formulations of the quadratic index of performance to be minimized such that the resulting solutions have desired features. Additionally, the problem formulation utilizes more completely the available information on the systems to be controlled and their environments. One of the measures for evaluating the effectiveness of the resulting problem formulations and solutions is their ability to accommodate a greater variety of problems and performance specifications. Another such measure is their ability to incorporate in the proposed solutions features that are known to be effective in practice. The resulting variety of problem formulations and solutions runs somewhat counter to the unifying nature of the standard LQR problem formulation and constitutes a recognition of its limitations in practice.

8.42.1 Disturbance Accommodation

Extensions of the standard LQR problem to accommodate unknown disturbance inputs have been proposed by Johnson[33, 34] and Davison and Ferguson.[35]

Johnson[33] considers a general class of disturbances and state-space equations:

$$\dot{x}(t) = A(t)x(t) + B(t)u(t) + B_w(t)w(t) \qquad (8.288)$$

$$y(t) = C(t)x(t) + D(t)u(t) + D_w(t)w(t) \qquad (8.289)$$

where $w(t)$ is a vector of disturbance inputs. The disturbance inputs are assumed to be described by linear time-varying differential equations that constitute a state-space model for the disturbances:

$$\dot{z}_\sigma(t) = A_\sigma(t)z_\sigma(t) + B_\sigma(t)x(t) + \sigma(t) \qquad (8.290)$$

$$w(t) = C_\sigma(t)z_\sigma(t) + D_\sigma(t)x(t) \qquad (8.291)$$

$z_\sigma(t)$ represents the state of the disturbance and $\sigma(t)$ is a vector of Dirac delta impulses occurring at unknown times. The terms including $x(t)$ in the preceding equations enable cases of state-dependent disturbances to be considered within this framework. The coefficient matrices are determined experimentally by examination of the records of the disturbances. This type of description of disturbances constitutes a waveform mode description and is applicable to a broad class of disturbances of practical interest that are not described well either by deterministic process models or by stochastic process models. Examples of such disturbances are piecewise linear, piecewise polynomial, or piecewise periodic signals.

The waveform mode description of disturbances is combined with the system state-space equations to provide a rather complete description of the system to be controlled and the inputs affecting

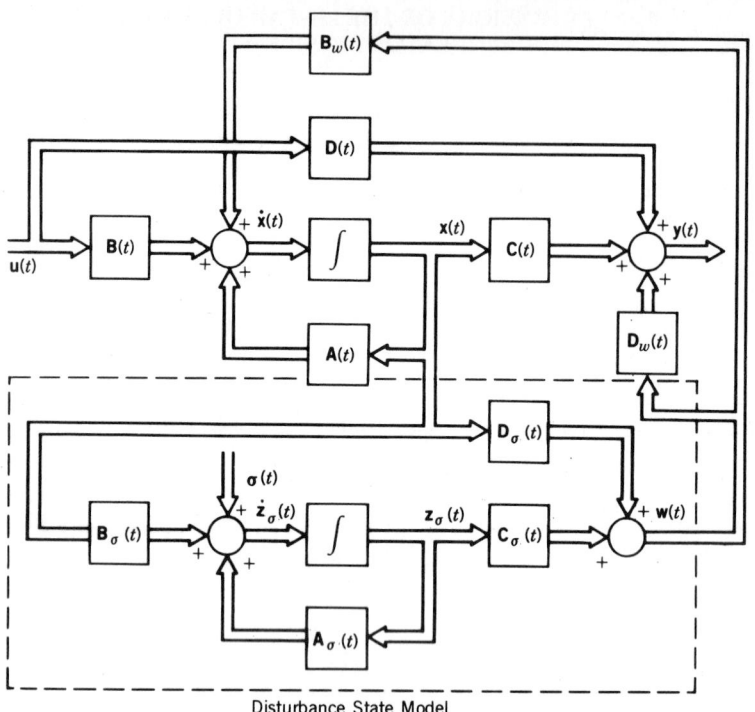

Fig. 8.172   Disturbance state model, continuous-time system.

its behavior. The exact design of the controller depends on the specific objectives used to govern the design. If one of the objectives of the control system design is to counteract as completely as possible the effects of the disturbance inputs, then the control input is considered to be composed of two parts:

$$\mathbf{u}(t) = \mathbf{u}_m(t) + \mathbf{u}_d(t) \tag{8.292}$$

where $\mathbf{u}_d(t)$ is the component used to counteract disturbance effects either completely or partially and $\mathbf{u}_m(t)$ is the component used to accomplish other objectives such as closed-loop pole placement. Alternatively, the objective of control system design may be the minimization of a quadratic index of performance. In either of these cases, the control law would require the feedback of the system state $\mathbf{x}$ as well as the disturbance state $\mathbf{z}_\sigma$ (Fig. 8.172). The state estimation methods described in Section 8.43 can be used to generate these state estimates from available measurements. The extension of this disturbance accommodation approach to discrete-time systems has also been considered by Johnson.[34]

The formulation of disturbance state models and their incorporation in controller design results in controller features familiar from more classical approaches to disturbance suppression. Examples are integral control for constant disturbances, notch filter control for sinusoidal disturbances and disturbance feedforward if some components of the disturbance inputs are measurable.[33] Hence, the disturbance accommodation controllers described here may be viewed as generalizations of classical solutions to disturbance suppression. A similar comment may be made concerning the mechanism for disturbance suppression inherent in the robust servomechanism structure described by Davison and Ferguson.[35] The robust controller structure is described later in this section.

8.42.2   Tracking Applications

Anderson and Moore,[27] Davison and Ferguson,[35] Trankle and Bryson,[36] and Tomizuka et al.[37–40] have considered extensions of the LQR formulation to accommodate tracking applications. Anderson and Moore[27] have considered the servomechanism problem where the linear system state equations are given by Eqs. (8.273) and (8.274) with $\mathbf{D}(t) = \mathbf{0}$ and the class of desired trajectories $\mathbf{y}_r$ is given by

$$\dot{\mathbf{x}}_r(t) = \mathbf{A}_r(t)\mathbf{x}_r(t) \tag{8.293}$$

$$\mathbf{y}_r(t) = \mathbf{C}_r(t)\mathbf{x}_r(t) \tag{8.294}$$

and a specified initial condition $\mathbf{x}_r(t_0)$. The index of performance to be optimized for the finite-time problem is

$$J = \int_{t_0}^{t_1} \left\{ \mathbf{x}^T \left[\mathbf{I} - \mathbf{C}^T(\mathbf{CC}^T)^{-1}\mathbf{C} \right]^T \mathbf{Q}_1 \left[\mathbf{I} - \mathbf{C}^T(\mathbf{CC}^T)^{-1}\mathbf{C} \right] \mathbf{x} + (\mathbf{y} - \mathbf{y}_r)^T \mathbf{Q}_2(\mathbf{y} - \mathbf{y}_r) + \mathbf{u}^T \mathbf{R}_2 \mathbf{u} \right\} dt$$

(8.295)

where the time dependencies of the vectors and matrices have been omitted for convenience. The matrices $\mathbf{Q}_1$ and $\mathbf{Q}_2$ are positive semidefinite matrices and $\mathbf{R}_2$ is a positive definite matrix. The weighting on the tracking error $\mathbf{y} - \mathbf{y}_r$ helps reduce it, whereas the weighting on the state $\mathbf{x}$ achieves a smooth response. The optimal control law involves linear feedback of the system state as well as feedforward of the state of the trajectory model:

$$\mathbf{u} = -\mathbf{K}(t)\mathbf{x}(t) - \mathbf{K}_r(t)\mathbf{x}_r(t)$$

(8.296)

where the gain matrices $\mathbf{K}(t)$ and $\mathbf{K}_r$ are linearly related to solutions of the matrix Riccati differential equations. Conditions for the time-invariant version of this servo problem to reduce to the standard infinite-time LQR problem have also been noted.[27]

A variation on the servomechanism problem already described is that of tracking where the desired trajectory $\mathbf{y}_r$ is known a priori rather than being defined by a model as in Eqs. (8.293) and (8.294). Anderson and Moore[27] have determined the optimal control law for the index of performance Eq. (8.295):

$$u = -\mathbf{K}(t)\mathbf{x}(t) - \mathbf{R}_2(t)^{-1}\mathbf{B}^T(t)\mathbf{b}(t)$$

(8.297)

where $\mathbf{b}(t)$ is the solution of a linear ordinary differential equation with $\mathbf{y}_r(t)$ as the forcing function, and $\mathbf{K}(t)$ is related, in the usual manner, to the solution of a matrix Riccati differential equation.

The control law, Eq. (8.297), incorporates information about future inputs over the entire interval of interest (t_0, t_1) and is said to have infinite preview control in addition to feedback control. A related problem is one where only finite preview of the desired trajectory is available; that is, at any time τ, $\mathbf{y}_r(t)$ is known for $\tau \le t \le \Delta T + \tau$, where ΔT is called the preview length. Tomizuka[37] has examined the continuous-time finite preview problem and determined the optimal control law for a quadratic index of performance over the entire interval of interest (t_0, t_1) where t_1 is greater than $t_0 + \Delta T$. The desired trajectory, not known from preview at any time t, is assumed to be modeled by a stochastic process. A discrete-time version of the problem is given by Tomizuka and Whitney.[38] Discrete-time finite preview of disturbance inputs in addition to the desired trajectory has also been considered by Tomizuka et al.[39, 40] The results indicate that preview control improves the control system performance, especially in the low-frequency range and that there exists a critical preview length beyond which preview information is less important.

8.42.3 Robust Servomechanism Control

Davison and Ferguson[35] have proposed a controller structure and formulated a design procedure for the robust control of servomechanisms. Robustness is defined here to imply asymptotic stability of the closed-loop system and asymptotic tracking of the desired trajectory for all initial conditions of the controller used and for all variations in the system model parameters that do not cause the controlled system to become unstable. The system equations are the time-invariant versions of Eqs. (8.288) and (8.289). The disturbance inputs $\mathbf{w}(t)$ are modeled by time-invariant versions of Eqs. (8.290) and (8.291) with no provision for either state-dependent disturbances $[\mathbf{B}_g(t) = \mathbf{0} = \mathbf{D}_g(t)]$ or impulsive inputs $[\boldsymbol{\sigma}(t) = 0]$. The desired trajectory $\mathbf{y}_r(t)$ is described by time-invariant versions of Eqs. (8.293) and (8.294).

Under certain specified conditions,[35] the robust servomechanism problem is assured of a solution. The resulting controller structure consists of a servocompensator and stabilizing compensator (Fig. 8.173), and the robust control input is given by

$$\mathbf{u} = -\mathbf{K}_\eta \boldsymbol{\eta} - \mathbf{K}_\theta \boldsymbol{\theta}$$

(8.298)

where $\boldsymbol{\eta}$ and $\boldsymbol{\theta}$ are the outputs of the servocompensator and stabilizing compensator respectively, and $\mathbf{K}_\eta$ and $\mathbf{K}_\theta$ are constant-gain matrices. The servocompensator is a dynamic controller with the trajectory error as input and its form and parameters are determined from the state-space models of the system and the disturbance and trajectory inputs. The servocompensator is a generalization of the integral controller from classical control theory. The stabilizing compensator has the function of stabilizing the augmented system consisting of the servocompensator and the system to be controlled. In general, the stabilizing compensator has a number of inputs as shown in Fig. 8.173. It is not

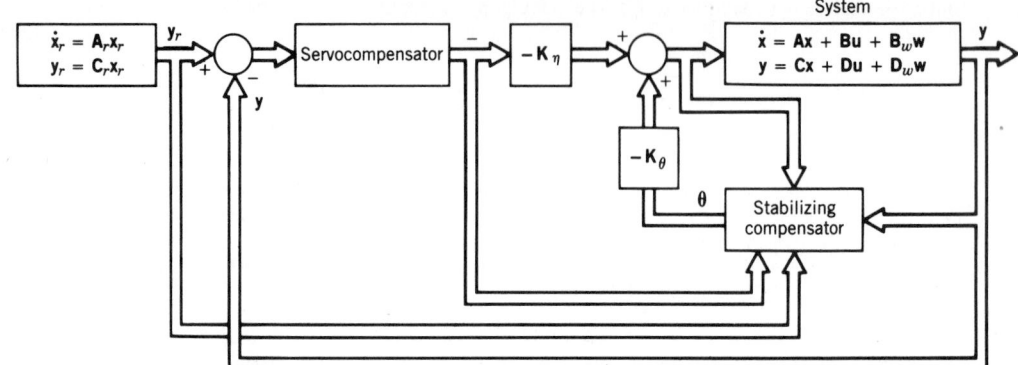

Fig. 8.173   Extension of LQR for the robust LTI servomechanism problem.

uniquely defined, however, and is usually chosen to have as simple a form as is possible given the performance requirements on the controlled system. Complete state feedback, if measurable, and observer-based controllers of the type described in Section 8.44 are among the more elaborate forms of the stabilizing compensator.

Once the structure of the stabilizing compensator is determined, the unknown controller parameters are determined by minimization of a quadratic index of performance. The index of performance is specified such that minimizing it gives a system with fast response and low interaction for MIMO systems. The optimum value of the index of performance is given in terms of the controller parameters. Since controller parameters are not known a priori, the optimal controller design reduces to a multiparameter optimization problem where the quantity being optimized is the quadratic index of performance. The parameter optimization can be constrained to allow the designer to handle closed-loop system damping constraints, controller gain constraints to avoid saturation effects, controller integrity constraints for sensor and actuator failures, and tolerance constraints to system parameter variations. When applied to example systems,[35] the robust control approach yields controller features commonly obtained from other frequency-domain-based design procedures, such as error integral control, phase-lead compensation, pole-zero cancellation, and low interaction for MIMO systems. Its ability to accommodate a variety of constraints on the controllers to ensure their practical utility makes the robust controller design approach a practically useful approach. Care is needed, however, to keep the resulting controller as simple as possible.

The recent extensions of the LQR problem described have addressed many of the limitations of state-space-based approaches to control system design noted by Horowitz[30] and others. As a result, the state-space approach is expected to be more useful in a greater variety of practical applications. It should be noted, however, that there are still no specific guidelines for the selection of weighting matrices in the quadratic indices of performance used by the extended versions of the LQR problem. Consequently, control system design using these methods would still involve considerable trial and error.

8.43   DESIGN OF LINEAR STATE ESTIMATORS

The optimal control laws for the standard LQR problem and the extensions described earlier require feedback of the entire system state. Pole placement algorithms require state feedback as well. Complete state measurement is not possible or practical in many instances. Therefore, the state variables often must be estimated from the measured output variables. Closed-loop dynamic realizations of such state estimators are described here. The property of reconstructibility (or equivalently, observability for LTI systems) is critical for the design of such estimators. It plays a role relative to state estimators that is very similar to the role played by state controllability relative to state feedback controller design. This similarity is the consequence of the duality of the two properties, referred to earlier in Section 8.37. Therefore, many of the results presented in this section parallel those of the preceding section on controller design.

8.43.1   The Observer

The structure of a closed-loop dynamic system, called an observer, to estimate the state $\mathbf{x}(t)$ of an LTV system described by Eqs. (8.213) and (8.214), from output measurements $\mathbf{y}(t)$ and input

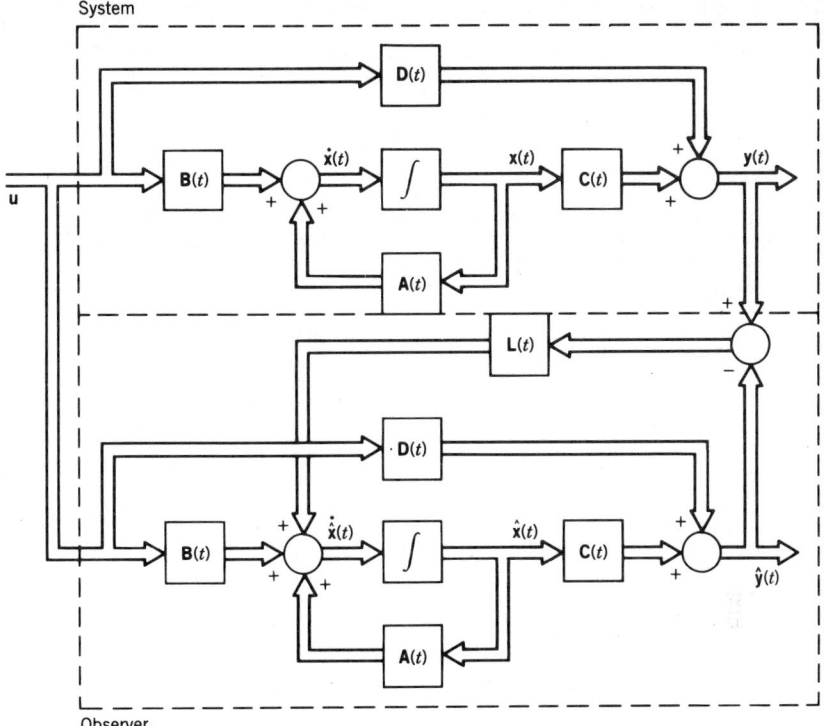

Fig. 8.174 Observer for LTV continuous-time system.

measurements $\mathbf{u}(t)$, was proposed by Luenberger[41] (Fig. 8.174):

$$\dot{\hat{\mathbf{x}}}(t) = \mathbf{A}(t)\hat{\mathbf{x}}(t) + \mathbf{B}(t)\mathbf{u}(t) + \mathbf{L}(t)[\mathbf{y}(t) - \mathbf{C}(t)\hat{\mathbf{x}}(t) - \mathbf{D}(t)\mathbf{u}(t)] \qquad (8.299)$$

where $\hat{\mathbf{x}}(t)$ is the estimated state and $\mathbf{L}(t)$ is a time-varying matrix of observer gains. Combination of Eqs. (8.299), (8.213), and (8.214) yields a dynamic equation for the state estimation error $\mathbf{e}(t)$:

$$\dot{\mathbf{e}}(t) \triangleq \dot{\mathbf{x}}(t) - \dot{\hat{\mathbf{x}}}(t) = [\mathbf{A}(t) - \mathbf{L}(t)\mathbf{C}(t)]\mathbf{e}(t) \qquad (8.300)$$

with the initial condition

$$\mathbf{e}(t_0) = \mathbf{x}(t_0) - \hat{\mathbf{x}}(t_0) \qquad (8.301)$$

Thus, if the observer is asymptotically stable

$$\lim_{t \to \infty} \mathbf{e}(t) = \mathbf{0} \qquad (8.302)$$

for all $\mathbf{e}(t_0)$. If the system under consideration is time invariant, the eigenvalues of the matrix $\mathbf{A} - \mathbf{LC}$ govern the transient behavior of the estimation error. These eigenvalues, referred to as observer poles, can be arbitrarily located in the complex plane by appropriate choice of the constant observer gain matrix $\mathbf{L}$, if and only if the system given by Eqs. (8.213) and (8.214) is completely reconstructible.[41] For LTI systems, reconstructibility is exactly equivalent to observability. Also, if the $\mathbf{L}$ matrix is to be real, the complex eigenvalues of $\mathbf{A} - \mathbf{LC}$ should be specified as conjugate pairs.

The discrete-time version of the observer is given in detail by Kwakernaak and Sivan[12] and parallels that for continuous-time systems closely.

For single-output systems, specification of the desired observer pole locations uniquely specifies the gain vector $\mathbf{L}$. Formulas for the gain vector $\mathbf{L}$ will be given here. They are obtained by analogy with Eqs. (8.272), (8.273) and (8.274). The object is to get the observer gain vector $\mathbf{L}$ so that the characteristic equation governing the estimation error transients has specified coefficients:

$$\beta_e(s) = \det(s\mathbf{I} - \mathbf{A} + \mathbf{LC})$$

$$= \det(s\mathbf{I} - \mathbf{A}^{\mathsf{T}} + \mathbf{C}^{\mathsf{T}}\mathbf{L}^{\mathsf{T}})$$

$$= s^n + \sum_{i=0}^{n-1} \beta_{ie} s^i = 0 \qquad (8.303)$$

where the matrix $\mathbf{A} - \mathbf{LC}$ is transposed so that the observer gain selection problem can be reduced to a form similar to the state feedback controller gain selection problem. Comparison of Eqs. (8.303) and (8.274) suggests that the following substitutions are necessary to convert algorithms for state feedback controller gain selection to observer gain selection.

CONTROLLER GAIN SELECTION		OBSERVER GAIN SELECTION	
$\mathbf{A}$	$\longrightarrow$	$\mathbf{A}^{\mathsf{T}}$	
$\mathbf{B}$	$\longrightarrow$	$\mathbf{C}^{\mathsf{T}}$	(8.304)
$\mathbf{K}$	$\longrightarrow$	$\mathbf{L}^{\mathsf{T}}$	

Using the preceding transformations, we get

$$\mathbf{L} = \beta_e(\mathbf{A}) \begin{bmatrix} \mathbf{C} \\ \mathbf{CA} \\ \vdots \\ \mathbf{CA}^{n-1} \end{bmatrix}^{-1} \begin{bmatrix} 0 \\ 0 \\ \vdots \\ 1 \end{bmatrix} \qquad (8.305)$$

for continuous-time systems. In the preceding equations,

$$\beta_e(\mathbf{A}) = \mathbf{A}^n + \sum_{i=0}^{n-1} \beta_{ie} \mathbf{A}^i \qquad (8.306)$$

The β_{ie}'s are the coefficients of the desired characteristic Eq. (8.303) for the observer. A similar equation describes the discrete-time observer characteristic equation. If CACSD packages[22-24] supporting only controller pole placement algorithms are available, the transformations of Eq. (8.304) are needed to select observer gains using these algorithms.

The observer of Eq. (8.299) is called a full-order observer since its dimensions are the same as those of the system whose states are being estimated. Reduction of the observer dimension can be achieved by using the fact that the output equation provides us with p linear equations in the unknown state, where p is the number of output variables. Therefore, the observer need only provide $n - p$ additional linear equations and thus need only be of dimension $n - p$. For time-invariant systems, the corresponding observer gain matrix can be chosen to place the $n - p$ observer poles at any desired locations in the complex plane, if the original systems of Eqs. (8.215) and (8.216) are completely observable. Equations for reduced-order observers for continuous-time systems are given by Kwakernaak and Sivan.[12] In the presence of measurement noise, the state estimates $x(t)$ obtained using reduced-order observers are more sensitive than those obtained using full-order observers.

For multioutput LTI systems, specification of the desired observer poles does not specify the gain matrix $\mathbf{L}$ uniquely. The additional freedom in the gain matrix selection can be used to assign the eigenvectors (or generalized eigenvectors) of the matrix $\mathbf{A} - \mathbf{LC}$ in addition to the eigenvalues. The corresponding procedure would be very similar to that used for gain matrix selection for multi-input systems and described in Section 8.40. The transformations of Eq. (8.304) can be used to adapt the controller gain selection procedure to observer gain selection. Alternatively, the observer gains can be chosen to design observers whose state estimates have low sensitivity to unmeasured disturbance inputs.[42]

Equations for full-order and reduced-order observers for discrete-time systems are given by Franklin and Powell.[26] In addition, a variation of these observers, called the current estimator, is possible for discrete-time systems if the computation time associated with the observer is short compared to the sampling interval.[26]

The observer designs described work well in the absence of significant levels of measurement noise or unknown disturbance signals. The sensitivity of the state estimates to measurement noise and unknown disturbance signals depends on the specified location of the observer poles. If these pole locations are too far into the left half of the complex s plane, the state estimates would be unduly sensitive to measurement noise and disturbance signals. In the limiting case, the observers can be shown to reduce to ideal differentiators. The observer pole locations should therefore be chosen to avoid such high sensitivities of the state estimates, but at the same time ensure that the estimation error transients decay more rapidly than the state variables being estimated. Specification of the observer poles in practice involves considerable trial and error in much the same manner as specification of closed-loop poles does for state feedback controller design. If some information is available concerning the disturbance signals affecting the system and the measurement noise, the observer gain matrix selection problem can be formulated as an optimal estimation problem.

8.43.2   The Optimal Observer

Optimization of observer design has been primarily performed assuming stochastic models for the disturbance inputs and measurement noise, though the effect of deterministic model errors on observer design is also important.[43] Consider the continuous-time system[12]

$$\dot{x}(t) = A(t)x(t) + B(t)u(t) + S(t)w_1(t) \qquad t \geq t_0 \tag{8.307}$$

$$y(t) = C(t)x(t) + w_2(t) \qquad t \geq t_0 \tag{8.308}$$

where $w_1(t)$ is the random disturbance input, $w_2(t)$ is the random measurement noise, and their joint probabilities are assumed to be known. The column vector $[w_1^T(t) \ w_2^T(t)]^T$ is assumed to be a white-noise process with intensity

$$V(t) = \begin{bmatrix} V_1(t) & V_{12}(t) \\ V_{12}^T(t) & V_2(t) \end{bmatrix} \tag{8.309}$$

that is, the expected value

$$E\left\{ \begin{bmatrix} w_1(t_1) \\ w_2(t_1) \end{bmatrix} \begin{bmatrix} w_1^T(t_2) & w_2^T(t_2) \end{bmatrix} \right\} = V(t_1)\delta(t_1 - t_2) \tag{8.310}$$

where $\delta(t_1 - t_2)$ is the Dirac delta function.

The initial state $x(t_0)$ is assumed to be a random variable uncorrelated with w_1 and w_2 and its probability given by

$$E[x(t_0)] = x_0 \quad \text{and} \quad E\left\{ [x(t_0) - x_0][x(t_0) - x_0]^T \right\} = Q_0 \tag{8.311}$$

The observer form is given by Eq. (8.299) and Fig. 8.174. The optimal observer problem consists of determining $K(\tau)$, $t_0 \leq \tau \leq t$ and the initial condition on the observer $\hat{x}(t_0)$ so as to minimize the expected value $E\{[x(t) - \hat{x}(t)]^T W(t)[\hat{x}(t) - x(t)]\}$, where $W(t)$ is a symmetric positive definite weighting matrix.

If the problem as stated is nonsingular

$$\det[V_2(t)] > 0 \qquad t \geq t_0 \tag{8.312}$$

and if the disturbance and measurement noise are uncorrelated

$$V_{12}(t) = 0 \tag{8.313}$$

the optimal observer gain matrix is given by Kalman and Bucy[44] as

$$K(t) = Q(t)C^T(t)V_2^{-1}(t) \qquad t \geq t_0 \tag{8.314}$$

where Q is a solution of the matrix Riccati equation:

$$\dot{Q}(t) = A(t)Q(t) + Q(t)A^T(t) + S(t)V_1(t)S^T(t)$$
$$- Q(t)C^T(t)V_2^{-1}(t)C(t)Q(t) \qquad t \geq t_0 \tag{8.315}$$

with the initial condition

$$Q(t_0) = Q_0 \tag{8.316}$$

and the observer initial condition

$$x(t_0) = x_0 \tag{8.317}$$

The resulting state estimator is called the Kalman–Bucy filter. The similarity of Eqs. (8.314) and (8.315) to the corresponding Eqs. (8.281) and (8.282), respectively, for the LQR problem is a result of the duality of the state estimation and state feedback control problems noted earlier. One difference, however, is that the Riccati equation for the optimal observer can be implemented in real time since Eq. (8.316) is an initial condition for Eq. (8.315). In contrast, for the finite-time LQR problem, Eq. (8.283) gives the terminal condition for the Riccati Eq. (8.282).

The steady-state properties of the optimal observer for linear time-varying and time-invariant systems parallel those of the optimal controller for the LQR problem and are described by Kwakernaak and Sivan.[12] If the time-varying system

$$\dot{\mathbf{x}}(t) = \mathbf{A}(t)\mathbf{x}(t) + \mathbf{S}(t)\mathbf{w}_1(t) \tag{8.318}$$

$$\mathbf{y}(t) = \mathbf{C}(t)\mathbf{x}(t) + \mathbf{w}_2(t) \tag{8.319}$$

is uniformly completely controllable by $\mathbf{w}_1(t)$ and uniformly completely reconstructible, the solution $\mathbf{Q}(t)$ of Eq. (8.315) converges to a steady-state solution $\mathbf{Q}_s(t)$ as $t_0 \rightarrow -\infty$ for any positive semidefinite $\mathbf{Q}_0$. The corresponding steady-state optimal observer

$$\dot{\hat{\mathbf{x}}}(t) = \mathbf{A}(t)\hat{\mathbf{x}}(t) + \mathbf{L}_s(t)[\mathbf{y}(t) - \mathbf{C}(t)\hat{\mathbf{x}}(t)] \tag{8.320}$$

where

$$\mathbf{L}_s(t) = \mathbf{Q}_s(t)\mathbf{C}^{\mathrm{T}}(t)\mathbf{V}_2^{-1}(t) \tag{8.321}$$

is exponentially stable. Also, if the system and the noise statistics are invariant, the matrix Riccati differential equation (8.315) becomes an algebraic equation as $t_0 \rightarrow -\infty$:

$$\mathbf{A}\mathbf{Q}_s + \mathbf{Q}_s\mathbf{A}^{\mathrm{T}} + \mathbf{S}\mathbf{V}_1\mathbf{S}^{\mathrm{T}} - \mathbf{Q}_s\mathbf{C}^{\mathrm{T}}\mathbf{V}_2^{-1}\mathbf{C}\mathbf{Q}_s = 0 \tag{8.322}$$

If the corresponding time-invariant system is completely controllable by the input $\mathbf{w}_1(t)$ and completely observable, Eq. (8.322) has a unique positive definite solution $\mathbf{Q}_s$ and the corresponding steady-state optimal observer of Eqs. (8.320) and (8.321) is asymptotically stable. Note that the measurable input $\mathbf{u}(t)$ has been omitted from Eqs. (8.318) and (8.319) for simplicity but does not change the substance of the results.

Interested readers are referred to Kwakernaak and Sivan[12] for a more complete consideration of the optimal observer for continuous-time and discrete-time systems. There is also extensive literature available on Kalman filters.[44–46]

8.44   OBSERVER-BASED CONTROLLERS

The observers described in the preceding section can be used to provide estimates of system state that, in turn, can be used to provide state feedback as described in Sections 8.40–8.42. The resulting controllers are dynamic compensators and are referred to as observer-based controllers.

The design of such observer-based controllers for LTI systems is simplified somewhat by the fact that their modes or eigenvalues satisfy the separation property, that is, the eigenvalues of the observer-controller are the same as the eigenvalues of the observer and the eigenvalues of the controller, the latter evaluated assuming perfect state measurement. For continuous-time LTI systems described by Eqs. (8.215) and (8.216), the control law given by Eq. (8.271), and the observer given by Eq. (8.299) with constant coefficient matrices, the observer-controller is given by Fig. 8.175 and the characteristic equation of the corresponding closed-loop system is[3]

$$\det(s\mathbf{I} - \mathbf{A} + \mathbf{B}\mathbf{K})\det(s\mathbf{I} - \mathbf{A} + \mathbf{L}\mathbf{C}) = 0 \tag{8.323}$$

A similar result can be shown to be true for discrete-time LTI systems.[26]

For LTI systems subjected to unmeasured randomly varying disturbance inputs and measurement errors, if the statistics of these signals are known, state estimators of the Kalman–Bucy type will be used. The resulting estimator-based controllers have eigenvalues that also satisfy a separation property. As a result of the separation property for controllers based on observers or Kalman–Bucy filters, the design of the controllers can be treated independently of the observer.

The use of observers or Kalman filters to provide state estimates for state feedback controllers does, however, impair overall controller performance. For instance, the transient response of such controllers is poorer than that of controllers using complete state feedback. Moreover, the properties relating to the gain and phase margins of optimal controllers for the LQR problem are not applicable

System

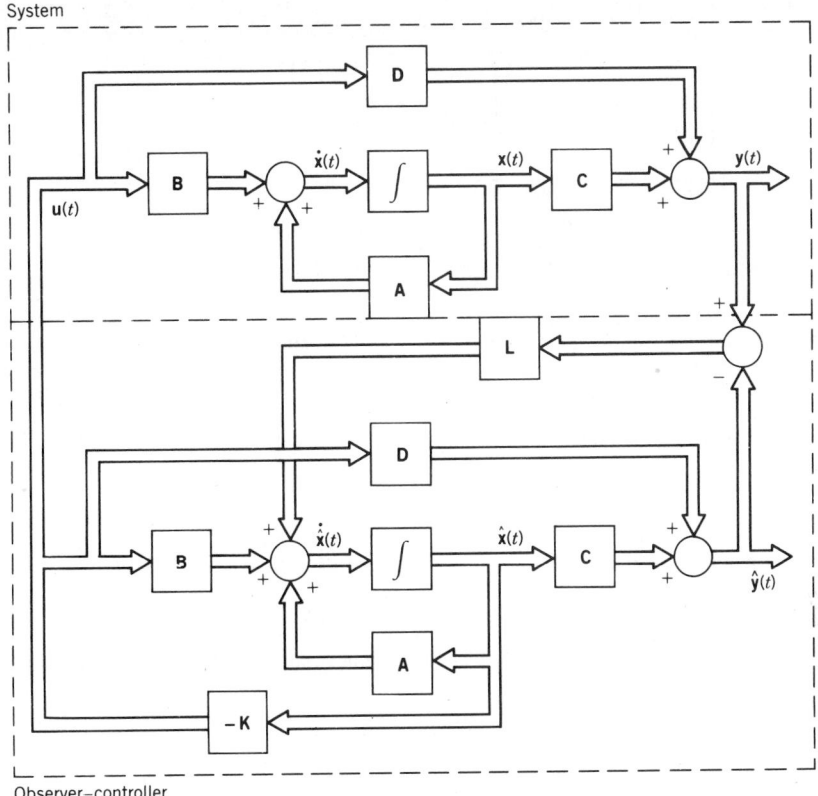

Observer–controller

Fig. 8.175 Observer-based controller for LTI continuous-time system.

if the controllers use estimated states for feedback rather than measured states.[28] There is a definite loss of controller robustness associated with the use of state estimators. The robustness of such controllers is more properly evaluated by considering them as dynamic compensators and using methods common to the frequency domain approach.[30]

Observer-based controllers for LTI systems can naturally be examined using transfer function or frequency-domain-based methods. Such a linking of time-domain- and frequency-domain-based controller designs offers a number of advantages. The state-space-based design approach leads to consideration of controller structures that are not obvious from a transfer function approach. In addition, the state-space approach alerts the designer to the problem of loss of controllability or observability via pole-zero cancellation. On the other hand, the transfer function approach can result in controllers than cannot be obtained by using observer-based controllers.[3] In addition, the consideration of observer-based controllers from a transfer function perspective helps evaluate the controllers to see whether proven and practical design guidelines are violated. Such guidelines invariably use transfer function terminology since they have evolved from years of experience using classical control techniques. If such guidelines are violated, the state-space-based controller design procedure can be modified appropriately.[47]

A recent example of a MIMO controller design method that has evolved from a combination of frequency domain methods and state-space methods is the linear-quadratic-Gaussian method with loop-transfer-recovery (LQG/LTR), developed by Athans.[48] The procedure relies on the fact that results and requirements relating to control system robustness to modeling errors are best presented in the frequency domain. The powerful controller and estimator structures resulting from LQR formulations of the control and state estimation problem are used. These structures are useful in this design method because their robustness and performance have been well studied, using frequency domain measures.[28] The resulting method relies upon designer expertise in formulating good performance specifications at the outset. The design method therefore avoids the main weakness of state-space methods—namely, the weak connection between performance measures used by these methods and performance measures of engineering significance. The computation of the controller is,

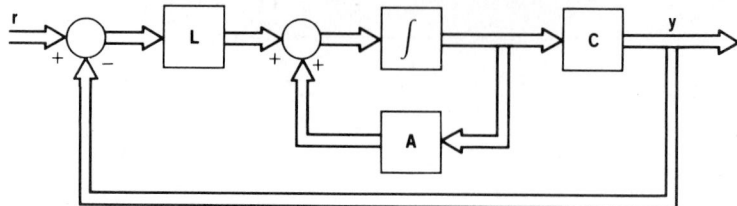

Fig. 8.176 Target feedback-loop block diagram.

however, straightforward in this method, since the controller structures are derived from well-established state-space methods and are well supported by commercial CACSD packages.[22-24]

The first step in the design method is the definition of the design plant model. This model includes not only the nominal model of the system to be controlled but also includes scaling of the variables and augmentation of the dynamics, such as the inclusion of integrators dictated by control objectives. The number of control inputs r and the number of outputs p are assumed to be equal in the following development. The model is also linear and time invariant and strictly proper, that is, the transmission matrix $\mathbf{D}$ in the system Eqs. (8.215) and (8.216) is zero. The transfer function matrix of the system is

$$\mathbf{H}(s) = \mathbf{C}(s\mathbf{I}_n - \mathbf{A})^{-1}\mathbf{B} \qquad (8.324)$$

Modeling inaccuracy is treated as follows. The actual transfer function matrix is given by

$$\mathbf{H}_A(s) = [\mathbf{I}_n + \mathbf{E}(s)]\mathbf{H}(s) \qquad (8.325)$$

where $\mathbf{E}(s)$ characterizes the modeling error. The maximum eigenvalue $\sigma_{\max}$ of the matrix $\mathbf{E}(j\omega)$ is assumed to be bounded by a known bound $e_m(\omega)$.

$$\sigma_{\max}[\mathbf{E}(i\omega)] < e_m(\omega) \qquad (8.326)$$

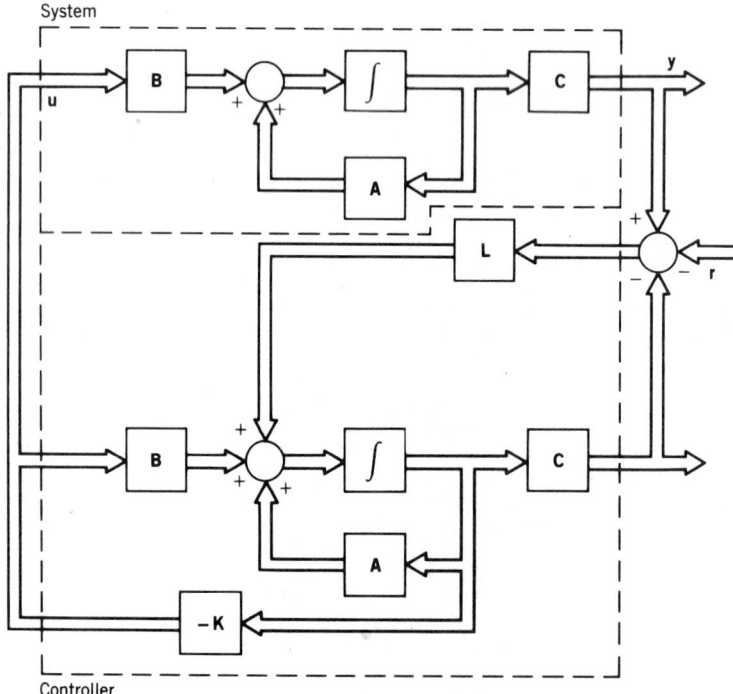

Fig. 8.177 Compensator structure for LQG/LTR method.

The second step in the design procedure is the specification of a target feedback loop that has satisfactory robustness, stability, and performance specifications. The target feedback loop is shown in Fig. 8.176. It is obviously not directly implementable since the control inputs u do not appear in the system. The matrix L is a constant matrix and is chosen as described later. It is the designer's task to experiment with different choices of L and evaluate whether the resulting system has satisfactory performance. This stage of the design therefore requires considerable trial and error.

Athans[48] has suggested using the steady-state Kalman–Bucy filter formulation with stationary system and noise characteristics. It should be noted, however, that the formulation is not used here to perform an estimation task. It is being used here to help calculate the matrix L because the resulting target feedback loop has well-known performance and robustness characteristics. At the minimum, selection of L as described by Athans[48] guarantees that the target feedback loop will not amplify disturbances entering the system at the output. In addition, the target feedback loop will not go unstable as long as the modeling uncertainty $e_m(\omega)$ in Eq. (8.326) is below 0.5.

Once the target feedback loop is chosen, the final step in the design procedure is the design of a compensator to enable the controlled system to approximate the behavior of the target feedback loop closely. Athans[48] has proposed the compensator structure shown in Fig. 8.177. The similarity of the controller to the observer-based controller in Fig. 8.175 is clear. The gain matrix K is computed via the solution of a version of the LQR problem. The loop transfer recovery (LTR) result, credited by Athans to other researchers, guarantees that for minimum-phase systems, the procedure described yields a controlled system behavior that approximates the target feedback loop behavior as closely as desired.

For non-minimum-phase systems, the design procedure remains the same. The only difference is that the final design may not approximate the behavior of the target feedback loop closely. In effect, this would result in additional design iterations to arrive at a satisfactory final design.

The LQG/LTR design procedure has been applied successfully to evaluate the feasibility of MIMO control for aircraft and helicopter flight control, jet engine control, and submersible control.

8.45   CONCLUSION

The state-space methods for dynamic systems analysis presented in Sections 8.33–8.38 offer a unified approach to the analysis of a great variety of dynamic systems. The primary emphasis, in these methods, on linear time-invariant systems is a reflection of the state of the literature on the subject and of the practice. Results for linear time-varying systems have been given in some of the standard texts.[2-4] In Sections 8.40–8.44, emphasis is on the synthesis of control systems, primarily for linear, time-invariant (LTI), single-input–single-output (SISO), and multiple-input–multiple-output (MIMO) systems.

The reader is referred to the following works for more exhaustive treatment of some of the topics not covered at great length here. The subject of multivariable control using state-space methods has been addressed at much greater length by Kailath[3] and others.[49] The application of state-space methods to nonlinear system analysis and control is treated at some length by Ramnath et al.[50] Optimal control problems other than the LQR formulation have been described in detail in a number of textbooks.[51,52] The subject of adaptive control refers to control situations where the controller parameters are adapted or adjusted as the behavior of the system being controlled changes. One approach to adaptive control, termed the model reference approach and employing state-space description, has been described at length by Landau.[53] Distributed-parameter systems are examples of systems with infinite-dimensional states. Application of state-space methods to these systems has been described by Tzafestas et al.[54] Time-delayed systems are also examples of systems with infinite-dimensional states. The analysis and control of such systems and of many of the other types of systems referred to in Sections 8.33–8.44 remain a subject of current research. For current research results in these areas, the reader is referred to journals such as the ASME *Journal of Dynamic Systems, Measurement and Control*, IEEE's *Transactions on Automatic Control*, the AIAA *Journal of Guidance, Control and Dynamics*, SIAM's *Journal on Control*, and *Automatica, The Journal of the International Federation of Automatic Control*.

REFERENCES FOR PART IV

1. Fortmann, T. E., and Hitz, K. L., *An Introduction to Linear Control Systems*, New York, Marcel-Dekker, 1977.
2. DeRusso, P. M., Roy, R. J., and Close, C. M., *State Variables for Engineers*, New York, Wiley, 1965.
3. Kailath, T., *Linear Systems*, Englewood Cliffs, N.J., Prentice-Hall, 1980.
4. Brogan, W. L., *Modern Control Theory*, Englewood Cliffs, N.J., Prentice-Hall, 1982.

5. Shinners, S. M., *Modern Control System Theory and Application*, Reading, Mass., Addison-Wesley, 1978.

6. Palm, W. J., III, *Modeling, Analysis and Control of Dynamic Systems*, New York, Wiley, 1983.

7. Kuo, B. C., *Digital Control Systems*, Champaign, Ill., SRL Publishing, 1977.

8. Kalman, R. E., and Bertram, J. E., "Control-System Analysis and Design via the Second Method of Lyapunov. I–Continuous Time Systems. II—Discrete-Time Systems," *Transactions of the ASME, Journal of Basic Engineering*, Vol. 82D, pp. 371–400, June 1960.

9. Kalman, R. E., "Mathematical Description of Linear Dynamical Systems," *SIAM Journal on Control*, Series A, Vol. 1, No. 2, pp. 153–192, 1963.

10. Kalman, R. E., "When Is a Linear Control System Optimal?" *Transactions of the ASME, Journal of Basic Engineering*, Vol. 86D, pp. 51–60, 1964.

11. Kalman, R. E., "On the General Theory of Control Systems," *Proceedings of the First International Congress on Automatic Control*, Butterworth's, England, 1960, pp. 481–493.

12. Kwakernaak, H., and Sivan, R., *Linear Optimal Control Systems*, New York, Wiley-Interscience, 1972.

13. Kreindler, E., and Sarachik, P., "On the Concepts of Controllability and Observability of Linear Systems," *IEEE Transactions on Automatic Control*, Vol. AC-9, No. 1, pp. 129–136, February 1964.

14. Kalman, R. E., Falb, P. L., and Arbib, M., *Topics in Mathematical System Theory*, New York, McGraw-Hill, 1969.

15. Gilbert, E. G., "Controllability and Observability in Multivariable Control Systems," *SIAM Journal on Control*, Series A, Vol. 2, No. 1, pp. 128–151, 1963.

16. Johnson, C. D., "Optimization of a Certain Quality of Complete Controllability and Observability for Linear Dynamical Systems," *Transactions of the ASME, Journal of Basic Engineering*, Vol. 91D, pp. 228–238, 1969.

17. Friedland, B., "Controllability Index Based on Conditioning Number," *ASME Transactions, Journal of Dynamic Systems, Measurement and Control*, pp. 444–445, Vol. 97, December 1975.

18. Muller, P. C., and Weber, H. I., "Analysis and Optimization of Certain Qualities of Controllability and Observability for Linear Dynamical Systems," *Automatica*, Vol. 8, pp. 237–246, 1972.

19. Silverman, L. M., and Meadows, H. E., "Controllability and Observability in Time-Variable Linear Systems," *SIAM Journal on Control*, pp. 64–73, Vol. 5, No. 1, 1967.

20. Horowitz, I. M., *Synthesis of Feedback Control Systems*, New York, Academic Press, 1963.

21. Rosenbrock, H. H., *Computer-Aided Control System Design*, New York, Academic Press, 1974.

22. Anonymous, *CTRL-C, A Language for the Computer-Aided Design of Multivariable Control Systems*, Systems Control Technology, Palo Alto, California, 1983.

23. Walker, R., Gregory, Jr., C., and Shah, S., "MATRIX$_x$: A Data Analysis, System Identification, Control Design and Simulation Package," *IEEE Control Systems Magazine*, pp. 30–36, December 1982.

24. Astrom, K. J., "Computer Aided Modeling, Analysis and Design of Control Systems—A Perspective," *IEEE Control Systems Magazine*, pp. 4–16, May 1983.

25. Wiberg, D. M., *State Space and Linear Systems*, Schaum's Outline Series, New York, McGraw-Hill, 1971.

26. Franklin, G. F., and Powell, J. D., *Digital Control of Dynamic Systems*, Reading, Mass., Addison-Wesley, 1980.

27. Anderson, B. D. O., and Moore, J. B., *Linear Optimal Control*, Englewood Cliffs, N.J., Prentice-Hall, 1971.

28. Safonov, M. G., *Stability and Robustness of Multivariable Feedback Systems*, Cambridge, Mass., MIT Press, 1980.

29. Perkins, W. R., and Cruz, Jr., J. B., "Feedback Properties of Linear Regulators," *IEEE Transactions on Automatic Control*, Vol. AC-16, No. 6, pp. 659–664, December 1971.

30. Horowitz, I. M., and Shaked, U., "Superiority of Transfer Function over State-Variable Methods in Linear Time-Invariant Feedback System Design," *IEEE Transactions on Automatic Control*, Vol. AC-20, No. 1, pp. 84–97, February 1975.

31. Rosenbrock, H. H., and McMorran, P. D., "Good, Bad or Optimal?" *IEEE Transactions on Automatic Control*, Vol. AC-16, No. 6, pp. 552–554, December 1971.

32. Tyler, Jr., J. S., "The Characteristics of Model Following Systems as Synthesized by Optimal Control," *IEEE Transactions on Automatic Control*, Vol. AC-9, No. 5, pp. 485–498, October 1964.

33. Johnson, C. D., "Theory of Disturbance-Accommodating Controllers," in *Control and Dynamic Systems, Advances in Theory and Applications*, C. T. Leondes (Ed.), Vol. 12, pp. 387–489, New York, Academic Press, 1976.

34. Johnson, C. D., "A Discrete-Time Disturbance-Accommodating Control Theory for Digital Control of Dynamical Systems," in *Control and Dynamic Systems, Advances in Theory and Applications*, C. T. Leondes (Ed.), Vol. 18, pp. 223–315, New York, Academic Press, 1982.

35. Davison, E. J., and Ferguson, I. J., "The Design of Controllers for the Multivariable Robust Servomechanism Problem Using Parameter Optimization Methods," *IEEE Transactions on Automatic Control*, Vol. AC-26, No. 1, pp. 93–110, February 1981.

36. Trankle, T. L., and Bryson, Jr., A. E., "Control Logic to Track Outputs of a Command Generator," *AIAA Journal of Guidance and Control*, Vol. 1, No. 2, pp. 130–135, March–April 1978.

37. Tomizuka, M., "Optimal Continuous Finite Preview Problem," *IEEE Transactions on Automatic Control*, Vol. AC-20, No. 3, pp. 362–365, June 1975.

38. Tomizuka, M., and Whitney, D. E., "Optimal Finite Preview Problems (Why and How Is Future Information Important?)" *ASME Transactions, Journal of Dynamic Systems, Measurement and Control*, Vol. 97, No. 4, pp. 319–325, December 1975.

39. Tomizuka, M., and Rosenthal, D. E., "On the Optimal Digital Share Vector Feedback Controller with Integral and Preview Actions," *ASME Transactions, Journal of Dynamic Systems, Measurement and Control*, Vol. 101, No. 2, pp. 172–178, June 1979.

40. Tomizuka, M., and Fung, D. H., "Design of Digital Feedforward/Preview Controllers for Processes with Predetermined Feedback Controllers," *ASME Transactions, Journal of Dynamic Systems, Measurement and Control*, Vol. 102, No. 4, pp. 218–225, December 1980.

41. Luenberger, D. G., "Observing the State of a Linear System," *IEEE Transactions on Military Electronics*, Vol. 8, pp. 74–80, April 1964.

42. Shah, S. L., Seborg, D. E., and Fisher, D. G., "Design and Application of Controllers and Observers for Disturbance Minimization and Pole Assignment," *ASME Transactions, Journal of Dynamic Systems, Measurement and Control*, Vol. 102, No. 1, pp. 21–27, March 1980.

43. Thau, F. E., and Kestenbaum, A., "The Effect of Modeling Errors on Linear State Reconstructors and Regulators," *ASME Transactions, Journal of Dynamic Systems, Measurement and Control*, Vol. 46, No. 4, pp. 454–459, December 1974.

44. Kalman, R. E., and Bucy, R. J., "New Results in Linear Filtering and Prediction Theory," *ASME Transactions, Journal of Basic Engineering*, Series D, Vol. 83, pp. 45–108, March 1961.

45. Mendel, J. M., and Gieseking, D. L., "Bibliography on the Linear-Quadratic-Gaussian Problem," *IEEE Transactions on Automatic Control*, Vol. AC-16, No. 6, pp. 847–869, December 1971.

46. Anderson, D. O., and Moore, J. B., *Optimal Filtering*, Englewood Cliffs, N.J., Prentice-Hall, 1979.

47. Bryson, Jr., A. E., "Some Connections Between Modern and Classical Control Concepts," *ASME Transactions, Journal of Dynamic Systems, Measurement and Control*, Vol. 10, No. 3, pp. 91–98, June 1979.

48. Athans, M., "A Tutorial on the LQG/LTR Method," Proceedings of the 1986 American Control Conference, Seattle, Washington, pp. 1289–1296, June 1986.

49. Sain, M., Ed., "Special Issue on Linear Multivariable Control," *IEEE Transactions on Automatic Control*, Vol. AC-26, No. 6, pp. 1–295, February 1981.

50. Ramnath, R. V., and Paynter, H. M., Ed., *Nonlinear System Analysis and Synthesis: Volume 2—Techniques and Applications*, Workshop/Tutorial Session at the Winter Annual Meeting of ASME, New York, December 1980.

51. Athans, M., and Falb, P., *Optimal Control*, New York, McGraw-Hill, 1966.

52. Sage, A. P., *Optimum Systems Control*, Englewood Cliffs, N.J., Prentice-Hall, 1968.

53. Landau, Y. D., *Adaptive Control. The Model Reference Approach*, New York, Marcel-Dekker, 1979.

54. Tzafestas, S. G., Ed., *Distributed Parameter Control Systems, Theory and Application*, Vol. 6, International Series on Systems and Control, Oxford, England, Pergamon Press, 1982.

CHAPTER 9
COMPUTER SCIENCE*

Jack L. Rosenfeld, Editor

Department of Computer Sciences
IBM Thomas J. Watson Research Center
Yorktown Heights, New York

Sheldon S. L. Chang, Editor

Department of Electrical Engineering
State University of New York
Stony Brook, New York

James E. Thornton

Network Systems Corporation
Brooklyn Park, Minnesota

Dennis J. McBride

Applied Research Department
IBM Thomas J. Watson Research Center
Yorktown Heights, New York

Earl E. Swartzlander, Jr.

TRW
Defense and Space Systems Group
Redondo Beach, California

George C. Feth

IBM Thomas J. Watson Research Center
Yorktown Heights, New York

Kishor S. Trivedi

Department of Computer Science
Duke University
Durham, North Carolina

Nikitas A. Alexandridis

Department of Digital Systems and Computers
National Technical University of Athens
Athens, Greece

David F. Bantz

Department of Computer Sciences
IBM Thomas J. Watson Research Center
Yorktown Heights, New York

Kenneth J. Thurber

Architecture Technology Corporation
Minneapolis, Minnesota

*Chapter 9 is adapted, with permission, from Sheldon S. L. Chang, Ed., *Fundamentals Handbook of Electrical and Computer Engineering*, Wiley, New York, 1982, Vol. I: Section 7; Vol. III: Section 1.

This chapter includes discussions of computer technology, design, and use. It is planned to serve those who (1) use computers to solve engineering problems, (2) design interfaces between computers and laboratory equipment or other devices, (3) design computers or parts of computers, (4) apply aspects of computer design to other engineering design problems, or (5) need to expand their knowledge of this important engineering discipline. We hope it will prove to be an indispensable reference for all such engineers, as well as students of computer engineering and computer science.

The full range of computer system capability is covered here, from functional units of microprocessors to collections of large mainframe computers. Stress is placed on microprocessors in the belief that this chapter will be used most widely by those interested in microprocessor-based systems.

Conventional Computer Organization

Before proceeding to summarize the sections in this chapter, we pause for a brief discussion of computer organization. The conventional computer organization shown in Fig. 9.1 is the starting point for most analyses of computers. There are three major blocks: the central processing unit (CPU), main memory, and input/output (I/O). The CPU performs the data processing. It controls the entire computer system. Main memory stores data that have been or will be processed. In conventional computers, main memory also holds instructions to the CPU: codes that describe the operations to be performed on the data. The input component brings data from devices such as keyboards and magnetic tape drives into main memory to be processed; the output component moves results from main memory to devices such as printers and transducers. The instructions are also brought from input devices to main memory. Some devices are used to store data rather than to serve as source input or sink output. These storage devices also transmit data to and from memory via the input/output control component.

The CPU comprises a control unit, an arithmetic-logic unit (ALU), and a number of registers. The control unit generates signals that control the operation of the entire system. The sequence of signals generated depends on what instruction is being executed. The registers are collections of flip-flops that hold control information or data. The ALU performs arithmetic—both for data processing requirements and for determining where required data and instructions can be found.

Memory is organized into words, each with the same number of bits of information and each with its own address. Consecutively executed instructions are normally fetched from consecutive addresses. This normal sequence may be modified by branches in the program. The capability to branch according to observed conditions gives the CPU the ability to do more than simply execute a fixed sequence of instructions. The instructions themselves specify where in memory or registers the operands are to be found. The control unit fetches the operands from their designated locations and stores the results appropriately.

Fig. 9.1 shows paths of data flow and control signals. Data and instructions move from input devices, through input control, and into main memory, all under the direction of the control unit in the CPU. Data words are moved from memory to registers. From there, they may make one or more passes through the ALU and back to the registers before being written back in memory. Results are moved from main memory to output devices. Instructions are brought from main memory into a register before being moved to the control unit for interpretation. The paths drawn with dashed lines in Fig. 9.1 represent alternative paths found in some computers.

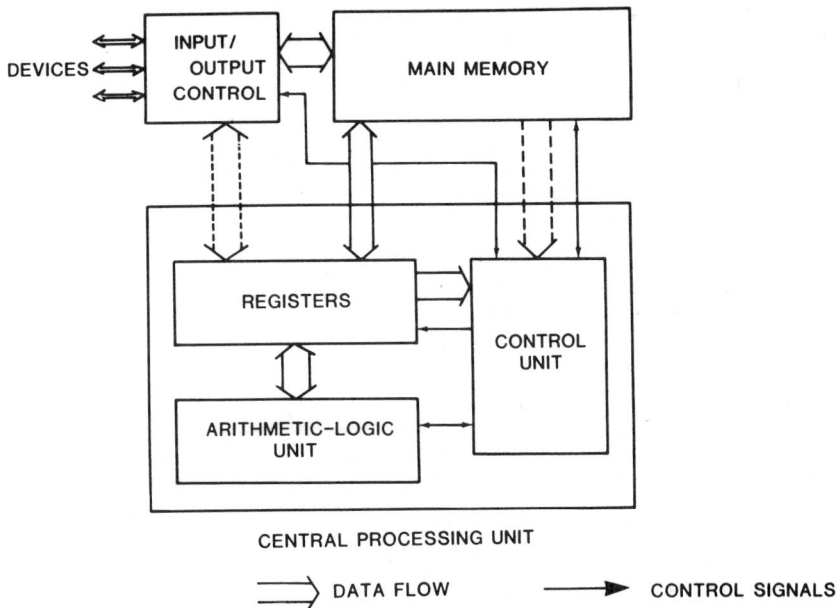

Fig. 9.1 Conventional computer organization.

Chapter Overview

The first sections, on architecture (Sections 9.1 and 9.2), present a complete discussion of computer architecture—the functional description of a computer's operation—from the point of view of the programmer. A single architecture can be implemented in many ways, using different technologies. The IBM System/370 is an example of a single architecture that has been implemented in many embodiments. These appear virtually identical to the programmer but have different costs and speeds. The sections discuss how registers and main memory can be addressed, how the instructions manipulate the information in the registers and main memory, and how the computer's "interrupt" system permits it to interact with external signals that signify the occurrence of events to which the computer must respond.

Section 9.3 discusses the preparation of sequences of CPU instructions to carry out specified functions. This article on computer programming is included in this section on computer design because it is important that the hardware designer understand how the computer instructions are used. Section 9.4 introduces the logic of switching devices—those building blocks from which all computer components are built. Techniques are given for realizing and minimizing logical functions. Section 9.5 describes the design of arithmetic-logic units and how they perform complex operations (e.g., floating-point multiplication) by executing series of simple operations such as addition, under the management of the control unit. Sections 9.6–9.8 present the electronic technologies and physical structures used to embody main memory, describe how main memory and attached storage devices are organized and how they can present the user with a "virtual memory" that is larger than the available main memory, and explore the concerns specific to microprocessor systems. A microprocessor is a CPU on a silicon chip. Some require more than one chip, and some include memory and I/O control on the chip. Section 9.9 contains an extensive analysis of control units. A frequently used control unit design uses very regular structures called "microprograms." (These must not be confused with programs for microprocessors, which are almost identical to programs for larger CPUs.)

Sections 9.10–9.12 include a general systems description of I/O: the interface among devices, controller, main memory, and CPU, and the interchange of control signals that takes place during an I/O operation. The technology of attached storage devices and techniques used for I/O in microprocessor systems are also analyzed. Section 9.13 describes buses, cables carrying data, and control signals between (and within) components of a computer. Alternative busing strategies are analyzed in this section.

A glossary follows.

9.1   ARCHITECTURE: REGISTERS, ADDRESSING, AND INSTRUCTIONS
James E. Thornton

9.1.1   Background

A computer architecture must be defined independently of any specific implementation since there are often several different implementations over a period of time. There are numerous ways to view an architecture, ranging from user application programs written in a high-level language to specific instruction sets, including those privileged ones used by the operating system. We choose the latter for this description. In order to provide enough understanding of the architecture, reference will be made to some implementation detail and occasionally to some alternative implementations.

During execution, a program views the architecture as a set of **registers**, a main memory with various **addressing modes**, and a processing unit **instruction set** (see the introduction to this chapter for a summary of computer organization). For very simple machines, there may be no privileged instructions or addressing modes. The machine executes programs for which all the resources of the machine are available (i.e., the entire main memory). For machines that are shared by many programs operating one at a time (multiprogramming), the architecture needs to provide privileged facilities for allocating and managing resources.

9.1.2   Registers and Local Memory

One distinguishing feature of a computer architecture is the register structure of the central processing unit. A register is a group of flip-flops or latches operating together to temporarily store data or control information.

All operations in a computer system result in modifications of bits in registers or main memory. An instruction may cause one or more operations to take place, generally in a specific sequence. At the completion of each instruction, the information stored in registers and main memory is available for subsequent instructions or other access.

Some processors make use of a single register for temporarily storing data. This register is typically called the **accumulator** (see Fig. 9.2). It operates in conjunction with the arithmetic-logic unit

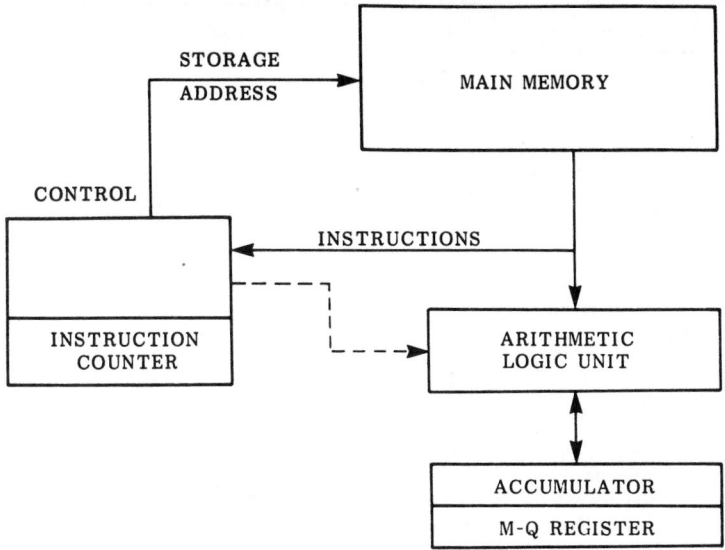

Fig. 9.2 Simple register configuration.

(ALU) component of the processing unit. The accumulator can, for example, receive a number from main memory and add it to the one already stored in it, clear its contents, and transfer its contents to main memory, among other properties. Most arithmetic operations store the results in the accumulator. In conjunction with the accumulator, an additional register is often used in order to implement multiply and divide operations. These operations usually require the additional register to hold partial results during the iterative execution of the instruction. This register has been called the Q register or the M-Q register since it holds the multiplier or quotient.

The length of the accumulator and other registers used to store data is determined by the largest number that needs to be held or processed. Since a fraction can be infinitely long, the choice of length is arbitrary. The choice sets the maximum integer size and determines the precision of the central processing unit. (See Section 9.3.6 for a discussion of scaling, which enables larger integers to be represented.) The relationship of the length of the accumulator to the length of a memory word is subject to varying implementation choices. The architecture defines data types (see later discussion) that may be shorter in length than a memory word or, alternatively, may require more than one memory word. For simplicity we may think of the accumulator length and memory word length as equal, but this is not necessary. The accumulator and memory, in any event, must accommodate the data types defined by the architecture.

Other registers are also utilized to store control information in the processing unit. These are generally not " visible" to the architecture but are described here to provide a description of execution of instructions. The implementation of these control registers may vary. The principal one is the **instruction counter**, which holds the main-memory address of the current instruction being executed. To begin a program, this register is set to the location of the first instruction in the program, either by an operator in a simple system or by the operating system. Subsequently, the register is incremented during the normal course of execution of each instruction and in preparation for obtaining the next instruction from main memory. Each instruction is brought from main memory into the processing unit and resides in a register (the instruction register) during the course of instruction execution.

In general, an instruction set may include instructions of different lengths for reasons to be explained later. The length in bits of an instruction is often a multiple of the length of a memory word since efficiency in storing programs in memory is a major objective.

The execution of a single instruction (see also Section 9.9.1) involves a number of elementary steps such as:

Increment instruction counter.

Read next instruction from main memory.

Interpret the instruction.

Read operand (if required).

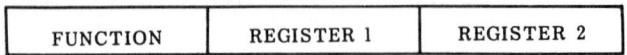

Fig. 9.3   Instruction format.

Perform the arithmetic function.

Store result (if required). (In branch-type instructions, the result may be a modification of the contents of the instruction counter.)

The accumulator represents all that is necessary to describe the registers of a simple architecture. The instruction set for such an architecture can also be quite simple. In general, an instruction will contain the function code, identifying the function to be performed, and an address in main memory. The address may refer to data to be fetched or stored or to a new location for the next instruction.

If main-memory access time were as rapid as that of registers, there would perhaps be no need to create a more complex architecture. Instructions would then execute as rapidly with operands from main memory as they do with operands from intimately located high-speed registers. However, historically, this factor plus other ease-of-use factors have combined to generate more registers and more architectural structure.

For instance, in the matter of addressing main memory for data, it is convenient to hold an index to be added to the base address of an array of data. The index provides a means to traverse the array without modifying the base address. In this case a more complex instruction may be added to the instruction set, or a program loop using simple instructions may be used. In either case the instruction itself now must contain some reference to the index register. In Fig. 9.3, the instruction contains three fields, identifying the function to be executed, the register containing the base address, and the register containing the index (in general, there may be more than one).

An architecture may utilize multiple base registers and multiple index registers for program efficiency or ease of use. For example, an architecture with only one base register may require reloading from main memory often to deal with multiple arrays of data. While this gets back to the implementation question of why registers are required at all, the architecture has been adjusted or enlarged to accommodate implementation efficiency. At this point, it is convenient to refer to the multiple index and base registers (as well as other registers used to store data) as a **local memory**. Local memory may be any size, and a field (or fields) of some instructions is used to refer to the desired word, just as a field in an instruction may refer to main memory. The local memory shown in Fig. 9.4 is that of the IBM System/370 architecture. For clarity, these registers are shown in two groups, one for indices, integers, and addresses, the other for floating-point numbers. These two groups may contain registers of different length, but this is not necessary. The arithmetic-logic unit utilizes the integer group of registers for fixed-point operations and the other register group for floating-point operations. In general, variable-field-length operations make references to main memory for data types that do not have a fixed length, such as alphanumeric text.

The registers remain an area where the user program can be entirely free of interaction with the operating system. If a user program is interrupted, the contents of these registers are stored in main memory and reloaded when the program is restarted. The operating system may be required to assist in saving the registers. Interaction with main memory, on the other hand, is subject to operating system control over the physical location and protection of a main-memory word. In Fig. 9.4 a virtual address translation is shown as one means of accomplishing this control. This is discussed in detail in the following section.

At this point, the nature of the information being stored or processed is not being examined. This is treated in sections that follow on the instruction set and data types.

9.1.3   Addressing

Increasing complexity of the central processing unit instruction set derives in a major way from the methods employed in addressing main memory and local memory. The address may be carried directly in the instruction or may be constructed by references to local-memory registers. Instructions may reference more than one address; however, we treat first the concepts of addressing memory and then the modes by which the addresses are constructed.

Main Memory

Traditional methods of addressing main memory use an address that identifies a specific word in real memory. (Reference is made to "real" memory here to differentiate from "virtual" memory, which is discussed later.) Differences in the implementation of main memory may require fetching or storing a

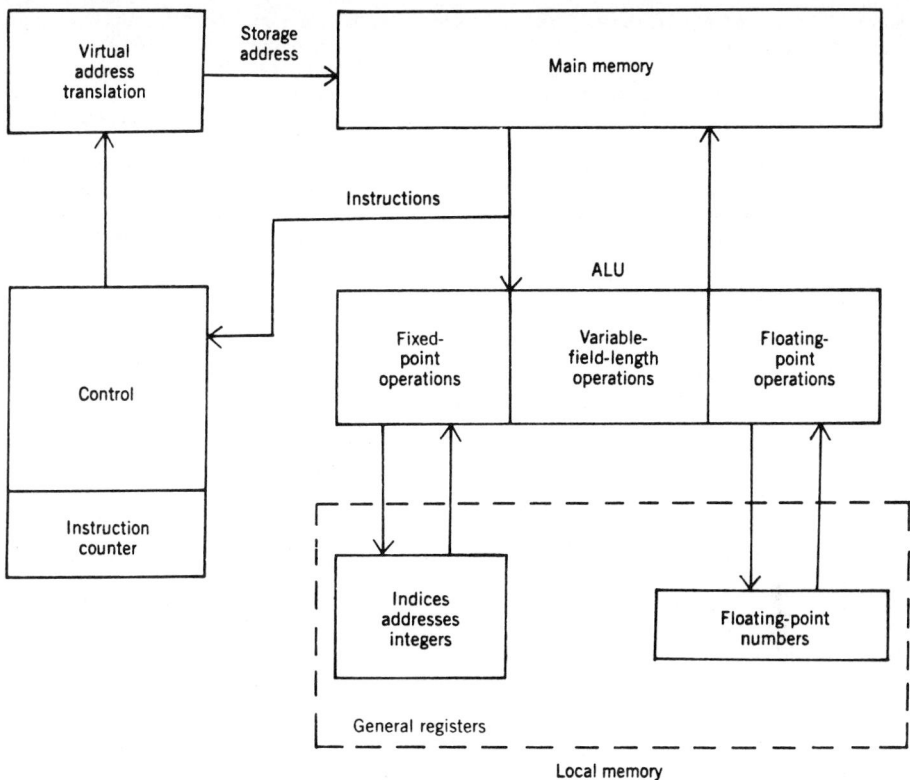

Fig. 9.4   General register configuration.

portion of a memory word or more than one memory word, depending on the data type called for in the instruction. The architecture does not view these implementation details.

Space in main memory may be allocated by the operating system (if there is one) for the user's program. In computers that do not employ virtual addressing, the user is not allowed to make memory references outside that allocated space. Such space is usually allocated in contiguous blocks.

Multiprogramming (jobs sharing memory, executing one at a time) requires that independent blocks of addresses be allocated to each program. As this complexity has grown, the user program has deferred to the operating system to make space available and to protect it from other programs coresident in main memory. In central processing units that do not employ virtual addressing, programs that require more space may be halted until more contiguous space is available for assignment. This operation may also entail relocation of blocks to utilize "holes" left in the address space by departing programs. Much of this difficulty is removed by simply having enough real main memory. Technology has responded admirably to this need over the years. Another reduction of difficulty would be to return to "monoprogramming" in the computer and relegate the problem to secondary storage only.

Virtual Memory.* The concept of virtual memory entirely separates the real address space from the user program address space. The user program may assume that it has nearly unlimited space available, limited only by the maximum address that can be generated using the instruction set address fields and index and address registers. These fields and register lengths have expanded gradually over the years, with a few architectures employing truly "single-level" storage of essentially unlimited space. These require very long addresses, one example being 48 bits. Earlier virtual storage address fields of 24 bits and even less place upper limits on the space available to the user program.

The virtual-memory concept is built on a system of fixed blocks of contiguous real-memory locations called **page frames**. The program or its data may be stored in main memory by assigning it a

---

*More detailed aspects of virtual memories are presented in Section 9.7.1.

number of independent page frames. The locations within each page frame must be contiguous in the real address realm (see Fig. 9.5). Page sizes are fixed (usually ranging from 512 to 4096 8-bit bytes) and may not overlap when stored in main memory or secondary storage. Pages from virtual address space containing programs and data are loaded by the operating system by locating them in secondary storage and moving them to main memory. On initially loading a program, page frames are assigned to it by the operating system, utilizing a table of the available page frames. As the program executes and calls for virtual addresses not yet loaded, it will be halted on each such memory reference while the operating system brings the appropriate page into main memory from secondary storage. If all page frames in main memory are in use, the operating system must also remove pages to allow new ones in. For this purpose, typically the page removed is the one least recently used.

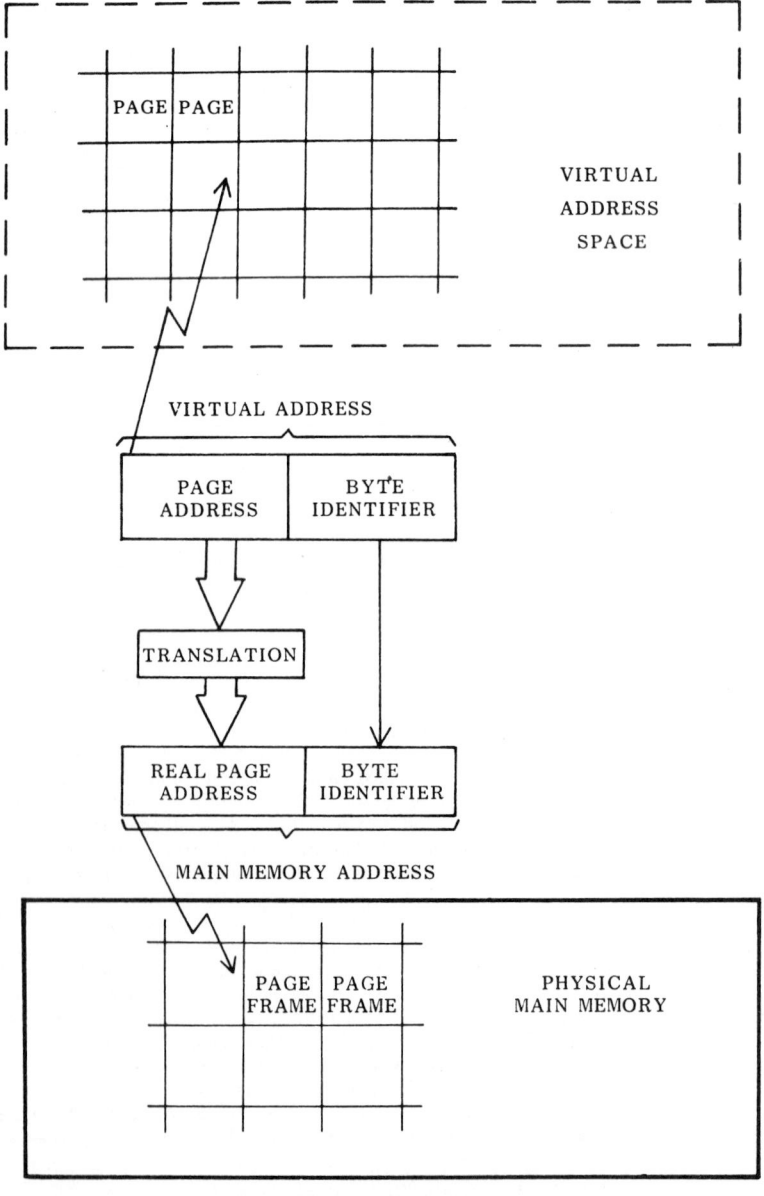

Fig. 9.5   Virtual memory.

The virtual address map (or page table) is typically held in a protected area of main memory. The virtual address translation unit (Fig. 9.4) may also include a small associative memory that contains most recently used entries in the page table.

Segments

Program and data areas are generally larger than a single page. It is convenient in some systems to deal with these larger groups of stored information in an ordered way within the virtual addressing system. The reasons may derive from the way pages are stored in secondary storage (it may be convenient to transfer multiple pages) or from the usefulness of a higher-level organization of virtual address space by the user.

A **segment** is merely a collection of contiguous pages in virtual address space. The virtual address translation still deals with pages, and there is no need for the pages to occupy contiguous page frames in main memory. If the virtual address is subdivided as in Fig. 9.6 to include a segment number, this provides a kind of two-level addressing. A segment map is searched first. On a match, the proper page map is identified for the page address search. Segmenting allows the operating system to control access to the more natural segment rather than each page, on the basis that segments are entire programs or data sets.

To summarize the virtual addressing schemes, a number of benefits accrue to the programmer and system designer. Transfers of pages between main memory and secondary storage are hidden from the application program, allowing the program to run on a variety of system configurations or utilizing a variable amount of main memory. The operating system allocates main-memory space in fixed increments and overlaps transfers between storage levels.

More details can be found in Section 9.7.1.

Local Memory

Registers making up local memory are either explicitly defined, such as the accumulator, or belong to a set of general registers. Addressing these registers is accomplished by certain fields within each instruction. In instruction sets with multiple addresses per instruction, each address requires an address field. The number of registers is thus limited by the size of the address field. A 4-bit field, for example, will address 16 general registers.

The general register organization is something of a compromise between single-address instructions and multiple-address instructions with full address fields. It avoids some of the extra instructions necessary for shuffling data, inherent in a single-address organization, yet it does not take space in the instruction for a full additional main-memory address. In contrast, shorter instructions can be used in which multiple addresses need only enough address bits to reference the local memory.

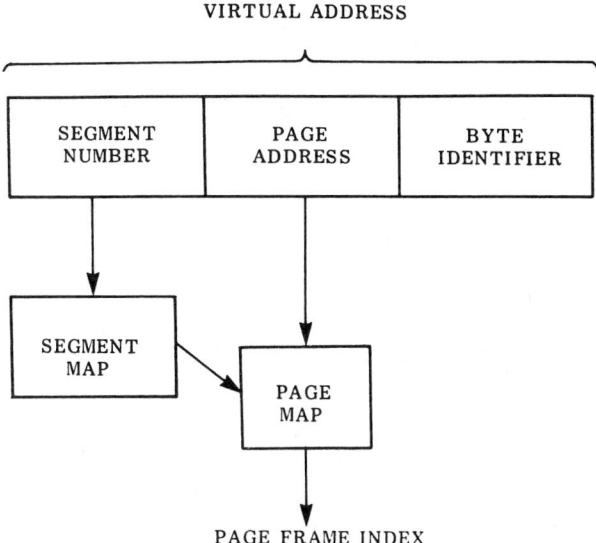

Fig. 9.6 Segment addressing.

Addressing Modes
There are a variety of useful ways to construct an address. These derive from the ease-of-use factors that have historically evolved in computer architecture. They also represent compromises in allocating fields of bits within an instruction. For example, it would be grossly inefficient to always carry full address fields, even in single-address instruction sets. Therefore, judicious use of general registers and local-memory addressing to produce main-memory addresses is preferable.

Base Register. Any of the general registers in local memory may be assigned as a base register. One convenient use of a base register is to point to a main-memory location representing the beginning of an array of data or a table. After the base register is loaded initially, subsequent instructions may construct main-memory addresses by adding a positive integer to the contents of the base register. The positive integer may be supplied directly from a field of bits in the instruction or from another local-memory word. When supplied from the instruction, this mode of addressing is referred to as **base-displacement addressing**. When supplied from local memory, the positive integer is normally considered an index. Remember that the resulting address is still only the logical address (or the virtual address for architectures utilizing virtual memory). This must undergo further translation, as required by the operating system, to the real main-memory address.

Accesses to several large arrays may take significant time if the base address has to be loaded each time. Therefore, it is convenient to maintain several base registers in those cases. For example, a move of data may use a base register for the beginning address of the source and another base register for the beginning address of the destination. An alternative is a single base register and two displacements.

Indexing. A base address may be indexed by another general register in constructing the main-memory address. In the case of a data move, the index is added to the source base address to obtain the first source word. The index is then added to the destination base address to obtain the first destination word. Following these two steps, the general register containing the index is incremented by a quantity equal to the word length, and the steps are repeated until the index reaches the length of the data field to be moved. This indexing operation would normally be executed in a program loop containing explicit instructions for constructing the memory addresses, reading, storing, incrementing the index, and either returning to the beginning or terminating the loop. As discussed in Section 9.1.5, this may require four to eight instructions.

This illustrates the value of an index for efficiency and for ease of use. The index serves the purpose of advancing the source and destination addresses in step as well as providing a test against the desired length.

Arrays in two dimensions may be addressed in a similar manner using two independent indices. The horizontal index is incremented a word at a time, with the assumption that the second row begins immediately following the last word of the first row as far as main memory is concerned. In the vertical direction, the index is incremented by a quantity equal to the length of a row. The indexing operation terminates at the end of either the row or the column.

The use of index registers in machines with simple instruction sets permits general treatment of such data types as vectors, arrays, and strings. Computer architectures may include complex instructions that apply to operations on entire vectors or arrays. These would implement entire loops, such as those above, in a single instruction, including the indexing, incrementing, terminating, and arithmetic function on the array.

Relative Addressing. Another form of constructing a main-memory address is relative addressing. This has some similarity to the base-displacement method. A positive or negative integer is carried in the instruction identifying a memory location in relation to the location of the current instruction. The integer is similar to the displacement field, except that it can be positive or negative. Rather than a general register for the base, the instruction counter is used. Thus an address can be calculated relative to the current instruction's address in main memory. This type of addressing generally does not count forward or backward by instruction but rather by memory word.

The principal value of relative addressing is for small displacement fields in the instruction, to conserve wasted instruction space. This makes it useful only in limited ranges of addresses. It is also complicated for hand assembly and debugging.

Indirect Addressing. In indirect addressing, the main-memory address is first calculated using any of the methods previously described. Under control of a single bit carried in the instruction, this address is used to read another address from main memory. This new address is the "indirect" address actually used by the instruction. Each time an indirect address step is taken, it requires an additional main-memory reference.

Through indirect addressing under the control of a single bit in the instruction, a larger block of seldom used base addresses can be kept in main memory, for example. Other innovative uses can be made, including some schemes that allow multiple levels of indirection. In these cases, a bit is carried in the main-memory word, together with the address it contains, identifying whether another level exists. One can construct decision trees in this manner.

Stack. A contrasting method of addressing main memory or local memory, especially for intermediate results, is the use of a **pushdown stack**.* These are generally employed instead of general registers and local memory. The stack operates as a last-in-first-out storage. A program may place intermediate results or input data into the stack without regard to the number of operands. They are simply pushed down as needed. Arithmetic operations may be executed on the last value in the stack (the top of the stack), with the intermediate result left there or removed to an explicit address in main memory.

A stack is a useful concept for computing expressions with variable numbers of operands and operators. This is quite different from the general register approach, in which the general registers are explicitly addressed by the instruction. Methods are available by which the correct intermediate result will arrive at the top of the stack in the correct order for solving the expression. The competitive issue between a stack architecture and a general register architecture reduces to the relative difficulty of managing the assignment of general registers versus the difficulty of maintaining the proper order of operands in the stack. Most user programs depend on the compiler to handle these problems.

Protection / Relocation

In a multiprogramming environment, there is a need to prevent one program from making references to main memory at addresses assigned to another program. The means to accomplish this are not visible to the user program but are under the control of the operating system. A primitive and inefficient means would be to carry a protection bit with each word in main storage. This permits two independent programs or groups of programs to be protected from each other.

A more general means to provide protection is to control access to entire program segments and data areas on a program-by-program basis. In an architecture that does not employ virtual storage addressing, this protection is generally applied to blocks of contiguous main-memory addresses. A user program requests a block of addresses from the operating system. In loading a user program, the operating system uses two control registers: a relocation register and a protection register. The contents of these registers represent the boundary addresses for the contiguous space assigned to the program. The registers are loaded by the operating system whenever the program is to be activated.

The user program is written as though its origin were location 0. The relocation register specifies the actual location of the program, and the protection register specifies the number of words allowed or the upper boundary address. To accomplish this, all storage addresses for main memory will be added to the contents of the relocation register and tested against the protection register. This function takes the same position in the central processing unit as the virtual translation unit does in a machine with virtual memory.

For an architecture employing virtual storage addressing, the relocation and protection function applies to pages rather than entire program segments and data fields. Each page is relocated independently, and each page may be protected independently. The relocation occurs simply by the translation from virtual to real address, as explained earlier. The virtual map containing the virtual address and its current real address may also be used to provide protection. To do this, each virtual map entry also contains a field of bits defined as a **lock**. In loading a page into main memory, the operating system assigns the lock bits. Thereafter, each program is provided with a key or set of keys that are loaded into a control register by the operating system in preparation for the program to be activated. One key must match the lock for every storage address being translated; otherwise, an abort occurs. An abort of this kind returns control to the operating system for action to be taken following an attempted unauthorized access. Each key may also be assigned a particular form of restriction; for example:

1. No restriction.
2. Read only as data.
3. Read only as program.
4. Write.

This fairly elegant type of protection allows many user programs to access main memory in a well-protected but flexible way. The operating system would, of course, be provided with a master key and would have a special key controlling its main memory usage. For architectures that do not employ virtual addressing, the entire contiguous block of addresses can be protected implicitly by not allowing accesses beyond the upper boundary.

*The use of a pushdown stack to save status upon interruption or subroutine entry is discussed in Section 9.3.8.

9.1.4 Data Representations

An architecture must generally deal with a variety of data types, including numeric and nonnumeric data. The possible data types include:

Bit	Integer
Bit string	Multiple-precision integer
Byte	Floating-point number
Byte string	Single precision
Character	Double precision
Character string	Vector
Decimal digit	Array
Decimal number	

The bit and bit string data types are identified with control and decision functions and may not be directly viewed by the architecture. However, the instruction set may have the capability to directly address to the bit level and may contain bit processing instructions, or it may require manipulation, such as shifting a data word left or right to bring a specific bit or bit string into alignment for processing.

The byte and character types are related in that they are interchangeable in some architectures. A byte is usually 8 bits in length and contains alphanumeric data, one symbol per byte. Characters have generally been in 6-bit lengths. Since characters may be stored or transmitted in 6-, 7-, or 8-bit formats, there is some variety to deal with. In general, a character set is intended to represent all of the symbols used to display, print, read, or store words or numbers, plus a collection of control characters used to control a variety of functions related to the human-machine interface. This printed page, for example, can be represented by a long string of characters that, if the page were printed by an automatic typewriter, would include characters for the spaces between words and carriage returns where needed. It should be noted that the 8-bit character representation allows more symbols to be added to the character set to account for advancing technology in the display and printing area. Two standards exist for interchange codes and specify the byte or character code representation: the ASCII (American Standard Code for Information Interchange) and EBCDIC (Extended Binary Coded Decimal Interchange Code).

Decimal numbers are a subset of the full character set, as represented in the 8-bit coded format. However, they may also be packed two to an 8-bit byte and treated separately from the character-handling logic. Decimal numbers are represented as sign-magnitude, as seen in the example in Fig. 9.7. Other representations are possible; this one is in use in the IBM System/370 architecture. The 4-bit sign representation is in the rightmost 4 bits of the rightmost byte of the field. The leftmost 4 bits of the rightmost byte contain the least significant digit of the number. This is convenient in implementations that execute decimal arithmetic a byte at a time. The sign is delivered to the ALU first, followed by the least significant digit, and so on.

Integers are stored as a sign bit and n bits of integer, generally limited to the length of a general register. There are three common methods of representing integers: the 1's- and 2's-complement notations and the sign-magnitude notation (see Section 9.5.4 for definitions of these notations). Integers are normally treated as fixed-point quantities with the radix point assumed just to the right of the least significant digit. In decimal arithmetic, the radix point is the decimal point.

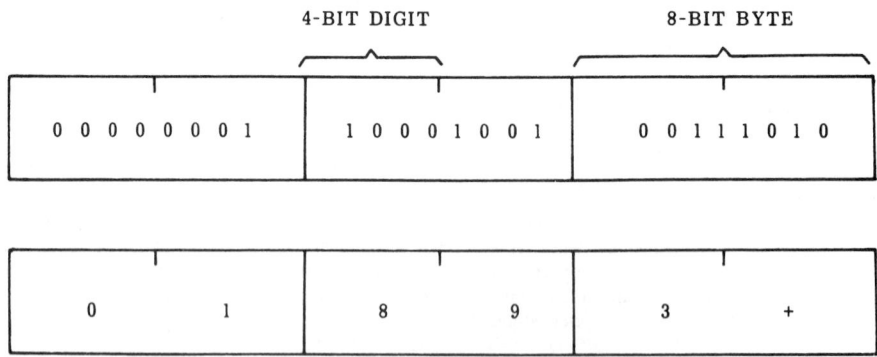

Fig. 9.7   Packed decimal format.

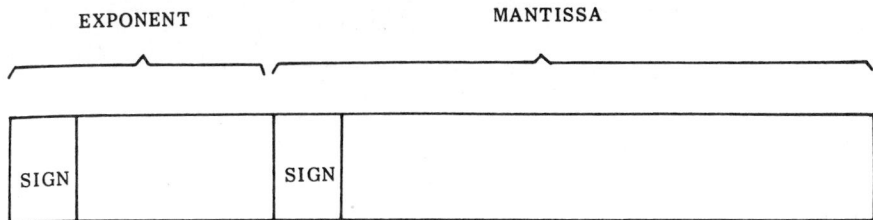

Fig. 9.8 Floating-point number.

Floating-point numbers take the form of a signed mantissa and a signed exponent, as in Fig. 9.8. Not all architectures handle floating-point numbers, just as some do not directly handle decimal numbers, character strings, and so on. For a fixed word size, machines have a much wider range of representable numbers using floating-point representation than fixed point, unless the numbers are scaled using a scaling algorithm, which effectively moves the radix point. A floating-point number can be expressed as mr^e, where m is the mantissa, r the radix, and e the exponent. Normally, the radix of the exponent is 2; however, some representations (IBM/370) use a higher radix in order to extend the number range. The mantissa is normally represented as a fraction with the radix point at the left end. Conversion from floating-point to integer form therefore requires realignment of the mantissa to shift the radix point to the right of the least significant digit.

Single-precision floating-point representations occupy one word in most common practice. Since floating-point operations generate results exceeding a single word size, the mantissa is generally normalized for single-precision computation. This means that the mantissa is shifted to the left as far as possible, eliminating any leading zeros, with an appropriate reduction in the value of the exponent. Any remaining bits of the mantissa to the right of the word are lost. In some "overflow" cases, a right shift is required. In multiple-precision computation, extra words are used to capture more significant bits of the results of floating-point operations. The representation of such numbers includes a signed exponent and signed mantissa for the word retained. Alternatively, the least significant bits of the mantissa could take up an entire word without the need for another exponent.

A **vector** is an ordered sequence of scalar numbers. Vectors are stored in main memory as a contiguous sequence of data words. A two-dimensional array can be represented in a linear fashion by a vector in which each row of the array follows immediately in sequence with the previous row. In accessing the array in the vertical, or column, direction the sequence of addresses is indexed by an integer equal to the length of a row. Vectors and arrays are treated in some architectures with complex instructions. In the more traditional architectures, they are handled by programs.

9.1.5 Instructions

The instruction set of the processing unit can range from very simple to very complex. Complexity is greater in architectures that are designed for greater specialization or ease of use to the programmer (or language compiler). We deal with the instruction set as viewed by the user programmer and the system programmer, although it should be understood that the two instruction sets could be grossly different. Conventional architectures add certain privileged instructions or properties to the basic set for the use of the systems programmer.

Two instruction sets are discussed here and are depicted in Figs. 9.9 and 9.10. The first is a simple instruction set used in the central processing unit of the Control Data 6600 (CYBER). The second is the more complex instruction set used in the IBM System/370.

Formats

It is convenient to organize the instruction set into classes by format. The CYBER formats are shown in Fig. 9.11. Instructions are either 15 or 30 bits in length and are packed in the 60-bit main-memory word. A 30-bit instruction may not cross word boundaries. The 15-bit instructions are three-address register transfer types that identify two source registers and a destination register. There are three separate sets of general registers in this architecture. They are **address**, **integer**, and **floating-point** registers. The 30-bit format contains an 18-bit immediate field that may be a source operand, as noted in the instruction set.

The System/370 formats are shown in Fig. 9.12. There are seven basic instruction formats that require from one to three half-words. (A half-word is 16-bits or 2 bytes.) This architecture is two-address, in general. The RR format provides for register transfer instructions in local memory; more complex instructions make use of additional general registers. In addressing main memory, the address is contained in a register designated by the R field in the instruction or is calculated from the

00	STOP	40	FLOATING PROD Xj × Xk → Xi
010	RETURN JUMP	41	ROUND FLT PROD Xj × Xk → Xi
011	READ ECS (Bj + K) WORDS FROM XO	42	FLT DP PROD Xj × Xk → Xi
012	WRITE ECS	43	FORM jK MASK IN Xi
02	GO TO K + Bi	44	FLT DIVIDE Xj by Xk → Xi
03	GO TO K if Xj (ZR, NZ, etc.)	45	ROUND FLT DIVIDE
04	GO TO K if Bi = Bj	46	
05	Bi ≠ Bj	47	SUM OF 1s in Xk → Xi
06	Bi ⩾ Bj		
07	Bi < Bj		
10	Xj → Xi	50	Aj + K → Ai
11	Xj AND Xk → Xi	51	Bj + K → Ai
12	Xi OR Xk → Xi	52	Xj + K → Ai
13	Xi ⊕ Xk → Xi	53	Xj + Bk → Ai
14	X̄k → Xi	54	Aj + Bk → Ai
15	X̄k AND Xj → Xi	55	Aj − Bk → Ai
16	X̄k OR Xj → Xi	56	Bj + Bk → Ai
17	X̄k ⊕ Xj → Xi	57	Bj − Bk → Ai
20	SHIFT Xi L jk places → Xi	60	Aj + K → Bi
21	SHIFT Xi R jk places → Xi	61	Bj + K → Bi
22	SHIFT Xk L Bj places → Xi	62	Xj + K → Bi
23	SHIFT Xk R Bj places → Xi	63	Xj + Bk → Bi
24	NORMALIZE Xk in Xi; CT → Bj	64	Aj + Bk → Bi
25	ROUND & NORMALIZE	65	Aj − Bk → Bi
26	UNPACK Xk → Xi; EXP → Bj	66	Bj + Bk → Bi
27	PACK Xk & Bj → Xi	67	Bj − Bk → Bi
30	FLOATING ADD Xj + Xk → Xi	70	Aj + K → Xi
31	FLOATING SUBT Xj + Xk → Xi	71	Bj + K → Xi
32	FLOATING DP ADD Xj + Xk → Xi	72	Xj + K → Xi
33	FLOATING DP SUBT Xj + Xk → Xi	73	Xj + Bk → Xi
34	ROUND FLT ADD Xj + Xk → Xi	74	Aj + Bk → Xi
35	ROUND FLT SUBT Xj + Xk → Xi	75	Aj − Bk → Xi
36	INTEGER Xj + Xk → Xi	76	Bj + Bk → Xi
37	INTEGER Xj − Xk → Xi	77	Bj − Bk → Xi

Fig. 9.9   CYBER instruction set.

following three binary numbers:

1. **Base address** is a 24-bit number contained in a general register specified by the 4-bit field, called the B field, in the instruction.
2. **Index** is a 24-bit number contained in a general register designated in a 4-bit field, called the X field, in the instruction.
3. **Displacement** is a 12-bit number contained in a field, called the D field, in the instruction.

Types (CYBER)

In the following subsections, the instruction set is classified by function as opposed to addressing method, data type, or format. The simpler CYBER instruction set is treated first, with selected instructions from System/370. (For more comprehensive explanation, see the Principles of Operations manuals for these machines.)

Data Manipulation. The CYBER architecture provides main memory references for *LOAD* and *STORE* through the use of specified address (A) registers as shown in Fig. 9.13. The eight instructions that define an A register as the destination (codes 50–57), calculate the main-memory address, which, when it is entered into the A register, initiates the memory reference. Five A registers (A_1–A_5) are assigned to the LOAD function and will cause a main-memory word to be loaded into the corresponding X register (X_1–X_5). Two A registers (A_6 and A_7) are assigned to the STORE function and will cause a word to be stored in main memory from the corresponding X register (X_6

STANDARD INSTRUCTION SET

ADD	MOVE
ADD LOGICAL	MOVE NUMERICS
AND	MOVE WITH OFFSET
BRANCH AND LINK	MOVE ZONES
BRANCH ON CONDITION	MULTIPLY
BRANCH ON COUNT	OR
BRANCH ON INDEX HIGH	PACK
BRANCH ON INDEX LOW OR EQUAL	SET CLOCK
CLEAR I/O	SET PROGRAM MASK
COMPARE	SET STORAGE KEY
COMPARE LOGICAL	SET SYSTEM MASK
COMPARE LOGICAL CHARACTERS	SHIFT LEFT
UNDER MASK	SHIFT LEFT LOGICAL
CONVERT TO BINARY	SHIFT RIGHT
CONVERT TO DECIMAL	SHIFT RIGHT LOGICAL
DIAGNOSE	START I/O
DIVIDE	START I/O FAST RELEASE
EXCLUSIVE OR	STORE
EXECUTE	STORE CHANNEL ID
HALT DEVICE	STORE CHARACTERS UNDER MASK
HALT I/O	STORE CLOCK
INSERT CHARACTER	STORE CONTROL
INSERT CHARACTERS UNDER MASK	STORE CPUID
INSERT STORAGE KEY	STORE MULTIPLE
LOAD	SUBTRACT
LOAD ADDRESS	SUBTRACT LOGICAL
LOAD AND TEST	SUPERVISOR CALL
LOAD COMPLEMENT	TEST AND SET
LOAD CONTROL	TEST CHANNEL
LOAD MULTIPLE	TEST I/O
LOAD NEGATIVE	TEST UNDER MASK
LOAD POSITIVE	TRANSLATE
LOAD PSW	TRANSLATE AND TEST
MONITOR CALL	UNPACK

FLOATING-POINT INSTRUCTIONS	DECIMAL-INSTRUCTIONS
ADD NORMALIZED	ADD DECIMAL
ADD UNNORMALIZED	COMPARE DECIMAL
COMPARE	DIVIDE DECIMAL
DIVIDE	EDIT
HALVE	EDIT AND MARK
LOAD	MULTIPLY DECIMAL
LOAD AND TEST	SHIFT AND ROUND DECIMAL
LOAD COMPLEMENT	SUBTRACT DECIMAL
LOAD NEGATIVE	ZERO AND ADD
LOAD POSITIVE	
MULTIPLY	
STORE	
SUBTRACT NORMALIZED	
SUBTRACT UNNORMALIZED	

Fig. 9.10   IBM System/370 instruction set.

and X_7). A LOAD or STORE address can thus be calculated by base-displacement methods, indexing, or direct address. (The reason for several registers is that, for some models of CYBER, a number of main-memory references can be executed in parallel. An architectural benefit of retaining key addresses is also a reason.)

The CYBER architecture also contains a group of instructions (10–17) that perform logical manipulation of binary data. These instructions operate on data in two registers on a bit-by-bit basis, transferring the result to a third register. Such operations as AND, OR, and exclusive OR are provided. These are useful for masking, merging, and selecting data contained within one 60-bit word.

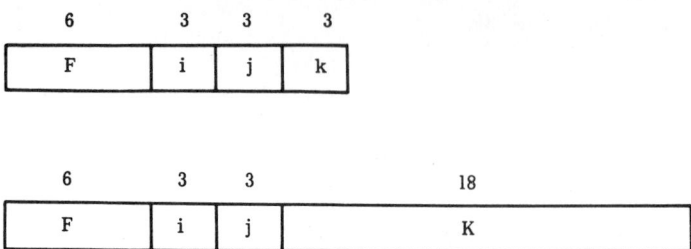

Fig. 9.11   CYBER instruction formats.

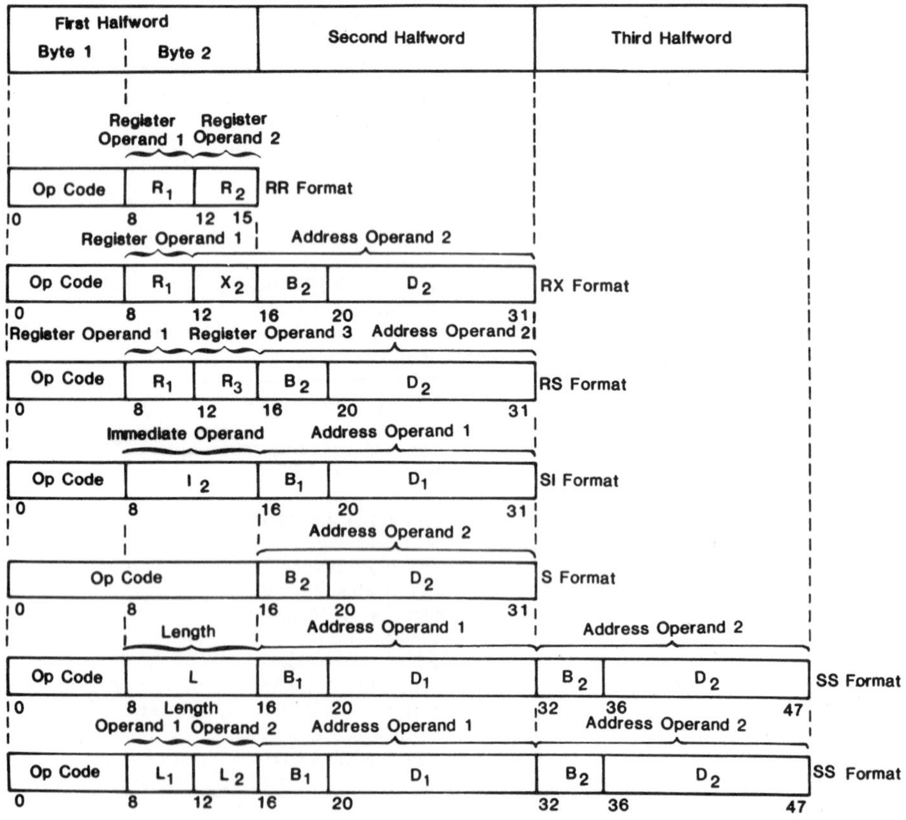

First Halfword		Second Halfword	Third Halfword
Byte 1	Byte 2		

Register Operand 1   Register Operand 2

| Op Code | R₁ | R₂ | RR Format |

0   8   12   15

Register Operand 1   Address Operand 2

| Op Code | R₁ | X₂ | B₂ | D₂ | RX Format |

0   8   12   16   20   31

Register Operand 1   Register Operand 3   Address Operand 2

| Op Code | R₁ | R₃ | B₂ | D₂ | RS Format |

0   8   12   16   20   31

Immediate Operand   Address Operand 1

| Op Code | I₂ | B₁ | D₁ | SI Format |

0   8   16   20   31

Address Operand 2

| Op Code | B₂ | D₂ | S Format |

0   8   16   20   31

Length   Address Operand 1   Address Operand 2

| Op Code | L | B₁ | D₁ | B₂ | D₂ | SS Format |

0   8   Length   16   20   32   36   47

Operand 1 Operand 2   Address Operand 1   Address Operand 2

| Op Code | L₁ | L₂ | B₁ | D₁ | B₂ | D₂ | SS Format |

0   8   12   16   20   32   36   47

Fig. 9.12   IBM System/370 instruction formats.

In CYBER, binary arithmetic instructions on integers can be executed with results transferred to any of the 24 general registers. Data representation is in 1's-complement form, as described in Section 9.5.4. The 50 through 77 series of instructions operate on integers of 18 bits in length (sign is extended in the 60-bit X registers). Two instructions (36 and 37) provide integer add and subtract on 60-bit numbers.

Four arithmetic operations can be performed on floating-point numbers with single and double precision and with results unrounded or rounded. There are floating ADD, SUBTRACT, MULTI-PLY, and DIVIDE (no double-precision for DIVIDE). The execution of ADD and SUBTRACT (see Section 9.5.5) requires an alignment of the mantissas first. This is accomplished by comparing

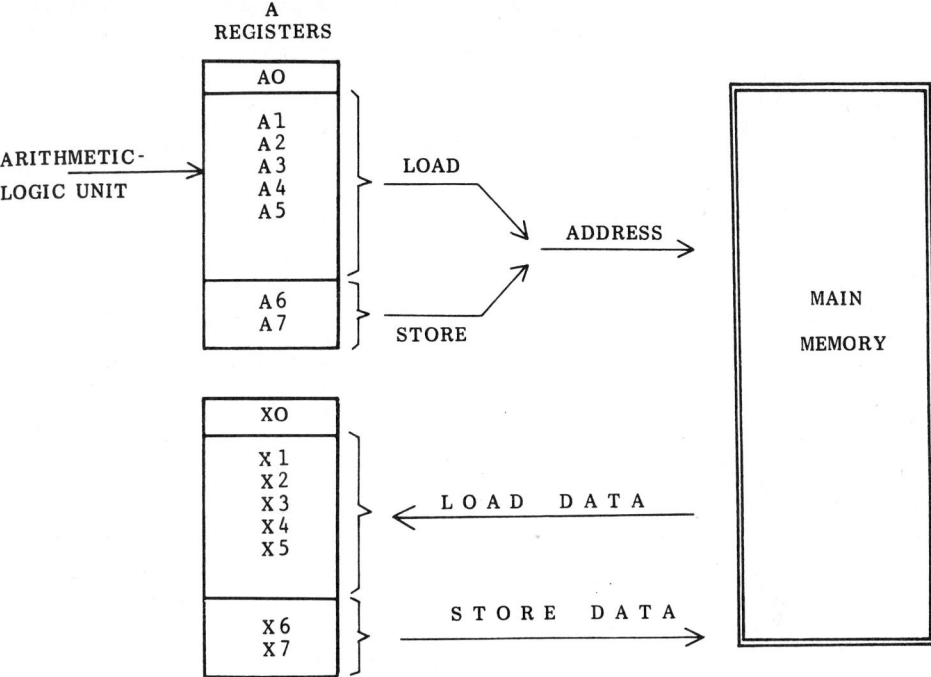

Fig. 9.13   CYBER load and store.

exponents, following which the mantissa of the operand having the smaller exponent is effectively shifted to the right the number of bits corresponding to the difference in exponents, prior to the add or subtract.

The floating-point ADD, SUBTRACT, and MULTIPLY instructions may produce result mantissas up to twice the length of the source operands' mantissas. This is handled in the CYBER architecture by unnormalized instructions that separately compute the two halves of the result. A normalized result is one in which the mantissa is positioned with its most significant nonzero digit shifted to the far left. In multiple-precision floating-point arithmetic, this shifting is not done, thereby retaining leading zeros and increasing precision. Single-precision floating-point instructions can optionally round the result.

Finally, the CYBER instruction set includes a set of SHIFT instructions (20–23). A 60-bit word in one of the X registers may be shifted left or right by a quantity contained in the j and k fields of the instruction. The left shift is circular with the leftmost bit returned to the rightmost bit location for each bit shifted. The right shift does not recirculate but allows the rightmost bit to be dropped for each bit shifted. Also on right shift, the sign bit is retained and also shifted right, effectively retaining the proper sign of the number being shifted. A quantity in a selected X register may also be shifted left or right by an amount contained in a B register. This is useful for scaling and for realigning unpacked floating-point numbers. The PACK and UNPACK instructions (26 and 27) utilize an X register for mantissa and a B register for exponent with an X register for the packed floating-point number. An X register floating-point quantity may also be normalized and optionally rounded. The NORMALIZE operation shifts the mantissa left until the most significant nonzero bit is positioned to the leftmost bit location of the mantissa. Each bit shift is accompanied by a reduction of the exponent, and the count of the number of bits shifted is placed in a B register.

Although the CYBER instruction set is simpler than the IBM System/370, many of the same kinds of instructions as those described here are found in both. A set of variable-field-length instructions will be described that also comes under the category of data manipulation, but these are not present in the CYBER architecture.

Instruction Sequencing. A program proceeds in sequence until a jump is encountered or until an interrupt occurs. In the latter case, the sequence is not modified, but the program will not be resumed until the interrupt is processed. Most machines allow interruptions to occur following completion of an instruction. The interrupt may be generated by an internal fault condition or an external condition requiring the operating system to respond. Control is therefore returned to a predetermined location in the operating system program. The operating system is then obliged to place the interrupted program in a state of suspension from which it can be resumed as if no interruption had occurred. In CYBER, this is significantly aided by a sequence called the EXCHANGE JUMP, which exchanges the contents of all the registers in local memory with an "exchange package" in main memory. In effect, the entire state of the central processing unit is exchanged, leaving it immediately available for the operating system program to proceed. When the suspended program is ready to be resumed, another EXCHANGE JUMP is executed, returning the CPU to the state that existed at the time of the interruption.

1. *Branching.* The sequence of instructions may be modified by several BRANCH instructions in the CYBER architecture. An UNCONDITIONAL BRANCH (02) will cause the instruction counter to be set to a new value equal to the sum of the immediate K field (18 bits) and the quantity contained in a specified B register. This can therefore be a form of base-displacement or indexed addressing for calculating the destination of the branch.

Conditional branching can be done conditional on the quantity in an X register (03) or on comparison between quantities in two B registers (04–07). In the former, the condition may be specified as zero, nonzero, positive, negative, and so on. In the latter, the condition may be specified as equal, not equal, greater than or equal, and less than, referring to the comparison between the specified registers.

In more complex instruction sets than CYBER, other data manipulation functions can be included in the branch beyond that of the comparison and address calculation. These could include an increment or decrement of an index or other arithmetic function with branching conditional on the result. Also, other architectures may branch conditional on previous results that may be stored in a control register or a normal operand location.

2. *Subroutine Calls.* Subroutine or procedure calls utilize an instruction that, in addition to an unconditional branch, provides the subroutine with arguments and a way of returning to the calling program, so that the calling program can resume execution properly. A subroutine is a program that is expected to be used often and that is more convenient to store in main memory once rather than repeating it "in line" with the normal program sequence. Such programs are very often written as "reentrant" subroutines, which means that neither the instructions nor the data are modified during execution of the subroutine in such a way as to prevent reuse of the subroutine. This must also hold true even if the subroutine is interrupted.

In the CYBER architecture, a subroutine is normally entered by a RETURN JUMP (010). This instruction is unconditional and causes a word to be stored in main memory at the address defined by the immediate K field in the instruction and then to branch to address K plus one. The word stored contains an unconditional jump back to an address, which is essentially the address of the next word in sequence of the originating program. This type of instruction allows the subroutine to be "called" by any program module with great flexibility. This is different from the EXCHANGE JUMP, which exchanges all the registers (see "Monitor Calls"). In the RETURN JUMP, the registers are left intact, and, by convention, some registers will represent input operands for the subroutine and others will carry the results.

3. *Monitor Calls.* A user program may contain instructions to invoke trace and other monitor services from an operating system utility. The MONITOR CALL is treated as if it were an interrupt (see Section 9.2.2). The user program is halted, and control is given to the operating system. The operating system utility will maintain records of such monitor calls and provide other program debugging services. In CYBER, the EXCHANGE JUMP (013) is used for the MONITOR CALL.

Variable-Field-Length Operations. There are no CYBER instructions for manipulating variable-field-length data. Such data may be strings of characters or bytes, decimal numbers, or vectors or linear arrays. Programs operating in the CYBER architectures deal with these data directly by means of sequences of instructions.

In contrast, the IBM System/370 contains a number of instructions that execute complex byte string and decimal numeric manipulation. These instructions are highlighted in this section, and the simpler instructions that are roughly comparable to CYBER are not discussed. Full treatment of all instructions in either machine may be found in their publications on Principles of Operations. Neither basic machine has instructions that directly manipulate vectors or arrays.

An example of a byte string operation is the MOVE (MVC) and MOVE LONG (MVCL). The MVC instruction uses the first 48-bit format (SS) in Fig. 9.12. This format contains a length field (L) and two operand address fields utilizing base-displacement addressing. The base addresses are contained in registers in local memory and specified by B1 and B2. The displacements are 12-bit immediate fields (D1 for the first operand and D2 for the second). The instruction moves the second

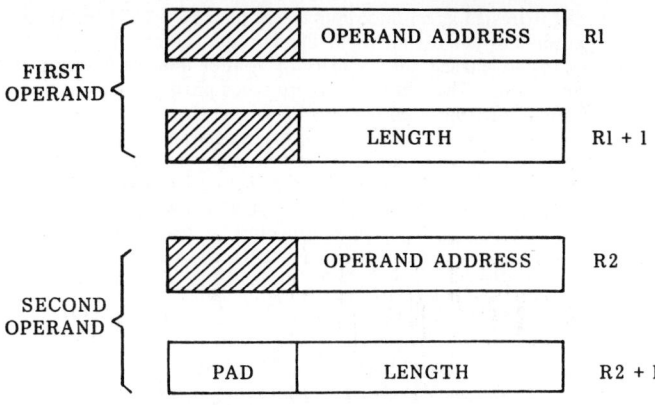

Fig. 9.14 Register use in MOVE LONG.

operand into the first operand location. Each operand field is processed from left to right (i.e., from lower byte address up). When the operands overlap, the result is obtained as if the operands were processed one byte at a time and each result byte were stored immediately after the necessary operand byte is fetched. This move instruction allows up to 256 bytes to be moved.

The MOVE LONG (MVCL) instruction utilizes the 16-bit format (RR) shown in Fig. 9.12. The R1 and R2 fields each specify an even-odd pair of general registers in local memory and must designate an even-numbered register (see Fig. 9.14). The R1 register contains a 24-bit address of the first operand. The R1 + 1 register contains a 24-bit length of the first operand. These are contained in 32-bit registers with the remaining bits ignored. Similarly, R2 and R2 + 1 registers contain the address and length, respectively, of the second operand, except that the remaining 8-bit field in register R2 + 1 contains a pad byte. The second operand is placed in the first operand location, provided that overlapping does not affect the final contents of the first operand location. The remaining low-order byte positions, if any, of the first operand location are filled with the padding byte. If the length of the second operand is longer than the first, the operation is halted at the limit of the first operand length.

Since the MVCL instruction can take an extended time, it is interruptible. When an interruption occurs, the contents of registers R1 + 1 and R2 + 1 are decremented by the number of bytes moved, and the contents of registers R1 and R2 are incremented by the same number, so that the instruction when reexecuted, resumes at the point of interruption.

This class of variable-field-length instructions illustrates the "long" and "short" methods used in System/370. The long requires use of four registers in local memory, whereas the short requires only two base registers.

A similar pair of instructions, CLC and CLCL, are used to COMPARE LOGICAL. The first operand is compared with the second operand, and the result is indicated in the condition code (a part of a control register). A following instruction will test the result. The COMPARE LOGICAL instruction treats all bits alike as part of an unsigned binary quantity, with all codes valid. These instructions also use formats SS and RR, respectively, and treat the two operands in a manner similar to the MOVE instructions. The short method (CLC) uses two base registers and can compare field lengths up to 256 bytes. The long method (CLCL) uses four general registers, similar to MOVE LONG (see Fig. 9.14). The instruction executes until a mismatch occurs between the two operands. At this point, the contents of R1 + 1 and R2 + 1 are decremented by the number of bytes that matched and the contents of registers R1 and R2 are incremented by the same number. The condition code can identify operands equal, first operand low and first operand high.

Other, more complex instructions are included in the System/370 architecture, which execute on variable-length fields of data. These include a number of DECIMAL instructions and text manipulation instructions such as EDIT and TRANSLATE.

Supervisory. Both the CYBER architecture and the System/370 architecture assume a supervisory program that coordinates the use of system resources and executes all I/O instructions, handles exceptional conditions, and supervises scheduling and execution of multiple programs. System/370 differentiates between MONITOR calls and SUPERVISOR calls, with the former being used for tracing the course of execution of a program and the latter being used for communication with the supervisory program. The CYBER architecture utilizes the CPU and other independent peripheral processing unit (PPUs) to execute the supervisory program.

A SUPERVISOR call is treated as an interruption that places the CPU in the supervisory state. While in this state, certain instructions are valid that are not allowed for user programs. These generally provide a means to change the contents of control and status registers and provide relocation and protection services. These instructions are called **privileged** (see Section 9.2.1) and will cause an exception condition interrupt if encountered in a user program.

A SUPERVISOR call may be made to the supervisory program for execution of I/O operations and other services. Main-memory management is not normally included in this service, since a memory reference to an address out of range, or to a page not currently in main memory, is handled by exception interrupt. The supervisory program maintains time-of-day clock and other timers accounting for the CPU time of individual user programs. A user program is restarted by the supervisory program by replacing appropriate contents of the general registers and control registers such that the program will resume where it left off. In CYBER, this is accomplished in one instruction, the EXCHANGE JUMP (01).

Input / Output. Section 9.10 discusses all aspects of I/O architecture. In this section, attention is restricted to the I/O channel method. The CYBER architecture utilizes a set of PPUs, each of which is a small programmable processor with private memory and access to main-memory. Part of the operating system function, particularly that of I/O, is executed in the PPUs. In fact, the CPU need not execute any I/O except to communicate with the PPUs through privileged locations in main-memory.

System/370 utilizes programmable data channels for the execution of I/O operations. The data channels provide the data path and control for I/O devices as they communicate with main memory (see Fig. 9.15). The data channel presents a standard interface to a control unit driving the I/O device. The interface passes commands and data to the control unit and receives responses, data, and external interrupts. The commands are part of a channel program that directs the control unit operation independently of the CPU.

The CPU initiates a channel program with a privileged instruction (START I/O) that is allowable only in the supervisory mode. When initiated, the channel fetches its own control word sequence (the channel program), governs the transfer of data and control signals, and counts record lengths. The channel executes the channel program and, on completion, interrupts the CPU. For this purpose, each command word contains command code, main-memory address, and a length where needed. This is a rudimentary single-address program structure that can branch in a limited way to deliver commands and data to the device control unit.

Thus, the principal method of operation is for the user program (or an operating system utility program) to prepare a channel program in main memory and call the operating system, which initiates the specified channel and returns control to the user program or to another user program.

More than one channel can execute concurrently, and certain types of channels can connect more than one device control unit and can multiplex channel programs on a single channel. Since the user program does not see this activity, the implementation may vary extensively and may include microprogrammed auxiliary control similar to the PPU concept. The programmable data channels each operate independently of the CPU and each other.

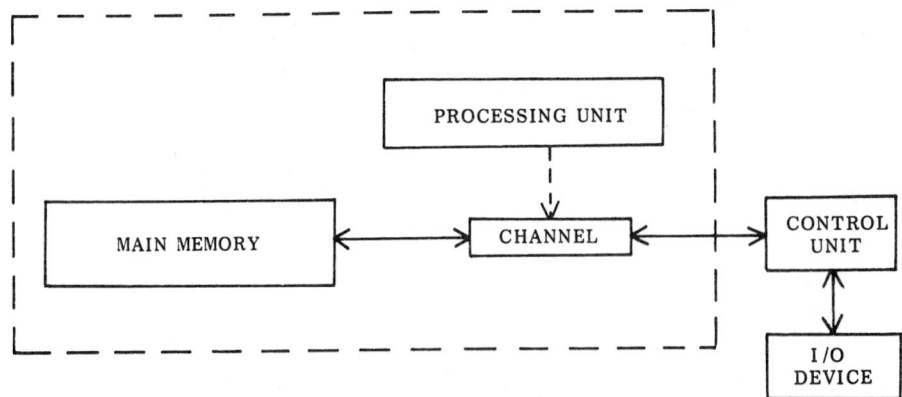

Fig. 9.15   Data channel.

9.2   ARCHITECTURE: INTERRUPT SYSTEMS AND SYSTEM STATES
Dennis J. McBride and Earl E. Swartzlander, Jr.

One of the most useful architectural features of a CPU is its ability to conditionally alter the instruction execution sequence. This means that conditions arising during program execution can be tested by branch instructions (see Section 9.1.5), and a new instruction address can be specified. Without this ability, the computer would be able to execute only a very restricted class of algorithms. Designating the types of conditions that are tested by branch instructions is an important architectural issue.

The ability of the CPU to do useful work is usually predicated on its ability to perform various I/O tasks as well, and the I/O data transfer process is typically controlled by an interrupt system. Consequently, the interrupt system is a fundamental part of the system architecture. The CPU hardware can be designed to alter the instruction execution sequence automatically upon the occurrence of interrupts. Either external or internal CPU events can cause interrupts. The currently executing program is usually suspended, rather than abnormally terminated, so that it can resume execution once the interrupt has been serviced (by the execution of a program that takes care of the situation that created the interrupt).

Finally, some CPU control mechanisms provide the ability to declare different subsets of the full instruction set to be valid at different times or to switch among several complete instruction sets. Specific instructions or hardware conditions cause appropriate changes from one instruction set to another. Traps (internally generated interrupts) are used to detect attempts to execute illegal instructions.

All three of these mechanisms are important architectural features. The branch conditions are essential for normal algorithmic processing. The interrupts and traps have implications at a higher architectural level. An attribute, often called the **system state**, is used to differentiate the various modes of CPU operation (i.e., normal versus interrupted, running versus stopped, etc.). Section 9.2.1 discusses the system states, and Section 9.2.2 discusses interrupt mechanisms.

9.2.1   System States

In the most general sense, the contents of all registers, memory locations, and other storage elements are necessary to describe an entire computer's state at any instant. The minimum information that must be retained (including next instruction address) in order to support a call-return control structure for interrupt handling, or for related activities such as subroutine nesting and multiprogramming, is collectively referred to as the **processor status**, or **machine status**. Even though the components of processor status may be generated in various parts of the CPU, they are often brought together in a single register to facilitate easy storage and retrieval when necessary. Typical names for this register are **status register**, or **program status word** (PSW). Memory locations may be reserved for storing additional status and control information.

The term *system state* carries a particular meaning, as defined by the architectural description of a given CPU. The commonly identified states that a CPU can assume are discussed in the following paragraphs.

Supervisor or Privileged State
For the case of large mainframe computers or minicomputers, the base architecture is usually designed with a software operating system in mind. A set of special, or privileged, instructions is made available for the operating system programmers to use (e.g., set CPU timer, insert storage key, load PSW). In the case of IBM System/370, these instructions can execute only while the system is in the **supervisor state**. This concept also exists in other machines. In the supervisor state, all instructions are valid, although some of these instructions may be configuration, model, or timing dependent. The supervisor state is reserved for software that handles interrupts and other exception conditions at the primitive hardware level, or CPU control level. Application programs do not have access to the supervisor state, except indirectly through special calls to "utility" programs within the supervisor.

Normal or Problem State
The normal or problem state is defined for application programs and higher levels of system software in systems that have a privileged class of instructions. Only a subset of the executable instructions is allowed in this state. An attempt to execute a privileged instruction usually generates an exception condition, or trap. The application software and much of the operating system software can therefore run at minimum risk to the vital CPU control mechanisms. In other words, the system is protected against misuse of instructions or accidental errors in most programs, with respect to the vital CPU control and hardware status operations.

A less sophisticated CPU, with a less complicated operating system, might not have these two modes or states. In this case, the application program must be able to handle I/O and special CPU control.

Wait State

Since a CPU may not have any instruction processing to do at a particular instant (e.g., because it might be waiting for I/O activity to execute), the system can enter the wait state, pending the arrival of an interrupt. The computer is not performing any useful work while in the wait state.

Stopped or Halted State

The stopped or halted state is normally provided as a console function, which allows the computer to be powered on but otherwise inactive. This function is desirable in the event of machine malfunction and for certain types of manual I/O operations: for the purpose of changing disks, tapes, printer paper, card decks, and so on. In large computing systems, however, much of this manual activity can be done by making an I/O device temporarily "unavailable." The system will either wait, as already described, or process other pending jobs if this kind of multitasking operation is supported by the operating system.

Other States or Modes

Other modes of operation are possible: for example, a **diagnostic** mode, which is not normally evident to application or operating system programmers. This would be used primarily by engineers and programmers who have special training for the purpose of architectural validation or hardware repair.

Another special mode of operation is **emulation**, whereby one machine type interprets instruction formats from a different machine architecture (see Section 9.9.6). This mode is sometimes made available to aid in installing a new computer, so that previously written programs for the previous computer can still be used.

Interrupt State

There are I/O device conditions as well as internal CPU exception conditions that demand immediate attention. All interrupts cause a fundamental change in the system state, so that the interrupt state is explicitly defined in most systems. The interrupt system itself is described in the following section.

9.2.2   Interrupt Handling

A computer can be viewed as a finite-state machine that moves from one state to another via the execution of a program. Interrupt mechanisms provide a well-defined way of altering the sequence of states in response to internal traps or outside asynchronous events (interrupts). There are many ways to handle interrupts, depending on the system requirements. The choice of a particular interrupt mechanism can have a significant impact on the throughput and flexibility of a system.

Polling and Nonpolling

One of the simplest ways to handle asynchronous events is the polling method. With each possible event, there is an associated flag that can be accessed by the program, to determine if service is required. (The flag is generated within the CPU in the case of traps, and externally for interrupts.) This method trades simple hardware for software complexity. It requires program memory space and also requires time for polling the flags (whether or not service is required). The polling method has low system throughput, high overhead time, and slow response time (see also Section 9.10.5).

In nonpolled interrupt handling systems, the asynchronous event generates an interrupt request signal, which is passed to the processor. The processor suspends execution of the current process and starts execution of an interrupt service routine. When the interrupt routine is completed, the processor resumes execution of the suspended process. This type of system is called an **interrupt-driven system** because it executes interrupt service routines that are initiated by interrupt requests.

Although this method requires more hardware, it has many advantages. Because the execution of interrupt service routines is transparent to the current process, the programmer does not need to write specialized polling routines. The response time is faster because no time is spent interrogating inactive components, which also increases the system throughput. There is less time overhead and less memory space required because only the service routine exists in memory, and no polling routines are required.

Level of Interrupt Handling

There are two levels on which interrupts may be handled. The first and most common is the machine-level interrupt. In this method, the presence of interrupt requests is checked at the start of each instruction fetch cycle. This guarantees that an interrupt is detected only when a machine instruction is complete and before a new instruction starts.

The second level of handling interrupts is at the microprogram level in which the microprogram can be interrupted at any time. This method has a faster response for servicing interrupt requests but requires that restrictions be placed on the microprogram and the hardware used to implement the interrupt mechanism.

Types of Interrupts

There are basically four types of interrupts, based on the relationship of the source of the interrupt to the processor:

1. *Intraprocessor* interrupts, or traps, are those asynchronous events that happen within the processor during the execution of a machine instruction. This group includes such things as attempting to divide by zero, overflow, accessing restricted memory, execution of a privileged instruction, and machine failure.

2. *Intrasystem* interrupts are created by system peripherals, such as disks, printers, and keyboards, that have completed operations or require service.

3. *Executive* interrupts are caused by the program that is currently executing. This provides a way for the current program to make a request of the operating system (executive). These requests include starting new tasks, allocating resources (disks, line printers, main memory), and communication with other tasks.

4. *Interprocessor* interrupts are used in multiprocessor systems as, for example, part of communication protocols.

Events for Interrupt Handling

When an interrupt occurs, a sequence of six events must take place. These events, which can be implemented in microcode or machine code, comprise, when integrated with the hardware, the interrupt mechanism. The sequence of events provides a smooth transfer from the current process environment to an interrupt servicing environment and back again. The sequence ensures that the processor status will be the same immediately after an interrupt is serviced as it was immediately before the interrupt occurred. The events may depend on the machine design and application.

1. *Interrupt Recognition.* This step is the recognition by the processor of an interrupt request due to activation of an interrupt request line or an internal mechanism. In this step, the processor can determine which device or CPU component made the request. The method used to determine which device to service is directly related to the interrupt structure of the machine. The different types of interrupt structures are discussed in more detail below.

2. *Status Saving.* The goal of this step is to make the interrupt sequence transparent to the interrupted process. Therefore, the processor saves the flags and registers that may be changed by the interrupt service routine so that they may be restored after the service routine is finished. The flags and registers that are saved automatically by the CPU are those that will be destroyed in the transfer of control from the current process to the interrupt service routine. It is then the responsibility of the service routine to save any other registers that it might change. The minimum set of flags and registers might include the instruction counter, overflow flag, sign flag, interrupt mask, and so on. The minimum set also includes any register or flag that needs to be saved but that the interrupt service routine cannot access.

3. *Interrupt Masking.* This step can overlap some of the other steps. For the first few steps of the sequence, all interrupts are masked out so that no other interrupt may be processed before the processor status is saved. Usually, the mask is then set to accept interrupts of higher priority. Some machines also allow the service routine, selectively, to enable or disable interrupts. There may be other variations to this step, depending on the application.

4. *Interrupt Acknowledgment.* At some point, the processor must acknowledge the interrupt being serviced, so that the interrupting device becomes free to continue its task. The processor can acknowledge in several different ways. One of the ways is to have an external signal line devoted to interrupt acknowledge. Another method relies upon the interrupting device recognizing an acknowledge when the cause of the interrupt is serviced by some CPU action. Some processor designs also use an acknowledge signal as a request for the interrupting device to place an identification code (ID) on the data bus (see the following paragraph).

5. *Interrupt Service Routine.* At this point the processor initiates the interrupt service routine. The address of the routine can be obtained in several ways, depending on the system architecture. The simplest is found in the polling method, in which one routine polls each device to find which one interrupted. Some designs require that the interrupting device put an address on the data bus so that the processor can store it in its program counter and branch to it. Other designs use an ID number derived from the priority of the interrupt, which is put through a mapping ROM or lookup table in memory in order to obtain the address of the service routine.

6. *Restoration and Return.* After the interrupt service routine has completed its processing, it restores all the registers it has changed, and the processor restores all the registers and flags that were saved at the initiation of the interrupt routine. If this is done correctly, the processor should have the same status as before the interrupt was recognized.

Interrupt Structures

The following is a list of the more common interrupt structures used with attached devices. As usual, there is a trade-off among hardware, software, and firmware. The particular structures vary in the way that the processor determines which device made the interrupt request.

Single Request, Multiple Poll. In this structure there is one request line that is shared among all the interrupting devices. When the processor recognizes an interrupt request, it polls all the devices to find the interrupting device. Priority is introduced via the order in which the devices are polled and can be dynamically reallocated by changing the polling sequence, if desired.

Single Request, Daisy-Chain Acknowledge. In this structure there is one request line, which is shared. When the processor receives an interrupt, it sends out a signal acknowledging the interrupt. The acknowledge signal is passed from I/O device to I/O device until the interrupting device receives the signal. At this point, the interrupting device identifies itself by putting an ID number on the data bus. This structure requires less software but has a static priority associated with the order of interconnection of the devices. There is also a time delay associated with the daisy-chain acknowledge structure because the interrupt acknowledge signal has to pass through several gate delays in each device, although this is much faster than polling.

Multiple Request. This structure features one request line (and one device) per priority level. The multiple-line structure gives the fastest response time since the interrupting device can be identified immediately. It also results in simpler interfaces in the peripheral units—in general, a single interrupt request flip-flop. This structure often has a mask bit associated with each priority level. The limitations associated with this method are that more hardware is required in its implementation and that there is a limit of one peripheral per priority level.

Multiple Request, Daisy-Chain Acknowledge. This structure combines the single request, daisy-chain acknowledge with the multiple-request structure. For each interrupt request line, there is an interrupt acknowledge line that is connected to a string of devices in a daisy-chain fashion. When the appropriate device receives the interrupt acknowledge, it puts an ID number on the data bus. The advantage of this structure is that many devices can be handled while maintaining short daisy chains. This gives a shorter access time than a single daisy chain, with less CPU hardware than an interrupt request line per device. The disadvantage is that each device must contain logic to pass on the acknowledge signal, which requires more hardware in each device.

Priority. In systems with asynchronous requests, two or more requests can arrive simultaneously. In order to handle this situation, there must be a prioritization mechanism to pick which request is serviced first. The two most common priority schemes are the static and the rotating approaches. In the static scheme, all the interrupt levels are ordered, from the lowest priority to the highest priority. This can be fixed in software or hardware and is usually permanent. In the rotating structure, the possible interrupt requests are logically arranged in a circle, with a pointer indicating the lowest-priority interrupt. The priority of each interrupt increases as one travels around the circle, with the highest-priority interrupt being adjacent to the lowest-priority interrupt. As an interrupt is serviced, the pointer is changed to point to the interrupt that was just serviced. This structure is advantageous when all interrupts have similar priority and service-rate requirements.

"Nesting" allows only higher-priority interrupts to interrupt an active interrupt service routine. Nesting requires inhibiting equal and lower-level interrupts. This requires that the interrupt structure hold the priority of the interrupt being serviced. This can be implemented with a register that holds the value as a binary encoded number or as a bit associated with each interrupt. Prioritization determines which interrupt request is serviced first when several requests are pending and control returns from the interrupt service routine for a higher-priority request.

However nesting is performed, all computers must have machine-level instructions to enable and disable interrupts and set and clear mask bits. Interrupt handlers can be written using these instructions to accomplish nesting of interrupts, although it can be implemented quite efficiently with microcode and hardware. In the simplest computers, the interrupt structure only prioritizes interrupts, leaving nesting to the software interrupt handlers.

9.3 PROGRAMMING BASICS
Jack L. Rosenfeld

This section deals with techniques for programming in **machine language**: designing sequences of CPU instructions. Because an understanding of how machine instructions are used is essential for processor architects and designers, this section is included in a chapter otherwise devoted to computer hardware. Machine language instructions are sequences of 0 and 1 bits that are interpreted by the control unit of the CPU and executed (see Section 9.9). Instructions in assembly language, which is closely related to machine language, contain alphanumeric symbols (more easily remembered by the programmer) for instruction operation codes, storage addresses, and, occasionally, the data itself. An assembler is a program that transforms a program in assembly language to its equivalent in machine language.

High-level languages (HLLs) such as PL/I, FORTRAN, COBOL, BASIC, and Pascal are provided for virtually all mainframes and minicomputers, and for an increasing number of microcomputers. The advantages of HLLs—ease of programming, of modifying, and of debugging, portability, and self-documentation—so greatly outweigh the disadvantages of somewhat reduced speed, increased storage occupancy, and limited flexibility that the programmer must have a compelling reason to eschew a HLL for machine language. In fact, optimizing compilers frequently produce better code than that created by average programmers in machine language.

The two most significant reasons for choosing machine language are unavailability of an appropriate HLL compiler and need to make the program being designed, or a part of it, especially efficient in its use of resources such as CPU time or storage occupied by the program or the data. If only a part of the program is critical, the remainder may well be programmed in a HLL.

The following sections deal with the basic aspects of machine instructions: accessing operands, branching, using index registers, invoking subroutines, and fixed-point arithmetic. A section on self-modifying code is included to strengthen the reader's comprehension of the concepts introduced. A brief section on "tricks of the trade" points out what facilities may be available to the alert programmer. The section concludes with a discussion on topics of concern to the programmer of a microprocessor system—how to cope with restrictions found in microprocessor systems—that programmers of larger systems generally may not be concerned with.

The program examples in this section are in a format resembling assembly language for many CPUs. The instruction set is typical of those of a number of CPUs and is designed to be illustrative but not complete. The addresses and values selected for the examples are completely arbitrary. Numbers rather than symbols are used for addresses in order to emphasize the difference between addresses and data values.

All CPU registers are assumed to have the width of memory words, and all instructions and numbers are also assumed to be one word wide. Cases in which the assumptions are unreasonable are dealt with. The form of the instructions is meant to enhance understanding—not to resemble any specific assembly language. One can learn the details of a specific computer's instruction set from the manufacturer's manual and from texts based on the more popular computers.

9.3.1 Operands

Operands of some instructions are explicit memory locations.* For example, the instruction

LOADA 123[†]

means load register A from memory location 123: that is, replace the current contents of A with the contents of memory at 123. It *does not* mean place the number 123 in A. In fact, LOADA has two operands: register A and memory location 123; the former is implicit, and the latter is explicit. Many other instructions also have implicit operands. For example, the instruction

CLEARA

means "clear register A," that is, replace the contents of A with the number 0. We now consider means of specifying explicit operands in memory. These are discussed in Section 9.1.3.

Addressing Memory Operands

A memory operand may be addressed directly, indirectly, or relatively. A **direct address** is an explicit memory address. The previous example, LOADA 123, has such a direct memory address.

In an architecture that permits **indirect addresses**, one generally finds one bit in the instruction that indicates that the operand is an indirect address. When that bit is 1, the contents of the memory location specified in the instruction are accessed; the data in that memory word are treated as an address; the contents of that indirect memory location are then used as the operand of the

---

*In this section the size of memory operands is treated, in general, as a fixed number of bits. In some architectures, each instruction implies the number of bits comprising its operand(s); for some designs, operands are multiples of the basic memory word size. For example, in IBM System/370, the basic memory unit is the 8-bit "byte." Operands are denoted byte, half-word (2 bytes), full-word (4 bytes), and double-word (8 bytes). The n-byte operand with address X comprises the bytes at locations X to $X + n - 1$.

[†] This is the notation for the machine language binary instruction. See Fig. 9.16 for an example.

instruction. Let us designate an instruction with the indirect address bit equal 1 by "IND" following the operation code mnemonic, such as

LOADA IND 123

Consider the following example:

Location	Instruction or Data
	LOADA IND 123
	:
123	257
	:
257	142

The result of executing the instruction LOADA IND 123 is to replace the contents of register A with the number 142. First, location 123 is accessed to fetch the address (257). Then the LOADA operation is executed with the contents of address 257 as operand, which means that 142 (the contents of memory location 257) replaces the previous contents of register A.

(Before proceeding with the next example, recall that successive instructions are normally fetched from successive main-memory locations. This sequence may be modified by branching instructions which are discussed in Section 9.3.2.)

Indirect addressing is useful for looking up values in a table. Consider the following example in which codes for characters may range from 0 to 255 and must be translated to other codes (e.g., from codes for lowercase characters to uppercase codes). The program segment starts with the source code in register A and ends with the translated code in register A.

Location	Instruction or Data		Comments
	ADDA	100	Generates address
	STORA	101	in table
	LOADA IND	101	
	:		
100	389		
101	—		Receives address
			in table: $389 + i$
	:		
389	Code for 0		
	:		
644	Code for 255		

The first instruction, ADDA 100, means add the operand (contents of location 100, which equals 389) to the current contents of register A and leave the sum in register A. The second instruction means place the contents of register A in memory location 101. If one started with i in register A, then $i + 389$ would now be found at location 101. The third instruction has its indirect addressing flag equal to 1, so the contents of memory location 101 are accessed, and then the contents of location $389 + i$ are loaded into register A. So if A originally contained the code i for a lowercase letter and location $389 + i$ contained the code for the same letter in uppercase, the result in register A would be the code for the uppercase letter. Section 9.3.4 describes how the same operation can be done with the use of an index register.

Some architectures permit more than one level of indirection. There could be an indirect addressing bit in each indirectly addressed location. Depending on that bit, the word is interpreted as the address of the memory word holding either the operand or the address of the operand. Obviously, an indefinite number of levels of indirection can be achieved, if the architecture permits it.

Relative addressing in an architecture implies that the address of the operand is determined by adding the number in the instruction to the address of the instruction itself. Instructions with relative addressing generally have short address fields and occupy less storage than instructions with direct addressing. Consider this code:

Location	Instruction or Data
100	35
	:
223	LOADA −123

In a CPU with relative addressing, the LOADA instruction would replace the contents of register A with the number 35, since the address of the instruction's operand is determined by adding −123 to 223 (the address of the instruction itself). Although relative addressing may be a nuisance to programmers, especially when they must insert or delete instructions from an existing sequence, an assembler can relieve most such problems.

An **immediate operand** is one found in the instruction itself; it is treated like data rather than the address where data can be found. For architectures permitting immediate operands, we designate an instruction with an immediate operand by "IMM" following the operation code. If the following sequence of three instructions is executed, the result in register A will be 155:

Location	Instruction or Data			Contents of Register A
	LOADA	IMM	25	25
	ADDA		25	125
	ADDA	IND	25	155
	⋮			
25	100			
	⋮			
100	30			

In some CPUs, the relative address is with respect to the address of the following instruction rather than the current instruction.

Register Operands
Registers may be explicit or implicit operands of an instruction. In all the instructions introduced so far, register A was an implicit operand; it either received data or was the source of data. Some architectures provide for a multiplicity of explicit operands in registers (a set of registers is also called local memory or a register file), main memory, or both: for example,

```
MOVE    A, B
ADD     C, 75
```

The first instruction replaces the contents of register A with those of register B. The second instruction adds the contents of memory location 75 to the contents of register C and places the sum in C.

The following code segment places the sum of the contents of locations 300 to 305 in register A.

```
LOAD    A, 300
ADD     A, 301
ADD     A, 302
ADD     A, 303
ADD     A, 304
ADD     A, 305
```

A better way to do this is with the use of an index register and branching.

9.3.2 Branching
Unconditional branching instructions, such as

```
BRA     75
```

cause the next instruction to be executed from an explicitly specified location (in this case, 75) rather than the next instruction in sequence. Conditional branching instructions specify what instruction is to be executed next if a specified condition is satisfied.

Conditions
Some of the conditions that may be tested are arithmetic result is positive, negative, zero, overflowed, underflowed (for floating-point numbers), or generated a carry. Other conditions that can be tested include parity errors, I/O status, protection violation, machine failure, and interrupt pending. Some architectures require that if a result is to be tested, it be done immediately after an arithmetic operation. In other architectures, certain control bits are set, depending on the instruction executed, and are not changed until another instruction that affects them is executed. For example, a "carry"

bit would be set by add and subtract instructions but not by store or branch instructions. If a condition is not to affect the flow of program control until after the pertinent control bit is no longer valid, the program must save the status of the control bit (perhaps by setting a storage word to 0 or 1, depending on the control bit status) and test that saved status at the appropriate time.

Branch Instructions

Conditional branch instructions generally specify the condition(s) tested and the location of the instruction to be executed next if the condition is satisfied. In some architectures, each condition is implied by the instruction. For example,

BRA NEG   200

executes the instruction at location 200 if the sign control bit is 1 (the result of the preceding arithmetic operation was negative) and the following instruction otherwise. Other general conditional branch instructions permit arbitrary combinations of the condition control bits to be tested. Some branch instructions have no explicit address, but skip the instruction after the branch instruction if the condition is satisfied or execute that instruction if it is not. There are variations to this implicit addressing mode.

At this point we have the repertoire of instructions necessary to sum a specified number of memory locations in sequence. Suppose that initially register A holds the address (ad) of the first number (N_1) in the sequence and B holds the count (ct) of numbers to be added:

Location	Instruction or Data			Comments
100	STORE		A, 200	Location 200 holds the address of the next number to be added
101	CLEAR		A	Sum is 0 so far
102	STORE		B, 201	Location 201 holds the count of numbers remaining
103	ADD	IND	A, 200	Add next number to sum
104	LOAD		B, 200	Increment address
105	ADD	IMM	B, 1	of next number
106	STORE		B, 200	
107	LOAD		B, 201	Decrement count of
108	SUB	IMM	B, 1	numbers remaining
109	BRA	NONZERO	102	Test count of numbers remaining
110	Next instruction			
	⋮			
200	—			Address of next number
201	—			Count of numbers left to accumulate

When the instruction at memory location 110 is executed, register A holds the sum of the ct numbers from ad to ad + ct − 1: $N_1 + N_2 + \cdots + N_{ct}$. The conditional branch instruction at memory location 109 tests the condition control bit set by the subtract instruction at 108. If the result of that subtraction is not zero (when fewer than ct numbers have been added), control branches back to the STORE instruction at location 102. The indirect add instruction at location 103 accesses location 200, which holds the *address* of the next number to be added, and adds the number. The STORE instruction at 106 replaces the address at 200 with the contents of register B, the incremented address. The STORE at 102 puts the decremented count at 201. Let us now satisfy ourselves that the program is correct. When the *indirect* ADD instruction at 103 is executed the first time, location 200 holds ad, 201 holds ct, and A holds 0. When the branch instruction at 109 is executed the first time, 200 holds ad + 1, 201 holds ct − 1, and A holds N_1. This table gives the contents *after* each iteration of the loop from 102 to 109:

Iteration	200	201	A
1	ad + 1	ct − 1	N_1
2	ad + 2	ct − 2	$N_1 + N_2$
⋮			
ct	ad + ct	0	$N_1 + \cdots + N_{ct}$

In comparing this with the code sequence at the end of Section 9.3.1, one finds that in the former, ct instructions are executed, whereas in the program just described, the same function requires 8ct + 2 instructions. In the former, the "program" occupies ct words of memory, and in the latter 10 words (plus two temporary storage words). The former program is dependent on the number of numbers to be added; the latter is not. The use of an index register to simplify this program is discussed in Section 9.3.4.

9.3.3 Self-Modifying Code

This subject is introduced to reinforce the reader's understanding of the meaning of machine instructions—not as a suggestion of acceptable programming technique. [Code that modifies itself is very risky to use since any error can easily be responsible for the code being incorrectly modified. Furthermore, such "bugs" are especially difficult to detect. In a microprocessor system in which code is written in read-only memory (ROM), self-modifying code is, obviously, impossible.] If, however, one uses a primitive architecture without indirect addressing or an index register, one might arrive at the following code segment to perform the operation of the code sequence in the previous example.

Location	Instruction or Data			Comments
400	ADD		A, 500	Initialize the instruction
401	STORE		A, 405	at 405
402	ADD		A, B	Initialize limit constant
403	STORE		A, 501	
404	CLEAR		A	
405	—			ADD A, $ad + i - 1$
406	LOAD		B, 405	Increment address field
407	ADD	IMM	B, 1	
408	STORE		B, 405	
409	SUB		B, 501	
410	BRA	NONZERO	405	
⋮				
500	ADD		A, 0	Constant
501	—			ADD A, $ad + ct$

To understand this, one must recall that each instruction is a word in main memory with a bit pattern that includes a code for the operation and may include a code for the register and a code for the address. Like any data word, it can be manipulated by other instructions. Let us assume that the instruction ADD A, 100 looks something like Fig. 9.16. One can increment the address field by loading the instruction word itself into a register, adding 1 to it, and storing it back. That is the basic principle in the loop from 405 to 410. Here it is assumed, as before, that initially register A holds ad and B holds ct.

The first instruction creates the instruction ADD A, ad in register A, and the second instruction places it at location 405. The third instruction, ADD A, B, adds the contents of registers A and B and places the result in A. In this case, the sum is the instruction ADD A, $ad + ct$, which is placed in location 501 by the following STORE instruction. The first time the instruction at 405 is executed, the contents of memory location ad (N_1) are placed in register A. The instructions from 406 to 408 increment the address in that instruction. The instruction at 409 subtracts the bit pattern for the instruction ADD A, $ad + ct$, which has been placed in 501, from the bit pattern for ADD A, $ad + 1$, leaving $-ct + 1$ in register B.

The following table gives the contents of location 405 and the A and B registers *after* each execution of the instruction at 410.

Iteration	405		B	A
1	ADD	A, $ad + 1$	$-ct + 1$	N_1
2	ADD	A, $ad + 2$	$-ct + 2$	$N_1 + N_2$
⋮				
ct	ADD	A, $ad + ct$	0	$N_1 + \cdots + N_{ct}$

ADD	A	100
01011010	0	00...001100100

Fig. 9.16 Assembly language instruction and its machine language equivalent.

Note that should the operation code of the ADD instruction correspond to a negative number, a subtract instruction would be used at 407, and the initializing instructions would be modified somewhat. If the address field were not in the low-order bits of the instruction, the code would have to be changed slightly.

Another example of the use of self-modifying code is to determine the proper address for a branch to take and place the address in the branch instruction before it is executed.

9.3.4 Index Registers

An additional register can eliminate difficulties associated with addressing arrays of data, as well as general counting-type operations. Index registers, in addition to participating in some arithmetic operations, participate in the calculation of storage addresses. Most instructions that address operands in main memory can have the contents of an index register added to the address in the instruction itself, the sum specifying the main-memory location of the operand. If we denote by "X" a bit in an instruction that is 1 when such an effective address calculation is to be made, then the notation

$$\text{LOAD X A, 100}$$

means: fetch the contents of memory at location (100 + contents of index register) and place it in register A.

With such a facility, the problem of adding ct numbers starting at ad can be solved as follows. Here XR denotes the index register. Assume that ad is initially in register A and ct in B.

Location	Instruction or Data			Comments
600	MOVE		XR, A	Move ad to index register
601	CLEAR		A	
602	ADD	X	A, 0	Effective address is ad + i − 1
603	ADD	IMM	XR, 1	Increment index register
604	SUB	IMM	B, 1	Decrement count
605	BRA	NONZERO	602	

The first instruction copies the address ad from A into the index register. The ADD instruction at 602 has an effective address of ad + 0 (the index register contents plus 0). This is incremented by 1 in the following instruction. The next instruction at 604 decrements the count, and the branch instruction loops back until the count has been reduced to 0. This code carries out the operation with 4ct + 2 instruction executions, considerably better than before.

In some CPUs, special instructions combine the incrementing and conditional branching operations; the IBM System/370 instruction Branch on Index High:

$$\text{BXH R1, R2, R3, a}$$

is such an instruction. The following description of BXH is somewhat simplified: the contents of register R2 are added to the contents of register R1, the sum being placed in R1; if the sum is greater than the contents of register R3, the next instruction is fetched from location a; otherwise, the instruction following the BXH is executed next. The sum of an array can be coded as follows, using this instruction:

Location	Instruction or Data			Comments
801	MOVE		C, A	Place ad in register C
802	ADD		A, B	Place ad + ct in
803	MOVE		XR, A	index register
804	LOAD	IMM	B, −1	Place −1 in B
805	CLEAR		A	
806	ADD	X	A, −1	Effective address: ad + ct − i
807	BXH		XR, B, C, 806	

Values have been chosen to decrement the index, for the sake of illustrating a different approach. The instructions from 801 to 804 load the registers with proper initial values. Since the B register holds −1 during the loop from 806 to 807, the BXH instruction always decrements the XR by 1. The comparand in register C is the constant address ad. The register contents are as follows each time *after* the BXH is executed:

Iteration	XR	A
1	ad + ct − 1	N_{ct}
2	ad + ct − 2	$N_{ct} + N_{ct-1}$
⋮		
ct	ad	$N_{ct} + \cdots + N_1$

The last time the ADD at 806 is executed, XR holds ad + 1. The effective address is ad since the XR contents are added to the address in the instruction (−1). The BXH instruction decrements XR to ad, and the branch back to 806 is not taken since ad is not higher than the comparand value in register C. This code has 2ct + 5 instructions executed to accumulate the ct values.

Some architectures have several registers that can be used as index registers; a field in the instruction is used to specify which one, if any, is used. In some architectures, the effective address is calculated by summing an address field, contents of an index register, and the contents of one or more other registers. (In IBM System/370, the additional register is called a base register.) This permits a great deal of flexibility.

9.3.5 Subroutine Calls

Virtually every architecture has at least one instruction to facilitate invoking a subroutine, and often an instruction to return control to the calling program. The primary feature for transferring control involves saving the address of the calling location. This is not essential; witness the following program, which calls a subroutine to sum an array of numbers. Assume that before the code is executed, the array address and count are loaded at 353 and 354, respectively.

	Location	Instruction or Data			Comments
	350	LOAD		XR, 352	Place address of next instruction (355) in XR
	351	BRA		798	Subroutine call
Calling sequence	352	355			Location of next instruction
	353	ad			Address of array
	354	ct			Count of numbers
	355	Next instruction			
	⋮				
	798	LOAD	X	A, −2	ad into register A
	799	LOAD	X	B, −1	ct into B
	800	STORE		XR, 901	Save return address
Subroutine	801				
					Code from previous example
	⋮				
	807				
	808	LOAD		XR, 901	Restore return address
	809	BRA	X	0	
	⋮				
	901	—			Holds return address

The LOAD at 350 places the address 355 in the index register. (The instruction LOAD IMM 355 would also be satisfactory.) It serves as both return address and address by which arguments can be located. The subroutine itself consists of the code from the last example in Section 9.3.4 plus a prologue (798 to 800) and an epilogue (808 and 809). In the prologue, ad and ct are loaded into registers A and B, respectively. (The effective address of the LOAD at 798 is 355 − 2, the address of the location holding ad.) The STORE at 800 saves the return address; it is restored in the index register by the LOAD at 808 and used by the *indexed* BRA at 809 to return control to location 355.

The subroutine can be written more efficiently. Arguments can be passed in registers rather than in a list, as shown. The example should be studied more to understand the techniques involved than to learn a model.

The preceding example would be slightly simpler with an instruction

BSY   a

that saves the address of the following memory word in an index register YR and branches to the subroutine at address a:

	Location	Instruction or Data		Comments
Calling sequence	⎡350	BSY	798	YR gets 351
	⎣351	400		Address of argument list
	352	Next instruction		
		⋮		
	400	ad		Argument list
	401	ct		
		⋮		
	⎡798	LOAD Y	XR,0	XR gets argument list address
	799	LOAD X	A,0	A gets ad
Subroutine	800	LOAD X	B,1	B gets ct
	801			
				Code from previous example
	807	⋮		
	⎣808	BRA Y	1	

(A different concept is illustrated here: an argument list separate from the calling sequence. This is preferred since it permits calling the subroutine from the same point, at different times with different arguments, without modifying the calling sequence itself.) The indexed LOAD at 798 uses YR as an index register to load XR from memory at 351. The BRA at 808 has an effective address of 352.

Architectural features helpful for subroutine execution are an instruction that combines the saving of the return address in some accessible location with a branch to the subroutine, an instruction that returns control to the calling program, an instruction that saves the contents of registers that will be used by the subroutine, an instruction that restores those registers, and means for accessing an argument list created by the calling program. Pushdown stacks, which facilitate subroutine execution, are included in many architectures, especially those of microprocessors. They are described in Section 9.3.8.

9.3.6 Fixed-Point Arithmetic

Floating-point arithmetic varies so much from architecture to architecture that it is difficult to generalize. Consequently, this section deals only with fixed-point numbers and decimal arithmetic.

Addition and Subtraction

Sections 9.1 and 9.5 treat binary numbers as having a fixed binary point either just beyond the least significant bit or between the most significant bit and the second most significant bit. In the former case, all numbers are integers; in the latter, all numbers X are fractions $(-1 \leq X < 1)$. In reality, adders (described in Section 9.5.3) treat all bit positions identically. The binary point is an imaginary concept helpful to the programmer. He or she may think of the bit pattern $b_{n-1} \ldots b_0$ in an n-bit register as representing a number $X = N \times S$, where N is the integer represented by the bits in the register ($N = \sum_{i=0}^{n-1} b_i 2^i$), and S is a scale factor. If $S = 1$, the number is an integer $(0 \leq X < 2^n)$ corresponding to the binary point to the right. If $S = 2^{-n}$, the number is a fraction $(0 \leq X < 1)$ corresponding to the binary point to the left. If $S = 2^{-n+1}$, the number $(0 \leq X < 2)$ is less than 2. In this case, the number can also be interpreted as a fraction in 2's-complement form $(-1 \leq X < 1)$ (see Section 9.5.4). Very small numbers may be handled by using $S = 2^{-a}$, very large numbers by $S = 2^a$, where a is large.

To illustrate the principle, we consider a program that adds a number of positive integers, the sum of which may exceed 2^n. It is preceded by code to initialize the index register and other registers for looping:

Location	Instruction or Data		Comments
78	CLEAR	A	A will hold scaled sum
79	CLEAR	B	B will hold overflow count
80	LOAD X	C,0	C receives N_i (effective address: ad $+ i - 1$)
81	SHIFT RIGHT	C,B	Scale down N_i
82	ADD	A,C	A holds $N_1 + \cdots + N_i$

Location	Instruction or Data		Comments
83	BRA NOTOVER	86	scaled down
84	SHIFT RIGHT	A, 1	Scale down $N_1 + \cdots + N_i$
85	ADD IMM	B, 1	Increment overflow count
86	BXH	XR, ..., 80	

The ADD instruction at 82 adds N_i to the sum $N_1 + \cdots + N_{i-1}$, just like the accumulating add in the previous examples. If the sum is at least 2^{n-1}, an overflow bit is set in a control register (see Section 9.5.5). This bit can be tested by a conditional branch instruction such as that at 83, which skips to 86 if the overflow bit is 0. When an overflow occurs, the sum is scaled down by one-half by a shift right instruction that shifts the contents of A right one position and introduces a 0 in the high-order bit. Such a SHIFT is at 84. After this has occurred, all numbers accumulated must also be scaled down by one-half before being added to the sum. If the overflow occurs a second time, the sum is scaled down by one-half again, and all numbers added in the future must be scaled down by one-fourth, and so on. The instruction at 85 increments, in register B, the count of number of times an overflow has occurred. The LOAD at 80 brings N_i into register C, and the SHIFT at 81 shifts this right the number of times specified in the B register (the overflow count). Note the difference between the two SHIFT instructions. At 81 the number of shifts of C is specified by a register contents; at 84 it is specified by an immediate operand. Other architectures may permit other shift specifications. The BXH instruction causes the loop to be executed and XR to be incremented (or decremented) an appropriate number of times. Upon completion, A holds I and B holds J, and the total is $I \times 2^J$. Of course, some precision may be lost in this process, and we have not considered rounding. Note that some shift instructions shift operands while preserving the sign of the number, some shift over two registers, and others shift in other ways.

Multiplication and Division
Fixed-point **multiplication** generally produces a product with the number of bits equal to the sum of the multiplier and multiplicand bits (e.g., two 16-bit operands result in a 32-bit product). The user must use either the entire product or the part that is known to be significant. Assume that the instruction

$$\text{MPY} \qquad \text{A, a}$$

creates the product of the contents of register A and of the memory location a and places the high-order n bits of the product in register A and the low-order n bits in B.

If it is known that $\Sigma x_i y_i$ is always less than 2^{n-1}, the following program calculates that scalar product. Assume that the vector of multiplicands is located in memory from ad to ad + ct − 1 and the multipliers from ad + 1000 to ad + 1000 + ct − 1. The following code sequence is preceded by code to initialize the index register and other registers for looping:

Location	Instruction or Data		Comments
32	CLEAR	C	
33	LOAD X	A, 0	Multiplicand into A from ad + i − 1
34	MPY X	A, 1000	Multiplier at ad + 1000 + i − 1
35	ADD	C, B	Accumulate sum in C
36	BXH	XR, ..., 33	

The LOAD in 33 brings the multiplicand from memory location ad + i − 1. The MPY instruction multiplies this by the multiplier. Note that the address field is 1000, so the effective address is 1000 greater than for the preceding instruction. The lower-order half of the product is in register B and accumulated in C by the ADD at 35. The high-order half is 0. Should the high-order half be significant, instead, the ADD instruction could accumulate the products from register A. Some precision would be lost. Alternative scaling is possible.

Division often involves a double-length dividend (DD) and a single-length divisor (DR), quotient (Q), and remainder (R). They satisfy the relation

$$DD = Q \times DR + R$$

A key requirement imposed by many architectures is that the user must scale the dividend and divisor so that the quotient fits in a single n-bit register. Assume that

$$\text{DIV} \qquad \text{AB, a}$$

divides the double-length dividend in registers A and B by the divisor at a in memory, leaving the quotient in A and the remainder in B.

Following is a program to convert from binary to decimal using the divide instruction. (The binary number is assumed less than 2^n.) This is based on the knowledge that a decimal number can be described as

$$(((\dots)10 + d_2)10 + d_1)10 + d_0$$

where d_i are the digits, d_0 being the least significant. The following code sequence is preceded by code to initialize the index register and other registers for looping:

Location	Instruction or Data			Comments
31	LOAD		B, 76	Load binary number
32	CLEAR		A	into AB
33	DIV	IMM	AB, 10	Divide by 10
34	STORE	X	B, 0	Remainder d_i
				stored at ad $- i$
35	MOVE		B, A	Quotient moved to B
36	BXH		XR, ..., 32	
:				
76	Binary number to be converted			

The first division at 33 yields a quotient

$$((\cdots)10 + d_2)10 + d_1$$

in A and a remainder d_0 in B. Each number d_i is the binary representation of a digit from 0 to 9. d_0 is stored in the last word of an array ending at ad by the STORE instruction. The quotient is moved to the A/B register by the instructions at 35 and 32. Successive iterations store d_i at ad $- i$. This loop is repeated ct times, corresponding to the maximum number of decimal digits possible for an n-bit number. Otherwise, the iterations can be ended when the quotient is 0.

Decimal Arithmetic

Each architecture has its own means of encoding decimal numbers (strings of codes for decimal digits) as well as instructions for manipulating them. Operands are generally variable length, with the maximum allowed length specified in the instruction or elsewhere. Arithmetic instructions generally have two such operands, the result replacing one of the operands. Other operations on strings in memory are also included in some architectures.

In CPUs without decimal addition instructions, there is often a helpful "decimal adjust" instruction. This is used following a normal binary addition of two words, each containing two or more binary-coded decimal strings. It uses control bits set by internal carries to adjust the sum to represent a binary-coded decimal number.

9.3.7 Tricks of the Trade

Each instruction set has some instructions that can be used effectively in ways that are not obvious initially. For example, the IBM System/370 architecture has no CLEAR instruction; however, the subtract instruction (denoted SR) can be used to clear any register by subtracting it from itself: SR R, R.

Another example concerns a microprocessor whose index register is twice as long as its memory word length. One way to move an address from locations 100 and 101 to 200 and 202 is

```
LOAD    A, 100
STORE   A, 200
LOAD    A, 101
STORE   A, 202
```

A better way is

```
LOAD    X, 100
STORE   X, 200
```

if the LOAD and STORE instructions for index registers automatically handle two-word addresses.

The latter method would normally be executed more rapidly than the former and occupy less memory space.

These techniques are learned from texts and manuals that discuss specific CPUs—sometimes even by experimenting with an instruction set.

9.3.8 Microprocessor Techniques

In most ways, microprocessor programs are similar to those for mainframe processors. (Incidentally, programs for microprocessors are not "microprograms"—operation sequences for CPU control units. The latter are described in Section 9.9.3.) On the other hand, there are certain areas in which dealing with microprocessors requires more ingenuity on the part of the programmer. This section deals with those areas.

Word Size

Frequently, microprocessor word sizes are smaller than mainframe or miniprocessor word sizes, and techniques to cope with these smaller sizes must be mastered. The basic approach to achieving precision with small word width is to combine a number of smaller words into a wider word. For example, if a word is n bits wide but the user requires kn bits of precision, he or she can represent numbers by k consecutive words of storage. The principle behind addition (and subtraction) of such numbers is to add the k corresponding pairs of n-bit words, starting with the least significant word pair, and adding the carry bit generated by the previous operation. The following example subtracts the three-word number at memory locations 200–202 from the one at 100–102 and places the difference at 300–302. The least significant words of the numbers are at 102 and 202:

```
LOAD                A, 102
SUB                 A, 202
STORE               A, 302
LOAD                A, 101
SUB      BOR        A, 201
STORE               A, 301
LOAD                A, 100
SUB      BOR        A, 200
STORE               A, 300
BRA      BORROW     1000
```

The BOR following two of the three SUB instructions means that when the second operand is subtracted from the first operand (e.g., contents of memory at 201 from the A register), the borrow bit generated by the previous operation is also subtracted. Often, the borrow and carry flags are identical. The last instruction of the sequence checks whether a borrow remains after subtraction of the most significant words. This implies a negative result, which is presumed to be handled by code at location 1000.

The subtract with borrow instructions use the borrow flag that was set by the previous SUB instruction. Should the intervening STORE and LOAD instructions affect the borrow flag, this code would not be correct. The programmer must always be aware of what instructions affect what status flags, if tests are not made immediately after the execution of the instructions setting the flags.

This type of software manipulation is similar to the functions performed on strings of data by hardware, in mainframes with rich instruction sets. Multiplication and division can also be performed in microprocessor software using similar techniques.

Addressing Modes

Consider a CPU with only one index register and an address field smaller than the address length. The simple task of moving a string of words from one set of locations to another is not so simple to program, in general. Although the second example in Section 9.3.6 illustrates how two arrays can be addressed by a single index register, that technique cannot be used in many instances. If the starting points of the arrays are w words apart, and if the address field in indexed instructions cannot accommodate numbers as large as w, a program using two index quantities must be used for the task:

Location	Instruction or Data		Comments
85	LOAD	XR, 201	Initialize pointer
86	STORE	XR, 204	to target array
87	LOAD	XR, 200	Initialize source pointer
88	LOAD X	A, 0	Register A gets N_i
89	INCREMENT XR		

Location	Instruction or Data		Comments
90	STORE	XR, 203	Source pointer
91	LOAD	XR, 204	Target pointer
92	STORE X	A, 0	
93	INCREMENT XR		
94	STORE	XR, 204	
95	LOAD	XR, 203	Source pointer
96	COMPARE XR	202	
97	BRA UNEQUAL	88	
	⋮		
200	ad1		Address of source array
201	ad2		Address of target array
202	ad1 + ct		Upper limit of source array
203	—		ad1 + i − 1
204	—		ad2 + i − 1

The instruction

INCREMENT XR

simply adds 1 to the index register contents. The instruction

COMPARE XR a

compares the index register with memory at location a. A control bit, which can be tested by the BRA UNEQUAL instruction, is set if the contents are equal. The index register is loaded with one address for accessing the source array at instruction 95 and with another address for accessing the target array at instruction 91. The preponderance of the program updates these two addresses and moves them between the index register and memory.

Note the assumption that the address of the word beyond the end of the source array can be found at location 202. If it is necessary to calculate this, and if addresses occupy two or more memory words, the addition must be performed as described in the previous section.

Stacks

Automatic pushdown stacks are used to save information when an interrupt occurs, when a subroutine is called, and when a program pushes data onto them. This stack is different from the stack described in Section 9.1.3. The stack described there is an integral part of the arithmetic-logic unit and participates in virtually every data manipulation instruction. In "stack" architectures, it is sometimes possible to combine into a single stack the pushdown stack that holds operands and the one that holds status information.

The principle behind the pushdown stack for storing status is easy to understand. An area of main memory is reserved (generally by the programmer) to hold the stack. The number of words occupied by the stack changes from time to time. A special register called the **stack pointer** (SP) is used to hold the address of the top of the stack, usually the next location available to receive data. When information is pushed onto the stack, either automatically or by program, the data are placed into adjacent memory locations at the SP address and above, and the address in the SP is decremented by the number of words pushed onto the stack. Conversely, when data are removed from the stack, the words are read from main memory at the locations below the stack pointer, placed in the designated registers, and the address in the SP is incremented appropriately. Instructions usually exist to set the SP and to manipulate its contents in other ways; these instructions vary from architecture to architecture.

In general, some CPU status information (instruction counter and other register contents, and control bits) is automatically pushed onto the stack when an interrupt is accepted by the CPU, and automatically restored when an instruction to return from the interrupt handler is executed. The information stored may be only the instruction counter, or all registers and control bits, depending on the architecture. In some architectures, a separate set of registers for the interrupt level is provided, in which case no saving and restoring is necessary. Depending on what registers the interrupt-handling program uses, some additional status information must be saved explicitly by the interrupt-servicing program, usually in the stack. Prior to returning to the interrupted program, the interrupt handler must restore the information explicitly saved. A return-type instruction restores automatically what was previously saved automatically. The interrupt handler may also inspect the stack if there is a need to determine something about the CPU status when the interrupt occurred. Enough memory must be reserved for the stack. If the stack is in the same memory space as instructions and data, there may be no means of protecting the stack from programs, or vice versa. The programmer must ensure that no interference occurs.

Subroutine calls are analogous to interrupts. The subroutine call instruction pushes the instruction counter contents, and perhaps more information, onto the stack. The return instruction pops (restores) all that was stored. It is up to the programmer explicitly to push anything more that he or she requires. The stack permits passing arguments to the subroutine either in the stack itself or in an argument list located at a known displacement from the instruction that passes control to the subroutine. In either case, the programmer must be aware of how much data is automatically pushed onto the stack, so that the information desired can be located.

The following two examples assume that (1) when a subroutine is called, 13 words of status are automatically saved; (2) the stack pointer is decremented by 13; (3) the return address, which is the address of the instruction following the branch-to-subroutine instruction, is located at the new stack pointer contents $+9$; (4) the index register contents are not saved automatically in the stack; and (5) two arguments must be passed to the subroutine. The following code illustrates passing the arguments in the stack:

	Location	Instruction or Data			Comments
	250	PUSH		A	First argument
Calling sequence	251	PUSH		B	Second argument
	252	BRASUB		305	
	253	POP		B	Clear stack upon
	254	POP		A	return
	⋮				
	305	PUSH		XR	Save index register
	306	MOVE		XR, SP	
	307	LOAD	X	B, 15	Load second argument
Subroutine	308	LOAD	X	A, 16	Load first argument
	⋮ Body of subroutine				
		POP		XR	Restore index register
		RETURN			

The two PUSH instructions at the beginning of the calling sequence push the arguments onto the stack, where the subroutine can fetch them. The BRASUB pushes CPU status onto the stack and loads the instruction counter with the subroutine address. The initial PUSH in the subroutine saves the index register contents on the stack, since the register is needed for what follows. The PUSH also decrements the stack pointer address. (Assume that the decrement is 1.) The MOVE instruction places the stack address in the index register so that it can be used to address data. The following indexed LOAD instruction addresses memory at the stack pointer plus 15 (1 for the XR, 13 for the CPU status, and 1 for the second argument pushed). The instruction at 308 has an address field of 16, corresponding to the first argument, which is found below the second. After the subroutine is executed, the index register is restored, and the RETURN from subroutine instruction is executed. When control returns to 253, two POP instructions are executed to clear the arguments from the stack. Note that they are popped in the opposite order from that in which they were pushed.

Now consider passing arguments in a list in the calling sequence:

	Location	Instruction or Data			Comments
	321	BRASUB		430	
	322	BRA		325	Branch past arguments
Calling sequence	323	Argument 1			
	324	Argument 2			
	325	Next instruction			
	⋮				
	430	PUSH		XR	Save XR
	431	MOVE		XR, SP	
	432	LOAD	X	XR, 10	XR gets address of arg1
	433	LOAD	X	A, 1	A gets arg1
Subroutine	434	LOAD	X	B, 2	
	⋮ Body of subroutine				
		POP			Restore XR
		RETURN			

Here the indexed LOAD at 432 in the subroutine fetches the word 10 words down in the stack (1 for the index register and 9 for the known displacement of the return address). In this case the return address is 322. This address is placed in the index register by the LOAD. The following two *indexed* LOADs bring the arguments into the A and B registers. The branch instruction at 322 exists to hop over the arguments when control returns to the calling program.

Note that XR is first loaded with the stack pointer and used to calculate the effective address of the place where the return address is stored. Then it is loaded with the address pushed onto the stack and used to calculate the effective address of the arguments. This indirect addressing sequence of loading the index register with a value used to calculate the effective address for another load of the index register is a common technique with processors having few index registers.

Although the stack pointer often can be manipulated like any other register, it may not be safely used if interrupts can occur, in which case, CPU status can be written at locations to which the pointer points.

Other Concerns

There are never enough registers in any processor. This is especially true for microprocessors. Some techniques for dealing with this have been observed already (e.g., saving a register's contents in memory while another value is loaded into the register). Another useful technique is "unwinding" a loop. This means repeating the code sequence for an operation the desired number of times rather than using a register to contain a count, then decrementing and conditionally branching after each time the sequence is executed. The code sequence of LOAD and ADDs in Section 9.3.1 is such an unwound loop. It requires no register for indexing or counting. Unwinding is clearly impractical when the sequence is repeated many times. This approach also represents a trade-off in program design: the greater speed of the unwound loop against the larger amount of storage required for the program.

The limited instruction sets of microprocessors means that operations done by hardware in some mainframes must be done in software in microprocessors. Obvious examples are instructions for floating-point arithmetic, variable-length decimal arithmetic, and fixed-point multiplication and division. The hardware algorithms described in Section 9.5 can be translated into software subroutines to perform these same operations. Other limitations are apt to prove irksome. For example, in a processor where addresses are double-word size, there may be no instruction to add a single-word number to a double-length index register. If one desires to add 7 to the index register, one may do this:

Location	Instruction or Data			Comments
13	STORE		XR, 38	Assume that this stores a double-word
14	LOAD		A, 39	Low-order half of
15	ADD	IMM	A, 7	XR contents
16	STORE		A, 39	
17	CLEARA			
18	ADD	CAR	A, 38	High-order half
19	STORE		A, 38	
20	LOAD		XR, 38	
⋮				
38	—			Hold contents of XR
39	—			and later 7 + C(XR)

The STORE at 13 places the double-word index register contents in storage at 38 and 39, so that they can be loaded into another register for addition. The ADD at 15 adds 7 to the low-order half of the index. The ADD CAR at 18 adds 0 (in register A) and the high-order half of the index and the carry from the addition of the low-order half. This is then put back to storage, from which the index register is loaded at instruction 20. Note that fewer instructions are required to increment the index register 7 times.

One major concern of the microprocessor programmer is that of handling interrupts (see Section 9.2). Because of the variety of architectures, it is impossible to generalize on useful programming techniques. In some systems, integral hardware exists to direct control to an interrupt handler specifically for the condition causing the interrupt. In other systems, the designer must either provide such hardware or the program must determine the condition and transfer control accordingly. Programming for Input/Output is often simpler in microprocessor systems (see Section 9.1) since devices can be addressed like memory locations. Nonetheless, "handshaking" protocols do exist and vary from system to system.

A final mention should be made of the occasional need to assemble a microprocessor program by hand. The need may arise due to the unavailability of an assembler or because it is faster to assemble a small program by hand. This entails assigning memory locations to all instructions and data words.

Then any symbols used for addresses must be converted to numerical addresses, and the binary codes for the instructions must be looked up.

9.4 SWITCHING THEORY AND LOGIC DESIGN
Sheldon S. L. Chang

This section is confined to design of logical (or digital) circuits that recognize only two voltage levels, $v < V_l$ and $v > V_h$, where $V_l < V_h$. We may denote either level as 0 and the other as 1. The finite difference $V_h - V_l$ is called the **noise margin**, as illustrated in Fig. 9.17. A logic circuit may have a multiple number of inputs and outputs and satisfies the requirement that if every input voltage stays in the 0 or 1 region, all its output voltages also stay in the 0 or 1 region, but not in between. Thus a logic circuit can be described by a special branch of mathematics that deals only with two numbers, **Boolean algebra**. Conversely, logic circuits are realizations of Boolean algebra.

There are two basic types of logic circuits. A **combinational circuit** has no internal memory. Its outputs z_i, $i = 1, 2, \ldots, m$, are functions of its inputs x_j, $j = 1, 2, \ldots, n$:

$$z_i = F_i(x_1, x_2, \ldots, x_n) \qquad i = 1, 2, \ldots, m \tag{9.1}$$

Equation (9.1) can be written in vector form as

$$\mathbf{z} = \mathbf{F}(\mathbf{x}) \tag{9.2}$$

where $\mathbf{z}$, $\mathbf{F}$, and $\mathbf{x}$ are vectors, and (9.2) is completely equivalent to (9.1).

A **sequential circuit** has memories in the form of internal states described by variables y_i, $i = 1, 2, \ldots, k$. The variables y_i change to Y_i with each application of inputs. Both the output variables $\mathbf{z}$ and new state $\mathbf{Y}$ are functions of the inputs $\mathbf{x}$ and present state $\mathbf{y}$:

$$\mathbf{z} = \mathbf{F}(\mathbf{x}, \mathbf{y}) \tag{9.3}$$
$$\mathbf{Y} = \mathbf{G}(\mathbf{x}, \mathbf{y}) \tag{9.4}$$

A sequential circuit as described here is called a **Mealy circuit**. If the outputs $\mathbf{z}$ depend only on the new state $\mathbf{Y}$, it is called a **Moore circuit**:

$$\mathbf{z} = \mathbf{f}(\mathbf{Y}) \tag{9.5}$$

Obviously, Eq. (9.5) is a special form of (9.3).

The property of two distinct voltage levels separated by a sufficiently wide noise margin is of fundamental importance; it makes logic circuits extremely simple and reliable so that they can be made into LSIs and VLSIs and serve as building blocks of a large computer complex.

9.4.1 Boolean Algebra

Definition
A Boolean variable can take on two values: 0 and 1. The **or** operator " $+$ " is defined by

$$0 + 0 = 0 \tag{9.6}$$
$$1 + 0 = 0 + 1 = 1 + 1 = 1 \tag{9.7}$$

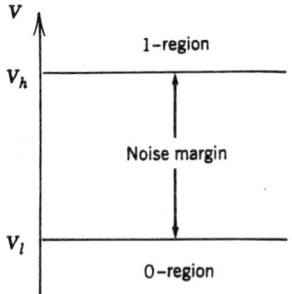

Fig. 9.17 Acceptable voltages.

The **and** operator "·" is defined by

$$1 \cdot 1 = 1 \tag{9.8}$$

$$1 \cdot 0 = 0 \cdot 1 = 0 \cdot 0 = 0 \tag{9.9}$$

The **not** operator "−" is defined by

$$\bar{0} = 1 \qquad \bar{1} = 0 \tag{9.10}$$

Equations (9.6)–(9.10) define Boolean algebra.

The operators + and · are also called **sum** and **product**, respectively.

When we work with symbols (or variables), it is customary to write AB instead of $A \cdot B$:

$$AB \triangleq A \cdot B$$

Duality. Duality is inherent in the definition of Boolean algebra. If we interchange the operators + and · and the values 0 and 1 simultaneously, all five defining equations remain valid. This property is called the **duality principle**.

The duality principle is useful in many ways. It helps us to verify and to remember the Boolean identities. Since an identity is valid no matter what values we assign to the Boolean variables, it follows that the dual of an identity is an identity.

Another use of the duality principle is in realizing the Boolean function by logic circuits. It provides insight on possible alternatives.

Boolean Expression and the Truth Table

A Boolean function is defined by its truth table or an expression of Boolean variables with operators · and + as connectives. The truth table is an exhaustive listing of all possible combinations of values of the variables and the resulting value of the expression. For instance,

$$F = AC + \bar{B}C \tag{9.11}$$

is a Boolean expression. As is customary in algebraic expressions, *the product takes precedence over the sum*, and (9.11) means

$$F = (AC) + (\bar{B}C) \tag{9.12}$$

The truth table of (9.12) is easily calculated and given in Table 9.1. *Two Boolean expressions are equal if their truth tables are the same.*

Basic Laws and Identities

In the following, the basic laws and identities are arranged in dual pairs. They can be easily verified by showing that the expressions on two sides of the equality sign have the same truth table.

1. Associative

$$(A + B) + C = A + (B + C) \tag{9.13}$$

$$(AB)C = A(BC) \tag{9.14}$$

2. Commutative

$$A + B = B + A \tag{9.15}$$

$$AB = BA \tag{9.16}$$

TABLE 9.1   Truth Table for Eq. (9.12)

A B C	F
0 0 0	0
0 0 1	1
0 1 0	0
0 1 1	0
1 0 0	0
1 0 1	1
1 1 0	0
1 1 1	1

3. Distributive

$$A(B + C) = AB + AC \tag{9.17}$$
$$A + BC = (A + B)(A + C) \tag{9.18}$$

4. DeMorgan's law

$$\overline{A + B} = \overline{A} \cdot \overline{B} \tag{9.19}$$
$$\overline{A \cdot B} = \overline{A} + \overline{B} \tag{9.20}$$

5. Single-variable identities

$$1 \cdot A = A \tag{9.21}$$
$$0 + A = A \tag{9.22}$$
$$1 + A = 1 \tag{9.23}$$
$$0 \cdot A = 0 \tag{9.24}$$
$$A + A = A \tag{9.25}$$
$$AA = A \tag{9.26}$$
$$A + \overline{A} = 1 \tag{9.27}$$
$$A\overline{A} = 0 \tag{9.28}$$
$$\overline{(\overline{A})} = A \tag{9.29}$$

6. Two-variable identities

$$A + AB = A \tag{9.30}$$
$$A(A + B) = A \tag{9.31}$$
$$A + \overline{A}B = A + B \tag{9.32}$$
$$A(\overline{A} + B) = AB \tag{9.33}$$

The two-variable identities can be easily proved from basic laws and single-variable identities:

$$
\begin{aligned}
A + AB &= A(1 + B) && \text{[from (9.17)]} \\
&= A \cdot 1 = A && \text{[from (9.23), (9.21)]} \\
A + \overline{A}B &= (A + AB) + \overline{A}B && \text{[from (9.30)]} \\
&= A + (A + \overline{A})B && \text{[from (9.13), (9.17)]} \\
&= A + B && \text{[from (9.27)]}
\end{aligned}
$$

An application of these identities is illustrated in the following example.

Example 9.1. Simplify the Boolean function F, which is given as

$$F = A(B + B\overline{C}) + AC + \overline{C}B + (\overline{C + A\overline{B}}) \tag{9.34}$$

SOLUTION. From Eq. (9.30),

$$A(B + B\overline{C}) = AB \tag{9.35}$$

From Eqs. (9.19), (9.20), (9.29), (9.17), and (9.16)

$$\overline{C + A\overline{B}} = \overline{C}(\overline{A} + B) = \overline{A}\overline{C} + B\overline{C} \tag{9.36}$$

Substituting Eqs. (9.35) and (9.36) into (9.34) gives

$$
\begin{aligned}
F &= AB + AC + \overline{C}B + \overline{A}\overline{C} + B\overline{C} \\
&= AB + AC + \overline{C}(\overline{A} + \overline{B} + B) \\
&= AB + AC + \overline{C} = AB + A + \overline{C} \\
&= A + \overline{C} \tag{9.37}
\end{aligned}
$$

9.4.2 Realization of Boolean Functions

A Boolean function can be realized in two ways:

1. A network of standard *gate circuits* or gates
2. A *programmable logic array* (PLA) or a *read-only memory* (ROM)

Realization by Gates

Each gate circuit represents one logic operation with a multiple number of inputs and one output variable. The output variable may be in its direct form, its negation, or both. Fig. 9.18 shows the names of the gates, their representative symbols, and the Boolean operations they perform.

The gates can be interconnected to represent literally a Boolean expression. For instance, the expression (9.34) is realized by the logic circuit of Fig. 9.19 in which the intermediate output variables are not marked, as they are easily readable from the circuit itself. In contrast, the expression of (9.37) is realized in Fig. 9.20. Although the circuit of Fig. 9.20 is considerably simpler, it performs the same function as that of Fig. 9.19.

An important property of the gates is that of *sufficiency*. A gate type is said to be sufficient if any arbitrary Boolean function can be realized by using entirely gates of the same type. It can be shown that any function can be realized with only *and*, *or*, and *not* operators. Then, the sufficiency of NAND and NOR gates can be established as follows.

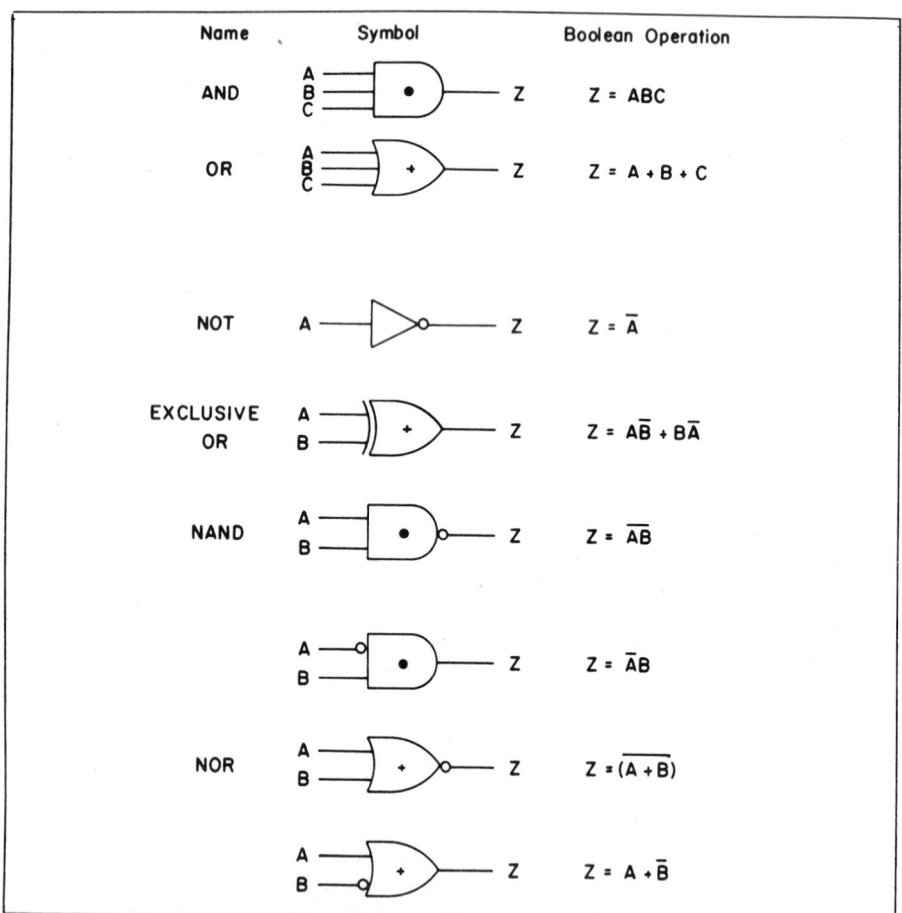

Fig. 9.18 Names and symbols of commonly used gates.

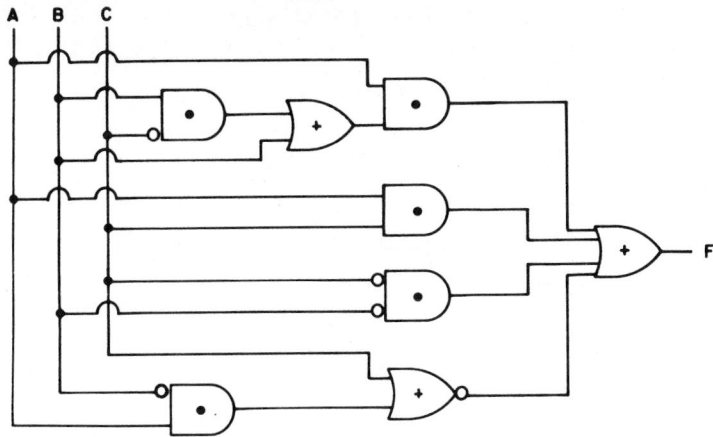

Fig. 9.19   Logic circuit.

Fig. 9.20   Logic circuit equivalent to Figure 9.19.

NAND Gate

$$\bar{A} = \overline{(A \cdot A)} \tag{9.38}$$

$$A + B = \overline{(\overline{A \cdot A}) \cdot (\overline{B \cdot B})} \tag{9.39}$$

$$AB = \overline{(\overline{AB})(\overline{AB})} \tag{9.40}$$

Equations (9.38)–(9.40) are realized in Fig. 9.21(a). Since every Boolean function is expressed in terms of "and," "or," and "not" operations, realization of the trio proves the sufficiency of NAND. Similarly, Fig. 9.21(b) illustrates realization of the trio by NOR.

Integrated-Circuit Logic Families
In integrated circuits it is convenient to interconnect one or two basic circuit types to realize a Boolean function. We refer to each type of basic circuit as a **logic family**.
   The following are important characteristics of a logic family:

1. *Fan-In and Fan-Out*. Fan-in is the number of inputs a gate can have. Fan-out is the maximum number of input terminals of the same type that its output can feed without being overloaded. High fan-ins and fan-outs offer greater flexibility of interconnection.
2. *Watts Loss per Gate*. For any given logic family, the power requirement per gate increases with the speed of response, which can be measured in terms of the gate's delay time.
3. *Delay Time*. The time interval between the arrival of the input signal to the time when the output reaches its new threshold level.

The details of various families of logic circuits are beyond the scope of this book.

9.4.3   Minimization of Boolean Functions

The complete equivalence of the two circuits shown as Figs. 9.19 and 9.20 illustrates the value of minimization. While the criterion of minimum number of gates and then minimum number of gate inputs does not necessarily correspond to minimum cost, it does weed out grossly wasteful designs. However, minimization with Boolean identities is a hit-and-miss method. The Karnaugh map, a systematic method useful for relatively small circuits, is discussed in the present section.

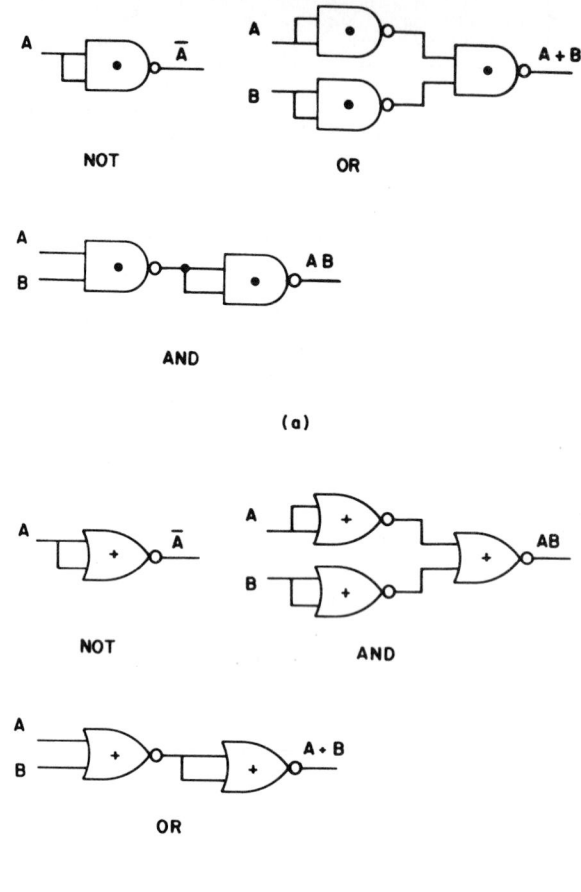

Fig. 9.21 Sufficiency of NAND and NOR: (*a*) NAND gates, (*b*) NOR gates.

Minterm and Maxterm: Standard Form of Boolean Functions

A Boolean function can be specified by either an algebraic expression or a truth table. In Section 9.4.1 the derivation of a truth table from a Boolean expression is shown. Conversely, a truth table can be represented by many equivalent Boolean expressions. There are two standard representations:

Sum of Product (S-of-P). Each "1" in the truth table output column is represented by a product of the input variables that defines its position. For instance, there are three 1's in Table 9.1, occurring at 001, 101, and 111. They are represented as $\overline{A}\,\overline{B}C$, $A\overline{B}C$, and ABC, respectively. The term $\overline{A}\,\overline{B}C$ gives a 1 at 001 and 0 elsewhere. Similarly, $A\overline{B}C$ and ABC give 1's at their respective positions. The function F can be represented as

$$F = \overline{A}\,\overline{B}C + A\overline{B}C + ABC \qquad (9.41)$$

Some standard definitions will now be introduced:

A **literal** is an input variable either in its direct or negated form: for example, A, $\overline{A}$, B, $\overline{B}$, C, and, $\overline{C}$ are all literals.

A **normal product term** is a product of literals that contains each input variable no more than once: for example, AB and $AB\overline{C}$ are normal products terms, but AAB is not. From Eqs.

(9.26) and (9.28) every product term can be reduced to a normal product term or 0:

$$ABA\overline{C} = AAB\overline{C} = AB\overline{C}$$
$$AB\overline{A}\,\overline{C} = A\overline{A}B\overline{C} = 0$$

A **minterm** is a product of literals with every input variable represented.
A **maxterm** is a sum of literals with every input variable represented.

It follows that the truth table of a minterm has only one 1 output and the rest are 0's, and the truth table of a maxterm has only one 0 and the rest are 1's. It is convenient to denote a minterm by $m(n)$, where n is the binary position of its 1. For instance, in a function of four variables, A, B, C, and D,

$$m(7) = \overline{A}BCD$$

The corresponding maxterm is

$$M(7) = \overline{m(7)} = A + \overline{B} + \overline{C} + \overline{D}$$

Equation (9.41) can be written as

$$F(A, B, C) = m(1) + m(5) + m(7) \qquad (9.42a)$$

Product of Sum (P-of-S). Each 0 in the truth table is represented by a maxterm, and the function is represented as a product of maxterms. For instance, the function of Table 9.1 can be represented as

$$F(A, B, C) = M(0)M(2)M(3)M(4)M(6) \qquad (9.43a)$$

Equations (9.42a) and (9.43a) are sometimes written in shorthand form as

$$F(A, B, C) = \sum m(1, 5, 7) \qquad (9.42b)$$

$$F(A, B, C) = \prod M(0, 2, 3, 4, 6) \qquad (9.43b)$$

The S-of-P and P-of-S Boolean expressions are called **second-order expressions** in the sense that from any input into the circuit to its output, there are no more than two gate levels. For a given technology, the delay time is proportional to the number of gate levels (between input and output), and a two-gate-level circuit is the fastest realization of a general Boolean function.

The terms in the S-of-P and P-of-S expressions are not necessarily minterms and maxterms. A term with $n - m$ literals, where n is the number of input variables and m is an integer less than n, is equivalent to 2^m minterms or maxterms.

Incompletely Specified Functions

In most applications, a logic circuit is part of a larger system, and some possible input combinations may never occur. Their corresponding outputs are then immaterial. Alternatively, the outputs under certain input combinations do not affect the larger system. In either case, we may regard the outputs for these input combinations as don't-cares. In an S-of-P expression, the don't-cares can be denoted by $d(n_1, n_2, \ldots)$. For instance, in

$$F(x, y, z, w) = \sum m(1, 3, 6, 8, 9, 12, 14) + d(4, 13) \qquad (9.44)$$

the term $d(4, 13)$ means that the values of F at $(0, 1, 0, 0)$ and $(1, 1, 0, 1)$ are not specified.

Minimization Criterion

An S-of-P (P-of-S) expression is *minimal* if (1) no other equivalent expression has fewer product (sum) terms, and (2) no other equivalent expression with the same number of terms has fewer literals.

The mathematical definition of "minimal," when interpreted in practical terms, means (1) a minimum number of gates and (2) a minimum total number of inputs to the gates among circuits that have a minimum number of gates.

Karnaugh Map

A **Karnaugh map** is a rearranged truth table in which the groupings of minterms into product terms with reduced numbers of literals are easily recognized. For instance, the incompletely specified

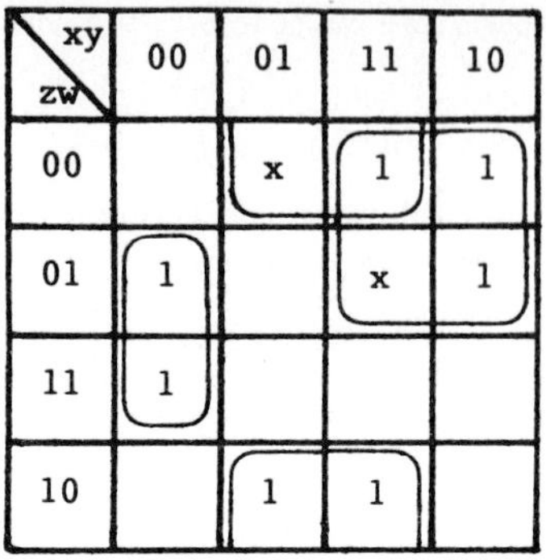

zw \ xy	00	01	11	10
00		x	1	1
01	1		x	1
11	1			
10		1	1	

Fig. 9.22 Karnaugh map.

function of Eq. (9.44) is represented in Fig. 9.22. The output values are rearranged in a 4 × 4 array
with x, y running horizontally and z, w running vertically. Both pairs follow the sequence

$$00, 01, 11, 10$$

so that between adjacent numbers, only one variable changes in value. The numbers at two ends, 00
and 10, are also adjacent.

The position of 1's and don't-cares "×" are marked in the 4 × 4 array, whereas the positions of
0's are left blank.

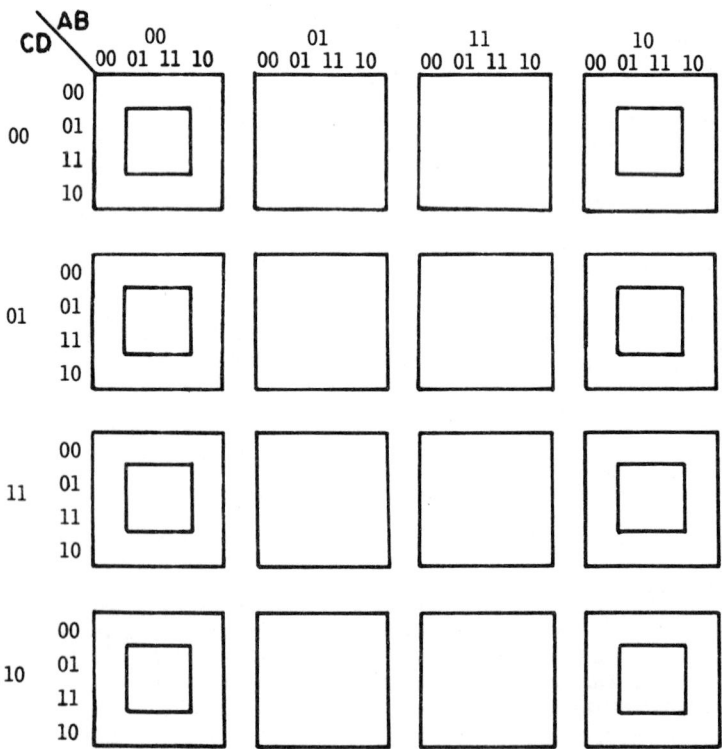

Fig. 9.23 The linked interior blocks represent five-cube $\overline{B}FH$. Each exterior block represents a
four-variable Karnaugh map for the variables EFGH.

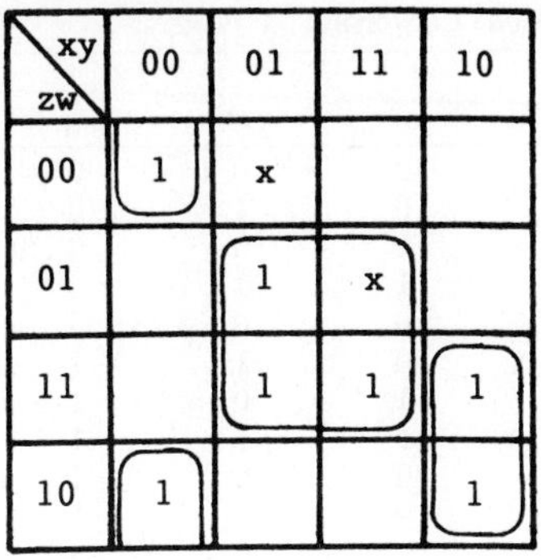

xy zw	00	01	11	10
00	1	x		
01		1	x	
11		1	1	1
10	1			1

Fig. 9.24   Inverse Karnaugh map.

Each unit cell, 1×2 block, 1×4 or 2×2 block, and so on, no matter where located, corresponds to a product term of $n, n - 1, n - 2, \ldots$ literals. In the literature these blocks are also called 0-cube, 1-cube, $\ldots, m$-cube, $\ldots$. For instance, the positions 4, 6, 12, and 14 form a 2-cube since the top and bottom rows are considered adjacent:

$$m(4,6,12,14) = y\bar{w}$$

Using the Karnaugh map, minimization is simplicity itself in principle. A set of blocks is to be found that covers all the 1's, possibly some $\times$'s, but none of the blank spaces. Among all possible ways, a minimal solution has the minimum number of largest blocks. For instance the Karnaugh map of Fig. 9.22 can be realized by either of the following:

$$
\begin{aligned}
F_1 &= m(1,3) + m(6,14) + m(8,9,12,13) \\
&= \bar{x}\bar{y}w + yz\bar{w} + x\bar{z} \\
F_2 &= m(1,3) + m(4,6,12,14) + m(8,9,12,13) \\
&= \bar{x}\bar{y}w + y\bar{w} + x\bar{z}
\end{aligned}
\tag{9.45}
$$

and F_2 is the minimal solution.

With six or more input variables, the minimal solution is not always easy to find. Figure 9.23 shows an eight-variable Karnaugh map and the location of a typical 5-cube: $\bar{B}FH$. There are 448 5-cubes alone, and they look different at different locations. The proficiency in recognizing high-order cubes in a Karnaugh map and to select a minimal solution can only improve with experience.

A minimal P-of-S expression for F can be found by first finding the minimal S-of-P expression of $\bar{F}$. For instance, the complement of the Karnaugh map of Fig. 9.22 is obtained by interchanging its blanks and 1's. The result is shown in Fig. 9.24. Its minimal S-of-P expression is

$$\bar{F}_3 = yw + \bar{x}\bar{y}\bar{w} + x\bar{y}z$$

Complementing this gives

$$F_3 = (\bar{y} + \bar{w})(x + y + w)(\bar{x} + y + \bar{z}) \tag{9.46}$$

9.4.4   Multiple-Output Logic Networks

Multiple-output combinational networks can be designed and realized with

1. Gates
2. Programmable logic array (PLA)
3. Read-only memory (ROM)

Example 9.2. A logic network is to be designed for converting a BCD coded word to an excess-3 coded word. The truth table is given in Table 9.2. The Karnaugh maps of F_1, F_2, F_3, and F_4 are

TABLE 9.2 BCD-to-Excess-3 Code Conversion

Decimal Digit	A	B	C	D	F_1	F_2	F_3	F_4
0	0	0	0	0	0	0	1	1
1	0	0	0	1	0	1	0	0
2	0	0	1	0	0	1	0	1
3	0	0	1	1	0	1	1	0
4	0	1	0	0	0	1	1	1
5	0	1	0	1	1	0	0	0
6	0	1	1	0	1	0	0	1
7	0	1	1	1	1	0	1	0
8	1	0	0	0	1	0	1	1
9	1	0	0	1	1	1	0	0

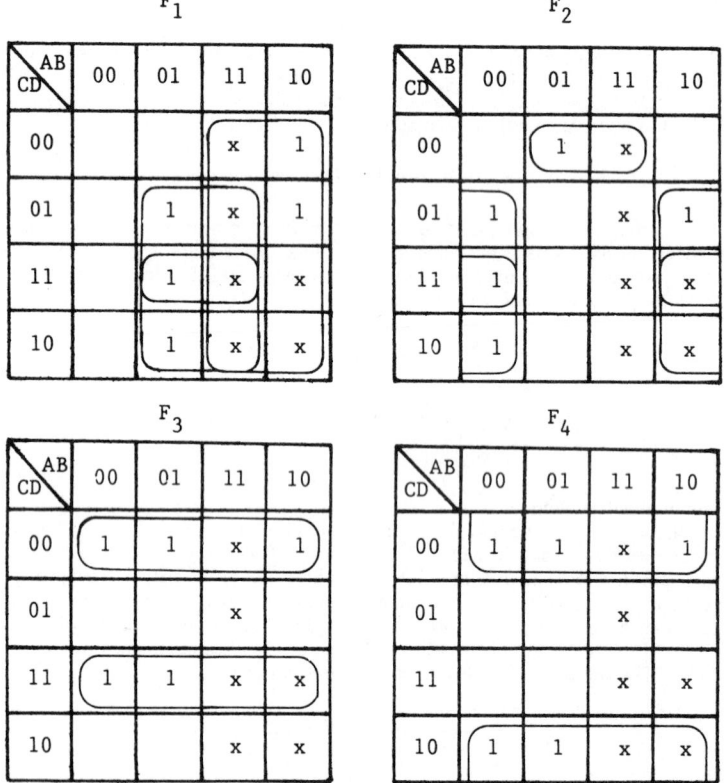

Fig. 9.25 Karnaugh maps for code converter.

illustrated in Fig. 9.25. The S-of-P expressions of the output functions are

$$F_1 = A + BD + BC$$
$$F_2 = B\overline{C}\,\overline{D} + \overline{B}C + \overline{B}D$$
$$F_3 = \overline{C}\,\overline{D} + CD$$
$$F_4 = \overline{D} \qquad\qquad\qquad (9.47)$$

Figure 9.26 shows the realization of Eq. (9.47) by NAND gates. Equation (9.47) can also be written in

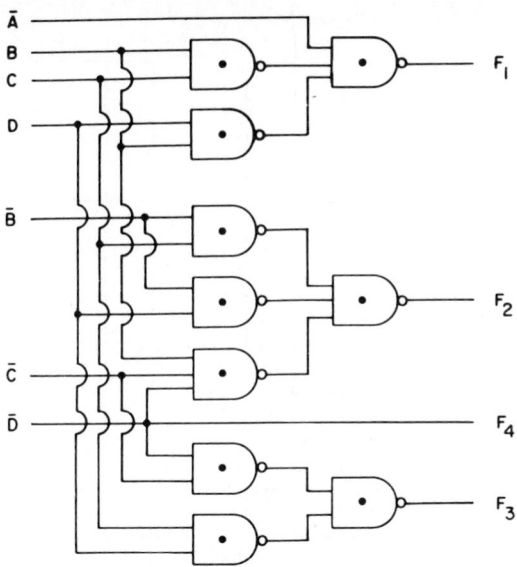

Fig. 9.26   Logic for code converter.

the form

$$F_1 = A + B(C + D)$$
$$F_2 = B(\overline{C + D}) + \overline{B}(C + D)$$
$$F_3 = (\overline{C + D}) + CD$$
$$F_4 = \overline{D} \tag{9.48}$$

Equation (9.48) can be realized with fewer gates than (9.47).

Programmable Logic Array
The programmable logic array (PLA) is a simple integrated circuit for realizing a multiple-input multiple-output logic network. Its circuit diagram is shown in Fig. 9.27(a). The input literals and output variables are represented by vertical lines as shown. Each product term is represented by a horizontal line or word line, which is connected to a selected number of input literals through diodes or switching elements. The second line is "high" if both B and C are "high." The product term represented by each word line is denoted at its right end. The word lines are in turn connected to the output lines through diodes or switching elements. As connected, the PLA realizes the code converter represented by Eq. (9.47). Figure 9.27(b) is a simpler way of representing the network of Fig. 9.27(a).

There are two types of PLAs. In a **mask programmable** PLA, the interconnections are permanently made at the time of manufacture. In a **field programmable** PLA (FPLA), all switching elements are present at the line intersections, but the connections are made by fusible links. To store data in the FPLA, the fuse links are selectively blown with voltage pulses supplied by special equipment called a FPLA programmer.

Example 9.3. A decoder with n input variables selects 1 out of 2^n lines. The truth table of a 3-bit decoder is given in Table 9.3.

Since each output function contains only one minterm, there is no need for minimization. Figure 9.28(a) shows realization of the decoder by AND gates and Fig. 9.28(b) shows realization by a decoding network that is in effect the first half of a PLA.

Read-Only Memory
A read-only memory (ROM) with n input lines and m output lines can store 2^n words of m bits each. With each input combination, one word is selected and appears at the output lines. A typical ROM circuit with $n = m = 4$ is shown in Fig. 9.29(a). Figure 9.29(b) shows the same circuit in a simpler

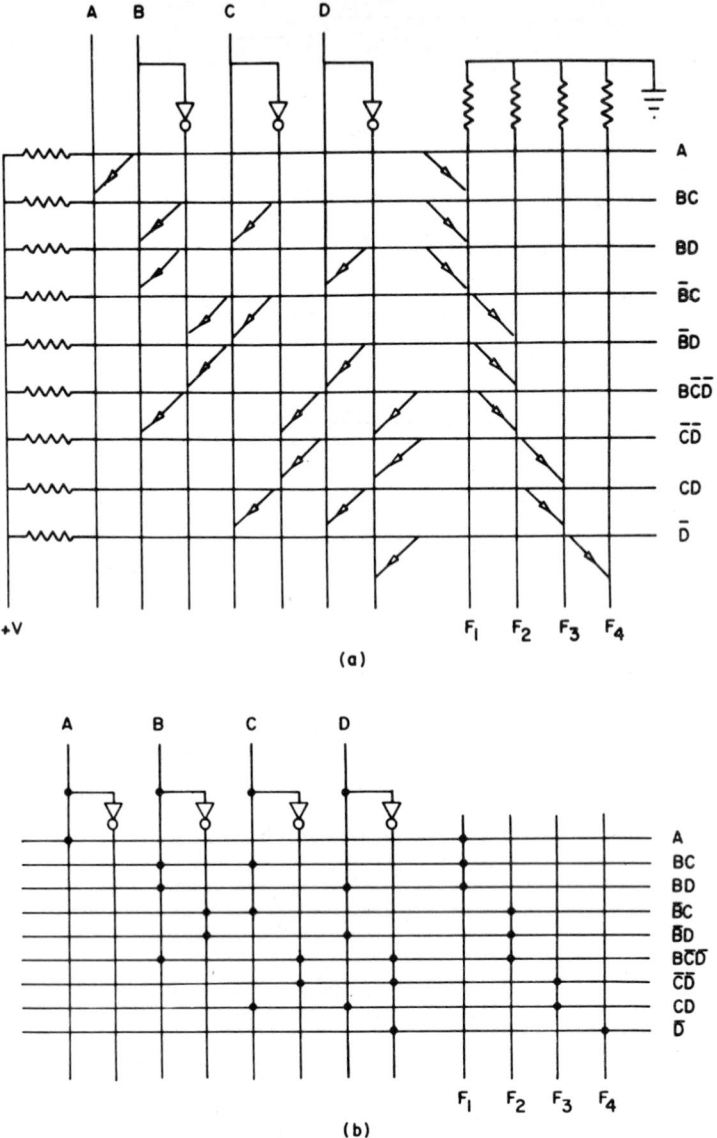

Fig. 9.27   PLA realization for code converter.

representation. According to the input binary address, a word line is selected by the decoder, and a 1 appears at each of the output lines connected to it. An output line and a word line are connected if there is a 1 at the corresponding intersection in the truth table. The ROM of Fig. 9.29 is a realization of the code converter of Table 9.2.

ROM is the most straightforward way of realizing a multiple-input multiple-output logic network, as its required storage is an exact copy of the truth table. However, it usually takes more electronics than the PLA because of the decoding operation and the greater number of word lines and connected intersections.

Three basic types of ROMs are available: (1) **mask programmable**, (2) **field programmable** (PROM), and (3) **erasable programmable** (EPROM). Types 1 and 2 are similar to their counterparts in PLA in construction. Instead of fusible links, a charge storage array is provided in EPROMs to enable or disable the switching elements. An EPROM is programmed with special equipment that

TABLE 9.3 3-Bit Decoder

A	B	C		F_0	F_1	F_2	F_3	F_4	F_5	F_6	F_7
0	0	0		1	0	0	0	0	0	0	0
0	0	1		0	1	0	0	0	0	0	0
0	1	0		0	0	1	0	0	0	0	0
0	1	1		0	0	0	1	0	0	0	0
1	0	0		0	0	0	0	1	0	0	0
1	0	1		0	0	0	0	0	1	0	0
1	1	0		0	0	0	0	0	0	1	0
1	1	1		0	0	0	0	0	0	0	1

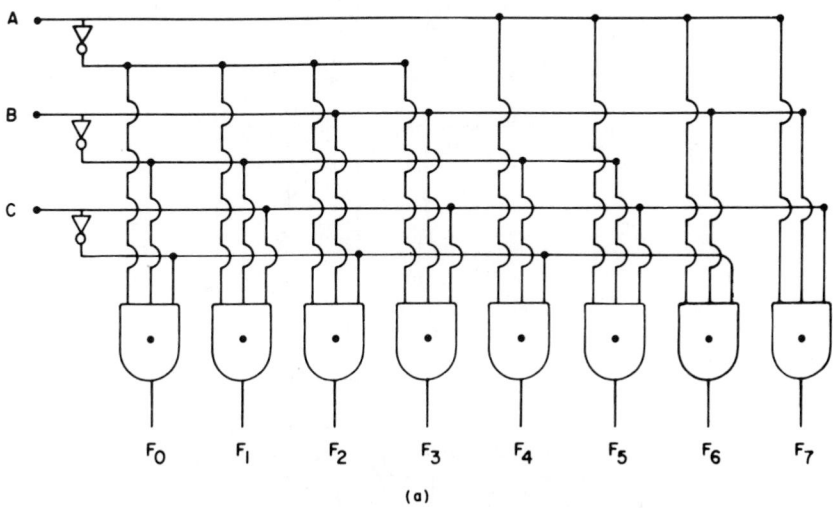

(a)

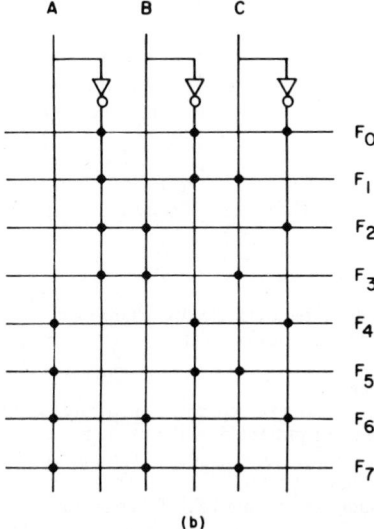

(b)

Fig. 9.28 (*a*) AND, NOT gates, (*b*) PLA decoder realization.

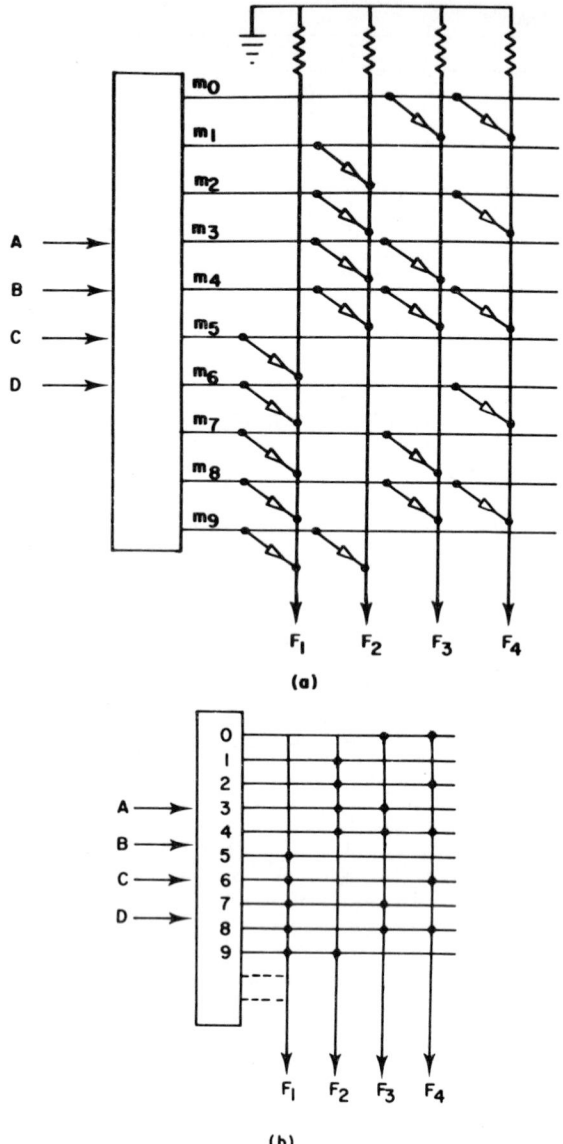

Fig. 9.29 ROM realization of BCD-to-excess-3 converter: (*a*) full representation, (*b*) simple representation.

applies appropriate voltage pulses to store electric charges in the array. The stored data are permanent until they are erased with ultraviolet light.

Full Adder and Parallel Adder

In this section we use adders of increasing complexity to illustrate the three methods of realizing logic networks. A 1-bit full adder can be represented by the truth table of Table 9.4, where A_i and B_i are the ith binary digits of the two numbers to be added, C_i is the carry bit from the $(i - 1)$th digit, S_i is the ith sum bit, and C_{i+1} is the new carry bit.

Figure 9.30 shows the Karnaugh maps from Table 9.4. The S-of-P expressions are

$$S_i = \bar{A}_i \bar{B}_i C_i + \bar{A}_i B_i \bar{C}_i + A_i \bar{B}_i \bar{C}_i + A_i B_i C_i \tag{9.49}$$

$$C_{i+1} = A_i B_i + A_i C_i + B_i C_i \tag{9.50}$$

TABLE 9.4   1-Bit Full Adder

	Input			Output	
A_i	B_i	C_i	S_i		C_{i+1}
0	0	0	0		0
0	0	1	1		0
0	1	0	1		0
0	1	1	0		1
1	0	0	1		0
1	0	1	0		1
1	1	0	0		1
1	1	1	1		1

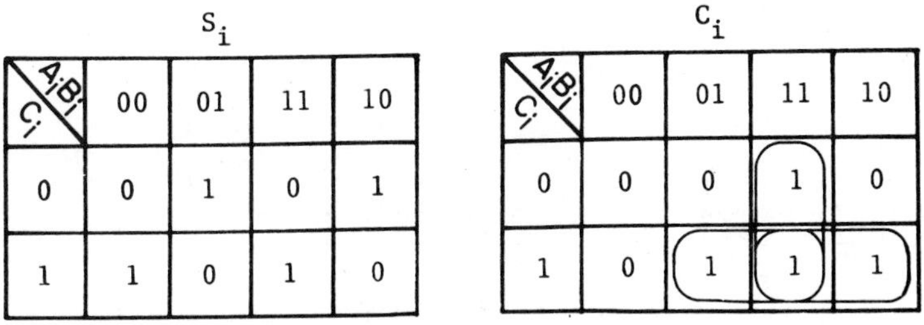

S_i

$\frac{A_iB_i}{C_i}$	00	01	11	10
0	0	1	0	1
1	1	0	1	0

C_i

$\frac{A_iB_i}{C_i}$	00	01	11	10
0	0	0	1	0
1	0	1	1	1

Fig. 9.30   Karnaugh maps of full adder.

Figure 9.31 shows three ways of realizing a full adder: (1) ROM realization of Table 9.4, (2) NAND realization of Eqs. (9.49) and (9.50), and (3) PLA realization of Eqs. (9.49) and (9.50).

To double the speed of addition, a 2-bit parallel adder can be used. The two binary numbers and the carry bit from previous addition are represented by A_1A_0, B_1B_0, and C_0, respectively. The sum and new carry bit are represented by S_1S_0 and C_2. See the discussion of carry lookahead adders in Section 9.5.3. For this problem, it is convenient to start with the Karnaugh maps shown in Fig. 9.32. The PLA realization of Fig. 9.32 and the ROM realization of its corresponding truth table are illustrated in Figs. 9.33 and 9.34, respectively. The hardware savings with PLA illustrates a general trend: The advantage of PLA over ROM widens as network complexity increases with more and more input and output variables.

9.4.5   Flip-Flops

Flip-flops are standard circuits for storing state information in a sequential circuit. Each flip-flop stores one binary variable. With N flip-flops, the stored binary number ranges from 0 to $2^N - 1$. Each stored binary number represents one state of the sequential circuit. Therefore, a maximum of 2^N states can exist in a circuit with N flip-flops.

S-R Flip-Flop
Figure 9.35 shows an S-R flip-flop with two NOR gates. There are two inputs to the circuit, S and R, and an output variable, Q. The truth table is shown in Table 9.5. With $S = R = 0$, Q can be either 0 or 1, and the input/output relation at each NOR gate is satisfied. The value of Q then represents the stored information. With S, R equal to 1,0 (or 0,1), only $Q = 1$ (or 0) is possible. The input condition $S = R = 1$ is to be avoided for the following reasons: (1) Both outputs, Q and $\bar{Q}$, are 0 and so represent an undefined state, and (2) if both S and R change to zero simultaneously, the outcome of Q is uncertain and may oscillate between 0 and 1.

A symbol representing an S-R flip-flop is shown in Fig. 9.35(b). We have marked the output terminals as Q and $\bar{Q}$ for clarity. Sometimes the output terminals Q and $\bar{Q}$ are not explicitly marked. Then the output terminal nearer S is meant to be Q. The input terminals are always marked to distinguish the type of flip-flop.

J-K Flip-Flop; D and T Flip-Flop
A J-K flip-flop is pulse operated as shown in Fig. 9.36. Its J, K inputs have no effect until the arrival of the clock pulse. There are delays in the AND circuits and in the S-R flip-flop, and these delays

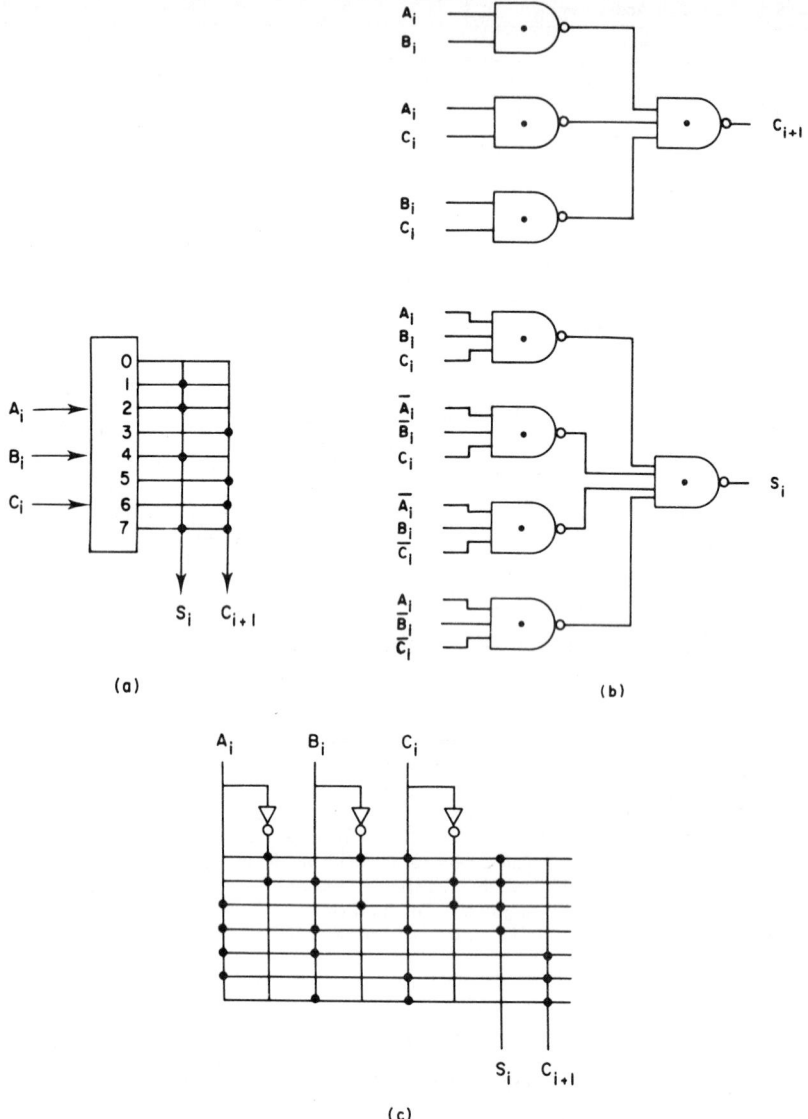

(a)   (b)

(c)

Fig. 9.31 Three ways of realizing full adder: (a) ROM, (b) basic two-level realization, (c) PLA.

play an indispensable part in the operation of a J-K flip-flop. Figure 9.37 shows timing diagrams of a J-K flip-flop. In Fig. 9.37(a), due to internal delays, the output variables Q and $\overline{Q}$ do not change until after the clock pulse. The flop-flop operates properly with its new state Q^+ given as

$$Q^+ = J\overline{Q} + \overline{K}Q \qquad\qquad (9.51)$$

where Q and Q^+ represent the output Q before and after a clock pulse, respectively. In Fig. 9.37(b), the internal delay is insufficient and the new value Q^+ is returned before the clock pulse is over. If the inputs are $J = 1$ and $K = 1$, the flip-flop can trigger another round of changes and yield erroneous results. Figure 9.38 shows a J-K flip-flop made entirely from NAND circuits. Its principle of operation is identical with that of Fig. 9.36.

There are two variations of the J-K flip-flop, as shown in Fig. 9.39. In Fig. 9.39(a) the J, K inputs are connected permanently to 1, and Q changes with each input pulse. It is called a T flip-flop.

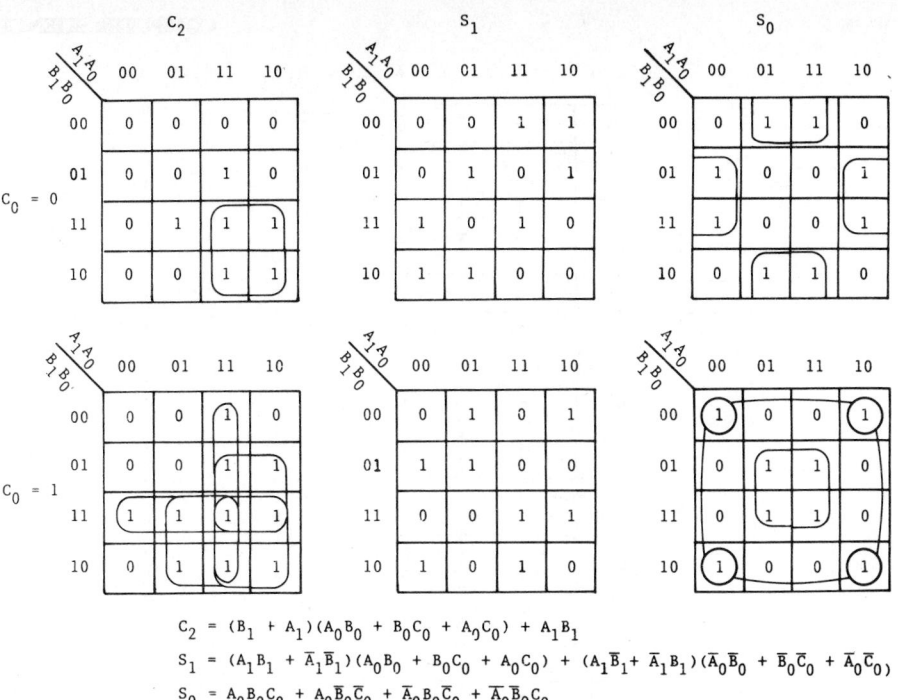

$$C_2 = (B_1 + A_1)(A_0 B_0 + B_0 C_0 + A_0 C_0) + A_1 B_1$$

$$S_1 = (A_1 B_1 + \overline{A}_1 \overline{B}_1)(A_0 B_0 + B_0 C_0 + A_0 C_0) + (A_1 \overline{B}_1 + \overline{A}_1 B_1)(\overline{A}_0 \overline{B}_0 + \overline{B}_0 \overline{C}_0 + \overline{A}_0 \overline{C}_0)$$

$$S_0 = A_0 B_0 C_0 + A_0 \overline{B}_0 \overline{C}_0 + \overline{A}_0 B_0 \overline{C}_0 + \overline{A}_0 \overline{B}_0 C_0$$

Fig. 9.32 Karnaugh maps of 2-bit adder.

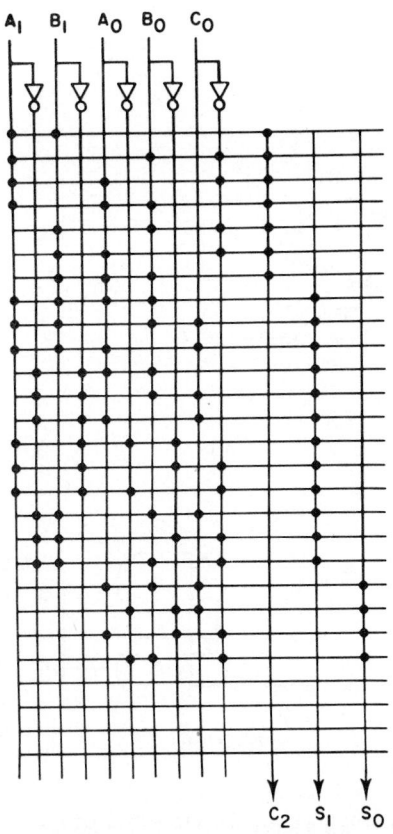

Fig. 9.33 PLA realization of a 2-bit adder.

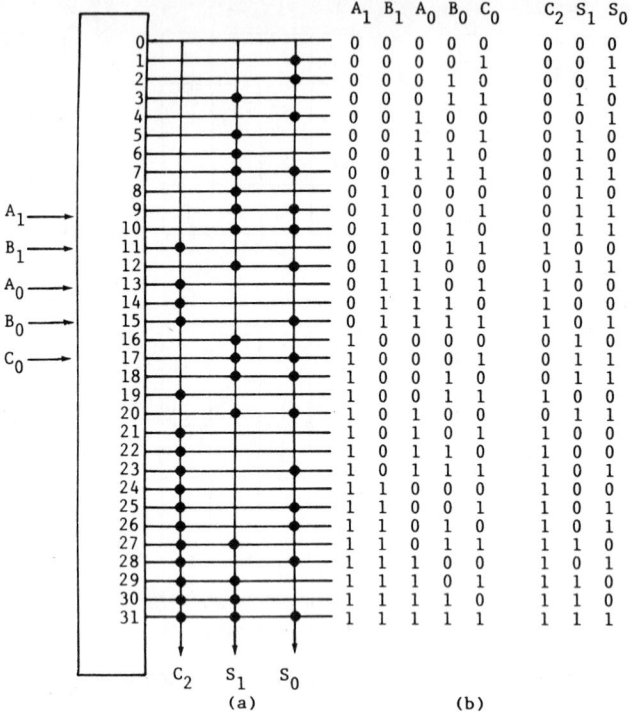

	A_1	B_1	A_0	B_0	C_0	C_2	S_1	S_0
0	0	0	0	0	0	0	0	0
1	0	0	0	0	1	0	0	1
2	0	0	0	1	0	0	0	1
3	0	0	0	1	1	0	1	0
4	0	0	1	0	0	0	0	1
5	0	0	1	0	1	0	1	0
6	0	0	1	1	0	0	1	0
7	0	0	1	1	1	0	1	1
8	0	1	0	0	0	0	1	0
9	0	1	0	0	1	0	1	1
10	0	1	0	1	0	0	1	1
11	0	1	0	1	1	1	0	0
12	0	1	1	0	0	0	1	1
13	0	1	1	0	1	1	0	0
14	0	1	1	1	0	1	0	0
15	0	1	1	1	1	1	0	1
16	1	0	0	0	0	0	1	0
17	1	0	0	0	1	0	1	1
18	1	0	0	1	0	0	1	1
19	1	0	0	1	1	1	0	0
20	1	0	1	0	0	0	1	1
21	1	0	1	0	1	1	0	0
22	1	0	1	1	0	1	0	0
23	1	0	1	1	1	1	0	1
24	1	1	0	0	0	1	0	0
25	1	1	0	0	1	1	0	1
26	1	1	0	1	0	1	0	1
27	1	1	0	1	1	1	1	0
28	1	1	1	0	0	1	0	1
29	1	1	1	0	1	1	1	0
30	1	1	1	1	0	1	1	0
31	1	1	1	1	1	1	1	1

C_2 S_1 S_0

(a)   (b)

Fig. 9.34 ROM realization of a 2-bit adder: (*a*) ROM, (*b*) truth table.

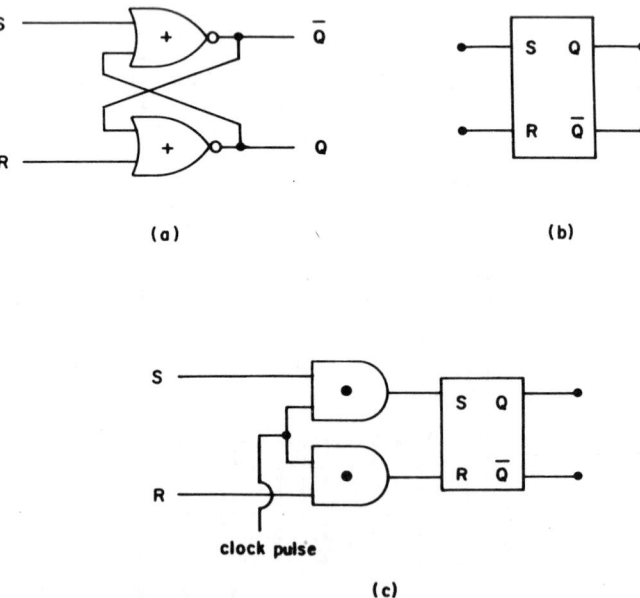

(a)   (b)

(c)

Fig. 9.35 *S-R* flip-flop: (*a*) circuit, (*b*) symbol, (*c*) clocked.

TABLE 9.5   Truth Table of *S-R* Flip-Flop

S	R	Q
0	0	0, 1
0	1	0
1	0	1
1	1	—

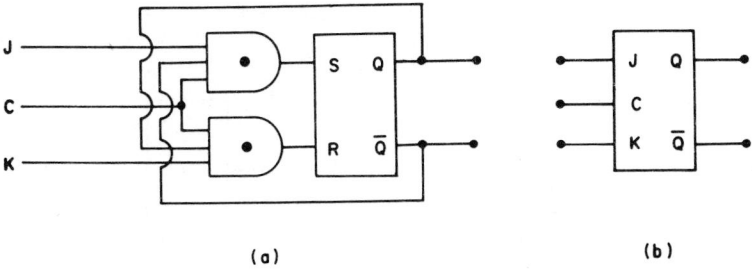

(a)   (b)

Fig. 9.36   Clocked *J-K* flip-flop: (*a*) circuit, (*b*) symbol.

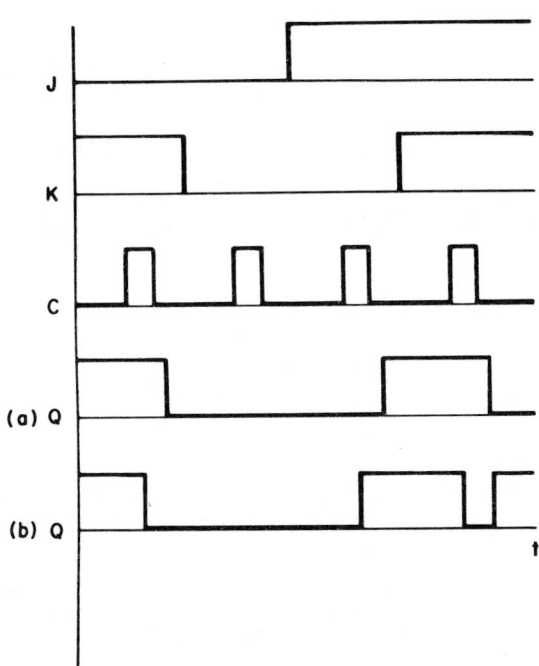

Fig. 9.37   Clocked *J-K* flip-flop timing diagram: (*a*) proper operation, (*b*) faulty operation caused by insufficient internal delay.

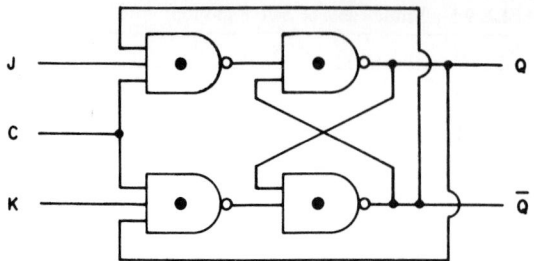

Fig. 9.38   All-NAND J-K flip-flop.

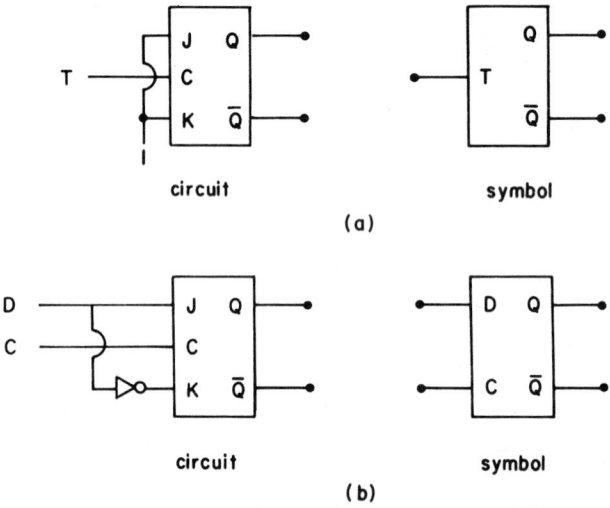

Fig. 9.39 D and T flip-flops: (a) T flip-flop, (b) D flip-flop.

In Fig. 9.39(b), $J = D$ and $K = \bar{D}$. Equation (9.51) gives

$$Q^+ = D\bar{Q} + DQ = D \qquad\qquad (9.52)$$

The next state is the same as the input D. It is called the D flip-flop.

Master-Slave Flip-Flops
Figure 9.40 illustrates the master-slave flip-flop concept, which ensures an adequate time delay between input and output signals. The first flip-flop (FF$_1$) changes state during the clock pulse.

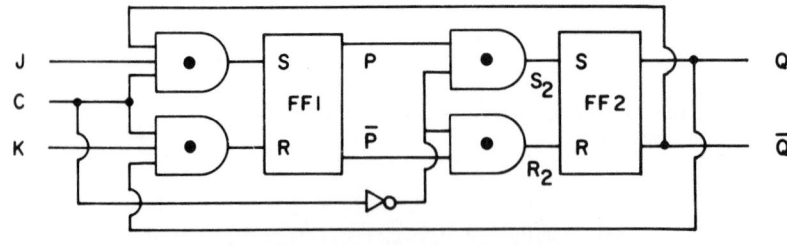

Fig. 9.40   Master-slave flip-flop.

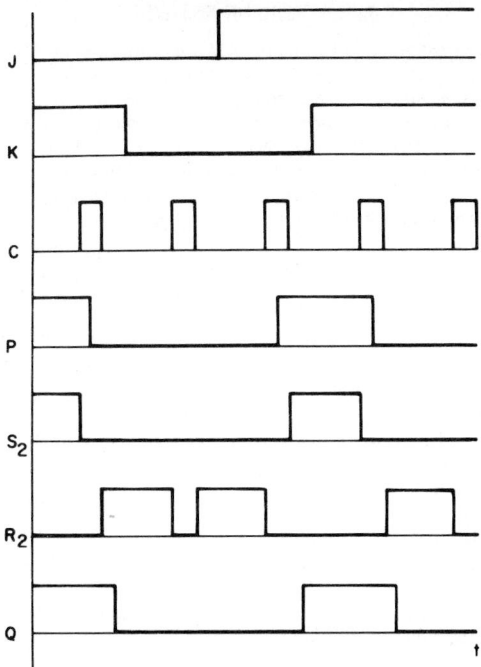

Fig. 9.41 Master-slave flip-flop timing diagram. Note that S_2 and R_2 are zero during the timing pulse.

However, $\bar{C} = 0$, and P and $\bar{P}$ are inhibited by the second set of AND circuits from reaching the SR inputs of FF_2. Only after the clock pulse is over and $\bar{C}$ becomes 1 again do the $P\bar{P}$ signals reach the SR inputs and switch FF_2 accordingly. The timing diagram is shown in Fig. 9.41.

Shift Register
One of the simplest but most useful sequential circuits is the shift register. It is made of a string of D flip-flops, as shown in Fig. 9.42. At each clock pulse, the stored data in each flip-flop advance to the next flip-flop. Mathematically, the shift register can be represented as

$$Z_k = \begin{cases} Z_{k-1} & k = 2, 3, \ldots, L \\ I & k = 1 \end{cases} \tag{9.53}$$

where I is the input data bit; Z_k, $k = 1, \ldots, L$, the output, and L the length of the register.

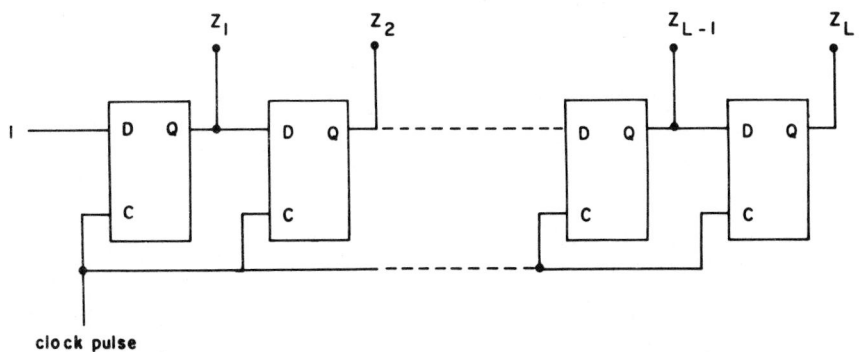

Fig. 9.42 Shift register.

9.4.6 State Minimization

The complexity of a sequential circuit usually increases with its number of states, and state minimization is an essential part of sequential circuit design.

Two sequential circuits A and B are equivalent if for each state in A there is an equivalent state in B, and vice versa, such that identical output sequences are generated if identical input sequences are applied to the two circuits. In other words, there is no way of distinguishing the two circuits from external tests. Minimization involves finding the one with a minimum number of states among all sequential circuits that are equivalent to the one at hand. Although many different ways can be used, the prevailing method starts with a given state table and then simplifies it by combining equivalent states.

Two states p and q are equivalent if their next states and outputs for any given input are the same.

$$z = F(x, p) = F(x, q) \tag{9.54}$$
$$Y = G(x, p) = G(x, q) \tag{9.55}$$

These conditions are sufficient for p and q to be equivalent. After one input, the system enters the same new state, and there is no way of telling from its output whether the system is initially in state p or in state q. We can discard q altogether and change any reference to q to a reference to p.

After all equivalent states are combined as described, there are two methods for further minimization or simplification:

1. The next-class tables
2. The implication chart

Both methods can be described as the step-by-step tightening of a necessary condition for equivalence until it becomes both necessary and sufficient.

Simplification by Combining Equivalent States
We shall start with a state table, which is one way of representing the defining equations (9.3) and (9.4) of a sequential circuit. Table 9.6 represents a typical sequence detector with eight states: a, b, c, d, e, f, g, and h. For any given present state, the next state and output values are listed in the same row. Each column corresponds to a value of the input variable x. For instance, if the present state and input are c and 1, respectively, we look for the third row under column $x = 1$. The next state is f, and the output is 0.

In Table 9.6, (a, e), (b, f), and (d, h) are equivalent pairs. Combining the equivalent pairs gives the data shown in Table 9.7 States a and c now have identical rows. Combining a and c, we obtain the results shown in Table 9.8, which represent a substantial simplification from Table 9.6. It will be shown later that Table 9.8 cannot be simplified further.

However, combining equivalent states does not always work. Other minimization techniques are described in the bibliography.

Incompletely Specified Sequential Circuits
In sequential circuits it is often true that a certain input does not occur at a given state; the output and the next state are then unspecified. It is also possible that outputs at a certain state and input conditions have no effect on the larger system; the outputs are then regarded as unspecified.

For any given state table S, there is a restricted class Σ of input sequences under which S is supposed to operate. If an input sequence does not belong or conform to Σ, the output of S is undefined. Our task of simplification is to find a (minimal) state table T such that with any input

TABLE 9.6 Sequence Detector 1101

Present State	Next State $x = 0$	1	Output $x = 0$	1
a	a	b	0	0
b	c	d	0	0
c	e	f	0	0
d	g	h	0	0
e	a	b	0	0
f	c	d	0	0
g	e	f	0	1
h	g	h	0	0

TABLE 9.7   Initial Simplification

Present State	Next State		Output	
	$x = 1$	1	$x = 0$	1
$a(e)$	a	b	0	0
$b(f)$	c	d	0	0
c	a	b	0	0
$d(h)$	g	d	0	0
g	a	b	0	1

TABLE 9.8   Simplified Sequence Detector

Present State	Next State		Output	
	$x = 0$	1	$x = 0$	1
$a(c, e)$	a	b	0	0
$b(f)$	a	d	0	0
$d(h)$	g	d	0	0
g	a	b	0	1

sequence $I \in \Sigma$ applied to both S and T, the outputs are identical for both S and T at places where the outputs from S are specified. T is then said to be a (minimal) cover of S.

Looking at it another way, if by specifying certain unspecified items (next states or outputs) in the state table of S, S and T can be made equivalent, then T is a cover of S. Similarly, two states are *compatible* if specification of unspecified items can make them equivalent.

If a is equivalent to b and b is equivalent to c, then a is equivalent to c. However, this does not necessarily hold for compatibility. If a is compatible with b, and b is compatible with c, a *may or may not be* compatible with c. The reason is that to make a equivalent to b, some unspecified items are specified in one way; to make b equivalent to c, the items are then specified in a different way. The two sets of additional specifications may very well be in conflict.

9.4.7   Design of Clock-Mode Sequential Circuits

In a **clock-mode** sequential circuit, it is assumed that the change of state in every part of the circuit is synchronized by an enabling clock pulse. The design procedure is as follows:

1. Work out a state table from the problem specification.
2. Minimize the state table.
3. Assign a binary number to each state. Each binary digit corresponds to the state of a flip-flop.
4. Realize the combinational circuit, which has a present state and input as its input variables, and a next state and output as its output variables. It can be done in three ways:
   a. gates
   b. PLA
   c. ROM

In the following we use the design of a sequence detector to illustrate the various steps.

State-Table Formulation

An important step in state-table formulation is to identify the states of the system from the system's function as described.

Example 9.4. Design a sequence detector that gives a 1 output after detection of the 1101 sequence, and a 0 output otherwise.

Solution 1. The state of the system can be identified by its three preceding inputs. If it is 110, a 1 input would bring an output of 1; otherwise, the output is 0. The eight states transfer into each other as the inputs are updated. A 101 state goes into a 010 state with a 0 input or a 011 state with a 1 input. The state transition and output specifications are given in Table 9.6, where $a, b, c, \ldots, h$ represent the states $000, 001, 010, \ldots, 111$, respectively.

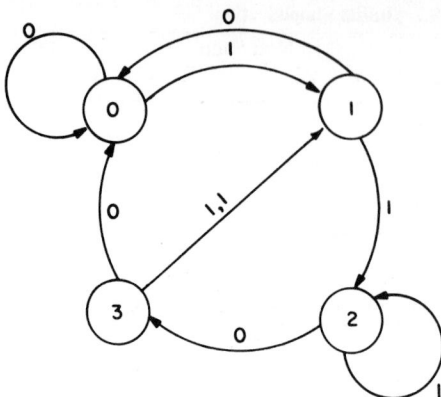

Fig. 9.43   Transition diagram of 1101 sequence detector.

TABLE 9.9   Two Methods of State Selection

Method 2	Method 1: Input Sequence	Name
0	000	a
	010	c
	100	e
1	001	b
	101	f
2	011	d
	111	h
3	110	g

Solution 2. Another way to select the states is to denote the number of usable correct digits. For instance, both 001 and 101 have one digit correct. We denote it as state 1. A 1 input would bring it to state 2. The complete transition diagram is shown in Fig. 9.43. Table 9.9 gives the correspondence between the two methods. If we identify the states 0, 1, 2, 3 with a, b, d, g of Table 9.8, it is readily verified that Table 9.8 and the transition diagram of Fig. 9.43 are one and the same.

Method 2 also gives a direct insight as to why Table 9.8 cannot be simplified further: the number of correct digits must increase step by step from 0 to 1 to 2, and to 3. Each step must be represented. The state-minimization procedure is discussed in Section 9.4.6.

State Assignment
The first step in realization is to assign binary numbers to the minimal set of states. We define the *distance* between two binary numbers as the number of digits that are different from each other. For instance, 101 has a distance of 1 from all three numbers 001, 111, and 100. A simple rule to follow is to assign linked states to binary numbers with distance of 1 as much as possible. It reduces the number of flip-flop state changes and generally results in the least logic requirement. In a critical design, it is advisable to try out two or three different assignments. In Table 9.10, the first column lists the four states of Fig. 9.43. The second and third columns list their new state assignments. Note that the assigned binary numbers are not in accord with the name sequence. The fourth column lists the two possible inputs for each state. The next three columns, Q_1^+, Q_2^+, and output Z, follow from Fig. 9.43. For instance, if the state is originally 1 or 01, a 1 input brings it to state 2 or 11, with an output 0. The meaning of the last four columns is discussed later.

Realization with ROM and PLA
The transition table is directly realized by a ROM with three address bits and three data bits, as shown in Fig. 9.44. The two D flip-flops hold Q_1^+ and Q_2^+ for one clock period before they appear at the output end as the new states Q_1 and Q_2.

TABLE 9.10   State Assignment and Transition Table

State	Q_1	Q_2	I	Q_1^+	Q_2^+	Z	S_1	R_1	S_2	R_2
0	0	0	0	0	0	0	0	×	0	×
	0	0	1	0	1	0	0	×	1	0
1	0	1	0	0	0	0	0	×	0	1
	0	1	1	1	1	0	1	0	×	0
2	1	1	0	1	0	0	×	0	0	1
	1	1	1	1	1	0	×	0	×	0
3	1	0	0	0	0	0	0	1	0	×
	1	0	1	0	1	1	0	1	1	0

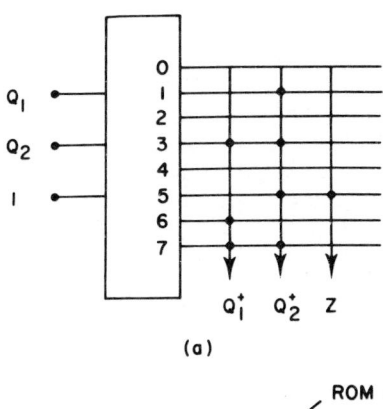

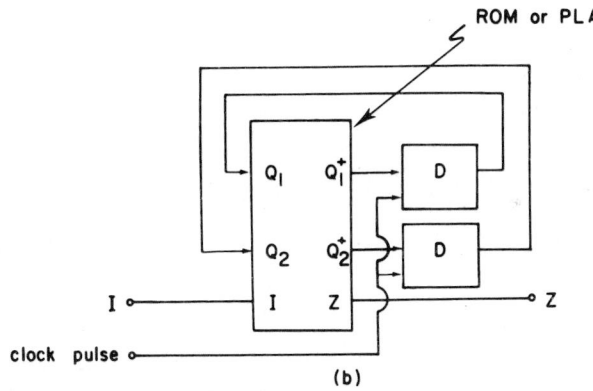

Fig. 9.44   ROM realization of sequence detector: (a) ROM; (b) circuit.

Alternatively, we can realize the transition table with a PLA. The Karnaugh maps of the three output variables are shown in Fig. 9.45. Correspondingly,

$$Q_1^+ = Q_1 Q_2 + Q_2 I$$
$$Q_2^+ = I$$
$$Z = Q_1 \overline{Q}_2 I \qquad\qquad (9.56)$$

Equation (9.56) is realized by the PLA of Fig. 9.46. The sequence detector is realized with two additional D flip-flops, as shown in Fig. 9.44(b).

Realization with S-R Flip-Flops

Realizations with basic S-R flip-flops are sometimes used to minimize the number of logic levels. The output variables from the combinational circuit are then S_1, R_1, S_2, R_2, and Z instead of Q_1^+, Q_2^+, and Z. The rule of assignment is as shown in Table 9.11. Table 9.11 is derived from truth table 9.5. It is readily verified that S and R as assigned bring about the specified state transition from Q to Q^+.

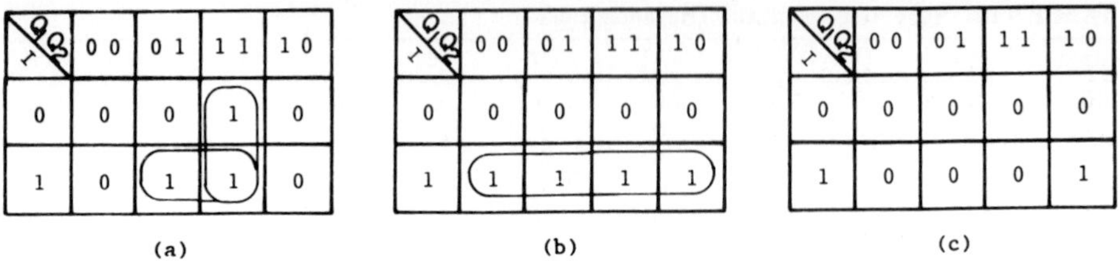

$Q_1 Q_2$ \ I	0 0	0 1	1 1	1 0
0	0	0	1	0
1	0	1	1	0

$Q_1 Q_2$ \ I	0 0	0 1	1 1	1 0
0	0	0	0	0
1	1	1	1	1

$Q_1 Q_2$ \ I	0 0	0 1	1 1	1 0
0	0	0	0	0
1	0	0	0	1

(a)   (b)   (c)

Fig. 9.45   Karnaugh maps of the variables Q_1^+, Q_2^+, Z: (a) Q_1^+, (b) Q_2^+, (c) Z.

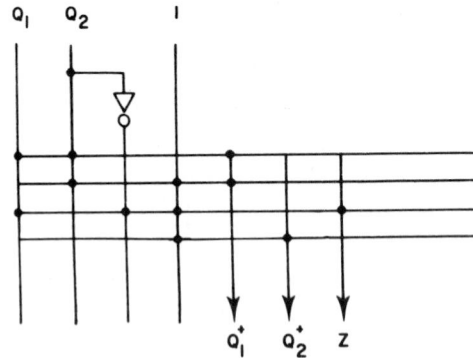

Fig. 9.46   PLA realization of sequence detector.

TABLE 9.11   Excitation Rule

Q	Q^+	S	R
0	0	0	$\times$
0	1	1	0
1	0	0	1
1	1	$\times$	0

With the aid of Table 9.11, Table 9.10 can be extended to give S_1, R_1, S_2, and R_2 for each row. The Karnaugh maps of S_1, R_1, S_2, R_2, and Z are called **excitation tables** and are shown in Fig. 9.47. Figure 9.48 gives a realization of the sequence detector by AND and NOT gates and S-R flip-flops.

Flowchart of Clock-Mode Sequential Circuit Design

Sequential circuit design is summarized in the flowchart of Fig. 9.49. With ROM realization, a minimum amount of design work is required. With PLA realization, a minimal set of product terms for the multiple outputs $Q_1^+, Q_2^+, \ldots, Z_1, Z_2, \ldots$ needs to be determined. With S-R flip-flops the number of output variables in the combinational part of the circuit is increased, with each Q replaced by S and R.

Both D flip-flops and S-R flip-flops can be used with either PLA or gate circuits. However, S-R flip-flops are rarely used with ROM because the increased number of combinational outputs means increased word length, which is expensive.

9.4.8   Pulse-Mode Sequential Circuits

Two important criteria for pulse-mode sequential circuits are the following:

1. Each pulse is sufficiently wide to trigger the flip-flops involved.
2. Various input pulses are sufficiently separated so that each one can be considered independently (in other words, in sequence).

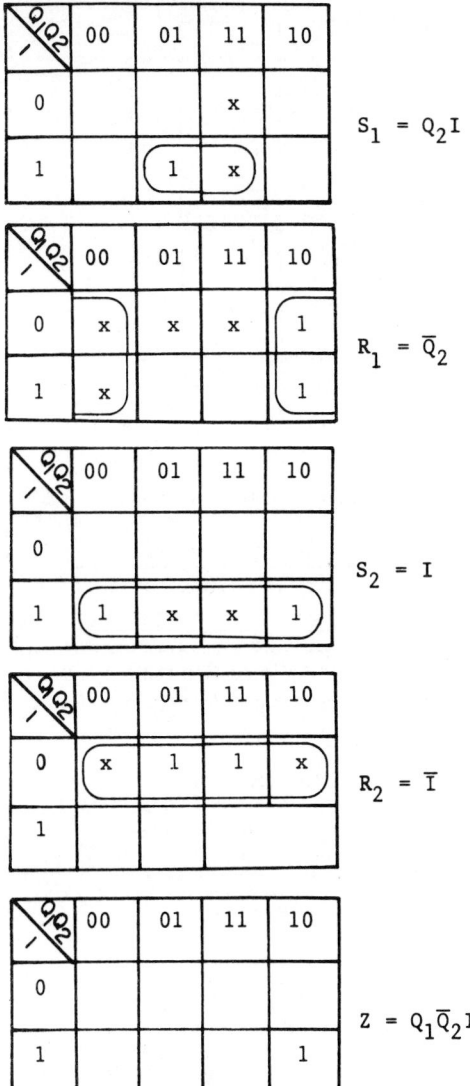

Fig. 9.47   Excitation tables and results.

With these two stipulations, there is no substantial difference between a clock-mode circuit and a pulse-mode circuit. As an example, the sequence detector problem is modified as follows. I and $\bar{I}$ are no longer regarded as complementary voltage levels but as independent pulses. A 1 input means a 1 pulse at I and nothing at $\bar{I}$. A 0 input means a 1 pulse at $\bar{I}$ and nothing at I. In between pulses, both I and $\bar{I}$ are zero. The sequence detector problem then becomes a pulse-mode sequential circuit problem. The same steps lead to the excitation table of Fig. 9.47 and its realization.

9.4.9   Level-Mode Sequential Circuits

There are certain interesting characteristics of level-mode sequential circuits that can be described in general terms:

1. *The Stable State.* In Eq. (9.4), the vectors x, y, and Y have various components that represent various input and state variables. For a clock-mode or a pulse-mode sequential circuit the state

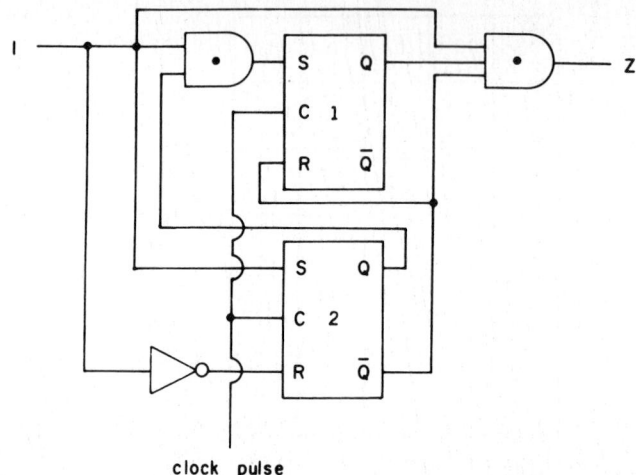

Fig. 9.48 Realization of sequence detector by gates and clocked *S-R* flip-flop.

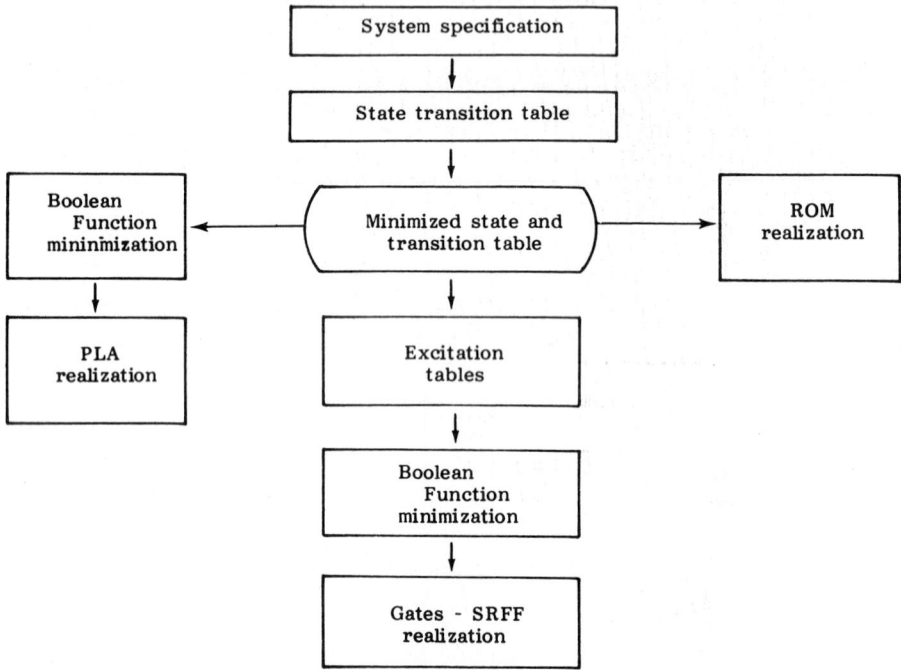

Fig. 9.49 Steps in designing (synchronized) sequential circuits.

variables stay put until the next pulse arrives. However, there is no such sharp demarcation line in a level-mode sequential circuit. As soon as the state variables change to their new values **Y**, the new states initialize another round of changes, unless

$$\mathbf{Y} = \mathbf{G}(\mathbf{x}, \mathbf{y}) = \mathbf{y} \tag{9.57}$$

If Eq. (9.57) is satisfied, **y** is called a **stable** state with input **x**.

2. *Multiple Stable States for the Same Input.* If for the same x, Eq. (9.57) allows more than one solution, then a multiple number of stable states exist. On the other hand, the existence of multiple

stable states is the main justification for using a level-mode sequential circuit. If there is only one stable state for each x, we might as well use a combinational circuit.

3. *Race and Cycle.* With a change in x, y goes through a number of transient states before arriving at a stable state. If the sequence of transient states is unique, we call it a **cycle**. If there are two or more possible sequences leading to the same stable state, we call it a **noncritical race**. If the possible sequences lead to different stable states, we call it a **critical race**.

4. *Fundamental Mode Circuit.* If two input variables change "simultaneously," a race results, as the priority of occurrence makes a difference in the sequence of transient states. To require the races to be noncritical for every combination of input changes is too severe a restriction. We specify instead that *the input variables can change only one at a time.* A circuit designed under this assumption is called a **fundamental mode circuit**.

5. *Hazards.* Similar to clock- and pulse-mode sequential circuits, a level-mode sequential circuit is realized by flip-flops and combinational circuits. The latter can introduce transient pulses due to uneven delays. For instance,

$$Y_1 = x_1 x_2 + \bar{x}_1 y_2 \tag{9.58}$$

With $x_2 = y_2 = 1$, Y_1 should remain 1 when x_1 switches from 1 to 0. However, due to delay in the NOT gate, if $\bar{x}_1$ remains 0 for a short duration, Y_1 is momentarily 0.

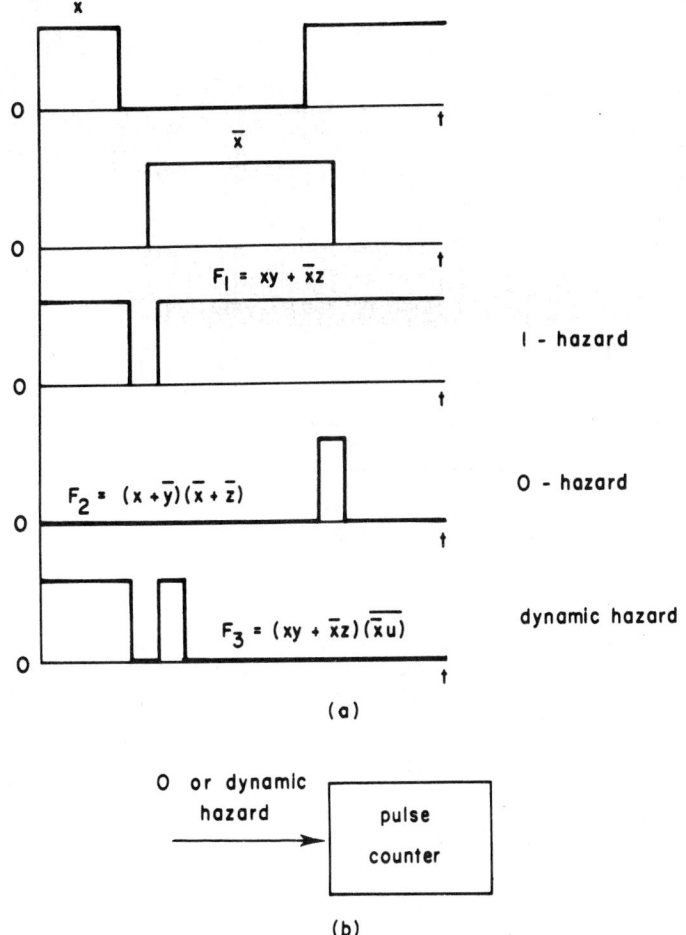

(a)

(b)

Fig. 9.50 Types of hazards: (*a*) hazards in combined circuits, (*b*) simple example of an essential hazard.

Figure 9.50 shows the common types of hazards and their technical names. The 1 hazard in Eq. (9.58) can be easily cured by adding another term:

$$y_1 = x_1 x_2 + \bar{x}_1 y_2 + x_2 y_2 \tag{9.59}$$

The Boolean expression (9.59) is equivalent to (9.58), but the term $x_2 y_2$ assures that the 1 hazard would not occur. Although such simple special cases are corrected easily, there is no general solution to the hazard problem. In a level-mode sequential circuit, hazards can cause erratic changes in the flip-flops, with the circuit ending up in a different stable state than the one for which it was originally designed.

Because of the critical races and hazards, the design problem of level-mode sequential circuits, even with the restriction to the fundamental mode, is still not completely solved. An excellent account of what can be done is given in Hill and Peterson's book listed in the bibliography.

9.5 ARITHMETIC-LOGIC UNIT
Earl E. Swartzlander, Jr.

The arithmetic-logic unit (ALU) is the component of a computer that performs arithmetic and logical operations. This section begins with an overview of ALU usage and then examines ALU functions, implementation, and algorithms.

9.5.1 ALU Usage

In a general-purpose computer, the ALU implements and performs all arithmetic operations as specified by the program sequence and is also used in generating memory addresses. This section briefly examines the concept of the basic stored program computer since it provides a perspective for how the ALU is used.

At a minimum, the ALU must compute the four basic arithmetic operations (add, subtract, multiply, and divide) as well as a number of logical operations (and, or, invert, etc.). With the exception of a few operations such as logical inversion and square root, ALU operations are dyadic (i.e., performed on two operands). Since data are fetched from the memory a word at a time, some form of data storage must be provided in conjunction with the ALU. A variety of approaches have been used, including accumulators, stacks, register files, and scratch pad memories. Section 9.1 examines these approaches in greater detail.

In implementing the ALU for minicomputers and microprocessors, it is frequently too complex to provide a dedicated hardware multiplier or divider. Instead, these functions are performed in firmware with a series of add and shift operations (for multiply) or subtract and shift operations (for divide). By contrast, large scientific computers use fully parallel logical arrays to implement multiplication directly. In the large machines, division is implemented either with special-purpose hardware or with an iterative procedure based on Newton's method. A final distinction between the smaller machines and large scientific computers is that the latter generally contain floating-point ALUs that directly compute both the exponent and the mantissa (or fraction) for arithmetic operations. In smaller machines, floating-point arithmetic is often implemented via a microprogrammed sequence of integer or fractional (micro) operations.

ALU Usage in Signal Processors

Special-purpose computers are becoming practical for applications in real-time signal processing. In this context the ALU is used strictly for data stream computation. This results in a somewhat different set of requirements for the ALU design. In basic terms, signal processing differs from general-purpose computing in that the operation sequences are usually not data dependent. Since the ALU is processing data only, higher instruction rates are achieved in signal processing.

In some signal-processing applications, it may be desirable to use an ALU with multiple data ports. This concept is shown in Fig. 9.51. On each memory cycle, a three-port data memory accesses two operands and stores a result. These three ports are directly connected to the inputs and output of the ALU. High ALU throughput is achieved without requiring high-speed data transfers.

9.5.2 ALU Functions

The most important characteristics of an ALU are the functional capabilities and the internal architecture. This section provides an overview of the capability and architecture of commercially available building-block ALU integrated circuits.

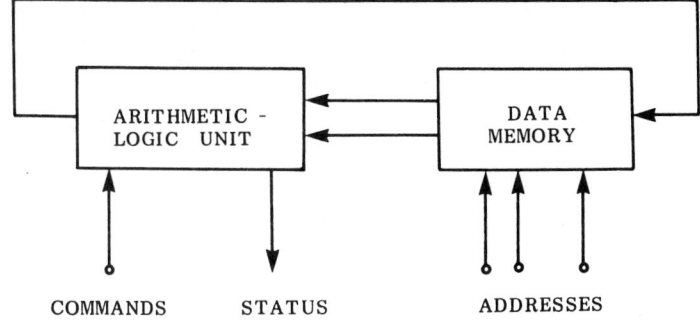

Fig. 9.51 Multiport memory/ALU interface.

Arithmetic and Logic Functions

Commercial ALU circuits are an extension of standard adder/subtractor circuits, where additional arithmetic and logic functions can be performed at little increase in complexity. From a 4-bit-wide building block one can implement arbitrary width ALUs. Consider a typical circuit. Inputs consist of two 4-bit operands (A_0, A_1, A_2, A_3 and B_0, B_1, B_2, B_3), a carry-in to the low-order stage, C_n, and a 4-bit function select code (S_0, S_1, S_2, S_3). The outputs consist of four data lines (F_0, F_1, F_2, F_3), a flag that indicates when all outputs are zero, a carry-out signal from the high-order stage, C_{n+4}, and carry lookahead signals, P and G (see Section 9.5.3 for a detailed discussion of carries). The arithmetic and logical functions, realized by the device include addition, subtraction, AND, OR, and so on. In most applications, only a few of the available functions are used.

The ALU device is a combinational circuit (i.e., it does not have internal data registers or latches). In many applications, external storage registers are needed to buffer the ALU operands. Recent integrated-circuit ALU designs have added internal registers. Bit-slice microprocessor designs are extensions of this concept that include complete sets of working registers.

Commercial ALU Circuits

Figure 9.52 shows an 8-bit adder/subtractor built with two 4-bit ALU devices. The add/subtract command is latched into a flip-flop so that both Q and $\overline{Q}$ outputs are available to generate the necessary add and subtract commands. (See Section 9.4.5.) The Q output is used as the least

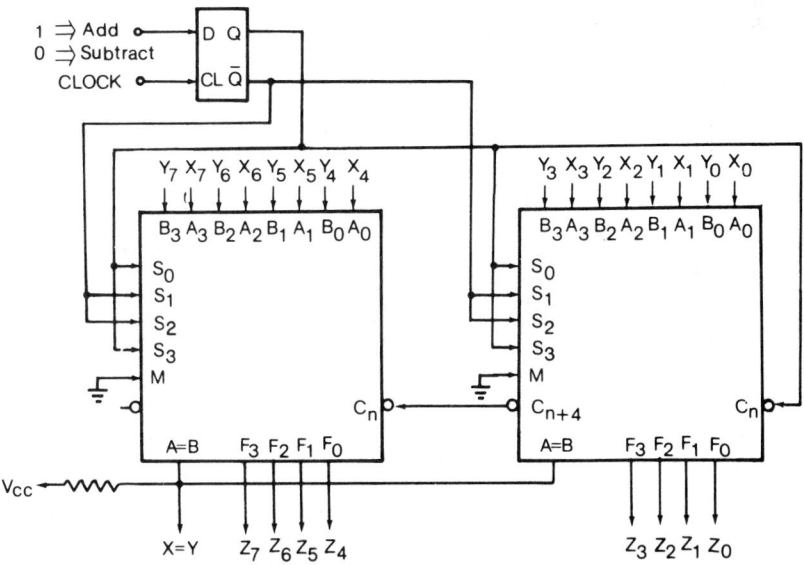

Fig. 9.52 Eight-bit adder/subtractor implemented with two 4-bit ALU devices using ripple carry.

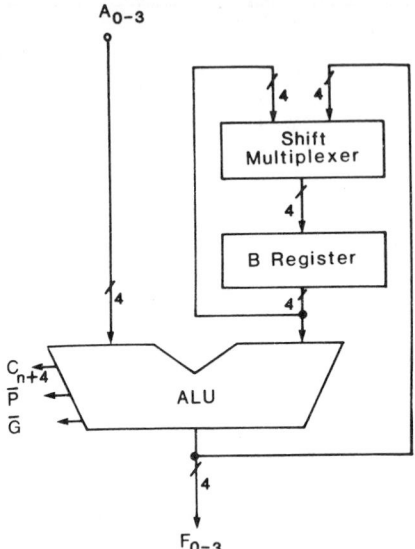

Fig. 9.53   Arithmetic-logic unit/accumulator.

significant carry input, so that a carry in is provided during subtraction. This implements 2's-complement subtraction, which is examined in greater depth in Section 9.5.5.

The functional block diagram of an ALU/accumulator is shown in Fig. 9.53. Basically, it consists of an ALU with an internal register for one input and a shift matrix between the ALU output and the register. The internal register is used as an accumulator when a sum of many terms is to be computed. The shift unit allows intermediate terms to be shifted up or down, as required, for rescaling arithmetic results.

In comparing ALU designs for a specific application, several interdependent specifications should be considered. These include operations to be performed, number of data ports, and storage (register) requirements. The operations required for the system must be realized with the available arithmetic and logical functions of the selected ALU. There are three data ports (two inputs and one output) for the combinational ALU designs or two (one input and one output) for ALUs that include internal storage. Three-port designs are preferred for large mainframe computers, high-speed signal processors, and other specialized high-speed applications where maximum speed is required; two-port architectures are well suited to general-purpose (i.e., minicomputer) applications.

Internal storage in the form of input registers, an accumulator, or register files reduces the need to access main memory and is also useful in storing temporary variables in signal-processing applications.

9.5.3   ALU Implementation

This section examines the implementation of ALUs of arbitrary size. The basic approach involves developing a module that is a slice of a wide-word ALU. It is an extension of the use of full adders to implement wide-word addition circuits. Figure 9.54 shows the truth table of a full-adder stage and the interconnection of n such stages to create a parallel adder for n-bit words. At the kth-bit position, the kth bits of the operands A_k and B_k and a carry signal from the preceding full adder, C_k, are used to generate the kth bit of the sum, S_k, and a carry to the $(k + 1)$st full-adder stage, C_k'.

The logical equations that describe the operation of the kth adder stage are

$$S_k = A_k \oplus B_k \oplus C_k \tag{9.60}$$

$$C_k' = A_k B_k + A_k C_k + B_k C_k \tag{9.61}$$

where A_k, B_k, and C_k are the inputs to the kth full-adder stage, and S_k and C_k' are the sum and carry outputs, respectively. This is called a **ripple carry adder** since the carry signals propagate (i.e., ripple) from the least significant stage to the most significant. This rippling occurs since the carry-out from each stage depends on the carry input to that stage (i.e., the carry-out of the preceding stage).

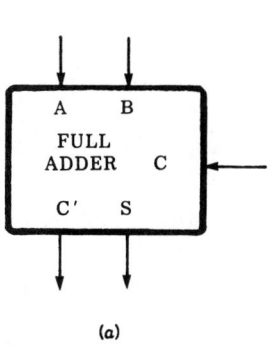

(a)

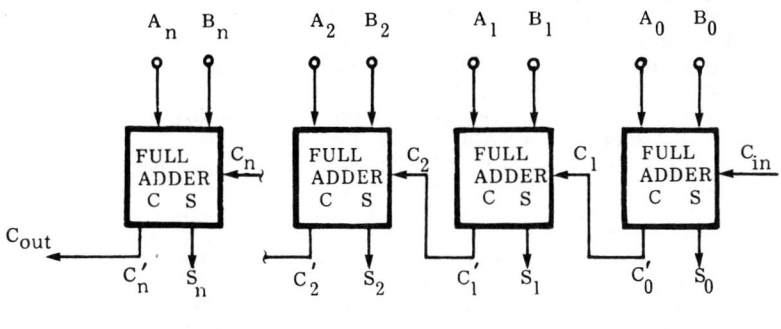

INPUTS			OUTPUTS	
A	B	C	S	C'
0	0	0	0	0
0	0	1	1	0
0	1	0	1	0
0	1	1	0	1
1	0	0	1	0
1	0	1	0	1
1	1	0	0	1
1	1	1	1	1

(b)

(c)

Fig. 9.54 Full-adder circuit and use: (a) circuit symbol, (b) truth table, (c) ripple carry adder.

Thus if the time from carry-in to carry-out requires a single delay, an n-bit ripple carry adder may require n delays. Although analysis has shown that the average maximum carry chain length is approximately $1 + \log_2 n$, the worst-case delay of n must be allowed.

One approach for circumventing the long, worst-case delay is the use of adders that generate signals indicating when carry propagation is complete. This technique is not widely used since it results in data-dependent timing, which greatly raises the system complexity.

A third adder approach is the carry lookahead adder. Here specialized carry logic determines the carry status by looking across a wide word. In each module, outputs indicate that a carry has been generated within that module or that a lower-output carry would be propagated across that module. To understand the concept, refer to Eq. (9.61), which describes ripple carry operation, and define $G_k \equiv A_k B_k$ and $P_k \equiv A_k + B_k$:

$$C'_k = G_k + P_k C_k \qquad (9.62)$$

This helps to explain the concept of carry generation and propagation: at a given stage, a carry is generated if G_k is true (i.e., $A_k = B_k = 1$), and a stage transmits (or propagates) an input carry to its output if P_k is true (i.e., either A_k or B_k is a 1). Extending this definition to a 4-bit-wide module, we begin with

$$S_k = A_k \oplus B_k \oplus C_k \qquad (9.63)$$

$$S_{k+1} = A_{k+1} \oplus B_{k+1} \oplus (G_k + P_k C_k) \qquad (9.64)$$

Equation (9.64) results from evaluating (9.63) for the $(k + 1)$st stage and substituting C_{k+1} $(\equiv C_k')$ from (9.62). Extending to subsequent stages, we obtain

$$S_{k+2} = A_{k+2} \oplus B_{k+2} \oplus (G_{k+1} + P_{k+1}G_k + P_{k+1}P_kC_k) \qquad (9.65)$$

$$S_{k+3} = A_{k+3} \oplus B_{k+3} \oplus (G_{k+2} + P_{k+2}G_{k+1} + P_{k+2}P_{k+1}G_k + P_{k+2}P_{k+1}P_kC_k) \qquad (9.66)$$

Recognize that a carry is emitted from stage $k + 2$ if (1) a carry is generated there, (2) a carry was emitted from stage $k + 1$ and propagated across stage $k + 2$, (3) a carry was emitted from stage k and propagated across both stages $k + 1$ and $k + 2$, and so on:

$$P_{\text{block}} = P_{k+3}P_{k+2}P_{k+1}P_k \qquad (9.67)$$

$$G_{\text{block}} = G_{k+3} + P_{k+3}G_{k+2} + P_{k+3}P_{k+2}G_{k+1} + P_{k+3}P_{k+2}P_{k+1}G_k \qquad (9.68)$$

$$C_{k+4} = G_{\text{block}} + P_{\text{block}}C_k \qquad (9.69)$$

Equation (9.69), which is Eq. (9.62) extended over the 4-bit block, shows that the carry of a 4-bit-wide block can be computed in only two gate delays [one to compute P_n and G_n for $n = k$ through $k + 3$ and one to evaluate (9.69) using (9.67) and (9.68) and the P_n and G_n values]. This compares with eight delays to compute the carry from a four-stage ripple carry adder [from Eq. (9.61) and Fig. 9.54].

Carry lookahead can be extended to larger blocks, but the number of inputs to the logical gates increases in direct proportion to the block size. Four-bit-wide blocks, for example, (1) can be combined to form adders for the most common word sizes (i.e., 16, 24, 32, 60, and 64 bits), (2) use a small number of package pins, and (3) are implemented with gates with at most four inputs. The various bit-slice ALU designs of Section 9.5.2 all provide P and G signals for carry lookahead according to these equations.

Figure 9.55 shows the interconnection of four 4-bit ALU circuits and a carry lookahead block to realize a 16-bit fast adder. The sequence of events that occur during an add operation is as follows: (1) apply A, B, and carry-in signals, (2) each 4-bit ALU computes P and G, (3) carry lookahead logic computes all carry outputs, and (4) each ALU computes the F outputs.

This process may be extended to larger adders by subdividing the large adder into 16-bit blocks. Each 16-bit-wide block is implemented as shown in Fig. 9.55. The easiest method to cascade the 16-bit block adders is with ripple carry: the carry-out of each adder is connected to the carry input of the next more significant adder. An alternative approach is to use a second level of carry lookahead, as exemplified for 64-bit word size in Fig. 9.56. This segments the words to be added into 16-bit blocks with carry lookahead between the carry lookahead blocks. The advantage is that with two-level carry lookahead, the delay for a 64-bit adder is only slightly greater than for the 16-bit single-level lookahead adder. With ripple carry between blocks, the delay is proportional to the number of 16-bit blocks.

The delay of each of these approaches is graphed in Fig. 9.57 for word sizes of 4 to 64 bits. At all sizes, carry lookahead produces a speed increase, but the improvement is most striking for wide-word (i.e., 32 to 64 bits wide) ALU designs. Because of the simplicity and performance gain achieved by use of carry lookahead logic, it is nearly universally accepted for ALU implementation.

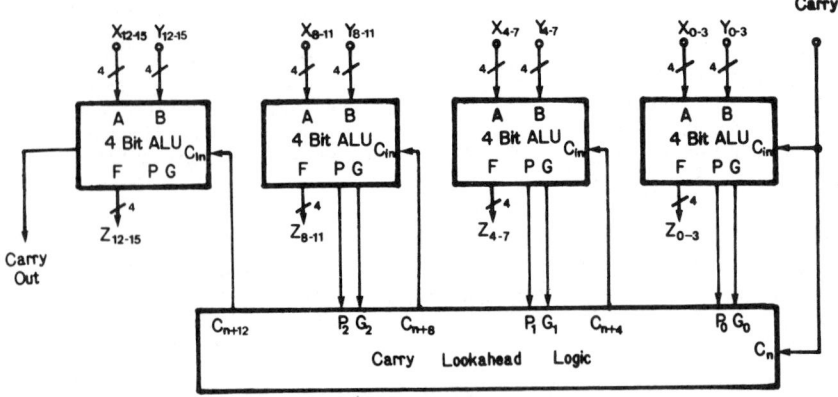

Fig. 9.55   Sixteen-bit carry lookahead adder.

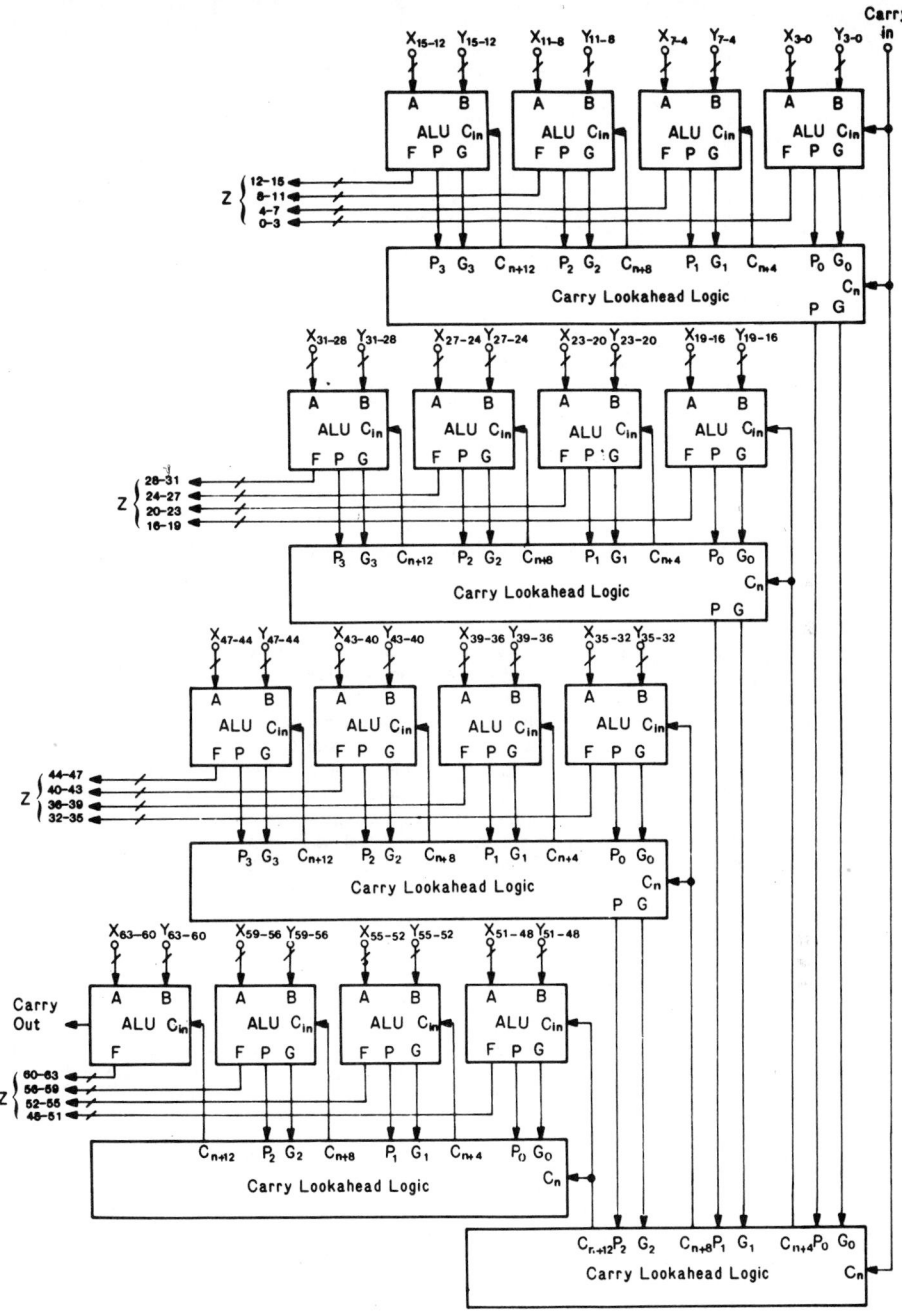

Fig. 9.56 Sixty-four-bit carry lookahead adder.

9.5.4 Representation of Numbers

Before considering algorithms for addition, subtraction, multiplication, and division, it is appropriate to consider the choice of number system. Fixed point and floating point are the two types in common use. Fixed-point systems use constant scaling (i.e., binary points that are fixed in a specific position) and are generally fractional systems that express numbers in the range $|X| \le 1$. Integers are often

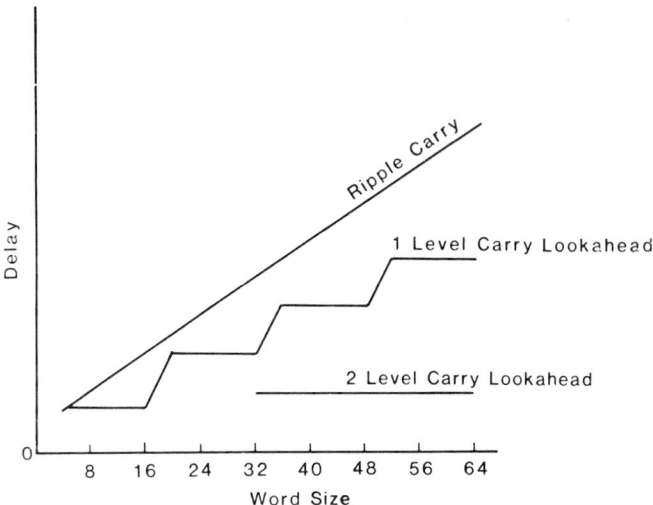

Fig. 9.57 ALU delay for various word sizes.

used in addressing, but the arithmetic is identical for all fixed-point number systems. Floating-point numbers consist of a fixed-point exponent (used to denote a scale factor) and a fixed-point fraction (i.e., the mantissa).

Fixed-point numbers and the mantissas of floating-point numbers are usually expressed as 2's-complement numbers. This choice has prevailed over sign-magnitude and 1's-complement forms because the 2's-complement form produces much simpler algorithms for addition and subtraction, the two most frequent arithmetic operations. Sign-magnitude produces a simpler multiplication algorithm, but the lower frequency of multiplication and the existence of satisfactory 2's-complement multiplication algorithms have resulted in nearly universal selection of the 2's-complement form. Positive 2's-complement fractions are represented by n bits $(a_{n-1}, a_{n-2}, \ldots, a_1, a_0)$:

$$A = \sum_{i=0}^{n-2} a_i 2^{i-n+1} \tag{9.70}$$

for example: $1/8 \leftarrow 0.00100\ldots$, $1/4 \leftarrow 0.0100\ldots$, and so on. The sign bit $a_{n-1} = 0$ (by definition), and $A < 1$. A negative 2's-complement number is represented by subtracting its absolute value from 2 (hence the name "2's complement"). Since the absolute value of the number is less than or equal to 1, the leading bit (a_{n-1}) is 1. A single notation will suffice to express both positive and negative 2's-complement numbers:

$$A = -a_{n-1} + \sum_{i=0}^{n-2} a_i 2^{i-n+1} \tag{9.71}$$

Note that $-1/8 \leftarrow 1.11100\ldots$, $-1/4 \leftarrow 1.1100\ldots$, and so on.

The largest and smallest positive numbers are $1 - 2^{-n+1}$ and 2^{-n+1}, expressed as $0.111\ldots11$ and $0.000\ldots01$, respectively; the most negative number is -1, expressed as $1.000\ldots00$; and there is a single representation for zero, $0.000\ldots00$. The 2's complement of a number is obtained by complementing each bit and adding a one to the least significant bit of the word.

The 1's-complement representation of a number is identical to the 2's-complement form for positive numbers. The representation for negative numbers, however, is the bit-by-bit complement of the corresponding positive value. For example, $-1/8 \leftarrow 1.11011\ldots$, $-1/4 \leftarrow 1.1011\ldots$, and so on. The most negative number is $2^{-n+1} - 1 \leftarrow 1.00\ldots$. There are two representations for zero: $0.00\ldots$ and $1.11\ldots$ (called minus 0).

The sign-magnitude representation is the same as the 1's- and 2's-complement forms for positive numbers. The representation for negative numbers uses a sign bit of 1 followed by the representation for the corresponding positive value. For example, $-1/8 \leftarrow 1.00100\ldots$, $-1/4 \leftarrow 1.0100\ldots$, and so on. There are two representations for zero: $0.00\ldots$ and $1.00\ldots$ (called minus 0).

In large computers, especially, decimal and hexadecimal (i.e., base 16) arithmetic is often performed. In both cases, data are represented by 4-bit characters. In decimal implementations, special arithmetic circuits perform the arithmetic; hex implementations use 4-bit ALU modules as described in Section 9.5.2.

9.5.5 Arithmetic Algorithms

In this section, fixed-point algorithms for addition/subtraction, multiplication, and division are examined. An introduction to arithmetic algorithms for floating-point number systems is also provided. Further information on ALU arithmetic is available in the literature.

Fixed-Point Addition / Subtraction

Two's-complement addition is performed as a normal addition of two n-bit binary numbers, including the sign bit. Subtraction is performed by summing the minuend and the 2's complement of the subtrahend.

Overflow is detected by comparing the carry signals into and out of the most significant adder stage (i.e., the stage that computes the sign bit). If the carries disagree, the arithmetic has overflowed and the result is invalid.

A circuit for a ripple carry adder/subtractor is shown in Fig. 9.58. It consists of a standard ripple carry adder (as shown in Fig. 9.54) with exclusive-OR gates to complement operand Y when performing subtraction, and a sign comparison circuit to check for overflow. It forms either $X + Y$ or $X - Y$ with 2's-complement arithmetic. In the case of $X + Y$, the mode selector causes the exclusive-OR gates to pass operand Y directly to the ripple carry adder. The carry into the least significant adder stage is a zero, so standard addition occurs. Overflow is detected by comparing the carry into and out of the most significant adder stage; as long as these carrys agree, overflow has not occurred. Subtraction is implemented by using the exclusive-OR gates to complement the bits of Y; the 2's complement is formed by setting the carry into the least significant adder stage.

Fixed-Point Multiplication

Multiplication is generally implemented either via a sequence of addition, subtraction, and shift operations or with a specialized circuit. In the latter case, which is described later, a parallel array of adders forms the product at speeds comparable to memory or ALU cycles. Booth's algorithm is widely used for 2's-complement multiplication since it is easy to implement and achieves high speed.

Booth's method for the multiplication of X and Y involves examining the ith bit of the multiplier X (x_i) starting with the least significant (i.e., $i = 0$) and assuming that $x_{-1} = 0$:

If $x_i = x_{i-1}$, shift the sum of partial products to the right (multiply by $\frac{1}{2}$).

If $x_i = 0$, $x_{i-1} = 1$, add Y to the existing sum of partial products and shift one place to the right.

If $x_i = 1$, $x_{i-1} = 0$, subtract Y by adding its 2's complement to the existing sum of partial products and shift one place to the right.

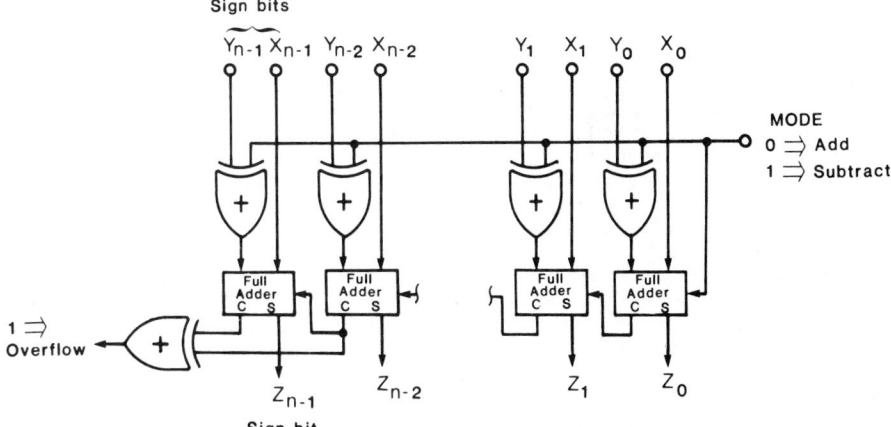

Fig. 9.58 Two's complement adder/subtractor.

TABLE 9.12   Example of Booth's Algorithm

$i\ X_i$	X_{i-1}	Sum of Partial Products		Operation Result before Shift
$X = \frac{5}{8}$ $(= 0.101)$				$Y = \frac{3}{4}$ $(= 0.110)$
0 1	0	0.000	$-Y$	1.010
1 0	1	1.1010	$+Y$	0.0110
2 1	0	0.00110	$-Y$	1.01110
3 0	1	1.101110	$+Y$	0.011110 $(= \frac{15}{32})$

Thus $\frac{5}{8} \times \frac{3}{4} = \frac{15}{32}$

$X = -\frac{5}{8}$ $(= 1.011)$				$Y = \frac{3}{4}$ $(= 0.110)$
0 1	0	0.000	$-Y$	1.010
1 1	1	1.1010	—	1.1010
2 0	1	1.11010	$+Y$	0.10010
3 1	0	0.010010	$-Y$	1.100010 $(= -\frac{15}{32})$

Thus $(-\frac{5}{8}) \times \frac{3}{4} = -\frac{15}{32}$

$X = \frac{5}{8}$ $(= 0.101)$				$Y = -\frac{3}{4}$ $(= 1.010)$
0 1	0	0.000	$-Y$	0.110
1 0	1	0.0110	$+Y$	1.1010
2 1	0	1.11010	$-Y$	0.10010
3 0	1	0.010010	$+Y$	1.100010 $(= -\frac{15}{32})$

Thus $\frac{5}{8} \times (-\frac{3}{4}) = -\frac{15}{32}$

$X = -\frac{5}{8}$ $(= 1.011)$				$Y = -\frac{3}{4}(= 1.010)$
0 1	0	0.000	$-Y$	0.110
1 1	1	0.0110	—	0.0110
2 0	1	0.00110	$+Y$	1.01110
3 1	0	1.101110	$-Y$	0.011110 $(= \frac{15}{32})$

Thus $(-\frac{5}{8}) \times (-\frac{3}{4}) = \frac{15}{32}$

No shift is performed on the last stage (i.e., when $i = n - 1$). All shift operations are arithmetic shifts (i.e., the sign bit is copied into the position to its right), and overflows in the addition process are ignored. Booth's multiplication algorithm is illustrated in Table 9.12, in which all sign combinations for the product of $\pm \frac{5}{8} \times \pm \frac{3}{4}$ are computed for 4-bit operands.

An alternative approach to multiplication involves the combinational generation of all bit products and their summation with an array of full adders. Since the speed of this approach is constrained only by the delay of cascade chains of logic gates, this approach is much faster than sequential/iterative processes. Figure 9.59 shows an example of the bit product terms that are summed to form the product of an n-bit number X and an m bit number Y (for arbitrary m and n). The terms that are added consist of all combinations of $x_i \cdot y_j$ for $0 \le i \le n - 2$ and $0 \le j \le m - 2$ and three rows of correction terms shown at the bottom of the matrix.

Although much effort has been expended in the development of efficient implementations of parallel or array multipliers, the details are beyond the scope of this section. It is sufficient for the user to understand the chip interfaces and control options.

A block diagram of a typical 16-bit single-chip multiplier constructed among these lines is shown in Fig. 9.60. It consists of an array of full adders that form and sum the bit products shown in Fig. 9.59 (shown as "asynchronous multiplier array"). Registers are provided for the two 16-bit operands, X and Y, and for the least significant and most significant halves of the product, LSP and MSP, respectively. Due to package pin limitations, the least significant half of the product, LSP, is output via the Y input port.

These devices perform 2's-complement arithmetic directly, with rounding (if desired), which is implemented by adding a one at the most significant bit of the LSP. This is especially useful for signal-processing applications where constant data precision is desired at all stages of the computation, and for single-precision floating-point operations.

Fixed-Point Division

Division is traditionally implemented as a shift, subtract, and compare operation, in contrast to the shift and add approach employed for multiplication. The comparison operation is significant: it results in a serial process that is not amenable to parallel implementation.

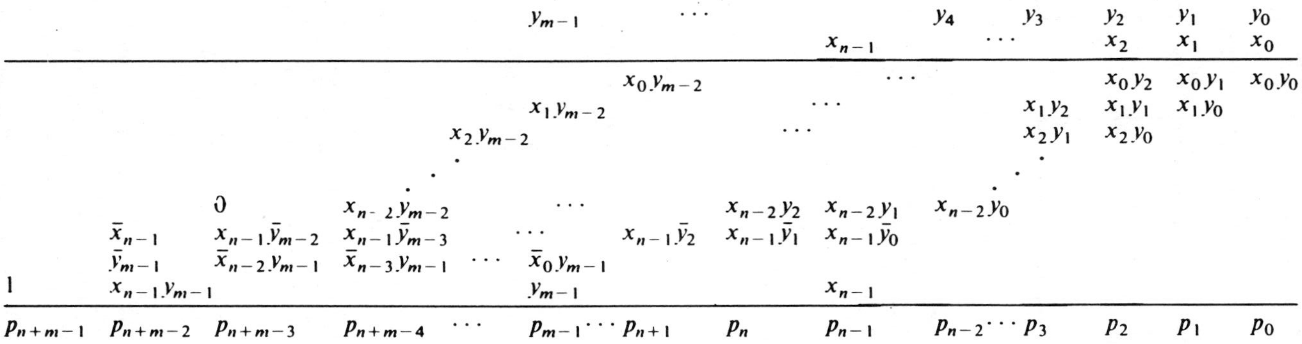

$$
\begin{array}{c}
\begin{array}{cccccccc}
 & & y_{m-1} & \cdots & y_4 & y_3 & y_2 & y_1 & y_0 \\
 & & x_{n-1} & \cdots & & & x_2 & x_1 & x_0 \\
\end{array}
\end{array}
$$

$$
\begin{array}{l}
x_0 y_0 \\
x_0 y_1 \quad x_1 y_0 \\
x_0 y_2 \quad x_1 y_1 \quad x_2 y_0 \\
\ \ x_1 y_2 \quad x_2 y_1 \\
\qquad\qquad\qquad \vdots \qquad\qquad x_{n-2} y_0 \\
x_0 y_{m-2} \\
x_1 y_{m-2} \\
x_2 y_{m-2} \qquad \cdots \\
\qquad\qquad \cdots \qquad x_{n-2} y_2 \quad x_{n-2} y_1 \\
\bar{x}_0 y_{m-1} \qquad\qquad x_{n-1}\bar{y}_2 \quad x_{n-1}\bar{y}_1 \\
\qquad y_{m-1} \qquad\qquad\qquad x_{n-1} \\
\bar{x}_{n-1} \\
\bar{y}_{m-1} \qquad\qquad x_{n-1}\bar{y}_{m-2} \qquad x_{n-2} y_{m-2} \qquad x_{n-2} y_2 \\
x_{n-1} y_{m-1} \qquad \bar{x}_{n-2} y_{m-1} \quad \bar{x}_{n-1} y_{m-3} \qquad x_{n-1}\bar{y}_0 \\
\qquad\qquad\qquad\ \ 0 \qquad\qquad \bar{x}_{n-3} y_{m-1}
\end{array}
$$

$$
\begin{array}{ccccccccccccc}
1 & p_{n+m-1} & p_{n+m-2} & p_{n+m-3} & p_{n+m-4} & \cdots & p_{m-1} & \cdots & p_{n+1} & p_n & p_{n-1} & p_{n-2} \cdots p_3 & p_2 & p_1 & p_0
\end{array}
$$

Fig. 9.59 Two's complement array multiplier bit product matrix.

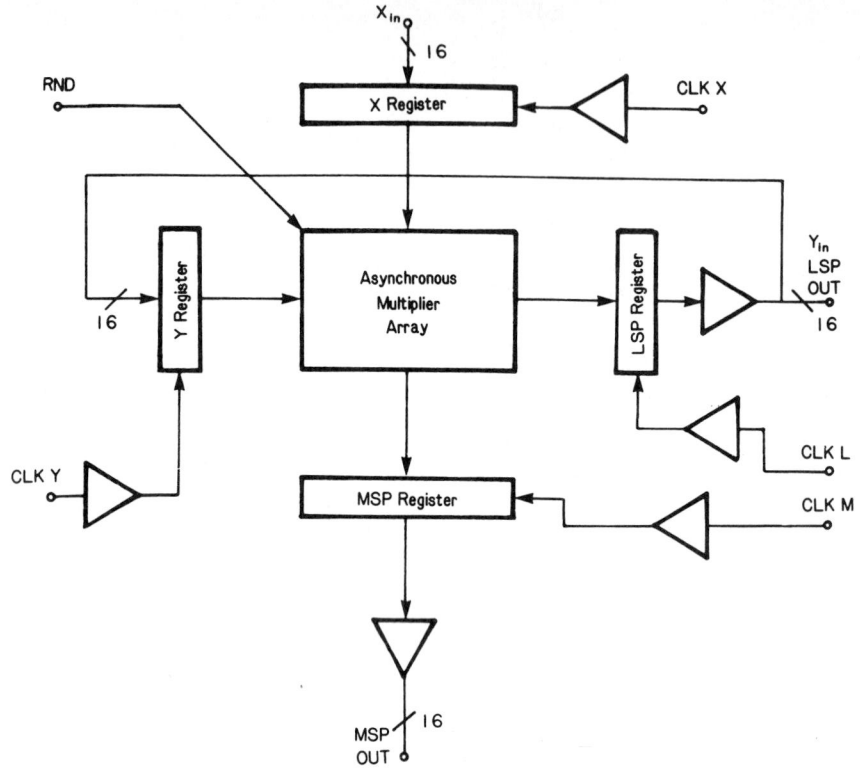

Fig. 9.60 Single-chip multiplier block diagram.

Traditional division is based on selecting digits of the quotient Q to satisfy the following equation:

$$R^{(i+1)} = rR^{(i)} - q_{n-j-1}D \quad \text{for } j = 0, 1, \ldots, n-1 \tag{9.72}$$

where D is the divisor, $R^{(0)}$ is the dividend (subject to the constraint $|R^{(0)}| < |D|$), q_{n-j-1} is the jth quotient digit to the right of the radix point, $R^{(j)}$ is the partial remainder after selection of the jth quotient digit, and r is the radix ($= 2$ for binary numbers). Nonrestoring division is the most widely used algorithm. In nonrestoring division, the quotient digits are constrained to be either $+1$ or -1. The digit selection and resulting partial remainder are given for the jth iteration by the following relations:

$$\text{If } 0 < R^{(j)} < D, \quad \text{then } q_{n-j-1} = 1 \quad \text{and } R^{(j+1)} = 2R^{(j)} - D$$

$$\text{If } 0 > R^{(j)} > -D, \quad \text{then } q_{n-j-1} = -1 \quad \text{and } R^{(j+1)} = 2R^{(j)} + D \tag{9.73}$$

This process continues either for a set number of iterations or until the partial remainder is smaller than a specified error criterion. The quotient resulting from this procedure is a coded signed digit number (comprised of digits of ± 1). It is converted into a conventional binary number by subtracting a number formed from the -1's from the number with the $+1$'s. The algorithm is illustrated in Table 9.13 where $\frac{5}{16}$ is divided by $\frac{3}{8}$.

A second division technique uses a form of Newton–Raphson iteration to derive a quadratically convergent approximation to the reciprocal of the divisor, which is then multiplied by the dividend to produce the quotient. This process often is faster than conventional division algorithms, especially in systems that include a fast multiplier. For this technique it is assumed that the divisor D is normalized (i.e., $\frac{1}{2} \le D < 1$). Since $1/D$ would be larger than 1, $Y = 1/2D$ is computed to avoid

TABLE 9.13 Example of Binary Nonrestoring Division

$R^{(0)}$ = dividend 0.0101 $(\leftarrow \frac{5}{16})$

D = divisor 0.0110 $(\leftarrow \frac{3}{8})$

$R^{(0)} > 0 \rightarrow q_1 = 1$	$2R^{(0)} \rightarrow$	0.1010
	$-D \rightarrow$	+1.1010
	$R^{(1)} \rightarrow$	0.0100
$R^{(1)} > 0 \rightarrow q_2 = 1$	$2R^{(1)} \rightarrow$	0.1000
	$-D \rightarrow$	+1.1010
	$R^{(2)} \rightarrow$	0.0010
$R^{(2)} > 0 \rightarrow q_3 = 1$	$2R^{(2)} \rightarrow$	0.0100
	$-D \rightarrow$	+1.1010
	$R^{(3)} \rightarrow$	1.1110
$R^{(3)} < 0 \rightarrow q_4 = -1$	$2R^{(3)} \rightarrow$	1.1100
	$+D \rightarrow$	+0.0110
	$R^{(4)} \rightarrow$	0.0010

Q = 0. 1 1 1 -1 (signed digit form)

$Q = P - N$, where P = 0.1110 and N = 0.0001

$P \rightarrow$	0.1110
$-N \rightarrow$	+1.1111
$Q \rightarrow$	0.1101 $\leftarrow \frac{13}{16}$

TABLE 9.14 Example of Division Using Newton–Raphson Iteration

Dividend: $\frac{5}{8}$

Divisor: $\frac{3}{4}$

	Error
$Y_0 = 1.457107 - 0.75 = 0.707107$	0.040440
$Y_1 = 2 \times 0.707107(1 - 0.75 \times 0.707107) = 0.664214$	0.002453
$Y_2 = 2 \times 0.664214(1 - 0.75 \times 0.664214) = 0.666658$	0.000009
$Q = 2 \times 0.625 \times 0.666658 = 0.833323$	

arithmetic overflow. The algorithm consists of three basic steps:

$$Y^{(0)} = 1.457107 - D \tag{9.74}$$

$$Y^{(i+1)} = 2Y^{(i)}[1 - DY^{(i)}] \qquad i = 0,1,\ldots,k \tag{9.75}$$

$$Q = 2NY^{(k)} \tag{9.76}$$

where i is the iteration count and N is the numerator.

With Newton–Raphson iterative division, the error decreases quadratically, so that the number of correct bits in each approximation is roughly twice the number of correct bits in the previous iteration. Thus from a 4-bit initial approximation, two iterations produce a reciprocal estimate accurate to 16 bits, and so on.

The efficiency of this process is dependent on the availability of a fast multiplier since each iteration of Eq. (9.75) requires two multiplications, a subtraction, and a doubling operation. The complete process for the initial estimate, two iterations, and the final quotient determination requires six addition operations and five multiplication operations to produce a 16-bit quotient. This is two to three times faster than a conventional nonrestoring divider if multiplication is roughly as fast as addition, a condition that is usually satisfied for systems that include single-chip parallel multipliers. Table 9.14 illustrates the Newton–Raphson iterative division process.

Floating-Point Arithmetic

Floating-point numbers consist of a pair of fixed-point numbers: a mantissa (or fraction) M, and an exponent (or characteristic) e. The value of a number V is given by

$$V = Mr^e \tag{9.77}$$

where r is the radix of the floating-point system. Many floating-point implementations are based on normalized representations of all numbers: this indicates that the most significant digit of the fraction is nonzero.

A typical 32-bit floating-point format has an 8-bit integer characteristic that ranges between 0 and 255. The value of the exponent is determined by subtracting a bias value from the characteristic. For this number system, a bias value of 128 would be used, so the exponent range is -128 to 127. The mantissa is a 24-bit 2's-complement fraction.

A flowchart for floating-point addition is shown in Fig. 9.61. The operands are (e_1, m_1) and (e_2, m_2), the result is (e_s, m_s). In ① the operand exponents or characteristics are compared; if they

$$S = \text{OPERAND 1} + \text{OPERAND 2}$$

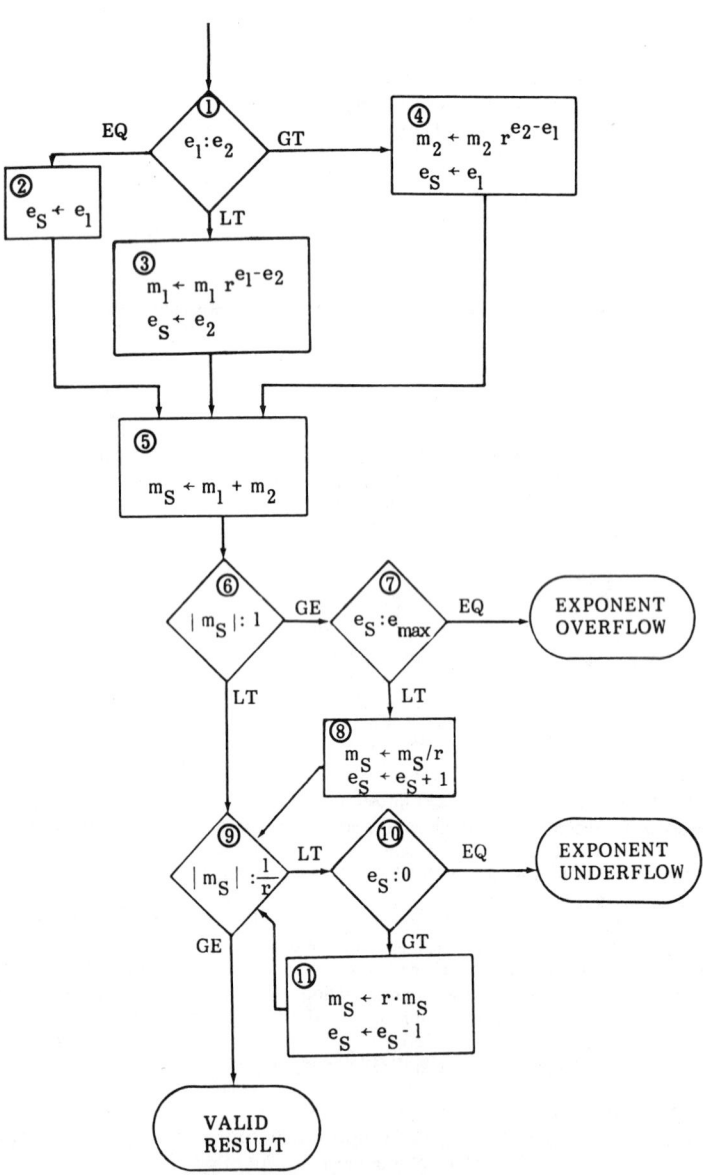

Fig. 9.61 Floating-point addition.

are unequal, the mantissa of the number with the smaller exponent is shifted right in ③ or ④ by the exponent difference, to properly align the mantissas. For example, to add 0.367×10^5 and 0.112×10^4, the latter would be shifted right one place and 0.367 added to 0.0112 to give a sum of 0.3782×10^5. The addition is performed in ⑤. Step ⑥ tests for overflow and shifts the result right (in ⑧) unless the downshift operation would cause exponent overflow. Similarly, the loop ⑨ – ⑪ scales small values upward to normalize the result unless exponent underflow would occur as a result.

Floating-point subtraction is implemented with a similar algorithm. Many refinements are possible to improve the speed of the addition and subtraction algorithms, but floating-point addition will, in general, be much slower than fixed-point addition.

The algorithm for floating-point multiplication is simpler. In the first step, the fixed-point double-precision product of the operand mantissas is computed. The exponent of the product is formed by summing the operand characteristics and subtracting the bias (otherwise, two biased exponents have been summed, yielding a doubly biased product exponent). Next, the computed exponent is tested for overflow, and the product mantissa is tested and normalized. Only a single left shift need be performed; if the operands were normalized, this will be adequate. Then the normalized product is rounded to single precision. Finally, the product exponent is checked for underflow (i.e., a result that is smaller than the minimum representable number).

In implementing floating-point multiplication, it is possible to perform the mantissa multiplication, normalization, and rounding in a single VLSI circuit. Such circuits must provide flags to indicate when overflow has occurred and when the product is normalized. Subsequent to the generation of these flags, a shifter is provided to shift the product down one bit position (to correct for overflow) or up one position (to normalize an unnormalized output).

Floating-point division is similarly implemented. A fixed-point divider computes the quotient of the mantissas, and a subtractor forms the quotient exponent. The quotient is normalized (if possible), and overflow and underflow tests are performed.

All floating-point algorithms may require rounding to produce a result in the correct format. This may be performed by adding a 1 at the most-significant-bit position which will be truncated. This causes a propagation of a carry into the bits that will be retained if the most significant bit of the "truncated" data is a 1. A variety of alternative rounding schemes have been developed over the years.

9.6 MAIN MEMORY AND HIGH-SPEED STORAGE: TECHNOLOGIES
George C. Feth

9.6.1 Overview

A memory may be characterized by the function that it performs, how it carries out the function, and the technology in which it is implemented. The basic function of memory is to retain information, as described in the introduction to Chapter 9.*

Read-Write and Read-Only Memory. Read-write (RW) memory is used to retain information that must be "stored" or "written" and later "retrieved," "fetched," or "read." **Read-only memory** (ROM; sometimes called **read-only storage**, ROS) is used to retain information that does not change, such as control programs or conversion tables. ROM has these advantages: it retains its information content when the computer is turned off, it costs less per bit than RW memory, and it is not susceptible to inadvertent changes of stored information.

There are several forms of memory in which the capability of writing is intermediate between RW and ROM. These are **programmable ROM** (PROM) or **write-once ROM** (WOROM), **erasable programmable ROM** (EPROM), and **electrically alterable ROM** (EAROM) or **read-mostly memory** (RMM). These are described in more detail, together with the technologies commonly used to implement them (primarily semiconductors), in Section 9.6.3.

Random-Access, Sequential-Access, and Content-Addressable Memory. In memory, the information usually is stored as fixed-length words that have a specific address; any such address can be selected at random for either reading or writing. This is called **random-access memory** (RAM). In contrast, a memory in which the bits or words are available only one after the other in a certain time sequence is called **sequential-storage**, or a **sequentially accessed**, memory (e.g., a shift register or magnetic tape). Another way of accessing a memory is by way of its contents: a **content-addressable memory** (CAM), or **associative memory**. Here the contents, or a selected portion of them for each

*Throughout Chapter 9, *memory* is used to refer to entities that store information that is immediately accessible to the CPU; *storage* is used to refer to entities that store information in a less rapidly available way (as well as for the more inclusive function of retaining information in general).

word, constitute its address, as in a directory or dictionary (see Section 9.7.2). An alternative means of realizing the same function is a **sequential search**, sequentially reading each word and comparing the selected portion of the contents with the desired value.

Destructive Read and Nondestructive Read. The reading of stored information from a memory cell may cause the original storage state to be altered, depending on the technology with which the memory is implemented. If reading causes destruction of the information (as with one-device dynamic-RAM semiconductor memory and ferrite-core memory), the read operation is termed **destructive read** (DR); if the original information is reliably retained (as with static semiconductor memory), it is termed **nondestructive read** (NDR). DR memories are provided with means to rewrite, or **regenerate**, the information that has been read, so functionally, the data are not lost.

Static and Dynamic: Volatility. In some memories, once the information is written, it is retained as long as the necessary power is supplied to the memory; this is termed a **static** memory. In other memories, even though the power is maintained, the physical quantity that represents the data gradually degrades unless it is refreshed periodically; such a memory is termed **dynamic**. Although the periodic refreshing does decrease the availability of the memory for reading and writing, long retention time and simultaneous refreshing of many bits keep the refresh duty cycle small (about 10%) practically.

Memories that can retain the stored information even if power is not maintained (e.g., magnetic bubbles and ferrite-core) are termed **nonvolatile**, in contrast to **volatile** memories, which lose information when the power is removed. However, even though a memory cell may be inherently nonvolatile, loss of information may occur unless special care is taken in the design of the memory system so that no spurious writing or destructive reading can occur during transients in the powering.

Criteria of Merit

The **criteria of merit** for memories include capacity, speed, cost, reliability, volatility, life, volume, weight, and power consumption. These are interrelated, and their relative importance depends on the application. Desirable characteristics of memories in a computing system are

Large *capacity*.

High *speed*, in order to interchange blocks of information rapidly between faster and slower memory units, between memory and attached storage, and between memory and processor, the latter in times of the order of a machine cycle. Memory access time, cycle time, and data rate are pertinent measures of performance and may have different values for write, read, or read-modify-write cycles.

Low *cost*, specifically, the cost per bit of memory should be much less than that of a logic gate. For large-capacity stores, the cost per bit is most important; for small capacity, the cost of the minimum capacity store (i.e., the entry cost) is often more important. (Capacity, speed, and cost are usually intimately related.)

Small *volume*, *weight*, and *power* consumption, which are especially important for aerospace applications and for portable equipment; power consumption and the attendant cooling requirements must be dealt with in any design. Power requirements increase with speed, and speed is related to physical size due to the propagation delay of the signals along cables, wires, or optical paths, as limited by the speed of light.

High *reliability* and long *life*, which may be influenced by environmental factors and design. Error-detection and error-correction codes (ECC) often are included to extend the inherent device capabilities to meet the system objectives.

Nonvolatility, which is mandatory at the largest-capacity storage level of the system but is not necessary at every level. Instead, the system may be designed to transfer information from volatile memories into a nonvolatile store whenever a threat is detected.

Of course, simultaneous attainment of all these attributes in a single memory is impractical; compromises need to be made. One means of doing this is via a **storage hierarchy**, which uses several levels of storage having a range of characteristics and employs an appropriate technology for each level (see Section 9.7). The relatively large capacity levels, having low cost per bit but slow access, are referred to as **attached storage**. For these, **magnetic recording** is a practical technology (i.e., magnetic tape and disk files; see Section 9.11). At the other extreme, maximum memory speed for interaction with the CPU is obtained by supplementing the **registers**, which are part of the CPU, by a very high speed but modest-capacity **cache memory**. This complements the **main memory**, which has larger capacity and lower cost per bit but slower speed. For cache (see Section 9.7.1) and main memory, **semiconductor technology** provides an attractive range of speed and cost; nonvolatility is not mandatory, and ECC is available for reliability.

Between the high-speed but rather expensive memory closely associated with the CPU and the very low-cost-per-bit but slow-access attached storage, there is a large disparity in speed, cost, and capacity. This motivates hardware approaches to provide intermediate levels, sometimes referred to as **backing stores**, or **file-gap fillers**, with capacity and cost intermediate between main memory and files. Such backing stores may be implemented with either RAM or sequential-storage technologies. RAM technologies are discussed in this section, attached storage in Section 9.11.

9.6.2   How a Memory Works

Three types of signals flow between the CPU and memory: *address*, *data*, and *control*, as indicated in Fig. 9.62. The address bits determine the sites in memory that are accessed, the controls effect the desired operation (e.g., write or read), and the data lines carry the information to and from the memory.

The **memory cell** is the site where binary information is stored. These cells are arranged in two- or three-dimensional arrays, and each cell is coupled to at least two external paths, which are used for selecting the cell and for data transfer. The details of the memory cell and its external connections vary with the technology used for the cells and are covered in Section 9.6.3.

Memory cycles are initiated by the processor; it issues an address and also provides control signals (possibly by way of a storage control unit). Common modes of operation are write, read, or read-modify-write. For a write operation, the data to be written are presented by way of the memory bus (see Section 9.13), and the address is supplied from the memory address register via the address bus. The memory address register and the memory data register may be considered part of either the CPU or the memory; in some systems, additional registers or buffers for either or both may be included within the memory, as depicted in Fig. 9.62. The address bits are decoded in the **address decoder**, which selects the appropriate set of **array drivers** to energize the addressed memory cells for writing. However, the output of the decoder is gated by a control signal that **enables** the write operation to proceed; the design must assure that the correct address and data signals are valid when the decoder receives the enable signal. The data to be written are provided via the data drivers and data lines (also commonly called **bit-sense** lines) to the selected cells, resulting in storing this information. For a read operation, the address is similarly presented, but now the driving of the

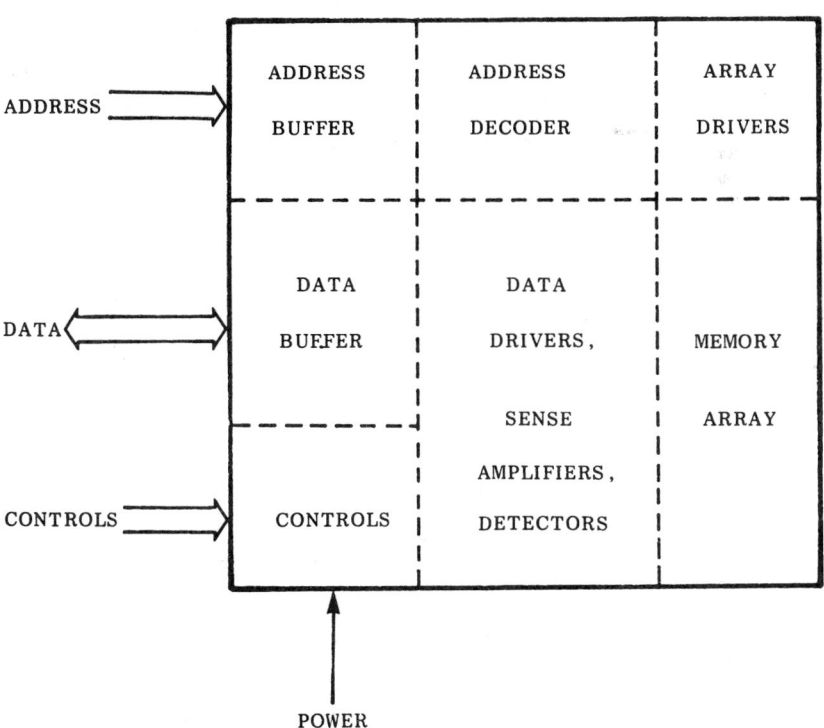

Fig. 9.62   Memory system.

selected memory cells causes them to apply signals representing their stored information onto their respective data lines; from there, the signals propagate to the **sense amplifier** and **detector** circuits, where they are amplified and detected (sensed). Then they set the data buffer and propagate to the memory bus. (If reading is destructive, then, as part of the read cycle, these sense amplifier and detector circuits will provide outputs also to the data lines to rewrite or **restore** the same data.) A read-modify-write cycle combines the read cycle at a given address, propagation of the detected data to the memory bus, presentation of new data on the memory bus by the processor, and writing the modified data at the same address.

A basic function of the control signals is to select the desired memory mode. The controls then generate and distribute appropriate timing and other control pulses to various parts of the memory. For instance, the same lines in the memory array are used for both reading and writing a given word. However, it is the combination of appropriate polarity, magnitude, and timing of the excitation of these lines that effects the desired mode, and this is governed by the controls.

The power required to maintain stored information is much less than that needed to read or write it at the required speed; hence, in many memory implementations, the supplied voltage(s) is reduced in the standby condition to a level where the data are just maintained. When a read or write operation is initiated, the full operating voltage is applied by the controls to that portion of the memory that is addressed, continuing for the duration of that memory cycle. In dynamic memories, the controls must initiate a refresh mode at intervals less than the minimum retention time. Some memory applications provide more than one port, so that data may be controlled to flow simultaneously to and from the memory via different ports.

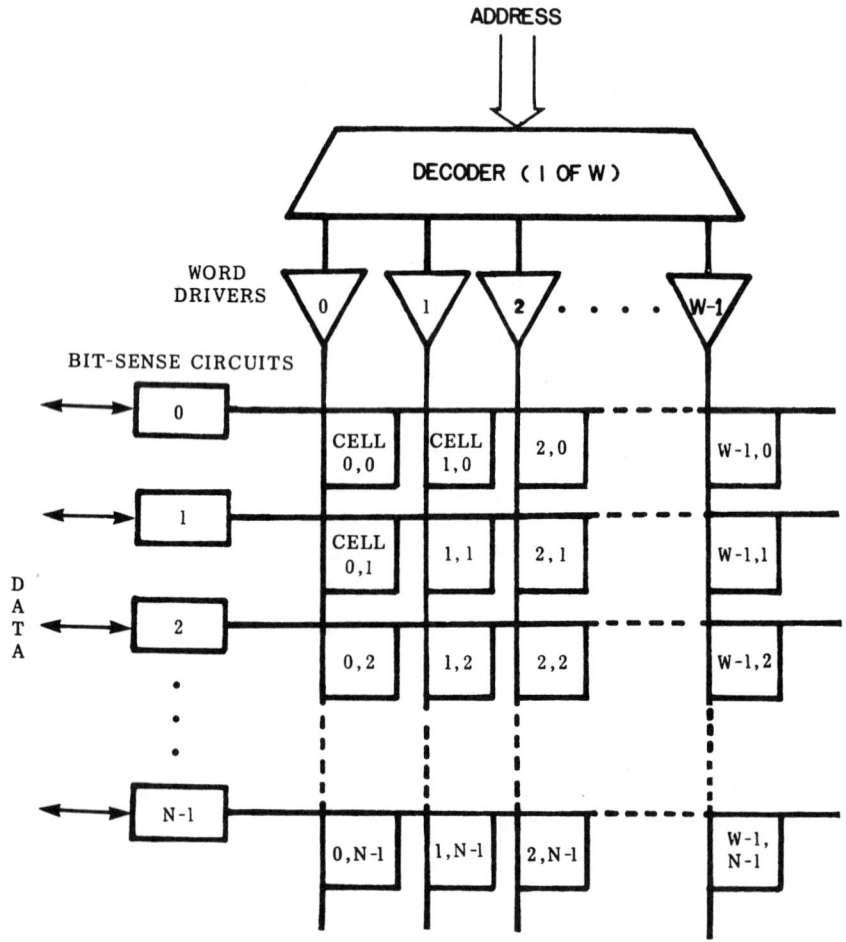

Fig. 9.63 Word-organized memory.

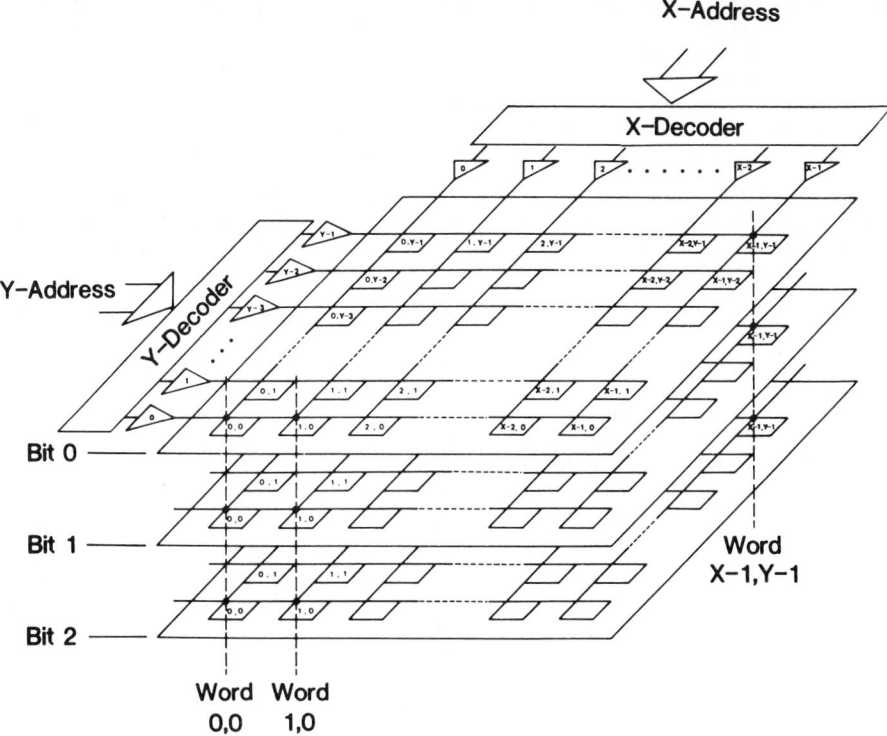

Fig. 9.64 Bit-organized memory.

Memories may be **organized** and **addressed** in a variety of ways. Usually, they are organized as some number of words, W, by some number of bits, N, where N may include both the data bits and bits for parity or for more sophisticated ECC. Memories are frequently composed of multiple subunits (e.g., basic storage modules, semiconductor chips, ferrite-core planes, etc.). One common memory organization is the **word-organized** memory (see Fig. 9.63). In such an organization, all the bits of the memory word are located along a physical **word line**, which passes through the memory array and is energized by the appropriate array driver when that word is selected. Each bit on the word line is also associated with a **data line**, or bit-sense line (which also links the other bits in the same position on other word lines in the array). When the word line is energized for writing or reading, all the memory bits on that line are written or read simultaneously, each bit storing the signal from its own data line for writing or coupling its stored information to its data line for reading. In contrast, the memory may be organized into multiple two-dimensional arrays of memory sites, each site being linked by a pair of orthogonal x and y conductors (see Fig. 9.64). One site in each array is selected by simultaneously energizing the appropriate x and y lines in each array. This is termed **coincident selection** [or XY or **two-dimensional selection** (2D)], the logical AND function of the signals on the x and y lines being performed in each memory cell. Additional lines are used to carry the signals from the cells to the sense amplifiers. In this scheme (called **bit-organized** in a ferrite-core memory or **bit per chip** in a semiconductor memory), each bit of the memory word comes from a different array. Variations of these organizations are possible; the relative advantages have to do with the number of devices and circuits required for driving and sensing and the speed of operation. Briefly, bit organization requires fewer supporting circuits (of the order of $X + Y + N$ compared to $X \cdot Y + N$ for word organization) but is slower because of the heavier loading on the bit circuitry.

9.6.3 Memory Technologies and Their Characteristics

Semiconductors are the primary technology for memory. This has come about because of the advances that have been made in silicon planar technology, making possible high levels of integration, inclusion of the decode, drive, sense, and control circuitry on the memory chip, and achieving high density, good yield, low cost, high speed, and low power. Semiconductors are superior to magnetic cores, which preceded semiconductors as the major memory technology, in essentially all aspects except nonvolatility.

Semiconductor Memories

There are two basic types of memory cell used in semiconductor RW memories; one uses the flip-flop, the other uses charge storage on a capacitor. The charge-storage cell can be implemented with as few as one active (switching) device plus capacitor (one-device or 1D cell); it is dynamic cell. The flip-flop cell requires at least the four devices of the flip-flop; it provides a static cell. The trade-offs between the flip-flop and capacitor switch cells involve density (hence cost), performance, and static versus dynamic operation. The dynamic, charge-storage cells, requiring fewer devices than the static, can be laid out more densely on a chip. (Since semiconductor processing is a batch operation, the greater the density, the more bits per batch and the lower the cost per bit.) However, the signal to be sensed is just the small amount of stored charge; hence sensing is more complex and slower than for the static cells, which inherently have gain within the cell. Moreover, the charge-storage cells require periodic refresh; however, this can be done for many cells simultaneously and uses a relatively insignificant fraction of memory time. Thus the cache and high-speed memory functions are ordinarily implemented with static memory technologies, whereas large-capacity main memories (especially when used in conjunction with a cache memory) and backing stores are usually designed using dynamic memory technologies. Semiconductor RW memories are volatile; however, at the system level they can be provided with uninterrupted power sources to maintain data during outages of utility power.

There are two major branches of semiconductor technology: **bipolar** and **metal-oxide semiconductor (MOS)**: each of these has further subbranches corresponding to particular device and circuit technologies. Bipolar memories are used primarily for high speed and are usually implemented with static cells. Bipolar dynamic memory is achieved by storing minority carriers in *pn* junctions; it offers lower cost and fairly high speed and is competitive with MOS memory.

Semiconductor main memory first became practical with MOS technology as sufficient density was achieved to lower the cost to levels competitive with ferrite-core memory at comparable performance. Both static and dynamic MOS memories are practical commercial products. Increasingly large levels of integration are being achieved as lithography capabilities for fine lines over large fields are improved, as chip dimensions are increased, and as device and cell design and technology capabilities are enhanced.

A common static MOS RAM cell is implemented with six devices (6D), four for the flip-flop and two as transmission gates to connect the cell to the data lines under control of the select line [see Fig. 9.65(*a*)]. A common dynamic MOS RAM cell is the (1D) charge-storage cell [see Fig. 9.65(*b*)]. The single MOS transistor connects the storage capacitor to its data line; it is switched from its gate, which is connected to a select line. (In a word-organized memory, the select line is the word line, whereas in a bit per chip organization, the select node itself may be driven from XY selection gates.) Sophisticated sensing techniques have been developed to detect the small stored charge of the cell when it is transferred to the capacitance of the data line. Retention times are of the order of milliseconds. (Other RAM cells have also been used.)

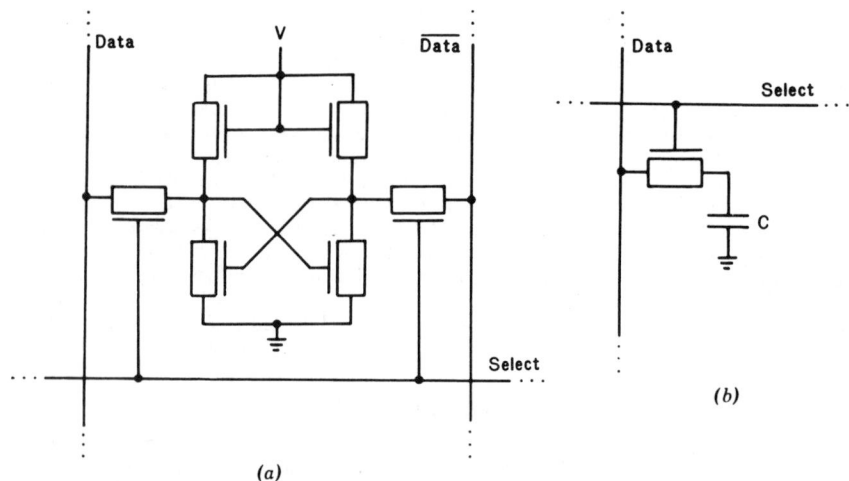

Fig. 9.65 Common MOS RAM cells: (*a*) 6D static RAM cell, (*b*) 1D dynamic RAM cell.

An important application of semiconductor memory is ROM. Semiconductor memory became practical just after microprogrammed control became widespread in mainframe computers; then, at the beginning of the microcomputer era, semiconductors were applied to implement ROMs. A ROM system is essentially the same as shown in Fig. 9.62, except that data flow only out from the memory, and no write or refresh controls or data drivers are used. The ROM cell must couple an electrical signal from the word drive line to the data line wherever a binary 1 is stored and nowhere else, and must prevent unwanted coupling or "sneak paths." This can be accomplished in any of the semiconductor technologies using a suitable diode or transistor in the memory cell to provide the coupling between word lines and data lines. Actually, the coupling device may be present in every cell, the distinction between the binary value of 1 or 0 of the stored information being determined only by whether the coupling device is electrically connected. In a ROM, this connection is preprogrammed into the mask that is used to fabricate the metal interconnections. In a PROM, a fusible link (e.g., of Nichrome or polycrystalline silicon) is included in each cell, and each cell can be selected and written once by burning out its fusible link. [This requires extra write circuitry (e.g., data drivers and higher current capability compared to a ROM) and is implemented in bipolar technology.] Such ROM and PROM memories are nonvolatile. The choice of the technology with which to implement a ROM is based on the required speed-cost-power trade-offs: bipolar for high speed, MOS for low cost, CMOS for low power.

Ferrite-Core Memories

Ferrite-core memory was the mainstay of main memory until semiconductors displaced it, and in fact, the main memory of a computing system is still sometimes referred to as "core memory." Ferrite-core memory can be designed to be nonvolatile, which is its only remaining advantage over semiconductor memory for most new applications.

Toroidal ferrite cores with pronounced magnetic hysteresis (called "square-loop" cores) are the memory cells. They are arranged in **core planes**, where they are threaded by the appropriate wires used for driving and sensing. The operation of the basic memory cell depends on the magnetic hysteresis characteristics of the core.

9.7 MAIN MEMORY AND HIGH-SPEED STORAGE: ORGANIZATION
Kishor S. Trivedi

The purpose of a memory system is to hold information in a cost-effective manner. The need for a large-capacity high-speed main memory cannot usually be fulfilled at an acceptable cost with a single memory technology. To meet cost-performance goals, it becomes necessary to synthesize a memory hierarchy combining a variety of technologies with differing characteristics. The technologies are ordered into levels on the basis of speed and cost (per byte), so that most frequently used information is kept in fast but expensive memory while the less frequently used information is stored in cheaper and slower memories. A properly designed hierarchy displays the favorable features of its constituent elements and behaves like a high-speed, low-cost, large-capacity storage device.

The levels of a hierarchy may be divided into two groups by an access gap (or **delay boundary**), as shown in Fig. 9.66. The memory levels above the delay boundary are directly addressable by the CPU, whereas the information in levels below the delay boundary can be accessed only after it is moved to one of the directly addressable memories. The lower speeds of memories below the delay boundary require that a process generating a request to such a memory be suspended and the CPU context be switched to another process until the information is moved to a directly addressable memory. The alternative of waiting for the request to be satisfied is usually too wasteful of CPU time.

The study of a memory hierarchy can be divided into the design issues and the management issues. The design problem consists of determining the number of levels in the hierarchy, the technology of each level, the capacity of each level, and the topology of the hierarchy. The management problem consists of determining where information is stored, how it can be located upon request, and when it should be moved. We restrict our discussion to management issues only; a discussion of design issues may be found in the literature.

The objective of a management strategy is to maintain currently used data in faster memory levels in order to minimize access time. This requires recognizing periods of inactivity of data and moving (or percolating) such data to slower and cheaper devices. The movement of data may be controlled by users by allowing them to input their knowledge of data and program usage; however, the current trend is toward automatically managed hierarchies to relieve the programmer of the burden of hierarchy management. Portions of the hierarchy may be managed completely in hardware (typically, the levels above the delay boundary), and portions may be managed by the operating system (typically, the levels below the delay boundary). Hardware control is more efficient, but it is inflexible and expensive. Software (operating system) control is flexible, but the overhead of managing the hierarchy may be large. A judicious mixture of the two techniques may be used to achieve the desired cost-performance goals.

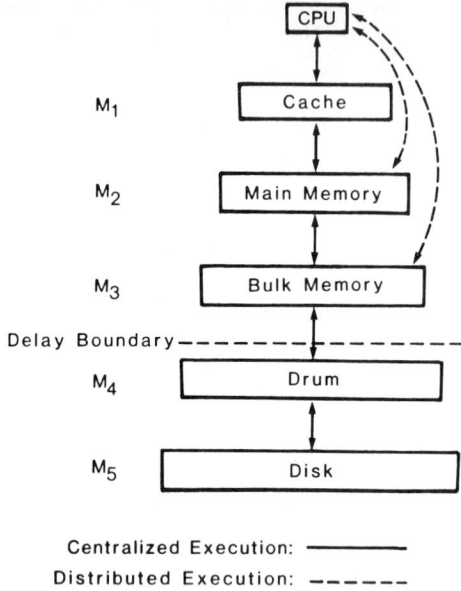

Fig. 9.66   Memory hierarchy.

Two techniques of managing directly addressable memories are available: centralized execution and distributed execution. In the **centralized mode** of execution, all information must be transferred into the fastest level before it can be referenced by the CPU. In the distributed mode of execution, the CPU is allowed to reference information in a slower, directly addressable level without transferring it into the fastest level. The more frequently referenced information should be stored in faster levels. The **distributed mode** of execution can reduce the traffic between the levels in the hierarchy. If the access frequencies are stationary, a preallocation of information across levels may be made. More commonly, however, access patterns are time dependent, and a promotion policy is based on a trade-off between the time to move the information to a faster level and the performance penalty of executing at the slower speed. Because of the difficulties involved in implementing such a promotion policy, the centralized mode of execution is usually preferred.

Next we consider a management strategy for the entire memory hierarchy, including the levels below the delay boundary. A staging hierarchy is a good example of an automatically managed hierarchy. Such a hierarchy has two or more levels of memory, denoted by $M_1, M_2, \ldots, M_H$, with capacities $C_1 \le C_2 \le \cdots \le C_H$ (bytes), and access times $t_1 \le t_2 \le \cdots \le t_H$. Information movement is between adjacent levels only. Movement between levels M_i and M_{i+1} occurs in blocks of size B_i (bytes), so that $B_1 \le B_2 \le \cdots \le B_{H-1}$. Each block of size B_i is composed of an integral number of blocks of size B_{i-1}, called its **descendants**. Similarly, each block of size B_i has a unique **parent** block of size B_{i+1}. All CPU references to the hierarchy are always satisfied at the fastest level by M_1 (i.e., centralized mode of execution). Movement of information within the hierarchy obeys the staging rule: whenever a block x is moved upward into a faster level M_i, its parent block x' must have been moved into level M_{i+1} (if it was not already there). Blocks may be moved up on demand or in anticipation. The usual approach is to use demand staging. The downward movement at any level occurs as a result of the application of a replacement algorithm. The usual replacement algorithm for each level is LRU (least recently used) or an approximation to it.

The management of memory hierarchies is explored further in the next section. Although the discussion is based on two-level hierarchies, the concepts generalize naturally to the multilevel hierarchies. The discussion of hierarchy management will reveal the fact that associative (or content-addressable) memories are needed for efficiency in implementation. Therefore, we discuss associative memories in the last section.

9.7.1   Two-Level Hierarchies

In the early days of computing, main memory was expensive and small. This implied that programs requiring larger memories had to be divided into a number of pieces, called **overlays**. The programmer was responsible for breaking the program into pieces and issuing input/output commands to transport overlays between the main and the secondary memory. This manual overlaying process tended to consume a large fraction of total program development time. The concept of virtual memory, which provides automatic overlaying, is now commonly employed, freeing the programmer from the tedious bookkeeping task. Such two-level hierarchies consisting of the main memory and the secondary storage are discussed later.

Another type of memory hierarchy evolved from the ubiquitous speed differential between the CPU and the main memory. This speed differential gave rise to program-accessible registers, instruction buffers, data buffers, and ultimately to the cache. Cache (or high-speed buffer) is a small, fast memory, typically implemented in bipolar technology, interposed between the CPU and the main memory. The most frequently used information (both instructions and data) is kept in the high-speed cache, so that the slower main memory is referenced only infrequently. Cache-to-main memory hierarchies are discussed later in this section.

Management of cache-based systems is usually performed by hardware for high speed; whereas, management of virtual memory systems is performed partly by software (as part of the operating system) for flexibility. Other than these differences, the two types of hierarchies are common in concept.

Virtual-Memory Systems

In order to classify various techniques of memory management, we make a distinction between the address space N of a program and the physical memory space M allocated to it. (Note that in a multiprogramming environment, the physical space M allocated to a program may be a small fraction of the total available memory space.) Since the program generates references to the address space N while the CPU can only make references to the locations in M, an address translation is required. Let f denote this address map, $f: N \rightarrow M$. We classify memory management techniques by the nature of the associated address map (see Fig. 9.67). The simplest scheme fixes f; hence it does not allow the program to be relocated in memory. The static relocation scheme allows the program to be relocated from one run to the next, but no relocation is allowed during a run of the program.

Schemes with static f are constraining for two reasons. In multiprogrammed systems, main memory is simultaneously shared among many programs. Programs may have to be moved around in memory to make space for other programs, or they may have to be moved in and out of memory. This implies that the program occupies different areas of main memory at different times; hence, f must be dynamic. Such a scheme is known as **dynamic relocation**. Another reason for a dynamic f occurs when the size of the address space N is larger than the size of the allocated memory space M.

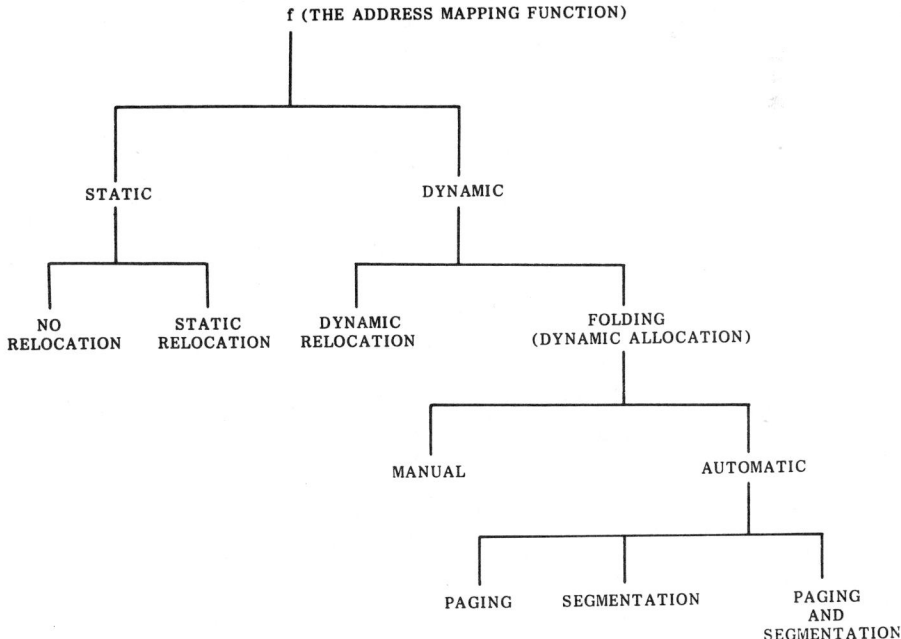

Fig. 9.67 Classification of memory management schemes.

PAGE FRAME NUMBER	SECONDARY STORAGE ADDRESS INDEX	FAULT	USAGE	DIRTY
4	4096	0	1	0
16	5192	0	0	1
-	377	1	-	-
-	6380	1	-	-
8	259	0	1	1
36	4000	0	1	0

2^{n-p} ENTRIES

$m-p$ BITS

Fig. 9.68   Example page table.

Different subsets of N will occupy the memory space M at different times, and we have a dynamic allocation of main memory. Such a scheme is also known as **folding**. The folding process may be carried out manually by the programmer, or it can be performed automatically by the system. Systems with automatic folding are said to provide a virtual memory that is potentially larger than the real memory space. The mapping mechanisms that implement virtual memory also provide the capability of dynamic relocation. Paging, segmentation, and a combination of the two are the three common schemes for providing a virtual memory.

Besides relieving the programmer of the difficult task of manual folding, a virtual-memory system provides several other advantages. Programs are unaffected by changes in the size of main memory, which would require reprogramming in a non-virtual-memory system. In a multiprogramming system, the use of virtual memory provides flexibility in memory allocation, since programs may be provided variable amounts of memory. This flexibility in memory allocation can be used to improve system efficiency.

Paging Systems.* In paging systems it is assumed that the entire address space N of the program resides in secondary storage. This address space is broken up into equal-size units called **pages**. Page sizes ranging from 512 to 4096 bytes are common. For our discussion, assume a page size of 2^p bytes. Main memory is broken into units of the same size called **page frames**. Assume that a main-memory real address is m bits long and a virtual address is n bits long. Thus the size of the address space is 2^n bytes (or 2^{n-p} pages), whereas the size of main memory is 2^m bytes (or 2^{m-p} page frames). Typically m is much smaller than n. Therefore, all pages cannot be in memory at once. A page table is used to keep track of the pages that are currently in memory. The page table is an array with an entry for each of the 2^{n-p} pages. Each entry includes several flags and pointers (see Fig. 9.68).

The mapping of an n-bit virtual address to an m-bit real address is performed by a hardware-implemented address translation mechanism. The most significant $(n-p)$ bits of the virtual address represent the index, in the page table, of the required page, identifying a page table entry, while the least significant p bits give the byte address within the page. See Fig. 9.5. If the **fault** flag of the page table entry is zero, the required page is in main memory, and a field of the entry gives the number of the main memory page frame on which the page resides. The real address is then obtained by concatenating the $(m-p)$-bit page frame number with the p-bit byte address.

If the fault flag of the page table entry is equal to 1, the required page is not in memory, and a hardware interrupt known as a **missing page fault** is generated. Since address mapping takes place on every memory reference, address mapping is implemented in hardware. A page fault is a less frequent event, however, and it triggers a complex sequence of actions. Therefore, an operating system routine called the **paging algorithm** handles the response to a page fault. The paging algorithm locates the required page on the secondary storage, using a field of the page table entry, decides where to place

*See also Section 9.1.3.

this page in main memory (this may require the replacement of a page currently resident in main memory), issues instructions to fetch the page from the secondary storage, and updates the page table entries as required. The program that generated the page fault is unable to execute until the page fetch is complete; therefore, the operating system schedules another program for execution, if one is available.

The page tables of all active users are usually kept in main memory. The address mapping scheme described is inefficient since every virtual address reference requires two accesses to main memory (one to consult the page table and the other to access the required byte). To solve this problem, a small associative memory (see Section 9.7.2) is used to hold the page table entries for the most recently used pages. Since the associative memory is capable of fast parallel search, the penalty of consulting the memory resident page table is eliminated whenever the required entry is found in the associative memory. Experiments have indicated that an associative memory that can hold 8 to 16 page table entries provides a high probability of finding the required entry in the associative memory.

When a new page needs to be fetched and there is no available page frame, one of the pages currently resident in main memory needs to be replaced. The most common replacement policy is to replace the least recently used (LRU) page. This algorithm replaces that memory page whose last reference was longest ago. This implies that the pages in main memory have to be ordered by their most recent usage, and this list needs to be updated at the time of every reference. Therefore, the implementation of a true LRU algorithm tends to be expensive. Fortunately, however, one can make good approximations to LRU with the help of one or more usage bits associated with each page.

When a particular page is selected for replacement, a page-out operation will normally be required. However, if the page was not modified (or written into), no page-out operation is needed since a valid copy exists in the secondary storage. To exploit this possibility, a flag known as the **dirty bit** is provided as part of each page table entry. Whenever the contents of a page frame are written into, this bit is set. The paging algorithm may examine appropriate dirty bits in making a replacement decision.

The complex sequence of actions triggered by a page fault usually requires the execution of several thousand instructions. This overhead can be reduced by either reducing the number of page faults or by decreasing the time to service a page fault. The latter can be achieved by a proper redesign of the paging algorithm or by increased use of hardware/firmware aids. To reduce the number of page faults, one can train programmers to produce better programs or use automatic methods of program transformations. Another way to reduce page faults is to increase the main-memory allocation of a program.

Increasing memory allocation of a program can reduce its page faults, but it may have a detrimental effect on system performance in a multiprogramming environment. For a given amount of real memory, increased allocation per program implies a reduction in the degree of multiprogramming, which in turn reduces the probability of finding a program ready to run when the current program incurs a page fault. On the other hand, a large degree of multiprogramming will reduce the memory allocation of each program and can result in an excessive number of page faults, producing very little useful activity. Such behavior of the system is called **thrashing**. To avoid thrashing it is important, therefore, to make accurate estimates of memory demands of programs and allocate main memory based on these estimates.

A practical scheme for main-memory allocation is to estimate the program's future memory requirements from the observation of its immediate past behavior. Furthermore, since we expect the memory requirements of a program to vary during execution, dynamic estimation of memory requirements is needed.

Next we consider the factors that the system designer has to consider when selecting the page size. For systems that use an electromechanical secondary storage device, a major consideration in the selection of the page size is the efficiency of the page transport operation (see Section 9.11.2). To reduce the latency overhead per byte of accessing the device, a large page size is required. With an electronic secondary storage device, however, a small page size is feasible.

Other considerations in the selection of the page size are its effect on the page fault rate and storage fragmentation. Programs are known to be more compressible when the page size is small; that is, they will not page fault heavily with a small allotment of main memory. A small page size is, therefore, desirable in a system with small memory size. A small page size is also desirable in a time-sharing system with a small time quantum. Page fault behavior is relatively insensitive to the page size with a large allotment of main memory.

Next consider the fragmentation problem. The size of the address space of a program may not be an integral multiple of the page size. The effect of rounding to obtain an integral number of pages is to leave unused words in the last page. Reducing the page size reduces this internal fragmentation of storage but also implies an increase in the size of the page table. The loss of storage due to page tables is known as table fragmentation. It has been shown that a page size close to 50 words is optimal. However, as pointed out earlier, such a small page size is feasible only when an electronic secondary storage device is used.

Segmentation.* Division of the address space into equal-size pages provides an efficient technique for storage management. However, the pages are physical units having no logical significance, and the address space is a linear array devoid of any visible logical structure. As a result, sharing programs or data between several concurrently active users tends to be cumbersome. Also, if a variable-size data structure is a part of the address space, a portion of the address space large enough to hold the maximum size must be allocated in advance. Segmentation avoids some of these difficulties of a paging system.

The technique of segmentation divides the address space into user-defined, variable-size units, called **segments**. The segments can be referenced by symbolic names, and these names need not be bound to the address space until the time of first reference. The address mapping process is similar to that for paging systems.

The advantages of segmentation stem from the fact that segment boundaries are usually natural program and data boundaries. Each program module (procedure or data) can be a named segment, and it can have unique access privileges (read, write, execute, etc.). Access control to the segments is naturally exercised as part of the address mapping process. Because of their logical independence, procedure segments can be recompiled and data segments can grow (or shrink) in size without affecting other segments. A segment can be given different names in different users' address spaces, providing a convenient method of sharing segments.

We have seen that paging provides a convenient mechanism for memory management, while segmentation provides a useful logical division of the address space. It is possible to combine the advantages of the two schemes.

Cache-Based Systems

The cache provides a cost-effective technique for speeding up mainframe computers and minicomputers. In such systems, the cache and the main memory are divided into equal-size units called **blocks**. There is an associative memory (Section 9.7.2) to perform the mapping between main-memory blocks and cache blocks. When the CPU generates a reference to a main-memory block, the mapping memory is searched using the main-memory address of the block as the key. If a match occurs, the required block is in cache (this is called a **hit**), and the execution proceeds at cache speed. If there is no match (this is called a **miss**), the required block must be fetched from the main memory.

Like the page fault rate of paging systems, the performance of a cache is measured by the **miss ratio**, which is defined to be the fraction of CPU references not found in cache. As in virtual memory systems, the mapping of a main-memory address into a cache address is performed in hardware. In addition, the response to a cache miss is also handled by hardware, which constitutes the main difference between a cache-based system and a virtual-memory system.

Important design parameters of such a system are the cache size, the cache organization, the block replacement algorithm, the block size, and the manner in which CPU write operations to memory are handled. We discuss each of these in the following paragraphs.

Of the parameters already described, performance is most sensitive to cache size. However, cache is expensive, and cost is a major consideration. Selection of optimal cache size to meet the desired cost-performance goals is an important system design problem.

The cache organization providing the smallest miss ratio, known as **fully associative cache**, allows any memory block to be assigned to any cache block. To perform the mapping from a CPU-generated main-memory address into a cache address, an associative mapping memory (or index directory) is needed to indicate which block from the memory is in which block frame of the cache. Also, a list must be maintained that orders the cache blocks by their replacement priority when space is needed. For each CPU request, the entire directory must be searched associatively, and the priority list must be updated. With current technology, fully associative cache is either slow or expensive or both.

An alternative cache organization is direct addressing, where the main memory is partitioned in such a way that all blocks in one partition can reside only in a unique cache block. The required block can now be accessed by indexing (if present in cache). This eliminates a search of the mapping memory altogether, but since many main-memory blocks will compete for a single cache slot, the miss ratio is likely to be large.

An intermediate type of cache organization, called the **set associative scheme**, is known to perform well at an acceptable cost. In this scheme, the cache is divided into a number of sets (or classes), each of which contains one or more blocks. Blocks within a set are searched associatively, whereas the required set is obtained by indexing. Since the replacement decision concerns the blocks in a set (which will be small), a true LRU algorithm is feasible (unlike the situation in paging systems). An important design consideration is the number of sets employed (or equivalently, the size of each set).

The size of a cache block is also an important factor. It has been observed that programs possess the property of sequentiality—a tendency of referencing successive addresses in the address space. A larger block size exploits the sequentiality property but implies that a smaller number of blocks are in

*See also Section 9.1.3.

cache. A larger block significantly reduces the miss ratio, especially in a smaller cache. However, a large block size implies a long time to fetch a block from main memory. The block fetch time is usually reduced by employing wide data paths between the cache and main memory. But this implies increased cost in bus hardware and control logic. For minicomputers, a small block size (e.g., 4 bytes) may be chosen to minimize cost. For larger systems, a block size of 32–64 bytes is common to provide enhanced performance.

So far, we have not made a distinction between read and write accesses to memory. When the CPU makes a write access, one strategy is to update both the main-memory location and its cache copy (if there is a hit) simultaneously. This strategy is called **write-through** (or **store-through**). Thus there is never a need to replace a block from cache; it can simply be overwritten without copying it into main memory. The alternative strategy is to update only the cache on a write hit. Now, when a modified cache block is replaced on a future miss, it must be recopied into main memory. This strategy is known as **write-back**. Experience suggests that only a small percentage (roughly 10%) of memory references are write accesses; hence write-through can be more efficient and is often used in practice. A related technique for speeding up cache operation is known as **load-through**. It allows portions of the block, as it is being fetched into cache, to be available to the CPU simultaneously, before the block transfer is completed.

As an example, we consider a typical cache organization, shown in Fig. 9.69. The 8K cache is divided into 256 32-byte blocks. This is organized into 128 sets (or columns), each set containing two

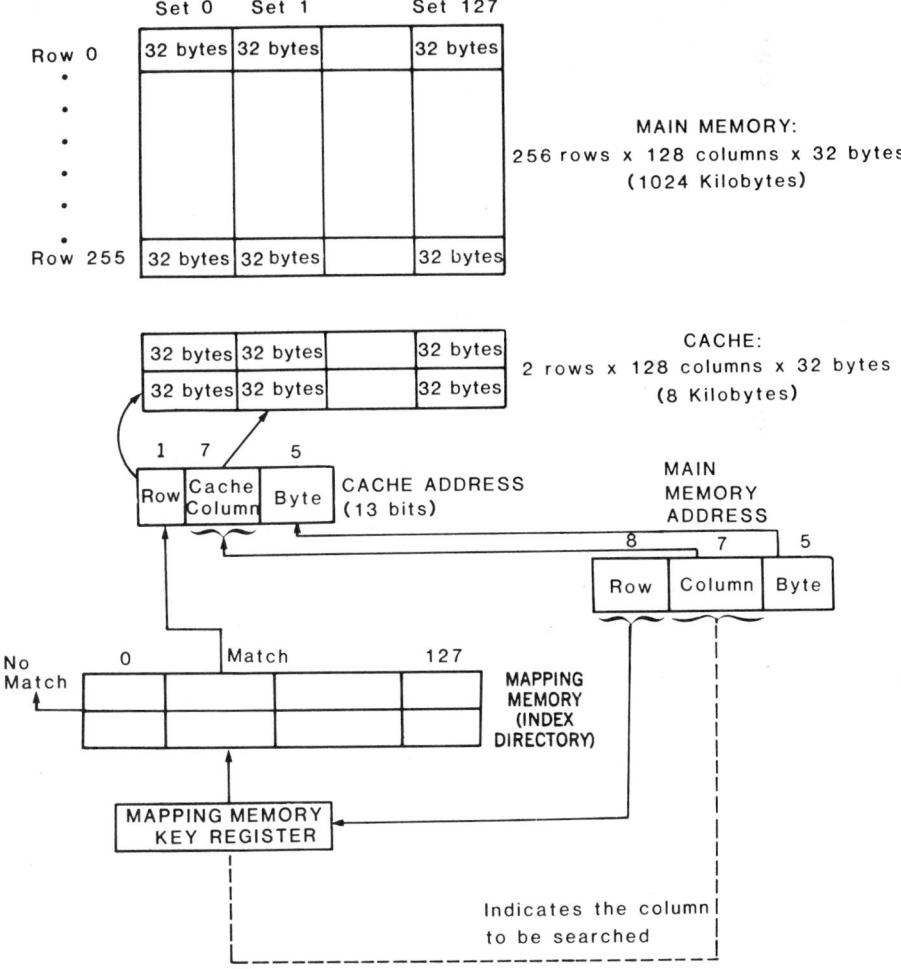

Fig. 9.69 Mapping process in cache main-memory hierarchy.

blocks. The main memory (assumed to hold 1024K bytes) is divided into 128 sets with 256 blocks per set (called rows). Blocks in set i of main memory may be placed only in set i of the cache.

A 1-million-byte main memory is addressed with a 20-bit address, whereas the 8K cache is addressed with a 13-bit address. (The associative mapping memory is also called the index directory.) The main-memory address is divided into several fields. The column field designates the mapping memory column to be used for search. Each of the two mapping memory entries indicates the main memory row whose block contents are currently held in the corresponding cache block. The row field of the effective address is associatively compared against the two entries in the index directory. If a match is found, the column field is used to access the corresponding cache block. If no match occurs, the requested block must be fetched from main memory, requiring a replacement of an existing cache block. Since the choice for replacement is restricted to one of the two blocks for the set involved, the LRU algorithm is easily implemented with a single usage bit.

If a cache-based system also possesses virtual memory, the main-memory address itself is obtained after a mapping from the CPU-generated virtual address. All CPU references are directed to a unit known as the **storage control unit**, which contains the virtual-memory address mapping hardware, the index directory of the cache, and cache hardware controls (e.g., the block replacement algorithm). The storage control unit also handles main-memory requests generated by the cache controls and by the I/O channel controls.

9.7.2 Associative Memories

Many computer applications require the search of items stored in a table. Searching a page table is an example discussed in Section 9.7.1. Searching a table is a common operation in a data base system. If a conventional random-access memory is used to store the table, an access can be performed by specifying the physical address of the item desired. If the physical address of the desired item is not known, a search procedure is required. The search procedure is a strategy for choosing a sequence of addresses, reading the content of memory at each address, and comparing the data read with the item being searched. The search procedure terminates successfully if a match is found, or unsuccessfully if the entire table is searched without a match. Many techniques have been developed to minimize the number of accesses while searching for an item in a random- or sequential-access memory.

The search time can be reduced considerably by using an associative memory or content-addressable memory (CAM). Such a memory can be accessed by using the contents of a subfield of the information being searched rather than by an address. The subfield chosen to address the memory is known as the key. Thus items stored in an associative memory may be thought of as pairs, (KEY, DATA). For example, if items of a page table are stored in such a memory, the page number (in the address space) will be the KEY, while the page frame number in which the said page is located and various flags will constitute the DATA.

An associative memory is searched in parallel for the required item. When a word is written in such a memory, no address is specified. The memory hardware is equipped to find an unused location and store the word. Upon reading, the content of the word, or a part thereof, is specified. The memory will locate all cells whose contents match the specified content and mark the cells for reading. The additional logic required for the parallel searching and matching circuits makes associative memories more complex and expensive compared to conventional memories. For this reason, the use of associative memories is limited to applications with a relatively small search space and requirements for a very short search time (e.g., address mapping hardware in virtual-memory systems and cache-based systems).

Figure 9.70 shows the block diagram of an associative memory. In this simple example, each entry in the storage array is a fixed-length (w-bit) word. Any subfield of the word may be chosen as a search field. The desired key is specified by the key register, which is compared simultaneously with the appropriate bits of all S words. Words matching the key emit match signals, which enter a select circuit. If several words match the key, the select circuit determines which word will be read out and placed in the output register. The select circuit may also read out all matching words in some predetermined sequence. Note that all words in the memory must simultaneously compare their keys with the input key; hence each word must have its own match circuit.

In the associative mapping memory of a virtual memory system, the $(n - p)$-bit index from the virtual address will be input to the key register. If a match occurs, the required page table entry is obtained in the output register. If no match occurs, the required page table entry must be obtained from the main-memory resident page table.

As another example, consider a set associative cache as in Fig. 9.69. Each column i of the index directory is an independent associative memory for mapping into set i of the cache. Thus the column field of the main-memory address indicates the specific associative memory to be used, and the row field is used as the search key. If no match occurs, the required block is not in cache. In case of a match, the cache row in which the required block resides is the output of the associative memory (no DATA field is necessary).

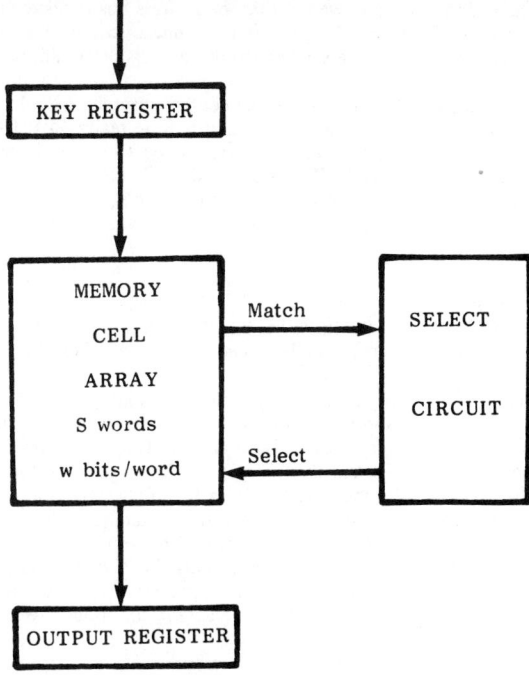

Fig. 9.70 Associative memory.

Since the cells of an associative memory have logical capabilities, such a memory is also referred to as a logic-in-memory device (or distributed logic memory). This can naturally be extended so that each cell may perform standard arithmetic and logic functions. In this case, it is more appropriate to refer to the resulting device as an associative processor.

9.8   MAIN MEMORY AND HIGH-SPEED STORAGE: MICROPROCESSOR SYSTEM CONSIDERATIONS
Dennis J. McBride

The term **microcomputer** indicates a computer that is small in terms of its size, power consumption, performance, cost, and other characteristics compared with other types of general-purpose computers. Some applications can take advantage of single-chip integrated-circuit microcomputers. Otherwise, the CPU, memory, and I/O functions are spread across some small number of chips. In this section we discuss aspects of system design wherein the CPU, memory, and I/O are separate entities and therefore need to be interfaced with each other. First is a discussion of selection signals, by which one memory location is chosen for reading or writing. Also, alternative means of using memory chips (and I/O devices) to fill the required address space are presented. Finally, we discuss other concerns, such as reserving memory for interrupt handlers, stack operations, and restrictions on address decoding.

9.8.1   Selection Signals

Consider the microcomputer system memory as an array of N memory locations, with each location M bits wide. For ease of design the width M is usually made to match the width of the CPU internal data path. Although this width is a reflection of the number of bits that can be operated on in parallel, it does not preclude the use of two or more contiguous memory locations to represent a data structure or an instruction.

Address Selection
In order to specify one memory location, the CPU generates an ordered set of binary logic signals that can be thought of as representing a positive binary integer. We call the conductors on which

these signals appear the address bus, and we say that the address bus holds the encoded representation of the memory location. The decoded representation appears within the memory subsystem to allow a read or write operation to only the single location that has been addressed.

Example 9.5. Memory integrated circuits are available that contain 64K locations, where each location is 1 bit wide. (We define 1K as equal 1024, so 64K is really 64×1024.) These memory devices thus require a 16-bit address bus as an input to describe a memory location uniquely. If the data path between the memory and the CPU is 8 bits wide, this address bus must be connected to each of eight identical memory devices in order to have a total configuration that is $64K \times 8$. We also refer to the data path as a bus; in this case, each memory device is responsible for a bit on one of the eight data bus paths.

Read and Write Selection

In addition to the address signals, it is necessary to supply a control signal to designate whether the operation is read or write. This is typically implemented as a single control line, and is often labeled R/W.

The timing and control for memory and I/O operations originate within the CPU. The technical specifications of a given microprocessor include this detailed timing information (see Section 9.9.7 for an example). It is usually presented as a set of timing diagrams. These diagrams illustrate (1) when the CPU address to memory is valid, (2) when the CPU places data signals on the data bus for a write operation, (3) when the CPU samples the data bus for a read operation, and (4) when the states of control signals change in order to effect the given operation. A unique sequence may exist for each type of CPU-memory interaction (instruction fetch, data fetch, data store). These details become important in system design, because memory devices must be chosen that match or exceed the performance required by the microprocessor. Timing diagrams are also used to present the performance of a particular memory device. Again, different diagrams are usually required for the read and the write operations.

For whatever memory operation is required, the microprocessor presents the address first. Some systems have a control signal that explicitly declares the time period of the valid address. After the memory location is specified by the address bus, the memory data appear on the data bus. For the read, or fetch, sequence the microprocessor simply waits one or more clock cycles after presenting a valid address and then samples the data bus, storing it in an internal CPU register. In some systems, the processor requires a "ready" signal from memory to indicate when the data bus may be sampled. This permits the use of a variety of memories (or other devices) with different cycle times. If the CPU is going to write data into memory, the address is presented first, with the R/W signal in its read state. This ensures that an arbitrary memory location will not be altered while the address bus is settling into its valid state. The CPU ignores the state of the data bus. The bus must be turned around, so to speak, by now declaring a write operation. This transition on the R/W control line tells the memory devices to deactivate their driver circuits. The contents of an internal CPU register can now be presented to memory, and will be saved in the designated location as the R/W line returns to its read state. The CPU is now allowed to change the state of the address bus to initiate the next operation.

Notice that this discussion assumes that the microprocessor and the memory are connected to a bidirectional data bus. Both sources must therefore be able to put their bus driver circuitry in a high (virtually infinite) impedance state. This is standard practice. Some memory devices, however, use separate signal paths for incoming and outgoing data. This permits faster memory performance. These devices can still be interfaced with a bidirectional bus, and bus interface devices are commonly available for this purpose.

Device Selection

The addressing range, or address space, as defined by the width of the address bus is usually not fully populated; part of the address space, perhaps even most of it, remains vacant. That is, physical memory locations do not exist for all addresses that can be generated by the CPU. There are various reasons for this. First, the majority of applications do not demand the maximum memory configuration. Second, the width of the address bus is defined by CPU architectural considerations, and not necessarily with a fully populated address space in mind. For example, some microprocessors have a 24-bit address bus, permitting a 16-megabyte address space. Only a modest proportion of large mainframe computers, or very specialized applications, can justify the associated cost of this quantity of memory. A third aspect, discussed later, is that some microprocessors do not have instructions devoted to I/O or for a dedicated I/O device address space. Part of memory address space must be reserved for I/O-device addresses. Frequently, the highest-order address bit is defined by the system designer as a memory versus I/O address indicator. This cuts the memory address space in half. In view of the infrequent need to populate all of address space, however, the assignment of I/O-device addresses within memory address space is not a burden. In fact, it can be an advantage.

Memory integrated circuits are available in standard widths of 1, 4, and 8 bits. The number of locations on a single device is chosen to be an integer power of 2. Clearly, the product of length times width is limited by the current state of technology. We therefore need a technique for expanding the width and length of main memory, to match our requirements, by using an array of standard memory devices.

When an array of memory devices is used to implement the memory space of a microcomputer system, means must be provided to select the appropriate one(s) of these to satisfy a specific memory access operation. Therefore, we require an additional control input to each memory integrated circuit. It is called a **device select**, or **chip enable signal**. When this is in its active state, the device can be addressed for read or write operations. Otherwise, the device virtually disappears from the data bus by assuming a high-impedance state. One can use this, for example, to provide a 2K location memory using two 1K location chips. Connect the 10 least significant bits of the address bus to both memory devices. The next most significant address bit is connected to the chip enable for one device, and the same address bit is inverted and then connected to the second device. Now only one device can be active at a time, so the data buses of the two devices can be common. One device populates locations 0 through 1023, and the other device occupies locations 1024 through 2047. This explicit control of the high-impedance state for memory devices gives very good expansion capability for ease of system design.

Some systems provide the minimum amount of decoding to select one of the memory devices at a time. This minimum decoding, however, causes a given actual location to be selected periodically throughout address space (e.g., in the preceding example, any memory word at address X from 0 to 2K will also be selected if the CPU generates address $X + 2nK$ for $n = 1, 2, \ldots$; however, location X is usually referred to in software only at its primary intended address). This technique of minimum decoding reduces costs for a minimum system but is otherwise not a recommended practice.

Bus Loading

We have several types of memory devices to choose from in configuring a total memory system. This means that many implementations are possible to realize a given total array size. Furthermore, the chip select technique makes practically any one of these implementations easy to carry out. We should now consider the second-order effects, however, since there may be an optimum approach. If the address bus is 16 bits wide, there are 64 regions of 1K locations each. In order to do a complete decoding for selection of 1K memory chips, we have to perform a 6-line-to-64-line decode of the 6 highest address bits. The 10 lowest-order address lines still go to all the memory devices. The extra decoding is explicitly required because there are 64 possible states that the 6 highest address bits can assume, so we use one of the 64 lines as a chip select for the lowest 1024 locations. We use a second line to select the next 1024 locations, and so on.

Let us now compare implementing a memory space with chips of 1K locations by 4 bits versus 4K $\times$ 1 chip organization, for an 8-bit data bus. For the 1K $\times$ 4 chips, two devices must be placed in parallel, each for 4 bits of the data bus. The chip select line for each region must simultaneously activate two memory devices, and each chip sees the 10 lowest-order address lines. For modest amounts of total memory, this design approach is reasonable. Now, a 4K $\times$ 1 device, with eight chips in parallel, needs two more address lines, but it uses three fewer data lines, per chip. It is a slightly more efficient package. Second, the remaining address lines are now only 4 in number compared to 6. Unique decoding can be accomplished via a 4-to-16 line decoder, as there are only 16 regions of 4K in the address space. Next, eight of these memory devices must be used in parallel to have an 8-bit bus. The address bus therefore feeds eight memory chips per 4K region. This was also the case before since the 1024 $\times$ 4 implementation used eight devices, stacked four high and two wide. Finally, notice that our new arrangement places only a single device load on each data bus line, whereas the previous design placed four loads on each data line. For maximum performance (highest speed) the data bus should have a minimum of loading. The general approach is therefore to use the single-bit-wide organization, with n bits in parallel for an n-bit data bus, when minimum loading is essential.

I / O Selection

The same combination of R/W and chip select signals used to select memory devices can activate a register whose width matches the data bus. This register can serve as an I/O device itself or be part of a device controller (see Section 9.10.6). This register can be thought of as a memory device that makes the state of each bit visible to the external environment. We now have an **output port** at a designated memory address. Similarly, the address bus can be decoded for a CPU read operation. Data from the external environment would be placed on the data bus when this location is referenced. We call this an **input port**. This technique can be used for any modest number and combination of input and output ports. The circuitry to perform unique decoding becomes a limiting factor, as does loading of the data bus by too many devices. Some of the decoding function can be implemented within each I/O peripheral unit, thus distributing the logic as well as the loading. Access timing for these I/O locations must match or exceed the performance required by the CPU. This technique is called

memory-mapped I / O, and it generally does not interfere with the implementation of modest amounts of main memory. The beauty of this technique from a programming point of view is that input- and output-device addresses are accessible via all the powerful addressing modes made available for memory operations. Microprocessors that do support a separate address space for I/O devices also have a weaker instruction set for accessing these locations. This is because the complexity of the instruction decode logic in the CPU is limited. The extra decoding for I/O instructions must necessarily subtract from the total function possible on a single chip. There are applications where the optimum design takes advantage of both techniques in order to minimize explicit decoding.

Refresh Control

Microcomputers that use a large array for main memory often use devices characterized as dynamic (see Section 9.6.3). These devices do not use bistable storage elements such as flip-flops or latches. They rely on the presence or absence of charge on a capacitor to represent each memory bit. Those bits that contain charge will eventually change state since the charge will leak away. These memory devices have a provision for comparing the state of a bit against a threshold, and then resetting the charge to maximum, or zero, but this happens only when the chip is accessed. The microcomputer system must guarantee that specific addresses will be reactivated more often than a specified period. This is much too cumbersome to implement in software, so special hardware must be designed to address specific locations of all memory devices according to the given time constraint. Devices are available that assist in this task, and some memory devices incorporate all or most of this special function. This makes them appear to be bistable when, in fact, they are dynamic. The term often applied to these devices is **quasistatic** since most of the dynamic behavior is transparent from the system designer's point of view. In any memory that is fundamentally dynamic, though, there is a probability that memory access by the CPU will be delayed because of a concurrent refresh operation at that same address. The details of timing and performance must take into account the worst-case access situation.

9.8.2   Organizing Address Space

If we look at the address bus of a microcomputer, we may find several different devices sharing the total address space. An arbitrary memory location might be vacant, or it could be within the address space of a RAM, ROM, or PROM, or it could be a memory-mapped I/O-device address. For a discussion of ROM, see Section 9.6.3. The memory locations that remain vacant and those that are populated by various devices are determined by the decoding logic, which in turn is defined by the system designer. All of these devices are wired to the data bus. Only one source may be active at a time. We have to understand the microprocessor architecture, though, to guarantee that any special-purpose memory locations required for instruction execution have been implemented. This varies from one architecture to another in terms of detail, but the concepts are general in nature.

Reserved Memory Locations

Most microprocessors have one or more interrupt mechanisms, which access specific memory locations. During the interrupt process the currently active instruction sequence is suspended. Information pertaining to the current state is saved, typically on a push-pop stack, so that instruction processing can resume where it was stopped after the interrupt-handling program has been executed. Except for any implications that a time delay might have, the interrupted program is not aware that it was ever disturbed. Meanwhile, the address of an interrupt-handling program is placed in the instruction counter, and interrupt processing occurs.

Most microprocessor architectures define a fixed address to be loaded into the instruction counter when an interrupt occurs. Some architectures allow the interrupting device to specify part of this address field. Generally, these fixed locations should contain branch instructions to other locations in main memory where the interrupt-servicing routines are found. After power is applied, and before the first interrupt is allowed to occur, these locations must be loaded (written into) with the entry addresses for interrupt processing. Frequently, two or more sequential locations are required to hold these entry addresses. These locations must therefore be populated, either with RAM or ROM, so that interrupt processing can be supported. Some microprocessors define several such sets of memory locations, one set for each distinct type of interrupt. These interrupt types might include external and internal, maskable or nonmaskable, single level or multilevel.

Next, let us consider some stack techniques (see Section 9.3.8). It is possible to maintain a small stack on the CPU chip itself. This type of stack is not part of main-memory address space. An arbitrarily large stack can be implemented by using an internal CPU register called the stack pointer. The address in this register is the location of the top of the stack. If the stack pointer is as wide as the address bus, any location can be the top of the stack. As interrupts occur or as explicit push instructions are executed, this register points to successively lower locations. Nesting of interrupts or push operations is thus limited according to the amount of contiguous address space that has been populated below the starting stack address.

If the stack pointer register is not as wide as the address bus, at least part of the address space that can be reached by this register must contain RAM.

Many microprocessors are designed to detect the initial application of power, at which time the instruction counter is reset to location zero or preset to a defined value. The instruction at this address is loaded into the instruction register, and processing begins. For ease of use, many microcomputers are designed with ROM in this part of address space. Some level of application support can thus be made immediately available upon which higher levels of function can be loaded into RAM from diskette, magnetic tape, and so on.

Decoding the Address Bus

The most straightforward decoding approach is to choose RAM and ROM chips with the same number of locations. They do not have to be the same width, as this is easily handled via parallel layout of the appropriate number of devices for a given width. The decoding for chip enable can now select banks of memory that all contain the same number of locations. This means that banks of memory can easily be relocated by interchanging the routing of the select lines between the decoder and the memory devices. This is very important when address space is not fully populated, especially for satisfying absolute address regions required by previously written software. Systems are often designed to have these address select lines brought out to some kind of circuit card connector and then back onto the card on their way to the memory devices. Relocation of memory addresses is particularly easy in this case, by rewiring or replacing the circuit card I/O connector. Equivalently, miniature switches can be placed directly on the circuit card, so that the absolute address of a block of memory is determined by the switch settings.

If enough address space is uncommitted, an I/O device address can be enabled with an unused memory chip enable, or bank select signal. The input or output port will be selected whenever any of the memory addresses within the decoded region are placed on the address bus. If, for example, external memory decoding has been tailored for devices with 4096 locations each, I/O selection using these same lines will cause each port to consume 4096 locations of the total address space. For this reason, a second level of address decoding is often made part of the system design, for I/O selection, to reduce the address space defined by each port. In the extreme, decoding can be used to guarantee that each port is assigned a unique address. But this requires additional circuitry, and the uniqueness must be justified.

9.9 CONTROL, TIMING, AND SYSTEM SIGNALS
Nikitas A. Alexandridis

In this section we describe the control and timing function of a digital processing system. Both intramodule and intermodule signals are analyzed. Signals required to carry out the instruction execution steps are described. We also illustrate various ways of implementing this control and timing function, all of them grouped into two major alternatives: the hard-wired sequential control implementation and the microprogrammed control implementation. Other features related to the control function are also described, such as operation overlap, instruction lookahead, pipelined operation, and microprogrammable control as a tool for emulation. Finally, LSI devices incorporated in the control section of a digital system are presented.

Two of the major components of any digital processing system are the arithmetic-logic unit (ALU) and the control section. They make up what is called the central processing unit (CPU) of a computer. The ALU is composed of properly interconnected data routing switches (that establish the paths for moving data around) and a data transformation unit (that carries out the arithmetic and logical operations). The control section directs the internal activities of the ALU and controls all modules of the system. During the system's operation, the CPU fetches machine language instructions (sometimes called "macroinstructions") from main memory. Then the operation code field of the instruction is interpreted by the circuitry in the control section (i.e., mapped into a series of control signals), and the operand(s) specified by the instruction is (are) routed to the ALU to be used in computations.

The function of the control section is to issue properly timed signals to control the data-handling operations within the processor module (**intramodule control**), to coordinate the operations in all other modules of the system, and to synchronize the intermodule communications and data transfers (**intermodule control**). Intermodule control is discussed in Section 9.9.7, where we cover microprocessor system signals. Some of the activities carried out in the control section include: use the instruction counter to address the instruction, fetch this instruction, modify the instruction counter to point to the "next" instruction, decode the fetched instruction, and generate the appropriate control signals needed for its execution. To carry out its tasks properly, the control section must contain at least the following three facilities:

1. Decision logic for generating the next control state. This next-state selection is done through a proper combination of the present-state information, the CPU module's external inputs, and

certain feedback lines either from within the module (e.g., flags and conditions from the ALU) or from other modules of the system (system status signals).

2. A facility to store the information that corresponds to the current control state.
3. Some means for translating this state into proper control signals to be issued either within the module or to other modules of the system.

There are differences in the requirements and characteristics between intra- and intermodule controls and their respective ways of implementing these control and timing functions.

Since almost all modern digital systems synchronize their events with single- or multiple-phase clocks, these control signals are issued in synchronism with precise clock pulses. We therefore confine our presentation to the synchronous (clocked) implementation, where it will be inferred that there exists such a timing mechanism or module to synchronize control state transitions and data-flow switching by supplying those precisely timed, properly shaped clock pulses to all parts of the system. The organization of this timing module is basically very simple, and it is discussed as an integral part of the control system.

9.9.1 Timing and Control Sequences

Timing

All actions of the logic circuits in a computer are initiated by a sequence or sequences of pulses generated by a clock. Usually, polyphase clocks are used. Figure 9.71 depicts the pulses of a single-phase and a three-phase clock that we use in this section. The time interval between two adjacent clock pulses of the single-phase clock is the **clock period**. In the polyphase clock, the adjacent pulses, with one pulse from each phase, form a cycle, called the **clock cycle**. The three pulses P_0, P_1, P_2 in a clock cycle of the clock in Fig. 9.71(b) can be used to control a three-step sequence.

Various physical implementations exist for generating multiple clock phases. Figure 9.72(a), (b), and (c) show three implementation examples (Fig. 9.72(c) and (d) are similar) for generating an eight-phase clock, the pulses of each phase applied to a separate wire. Figure 9.72(d) shows a register and decoder implementation that we use in Section 9.9.2 in the design of the control section. At the beginning of each clock pulse, the corresponding phase P_i will be activated, depending on the binary value contained in register C.

Microoperations

Digital systems and computers are required to perform complex operations that are effected by executing sequences of primitive operations called **microoperations**. Microoperations are carried out by properly connecting the system's basic hardware resources, such as registers, ALU components, and data links in a variety of combinations.

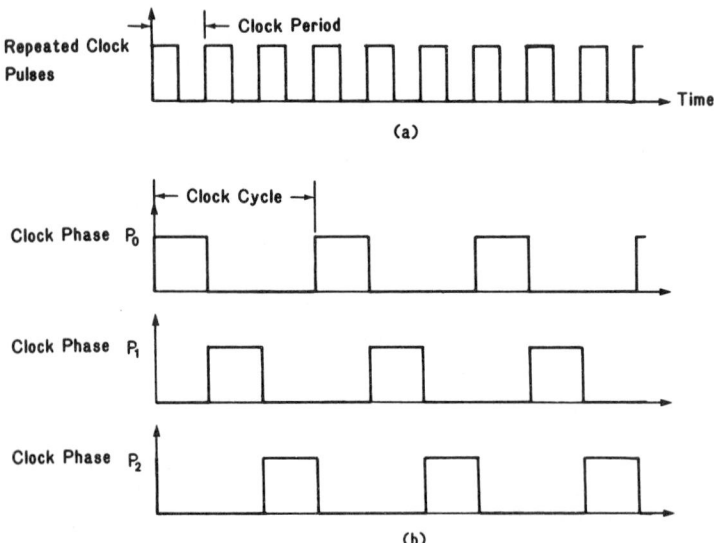

Fig. 9.71 Clocks: (a) single-phase clock, (b) three-phase clock.

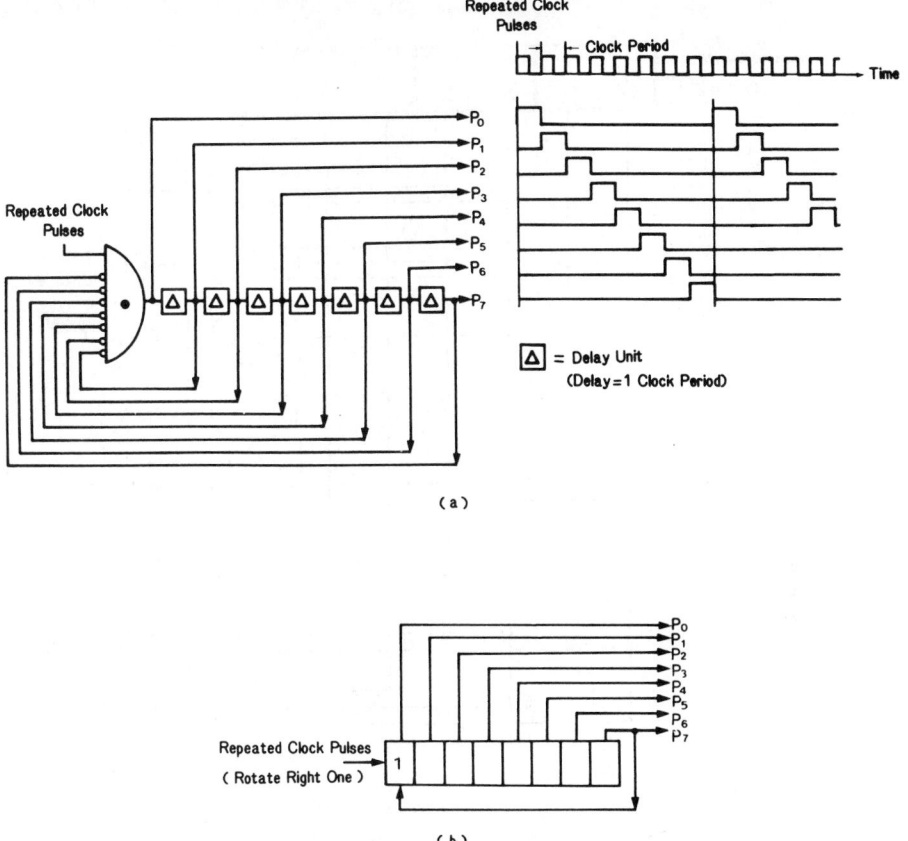

(a)

(b)

Fig. 9.72 Implementations for generating eight clock phases: (a) pulse distribution, (b) ring counter.

Microoperations may be classified into set-constant, transfer, unary, and binary microoperations. The most fundamental of them is the set-constant microoperation, which sets a register to a constant number value. The transfer microoperation transfers the contents of one or two registers to another register. In the unary and binary microoperations, the ALU receives information from one or two registers over data paths, and transforms the input into a result, which is usually gated to another register. Examples of unary microoperations include the shift, rotate, and count operations. Binary microoperations include logical (e.g., AND and OR) and arithmetic (e.g., ADD and SUBTRACT) operations.

In this section we also use complex conditional microoperations of the type IF-THEN or IF-THEN-ELSE, which are carried out under certain conditions. For example, the microoperation IF (STR ≠ 0)THEN (IC ← (IC) + 1) ELSE (IC ← (IC) + 2) indicates that the instruction counter will be incremented by one if STR ≠ 0; otherwise, the microoperation of incrementing the instruction counter by two will be executed.

In carrying out these microoperations, proper control signals must be issued to activate data paths and to select source and destination registers, as well as a transformation unit. These control signals must be precisely timed so that all actions will take place at the appropriate instant.

Some microoperations are more complex than others. For example, a transfer microoperation is simpler than a unary or binary microoperation and requires only one control signal. Even for this simplest microoperation, however, the transfer is not done instantaneously; a finite amount of time is required. A clock period larger than this amount of time is usually chosen so that the microoperation can be completed within the clock period. Figure 9.73 shows graphical representations of the "transfer contents of register A to register B" microoperation. Here, two 4-bit registers are connected through a 4-bit data path containing an equal number of control gates. The duration of the clock period of P has been properly chosen to allow enough time for the execution of the microoperation.

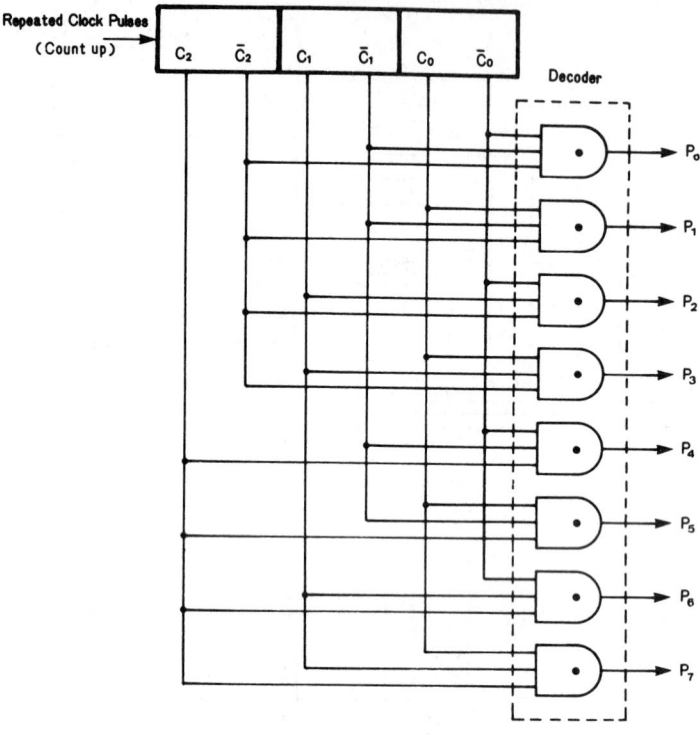

(c)

(d)

Fig. 9.72 Continued: (c) binary counter, (d) register and decoder combination.

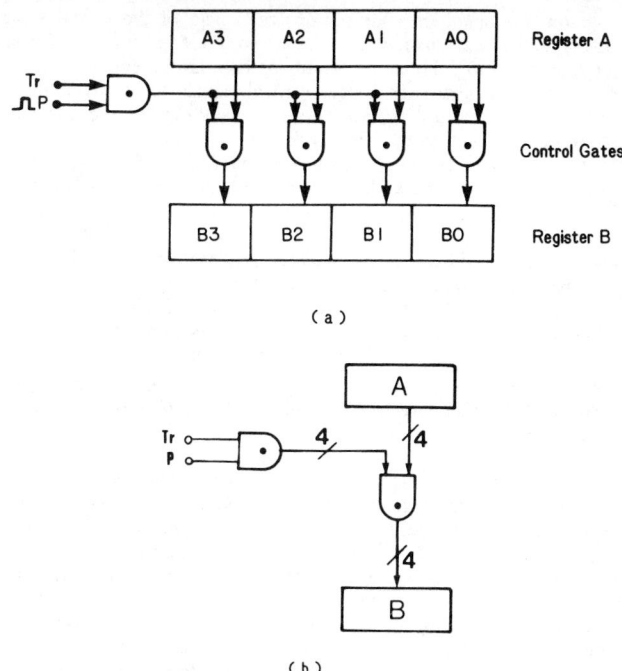

(a)

(b)

Fig. 9.73 Two graphic representations for the register transfer microoperations T_r and $P: B \leftarrow (A)$.

Symbolically, this transfer microoperation is expressed as*

$$T_r \text{ AND } P: B \leftarrow (A)$$

that is, the contents of register A are transferred to register B when both the command signal T_r and the pulse P of the single-phase clock are present. The logical-AND operation of T_r and P constitutes the **control signal** required for this microoperation. In general, such a control signal may initiate one or more microoperations. For example, the statement above should be changed to

$$T_r \text{ AND } P: B \leftarrow (A), T_r \leftarrow 0$$

if we do not want the transfer to be taking place each time a clock pulse P appears. Here the second microoperation resets the command signal T_r, and the logical value of the control signal T_r AND P is no longer 1.

Instruction Sequencing
During the normal sequencing of a stored-program digital system, the instructions are executed one after the other following the order in which they are stored in main memory. In general, each machine language instruction is carried out by executing a sequence of one or more microoperations. For each instruction, the fetch sequence is carried out first, followed by the execution sequence. During the fetch sequence, the instruction addressed by the instruction counter is fetched from main memory

*The following notation is used in this section:

Notation	Meaning
X	Register X
(X)	Contents of register X
M	Memory
M^x	Memory location specified by *contents* of register X; note that this is not consistent with the preceding, but it is less cumbersome than $M^{(x)}$
(M^x)	Contents of memory at location specified by contents of register X
$Y \leftarrow (X)$	Contents of Y are replaced by contents of X
$Y \leftarrow k$	Contents of Y are set to k

into the control section, the operation code (or op code) field of the instruction is decoded, the operand address is computed and transferred to the memory address register (if necessary), and the instruction counter is modified to point to the "next" instruction in memory. Having decoded the op code, the control section must now issue proper control signals to carry out the microoperations necessary to execute this instruction. The time taken to carry out the execution cycle of a simple instruction (e.g., ADD) corresponds to the **CPU cycle time**.

Like the CPU, the main memory operates on a **memory cycle**. A memory cycle is used for a memory read or write operation. Each such operation is carried out as a sequence of activities as follows:

1. *For a memory read.* The address is first transferred to the memory address register (MAR); a memory read operation is initiated; the addressed word is read out of memory into the memory data register (MDR); and the word is then restored (written back) at the same location (assuming that reading is a destructive operation). The **memory access time** is that part of the memory cycle time during which the read operation is performed.

2. *For a memory write.* The address is first transferred to the MAR; a memory write operation is initiated; reading of the addressed word into the MDR is inhibited; the word in the MDR is written into the memory.

In order to be more specific in our description, we shall use the simple system shown in Fig. 9.74. Although the CPU cycle time is much faster than the memory cycle time, for simplicity we assume here that the memory cycle time is equal to the CPU cycle time, and that both are chosen to coincide with the clock cycle defined in Fig. 9.71, composed of three clock phases, P_0, P_1, P_2.

This system has a memory of 1024 words by 18 bits (with a memory address register and memory data register), an ALU with an 18-bit accumulator, and a control section containing a 6-bit op code register, a 2-bit register I (for indexing and indirection), a 10-bit instruction counter, one 10-bit index register, and the necessary circuitry (gate signal generator) for issuing appropriate control signals to the rest of the system. The control section, the ALU, the registers, and the memory module of this system are interconnected via appropriate buses. Figure 9.74 also shows some interconnections and control gates that, when activated by their **control gate signals** G_i, initiate respective microoperations (discussed in detail in Section 9.9.2). Each instruction is composed (from left to right) of a 6-bit

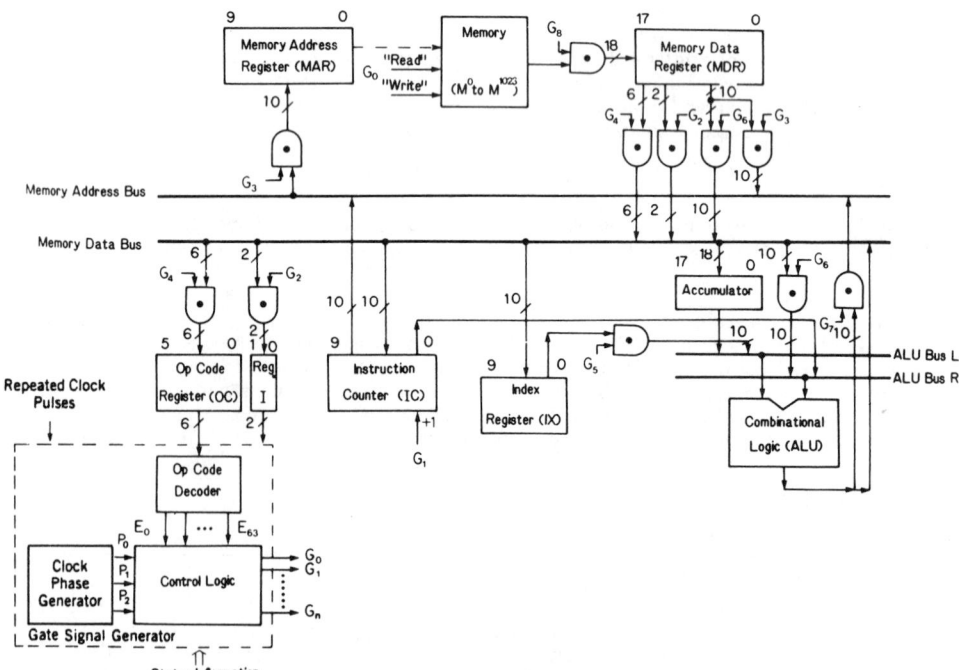

Fig. 9.74 Configuration of a simple, stored program, digital system.

TABLE 9.15   Instruction Subset

Command Signal	Mnemonic Name	Function	Execution
E_0	HLT	Halt	STR $\leftarrow$ 0 (the START flag is reset)
E_1	ADD	Addition	ACC $\leftarrow$ (ACC) + (M^{adr})
$\vdots$			
E_{10}	LDA	Load accumulator	ACC $\leftarrow$ (M^{adr})
E_{11}	STA	Store accumulator	M^{adr} $\leftarrow$ (ACC)
$\vdots$			
E_{20}	JMP	Jump (unconditional)	IC $\leftarrow$ adr
E_{21}	BAZ	{ Branch if accumulator zero	{ If (ACC) $\neq$ 0, then IC $\leftarrow$ (IC) + 1 { If (ACC) = 0, then IC $\leftarrow$ adr
$\vdots$			

operation code field, a 1-bit index tag, a 1-bit indirect addressing tag, and a 10-bit address field (adr). We assume single-address instructions and no overlap between the instruction fetch and execute sequences.

Decoding of the 6-bit-wide op code may activate one of up to 64 different E_i **command signals,** each command signal initiating a different set of microoperations required to execute the instruction. Each command signal E_i remains valid for the whole duration of the respective instruction execution sequence. As we will see later, the 6-bit combination 111111 is not assigned to an instruction op code but, instead, is reserved for setting the op code register to a value that generates the special command signal E_{63}. If $E_{63} = 1$, this implies that the system is in its instruction fetch sequence; if $E_{63} = 0$, this implies that the system is in its instruction execution sequence. It is assumed that the command signal E_{63} also remains valid for the whole duration of the respective sequence.

The gate signal generator in Fig. 9.74 combines the command signals, the timing pulses from the clock, and other system state information (such as ACC = 0) and generates gate signals to initiate the respective microoperations. The details of this gate signal generator are discussed in Section 9.9.2. Table 9.15 shows a subset of the system's instructions, their mnemonic names, and the symbolic representation of their executions.

Microoperation Sequencing

The role of the control section is to provide the control signals for the rest of the system and to determine what the next control signals will be. The *next* control signals generally depend on the results of the present microoperation(s) and on the status of certain system flags that are generated.

The processing of each instruction always starts with the fetch sequence for extracting the instruction from memory. This instruction fetch sequence is the same for all instructions (i.e., it is carried out by the execution of the same sequence of microoperations). (In this section we disregard both index and indirect addressing tags. We will use these tags in Section 9.9.2 when we give the detailed implementation of the hard-wired sequential control section.) The instruction execution sequence varies from instruction to instruction depending on the specific instruction to be executed.

For the simple system in Fig. 9.74, the following activities must be carried out during the instruction fetch sequence:

MAR $\leftarrow$ (IC). The contents of the instruction counter are transferred to the memory address register.

Read $\leftarrow$ 1. Memory fetch request is signaled.

MDR $\leftarrow$ (M^{MAR}). The memory word located at the address in MAR is transferred to the memory data register.

IC $\leftarrow$ (IC) + 1. The instruction counter is incremented by one to point to the next instruction.

MAR $\leftarrow$ (MDR_{9-0}). The address part (the rightmost 10 bits) of the MDR is transferred to the MAR register.

OC $\leftarrow$ (MDR_{17-12}). The op code part (the leftmost 6 bits) of the MDR is transferred to the op code register.

Before we continue to define the control signals required to sequence the activities outlined, we pause momentarily to explain the control signals needed for accessing the (main) memory. Since we assumed that the memory cycle coincides with the clock cycle, and since the clock cycle is made up of three clock phases, we must allocate all memory activities (discussed earlier) into three major steps, each step controlled by one clock phase. One way of doing this for the memory read and write cycles is as follows:

1. *Memory read cycle.* The transfer of the address to the memory address register occurs at clock phase P_2 of the preceding clock cycle; the signaling to initiate a memory read operation and the reading out of the addressed word into the memory data register both occur at the same clock phase: phase P_0 of the current clock cycle. Thus we have

$$P_2: \text{MAR} \leftarrow (\text{IC}) \text{ or MAR} \leftarrow (\text{MDR}_{9-0}) \qquad \text{Beginning of memory cycle}$$
$$P_0: \text{MDR} \leftarrow (\text{M}^{\text{MAR}}), \text{Read} \leftarrow 1$$
$$P_1: \qquad\qquad\qquad\qquad\qquad\qquad\qquad\qquad \text{End of memory cycle}$$

2. *Memory write cycle.* The transfer of the address to the memory address register occurs at clock phase P_2 of the preceding clock cycle; the word to be stored in memory is placed into the memory data register at clock phase P_0 of the current clock cycle; the signaling to initiate a memory write operation and the word in the memory data register written into the memory both occur at the same clock phase P_1. Thus we have

$$P_2: \text{MAR} \leftarrow (\text{IC}) \text{ or MAR} \leftarrow (\text{MDR}_{9-0}) \qquad \text{Beginning of memory cycle}$$
$$P_0: \text{MDR} \leftarrow (\text{ACC})$$
$$P_1: \text{M}^{\text{MAR}} \leftarrow (\text{MDR}), \text{Write} \leftarrow 1 \qquad\qquad \text{End of memory cycle}$$

Let us now return to define the control signals required to execute the instruction fetch sequence. Again the three phases of the clock must be used, and all the activities of the fetch sequence must be grouped into three major steps. The command signal E_{63} initiates the fetch sequence. This E_{63} can be generated by putting six 1's in the op code register of Fig. 9.74 (i.e., OC $\leftarrow$ 63) at the end of each instruction execution sequence. Then, assuming that MAR $\leftarrow$ (IC) was done during clock phase P_2 of the preceding clock cycle, we have the following for the instruction fetch sequence:

$$E_{63} \text{ AND } P_0: \text{MDR} \leftarrow (\text{M}^{\text{MAR}}), \text{Read} \leftarrow 1$$

$$E_{63} \text{ AND } P_1: \text{IC} \leftarrow (\text{IC}) + 1$$

$$E_{63} \text{ AND } P_2: \text{OC} \leftarrow (\text{MDR}_{17-12}), \text{MAR} \leftarrow (\text{MDR}_{9-0})$$

It is noticed that more than one microoperation may be executed during a single clock phase, initiated by a control signal generated by the logical-AND operation of E_{63} with the respective clock phase P_i. Also note that the operand address is placed into MAR during P_2, to be ready for the succeeding instruction execution sequence, regardless of whether the operand will be stored (or fetched) by the instruction.

The sequence chart for our simplified system is given in Fig. 9.75, showing the respective microoperation sequences for the instruction subset of Table 9.15. With the powering-on of the system, flag STR is set to zero, and the system is in the wait state, executing the wait loop at the lower left corner of Fig. 9.75. When the START button is pressed, flag STR is set to 1, and the computer begins execution of the fetch sequence, executing the first instruction from memory location 0. Transitions from one state to another take place at the end of every clock phase. It is observed that the fetch sequence and its microoperations are the same for each instruction and that three clock phases are required to traverse through its three states. At the end of the fetch sequence, the control section interprets the operation code field of the instruction that has arrived from main memory. This usually results in a sequence of changes of the system's state and causes stimuli to be sent to the other modules of the machine. Depending on the command signal E_i generated from decoding the instruction, a separate and unique sequence of steps will be followed during the instruction execution sequence. It has been assumed that the reading part of the memory cycle occurs during clock phase P_0 and the writing part during clock phase P_1, and that for all instructions, their execution sequence is complete at the end of clock phase P_2. For the HLT instruction, the START flag is reset to zero (STR $\leftarrow$ 0) and the system goes into the wait loop, from which it exits when the START flag is (manually) set to 1 (i.e., STR $\leftarrow$ 1).

For example, consider the execution sequence for instruction ADD, which, when decoded, generates the command signal E_1. A sequence of three control signals is required to carry out the three steps of the instruction shown in Fig. 9.75, as follows:

$$E_1 \text{ AND } P_0: \text{MDR} \leftarrow (\text{M}^{\text{MAR}}), \text{Read} \leftarrow 1$$

$$E_1 \text{ AND } P_1: \text{ACC} \leftarrow (\text{ACC}) + (\text{MDR})$$

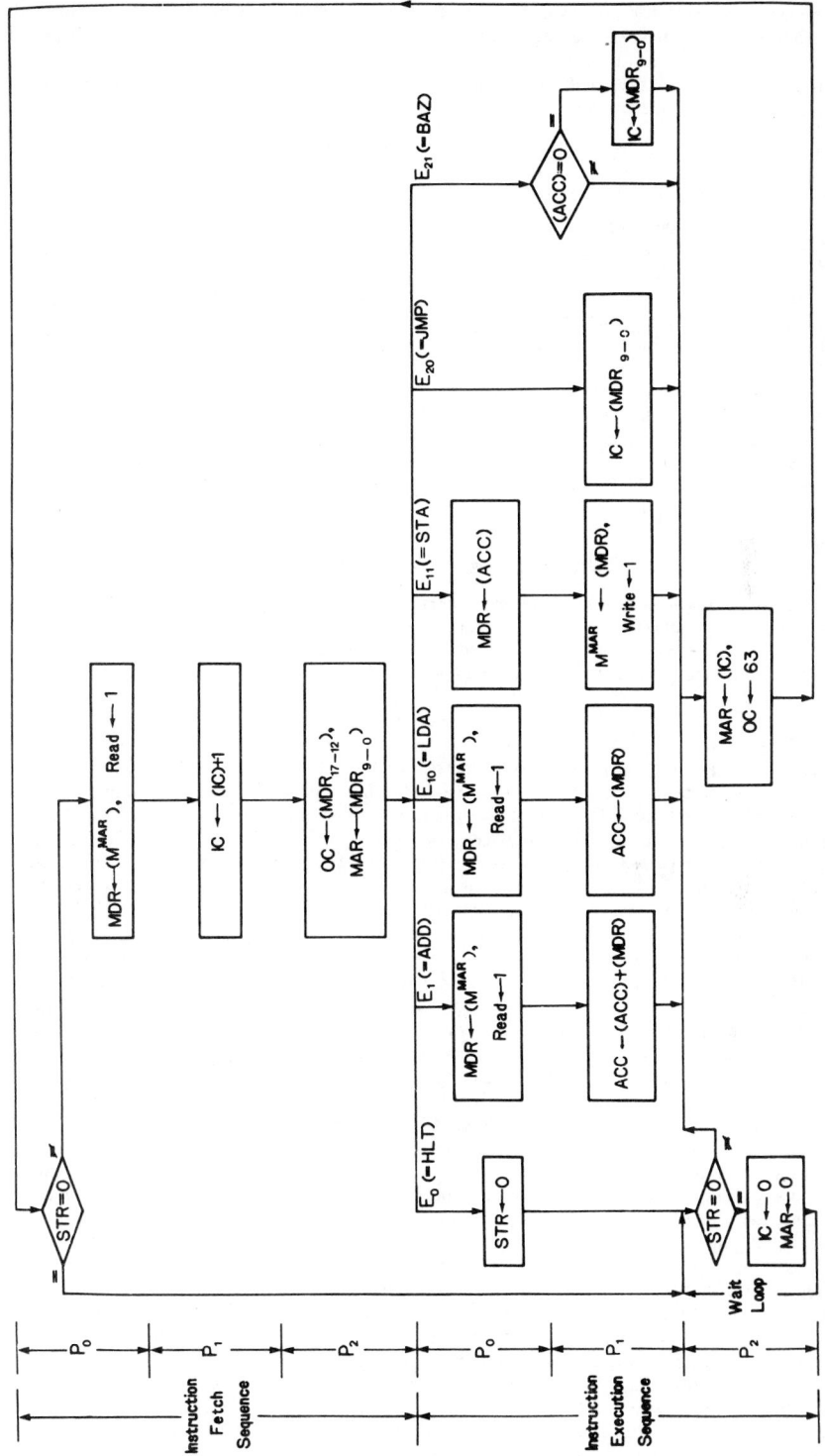

Fig. 9.75 Sequence chart with the respective microoperations for the instruction subset of Table 9.15 (no indexing, no indirect addressing).

Here we assume that both simple operations (such as a register transfer) and complex operations (such as addition) are carried out within one clock phase. The microoperation $OC \leftarrow 63$ terminates the instruction execution sequence and causes command signal E_{63} to be issued, to initiate a new fetch sequence for the next instruction.

For the execution of the conditional transfer instruction E_{21} (= BAZ), the following control signals are required for the two steps:

Step 1. E_{21} AND P_1 AND [(ACC) = 0]: $IC \leftarrow (MDR_{9-0})$
E_{21} AND P_1 AND [(ACC) $\neq$ 0]: no action
Step 2. E_{21} AND P_2: $MAR \leftarrow (IC)$, $OC \leftarrow 63$

Of course, it must be appreciated that this is an overly simplified design. Only a subset of very simple instructions has been used, and the CPU cycle time and memory cycle time were chosen to be the same and both equal the three-phase clock cycle. It has also been implied here that the control section, the ALU, and the memory module are working together synchronously. In real situations, many more complicated instructions and clocking schemes are used. Longer instructions also exist (e.g., multiply, divide) that require iterative executions of sequences of microoperations (Section 9.5.5).

To summarize, the control section must contain capabilities for issuing several control signals, properly synchronized with a timing source, to ensure that the microoperations are executed at the proper times and sequence. There are two major ways of generating the control signals to activate the system's microoperations: through hard-wired sequential logic implementation (discussed in Section 9.9.2) and through microprogrammed implementation (discussed in Section 9.9.3).

9.9.2 Hard-wired Sequential Control

In this section we discuss the techniques and implementations for hard-wired sequential control. The hard-wired implementation of the control section is a complex system of discrete logic elements and sequential circuits, which are often interrelated and synchronized by an elaborate timing generator, that ensures that the microoperations required for the instruction occur at the right times in the correct sequence.

Next we present the detailed hardware implementation of the control section for the simple system of Fig. 9.74. We again use the instruction fetch sequence as an example, but augmented here to include the handling of indexing and indirection. We show the respective microoperations involved, and describe the control circuit that will issue the appropriate signals to sequence through this set of microoperations. The concepts we present through this example can also be applied to implement the instruction execution sequence.

The detailed hardware configuration of the gate signal generator is given in Fig. 9.76(a). It is designed using sequential logic, and it is composed primarily of the next-control-state generator and the control matrix. The next-control-state generator is similar to the circuit of Fig. 9.72(d), using a combination of a 3-bit register C and a decoder. The eight outputs T_i from the decoder give eight single-step control state signals, T_0 to T_7. Only one of the eight outputs is active at any one time, according to the value set in register C. Its signal will activate the respective microoperation(s) to be executed during that control state. Each control state signal T_i occurs at the beginning of the clock period, lasts for one period, and all microoperations activated will complete before the end of that clock period. The control matrix receives as inputs the control state signals T_0-T_7, the command signals E_0-E_{63} from the op code decoder, the index and indirect tags I_1 and I_0 from register I, and some other status information, such as (ACC) = 0. For each state T_i, the control matrix generates: (1) the gate signals required to initiate the corresponding microoperation(s) to be executed during this state, and (2) appropriate signals to the next-control-state generator to set register C to the proper value of the next state.

The instruction fetch sequence we discuss here is more complicated than that described in Section 9.9.1, since we are now also making use of the index and indirect addressing tags of the 18-bit instruction of our system. (For indirect addressing, see also the discussions in Sections 9.1.3 and 9.3.1.) Each instruction being fetched must now be examined further to see whether it specifies indexing or indirect addressing, as shown in the flowchart of Fig. 9.77.

Therefore, the instruction fetch sequence of Fig. 9.75 has now been enlarged to that given in Table 9.16. (For simplicity, we disregard here the examination of STR = 0.) It is noticed that eight control states T_0-T_7 are required for the instruction fetch sequence. The second and third columns in Table 9.16 show the control signals issued and the corresponding microoperation(s) executed during each control state T_i. The fourth and fifth columns in Table 9.16 signify the outputs of the control matrix of the gate signal generator. The gate signals G_i listed in the fourth column activate the respective control gates to execute the microoperations. The signals generated by the control matrix to define the next control state of the system are in the fifth column, and the next control states in the sixth column.

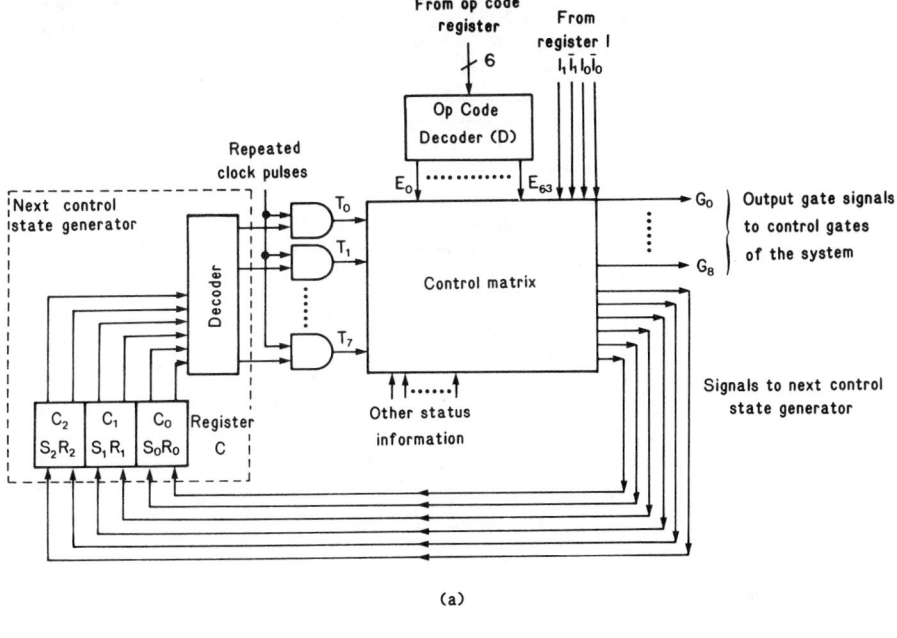

(a)

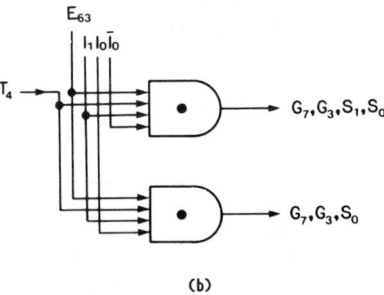

(b)

Fig. 9.76 (*a*) Hardware implementation of the gate signal generator. (*b*) Section of the control matrix for generating the signals needed during control state T_4 (see also Table 9.16).

The operation of Table 9.16 is as follows. At the end of each instruction execution sequence, (1) the operation code register is set to six 1's (i.e., OC ← 63), which, when decoded, will issue command signal E_{63} to initiate the instruction fetch sequence; this E_{63} will remain on during the whole fetch sequence, and (2) register C in the next-control-state generator is set to zero (by triggering its three reset inputs) to initiate the first state T_0 of the instruction fetch sequence.

During control state T_0, the control matrix generates the gate signals G_8 and G_0 to execute the two microoperations that read the instruction out of memory into MDR. The control matrix also issues appropriate signals (here, $S_0 = 1$) to advance the next-control-state generator to the next control state T_1.

During control state T_1, the instruction counter is incremented (by opening gate G_1), and register C is set to the value 010 to advance to the next control state T_2.

During control state T_2, the index and indirect addressing tags of the instruction are examined (by opening gates G_2), the address field of the instruction is transferred to MAR (by opening gates G_3) to be ready for the instruction execution sequence in case neither indexing nor indirect addressing is specified, and the control matrix generates the inputs to register C (here, $S_0 = 1$) to advance to the next control state T_3.

If it is concluded during control state T_3 that neither indexing nor indirect addressing was specified by the instruction (i.e., $I_1 = I_0 = 0$), the system advances to the last control state T_7 of the instruction fetch sequence. If the instruction specifies either indexing only or both indexing and

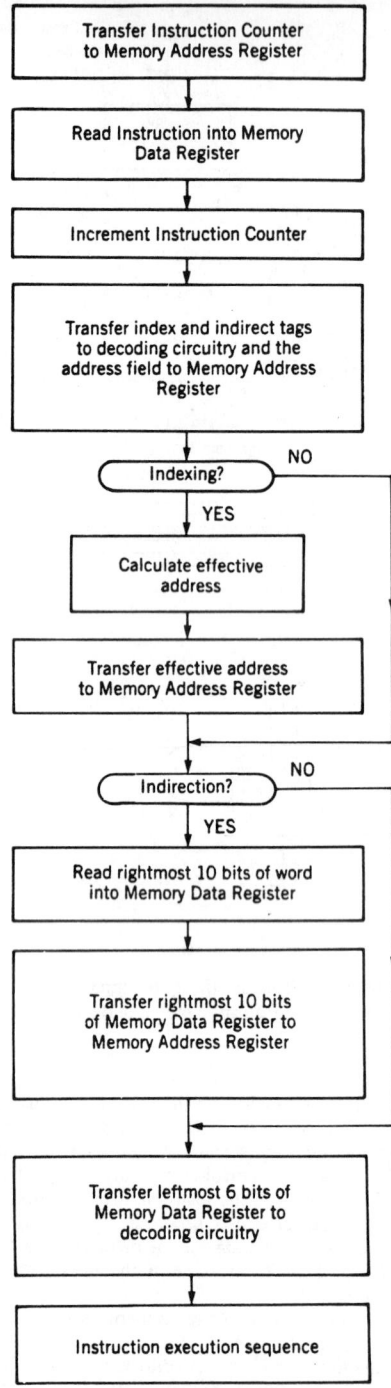

Fig. 9.77 Flowchart for indexing and/or indirect addressing (single level) during the instruction fetch.

TABLE 9.16 Gate Signals and Inputs to the Next-Control-State Generator for an Instruction Fetch Sequence (Including Indexing and / or Indirect Addressing)

Control State	Control Signal	Microoperation	Control Gate Signal	Inputs to Register C	Next Control State	Explanation
T_0	E_{63} AND T_0:	MDR ← (M^{MAR}),	G_8	—		Instruction read out into MDR.
		Read ← 1,	G_0	—		
		C ← (C) + 1		$S_0 = 1$	T_1	Advance to next state T_1.
T_1	E_{63} AND T_1:	IC ← (IC) + 1,	G_1			Increment instruction counter and
		C ← (C) + 1		$S_1 = 1, R_0 = 1$	T_2	advance to next state T_2.
T_2	E_{63} AND T_2:	I ← (MDR$_{11-10}$),	G_2			Index and indirect tags transferred to register I.
		MAR ← (MDR$_{9-0}$),	G_3			Address field of instruction (10 bits) transferred to MAR.
		C ← (C) + 1		$S_0 = 1$	T_3	Advance to next state T_3.
T_3	E_{63} AND T_3 AND $\bar{I}_1$ AND $\bar{I}_6$:	C ← 7		$S_2 = 1$	T_7	Neither indexing nor indirection. Go to T_7.
	E_{63} AND T_3 AND I_1:	ALU ← (IX),	G_5			Either indexing only, or indexing followed by
		ALU ← (MDR$_{9-0}$),	G_6			indirect addressing.
		C ← (C) + 1		$S_2 = 1, R_1 = 1, R_0 = 1$	T_4	Advance to next state T_4.
	E_{63} AND T_3 AND $\bar{I}_1$ AND I_6:	C ← 5		$S_2 = 1, R_1 = 1$	T_5	Only indirection. Go to T_5.
T_4	E_{63} AND T_4 AND I_1 AND $\bar{I}_6$:	MAR ← (ALU),	G_7, G_3			Perform indexing only. Go to T_7.
		C ← 7		$S_1 = 1, S_0 = 1$	T_7	
	E_{63} AND T_4 AND I_1 AND I_6:	MAR ← (ALU),	G_7, G_3			After indexing go for indirect addressing.
		C ← (C) + 1		$S_0 = 1$	T_5	Advance to next state T_5.
T_5	E_{63} AND T_5	MDR ← (M_{9-0}^{MAR}),	G_8^*			Start indirect addressing (G_8^* reads the
		Read ← 1,	G_0			rightmost 10 bits of the word into MDR).
		C ← (C) + 1		$S_1 = 1, R_0 = 1$	T_6	Advance to next state T_6.
T_6	E_{63} AND T_6:	MAR ← (MDR$_{9-0}$),	G_3			End of indirect addressing. Advance to
		C ← (C) + 1		$S_0 = 1$	T_7	next state T_7.
T_7	E_{63} AND T_7:	OC ← (MDR$_{17-12}$),	G_4			Op code field of the instruction (6 bits) is
		C ← 0		$R_2 = 1, R_1 = 1, R_0 = 1$	T_0 of execute sequence	transferred to op code register to be decoded. Advance to T_0, the first state of the instruction execution sequence.

indirect addressing (i.e., $I_1 = 1$), the indexing operation starts by transmitting the contents of the index register and the address field of the instruction to the ALU (by opening gates G_5 and G_6) where they are added, and the system advances to the next control state T_4. If it is concluded that only indirect addressing was specified (i.e., $I_1 = 0$, $I_0 = 1$), the system goes to control state T_5 (and then to T_6) to perform the indirection.

If indexing only were specified (i.e., $I_1 = 1$ and $I_0 = 0$), then, during control state T_4, the indexed effective address is transferred from ALU (assuming intermediate latching in the ALU) to the MAR, and the system advances to the last control state T_7 of the instruction fetch sequence. If both indexing and indirect addressing were specified (i.e., $I_1 = I_0 = 1$), then, during control state T_4, the indexed effective address is again transferred to MAR, but now the system advances to the next control state T_5 (and then T_6) to perform indirect addressing. Notice that two control states are allocated here for the more complex operations (such as indexing).

The last control state of the instruction fetch sequence is T_7, during which the op code field of the instruction is transferred to the op code register to be decoded. When decoded, it will issue a command signal E_i ($E_i \neq E_{63}$) that will indicate the beginning of the instruction execution sequence. This command signal E_i will remain on during the whole instruction execution sequence and will be used by the control matrix to issue appropriate gate signals needed to execute the instruction. Also at T_7, the next-control-state generator is signaled to advance to control state T_0, the first control state of the instruction execution sequence.

Figure 9.76(b) shows, as an example, the part of the control matrix involved during control phase T_4. As noted from Table 9.16, depending on its inputs I_1 and I_0, the control matrix will either generate gate signals G_7 and G_3 to initiate microoperation MAR ← (ALU) and signals $S_1 = 1$ and $S_0 = 1$ to set register C to 7, or it will generate gate signals G_7 and G_3 to initiate the same microoperation and the signal $S_0 = 1$ to set register C to 5.

Of course, it must be appreciated that this has been an oversimplified, nonoptimal, partial design. Nothing has been said about the details of the instruction execution sequence (which becomes very complicated for complex instructions such as multiply and divide), about effective address calculations through multiple index and base registers, about ways of saving return addresses for subroutine jumps, and so on. Also, no attempt was made to minimize the number of control states required or the hardware involved for the instruction fetch sequence given in Table 9.16.

To summarize: The instruction fetch sequencing of Table 9.16 (with indexing and indirection) requires eight control states T_0–T_7 used to control the execution of the 14 microoperations that are necessary. Each microoperation is initiated by the opening of the respective control gate in the fourth column of the table. The second column shows the control state and test conditions examined for the execution of these microoperations. The gate signal generator of Fig. 9.74 makes use of sequential logic control, and its register C generates the control sequence according to the inputs it receives from the control matrix. Each value of register C corresponds to one control state T_i assigned to command one or more microoperations (see Table 9.16).

9.9.3 Microprogrammed Control

Instead of being embodied with sequential logic, a control unit may also be realized by means of a microprogram stored in a fast memory called a **control store**. The functional operation of the system and the data flow within a module or between modules are usually considered to be independent of whether control is hard-wired or microprogrammed.

A simple microprogrammed control section has the configuration shown in Fig. 9.78. It typically consists of a control store with its control store address register and microinstruction register (the equivalent of the memory data register), a control store sequencer, and some remaining circuitry such as selection and control logic, registers, and so on. The control store (also called microprogram control memory or microprogram memory) is generally several times faster than the system's main memory; it contains "microprograms" composed of "microinstructions" and generates control signals timed to control the rest of the system. Each **microinstruction** defines the state of the control store's output control lines, and the execution of these microinstructions specifies the steps through which the machine will sequence to exercise control over the operation of all modules. The **control store sequencer** (also called the next-control-store-address generation circuit, microprogram control unit, or microprogram sequencer) provides the machine instruction decode logic and specifies the "next microaddress" generation scheme for sequencing the execution of these microinstructions. Program instructions are fetched from the system's main memory under the direction of microinstructions read from the control store. The op code of each instruction is interpreted by the control store sequencer (i.e., mapped into a control store address), and the instruction is then executed as a **microprogram** (i.e., a sequence of microinstructions).* Each microinstruction fetched from the control store and placed into the MIR activates a number of control gates G_i. If each bit in the microinstruc-

---

*"Microprograms" should not be confused with programs for microcomputers.

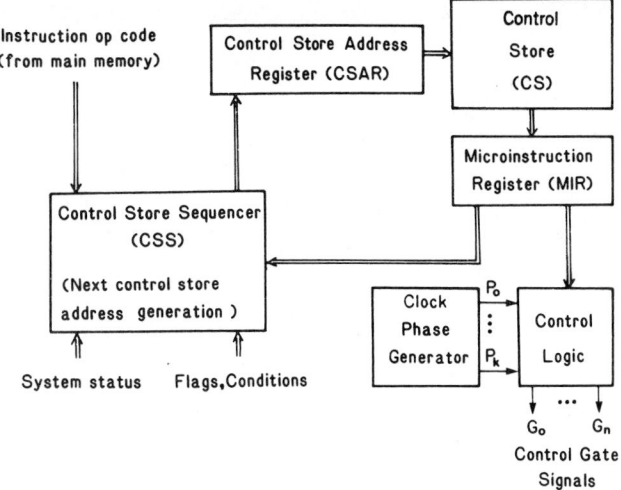

Fig. 9.78 Simple microprogrammed implementation of the control function.

tion is assigned to one microoperation, then, since more than one bit can be 1, more than one microoperation can occur simultaneously. Therefore, the execution of each microinstruction generates one or more control signals to discrete control gates of the system, each such signal initiating the respective microoperation. Since each microinstruction can generate a multitude of control signals, the microprogram executed generates the sequence of control signals required to execute the machine instruction. The sequencing of the microinstructions in a microprogram is established by the control store sequencer. As we see later, there are various ways of doing this; for example, by incrementing the control store address register, by transferring an address field of the microinstruction to the control store address register, or by using branch condition bits in the microinstruction to examine external conditions before generating the proper next microinstruction address. Detailed descriptions of the control store sequencer and the control store will be given.

Timing and Sequencing Considerations

Microinstruction Execution Timing Schemes. We assume that microinstructions are fetched serially from control store; the next microinstruction to be executed does not begin until the execution of the current microinstruction terminates. The control store is a read-only memory. The control store operates on a cycle time called **control store cycle time**, or **control cycle**, of the microprogrammed control unit, or **microcycle**. It corresponds to the time required to transfer the next microinstruction address to CSAR plus the time required for the microinstruction to be read into the MIR (i.e., the access time). During each microcycle, a microinstruction is fetched from control store, and the microoperations it specifies are then executed. Under this description then, a two-phase clock (or two-phase microcycle) would suffice for the microprogrammed control unit. During the first phase, the address of the next microinstruction is transferred to CSAR, and, at the same time, the microoperations activated by the microinstruction control bits in the MIR register are carried out. During the second phase, the next microinstruction is read out of control store into register MIR. However, some microoperations may be of the conditional type—for example, for the BAZ instruction in Fig. 9.75, we have the conditional microoperation IF $((ACC) = 0)$ THEN $(IC \leftarrow (MDR_{9-0}))$ ELSE $(MAR \leftarrow (IC), OC \leftarrow 63)$. Since this condition may not be known until the CPU microoperations are executed, the next microinstruction address transfer to CSAR and the execution of the CPU microoperations cannot occur at the same clock phase. Therefore, a third phase is used for the microcycle. Now, during one phase, the address of the next microinstruction is transferred to CSAR; during the next phase, the microinstruction is read into MIR; and during the third phase, the microinstruction is executed (i.e., the respective microoperations are performed). The events involved during each microcycle are shown in Fig. 9.79, and they are as follows:

$$P_2: MIR \leftarrow (CS^{CSAR}) \qquad \text{Fetch next microinstruction}$$
$$P_0: \qquad\qquad\qquad\qquad \text{Execute microinstruction}$$

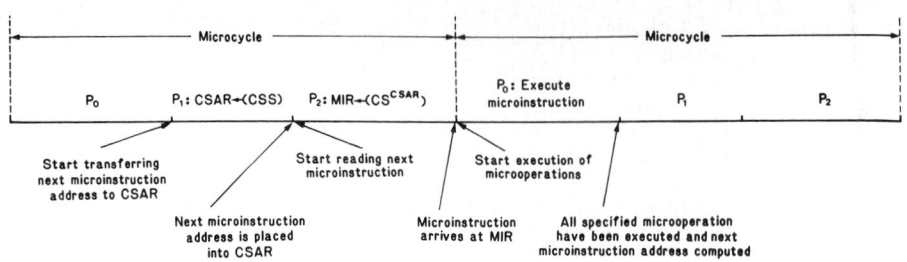

Fig. 9.79   Events involved during a microcycle.

P_1: CSAR ← (CSS)   (Conditions have been examined)
                                       Set next microinstruction
                                       address into CSAR

Depending on the way the control bits of the microinstruction activate microoperations, there can be a monophase or a polyphase implementation. In the **monophase** implementation, the execution of each individual microinstruction requires only one clock phase. In other words, all the microoperations that this microinstruction initiates are completed within this clock phase duration P_0 as shown in Fig. 9.80(a). In the **polyphase** implementation, the execution of each individual microinstruction requires more than a single clock phase. The control signals of a single microinstruction are issued in sequence over a number of clock phases (Fig. 9.80(b)), sequencing accordingly the respective microoperations. Furthermore, polyphase implementations are characterized as synchronous or asynchronous. In the **synchronous polyphase**, the number of clock cycles required for the microinstruction execution remains constant (i.e., same number of clock phases for every microinstruction). In the **asynchronous polyphase**, the number of clock phases required to execute a microinstruction depends on the complexity of its microoperations.

Microinstruction Sequencing. Another major consideration involves the way of sequencing (i.e., fetching and executing) these microinstructions. The two aspects of microinstruction sequencing have to do with timing of microinstruction fetches (serial or parallel, depending on the nonoverlap or complete overlap of the execution of the current microinstruction with the fetching of the next microinstruction) and with methods used to take care of conditional branch situations in the microprogram and to generate the next-control-store-address. So far we have discussed **serial** microinstruction sequencing, where the next microinstruction is not fetched from the control store before the current microinstruction has completed its execution. **Parallel** microinstruction sequencing, on the other hand, allows the next microinstruction to be fetched concurrently with the execution of the current microinstruction (Fig. 9.81(a)). This, of course, yields speed advantages, and parallel implementation runs almost twice as fast as the serial implementation. The parallel case, however, presents some problems, the most important one having to do with properly treating conditional branches. The microcycle for the next microinstruction cannot start before the current microinstruction has been executed (or its op code decoded) and the correct next address has been extracted. Several techniques are available to deal with such problems. One approach to deal with conditional branches is the **serial-parallel** implementation shown in Fig. 9.81(b). Here the microcycle for the next microinstruction follows the execution of the current microinstruction if the current microinstruction contains a conditional branch microoperation.

Next-Control-Store Address Generation. As shown in Fig. 9.78, the logic to generate the next control address is included in the control store sequencer. This sequencing logic may be as simple as a number of logic gates or as complex as a special LSI device.

In the simplest implementation, the control store sequencer and the control store address register may be replaced by a control store address counter (Fig. 9.82(a)). The instruction op code from main memory goes directly into the control store address counter. The address of the next microinstruction is generated simply by incrementing the control store address counter by one at the end of each microinstruction execution. The microinstruction itself does not contain a "next address" field. This counter technique permits only sequential control; it provides no means for altering—conditionally or unconditionally—the established flow of control.

Control store sequencers can be used for sequencing with branching capabilities, if they may load a new number that is present on their data input lines. For example, for each microinstruction, the counter shown in Fig. 9.82(a) may be loaded with the next address from a field in the microinstruc-

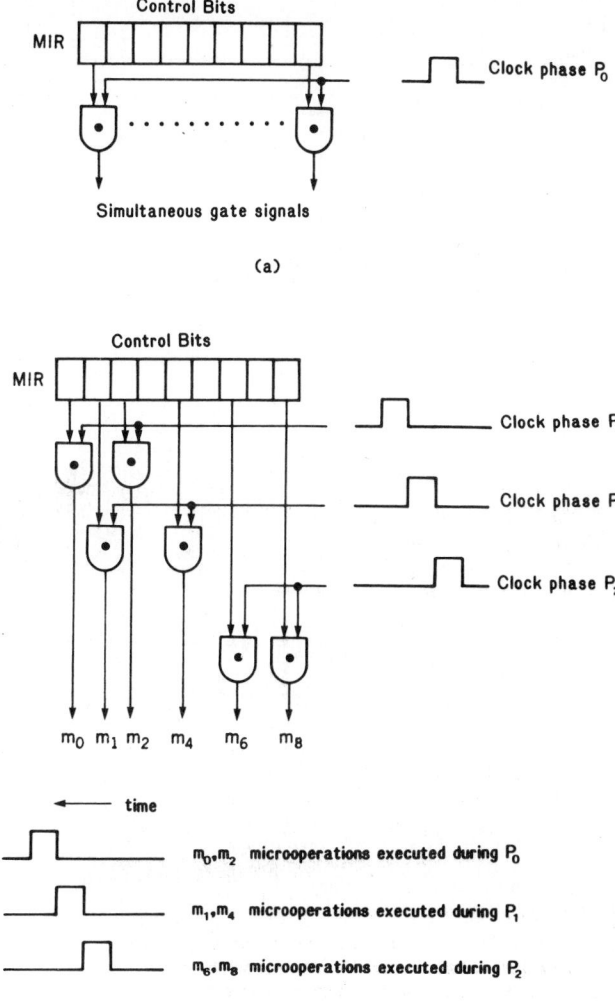

Fig. 9.80 (*a*) Monophase implementation: all microoperations initiated and completed within one clock phase P_0, (*b*) polyphase implementation: during every clock phase a pair of microoperations is executed, but these three pairs are executed during three different clock phases P_0, P_1, P_2.

tion currently executing. The microinstructions now become wider, but they offer additional capability for performing unconditional jumps in the microprogram.

In most situations, however, control store sequencers are also required to facilitate conditional branching capabilities in the microprogram. The control store sequencer will have more decision-making capability if it can examine—before branching—condition or status signals that originate from various parts within the system. There are many possible alternative implementations available. We present here only one of them, shown in Fig. 9.82(*b*). This control store sequencer is composed of a decoder and decision logic box. The decoder decodes the condition code field bits of the microinstruction. The decision logic combines the decoder outputs and the condition signals originating from various parts of the system and decides whether it should advance to the next microinstruction in the microprogram (i.e., it signals the +1 input of the control store address counter to advance by one), or whether it should branch to the address specified by the current microinstruction (by enabling the LD input of the control store address counter to parallel-load the next address).

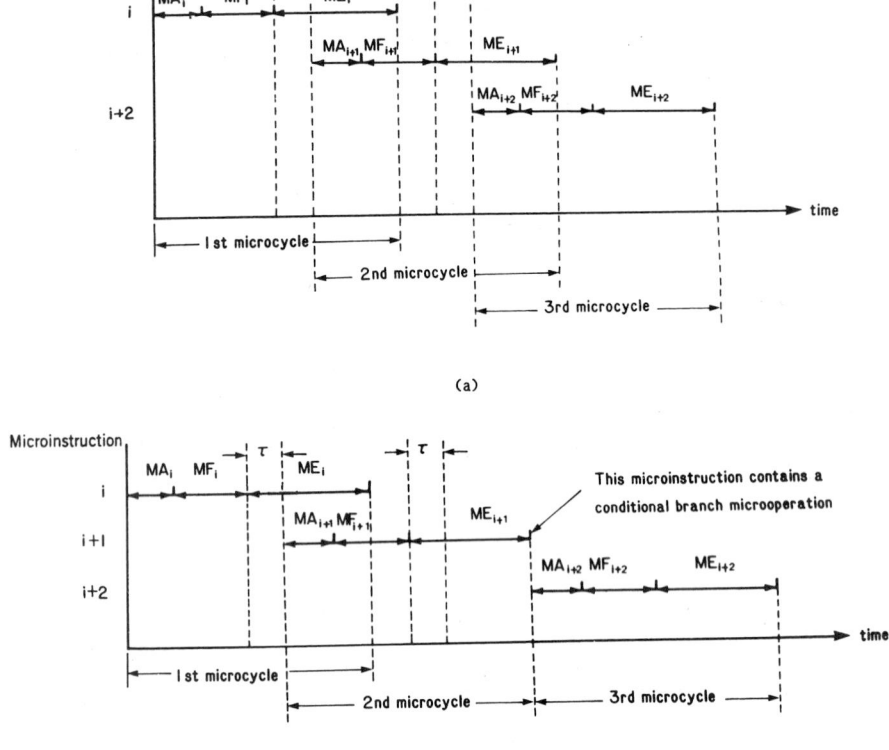

Fig. 9.81 (a) Parallel implementation, (b) combined serial-parallel implementation, where microinstruction $i + 1$ contains a conditional branch microoperation. MA, transfer microinstruction address to CSAR; MF, fetch microinstruction into MIR; ME, execute microinstruction; τ, time to decode microinstruction op code (to assure no conditional branches).

For example, if the branch condition code field of the microinstruction is 3 bits wide, the decoder specifies eight condition codes. Assume that a branch condition code field having the values of 000, 001, or 010 indicates normal advancing to the next microinstruction; a branch condition code of 011 indicates branch if (ACC) = 0; a code of 100 indicates branch if (IX) = 0; a 101 indicates branch if (IC) = 0; and codes 110 or 111 indicate branch if ACC overflows.

Microprogrammed Control Unit
The hard-wired control implementation for the configuration of Fig. 9.74 was realized by the sequential logic network of the type shown in Fig. 9.76, associated with a control register C that generated the control sequence. Each value in register C was assigned to command one or more microoperations to be executed during the same control state. For our simple system of Fig. 9.74, three control states would have been required, corresponding to the three clock phases used: P_0, P_1, P_2.

In the microprogrammed control implementation in the configuration of Fig. 9.78, the control sequence is carried out by the microprogram stored in the control store. Under such an implementation, the sequence chart of Fig. 9.75 will now be converted to that of Fig. 9.83 (where, for the sake of simplicity, we do not show the read signaling of the control store). One microinstruction is used for the instruction fetch sequence and one microinstruction for each of the execution sequences of the six instructions. We assume a polyphase timing scheme, where each microinstruction requires three clock phases, P_0, P_1, P_2, and no indexing or indirect addressing is considered. It must be noticed that microoperation OC $\leftarrow$ (MDR$_{17-12}$) of Fig. 9.75 has now been modified to CSAR $\leftarrow$ (MDR$_{17-12}$) and

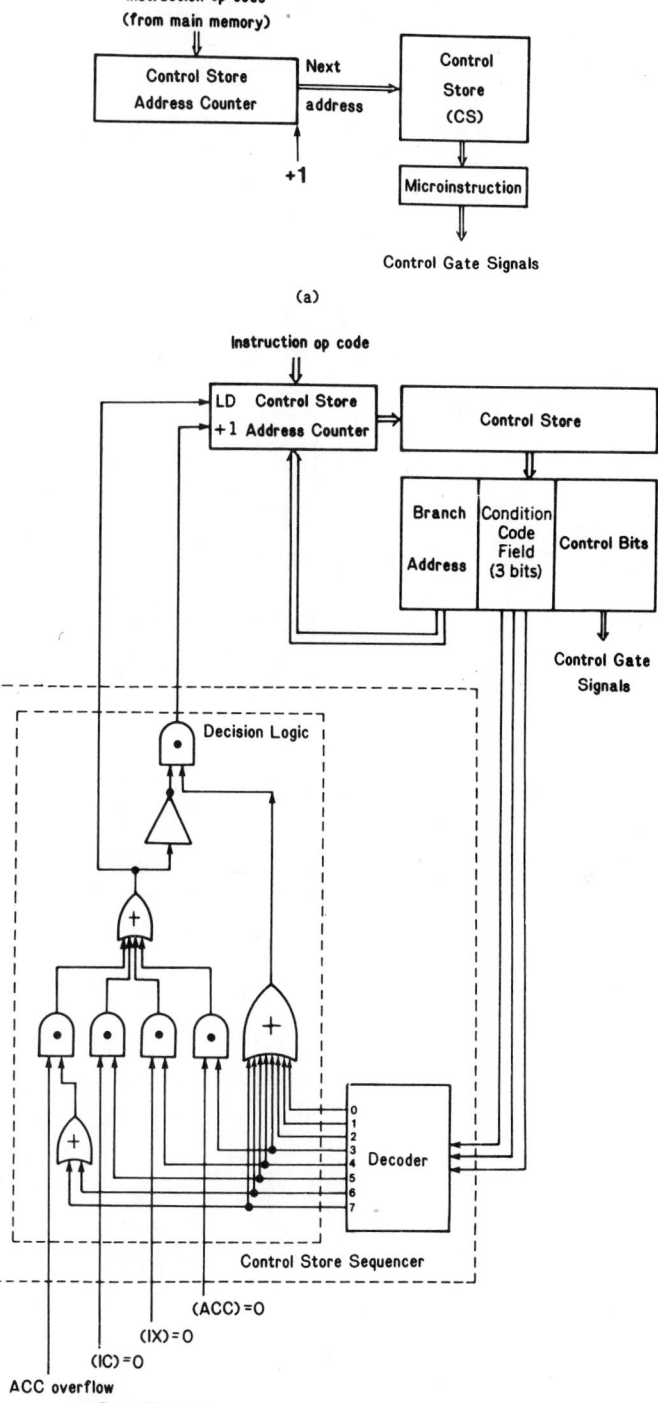

Fig. 9.82 Alternative implementations of control store sequencers: (*a*) using a control store address counter, (*b*) providing conditional branching capability based on external condition signals.

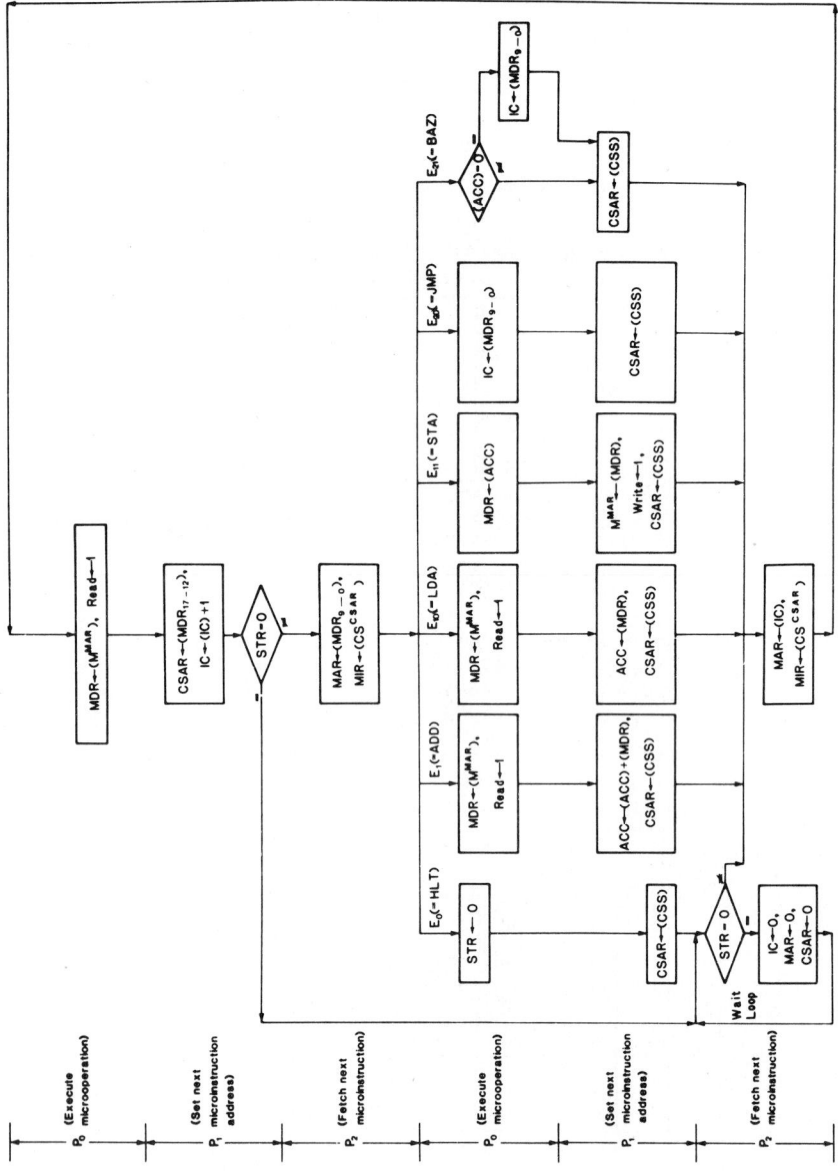

Fig. 9.83 Sequence chart of Fig. 9.75 for the microprogrammed implementation of Fig. 9.78 (no indexing and no indirect addressing examined).

TABLE 9.17　Assigning the 18 Microoperations of Fig. 9.83 to Control Bits of the Microinstruction

Control Bit	Clock Phase	Microoperation
MIR_{10}		If $MIR_{10} = 1$ then fetch sequence; otherwise execute sequence
MIR_9	P_0	$MDR \leftarrow (M^{MAR})$, Read $\leftarrow 1$
MIR_8	P_1	$CSAR \leftarrow (MDR_{17-12})$, $IC \leftarrow (IC) + 1$
MIR_7	P_2	IF $(STR \neq 0)$ THEN $(MAR \leftarrow (MDR_{9-0})$, $MIR \leftarrow (CS^{CSAR}))$ ELSE $(IC \leftarrow 0, MAR \leftarrow 0, CSAR \leftarrow 0)$
MIR_6	P_1	$CSAR \leftarrow (CSS)$
	P_2	$MAR \leftarrow (IC)$, $MIR \leftarrow (CS^{CSAR})$
MIR_5	P_0	$STR \leftarrow 0$
MIR_4	P_1	$ACC \leftarrow (ACC) + (MDR)$
MIR_3	P_1	$ACC \leftarrow (MDR)$
MIR_2	P_0	$MDR \leftarrow (ACC)$
	P_1	$M^{MAR} \leftarrow (MDR)$, Write $\leftarrow 1$
MIR_1	P_0	$IC \leftarrow (MDR_{9-0})$
MIR_0	P_0	IF $(ACC = 0)$ THEN $(IC \leftarrow (MDR_{9-0}))$

that two more microoperations

$$CSAR \leftarrow (CSS) \qquad \text{Set "next" microinstruction address}$$

$$MIR \leftarrow (CS^{CSAR}) \qquad \text{Fetch next microinstruction}$$

are now required due to the existence of the control store. With the power-on of the system, the STR flag is set to zero and the system is in the wait state, executing the wait loop at the lower left corner of Fig. 9.83. The microinstruction that corresponds to the instruction fetch sequence is assumed located in control store location zero. When the START button is pressed, the STR flag is set to 1, the first microinstruction is read out from control store location zero, and the computer operation is started.

The sequence chart of Fig. 9.83 for the microprogrammed control implementation involves a total of 18 microoperations. If each control bit in the microinstruction is assigned to one microoperation, the microinstruction will require an 18-bit-long control field. An additional nineteenth bit will be required to indicate instruction fetch or execution sequence. Since more than one microoperation occurs simultaneously during each clock phase, more than one bit of each microinstruction will be 1.

For our specific example here, two techniques can be used to reduce the microinstruction width. For the seven microinstructions required, we can do the following:

1. We can group the microoperations that always occur at the same time under one control signal. For example, the two microoperations $MDR \leftarrow (M^{MAR})$ and Read $\leftarrow 1$ that occur at phase P_0 can be grouped under one control signal; similarly, the two microoperations $IC \leftarrow (IC) + 1$ and $CSAR \leftarrow (MDR_{17-12})$ that occur at P_1, the two microoperations $M^{MAR} \leftarrow (MDR)$ and Write $\leftarrow 1$ that occur at P_1, and the two microoperations $MAR \leftarrow (IC)$ and $MIR \leftarrow (CS^{CSAR})$ that occur at P_2. It must be noted, however, that such a reduction gives up the use of each individual microoperation if such a use occurs later.

2. In addition, we can group those microoperations that can take place in the different phases of the same clock cycle under one control bit. For example, the following microoperations can be grouped under one control bit: microoperation $CSAR \leftarrow (CSS)$ at clock phase P_1 and microoperations $MAR \leftarrow (IC)$ and $MIR \leftarrow (CS^{CSAR})$ both at clock phase P_2. Again this reduction limits the future use of the individual microoperations.

Therefore, the microoperations can now be assigned to a smaller number of control bits, the bits MIR_{10} to MIR_0 of the microinstruction, as shown in Table 9.17.

Let us assume that the seven microinstructions are stored in control store locations 0–6. Then with the control bit assignments of Table 9.17, the microprogram in the control store of our simple system would be that of Fig. 9.84.

Microinstruction Format
Designing the microinstruction word is not a simple task. The factors affecting the number of bits and how they are arranged in a microinstruction word are the number of different microoperations the system can perform, the desired degree of parallelism in executing these microoperations, the way of sequencing the microinstructions of the microprogram, and the flexibility to be retained.

Control
Store
Location

Code	MIR$_{10}$	MIR$_9$	MIR$_8$	MIR$_7$	MIR$_6$	MIR$_5$	MIR$_4$	MIR$_3$	MIR$_2$	MIR$_1$	MIR$_0$
0	FETCH	1	1	1	1	O	O	O	O	O	O	O
1	HLT	O	O	O	O	1	1	O	O	O	O	O
2	ADD	O	1	O	O	1	O	1	O	O	O	O
3	LDA	O	1	O	O	1	O	O	1	O	O	O
4	STA	O	O	O	O	1	O	O	O	1	O	O
5	JMP	O	O	O	O	1	O	O	O	O	1	O
6	BAZ	O	O	O	O	1	O	O	O	O	O	1

Fig. 9.84 Microprogram of our simple system (with the control bit assignments of Table 9.19 and no indexing and no indirect addressing).

In its simpler form, a microinstruction usually has two function parts: (1) those bits that correspond to the control signal patterns that control the microoperations to be carried out and (2) those bits that specify the address of the next microinstruction to be executed.

Fixed and Variable Formats
The microinstruction's format may be fixed or variable. In the fixed format, each bit in the microinstruction is always interpreted the same way, whereas in the variable format, the meaning of the bits or fields changes according to certain system states. The variable format leads to shorter microinstructions, but it also requires more complex decoding logic.

Horizontal and Vertical Microinstructions
Horizontal microinstructions represent several microoperations that in general are executed concurrently. In the extreme, each horizontal microinstruction controls all the hardware resources of the system. Each control bit corresponds to a distinct microoperation: 1 appearing in a bit position is used solely to activate a separate control line or point in the system, indicating that the corresponding microoperation is to be executed. This formatting was used earlier. Because these microinstructions control multiple resources simultaneously, they usually have more control bits than the other implementations we will see. Lengths of 48 bits or more are common. Such an organization has the advantage of utilizing hardware more efficiently, and it requires smaller numbers of microinstructions per microprogram. However, it is rarely used in organizing control stores since for any reasonable size machine, the microinstruction lengths become very large. This scheme is of historical interest since it corresponds to Wilkes's original model* where each microoperation was the opening of a hardware gate.

The **vertical** microinstruction format resembles the classical instruction formats comprised of one microoperation code (mop code), one or more operands, and some other fields (such as a condition code field for conditional branches). Each vertical microinstruction generally represents a single microoperation (e.g., a transfer or a data transformation operation), while operands may specify the data sink and source. For example, for the simple microprogrammed implementation discussed in the preceding section, the vertical microinstructions would have a one-to-one correspondence with the microoperations shown in Fig. 9.83. Since vertical microinstructions resemble instructions in format and operation, it is easier to write vertical microprograms than their horizontal counterparts, and this ability enhances a system's microprogrammability. This scheme also requires relatively short word lengths, in the range 16 to 64 bits. On the other hand, vertical microinstructions require more complex decoding logic and do not take full advantage of parallelism in the system's microarchitecture.

*M. V. Wilkes and J. B. Stringer, "Micro-programming and the Design of the Control Circuits in an Electronic Digital Computer," *Proc. Camb. Philos. Soc.*, **49**, 230–238 (Apr. 1953).

Microprogram execution is slower since only one microoperation is done at a time. Vertical control store organizations are also relatively inflexible because the addition of a new microinstruction type may require a change in the decoder logic. Since a number of simple microoperations can now be combined into one more complex microoperation, the dividing line between horizontal and vertical microinstructions is difficult to define exactly.

Reducing the Number of Control Bits

One way of reducing microinstruction length, and consequently control store width, is to group these microoperations that always occur at the same time under one control signal. Thus only one control bit is used to control several microoperations. Such a grouping was used earlier.

Another way is to group together in one encoded control field the mutually exclusive microoperations, those microoperations of which only one occurs at one time. No information is lost with such an encoding scheme. If each microoperation corresponds to the activation of a single control line and if, for example, 2^n mutually exclusive lines exist, an n-bit field in the microinstruction is required. These n bits, through a decoder, regenerate these 2^n control signals when the microinstruction is executed. Two such groups of microoperations that are not mutually exclusive must be controlled by two separate encoded fields in the microinstruction. There may be several ways of performing this microinstruction encoding, and the decision to group microoperations together in one or more separate encoded fields depends on their potential for being executed in parallel. The use of separate control fields for each independent functional unit, such as ALU, shifters, and so on, allows simultaneous use of system resources, as long as they do not conflict. Generally, for the encoded formats, a number of output decoders are required for the various fields of the microinstruction. However, since the widths of these fields are usually small, these decoders are less complex than those required for the highly encoded vertical microinstructions.

In choosing the microinstruction format, the designer has the task of balancing control store size, complicated clocking schemes and speeds, branching flexibility, and the complexity of output circuitry required to handle microinstruction control bits or to decode encoded microinstruction fields. In general, microinstruction characteristics are strongly influenced by the data flow design of the processing section and by the machine language instruction set the system supports.

Organization of the Control Store

A central issue in the design of a microprogrammed control section is the organization of its control store, which contains the microinstructions that constitute the microprograms to be executed. Advances in memory technology have resulted in the availability of large, fast, inexpensive control stores and in the appearance of several types of control store structures.

Functional Characteristics. Most microprogrammed computers have several memory facilities available: main memory, control store, and local memory, the latter consisting primarily of functional registers. The control store usually has one level, although there are machines that also contain a second level, the nanostore. Usually, control store is distinct from main memory, although there are exceptions to this, in which microprograms are executed from a protected section of main memory. Between the microinstructions in the control store and the control gates that they activate, there exists at least a buffer register (the microinstruction register, MIR) and there may also exist some decoding logic.

One characteristic of the control store is whether it is writeable. ROM-based nonalterable control stores are usually provided (written) by the manufacturer, are usually tailor-made, and depend on the system's architecture. On the other hand, writeable control store (WCS), realized using fast random-access memories (RAMs), has the advantage that it provides user microprogrammability, allowing a user to create the system's architecture. Advances in technology will continue to produce larger and faster RAMs, so their utilization for the implementation of control stores will allow greater freedom and flexibility in system designs. Users will then have the ability of defining the instruction set best suited to the specific application. They can also implement certain critical routines (e.g., specialized I/O and interrupt-handling procedures, floating-point routines, system tables, etc.) as microprograms, thus improving system performance for a given application. Dynamically microprogrammable systems can also be implemented with WCS, whose microprograms can be modified electronically during operation. Between these two extremes, various technologies also permit intermediate characteristics employing, for example, a facility to switch from one ROM to another.

Speed. An important characteristic of the control store is its speed, described by its **access time**, which corresponds to the time required to read a microinstruction from the control store into the MIR. This speed is very important and is generally the limiting factor in the overall speed of the system. On the other hand, the control store should be fast enough so that the degradation it may cause (relative to hard-wired control) to the system's speed will be compensated for by the advantages of microprogramming. If user microprogrammability is to be allowed, to make attractive the implementation of functions in firmware rather than in software, the ratio of main-memory cycle time to control store cycle time is usually of the order of 10 : 1.

Size. The size of the control store may be a limiting factor to what applications can be microprogrammed on the system. The number of microinstruction words in the control store depends primarily on the instruction repertoire.

Control Store Organizations. In the simplest and most common structure, there is one microinstruction in each control store location. A different structure may have two microinstructions per location, thus reducing the number of control store references. A third alternative divides the control store into blocks (sometimes called pages), and the addressing scheme used addresses either within the same block as the current microinstruction or to another block. This shortens the addresses when addressing is done within the same block. Another scheme is the split control store, which comprises two separate storage units that have different word lengths. The shorter-word-length unit contains microinstructions that move data or initiate the execution of a microinstruction that resides in the other storage unit. The longer microinstructions that reside in the other unit can exercise more direct control over machine resources. A final scheme involves structuring the control store into multilevel organizations; in the two-level case, the second level contains longer microinstructions called **nanoinstructions** and the unit that contains them is called the **nanostore**.

Hard-wired versus Microprogrammed Control

The speed of operation of a hard-wired control is limited by the speed of logic circuitry. The speed of operation of a microprogrammed implementation is limited by the speed of the control store. Since control stores operate more slowly than the logic circuitry, microprogram control does not give the fastest speed of operation. It is also generally agreed that for a well-defined application, the hard-wired implementation approach results in a somewhat smaller control section than that of the microprogrammed approach. However, changing the visible machine and its instruction set requires rewiring, new layouts, and so on, whereas in the microprogrammed version, this involves only changes or additions to the microprogram. The microprogrammed approach also provides flexibility and coherence to the control design, and a level of diagnosability not attainable with a nonmicroprogrammed approach. Although the hard-wired control can be faster, as the complexity of the system increases, it becomes more expensive since its cost is somehow proportional to the complexity of the control. The cost of microprogrammed implementation is essentially determined by the cost of the control store, and ROMs, PLAs, and other LSI devices used today as control stores continuously show a dramatic reduction in prices. At some level, however, there is always the need for some hardwired control (e.g., in decoding the outputs of the control store, for sequencing the microprogrammed control, etc.). Even with microprocessors and microcomputers, some of them have microprogrammed control and some do not.

Execution of a COMPARE Instruction in a Microprogrammed System

In order to make clearer the design requirements of a microprogrammed control unit, we now present an example of defining the microoperations and control signals involved in the execution of the following simple COMPARE instruction:

$$IF(ACC) = (B) \quad THEN \ IC \leftarrow (IC) + 2$$

$$ELSE \ IC \leftarrow (IC) + 1$$

The details of the system to be used are shown in Fig. 9.85. The system uses 2's-complement notation, and the parallel adder can add two operands, with input carry G_{20} being 0 or 1. The ALU section has zero detect logic that issues a 1 when the output of the adder is all zeros; otherwise, it issues a 0. A temporary register B is also used.

The necessary microoperations are grouped and coded in assigned fields of the microinstruction with one code of the field usually designating one microoperation. The leftmost 6 bits of the microinstruction are the field NMA, which signifies the next microinstruction address. The next 2-bit field LIN is chosen to designate the left input to the adder, as shown in Table 9.18. This table also shows the corresponding control gate signals.

The next 2-bit field RIN is used to designate the right input to the adder, as shown in Table 9.19. (B') is the 1's complement of (B).

The next 1-bit field ADC is used to designate the microoperations of the parallel adder as follows. When ADC = 0 (i.e., $G_{20} = 0$) the addition is performed with no carry-in; if ADC = 1 (i.e., $G_{20} = 1$) the addition is performed with carry-in. (See Section 9.5.4 regarding 2's-complement addition.)

The 2-bit field DBIN designates connections to the data bus, as shown in Table 9.20.

The next 3-bit field DBOUT designates the destination of the information on the data bus, as shown in Table 9.21.

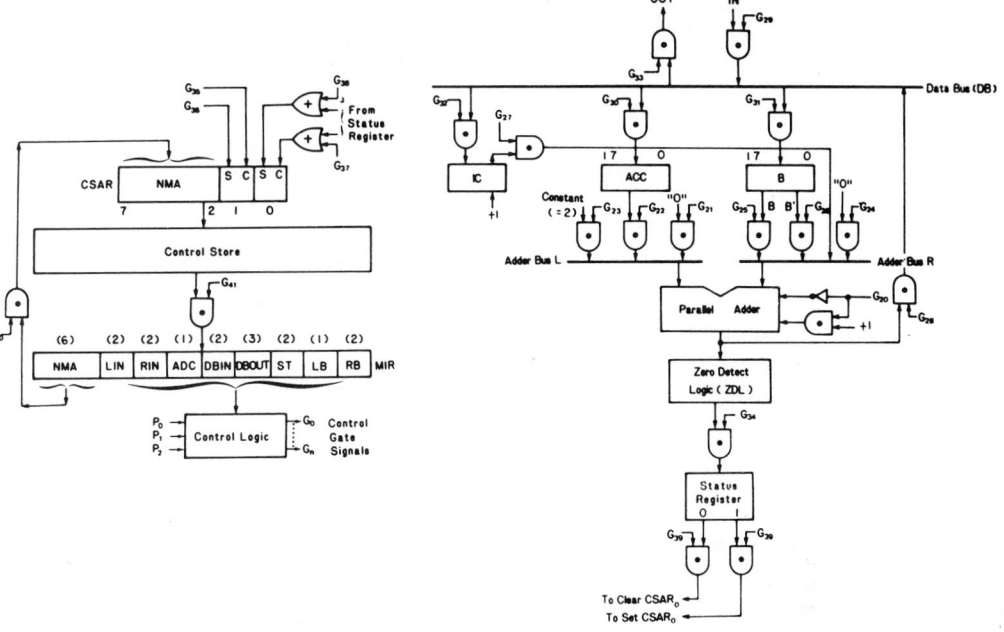

Fig. 9.85 Part of the microprogrammed system and the control gates involved in the execution of the COMPARE instruction.

TABLE 9.18   Field LIN

Code	Microoperation	Control Gate Signal
00	Adder ← "0"	G_{21}
01	Adder ← (ACC)	G_{22}
10	Adder ← "2"	G_{23}
11	Not used	—

TABLE 9.19   Field RIN

Code	Microoperation	Control Gate Signal
00	Adder ← "0"	G_{24}
01	Adder ← (B)	G_{25}
10	Adder ← (B')	G_{26}
11	Adder ← (IC)	G_{27}

TABLE 9.20   Field DBIN

Code	Microoperations	Control Gate Signal
00	No operation	—
01	DB ← Adder	G_{28}
10	DB ← IN	G_{29}
11	Not used	—

TABLE 9.21   Field DBOUT

Code	Microoperation	Control Gate Signal
000	No operation	—
001	ACC ← (DB)	G_{30}
010	B ← (DB)	G_{31}
011	IC ← (DB)	G_{32}
100	OUT ← (DB)	G_{33}
101		—
110 }	Not used	—
111		—

TABLE 9.22   Field ST

Code	Microoperation	Control Gate Signal
0	No operation	—
1	IF (ZDL = 0) THEN (STATUS ← 0) ELSE (STATUS ← 1)	G_{34}

TABLE 9.23   Field LB

Code	Microoperation	Control Gate Signal
0	$CSAR_1 \leftarrow 0$	G_{35}
1	$CSAR_1 \leftarrow 1$	G_{36}

TABLE 9.24   Field RB

Code	Microoperation	Control Gate Signal
00	$CSAR_0 \leftarrow 0$	G_{37}
01	$CSAR_0 \leftarrow 1$	G_{38}
10	$CSAR_0 \leftarrow$ (STATUS)	G_{39}
11	Not used	—

Storing of the result of the zero detect logic ZDL in the 1-bit register STATUS is designated by the 1-bit field ST, as shown in Table 9.22.

The control store in Fig. 9.85 has an 8-bit control store address register (CSAR). Its six high-order bits, $CSAR_{7-2}$, are provided by the next microinstruction address field NMA of the microinstruction in MIR. The two low-order bits $CSAR_1$ and $CSAR_0$ are determined by fields LB and RB, as shown in Tables 9.23 and 9.24.

The comparison in this example is performed by subtracting the number in register B from the accumulator; then the examination of the value of the zero detection logic will determine whether IC will be incremented by one or by two.

As shown in Fig. 9.86, this COMPARE instruction requires three microinstructions. The first microinstruction, in location YYYYYYYY, performs the comparison already noted above; the subtraction is carried out by adding the 2's-complement of the number in register B to the number in the accumulator. Either the microinstruction in location ZZZZZZ00 [i.e., IC ← (IC) + 1] or the microinstruction in location ZZZZZZ01 [i.e., IC ← (IC) + 2] will be executed next, depending on the result of the first microinstruction. Finally, no matter which one of the two is executed, the sequence eventually reaches the microinstruction in location ZZZZZZ11.

As before, each microinstruction in the control store is processed under the control of a three-phase clock, P_0, P_1, P_2. Assume that during the previous microcycle a control signal G is provided to start the sequence with a proper address in CSAR as follows:

$$P_0 : G \leftarrow 0$$
$$\bar{G} \text{ AND } P_1 : CSAR \leftarrow YYYYYYYY, G \leftarrow 1$$
$$G \text{ AND } P_2 : MIR \leftarrow (CS^{CSAR})$$

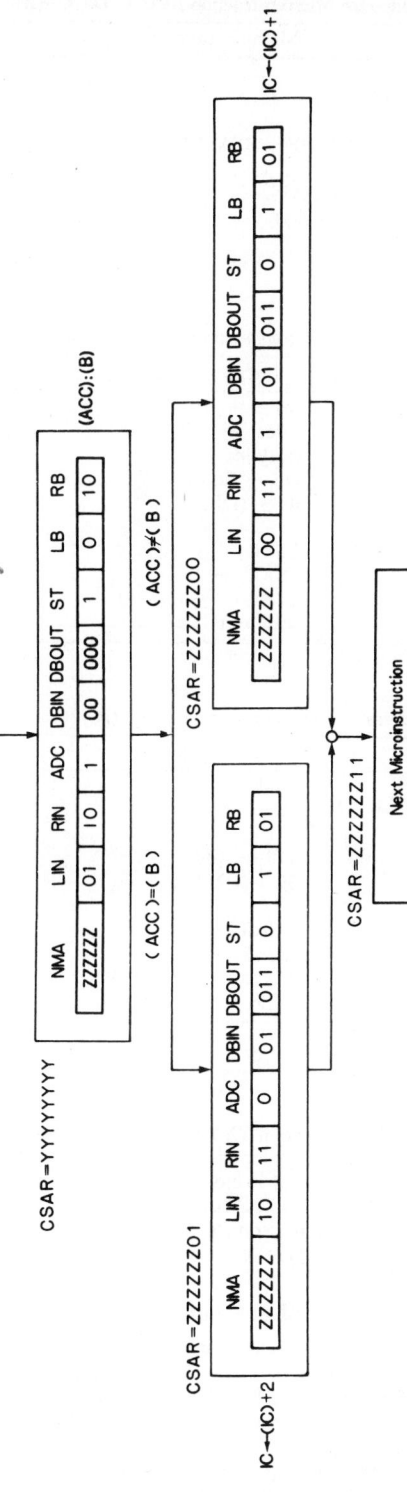

Fig. 9.86 Microinstruction address assignments and microinstruction structure for execution of the COMPARE instruction.

TABLE 9.25   Execution Sequence for Microinstruction ACC ← (ACC)-(B)

Timing	Microoperation	Control Gate Signal
(LIN = 1) AND P_0	Adder ← (ACC)	G_{22}
(RIN = 2) AND P_0	Adder ← (B′)	G_{26}
(ADC = 1) AND P_0	carry-in	G_{20}
(ST = 1) AND P_0	IF (ZDL = 0) THEN (STATUS ← 0) ELSE (STATUS ← 1)	G_{34}
(LB = 0) AND P_1	$CSAR_1$ ← 0	G_{35}
(RB = 2) AND P_1	$CSAR_0$ ← (STATUS)	G_{39}
G AND P_1	$CSAR_{7-2}$ ← (NMA)	G_{40}
G AND P_2	MIR ← (CS^{CSAR})	G_{41}

TABLE 9.26   Execution Sequence for Microinstruction IC ← (IC) + 1

Timing	Microoperation	Control Gate Signal
(LIN = 0) AND P_0	Adder ← 0·	G_{21}
(RIN = 3) AND P_0	Adder ← (IC)	G_{27}
(ADC = 1) AND P_0	carry-in	G_{20}
(DBIN = 1) AND P_0	DB ← Adder	G_{28}
(DBOUT = 3) AND P_0	IC ← (DB)	G_{32}
(ST = 0) AND P_0	—	—
(LB = 1) AND P_1	$CSAR_1$ ← 1	G_{36}
(RB = 1) AND P_1	$CSAR_0$ ← 1	G_{38}
G AND P_1	$CSAR_{7-2}$ ← (NMA)	G_{40}
G AND P_2	MIR ← (CS^{CSAR})	G_{41}

TABLE 9.27   Execution Sequence for Microinstruction IC ← (IC) + 2

Timing	Microoperation	Control Gate Signal
(LIN = 2) AND P_0	Adder ← 2	G_{23}
(RIN = 3) AND P_0	Adder ← (IC)	G_{27}
(ADC = 0) AND P_0	—	G_{20}
(DBIN = 1) AND P_0	DB ← Adder	G_{28}
(DBOUT = 3) AND P_0	IC ← (DB)	G_{32}
(ST = 0) AND P_0	—	—
(LB = 1) AND P_1	$CSAR_1$ ← 1	G_{36}
(RB = 1) AND P_1	$CSAR_0$ ← 1	G_{38}
G AND P_1	$CSAR_{7-2}$ ← (NMA)	G_{40}
G AND P_2	MIR ← (CS^{CSAR})	G_{41}

Then, for the first microinstruction in location YYYYYYYY, which compares the contents of register B and the accumulator, the sequence of microoperations executed and the control gate signals emitted are shown in Table 9.25. The sequence of microoperations executed and the control gate signals emitted for the microinstruction in location ZZZZZZ00 [i.e., IC ← (IC) + 1] are shown in Table 9.26, and the microoperations executed and the control gate signals emitted for the microinstruction in location ZZZZZZ01 [i.e., IC ← (IC) + 2] are shown in Table 9.27.

9.9.4   LSI Control Elements

So far, we have seen some of the hardware elements that support microprogrammed control: control stores (usually ROMs, but sometimes RAMs) to hold the microinstructions that sequence control; circuits (counters, decision logic gates, etc.) for testing conditions and status signals, extracting microinstruction fields, and generating the next microinstruction address; and various other elements, such as registers and decoders.

In this section we discuss two LSI components in detail—the programmable logic array (PLA) and the control store sequencer—and their use in the control function.

Programmable Logic Arrays

Principles of Operation. The principles of PLA operation are covered in detail in Section 9.4.2.
Simply stated, the PLA may be viewed as an LSI implementation of the classic logical sum-of-products structure. Figure 9.87(a) gives a block diagram of the PLA, and Fig. 9.87(b) presents more details of its internal structure. The first-level AND array performs the logical AND for selected combinations of the input signals. These AND gates produce logical product terms (or P terms), $P_1, P_2, \ldots, P_m$. The second level is considered to be composed of OR gates that, by logically summing some or all of the P terms produced by the first level, produce the final outputs (or sum terms), $F_1, F_2, \ldots, F_k$. One may go directly from the logic equations (preferably in their minimal form) to the programming of the AND and OR arrays of the PLA. Since the PLA can be programmed to handle only the useful combinations of the input variables, it is advantageous to use PLAs when there is a large number of inputs and only a small subset of the possible combinations of these inputs is required (used).

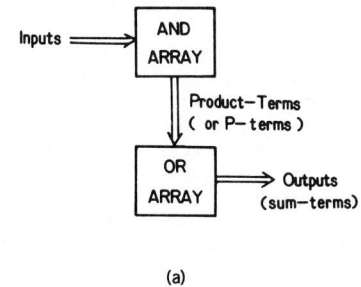

(a)

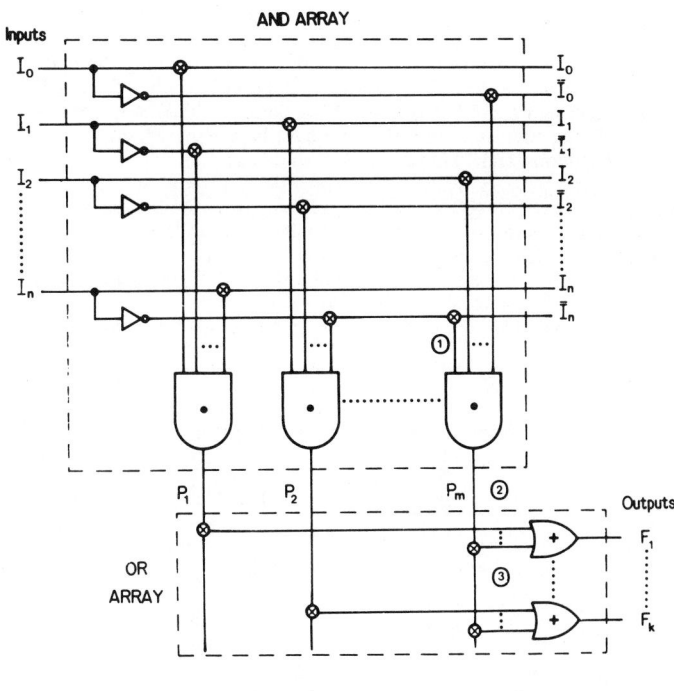

(b)

Fig. 9.87 PLA: (a) block diagram, (b) details of its internal structure. ① up to 2^n inputs/gate; ② P_i's are product terms; ③ up to m inputs/gate; ⊗ programmable connection.

In sequential logic implementations, the PLA can be used to replace the combinational part of the circuit. A new type of PLA, the "sequential PLA," includes latching flip-flops at its output (i.e., storage elements in the feedback path to store the state of the circuit) and can lead directly to an easier realization of clocked sequential logic or control-logic circuits.

PLAs Used in the Control Function. In the hard-wired sequential implementation of the control function discussed in Section 9.9.2, the PLA can be used to realize the gate signal generator (see Fig. 9.76). The control matrix especially, which is of AND-OR nature, can be easily replaced by a PLA.

In the microprogrammed implementation, the PLA can be effectively used at various places in the control section:

1. The PLA can be used as an *instruction op code decoder*, to translate the instruction op code in the instruction register to the control store starting address. In simple machines, this address mapping can be done by a decoder or a ROM. But in more complex cases where the instruction is composed of a number of fields and has a variable format, PLAs can be used to interpret a large number of bits at once, ignoring or interpreting some fields based on the contents of other fields. For example, the control store address can be the output of two PLAs, one PLA used as an instruction interpreter, the second PLA used as an addressing mode interpreter.

2. Another application of the PLA is for *control store patching*. Quite often, after the final design of the control section, there may be cases where the microprogram for the system needs revisions, or faults may appear in some locations of a large, otherwise functional control store. Rather than replacing the control store with a new one, one may use a PLA. This PLA will decode the address locations the engineer wants to change or will detect attempts to access a faulty location. In such cases, it can disable the control memory and either generate new data or enable a small "backup" ROM to which it redirects addresses translated from those originally intended for the control store. Figure 9.88 shows such an implementation where, for example, a PLA with i inputs, j P terms, and k outputs can detect j addresses in a 2^i control store address space, generate an enable/disable signal, and generate j addresses to a small 2^{k-1}-word backup ROM.

3. The PLA can also be used, although not generally, *to store the microprogram instructions*.

4. Finally, another example of applying PLAs in the microprogrammed control section is for *generating the branch addresses* for the modified counter implementation scheme. In such an implementation, the branch addresses may be encoded prior to storing the microprogram in the appropriate next address bits of the microinstructions and decoded with a PLA. If for example, there are 32 branch addresses in a control store of 1024 locations, the PLA implementation would require only 5 bits to encode the addresses; 10 bits would have been required, otherwise. This PLA would require 5 inputs, 32 P terms, and 10 outputs.

LSI Sequencers

The main purpose of any control store sequencer is to present an address to the control store so that a microinstruction may be fetched and executed. The "next address" logic part of the sequencer determines the source of the address to be loaded into the control store address register/counter.

Today's commercial LSI control store sequencers have much more than the capabilities mentioned so far. They receive many more control signals and system status bits, and they include everything

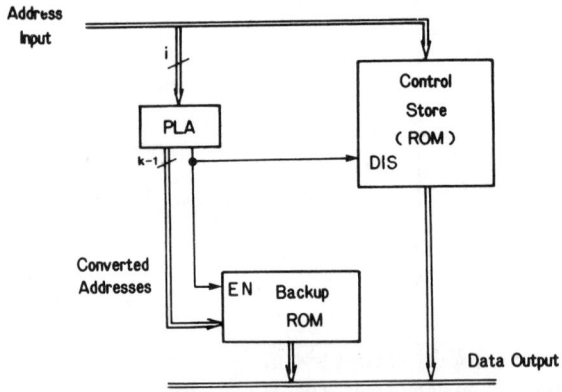

Fig. 9.88 PLA used for control store patching.

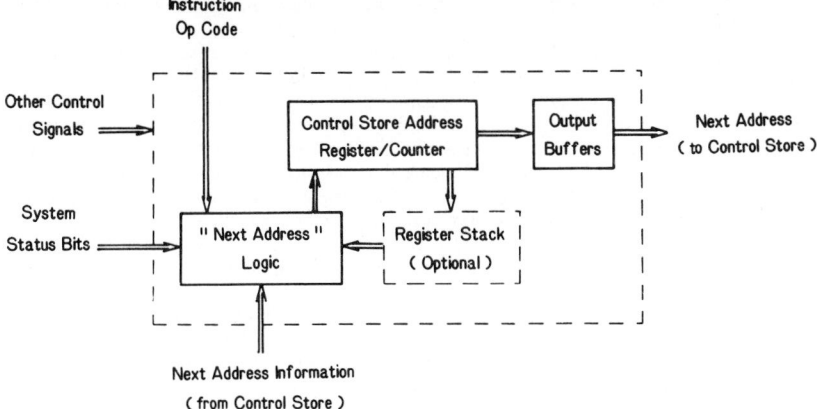

Fig. 9.89 General block diagram of an LSI control store sequencer.

required for address-incrementing functions and complex multiway conditional microbranching. They have on-chip loop counters for repeated microprogram loops and may also include on-chip hardware stacks and pointers, a useful feature for saving the return-to addresses of nested microsubroutines. A typical block diagram of a commercial sequencer is shown in Fig. 9.89, where the "next address" logic is now quite complex. The choice of the address source is guided by the next address information bits the sequencer receives; these bits can generally specify one of the following: increment, conditional skip next microinstruction, conditional branch to a microsubroutine, and others.

Available LSI control store sequencers may be divided into two general categories: those with a fixed number of memory locations within their addressing capability and those that are cascadable bit-slice devices. While bit-slice sequencers are easily expandable, the fixed-address sequencers require additional circuits and paging techniques to extend their memory access range.

Another difference among the available sequencers is that some of them provide on-chip address output latches, whereas others do not have such latches. The latter category allows more flexibility but requires the designer to add his or her own latches.

9.9.5 Lookahead and Pipelining

The concepts of operation overlap, instruction lookahead, and processor pipelining are significant architectural features used to increase the average speed of the instruction set and the overall throughput rate of the system. Such concepts are found not only in larger mainframe CPUs but also in minicomputer CPUs and even in 16-bit microprocessors.

Instruction Lookahead

Through instruction **prefetching**, or **lookahead**, an effective overlap is achieved among instruction fetching, decoding, and execution (data handling and transformation) times. The delay involved between the initiation of a memory read and the arrival of the word at its destination [memory access time delay, memory busy time (servicing a higher-priority request), and delays in the word transfer path] is usually relatively long. Lookahead mechanisms that fetch a number of instructions and operands from main memory in advance of execution by the ALU section of the system can effectively reduce the average delay. This technique "looks" several instructions ahead of the one currently being executed. Buffers or scratch registers are used within the CPU as speed-matching devices (1) in the control section of the CPU to hold the instructions fetched in advance or to hold several instructions to improve performance for small loops, and (2) interposed between the ALU and memory to hold prefetched operands and to avoid storing and refetching data used by successive instructions. Through this lookahead technique, if various references are initiated early enough, the operands are usually available in the ALU buffer by the time the ALU is ready for them.

If we assume that processing for each instruction is composed of three subtasks:

F = instruction fetch and decoding

A = address calculation and operand fetches

E = instruction execution

the instruction lookahead is similar to Fig. 9.81(a), where F of one instruction, A of the prior instruction, and E of the one before that may occur simultaneously, yielding a threefold improvement over serial processing.

This lookahead concept attempts to have instructions and operands arrive from memory at a rate approximately equal to the rate at which the CPU can utilize them. However, some interlock problems may arise. First, if an instruction depends on previous ALU results, it should not be started before these results are available. Second, if a conditional-branch-type instruction is encountered, and the previous instruction that sets those conditions has not been completed, the control section will not know whether to prefetch the instruction following the conditional branch or the instruction at the target of the branch. Third, problems also arise when fetching operands that have not reached storage or when attempting to modify an instruction by a preceding store operation (i.e., treating instructions as data, as in Section 9.3.3).

Several techniques have been used to take care of such deadlocks:

1. There are three basic ways for handling the problem of conditional branching on ALU results:
   a. The control section can stop the flow of instructions until the ALU has completed the preceding operation and the result is known, and then fetch the next instruction. Whether the branch is taken or not, this always causes a delay.
   b. Based on past experience with the program, make a "guess" which way the branch is going to go before it is taken, follow this path, and continue to prepare instructions; if the guess proves later to be wrong, the prepared instructions must be discarded and the correct path taken instead.
   c. The control section can be simultaneously fetching the instructions following the conditional branch as well as instructions at the target of the branch; when the ALU generates the condition(s), the control section can decide which of the two groups of prepared instructions to use.
2. Problems similar to that of treating instructions as data can be handled by the control section if, as soon as an instruction is loaded, the control section tests the instruction to see whether it is a *store-type* instruction, in which case the fetch sequence must wait until the effective address has been prepared to see whether it is going to modify a successive instruction.
3. The problems of fetching operands that have not reached storage yet can be handled, if the result of the ALU can return first to the operand's buffer, before being sent to storage.

Prefetching and lookahead have an impact on the control function, since it must implement this technique and take care of all deadlocks that may arise during system operation. Incorporating additional scratch registers and buffers in the system increases the number of gates requiring control. In the microprogrammed control version, this increases the number of control bits in the microinstruction, and the width and cost of the control store. It also requires proper microprograms to achieve this instruction overlap, incorporating sophisticated algorithms for determining the next step (especially in the case of conditional branches).

Pipelining

The lookahead operation we just discussed can be used for matching the speeds of the ALU and main memory by greatly reducing the time spent by the ALU waiting for an operand.

The effect of the speed of the ALU itself on overall system performance is also very important. To increase the ALU speed, an "assembly-line technique" called **pipelining** has been effectively used. The basic philosophy is to divide a computational task into several sequential subtasks, each being handled by a separate stage or station of the ALU. All stages operate simultaneously with independent sets of data; temporary storage buffers exist between stages; individual computation advances from one stage to the next; and results are closer to completion as the end of the pipeline is approached. This pipelined ALU accepts new inputs before the previously accepted inputs have been completely processed and put out from the pipeline. When one subtask result leaves a stage, the logic associated with that stage becomes free and can accept new data (results) from the previous stage. The rate at which inputs are accepted to the ALU is chosen in relation to the time required to get through one stage, with the main goal of keeping all portions of the pipeline fully utilized. Once the pipeline is full, the output rate will match the input rate.

The pipelined ALU can be effectively used for executing complex repetitive-type instructions (such as multiply/divide) and floating-point instructions. As far as a single individual instruction goes, the time required for its execution via a pipelined implementation may be the same or slightly longer than with a parallel ALU used iteratively. However, the fact that several steps of the instruction execution are carried out concurrently in the pipeline increases the number of instructions that may be completed per second, thus increasing significantly the overall throughput capability of the system.

Again, several requirements exist to properly control the operation of such a pipelined ALU. The control function for pipelines is considerably more complex than for more conventional structures.

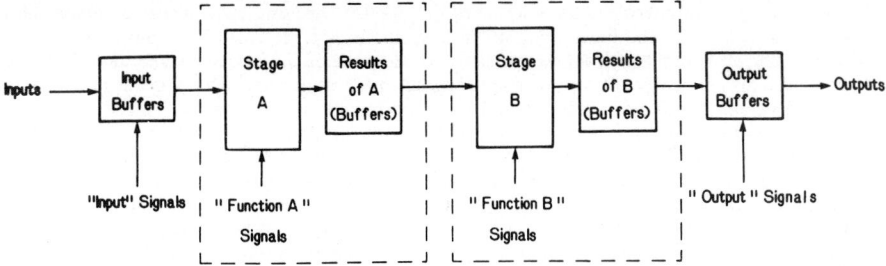

Fig. 9.90   Simple two-stage pipeline data flow.

	P_i	P_{i+1}	P_{i+2}	P_{i+3}	P_{i+4}	P_{i+8}	P_{i+6}	P_{i+7}	P_{i+x}
datum$_j$	Input	Function A	Function B	Output					
datum$_{j+1}$		Input	Function A	Function B	Output				
datum$_{j+2}$			Input	Function A	Function B	Output			
datum$_{j+3}$				Input	Function A	Function B	Output		
datum$_{j+4}$					Input	Function A	Function B	Output	
datum$_{j+5}$						Input	Function A	Function B	Output

Fig. 9.91   Timing diagram of the operations performed by a two-stage pipeline during every clock phase P_i.

The control section must determine what data should be input to the pipeline, the specific operation of each stage (i.e., what inputs it should receive from the previous stage, what functions it should perform on them, what results it should pass to the next stage), and what outputs should be sequentially generated at the end of the pipeline.

For example, consider the simple two-stage pipeline of Fig. 9.90. Figure 9.91 shows the timing diagram of the operations performed by this two-stage pipeline, for every clock phase P_i. Inputs are gated to the pipeline by the "input" signals during clock phase P_i. During clock phase P_{i+1}, the function A control signals are issued to select the inputs to be gated to stage A and to specify the function to be performed by stage A;* at the same time, since the input buffers of the pipeline become available, the "input" signals may be issued again to fill the buffers with the next data to be fed into the pipeline. During clock phase P_{i+2}, three things happen simultaneously.

From P_{i+3} on, during every clock phase, an output leaves the pipeline at almost twice the rate possible without pipelining. In general, for a k-stage pipeline, $k + 2$ types of control signals would be simultaneously issued during each clock phase. Inputs are gated into the pipeline during each clock phase, and similarly, during each clock phase, outputs are being produced by the pipeline at nearly k times the rate of a system without parallelism. All these operations, of course, are transparent to the user and the programmer of the system.

9.9.6   Emulation

Often, a computer user may purchase a new computer and wish to convert old programs to run on the new machine. Or a user may wish to convert application programs written for one machine to run on another machine that exists in the local environment, to permit the user to take advantage of all the additional features of his or her machine.

There are several approaches for performing this conversion. At the one end of the spectrum lies the reprogramming effort, where the user reprograms all problems in the new machine language. The next point in the spectrum is the recompiling process; if the programs are all written in a high-level language, guaranteeing absolute machine independence, the task of conversion is reduced to recompiling these programs for the new system. The next approach involves substituting for each machine instruction in the source program for the old system, one or more instructions of the new machine that execute the same operation, then assembling the resulting code and executing it. This is an inefficient approach of converting one program to another and presents serious problems since the

---

*We assume that during one clock phase, the operation of a stage is carried out and the results are properly placed in its output buffers.

architectures may be very different. Another approach used to perform this conversion is (software) **simulation**, through which the functioning of the one system is represented by another. Simulators usually run their programs significantly more slowly than on the original machine, and their degree of efficiency depends on the detail to which the original machine is simulated. On the other hand, simulation does not require any hardware, and it is conceptually simple to design.

The final point on the program conversion scale is **emulation**. Microprogramming techniques may be used here to accomplish this step of executing programs in the old machine language whose instructions are properly interpreted in the new system. This new physical machine, as defined by its hardware, its microinstructions, and their actions, is called the **host** machine. The old machine emulated by sets of microprogrammed routines is called the **image**, **virtual**, or **target** machine. Thus emulation is a combined hardware-software approach to simulation. It can be done by either adding hardware facilities in the host system for handling special features of the target machine or by providing these facilities through microprogramming techniques. An emulator is basically an extension of the host machine's architecture, hardware, and software, to include the microarchitecture of the target machine. It is relatively easy for a simple host system to emulate more than one target machine. Since it executes the instruction set of the target machine at the control level rather than in software, the emulator is much faster than a simulator. In general, emulators offer a one-to-one throughput compared to the target machine, while simulators are normally 5 to 10 times slower. Usually, a host machine used for emulation will have (at least) two types of microprograms: one for its own conventional machine language and one or more (called the emulators) for the target machines' languages. Means should be provided to select which one of such microprogram types the host executes.

The emulator must:

1. Map the components of the target machine (i.e., main memory, registers, I/O subsystem, and other resources that are addressable by the source program) into those of the host machine.
2. Interpret the machine language instructions of the target machine.
3. Identify the respective microroutines in the host machine that will execute the target machine instructions.
4. Provide some means for efficient linkage and interfacing between host machine mode of operation and emulation mode of operation.

All activities of the emulator are governed by an **emulator control program** discussed later.

Mapping the Components of the Target Machine to Those of the Host Machine

In general, the smaller the mismatch between the host microarchitecture and the target machine architecture, the easier the construction of an efficient emulator. When the differences are significant, they may impose real restrictions and implementation shortcomings, resulting in certain features of the target machine not being supported by the host.

When mapping the target main memory to the host memory, the following considerations must be taken into account: main-memory capacities for both the target and the host machines, addressing modes and address boundary alignments, lengths and formats of memory words, how many host bits to be used for emulating a target word or byte, and other factors. This mapping function must ensure that a minimum addressable memory cell of the target machine will be contained in the smallest possible host machine memory cell, and that address translations from target address to host address and vice versa should not be time consuming. If, after such mapping, there is still surplus host memory available, this can be effectively used to temporarily store I/O-type information (for the case where the target machine performs I/O in concurrent mode) or to save some of the target machine's registers containing privileged data that the target machine user cannot use.

If the host machine has at least as many registers, counters, and so on, as the target machine, of equal or larger length, they can be used effectively to emulate those of the target machine. If the number of the host machine registers is less than those of the target machine to be emulated, a secondary register file might need to be created in the host's main memory. The most important and frequently used registers of the target machine (e.g., the accumulator, the instruction counter, the instruction register, etc.) can be assigned to the main registers of the host, while the rest of the target registers (e.g., index registers, overflow and indicators, etc.) can be assigned to the secondary file in host's memory. A page map mechanism (see Section 9.7.1) of the emulator control program can designate which target registers are in the host's register file and which are in its secondary file at any time. This, of course, introduces considerable overhead and requires some sort of scheduling to determine which register should be paged out to make room for the incoming register. It must also be noted here that mapping the condition codes generated by the host's hardware to convert them to target machine condition codes is not a trivial effort.

Mapping of I/O devices is simpler than main-memory mapping. A static one-to-one mapping of target I/O device registers into host I/O subsystem registers or host main-memory locations may be done. We shall see below how these are handled by an I/O instruction or an I/O interrupt.

Interpreting and Executing Target Instructions

Having defined the mapping of the target machine microarchitecture into the host machine, the designer must then develop the firmware for interpreting target instructions. A number of micro-routines must be written in the microprogrammable host, such as (1) a start microroutine, for initializing the emulator prior to the processing of any target instructions; (2) a fetch microroutine, for fetching the next target instruction pointed to by the instruction counter, and one or more microroutines to perform target instruction format parsing, effective operand address computation, and op code decoding (these microroutines are repeated for every target instruction in the program to be converted, before branching off to the corresponding instruction module for executing it); and (3) the various instruction modules containing the microroutine(s) that execute the individual target instructions.

An emulator control program must also be designed to sequence the activities previously described, and contain proper microroutines for handling exceptions and error situations (error handler), interrupts (interrupt handler), and I/O microroutines (I/O handler) to be used by the I/O instructions as well as the interrupt handler, as described later. The emulator control program may also perform data conversion and transmission, as well as the linkage between the CPU and I/O.

When the target instruction to be executed is an I/O instruction, control is passed to the emulator control program. The control program searches a table of address pairs to match the I/O device register address; the second address of the pair is the entry point into the I/O device handler. Also passed to the I/O device handler microroutine is a read/write flag. The device handler responds to the request by accessing the I/O devices directly. When the I/O is done in a concurrent mode, part of the host's surplus main memory may be used to store the I/O information, to be used by the I/O handler for carrying out this data transfer activity.

In an analogous fashion, the appearance of an interrupt transfers control to an interrupt handler in the emulator control program. This handler may use the I/O microroutines. It identifies the interrupt type and source, saves all pertinent information into a priority-ordered queue, and sets a flag. The interrupt-handler microroutine is executed after the execution of one instruction and prior to fetching the next instruction. When the emulator is restarted, the instruction fetch microroutine interrogates this flag, and control is conditionally passed to the next stage of processing.

9.9.7   Microprocessor System Signals

So far in Section 9.9, we have discussed primarily intramodule control and timing signals. We have chosen as an example the most complicated module of a digital system—the CPU processor module. Therefore, the control and timing signals we mentioned corresponded mainly to those required by the ALU to carry out the instruction execution microsequences. We also mentioned signals to memory to fetch instructions and data, to move addresses, to increment counters, and so on. System-wide control and timing signals are also required to synchronize the operations of the separate and different modules and to facilitate their intercommunication activities over the buses.

As a typical modular system, we consider a microprocessor system whose general block diagram is given in Fig. 9.92. Note that we are now talking about systems controlled by a microprocessor, not about a microprogrammable control unit. Before describing the microprocessor system signals, we explain the operation of a microprocessor and the cycles it follows for fetching and executing an instruction. We then discuss the timing and synchronization requirements of a microprocessor-based system. This will help us understand better the role and function of some of the microprocessor system signals. The rest of the system signals are covered later.

Microprocessor Systems

Operation. The microprocessor of a system,* like the CPU of a general-purpose computer, communicates with the memory modules and the I/O modules by sending addresses to them and sending them data or receiving data from them (see Sections 9.12.1 and 9.13). Thus two of the interconnecting buses shown in Fig. 9.92 are devoted to addresses and data. In the most straightforward design, the widths of these buses are selected to correspond to the widths of the system addresses and data. In this case, for an 8-bit data word microprocessor capable of directly addressing a 64 K-byte memory space, an 8-bit data bus and a 16-bit address bus would be required. Alternative design techniques may also be used, incorporating bus multiplexing (see Section 9.13.6). The

---

*Some microprocessors are made up of a number of chips. We discuss here the single-chip microprocessor, for reasons of simplicity.

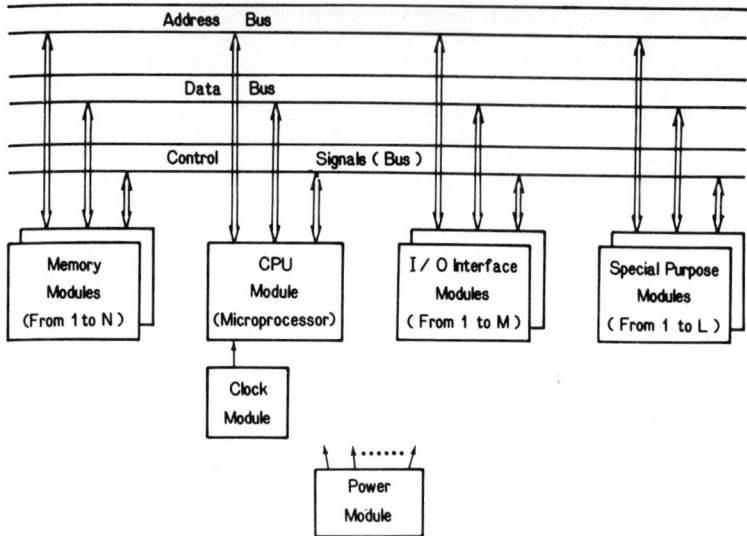

Fig. 9.92 Modular microprocessor-based digital system.

following key questions arise with regard to the use of the buses: Which module should use a bus at any time? If all modules attempt to use a bus, should it be allocated to the module that first requested its use? Should there be a more orderly sequence for allocating a bus to the proper module at a given time? Who will determine to which module to allocate the bus at any given time?

Since the microprocessor is the major component of such a digital system, it has been assigned this task of determining the bus allocation. Knowing the system state at all times, it provides all other modules of the system with properly timed control signals to inform them of the state of the microprocessor and what information is on the bus at all times. These signals then synchronize the rest of the system to the microprocessor.

Microprocessor Cycles. The system-wide signals issued by the microprocessor provide the following functions: synchronization, microprocessor system scheduling (interrupt and I/O), and other facilities, such as clock and reset.

We use an 8-bit microprocessor (Intel 8080) to discuss the microprocessor system signals shown in Fig. 9.93. This microprocessor chip is driven by an external two-phase clock on a separate chip (input

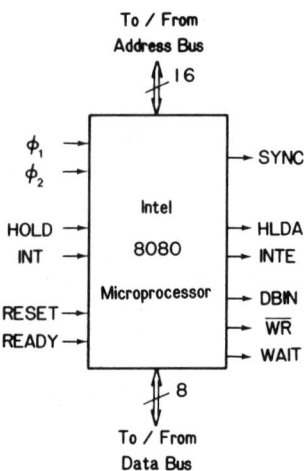

Fig. 9.93 Intel 8080 microprocessor system signals.

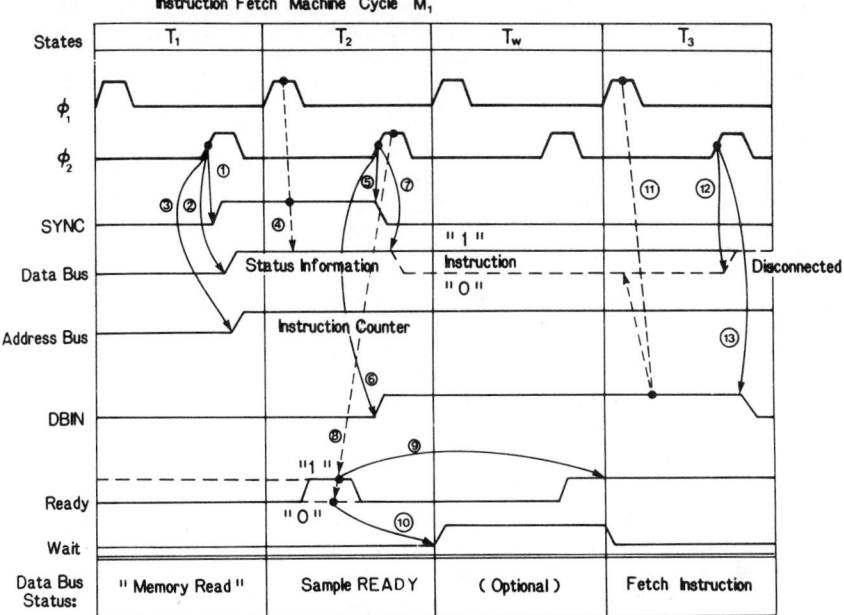

Fig. 9.94 Timing relationships for fetching a single-byte instruction at machine cycle M_1.

pins ϕ_1 and ϕ_2), and all processing activities are referred to the period of this clock.

The execution time of an instruction (fetching and execution) may consist of one to five **machine cycles**, denoted $M_1, M_2, \ldots, M_5$. The beginning of every machine cycle is specified by the SYNC signal issued by the microprocessor, as explained in the next section. The 8080, like many other microprocessors, can issue only one memory address per machine cycle, each address accessing a byte. Since the instruction length may be 1, 2, or 3 bytes long, fetching an instruction may require from one (M_1) to three machine cycles (M_1, M_2, M_3). The additional number of machine cycles required to execute the instruction varies from instruction to instruction. Some instructions do not reference memory or a peripheral device (i.e., no address is issued during their execution); therefore, no additional machine cycles are required. Others may require the extra M_4 and even M_5 machine cycles since their execution may send data to memory or to a peripheral device or receive data from them.

Each machine cycle consists of three to five clock periods, called **states** and denoted $T_1, T_2, \ldots, T_5$. See Fig. 9.94. States are the smallest unit of processing activity. The clock phase ϕ_1 subdivides each machine cycle into states. A state lasts one clock period of ϕ_1 (i.e., the time between two consecutive rising edges of ϕ_1). With each clock pulse ϕ_1, a transition occurs from one state to the next.

System Timing and Synchronization

To understand better the timing and synchronization requirements of the microprocessor system, we examine in detail the fetching and execution of the single-byte instruction

$$\text{``ADD M'': ACC} \leftarrow (\text{ACC}) + (M^{H,L})$$

This instruction specifies addition between the contents of the 8-bit accumulator with the contents of a memory location (one byte) whose 16-bit address is specified by the two bytes previously loaded in registers H and L in the microprocessor. The 8-bit sum is stored in the accumulator.

Instruction Fetching. The instruction fetching is always started at state T_1 of machine cycle M_1. In this simplest case of a single-byte instruction, memory cycle M_1 consists of three states, T_1, T_2, T_3. The timing relationships for the machine cycle M_1 during which the instruction fetching is performed is given in Fig. 9.94.

State T_1. The microprocessor communicates with the other system modules through a 16-bit address and an 8-bit data bus. In state T_1, the rising edge of clock phase ϕ_2 initiates the SYNC signal.

This dependency is shown in Fig. 9.94 with the number ① arrow from the rising edge of the ϕ_2 pulse to the SYNC signal. This SYNC signal issued by the microprocessor is used simply to inform all system modules of the first state T_1 of each machine cycle M_i. During this state T_1, the microprocessor uses the 8-bit data bus to transmit an 8-bit status word, through which it notifies all other modules of the system of the type of machine cycle (e.g., instruction fetch, memory read, memory write). The transmission of this status word on the data bus during T_1 is also initiated by the rising edge of ϕ_2 (arrow number ② in Fig. 9.94). Furthermore, the rising edge of ϕ_2 during T_1 causes the microprocessor to place the contents of the instruction counter on the address bus (arrow ③). The address bus lines hold their values stable until the first clock ϕ_2 after state T_3. This gives ample time for the microprocessor to read the instruction byte returning from memory.

State T_2. External logic reads and examines the status word when both ϕ_1 and SYNC are "1" (dotted time marker ④). Assume that during state T_2 the external signal READY input to the microprocessor is "1." (We shall see what happens when READY $= 0$.) This means that memory had enough time to access the instruction byte, and at the same time send the signal READY $= 1$ to the microprocessor. The rising edge of ϕ_2 ends the SYNC signal (arrow ⑤), initiates the signal DBIN (arrow ⑥), and removes the status word from the data bus (arrow ⑦). The microprocessor issues this DBIN signal to notify the rest of the system that the data bus is in an input state and can therefore be used by memory to place information on it. Immediately thereafter, memory places the accessed instruction byte on the data bus.

State T_3. Since READY was 1, the system enters state T_3 (arrows ⑧ and ⑨). The specific operations performed during state T_3 depend on the type of machine cycle M_i. Since here it is the M_1 cycle of the instruction fetch, the microprocessor recognizes that the word found on its data bus is an instruction, and reads it during state T_3 (dotted time marker ⑪). The rising edge of ϕ_2 ends the period that the data bus carries information, and the data bus is considered now as being "disconnected" from the microprocessor (arrow ⑫). Finally, the rising edge of ϕ_2 also ends the signal DBIN (arrow ⑬).
Since the microprocessor is usually faster than most of the external memories, it can remain idle until the memory has had time to supply the instruction. If, after the microprocessor has sent an address to memory, the microprocessor receives a "1" on its READY input, it will proceed to state T_3 and read the information found on the data bus. If, however, the memory did not have enough time to access the instruction, it would not send READY $= 1$ to the microprocessor. In that case, instead of advancing to the next state, T_3, the microprocessor enters a "wait" state, T_w, after finishing its T_2 state. State T_w, which lies between states T_2 and T_3, starts—like all other states—with the rising edge of ϕ_1. At the same time, the microprocessor notifies the rest of the system that it is entering state T_w by issuing a "1" on its output pin WAIT (arrow ⑩). This wait period can last for an indefinite number of clock periods, as long as READY $= 0$. Then the microprocessor enters state T_3 with the next rising edge of clock ϕ_1. With the READY signal, the microprocessor can be delayed so that it will always stay in synchronism with other, slower modules of the system. This input pin READY can also be used to let the microprocessor execute the instructions step by step, something that is very useful for debugging purposes.

Instruction Execution. Thus the one-byte instruction "ADD M" has been fetched into the microprocessor during the first three states T_1, T_2, T_3 of the first machine cycle M_1. Since it specifies adding the contents of the ACC and the contents of a memory location, a new reference to memory is required to fetch the operand. This will be done during the first three states, T_1, T_2, T_3, of the second machine cycle M_2. The microprocessor issues the data address (the contents of its two registers H and L) to memory during T_1, the memory accesses the operand and places it on the data bus during the state T_2, and the microprocessor reads this operand byte and places it in its temporary register during T_3. The timing diagram during these three states for fetching the data is similar to that of Fig. 9.94. Finally, during state T_4 of machine cycle M_2, the contents of the ACC and of this temporary register are added in the ALU and the result is placed in the ACC. Thus the fetching and execution of this single-byte instruction "ADD M" requires two machine cycles and a total of seven states.

Other System Signals
We now describe some of the other system signals shown in Fig. 9.93 that have not been discussed so far.

RESET. This input signal to the microprocessor is used to initialize the system after a power-down condition or to reinitialize the microprocessor. Usually, after the RESET signal has been applied, the instruction counter is loaded with the contents of a standard starting address.

INT. The microprocessor recognizes an interrupt request—placed on this line by an external I/O interface device—at the end of the current instruction or while halted. If the interrupt is accepted, the microprocessor sends during state T_1 the INTA (Interrupt Acknowledge) signal, which is a specific bit of the status word placed on the data bus lines, and enters the "interrupt cycle."

INTE. Most microprocessors have two instructions, the EI (Enable Interrupts) and DI (Disable Interrupts), that place an internal Interrupt Enable flip-flop in the state 1 or 0, respectively, and specify whether the microprocessor is allowed to accept interrupts or not. If this Interrupt Enable flip-flop is reset, the microprocessor will not honor the interrupt request. The INTE (Interrupt Enable) signal indicates the contents of this internal flip-flop. It is automatically reset (disabling further interrupts) at time T_1 of the instruction fetch cycle (M_1) when an interrupt is accepted and is also reset by the RESET signal. The INTE signal is used by I/O devices to decide whether or not they can interrupt the microprocessor (by sending a signal to its INT input).

HOLD and HLDA. The HOLD input signal requests the microprocessor to enter the HOLD state. The microprocessor acknowledges being in the HOLD state by placing a signal on its HLDA (Hold Acknowledge) pin, indicating to the other modules of the system that its data bus and address bus are in their high-impedance state (i.e., "disconnected" from the microprocessor). Therefore, an external module can gain control of these buses as soon as the microprocessor has completed its use of these buses for the current machine cycle. These HOLD and HLDA signals are usually used by certain I/O modules that want to perform high-speed data exchange with the system.

WR. This signal is issued by the microprocessor to indicate memory write or I/O control operation and used by memory modules and I/O devices. (Signals are often denoted by $\overline{X}$ rather than X.) Data to be written in memory or to be output remain stable on the data bus as long as $\overline{WR} = 0$.

To summarize, then, in order to orchestrate and synchronize the operation of all the system modules, timing, control, and status signals are required. The timing signals input to the microprocessor (in our case clock phases ϕ_1 and ϕ_2) are generated by an external clock generator. The timing signal used by all other modules of the system is the SYNC signal issued by the microprocessor during state T_1 of each machine cycle M_i. It was also mentioned that, during state T_1 of each machine cycle M_i, the information the microprocessor places on the data bus is an 8-bit status word. This status word defines the type of the machine cycle (e.g., instruction fetch, memory read, memory write, input operation, and output operation). Since this status word is available on the data bus for only one ϕ_2 clock period, external latching logic must be used to store it, in order to use it appropriately during subsequent states of the machine cycle.

9.10 INPUT / OUTPUT AND DEVICES: GENERAL CONSIDERATIONS
David F. Bantz

9.10.1 Definitions

Input/output (I/O) refers to the exercise of control by a processor over its attached peripheral devices and the exchange of data between the processor and those devices. Figure 9.95 is a block diagram of the general form of an I/O system. Devices are attached to **device controllers**, which contain device-specific logic. Device controllers are sometimes referred to as "control units." Two forms of device controller attachment are shown in the figure: to a memory bus and to an I/O bus. The system design may allow device controller attachment directly to the memory bus or may contain a subsystem to control I/O, and an I/O interface that exchanges signals with an I/O bus.

A **unit data transfer** is that amount of data that can be transferred with one exchange of signals, either on the memory or I/O buses. The width of the buses determines how many bits of data are transferred in a unit data transfer. Typically, from 1 to 16 bits are transferred. A **block** of data consists of an integral number of units of data, requiring multiple unit data transfers.

9.10.2 Device Controller Attachment

In some I/O system designs, device controllers are attached directly to the memory bus. Device controller A in Fig. 9.95 is attached to the memory bus. In these systems the processor may move data between device controller registers and processor registers using the same mechanism that the processor uses to exchange data with its own memory. Alternatively, the device controllers may access processor memory directly to retrieve or deposit data. If the device controller can move a block of data between the device and memory, it is capable of the direct memory access function, described in more detail in Section 9.12.1. The programming of DMA is described in Section 9.10.5.

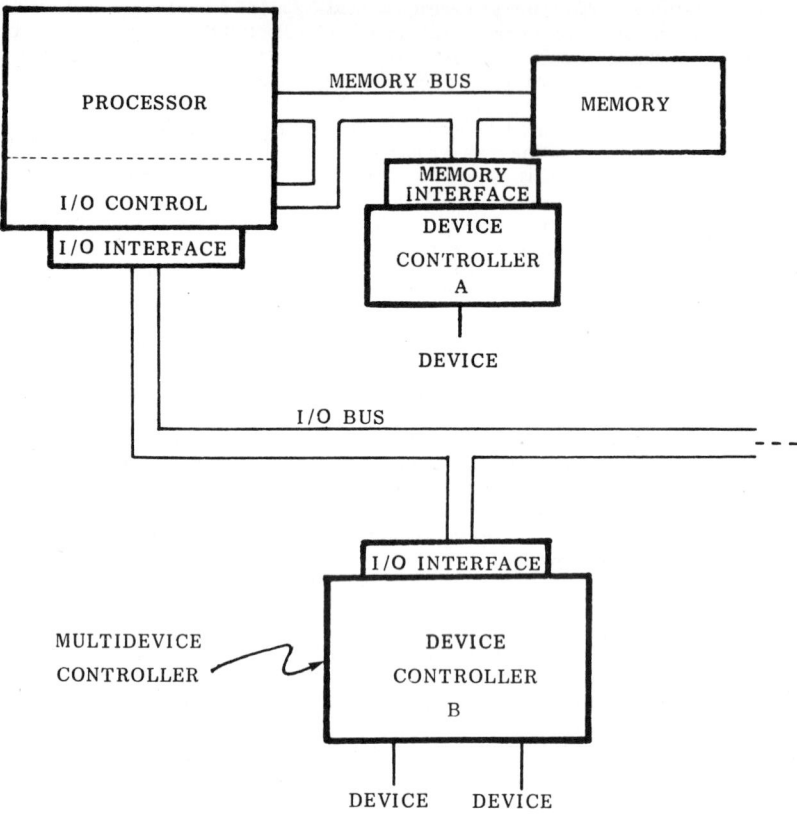

Fig. 9.95 I/O system.

The processor may communicate control information with the device controller by moving data between processor and device controller registers. Certain portions of the memory address space may be recognized as designating the control registers in the device controller. This technique is known as **memory mapping** the device controller registers (see also Section 9.8.1). The program in the processor first sets these registers by executing "store" instructions to the designated addresses and then initiates the data transfer with an access to a designated address. Status (see Section 9.10.4) can be retrieved from the device controller by executing a **load** instruction whose operand address is recognized by the device controller status register.

A limitation of memory bus attachment of device controllers for large computer systems is the difficulty of satisfying simultaneously the physical and electrical requirements of an I/O interface and a memory interface. Memory-to-processor data rates are often an order of magnitude greater than I/O transfer rates, and physical distances are small. In large systems, I/O systems must allow tens or hundreds of meters between the device controllers and the processor, and widely varying device controller response times must be accommodated.

An alternative to using the memory bus for I/O is to devote some processor hardware to I/O control, which allows transfer of data between a separate I/O interface and either memory or processor registers. This allows separate optimization of the memory and I/O interfaces. The I/O interface is attached to a bus, called the **I/O bus**, which communicates signals between the processor I/O interface and the I/O interfaces of all the device controllers. Device controller B in Fig. 9.95 is attached to an I/O bus.

Processor hardware capable of communicating blocks of data between memory and an I/O interface is called an **I/O channel**. When I/O channels are used, the processor must have a set of special instructions to control the channel. These instructions interrogate the state of the channel, transmit parameters of an I/O transfer, and start and stop transfers. I/O channels are described further in Section 9.10.5.

9.10.3   I / O Path Alternatives

The I/O path connects the processor I/O interface with those of the device controllers. The most common form of the I/O path is a bus, controlled by the processor I/O interface. For an example of a protocol constraining the exchange of signals on a bus, see Section 9.13.4. In large computer systems with many device controllers, physical cabling restrictions and cost may become limiting factors in the I/O bus design. Signal-to-noise ratios and propagation delays limit the length of a bus and the number of device controllers that can share the bus. One solution is a loop topology. A commonly adopted solution is to provide multiple I/O buses, each controlled by a separate I/O channel.

The connection of a device controller to a particular I/O bus may be static through cabling or may be dynamic. In the dynamic case, device controllers may be switched from one I/O bus to another, manually or under processor control. Often, device controllers are designed to attach to two I/O buses so that the device controller and its attached devices can be shared by two processors. Dynamic connections also allow a system to continue functioning with all its device controllers in the event of the failure of one I/O bus.

9.10.4   I / O Protocols

Communication between a processor and a device controller follows a structured form: a protocol. I/O protocols prescribe what are the permitted actions and necessary reactions while a processor and a device controller are exchanging control information and data. Protocols are effectively described by multiple interacting state diagrams and timing charts. An example of a protocol described as a timing chart is given in Fig. 9.96 and is described later in this section. Compare this with the protocol shown in Fig. 9.94.

Sessions
A **session** is a period of time during which there is an interchange of control information and data between the processor and a device controller. Sessions do not necessarily consist of continuous signaling but are characterized by retained information (**status**) in both the processor and the device controller. For example, if a magnetic tape drive has been commanded to rewind by a processor and its device controller has accepted the command, the processor expects to be notified, sooner or later, of the completion of the rewind operation. Furthermore, once rewound, the magnetic tape drive is expected to stay at the load point; both the processor and the device controller retain this information.

Device controllers may be **off-line**, meaning that no data communication between the device controller and the processor is possible, or **on-line** and thus capable of exchanging data with the processor. Off-line device controllers are logically removed from the system, and if their design permits may be powered down or physically disconnected for maintenance or replacement. On-line device controllers are expected to be capable of entering a session with a processor.

Device Controller Selection
A processor enters a session with a device controller through the process of **selection**. In general, selection consists of the transmission of a device controller address by the processor together with a command. The device controller must respond with either a positive or diagnostic negative response to the command; if the response is positive, a session is established.

Status
Status is information retained in the device controller that characterizes the state of the device controller and its attached devices. Especially for device controllers that serve several devices (e.g., device controller B in Fig. 9.95), there is an important distinction between the **device** status (or the status of some portion of the device controller that is dedicated to a device) and the **controller** status. Separate status indications should be available for the controller and for each device connected to the controller. Codes for device and device controller status information are, in general, specific to a particular unit, but several general classes of status exist. Examples of controller status are "off-line," "temporarily unavailable," "busy," and "command reject." Examples of device status are "busy," "not operational," and "data error." Status must be sufficiently detailed that the processor operating system may take recovery action, but not so detailed that many processor instructions are needed to analyze the status. Different device controllers should encode status in the same way, whenever possible.

Commands
Commands are control information sent by the processor to a device controller to invoke a device controller function. Commands are, in general, specific to a device or device controller. However,

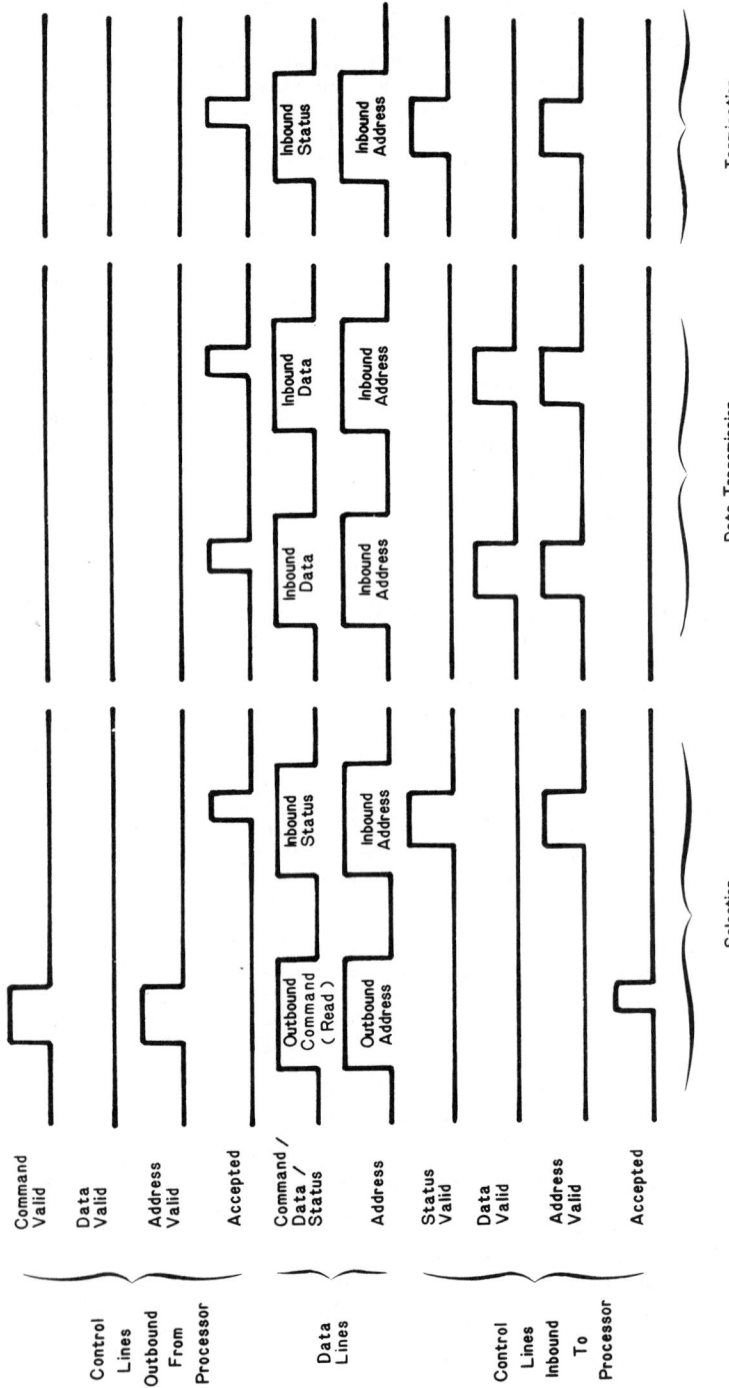

Fig. 9.96 Example session.

there are several general classes of such commands. For each device and device controller, some command from the processor should cause the device controller to return status. Similarly, some command should force a device and device controller to be reset to some known state. One command should cause the device controller to disconnect from the I/O interface, to facilitate maintenance and isolate faults. Another command should cause the device controller to return an identification of the type of device controller and number and type of devices to facilitate automatic system configuration.

Commands requesting data transmission must specify the direction of that transmission but may *imply* which or how much data is to be transferred. For example, a command to read a card from an 80-column card reader implies that all 80 columns of the next card are to be read.

Data Transmission

One major variation in data transmission concerns whether the I/O bus is shared or used exclusively by a device controller during data transmission and—if used exclusively—for how much data. IBM terminology distinguishes among selector, multiplexer, and block multiplexer channels. A **selector** channel dedicates the I/O bus to a device controller for the duration of a session. A **multiplexer** channel allows sharing of the I/O bus by several sessions, switching the path for each unit data transmission. A **block multiplexer** channel allows several simultaneous sessions but dedicates the path during the transmission of a block of data.

When the I/O bus is shared, the channel must be able to resolve contention for the path, and device controllers must be able to notify the processor of a data transmission request. The channel must be able to identify which of the device controllers wishes to exchange data. One or more units of data may follow. The last unit of data may be flagged, or a count may be prefixed to the data, or the count may be implicit.

It may be necessary during a data transmission sequence for either the processor or the device controller to terminate the sequence prematurely. This is referred to as an **abort**. When the device controller aborts, the signaling of the abort can take the form of an unsolicited error status transmission by the device controller. When the processor aborts, this may take the form of a missing or negative acknowledgment to a full block of data or a special signal to be interpreted by the device controller.

Normal termination of a data transmission sequence may be inferred or may be explicitly signaled by the processor or the device controller. One method for a device controller to terminate or acknowledge an inferred termination is to present normal status to the processor. This termination may leave either the device controller or device in a state incapable of accepting new commands since the completion of data transmission does not necessarily imply readiness to accept a new command. For this reason, some systems signal "device controller available" status and "device available" status some time after the completion of data transmission.

Processor Interruptions

The processor program involved in I/O may continually interrogate (**poll**) a device controller for a change in status or may prepare an interrupt-handling program to be invoked automatically when a device controller changes status. Polling is described further in Section 9.10.5 and interrupt system implementation in Section 9.2. Interrupt system programming considerations are discussed in Section 9.10.5. Interrupts may occur to signal a change in status of a device controller from off-line to on-line. During a session, interrupts can be used to signal the rejection of a command by a device controller, the completion of the transfer of a block of data, or an error during the I/O operation.

Example of a Session

Figure 9.96 describes a complete session (in simplified form) between the processor and a device controller in the form of a **timing chart**. Timing charts show the timing relationships between signals. In Fig. 9.96, three sets of signal lines are shown: control signals outbound from the processor, control signals inbound to the processor from the device controller, and the signals COMMAND/ DATA/ STATUS and ADDRESS, which may originate at either the processor or the device controller. The direction of signal propagation on this latter set of lines is determined by the outbound and inbound control lines. For example, if the outbound COMMAND VALID line is 1, the processor has placed a command on the COMMAND/DATA/STATUS lines, and the signals are outbound from the processor.

Both the outbound and inbound lines have an ACCEPTED line, driven by the processor and the device controller, respectively. If the direction of signals on COMMAND/DATA/STATUS is outbound, the device controller signals its acceptance of those data on the inbound ACCEPTED line; if the device controller is placing signals on COMMAND/DATA/STATUS, the processor signals its acceptance of those signals on the outbound ACCEPTED line.

Figure 9.96 is divided into three major sections: selection, data transmission, and termination. During the selection sequence the processor sends a READ command, together with a device controller address. The device controller responds with its own address and with its current status.

The purpose of the inbound device controller address is twofold: to ensure that no error in address recognition was made by the device controller, and to be compatible with the case where multiple sessions can be active at the same time ("multiplexer" channel). Once the processor has signaled its acceptance of this status, a session is established. During data transmission, the device controller sends its address and data inbound, and the processor signals its acceptance for each unit data transfer. During the termination sequence, the device controller sends completion status inbound together with the device controller address. Once the processor signals acceptance, the session is complete. The I/O channel may generate an interrupt to the processor on receipt of termination status from the device controller.

In some I/O bus designs, there are limits on the time intervals between changes in signal levels. For example, in Fig. 9.96 a design might require that the response by the device controller to an outbound command should occur not more than 1 μs after the fall of the inbound ACCEPTED signal. This is indicated on a timing chart by drawing an arrow from the trailing edge of the pulse on the inbound ACCEPTED line to the leading edge of the inbound status and address on COM-MAND/DATA/STATUS and ADDRESS, respectively, and marking the arrow with a description of the timing constraint. Both minimum, maximum, and nominal timing requirements are found in such descriptions.

Another example of bus handshaking appears in Section 9.9.7.

9.10.5   Processor I / O Considerations

This section concerns the processor hardware for the I/O interface and programming considerations for I/O. Different techniques for processor I/O can be distinguished by the amount of processor intervention during an I/O transfer.

Polling I / O
When a processor performs I/O by repetitively interrogating a device controller and transferring data whenever possible, the mode of I/O is called **polling**. Polling is done to determine a change in device controller status when hardware is incapable of automatically alerting the processor to that change. Polling requires a minimum of processor hardware but requires that the processor possibly execute many unproductive instructions, interrogating device controller status that does not change. Polling need not be continuous; a program may poll several device controllers in turn. In general, a program determines a **polling schedule**, which establishes how often to poll and which device controllers are to be polled. Continuous polling is known as "busy-waiting."

Interrupt-Driven I / O
If a processor can be alerted automatically to a change in device controller status, many of the instructions executed by a polling program are not needed, and the processor can perform other work during I/O. Interrupts provide a limited capability for processor alerting. Usually, the interrupt does not provide all the status change information, so the processor must subsequently read device status.

The program to perform interrupt-driven I/O must first set the address of the I/O interrupt-handling routine into a processor register or memory location, so that when an I/O interrupt occurs, the processor will automatically invoke that routine. The program can then initiate the I/O operation. When the device controller needs service by the processor, it generates an interrupt request. When the processor recognizes the interrupt, it must first save the state of the interrupted program and then begin execution of the interrupt-handling routine. This routine then determines the cause of the interrupt, usually by interrogating each device controller capable of generating the interrupt. In order that the processor not receive multiple responses to an interrogation, the processor must choose a specific device controller to respond to an interrogation request. This is done by prefixing or accompanying the interrogation request with an address—a number unique to the particular device controller. When the interrupt-handling routine finishes, the state of the interrupted program must be restored and its execution resumed.

Once the I/O initiation is complete, the initiating program continues to run (with momentary interruptions) while the I/O transfer continues. This is known as **immediate return** or **asynchronous** program execution. At some point in its execution, the initiating program should check to see whether the I/O transfer completed correctly: this can be done by busy-waiting on a flag set by the I/O interrupt handler, or through the facilities of a multitasking monitor. Interrupt-driven I/O still requires processor intervention for each unit of data transfer, but the processor executes interrupt-handler instructions only in response to a status change in the device controller.

Direct Memory Access I / O
In interrupt-driven I/O, a major portion of the instructions executed are for the purpose of detecting and responding to a change in device controller status, and for transferring data. Additional hardware can be provided in memory bus I/O systems (discussed earlier) to handle data transmission without

the need to execute instructions in the processor. This hardware, called **direct memory access** (DMA), enables the transmission of data between the device controller and main memory, terminating the transfer when a predetermined number of units of data have been transmitted. DMA implementation is treated in detail in Section 9.12.1.

DMA hardware consists of a count register that may be decremented and tested for zero, a memory address register that may be incremented, and sequencing logic to request data transfers on the memory bus. Once the I/O transfer is initiated, the DMA sequencing logic contends for the memory bus and, when access is granted, transmits the data between the device controller and main memory. Use of the memory bus implies that a memory address and a function (READ, WRITE) be transmitted by the DMA as well. The DMA is usually capable of generating an interrupt on completion of the I/O transfer. The sharing of the memory bus between the processor and the DMA, where the DMA has priority, is referred to as **cycle stealing**. The DMA logic is often implemented in the device controller. Note that I/O is asynchronous with the execution of the I/O invoking program, except for memory cycles stolen by the DMA. Provided that the DMA hardware is sufficiently fast, the maximum data transfer rate of DMA I/O is limited only by the rate at which main-memory cycles are available to the DMA.

To transfer data via DMA, the processor must first prepare an interrupt-handler routine to be invoked at the completion of the data transfer. The processor must set the starting address of the I/O data buffer area of main memory into a DMA register, together with a count of the number of unit data transfers required. The processor must then enable the DMA, often by setting a register in the DMA logic.

Channel I / O

In systems with a separate I/O bus, the hardware necessary to handle the transmission of blocks of data between memory and device controllers is called a **channel** or an **input / output processor** (IOP). The purpose of a channel or IOP is to reduce to a minimum the number of instructions executed by the main processor in support of I/O. Figure 9.97 shows the relationships among the channel, the processor, and memory. The channel fetches **channel command words** (CCWs), which specify the address and length of I/O data buffers in main memory, from an area of memory called the **channel program**. The processor sends the address of the first CCW in the channel program to the channel at

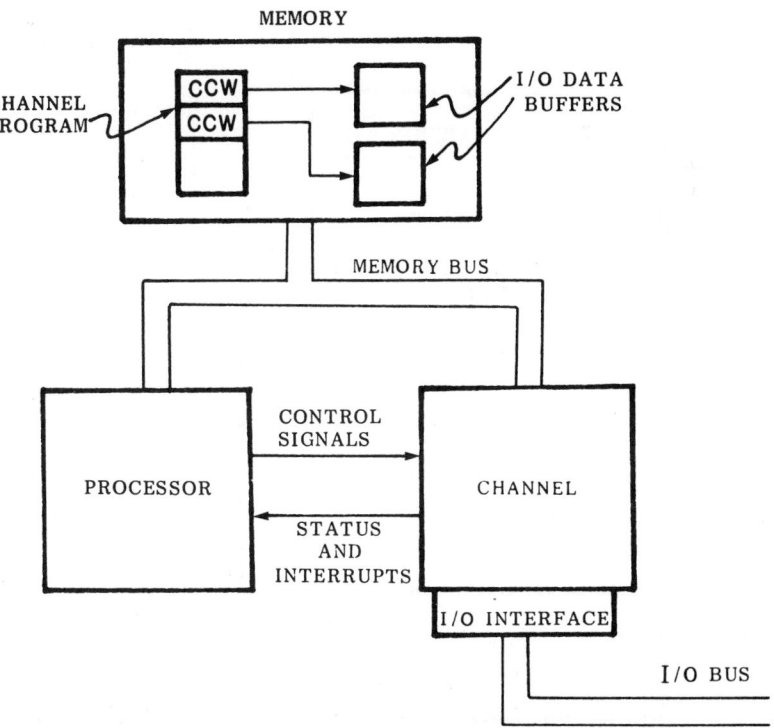

Fig. 9.97 Relationship of the channel to the processor and memory.

the time of initiation of an I/O transfer. Once the channel has been started, the channel uses the CCWs, one after the other, to specify the transfer of blocks of data between a device controller and the I/O data buffers. Each CCW may specify that an interrupt be generated to the processor when an I/O data buffer has been filled or emptied.

Another alternative in the design of an IOP is to provide it with a more general-purpose instruction set and with its own memory for channel programs. This allows the transfer of more complex data structures and permits "intelligent" preprocessing of I/O data.

I / O Control Software

Operating systems usually contain a component responsible for managing the I/O subsystem, recovering from errors, and providing a more convenient way for user programs to perform I/O operations. This component, the input/output control system (IOCS), can easily become the most complex and sophisticated software subsystem of the operating system.

9.10.6   Device Controller Design

The device controller serves several purposes: it must conform to the I/O interface and execute the "universal" commands discussed in Section 9.10.4, it must conform to the device interface, and it must control the device in response to a device-specific command. A general form of a device controller is illustrated in Fig. 9.98.

I / O Interface

The I/O interface portion of a device controller connects to the I/O bus. In Fig. 9.98 the I/O interface is shown as having a signal conversion portion and a handshaking portion. The purpose of the signal conversion portion is to transmit signals to the I/O bus and to receive signals from it. This may involve driver/receiver circuits and protection from potentially dangerous transients. Logic and switches to allow orderly disconnection of the device controller from the I/O bus for maintenance also reside in this section. The handshaking portion responds to and generates the signal sequences defined for the I/O bus. If the I/O bus signaling protocol has timing restrictions, these restrictions apply to the handshaking portion. The signal conversion and handshaking portions of the I/O interface act together to exchange commands, status, and data with the I/O bus.

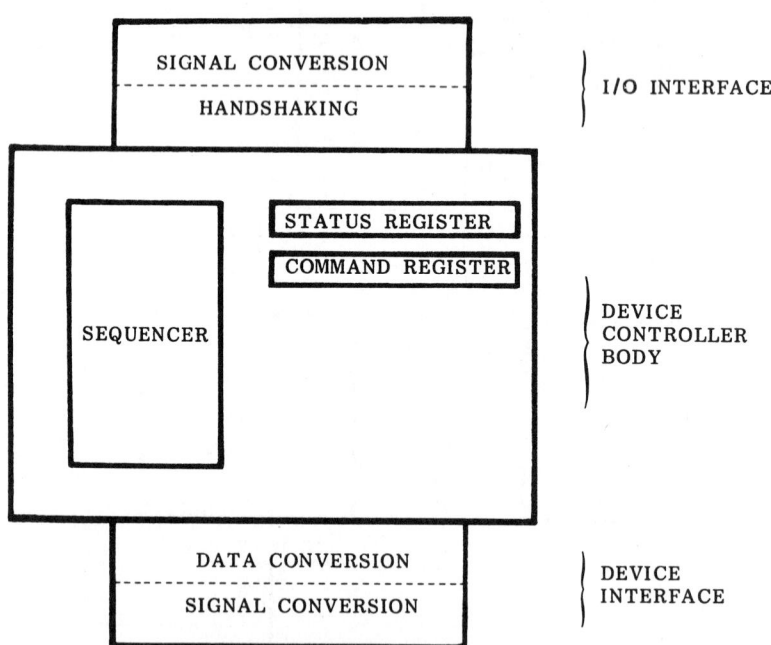

Fig. 9.98   Device controller.

Device Controller Body

This section of a device controller contains a sequencer to implement the I/O protocol and registers to hold the current command and status. Some device controller designs, especially disk controllers, may contain data buffers in the device controller body. For example, one portion of the sequencer might behave as shown in Fig. 9.96 and described in Section 9.10.4. After selection, the command register would contain the command for the device controller to execute. Certain commands can be executed by the device controller body without affecting the device. For example, if the command specified that the device controller be placed off-line, the sequencer would enter a state in which no subsequent commands could be accepted and only "off-line" status returned. Other commands require interaction with the device: for example, those that specify transfer of data. The sequencer would then exchange control and data signals with the device through the device interface to perform the desired function.

If the I/O protocol is simple, the sequencer may be realized directly as a finite-state machine, but a more general and flexible implementation partitions the sequencer functions into those that are done during a unit data transfer (and must be fast) and those that are concerned with the interpretation of commands and exceptional conditions within the device. The data transfer sequencing functions are done by specialized finite-state machines, as for DMA. The more complex but less time-critical sequencing is done by a small processor, typically with limited RAM to hold status and command information and with a ROM program store. For slow transfers, the data transfer sequencing can also be done by this small processor.

Device Interface

Among the devices that may be found attached to device controllers are:

Open reel and cassette magnetic tape transports
Rigid disk drives
Serial and line printers
Graphic and video display terminals
Punched card and tape readers and punches

Current, detailed information on these devices and their interfaces is available from their manufacturers and from the trade press.

The device interface varies greatly from device to device, but the general requirement of a device controller is to conform to the electrical, mechanical, timing, and sequencing restrictions imposed by the device. In Fig. 9.98 the device interface is shown as having a signal conversion portion and a data conversion portion. Especially for electromechanical devices, the protection of signal lines from dangerous electrical transients by the signal conversion portion is essential, and wide variations in timing must be accommodated. The data conversion portion reformats the data transferred between the device and the device controller body. For example, for a disk device, the data will have to be serialized or deserialized. If the data are recorded in encoded form, the code conversion is done in the data conversion portion of the device interface. Some specific examples of device interfaces are given in Section 9.12.5.

9.11 INPUT / OUTPUT AND DEVICES: ATTACHED STORAGE DEVICES

George C. Feth

9.11.1 Overview

Attached storage is used to provide the large-capacity, nonvolatile storage that is required in computing systems. The most common attached-storage devices are disk, drum, and magnetic tape files; these are described in more detail later in this section. Although such files are commonly called "devices," they are really rather complex systems or subsystems. Several kinds of storage devices may be integrated into a storage hierarchy (see Section 9.7), to optimize the capacity-speed-cost trade-off.

In the majority of attached storage devices, the data are stored sequentially along paths in a continuous two-dimensional **storage medium, transducers** are used to write and read the data, and relative **mechanical motion** between medium and transducers is used to position the transducers and the storage sites. This is in contrast to random-access memory, where individual memory sites are identifiable entities, not part of a continuum, and are selected electronically without mechanical motion.

Magnetic recording continues to be the major technology for storage; however, it is not the sole possibility. **Electron-beam** or **optical** transducers can be used together with the appropriate media, and have been used in some write-once, read-only stores (WOROS). In addition, solid-state **magnetic bubbles, charge-coupled devices (CCD)**, and **electron-beam accessed MOS (EBAM)** have been developed.

The **criteria of merit** for attached stores are primarily capacity and cost per bit. Nonvolatility is essential, at least in the largest capacity store in a system. Speed is also important, primarily in terms of the time required to transfer a block of data. In data-base-type systems with many concurrent users, an additional criterion is the access rate: the number of accesses per second that the system can provide. Cost has two components: a more or less fixed cost due to the controls and ancillary equipment not directly related to the capacity, and a variable cost, which increases with the storage capacity. For small systems, the cost of a minimum-capacity system, the "entry-level" cost, may be most important. To achieve the very lowest cost per bit, **off-line** storage is used; that is, the storage medium is removed from the file and stored elsewhere. For example, a tape reel is removed from a tape drive or a disk cartridge from a disk drive and stored in racks "off-line." The cost per bit of off-line storage is primarily the cost of the medium and its carrying assemblage; whereas for **on-line** storage (i.e., storage that is not removed from its file), the true cost per bit is more nearly the total cost of the file as well as its dedicated control system divided by the total storage capacity. Reliability, as well as nonvolatility, is essential to assure that data are not irretrievably lost due to malfunction. To this end, error detection and error correction coding is used with many storage devices so that errors due to noise or dropouts, so-called soft errors, can be corrected. Uncorrectable errors, or hard errors, must be drastically limited. Practically, hard-error rates of the order of 1 in 10^{11} or more bits can be achieved; soft-error rates may be one or two orders of magnitude larger.

9.11.2 Magnetic-Recording Storage Devices and Technology

A magnetic-recording storage device includes, in general, the storage medium, the transducers or **heads** for transferring information into and out of the storage medium, mechanisms to move the storage medium, and in some devices, to position the transducers relative to the storage medium, and electronics for reading and writing, for control and for signal and data processing. Signals representing data and control flow to and from the file (see Fig. 9.99). The physical distinctions between the different kinds of devices—tape, disk, and drum—have to do primarily with the forms of the storage medium and the mechanical aspects of the devices, which in turn affect the storage capacity, cost, and performance of the files.

In a magnetic-tape file, the medium is magnetic tape, and information is stored sequentially in records, one after the other, along the tape (see Fig. 9.100). The long, flexible, tape medium is stored in tape reels, cartridges, or cassettes. In a magnetic drum, the recording medium is a coating on the surface of a cylinder, the information is recorded in circular **tracks** around the cylinder and one or more read-write transducers is provided for each track (see Fig. 9.101). These are commonly called **head-per-track** or **fixed-head** files. In order to select the desired track, the path between the read-write amplifier and the appropriate head is switched electronically. The medium may also be in the form of a disk, rather than a drum; the tracks are recorded in a band of concentric circles, or **race**, near the outer periphery of the disk; this is called a **fixed-head disk** file. Some disk files are built so that the

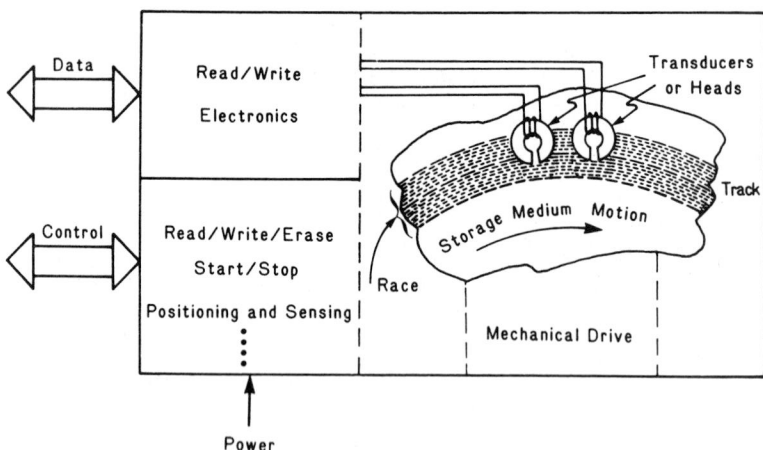

Fig. 9.99 Attached store.

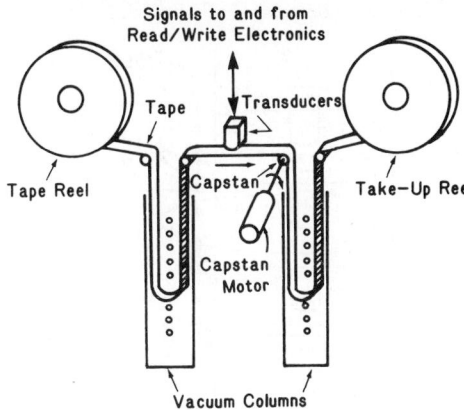

Fig. 9.100 Magnetic tape drive.

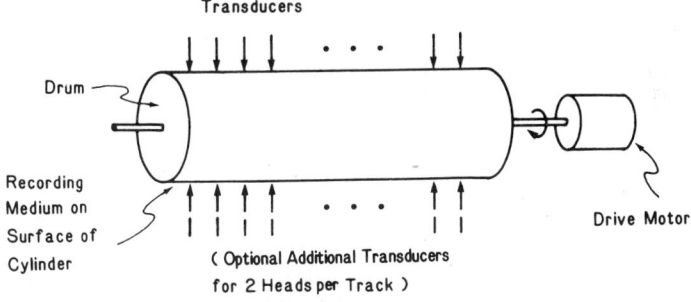

Fig. 9.101 Magnetic drum.

head can be moved radially over the disk from track to track, so that a single set of recording heads can **access** any of the recorded track positions in the race; these are referred to as **moving-head disk files** (see Fig. 9.102). Disks can be made to be recorded on one or both sides, and multiple disks can be mounted coaxially along the same shaft or **spindle**. One set of transducers is used for each disk side or surface, the transducers are mounted on arms that extend between the disks, and the arms are mechanically attached to the same structure, or ganged, so that they move radially together. This structure is moved by an access mechanism to position the heads on the desired track. One head and disk side are commonly used in a feedback control system for positioning the heads. The set of tracks that are simultaneously beneath the transducers on the various disks is called a **cylinder**; electronic switching is used to access the single desired head within a cylinder. In disk files, the heads can be used to access the desired track location directly, and these files are commonly called **direct-access storage devices (DASD)**, in contradistinction to tape files, where the records are sequentially accessed. Disks may be flexible or rigid, and they may be designed to be removed from the spindle for storage off-line. The arms holding the heads may be designed to be retractible so that the disks can be removed, or the heads and disks may be designed into the same sealed cartridge, which is removed as a unit. More than one spindle may be driven from the same motor or disk **drive unit**. A **control unit** is frequently designed separately from the disk or tape storage unit but for operation with that specific storage device. The control unit may be designed to handle numerous storage units, so that its cost is distributed over more storage capacity.

Magnetic-Recording Technology

The magnetic-recording medium is central to the magnetic-recording technology. Storage media consist of a layer of ferromagnetic material coated onto a supporting substrate. The information can be visualized as being stored in the ferromagnetic layer in the form of little magnets having magnetic flux in one of two antiparallel directions.

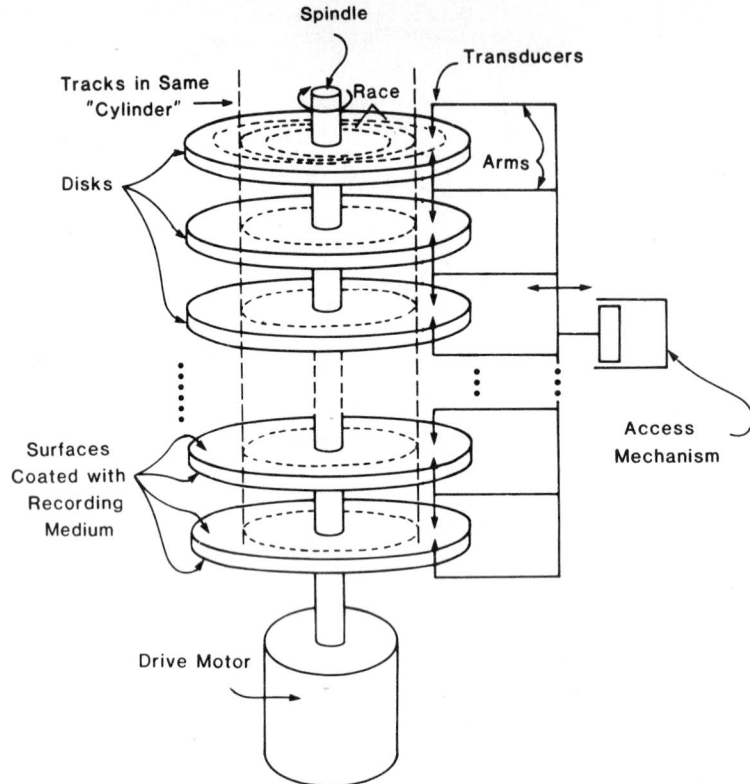

Fig. 9.102 Moving-head disk.

Magnetic-recording heads (see Fig. 9.99) are the transducers for changing and sensing the recorded flux. **Write heads** are used to change or write the magnetization to the intended direction corresponding to the data; **read heads** are used to sense the change of direction of magnetization; **erase heads** may be used to obliterate previously stored information by driving the magnetization to saturation in one direction throughout its depth (although erasing is not mandatory with saturation recording). A head contains a **ferromagnetic path** that is closed except for a fine air gap that separates the two pole pieces in the face of the head; this face, containing the pole pieces and gap, is the portion of the head that is brought into close proximity to the storage medium. A winding of one or more turns of a conductor links the ferromagnetic structure. In the write head, the current passing through the winding produces the drive field in the vicinity of the gap, and this field writes the desired magnetization state in the medium, aligning the magnetization of the medium with that of the drive field as the medium passes the gap. In the read head, the flux that enters the two pole pieces changes when the transition region of the recorded magnetization passes the gap. In an **inductive** read head the resulting change of flux linkages induces a voltage in the winding, the sense of the voltage corresponding to the sense of the flux change in the transition. Read heads can also be implemented using other magnetic effects. A single head, possibly with multiple windings, can be used for write and read (and erase). However, in many instances it is preferable to separate the functions. Track-to-track spacing, or track density, together with bit density determine overall magnetic-recording density. Track density is limited by signal and noise considerations. The signal magnitude is proportional to the overlap of the read head with recorded track width. Crosstalk is produced by the proximity of neighboring tracks and is affected by inaccuracies of the relative positioning of head and medium, specifically the repeatability of the positioning.

Very crucial aspects of the recording process, in addition to the magnetic properties of the medium, are the gap dimension in the head, the head-to-medium spacing, and the thickness of the medium. All three of these dimensions must be considerably smaller than the shortest magnet to be recorded. Often, on flexible media, **contact recording** is used, the head touching the medium to minimize the head-to-medium spacing; even so, there is some minimum value of effective spacing. Moreover, such a system is prone to wear and is susceptible to surface irregularities of the medium.

The electronic circuitry for magnetic recording is used to drive the write heads, to amplify and detect the readback data signals, and to convert between binary data and the signal waveforms involved with both writing and reading, in accord with the selected recording technique.

Several recording techniques are in common use; they were developed to achieve high packing density, yet reliable data recovery. **Longitudinal recording** is usually used (i.e., with the recorded magnetization parallel to the direction of motion between the medium and head, in contrast to transverse recording or vertical recording). **Saturation recording** is common, driving the flux to saturation in one or the other direction, in contrast to linear recording. One such method uses a non-return-to-zero (NRZ) technique. In such a system, the write head is driven with either positive or negative current to record one or the other direction of saturation magnetization in the medium. The direction of magnetization corresponds to the binary data, changes in magnetization and the corresponding induced voltage in the read head occurring only when there is a change in value of the binary data, the polarity of the sensed signal indicating whether the transition is from binary 0 to 1, or vice versa. Another recording scheme uses a change of the magnetization direction for each bit that is a 1, with no change for a 0, so-called non-return-to-zero-invert (NRZI) recording. Other schemes employ a single reversal of direction of magnetization within a bit cell for one binary state, and two reversals within the same cell interval for the other state; this is effectively a frequency modulation technique.

Actually, the signals that are recorded are not the original binary data issued by the data stream of the computer but are encoded in order to serve several special purposes (e.g., self-timing, formatting, baseline independence). Error detection and error correction features can also be included in the encoding. Such encoding and decoding require suitable circuitry for buffering the data and for the required logical manipulations and transformations.

Additional circuitry is used for controls in the storage devices. In devices with many transducers but only a single data stream, such as fixed-head files or moving-head files with more than one recording head, the transducer to be activated is switched electronically. In moving-head files, the head positioning is controlled electronically through an access mechanism; often, a feedback control system is used to position the heads to the desired cylinder. In a tape or removable-disk file, the position of the tape and the presence or absence of the medium may be electronically sensed. In some disks and drums, there are several blocks of data, or records, per revolution of the medium, and **rotational position sensing** is included to sense what sector is approaching the transducer, so that the appropriate transducer can be selected to read or write in that sector.

Control Units and Data Formats

Generally, the device control unit that provides the interface between an attached storage device and the I/O bus (see Section 9.10.2) performs the following functions:

Converting the parallel data words from the I/O bus into a serial by bit data stream to the device and deserializing for data flow in the opposite direction.

Controlling the mechanical motion of access arms, tape-drive capstans, and so on.

Selecting the proper read or write head.

Monitoring the status of the device and communicating this to the processor.

Keeping track of the rotational position.

Formatting the data into records, blocks, sectors, and so on.

Keeping count of the records read.

Searching for a desired record by comparing data on the device with data in main memory.

Adding error-checking and error-correcting codes (ECCs) to written data.

Checking ECCs in data from the device.

The format of the data on the device varies from device to device and from manufacturer to manufacturer. Data on tape are generally written in blocks or records with gaps between them. Special codes may be recorded to denote the end of each file or similar division. On fixed-head files and moving-head files, data are commonly divided into blocks or records that are usually shorter than a total track or revolution of the medium. Tracks may be divided further into sectors. There is generally a code to mark the start of each track or sector. Special fields within each record, perhaps written by the control unit and removed by it when data are transmitted to the processor, may be added to the data in order to identify the track and cylinder, to index the record, and to include ECC.

Moving-Head Disk Files (MHF)

MHFs are used to store large amounts of data that must be retrieved quickly. The time to access and transfer a random block of data is the sum of three components: the time required for the head to move to the specified track and settle there—**seek time**; the time for the disk to rotate so that the beginning of the block reaches the head—**latency time**; and the time for the entire block to pass beneath the head—**block transmission time**. The time required to electronically switch between heads is negligible in comparison. Since actual read or write time is but a small part of the overall block acquisition time, the read-write circuitry can be used to handle several drives. This is done by maintaining a queue of requests for all the drives and setting the accessing in motion for each drive as it is freed up or as a request for that drive arrives. When the head is at the sought cylinder (and approaches the appropriate sector in a disk that has a rotational position-sensing feature), the pertinent head is selected electronically. Sometimes there is blocking due to the electronics being busy, so individual requests are delayed. The **access rate**, the number of data blocks that can be accessed in a given time, depends on the number of independent electronic channels, the number of access mechanisms, the block size, and the distribution or clustering of requests in the file.

MHFs span a large range of applications. At the one end are the small-diameter diskettes with storage capacity starting in the fractional megabyte range, costing several hundred dollars, and having block acquisition times of tenths of a second; these are commonly used with minicomputers and microcomputers. At the other extreme are the large disk systems with capacities of hundreds of megabytes per spindle. For even lower cost per bit, removable media—disk packs or cassettes—are used to store data off-line at a small fraction of the cost of the storage system. However, for the highest density recording, often only fixed media are available because of the sophistication required for the fine dimensions and tolerances and the required degree of cleanliness.

Drums and Fixed-Head Files (FHF)
Magnetic recording drums were the early form of fixed-head file, and FHFs are still often loosely called drums, even though disks may be used as the recording surfaces in order to increase the capacity per drive. FHFs are used for the fastest access requirements in order to eliminate the seek time encountered in MHFs, this being the largest component of their acquisition time. FHFs are often used for swapping data with main memory in virtual memory systems (see Section 9.7.1) and for storage of frequently used parts of the operating system. The data are arranged in sectors, and access requests are queued for each sector in order to increase the throughput. The cost penalty for this less-than-order-of-magnitude improvement in speed over MHFs is often well over an order-of-magnitude increase in cost per bit.

FHFs with lower cost, albeit lower capacity and performance, are available also for use with smaller mainframe computers and minicomputers. FHFs find application in militarized systems, where they have the advantage of being more rugged than MHFs.

Tape Files
Magnetic tape files are used for the lowest cost per bit and largest capacity read-write storage. Data are recorded sequentially on the tape and are commonly grouped in blocks called **records**. Tape drives are usually designed to reverse direction for rewinding and for backing up to preceding records; this requires start/stop capability. Gaps in the recording are left between records so that the tape motion can be brought up to speed in the interrecord gap in start/stop operation. All but the lowest-performance files record in several tracks in parallel on the tape, six and eight data tracks being common. An additional track for parity checking is often used, and additional bits may be added to the record length, as well, for error detection and correction. The bits that were written simultaneously may not all reach the read heads simultaneously, due, for example, to nonuniform stretching of the tape medium or different alignment of the heads on the writing and on the reading tape drives. This results in skew of the sensed data. De-skewing buffers are used to realign the data for reading.

In most systems, the tape is stored off-line and is accessed manually by a machine operator. However, automated tape files, in which the tape handling and storage are performed without human intervention, have been developed for mass storage, and a random-access time of the order of 10 seconds for storage capacity of the order of 10^{12} bits is achieved.

9.12 INPUT/OUTPUT AND DEVICES: MICROPROCESSOR SYSTEM I/O
David F. Bantz

Microprocessor I/O systems differ from large- and medium-scale computer I/O systems primarily in that a limited number of pins on the microprocessor package itself are available for I/O. A separate I/O interface is generally not feasible: virtually every microprocessor communicates with its I/O attachments over the same bus that is used to attach the microprocessor's memory. In some microprocessors, memory access instructions are used to set and read I/O attachment registers; in others, a special class of I/O instructions performs this function, although the path to the I/O attachment is the memory bus. Some microprocessors have a bit-serial I/O facility, intended for the attachment of slow devices.

9.12.1   Memory Bus I / O

Microprocessors differ in whether bus cycles for I/O are distinguished from bus cycles for memory access or not. In some microprocessors, a signal line is available to designate a particular bus cycle as having resulted from the execution of an I/O instruction. The attachments to these microprocessors can use this signal line to determine whether the address portion of the bus contains a memory address or an I/O attachment address. In microprocessors lacking such a signal line, the registers in the I/O attachment are accessed as memory locations by the microprocessor; only the contents of the address bus distinguish an I/O operation from a memory access operation. This form of memory bus I/O is referred to as **memory mapped** (see Section 9.8.1). Microprocessors using memory-mapped I/O have no explicit I/O instructions. Any processor with a memory bus can be adapted to do memory-mapped I/O.

Disadvantages of having a signal that distinguishes I/O bus cycles from memory cycles include the extra operation codes and processor circuitry and the extra package pin needed for the signal. A disadvantage of memory-mapped I/O is that a portion of the processor address space (the range of valid addresses) must be dedicated to I/O registers.

Address Decoding

Each I/O attachment must contain circuitry to recognize its address. In order to simplify address recognition circuitry, one may design circuitry to examine only a subset of the address lines. For example, the circuitry may be designed to select the register if the most significant bit of the address is true. Then for 16-bit addresses, any address in the range 8000 to FFFF (hexadecimal) will select the register. In general, if the address recognition circuitry has N gate inputs connected to the high-order bits of an address bus of M bits, a contiguous block of 2^{M-N} addresses is dedicated to the register.

Multiplexed Addressing

In some systems, a single bus is multiplexed between address and data information. The implication for I/O attachments is that the address information must be latched into a register to enable valid address recognition. Alternatively, the address may be recognized at the time it is presented by the microprocessor, and only the signal representing address recognition need be latched. This indication must be reset before the next address is presented by the microprocessor.

Data Transmission

Once the address of an I/O attachment has been recognized, the attachment must either accept or present data to the microprocessor. Some microprocessors have a bus signal that indicates, when 1, that data are to be presented, and indicates, when 0, that data are to be accepted. Other microprocessors use two separate lines for this purpose. Microprocessors usually generate a signal that indicates that data are valid on the data lines of the bus, or that data have been captured by the microprocessor from the data lines. Most often, this signal is a transition from logical 1 to 0, or vice versa. Thus to present data to the microprocessor, the I/O attachment must recognize its address, sense that data are to be presented, and remove the data from the data lines after detecting the signal that the data have been captured. To accept data from the microprocessor, the I/O attachment must recognize its address, sense that data are to be accepted, and capture the data when the transition occurs that signals that the data are valid on the data lines.

Example

Figure 9.103 details the logic necessary to connect an external register to the Intel 8085 bus, which has a signal line distinguishing bus cycles for I/O (IOCYCLE). The register is selected by the RECO output from the address recognition circuitry. This circuitry compares a prewired address with the contents of A8–A15 (high-order address bits), which contains the I/O attachment address during an I/O operation; when the comparison succeeds, when IOCYCLE designates an I/O bus cycle, and when ADDRESS LATCH ENABLE makes a positive transition, the RECO output becomes true. The next positive transition on ADDRESS LATCH ENABLE sets RECO false. If the register is selected and NOT READ is 0, indicating a bus read operation, the OUTPUT ENABLE control input to the register is 1, causing the register to place its contents on DATA. During an I/O write, INPUT ENABLE is 1, and when STROBE makes a negative transition, the data on DATA are latched into the register. The 8085 bus timing ensures that IOCYCLE is 1, NOT WRITE is 0, and CLK makes a negative transition only while data are valid on AD0–AD7 (the multiplexed address/data bits).

A timing diagram for the I/O read operation is shown in Fig. 9.104. Note that the signal OUTPUT ENABLE can be 1 only when the signal NOT READ is 0, and when the I/O attachment has been recognized.

Direct Memory Access (DMA)

When data are transferred between the microprocessor and an I/O attachment under direct control of the microprocessor program, each byte or word of data requires the execution of several

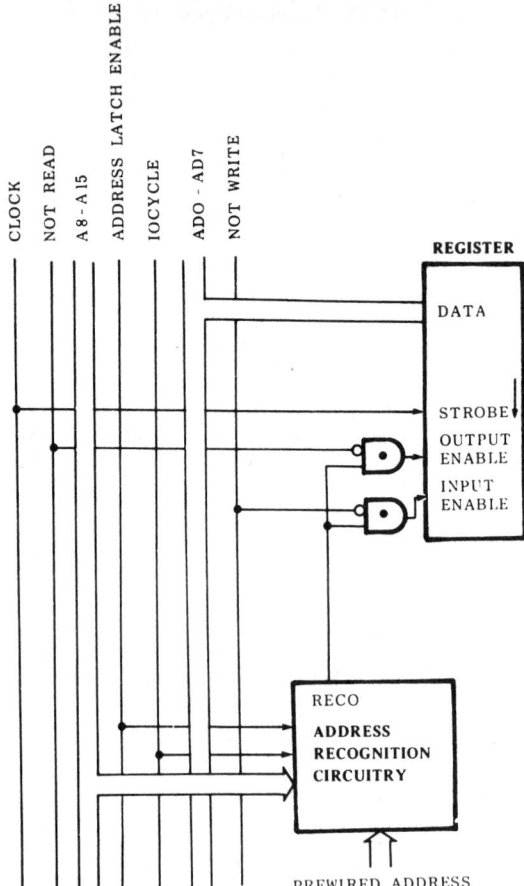

Fig. 9.103   Intel 8085 I/O interface.

instructions to transfer. Some I/O devices require higher data transfer rates than are achievable with this technique. For these devices, the I/O attachment may use direct memory access (DMA). DMA allows the direct transfer of data between an I/O attachment and the microprocessor memory. The microprocessor first initializes the DMA controller circuitry by storing a count and a starting memory address in its registers. Once started, a DMA transfer proceeds without further microprocessor intervention, except that an interrupt (see Section 9.12.3) may be generated upon completion of the DMA operation.

I/O attachments using DMA (see Section 9.10.5) incorporate circuitry similar to that of Fig. 9.105. Circuitry to allow the microprocessor to set the COUNTER and ADDRESS COUNTER registers is not shown. The signal BUS CYCLE is assumed to define the interval of time during which addresses are presented and data are exchanged on the bus. The DMA controller connects to the I/O attachment with the lines TRANSMIT REQUEST and REQUEST GRANTED.

When the I/O attachment wishes to use a bus cycle, it raises the line TRANSMIT REQUEST. If the DMA count register is nonzero, the signal is placed on the BUS REQUEST line to the processor. The processor hardware periodically examines this signal, and when it is 1 the processor waits until the end of the current bus cycle, stops, places its address and data line drivers in the high-impedance state, and raises the line BUS GRANT. The processor is effectively isolated from the bus during bus cycles granted to the DMA controller. When BUS GRANT is sensed 1 by the DMA controller, it places the contents of its ADDRESS COUNTER register on the ADDRESS lines and signals the I/O attachment on REQUEST GRANTED that it may use the current bus cycle for transmission of data. The I/O attachment itself may drive the bus lines that determine the direction of data transfer, or additional circuitry in the DMA controller may drive these lines. As long as TRANSMIT REQUEST is held at 1, consecutive bus cycles may be used by the I/O attachment.

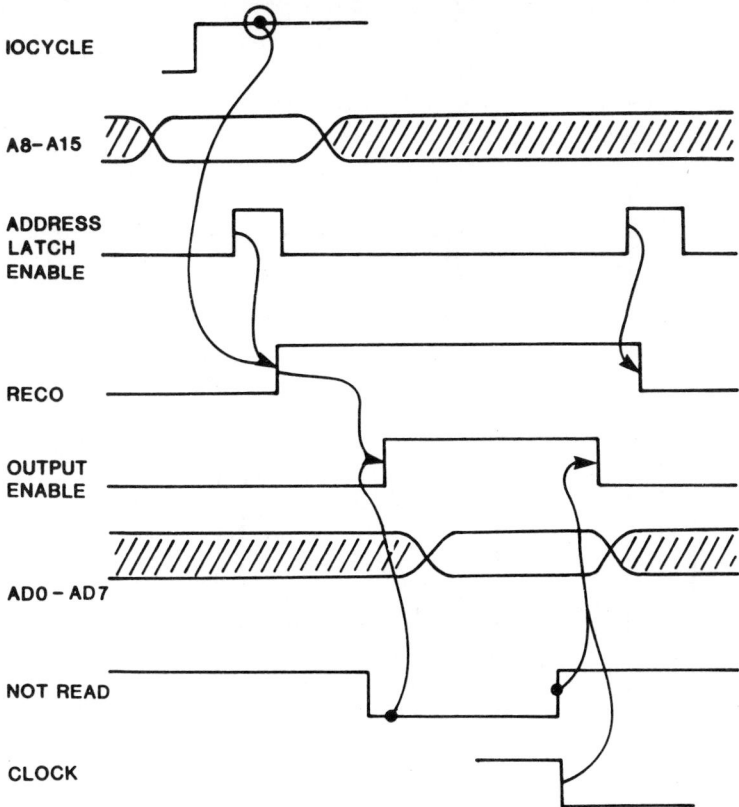

Fig. 9.104   Intel 8085 I/O interface timing for I/O read operation.

If several I/O attachments, each with its own DMA channel, wish to use the bus simultaneously, hardware must be provided to resolve the contention among the various channels. This resolution may be either on a priority basis or may grant bus cycles to competing DMA channels on a round-robin basis.

The circuitry of Fig. 9.105 is capable of using successive bus cycles ("burst mode") or using bus cycles intermittently. The choice depends on the data transfer rate of the I/O attachment. Often, the microprocessor must use several bus cycles in preparation for relinquishing the bus by generating BUS GRANT, and must use several bus cycles after regaining the bus. These cycles are unproductive in that they do not contribute to instruction execution or data transfer. Therefore, DMA transfers that use consecutive bus cycles make more efficient use of the bus.

9.12.2   Non-Memory-Bus I/O

Some microprocessors have a serial I/O bus separate from their memory bus. Here, separate serial input and output channels allow transmission between an I/O attachment and the microprocessor. The I/O attachment address is signaled serially by the processor, followed by outbound or inbound data. The I/O attachment uses a separate clock line driven by the microprocessor to control the rate at which data are captured or signaled. A single instruction causes a transfer; the microprocessor cannot execute the next instruction until the transfer is complete. This serial I/O facility is usually intended for the attachment of I/O devices with low data rates.

9.12.3   Interrupts

To determine whether a particular I/O attachment requires data transmission, the microprocessor must interrogate the I/O attachment. This usually repetitive interrogation is called **polling**. The ability of an I/O attachment to interrupt the execution of a program in the microprocessor can be used to signal the requirement for data transmission more efficiently than polling.

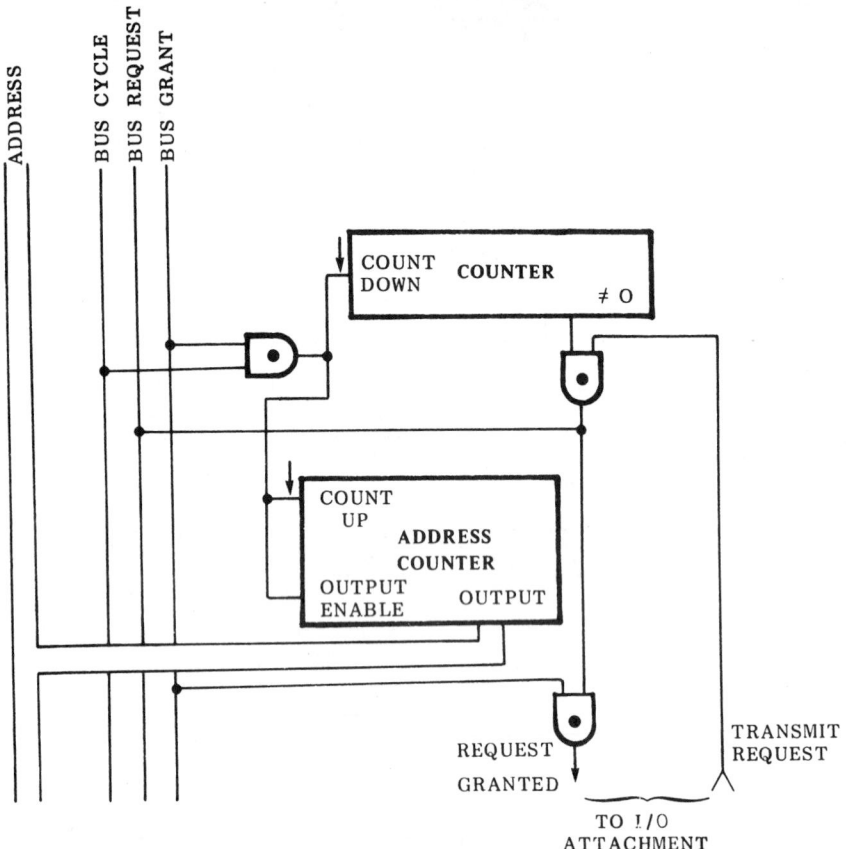

Fig. 9.105   DMA controller.

Microprocessor interrupt systems usually depend on a logic circuit (attached to the micro-processor like an I/O attachment) devoted to controlling the interrupts on behalf of the processor. This alternative is most often chosen in preference to dedicating a pin of the microprocessor package to each interrupt level. One pin signals the microprocessor that an interrupt has occurred; a second pin may be used for a high-priority interrupt to signal error conditions. The interrupt controller resolves contention among the individual interrupts on a priority basis, encoding the number of the highest-priority interrupt into an internal "status" register. When the "status" register is read by the microprocessor, its contents are automatically transferred to an internal "level" register. Only an interrupt whose priority exceeds the current contents of the "level" register will cause a new interrupt to the microprocessor. Finally, the interrupt controller contains a "mask" register, which can be set by the microprocessor and which is ANDed with the individual interrupts. This facility is valuable since it allows the microprocessor program to defer the recognition of selected interrupts without masking them all.

9.12.4   General Parallel and Serial Interface Systems

Integrated-circuit manufacturers, recognizing the need to implement I/O attachments, supply a wide variety of packaged components that attach directly to microprocessor buses and present a general interface to a device. This interface may allow either serial or parallel transfer of data to the device and often provides the ability to generate an interrupt signal to the microprocessor when a transfer is complete.

Parallel Interface
This family of integrated circuits, sometimes called **parallel interface adapters** (PIAs), is designed to simplify the attachment of devices requiring parallel data transfer or devices requiring registers

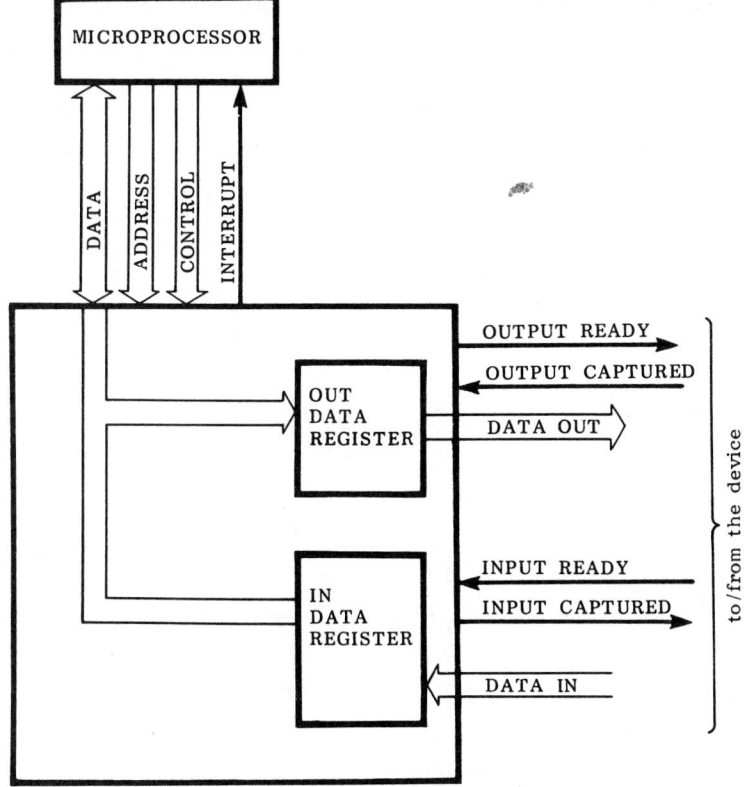

Fig. 9.106   Parallel interface.

containing control information set by the microprocessor. An example of such a parallel interface device is shown in Fig. 9.106. The signals between the parallel interface and the microprocessor, on the memory bus, are the data, address, and control signals required for any I/O attachment. The signals between the parallel interface and the device carry data (DATA IN, DATA OUT) and control (OUTPUT READY, OUTPUT CAPTURED, INPUT READY, and INPUT CAPTURED). The control signals are sometimes called **handshaking** signals. For example, when the microprocessor sets data into the OUT DATA REGISTER, the handshaking line OUTPUT READY signals the device that data are ready for it. When the device has captured the data, it signals on OUTPUT CAPTURED, which may generate an interrupt to the microprocessor. The microprocessor program can then supply more data or respond in some other way. Similarly, when the device has data for the microprocessor, it signals on the INPUT READY line. The parallel interface may then generate an interrupt to the microprocessor, which would then read the data now present in the IN DATA REGISTER. The act of reading that register causes the parallel interface to generate the signal INPUT CAPTURED to the device.

Serial Interface
This family of integrated circuits is designed to simplify the attachment of devices requiring serial data transfer. An example of such a serial interface device, called a **universal synchronous asynchronous receiver transmitter (USART)**, is shown in Fig. 9.107. The signals between the microprocessor and the serial interface (on the memory bus) are those required by an I/O attachment. The signals between the serial interface and the device consist of status signals (TRANSMIT READY, RECEIVE READY), clock signals (TRANSMIT CLOCK, RECEIVE CLOCK), and data (TRANSMIT DATA, RECEIVE DATA). When the device has data for the microprocessor, it signals those data, serial by bit, on the RECEIVE DATA line in synchronism with a clock signal on RECEIVE CLOCK. Usually, the data are prefixed by a 1 bit (the "start" bit), indicating the start of data, and the data are from 5 to 8 bits long. A 1 bit (the "stop" bit) may be suffixed to the data. After all bits of the data have been received by the serial interface, the stop and start bits are stripped and an interrupt may be generated

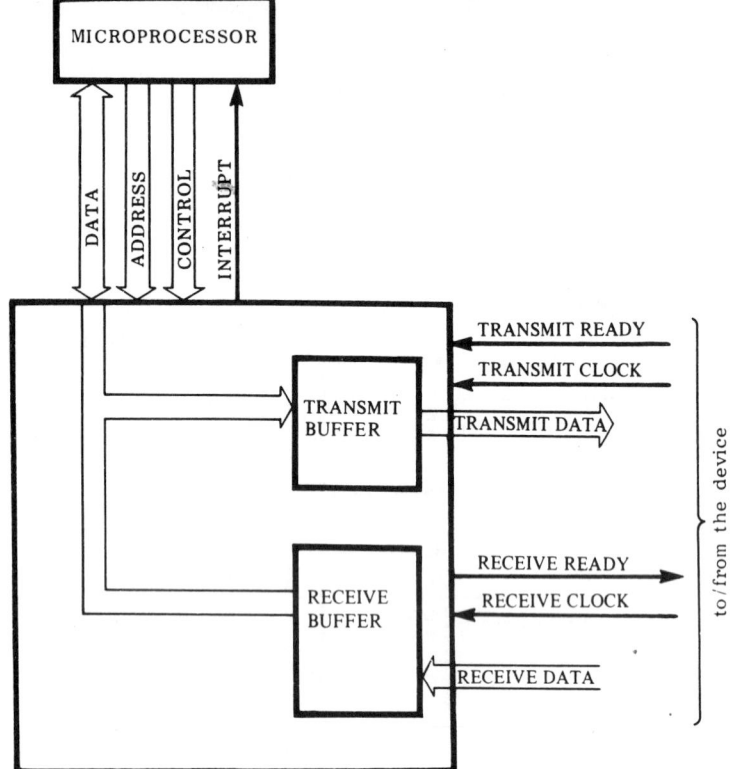

Fig. 9.107   Serial interface.

to the microprocessor to indicate that data may be read from the RECEIVE BUFFER. The mode of data transmission in which one 5- to 8-bit character is transmitted at a time is called **start / stop**.

When data are to be sent to the device, the microprocessor sets the data in the TRANSMIT BUFFER. The data are then serialized, prefixed with the start bit and suffixed with the stop bit, and sent to the device on TRANSMIT DATA in synchronism with a clock on TRANSMIT CLOCK.

Other Attachments
Another form of general-purpose attachment is the IEEE 488 General-Purpose Interface Bus,* a standard for the connection of laboratory instrumentation. Integrated-circuit manufacturers supply interface devices that connect to a microprocessor bus on one side and to the IEEE 488 bus on the other.

9.12.5   Device Attachment Examples
This section contains a number of examples of specific I/O attachments for microprocessors.

Keyboard Attachment
Figure 9.108 shows how a keyboard can be constructed as an array of contacts connected to a parallel interface. The DATA OUT signals are used to select a row of contacts for sensing; the DATA IN signals give the state of all contacts in the row. The microprocessor sets only one bit to 0 at a time in the OUT DATA REGISTER; otherwise, several rows might be selected simultaneously. All inputs to the IN DATA REGISTER are tied to a logical 1 and will remain 1 if no contact is closed. Only those contacts tied to an OUT DATA REGISTER line that is 0 can affect the corresponding IN DATA REGISTER bit.

*IEEE Standard Digital Interface for Programmable Instrumentation, IEEE Std 488-1975, IEEE, New York.

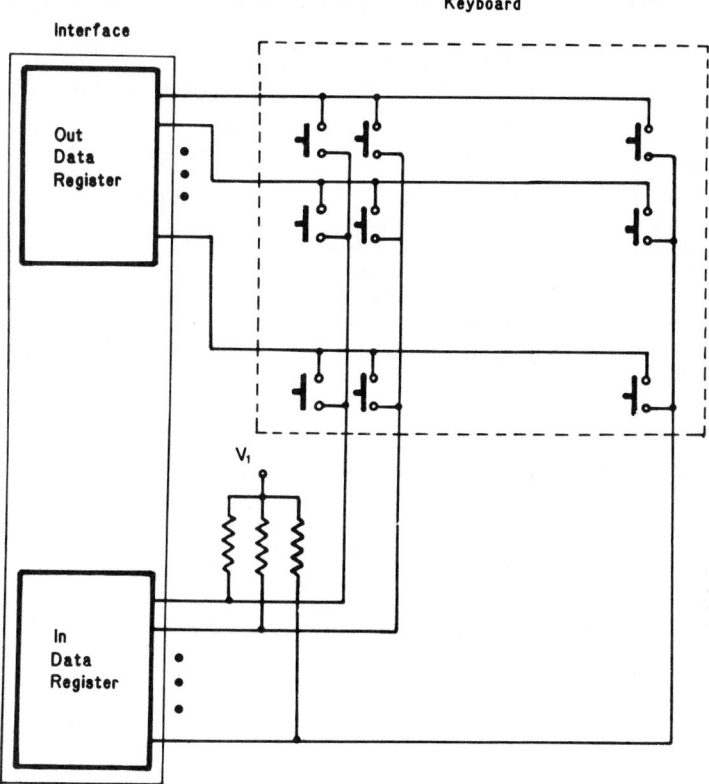

Fig. 9.108   Nonencoded keyboard.

Since the contacts may open and close rapidly for up to 10 ms after a key is depressed, the state of a contact should be sampled several times before the microprocessor program decides that the key has been depressed. This process is called **debouncing**. As peak keying rates may exceed 20 keystrokes per second, the keyboard scanning rate should be at least 40 scans per second to allow for at least two scans per key per keystroke.

Communications Attachment

When microprocessors are used in communications systems, more complex and efficient modes of transmission than start/stop (Section 9.12.4) must often be supported. Higher-data-rate devices require a **synchronous**, or continuous, mode of transmission that transmits a continuous sequence of transitions, allowing the receiver to maintain clock synchronization. Messages are transmitted according to a link protocol, where a message consists of a destination address, control information, a block of characters, and an error control unit.

Typical of the devices that can be used to implement these communications protocols is an advanced data link controller (ADLC). The general form of this device is similar to that of the serial interface (Section 9.12.4). Among the differences are the multistage transmitter and receiver buffers, capable of buffering several characters, allowing the processor more time to respond to an interrupt. The ADLC computes a 16-bit polynomial check code, which is automatically appended to transmitted messages and stripped from received ones. Provision is made for the ADLC to work with DMA (see Section 9.12.1) so that data link rates too fast to be processed by the microprocessor with interrupts can be supported. This device must be serviced continually once a message is begun; if not, a data overrun or underrun condition will cause a portion of a message to be lost. This is a real-time device whose internal buffering is minimal and whose data rate is determined by the link speed—not by the microprocessor program.

Display Attachment

Alphanumeric CRT displays are often designed with an internal microprocessor. This processor can be used to scan, debounce, and encode the keyboard; to provide the interface, either to a computer or

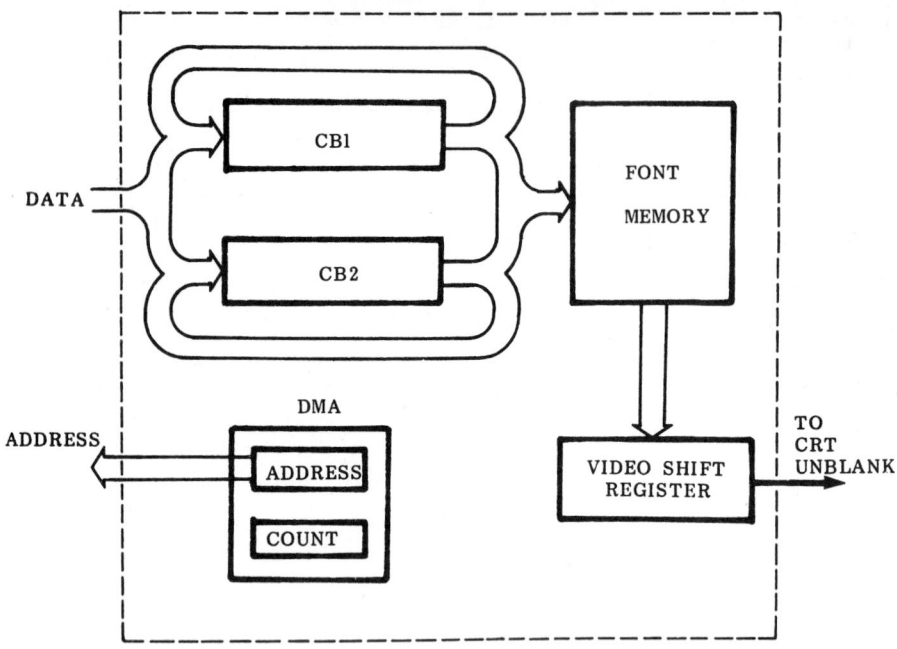

Fig. 9.109   Character generator.

to a communications line; and to perform editing functions in response to keyboard entries. Earlier we described the I/O attachment of a keyboard and communications. This section describes the attachment of the display.

The display considered here is a raster CRT whose beam moves horizontally left to right across its screen. The beam is then shut off, moved downward by one beam width, and retraced to the left. The beam is then moved horizontally from left to right. This process is repeated many times per second. When the beam reaches the bottom of the screen, it is retraced to the top. To display characters with this beam, a signal must be supplied to an unblanking circuit as the beam moves horizontally left to right.

Figure 9.109 shows the circuitry necessary to convert a buffer of character codes from the microprocessor memory into the serial bit stream of data necessary to unblank the beam. This circuitry is called a **character generator**. When the beam begins its scan at the top of the screen, the character buffer CB1 contains the codes of all the characters to be displayed on that scan line. The code for the leftmost character is in the rightmost end of the character buffer and is used to address a memory (the FONT ROM) containing the bit image of the character. A horizontal slice of the character is then accessed from the FONT ROM and placed in the VIDEO SHIFT REGISTER, which serializes the slice to the unblanking amplifier of the CRT. As the horizontal slice of the first character image is displayed, the character buffer is shifted circularly one position to the right, bringing the code for the next character to be displayed to the right end of the character buffer. That code is now in position to address the FONT ROM. In this way the character codes are converted to their serial video representation.

While one of the character buffers is being shifted to access the FONT ROM and create the character images, the other is being filled from microprocessor memory using the DMA controller. Successive character codes are retrieved from microprocessor memory and stored in the character buffer. At the bottom of the screen, the DMA address register and counter must be reset so that codes for characters at the top of the screen are the next to be accessed. This process of alternately filling the two character buffers continues for as long as an image is required on the display screen.

9.13 BUSES
Kenneth J. Thurber

A bus is a communication device used to interconnect components in a digital system by providing a data highway between system elements. There are many forms of buses; however, the general context of a computer bus is that of a digital structure designed to support high-speed traffic (rates of many megabits per second) over short distances (millimeters to kilometers), in contrast to more general circuits designed to handle low-bandwidth/long-distance traffic (e.g., phone lines). Typically, the bus consists of a set of lines that carry information, such as addresses (memory location, device ID, etc.), control signals, and data. The primary function of a bus is to provide the system with a means of device interconnection that supports the allowable communication modes.

The operation of a bus may be viewed as a sequence of functions: the devices that use the bus make requests for allocation of the bus, the bus requests are arbitrated and a device selected, the bus is allocated to the device, the data are transferred, the transmission errors are resolved, and the bus is released for allocation. Arbitration may be performed implicitly (i.e., it may be designed into other system functions). For example, in an intra-ALU (arithmetic-logic unit) bus, one central unit may make all decisions. In other cases, the arbitration may be an explicit function provided by a unit or series of units. Clearly, many of these functions may be overlapped; for example, it is quite common to develop a bus that is simultaneously arbitrating for bus usage at time t_1, is transferring data for devices that were selected during the arbitration at $t_1 - t$, and is providing error control for data that were transferred at time $t_1 - 2t$ by a third set of devices.

All bus operations and concepts will be functionally described as explicit implementations. In an actual design, some functions would be combined or even deleted via some implicit design decision intended to simplify the hardware.

9.13.1 Terminology and Definitions

A number of terms are commonly used to describe design parameters of digital bus structures. The major terms necessary for understanding the remainder of this section are defined.

There are two general forms of buses: **dedicated** and **nondedicated**. A dedicated bus may be dedicated to a pair of devices or to the performance of a particular function. Nondedicated buses are shared between multiple devices or functions or both.

The **arbitration**/control mechanism consists of the algorithm and hardware structure necessary to resolve which device (among requesting devices) is to be allocated the bus for the next communication period. Arbitrators may be **centralized** (the arbitration mechanism is at a single location) or **decentralized** (the arbitration mechanism is physically or logically dispersed among the devices).

After a device has gained control of a bus for communication, the device must be connected to the bus to permit the actual transfer of information. The actual transmission occurs between a source device and a destination device. If prearranged timing agreements between the devices exist, the transmission is called **synchronous**; if not, it is called **asynchronous**. Asynchronous transmissions may require interlocking transmissions between source and destination devices; this procedure is known as **handshaking** or request/acknowledge asynchronous communication.

The **length** of a bus is its physical distance from beginning to end. In a loop, length can be measured as either the maximum distance between any pair of devices or the total of all distances between pairs. Generally, loop length is the total of distances. The number of circuits (wires) in the bus is called the number of lines or **bus width**. The transmission capacity of the bus in bits (or bytes, words, etc.) per second is known as bus **bandwidth**. Bus **latency** is the time necessary to effect a transaction on the bus. A **transaction** consists of all steps from the initial source device request through all communication steps until the source device ascertains that the destination has correctly received the data.

9.13.2 Types of Buses

At least two distinct views of a bus exist: **hierarchical** and **functional**. The hierarchical approach notes that buses occur repeatedly in a computer system. They are used to interconnect registers on a chip, to connect chips together inside a computer, and to interconnect computers. This view notes that buses at different levels in a system differ in implementation rather than kind (i.e., all buses perform the same basic functions); however, for hardware efficiency reasons, functions that may be explicit in one bus may be implicit in another bus.

The hierarchical approach categorizes the bus as dedicated or nondedicated. Further, these terms may apply to either function or physical characteristics. That is, a bus may be physically dedicated to a pair of devices while being used for a variety of functions performed between the devices.

The function approach generally classifies buses into groups. The most common grouping breaks out bus categories as follows:

1. CPU–memory
2. Input/output controller–memory
3. CPU–input/output controller
4. Input/output controller–peripheral device
5. Input/output controller–external communication lines
6. CPU–CPU

Inside a central processor there are also buses. Some of these are used to interconnect register structures (interregister buses), and some are used to connect various functional units in the ALU (intra-ALU buses). Generally, these buses are very simple. There may be implicit arbitration done by the microprogrammer, who specifies the use of the bus via the microprogram sequences; no error control other than bus parity causing a hardware fault; and so on.

9.13.3   Arbitration / Control Units

One of the key functions performed by the bus structure is to allocate the bus. In general, a bus may be viewed as a shared data highway in which users (devices) compete on the basis of some performance measure (e.g., priority) to use the data highway. One integral portion of this concept is that of the bus control unit/bus master or arbitrator. The function of the arbitrator is to accept and process requests to use the bus. The control unit then selects the next device to receive the bus and allocates the bus to the appropriate device. This function may be either centralized (i.e., there exists a specific unit, sometimes called the bus master) or decentralized (there are multiple units capable of being the bus master). One way of building a decentralized controller is to replicate the bus controller at every device location. This may improve fault tolerance, but also increase cost, size, power, and weight. This same type of function, arbitration, is performed at other parts in the computer system; for example, at the memory multiport (a multiport is a priority sorting device that determines which of several requestors is allowed access to a specific memory bank). In fact, a memory multiport may be viewed as one form of a bus structure in which all arbitration for a given memory bank is centralized in the memory multiport front end.

An example of a centralized bus control unit operation for independent requests is shown in Fig. 9.110. In this case, each device has a separate pair of bus request and bus grant lines, which it uses for bus controller communication. When a device requires use of the bus, it generates a bus request signal, which is sent to the controller. The controller selects the next device to receive service via an algorithm and generates a bus-granted signal to the selected device. The selected device lowers its request and generates a bus-assigned signal, indicating bus busy to all other devices. After the transmission completes, the device lowers the bus-assigned line; the bus controller lowers bus granted, and then selects the next device to use the bus.

From the perspective of the bus controller, at the start of the arbitration algorithm, all pending bus requests appear to be presented simultaneously to the bus controller; thus the overhead time required for allocating the bus can be short—namely, the time to compute the algorithm. The controller can use any desired algorithm, such as a fixed device priority scheme, a programmable (modifiable) device priority scheme, or a circular allocation (round-robin) scheme. It is also easy to disable requests from a device that is suspected to have failed.

The major disadvantage of independent requests is the potentially large number of lines and connectors. Furthermore, the complexity of the allocation algorithm will be reflected in the bus controller hardware.

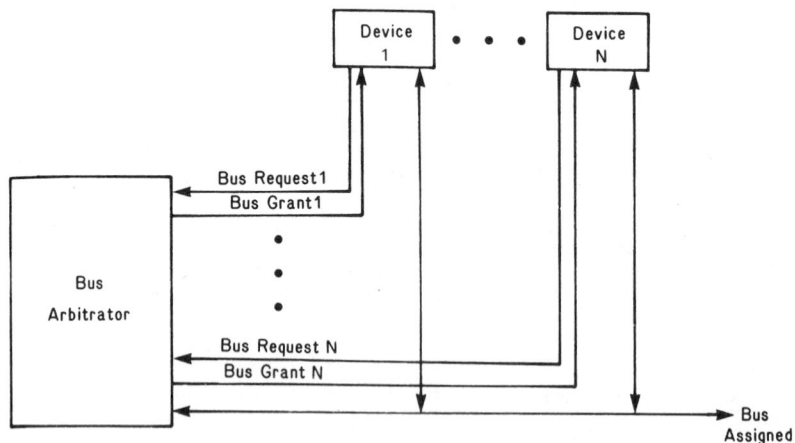

Fig. 9.110   Centralized independent requests.

9.13.4 Communication

There are two main communication strategies: synchronous communication (explicit timing agreements exist between devices) and asynchronous communication (no explicit agreements). Generally, communication design depends on timing trade-offs that specify the allowable sequencing of specific control lines (e.g., the relationship between data ready and data accept lines).

In the digital data communications environment, in general, and the computer network environment, in particular, paths are usually the point-to-point links characteristic of common-carrier systems. Sharing these paths, line sharing, involves the use of units known as "multiplexers" and "concentrators" for connection of (other) nodes to the path (multiplexing is discussed in Section 9.13.6). Synchronous devices are normally defined to be those that exchange data based on advance agreements as to timing (i.e., without synchronization signals). Frequently, the devices have either a common clock or separate clocks that are "implicitly" synchronized (i.e., within a specified tolerance of one another). One of the primary techniques of implicit systems is to have the sender and receiver of information both contain local clocks of a specified accuracy. A series of special signals is sent between devices to allow them to synchronize these clocks. Then transmissions can occur at a predefined rate, as long as the accuracy of the clocks remains within a specified tolerance. Periodically, the clocks will have to be resynchronized.

As an example of a form of synchronous communication, consider Fig. 9.111. This figure illustrates a logical clocking mechanism in which local clocks are synchronized and used to derive the data from an encoded waveform. In the example shown, the receiver is assumed to be synchronized onto the time standard of the sender. The sender sends a signal that is encoded as follows. There is a transition at the beginning of every slot (t between two successive clock pulses). During the time slot, a 0 is represented by no change in the signal, and a 1 is represented by a change (either high to low or low to high). This coded information is transmitted on the bus. The receiver derives the data by detecting when the signal changes level near the center of the time slot (data = 1) and when it does not (data = 0). Since there is always a transition at the beginning of every time slot, the receiver can maintain synchronization with the sender's clock.

As a detailed asynchronous communication example, asynchronous request/acknowledge communication will be illustrated (see Section 9.9.7). The request/acknowledge method of asynchronous communication is separated into three cases: noninterlocked, half interlocked, and fully interlocked.

Figure 9.112(a) illustrates the noninterlocked case. The source puts data (D) on the bus and raises data ready (DR), and the destination stores D and responds with data accept (DA), which causes the source to drop DR and to place new D on the lines. If an error is found in the data, the receiving device raises a data error line instead of DA. This signal interchange provides error control and permits operation between devices of any speeds. The price paid is speed and added logic complexity

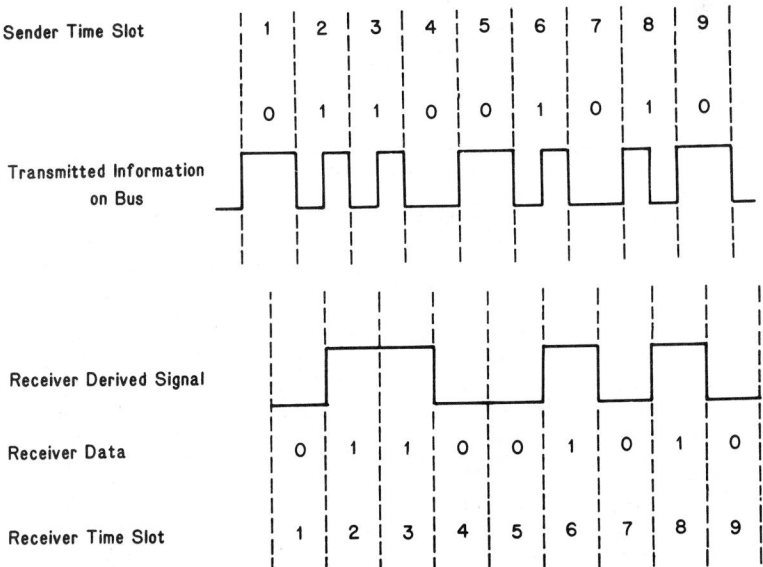

Fig. 9.111 Biphase synchronous communication.

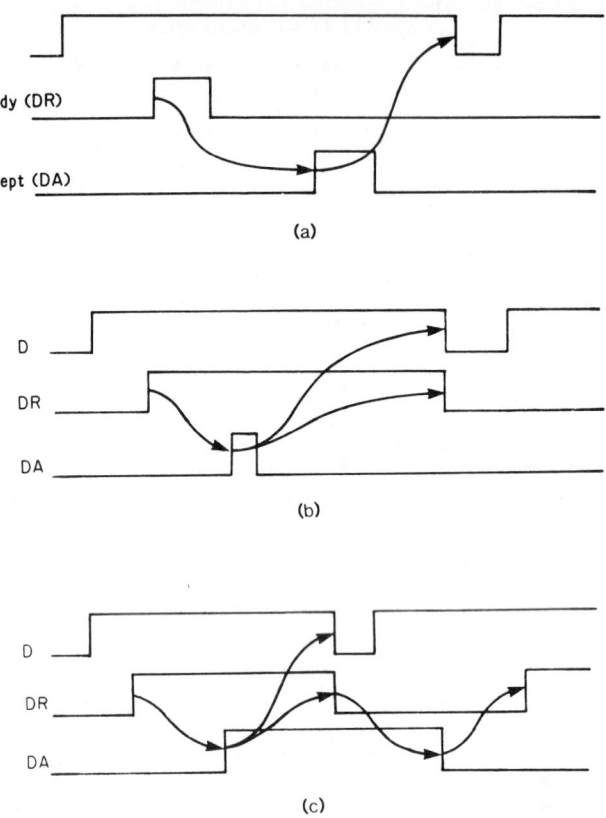

Data (D)

Data Ready (DR)

Data Accept (DA)

(a)

D

DR

DA

(b)

D

DR

DA

(c)

Fig. 9.112 Asynchronous request/acknowledge communication: (*a*) noninterlocked, (*b*) half-interlocked, (*c*) fully interlocked.

compared to a noninterlocked, one-way command interface, which would simply send the data and hope it would arrive. The exact timing is a function of the implementation. There are two lines, DA and DR, susceptible to noise, and twice as many bus length delays as in a one-way noninterlocked interface.

Improper ratios of bus propagation time and communication signal pulse widths could allow a second DR to come and go while DA is still high in response to the first DR. This could hang up the entire bus, because the destination would never send a second DA, having missed the second DR. Therefore, the second transmission would never be completed.

This can be avoided by making DR remain up until DA (or data error) is received by the source, as seen in Fig. 9.112(*b*). In this half-interlocked interface, if DR comes up while DA is still high, the second transfer will only be delayed. When the source of the first transmission sees DA, it "drops" DR. The second transmission holds up until this event, ensuring a proper sequence of interlocks. The variable width of DR tends to protect it from noise. In Fig. 9.112(*a*) a noise pulse could be interpreted as DR, but not in the half-interlocked case. There is no speed penalty and little hardware cost associated with these improvements over the noninterlocked case.

Assume that the source generates outputs from a buffer register, which has a cycle time called the buffer reload period. One more potential timing error is possible if DA extends over the source buffer reload period and masks the leading edge of the next DR (i.e., DR's leading edge is sent before DA is dropped). Figure 9.112(*c*) shows how this can be avoided with a fully interlocked interface, where a new DR does not occur until the trailing edge of the old DA (or data error) signal has been received. Both communication signals are now comparatively noise immune. The device logic is more complex. The major disadvantage is that the bus delays for the fully interlocked case have doubled over the half-interlocked case, thus halving the maximum possible transfer rate.

Other examples of bus handshaking appear in Sections 9.9.7 and 9.10.4.

9.13.5   Error Control

The function of error control is to provide means for checking, detecting, and correcting transmission errors. The most basic error control concepts are:

1. VRC (vertical-redundancy checks): parity or M-out-of-N codes
2. LRC (longitudinal-redundancy checks): for example, check-sum procedures (cyclic error checking or polynomial codes)

In the VRC technique, a parity bit or other encoding is added to each character transmitted. LRC adds parity bits to each message. The M-out-of-N code technique uses a special code set in which characters are mapped onto the code assignments in such a manner that all N-bit characters contain M bits set to the value 1. Check-sum procedures compute a correctness value based on either a cyclic error code or a polynomial error code. Of course, combinations of these techniques could be used.

Some techniques (polynomial codes) provide for both error detection and error correction capabilities. If an error-correcting technique is selected, enough redundant information is added to each transmission so that not only can transmission errors be detected, but they can also be corrected automatically by the destination device. This approach is expensive. Encoding and decoding redundant data require complex computational capabilities and use significant transmission bandwidth.

The error-detecting approach is efficient in terms of bandwidth utilization for systems with a small percentage of bad transmissions. The concept requires retransmission of any erroneous incoming information. The destination typically must return an acknowledgment (ACK) signal to the source before some specified event occurs at the source (e.g., a clocked time-out period is completed), or the information will be retransmitted. After a specified number of transmissions have failed to complete (generate a received ACK within the allowable clock time-out period), the line or destination device will be declared inoperable.

9.13.6   Signal Lines

The width of a bus affects many aspects of the system, including cost, reliability, and throughput. Signals have been discussed as if separate lines exist dedicated to specific functions. In an actual implementation, this may not be true. Techniques exist designed to minimize/combine the number of lines associated with functions, such as device addressing lines, data lines, and control. The objective is to achieve the smallest number of lines consistent with the necessary types and rates of communication.

Bus lines require drivers, receivers, cable, connectors, and power, all of which tend to be costly compared to logic. Physically and electrically, the bus can be viewed as drivers, connectors, cabling, connectors, and receivers—in that order. The connectors physically connect the cable ends into the electrical system. The drivers and receivers convert the signals into a form suitable for transmission through the cable. Clearly, media other than cable are available and would use different connectors, drivers, and receivers. Because of media, distance, and speed, conventional logic generally cannot be used to drive the bus; thus special circuits and interfaces must be designed. Connectors occupy a significant amount of physical space and are also among the least reliable components in the system. Reliability is often diminished even further as the number of lines increases, because of the additional signal switching noise.

Line combination, serial/parallel conversions, and multilevel encoding are some of the fundamental techniques for reducing bus width. These are discussed in the next three paragraphs.

Combination is a method of reducing the number of lines based on function and direction of transmission. Complementary pairs of simplex lines might be replaced with single half-duplex lines. Instead of dedicating individual lines to separate functions, a smaller number of time-shared lines might be more cost effective, even if extra logic is involved. This includes the performance of bus control functions with coded words on the data lines.

Serial/parallel trade-offs are frequently employed to balance bus width against system cost and performance. Transmitting fewer bits at a time saves lines, connectors, drivers, and receivers, but adds conversion logic at each end. It may also be necessary to use higher-speed circuits (thus more expensive) to maintain effective throughput. Serial/parallel converters at each end of the bus can be augmented with buffers that absorb traffic fluctuations and permit the use of a slower bus, relieving the device of having to accept each input by allowing inputs to be buffered automatically into larger quantities of information before requiring device attention. Independent of bus width considerations, this concept can minimize communication delays due to busy destination devices. Bit-serial transmission generally is the slowest, requires the most buffering and the least line hardware, produces the smallest amount of noise because only one line is changing state at a time, and is the most applicable approach in cases with long lines. Parallel transmission, although faster, uses more line hardware, generates greater noise, and is more cost effective over shorter distances.

Multilevel encoding converts digital data into analog signals on the bus. It is occasionally used to increase bandwidth by sending parallel data over a single line, but there are numerous disadvantages, such as complexity, line voltage drops, and lack of noise immunity.

Time-division multiplexing (TDM) provides the capability to share the set of lines in time rather than by space or electrical implementation. At time T_1, the line set may appear (be interpreted) as control lines; at time T_2, as data lines; and so on. Whether the interpretation is performed periodically or at an arbitrary time, under control of a set of specification lines, differentiates the concept of pure time division multiplexing from the concept of statistical multiplexing (STDM). Statistical multiplexing requires more lines (interpretation codes) but promises more efficient use of bandwidth.

TDM can be viewed as dividing the system bus bandwidth among a set of devices. A device wishing to transmit simply waits until it is assigned time, and transmits information of the form: data, destination. If a device does not transmit, its slot goes unused, wasting bandwidth. However, there is no need for an arbitration function. (Another perspective is that polling is used; thus the source device is known.) STDM allows any device to use any slot. Devices compete for slots; thus all slots could be used. However, arbitration must be provided for each slot allocation, and devices must know the source. Typically, the transmitted information is: source, data, destination.

9.13.7 Conclusion

Buses are the cornerstone of digital communications. They are a hierarchical concept that provides the designer the ability to interconnect functional elements (e.g., ALU to registers) or complete systems. Buses differ in implementation, with a set of common functions being implemented via either implicit or explicit functional capabilities rather than in kind. This section summarized the most important trade-offs in the area of bus design. The functional and physical parameters discussed will provide a designer the capability to exploit fully contemporary hardware technology via a state-of-the-art interconnection system specialized to a specific requirement.

9.14 GLOSSARY
Jack L. Rosenfeld

Access Time   *See* Cycle Time.

Accumulator   Register in CPU used to hold data and results of arithmetic and logic operations (e.g., accumulation); occasionally, this is the only such register.

Adder   *See* ALU; Carry.

Address   Number specifying location of data in memory, on device, etc.

Address Space   Set of all addresses that can be generated by CPU.

ALU (Arithmetic-Logic Unit)   Combinatorial logical circuitry in CPU that performs arithmetic and logic functions; main component is generally an "adder"—a combinatorial circuit that performs binary (perhaps decimal) addition of input words; ALU sometimes contains buffer registers; *see* Carry.

Arbitration   Process of deciding which one of contending requesters (usually devices attempting to use a bus or to interrupt a CPU) may use the requested facility.

Architecture   Characteristics of a computer system visible to the programmer (e.g., instruction set, registers, interrupt structure).

Arithmetic-Logic Unit   ALU.

Array Processor   Processor executing identical operations on multiple elements within an array, allowing interactions between results and operations on adjacent array elements.

ASCII   *See* Code.

Assembler   *See* Machine Language.

Assembly Language   *See* Machine Language.

Associative Memory = Content-Addressable Memory   Memory in which selection of a word is based on the information stored in a portion of the word, rather than its location; "key field" is that part of a memory word containing unique identifier; used for address translation and other special purposes.

Associative Processing   Selecting data elements for processing by value—not by location within memory.

Asynchronous   *See* Synchronous.

Attached Storage = Secondary Storage   *See* Memory.

Attached Support Processor Computer dedicated to a specialized function in a multicomputer organization (e.g., input/output and array processing).

Availability Fraction of time a system is able to perform its required function.

Backing Store Memory subsystem with larger capacity slower access, and lower cost per bit than main memory; usually used for storing information needed less frequently than information kept in main memory; generally not accessed like main memory.

Base Register CPU register holding base address; *see* Effective Address.

Bidirectional *See* Tristate.

Bit = Binary Digit 0 or 1.

Bit Slice Slice.

Board *See* Packaging Hierarchy.

Borrow *See* Carry.

Branch Instruction CPU instruction causing control to jump to another instruction, generally out of the normal sequence; "conditional branch" instruction causes control to jump only if specified conditions are satisfied (e.g., if a result is 0).

Break Point (for debugging) Point in program under test at which control is caused to pass to the debugging system.

Buffer Storage element, between data transmitter and receiver that are not synchronized, used to hold data temporarily; "memory buffer" is area in memory device to hold block of several words written by transmitter before being read by receiver; similarly, "buffer register" is device to hold a word of information; *see* Stack.

Bus Communication hardware interconnecting components in a digital system; "memory bus" connects main memory, CPU, and I/O control; "I/O bus," controlled by I/O channel, transmits data and control information between I/O channel and attached device controllers.

Byte Data word, generally 8 bits wide.

Cache Very high speed buffer memory interposed between CPU and main memory, holding recently referenced information; used to improve system performance.

CAM (Content-Addressable Memory) Associative memory.

Card *See* Packaging Hierarchy.

Carry (in an adder) Signal from one stage to the next-higher-order stage denoting a sum greater than or equal to the radix; "carry propagation" is passing carry from each stage to the next-higher stage; a "ripple-carry" adder propagates carries between neighboring stages; "carry lookahead" is determination of carry from a stage according to inputs to that stage and several lower-order stages, but without carry propagation, thus increasing the speed; "borrow" is like carry for subtraction process; *see* Full Adder.

Central Processing Unit CPU.

Channel I/O channel.

Character Symbol used to represent data (e.g., alphabet, numerals).

Characteristic *See* Floating Point.

Chip *See* Packaging Hierarchy.

Clock Electronic circuitry for generating periodic timing signals.

Code Bit pattern used to represent computer instruction, data character, etc.; "EBCDIC" (Extended Binary Coded Decimal Interchange Code) uses 8 bits to represent alphanumeric and control characters; "ASCII" (American National Standard Code for Information Interchange) is a different code, using 7 bits plus parity bit.

Compiler *See* High-Level Language.

Complement The opposite; for example, binary bits 0 and 1 are complements.

Computer System for performing sequences of arithmetic and logical operations specified by a program; includes CPU, memory, and I/O.

Conditional Branch *See* Branch Instruction.

Console Device by which operator or user interacts with a computer; generally has input means (e.g., keyboard and switches) and output means (e.g., cathode-ray-tube display and lights).

Content-Addressable Memory Associative memory.

Control Program = Supervisor *See* Operating System.

Control Store *See* Microprogrammed Control.

Control Store Sequencer Logic unit that provides the machine instruction decode logic and specifies the "next-control-store-address" generation scheme for fetching microinstructions from a control store; *see* Microprogrammed Control.

Control Unit   Part of a computer's CPU that issues timing and control signals to direct the activity of the entire system; device controllers are sometimes called control units.

Coupling (in a multiple-processor system)   The nature of physical interconnection or level of logical interaction; "loosely coupled" systems interact in the same manner as they perform input/output operations; "tightly coupled" systems share directly addressable, executable main memory, interacting by sharing each other's address space.

CPU (Central Processing Unit)   Section of a computer system that includes ALU, control unit, and registers.

Cross-Assembler   Assembler that runs on one computer and produces code to run on another computer; *see* Machine Language.

Cross-Compiler   Compiler that runs on one computer and produces machine language to run on another computer; *see* High-Level Language.

Cycle Stealing   Mode of operation in which I/O channel or DMA controller has priority over CPU with regard to requests for memory access cycles.

Cycle Time (in a memory system)   Total time for a memory operation, from initiation until memory is ready for the next such operation; "access time" is the part of memory read cycle time from initiation until data are available at memory output; (for CPU) interval during which basic operations occur.

Cylinder   *See* Track.

DASD (Direct-Access Storage Device)   Storage device such as disk or drum in which access time is relatively independent of location of data, in comparison with, for example, magnetic tape unit; *compare* Sequential-Access Memory.

Data Channel   I/O channel.

Data-Flow Processing   Machine organization in which successive operations on data are executed as soon as the data are available, rather than at predetermined time.

Destructive Read   Memory read operation in which original information is not preserved (e.g., for charge storage cells and magnetic cores); means are generally provided for regeneration of the information.

Device Controller = I/O Device Controller (sometimes called **Control Unit**)   Hardware that attaches peripheral devices to I/O interface, generating control signals to the device, serializing and deserializing data, monitoring status, and performing other functions.

Direct-Access Storage Device   DASD.

Direct Addressing   Simplest CPU addressing method, in which unmodified address specifies memory location of operand; *see also* Indirect Addressing; Relative Addressing.

Direct Memory Access   DMA.

Disk   Magnetic disk.

Diskette = Floppy Disk   Flexible disk for magnetic recording; drives for diskettes are generally less expensive and slower than drives for stiff disks.

Displacement   *See* Effective Address.

Division   *See* Restoring Division.

DMA (Direct Memory Access)   Transmission of data on memory bus directly between memory and a device controller; a "DMA controller" manages a sequence of DMA transmissions for a peripheral device.

Dynamic Allocation = Folding   Process in which different subsets of programs and data in the address space occupy different areas of main memory at different times; "paging" is one technique for dynamic allocation.

Dynamic Memory   Memory device in which information must be regenerated periodically, generally due to loss of charge; *see* Static Memory.

Dynamic Redundancy   *See* Static Redundancy.

Dynamic Relocation   Process in which programs and data may occupy different areas of main memory at different times.

EAROM   Electrically alterable read-only memory.

EBCDIC (Extended Binary Coded Decimal Interchange Code)   *See* Code.

ECC   Error Checking and Correction.

Effective Address   Address used to access main memory, the sum of several quantities: a "base address" is generally the address of the start of a block of storage; an "index" is usually the distance between an element of an array and the origin of the array; a "displacement" is the difference in addresses between the start of a block and an element of interest (e.g., a single word or the origin of an array).

Embedded Microprocessor System Microprocessor, memory, and I/O, all packaged into a specific product.

Emulation Use of microcode to simulate an architecture; hardware on which microcode is executed is called "host"; architecture simulated is called "target" architecture; host may or may not be designed specifically to emulate target; host may be capable of switching emulation among two or more targets; *see* Microprogrammed Control.

Enable Signal Signal sent to a device to cause it to become active.

EPROM Erasable programmable read-only memory.

Error Checking and Correction (ECC) Process by which additional checking bits are added to data by a transmitter so that if any of the bits (data or checking) are received incorrectly by a receiver, the error can be detected and the erroneous bits corrected.

Exponent *See* Floating Point.

Failure Event said to occur when system does not perform its service as specified.

FAMOS Floating gate avalanche injection metal-oxide semiconductor.

Fault *See* Paging.

Fault Tolerant Able to function satisfactorily despite occurrence of faults.

FHF Fixed-head file.

FIFO (First-In First-Out) Rule for specifying order of processing data, requests, etc.; *see* LIFO; Stack.

File A large unit of information on a secondary storage device; also, a secondary storage device itself; *see* Memory.

Firmware Microcode; *see* Hardware.

Fixed-Head File (FHF) Magnetic disk or drum storage device in which reading and recording transducers do not move; *compare* Moving-Head File.

Fixed Point Representation of numbers and the associated arithmetic in which radix point (e.g., decimal or binary point) is considered to be in the same position for all such numbers; *see* Floating Point.

Floating Point Representation of numbers and the associated arithmetic using a movable radix point (e.g., decimal or binary point); representation (e, f) of a number N comprises a fraction or "mantissa" part f and an "exponent" or "characteristic" part e ($N = f \times r^e$, where r is the radix); occasionally, a bias quantity is subtracted from the characteristic to give the true exponent; *see* Fixed Point; Radix.

Floppy Disk Diskette.

FLOPS (Floating Operations Per Second) Measure of CPU performance.

Folding Dynamic allocation.

Fragmentation Creation of partitions in memory with alternating used and unused sections, due to a succession of storage allocations.

Full Adder One stage of a binary adder with two operand inputs and a carry input, which produces a sum and a carry output; *see* Carry.

Gate Array = Masterslice Logic integrated circuit in which the primitive logic circuits (AND, OR, etc.) are in a fixed pattern, and different functions are implemented by creating interconnections among the logic circuits.

Handshaking Means of coordinating two units, in which signals are exchanged in a prescribed sequence.

Hardware The physical components of a computer system; "software" is the collection of programs executed on the system; "firmware" is the microcode for a CPU that has microprogrammed control; *see* Microprogrammed Control.

Hard-wired Control Implementation of a control unit by means of discrete logic elements and sequential circuits; *see* Microprogrammed Control.

Head (in a magnetic-recording system) Transducer that converts electrical signals to magnetic, or vice versa; also, physical module holding read, write, and/or erase transducers.

Hexadecimal Base 16 representation of numbers.

Hierarchy *See* Memory Hierarchy; Packaging Hierarchy.

High-Level Language (HLL) Programming language using statements close to user's normal expressions, such as $A = B \times (C + D)$ to express an arithmetic operation and IF $A = B$ THEN $I = I + 1$ to represent a control function; a "compiler" is a program that translates from an HLL to machine language; *see* Machine Language.

Hit *See* Miss.

HLL High-level language.

Horizontal Microinstruction Microinstruction that encodes several microoperations to be executed concurrently; *see* Microprogrammed Control; Vertical Microinstruction.

Host *See* Emulation.

Index *See* Effective Address.

Index Register CPU register to hold index; automatically added to address of origin of array to calculate address of array element in main memory; "indexed addressing" is this mode of calculating address; index register also used for counting iterations; *see* Effective Address.

Indexed Addressing *See* Index Register.

Indirect Addressing Mode for CPU to access operand, in which access is first made to location holding address of operand rather than operand itself; *see also* Direct Addressing.

Instruction A command to be executed by a computer's CPU; a "program" is a sequence of such commands (perhaps in a high-level language) for performing a useful function.

Instruction Counter = Program Counter CPU register holding address in main memory of instruction currently being executed (or next instruction).

Instruction Register CPU register holding instruction being executed.

Interrupt Signal to CPU causing it to suspend execution of current program and execute another program ("interrupt handler") designed to respond to the source of the signal; a "trap" is an interrupt generated internally to the CPU; *see* Priority Interrupt Device.

I/O (Input/Output) Subsystem of computer for reading in and writing out data and programs.

I/O Channel = Data Channel = Channel Unit to control transmission of blocks of data between I/O interface of a CPU and memory; *see* Bus.

I/O Device Controller Device controller.

I/O Interface Connection between device controllers and channel or CPU; usually completely specified in terms of a description of physical connection and a description of all permissible signal sequences with their timing constraints.

K Abbreviation for 1024 (2^{10}); often used in referring to memory size; "M" is abbreviation for 1,048,576 ($2^{20} = 1K \times 1K$).

Key Field That part of a block of data that contains identification of the block; *see also* Associative Memory.

Latency Interval between request for data and availability of data, often applied to rotation time of magnetic disk or drum device.

LIFO (Last-in First-Out) Rule specifying order of processing data requests, etc.; *see* FIFO; Stack.

Local Memory Small memory integrated in CPU to hold intermediate results.

Logic Analyzer Laboratory instrument for storing and observing sequences of digital signals; *see* Microprocessor Analyzer.

Lookahead "Instruction lookahead" is a technique to overlap instruction fetching, decoding, and execution times by fetching instructions and operands in advance of execution; *see also* Carry.

Loose Coupling *See* Coupling.

M *See* K.

Machine Language Programming language consisting of instructions ("machine-level" instructions) for a computer's CPU, encoded in binary form and interpreted by the control unit; "assembly language" is the same instruction set encoded in alphanumeric form with added instructions understood by the assembler; an "assembler" is a program that translates a program written by a programmer in assembly language to its machine language equivalent, for execution by the CPU; *see* High-Level Language.

Magnetic Disk = Disk Storage device using one or more disks coated with magnetic material as the storage medium.

Magnetic Drum Storage device using cylinder coated with magnetic material as the storage medium.

Magnetic Tape Drive Storage device using magnetic tape as the storage medium.

Mainframe CPU and memory of large computer system.

Main Memory *See* Memory.

Mantissa *See* Floating Point.

Mask Bit pattern designating which bits of a word are to be examined or processed (e.g., in selecting which of several external interrupt signals to accept).

Masterslice Gate array.

MDS Microprocessor development system.

Memory   Computer component for storing data and instructions; "main memory" in this handbook section refers to memory directly addressed by CPU; "attached storage" or "secondary storage" refers to slower memory that can be accessed only by first moving contents to main memory; "file" is sometimes used to refer to attached storage devices (e.g., tape file, disk file, fixed-head file).

Memory Address Register (MAR)   Register in CPU or main memory holding address for memory word being accessed.

Memory Hierarchy   Assemblage of storage devices with different speeds, costs, and sizes linked together so as to appear to have the speed of the fastest device and the size of the largest device; a "staging" hierarchy permits movement of data between adjacent members of the hierarchy only, although several such movements may be performed in sequence.

Memory-Mapped I/O   Principle of CPU accessing device controllers attached directly to memory bus as if they were memory locations.

MHF   Moving-head file.

Microcode = Microprogram = Firmware   *See* Microprogrammed Control. (*Note*: microcode is *not* code for a microcomputer.)

Microcomputer   *See* Microprocessor.

Microinstruction   *See* Microprogrammed Control.

Microinstruction Register (MIR)   Register receiving microinstructions from control store in microprogrammed control system; *see* Microprogrammed Control.

Microoperation   Primitive operation of a computer system (e.g., open gates between two registers, start memory cycle).

Microprocessor   CPU implemented on a single semiconductor chip or a small number of chips; a "microprocessor system" is controlled by one or more microprocessors; a "microcomputer" comprises a microprocessor, main memory, and I/O.

Microprocessor Analyzer   Laboratory instrument for storing sequences of microprocessor signals and displaying them as sequences of instructions and addresses; *see* Logic Analyzer.

Microprocessor Development System (MDS)   Microcomputer with software and hardware for developing programs for microprocessor systems; has facilities for writing, testing, and storing programs.

Microprogram = Microcode   *See* Microprogrammed Control. (*Note*: a microprogram is *not* a program for a microcomputer.)

Microprogrammed Control   Implementation of the control unit of a CPU using information found in a storage unit (generally read-only) called the "control store"; "microinstructions" are the binary words in the control store, containing encodings of the microoperations to be performed (microinstructions may specify that several microoperations occur simultaneously); "microcode" or "microprogram" is the collection of all microinstructions in the control store, analogous to a machine language program in main memory; *see* Control Store Sequencer; Hardwired Control; Microoperation.

Microprogram Sequencer   Control store sequencer.

MIPS (Millions of Instructions per Second)   Unit of CPU performance.

MIR   Microinstruction register.

Miss (in a paging or cache system)   Reference to a block not currently found in main memory or the cache; a "hit" is a reference to a block currently in main memory or the cache; *see* Paging.

Module   *See* Packaging Hierarchy.

Moving-Head File (MHF)   Magnetic disk storage device in which reading/recording transducers are on access arms that move to access tracks at different radii; *see* Fixed-Head File.

MTBF   Mean time between failures.

MTFF   Mean time before first failure.

Multiplex   Share a facility; in "time multiplexing" (e.g., of a bus) each user is assigned a time interval for using the facility; in "frequency multiplexing," each user uses a different frequency range so all can transmit simultaneously; "space multiplexing" permits use of different physical parts simultaneously; a "multiplexer" is a device that controls multiplexing; a multiplexer is also a device with more than one input and one output in which any one of the inputs at a time can be connected to the output.

Multiport Memory   *See* Port.

Multiprocessor   Computer system with two or more CPUs that are controlled by a single operating system and that communicate with each other.

Nondestructive Read   Opposite of "destructive read."

Nonrestoring   *See* Restoring Division.

Non-Return-to-Zero (NRZ)   *See* Return-to-Zero.

Nonvolatile   Opposite of "volatile."

NRZ (Non-Return-to-Zero)   *See* Return-to-Zero.

Off-Line   *See* On-line.

One's Complement   Binary representation of numbers in which negative number $-n$ is represented by the bit-by-bit complement of the binary representation of n; *see also* Sign-Magnitude; Two's Complement.

On-Line   Physically connected and capable of performing normal functions; "off-line" means not capable of performing normal functions: not logically connected, but may be physically connected.

Op Code   Operation code.

Operand   Data manipulated by a computer instruction.

Operating System (OS)   Software to manage all the resources of a computer: storage, processing time, system programs, I/O, etc.; the "control program" or "supervisor" is the part of the operating system that manages the execution of other programs.

Operation Code (Op Code)   Part of computer instruction that specifies what operation is to be performed.

OS   Operating system.

Packaging Hierarchy   Set of physical structures used to construct computer system; in contemporary systems, the most common hierarchy is: "chip," a semiconductor entity on which integrated circuits are fabricated; "module," which houses one or more chips, provides protection for them, and provides interconnection between them and the card; "card," composed of laminated insulation and conductor layers, which provides support and interconnection structures for many modules; cards may be plugged orthogonally into "boards," which provide pluggable connections and wiring structures for the cards.

Page   *See* Paging.

Page Fault   *See* Paging.

Page Frame   *See* Paging.

Page Table   *See* Paging.

Paging   Technique for implementing virtual memory; address space and main memory are divided into equal, fixed-size blocks called "pages" and "page frames," respectively; "paging algorithm" is used to determine what pages to move between secondary storage and main memory, and when to move them; a "page table" records status of pages and page frames; when a page referred to by the program is not in main memory, a "page fault" interruption occurs, and the paging algorithm is used to determine what action to take; *see* Miss; Virtual Memory.

Parameter   *See* Subroutine.

Partition   Grouping of logical or physical objects for the purpose of defining a higher-order function or packaging part; an example of a logical partition may be a group of lower-order logical elements such as a binary adder, storage cells, and gates to form a "bit slice"; a physical partition may be defined by all the chips to be included in a single module or all logic circuits on a single chip, etc.

Peripheral Device   Input and/or output device attached to a computer.

Pipeline   A type of ALU design for improved performance in which an arithmetic operation is performed on operands by automatically moving them through a set of individual units, each unit performing a separate part of the arithmetic operation; more than one set of operands are generally in the pipeline simultaneously.

PLA (Programmable Logic Array)   Combinatorial logic device, usually consisting of AND gates feeding OR gates, in a physical array; used as a ROM; the logic function is generally established ("programmed") at time of manufacture.

Poll   To interrogate status repeatedly (e.g., to test status of devices one after the other).

Pop   *See* Stack.

Port   Location where data can enter or exit; "multiport memories" permit simultaneous entrance and/or exit of data.

Printer   Computer output device that prints information on paper; a "line printer" prints one line at a time; a "serial printer" prints a character at a time.

Priority Interrupt Device   Semiconductor device used in small computer systems to select one of several input interrupt signals on the basis of some priority algorithm, and generate interrupt signal to CPU if specified conditions are satisfied.

Program *See* Instruction; also, to write a specified bit pattern in a permanent or semipermanent storage device (e.g., ROM, PROM, PLA).

Program Counter Instruction counter.

Programmable Logic Array PLA.

Programmable Read-Only Memory PROM.

Programmer Person who writes programs; also, device for writing specified bit patterns in a permanent or semipermanent storage device (e.g., PROM programmer).

PROM (Programmable Read-Only Memory) Memory that can be written selectively ("programmed") only once—after that, it behaves like ROM.

Protection "Storage protection" is prevention of access to designated areas of memory by unauthorized programs.

Protocol Rules for communication; specifically, the sequence of signals that must be interchanged and the actions that must be taken to complete a communication.

Push *See* Stack.

Pushdown Stack *See* Stack.

Radix Base of a number system (e.g., 2 for binary numbers, 10 for decimal, 16 for hexadecimal).

RAM (Random-Access Memory) Memory system in which words can be accessed in any sequence and access time is independent of location; memory capable of both read and write operations; *see* ROM.

Random-Access Memory RAM.

Read-Only Memory ROM.

Read-Only Store (ROS) ROM.

Record Basic block of data on input or output device.

Refresh (in dynamic memory) To restore the original physical quantity (e.g., charge) that represents stored information.

Register Set of flip-flops to hold information; usually, all flip-flops of the register (or a subset of them) are set simultaneously, and information is read out from all (or a subset) simultaneously.

Register File Group of registers in CPU.

Relative Addressing Mode for CPU to generate address by adding displacement to address of current instruction; *see also* Direct Addressing.

Reliability Probability of being able to perform specified function.

Relocation Process of moving program or data to a new main memory location for execution there.

Replacement Algorithm (in a paging system) A rule for selecting which memory page frame is to be replaced by new page from secondary storage; *see* Paging.

Restoring Division Division algorithm involving making trial divisions and restoring original dividend when trial fails; "nonrestoring division" does not restore when trial fails.

Return-to-Zero Method of magnetic recording or digital communication in which signal returns to 0 between excursions to positive or negative limits; "non-return-to-zero" is method in which signal is always at either positive or negative limit.

Ripple-Carry Adder *See* Carry.

ROM (Read-Only Memory) = ROS Memory whose information contents cannot be changed after manufacture.

ROS (Read-Only Storage) ROM.

Search (for input device) To scan data until the desired information is located.

Secondary Storage = Attached Storage *See* Memory.

Seek (for moving-head file) To move access arm(s) from one track position to another.

Segment *See* Segmentation.

Segmentation Technique for implementing virtual memory in which address space is divided into variable-size blocks called segments.

Self-Modifying Code Program that modifies parts of its own instructions.

Sequencer Control store sequencer.

Sequential-Access Memory Memory system in which words can be read only in the order in which they were written (or reverse order); (e.g., magnetic tape); *compare* DASD.

Serviceability Ability to be repaired or maintained.

Sign-Magnitude Binary representation of numbers with one bit for plus or minus sign and other bits for the magnitude; *see also* One's Complement; Two's Complement.

Slice   Building-block section of ALU, registers, memory, or other components, performing a function for *m* bits of a word; generally, several of these (*n*) are used in parallel to provide the function for *n* × *m* bits; provides flexible means of designing CPUs or other subsystems to satisfy special requirements.

Software   *See* Hardware.

Source-Sink I / O   Input or output between system and original source of data or final recipient of data, as opposed to I/O with attached storage device intended for temporary storage.

Stack   Mechanism, implemented by hardware, software, or both, for storing a variable number of units of information; a "pushdown" or "LIFO" stack has a single port, and the order of retrieval is the inverse of the order of storing [i.e., the item most recently written ("pushed") onto the stack is the first item read ("popped") from it]; a "FIFO" mechanism has a port for writing and another for reading, and the order of retrieval is the same as the order of storing (i.e., the first item written is the first item read); *see* Buffer.

Staging   *See* Memory Hierarchy.

Static Memory   Memory device that retains information as long as power is applied to it; *see* Dynamic Memory.

Static Redundancy   Approach to system reliability using redundant components, with a permanent system configuration (e.g., triple-modular redundancy); "dynamic redundancy" is an approach to reliability using redundant components with a configuration that is changed according to the errors detected.

Storage   *See* Memory.

Store-Through = Write-Through   (in cache system)   Process of writing both in the cache and the corresponding location in main memory when a memory write operation is called for, to avoid subsequently writing entire cache block to memory.

String Processor = Vector Processor   Computer system designed specifically to optimize execution of the same arithmetic or logical operation on a continuous series of data elements; often uses ALU with pipeline.

Subroutine   Program sequence invoked ("called") from one or more points in another program (the "calling" program); generally returns control to instruction immediately after point of invocation; the "parameters" of the subroutine are operands passed to it by the calling program and manipulated by the subroutine.

Supervisor = Control Program   *See* Operating System.

Synchronous   Associated with clock having fixed cycle time; "asynchronous" is not associated with clock, so asynchronous events can occur at any time.

Target   *See* Emulation.

Three-State   Tristate.

Throughput   Rate of work completed by a computer system; *see* MIPS.

Tight Coupling   *See* Coupling.

Time Sharing   (in computer system)   Permitting more than one user to use system concurrently by causing the CPU to execute a part (or all) of each user's program, one at a time.

TMR   Triple-modular redundancy.

Track   Circular path on disk or drum storage medium, or linear path on magnetic tape; contains data recorded by single head, without motion of access mechanism; (in moving-head file) "cylinder" is the set of tracks that can be accessed for a single position of the access mechanism.

Transaction   (for communication along a bus)   All steps required to complete transmission of a unit of data.

Trap   *See* Interrupt.

Triple-Modular Redundancy (TMR)   Approach to high reliability in which three copies of system modules are provided, all fed with same inputs, and in which the three module outputs are compared by "voters" that select final outputs according to majority vote.

Tristate = Three-State   Logic circuit capability to force a wire to one of two different voltages to represent two different logic values, or to assume a high-impedance state so that another circuit can place a voltage on the line; a "bidirectional" line is one with tristate drivers at both ends.

Two's Complement   Binary representation of numbers in which a negative number $-n$ is represented by the binary representation of $2 - n$; *see also* One's Complement; Sign-Magnitude.

UART (Universal Asynchronous Receiver / Transmitter)   Device for communicating along a serial bus one character at a time, with facilities for serializing and deserializing data.

UPS   Uninterruptible power system.

USART (Universal Synchronous / Asynchronous Receiver / Transmitter)   Device able to operate like either UART or USRT.

USRT (Universal Synchronous Receiver / Transmitter)   Device similar to UART, except for transmitting or receiving streams of characters rather than one at a time.

Variable Field Length (for instructions and data)   Having more than one possible length.

Vector Processor   String processor.

Vertical Microinstruction   Microinstruction with format similar to machine instruction format; *see* Horizontal Microinstruction; Microprogrammed Control.

Virtual Memory   System whereby main memory is made to appear larger than it actually is by automatically moving blocks of memory from secondary storage as needed; generally, implemented by a combination of hardware and software; *see* Paging.

Volatile (in memory system)   Losing information when power is removed or interrupted.

Word   Group of characters or bits treated as a unit for purposes of reading or writing to memory and performing arithmetic or logic operations; multiple-word units and fractional-word units, however, are often encountered.

WOROM   Write-once read-only memory.

Write-Through   Store-through.

BIBLIOGRAPHY FOR ARCHITECTURE: REGISTERS, ADDRESSING, AND INSTRUCTIONS

Alexandridis, N. A., *Microprocessor System Design Concepts*, Rockville, Md., Computer Science Press, 1984.

Baer, J., *Computer Systems Architecture*, Rockville, Md., Computer Science Press, 1980.

Bell, C. G., and Newell, A., "Parallel Operation in the Control Data 6600," in *Computer Structures*, New York, McGraw-Hill, 1971, pp. 470–476, 489–503.

Chang, S. S., Ed., *Fundamentals Handbook of Electrical and Computer Engineering*, Vol. 3, New York, Wiley, 1985.

Foster, C. C., and Iberall, A. R., *Computer Architecture*, 3rd ed., New York, Van Nostrand-Reinhold, 1985.

Hamacher, V. C., Vranesic, Z. G., and Zaky, S. G., *Computer Organization*, 2nd ed., New York, McGraw-Hill, 1984.

Hayes, J. P. *Computer Architecture and Organization*, New York, McGraw-Hill, 1978.

Hwang, K., and Briggs, F. A., *Computer Architecture and Parallel Processing*, New York, McGraw-Hill, 1984.

IBM System / 370 Principles of Operation, IBM Publication GA 22-7000, IBM Corp., White Plains, N.Y.

Iliffe, J. K., *Advanced Computer Design*, Prentice-Hall International, 1982.

Kogge, O., *The Architecture of Pipelined Computers*, New York, McGraw-Hill, 1981.

Kuck, D. J., *The Structure of Computers and Computations*, New York, Wiley, 1978, pp. 37, 254, 284.

Lorin, H., *Introduction to Computer Architecture and Organization*, New York, Wiley-Interscience, 1982.

Meyer, T. H., *Computer Architecture and Organization*, Beaverton, Or., Dilithium Press, 1982.

Peatman, J. B., *Microcomputer-Based Design*, New York, McGraw-Hill, 1977.

Prasad, N. S., *Architecture and Implementation of Large Scale IBM Computer Systems*, Wellesley, Mass., QED Information Sciences, 1981.

Short, K., *Microprocessors and Programmed Logic*, Englewood Cliffs, N.J., Prentice-Hall, 1981.

Siewiorek, D. P., Bell, C., G., and Newell, A. P., *Computer Structures: Principles and Examples*, New York, McGraw-Hill, 1982.

Stallings, W. E., *Reduced Instruction Set Computers*, Washington, D.C. Computer Society Press, 1986.

Stone, H., Ed., *Introduction to Computer Architecture*, 2nd ed., Chicago, Ill., SRA, 1980.

Thornton, J. E., *Design of a Computer, The Control Data 6600*, Glenview, Ill., Scott, Foresman, 1970.

Toy, W., and Zee, B., *Computer Hardware/Software Architecture*, Englewood Cliffs, N.J., Prentice-Hall, 1986.

BIBLIOGRAPHY FOR ARCHITECTURE: INTERRUPT SYSTEMS AND SYSTEM STATES

Alexandridis, N. A., *Microprocessor System Design Concepts*, Rockville, Md., Computer Science Press, 1984.

Chang, S. S., Ed., *Fundamentals Handbook of Electrical and Computer Engineering*, Vol. 3, New York, Wiley, 1983.

Foster, C. C., and Iberall, A. R., *Computer Architecture*, 3rd ed., New York, Van Nostrand-Reinhold, 1985.

Peatman, J. B., *Microcomputer-Based Design*, New York, McGraw-Hill, 1977.

Stone, H. B., *Introduction to Computer Organization and Data Structure*, New York, McGraw-Hill, 1972.

Tanenbaum, A. S., *Structured Computer Organization*, Englewood Cliffs, N.J., Prentice-Hall, 1976.

BIBLIOGRAPHY FOR PROGRAMMING BASICS

Chang, S. S., Ed., *Fundamentals Handbook of Electrical and Computer Engineering*, Vol. 3, New York, Wiley, 1983.

BIBLIOGRAPHY FOR SWITCHING THEORY AND LOGIC DESIGN

Chang, S. S., Ed., *Fundamentals Handbook of Electrical and Computer Engineering*, Vol. 3, New York, Wiley, 1983.

Dietmeyer, D. L., *Logic Design of Digital Systems*, Boston, Allyn and Bacon, 1978.

Hill, F. J., and Peterson, G. R., *Introduction to Switching Theory and Logic Design*, New York, Wiley, 1974.

Krutz, R. L., *Microprocessors and Logic Design*, New York, Wiley, 1980.

Mead, C., and Conway, L., *Introduction to VLSI Systems*, Reading, Mass., Addison-Wesley, 1980.

Ross, Jr., C. H., *Fundamentals of Logic Design*, St. Paul, Minn., West Publishing, 1979.

Weste, N., and Eshraghian, K., *Principles of CMOS VLSI Design, a Systems Perspective*, Reading, Mass., Addison-Wesley, 1985.

BIBLIOGRAPHY FOR ARITHMETIC-LOGIC UNIT

Alexandridis, N. A., *Microprocessor System Design Concepts*, Rockville, Md., Computer Science Press, 1984.

Baer, J., *Computer Systems Architecture*, Rockville, Md., Computer Science Press, 1980.

Chang, S. S., Ed., *Fundamentals Handbook of Electrical and Computer Engineering*, Vol. 3, New York, Wiley, 1983.

Foster, C. C., and Iberall, A. R., *Computer Architecture*, 3rd ed., New York, Van Nostrand-Reinhold, 1985.

Hwang, K., *Computer Arithmetic: Principles, Architecture, and Design*, New York, Wiley, 1979.

Mano, M., *Computer Systems Architecture*, 2nd ed., Englewood Cliffs, N.J., Prentice-Hall, 1985.

Meyer, T. H., *Computer Architecture and Organization*, Beaverton, Or., Dilithium Press, 1982.

Peatman, J. B., *Microcomputer-Based Design*, New York, McGraw-Hill, 1977.

Swartzlander, Jr., E. E., *Computer Arithmetic*, Stroudsburg, Penn., Dowden, Hutchinson & Ross, 1980.

Weste, N., and Eshraghian, K., *Principles of CMOS VLSI Design, a Systems Perspective*, Reading, Mass., Addison-Wesley, 1985.

BIBLIOGRAPHY FOR MAIN MEMORY AND HIGH-SPEED STORAGE: TECHNOLOGIES

Alexandridis, N. A., *Microprocessor System Design Concepts*, Rockville, Md., Computer Science Press, 1984.

Chang, S. S., Ed., *Fundamentals Handbook of Electrical and Computer Engineering*, Vol. 3, New York, Wiley, 1983.

Hnatek, E. R., *A User's Handbook of Semiconductor Memories*, New York, Wiley, 1977.

Luecke, G., Mize, J. F., and Carr, W. N., *Semiconductor Memory Design and Applications*, New York, McGraw-Hill, 1973.

Matick, R. E., *Computer Storage Systems and Technology*, New York, Wiley, 1977.

Middelhoek, S., George, P. K., and Dekker, P., *Physics of Computer Memory Devices*, New York, Academic Press, 1976.

Proebster, W. F., Ed., *Digital Memory and Storage*, Braunschweig, Vieweg, 1978.

BIBLIOGRAPHY FOR MAIN MEMORY AND HIGH-SPEED STORAGE: ORGANIZATION

Alexandridis, N. A., *Microprocessor System Design Concepts*, Rockville, Md., Computer Science Press, 1984.

Bell, C. G., Mudge, J. C., and McNamara J. E., Eds., *Computer Engineering: A DEC View of Hardware System Design*, Bedford, Mass., Digital Press, 1978, pp. 263–267.

Chang, S. S., Ed., *Fundamentals Handbook of Electrical and Computer Engineering*, Vol. 3, New York, Wiley, 1983.

Hellerman, H. and Conroy, T. E., *Computer System Performance*, New York, McGraw-Hill, 1975.

Matick, R. E., *Computer Storage Systems and Technology*, New York, Wiley-Interscience, 1977.

Organick, E. I., *The Multics Systems: An Examination of Its Structure*, Cambridge, Mass., MIT Press, 1972.

Peatman, J. B., *Microcomputer-Based Design*, New York, McGraw-Hill, 1977.

Thurber, K. J., *Large Scale Computer Architecture*, Rochelle Park, N.J., Hayden, 1976.

BIBLIOGRAPHY FOR MAIN MEMORY AND HIGH-SPEED STORAGE: MICROPROCESSOR SYSTEM CONSIDERATIONS

Alexandridis, N. A., *Microprocessor System Design Concepts*, Rockville, Md., Computer Science Press, 1984.

Chang, S. S., Ed., *Fundamentals Handbook of Electrical and Computer Engineering*, Vol. 3, New York, Wiley, 1983.

Hilburn, J. L., and Julich, P. M., *Microcomputers/Microprocessors Hardware, Software, and Applications*, Englewood Cliffs, N.J., Prentice-Hall, 1976.

Ogdin, C. A., *Microcomputer Design*, Englewood Cliffs, N.J., Prentice-Hall, 1978.

Peatman, J. B., *Microcomputer-Based Design*, New York, McGraw-Hill, 1977.

BIBLIOGRAPHY FOR CONTROL, TIMING AND SYSTEM SIGNALS

Agrawala, A. K., and Rauscher, T. G., *Foundations of Microprogramming: Architecture, Software, and Applications*, New York, Academic Press, 1976.

Alexandridis, N. A., *Microprocessor System Design Concepts*, Rockville, Md., Computer Science Press, 1984.

Buchholz, W., Ed., *Planning a Computer System*, New York, McGraw-Hill, 1962.

Chang, S. S., Ed., *Fundamentals Handbook of Electrical and Computer Engineering*, Vol. 3, New York, Wiley, 1983.

Chu, Y., *Computer Organization and Microprogramming*, Englewood Cliffs, N.J., Prentice-Hall, 1972.

Foster, C., and Iberall, A. R., *Computer Architecture*, 3rd ed., New York, Van Nostrand-Reinhold, 1985.

Husson, S. S., *Microprogramming: Principles and Practices*, Englewood Cliffs, N.J., Prentice-Hall, 1970.

Klingman, E. E., *Microprocessor Systems Design*, Englewood Cliffs, N.J., Prentice-Hall, 1977.

Langdon, Jr., G., *Computer Design*, San Jose, Calif., Computeach Press, 1982.

McGlynn, D. R., *Microprocessors*, New York, Wiley, 1976.

Peatman, J. B., *Microcomputer-Based Design*, New York, McGraw-Hill, 1977.

Short, K. L., *Microprocessors and Programmed Logic*, Englewood Cliffs, N.J., Prentice-Hall, 1981.

Tanenbaum, A. S., *Structured Computer Organization*, Englewood Cliffs, N.J., Prentice-Hall, 1976.

BIBLIOGRAPHY FOR INPUT / OUTPUT AND DEVICES: GENERAL CONSIDERATIONS

Alexandridis, N. A., *Microprocessor System Design Concepts*, Rockville, Md., Computer Science Press, 1984.

Chang, S. S., Ed., *Fundamentals Handbook of Electrical and Computer Engineering*, Vol. 3, New York, Wiley, 1983.

Foster, C. C. and Iberall, A. R., *Computer Architecture*, 3rd ed., New York, Van Nostrand-Reinhold, 1985.

Mano, M., *Computer Systems Architecture*, 2nd ed., Englewood Cliffs, N.J., Prentice-Hall, 1982.

Peatman, J. B., *Microcomputer-Based Design*, New York, McGraw-Hill, 1977.

BIBLIOGRAPHY FOR INPUT / OUTPUT AND DEVICES: ATTACHED STORAGE DEVICES

Chang, S. S., Ed., *Fundamentals Handbook of Electrical and Computer Engineering*, Vol. 3, New York, Wiley, 1983.

Matick, R. E., *Computer Storage Systems and Technology*, New York, Wiley, 1977.

Middlehock, S., George, P. K., and Dekker, P., *Physics of Computer Memory Devices*, New York, Academic Press, 1976.

Proebster, E., Ed., *Digital Memory and Storage*, Braunschweig, Vieweg, 1978.

BIBLIOGRAPHY FOR INPUT / OUTPUT AND DEVICES: MICROPROCESSOR SYSTEM I / O

Alexandridis, N. A., *Microprocessor System Design Concepts*, Rockville, Md., Computer Science Press, 1984.

Chang, S. S., Ed., *Fundamentals Handbook of Electrical and Computer Engineering*, Vol. 3, New York, Wiley, 1983.

Mano, M., *Computer Systems Architecture*, 2nd ed., Englewood Cliffs, N.J., Prentice-Hall, 1982.

Peatman, J. B., *Microcomputer-Based Design*, New York, McGraw-Hill, 1977.

BIBLIOGRAPHY FOR BUSES

Alexandridis, N. A., *Microprocessor System Design Concepts*, Rockville, Md., Computer Science Press, 1984.

Chang, S. S., Ed., *Fundamentals Handbook of Electrical and Computer Engineering*, Vol. 3, New York, Wiley, 1983.

IEEE Standard Microcomputer System Bus, IEEE Std 796-1983, IEEE, New York, 1983.

IEEE Standard Digital Interface for Programmable Instrumentation, IEEE Std 488-1975, IEEE, New York, 1975.

Levy, J. L., "Buses, the Skeleton of Computer Structures" in *Computer Engineering: A DEC View of Hardware System Design*, C. G. Bell et al., Eds., Digital Press, Bedford, Mass., 1978.

Peatman, J. B., *Microcomputer-Based Design*, New York, McGraw-Hill, 1977.

Thurber, K. J., and Masson, G. M., *Distributed-Processor Communication Architecture*, Lexington, Mass., Lexington Books (D.C. Heath), 1979.

Thurber, K. J., *Distributed Process Architecture: A Tutorial*, Long Beach, Calif., IEEE Computer Society, 1979.

CHAPTER 10

ENGINEERING THERMODYNAMICS AND HEAT TRANSFER

Bernard D. Wood

Department of Mechanical and Aerospace Engineering
Syracuse University
Syracuse, New York

Frank P. Incropera

Department of Mechanical Engineering
Purdue University
West Lafayette, Indiana

10.1 ENGINEERING THERMODYNAMICS
Bernard D. Wood

List of Symbols

English Alphabet

A	Availability	g	Gravitational acceleration
B	Darrieus function	g_c	Dimensional constant; $1.0 \text{ kg/N} \times \text{m/s}^2$
C	Specific heat		or $32.174 \text{ lb/lbf} \times \text{ft/sec}^2$
CR	Compression ratio	H	Enthalpy
d	Exact differential	K	Equilibrium constant
E	Energy	KE	Kinetic energy
F	Helmholtz function; force	LW	Lost work
G	Gibbs function	$\mathcal{M}$	Molecular weight

English Alphabet   *(Continued)*

m	Mass		T	Temperature
n	Number of moles; exponent in		U	Internal energy
	polytropic expression		V	Volume
P	Pressure		$\mathscr{V}$	Velocity
PE	Potential energy		W	Humidity ratio
$\mathscr{Q}$	Heat Transfer		$\mathscr{W}$	Work
$\mathscr{R}$	Gas constant per unit mass		x	Dryness fraction ("quality")
$\overline{\mathscr{R}}$	Gas constant per mole		Z	Compressibility factor
S	Entropy		z	Elevation

Greek and Other Symbols

β	Volume expansivity		μ	Joule–Thomson coefficient; chemical
γ	Ratio of specific heats (C_p/C_v)			potential; degree of saturation
Δ	Finite difference		ν	Stoichiometric coefficient
δ	Inexact differential		ρ	Density (inverse of specific volume)
ε	Degree of reaction		τ	Torque
ε'	Effectiveness		ϕ	Constant-pressure entropy function;
ε''	Second-law efficiency			availability function
η	Efficiency		ω	Speed of rotation
θ	Temperature, ideal-gas scale		f	Fugacity
κ	Compressibility			

Subscripts

cy	Cycle		oa	Overall
e	Exit conditions		P	Products of a reaction
f	Saturated liquid conditions		R	Reactants; reduced value
fg	Change from saturated liquid to		r	Relative value
	saturated vapor		th	Thermal
g	Saturated vapor conditions		u	Useful
H	Higher		v	Vapor component
i	Inlet conditions		vs	Saturated vapor component
L	Lower			
0	Stagnation value; arbitrary base value;			
	"zero pressure" value;			
	environment conditions			

Note: Lowercase letters frequently indicate specific values of an extensive property that is represented by a capital letter, e.g., $h = H/m$ kJ/kg   or   Btu/lb.

10.1.1   Basic Principles

Introduction

Thermodynamics is concerned primarily with the transfer of energy, the conversion of energy from one form to another, and the storage of energy. Because energy transfer to or from a system changes the properties of the system, thermodynamics is concerned with property values and phase changes of the substances comprising the system. One might limit the term *thermodynamic property* to properties such as internal energy, entropy, and the Gibbs function for which temperature is a component of the property values. But changes in what might be called *mechanical properties*, such as position, velocity and stress, also imply energy changes. Therefore, we cannot make a clear distinction between thermodynamics and mechanics except in the limited situations in which either the thermal or the mechanical aspects become essentially negligible. The topics to be dealt with in this section are those in which thermal effects predominate and those with which engineers normally are concerned.

Engineering thermodynamics, as dealt with in this chapter, will be limited largely to the classical approach. Classical or macroscopic thermodynamics is based on observable and measurable phenomena. Its "laws" are established by the failure of repeated attempts to disprove the generalizations based on innumerable observations. By contrast, microscopic thermodynamics depends on assumptions concerning the characteristics of molecules and the laws of mechanics applied to a sufficiently large number of them so that the rules of statistics may be invoked. The implications drawn from statistical thermodynamics are only valid to the extent that the results can be confirmed by direct or indirect macroscopic observations. Still, the statistical mechanics approach to thermodynamics is very valuable in that it provides a basis for mathematical extrapolation beyond the limits of actual observation.

Definition of Terms

In the discipline we call thermodynamics, numerous terms have specific limited meanings that may or may not conform to their usage elsewhere. Figure 10.1 illustrates some of these terms and establishes a sign convention that will be used throughout Section 10.1.

A thermodynamic analysis must start with the identification of the system to be considered. The **system boundaries** may be real or imaginary and are chosen for convenience. The system may expand or contract but the boundaries remain identifiable. All things with which the system exchanges either energy or mass make up the **surroundings**. Arbitrarily, *heat transfer to* the system is considered positive, and *work done by* the system is considered positive. (In some disciplines such as chemical thermodynamics, both heat transfer to and work done on the system are considered to be positive). When an energy balance is undertaken, the sign convention, whichever is chosen, must be consistent throughout. A **closed system** does not exchange mass with the surroundings. Mass transfer *to* an **open system** is designated by m_i and *from* the system by m_e, the subscripts standing for inlet and exit. If these two are not equal over some time period, obviously the mass of the system is changed. An **isolated system** exchanges neither mass nor energy with surroundings. The system plus its surroundings, in effect, become an isolated system and are sometimes referred to as the **universe** of the particular analysis.

The thermodynamic **state point** of a system is established by its **property values**. Properties therefore are **point functions**, the value of each being some function of the values of others at the same state point. The functional relationships, whether simple or complex, among the various property values imply that these relationships can be dealt with according to the rules of mathematics. Therefore, an infinitesimal change in a property value is an **exact differential**, written dx.

A **quasistatic process** is one in which the property values change progressively from one equilibrium state point to another. That sequence of state points is the thermodynamic **path** of the process. Because the amount of work done or heat transferred during that process will depend on the path followed, work and heat transfer are **path functions**. An infinitesimal amount of work or heat transfer is therefore an **inexact differential**, written $\delta \mathscr{W}$ or $\delta \mathscr{Q}$.

Property values such as pressure and temperature that are independent of the amount of mass being considered are **intensive properties**. Those, such as volume and internal energy, that do depend on the mass are **extensive properties**. For example, for a substance uniform throughout, one half the mass would occupy one half the volume of the whole, while the temperature value, at equilibrium, would be the same in the two halves.

A completely **reversible process** is one that can be returned to its beginning state point along the path of the original process and leave no change in either the system or the surroundings. (This is an ideal process, never completely achievable.) A system may be **internally reversible** in that the system's property values of the original path are retraced despite the fact that some change will be found in the surroundings.

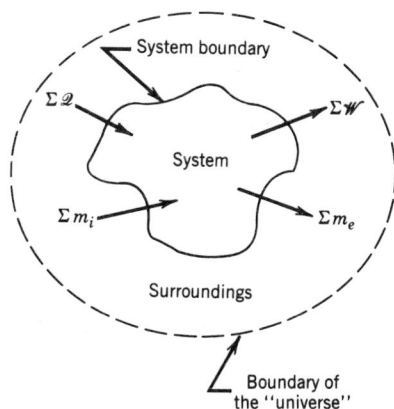

Fig. 10.1 Definition of terms relating to a system and its surroundings.

Units of Energy and Power

All forms of energy must have the same basic units. **Work** is one form of energy and is defined as the product of a force times the collinear distance through which that force is applied. In SI units, the basic unit of energy therefore is the newton-meter (N-m). One newton-meter is a joule (J). In the English engineering system, the basic unit of energy is the foot-pound force (ft-lbf). Other units of energy used particularly in heat transfer are the calorie and the British Thermal Unit (Btu). Conversions from one system of units to another are given in Chapter 1.

Power is the rate of energy transfer. In SI units, one watt (W) is defined as one newton-meter per second (N-m/s). One kilowatt (kW) then is one thousand newton-meters per second. In the English engineering system, 1 horsepower (hp) is defined as 550 foot-pounds force per second. One horsepower is approximately 0.7457 kW.

The concept of force is derived from Newton's second law of motion, which states that force is proportional to the product mass times acceleration:

$$F = \frac{ma}{g_c}$$

One newton (N) is defined as the force that will accelerate one kilogram (the basic unit of mass in SI) at the rate of one meter per second per second (the basic units of distance and time). The dimensional constant (g_c) therefore has the value unity and the units kg-m/N-s^2.

In the English engineering system of units, one pound force is defined as causing one pound mass to accelerate at the rate equivalent to "standard" gravitational acceleration, which is 9.80665 m/s^2, or approximately 32.174 ft/sec^2. Thus, the dimensional constant in this system is 32.174 lb-ft/lbf-sec^2.

Work ($\mathscr{W}$)

From the definition of work as the product of a force and the distance through which the force acts, the work done between two thermodynamic state points 1 and 2 is $\int_1^2 F\,dx$. Force and distance, however, are not convenient thermodynamic properties; pressure and volume are. Pressure is force divided by area, and area times distance is change of volume. Therefore, $\mathscr{W} = \int_1^2 PA\,dx = \int_1^2 P\,dV$. Note that when dV is the change of volume of the thermodynamic system, a positive change in volume indicates work done by the system on the surroundings, which is positive work in our convention. $P\,dV$ work is known as **displacement work** ($\mathscr{W}_D$), which can be calculated if there is a known relationship between pressure and volume. The integration of $F\,dx$ or $P\,dV$ implies that the force or pressure, while perhaps varying, is continuously resisted by an equal and opposite force or pressure. That is to say, a **quasistatic process** is implied.

When a flowing fluid crosses a system boundary against a resisting pressure, **flow work** ($\mathscr{W}_F$) is done. In a quasistatic situation, the resisting pressure is equal to the pressure of the flowing fluid. The work done in forcing a volume V across the surface is the product PV. For a specific volume ($v = V/m$), the flow work per unit mass is Pv N-m/kg or ft-lbf/lb, in the SI or the English engineering system, respectively.

Work done by or on a system is frequently transmitted by a rotating shaft. The **shaft work** ($\mathscr{W}_{sh}$) done in rotation through one radian is equal to the torque (τ), which is the product of the tangential force times the radius at which it acts. Therefore, in one revolution at uniform torque, the shaft work is $2\pi\tau$. The **shaft power**, then, is $\tau\omega$ or $2\pi\tau(\text{rpm}/60)$, where ω is speed of rotation in radians per second and rpm is revolutions per minute. Because shaft power from an engine historically was measured by means of a "brake" arrangement, shaft power is often called **brake power**. The more general term, shaft power, will be used here, whether positive or negative.

Heat Transfer ($\mathscr{Q}$)

Energy crossing a system boundary because of a temperature difference is **heat transfer** ($\mathscr{Q}$). The basic units of this form of energy are the same as for work, and the units for the rate of heat transfer ($\dot{\mathscr{Q}}$) are the same as for power. Because heat transfer cannot be continuously converted entirely to work (see the second law of thermodynamics) whereas work can be entirely dissipated in heat transfer, work is referred to as a high-grade form of energy and heat transfer as low-grade energy.

Efficiency and Coefficient of Performance

Both of these terms can be defined as "the desired effect divided by the energy expended to achieve that effect." Efficiency and coefficient of performance values, therefore, are only meaningful in the context in which they are used.

For a power-producing cycle as shown schematically in Fig. 10.2, the desired effect is the net work delivered ($\sum \mathscr{W}$) and the energy expended is the net heat transfer ($\sum \mathscr{Q}_H$) to the system from the higher temperatures (T_H). From the first law of thermodynamics,

$$\sum \mathscr{W} = \sum \mathscr{Q}_H - \sum \mathscr{Q}_L$$

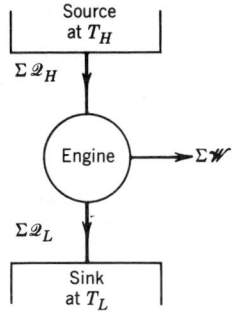

Fig. 10.2 Thermal efficiency of a power cycle:

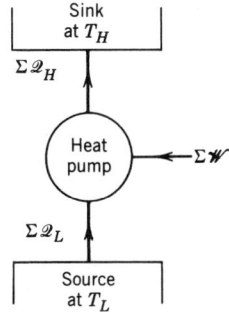

Fig. 10.3 Coefficient of performance of a vapor compression refrigeration or heat pump cycle.

Thus the thermal efficiency is

$$\eta_{th} = \frac{\Sigma \mathscr{W}}{\Sigma \mathscr{Q}_H} = \frac{\Sigma \mathscr{Q}_H - \Sigma \mathscr{Q}_L}{\Sigma \mathscr{Q}_H} = 1 - \frac{\Sigma \mathscr{Q}_L}{\Sigma \mathscr{Q}_H}$$

For a vapor compression refrigeration or heat pump cycle, as in Fig. 10.3, the energy supplied is shaft work. The desired effect for refrigeration is heat removal from a low-temperature source $\Sigma \mathscr{Q}_L$ at T_L. For a similar device operating as a heat pump, the desired effect is heat delivery to a sink, $\Sigma \mathscr{Q}_H$ at T_H. Again $\Sigma \mathscr{W} = \Sigma \mathscr{Q}_H - \Sigma \mathscr{Q}_L$. Thus, the coefficient of performance (COP) is follows:

As a refrigerator:

$$COP_R = \frac{\Sigma \mathscr{Q}_L}{\Sigma \mathscr{W}} = \frac{\Sigma \mathscr{Q}_L}{\Sigma \mathscr{Q}_H - \Sigma \mathscr{Q}_L}$$

As a heat pump:

$$COP_{HP} = \frac{\Sigma \mathscr{Q}_H}{\Sigma \mathscr{W}} = \frac{\Sigma \mathscr{Q}_H}{\Sigma \mathscr{Q}_H - \Sigma \mathscr{Q}_L}$$

In an adiabatic expansion process, the maximum work that could be delivered would be the work of isentropic expansion. In an adiabatic compression process, the minimum work required would be the work of isentropic compression (see Fig. 10.12). Thus, between two pressures P_H and P_L, the isentropic efficiency of expansion is

$$\eta_{E(s)} = \frac{\text{actual work}}{\text{isentropic work}} = \frac{\mathscr{W}_a}{\mathscr{W}_s}$$

For compression, the isentropic efficiency is

$$\eta_{C(s)} = \frac{\text{isentropic work}}{\text{actual work}} = \frac{\mathcal{W}_s}{\mathcal{W}_a}$$

Basic Laws of Thermodynamics

The laws of thermodynamics have been stated in many different ways, the formulation depending primarily on the intended application. It can always be shown, however, that one statement is equivalent to another in that a violation of any one would result in a violation of the others. The statements given here are the most useful for the material of Section 10.1.

Zeroth Law. The zeroth law precedes and is basic to the concept of a **temperature scale**. It states:

If two bodies are each in thermal equilibrium with (at the same temperature as) a third body, they are in thermal equilibrium with (at the same temperature as) each other.

First Law. The first law of thermodynamics is essentially a law of conservation of energy and assumes a conservation of mass. It was first recognized as a result of energy balances applied to **thermodynamic cycles**:

For a closed system executing a thermodynamic cycle, the cyclic integral of work done by the system is exactly equal to the cyclic integral of heat transfer to the system.

$$\oint \mathcal{Q} = \oint \mathcal{W}$$

The implication of this statement for a closed, stationary system following any process from state point 1 to state point 2 is the existence of a property called **internal energy** (U) such that

$$U_2 - U_1 = \int_1^2 \delta \mathcal{Q} - \int_1^2 \delta \mathcal{W}, \quad \text{and} \quad dU = \delta \mathcal{Q} - \delta \mathcal{W}$$

When the first law of thermodynamics, which embodies the concepts of conservation of mass and energy, is applied to an **open system** between time 1 and time 2, it follows that

From the beginning to the end of a particular time period, the difference between the sums of all the energy values associated with the masses within the control volume of a system must be accounted for by the net sum of all the energies associated with all the masses crossing the control surface, plus the net heat transfer to the system, minus the net sum of all the forms of work done by the system.

Mathematically, this may be stated as

$$\sum m_2 \left(u_2 + \frac{\mathcal{V}_2^2}{2g_c} + \frac{g}{g_c} z_2 \right) - \sum m_1 \left(u_1 + \frac{\mathcal{V}_1^2}{2g_c} + \frac{g}{g_c} z_1 \right)$$

$$= \left[\sum m_i \left(u_i + \frac{\mathcal{V}_i^2}{2g_c} + \frac{g}{g_c} z_i \right) + \sum m_e \left(u_e + \frac{\mathcal{V}_e^2}{2g_c} + \frac{g}{g_c} z_e \right) \right]$$

$$+ \sum \mathcal{Q} - \left[\sum \mathcal{W}_{sh} + \int_1^2 P \, dV + \sum m_e (P_e v_e) - \sum m_i (P_i v_i) \right]$$

where internal energy (u) kinetic energy $(\mathcal{V}^2/2g_c)$; potential energy $[(g/g_c)z]$; heat transfer $(\mathcal{Q})$; and shaft work, displacement work, and flow work are the energies of interest. For each identified mass entering the system (m_i) or leaving the system (m_e), the flow work term (Pv) may conveniently be grouped with the internal energy term (u) as **specific enthalpy** (h), which is defined as $h = u + Pv$.

For an open system in which there is a uniform state throughout at time 1 and again at time 2, and for which the boundaries are fixed ($dV = 0$), the first law may be simplified to

$$m_2 \left(u_2 + \frac{\mathscr{V}_2^2}{2g_c} + \frac{g}{g_c} z_2 \right) - m_1 \left(u_1 + \frac{\mathscr{V}_1^2}{2g_c} z_1 \right)$$

$$= \sum m_i \left(h_i + \frac{\mathscr{V}_i^2}{2g_c} + \frac{g}{g_c} z_i \right) - \sum m_e \left(h_e + \frac{\mathscr{V}_e^2}{2g_c} + \frac{g}{g_c} z_e \right)$$

$$+ \sum \mathcal{Q} - \sum \mathscr{W}_{sh}$$

When there is **steady state** within the control volume (no change in energy values) and **steady flow** ($m_i = m_e$) as well as **fixed boundaries** ($dV = 0$), this reduces to

$$\sum m_i \left(h_i + \frac{\mathscr{V}_i^2}{2g_c} + \frac{g}{g_c} z_i \right) + \sum \mathcal{Q} = \sum m_e \left(h_e + \frac{\mathscr{V}_e^2}{2g_c} + \frac{g}{g_c} z_e \right) + \sum \mathscr{W}_{sh}$$

Finally, when there is steady state, steady flow, fixed boundaries, and only one inlet and one exit so that $m_i = m_e = m$, then mass can be eliminated, and the first law becomes

$$\left(h_i + \frac{\mathscr{V}_i^2}{2g_c} + \frac{g}{g_c} z_i \right) + q = \left(h_e + \frac{\mathscr{V}_e^2}{2g_c} + \frac{g}{g_c} z_e \right) + \sum \omega_{sh}$$

where changes in kinetic and potential energies are negligible, this reduces to

$$(h_i - h_e) + q = \omega_{sh}$$

The sum $(h_i - h_e) + q$ in the preceding equations can be grouped when all the heat transfer (q) is actually delivered to the working fluid. The definition of enthalpy is $h = u + pv$. Thus

$$dh = du + p\,dv + v\,dp$$

From the first law applied to a unit mass for which all the work is displacement work,

$$\delta q = du + P\,dv$$

Integrating dh between i and e gives

$$h_e - h_i = \int_i^e (du + P\,dv) + \int_i^e v\,dP$$

$$= q + \int_i^e v\,dP$$

so that

$$(h_i - h_e) + q = -\int_i^e v\,dP$$

This may be substituted into the first law equations for an open system with steady state and steady flow.

Second Law. Three statements of the second law of thermodynamics will be useful in this section. It can be shown that they are in fact equivalent to one another because a violation of any one statement will result in a violation of the other two.

The Kelvin–Planck Statement. No system operating continuously in a cycle can receive heat transfer from a single source at a uniform temperature and convert all of that energy to useful work.

The Clausius Statement. No system can cause net heat transfer to flow from a source at some temperature to a sink at a higher temperature spontaneously, that is, without some other effect being required.

The Caratheodory Statement. In the vicinity of any equilibrium state point of a system, there are other state points that cannot be reached by a reversible adiabatic path.

Consequences or Corollaries of the Second Law. As a consequence of the second law of thermodynamics, a number of secondary principles must follow. In fact, some of these, discussed next, could serve as satisfactory statements of the second law from which the preceding statements would follow.

1. The existence of the property internal energy followed from the first law. The second law dictates that there is a property, entropy (S), defined as $dS = (\delta \mathcal{Q}/T)_{\text{rev}}$. The subscript rev indicates that this definition is restricted to a reversible process. Entropy is the only thermodynamic property defined with such a restriction.

2. The thermal efficiency of any working cycle is less than unity, $\eta_{\text{th}} < 1.0$.

3. The **Carnot cycle** is an ideal reversible cycle composed of two isothermal processes and two adiabatic processes. All the heat transfer $\mathcal{Q}_H$ is at the higher temperature T_H, and all the heat transfer $\mathcal{Q}_L$ is at the lower temperature T_L. The thermodynamic temperature scale (discussed later) is defined by the equality

$$\frac{T_L}{T_H} = \frac{\mathcal{Q}_L}{\mathcal{Q}_H}$$

where $\mathcal{Q}_H$ and $\mathcal{Q}_L$ are the two heat transfer quantities in a Carnot cycle.

4. The **maximum possible thermal efficiency** of a working cycle between a source at T_H and a sink at T_L is the efficiency of a reversible cycle receiving heat only at T_H and rejecting heat only at T_L. The Carnot cycle is an example of such a reversible cycle. Its thermal efficiency is $\eta_{\text{th}} = 1 - (T_L/T_H)$.

5. All reversible power cycles receiving heat at T_H and rejecting heat at T_L will have the same thermal efficiency. Other such cycles are the ideal Stirling cycle with perfect regeneration and the ideal Ericsson cycle with perfect regeneration.

6. For a refrigeration system or heat pump, the **maximum possible coefficient of performance** (COP) is achieved by a reversible cycle such as the Carnot cycle, receiving heat only at T_L and rejecting heat only at T_H.

$$\text{Maximum COP}_R = \frac{T_L}{T_H - T_L}, \qquad \text{maximum COP}_{\text{HP}} = \frac{T_H}{T_H - T_L}$$

7. The **inequality of Clausius** states that for any cycle, $\oint(\delta \mathcal{Q}/T) \leq 0$, and consequently, for any process, $dS \geq \delta \mathcal{Q}/T$. The equalities hold only for a completely reversible cycle or process.

8. The **entropy principle**, or the principle of increase of entropy, states that, for a system plus its surroundings, $\Sigma \, dS \geq 0$. Again, the equalities hold only for reversible situations.

Third Law. As with the second law of thermodynamics, there are several statements of the third law. All of these are essentially equivalent:

1. The entropy change associated with any isothermal reversible process of a condensed system approaches zero as the temperature approaches zero.
2. It is impossible to reduce the entropy of a system to its zero-point value by any finite number of operations.
3. It is impossible to reduce the temperature of a system to absolute zero in any finite number of operations.

This last statement can be called the "principle of unattainability of absolute-zero temperature."

The Thermodynamic Temperature Scale

As already noted, the thermodynamic temperature scale is defined in terms of the two heat transfer quantities in a Carnot cycle such that

$$\frac{T_L}{T_H} = \frac{\mathcal{Q}_L}{\mathcal{Q}_H}$$

Only one point need be established on this scale, and that is the triple point of pure water, defined as 273.16 K. The size of one Kelvin degree is the same as the size of one Celsius degree. The ice point of water is 0.01° below the triple point of water. Thus the ice point, previously defined as zero on the

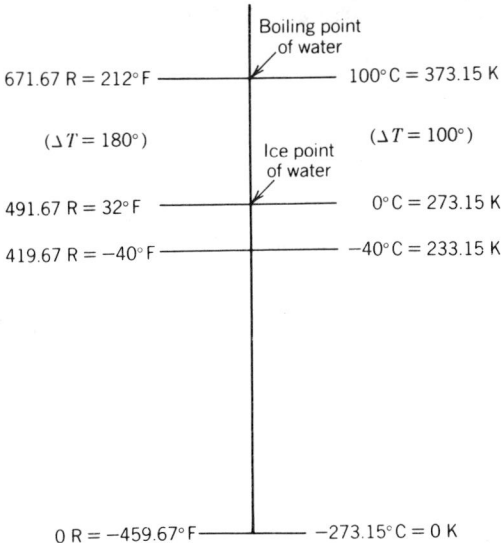

Fig. 10.4 Exact equivalents on the Kelvin, Celsius, Rankine, and Fahrenheit temperature scales.

Celsius scale, is at 273.15 K. The four temperature scales, Kelvin and Celsius, Rankine, and Fahrenheit, are compared with exact equivalents shown in Fig. 10.4. As one Celsius degree is equivalent to 1.8 Fahrenheit degree, so one Kelvin degree is equivalent to 1.8 Rankine degree, exactly.

The fixed point on the Kelvin scale (273.16 K) was established to make the thermodynamic temperature scale compatible with the familiar ideal-gas temperature scale. That scale is defined as

$$\theta = 273.16 \lim_{P_3 \to 0} \frac{P}{P_3} \quad \text{(at constant volume)}$$

where P is the gas pressure at temperature θ and P_3 is the pressure of the same gas in the same volume but at the triple-point temperature of water, and P_3 is very low. For an ideal gas (see Section 10.1.3), the pressure ratio P/P_3 would be zero at temperature θ equal to absolute zero.

10.1.2 Thermodynamic Properties of a Pure Substance

Introduction
A pure substance is one of constant uniform composition. For a given mass of such a substance, the values of any two **independent** properties will establish a state point. Thus all other properties are **dependent**, that is, their property values are given by (are a function of) these two. Consequently, for a pure substance, changes in property values along any path and functional relationships among the various thermodynamic properties can be evaluated by the application of the normal rules of mathematics.

Changes in property values of a given mass of a pure substance are governed by the first law of thermodynamics applied to this "closed system," in which the only work done is displacement work. In a quasistatic process, which implies that the pressure of the system is balanced by an equal and opposite external pressure, the displacement work is the integral of $P\,dV$. Also, when there is heat transfer to this mass, the property changes are internally reversible, so that $\delta \mathcal{Q} = T\,dS$. As a result, work and heat transfer, which are path functions, are reflected in changes in property values. From these basic facts, the relationships among the various thermodynamic properties are developed based on their respective definitions.

Internal Energy
From the first law,

$$dU = \delta \mathcal{Q} - \delta \mathcal{W}$$

Thus,

$$dU = T\,dS - P\,dV$$

From this equation and the rules of mathematics, it follows that

$$T = \left(\frac{\partial U}{\partial S}\right)_V \qquad P = \left(\frac{\partial U}{\partial V}\right)_S \qquad \left(\frac{\partial T}{\partial V}\right)_S = -\left(\frac{\partial P}{\partial S}\right)_V \qquad \text{(Maxwell's Eq. I)}$$

Enthalpy

By definition, $H = U + PV$. Thus,

$$dH = dU + P\,dV + V\,dP = T\,dS + V\,dP$$

It follows that

$$T = \left(\frac{\partial H}{\partial S}\right)_P \qquad V = \left(\frac{\partial H}{\partial P}\right)_S \qquad \left(\frac{\partial T}{\partial P}\right)_S = \left(\frac{\partial V}{\partial S}\right)_P \qquad \text{(Maxwell's Eq. II)}$$

Helmholtz Function or Helmholtz "Free Energy"

By definition, $F = U - TS$. Thus,

$$dF = dU - T\,dS - S\,dT = -P\,dV - S\,dT$$

It follows that

$$P = -\left(\frac{\partial F}{\partial V}\right)_T \qquad S = -\left(\frac{\partial F}{\partial T}\right)_V \qquad \left(\frac{\partial S}{\partial V}\right)_T = \left(\frac{\partial P}{\partial T}\right)_V \qquad \text{(Maxwell's Eq. III)}$$

Gibbs Function or Gibbs "Free Energy"

By definition, $G = H - TS$. Thus,

$$dG = dH - T\,dS - S\,dT = V\,dP - S\,dT$$

It follows that

$$V = \left(\frac{\partial G}{\partial P}\right)_T \qquad S = -\left(\frac{\partial G}{\partial T}\right)_P \qquad \left(\frac{\partial V}{\partial T}\right)_P = -\left(\frac{\partial S}{\partial P}\right)_T \qquad \text{(Maxwell's Eq. IV)}$$

Specific Heats

The specific heat of a substance is a property of that substance, and its value is a function of other property values. In addition, a process is implied, so that the value of specific heat must depend on the implied process. Two particular specific heats are of interest.

Specific Heat at Constant Volume. By definition

$$C_V = \left(\frac{\partial u}{\partial T}\right)_V$$

From the first law applied to a unit mass of a given substance

$$du = \delta q - P\,dv = T\,ds - P\,dv$$

Thus at constant volume ($P\,dv = 0$), and in the absence of any other work term,

$$du = T\,ds \qquad C_V = \left(\frac{\partial q}{\partial T}\right)_V$$

This is the historic concept of specific heat as related to heat transfer.

Specific Heat at Constant Pressure. By definition

$$C_P = \left(\frac{\partial h}{\partial T}\right)_P$$

From the definition of enthalpy

$$dh = du + P\,dv + v\,dP = \delta q + v\,dP = T\,ds + v\,dP$$

Again it is seen that the historic concept of specific heat, as related only to heat transfer, is compatible with the definition of C_P in the absence of work other than $P\,dv$. When $dP = 0$, then $dh = T\,ds$, and $C_P = (\partial q/\partial T)_P$.

Volume Expansivity (β)
Volume expansivity is the rate of change of volume with temperature at constant pressure:

$$\beta = \frac{1}{V}\left(\frac{\partial V}{\partial T}\right)_P$$

Compressibility (κ)
Compressibility is the rate of change of volume with pressure in a compression or expansion process.

Isothermal Compressibility: $\qquad \kappa_T = -\frac{1}{V}\left(\frac{\partial V}{\partial P}\right)_T$

Isentropic Compressibility: $\qquad \kappa_S = -\frac{1}{V}\left(\frac{\partial V}{\partial P}\right)_S$

T ds Equations
Entropy (S), being a property, can be considered as a function of any two others, for example, temperature (T), volume (V), or pressure (P). When this is stated mathematically and the appropriate Maxwell equation and definition of specific heat are substitute for the partial derivatives, three $T\,ds$ equations result:

$$T\,ds = C_v\,dT + T\left(\frac{\partial P}{\partial T}\right)_V dv$$

$$= C_P\,dT - T\left(\frac{\partial v}{\partial T}\right)_P dP$$

$$= C_P\left(\frac{\partial T}{\partial v}\right)_P dv + C_V\left(\frac{\partial T}{\partial P}\right)_V dP$$

Energy Equations
Similarly, from the first-law property relationship with appropriate substitutions from the Maxwell equations, two energy equations are derived:

$$\left(\frac{\partial U}{\partial V}\right)_T = T\left(\frac{\partial P}{\partial V}\right)_V - P$$

$$\left(\frac{\partial U}{\partial P}\right)_T = -T\left(\frac{\partial V}{\partial P}\right)_P - P\left(\frac{\partial V}{\partial P}\right)_T$$

Specific Heat Relationships
When the first and second energy equations are combined mathematically, and the definitions of volume expansivity and compressibility are inserted, two important specific-heat relationships result:

$$C_P - C_V = -T\left(\frac{\partial V}{\partial T}\right)_P^2\left(\frac{\partial P}{\partial V}\right)_T = \frac{TV\beta^2}{\kappa_T}$$

$$\gamma = \frac{C_P}{C_V} = \frac{\kappa_T}{\kappa_S}$$

The Thermodynamic Surface
Because the value of any one property of a pure substance is dependent on the values of any two other independent properties, the thermodynamic conditions at which a unit mass (1 kg or 1 lb) of a particular pure substance can exist are strictly limited. For instance, the pressure value is a function of the temperature and the volume values. Thus, on pressure-volume-temperature coordinates, the possible state points for a unit mass form a **thermodynamic surface** as shown in Fig. 10.5. The shape of the surface is for a substance that expands on melting. (Water, because of its unusual crystal structure as a solid, is one of a very few naturally occurring simple substances that expand when solidified). The coordinates are logarithmic in Fig. 10.5 because of the wide range of values between significant points for most substances.

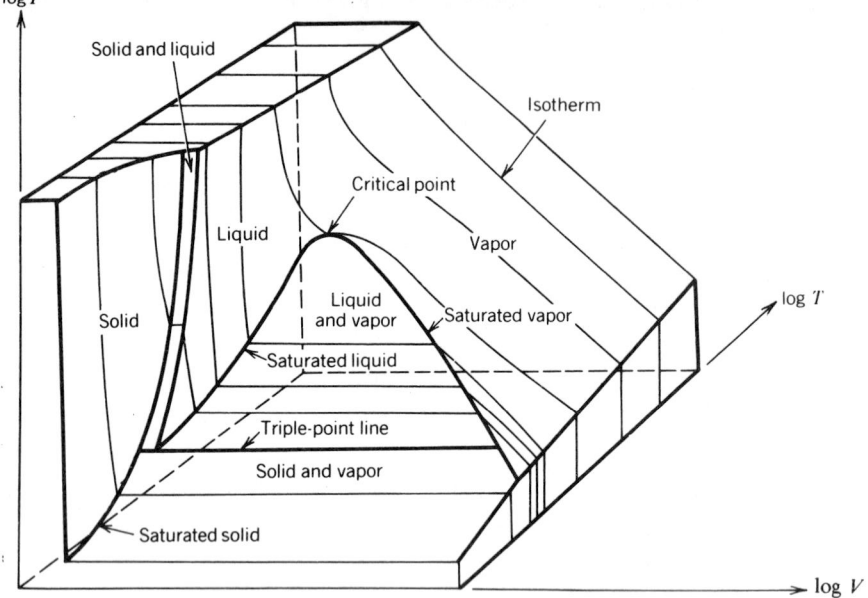

Fig. 10.5 The thermodynamic surface for an ordinary substance.

The thermodynamic conditions at which two or more phases (solid, liquid, and vapor) can exist in equilibrium are **saturation conditions**. The pressure at which a phase change occurs establishes the temperature (or vice versa). Thus, saturation conditions form lines rather than areas on pressure-temperature coordinates. Because a phase change generally implies a volume change, saturation conditions do form surfaces on pressure-volume or temperature-volume coordinates.

A pressure and temperature at which three phases can exist in equilibrium is a **triple point**, which is a line on P-v or T-v coordinates. The substance can exist as a liquid and a vapor in equilibrium only in the region below the **critical-point** values of pressure and temperature. Above the critical point, the two phases cannot be distinguished.

When thermodynamic property values at saturation conditions are tabulated, it is conventional to use the subscript f to indicate saturated liquid conditions, the subscript g to indicate saturated vapor conditions, and the subscript fg to indicate the change from liquid to vapor. For instance, $v_f + v_{fg} = v_g$. When liquid and vapor are in equilibrium, the mass fraction in the vapor phase is the **dryness fraction**, sometimes called **quality**, conventionally designated as x. Thus, the specific volume (for instance) of the mixture would be

$$v_{\text{mix}} = (1 - x)v_f + xv_g = v_f + xv_{fg}$$

Similarly,

$$h_{\text{mix}} = h_f + xh_{fg}$$
$$s_{\text{mix}} = s_f + xs_{fg} \ldots$$

The line identified *saturated vapor* in Fig. 10.5 shows the P-v-T conditions at which dry, saturated vapor can exist. Above and to the right of the liquid-vapor region, the vapor is **superheated**. That is, the vapor is at a temperature higher than the saturation temperature corresponding to its pressure.

10.1.3 The Ideal Gas

Definition

An **ideal gas** is defined simply by its equation of state, which may be written as

$$Pv = \mathcal{R}T = \frac{\bar{\mathcal{R}}}{\mathcal{M}}T \quad \text{or} \quad PV = m\mathcal{R}T = n\bar{\mathcal{R}}T$$

where P = absolute pressure, N/m² or lbf/ft²

v = specific volume, m³/kg or ft³/lb

T = absolute temperature, Kelvin (K) or Rankine (R) degrees

$\mathcal{M}$ = molecular weight of gas

m = mass, kg or lb

n = number of moles = $m/\mathcal{M}$

$\mathcal{R}$ = individual gas constant, J/kg-K or ft-lbf/lb-R

$\bar{\mathcal{R}}$ = universal gas constant = $\mathcal{M}\mathcal{R}$

= 8.31434×10^3 J/kg-mol-K

= 1545 ft-lbf/lb-mol-R

= 1.9859 Btu (IST)/lb-mol-R

Note that the numerical value of the molecular weight ($\mathcal{M}$) is the same in all systems of units. For instance, the molecular "weight" of oxygen (O_2) is 31.999 whether expressed as kilograms per kilogram-mole (kg-mol), grams per gram-mole (g-mol), or pounds per pound-mole (lb-mol). Thus, the specific volume of oxygen at a pressure of 1 atm (1.01325×10^5 Pa or 2116.22 lb/ft²) and 15°C (59°F) is

$$v = \frac{(8.3143 \times 10^3)(273.15 + 15)}{(31.999)(1.01325 \times 10^5)} = 0.7389 \text{ m}^3/\text{kg}$$

$$= \frac{(1545)(459.67 + 59)}{(31.999)(2116.22)} = 11.834 \text{ ft}^3/\text{lb}$$

Property Relationships for an Ideal Gas

By a combination of the ideal-gas equation of state ($Pv = \mathcal{R}T$) and the appropriate energy and specific-heat relationships already developed, it can be shown that the following statements can be made *for an ideal gas*:

1. Internal energy (u), enthalpy (h), specific heat at constant pressure (C_p), and specific heat at constant volume (C_v) are all functions of temperature only. That is, these terms are not affected by pressure or specific volume.

2. The difference between the two specific heats is the gas constant $\mathcal{R}$. Thus

$$C_p - C_v = \mathcal{R}$$

$$\gamma = \frac{C_p}{C_v} = 1 + \frac{\mathcal{R}}{C_v}$$

$$C_v = \frac{\mathcal{R}}{\gamma - 1} \quad \text{and} \quad C_p = \frac{\gamma}{\gamma - 1}\mathcal{R}$$

3. For a polytropic process in which PV^n = a constant,

$$\frac{T_1}{T_2} = \left(\frac{v_2}{v_1}\right)^{n-1} = \left(\frac{P_1}{P_2}\right)^{(n-1)/n}$$

Here, n has any value appropriate to the process. For a reversible adiabatic process, which is an **isentropic process**, the value of n becomes γ. The value of γ is the ratio C_p/C_v, where both are constant, or it is the logarithmic mean value of C_p/C_v between T_1 and T_2.

4. The change of entropy for an ideal gas is given by either of the following equations:

$$s_2 - s_1 = \int_1^2 C_v \frac{dT}{T} + \mathcal{R}\int_1^2 \frac{dV}{V}$$

$$= \int_1^2 C_p \frac{dT}{T} - \mathcal{R}\int_1^2 \frac{dP}{P}$$

In these two equations the first term on the right-hand side is a function of temperature only, and the last term is easily calculated as $\mathcal{R} \ln(V_2/V_1)$ or $\mathcal{R} \ln(P_2/P_1)$, respectively. Clearly, then, at constant volume or constant pressure the change of entropy depends only on the temperature limits and the relationship between temperature and either C_v or C_p, respectively, for a particular gas.

5. Insofar as real gases follow the ideal-gas relationship with reasonable accuracy, the temperature-dependent terms can be tabulated against temperature only, which is done in "gas tables." The terms conventionally tabulated are

T = absolute temperature

$h = \int_{T_0}^{T} C_p \, dT$

P_r = relative pressure = P/P_0 in an isentropic process

$u = \int_{T_0}^{T} C_v \, dT$

v_r = relative volume = v/v_0 in an isentropic process

ϕ = constant-pressure entropy function = $\int_{T_0}^{T} C_p(dT/T)$

The Compressibility Factor (Z)

An indication of the deviation of **real-gas** relationships from the simple ideal-gas relationship is given by the **compressibility factor** (Z) defined as

$$Z = \frac{Pv}{\mathscr{R}T}$$

A value of unity, indicating ideal gas, is approached for all real gases as the pressure approaches zero.

It has been found that the value of the compressibility factor depends on the value of the reduced pressure (P_R) and the value of the reduced temperature (T_R) where

$$P_R = \frac{P}{P_c} \quad \text{and} \quad T_R = \frac{T}{T_c}$$

P_c and T_c being the critical-point values of pressure and temperature for the gas considered. In other words,

$$Z = \frac{Pv}{\mathscr{R}T} = f(P_R, T_R)$$

and that functional relationship is very similar for most substances over a wide range of values for P_R and T_R. This is illustrated in various references by means of a "compressibility chart" that is useful when tabulated data are not available.

10.1.4 Nonreactive Mixtures and Solutions

Mixtures of Ideal Gases

Just as it is an idealization to assume that actual gases follow the equation of state of the ideal gas ($PV = n\mathscr{\bar{R}}T$), some error is introduced in the assumption that each gas in a mixture of two or more gases will act as if it alone filled the space at some "partial pressure" or that it occupied some "partial volume" subjected to the "total pressure." Nevertheless, if all components do approximate the ideal-gas model, these assumptions will provide a good approximation for the mixture. Two models are commonly employed.

The **Dalton rule** of partial pressures assumes that each individual gas (i) in the mixture occupies the total volume (V) at a pressure P_i consistent with the number of moles (n_i) of that gas present at the uniform temperature (T), so that

$$P_i V = n_i \mathscr{\bar{R}} T$$

It assumes that the total pressure is the sum of all the individual pressures, that the total number of moles is the sum of the individual number of moles, and that the partial pressures for each component are proportional to the number of moles of each in the mixture:

$$P = \sum P_i \qquad n = \sum n_i \qquad \frac{P_i}{P} = \frac{n_i}{\sum n_i}$$

It follows that, in the Dalton model,

$$\frac{PV}{\mathscr{\bar{R}}T} = \sum \frac{P_i V}{\mathscr{\bar{R}}T} = \frac{V}{\mathscr{\bar{R}}T} \sum P_i$$

The **Amagat rule** of partial volumes, by contrast, assumes that each component gas can be treated as if it occupies an individual volume proportional to the number of moles (n_i) of that gas and subjected to the total pressure (P) at the uniform temperature (T), so that

$$PV_i = n_i \bar{\mathcal{R}} T$$

and that

$$V = \sum V_i \qquad n = \sum n_i \qquad \frac{V_i}{V} = \frac{n_i}{\sum n_i}$$

It follows that, in the Amagat model,

$$\frac{PV}{\mathcal{R}T} = \sum \frac{PV_i}{\mathcal{R}T} = \frac{P}{\mathcal{R}T} \sum V_i$$

The two idealizations, Dalton and Amagat, are equally useful, one or the other being more convenient in particular situations.

Calculations of Mass Fractions in a Mixture of Gases

The reliability of the assumptions described for the Dalton and Amagat rules is illustrated by the fact that their application in a conversion from the generally accepted molecular fractions of the components of "dry air" to mass fractions provides values good to approximately five significant figures. The resulting values are approved, for instance, by the American Society of Heating, Refrigerating, and Air Conditioning Engineers. Table 10.1 explains the procedure in which the partial "molecular weights" are calculated as the product of the individual molecular weights multiplied by the molecular fraction in the definition of "dry air."

Note that for any gas, the mass is directly proportional to the mass of its individual molecules because, according to **Avogadro's rule**, a kilogram mole of any gas has the same number of molecules (6.022169×10^{26}) as any other gas.

The **Gibbs–Dalton** formulations for internal energy, enthalpy, and entropy per mole of a mixture of gases are simply the summation of values for each component at the common temperature, taken in proportion to the molecular fraction of each in the whole:

$$\bar{u} = \frac{\sum n_i \bar{u}_i}{\sum n_i} \qquad \bar{h} = \frac{\sum n_i \bar{h}_i}{\sum n_i} \qquad \bar{s} = \frac{\sum n_i \bar{s}_i}{\sum n_i}$$

Whether the Dalton or the Amagat model is used to calculate internal energy values or enthalpy values does not matter because these properties are not dependent on pressure or volume for an ideal gas. Entropy, however, does depend on pressure, and the individual entropy values must be calculated at the partial pressure of each, the Dalton model.

When gases are mixed so that each assumes only its partial pressure, which is less than its original pressure, there is an increase in entropy for each. Recall that for an ideal gas changing from state point 1 to state point 2, the entropy change is

$$\bar{s}_2 - \bar{s}_1 = \int_1^2 \bar{C}_P \frac{dT}{T} - \bar{\mathcal{R}} \int_1^2 \frac{dP}{P}$$

TABLE 10.1 Dry Air Components

Substance	Molecular Weight Based on C = 12 (Definition)	Molecular Fraction in Dry Air ‘ (Definition)	Partial Molecular Weight in Dry Air (Calculated)	Mass Fraction in Dry Air (Calculated)
Nitrogen (N_2)	28.0134	0.78084	21.87398	0.75520
Oxygen (O_2)	31.9988	0.209476	6.70298	0.23142
Argon (A)	39.948	0.00934	0.37311	0.01288
Carbon Dioxide (CO_2)[a]	44.00995	0.000314	0.01382	0.00048
Other[a]	—	0.00003	0.00061	0.00002
Dry air		1.0	28.9645	1.0

[a] Variable.

Thus, at constant temperature,

$$\Delta \bar{s}_i = +\bar{\mathscr{R}} \ln \frac{P}{P_i} = +\bar{\mathscr{R}} \ln \frac{\sum n_i}{n_i}$$

This "entropy of mixing" is always positive.

Mixture of a Gas and a Vapor (Psychrometrics)

When one of the components in a mixture is close to or at the saturation temperature for its partial pressure, it must be considered as a vapor, and the terms *humidity ratio*, *degree of saturation*, *relative humidity*, and *dew point* have relevance. When one of the components condenses, the mass ratio of the gas-vapor mixture changes. The most common example of a mixture of a gas and a vapor is "moist air" where the components of the dry air can be assumed constant and acting as a single gas, and the mass of water vapor present per unit volume cannot be larger than the density for saturation conditions at the existing temperature. The term *saturation* has exactly the same significance here as in the discussion of thermodynamic properties of a pure substance. The thermodynamic properties of moist air are called **psychrometric properties**. The **humidity ratio** (W) of a mixture of a gas and a vapor is simply the mass ratio.

$$W = \frac{\text{mass of vapor}}{\text{mass of gas}}$$

In psychrometrics

$$W = \frac{\text{mass of water vapor}}{\text{mass of dry air}}$$

Degree of saturation (μ) is the ratio of the existing humidity ratio (W) to the humidity ratio that would exist at saturation (W_s) for the same temperature (T) and total pressure (P). Thus

$$\mu = \frac{W}{W_s}(T, P)$$

Relative humidity (RH) is a term used most frequently in connection with the properties of moist air. It is applicable, however, to any mixture of gas and vapor, or in fact to any vapor alone. Relative humidity is the existing mole fraction of vapor in a given volume of the mixture divided by the mole fraction of vapor for saturation conditions of the vapor at the same temperature and total pressure. This is the same as the ratio of the moles or mass of vapor to the moles or mass of vapor for saturation, per unit volume, at the same temperature:

$$\text{RH} = \frac{n_v / \sum n}{n_{vs} / \sum n}(T, P)$$

$$= \frac{n_v}{n_{vs}}(T, V) = \frac{m_v}{m_{vs}}(T, V)$$

When the vapor is at a low value of partial pressure so that it acts as an ideal gas ($PV = n\bar{\mathscr{R}}T$),

$$\text{RH} \simeq \frac{P_v}{P_{vs}}$$

The relationship between relative humidity and degree of saturation, when ideal-gas conditions exist, is

$$\text{RH} \simeq \frac{\mu}{1 - (1 - \mu)(P_{vs}/P)}$$

The **dew-point temperature** of a mixture of gas and vapor is the saturation temperature of the vapor component corresponding to its partial pressure in the mixture.

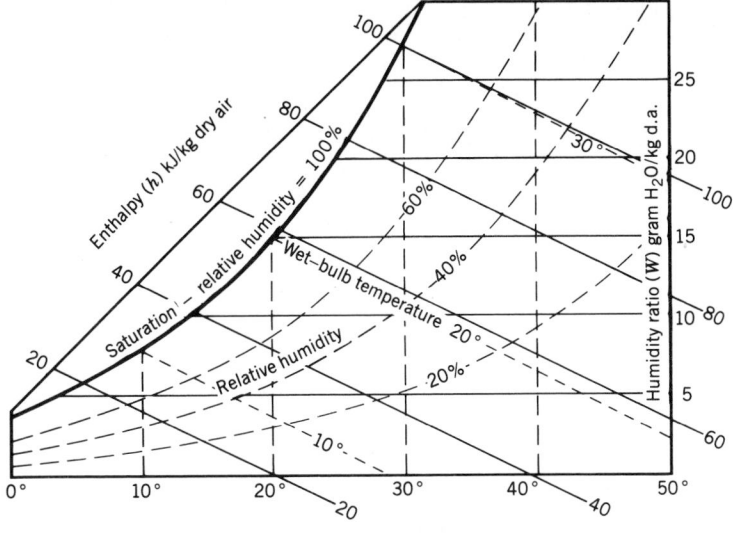

Fig. 10.6 Outline of psychometric chart based on ASHRAE Chart No. 1 (normal temperature range, sea level).

The Psychrometric Chart

The psychrometric chart is explained briefly here as an example of the graphical presentation of the thermodynamic properties of a mixture of nonreactive, simple substances, in this case a gas and a vapor.

The components of "dry air", defined in Table 10.1, can be assumed to act as a single gas because the component fractions do not change over the wide range of atmospheric or air-conditioning temperatures and pressures. The vapor component, which is water, does change, however. Both the dry air and the water vapor act very nearly as ideal gases because the individual vapor pressures are all very low compared to their respective critical pressures. A properly constructed chart takes into account the slight but significant variations in properties from the simple, ideal-gas relationships. The variations and corrections are detailed in the American Society of Heating, Refrigerating, and Air Conditioning Engineers (ASHRAE) *Handbook of Fundamentals* and the references are cited there.

Figure 10.6 is an outline of the psychrometric chart. For a given total pressure, two independent property values establish a state point; all the other properties are dependent. All the properties shown have been defined except **wet-bulb temperature**, more properly called the **temperature of adiabatic saturation**. It is easily shown that degree of saturation (μ) and relative humidity (RH) are a function of the actual temperature of the mixture, called here the **dry-bulb temperature**, and the temperature of adiabatic saturation.

Adiabatic Combination or Separation of Two-Component Mixtures

Figure 10.7 shows two streams (1 and 2) being brought together with no heat transfer, and it shows the combination of the two leaving as a single stream (3). Each stream is a mixture of two substances, A and B, with different mixture ratios (x) and different enthalpy values (h).

The mixture ratio may be defined as the mass ratio (m_A/m_B) or as the mass fraction ($m_A/\Sigma m$). Enthalpy may be given as enthalpy per unit mass of A (kJ/kg of A) or per unit mass of the mixture (kJ/kg A plus B). In either situation, an energy and mass balance for the system, assuming either steady flow or batch mixing, will yield a straight-line relationship for points 1, 2, and 3 on a plot of enthalpy versus mixture strength. Furthermore, the position of point 3 will be such that

$$\frac{x_2 - x_3}{x_3 - x_1} = \frac{h_2 - h_3}{h_3 - h_1} = \frac{m_1}{m_2}$$

The same equation would apply if the arrows on the mixing diagram were reversed, implying a separation of stream 3 into two streams, 1 and 2.

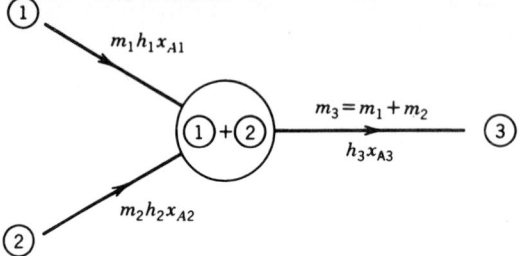

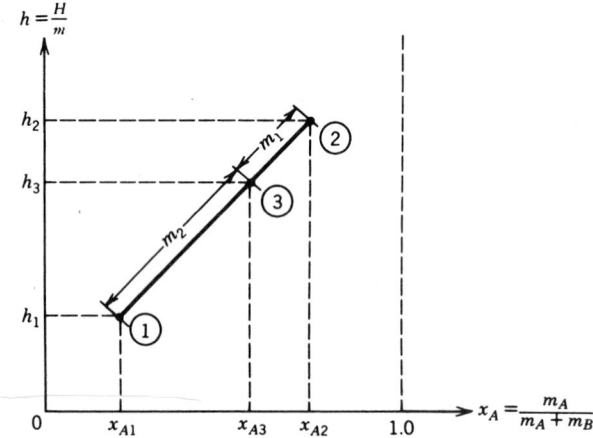

Fig. 10.7   Adiabatic combination of two different two-component streams.

The psychrometric chart is an example of an enthalpy-mixture strength $(h - W)$ diagram. It is true that when two streams of moist air with different state points (1 and 2) are mixed adiabatically in an air-conditioning system, the state point of the emerging stream is to be found on a straight line between point 1 and point 2 in a position dictated by the mass rates of the two streams. See Fig. 10.8.

When the mixing process is not adiabatic ($\mathcal{Q} \neq 0$), the end point on Fig. 10.7 or 10.8 can be found by assuming the heat transfer to occur after the mixing. Then point 4 is found on a constant x or constant W line above or below, to the right or the left, of point 3, such that

$$h_4 - h_3 = \frac{\mathcal{Q}}{\sum m}$$

Evaporation and Condensation of a Two-Component Mixture

A more complicated situation obtains when two substances (A and B) that have different evaporation temperatures for the prevailing pressure are mixed and there is a change of phase from liquid to vapor or vapor to liquid. This is illustrated on temperature-mixture strength and on enthalpy-mixture strength coordinates in Fig. 10.9. In the illustration, A and B might be ammonia (NH_3) and water (H_2O) or nitrogen (N_2) and oxygen (O_2). The diagrams are not to scale, but the shapes are consistent with the respective properties: At a given pressure below the critical points of these substances, the saturation temperature of ammonia is lower than that of water, and its latent heat of vaporization is less than that of water; the same is true for nitrogen compared to oxygen.

These two mixtures NH_3/H_2O and N_2/O_2 are chosen for illustration because the former is a common mixture used in absorption refrigeration systems and the latter are the main components of air. In both examples, changes in the pressure and temperature are employed to separate one component from another.

The saturation lines in Fig. 10.9(a) show that the vapor at any temperature has a higher concentration of A than does the liquid with which it is in equilibrium. Therefore, when a subcooled

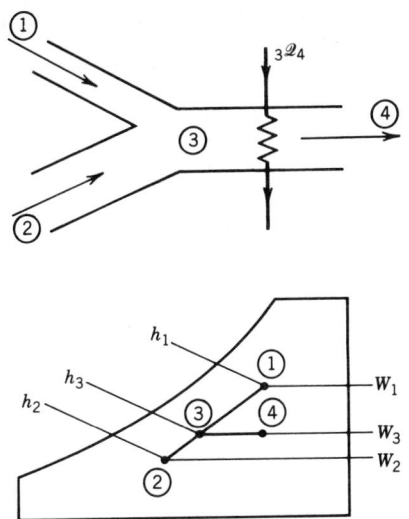

Fig. 10.8 Mixing of two streams of moist air with subsequent positive heat transfer. $m_2 = 2m_1$.

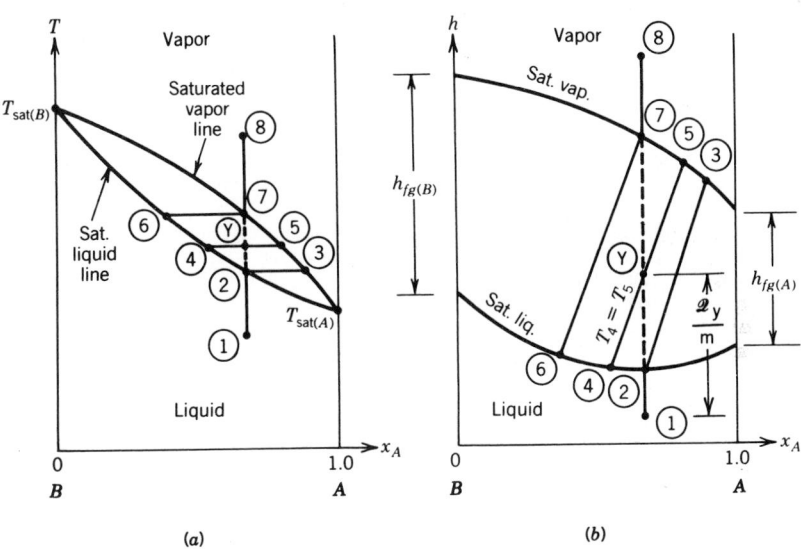

Fig. 10.9 Constant-pressure evaporation of a two-component mixture.

liquid at concentration x_1 is heated to saturation, component A vaporizes preferentially. As vaporization proceeds, the liquid concentration moves toward point 4 while the vapor concentration moves toward 5. On both Figs. 10.9(a) and 10.9(b), the line 4-5 is an isotherm, reached at point Y when energy equal to $\mathcal{Q}_y$ has been added. As was shown for mixtures in general, the ratio mass of vapor to mass of liquid (m_5/m_4) is indicated by the division of the line on h-x coordinates.

$$\frac{m_5}{m_4} = \frac{h_y - h_4}{h_5 - h_y} = \frac{x_y - x_4}{x_5 - x_y}$$

Finally, if no vapor has left the system, the last liquid to vaporize produces vapor at point 7 where concentration is the same as it was at point 1. Further heating produces superheated vapor at point 8.

Condensation of vapor from point 8 would simply reverse the whole procedure assuming that no liquid had been drained off during the process.

At some mixture strength between 0 and 1.0, the latent heat of vaporization (h_{fg}) is shown to be larger than the weighted average between the values for A and B at the prevailing pressure. This is explained by the fact, for the two examples chosen, that there is a tendency for A to dissolve in B spontaneously so that energy is released as the two combine. This additional "heat of solution" must be supplied to separate the two again. This is an example of the generalization that a spontaneous reaction moves toward a lower energy level.

Various substances in solution exhibit **azeotropic concentration** values, over certain temperature and pressure ranges, at which the solution will change phase as a single substance, maintaining the same concentration value through the evaporation or condensation process. The azeotropic temperature may be higher or lower (and the corresponding pressure lower or higher) than the saturation value for either component separately. The azeotropic mixture strength (x value) is a function of temperature and pressure. At sufficiently low pressures the azeotropic point disappears for all mixtures.

Partial Molal Properties

The value of any extensive property varies directly with the mass or the number of moles of the substance being considered. In a mixture of different substances the value of the extensive property is the sum of the products of the partial molal value of each times the number of moles of each. For instance, for a mixture of several substances the total volume at a particular temperature and pressure is the sum of the volumes of the components in the mixture:

$$V = \sum n_i \left(\frac{\partial v}{\partial n_i} \right)_{P, T, n_j}$$

where $\partial V / \partial n_i$ is the **partial molal volume** of component i at the prescribed temperature (T) and pressure (P) when all other components (j) are at prescribed values in the mixture.

For ideal gases, in accordance with the Amagat model, the partial molal volume is the same as the specific volume (mole basis) for each component. For nonideal substances the volume of the mixture is typically less than the sum of the volumes of the components separately at the same temperature and pressure.

A solution may be in the solid, liquid, or vapor phase. An **ideal solution** is one in which the total volume is the same as the sum of the volumes of the separate parts at the same temperature and pressure. An ideal mixture of ideal gases is a special case of the ideal solution.

Other extensive thermodynamic properties such as internal energy (U), enthalpy (H), Gibbs function (G), and Helmholtz function (F), when related to a mixture or solution, must be expressible in partial molal form, as was illustrated previously for volume (V).

Mixtures with Variable Composition

When a small amount of one or more components in a mixture is added or removed, each extensive property value is changed in proportion to the number of moles added or removed, and in proportion to the partial molal values of that property for those components. Consider the internal energy (U) of a system of fixed mass:

$$dU = T \, dS - P \, dV$$

When a relatively small amount of any one component is added,

$$dU = T \, dS - P \, dV + \left[\left(\frac{\partial U}{\partial n_i} \right)_{S, V, n_j} \right] dn_i$$

If there is a change in the number of moles of several components,

$$dU = T \, dS - P \, dV + \sum \left[\left(\frac{\partial U}{\partial n_i} \right)_{S, V, n_j} \right] dn_i$$

Similarly, the changes in the other energy terms are

$$dH = T\,dS + V\,dP + \sum \left(\frac{\partial H}{\partial n_i}\right)_{S,P,n_j} dn_i$$

$$dG = -S\,dT + V\,dP + \sum \left(\frac{\partial G}{\partial n_i}\right)_{T,P,n_j} dn_i$$

$$dF = -S\,dT - P\,dV + \sum \left(\frac{\partial F}{\partial n_i}\right)_{T,V,n_j} dn_i$$

It can be shown that each of the partial energy properties in the preceding equations has the same value. Each is given the name **chemical potential** (μ).

$$\mu_i = \left(\frac{\partial U}{\partial n_i}\right)_{S,V,n_j} = \left(\frac{\partial H}{\partial n_i}\right)_{S,P,n_j}$$

$$= \left(\frac{\partial G}{\partial n_i}\right)_{T,P,n_j} = \left(\frac{\partial F}{\partial n_i}\right)_{T,V,n_j}$$

The chemical potential (μ), defined here for a nonreactive mixture, becomes very significant as a "driving force" in a reactive system. It is an intensive property.

Of the four different expressions for μ_i only the one concerning the Gibbs function (G) is a partial molal property, because for that one it is specified that temperature (T) and pressure (P) do not change.

Equilibrium Between Phases

In Section 10.1.2, the Gibbs function was defined as $G = H - TS$ and it was shown that $dG = V\,dP - S\,dT$. Since a pure substance or an ideal nonreactive mixture that changes from phase 1 to phase 2 at constant pressure does so also at constant temperature, $dG = 0$.

It is noted previously that the chemical potential of one component in a mixture (μ_i) is the partial molal value of the Gibbs function, $(\partial G/\partial n_i)_{T,P,n_j}$. During a phase change from phase 1 to phase 2, the decrease in one ($-dn_1$) is the increase in other ($+dn_2$). At constant T and P

$$dG = \mu_1\,dn_1 + \mu_2\,dn_2 = 0 \quad \text{and} \quad \mu_1 = \mu_2$$

Thus, the specific value of the Gibbs function is the same for the two phases:

$$\bar{g}_1 = \bar{g}_2$$

This remains true as the pressure and corresponding temperature change:

$$d\bar{g}_1 = d\bar{g}_2$$

$$v_1\,dP - s_1\,dT = v_2\,dP - s_2\,dT$$

$$\frac{dP}{dT} = \frac{s_2 - s_1}{v_2 - v_1} = \frac{h_2 - h_1}{T(v_2 - v_1)}$$

This is the Clausius–Clapeyron equation, applicable to changes between any two phases. If the change is from liquid to vapor, and the conventional subscripts are employed,

$$\frac{dP}{dT} = \frac{h_{fg}}{T_{sat}v_{fg}}$$

which is the slope of the saturation line on the P-T coordinates. At a triple point, where three phases are in equilibrium,

$$g_1 = g_2 = g_3$$

10.1.5 Reactive Systems

Reactive systems are those in which there is a change in the chemical composition of the components. An example is the combustion of fuel with oxygen alone or with air.

Stoichiometry

The **stoichiometric mixture** of fuel and oxidizer is the mixing proportion that is just sufficient to produce complete oxidation of the components of the fuel with no excess of either fuel or oxidizer. This may also be called the **chemically correct mixture** strength, the **theoretically correct mixture** strength, or the **combining proportions**.

For example, a hydrocarbon fuel with the chemical formula $C_n H_m$ will require sufficient oxygen (O_2) to produce only carbon dioxide (CO_2) and water (H_2O) molecules with zero excess oxygen. The reaction equation would be

$$C_n H_m + \left(n + \tfrac{1}{4}m\right)O_2 \rightarrow nCO_2 + \tfrac{1}{2}mH_2O$$

Example 10.1. The stoichiometric mixture strength of the hydrocarbon fuel octane with oxygen would be

$$(C_8H_{18}) + 12.5O_2 \rightarrow 8CO_2 + 9H_2O$$

On a mole basis,

$$1.0 + 12.5 \rightarrow 8 + 9$$

On a mass basis (approximately),

$$1.0(8 \times 12 + 18 \times 1) + 12.5(32) \rightarrow 8(12 + 32) + 9(2 + 16)$$
$$114 + 400 \rightarrow 352 + 162$$

Note that there must be a mass balance, but that the number of moles will not necessarily balance. Also, if both fuel and oxidizer could be considered as ideal gases, which is not so in this example at normal temperature, a mole basis would also be a volume basis.

In this example the fuel-to-oxidizer ratio, mass basis, is $114/400 = 0.285$.

Frequently, of course, the oxygen is supplied in air. It is conventional in this type of calculation to consider that air is 21.0% oxygen and 79.0% nitrogen on a mole or volume basis, the other inert constituents of air being grouped with the nitrogen. (Compare the constituents of standard dry air as given in Table 10.1). The ratio 79/21 is approximately 3.76.

The stoichiometric or chemically correct mixture proportions for octane and air would be

$$C_8H_{18} + 12.5O_2 + (3.76)(12.5)N_2 \rightarrow 8CO_2 + 9H_2O + (3.76)(12.5)N_2$$

On a mole basis,

$$1 + 12.5 + 47 \rightarrow 8 + 9 + 47$$

On a mass basis,

$$114 + 400 + 47(28) \rightarrow 1830$$

The stoichiometric fuel-to-air ratio then is $114/1716 = 0.0664$, and the air-to-fuel ratio would be $15.05/1.0$, on a mass basis.

A mixture different from the stoichiometric proportions is either **rich** (insufficient oxygen for complete combustion) or **lean** (an excess of oxygen). For the example of octane in air, a 10% lean mixture or a mixture with 10% excess air would have an air-to-fuel ratio of $1.10 \times 15.05 = 16.56/1.0$. A mixture of 10% rich would have a fuel-to-air ratio of $1.10 \times 0.0664 = 0.0730/1.0$. Unless otherwise stated, fuel-to-air and air-to-fuel ratios should be given on a mass basis.

Enthalpy of Formation

For most tables and charts giving thermodynamic properties of pure substances, the enthalpy is given the value zero at some arbitrary temperature chosen for convenience, for example, the ice point of water for the Steam Tables and $-40°C$ for most refrigerants. For reacting systems, where substances previously dealt with as having fixed chemical formulas will combine with others or perhaps dissociate, the energy transfers must involve the energy supplied or released in the original formation of the substance. It is conventional to take the value of enthalpy as equal to zero at the **standard** temperature and pressure, 25°C (298 K) and 1.0 bar (10^5 Pa or 0.1 MPa) for an element or for a particular stable molecular structure. Thus, solid carbon (C), diatomic hydrogen (H_2), diatomic oxygen (O_2), and so on, all have the same value (zero) at that standard temperature and pressure.

The **enthalpy of formation** is the energy released when two elements or molecules at 25°C and 1.0 bar, at which their enthalpy values are zero, combine to produce another molecular structure delivered at the same temperature and pressure in a steady-state–steady-flow process. Consider the formation of water from hydrogen and oxygen:

$$\left(\sum n_i h_i\right)_{25°C,\,1\,bar} + \mathcal{Q} = \left(\sum n_e h_e\right)_{25°C,\,1\,bar}$$

$$H_2 + \tfrac{1}{2}O_2 + \mathcal{Q} = H_2O$$

When the water is in the vapor phase, the value of $\mathcal{Q}$ is approximately -2.148×10^5 kJ/kmol of water. That value, then, is the enthalpy of formation per mole ($\bar{h}_f^\circ$) for water vapor, $H_2O(g)$. The superscript (°) indicates standard conditions, 25°C and 1.0 bar.

If the product were to be $H_2O(l)$, which is water in the liquid phase, then ($\bar{h}_f^\circ$) would be -2.858×10^5 kJ/kmol. The difference is the latent heat of vaporization for water, per kg-mol ($\bar{h}_{fg}$) at 25°C. This example was taken, because of our familiarity with the properties of water, to illustrate the importance of phase and to note that, in fact, water would not be vapor at these conditions. This conventional adjustment of the $\bar{h}_f^\circ$ value simplifies reaction calculations for various substances when the products might be found in one phase or in another.

Enthalpy values at 1.0 bar and at temperatures above or below 298 K are tabulated as

$$\bar{h}_T^\circ - \bar{h}_f^\circ = \int_{298}^{T} \bar{C}_p \, dT$$

This assumes the products are ideal gases for which the specific heat at constant pressure is a function of temperature only. Conventionally, entropy values are tabulated at 1.0 bar pressure as absolute values. That is, entropy is zero at 0 K in accordance with the third law of thermodynamics.

Heat of Reaction

When several substances (reactants) combine chemically to produce several different substances (products), the first law of thermodynamics can be applied to the process. For a **closed system** at constant volume,

$$\sum U_R + \mathcal{Q}_V = \sum U_P + \mathcal{W}$$

For an **open system** at constant pressure,

$$\sum H_R + \mathcal{Q}_P = \sum H_P + \mathcal{W}$$

In the summation of internal energies or enthalpies for both products and reactants, the energy of formation at standard temperature and pressure as well as changes above or below those conditions must be taken into account. The **heat of reaction** (T, P) would be the value of $\mathcal{Q}$ when the work ($\mathcal{W}$) is zero.

This is illustrated in Fig. 10.10. For the open system, $\sum H_R = \sum n_i \bar{h}_i^\circ)_R$ and $\sum H_P = \sum n_i \bar{h}_i^\circ)_P$ are both plotted as a function of temperature (T). All the components supplied and all the products including unburned fuel and excess oxidizer as well as inert components must be included in the summations. For simplicity, it is assumed that all components are ideal gases so that pressure changes are ignored. For the conditions shown (arbitrarily selected) the temperature (T_R) at which the reactants are supplied is greater than the standard temperature (298 K), and the temperature of the products (T_P) is higher than T_R. The value of $\mathcal{Q}$ for the conditions shown is negative, which indicates heat transfer from the system; this is the most commonly expected situation.

For what is generally referred to as the **heating value** of a fuel, as determined by a flow calorimeter at constant pressure, the products are assumed to leave at $T_P = T_R$, and the $\mathcal{Q}_P$ value is the vertical distance between the two $\sum H$ lines at that temperature, which may or may not be standard temperature.

For a constant-volume reaction, as for the heating value determined in a closed calorimeter, the diagram is similar, as shown, but $\mathcal{Q}_V = \sum U_P - \sum U_R$ is the significant term.

Adiabatic Flame Temperature

When there is zero heat transfer, the temperature of the products that satisfies the energy balance equation with $\mathcal{Q} = 0$ is the **adiabatic flame temperature** as shown in Fig. 10.11. Unless otherwise specified, this is given for an open system where $\sum H_R = \sum H_P$.

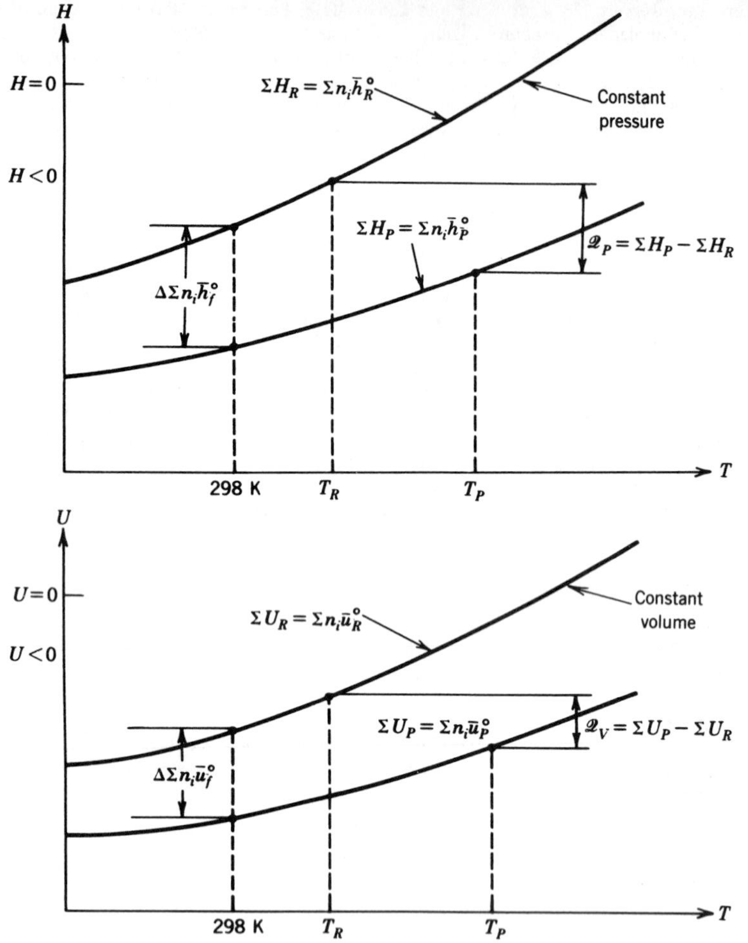

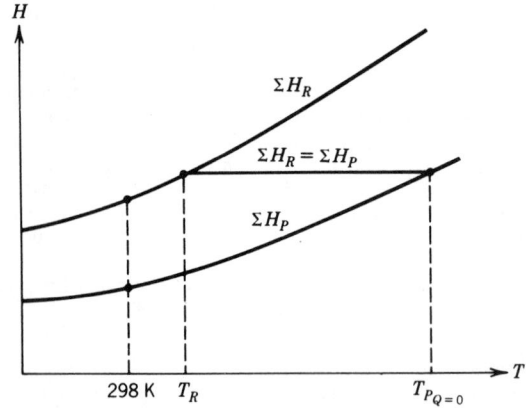

Fig. 10.10   Heat of formation and heat of reaction at constant pressure and at constant volume.

Fig. 10.11   Adiabatic flame temperature at constant pressure.

The maximum adiabatic flame temperature is achieved for the stoichiometric mixture proportions. It is possible, therefore, to control the combustion temperature by an adjustment of the mixture ratio of the reactants. Either a lean mixture (excess air) or a rich mixture (excess fuel) will produce an adiabatic flame temperature less than the maximum because part of the energy released must raise the temperature of these inert components.

Note that dissociation of products at the higher temperatures has not been considered in these definitions. That will be discussed later.

Chemical Equilibrium

In the foregoing discussions of stoichiometric mixtures and adiabatic flame temperatures, it has been assumed that reactions have gone to completion. In fact, reactions will proceed in one direction only until an equilibrium situation is reached, at which the driving forces (chemical potentials) are equal in both directions.

Consider the reactions involving four constituents described by the equation

$$\nu_A A + \nu_B B \rightleftharpoons \nu_C C + \nu_D D$$

The values of ν_i are the individual stoichiometric coefficients for each, in other words the number of moles of each that would produce a balance for all elements in the equation. On the other hand, the number of moles of each (n_i) need not be the correct combining proportions. As the reaction proceeds to the right or to the left, the changes in the number of moles of each (dn_i) will be of importance. Also, the total number of moles present will determine partial pressures, which are significant.

The **degree of reaction** will be ε, and the **degree of dissociation** will be $1 - \varepsilon$, where reaction means movement toward the right and dissociation is movement toward the left, as defined by the following equations:

$$dn_A = -\nu_A \, d\varepsilon \qquad dn_B = -\nu_B \, d\varepsilon$$
$$dn_C = +\nu_C \, d\varepsilon \qquad dn_D = +\nu_D \, d\varepsilon$$

For simplicity, in the following discussion it will be assumed that each constituent acts as an ideal gas so that its mole fraction ($y_i = n_i/\Sigma n$) is the ratio of its partial pressure to the total pressure (P_i/P). Where this is not so, the fugacity of the components must be substituted for pressure (discussed later).

Equilibrium is achieved in any reaction when the Gibbs function values are the same on both sides of the reaction equation. The value of the Gibbs function per mole for each component assumed to be an ideal gas is

$$\bar{g}_i = \bar{h}_i - T\bar{s}_i = \mu_i$$

and

$$\bar{s}_i = \int_0^T \bar{C}_p \frac{dT}{T} - \mathscr{R} \ln \frac{P}{P^\circ}$$

The change in the Gibbs function for the mixture, at a given temperature and total pressure, depends on the change in the degree of reaction, so that

$$dG^\circ_{T,\,P} = d\varepsilon \sum \nu_i \left[\bar{g}_i^\circ + \mathscr{R}T \ln y_i \left(\frac{P}{P^\circ} \right) \right]$$

$$= d\varepsilon \left[\Delta G^\circ + \sum \nu_i \mathscr{R}T \ln \left(\frac{P_i}{P^\circ} \right) \right]$$

where

$$\Delta G^\circ = -\nu_A \mu_A - \nu_B \mu_B + \nu_C \mu_C + \nu_D \mu_D$$

and

$$\sum \nu_i \mathscr{R}T \ln P_i = \mathscr{R}T \ln \left[\frac{y_C^{\nu_C} y_D^{\nu_D}}{y_A^{\nu_A} y_B^{\nu_B}} \right] \left(\frac{P}{P^\circ} \right)^{\nu_C + \nu_D - \nu_A - \nu_B}$$

$$= \mathscr{R}T \ln K$$

As defined by these equations, K is the **equilibrium constant**. At equilibrium, $dG_{T,P}^\circ = 0$, so that for equilibrium,

$$\ln K = -\frac{\Delta G^\circ}{\mathscr{R}T}$$

The equilibrium constant (K) is a function of temperature as is the **heat of reaction** (ΔH°) defined as

$$\Delta H^\circ = \nu_C \bar{h}_C^\circ + \nu_D \bar{h}_D^\circ - \nu_A \bar{h}_A^\circ - \nu_B \bar{h}_B^\circ$$

The differential of $\ln K$ with respect to temperature at constant pressure produces the **van't Hoff equation**, or van't Hoff isobar:

$$\frac{d}{dT} \ln K = \frac{\Delta H^\circ}{\mathscr{R}T^2}$$

In the defining equation for K, the ratio P/P° can be written simply as P if its units are bars and $P^\circ = 1$ bar, as it conventionally is.

As noted previously, the development was presented with the assumption that all constituents in the reaction equation are ideal gases so that enthalpy is a function of temperature but not pressure. Where that is not so, **fugacity** (f) must be substituted for pressure. Fugacity has been referred to as a "pseudo-pressure" defined as follows: For an ideal gas

$$Pv = \mathscr{R}T \quad \text{and} \quad dg_T = v\,dP_T = \left(\frac{\mathscr{R}T}{P}\right) dP_T = \mathscr{R}T\,d(\ln P)_T$$

For a real gas

$$Pv = Z\mathscr{R}T \quad \text{and} \quad dg_T = Z\mathscr{R}T\left(\frac{dP_T}{P}\right) = Z\mathscr{R}T\,d(\ln P)_T$$

$$= \mathscr{R}T\,d(\ln f)_T$$

in which the fugacity (f) is such that the ratio f/P approaches unity as P and f approach zero.

When it is necessary to use fugacity rather than pressure in the determination of the equilibrium constant (K), the development of the defining equation is more complicated and results in the following definition in which P° is unity and $P/P^\circ = P$:

$$K_f = \left[\frac{y_C^{\nu_C} y_D^{\nu_D}}{y_A^{\nu_A} y_B^{\nu_B}}\right] P^{\nu_C + \nu_D - \nu_A - \nu_B} \left[\frac{(f/P)_C^{\nu_C} (f/P)_D^{\nu_D}}{(f/P)_A^{\nu_A} (f/P)_B^{\nu_B}}\right]$$

The change in the equilibrium constant (K_f) with temperature is again given by the van't Hoff equation. Thus

$$\frac{d}{dT} \ln K_f \approx \frac{\Delta H^\circ}{\mathscr{R}T^2}$$

For a more detailed discussion of fugacity, see a reference text such as those by Denbigh, Zemansky or Van Wylen and Sonntag, listed in the bibliography.

Note that dissociation, movement toward the left in the reaction equation, includes dissociation of polyatomic molecules into monatomic molecules; it also includes ionization of molecules, which is the separation of negatively charged electrons from molecules to leave positively charged ions. These two phenomena occur at relatively high temperatures.

10.1.6 Thermodynamic Processes

The processes examined here are discussed in terms of the changes in thermodynamic property values of the working substance. For simplicity, changes in mechanical energies (kinetic and potential) are not considered, but must be included where applicable. When kinetic energy changes are not negligible in the open system, stagnation enthalpy ($h_0 = h + \mathscr{V}^2/2g_c$) should be substituted for h in the following equations.

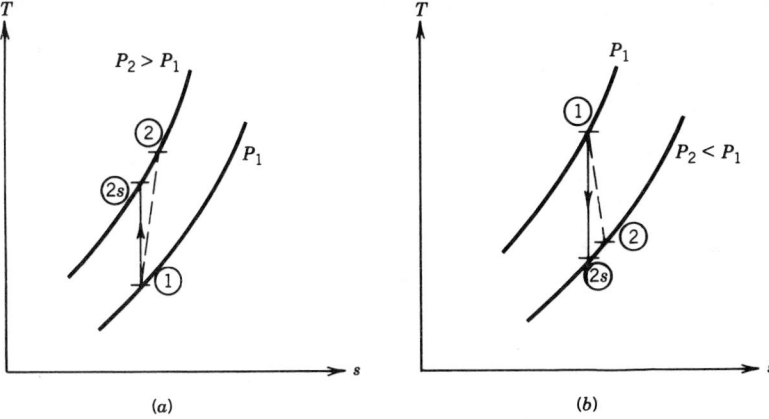

Fig. 10.12 Adiabatic compression and expansion of a working fluid from P_1 to P_2. (a) Compression; (b) expansion.

The Adiabatic Process

Adiabatic means simply "with no heat transfer," that is, $\mathcal{Q}$ is equal to zero or negligible compared to other energy terms involved. For a reversible adiabatic process, the change of entropy is zero. The principle of increase of entropy dictates that, for an adiabatic process, the change of entropy is zero or positive.

The work of adiabatic compression or expansion has been shown to be the change of internal energy for a closed system and the change of enthalpy for a simple steady-state–steady-flow open system. The temperature-entropy diagrams of Fig. 10.12 show that, with irreversibilities, the energy changes and therefore the work involved must be greater than isentropic for compression and less than isentropic for expansion. The **efficiency of compression** (η_c) then is the ratio ideal work–actual work, and the **efficiency of expansion** (η_E) is the ratio actual work–ideal work. For a closed system

$$\eta_{cs} = \frac{u_{2s} - u_1}{u_2 - u_1}$$

and for an open steady-state–steady-flow system

$$\eta_{cs} = \frac{h_{2s} - h_1}{h_2 - h_1}$$

Similarly, the efficiency of adiabatic expansion for a closed system is

$$\eta_{ES} = \frac{u_1 - u_2}{u_1 - u_{2s}}$$

and for open steady-state–steady-flow

$$\eta_{ES} = \frac{h_1 - h_2}{h_1 - h_{2s}}$$

The Isothermal Process

Isothermal means simply "constant temperature." For an ideal gas this implies constant internal energy and constant enthalpy, but in general u and h are functions of pressure as well. The temperature-entropy diagrams of Fig. 10.13 show the ideal isothermal process [10.13(a)] for an ideal gas and [10.13(b)] for a real gas over a large pressure range where enthalpy is not a function of temperature alone.

For a closed system, the first law gives the work of compression or expansion as

$$\mathcal{W}_D = \mathcal{Q} + (U_1 - U_2)$$

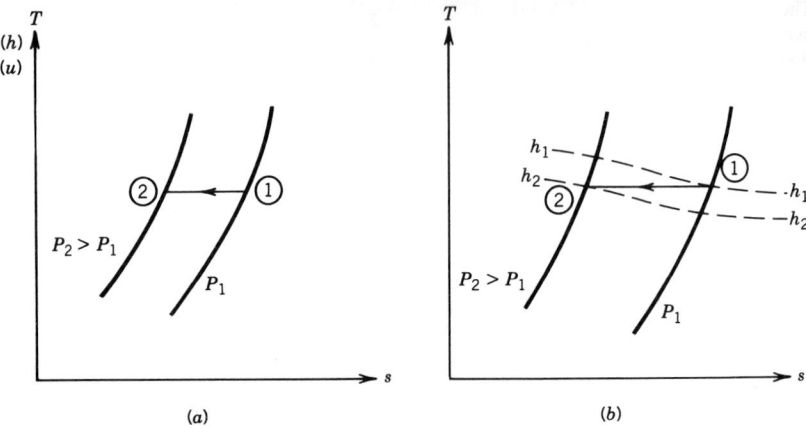

Fig. 10.13   Isothermal compression. (*a*) Ideal gas; (*b*) real gas, large pressure range.

For the simple steady-state–steady-flow system, the shaft work for compression or expansion is

$$\mathcal{W}_{sh} = \mathcal{Q} + (H_1 - H_2)$$

In both cases, for the ideal process

$$\mathcal{Q} = T(S_2 - S_1)$$

For the particular case of the ideal gas, $\mathcal{Q} = \mathcal{W}$. Further, the displacement work of the closed system is equal to the shaft work of the steady-state–steady-flow system when $PV = m\mathcal{R}T$. In other words, for an ideal gas undergoing a quasistatic isothermal process,

$$\mathcal{Q} = T\,dS = m\mathcal{R}\ln\frac{v_2}{v_1} = m\mathcal{R}\ln\frac{P_1}{P_2} = \int_1^2 P\,dV = -\int_1^2 V\,dP$$

The actual work cannot be expressed simply in terms of thermodynamic property changes, but generally it can be measured as the input or output work for compression or expansion. The efficiency of isothermal compression is

$$\eta_{CT} = \frac{\text{ideal work}}{\text{actual work}}$$

and the efficiency of isothermal expansion is

$$\eta_{ET} = \frac{\text{actual work}}{\text{ideal work}}$$

The Polytropic Process

The polytropic process (literally several things changing) is one that follows the relationship $PV^n = $ constant, for any value of n except 1.0, zero, and infinity. For a quasistatic process, the integration of $P\,dV$ and of $-V\,dP$ with PV^n constant gives the displacement work or the shaft work, respectively:

$$\mathcal{W}_D = \int_1^2 P\,dV = \frac{P_2 V_2 - P_1 V_1}{1 - n}$$

and

$$\mathcal{W}_{sh} = -\int_1^2 V\,dP = \frac{n}{1 - n}(P_2 V_2 - P_1 V_1)$$

The three values of n that are exceptions to the preceding indicate processes in which certain properties are constant. If the exponent n is equal to 1.0, then the product PV is constant. For that process

$$\int_1^2 P \, dV = -\int_1^2 V \, dP = P_1 V_1 \ln \frac{V_2}{V_1} = P_1 V_1 \ln \frac{P_1}{P_2}$$

When the exponent is equal to zero, the pressure is constant, and

$$\int_1^2 P \, dV = P(V_2 - V_1)$$

$$\int_1^2 V \, dP = 0 \quad \text{and} \quad \mathcal{Q} = H_2 - H_1$$

When n is equal to infinity, the volume is constant. Thus

$$\int_1^2 P \, dV = 0 \quad \text{and} \quad -\int_1^2 V \, dP = V(P_1 - P_2)$$

The Throttling Process

When a flowing fluid is allowed to expand through a restriction such as a valve or orifice from a higher to a lower pressure without doing any useful work, the process is called throttling. The first law for steady-state–steady-flow in which there is no heat transfer and negligible change in kinetic and potential energies shows that the enthalpy value of exit will be the same as the inlet enthalpy. Such a situation is shown in Fig. 10.14.

For an ideal gas, zero change in enthalpy would imply zero change in temperature. For a real gas or vapor, throttling usually implies a change of temperature that may be either positive or negative. This change of temperature with pressure at constant enthalpy is called the **Joule–Thomson** or the **Joule–Kelvin coefficient** (μ) defined as

$$\mu = \left(\frac{\partial T}{\partial P}\right)_h$$

It can be shown that this coefficient has two components, either of which may predominate:

$$\mu = \frac{1}{c_P}\left[-\left(\frac{\partial u}{\partial P}\right)_T - \left(\frac{\partial Pv}{\partial P}\right)_T\right]$$

Figure 10.15 shows the change in the sign of μ from positive to negative for a typical real gas over a wide range of conditions. Each substance has a maximum **inversion temperature** above which the sign of μ is negative only.

The familiar change of temperature with pressure as a liquid at point 1 is throttled to a pressure low enough for the end point (2) to be in the liquid-vapor region is shown in Fig. 10.16. This is a basic process in any vapor compression refrigeration process. Also shown on both P-h and h-s coordinates is throttling from a slightly wet vapor at point 3 to superheated vapor at point 4. This process is the basis of the so-called **throttling calorimeter**. In that device, a sample of vapor (commonly steam) is drawn from a line at a reasonably high pressure and allowed to expand to approximately atmospheric pressure. In the superheat region, the pressure and temperature values at point 4 establish the enthalpy value that in turn indicates the dryness fraction at point 3.

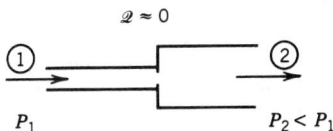

Fig. 10.14   Fluid flow throttled through an orifice: $V^2/2g_c \approx V_2^2/2g_c$.

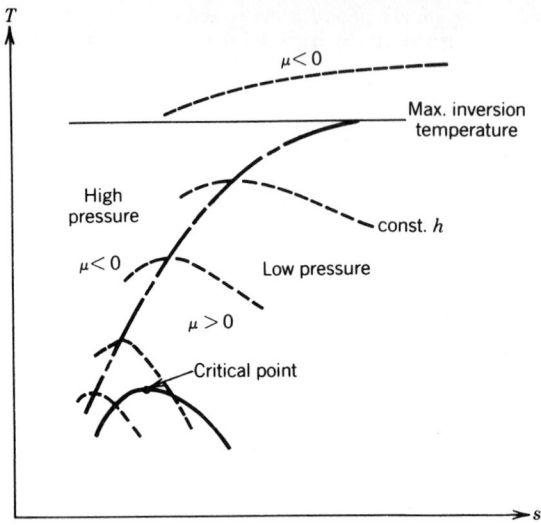

Fig. 10.15   The Joule–Thomson (Joule–Kelvin) coefficient on T-s coordinates.

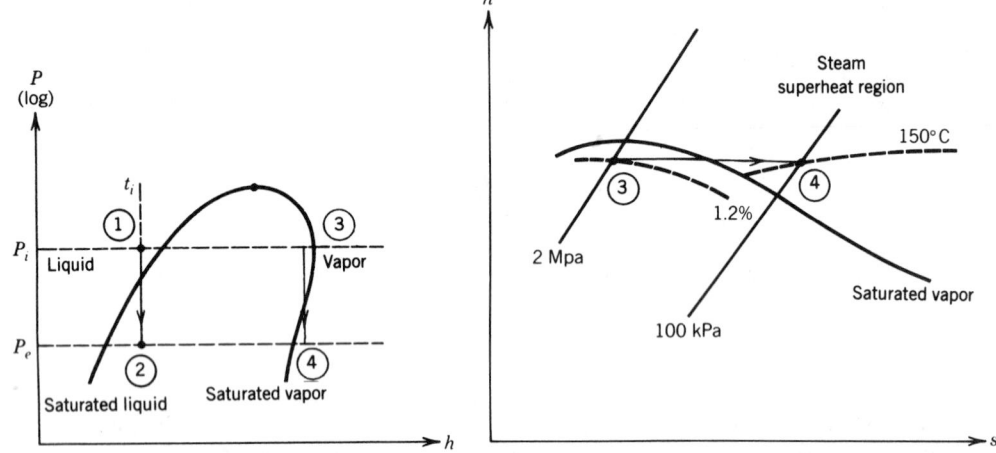

Fig. 10.16   Throttling processes: 1-2, from liquid to liquid-vapor region (as in refrigeration cycle); 3-4, from slightly wet vapor to superheat region (as in throttling calorimeter).

10.1.7   Ideal Power Cycles

A cycle is any series of thermodynamic processes that returns a working fluid to its original state point. This is the most common manner in which heat transfer from a high-temperature source is utilized to deliver power continuously. (A less common alternative is one or another direct energy conversion device such as a thermoelectric generator or a fuel cell.)

The Carnot, Stirling, and Ericsson Cycles

Two corollaries of the second law of thermodynamics (see Section 10.1.1) state that the maximum possible thermal efficiency of a working cycle between a source at T_H and a sink at T_L is the efficiency of a reversible cycle receiving heat only at T_H and rejecting heat only at T_L, and that all reversible cycles so constrained will have the same thermal efficiency. That thermal efficiency is

$$\eta_{\text{th}} = 1 - \frac{|\mathcal{Q}_L|}{|\mathcal{Q}_H|} = 1 - \frac{T_L}{T_H}$$

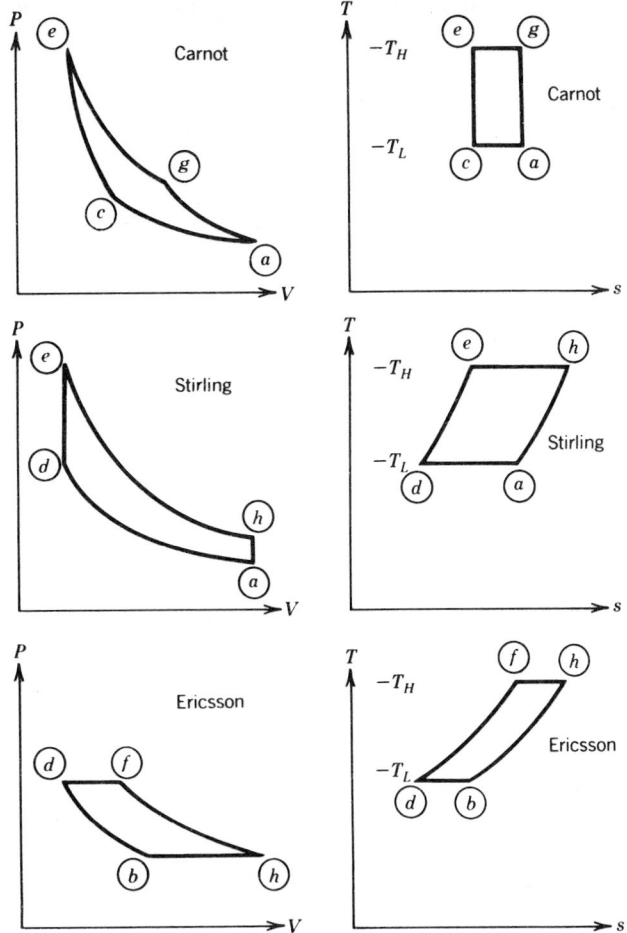

Fig. 10.17 Gas power cycles—Carnot, Stirling, and Ericsson—between T_H and T_L and between V_a and V_d.

Three such cycles are shown in Fig. 10.17 on P-v and on T-s coordinates with the assumption that the working fluid is an ideal gas. All three are between the same temperature limits, T_H and T_L, and between the same volume limits, $v_d = v_e$ and $v_a = v_h$.

The Carnot cycle is composed of two reversible adiabatic (isentropic) processes, c-e and g-a, and two isothermal processes, e-g at T_H and a-c at T_L.

The Stirling cycle has its heat supply on the isothermal line e-h at T_H and its heat rejection on the isothermal line a-d at T_L. Between these two limiting temperatures there is perfect **regeneration** from the constant-volume line h-a to the constant-volume line d-e. That is to say, the negative heat transfer quantity $\int_h^a C_V\, dT$ is not removed from the system but is transferred internally and is equal to $\int_d^e C_V\, dT$.

The Ericsson cycle is similar except that regeneration takes place from the constant-pressure line h-b to the constant-pressure line d-f, the regenerative heat transfer quantity being $\int_h^b C_p\, dT$ equal to $\int_d^f C_p\, dT$. Heat transfer quantities from and to the surroundings in this cycle are along the lines f-h and b-d.

Air-Standard Cycles

Because so many gas cycles use air as the major part of the working fluid, ideal values of thermal efficiency are often calculated for various cycles with the assumption that the working fluid is a hypothetical frictionless substance with the thermodynamic properties of real air at standard conditions: 15°C and 1.0 atm pressure. It is assumed that the specific heat values remain constant and that γ, which is C_p/C_v, is constant at 1.40. Note that γ is significant in the expression for thermal efficiency for the gas power cycles that follow.

The Otto, Diesel, and Brayton (Joule) Cycles

These three **gas power cycles** are shown in Fig. 10.18 on P-v and on T-s coordinates. Each has an isentropic compression process, line a-b.

The ideal Otto cycle is composed of two isentropic processes, a-b and c-d, and two constant-volume processes b-c and d-a. This is the cycle attempted by most high-speed internal combustion engines, which are, of course, open cycles. Air alone or an air and fuel mixture is compressed from a to b, energy is supplied ideally at constant volume during combustion from b to c, and the products of combustion expand from c to d. Exhaust gases are released to the atmosphere in place of the constant-volume line d-a.

Ideally, the thermal efficiency of this cycle is

$$\eta_{th} = 1 - \frac{T_a}{T_b} = 1 - \left[\frac{1}{CR}\right]^{\gamma - 1}$$

where CR is the compression ratio, V_a/V_b. Because the specific heat of real air increases with temperature above the value at 15°C, γ is not in fact constant, but decreases below 1.40. Components of the working fluid other than air also have values of γ less than 1.40. Thus the thermal efficiency of the air-standard cycle, which is a function of compression ratio only, is the maximum attainable for this open cycle.

The Diesel cycle differs from the Otto cycle in that the heat supply is at constant pressure along the line b-c. The thermal efficiency thus depends on the isentropic expansion ratio V_d/V_c as well as on

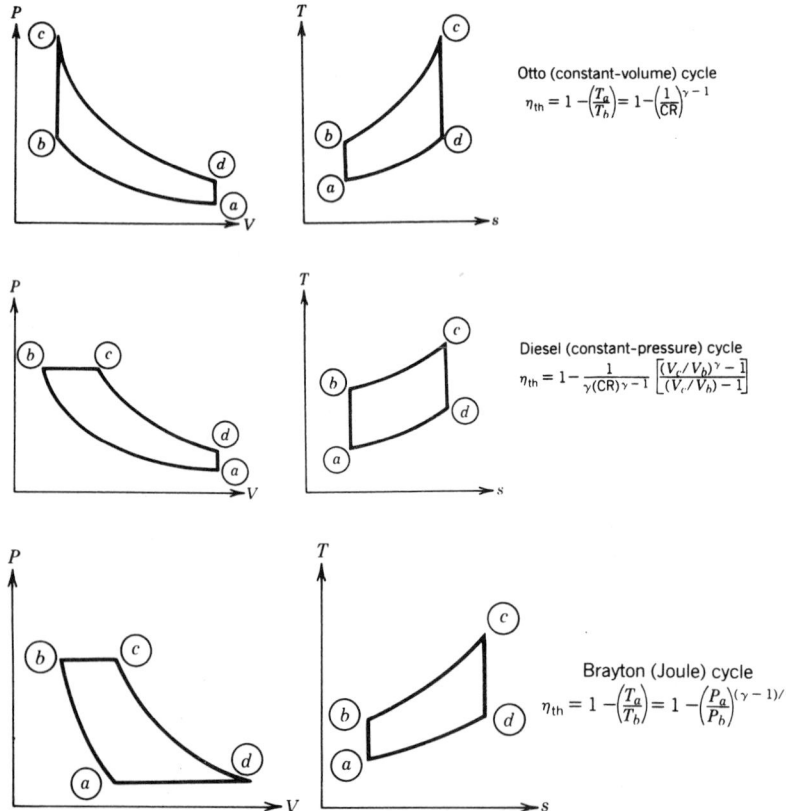

Otto (constant-volume) cycle
$$\eta_{th} = 1 - \left(\frac{T_a}{T_b}\right) = 1 - \left(\frac{1}{CR}\right)^{\gamma - 1}$$

Diesel (constant-pressure) cycle
$$\eta_{th} = 1 - \frac{1}{\gamma(CR)^{\gamma - 1}}\left[\frac{(V_c/V_b)^{\gamma} - 1}{(V_c/V_b) - 1}\right]$$

Brayton (Joule) cycle
$$\eta_{th} = 1 - \left(\frac{T_a}{T_b}\right) = 1 - \left(\frac{P_a}{P_b}\right)^{(\gamma - 1)/\gamma}$$

Fig. 10.18 Gas power cycles—Otto, Diesel, and Brayton (Joule), each with isentropic compression a-b.

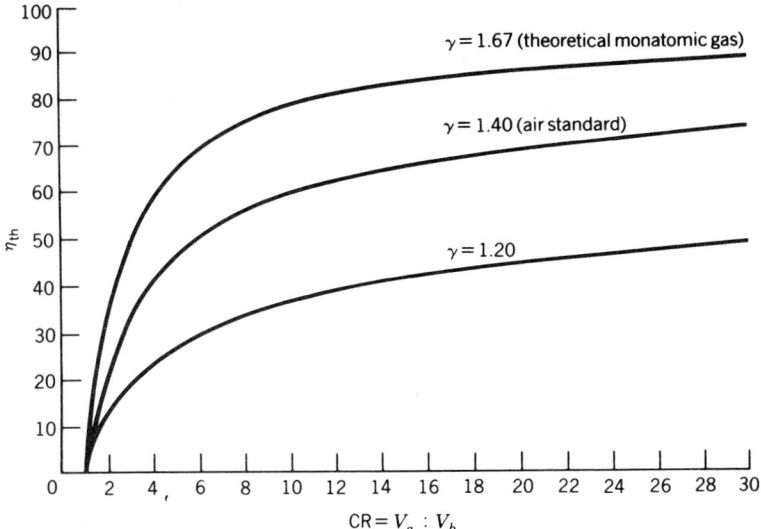

Fig. 10.19   Thermal efficiency values for ideal Otto and Brayton cycles as a function of compression ratio and γ.

the compression ratio V_a/V_b. This reduces the thermal efficiency:

$$\eta_{th} = 1 - \frac{1}{\gamma(CR)^{\gamma-1}}\left[\frac{(V_c/V_b)^\gamma - 1}{(V_c - V_b) - 1}\right]$$

The Brayton cycle is called also the **Joule cycle**. It is composed of two isentropics, *a-b* and *c-d*, and two constant-pressure lines. Its thermal efficiency takes exactly the same form as that for the Otto cycle, but is more usefully expressed as a function of pressure ratio rather than volume ratio:

$$\eta_{th} = 1 - \frac{T_a}{T_b} = 1 - \left[\frac{P_a}{P_b}\right]^{(\gamma-1)/\gamma}$$

As a power cycle in practice, it is most often found as a gas turbine cycle with net power delivered either as shaft power or as kinetic energy in aircraft propulsion.

The efficiency for the Otto and the Brayton cycles is plotted in Fig. 10.19 as a function of isentropic compression ratio (V_a/V_b) for three values of γ.

The efficiency of the open Brayton cycle can be improved, at the lower pressure ratios, by **regeneration**. This is shown in Fig. 10.20 where the compressed air is preheated before the combustion chamber by the turbine exhaust gas. Both heat supply (q_H) and heat rejection (q_L) are reduced by the amount of heat transfer in the regenerator.

With **perfect regeneration**, $T_e = T_d$ and $T_f = T_b$. The thermal efficiency of this ideal cycle becomes

$$\eta_{th} = 1 - \left(\frac{T_a}{T_b}\frac{T_b}{T_d}\right)$$

That is an improvement over the simple cycle without regeneration as long as T_b is less than T_d. As the pressure ratio is increased, T_b becomes higher and T_d becomes lower, and the possibility of regeneration disappears.

The Atkinson Cycle
A limiting factor in the Otto cycle is its failure to allow the expansion process to proceed to the lowest possible pressure. In the Atkinson cycle, the expansion process *c-d* is extended to $P_d = P_a$ so that the thermal efficiency is improved. See Fig. 10.21. The cycle is accomplished in practice in the combina-

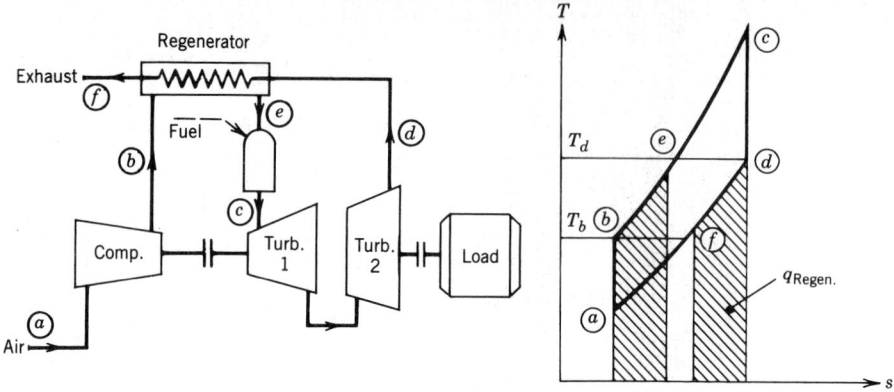

Fig. 10.20   Open Brayton cycle with regeneration.

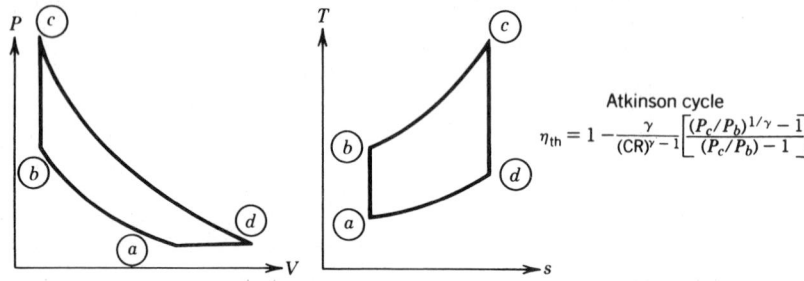

Atkinson cycle

$$\eta_{th} = 1 - \frac{\gamma}{(CR)^{\gamma} - 1}\left[\frac{(P_c/P_b)^{1/\gamma} - 1}{(P_c/P_b) - 1}\right]$$

Fig. 10.21   The Atkinson cycle.

tion of an internal combustion reciprocating engine (possibly free-piston) with a gas turbine through which the exhaust gases expand to atmospheric pressure.

The Simple Rankine Cycle

The Rankine cycle is the basis of all **vapor power cycles**. (In contrast to a gas cycle in which the working fluid remains in one phase, in a vapor cycle the working fluid is converted from liquid to vapor in one process and condensed from vapor to liquid in another.)

Figure 10.22 shows the four basic components of the Rankine cycle: boiler or steam generator, turbine, condenser, and liquid pump. In large central power plants, the working fluid is almost invariably water. Consequently the accompanying T-s and h-s diagrams show the shape of the liquid-vapor region for water, which has a relatively low temperature and high pressure at the critical point. Therefore the saturated vapor at point c is usually superheated to the highest practical temperature at d before expansion through the turbine.

In a conventional power plant the source of energy in the boiler is combustion of coal, oil, or natural gas with air. In a nuclear power plant the source is radioactive "fuel." The cycle for the working fluid is essentially the same except that top temperatures and pressures may be different for safety reasons. The system for a large plant will contain many more components to improve thermal efficiency, and those involving internal heat exchange or regeneration will alter the shape of the thermodynamic cycle.

In the simple cycle shown, all energy supply (q_H) is in the boiler, and all heat rejection (q_L) is in the condenser. Isentropic expansion is shown through the turbine for this ideal cycle. The pump work is very small compared to the other energy values so that $h_b \approx h_a$. Thus, the thermal efficiency is

$$\eta_{th} = \frac{\Sigma w}{q_H} = \frac{(h_d - h_e) - (h_b - h_a)}{h_d - h_b}$$

$$\approx 1 - \frac{h_e - h_a}{h_d - h_a}$$

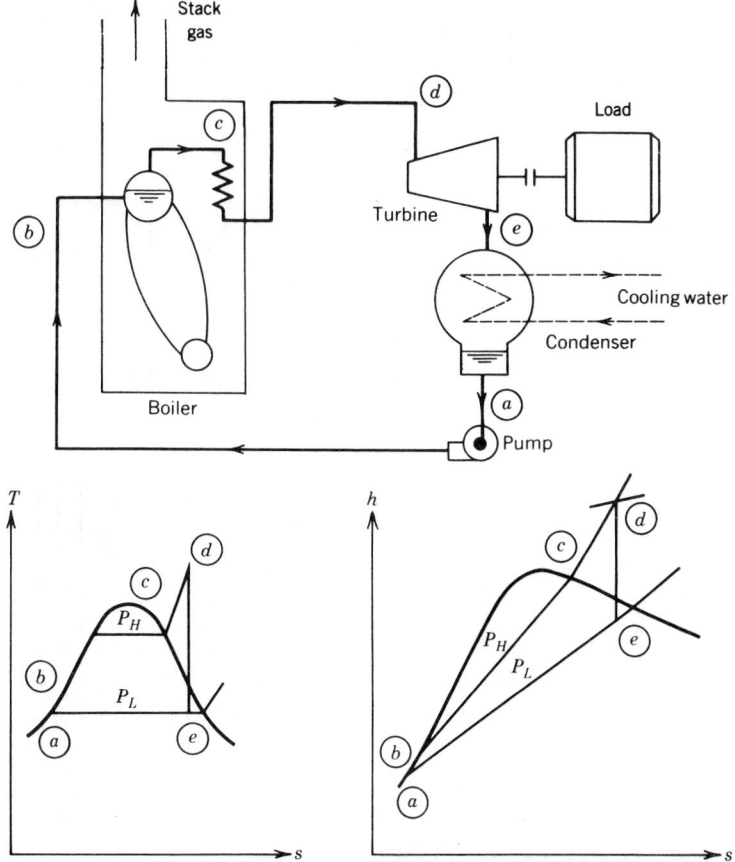

Fig. 10.22   The simple, ideal Rankine power cycle.

Combined Cycles

Innumerable possibilities exist for utilizing the heat rejection from one power cycle as the heat input to another. Two of these are shown schematically in Figs. 10.23 and 10.24.

In Fig. 10.23, an open gas turbine system operating on the Brayton cycle is **topping** a simple Rankine cycle. In order to limit the maximum temperature an inlet to the gas turbine, the fuel-to-air ratio in the combustion chamber is very lean. The hot exhaust gases therefore contain excess oxygen available for further combustion in the boiler of the Rankine cycle. While the hot exhaust gases alone might provide sufficient energy for the boiler, additional fuel might be supplied to raise the temperature. With an auxiliary air fan at the boiler and a bypass line for the gas turbine exhaust, either cycle could operate independently at part load.

Figure 10.24 shows a Rankine cycle **bottoming** an internal combustion engine. Here the hot exhaust gases from the engine, which contain very little excess air, are simply cooled in the boiler of the lower cycle. The working fluid of the Rankine cycle is generally not water because the top temperature is limited by the exhaust gas temperature, and it is desirable to keep the pressure in the condenser above atmospheric for simplicity.

Relative Efficiency and Cycle Efficiency

Ideal cycles, other than the Carnot, Stirling, and Ericsson cycles, will have efficiencies less than the maximum between the top and bottom temperatures (T_H and T_L) if there is any heat transfer to the system at a temperature below T_H or any heat transfer from the system above T_L. The relative efficiency (η_r) compares an ideal cycle, such as the Otto or the Rankine, with the best possible:

$$\eta_r = \frac{\eta_{th(ideal)}}{\eta_{th(Carnot)}}$$

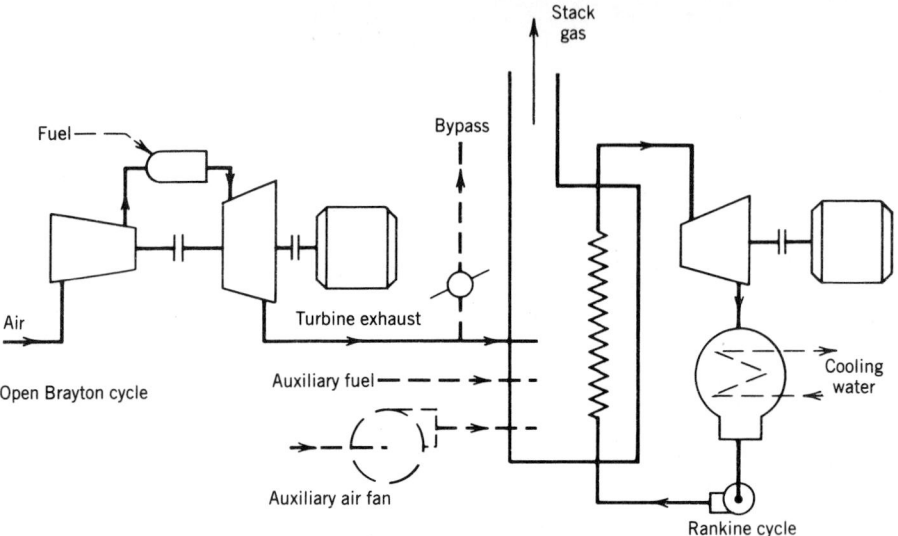

Fig. 10.23 Open gas turbine system topping a vapor power system.

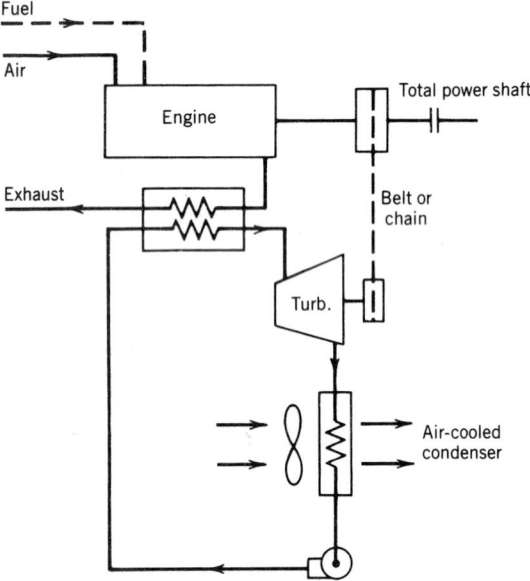

Fig. 10.24 Vapor power cycle bottoming a reciprocating internal combustion engine.

The cycle efficiency (η_{cy}) compares the efficiency of an actual cycle with that of the ideal cycle presumably being attempted:

$$\eta_{cy} = \frac{\eta_{th(actual)}}{\eta_{th(ideal)}}$$

When these two efficiency ratios are combined,

$$\eta_{th(actual)} = \eta_{th(Carnot)}\eta_r\eta_{cy}$$

10.1.8   Refrigeration and Heat Pump Cycles

One statement of the second law of thermodynamics confirms our experience that energy will not flow spontaneously by heat transfer from a lower to a higher temperature level. This energy transfer can be accomplished, however, if some energy in the form of work or its equivalent is supplied to a system. That system, shown symbolically in Fig. 10.3, can be considered as a refrigerator where the purpose is to produce a cooling effect at the lower temperature (T_L) or as a heat pump if the purpose is to provide heat transfer to the higher temperature (T_H).

The Vapor Compression Cycle

The most common cycle for refrigeration or heat pumping is one in which a fluid evaporating at a low pressure provides a cooling effect at the corresponding low saturation temperature. The resulting vapor is then compressed adiabatically to a higher pressure at which it is condensed in order to reject energy by heat transfer at the corresponding higher saturation temperature. This system is shown schematically in Fig. 10.25 along with both *T-s* and *P-h* diagrams for a typical working fluid or "refrigerant."

The vapor leaving the evaporator is shown to be slightly superheated to prevent the possibility of liquid entering the compressor. The condensate leaving the condenser is shown slightly subcooled to avoid vaporization before it passes through the throttling valve. It is in the throttling process that the state point *f* is reached, close to the saturated liquid line at low pressure, from which point evaporation procedes to produce the cooling effect.

In the system shown the refrigerant receives energy (q_L) equal to $h_a - h_f$ per unit mass in the process of cooling fluid *A*. Work (w) equal to $h_b - h_a$ per unit mass of refrigerant is supplied by an

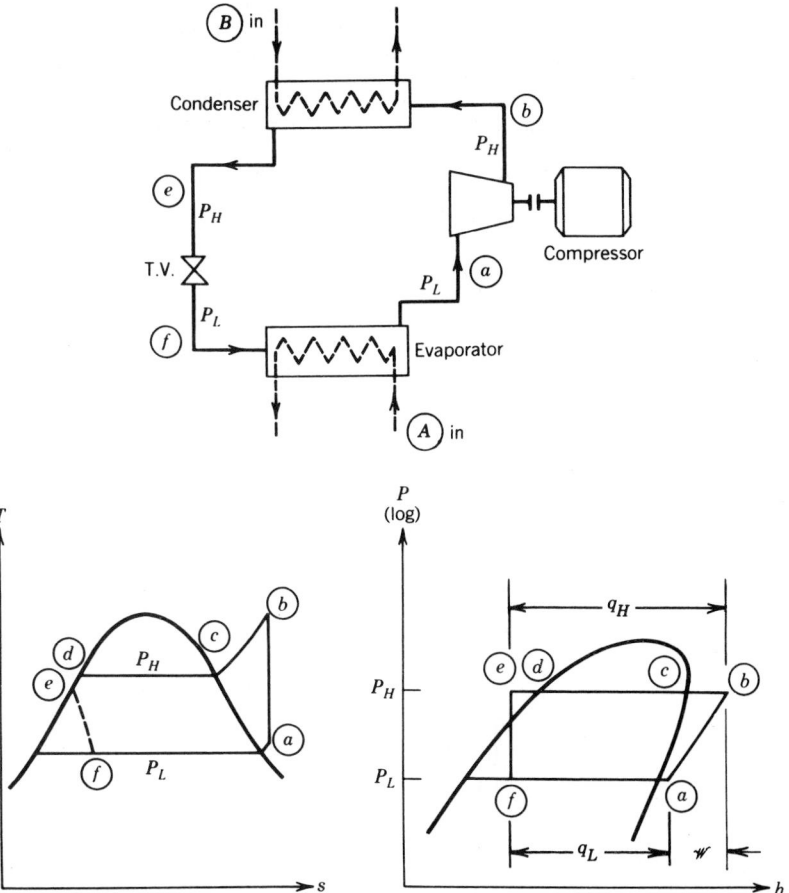

Fig. 10.25   A simple, ideal vapor compression refrigeration cycle.

electric motor or engine. In the condenser, fluid B carries away q_H equal to $h_b - h_e$ per unit mass of refrigerant. In the throttling process there is no work done and negligible heat transfer so that $h_f = h_e$.

As a refrigerator, the coefficient of performance (COP) for this system is

$$\text{COP}_R = \frac{q_L}{w} = \frac{h_a - h_f}{h_b - h_a}$$

As a heat pump, the coefficient of performance would be

$$\text{COP}_{HP} = \frac{q_H}{w} = \frac{h_b - h_e}{h_b - h_a}$$

For the simple, ideal system shown

$$\text{COP}_{HP} = \text{COP}_R + 1.0$$

Absorption Refrigeration Systems

An alternative to vapor compression refrigeration is one form or another of the absorption system. In this, a refrigerant vapor is driven out of a liquid solution by heat transfer to it. The vapor is then condensed and throttled to a lower pressure as in a vapor compression cycle. After evaporation at the low pressure to produce the refigeration effect, the resulting vapor is redissolved into the liquid solution. In that process, the latent heat of condensation as well as the heat of solution must be rejected to the surroundings by cooling water or air. The liquid solution can then be pumped back to a chamber, sometimes called a "generator," at higher pressure where vapor is again driven out of solution (generated) by heat transfer. The primary advantage of this system compared to a vapor compression system is the substitution of a relatively small amount of liquid pump work for the relatively large amount of vapor compression work.

A symbolic representation of such a system is shown in Fig. 10.26. The total heat rejection to ambient conditions $(\Sigma \mathscr{Q}_A)$ is the sum of the energy supplied from a high-temperature source $(\mathscr{Q}_H)$, the refrigeration effect produced, $(\mathscr{Q}_L)$, and the relatively small pump work. The overall coefficient of performance for this refrigeration system, ignoring the pump work, is

$$\text{COP}_{oa} = \frac{\mathscr{Q}_L}{\mathscr{Q}_H}$$

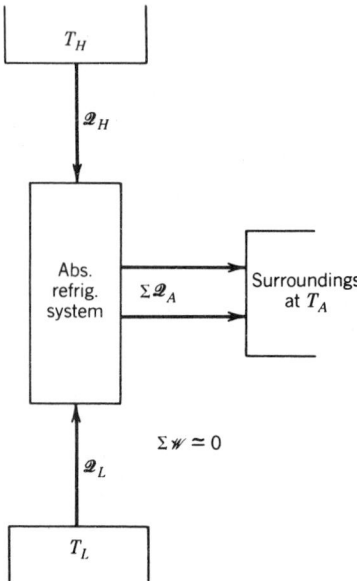

Fig. 10.26   Symbolic representation of an absorption refrigeration system.

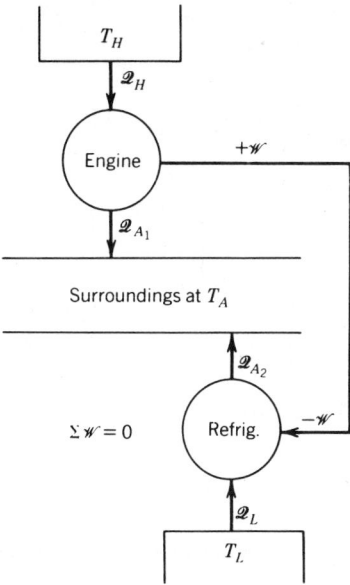

Fig. 10.27 Combined engine plus refrigeration system. Overall COP $= \mathcal{Q}_L / \mathcal{Q}_H$.

While the overall COP is small, the energy supply is "low-grade" heat transfer rather than "high-grade" expensive work.

Maximum Possible Overall COP

While no working cycle and no refrigeration cycle can be identified in the absorption refrigeration system described, there is in fact the equivalent of the two thermodynamically.

Figure 10.27 shows this equivalent combined cycle: an engine receiving energy from a source at T_H and rejecting energy to ambient conditions at T_A, the sole purpose of which is to drive a refrigeration machine producing a cooling effect at the low temperature, T_L, and also rejecting energy at T_A.

The principle of increase of entropy shows that the maximum possible overall coefficient of performance for the combined system is the product of the maximum thermal efficiency for the engine times the maximum COP for the refrigerator between their respective temperature levels.

$$\Sigma \, \Delta S = \Delta S_H + \Delta S_L + \Delta S_A$$
$$= -\frac{\mathcal{Q}_H}{T_H} - \frac{\mathcal{Q}_L}{T_L} + \frac{\mathcal{Q}_H + \mathcal{Q}_L}{T_A} \geq 0$$

Thus

$$\frac{\mathcal{Q}_L}{\mathcal{Q}_H} \leq \left(\frac{T_H - T_A}{T_H} \right) \left(\frac{T_L}{T_A - T_L} \right)$$

The apparently low value of COP for an absorption system should therefore be compared to a vapor compression machine along with the power system driving it, rather than to the refrigeration machine alone.

Gas Cycle Refrigeration

Any ideal reversible cycle used for power could theoretically be reversed to operate as a refrigerator or heat pump. In fact only the Stirling gas cycle and the Brayton gas cycle are so employed. The Stirling cycle can be approached in a reciprocating machine, and with very effective regeneration can reach extremely low temperatures in one stage. It will not be illustrated here.

The Brayton cycle, accomplished with separate compressor and turbine on a common shaft will produce refrigeration in a simple cycle. When internal heat transfer (regeneration) is added, lower temperatures can be achieved. Such a system is shown schematically in Fig. 10.28.

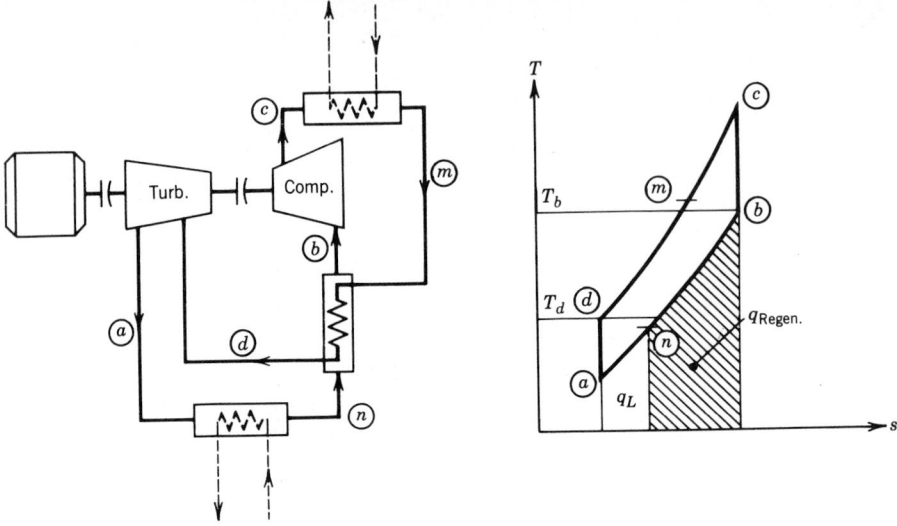

Fig. 10.28   Closed Brayton cycle with regeneration for low-temperature refrigeration.

Minimum Work Required to Cool a Finite System

The refrigeration device shown in Fig. 10.29 is employed to cool some finite body or system from temperature $T_1 = T_A$ to some lower temperature T_2. The change of entropy of the finite system is $S_2 - S_1$. The change of entropy of the surroundings at T_A, to which is delivered the sum of heat transfer from the system plus the work done to drive the refrigerator must be $({}_1\mathcal{Q}_2 + \mathcal{W})/T_A$. The entropy principle dictates that $\Sigma \Delta S$ be equal to or greater than zero.

$$\Sigma \Delta S = \frac{{}_1\mathcal{Q}_2 + \mathcal{W}}{T_A} + (S_2 - S_1)$$

Therefore,

$$\mathcal{W} \geq T_A(S_1 - S_2) - {}_1\mathcal{Q}_2$$

These energy quantities are shown as areas on the T-S diagram.

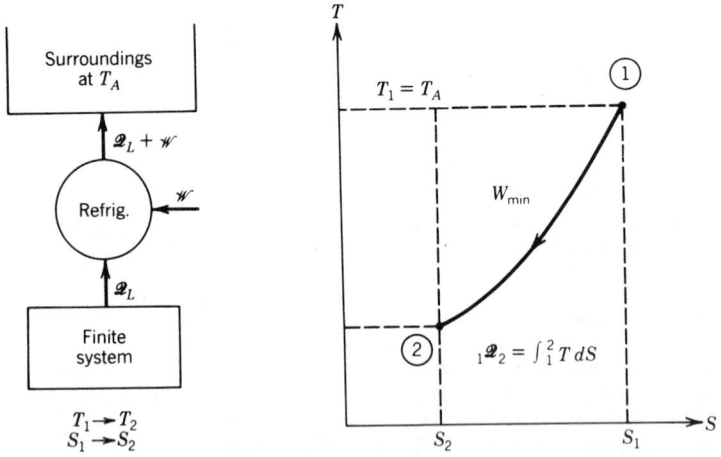

Fig. 10.29   The minimum work required to cool a finite body.

10.1.9 Availability, Effectiveness, and Second-Law Efficiency

Because a working cycle receiving energy from some source above ambient temperature must eventually reject heat at ambient temperature, there is a limit to the amount of useful work that can possibly be obtained from the heat transfer received. Also, energy in any system can be converted to useful work only until that system comes into mechanical and thermal equilibrium with the surroundings at ambient temperature. The maximum possible useful work obtainable in any situation is called the **available energy** or the **availability** of that system in that environment. In recent years, some authorities have used the term **exergy** in place of available energy.

Maximum output of work (or minimum work required) depends on all cycles and processes being reversible. Any **irreversibility** results in **lost work**. Either term is the difference between the available energy and the work actually done (or the difference between the work actually required and the minimum possible work required). Lost work is not a loss in the usual sense; it is a lost opportunity to harness some part of the available energy (or the expenditure of more than the minimum work required).

Kinetic and potential energies can be converted directly to useful work until the velocity and elevation, respectively, of the mass of the system become zero relative to the environment of the system. Those terms, then, are part of the available energy of the system.

Nonuseful work must be subtracted from the maximum work to give the maximum useful work. A distinction is made here between the terms surroundings and environment. The *surroundings* include everything with which the system interacts, including other systems on which useful work can be done. The *environment* is the atmosphere or other medium around the system with which the system exchanges energy by heat transfer. If the system expands, pushing against the environment, the work done is not generally useful. The pressure of the environment will be designated P_0. For a finite change in system volume ΔV, the nonuseful work is $P_0 \, \Delta V$ and may be positive or negative depending on the sign of ΔV.

Available Energy and Lost Work in Various Situations

For each of the four situations described here, the environment is at temperature T_0 and at pressure P_0. See Fig. 10.30. Note that, for each system plus its environment, the sum of all entropy changes ($\sum \Delta S$) must be greater than, or in the limit equal to zero.

1. Available fraction of heat transfer ($\mathcal{Q}_H$) from an infinite source at T_H:

$$A = \mathcal{Q}_H \left(1 - \frac{T_0}{T_H} \right) \qquad \sum \Delta S = -\frac{\mathcal{Q}_H}{T_H} + \frac{\mathcal{Q}_0}{T_0}$$

$$\mathcal{W} = \mathcal{Q}_H - \mathcal{Q}_0 \qquad \mathrm{LW} = \mathcal{W} - A = T_0 \sum \Delta S$$

2. Minimum work to maintain a temperature $T_L < T_0$:

$$A = \mathcal{W}_{\min} = \mathcal{Q}_L \left(\frac{T_0}{T_L} - 1 \right) \qquad \sum \Delta S = -\frac{\mathcal{Q}_L}{T_L} + \frac{\mathcal{Q}_0}{T_0}$$

$$\mathcal{W} = \mathcal{Q}_0 - \mathcal{Q}_L \qquad \mathrm{LW} = \mathcal{W} - A = T_0 \sum \Delta S$$

3. Available energy from a closed, finite system changing from state point 1 to state point 2 while receiving heat transfer ($\mathcal{Q}_0$) from the environment. According to the first law, the work of the system must be

$$_1\mathcal{W}_2 = \mathcal{Q}_0 - (E_2 - E_1)$$

where

$$E = U + \frac{m}{g_c} \left(\frac{\mathcal{V}^2}{2} + gz \right)$$

Of this work done, $P_0(V_2 - V_1)$ is not useful. Also, $S_2 - S_1 \geq \mathcal{Q}_0/T_0$, so that $\mathcal{Q}_0 \leq T_0(S_2 - S_1)$. Consequently, the maximum possible *useful work* between conditions 1 and 2, which is the availabil-

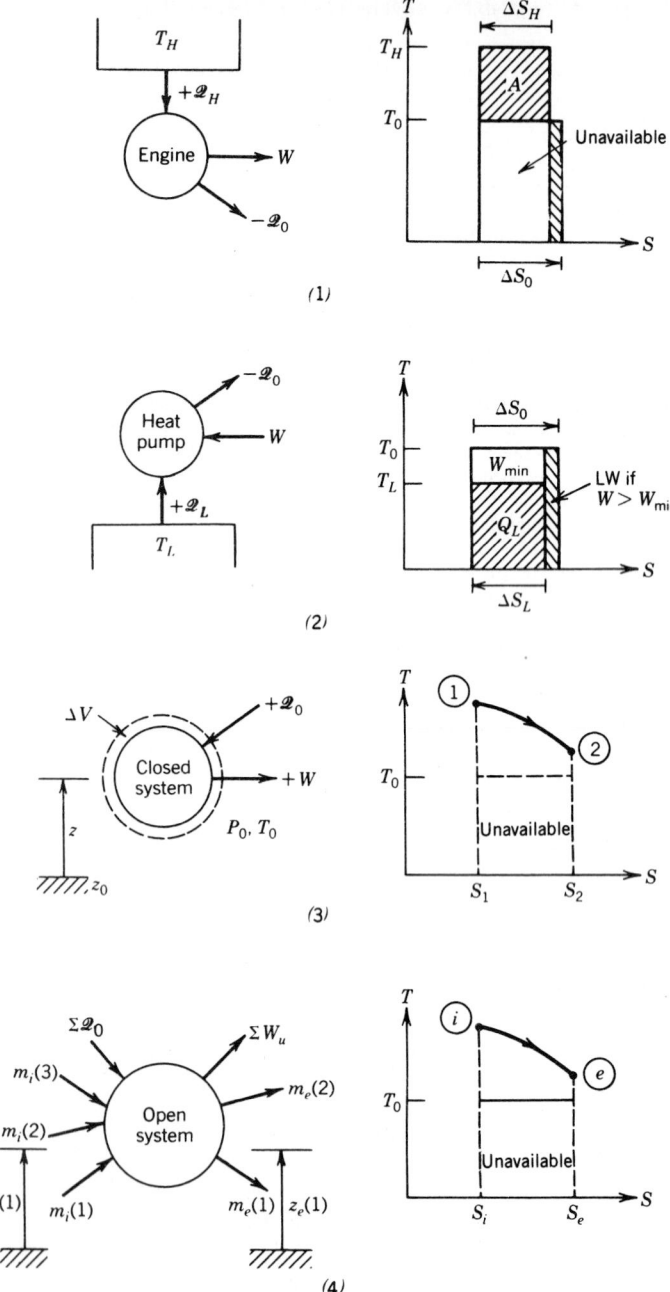

Fig. 10.30 Available energy in four situations: (1) engine with infinite source at T_H; (2) minimum work to maintain $T_L < T_0$; (3) closed finite system between state points 1 and 2; (4) open system, steady-state, steady flow.

ity change between 1 and 2, is

$$_1A_2 = T_0(S_2 - S_1) - (E_2 - E_1) - P_0(V_2 - V_1)$$
$$= (E_1 + P_0V_1 - T_0S_1) - (E_2 + P_0V_2 - T_0S_2)$$
$$= \phi_1 - \phi_2$$

where ϕ is the *availability function*.

When the closed, finite system undergoes no change in volume or position,

$$_1A_2 = (U_1 - T_0S_1) - (U_2 - T_0S_2)$$

In certain systems, such as a closed calorimeter for the determination of "heat of reaction," there is heat transfer to the environment and the system is in thermal equilibrium with the environment before and after the process. Then T_1 and $T_2 = T_0$, and

$$_1A_2 = (U_1 - T_1S_1) - (U_2 - T_2S_2) = (F_1 - F_2)_{T_0V}$$

where F is the **Helmholtz function** sometimes called the "nonflow" or "batch" energy function because of its significance in the situation described. When the Helmholtz function was defined in Section 10.1.2, it was noted that

$$dF = -P\,dV - S\,dT$$

For a simple, nonreacting substance with no change in volume or temperature, $dF = 0$, and change of availability has no significance.

4. Available energy from steady flow through an open system at steady state. The first law for this system gives the *useful work* (which is the *shaft work*) as

$$\Sigma \mathcal{W}_u = \Sigma m_i \left(h_i + \frac{\mathcal{V}_i^2}{2g_c} + \frac{g}{g_c}z_i \right) + \Sigma \mathcal{Q}_0 - \Sigma m_e \left(h_e + \frac{\mathcal{V}_e^2}{2g_c} + \frac{g}{g_c}z_e \right)$$
$$= (\Sigma H_i - \Sigma H_e) + \Sigma \mathcal{Q}_0 - \Sigma \Delta KE - \Sigma \Delta PE$$

Again, from the entropy principle, $\Sigma \mathcal{Q}_0 \leq T_0(S_0 - S_i)$. Consequently, the maximum possible useful work, which is the availability, would be

$$_iA_e = (\Sigma H_i - \Sigma H_e) - T_0(S_i - S_e) - \Sigma \Delta KE - \Sigma \Delta PE$$

Frequently there is only one inlet and one outlet and ΔKE and ΔPE are negligible. Then

$$_iA_e = (H_i - T_0S_i) - (H_e - T_0S_e) = B_i - B_e$$

where B is the **Darrieus function**.

To generalize to more than one inlet, more than one exhaust, and significant changes in kinetic and potential energies,

$$\Sigma_i A_e = (\Sigma B_i - \Sigma B_e) - \Sigma KE - \Sigma PE$$

In certain systems, such as fuel cell or an open "calorimeter" for determining the energy of reaction of a mixture of fluids, both the inlet and exhaust streams are in equilibrium with the environment at T_0 and P_0. When ΔKE and ΔPE can be ignored, the availability of the process is

$$_iA_e = (H_i - T_iS_i) - (H_e - T_eS_e) = (G_i - G_e)_{T_0P_0}$$

where G is the **Gibbs function**, sometimes called the steady-state–steady-flow energy function. In Section 10.2.2 it was noted that

$$dG = V\,dP - S\,dT$$

Thus, the availability for a nonreacting open system with $T_i = T_e$ and $P_i = P_e$ would be zero.

Effectiveness
Since only a part of the energy associated with or delivered to a system is available for conversion to useful work, the value of thermal efficiency (η_{th}) does not truly indicate the effectiveness of a device

used to harness the energy. The concept of availability does provide a meaningful comparison. The **effectiveness** (ε') of a device or a system is defined as

$$\varepsilon' = \frac{\text{useful work actually performed}}{\text{energy available for useful work}}$$

Note that this term evaluates the performance of a device without consideration of whether an alternative system might have been more effective.

Consider the adiabatic expansion of a fluid through a turbine, as shown in Fig. 10.31. The adiabatic expansion efficiency of the turbine in the process from i to e is

$$\eta_{E(s)} = \frac{h_i - h_e}{h_i - h_{e(s)}} = \frac{h_i - h_e}{\left(h_i - h_e\right) + \left(h_e - h_{e(s)}\right)}$$

The turbine *loss* would be $h_e - h_{e(s)}$, which is the difference between the actual turbine work and the isentropic work.

The available energy in this process is the change in the Darrieus function (b kJ/kg), so that the effectiveness is

$$\varepsilon' = \frac{h_i - h_e}{\phi_i - \phi_e} = \frac{h_i - h_e}{b_i - b_e}$$

$$= \frac{h_i - h_e}{\left(h_i - T_0 S_i\right) - \left(h_e - T_0 S_e\right)}$$

$$= \frac{h_i - h_e}{\left(h_i - h_e\right) + T_0\left(s_e - s_i\right)}$$

The *lost work* in this process is $T_0(s_e - s_i)$. Because $T_0(s_e - s_i)$ is less than $h_e - h_{e(s)}$, the effectiveness is greater than the efficiency. This recognizes the fact that, although the work done was less than the work of isentropic expansion, the loss was retained by the fluid; a part of this loss (above T_0) is still theoretically available for work production, as in a subsequent stage of a multistage turbine.

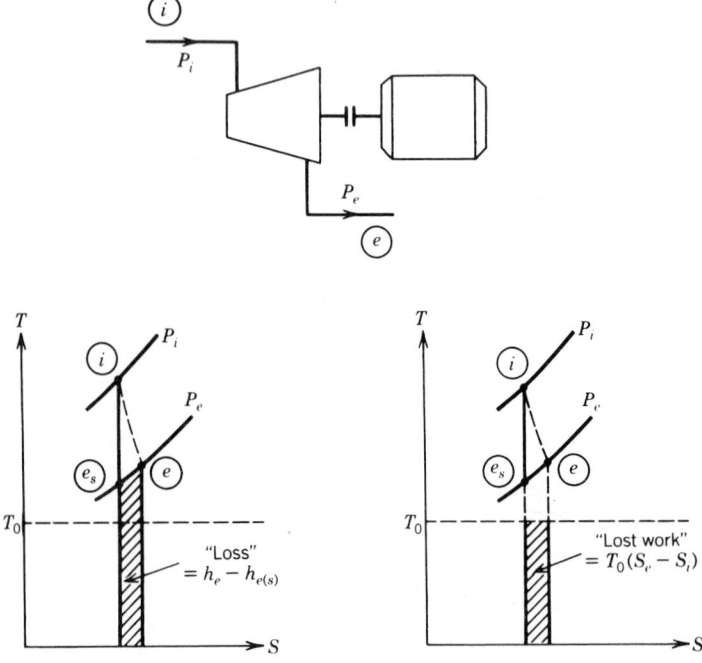

Fig. 10.31 Comparison between loss and lost work in turbine expansion.

Second-Law Efficiency

The term *second-law efficiency* (ε'') goes beyond the term *effectiveness* (ε') in the application of the concept of availability. While effectiveness evaluates how well a system or device has utilized the available energy, the second-law efficiency considers whether some other system might accomplish the same task with a smaller expenditure of available energy. It therefore evaluates the *task*:

$$\varepsilon'' = \frac{\text{heat or work usefully transferred}}{\begin{array}{c}\text{maximum possible heat or work} \\ \text{transfer for the same energy input}\end{array}}$$

$$= \frac{\text{minimum availability required}}{\text{actual availability expended}}$$

A simple example of the significance of the second-law efficiency is found in the task of heating a building. The first-law efficiency of a furnace would be the heat actually delivered to the building divided by the heating value of the fuel used. This efficiency might be very high. The second-law evaluation of the task would consider possible alternatives to the furnace. The fuel consumed by an engine or steam power plant used to drive a heat pump that would in turn deliver heat to the building might show that the furnace consumed much more fuel than that supplied to the combined power and heat pump system accomplishing the same task. The second-law efficiency would then be lower than the first-law value and more significant.

Of course a complete economic evaluation of alternative methods of accomplishing a particular task must consider the cost in money, materials, and energy to provide alternative equipment.

When the sole purpose of a device is the production of useful work, then the amount of work actually produced is the minimum availability required. In that case, the second-law efficiency is the same as the effectiveness.

BIBLIOGRAPHY

Benson, Rowland S., *Advanced Engineering Thermodynamics*, 2nd ed., Oxford, Pergamon, 1977.

Denbigh, Kenneth, *The Principles of Chemical Equilibrium*, 3rd ed., London, Cambridge University Press, 1971.

Hatsopoulos, G. N., and Keenan, J. H., *Principles of General Thermodynamics*, New York, Wiley, 1965.

Keenan, Joseph H., *Thermodynamics*, New York, Wiley, 1941.

Kestin, Joseph, *A Course in Thermodynamics*, Waltham, Mass., Blaisdell, Vol. 1, 1966; Vol. II, 1968.

Lee, J. F., and Sears, F. W., *Thermodynamics*, 2nd ed., Addison-Wesley, Reading, Mass., 1963.

Reynolds, W. C., and Perkins, H. C., *Engineering Thermodynamics*, 2nd ed., New York, McGraw-Hill, 1977.

Roberts, J. K. and Miller, A. R., *Heat and Thermodynamics*, 4th ed., London, Blackie, 1951.

Van Wylen, G. J. and Sonntag, R. E., *Fundamentals of Classical Thermodynamics*, 3rd ed., New York, Wiley, 1985.

Wilbur, Leslie C. (Ed), *Handbook of Energy Systems Engineering*, New York, Wiley, 1985.

Wood, Bernard D., *Applications of Thermodynamics*, 2nd ed., Addison-Wesley, Reading, Mass., 1982.

Zemansky, Mark W., *Heat and Thermodynamics*, 5th ed., New York, McGraw-Hill, 1957.

Zemansky, M. W., Abbott, M. M., and Van Ness, H. C., *Basic Engineering Thermodynamics*, 2nd ed., New York, McGraw-Hill, 1975.

10.2 HEAT TRANSFER
Frank P. Incropera

List of Symbols

English Alphabet

A	Area	c_p	Specific heat
b	Film width	D	Diameter
Bi	Biot number	E	Emissive power

English Alphabet (Continued)

f	Friction factor		q	Heat rate
F_{ij}	View factor		q''	Heat flux
Fo	Fourier number		$\dot{q}$	Heat generation rate per unit volume
G	Irradiation		r	Radius
g	Gravitational acceleration		R_t	Thermal resistance
h	Convection coefficient		Ra	Rayleigh number
h_{fg}	Latent heat of vaporization		Re	Reynolds number
J	Radiosity		S	Conduction shape factor
k	Thermal conductivity		t	Time
L	Length or thickness		T	Temperature
$\dot{m}$	Mass flow rate		u_m	Mean velocity
Nu	Nusselt number		U	Overall heat transfer coefficient
P	Perimeter		V	Volume; velocity
Pr	Prandtl number		x	Longitudinal coordinate

Greek Symbols

α	Thermal diffusivity; absorptivity		μ	Dynamic viscosity
β	Thermal expansion coefficient		ν	Kinematic viscosity
Γ	Flow rate per unit width		σ	Surface tension; Stefan–Boltzmann constant
δ	Boundary layer thickness; film thickness			
ε	Fin effectiveness; emissivity		ρ	Density; reflectivity
η	Fin efficiency		θ	Reduced temperature $(T-T_f)$
λ	Wavelength			

Subscripts

abs	Absorbed		l	Liquid
b	Fin base condition; blackbody		m	Mean
c	Contact; cross-sectional; critical		o	Outlet; outer
cond	Conduction		rad	Radiation
conv	Convection		ref	Reflected
e	Excess		s	Surface
f	Fluid; fin		sat	Saturated
fd	Fully developed		t	Thermal
h	Hydrodynamic		v	Vapor
i	Initial; inlet; inner			

10.2.1 Introduction

In any engineering system that is characterized by temperature differences, **heat transfer** will occur in the direction of decreasing temperature. Since it is the **internal thermal energy** of matter that is being transferred, the process could be **transient**, causing the temperature (thermal energy) of hotter regions to decrease with time, while the temperature (thermal energy) of the colder regions increased. This condition would prevail if energy were not being supplied to and extracted from the hot and cold regions, respectively, by other means. The process would continue until the temperatures equalized, at which point **thermal equilibrium** would be established and there would no longer be any heat transfer. Alternatively, if thermal energy were supplied to and extracted from the hot and cold regions, respectively, at the same rate at which heat transfer occurred between the two regions, the process would be **steady** and temperatures would be invariant with time.

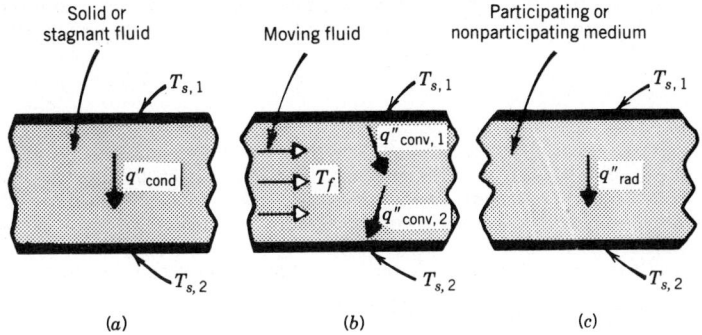

Fig. 10.32 Heat transfer between isothermal surfaces maintained at temperatures $T_{s,1} > T_{s,2}$: (a) conduction, (b) convection ($T_{s,1} > T_f > T_{s,2}$), (c) thermal radiation.

Modes of Heat Transfer

There are three ways in which heat may be transferred in the direction of decreasing temperature. Termed **modes**, they include heat transfer by *conduction*, *convection*, and *thermal radiation* (Fig. 10.32). If temperature gradients are induced in a solid or stagnant fluid layer by maintaining its surfaces at different temperatures, heat is transferred through the layer by conduction. In the presence of a temperature gradient, it is the random motion of molecules in the medium that provides for the transfer of thermal energy. The basic equation that governs this transfer is termed **Fourier's law**, and its vector form is

$$\mathbf{q}'' = -k\nabla T \tag{10.1}$$

The heat flux vector, which represents the rate at which heat is transferred per unit area perpendicular to the direction of transfer (W/m^2), is proportional to the gradient of the scalar temperature field ($°C/m$), where the proportionality constant k is a property of the medium termed the **thermal conductivity**. The minus sign is a consequence of the fact that heat is transferred in the direction of decreasing temperature.

In practice, fluids are seldom stationary and, if temperature variations are imposed on a moving fluid, thermal energy is transferred by the bulk, as well as the random, motion of the fluid. A common example involves motion of a fluid at one temperature, T_f, over a solid surface of different temperature, T_s. Heat transfer is said to occur by convection, and the rate of heat transfer per unit surface area is expressed in terms of **Newton's law of cooling**

$$q'' = h(T_s - T_f) \tag{10.2}$$

where the convection coefficient h has units of $W/m^2 \ °C$. For the conditions of Fig. 10.32(b), there is heat transfer by convection from surface 1 to the fluid and from the fluid to surface 2. Convection is strongly influenced by *velocity* and *thermal boundary layers* that develop in the fluid adjoining a surface.

Thermal radiation refers to energy which is released (emitted) in the form of electromagnetic waves by matter that is at a finite absolute temperature. Emission is an energy conversion process, whereby thermal energy of the matter is converted to electromagnetic energy of the waves. If the waves are intercepted and *absorbed* by matter, there is then a conversion from electromagnetic energy back to thermal energy. In this manner, there is radiant energy exchange between the two surfaces of Fig. 10.32(c). Both surfaces emit and absorb radiation, but if $T_{s,1} > T_{s,2}$, there is a net transfer from surface 1 to 2. This exchange can, of course, be influenced by a *participating* intervening medium. An example would be a high-temperature combustion gas that absorbs some of the radiation emitted by the surfaces, while emitting radiation that can be absorbed by the surfaces. If, instead, the surfaces were separated by air, the medium would be *nonparticipating* (neither emitting nor absorbing) and would hence have no influence on radiation exchange between the two surfaces.

10.2.2 Conduction

The Heat Equation

Fourier's law indicates that heat transfer by conduction in a solid may be computed if the temperature distribution in the solid is known. The temperature distribution may be determined by solving a differential equation that is based on the requirement that energy must be conserved for an infinitesimally small control volume located anywhere in the solid. Termed the **heat equation**, it is of the form

$$k \nabla^2 T + \dot{q} = \rho c_p \frac{\partial T}{\partial t} \tag{10.3}$$

where constant properties (k, ρ, c_p) are assumed and, from left to right, the terms account for net energy transfer by conduction, thermal energy generation, and thermal energy storage. The *energy generation* term $\dot{q}$, represents a thermal energy source due to ohmic (electric) heating, chemical reactions, or nuclear reactions. Each term has the units of energy rate per unit volume. Exact solutions to Eq. (10.3) may be obtained for several simplified, yet important, conditions.

One-Dimensional, Steady Conduction

Common geometries for which one-dimensional conduction may often be assumed include the large plane wall, the *long* cylindrical tube, and the spherical shell (Fig. 10.33). Assuming steady-state conditions with no thermal energy generation, the appropriate forms of the heat equation are listed in Table 10.2, along with the corresponding temperature distributions. The temperature distributions are obtained by solving the heat equations subject to prescribed uniform temperatures, $T_{s,1}$ and $T_{s,2}$, at the inner and outer surfaces, where $\Delta T = T_{s,1} - T_{s,2}$. Once the temperature distribution is known, Fourier's law may be applied to obtain the heat flux, q'', which may then be multiplied by the heat transfer area to obtain the heat rate. In each case a thermal resistance may be defined as the ratio of the driving potential (ΔT) to the current (heat) flow (q):

$$R_t \equiv \frac{\Delta T}{q} \tag{10.4}$$

From Table 10.2 it is evident that both q'' and q are constant (independent of x) for one-dimensional, steady-state conduction in a plane wall without generation. However, although q is constant (independent of r) for the cylindrical and spherical walls, q'' decreases with increasing r and r^2, respectively.

Frequently, conduction systems are comprised, not of a single wall but, of a *composite wall* in which solids of different thermal conductivity are joined in series or parallel. Moreover, temperatures may not be known for the outer surfaces of the composite but, instead, may be known for fluids passing over the surfaces. A representative case is shown in Fig. 10.34, where two solid materials of different thermal conductivity are joined and heat transfer through the composite results from the flow of a hot fluid through the inner tube and a cold fluid over the outer surface. The rate of heat transfer in the radial direction is a constant whose value depends on the overall temperature difference, $\Delta T = T_{f,1} - T_{f,3}$, and the total thermal resistance, R_{tot}. This resistance would include contributions due to conduction in A and B, as well as an interfacial contact resistance between A and B and convection resistances at the inner and outer surfaces.

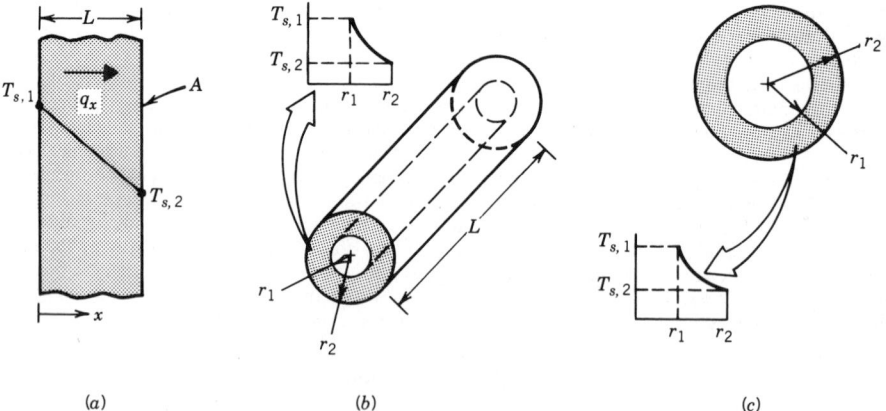

(a)   (b)   (c)

Fig. 10.33 One-dimensional conduction systems: (*a*) plane wall of area A, (*b*) long cylindrical tube of length L, (*c*) spherical shell.

TABLE 10.2 One-dimensional, Steady-state Solutions to the Heat Equation with No Generation

	Plane Wall	Cylindrical Wall	Spherical Wall
Heat equation	$\dfrac{d^2T}{dx^2} = 0$	$\dfrac{1}{r}\dfrac{d}{dr}\left(r\dfrac{dT}{dr}\right) = 0$	$\dfrac{1}{r^2}\dfrac{d}{dr}\left(r^2\dfrac{dT}{dr}\right) = 0$
Temperature distribution	$T_{s,1} - \Delta T\dfrac{x}{L}$	$T_{s,2} + \Delta T\dfrac{\ln(r/r_2)}{\ln(r_1/r_2)}$	$T_{s,1} - \Delta T\left[\dfrac{1 - (r_1/r)}{1 - (r_1/r_2)}\right]$
Heat flux (q'')	$k\dfrac{\Delta T}{L}$	$\dfrac{k\,\Delta T}{r\ln(r_2/r_1)}$	$\dfrac{k\,\Delta T}{r^2[(1/r_1) - (1/r_2)]}$
Heat rate (q)	$kA\dfrac{\Delta T}{L}$	$\dfrac{2\pi Lk\,\Delta T}{\ln(r_2/r_1)}$	$\dfrac{4\pi k\,\Delta T}{(1/r_1) - (1/r_2)}$
Thermal resistance ($R_{t,\mathrm{cond}}$)	$\dfrac{L}{kA}$	$\dfrac{\ln(r_2/r_1)}{2\pi Lk}$	$\dfrac{(1/r_1) - (1/r_2)}{4\pi k}$

In general, contact and convection resistances are defined as

$$R_{t,c} = \frac{R''_{t,c(i)}}{A_i} \tag{10.5}$$

$$R_{t,\mathrm{conv}} = \frac{1}{h_i A_i} \tag{10.6}$$

where A_i is the contact or convection surface area, $R''_{t,c}$ is the contact resistance for a unit area (m²-°C/W), and h_i is the surface convection coefficient. The value of $R''_{t,c}$ depends on the types of adjoining materials, their surface roughnesses, and the interface pressure.[1] To determine the heat rate for one-dimensional transfer in a composite system with multiple conduction, contact, and convection resistances, it is only necessary to divide the overall temperature difference by the total resistance:

$$q = \frac{\Delta T}{R_{\mathrm{tot}}} = \frac{\Delta T}{\Sigma(R_{t,\mathrm{cond}} + R_{t,c} + R_{t,\mathrm{conv}})} \tag{10.7}$$

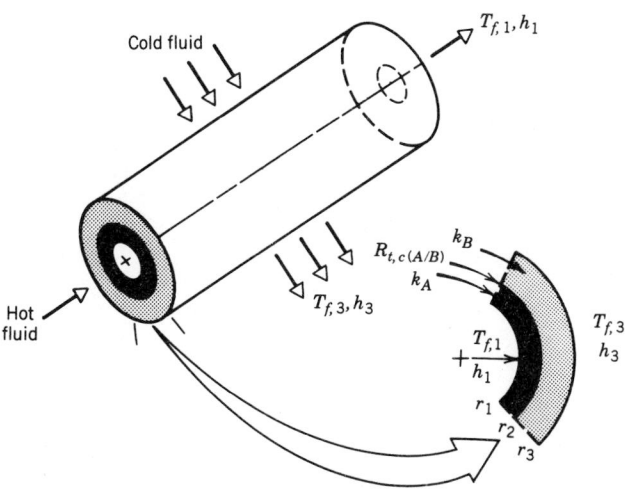

Fig. 10.34 Two-layer cylindrical composite with convection at the inner and outer surfaces and an interfacial contact resistance.

Alternatively, the heat rate may be expressed in terms of an *overall heat transfer coefficient* U_i

$$q = U_i A_i \Delta T \tag{10.8}$$

where U_i depends on the choice of surface area A_i and is related to the total resistance

$$U_i A_i = R_{tot}^{-1} \tag{10.9}$$

If thermal energy is generated within a plane, cylindrical or spherical wall, the appropriate form of the heat equation is obtained by simply adding $\dot{q}/k$ to the left-hand side of the differential equations listed in Table 10.2. Subject to appropriate boundary conditions, each equation may readily be solved for the temperature distribution, which may then be used with Fourier's law to determine the conduction heat flux and rate.[2] With generation, the heat rate is no longer a constant, independent of x or r, in which case it may no longer be related to the temperature difference through the conduction resistance.

Extended Surfaces

In thermal system design, there is often the need to *enhance* heat transfer between a wall and an adjoining fluid. An important option for effecting such enhancement involves using **extended surfaces** or **fins**, which increase the surface area available for convection. Common examples are shown in Fig. 10.35. The pin and triangular fins are mounted to a plane wall, and the annular fin is attached to a circular tube. The pin has a uniform circular cross section, while the triangular and annular fins have rectangular cross sections that vary with x and r, respectively.

Assuming steady one-dimensional conduction, negligible surface radiation, and a uniform convection coefficient, the fin temperature distribution is governed by the following differential equation[2]

$$\frac{d^2\theta}{dx^2} + \left(\frac{1}{A_c} \frac{dA_c}{dx} \right) \frac{d\theta}{dx} - \left(\frac{1}{A_c} \frac{h}{k} \frac{dA_s}{dx} \right) \theta = 0 \tag{10.10}$$

where $\theta(x)$ is the temperature at any point along the fin relative to the fixed fluid temperature ($\theta = T - T_f$) and x may be replaced by r for an annular fin. If the fin is of uniform cross-sectional area ($dA_c/dx = 0$), the change in the surface area with x is simply equal to the perimeter ($dA_s/dx = P$) and the *fin equation* reduces to

$$\frac{d^2\theta}{dx^2} - m^2\theta = 0 \tag{10.11}$$

where

$$m^2 = \frac{hP}{kA_c} \tag{10.12}$$

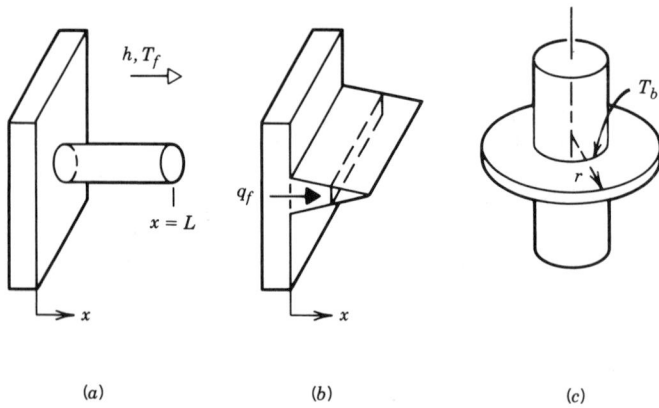

(a)   (b)   (c)

Fig. 10.35 Fin configurations: (a) pin fin, (b) triangular fin, (c) annular fin.

Equation (10.10) or (10.11) may be solved subject to boundary conditions imposed at the base $(x = 0)$ and the tip $(x = L)$ of the fin. Typically, the base temperature T_b is prescribed, in which case $\theta(0) = \theta_b$, while the tip condition may correspond to a prescribed temperature, an adiabatic surface, or convective transport. In this third case, heat transfer by conduction to the tip is balanced by convection from the tip:

$$-k\frac{d\theta}{dx}\bigg|_{x=L} = h\theta(L) \tag{10.13}$$

Using Eq. (10.13) as a boundary condition in solving Eq. (10.11), the fin temperature distribution is

$$\frac{\theta}{\theta_b} = \frac{\cosh m(L-x) + (h/mk)\sinh m(L-x)}{\cosh mL + (h/mk)\sinh mL} \tag{10.14}$$

Applying Fourier's law at $x = 0$, the corresponding fin heat transfer rate is

$$q_f = \frac{\sinh mL + (h/mk)\cosh mL}{\cosh mL + (h/mk)\sinh mL}(hPkA_c)^{1/2} \tag{10.15}$$

Solutions to Eq. (10.11) for other tip conditions, as well as solutions to Eq. (10.10) for fins of nonuniform cross section, have been obtained.[2-4] Often the results are presented in terms of fin performance parameters such as the effectiveness

$$\varepsilon_f = \frac{q_f}{hA_{c,b}\theta_b} \tag{10.16}$$

and the efficiency

$$\eta_f = \frac{q_f}{q_{max}} \tag{10.17}$$

The effectiveness assesses pin performance by comparing fin heat transfer to the heat transfer that would occur without the fin. Values should be well in excess of unity. The efficiency compares actual fin heat transfer to the maximum possible transfer, which would occur if the entire fin were at the base temperature.

Two-Dimensional, Steady Conduction

In many cases, conduction analyses are oversimplified by invoking a one-dimensional approximation, and it is necessary to account for multidimensional effects. The heat equation, Eq. (10.3), becomes a partial differential equation that typically must be solved numerically, using approximate finite-difference or finite-element techniques. However, for certain simple geometries and boundary conditions, exact (analytical) solutions may be obtained.[5-7]

A simple, approximate procedure for determining the conduction heat rate in two-dimensional geometries with isothermal boundaries uses the **conduction shape factor** S, such that

$$q = kS(T_1 - T_2) \tag{10.18}$$

The shape factor has been determined for many geometries,[8-10] and representative results are shown in Table 10.3.

Transient Conduction

Lumped Capacitance Analysis. Transient conduction occurs when a solid experiences a change in its surface thermal condition as, for example, when a hot metal billet is quenched in a liquid bath (Fig. 10.36). If the solid is initially at a uniform temperature T_i, its temperature will decrease with time, until thermal equilibrium with the fluid is eventually reached. The reduction in temperature is due to convection heat transfer from the solid to the fluid.

Under conditions for which the resistance to heat transfer by conduction in the solid is small relative to heat transfer by convection from the solid to the fluid, spatial temperature variations within the solid may be neglected during the transient process. Hence, the solid may be characterized by a single temperature $T(t)$ whose value changes with time. The necessary conditions correspond to

TABLE 10.3   Conduction Shape Factors

Geometry	Schematic	Shape Factor
Isothermal sphere at T_1 in a semiinfinite medium with surface at T_2.		$\dfrac{2\pi D}{1 - D/4z}$
Isothermal cylinder of length L at T_1 in a semiinfinite medium with axis parallel to surface at $T_2 (L \gg D)$.		$\dfrac{2\pi L}{\cosh^{-1}(2z/D)}$
Isothermal cylinder of length L at T_1 centered in a square solid rod of same length with surface at $T_2 (L \gg W)$.		$\dfrac{2\pi L}{\ln(1.08W/D)}$
Isothermal cylinder of length L at T_1 in a semiinfinite medium with axis normal to surface at T_2 $(L \gg D)$.		$\dfrac{2\pi L}{\ln(4L/D)}$

satisfaction of the following inequality

$$\text{Bi} = \frac{h(V/A_s)}{k} < 0.1 \tag{10.19}$$

where the Biot number Bi is interpreted as a ratio of conduction to convection resistances.

If Bi < 0.1, the transient response may be determined from a **lumped capacitance analysis** for which the heat loss by convection is equated to the rate of change of thermal energy for the solid. It

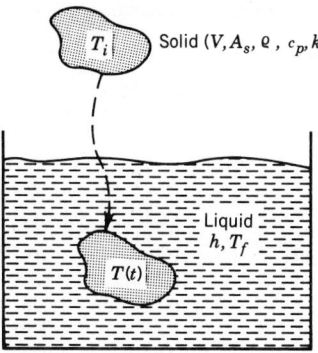

Fig. 10.36   Quenching of hot billet in liquid bath.

follows that

$$\frac{T - T_f}{T_i - T_f} = \exp\left(-\frac{hA_s}{\rho V c_p}t\right) \tag{10.20}$$

in which case the temperature of the solid, relative to T_f, decays exponentially to zero.

    One-Dimensional Conduction in Finite Solids. If Bi > 0.1, temperature gradients within the solid are significant and variations of temperature with position, as well as time, should be considered. Simple, yet important, cases for which exact solutions to the heat equation have been obtained correspond to one-dimensional conduction in a plane wall, a long (infinite) cylinder, and a sphere (Fig. 10.37). In each case, the solid is initially at a uniform temperature, $T(x, 0) = T_i$ or $T(r, 0) = T_i$, and thermal change is induced by allowing for convection heat transfer to or from a fluid of different temperature. The subsequent thermal response, $T(x, t)$ or $T(r, t)$, is determined by solving the appropriate form of the heat equation subject to imposed symmetry (at $x = 0$ or $r = 0$) and convection (at $x = L$ or $r = r_0$) conditions.

    Although the foregoing solutions take the form of infinite series, simple one-term approximations may be used when the dimensionless time, termed the Fourier number, Fo, exceeds 0.2. Hence, for the plane wall and Fo $= \alpha t/L^2 > 0.2$,

$$\frac{T - T_f}{T_i - T_f} \approx C_1 \exp\left(-\zeta_1^2 \mathrm{Fo}\right)\cos(\zeta_1 x^*) \tag{10.21}$$

where $x^* = x/L$ and the constants C_1 and ζ_1 depend on Bi $= hL/k$ (Table 10.4). Similarly, when

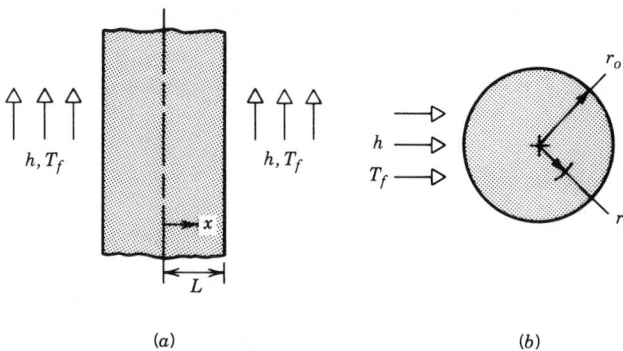

(a)   (b)

Fig. 10.37   Transient, one-dimensional conduction: (a) plane wall of thickness $2L$ with symmetric convection, (b) long cylinder or sphere of radius r_o.

TABLE 10.4 Constants Used in One-term Approximations to Series Solutions for Transient, One-dimensional Conduction

	Plane Wall		Infinite Cylinder		Sphere	
Bi	ζ_1 (rad)	C_1	ζ_1 (rad)	C_1	ζ_1 (rad)	C_1
0.02	0.1410	1.0033	0.1995	1.0050	0.2445	1.0060
0.04	0.1987	1.0066	0.2814	1.0099	0.3450	1.0120
0.06	0.2425	1.0098	0.3438	1.0148	0.4217	1.0179
0.08	0.2791	1.0130	0.3960	1.0197	0.4860	1.0239
0.10	0.3111	1.0160	0.4417	1.0246	0.5423	1.0298
0.15	0.3779	1.0237	0.5376	1.0365	0.6608	1.0445
0.20	0.4328	1.0311	0.6170	1.0483	0.7593	1.0592
0.25	0.4801	1.0382	0.6856	1.0598	0.8448	1.0737
0.30	0.5218	1.0450	0.7465	1.0712	0.9208	1.0880
0.4	0.5932	1.0580	0.8516	1.0932	1.0528	1.1164
0.6	0.7051	1.0814	1.0185	1.1346	1.2644	1.1713
0.8	0.7910	1.1016	1.1490	1.1725	1.4320	1.2236
1.0	0.8603	1.1191	1.2558	1.2071	1.5708	1.2732
2.0	1.0769	1.1795	1.5995	1.3384	2.0288	1.4793
3.0	1.1925	1.2102	1.7887	1.4191	2.2889	1.6227
4.0	1.2646	1.2287	1.9081	1.4698	2.4556	1.7201
6.0	1.3496	1.2479	2.0490	1.5253	2.6547	1.8338
8.0	1.3978	1.2570	2.1286	1.5526	2.7654	1.8921
10.0	1.4289	1.2620	2.1795	1.5677	2.8363	1.9249
20.0	1.4961	1.2699	2.2881	1.5919	2.9857	1.9781
40.0	1.5325	1.2723	2.3455	1.5993	3.0632	1.9942
100.0	1.5552	1.2731	2.3809	1.6015	3.1102	1.9990

$Fo = \alpha t / r_o^2 > 0.2$

$$\frac{T - T_f}{T_i - T_f} \approx C_1 \exp\left(-\zeta_1^2 Fo\right) J_0\left(\zeta_1 r^*\right) \tag{10.22}$$

for the infinite cylinder and

$$\frac{T - T_f}{T_i - T_f} \approx C_1 \exp\left(-\zeta_1^2 Fo\right) \frac{\sin(\zeta_1 r^*)}{\zeta_1 r^*} \tag{10.23}$$

for the sphere, where $r^* = r/r_o$, J_0 is the zero-order Bessel function of the first kind, and C_1 and ζ_1 are listed in Table 10.4. Convenient graphical representations of the foregoing equations are provided in most introductory heat transfer texts.[2]

One-Dimensional Conduction in Semiinfinite Solids. Another simple geometry for which analytical solutions have been obtained is the **semiinfinite solid**. Such a solid has a single, identifiable surface and extends to infinity with increasing distance from the surface. If its temperature is initially uniform and a different surface temperature is imposed at $t = 0$ [Fig. 10.38(a)], its thermal response is given by

$$\frac{T(x, t) - T_s}{T_i - T_s} = \text{erf}\left[\frac{x}{2(\alpha t)^{1/2}}\right] \tag{10.24}$$

This expression may be used with Fourier's law to determine the conduction heat flux, and evaluating the result at the surface it follows that

$$q_s''(t) = \frac{k(T_s - T_i)}{(\pi \alpha t)^{1/2}} \tag{10.25}$$

Equations (10.24) and (10.25) could be used to predict conditions for the early stages of the transient

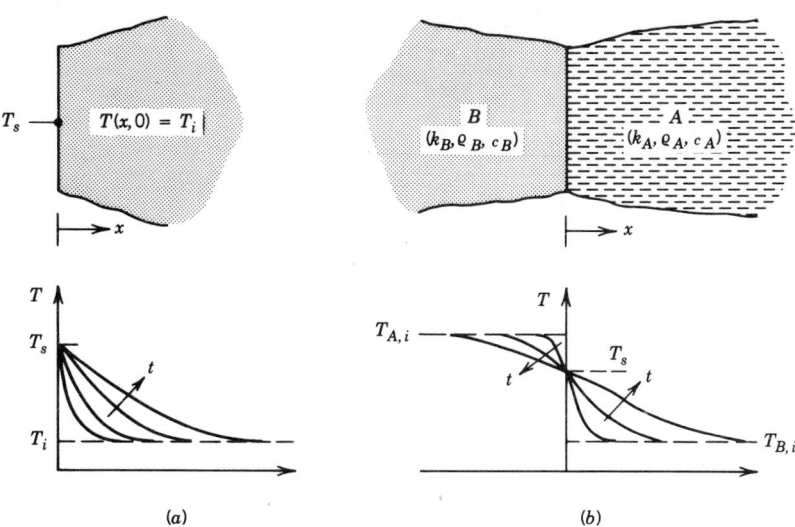

Fig. 10.38 Conduction in a semiinfinite solid: (*a*) imposed surface temperature, (*b*) contact between two semiinfinite solids at different initial temperatures.

in a finite solid wall, up to the point in time at which thermal effects penetrate to the opposite surface. The equations may also be used to predict the thermal response of two semiinfinite solids, which are initially at different uniform temperatures and are brought into contact [Fig. 10.38(*b*)]. Equation (10.25) may be written for each solid, and from conservation of energy, the results may be equated to yield the following expression for the surface (contact) temperature

$$T_s = \frac{(k\rho c)_A^{1/2} T_{A,i} + (k\rho c)_B^{1/2} T_{B,i}}{(k\rho c)_A^{1/2} + (k\rho c)_B^{1/2}} \qquad (10.26)$$

The surface temperature is independent of time and may be used with Eqs. (10.24) and (10.25) to obtain temperatures in each of the solids and the surface heat flux.

10.2.3 Convection

Convection heat transfer is due to fluid motion over a surface of different temperature. Different types of convection phenomena are associated with whether the flow is forced or natural, external or internal, and single phase or multiphase. In the next three sections we consider convection heat transfer without phase change, and in the fourth section we consider phase-change phenomena corresponding to boiling and condensation.

Forced Convection: External Flow
In external flow, boundary layer development on a surface occurs without constraints imposed by neighboring surfaces. Accordingly, there will always exist a region of flow outside the boundary layer in which velocity and temperature gradients are negligible. Common examples include the flat plate in parallel flow, the cylinder in cross flow, and the tube bank in cross flow (Fig. 10.39). In each case a portion, or all, of the surface is presumed to be at a uniform temperature T_s, which differs from the free-stream fluid temperature T_∞ (for external flows $T_f = T_\infty$).

For the flat plate in parallel flow [Fig. 10.39(*a*)], conditions depend on whether the flow is laminar or turbulent and on whether the heat transfer section for which $T_s \neq T_\infty$ is preceded by an *unheated section* for which $T_s = T_\infty$. With an unheated section of length ξ, the local dimensionless convection coefficient, or Nusselt number ($\text{Nu}_x = h_x x/k$), on the heated section is

$$\text{Nu}_x = \frac{\text{Nu}_x|_{\xi=0}}{\left[1 - (\xi/x)^m\right]^n} \qquad (10.27)$$

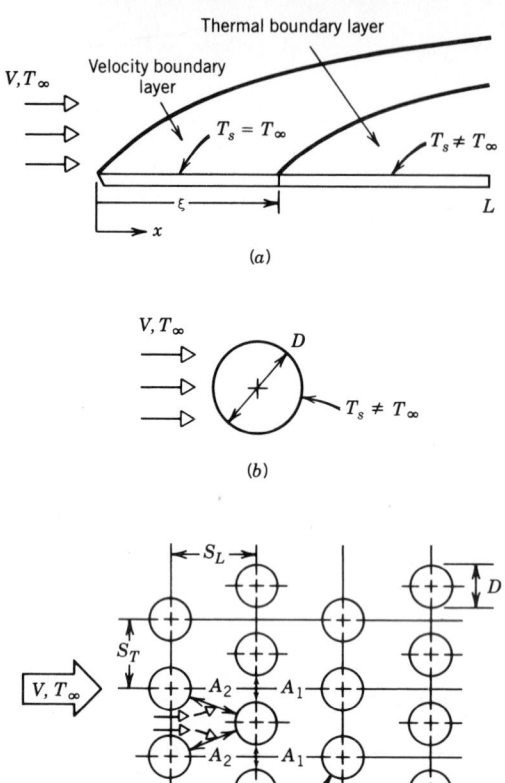

Fig. 10.39 Common external flow geometries: (*a*) flat plate in parallel flow, (*b*) cylinder in cross flow, (*c*) tube bank in cross flow.

where ($m = \frac{3}{4}$, $n = \frac{1}{3}$) and ($m = \frac{9}{10}$, $n = \frac{1}{9}$) for laminar and turbulent flow, respectively. The local Nusselt number corresponding to no unheated section, $Nu_x|_{\xi=0}$, is given by

$$Nu_x|_{\xi=0} = 0.332 \, Re_x^{1/2} \, Pr^{1/3} \tag{10.28}$$

$$Nu_x|_{\xi=0} = 0.0296 \, Re_x^{4/5} \, Pr^{1/3} \tag{10.29}$$

for laminar and turbulent flow, respectively, where $Re_x = Vx/\nu$ is the Reynolds number of the flow and Pr, the Prandtl number, is a fluid property. Laminar flow will exist for Reynolds numbers less than a critical value $Re_{x,c}$, while turbulent flow may be assumed for $Re_x > Re_{x,c}$. The critical Reynolds number depends on surface roughness and the turbulence level of the free stream, with values typically in the range $10^5 < Re_{x,c} < 3 \times 10^6$.

The average convection coefficient over a surface of length L may be obtained by integrating local values over the surface. For $\xi = 0$, it follows that, if $Re_L < Re_{x,c}$,

$$\overline{Nu}_L = 0.664 \, Re_L^{1/2} \, Pr^{1/3} \tag{10.30}$$

and, if $Re_L > Re_{x,c}$,

$$\overline{Nu}_L = \left[0.037\left(Re_L^{4/5} - Re_{x,c}^{4/5}\right) + 0.664 \, Re_{x,c}^{1/2}\right] Pr^{1/3} \tag{10.31}$$

In using the foregoing expressions, fluid properties should be evaluated at the mean temperature, $(T_s + T_\infty)/2$.

Numerous correlations have been proposed for heat transfer between a long cylinder and a fluid flowing normal to the cylinder's axis. One such correlation for the average coefficient, which applies over a wide range of conditions, is of the form[11]

$$\overline{\mathrm{Nu}}_D = 0.3 + \frac{0.62\,\mathrm{Re}_D^{1/2}\,\mathrm{Pr}^{1/3}}{\left[1 + (0.4/\mathrm{Pr})^{2/3}\right]^{1/4}}\left[1 + \left(\frac{\mathrm{Re}_D}{282{,}000}\right)^{5/8}\right]^{4/5} \tag{10.32}$$

where $\overline{\mathrm{Nu}}_D = \bar{h}D/k$ and $\mathrm{Re}_D = VD/\nu$. Due to boundary layer transition and separation phenomena, local coefficients exhibit sharp variations about the average. All properties are evaluated at $(T_s + T_\infty)/2$.

Many engineering applications involve a bank of cylinders or tubes in cross flow. The tube rows are either *staggered* [Fig. 10.39(c)] or *aligned* in the longitudinal (fluid flow) direction. For 20 or more tube rows in the flow direction, the average Nusselt number is given by[12]

$$\overline{\mathrm{Nu}}_D = C\,\mathrm{Re}_{D,\mathrm{max}}^m\,\mathrm{Pr}^{0.36}\,(\mathrm{Pr}/\mathrm{Pr}_s)^{1/4} \tag{10.33}$$

where $\mathrm{Re}_{D,\mathrm{max}} = V_{\mathrm{max}} D/\nu$ is based on the maximum flow velocity in the tube bank. For the staggered arrangement of Fig. 10.39(c), the maximum velocity will exist between adjoining tubes of a row (A_1) or along the diagonal between neighboring tubes of adjoining rows (A_2). Values of C and m depend on the tube bank geometry and $\mathrm{Re}_{D,\mathrm{max}}$. For a staggered array with $10^3 < \mathrm{Re}_{D,\mathrm{max}} < 2 \times 10^5$, $m = 0.60$ and $C = 0.35(S_T/S_L)^{0.2}$ or $C = 0.40$ for $S_T/S_L < 2$ or $S_T/S_L > 2$, respectively. For $2 \times 10^5 < \mathrm{Re}_{D,\mathrm{max}} < 2 \times 10^6$, $C = 0.022$, and $m = 0.84$. For an aligned array $C = 0.27$ and $m = 0.63$ if $10^3 < \mathrm{Re}_{D,\mathrm{max}} < 2 \times 10^5$, while $C = 0.021$ and $m = 0.84$ if $2 \times 10^5 < \mathrm{Re}_{D,\mathrm{max}} < 2 \times 10^6$. Except for Pr_s, all fluid properties are evaluated at $(T_s + T_\infty)/2$. The average Nusselt number decreases with decreasing row number below 20, but even for as few as 4 rows it remains within 10% of the value calculated from Eq. (10.33).

Forced Convection: Internal Flow
An internal flow is completely confined by a surface, in which case hydrodynamic and thermal boundary layers are unable to develop without eventually being constrained. Hydrodynamic boundary layer development for laminar flow in a circular tube is shown in Fig. 10.40. Entering with a uniform velocity profile, fluid is decelerated to zero velocity at the wall and viscous forces cause a hydrodynamic boundary layer of thickness δ to develop with increasing x. This development occurs at the expense of a shrinking inviscid region, in which the fluid is accelerated, until boundary layer merger occurs at the centerline. Shortly downstream of this point, a parabolic velocity profile is achieved. The profile is invariant with further increases in x, and *fully developed hydrodynamic conditions* are said to exist. In laminar flow the length of the *hydrodynamic entry region* is given by

$$\frac{x_{\mathrm{fd},h}}{D} \approx 0.05\,\mathrm{Re}_D \tag{10.34}$$

where $\mathrm{Re}_D = \rho u_m D/\mu$ and $u_m = \dot{m}/\rho A_c$ is the mean velocity.

Similarly, if the fluid enters at a uniform temperature $T_{m,i}$ and either a uniform surface temperature, $T_s \neq T_{m,i}$, or a uniform surface heat flux q_s'' is maintained, a thermal boundary layer will develop. Development of this boundary layer corresponds to a decay in the local Nusselt number from infinity at $x = 0$ to a *fully developed* value that is independent of x for $x > x_{\mathrm{fd},t}$. For laminar

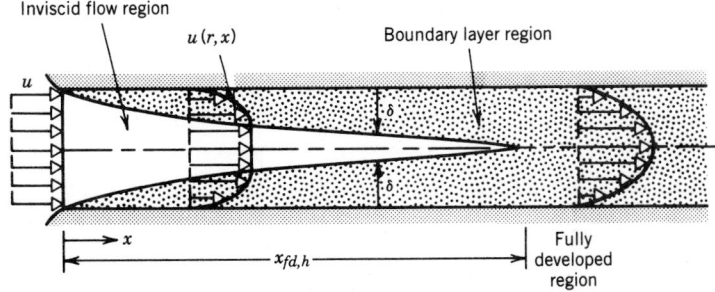

Fig. 10.40 Laminar boundary layer development in a circular tube.

flow in a circular tube, the length of the *thermal entry region* is given by

$$\frac{x_{fd,t}}{D} \approx 0.05 \, Re_D \, Pr \qquad (10.35)$$

For turbulent flow, the hydrodynamic and thermal entry lengths are approximately equal to each other and independent of Reynolds number. To a first approximation,

$$\frac{x_{fd,h}}{D} \approx \frac{x_{fd,t}}{D} \approx 10 \qquad (10.36)$$

Flow in a circular tube may be assumed to be turbulent for $Re_D > Re_{D,c} \approx 2300$.

For internal flows, the *mixed-mean* fluid temperature is the reference temperature ($T_m = T_f$) and, from energy balance considerations, its variation with x along a duct of uniform perimeter P can be expressed as

$$T_m(x) = T_{m,i} + \frac{q_s'' P}{\dot{m} c_p} x \qquad (10.37)$$

for a uniform surface heat flux, or as

$$\frac{T_s - T_m(x)}{T_s - T_{m,i}} = \left(-\frac{Px}{\dot{m} c_p} \bar{h} \right) \qquad (10.38)$$

for a uniform surface temperature. The total heat rate for a duct of length L is

$$q = q_s'' A_s \qquad (10.39)$$

for a uniform surface heat flux, or

$$q = \bar{h} A_s \frac{\Delta T_o - \Delta T_i}{\ln(\Delta T_o / \Delta T_i)} \qquad (10.40)$$

for a uniform surface temperature, where $A_s = PL$, $\Delta T = T_s - T_m$, and $T_{m,o}$ is the mean fluid temperature at $x = L$.

To use Eq. (10.38) or (10.40), the average convection coefficient from the tube inlet to x (or L) must be known. In turbulent flow, x_{fd} is often much less than L, and it is reasonable to assume that $\bar{h} \approx h_{fd}$. The most accurate correlation for the Nusselt number in fully developed turbulent flow is attributed to Petukhov[13] and has the form

$$Nu_D = \frac{(f/8) Re_D \, Pr}{1.07 + 12.7(f/8)^{1/2}(Pr^{2/3} - 1)} \left(\frac{\mu}{\mu_s} \right)^n \qquad (10.41)$$

where $n = 0.11$ for $T_s > T_m$ and $n = 0.25$ for $T_s < T_m$. All properties, except μ_s, are evaluated at T_m, or at $(T_{m,i} + T_{m,o})/2$ when using Eq. (10.41) to calculate Nu_D for the entire tube length. For smooth tubes the friction factor can be evaluated from

$$\frac{1}{f} = (1.82 \log_{10} Re_D - 1.64)^2 \qquad (10.42)$$

For laminar fully developed flow, a solution to the appropriate form of the energy equation for uniform surface temperature yields

$$Nu_D = 3.66 \qquad (10.43)$$

However, due to the long entry regions that typically characterize laminar flows, it is seldom reasonable to use Eq. (10.43) as an approximation for the average Nusselt number over the entire tube length. Instead, the following correlation due to Hausen[14] may be used:

$$\overline{Nu}_D = 3.66 + \frac{0.0668(D/L) Re_D \, Pr}{1 + 0.04[(D/L) Re_D \, Pr]^{2/3}} \qquad (10.44)$$

Free Convection

In forced convection, fluid motion is driven by an external agent such as a fan or pump. In free convection motion is driven by a *buoyancy force* that results from the presence of density variations in the fluid and a *body force* that is proportional to density. Typically, the body force is gravitational and the density variations are due to temperature gradients in the fluid.

Free convection problems may be grouped according to whether they involve an object submerged in an infinite, quiescent fluid or the walls of a fluid enclosure. Examples of the first group include vertical plates or horizontal plates and cylinders [Figs. 10.41(a), (b), and (c)], while examples of the second group include rectangular cavities or annular spaces [Figs. 10.41(d) and (e)]. In each case fluid motion is driven by buoyancy forces arising from imposed temperature differences.

Free convection boundary layer development on a heated (or cooled) plate has been studied extensively, and, over a wide Rayleigh number range, $0.1 < \text{Ra}_L < 10^{12}$, measured, as well as predicted, average convection coefficients are well correlated by the expression[15]

$$\overline{\text{Nu}}_L = \left\{ 0.825 + \frac{0.387\,\text{Ra}_L^{1/6}}{\left[1 + (0.492/\text{Pr})^{9/16}\right]^{8/27}} \right\}^2 \tag{10.45}$$

where L is the plate length. The *Rayleigh number* is indicative of the strength of buoyancy forces relative to viscous forces in the fluid and is defined as $\text{Ra}_L = g\beta\,\Delta T L^3/\alpha\nu$. For a heated horizontal plate facing upward (or a cooled plate facing downward), the following correlation is recommended[16]

$$\overline{\text{Nu}}_L = C\,\text{Ra}_L^m \tag{10.46}$$

where $C = 0.54$, $m = \frac{1}{4}$ for $10^5 < \text{Ra}_L < 2 \times 10^7$ and $C = 0.14$, $m = \frac{1}{3}$ for $2 \times 10^7 < \text{Ra}_L < 3 \times$

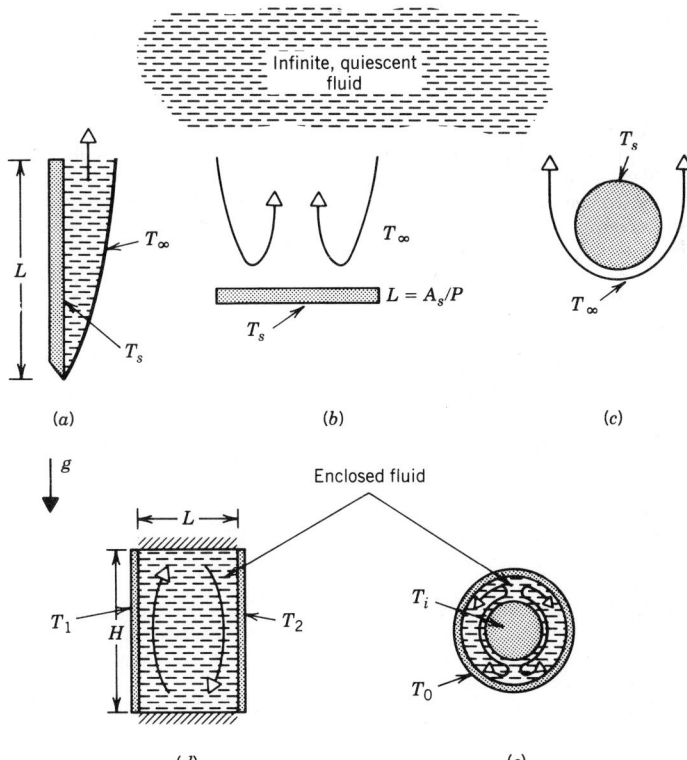

Fig. 10.41 Free convection heat transfer from (a) vertical plate, (b) horizontal plate (upper surface), and (c) horizontal cylinder to infinite, quiescent fluid ($T_\infty < T_s$). Free convection heat transfer between (d) vertical walls of an enclosure ($T_1 > T_2$) and (e) concentric cylinders ($T_i > T_o$).

10^{10}. The characteristics length is the ratio of the plate surface area to the plate perimeter, $L = A_s/P$. For the long horizontal cylinder, the following expression satisfactorily correlates average Nusselt number data for $10^{-5} < \text{Ra}_D < 10^{12}$ (Ref. 17)

$$\overline{\text{Nu}}_D = \left\{ 0.60 + \frac{0.387\,\text{Ra}_D^{1/6}}{\left[1 + (0.559/\text{Pr})^{9/16}\right]^{8/27}} \right\}^2 \tag{10.47}$$

In each of the preceding expressions fluid properties should be evaluated at the mean temperature $(T_s + T_\infty)/2$.

For free convection in the vertical cavity of Fig. 10.41(d), the average heat flux is $q'' = \overline{h}(T_1 - T_2)$ and an appropriate correlation for height-to-width ratios in the range $2 < H/L < 10$, $\text{Pr} < 10^5$, and $\text{Ra}_L < 10^{10}$ is[17]

$$\overline{\text{Nu}}_L = 0.22 \left(\frac{\text{Pr}}{0.2 + \text{Pr}} \text{Ra}_L \right)^{0.28} \left(\frac{H}{L} \right)^{-1/4} \tag{10.48}$$

The heat transfer per unit length of the concentric cylinders [Fig. 10.41(e)] may be expressed as

$$q' = \frac{2\pi k_{\text{eff}}}{\ln(D_o/D_i)}(T_i - T_o) \tag{10.49}$$

where the *effective conductivity* is[18]

$$\frac{k_{\text{eff}}}{k} = 0.386 \left\{ \frac{\text{Pr}\,\text{Ra}_L}{0.861 + \text{Pr}} \frac{[\ln(D_o/D_i)]^4}{L^3 \left(D_i^{-3/5} + D_o^{-3/5} \right)^5} \right\}^{1/4} \tag{10.50}$$

and $L = (D_o - D_i)/2$. In using Eqs. (10.48) and (10.50), fluid properties should be evaluated at the mean enclosure temperature.

Boiling and Condensation

Previous sections have considered convection heat transfer for a single-phase fluid. However, many important problems involve heat transfer with phase change between the liquid l and vapor v states, as in boiling and condensation. Due to latent heat effects and large buoyancy forces resulting from the density difference $(\rho_l - \rho_v)$, heat transfer coefficients and rates can substantially exceed values commonly associated with single-phase forced convection.

The most widely studied boiling condition is one for which a solid is submerged in a saturated liquid and maintained at a temperature in excess of the saturation temperature $(T_s > T_{\text{sat}})$. Termed *pool boiling*, its key features are revealed by the representative boiling curve of Fig. 10.42. For excess temperatures, $\Delta T_e = T_s - T_{\text{sat}}$, less than approximately 5°C, vapor bubble production is minimal, and fluid motion is determined primarily by single-phase free convection. At the onset of *nucleate boiling* (ONB), isolated bubbles form at discrete nucleation sites and ascend from the surface. Bubble production increases with increasing ΔT_e, causing significant fluid mixing near the surface and a large increase in the heat flux. There comes a point, however, at which bubble formation is so pronounced that liquid motion to the surface is inhibited and a maximum heat flux is reached. If ΔT_e is increased beyond this point, vapor patches begin to form on the surface, causing a reduction in q_s''. This *transition boiling* regime is terminated when the surface is completely covered by a vapor blanket and a minimum heat flux is reached (the *Leidenfrost point*). As ΔT_e is increased beyond this point (film boiling), q_s'' again begins to increase due to increasing convection and radiation across the vapor blanket.

The foregoing discussion presumes that control over ΔT_e is maintained as the boiling curve is traversed. In practice, however, it is the heat flux, q_s'', that is usually controlled. In that case, it is evident that only a slight increase in q_s'' beyond q_{max}'' will cause a dramatic rise in ΔT_e. Termed the *boiling crisis*, it could induce thermal failure of the solid.

Engineering systems are typically operated in the nucleate boiling regime, where for pool boiling

$$q_s'' = \mu_l h_{fg} \left[\frac{g(\rho_l - \rho_v)}{\sigma} \right]^{1/2} \left(\frac{c_{p,l} \Delta T_e}{C_{s,f} h_{fg} \text{Pr}_l^n} \right)^3 \tag{10.51}$$

Termed the *Rohsenow correlation*, values of $C_{s,f}$ and n depend on the fluid/surface combination and

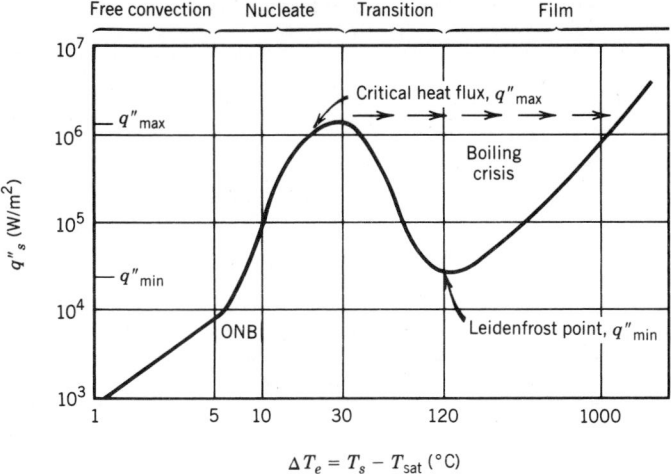

Fig. 10.42 Typical boiling curve for water at one atmosphere.

are, for example, equal to 0.013 and 1.0 for water on a polished copper surface.[19] The maximum or *critical* heat flux is given by the expression[20]

$$q''_{\max} = 0.149 h_{fg} \rho_v \left[\frac{\sigma g (\rho_l - \rho_v)}{\rho_v^2} \right]^{1/4} \tag{10.52}$$

Condensation occurs when a surface is maintained below the saturation temperature of an adjoining vapor ($T_s < T_{sat}$). Condensate may form on the surface as droplets or as a continuous film. Film condensation is favored for clean, uncontaminated surfaces and has been studied extensively for a vertical plate (Fig. 10.43). The film forms at the top of the plate, and its thickness δ and flow rate per unit width Γ increase with increasing x due to continuous condensation at the liquid-vapor interface. Film conditions may be laminar (wavefree or wavy) and turbulent, depending on the Reynolds number,

$$\text{Re}_\delta = \frac{4\Gamma}{\mu_l} = \frac{4\dot{m}}{\mu_l b} \tag{10.53}$$

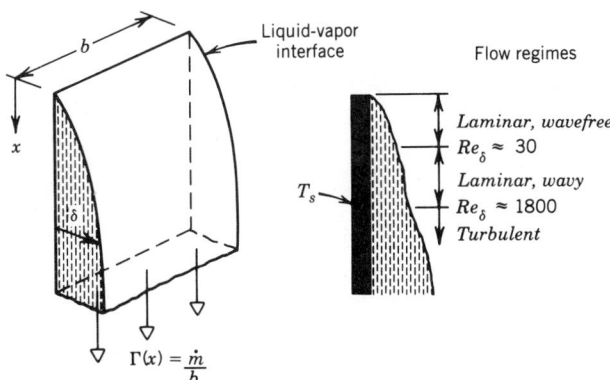

Fig. 10.43 Film condensation on a vertical plate. Adapted from Ref. 2 with permission.

and the average convection coefficient for a plate of length L is[2]

$$\frac{\bar{h}_L\left(\nu_l^2/g\right)^{1/3}}{k_l} = \begin{cases} 1.47\,\mathrm{Re}_\delta^{-1/3} & (\mathrm{Re}_\delta < 30) \\[2mm] \dfrac{\mathrm{Re}_\delta}{1.08\,\mathrm{Re}_\delta^{1.22} - 5.2} & (30 < \mathrm{Re}_\delta < 1800) \\[2mm] \dfrac{\mathrm{Re}_\delta}{8750 + 58\,\mathrm{Pr}^{-0.5}\left(\mathrm{Re}_\delta^{0.75} - 253\right)} & (\mathrm{Re}_\delta > 1800) \end{cases} \tag{10.54}$$

10.2.4 Radiation

Thermal radiation may be viewed as electromagnetic waves emitted by matter that is at a finite absolute temperature. An important feature of the process is that the rate of radiant energy propagation may depend on direction and will always depend on wavelength. To simplify matters our discussion will presume directional independence but, of necessity, will consider wavelength (*spectral*) effects. The most common applications involve radiation exchange at a surface or between surfaces separated by a nonemitting, nonabsorbing gas.

Blackbody Radiation

The blackbody is an idealization that, as the perfect emitter and absorber, provides a standard against which all other radiating matter may be compared. Regardless of wavelength, a blackbody absorbs all incident radiation and, for a prescribed temperature, no surface can emit more energy than a blackbody. The spectral distribution of the radiant energy flux (the spectral emissive power) emitted by a blackbody is shown in Fig. 10.44. The spectral emissive power varies continuously with wavelength and, at all wavelengths, its value increases with increasing temperature. An important feature of the distribution is the existence of a maximum whose wavelength depends on temperature according to Wien's law

$$\lambda_{\max} T = 2898 \ \mu\mathrm{m\text{-}K} \tag{10.55}$$

Hence, with increasing temperature, proportionately more radiation is emitted at smaller wavelengths.

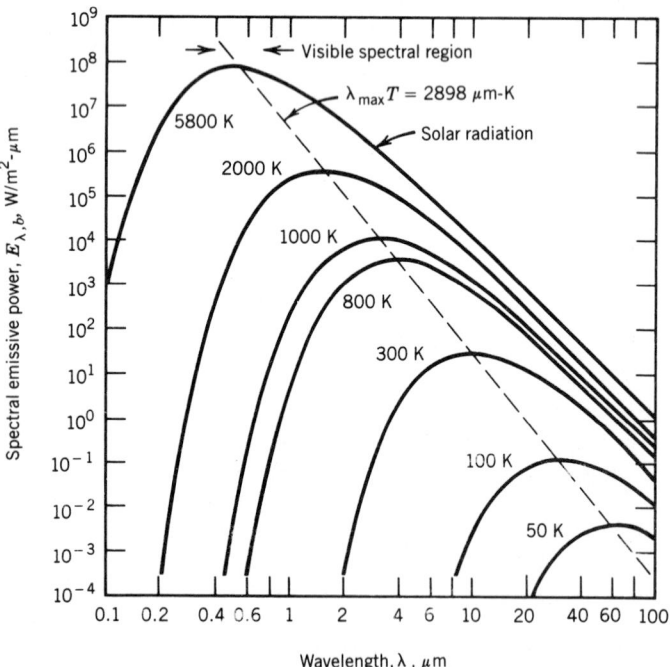

Fig. 10.44 Spectral distribution of blackbody radiation. Used with permission from Ref. 2.

In equation form the spectral distribution of blackbody radiation is given by the Planck function

$$E_{\lambda,b}(\lambda,T) = \frac{C_1}{\lambda^5[\exp(C_2/\lambda T) - 1]} \tag{10.56}$$

where $C_1 = 3.742 \times 10^8$ W-μm^4/m^2 and $C_2 = 1.439 \times 10^4$ μm-K. Equation (10.56) may be multiplied by $d\lambda$ and integrated from any λ_1 to λ_2 to obtain the emissive power for a finite spectral region. The result could also be divided by the integral over all possible wavelengths ($0 \leq \lambda \leq \infty$) to obtain the fraction of total blackbody emission from λ_1 to λ_2. The expression for the *total* blackbody emissive power is termed the *Stefan–Boltzmann law*:

$$E_b = \int_0^\infty E_{\lambda,b} \, d\lambda = \sigma T^4 \tag{10.57}$$

where $\sigma = 5.670 \times 10^{-8}$ W/m^2-K^4. The ratio $\int_0^\lambda E_{\lambda,b} \, d\lambda / \sigma T^4$ has been conveniently tabulated as a function of λT (Ref. 2).

Radiation Exchange at a Surface

Total emission from any real surface may be computed if the total emissivity ε and the temperature T of the surface are known. By definition

$$E = \varepsilon E_b = \varepsilon \sigma T^4 \tag{10.58}$$

Although total emissivities are tabulated for many surface materials,[2] it is important to recognize that ε may depend strongly on the surface temperature and that, at times, it may be necessary to determine this dependence by computing ε from knowledge of how its spectral component varies with λ (Refs. 2, 21).

In addition to emitting radiation, a surface may also receive radiation from other sources, which it will *absorb*, *reflect*, or *transmit*. If the surface is opaque, there is no transmission and the total incident radiation, termed the irradiation G, is simply absorbed or reflected. Surface properties are introduced to determine the response to incident radiation where, by definition, the total absorptivity α is used to compute the rate of radiant energy absorption per unit surface area

$$G_{abs} = \alpha G \tag{10.59}$$

and the total reflectivity ρ is used to calculate the rate at which radiation is reflected

$$G_{ref} = \rho G \tag{10.60}$$

Although tabulations of α and ρ are available in the literature, it is important to note that values for a particular surface can vary over a wide range, depending on the spectral distribution of the irradiation. In engineering calculations, it is common to assume *gray surface* behavior for which $\alpha = \varepsilon$. However, there are several important restrictions that must be satisfied before the equality may be used.[2] If the surface is opaque and α is known, the reflectivity may be obtained from $\rho = 1 - \alpha$.

In general, heat transfer at a surface may be influenced by conduction, convection, and radiation. As shown in Fig. 10.45, for an opaque surface the corresponding surface energy balance is

$$q''_{cond} = q''_{conv} + E - \alpha G \tag{10.61}$$

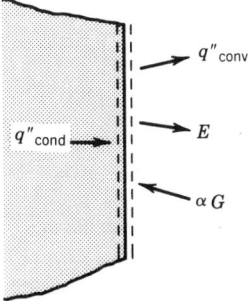

Fig. 10.45 Surface energy balance.

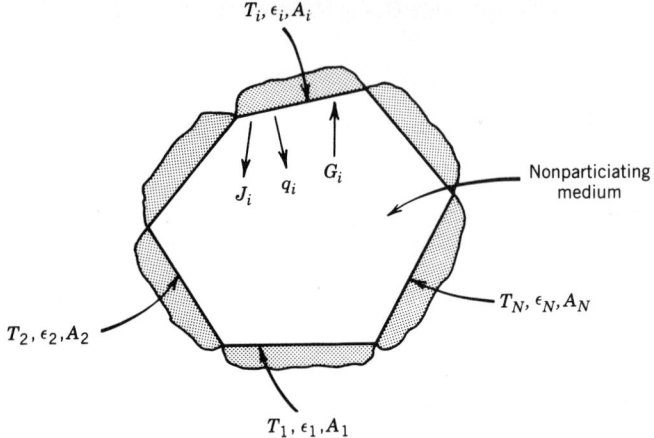

Fig. 10.46　Radiation exchange in an N surface enclosure.

Note that only the absorbed, and not the reflected, irradiation contributes to the thermal energy budget of the surface.

Radiation Exchange Between Surfaces

Problems frequently arise for which there is radiation exchange between two or more surfaces that form an enclosure and are separated by a nonparticipating medium (Fig. 10.46). If conduction and convection effects may be neglected, the net rate at which heat is transferred from any surface i may be expressed as

$$q_i = A_i(J_i - G_i) \tag{10.62}$$

where the radiosity J_i is the rate at which radiation leaves the surface by both emission and reflection. By assuming an opaque, gray surface, this expression may be cast in the form

$$q_i = \frac{E_{bi} - J_i}{(1 - \varepsilon_i)/\varepsilon_i A_i} \tag{10.63}$$

Hence, if J_i is known, q_i could be obtained from knowledge of T_i, which fixes E_{bi} from Eq. (10.57), or vice versa. The radiosity of each surface is determined by simultaneously solving a system of N equations, which are obtained by applying

$$q_i = \sum_{j=1}^{N} A_i F_{ij}(J_i - J_j) \tag{10.64}$$

or

$$\frac{E_{bi} - J_i}{(1 - \varepsilon_i)/\varepsilon_i A_i} = \sum_{j=1}^{N} \frac{J_i - J_j}{(1/A_i F_{ij})} \tag{10.65}$$

to each surface according to whether q_i or T_i, respectively, is known. The configuration, or view, factor F_{ij} is a geometrical quantity that provides the fraction of radiation leaving surface i that is intercepted by surface j. Results are available for many common geometries.[2, 21, 22]

10.2.5　Summary

Section 10.2 has provided an overview of some of the basic elements of conduction, convection, and radiation heat transfer. However, because of space constraints, presentations have been highly abbreviated and many important topics have been completely omitted. Examples of such topics include numerical methods of solving the differential equations of conduction and convection, combined forced and free convection, two-phase flow, heat exchangers, gaseous radiation, and multimode (combined conduction, convection or radiation) effects. For the reader with serious interests in heat transfer, this section should therefore be viewed as no more than a starting point. For a more comprehensive, general treatment of the subject, one of the many undergraduate textbooks, such as that by Incropera and DeWitt,[2] or the recently published *Handbook of Heat Transfer*[23] should be consulted. Detailed expositions of specific topics are also available, as, for example, for free and mixed convection,[24] heat exchangers,[25,26] and radiation.[21]

REFERENCES

1. Kraus, A. D., and Bar-Cohen, A., *Thermal Analysis and Control of Electronic Equipment*, Chap. 9, New York, McGraw-Hill, 1983.

2. Incropera, F. P., and DeWitt, D. P., *Introduction to Heat Transfer*, New York, Wiley, 1985.

3. Schneider, P. J., *Conduction Heat Transfer*, Reading, Mass., Addison-Wesley, 1955.

4. Kern, D. Q., and Kraus, A. D., *Extended Surface Heat Transfer*, New York, McGraw-Hill, 1972.

5. Arpaci, V. S., *Conduction Heat Transfer*, Reading, Mass., Addison-Wesley, 1966.

6. Myers, G. E., *Analytical Methods in Conduction Heat Transfer*, New York, McGraw-Hill, 1971.

7. Ozisik, M. N., *Heat Conduction*, New York, Wiley, 1980.

8. Sunderland, J. E., and Johnson, K. R., *Trans. ASHRAE*, **10**, 237–241 (1964).

9. General Electric Co., *Heat Transfer Data Book*, Section 502, Schenectady, N.Y., 1973.

10. Hahne, E., and Grigul, U., *Int. J. Heat Mass Transfer*, **18**, 751–769 (1975).

11. Churchill, S. W., and Bernstein, M., *J. Heat Transfer*, **99**, 300–306 (1977).

12. Zhukauskas, A., in *Advances in Heat Transfer*, J. P. Hartnett and T. F. Irvine, Jr. (Eds.), Vol. 8, 93–160, New York, Academic Press, 1972.

13. Petukhov, B. S., in *Advances in Heat Transfer*, J. P. Hartnett and T. F. Irvine, Jr. (Eds.), Vol. 6, 504–564, New York, Academic Press, 1970.

14. Hausen, H., *Z. VDI. Beih. Verfahrenstech* **4**, 91 (1943).

15. Churchill, S. W., and Chu, H. H. S., *Int. J. Heat Mass Transfer* **18**, 1323–1329 (1975).

16. Lloyd, J. R., and Moran, W. R., ASME Paper 74-WA/HT-66, 1974.

17. Catton, I., *Proc. Sixth Int. Heat Transfer Conf.* **6**, 13–31 (1978).

18. Raithby, G. D., and Hollands, K. G. T., in *Advances in Heat Transfer*, Vol. 11, 265–315, New York, Academic Press, 1975.

19. Rohsenow, W. M., *Trans. ASME* **74**, 969–975 (1952).

20. Lienhard, J. H., Dhir, V. K., and Riherd, D. M., *J. Heat Transfer*, **95**, 477–482 (1973).

21. Siegel, R., and Howell, J. R., *Thermal Radiation Heat Transfer*, 2nd ed., New York, McGraw-Hill, 1981.

22. Howell, J. R., *A Catalog of Radiation Configuration Factors*, New York, McGraw-Hill, 1982.

23. Rohsenow, W. M., Hartnett, J. P., and Ganic, E. N., *Handbook of Heat Transfer Fundamentals*, 2nd ed., New York, McGraw-Hill, 1985.

24. Kakac, S., Aung, W., and Viskanta, R., *Natural Convection: Fundamentals and Applications*, New York, Hemisphere Publishing, 1985.

25. Kakac, S., Bergles, A. E., and Mayinger, F., *Heat Exchangers*, New York, Hemisphere Publishing, 1981.

26. Kakac, S., Shah, R. K., and Bergless, A. E. (Eds.), *Low Reynolds Number Flow Heat Exchangers*, New York, Hemisphere Publishing, 1983.

CHAPTER 11
ELECTROMAGNETICS AND CIRCUITS

Albert Rosa

Department of Engineering
University of Denver
Denver, Colorado

This section presents a brief insight into the principles of electromagnetics as required for understanding the underlying fundamentals of the behavior of linear circuits. An introductory development of linear active and passive electric circuits emphasizing modern interfacing concepts as well as traditional dc and ac circuit analysis completes the section.

11.1 SYSTEM OF UNITS

The rationalized system of meter-kilogram-second (MKS) units, often referred to as the Giorgi system, is used throughout this section unless otherwise indicated. The so-called MKS system is formed through the use of practical units for the electromagnetic quantities in addition to the meter as a unit of length, the kilogram as a unit of mass, and the second as a unit of time, in order to convert the practical system into an "absolute" system. Historically, several systems of units preceded the MKS system. The electrostatic and electromagnetic systems were both "absolute" systems based on the centimeter as a unit of length, the gram as a unit of mass, and the second as a unit of time. The electrostatic unit of charge, the statcoulomb, is defined on the basis of Coulomb's law for the force that exists between two charges with a given distance between them and in a homogeneous medium. On the basis of this law, the electrostatic unit of charge is related to the fundamental units of length, mass and time. The electromagnetic unit of charge follows directly from the electromagnetic unit of current, which is defined by Ampere's law for the force between two current elements with a given distance of separation. On this basis the electromagnetic unit of charge, the abcoulomb, was a much larger unit of charge than the statcoulomb, the corresponding unit quantity of charge in the electrostatic system. For practical usage, certain units that constitute the practical system were found convenient. These units are the unit of resistance, the ohm; the unit of potential, the volt; the unit of current, the ampere; the unit of quantity of charge, the coulomb; the unit of capacitance, the farad; the unit of inductance, the henry; the unit of power, the watt; and the unit of energy, the watt-second or joule. Working backward from these practical units of length, mass and time associated with the MKS system, and substituting all of these in Ampere's law, it was found possible to define the permeability of free space μ_0. Substituting these units, or whichever ones are appropriate, into Coulomb's law, it was found possible to define the permittivity of free space ε_0. In this way, the practical system has been extended into an "absolute" system, thus giving rise to the MKS system of units. Since the MKS system as here used includes the coulomb as a unit charge, it is sometimes designated the MKSC system. By occasional reference to Chapter 1 on physical units and standards, it is possible for the reader to transfer from one system of units to another. Standard symbols for quantities are given in Table 1.3 of Chapter 1. When vector relations are involved, the symbol is in boldface type.

11.2   ELECTRON THEORY

11.2.1   Electrical Entities and Their Properties

Electrical Entities
In the study of electromagnetics and circuits, several electrical entities must be understood. These are the negative electron and positive particles such as the proton and hole.

The Electron. The electron is the smallest quantity of negative $(-)$ electricity that can exist by itself. From empirical data, the following assumptions regarding the electron can be justified:

1. No charge of electricity can be produced that is not an integral multiple of the charge carried by a simple electron. The value of this charge is

$$e = -q \qquad q_E = 1.602 \times 10^{-19} \text{ C} \tag{11.1}$$

2. The mass of an electron may be defined as the force required to give it unit acceleration, that is, $f = ma$, providing the resulting electron velocities are less than $\frac{1}{10}$ the velocity of light. Experiments indicate that the "rest" mass of an electron, thus defined, is

$$m_0 = 9.1066 \times 10^{-31} \text{ kg} \tag{11.2}$$

The actual mass of a free electron increases with increasing velocity as a result of relativistic analysis as

$$m = \frac{m_0}{\left[1 - (v/c)^2\right]^{1/2}} \tag{11.3}$$

Electrons in a crystal, such as a semiconductor, are not treated as "free" electrons. Rather, these particles interact with the crystal lattice resulting in a mass quite different than the electron's "free" mass. In order to apply electrodynamic equations to electrons in crystals an altered or "effective" mass (m^*) must be used. The effective mass varies from material to material. While effective mass is quite difficult to calculate, handbooks of semiconductor properties list empirical values measured using techniques such as cyclotron resonance. For example, for n-type germanium, the effective mass of an electron m_n^*, is 0.55 its rest mass, while for silicon, it is $1.1m_0$.

3. An electron having a charge e and moving with velocity v produces the same magnetic field as an elementary length ds of a conduction current of strength i, where $ev = i\,ds$.

4. Every neutral atom of matter contains at least one electron that is held in position by forces analogous to elastic forces; that is, an electron may oscillate within the atom or may be displaced by an impressed electrostatic field. The electrons are always of the same nature irrespective of the substance in which they exist.

5. An electron may be forced from the atom by the influence (mutual repulsion) of a free electron moving at a high velocity in its immediate vicinity. This action is usually spoken of as a bombardment, or collision, although it is not necessary to assume that the free electron hits the atom.

6. When a neutral atom loses an electron, this atom manifests the properties of a positively charged body and is called a positive ion.

7. When a neutral atom gains an electron, this atom manifests the properties of a negatively charged body and is called a negative ion.

8. In every substance, at temperatures above absolute zero, there exists a probability of finding free electrons in the substance. For metals, the number of free electrons is relatively constant versus temperature and quite high, accounting for the relative ease with which metals conduct electricity. In n-type semiconductors where electrons are the principal carriers of electricity, the number can vary over many orders of magnitude with temperature and with the kind and amount of impurity. It is their control of electrical properties that accounts for much of the great usefulness of semiconductors. Under certain circumstances, for example, a gas at ordinary pressures, a free electron may attach itself to an atom or molecule or the free electron may pass from one substance to another causing a flow of electricity.

The Positive Particles. Experiments can also justify the following assumptions regarding positively charged particles or entities:

1. The smallest positive charge that it has been possible to produce is numerically equal to that of an electron, namely, 1.602×10^{-19} C.

2. This smallest possible positive charge when associated with a particle having a mass of the same order of magnitude as that of a hydrogen atom is called a **proton**. The mass of a proton is 1.6734×10^{27} kg.

3. This smallest possible positive charge when associated with a particle having a mass of the same order of magnitude as that of the negative electron is called a positive electron, or **positron**.

4. In semiconductors there exist electron vacancies that may be free to move about the semiconductor crystal. These vacancies are known as **holes**. While a hole is not a particle, its behavior is identical to that of an electron but with positive charge, that is,

$$h = -e = +q_H = 1.602 \times 10^{-19} \text{ C} \tag{11.4}$$

In p-type semiconductors, the principal carrier of electricity is the hole. Its "mass" is of the same order as that of an electron in the same material and is also referred to as its effective mass, m_p^*. It too varies from material to material; for example, in germanium the effective mass of a hole is $0.37m_0$, and in silicon, it is $0.59m_0$, where m_0 is the rest mass of the electron.

5. The term **ion** is usually reserved to designate any charged body having a mass of the order of magnitude of that of a molecule or atom. In this sense a positive or negative electron is not an ion, but the proton is an ion. If, however, the electron becomes attached to an atom or molecule, then this combination forms an ion. In the discharge or electricity through gases at low pressure, the negative electron, although it is not attached to an atom or molecule, is sometimes called an ion.

11.2.2   Emission Phenomena

Thermionic Emission of Electrons
When a metal is heated to a high temperature in a high vacuum, it gives off electrons freely. The amount of the electron emission is subject to the temperature of the metal or cathode and to the electric field at its surface, as due, for example, to a neighboring anode.

The theory assumes that in accordance with the kinetic theory of matter the electrons within the metal are in a constant state of motion, and that of those nearest the surface a certain proportion escape, a few at high, the greater number at lower, velocities in accordance with Maxwell's distribution law. When an electron leaves, the surface of the metal becomes positively charged, thus putting a retarding force on the electron and tending to make it return. The distance traversed by the electron is greater the greater its initial velocity, and in the presence of a neighboring metal charged surface it may not return at all. The higher the temperature of the heated metal or cathode, the greater the velocities of the electron leaving it, and the greater the volume of electron discharge.

An electron can escape from a metal when, and only when, its kinetic energy is greater than or equal to the work that must be done in escaping; that is, when

$$\tfrac{1}{2}mv^2 \geq w_0 \tag{11.5}$$

where m is the mass of the electron, v is the component of its velocity normal to the emitting surface, and w_0 is the electron affinity or internal work of electron evaporation of the emitting substance.

With a given cathode temperature there is a maximum electron current beyond which an increase in anode voltage is ineffective. Under these conditions all the electrons emitted from the cathode are drawn to the anode. This condition is known as **voltage saturation** or **temperature limited**.

This electron current follows the law

$$i = Nq_E = AT^\lambda e^{-(w_0/kT)} \tag{11.6}$$

where w_0 is the work function in electron volts, i is the saturation current per square centimeter, q_E is the charge per electron, A is a constant of the material, T is the absolute temperature in degrees Kelvin. λ is a number not much different from unity, e is the base of the Napierian system of logarithms, and k is the Boltzmann constant, 8.620×10^{-5} electron volts per degree Kelvin. As far as the individual electron is concerned, this is a statistical law. The value of the current given in Eq. (11.6) is independent of the potential of the anode.

With a given anode voltage there is a maximum electron current beyond which an increase in cathode temperature is ineffective. When the cathode is first heated, electrons are emitted with varying velocities. If the anode voltage is zero, the first few of these electrons experience no external force except a slight attraction toward the cathode. These first few electrons ultimately land on the anode or on the walls of the container. The succeeding electrons experience a repulsion due to the electrons, previously emitted but still near the cathode, in the space between the anode and cathode; and if the energies of the newly emitted electrons are small, their motion will be retarded or even reversed. There is thus built up an electric field of force tending to decelerate an electron emerging from the cathode and to make it return to the cathode. If there is no removal of electrons from the space surrounding the cathode, this electric field increases in strength until a condition of equilibrium is reached, that is, until as many electrons are entering the cathode as are leaving it. The charge due to the presence of these electrons constitutes the **space charge**. From this it is evident that for a given anode voltage, an increase in cathode temperature beyond a certain value will serve only to increase the space charge without increasing the current to the anode. This condition is known as **temperature saturation** or **space-charge limited**.

Electron Tubes

A glass or metal bulb, either evacuated or gas filled, and containing two or more electrodes, is called an **electron tube**. Electrons (and positive ions) are produced in such tubes by thermionic emission, by photoelectric emission (q.v.), or by bombardment and are caused to move by the action of an electric field. This electron flow is made to serve various purposes depending on its magnitude, velocity, and method of control. Two-electrode tubes can be used for rectification since current will pass more readily in one direction than the others. Tubes with three or more electrodes are important since the voltage signal at one or more electrodes can exert great influence to obtain a much higher level of signal at another electrode.

Cathode Rays. At the occurrence of an electric discharge in an electron tube and at a certain value of the pressure of the gas contained in the tube, it is noticed that a greenish fluorescence occurs on the walls of the tube. By placing solid bodies in the tube, it appears that the fluorescence is due to something proceeding in straight lines perpendicularly from the surface of the cathode. The name **cathode rays** has been given to the agent that produces the fluorescence. These rays are deflected by both electric and magnetic fields and convey a negative electric charge to an insulated conductor. They consist of a stream of rapidly moving electrons. The amount of the deflection of the rays, under the action of magnetic and electrostatic fields, can be calculated or observed.

11.3   DC CIRCUITS (TIME INDEPENDENT CIRCUITS)

11.3.1   Basic Circuit Concepts

Circuit Variables

The underlying physical quantities in the study of electronic systems are two basic variables—**charge** and **energy**. Of the two, charge is electric in character. The concept of electric charge explains the very strong electrical forces that occur in nature. To explain both attraction and repulsion, we say that there are two kinds of charge—positive and negative. Like charges repel whereas unlike charges attract. The symbol q is used to represent charge. In the MKS system charge is measured in coulombs. The smallest quantity of charge in nature is an electron's charge ($q_E = 1.6 \times 10^{-19}$ C).

Electric charge is a rather cumbersome variable to work with in practice. Moreover, in most situations the charges are moving, and so we find it more convenient to measure the amount of charge passing a given point per unit time. To do this, we define a signal variable i called **current** as follows:

$$i = \frac{dq}{dt} \tag{11.7}$$

Current is a measure of the flow of electric charge. It is the time rate of charge passing a given point. The physical dimensions of current are coulombs per second. In the MKS system, the unit of current is the ampere. That is,

$$1 \text{ C/s} = 1 \text{ A}$$

since there are two types of electric charge, there is a bookkeeping problem associated with the direction of current flow. In electrical engineering it is customary to define the direction of current as the direction of the net flow of *positive* charges.

The concept of **voltage**, a second signal variable, is associated with the change in energy that would be experienced by a charge as it passes through a circuit. The symbol w is commonly used to represent energy. In the MKS system, energy carries the units of joules. If a small charge dq were to experience a change in energy dw in passing from point A to point B, then the voltage v between A and B would be defined as the change in energy per unit charge or

$$v = \frac{dw}{dq} \qquad (11.8)$$

Voltage does not depend on the path followed by the charge dq in moving from point A to point B. Furthermore, there can be a voltage between two points even if there is no charge motion, since voltage is a measure of how much energy would be involved if a charge dq were moved. The dimensions of voltage would be joules per coulomb. In the MKS system, the unit of voltage is the volt. That is,

$$1 \text{ J/C} = 1 \text{ V}$$

A third signal variable is **power** which is defined as the time rate of change of energy:

$$p = \frac{dw}{dt} \qquad (11.9)$$

The dimensions of power would be joules per second, which in the MKS system is called a watt. In electrical situations, it is useful to have power expressed in terms of current and voltage. This is done by writing Eq. (11.9) as

$$p = \frac{dw}{dq} \times \frac{dq}{dt} = vi \qquad (11.10)$$

Thus the electric power associated with a situation is determined by the product of current and voltage.

Figure 11.1(a) shows the interrelation between the two basic variables, charge and energy, and the three signal variables, current, voltage, and power. Charge and energy, like mass, length, and time, are taken as basic assumed concepts.

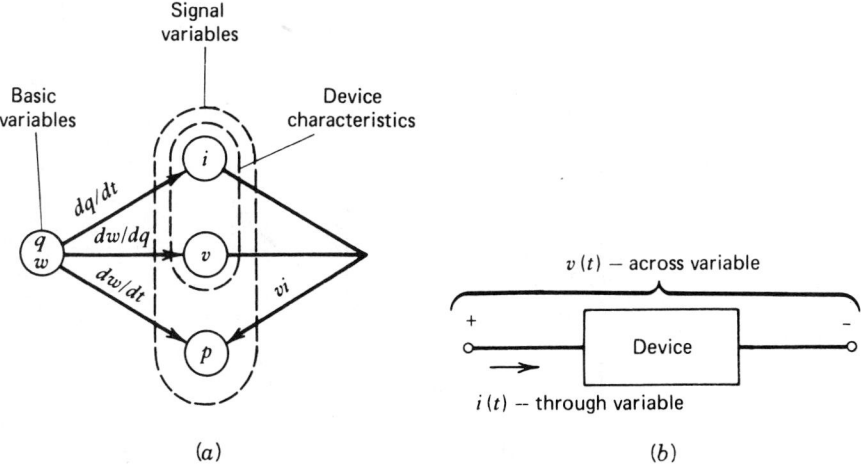

(a)   (b)

Fig. 11.1 (*a*) Flow diagram of circuit variables. (*b*) Assignment of reference direction to current and reference polarity to voltage in a two-terminal device. Reproduced, with permission, from R. E. Thomas and A. J. Rosa, *Circuits and Signals: An Introduction to Linear and Interface Circuits*, New York, Wiley, 1984.

While the roots of electrical engineering are intimately entwined with the basic variables, the engineer only occasionally becomes concerned about them, but works instead with the signal variables. The reason for this choice of working parameters is that current and voltage are relatively easy to measure, and therefore convenient to use to represent data. Since the processing of data is the fundamental reason for studying electronic systems, it follows that current and voltage are our working variables.

It is essential that the reader recognize that current and voltage are not the same thing. Current is a measure of the time rate of charge passing a point. Since it is a measure of flow, we think of current as a *through* variable. Voltage is a measure of the net change in energy involved in moving a charge from one point to another. Note that the concept of voltage involves two points. Voltage is not measured at a point but rather between two points. We therefore call voltage an *across* variable.

Figure 11.1(*b*) shows a notation used for assigning reference directions to current and voltage. The reference mark for current (an arrow below the wire) does *not* indicate the actual direction of current flow. The actual direction of flow may be reversing a million times per second. However, when the actual direction coincides with the reference direction, we say that the current is positive. When the opposite occurs, we say that the current is negative. Thus, if the net flow of positive charge in Fig. 11.1(*b*) is to the right, we say that the current $i(t)$ is positive. Conversely, if the current $i(t)$ is positive, then we know the net flow of positive charge is to the right.

Similarly, the voltage reference marks (+ and − symbols) do not imply that the potential at the positive terminal is always higher than the potential at the negative terminal. However, when this is true, we will say that the voltage across the device is positive. When the opposite is true, we say that the voltage is negative.

Element Constraints

The device shown as a rectangular box in Fig. 11.1(*b*) is representative of a whole family of two-terminal components that engineers connect together to form circuits. Since there are many different devices that possess different, and at times divergent, current-voltage (*i-v*) characteristics, engineers have developed models representing the different devices. An electric device is a component that is viewed as an entity. To study a device is to study the *i-v* characteristics of that device. Most devices are nonlinear, that is, their *i-v* relationship is nonlinear. The study of circuits containing several nonlinear devices would require enormous amounts of effort to analyze if it were not for the model. A model, which usually takes the form of an equation or graphical relation, is a linearized version or approximation of the more complex nonlinear *i-v* relation. To distinguish between circuit devices (the real thing) and circuit models (an approximate stand-in), we call the model a circuit element. Thus circuit devices appear as hardware, and are listed in manufacturers' catalogs. Circuit elements, or models, appear in schematics and in books on circuit analysis and electronics.

The first device in our study is the resistor shown in Fig. 11.2(*a*). The current-voltage characteristic of this device is quite nonlinear, as shown by the dashed line in Fig. 11.2(*b*). To write a reasonably accurate mathematical expression for this device would require a cubic equation. Such a complex relation would make analysis of circuits containing resistors very difficult.

If we look closely at the *i-v* relation, we can see that if we limit the range over which we use the device, we can approximate, with only minimal error, its characteristics with a straight line. This limit

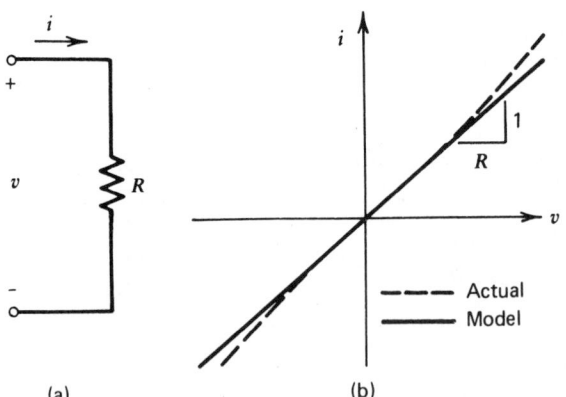

(a)   (b)

Fig. 11.2 The linear resistor. (*a*) Circuit symbol. (*b*) *i-v* characteristics. Reproduced, with permission, from R. E. Thomas and A. J. Rosa, *Circuits and Signals: An Introduction to Linear and Interface Circuits*, New York, Wiley, 1984.

is usually described as a power rating of the resistor and in general is not exceeded in designing circuits that employ resistors. In sum, if we stay within the device's prescribed limits, we can obtain our first circuit element, the linear resistor. The equations that describe this element are

$$v = Ri \quad \text{or} \quad i = Gv \tag{11.11}$$

where

$$G = \frac{1}{R} \tag{11.12}$$

These equations are collectively known as Ohm's law. The parameter R is called resistance and has the unit ohms (Ω). The parameter G is called conductance, with the unit mho ($\mho$). The relationships expressed in these equations can also be presented graphically, as the solid line in Fig. 11.2(b).

The power associated with the resistor can be found from $p = vi$ and Eq. (11.11):

$$p = Ri^2 = \frac{v^2}{R} \tag{11.13}$$

Since the parameter R is positive, these equations tell us that the power is always nonnegative. This means that the resistor always absorbs power.

Electronic circuits require power to operate. In electronics, there are two types of power sources: **voltage sources** and **current sources**. In addition, each of these sources can be either constant (non-time-varying) or time-varying signal sources. The circuit symbols and the i-v characteristics of each of these four sources are shown in Fig. 11.3. It is noted that there is no separate symbol for a constant-current source, whereas there is a symbol for a constant-voltage source (a battery).

The element equations for the current source are

$$i = i_s \quad \quad v \text{ varies} \tag{11.14}$$

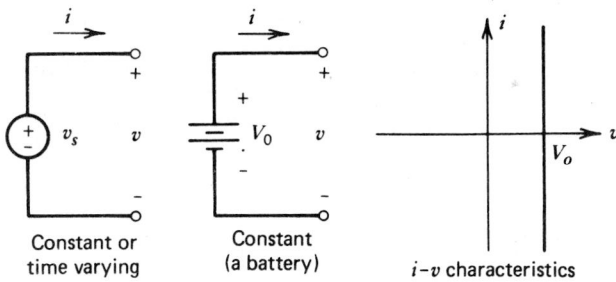

Voltage Sources

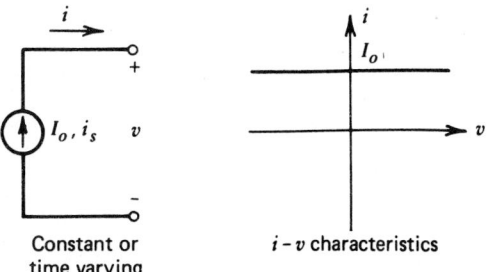

Current Source

Fig. 11.3 Ideal sources. Reproduced, with permission, from R. E. Thomas and A. J. Rosa, *Circuits and Signals: An Introduction to Linear and Interface Circuits*, New York, Wiley, 1984.

In words, the current source supplies i_s amperes out of its positive terminal and into its negative terminal, and will furnish whatever voltage is required by the circuit to which it is connected. The signal voltage source is described by

$$v = v_s \qquad i \text{ varies} \tag{11.15}$$

This means that the voltage source produces v_s volts across its terminals and will supply whatever current may be required by the circuit to which it is connected. Sources are often called forcing functions and are said to force or drive the circuit.

Connection Constraints

A **circuit** is any collection of devices connected at their terminals. The laws governing circuit behavior are based on Kirchhoff's laws. They tell us that the process of interconnecting devices to form a circuit forces the device currents and voltages to behave in certain ways. These constraints are based only on the circuit connections and not on the specific devices in the circuit. For this reason, we call the equations derived from Kirchhoff's law **connection constraints**.

The junction of two circuits is called an **interface** and represents not only a physical electrical connection, but also often is a contractual boundary between the different manufacturers of the individual circuits.

While it is customary to designate a juncture of two or more elements as a node (a **node** is a point at which two or more devices are connected together), it is important to realize that a node is not confined to a point but includes all the wire from the point to each element.

Kirchhoff's first law, known as Kirchhoff's current law (KCL), states: *The algebraic sum of the currents entering a node is zero at every instant.* In forming the algebraic sum of currents, we must take into account the current reference directions associated with the devices. If the current reference direction is into the node, we assign a positive sign in the algebraic sum to the corresponding current. If the reference direction is away from the node, we assign a negative sign. There are two signs associated with each current in the application of KCL. The first is the sign given to a current in writing a KCL connection equation. This sign is determined by the orientation of the current reference direction relative to a node. The second sign is determined by the actual direction of current flow relative to the reference direction.

The second of Kirchhoff's circuit laws, Kirchhoff's voltage law (KVL), states: *The algebraic sum of all of the voltages around a loop is zero at every instant.* A loop is a sequence of devices that forms a closed path. In writing the algebraic sum of voltages, we must account for the assigned reference marks. When we traverse a loop, if we go from a $+$ to $-$ reference mark, we use a positive sign in the sum for the corresponding voltage. If the opposite occurs, then we use a minus sign. There are two signs associated with each voltage. The first is the sign determined by the actual polarity of a voltage relative to its assigned reference polarity. The second is the sign given the voltage in a KVL connection equation. This second sign is determined by the reference polarity relative to the direction of traversing the loop.

When two or more elements are connected between two common nodes the elements are said to be connected **in parallel**, as a result, all the voltages across them are equal.

Two elements are said to be connected **in series** if they share one common node to which no other element is connected. A consequence of the series connection is that the current through each element must be equal. Any number of elements may be connected in series, as a result, the same current flows through all.

Combined Constraints

The usual goal of circuit analysis is to determine the currents or voltages at various places in a circuit. This analysis is based on sets of equations or constraints of two distinctly different types. The element constraints are based on the models of the specific devices connected in the circuit. The connection constraints are based on Kirchhoff's laws and the circuit connections. The element equations are independent of the circuit in which the device is connected. Likewise, the connection equations are independent of the specific devices that may be in the circuit. But taken together, the element and connection equations provide the data needed to analyze a circuit.

The circuit in Fig. 11.4 can be used to illustrate the formulation of these equations. The element equations for this circuit are

$$v_A = V_0 \qquad v_1 = R_1 i_1 \qquad v_2 = R_2 i_2 \tag{11.16}$$

These equations describe the three devices in the circuit. They do not depend on how the devices are

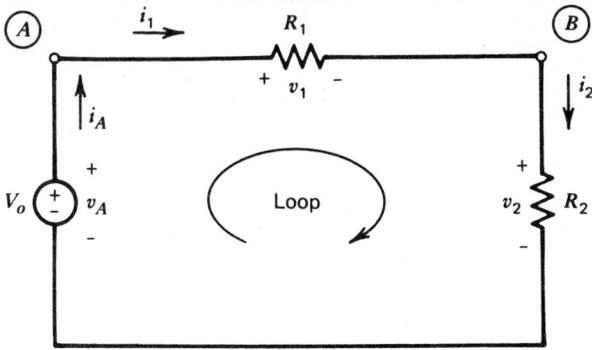

Fig. 11.4 Circuit used to demonstrate Kirchhoff's law. Reproduced, with permission, from R. E. Thomas and A. J. Rosa, *Circuits and Signals: An Introduction to Linear and Interface Circuits*, New York, Wiley, 1984.

connected in the circuit. The connection equations are obtained from Kirchhoff's laws:

Node (A): $i_A - i_1 = 0$

Node (B): $i_1 - i_2 = 0$ (11.17)

Loop: $-v_A + v_1 + v_2 = 0$

These equations are independent of the specific devices in the circuit. They depend only on the circuit connections.

Suppose $V_0 = 10$ V, $R_1 = 2000$ Ω, and $R_2 = 3000$ Ω for the circuit in Fig 11.4. Find all the device currents and voltages. Substituting the element equations into the KVL connection equation produces

$$- V_0 + R_1 i_1 + R_2 i_2 = 0 \qquad (11.18)$$

This equation can be used to solve for i_1 because the second KCL connection equation requires that $i_2 = i_1$, since R_1 and R_2 are connected in series:

$$i_1 = \frac{V_0}{R_1 + R_2} = \frac{10}{2000 + 3000} = 2 \text{ mA} \qquad (11.19)$$

By finding this current, we have determined every device current, because the KCL connection equations collectively require that

$$i_A = i_1 = i_2 \qquad (11.20)$$

since all three elements are connected in series. Substituting all of the known values into the element equations gives

$$v_A = 10 \text{ V} \qquad v_1 = R_1 i_1 = 4 \text{ V} \qquad v_2 = R_2 i_2 = 6 \text{ V} \qquad (11.21)$$

We have found every device voltage and current. Note the analysis strategy used here. We first determined all of the device currents and then found the voltages from these values using the element constraints.

Equivalent Circuits

Very often in the analysis of circuits, it is to the engineer's benefit to replace a portion of the circuit with a simpler equivalent one. The underlying basis for two circuits to be equivalent is contained in their *i-v* relation, as stated by the following definition: *Two circuits are said to be equivalent if they have identical i-v characteristics at a specified pair of terminals.*

Figure 11.5 summarizes some of the more common equivalent circuits encountered in linear circuit analysis.

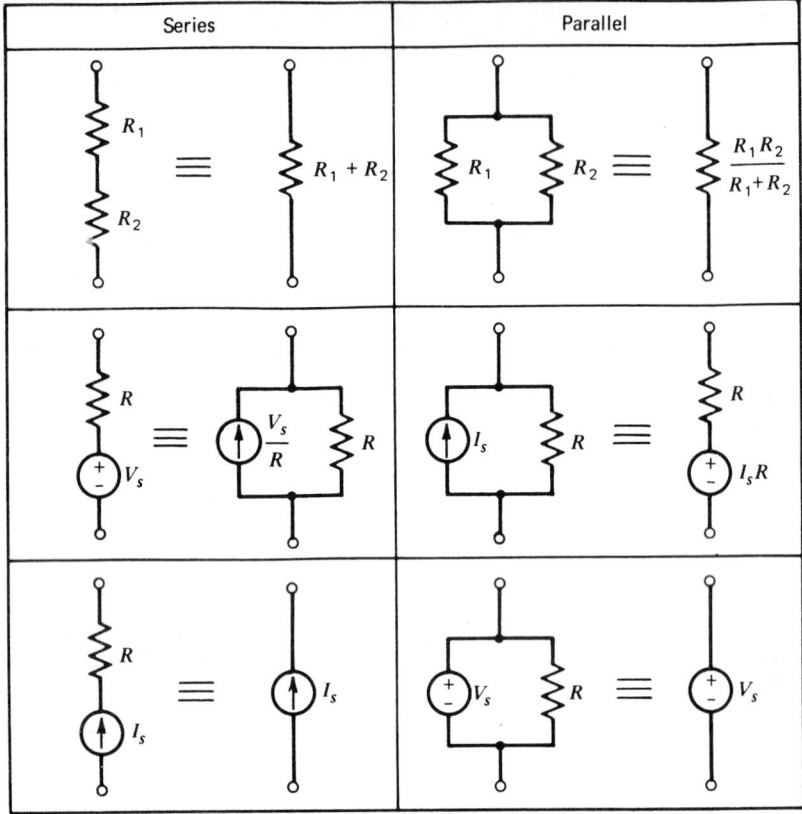

Fig. 11.5 Some two-terminal equivalent circuits. Reproduced, with permission, from R. E. Thomas and A. J. Rosa, *Circuits and Signals: An Introduction to Linear and Interface Circuits*, New York, Wiley, 1984.

Voltage and Current Division

No discussion of series and parallel circuits can be complete without the inclusion of voltage and current division. These two tools find wide application in the analysis of circuits.

Suppose we have a circuit consisting of a voltage source in series with three resistors, as shown in Fig. 11.6. Suppose that the voltage across R_2 is desired. The following analysis will yield the correct result, and a very useful analytic tool.

The application of KVL around the loop in Fig. 11.6 yields

$$v_s = v_1 + v_2 + v_3 \tag{11.22}$$

Using Ohm's law and the fact that the same current i flows through all three resistors since they are connected in series, we find that

$$v_s = R_1 i + R_2 i + R_3 i \tag{11.23}$$

Solving for i,

$$i = \frac{v_s}{R_1 + R_2 + R_3} \tag{11.24}$$

Now to find v_2 we again use Ohm's law and substitute for i:

$$v_2 = iR_2 = \frac{v_s R_2}{R_1 + R_2 + R_3} \tag{11.25}$$

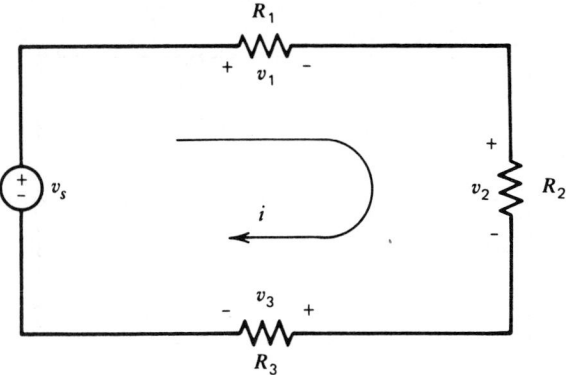

Fig. 11.6 A voltage divider circuit. Reproduced, with permission, from R. E. Thomas and A. J. Rosa, *Circuits and Signals: An Introduction to Linear and Interface Circuits*, New York, Wiley, 1984.

Finding the result for v_1 and v_3 is similar, and we obtain

$$v_1 = \frac{v_s R_1}{R_1 + R_2 + R_3}, \qquad v_3 = \frac{v_s R_3}{R_1 + R_2 + R_3} \tag{11.26}$$

In looking over our results, we observe an interesting phenomenon. In each case the voltage across the desired resistor is equal to the value of its resistance divided by the total resistance in the circuit times the total voltage across the series circuit. In other words, each resistor extracts its fraction of voltage so that the sum of the voltages across all the resistors in the loop equals the total voltage applied. Once we recognize this phenomenon, we need not repeat the analysis! Whenever we have a number of resistors in series with a voltage source, we can immediately use the voltage division rule to find the desired voltages.

An important device in electrical engineering is the **potentiometer**. This device makes use of voltage (*potential*) division to *meter* out a desired fraction of the input voltage. Figure 11.7 shows the circuit symbol of a potentiometer and a typical application.

The voltage v_{out} can be varied by simply turning the knob on the potentiometer to move the wiper arm contact. By using the voltage divider rule, v_{out} can be found as

$$v_{out} = v_s \frac{R_{total} - R_1}{R_{total}} \tag{11.27}$$

When we make R_1 zero by moving the wiper all the way to the top, we get $v_{out} = v_s$. If we move the

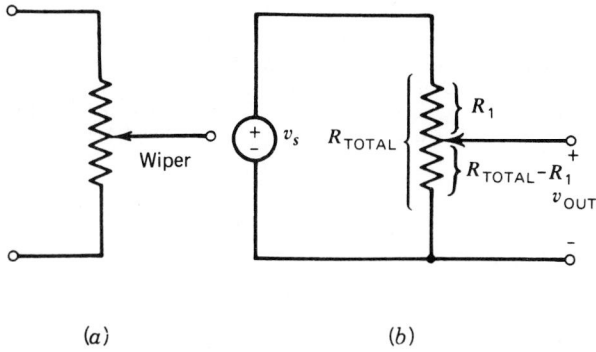

(a)   (b)

Fig. 11.7 The potentiometer. (*a*) Circuit symbol. (*b*) Application. Reproduced, with permission, from R. E. Thomas and A. J. Rosa, *Circuits and Signals: An Introduction to Linear and Interface Circuits*, New York, Wiley, 1984.

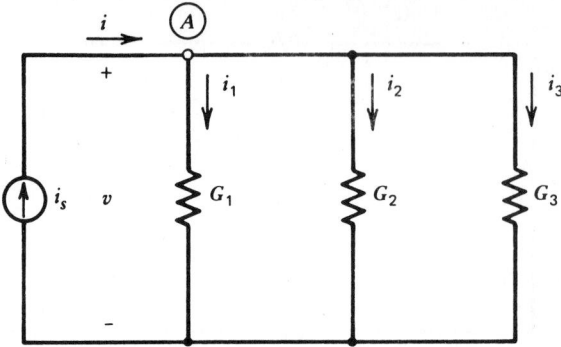

Fig. 11.8 A current divider. Reproduced, with permission, from R. E. Thomas and A. J. Rosa, *Circuits and Signals: An Introduction to Linear and Interface Circuits*, New York, Wiley, 1984.

wiper all the way to the bottom, we make R_1 equal to R_{total}, hence $v_{\text{out}} = 0$. We can vary the output voltage all the way from v_s, the applied voltage, to zero.

Applications for potentiometers are almost endless—control of volume, control of voltage output, and fine adjustments for alignment of circuits are just a few.

The dual of voltage division is current division. Since it is the dual, we deal with conductances and current sources connected in parallel. Using a similar analysis we find for the circuit of Fig. 11.8:

$$i_1 = \frac{G_1 i_s}{G_1 + G_2 + G_3} \qquad i_2 = \frac{G_2 i_s}{G_1 + G_2 + G_3} \qquad i_e = \frac{G_3 i_s}{G_1 + G_2 + G_3} \qquad (11.28)$$

We see that the total current delivered to the parallel network is *divided* among the conductances G_1, G_2, and G_3 in proportion to their value compared with the total of the conductances in parallel.

Example 11.1. The D'Arsonval galvanometer is a device used to measure direct currents and voltages. In simple terms, a coil of wire is mounted between a permanent magnet in such a manner that it is free to rotate. As a current flows through the coil, a torque is created that causes the coil to turn. A pointer or needle is attached to the coil and as the current is increased the pointer deflection is linearly proportional to the current. See Fig. 11.9(*a*).

D'Arsonval movements are rated as to the amount of current necessary to achieve full-scale deflection. Depending on the actual construction, the usual range of current necessary to achieve full-scale deflection is 1 μA to 1 mA.

Clearly scientists and engineers require instruments that can measure a much wider range of values. To get around this limitation, meter designers employ a shunt resistance R_s, as shown in Fig. 11.9(*b*). Suppose it was desired to measure 10 A full-scale, and the D'Arsonval movement we had available provided a full-scale reading with 10 μA. What shunt resistance will be required if the coil resistance is 20 Ω?

Consider the model of an ammeter shown in Fig. 11.9(*b*). Essentially the task at hand is to shunt most of the current around the actual meter movement, letting only (and exactly) 10 μA flow through the meter movement when 10 A is flowing through the newly designed meter. In Fig. 11.9(*b*), the actual current desired to be measured is indicated by i_m, the resistance of the meter movement R_m, and the full-scale current of the movement I_{fs}. R_s is the shunt resistance for which we must design. Once we show the model, it should become clear that the problem can be solved using current division, that is,

$$I_{fs} = \frac{R_s I_m}{R_m + R_s} \qquad 10^{-5} = \frac{R_s(10)}{20 + R_s} \qquad (11.29)$$

which yields $R_s = 20\ \mu\Omega$.

Ground—The Reference Node

Voltage is an across variable that is defined and measured between two nodes. It is often convenient to identify one of the nodes as the reference node, commonly called ground, and to measure (and define) the voltage at all other nodes with respect to this reference node. This concept should not be new to us. For example, the concept of elevation is similar. If one asks for the elevation of a particular

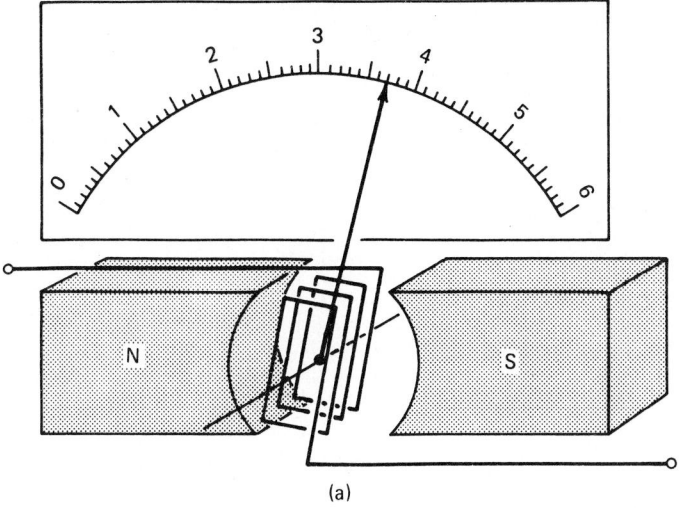

(a)

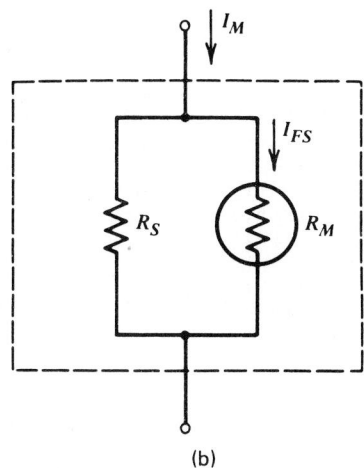

(b)

Fig. 11.9 Simple representation of a D'Arsonval meter movement and model. (*a*) Representation of a D'Arsonval galvanometer. (*b*) Model of an ammeter. Reproduced, with permission, from R. E. Thomas and A. J. Rosa, *Circuits and Signals: An Introduction to Linear and Interface Circuits*, New York, Wiley, 1984.

mountain, one usually expects to obtain the number of feet or meters between the top of the mountain and a reference—usually mean sea level. In electric circuits which node we select as reference is not always as obvious as mean sea level is in measuring elevations. Hence we denote the selected reference node as ground by using one of the "ground" symbols shown in Fig. 11.10. The term **node voltage** can then be used to define the voltage of any node in the circuit with respect to the selected reference node. It should be noted that, in general, voltages measured with respect to the reference can be positive (greater than the reference) or negative (less than the reference)—just as Death Valley has a minus elevation indicating that it is below mean sea level. Once a reference node has been chosen, it can be used to advantage to reduce complex circuit drawings, or schematics. To do this, every element connected to the reference node or ground is shown connected to the ground symbol. All elements connected to the ground symbol are assumed to be connected together at the end tied to that symbol, even though they may be located far apart on the schematic. The result is a simplified drawing with fewer interconnecting wires shown.

Fig. 11.10 Ground symbols. Reproduced, with permission, from R. E. Thomas and A. J. Rosa, *Circuits and Signals: An Introduction to Linear and Interface Circuits*, New York, Wiley, 1984.

11.3.2 Circuit Theorems

Linearity

The hallmark feature of **linear** circuits is that signal outputs are linear functions of its inputs. Mathematically we say that a function is linear if it has two properties:

$$f(AX) = Af(X) \quad \text{(homogeneity)}$$

and   (11.30)

$$f(X_1 + X_2) = f(X_1) + f(X_2) \quad \text{(additive)}$$

where A is a constant. In terms of circuits the homogeneity property indicates that the output of a linear circuit is proportional to the input. The additive property means that the output due to two or more inputs can be found by adding the outputs obtained when each input is applied separately. In circuit analysis the first property is called **proportionality** and the additive property is called **superposition**.

Proportionality

If we consider the *i-v* characteristics of a linear resistor $v = iR$, then it is clear that if the current (input) is doubled, the voltage (output) is doubled. However, the power delivered is $p = i^2R$, hence doubling the current quadruples the power. Thus we observe that a circuit can be linear only in the current and voltage signal variables, but not power. Power is proportional to the product of current and voltage, and is inherently nonlinear, even though the circuit may be linear.

The significance of linearity is that for resistive circuits we can write input-output relationships as

$$y = Kx \quad\quad (11.31)$$

where x is the input (current or voltage), y is an output (current or voltage), and K is a constant. We already have seen several examples of this relationship in voltage and current divider relationships [Eqs. (11.25)–(11.28)].

Superposition

The principle of superposition is a very useful computational and conceptual tool that finds many applications in the analysis of linear circuits with several inputs. The input-output relationships of linear circuits have the additive property of linear functions. For linear, resistive circuits this means that we can always write any output y as

$$y = K_1x_1 + K_2x_2 + K_3x_3\ldots \quad\quad (11.32)$$

where $x_1, x_2, x_3, \ldots$, are circuit inputs, and $K_1, K_2, K_3, \ldots$, are constants that depend on the circuit. Briefly stated, for a linear, resistive circuit the output is a linear combination of the various inputs.

A circuit analysis technique based on superposition proceeds as follows:

1. Turn off* all input signal sources except one and find the output due to that input acting alone.
2. Repeat step 1 successively for each input signal source.
3. The output when all input sources are on is then found by simply adding the responses due to each source acting alone.

Example 11.2. Determine the output of the circuit in Fig. 11.11(a) using superposition. To find the output due to the voltage source, we turn off the current source (replace it by an open circuit) and obtain the circuit in Fig. 11.11(*b*).

---

*To turn off a voltage source we set its voltage to zero ($V_s = 0$), which is the equivalent of replacing it with a short circuit. Similarly, turning off a current source ($I_s = 0$) is the equivalent of replacing it with an open circuit.

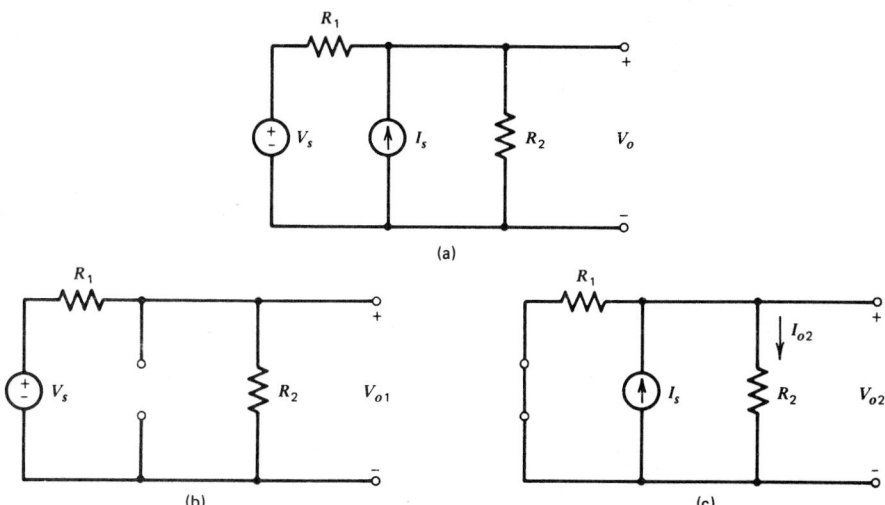

Fig. 11.11 Circuit for Example 11.2, analysis using superposition: (*a*) circuit; (*b*) current source turned OFF; (*c*) voltage turned OFF. Reproduced, with permission, from R. E. Thomas and A. J. Rosa, *Circuits and Signals: An Introduction to Linear and Interface Circuits*, New York, Wiley, 1984.

By voltage division,

$$V_{01} = \frac{R_2}{R_1 + R_2} V_s \tag{11.33}$$

Next we turn off the voltage source (replace it by a short circuit) and obtain the circuit in Fig. 11.11(*c*). By current division,

$$I_{02} = \frac{R_1}{R_1 + R_2} I_s \tag{11.34}$$

But by Ohm's law,

$$V_{02} = I_{02} R_2 = \frac{R_1 R_2}{R_1 + R_2} I_s \tag{11.35}$$

Using superposition we then write

$$V_0 = V_{01} + V_{02} = \frac{R_2}{R_1 + R_2} V_s + \frac{R_1 R_2}{R_1 + R_2} I_s \tag{11.36}$$

Thevenin and Norton Equivalent Circuits

An interface is a connection between two or more circuits that perform different functions. For the two-terminal interface shown in Fig. 11.12(*a*), we normally think of one circuit as the source *S* and the other as the load *L*. That is, we think of signals as being produced by the source circuit and delivered to the load.

A powerful tool for dealing with interfaces is the concept of the **Thevenin** and **Norton equivalent circuits** shown in Fig. 11.12(*b*) and 11.12(*c*), respectively. Stated formally: *Given a two-terminal interface in which the source circuit is linear, then the same interface signals will exist if the source is replaced by its Thevenin or Norton equivalent circuit.* Note that the equivalences require that the source circuit be linear, but place no restriction on the nature of the load circuit. The load may be linear or nonlinear.

The Thevenin equivalent consists of a voltage source (V_T) in series with a resistor (R_T). The Norton equivalent is a current source (I_N) in parallel with a resistor (R_N). Since either equivalent

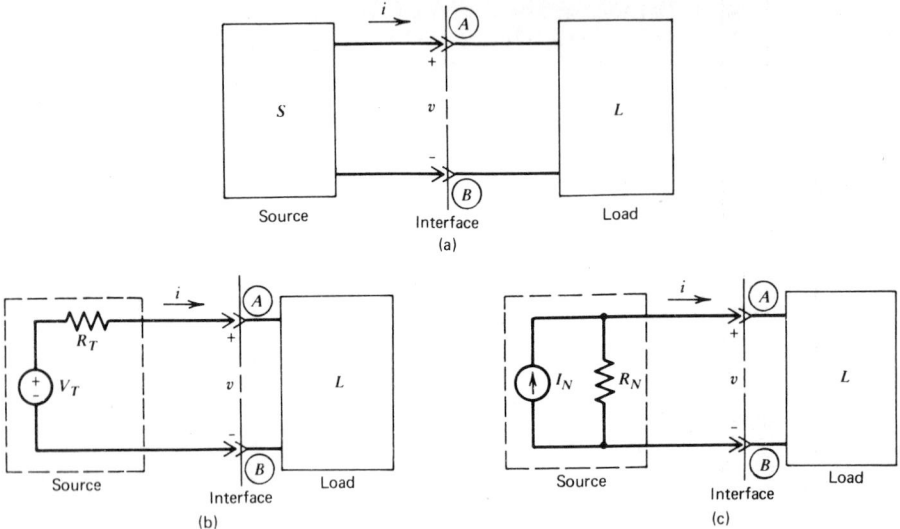

Fig. 11.12 Source-load equivalent circuits: (a) two-terminal interface, (b) Thevenin equivalent, (c) Norton equivalent. Reproduced, with permission, from R. E. Thomas and A. J. Rosa, *Circuits and Signals: An Introduction to Linear and Interface Circuits*, New York, Wiley, 1984.

leaves the interface signals unchanged, the two circuits must be equivalent to each other. Using KVL and Ohm's law in the Thevenin equivalent yields the i-v characteristics at terminals Ⓐ and Ⓑ as

$$v = V_T - iR_T \tag{11.37}$$

Applying KCL and Ohm's law to the Norton equivalent yields the i-v characteristics at terminals Ⓐ and Ⓑ as

$$i = I_N - \frac{v}{R_N} \quad \text{or} \quad v = I_N R_N - iR_N \tag{11.38}$$

Since two equivalent circuits must have identical i-v characteristics, if follows by comparing Eqs. (11.37) and (11.38) that

$$R_N = R_T$$
$$I_N R_N = V_T \tag{11.39}$$

In essence the Thevenin and Norton equivalents are related by the source transformation relationships shown in Fig. 11.5. This means that we do not need to find both equivalent circuits independently. Once one is found, the other can be determined by source transformation. Put differently, the two equivalent circuits involve four parameters (V_T, R_T, I_N, R_N), but Eqs. (11.39) provide two relations between the parameters. Hence only two conditions are required to determine both equivalent circuits.

These two conditions are easily obtained by using open-circuit and short-circuit loads. That is, if the actual load is disconnected from the source, then an open-circuit voltage v_{oc} appears between terminals Ⓐ and Ⓑ. Applying the same condition to the Thevenin equivalent reveals that $V_T = v_{oc}$. Similarly, disconnecting the load and connecting a short circuit causes a current i_{sc} to flow. Applying the same connection to the Norton equivalent indicates that $I_N = i_{sc}$. In summary, if we find the open-circuit voltage and the short-circuit current, we can determine the Thevenin and Norton equivalent circuit parameters as

$$V_T = v_{oc}$$
$$I_N = i_{sc} \tag{11.40}$$
$$R_N = R_T = \frac{v_{oc}}{i_{sc}}$$

An important use of Thevenin and Norton equivalent circuits is the graphical analysis of circuits containing a two-terminal nonlinear element. An interface is defined at the terminals of the nonlinear element and the linear part of the circuit is reduced to a Thevenin equivalent as indicated in Fig. 11.13(a). The i-v characteristics of the Thevenin equivalent are constrained to be

$$i = \frac{-1}{R_T}v + \frac{V_T}{R_T} \tag{11.41}$$

This i-v characteristic is a straight line in the i-v plane, as shown in Fig. 11.13(b). The straight line is called the load line.

As also shown in Fig. 11.13(c), the nonlinear device has some i-v characteristic determined by its physical makeup. Mathematically this characteristic can be written as

$$i = f(v) \tag{11.42}$$

The problem is to solve Eqs. (11.41) and (11.42) simultaneously. This can be done by numerical iteration if the $f(v)$ is known explicitly, but often a graphical solution is adequate.

If we graphically superimpose the source load line on the i-v characteristics of the nonlinear element, the point (or points) of intersection represent the values of i and v that satisfy both the source and load constraints. The point of intersection is called the **operating point**, or **Q point**, in the terminology of electronics.

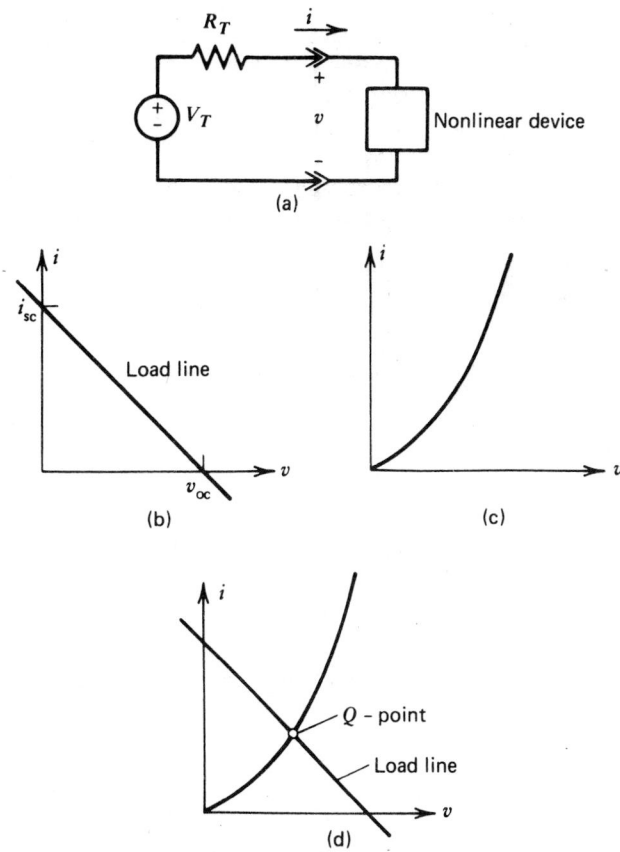

(a)

(b)

(c)

(d)

Fig. 11.13 Using Thevenin to find current through and voltage across a nonlinear element: (a) circuit, (b) source i-v characteristics, (c) nonlinear device characteristics, (d) source and device i-v characteristics showing Q point. Reproduced, with permission, from R. E. Thomas and A. J. Rosa, *Circuits and Signals: An Introduction to Linear and Interface Circuits*, New York, Wiley, 1984.

Maximum Signal Transfer

With the advent of integrated circuits engineers spend a considerable portion of their efforts on interfacing standard, commercially available building blocks called **integrated circuits** (ICs) or **chips**. In this context interfacing means interconnecting building blocks so that they operate together in some desired fashion. One of the constraints on achieving desired circuit performance is the maximum signal levels that can be delivered at an interface. Simply put, given a *fixed* source and an *adjustable* load, what are the maximum values of the signals available at an interface?

For simplicity we shall treat the case in which both the source and load circuits are linear. The source can be represented by its Thevenin equivalent and the load by an equivalent resistance R_L, as shown in Fig. 11.14.

By voltage division the voltage at the interface is

$$v = \frac{R_L}{R_L + R_T} V_T \tag{11.43}$$

For a fixed source and a variable load, the voltage will be a maximum if R_L is very large compared with R_T. Ideally R_L should be infinite, in which case

$$v_{max} = V_T = V_{oc} \tag{11.44}$$

The current delivered at the interface is

$$i = \frac{V_T}{R_L + R_T} \tag{11.45}$$

Again, for a fixed source and a variable load, the current will be a maximum if R_L is very small compared with R_T. Ideally R_L should be zero, in which case

$$i_{max} = \frac{V_T}{R_T} = I_N = i_{sc} \tag{11.46}$$

The power delivered at the interface is the product vi. By using Eqs. (11.43) and (11.45) we can write

$$p = vi$$

$$= \frac{R_L V_T^2}{(R_L + R_T)^2} \tag{11.47}$$

For the constraint of a fixed source and a variable load, we find the maximum available power by differentiating Eq. (11.47) with respect to R_L.

$$\frac{\partial p}{\partial R_L} = \frac{\left[(R_L + R_T)^2 - 2R_L(R_L + R_T)\right] V_T^2}{(R_L + R_T)^4}$$

$$= \frac{R_T^2 - R_L^2}{(R_L + R_T)^4} \tag{11.48}$$

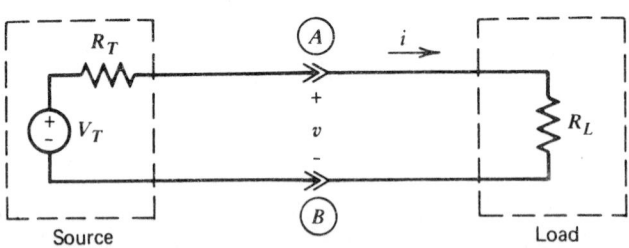

Source   B   Load

Fig. 11.14 Circuit for determining maximum power transfer relationships. Reproduced, with permission, from R. E. Thomas and A. J. Rosa, *Circuits and Signals: An Introduction to Linear and Interface Circuits*, New York, Wiley, 1984.

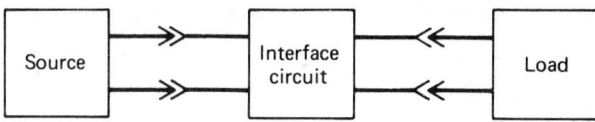

Fig. 11.15 Interface circuit. Reproduced, with permission, from R. E. Thomas and A. J. Rosa, *Circuits and Signals: An Introduction to Linear and Interface Circuits*, New York, Wiley, 1984.

Situation	Source R	Load R
I	Variable	Variable
II	Variable	Fixed
III	Fixed	Variable
IV	Fixed	Fixed

Fig. 11.16 Interface design situations. Reproduced, with permission, from R. E. Thomas and A. J. Rosa, *Circuits and Signals: An Introduction to Linear and Interface Circuits*, New York, Wiley, 1984.

Clearly the derivative will be zero (maximum power) if $R_T = R_L$. Thus for maximum power transfer the source and load resistances should be equal. When this occurs the source and load are generally said to be *matched*.

By substituting the matched condition ($R_T = R_L$) back into Eq. (11.47) we find the maximum power available to be

$$p_{max} = \frac{V_T^2}{4R_T} = \frac{I_N^2 R_T}{4} \qquad (11.49)$$

Principles of Signal Transfer

We have discussed the maximum signal levels that are available at the output of a fixed source circuit. These results place bounds on what is achievable at an interface but do not describe all of the situations normally encountered in circuit design. Usually we are confronted with a situation in which one or more of the signal levels at the interface is prescribed and either the source or the load, or both, must be adjusted to achieve a workable design. Sometimes it is necessary to design an interface circuit that is inserted between the source and load, as shown in Fig. 11.15.

The interface circuit may be purely resistive, active such as op-amps, or made up of impedance devices such as inductors, capacitors or transformers.

Figure 11.16 lists the four canonic situations in which the interface design can arise. Situation I can be encountered when designing with discrete circuit elements. Generally we can pick both the source and the load, but the design problem may involve iteration since the source and load interact. Situation II can arise when we attempt to deliver signals to an existing load or receiver. Generally in this situation it is desirable to reduce the source resistance to a minimum, ideally to zero. There are, of course, limits on the achievement of this goal, but we will see that the op-amp can be used to approach the ideal. Situation III occurs fairly commonly. Certain types of transmission lines and laboratory signal sources have standard source resistances such as 50, 75, 300, and 600 Ω. In such situations the load must be selected to achieve the desired interface objective.

The last situation, situation IV, occurs all too frequently. We are often faced with the problem or interconnecting standard IC building blocks. The problem usually involves designing an interface circuit that will allow the source and load to interact in harmony. The engineer must consider a wide range of devices such as bridges, pads or potentiometers, op-amp buffers, or tristate logic—whatever will work must be considered. However, simplicity and cost are major factors that must also influence the design decision.

11.3.3 Active Circuits

Linear Dependent Sources

One of the most important signal-processing functions available to electrical engineers is signal **amplification**. Circuits consisting only of linear resistors cannot produce output signal voltages, currents, or, most important, powers that are larger than their signal inputs. To obtain signal amplification we need active devices such as vacuum tubes, transistors, and operational amplifiers

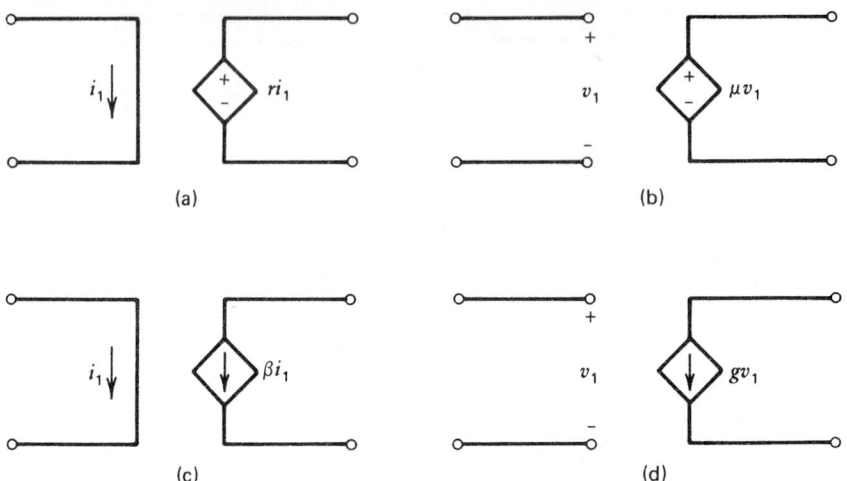

Fig. 11.17 Dependent sources. (*a*) Current-controlled voltage source. (*b*) Voltage-controlled voltage source. (*c*) Current-controlled current source. (*d*) Voltage-controlled current source. Reproduced, with permission, from R. E. Thomas and A. J. Rosa, *Circuits and Signals: An Introduction to Linear and Interface Circuits*, New York, Wiley, 1984.

(op-amps). Simple models of these active devices require a new set of circuit elements called **dependent** or **controlled** sources.

Active devices are usually very nonlinear. However, many can be made to operate in a linear mode. When active devices are operating in a linear mode, we can model their characteristics by using one or more of the four dependent source elements shown in Fig. 11.17. The dominant feature of a dependent source is that the strength or magnitude of the source is proportional to—that is, controlled by—a voltage or current appearing elsewhere in the circuit.

Several matters of notation and symbols should be mentioned. Each of the controlled sources is characterized by a single parameter, either μ, β, r, or g. These parameters are often somewhat loosely called the **gain** of the controlled source. Strictly speaking, the parameters μ and β are dimensionless quantities called the open-circuit voltage gain and the short-circuit current gain, respectively. The parameter r has the dimensions of ohms and is called the transresistance, a contraction of transfer resistance. The parameter g is called transconductance and has the dimensions of mhos.

Dependent sources are linear elements that are used in circuit analysis. But in some respects they are conceptually different than the other elements we have used. The resistance element is an ideal model of a set of devices called resistors. The ideal switch is a model of a component called a **switch**. But you will not find controlled sources in parts lists and catalogs. Controlled sources in combination with other elements are used to model real devices such as transistors and op-amps.

A word of caution: We use the diamond symbol to indicate a linear dependent source. Many others do not employ this symbol, but instead use the circle symbol as for the independent sources. One should realize that it is the *gain* parameter that really identifies a source as dependent.

Analysis of Circuits with Dependent Sources

The analysis of circuits containing controlled sources is fundamentally the same as the passive circuits. Kirchhoff's laws still apply and so the connection constraints are the same. The element constraints are still independent of the connections. But we have a new set of elements so that the combined constraints yield circuit behavior that is significantly different than passive circuits.

A simple example of a circuit using a voltage-controlled source is shown in Fig. 11.18(*a*). Applying Kirchhoff's voltage law around loop *I* yields

$$v_s = R_s i + v_1 \tag{11.50}$$

But $i = 0$ since there is an open circuit, and hence $v_s = v_1$. The controlled voltage source output therefore is equal to μv_s. Since v_0 is in parallel with the controlled source, we have

$$v_0 = \mu v_s \tag{11.51}$$

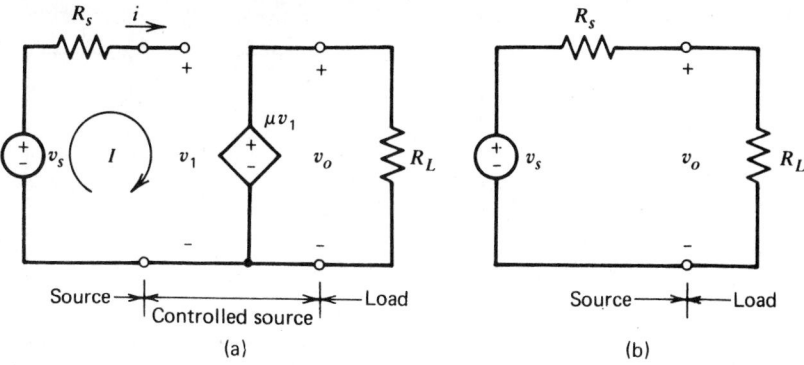

Fig. 11.18 Comparison of active and passive circuits: (*a*) active circuit; (*b*) passive circuit. Reproduced, with permission, from R. E. Thomas and A. J. Rosa, *Circuits and Signals: An Introduction to Linear and Interface Circuits*, New York, Wiley, 1984.

The output is directly proportional to the input. If μ is greater than 1, then we say that the circuit *amplified* the input, and we call the circuit an **amplifier**. If μ is less than 1, we say that the circuit *attenuated* the input and we call the circuit an **attenuator**.

The real value of linear active circuits can be demonstrated by performing a similar analysis of the passive circuit in Fig. 11.18(*b*). Application of the ever-so-familiar voltage division relationship yields

$$v_0 = \frac{R_L v_s}{R_s + R_L} \qquad (11.52)$$

If we compare the results of our analysis of both circuits [Eqs. (11.51) and (11.52)], we can make some important observations.

First of all, the output of the circuit with the dependent source does not depend on either the source or load resistance; rather, it depends only on the gain μ. This means that the output voltage is not limited by the maximum signal transfer conditions derived earlier. In fact, if the gain μ is greater than 1, the output voltage can be greater than the input voltage. In effect the controlled source provides unilateral signal transfer that isolates the source and load. This eliminates the source-load interaction that led to the maximum signal transfer conditions.

To illustrate this further consider the power delivered to the load in the controlled source circuit. Using Eq. (11.51),

$$P_L = \frac{(V_0)^2}{R_L} = \frac{(\mu V_s)^2}{R_L} \qquad (11.53)$$

The load power is not dependent on the source resistance, and hence is not limited by the maximum power transfer condition ($R_s = R_L$). In fact, the power delivered by the source is zero since it is connected to an open circuit and it is necessarily zero. The circuit produces an output power with zero input power. This apparent contradiction can be resolved by realizing that a dependent source is a *model* of devices that requires an external energy source to operate. We assume that the external supply can indeed provide whatever power the independent source requires. In real devices this is not always the case, and the design engineer must ensure that the limits of the external supply are not exceeded. In our discussion of op-amps we shall consider at least some limitations of the external supply.

Example 11.3. Determine the equivalent resistance R_{eq} between terminals Ⓐ and Ⓑ of the circuit of Fig. 11.19(*a*). The first observation one should make is that this is a **feedback** circuit. Feedback occurs when there is a path from the dependent source back to its input control circuit. The circuit in Fig. 11.19(*a*) has no excitation from independent sources, and hence the dependent current source is off and acts as an open circuit. As a result it intuitively appears that the equivalent resistance between Ⓐ and Ⓑ should be simply R_L.

However, the effect of the dependent source has not been accounted for since the circuit has no excitation. To provide excitation we connect the test source shown in Fig. 11.19(*b*) and analyze the

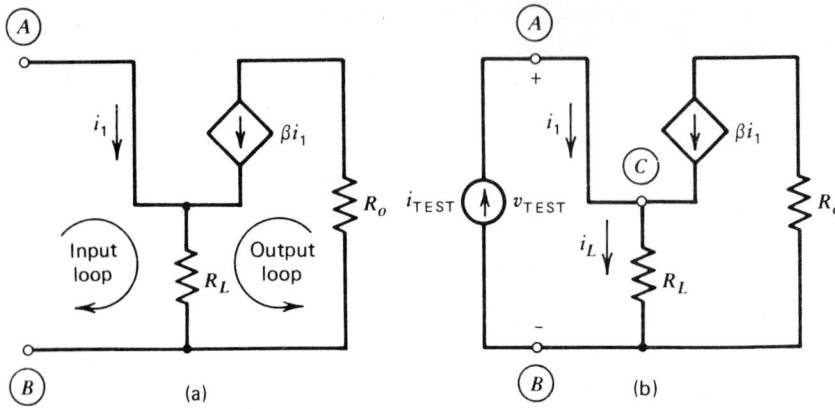

Fig. 11.19 Analysis of dependent source circuit with feedback. (*a*) Active circuit with feedback. (*b*) Circuit driven by test source. Reproduced, with permission, from R. E. Thomas and A. J. Rosa, *Circuits and Signals: An Introduction to Linear and Interface Circuits*, New York, Wiley, 1984.

circuit response. By KCL at node $\textcircled{C}$:

$$i_L = i_1 + \beta i_1$$
$$= (1 + \beta)i_1 \tag{11.54}$$

By KCL at node $\textcircled{A}$ $i_1 = i_{\text{test}}$, and hence

$$i_L = (1 + \beta)i_{\text{test}} \tag{11.55}$$

But by Ohm's law,

$$v_{\text{test}} = R_L i_L \tag{11.56}$$

Hence

$$v_{\text{test}} = (1 + \beta)R_L i_{\text{test}} \tag{11.57}$$

Now the principle of equivalence says that two circuits are equivalent if they have the same *i-v* characteristics. Since equivalent resistance means

$$v_{\text{test}} = R_{\text{eq}} i_{\text{test}} \tag{11.58}$$

by comparing Eqs. (11.57) and (11.58) we conclude that

$$R_{\text{eq}} = (1 + \beta)R_L \tag{11.59}$$

This result is significantly different from our initial intuitive estimate of $R_{\text{eq}} = R_L$. Moreover the circuit in question is a model of a transistor circuit and the gain parameter β would typically be on the order of 100. So not only is our intuition wrong, but it is wrong by two orders of magnitude! The reason for this spectacular result is the unique placement of the feedback resistor R_L that ties the input and output loops together. This permits both the input current i_1 and the output current βi_1 to add dramatically, thereby increasing the effect of R_L on the circuit.

Active circuits with feedback may have Thevenin resistances that are many orders of magnitude larger or smaller than the apparent input resistance with no excitation. As a result, to find the Thevenin equivalent of an active circuit we should always find the open-circuit voltage and short-circuit current since both of these calculations require that the circuit be driven by independent sources.

The Operational Amplifier
The operational amplifier (op-amp) is the premier linear active device made available by IC technology. The device itself is a complex array of transistors, resistors, diodes, and capacitors, all fabricated and interconnected on a tiny piece of silicon commonly called a chip. The completed

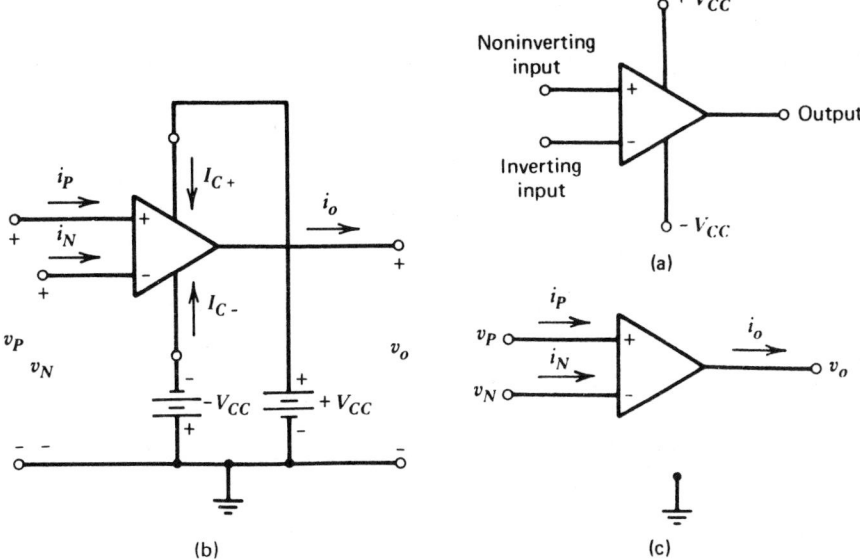

Fig. 11.20 (*a*) Op-amp circuit symbol. (*b*) Complete set of variables for op-amp currents and voltages. (*c*) Signal variables only. Reproduced, with permission, from R. E. Thomas and A. J. Rosa, *Circuits and Signals: An Introduction to Linear and Interface Circuits*, New York, Wiley, 1984.

device is then packaged and sold as a single unit. Such devices are available at very low unit costs and have many applications in both linear and nonlinear circuits.

In spite of its complexity, the op-amp can be modeled by rather simple *i-v* characteristics. In other words, we do not need to concern ourselves with what is going on inside the package; rather we can treat the op-amp as a circuit element with certain terminal characteristics. In what follows we learn how to analyze and design op-amp circuits using what is called its *ideal* model.

Before we develop this model there are certain matters of notation and nomenclature that must be discussed. The op-amp to be presented is a five-terminal device as shown in Fig. 11.20(*a*). The + and − symbols in the triangle identify the input terminals and are a shorthand notation for the noninverting and inverting input terminals, respectively. These symbols simply identify the two input terminals and have nothing to do with the polarity of the voltages applied. The other terminals are the output and the plus and minus supply voltage, usually labeled $\pm v_{cc}$. While some op-amps have more terminals than these, five are always present and are the only ones we shall use in this discussion.

It is especially important to note that two terminals marked $\pm V_{cc}$. These two terminals usually are not included in op-amp circuits so as to keep the diagrams uncluttered; however, they are always there. It is through these terminals that the power for the operation of the op-amp must flow. The supply voltages, as $\pm V_{cc}$ are often called, provide the power for amplification and determine the upper and lower voltage limits on the op-amp output. In general these voltages mark the boundary between linear and nonlinear operation of the op-amp.

Figure 11.20(*b*) shows a complete set of voltage and current variables for the op-amp, and Fig. 11.20(*c*) shows the abbreviated set of signal variables we will use. All voltages are defined with respect to a common reference node, usually ground. Figure 11.20(*c*) shows the shorthand notation we use to describe the signal voltages. The voltage symbol is written beside the terminal, and we describe these voltages as the output voltage, or the voltage at the inverting input, and so forth. By such terminology we really mean the voltage between the output terminal and the reference node, or the voltage between the inverting input terminal and ground.

The reference directions for the currents are the traditional ones, but the use of the abbreviated set of signal variables can cause conceptual problems. If we were to write a global KCL equation for the complete set of variable in Fig. 11.20(*b*), it would read

$$i_0 = I_{c+} + I_{c-} + i_P + i_N \tag{11.60}$$

which is a correct application of Kirchhoff's current law. A similar equation or the signal variables in

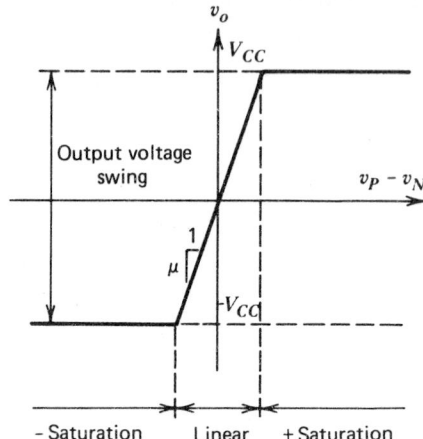

Fig. 11.21 Op-amp transfer characteristics. Reproduced, with permission, from R. E. Thomas and A. J. Rosa, *Circuits and Signals: An Introduction to Linear and Interface Circuits*, New York, Wiley, 1984.

Fig. 11.20(*c*) reads

$$i_0 = i_P + i_N \tag{11.61}$$

this should result in an incorrect analysis since the supply voltage currents have been ignored.

The dominant feature of the op-amp is its transfer characteristics shown in Fig. 11.21. These characteristics provide the relationships between the two input voltages (v_P, the noninverting input, and v_N, the inverting input) and the output voltage (v_0). The transfer characteristic is divided into linear and saturation ranges. In the linear range the output is proportional to the difference between the two inputs, and consequently the op-amp is called a **differential amplifier**. The slope of the line in the linear region is called the op-amp gain or open-loop gain, and is denoted as μ. Thus in the linear range the input-output relation can be written as

$$v_0 = \mu(v_P - v_N) \tag{11.62}$$

The open-loop gain of an op-amp is very large, usually greater than 10^5. As long as the net input ($v_P - v_N$) is small, the output will be proportional to the input. However, when μ times the net input lies outside the range defined by the supply voltages $\pm V_{cc}$, the output is limited by the supply voltages (less some small internal losses). When this occurs, the op-amp is said to be saturated, and the output is no longer proportional to the input but is determined by the supply voltages.

There are applications such as the **comparator** or **Schmidt trigger** where the op-amp is intentionally driven into the saturation regions. Such nonlinear uses of the op-amp find applications in many electronic circuits. However, in our treatment for now, we shall restrict the use of the op-amp to its linear region. In general we wish to analyze and design circuits that do not saturate for the given input(s).

A controlled source model of the op-amp is shown in Fig. 11.22. In addition to the open-loop gain μ, this model includes an input resistance (R_i) and an output resistance (R_0). Representative values for these parameters are given in Fig. 11.23(*a*), along with the values for the ideal op-amp.

The controlled source model can be used to develop the *i-v* characteristics of the ideal model. We have restricted our treatment to the linear range of operation. This means that

$$-V_{cc} \le v_0 \le +V_{cc} \tag{11.63a}$$

By using Eq. (11.62), we can write this bound as

$$-\frac{V_{cc}}{\mu} \le (v_P - v_N) \le +\frac{V_{cc}}{\mu} \tag{11.63b}$$

The supply voltage $\pm V_{cc}$ is typically ± 15 V while μ is a very large number, usually 10^5 or greater. Consequently $v_P \simeq v_N$ for linear operation. For the ideal op-amp, the open-loop gain is infinite ($\mu = \infty$) and this fuzzy equality becomes an exactitude. Moreover the input resistance of the ideal

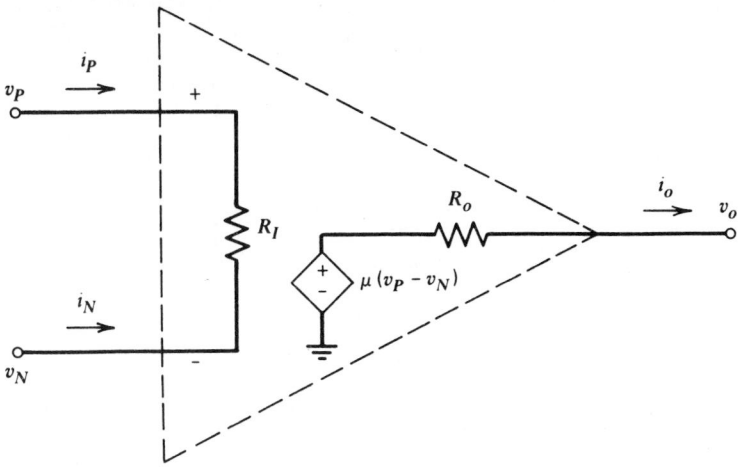

Fig. 11.22 Controlled source model of linear op-amp. Reproduced, with permission, from R. E. Thomas and A. J. Rosa, *Circuits and Signals: An Introduction to Linear and Interface Circuits*, New York, Wiley, 1984.

Parameter	Name	Typical Values	Ideal OP AMP Values
μ	Open-loop gain	10^5–10^7	∞
R_I	Input resistance	10^6–10^{13}	∞
R_o	Output resistance	10–100	0
$\pm V_{CC}$	Supply voltages	± 15 V	± 15 V

(a)

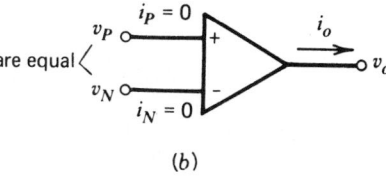

(b)

Fig. 11.23 (*a*) Op-amp parameters. (*b*) Ideal op-amp characteristics. Reproduced, with permission, from R. E. Thomas and A. J. Rosa, *Circuits and Signals: An Introduction to Linear and Interface Circuits*, New York, Wiley, 1984.

op-amp model is also infinite so no current is drawn at either input. In sum, the *i-v* characteristics of the ideal op-amp are

$$v_P = v_N$$
$$i_P = i_N = 0 \qquad\qquad (11.64)$$

These characteristics are illustrated on the op-amp circuit symbol in Fig. 11.23(*b*).

At first glance the element constraints of the ideal op-amp appear to be fairly useless. They actually look more like connection constraints and are totally silent about the output quantities (v_0 and i_0), which are usually the signals of interest. The answer to this dilemma is that in linear applications feedback is always present. That is, for the op-amp to operate in a linear mode it is

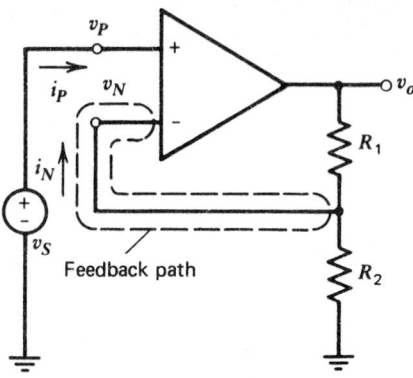

Fig. 11.24 Noninverting op-amp circuit. Reproduced, with permission, from R. E. Thomas and A. J. Rosa, *Circuits and Signals: An Introduction to Linear and Interface Circuits*, New York, Wiley, 1984.

necessary that there be some feedback paths from the output to one or both of the inputs. These feedback paths allow us to analyze op-amp circuits using the ideal op-amp constraints.

To illustrate this process let us determine the input-output characteristics of the circuit in Fig. 11.24(a). This circuit has a feedback path from the output to the inverting input via a voltage divider. Since the ideal op-amp draws no current at either input ($i_P = i_N = 0$), we can use the voltage division rule to determine the voltage at the inverting input:

$$v_N = \frac{R_2}{R_1 + R_2} v_0 \qquad (11.65)$$

The input source connection at the noninverting input requires that

$$v_P = v_s \qquad (11.66)$$

But the ideal op-amp constraint demands that $v_P = v_N$; hence we can equate the right sides of Eqs. (11.65) and (11.66) to obtain the input-output relationship of the circuit as

$$v_0 = \frac{R_1 + R_2}{R_2} v_s \qquad (11.67)$$

The analysis strategy is to use the input signal source constraint together with the feedback path to determine the two op-amp input voltages. The ideal op-amp constraint requires that these voltages be equal, and this equality is used to determine the overall circuit input-output relationship.

The circuit in Fig. 11.24 is called the **noninverting amplifier**. The input-output relationship is of the form $v_0 = Kv_s$, which reminds us that the circuit is linear. The constant

$$K = \frac{R_1 + R_2}{R_2} \qquad (11.68)$$

is called the closed-loop gain since it includes the effect of the feedback path. In discussing op-amp circuits it is necessary to distinguish between two types of gains. The first is the *open-loop* gain provided by the op-amp. This gain is a very large number and ideally it is infinite. Then there is the *closed-loop* gain of the op-amp circuit, which includes a feedback path. This gain must be much smaller and is determined by the elements in the feedback path. For example, the closed-loop gain in Eq. (11.68) is really the voltage division rule upside down. Thus the feedback converts the very high but imprecisely known open-loop gain into a much smaller but precisely known closed-loop gain.

Basic Linear Op-amp Circuits

Analog signal-processing systems are often constructed using interconnections of relatively simple op-amp circuits in a building block fashion. In this section we introduce a basic set of circuits that can be used as these building blocks. We have already introduced one of these circuits—the noninverting amplifier discussed in the preceding section. Other basic circuits are the **inverting amplifier**, the **summer**, and the **subtractor**, shown in Fig. 11.25.

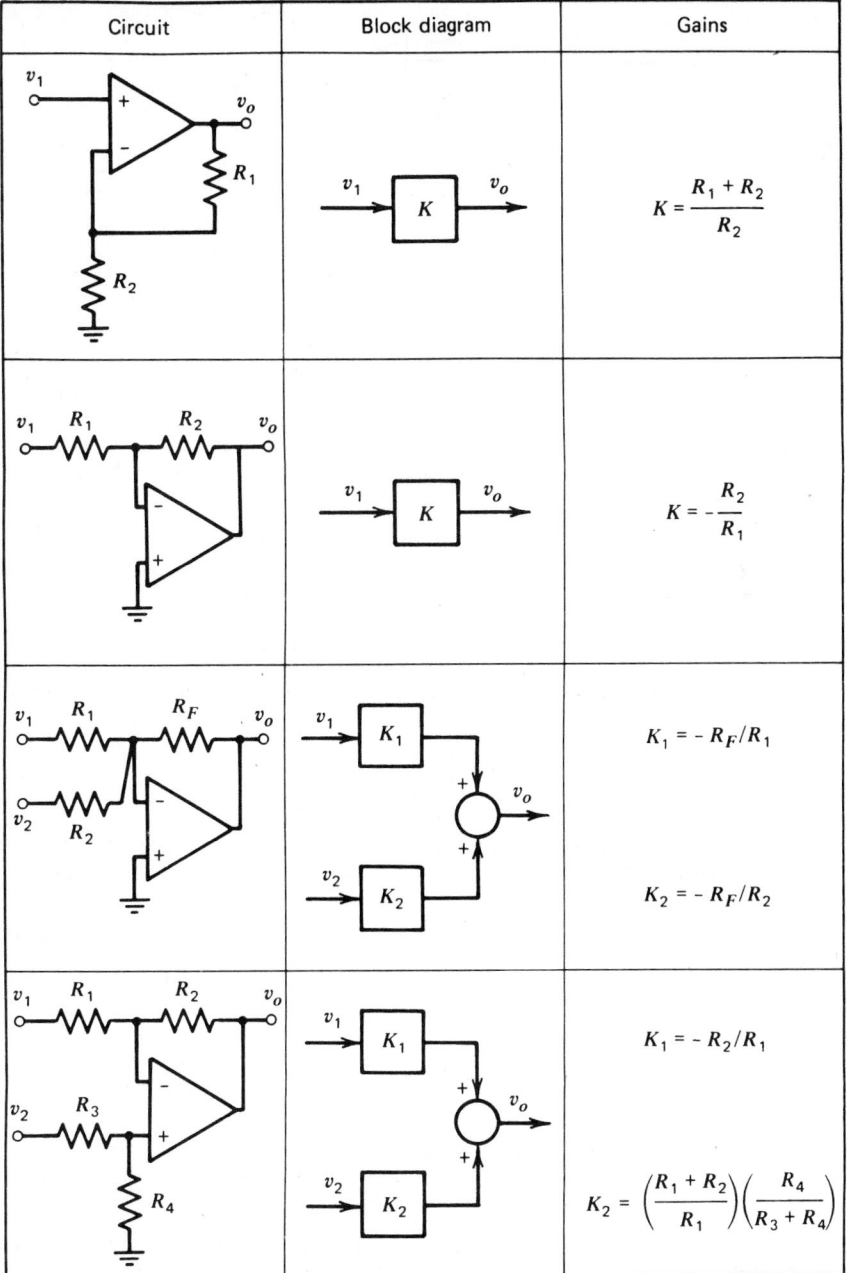

Circuit	Block diagram	Gains

Fig. 11.25 Block diagram representation of some basic linear op-amp circuits. Reproduced, with permission, from R. E. Thomas and A. J. Rosa, *Circuits and Signals: An Introduction to Linear and Interface Circuits*, New York, Wiley, 1984.

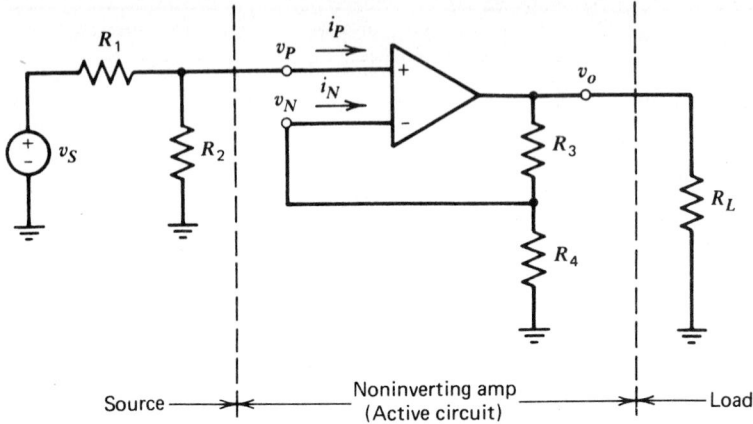

Fig. 11.26 Op-amp circuit for Example 11.4. Reproduced, with permission, from R. E. Thomas and A. J. Rosa, *Circuits and Signals: An Introduction to Linear and Interface Circuits*, New York, Wiley, 1984.

The key to using the building block approach is to recognize the feedback pattern and to isolate the basic circuit as a building block. The following example illustrates this process.

Example 11.4. Determine the input-output relationship of the circuit in Fig. 11.26.
　　When the circuit is partitioned as shown, we should recognize the noninverting amplifier. Since this circuit has zero output resistance, the load has no effect on the output voltage. Hence we can write, by inspection,

$$v_0 = \frac{R_3 + R_4}{R_4} v_P \tag{11.69}$$

since R_3 and R_4 form the feedback path. Because the input current at the noninverting terminal is zero, we can determine v_P using voltage division:

$$v_P = \frac{R_2}{R_1 + R_2} v_s \tag{11.70}$$

Thus the overall circuit input-output relationship is

$$v_0 = \frac{R_3 + R_4}{R_4} \frac{R_2}{R_1 + R_2} v_s \tag{11.71}$$

Example 11.5. Design an inverting summer to implement the input-output relationship

$$v_0 = -(5v_1 + 13v_2)$$

Recognizing the summer relationship from Fig. 11.25, the design problem is to select input and feedback resistors such that

$$\frac{R_F}{R_1} = 5 \quad \text{and} \quad \frac{R_F}{R_2} = 13 \tag{11.72}$$

One solution is to select $R_F = 65$ kΩ, which requires $R_1 = 13$ kΩ and $R_2 = 5$ kΩ. This design is shown in Fig. 11.27(*a*). There is nothing particularly wrong with this design, except that the resistance values are not the standard values and available only in precision resistors. An alternative in Fig. 11.27(*b*) is to select $R_F = 100$ kΩ, which leads to $R_1 = 20$ kΩ and $R_2 = 7.69$ kΩ. The first two values are standard sizes and only R_2 would require special consideration.

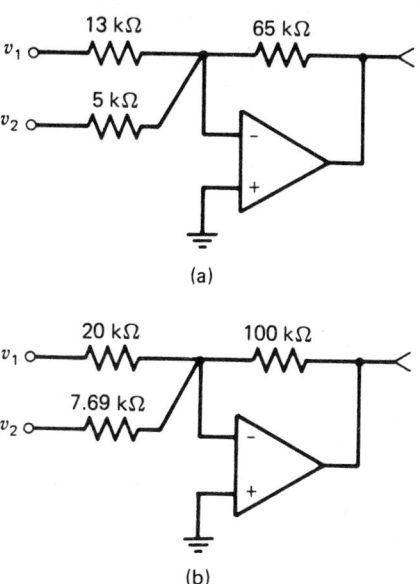

Fig. 11.27 Possible design solutions for Example 11.5: (*a*) first design, (*b*) alternative design. Reproduced, with permission, from R. E. Thomas and A. J. Rosa, *Circuits and Signals: An Introduction to Linear and Interface Circuits*, New York, Wiley, 1984.

The four basic op-amp circuit building blocks shown in Fig. 11.25 are often conveniently represented in block diagram form. The reason is that much of analog system analysis and design take place at the *functional* or *block diagram* level. A major advantage of op-amp circuits is that there is a near one-to-one correspondence between the block diagrams and the circuits that actually implement the system function. The amplifier or gain block of Fig. 11.28(*a*) says that the output is obtained by multiplying the input by the block gain, that is, $v_0 = K v_s$. The summing symbol is almost self-explanatory. It indicates that the output is simply $v_s = v_1 + v_2$. The amplifier block and the summing symbol can then be combined to produce a weighted summer as shown in Fig. 11.28(*c*).

One of the important advantages of op-amp circuits is that we can connect them in cascade without worrying about the loading effects of one circuit on the other, provided we take a few simple precautions. There are several reasons for connecting circuits in cascade; the main ones are to achieve more complex input-output relationships and to obtain higher gains without sacrificing bandwidth.

An example of a cascade connection is shown in Fig. 11.29. From a block diagram point of view we would say that the overall gain of the circuit is simply K_1, times K_2 times K_3, where these individual circuit gains were calculated with no load at the output. The input-output equivalent circuit of the cascade reminds us that the op-amp circuits have essentially zero output resistance, so that generally it is legitimate to consider the individual circuits independently. However, this is strictly true only if the load presented by the succeeding stage is reasonable, that is, within the power-handling capability of the op-amp. For this and other reasons, it is important to know the input resistance of an op-amp circuit.

Figure 11.30 shows the circuit diagram for determining the input resistance for the two basic op-amp circuit configurations. To obtain input resistance we conceptually apply a test signal and determine R_{in} from equivalence.

$$R_{\text{in}} = \frac{v_{\text{test}}}{i_{\text{test}}} \qquad (11.73)$$

For the noninverting circuit the test current is necessarily zero of the ideal op-amp model. Therefore apparently

$$R_{\text{in}} = \infty \quad \text{(noninverting amplifier)} \qquad (11.74)$$

If we were to use the detailed controlled source model of the op-amp shown in Fig. 11.22, we would

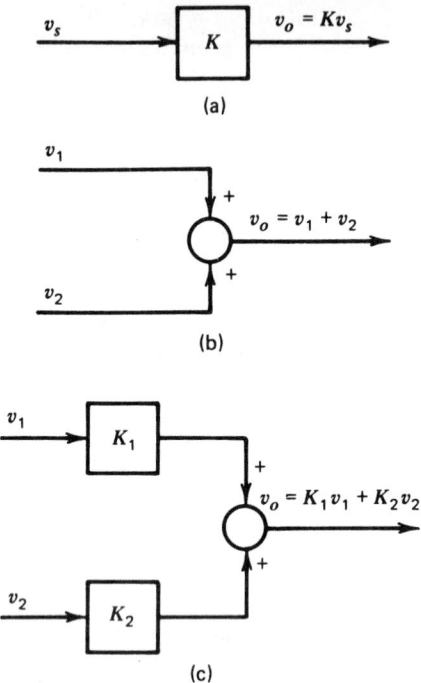

Fig. 11.28 Block diagram symbols: (*a*) amplifier block, (*b*) summer, (*c*) combined functions. Reproduced, with permission, from R. E. Thomas and A. J. Rosa, *Circuits and Signals: An Introduction to Linear and Interface Circuits*, New York, Wiley, 1984.

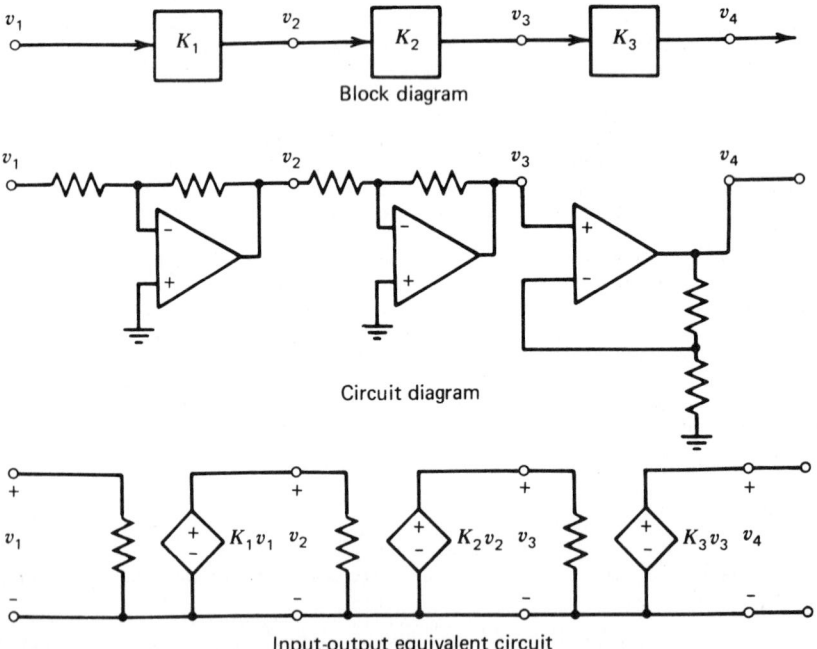

Fig. 11.29 Cascade connection. Reproduced, with permission, from R. E. Thomas and A. J. Rosa, *Circuits and Signals: An Introduction to Linear and Interface Circuits*, New York, Wiley, 1984.

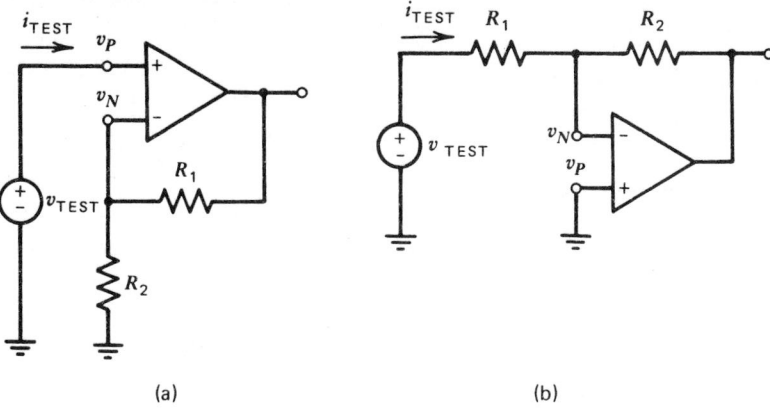

Fig. 11.30 Input resistance test for the basic op-amp circuits: (*a*) noninverting, (*b*) inverting. Reproduced, with permission, from R. E. Thomas and A. J. Rosa, *Circuits and Signals: An Introduction to Linear and Interface Circuits*, New York, Wiley, 1984.

find that the actual input resistance is approximately $\mu R_1/K$, where μ is the open-loop gain, K is the closed-loop gain, and R_1 is the op-amp input resistance. For any reasonable closed-loop gain this is a resistance of the order of 10^{10} to 10^{15} Ω. Thus for all practical purposes the noninverting amplifier has an infinite input resistance.

For the noninverting amplifier the input current is

$$i_{\text{test}} = \frac{v_{\text{test}} - v_N}{R_1} \tag{11.75}$$

For the ideal op-amp model $v_P = v_N$, and since the noninverting input is grounded $v_N = 0$. Thus the input resistance is

$$R_{\text{in}} = R_1 \quad (\text{inverting amplifier}) \tag{11.76}$$

Again if we were to use the complete controlled source model of the op-amp this result would be modified very slightly by a parallel resistance of the order of $\mu R_1/K$, a resistance whose value is of the order of 10^{10} to 10^{15} Ω. Thus for all practical purposes the input resistance of the inverting amplifier is R_1.

The impact of this is that in designing inverting amplifiers we must pick R_1 with two goals in mind. The first is to achieve the specified gain, and the second is to minimize the load presented in the preceding stage. This is particularly true when the preceding state is *not* the output of an op-amp circuit. It is particularly important to control the input resistance of an inverting amplifier when interfacing with signal transducers at the input to an amplifier cascade.

Interfacing with Op-amps

Designing circuits to interface with each other is greatly simplified by using operational amplifiers. This simplification is primarily attributable to the isolation properties of the op-amps, which enable an engineer to design one circuit without worrying about the interaction of connecting circuits. A drawback of op-amp interfacing is that supply voltages must be provided to operate the amplifier. Often, the low cost of op-amps and the ease of designing using op-amps will outweigh any problems associated with providing supply voltages. Furthermore, in many applications, the supply voltages are already available to drive other parts of the system.

To understand the utility of op-amps in interfacing, consider the circuit of Fig. 11.31(*a*). The problem is to provide a constant voltage to a load that can vary over several orders of magnitude. What we really want is an ideal voltage source, but signal sources usually do not cooperate in this regard. Thus if we are limited to a real voltage source, it is obvious that if R_L is changed, so too must R_s or V_s be changed if we want to maintain a constant voltage. This could become very inconvenient since V_s or R_s is usually not accessible. The circuit of Fig. 11.31(*b*), on the other hand, readily provides a constant voltage regardless of which R_L is selected. In reality there is a maximum current that the op-amp can deliver, and once this maximum is reached, any further reduction of R_L will

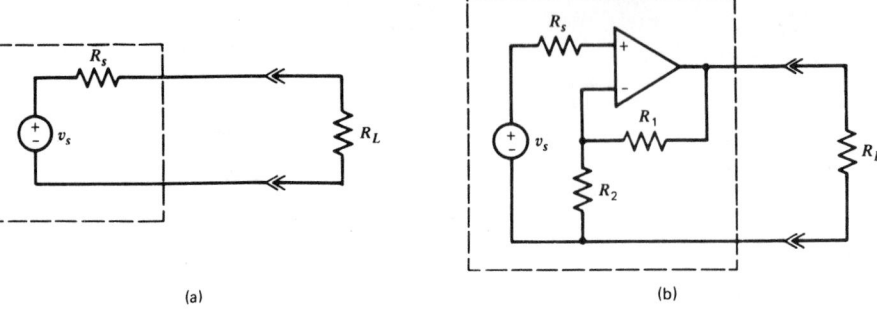

(a)   (b)

Fig. 11.31 Source-load interface with an op-amp circuit: (*a*) real voltage source, (*b*) ideal voltage source. Reproduced, with permission, from R. E. Thomas and A. J. Rosa, *Circuits and Signals: An Introduction to Linear and Interface Circuits*, New York, Wiley, 1984.

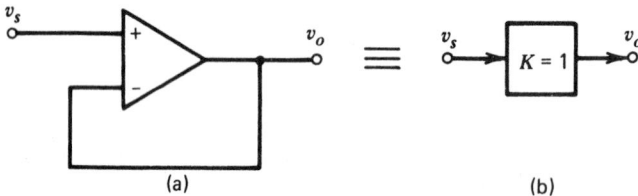

(a)   (b)

Fig. 11.32 Op-amp voltage follower: (*a*) circuit representation, (*b*) block diagram. Reproduced, with permission, from R. E. Thomas and A. J. Rosa, *Circuits and Signals: An Introduction to Linear and Interface Circuits*, New York, Wiley, 1984.

cause the voltage to drop. However, for any reasonable range of load, the op-amp will look like an ideal voltage source.

A most important interfacing tool, the voltage follower, is shown in Fig. 11.32. This is a special case of the noninverting amplifier. In this case R_1 is made equal to zero by using a short circuit in place of R_1. If R_1 is zero, the gain is one, regardless of what R_2 equals. Therefore R_2 can be removed from the circuit without effect.

The result is that

$$\frac{v_0}{v_s} = 1 \tag{11.77}$$

To illustrate an application for a voltage follower, consider the problem of measuring the voltage across a 1-MΩ resistor with a typical voltmeter of 1 MΩ input resistance. This problem is shown in Fig. 11.33. The circuit in Fig. 11.33(*a*), the 1-MΩ resistor in the voltmeter is in parallel with the 1 MΩ that is being measured. The result is a 500-kΩ resistor in series with the rest of the circuit. By voltage division the voltage measured by the voltmeter is

$$v_{\text{meter}} = \frac{500\ \text{k}\Omega}{500\ \text{k}\Omega + 1\ \text{M}\Omega} 10 = 3.3\ \text{V} \tag{11.78}$$

This 3.3 V is not the 5.0 V expected. By inserting a voltage follower between the meter and the voltage measured, the current flowing in the circuit will no longer split as i_p is zero. Hence we measure the true voltage of 5 V across the 1-MΩ resistor.

Example 11.6. Another very important application of op-amps is in interfacing digital systems with analog systems. A digital-to-analog (D/A) converter requires the following input-output relationship:

$$v_0 = 8v_1 + 4v_2 + 2v_3 + v_4 \tag{11.79}$$

A simple way to implement this function is to use a summer with properly weighted resistors. Figure

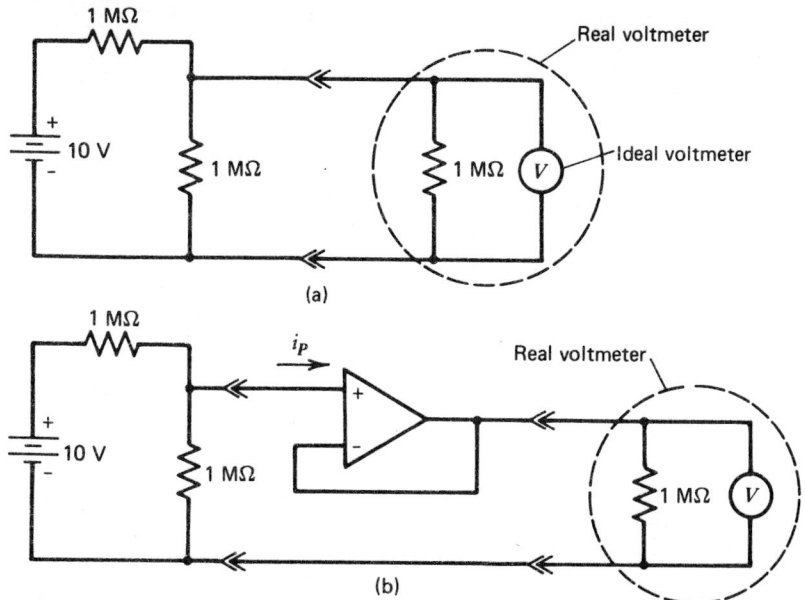

Fig. 11.33 Voltage follower application: (*a*) voltmeter connected directly, (*b*) follower added. Reproduced, with permission, from R. E. Thomas and A. J. Rosa, *Circuits and Signals: An Introduction to Linear and Interface Circuits*, New York, Wiley, 1984.

11.34(*a*) shows a binary-weighted resistance D/A converter. Figure 11.34(*b*) shows the same function in block diagram form.

The output of the op-amp will be the binary-weighted sum of the digital input, scaled or amplified by $-R_F/R$. That is,

$$v_0 = \frac{-R_F}{R}(8v_1 + 4v_2 + 2v_3 + v_4) \tag{11.80}$$

which if we pick $R_F/R = 1$ is what is desired.

The problem of interfacing linear transducers is a very common one. A generalized block diagram approach to this interfacing problem is shown in block diagram form in Fig. 11.35. The transducer converts some physical parameter to a voltage that is proportional to the physical parameter being measured. The output of the transducer is usually very weak and has to be greatly amplified. To achieve a desired output, the amplified transducer output often has to be offset, a process known as adding **bias**. Example 11.7 will demonstrate the ease of designing an interface for a linear transducer using op-amps.

Example 11.7. In a laboratory experiment it is desired to easily measure the amount of light incident on a certain photocell when the incident light is between 5 and 20 lumens (lm). The output is to be displayed on a 0- to 10-V voltmeter. Design the interface if the photocell output is as shown in the graph of Fig. 11.36(*a*). Furthermore, it is desired that when 5 lm is incident that the meter indicates 0 V and for 20 lm it indicates 10 V.

The output of the transducer will change (linearly) (0.5 − 0.2 mV = 0.3 mV) for a light intensity change between 5 and 20 lm. This 0.3-mV change must produce a (10 − 0)-V change in the meter. Thus our signal needs to be amplified by K:

$$K = \frac{\text{desired range}}{\text{available range}}$$

$$= \frac{(10 - 0)\text{ V}}{(0.5 - 0.2) \times 10^{-3}\text{ V}} = 3.3 \times 10^4 \tag{11.81}$$

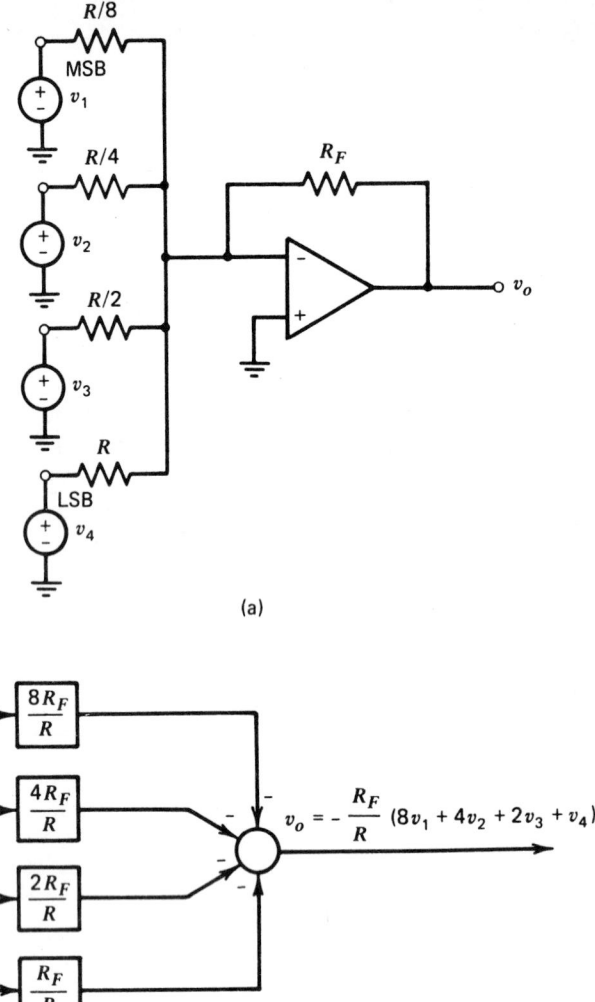

(a)

(b)

Fig. 11.34 A binary-weighted resistance D/A converter: (*a*) circuit, (*b*) block diagram. Reproduced, with permission, from R. E. Thomas and A. J. Rosa, *Circuits and Signals: An Introduction to Linear and Interface Circuits*, New York, Wiley, 1984.

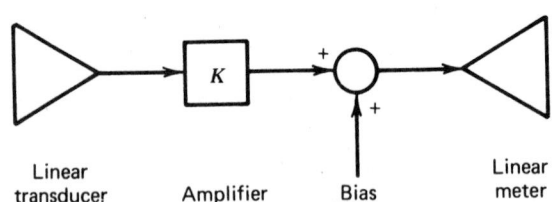

Fig. 11.35 Block diagram of a linear transducer interface. Reproduced, with permission, from R. E. Thomas and A. J. Rosa, *Circuits and Signals: An Introduction to Linear and Interface Circuits*, New York, Wiley, 1984.

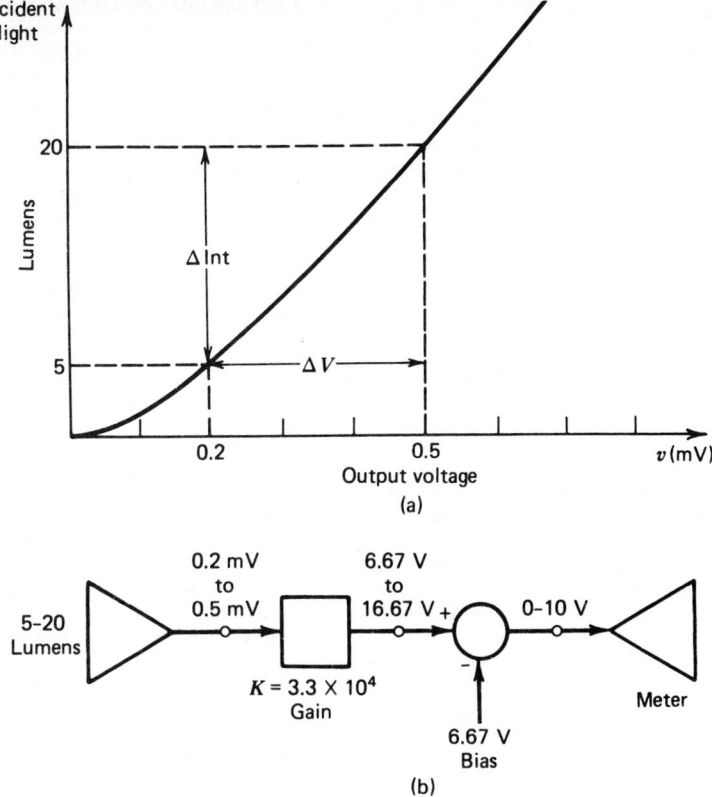

Fig. 11.36 Linear transducer op-amp interface example: (*a*) transducer characteristics, (*b*) block diagram. Reproduced, with permission, from R. E. Thomas and A. J. Rosa, *Circuits and Signals: An Introduction to Linear and Interface Circuits*, New York, Wiley, 1984.

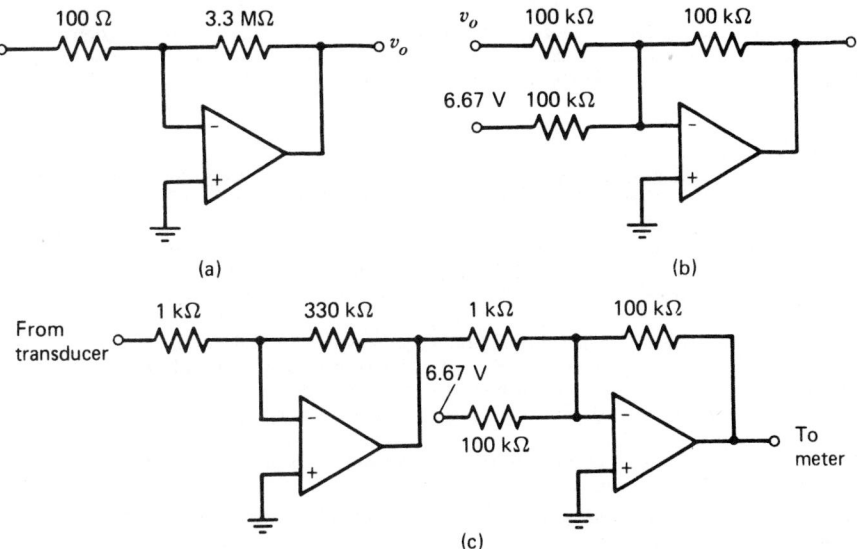

Fig. 11.37 Solution to a linear transducer design example: (*a*) scalar multiplier, (*b*) bias summer, (*c*) final design. Reproduced, with permission, from R. E. Thomas and A. J. Rosa, *Circuits and Signals: An Introduction to Linear and Interface Circuits*, New York, Wiley, 1984.

If we multiply the transducer's output voltage range by the gain, we obtain an output range of 6.67–16.67 V instead of the desired 0- to 10-V spread. However our meter is only good from 0 to 10 V. To satisfy this interface requirement we must subtract 6.67 V from the amplifier's output. The constant 6.67 V is the bias. The block diagram solution is shown in Fig. 11.36(b). To realize the solution using op-amps, we design each piece of the interface independently and then cascade them. Figure 11.37(a) shows a suitable scalar multiplier of gain 3.3×10^4. Figure 11.37(b) shows the subtractor or bias operation. Be aware that the designer did not actually use a subtractor but a summer, taking advantage of the inverting properties of the scalar multiplier and summer used. The total circuit is shown in Fig. 11.37(c).

Notice that to achieve the 33,000 gain would require a relatively small input resistor and a very large, untypical resistor in the feedback. In the final design the gain of the multiplier was reduced by 100 to permit use of better resistor values. The missing gain was then made up in the summer circuit.

11.3.4   General Circuit Analysis

Device and Connection Analysis
To develop general methods of circuit analysis we must first understand the foundations on which all methods of circuit analysis are based. The behavior of a circuit, an interconnection of devices, is fundamentally rooted in constraints of two types. First are the connection constraints embodied in Kirchhoff's laws. These constraints depend only on the manner in which the circuit is interconnected and not on the nature of the individual devices. Device individuality is captured by its *i-v* characteristics, which, in turn, do not depend on the nature of the circuit connections. In sum then, equilibrium in a circuit is the result of the balancing of two independent types of constraints: (1) connection constraints (Kirchhoff's laws) and (2) device constraints (*i-v* characteristics).

This observation provides a way to formulate a system of independent, linear equations that completely describe the circuit behavior. We first identify a current and voltage variable with every element in the circuit. We then write the connection constraints in terms of these variables using Kirchhoff's laws. The device constraints also can be expressed in terms of these variables using the element *i-v* characteristics. Collectively these steps provide a complete description of the circuit. More formally, for a circuit with N nodes and E elements we need to complete four steps:

1. Identify a current and a voltage variable with every element in the circuit.
2. Write KCL connection constraints in terms of the element currents at $N - 1$ nodes.
3. Write KVL connection constraints in terms of the element voltages around $E - N + 1$ loops.
4. Write device constraints in terms of the element currents and voltages using the element *i-v* characteristics.

These steps lead to $N - 1$ KCL connection equations, $E - N + 1$ KVL connection equations, and E element equations. Collectively this leads to a total of $(N - 1) + (E - N + 1) + E = 2E$ independent equations, which is precisely the number needed to solve for all of the variables identified in step 1.

Node Analysis
For most electronic circuits, particularly those containing active elements, the most convenient solution variables are the node voltages. To define a set of node voltages we first select a reference node. The node voltages are then defined as the voltages at the remaining $N - 1$ nodes with respect to the selected reference node. The reference node is indicated by the ground symbol, and the node voltage is identified by a voltage symbol written adjacent to the remaining nodes. This notation means the positive reference mark is located at the node in question while the negative mark is at the reference node.

To use these variables to formulate equilibrium equations for a circuit, we proceed as in device and connection analysis, except that the KVL connection equations are not explicitly written down. Instead the KVL constraints are used to express the device constraints in terms of the node voltages. More formally, to develop a set of node voltage equations we need four steps:

1. Select a reference node and identify a node voltage at each of the remaining $N - 1$ nodes. Identify a current with every element in the circuit.
2. Write KCL connection constraints in terms of the element currents at the $N - 1$ nonreference nodes.
3. Use KVL and the element *i-v* characteristics to express the element currents in terms of the node voltages.
4. Substitute the device constraints from step 3 into the KCL connection constraints from step 2 and arrange the resulting $N - 1$ equations in a standard form.

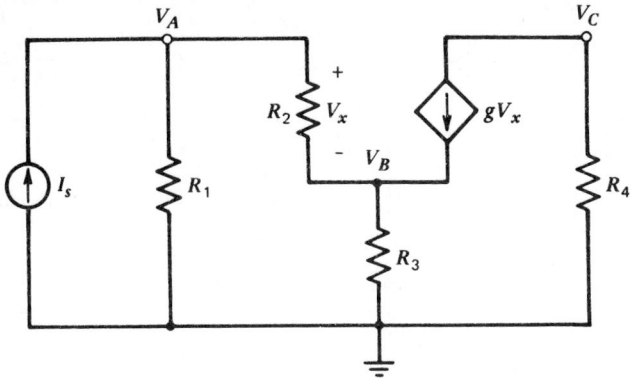

Fig. 11.38 Dependent source circuit for Example 11.8. Reproduced, with permission, from R. E. Thomas and A. J. Rosa, *Circuits and Signals: An Introduction to Linear and Interface Circuits*, New York, Wiley, 1984.

This process leads to $N-1$ linear equations in $N-1$ unknown node voltages. Thus the node voltage formulation leads to a significant reduction in the number of linear equations that must be manipulated simultaneously, particularly if the circuit contains a large number of devices connected in parallel. The following example illustrates the node voltage analysis process.

Example 11.8. The circuit in Fig. 11.38 contains an independent current source and voltage-controlled dependent current source. We shall treat the dependent source temporarily as if it were an independent source, and write node voltage equations by inspection. Starting with node (A), the sum of conductances connected is $G_1 + G_2$. The conductance between (A) and (B) is G_2, and between (A) and (C) there is none. Further, I_s is injected into node (A), and hence it would appear as a positive current on the right-hand side of the equation. Consequently the node (A) equation is

$$(G_1 + G_2)V_A - G_2V_B - 0V_C = I_s \qquad (11.82)$$

Similarly at node (B) the sum of conductances is $G_2 + G_3$. Between (B) and (A) the conductance is again G_2, while there is not conductance between nodes (B) and (C). Now treating the dependent source as if it were independent, node (B) is

$$(G_2 + G_3)V_B - G_2V_A - 0V_C = gV_x \qquad (11.83)$$

By inspection, node (C) is

$$G_4V_C - 0V_A - 0V_C = -gV_x \qquad (11.84)$$

We have derived a set of symmetrical node equations by treating the dependent source as independent. By KVL we can write

$$V_x = V_A - V_B \qquad (11.85)$$

When this result is substituted into the node equations and all unknowns moved to the left side, there results:

Node (A):
$$(G_1 + G_2)V_A - G_2V_B = I_s$$

Node (B):
$$-(G_2 + g)V_A + (G_2 + G_3 + g)V_B = 0 \qquad (11.86)$$

Node (C):
$$gV_A - gV_B + G_4V_C = 0$$

We now have three equations in three unknowns, including the effect of the dependent source.

In developing the formulation of node equations, we have thus far tacitly assumed that the circuit contains only independent or dependent *current* sources. It might appear that inclusion of *voltage*

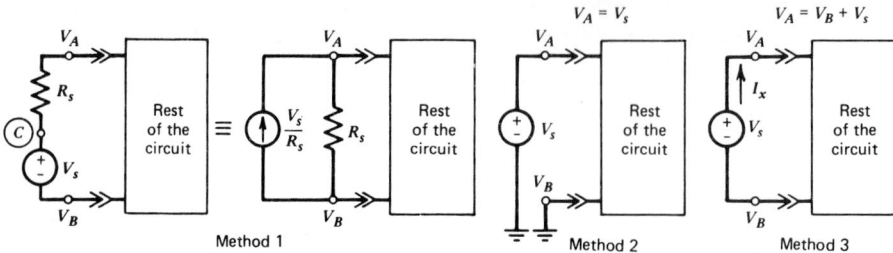

Fig. 11.39 Methods of handling voltage sources in the formulation of node equations. Reproduced, with permission, from R. E. Thomas and A. J. Rosa, *Circuits and Signals: An Introduction to Linear and Interface Circuits*, New York, Wiley, 1984.

sources (dependent or independent) would complicate node analysis since the current through a voltage source is a dependent variable. In fact, the opposite is often the case. The presence of voltage sources often simplifies node analysis.

There are basically three ways to deal with voltage sources, as shown in Fig. 11.39. If there is a resistor in series with the voltage source, then method 1 applies. The series combination of voltage source and resistor can be replaced at terminals Ⓐ and Ⓑ by an equivalent current source using source conversion. We can then develop node equations in the usual way. It is evident that this method eliminates node Ⓒ from the circuit, which simplifies the analysis.

If there is no resistor in series with the source, then we can use method 2 in Fig. 11.39. Here we have selected node Ⓑ as the reference node. As a result the node voltage V_A is no longer an unknown since it must be equal to the input V_s. We now do not need to develop a node voltage equation at node Ⓐ since its voltage is known. We then write node equations at the remaining nodes in the usual way, but as a final step we move all terms involving V_A to the right side since it is a known input and not an unknown response. This reduces the number of node equations as one is not needed at node Ⓐ.

If the circuit contains two or more voltage sources that do not share a common node, and if none are connected in series with a resistor, then we must use method 3 shown in Fig. 11.39. In this case we identify an unknown current I_x with the voltage source. This adds one additional unknown to our problem. However, we can also write a KVL constraint

$$V_A - V_B = V_s \qquad (11.87)$$

which adds an additional equation.

In developing the resulting equations we either modify the circuit (method 1), do not write node equations at all nodes (method 2), or introduce an unknown current to the list of variables (method 3). It should be noted that the three methods are not mutually exclusive. In a complicated circuit we might appeal to all three methods.

We have seen in node analysis that the node voltages are a very useful set of analysis variables. Many large computer circuit analysis programs such as ECAP and SPICE are based on node equations since the formulation process is very algorithmic. But node voltages are also very useful in the laboratory. To evaluate a circuit, we connect one terminal of a voltmeter to ground (the reference node) and probe the remaining nodes with the other voltmeter terminal, thereby measuring V_A, V_B, V_C, and so on. Thus node voltages are useful for both analysis and experimentation.

Mesh Analysis

Mesh currents are an alternative set of solution variables that are useful for the analysis of circuits containing many elements connected in series, and hence there are many nodes. We restrict our development of mesh analysis to planar circuits, that is, circuits that can be drawn on a flat surface without crossovers. Such circuits can always be drawn in the "window-pane" form illustrated in Fig. 11.40.

Mesh currents are depicted as flowing through the elements around the perimeter of the mesh. A mesh current is then defined by labeling the meshes (I_1, I_2, I_3, etc.) and assigning a reference direction to each current. Mesh currents are all assumed to flow in a clockwise sense, but there is no fundamental reason for this, except perhaps tradition. We then think of these mesh currents as flowing around their respective meshes, as indicated in Fig. 11.40.

We should emphasize that we are not describing the physical process that takes place in a circuit. The electrons do not somehow get assigned to I_1 or I_2, and so on. We are here defining a set of

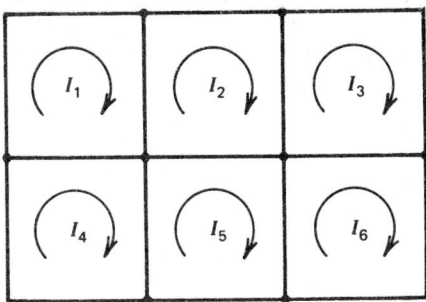

Fig. 11.40 Meshes in planar circuits. Reproduced, with permission, from R. E. Thomas and A. J. Rosa, *Circuits and Signals: An Introduction to Linear and Interface Circuits*, New York, Wiley, 1984.

variables that can be used to determine the voltage across and current through the circuit elements. In other words, mesh currents are variables of convenience that are used in analysis, but are only somewhat abstractly related to the physical process.

To use mesh currents to formulate circuit equations, we proceed as in device and connection analysis, except that the KCL connection constraints are not explicitly written down. Instead the KCL constraints are used to express the device constraints in terms of the mesh currents. Stated formally, there are four steps in mesh analysis:

1. Identify a mesh current with every mesh and identify a voltage with every circuit element.
2. Write KVL connection constraints in terms of the element voltages around every mesh.
3. Use KCL and the element *i-v* characteristics to express the element voltages in terms of the mesh currents.
4. Substitute the device constraints from step 3 into the connection constraints from step 2 and arrange the resulting equations in a standard form.

If there are current sources (independent or dependent), then in a dual fashion to what we did in node analysis, we can identify three possibilities.

1. If the current source is connected in parallel with a resistor, then it can be converted to an equivalent voltage source by source conversion.
2. If only one mesh current flows through a current source, then that mesh current is no longer an unknown but is determined by the source current. We can write mesh equations around the remaining mesh in the usual way and move the known mesh current to the source side of the equations in the final step.
3. If two mesh currents flow through a current source, then their difference is fixed, and we must then introduce an unknown voltage across the current source as one of our solution variables.

These approaches are not mutually exclusive (we might apply all three in a very complicated circuit) and are the duals of the three methods used in developing modified node equations.

Example 11.9. The circuit in Fig. 11.41 is a small-signal model of a transistor circuit called an **emitter follower**. The given circuit contains a voltage-controlled current source that can be converted to a voltage-controlled voltage source using source conversion, as shown in the modified circuit. We now temporarily treat the dependent voltage source as an independent source and write two mesh equations by inspection:

Mesh ① : $$(R_1 + R_2 + R_3)I_1 - R_3 I_2 = V_s - gR_3 V_x \qquad (11.88)$$

Mesh ② : $$(R_3 + R_4)I_2 - R_3 I_1 = gR_3 V_x \qquad (11.89)$$

We observe that the control voltage V_x can be written in terms of mesh currents as

$$v_x = R_2 I_1 \qquad (11.90)$$

When this result is substituted into mesh equations and all unknowns shifted to the left side of the

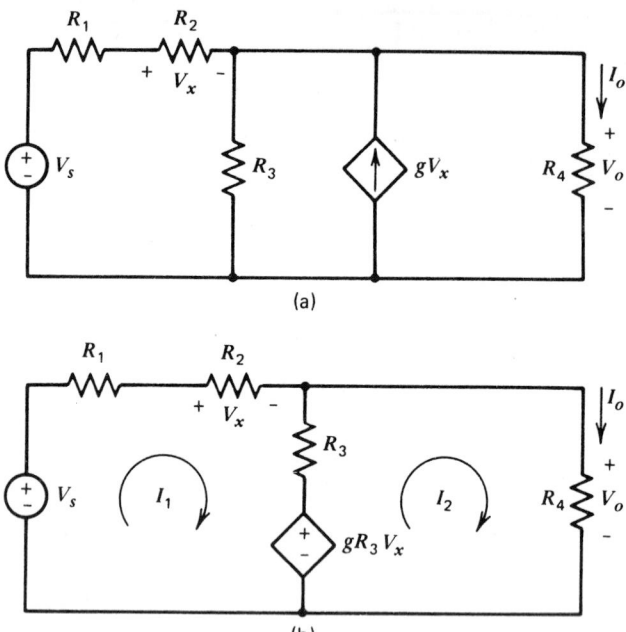

(a)

(b)

Fig. 11.41 Dependent source circuit for applying mesh analysis: (a) given circuit, (b) modified circuit. Reproduced, with permission, from R. E. Thomas and A. J. Rosa, *Circuits and Signals: An Introduction to Linear and Interface Circuits*, New York, Wiley, 1984.

resulting equations, we get

$$(R_1 + R_2 + R_3 + gR_2R_3)I_1 - R_3I_2 = V_s$$
$$-(R_3 + gR_2R_3)I_1 + (R_3 + R_4)I_2 = 0 \qquad (11.91)$$

The controlled source is a unilateral element and therefore the resulting mesh equations are not symmetrical.

Turning to the solution phase we solve first for the indicated output current:

$$I_0 = I_2 = \frac{(R_3 + gR_2R_3)V_s}{R_1R_3 + R_1R_4 + R_2R_3 + R_2R_4 + R_3R_4 + gR_2R_3R_4} \qquad (11.92)$$

And finally,

$$V_0 = I_0R_4$$
$$= \frac{(R_3R_4 + gR_2R_3R_4)V_s}{R_1R_3 + R_1R_4 + R_2R_3 + R_2R_4 + R_3R_4 + gR_2R_3R_4} \qquad (11.93)$$

If g is a very large number, then the last terms in the numerator and denominator dominate and

$$V_0 \approx \frac{gR_2R_3R_4}{gR_2R_3R_4}V_s = V_s \qquad (11.94)$$

Hence the name voltage follower is often used for the emitter follower circuit since the output is the same as (follows) the input.

11.4   ELECTROSTATICS AND ELECTROMAGNETICS

11.4.1   Electrostatics, Potentials, and Currents

Coulomb's Law

Coulomb observed that an electric charge in the vicinity of another electric charge had a force exerted on it. If the charges were like charges, the force on each charge was such as to move the charges apart; for unlike charges the forces were such as to move the charges together. Coulomb's formulation of the results of his experiments produced the following basic relation known as Coulomb's law:

$$F = \frac{q_1 q_2}{4\pi\varepsilon r^2} \ \text{N} \tag{11.95}$$

In this law F is the magnitude of the force in newtons; q_1 and q_2 are charges in coulombs separated a distance r meters; ε is the absolute dielectric constant of the medium, also called its permittivity. The forces on the two charges have the same magnitude but opposite directions. Either charge may be considered as in the electric field of the other.

Electric Fields of Force

In any portion of a substance in which the electricity is acted upon by a force tending to move it, there is said to be an **electric field**. An electric field is also said to exist in any region of free space where a charge, if placed there, would have a force exerted upon it tending to move it.

Intensity of an Electric Field. The *intensity* of an electric field **E** at any point is defined as the force exerted on a unit positive charge at this point by the charges producing the field. The direction of the field intensity is defined as the direction of the force acting on a positive charge at this point. A positive charge then moves or tends to move in the direction of the field, and a negative charge moves or tends to move in the opposite direction.

The intensity of electric field can be obtained from Coulomb's law by setting one charge, say q_2 equal to 1 C (unit charge). Then

$$\mathbf{E} = \frac{q}{4\pi\varepsilon r^2} \quad \text{(N/C or V/M)} \tag{11.96}$$

The unit of electric field intensity is expressed as so many volts per meter. Alternatively, it may be expressed as a force per unit charge, such as so many newtons per coulomb.

Lines of Electric Force and Lines of Electric Intensity. A line drawn in an electric field in such a manner that its direction at each point coincides with the direction of the field at that point is called a **line of electric field**. Any number of such lines may be drawn in an electric field, but no two of these lines can intersect. The density of these lines, that is, the number per unit area perpendicular to their direction, may be chosen arbitrarily to represent the value of the field intensity at this area, and when so drawn they are called **lines of electric intensity**, as distinguished from flux. The term *line of electric force* is used to designate the direction of the field of any point; in any statement involving the density of these lines the proper terms will be employed.

Electric Potential Difference

The electric potential difference between two points 2 and 1 (see Fig. 11.42) is the work per unit positive charge in moving the unit positive charge from point 1 to point 2 against the electric field

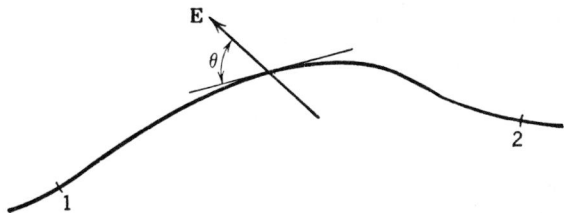

Fig. 11.42   Calculation of potential difference between two points in the presence of an electric field.

intensity, that is,

$$V_2 - V_1 = \int_1^2 - (E\cos\theta)\,dl \qquad \text{(N-m/C or J/C or V)} \qquad (11.97)$$

where dl represents an elementary length of the path, and $-E\cos\theta$ the component of force to overcome the field intensity E along dl.

If $V_2 - V_1$ is positive, then point 2 is said to be at a higher electric potential than point 1. If $V_2 - V_1$ is negative, point 2 is at a lower potential than point 1.

A point may be designated as zero electric potential for reference purposes; this designation may be entirely arbitrary, although in some physical cases it is often both logical and convenient to consider a special point as being at zero potential. The electric potential at a point is the difference of potential between the point and a reference point that has been designated as being at zero electric potential.

Voltage or Potential Gradient
Equation (11.97) can be rewritten as

$$V_2 - V_1 = \int_1^2 dV = \int_1^2 - E\cos\theta\,dl = \int_1^2 - \mathbf{E}\,dn \qquad (11.98)$$

where dn represents the differential element taken perpendicular to electric equipotential surfaces. Differentiation of Eq. (11.98) with respect to dn results in

$$\mathbf{E} = \frac{-dV}{dn} \qquad (11.99)$$

The boldface type indicates that vector quantities are involved. Now dV/dn is called the potential or voltage gradient; it is the maximum value of the directional derivative of voltage with respect to distance. Symbolically ∇V represents gradient of V, thus

$$\mathbf{E} = -\operatorname{grad} V = -\nabla V \qquad (11.100)$$

Flow of Electricity
Whenever an electric field is set up in a substance by any means whatever, a displacement of the electricity in that substance always takes place, the nature of the displacement depending on the nature of the substance. The positive electricity within the substance is displaced or orientated in the direction of the field intensity and the negative electricity in the opposite direction, until an opposing force is set up that just balances the forces due to the impressed field. In metallic conduction, the flow of electrons in a direction opposite to the field constitutes the electric current. In electrolytes there is a migration of positive ions in the direction of the field, and of electrons opposite to the direction of the field. In semiconductors, free electrons move in direction opposite the field, while free holes move in direction of the field.

The displacement of charge carriers within a substance can be measured directly only in semiconductors using the Hall effect. Two effects that always result when electricity is displaced are (1) a magnetic field is established around the path along which the displacement takes place and (2) heat is developed in the path of the displacement. The magnetic field produced by a displacement or flow of electricity is usually taken as the measure of the rate of flow, that is, of the quantity of electricity displaced per unit time through a surface perpendicular to the direction of the displacement. This rate of flow is called the electric current.

Conduction Current and Displacement Current
Experience shows that, when an electric field is established in any substance, the total equivalent electric current set up depends on (1) the value of the field intensity, (2) the rate of change of the field intensity, and (3) the nature of the substance in which the field is established. The current density $\mathbf{J}$ at any point at any instant may in general be expressed by the relation

$$\mathbf{J} = \sigma\mathbf{E} + \frac{d}{dt}\varepsilon\mathbf{E} \qquad (11.101)$$

where $\mathbf{E}$ is the field intensity, σ and ε are coefficients depending on the chemical nature and physical condition of the substances at the point in question. σ is the conductivity of a substance in mho-meters. The total current density may then be considered as the sum of the two components having, respectively, the densities

$$J_c = \sigma E \quad \text{and} \quad J_d = \frac{d}{dt}(\varepsilon E) \tag{11.102}$$

The first of these components, J_c, is called the conduction current density; the second, J_d is commonly called the displacement current density. The conduction current density is the only appreciable component in substances usually classed as conductors although it must often be accounted for in semiconductors, and the displacement current density is appreciable only in substances ordinarily classed as semiconductors and dielectrics. The conduction current density in a dielectric is usually small, though measurable; it is frequently called the **leakage current density**. When the electric field in a dielectric is rapidly varying, the displacement current density may be many times greater than the conduction or leakage current density through the dielectric. The ability to control the displacement current in semiconductors is responsible for much of their usefulness.

Conductivity and Resistivity

The quotient of the density, J_c of the conduction current by the field intensity E, that is, the coefficient σ in the expression $J_c = E$ is called the conductivity or specific conductance of the substance at the point in question. Since in an ordinary conductor the displacement current density is insignificant, the conductivity of an ordinary conductor is also equal to the density of the total current divided by the field intensity; that is, for a conductor

$$J = \sigma E \, \text{A/m}^2 \tag{11.103}$$

where J represents the density of the total current. Experience shows that for a given conductor at constant temperature (and also at constant pressure, in a gas) this coefficient (σ) is a constant irrespective of the strength, distribution, or time variation of the current. The value of σ for a dielectric, however, is not in general a constant but depends on the time variation of the field intensity. The conductivity of a semiconductor is readily varied over many orders of magnitude by a process called **doping**. It is this ability to control a semiconductor's electrical properties that permits devices such as transistors to be fabricated.

This relation between J and E may also be written

$$E = \rho J \tag{11.104}$$

where ρ is the reciprocal of the conductivity σ. The constant ρ is called the resistivity or specific resistance of the substance. Equation (11.104) is referred to as the generalized form of Ohm's law.

11.4.2 Dielectric Flux

Dielectric Flux and Dielectric Flux Density

The displacement current density through a dielectric at any point depends on the rate of change of the electric field intensity and on the nature of the dielectric. The density of this displacement current at any point may be expressed by the relation

$$J_d = \frac{d}{dt}(\varepsilon E) \tag{11.105}$$

and

$$\varepsilon = \varepsilon_0 \varepsilon_r$$

where E is the field intensity, ε_0 is the dielectric constant or permittivity of free space, and ε_r is the relative dielectric constant of the material.

The quantity εE whose rate of change is equal to the density of the displacement current, is called the **dielectric flux density** and may be represented by the symbol D. Then

$$D = \varepsilon E \tag{11.106}$$

The direction of the dielectric flux density is arbitrarily chosen to be the same as that of the electric field intensity E. Through any surface of area A at each point of which the dielectric flux density has a constant value D and is perpendicular to that surface, there is said to exist a dielectric flux equal to D/A. The total dielectric flux through a surface may be represented by the symbol ψ.

In general, the total dielectric flux through any surface is

$$\psi = \int D \cos \alpha \, ds \tag{11.107}$$

where ds represents any elementary area of this surface, $D \cos \alpha$ the component of flux density normal to the surface at ds, and ψ the sum of all the products $D \cos \alpha \, ds$ for that surface. The total displacement current through this surface is then

$$i_d = \frac{d\psi}{dt} \tag{11.108}$$

Lines of Dielectric Flux

The electric flux through any surface may be represented by lines drawn in the same direction as the lines of electric intensity, but of such a density that their number per unit area perpendicular to their direction at any point is equal to the dielectric flux density at this point. The number of these lines cutting any surface is then equal to the total dielectric flux through this surface. The ratio of the number of flux lines through any surface to the number of lines of electric intensity through that surface is equal to the dielectric coefficient of the substance in which the field exists.

Electric Charge and Dielectric Flux

Within any substance of uniform structure throughout, the dielectric flux lines are continuous lines; that is, the number of these lines coming up to one side of a surface within such a substance is equal to the number of these lines leaving the other side of that surface. Experience shows that it is impossible to produce an appreciable dielectric flux in those substances ordinarily classed as conductors; hence dielectric flux lines cannot pass through a good conductor, but terminate at its surface. Since every dielectric is a conductor to at least a slight extent, not all the dielectric flux lines coming up through one dielectric to the surface of contact between this dielectric and another pass through the second dielectric, but some of them terminate at this surface.

Experience shows that, to establish an electric field in the dielectric around a conductor, electricity must be conducted through the conductor to the surface of contact between the conductor and the dielectric. For example, consider a good conductor in contact with a perfect dielectric (Fig. 11.43); a momentary conduction current must flow through the conductor along the stream lines of the conduction current, represented by the dotted lines. While the field is being established (and therefore varying), a displacement current is set up in the dielectric requiring an equal conduction current in the conductor, and consequently the rate of change of the dielectric flux (ψ) established in the dielectric must be equal to the conduction current (i) flowing up to this surface through the conductor; that is,

$$\frac{d\psi}{dt} = i$$

or   (11.109)

$$\psi = \int i \, dt \; C$$

This relation is a general one, namely, the total dielectric flux from any area A in the surface of a conductor is equal to the total charge on this area. Hence every flux line originates at a positively charged conducting surface and terminates at a negatively charged conducting surface, a line connecting each unit positive to each unit negative charge.

The quantity of electricity conducted through a conductor when a momentary current is established through it can be measured readily by means of a ballistic galvanometer, and consequently the dielectric flux may readily be determined.

Dielectric flux may be expressed in the same units as electric charge, namely, coulombs, statcoulombs, or abcoulombs.

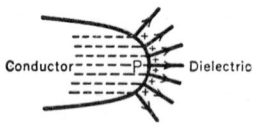

Fig. 11.43   Conductor–dielectric interface.

Surface Density of Charge

When there is no current in a conductor, there can be no electric field [see Eq. (11.103)]; and therefore the surface of a conductor in which no current is flowing is always an equipotential surface. Hence the lines of electrostatic intensity, in the surrounding dielectric, and therefore the dielectric flux lines also, must leave or enter this surface in a direction perpendicular to it. The dielectric flux density in the dielectric just outside a conducting surface in which there is no electric current is perpendicular to this surface and has the magnitude

$$D = \sigma_s \tag{11.110}$$

where σ_s is the charge per unit area of the surface at this point, or the **surface density** of the charge.

11.4.3   Magnetizing Flux

Magnetic Field of Force

Any region in which a magnetic substance (e.g., a piece of soft iron), when placed therein, becomes magnetized is said to be a **magnetic field**. A magnetic field exists in and around every magnetized substance and around every electric current. The direction of the magnetic field at any point P is arbitrarily chosen as the direction in which a small magnetic needle point would point when placed at P without disturbing appreciably the existing conditions.

Magnetic Flux

Consider a small closed turn of wire (Fig. 11.44) placed in a magnetic field with its plane perpendicular to the direction of the field. Experience shows that when such a turn of wire is removed from the field in any manner whatever (the coil remaining short-circuited on itself or forming part of a closed circuit), or when the magnetic field is caused to disappear in any manner whatever, a momentary electromotive force (emf) or voltage is set up or *induced* in this coil, which in turn causes a momentary electric current to flow through the coil. This emf exists only while the coil is moving across the field or while the field through the coil is varying.

The time integral of the induced emf when the coil is removed entirely from the magnetic field is taken as the measure of the magnetic flux existing through the coil when in its original position. That is, calling v the emf induced in the coil at any instant by its motion through the field, and t the time during which the emf exists in the coil, then the magnetic flux through the coil when in its original position is

$$\phi = \int_0^t v \, dt \tag{11.111}$$

When v is in volts and t is in seconds, ϕ is in webers. This quantity ϕ is readily measured by means of a ballistic galvanometer.

Magnetic Flux Density

Experience shows that the magnetic flux through any closed loop, such as the turn of wire described above, depends on the area inclosed by this loop. The magnetic flux per unit area through any surface perpendicular to the direction of the field is defined as the *magnitude* of the magnetic flux density at this surface and is represented by the symbol B. By the magnitude of the flux density at any point is meant the magnitude of the flux density at any infinitely small surface drawn perpendicular to the field at this point. The *direction* of the magnetic flux density at any point is the same as that in which a magnetic needle would point if placed at this point; that is, the direction of the flux density and the direction of the magnetic field are the same. The vector having this defined magnitude and direction is called the flux density and is usually designated by the symbol **B**. When the flux density has the same

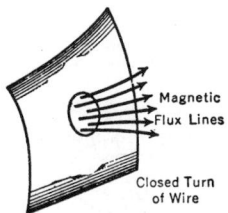

Fig. 11.44   Induction of an emf in a closed coil by an applied magnetic field.

value **B** at every point of a surface of area A and is perpendicular to this surface, then the total flux through this surface is

$$\phi = BA \text{ Wb} \tag{11.112}$$

The total magnetic flux across any surface S may in general be expressed mathematically by the surface integral

$$\phi = \oint (B \cos \alpha)\, ds \text{ Wb} \tag{11.113}$$

where ds represents any elementary area of this surface and $(B \cos \alpha)$ the component of the flux density perpendicular to ds. Magnetic flux density in the MKS system is expressed in webers per square meter.

Magnetic Flux Lines

Magnetic flux can be represented by lines so drawn in the field that their direction coincides at each point with the direction of the field at that point, and of such a number that their density at each point (number per unit area perpendicular to their direction) is equal to the magnetic flux density at that point. Such lines are called **magnetic flux lines**. Experience shows that lines thus drawn in a magnetic field always form closed loops; that is, a magnetic flux line has no ends. As a consequence the total magnetic flux coming up to any surface in a magnetic field is always equal to the total flux leaving that surface.

11.4.4   Electromagnetism

Magnetic Fields Due to Electric Currents

Experience shows that every filament of electric current is always accompanied by a magnetic field the flux lines of which link the filament of current. That is, the flux lines thread the loops formed by the current filament (see Fig. 11.45).

Right-handed Screw Law

The direction of the current flowing around any electric circuit and the direction in which the flux lines due to that current thread this circuit are related to each other in the same manner as the direction of motion of a point on the edge of the head of a right-handed screw placed at the center of the circuit and the direction of advance of the screw. Or, if one faces the electric circuit looking in the direction of the flux lines threading it, the current producing these lines is in the clockwise direction around the circuit. The relative direction of the current and its magnetic flux may be briefly described by saying that the current is in the right-handed screw direction with respect to the flux that it produces.

Induced emf or Voltage

The measure of magnetic flux is based on the experimental fact that, whenever the magnetic field threading an electric circuit changes, an emf is induced in that circuit. When the circuit is formed by a single turn of wire, this induced emf is, from our definition, equal to the rate of change of this flux with respect to time; that is, $v = d\phi/dt$. When the circuit is in the form of a coil each turn of which links the flux, the emf induced in each turn is equal to $d\phi/dt$, where ϕ is the flux that links that particular turn. When each turn links the same number of flux lines, then the total induced emf in a coil of N turns is

$$v = N\frac{d\phi}{dt} \text{ V} \tag{11.114}$$

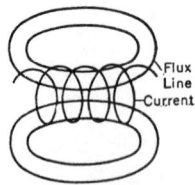

Fig. 11.45   Magnetic flux lines caused by a current flowing in a coil of wire.

When the change in flux is due to a motion of a circuit or part of a circuit through a magnetic field, the induced emf in any conductor is also equal to the negative of the number of flux lines that cut across this conductor per unit time.

Magnetic Linkages

The condition that each turn of a coil be linked by the same flux ϕ seldom exists; some of the flux lines usually link only part of the turns. In general, the total voltage is

$$v = -\frac{d}{dt}\left(n_1\phi_1 + n_2\phi_2 + \cdots + n_n\phi_n\right) \qquad (11.115)$$

where $\phi_1, \phi_2, \ldots$, represent the fluxes linking the various numbers of turns, $n_1, n_2, \ldots$, respectively. The sum $(\lambda_1 + \lambda_2 + \cdots + \lambda_n)$ may be called the total number of **magnetic linkages**, or **flux linkages**, and may be conveniently represented by the symbol λ, namely,

$$\lambda = \lambda_1 + \lambda_2 + \cdots + \lambda_n \text{ weber-turns} \qquad (11.116)$$

and the total induced emf may then be written

$$v = \frac{d\lambda}{dt} \text{ V} \qquad (11.117)$$

When all the N turns link the same flux, ϕ, then $\lambda = N\phi$.

Direction of Induced emf or Voltage

The direction of the induced emf or voltage around a circuit is found to be in the left-handed screw direction with respect to the increase of flux; namely, if one faces the circuit looking in the direction of the increase of flux, the induced emf is in the counterclockwise direction. The current that would be set up by this emf, however, would produce a flux linking the circuit in the right-handed screw direction. Hence a change in the magnetic flux through an electric current always sets up a voltage that tends to produce a current around this circuit in such a direction as to set up an opposing flux. This fact may be expressed mathematically by writing a minus sign before $d\phi/dt$ as in Eq. (11.114), that is, by putting

$$v = -N\frac{d\phi}{dt} \qquad (11.118)$$

The value of $(-N\,d\phi/dt)$ is then the voltage induced in the circuit in the right-handed screw direction with respect to the increase of flux. Or stated in other words $(-N\,d\phi/dt)$ represents the rise of electric potential and $N(d\phi/dt)$ represents the drop of potential around the circuit in the right-handed screw direction with respect to the increase of flux.

11.4.5   Magnetizing Force and the Magnetic Circuit

Simple Magnetic Circuit

In certain simple magnetic circuits the treatment is mathematically analogous to that of simple electric circuits. An expression similar to Ohm's law is employed. Consider a closed magnetic circuit such as a uniformly wound torus (Fig. 11.46).

For this circuit we may write

$$\mathcal{F} = \mathcal{R}\phi \qquad (11.119)$$

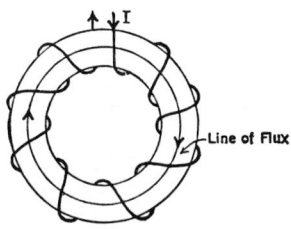

Fig. 11.46   A uniformly wound focus electromagnet.

where $\mathscr{F}$ is called the **magnetomotive force** (mmf) and is analogous to electromotive force; $\mathscr{R}$ is a factor of proportionality called the **reluctance** and is analogous to resistance; and ϕ, the flux, is analogous to electric current. It must be remembered in considering this analogy that an unvarying flux is thought of as a static condition, whereas electric current is defined as the motion of electric charge. Therefore, there is no true *physical* analogy. Equation (11.119) is sometimes called Ohm's law for magnetic circuits.

Magnetomotive Force

The mmf $\mathscr{F}$, of Eq. (11.119), may also be expressed

$$\mathscr{F} = NI \tag{11.120}$$

where $\mathscr{F}$ is in amperes, N is the total number of turns of the coil, and I is the current in amperes flowing through the coil. Obviously $\mathscr{F}$ is in direct proportion to the ampere-turns and dimensionally is the same as the ampere. In general, mmf may be expressed

$$\mathscr{F} = \sum NI \tag{11.121}$$

where $\sum NI$ represents the algebraic summation of all current-turns linking the magnetic circuit. The mmf is taken as positive when the current links the flux lines in the right-handed screw direction, and negative when the current links the flux lines in the left-handed screw direction.

Work Done by a Varying Magnetic Flux

Consider a coil A (Fig. 11.47) of N turns of wire, and let each of these N turns be linked by a flux ϕ due to some external agent, for example, another coil B in which an electric current is flowing. Let the flux ϕ through A due to B be increasing at any instant at the rate $d\phi/dt$ in the left-handed screw direction with respect to the current I in A at this instant. Then there is induced in A at this instant a voltage in the direction of I equal to $v = N(d\phi/dt)$, and therefore the electric power developed in A at this instant is p. This power is transmitted to the coil A as a result of the varying flux through it; hence the power

$$p = NI\frac{d\phi}{dt} \tag{11.122}$$

may be looked upon as the magnetic power input, this power being converted within the coil into electric power.

Magnetizing Force or Magnetic Field Intensity

Experience shows that the magnetic flux density produced at any point by a given mmf depends (a) on the position of the point with respect to the source of the mmf and (b) on the nature of the substances through which this mmf produces the magnetic flux. These facts lead to the conception of the flux density at any point in a magnetic field as being due to a "magnetizing force" H at that point, this magnetizing force H depending solely on the mmf producing the field and the distribution of the flux lines, as distinguished from the flux density B, which depends not only on these two items but also upon the nature of the medium at the point in question.

The magnetizing force (also called the magnetic field intensity) at successive points along any closed path in a magnetic field may be defined by the relation that its line integral around such a path is equal to the total mmf acting around this path, namely,

$$\sum NI = \oint (H\cos\theta)\, dl \tag{11.123}$$

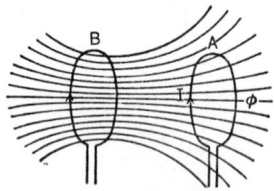

Fig. 11.47   Inductive coupling of power between two coils.

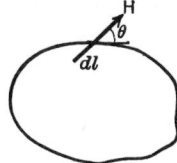

Fig. 11.48   Calculation of H around a closed path.

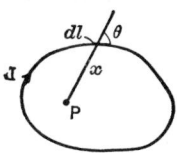

Fig. 11.49   Magnetizing force due to an element of current filament.

where dl represents any elementary length of this path (see Fig. 11.48), $H\cos\theta$ the value of the component of the magnetizing force at dl in the direction of dl, and $\sum NI$ the total number of current turns linked by the path.

When the path coincides in direction with the magnetizing force at each point, the total mmf acting around this path may be written

$$\sum N = \oint H\,dl \qquad\qquad (11.124)$$

In general **B** and **H** are in the same direction.

Lines of Magnetizing Force

The magnetizing force at any point in a magnetic field may be represented by lines drawn in the field in a direction such as to coincide with the direction of the magnetizing force at each point, and of such a number per unit area perpendicular to their direction that their density at each point gives the value of the magnetizing force at that point. Such lines are called **lines of magnetizing force**, or **lines of magnetic field intensity**. In general the lines of magnetizing force and the magnetic flux lines coincide in direction, but their densities are different. In nonmagnetic substances the flux lines and lines of magnetizing force measured in cgs units coincide in both number and direction.

Magnetizing force is of the nature of mmf per unit length. The MKS unit of magnetizing force is called an ampere per meter.

When the medium surrounding a filament of electric current is of a uniform magnetic nature throughout, the magnetizing force at any point may be calculated from the shape and distribution of the stream lines of the current, irrespective of whether the medium is nonmagnetic or highly magnetic.

Magnetizing Force at Any Point Due to an Element of a Current-Filament (Fig. 11.49). Consider any closed filament of electric current and let the surrounding medium be uniform in its magnetic properties throughout the region in which the magnetic field produced by this current exists. It can be shown that each elementary length dl of this filament may be considered as contributing to the magnitude of the magnetizing force **H** at any point P in this region an amount

$$dH = \frac{(I\sin\theta)\,dl}{4\pi x^2} \qquad (A/m) \qquad\qquad (11.125)$$

where I is the current flowing along this filament, x the distance from P to dl, and θ the angle between x and dl. The direction of dH is perpendicular to the plane determined by x and dl. The total magnetizing force at P is then the vector sum or vector integral of dH for all the elementary lengths into which the filament is divided.

Magnetizing Force Due to a Straight Wire (Fig. 11.50). Applying Eq. (11.125) to the case of a straight wire of circular cross section carrying a current I, the magnitude of the magnetizing force at

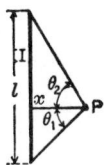

Fig. 11.50   Magnetizing force due to a current carrying wire.

any point P due to a length l of this wire is

$$H = \frac{I}{4\pi x}(\sin\theta_1 + \sin\theta_2) \quad (\text{A/m}) \tag{11.126}$$

where x is the perpendicular distance from P to the wire, and θ_1 and θ_2 the angles designated in Fig. 11.50.

If the wire is very long compared with x, this becomes

$$H = \frac{I}{2\pi x} \quad (\text{A/m}) \tag{11.127}$$

This formula also holds approximately for any point outside a wire of any shaped cross section, provided x is large compared with the maximum diameter of this section. For a point inside a long wire of circular cross section of radius a, the magnetizing force is also given by Eq. (11.127) when I is taken to represent that part of the current inside the circle through P concentric with the axis of the wire. When the current density is uniform over the cross section, as is usual, the magnetizing force inside the wire is

$$H = \frac{xI}{2\pi a^2} \quad (\text{A/m}) \tag{11.128}$$

Magnetizing Force on the Axis of a Circular Coil of N Turns. Let I be the current, r the mean radius of the coil, and x the distance of the point from the center of the circle; then

$$H = \frac{NIr^2}{2(r^2 + x^2)^{3/2}} \quad (\text{A/m}) \tag{11.129}$$

Magnetizing Force Due to a Solenoid. A solenoid is a helical coil of wire, each turn having the same radius. Let N = total number of turns, I = current in amperes, r = mean radius of the helix in meters, and l = length of helix in meters. Then at any point on the axis of the helix (inside or outside), at a distance of x meters from its center, the magnetizing force is

$$H = \frac{NI}{2l}\left[\frac{0.5l + x}{r^2 + (0.5l + x)^2} + \frac{0.5l - x}{r^2 + (0.5l - x)^2}\right] \quad (\text{A/m}) \tag{11.130}$$

This formula holds only when the thickness of the winding is small compared with the mean radius r. When l is large compared with r, this reduces to

$$H = \frac{NI}{l} \quad (\text{A/m}) \tag{11.131}$$

For all points inside the solenoid (whether on the axis or not) at a distance from the ends (large compared with r), that is, inside the central portion of a long solenoid, the field is uniform over the cross section of the solenoid and its value is given by Eq. (11.131).

Magnetizing Force Inside a Torus (Fig. 11.51). When a torus is uniformly wound with an insulated wire so that the turns of the wire are close together and cover the entire surface of the torus, the magnetic field is confined entirely within the space inclosed by these turns, and therefore, when the core on which the wire is wound is of uniform magnetic material throughout, both the lines of magnetizing force and the flux lines must be concentric circles, as shown in the figure. The

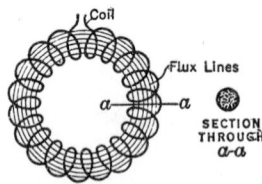

Fig. 11.51 Magnetizing force inside a torus.

magnetizing force will have the same value at every point on the circumference of any one of these circles, and, therefore, from Eq. (11.124) the value of H at any point P within the core is

$$H = \frac{NI}{l} \quad (A/m) \tag{11.132}$$

where N is the total number of turns on the core, I the current in each turn, and l the length of the circumference through P. Unless the ring has a large radius compared with the radius of the cross section of the core, H will not be uniform over this section, since l for the various points in the cross section will differ considerably.

It should be noted that the value of H is independent of the material of the core provided only that the core is of uniform material throughout. That is, Eq. (11.132) applies to an iron core as well as to an air or wood core, provided the iron is uniform throughout and there is no air gap across the path of the flux lines. Even a mechanically perfect contact between two pieces of iron of the same kind, however, is sufficient to make this formula useless.

Magnetic Permeability
In free space, magnetic flux density is related to magnetic field intensity by the defining formula

$$\mathbf{B} = \mu_0 \mathbf{H} \tag{11.133}$$

where μ_0, the permeability of free space, has the value $4\pi \times 10^{-7}$ H/m. In other materials, the magnetic flux density at a point is related to the magnetic intensity at the same point by

$$\mathbf{B} = \mu \mathbf{H} \quad \text{where} \quad \mu = \mu_0 \mu_r \tag{11.134}$$

The relative permeability μ_r is a numeric comparing a given material with free space. The magnetic permeability μ for a medium is the product of the relative permeability of that medium and the permeability of free space.

The relative permeability of most substances differs inappreciably from unity. Aside from iron, steel, nickel, cobalt, and synthetic magnetic alloys, all the materials used in electrical engineering have a relative permeability that may be considered equal to that of air. The relative permeability of nonmagnetic substances is for all practical purposes a constant irrespective of the flux density. A physical picture of relative permeability may be gained from the following. A coil carrying an unvarying current produces a magnetic field in the air surrounding it. If a substance is placed, for example, inside this coil without changing the amount or distribution of the flux inside the coil, this substance has a relative permeability of unity. If the flux inside the coil is altered in value or distribution, the relative permeability of the introduced substance differs from unity.

Difference of Magnetic Potential
Consider any two points 1 and 2 in a magnetic field (Fig. 11.52) and let the path between them from 1 to 2 pass through an electric circuit producing an mmf in the direction from 1 to 2; then the expression

$$U_{12} = \int_1^2 (H \cos \theta) \, dl - \mathscr{F}_{12} \; A \tag{11.135}$$

is called the *drop of magnetic potential* from 1 to 2. From the definition of magnetizing force [Eq. (11.123)], it follows that around any closed circuit the drop of magnetic potential is always zero. A magnetomotive force $\mathscr{F}_{12}$ is, therefore, equivalent to a rise of magnetic potential from 1 to 2.

When there is no source of mmf between 1 and 2 and the path coincides with a line of magnetizing force, the drop of magnetic potential is

$$U_{12} = \int_1^2 H \, dl \; A \tag{11.136}$$

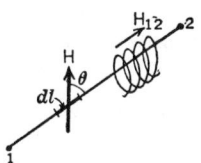

Fig. 11.52   Illustration of drop of magnetic potential.

Magnetic potential difference is of the same nature as mmf and may, therefore, be expressed in the same units, namely amperes.

Magnetic Reluctance

To establish a magnetic flux ϕ through a given portion of a substance that is not itself linked by a source of mmf, a difference of magnetic potential must always be established between the end surfaces of this substance. Let U be the magnetic potential drop established from one surface to the other; then, the quotient

$$\mathcal{R} = \frac{U}{\phi} \quad \text{(A/Wb)} \tag{11.137}$$

is defined as the magnetic reluctance of the given portion of the substance. It should be noted that this definition is meaningless except when applied to a portion of a substance of which the end surfaces are magnetic equipotential surfaces and through every cross section of which the same flux passes.

Magnetic Permeance

The reciprocal of magnetic reluctance is called **magnetic permeance**. Magnetic permeance is analogous to electric conductance, except that it is not a factor that affects the dissipation of energy in a substance. It does, however, enter into the expression for the energy stored in a magnetic field in the same way that the electrostatic capacity of a dielectric is a determining factor in the expression for the energy stored in the electric field.

Kirchhoff's Laws for the Magnetic Circuit

As already noted in Eq. (11.113), the total magnetic flux coming up to any surface in a magnetic field is always zero, provided a flux leaving a surface is considered as a negative flux coming up to that surface. This face may be represented by the formula

$$\sum \phi = 0 \tag{11.138}$$

for every surface in the field. Similarly, from the definition of magnetic potential drop, it follows that the total magnetic potential drop around any closed circuit is zero, or that the total mmf acting around any closed circuit is equal to the sum of the reluctance drops around that circuit, which may be represented by the formula

$$\sum \mathcal{F} = \sum \mathcal{R} \phi \tag{11.139}$$

These two equations are identical in form with those representing Kirchhoff's laws for the electric circuit. They are, however, not so easy to use for practical calculations, for the magnetic flux is not confined to approximately geometrical lines like the currents in a network of insulated wires, but in general fills all space surrounding the coils that establish the mmf's; also, when there is iron or other magnetic material in the circuit, the permeability depends on the flux density and the previous history of the iron. Only in the special case of a uniformly wound circular ring or toroid are the lines of induction confined entirely to an iron circuit; in general a certain number also exist in the air and in whatever other substances are in the vicinity of the iron circuit.

11.4.6 Basic Electromagnetic Theory

Wave propagation represents a major electromagnetic phenomenon that finds unlimited application in radio waves, light, microwave circuits, lasers, radar, antennas, transmission lines, and other applications too numerous to mention.

The basis for electromagnetic theory stems from James Clerk Maxwell's discovery of the theory of light. Using Faraday's view of fields of force, Maxwell wrote a classic paper in his search for understanding electromagnetic action. His *General Equations of the Electromagnetic Field* developed the light general relationship that described in a compact mathematical language the sum knowledge of electromagnetic action. As ultimately developed, the eight general relationships were combined into four classic equations known as Maxwell's equations first published in his *Treatise on Electricity and Magnetism*.

The first equation is known as the differential form of Faraday's law:

$$\nabla \times \mathbf{E} = \frac{-\partial \mathbf{B}}{\partial t} \tag{11.140}$$

where $\mathbf{E}$ is the electric field and $\mathbf{B}$ the magnetic flux density or simply the magnetic field. This equation describes the electrical field produced by a time-varying magnetic field.

The second equation is a generalization of Ampere's circuital law also known as the Biot–Savart law to which Maxwell added the displacement current term $\partial \mathbf{D}/\partial t$ to account for time-variant fields and currents:

$$\nabla \times \mathbf{H} = \mathbf{J} + \frac{\partial \mathbf{D}}{\partial t} \qquad (11.141)$$

where $\mathbf{H}$ is the magnetic field intensity, $\mathbf{J}$ the conduction current density, and $\mathbf{D}$ the displacement current density. This equation relates the magnetic field intensity produced by a current density in both static and time-varying fields.

The third equation is the differential form of Gauss's law and derived from Coulombs's and Gauss's law

$$\nabla \cdot \mathbf{E} = \frac{\rho}{\varepsilon} \qquad (11.142)$$

where $\mathbf{E}$ is the electric field, ρ the charge density, and ε the permittivity of the medium. This equation states that the electric field that diverges from a closed surface is a function of the charge density enclosed by that surface.

The final equation is the magnetic dual of the third equation. It states that magnetic lines of flux form a system of closed loop, and nowhere do they terminate on a magnetic monopole—unlike electric lines of flux.

$$\nabla \cdot \mathbf{B} = 0 \qquad (11.143)$$

where $\mathbf{B}$ is the magnetic field.

A fifth equation is often attributed to Maxwell and listed with the other four. This relationship, known as charge conservation, states that the amount of charge diverging away per second from a volume element must equal the time rate of decrease of the enclosed charge, that is,

$$\nabla \cdot \mathbf{J} = \frac{-\partial \rho}{\partial t} \qquad (11.144)$$

where $\mathbf{J}$ and ρ were previously defined.

As noted in Eqs. (11.103), (11.106), and (11.134), there are relations between flux density and field intensity, and between current density and electric field intensity.

The parameters σ, ε and μ (conductivity, permittivity, and permeability) in anisotropic media will depend on the direction and magnitude of the accompanying vector. Under these conditions these are nonlinear dyadic quantities.

The wave equation, which gives rise to wave propagation phenomenon, is readily found by taking the curl of Eq. (11.140) and substituting Eq. (11.141) as appropriate on the right side of the new equation and using Eqs. (11.142), (11.106), and (11.134). Doing this we obtain

$$\nabla^2 \mathbf{E} + k^2 \mathbf{E} = \frac{\partial \mu \mathbf{J}}{\partial t} + \frac{\nabla \rho}{\varepsilon} \qquad (11.145)$$

where $k = \omega\sqrt{\mu\varepsilon}$ and is called the wave number. The wave number is equal to ω divided by the velocity of propagation of electromagnetic waves in a medium with parameters μ and ε. k is also equal to $2\pi/\lambda$ where λ is the wavelength of a plane wave in the same medium.

In a source-free region Eq. (11.145) can be simplified into the vector Helmholtz equation:

$$\nabla^2 \mathbf{E} + k^2 \mathbf{E} = 0 \qquad (11.146)$$

The study of wave propagation usually involves the interaction of the electromagnetic wave with a different media than the one it is propagating in. In general the solution of problems dealing with electromagnetic waves requires that Maxwell's equation be satisfied as well as the boundary conditions at the interface between the two media. The mathematical solution of these problems are greatly simplified if an appropriate coordinate system is chosen. In dealing with boundary conditions, the components of an impinging field on a boundary must meet the particular constraints of the boundary conditions. These relations are especially important since their solution often dictates the nature of the field equations that are operable within the boundary configuration—inside a waveguide, for example.

At a boundary the normal and tangential components of fields on both sides must match. The particular boundary conditions are as follows:

For an electric field traveling from medium 1 to medium 2, we obtain:

$$\varepsilon_2 E_{n2} - \varepsilon_1 E_{n1} = \sigma \tag{11.147}$$

$$E_{t1} = E_{t2} \tag{11.148}$$

For a magnetic field traveling from medium 1 to medium 2, we obtain:

$$\mu_1 H_{n1} = \mu_2 H_{n2} \tag{11.149}$$

$$H_{t1} = H_{t2} + J_\nu \delta \tag{11.150}$$

where the subscript n relates to the normal component of the field and the subscript t relates to the tangential component. J_ν is the current density at the surface of each medium at the interface. δ is associated with the skin depth, the actual penetration of the fields and current into a medium. The skin depth is equal to $1/\sqrt{\pi f \mu \sigma}$ and decreases with increasing frequency. For a good conductor, such as copper at 10 GHz, the skin depth is only 0.6 μm.

For the special case of the interface between an ideal conductor and air or vacuum, such as for an antenna or waveguide, the boundary conditions simplify to

$$\varepsilon_0 E_n = \sigma \tag{11.151}$$

$$H_n = 0 \tag{11.152}$$

$$E_t = 0 \tag{11.153}$$

$$H_t = J_\nu \delta \tag{11.154}$$

The time-average amount of energy stored in an electric field contained in volume V is given by

$$W_e = \frac{1}{4} \operatorname{Re} \iiint_V \mathbf{D} \cdot \mathbf{E}^* \, dV \tag{11.155}$$

likewise for a magnetic field

$$W_m = \frac{1}{4} \operatorname{Re} \iiint_V \mathbf{B} \cdot \mathbf{H}^* \, dV \tag{11.156}$$

where the asterix (*) designates the complex conjugate value.

The time-average complex power flow across a surface S is given by

$$P = \frac{1}{2} \iint_S \mathbf{E} \times \mathbf{H}^* \, d\mathbf{a} \tag{11.157}$$

The integrand of Eq. (11.157) is the Poynting vector $\mathbf{S}$, an energy transport quantity.

$$\mathbf{S} = \tfrac{1}{2} \mathbf{E} \times \mathbf{H}^* \tag{11.158}$$

where $\mathbf{E}$ and $\mathbf{H}^*$ are assumed to be time-varying fields.

Even though the Poynting vector relates to integration over a closed surface, that is, the loss of electromagnetic energy from an enclosed volume, the Poynting vector is often calculated for a nonclosed surface as a measure of the flow of electromagnetic energy from that surface. A typical application would have one calculate the Poynting vector at a traverse section of a waveguide. If correctly applied, the resulting answer gives the amount of microwave power flowing in the waveguide.

11.5   AC CIRCUITS (TIME-DEPENDENT CIRCUITS)

11.5.1   Signal Waveforms

Introduction

Electrical engineers normally consider a signal to be an electric current or voltage. A signal is generally a time-varying quantity. The pattern of time variation is called a **waveform**. More formally, *a waveform is an equation or graph that describes the signal characteristics as a function of time*. Some examples of signal waveforms are shown in Fig. 11.53.

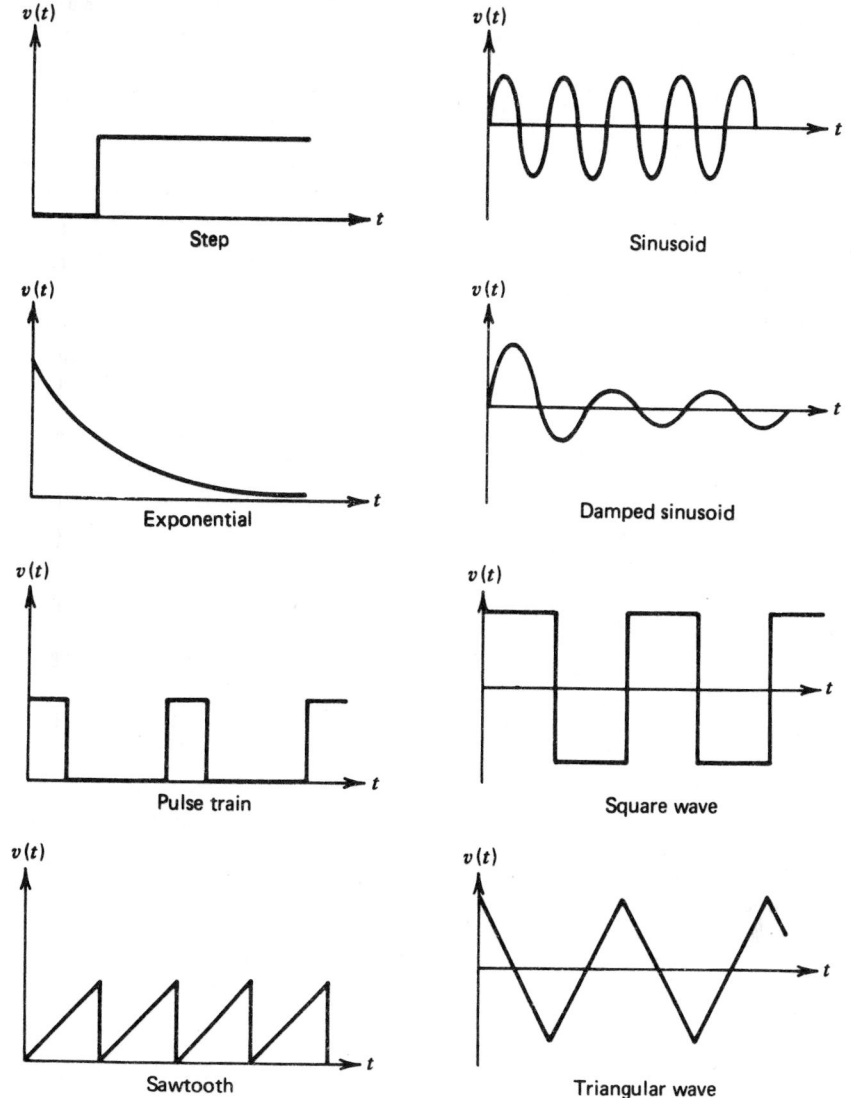

Fig. 11.53 Some example waveforms. Reproduced, with permission, from R. E. Thomas and A. J. Rosa, *Circuits and Signals: An Introduction to Linear and Interface Circuits*, New York, Wiley, 1984.

Most of the waveforms of interest can be constructed using just three basic signal models: the step function, the exponential function, and the sinusoidal function.

There are two matters of notation and convention that must be discussed before we continue. First, quantities that are constant (non-time-varying) are usually represented by uppercase letters (V_A, I, T_0, A_1) or lowercase letters in the early part of the alphabet (a, b, c, d_0). Time-varying quantities are represented by lowercase letters that are not in the early part of the alphabet. This time variation is expressly indicated when we write a term such as $v_1(t)$, $i_A(t)$, or $u(t)$. The time variation is usually implied when such terms are written as v_1, i_A, or u.

Second, the signal variables in a circuit are normally provided with + and − reference marks. It is important to remember that these reference marks do not indicate the polarity of a voltage or the direction of flow of a current. The signal waveforms in a circuit can assume both positive and negative values. The purpose of these marks is to provide a reference against which the actual value of the signal waveform can be compared. When the actual polarity of a voltage or direction of current flow coincides with the reference direction, then we say that the signal is positive. When the opposite occurs, then the signal is considered negative.

The Step Waveform

To develop the general step function waveform, let us first consider the unit step function defined by the relationship

$$u(t) = \begin{cases} 0 & \text{for } t < 0 \\ 1 & \text{for } t > 0 \end{cases} \tag{11.159}$$

The unit step function is zero when its argument (t) is negative, and it is unity when its argument is zero or positive. Mathematically the unit step function $u(t)$ has a jump or discontinuity at $t = 0$.

It is not possible to generate a true step function since no physical variable can undergo a jump change in zero time. Practically speaking it is possible to generate very good approximations to the step function. What is required is that the finite switching time be very short compared with the time the variable remains in its new state.

On the surface, it might appear that the step function is not a very exciting waveform or, at best, only a source of temporary excitement. However, we shall see that the step waveform is a versatile signal and is used to construct models of a wide range of interesting waveforms. First of all, we can multiply $u(t)$ by a real constant A to obtain

$$Au(t) = \begin{cases} 0 & \text{for } t < 0 \\ A & \text{for } t > 0 \end{cases} \tag{11.160}$$

This constant A is called the **amplitude** of the waveform. In addition, we can shift the time at which the step occurs by replacing (t) by ($t - T_s$). Since the definition of the step function indicates that the jump occurs when the argument is zero, the function $u(t - T_s)$ takes on the values

$$u(t - T_s) = \begin{cases} 0 & \text{for } t - T_s < 0 \text{ or } t < T_s \\ 1 & \text{for } t - T_s > 0 \text{ or } t > T_s \end{cases} \tag{11.161}$$

Thus we can delay or advance the time at which the step occurs by changing the argument. The constant T_s is called the step time.

The general step waveform is then written as $Au(t - T_s)$, and we see that it takes two constants to define the signal. The constant A determines the amplitude of the step and normally carries the units of volts, amperes, or watts in an electrical context. The constant T_s determines the time at which the step occurs and carries the units of time, usually seconds. Either A or T_s or both can be negative.

Figure 11.54 shows the effects on the step waveform of changing the amplitude A and the step time T_s. By changing A and T_s and adding several step functions, we can represent a number of

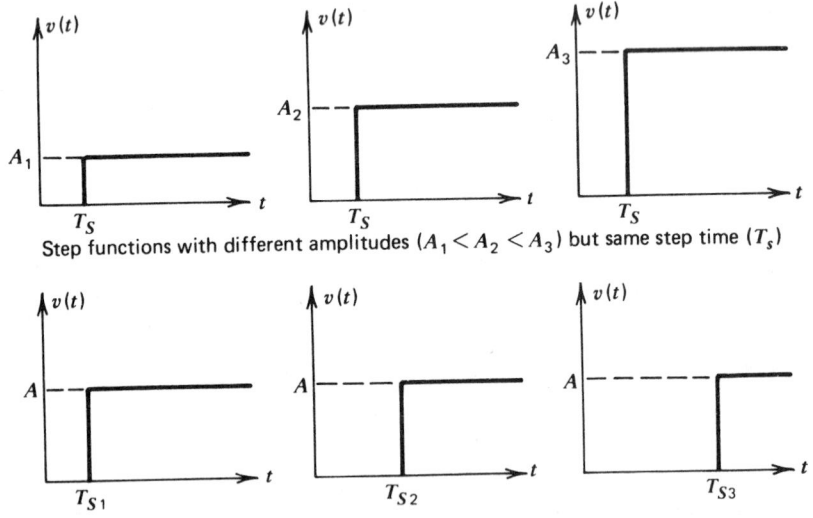

Step functions with different amplitudes ($A_1 < A_2 < A_3$) but same step time (T_s)

Step functions with the same amplitude (A) but different step times ($T_{S1} < T_{S2} < T_{S3}$)

Fig. 11.54 Effects of amplitude and step time on the general waveform, $Au(t - T_s)$. Reproduced, with permission, from R. E. Thomas and A. J. Rosa, *Circuits and Signals: An Introduction to Linear and Interface Circuits*, New York, Wiley, 1984.

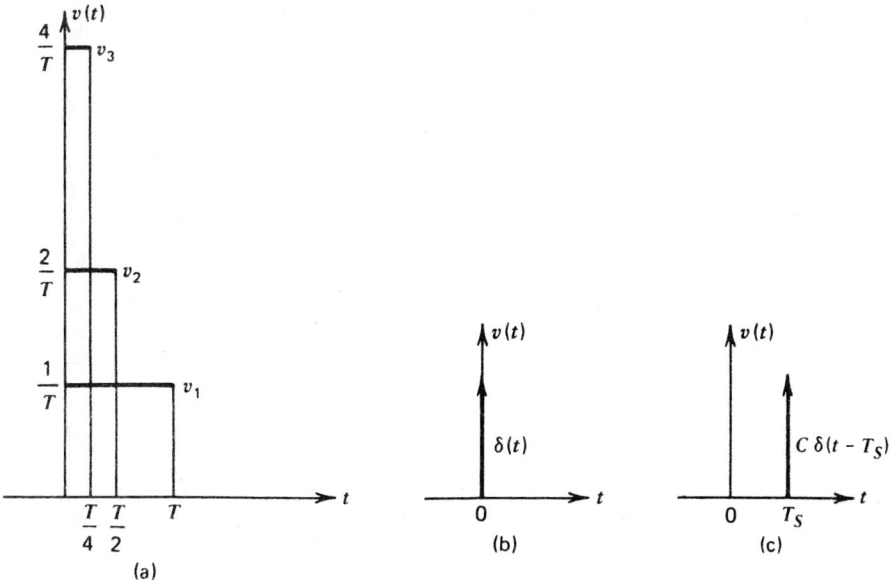

Fig. 11.55 Impulse waveforms: (*a*) pulse waveforms, (*b*) unit impulse, (*c*) general impulse. Reproduced, with permission, from R. E. Thomas and A. J. Rosa, *Circuits and Signals: An Introduction to Linear and Interface Circuits*, New York, Wiley, 1984.

important waveforms. An important application of step functions is in obtaining an understanding of the **impulse** function. Consider the rectangular pulse labeled v_1 in Fig. 11.55(*a*). Such a pulse can be written as

$$v_1(t) = \frac{1}{T}[u(t) - u(t - T)] \tag{11.162}$$

This pulse has an amplitude that is inversely proportional to its duration. If we halve the duration and double the amplitude, we obtain the pulse v_2. By repeating the process of halving and doubling, we obtain v_3. For all three pulses we observe that if we integrate with respect to time (find the area under the waveform), we obtain unity. Now consider the special function obtained by carrying this process to the limit. The duration approaches zero, the amplitude becomes infinite, but the area remains unity. Such a function finds wide application in circuit and system analysis, and goes by the name impulse or Dirac delta function. Its mathematical symbol is $\delta(t)$, and its graphical representation is shown in Fig. 11.55(*b*). More formally, the unit impulse is defined as

$$\delta(t) = 0 \quad \text{for } t \neq 0 \qquad \delta(t) = \infty \quad \text{for } t = 0$$

$$\int_{-\infty}^{t} \delta(x)\, dx = u(t) \tag{11.163}$$

This latter expression suggests that the impulse is the derivative of a step function:

$$\frac{du(t)}{dt} = \delta(t) \tag{11.164}$$

This relationship cannot be justified using elementary mathematics since the function $u(t)$ has a discontinuity at $t = 0$, and therefore its derivative does not exist at that point in the usual sense. However, the concept can be justified using the theory of distributions as shown in advanced texts on circuits and systems. The general impulse is written as $C\delta(t - T_s)$ and its graphical representation is shown in Fig. 11.55(*c*). Notice that the magnitude of the impulse is represented by its area and not its amplitude, which is infinite.

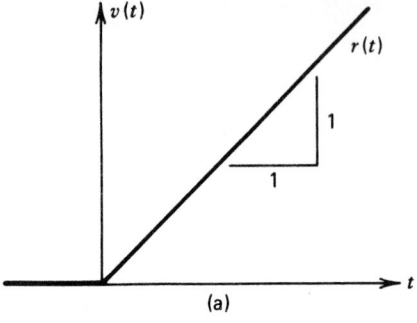

(a)

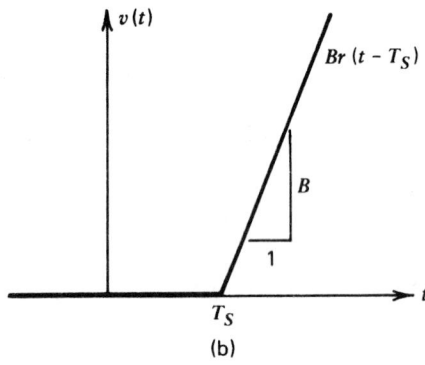

(b)

Fig. 11.56 Ramp waveforms: (a) unit ramp waveforms, (b) general ramp waveforms. Reproduced, with permission, from R. E. Thomas and A. J. Rosa, *Circuits and Signals: An Introduction to Linear and Interface Circuits*, New York, Wiley, 1984.

The unit ramp waveform can be derived from the step function by integration:

$$r(t) = \int_{-\infty}^{t} u(x)\, dx$$

or   (11.165)

$$r(t) = tu(t)$$

For negative times the ramp waveform is zero and for positive time it is simply equal to t. The unit ramp waveform is shown in Fig. 11.56(a). Notice that its slope is unity. The general ramp waveform can then be written as $Br(t - T_s)$ and is shown graphically in Fig. 11.56(b). The general ramp is zero until $t = T_s$, and then increases with a constant slope of B. Thus the magnitude of a ramp waveform is measured by its slope. The ramp is useful in representing triangular and sawtooth waveforms used as timing or sweep signals in both digital and analog circuits.

The impulse, step, and ramp form a set of signals that are commonly referred to as singularity functions. The three waveforms can be related by integration as

$$u(t) = \int_{-\infty}^{t} \delta(x)\, dx$$

$$r(t) = \int_{-\infty}^{t} u(x)\, dx$$ (11.166)

or by differentiation as

$$\delta(t) = \frac{du(t)}{dt}$$

                                                                                          (11.167)

$$u(t) = \frac{dr(t)}{dt}$$

These waveforms are often used as standard forcing functions or inputs in the study of circuits and systems.

The Exponential Waveform

The waveform defined by the relationship

$$v(t) = Au(t)\exp\left(-\frac{t}{T_c}\right) \tag{11.168}$$

is an exponentially decreasing function. A plot of this waveform is shown in Fig. 11.57. Because of the step function in the definition, the waveform is zero for negative time and jumps to an amplitude of A at $t = 0$. Thereafter the waveform monotonically decays toward zero as time approaches infinity. The two parameters required to define the waveform are the *amplitude* (A) and the *time constant* (T_c). The parameter A has the same units as the signal quantity (usually volts, amperes, or watts) and represents the initial ($t = 0$) amplitude of the waveform. The time constant carries the units of time and determines the rate at which the waveform decays.

The time constant is of special significance since it determines the decay rate. For $t = T_c$, $v(T_c) = Ae^{-1}$, which is approximately 0.368 A. Therefore an exponential waveform decays to about 37% of its initial amplitude in a time span of one time constant. At $t = 5T_c$, the value of the waveform is Ae^{-5}, which is approximately 0.00674 A. Thus an exponential signal decays to less than 1% of its amplitude in a time span of five time constants. While theoretically an exponential waveform endures forever, practically speaking the amplitude becomes negligible after about $5T_c$. For this reason the *signal duration* is often taken as $5T_c$. This definition is somewhat arbitrary and is based on practical considerations.

The exponential waveform has two important properties regarding the rate at which the signal decays.

Decrement Property. The value of the waveform for $t > 0$ is given by

$$v(t) = A \exp\left(-\frac{t}{T_c}\right) \tag{11.169}$$

Note that the step function $u(t)$ has been omitted since its value is unity for $t > 0$. After an additional time period Δt has gone by, the amplitude is

$$v(t + \Delta t) = A \exp\left(-\frac{t + \Delta t}{T_c}\right) \tag{11.170}$$

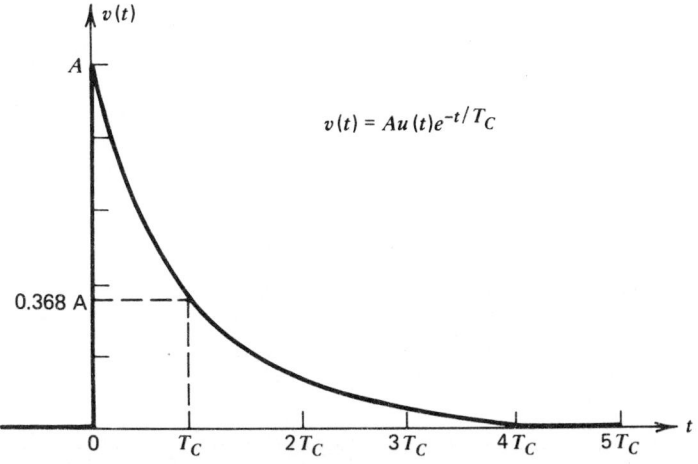

Fig. 11.57 Exponential waveform. Reproduced, with permission, from R. E. Thomas and A. J. Rosa, *Circuits and Signals: An Introduction to Linear and Interface Circuits*, New York, Wiley, 1984.

The ratio of these two amplitudes is

$$\frac{v(t + \Delta t)}{v(t)} = \frac{A \exp[-(t + \Delta t)/T_c]}{A \exp(-t/T_c)} = \exp\left(-\frac{\Delta t}{T_c}\right) \tag{11.171}$$

This ratio is independent of the starting time and the amplitude A. In other words, in any fixed time period Δt, the fractional decrease in the amplitude of an exponential waveform depends only on the time constant. The decrement property can be expressed as equal percent change in equal time intervals. For example, if an exponential waveform decreases to one-half of its initial amplitude in, say, 15 ms, then it will be reduced by the same 50% factor in every subsequent 15-ms interval.

Slope Property. The slope of the exponential waveform (for $t > 0$) is found by taking the derivative of Eq. (11.169) with respect to time:

$$\frac{dv(t)}{dt} = \frac{-A}{T_c} \exp\left(-\frac{t}{T_c}\right) = \frac{-1}{T_c} v(t) \tag{11.172}$$

The slope of the exponential waveform is inversely proportional to the time constant. This means that small time constants lead to large slopes or rapid decays, while large time constants describe signals with shallow slopes and long decay times. All of this is summed up in the fact that the derivative of an exponential signal is itself an exponential with the same time constant but a different amplitude.

The exponential waveform is a good representation of a number of natural phenomena occurring in the physical and biological sciences. The decay in the pressure of a punctured tire and the transfer of nutrients through a cell membrane are but two examples. The term **half-life** is often used to describe the exponential decay of physical phenomena.

The Sinusoidal Waveform

The cosine and sine functions arise frequently in science and engineering, and the corresponding time-varying signal is especially hallowed by electrical engineers. In contrast to the step and exponential waveforms, the sinusoid extends indefinitely in time in both the positive and negative directions. It has neither a beginning nor an end! We admit that a real signal must have been turned on at some time in the past. We acknowledge that in all likelihood it will be turned off at some time in the future. While all real signals are surely finite in duration, the eternal sine wave turns out to be both a very convenient artifice and a good approximation.

Since this model of the sinusoid is infinite in extent, it turns out to be an endless repetition of the same old thing—a periodic oscillation between positive and negative values. Since it is unendingly repetitive, we need not examine the entire signal but only a representative segment called a cycle. Some cycles of sinusoidal waveforms are shown in Fig. 11.58. Since the signal does not "start" somewhere, the location of the time origin ($t = 0$) is somewhat uncertain, or at least arbitrary.

If the time origin is located as in Fig. 11.58(a), then the most natural way to describe the sinusoid is

$$v(t) = a \cos\frac{2\pi t}{T_0} \tag{11.173}$$

On the other hand, if the time origin is located as in Fig. 11.58(b), then the natural expression for the waveform is

$$v(t) = b \sin\frac{2\pi t}{T_0} \tag{11.174}$$

In either case, the **period** T_0 is the interval of time over which the waveform repeats itself and carries the units of time.

The general case is shown in Fig. 11.58(c). This waveform can be expressed as a linear combination of the cosine and sine functions. This form is often referred to as its **rectangular form**.

$$v(t) = a \cos\frac{2\pi t}{T_0} + b \sin\frac{2\pi t}{T_0} \tag{11.175}$$

The constants a and b are called the **Fourier coefficients** of the sinusoid and can be defined as

$$a = v(0) \qquad b = v\left(\frac{T_0}{4}\right) \tag{11.176}$$

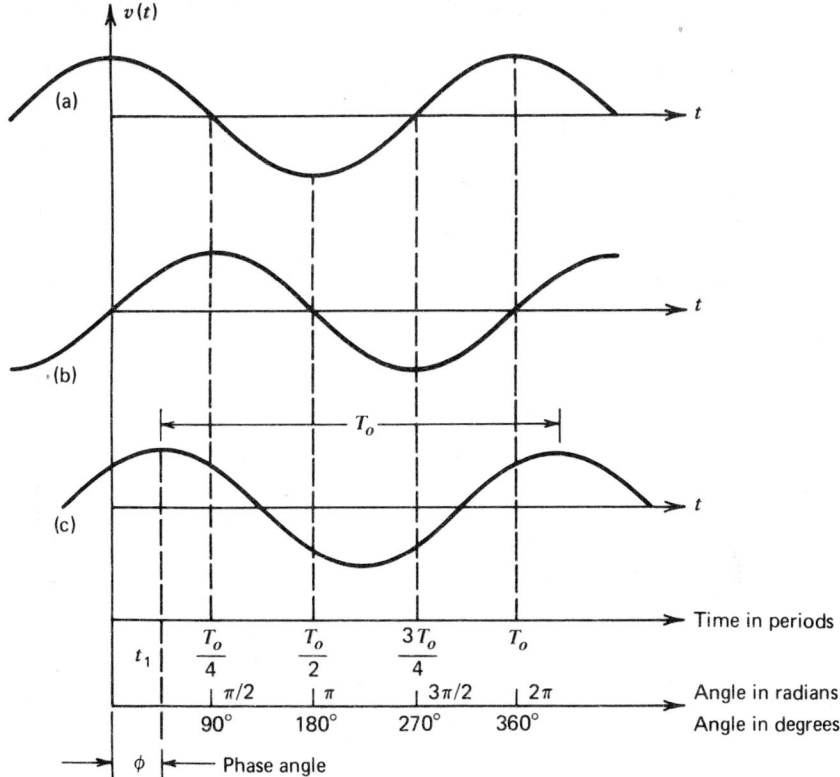

Fig. 11.58 Representative cycles of a sinusoidal waveform. Reproduced, with permission, from R. E. Thomas and A. J. Rosa, *Circuits and Signals: An Introduction to Linear and Interface Circuits*, New York, Wiley, 1984.

The Fourier coefficients have the same units as the waveform (volts or amperes) and either a or b or both can be negative.

An alternative or **polar form** representation of the general sinusoid is

$$v(t) = A \cos\left(\frac{2\pi t}{T_0} - \phi\right) \tag{11.177}$$

In this form the constant A is called the *amplitude* and carries the units of the waveform $v(t)$. The maximum value of the waveform is $+A$ and the minimum value is $-A$. The parameter T_0 is the period as before, and ϕ is called the phase angle.

The term *phase angle* comes from the circular interpretation of the cosine and sine functions. We think of the period as being divided into $360°$ or 2π radians, as indicated at the bottom of Fig. 11.58. From this viewpoint, and since we are referring our angle to the cosine, the phase angle is the "angle" between $t = 0$ and the first positive peak (the place where a cosine begins) after the origin (t_1). The phase angle can be expressed in units of degrees or radians, but note that the term $2t/T_0$ has the units of radians in this sense. Therefore care must be used in those situations where we must perform the addition operation implied by the expression ($2\pi t/T_0 - \phi$) to ensure that both terms have the same units. We can express the relation between ϕ and t_1 as

$$\phi = \frac{t_1}{T_0}(360°) = \frac{t_1}{T_0}(2\pi) \tag{11.178}$$

Since we have two representations of the general sinusoid, we are naturally led to inquire into the relationship between them. One way to derive these relationships is to apply the definitions in Eq.

(11.176) to the representation of the sinusoid in Eq. (11.177). Substituting $t = 0$ into Eq. (11.177) yields

$$a = v(0) = A \cos(-\phi)$$

or   (11.179)

$$a = A \cos(\phi)$$

Likewise, substituting $t = T_0/4$ into Eq. (11.177) yields

$$b = v\left(\frac{T_0}{4}\right) = A \cos\left(\frac{\pi}{2} - \phi\right)$$

or   (11.180)

$$b = A \sin\phi$$

The converse problem also arises—that is, given the Fourier coefficients a and b, find the constants A and ϕ. These relationships can be derived by squaring Eqs. (11.179) and (11.180), and adding the result to obtain

$$A = \sqrt{a^2 + b^2} \qquad (11.181)$$

Likewise, dividing Eqs. (11.179) and (11.180) and solving for ϕ yields

$$\phi = \tan^{-1} \frac{b}{a} \qquad (11.182)$$

It is also customary to describe the time variation of the sinusoid in terms of frequency. Frequency is defined as the number of periods per unit time, or cycles per unit time. Clearly the period T_0 is the number of seconds per cycle; hence the number of cycles per second is

$$f_0 = \frac{1}{T_0} \qquad (11.183)$$

where f_0 represents the **cyclic frequency** or, as it is more commonly called, simply the frequency. The unit of frequency (cycles per second) is called a hertz (Hz). The frequency can also be expressed as an angular quantity in radians per second. This method of representing frequency is called **angular frequency** ω_0, where

$$\omega_0 = 2\pi f_0 = \frac{2\pi}{T_0} \qquad (11.184)$$

since there are 2π radians per cycle.

In working with signals, engineers prefer to express frequency in cyclic fashion—for example, we tune our radios to 690 kHz (AM) or 101 MHz (FM). However, when working with systems, engineers often design and analyze using radian frequency.

The sinusoid has three important properties.

Periodic Property. Waveforms that are endlessly repetitious are called **periodic**. The definition of a periodic waveform can be formalized in the statement: *A waveform is said to be periodic if*

$$v(t + T_0) = v(t) \qquad (11.185)$$

for all values of t. The constant T_0 is called the period of the waveform if it is the smallest nonzero interval for which this equation is true. The sawtooth, square, and triangular waves as well as the sinusoid are all examples of periodic waveforms. Signals that are not periodic are called **aperiodic**.

Additive Property. If two sinusoids with the same frequency are added, we get a sinusoid with different amplitude parameters but the same frequency.

Derivative and Integral Properties. The sinusoid maintains its waveshape when it is differentiated or integrated. These operations change the amplitude and phase angle of the sinusoid but do not change the frequency. The fact that the frequency is unchanged under differentiation and integration is a unique property of the sinusoid. No other periodic waveform has this shape-preserving property.

Composite Waveforms

The three basic waveforms, the *step* function, the *exponential* function and the *sinusoidal* function, are the constituents of almost all of the other signals and form the basis for most of the signals used in electrical engineering. By adding and multiplying these basic waveforms we can generate most of the signals used in signal and system analysis.

Consider an exponential rise. It is obtained by taking the difference between a step function and an exponential. This gives the waveform shown in Fig. 11.59(*a*). Mathematically the waveform is obtained by writing

$$v_1(t) = Au(t)$$

$$v_2(t) = Au(t)\exp\left(-\frac{t}{T_c}\right) \tag{11.186}$$

and forming the exponential rise as

$$v_3(t) = v_1(t) - v_2(t) = Au(t)\left[1 - \exp\left(-\frac{t}{T_c}\right)\right] \tag{11.187}$$

As time becomes large the waveform approaches a final value of A. Practically speaking the amplitude is within less than 1% of this final value at $t = 5T_c$. At $t = T_c$, $v_3(T_c) = A(1 - 0.368) = 0.632A$. Thus the waveform rises to about 63% of its final value in a time span of one time constant. This waveform is also often called a "charging exponential" and represents the behavior of signals that occur in the charging of circuits that contain memory elements (capacitors or inductors).

The next composite waveform, the *damped ramp*, is obtained by multiplying a ramp an an exponential. Let

$$v_1(t) = r(t)$$

$$v_2(t) = Au(t)\exp\left(-\frac{t}{T_c}\right) \tag{11.188}$$

for $t > 0$ this waveform can be written as

$$v_3(t) = At\exp\left(-\frac{t}{T_c}\right) \tag{11.189}$$

As time becomes large the ramp term increases without bound while the exponential decays to zero. Since the composite waveform is the product of these terms, it is important to determine which effect predominates. A single application of L'Hôpital's rule will convince the reader that the exponential wins the race. That is, the exponential decay is stronger than the unbounded behavior of the ramp. For this reason the waveform has the shape shown in Fig. 11.59(*b*).

The *damped sinusoid* of Fig. 11.59(*c*) is realized by multiplying an exponential and a sinusoid. Let

$$v_1(t) = Au(t)\exp\left(-\frac{t}{T_c}\right)$$

$$v_2(t) = \cos(\omega t - \phi) \tag{11.190}$$

Then define

$$v_3(t) = v_1(t)v_2(t)$$

$$= Au(t)\exp\left(-\frac{t}{T_c}\right)\cos(\omega t - \phi) \tag{11.191}$$

We can view this signal as a sinusoid whose amplitude is not constant but decays with time. The decay is provided by the exponential, and although theoretically the waveform does not reach zero in finite time, practically the signal duration is about $5T_c$. This waveform occurs in the pulse response of amplifiers and is given the descriptive name *ringing*, which is a negative word among audio purists.

Partial Waveform Descriptors

A waveform is an equation or graph that gives a complete description of the signal. However, we often work with quantities such as peak-to-peak value and root-mean-square (rms) value that are only

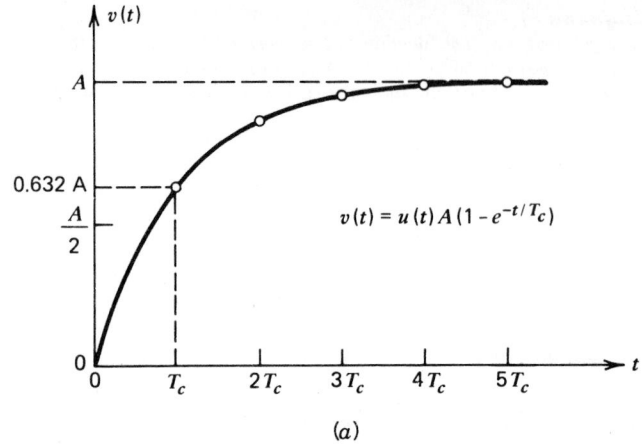

$$v(t) = u(t)A(1 - e^{-t/T_c})$$

(a)

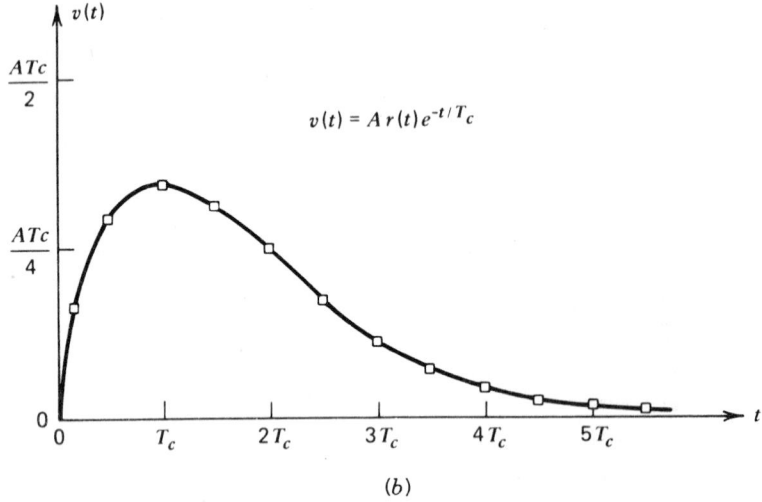

$$v(t) = Ar(t)e^{-t/T_c}$$

(b)

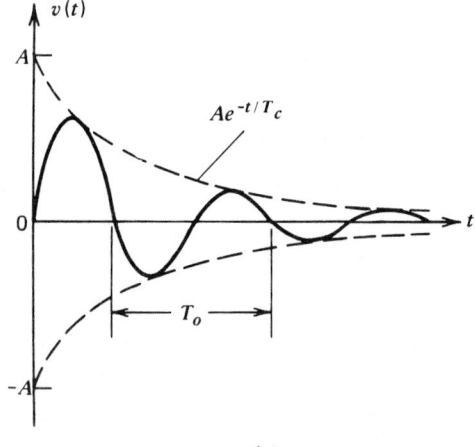

(c)

Fig. 11.59 (*a*) Exponential rise. (*b*) Damped ramped waveform. (*c*) Damped sinusoidal waveform. Reproduced, with permission, from R. E. Thomas and A. J. Rosa, *Circuits and Signals: An Introduction to Linear and Interface Circuits*, New York, Wiley, 1984.

partial descriptions of the waveform that highlight certain characteristics but do not completely specify the signal. For example, the rms value is a measure of the average power carried by the signal, while peak-to-peak value is an indication of the total excursion of the waveform. In other words, partial descriptors bring out important attributes such as the amount of energy or data that can be carried by a waveform.

Generally a waveform $v(t)$ varies between two extreme values, which we will denote as V_{max} and V_{min}. The peak-to-peak (V_{PP}) describes the total excursion and is defined as

$$V_{PP} = V_{max} - V_{min} \qquad (11.192)$$

This definition means that V_{PP} is always positive even if V_{max} and V_{min} are both negative. The peak value, V_p, is the maximum absolute value of the waveform. That is, V_P is $|V_{max}|$ or $|V_{min}|$, whichever is larger. The peak value is always a positive number that indicates the maximum excursion of the waveform. The waveforms in Fig. 11.60 illustrate the definitions of these two descriptors.

The peak and peak-to-peak values describe the hills and valleys of a waveform. The average value, on the other hand, involves using integration to level the hills and valleys into a flat plane. Basically, the average value is the area under the waveform over some period of time T, divided by that time period. Mathematically we define average value over the time T as

$$V_{ave} = \frac{1}{T}\int_0^T v(t)\, dt \qquad (11.193)$$

For periodic signals the averaging interval is taken to be one period (T_0). Technically this would be called the one-cycle average. But for a periodic waveform it is simply called the average value without any qualifiers about the averaging interval, which is understood to be one period.

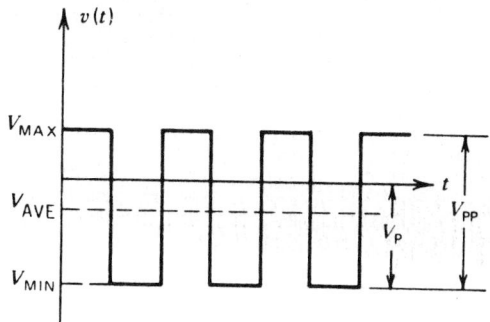

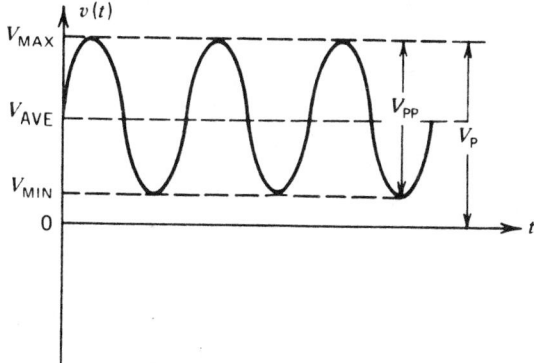

Fig. 11.60 Peak, peak-to-peak, and average values. Reproduced, with permission, from R. E. Thomas and A. J. Rosa, *Circuits and Signals: An Introduction to Linear and Interface Circuits*, New York, Wiley, 1984.

For many periodic waveforms the integration in Eq. (11.193) can be simplified by observing that the net area under the waveform is equal to the area above the time axis minus the area below the time axis. For example, the sinusoid in Fig. 11.60 obviously has a positive average value since the area is entirely above the axis. The pulsetrain in Fig. 11.60 clearly has a negative average value.

The average value measures the offset or asymmetry of the waveform with respect to $v = 0$; that is, it indicates how completely the hills fill in the valleys in the leveling process. It is also called the direct-current or dc component of the waveform.

The rms value of a waveform is a measure of the average power carried by the signal. To derive this measure we first note that the instantaneous power delivered to a resistor by a voltage $v(t)$ is

$$p(t) = \frac{1}{R}[v(t)]^2 \tag{11.194}$$

The average power delivered to the resistor in a time span T is defined as

$$P_{\text{ave}} = \frac{1}{T}\int_0^T p(t)\, dt \tag{11.195}$$

Combining Eqs. (11.194) and (11.195) then yields

$$P_{\text{ave}} = \frac{1}{R}\left\{\frac{1}{T}\int_0^T [v(t)]^2\, dt\right\} \tag{11.196}$$

The quantity inside the curly brackets in this equation is the average value of the square of the waveform over the time interval T. This term is a partial descriptor related to the average power carried by the waveform. The units of this term are volts squared. It is customary to use the square root of this term to define the descriptor.

$$V_{\text{rms}} = \sqrt{\frac{1}{T}\int_0^T [v(t)]^2\, dt} \tag{11.197}$$

The resulting partial descriptor is called the rms value since it is obtained by taking the square *root* of the average (*mean*) of the *square* of the waveform. For periodic signals the averaging interval is taken as one cycle since such a waveform duplicates itself every T_0 seconds. Thus V_{rms} for a periodic waveform can be obtained from Eq. (11.197), except that T is replaced by T_0.

We can now write the average power in terms of V_{rms} as

$$P_{\text{ave}} = \frac{1}{R}V_{\text{rms}}^2 \tag{11.198}$$

The average power is proportional to the square of the rms value of the signal. If the waveform amplitude is doubled, the rms value is doubled, and the average power is quadrupled. If the purpose is to transfer power, then the signal level should be as high as possible. Commercial electric power systems often use transmission voltages on powerlines in the range of hundreds of kilovolts (rms) to transfer large amounts of power.

11.5.2 The Capacitor and the Inductor

The Capacitor
A **capacitor** is a two-terminal energy storage device. Capacitors employ the physical property of *action at a distance*. Whenever charges are separated, a coulomb force of attraction or repulsion, **F**, exists between them. The magnitude of this force takes the following form:

$$F = \frac{q_1 q_2}{4\pi\varepsilon r^2} \tag{11.199}$$

where q_1 and q_2 are the value of the separated charges in coulombs, ε is the permittivity of the medium separating the charge, and $\varepsilon = \varepsilon_0\varepsilon_r$ where ε_r is the relative dielectric constant and ε_0 is the permittivity of free space (8.85×10^{-12} N/m), and r is the separation between the charges. From this equation, it should be apparent that physical geometry will play a part in our determination of capacitance.

The simplest vehicle for collecting and separating charge is a pair of parallel plates. Since two separated parallel plates with charge on them will exhibit a force according to Eq. (11.199), it is often convenient to measure this force in terms of an electric field **E** equal to the force per unit charge, that is,

$$\mathbf{E}(t) = \frac{\mathbf{F}}{q(t)} \qquad (11.200)$$

From Gauss's law (which states that the electric flux passing through any closed surface is equal to the total charge enclosed by that surface), one can relate the electric field to the geometry of the parallel plates. If the separation d is small compared with the dimension of the plates, then the electric field can be written as

$$E(t) = \frac{q(t)}{\varepsilon A} \qquad (11.201)$$

where $E(t)$ is the one-dimensional electric field in the direction perpendicular to the plates, ε is the permittivity of the dielectric, A is the area of the plates, and $q(t)$ the magnitude of the charge on each plate. With d small compared with the dimensions of the plates, the electric field can be assumed to be uniform. We can then write

$$E(t) = \frac{v_c(t)}{d} \qquad (11.202)$$

Substituting into Eq. (11.201) and solving for q, we find

$$q(t) = v_c(t)\left[\frac{\varepsilon A}{d}\right] = v_c(t)C \qquad (11.203)$$

The quantity in the brackets of Eq. (11.203) is called the **capacitance** C of the capacitor.

Capacitance is measured in farads (F) and depends only on the physical properties of the dielectric and the dimensions of the plates.

In practice it is difficult to measure charge. Rather it is customary to measure the time rate of change of charge, or current. If we take the time derivative of Eq. (11.203), and consider C to be a constant, we get

$$i_c(t) = C\frac{dv_c(t)}{dt} \qquad (11.204)$$

Equation (11.204) represents a mathematical model of an ideal element that at times might break down when working with real devices. Figure 11.61(a) shows the ideal model defined by Eq. (11.204). Figure 11.61(b) shows a less ideal model of a capacitor with many of its parasitic properties.

In Eq. (11.204) we express the capacitor current in terms of the capacitor voltage. To find a representation of the capacitor voltage in terms of current, we must perform an integration. Selecting the limits of integration requires some thought. In general, we can assume that at some time long ago, possibly when the capacitor was first built, there was no voltage across the capacitor. Hence we can

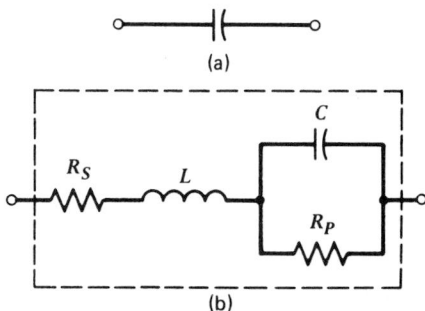

(a)

(b)

Fig. 11.61 Ideal and parasite model of a capacitor: (a) ideal, (b) extended model. Reproduced, with permission, from R. E. Thomas and A. J. Rosa, *Circuits and Signals: An Introduction to Linear and Interface Circuits*, New York, Wiley, 1984.

reasonably assume the voltage at $t = -\infty$ to be zero. How much voltage there is at some other time will depend on how i behaved since $t = -\infty$ to the time t that we want to determine v_c. Therefore we can write

$$\int_0^{v_c(t)} dv_c(t) = \frac{1}{C} \int_{-\infty}^t i_c(t)\, dt = v_c(t) \tag{11.205}$$

This expression states that the voltage at any time t is proportional to the integral of the current flowing through the capacitor from time immemorial to time t. This integral may be easy to interpret, but it is difficult to evaluate since we usually have no idea of how the current behaved from time immemorial to when we were ready to use it in our circuit. Hence we need to modify the expression so we can use it.

A way around our problem is to divide our integral into two parts. The first part represents all the time from time immemorial to when we want to start our problem. The second part is from the time we start our problem to the present. The exact time we choose to divide our problem is not important, but we can simplify our calculation if we let that time be $t = 0$. In general, we will choose the occurrence of some event as $t = 0$; for example, it can be the instant a switch is thrown or the start of a particular clock pulse. Such epoch-making events occur quite frequently in circuits and offer good landmarks from which to measure the behavior of energy storage devices. Using this technique we can write Eq. (11.206) as

$$v_c(t) = \underbrace{\frac{1}{C} \int_{-\infty}^0 i_c(t)\, dt}_{v_c(0)} + \frac{1}{C} \int_0^t i_c(t)\, dt \tag{11.206}$$

The first term on the right of Eq. (11.206) can be evaluated and is equal to a constant $v_c(0)$. It represents the value of the voltage across the capacitor at time $t = 0$ due to the current flowing through it from time immemorial to $t = 0$. It may or may not be possible to calculate the value of $v_c(0)$, but it can always be measured. Regardless of how the current is changing, there always is only one value of $v_c(t)$, at any instant of time including $t = 0$. Hence Eq. (11.206) is a sum of two terms: an integral that represents the positive time variation of the voltage and a constant $[v_c(0)]$ that represents the *initial condition* of the capacitor.

Let us pause and reflect on Eq. (11.206) for a moment. At $t = 0$ the left side of Eq. (11.206) clearly is equal to $v_c(0)$. At $t = 0^+$, an instant after $t = 0$, $v_c(0^+)$ is given as

$$v_c(0^+) = \frac{1}{C} \int_0^{0^+} i_c(t)\, dt + v_c(0) \tag{11.207}$$

For any finite value of $i_c(t)$ (any realizable signal), the integral from $t = 0$ to $t = 0^+$ is zero. This means that

$$v_c(0^+) = v_c(0) \tag{11.208}$$

The implication of Eq. (11.208) is that *the voltage across a capacitor cannot change instantaneously*. To support our implication, suppose for a moment that it was possible to change the voltage across a capacitor instantaneously. This means that $dv_c(t)/dt$ is infinite. But from Eq. (11.204) this would mean that the current $i_c(t)$ would be infinite—a physical impossibility. Thus we can say that the voltage across a capacitor cannot change instantaneously since it would require an infinite current. This simple but important realization is called the **continuity** relation for capacitors and is usually written as

$$v_c(t^+) = v_c(t^-) \tag{11.209}$$

Power, Energy, and Memory
Consider an ideal capacitor. The relationship between current, voltage, and power is given by

$$p_c(t) = i_c(t) v_c(t) \tag{11.210}$$

which can be rewritten using the capacitor's *i-v* characteristics as

$$p_c(t) = C v_c(t) \frac{dv_c(t)}{dt} \tag{11.211}$$

From Eq. (11.211) we can see that the power can be either positive or negative since the time rate of change of voltage can be positive or negative regardless of the absolute value of voltage. This is in stark contrast to a resistor where the power can only be positive. Recall that positive power implies power *delivered* to an element, while negative power suggests power *supplied* by that element. The ability to supply power implies the ability to store energy. Let's focus our attention on this special ability.

The power expression of Eq. (11.211) is a perfect derivative:

$$p_c(t) = Cv_c(t)\frac{dv_c(t)}{dt} = \frac{d}{dt}\left[\frac{1}{2}Cv_c^2(t)\right] \qquad (11.212)$$

Since power is the time rate of change of energy, the quantity inside the brackets must represent the energy stored in the electric field of the capacitor, that is,

$$w_c(t) = \frac{1}{2}Cv_c^2(t) \qquad (11.213)$$

The amount of energy stored at any instant of time is dependent only on the instantaneous value of the voltage. Because of the $v_c^2(t)$ term, the energy stored is always positive. Looking back at Eq. (11.211) we can see further justification for the fact that the voltage across a capacitor cannot change instantaneously [Eq. (11.209)]. An instantaneous change would mean that $dv_c(t)/dt$ is infinite, which, in turn, would mean that infinite power would be required, an impossible situation. The important idea here is that the capacitor can store energy, and that the energy is proportional to the square of the voltage across it. Voltage, then, is the *state* variable for the capacitor.

Example 11.10. The voltage across the $\frac{1}{2}F$ capacitor varies as shown in Fig. 11.62(a). Find the current through, the energy stored in, and the power delivered to or supplied by the capacitor.

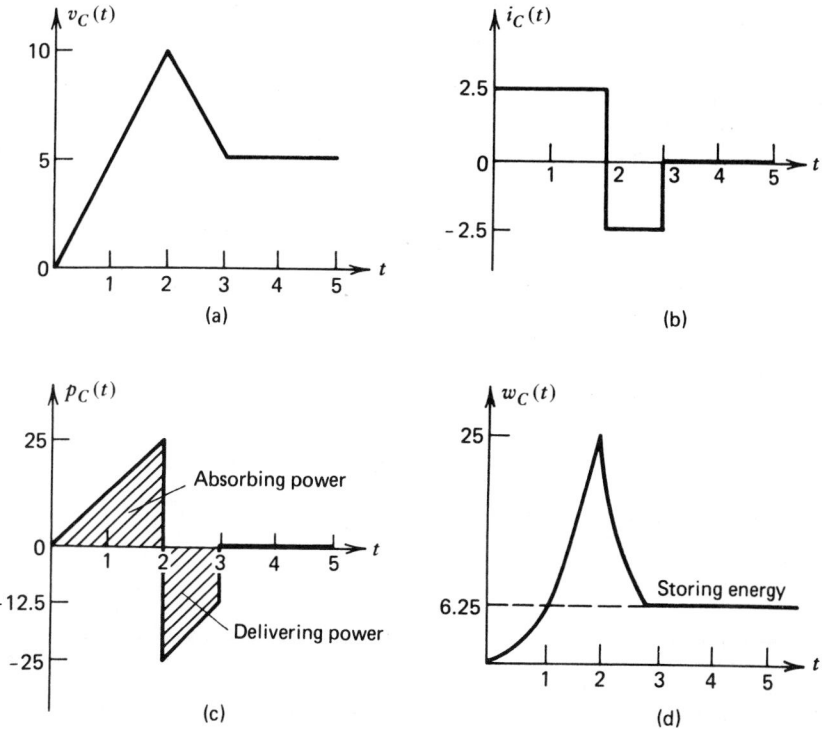

Fig. 11.62 Waveforms showing the relationships between i, v, p, and in a capacitor: (a) voltage, (b) current, (c) power, (d) energy. Reproduced, with permission, from R. E. Thomas and A. J. Rosa, *Circuits and Signals: An Introduction to Linear and Interface Circuits*, New York, Wiley, 1984.

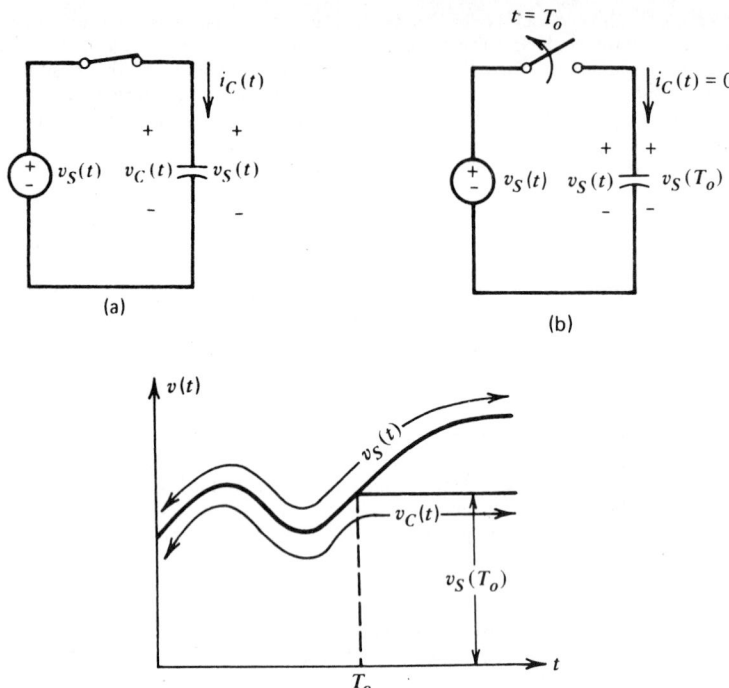

Fig. 11.63 Demonstration of capacitor memory: (*a*) switch closed, track mode; (*b*) switch open, hold mode. Reproduced, with permission, from R. E. Thomas and A. J. Rosa, *Circuits and Signals: An Introduction to Linear and Interface Circuits*, New York, Wiley, 1984.

The current through the capacitor is found by using Eq. (11.204). The power is simply given by the product of the voltage curve and the current curve. The energy is found by either integrating the power curve or by finding $\frac{1}{2}Cv_c^2(t)$ point by point. The resulting graphs are shown in Fig. 11.62(*b–d*).

Let us return to the concept that capacitors can store and *remember* a particular voltage. Consider the circuit of Fig. 11.63(*a*). The voltage across the capacitor must *track* the voltage supplied by the source since they are connected in parallel.

Suppose at some time $t = T_0$ the switch is opened [Fig. 11.63(*b*)]. The source will continue to vary and produce a voltage $v_s(t)$. However, the capacitor is disconnected from the circuit and, lacking a path for the charges on its plates to recombine, will *hold* the voltage it had across it at the time the switch was opened, that is, $v_s(T_0)$. Another way to look at this is that at $t = T_0$, the current is forced to zero, $i_c(t) = 0$, since the switch is open. If $i_c(t) = 0$, then the voltage across the capacitor is a constant, that is,

$$i_c(t) = 0 = C\left[\frac{dv_c(t)}{dt}\right] \tag{11.214}$$

but

$$0 = C\left[\frac{d(\text{const.})}{dt}\right] \tag{11.215}$$

Our ideal capacitor theoretically will hold or *remember* the voltage applied at the instant the switch was opened forever. In reality, because of parasitic effects, real capacitors gradually will lose energy, and hence the stored value gradually will decrease. For this reason, if the actual value remembered is to be retained for a long period of time, it occasionally must be refreshed. This concept of memory and energy storage is a fundamental property of digital electronic circuits used for data computation in computers and ultimately provides a limitation to the speed with which these computations can occur. (See, for example, MOS devices in electronic texts such as Millman, *Microelectronics*, McGraw-Hill, New York, 1979.)

The Inductor

Much like the capacitor, **inductors** are two-terminal energy storage devices that exhibit the *action at a distance* phenomenon. A compass placed near a wire carrying current would be deflected to align with a magnetic force surrounding the wire. The amount of force exerted by each magnetic pole is called the magnetic field density **B**. This magnetic field density **B** is proportional to several key parameters, including the amount of current i the wire is carrying and the distance r from the wire. Mathematically,

$$\mathbf{B} \alpha \frac{\mu i}{r} \qquad\qquad (11.216)$$

where μ is the magnetic permeability of the medium surrounding the conductor.

In dealing with inductors it is often more convenient to talk about a quantity called magnetic flux, ϕ. This new parameter is related to **B** as originally derived in Eq. (11.113) and is repeated here as

$$\phi = \oint B \cos \alpha \, ds \qquad\qquad (11.217)$$

where s is the surface of integration and α is the angle between the surface of integration and the field density **B**. This magnetic flux can be said to be represented by lines of flux surrounding the current-carrying wire. If the wire is wound into a coil, the lines of flux become concentrated along the axis of the coil. The more turns N the coil has, the more flux.

The relationship among flux, current, and turns is given by

$$\phi(t) = kNi_L(t) \qquad\qquad (11.218)$$

where k is a constant of proportionality. This flux, in turn, will tie together or link all the turns of the inductor. The more turns, the more linkage. This total linkage is called the **flux linkage** and has the symbol λ and the units of weber (Wb). Flux linkage is another basic variable like charge and energy and is the dual of charge. We shall develop the i-v relations for the inductor in a manner dual to the capacitor. Flux linkages and flux are related by

$$\lambda(t) = N\phi(t) \qquad\qquad (11.219)$$

so that by substitution we get

$$\lambda(t) = [kN^2]i_L(t) = Li_L(t) \qquad\qquad (11.220)$$

The quantity in the brackets in Eq. (11.220) is called the **inductance** L of the coil.

The unit of inductance is the henry (H) and depends entirely on the physical properties of the coil.

As with trying to measure charge, trying to measure flux linkages is difficult. Rather it is customary to measure the time rate of change of flux linkages or voltage. If we take the time derivative of Eq. (11.220) and consider L a constant, we get

$$v_L(t) = L\frac{di_L(t)}{dt} \qquad\qquad (11.221)$$

Once again the reader is cautioned that Eq. (11.221) represents a mathematical model of an ideal device. Real inductors are far less close to their ideal models than capacitors or resistors. For example, we assume that inductor coils are wound with zero-resistance wire, but in practice this is never the case. Figure 11.64(a) shows the model for an ideal inductor while Fig. 11.64(b) shows a more realistic model for a real inductor.

In Eq. (11.221) we represent the inductor voltage in terms of the inductor current. To find a representation of the inductor current in terms of its voltage, we must integrate Eq. (11.221). Using analogous arguments as for the capacitor in selecting limits of integration, we can write

$$\int_0^{i_L(t)} di_L(t) = \frac{1}{L}\int_{-\infty}^t v_L(t) \, dt = i_L(t) \qquad\qquad (11.222)$$

This equation states that the current at any time t is proportional to the integral of the voltage across the inductor from time immemorial to time t. Just as with the capacitor, we divide our integral

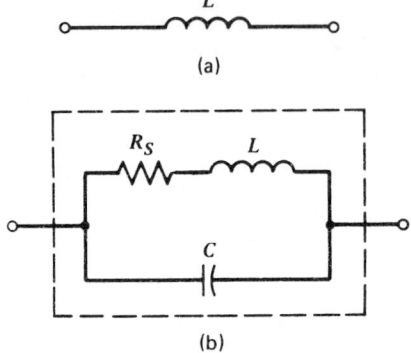

Fig. 11.64 Ideal and parasitic model of an inductor: (*a*) ideal, (*b*) extended model. Reproduced, with permission, from R. E. Thomas and A. J. Rosa, *Circuits and Signals: An Introduction to Linear and Interface Circuits*, New York, Wiley, 1984.

into two parts using $t = 0$ as our benchmark. The result then is

$$i_L(t) = \underbrace{\frac{1}{L}\int_{-\infty}^{0} v_L(t)\, dt}_{i_L(0)} + \frac{1}{L}\int_{0}^{t} v_L(t)\, dt \tag{11.223}$$

The first term on the right can be evaluated and is equal to a constant, $i_L(0)$. It represents the value of the current flowing through the inductor at time $t = 0$ due to the voltage impressed across it from time immemorial to $t = 0$. Like the voltage across the capacitor, it may or may not be possible to calculate this value of current, but it can always be measured. There is only one value of current $i_L(t)$ flowing through the inductor at any instant of time including $t = 0$. Hence we can rewrite Eq. (11.223) as a sum of two terms: an integral that represents the positive time variation of the current and a constant that represents the initial condition of the inductor.

Following a similar argument as for the capacitor, we should realize that *the current through an inductor cannot change instantaneously*. That is,

$$i_L(t^+) = i_L(t^-) \tag{11.224}$$

This simple but important relation is called the **continuity** relation for inductors.

An inductor, like the capacitor or any other element, has the following relationship among voltage, current, and power:

$$p_L(t) = i_L(t) v_L(t) \tag{11.225}$$

which can be rewritten using the inductor's *i-v* relation as

$$p_L(t) = L i_L(t) \frac{d i_L(t)}{dt} \tag{11.226}$$

As with the capacitor, the power can be either positive or negative, which means that an inductor can absorb or provide power. The power relation in Eq. (11.226) is a perfect derivative, that is,

$$p_L(t) = \frac{d}{dt}\left[\frac{1}{2} L i_L^2(t) \right] \tag{11.227}$$

Since power is the time rate of change or energy, the quantity inside the bracket represents the energy stored in the magnetic flux of the inductor, hence,

$$w_L(t) = \tfrac{1}{2} L i_L^2(t) \tag{11.228}$$

The amount of energy stored at any instant of time is dependent only on the instantaneous value of the current. The energy stored is always positive. Like the capacitor, an inductor can store energy, except it stores it in its magnetic rather than electric field. An inductor can remember the last current to flow through it. Current then is the *state* variable for the inductor.

Inductors are the dual of capacitors. This means that the *i-v* characteristics of one element give rise to similar *v-i* characteristics of the other. That is, inductors and capacitors give rise to similar looking equations except that the roles of current and voltage are interchanged. One may be tempted to believe that inductors and capacitors can be interchanged freely in circuits. In some ways this is true—especially in theory. In practice, however, real inductors tend to be lossy and heavy as discrete components, and are rarely produced on ICs because of severe limitations in building them on a planar surface. The end result is that capacitors are the principal energy storage elements used in electronics. Inductors are used in special applications where both types of energy storage elements are necessary, such as in resonant circuits; in applications where interaction between two or more inductors is desired, such as in transformers; and in other applications where size does not matter, such as in power applications.

Op-amp Circuits with Energy Storage Devices

The *i-v* relationships of capacitors and inductors when combined with the special properties of operational amplifiers give rise to several important applications.

Consider the circuit of Fig. 11.65. We began our analysis by writing a KCL equation at node (A):

$$i_R(t) + i_c(t) = i_N(t) \tag{11.229}$$

and the element equations are

$$i_c(t) = C\frac{d[v_0(t) - v_N(t)]}{dt} \tag{11.230}$$

and

$$i_R(t) = \frac{1}{R}[v_s(t) - v_N(t)] \tag{11.231}$$

Finally, if we use the properties of an ideal op-amp,

$$v_N(t) = v_P(t) \qquad i_N(t) = i_P(t) = 0 \tag{11.232}$$

we can substitute all these element constraints into our original KCL equation. In doing so we should also note that $v_P(t)$ is connected to ground and hence is zero, making $v_N(t)$ equal to zero as well. We get

$$\frac{v_s(t)}{R} + C\frac{dv_0(t)}{dt} = 0 \tag{11.233}$$

which we can rewrite to solve for $v_0(t)$ in terms of $v_s(t)$ as

$$v_0(t) = -\frac{1}{RC}\int_{-\infty}^{t} v_s(t)\, dt \tag{11.234}$$

The output of our op-amp circuit is proportional to the integral of the input. We call this circuit an **integrator**. Note that the proportionality constant $(-1/RC)$ has the units of inverse time

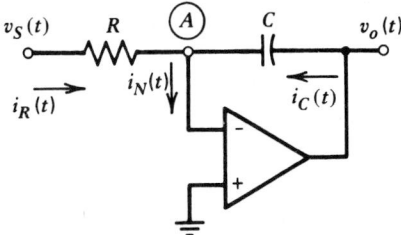

Fig. 11.65 An op-amp integrator. Reproduced, with permission, from R. E. Thomas and A. J. Rosa, *Circuits and Signals: An Introduction to Linear and Interface Circuits*, New York, Wiley, 1984.

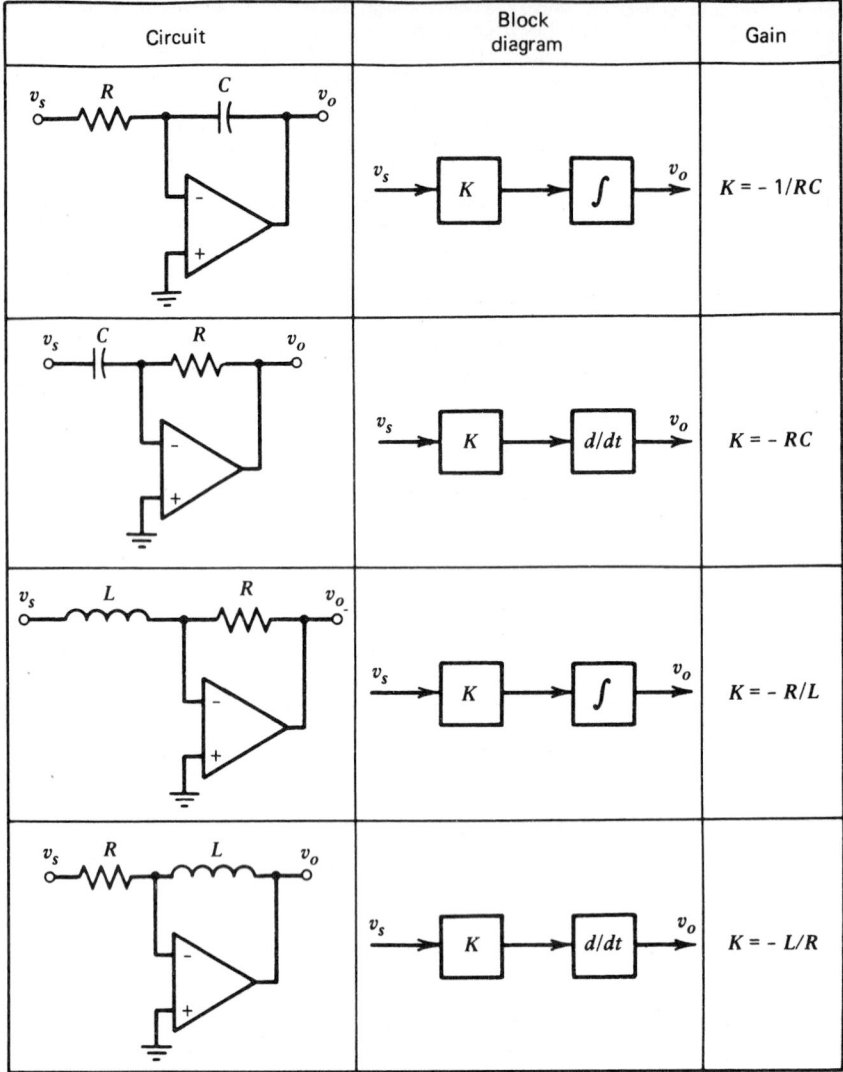

Fig. 11.66 Block diagram representation of op-amp integrators and differentiators. Reproduced, with permission, from R. E. Thomas and A. J. Rosa, *Circuits and Signals: An Introduction to Linear and Interface Circuits*, New York, Wiley, 1984.

(second^{-1}). By following a similar analysis with R and C interchanged we get

$$v_0(t) = -RC\frac{dv_s(t)}{dt} \qquad (11.235)$$

This circuit performs as a **differentiator**. From the concept of duality two similar relations evolve if we replace C by L.

We have introduced two more building blocks to our collection of functional mathematical operations. Along with the scalar multipliers and summers introduced earlier, Fig. 11.25, we now add the mathematical operations of integration and differentiation. The operational amplifier received its name from its ability to perform so many varied mathematical operations. These two new building blocks are collected and represented along with circuit, block diagram, and gains in Fig. 11.66.

This entire collection of op-amp realizable mathematical operations forms the basis of the so-called analog computer that finds extensive application in simulating systems described by differential or integrodifferential equations.

Example 11.11. Using the building blocks contained in Fig. 11.25 and Fig. 11.66, devise a solution to the following differential equation. Assume all initial conditions to be zero:

$$\frac{d^2y}{dx^2} + 6\frac{dy}{dx} + \frac{1}{2}y = x \qquad (11.236)$$

The solution to our equation is y. Hence we begin by solving our equation for y.

$$y = 2x - 2\frac{d^2y}{dx^2} - 12\frac{dy}{dx} \qquad (11.237)$$

Now we must realize that y is the output of a summer that has all the terms on the right as inputs [see Fig. 11.67(a)]. If we have y, and we differentiate it, we obtain dy/dx, and if we then multiply this output by -12, we have one of the necessary inputs to our summer. Continuing this process, if we differentiate dy/dx, we get our second derivative, which we can scale by -2 to obtain our second input. Figure 11.67(b) shows the result of our block diagram analysis. From this block diagram we can then design each individual block, being careful to consider the gains and inverting properties of each block.

Practical realities make our approach to this problem unrealistic. Differentiators are rarely used as they accentuate noise, and noise is always present. An alternative design uses integrators in lieu of differentiators as shown in Fig. 11.67(c).

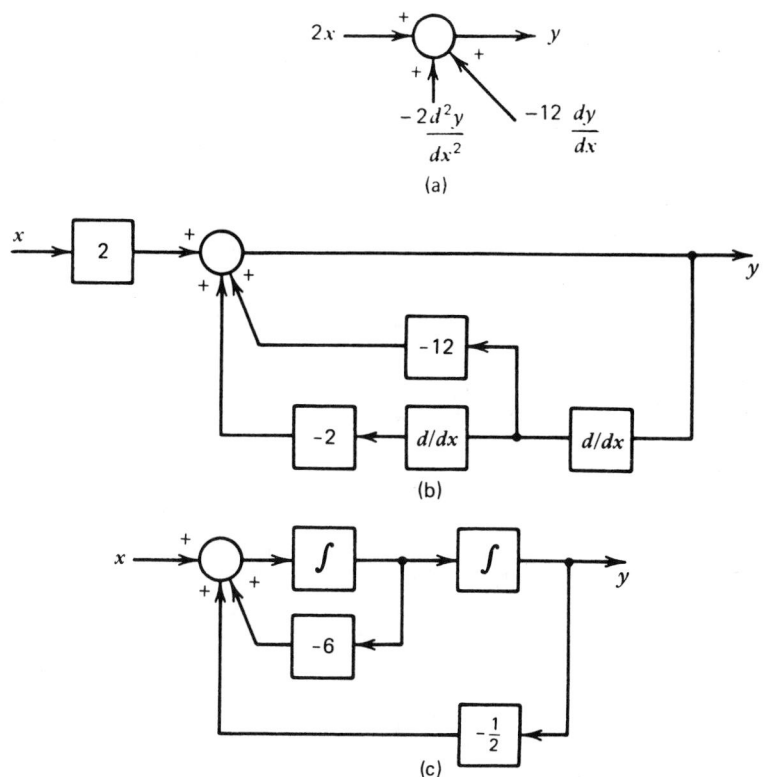

(a)

(b)

(c)

Fig. 11.67 Example use of block diagrams to solve differential equations: (a) summing operation, (b) solution using differentiators, (c) solution using integrators. Reproduced, with permission, from R. E. Thomas and A. J. Rosa, *Circuits and Signals: An Introduction to Linear and Interface Circuits*, New York, Wiley, 1984.

Linearity and Equivalent Circuits

Capacitors and inductors, like resistors, are linear elements. Circuits composed of R's, L's, and C's are linear circuits, and all the analysis tools applicable to linear circuits (Thevenin, superposition, etc.) apply.

One of the first and simplest applications of the linearity property is in establishing a rule for equivalent circuits. There are many occasions where two or more capacitors are connected in series or parallel. Analysis of such circuits would be simplified if those capacitors could be combined and replaced by a simpler equivalent.

If an analysis of the effects of combining capacitors in parallel is conducted, the net result is that the parallel connection of two or more capacitors produces an equivalent capacitance that is the sum of the individual capacitances. For the series connection the capacitance reciprocals are added to produce the reciprocal of the equivalent capacitance. A similar analysis can be performed for the inductor. But since the inductor is the dual of the capacitor, the equivalence rules are interchanged. That is, for the series connection of inductors the equivalent inductance is found by adding the individual inductances, while for the parallel connection the reciprocals are added. These observations on combining capacitors or inductors in series of parallel are summarized in Fig. 11.68.

Element	Series	Parallel
Capacitors	$\dfrac{1}{C_{EQ}} = \dfrac{1}{C_1} + \dfrac{1}{C_2} + \cdots + \dfrac{1}{C_N}$	$C_{EQ} = C_1 + C_2 + \cdots + C_N$
Inductors	$L_{EQ} = L_1 + L_2 + \cdots + L_N$	$\dfrac{1}{L_{EQ}} = \dfrac{1}{L_1} + \dfrac{1}{L_2} + \cdots + \dfrac{1}{L_N}$

Fig. 11.68 Equivalent circuits. Reproduced, with permission, from R. E. Thomas and A. J. Rosa, *Circuits and Signals: An Introduction to Linear and Interface Circuits*, New York, Wiley, 1984.

Mutual Inductance

It was earlier shown that a current flowing through a coiled wire produces magnetic flux lines that form closed-loops about the conductor and that the number of turns in the coil enhances that flux. Furthermore if the flux lines or linkages were changing with respect to time, a voltage would be generated across the coil of wire. Suppose that a second coil of wire was brought close to the first coil. It is reasonable to assume that the flux linkages of the first coil would cut the second coil and generate a voltage across it. In this section we will look at the effect of one coil on another, called **mutual inductance**.

Consider the coils of wire in close proximity as shown in Fig. 11.69.

Suppose a current $i_1(t)$ flows in the first coil of turns N_1. The amount of flux linkages produced by this current is given by Eq. (11.220)

$$\lambda_1(t) = k_1 N_1^2 i_1(t) \tag{11.238}$$

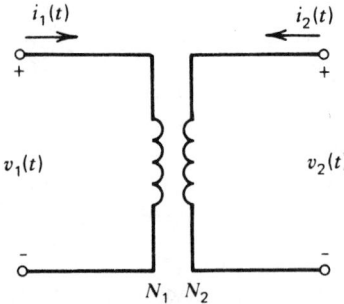

Fig. 11.69 Mutually coupled coils. Reproduced, with permission, from R. E. Thomas and A. J. Rosa, *Circuits and Symbols: An Introduction to Linear and Interface Circuits*, New York, Wiley, 1984.

The voltage produced across this coil is given by Eq. (11.221):

$$v_1(t) = k_1 N_1^2 \frac{di_1(t)}{dt} = L_1 \frac{di_1(t)}{dt} \tag{11.239}$$

The constants $K_1 N_1^2$ are called the self-inductance, or simply the inductance, and are referred to by the symbol L_1.

Now suppose that the second coil was close enough so that the flux linkages produced by i_1 intercepted it. A voltage would be generated in the second coil that would be proportional to several things. First of all, it would be proportional to the current in the first coil, $i_1(t)$. Then it would be proportional to the turns in the first coil, N_1. It also would be proportional to the turns in the second coil, N_2. And finally it would be proportional to the coupling between the two coils and a number of secondary effects, k_m. We can write

$$v_2(t) = N_1 N_2 k_m \frac{di_1(t)}{dt} = M \frac{di_1(t)}{dt} \tag{11.240}$$

We call the constant $N_1 N_2 k_m$ the mutual inductance and give it the symbol M. The units of M, just like L_1, are expressed in henrys. The same reasoning could be followed if the current were flowing in coil 2. There would be a self-induced voltage and one voltage induced in coil 1 via mutual inductance. That is, we would obtain two terms similar to Eqs. (11.239) and (11.240),

$$v_2(t) = k_2 N_2^2 \frac{di_2(t)}{dt} = L_2 \frac{di_2(t)}{dt} \tag{11.241}$$

and

$$v_1(t) = N_1 N_2 k_m \frac{di_2(t)}{dt} = M \frac{di_2(t)}{dt} \tag{11.242}$$

Now suppose *both* currents were flowing as in Fig. 11.70(a). We could combine the results of Eqs. (11.239)–(11.242) using superposition to obtain the terminal voltages for each coil:

$$v_1(t) = L_1 \frac{di_1(t)}{dt} + M \frac{di_2(t)}{dt}$$

and

$$\tag{11.243}$$

$$v_2(t) = M \frac{di_1(t)}{dt} + L_2 \frac{di_2(t)}{dt}$$

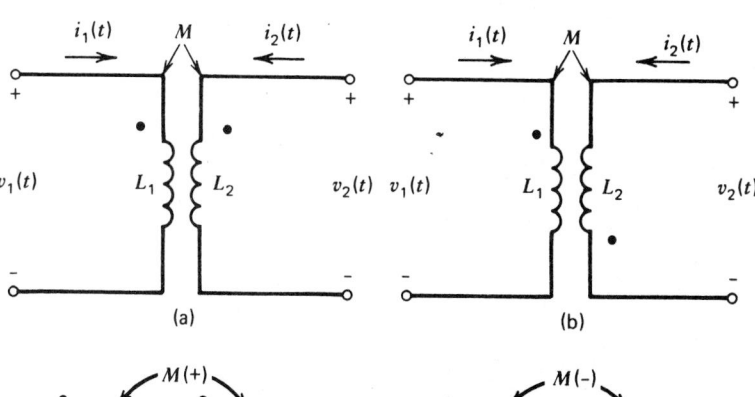

Fig. 11.70 Coupled coils showing the dot convention. Reproduced, with permission, from R. E. Thomas and A. J. Rosa, *Circuits and Signals: An Introduction to Linear and Interface Circuits*, New York, Wiley, 1984.

Up to now we have assumed that the two coils were wound with *positive* polarity. This may not be the case. The way a coil is wound—clockwise or counterclockwise—can make a difference to the polarity of the mutual voltages generated. In Fig. 11.70(*a*) it was assumed that both coils were wound in the same direction. To show that this was true we added two dots to the top of each coil. The result then is Eq. (11.243). Suppose, however, that one coil was wound differently than the other. This would result in a *negative* polarity. We would indicate this winding by the dot convention shown in Fig. 11.70(*b*). What the dots refer to is the positive polarity for the *mutually* induced voltage not the self-induced voltage. When the windings are as depicted in Fig. 11.70(*b*), Eq. (11.243) must be written as

$$v_1(t) = L_1 \frac{di_1(t)}{dt} - M \frac{di_2(t)}{dt}$$

and   (11.244)

$$v_2(t) = -M \frac{di_1(t)}{dt} + L_2 \frac{di_2(t)}{dt}$$

Let us try to realize why this is so. Recall that whenever a current flows in a conductor, a magnetic field is generated. This field is oriented in a particular direction in accordance with what is often referred to as the *right-hand rule*. Hence if we looked at a coil head on and the windings were clockwise, the north pole of the resultant field would point away from us. If the windings were counterclockwise, the north pole would point toward us. It seems reasonable to assume that the voltages generated by flux linkages created by currents running in opposite directions would have opposite polarities. The dots are a simple way to keep track of these polarities. If both the flux linkage generating current and the mutually induced current enter dotted terminal, the mutual inductance term is positive. If one of the currents exits a dotted terminal while the other enters one, the mutual inductance term is negative. If both currents exit dotted terminals, the mutual inductance term is once again positive. Figures 11.70(*c*) and 11.70(*d*) are another way to look at additive and subtractive mutual inductance. We have assumed that the mutual inductance M is the same regardless of which current is inducing the voltage. In some rare cases this will not be true, and two different M's will have to be identified. Under these conditions Eq. (11.243) will have to be written as

$$v_1(t) = L_1 \frac{di_1(t)}{dt} + M_{12} \frac{di_2(t)}{dt}$$

and   (11.245)

$$v_2(t) = M_{21} \frac{di_1(t)}{dt} + L_2 \frac{di_2(t)}{dt}$$

Consider three coils that are mutually coupled as shown in Fig. 11.71. Let's find a set of equations describing $v_1(t)$, $v_2(t)$, and $v_3(t)$ in terms of the currents and the inductances.

Following the example used to develop Eqs. (11.243) and (11.244) we can write the results as

$$v_1(t) = L_1 \frac{di_1(t)}{dt} + M_1 \frac{di_2(t)}{dt} - M_3 \frac{di_3(t)}{dt}$$

$$v_2(t) = M_1 \frac{di_1(t)}{dt} + L_2 \frac{di_2(t)}{dt} - M_2 \frac{di_3(t)}{dt} \qquad (11.246)$$

$$v_3(t) = M_3 \frac{di_1(t)}{dt} - M_2 \frac{di_2(t)}{dt} + L_3 \frac{di_3(t)}{dt}$$

The Ideal Transformer

A most important use for coupled coils is the **transformer**. Transformers are physically designed so that a small amount of current can generate a lot of flux linkages. Furthermore almost 100% of this flux linkage couples both coils. In our brief study of the transformer we discuss the ideal model. The transformer is a very efficient device. Over 99% of the power input to the transformer comes out! Transformers are ubiquitous; they find applications in all kinds of power and instrumentation systems. Their ubiquitousness derives from their ability to simplify interfacing circuits.

In our development of the ideal transformer we make two idealizations. The first assumes that all the flux linkages produced by the first coil link the second coil as well, that is, $k_1 = k_2 = k_m \approx 1$. This really is not a far-fetched idealization since in practical iron-core-type transformers $k_1 \approx k_2 \approx$

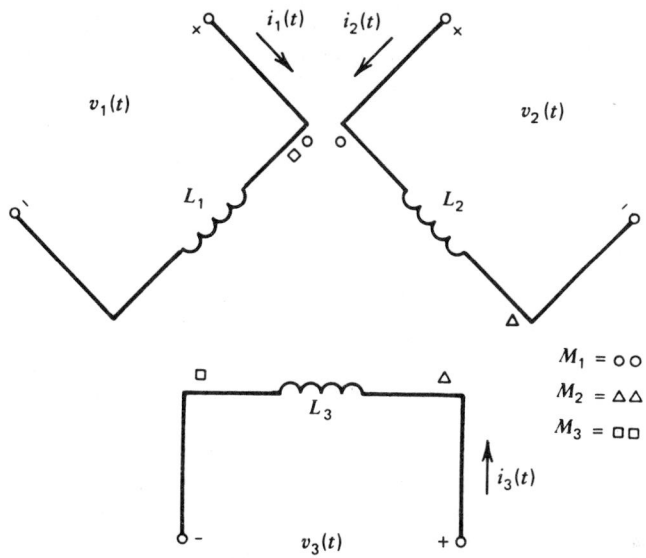

Fig. 11.71 Three mutually coupled coils. Reproduced, with permission, from R. E. Thomas and A. J. Rosa, *Circuits and Signals: An Introduction to Linear and Interface Circuits*, New York, Wiley, 1984.

$k_m \simeq 0.99$. This idealization allows us to rewrite Eq. (11.243) as

$$v_1(t) = N_1^2 \frac{di_1(t)}{dt} + N_1 N_2 \frac{di_2(t)}{dt}$$

$$v_2(t) = N_1 N_2 \frac{di_1(t)}{dt} + N_2^2 \frac{di_2(t)}{dt} \qquad (11.247)$$

Factoring N_1 from the first equation and N_2 from the second and finding the ratio of $v_1(t)$ to $v_2(t)$ yields

$$\frac{v_1(t)}{v_2(t)} = \frac{N_1}{N_2} \equiv n \qquad (11.248)$$

where n is called the **turns ratio**.

Thus our first idealization leads to the statement that the ratio of the voltages across the coils is directly proportional to the turns ratio.

The second idealization we must make is that negligible power is lost in the transformer. That essentially means that the instantaneous power input to the transformer equals the instantaneous power output or

$$p_1(t) = p_2(t) \quad \text{or} \quad v_1(t)i_1(t) = -v_2(t)i_2(t) \qquad (11.249)$$

The minus sign simply takes into account our referenced direction for the current $i_2(t)$. The transformer *supplies* power at its output, hence the negative sign on the right side of the equation. If we manipulate Eqs. (11.248) and (11.249), we can generate the remaining ideal transformer relation

$$\frac{v_1(t)}{v_2(t)} = -\frac{i_2(t)}{i_1(t)} = n \qquad (11.250)$$

In summary, Eqs. (11.248)–(11.250) contain the properties of the ideal transformer.

Some vocabulary used with transformers is as follows: The input is generally referred to as the *primary*, while the output is called the *secondary*. If the number of turns of the secondary is larger than the number of turns in the primary, that is, $N_2 > N_1$, then $n < 1$. This means that a voltage

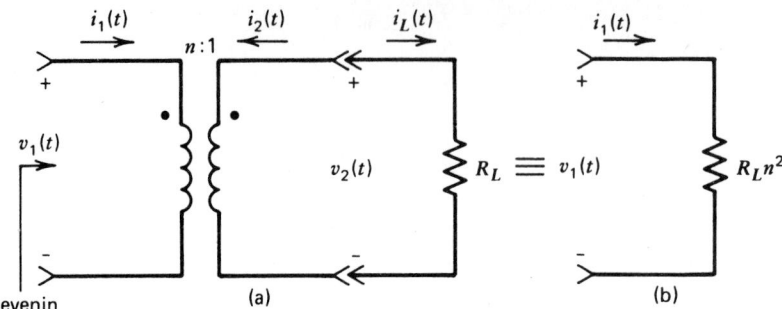

Fig. 11.72 Transformer used as an interfacing tool: (*a*) transformer circuit, (*b*) equivalent. Reproduced, with permission, from R. E. Thomas and A. J. Rosa, *Circuits and Signals: An Introduction to Linear and Interface Circuits*, New York, Wiley, 1984.

$v_2(t)$ will be greater than $v_1(t)$. Thus the voltage is stepped up. Such a transformer is often referred to as a **step-up transformer**. If the opposite is true, then $n > 1$ and the transformer is called a **step-down transformer**.

A word of caution is in order in working with circuits containing transformers. Our two idealizations mean that *no* current is needed to produce flux linkages and that all the inductances, both self and mutual are infinite. Do not try to write coupled-coil equations such as Eqs. (11.243) and (11.244) if the transformer is to be treated as ideal.

The real advantage of transformers is in their ability to ease the interfacing of circuits. Consider the circuit of Fig. 11.72.

The transformer will alter the way in which the load R_L appears at the input. What we wish to find then is the Thevenin circuit presented at the input. To do this we note that

$$v_2(t) = i_L(t) R_L \tag{11.251}$$

but

$$i_2(t) = -i_L(t) \tag{11.252}$$

so that

$$v_2(t) = -i_2(t) R_L \tag{11.253}$$

Now using the ideal transformer equation, we let

$$v_2(t) = \frac{v_1(t)}{n} \tag{11.254}$$

and

$$i_2(t) = -i_1(t) n \tag{11.255}$$

Substituting, we find that

$$\frac{v_1(t)}{n} = -[-i_1(t)n] R_L$$

or
$$\tag{11.256}$$

$$\frac{v_1(t)}{i_1(t)} = n^2 R_L$$

Hence the load R_L will appear to be n^2 times what it actually is. For example, if R_L is 1 kΩ and $n = 10$, R_L will seem to be 100 kΩ at the input. Transformers are used frequently for matching a source to a load for maximum power transfer.

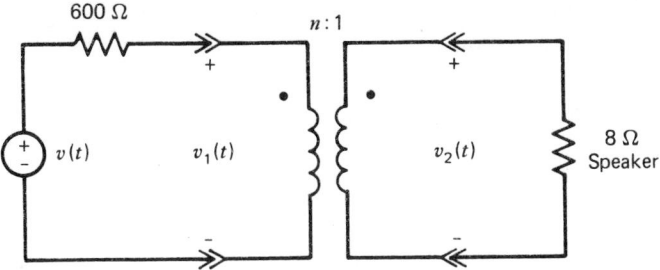

Fig. 11.73 "Stereo amplifier" output transformer circuit for Example 11.12. Reproduced, with permission, from R. E. Thomas and A. J. Rosa, *Circuits and Signals: An Introduction to Linear and Interface Circuits*, New York, Wiley, 1984.

Example 11.12. The output stage of a stereo amplifier has a resistance of 600 Ω (Fig. 11.73). The speaker (the load) has a resistance of 8 Ω. Match the two for maximum power transfer.

We recall that for maximum power transfer R_L must equal R_s. We could use a resistive interface circuit, but while maximum power would be transferred from the source, much of what is transferred would be lost in the resistive network, not in the load. If we use a transformer, essentially all the transferred power will be delivered to R_L, provided we use a transformer with the right turns ratio.

What we wish to do is make our load appear to the source as 600 Ω. We saw from Eq. (11.256) that a load will appear to be n^2 times larger than it actually is at the primary. Therefore we must select a transformer with a turns ratio equal to the following:

$$600 = n^2 8 \qquad (11.257)$$

$$n = \sqrt{\frac{600}{8}} = 8.66 \qquad (11.258)$$

11.5.3 Classical Analysis of AC Circuits

Circuits With a Single Energy Storage Device

The equilibrium in any circuit is always the result of constraints of two types: (1) connection constraints (KVL and KCL) and (2) device constraints (element equations). The major difference with energy storage device circuits is that the equilibrium is dynamic and the formulation process leads to differential equations rather than the algebraic equations that characterize resistive circuits. Nonetheless the formulation methods studied in the context of resistive circuits are quite general, and can easily be extended to the study of circuits with inductors and capacitors.

The simplest of these circuits contain a single capacitor or a single inductor as illustrated in Fig. 11.74. The resistors and sources can always be replaced by a Thevenin or Norton equivalent, which leads to the simple series and parallel circuits shown. To formulate the differential equation of the resistor-capacitor (RC) circuit, Fig. 11.74(a), we note that the Thevenin equivalent source circuit is governed by the constraint

$$v + R_T i = v_T \qquad (11.259)$$

while the capacitor is characterized by the i-v constraint

$$i = C\frac{dv}{dt} \qquad (11.260)$$

Substituting the element constraint into the source constraint yields

$$R_T C\frac{dv}{dt} + v = v_T \qquad (11.261)$$

Mathematically this result is called a *first-order, linear differential equation*. The dependent variable (v) is the voltage across the capacitor. Following similar logic we can obtain the dual relationship for

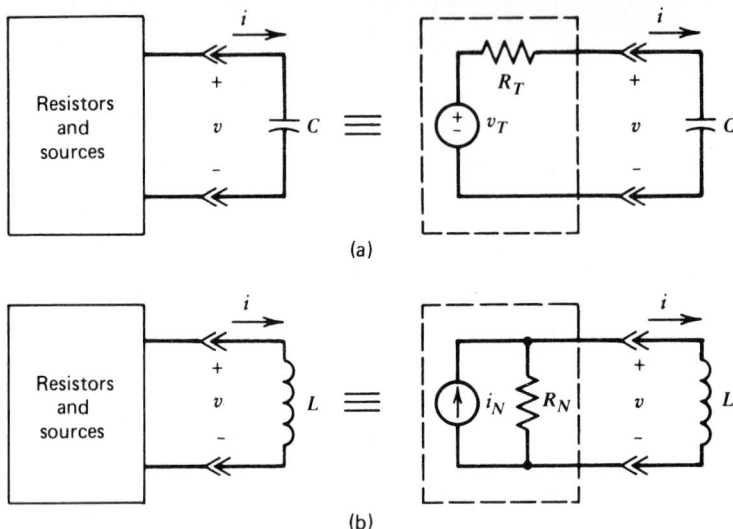

Fig. 11.74 General single energy storage device circuits: (a) resistor-capacitor circuits, (b) resistor-inductor circuits. Reproduced, with permission, from R. E. Thomas and A. J. Rosa, *Circuits and Signals: An Introduction to Linear and Interface Circuits*, New York, Wiley, 1984.

the *RL* circuit of Fig. 11.74(b). That is,

$$G_N L \frac{di}{dt} + i = i_N \tag{11.262}$$

which is likewise a linear, first-order differential equation with current as the dependent variable. In the present context this means we do not need to study both equations in detail. To have studied one is to have studied the other. Therefore, in what follows, we concentrate on the *RC* circuit.

In dealing with *RC* circuits we shall find that the *response* (solution of the differential equation) depends on three kinds of things:

1. The form of the input signal $v_T(t)$
2. The values of R_T and C
3. The initial condition

This last dependence is something new. The response of dc circuits depends only on the input signals and the circuit parameters. But here we have those dependencies and a third one, which is related to what has happened in the past, that is, memory due to energy storage. Put differently, the circuit can have a nonzero response even if there is no input signal.

To highlight this difference we first find what is called the *zero-input* response of the circuit. Setting $v_T = 0$ in Eq. (11.261) produces

$$R_T C \frac{dv}{dt} + v = 0 \tag{11.263}$$

Mathematically this is called a *homogeneous equation*. The equation requires that a linear combination of the voltage and its derivative be zero. This observation suggests that we try a solution in the form of an exponential,

$$v(t) = Ke^{st} \tag{11.264}$$

where K and s are constants to be determined. Substituting this trial solution into Eq. (11.263) results in

$$Ke^{st}(R_T Cs + 1) = 0 \tag{11.265}$$

which yields

$$R_T Cs + 1 = 0 \qquad (11.266)$$

The result is called the circuit's *characteristic equation* and has the root $s = -1/R_T C$. Thus the response is of the form

$$v(t) = K \exp\left(-\frac{t}{R_T C}\right) \qquad (11.267)$$

The remaining constant K can be evaluated by knowing the value of $v(t)$ at one particular time. In circuit analysis this time is usually taken to be $t = 0$, and the value of $v(0)$ is denoted as V_0. Thus

$$v(0) = Ke^{-0} = V_0 \qquad (11.268)$$

Therefore $K = V_0$ and we obtain the zero-input response as

$$v(t) = V_0 \exp\left(-\frac{t}{R_T C}\right) \qquad (11.269)$$

The zero-input response of the RC circuit is an exponential. The time constant of the waveform is $T_c = R_T C$ and the initial amplitude is V_0. In sum, the zero-input response of the capacitor circuit is determined by two quantities: (1) the circuit time constant and (2) the initial value of the voltage across the capacitor. In the case of a RL circuit, the zero-input response is determined by (1) the circuit time constant L/R and (2) the initial value of the current through the inductor.

Step Response of Circuits With a Single Energy Storage Device

The concept of step response is one of the fundamental notions in circuit and system theory. By step response we mean the response of a circuit or system to a step function input. The step function is one of the two premier test inputs used to study the dynamics of linear systems. The other standard input is the sinusoid, which we study later.

In developing the step response of a circuit with a single energy storage device, we treat the general RC circuit in detail. Its results are directly applicable to its dual, the RL circuit. If the input to the general RC circuit in Fig. 11.74 is a step function $v_T(t) = Au(t)$, then the circuit differential equation [Eq. (11.261)] becomes

$$R_T C \frac{dv}{dt} + v = A \qquad \text{for } t > 0 \qquad (11.270)$$

There are a number of ways to solve this equation, including separation of variables and integrating factors. However, because the circuit is linear, we chose an approach that divides the solution into two parts as

$$v(t) = v_N(t) + v_F(t) \qquad (11.271)$$

In this expression $v_N(t)$ is called the *natural response* and is the general solution of the homogeneous equation (input set to zero). The component $v_F(t)$ is called the *forced response* and is a particular solution of Eq. (11.270). To find the natural response, the input is set to zero. That is,

$$R_T C \frac{dv_N}{dt} + v_N = 0 \qquad (11.272)$$

But this is the same equation used to obtain the zero-input response [Eq. (11.263)], so we know that the natural response takes the form

$$v_N(t) = K \exp\left(-\frac{t}{R_T C}\right) \qquad (11.273)$$

This is a general solution of the homogeneous equation since it contains an arbitrary constant K. But we cannot evaluate K at this point from the initial conditions as we did for the zero-input response. The initial condition applies to the *total* response (natural plus forced), and we still have to determine the forced response.

Turning now to the forced response, we seek a particular solution of the equation

$$R_T C \frac{dv_F}{dt} + v_F = A \qquad t > 0 \tag{11.274}$$

It seems clear by inspection that a particular solution is $v_F = A$, since

$$\frac{dv_F}{dt} = \frac{dA}{dt} = 0 \tag{11.275}$$

so that

$$R_T C(0) + A = A \tag{11.276}$$

and Eq. (11.274) reduces to the identity $A = A$.

Combining the forced and natural responses we obtain

$$v(t) = v_N(t) + v_F(t)$$
$$= K \exp\left(-\frac{t}{R_T C}\right) + A \tag{11.277}$$

This result is a general solution for the step response since it satisfies Eq. (11.270) and contains an arbitrary constant K. This constant now can be evaluated if we make use of the initial condition as

$$v(0) = V_0 + Ke^{-0} + A = K + A \tag{11.278}$$

which means that $K = (V_0 - A)$. If we substitute this into our total solution, the step response can be expressed as

$$v(t) = (V_0 - A)\exp\left(-\frac{t}{R_T C}\right) + A \tag{11.279}$$

A general plot of this response waveform is shown in Fig. 11.75.

It is interesting to contrast the difference between the zero-input response and the step response. If left to its own devices (zero input), the capacitor voltage simply exponentially decays to zero with a time constant determined by the circuit. When the circuit is driven by a step function, the response departs from the initial condition and asymptotically approaches a constant value determined by the amplitude of the step input. That is, if we evaluate the response $v(t)$ at $t = 0$, the time we start our event, and $t = \infty$, a time long after our circuit has quieted down, we obtain

$$v(0) = (V_0 - A)e^{-0} + A = V_0 - A + A = V_0 \tag{11.280}$$

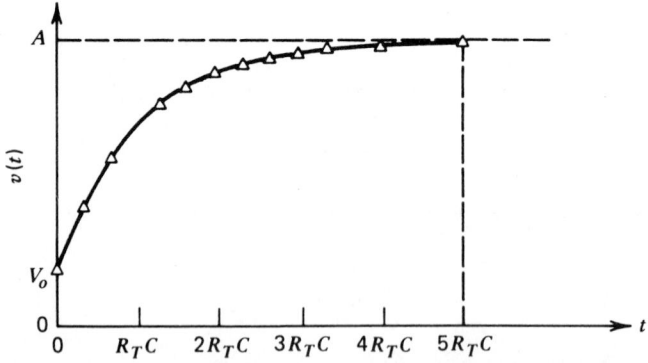

Fig. 11.75 Step response of the general RC circuit. Reproduced, with permission, from R. E. Thomas and A. J. Rosa, *Circuits and Signals: An Introduction to Linear and Interface Circuits*, New York, Wiley, 1984.

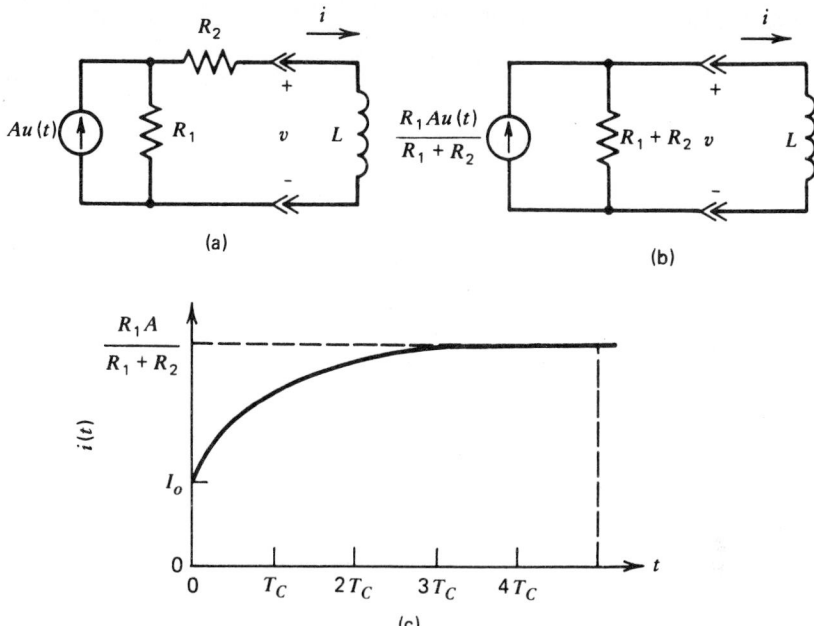

Fig. 11.76 Circuits and pulse response for Example 11.13: (*a*) given circuit, (*b*) Norton equivalent, (*c*) step response. Reproduced, with permission, from R. E. Thomas and A. J. Rosa, *Circuits and Signals: An Introduction to Linear and Interface Circuits*, New York, Wiley, 1984.

and

$$v(\infty) = (V_0 - A)e^{-\infty} + A = (V_0 - A)0 + A = A \qquad (11.281)$$

The path between these two values is an exponential rise waveform whose time constant is none other than the circuit's time constant. In summary then, the step response of an RC or RL circuit depends on three things:

1. The circuit time constant (RC or L/R)
2. The initial ($t = 0$) capacitor voltage or inductor current
3. The amplitude of the step function applied at $t = 0$

Example 11.13. The circuit in Fig. 11.76(*a*) is a single energy storage device RL network driven by a step function current input. Determine the step response.

To solve this problem we must first find the Norton equivalent of the source-resistor circuit. The short-circuit current at the interface is by current division,

$$i_{sc}(t) = \frac{R_1}{R_1 + R_2} Au(t) \qquad (11.282)$$

Looking back into the source-resistor circuit with the current source off (replaced by an open circuit), we see a resistance that is simply R_1 and R_2 in series. Hence

$$R_N = R_1 + R_2 \qquad (11.283)$$

The Norton equivalent just derived is then shown in Fig. 11.76(*b*). The time constant of the circuit is

$$T_c = G_N L = \frac{L}{R_1 + R_2} \qquad (11.284)$$

The differential equation describing the step response is

$$\frac{L}{R_1 + R_2} \frac{di}{dt} + i = \frac{AR_1}{R_1 + R_2} \qquad t > 0 \tag{11.285}$$

The forced and natural responses are then

$$i_F = \frac{AR_1}{R_1 + R_2}$$

$$i_N = K \exp\left[-\frac{(R_1 + R_2)t}{L}\right] \tag{11.286}$$

Combining these responses yields

$$i(t) = K \exp\left[-\frac{(R_1 + R_2)t}{L}\right] + \frac{AR_1}{R_1 + R_2} \tag{11.287}$$

The constant K can be evaluated from the initial condition as

$$i(0) = I_0 = K + \frac{R_1 A}{R_1 + R_2} \tag{11.288}$$

which implies that

$$K = I_0 - \frac{AR_1}{R_1 + R_2} \tag{11.289}$$

Finally we can write down the step response in all its glory as

$$i(t) = I_0 - \frac{AR_1}{R_1 + R_2} \exp\left[-\frac{(R_1 + R_2)t}{L}\right] + \frac{AR_1}{R_1 + R_2} \tag{11.290}$$

The general form of this response waveform is shown in Fig. 11.76(b). Notice that the final value of the inductor current is determined by the amplitude of the Norton equivalent step function and not the amplitude of the current step function in the original circuit shown in Fig. 11.76(a).

Initial and Final Conditions

If we review the step responses of the last section carefully, we find that the capacitor voltage of an RC circuit can be written as

$$v_c(t) = (\text{IC} - \text{FC})\exp\left(-\frac{t}{T_c}\right) + \text{FC} \tag{11.291}$$

Similarly, the inductor current in an RL circuit has a step response, which is

$$i_L(t) = (\text{IC} - \text{FC})\exp\left(-\frac{t}{T_c}\right) + \text{FC} \tag{11.292}$$

In both of these expressions IC stands for the initial condition ($t = 0$) and FC for the final condition ($t = \infty$). Thus to determine the step response of any single energy storage device circuit, we need to find three quantities: IC, FC, and T_C. We know how to determine the time constant of a circuit. Clearly it would be useful if we had a simple way to determine IC and FC.

The key to accomplishing this goal is to observe that as time approaches infinity both responses approach a constant value (FC). Thus for large time $v_c = \text{FC}$ and hence

$$i_c = C\frac{dv_c}{dt} = 0 \tag{11.293}$$

In words, for constant inputs a capacitor eventually (after five time constants or so) acts as an open circuit ($i_c = 0$). By similar reasoning the inductor current approaches a constant value (FC) for large time, and hence

$$v_L = L\frac{di_L}{dt} = 0 \tag{11.294}$$

and we see that for constant inputs an inductor eventually (five time constants more or less) acts as a short circuit ($v_L = 0$).

These observations allow us to determine the final value of the step response directly from the circuit. For an *RC* circuit we replace the capacitor by an open circuit, and then analyze the resulting dc circuit to determine the final value of the capacitor voltage. For the *RL* circuit we replace the inductor by a short circuit and use dc analysis methods to find the final value of the inductor current.

In addition, it turns out that we can use this method to determine the initial condition in certain types of situations. One common case is a circuit containing a switch that remains in one state, say closed, for a period of time that is long compared with the circuit time constant. If the circuit input is constant, then the capacitor voltage (or inductor current) will approach a constant value, which is the final condition for the circuit with the switch closed. If the switch is now opened at a time that we choose to call $t = 0$, then a transient will ensue in which the capacitor voltage (or inductor current) is driven to a new final condition. But note, the final condition that existed for $t < 0$ is the initial condition at $t = 0$ when the transient begins. The reason for this is the continuity conditions for capacitors and inductors

$$v_c(0^+) = v_c(0^-) \quad \text{and} \quad i_L(0^+) = i_L(0^-) \tag{11.295}$$

Here $x(0^-)$ implies the value of the variable x, a whisker before the epoch-making event occurs. $x(0^+)$, on the other hand, is the value of the variable an instant after the event occurred. The event is said to occur in zero time (instantaneously).

The switching action thus cannot cause an instantaneous change in the value of the *capacitor voltage* or the *inductor current* at $t = 0$. What this means in the present context is that

$$FC(0^-) = IC(0^+) \tag{11.296}$$

In other words, the switch is an epoch-making-event maker. The final conditions of the state variables from the previous epoch are their initial conditions for the subsequent epoch.

The usual way to state such analysis problems is to say that a switch has been (say) closed for a "long time." In this context a "long time" means something more than five time constants. Since electric circuits rarely have time constants that exceed a few hundred milliseconds, we do not need great patience to wait the requisite "long time" before opening the switch.

Example 11.14. Consider the op-amp circuit in Fig. 11.77. The switch has been in position A for a long time. At $t = 0$ it is moved to position B. Determine the output voltage $v_0(t)$.

The differential equation of this circuit with the switch in position A is

$$R_2C\frac{dv_0}{dt} + v_0 = -\frac{R_2}{R_1}V_s \tag{11.297}$$

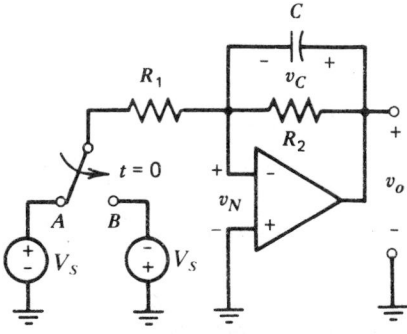

Fig. 11.77 Op-amp circuit for Example 11.14. Reproduced, with permission, from R. E. Thomas and A. J. Rosa, *Circuits and Signals: An Introduction to Linear and Interface Circuits*, New York, Wiley, 1984.

Therefore we know the circuit time constant is R_2C. The initial value circuit is a simple inverting amplifier, since C is replaced with an open circuit, we can write

$$v_0(0^-) = v_0(0^+) = \text{IC} = -\frac{R_2}{R_1}V_s \tag{11.298}$$

In the final value circuit the polarity of the input source is reversed, but the configuration is still an inverting amplifier, and hence

$$v_0(\infty) = \text{FC} = -\frac{R_2}{R_1}V_s = -\text{IC} \tag{11.299}$$

Thus the step response is

$$v_0(t) = (\text{IC} - \text{FC})\exp\left(-\frac{t}{T_c}\right) + \text{FC} \tag{11.300}$$

$$= \frac{R_2V_s}{R_1}\left[1 - 2\exp\left(-\frac{t}{R_2C}\right)\right] \quad t > 0 \tag{11.301}$$

Since v_0 is not a state variable, we might wonder if there could be a discontinuity at $t = 0$. But a KVL equation around the perimeter of the op-amp in the circuit shows that

$$-v_0 + v_c + v_N = 0 \tag{11.302}$$

But $v_N = 0$ since the op-amp input constraint requires $v_N = v_P$, and $v_P = 0$ because the noninverting input is grounded. Thus from the KVL equation $v_0 = v_c$. In words, the output voltage equals the capacitor voltage, which is the state variable, and there cannot be a jump at $t = 0$.

Sinusoidal Response of Circuits With a Single Energy Storage Device

The response of linear circuits to sinusoidal inputs is one of the central themes of electrical engineering. In this introduction to the concept we treat the general RC circuit by means of its differential equation. Later we shall see that sinusoidal response can be found by means of the Laplace transformation, and ultimately from the circuit itself using the concept of a phasor. But for now, to attain an initial physical insight we concentrate on the use of the classical method of finding solutions to differential equations.

If the input to the general RC circuit in Fig. 11.74 is a sinusoid, then the circuit differential equation [Eq. (11.261)] can be written as

$$R_TC\frac{dv}{dt} + v = u(t)A\cos\omega t \tag{11.303}$$

Now the input is a sinusoid that is initiated at $t = 0$ through some action such as the closing of a switch. We seek a solution function $v(t)$ that satisfies Eq. (11.303) for $t > 0$ and that meets a prescribed initial condition $v(0) = V_0$.

As with the step response, we find the solution in two parts: natural response and forced response. The natural response is of the form

$$v_N(t) = K\exp\left(-\frac{t}{R_TC}\right) \tag{11.304a}$$

It represents what the circuit naturally tends to do with zero input, and thus is called the *natural response*.

The forced response is a particular solution of Eq. (11.304). Since the input is a sinusoid, we are naturally led to try a solution in the form of a sinusoid:

$$v_F = a\cos\omega t + b\sin\omega t \tag{11.304b}$$

In this expression the Fourier coefficients a and b are unknown. The technique we are using is known mathematically as the *method of undetermined coefficients*, in this case, undetermined *Fourier coefficients*. To show that Eq. (11.305) is indeed a solution, we substitute into Eq. (11.303)

$$R_TC\frac{dv_F}{dt} + v_F = A\cos\omega t \quad t > 0 \tag{11.305}$$

or

$$R_T C(-\omega a \sin \omega t + \omega b \cos \omega t) + a \cos \omega t + b \sin \omega t = A \cos \omega t \tag{11.306}$$

Now gather all sine and cosine terms on one side of the equation:

$$(a + R_T C \omega b - A)\cos \omega t + (b - R_T C \omega a)\sin \omega t = 0 \tag{11.307}$$

This equation can be valid for all $t > 0$ only if the coefficients of the cosine and sine terms are identically zero. From this observation we obtain two linear equations in a and b:

$$\begin{aligned} a + R_T C \omega b &= A \\ - R_T C \omega a + b &= 0 \end{aligned} \tag{11.308}$$

which yield solutions

$$a = \frac{A}{1 + (\omega R_T C)^2} \qquad b = \frac{\omega R_T C A}{1 + (\omega R_T C)^2} \tag{11.309}$$

The undetermined Fourier coefficients are now determined. That is, a and b have been expressed in terms of known circuit parameters ($R_T C$) and known input signal parameters (ω and A).

We now combine the forced and natural responses as

$$v(t) = K \exp\left(-\frac{t}{R_T C}\right) + \frac{A}{1 + (\omega R_T C)^2}(\cos \omega t + \omega R_T C \sin \omega t) \tag{11.310}$$

The constant K can now be evaluated from the prescribed initial condition

$$v(0) = V_0 = K + \frac{A}{1 + (\omega R_T C)^2} \tag{11.311}$$

which yields

$$K = V_0 - \frac{A}{1 + (\omega R_T C)^2} \tag{11.312}$$

We can now write down the function $v(t)$ that satisfies the differential equation and the given initial condition:

$$v(t) = \underbrace{\left[V_0 - \frac{A}{1 + (\omega R_T C)^2}\right]\exp\left(-\frac{t}{R_T C}\right)}_{\text{Natural response}}$$

$$\underbrace{+ \frac{A}{1 + (\omega R_T C)^2}(\cos \omega t + \omega R_T C \sin \omega t)}_{\text{Forced response}} \tag{11.313}$$

This result is worthy of several comments. The natural response is a decaying exponential whose time constant is none other than the circuit time constant. This component of the response essentially vanishes after about five time constants. Thereafter there remains the forced response, which is a sinusoid with the same frequency (ω) as the input, but with a different amplitude and phase angle. The amplitude of the forced response is proportional to the amplitude of the input (A), reminding us that the circuit is linear and has the proportionality property. However, the amplitude and phase angle of the forced response are *not* linear functions of the *input frequency*. In other words, the circuit will respond to different input frequencies in quite different ways. This frequency-selective character-istic of linear circuits gives rise to a signal-processing method called **filtering**.

In the terminology of electrical engineering, the forced response is usually called the **sinusoidal steady-state response**.

Circuits with Two Energy Storage Devices

In this section we shall discuss circuits that contain at least two energy storage elements that cannot be combined to produce a single equivalent energy storage element. Although the number of such circuits is endless, we concentrate our attention on two classical, canonical forms, the series RLC circuit and the parallel RLC circuit.

The general series RLC circuit is shown in Fig. 11.78. The inductor and capacitor are connected in series, and the source-resistor circuit can be reduced to its Thevenin equivalent (R_T and v_T). For reasons that will become apparent, the most convenient solution variable is the capacitor voltage. To isolate v_C we can write

$$v_L + v_C + R_T C \frac{dv_C}{dt} = v_T \tag{11.314}$$

where $C\, dv_C/dt$ is the current through the Thevenin resistor R_T. To eliminate the inductor voltage we note that the same current flows through the inductor, hence,

$$v_L = LC \frac{d^2 v_C}{dt^2} \tag{11.315}$$

When this result is substituted into Eq. (11.314), there results

$$LC \frac{d^2 v_C}{dt^2} + R_T C \frac{dv_C}{dt} + v_C = v_T \tag{11.316}$$

This is a *second-order, linear differential equation* in which the dependent variable is the capacitor voltage. Once we have solved this equation for v_C, we can work backward through our analysis to determine every other voltage or current in the circuit if need be. In fact, we could have derived a differential equation in which the dependent variable is any one of the other signal variables. If we had done so, the left side of the equation would have the same form as Eq. (11.316), but the right side would involve derivatives of the input v_T. Since this could be a bit awkward at times, we choose the capacitor voltage as our solution variable.

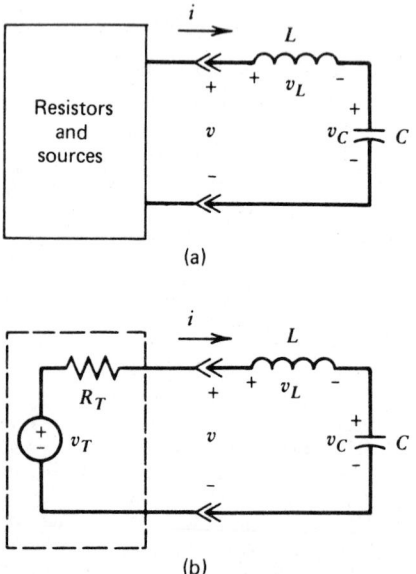

(a)

(b)

Fig. 11.78 The series RLC circuit: (*a*) given circuit, (*b*) equivalent circuit. Reproduced, with permission, from R. E. Thomas and A. J. Rosa, *Circuits and Signals: An Introduction to Linear and Interface Circuits*, New York, Wiley, 1984.

In dealing with Eq. (11.316) we shall find that the solution depends on three things:

1. The nature of the input signal (v_T).
2. The values of the circuit parameters (R_T, L, C).
3. The values of the capacitor voltage and the inductor current at $t = 0$, which will be denoted as V_0 and I_0, respectively.

The first two categories are not surprising since these things also influence the response of resistive circuits. The last category points out that the circuit can store energy and can remember a voltage or a current. That is, its response is influenced by what has happened in the past, as represented by the energy stored in the capacitor ($\frac{1}{2})CV_0^2$ and the inductor ($\frac{1}{2})LI_0^2$ at $t = 0$, the time at which we initiated our solution.

Among the consequences of energy storage is that the circuit can have a response even though the input signal is identically zero for $t > 0$. To highlight this characteristic we first deal with the zero-input response. That is, we make $v_T = 0$. Equation (11.316) then becomes

$$LC\frac{d^2 v_C}{dt^2} + R_T C\frac{dv_C}{dt} + v_C = 0 \qquad (11.317)$$

This is a homogeneous differential equation. In words, it requires that a linear combination of a dependent variable and its first and second derivative somehow sum to zero for all $t > 0$. This suggests, as it did for circuits with a single energy storage device, that we try a solution in the form

$$v_C(t) = Ke^{st} \qquad (11.318)$$

where s can be real, imaginary, or complex. When this trial solution is inserted in Eq. (11.317) there results

$$Ke^{st}(LCs^2 + R_T Cs + 1) = 0 \qquad (11.319)$$

The possibility that $K = 0$ is called the *trivial solution*. The quantity e^{st} cannot be zero for all $t > 0$. Hence the general condition is that

$$LCs^2 + R_T Cs + 1 = 0 \qquad (11.320)$$

This is the circuit's characteristic equation. It is a quadratic because the circuit contains two noncombinable energy storage elements.

Thus the presence of two energy storage elements leads to a quadratic characteristic equation that necessarily has two roots:

$$s_1, s_2 = \frac{-R_T C \pm \sqrt{(R_T C)^2 - 4LC}}{2LC}. \qquad (11.321)$$

Apparently then there are two possible solutions: $K_1\exp(s_1 t)$ and $K_2\exp(s_2 t)$. But we seek the general solution, and we also know that we must satisfy two initial conditions (V_0 and I_0). Consequently the general solution of the homogeneous equation must be

$$v_C(t) = K_1\exp(s_1 t) + K_2\exp(s_2 t) \qquad (11.322)$$

We can evaluate the constants K_1 and K_2 from the initial conditions. If we let $t = 0$, we can find the first equation:

$$v_C(0) = V_0 = K_1 + K_2 \qquad (11.323)$$

To obtain a second equation we must use the initial value of the inductor current. Looking back to Fig. 11.78 we see that

$$i_L(0) = C\frac{dv_C(t)}{dt}\bigg|_{t=0} = Cv'_C(0) = I_0 \qquad (11.324)$$

Hence,

$$v_C'(0) = \frac{I_0}{C} \tag{11.325}$$

Thus the initial inductor current specifies the value of the first derivative of the capacitor voltage at $t = 0$. To use this result we differentiate Eq. (11.322) and substitute the result found in Eq. (11.325) and we obtain

$$v_C'(0) = \frac{I_0}{C} = K_1 s_1 + K_2 s_2 \tag{11.326}$$

Equations (11.323) and (11.326) provide two equations in two unknowns:

$$K_1 + K_2 = V_0$$
$$s_1 K_1 + s_2 K_2 = \frac{I_0}{C} \tag{11.327}$$

whose solutions are

$$K_1 = \frac{s_2 V_0 - I_0/C}{s_2 - s_1} \quad \text{and} \quad K_2 = \frac{-s_1 V_0 + I_0/C}{s_2 - s_1} \tag{11.328}$$

Putting these results back into Eq. (11.322) yields the zero-input response as

$$v_C(t) = \frac{s_2 V_0 - I_0/C}{s_2 - s_1} \exp(s_1 t) + \frac{-s_1 V_0 + I_0/C}{s_2 - s_1} \exp(s_2 t) \tag{11.329}$$

This result is indeed the general form of the zero-input response of the series RLC circuit. The response depends on two initial conditions (V_0 and I_0), and on the circuit parameters (R_T, L, and C) since the roots s_1 and s_2 derive from the characteristic equation. However, there is more here than meets the eye. The nature of the roots can dramatically change the basic form of the response.

Looking back at Eq. (11.321) defining the roots

$$s_1, s_2 = \frac{-R_T C \pm \sqrt{(R_T C)^2 - 4LC}}{2LC} \tag{11.330}$$

we find that there are three distinct possibilities:

CASE A: If $(R_T C)^2 - 4LC > 0$, the radicand is positive and there will be two real, unequal roots ($s_1 \neq s_2$).

CASE B: If $(R_T C)^2 - 4LC = 0$, the radicand vanishes and there will be two real, equal roots ($s_1 = s_2$).

Case C: If $(R_T C)^2 - 4LC < 0$, the radicand will be negative, which leads to two complex conjugate roots ($s_1 = s_2^*$).

Let us consider each of these cases in detail. For case A the roots are *real* and *unequal*. Using the notation

$$s_1 = \alpha_1 \quad \text{and} \quad s_2 = \alpha_2 \tag{11.331}$$

we can write the general form of the response from Eq. (11.329) as

$$v_C(t) = \frac{(\alpha_2 V_0 - I_0/C)\exp(\alpha_1 t) + (-\alpha_1 V_0 + I_0/C)\exp(\alpha_2 t)}{\alpha_2 - \alpha_1} \tag{11.332}$$

There is nothing particularly surprising here. The response is called the *overdamped* case and simply consists of two exponential waveforms. So to speak, the response has two time constants. The two time constants can be greatly different, or nearly equal, but they cannot be exactly equal because we would then have case B.

For case B the roots are *real* and *equal*. Using the notation

$$s_1 = s_2 = \alpha \qquad (11.333)$$

the general form in Eq. (11.329) becomes

$$v_C(t) = \frac{(\alpha V_0 - I_0/C)\exp(\alpha t) + (-\alpha V_0 + I_0/C)\exp(\alpha t)}{\alpha - \alpha} \qquad (11.334)$$

We immediately see a problem with this result since the denominator vanishes. However, a careful examination of the numerator shows that it vanishes as well. Our solution reduces to the indeterminant form 0/0! To investigate this further we let

$$s_1 = \alpha \quad \text{and} \quad s_2 = \alpha + x \qquad (11.335)$$

and explore the situation as x approaches zero. Using this notation in Eq. (11.329) produces

$$v_C(t) = \exp(\alpha t)\left[V_0 + (\alpha V_0 - I_0/C)\frac{1 - \exp(xt)}{x} \right] \qquad (11.336)$$

In this form we see that our problem comes from the term $(1 - e^{\alpha t})/x$, which is indeed indeterminant as x approaches zero. But the application of l'Hôpital's rule reveals

$$v_C(t) = e^{\alpha t}\left[V_0 - (\alpha V_0 - I_0/C)t \right] \qquad (11.337)$$

This special form of the response is called the *critically damped* case. A special form is required because the two roots are identical.

In case C we have *complex conjugate* roots, which we denote as

$$s_1 = \alpha - j\beta \quad \text{and} \quad s_2 = \alpha + j\beta \qquad (11.338)$$

Inserting these roots directly into Eq. (11.329) yields

$$v_C(t) = e^{\alpha t}\left[V_0\frac{e^{j\beta t} + e^{-j\beta t}}{2} - \frac{(\alpha V_0 - I_0/C)}{\beta}\frac{e^{j\beta t} - e^{-j\beta t}}{j2} \right] \qquad (11.339)$$

or in sinusoidal form as

$$v_C(t) = e^{\alpha t}\left[V_0\cos\beta t - \frac{(\alpha V_0 - I_0/C)}{\beta}\sin\beta t \right] \qquad (11.340)$$

Thus, even though the roots are complex, we have ended up with a real function of time. The real part of the roots (α) ends up as the damping term in the exponential, while the imaginary part (β) ends up as the frequency of oscillation of the resulting damped sinusoidal waveform. This case is referred to as the *underdamped* response.

Example 11.15. The circuit in Fig. 11.79(a) is a series *RLC* circuit in which the switch is closed at $t = 0$. The initial capacitor voltage is 15 V and the inductor current is initially zero. The values of the capacitance and inductance are fixed. Find the roots of the characteristic equation for three different values of resistance: 8.5, 4, and 1 kΩ.

From Eq. (11.317) we know that the trial solution $v_C = Ke^{st}$ yields the following characteristic equation

$$LCs^2 + RCs + 1 = 0 \qquad (11.341)$$

What we need to do is to find the roots of this equation for the three values of resistance given in the foregoing. For the first value (8.5 kΩ) the characteristic equation is

$$s^2 + 8500s + 4 \times 10^6 = (s + 500)(s + 8000) = 0 \qquad (11.342)$$

The solution contains *two real, unequal* roots ($s_1 = -500$ and $s_2 = -8000$), which is an example of case A. For the second value of R, 4 kΩ, the equation becomes

$$s^2 + 4000s + 4 \times 10^6 = (s + 2000)^2 = 0 \qquad (11.343)$$

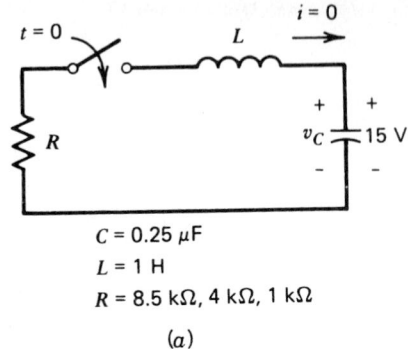

$$C = 0.25 \ \mu F$$
$$L = 1 \ H$$
$$R = 8.5 \ k\Omega, 4 \ k\Omega, 1 \ k\Omega$$

(a)

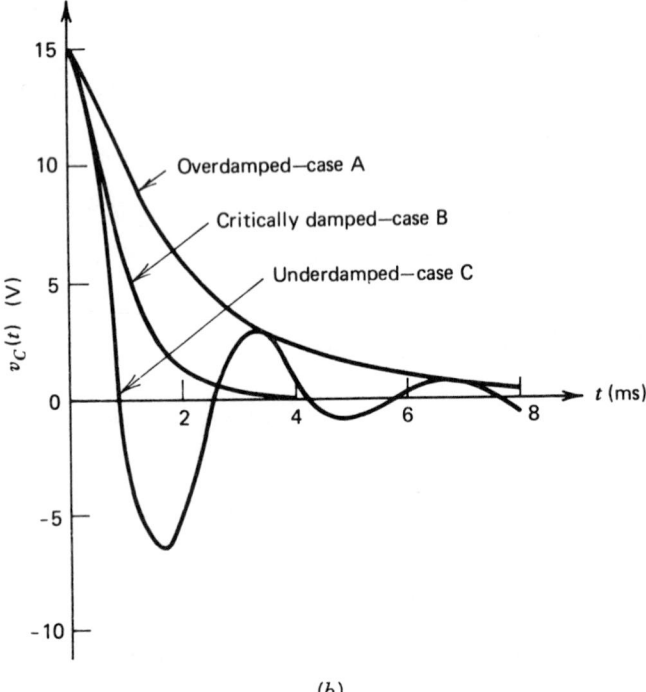

(b)

Fig. 11.79 Possible responses for the series *RLC* circuit (Example 11.15). Reproduced, with permission, from R. E. Thomas and A. J. Rosa, *Circuits and Signals: An Introduction to Linear and Interface Circuits*, New York, Wiley, 1984.

For this value of resistance there are *two real, equal* roots ($s_1 = s_2 = -2000$), which illustrates case B. For the final value of 1 kΩ, the characteristic equation is

$$s^2 + 1000s + 4 \times 10^6 = 0 \tag{11.344}$$

whose roots are complex and conjugate ($s_1, s_2 = -500 \pm 500\sqrt{-15}$). The quantity under the radical is negative, illustrating the case C situation.

The initial conditions were given as $V_0 = 15$ V and $I_0 = 0$. We are now in a position to apply the results derived for each of the types of responses.

CASE A: For $R = 8.5$ kΩ, we found

$$v_C(t) = 16e^{-500t} - e^{-8000t} \tag{11.345}$$

This is the overdamped response and consists of two exponential waveforms.

CASE B: For $R = 4$ kΩ, we found

$$v_C(t) = 15(1 + 2000t)e^{-2000t} \tag{11.346}$$

This special form of the response occurs because the roots are equal and is called the critically damped case.

CASE C: For $R = 1$ kΩ, we found

$$v_C(t) = e^{-500t}(15 \cos 500\sqrt{15}\,t + 7500 \sin 500\sqrt{15}\,t) \tag{11.347}$$

This is a damped sinusoidal response characteristic of the underdamped case.

All three of these responses are shown in Fig. 11.79(b). All start out at 15 V (the initial condition) and eventually decay to zero. However, the nature of the response waveforms is dramatically different. The overdamped response (case A) is relatively sluggish and unexciting. The critically damped response (case B) decays rapidly to zero but does not undershoot its final value. The underdamped response passes through zero rapidly but undershoots the mark, and eventually decays to zero in a sequence of damped oscillations.

All of the results derived for the series RLC circuit apply to the parallel RLC case with the appropriate duality replacements. In particular, the concept of overdamped, critically damped, and underdamped response applies to both cases.

Step Response of Circuits Containing Two Energy Storage Devices

The step response is an important characterization of linear circuits, and so we are naturally led to investigate this matter for circuits with two energy storage devices. In this introductory development we use the series RLC circuit to demonstrate the results employing classical differential equation methods. As we now know, similar results would be obtained for the parallel RLC circuit and, in fact, for any circuit with two energy storage devices.

If the input to a series RLC circuit is a step function $v_T(t) = Au(t)$, then the differential equation is given by Eq. (11.316). We seek a solution of this equation for $t > 0$, subject to the initial conditions V_0 and I_0, which are the initial capacitor voltage and inductor current, respectively. As we found with circuits with only one energy storage device, this can be done by dividing the solution into forced and natural components. The natural response is the general solution of the homogeneous equation (input set to zero), while the forced response is a particular solution of

$$LC\frac{d^2v_{CF}}{dt^2} + R_TC\frac{dv_{CF}}{dt} + f_{CF} = A \qquad t > 0 \tag{11.348}$$

It is not hard to see that $v_{CF} = A$ is a particular solution since dA/dt and d^2A/dt^2 are both zero.

Turning now to the natural response, we seek a general solution of the homogeneous equation. But we know that the natural response will take one of the three possible forms of the zero-input response: overdamped, critically damped, or underdamped. To highlight these possibilities we introduce two new parameters, ω_0 and ζ, and define them as

$$\frac{1}{\omega_0^2} \equiv LC \quad \text{and} \quad \frac{2\zeta}{\omega_0} \equiv R_TC \tag{11.349}$$

The parameter ω_0 is called the *undamped natural frequency*, or simply the *natural frequency*, and ζ is called the *damping ratio*. In terms of these two parameters the homogeneous equation is

$$\frac{1}{\omega_0^2}\frac{d^2v_C}{dt^2} + \frac{2\zeta}{\omega_0}\frac{dv_C}{dt} + v_C = 0 \tag{11.350}$$

The trial solution of the form $v_C = Ke^{st}$ then leads to the characteristic equation

$$s^2 + 2\zeta\omega_0 s + \omega_0^2 = 0 \tag{11.351}$$

whose roots are

$$s_1, s_2 = -\zeta\omega_0 \pm \omega_0\sqrt{\zeta^2 - 1} \tag{11.352}$$

We can now begin to see the virtue of these two parameters. The radicand defining the roots depends only on ζ, the damping ratio. We can write down the three possible types of roots in terms of that parameter.

CASE A: For $\zeta > 1$, the radicand is positive and there are two unequal, real roots

$$\alpha_1 = -\zeta\omega_0 + \omega_0\sqrt{\zeta^2 - 1}$$

and   (11.353)

$$\alpha_2 = -\zeta\omega_0 - \omega_0\sqrt{\zeta^2 - 1}$$

and the natural response will be of the form

$$v_{CN}(t) = K_1\exp(\alpha_1 t) + K_2\exp(\alpha_2 t) \tag{11.354}$$

CASE B: For $\zeta = 1$, the radicand vanishes and there are two real, equal roots

$$s_1 = s_2 = -\zeta\omega_0 \tag{11.355}$$

and the natural response has the form

$$v_{CN}(t) = (K_1 + K_2 t)\exp(-\zeta\omega_0 t) \tag{11.356}$$

CASE C: For $\zeta < 1$, the radicand is negative leading to two complex, conjugate roots $s_1, s_2 = \alpha \pm j\beta$, where

$$\alpha = -\zeta\omega_0 \quad \text{and} \quad \beta = \omega_0\sqrt{1 - \zeta^2} \tag{11.357}$$

and the natural response has the form

$$v_{CN}(t) = e^{\alpha t}(K_1\cos\beta t + K_2\sin\beta t) \tag{11.358}$$

In other words, for $\zeta > 1$ we have the overdamped response, for $\zeta = 1$ we have the critically damped response, and for $\zeta < 1$ we have the underdamped response.

We can now at least write down the general form of the step response of the series RLC circuit:

$$v_C(t) = A + v_{CN}(t) \tag{11.359}$$

where $v_{CN}(t)$ takes one of the three forms listed. The appropriate form is determined by the value of the damping ratio ζ. The arbitrary constants in each form can be evaluated from the initial conditions applied to Eq. (11.359). In other words, it takes three kinds of parameters to determine the step response:

1. The amplitude of the input step function (A).
2. The values of the initial conditions (V_0 and I_0).
3. The natural frequency and damping ratio (ζ and ω_0).

In this regard the natural frequency and damping ratio play the same role for circuits with two energy storage devices that the time constant plays for circuits with only one. That is, they are parameters that completely describe the form of the natural response, just as the circuit time constant completely defines the form of the natural response for a circuit with one energy storage device. That it should take two parameters to do this should not be surprising since the circuit has two energy storage elements.

The circuit in Fig. 11.80(a) is a series RLC circuit with zero initial conditions and a step function input. The differential equation for the circuit is

$$LC\frac{d^2v_C}{dt^2} + RC\frac{dv_C}{dt} + v_C = A \qquad t > 0 \tag{11.360}$$

Upon inserting the numerical values given in the figure,

$$10^{-6}\frac{d^2v_C}{dt^2} + \frac{10^{-3}}{2}\frac{dv_C}{dt} + v_C = 10 \tag{11.361}$$

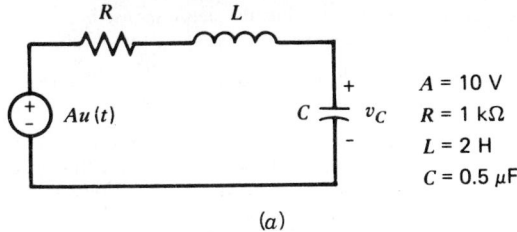

$A = 10$ V
$R = 1$ kΩ
$L = 2$ H
$C = 0.5$ μF

(a)

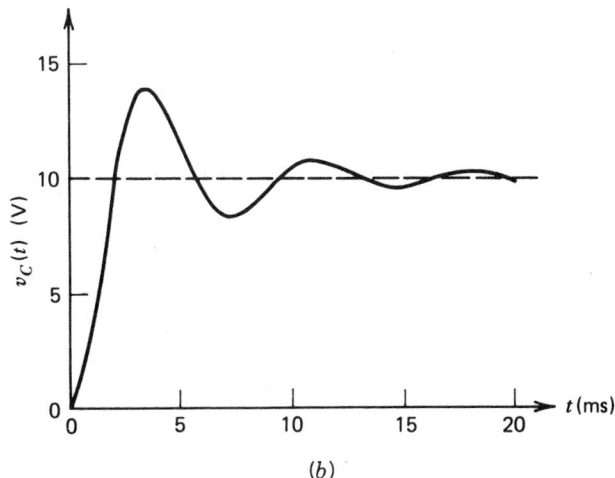

(b)

Fig. 11.80 (a) Two energy storage device circuits RLC series circuit. (b) Step response. Reproduced, with permission, from R. E. Thomas and A. J. Rosa, *Circuits and Signals: An Introduction to Linear and Interface Circuits*, New York, Wiley, 1984.

from which it is apparent that $v_{CF} = 10$ V. Comparing this equation to the standard form yields

$$\frac{1}{\omega_0^2} = 10^{-6} \quad \text{and} \quad \frac{2\zeta}{\omega_0} = \frac{10^{-3}}{2} \tag{11.362}$$

Hence $\omega_0 = 10^3$ and $\zeta = 0.25$.

Since the damping ratio is less than unity, we now know that the response is underdamped. Thus

$$\alpha = -\zeta\omega_0 = -250$$

and
$$\tag{11.363}$$

$$\beta = \omega_0\sqrt{1 - \zeta^2} = 250\sqrt{15}$$

and the step response has the form

$$v_C(t) = 10 + e^{-250t}\left(K_1\cos 250\sqrt{15}\,t + K_2\sin 250\sqrt{15}\,t \right) \tag{11.364}$$

which is the underdamped form. The constants K_1 and K_2 can be determined now from the initial conditions (both zero):

$$v_C(0) = 10 + K_1 = 0 \tag{11.365}$$

$$v_C'(0) = -250K_1 + 250\sqrt{15}\,K_2 = 0$$

which yields $K_1 = -10$ and $K_2 = -10/\sqrt{3}$. Thus the step response is

$$v_C(t) = 10 - e^{-250t}\left(10\cos 250\sqrt{15}\,t + \frac{10}{\sqrt{3}}\sin 250\sqrt{15}\right) \qquad (11.366)$$

This response is plotted in Fig. 11.80(b). The waveform is zero at $t = 0$ as required by the initial condition, and eventually settles down on the final value of 10 V. The final value is the amplitude of the forced response since the natural response decays to zero. However, between zero and the steady-state value the response initially overshoots the mark, then undershoots, and gradually decays to the steady response. This damped sinusoidal behavior is, of course, the result of the underdamped natural response.

11.5.4   Transform Analysis of AC Circuits

Signal Waveforms and Transforms
A transformation is the process of changing the form of data in accordance with a specified rule. In the present context, transformations are used to obtain alternative representations of a signal or a system. They are used because they often simplify certain signal-processing operations or because they provide a different perspective that can be quite useful or even essential. Modern signal processing uses a host of transformation methods, including Fourier transforms, fast Fourier transforms (FFT), discrete Fourier transforms (DFT), Z transforms, and Laplace transforms. In every case, these methods involve a specific transformation rule, make certain analysis techniques more manageable, and provide a useful viewpoint for system design.

In this context, we concern ourselves only with the Laplace transformation. Our initial exposure to Laplace ($\mathscr{L}$) transforms will follow the pattern shown in Fig. 11.81. The process begins with a linear circuit and the differential equation that characterize its behavior. This equation is transformed into the complex frequency domain, where it is changed into an algebraic equation. Simple algebraic techniques are then used to obtain the solution in the frequency domain. The inverse Laplace ($\mathscr{L}^{-1}$) transform changes the s-domain solution back into the time domain, yielding the solution of the original differential equation. The figure also points out that there is another route to the solution using classical techniques such as we found in Section 11.5.3. For simple circuits, this route may be easier and more direct. But for more complex circuits, the advantages of the Laplace transformation can be quite significant.

An advantage of the transform approach is that solving a differential equation becomes a purely algebraic process. However, Laplace transforms are more than just another way to solve differential equations. They offer a new and different perspective toward circuit behavior. One must learn to think about signals and systems in both the familiar time domain and the arena of Laplace transforms called the *complex frequency domain*.

Symbolically we represent the Laplace transformation operation as

$$\mathscr{L}[v(t)] = V(s) \qquad (11.367)$$

In words, this expression says that $V(s)$ is the Laplace transform of the waveform $v(t)$. The transformation operation involves two domains: (1) the time domain in which the signal is characterized by its waveform $v(t)$ and (2) the complex frequency domain in which the signal is represented by its transform $V(s)$. The variable s is called the complex frequency variable and, being complex, is equal to $\sigma + j\omega$, where σ is the attenuation constant and ω is the radian frequency. In sum, a signal can uniquely be expressed as a waveform or a transform. Collectively $v(t)$ and $V(s)$ are called a *transform pair*, meaning alternative representations of the same thing, the signal. Note that a lowercase letter is used to represent the signal waveform and an uppercase letter for its transform.

The Laplace transform of a signal waveform $v(t)$ is defined as

$$V(s) = \int_{0-}^{\infty} v(t)e^{-st}\,dt \qquad (11.368)$$

This definition involves an improper integral (the upper limit is infinite), and so we must discuss the conditions under which the integral converges. Briefly, the integral will exist if the waveform $v(t)$ is piecewise continuous and of exponential order. Piecewise continuous means that in any finite interval, $v(t)$ has a finite number of steplike discontinuities. Exponential order means that there exists constants K and b such that $|v(t)| < Ke^{bt}$ for all t greater than some value T. Fortunately essentially all signals encountered in real systems meet these two conditions.

Equation (11.368) used a lower limit denoted $t = 0-$ to indicate a time just a whisker before $t = 0$. This is done because in dealing with system response we divide all of history into two eras as a result of an epoch-making event that occurs at $t = 0$, the movement of a switch, for example.

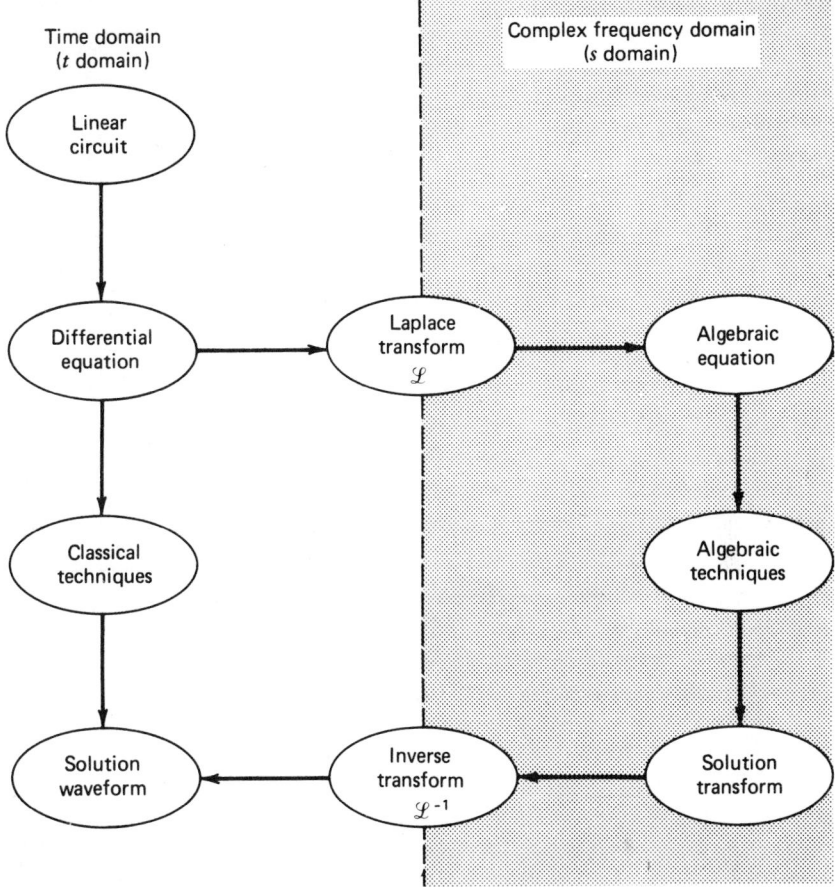

Fig. 11.81 Laplace transformation pattern. Reproduced, with permission, from R. E. Thomas and A. J. Rosa, *Circuits and Signals: An Introduction to Linear and Interface Circuits*, New York, Wiley, 1984.

Occasionally the epoch-making event leads to a discontinuity in $v(t)$ at $t = 0$, that is, $v(0-) \neq v(0)$. To capture this discontinuity in the integration process, we set the lower limit at $t = 0-$, just prior to the event. Fortunately, in most situations, there is no discontinuity, so we will not distinguish between $t = 0-$ and $t = 0$, unless it is crucial.

Finally, it is important to note at the outset that the Laplace transformation that we shall deal with inherently involves waveforms that are *identically zero for all negative t*. This is not obvious from the definition of the direct transformation in Eq. (11.368) since the integral does not involve the values of $v(t)$ for negative t. However, the inverse transformation, which we study in a later section, always produces a waveform that is zero for $t < 0$. Hence to preserve *uniqueness* of the transformation we must assume that all waveforms are identically zero for $t < 0$.

Equally fortunate is the fact that the number of different signal waveforms encountered in linear systems is relatively small. The list includes the basic waveforms, that is, step, exponential, sinusoid, as well as a number of composite waveforms such as the exponential rise, damped ramp, impulse, ramp, and damped sinusoid. Since the total number is relatively small, we do not need to appeal to the definition repeatedly in using transform methods. Once the transform of a waveform has been found, it can be cataloged in a table of transform pairs for future reference and use. Table 11.1 gives some important pairs of transforms. More extensive tables are available in other references such as in *Standard Mathematical Tables*, published yearly by the CRC Press.

The foremost feature of the Laplace transformation is that it is a linear operation. If

$$\mathscr{L}\{v_1(t)\} = V_1(s) \quad \text{and} \quad \mathscr{L}\{v_2(t)\} = V_2(s) \qquad (11.369)$$

TABLE 11.1   Basic Laplace Transform Pairs

Signal	Waveform	Transform
Impulse	$\delta(t)$	1
Step or constant	$u(t)$	$\dfrac{1}{s}$
Ramp	$r(t) = tu(t)$	$\dfrac{1}{s^2}$
Exponential	e^{-at}	$\dfrac{1}{s+a}$
Damped ramp	te^{-at}	$\dfrac{1}{(s+a)^2}$
Sinusoid	$\cos \beta t$	$\dfrac{s}{s^2 + \beta^2}$
Sinusoid	$\sin \beta t$	$\dfrac{\beta}{s^2 + \beta^2}$
Damped sinusoid	$e^{-at}\cos \beta t$	$\dfrac{s + \alpha}{(s + \alpha)^2 - \beta^2}$
Damped sinusoid	$e^{-at}\sin \beta t$	$\dfrac{\beta}{(s + \alpha)^2 - \beta^2}$

and if A and B are constants, then

$$\mathcal{L}\{Av_1(t) + Bv_2(t)\} = \int_0^\infty [Av_1(t) + Bv_2(t)]e^{-st}\,dt$$

$$= A\int_0^\infty v_1(t)e^{-st}\,dt + B\int_0^\infty v_2(t)e^{-st}\,dt \qquad (11.370)$$

Hence

$$\mathcal{L}\{Av_1(t) + Bv_2(t)\} = AV_1(s) + BV_2(s) \qquad (11.371)$$

The linearity property is a very important feature. Among its many implications is that Kirchhoff's laws are valid in both the time domain and the complex frequency domain.

Since memory element i-v relationships involve the time-domain operations of integration and differentiation, their frequency-domain equivalents are important. For integration, we write as follows:
If

$$\mathcal{L}\{v(t)\} = V(s) = \int_0^\infty v(t)e^{-st}\,dt \qquad (11.372)$$

then

$$\int_0^t v(t)\,dt = \int_0^\infty \left\{\int_0^t v(t)\,dt\right\}e^{-st}\,dt \qquad (11.373)$$

This expression can be integrated by parts using

$$y = \int_0^t v(t)\,dt \qquad dx = e^{-st}\,dt \qquad (11.374)$$

Hence

$$dy = v(t) \qquad x = \frac{-e^{-st}}{s} \qquad (11.375)$$

TABLE 11.2   Basic Laplace Transformation Properties

Property	Time Domain	Frequency Domain
Independent variable	Time t	Complex frequency $s = \sigma + j\omega$
Signal representation	Waveform $v(t)$	Transform $V(s)$
Linearity property	$Av_1(t) + Bv_2(t)$	$AV_1(s) + BV_2(s)$ A and B are constants
Integration property	$\int_0^t v(t)\, dt$	$\dfrac{V(s)}{s}$
Differentiation property	$\dfrac{dv(t)}{dt}$	$sV(s) - v(0-)$
	$\dfrac{d^2v}{dt^2}$	$s^2V(s) - sv(0-) - v'(0-)$
	$\dfrac{d^3v}{dt^3}$	$s^3V(s) - s^2v(0-) - sv'(0-) - v''(0-)$

Using these factors reduces Eq. (11.373) to

$$\mathcal{L}\left\{\int_0^t v(t)\, dt\right\} = \left[\frac{-e^{-st}}{s}\int_0^t v(t)\, dt\right]_0^\infty + \frac{1}{s}\int_0^\infty v(t)e^{-st}\, dt \qquad (11.376)$$

The first term on the right in Eq. (11.376) is zero. It vanishes at the lower limit because $\int_0^0 f(t)\, dt = 0$. It vanishes at the upper limit if $\sigma = \text{Re}\{s\}$ is sufficiently large so that $e^{-st} \to 0$ as $t \to \infty$. Using Eq. (11.372) in the second term yields

$$\mathcal{L}\left\{\int_0^t v(t)\, dt\right\} = \frac{V(s)}{s} \qquad (11.377)$$

Integration of a waveform $v(t)$ in the time domain can be accomplished by the algebraic process of dividing its transform $V(s)$ by s in the frequency domain.

By following a similar process, the differentiation operation is transformed into the frequency domain as follows:

$$\mathcal{L}\left\{\frac{dv}{dt}\right\} = sV(s) - v(0-) \qquad (11.378)$$

Time differentiation of the waveform $v(t)$ transforms into multiplication of $V(s)$ by s to within an additive constant $v(0-)$. Note that in this case we must use $t = 0-$ to account for any discontinuities in $v(t)$ that occur at $t = 0$. If none exist, the difference between 0 and $0-$ can be ignored. By repeated application of the differentiation rule in Eq. (11.378) we obtain

$$\mathcal{L}\left\{\frac{d^2v}{dt^2}\right\} = s^2V(s) - sv(0-) - v'(0-) \qquad (11.379)$$

$$\mathcal{L}\left\{\frac{d^3v}{dt^3}\right\} = s^3V(s) - s^2v(0-) - sv'(0-) - v''(0-) \qquad (11.380)$$

where $v' = dv/dt$ and $v'' = d^2v/dt^2$.

The fact that time integration and differentiation change into algebraic processes in the s domain is a hallmark feature of the Laplace transformation.

Table 11.2 summarizes some of the basic Laplace transform properties.

Inverse Laplace Transforms

So far, we have been concerned with the process of obtaining the transform of a signal from its waveform. But, as the pattern in Fig. 11.81 shows, we eventually will be confronted with the need to perform the inverse process. The process of recovering the waveform from a transform is called the **inverse Laplace transformation**. The important feature of this process is that the Laplace transformation is unique. Symbolically, we can state the uniqueness property as follows:

If

$$\mathcal{L}\{v(t)\} = V(s)$$

then

$$\mathcal{L}^{-1}\{V(s)\}(=)u(t)v(t)$$

The symbol $\mathcal{L}^{-1}$ stands for the inverse Laplace transform. The notation $(=)$ means equal almost everywhere. The only points at which equality may not hold are at the discontinuities of $v(t)$.

Finally, note that the waveform recovered by the inverse transformation is zero for $t < 0$. The mathematical justification for this is beyond the scope of our treatment. But to preserve uniqueness, that is, that one and only one signal waveform for each signal transform exists and vice versa, we must assume that all waveforms involved are zero for negative time. This is a very important point. For example, when we write

$$V(s) = \mathcal{L}\{\sin \beta t\} = \frac{\beta}{s^2 + \beta^2} \tag{11.381}$$

and

$$v(t) = \mathcal{L}^{-1}\left\{\frac{\beta}{s^2 + \beta^2}\right\} = \sin \beta t \tag{11.382}$$

we do *not* mean an eternal sine wave, but a waveform $v(t) = u(t)\sin \beta t$ that starts at $t = 0$. The step function required is not always shown explicitly. However, an implied step function must be understood, otherwise the process is not unique.

The upshot of all this is that a table of Laplace transform pairs can be used in either direction: waveform to transform or transform to waveform. For example, if we are given the transform

$$V(s) = \frac{10}{s + 2} \tag{11.383}$$

and need to determine the corresponding waveform, we scan the transform column in Table 11.1 and discover that $V(s)$ is of the form of $1/(s + a)$. This is the transform of the exponential waveform. Using this observation, together with the linearity property, we conclude that the corresponding waveform is

$$v(t) = 10u(t)e^{-2t} \tag{11.384}$$

This is an exponential signal with an amplitude of 10 and a time constant of 0.5. Note that these data are contained in the transform. In other words, the transform and the waveform are equivalent representations of the exponential signal. Stated differently, the data needed to describe the signal are available in either the transform or the waveform. The Laplace transformation alters the form of the data, but not the data themselves.

It is a simple matter to go from a transform to a waveform if we can find $V(s)$ in a table of pairs. Unfortunately it does not take a very complicated signal (or system) before we exceed the capability of Table 11.1, or even the more extensive tables available in the literature. However, all is not lost, because the general method of performing the inverse transformation is based on decomposing $V(s)$ into a linear combination of terms, each of which is available in Table 11.1. A most useful method is that of partial fraction expansion.

Step Response of Circuits with a Single Energy Storage Device
In using the Laplace transformation to study single energy storage device circuits we can either (1) transform the circuit differential equation or (2) transform the circuit itself. In this section, we use the differential equation approach. The latter viewpoint is developed later.

The RC circuit in Fig. 11.82 is driven by a constant source and a switch that closes at $t = 0$. For the purpose of analysis, we can replace the switch and constant source by a voltage source that delivers a signal $V_i u(t)$, a step function. In sum, the circuit input signal is a step function, and so the resulting circuit response is called the **step response**.

From our previous study of circuits with a single energy storage device, we can predict that the step response will depend on three things:

1. The amplitude of the input step function (V_i)
2. The circuit time constant (RC)
3. The voltage across the capacitor at $t = 0$

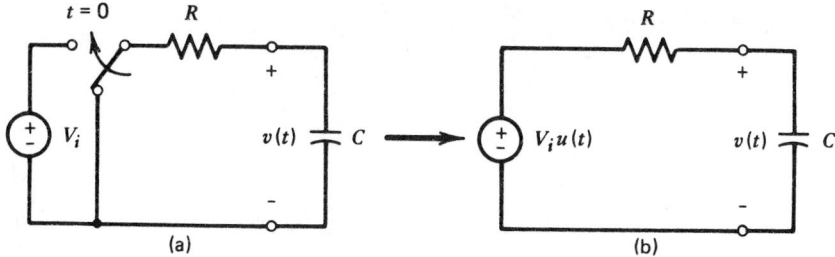

Fig. 11.82 Transformation of a single energy storage device RC circuit: (a) original circuit, (b) equivalent step function. Reproduced, with permission, from R. E. Thomas and A. J. Rosa, *Circuits and Signals: An Introduction to Linear and Interface Circuits*, New York, Wiley, 1984.

We also know that the response is governed by the circuit differential equation

$$RC\frac{dv}{dt} + v = V_i u(t) \qquad (11.385)$$

with the initial condition $v(0) = v_C(0)$.

Our objective is to determine the waveform $v(t)$ that satisfies this differential equation and the initial condition. Simply put, we must determine the step response. To achieve this objective, we follow the pattern shown in Fig. 11.81 by transforming the differential equation into the s domain, solving for the transform of the response, and then performing the inverse transformation to obtain the response waveform $v(t)$.

To begin this process, we transform Eq. (11.385) as

$$\mathscr{L}\left\{RC\frac{dv}{dt} + v\right\} = \mathscr{L}\{V_i u(t)\} \qquad (11.386)$$

$$RC\mathscr{L}\left\{\frac{dv}{dt}\right\} + \mathscr{L}\{v(t)\} = V_i\mathscr{L}\{u(t)\} \qquad (11.387)$$

$$RC[sV(s) - v(0)] + V(s) = \frac{V_i}{s} \qquad (11.388)$$

To achieve this transformation, we have used the linearity property, the differentiation property, and the transform pair for a unit step function. The beauty of this result is that it is now an algebraic equation in $V(s)$, the transform of the response we seek.

We now algebraically separate $V(s)$ in Eq. (11.388):

$$(RCs + 1)V(s) = \frac{V_i}{s} + RCv(0) \qquad (11.389)$$

Solving for $V(s)$ yields

$$V(s) = \frac{V_i/RC}{s(s + 1/RC)} + \frac{v(0)}{(s + 1/RC)} \qquad (11.390)$$

The function $V(s)$ in this equation is the transform of the waveform $v(t)$ that satisfies the differential equation and meets the initial conditions. Note that the initial condition first appeared explicitly as a result of applying the differentiation rule to obtain Eq. (11.388).

To recover the waveform $v(t)$, we must apply the inverse transformation to Eq. (11.390). The partial fraction expansion of the first term on the right is

$$\frac{V_i/RC}{s(s + 1/RC)} = \frac{k_1}{s} + \frac{k_2}{s + 1/RC} \qquad (11.391)$$

The coverup algorithm yields the residues as

$$k_1 = \frac{V_i/RC}{s + 1/RC}\bigg|_{s=0} = V_i \qquad k_2 = \frac{V_i/RC}{s}\bigg|_{s=-1/RC} = -V_i \qquad (11.392)$$

Hence Eq. (11.390) can be written as

$$V(s) = \frac{V_i}{s} - \frac{V_i}{s + 1/RC} + \frac{v(0)}{s + 1/RC} \qquad (11.393)$$

Each term in this expansion is now recognizable. The first term on the right is a step function, and each of the next two terms is an exponential:

$$v(t) = V_i + [v(0) - V_i]e^{-t/RC} \qquad t > 0 \qquad (11.394)$$

This is a familiar result from our previous study of RC circuits with a single effective capacitor. The first term on the right is the forced response due to the input force, while the second is the natural response due to the characteristic of the circuit itself. The complete response, as predicted, depends on three parameters: the input amplitude V_i, the time constant RC, and the initial condition $v(0)$. Note, the Laplace transformation method yields the complete response (forced plus natural) in one process. The pattern is the one outlined in flow graph form in Fig. 11.81.

Response to General Inputs
It is comforting to discover that the Laplace transformation yields results that agree with those obtained by classical methods. However, the Laplace transformation is more than just another way to find the step response of a circuit. It can also be used to find the response attributable to a wide variety of input signals.

For example, the RC circuit in Fig. 11.83 is driven by an input signal denoted $v_A(t)$, where the input signal is not necessarily a step function, but any time-varying waveform. The differential equation describing the voltage across the capacitor is

$$RC\frac{dv}{dt} + v = v_A(t) \qquad (11.395)$$

The only difference here is that the right side of the differential equation, the driving force, can be any time-varying waveform. The classical methods of dealing with this situation involve such concepts as variation of parameters, integrating factors, and undetermined coefficients.

However, with the Laplace transform method we can proceed without specifying the exact nature of the input signal. We first transform Eq. (11.395) into the s domain as

$$RC[sV(s) - v(0)] + V(s) = V_A(s) \qquad (11.396)$$

The only assumption necessary is that the input $V_A(t)$ be Laplace transformable, a condition met by

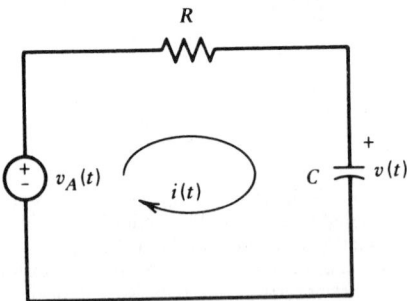

Fig. 11.83 Single energy storage device RC circuit with a general input signal. Reproduced, with permission, from R. E. Thomas and A. J. Rosa, *Circuits and Signals: An Introduction to Linear and Interface Circuits*, New York, Wiley, 1984.

all signals of interest. We now solve for the response $V(s)$ as

$$V(s) = \frac{V_A(s)/RC}{s + 1/RC} + \frac{v(0)}{s + 1/RC} \qquad (11.397)$$

The function $V(s)$ is the transform of the response of the circuit due to any input signal $v_A(t)$. Notice that we can proceed this far in the solution process without knowing the exact form of the input signal. In a sense we have found the general solution—in the s domain, of course—of the differential equation [Eq. (11.395)] for any input signal. All of the necessary ingredients are present in Eq. (11.397): the input signal $V_A(s)$, the circuit time constant RC, and the initial condition $v(0)$. However, to recover the waveform $v(t)$, we must have a particular input in mind.

Transformed Circuits

So far we have used the Laplace transformation to change waveforms into transforms, and to change circuit differential equations into algebraic equations. These operations are useful and provide insight into the nature of the s domain. However, the utility of the Laplace transformation is not just solving differential equations, but providing an alternative representation of signals and systems. The real advantage of the Laplace transformation shows up when we transform the circuit itself and analyze its behavior in the s domain.

The pattern of s-domain circuit analysis is outlined in Fig. 11.84. We begin with a circuit described in the t domain in the usual way. We transform the circuit into the s domain, write the circuit equilibrium equations directly in that domain as algebraic equations, and then solve for current or voltage transforms using algebraic techniques. If necessary, we can obtain current or voltage waveforms by performing the inverse Laplace transformation. The illustration also points out that there is another route to the solution waveform, using the circuit differential equation and classical time-domain techniques.

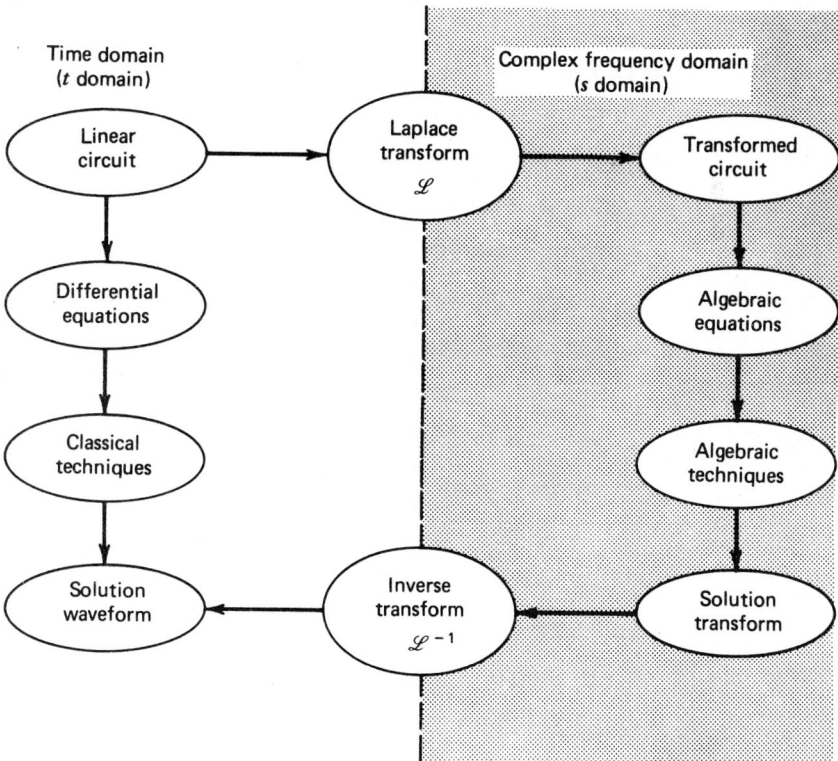

Fig. 11.84 Pattern of s-domain circuit analysis. Reproduced, with permission, from R. E. Thomas and A. J. Rosa, *Circuits and Signals: An Introduction to Linear and Interface Circuits*, New York, Wiley, 1984.

From the flow diagram in Fig. 11.84 it may appear that s-domain circuit analysis is just a way to circumvent the classical differential equation technique. But it is really more than this. It provides a new viewpoint, a different way of thinking about signals and circuits. The solution transform is actually just another representation of a signal. The solution transform contains the same data as the solution waveform, otherwise we could not recover the waveform from the transform by using the inverse transformation. We can now begin to think of the solution transform as *the* solution, to think of circuit behavior in terms of signal transforms rather than signal waveforms. In sum, we can think about circuits in the s domain.

We have seen many times that the behavior of a circuit is rooted in an equilibrium established by constraints of two types: (1) connection constraints and (2) device constraints. The connection constraints are represented mathematically by equations obtained using KCL and KVL. The device constraints are mathematically expressed by the *i-v* relationships of the elements used to model the devices in the circuit. In other words, connection equations and element equations are the foundation of circuit analysis. How are we to transform circuits? We must see how connection and element equations are altered by the Laplace transformation.

The connection equations are based on Kirchhoff's laws. A typical current-law equation in the time domain would be

$$i_1(t) + i_2(t) + i_3(t) + i_4(t) = 0 \tag{11.398}$$

In words, this equation says that the sum of current waveforms at a node is zero for all values of t. If we take the Laplace transform of this equation, then because of the linearity property of the transformation, we obtain

$$I_1(s) + I_2(s) + I_3(s) + I_4(s) = 0 \tag{11.399}$$

What this equation says is that the sum of current transforms is zero for all values of s. Clearly this idea generalizes to any number of currents at any node. Just as clearly, it applies to Kirchhoff's voltage law as well. In sum, KCL and KVL are not altered by the Laplace transformation. They apply in either the t domain to waveforms or the s domain to transforms. Hence the connection constraints in a circuit are the same in the s domain.

Turning now to the device constraints, we deal first with the signal voltage source in Fig. 11.85. The *i-v* relationships for this element are

$$v(t) = v_s(t) \quad \text{and} \quad i(t) \text{ varies} \tag{11.400}$$

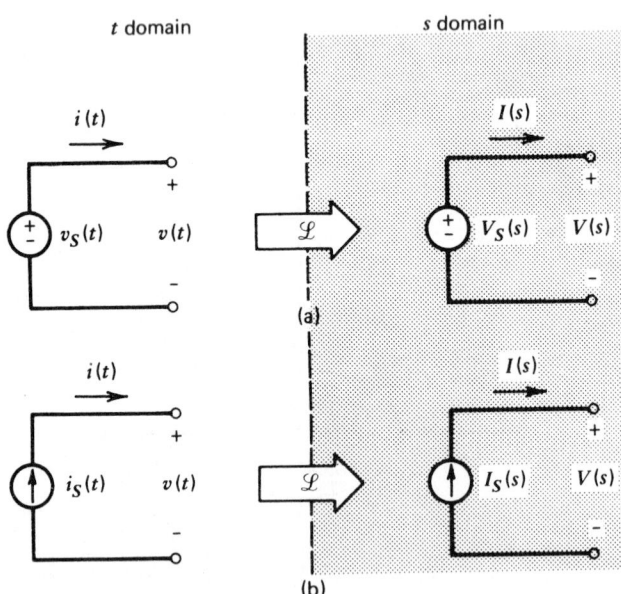

Fig. 11.85 s-domain models of signal sources: (*a*) voltage sources, (*b*) current sources. Reproduced, with permission, from R. E. Thomas and A. J. Rosa, *Circuits and Signals: An Introduction to Linear and Interface Circuits*, New York, Wiley, 1984.

This says that a voltage source produces a prescribed waveform at its terminals and can deliver any current waveform that may be demanded by the circuit to which it is connected. Taking the Laplace transform of this relationship produces

$$V(s) = V_s(s) \quad \text{and} \quad I(s) \text{ varies} \tag{11.401}$$

This relationship says that the voltage source produces a prescribed transform and can deliver whatever current transform may be required. Clearly the same idea applies to the current source in Fig. 11.85. In sum, signal sources behave in exactly the same way in the s domain, except that we think of them as producing a transform rather than a waveform.

Next we consider the three linear passive circuit elements shown in Fig. 11.86. In the time domain the i-v relationships are

Resistor: $\qquad\qquad\qquad v_R(t) = R i_R(t)$

Inductor: $\qquad\qquad\qquad v_L(t) = L\dfrac{di_L(t)}{dt}$ $\qquad\qquad\qquad$ (11.402)

Capacitor: $\qquad\qquad\qquad v_C(t) = \dfrac{1}{C}\displaystyle\int_0^t i_C(t)\,dt + v_C(0)$

These relationships can be transformed into the s domain using the linearity, differentiation, and

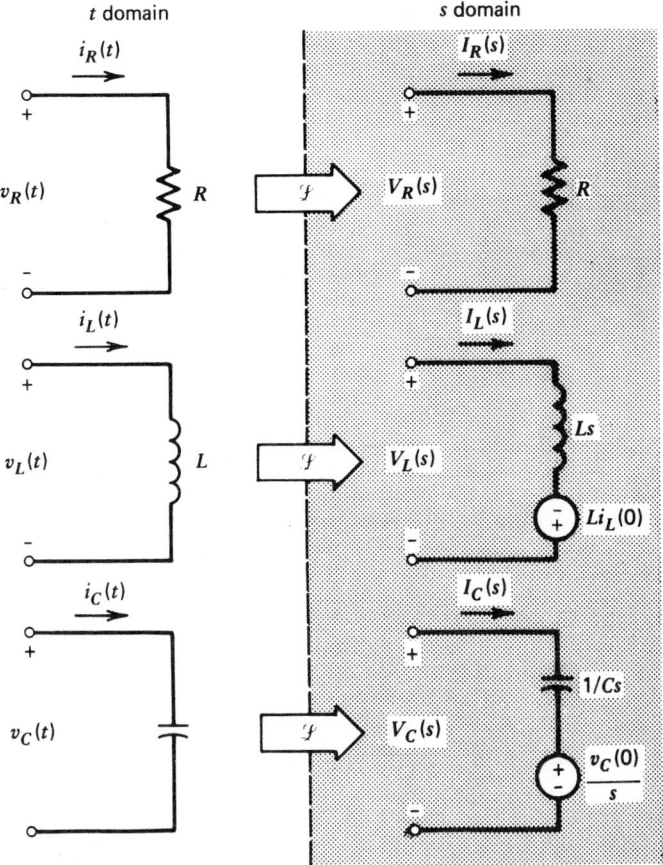

Fig. 11.86 s-domain models for passive elements used in meshed circuit analysis. Reproduced, with permission, from R. E. Thomas and A. J. Rosa, *Circuits and Signals: An Introduction to Linear and Interface Circuits*, New York, Wiley, 1984.

integration properties of the Laplace transformation:

Resistor: $\qquad\qquad\qquad\qquad V_R(s) = RI_R(s)$

Inductor: $\qquad\qquad\qquad\qquad V_L(s) = (Ls)I_L(s) - Li_L(0)$ $\qquad\qquad$ (11.403)

Capacitor: $\qquad\qquad\qquad\qquad V_C(s) = \dfrac{1}{Cs}I_C(s) + \dfrac{1}{s}v_C(0)$

As might be expected, the *i-v* relationships in the *s* domain for all three elements turn out to be linear algebraic equations. In particular we see that for the resistor Ohm's law is the same in the *s* domain.
These three element equations lead to the *s* domain circuit models in Fig. 11.86. The initial conditions associated with the two energy storage elements are modeled as voltage sources connected in series with the elements. The element symbols in the *s* domain stand for what is called the element *impedance*. The concept of impedance is based on writing Eqs. (11.403) in the form

Resistor: $\qquad\qquad\qquad\qquad V_R = Z_R I_R$

Inductor: $\qquad\qquad\qquad\qquad V_L = Z_L I_L - Li_L(0)$ $\qquad\qquad$ (11.404)

Capacitor: $\qquad\qquad\qquad\qquad V_C = Z_C I_C + \dfrac{v_C(0)}{s}$

The symbol Z stands for the element impedance, which can be defined as the proportionality factor in the *s*-domain relationship between the current transform and the voltage transform. In comparing Eqs. (11.403) and (11.404) we identify the three element impedances as

Resistor: $\qquad\qquad\qquad\qquad Z_R = R$

Inductor: $\qquad\qquad\qquad\qquad Z_L = Ls$ $\qquad\qquad$ (11.405)

Capacitor: $\qquad\qquad\qquad\qquad Z_C = \dfrac{1}{Cs}$

Impedance is inherently an *s*-domain concept since it is based on a proportionality between a current transform and a voltage transform. It is a generalization of the concept of resistance, and hence we have the name *impedance*. The impedance of a resistor is a constant, its resistance. The impedances of the two energy storage devices are not constants but depend on the complex frequency variable *s*.
Alternatively the *i-v* relationships in Eq. (11.403) can be solved for the current transform.

Resistor: $\qquad\qquad\qquad\qquad I_R(s) = GV_R(s)$

Inductor: $\qquad\qquad\qquad\qquad I_L(s) = \dfrac{1}{Ls}V_L(s) + \dfrac{1}{s}i_L(0)$ $\qquad\qquad$ (11.406)

Capacitor: $\qquad\qquad\qquad\qquad I_C(s) = (Cs)V_C(s) - Cv_C(0)$

This form of the *i-v* relationships leads to the *s*-domain circuit models in Fig. 11.87. Here the initial conditions for the energy storage devices appear as current sources in parallel with what is called the element *admittance* Y. The concept of admittance is a generalization of conductance and can be defined as the reciprocal of impedance ($Y = 1/Z$). The three-element admittances are

Resistor: $\qquad\qquad\qquad\qquad Y_R = \dfrac{1}{Z_R} = \dfrac{1}{R} = G$

Inductor: $\qquad\qquad\qquad\qquad Y_L = \dfrac{1}{Z_L} = \dfrac{1}{Ls}$ $\qquad\qquad$ (11.407)

Capacitor: $\qquad\qquad\qquad\qquad Y_C = \dfrac{1}{Z_C} = Cs$

Alternatively we see that the element admittances are the proportionality factor in the linear *i-v* characteristics relating transform voltage to transform current, Eq. (11.406). In any event, in the *s* domain the passive elements can be represented by impedances, with initial-condition voltage sources in series with the energy storage elements, or by admittances with initial-condition current sources in parallel.
Circuit analysis in the *s* domain closely parallels *t*-domain analysis since KCL and KVL are unchanged. To transform a circuit into the *s* domain, we replace each element by its *s*-domain model.

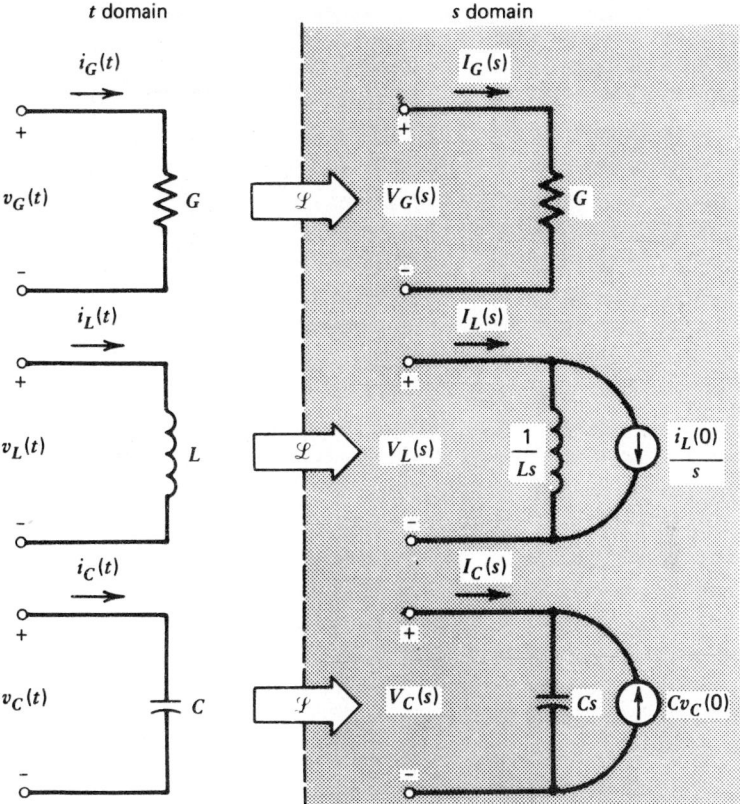

Fig. 11.87 s-domain model for passive elements as used in node voltage analysis. Reproduced, with permission, from R. E. Thomas and A. J. Rosa, *Circuits and Signals: An Introduction to Linear and Interface Circuits*, New York, Wiley, 1984.

For sources and resistors this is really no change at all. For capacitors and inductors we replace the element by its impedance in series with an initial-condition voltage source or by its admittance in parallel with an initial-condition current source. Note that in the *s* domain the circuit initial conditions appear as sources in the circuit, rather than as boundary conditions on the solution of the circuit differential equation. The *i-v* relationships in the *s* domain have an Ohm's-law-like character involving the impedance or admittance of the element. All of these features make *s*-domain circuit analysis an algebraic process similar to the analysis of resistive circuits in the *t* domain.

Example 11.16. Figure 11.88(*a*) shows a series *RL* circuit. Figure 11.88(*b*) shows the circuit transformed into the *s* domain. The sum of voltage transforms around the one loop in this circuit is

$$-\frac{V}{s} + V_R(s) + V_L(s) = 0 \tag{11.408}$$

The voltages $V_R(s)$ and $V_L(s)$ can be written in terms of the mesh current $I(s)$ using the *s*-domain element equations

$$V_R(s) = RI(s) \quad \text{and} \quad V_L(s) = LsI(s) - Li_L(0) \tag{11.409}$$

Substituting these relationships into the foregoing KVL equation and collecting terms yields

$$(Ls + R)I(s) = \frac{V}{s} + Li_L(0) \tag{11.410}$$

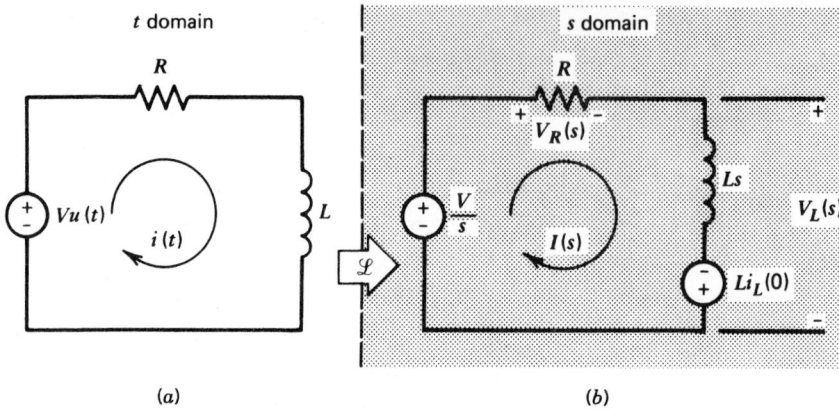

Fig. 11.88 RL circuit used in Example 11.16. (a) t-domain circuit, (b) s-domain equivalent. Reproduced, with permission, from R. E. Thomas and A. J. Rosa, *Circuits and Signals: An Introduction to Linear and Interface Circuits*, New York, Wiley, 1984.

Solving for $I(s)$,

$$I(s) = \frac{V/L}{s(s + R/L)} + \frac{i_L(0)}{s + R/L} \tag{11.411}$$

This is the transform of the mesh current for a step function input. If necessary, we can determine the corresponding waveform. First, we expand the transform by partial fractions:

$$I(s) = \frac{V/R}{s} - \frac{V/R}{s + R/L} + \frac{i_L(0)}{s + R/L} \tag{11.412}$$

Performing the inverse transformation yields

$$i(t) = \frac{V}{R} + \left[i_L(0) - \frac{V}{R} \right] e^{-Rt/L} \qquad t > 0 \tag{11.413}$$

The first term on the right is the forced response and the second term the natural response.

Transfer Functions

An important application of circuit analysis is the processing of a signal in its passage from input to output. In the s domain the signal processing involved is described by a rational function of the complex frequency variable called a **transfer function**:

$$\text{Transfer function} = \frac{\text{zero-state response transform}}{\text{input signal transform}} = T(s) \tag{11.414}$$

Note that the formal definition applies only to the zero-state response and implies that the circuit has only one input. Both of these conditions simplify the process of finding and using transfer functions.

The pattern of signal transfer from input to output is illustrated in Fig. 11.89. There is a single input signal, either a current I_1 or voltage V_1, and a single output, either a voltage V_2 or a current I_2. Since the input and output signals can take one of two possible forms, it is possible to define four different kinds of transfer functions (V_2/V_1, I_2/I_1, V_2/I_1, I_2/V_1). We shall concentrate our attention on the two dimensionless functions:

$$\text{Voltage transfer function} = T(s) = \frac{V_2(s)}{V_1(s)}$$

$$\text{Current transfer function} = T(s) = \frac{I_2(s)}{I_1(s)} \tag{11.415}$$

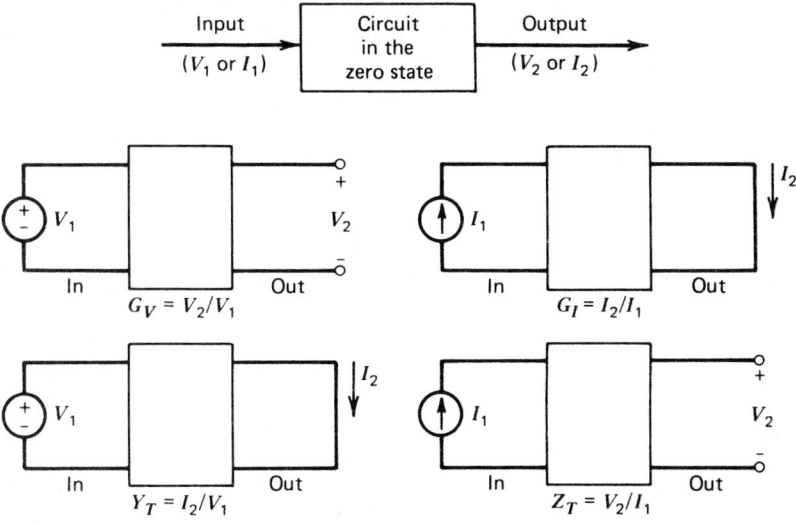

Fig. 11.89 Signal transfer through a circuit. Reproduced, with permission, from R. E. Thomas and A. J. Rosa, *Circuits and Signals: An Introduction to Linear and Interface Circuits*, New York, Wiley, 1984.

The distinguishing feature of a transfer function is that it involves an input at one place in a network and a response that occurs somewhere else. In this regard the transfer function $T(s)$ is the s-domain generalization of the input-output relationship $y = kx$, which we studied for resistive circuits. As such we see that it is the result of the circuit being linear. Strictly speaking, only linear circuits have transfer functions, although it is fairly common for practicing engineers to apply the term loosely (and usually incorrectly) to nonlinear circuits.

Transfer functions always turn out to be rational functions: quotients of polynomials in the complex frequency variable of the form $r(s)/q(s)$. In this regard they look just like signal transforms. However, except for the very special and very important case when the input is an impulse $\delta(t)$, transfer functions are not, in general, signal waveforms. They are s-domain characterizations of the zero-state signal transfer through the circuit.

Impulse and Step Response

We can use the circuit transfer function to predict the response of circuits. The definition of this function is the ratio of the zero-state response to the input signal transform. As such, the transfer function is an s-domain description of the zero-state characteristics of a circuit. However, to use the transfer function to predict responses, we write the defining relationship as

$$\left(\begin{array}{c}\text{Zero-state}\\\text{response}\end{array}\right) = \left(\begin{array}{c}\text{Transfer}\\\text{function}\end{array}\right) \times \left(\begin{array}{c}\text{Input signal}\\\text{transform}\end{array}\right) \qquad (11.416)$$

$$Y(s) = T(s) \times X(s)$$

where $T(s)$ is the circuit transfer function, $X(s)$ is the input signal transform, and $Y(s)$ is the zero-state response or output signal.

The block diagram representation of the input-output relationship is shown in Fig. 11.90.

This viewpoint indicates that the transfer function is the rational function, which is multiplied by the input to obtain the output. Since this input-output relationship involves transferring the signal through the circuit, it is reasonable to call the circuit's contribution the transfer function. Likewise it is reasonable to think of the input as the excitation or cause and the output as the response or effect. The input-output or cause-effect relationship is imbedded in the transfer function. A major role of circuit analysis is to deal with the problem of finding the output when given the transfer function and the input signal. Contrast this with the problem of design. In design the excitation or input is known and the output desired for that input is known. The designer's task is to produce a circuit, usually within some equipment, component, and monetary constraints, that has a suitable transfer function.

While the transfer function is relatively easy to come by, it is clear that we cannot find the circuit response until we are given an input signal. Here we encounter a central paradox of circuit analysis.

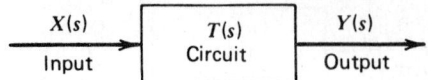

Fig. 11.90 Block diagram of the input, output, and transfer function. Reproduced, with permission, from R. E. Thomas and A. J. Rosa, *Circuits and Signals: An Introduction to Linear and Interface Circuits*, New York, Wiley, 1984.

In practice the input signal is a carrier of information and is therefore unpredictable. We could spend a lifetime studying a circuit for various inputs and still not treat all possible signals that might be encountered in practice. What we must do is calculate the responses due to certain simple test signals. Although these test signals may never occur as real input signals, their responses tell us enough to begin to understand how the circuit will react to other signals.

The two premier test signals used are the *pulse* and the *sinusoid*. These waveforms have the recommendation that they are relatively easy to describe mathematically and can be easily generated in the laboratory when we need to test our mathematical predictions against the reality of hardware. Moreover the pulse is in some sense representative of the signals produced by digital transducers such as keyboards and teletypewriters. Likewise the sinusoid is characteristic of the signals produced by the human voice and musical instruments. Furthermore the study of linear systems is the study of the differential equations that represent those systems. Differential equations have two-part solutions, the transient and the forced or steady-state response. A circuit's transient behavior is alluded to by its pulse response and its forced behavior by its sinusoidal steady-state response.

The study of pulse response of a circuit divides into the two extreme cases. When the pulse is very short compared with the response time of the circuit, the input can be modeled as an impulse. This all-or-nothing pulse is so brief that the circuit barely has time to begin to respond before the input returns to zero. In essence we get a zero-input response because the sluggish circuit sees essentially no input except for one brief, but spectacular excursion.

At the other extreme is the more sedate "long" pulse whose duration exceeds the inherent response time of the circuit. Here the circuit has more than enough time to respond and to reach a steady output before the input changes again. The leading edge of this pulse can be treated as a step function as far as the circuit is concerned.

The two extreme models for a pulse input are the step function $u(t)$ and the impulse $\delta(t)$. We first introduce some notation and then show that there is a simple set of relationships between the two responses.

We first write the transfer function as

$$T(s) = \frac{b_m s^m + b_{m-1} s^{m-1} + \cdots + b_1 s + b_0}{s^n + a_{n-1} s^{n-1} + \cdots + a_1 s + a_0} \tag{11.417}$$

The natural poles, or natural frequencies, are the roots of the denominator polynomial in this expression. Now, if the input is a unit impulse, then

$$x(t) = \delta(t)$$

Hence   (11.418)

$$X(s) = 1$$

The response due to a unit impulse input, hereafter called simply the *impulse response* is

$$Y(s) = T(s) \times 1 = T(s) \tag{11.419}$$

The remarkable result is that the impulse response transform is the transfer function itself! Only in this very special, very important, case of an impulse input, can we treat a transfer function as a signal transform. To avoid possible confusion between transfer functions $T(s)$ and signal transforms, we denote the impulse response as $H(s)$, the signal transform, and $h(t)$, the corresponding signal waveform. Using this notation, we then write

$$
\begin{array}{cc}
 & \text{IMPULSE RESPONSE} \\
\text{TRANSFORM} & \text{WAVEFORM} \\
H(s) = T(s) & h(t) = \mathscr{L}^{-1}\{H(s)\}
\end{array}
\tag{11.420}
$$

When the input to the circuit is a unit step function, then

$$x(t) = u(t) \tag{11.421}$$

Hence

$$X(s) = \frac{1}{s}$$

The corresponding response, hereafter called simply the step response, can be written as

$$Y(s) = T(s)\frac{1}{s} \tag{11.422}$$

The step responses will be denoted as $G(s)$ and $g(t)$, respectively, and hence we write

$$\begin{array}{ccc} & \text{STEP RESPONSE} & \\ \text{TRANSFORM} & \text{WAVEFORM} & (11.423) \\ G(s) = \dfrac{T(s)}{s} & g(t) = \mathscr{L}^{-1}\{G(s)\} & \end{array}$$

Using this notation, we now show that there are simple relationships between the impulse and step responses. First, combining Eqs. (11.420) and (11.423), we write

$$G(s) = \frac{T(s)}{s} = \frac{H(s)}{s} \tag{11.424}$$

In words, the step response transform is obtained by dividing the impulse response transform by s. But division by s in the s domain corresponds to integration in the time domain. Hence

$$g(t) = \int_0^t h(t)\,dt \tag{11.425}$$

By the fundamental theorem of calculus then,

$$h(t)(=)\frac{dg(t)}{dt} \tag{11.426}$$

where the symbol $(=)$ means equal almost everywhere. The almost-everywhere restriction excludes only those points at which $g(t)$ has a jump or discontinuity. Thus the step response waveform is the integral of the impulse response waveform, and conversely, the impulse response waveform is the derivative of the step response waveform.

In sum, the five quantities $T(s)$, $H(s)$, $h(t)$, $G(s)$, and $g(t)$ are all related. If we have any one of these quantities, we can obtain all of the others by relatively simple mathematical operations. A summary of the relationships is given in Fig. 11.91.

Step Response Partial Descriptors

A one-pole or two-pole transfer function contains all of the data needed to construct a complete description of its step response waveform. Often we do not need a complete description but work with partial descriptors of the response. The step responses of devices, circuits, and systems are often specified in terms of these partial descriptors.

The step responses of many circuits and systems have waveforms similar to that shown in Fig. 11.92. The response is initially driven toward its final value, which it overshoots, then undershoots, and eventually settles down on. The initial rise is often described in terms of **rise time** (T_R):

$$T_R = T_2 - T_1 \tag{11.427}$$

where T_1 is the time at which $g(t) = 0.1g(\infty)$ and T_2 is the time at which $g(t) = 0.9g(\infty)$. The constants 0.1 and 0.9 are the most common values, although other values sometimes are used. Rise time is an important descriptor of the speed of response of a system. Rise times range from fractions of nanoseconds for high-speed digital computers to several seconds for large electromechanical systems.

For some systems the time between $t = 0$ and $t = T_1$ is large compared with the rise time. In such situations an additional parameter called **delay time** T_D is often specified. Delay is defined as the time

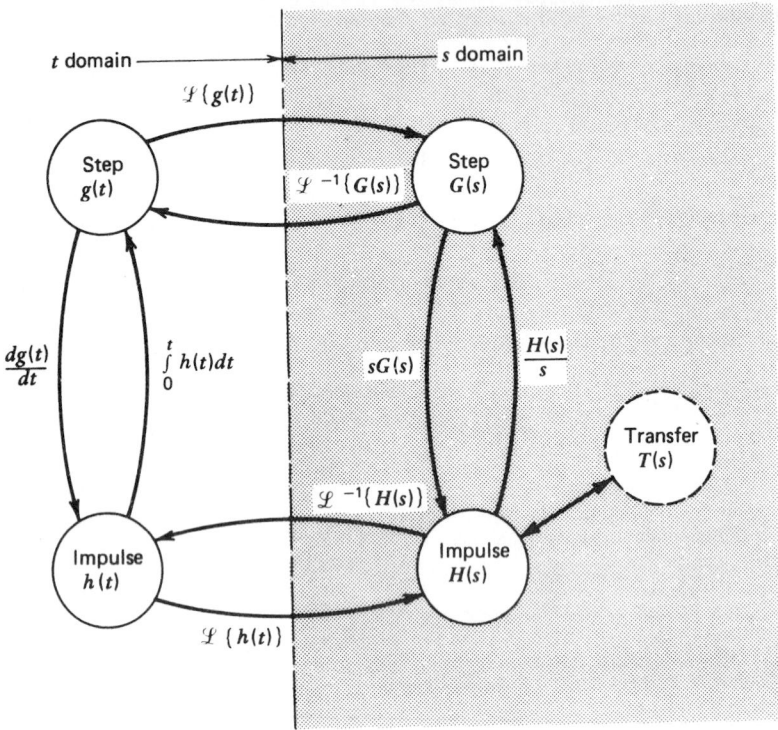

Fig. 11.91 Relationships between step and impulse responses and the transfer function in the t and s domains. Reproduced, with permission, from R. E. Thomas and A. J. Rosa, *Circuits and Signals: An Introduction to Linear and Interface Circuits*, New York, Wiley, 1984.

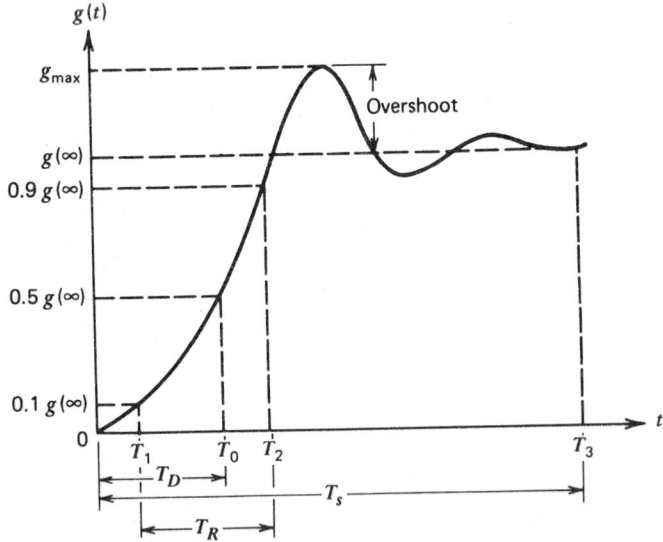

Fig. 11.92 Representative step response showing time (T_R), delay time (T_p), settling time (T_s), and overshoot. Reproduced, with permission, from R. E. Thomas and A. J. Rosa, *Circuits and Signals: An Introduction to Linear and Interface Circuits*, New York, Wiley, 1984.

interval between $t = 0$ and the time at which $g(t) = 0.5g(\infty)$. In other words, it is the time required for the response to get halfway to its final value. This is indicated as T_0 in Fig. 11.92.

A distinguishing feature of many step responses is **overshoot**. Quantitatively the percent overshoot is defined as

$$P_o = \text{percent overshoot} = \frac{g_{max} - g(\infty)}{g(\infty)} \times 100\% \qquad (11.428)$$

Percent overshoot is not defined if $g(\infty) = 0$.

The **settling time** (T_s) is a measure of how long it takes for the response to "settle down." Mathematically T_s is the time interval between $t = 0$ and the time at which the response remains within a certain small percentage of its final value, T_3 in Fig. 11.92. Typical values of this small percentage are 1, 3, and 5%.

The parameters rise time, delay time, percent overshoot, and settling time often describe step responses in sufficient detail for specification and design. The reader should remember that the definitions given here cannot be blindly applied to every type of step response. As noted, overshoot is not defined if the final value is zero. In such cases rise time is defined as the time interval between $g(t) = 0.1g_{max}$ and $g(t) = 0.9g_{max}$; that is, the time between reaching 10 and 90% of the peak value of the response.

11.5.5 Sinusoidal Response

The Sinusoidal Steady State

The unique properties of the sinusoidal signal make it a very common and useful waveform for testing and describing circuits and systems. The two key properties are that the sum of two sinusoids of the same frequency produces a sinusoid, and that a sinusoid may be integrated or differentiated any number of times and the result remains a sinusoid. No other periodic waveform has these special properties.

The upshot is that if a linear, stable circuit is driven by a sinusoid, then ultimately all currents and voltages in the circuit are sinusoids of the same frequency.

To deal with the sinusoidal response mathematically, we first consider the case in which the circuit is described by a transfer function $H(s)$. Figure 11.93 shows the pattern of interaction. There is an input $x(t)$ or $X(s)$ and a resulting response $y(t)$ or $Y(s)$. In the s domain, the connection between the responses is the transfer function

$$Y(s) = H(s)X(s) \qquad (11.429)$$

In the present case the input is a sinusoid, which we write in general form as

$$x(t) = X_0 \cos(\omega t + \phi) = X_0(\cos \omega t \cos \phi - \sin \omega t \sin \phi) \qquad (11.430)$$

In this form we recognize $\cos \omega t$ and $\sin \omega t$ and can write the input transform as

$$X(s) = X_0 \frac{s}{s^2 + \omega^2} \cos \phi - \frac{\omega}{s^2 + \omega^2} \sin \phi = X_0 \frac{s \cos \phi - \omega \sin \phi}{s^2 + \omega^2} \qquad (11.431)$$

Substituting this for $X(s)$ in Eq. (11.429) we get

$$Y(s) = X_0 \left[\frac{s \cos \phi - \omega \sin \phi}{(s - j\omega)(s + j\omega)} \right] H(s) \qquad (11.432a)$$

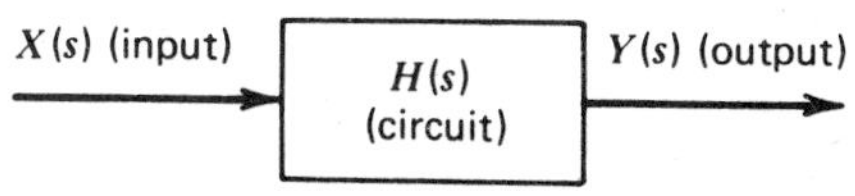

Fig. 11.93 Block diagram of the input, output, and transfer function. Reproduced, with permission, from R. E. Thomas and A. J. Rosa, *Circuits and Signals: An Introduction to Linear and Interface Circuits*, New York, Wiley, 1984.

We now expand the response transform by partial fractions as

$$Y(s) = \underbrace{\frac{k}{s - j\omega} + \frac{k^*}{s + j\omega}}_{\text{Forced response}} + \underbrace{\frac{k_1}{s - p_1} + \frac{k_2}{s - p_2} + \cdots + \frac{k_n}{s - p_n}}_{\text{Natural response}} \qquad (11.432b)$$

where $p_1, p_2, \ldots, p_n$ are the roots of the transfer function $H(s)$, and hence do not depend on the nature of the input. We now perform the inverse transformation and obtain the response waveform as

$$y(t) = ke^{j\omega t} + k^*e^{-j\omega t} + k_1\exp(p_1 t) + k_2\exp(p_2 t) + \cdots + k_n\exp(p_N t) \qquad (11.433)$$

This form gives a complete accounting of the response. There are two response terms arising from the forcing function $X(s)$ and N terms from the transfer function $H(s)$.

At this point we introduce a key assumption or condition. If the circuit is stable, then all of the natural roots have negative real parts. Consequently all of the components of the natural response ultimately decay to zero, and eventually only the forced component of the response persists. It is important to realize that in electric circuits this process does not take a great deal of time. If we apply a sinusoidal input to a stable electronic circuit, then after a few hundred milliseconds at most, only the forced component of the response will be observable.

The persistent response is called the **sinusoidal steady-state response**. The term *steady state* is perhaps unfortunate since it implies a steady or constant response, when in fact the forced response is an ever-changing sinusoid. Nonetheless we use the steady-state (ss) terminology, and we can now write it as

$$y_{ss}(t) = ke^{+j\omega t} + k^*e^{-j\omega t} \qquad (11.434)$$

In this form we see that the steady-state response is the sum of two conjugate terms. That is, $(ke^{+j\omega t})^* = k^*e^{-j\omega t}$. The sum of two conjugate terms is twice their real parts. Hence we can express the steady-state response as

$$y_{ss}(t) = \text{Re}\{2ke^{j\omega t}\} \qquad (11.435)$$

where Re{ } stands for the real part and k is the residue of the roots of the forcing function. That is,

$$2k = X_0|H(j\omega)|\exp[j(\phi + \phi_H)] \qquad (11.436)$$

Thus the *steady-state component is a sinusoid of the same frequency as the input but with a different amplitude and phase angle*. The input-output relationships can be summarized in the statements:

Output amplitude = (input amplitude) × [magnitude of $H(j\omega)$]

Output phase = (input phase) + [angle of $H(j\omega)$]

Output frequency = input frequency

Figure 11.94 presents these statements in an input-output format showing the effect of the circuit on the steady-state output. In the case of frequency there is no effect, since the input and output have the same frequency—but the amplitude and phase of the response are altered by the function $H(j\omega)$.

Phasors

The concept of a **phasor** was originally put forth by the American engineer Charles Steinmetz at the International Electrical Congress in 1893. Steinmetz also popularized the phasor by demonstrating its many applications so that by the early twentieth century it was universally used in the study of ac circuits and systems. Simply put, a phasor is an alternative way to represent a sinusoid.

Thus if the sinusoidal signal $v(t)$ is written as

$$v(t) = A\cos(\omega t + \phi) \qquad (11.437)$$

its phasor representation is

$$\mathbf{V} = Ae^{j\phi} \qquad (11.438)$$

Alternatively, if the waveform is written as

$$v(t) = a\cos(\omega t) + b\sin(\omega t) \qquad (11.439)$$

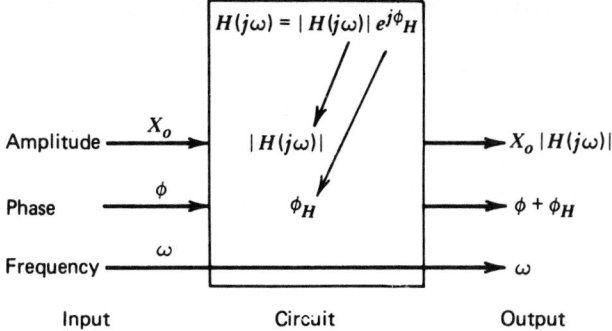

Fig. 11.94 Signal transfer in the sinusoidal steady state. Reproduced, with permission, from R. E. Thomas and A. J. Rosa, *Circuits and Signals: An Introduction to Linear and Interface Circuits*, New York, Wiley, 1984.

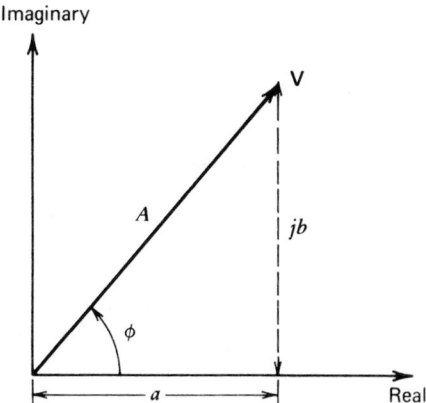

Fig. 11.95 Graphical interpretation of the phasor representation. Reproduced, with permission, from R. E. Thomas and A. J. Rosa, *Circuits and Signals: An Introduction to Linear and Interface Circuits*, New York, Wiley, 1984.

then its phasor representation is

$$\mathbf{V} = a + jb \qquad (11.440)$$

The relationship between these two representations is shown in Fig. 11.95.

There are two things about the phasor that require emphasis. First, phasors are indicated by boldface symbols to distinguish them from other signal representations, such as $v(t)$ and $V(s)$. Second, the phasor itself does not contain any information about the frequency of the sinusoid. Since all signals in the sinusoidal steady state are sinusoids of the same frequency, this information is not crucial. On the other hand, a frequency *must* be given or implied in any particular situation. Thus waveforms and phasors are simply two different and essentially equivalent ways to represent the eternal sinusoid. All of the data needed to construct the phasor are contained in the waveform, and conversely, all of the data (except frequency) required to define the waveform are contained in the phasor. There is a one-to-one correspondence between a sinusoidal waveform sum and the corresponding phasor sum. Any constraint that applies to the sum of waveforms also applies to the sum of phasors. Among the several implications of this is that Kirchhoff's laws apply to both waveforms and phasors.

In the sinusoidal steady state we can write the general input-output relationship of Eq. (11.429) in phasor form as

$$\mathbf{Y} = H(j\omega)\mathbf{X} \qquad (11.441)$$

That is, in the sinusoidal steady state the transfer function provides a relationship between the input phasor and the output phasor. To see that the relationship in Eq. (11.441) is true, we use the notation

$$\mathbf{Y} = Y_0 e^{j\theta}$$
$$\mathbf{X} = X_0 e^{j\phi} \qquad (11.442)$$
$$H(j\omega) = |H(j\omega)|\exp(j\phi_H)$$

Inserting these expressions into Eq. (11.441) yields

$$Y_0 e^{j\theta} = |H(j\omega)| X_0 \exp(j\phi + \phi_H) \qquad (11.443)$$

Now two phasors can be equal only if they have the same amplitude and phase angle. Hence Eq. (11.443) tells us that

$$\text{Output amplitude} = Y_0 = |H(j\omega)| X_0$$
$$\text{Output phase} = \theta = \phi + \phi_H$$
$$\text{Output frequency} = \omega$$

But these are the same input-output rules obtained using Laplace transforms. Thus we can think of the input and the output as phasors, and the transfer function as the connection between the two.

Phasor Circuit Analysis

Phasor circuit analysis is the analysis of circuits in the sinusoidal steady state in which the signals are represented as phasors. We have seen that circuit analysis is based on an equilibrium established by constraints of two types: (1) connection constraints (Kirchhoff's laws) and (2) device constraints (element equations). To deal with phasor circuit analysis we must first see how these constraints are written in phasor form.

In the sinusoidal steady state the application of Kirchhoff's voltage law around a loop in the circuit would lead to an equation of the form

$$V_1\cos(\omega t + \phi_1) + V_2\cos(\omega t + \phi_2) + V_3\cos(\omega t + \phi_3) = 0 \qquad (11.444)$$

But as we saw in the preceding section, there is a one-to-one correspondence between waveform sums and phasor sums. Hence it must be also true that

$$\mathbf{V}_1 + \mathbf{V}_2 + \mathbf{V}_3 = 0 \qquad (11.445)$$

Clearly this result would apply for any number of voltages and to sinusoidal currents in applying Kirchhoff's current law as well. In sum, the connection constraints are unchanged. Kirchhoff's laws apply to waveforms and phasors.

The general element equation in phasor form is written as

$$\mathbf{V} = Z\mathbf{I} \quad \text{or} \quad \mathbf{I} = Y\mathbf{V} \qquad (11.446)$$

which is the phasor version of Ohm's law, where Z is the element impedance and Y its admittance. The impedances and admittances for the resistor, inductor, and capacitor are

$$Z_R = R \quad \text{or} \quad Y_R = \frac{1}{R}$$
$$Z_L = j\omega L \quad \text{or} \quad Y_L = \frac{1}{j\omega L} \cdot \qquad (11.447)$$
$$Z_C = \frac{1}{j\omega C} \quad \text{or} \quad Y_C = j\omega C$$

We can think of phasor circuit analysis in terms of the flow diagram in Fig. 11.96. We begin in the time domain with a circuit that is operating in the sinusoidal steady state. We transform the circuit into the frequency domain, which means that all signals become phasors and the circuit elements are described by their impedances with s replaced by $j\omega$. We can then employ our lexicon of algebraic analysis techniques to solve for the phasor response. This phasor response can then be taken back into the time domain, if need be, to obtain the response waveform.

Example 11.17. Determine $i(t)$, $i_C(t)$, and $i_R(t)$ for the circuit of Fig. 11.97(a).

From the data given we assume that the circuit is operating in the sinusoidal steady state. Figure 11.97(b) shows the frequency-domain version of the circuit with the desired responses shown as

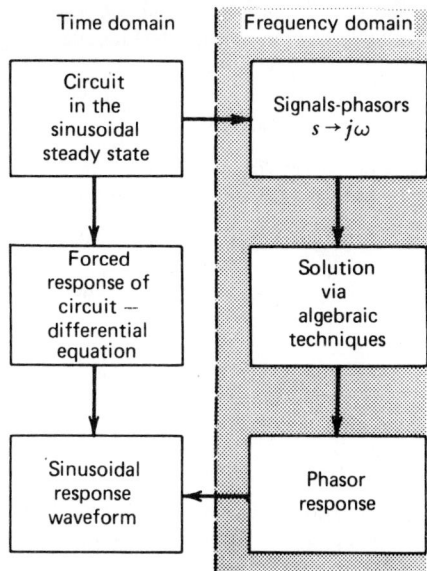

Fig. 11.96 Pattern of circuit analysis using phasors. Reproduced, with permission, from R. E. Thomas and A. J. Rosa, *Circuits and Signals: An Introduction to Linear and Interface Circuits*, New York, Wiley, 1984.

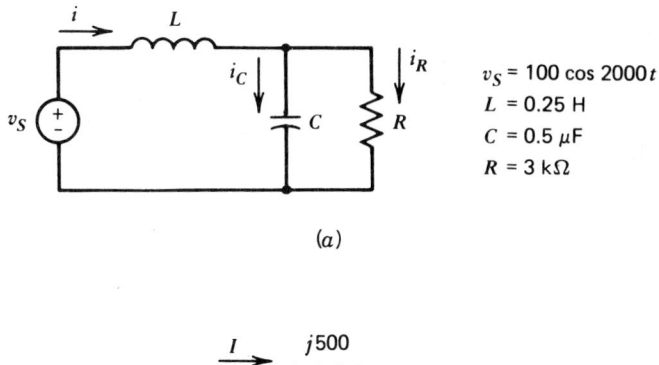

(a)

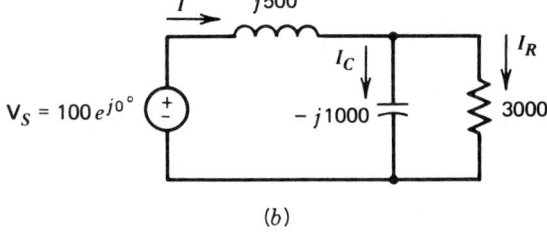

(b)

Fig. 11.97 Circuit (*a*) and solution (*b*) for Example 11.17. Reproduced, with permission, from R. E. Thomas and A. J. Rosa, *Circuits and Signals: An Introduction to Linear and Interface Circuits*, New York, Wiley, 1984.

phasors. To solve for these responses we shall use circuit reduction. The resistor and capacitor are connected in parallel and can be replaced by an equivalent impedance Z_{eq1}:

$$Z_{eq1} = \frac{Z_R Z_C}{Z_R + Z_C} = 300 - j900 \qquad (11.448)$$

In this circuit the equivalent impedance Z_{eq1} is connected in series with the inductor and the combination can be replaced by an equivalent input impedance Z_{eq2}:

$$Z_{eq2} = Z_{eq1} + Z_L = 300 - j400 \qquad (11.449)$$

We now have an equivalent impedance at the input terminals and the current $\mathbf{I}$ is found as

$$\mathbf{I} = \frac{\mathbf{V}_s}{Z_{eq2}} = \frac{100 + j0}{300 - j400} = 0.2 \exp(j53.1°) \qquad (11.450)$$

We can write the corresponding current waveform as

$$i(t) = 0.2 \cos(2000t + 53.1°) \qquad (11.451)$$

To find the other required current responses we observe from Fig. 11.97(b) that the current $\mathbf{I}$ flows through the inductor and upon reaching the parallel combination of the resistor and capacitor divides into two components. Therefore by current division we have

$$\mathbf{I}_R = \frac{Z_C}{Z_R + Z_C} \mathbf{I} = \frac{-j1000}{3000 - j1000} 0.2 \exp[j53.1°]$$

$$= 0.0632 \exp(-j18.5°)$$

and $\qquad\qquad\qquad\qquad\qquad\qquad\qquad\qquad\qquad\qquad\qquad\qquad (11.452)$

$$\mathbf{I}_C = \frac{Z_R}{Z_R + Z_C} \mathbf{I} = \frac{3000}{3000 - j1000} 0.2 \exp[j53.1°]$$

$$= 0.190 \exp(j71.5°)$$

Hence the two waveforms are

$$i_R(t) = 0.0632 \cos(2000t - 18.5°)$$
$$i_C(t) = 0.190 \cos(2000t + 71.5°) \qquad (11.453)$$

One of the important consequences of linearity is superposition. Superposition can be applied in phasor circuit analysis if we observe certain precautions. If all of the sources have the *same* frequency, then the superposition (summation) can be made in phasor form. That is, we can find the phasor response due to each source separately and obtain the response due to all sources by simply adding the individual phasors. However, if the sources have *different* frequencies, we cannot simply add the individual phasors to find the total response. Remember that the phasor contains no information about the frequency of the sinusoid it represents. We cannot add two phasors unless they have the same frequency. All is not lost since we can still use superposition and phasor analysis in the multiple-frequency case. However, we must use caution when finding the individual responses, taking due account of the changes in element impedances because of the different frequencies. But then the individual phasor response for each frequency must be converted into waveforms and superposition applied in the time domain. The multiple-frequency case gives rise to what is termed *frequency response*.

In more complicated circuits these direct methods of analysis could prove cumbersome, and we would need to appeal to a general method such as node analysis or mesh analysis. The formulation of node-voltage or mesh-current equilibrium equations in phasor form follows exactly the same pattern that we have seen before. A typical node equation would be of the form

$$Y_1\mathbf{V}_1 - Y_2\mathbf{V}_2 - Y_3\mathbf{V}_3 = \mathbf{I}_{s1} \qquad (11.454)$$

where the phasors $\mathbf{V}_1$, $\mathbf{V}_2$, and $\mathbf{V}_3$ are the node voltages, Y_1 is the sum of admittances connected to node ①, Y_2 is the admittance connected between nodes ① and ②, Y_3 is the admittance connected between nodes ① and ③, and finally, $\mathbf{I}_{s1}$ is the sum of the current source inputs connected to node

①. Similarly a typical mesh equation would be

$$Z_1 I_1 - Z_2 I_2 = Z_3 I_3 = V_{s1} \tag{11.455}$$

where I_1, I_2, and I_3 are the mesh currents, Z_1 is the sum of impedances in mesh ①, Z_2 is the impedance common to meshes ① and ②, Z_3 is the impedance common to meshes ① and ③, and V_{s1} is the sum of voltage source inputs in mesh ①.

Impedance and Admittance

The concepts of impedance and admittance play key roles in the behavior of circuits in the sinusoidal steady state. The impedances of the three passive elements were given in Eq. (11.446). The impedance of a resistor is purely real while the impedances of energy storage devices are purely imaginary. When we combine these elements in a circuit to obtain an equivalent impedance, we find that it can be written as

$$Z = R + jX \tag{11.456}$$

In this rectangular form we call the real part of the impedance R *resistance* and the imaginary part X *reactance*.

For example, the series RLC circuit in Fig. 11.98(a) has an equivalent impedance of

$$Z = Z_R + Z_L + Z_C = R + j\left(\omega L - \frac{1}{\omega C}\right) \tag{11.457}$$

and we can identify the resistance and reactance as

$$R = R \quad \text{and} \quad X = \left(\omega L - \frac{1}{\omega C}\right) \tag{11.458}$$

In this case it is easy to see that the real part of the input impedance comes from the resistor and the reactive component from the inductor and capacitor. The real part is always positive, but the reactive part can be either positive or negative. In fact we see that at one frequency the reactive part can be

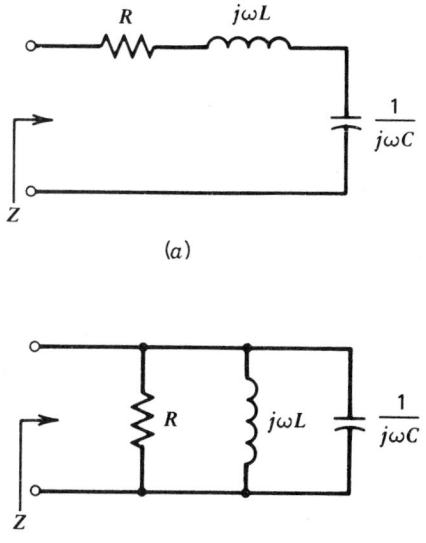

(a)

(b)

Fig. 11.98 (a) A series RLC circuit. (b) A parallel RLC circuit. Reproduced, with permission, from R. E. Thomas and A. J. Rosa, *Circuits and Signals: An Introduction to Linear and Interface Circuits*, New York, Wiley, 1984.

made zero. Specifically at the frequency

$$\omega_0 = \frac{1}{\sqrt{LC}} \tag{11.459}$$

the reactance vanishes and the input impedance is purely resistive. When this occurs, the circuit is said to be in **resonance** and the frequency at which it occurs is called the **resonant frequency**.

Admittance is the reciprocal of impedance ($Y \equiv 1/Z$), and the three passive elements have admittances that were also given in Eq. (11.447). When we combine elements to produce an equivalent admittance, we find that it can be written as

$$Y = G + jB \tag{11.460}$$

In this form we call the real part of the admittance G *conductance* and the imaginary part B *susceptance*.

An example is the parallel RLC circuit in Fig. 11.98(b). The equivalent input admittance of this circuit is

$$Y = Y_R + Y_L + Y_C = \frac{1}{R} + j\left(\omega C - \frac{1}{\omega L}\right) \tag{11.461}$$

and we identify the conductance and susceptance as

$$G = \frac{1}{R} \quad \text{and} \quad B = \left(\omega C - \frac{1}{\omega L}\right) \tag{11.462}$$

The conductance is always positive and is contributed by the resistance. The susceptance comes from the inductor and capacitor and can be either positive or negative. In fact the susceptance vanishes at the frequency

$$\omega_0 = \frac{1}{\sqrt{LC}} \tag{11.463}$$

Thus the frequency at which *either* the reactance *or* the susceptance is zero is called the resonant frequency.

The real and imaginary parts of an impedance or admittance generally depend on frequency. Therefore the circuit will respond to different input frequencies in different ways. In particular, when the input frequency corresponds to a resonant frequency, one at which the imaginary part is zero, the circuit acts as if it were made up of pure resistances. For the reactances to cancel to produce resonance, the circuit must have at least one inductor and one capacitor. As a consequence passive RC or RL circuits cannot exhibit resonance.

In addition to the resonant frequencies, we are often interested in the circuit's response at zero and infinite frequencies. Figure 11.99 contains a summary of the impedances and admittances of the circuit elements. First note that the impedance of the resistor does not depend on frequency. However, the other two elements have zero impedance or admittance at zero and infinite frequencies. Now since $V = ZI$, zero impedance means $V = 0$ regardless of I. But this condition describes a short circuit. Thus the inductor acts as a short circuit at zero frequency and the capacitor as a short circuit at infinite frequency.

Conversely, since $I = YV$, zero admittance means the $I = 0$ regardless of V. But $I = 0$ for any V describes an open circuit. Hence the capacitor acts as an open circuit at zero frequency and the inductor at infinite frequency. These results are very important because capacitors and inductors

	At $\omega = 0$		At $\omega = \infty$	
Resistor	$Z_R = R$	$Y_R = G$	$Z_R = R$	$Y_R = G$
Inductor	$Z_L = 0$	$Y_L = \infty$	$Z_L = \infty$	$Y_L = 0$
Capacitor	$Z_C = \infty$	$Y_C = 0$	$Z_C = 0$	$Y_C = \infty$

Fig. 11.99 Behavior of R, L, C at $\omega = 0$, and $\omega = \infty$. Reproduced, with permission, from R. E. Thomas and A. J. Rosa, *Circuits and Signals: An Introduction to Linear and Interface Circuits*, New York, Wiley, 1984.

are often used in circuits to short-circuit or block either very high or very low frequencies. To study a circuit at zero or infinite frequency we can replace the capacitors and inductors by short or open circuits and analyze the remaining resistive circuit. In this way we can begin to see how to obtain circuits that are frequency selective.

Power in the Sinusoidal Steady State

To investigate power in the sinusoidal steady state we must return to the time domain since power, as we shall see, cannot be represented as a phasor. If we write the voltage and current at the input to a circuit as

$$v(t) = V_0 \cos(\omega t + \phi_V)$$
$$i(t) = I_0 \cos(\omega t + \phi_I)$$
(11.464)

Then the power delivered at the input is

$$p(t) = [v(t)][i(t)]$$
$$= V_0 I_0 \cos(\omega t + \phi_V) \cos(\omega t + \phi_I)$$
(11.465)

we can write the power as

$$p(t) = \frac{V_0 I_0}{2} \cos(\phi_V - \phi_I) + \frac{V_0 I_0}{2} \cos(2\omega t + \phi_V + \phi_I)$$
(11.466)

The instantaneous power in the sinusoidal steady state turns out to be the sum of two terms. The first term is a constant and the second is a sinusoid whose frequency is twice that of the current and voltage. In dealing with power we are usually interested in its average value. Since for the sinusoid the average value is zero, the double-frequency term in Eq. (11.466) makes zero contribution to the average power and therefore

$$P_{ave} = \frac{V_0 I_0}{2} \cos(\phi_V - \phi_I)$$
(11.467)

The average power is proportional to the cosine of the difference between the voltage and current phase angles. It is convenient to represent this phase difference as the angle

$$\theta = \phi_V - \phi_I$$
(11.468)

and then to write Eq. (11.467) as

$$P_{ave} = \frac{V_0 I_0}{2} \cos \theta$$
(11.469)

When written in this form, the term $\cos \theta$ is called the **power factor**.

It is also useful to relate the average power to the input impedance. We can express the phasors of the input current and voltage in Eq. (11.464) as

$$\mathbf{V} = V_0 \exp(j\phi_V) \quad \text{and} \quad \mathbf{I} = I_0 \exp(j\phi_I)$$
(11.470)

These two phasors are related by an input impedance, which we write in polar form as

$$Z = |Z| \exp(j\phi_Z)$$
(11.471)

But since $\mathbf{V} = Z\mathbf{I}$ we can write

$$\frac{\mathbf{V}}{\mathbf{I}} = \frac{V_0 \exp(j\phi_V)}{I_0 \exp(j\phi_I)} = \frac{V_0}{I_0} \exp[j(\phi_V - \phi_I)] = |Z| \exp(j\phi_Z)$$
(11.472)

and we find that the difference between the voltage and current phase angles equals the angle of the impedance. But writing the impedance in rectangular form as $Z = R + jX$, we see that the cosine of ϕ_Z, and hence the power factor, is

$$\cos \phi_Z = \cos \theta = \frac{R}{|Z|}$$
(11.473)

substitution back into Eq. (11.469) we get

$$P_{ave} = \frac{V_0 I_0}{2} \frac{R}{|Z|} = \frac{I_0^2}{2} R = I_{rms}^2 R \qquad (11.474)$$

Since $V_0 = |Z| I_0$, and $I_0 / \sqrt{2} = I_{rms}$.

Equation (11.474) is another way to determine average power, and it points out that average power is proportional to the input resistance. If we apply this idea to the individual circuit elements, we see that the resistor absorbs power, but that the inductor and capacitor cannot absorb power since they are purely reactive and have no resistance. This does not mean that the instantaneous power drawn by the reactive elements is zero since, as Eq. (11.466) points out, there is a double-frequency sinusoidal term. Thus the reactive elements extract power from the circuit on one-half of the cycle but promptly return exactly the same amount of power on the next half-cycle, thus drawing no power on the average.

For maximum power transfer it can be shown that Z_L, the equivalent load impedance, must be the complex conjugate of the source impedance, Z_S.

Three-Phase Power

One of the major applications of phasor analysis is in the study of ac electric power systems. Although the loads on power systems vary during any one day, these variations are extremely slow compared with the period of the sinusoid involved. Consequently the analysis of a power system can be carried out using steady-state concepts and the phasor.

The predominant form for generating and distributing electric power is the three-phase system shown in Fig. 11.100. The system has three lines (A, B, C), which are used to transmit power from the source to the loads. The figure also shows a fourth line, labeled N, for neutral. In a balanced system, the only type we consider, this fourth line carries no current or power, but does serve as a reference for defining voltages. The three-phase source in the illustration is modeled as three independent sources, although a three-phase generator is a single piece of equipment with three separate windings. Similarly the loads are modeled as three separate impedances, although a three-phase load may be housed within a single container.

The terminology Y connected and Δ connected refers to the two different ways the source and loads can be electrically arranged. In almost all practical systems the source is Y connected while the loads can be either Y or Δ, although Δ is more common.

To identify voltages in the system we use a special double-subscript notation. The reason is that there are at least six voltages to deal with; the three line-to-line voltages and the three line-to-neutral voltages. Consequently we use two subscripts to define the points across which the voltage is defined. For example, V_{AN} means the voltage of line A with respect to neutral (N), with the implied plus reference mark at A and the implied minus at N. The three line-to-neutral voltages, called the **phase**

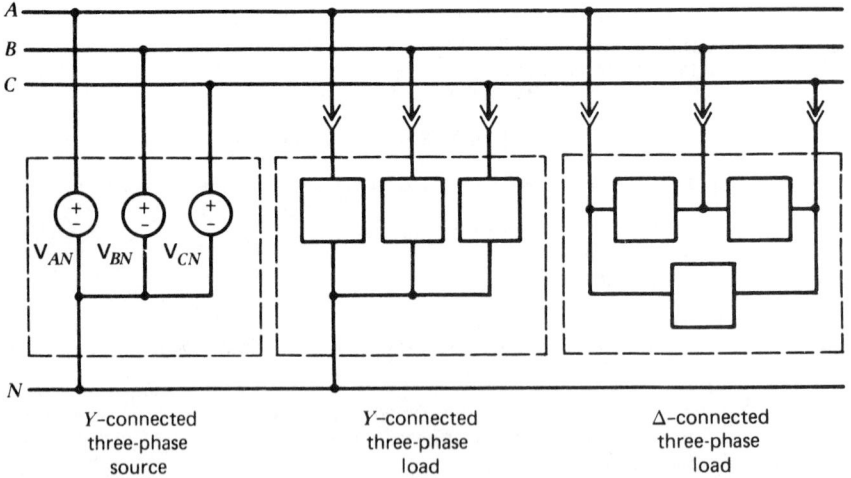

| Y-connected three-phase source | Y-connected three-phase load | Δ-connected three-phase load |

Fig. 11.100 Three-phase power system. Reproduced, with permission, from R. E. Thomas and A. J. Rosa, *Circuits and Signals: An Introduction to Linear and Interface Circuits*, New York, Wiley, 1984.

voltages, are identified as V_{AN}, V_{BN}, and V_{CN}. The line-to-line voltages, called simply the **line voltages**, are identified as V_{AB}, V_{BC}, and V_{CA}. In this notation it follows that $V_{BN} = -V_{NB}$, and consequently $V_{AB} = V_{AN} + V_{NB} = V_{AN} - V_{BN}$.

A balanced three-phase source produces phase voltages that obey the following two constraints

$$|V_{AN}| + |V_{BN}| + |V_{CN}| = V_P$$
$$V_{AN} + V_{BN} + V_{CN} = 0 \qquad (11.475)$$

That is, the phase voltages all have the same amplitude (V_P), and they sum to zero. There are two possible arrangements that satisfy these constraints:

POSITIVE SEQUENCE	NEGATIVE SEQUENCE	
$V_{AN} = V_P \angle 0°$	$V_{AN} = V_P \angle 0°$	
$V_{BN} = V_P \angle -120°$	$V_{BN} = V_P \angle -240°$	(11.476)
$V_{CN} = V_P \angle -240°$	$V_{CN} = V_P \angle -120°$	

These two cases are called the positive phase sequence and the negative phase sequence, respectively, and Fig. 11.101 shows the corresponding phasor diagrams. It is apparent that either case involves three phasors of equal length, which are all separated by a phase angle of 120°, so that the sum of any two exactly cancels the third. It is also apparent that we can convert one phase sequence into the other by simply interchanging the labels on lines B and C. Thus there is no conceptual difference between the two sequences. In what follows we always use the positive phase sequence.

The reader is cautioned that although there is no conceptual difference, this is not the same as saying that the phase sequence is unimportant. It turns out that three-phase motors run in *one* direction when the positive sequence is applied, and in the *opposite* direction for the negative sequence.

The relationship between the line voltages and the phase voltages can be obtained by manipulating the phase voltage phasors:

$$V_{AB} = V_{AN} - V_{BN} = \sqrt{3}\, V_P \angle 30°$$

$$V_{BC} = \sqrt{3}\, V_P \angle -90° \qquad (11.477)$$

$$V_{CA} = \sqrt{3}\, V_P \angle -210°$$

Figure 11.102 shows the phasor diagram of these results. The line voltages are all of equal amplitude and displaced by 120°, and hence obey the same constraints as the phase voltages. If we denote the amplitude of the line voltages as V_L, then

$$V_L = \sqrt{3}\, V_P \qquad (11.478)$$

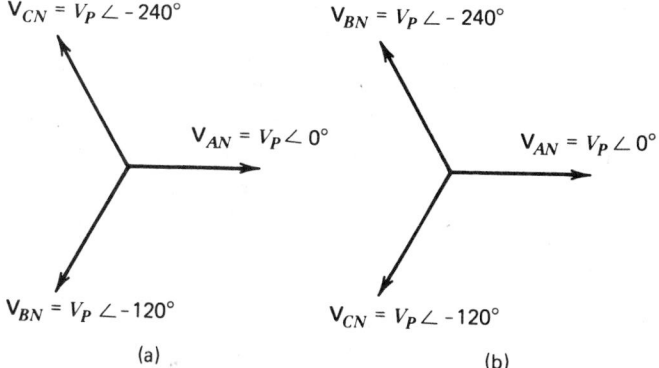

(a)   (b)

Fig. 11.101 Two possible phase sequences: (*a*) positive, (*b*) negative. Reproduced, with permission, from R. E. Thomas and A. J. Rosa, *Circuits and Signals: An Introduction to Linear and Interface Circuits*, New York, Wiley, 1984.

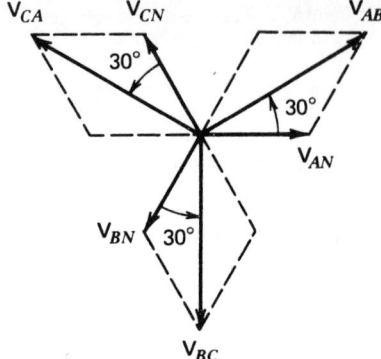

Fig. 11.102 Phasor diagram showing the relationships between phase voltages and line voltages. Reproduced, with permission, from R. E. Thomas and A. J. Rosa, *Circuits and Signals: An Introduction to Linear and Interface Circuits*, New York, Wiley, 1984.

Thus the line voltage amplitude is $\sqrt{3}$ greater than the phase voltage amplitude. This ratio appears in descriptions such as 120/208 V three phase. The 120 is the phase voltage and 208 is the line voltage.

The voltages are also normally expressed in terms of rms values rather than peak values. For sinusoids the relationship between peak and rms is $\sqrt{2}$, so that the equation for the average power can be written as

$$P = \frac{|V|}{\sqrt{2}} \frac{|I|}{\sqrt{2}} \cos = V_{\text{rms}} I_{\text{rms}} \cos \theta \qquad (11.479)$$

Since we shall make frequent use of this equation to determine power, it is more convenient to use the rms value. In dealing with three-phase power systems, phasor magnitudes always represent the rms value of a current or voltage rather than the peak value.

The phase reference for all voltages and currents is taken to be the line A phase voltage, that is, $\mathbf{V}_{AN} = V_P \angle 0°$. This practice can be seen in Figs. 11.101 and 11.102. If we report that a current is $14.7 \angle - 137°$, this means that the amplitude of the current is 14.7A rms and it is displaced from line A phase voltage by $-137°$.

We now turn to the analysis of a balanced three-phase circuit shown in Fig. 11.103. The load is balanced; that is, the impedances in the legs of the Y-connected load are equal. Our goal is to determine the source-load interaction at the interface in terms of the three line currents $\mathbf{I}_A$, $\mathbf{I}_B$, and

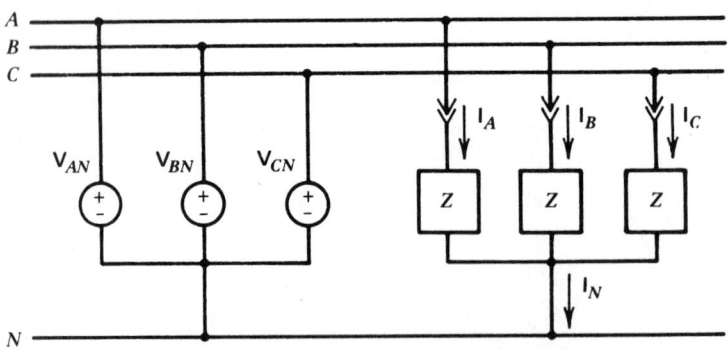

Fig. 11.103 Y-connected source and load. Reproduced, with permission, from R. E. Thomas and A. J. Rosa, *Circuits and Signals: An Introduction to Linear and Interface Circuits*, New York, Wiley, 1984.

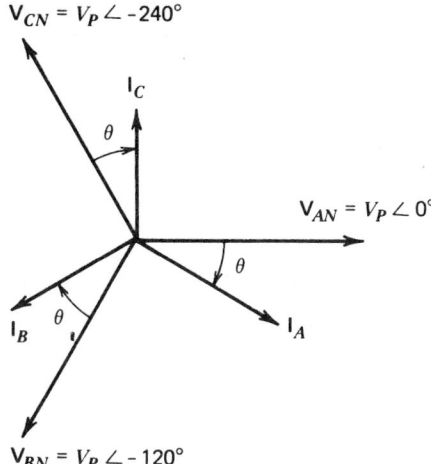

Fig. 11.104 Phasor diagram relating line currents and phase voltages in a balanced three-phase system. Reproduced, with permission, from R. E. Thomas and A. J. Rosa, *Circuits and Signals: An Introduction to Linear and Interface Circuits*, New York, Wiley, 1984.

I_C, and the total power delivered to the load:

$$\mathbf{I}_A = \frac{\mathbf{V}_{AN}}{Z} = \frac{V_P\angle 0°}{|Z|\angle\theta} = \frac{V_P}{|Z|}\angle -\theta$$

$$\mathbf{I}_B = \frac{\mathbf{V}_{BN}}{Z} = \frac{V_P\angle -120°}{|Z|\angle\theta} = \frac{V_P}{|Z|}\angle -120° -\theta \qquad (11.480)$$

$$\mathbf{I}_C = \frac{\mathbf{V}_{CN}}{Z} = \frac{V_P\angle -240°}{|Z|\angle\theta} = \frac{V_P}{|Z|}\angle -240° -\theta$$

Figure 11.104 shows the phasor diagram relating the line currents and phase voltages. The three line currents are of equal amplitude and symmetrically disposed at 120° intervals. This means that $\mathbf{I}_A + \mathbf{I}_B + \mathbf{I}_C = 0$. If we apply KCL at the neutral connection of the load in Fig. 11.103, we conclude that $\mathbf{I}_N = 0$. In a balanced three-phase system no current flows in the neutral wire. The neutral wire could be replaced by any impedance, including infinity. In other words, the neutral wire can be disconnected without upsetting the system balance. In real systems the neutral wire may or may not be present.

We now turn to the more common balanced Δ load shown in Fig. 11.105. Our objective remains to determine the interface line currents and the total load power. But since the impedances here are

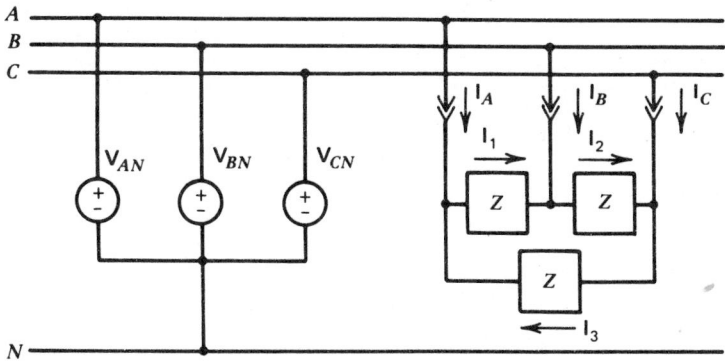

Fig. 11.105 Y-connected source and a Δ-connected load. Reproduced, with permission, from R. E. Thomas and A. J. Rosa, *Circuits and Signals: An Introduction to Linear and Interface Circuits*, New York, Wiley, 1984.

connected from line to line, we must go through the intermediate step of first determining the phase currents I_1, I_2, and I_3 from the phase impedance and the line voltages:

$$I_1 = \frac{V_{AB}}{Z} = \frac{V_L \angle 30°}{Z} = \frac{0.866V_L + j0.5V_L}{Z}$$

$$I_2 = \frac{V_{BC}}{Z} = \frac{V_L \angle -90°}{Z} = \frac{-jV_L}{Z} \qquad (11.481)$$

$$I_3 = \frac{V_{CA}}{Z} = \frac{V_L \angle -210°}{Z} = \frac{-0.866V_L + j0.5V_L}{Z}$$

The phase currents are of equal amplitude (the system is balanced) and disposed at 120° intervals. Applying KCL at the connection points of the impedance, we obtain the line currents:

$$I_A = I_1 - I_3 = \frac{\sqrt{3}\,V_L}{Z} = \frac{\sqrt{3}\,V_L}{|Z|} \angle -\theta$$

$$I_B = I_2 - I_1 = \frac{-0.866V_L - j1.5V_L}{Z} = \frac{\sqrt{3}\,V_L}{|Z|} \angle -120° -\theta \qquad (11.482)$$

$$I_C = I_3 - I_2 = \frac{-0.866V_L + j1.5V_L}{Z} = \frac{\sqrt{3}\,V_L}{|Z|} \angle -240° -\theta$$

The power factor angle of the line currents is the angle of the impedance Z. This was true for the Y load [see Eq. (11.480)] because the impedances were connected from line to neutral. But it is also true here even though the impedances are connected from line to line. The power factor of the load is determined by its impedance and it matters not whether the load is Y or Δ connected. Finally, by comparing Eqs. (11.481) and (11.482), we find that

$$I_L = \sqrt{3}\,I_P \qquad (11.483)$$

The line current amplitude in a Δ load is $\sqrt{3}$ greater than the phase current.

Three-phase loads are often described in terms of the power they draw at the interface, rather than their impedances. In such cases we still have the goal of determining the line currents. For a given total power P_T and power factor $(\cos \theta)$, in a balanced system we can obtain the amplitude of the line current from the phase-power relationship as

$$I_L = \frac{P_T/3}{V_P \cos \theta} \qquad (11.484)$$

This yields the amplitude of the line current but not its phase angle. The phase angle is provided by specifying that the power factor is **lagging**, or **leading**. This means that the phase A line current is

$$I_A = I_L \angle \pm \theta \qquad (11.485)$$

where the plus sign applies for leading power factors and the minus for lagging.

Frequency Response
We have seen that signals can be described in terms of their frequency content. Frequency-domain signal processing is performed when a circuit selectively affects the frequencies contained in the input signal to produce the desired output. We found that for an input signal $A \cos(\omega t + \phi)$ to a circuit with a transfer function $H(s)$, the sinusoidal steady-state output was given by Eq. (11.436), which we could wire as

$$y_{ss} = A|H(j\omega)|\cos[\omega t + \phi + \angle H(j\omega)] \qquad (11.486)$$

which leads to the input-output rules:

$$\text{Output amplitude} = A|H(j\omega)|$$
$$\text{Output phase} = \phi + \angle H(j\omega)$$
$$\text{Output frequency} = \omega$$

In sum, the output amplitude is modified by the magnitude of the transfer function evaluated at the

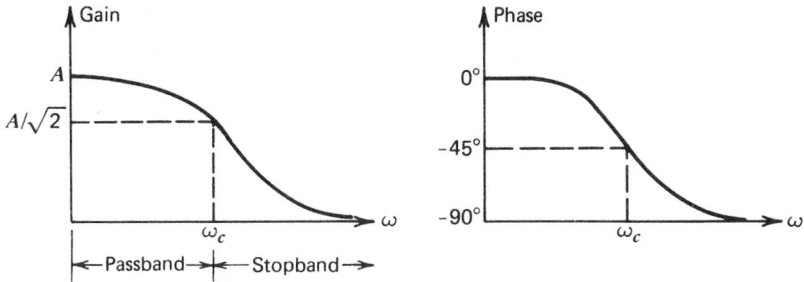

Fig. 11.106 Typical frequency response plots. Reproduced, with permission, from R. E. Thomas and A. J. Rosa, *Circuits and Signals: An Introduction to Linear and Interface Circuits*, New York, Wiley, 1984.

frequency of the input. Likewise the output phase angle is altered by the angle of the transfer function evaluated at the same frequency. To study the frequency selectivity characteristics of a circuit we need to look at its transfer function for all frequencies of interest.

The effect of the circuit on the frequency response is represented by the functions $|H(j\omega)|$ and $\angle H(j\omega)$, called the gain and the phase, respectively. The gain and phase functions are the frequency-dependent relationships that are used to describe the *frequency response* of the circuit. In effect then, if we evaluate a transfer function $H(s)$ at $s = j\omega$, then the magnitude of $H(j\omega)$ is the circuit gain as a function of frequency and the angle is its phase shift. The gain and phase functions of a circuit can be expressed mathematically or graphically. When presented graphically the result is called a frequency response plot, such as the example in Fig. 11.106.

A system of terminology based on the gain of the circuit is used to describe its frequency response. For example, Fig. 11.106 shows that the gain is essentially constant if the input frequency is below ω_c. The range of frequencies over which this occurs is called a **transmission band**, or **passband**. Conversely, for frequencies above ω_c, the gain falls off so that the output signals in this range are smaller than those in the passband. The range of frequencies over which this occurs is called an **attenuation band**, or **stopband**.

The frequency associated with the boundary between a transmission band and an adjacent attenuation band is called the **cutoff frequency** (ω_c in Fig. 11.106). As Fig. 11.106 shows, the transition from the passband to the stopband is gradual, and therefore the precise location of the cutoff frequency is a matter of definition. The most widely used definition is that cutoff occurs when the gain has decreased by a factor of $1/\sqrt{2} = 0.707$ from the maximum gain in the passband. Again this definition is arbitrary since there is no sharp boundary between a transmission band and the adjacent attenuation band. However, the definition is motivated by the fact that the power carried by a waveform (current or voltage) is proportional to the square of its amplitude. If the amplitude is reduced by a factor of $1/\sqrt{2}$, then the power carried is reduced by a factor of one-half. For this reason the cutoff frequency is also called the **half-power frequency**.

Additional frequency response terminology is based on the four prototype gain characteristics shown in Fig. 11.107. A *low-pass* gain has single transmission band extending from zero frequency (dc) to some cutoff frequency ω_c. Conversely a *high-pass* characteristic has a single transmission band extending from some cutoff frequency ω_c to infinite frequency. A *bandpass* gain has a single transmission band with two cutoff frequencies ω_{c1}, ω_{c2}, neither of which is zero or infinite. Finally, the *bandstop* gain has a single attenuation band with two cutoff frequencies ω_{c1}, ω_{c2} neither of which is zero or infinite. Figure 11.107 also shows how each of the four prototypes alters the input spectrum to produce distinctly different outputs.

Frequency response plots are almost always made using a logarithmic scale for the frequency variable and the gain. In part this is done because the frequency ranges of interest span many orders of magnitude. But a logarithmic frequency scale also tends to compress the data and highlight dominant features. The use of a logarithmic frequency scale involves some special terminology. Any frequency range whose end points have a 2 : 1 ratio is called an **octave**. Likewise any range whose end points have a 10 : 1 ratio is called a **decade**. For example, the frequency range from 10 to 20 Hz is one octave, as is the range from 20 to 40 MHz. The standard UHF (ultra high-frequency) band extends from 0.3 to 3 GHz, or one decade. The common audio range (20 Hz to 20 kHz) is then three decades (20 : 200, 200 : 2000, 2000 : 20,000). Frequency response plots that use a logarithmic scale for both gain and frequency axes are called **Bode plots**. These plots highlight dominant features, reveal certain symmetries, and open the way to new insights into the relationships between poles and zeros and frequency response.

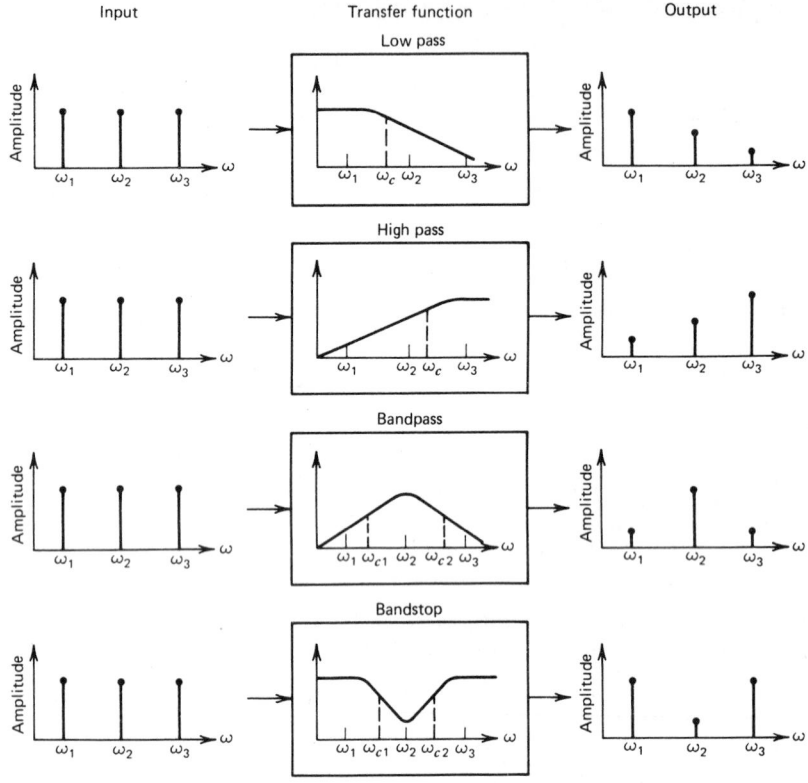

Fig. 11.107 Four prototype filter gain characteristics with resulting effect on constant input. Reproduced, with permission, from R. E. Thomas and A. J. Rosa, *Circuits and Signals: An Introduction to Linear and Interface Circuits*, New York, Wiley, 1984.

BIBLIOGRAPHY

Ampere, A. M., "Sur l'etat magnetique des corps qui transmettant un courant d'electricite," *Ann. Chem. Phys.* **XVI**, Paris (1821).

Attwood, S. S., *Electric and Magnetic Fields*, New York, Wiley, 1949.

Bobrow, L. S., *Elementary Linear Circuit Analysis*, New York, Holt, Rinehart and Winston, 1981.

Bohr, N., "On the Constitution of Atoms and Molecules," *Philos. Mag.* **26**, 476 (1913).

Burr-Brown, *Handbook of Operational Amplifier Applications*, Arizona, Burr-Brown Research Corp., 1963.

Dahl, O. G. C., *Electric Circuits, Theory and Applications*, New York, McGraw-Hill, 1928.

Faraday, M., *Experimental Researches in Electricity*, 3 vols., London, R. and J. E. Taylor, 1839–1855.

Fink, D. G., and J. M. Carroll, *Standard Handbook for Electrical Engineers*, 10th ed., New York, McGraw-Hill, 1968.

Hague, B., *Electromagnetic Problems in Electrical Engineering*, London, Oxford University Press, 1929.

Hayt, W. H., Jr., *Engineering Electromagnetics*, 4th ed., New York, McGraw-Hill, 1981.

Hayt, W. H., Jr., and J. E. Kemmerly, *Engineering Circuit Analysis*, 3rd ed., New York, McGraw-Hill, 1978.

Heaviside, O., *Electromagnetic Theory*, London, Electrician Publishing, 1893.

Hertz, H., *Electric Waves*, Trans. by D. F. Jones, London, 1893, Reprint Dover, New York, 1962.

Hodgman, C. D. (Ed.), *Mathematical Tables*, from *Handbook of Chemistry and Physics*, Ohio, The Chemical Rubber Publishing Co., pp. 294–301, 1963.

BIBLIOGRAPHY

Jeans, J. H., *Mathematical Theory of Electricity and Magnetism*, Cambridge, University Press, 1908.

Jordan, E. C., and K. G. Balmain, *Electromagnetic Waves and Radiating Systems*, 2nd ed., Englewood Cliffs, N.J., Prentice-Hall, 1968.

Kittel, C., *Introduction to Solid State Physics*, 5th ed., New York, Wiley, 1976.

Kraus, J. D., *Electromagnetics*, 3rd ed., New York, McGraw-Hill, 1984.

Lorentz, H. A., *The Theory of Electrons*, Leipzig, B. G. Teubner, 1909.

McGillem, C. C., and Cooper, G. R., *Continuous and Discrete Signal and System Analysis*, New York, Holt, Rinehart and Winston, 1974.

Maxwell, J. C., *A Treatise on Electricity and Magnetism*, 2 Vols., London, Oxford University Press, 1904.

Maxwell, J. C., *Elementary Treatise on Electricity*, W. Garnett (Ed.), London, Oxford University Press, 1888.

Millikan, R. A., *The Electron*, Chicago, University of Chicago Press, 1924.

National Bureau of Standards (U.S.), *Monograph 56*, August, 1963.

Nilsson, J. W., *Electric Circuits*, Reading, Mass., Addison-Wesley, 1983.

Ohm, G. S., *The Galvanic Circuit Investigated Mathematically*, Transl. W. Francis, New York, Van Nostrand, 1891.

Oliver, B. M., and J. M. Gage, *Electronic Measurements and Instrumentation*, New York, McGraw-Hill, 1971.

Pender, H., and S. R. Warren, Jr., *Electric Circuits and Fields*, New York, McGraw-Hill, 1943.

Priestley, J., *History and the Present State of Electricity with Original Experiments*, London, 1767, Johnson Reprint, New York, 1966.

Russell, A., *The Theory of Alternating Currents*, Cambridge, University Press, 1914.

Shockley, W., *Electrons and Holes in Semiconductors*, Bell Laboratories Series, New Jersey, D. Van Nostrand Company, 1950.

Startling, S. G., *Electricity and Magnetism for Advanced Students*, London, Longmans Green, 1924.

Steinmetz, C. P., *Transient Electric Phenomena and Oscillations*, New York, McGraw-Hill, 1920.

Taylor, B. N., Parker, W. H., and Langenberg, D. N., "Determination of e/h, Using Macroscopic Quantum Phase Coherence in Superconductors: Implications for Quantum Electrodynamics and the Fundamental Physical Constants," *Rev. Mod. Phys.*, **41**, 375, (July 1969).

Thomas, R. E., and Rosa, A. J., *Circuits and Signals*: *An Introduction to Linear and Interface Circuits*, New York, Wiley, 1984.

Thompson, J. J., *Elements of Electricity and Magnetism*, New York, MacMillan, 1921.

Thompson, S. P., *Elementary Lessons in Electricity and Magnetism*, New York, MacMillan, 1915.

Thompson, S. P., *LaPlace Transformations*, New York, Prentice-Hall, 1950.

Van Valkenburg, M. E., *Network Analysis*, Englewood Cliffs, N.J., Prentice-Hall, 1955.

Van Vleck, J. H., *The Theory of Electric and Magnetic Susceptibilities*, London, Oxford University Press, 1932.

Wilson, A. H., *Theory of Electricity and Magnetism*, London, MacMillan, 1897.

CHAPTER 12

ELECTRONICS*

Velio A. Marsocci, Editor

Department of Electrical Engineering
State University of New York
Stony Brook, New York

Narenda Mohan

Department of Electrical Engineering
University of Minnesota
Minneapolis, Minnesota

Robert L. Kustom

Argonne National Laboratory
Argonne, Illinois

Kenneth Sohn

Department of Electrical Engineering
New Jersey Institute of Technology
Newark, New Jersey

Denny D. Tang

IBM Thomas J. Watson Research Center
Yorktown Heights, New York

Andres Fortino

Department of Electrical Engineering
Temple University
Philadelphia, Pennsylvania

Sheldon S. L. Chang

Department of Electrical Engineering
State University of New York
Stony Brook, New York

Harry N. Norton

Jet Propulsion Laboratory
California Institute of Technology
Pasadana, California

Yusuf Ziya Efe

Department of Electrical Engineering
The Cooper Union College
New York, New York

Kenneth Short

Department of Electrical Engineering
State University of New York
Stony Brook, New York

David F. Bantz

Department of Computer Science
IBM Thomas J. Watson Research Center
Yorktown Heights, New York

Emory J. Harry

Tektronix, Inc.
Beaverton, Oregon

Robert E. Metzler

Tektronix, Inc.
Beaverton, Oregon

*Chapter 12 is adapted, with permission, from Sheldon S. L. Chang, Ed., *Fundamentals Handbook of Electrical and Computer Engineering*, Wiley, New York, 1982, Vol. 1: Section 6, 7; Vol. 2: Section 5; Vol. 3: Section 1.

12.1   INTRODUCTION
Velio A. Marsocci

Not too many decades ago the task of producing an overview of the subject area of electronics would have focused on decisions involving how many of the details relative to the subject should be included rather than on the topics to be selected. The electronics engineer rarely, if ever, was exposed to the details of electron ballistics as they pertained to vacuum-tube devices; the circuit analyses usually assumed that the frequency limitations were imposed by the design of the circuitry external to the device, and by the interelectrode capacities of the vacuum device. The effects produced by electron transport were mentioned but not brought into the usual circuit analyses. Thus, the electronic devices were representable by equivalent circuits composed of discrete circuit elements. A reasonably good coverage of the subject of electronics, as referred to at that time, could have been presented by a summary describing the characteristics of a few types of vacuum tubes and of some of the conventional audio and radio-frequency amplifier circuits, and possibly some microwave circuitry. Nonlinear circuits, which include digital circuits, would have been given a small space in the work. In fact, if the text were aimed at engineers in general, and not necessarily only electrical engineers, it was most likely material on digital electronics would have been entirely omitted.

The situation has changed considerably when the aforementioned task of selecting subject matter is to be implemented in the context of present-day electronics. The complications that enter arise from several causative factors. The usage of the word *electronics* has evolved to be the generic term for all hardware at the device, circuit, and overall systems levels; whereas in the past, electronics was understood to mean only the active devices and only the circuits employing such elements. In addition, the active as well as the passive electronic devices are now overwhelmingly of the solid-state type, and the details of the underlying physical mechanisms that produce their characteristics can no longer be separated from the analyses and the design of systems utilizing such devices as components. The added complication to the problem by the pervasion of digital systems into the arena is manifest and the subject is, of itself, vast in the coverage required for its description.

The present handbook is meant to provide information concerning the broad spectrum of engineering principles, and therefore this chapter on electronics must necessarily be very limited. The selection of material is, therefore, such that it can provide only a brief review of the fundamentals of certain basic topics of interest; in order to pursue any of the subjects presented to greater depths, the electronics/electrical engineer is directed, as a starting point, to the bibliographies at the end of the chapter. The viewpoint that guided the selection of material took, as a prime consideration, the topics that would be oriented toward the manner in which electronic hardware would be most likely utilized by nonelectrical as well as electrical engineers. It was noted that many engineers, primarily nonelectrical, encounter electronic systems mainly in the context of instrumentation. Therefore, it was decided to arrange the presentation of the material herein so that attention was given both to the description of the electronic subsystems, that is, sensors, operational amplifiers, microprocessors, and so on, and to the manner in which these influence the characteristics, and are incorporated into the design, of electronic instrumentation.

The present chapter may be viewed as consisting of two parts. The first part, comprising Sections 12.2–12.7, contains material describing the basic characteristics of modern electronic devices and of the electronic networks at the individual circuit level. The second part, Sections 12.8–12.12 are oriented toward the description of electronic subsystems with emphases on operation amplifiers and on microprocessors. These are arranged, as previously mentioned, in a manner that describes their inclusion and their function in electronic instrumentation.

It was further decided at the outset that a selected but brief bibliography would be provided by the author of each of the sections. That is, the intent is to instruct the designer and to provide orientation for the individual who may be reading into a particular subject field for the first time, and not to provide the reader with a large and comprehensive bibliography and list of references. Given the rate of generation of new publications, these latter tend to become obsolete with great rapidity.

Needless to say, any systems or circuits, as described in this chapter, are presented only as illustrative material to enhance the discussion given by the author. There has been no attempt made in the material to offer a design that is complete in the sense that all aspects of testing, packaging, and certifying for safety against hazards have been assured. The authors, the editors, and the publishers are not responsible for any harm or injury caused by the use of the material described herein to design, to construct, to test, or to give instructions to others, with regard to any devices, circuits, subsystems, or systems.

12.2 POWER ELECTRONICS I
Narenda Mohan

Power electronics deals with the use of power semiconductor devices for the control and conversion of large amounts of electrical power. The effective voltage, effective current, frequency, and power factor are some of the electrical waveform parameters that are controlled. Generally, control of the electrical waveform is desired simply as a means of controlling the output, which may be nonelectrical: for example, the speed of a motor driving a pump. The most important distinction between power electronics and small-signal electronics is the requirement of high power conversion efficiency because of the cost of wasted energy, and the difficulty in removing the dissipated heat may be very significant. This requires that power semiconductor devices must operate as a switch–either in the fully "on" or "off" state—rather than operating in their linear region.[1]

12.2.1 Power Semiconductor Devices

All semiconductor switches carry current in only one direction, known as the forward direction. These switches can be divided into three categories based on their external operating characteristics:

1. Diodes. These conduct current when the forward polarity voltage is applied.
2. Thyristors or Silicon-Controlled Rectifiers. These begin to conduct when a forward polarity voltage and a gate control signal are applied, and continue to conduct until the current is brought to zero under the influence of the external electrical circuit.
3. Power Transistors (Bipolar Junction and Field-Effect Transistors), Gate Turn-Off Thyristors, or Forced Commutated Thyristors. In these the forward conduction can be initiated and interrupted by means of control signals.

Only silicon-controlled rectifiers will be described.

Silicon-Controlled Rectifiers

Silicon-controlled rectifiers (SCRs) contain three internal *pn* junctions, as shown in Fig. 12.1(*a*). Probably the best way of understanding SCR operation is to consider its equivalent circuit, shown in Fig. 12.1(*c*). With a forward voltage applied (anode positive with respect to the cathode), the SCR is switched on by applying a positive gate pulse with respect to the cathode. As the gate current flows into *G*, the *npn* transistor is turned on, causing the base current of the *pnp* transistor to flow, thereby turning it on. The collector current of the *pnp* transistor goes into the base of the *npn* transistor; thereby the two transistors are latched into conduction. Once the SCR begins to conduct, the gate loses its control function and the gate pulse is removed to minimize the internal power dissipation. The SCR turns off only if the current through it is decreased to zero under the influence of the external circuit in which it is connected.

Figure 12.2(*a*) shows the static $V - I$ characteristic, and Fig. 12.2(*b*) shows the gate triggering requirements. Care should be taken not to exceed the maximum voltage and the power capabilities of the gate. The gate should be operated in the region shown crosshatched when the triggering is not required. For "hard firing," which assures SCR triggering, the gate signals should be in the upper right part of the next region in Fig. 12.2(*b*).[1] The ratings and the gate triggering circuits of various SCRs are discussed in Refs. 2 and 3.

Triacs

Triacs are two back-to-back connected thyristors integrated into a single device as shown in Fig. 12.3. Triacs are widely used as switches for moderate power levels because of their low cost and the need for only one triggering circuit instead of two for the back-to-back-connected SCRs. One important difference between the use of a triac and a pair of SCRs in an ac circuit is that each SCR has an entire half-cycle to turn off, whereas the triac must turn off during the brief instant while the load current is passing through zero (i.e., one-half of the device must block forward current immediately after the other half has ceased conduction). Because of this difficult requirement, it has not been possible to build triacs with voltage, current, or frequency ratings as high as those of SCRs.[2,3]

12.2.2   Phase-Controlled Rectifiers and ac-Line-Voltage-Commutated Inverters

Converter is a general term and refers to the switch arrangement capable of converting ac and dc quantities. The term **rectifier** is specifically used for ac-to-dc conversion and the term **inverter** for dc-to-ac conversion. The gate control of SCRs, which generally make up these converters, makes it possible to operate the same circuit either in the rectifier or in the inverter mode. When operated as a rectifier, the converter supplies power to the dc side, and when operated as an inverter, the dc source supplies power to the ac system.

Single-Phase Full-Wave Converter

Figure 12.4 shows a single-phase converter that makes use of a transformer with a center-tapped secondary winding. The transformer turns ratio is selected to yield the desired dc voltage range from the ac voltage of a specified magnitude. In the following analysis, the SCRs, the transformer, and the ac voltage source are assumed to be ideal, and the smoothing inductor L_d is sufficiently large to maintain continuous dc current into the battery-type dc load.

The gate triggering of each of the two SCRs is delayed by α degrees (called the firing angle) with respect to the instant when the forward polarity voltage first appears across it in each cycle of operation. The average voltage across L_d must be zero and the battery voltage E_d depends on the

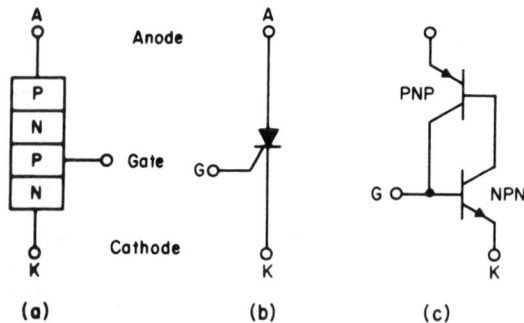

Fig. 12.1   (*a*) SCR with three *pn* junctions, (*b*) device symbol, (*c*) equivalent circuit.

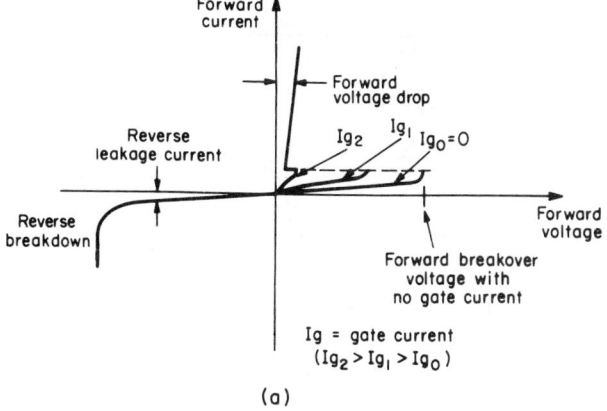

(a)

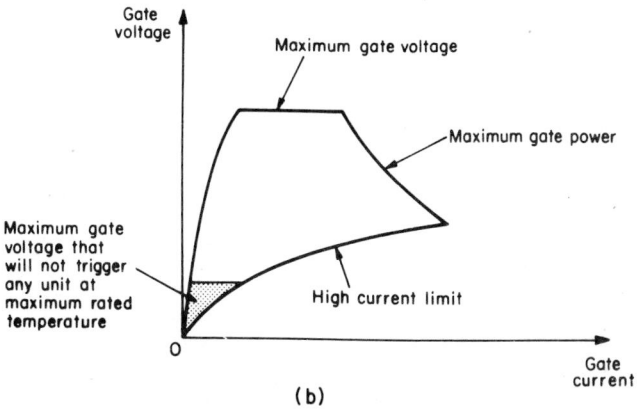

(b)

Fig. 12.2 (a) SCR V-I characteristic, (b) gate trigger characteristic of an SCR.

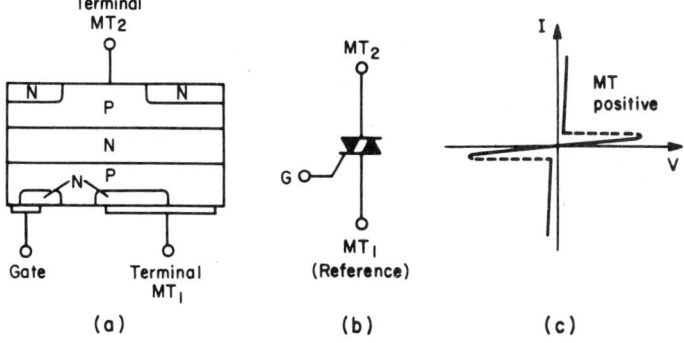

(a) (b) (c)

Fig. 12.3 (a) Simplified triac structure, (b) triac symbol, (c) triac V-I characteristic.

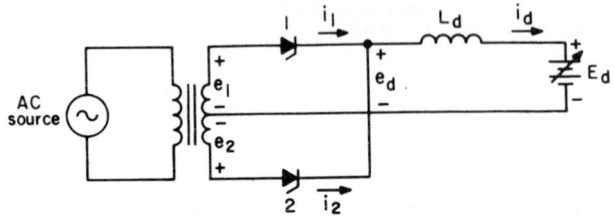

Fig. 12.4   Single-phase full-wave converter.

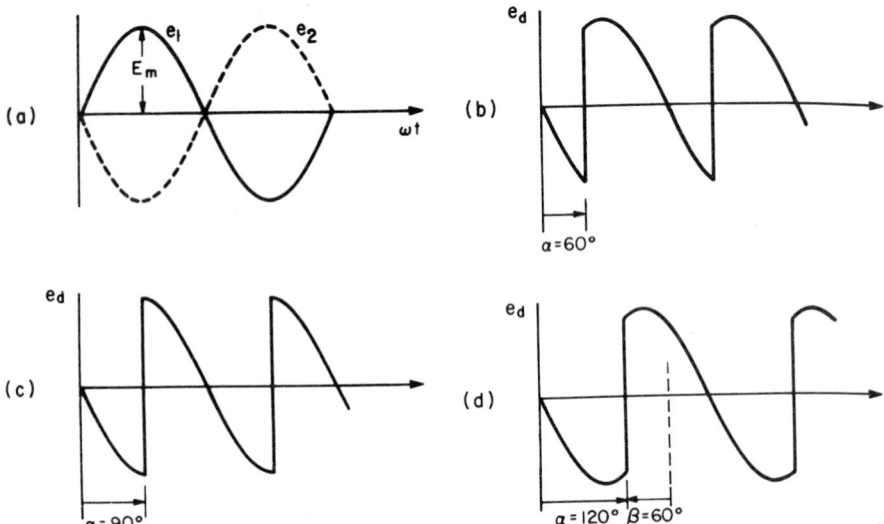

Fig. 12.5   Single-phase full-wave converter voltage waveforms: (*a*) transformer secondary voltages, (*b*) e_d with $\alpha = 60°$, (*c*) e_d with $\alpha = 90°$, (*d*) e_d with $\alpha = 120°$ or $\beta = 60°$.

firing angle α. An arbitrary and nonzero magnitude of dc current i_d is assumed to flow. The average output voltage is given as

$$E_d = \frac{2E_m}{\pi}\cos\alpha$$

Figure 12.5 shows the voltage waveforms for three values of α. For α equal to 90°, the average dc output is zero, and this is the changeover point from the rectifier to the inverter operation. Figure 12.5(*d*) shows inverter operation with $\alpha = 120°$ and $\beta = 180° - \alpha = 60°$. Ideally, β can be increased up to 180°; however, in practical circuits, $\beta = 180° - \alpha$ should be large enough to allow sufficient time between the instant the current through a SCR goes to zero and the instant when a forward voltage appears across it. This interval should be greater than the turn-off or recovery time of the SCR, which permits it to block a forward polarity voltage.

12.3   POWER ELECTRONICS II
Robert L. Kustom

12.3.1   Thyratrons*

Thyratrons are popularly used as accurately timed switching devices for medium- and high-power short-pulse applications. They consist of three- or four-element low-pressure, gas-filled tubes. The

*Work supported by the U.S. Department of Energy.

elements consist of the cathode, which is a primary electron emitter; the anode, to which a positive voltage is applied relative to the cathode; and one or two grids used to control the start of conduction.

The cathode emits electrons that are accelerated by the anode potential when the grid is not acting as a shield. In ordinary operation, the voltage condition on the grid, or grids, is applied to form a shield against acceleration of the electrons until switching action is desired. Depending on the actual design of the grid structure, a negative or positive grid bias is required to shield the cathode from the anode. Sufficient reduction of the negative bias on a negative grid thyratron or sufficient increase of the positive bias on a positive grid thyratron starts conduction of electrons beyond the grid into the anode field. The electrons create a column of electrons and gas ions through collisions with the gas molecules. A sheath of ions forms around the grid, preventing any potentials from penetrating into the main body of the discharge. Once the discharge starts, the grid provides no further control over the discharge behavior. The thyratron does not recover its insulation capabilities until current in the anode circuit stops and sufficient time elapses at zero current for ion recombination to occur. A potential drop of about 30 to 100 V occurs between the plate and cathode during conduction.

Two-grid, four-element thyratrons are used to minimize variations in firing time. The grids can be operated in a variety of ways. One technique is to operate the first grid continuously ionized between it and the cathode. A keep-alive current of a few tens of milliamperes is adequate. The second grid is used for the gating function with a pulse driven over its negative bias. The two grids could be pulsed successively with a delay of a few microseconds between leading edges of the pulses or the grids may be driven by a single trigger source.

Hydrogen and deuterium gas are popularly used in high-voltage, high-power thyratrons because of their ability to provide fast switching times, short recovery times, and long life.

Pulsed Power Circuits with Thyratrons

A very simple pulsed power discharge circuit using a thyratron is shown in Fig. 12.6. The capacitor is slowly charged through the charging choke and the load resistor by a high-voltage, low-current power supply, while the thyratron is nonconducting. A gate pulse ionizes the thyratron, allowing the capacitor to discharge through the resistor, normally at a current many times higher than the charging current. When the capacitor is discharged and current stops, the tube deionizes, restoring its insulation value. The recombination action takes a finite time, normally called the **recovery time**. The choke in the power supply circuit is added to retard the flow of power supply current into the thyratron at the end of the capacitor discharge. The tube could be kept in an ionized state by the power supply current if a preventive mechanism is not employed.

A typical fast-discharge circuit is shown in Fig. 12.7. A delay line with characteristic impedance Z_0 is attached to the anode of the thyratron and a transmission line with characteristic impedance Z_0

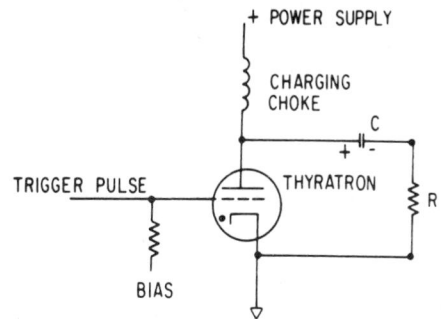

Fig. 12.6 Simple capacitor discharge circuit using a thyratron.

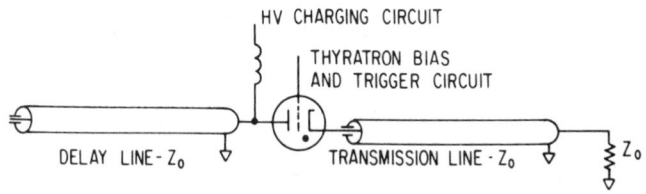

Fig. 12.7 Delay-line discharge circuit for fast, shaped pulses.

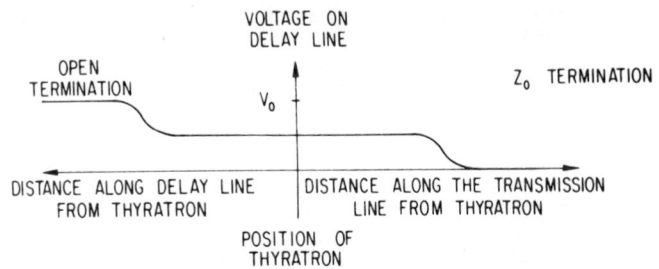

Fig. 12.8 Voltage distribution on the delay line and transmission line at an arbitrary time before the voltage wave on the delay line reaches the unterminated end of the line.

is attached to the cathode. The transmission line is terminated in a load resistance Z_0. The charging circuit charges the delay line to a desired voltage V_0. A trigger pulse on the grid drives the thyratron into conduction, connecting the charged delay line into the uncharged transmission line. This effectively acts as though a resistor of value Z_0 has been connected into the thyratron end of the charged delay line. A voltage wave of $+V_0/2$ travels along the transmission line toward the terminating resistor. A voltage wave of $-V_0/2$ travels along the delay line toward the unterminated end of the line. Behind the wave front the voltage has been reduced to $V_0/2$, whereas it is still V_0 before the wave front. This is graphically displayed in Fig. 12.8 at some arbitrary time before the traveling wave on the delay line reaches the open termination. When the $-V_0/2$ voltage wave on the delay line reaches the open termination, it is reflected backward down the delay line toward the thyratron. The reflection coefficient k_r is

$$k_r = \frac{Z_t - Z_0}{Z_t + Z_0} \tag{12.1}$$

where Z_0 is the characteristic impedance of the transmission line and Z_t is the termination impedance of the transmission line. Since Z_t is infinite at the open end of the delay line, the reflection coefficient equals unity. The $-V_0/2$ wave is therefore reflected back down the delay line, reducing the voltage to zero in its wake. Eventually, the $-V_0/2$ wave reaches the transmission line.

The $+V_0/2$ wave traveling on the transmission line is absorbed without reflection by the terminating resistor since the resistance is equal to Z_0, making k_r equal to zero. A $+V_0/2$ voltage appears across the terminating resistor for a period equal to the ratio of twice the length of the delay line divided by the electromagnetic-wave propagation velocity of the delay line. The pulse length τ across the load is

$$\tau = 2l_d\sqrt{LC} \tag{12.2}$$

where l_d is the length of the delay line, L the inductance of the delay line per unit length, and C the capacitance of the delay line per unit length.

In a fast-rise-time circuit of this type, stray capacitances and inductances must be minimized. The thyratron housing must be designed to help match the characteristic impedance of the delay line and transmission line.

The delay time in achieving anode conduction is an important thyratron parameter. The rise time of the control grid trigger pulse must be as rapid as possible to avoid triggering delays. The thyratron has an inherent delay known as the **anode delay time**, often defined as the time interval between the point where the rising slope of the control grid trigger pulse reaches 25% of the maximum unloaded pulse amplitude and the point where anode conduction takes place. This delay time is caused in part by the gas dynamics within the tube and in part by the stray inductances associated with the tube housing and the interconnection to the delay line. The latter is roughly 1.4 times the ratio of stray inductances L_s and Z_0. The total delay time t_d is approximately

$$t_d \simeq \sqrt{t_{\text{inh}} + 1.4(L_s/Z_0)} \tag{12.3}$$

where t_{inh} is the inherent time delay of the tube.

The rise times that can be typically achieved are on the order of tens of nanoseconds for multimegawatt instantaneous powers. An oscillograph of a fast discharge pulse is shown in Fig. 12.9. The horizontal calibration is 100 ns per division, and the vertical calibration is 10 kV per division.

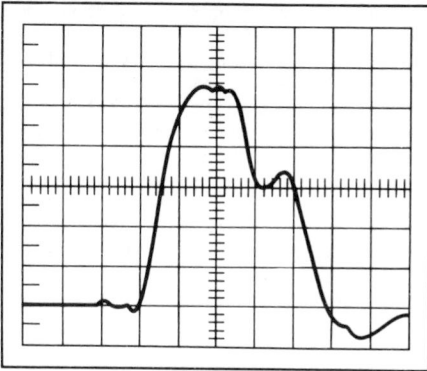

Fig. 12.9   Oscillograph of a fast-discharge pulse.

This particular pulse was developed by discharging a 14-Ω delay line into a 14-Ω transmission line by a deuterium thyratron with an 80-kV rating.

It should be noted that the cathode and grid of the thyratron in Fig. 12.7 rise to a voltage of $V_0/2$ during conduction of the delay line pulse. This means that the cathode heater circuit, grid bias, and trigger pulse must be capable of withstanding the high-voltage pulse without arcing to ground or distorting the pulse due to the stray impedances of the insulation networks.

Parallel and Series Operation and Multigap Thyratrons
Parallel operation of thyratrons cannot be achieved without some form of impedance in series with each tube. Further, the larger variations in anode delay times for three-element thyratrons make them very difficult to use in parallel operation. Because anode delay variations are considerably reduced with four-element devices, reliable parallel operation can be achieved. In practice, a limit of about six parallel tubes is typical. Since the addition of anode impedances either degrades rise time or degrades impedance matching, fast pulse circuits rarely can employ parallel tubes without dividing the load circuit so that individual sections can be driven by individual tubes.

Series operation of thyratrons is complicated by the necessity of insulating the cathode heaters, grid bias, and trigger circuits from ground at the intermediate voltage levels of series-connected tubes. Figure 12.10 shows two series-connected thyratrons with the necessary voltage balancing impedances to avoid uneven voltage distributions. The capacitor C_2 is required across thyratron T_2 because the isolating network for the heater circuit of T_2 results in a stray capacitance to ground that effectively adds a capacitor C_1 across T_1. The value of C_2 must be adjusted to compensate for T_1 and any other stray impedances that might result from intermediate position of the cathode network of T_2. The resistors are needed to balance the dc voltage across the tubes that might occur because of variations

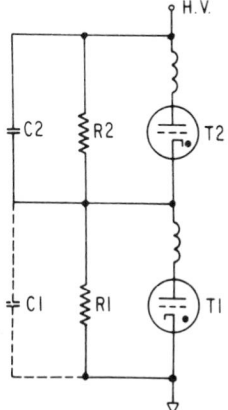

Fig. 12.10   Two series-connected thyratrons.

in the leakage resistances of the tubes. The values of R_1 and R_2 should be made as large as possible to avoid power supply drain.

It is possible to avoid a second trigger network for T_2 by capacitively coupling the anode of T_1 to the grid of T_2. As the voltage on T_1 collapses, the capacitor network would force T_2 into conduction.

Multigap thyratrons are commercially available. These devices basically create virtual cathodes by making the anode of T_1 the cathode of T_2. The anode plate is built with an annular hole so that when T_1 ionizes, some of the electrons escape past the anode of T_1 and ionize T_2. Two-, three-, and four-gap devices are available using this technique. Each gap is individually able to hold as much as 40 kV.

Modern high-power thyratrons can operate at voltages as high as 160 kV, with currents as high as 10,000 A and instantaneous output powers in excess of 300 MW. The power rating for the high-power thyratrons used in repetitive pulse operation is often given in terms of an average current and an anode heating factor that is the product of volt-amperes and pulses per second. Single-shot ratings are given in ampere-seconds.

12.3.2 Ignitrons

The ignitron is a mercury-arc tube with a pool-type cathode in which an ignitor rod is inserted. The main power flow occurs between the anode and the pool-type cathode and is initiated by a pulsed current that flows through the ignitor rod into the mercury pool. The ignitor rod is made of a high-resistance semiconductor so that a hot cathode spot is formed on the pool of mercury when the ignitor pulse is applied. If the anode is positive at the time of the ignitor pulse, an anode-cathode arc is formed, and the tube conducts high current with a relatively low voltage between the anode and cathode. Conduction continues until the current stops flowing and recombination of the mercury gas occurs.

Ignitor pulses for relatively large pulsed tubes require on the order of a few hundred to a few thousand volts and a few joules of energy. Anode delay times can be reduced to about 1 μs or slightly less.

Ignitrons must be mounted vertically during operation and are very sensitive to vibration. They require heat treatment after shipping or handling because the mercury splashes over the interior walls, leaving a residue that will degrade voltage-holding capabilities. The heat treatment consists of heating the anode stud to about 100°C for 2 h after the tube is mounted in its operating position. Voltage conditioning is often required as well as heat conditioning. A circuit consisting of an adjustably charged high-voltage capacitor and a current-limiting resistor is used to raise the holding voltage across the tube. A typical voltage-conditioning circuit is shown in Fig. 12.11.

Pulsed Power Circuits Using Ignitrons

Ignitrons are limited in their application to relatively few circuits, primarily pulsed power circuits with very high di/dt and very large energy transfer values and high-power "crowbar"-type circuits. Modern power convertor circuits almost exclusively use silicon-controlled rectifiers (SCRs) and most pulsed power circuits use SCRs as well. Thyratrons are used where rise times on the order of tens of nanoseconds are required. Modern circuits use ignitrons where the di/dt rise times exceed the capabilities of large SCR devices and the energy transfer requirements exceed the capabilities of modern thyratrons.

The energy discharge circuits are similar to that of the simple capacitor discharge circuit shown in Fig. 12.6 with the thyratron replaced by an ignitron.

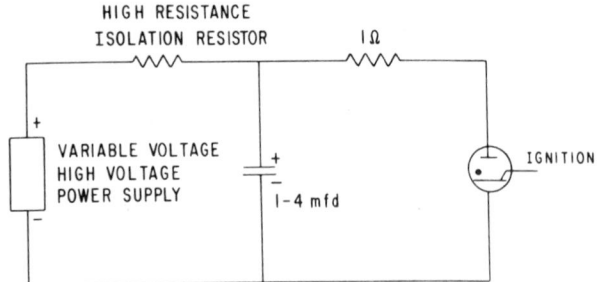

Fig. 12.11 Typical ignitron voltage-conditioning circuit.

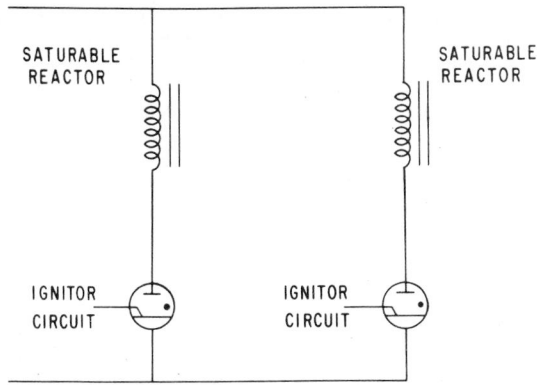

Fig. 12.12   Circuit for parallel operation of operation of ignitrons.

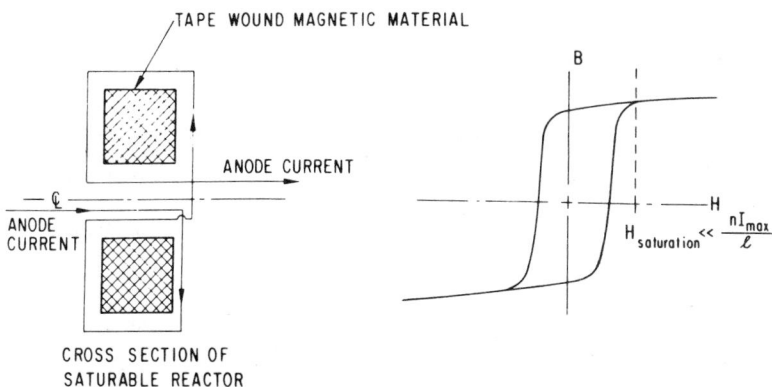

Fig. 12.13   Saturable reactor for parallel operation of ignitrons.

Ignitrons for power-switching applications have anode voltage ratings between 15 and 50 kV and current ratings between 30 and 100 kA. Energy-switching capabilities vary between 10 and 50 kJ/min.

Series and Parallel Operation of Ignitrons
Parallel operation of ignitrons encounters the same difficulty as does parallel operation of thyratrons. Two tubes do not have identical anode delay times or anode-to-cathode potential drops; therefore, series impedances are needed in series with the anodes to balance the current flow between different tubes, and a technique is needed to eliminate time differences between anode delay times.

A form of saturable reactor is frequently used in ignitron circuits for balancing anode firing times. A typical circuit for parallel operation is shown in Fig. 12.12. A cross section of a typical saturable reactor and the desired $B - H$ curve for the magnetic material are shown in Fig. 12.13. The cores are tape wound, using appropriately thin material for the rise times involved. The current at which the magnetomotive force achieves saturation of the core is small compared to the final anode current. During the early period, shortly after ionization of the tube, the reactor limits the flow of current and supports a fairly high voltage. The reactor voltage prevents the collapse of the anode voltage due to the ionization of one of the tubes, an action that would otherwise impede the ionization of the other tube. The voltage across the reactor, v, is

$$v = L\frac{di}{dt} + i\frac{dL}{dt} = N\frac{\Delta\phi}{\Delta t} \qquad (12.4)$$

where L is the inductance of the reactor, i the anode current of the ignitron, N the number of turns in the reactor, $\Delta\phi$ the change in flux from zero to saturation level, and Δt the rise time of the anode current. The voltage across the reactor collapses to zero when the saturation point of the magnetic material is reached. Thus if the time jitter in ionization times of parallel ignitrons is expected to be Δt and a minimum voltage of v is required across the anodes of those ignitrons not yet ionized, then the total flux change, $N\Delta\phi$, must be equal to $v\,\Delta t$. This requirement defines the cross-sectional area of the core. The number of turns normally must be kept small because the copper adds resistance in series with the ignitron. To some degree the resistance benefits operation because it is needed to balance the currents between the ignitrons after the reactors saturate because of variations in the anode voltage during conduction. However, because of the very large currents handled by the ignitrons, an unacceptably large resistance will result rather quickly if too many turns are used. It is also difficult to wind many turns when the conductors must handle between 20 and 100 kA.

Series operation of ignitrons must be handled in a similar manner to the series connection of thyratrons. The cold cathode of the ignitron reduces the stray capacitance to ground relative to that of a thyratron; however, balancing networks are still needed to avoid voltage unbalances.

12.3.3   High-Vacuum Electron Tube

The high-vacuum electron device consists of a thermionic-electron-emitting cathode, an anode, and in some devices anywhere from one to three control grids in a high-vacuum enclosure. In most ordinary modern applications, semiconductor devices have replaced the use of high-vacuum devices. The exceptions are applications requiring extremely high voltage or high power or applications in extremely hostile atmospheres, such as high-temperature or high-radiation areas.

The cathode can be a directly heated type, in which the heater filament is also the electron-emitting surface, or it can be an indirectly heated type, in which the electron-emitting surface is a thin metal shell closely surrounding the filament. Since the electrons are only emitted from the hot cathode surface, substantial current conduction occurs only when the anode potential is positive relative to the cathode.

A two-element vacuum device without control grids is called a **diode**. A typical voltage-current characteristic for a vacuum diode is shown in Fig. 12.14. The portion of the characteristic between $e_b = 0$ and the saturation point of current is dominated by space charge effects at the cathode. After the saturation point, the current is emission limited. Operation at higher temperature increases the saturation current with only slight increases in the space-charge-limited region.

Compared to thyratrons and ignitrons, a vacuum diode has an extremely high voltage drop between the anode and cathode. Further, as indicated previously, the advent of semiconductor diodes has virtually eliminated the use of vacuum diodes for most ordinary applications.

The high-vacuum triode has a slotted or semitransparent grid structure positioned between the anode and cathode. The electric field existing on the cathode is generated by the voltage on the grid and the voltage on the anode by virtue of the semitransparent nature of the grid. Most of the electrons accelerated away from the cathode by this electric field pass through the openings in the grid structure and onto the anode. Some, however, strike the grid, causing additional heating. Similarly, on

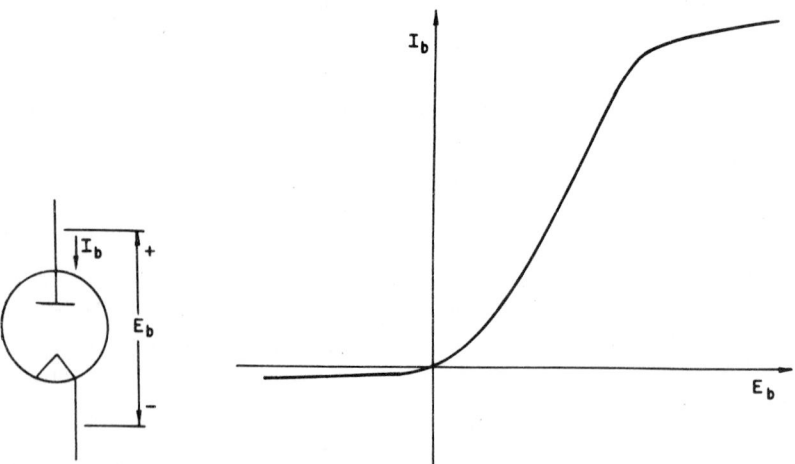

Fig. 12.14 Typical voltage-current characteristics for a vacuum diode: $E_{g6} > E_{g5} > E_{g4} > E_{g3} > E_{g2} > E_{g1}$.

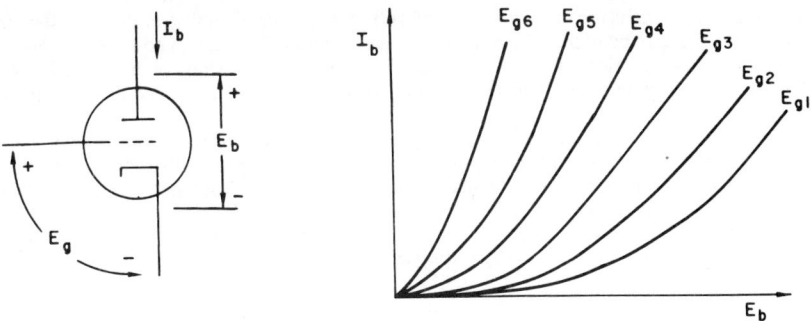

Fig. 12.15 Typical voltage-current characteristics for a vacuum triode.

striking the anode the electrons surrender most of their kinetic energy to heating the anode. The grids and anodes of high-power tubes are cooled either by specially designed convection cooling structures, fan-cooled structures, for water-cooled structures, depending on the power level of the device.

A typical voltage-current characteristic for a vacuum-tube triode is shown in Fig. 12.15. Several $E_b - I_b$ curves for fixed values of grid voltage, E_{g1}, E_{g2}, E_{g3}, E_{g4}, E_{g5}, and E_{g6}—where the grid voltage E_{g1} is the largest negative value—E_{g2} is less negative than E_{g1}, and E_{g3} is less negative than E_{g2}, and so on. Normally, the grid voltage remains negative over the full operating range of the triode, although the geometrical construction of some tubes requires the grid voltage to extend into the positive range.

The operating point on the triode characteristics can be found graphically. Referring to Fig. 12.16, the plate voltage E_b and the plate current I_b must simultaneously satisfy the expression

$$E_b = V_b - I_b R_L \tag{12.5}$$

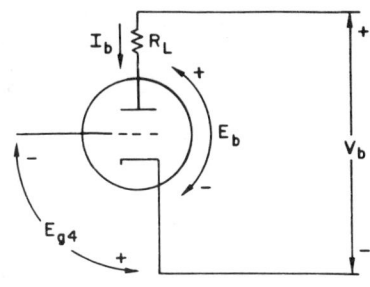

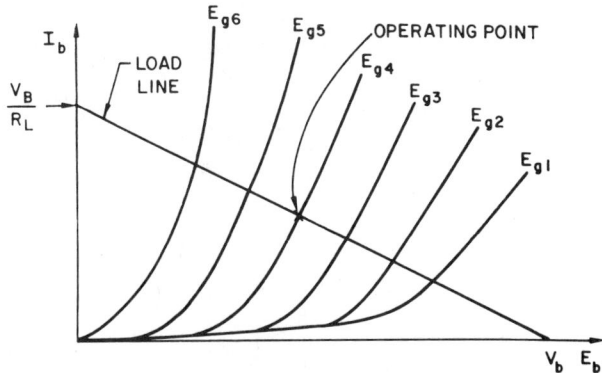

Fig. 12.16 Graphical determination of the operating point for a vacuum triode.

and the voltage and current for the given value of grid voltage corresponding to the voltage-current characteristics of the specific triode. The straight line described by Eq. (12.5), known as the load line, is drawn on the triode characteristics as shown in Fig. 12.16. The operating point for the triode is found from the intersection of the load line and the voltage-current characteristic line for the given value of grid voltage.

If the grid voltage changes, the operating point will also change, but the values of plate voltage and current will correspond to points lying on the load line. This is graphically shown in Fig. 12.17. The circuit in Fig. 12.17 has a relatively low-frequency sine-wave voltage in series with a dc voltage of value E_{g4}. Thus the grid voltage is

$$E_g = v_s \sin \omega t - E_{g4} \tag{12.6}$$

where

$$v_s = E_{g4} - E_{g5} \cong E_{g3} - E_{g4} \tag{12.7}$$

A point-by-point plot of the voltage and current shows that the plate voltage E_b varies almost

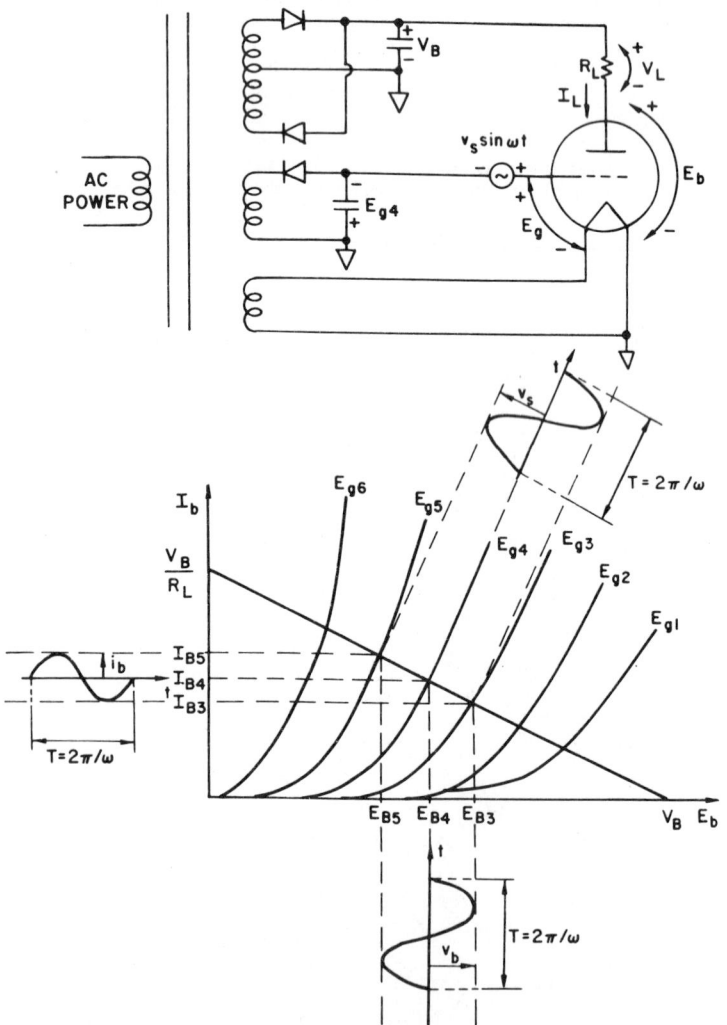

Fig. 12.17 Effect of sinusoidal variation of grid voltage on operating point of triode.

sinusoidally by the amount

$$E_b = E_{b4} - v_b \sin \omega t \qquad (12.8)$$

where

$$v_b \approx E_{b4} - E_{b5} \approx E_{b3} - E_{b4} \qquad (12.9)$$

the plate current I_b varies almost sinusoidally by the amount

$$i_b = I_{b4} + i_b \sin \omega t \qquad (12.10)$$

where

$$i_b \approx I_{b4} - I_{b3} \approx I_{b5} - I_{b4} \qquad (12.11)$$

and the load voltage V_L varies almost sinusoidally by the amount

$$V_L = I_{b4} R_L + i_b R_L \sin \omega t \qquad (12.12)$$

The amounts by which Eqs. (12.10) and (12.12) fail to replicate sine waves depend on the linearity of the triode characteristics and the size of the signal relative to the saturation and cutoff levels, respectively. Nonlinearities in the tube characteristics obviously lead to the generation of higher harmonic sine components.

It is possible to treat analysis of various circuit designs incorporating vacuum-tube triodes by developing an equivalent circuit using only the small-signal components v_s, v_b, and i_b of Eqs. (12.6), (12.8), (12.10), and (12.12). The most common form of the equivalent circuit is shown in Fig. 12.18. The constant μ, known as the **amplification factor** is the ratio of the increments ΔE_b to ΔE_g at constant plate current. The constant r_p, known as the **plate resistance**, is the ratio of the increments ΔE_b to ΔI_b at constant grid voltage. The dc values I_{B4}, E_{B4}, and E_{g4} in Eqs. (12.6), (12.8), (12.10), and (12.12) are known as the **bias** or **quiescent conditions** for the tube.

An idealized cross section of a triode is shown in Fig. 12.19. Although actual tube geometries can vary widely from that shown in Fig. 12.19, the basic elements are adequately illustrated by the idealized cross section. It is clear that a physical geometry as shown will have capacitances existing among all three elements. Thus at high-frequency operation, the equivalent circuit of Fig. 12.18 must be modified to include the capacitances. This is shown in Fig. 12.20.

In addition to the triode, common vacuum tubes are the tetrode, which has two grids, and the pentode, which has three grids. The common use of the second grid, located radially outward from the control grid in these tubes (or in the relatively same geometric position in other geometries), is as a screen between the control grid and the anode. The screen grid is normally operated with a constant, positive voltage with respect to the cathode at about two-thirds the value of the dc supply voltage used for the plate circuit of the tube. A typical voltage-current characteristic is shown in Fig. 12.21.

The effect of the screen grid is to reduce the normal control grid-plate capacitance C_{pg}, shown in Fig. 12.20, to extremely low value. The value of C_{pg} is a triode can have a significantly negative effect on amplifier performance at fairly moderate frequencies; thus the addition of a screen grid can greatly extend the utility of vacuum tubes.

A problem does arise with the screen grid at relatively low plate voltages, as can be seen from the erratic nature of the voltage-current curves at plate voltages lower than the vertical dashed line in Fig. 12.21. This effect is caused by the acceleration of secondary electrons from the anode to the screen grid when the plate voltage decreases below the screen voltage. The common use of the third grid of the pentode is to suppress this effect. This is accomplished by connecting the grid geometrically

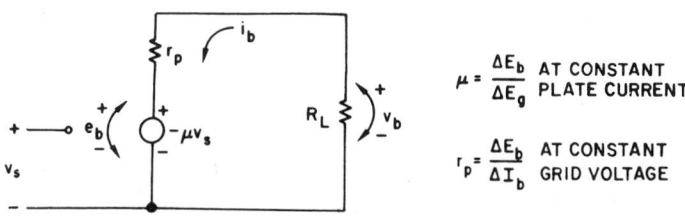

Fig. 12.18   Small-signal, low-frequency equivalent circuit for a vacuum triode.

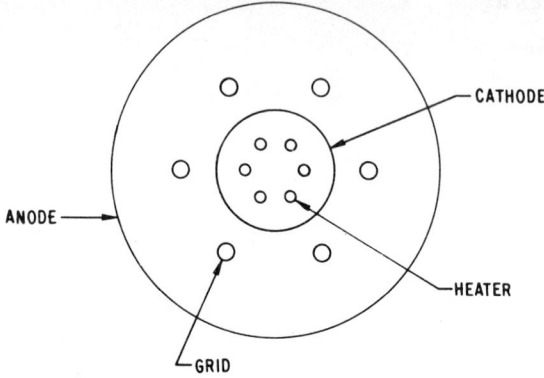

Fig. 12.19 Cross section of an idealized high-vacuum triode.

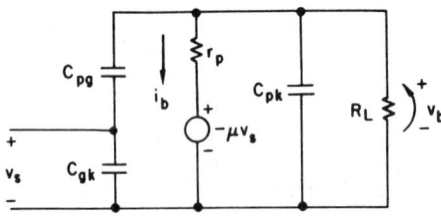

Fig. 12.20 Small-signal equivalent circuit for a vacuum triode, including interelement capacitance.

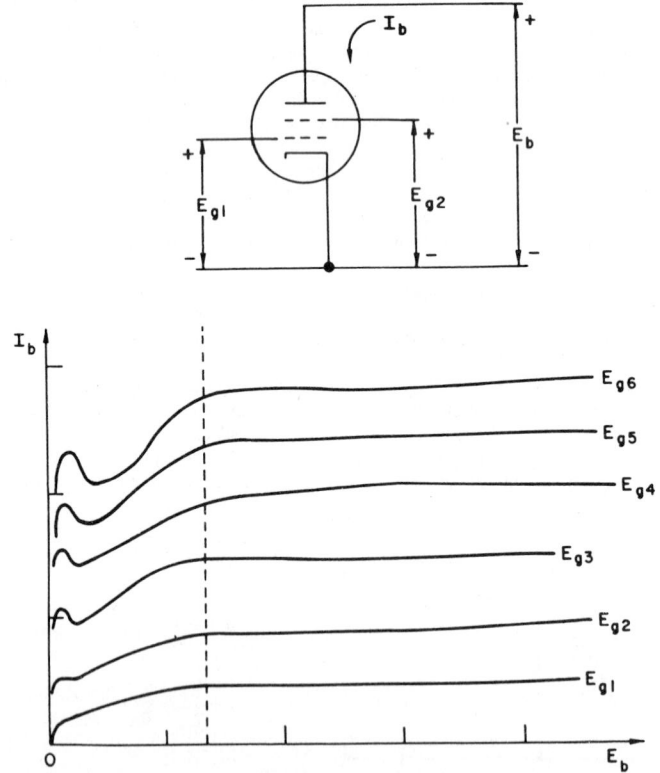

Fig. 12.21 Typical tetrode voltage-current characteristic curves when E_{g2} is held at a constant dc voltage.

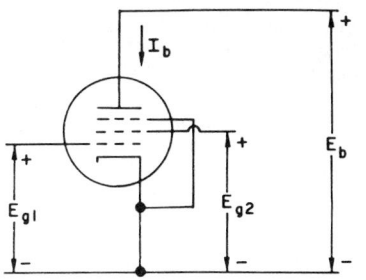

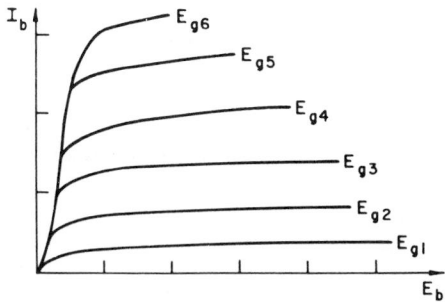

Fig. 12.22 Typical voltage-current characteristics for a vacuum pentode.

closest to the anode electrically to the cathode. Thus electrons are accelerated from the cathode through the control grid by the screen grid potential and are transported through the semitransparent screen and suppressor grids to the anode. The suppressor grid maintains sufficient retarding electrical field over the surface of the anode to prevent movement of secondary electrons away from the anode.

A typical voltage-current characteristic for a pentode is shown in Fig. 12.22.

Both the tetrode and pentode are better characterized by a current source than by a voltage source in the small-signal equivalent circuit. This is a result of the very nearly constant current as a function of plate voltage when measured at constant control grid voltage. The ideal equivalent circuit for a tetrode and pentode is shown in Fig. 12.23. The parameter g_m for the current source is called the **plate-grid transconductance** and is the following incremental ratio:

$$g_m = \frac{\Delta I_B}{\Delta E_{g1}} \tag{12.13}$$

for constant plate voltage and screen grid voltage. The value of C_{in} is the sum of the control grid-to-cathode capacitance and the control grid-to-screen grid capacitance for either a tetrode or a pentode. The value of C_{out} is the sum of the plate-to-screen grid capacitance and the plate-to-cathode capacitance for a tetrode and is the sum of these two capacitances and the plate-to-suppressor grid capacitance for a pentode. The capacitance C_{pg1}, shown dashed in Fig. 12.23, is the plate-to-control grid capacitance. In the ideal case, this is the capacitance that has been reduced to virtually zero;

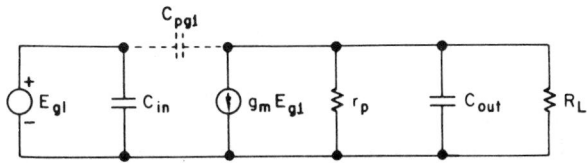

Fig. 12.23 Small-signal equivalent circuit for a tetrode or pentode.

however, when tuned circuits are employed in the input and output networks, this capacitance can become a dominant value, relatively speaking, even though the reason for using tetrodes and pentodes was to reduce this capacitance to a negligible level.

Typical Amplifier Circuits and Their Characteristics

The different types of circuits that can be built using vacuum tubes are almost numberless. They include amplifiers, oscillators, modulators, demodulators, detectors, frequency multipliers, mixers, noise sources, and many others. Space does not permit considering any significant number of these possible circuit configurations. This section briefly describes the characteristics of low-frequency common-cathode, common-grid, and common-plate amplifiers. The reader is referred to the many excellent reference books listed at the end of the section for more details on the possible circuits and their analyses.

The three basic amplifier connections and their small-signal equivalent circuits are shown in Fig. 12.24. The impedance Z_{gk} is the sum of leakage resistance between the grid and the cathode and the grid-to-cathode capacitive reactance.

The voltage gain, power gain, input impedance, and output impedance are common parameters of interest in amplifier circuits. Comparatively, the common-cathode amplifier has a relatively high input impedance, high voltage gain, high power gain, and a medium-level output impedance. The grounded-grid amplifier has a relatively low input impedance, high voltage gain, medium power gain, and high output impedance. The cathode-follower amplifier has a very high input impedance, near-unity voltage gain, medium power gain, and very low output impedance. The algebraic expressions for voltage gain, input impedance, and output impedance are listed in Table 12.1 without derivation.

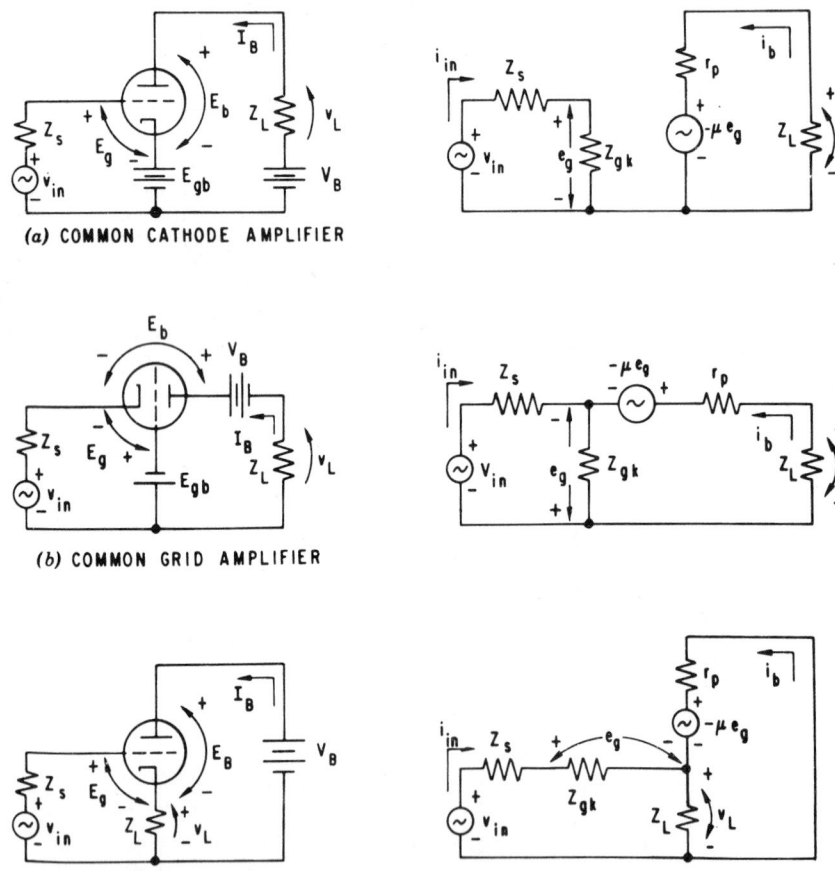

(a) COMMON CATHODE AMPLIFIER

(b) COMMON GRID AMPLIFIER

(c) COMMON PLATE (Cathode Follower) AMPLIFIER

Fig. 12.24 Basic amplifier connections and their small-signal equivalent circuits.

TABLE 12.1 Algebraic Expressions for Voltage Gain, Input Impedance, and Output Impedance for Three Basic Amplifier Connections

	Common Cathode	Common Grid	Cathode Follower
Voltage gain, $\dfrac{v_L}{v_{\text{in}}}$	$-\mu\left(\dfrac{Z_L}{r_p+Z_L}\right)\dfrac{Z_{gk}}{Z_s+Z_{gk}}$	$\dfrac{(\mu+1)Z_L}{r_p+Z_L+(\mu+1)Z_s}$	$\dfrac{(\mu Z_{gk}+r_p)(Z_L/r_p)}{Z_s+Z_{gk}+(\mu Z_{gk}+r_p)(Z_L/r_p)}\simeq\dfrac{\mu}{\mu+r_p/Z_L}$
Input impedance, $\dfrac{V_{\text{in}}}{I_{\text{in}}}$	Z_s+Z_{gk}	$Z_s+\dfrac{r_p+Z_L}{\mu+1}$	$Z_s+Z_{gk}+(\mu Z_{gk}+r_p)(Z_L/r_p)$
Output impedance, $-\dfrac{v_L(Z_L=\infty)}{i_b(Z_L=0)}$	r_p	$\dfrac{(Z_s+r_p)Z_{gk}}{Z_s+Z_{gk}}$	$r_p\left(\dfrac{Z_s+Z_{gk}}{Z_s+Z_{gk}+\mu Z_{gk}+r_p}\right)\dfrac{\mu Z_{gk}+r_p}{\mu Z_{gk}}\simeq r_p\left(\dfrac{1}{\mu+1}\right)$

High-Power Tuned Circuit Amplifiers

As already indicated, the most common modern application of high-vacuum electron tubes is in high power, many kilowatts to several megawatts, and high frequency, 1–400 MHz. In almost all of these applications, some form of tuned circuit, either a lumped LC parameter network or a transmission line or cavity resonator, is used in the output circuit, and often also in the input circuit as well, to neutralize capacitive loading effects. Furthermore, the small-signal condition under which the equivalent circuit described earlier is applicable is not normally satisfied in high-power applications.

The operating conditions shown in Fig. 12.17 are generally known as *class A* operation. Plate current flows throughout the whole radio-frequency (RF) cycle, a situation that remains true as long as the input signal does not drive the triode to the cutoff point on the high-voltage end of the load line. The primary shortcoming of class A operation is the low power efficiency as defined by the ratio of RF power delivered to the load relative to the DC power delivered by the plate power supply. This parameter is known as the **plate efficiency**, η. A rough estimate of the maximum plate efficiency in class A operation can be made by assuming that cutoff occurs along the abscissa and saturation occurs along the ordinate. In the idealized case, the bias point would be selected at $V_B/2$ and $I_B = V_B/2R_L$. The peak RF values would also be $V_B/2$ and $I_B = V_B/2R_L$, causing the amplifier to operate over the total load line from V_B and $I_B = 0$ to $V_B = 0$ and $I_B = V_B/R_L$. The average RF power delivered to the load is

$$P_{rf} = \frac{1}{T}\int_0^T i_b^2 R_L \, dt$$

$$= \frac{1}{T}\int_0^T \frac{V_B^2}{4R_L^2}(R_L)\sin^2\omega t \, dt$$

$$= \frac{V_B^2}{8R_L} \tag{12.14}$$

The power delivered by the dc power supply is the product of V_B and $V_B/2R_L$; therefore,

$$\eta = \frac{P_{rf}}{P_{dc}} \times 100$$

$$= 25\% \tag{12.15}$$

Two important comments should be added: (1) the value of 25% is a theoretical maximum that cannot quite ever be achieved because saturation and cutoff do not occur exactly along the ordinate and abscissa; and (2) the maximum efficiency occurs only when the triode is driven at maximum power. As the RF drive is reduced, plate efficiency is reduced as well because the dc power supply always delivers the power at the bias level equal to $V_B^2/2R_L$. Therefore, plate efficiency in a class A amplifier varies between 0 and almost 25%, at best.

Low plate efficiency translates into unnecessarily large, bulky vacuum tubes consuming large internal power that somehow must be cooled to avoid overheating and into very high power dc plate supplies relative to the amount of RF delivered to the load. In high-power applications, a tuned circuit that includes the output capacitance of the tube is added onto the plate to resonate at the frequency of the input signal. A typical tuned plate amplifier circuit is shown in Fig. 12.25. The bias point is indicated on the voltage-current characteristics for this triode in Fig. 12.25. Since this bias point is either at, or very nearly at, the cutoff point of the triode, plate current only flows for the positive cycle of the input signal voltage. This type of operation is known as *class B* operation.

The resonant circuit in the output section of the triode is excited by the fundamental component of the half-sine-wave current of the triode. A full-wave voltage is generated across the tube and the load resistance by the resonant circuit; a full-wave current is also generated in the load by the resonant circuit. Some of the voltage and current waveforms for a class B amplifier are shown in Fig. 12.26. A harmonic analysis of the plate current waveform shown in Fig. 12.26 is given by

$$i_b = \frac{I_{bm}}{\pi} + \frac{I_{bm}}{2}\cos\omega t - \sum_{n=2}^{\infty}\frac{2I_{bm}}{\pi(n^2-1)}\cos\frac{n\pi}{2}\cos n\omega t \tag{12.16}$$

All the harmonics for $n > 1$ are shunted by the capacitor, and the dc current I_{bm}/π is bypassed by the inductor. The load current through R_L has the negative value of the fundamental frequency component, $I_{bm}/2$, in Eq. (12.16). The load voltage is

$$v_L = -\frac{I_{bm}R_L}{2}\cos\omega t \tag{12.17}$$

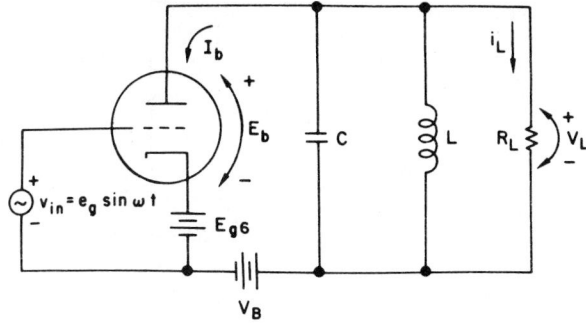

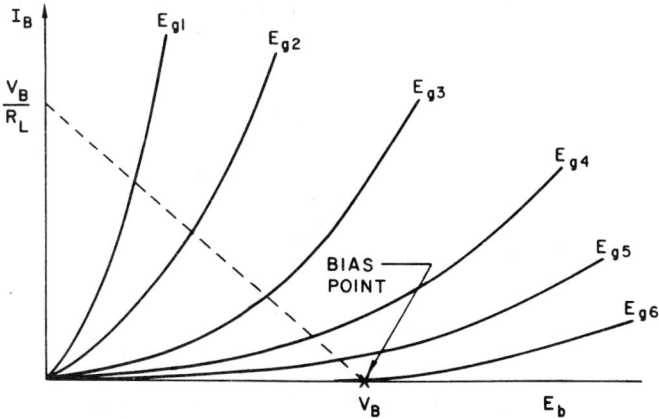

Fig. 12.25 Tuned plate triode amplifier biased for class B operation.

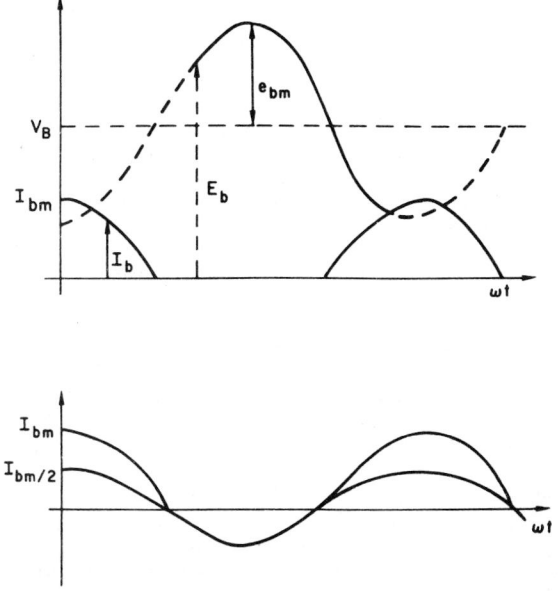

Fig. 12.26 Current and voltage waveforms for a class B tuned amplifier.

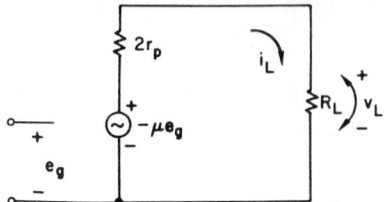

Fig. 12.27 Equivalent circuit for a class B tuned amplifier.

During the half-cycle when the triode is conducting, the ordinary voltage source, $-\mu e_g$, drives the plate current I_b through the plate resistance r_p and the load voltage v_L. The load voltage is described by Eq. (12.17); therefore;

$$-\mu e_g \cos \omega t + I_{bm} r_p \cos \omega t - v_L = 0$$

$$I_{bm}\left(r_p + \frac{R_L}{2}\right) = \mu e_g$$

$$I_{bm} = \frac{2\mu e_g}{2r_p + R_L} \tag{12.18}$$

The load current has been shown to be one-half the magnitude of I_{bm}, so

$$i_L = \frac{\mu e_g}{2r_p + R_L} \cos \omega t \tag{12.19}$$

The equivalent circuit for a class B tuned plate amplifier is similar to a class A amplifier except that the plate resistance is doubled. This is shown in Fig. 12.27.

The RF power delivered to the load is equal to $I_{bm}^2 R_L/4$. The power delivered by the dc power supply is equal to the product of the supply voltage V_b and the dc current term in the Fourier expansion of Eq. (12.16); that is,

$$\eta = \frac{I_{bm}^2 R_L/4}{V_b I_{bm}/\pi}$$

$$= \frac{\pi}{4}\frac{I_{bm} R_L}{V_b} \tag{12.20}$$

The maximum value that I_{bm} can have is V_B/R_L. Thus the efficiency at maximum output current is 78.5%, which is three times higher than class A operation. The efficiency at less than maximum output current decreases linearly with current. The efficiency in class B operation remains much higher at all levels, not only because the peak level is three times higher, but also because the efficiency is proportional to plate current; whereas in class A operation, the efficiency is proportional to the square of the plate current.

Another common connection for high-power operation is the push-pull circuit shown in Fig. 12.28. The grid voltage on one triode increases, while the other decreases in equivalent amounts. This action produces a fundamental component of current that flows through the two tubes and through the transformer, but not through the power supply because the signs of the tube currents are such that they cancel each other in the power supply line. The equivalent circuit for the push-pull connection is also shown in Fig. 12.28. It is possible to operate push-pull amplifiers in class B without tuned circuits because of the complementary effect of the two tubes. This circuit will not be analyzed in detail. Interested readers are referred to the bibliography for this section at the end of the chapter.

Class A and class B operations are not the only designated classes of operation. The definition for the complete list are as follows:

CLASS A OPERATION. Plate current flows throughout the whole cycle.

CLASS AB OPERATION. Plate current flows for appreciably more than one-half cycle, but less than the full cycle.

CLASS B OPERATION. Plate current flows for approximately one-half cycle.

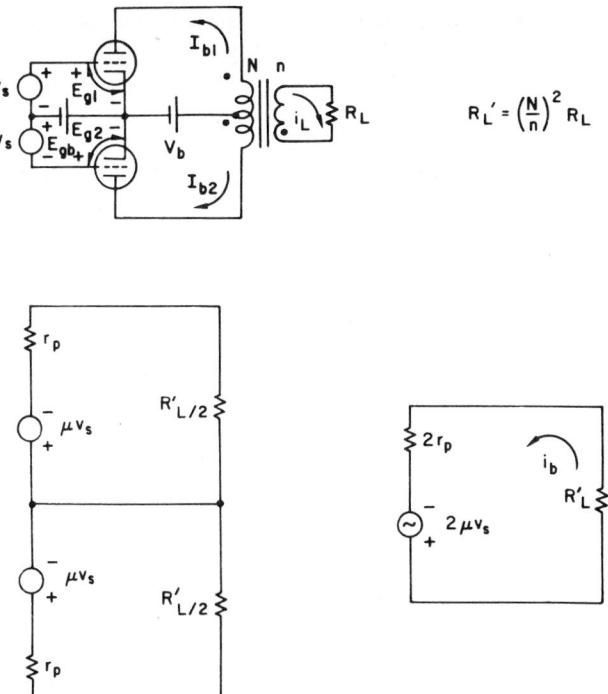

Fig. 12.28 Typical circuit and equivalent circuit for a push-pull connection.

CLASS C OPERATION. The grid bias is much greater than the cutoff value, so that plate current is zero in the tube when no signal is applied. Plate current flows for appreciably less than one-half cycle.

If the suffix 1 is added to the classification, it denotes that no grid current flows during the cycle; and if the suffix 2 is added to the classification, it denotes that grid current does flow at some time during the cycle.

12.4 OSCILLOSCOPES
Kenneth Sohn

12.4.1 Cathode-Ray Tube

Although vacuum tubes have vanished from most electronic system design today, the cathode-ray tube still plays an important role in television, radar, and oscilloscope. The operation of this device can be easily analyzed using ballistic theory. The purpose of the tube is to display an electrical signal. This is accomplished by controlling a beam of electrons produced at one end of the tube so that by the proper acceleration and deflection of the beam (to convert the electron-beam energy to light energy), different electrical signals can be displayed on the screen. All types of cathode-ray tubes have basically four parts:

1. A source of electrons
2. Control elements, that focus and accelerate the electron beam toward the screen
3. Deflection plates
4. Phosphorescent screen

Typical construction of a cathode-ray tube is shown in Fig. 12.29.

The first two sections consist of the source of the electron beam and focus area. These two sections are also called the **electron gun**. After leaving the focusing region, the electronic beam is then passed

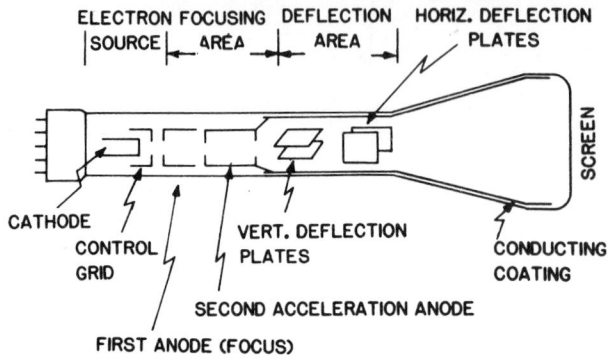

Fig. 12.29 Basic electrostatic cathode-ray tube.

through two pairs of deflection plates, positioned at right angles to each other. One pair is intended for horizontal deflection and the other for vertical deflection. Although the accelerating anode is a basic element of the focusing area, it is usually extended beyond the gun area into the cone of the envelope. This is done by coating a conducting material inside the glass envelope. After the electrons leave the deflection plate regions, they are accelerated toward the phosphorescent screen until they strike the screen. The function of the screen is to convert the electrical energy into light energy. This screen will emit light upon bombardment by the electrons; the color of the light depends on the type of phosphor used. After removal of the electron beam, the screen will continue to glow for some time. The color of the light, after the removal of the beam, may be different from the color before the removal of the beam. The "persistence" of the after-glow varies for different types of phosphor material used. The degree of persistence is usually divided into three classifications: short, medium, and long. Recently, a variable-persistence screen has become available. Its greatest advantage is that it reduces flicker in very lower frequency waveforms (output of spectrum analyzers, medical instrumentations, and many industrial transducers). Controlling the amount of persistence of the CRT screen, the display can be viewed easily.

Variable-persistence CRTs involve placing a mesh of very fine wires, which reduces the effect of the electrostatic field created by the acceleration, on the deflection of the beams. There is some loss in resolution of the CRT as the beam passes through the grid.

A fiber-optic screen has been developed by several manufacturers during the last 15 years. Since the tiny fiber-optic filaments can carry the light from the back of the face to the front with very little attenuation, a bright trace (display) with no parallax is possible.

Displaying more than one trace is very important in such applications as phase measurements and medical instrumentation. There are many different schemes used today to accomplish this multitrace. One method is to have more than one gun; some manufacturers use as many as four guns.

Dual traces are also provided by a single gun with a time-sharing sweep (chopped presentation).

Every oscilloscope has a **graticule**, which is often, a piece of plexiglass with a ruled scale inscribed on one of its surfaces. The graticule is placed in front of the CRT screen. This can cause parallax errors. Some of the newer CRTs have a grid etched on the CRT face plate, which eliminates most parallax error. Newer CRTs have their graticules etched directly on the inside surface of the face plate, on the same side as the phosphor coating. This type of construction is almost free of parallax error.

Parallax error is an optical phenomenon that results when the viewer's eye is not at right angles to the surface of the CRT screen. Therefore, one should view the CRT screen in line with a point on the graticule and a point on the waveform on the CRT display.

12.4.2 Real-Time Oscilloscopes

An **oscilloscope** is an instrument that displays physical information on a cathode-ray tube; it allows visual measurement of the input quantity. Although this input quantity is usually voltage, other quantities, such as current, temperature, pressure, speed, and strain can be translated into voltages by means of suitable transducers, then displayed on the oscilloscope. There are fundamentally two different types of oscilloscope: the **real-time display** oscilloscope and the **real-time sampling display**, or **equivalent-time sampling** oscilloscope.

A typical real-time oscilloscope is depicted in Fig. 12.30. Input signals for vertical deflection on the CRT are applied at the input of the preamplifier. This stage controls input coupling, vertical deflection factor, and dc balance. Input signals applied to the vertical input may be ac coupled or dc

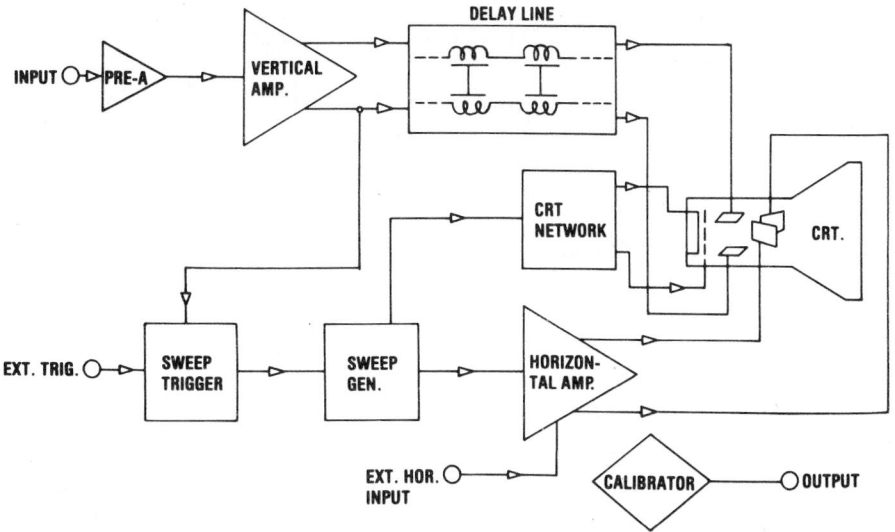

Fig. 12.30 Basic diagram of an oscilloscope.

coupled to the input attenuator. The effective overall vertical deflection factor is determined by the VOLTS/DIV. switch. Dc balance adjustment in the preamplifier is used to set the same dc voltage level throughout the full rotation of variable VOLTS/DIV. control; thus the image on the CRT will not shift. The vertical amplifier provides sufficient voltage gain for the vertical deflection. This amplifier circuit also provides adjustment of variable deflection factors, vertical positions, and samples of the vertical signal to the sweep trigger for the purpose of internal triggering.

The double-ended output from the vertical amplifier is connected to a delay line. The delay line delays the vertical amplifier signal to the vertical deflection plates until the trigger and time-base circuits have had sufficient time to get the unblanking and horizontal sweep started. This is an absolutely required circuit in order to display the entire signal (waveform); otherwise, the leading edge of the waveform will not be displayed.

The sweep trigger circuit provides trigger pulses in order to start the sweep generator. These trigger pulses can be derived from either the vertical amplifier or an external signal connected to the EXT. TRIG. INPUT. In this sweep trigger circuit, controls for selecting trigger level, slope, coupling, source, and mode are also provided.

Sweep Generator
The **sweep generator** (time-base generator) generates a periodic sawtooth voltage. This sawtooth voltage is then applied to the horizontal amplifier. The horizontal amplifier circuit, in turn, provides the output signal to the CRT horizontal deflection plates. In the EXT. HORIZ. position, the horizontal deflection signal is supplied by an external signal. This positive-going sawtooth waveform is linear; that is, it is a ramp function. The rate of rise is set by the TIME/DIV. control. The time-base amplifier (horizontal amplifier) also provides both positive and negative sawtooth waveforms for the two horizontal deflection plates. The cathode-ray beam is swept horizontally to the right through a given amount of graticule divisions during each unit of time. In order to display a steady waveform on the CRT screen, the horizontal sweep waveform must be synchronized with the waveform being displayed. This is accomplished by starting the sweep waveform (time-base sawtooth waveform) at a selected point on the displayed waveform. Figure 12.31(a) illustrates a stable and Fig. 12.31(b) an unstable display of the same signal.

The CRT network provides the high voltage and control circuits necessary to operate the cathode-ray tube. The high voltage is derived from a high-frequency oscillator that operates at approximately 25 kHz. Controls for intensity, focus, and astigmatism adjustments are provided in this network. A rectangular pulse generated in the time-base generator is used as an unblanking signal connected to the grid of the cathode-ray tube. This unblanking signal unblanks the CRT so that a trace can be displayed.

Trigger Level
On some waveforms being displayed, one may want to display the signal waveform starting on "+" slope or on "−" slope of the waveform (Fig. 12.32). The slope control allows this selection. The

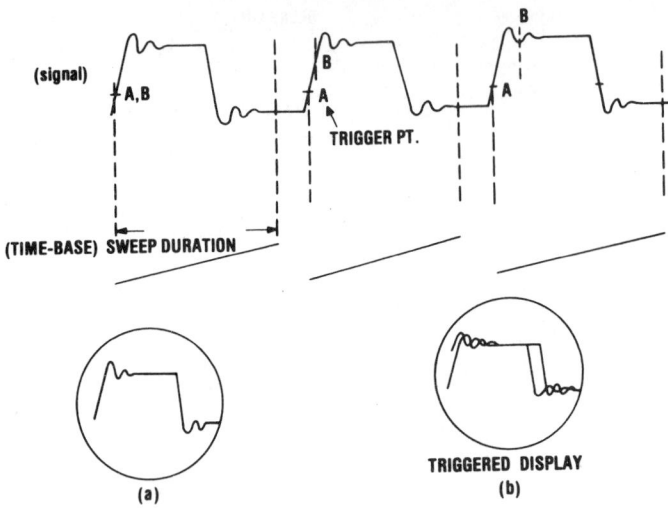

Fig. 12.31 (*a*) Correctly triggered display, (*b*) incorrect slope triggering.

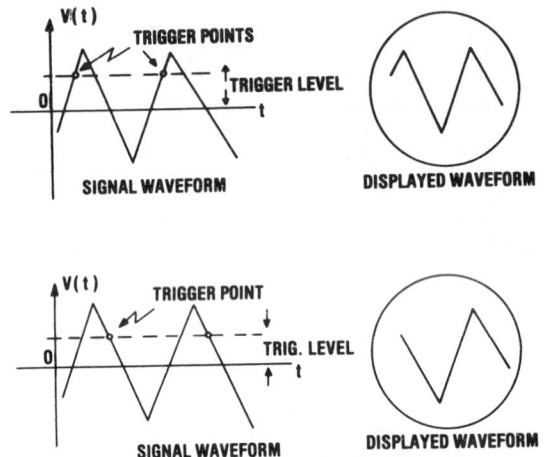

Fig. 12.32 (*a*) Positive slope triggering, (*b*) negative slope triggering.

trigger level selects a particular signal amplitude point at which the sweep begins. The mode control allows one to choose either ac coupling or dc coupling for specific advantages.

Calibration and Measurement
The calibration circuit generates a standard square-wave voltage. The square-wave signal is used to calibrate the vertical deflection factor as well as to adjust the compensated probe for the vertical input.

In some cases the input impedance of the oscilloscope can produce undesirable loading on the circuit under test. This loading will cause the display on the CRT screen to be other than an actual waveform in the circuit under test. A compensated probe (passive or active) can be used to reduce the loading effect. The probe includes a resistor R_1 shunted by a capacitor C_1 (adjustable) (Fig. 12.33). This combination is connected with the oscilloscope input impedance that is a parallel combination of C_{in} and R_{in}. The perfectly compensated probe requires the condition: $R_1 C_1 = R_{in} C_{in}$. The compensated probe can be adjusted with the calibrating signal (Fig. 12.34).

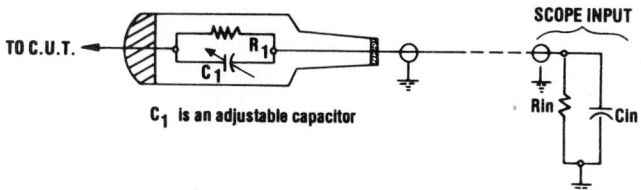

Fig. 12.33 Passive-compensated probe.

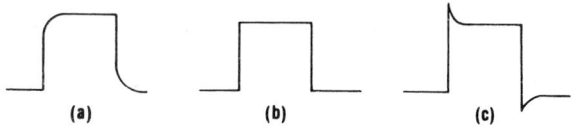

Fig. 12.34 Response of compensated probe to square wave: (*a*) undercompensation, (*b*) perfect compensation, (*c*) overcompensation.

Storage Oscilloscope

The real-time oscilloscope performs more than simple display of the input phenomena. Because of the CRT screen's persistence, it allows us to integrate and display with some delay time. The storage CRT screen stores transient waveforms (nonrecurrent waveforms) and displays them as though the input waveform was periodic. However, the signal-processing capabilities often create a new problem for oscillo-photorecordings when the screen persistence interferes with the observation of the transient waveforms. Some oscilloscope manufacturers solve these difficulties by digital processing of the input waveforms. For example, a low-frequency waveform may be recorded in real time and displayed in a faster time to eliminate the flickering. Using today's highly developed solid-state memory technology, such processing should be a relatively simple task.

A processor can be interfaced between a sample probe and a general-purpose oscilloscope, or inserted between the sample probe and a sampling scope. The modified real-time oscilloscope, which is used as a sampling scope, is shown in Fig. 12.35.

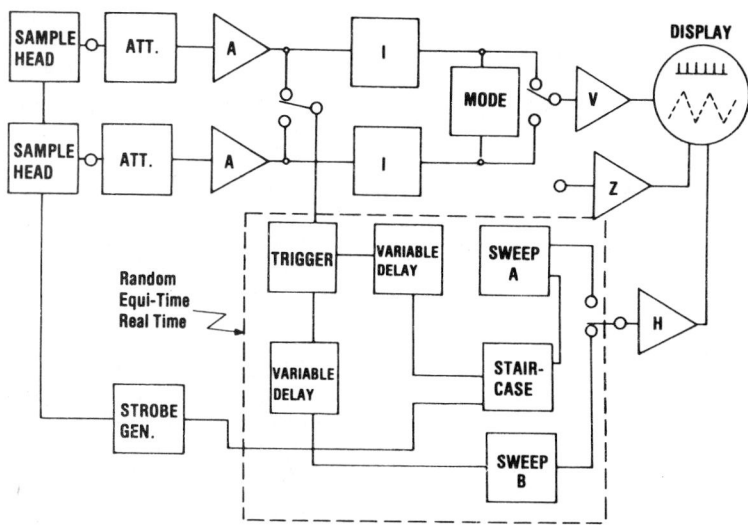

Fig. 12.35 Dual-trace scope modified to operate as a sampling scope. (Adapted from Ref. 4.)

12.4.3   Sampling Concepts[4]

Sampling in oscilloscopes is to solve the limitations of gain-bandwidth in most oscilloscopes. When a 100-ps segment of a 1-GHz signal is sampled and stretched 100 times to 10 ns, the 1-GHz waveform can be displaced on a 50-MHz-bandwidth oscilloscope. If equivalent time sampling is used to lower the frequency by 100 times, then a 100-ns sweep will display one cycle corresponding to 10 MHz. From Table 12.2,[4] one can see that if random sampling at 10 kHz is used, a sample occurs every 1000 cycles. Random sampling at 10 kHz would require 100 ms to synthesize a cycle.

Table 12.2 shows the number of samples taken for various sweep rates if a 100-kHz sample is at free run. With a 50-MHz scope and with sampling at 50 MHz, one is able to view a 2-μs/cm waveform at 100 samples/cm in real time. Figure 12.36 illustrates three different types of sampling scopes and displays. In Fig. 12.36(a), a free-running sampler presents a dotted tracing of the train of pulses. In Fig. 12.36(b), an equivalent-time sample is used. A 1-GHz waveform is sampled at 100 kHz in real time, and there is no phase lock between the 1-GHz signal and the sampler; however, two signals occasionally will momentarily be in phase. Thus a recognizable waveform will be displayed on the CRT screen. Equivalent-time sampling requires occasional phase locking between the signal and the sampling signal. In Fig. 12.36(c), random sampling I uses periodic sampling but of random phase. The synchronizing trigger starts a fast ramp that is stopped when a sample pulse occurs within a sweep window. The heights of the signal and the ramp determine the position of the CRT spot for that sample. Since there is no phase lock involved, the sampling is random in nature.

12.4.4   Applications

1. *Period and Frequency Measurements.* Set up an oscilloscope and obtain a stable waveform using the trigger controls. Now set the VOLTS/DIV. control so that the peak to peak is approximately two-thirds of the CRT screen. Also set the SEC/DIV. control to obtain approximately one or two cycles of the display. Next, adjust the vertical position and the horizontal position controls until the display is centered, as shown in Fig. 12.37. To calculate the period, count the number of divisions for the complete cycle. Then, using the following formula, the period is determined:

$$\text{Period } (T) = \text{number of divisions} \times \text{SEC/DIV. setting}$$

Once you have determined the period T, the frequency is

$$\text{Frequency} = \frac{1}{\text{period}}$$

2. *Rise-Time Measurement.* The rise time is illustrated in Fig. 12.38. Set up an oscilloscope and obtain a single pulse, as shown in Fig. 12.38, using the SEC/DIV. control. Then adjust the time VOLTS/DIV. control and its VAR (variable) control until the display of the waveform is five graticule divisions. This is the distance between the 0 and 100% points on most oscilloscopes. One-half division of the graticule division is 10%. To calculate the rise time T_r, count the number of horizontal divisions between the 10 and 90% points. Then, with the following formula, T_r is found.

$$T_r = \text{horizontal distance (division)} \times \text{SEC/DIV. setting}$$

TABLE 12.2   Sampling Rate Necessary for 100 and 1000 Dots per Centimeter of Display for Representative Sweep Rate

Sweep Rate	Display Cycles/10 cm (Hz)	Samples/cm for 100 kHz Sample Rate	Time for 1000 Samples (100-kHz rate)	Samples/s (Hz) for Samples/cm	
1.0 s/cm	0.1	10^5	In excess	100	1,000
0.1 s/cm	1.0	10^4	In excess	1,000	10,000
10 ms/cm	10	10^3	In excess	10,000	100,000
1.0 ms/cm	100	10^2	10 ms	100,000	1×10^{-6}
100 μs/cm	1,000	10	10 ms	1×10^{-6}	10×10^{-6}
10 μs/cm	10,000	1	10 ms	10×10^{-6}	100×10^{-6}
1.0 μs/cm	100,000	10^{-1}	10 ms	100×10^{-6}	—
100 ns/cm	1×10^{-6}	10^{-2}	10 ms	—	—
10 ns/cm	10×10^{-6}	10^{-3}	10 ms	—	—
1 ns/cm	100×10^{-6}	10^{-4}	10 ms	—	—

Source. Adapted from Ref. 4.

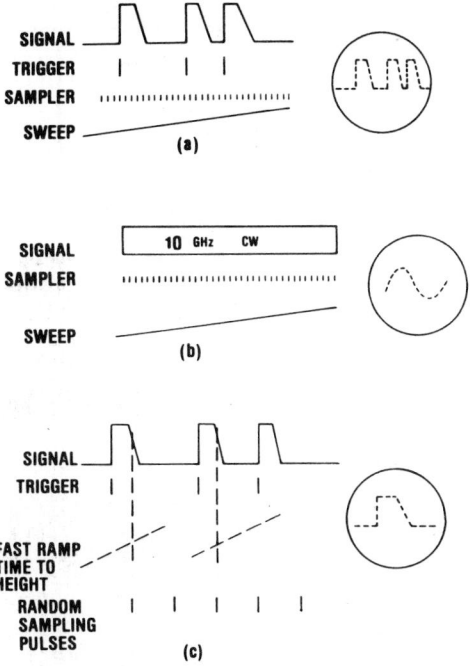

Fig. 12.36 (*a*) Real-time sampling. Random-phase sampling: dotted trace of pulses with high dot density. (*b*) Equivalent-time sampling. A periodic wave with long coherence length can be sampled by a periodically occurring sampling signal if the relative phase shift between the two is not too great. (*c*) Equivalent random sampling. This method of sampling is due to McQueen[5] {2} and Frye and Nahman.[6] (Adapted from Ref. 4.)

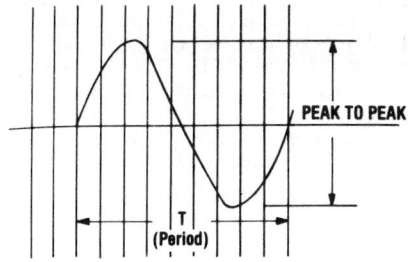

Fig. 12.37 Measuring the period of a waveform.

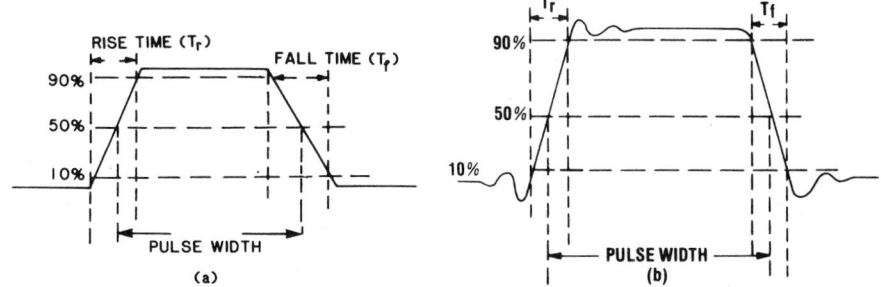

Fig. 12.38 Rise- and fall-time measurements: (*a*) ideal pulse, (*b*) practical pulse.

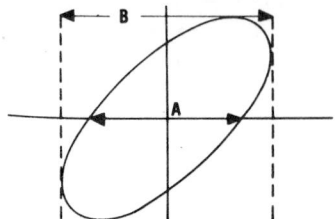

Fig. 12.39   Lissajous figure.

3. *X-Y Phase Measurements*. When an *X-Y* phase measurement technique is used, what is observed is not an individual waveform but a Lissajous figure. When a single-channel oscilloscope is used, both *X* and *Y* inputs must be calibrated for the same sensitivity before the two inputs are applied: one a X INPUT, the other at Y INPUT. Next, adjust the *X*- and *Y*-position control to center the Lissajous figure as shown in Fig. 12.39. Then count the number of horizontal divisions between two points on the Lissajous figure with the widest horizontal separation (distance *B* in Fig. 12.39). Next, count the number of horizontal divisions between the intersections of the Lissajous figure and the center horizontal graticule line (see Fig. 12.39). The phase angle between the *X* and *Y* inputs is calculated by

$$\theta = \sin^{-1}\frac{A}{B}$$

12.5   BIPOLAR TRANSISTORS AND INTEGRATED CIRCUITS
Denny D. Tang

Bipolar transistors have been fabricated on semiconducting germanium, silicon, and gallium arsenide. Among these, GaAs is the most attractive material for building high-performance transistors because it has the highest electron mobility and largest band gap, which permits operation over the widest temperature range. However, for many years, silicon has been the most important semiconductor material, owing to the earlier success of the silicon planar technology while GaAs technology has yet to mature, and owing to its wider band gap than its rival germanium.

The advance in planar technology has helped to increase the power and frequency capabilities of bipolar transistors as well as their reliability, and more importantly, has led to the creation of integrated circuits. Integrated circuits are categorized into analog IC and digital IC by the area of their applications. Analog circuits deal with signals of continuous magnitude, where a transistor performs signal amplification. Special design techniques have been developed for analog ICs due to the lack of inductive circuit components, limited range, large tolerance of the component values on the IC. However, integrated circuits offer an advantage not found in a discrete component: good tracking in component characteristics and temperature coefficient. Such advantages have been utilized and have led to the development of differential and operational amplifiers and analog multipliers, which become the basic building blocks of the analog IC.

Digital circuits deal with digital signals with only two states, where the transistor acts as a switch, switching from conducting state to nonconducting state, and vice versa. The nature of the digital circuit makes it immune to the processing tolerance of the integrated circuit and makes it most suitable for integration. Both bipolar IC and MOSFET IC share the digital IC market, while bipolar circuits dominate the high-speed applications. Spurred by the enormous market demand and as a result of the advances in processing technology, the transistor size is miniaturized and the number of transistors on a digital IC, or the level of integration, increases continuously.

This section is written for readers with a background in basic one-dimensional transistor theory and switching characteristics of transistors. Section 12.5.1 covers microwave and power transistors with special emphasis on the frequency and power-handling capabilities. The first half of Section 12.5.2 provides a general discussion on the topic of analog IC, including the design techniques, the basic building blocks, and their applications. Digital integrated circuits are covered in the second half of where memory, logic, and LSI design approaches are discussed.

12.5.1   Microwave Transistors and Power Transistors

Although there is no well-defined boundary between power transistors and microwave transistors, power gain and efficiency are usually the prime considerations for a power transistor, whereas cutoff frequency and noise figure are the prime considerations for a microwave transistor.

Microwave Transistors

Frequency Response. Transistors operated in the high-frequency region, where the range goes up into GHz, are considered as microwave transistors. At high frequencies, the transistor gain decreases with increasing frequency due to the transit-time effect. The general form of the transistor power gain as a function of frequency is

$$G(f) \approx \frac{G_0}{\left[1 + G_0^2 (f/f_{max})^4\right]^{1/2}} \tag{12.21}$$

$$G_0^2 \left(\frac{f}{f_{max}}\right)^4 \gg 1$$

Then

$$G = Z \left(\frac{f_{max}}{f}\right)^2 \tag{12.22}$$

and the gain falls off at 6 dB per octave, where f_{max} is the maximum frequency of oscillation and is defined as the frequency at which the transistor unilateral power gain equals unity. The proportionality constant Z, which accounts for the transistor parasitic elements, typically ranges from 0.2 to 0.3.

At a power gain of 1, the maximum frequency of oscillation f_{max} can be defined as[7]

$$f_{max} = \left(\frac{\alpha_0 f_T}{8 \pi r_b' C_c}\right)^{1/2} \tag{12.23}$$

where α_0 is the common-base dc gain, f_T the transistor current gain-bandwidth product, and $r_b' C_c$ the collector-base time constant, which is the product of the base resistance and the collector capacitance.

Figure 12.40 shows the state-of-the-art performance of bipolar microwave transistors.[8] Maximization of f_{max} requires minimizing the $r_b' C_c$ and maximizing f_T. The emitter and base geometry of a typical microwave and power transistor is shown in Fig. 12.41(a). The interdigital structure of the emitter and base stripes are designed to reduce r_b'. For a given emitter strip width S, length l, and

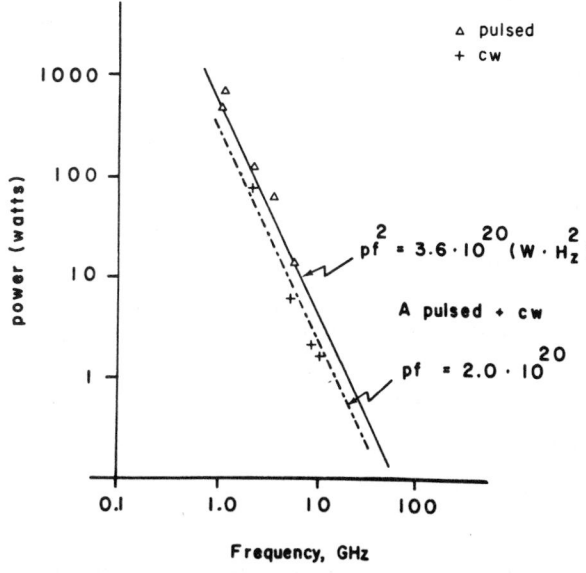

Fig. 12.40

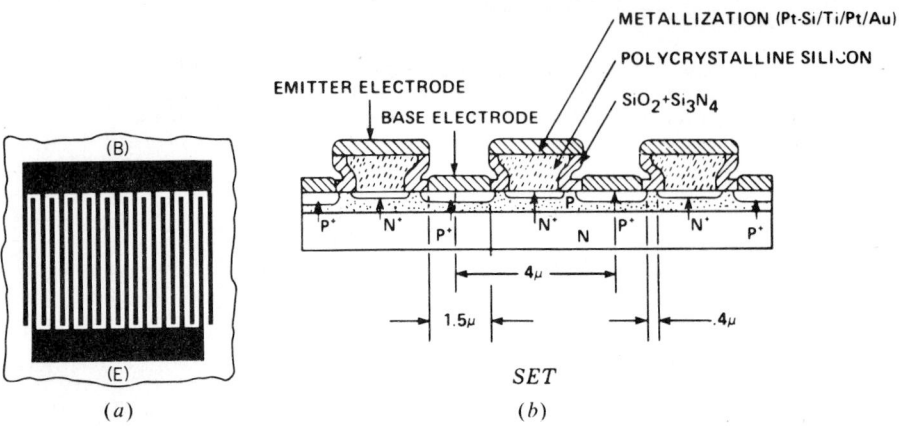

Fig. 12.41 (a) Interdigital emitter-base geometry of a high-frequency bipolar transistor, (b) step electrode transistor (SET) (from T. Sakai, ISSCC Dig. Tech. Paper, 1977. p. 196; reprinted by permission (After Ref. 9.)

periodicity of emitter strip of spacing d,

$$r_b' \propto R_{sb} \frac{S}{l} \quad \text{and} \quad C_c \propto C_0 l d$$

Then

$$f_{max} \propto \left(\frac{1}{dS} \frac{\alpha_0 f_T}{8\pi R_{sb} C_0} \right)^{1/2} \tag{12.24}$$

where R_{sb} is the sheet resistance of the base and C_0 the capacitance per unit area.

Reducing the width of the emitter S and the spacing d between the emitter stripes (or the extrinsic base area) is the most effective approach to achieve the reduction of $r_b' C_c$. Figure 12.41(b) shows an advanced bipolar transistor structure called step-electrode transistor (SET),[9] the spacing between the emitter and base is reduced to 0.4 μm by an advanced bipolar process. The cutoff frequency f_T of the transistor is directly related to the impurity profile of the transistor and the operating condition. Let us consider a transistor built on an epitaxial collector, as shown in Fig. 12.42.

The cutoff frequency is given by

$$f_T = \left\{ 2\pi \left[\frac{kT(C_{te} + C_{tc} + C_p)}{qI_c} + \frac{W_B^2}{ND_B} + \frac{Xc}{2v_s} \right] \right\}^{-1} \tag{12.25}$$

where C_{te}, C_{tc}, and C_p are depletion layer capacitance of the emitter and collector, and the parasitic capacitance at the base node, respectively. N is the field-assisted diffusion factor, ranging from 2 to 60; D_B is the diffusion coefficient; and v_s is the carrier saturation velocity ($\sim 8 \times 10^6$ cm/s for silicon). At low current, f_T is limited primarily by the discharging and charging time of the depletion capacitances C_{te} and C_{tc}. As I_c increases, the limiting component of f_T is contributed by the stored changes, given in the second and third terms. To increase the cutoff frequency, the transistor should have a very narrow base thickness, which is one of the most important dimensions, a narrow collector region, and should be operated at a high current level. As the collector width decreases, however, there is a corresponding decrease in breakdown voltage. Compromises must be made for high-frequency and high-power operation. Since the collector current drifts through the collector junction, the maximum current density at the collector is limited by the collector doping density as given by

$$J_c \approx q\mu_c N_c E_c = \frac{q\mu_c N_c (V_{co} + \phi_{bi})}{W_c} \tag{12.26}$$

where μ_c N_c, W_c, and E_c are the mobility, impurity doping density, thickness of epitaxial layer, and

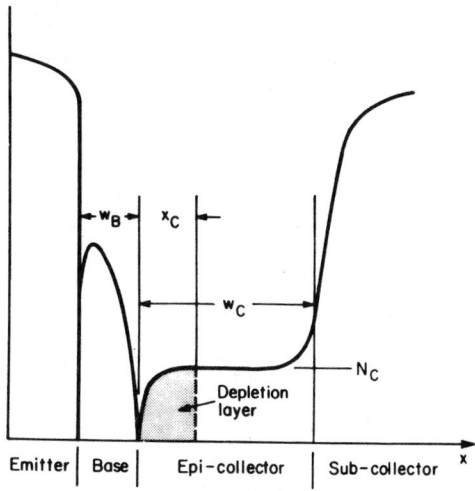

Fig. 12.42   Schematic impurity profile of an epicollector transistor.

electric field in the collector epitaxial layer, respectively; V_{co} is the applied reverse-bias voltage, and ϕ_{bi} the built-in potential of the collector junction. When collector current density is greater than that given by Eq. (12.26), the carrier density is greater than the doping density of the collector and upsets the charge neutrality in the collector region, resulting in relocation of the high-field region from the junction into the collector region and widening of the base region, which is called the Kirk effect[10] or **base stretching**. Thus the f_T of the transistor peaks at this current density.[11]

For a transistor operating above 8 GHz, f_T needs to be approximately 10 GHz. To get f_{max} to the required 25-GHz levels, it is necessary to get the $r_b'C_c$ time constant down to 0.7 ps, which requires micrometer and submicrometer emitter stripe widths and a 100-nm or thinner base width. Detailed design considerations of microwave transistors are given in Ref. 12.

Noise Figure. The ratio of total mean-square noise voltage at the output of the transistor to mean-square noise voltage at the input resulting from thermal noise in source resistance R_g. At lower frequencies the dominant noise source in a transistor is due to the surface effect, which gives rise to the $1/f$ noise spectrum. At medium and high frequencies the noise figure (NF) is given by[13]

$$ \text{NF} = 1 + \frac{r_b}{R_g} + \frac{r_e}{2R_g} + \frac{(1 - \alpha_0)\left[1 + (1 - \alpha_0)^{-1}(f/f_\alpha)^2\right](R_g + r_b + r_e)^2}{2\alpha_0 r_e R_g} \qquad (12.27) $$

where R_g is the generator resistance, f_α is the α cutoff frequency, and r_b and r_e are base and emitter resistance, respectively. From Eq. (12.27) it can be shown that at medium frequencies where $f \ll f_\alpha$, the noise figure is essentially a constant determined by r_b, r_e, $1 - \alpha_0$, and R_g. There is an optimum termination R_g that can be calculated from the condition $d(\text{NF})/dR_g = 0$. For a low-noise design, a low value of $1 - \alpha_0$ (i.e., a high β_0) is very important. At high frequencies beyond the "corner" frequency $f = \sqrt{1 - \alpha_0} f_\alpha$, the noise figure will increase approximately as f^2.

Figure 12.43 shows the noise figures of silicon bipolar transistors;[12] the noise figure decreases at the high-frequency region as the emitter stripe width is narrowed (thus the r_b decreases).

Although transistor power gain falls off at 6 dB per octave at high frequencies, the frequency characteristics of packaged transistors can be greatly altered by the parasitic reactance of the package. Careful design of the package and impedance-matching circuit is required to obtain the desired performance. Figure 12.44 shows a complete amplifier circuit[14] built on a dielectric board on which TEM-mode microstrip circuits of various shapes are utilized to match the input and output impedance of the transistor to the source and load, respectively, so that wideband amplification can be realized. In this figure , Q_1 is the microwave transistor, and the rest are the dc biasing circuits.

Power Transistors
Although at higher frequencies the power output of transistors varies approximately as $1/f^2$, at lower frequencies the limitation on power output is due mainly to the thermal effect. As the power

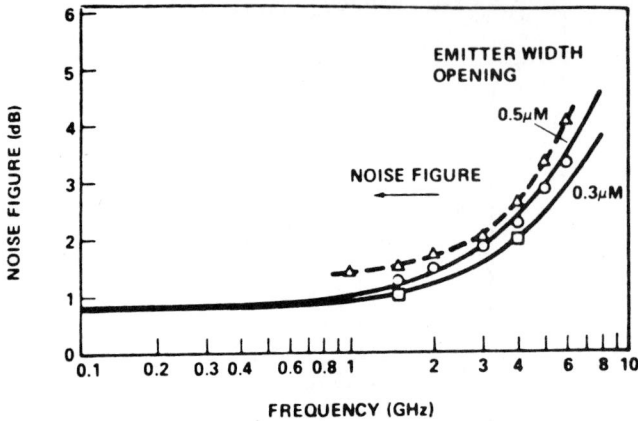

FREQUENCY (GHz)

Fig. 12.43 Typical noise figure of Si-microwave transistor as a function of frequency (solid line HP 505, 9HXTR-6101). [From T. Hsu and C. P. Snapp, *IEEE Trans. Electron Devices*, **ED-55**, 723 (1978). Reprinted by permission.]

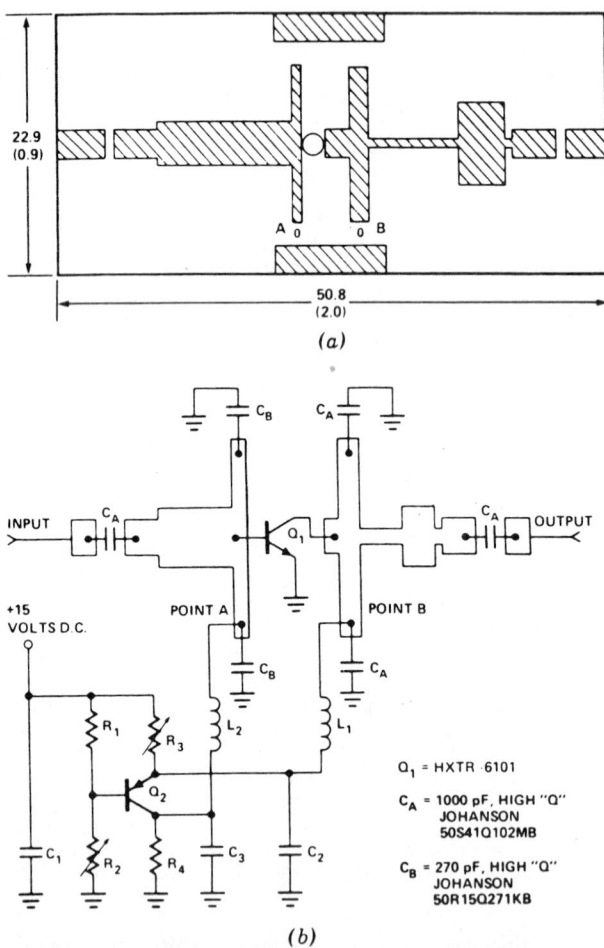

Fig. 12.44 (*a*) RF circuit board layout, (*b*) complete bipolar transistor amplifier schematics. (HP Application Note 967.)

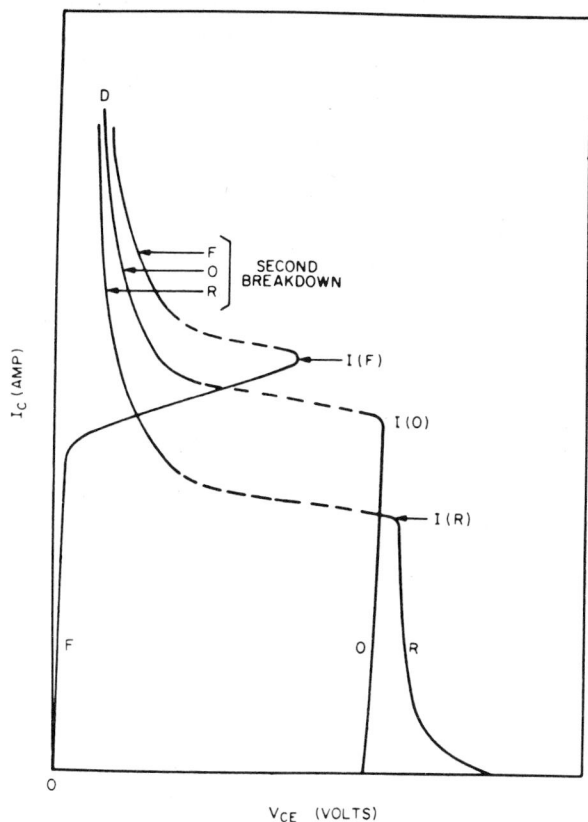

Fig. 12.45 Collector-current versus collector-emitter voltage under second breakdown condition. *F*, 0, and *R* indicates forward-, zero-, and reverse-base drive, respectively. [From H. A. Schaft, *Proc. IEEE*, **55**, 1272 (1967). Reprinted by permission.]

increases, the junction temperature T_j increases. The maximum T_j is limited by the temperature at which the base region becomes intrinsic. About T_j the transistor action ceases, since by then the collector is effectively short-circuited to the emitter. To improve transistor performance, one must improve the encapsulation sufficiently to provide an adequate heat sink for efficient thermal dissipation, and must use materials with large band gaps that will allow higher-temperature operation.

Second breakdown[15] is another phenomenon that limits the power capability of the transistor. This phenomenon differs from avalanche breakdown by the time delay of its occurrence and its voltage snapback $I - V$ characteristics as shown in Fig. 12.45. The initiation of instability is mainly due to the temperature effect. When a pulse with a power of $P = I_c V_{CE}$ is applied to a transistor, the transistor heats up, and hot spots are usually located at the center of the device. Within the hot spot, the intrinsic carrier concentration n_i increases, as temperature rises, and when n_i is greater than the collector doping the collector series resistance decreases and the instability triggers the collapse of the voltage across the junction. As a result, the resistance of the breakdown spot becomes drastically reduced and current constriction occurs that eventually leads to melting of the spot, causing permanent damage to the transistor.

12.5.2   Bipolar Integrated Circuits

Components

Bipolar integrated circuits are exclusively fabricated on silicon substrate with planar technology. Junction transistors (mostly *npn* transistors), *pn*-junction diodes, resistors, capacitors, and Schottky barrier diodes (SBDs) are the basic circuit components of bipolar integrated circuits. Figure 12.46(*a*) shows the top and cross-sectional views of a typical structure of an integrated *npn* transistor and a resistor before the metal layer on the surface is processed.[16] These devices are fabricated on an n^-

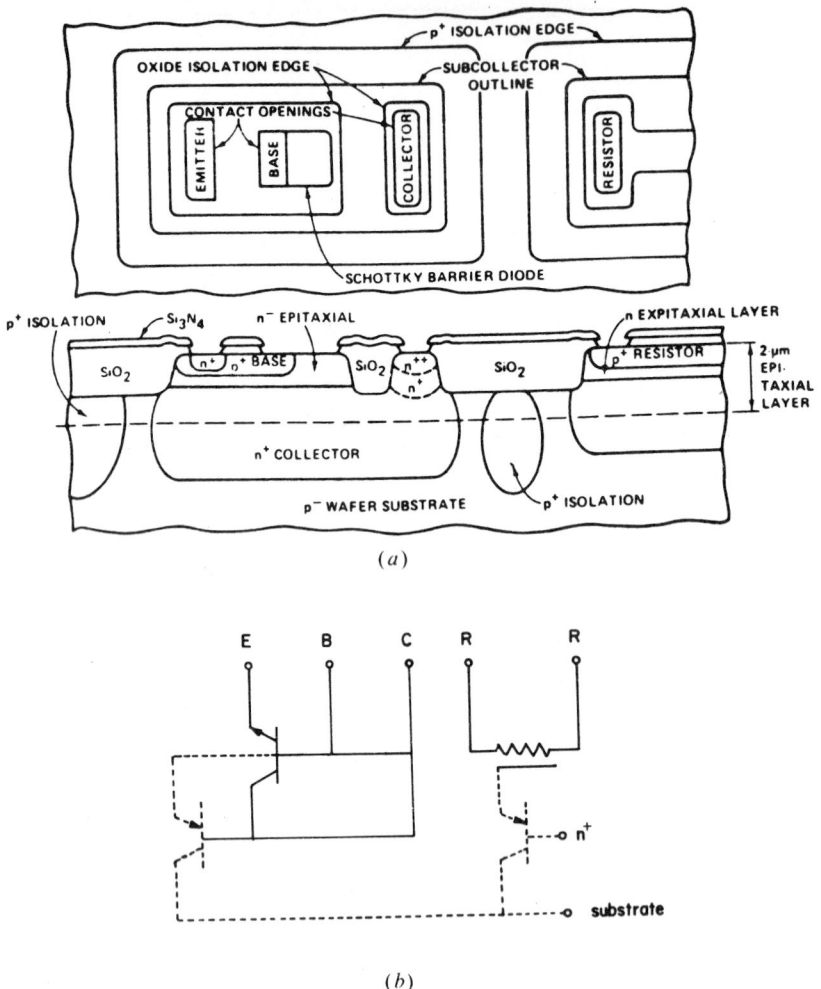

Fig. 12.46 (*a*) Top and cross-sectional views of a typical Schottky diode clamped transistor and resistor fabricated using a 2.0-μm-thick epitaxial layer bipolar process. [From R. J. Blumberg, *IEEE J. Solid State Circuits*, **SC-14**, 818 (1979), reprinted by permission]; (*b*) equivalent circuit.

epitaxial layer on the p^- substrate patterned with n^+ sublayer islands, where the n^+ sublayer is called the n^+ collector in Fig. 12.46(*a*) and serves to reduce the series resistance of the collector measured from the epitaxial collector of the *npn* transistor to the collector contact on the surface. The Schottky barrier diode is located over the region where metal contacts to the n^- epitaxial layer and serves as a "clamp" to prevent the transistor from deep saturation. Unlike a *pn*-junction diode, a Schottky barrier diode is a majority carrier device with no minority carrier storage time constant associated to its switching characteristics; therefore, its switching speed is faster than that of a *pn*-junction diode.

Lateral *pnp* transistors are frequently built together with *npn* transistors on the same integrated circuit for certain circuit applications. Figure 12.47 shows two *pnp* structures; one is called an epi-base *pnp* transistor, the other is called a double-diffused lateral *pnp* transistor. The former can be built with the same process as that used for a *npn* transistor, where the base of the *npn* transistor is used as the emitter and the collector of the *pnp* transistor. The latter is built by diffusion of *n*-type and *p*-type dopants through a common opening into a *p*-type epitaxial layer, and as a result, it has a much thinner base width.

In Fig. 12.46(*a*) the thick oxide and the underlying p^+ region surrounds the transistor and the resistor and serves as an isolation called oxide isolation. The p^+ region under the thick oxide is

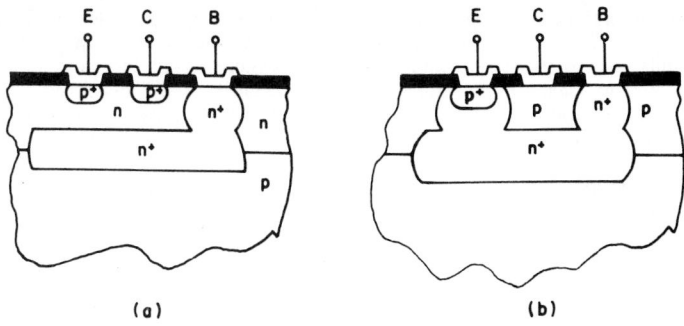

Fig. 12.47 Lateral *pnp* transistors: (*a*) epi-base transistor, (*b*) double-diffused transistor.

formed to prevent shorting between neighboring n^+ sublayer islands, which is caused by inversion of the p^- region under the oxide and is called channel stop. LOCOS (local oxidation of silicon), isoplanar, OXIM (oxide-isolation-monolithic) are the names of the process that fabricates bipolar integrated circuits with oxide isolation. There are other kinds of isolation. One is diffusion isolation in which a p^+ or n^+ ring is diffused from the surface deep down to the p^- substrate around the active device. The latter is called collector diffusion isolation (CDI) and serves as the collector contact of the *npn* transistor at the same time. The diffusion-isolated device is larger and has more sidewall capacitance than the oxide-isolated device; thus the oxide-isolated device is more suitable for large-scale integration.

The equivalent circuit of the integrated circuit shown in Fig. 12.46(*a*) is given in Fig. 12.46(*b*). The actual circuit consists not only of the transistor and the resistor but also of the parasitic capacitance (associated with the substrate and isolation) and the parasitic *pnp* transistor formed by the base (p), the collector (n), and the substrate (p) (with very low current gain). Such parasitics inherently exist in the integrated device structure, and to some extent they all affect the electrical characteristics and performance of the integrated circuit.

The typical range of parameters of the integrated components is shown in Table 12.3. The tolerance is the variation in value of the parameters over the wafer, while matching means the

TABLE 12.3   Typical Performance of Integrated Components

	Symbol	Typical Value	Tolerance	Temperature Coefficient
npn transistors and diodes				
Current amplification factor	h_{FE}	30–100	$\pm 50, -30\%$	$+0.5\%/\,°C$
Matching of h_{FE} between identical transistors in close proximity	Δh_{FE}	—	$\pm 10\%$	$+10.5\%/\,°C$
Base-emitter diode forward voltage drop (low current)	V_{BE}	0.7 V	$\pm 3\%$	$-2\ \text{mV}/\,°C$
Matching of V_{BE} between identical transistors in close proximity	ΔV_{BE}	—	$\pm 2\ \text{mV}$	$\pm 10\ \mu\text{V}/\,°C$
Base-emitter diode reverse breakdown voltage	BV_{EBO}	6–8 V	$\pm 5\%$	$\pm 3\ \text{mV}/\,°C$
Collector-base breakdown voltage	BV_{CBO}	< 45 V	$\pm 30\%$	—
Collector-substrate breakdown voltage	BV_{CS}	< 60 V	—	—
Lateral *pnp* transistors				
Current amplication factor	h_{FE}	0.5–20	$+200, -50\%$	$\pm 0.5\%/\,°C$
Base-collector breakdown voltage	BV_{CBO}	< 45 V	—	—
Resistors				
Resistance of diffused resistors (base layer)	R	100 Ω–20 kΩ	$\pm 20\%$	$\pm 0.15\%/\,°C$
Resistance of deposited resistors	R	100 Ω–100 kΩ	$\pm 18\%$	$\pm 0.10\%/\,°C$
Matching between identical resistors in close proximity	ΔR	—	$\pm 3\%$	$\pm 0.005\%/\,°C$
Pinch resistors				
Sheet resistance (small applied voltage)	R_S	5–10 kΩ/□	$+100, -50\%$	$+0.3-0.5\%/\,°C$
Matching of identical pinch resistors in close proximity	ΔR	—	$\pm 5\%$	—

variation of parameters when the parameter is measured from the neighboring devices. Clearly, only a limited range is covered by each parameter. The circuit designer must avoid circuit components that are not integrable, such as transformer and coils, and select components and circuits that occupy a small surface area to realize the advantage of the integrated circuit. Furthermore, the circuit performance must be considered under worst-case conditions, when parameters vary with location on the circuit chip and temperature.

Bipolar Analog Circuits[17]

In the analog circuit applications, there are certain inherent limitations of monolithic device structures, such as poor absolute value tolerances (see Table 12.3) and large temperature coefficients, limited choice of component values and compatible device types, and last but not least, lack of monolithic inductors. On the other hand, IC fabrication methods offer a number of unique and powerful advantages to the circuit designer. Some of these are the availability of a large number of active devices, good matching and thermal tracking of component values, and the control of the device layout and the geometries during chip design. By making efficient use of these inherent advantages associated with monolithic IC structures, it is often possible to come up with designs that can greatly exceed the performance of similar discrete component circuits.

Various design techniques to overcome limitations in the range of the parameters of the integrated devices are given as following. Impedance transformation, which is used to obtain large value of capacitance and resistance from smaller ones, is realized with transistors as shown in Fig. 12.48(a) and (b), in which the value of components is multiplied by the current gain of the transistors. Thus impedance transformation enables us to use smaller components, which occupy a smaller chip area. The current source in Fig. 12.48(c) is realized with a transistor and a small-value resistor and performs better than a large resistor and provides an economical way to obtain better performance.

A feedback amplifier is an example of the advantage of using resistor tracking of the integrated circuit. In a feedback amplifier [see Fig. 12.48(d)], the gain of the amplifier is equal to the ratio of the feedback resistors, not the absolute value of the resistor, provided that the gain of the transistor is much greater than the gain of the amplifier. A differential amplifier is another important example. Since the two transistors (see Fig. 12.49) and load resistors ideally should be identical to realize high

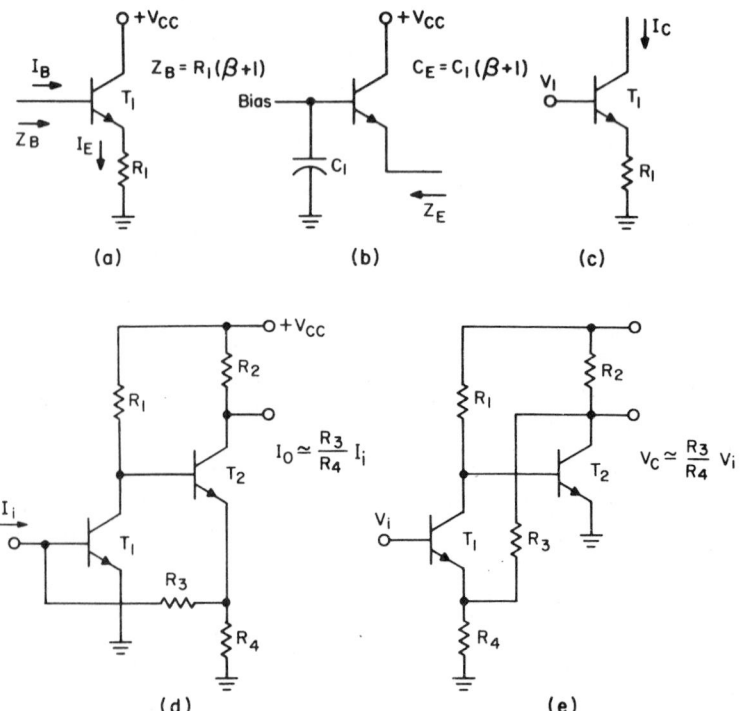

Fig. 12.48 (a) Multiplication of resistance with the gain of a transistor, (b) multiplication of capacitance with the gain of a transistor, (c) constant-current source, (d) shunt series feedback pair, (e) series shunt feedback pair.

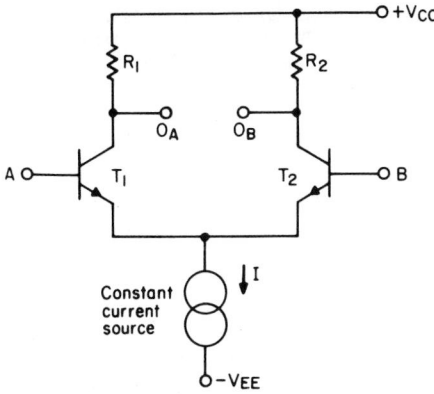

Fig. 12.49 Basic differential amplifier circuit.

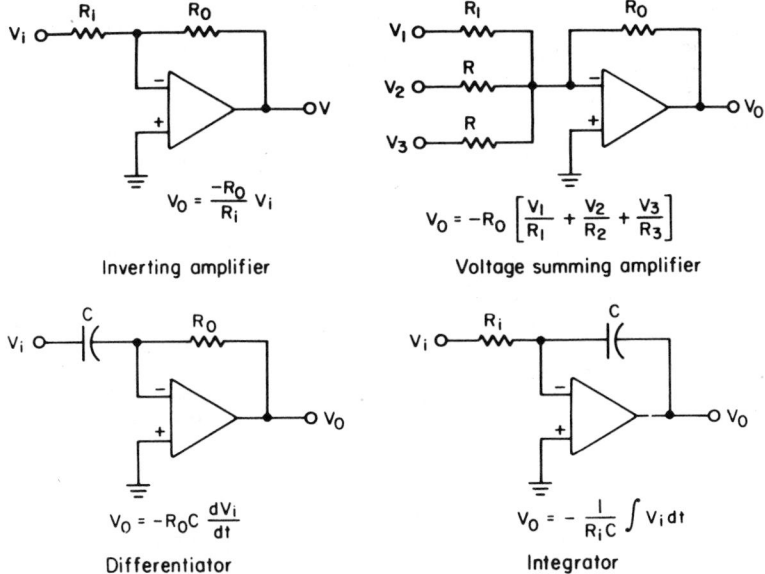

Inverting amplifier
$$V_O = \frac{-R_O}{R_I} V_i$$

Voltage summing amplifier
$$V_O = -R_O \left[\frac{V_1}{R_1} + \frac{V_2}{R_2} + \frac{V_3}{R_3} \right]$$

Differentiator
$$V_O = -R_O C \frac{dV_i}{dt}$$

Integrator
$$V_O = - \frac{1}{R_i C} \int V_i\, dt$$

Fig. 12.50 Applications of the operational amplifier: (*a*) inverting amplifier, (*b*) voltage-summing amplifier, (*c*) differentiator, (*d*) integrator.

common-mode rejection, the tracking of transistor and resistor temperature coefficient and other characteristics obtainable from transistors in close proximity on the integrated circuit chip provides better matching than that of any discrete devices.

An operational amplifier (op-amp) consists of several stages of differential amplifiers of great stability. It has a high input impedance, low output impedance, and a large amount of gain. The voltage gain of the op-amp is determined by the external feedback resistors, as shown in Fig. 12.50(*a*). Furthermore, it can be used as a summing amplifier with more than one input, as shown in Fig. 12.50(*b*), a differentiator as in 12.50(*c*), and an integrator as in 12.50(*d*). These and other applications, such as an analog multiplier, modulator,[16] analog-to-digital (A/D) converter, and digital-to-analog (D/A) converter,[18,19] of the operational amplifier make it the basic analog system.

Bipolar Digital Circuits
Bipolar logic and memory circuits are covered in this section. There are many bipolar logic gates and they are described in the following section. Propagation delay (switching delay) versus power dissipation and noise margin are the figures of merit of a logic gate. In a memory circuit, cell size,

access time, and write time are the three major considerations of memory chips. Various types of bipolar memory are discussed next. As the number of circuits on the integrated-circuit chip increases, the design work becomes astronomical. The design approach of a large-scale-integration (LSI) circuit is also discussed.

Logic Gates. Many approaches to the logic gate are possible and have been used in the past. They are direct-coupled-transistor logic (DCTL), resistor-transistor logic (RTL), diode-transistor logic (DTL), transistor-transistor logic (TTL), emitter-coupled logic (ECL), merged-transistor logic (MTL), and Schottky-transistor logic (STL). The most commonly used logic gates today are TTL and ECL, and recently, MTL and STL have received increased attention. We discuss these four circuits in more detail in this section, in the sequence STL, TTL, MTL, and ECL.

1. *STL.*[20] The three Schottky barrier diodes couple the input to the gate as shown in Fig. 12.51(*a*). T_1 is the switching transistor and its base current flows through R_1. If one of the three inputs *A*, *B*, or *C* is at the ground potential, all the current that flows through R_1 is sunk by the Schottky barrier diode instead of flowing into the base of the transistor, since the Schottky barrier diode turns on at a voltage lower than the emitter-base diode of the transistor. Thus transistor T_1 switches off and the output voltage is pulled up to V_{cc}. Only when all three inputs are at V_{cc}, so that none of the Schottky barrier diodes conduct, does the transistor switch on, due to the availability of base current, and the output voltage drops to V_{CE}(sat), the saturation voltage. Thus STL is a NAND gate and the logic operations are performed by the Schottky barrier diodes, and the transistor acts as an inverter. Alternatively, Schottky barrier diodes can be connected to the collector of the transistor as shown in Fig. 12.51(*c*). The latter is more compact in size and a NOR function is realized by the use of an STL NOT arrangement at the input. Figure 12.51(*b*) shows the layout of a gate of this form.

2. *TTL.* TTL is the most popularly used logic circuit. Figure 12.52 shows the circuit. Notice that the difference between the TTL and the STL is that the Schottky diodes at the input of the gate are replaced with a multiple-emitter transistor T_2. If all the inputs *A*, *B*, and *C* are open circuit or at high potential, the current that flows through R_1 and the base-collector diode of the gating transistor T_2 turns on the switching transistor T_1, resulting in a low output voltage, which is equal to V_{CE}(sat). If any input potential is low, so that a large current flows through resistor R_1 and the emitter-base diode of T_2, the base voltage of T_2 is lowered and no current flows through the collector-base diode of T_2. Due to a lack of base current, T_1 switches off and the potential of the output node rises to V_{CC}. The resistor R_3 at the base of T_1 provides a base discharging current path and thus reduces the turn-off time.

When the gate is driving a large capacitive load, the *RC* network formed by R_L of the gate and the capacitance of the load introduces an exponential rise of voltage at the output, thus introducing a time delay. The switching speed of a logic circuit driving a capacitive load can be improved by introducing additional transistors in the output stage with a push-pull driver circuit, as shown in Fig. 12.53. In this circuit, transistor T_3 is connected as an emitter follower between the load resistor and the load. Its gain is used to produce the necessary high current during the turn-on switching transient.

In STL and TTL the switching transistors of these circuits are allowed to saturate. As soon as a transistor becomes saturated, a significant amount of stored charge is stored in the epi region of the collector-base junction, resulting in a time delay when the transistor is turned off. The most common technique to prevent such deep saturation from occurring is to add a Schottky barrier diode, called a clamp, between the collector and the base of the transistor, as shown in Fig. 12.54. The V_{CE} of the clamped transistor remains about 0.4 V when switched on.

3. *MTL* (I^2L).[21] A merged-transistor-logic (MTL) gate is also called an integrated-injection-logic (I^2L) gate. Unlike other bipolar logic gates, the MTL gate uses a lateral *pnp* transistor instead of a resistor as a current source to supply current to the switching *npn* transistors. As shown in Fig. 12.55, the *npn* transistors are "merged"; they shared a common base region and a common emitter region, the *n*-epi region. The *npn* transistors are operated "upward"; that is, when they are switched on, they sink current from the top n^+ collector node to the n^+ ground. The lateral *pnp* transistor is merged together with the *npn* transistors; its base region is the emitter of the *npn* transistors and its collector is the base of the *npn* transistors. When the base node *B* is open circuit, the current supplied by the *pnp* transistor turns on the *npn* transistors. Conversely, when the collector current of the *pnp* transistor is completely sunk by the *npn* transistor of the previous stage through node *B*, the *npn* transistors turn off due to a lack of base current. This logic gate configuration does not need any isolation in the n^+ sublayer and the circuit size is smallest in the bipolar logic circuit family due to its merged structure. It is also the circuit that dissipates the lowest power in the bipolar circuit family. This logic gate has a single input port and multiple output ports, the collectors of *npn* transistors. Logic function is realized by dotting the output of the previous stages together. The circuit switching speed of MTL is limited by the stored charge in the base-epi junction. An MTL gave fabricated on thin epitaxial film has shown a gate switching delay of less than 1 ns, comparable with the TTL gate.

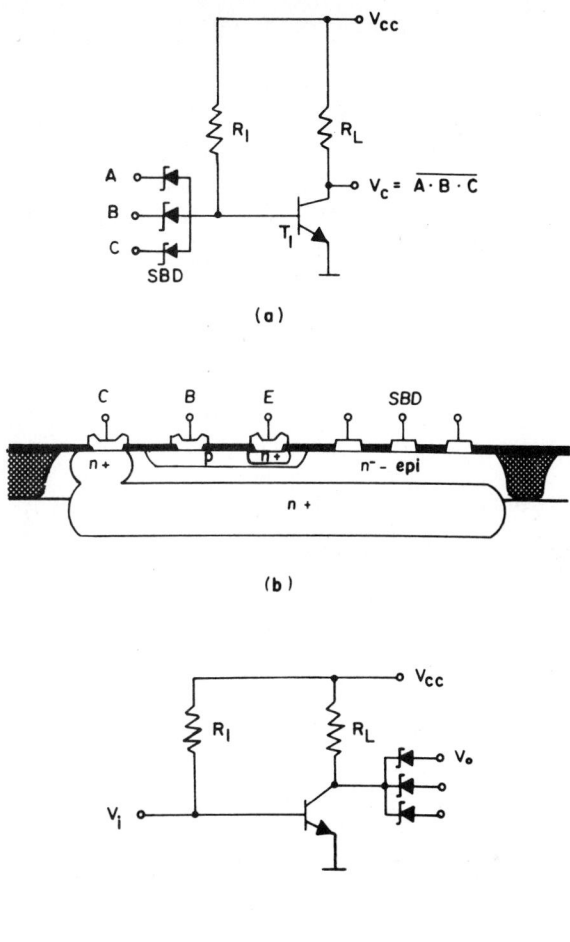

Fig. 12.51 STL circuits: (a) NAND circuit with SBD at its inputs, (b) cross-sectional view of (c), (c) STL NOT circuit with SBD at its output.

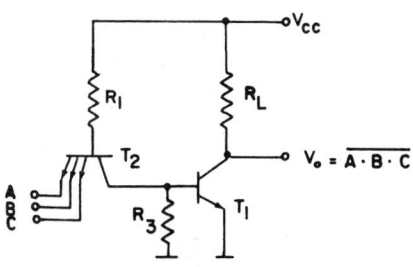

Fig. 12.52 TTL circuit diagram.

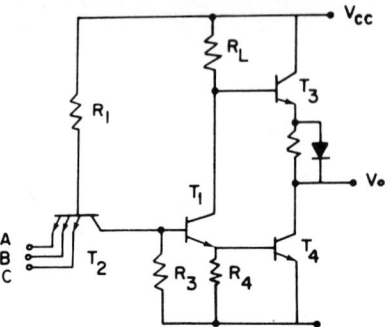

Fig. 12.53 TTL with push-pull driver.

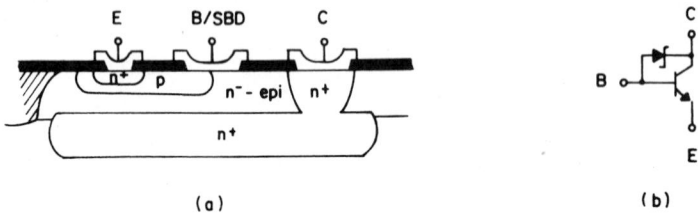

(a)   (b)

Fig. 12.54 Schottky barrier-clamped *npn* transistor: (*a*) cross section of view, (*b*) circuit diagram.

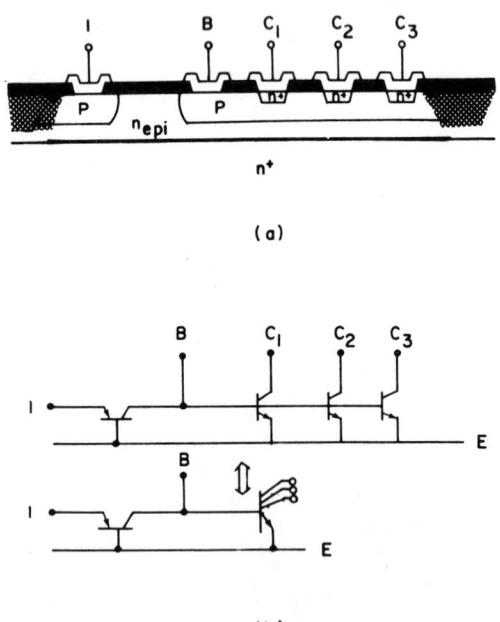

(a)

(b)

Fig. 12.55 (*a*) Cross section of view of MTL gate, (*b*) circuit diagram.

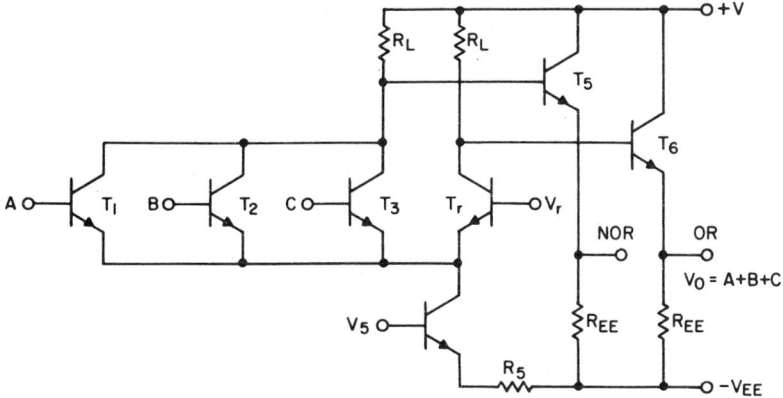

Fig. 12.56 Current mode logic (CML) circuit with level shifting.

4. *ECL/CML.* The logic in which transistors are not operated in the saturation mode is the emitter-coupled-logic circuit. Figure 12.56 shows the circuit. Transistors T_r is biased at a reference voltage V_r, which is at the midpoint between the high and the low logic voltage level at the input of the gate (i.e. at the base of the input transistors T_A, T_B, T_C). The emitters of transistor T_r and the input transistors are coupled together and connected to a current source formed by transistors T_s and R_s. When all the input voltages are at the low level, the input transistors are off and the reference transistor is on, resulting in a high potential at the collector node N_1 and a low potential at N_2. Notice that the resistor value R_L is small enough such that these switching transistors stay out of saturation. When any one of the input voltages is high, the input transistor is on the transistor T_r is off, resulting in a high potential at N_2 and a low potential at N_1; the current switches from the right load resistor to the left load resistor, hence the name current switch. These are two outputs from this gate, and the gate performs functions of OR and NOR.

Transistors T_5 and R_{EE} form an emitter-follower circuit that increases the output current driving capabilities of the gate and also serves to downshift the voltage level at N_2 by one $V_{BE(ON)}$. This output is the inverse (out of phase) of the input signal. Similarly, the transistors T_6 and R_{EE} on the other side form another emitter-follower circuit, which gives a noninverting output. Current-mode logic (CML) gates are ECL gates without the two emitter followers and thus dissipate less power.

This gate uses a large number of transistors and three different power supplies. It occupies a large silicon area and dissipates a large amount of power. However, due to the nonsaturate operation of the transistors and its capability to provide both noninverting and inverting outputs, it is the most powerful and fastest logic gate available. The state-of-the-art switching speed of this gate is of the order of 0.1 ns per gate delay.

5. *Power-Delay Characteristics of Bipolar Switching Circuit and Design Optimization Techniques.*[22] Besides the area of the gate, the power dissipation and switching delay are the most important figures of merit of logic gates. All bipolar logic gates show a common power-delay characteristic. Figure 12.57 shows a typical switching delay of the bipolar logic gate as a function of power dissipation. There are three distinct regions. To understand the power-delay characteristics, consider the bipolar transistor in a gate as a current source being switched on and off. Let us use MTL in Fig. 12.55 as an example. The *pnp* transistor is represented by a constant-current source. As shown in Fig. 12.58, the collectors of T_1 and T_2 are dotted and connected to the base of T_3. The transitors T_1 and T_2 are initially off, and then TR_1 is switched on to sink the base current of T_3. When the gates are operated at the low-power dissipation region, region I of Fig. 12.57, the operating current from the current source is small. The emitter-base junction is biased in such a way that the depletion-layer capacitance of the junction is larger than the diffusion capacitance. The switching delay is determined by the charging and discharging time of the node capacitance at the collector of switching transistor T_1, which is node B, and the delay is inversely proportional to the operating current, which can be written as

$$\tau_d \propto \frac{\sum \overline{C}}{I} V_l \qquad (12.28)$$

where $\sum \overline{C}$ is the sum of the average depletion-layer capacitances at node B and V_l is the voltage of the logic swing. The power dissipation is roughly proportional to the current since the power supply

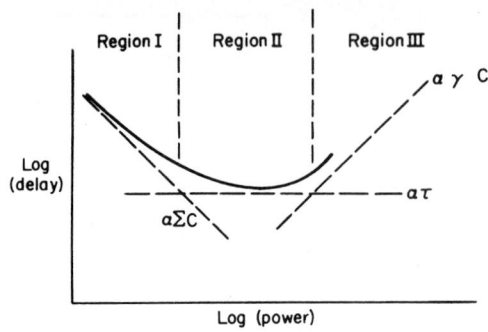

Fig. 12.57 Power-delay characteristics of bipolar logic gate.

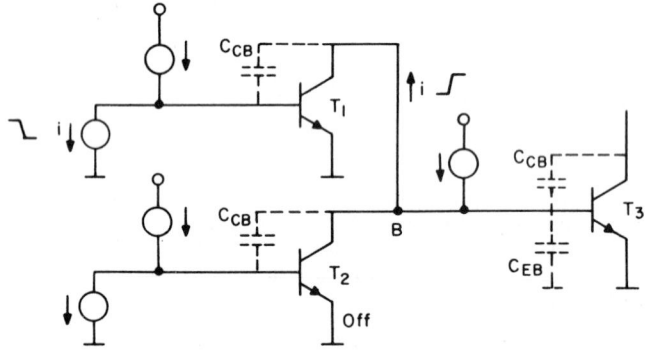

Fig. 12.58 Node capacitance at the input of an MTL gate.

voltage or the emitter-base voltage of the *pnp* transistor, which acts as a current source, varies slowly. Thus in this region the power-delay product

$$P\tau_d \propto I\tau_d \approx \left(\sum \overline{C} \right) V_l \tag{12.29}$$

is nearly constant. In region II on Fig. 12.57 the operating current of the transistor is large enough so that the diffusion capacitance of the junctions becomes dominant. Since the diffusion capacitance associated with the stored charge in the junction is proportional to the current as given by

$$C_{\text{diff}} \propto \frac{\tau_s I}{V_T} \tag{12.30}$$

where τ_s and V_T are the stored-charge time constant and thermal voltage kT/q, respectively, the switching delay is constant and independent of the operating current as given by

$$\tau_d \propto \frac{C_{\text{diff}}}{I} V_l \propto \left(\frac{\tau_s}{V_T} \right) V_l \tag{12.31}$$

When the operating current further increases, not only the diffusion capacitance but also the series resistance become dominant delay components: namely, the *RC*-time constant of $r_b C_d$, where r_b is the base resistance of the transistor. This time constant increases with current since C_d increases rapidly with current, especially when the transistor is driven into the high injection state. As a result, the switching delay in region III increases with current.

The optimal operating point of the switching circuit is around the border of regions I and II, where the power dissipation is low and the switching speed of the logic gate is near maximum.

In gates other than MTL, a resistor is connected to the collector of the switching transistor for pull-up purposes, and more parameters are involved in the power-delay characteristics. Although

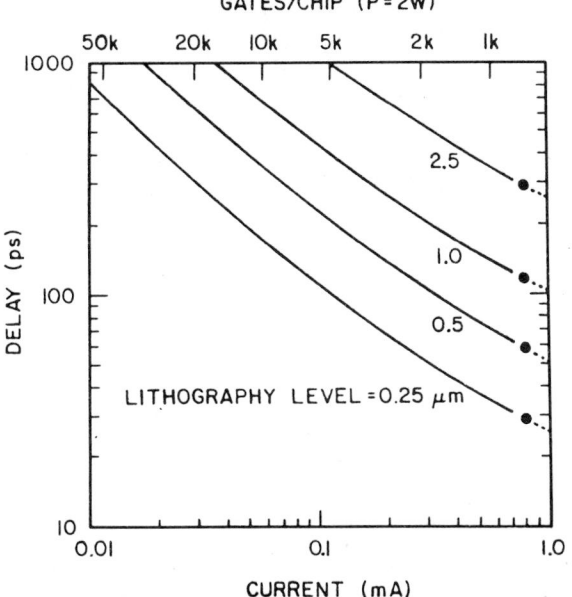

Fig. 12.59 Delay versus operating current of ECL gate. The top scale shows a maximum number of circuits on a chip that dissipates 2 W. (From Ref. 9.)

quantitatively the power-delay characteristics of each gate are different, the general characteristics remain the same. To reduce the power dissipation of the gate, the depletion-layer capacitance should be reduced, meaning that the doping density near the junction should be reduced. To increase the speed of the circuit, the stored-charge time constant should be decreased and the transistor should be operated in the higher current region, region II of Fig. 12.57. Thus the base region of the transistor should be thin, the doping gradient at the junction should be sharp, and the collector doping should be high to allow the transistor to be operated at high current density. The transistor structure should be designed according to the intended collector current density to obtain optimal power-delay characteristics. Figure 12.59 shows the switching delay of an ECL gate designed with various lithographic line widths. The smaller the lithographic line width, the narrower the emitter width of the transistor and the smaller the transistor that can be made. A bipolar circuit scaling scheme[22] has been devised to establish a proper device impurity profile as a function of lithographic line width. The doping concentration of the collector region, the base region, the base width, and the operating current are the primary parameters to be scaled according to the emitter width, whose minimum width is limited by the lithographic line width. Table 12.4 summarizes the transistor structure optimized for ECL circuit applications.

TABLE 12.4   Transistor Parameters[a]

	n^+pnn^+		n^+pn^+	
	Basis	Scaled	Basis	Scaled
a (μm)	2.5	0.23	2.5	0.32
W_b (μm)	0.16	0.023	0.25	0.038
W_{epi} (μm)	2.5	0.23	—	—
N_b (cm^{-3})	1.5×10^{17}	7.8×10^{18}	1.9×10^{17}	8×10^{18}
N_c (cm^{-3})	8.8×10^{15}	1.0×10^{18}	—	—
r_{db} (k$\Omega/\square$)	10.6	6.2	6.1	3.7
J (kA/cm^2)	4.4	500.	8.4	500.

Source. From Ref. 22.
[a]a, emitter stripe width; W_b, base width; W_{epi}, width of epitaxial layer; N_b, base doping density; N_c, collector doping density; r_{db}, base sheet resistance; J, current density.

Memory Cells

1. *ROM (Read-Only Memory).*[23] Memory cells in a ROM are generally arranged in a two-dimensional rectangular matrix. The matrix and the associated circuitry are shown in Fig. 12.60. Each cell in the matrix is accessed through parallel binary address (*W* lines) and parallel binary output (*B* lines). The *W* and *B* lines are commonly referred to as word and bit lines, respectively. The Schottky barrier diode or transistor is placed between the word address lines and the specific bit lines, which is indicated in Fig. 12.60 as the coupling cell. The separate multiemitter transistors within the ROM storage array consist of long base stripes into which emitters have been selectively programmed using a photolithographic mask. Figure 12.61 shows an emitter contact mask programmed to form transistor emitters at specific points along the base stripe. Similar techniques can be used for the Schottky barrier diode array.

If one uses a fusible metalization (such as Nichrome) with bipolar circuits, the connections to the individual emitters for separate memory cells can be open circuited in a controllable way and the memory may be programmed after packaging.

2. *RAM (Random-Access Memory).* There are two basic types of bipolar RAM cells, the static cell and the dynamic cell. The **static** cell has two bistable states, and each of the states remains unchanged unless an external voltage current is applied to force the cell to change from one state to another state. The **dynamic** memory cell has one nonequilibrium state that dissipates itself and gradually turns into the equilibrium state, the other state.

a. *Static Cell.* A static cell consists of a flip-flop and a coupling circuit as shown in Fig. 12.62. Transistors T_1 and T_2 and resistors form a flip-flop, and the state of the flip-flop determines the information stored in the memory cell. Transistors T_3 and T_4 are the coupling transistors for sensing and writing purposes. To sense the state of the cell, equal amounts of currents are sunk through the emitters of T_3 and T_4 and the voltage of those emitters is detected. In this case, the potential of the $\bar{B}$ node is higher than that of the B node since the base potential of T_4 is higher than that of T_3. To change the state of the flip-flop, one has to switch around the potential of the two collector nodes, C_1

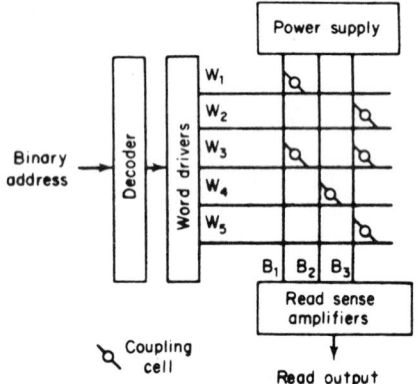

Fig. 12.60 Diagram of a ROM with memory arranged in a rectangular matrix.

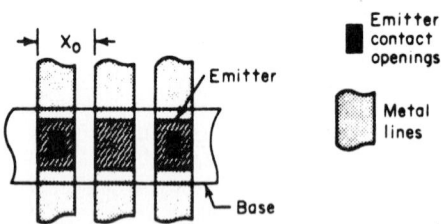

Fig. 12.61 Emitter contact mask programmed to form transistor at specific point in bipolar circuit.

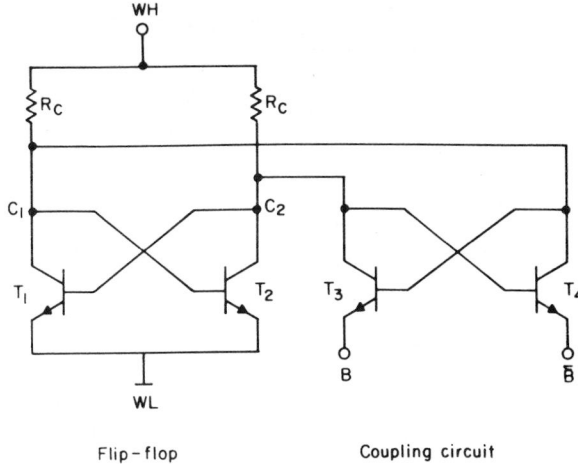

Flip–flop   Coupling circuit

Fig. 12.62 Basic bipolar static memory cell.

and C_2. This is accomplished by sinking a large amount of current from the high potential node, C_2 in this case. The current sinks through the load resistor of T_2 and through T_3 lowers the potential of node C_2 and thus turns off T_1. When T_1 is off, the potential at its collector rises and in turn switches on T_2. When T_2 is on, the potential at its collector C_2 drops further down and the cell completes its change of state.

When the cell is in the standby state, no current flows out of the coupling transistors T_3 and T_4. Thus the value of the resistors R_c is determined by the standby current dissipated by the cell. Additional Schottky diodes are often placed in shunt with the load resistor to keep the transistor out of saturation and to allow a large amount of current flow from the power supply to the coupling transistors when the cell is read. Such a cell is called an **ECL cell**. In integrated circuits the coupling transistors are merged into the flip-flop transistors; that is, T_2 and T_3 are placed as one transistor with two emitters, as are T_1 and T_4. Figure 12.63 shows the layout of a memory cell and the circuit diagram of this kind of memory cell. This memory cell has been the highest-speed cell. Figure 12.64 shows a typical memory circuit organization. A cell is selected by raising the word line potential in the X direction and sinking current into the selected pair of bit lines in the Y direction through a bit selection switch, and the differential potential of the selected bit lines is then detected.

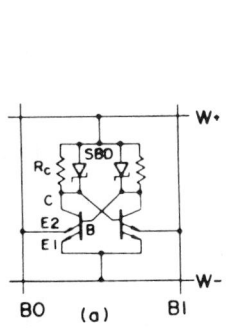

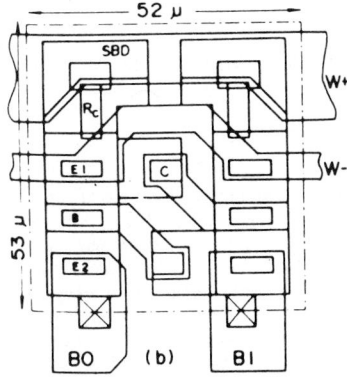

Fig. 12.63 Multiemitter memory cell with SBD clamping: (a) simplified cell circuit, (b) cell pattern layout. [From Ref. 24; reprinted by permission.]

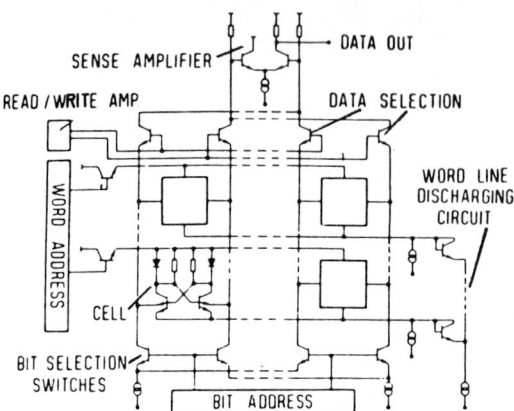

Fig. 12.64   Typical functional block of RAM.

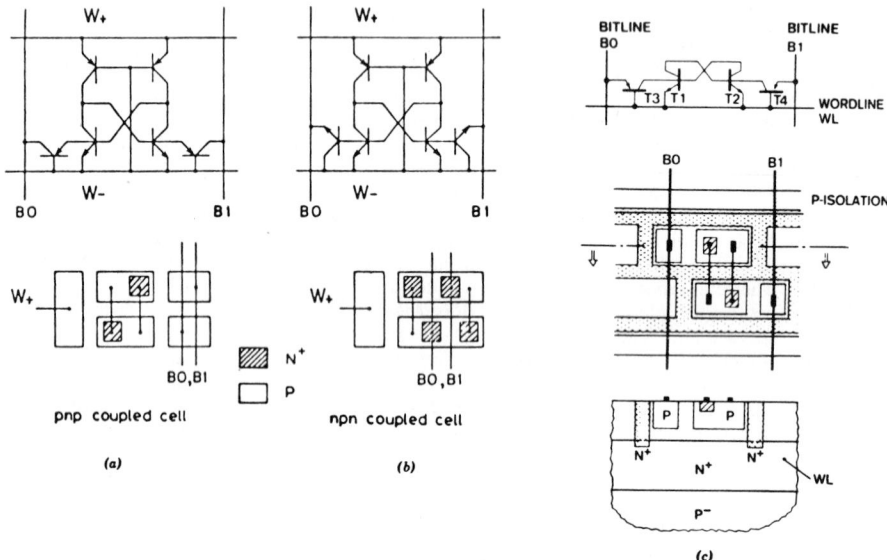

Fig. 12.65   Third vertical of MTL memory cell: (*a*) *pnp* coupled, (*b*) *npn* coupled, (*c*) injector sensed.

When designing a static RAM cell of small standby current, load resistors of large resistance are required. Since in integrated circuits, the reproducible maximum sheet resistivity is limited, a larger resistor means a larger resistor size and that increases the cell size. This undesired large resistor can be replaced by a current source such as a *pnp* transistor[25] operated in the common-base mode. Figure 12.65 shows three versions of a *pnp*-load memory cell called the **MTL cell**. The resistance of the *pnp* emitter diode (kT/qI) is very large at low emitter-base bias. Thus the standby current is small. When the cell is selected by raising the potential of the injector (the emitter of the *pnp* transistor), the current supplied to the cell increases exponentially. This large amount of current sinks to the bit line of the "on" transistor side; thus the cell can be accessed with a "read" current much greater than the standby current. A large read current allows the cell to be read rapidly. Thus the *pnp* transistor effectively acts as a variable-load resistor. It is merged into the *npn* transistor; thus the cell occupies an area much smaller than that of the resistor-load cell, resulting in a dense memory circuit.

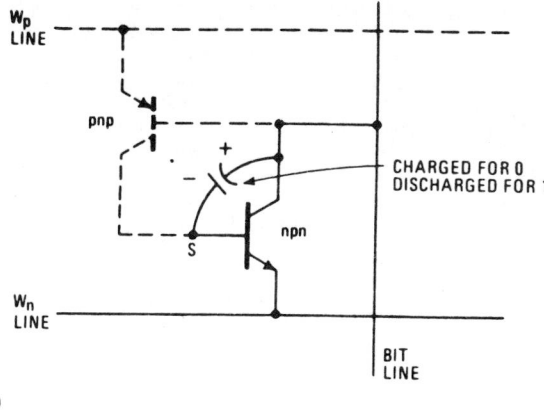

(a)

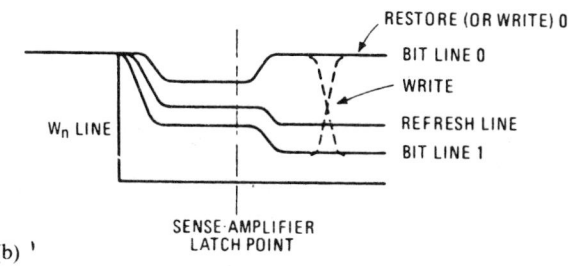

(b)

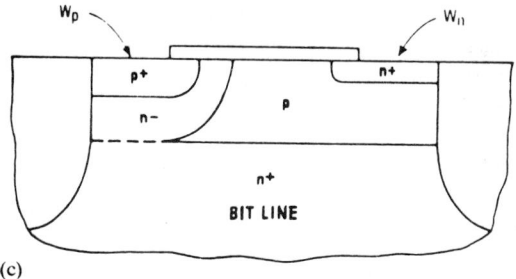

(c)

Fig. 12.66 Dynamic RAM cell: (*a*) circuit schematic, (*b*) potential waveform of word line when read, (*c*) cross-sectional view of the cell.

b. *Dynamic Cell.*[26] The dynamic cell, also called the I^3L **cell**, consists of a *pnp* transistor and an *npn* transistor with one base node connected to the collector of the other, as shown in Fig. 12.66. This cell is smaller in area than the static cell since it consists of fewer transistors. The two states of the cell are high potential (1) and low potential (0) at the base node S of the *npn* transistor. During standby condition, the potential of word lines W_n is higher than that of W_p, and the emitter-base diode of *npn* transistor is reverse biased. To read the cell, the potential of the lower word line W_n is lowered to that of W_p, and the bit line is floating. If the cell stores 0, meaning that the collector-base diode of the *npn* transistor is reverse biased and the storage node S is at a low potential, this sudden pull-down of the word line W_n will not forward bias the emitter-base junction of the *npn* transistor; therefore, very little signal is coupled capacitively from W_n to the bit line. If the cell stores 1, meaning 0 voltage across the collector-base junction of the *npn* transistor, the sudden pull-down of line W_n (about 1 V below its standby potential) causes the emitter-base junction of the *npn* transistor to become forward biased and drives the *npn* transistor into the active region. The collector-base capacitance then acts as a feedback capacitor coupling the W_n potential to the bit line through a capacitor with capacitance C_{CB} multiplied by the current gain of the *npn* transistor. Thus the bit line potential is pulled down. By sensing the potential of the bit line, the state of the cell can be read. To write the cell, the potential of W_n is lowered and the bit line potential is set either high for writing 0 or low for writing 1. Then the potential of W_n is raised above that of W_P. Afterward, the cell is in the standby state. Notice that the state 1 of the cell is not an equilibrium state since the emitter-base junction of the *pnp* transistor is reverse biased during standby. The leakage current through this junction gradually charges up the collector-base junction of the *npn* transistor and eventually changes the cell into the 0 state. Thus this kind of dynamic cell has to be rewritten or "refreshed" periodically to maintain the state 1.

For more details about bipolar ROM and static RAM chip and organization, the reader is referred to Ref. 23.

12.6 MOS TECHNOLOGY
Andres Fortino

TERMS

N_A, N_D = acceptor and donor concentrations

ϕ_s = surface potential at the insulator semiconductor interface

N_i = intrinsic carrier concentration (1.45×10^{10} cm^{-3} at 300 K)

ε_s = permittivity of silicon (semiconductor)

ε_{ox} = permittivity of oxide (insulator)

$q = 1.6 \times 10^{-19}$ C (magnitude of the charge of an electron)

k = Boltzmann constant

T = temperature (K)

ϕ_{ms} = metal-to-semiconductor work function difference

Q_{ss} = surface charge at the insulator-semiconductor interface

C_0 = capacitance per unit area of the insulator

ϕ_{sox} = semiconductor-to-oxide work function difference

ϕ_{mox} = metal-to-oxide work function difference

$\mathscr{E}_g$ = semiconductor valence to conduction band energy gap

ϕ_F = bulk Fermi potential

L = IGFET channel length

W = IGFET channel width

L_B = extrinsic Debye length = $(\varepsilon_s kT/q^2 N_B)^{1/2}$

N_B = bulk doping

Metal-oxide-semiconductor (MOS) technology is of key importance in the production of integrated circuits composed of surface devices. The MOS capacitor structure provides a useful vehicle to study surfaces and is a key process tool. One may substitute for the oxide insulator any variety of other single insulators or layered structures. In this case a more general generic label for these devices would be MIS (metal-insulator-semiconductor). Within this family of devices one includes the CCD (change-coupled device), a surface-controlled serial storage element of minimal area. The workhorse device of the technology is the MOSFET (metal-oxide-semiconductor field-effect transistor), more generally referred to as an IGFET (insulated-gate field-effect transistor). To understand the operation and design of these devices it is necessary to have a thorough knowledge of electric field control of surface conditions, rectification in a *pn* junction, and their combined action in a FET. Successful manufacture of these devices requires a thorough knowledge of chemical processes. A great deal of know-how for a processing line is developed by each manufacturer through experience and is proprietary, but the general principles, together with some process data, are available publicly and enable one to understand and design processes. Advances in computer technology have also advanced the art of design of the devices and processes. Computer-aided design (CAD) is a well-developed field in support of this technology.

12.6.1   Device Operation

MOS System

The MOS structure shown in Fig. 12.67 is a surface-controlled device: the MOS capacitor.[27] The variation of mobile charge densities (electrons and holes) and fixed charge centers (doping ions) with depth into the material as a function of applied potential is shown in Fig. 12.68. Four regions of

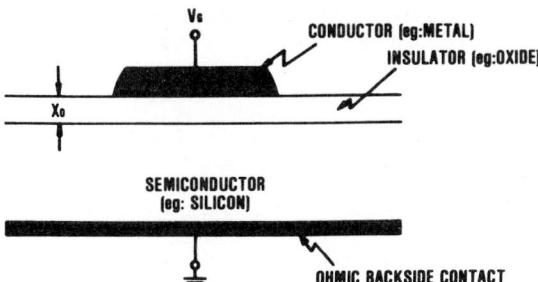

Fig. 12.67 Metal-oxide-semiconductor capacitor.

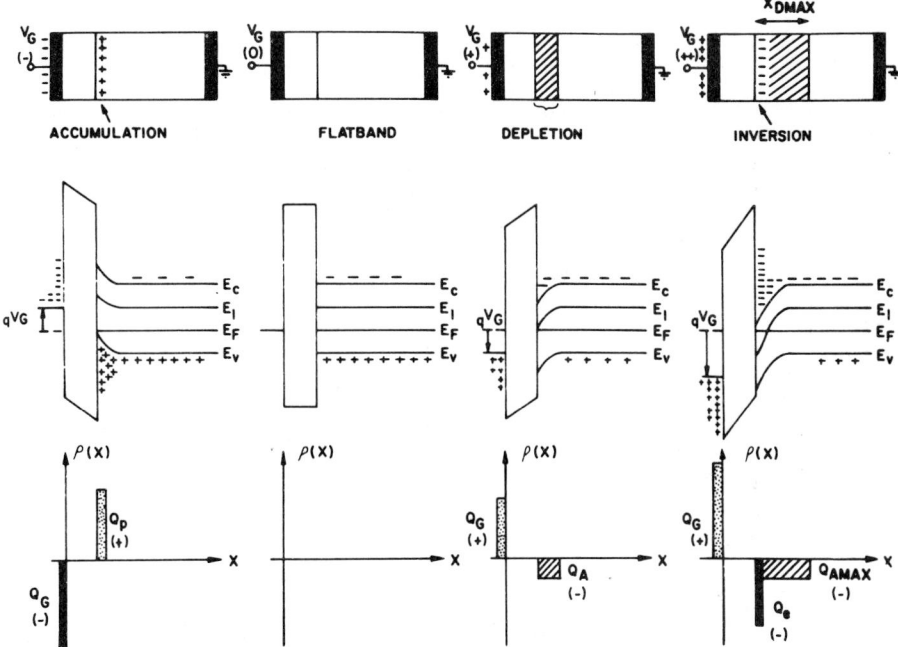

Fig. 12.68 Charge accumulation and energy-band diagrams of an ideal *p*-type MOS capacitor under varying bias conditions: (*a*) accumulations, (*b*) flatband, (*c*) depletion, (*d*) inversion.

operation are evident: accumulation, depletion, inversion, and deep depletion. The semiconductor in this case is uniformly doped.

Accumulation occurs when the correct bias is applied to the gate to attract majority carriers to the oxide-semiconductor interface (negative for a *p*-type semiconductor, attracting holes as shown). If the bias is made zero or is applied in the opposite direction, majority carriers are rejected from the surface region and fixed doping ionic centers are left uncovered as fixed space charge, causing the condition termed **depletion**. As a higher bias is applied in this direction, minority carriers begin to accumulate at the surface. Once their density is higher than the background doping, a region called an **inversion layer** is formed, which is, by field effect, of opposite conductivity type compared to the bulk. This process is depicted in Fig. 12.68.

Deep Depletion

Accumulation, depletion, and inversion are the three stable dc regions of operation. Pulsing the metal gate in the direction of inversion will cause a condition termed **deep depletion**. A transient condition occurs in which the minority carriers accumulate at the surface over a finite time interval. A measure of the carrier lifetime is the speed at which this deep depletion condition degrades into inversion. This is normally done in the dark via measuring capacitance change as a function of time.[28]

Capacitance

The relationship between capacitance and applied voltage at 1 MHz and at 100 Hz is illustrated in Fig. 12.69 for a *p*-type semiconductor MOS capacitor.[28] The relationship between capacitance and voltage for a uniformly doped device is given in Table 12.5. For *n*-type substrates, the *C – V* (capacitance-voltage) relationship is reversed, with the minimum capacitance (C_{min}) occurring for negative biases. This minimum capacitance, C_{min}, occurs at high frequency due to the finite lifetime of the minority carriers in the inversion layer. At high frequencies (> 100 Hz), these carriers are not fast enough to move through the depletion layer back to inversion layer to follow the ac signal. Thus for the inversion bias condition, the inversion charge remains fixed by the dc bias level. It is the edge of the depletion region that is modulated by the ac signal. During the transient conditions of pulsing the device into deep depletion, since no inversion layer has been formed, all of the silicon capacitance is due to space charge modulation. This results in a reduced total capacitance, as shown in Fig. 12.69.

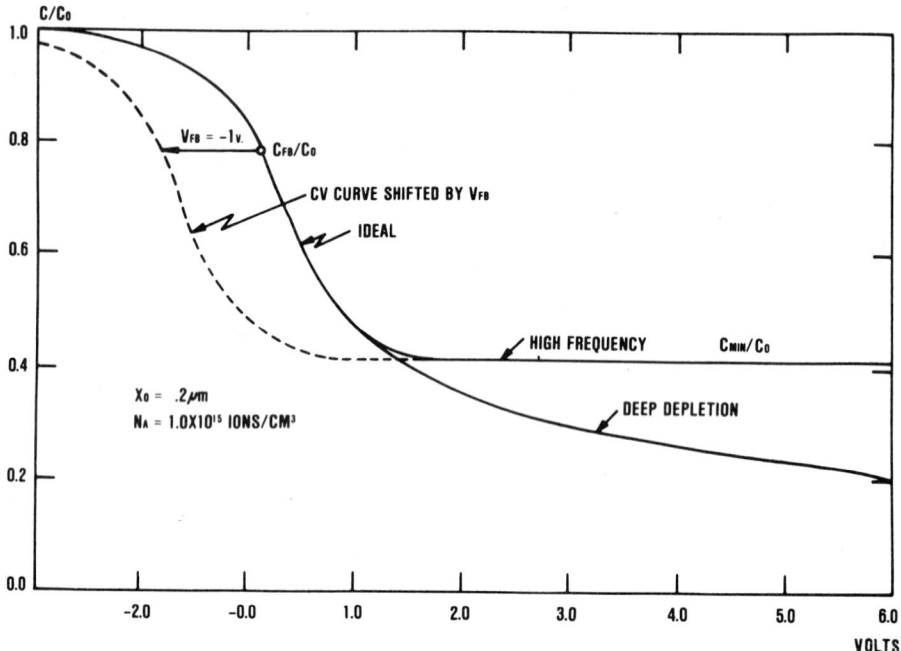

Fig. 12.69 Normalized capacitance-voltage relationship for a *p*-type MOS capacitor ($N_a = 1 \times 10^{15}$ ions/cm³ with an oxide thickness x_0 of 2200 nm). High-frequency and deep depletion characteristics are observable, as is the typical flatband voltage shift (dashed line).

TABLE 12.5 Minimum Capacitance C_{min} (Normalized to Maximum) versus Uniform Doping Level for an MOS Capacitor for Various Oxide Thicknesses

Oxide Thickness T_{ox}(Å)	Doping Level N_A (cm⁻³)						
	10^{14}	5×10^{14}	1×10^{15}	5×10^{15}	10^{16}	10^{17}	10^{18}
500	0.066	0.112	0.148	0.266	0.333	0.593	0.812
1000	0.110	0.202	0.258	0.420	0.499	0.745	0.896
5000	0.367	0.559	0.634	0.784	0.833	0.936	0.997

Flatband

Flatband is defined as that voltage applied to the gate that produces zero energy band bending at the surface. It is given simply by

$$V_{FB} = \phi_{ms} - \frac{Q_{ss}}{C_0} \tag{12.32}$$

where the work function difference may be found from the chart in Fig. 12.70. The diagram in Fig. 12.71, shows all the components of ϕ_{ms} with respect to the vacuum level:

$$\phi_{ms} = \phi_{mox} - \phi_{sox} - \frac{\mathscr{E}_g}{2q} - \phi_F \tag{12.33}$$

The surface charge Q_{ss} is due to many factors.[29, 30] It is normally positive trapped interface charge and shifts the $C - V$ characteristic negative. Trapped electrons at the interface will cause a positive shift. Various processing sequences have a varying effect on this parameter.[31]

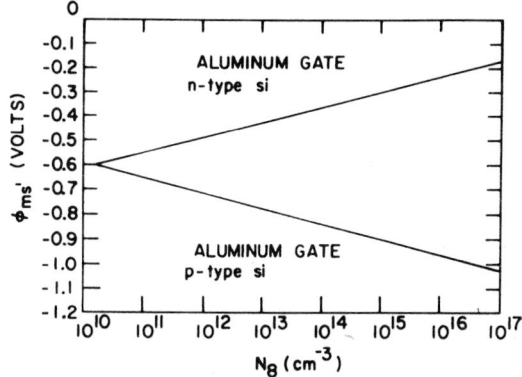

Fig. 12.70 Aluminum-silicon work function difference for *n*- and *p*-type silicon as a function of doping levels (From Ref. 28.)

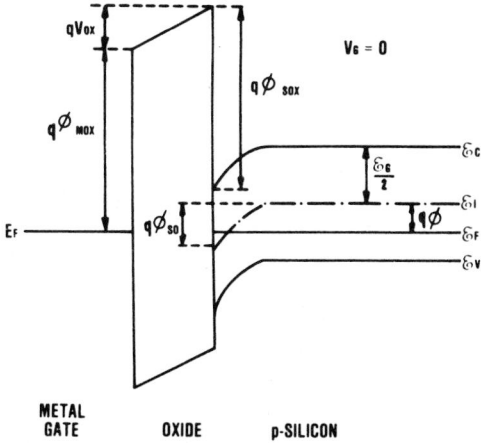

Fig. 12.71 Energy-band diagram of a metal-oxide *p*-type silicon capacitor, showing all pertinent energies in calculation of the flatband. (After Ref. 27.)

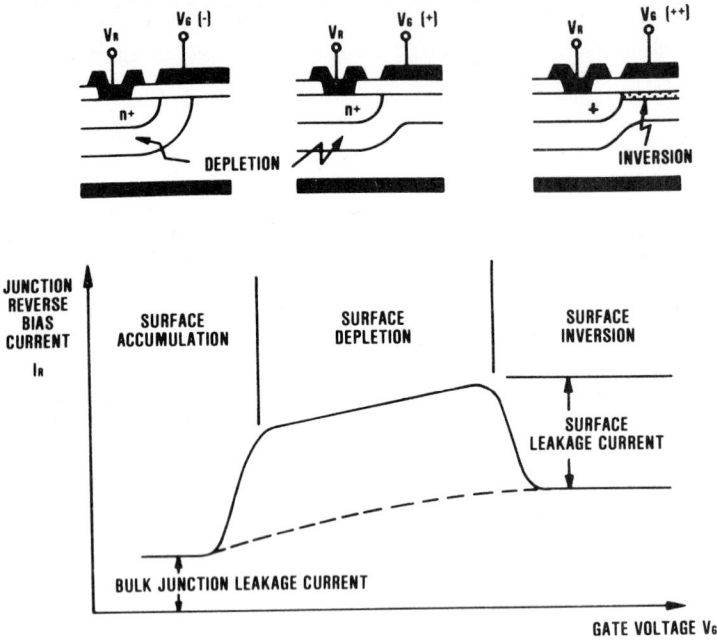

Fig. 12.72 Gate-controlled dial characteristic showing surface accumulation, depletion, and inversion under the metal gate.

pn Junction

In MOS technology the *pn* junction plays many roles: (1) as sources and drains of IGFETs, (2) as input diodes for protection of gates from over voltage, (3) as a charge injector for charge-coupled devices (CCDs), and (4) as increased capacitance for memory. The theory of operation of the *pn* junction has been described elsewhere in this book.

There are a few characteristics of junctions that are key, and we touch on them briefly here. Junction leakage requirements, especially for memory applications, are very stringent (nominally less than a few pA/cm^2). Junction breakdown characteristics must be compatible with the supply voltage levels used. Also, the resistance of diffused lines is required to be as low as possible.

Gate-Controlled Diode

The gate-controlled diode is a special type of *pn* junction fabricated with a MOS gate guard ring employed for process monitoring.[28] It is used to measure both bulk and surface leakage currents and to monitor junction quality in MOS processing. The structure, typical measurements, and the type of extracted data are illustrated in Fig. 12.72.

12.6.2 IGFET

The IGFET is a combination of a surface control device, the MOS capacitor, and a *pn* junction. The junctions act as the contacts of a switch and the establishment of a conducting layer at the surface between the junction by field effects acts as the connection. Figure 12.73 shows a typical IGFET cross section with a corresponding top view.

For a *p*-type doped semiconductor substrate and *n*-type source and drain diffusions, the IGFET is labeled an *n* channel since when a channel is established between source and drain, the inversion region is *n* type. The reverse is true for a *p*-channel IGFET. Since the carriers of the current for *n*-channel devices are electrons, which have the advantage of higher mobility in silicon over holes, they are faster and thus *n*-channel technology is more common and desirable. *p*-channel technology commonly appears in situations such as CMOS (complementary MOS) devices.[32]

Principles of Operation

There are three distinct regions of operation for the typical long-channel IGFET. Let all voltages be referenced to the source terminal (see Fig. 12.74). For all drain voltages, if the gate voltage is less than what is needed to create an inversion region, no channel exists and almost no current flows. This subthreshold region of operation has two characteristics. First, the current has an exponential

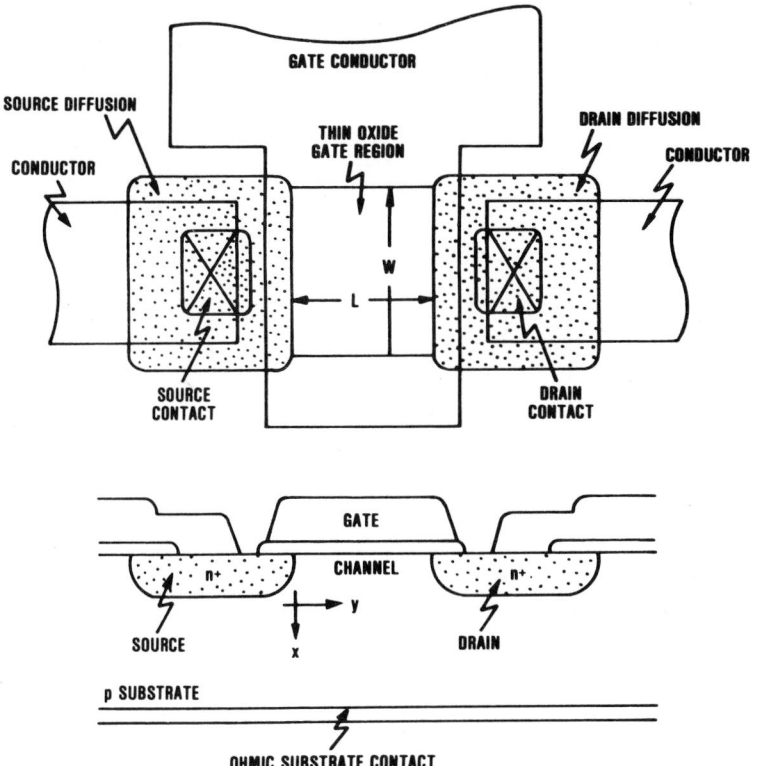

Fig. 12.73 Cross section and top view of a *p*-type IGFET.

relationship to the drain bias and is very small:

$$I_s = \Omega W L_B q D_{\text{eff}} \frac{n_i}{L} \left(1 - e^{-v_{ds}} \right)$$ (12.34)

(where Ω is a function of substrate bias; see Troutman and Chakravarti).[33] Second, the gate voltage needed to create a channel, defined as inversion (the surface potential being at least $2\phi_F$), is

$$V_T = \left(\frac{-Q_{\text{ss}} - Q_{\text{sd,max}}}{\varepsilon_{\text{ox}}} \right) x_0 + \phi_{\text{ms}} + 2\phi_F$$ (12.35)

For device design and analysis purposes, another operational definition of threshold may be that gate voltage for which the subthreshold channel current is $40nA(w/L)$, which corresponds approximately to a surface band bending of $\phi_s = 2\phi_F$. Subthreshold characteristics and the definition of threshold above are illustrated in Fig. 12.74.

Once past threshold, for small drain-to-source biases, there exists an "ohmic" region of operation derived below. For large drain and gate biases along threshold, the device is in saturation, where increases in drain bias increase the channel current very little.

Consider the IGFET shown in Fig. 12.75 with an applied gate bias above threshold so that a channel has formed. The voltage drop across an infinitesimal section of the channel is

$$dV = I_D \, dR = - \frac{I_D \, dy}{W \mu_n Q_n(y)}$$ (12.36)

where $Q_n(y)$ is the charge density of electrons in the channel per unit area and I_D is the drain current. The charge in the silicon surface directly under the gate is made up of mobile channel

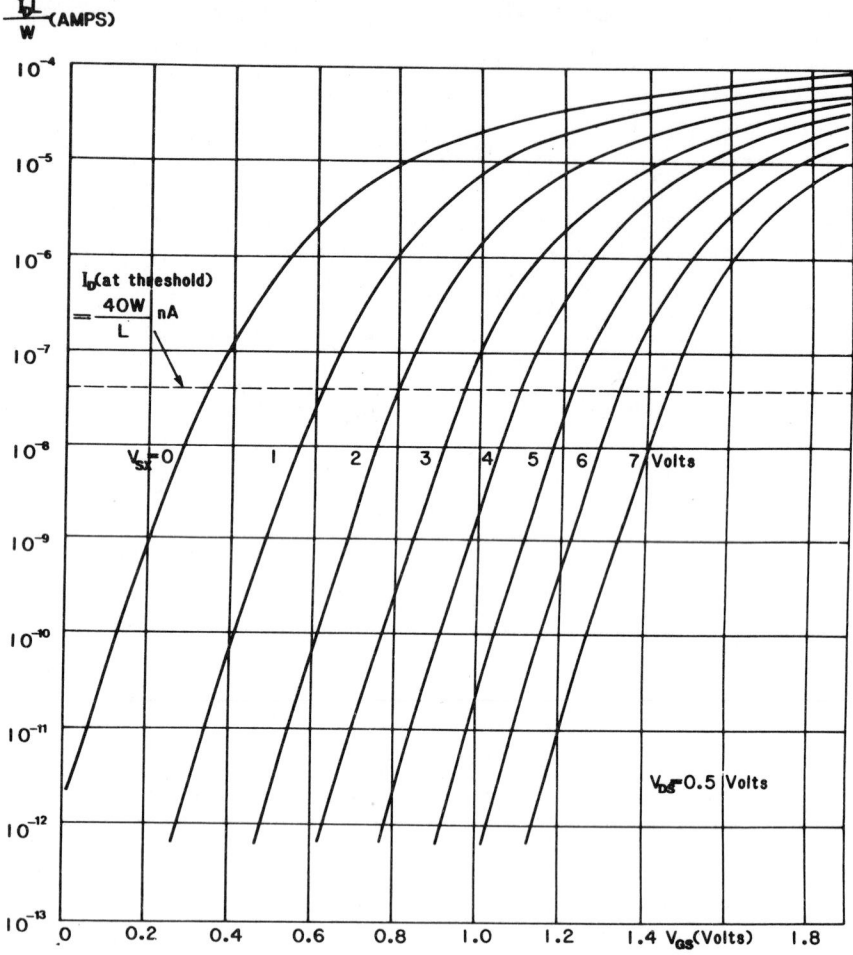

Fig. 12.74 Subthreshold current characteristic for a *p*-type IGFET as a function of substrate-to-source bias. Drain bias is typical for this measurement at 0.5 V. The device width-to-length ratio is $W/L = 10$, which yields for this device a threshold-defining subthreshold current of 40 nA.

electrons and ionized acceptor ions in the depletion region.

$$Q_s(y) = Q_n(y) + Q_B(y) \tag{12.37}$$

This charge is reflected in the gate conductor, with the insulator acting as a capacitor. The voltage drop across this insulator capacitance is then

$$V_G - V_{FB} - \phi_s(y) = -\frac{Q_s(y)}{C_0} \tag{12.38}$$

where $\phi_s(y)$ is the surface potential in the silicon along the channel. Combining the two equations, we obtain

$$Q_n(y) = -[V_G - V_{FB} - \phi_s(y)]C_0 - Q_B(y) \tag{12.39}$$

Since an inversion layer exists, the surface potential is given by

$$\phi_s(y) = V(y) + 2\phi_F \tag{12.40}$$

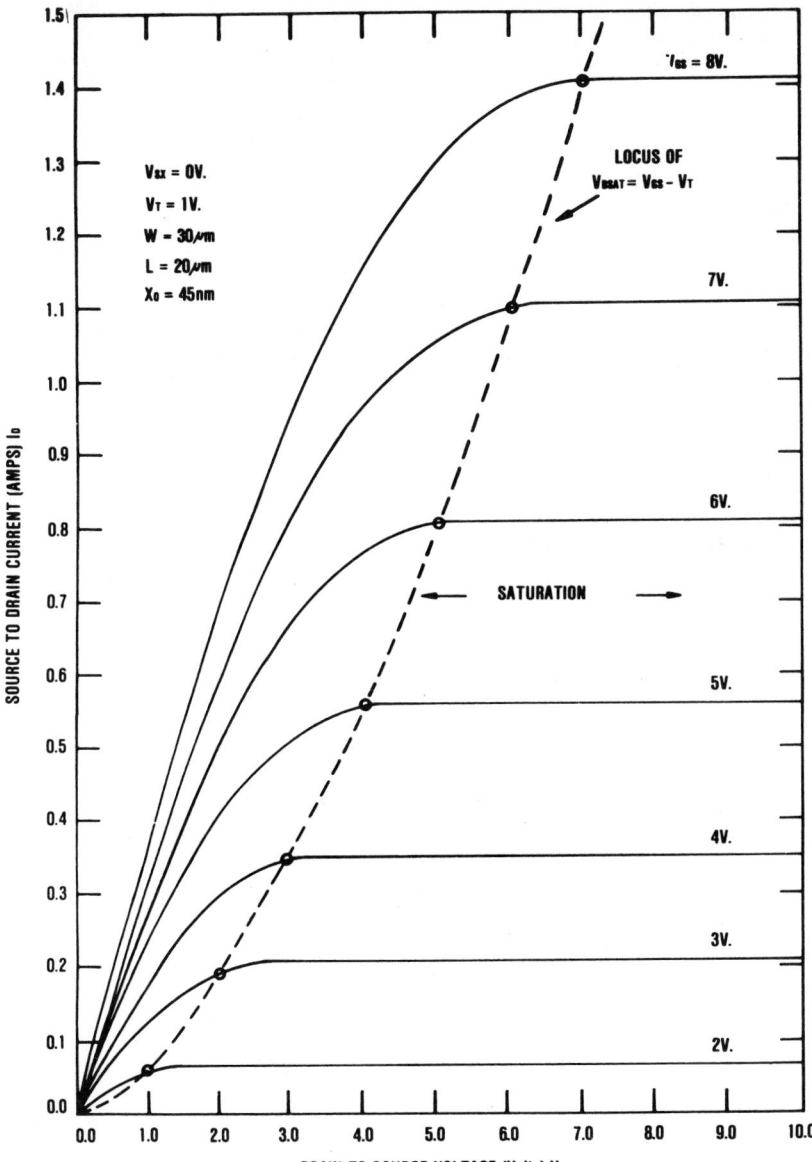

Fig. 12.75 Active-region drain bias versus drain-to-source current characteristics for an IGFET ($V_{SX} = 0$ V, $V_T = 1$ V, $W = 30$ μm, $L = 20$ m, $x_0 = 45$ nm).

where $V(y)$ is the voltage drop from drain to source, which becomes the reverse bias of the elemental section of channel (at y) with respect to the substrate. The depletion region charge Q_B is given by

$$Q_B(y) = -qN_A X_{D\,\text{max}}(y) = -\sqrt{2\varepsilon_s\varepsilon_0 qN_A[V(y) + 2\phi_F]} \qquad (12.41)$$

We may combine these equations, solve for I_D, and integrate from $y = 0$ ($V = 0$) at the source, to $y = L$ ($V = V_D$) at the drain and obtain.[34]

$$I_D = \frac{W}{L}\mu_n C_0\left\{\left(V_G - V_{FB} - 2\phi_F - \frac{V_D}{2}\right)V_D - \frac{2}{3C_0}(2\varepsilon_s\varepsilon_0 qN_A)^{1/2}\left[(V_D + 2\phi_F)^{3/2} - (2\phi_F)^{3/2}\right]\right\}$$

$$(12.42)$$

This equation is valid for drain biases less then a saturation value, $0 \leq V_D \leq V_{D\,\text{sat}}$. For $V_D > V_{D\,\text{sat}}$ the current remains constant. Figure 12.75 depicts typical n-channel current-voltage characteristics. For very small drain biases $V_D \ll 2\phi_F$ (typically 0.5 V), the current expression reduces to

$$I_D \simeq \frac{W}{L}\mu_n C_0 (V_G - V_T)V_D \qquad (12.43)$$

where V_T is defined in Eq. (12.35). This is the linear region of operation where the conductance of the channel is

$$g = \frac{W}{L}\mu_n C_0 (V_G - V_T)$$

For increasing drain biases, a drain-to-channel depletion region begins to form, pinching off the channel inversion region, as illustrated by the sequence of diagrams in Fig. 12.76. At this point, the current becomes a diffusion current through the drain-channel depletion region. The saturation drain

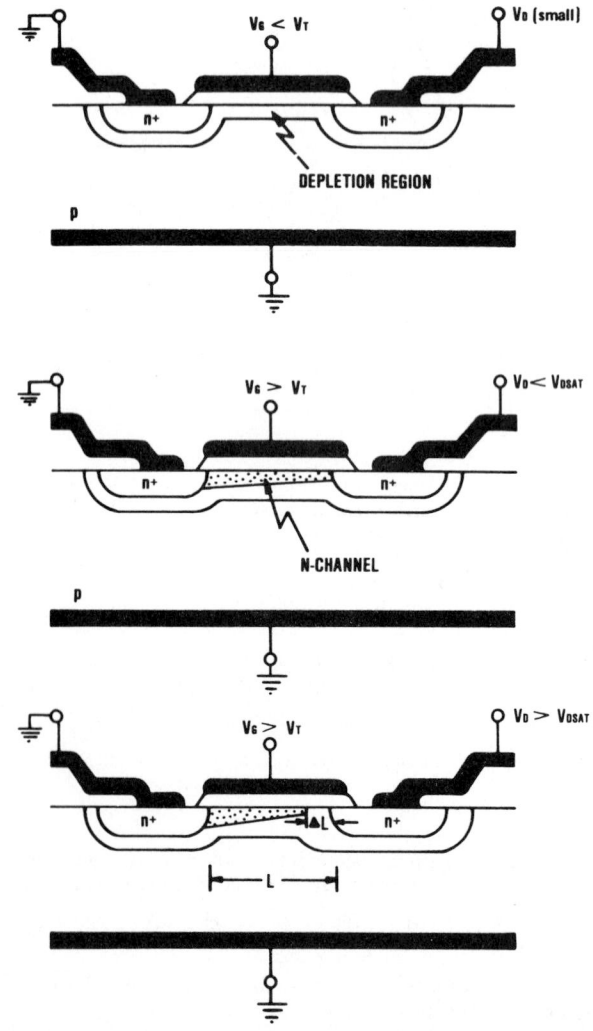

Fig. 12.76 Depleted region with an inversion region for three modes of long-channel IGFET operation: (a) below threshold, (b) active, (c) saturation.

voltage may be obtained from the definition of pinching: $Q_n(L) = 0$. Using this in Eq. (12.39), we obtain

$$Q_n(L) = 0 = V_{D\text{ sat}} + \frac{1}{C_0}\left[2\varepsilon_s\varepsilon_0 q N_A(V_{D\text{ sat}} + 2\phi_F)\right]^{1/2} + 2\phi_F - V_G - V_{\text{FB}} \qquad (12.44)$$

Solving for $V_{D\text{ sat}}$ and making the reasonable approximation that the oxide thickness X_0 is small compared to the depletion region, we obtain

$$V_{D\text{ sat}} \simeq V_G - V_{\text{FB}} - 2\phi_F = V_G - V_T \qquad (12.45)$$

Using this value of V_D in Eq. (12.41) and assuming the gate voltage is sufficiently high ($V_G - V_{\text{FB}} - 2\phi_F \gg 1$), we obtain the saturation current

$$I_{D\text{ sat}} = \frac{W}{L}\mu_n C_0(V_G - V_T)^2 \qquad (12.46)$$

Four-Terminal Operation

The preceding derivations were made for source and substrate at the same potential. It is common to operate the devices with a fixed substrate bias not necessarily zero with respect to the source.[34] For example, the source of the load device in an active load inverter circuit is not always at ground potential but floats with the input bias.

In this situation, the drain-current relationships derived still apply but the threshold voltage is no longer fixed; it varies as a function of source-to-substrate bias (V_{ss}):

$$V_T = \frac{-Q_{\text{ss}} - Q_{B\max}\left[(2\phi_F - V_{\text{ss}})/2\phi_F\right]^{1/2}}{\varepsilon_{\text{ox}}} + \phi_{\text{ms}} + 2\phi_F \qquad (12.47)$$

This formula applies for an IGFET fabricated on a uniformly doped substrate. Typical substrate sensitivity characteristics (threshold V_t versus source-to-substrate voltage V_{ss}) are displayed in Fig. 12.77. This characteristic is useful in the design of circuits since it gives the variation of threshold (and thus channel current) as a function of source-to-substrate bias. In the design of the IGFET, a nearly flat substrate sensitivity curve is sought. From a study of Eq. (12.47) one may observe that the slope of V_T versus V_{ss} is proportional to $N_A^{1/2}$. Thus for the desired effect, high-resistivity substrates (low N_A) are employed. This in turn lowers the overall threshold, as seen by comparing curves A and B in Fig. 12.77. Normally, the flatband is negative (the oxide charge Q_{ss} is commonly positive), tending to depress the threshold as well. Circuit requirements for enhancement devices (devices to which a gate bias must be applied to turn them on) are such that high thresholds are necessary.

To meet both requirements, additional doping is introduced at the surface of the silicon under the gate. This is normally done by ion implanting a species of the same doping type as the substrate. If the ion charge distribution is designed shallow enough, it acts essentially as a sheet charge, raising the threshold by the average amount of charge divided by the oxide capacitance.[35]

The thickness of the insulator also has a direct effect on the threshold (lower thresholds for thinner insulators). For this reason, as the devices are shrunk horizontally as well as vertically in the drive toward more integration,[36] techniques such as ion implantation are employed to maintain threshold levels high.

Enhancement and Depletion IGFETs

As already discussed, devices to which a gate bias must be applied to turn them on (i.e., for channel current to flow) are considered enhancement IGFETs. In the n-channel case, enhancement devices have a positive threshold; p channels have a negative threshold.

Another class of very useful devices are those that are turned on even before a gate bias is applied. Such devices are called **depletion IGFETs** and are very useful as dynamic loads for inverter circuits, for example. The four types of devices and typical characteristics are illustrated in Fig. 12.78.

Scaling and Geometry Effects

To achieve higher speed and greater packing density of MOS circuits, the geometry of the devices has been progressively scaled downward. The scaling of the device parameters, to a first order, has been studied by many workers[37-39] and may be succinctly illustrated as in Table 12.6. As one proceeds with scaling, several second-order effects show up that must be accounted for in the design and either minimized, or in some cases, taken advantage of.

In the saturation region, as the drain bias is increased, the depletion around the drain junction is increased. In the channel region, this causes an effective shortening of the length of channel that is

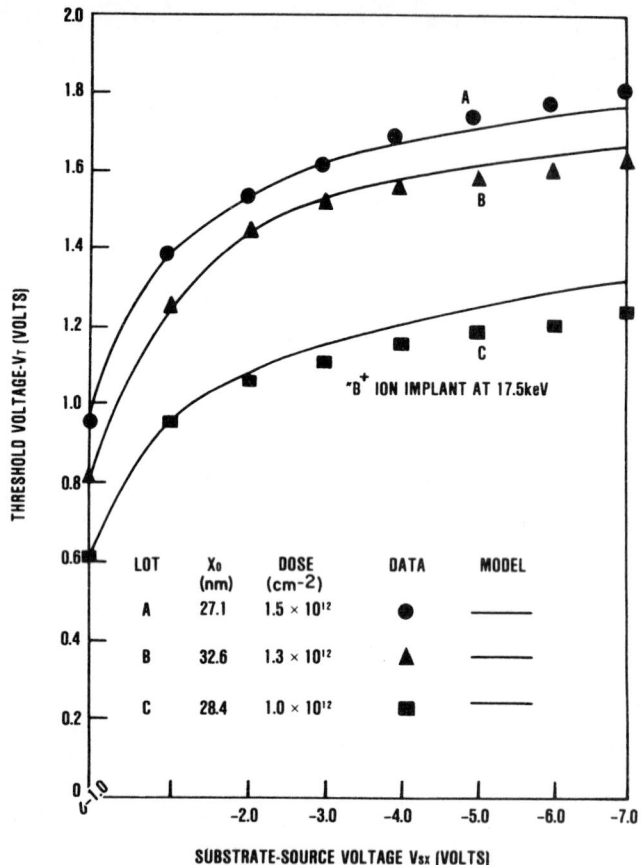

Fig. 12.77 Substrate sensitivity characteristics for an ion-implanted enhancement p-channel IGSETs, demonstrating comparison of measurement to model. (From Ref. 33.)

turned on, and the saturation current is reduced by a factor $\Delta L/L$.[34]

$$I'_{D\,\text{sat}} \simeq \frac{I_{D\,\text{sat}}}{1 - \Delta L/L} \tag{12.48}$$

where ΔL is the reduction of channel length due to drain bias increase, given approximately by

$$\Delta L \simeq \frac{2\varepsilon_s \left(V_D - V_{D\,\text{sat}}\right)^{1/2}}{qN_A} \tag{12.49}$$

It may be seen from Eq. (12.48) that for long channel lengths (long compared to possible ΔL's), this modulation of the effective channel length by the spreading of the drain depletion region has little effect on channel current. On the other hand, for short channel lengths this has a pronounced effect of changing the slope of the I_D versus V_D curves in saturation from zero to a finite resistance. This phenomenon, together with a diagram depicting operation of a short-channel IGFET beyond pinch-off, is illustrated in Fig. 12.79.

Another phenomenon also attributed to shrinking device size is termed the short-channel effect,[40] and its companion, the narrow-channel effect.[41] The short-channel effect has basically two components: the proximity effect and the drain bias effect.[42] The first is due to the fact that the channel length is decreased; the proximity of the drain to the source causes the surface potential in the channel to drop. This makes it easier to turn on the device, thus effectively lowering the threshold voltage. Even at very low drain voltages this effect is very pronounced for very short channel devices, especially those of channel length comparable to the bulk Debye length, L_B.

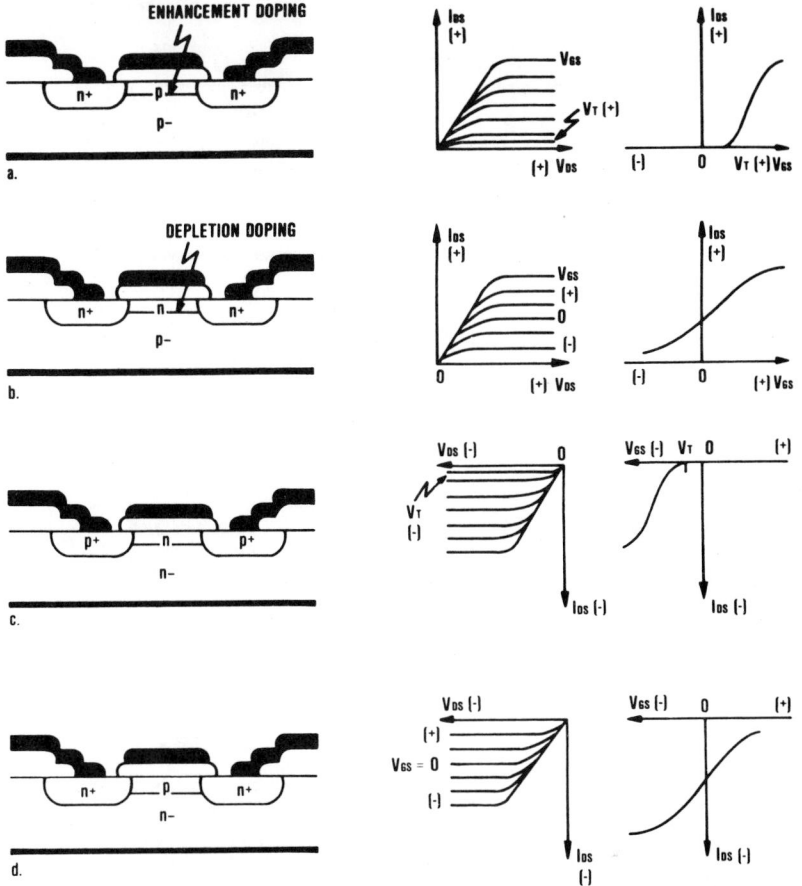

Fig. 12.78 Typical cross sections and characteristics for (*a*) *p*-channel enhancement, (*b*) *p*-channel depletion, (*c*) *n*-channel enhancement, and (*d*) *n*-channel depletion IGFETs.

TABLE 12.6   VLSI Parameter Dependence on Minimum Feature Size L

Parameter	Determining Equation	Dependence on L	Comment
$V = V_G - V_T$	$\sim LE$	L	Scale independent of electric field E
t = transit time	$\sim \dfrac{L}{\mu E}$	L	Time to traverse channel
C = capacitance	$\sim \dfrac{LW}{D}$	L	Assuming that $D \sim L$
$I_{D\,\text{sat}}$	$\sim JL$	L	Scale independence of surface current density J
P_{dc} = dc supply power	$\sim IV$	L^2	Per gate
E_{sw} = energy loss per switching	$\sim CV^2$	L^3	Per gate

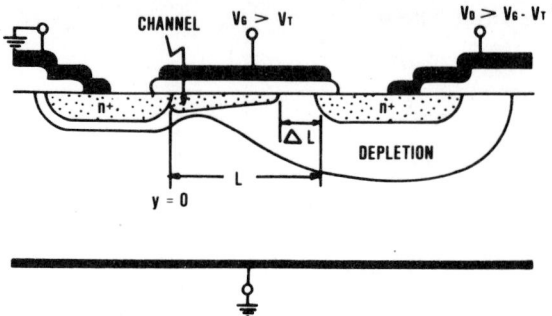

Fig. 12.79 Effect of drain bias on the electrical shortening of channel length in saturation: channel-length modulation. (From Ref. 34.)

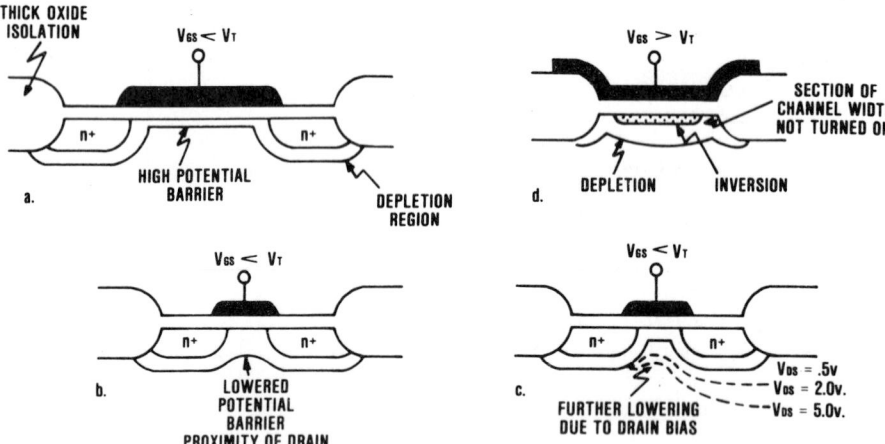

Fig. 12.80 Narrow-channel and short-channel effects: (*a*) long-channel EGFET, (*b*) effect of drain bias on threshold lowering in a short-channel IGFET, (*c*) same effect caused by the proximity of two diffusions, (*d*) cross section of IGFET along channel rather than across it, showing increase in threshold by incomplete channel formation at sidewalls. (From Refs. 40–42.)

For increasing drain voltages, the spread of the depletion region into the channel is concurrent with a lowering of the surface potential barrier. As a consequence a lowering of the threshold occurs. The combination of the drain bias effect and the proximity effect causes a lowering of the threshold for short-channel devices compared to long-channel devices of the same chip. Figure 12.80 illustrates the results of the two effects. This presents a problem that circuit designers must work around and device designers must minimize. Often, properly designed threshold-adjusted implanted profiles will minimize this effect.

A complementary problem occurs when the device width is reduced. In this case, due to the thick oxide sidewalls of the device, the section of the channel near the sidewalls is not fully inverted and thus carries less current. This causes an apparent threshold increase, since now the gate has to be driven harder to turn it on fully. Figure 12.80 illustrates this narrow-channel effect. For channels much wider than the small edge regions that fail to turn on, this is not noticeable. Thus, for chips with devices that have a variety of channel widths, this threshold shift may also be of concern and must be minimized by the device designer.

Thick Oxide IGFET

For most MOS fabrication technologies and device designs, electrical isolation of diffused lines is done by some form of thick oxide. If a conductor passes over this insulator separating the diffusions and has a sufficient voltage to turn on the surface underneath, conduction between the diffusions will occur in places where it was not intended. To minimize this leakage current, the insulator is made

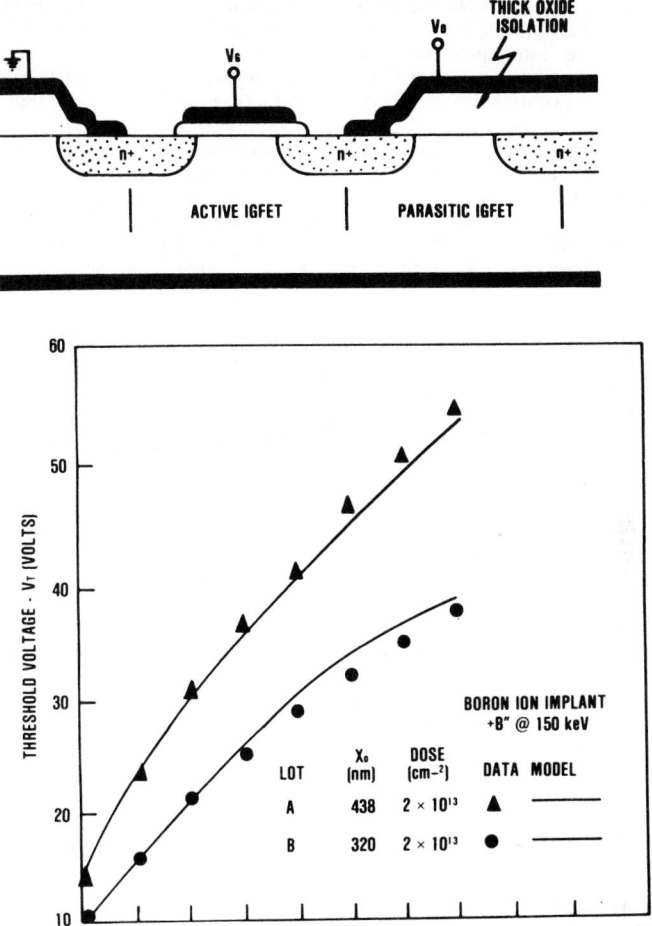

Fig. 12.81 Thick oxide (parasitic) IGSET cross section and typical ion-planted IGSET substrate sensitivity characteristics.

very thick and usually a heavy ion implant dose is introduced at the surface to increase the threshold voltage. Normally, the threshold of this parasitic IGFET is made several times the largest voltage available on the chip, thus ensuring very low leakage currents. A typical thick oxide IGFET and substrate sensitivity characteristic are illustrated in Fig. 12.81.

12.6.3 Processing

Traditionally, the simplicity of MOS processes was a mark in favor of MOS over bipolar technology.[37] Not so today, as VLSI has increased the level of complexity in the MOS world. Process complexity may be gauged roughly by the number of mask levels needed to achieve a certain level of functionality. A reasonable dividing line in processing, from starting silicon wafer to encapsulated chip, is the final conductive layer, which makes the device electrically operational. At this point in the process the device may be tested electrically. Past this point, passivation, connections to a package, and encapsulation are done.

The discussions that follow are descriptions of representative types of processes with steps leading up to a working device. Passivation and encapsulation technologies, although they must be compatible with the particular MOS process in question, are not discussed.

Basically, several key device features must be provided for by a process. The order in which these features appear and the method by which they are introduced vary and are constantly evolving. This is a key area of intense innovation in the semiconductor field.

Two basic processes are discussed in detail: the metal gate, thick oxide isolation IGFET; and the polysilicon gate, recessed oxide isolation IGFET. Advances on the latter process leading to VLSI are also introduced.

Although *p*-channel technology was the first to be productive in the marketplace, the advantages of speed of *n*-channel has caused this technology to capture the MOS world.[43] A special case is the CMOS (complementary MOS) technology,[32] where both *p*-channel and *n*-channel IGFETS are integrated side by side, providing a superior inverter configuration. This technology has increased process complexity when compared to the typical MOS process but is worth it in certain circuit applications. The process descriptions that follow involve exclusively *n*-channel devices.

Metal Gate IGFET

Processing of this device is basically done with four masks and is illustrated in Fig. 12.82. Starting with a bare substrate, a thick oxide is thermally grown. A mask is used to etch the oxide and define the source and drain and connecting diffusions. A diffusion is made at high temperatures, typically of phosphorus. The oxide is then removed and regrown to the proper thickness for isolation. In some processes the masking oxide is left in place for isolation, but this is not as typical as removal and regrowth. Then the gate region is etched through the thick oxide to the underlying silicon, using a second mask. Thermal oxidation of the gate insulator is then performed. Thicknesses range from 0.05 to 0.10 μm. A third mask is used to define contact openings to the diffusions. Finally, metal (usually aluminum) is blanket evaporated onto the wafer and by using a final mask, is etched to provide all necessary conductors. The aluminum is annealed and the device is complete and ready for preliminary electrical test. The final device cross section is illustrated in Fig. 12.82.

Polysilicon Gate IGFET

There are several drawbacks to the metal gate process technology compared to the use of polysilicon gate:

1. The gate is not aligned to the edges of the diffusion, causing excess gate to diffusion capacitance, slowing down most circuits.

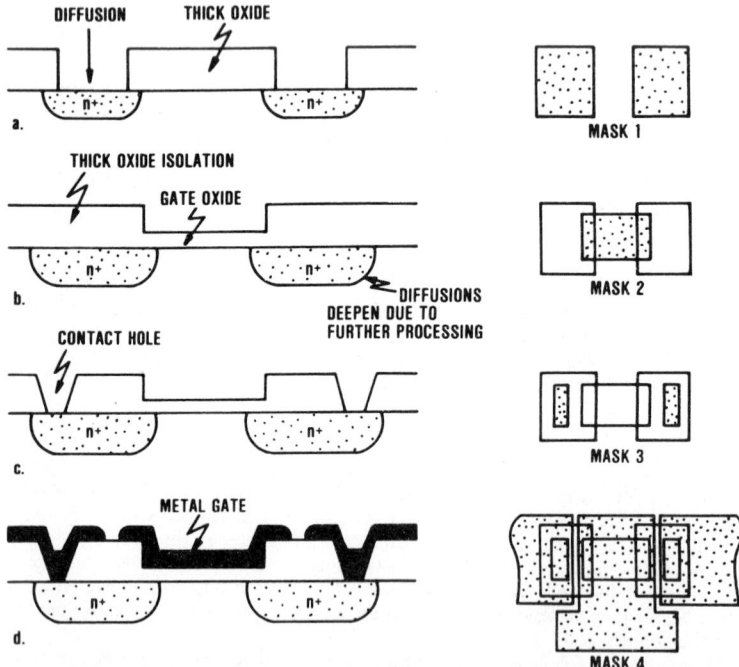

Fig. 12.82 Typical metal gate IGFET process sequence up to metal definition, showing device cross sections and top views.

2. Aluminum is volatile at the elevated temperatures of oxidation and diffusion, preventing further processing at those temperatures once the aluminum is deposited.

3. The limitation cited implies that only two levels of interconnect (diffusions and the metal) are possible at the device definition mask tolerances.

The use of polysilicon gates gets around these limits by depositing and defining the gate first and then diffusing the source and drains. This is possible since polysilicon can withstand elevated process temperatures. Several layers of polysilicon independently defined are possible in a process. These are isolated typically by thermal oxides and do not preclude the use of final metal layer, which is normally used.

Although thick oxide isolation may be employed, as already defined, a recessed oxide scheme is more advantageous and is widely practiced. To produce a recessed oxide, a thin layer of oxide is grown and a layer of silicon nitride is deposited at the beginning of the process. Both are etched to expose the silicon where the recessed oxide isolation is to be grown. During thermal oxidation, the silicon nitride prevents the penetration of oxygen to the Si-SiO$_2$ surface; thus no oxidation takes place there. Figure 12.83 depicts the cross section after thermal oxidation. Normally, to increase the threshold of the parasitic IGFET, an ion implant of boron is introduced before the oxidation. Subsequently, the oxide-nitride (SiO$_2$-Si$_3$N$_4$) mask is etched and the gate oxide is thermally grown (nominally 0.03–0.07 μm for LSI). At this point a threshold-adjusted implant may be introduced. A layer of doped polysilicon is deposited and etched with the second mask to define the gate conductors. Using the polysilicon as a mask, the gate oxide is etched in the source and drain regions (see Fig. 12.83). The junctions are formed by various methods, including diffusions and implantation. Reoxidation of the silicon passivates the junctions and the polysilicon (this contact is usually made over the isolation oxide). Evaporation of a metal, normally aluminum or a suitable alloy, etching it using a fourth mask, and annealing complete the process.

One may introduce several more levels of interconnect than the three already outlined. After polysilicon definition and before etching for diffusions, another level of polysilicon may be deposited

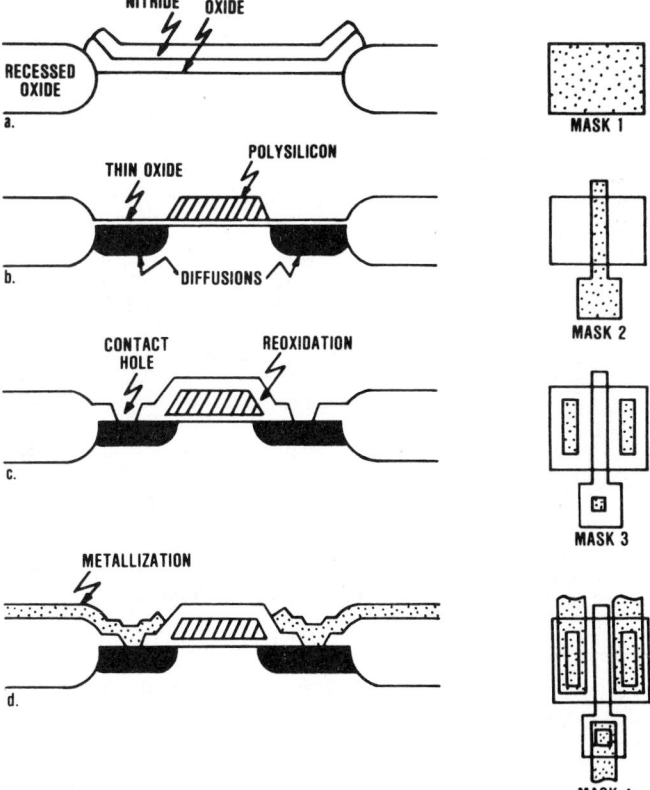

Fig. 12.83 Self-aligned polysilicon gate IGFET process sequence typical of LSI and VLSI technologies.

and defined. The process may be repeated several times with care that the resultant topography is not excessively steep for proper metal coverage.

Advances in Processing

The several layers of polysilicon that are available make possible efficient fabrication of CCDs (charge-coupled devices),[44] which depend on overlapping electrodes. One drawback of polysilicon is that even at the highest doping possible, the conducting lines have a very high resistance compared to a metal line. One solution currently being investigated is silicide structures, such as WuSi, which reduce the resistance, yet can withstand the high processing temperatures.[45]

The use of ion implantation has spread from resistors to FET thresholds[46] to the point where currently most of the doping on the wafer is done by implantation, including junctions.

Noble gas implants such as argon in the backside of the wafer are also frequently used to improve the silicon quality at the surface and reduce leakage due to crystal damage.[47]

12.7 SWITCHING CIRCUITS
Sheldon S. L. Chang

12.7.1 Boolean Algebra

Definition

A Boolean variable can take on two values: 0 and 1. The **or** operator " + " is defined by

$$0 + 0 = 0 \tag{12.50}$$

$$1 + 0 = 0 + 1 = 1 + 1 = 1 \tag{12.51}$$

The **and** operator " · " is defined by

$$1 \cdot 1 = 1 \tag{12.52}$$

$$1 \cdot 0 = 0 \cdot 1 = 0 \cdot 0 = 0 \tag{12.53}$$

The **not** operator "–" is defined by

$$\bar{0} = 1 \qquad \bar{1} = 0 \tag{12.54}$$

Equations (12.50)–(12.54) defined Boolean algebra.

The operators + and · are also called **sum** and **product**, respectively.

When we work with symbols (or variables), it is customary to write AB instead of $A \cdot B$:

$$AB \triangleq A \cdot B$$

Duality. Duality is inherent in the definition of Boolean algebra. If we interchange the operators + and · and the values 0 and 1 simultaneously, all five defining equations remain valid. However Eqs. (12.50)–(12.53) are interchanged. This property is called the **duality principle**.

The duality principle is useful in many ways. It helps us to verify and to remember the Boolean identities. Since an identity is valid no matter what values we assign to the Boolean variables, it follows that the dual of an identity is an identity.

Another use of the duality principle is in realizing the Boolean function by logic circuits. It provides insight on possible alternatives.

Boolean Expression and the Truth Table

A Boolean function is defined by its truth table or an expression of Boolean variables with operators · and + as connectives. The truth table is an exhaustive listing of all possible combinations of values of the variables and the resulting value of the expression. For instance,

$$F = AC + \bar{B}C \tag{12.55}$$

is a Boolean expression. As is customary in algebraic expressions, the product takes precedence over the sum, and (12.55) means

$$F = (AC) + (\bar{B}C) \tag{12.56}$$

The truth table of (12.56) is easily calculated and given in Table 12.7. Two Boolean expressions are equal if their truth tables are the same.

TABLE 12.7 Truth Table for Eq. (12.56)

A B C	F
0 0 0	0
0 0 1	1
0 1 0	0
0 1 1	0
1 0 0	0
1 0 1	1
1 1 0	0
1 1 1	1

Basic Laws and Identities

In the following, the basic laws and identities are arranged in dual pairs. They can be easily verified by showing that the expressions on two sides of the equality sign have the same truth table.

1. Associative

$$(A + B) + C = A + (B + C) \tag{12.57}$$
$$(AB)C = A(BC) \tag{12.58}$$

2. Commutative

$$A + B = B + A \tag{12.59}$$
$$AB = BA \tag{12.60}$$

3. Distributive

$$A(B + C) = AB + AC \tag{12.61}$$
$$A + BC = (A + B)(A + C) \tag{12.62}$$

4. DeMorgan's law

$$\overline{A + B} = \overline{A} \cdot \overline{B} \tag{12.63}$$
$$\overline{A \cdot B} = \overline{A} + \overline{B} \tag{12.64}$$

5. Single-variable identities

$$1 \cdot A = A \tag{12.65}$$
$$0 + A = A \tag{12.66}$$
$$1 + A = 1 \tag{12.67}$$
$$0 \cdot A = 0 \tag{12.68}$$
$$A + A = A \tag{12.69}$$
$$AA = A \tag{12.70}$$
$$A + \overline{A} = 1 \tag{12.71}$$
$$A\overline{A} = 0 \tag{12.72}$$
$$\overline{(\overline{A})} = A \tag{12.73}$$

6. Two-variable identities

$$A + AB = A \tag{12.74}$$
$$A(A + B) = A \tag{12.75}$$
$$A + \overline{A}B = A + B \tag{12.76}$$
$$A(\overline{A} + B) = AB \tag{12.77}$$

The two-variable identities can be easily proved from basic laws and single-variable identities:

$$A + AB = A(1 + B) \qquad\qquad [\text{from } (12.61)]$$
$$= A \cdot 1 = A \qquad\qquad [\text{from } (12.67), (12.65)]$$
$$A + \bar{A}B = (A + AB) + \bar{A}B \qquad\qquad [\text{from } (12.74)]$$
$$= A + (A + \bar{A})B \qquad\qquad [\text{from } (12.61)]$$
$$= A + B \qquad\qquad [\text{from } (12.71)]$$

An application of these identities is illustrated in the following example.

Example 12.1. Simplify the Boolean function F, which is given as

$$F = A(B + B\bar{C}) + AC + \overline{CB} + \left(\overline{C + A\bar{B}}\right) \qquad (12.78)$$

SOLUTION. From Eq. (12.74),

$$A(B + B\bar{C}) = AB \qquad (12.79)$$

From Eqs. (12.63), (12.64), (12.73), (12.61), and (12.60),

$$\overline{C + A\bar{B}} = \bar{C}(\bar{A} + B) = \overline{AC} + B\bar{C} \qquad (12.80)$$

Substituting Eqs. (12.79) and (12.80) into (12.78) gives

$$F = AB + AC + \overline{CB} + \overline{AC} + B\bar{C}$$
$$= AB + AC + \bar{C}(\bar{A} + \bar{B} + B)$$
$$= AB + AC + \bar{C} = AB + A + \bar{C}$$
$$= A + \bar{C} \qquad (12.81)$$

12.7.2   Realization of Boolean Functions

A Boolean function can be realized in two ways:

1. A network of standard gate circuits or gates
2. A programmable logic array (PLA) or a read-only memory (ROM)

Realization by Gates

Each gate circuit represents one logic operation with a multiple number of inputs and one output variable. The output variable may be in its direct form, its negation, or both. Table 12.8 shows the names of the gates, their representative symbols, and the Boolean operations they perform.

The gates can be interconnected to represent literally a Boolean expression. For instance, the expression (12.78) is realized by the logic circuit of Fig. 12.84 in which the intermediate output variables are not marked, as they are easily readable from the circuit itself. In contrast, the expression of (12.81) is realized in Fig. 12.85. Although the circuit of Fig. 12.85 is considerably simpler, it performs the same function as that of Fig. 12.84.

An important property of the gates is that of sufficiency. A gate type is said to be sufficient if any arbitrary Boolean function can be realized by using entirely gates of the same type. It can be shown that any function can be realized with only *and*, *or*, and *not* operators. Then, the sufficiency of NAND and NOR gates can be established as follows.

$$\bar{A} = \left(\overline{A \cdot A}\right) \qquad (12.82)$$

$$A + B = \overline{\left(\overline{A \cdot A}\right) \cdot \left(\overline{B \cdot B}\right)} \qquad (12.83)$$

$$AB = \left(\overline{\overline{AB}}\right) \qquad (12.84)$$

Equations (12.82)–(12.84) are realized in Fig. 12.86(*a*). Since every Boolean function is expressed in terms of "and," "or," and "not" operations, realization of the trio proves the sufficiency of NAND. Similarly, Fig. 12.86(*b*) illustrates realization of the trio by NOR.

TABLE 12.8   Names and Symbols of Commonly-Used Gates

Name	Symbol	Boolean Operation
AND		$Z = ABC$
OR		$Z = A + B + C$
NOT		$Z = \overline{A}$
EXCLUSIVE OR		$Z = A\overline{B} + B\overline{A}$
NAND		$Z = \overline{AB}$
		$Z = \overline{A}B$
NOR		$Z = \overline{(A + B)}$
		$Z = A + \overline{B}$

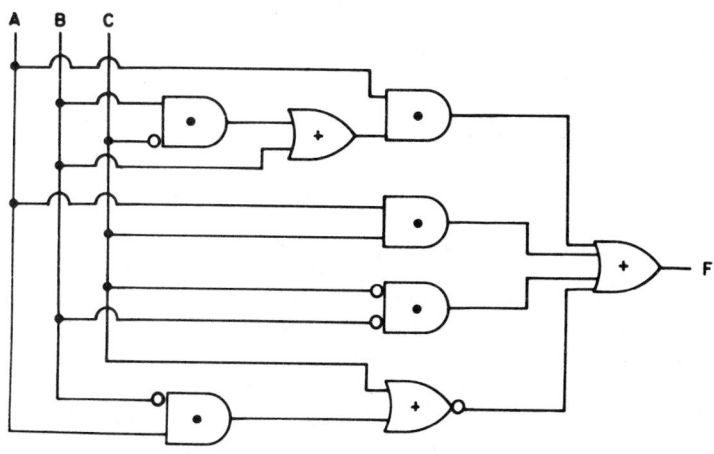

Fig. 12.84

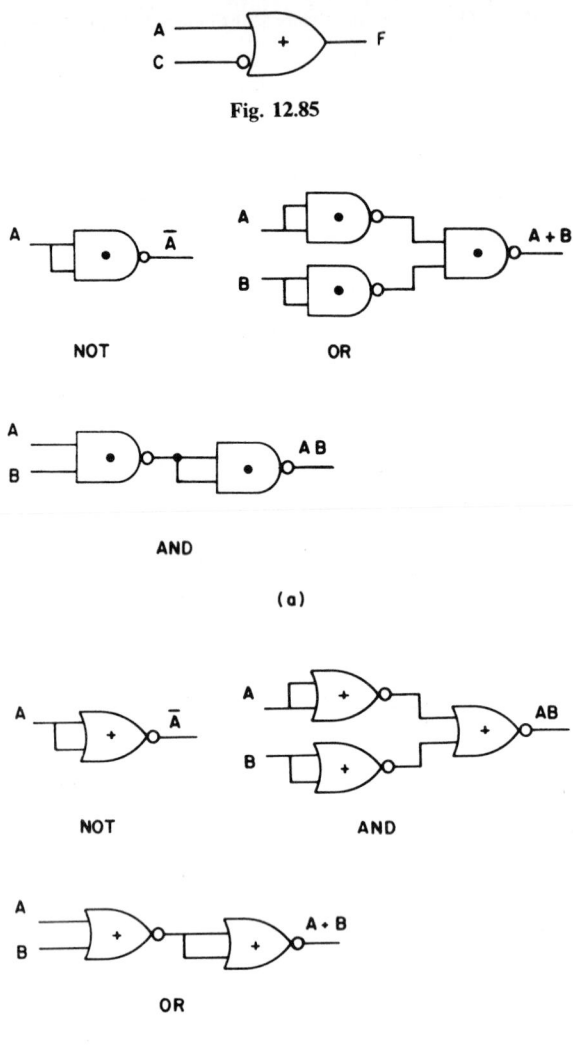

Fig. 12.85

Fig. 12.86 Sufficiency of NAND and NOR: (*a*) NAND gates, (*b*) NOR gates.

Integrated-Circuit Logic Families

In integrated circuits it is convenient to interconnect one or two basic circuit types to realize a Boolean function. We refer to each type of basic circuit as a **logic family**.

The following are important characteristics of a logic family:

1. *Fan-In and Fan-Out*. Fan-in is the number of inputs a gate can have. Fan-out is the maximum number of input terminals of the same type that its output can feed without being overloaded. High fan-ins and fan-outs offer greater flexibility of interconnection.

2. *Watts Loss per Gate*. For any given logic family, the power requirement per gate increases with the speed of response, which can be measured in terms of the gate's delay time.

3. *Delay Time*. The time interval between the arrival of the input signal to the time when the output reaches its new threshold level.

Figure 12.87 gives a summary of the logic families. We normally associate a high voltage level with 1, and low voltage with 0. In Fig. 12.87(*a*) "high"s at both inputs *A* and *B* are required to saturate

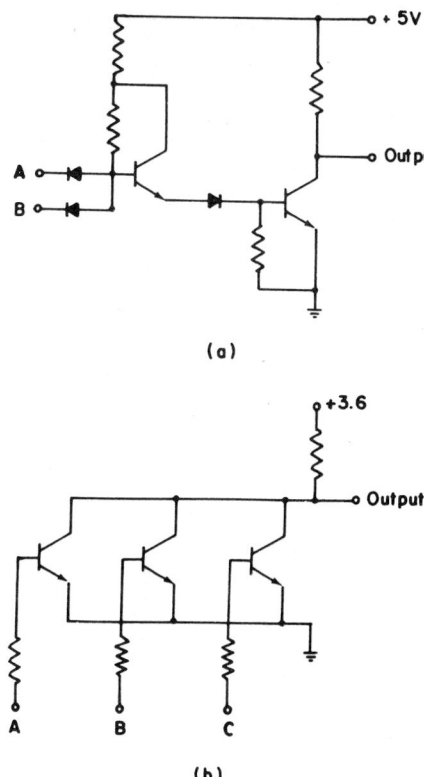

(a)

(b)

Fig. 12.87   IC logic families: (*a*) DTL, (*b*) RTL, *(Continued on the next page)*

the transistor base current and thereby cause a "low" output Z. Its logic expression is

$$Z = \overline{A \cdot B} \tag{12.85}$$

The CC-CE transistor connection is to increase the input impedance and thereby improve its fan-in and fan-out characteristics. In the circuit in Fig. 12.87(*b*), a "high" at any input terminal is sufficient to saturate its transistor and brings the output voltage down near ground level. Its logic expression is

$$Z = \overline{A + B + C} \tag{12.86}$$

In Fig. 12.87(*c*) the output at Z follows the voltage V with a much lower output impedance. It reduces delay time and improves fan-out capability. Any low at the input would draw base current away from Q_2 and cause it to cut off. The logic expression is

$$Z = \overline{A \cdot B \cdot C} \tag{12.87}$$

In all the preceding circuits, the transistors are either saturated or cut off. Considerable delay is caused by draining away accumulated charge in the base region to pull a transistor out of saturation. There are two methods of improvement. Moderate improvement can be obtained by using Schottky transistors in Fig. 12.87(*c*). A Schottky transistor has a diode clamp bridged across the base and the collector. It prevents the diode from being saturated. A greater improvement is obtained by keeping the transistor out of saturation, as shown in Fig. 12.87(*d*). The circuit is an emitter-coupled pair with two emitter followers as the output stage. All transistors operate in the active region. This circuit gives fastest response but requires the highest power consumption.

The circuit of a CMOS inverter is shown in Fig. 12.87(*e*). With high input, Q_1 is off and Q_2 is on. The voltage output is 0. With low input, Q_1 is on and Q_2 is off. The voltage output is V_+. The circuit does not consume power except at the switching instant. By making series and parallel connections of

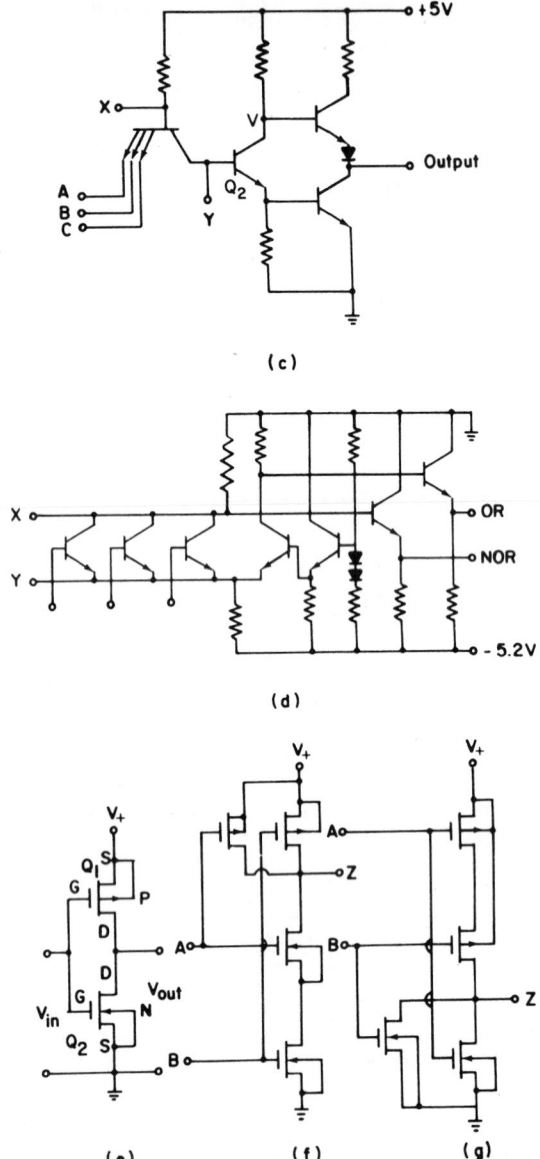

Fig. 12.87 continued IC logic families: (c) TTL, (d) ECL, (e) CMOS, (f) CMOS NAND, (g) CMOS NOR.

the top and bottom transistors, we obtain the NAND and NOR circuits of Fig. 12.87(f) and (g), respectively.

Table 12.9 gives a summary of the characteristics of integrated-circuit logic families. PLA and ROM are not included. They are most cost effective in realizing multiple output logic networks.

12.7.3   NMOS and VLSI

In LSI and VLSI, the basic transistor is an *n*-channel metal-oxide insulated-gate field-effect transistor (NMOS). There are two important advantages of the NMOS over its counterpart PMOS:

    1. Electron mobility in silicon is about twice the hole mobility. Thus *n*-channel transistors have greater gain and speed than do *p*-channel devices with the same geometry.

TABLE 12.9 Basic Characteristics of Integrated-Circuit Logic

Type Characteristic	DTL	RTL	TTL	STTL	ECL	CMOS
Positive Logic function of basic gate	NAND	NOR	NAND	NAND	OR/NOR	NAND or NOR
Maximum fan-in without expansion	10	5	8	8	5	8
Typical fan-out	8–10	4	8–10	8–10	20–25	Unlimited
Typical power dissipation per gate	10 mW	12 mW	1–25 mW	2–20 mW	High	0.01 μW static $\approx$ 1 mW at 1 MHZ
Typical gate delay (ns)	30	20	6–33	3–10	1–2	25–35
Noise performance	Good	Fair	Fair to medium	Fair to medium	Fair	Very good
Cost	Low	Low	Low	Medium to high	High	Medium to high
Availability of complex functions	Fair	Fair to medium	Excellent	Medium	Fair	Medium, growing

Source: Reprinted from F. J. Hill and G. R. Peterson, *Switching Theory and Logical Design*, by permission from John Wiley & Sons, Inc., 1974.

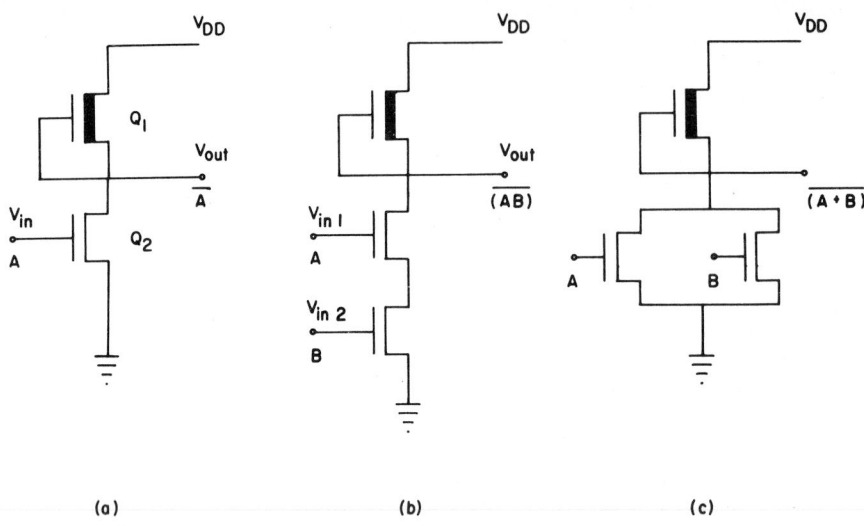

Fig. 12.88   NMOS logic: (*a*) NOT, (*b*) NAND, (*c*) NOR.

2. The threshold voltages of *n*-channel transistors tend to be low, and a lower supply voltage is possible.

Large-scale integration has important advantages in speed, supply power, and switching energy. In a first approximation, delay is proportional to a critical linear dimension L_0, while dc supply power and switching energy are proportional to L_0^2. Thus with progressively decreasing L_0 due to improved technology, NMOS gates will gain ascendancy as the basic building blocks of logic circuits.

Figure 12.88 illustrates NMOS logic circuits. In the inverter circuit [12.88(*a*)] the top transistor, Q_1, operates in a depletion mode with a negative threshold voltage. It conducts with gate source voltage equal to zero and behaves very much like a (nonlinear) resistor. The lower transistor, Q_2, operates in an enhancement mode with positive threshold voltage V_{th}. If $V_{in} < V_{th}$, Q_2 is cut off and

$$V_{out} = V_{DD} > V_1 > V_{th}$$

where V_1 is a voltage between V_{th} and V_{DD}. It defines the 1 level. If $V_{in} > V_1$, Q_2 conducts sufficiently to pull down V_{out} below V_{th}:

$$V_{out} < V_{th}$$

The correspondences between 0 and 1 levels and the voltage V are as follows:

$$0 \Leftrightarrow V < V_{th}$$
$$1 \Leftrightarrow V > V_1$$

NMOS NAND, and NOR circuits are illustrated in Fig. 12.88(*b*) and (*c*), respectively. For a more detailed discussion of the electronics of these devices, the reader is referred to Sections 12.4 and 12.5.

12.7.4 Flip-Flops

Flip-flops are standard circuits for storing state information in a sequential circuit. Each flip-flop stores one binary variable. With N flip-flops, the stored binary number ranges from 0 to $2^N - 1$. Each stored binary number represents one state of the sequential circuit. Therefore, a maximum of 2^N states can exist in a circuit with N flip-flops.

S-R Flip-Flop

Figure 12.89 shows an S-R flip-flop with two NOR gates. There are two inputs to the circuit, S and R, and an output variable, Q. The truth table is shown in Table 12.10. With $S = R = 0$, Q can be either 0 or 1, and the input/output relation at each NOR gate is satisfied. The value of Q then represents the stored information. With S, R equal to 1, 0 (or 0, 1), only $Q = 1$ (or 0) is possible. The

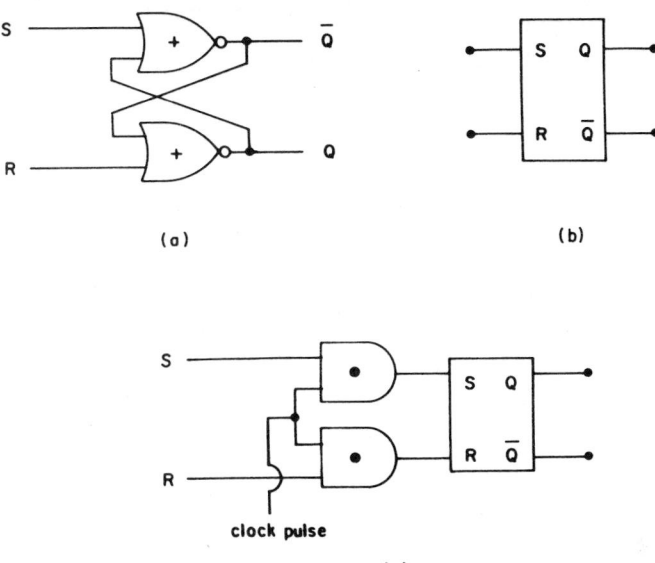

(a)   (b)

clock pulse

(c)

Fig. 12.89   *S-R* flip-flop: (*a*) circuit, (*b*) symbol, (*c*) clocked.

TABLE 12.10   Truth Table of *S-R* Flip-Flop

S	R	Q
0	0	0, 1
0	1	0
1	0	1
1	1	—

input condition $S = R = 1$ is to be avoided for the following reasons: (1) Both outputs, Q and $\overline{Q}$, are 0 and so represent an undefined state, and (2) if both S and R change to zero simultaneously, the outcome of Q is uncertain and may oscillate between 0 and 1.

A symbol representing an *S-R* flip-flop is shown in Fig. 12.89(b). We have marked the output terminals as Q and $\overline{Q}$ for clarity. Sometimes the output terminals Q and $\overline{Q}$ are not explicitly marked. Then the output terminal nearer S is meant to be Q. The input terminals are always marked to distinguish the type of flip-flop.

J-K Flip-Flop; D and T Flip-Flop
A *J-K* flip-flop is pulse operated as shown in Fig. 12.90. Its *J, K* inputs have no effect until the arrival of the clock pulse. There are delays in the AND circuits and in the *S-R* flip-flop, and these

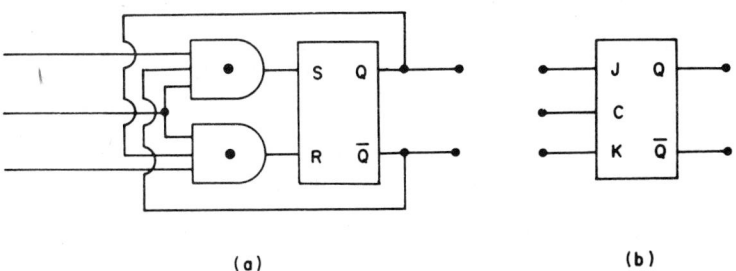

(a)   (b)

Fig. 12.90   Clocked *J-K* flip-flop: (*a*) circuit, (*b*) symbol.

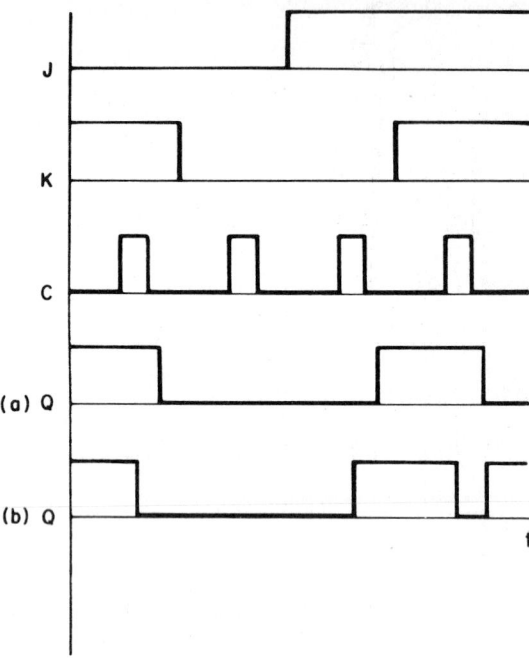

Fig. 12.91 Clocked *J-K* flip-flop timing diagram: (*a*) proper operation, (*b*) faulty operation caused by insufficient internal delay.

delays play an indispensable part in the operation of a *J-K* flip-flop. Figure 12.91 shows timing diagrams of a *J-K* flip-flop. In Fig. 12.91(*a*), due to internal delays, the output variables Q and $\overline{Q}$ do not change until after the clock pulse. The flip-flop operates properly with its new state Q^+ given as

$$Q^+ = J\overline{Q} + \overline{K}Q \tag{12.88}$$

where Q and Q^+ represent the output Q before and after a clock pulse, respectively. In Fig. 12.91(*b*), the internal delay is insufficient and the new value Q^+ is returned before the clock pulse is over. If the inputs are $J = 1$ and $K = 1$, the flip-flop can trigger another round of changes and yield erroneous results. Figure 12.92 shows a *J-K* flip-flop made entirely from NAND circuits. Its principle of operation is identical with that of Fig. 12.90.

There are two variations of the *J-K* flip-flop, as shown in Fig. 12.93. In Fig. 12.93(*a*) the *J, K* inputs are connected permanently to 1, and Q changes with each input pulse. It is called a *T* flip-flop. In Fig. 12.93(*b*), $J = D$ and $K = \overline{D}$. Equation (12.88) gives

$$Q^+ = D\overline{Q} + DQ = D \tag{12.89}$$

The next state is the same as the input D. It is called the *D* flip-flop.

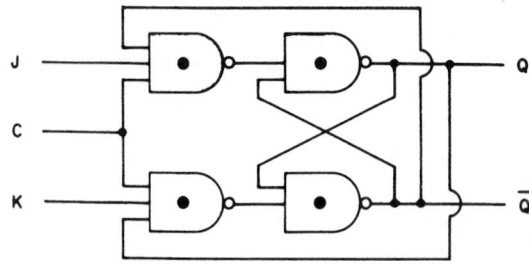

Fig. 12.92 All-NAND *J-K* flip-flop.

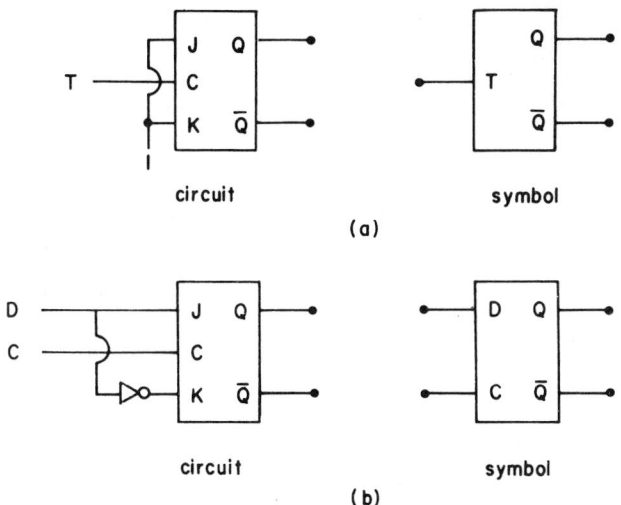

Fig. 12.93   *D* and *T* flip-flops: (*a*) *T* flip-flop, (*b*) *D* flip-flop.

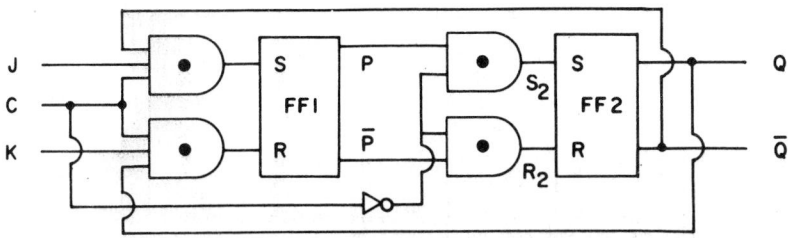

Fig. 12.94   Master-slave flip-flop.

Master-Slave Flip-Flops

Figure 12.94 illustrates the master-slave flip-flop concept, which ensures an adequate time delay between input and output signals. The first flip-flop (FF_1) changes state during the clock pulse. However, $\bar{C} = 0$, and P and $\bar{P}$ are inhibited by the second set of AND circuits from reaching the SR inputs of FF_2. Only after the clock pulse is over and $\bar{C}$ becomes 1 again do the $P\bar{P}$ signals reach the SR inputs and switch FF_2 accordingly. The timing diagram is shown in Fig. 12.95.

Shift Register

One of the simplest, but most useful, sequential circuits is the shift register. It is made of a string of *D* flip-flops, as shown in Fig. 12.96. At each clock pulse, the stored data in each flip-flop advance to the next flip-flop. Mathematically, the shift register can be represented as

$$Z_k = \begin{cases} Z_{k-1} & k = 2,3,\dots,L \\ I & k = 1 \end{cases} \tag{12.90}$$

where I is the input data bit; Z_k, $k = 1,\dots,L$, the output, and L the length of the register.

12.8   INSTRUMENTATION
Velio A. Marsocci

12.8.1   Introduction

The remaining sections in this chapter have been chosen with two purposes in mind: coverage of the subject areas in their own right and to provide information in an order that gives the reader a

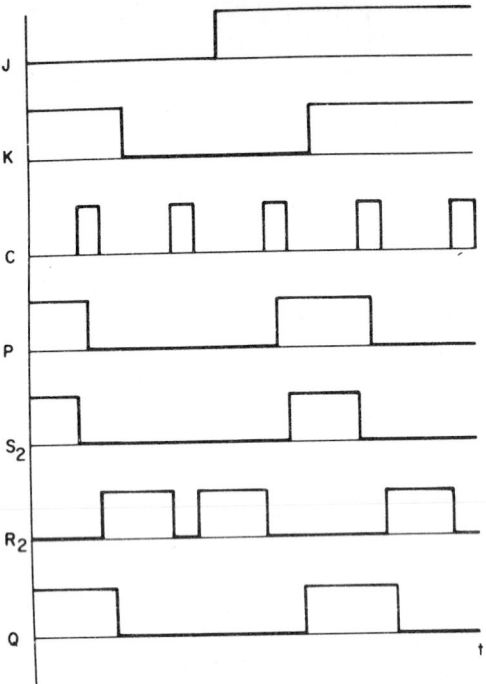

Fig. 12.95 Master-slave flip-flop timing diagram. Note that S_2 and R_2 are zero during the timing pulse.

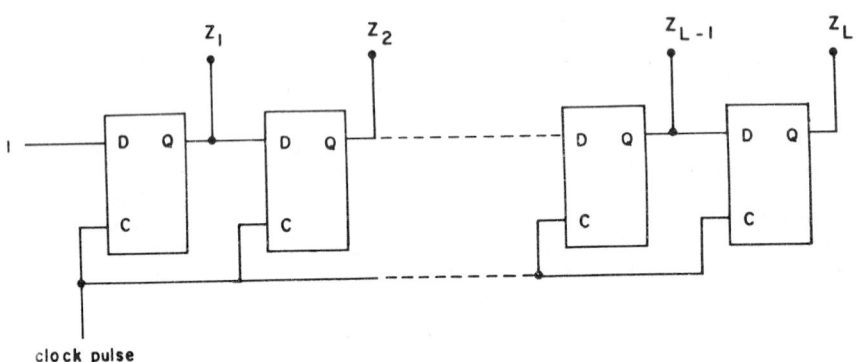

Fig. 12.96 Shift register.

perspective as to how electronic subsystems are utilized in the architecture of instrumentation hardware. Thus, any of the sections that follow may be read independently of the others, or they may be combined in whatever groupings are required to satisfy the needs of the reader.

Any attempt to place boundaries on the definition of that category of electronic equipment that we might wish to include as a part of **instrumentation** invariably fails. The selection of specific systems that might exemplify such a definition is extremely vast and is subject to the vagaries of how different individuals might wish to include, or not, a particular system in the definition. In the broadest sense, the kinds of systems that might be categorized as instrument systems may range from a benchtop voltmeter to the large overall system required to send a vehicle into space to make observations of a distant planet and to relay the results to an Earth station. In effect, for the latter case, the entire assemblage of the space vehicle, and the control station, as well as the computer system that acts as part of the control loop, all form a vast system whose essential purpose is to make a measurement and, after some predetermined form of data reduction, to produce appropriate readout corresponding to the measured data.

It may well be that the closing statement in the preceding sentence is the simplest approach to the determination of what kind of electronic hardware belongs to the category of instrumentation: that is, an instrument system is one that performs a measurement and displays the result of the measurement in some convenient form. Of course, there exist a large number of systems that may fit this category but are not electronic in nature. This section is concerned only with instrument systems whose subsystems are predominantly of the electronic variety. This is in keeping with the general nature of the subject of this chapter.

To provide some orientation to the discussion of instrument systems, it might be convenient to establish that the basic parameters and the hardware associated with any physical measurement must include[48,49] the measured input parameter (measured quantity), a reference input parameter (reference quantity), a difference detector, and an output parameter (output quantity). It is understood that each of the dynamical parameters (i.e., measured, reference, output) must be present in the form of pairs of quantities (e.g., voltage-current, force-displacement, etc.) such that the product of the two in each pair is energy or power. In addition, the reference parameter must possess the same characteristics or property of the measured parameter. Some simple examples of the reference parameter include an operating point, an initial condition, and a zero setpoint. The difference detector may take a variety of forms, from that of a simple differential input amplifier to a complicated electromechanical servo system, together with other attendant subsystems necessary for signal processing and display.

A generic instrument system may be envisioned as depicted in Fig. 12.97. The diagram indicates blocks that represent the typical functional subsystems associated with electronic instrumentation. The diagram is meant to be all-inclusive and any particular instrument may, or may not, contain one or more of the subsystems as shown. Further, there may not be unanimous agreement as to which block in the diagram a specific portion of a particular instrument system may belong. For example, there could arise an argument as to whether some amplifier in the system more properly belongs in the block labeled "amplification" or that labeled "signal processing," or whether some other subsystem is more properly a part of "data reduction" rather than "signal processing." Such decisions are not pertinent in the present context of an attempt to discuss the general characteristics of the instrument system, and are decisions that are best made by those who are to be involved with the details of the design at the subsystem and the circuit levels. In many cases, such a problem may be one more influenced by semantics than by any pressing design requirements.

However, it should be recognized that since the diagram is meant as a general representation, the blocks labeled "data reduction," "processing," and so on, may, in a particular system, include analog-to-digital and digital-to-analog converters as well as some level of computerization. The latter may extend from a small built-in microprocessor to a situation where the instrument is operating on-line with a large computer system that performs a complicated data reduction routine as well as exerting a controlling function for the instrument. It should also be recognized that the various connections between the subsystems may not be hardwire links but those implemented by telemetry.

The individual who is responsible for the overall project concerned with design of the instrument system will, in conjunction with others associated with the project, also be concerned with the question of which functional subsystems are to be included in the particular implementation and of how these may best be interfaced. Implicit in such decisions will be those related to which portions of the system will be analog types, which will be digital, and where the appropriate conversions are to be made. In such thinking, it must be kept in mind that the system must satisfy the requirements of essentially two related viewpoints. One is that the instrument system must process and transmit information and the other is that the instrument must, coincidently, be considered as a processor and a transmitter of energy.[49] Depending on the location of the interface point it may, for example, be desired to transmit a signal with maximum information content associated with either a minimal or maximal level of energy transfer.

These decisions, which eventually determine the accuracy and the precision of the instrument system, cannot be made independently of how the subsystems are to be designed. The kinds of

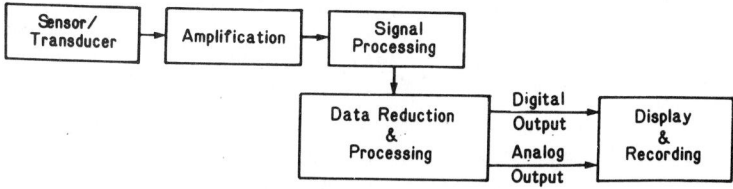

Fig. 12.97 General instrument system.

subsystems chosen, and how they are interfaced, will determine the number of data domain conversions through which the signal representing the measured parameter will be processed. The data domains are defined[50] as the property (i.e., voltage, current, force, displacement, time, pulse height, analog, digital, etc.) of the parameter used to represent the measured quantity. Each functional subsystem of the instrument acts as a domain converter and contributes to the inherent error that a final choice of design will impose on the instrument package. A very useful graphical method, which aids in comparing several alternative design choices from the viewpoint of data-domain conversions, is described in the literature.[50]

Obviously, a full treatment of instrument system theory and design, from the individual circuit to the overall system levels, could not be encompassed in a single section of a handbook. Therefore, only those aspects of electronic systems that are of particular relevance to electronic instrumentation are discussed in this section.

It is with the foregoing thoughts in mind that the structuring of the presentation of the subject matter in this section has been made. The present state of sensing technology still recognizes that almost all of the dynamical physical parameters, which come under measurement, are analog in nature, except for certain types of measurements associated with phenomena at the atomic or subatomic levels. There has not, as yet, been devised a transducer/sensor device that can convert the value of such measurands directly into a digital output. Thus the input end of most of the typical instrumentation is of an analog nature and, usually, is an electronic amplifier designed with regard to sensitivity, input impedance, and noise considerations. It then becomes a matter of design decision where, if at all, in the instrument system the signal is converted into the digital domain.

The ordering of the written material in this section was made to correspond, for the most part, with the sequence in which the hardware being discussed occurs in a typical instrument system, from the input to the output. Thus the category of sensor/transducers, which are certainly devices entirely associated with instrumentation, is the first subject covered in the sequence. This is followed by the coverage of material pertaining to instrumentation amplifiers, and so on through the gamut of subsystems. The latter parts of this section are concerned with essentially complete instrument packages that are designed for use in a specific subject field of measurement (i.e., electronic testing, biological/physiological, and clinical measurements). Here again, no attempt is made for a complete coverage of instrument systems; the constraints of space required the authors to make a selection of those instruments that are most prevalently encountered within each category.

The entry of microprocessors in the implementation of instrument systems has been so dramatic that, obviously, no discussion on the subject would be complete without some attention given to those subsystems. Thus an article on the role of microprocessors in instrumentation is included here. Finally, the section culminates in a discussion of testing instruments as they apply to telemetry systems.

12.9   SENSORS AND TRANSDUCERS
Harry N. Norton

12.9.1   Sensing Devices in Electronic Measuring Systems

Electronic measuring systems are used to provide information. This information is either used as such (e.g., for analyses, for determinations, for verification of performance), or it is used as a basis for exerting control. The control operations can be open-loop or closed-loop. Open-loop control requires someone who evaluates the information and then effects some sort of change in a system. When the control is closed-loop, the measuring system operates as a part of an automatic control system. In all such systems the information originates in a sensing device that provides an electrical signal in response to variations in a quantity, property, or condition that is being measured (the **measurand**).*
Such sensing devices are generally called **transducers** (devices that provide a usable output in response to a specified measurand). In conjunction with certain measurands, they are more commonly referred to as **sensors**, or sometimes as **detectors**; in process control applications they are still frequently called **transmitters**.

A transducer contains, as a minimum, two major elements: the part that responds directly to the measurand (**sensing element**) and the electrical portion in which the output originates (**transduction element**). For example, a strain-gage pressure transducer typically employs a diaphragm as sensing element and a strain-gage bridge as transduction element. In many transducers the sensing and transduction functions are provided by the same element [e.g., in a resistive temperature transducer (such as a platinum-wire thermometer or temperature sensor) the resistive winding responds directly to temperature and originates the output signal by changing its resistance as the temperature

---

*Terminology used in this section is generally in conformance with Standard MC6.1, "Electrical Transducer Nomenclature and Terminology," © Instrument Society of America, 1975.

changes]. Transducers may contain additional elements, such as those used for integral signal conditioning or excitation-power conditioning, or for limiting or conditioning the measurand in a known manner.

Transducers (and sensors) can be classified in two major categories: those that require an external electrical voltage or current (**excitation**) for their proper operation (**non-self-generating** or **passive** transducers) and those capable of providing an output signal without applied excitation (**self-generating** transducers). This classification is determined by the **transduction principle** employed.

12.9.2   Nomenclature

The nomenclature used to describe a transducer or sensor is based primarily on the measurand. Thus we classify transducers as, for example, acceleration transducers, force transducers, pressure transducers, torque transducers, light sensors, or nuclear radiation sensors (detectors). Secondarily, a transducer is described by its electrical transduction principle. Hence a transducer is described more completely as, for example, a piezoelectric acceleration transducer, a potentiometric pressure transducer, a strain-gage torque transducer, a photovoltaic light sensor, or an ionizing nuclear radiation transducer. An additional modifier can be added to restrict the measurand, for example, differential pressure, angular speed, or ultraviolet light, with the following serving as corresponding examples: strain-gage differential-pressure transducer, electromagnetic angular-speed transducer, and photoconductive ultraviolet-light sensor. Another modifier may be added to the nomenclature to denote the type of sensing element used, or special features or provisions. Examples including such modifiers are: toothed-rotor electromagnetic angular-speed transducer, dual-output reluctive pressure transducer, triaxial piezoelectric linear acceleration transducer, and rotary transformer strain-gage torque transducer.

For any given application the nomenclature must also include the measuring range of the transducer (i.e., the upper and lower limits of measurand values intended to be measured by the transducer). Range (and appropriate units) complete the nomenclature, for example, "0 to 2.5 cm potentiometric displacement transducer," "−50 to +150°C resistive temperature transducer," or "±50g triaxial, amplifying, piezoelectric acceleration transducer."

A number of special terms have been applied to transducers for specific measurands. These include "rate gyro" for gyroscopic attitude rate sensor, "accelerometer" for acceleration transducer, "load cell" for force transducer, "tachometer" for angular-speed transducer, "strain gage" for resistive strain sensor, "photodetector" for light sensor, and "flowmeter" for flow-rate transducer. Most of these terms are well understood by workers in specific fields. The use of some other terms, such as "pickup" or "gage" for transducers in general and "cell" for transducers other than force transducers (load cells), should be avoided.

12.9.3   Transduction Principles

Self-Generating Transduction
Piezoelectric transduction elements [Fig. 12.98(a)] convert measurand changes into changes in the electrostatic charge (Q) or voltage (E) generated by certain crystals when mechanically stressed (**piezoelectric effect**). The stress is developed by having the sensing element apply either tension and compression forces, shear forces, or bending forces to the crystal. Some natural crystals, notably quartz, exhibit this effect. Ceramic crystals, which are artificially polarized after the ceramic mixture is baked, can also exhibit this effect. The electrodes are typically metallic film deposited or fired onto the appropriate crystal surfaces. Piezoelectric elements are characterized by a very high output impedance.

Photovoltaic transduction elements convert measurand changes into changes of an electromotive force (voltage) generated when a junction between certain dissimilar materials is illuminated (**photovoltaic effect**) [see Fig. 12.98(b)]. The two materials can both be semiconductors, or one of them can be a metal. Most frequently the junction is a *pn* junction, a junction between *n* and *p* semiconductor materials. The junction is referred to as a **homojunction** when both semiconductors are of the same basic material and as a **heterojunction** when they are not. When used as light sensors, photovoltaic elements respond directly to the measurand. When used in sensors for other measurands (e.g., displacement, angular speed) a sensing element is used additionally.

Electromagnetic transduction elements convert measurand changes into changes of the electromotive force induced in a conductor by a change in magnetic flux. Such transduction elements are made up, typically, of a stationary electromagnet and one or more portions of ferromagnetic material that pass by the pole piece of the electromagnet in close proximity so that the flux of the moving portions interacts with (cuts) the flux around the electromagnet. The principle is illustrated in Fig. 12.98(c).

Thermoelectric transduction elements convert measurand changes into changes of the electromotive force that is generated by a temperature difference between the junctions of two selected dissimilar conductive materials [Fig. 12.98(d)]. The effect employed by such elements is the **Seebeck**

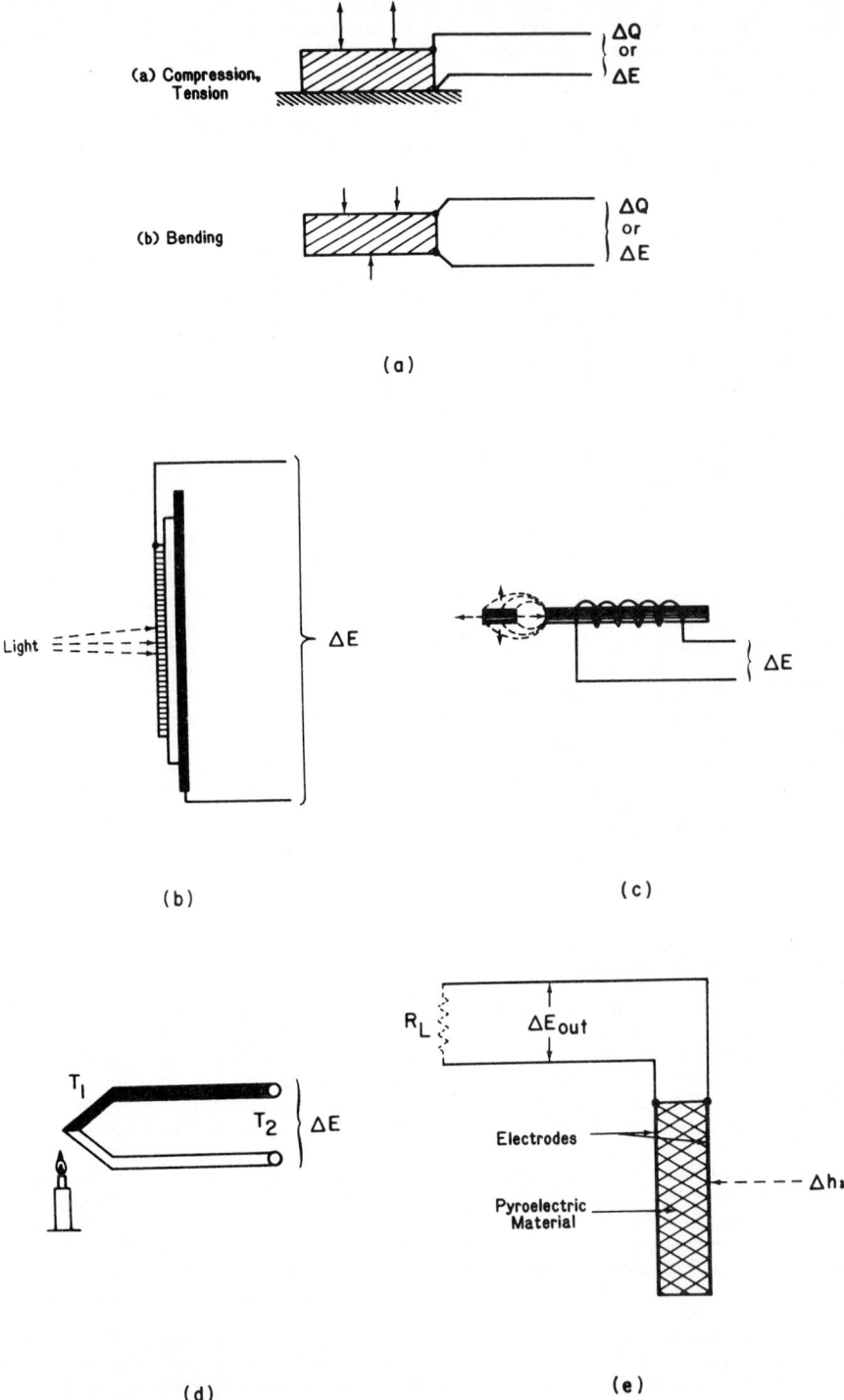

Fig. 12.98 Self-generating transduction system elements: (*a*) piezoelectric transduction, (*b*) photovoltaic transduction, (*c*) electromagnetic transduction, (*d*) pyroelectric transduction. (From Harry N. Norton, *Sensor and Analyzer Handbook*, © 1982, pp. 20–24. Reprinted by permission of Prentice-Hall, Inc., Englewood Cliffs, New Jersey.)

effect. This type of element is commonly used for temperature measurement and is referred to as **thermocouple**, or, when multiple junctions are used, as **thermopile**.

Pyroelectric transduction elements convert changes in the measurand into changes in the potential across the surfaces of certain materials, due to charge separation, caused by heating. As illustrated in Fig. 12.98(e), changes in the output voltage are generated by changes in incident photon energy ($h\nu$), usually in the infrared portion of the electromagnetic spectrum. The **pyroelectric effect** was first observed when piezoelectric crystals were found to produce spurious signals when exposed to temperature transients.

Non-Self-Generating Transduction

Passive transduction elements (see Fig. 12.99) require a source of excitation current or voltage (E_X) for their proper operation. Many of these transduction elements are used in conjunction with a mechanical sensing element.

Capacitive transduction [Fig. 12.99(a)] is the conversion of measurand changes into changes of capacitance. A capacitor consists of two electrodes (**plates**) separated by an insulator (**dielectric**). The capacitance changes can, therefore, be produced either by changing the proximity between the two plates (or two sets of plates), or by changing the dielectric between two stationary plates (different insulators have different dielectric constants). The motion of one plate can be caused by the motion of, for example, the shaft of a displacement transducer, of a seismic mass in an accelerometer, or of a pressure-sensing diaphragm or Bourdon tube. The change in dielectric can be effected by, for example, replacing air between the plates with liquid (as is done in some liquid-level sensor designs) or by moving an insulating sleeve (with a dielectric constant different from that of air) between the plates. The capacitance changes can be converted into ac voltage changes by connecting the element into an impedance bridge. They can also be converted into frequency changes (of an ac voltage) by using the capacitive element as the frequency-controlling element in an oscillator circuit. Either form of signal can then be converted into a dc voltage change if required.

Inductive transduction is the conversion of measurand changes into changes of the self-inductance of a single coil [Fig. 12.99(b)]. This can be effected by moving the ferromagnetic core in and out of a coil winding. It can also be caused by setting up eddy currents in a metallic material brought into the proximity of the pole piece of a winding with a stationary core. The inductance changes can be converted into voltage changes by connecting the element as one arm of an impedance bridge, usually an inductance bridge having another arm formed by a coil having the same characteristics as the transduction coil, and often exposed to the same temperature by close physical colocation, but not affected by the measurand.

Reluctive transduction is the conversion of measurand changes into ac voltage changes by a change in the reluctance path between two or more coils or separated portions of one coil, when ac excitation is applied to the coil(s). Included among reluctive transducers are those employing differential-transformer, inductance-bridge [see Fig. 12.99(c)], and synchro elements as well as elements in which the reluctance path is changed by other means such as stress acting on a ferromagnetic material. The differential transformer is the most commonly used type of reluctive transduction element, and a variety of winding configurations and connection schemes are used besides those shown in the illustration. In its typical operation, the coupling from the primary coil to one of the differential transformer's secondary coils increases, while it decreases to the other secondary coil. The coils are so configured that the output voltage is zero when the moving ferromagnetic core is centered. Core displacement then results in amplitude as well as phase changes in the output voltage. The core displacement can be rectilinear (linear variable differential transformer, LVDT) or angular (rotary variable differential transformer, RVDT).

Resistive transduction [Fig. 12.99(d)] is the conversion of measurand changes into resistance changes. The resistance changes are then converted into current or voltage changes by connecting the transduction element into a suitable circuit (e.g., voltage-divider, Wheatstone bridge) to which excitation is supplied. The resistance changes occur in a number of different ways, and different effects are employed in conjunction with different measurands, as indicated in the illustration. When a conductor is heated or cooled its resistance will change; the magnitude of the change is governed by the resistivity of the conductor and its temperature coefficient of resistance. The same holds true for certain semiconductors whose temperature coefficient, however, can be negative or positive, whereas it is normally only positive for conductors. This effect, which can be called **thermoresistive**, is widely employed in temperature sensors using either a conductor (e.g., platinum wire or film) or a semiconductor (e.g., thermistor, doped silicon, or germanium).

Another form of resistive element is one employing the **piezoresistive** effect, the change in resistance of a conductor or semiconductor due to mechanical strain. The most common use of this effect is in the **strain gage**, a thin conductor (metal wire, foil, or film) or semiconductor that is attached to a surface undergoing strain, so that the strain gage changes its resistance with elongation and contraction. A simple form of resistive element, no longer in common use, is the **rheostat**; the motion derived from a sensing element moves a sliding contact over a resistance winding, thereby

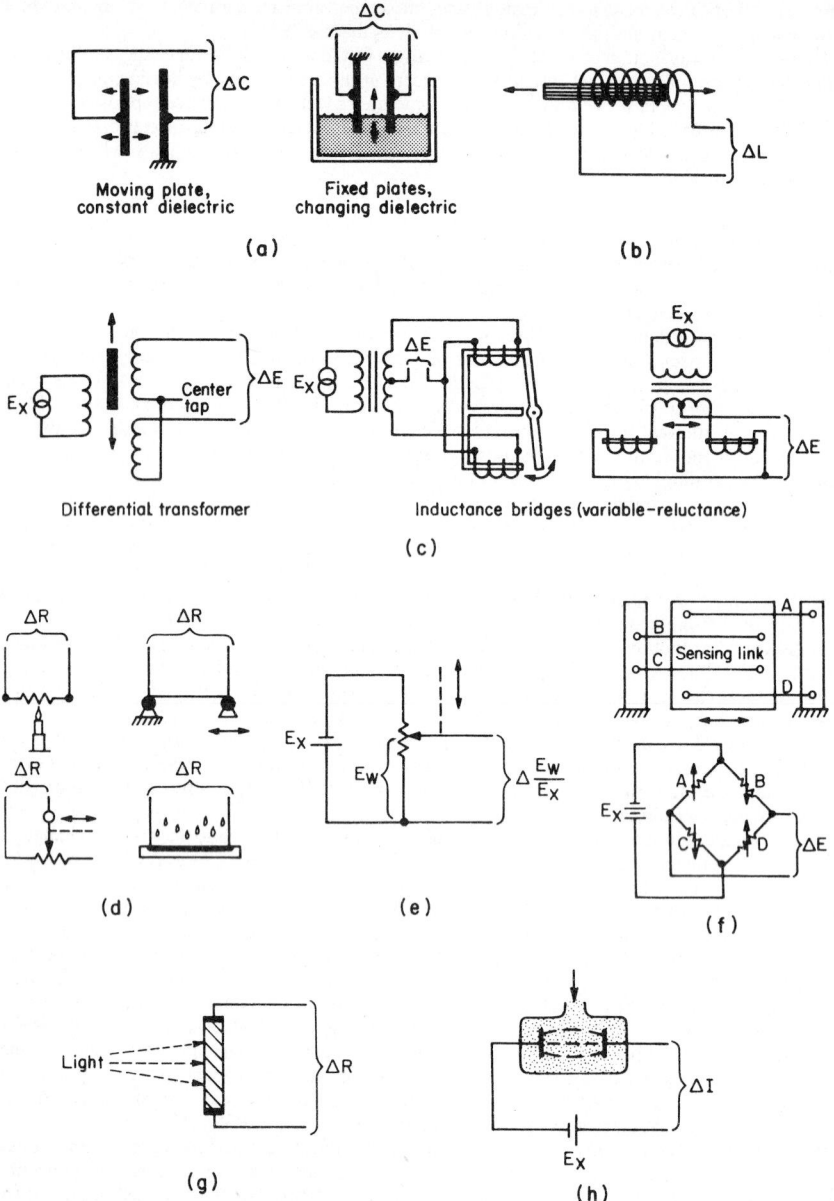

Fig. 12.99 Passive transduction elements: (*a*) capacitive transduction, (*b*) inductive transduction, (*c*) reluctive transduction, (*d*) resistive transduction, (*e*) potentiometric transduction, (*f*) strain-gage transduction, (*g*) photoconductive transduction, (*h*) ionizing transduction. (From Harry N. Norton, *Sensor and Analyzer Handbook*, © 1982, pp. 20–24. Reprinted by permission of Prentice-Hall, Inc., Englewood Cliffs, New Jersey.)

changing the resistance between the terminal of the winding and the sliding contact. Another form of resistive element is one containing a hygroscopic salt (or carbon powder), which changes its resistance with humidity.

Potentiometric transduction is the conversion of measurand changes into voltage-ratio changes [Fig. 12.99(*e*)] by a change in the position of a movable contact on a resistance element across which excitation is applied. The resistance elements are usually made by winding thin insulated wire on a coil form or mandrel, then removing the insulation along the portion to be contacted by the sliding contact (wiper); they can also be made of metal film or conductive plastic. The motion of a mechanical sensing element, linked to the wiper, causes wiper motion along the resistance element. Curved, annular, and sometimes helical resistance elements are used when the motion is rotary rather than rectilinear.

Strain-gage transduction is the conversion of a changing measurand into a voltage change due to the resistance changes of strain gages connected into a Wheatstone bridge (**strain-gage bridge**) circuit across which excitation is applied. A bridge circuit with four active arms (four strain gages subjected to mechanical strain) is shown in Fig. 12.99(*f*). Sometimes only two arms are active; one-active-arm bridges are rare. Strain variations in the mechanical member to which the strain gages are attached (or of which they are part) are caused by the action of a mechanical sensing element.

Photoconductive transduction (really a form of resistive transduction) is the conversion of measurand changes into changes of the conductance of a semiconductor material due to changes in the amount of illumination incident upon the material. Photoconductors are widely used as light sensors [Fig. 12.99(*g*)] but are also used, in conjunction with various sensing elements, for measurands other than light intensity.

Ionizing transduction is the conversion of measurand changes into a change in ionization current [Fig. 12.99(*h*)], such as through a gas between two electrodes. Among the means for changing an ionization current are changing the density of a gas (air, in the case of some types of vacuum sensors) or by the production of ions by the action of charged particles; the latter can occur in a sealed volume of gas, in certain crystals, or in semiconductors.

12.9.4 Sensing Elements and Their Uses in Transducers and Sensors

Sensing elements are combined with transduction elements primarily in transducers and sensors for the measurement of mechanical quantities (quantities in solid and fluid mechanics). Most sensing elements produce either a displacement (change in position) or a deformation resulting in stress, either of which can then be converted into output signals by a transduction element. In transducers for most mechanical measurands it is usually possible to combine a given sensing element with any of several types of transduction elements. However, some of these combinations are more practical or easier to produce than others, or they lend themselves better to transducers for some measurement applications than to others. Commonly used sensing elements and their typical use in transducers and sensors will be described.

Acceleration, Shock, and Vibration

These quantities are measured by **accelerometers**. The sensing element of an accelerometer is a **seismic mass**, typically a metal disk or bar, restrained by a spring. As acceleration forces act on the mass the mass displaces and is returned to its original position by the spring (**spring-mass system**). Instead of a measurable displacement, the mass can also apply mechanical stress to a transduction element. The transduction element itself (e.g., a piezoelectric crystal) can provide the function of a spring. Piezoelectric accelerometers are widely used for shock and vibration measurements where their high natural frequency provides a flat frequency response into the tens of kilohertz. However, such accelerometers cannot be used for steady-state accelerations. Some semiconductor-strain-gage accelerometers, which do have "dc response," also have fairly high natural frequencies. The most accurate transducers, among commonly used accelerometers, are those operating in a servo mode. **Servo accelerometers** employ closed-loop control to counteract the effects of the acceleration force. Their output is a function of the restoring force needed to attain a balance condition.

Attitude

Attitude transducers employ a **gyroscope** as sensing element when they are inertially referenced (**free gyro**) and some form of **pendulum** when they are gravity referenced. A **rate gyro** is a sensor that provides an output proportional to the time rate of change of attitude.

Displacement and Position

The sensing element of a displacement transducer, used to determine position or change in position, is a **shaft** with a suitable coupling. The shaft is attached to the movable portion of a transduction element, such as a wiper in a potentiometric element or a core in a reluctive (e.g., differential transformer) element. The motion can be linear or angular (rotary). The shaft can also move a strip or disk, carrying a conductive/nonconductive, magnetic/nonmagnetic, or transparent/opaque code pattern, which, by means of an appropriate reading head, provides a digital representation of displacement in the transducer's output (**linear** or **angular encoder**).

Force and Mass

Force transducers, which are also used for mass determinations ("weighing") in the presence of gravity, employ **bending beams**, **proving rings** or **frames**, or **columns** as sensing elements. The force to be measured is applied either to the free end of a cantilever beam or to the center of a beam supported at both ends, or to the top of a proving ring or frame (supported on the bottom), or to the top of a solid or hollow column. The applied force is usually sensed as stress induced in the sensing element; hence strain-gage transduction is most frequently used. Piezoelectric transduction, with the crystal responding to force-induced stresses, is used when high-frequency response is required.

Torque

The sensing element of torque transducers is the **torsion bar**, a thinned-down section of a rotating shaft. One end of the torsion bar deflects angularly with respect to the opposite end when a torque is applied. Torsion bars can be round or square, or machined into special variants of the square configuration that is preferred for strain-gage transduction, which utilizes stresses induced in the torsion bar in response to an applied torque. Photoelectric torque sensors utilize the (very small) angular displacement by having a coded disk attached to each end of the torsion bar and sensing the changes in shading between the two disks, through which a light beam is passed, as the mutual alignment between two transparent/opaque code patterns changes.

Speed and Velocity

Speed is a scalar quantity (magnitude only), whereas velocity is a vector quantity (magnitude and direction). Linear-velocity transducers are usually of the electromagnetic type (self-generating), are used to measure oscillatory velocity, and employ either a **moving magnet** as the sensing element in conjunction with a stationary coil or a **moving coil** in conjunction with a stationary magnet.

Angular speed (or velocity, if the output also indicates direction) transducers are commonly referred to as **tachometers**. Electric generators (dc tachometer generator, ac induction tachometer, ac permanent-magnet tachometer) provide an analog output proportional to rotary speed. A frequency output (which can easily be converted into a digital signal) is provided by electromagnetic and photoelectric tachometers. One or more teeth of ferromagnetic material on a rotor, attached to the shaft whose rotary speed is to be measured, induce a pulse in an electromagnetic transduction element whenever the passing of a tooth cuts the electromagnet's flux. Hence n pulses are obtained per revolution, where n is the number of teeth. Similarly, a disk coded with an alternatingly transparent/opaque pattern can be used to interrupt a light beam passing through the disk or one or more reflecting spots on the rotating member can reflect a light beam toward a light sensor.

Stress and Strain

Strain is measured (and stress can be calculated from the strain measurement) by resistive sensors called strain gages. Semiconductor strain gages exist besides the more commonly used metal-foil gages (which are usually produced by photoetching techniques); metal-wire gages are also still in use. A typical metal-foil gage is shown in Fig. 12.100. Many other patterns are available besides the basic pattern illustrated. They include multiple gages (rosettes) used for simultaneous strain measurements in two or more directions. A strain gage becomes a strain sensor only when it is properly bonded to the measured surface. Strain gages can be very small, down to less than 2×1.5 mm in overall size.

Flow

Although sensing elements such as a spring-restrained plug against a shaped constriction or a deflecting vane in the flow stream have been used in past designs, the use of mechanical flow-sensing elements is now essentially limited to the **turbine**, a bladed rotor that turns at a speed nominally proportional to the volume rate of flow. The turbine operates in conjunction with an electromagnetic, sometimes a photoelectric transduction element, in a manner similar to that described for tachometers. Other flowmeters use transduction elements in specially adapted forms. **Thermal** flowmeters utilize either the changes in heat transfer from a heat source to a temperature sensor by the measured fluid, or, as in the **hot-wire** or hot-film **anemometer** (the term applies when gas is the measured fluid), where the amount of current needed to maintain the temperature of a resistive element, cooled in proportion to flow rate, is a function of flow. **Magnetic** flowmeters use the interaction of a moving conductive liquid (many liquids, including water, are more or less conductive) with a magnetic field to provide an output. This is a form of electromagnetic transduction in which the fluid acts as the moving conductor. Another design sets the fluid into a swirling motion and detects the frequency of oscillation, which is proportional to flow, by a transduction element such as a rapidly responding temperature sensor. **Ultrasonic** techniques have also been applied to flow measurement; the difference in the propagation velocity of a pulse of ultrasound is indicative of flow rate.

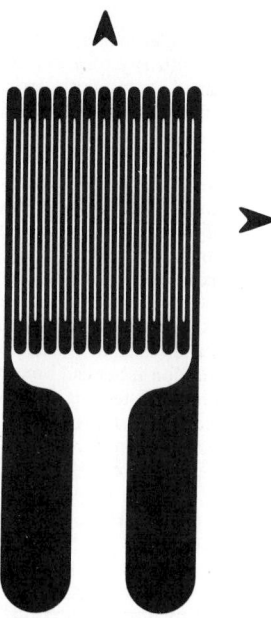

Fig. 12.100 Typical metal-foil strain gage.

Humidity and Moisture

Sensors for these quantities fall into one of three categories: hygrometers, psychrometers, and dew-point sensors. **Hygrometers** measure humidity directly. **Psychrometers** use the comparison between readings of a temperature sensor ("dry bulb") with another temperature sensor wrapped in a water-saturated wick or sleeve ("wet bulb") to determine humidity by referring to a table. **Dew-point sensors** measure the temperature of a surface, which is artificially cooled, at the instant when moisture is first precipitated on it. Resistive hygrometers employ a film of a hygroscopic salt or carbon powder, which changes its resistance with humidity. Other sensing/transduction elements include the aluminum oxide element, which changes its impedance with the vapor pressure of water in the measured fluid. Psychrometric humidity sensors employ two resistive (conductive or semiconductive) temperature sensors. Dew-point sensors measure the temperature of the cooled surface, typically with a resistive sensor, while simultaneously determining (e.g., photoelectrically) the point at which condensation occurs on the surface.

Liquid Level

Liquid-level measurements, including the level of slurries and granular solids, in tanks and vessels, are of two types: continuous measurement or point sensing. The latter refers to one or more discrete determinations of level. A variety of sensing elements and transduction methods are used. The only commonly used mechanical sensing element is a float, which rides on the surface of the liquid and whose up-and-down motion actuates a transduction element either continuously (e.g., potentiometric) or at a point (e.g., a magnetic switch). Other sensing methods employ sensing/transduction elements. Among these is the capacitive element, typically two concentric hollow cylinders, each one acting as an electrode, with the rising and falling liquid acting as variable dielectric. Resistive transduction utilizes the resistance path formed by a conductive liquid between two electrodes, or between one electrode and the tank wall. Another form is the hot-wire element, used as a point sensor, which is cooled when it comes in contact with the liquid. Electrooptical point sensors use either the reflectance method, where a change in the index of refraction due to presence or absence of liquid permits or stops light from being reflected to a light sensor, or the transmittance method, where the presence of liquid between light source and sensor changes the amount of light sensed. Ultrasonic methods are employed for continuous as well as point sensing. A popular method employs a differential-pressure transducer to measure the pressure difference between the top and the bottom of a tank, which varies with the height of liquid (head) above the bottom. Oscillating elements, whose oscillation is damped by the presence of liquid, are also used as point sensors.

Pressure

The most commonly used pressure-sensing elements are the diaphragm, capsule, Bourdon tube, and straight tube. These can be combined with any of a number of different transduction elements. A **diaphragm** is a thin circular plate fastened continuously around its edge. It can be flat or corrugated (with axially concentric corrugations that increase the effective area and hence the deflection of its center). Pressure is applied to the outside of the diaphragm. The inside, where deflection is transduced, is either evacuated and sealed (in absolute-pressure transducers), or vented to ambient pressure (in gage-pressure transducers), or ported to another source of pressure (in differential-pressure transducers, when the other fluid cannot harm the transduction element).

A **capsule** ("aneroid") consists of two corrugated diaphragms, formed into shells of opposite curvature and sealed to each other around their edge. When pressure is applied into the capsule, the capsule expands and provides an axial displacement. Two or more capsules can be used in a transducer to multiply the deflection. The action of multiple capsules is similar to that provided by **bellows**, which are still sometimes used as pressure-sensing elements. A **Bourdon tube** is a curved or twisted tube, elliptical in cross section and sealed at one end. When pressure is applied into the tube it will tend to straighten (or untwist). Bourdon tubes can be curved into a C or U shape, or they can be wound into a spiral or into a helix to increase the angular displacement of their sealed tip. A **straight tube** is a short cylinder, sealed at one end. When pressure is applied into the tube, the tube walls are stressed (hoop stress). The stress can be transduced by strain gages or, if the tube is set into oscillation, will change the resonant frequency of the tube, and such changes can be used to produce a frequency-output signal.

Capacitive, inductive, reluctive, potentiometric, strain-gage, and piezoelectric transduction are used, to varying degrees, in conjunction with these sensing elements. Closed-loop null-balance or force-balance servo operation is used in some designs. Piezoelectric and some types of capacitive pressure transducers are usable for **sound-pressure** (acoustic) measurements and are then often referred to as "microphones" or, when used underwater, as "hydrophones."

Vacuum

Vacuum sensors are used to measure pressures substantially below 1 standard atmosphere. Although one capacitive pressure transducer design is usable for medium-range vacuum measurements, vacuum sensors are generally quite different from pressure transducers. The **Pirani gage** is a resistive sensor employing a heated wire whose heat is reduced in proportion to heat conducted away by residual gas in a chamber. The resistance change in the heated filament is indicative of vacuum. A similar device is the **thermocouple gage**, which uses a thermocouple to measure the changes in filament temperature due to changes in vacuum. The **ionization gage** is similar in construction to a triode-type vacuum tube. The residual gas within the (usually glass) envelope is ionized, positive ions are collected at the anode, and the ionization current is a measure of vacuum. The addition of a magnetic field to an ionizing vacuum sensor, such as the **hot-cathode magnetron gage** (Lafferty gage) or the **cold-cathode magnetron gage** (Philips gage or Penning gage), greatly extends the measuring range into the high and ultrahigh vacuum (a "high" vacuum is a very low pressure).

Temperature

Temperature sensors use either thermoelectric sensing/transduction elements (thermocouples, thermopiles; see Section 12.8.4) or resistive sensing/transduction elements (resistance thermometers; see Section 12.8.4). They can be configured as immersion probes or surface sensors and a number of subcategories of these configurations are used. Temperature can also be sensed remotely (noncontacting) by **radiation pyrometers**, whose operation is usually based on Planck's radiation formula.

12.10   INSTRUMENTATION AMPLIFIERS
Yusuf Ziya Efe

12.10.1   Operational Amplifier

Operational amplifiers were introduced in the early 1940s. They were designed using bulky, power-hungry vacuum tubes. They were used primarily to accomplish addition, subtraction, and other mathematical operations, hence the name operational amplifier (op-amp). The use of operational amplifiers began to expand toward the present spectrum of applications in the early 1960s, as monolithic integrated circuits (ICs) developed. Op-amps have gained wide acceptance as versatile, predictable, and reliable inexpensive building blocks.

The op-amp is a direct-coupled high-gain amplifier, which also has provision for external feedback. Through the external feedback circuitry the response of the amplifier can be controlled virtually independent of its internal parameters. In this respect, a detailed analysis of the internal

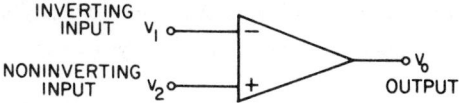

Fig. 12.101 Symbol for an operational amplifier.

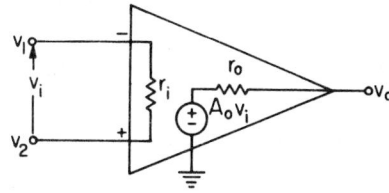

Fig. 12.102 Equivalent circuit for an op-amp.

circuitry of an op-amp is usually ignored. Figure 12.101 is widely used to represent a conventional op-amp.

Input terminals, labeled $(-)$ and $(+)$, are usually called **inverting** and **noninverting** terminals, respectively. These terminals are also called **differential input terminals** because output voltage v_o depends on the difference in voltage between them. That is,

$$v_o = A_o(v_2 - v_1) \qquad (12.91)$$

where A_o is the open-loop voltage gain of the op-amp. In some cases the noninverting input is not externally available, but it is preferable for a differential input to be provided.

The equivalent-circuit model of an op-amp consists of an input resistance r_i connected between the two input terminals, and the output circuit consists of a controlled source $A_o v_i$ in series with an output resistance r_o, as shown in Fig. 12.102.

The voltage gain A_o of the op-amp is usually very large; typically, $A_o > 100{,}000$. The input resistance is much larger than 100 kΩ. The output resistance, on the other hand, is about 100 Ω or less and it may be neglected for many applications. It is important to note that these quantities are defined at dc operations.

12.10.2 Ideal Operational Amplifier

The ideal op-amp possesses the following characteristics:

1. The open-loop voltage gain is infinite: $A_o = \infty$.
2. The input resistance is infinite: $r_i = \infty$.
3. The output resistance is zero: $r_o = 0$.
4. The bandwidth is infinite: $BW = \infty$.
5. The input offset voltage is zero; that is, $v_o = 0$ if $v_i = v_2 - v_1 = 0$.
6. The output voltage, $v_o = 0$, when the input voltage, $v_i = v_2 - v_1 = 0$. This means that the offset voltage is zero.
7. Insensitivity of op-amp parameters to temperature and power supply variations.

From these ideal characteristics, two very important additional properties can be deduced.

8. Since the input resistance is infinite, there will be no current flowing into the amplifier input terminals.
9. When feedback is employed, the differential input voltage reduces to zero.

None of these ideal parameters have been achieved by any op-amp, although manufacturers of op-amps continually improve their products. However, many of these parameters are sufficiently close to the ideal that the difference is nearly negligible in most practical applications. For example, input bias currents are in the range of 5 pA for FET input amplifiers, while input resistances are guaranteed a minimum of 10^{12} Ω. Offset voltage has been reduced to less than 1 mV in many cases.

12.10.3 Basic Operational Amplifier Configurations

Operational amplifiers can be connected in three basic amplifying circuits:

1. The inverting configuration
2. The noninverting configuration
3. The differential amplifier configuration

Practically all other op-amp circuits are based in some way on these three basic configurations.

Inverter

Figure 12.103 shows the basic inverting amplifier configuration. Observe that the signal is applied to the $(-)$ inverting terminal through R_1. The output is fed back to this input terminal (feedback) through R_2.

Applying Kirchhoff's current law (KCL) at the node v_x,

$$\frac{v_i - v_x}{R_1} + \frac{v_o - v_x}{R_2} = i_b \qquad (12.92)$$

Since the input resistance of an ideal op-amp is infinite, the current flowing into the op-amp is zero, $i_b = 0$. For a practical op-amp the input resistance is very large; typically, it is larger than 100 kΩ. Hence the current flowing into the amplifier would almost always be approximated to zero (i.e., $i_b \approx 0$). Thus $i_1 \approx i_2$. This means that the voltage across the input resistance r_i in Fig. 12.102 is zero or can be approximately equal to zero. That is, the $(-)$ input terminal is considered to be internally connected to ground. It is for this reason that the amplifier configuration in Fig. 12.103 is said to have a **virtual ground** at its input. With these approximations Eq. (12.92) becomes

$$\frac{v_i}{R_1} + \frac{v_2}{R_2} \approx 0 \qquad (12.93)$$

Thus the voltage gain with feedback, called **closed-loop gain**, is found to be

$$A_{cl} = \frac{v_o}{v_i} = -\frac{R_2}{R_1} \qquad (12.94)$$

Note that the closed-loop gain A_{cl} of the op-amp circuit depends only on the external passive components R_1 and R_2. The negative sign in Eq. (12.94) indicates the inverting property of the circuit. It is also important to observe that the input resistance seen by the signal source v_i is

$$R_i = \frac{v_i}{i_1} = R_1 \qquad (12.95)$$

because $v_x = 0$ for an ideal op-amp.

Practical Inverting Op-Amp

1. *Gain.* Equation (12.94) is valid only if the voltage gain of the op-amp is infinite (i.e., the op-amp is ideal). In a practical op-amp, however, $A_o \neq \infty$, $r_i \neq \infty$, and $r_o \neq 0$. In this case the equivalent circuit for a practical inverting op-amp circuit is as shown in Fig. 12.104. Applying the KCL at the

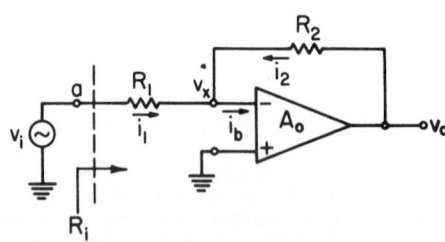

Fig. 12.103 Inverting amplifier.

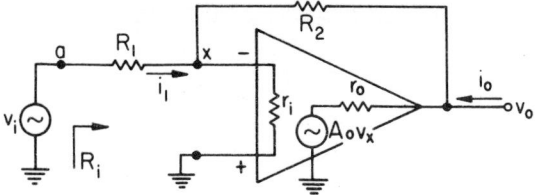

Fig. 12.104 Practical inverting op-amp equivalent circuit.

node x,

$$\frac{v_i - v_x}{R_1} + \frac{v_o - v_x}{R_2} = \frac{v_x}{r_i} \qquad (12.96)$$

Using the fact that $A_o = v_o/v_x$ and $A_{cl} = v_o/v_i$ and manipulating Eq. (12.96), the overall closed-loop voltage gain of the amplifier is found to be

$$A_{cl} = \frac{G_1 A_o}{G_1 + G_2(1 - A_o) + g_i} \qquad (12.97)$$

where $G_1 = 1/R_1$, $G_2 = 1/R_2$, and $g_i = 1/r_i$.

Observe that if $A_o = \infty$, Eq. (12.97) reduces to (12.94). Also note that A_{cl} is not only a function of R_1 and R_2, but is also influenced by the input resistance r_i and the open-loop gain A_o.

2. *Input and Output Resistances.* The input resistance between the terminal labeled a and the ground is found to be

$$R_i = R_1 + r_i \| R_t \qquad (12.98)$$

where

$$R_t = \frac{R_2 + r_o}{1 + A_o}$$

The output resistance R_o seen between the output terminal and ground is found to be

$$R_o = \frac{v_o}{i_o} = 1 \bigg/ \left[\frac{1 + R_1 A_o/(R_1 + R_2)}{r_o} + \frac{1}{R_1 + R_2} \right] \qquad (12.99)$$

Note that when $A_o = \infty$ or $r_o = 0$, the output resistance reduces to zero, as expected.

Noninverter

Figure 12.105 illustrates a noninverting op-amp configuration. The input voltage is applied to the (+) input terminal. A fraction of the output voltage is applied to the (−) input terminal through the voltage divider formed by R_1 and R_2. Since no input current i_b flows into either input terminal ($r_i = \infty$ for the ideal op-amp) and since $v_i - v_x = 0$,

$$v_i = v_x = \frac{R_1}{R_1 + R_2} v_o \qquad \text{(voltage divider)}$$

and

$$A_{cl} = \frac{v_o}{v_i} = 1 + \frac{R_2}{R_1} \qquad (12.100)$$

Hence the overall voltage gain A_{cl} of a noninverting op-amp must always be greater than or equal to unity, $1 \le A_{cl} < \infty$. $A_{cl} = 1$ when $R_2 = 0$. Under this condition R_1 may be deleted, and the circuit is called a **voltage follower**.

The input resistance of the noninverting amplifier is $R_i = v_i/i_i$. Since $i_i = 0$ for an ideal op-amp, $R_i = \infty$.

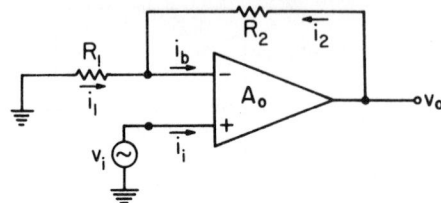

Fig. 12.105   Noninverting op-amp configuration.

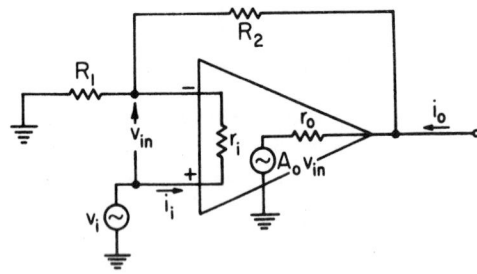

Fig. 12.106   Practical noninverting op-amp equivalent circuit.

Practical Noninverting Op Amp

1. *Gain.* Equation (12.100) is valid only if the op-amp is ideal. In a physical op-amp, however, $A_o \neq \infty$, and $r_o \neq 0$. In this case the equivalent circuit becomes as shown in Fig. 12.106.

The closed-loop gain of the real op-amp is $A_{cl} = v_o/v_i$. This, in most cases, is very close to that given in Eq. (12.100).

2. *Input and Output Resistances.* The input resistance defined as $R_i = v_i/i_i$ is found to be

$$R_i = \frac{A_o r_i R_1}{R_1 + R_2} \tag{12.101}$$

Note that as r_i or A_o approach infinity, $R_i \to \infty$ as expected.

The output resistance R_o defined by $R_o = v_o/i_o$ is found to be

$$R_o = 1 \Bigg/ \left[\frac{R_1 + R_2 + r_i A_o}{r_o(R_1 + R_2)} + \frac{1}{R_1 + R_2} \right] \tag{12.102}$$

Note that when A_o or r_i approach infinity, $R_o \to 0$, as expected.

Differential Amplifier

This is a combination of the two previous configurations. Note from Fig. 12.107 that the amplifier has signals applied to both input terminals.

Using superposition, let the output component due to v_1 be v_{o1}, and due to v_2 be v_{o2}. Using the gain equation for the inverting amplifier, the output v_{o1} will be

$$v_{o1} = -\frac{R_{12}}{R_{11}} v_1 \tag{12.103}$$

Using the gain equation for the noninverting amplifier, the output v_{o2} will be

$$v_{o2} = +\frac{R_{22}}{R_{21}} v_2 \tag{12.104}$$

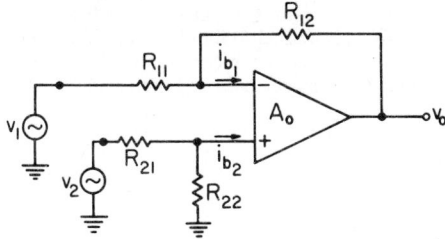

Fig. 12.107   Differential amplifier configuration.

Since the total output will be equal to the sum of v_{o1} and v_{o2},

$$v_o = v_{o1} + v_{o2} = \frac{R_{22}}{R_{21}} v_2 - \frac{R_{12}}{R_{11}} v_1 \qquad (12.105)$$

For simplicity, let $R_{22} = R_{12} = R_2$ and $R_{21} = R_{11} = R_1$; then from Eq. (12.105), the gain of the differential amplifier is found to be

$$A_{cl} = \frac{v_o}{v_2 - v_1} = \frac{R_2}{R_1} \qquad (12.106)$$

This is the gain of the amplifier for differential mode signals (i.e., $v_1 \neq v_2$). This configuration is unique in the sense that it tends to reject a signal common to both input terminals. This **common-mode rejection** is dominated by the accuracy of the resistors R_1 and R_2. Other errors would arise from the offset voltage, input offset current, input bias current, and so on.

Input Resistance. One can easily show that the input resistances for inverting and noninverting terminals are not the same and are given by

$$R_i \, [\text{for } (-) \text{ input}] = R_{11} \qquad (12.107)$$
$$R_i \, [\text{for } (+) \text{ input}] = R_{21} + R_{22} \qquad (12.108)$$

Ideally, the output of an ideal differential amplifier should be given by

$$v_o = A_d (v_2 - v_1) \qquad (12.109)$$

where A_d is the gain of the differential amplifier given in Eq. (12.106). That is, any signal common to both inputs should have no effect on v_o. However, this is not the case for a practical differential amplifier, as seen from Eq. (12.105). The output signal depends on both **difference signal** v_d and **common-mode signal** v_c defined as

$$v_d = v_2 - v_1 \quad \text{and} \quad v_c = \frac{v_1 + v_2}{2} \qquad (12.110)$$

or

$$v_2 = v_c + \frac{v_d}{2} \quad \text{and} \quad v_1 = v_c - \frac{v_d}{2} \qquad (12.111)$$

Using v_1 and v_2 with

$$A_1 = -\frac{R_{12}}{R_{11}} \quad \text{and} \quad A_2 = \frac{R_{22}}{R_{21}}$$

in Eq. (12.105), we have

$$v_o = A_d v_d + A_c v_c \qquad (12.112)$$

where $A_d = (A_2 - A_1)/2$ is the voltage gain for difference signal and $A_c = A_1 + A_2$ is the gain for the common-mode signal. Ideally, it is desired to have A_d very large and A_c should be zero.

The quantity used to measure how much the common-mode signal is suppressed relative to the input difference voltage is called the **common-mode rejection ratio** (CMRR). It is defined as

$$\text{CMRR} = \left| \frac{A_d}{A_c} \right| \quad \text{or} \quad \text{CMRR}_{\text{dB}} = 20 \log \left| \frac{A_d}{A_c} \right| \tag{12.113}$$

Using Eq. (12.113) in (12.112) yields

$$v_o = A_d v_d \left(1 + \frac{1}{\text{CMRR}} \frac{v_c}{v_d} \right) \tag{12.114}$$

It can be observed from Eq. (12.114) that the amplifier should be designed such that CMRR $\gg$ v_c/v_d to be able to reject common-mode signal effectively. CMRR is an inherent parameter in all op-amps, and it is a function of the internal circuitry. Its value is provided in data sheets stating the minimum or typical values at dc or CMRR versus frequency.

12.10.4 Applications of the Basic Configurations

Applications of the basic configurations are countless. Only several commonly used circuits are discussed here, among them the inverting summing amplifier, audio mixer, and inverting averaging amplifier.

Inverting Summing Amplifier

The amplifier shown in Fig. 12.108 is a variation of the inverting amplifier. The output voltage is proportional to the linear sum of the input voltages. The output voltage can be determined by noting that the feedback causes a virtual ground, $v_{\text{in}} = 0$, across the op-amp input terminals and also noting that $i_b = 0$.

Since the voltage at the node x, $v_x = v_{\text{in}} = 0$, applying the KCL at the node x, we have

$$i_{11} + i_{12} + i_{13} = -i_2$$

or

$$\frac{v_1}{R_{11}} + \frac{v_2}{R_{12}} + \frac{v_3}{R_{13}} = -\frac{v_o}{R_2}$$

or

$$v_o = -\left(\frac{R_2}{R_{11}} v_1 + \frac{R_2}{R_{12}} v_2 + \frac{R_2}{R_{13}} v_3 \right) \tag{12.115}$$

Observe that each input is multiplied by a different scale factor and then added. The gains of the individual channels are scaled independently by R_{11}, R_{12}, R_{13}, and so on. Also, R_{11}, R_{12}, R_{13}, and so on, are the input resistances of the respective channels.

Audio Mixer

Isolation is an important characteristic of the configuration in Fig. 12.108. This is the result of each signal source seeing virtual ground potential at the summing node. Thus the input signals v_1, v_2, and

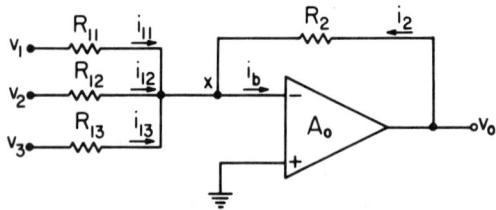

Fig. 12.108 Inverting summing amplifier.

v_3 do not interact. This is a very desirable feature, particularly for an audio mixer. For example, if v_1, v_2, and v_3 were replaced by microphones, the signals from each microphone are added or mixed at every instant. Hence the intensity of each signal can be controlled independently from others by adjusting the resistance in the signal path.

Inverting Averaging Amplifier

The output of an averaging amplifier is proportional to the average of all input voltages. That is, if three input voltages v_1, v_2, and v_3 are applied to the inputs in Fig. 12.108, the averager will add these inputs and divide the sum by 3. The output should be

$$v_o = -\frac{v_1 + v_2 + v_3}{3} \tag{12.116}$$

The negative sign in Eq. (12.116) comes from the fact that the amplifier is an inverting amplifier. To achieve this averaging process, the input resistances R_{11}, R_{12}, and R_{13} and the feedback resistance R_2 must be properly adjusted. For example, to realize Eq. (12.116), the following selection can be made. Let $R_{11} = R_{12} = R_{13} = R$ and $R_2 = R/3$. The reader can justify this by substituting these resistance values in Eq. (12.115).

12.10.5 Operational Amplifier Characteristics and Noise

The performance of an op-amp is generally characterized by a range of performance limits obtained from testing a large number of units. These error-causing characteristics of an op-amp can be divided into two categories:

1. The dc parameters
2. The ac parameters

DC Parameters

The following parameters add error components to the dc response of an op-amp.

Input Offset Voltage, V_{os}. An ideal op-amp produces a zero output voltage for a zero differential input. However, unavoidable imperfections in circuit components used within real amplifiers cause a voltage at the output when the input voltage is zero. This voltage is called the **output offset voltage**. The differential input voltage required between input terminals to obtain zero output is called the **input offset voltage**, V_{os}.

This dc voltage can be conveniently represented by a single dc source located at the input of the op-amp as shown in Fig. 12.109, noting, of course, the proper polarity. The voltage source V_{os} may also be placed in the path of the inverting terminal.

When the op-amp is operated without feedback as in Fig. 12.109(a) with $v_i = 0$, the effective input voltage to the amplifier is V_{os}. This input voltage, however, is usually enough to drive the amplifier into saturation, producing an output voltage close to the power supply voltage. However, if the op-amp is used in a feedback configuration shown in Fig. 12.109(b), the amplifier acts as an inverting amplifier for v_i and as a noninverting amplifier for the offset voltage V_{os}. Thus, using the relations developed previously and applying superposition, the output voltage can be written as

$$v_o = \left(1 + \frac{R_2}{R_1}\right) V_{os} - \frac{R_2}{R_1} v_i \tag{12.117}$$

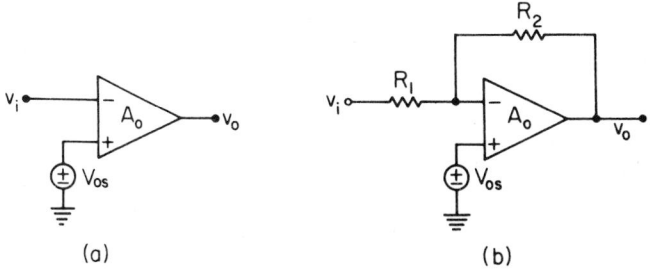

Fig. 12.109 Input offset voltage in op-amp.

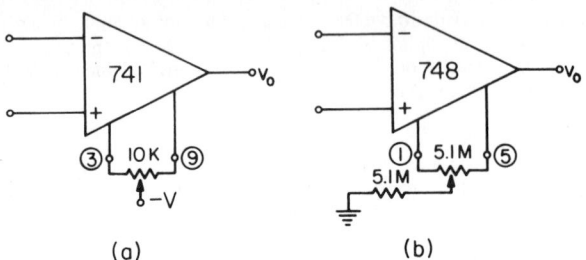

Fig. 12.110 Typical voltage offset nulling circuits. Numbers in circles indicate pin numbers.

When $v_i = 0$, the output offset voltage is found to be

$$v_o = \left(1 + \frac{R_2}{R_1}\right) V_{os} \tag{12.118}$$

where $(1 + R_2/R_1)$ is called **noise gain**.

Under normal operating conditions V_{os} is constant and is independent of the circuit configuration and the amount of feedback used. However, output offset voltage depends on the amount of feedback. Typically, the input offset voltage is less than 5 mV for bipolar-input op-amps and 10 mV or more for FET input op-amps.

In many applications the consequences of the offset voltage may be considered to be negligible. In other applications however, it may cause serious errors (i.e., in calibration of electronic instruments, logarithmic amplifiers, etc.). The effect of the input offset voltage can be reduced by adding a small dc voltage source at the input and adjusting its magnitude and polarity to give a zero output voltage when $v_i = 0$. This process is called **nulling the output offset voltage**. However, in many op-amps special terminals are provided to null the output offset voltage by adjusting an external potentiometer. Some examples are shown in Fig. 12.110.

Although input offset voltage may be nulled at room temperature by using nulling circuits, a change in temperature alters the state of balance, which induces an **input offset voltage drift**. This drift is amplified by a factor of noise gain, that is,

$$\Delta v_o = \left(1 + \frac{R_2}{R_1}\right) \Delta V_{os} \tag{12.119}$$

A technique to minimize the input offset voltage is shown in Fig. 12.111. A null potentiometer is set for zero output when the input is zero. An interesting feature of the circuit is that the performance is relatively unaffected by supply voltage variation. Drift is specified for offset voltage in $\mu V/^\circ C$ and usually ranges from 5 to 50 $\mu V/^\circ C$. It may differ at different temperatures and may even reverse (i.e., it may be either positive or negative). For this reason, manufacturers specify either a maximum drift or an average value for the drift. Sometimes a plot of drift versus temperature is provided.

Universal Balancing Technique. The internal offset nulling techniques presented previously can be used only if the appropriate terminals are provided by the manufacturers. Unfortunately, these methods cannot be applied for all op-amps, simply because offset nulling terminals may not be available to users. The universal external offset nulling techniques for both inverting and noninverting amplifier configurations are shown in Fig. 12.112. Some typical values for resistances are also indicated in the circuits.

The offset voltage adjustment range, OVAR, for the inverting amplifier configuration in Fig. 12.112(a) is

$$\text{OVAR} \simeq \frac{R_4}{R_3} V \tag{12.120}$$

and for the noninverting amplifier configuration in Fig. 12.112(b) is

$$\text{OVAR} = \frac{R_2 R_3}{R_4(R_1 + R_2)} V \tag{12.121}$$

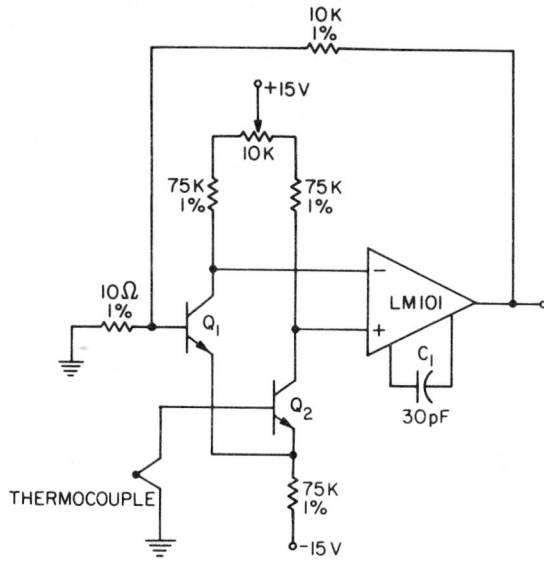

Fig. 12.111 Dc op-amp using the drift compensation technique. (Courtesy of National Semiconductor Corp.)

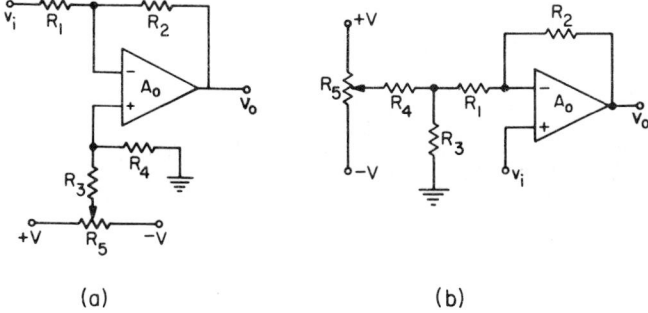

(a)   (b)

Fig. 12.112 Universal offset voltage nulling circuits: (*a*) inverting op-amp, (*b*) noninverting op-amp.

The methods shown here allow offset voltage balancing without regard to the internal circuitry of the op-amp (i.e., they can be used with any op-amp circuits).

Input Bias Current, I_b. Until now it has been assumed that input currents to an op-amp are zero. In reality, however, the input terminals conduct a small value of dc current to bias the internal transistors. These bias currents I_{b_1} and I_{b_2} can be modeled as shown in Fig. 12.113.

By definition, the **input bias current** of an op-amp is the average of two input currents when the output is nulled to zero volts, that is,

$$I_b = \frac{I_{b_1} + I_{b_2}}{2}$$

(12.122)

The typical input bias current ranges from picoamperes to nanoamperes. In general, FET-input amplifiers have lower input bias currents than do bipolar monolithic transistor input op-amps. As with input offset voltage, the input bias current is a dc parameter and may not affect the ac performance of the op-amp.

$$R_b = R_1 \| R_2 = \frac{R_1 R_2}{R_1 + R_2}$$

(12.123)

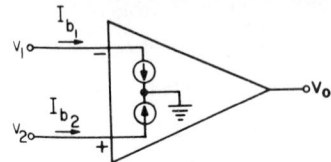

Fig. 12.113 Model for input bias current for op-amp.

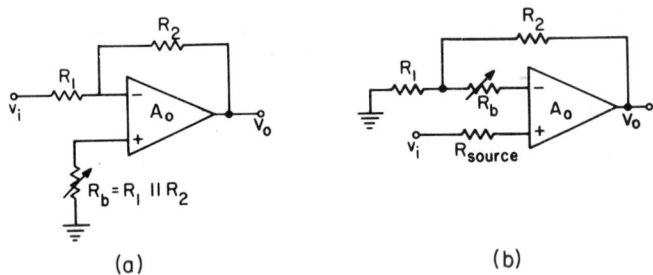

(a)   (b)

Fig. 12.114 Cancellation of effect of input bias current on output: (a) for inverting op-amp, (b) for noninverting op-amp.

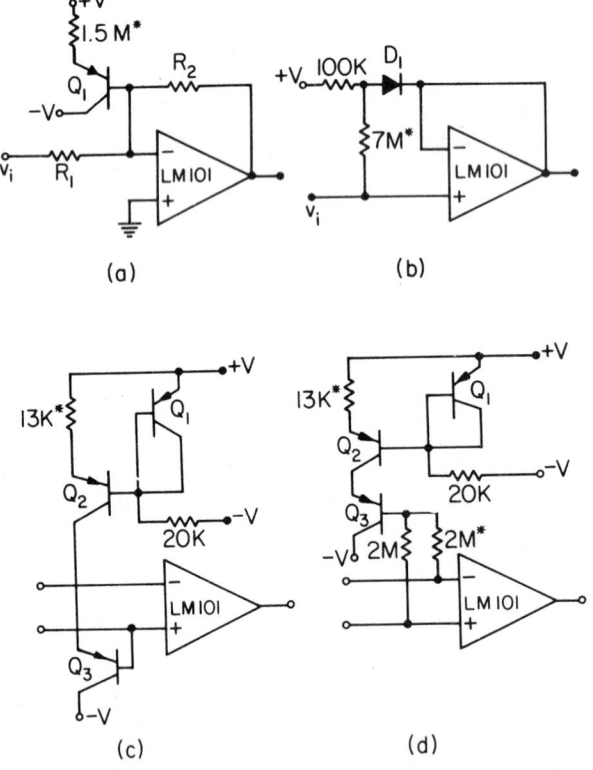

(a)   (b)

(c)   (d)

Fig. 12.115 Canceling the effect of changes in I_b with temperature for various amplifier circuits: (a) inverting amplifier, (b) voltage follower, (c) noninverting amplifier, (d) differential input amplifier. (Courtesy of National Semiconductor Corp.)

The voltage across the resistance R_b will tend to cancel the effect of I_b on the output V_o. The error at the output due to I_b is often larger than that caused by the input offset voltage. This error depends on the size of the resistance R_b. If $R_b > V_{os}/I_b$, the error at the output caused by I_b is larger than that caused by V_{os}.

The circuit in Fig. 12.114(b) is recommended for a noninverting amplifier. The idea behind these two cases in canceling the effect of the input bias current on the output is that the source resistance seen by both op-amp input terminals must be identical.

As with the input offset voltage, the input bias current varies with temperature. However, if the resistances seen by both inputs are made identical, input bias-current changes with temperature may be neglected.

In practice, as recommended by manufacturers, the range of the compensating resistance R_b should have a maximum value of about three times the source resistance. Also, the equivalent parallel resistance of R_1 and R_2 should be less than one-third the input resistance.

In some cases the input resistance of the source is not known. In this case, other compensation schemes must be used. Several of these techniques are shown in Fig. 12.115.

Input Offset Current, I_{os}. The difference in magnitudes between I_{b2} and I_{b1} is called the **input offset current**, I_{os}, in Fig. 12.113:

$$I_{os} = I_{b1} - I_{b2} \tag{12.124}$$

This current is specified by manufacturers in data sheets for a zero output voltage and at room temperature (25°C). The difference in Eq. (12.124) occurs due to the fact that the transistors in the first internal differential amplifier may not be matched properly, requiring different bias currents. The typical I_{os} is less than 25% of input bias current I_b defined in Eq. (12.122).

To minimize error in the output voltage due to offset current, a resistor R_b shown in Fig. 12.114 is connected, which is computed by Eq. (12.123).

12.10.6 Instrumentation Amplifiers

The instrumentation amplifier is a high-accuracy differential op-amp that can faithfully amplify low-level signals in the presence of high common-mode noise. Other features of the instrumentation amplifier are high input impedance, low offset and drift, low nonlinearity, stable gain, and low output impedance. It is commonly used for preamplification of small differential signals superimposed on high common-mode voltages, signal conditioning, signal translation, and so on. Although it can be constructed by using two or three op-amps and several precision resistors, instrumentation amplifier modules are also available and are usually preferred over the user-assembled variety because the former offer optimized and specified performance in a compact package. Some examples are the AD521, AD522, and 606.

The basic differential amplifier in Fig. 12.107 cannot be used as an instrumentation amplifier for the following reasons:

1. Input impedance seen from each terminal are different, as indicated in Eqs. (12.107) and (12.108).
2. If high gain is needed, the values of R_{22} and R_{12} in Fig. 12.107 are usually chosen to be large, which causes large output offset errors.

Several instrumentation amplifiers with specific characteristics are presented in the following paragraphs.

High Input Impedance Instrumentation Amplifier

An instrumentation amplifier with a high input impedance using two op-amps is shown in Fig. 12.116. High input impedance is achieved by applying signals v_1 and v_2 into the noninverting terminals of the op-amps since noninverting amplifiers present high input impedance.

To illustrate the operation of the circuit, let us first neglect the resistance R_G. The A_1 amplifies the signal v_1 by a factor of $(1 + R_2/R_1)$. The output of A_1 is further amplified by A_2 and the output due to V_1 will be

$$v_{o1} = -\frac{R_4}{R_3}\left(1 + \frac{R_2}{R_1}\right)v_1 \tag{12.125}$$

The signal v_2 will be amplified by A_2, and its contribution to the output is

$$v_{o2} = \left(1 + \frac{R_4}{R_3}\right)v_2 \tag{12.126}$$

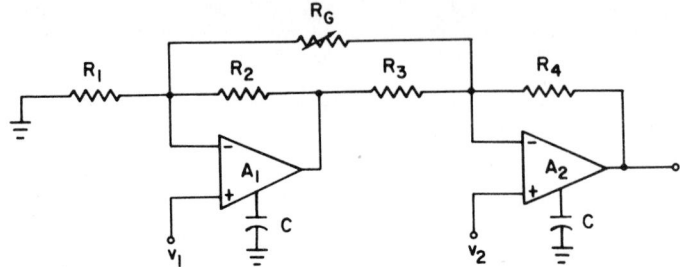

Fig. 12.116 High input impedance instrumentation amplifier.

Hence, using superposition, the output when both v_1 and v_2 are applied will be

$$v_o = -\frac{R_4}{R_3}\left(1 + \frac{R_2}{R_1}\right)v_1 + \left(1 + \frac{R_4}{R_3}\right)v_2 \qquad (12.127)$$

If the circuit is balanced by selecting resistances so that $R_1 R_3 = R_2 R_4$, Eq. (12.127) reduces to

$$v_o = \left(1 + \frac{R_1}{R_2}\right)(v_2 - v_1) \qquad (12.128)$$

When the resistance R_G is included in the circuit, a **summing** current $(v_2 - v_1)/R_G$ will be added to each summing modes and Eq. (12.128) will become

$$v_o = \left(1 + \frac{R_1}{R_2} + 2\frac{R_1}{R_G}\right)(v_2 - v_1) \qquad (12.129)$$

Using an LM108 with $R_2 = R_3 = 1\,\text{k}\Omega$ and $R_1 = R_4 = 100\,\text{k}\Omega$, Eq. (12.125) yields $v_{o1} = -101v_1$ and (12.126) yields $v_{o2} = +101v_2$. However, if a common-mode signal appears at both inputs (i.e., $v_1 = v_2$), the output will be zero.

The critical problem in this circuit is that the circuit is very sensitive to resistor mismatch. The advantage of this circuit is that the affect of the strong input capacitances is considerably smaller at high frequencies and may be neglected. The gain of the amplifier can be adjusted by R_G quite easily without affecting the matching.

Variable-Gain Differential Input Instrumentation Amplifiers
The instrumentation amplifier with differential input terminals, high input impedance, high CMRR, and a provision to vary the gain is shown in Fig. 12.117. High input impedance at both inputs is

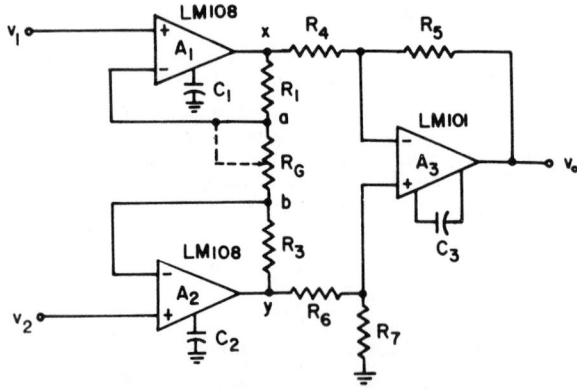

Fig. 12.117 Variable-gain, differential-input instrumentation amplifier.

guaranteed by using op-amps A_1 and A_2 in noninverting configurations. When the differential amplifier A_3 is balanced high, CMRR is obtained. When the input $v_2 = 0$, the voltages at a, b, and x are

$$v_a = v_1 \qquad v_b = 0 \qquad v_x = v_{o2} = \left(1 + \frac{R_1}{R_G}\right)v_1 \qquad (12.130)$$

When the input $v_1 = 0$, the voltages at the same points are

$$v_a = 0 \qquad v_b = v_2 \qquad v_x = v_{o1} = -\frac{R_1}{R_G}v_2 \qquad (12.131)$$

Hence, using superposition, the voltage at x due to both inputs v_1 and v_2 is

$$v_x = \left(1 + \frac{R_1}{R_G}\right)v_1 - \frac{R_1}{R_G}v_2 \qquad (12.132)$$

Similarly, the voltage at y is found to be

$$v_y = \left(1 + \frac{R_3}{R_G}\right)v_2 - \frac{R_3}{R_G}v_1 \qquad (12.133)$$

Let the op-amps A_3 be balanced (i.e., $R_5/R_4 = R_7/R_6$); also let $R_1 = R_3$. Then the output v_o is found to be [see Eq. (12.106)]

$$v_o = \frac{R_5}{R_6}\left(1 + \frac{2R_1}{R_G}\right)(v_2 - v_1) \qquad (12.134)$$

The gain of the complete amplifier can be varied by a single resistor R_G, as indicated by the dashed line in Fig. 12.117.

A mismatch in R_1 and R_3 does not degrade the CMRR of the instrumentation amplifier, but it introduces a gain error. To reduce the output offset voltage, the gain of A_3 is kept low (i.e., less than 10). To ensure a high CMRR, the balancing $R_5/R_4 = R_7/R_6$ should be chosen accurately and precision op-amps must be used for A_1 and A_2. Therefore, 1% resistors must be used for all resistors except R_G.

Linear-Gain Controlled Instrumentation Amplifier
Although the gains of instrumentation amplifiers in Figs. 12.116 and 12.117 can be controlled by varying a single resistor R_G, the variation of the gains is not linear with respect to R_G, as evidenced in Eqs. (12.129) and (12.134). On the other hand, it is desirable in many applications to adjust the gain of the amplifier linearly. This can be done by the circuit illustrated in Fig. 12.118, which is a modification of the circuit in Fig. 12.117. In this case the input op-amps A_1 and A_2 are at unity gain. The gain of the circuit is controlled by the additional amplifier A_4.

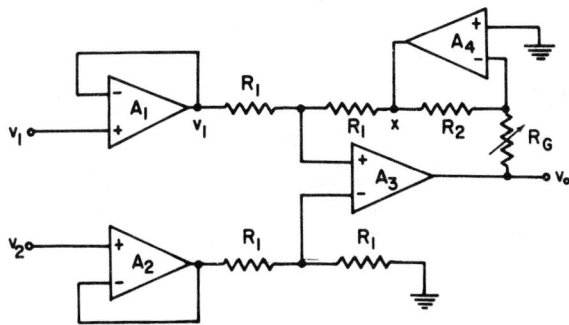

Fig. 12.118 Linear gain-controlled instrumentation amplifier.

The output voltage at x must be equal to $v_1 - v_2$, since the op-amp A_3 is in differential configuration [i.e., see Eq. (12.107)]. Then the output voltage v_o is found to be

$$v_o = -\frac{R_G}{R_2}(v_1 - v_2) \tag{12.135}$$

Hence the gain of the circuit is adjusted linearly by the control potentiometer R_G. However, the circuit in Fig. 12.118 has some limitations. For example, the output resistance of A_4 causes a mismatching in the resistors around the differential amplifier A_3. This of course degrades the CMRR. At low frequencies the effect of this output resistance is minimized by feedback; however, it has significant effect at high frequencies. Also, voltage offset and drift errors are increased. This is due to the fact that op-amps A_1 and A_2 are in unity-gain configuration; hence the input signal is not amplified before it is added to the offset voltage of A_3.

Signal-Conditioning Circuits

An objective of many op-amp circuits is to provide a linear relation between input and output signals. In this section some of the linear applications of op-amps (integrator, differentiator, etc.) and their limitations will be presented. An integrator is considered to be a low-pass filter and a differentiator is a high-pass filter. Although these operations are mathematically related to each other, problems associated with each circuit are completely different. For example, the main problems in differentiator circuits are noise and instability, while the integrator circuits are prone to dc drift and offset. Nonlinear applications of op-amps are presented later in this section.

Differentiator. An ideal differentiator produces an output voltage proportional to the derivative of the input voltage. The basic differentiator circuit is shown in Fig. 12.119(a).

The input current to the ideal op-amp is zero and the feedback through R_f maintains a virtual ground at the inverting input terminal. Therefore, the current through the capacitor C_f and the output voltage are

$$i(t) = C_1\frac{dv_i(t)}{dt} \quad \text{and} \quad v_o(t) = -i(t)R_f \tag{12.136}$$

Hence

$$v_o(t) = -R_fC_1\frac{dv_i(t)}{dt} \tag{12.137}$$

The magnitude of the gain of this ideal differentiator is then

$$|A_{cl}(f)| = \left|\frac{V_o(f)}{V_i(f)}\right| = 2\pi R_f C_1 F$$

or

$$|A_{cl}(f)| = \frac{f}{f_d} \quad \text{where} \quad f_d = \frac{1}{2\pi R_f C_1} \tag{12.138}$$

This is plotted in Fig. 12.119(c), which intersects the open-loop-gain frequency response at D. Note that the change in slope of both curves at D is 12 dB/octave. Hence the differentiator in Fig. 12.119(a) has a tendency toward instability. Another problem with this basic differentiator is that the input impedance of the circuit is a pure capacitance, which is not acceptable for most signal sources. Also, high-frequency noise would obscure the differentiated signal.

The modified circuit in Fig. 12.119(b) is preferred as a means of eliminating the problems stated previously. R_2 is included to the circuit to prevent the input bias current from producing a dc offset at the op-amp output, as discussed earlier. Capacitor C_2 is needed to bypass the thermal noise of R_2 to ground. The use of capacitor C_f is optional. Without C_f and ignoring C_2 and R_2, the closed-loop-gain frequency response of the circuit in Fig. 12.119(b) is found to be

$$A_{cl}(f) = \frac{-j(f/f_d)}{1 + j(f/f_1)} \tag{12.139}$$

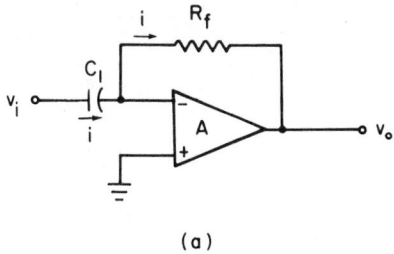

(a)

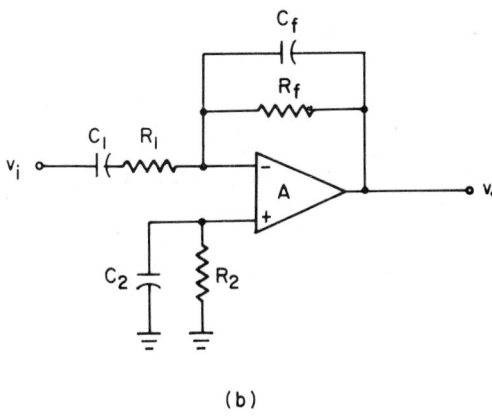

(b)

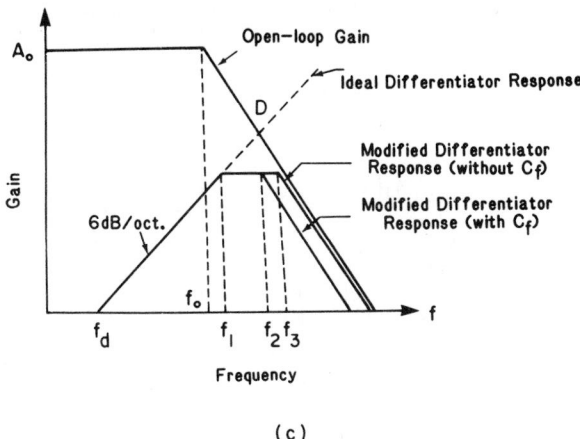

(c)

Fig. 12.119 (a) Basic differentiator circuit, (b) modified differentiator, (c) frequency responses of the ideal practical differentiators.

where

$$f_d = \frac{1}{2\pi R_f C_1} \quad \text{and} \quad f_1 = \frac{1}{2\pi R_1 C_1} \tag{12.140}$$

The plot of $|A_{cl}(f)|$ in Eq. (12.139) is shown in Fig. 12.119(c). Note that true differentiation is achieved for frequencies below f_1. Beyond this frequency the circuit acts as a voltage amplifier.

When C_f is included, the closed-loop-gain frequency response is found to be

$$A_{cl}(f) = \frac{-j(f/f_d)}{[1 + j(f/f_1)][1 + j(f/f_2)]} \tag{12.141}$$

where $f_2 = 1/2\pi R_f C_f$ and f_1 and f_d are given in Eq. (12.140). The plot of $A_{cl}(f)$ in Eq. (12.141) is also shown in Fig. 12.119(c). Note that the inclusion of R_1 and C_f introduced two real poles, one due to R_1 and the other due to C_f. This provides a stable system and reduces the high-frequency noise. The poles are placed at sufficiently high frequencies to prevent significant phase-shift error in the signal frequency range.

Integrator. An ideal integrator produces an output voltage proportional to the integral of the input voltage. The basic integrator circuit is shown in Fig. 12.120(a). The feedback around the op-amp is provided by the capacitor C_f, which maintains a virtual ground at the inverting input of the op-amp. Thus the voltage across C_f is equal to the output voltage v_o and is given by

$$v_o(t) = -\frac{1}{C_f} \int_0^t i(\tau)\, d\tau + v_o(0)$$

where $i(t) = v_i(t)/R_1$. Hence

$$v_o = -\frac{1}{R_1 C_f} \int_0^t v_i(\tau)\, d\tau + v_o(0) \tag{12.142}$$

The output voltage is proportional to the integral of the input voltage with respect to time. The initial output voltage $v_o(0)$ can be set to the desired value with the aid of simple additional circuitry to be discussed.

The magnitude of the gain of this ideal integrator in the frequency domain is then

$$|A_{cl}(f)| = \left| \frac{V_o(f)}{V_i(f)} \right| = \frac{1}{2\pi R_1 C_f f}$$

or

$$|A_{cl}(f)| = \frac{f_1}{f} \quad \text{where} \quad f_1 = \frac{1}{2\pi R_1 C_f} \tag{12.143}$$

This is plotted in Fig. 12.120(c).

The finite gain and bandwidth affect the response of the integrator. To see this, consider the op-amp whose open-loop-gain frequency response is approximated by a single pole located at f_o and a low-frequency gain of A_o, as shown in Fig. 12.120(c). If $A_o \gg 1$ and $A_o R_1 C_f \gg 1/f_o$, the resulting integrator response is found to be

$$A_{cl}(f) = \frac{-A_o}{[1 + j(1/A_o f_o)][1 + j(fA_o/f_1)]} \tag{12.144}$$

where $f_1 = 1/2\pi R_1 C_f$. The magnitude of $A_{cl}(f)$ in Eq. (12.144) is also plotted in Fig. 12.120(c). Note that the response of the real integrator departs from the ideal response at frequencies $f_2 = f_1/A_o$ and $f_3 = A_o f_o$.

The sources of error in an integrator are input offset voltage V_{os} and the input offset current I_{os}. Because of these dc errors, the output of the integrator consists of two components:

$$v_o(t) = \underbrace{-\frac{1}{R_1 C_f} \int v_i(t)\, dt}_{\text{signal term}} + \underbrace{\frac{1}{R_1 C_f} \int V_{os}\, dt + \frac{1}{C_f} \int I_{os}\, dt + V_{os}}_{\text{error terms}} \tag{12.145}$$

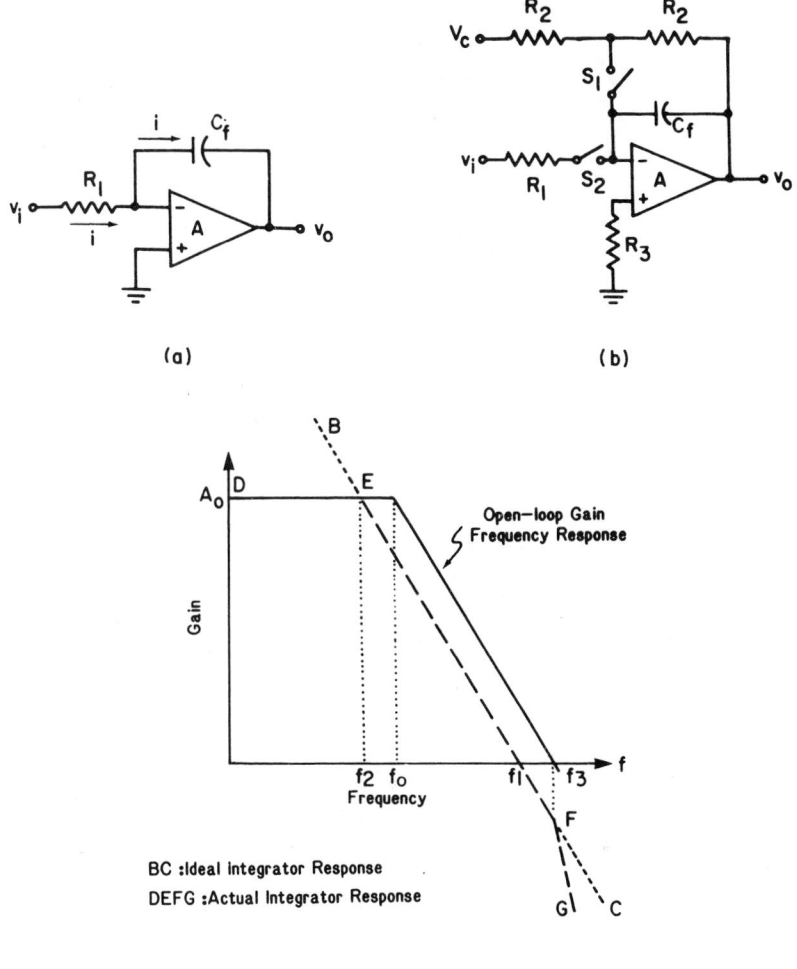

Fig. 12.120 (a) Basic integrator, (b) integrator with manual reset, (c) frequency responses of ideal and actual integrators: BC, ideal integrator response; DEFG, actual integrator response.

The ramp voltage errors caused by V_{os} and I_{os} in the second and third terms in Eq. (12.145) will continue to increase until the amplifier reaches its saturation voltage or the limit set by some external circuitry. If the resistance R_3 is not included in Fig. 12.120(b), the error causes by I_{os} will be replaced by a larger error due to input bias current I_B.

The integration process in an integrator can be initiated and terminated by the simple switching circuit shown in Fig. 12.120(b). For example, when switch S_1 is closed, capacitor C_f is charged and the output voltage rises to the negative of V_C (reset mode). If switch S_1 is opened and S_2 is closed, the circuit begins integration of the input signal $v_i(t)$ beginning at the value $-V_c$ (integrate mode). If both switches are held open, the output voltage will hold its latest value (hold mode).

For a good integrator, the feedback capacitor must be chosen with a dielectric leakage current less than the bias current of the op-amp. Selection of the op-amp is also critical. For long-term integration chopper-stabilized amplifiers are used because of their excellent long-term dc stability. Op-amps with bipolar transistor input stages are used in very short term integration (i.e., in special waveform generators). FET amplifiers are used for medium-length integration because of their low bias current.

Bridge Amplifiers

Bridge amplifiers are commonly used in amplifying the output signal from a transducer bridge such as a strain gage. The high-impedance instrumentation amplifiers discussed previously are most suitable

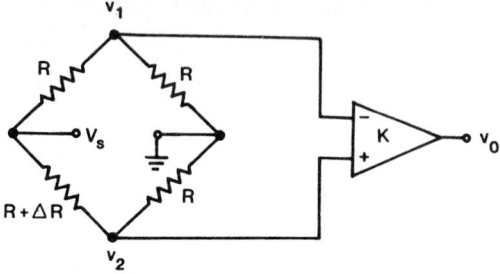

Fig. 12.121 Bridge amplifier with one active bridge arm.

for bridge amplifiers. A bridge amplifier with one active bridge arm (i.e., a sensor) is shown in Fig. 12.121.

A simple analysis of this circuit yields the following equations:

$$v_1 = \frac{R}{2R + \Delta R} V_s \qquad v_2 = \frac{V_s}{2} \qquad (12.146)$$
$$V_o = K(v_2 - v_1) \qquad (12.147)$$

where K is the gain of the instrumentation amplifier and R is a change in resistance in response to the measured physical quantity (i.e., pressure, force, temperature, etc.).

Let $\delta = \Delta R/R$; then manipulation of Eqs. (12.146) and (12.147) yields

$$v_2 - v_1 = -\frac{V}{4}\frac{\delta}{1 + \delta/2}$$
$$v_o = -\frac{KV}{4}\frac{\delta}{1 + \delta/2} \qquad (12.148)$$

If $\delta \ll 1$ (i.e., deviation ΔR is much smaller than bridge resistance R), Eq. (12.148) can be approximated by

$$v_o = -\frac{KV}{4}\delta \qquad (12.149)$$

Hence for a small percentage change in the element, the output signal is a linear function of the variation of the active element. For large variations the exact relation in Eq. (12.148) must be used.

Bridge amplifiers can also be constructed using only one op-amp rather than the fully developed differential instrumentation amplifier. Several circuits will be presented.

Bridge Current Amplifier. A bridge current amplifier is shown in Fig. 12.122. The differential output of the bridge is applied to the inputs of the op-amp. This forces the differential output of the bridge to be zero since for a high-gain op-amp, $v_2 - v_1 = 0$. Therefore, the bridge effectively operates

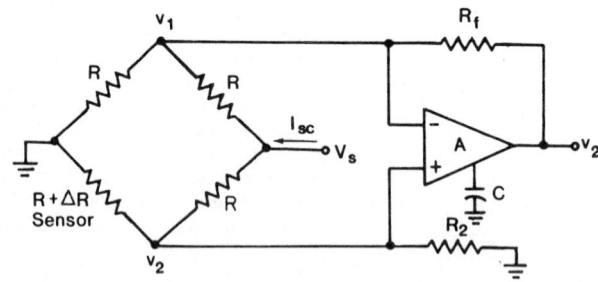

Fig. 12.122 Full-bridge current amplifier.

under short-circuit conditions, and the op-amp causes the power supply to act as a constant-current source. The value of the short-circuit current I_{sc} is determined by $I_{sc} = 2V_s/R$.

The following equations describe operation of the circuit:

$$v_2 = \frac{R_p}{R_p + R} V_s \tag{12.150}$$

where

$$R_p = (R + \Delta R) \| R_2 \tag{12.151}$$

Applying KCL at the node v_1,

$$\frac{v_o - v_1}{R_f} + \frac{V_s - v_1}{R} - \frac{v_1}{R} = 0$$

or

$$v_1 = \frac{Rv_o + R_f V_s}{R + 2R_f} \tag{12.152}$$

For a high-gain op-amp, $v_2 - v_1 = 0$. Thus Eqs. (12.150)–(12.152) yield

$$v_o = \frac{RR_2 + \delta R_2 R_f - RR_f}{[(2 + \delta)/(1 + \delta)]RR_2 + R^2} V_s \tag{12.153}$$

Note that the output is a nonlinear function of the variation of the active bridge element. If $\delta \ll 1$, $R_f \gg R$, and $R_f = R_2$, then Eq. (12.153) reduces to

$$v_o \simeq \frac{VR_f}{2R} \delta \tag{12.154}$$

Observe that under these conditions the output is a linear function of δ. However, it should be also noted that the bridge resistance R appears in Eq. (12.154). This means that the bridge elements must be insensitive to temperature to achieve a stable operation with temperature. The op-amp should have reasonably good CMR. To get high gains, the feedback resistor R_f must be large. However, large values of gains and R_2 increase the output offset error, as evidenced in the following equation:

$$\text{Output offset} = \left(1 + 2\frac{R_f}{R}\right) V_{os} \pm R_f I_{os} \tag{12.155}$$

where V_{os} is the input offset voltage and I_{os} is the input offset current of the op-amp.

Whenever the rejection of common-mode noise signals is not a problem, the half-bridge current amplifier in Fig. 12.123 can be used.

The output of the bridge is connected directly to the input of the op-amp. Since the other input is held at ground potential, $v_1 = 0$. Hence the op-amp responds to the short-circuit output current I_{sc}. The output voltage is then found to be

$$v_o = -R_f I_{sc} = -R_f \left(\frac{V_s}{R} - \frac{V_s}{R + \Delta R}\right)$$

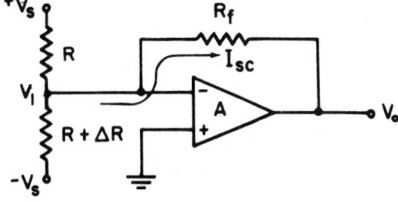

Fig. 12.123 Half-bridge current amplifier.

Manipulating this, we have

$$v_o = -\frac{R_f}{R}\frac{\delta}{1+\delta}V_s \qquad (12.156)$$

If $\delta < 1$, then

$$v_o \simeq -\frac{R_f V_s}{R}\delta \qquad (12.157)$$

Hence the output of the op-amp is a linear function of the variation of the active element if $\delta \ll 1$. The main disadvantage of the half-bridge amplifier is its inability to reject noise, which is normally accomplished by differential input bridge amplifiers. Therefore, noise and ripple of the power supply V_s must be kept very low. Additionally, since the output is a function of the bridge elements, these elements must be insensitive to the environmental factors. The output offset voltage of the half-bridge amplifier is given by

$$\text{Output offset} = \left(1 + 2\frac{R_f}{R}\right)V_{os} \pm I_B R_f \qquad (12.158)$$

where I_B is the input bias current. For lowest possible drift and offset errors, the op-amp can be chopper stabilized.

Inverting Bridge Amplifier. Another bridge amplifier using a single op-amp in inverting mode is shown in Fig. 12.124. Since the amplifier operates single ended, the op-amp can be chopper stabilized to have lowest drift and offset errors.
The output voltage of the op-amp is found to be

$$v_o = V_s\left(1 + \frac{R_f}{R_1}\right)\frac{\Delta R}{2(2R + \Delta R)} \qquad (12.159)$$

or

$$v_o = \frac{V_s}{4}\left(1 + \frac{R_f}{R_1}\right)\frac{\delta}{1 + \frac{\delta}{2}} \qquad (12.160)$$

Hence if $\delta \ll 1$,

$$v_o \simeq \frac{V_s}{4}\left(1 + \frac{R_f}{R_1}\right)\delta \qquad (12.161)$$

The advantage of this circuit is that the output of the op-amp does not depend on the bridge elements. In contrast to the previous circuits in which the op-amp places a short across the bridge output, the output voltages in this case are proportional to the open-circuit voltage of the bridge. This is due to the fact that the input to the op-amp draws negligible signal current.

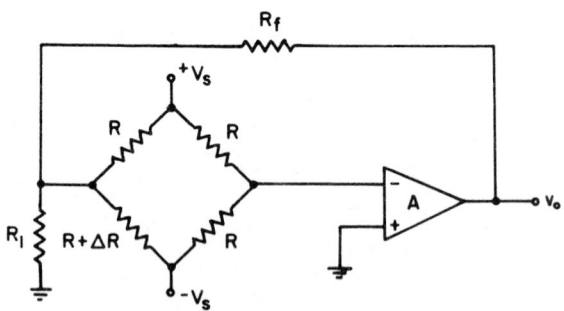

Fig. 12.124 Inverting bridge amplifier.

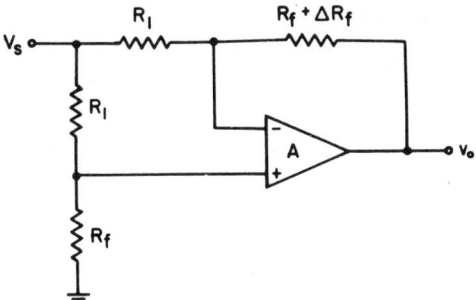

Fig. 12.125 Wide-deviation bridge amplifier.

A further advantage of this circuit is that this type of bridge amplifier can be very accurate. It is commonly used to detect very small bridge signals. The major disadvantage is that a floating bridge power supply is required. The output offset voltage in this case is determined by

$$\text{Output offset} = \left(1 + \frac{R_f}{R_1}\right)(V_{\text{os}} - RI_B) - R_f I_B \qquad (12.162)$$

where I_B is the input bias current of the op-amp.

Wide-Deviation Bridge Amplifier. The outputs of the op-amps in the previous bridge amplifiers are, in general, nonlinear unless the variation of the active element is very small. In case of large variations the circuit shown in Fig. 12.125 can be used. As seen from Eq. (12.163), the output of the op-amp is directly proportional to the δ. Hence no approximation is needed for linear operation:

$$v_o = -V_s \frac{R_f}{R_f + R_1} \delta \qquad (12.163)$$

The bridge resistors R_1 must be well matched for a proper operation. Also, the active element $R_f + \Delta R_f$ must be matched to the value of R_f when the bridge is at null. Because of the difficulty in calibration, the previous bridge circuits are used mainly for the case of small fractional changes in the active element.

Voltage Regulators
One of the common building blocks in the linear IC family is the voltage regulator. Many regulators are fixed in output voltage. These are regulators whose output voltages are **standard** (i.e., ± 5 V, ± 12 V, ± 15 V, etc.) and are used for various logic and analog device families. By the addition of a single resistor, these fixed regulators can be transformed into stable current sources. Furthermore, these units are available with maximum load currents in the range of several milliamperes to amperes, and their small size makes them ideal for local or on-card regulation.

The first and foremost function of any voltage regulator, either discrete or IC, is to maintain the specified voltage at its output. This should be expected over a specified range of load currents. There are, however, factors such as temperature and input voltage changes that tend to have an adverse effect. To compound the situation, there is a degree of interaction between the aforementioned performance-degrading factors.

Basic Voltage Regulators. Simply, a voltage regulator consists of an op-amp, voltage references (i.e., zener diodes), and a series-pass element. External components are usually required in a majority of regulators. These are included to improve transient response and to prevent oscillations.

Series Voltage Regulator
The circuit in Fig. 12.126 is a typical series voltage regulator in which the series-pass element (the transistor Q) acts as a variable resistor. In this case the series-pass element dissipates the excess voltage ($V_{\text{in}} - V_{\text{out}}$). In many cases of high-power requirements (high input/output voltage differential and/or large load currents) an external series-pass transistor may be utilized.

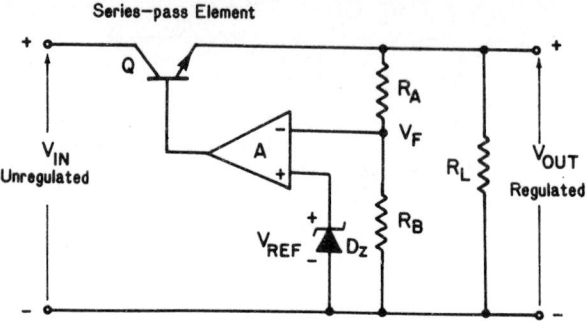

Fig. 12.126   Basic series voltage regulator.

The voltage reference is derived from the zener diode D_z. This is applied to the noninverting terminal of the op-amp. When

$$V_F = \frac{R_A}{R_A + R_B} V_{\text{out}} \tag{12.164}$$

is less than V_{ref}, the op-amp drives Q until a voltage at V_F equal to V_{ref} is obtained. If the voltage at V_F is greater than V_{ref}, the inverting input voltage V_F will drive the transistor Q to obtain a lower voltage at the output. A balance is obtained when

$$V_F = \frac{R_A}{R_A + R_B} V_{\text{out}} = V_{\text{ref}} \tag{12.165}$$

At this point both the inverting and noninverting inputs will be approximately equal. The desired output will then be

$$V_{\text{out}} = \frac{R_A + R_B}{R_A} V_{\text{ref}} \tag{12.166}$$

Clearly, the output will be determined by the ratio of R_A and R_B.

Series-pass regulators are inherently low in efficiency. This is, of course, due to the power dissipated (wasted) by the series-pass transistor. The regulation of dc voltage is accomplished efficiently by a nondissipative or switching type of regulator.

Switching Voltage Regulators

Switching regulators use a high-frequency switch to turn the series-pass transistor on or off. The higher the frequency, the smaller the components for a specified output power capability. The main drawbacks of the switching regulator are usually a larger number of external components and an inductor. The inductor, however, can be made small if high frequencies are used in the design. Figure 12.127 shows a typical switching regulator circuit.

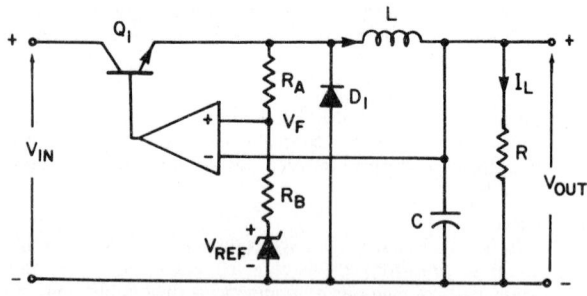

Fig. 12.127   Basic switching voltage regulator.

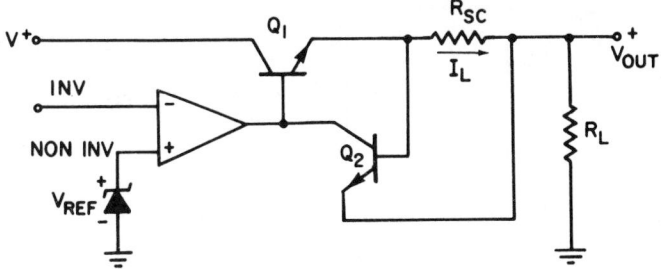

Fig. 12.128 Short-circuit protection in a voltage regulator.

In this circuit configuration, regulation is accomplished by controlling the duty cycle of Q_1. The transistor is switched either on or off so that its operation never lies in its linear region. Hence power dissipation is at a minimum. Diode D_1 is inserted so that it will conduct when Q_1 is off. When Q_1 is on, D_1 will be reverse biased and will therefore not conduct. Load current I_L, which flows through L, charges the capacitor C. When V_{out} reaches the value of V_F, the comparator will turn Q_1 off. The inductor current will begin to decrease and V_{out} will correspondingly decrease. When V_{out} reaches a value of slightly less than V_{ref}, the comparator will switch Q_1 back to on. The voltage output will be a function of the duty cycle of the switching circuit

$$v_o = \frac{t_{\text{on}}}{t_{\text{on}} + t_{\text{off}}} V_{\text{in}} \tag{12.167}$$

where t_{on} and t_{off} are the turn-on and turn-off times, respectively.

Short-Circuit Protection. In many voltage regulators a transistor is added for current-limiting purposes. Figure 12.128 is a simplified schematic of such an arrangement.

The base emitter of Q_2 will become forward biased at a particular level of I_L. This is due to the external current-sensing resistor R_{sc}. When this occurs, the collector of Q_2 will sink most of the available current from the op-amp (error amplifier) whose output is a current source. This in turn will tend to cut off the output stage and limit the output current.

Families of IC Regulators. Many classes of voltage regulators are available commercially. There are the **single-voltage** types, such as the National LM320, LM340, Fairchild MA7800, and Lambda 1400. Another class is the **voltage-adjustable** types, typical of which are the μA723, LM105, and MC1569. Furthermore, there are **dual tracking** devices, which provide both negative and positive regulated voltages, which can be varied (trimmed) to give a desired output.

Fixed output voltage regulators (three-terminal regulators) are adequate when standard output voltages can be used. They have an output accuracy of $\pm 4\%$ and will deliver from 0.5 A to several amperes. The fixed regulators need a minimum number of components. Usually, one or two capacitors are needed. Even these fixed regulators, by adding external components, can be converted to give variable outputs at higher power levels. The μA723 will be used in the examples.

1. *Positive Voltage Regulators.* A basic positive voltage regulator gives a regulated output that is positive with respect to ground. Such a circuit is shown in Fig. 12.129. Generally, the value of R_A and R_B sets a division ratio of the output with respect to the reference voltage. The basic equations that govern the values of R_A and R_B and other parameters are provided.

1. For minimum temperature drift choose

$$R_C = R_A \| R_B \tag{12.168}$$

R_C may be eliminated by shorting the V_{ref} and NON INV terminals for a minimum component count.

2. Current limiting:

$$I_{\text{sc}} = \frac{V_{\text{sense}}}{R_{\text{sc}}} \simeq \frac{0.65}{R_{\text{sc}}} \tag{12.169}$$

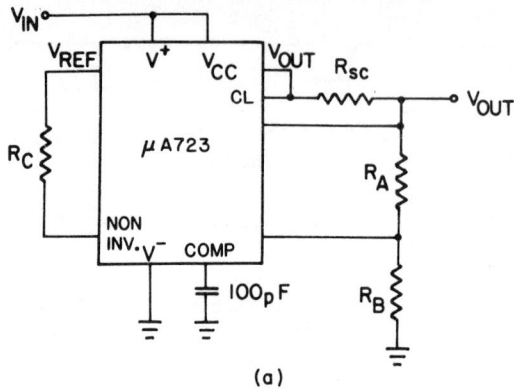

(a)

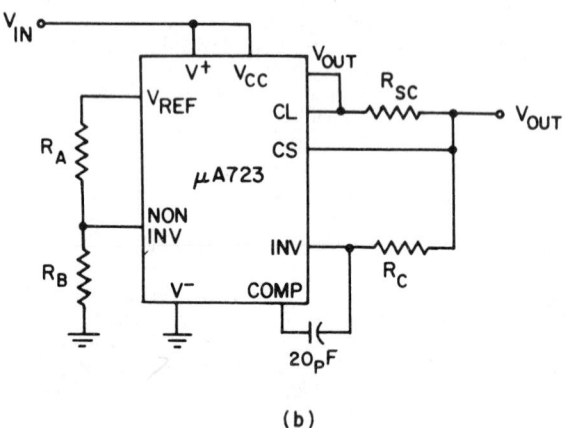

(b)

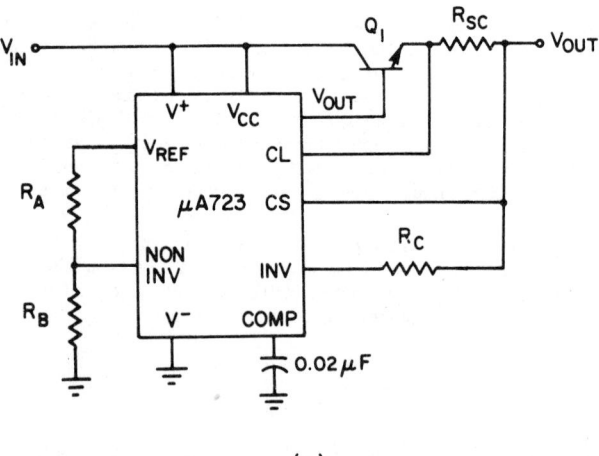

(c)

Fig. 12.129 Various voltage regulators using A723: (a) basic high-voltage regulator, 7 V, V_{out} = 37 V; (b) basic low-voltage regulator, 2 V, V_{out} = 7 V; (c) basic—5-V, 5-A voltage regulator.

3. Output voltage:

$$V_{out} = \frac{R_A + R_B}{R_B} V_{ref} \quad [\text{for Fig. 12.129}(a)] \tag{12.170}$$

$$V_{out} = \frac{R_B}{R_A + R_B} V_{ref} \quad [\text{for Fig. 12.129}(b)] \tag{12.171}$$

where $V_{ref} \simeq 7.15$ V (typical).

Consider that a voltage regulator with an output voltage $V_{out} = 5$ V will be designed. It is desired that current be limited to $I_{sc} = 100$ mA. The unregulated input voltage in $V_{in} = 30$ V (V_{in} can be as high as 40 V for the MA723).

The values of R_A and R_B should be such that R_A (and R_B) ≤ 15 kΩ so that V_{ref} will not be appreciably loaded. Although a current of at least 5 mA can be drawn from V_{ref}, excessive current drain will cause poor regulation at low V_{in} and high dissipation at high V_{in}, which are undesirable. Choose $R_B = 10$ kΩ; then from (12.171) we have $R_A \simeq 4$ kΩ ($V_{ref} = 7.15$ V. $V_{out} = 5$ V). Using Eqs. (12.168) and (12.169), we have $R_C = 4$ kΩ||10 k$\Omega = 2.9$ kΩ and $R_{sc} = 0.65/0.1 = 6.5$ Ω.

2. *Negative Voltage Regulators.* A negative voltage regulator gives a regulated output that is negative with respect to ground. Such a negative voltage regulator is shown in Fig. 12.130. This configuration makes use not only of its V_{ref} but an internal 6.2-V zener as well. Contrary to positive regulators, the inverting and noninverting inputs are reversed and the external pass transistor Q_1 acts as a level-shifted emitter follower from V_{out}. A good regulation is achieved since the regulator is driven from its own output. Regulation is controlled by the h_{FE} of Q_1 and the load regulation of the IC.

The resistor R_E should be of a sufficient value to drive the maximum load current required, through Q_1 at $V_{in}(\text{min})$. Yet the value of R_E should limit the current through the zener [at a minimum load and $V_{in}(\text{max})$] to 10 mA. This places a constraint on the lower limit of the h_{FE} of Q_1. For large current requirements, Q_1 may be replaced with a Darlington device. A limitation of this circuit is the inability to provide an output voltage of less than 9 V. The output voltage for the regulator in Fig. 12.130 is given by

$$V_{out} = -V_{ref}\frac{R_C(R_A + R_B)}{R_A(R_C + R_D)} \tag{12.172}$$

For negative output voltages less than 9 V, the V^+ and V_{cc} terminals must both be connected to a positive supply such that the voltage between V^+ and V^- is greater than 9 V; also the connection between x and y should be removed. The inputs to the internal amplifier should never be more positive than V_{out}. As V_{ref} is approximately 7.2 V, this should be reduced to below 5 V. If we let $R_C = R_D$, the input to the inverting terminal will be 3.6 V and will thus be satisfactory.

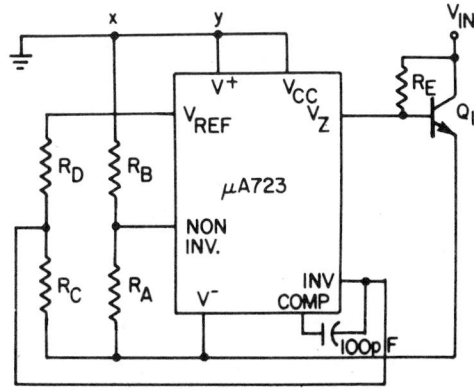

Fig. 12.130 Negative voltage regulator.

12.10.7   Nonlinear Operational Amplifier Applications

Examples and applications considered thus far have involved only linear elements, and it is assumed that the operational amplifier remains in its linear operating region. For many analog applications, linear relationships are sought. Although many devices are nonlinear, they are often linear enough over certain limited ranges to be useful. However, many nonlinear devices are used for both linear and nonlinear signal processing. These include multiplication/division, logarithmic operations, and square-law devices.

   If nonlinear elements are used in the feedforward or feedback paths of amplifiers, entire families of linear and nonlinear circuits can be constructed. Basic building blocks of mathematical functions are then used for myriad signal-processing applications.

Comparators
Comparators are used, as the name implies, to compare a signal with respect to a reference signal. The input and reference signals can be either voltage or current signals. Comparators are an essential part of analog-digital interfacing circuits. An ideal comparator and its transfer characteristic are shown in Fig. 12.131. The signal is applied to the inverting terminal (inverting comparator). Hence

$$v_o = \begin{cases} V_{o,\,\text{sat}} & \text{for } v_i < V_R \\ 0 & \text{for } v_i = V_R \\ -V_{o,\,\text{sat}} & \text{for } v_i > V_R \end{cases} \qquad (12.173)$$

If v_i and V_R are interchanged in Fig. 12.131(a), a noninverting comparator results. The reference voltage V_R can be any voltage provided that it does not exceed the maximum common-mode range.

   The op-amp in Fig. 12.131(a) is operated in an open-loop condition. Therefore, the slope of the line ab is infinite for an ideal op-amp and A_o for a practical op-amp. The output is limited by the saturation levels $+V_{o,\,\text{sat}}$ and $-V_{o,\,\text{sat}}$. These saturation voltages can be varied by the power supply voltages applied to the op-amp.

   Since comparator circuits normally operate without negative feedback, there is no need for frequency compensation. Moreover, removal of frequency compensation increases the high-frequency gain that means that the sensitivity of the op-amp is increased.

   It is usually desired to control the swing of the output independently of the power supply voltages. Two such circuits and their transfer characteristics are shown in Fig. 12.132. In these cases the breakdown voltage of the zener diode(s) limit the output voltage. The dotted line in the transfer characteristic in Fig. 12.132(a) indicates the realistic behavior in which the principal deviation from ideal is a forward drop of 0.6 V when the diode is forward biased.

   Regenerative Comparator (Schmitt Trigger). The comparators presented previously share the disadvantage that noise in the input signal causes serious output problems, particularly with very low frequency ac and dc signals. Also, for the slowly varying input signals, the output will change slowly. This becomes a serious disadvantage when the output of the comparator is used to trigger a logic circuit requiring fast trigger pulses. Both of these problems—noise immunity and speed—can be improved by introducing the positive feedback as shown in Fig. 12.133.

   When $v_1 < v_x$ so that $v_o = v_{o,\,\text{sat}}$, R_2 will feed back a signal and the new reference signal at x will be

$$v_x = V_1 = \frac{R_2 V_R}{R_1 + R_2} + \frac{R_1 V_{o,\,\text{sat}}}{R_1 + R_2} \qquad (12.174)$$

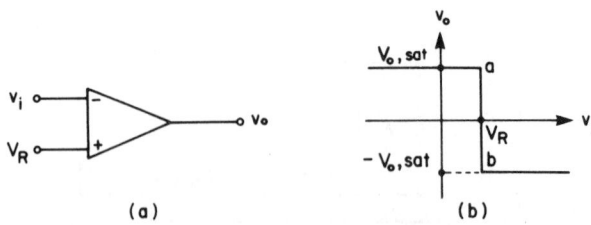

Fig. **12.131**   (a) Basic comparator, (b) its transfer characteristic.

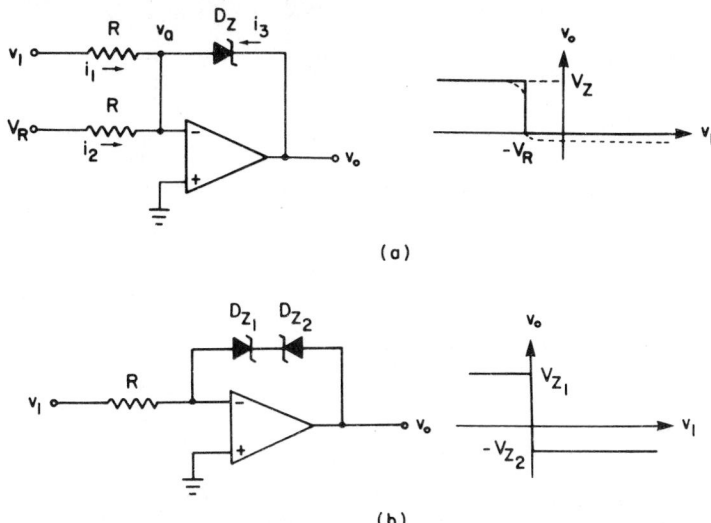

(a)

(b)

Fig. 12.132 Comparators with feedback limiters.

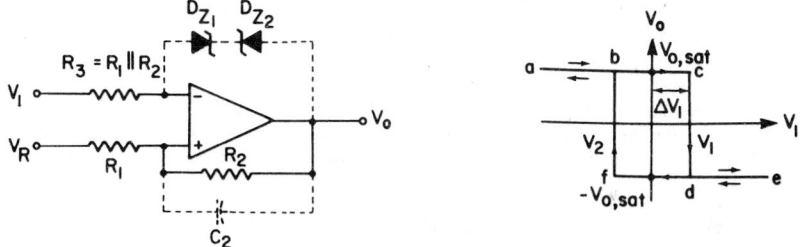

Fig. 12.133 Comparator with positive feedback and its transfer characteristic.

If v_i is increased further, v_o remains constant at $V_{o,\text{sat}}$. When $v_i = V_1$ the output regeneratively switches to $v_o = -V_{o,\text{sat}}$ and remains at this value as long as $v_i > V_1$. This portion of the transfer characteristic corresponds to the portion *abcde* in Fig. 12.133(b).

If $v_i > V_1$ so that $v_o = -V_{o,\text{sat}}$ and v_i is decreased,

$$v_x = V_2 = \frac{R_2 V_R}{R_1 + R_2} - \frac{R_1 V_{o,\text{sat}}}{R_1 + R_2} \tag{12.175}$$

At $v_i = V_2$ the output regeneratively switches to $v_o = +V_{o,\text{sat}}$, as indicated in Fig. 12.133(b) by the portion *edfba*. Note that $V_2 < V_1$, and the difference between V_1 and V_2 is called **hysteresis**, V_H:

$$V_H = V_1 - V_2 = \frac{2R_2 V_{o,\text{sat}}}{R_1 + R_2} \tag{12.176}$$

Note that because of the hysteresis, the circuit triggers at a higher voltage for increasing than for decreasing signals. Hence the circuit in Fig. 12.133 is more immune to noise than are the previous circuits. Also, the change of output in this circuit is almost instantaneous.

The output swing of a Schmitt trigger can also be controlled by zener diodes as indicated by dashed lines in Fig. 12.133(a). In this case the same formulas are used to determine V_1 and V_2 in Eqs. (12.174) and (12.175) by replacing $V_{o,\text{sat}}$ in (12.174) by V_{z_1} and in (12.175) by V_{z_2}. Also, a capacitor C_2 shown by dashed lines in Fig. 12.133, may be used to achieve maximum switching speed.

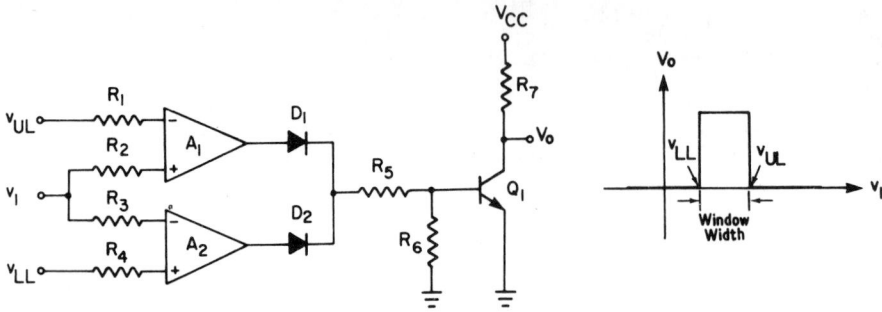

Fig. 12.134 Window comparator and its transfer characteristic.

Zero-Crossing Detector. If V_R is set equal to zero, the output will respond every time the input passes through zero. Such a comparator circuit is called a **zero-crossing detector**.

Window Comparator. Window comparators, also known as **double-ended limit detectors**, are used to detect whether a given voltage is within prescribed voltage limits. This can be accomplished by simply combining the outputs of two comparators, one indicating greater than a lower limit, v_{LL}, and the other indicating less than an upper limit v_{UL}. If the applied voltage v_1 is within the range (i.e., $v_{LL} \leq v_1 \leq v_{UL}$), the output will be true (i.e., high). If the applied voltage is not within the range, the output of one of the comparators is not true (i.e., low). A window comparator is shown in Fig. 12.134.

The output $v_o = V_{CC}$ if $v_{LL} \leq v_1 \leq v_{UL}$. If $v_1 < v_{LL}$, then A_2 turns the transistor Q_1 on through D_2. If $v_1 > v_{UL}$, then A_1 turns Q_1 on through D_1. If $v_{LL} \leq v_1 \leq v_{UL}$, both A_1 and A_2 are low; hence Q_1 remains off and $v_o = V_{CC}$.

Logarithmic Circuits
If a nonlinear device is connected in the feedback path of an op-amp, the relation between input and output would be nonlinear. Furthermore, the output would be a mathematical relation that is a function of the nonlinear device. Consider the circuit shown in Fig. 12.135, and the diode voltage-current relationship,

$$i_D = I_o\left(e^{qv_D/kT} - 1\right) \simeq I_o e^{qv_D/kT} \tag{12.177}$$

for $v_D \gg kT/q$, where

k = Boltzmann's constant, 1.380×10^{-23} J/K
q = electron charge, 1.602×10^{-19} C
T = absolute temperature (K)
I_o = reverse saturation current for diode
v_D = voltage across the diode

Since the voltage at the inverting input of the op-amp is approximately equal to zero, $v_D = v_o$. Also, if the input current to the amplifier is negligibly small, $i_1 = i_D$. Thus

$$-\frac{v_I}{R_1} = I_o\left(e^{qv_o/kT} - 1\right) \tag{12.178}$$

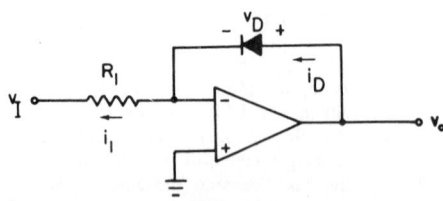

Fig. 12.135 Basic logarithmic amplifier.

Assume that the circuit is operating with a positive input voltage. The maximum negative value that i_D can have is $-I_o$. If $v_I/R_1 > I_o$, the current through the reverse-biased diode cannot balance the current i_1. Hence the output voltage of the op-amp will be driven negative until it is saturated. This means that the voltage at the inverting terminal will not be near ground. This is because the diode can conduct a limited current in the reverse direction. However, if the circuit is operated with a negative input voltage, this problem does not exist since the diode will be forward biased in this case. For $v_o = v_D \gg kT/q$, Eq. (12.178) reduces to

$$-\frac{v_I}{R_1} \simeq I_o e^{qv_o/kT}$$

Hence

$$v_o \simeq \frac{kT}{q}\ln\left(\frac{-v_I}{R_1 I_o}\right) \quad \text{or} \quad v_o \simeq \frac{2.3kT}{q}\log_{10}\left(\frac{-v_I}{R_1 I_o}\right) \tag{12.179}$$

Thus the output voltage is proportional to the logarithm of the magnitude of the input voltage for negative inputs. Diodes are often impractical for use in log converters. This is because they have a limited range of true logarithmic characteristic within their specified operating current rating. At high diode currents the ohmic resistance causes a significant error:

$$v_o = \frac{kT}{q}\ln\left(\frac{-v_I}{R_1 I_o}\right) + I_D R_B \tag{12.180}$$

where R_B is the bulk resistance.

Log Amplifier Using Transistors. Operation of diode log converters is not practical over two decades for true log characteristics. By using a transistor, connected in place of a diode as shown in Fig. 12.136, a greater dynamic range is achieved. In this configuration the input current or voltage will determine the collector current. Essentially, $i_1 = -i_C$ and the collector of Q_1 is virtually grounded. As the base is grounded, base and collector are at the same potential. Then v_o is actually equal to v_{BE} of Q_1 and the op-amp supplies the needed current i_E. For $h_{FE} \gg 1$, $i_E \simeq -i_C$ and the Ebers–Moll equation for a bipolar transistor model gives

$$i_C = \alpha_N I_{ES}\left(e^{-qv_{EB}/kT} - 1\right) \tag{12.181}$$

where $v_{EB} = v_o = $ emitter-base voltage, α_N is the normal (forward) current gain of the transistor, I_{ES} is the reverse saturation current of the emitter-base junction. For $v_{BE} < 100$ mV, Eq. (12.181) can be approximated by

$$i_C \simeq \alpha_N I_{ES} e^{-qv_{EB}/kT} \tag{12.182}$$

The op-amp holds $i_C \simeq i_1 = v_I/R_1$; then

$$v_o = -\frac{kT}{q}\ln\left(\frac{v_I}{R_1 \alpha_N I_{ES}}\right) \quad \text{or} \quad v_o = \frac{-2.3kT}{q}\log_{10}\left(\frac{v_I}{R_1 \alpha_N I_{ES}}\right) \tag{12.183}$$

Thus the output voltage is proportional to the logarithm of the input voltage. For some transistors, $I_{ES} < 10^{-14}$ A. This is orders of magnitude better than the I_o of diodes. Thus an expanded range of logarithmic operation (6 to 8 decades) can be achieved by using the circuit in Fig. 12.136.

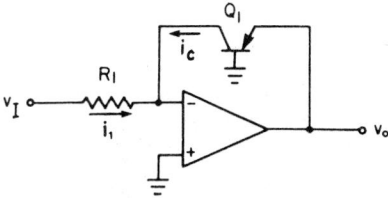

Fig. 12.136 Logarithm converter using a transistor-transistor model.

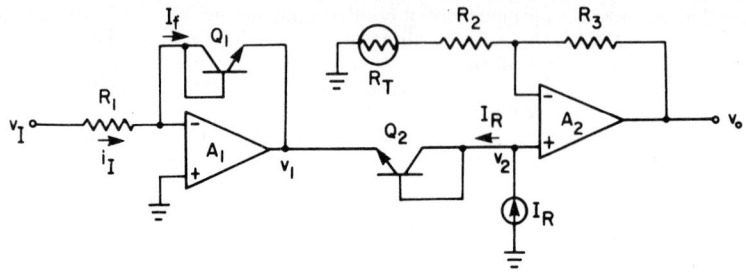

Fig. 12.137   Temperature-compensated logarithm converter.

Temperature Compensation. It should be noted from Eq. (12.183) that the output voltage v_o is temperature dependent due to the scale factor kT/q (which is approximately $T/11{,}000$) and the reverse saturation current I_{ES}. Both of these temperature effects can be reduced by using the circuit in Fig. 12.137, where R_T is a thermistor. The current source I_R forces a constant current through Q_2, which induces a voltage v_{CEQ_2}. Thus the voltage v_2 is

$$v_2 = v_1 + v_{CEQ_2} = -\frac{kT}{q}\left[\ln\frac{v_I}{\alpha_N R_1} - \ln I_{ESQ_1} - \ln I_R + \ln I_{ESQ_2}\right] \qquad (12.184)$$

If Q_1 and Q_2 are matched, then $I_{ESQ_1} = I_{ESQ_2}$ (i.e., the saturation current terms cancel each other); therefore,

$$v_2 = -\frac{kT}{q}\ln\left(\frac{v_I}{\alpha_N R_1 I_R}\right) \qquad (12.185)$$

Then the output voltage is found to be

$$v_o = \left(1 + \frac{R_3}{R_2 + R_T}\right)v_2$$

or

$$v_o = -\frac{kT}{q}\left(\frac{R_2 + R_3 + R_T}{R_2 + R_T}\right)\ln\left(\frac{v_I}{\alpha_N R_1 I_R}\right) \qquad (12.186)$$

Letting

$$K_1 = \frac{kT}{q}\left(\frac{R_2 + R_3 + R_T}{R_2 + R_T}\right) \qquad K_2 = \frac{1}{\alpha_N R_1 I_R}$$

Eq. (12.186) becomes

$$v_o = -K_1 \ln K_2 v_I \qquad (12.187)$$

Thus the remaining temperature sensitive factor, kT/q, can be compensated in the output amplifier by making its gain temperature sensitive. This is the purpose of R_T, which is a temperature-sensitive resistor (thermistor).

Log-Ratio Circuits
A variation of the logarithmic amplifier is the log-ratio circuit shown in Fig. 12.138. Two input signals v_1 and v_2 are converted to temperature-sensitive logarithmic voltages v_3 and v_4 by Q_1 and Q_2 and op-amps A_1 and A_2. Amplifier A_3 is a differential amplifier with a gain of R_3/R_2. The output voltage v_o is found to be

$$v_o = \frac{R_3}{R_2}(v_4 - v_3) = \frac{R_3}{R_2}\frac{kT}{q}\ln\frac{v_1}{v_2} \qquad (12.188)$$

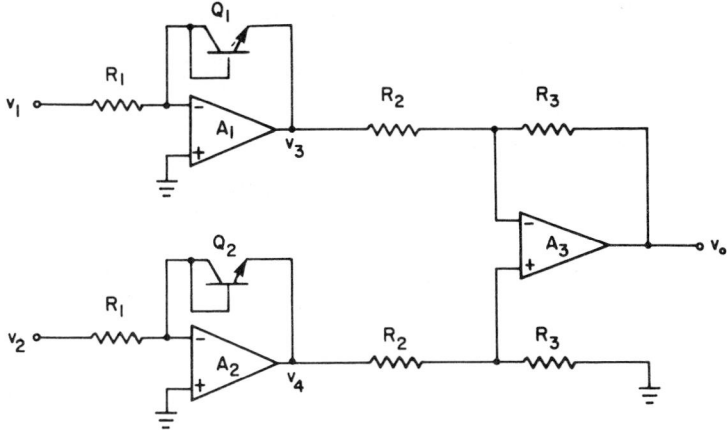

Fig. 12.138   Log-ratio amplifier circuit.

Subtraction of v_3 from v_4 tends to cancel the temperature-sensitive terms $(kT/q)(\ln I_{ES})$. If R_3/R_2 is made to be a temperature-compensating factor, the effect of kT/q can also be canceled.

Antilogarithm Circuits

If the positions of the input resistor and log device in Fig. 12.136 are exchanged, the circuit shown in Fig. 12.139 can be used as an inverse log (antilog) amplifier. There are, however, some serious drawbacks to this circuit configuration. For example, it has very poor temperature stability due to the kT/q factor of Q_1. Also, we want an output v_o such that

$$v_o = K_1 e^{K_2 v_I} \tag{12.189}$$

where K_1 and K_2 are constants. However, the circuit in Fig. 12.139 will not give $v_o = K_1$ for $v_I = 0$. If the op-amp is carefully offset trimmed, $v_o = 0$ for $v_I = 0$. Hence if the op-amp is properly biased, v_o can be made equal to K_1 for $v_I = 0$. It must then be ensured that the operation will not be affected by the kT/q factor of Q_1 (i.e., some bias must be applied). The current generator I_R in Fig. 12.140

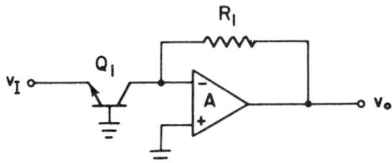

Fig. 12.139   Antilog circuit.

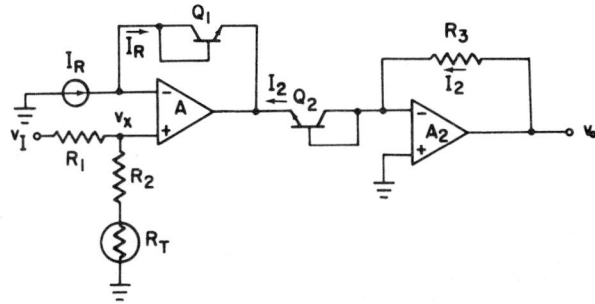

Fig. 12.140   Temperature-compensated antilog circuit.

serves this purpose. Hence

$$v_x = \frac{R_2 + R_T}{R_1 + R_2 + R_T} v_I \quad \text{(voltage divider)} \tag{12.190}$$

Also, neglecting I_{ES} and assuming that $\alpha_N \simeq 1$,

$$v_1 \simeq \frac{R_2 + R_T}{R_1 + R_2 + R_T} v_i - \frac{kT}{q} \ln I_R \tag{12.191}$$

Since $v_1 = -v_{CEQ_2}$,

$$v_1 \simeq -\frac{kT}{q} \ln I_2 \tag{12.192}$$

Combining Eq. (12.191) and (12.192), and using $v_o = I_2 R_3$, it follows that

$$v_o = R_2 I_R \ln^{-1} \left(-\frac{q}{kT} \frac{R_2 + R_T}{R_1 + R_2 + R_T} v_I \right) \tag{12.193}$$

or

$$v_o = K_1 e^{K_2 v_I}$$

where

$$K_1 = R_3 I_R \quad \text{and} \quad K_2 = -\frac{q}{kT} \frac{R_2 + R_T}{R_1 + R_2 + R_T}$$

Again, a pair of well-matched transistors Q_1 and Q_2 should be used for temperature tracking to cancel the effect of I_{ESQ_1} and I_{ESQ_2}. The temperature-sensitive resistance R_T is used to eliminate the effect of q/kT.

Multipliers and Dividers
Multipliers and dividers are another frequently used nonlinear application of op-amps. The output of the device is proportional to the product or quotient of the input signals.

Multiplication	$v_o = K_m v_1 v_2$	(12.194)

Division	$v_o = \dfrac{K_D v_1}{v_2}$	(12.195)

where K_m and K_D are gain factors.

The operating range of a multiplier (or a divider) is defined in terms of the polarity of its inputs and outputs as follows: If the output and inputs can have either positive or negative polarity, the device is designated as a **four-quadrant device**. If the output is bipolar to only one type of input polarity, it is designated as a **two-quadrant device**. If all signals are of a single polarity, it is called a **one-quadrant device**.

For a two-input multiplier, quadrature is illustrated in Fig. 12.141.

In the log circuits previously presented, inputs must be of one polarity. To reverse polarity of inputs (and hence output), substituting *npn* for *pnp* or diode reversal is imperative.

Many nonlinear functions, such as $X \cdot Y$, X/Y, and $X^{\pm n}$, are easily generated with the use of log circuits. Multiplication becomes addition and division becomes subtraction. Furthermore, exponents, fractional or otherwise, merely become gain coefficients of log amplifiers. Multipliers other than those using log circuits are also available. The most commonly used methods are quarter-square, time division, transconductance, and averaging.

Logarithmic Multiplier and Divider. The basic building-block approach to multiplication and division is the log and inverse log (antilog) method illustrated in Fig. 12.142.

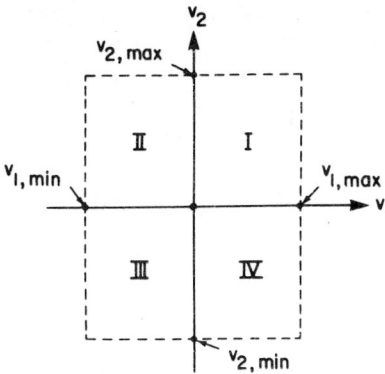

Fig. 12.141 Operating quadratures.

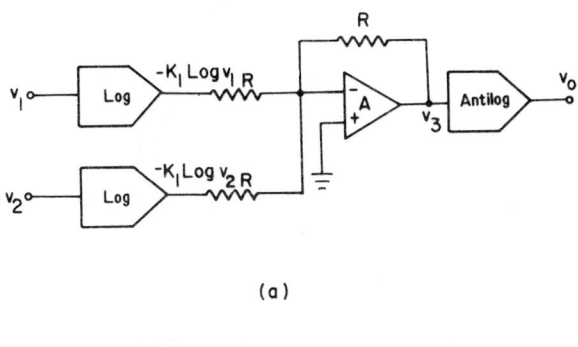

(a)

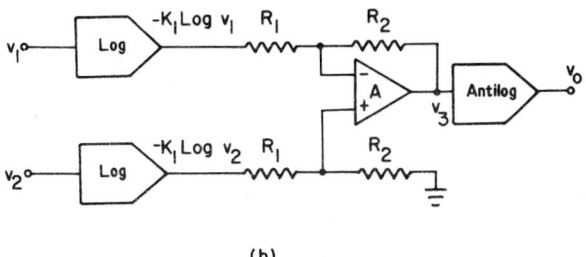

(b)

Fig. 12.142 Multiplication and division by log method: (a) logarithmic multiplier, (b) logarithmic divider.

For the multiplier in Fig. 12.142(a) it is only necessary to take the log of each input, v_1 and v_2, sum these logs, and then take the antilog as indicated. That is,

$$v_3 = +K_1(\log v_1 + \log v_2) = K_1\log(v_1 v_2) \tag{12.196}$$

$$v_o = K_2\log^{-1}\left(\frac{v_3}{K_1}\right) = K_2 v_1 v_2 \tag{12.197}$$

For the divider shown in Fig. 12.142(b) the logs of the two input voltages are subtracted, then the antilog is taken. Note that v_3 is actually the log ratio of v_1 and v_2. Thus

$$v_3 = K_1\log\left(\frac{v_1}{v_2}\right) \tag{12.198}$$

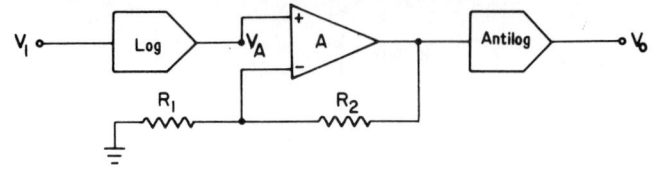

(a)

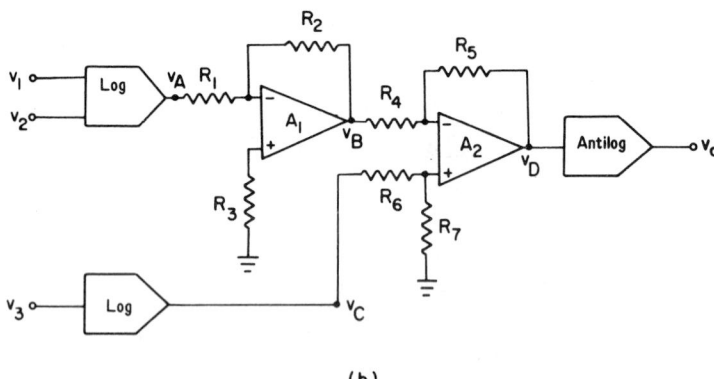

(b)

Fig. 12.143 Circuits raising (*a*) a signal to power, (*b*) ratio of two signals to a power and to multiplying the result.

and

$$v_o = K_2 \log^{-1}\left(\frac{v_3}{K_1}\right) = K_2 \frac{v_1}{v_2} \qquad (12.199)$$

It is interesting to note that if the inputs v_1 and v_2 are tied together,

$$v_o = K v_1^2 = K v_2^2 \qquad (12.200)$$

This is useful as a squaring circuit.

Raising to a Power. Using log and antilog blocks, any input signal can be raised to any power. Consider the circuit shown in Fig. 12.143. Assume that all K's in the previous equations are normalized (i.e., $K_1 = K_2 = K = 1$). Let $m = 1 + R_2/R_1$; then

$$v_A = \log v_1 \quad \text{and} \quad v_B = \left(1 + \frac{R_2}{R_1}\right) \log v_1 = m \log v_1$$

or

$$v_o = v_1^m \qquad (12.201)$$

In Fig. 12.143(*b*), a multiplier/divider circuit is combined with a power-raising circuit. The following equations describe the operation of this circuit:

$$v_A = \log v_1 - \log v_2 \qquad (12.202)$$

$$v_B = -\frac{R_2}{R_1} v_A = -\frac{R_2}{R_1}(\log v_1 - \log v_2) \qquad (12.203)$$

$$v_C = \log v_3 \qquad (12.204)$$

$$v_D = -\frac{R_5}{R_4} v_B + \frac{R_7}{R_6} v_C \qquad (12.205)$$

Letting $m = R_2/R_1$, $m_1 = (R_5/R_4)m$, and $R_6 = R_7$ and manipulating Eq. (12.205), it can be shown that

$$v_o = v_3 \left(\frac{v_1}{v_2} \right)^{m_1} \tag{12.206}$$

Hence the input signal v_3 is multiplied by the (m_1)th power of the ratio of v_1 and v_2.

The performance of a practical log-antilog multiplier is excellent for many applications. The primary sources of static errors are offset errors caused by input offset current and voltage, which may be reduced by properly nulling the op amps. Also, resistance tolerance causes an error in the scale factors. This error can be reduced by using precision resistors.

Nonlogarithmic Multipliers and Dividers. Not all multiplication schemes need be of the log-antilog category. When the order of magnitude of the multiplier and multiplicand are limited, alternative circuits may be employed. One alternative is to use an op-amp as a controlled-gain amplifier. Making use of the drain-to-source resistance R_{ds} of an FET, Fig. 12.144 illustrates a simple two-quadrant multiplier. Depending on the FET chosen, its R_{ds} can be linear over a limited range of v_{gs} (gate-to-source voltage). From the circuit

$$v_o = -\frac{R_1}{R_{ds}} v_1 \tag{12.207}$$

It is known that, for an FET, the drain-to-source resistance is

$$R_{ds} = \frac{r_0}{1 - kv_{gs}} = -\frac{r_0}{kv_{gs}} \tag{12.208}$$

where r_0 is the drain resistance at zero gate bias and k is a constant dependent on the FET type. If Q_1 is carefully chosen, then for v_o we have

$$v_o = -\frac{R_1}{R_{ds}} v_1 = \frac{kR_1}{r_0} v_1 v_{gs} \tag{12.209}$$

Hence the signals at v_1 and v_{gs} are multiplied together with a scale factor of $(kR_1)/r_0$.

For performing division, Q_1 and R_1 are interchanged.

Transconductance Multiplier. Most of the popular IC multipliers operate on the principle of variable transconductance. A simple circuit that performs two-quadrant multiplication using this technique is shown in Fig. 12.145.

The circuit operates by varying the emitter current of a perfectly matched pair of transistors Q_1 and Q_2. Then A_2, Q_3, and R_3 serve as a voltage-controlled current source; hence the current I_E is proportional to the input signal v_2. The differential output from Q_1 and Q_2 is applied to A_1, and the output v_o can be shown to be

$$v_o = \left(\frac{q}{2kT} I_E v_1 \right) R_{10} \tag{12.210}$$

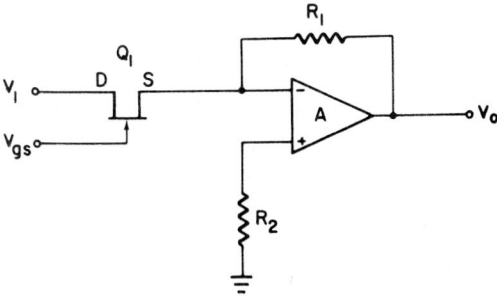

Fig. 12.144 Two-quadrant multiplier.

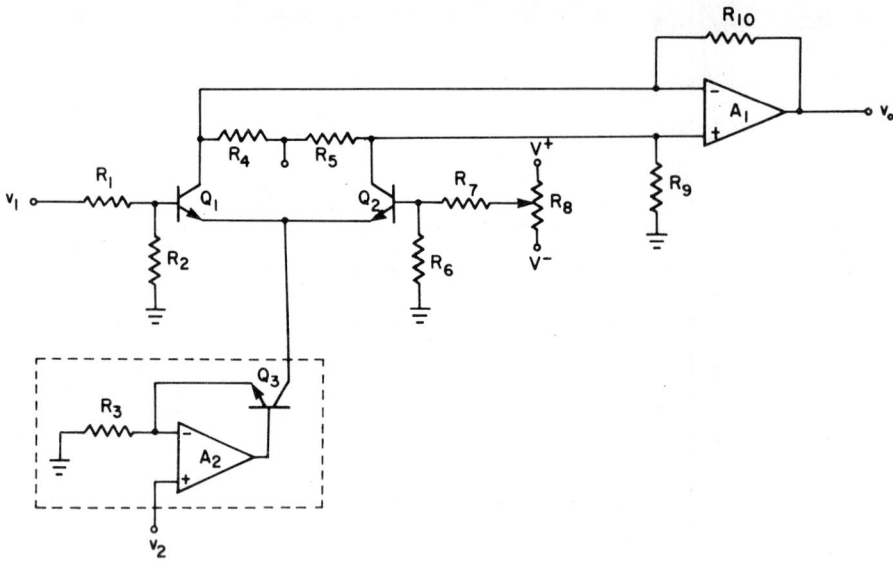

Fig. 12.145   Variable transconductor multiplier.

where

$$I_E = \frac{v_2}{R_3} \tag{12.211}$$

Hence

$$v_o = \frac{q}{2kT}\left(\frac{R_{10}}{R_3}\right)v_1 v_2 = K v_1 v_2 \tag{12.212}$$

Resistors R_1 and R_2 form a voltage divider. The divider is utilized because the output of the bridge configuration is nearly linear with an input of only ± 10 mV applied directly to the base of Q_1. Hence scaling is used to allow larger input voltage for v_1. The configuration shown is very sensitive to temperature. Both the scale factor K and the dc level tend to drift. The drift in K is caused primarily by the change in temperature, and the unavoidable mismatch in Q_1 and Q_2 causes drift in the dc level. Transconductance multipliers are well suited in communications circuits, where low-level signals are usually encountered. This basic two-quadrant multiplier circuit can be extended to operate in four quadrants. This is accomplished by adding a second differential pair.

Quarter-Square Multipliers.   The quarter-square multiplier relies on the basic relationship

$$\frac{(X+Y)^2}{4} - \frac{(X-Y)^2}{4} = XY \tag{12.213}$$

Although this approach has the advantage of moderately high bandwidth and low error, the high component count makes the cost of fabrication relatively high. With the exception of op-amps, all the components are usually discrete. Figure 12.146 is a block diagram of a quarter-square multiplier. Generally, the absolute value and squaring operations are often implemented by diode networks.

Diode function generating networks are used to approximate squaring operations. This is achieved by piecewise linear approximation utilizing bias sources and several diodes. Often, many segments are used in the diode network for a more precise approximation.

Time-Division Multiplier.   Time-division multipliers (also known as **pulse height-width multipliers**), despite low bandwidth, offer high precision and low error. They are particularly suited in environments where errors due to temperature might otherwise be appreciable.

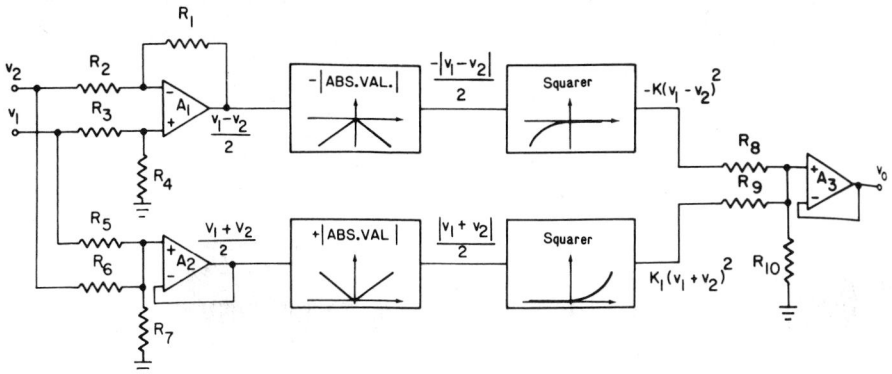

Fig. 12.146 Block diagram of a quarter-square multiplier.

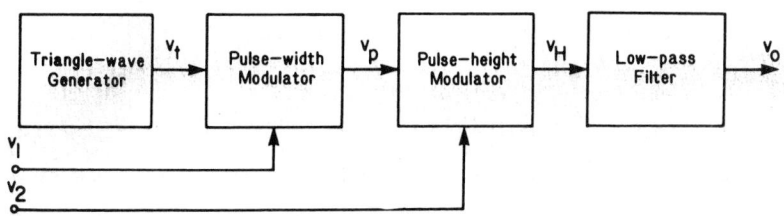

Fig. 12.147 Block diagram of a time-division multiplier.

Pulse height-width (PHW) multipliers are based on the principle that one input voltage controls the duty cycle of a pulse train, while another voltage is made to be proportional to the amplitude of the same pulse train. The waveform is then passed through a low-pass filter so that short-term variations are removed. The output will be proportional to the product of the two input signals.

Figure 12.147 is a block diagram of a PHW multiplier where v_1 and v_2 are voltages to be multiplied. The triangular-wave generator produces v_t, a triangular wave with period T and amplitude V_t. This waveform is then applied to the pulse width modulator, whose function is to compare V_t with v_1. This is accomplished through internal comparator circuitry.

From Fig. 12.148, it can be verified that T_1 and T_2 are a function of v_1. Since

$$T = T_1 + T_2$$

then

$$\frac{T_1}{T} = \frac{v_1}{V_t} \tag{12.214}$$

That is, as v_1 increases, so does the pulse width. The output of the pulse-width modulator then is v_p: a constant-amplitude pulse with magnitude V_p, period T, and duty cycle T_1/T.

In turn, v_p is applied to the pulse height modulator. Its output is a pulse with period T, duty cycle T_1/T, and amplitude V_H. However, V_H is made to be a function (linear) of v_2, that is,

$$v_H = Kv_2 \tag{12.215}$$

If the output is averaged per cycle (i.e., for every period T), the average output or v_o will be

$$v_o = \frac{T_1}{T} v_H \tag{12.216}$$

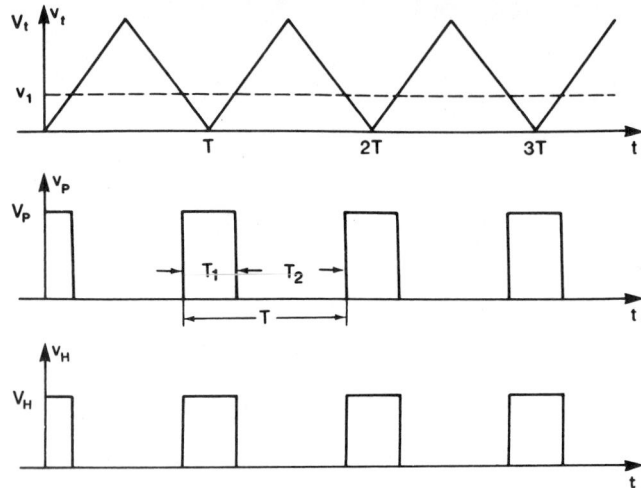

Fig. 12.148 Modulated signals v_p and v_H in pulse height-width multiplier.

TABLE 12.11   Multiplier Comparison Chart

Technique	Bandwidth	Accuracy (%)	Comments
Quarter-square	5 MHz	0.25	Poor linearity, error ripple significant for low-level signals, high cost, large component count
Time division	1 kHz	0.01	Narrow bandwidth, useful for near-static applications
Transconductance	10 MHz	2.0	Temperature stability a factor, low signal level, applications oriented
Log-antilog	100 kHz	0.2	Simple for one quadrant operations

or

$$v_o = \frac{v_1}{V_t} K v_2 = \frac{K}{V_t} v_1 v_2 \qquad (12.217)$$

If the constant $K = V_T$, then

$$v_o = v_1 v_2 \qquad (12.218)$$

The pulse height-width multipliers have good static accuracy; however, they have severe dynamic limitations due to the phase shift inherent in the output low-pass filter. Also, the typical -3-dB bandwidth is less than 1 kHz. This technique of multiplying two analog signals is used where high static accuracy is required.

A comparison of multiplier techniques presented with their respective characteristics is given in Table 12.11.

12.10.8   Sample-and-Hold Circuits

A sample-and-hold circuit is a device that

1. Samples (tracks) an input signal and transmits it to the output.
2. Holds the last sampled value until the input is sampled again.

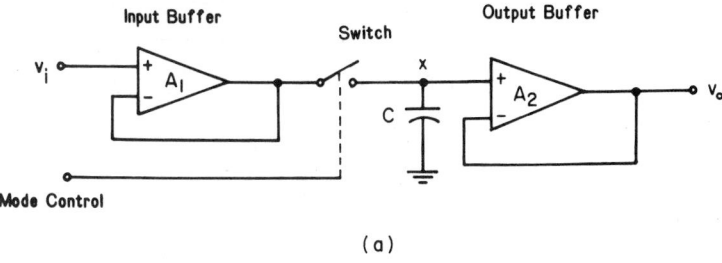

(a)

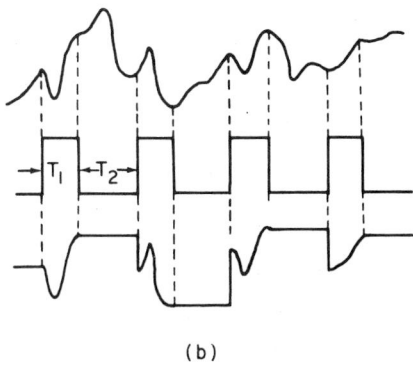

(b)

Fig. 12.149   (a) Basic sample-and-hold circuit, (b) input and output waveforms.

The operation of a basic sample-and-hold circuit is illustrated in Fig. 12.149.

The input buffer amplifier A_1 is a high-output-current op-amp, so that in the **sample mode** (S is closed) it charges the capacitor C quickly. Thus the voltage at x will follow the input voltage during the sampling interval T_1. During the **hold mode**, T_2, the input amplifier will no longer charge the capacitor. Hence the voltage at x is maintained until the next sampling interval. The output buffer A_2 holds the transferred value of v_x at the output v_o for the entire time T_2. To charge the capacitor quickly and retain its charge with minimal droop between sampling times, the R_{off}/R_{on} ratio must be very high. There is a tendency for capacitor C to be discharged by the input currents of amplifier A_2. Therefore, for minimal droop, A_2 must have very low bias and offset currents.

There are numerous applications of sample-and-hold circuits, particularly in high-speed, multiplexed data acquisition circuits. However, a frequent use is in sampled data systems in order to diminish aperture time with analog-to-digital (A/D) converters.

Figure 12.150 is an example of a typical sample-and-hold circuit. Input buffer A_1 offers a low output impedance for charging and discharging capacitor C_2, which holds the sampled charge. FET Q_2 acts as a switch that is enabled by applying V^+, and diode D_1 is reverse biased, which turns Q_2 on because of a zero gate bias. The loop of $A_1 - C_2$ is therefore closed and v_c will track v_{in}. Negative feedback to A_1 will balance any offsets across Q_1 and Q_2.

When the sample enable signal is brought to V^-, diode D_1 becomes forward biased and brings the gate of Q_2 toward V^-. Because of this, the channel resistance of Q_2 will be extremely high and the feedback loop is broken. Capacitor C_2 will, however, continue to hold its charge as FET Q_1 acts as a buffer. For interfacing to a low impedance, A_2 is used as a voltage follower.

Low-leakage devices must be used for Q_1 and Q_2 as the basic relation of

$$\frac{\Delta v_c}{\Delta t} = \frac{I_c}{C_2} \tag{12.219}$$

holds for the charge across capacitor C_2. Some improvement may be possible employing a similar circuit configuration using MOS devices for Q_1 and Q_2.

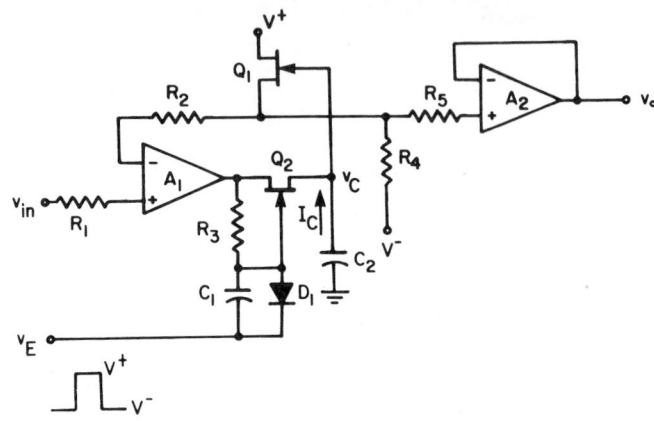

Fig. 12.150 Typical sample-and-hold circuit.

12.10.9   Peak Detector

On occasion an instrument design requires a circuit that will find and hold the maximum value of an input waveform. Figure 12.151 illustrates the requirement of such a circuit. The solid line of the graph represents a continuously varying input signal voltage as a function of time. The dashed line represents the output of a peak detection circuit. Figure 12.152 is an example of a peak detection circuit. This circuit produces an output voltage v_o that increases only when the slope of the input voltage v_i is both increasing and greater than any previous v_i. When the slope of v_i is either zero or decreasing, the value of v_o will remain at the highest previous level of v_i. If perfect op-amps, diodes, and capacitors are used, the output v_o will remain at the highest previous level of v_i until the circuit is reset by switch S_1.

Amplifier A_1 acts as an input buffer. The output of A_1 is equal to the input voltage and is used to charge C_1. The high reverse resistance of diodes D_1 and D_2 prevents the op-amp's low output impedance from discharging the capacitor.

Amplifier A_2 serves as an output buffer and is connected as a voltage follower. This is needed to isolate the capacitor C_1 from the output. In very critical applications certain precautions must be taken to maintain the charge on C_1. First, a high-quality capacitor should be used, such as a polystyrene or other low-leakage type. Second, amplifier A_2 should be of the FET input type, which has very high input impedance and, more important, low input current. Since the voltage discharge rate is

$$\frac{\Delta v_c}{\Delta t} = \frac{I}{C_1} \tag{12.220}$$

the lower the input current, the lower the drift in voltage across C_1.

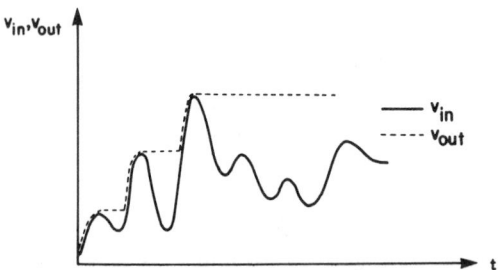

Fig. 12.151 Input and output signals of a peak detector.

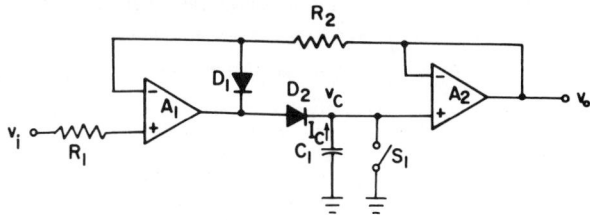

Fig. 12.152 Peak detector circuit.

Resistor R_2 is included in order to allow A_1 to be clamped in the off state by diode D_1. This results in faster recovery. Switch S_1 presents the only low-impedance path to ground and is used to discharge C_1.

While this circuit involves a single quadrant operation, the polarity of operation can be reversed if D_1 and D_2 are reversed.

12.10.10 Analog-to-Digital and Digital-to-Analog Converters

Digital (binary) signals are represented by signals that make a sudden transition between two values. Analog signals may instead assume any continuous range of values. There are at times advantages for processing signals in either the digital or the continuous domain.

Generally, digital-to-analog (D/A) converters are somewhat simpler than analog-to-digital (A/D) converters. Furthermore, since some A/D processors use D/A converters as part of the overall circuitry, D/A converters will be considered first.

Digital-to-Analog (D/A) Converter

A D/A converter takes a digital word at its input and converts it to an equivalent analog voltage. D/A converters can be designed to accept the digital word in a variety of codes such as BCD (binary-to-decimal) code or the binary code. The commonly used **binary-weighted-resistor** D/A converter shown in Fig. 12.153 will be considered.

The switches S_0 through S_{N-1} select either of two binary logic states, $v(0)$ or $v(1)$ [i.e., $S_i = v(0)$ or $v(1)$ for $i = 0, 1, 2, \ldots, N - 1$]. Thus the binary word $b_{N-1}b_{N-2}\ldots b_2b_1b_0$ is represented by $S_{N-1}S_{N-2}\ldots S_2S_1S_0$, where S_i represents the **states** of the switches. As there are N switches, this

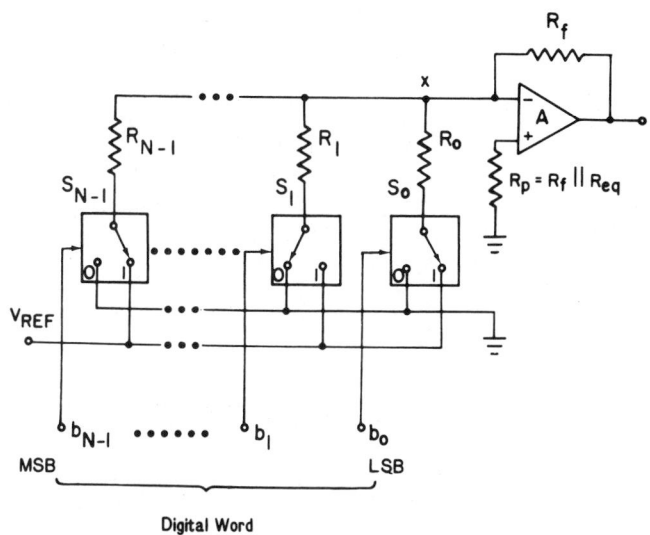

Digital Word

Fig. 12.153 Binary-weighted resistors, D/A converter.

represents an N-bit D/A converter. The values of resistances are determined by

$$R_k = \frac{R}{2^k} \qquad k = 0, 1, \ldots, N-1 \tag{12.221}$$

Observe that the values of resistances are inversely proportional according to the positional weight of each bit in an N-bit digital word. Binary-weighted currents set by the binary-weighted resistor network and the reference voltage V_{ref} are summed by the op-amp A. The op-amp operates as a low-output-impedance current-to-voltage converter. The output voltage v_o is

$$v_o = -\left(\frac{R_f V_{ref}}{R_{N-1}} S_{N-1} + \frac{R_f V_{ref}}{R_{N-2}} S_{N-2} + \cdots + \frac{R_f V_{ref}}{R_0} S_0 \right) \tag{12.222}$$

Using Eq. (12.221) in (12.222), we have

$$v_o = -\frac{R_f V_{ref}}{R} \left(S_{N-1} \cdot 2^{N-1} + S_{N-2} \cdot 2^{N-2} + \cdots + S_0 \cdot 2^0 \right) \tag{12.223}$$

Thus the output of the op-amp is proportional to the binary number represented by $S_{N-1} S_{N-2} \cdots S_1 S_0$, the weight for the most significant bit (MSB) being $(R_f V_{ref} \cdot 2^{N-1})/R$ and for the least significant bit (LSB) being $(R_f V_{ref} \cdot 2^0)/R$.

The main source of error is due to the voltage and current offsets. This can be compensated by inserting a resistor R_p between the noninverting terminal of the op-amp and the ground. The value of R_p is given by

$$R_p = R_f \| R_{eq} = \frac{R_f R_{eq}}{R_f + R_{eq}} \tag{12.224}$$

where R_{eq} is the equivalent resistance of the network to the left of the point x and is shown to be

$$R_{eq} = \frac{R}{2^N - 1} \tag{12.225}$$

The voltage error at the output due to offset current and offset voltage is given by

$$\text{Error} = R_f \left(I_{B_1} - I_{B_2} \right) + V_{os} \left(1 + \frac{R_f}{R_{eq}} \right) \tag{12.226}$$

Other sources of error are drift with temperature, the closed-loop-gain error (which may be neglected for a 12-bit, or less, D/A converter when the open-loop gain is 80 dB or higher). The slew rate and settling time of the op-amp are extremely important in high-speed D/A.

A major problem in commercially packaging such a circuit with large bit conversion is the difficulty in obtaining the proper resistances in the weighted-resistors network. For a large number of bits, the range of resistors R_0 through R_{N-1} becomes quite large. As an example, consider an $N = 12$-bit D/A converter with $R = 5$ kΩ. Then the resistor corresponding to the LSB will be $5 \times 2^{12-1} = 10.24$ MΩ. This entire wide range (from 5 kΩ to 10.24 MΩ) would require a precise ratio among all resistors (R_0 through R_{11}). This is a very difficult process in monolithic fabrication.

For a large number of bits a different approach should be considered. The commonly used $R, 2R$ ladder network illustrated in Fig. 12.154 solves this problem at the expense of two resistors per bit. The analysis of the circuit is similar to that of the previous circuit. The equivalent resistance is $3R$, the weight for the MSB is $-(V_{ref}/2)$ and the weight for the LSB is $-(V_{ref}/2^N)$. Other resistive network schemes also exist. They all produce a current proportional to the weight of each binary bit of a digital word.

Analog-to-digital (A/D) Converters
An A/D converter is used to convert an analog voltage into an equivalent digital word. There exist several methods of converting analog data to digital data. These methods can be grouped into two general categories:

1. A/D converters based on the **time-measurement principle** (i.e., dual-slope integrating A/D converters)
2. A/D converters based on **comparison methods** (i.e., successive-approximation A/D converters)

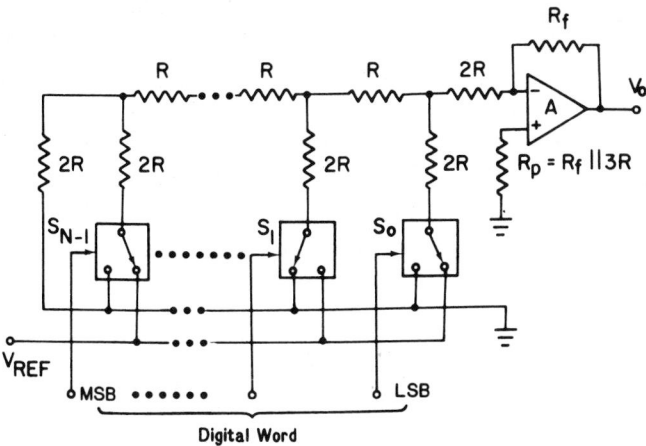

Fig. 12.154 D/A converter using an $R, 2R$ resistive ladder network.

Dual-Slope Integrating A/D Converters. The dual-slope conversion method makes use of the principle of indirectly measuring an unknown analog voltage by converting it to a time period. It consists of an integrator, a comparator (zero-crossing detector), an output counter, some control logic, and a clock. The block diagram of a 4-bit dual-slope A/D converter is shown in Fig. 12.155.

To start conversion, the 5-bit binary counter is first reset to zero (00000). The resulting zero at bit 4 opens S_2 and closes S_1, which connects the unknown analog voltage to the integrator. If the time constant of the integrator is $\tau = R_1C_1$, the integrator output v_1 is

$$v_1 = \left(-\frac{v_i}{\tau}\right)t \qquad (12.227)$$

the integration of v_1 continues for a fixed period of time T_1 and the counter counts up to 01111. On the next clock pulse the counter switches to 10000. It is assumed that v_i is always positive and is also constant during T_1, which can be achieved with a sample-and-hold circuit. This is shown in Fig. 12.155(b). The 1 in bit 4 causes the switch control to open S_1 and to close S_2. The output of the

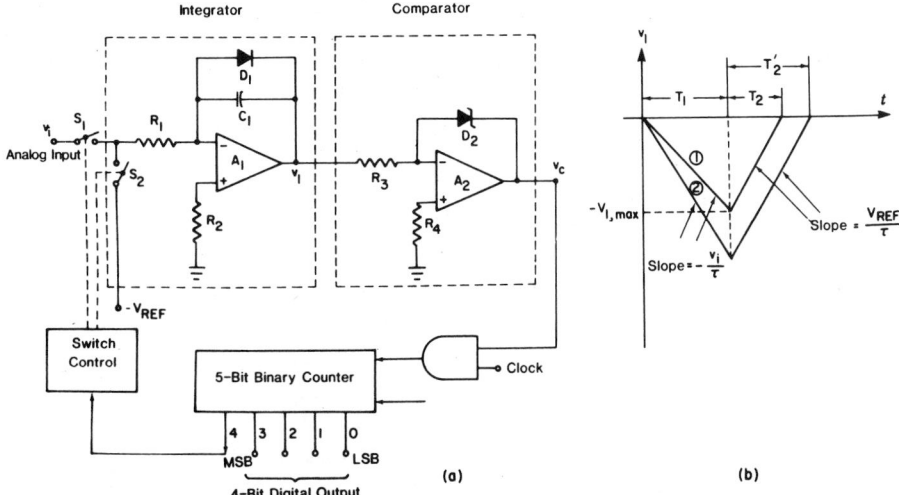

Fig. 12.155 Dual-slope integrator, A/D converter: (a) block diagram, (b) timing diagram.

integrate now starts to integrate $-V_{\text{ref}}$ and

$$v_1 = \left(\frac{V_{\text{ref}}}{\tau}\right)t \tag{12.228}$$

The counter continues to count up until the output of the integrator v_1 becomes zero (let the time required for this be T_2). At this instant the comparator output v_c goes to zero state and the AND gate prevents clock pulses from going to the counter. From Fig. 12.155(b) we can write

$$-V_{1,\,\max} = \frac{v_i}{\tau}T_1 = \frac{V_{\text{ref}}}{\tau}T_2 \tag{12.229}$$

or

$$T_2 = \frac{T_1}{V_{\text{ref}}}v_i \tag{12.230}$$

Observe that T_2 is proportional to the input analog signal v_i. Since T_2 is related to the number of counts in the binary counter, the 4-bit output digital word is equivalent to the input analog signal. When a higher input voltage is applied to the integrator [curve 2 in Fig. 12.155(b)] the output of the integrator at the end of T_1 is also larger. However, since the reference voltage V_{ref} remains the same, the slope during the integration of V_{ref} will also remain the same. Thus the time necessary for the output of the integrator to reach zero increases to T_2'. Therefore, higher input analog voltages give proportionally longer times T_2.

Observe that the tolerances of R_1 and C_1 do not affect the accuracy of the A/D converter. The advantages of the dual-slope integrating A/D converter are high noise rejection and excellent stability with both time and temperature. However, the voltage and current offsets of the integrator would limit the accuracy, and drift in V_{ref} also degrades the desired accuracy.

Examples of commercially available dual-slope integrating A/D converters are ADC1100 (11 bits, 42 ms), ADC14I (14 bits, 40 ms), AN2313 (10 bits, 6.7 ms), ADCE10B (10 bits, 1.25 ms). As this process of conversion is inherently slow, application is found mostly in voltmeters and other noncritical processing systems.

Successive-Approximation A/D Converters. This type of A/D converter approximates the analog input by successively trying a digital "1" in each successive bit of a feedback D/A converter. The block diagram of such an A/D converter is shown in Fig. 12.156. The successive-approximation approach is considerably faster than the dual-slope integrating A/D converter and other methods commercially available. Its accuracy is determined primarily by the accuracy of the internal D/A converter. It is important to note that the conversion process is not continuous (i.e., it begins with a START pulse and ends at the end of n clock periods), and the sample value of the analog input must be held constant during conversion. Therefore, the converter must be preceded by a sample-and-hold (S/H) circuit.

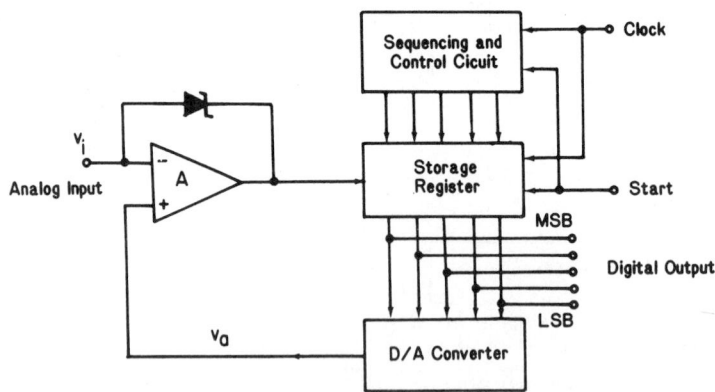

Fig. 12.156 Block diagram of a successive-approximation AID converter.

The conversion cycle begins with a START pulse that resets the flip-flops of the storage register and the flip-flops in the D/A converter. In this type of A/D converter each bit of the digital word is determined one at a time starting with the MSB as follows. During the first clock cycle, a 1 is tried for the MSB, which is equivalent to one-half the converter's full-scale (FS) range. If $v_i > v_a$, a logic 1 is retained in the MSB position in the storage register. If $v_i < v_a$, a logic 0 is placed. During the second clock cycle a logic 1 is applied to the bit next to the MSB. If the analog input is greater than the D/A output, the logic 1 at this bit is stored; otherwise, a logic 0 is stored in that bit position. This adds $\frac{1}{4}$ FS to the D/A output. The output of the D/A is not set to either $\frac{1}{4}$FS or $\frac{3}{4}$FS, depending on the previous comparison. This process is repeated for each bit. The final state of all the bits is the digital word equivalent to the analog input.

Conversion speed of the successive-approximation A/D converter is primarily determined by the D/A converter and the comparator. The accuracy of the converter can be no better than the accuracy of the internal D/A converter.

Other types of A/D converters are also available, such as the counter ramp A/D converter, voltage-to-frequency converters, and parallel A/D converters.

Some examples of commercially available successive-approximation A/D converters are ADC1103 (8 bits, 1.0 μs) and AD7570 (10 bits, 20 μs). These fast A/D converters are used mostly for real-time processing, where high speed is an utmost requirement.

12.10.11   Function Generators

Operational amplifiers, comparators, integrators, and associated circuitry can be used to generate specific waveforms. Many of these circuits are brought together in integrated-circuit form and are capable of generating sine, triangle, pulse, sawtooth, and square waves.

The approaches to generating any one single waveform are many. A few typical techniques will be described. The upper limit of usable speed is determined by the response time and slew rate of the op-amp employed. Therefore, to achieve maximum speed, compensation should not be used. Also, optimum performance of a function generator is based on the comparator.

Square-Wave Generator
The square-wave generator shown in Fig. 12.157(a) produces a waveform shown in Fig. 12.157(b). This circuit is also known as the **astable multivibrator**, as it has two quasistable states. That is, the output v_o remains in one state for a time T_1 and then changes to the second state abruptly for a time T_2. Hence the period of the square wave is $T = T_1 + T_2$.

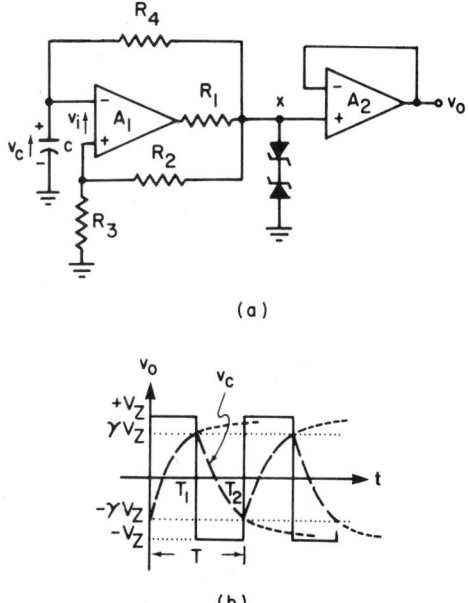

(a)

(b)

Fig. 12.157   (a) Square-wave generator, (b) output waveform.

Note from the circuit that a fraction of the voltage at point x is fed back to the noninverting input of the op-amp A_1. The fraction is determined by the voltage divider formed by R_2 and R_3 and is

$$\gamma = \frac{R_3}{R_2 + R_3} \qquad (12.231)$$

The voltage at x is also connected through the resistor R_4 to the inverting terminal of the op-amp, which charges and discharges capacitor C. Therefore, one can determine the differential input voltage v_i as

$$v_i = v_c - \gamma v_x \qquad (12.232)$$

When $v_i > 0$, the $v_x = -V_{z_1}$ and if $v_i < 0$, then $v_x = +V_{z_2}$. Hence the capacitor C will be charged exponentially toward V_{z_2} through the integrator formed by $R_4 C$. The voltage at x will remain constant at $v_x = V_{z_2}$ until $v_c = \gamma v_x = \gamma V_{z_2}$. When $v_c > \gamma v_x$, the output reverses itself abruptly such that $v_x = -V_{z_1}$. The capacitor now discharges exponentially toward $-V_{z_1}$. Since the op-amp A_2 is merely a voltage follower used as a buffer, $v_o = v_x$. The output voltage v_o and the capacitor voltage v_c are shown in Fig. 12.157(b) for $V_{z_1} = V_{z_2} = V_z$. For $0 < t < T/2$ it can be shown that

$$v_c(t) = V_z \left[1 - (1 + \gamma) e^{-t/R_4 C} \right] \qquad (12.233)$$

At the point of transition (positive to negative) $t = T_1 = T/2$, $v_c = \gamma v_x = \gamma V_z$, and it is found that

$$T = 2 R_4 C \ln \left(\frac{1 + \gamma}{1 - \gamma} \right) \qquad (12.234)$$

Observe that the frequency of the square wave $f = 1/T$ is independent of V_z. However, waveform symmetry, T_1 and T_2, is a function of V_{z_1} and V_{z_2}, and also depends on the op-amp.

The square-wave generator shown in Fig. 12.157(a) is excellent for fixed-frequency applications in the audio-frequency range. The frequency may be trimmed by varying R_4. Frequency stability depends primarily on C and zener diodes. The selection of A_1 is not critical for the frequency range 10 Hz to 10 kHz. At low frequencies, the op-amp bias current and the input impedance become significant. At high frequencies, the op-amp slew rate and the delay time of the op-amp, when coming out of saturation, are important. Hence to expand the range of frequency, the op-amp A_1 must be selected carefully. If the output of the op-amp, when saturated, is constant and symmetrical, then R_1 and zener diodes may be omitted.

Pulse Generators

Pulse waveforms are commonly used for timing and sampling. A pulse generator circuit is similar to the square-wave generator. As seen in Fig. 12.158, the resistor R_4 in Fig. 12.157(a) in the negative feedback loop is replaced by a resistance-diode network.

When the output is positive, D_1 conducts and the capacitor C is charged through R_{41}. When the output is negative, D_2 conducts and the capacitor C is charged through R_{42}. If $R_{41} < R_{42}$, then $T_1 < T_2$. Hence positive-going pulses are obtained. If the diodes are reversed or if $R_{42} < R_{41}$, then

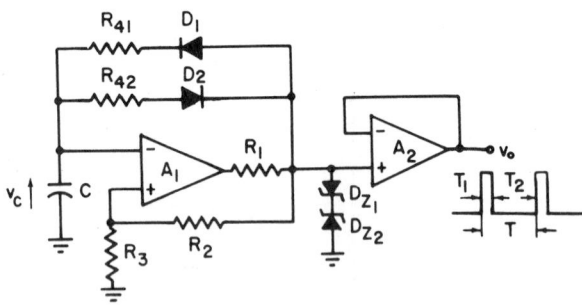

Fig. 12.158 Pulse generator.

negative-going pulses are obtained. The width of the pulses is determined by

$$T_1 = R_{41}C \ln\left(\frac{1 + \gamma}{1 - \gamma}\right) \qquad T_2 = R_{42}C \ln\left(\frac{1 + \gamma}{1 - \gamma}\right) \tag{12.235}$$

The period of the pulse train is $T = T_1 + T_2$.

Triangular-Wave Generator

A triangular wave is obtained when a square wave is integrated. In Fig. 12.157 this is achieved by R_4 and C. When the capacitor voltage v_c integrates up to γV_z, the comparator reverses the slope of the integration voltage. The result is a triangular wave. However, as seen from Fig. 12.157(b), the slope of v_c is quite nonlinear, being rather exponential. The linearity can be improved by using only the initial portion of the voltage v_c, which can be achieved by making γ very small. However, better triangular-wave linearity can be achieved by maintaining a constant capacitor charging current. This results in a constant rate of change of voltage with time.

In the circuit shown in Fig. 12.159, the FETs form a floating constant-current source that supplies both polarities of charging current. For a given output voltage polarity, one gate will be forward biased while the other gate will be reverse biased. When the output voltage polarity reverses, the roles of the FETs will be interchanged, which also changes the polarity of the charging current. The period of oscillation is determined by

$$T = 4\gamma V_z C \tag{12.236}$$

Observe that the period is a function of V_z, while in the previous case the period of v_c was independent of V_z. To ensure symmetry the FETs must be matched carefully. Also, to operate as current sources, the gate-drain voltage of FETs must be large enough to induce pinch-off.

To achieve a better control and greater precision a separate integrator may be used as shown in Fig. 12.160.

The integrator formed by A_2, R_f, and C_f integrates the voltage difference $v_s - V_s$, in which the polarity of v_s is changed periodically. Hence the voltage integrated will be increased or decreased by the amount of V_s. This in turn increases or decreases the integration rate. Therefore, V_s can be used to control symmetry. The midpoint of the triangular wave is adjusted by V_{os}, as indicated in Fig. 12.160. The period of the triangular wave is found to be

$$T = \frac{2V_{pp}V_z}{V_z^2 - V_s^2} R_f C_f \tag{12.237}$$

where V_{pp}, the peak-to-peak voltage of the wave, is given by

$$V_{pp} = 2V_z\left(1 + \frac{R_3}{R_2}\right) \tag{12.238}$$

Observe that the amplitude of the triangular wave is adjusted by the ratio R_3/R_2 and V_z. When V_{pp} is fixed, the frequency of the oscillation is controlled by varying R_f.

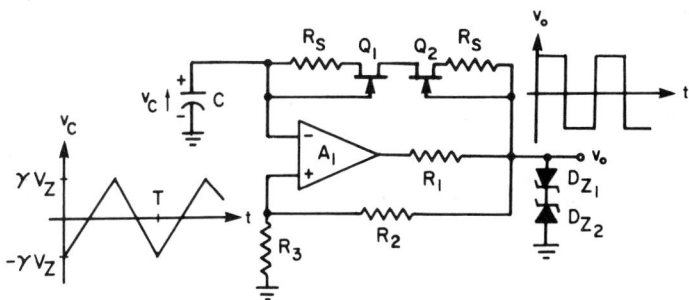

Fig. 12.159   Triangular-wave generator with improved linearity.

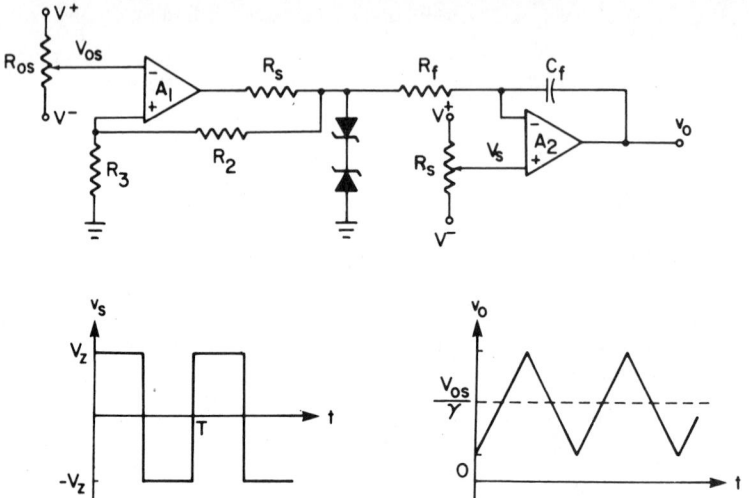

Fig. 12.160 Precision triangular-wave generator with a specific integrator.

Sawtooth Generators

Sawtooth generators are similar to triangular-wave generators in that a linear ramp is generated as shown in Fig. 12.161(b). A sawtooth waveform is used in sweep testing and display application. To obtain such a waveform, a pulse train is integrated. In this case, T_1 and T_2 is given by Eq. (12.235).

Another approach to obtaining a sawtooth waveform is shown in Fig. 12.162. If I is a constant-current source,

$$v_c = \frac{I}{C}t \tag{12.239}$$

which represents a linear ramp function. Voltage v_c is then applied to a comparator A_1. When v_c reaches an amplitude predetermined by V_{ref}, the comparator triggers the monostable, which acts as a driver for Q_1. This in turn discharges capacitor C. The cycle then repeats itself. Amplifier A_2 acts as a buffer. The time T_1 is the time required for voltage v_c to reach V_{ref}. Time T_2 is the time needed to discharge the capacitor. The pulse width of the monostable should then be sufficient so that v_c will return to the desired starting point (nominally zero). The repetition rate (or frequency) then depends on V_{ref}, current source I, the on-resistance R_{ds} of Q_1, as well as temperature instabilities and the capacitor C, that is,

$$T_1 \simeq \frac{C}{I}V_{ref} \qquad T_2 \simeq 4R_{ds}C \tag{12.240}$$

Several function generators capable of generating a square wave, triangular wave, sine wave, and so on, are available commercially (i.e., Intersil 8038, Signetics 566, Exar 2206, etc.).

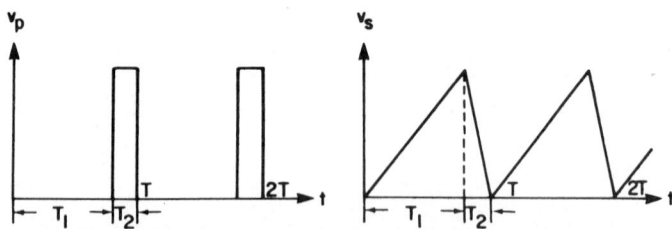

Fig. 12.161 (a) Pulse train, (b) sawtooth waveform obtained from the pulse train in (a).

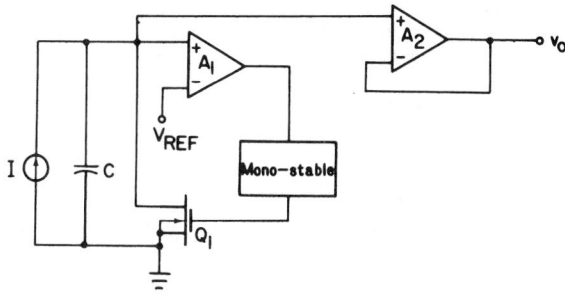

Fig. 12.162 Sawtooth generator.

12.11   MICROPROCESSORS IN INSTRUMENTATION
Kenneth Short

The advent of the microprocessor has allowed the logic and computational power of a digital computer to be economically included in an instrument. The effect on the instrument user is the availability of instruments with improved performance and a host of new features. These features would not be economically feasible in a hard-wired logic counterpart of an instrument containing a microprocessor. Table 12.12 lists some of the features that can be economically included in instruments that incorporate a microprocessor. Additionally, the use of microprocessors has resulted in instruments that combine the functions provided by several conventional instruments into a single instrument and in instruments that implement entirely new measurement functions.

The effect of microprocessors on the instrument designer is the emergence of a new and somewhat general structure for instrument design along with a significantly wider range of approaches to implementing the various functions required in a particular instrument. An instrument that is designed around either a microprocessor or a microcomputer will be referred to as a microprocessor-based instrument.

An example of a microprocessor-based instrument, with digital input, is a universal counter capable of frequency, period, time interval, ratio, and totalize measurements. Pushbuttons and knobs on the front panel allow the operator to select the desired instrument function, input attenuation, trigger level, and sensitivity. A seven-segment LED display indicates the value of the result as a mantissa and exponent. LED annunciators indicate the proper engineering units.

When the instrument is turned on, a power-up reset and self-check style is automatically entered. During this cycle the microprocessor performs a check sum on its internal program in read-only memory (ROM) and checks its read-write memory (RWM). Also, a partial check of other LSI devices and the I/O ports of the system is performed. Any failure is indicated by a code on the display.

After completion of the self-check the microprocessor scans the front panel controls to determine the selected mode of operation and configures the internal logic of the system appropriately. The measurement begins when the input signal is detected. The counter uses a reciprocal counting technique to provide maximum resolution when the frequency mode of operation has been selected and the input frequency is below its internal clock, f_c. In effect, the period of the input waveform is

TABLE 12.12   Some Features That Can Be Economically Included in a Microprocessor-Based Instrument

1.	Self-test
2.	Fault isolation
3.	Automatic calibration
4.	Automatic ranging
5.	Zero and gain error correction
6.	Mathematical and statistical computation
7.	Decision-making capability
8.	Repeated measurements
9.	Output or display of data in engineering units
10.	Interfaces to computer-type peripherals
11.	Interface to IEEE 488 bus

measured and the microprocessor computes and displays the frequency. For input frequencies above f_c the counter automatically switches to a conventional frequency counting technique.

A microprocessor-based universal counter might optionally allow its operating mode and initiation of measurement to be programmed remotely through a machine interface. Results of the measurement would also be transmitted over the machine interface. An instrument with a machine interface is readily incorporated with other instruments into an automated measurement system.

12.11.1 Generalized Microprocessor-Based Instrument

A simplified block diagram of a general microprocessor-based instrument, utilizing a single microprocessor, is shown in Fig. 12.163. This figure shows four major subsystems interconnected by a common bus: the microcomputer, the measurand interface, the operator interface, and the machine interface. All microprocessor-based instruments contain the first of these subsystems. A microprocessor-based instrument may contain either an operator interface or machine interface or both. The microcomputer subsystem consists of the microprocessor, read-only memory, read-write memory, and possibly I/O ports. Depending on the physical partitioning of the system hardware, the I/O ports may be associated with the various interfaces as implied in Fig. 12.163. The measurand interface provides the interface between the measurands of interest and the system bus. Typically, these measurands are analog in nature; thus the measurand interface includes analog signal conditioning, analog multiplexing, and A/D conversion. The operator interface provides the switches, pushbuttons, and other controls manipulated by an operator to control the measurement operation. The operator interface also provides the visual displays by which the operator monitors the measurement process and reads the measurement results. Information similar to that provided by and presented to the operator may be provided by or presented to an electronic device via the machine interface. The interfaced device may be a computer-type peripheral, which provides hard copy or mass storage of the results, or a controller that provides setup information to the instrument and receives the measurement results.

An application program in ROM is executed by the microprocessor and controls the operation of the measurand, operator, and machine interfaces. The microprocessor also implements the computations or data transformation desired. In a microprocessor-based instrument, the interface subsystems are optimized for control by an interaction with the microprocessor. This optimization is enhanced by the existence of a wide range of analog and digital LSI devices designed for use with a microprocessor. In complex instruments additional microprocessors may exist as part of the measurand, operator, or machine interfaces.

The components that comprise the four major subsystems may be physically partitioned in a number of ways. In small systems, all the components may be contained on a single printed-circuit board. The components are interconnected by the microprocessor's system bus, which includes an address, data, and control buses. Other more complex systems may distribute the components among several printed-circuit boards or cards, with each board providing a functionally modular subsystem. The printed-circuit boards are then interconnected via a backplane bus in the instrument mainframe. This structure allows features to be added to an instrument by simply adding printed-circuit boards to the mainframe.

Additionally, this structure allows implementation of virtual instruments that use a common mainframe, and common microcomputer, operator interface, and machine interface printed-circuit boards. The addition of a specialized measurand interface board(s) and appropriate software creates the desired instrument.

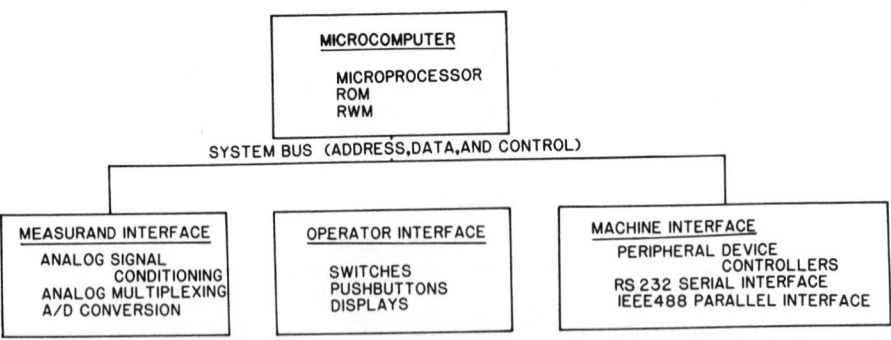

Fig. 12.163 Block diagram of the major subsystems of a generalized microprocessor-based instrument.

When power is applied to a microprocessor-based instrument, a power-on circuit resets the microprocessor, and it begins to immediately execute its application program stored in ROM. Additionally, the reset pulse from the power-on reset circuit is used to preset or clear any storage elements in the measurand, operator, or machine interfaces that require such initialization.

The application program first initializes any programmable LSI devices used to implement the measurand, operator, or machine interfaces. Next, a self-check routine may be executed that tests the operability of the instrument system. If the self-check is completed successfully, the system is then ready to accept commands from the operator or machine interface in order to determine its mode of operation, measurement function, and the format of the data to be displayed.

Once a measurement is initiated, the software controls and sequences the measurement. The microprocessor controls the acquisition of data from the measurand and stores it in its memory for processing. Once the data have been acquired, the microprocessor can carry out any digital transformations on the data that may be desired. Such computations include correcting data for offset and gain errors, filtering data digitally, linearizing input data, ratio taking, sorting, and computing desired results.

Once the desired results have been computed, the microprocessor formats them for output in engineering units. The results are then displayed at the operator interface or transferred to another electronic system through the machine interface.

12.11.2 Microcomputer Subsystem

Several alternatives exist for implementing the microcomputer subsystem. The initial choice is between the use of a single-chip microprocessor, single-chip microcomputer, or bit-slice microprocessor. A single-chip microprocessor contains an arithmetic-logic unit, control unit, general-purpose registers, and usually a clock on a single LSI chip. ROM and RWM are provided by additional IC devices.

The particular microprocessor chosen determines the signals that comprise the address, data, and control buses that provide the electrical interface between the microprocessor and ROM, RWM, and I/O ports. The entire operation of the microprocessor-based system is controlled by the execution of the instructions that comprise the application program that is sorted in ROM.

A number of types of IC memory devices may be used to provide the system's ROM. These include mask-programmed ROMs, and programmable ROMs or PROMs, which are programmed once with the bit patterns that comprise the application program and cannot be subsequently modified. Masked ROMs are programmed during integrated-circuit fabrication, whereas PROMs are programmed after their manufacture using a specialized PROM programmer.

Erasable programmable ROMs or EPROMs, and electrically alterable ROMs or EAROMs, can be reprogrammed to allow modification of the application program as often as is desired. These reprogrammable ROMs are used for system development.

All the aforementioned ROMs are nonvolatile; the contents of these memories are not altered due to interruption of the system's operating power. Thus an application program in ROM is always available for execution as soon as the system is powered up. In addition to the application program, any other fixed data needed by the system is stored in ROM. This includes tables of constants and fixed messages sent to the system operator.

The RWM is volatile; its contents are altered if the memory operating power is interrupted. The microprocessor uses the RWM for storage of data that has been input and is awaiting processing and the results of processing that are awaiting output. In addition, intermediate results of processing are temporarily stored in RWM. A portion of RWM is also used as a first-in-first-out (FIFO) stack to provide storage for subroutine return addresses.

Transfer of data between the microcomputer subsystem and other subsystems is accomplished via ports. The structure shown in Fig. 12.163 shows the various subsystems interconnected by a common system bus. In this type of structure the I/O ports are associated with the interface subsystems. Input ports provide a pathway for digital data to the microcomputer system and output ports provide a pathway for digital data from the microcomputer system. In its simplest form, a port is a register. An input port requires in addition to the register that holds the data to be input, a three-state buffer between the register and data bus, as shown in Fig. 12.164. When the input port is addressed by the microprocessor, the address decoding logic enables the three state outputs of the buffer, allowing the contents of the register to drive the data bus and the microprocessor to input the data. An output port need only consist of a register as shown in Fig. 12.165. To output data to the register, the microprocessor places the address of the register on the address bus, and the data to be output on the data bus, and when the data are stable, the microprocessor generates a write strobe. The address decoder associated with the output port generates a strobe to clock the data into the output register. More complex programmable LSI input and output ports have their direction (input or output) determined by a command that is sent to the programmable I/O port when it is initialized.

The circuitry associated with many LSI input and output devices contains the input or output port as an integral part of the device.

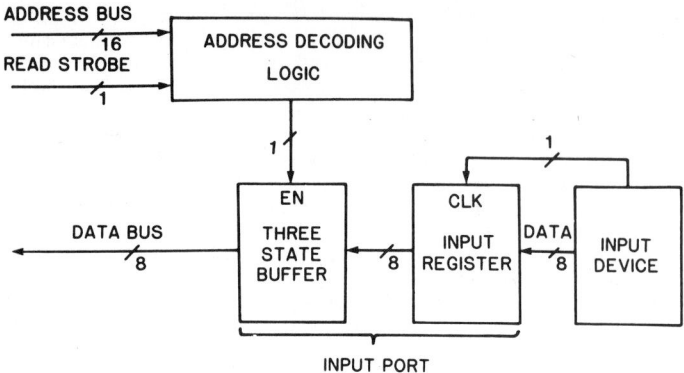

Fig. 12.164 Block diagram of hardware to input data to the system bus.

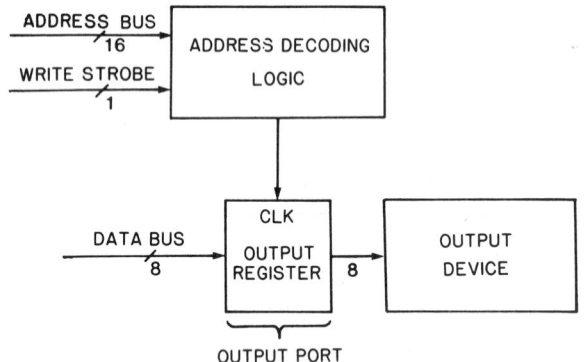

Fig. 12.165 Block diagram of hardware to output data from the system bus.

In a microprocessor-based instrument, the system bus interconnects all the major subsystems: microprocessor, memory, and I/O devices (via I/O ports). The actual signals that comprise a system bus depend on the microprocessor chosen. However, most LSI peripheral devices are designed to be compatible with most microprocessors of a particular word length: 4, 8, or 16 bits. Efforts have been made to standardize timing requirements of peripheral devices to enhance their compatibility further.*

For small systems all the LSI components may be contained on a single printed-circuit board. A single bus structure that interconnects the IC components on the board is all that is required. For larger systems a number of printed-circuit boards may be required. Partitioning of the ICs among the various boards is usually done in such a manner that each printed-circuit board implements a distinct subfunction of the entire system function. The interconnection of these boards is accomplished via a bus, which may differ somewhat from that of the microprocessor itself. A number of standard buses exists (e.g., STD BUS† and MULTIBUS‡). Each standard defines the signals that comprise the bus, their timing, and relationship to each other. These bus structures usually allow any microprocessor to be used as long as any additional logic that is required to provide signals compatible with the bus standard is included on the board containing the microprocessor.

Single-chip microcomputers contain the microprocessor ROM, RWM, and I/O ports on a single LSI circuit. The amount of ROM and RWM, and the number of input/output lines provided by a single-chip microcomputer is limited. However, for limited complexity applications where the ROM, RWM, and I/O lines available in a particular single-chip microcomputer are sufficient, the advantage

*Series 8000 Microprocessor Family Handbook; The MICROBUS, Chap. 2, and MICROBUS Electrical Specifications, App. A, National Semiconductor, 2900 Semiconductor Drive, Santa Clara, Calif. 95051, 1978.
†Series 7000 STD BUS Technical Manual, Pro-Log Corporation, Monterey, Calif. 93940, 1979.
‡Intel MULTIBUS Specification, Intel Corporation, 3065 Bowers Ave., Santa Clara, Calif. 95051, 1978.

of reduced parts count and greater reliability are significant. Thus single-chip microcomputers are preferred in small dedicated systems that do not provide for system expansion.

Bit-slice microprocessors are high-performance microprocessors composed of a number of ICs. Each of the constituent ICs is fabricated using a high-speed technology such as Schottky T^2L, I^2L, or ECL. Typically, the bit-slice microprocessor's control and sequencing logic is contained on one IC and its microprogram and application program on additional ICs. The registers and arithmetic logic unit are partitioned over several identical ICs. Each IC provides a 2-, 4-, or 8-bit slice of each of the microprocessor's registers and arithmetic and logic circuitry. The slices are cascaded to provide the desired word length for the microprocessor. The high performance provided by bit-slice microprocessors is not required in most instruments. This fact, together with the increased parts count and increased design effort required for their application, precludes consideration of the use of bit-slice microprocessors in all but the highest-performance instrument.

Reduced to its simplest, the microprocessor implements the instrument's functions by transferring data among the registers that comprise the microprocessor, memory, and I/O ports and transforming data in the microprocessor. The microprocessor provides the logic and arithmetic capability for the instrument. Once reset, the microprocessor sequentially fetches instructions from its application program in ROM and executes them. This sequential nature of operation can be altered by the execution of a branch instruction that causes control to be transferred to a nonsequential instruction. Conditional branch instructions allow control to be transferred based on the value of data input to the microprocessor or the results of a logic or arithmetic computation.

The application program is written as a main program and a number of subroutines. The subroutines are constructed in a modular fashion and implement a particular software subfunction.

One or more subroutines implement the instrument's initialization when power is turned on. A number of programmable LSI devices are typically used in the interfaces to provide I/O ports, keyboard scanning, display refreshing, serial communications, and peripheral device control. These devices are somewhat general in nature and have a number of modes of operation and formats. Such devices contain a command register(s) that must be loaded to configure the device for its particular mode of operation in the instrument. The initialization subroutines load each programmable device with the necessary command words from ROM.

Other subroutines provide a self-check to test the system's operability before it can be used. These tests usually assume that the microprocessor and the part of its ROM containing the self-check programs are operating properly. Each ROM IC in the system is checked by using some type of checksum technique. A simple technique requires that the last word of each ROM is programmed with a value so that the sum of all the words of the ROM is zero or some other predefined value. The self-check subroutine adds all the words of the ROM and checks that the sum is the desired value.

Each word of RWM is checked to see that it can be written and read. Any number of techniques may be used; one simple test writes words to the memory to form a checkerboard of 0's and 1's and then tries to read them back. If successful, it then writes a reversed pattern and tries to read it back. Another popular test is the "walking 1's and 0's test." The routine checks to see if a single 1 can be written to and read from a bit position with all other bit positions 0. This is repeated for each bit position, in turn, creating a "walking 1." The same test is performed with a single 0 and all 1's, a "walking 1."

I/O ports and programmable LSI devices are tested to whatever extent is possible and practical depending on the nature of the device. Once the self-check has been successfully completed, the instrument is ready to carry out its normal functions.

Other functions that are carried out in software will be considered as appropriate in discussing the interface subsystems. The data transferred to or from the I/O ports are digital, and transformations between the analog and digital domains occur in the interface systems.

12.11.3 Measurand Interface

The measurand interface has as its inputs the analog or digital measurands and provides at its output digital data appropriate for input to the microcomputer subsystem. For digital measurands the measurand interface typically implements only threshold detection, level conversion, or pulse shaping to provide a digital signal that can be input through a three-state buffer controlled by the microprocessor.

Analog signals must be converted to a binary word in order to be input by the microprocessor. This conversion is accomplished by an analog-to-digital (A/D) converter that provides a binary word as its output. Analog signal conditioning is typically required preceding the A/D converter input, so that the range of the analog signal at the A/D converter's input matches the input range of the A/D converter. Thus the A/D converter provides a binary output of maximum possible resolution. Analog signal conditioning may be as simple as an amplifier that matches the output range of the measurand transducer to the input range of the A/D converter. Depending on the measurand's characteristics and the transducer employed, other signal-conditioning operations such as buffering or filtering may be required.

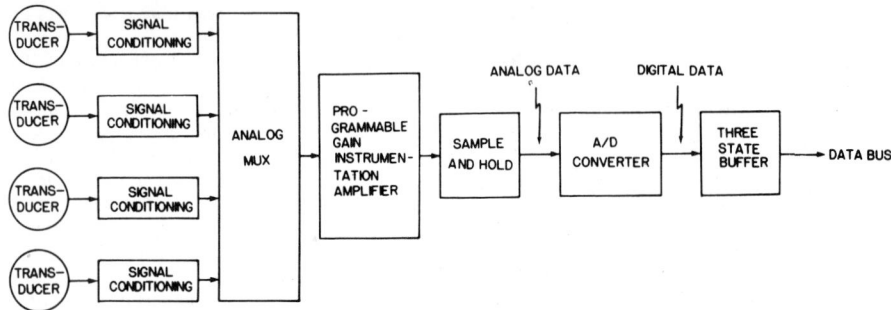

Fig. 12.166 Generalized measurand interface for the acquisition of analog data.

A generalized measurand interface for the acquisition of analog data is shown in Fig. 12.166. A transducer converts the analog measurand to a voltage for input to the analog multiplexer. The transducer may have signal-conditioning circuitry associated with it to provide the appropriate analog voltage range. The multichannel analog multiplexer receives a command (bit pattern) from the microprocessor to select the input channel.

The output from the analog multiplexer is amplified by a programmable-gain instrumentation amplifier. The instrumentation amplifier's gain is programmed by a command from the microprocessor. A sample-and-hold circuit, also controlled by the microprocessor, allows the microprocessor to control the sample time and provides a constant analog output for conversion by the A/D converter. The A/D converter provides the digital output to the microprocessor via three-state buffers to the system data bus. When the number of bits provided by the A/D converter exceeds the number of bits provided by the data bus, the three-state buffers are arranged so that the A/D output is transferred as multiple words to the microprocessor.

The data acquisition system of Fig. 12.166 can be implemented using integrated circuits. A/D converters that are microprocessor compatible simplify this task by providing the necessary three-state buffers for interfacing to the system bus as part of the A/D converter. Instruments that utilize a standard bus can take advantage of premanufactured data acquisition boards from various manufacturers. A typical data acquisition board has 16 channels of single-ended input or 8 channels of differential input. Its instrumentation amplifier has a software programmable gain from 1 to 1024, and its A/D converter provides a 12-bit output.

The microprocessor can readily convert the digitized transducer output from the measurand interface to a value representing the physical parameter measured or some derived quantity. The utilization of a microprocessor in an instrument allows a wider choice of devices as transducers. A transducer can be utilized in a microprocessor-based system provided that its output is either very stable or very linear.

If a transducer's output is stable but not linear, linearization can be accomplished by the microprocessor. Software linearization may be accomplished by a mathematical transformation of the digitized input or by a table-lookup procedure. For a transducer that is very linear, only short-term stability is required for it to be used with a microprocessor-based instrument. The microprocessor can provide automatic calibration, removing gain and zero errors. This is accomplished by using the multiplexer to allow measurement of a zero reference, a local transfer standard or internal reference, and the measurand. The microprocessor then computes the corrected value of the measurand from these three measurements.

The use of a programmable-gain instrumentation amplifier, together with the multiplexer, allows maximum resolution to be obtained from measurands with widely different output ranges without requiring individual analog signal conditioning for each channel in order to match the input range of the A/D converter. The gain setting for the instrumentation amplifier, for each channel selected, can be permanently stored in memory. Alternatively, the gain setting can be computed dynamically by the microprocessor providing increased resolution for lower values of the input measurand, at the price of increased conversion time.

In addition to linearization, calibration, and autoranging other front-end processing operations, such as digital filtering, may be accomplished.

12.11.4 Operator Interface

Conventional instruments provide operator control via front-panel switches. These switches directly implement the desired switching or selection function.

Several approaches are used on microprocessor-based instruments. Keypads or key switches on the front panel provide "soft keys" that do not directly implement any switching but are scanned by the microprocessor or keyboard scanning hardware in the operator interface. The microprocessor translates a keystroke or sequence of keystrokes into the desired switching or function selection operation. The microprocessor controls hardware in the various interfaces to implement commands entered through the soft keys.

Use of an LSI keyboard scanner relieves the microprocessor from directly scanning the keyboard's key matrix. The LSI keyboard scanner independently scans the keyboard and detects each key closure. In response to the detection of a key closure, the keyboard scanner sets a data available flag. This flag can be polled periodically by the microprocessor or can be used to interrupt the microprocessor directly. The code corresponding to the key pressed can then be read by the microprocessor from an input register in the keyboard scanner.

When mechanical switching is a necessity, the microprocessor can light LEDs on the front panel to indicate which button is to be depressed by the operator, thus providing microprocessor-directed control.

Conventional instruments use meter movements and various annunciator and numeric displays on the front panel to provide visual outputs to the operator. Microprocessor-based instruments primarily use numeric displays. However, alphanumeric displays that provide messages as well as measurement results to the operator are also used.

Multiplexed multidigit seven-segment displays can be refreshed by an LSI display driver. These devices contain a word of memory for each digit. The microprocessor writes the information to be displayed to this memory buffer. The display driver independently controls the multiplexing of information from the memory buffer to the display, providing the necessary refresh rate. The display driver has the current-handling capability to drive the seven-segment displays directly.

The conventional type of operator interface in the form of a front panel may not exist at all in a microprocessor-based instrument. That is, the instrument may have no front-panel controls or displays at all. For such an instrument the operator interacts with the instrument via a CRT terminal or similar device, or an instrument controller connected through the machine interface.

12.11.5   Machine Interface

The machine interface allows external devices to be connected to the microprocessor-based instrument. These devices may be either computer-type I/O peripherals such as printers, plotters, keyboards, CRT terminals, and magnetic tape, or flexible disk storage. The machine interface is also used to connect the microprocessor-based instrument to a computer or controller. The computer or controller provides setup information to the instrument, controls its overall operation, and receives the instrument's measurement results. Use of a separate computer or controller to supervise an instrument's operation allows a number of instruments and peripheral devices to be readily interconnected to form an automated instrumentation system.

Conventional instruments typically provide dedicated parallel outputs for interfacing to other electronic devices. These nonstandard interfaces differ from one instrument to another. Thus additional custom hardware interfaces are required to make these instruments' remote inputs and outputs compatible to devices to which they were being interfaced.

Substantial standardization for serial interfaces is provided by the use of the RS232C standard.* LSI universal asynchronous receiver/transmitter (UART) circuits provide an economical building block for serial interfaces. Standardization for parallel interfacing of instruments is provided by the IEEE-488 standard.

12.11.6   IEEE 488 General-Purpose Interface Bus

IEEE standard 488 is a standard for a digital interface for programmable instrumentation.[†] This standard specifies the mechanical, electrical, and functional requirements of a digital interface between instruments interconnected to form an automated system.

The interface is specified in essentially device- and system-independent terms. The interface consists of a passive interconnecting bus and interface logic contained in each instrument. The standard defines the bus's physical connector, the function of the interconnecting wires, logic conventions, and the format and timing of control and data signals.

A wide variety of commercial instruments from different manufacturers, each containing the 488 interface, can be readily interconnected to form an automated measurement system. Commercial

---

*EIA Standard RS232-C Interface Between Data Terminal Equipment and Data Communication Equipment Employing Serial Binary Data Interchange, Electronic Industries Association, 2001 Eye St., N.W., Washington, D.C. 20006.
[†]IEEE Standard 488-1978, IEEE Standard Digital Interface for Programmable Instrumentation. The Institute of Electrical and Electronic Engineers, Inc., 345 East 47th Street, New York, N.Y. 10017.

devices containing a 488 interface include measurement, display, stimulus, storage, and processor/controller devices. Special-purpose instruments that contain a 488 bus interface can easily be integrated into automated systems with commercial devices.

As many as 15 instruments containing IEEE 488 interfaces may be connected to a common bus. The maximum length of cable used to make the interconnection is 20 m. Devices connected to the IEEE bus must have the capability to implement one or more of the following interface functions: talker, listener, or controller. A talker (or device having the talker capability) is a device that can send data over the bus. A listener is a device that can receive data from the bus. The controller designates which device is to talk and which is to listen. A controller can also send special types of data (address and commands) to all devices connected to the bus. The controller can also receive status data from other devices. Typically, a system contains one controller, usually a programmable calculator or microcomputer, and several talkers and listeners. The controller assigns only one talker to be active at a time but several listeners may be active simultaneously. The controller uses a group of commands, known as interface messages, to direct the other instruments on the bus in carrying out their functions of talking and listening.

The bus structure consists of 16 signal lines divided into three groups: data input/output bus group, data byte transfer and control group, and general bus management group (see Fig. 12.167). The data input/output bus (DIO1-DIO8) is an 8-bit bidirectional, multipurpose data bus. Information is transferred between devices on the bus in a bit parallel-byte serial manner using these eight lines. This information includes universal commands, addresses, program data to set up an instrument, measurement results, and status data. The type of data on DIO1-DIO8 is indicated by ATN. If the controller has made ATN true, the DIO lines contain addresses or universal commands of interest to all devices. When the controller has ATN false, the device previously addressed to talk and those previously addressed to listen exchange device-dependent data on the DIO lines.

The data byte transfer control group consists of three lines (DAV, NRFD, NDAC) used to provide handshaking for asynchronous data transfers. The talker uses DAV (data valid) to indicate the presence of valid data on the DIO lines. Several devices may have been addressed to listen. Each device indicates its acceptance of the data by asserting DAC true and RFD false. The NRFD and NDAC outputs of each device are open collector and wired together to form logical AND (wired OR) functions. When the talker sees the NDAC line false, it sets DAV false and removes the data byte from the bus. When ready to receive a new data byte, each listener asserts RFD true and DAC false, thus completing the transfer of a single byte of data. The effect of wire-ORing all devices' NRFD outputs and NDAC outputs is that data are transferred asynchronously at the rate of the slowest device addressed to listen. The bus specification allows transfers up to the rate of 1 M byte/s.

The bus management group consists of five control lines. Interface Clear (IFC) is used by a controller to place the interface system in a known quiescent state. The Attention (ATN) line is, as previously mentioned, used by a controller to indicate that information on the data lines is an interface control message or device-dependent data. The Remote Enable (REN) line is used to select between local front-panel control and remote control via the bus. Devices use the Service Request (SRQ) line to indicate to a controller that the device requires attention. The End or Identify (EOI) line has two functions. It is used by a talker to indicate the last byte of a multibyte data transfer. Or, the controller can assert this line concurrently with the ATN line to execute a parallel poll, obtaining a single status bit from up to eight devices on the DIO lines.

The total functional capability of the IEEE 488 interface is divided into 10 interface functions: five basic functions (Source Handshake, Talker, Listener, Acceptor Handshake, Controller) and five

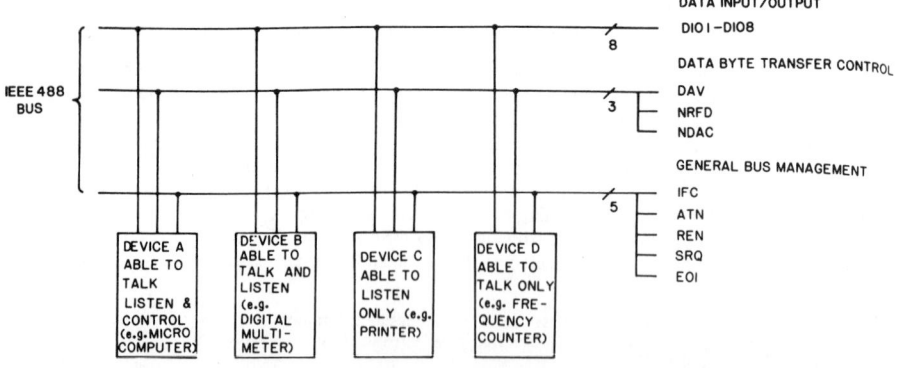

Fig. 12.167   IEEE 488 bus.

supplementary functions (Clear, Trigger, Remote/Local, Service Request, Parallel Poll). Almost all instruments will contain either a talker or a listener function or both. Only a few instruments will contain a controller function.

IEEE standard 488 uses state diagrams to describe each of the interface functions. The use of state diagrams provides a method of specifying the interface functions independent of their method of implementation. Several methods are available for implementing the interface function logic.

For some simple instruments where only a few of the interface functions need to be implemented and where the microprocessor is not totally occupied with implementing the device's functions, the microprocessor can directly implement the interface function state diagrams in software.*

Other instruments may require a hardware-intensive approach to implementing the interface function. Such is the case when high-speed transfers are required or when the microprocessor's capability is almost totally utilized in implementing the device's instrumentation function.† When the instrument requires only a few interface functions, they can be implemented economically using SSI and MSI circuits. Alternatively, particularly in cases where a number of, or all of the interface functions are to be implemented, LSI circuits specifically designed to implement IEEE 488 interface functions can be used.

All 10 of the functions defined by the IEEE 488 standard can be implemented with a single LSI general-purpose interface bus IC. These ICs are designed to interface between a microprocessor system bus and the IEEE 488 bus and provide the talker, listener, and controller function logic. Bus transceiver ICs are required between the 488 interface IC and the IEEE 488 bus to provide the needed buffering and drive.

Two types of messages are sent on the bus interface: control messages and device-dependent messages. The interface control messages are used to control the interaction of devices on the bus and include addresses, universal commands, and status data. The coding scheme for these messages is defined by the standard and is based on the ASCII code.

Device-dependent messages may be sent in any code, although the ASCII code is often used. For two devices on the bus to communicate, the protocol coding of device-dependent messages must be compatible. Numeric data are often transmitted in ASCII using FORTRAN exponential, floating-point, or integer format.

The microprocessor in a microprocessor-based system can translate from internal numeric codes to the code and format used to transfer numeric data on the bus. Additionally, the microprocessor must interpret device-dependent commands sent as data and cause the appropriate actions to take place within the instrument. A draft document provides recommended code and format guidelines for device-dependent messages.‡

12.11.7 Software Development

The design of a microprocessor-based instrument requires, in addition to the hardware design, the development of the application program, which is executed by the microprocessor to control the instrument. This application program must be translated to a sequence of binary words that can be programmed into a ROM.

The application program can be written in any one of the several different types of languages: machine language, assembly language, or a high-level language. Machine language is the simplest conceptually, but the most difficult in a practical design. Here the program is written directly in binary form, with each of the microprocessor's instructions consisting of one or more binary words. Often these binary words are written in shorthand notation as octal or hexadecimal numbers. Machine language programming has several drawbacks, including being tedious and error prone and requiring that all addresses and constants be specified as absolute values.

Assembly language programming represents instructions using easily remembered mnemonics and allows addresses and constants to be specified symbolically. These features of assembly language ease the programming effort considerably. Translation of a program written in assembly language to its binary equivalent requires another program called an assembler. The assembler takes the application program written in assembly language and referred to as the source program, and translates it to its binary equivalent, referred to as object code.

A one-to-one correspondence exists between the assembly language instructions and the machine language instructions to which they are translated. Thus the programmer still deals directly with the architecture of a specific microprocessor, specifying all transfers of data among registers and transformations on data in registers.

*R. YOUNG, "Implementing an IEEE-488 Bus Controller with Microprocessor Software," *IEEE Trans. Ind. Electron. Control Instrum.*, **IECI-27**(1), 10–15 (Feb. 1980).
†D. C. LOUGHRY and M. S. ALLEN, "IEEE Standard 488 and Microprocessor Synergism," *Proc. IEEE*, **66**(2), 162–172 (Feb. 1978).
‡Draft Code and Format Conventions for use with IEEE Standard 488-1978 Digital Interface for Programmable Instrumentation, The Institute of Electrical and Electronics Engineers, Inc., June 1979.

Programming effort is further reduced by using a suitable high-level language. Use of a high-level language relieves the programmer from dealing directly with the microprocessor's architectural structure. High-level languages use instructions of a more English-like nature. A programmer can usually write an application program faster in a high-level language than in assembly language. In addition, because of the nature of the instructions in a high-level language, it is somewhat self-documenting.

A program written in a high-level language is converted to object code by another program called a compiler. A number of high-level languages exist for use with microprocessors, including BASIC, FORTRAN, Pascal, C, and variants of PL/1. The use of a particular high-level language with a particular microprocessor requires a compiler for that particular language and particular microprocessor.

In addition to an assembler or a compiler, software development requires other developmental software, such as an editor and linker and a computer on which to run them. The editor is used for text manipulation to create the source program file. Large source programs consist of a main program and a number of subroutines. The subroutines can be created and assembled or compiled separately. The linker program combines the main program and subroutines into a single program. Some linkers can combine programs written in several different languages into one program. This allows the programmer to use whatever language is most appropriate for the implementation of each subroutine or task.

The computer used to run the development software can be a large computer, minicomputer, or a microcomputer. An assembler or compiler run on one machine to generate code for another machine is a cross-assembler or cross-compiler.

Microprocessor development systems are microcomputers designed for software development for microprocessor-based systems. These development systems include the microcomputer, a keyboard for data entry, a CRT for display, and flexible or hard disks, or magnetic tapes for mass storage. In addition, the necessary system software, including editor, assembler, compilers, and linker, is available.

12.12   INPUT/OUTPUT AND DEVICES: MICROPROCESSOR SYSTEM I/O
David F. Bantz

Microprocessor I/O systems differ from large- and medium-scale computer I/O systems primarily in that a limited number of pins on the microprocessor package itself are available for I/O. A separate I/O interface is generally not feasible: virtually every microprocessor communicates with its I/O attachments over the same bus that is used to attach the microprocessor's memory. In some microprocessors, memory access instructions are used to set and read I/O attachment registers; in others, a special class of I/O instructions performs this function, although the path to the I/O attachment is the memory bus. Some microprocessors have a bit-serial I/O facility, intended for the attachment of slow devices.

12.12.1   Memory Bus I/O

Microprocessors differ in whether bus cycles for I/O are distinguished from bus cycles for memory access or not. In some microprocessors, a signal line is available to designate a particular bus cycle as having resulted from the execution of an I/O instruction. The attachments to these microprocessors can use this signal line to determine whether the address portion of the bus contains a memory address or an I/O attachment address. In microprocessors lacking such a signal line, the registers in the I/O attachment are accessed as memory locations by the microprocessor; only the contents of the address bus distinguish an I/O operation from a memory access operation. This form of memory bus I/O is referred to as **memory mapped** (see Section 9.8.1). Microprocessors using memory-mapped I/O have no explicit I/O instructions. Any processor with a memory bus can be adapted to do memory-mapped I/O.

Disadvantages of having a signal that distinguishes I/O bus cycles from memory cycles include the extra operation codes and processor circuitry and the extra package pin needed for the signal. A disadvantage of memory-mapped I/O is that a portion of the processor address space (the range of valid addresses) must be dedicated to I/O registers.

Address Decoding

Each I/O attachment must contain circuitry to recognize its address. In order to simplify address recognition circuitry, one may design circuitry to examine only a subset of the address lines. For example, the circuitry may be designed to select the register if the most significant bit of the address is true. Then for 16-bit addresses, any address in the range 8000 to FFFF (hexadecimal) will select the register. In general, if the address recognition circuitry has N gate inputs connected to the high-order bits of an address bus of M bits, a contiguous block of 2^{M-N} addresses is dedicated to the register.

Multiplexed Addressing
In some systems, a single bus is multiplexed between address and data information. The implication for I/O attachments is that the address information must be latched onto a register to enable valid address recognition. Alternatively, the address may be recognized at the time it is presented by the microprocessor, and only the signal representing address recognition need be latched. This indication must be reset before the next address is presented by the microprocessor.

Data Transmission
Once the address of an I/O attachment has been recognized, the attachment must either accept or present data to the microprocessor. Some microprocessors have a bus signal that indicates, when 1, that data are to be presented, and indicates, when 0, that data are to be accepted. Other microprocessors use two separate lines for this purpose. Microprocessors usually generate a signal that indicates that data are valid on the data lines of the bus, or that data have been captured by the microprocessor from the data lines. Most often, this signal is a transition from logical 1 to 0, or vice versa. Thus to present data to the microprocessor, the I/O attachment must recognize its address, sense that data are to be presented, and remove the data from the data lines after detecting the signal that the data have been captured. To accept data from the microprocessor, the I/O attachment must recognize its address, sense that data are to be accepted, and capture the data when the transition occurs that signals that the data are valid on the data lines.

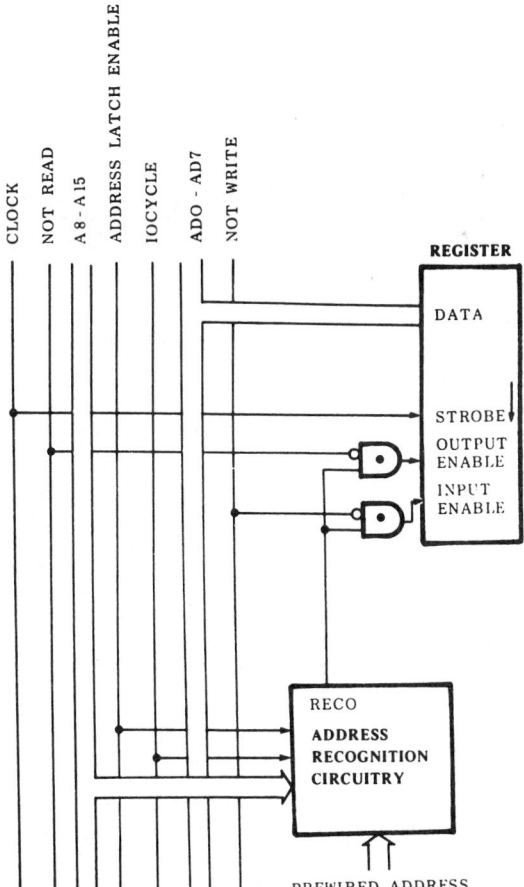

Fig. 12.168   Intel 8085 I/O interface.

Example

Figure 12.168 details the logic necessary to connect an external register to the Intel 8085 bus, which has a signal line distinguishing bus cycles for I/O (IOCYCLE). The register is selected by the RECO output from the address recognition circuitry. This circuitry compares a prewired address with the contents of A8–A15 (high-order address bits), which contains the I/O attachment address during an I/O operation; when the comparison succeeds, when IOCYCLE designates an I/O bus cycle, and when ADDRESS LATCH ENABLE makes a positive transition, the RECO output becomes true. The next positive transition on ADDRESS LATCH ENABLE sets RECO false. If the register is selected and NOT READ is 0, indicating a bus read operation, the OUTPUT ENABLE control input to the register is 1, causing the register to place its contents on DATA. During an I/O write, INPUT ENABLE is 1, and when STROBE makes a negative transition, the data on DATA are latched into the register. The 8085 bus timing ensures that IOCYCLE is 1, NOT WRITE is 0, and CLK makes a negative transition only while data are valid on AD0–AD7 (the multiplexed address/data bits).

A timing diagram for the I/O read operation is shown in Fig. 12.169. Note that the signal OUTPUT ENABLE can be 1 only when the signal NOT READ is 0, and when the I/O attachment has been recognized.

Direct Memory Access (DMA)

When data are transferred between the microprocessor and an I/O attachment under direct control of the microprocessor program, each byte or word of data requires the execution of several instructions to transfer. Some I/O devices require higher data transfer rates than are achievable with this technique. For these devices, the I/O attachment may use direct memory access (DMA). DMA allows the direct transfer of data between an I/O attachment and the microprocessor memory. The microprocessor first initializes the DMA controller circuitry by storing a count and a starting memory address in its registers. Once started, a DMA transfer proceeds without further microprocessor

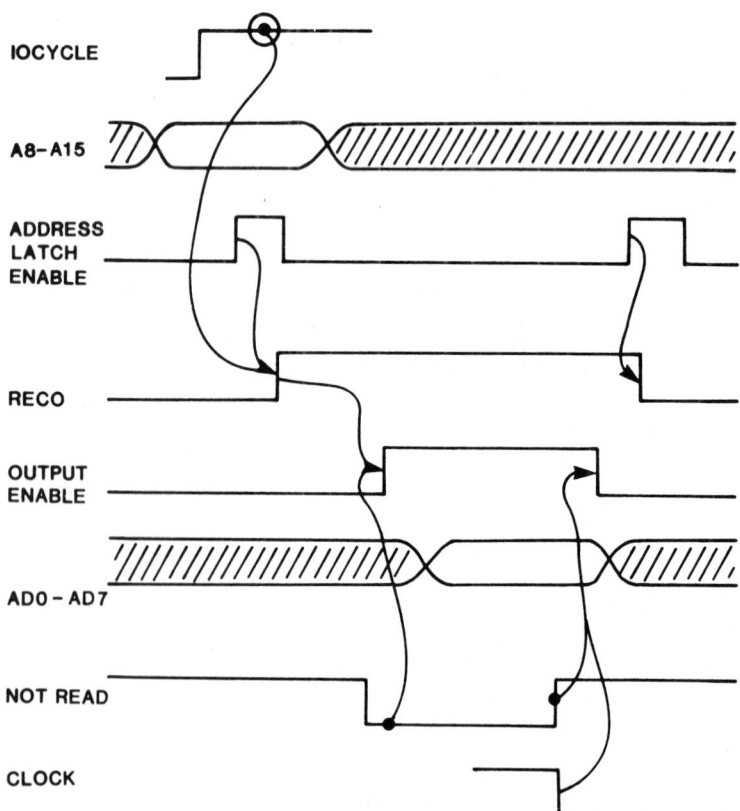

Fig. 12.169   Intel 8085 I/O interface timing for I/O read operation.

intervention, except that an interrupt (see Section 12.11.3) may be generated upon completion of the DMA operation.

I/O attachments using DMA (see Section 9.10.5) incorporate circuitry similar to that of Fig. 12.170. Circuitry to allow the microprocessor to set the COUNTER and ADDRESS COUNTER registers is not shown. The signal BUS CYCLE is assumed to define the interval of time during which addresses are presented and data are exchanged on the bus. The DMA controller connects to the I/O attachment with the lines TRANSMIT REQUEST and REQUEST GRANTED.

When the I/O attachment wishes to use a bus cycle, it raises the line TRANSMIT REQUEST. If the DMA count register is nonzero, the signal is placed on the BUS REQUEST line to the processor. The processor hardware periodically examines this signal, and when it is 1 the processor waits until the end of the current bus cycle, stops, places its address and data line drivers in the high-impedance state, and raises the line BUS GRANT. The processor is effectively isolated from the bus during bus cycles granted to the DMA controller. When BUS GRANT is sensed 1 by the DMA controller, it places the contents of its ADDRESS COUNTER register on the ADDRESS lines and signals the I/O attachment on REQUEST GRANTED that it may use the current bus cycle for transmission of data. The I/O attachment itself may drive the bus lines that determine the direction of data transfer, or additional circuitry in the DMA controller may drive these lines. As long as TRANSMIT REQUEST is held at 1, consecutive bus cycles may be used by the I/O attachment.

If several I/O attachments, each with its own DMA channel, wish to use the bus simultaneously, hardware must be provided to resolve the contention among the various channels. This resolution may be either on a priority basis or may grant bus cycles to competing DMA channels on a round-robin basis.

The circuitry of Fig. 12.170 is capable of using successive bus cycles ("burst mode") or using bus cycles intermittently. The choice depends on the data transfer rate of the I/O attachment. Often, the

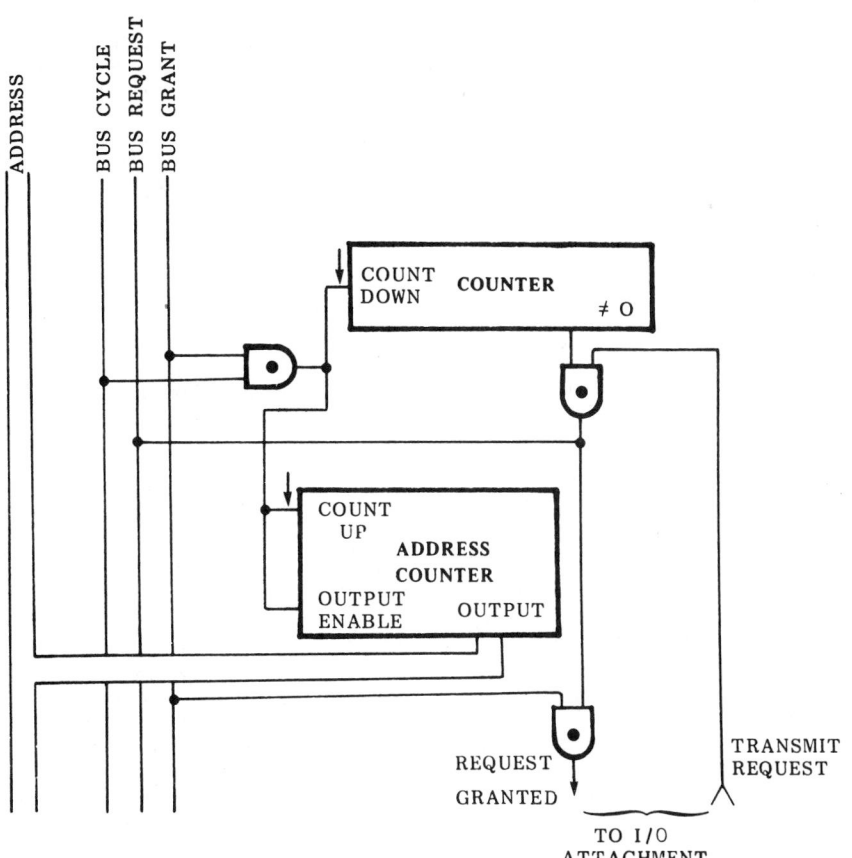

Fig. 12.170 DMA controller.

microprocessor must use several bus cycles in preparation for relinquishing the bus by generating BUS GRANT, and must use several bus cycles after regaining the bus. These cycles are unproductive in that they do not contribute to instruction execution or data transfer. Therefore, DMA transfers that use consecutive bus cycles make more efficient use of the bus.

12.12.2   Nonmemory Bus I/O

Some microprocessors have a serial I/O bus separate from their memory bus. Here, separate serial input and output channels allow transmission between an I/O attachment and the microprocessor. The I/O attachment address is signaled serially by the processor, followed by outbound or inbound data. The I/O attachment uses a separate clock line driven by the microprocessor to control the rate at which data are captured or signaled. A single instruction causes a transfer; the microprocessor cannot execute the next instruction until the transfer is complete. This serial I/O facility is usually intended for the attachment of I/O devices with low data rates.

12.12.3   Interrupts

To determine whether a particular I/O attachment requires data transmission, the microprocessor must interrogate the I/O attachment. This usually repetitive interrogation is called **polling**. The ability of an I/O attachment to interrupt the execution of a program in the microprocessor can be used to signal the requirement for data transmission more efficiently than polling.

Microprocessor interrupt systems usually depend on a logic circuit (attached to the microprocessor like an I/O attachment) devoted to controlling the interrupts on behalf of the processor. This alternative is most often chosen in preference to dedicating a pin of the microprocessor package to each interrupt level. One pin signals the microprocessor that an interrupt has occurred; a second pin may be used for a high-priority interrupt to signal error conditions. The interrupt controller resolves contention among the individual interrupts on a priority basis, encoding the number of the highest-priority interrupt into an internal "status" register. When the "status" register is read by the microprocessor, its contents are automatically transferred to an internal "level" register. Only an interrupt whose priority exceeds the current contents of the "level" register will cause a new interrupt to the microprocessor. Finally, the interrupt controller contains a "mask" register, which can be set by the microprocessor and which is ANDed with the individual interrupts. This facility is valuable, since it allows the microprocessor program to defer the recognition of selected interrupts without masking them all.

12.12.4   General Parallel and Serial Interface Systems

Integrated-circuit manufacturers, recognizing the need to implement I/O attachments, supply a wide variety of packaged components that attach directly to microprocessor buses and present a general interface to a device. This interface may allow either serial or parallel transfer of data to the device and often provides the ability to generate an interrupt signal to the microprocessor when a transfer is complete.

Parallel Interface

This family of integrated circuits, sometimes called **parallel interface adapters** (PIAs), is designed to simplify the attachment of devices requiring parallel data transfer, or devices requiring registers containing control information set by the microprocessor. An example of such a parallel interface device is shown in Fig. 12.171. The signals between the parallel interface and the microprocessor, on the memory bus, are the data, address, and control signals required for any I/O attachment. The signals between the parallel interface and the device carry data (DATA IN, DATA OUT) and control (OUTPUT READY, OUTPUT CAPTURED, INPUT READY, and INPUT CAPTURED). The control signals are sometimes called **handshaking** signals. For example, when the microprocessor sets data into the OUT DATA REGISTER, the handshaking line OUTPUT READY signals the device that data are ready for it. When the device has captured the data, it signals on OUTPUT CAPTURED, which may generate an interrupt to the microprocessor. The microprocessor program can then supply more data or respond in some other way. Similarly, when the device has data for the microprocessor, it signals on the INPUT READY line. The parallel interface may then generate an interrupt to the microprocessor, which would then read the data now present in the IN DATA REGISTER. The act of reading that register causes the parallel interface to generate the signal INPUT CAPTURED to the device.

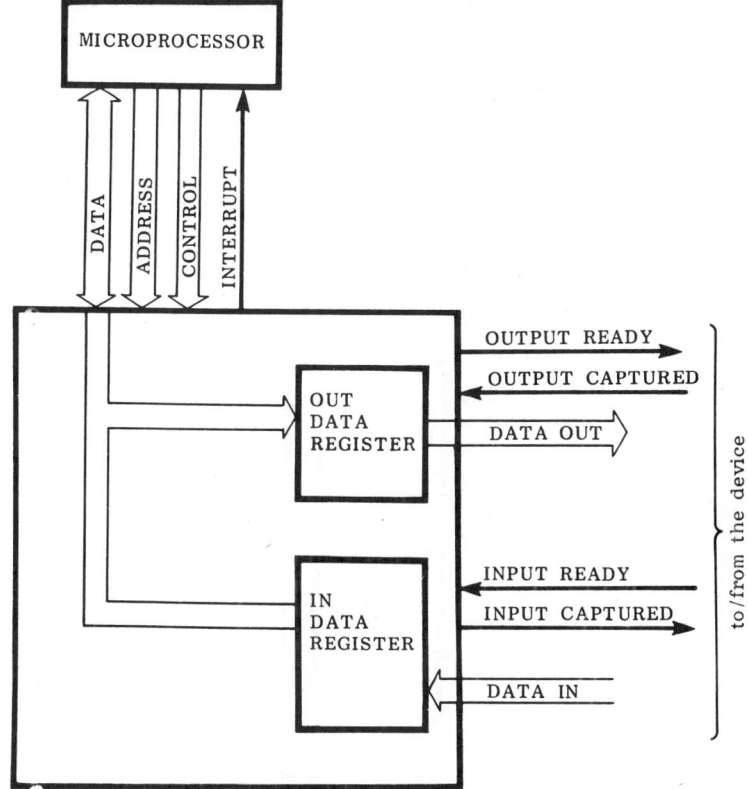

Fig. 12.171 Parallel interface.

Serial Interface

This family of integrated circuits is designed to simplify the attachment of devices requiring serial data transfer. An example of such a serial interface device, called a **universal synchronous asynchronous receiver transmitter (USART)**, is shown in Fig. 12.172. The signals between the microprocessor and the serial interface (on the memory bus) are those required by any I/O attachment. The signals between the serial interface and the device consist of status signals (TRANSMIT READY, RECEIVE READY), clock signals (TRANSMIT CLOCK, RECEIVE CLOCK), and data (TRANSMIT DATA, RECEIVE DATA). When the device has data for the microprocessor, it signals those data, serial by bit, on the RECEIVE DATA line in synchronism with a clock signal on RECEIVE CLOCK. Usually, the data are prefixed by a 1 bit (the "start" bit), indicating the start of data, and the data are from 5 to 8 bits long. A 1 bit (the "stop" bit) may be suffixed to the data. After all bits of the data have been received by the serial interface, the stop and start bits are stripped and an interrupt may be generated to the microprocessor to indicate that data may be read from the RECEIVE BUFFER. The mode of data transmission in which one 5- to 8-bit character is transmitted at a time is called **start / stop**.

When data are to be sent to the device, the microprocessor sets the data in the TRANSMIT BUFFER. The data are then serialized, prefixed with the start bit and suffixed with the stop bit, and sent to the device on TRANSMIT DATA in synchronism with a clock on TRANSMIT CLOCK.

Other Attachments

Another form of general-purpose attachment is the IEEE 488 General-Purpose Interface Bus,* a standard for the connection of laboratory instrumentation. Integrated-circuit manufacturers supply interface devices that connect to a microprocessor bus on one side and to the IEEE 488 bus on the other.

*IEEE Standard Digital Interface for Programmable Instrumentation, IEEE Std 488-1975, IEEE, New York.

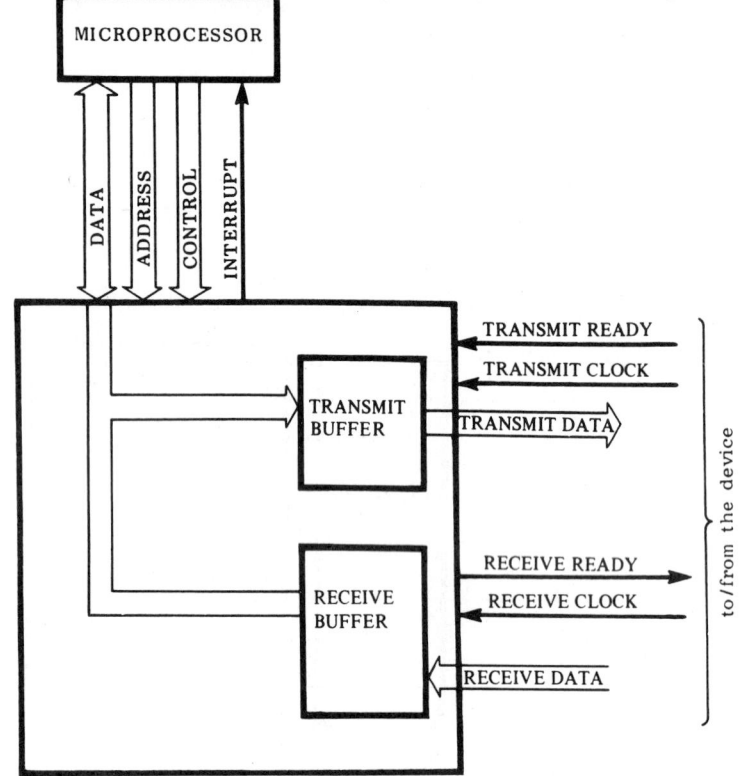

Fig. 12.172 Serial Interface.

12.12.5 Device Attachment Examples

This section contains a number of examples of specific I/O attachments for microprocessors.

Keyboard attachment

Figure 12.173 shows how a keyboard can be constructed as an array of contacts connected to a parallel interface. The DATA OUT signals are used to select a row of contacts for sensing; the DATA IN signals give the state of all contacts in the row. The microprocessor sets only one bit to 0 at a time in the OUT DATA REGISTER; otherwise, several rows might be selected simultaneously. All inputs to the IN DATA REGISTER are tied to a logical 1 and will remain 1 if no contact is closed. Only those contacts tied to an OUT DATA REGISTER line that is 0 can affect the corresponding IN DATA REGISTER bit.

Since the contacts may open and close rapidly for up to 10 ms after a key is depressed, the state of a contact should be sampled several times before the microprocessor program decides that the key has been depressed. This process is called **debouncing**. As peak keying rates may exceed 20 keystrokes per second, the keyboard scanning rate should be at least 40 scans per second to allow for at least two scans per key per keystroke.

Communications Attachment

When microprocessors are used in communications systems, more complex and efficient modes of transmission than start/stop (Section 12.11.4) must often be supported. Higher-data-rate devices require a **synchronous**, or continuous, mode of transmission that transmits a continuous sequence of transitions, allowing the receiver to maintain clock synchronization. Messages are transmitted according to a link protocol, where a message consists of a destination address, control information, a block of characters, and an error control unit.

Typical of the devices that can be used to implement these communications protocols is an advanced data link controller (ADLC). The general form of this device is similar to that of the serial interface (Section 12.11.4). Among the differences are the multistage transmitter and receiver buffers,

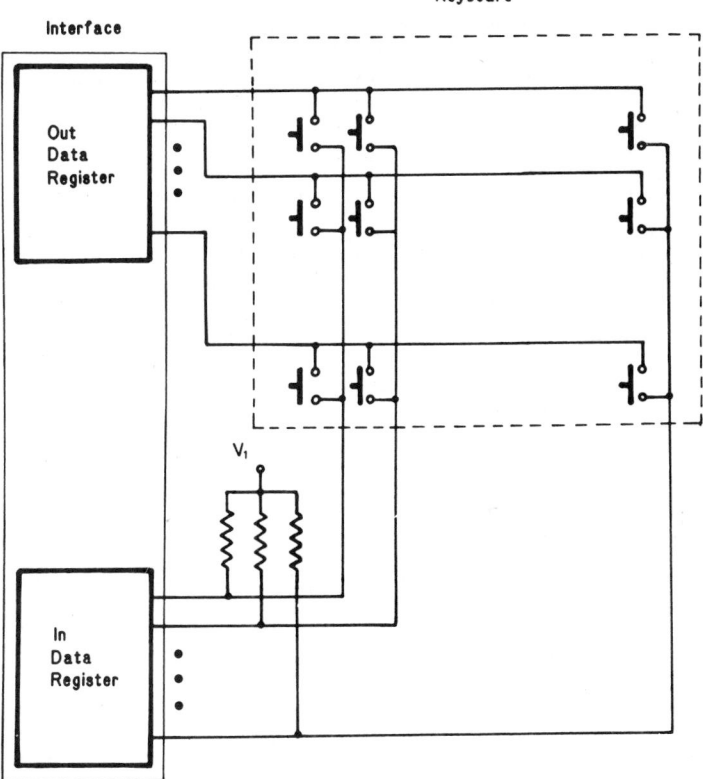

Fig. 12.173   Nonencoded keyboard.

capable of buffering several characters, allowing the processor more time to respond to an interrupt. The ADLC computes a 16-bit polynomial check code, which is automatically appended to transmitted messages and stripped from received ones. Provision is made for the ADLC to work with DMA (see Section 12.11.1) so that data link rates too fast to be processed by the microprocessor with interrupts can be supported. This device must be serviced continually once a message is begun; if not, a data overrun or underrun condition will cause a portion of a message to be lost. This is a "real-time" device whose internal buffering is minimal and whose data rate is determined by the link speed not by the microprocessor program.

Display Attachment

Alphanumeric CRT displays are often designed with an internal microprocessor. This processor can be used to scan, debounce, and encode the keyboard; to provide the interface, either to a computer or to a communications line; and to perform editing functions in response to keyboard entries. Earlier we described the I/O attachment of a keyboard and communications. This section describes the attachment of the display.

The display considered here is a raster CRT whose beam moves horizontally left to right across its screen. The beam is then shut off, moved downward by one beam width, and retraced to the left. The beam is then moved horizontally from left to right. This process is repeated many times per second. When the beam reaches the bottom of the screen, it is retraced to the top. To display characters with this beam, a signal must be supplied to an unblanking circuit as the beam moves horizontally left to right.

Figure 12.174 shows the circuitry necessary to convert a buffer of character codes from the microprocessor memory into the serial bit stream of data necessary to unblank the beam. This circuitry is called a **character generator**. When the beam begins its scan at the top of the screen, the character buffer CB1 contains the codes of all the characters to be displayed on that scan line. The code for the leftmost character is in the rightmost end of the character buffer and is used to address a

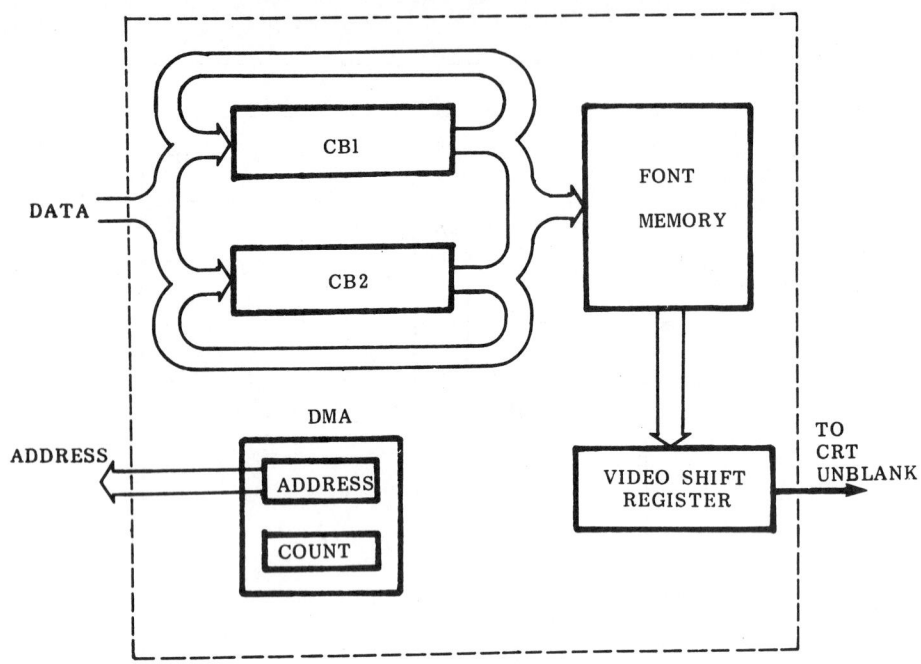

Fig. 12.174   Charactor generator.

memory (the FONT ROM) containing the bit image of the character. A horizontal slice of the character image is then accessed from the FONT ROM and placed in the VIDEO SHIFT REGISTER, which serializes the slice to the unblanking amplifier of the CRT. As the horizontal slice of the first character is displayed, the character buffer is shifted circularly one position to the right, bringing the code for the next character to be displayed to the right end of the character buffer. That code is now in position to address the FONT ROM. In this way the character codes are converted to their serial video representation.

While one of the character buffers is being shifted to access the FONT ROM and create the character images, the other is being filled from microprocessor memory using the DMA controller. Successive character codes are retrieved from microprocessor memory and stored in the character buffer. At the bottom of the screen, the DMA address register and counter must be reset so that codes for characters at the top of the screen are the next to be accessed. This process of alternately filling the two character buffers continues for as long as an image is required on the display screen.

12.13   MEASUREMENT INSTRUMENTS
Emory J. Harry and Robert E. Metzler

12.13.1   Oscilloscopes

The modern oscilloscope is the most versatile test and measurement instrument and has become one of the most universally used. Other instruments—counters and digital multi-meters (DMMs)—can measure a single parameter more accurately under most conditions, but an oscilloscope gives a visual display, in graphic form, of the entire input waveform. Virtually any phenomenon that can be transduced to a voltage can be displayed on an oscilloscope. From this display, the user can determine characteristics or measure parameters of the input that are not apparent or are impossible to measure with other types of instruments.

Oscilloscopes use a cathode-ray tube (CRT), which employs an almost massless electron beam to draw a graph on an internal screen of phosphor. The phosphor emits light where the electron beam strikes it. Because the electron beam has so little mass, it can be deflected electrostatically very rapidly, giving the oscilloscope very high bandwidth as compared to other instruments that are capable of displaying a graph of an input waveform (i.e., oscillographs, chart recorders).

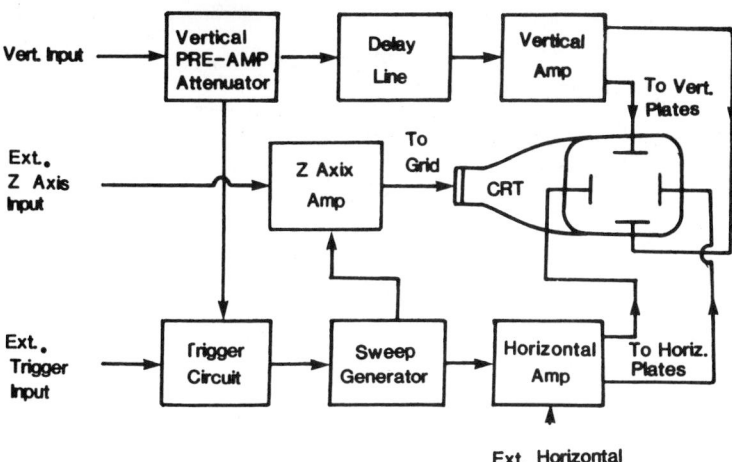

Fig. 12.175   Block diagram of a cathode-ray oscilloscope.

Block Diagram

Figure 12.175 is a simplified block diagram of an oscilloscope showing the major functional blocks.

Cathode-Ray Tube (CRT). A CRT is the heart of an oscilloscope. It performs the essential function of converting the input signal into a visual image.

Vertical Preamplifier. An input signal connected to the vertical input is conditioned or standardized by the vertical preamplifier and attenuator. This permits oscilloscopes to handle a wide range of input voltages, from several hundred volts to a few millivolts; in specialized oscilloscopes, to thousands of volts or just a few tens of microvolts.

Delay Line. The vertical delay line passes the input signal from the preamplifier to the vertical output amplifier, delaying the signal so that the signal (sawtooth) from the horizontal amplifier arrives at the CRT prior to the signal from the vertical output amplifier. Inexpensive oscilloscopes do not always have delay lines, in which case the vertical signal arrives at the CRT before or at the same time as the horizontal signal. As a result, the first part of the input waveform cannot be seen on the CRT.

Vertical Output Amplifier. The vertical output amplifier amplifies the signal to sufficient amplitude to drive the vertical deflection plates in the CRT.

Trigger. The trigger block picks off a portion of the input signal from the vertical amplifier or takes the external trigger input signal and from this signal generates a pulse that is used to start the time base or sweep generator.

Time Base. The time base generates a wide range of sweeps or sawtooth waveforms that deflect the beam horizontally at a linear rate. The sweep speeds will vary from several seconds to as fast as a few hundred picoseconds per division on the CRT of the fastest scopes. The sweep (sawtooth waveform) from the time base is started by a pulse from the trigger block; thus the sweep is synchronized with the vertical input signal.

Horizontal Amplifier. The horizontal amplifier amplifies the sawtooth from the time base so that it can drive the horizontal deflection plates in the CRT. In more specialized applications, the horizontal amplifier amplifies an externally applied signal so that it can drive the horizontal deflection plates directly.

Z-Axis Amplifier. The time base generates a signal that is applied to the Z-axis amplifier. The Z-axis amplifier amplifies this signal, which is then applied to the grid of the CRT, turning the beam of electrons on and off. The beam is turned on while it is being swept from left to right across the CRT face (generating the X axis of a graph). While the sawtooth is returned to zero to wait for the next sweep, the electron beam is turned off.

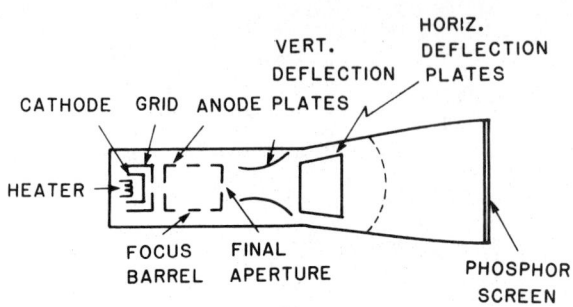

Fig. 12.176 Monoaccelerator cathode-ray tube.

Oscilloscope CRTs

Oscilloscope CRTs come in a variety of types, depending on their intended application.

Monoaccelerators. The simplest oscilloscope CRT is a monoaccelerator, which is adequate for most low-bandwidth applications. In the monoaccelerator (Fig. 12.176) the cathode is heated by the heater and, as a result, emits electrons. The grid controls the movement of these electrons from the cathode—which is at a high negative potential, 1 to 5 kV—to the anode. The grid and the anode both have small holes or apertures that allow the beam, which has considerable energy due to the high potential, to pass through. The anode is followed by a focusing element that focuses the beam down. It then passes through the final aperture element. The beam now passes through the vertical deflection plates, where it is deflected vertically an amount proportional to the amplitude of the vertical input signal. After the vertical plates, the beam passes through the horizontal plates, which deflect it horizontally, usually at a linear rate determined by the sweep generator. After passing the horizontal plates, the beam of electrons strikes the phosphor screen, which emits an amount of light proportional to the amount of energy in the beam.

Post-Deflection Acceleration (PDA). For higher-performance oscilloscopes that have more vertical bandwidth and faster sweep speeds, monoaccelerators are not adequate. They do not have enough writing rate at high deflection speeds. (As the deflection speed increases, fewer electrons per unit of time are striking any given area of phosphor and, therefore, the amount of light output drops.) In wider-bandwidth oscilloscopes, to obtain more light output (higher writing rate), PDA tubes are used. PDA tubes employ a high positive voltage—on the order of 10 kV—beyond the deflection plates to increase the beam velocity and energy and, therefore, light output.

Phosphors. Phosphor is deposited on the inside face of the CRT because it has the capacity of converting the kinetic energy in the electron beam to light. Two phenomena occur when an electron beam strikes phosphor. While the beam is striking the phosphor, it fluoresces (emits light). When the beam is removed, the phosphor will still emit light for some period of time—it phosphoresces.

There are a large number of different phosphors that have different wavelengths (different colors) and different phosphorescent decay times (how long they emit light after the excitation is removed). Different applications require different phosphors. The three major phosphor types used in oscilloscope applications are P7, P11, and P31. They have the following characteristics:

Phosphor Type	Fluorescence	Phosphorescence	Persistence	Relative Luminance	Relative Writing Rate
P7	Blue-white	Yellow-green	Long	45	95
P11	Blue-violet	Blue	Medium	25	100
P31	Green	Green	Short	100	75

P7 is used primarily in medical or other applications that require long persistence. P11 is used primarily in photographic applications because the color of its light output is close to the wavelength or color where photographic film is most sensitive. P31 is used in the vast majority of oscilloscope applications because the output color is in the area where human eyes are most sensitive. It has medium to short persistence, which avoids multiple images when the waveform is moving, high luminance (light output), relatively high writing rate, and is resistant to burning.

Storage CRTs. The longest persistence phosphor available gives only a few seconds of viewing time at best. Some applications require that the input waveform be stored or displayed on the CRT for an indefinite period of time. For example, a single-shot event (one that occurs only once) must be stored to be evaluated.

There are two basic types of storage CRTs: bistable storage and variable-persistence storage. A third type of CRT that employs most of the advantages of both is called a charge transfer CRT.

1. Bistable Storage. The first practical, commercially available storage oscilloscopes employed bistable storage CRTs. The operation of storage CRTs is based on the principle of secondary emission.

The phosphor particles are loosely deposited on the face of the CRT, as shown in Fig. 12.177, in such a way that they are insulated from each other. A material is added to the phosphor that has the desired characteristic of emitting a large number of secondary electrons when struck with a high-velocity, high-energy beam of electrons. As the writing beam sweeps across the phosphor, the phosphor particles struck by it emit secondary electrons and take on a positive charge. Because the particles are insulated from each other, this charge does not migrate from one particle to the next across the face of the CRT. In other words, the beam leaves a replica of its path in phosphor particles that have a positive charge. There is a second source of low-energy electrons in a storage CRT from flood guns mounted in front of the deflection plates. The flood guns flood the entire phosphor screen with low-velocity electrons. These electrons are attracted by the phosphor particles, which have a positive charge and have sufficient energy to cause these particles to emit a visible amount of light. The rest of the phosphor screen, which does not have a positive charge, repels the flood-gun electrons, or those that are not repelled do not strike the phosphor with sufficient energy to cause it to emit a significant amount of light. The result is illumination of the path of the writing beam that will last as long as the flood guns are on.

Bistable storage is the simplest and lowest-cost form of storage. Its major disadvantage is that it offers no half tones—the stored trace is either written or not—thus the name, bistable.

2. Variable-Persistence Storage CRTs. Variable-persistence CRTs work much like bistable CRTs. The major difference is that in a variable-persistence tube, there is a fine wire mesh with a dielectric material deposited on it a few millimeters in front of the phosphor that acts as the stored target. The positively charged portion of the mesh that has been struck by the high-energy beam allows electrons to pass through and strike the phosphor while the rest of the mesh repels the low-energy electrons from the flood guns.

Variable persistence has several advantages over bistable storage: half tones, higher nonstorage writing rate, and persistence that can be varied. Its major disadvantage is cost.

3. Charge Transfer Storage CRTs. The most recent development in storage CRTs is the charge transfer CRT. The storage target, whether it is the phosphor or a mesh, must have two properties: good secondary emission, and insulation to prevent the charge from migrating from one phosphor particle to the next. In either bistable or variable persistence tubes, these characteristics are traded off against each other. The charge transfer tube employs an extra mesh. The first mesh is optimized for good secondary emission, but cross migration takes place rather rapidly; the second mesh is optimized for good insulation. When a trace or waveform is stored on the first mesh, it is immediately transferred to the second mesh. This is accomplished by controlling the mesh voltages so that the flood-gun electrons do not pass through the first mesh but form a cloud in front of the mesh. A high-amplitude, short-duration voltage is then applied to the second mesh. At those places where a pattern is written on the first mesh, this causes electrons from the cloud behind the mesh to be pulled through the first mesh with sufficient energy to transfer the pattern or waveform to the second mesh. After the transfer of this charge takes place, the CRT operates just like a normal variable persistence CRT. Because each mesh is optimized only for the function it performs, the charge transfer storage CRT can achieve a much faster writing rate than either of the other two types of storage. Its major disadvantages are cost and complexity.

Dual Beam. It is often desirable to display two waveforms simultaneously for purposes of comparison. This can be accomplished in two ways. A single beam can be multiplexed or a CRT can

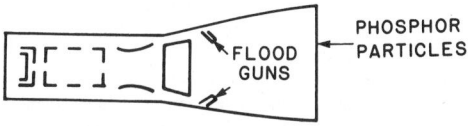

Fig. 12.177   Bistable storage cathode-ray tube.

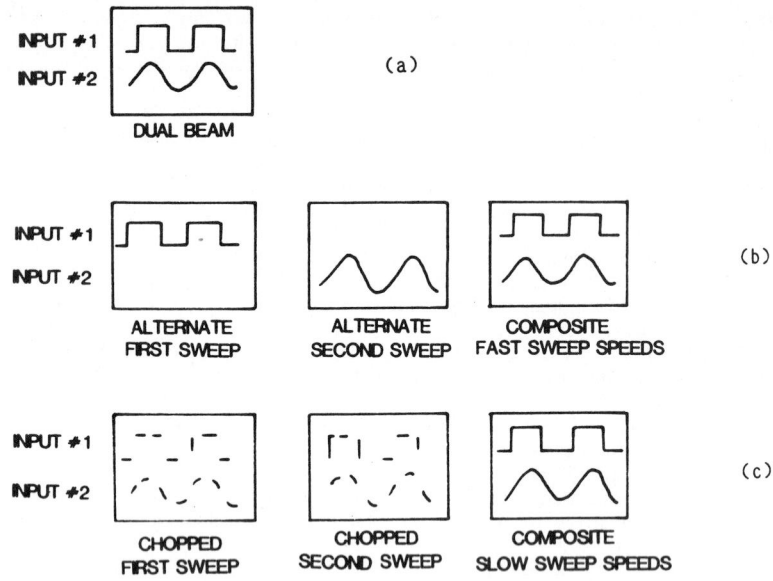

Fig. 12.178 Two input waveforms displayed as: (a) dual beam, (b) alternate, (c) chopped.

be used that has two separate beams and deflection systems that are both directed at a common screen.

Dual-beam oscilloscopes that employ two completely separate cathodes and deflection systems, shown in Fig. 12.178(a), are the most versatile. They are essentially two separate oscilloscopes that share a common CRT faceplate.

Many dual-beam instruments have completely separate vertical deflection systems but employ a common horizontal set of deflection plates. These instruments are not as versatile as those that employ two completely separate sets of horizontal plates, one for each beam, but they are more economical.

Some dual-beam oscilloscopes use beam splitters. Beam splitters split the electron beam from a single cathode and direct half of the beam to each of the two separate deflection structures. This type of instrument will also often only have a single set of horizontal plates; this approach is taken primarily for economic reasons.

Vertical Inputs

Dual Trace. For most applications, dual trace—the term used to describe the multiplexing of a single beam—will accomplish everything that dual beam accomplishes, and it is significantly less expensive. It is inexpensive enough and versatile enough that almost all medium- to high-performance oscilloscopes offer dual-trace capability.

The multiplexing is done in two ways. The two vertical inputs are either alternated between—each input is displayed on alternate sweeps of the time base (Fig. 12.178(b)); or chopped—the two inputs are chopped between at a fixed rate (Fig. 12.178(c)). The following drawings show two input waveforms displayed dual beam, alternately, and chopped.

Virtually all dual-trace oscilloscopes offer both alternate and chopped modes. One of the two is selected depending on the required sweep speed. At slow sweep speeds, the chopped mode is used. The chopping frequency is usually about 1 MHz, so at slow sweep speeds (below about 0.1 ms per division) the missing pieces of the displayed waveform are so small they cannot be seen—the displayed waveforms appear continuous. At faster sweep speeds, the alternate mode is selected. At these faster sweep speeds, each of the two inputs is being alternately displayed, but it occurs so rapidly that they appear to be simultaneous.

The major shortcoming in multiplexing between inputs is in single-sweep applications, where the events to be displayed only occur once. If the inputs are multiplexed, the portion of the waveform that is of interest may occur just as the opposite input is being displayed. Stable triggering is also sometimes a problem when the two inputs are nonsynchronous. Triggering is discussed later.

Differential Mode. Many oscilloscopes provide a mode that allows the two vertical inputs to be algebraically added or subtracted. The most common use for this is differential measurements—algebraic addition with one of the two inputs inverted. In this mode, only the difference between the two inputs is displayed; any signal that is common to both (common mode) is not displayed.

Input Coupling. Almost all oscilloscope vertical inputs can be either ac or dc coupled. Ac coupling will result in the loss of low-frequency response. However, ac coupling is useful in blocking the dc component of an input waveform so that the ac component can be displayed at the desired sensitivity without regard for the size of the dc component.

Some oscilloscopes have dc offset or slideback capability, which allows the user to add a variable dc voltage to the input. This permits the user to cancel a dc component.

Triggering

The trigger circuitry in an oscilloscope allows the user to select the portion of the input waveform on which triggering is desired. It then selects or triggers on this same point on each successive sweep so that the displayed waveform is stable.

Level and Slope. Virtually all oscilloscopes allow the user to select the desired trigger level (voltage amplitude) and the desired slope or polarity of the triggering point.

Coupling. Both ac and dc trigger coupling are usually offered. The ac coupling allows the user to block the dc component of the waveform and select the level on the ac portion of the waveform that is desired.

Some oscilloscopes also offer high-frequency reject and low-frequency reject—two filters that allow the user to trigger on either the low-frequency or the high-frequency components of an input waveform.

Source. Most oscilloscopes offer both internal and external triggering. Internal triggering operates from the vertical input signal, and external triggering operates from the signal that is connected to a separate "external trigger" input connector. Often, "line triggering" is also offered, which selects as a trigger a sample of the incoming power line voltage. In multichannel oscilloscopes, the trigger source selection should allow the user to select either input channel as a trigger source. Composite triggering, a triggering signal that is the composite of the two vertical inputs, is also available on some oscilloscopes.

Single Sweep. Some oscilloscopes offer single sweep, which is useful in displaying events that occur only once or occur randomly. In the single-sweep mode, the sweep generator will sweep once when triggered but will not sweep again until it is reset or armed by the operator.

Automatic Triggering. If there is no trigger signal present, or the trigger control is set at a level beyond the level of the trigger signal, no triggering pulse is generated to cause the time base to operate so as to generate a sweep—there is no display on the CRT. To avoid the uncertainty caused by the lack of a sweep, most oscilloscopes have automatic modes. The automatic modes cause the time base to free run (generate sweeps) in the absence of triggers so that there is always a trace on the screen.

Peak-to-Peak Automatic Triggering. There are a few oscilloscopes available that offer what is called peak-to-peak automatic triggering. This is a form of automatic triggering circuit that samples the peak-to-peak amplitude of the input trigger signal and automatically sets the limits of the trigger level control equal to this peak-to-peak value. This results in the trigger level control always having the optimum resolution. It also results in a trigger always being generated regardless of the input amplitude and setting of the trigger level control.

Time Base

The time base, when triggered, generates a sawtooth waveform that is applied to the horizontal deflection plates. This waveform results in the electron beam being deflected across the CRT at a linear rate, thus establishing the time per division on the X axis of the graph displayed on the face of the CRT. The range of sweep speeds will vary with the vertical bandwidth and overall performance of the oscilloscope—higher-performance instruments require faster sweep speeds. High-performance oscilloscope sweep speeds will start at 2 to 5 per division and increase in a calibrated 1-2-5 sequence to, in very high performance instruments, less than 1 ns per division. The accuracy of the sweep in most oscilloscopes is about 3%.

Magnifier. The fastest sweep speeds in most oscilloscopes are obtained through the use of a magnifier, usually times 5 or 10, that increases the gain of the horizontal amplifier by a factor of 5 or

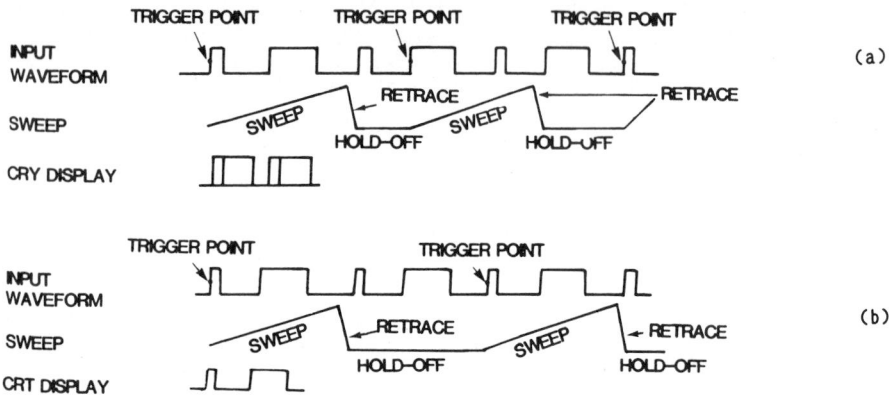

Fig. 12.179 (*a*) Erroneous trigger selection, (*b*) proper trigger selection.

10. This magnification can be used at other than the fastest sweep speeds to increase the displayed resolution by magnifying the desired portion of the sweep. A ×10 magnifier in conjunction with the horizontal position control will allow any single horizontal division to be magnified to 10 divisions. Magnifiers decrease the horizontal accuracy to approximately 5%.

Variable Hold-off. The trigger circuit can sometimes, particularly on complex input waveforms, select a triggering point that results in an erroneous, misleading display. Figure 12.179(*a*) is an example of an erroneous trigger being selected; as a result of the time lost during retrace and hold-off, the trigger circuit alternately selects narrow and wide input pulses on which to trigger. Some oscilloscopes, particularly those designed for digital applications, offer a feature called variable hold-off. It allows the user to continuously increase the hold-off time—the time that the sweep generator is prevented from sweeping, normally just long enough to allow it to stabilize. The user can thus avoid undesired triggers by increasing the hold-off time until the proper display is displayed on the CRT, as shown in Fig. 12.179(*b*).

Sweep Switching. Some higher-performance, single-beam oscilloscopes offer two independent time bases. By alternately sweeping with one time base and then the other, a single waveform can be displayed at two different sweep speeds or the time bases can be independently triggered, allowing two vertical inputs to be displayed, each at a different sweep speed. A few high-performance oscilloscopes also offer the ability to chop between the two time bases. Horizontal chopping, like vertical chopping, is most often used at the slower sweep speeds. At slow sweep speeds, alternating horizontally can result in a flickering display.

Delayed Sweep. Many medium- to high-performance oscilloscopes that have two sweep generators offer a horizontal mode called delayed sweep, which offers much higher magnifications than is possible with a magnifier and at no loss of accuracy. Used properly, the delayed sweep can even result in higher accuracy than is normally available.

In this mode, the first sweep generator (called the delaying sweep) is used to delay, from its trigger, the second sweep generator (called the delayed sweep). The delaying sweep sawtooth is compared, in a comparator, to a dc voltage determined by the position of a front-panel knob or dial called "delay" or "delay time" or "delay time multiplier." When the delaying sweep runs up to the point where its voltage is equal to the voltage set by the delay-time multiplier dial, a trigger is generated that starts or enables the delayed sweep. (The delayed sweep can either sweep immediately after the delay or wait for its own trigger.) The result, if the delayed sweep is directed to the horizontal deflection plates of the CRT, is that the portion of the input waveform that occurs after the delay will be displayed at a time per division set by the delayed sweep. Normally, the delayed sweep is set to run faster than the delaying sweep; the amount of magnification is determined by the ratio of the two sweep speeds. If the delaying sweep is set to 1 ms per division and the delayed sweep to 1 μs per division, the magnification is 1000. Oscilloscopes that incorporate a delayed sweep mode usually also incorporate an intensified sweep mode that intensifies the portion of the waveform that will be displayed in the delayed sweep mode. This is particularly useful, at the high magnifications that are possible, in selecting the portion of the waveform to be magnified. Figure 12.180 demonstrates the operation of these two modes. Waveform 1 is a hypothetical input signal. The fourth set of pulses, three pulses, are to be magnified. Waveform 2 is the delaying sweep trigger. Waveform 3 is the

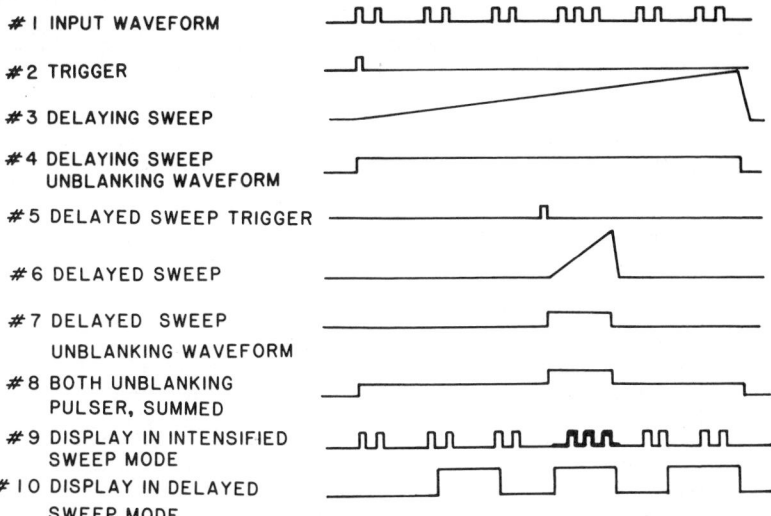

#1 INPUT WAVEFORM

#2 TRIGGER

#3 DELAYING SWEEP

#4 DELAYING SWEEP
UNBLANKING WAVEFORM

#5 DELAYED SWEEP TRIGGER

#6 DELAYED SWEEP

#7 DELAYED SWEEP
UNBLANKING WAVEFORM

#8 BOTH UNBLANKING
PULSER, SUMMED

#9 DISPLAY IN INTENSIFIED
SWEEP MODE

#10 DISPLAY IN DELAYED
SWEEP MODE

Fig. 12.180   Operation of delayed sweep mode and of intensified sweep mode.

delaying sweep. Waveform 4 is the delaying sweep unblanking waveform that the sweep generator sends to the CRT to unblank, or turn on, the CRT during the sweep. Waveform 5 is the trigger from the comparator that starts the delayed sweep. Waveform 6 is the delayed sweep. Waveform 7 is the delayed sweep unblanking pulse. Waveform 8 is the two unblanking pulses summed.

In the intensified sweep mode, waveform 1 goes to the vertical deflection plates, waveform 2 starts the delaying sweep (3), which goes to the horizontal plates, and waveform 8 goes to the Z axis of the CRT. The result is that the vertical input is displayed at a sweep speed determined by the delaying sweep generator, but the portion of the waveform to be magnified—the three pulses—is intensified, as shown in waveform 9. This intensified portion of the sweep can be positioned with the delay-time multiplier dial.

In the delayed sweep mode, waveform 1 goes to the vertical deflection plates of the CRT, waveform 6 goes to the horizontal deflection plates of the CRT, and waveform 7 goes to the Z axis of the CRT. Waveform 10 is the display on the CRT.

A few oscilloscopes incorporate both sweep switching and delayed sweep. This permits the display on the CRT of both the intensified and delayed modes simultaneously.

Mixed Sweep. Mixed sweep is a variation of delayed sweep that offers some of the advantages of the combination of delayed sweep and sweep switching, but with a single trace.

In this mode, the waveform that is first directed to the horizontal deflection plates of the CRT is the delaying sweep, but at the point in time determined by the delay-time multiplier, switching takes place that directs the delayed sweep to the horizontal deflection plates. The switching is accomplished by comparing the instantaneous amplitudes of the two sweeps. When the delayed sweep amplitude reaches the instantaneous level of delaying sweep, the switching takes place. The resulting display, using the same vertical input as before, is shown in Fig. 12.181.

External Horizontal

It is sometimes useful to deflect the beam horizontally with other than a sawtooth. For example, two sine waves can be applied to the vertical and horizontal deflection plates and a Lissajous pattern results from which one can determine the phase relationship of the two sine waves.

Most oscilloscopes offer an external horizontal input (input directly to the horizontal amplifier) for these applications.

Fig. 12.181   Mixed sweep mode.

External Z-Axis
It is also sometimes useful to control or modulate the intensity of the beam. For these applications, many oscilloscopes offer an external input to the Z-axis amplifier.

Probes
The signal to be measured can be significantly altered by the connection between the source of the signal and the oscilloscope. For example, just a few feet of coaxial cable can add 50 to 100 pF of capacitance, which is intolerable in many applications. The oscilloscope input impedance itself, which is typically 1 MΩ and 10–50 pF (some wide-bandwidth oscilloscopes have 50-Ω inputs), can have a major effect. There is a wide variety of oscilloscope probes available designed to facilitate this connection and increase the flexibility of the oscilloscope.

Passive Voltage Probes. The most common probe is the 10× passive voltage probe. It is essentially a compensated 10× voltage divider. These probes have a 10 MΩ and 5 to 20 pF input impedance. They come with a wide variety of hardware for connecting to wires, test points, ICs, and so on. Besides offering input resistance that is 10 times as high and several times lower input capacitance, they are very convenient to use. They also allow higher input voltages—typically two or three times higher than the oscilloscope—but decrease the maximum sensitivity by 10 times.

Passive voltage probes are also available that are increased 100 and 1000 times. They have even lower input capacitance—typically 2–3 pF—and higher input resistance—100 and 1000 MΩ, respectively. However, their application is more limited because of the 100 and 1000 times decrease in sensitivity.

Very wide bandwidth passive probes are also available for oscilloscopes with 50-Ω inputs. They have even lower input capacitance (on the order of 1 pF) but have relatively low input resistance, 500 Ω for 10× and 5 kΩ for 1000× .

Active Voltage Probes. There are two types of active voltage probes available. The first is optimized for wide bandwidth, high input resistance, and low input capacitance. They have capacitance on the order of 2 or 3 pF and input resistance from 100 kΩ to 100 MΩ. The second type of active voltage probe is optimized for high CMRR (common-mode rejection ratio) and differential applications in general. If two passive voltage probes are used differentially, high CMRR is not practical. One percent attenuator resistors in the probe would reduce the CMRR to approximately 50 : 1, but active differential probes can have CMRRs of 10,000 : 1 at dc.

Passive Current Probes. Passive current probes are available that convert an input current to voltage. These probes are essentially transformers and their insertion impedance is on the order of 0.5–1 Ω shunted by 5–0 μH. They are very useful in some applications, and in fact, make some measurements possible that would not be possible with a voltage probe.

Active Current Probes. There are two basic types of active current probes. The first is simply an amplifier with equalization to extend the bandwidth and sensitivity of passive current probes. Passive current probes, because they are basically transformers, have limited low-frequency response. They roll off in the range 100 Hz to 1 kHz. When they are used in conjunction with these amplifiers, the low-frequency response is extended by about a factor of 10, to 10–100 Hz. The second type of active current probe is the Hall effect probe, which has a bandwidth of dc to about 50 MHz. The Hall effect probe combines a current transformer with a Hall effect drive. The Hall effect drive extends the low-frequency response to dc, which is the major advantage of the probe.

High-Voltage Probes. Probes are available that permit an oscilloscope to measure voltages up to 40 kV. They have reasonable bandwidth, relatively low capacitance, and high input resistance, but should be used carefully because the maximum voltage that they can measure significantly decreases with frequency.
Note: The specifications of all probes must be examined carefully due to the derating of input voltage or current as a function of frequency.

12.13.2   Digital Multimeters

Functional Converters

A/D Converters. The A/D (analog-to-digital) converter is the heart of any DMM—a DMM being basically an instrument that converts an analog input to a digital presentation, either a front-panel readout or digital output signal. There are a variety of different types of A/D converters used in DMMs, each with advantages and disadvantages. The most important trade-offs are cost, accuracy, speed, and noise immunity. The two most common types are dual-slope integrating and

TABLE 12.13   Comparison of Dual-Slope and Successive-Approximation A / D Converters

Parameter	Dual-Slope Integration	Successive Approximations
Cost	Low	Medium
Accuracy	Medium	High
Speed	Medium	High
Noise immunity	Good	Poor

successive approximations. Most benchtop DMMs employ dual-slope integration or a variation, while most "systems" DMMs employ successive approximations. Table 12.13 demonstrates why. Dual-slope integration DMMs have reasonably good accuracy, speed, and noise immunity, particularly at multiples of line frequency, at low cost; successive approximations have high accuracy and high speed.

Dc voltages are applied directly to the input of the A/D converter after being ranged either automatically or with a range switch. Any other analog input, ac voltage, current, ohms, temperature, and so on, must be converted to a dc voltage.

AC Converter. An ac converter must produce a dc voltage output numerically equal to the ac voltage input. The simplest ac converters are rectifier-filter combinations that respond to the average value of the ac input signal. They customarily are calibrated in terms of the rms value with a sinusoidal waveform. For any distorted or other non-sine-wave forms, average-responding converters will indicate considerable error. True rms ac converters produce dc outputs equal to the rms value of the input, independently (within limits) of waveshape.

Current Converters. A simple shunt resistance serves as the current-to-voltage converter, both dc and ac, in multimeters. Switchable decade values are typically provided. Thus for a basic DVM full-scale value of 200 mV, a 0.1-Ω shunt gives a 2-A range; 1.0 Ω provides a 200-mA range, and so on.

Ohms Converters. Resistance ranges are added to the basic DVM by providing a constant-current source to force selectable decade values of current through the unknown resistance. The DVM then measures the resulting voltage drop. A 1-mA current will produce a 200-Ω full-scale resistance range when monitored by a 200-mV full-scale voltmeter, 100 μA gives a 2000-Ω range, and so on.

Other Functions. Five functions are relatively standard on DMMs: dc volts, ac volts, dc current, ac current, and ohms. Other functions available include decibel measurements of ac signals, surface temperature measurements with an accessory thermally sensitive probe, capacitance measurements, and conductance measurements.

Resolution and Accuracy

Fractional Digits and Overrange. Most digital voltmeter manufacturers use "fractional digit" terminology to describe the full-scale reading capability of their instruments. Thus a $4\frac{1}{2}$-digit meter has four digits with full zero-through-nine display capability and a fifth (most significant) digit that can display only zero or one. The maximum reading is thus 19,999 counts. Typically, the ranges will be labeled 2 V, 20 V, 200 V, and so on, on the instrument panel, but the true full-scale value is one count less. Some older instruments may still label the ranges 1 V, 10 V, and so on, and refer to the additional coverage as overrange. With that terminology, the 19,999 display capability would be called 100% overrange and a 1199 ... maximum would be called 20% overrange.

Resolution and Sensitivity. Resolution refers to a change of one unit in the least significant digit on the most sensitive range. A $4\frac{1}{2}$-digit meter on its 200 (199.99)-mV range has 0.01-mV resolution. A $5\frac{1}{2}$-digit meter has 10× better resolution. Upranging to the next least sensitive range causes a 10× reduction in resolution.

Accuracy. Typical accuracy statements are phrased as "$\pm X\%$ of reading $\pm Y\%$ of full scale" or sometimes "$\pm X\%$ of reading $\pm Y$ counts." Any digital meter has a minimum ambiguity of ± 1 count. Noise and other short-term variations can increase this specification from the theoretical one-count minimum. Economical instrument design requires a reasonable balance between the "$X\%$ of reading" basic accuracy and the instrument's number of digits. Typical $3\frac{1}{2}$-digit instruments have basic dc accuracies of 0.1 to 0.3%; $4\frac{1}{2}$-digit meters are usually 0.02 to 0.05%, and so on.

Ranging
Multimeters offer multiple ranges within each function in addition to multiple functions. In dc and ac voltage modes, high-impedance-resistive attenuators (compensated for wide ac bandwidth) are switched in ahead of the basic A/D converter in order to read higher voltages. Current and resistance range switching was discussed. Some multimeters add an autorange feature, whereby they sense the display reading and automatically switch ranges to obtain maximum resolution while avoiding display overflow.

Noise Rejection
Good DMM design controls layout, shielding, and capacitive balance to ground to reject noise. A balanced or true differential input provides high common-mode rejection ratios (CMRR) for the case of noise signals present between either input lead and ground. Furthermore, in dc modes additional rejection of noise pickup at power-line frequencies and their multiples can be provided by selecting A/D converter gate periods or integration times to be an integral number of power-line cycles. Most DMMs select times to give this high normal-mode rejection ratio (NMRR) for both 50- and 60-Hz power frequencies.

Probes and Accessories
In addition to common test leads, high-voltage probes are available that attenuate input voltages by a factor of 1000. RF probes detect high-frequency signals—often to 1 GHz—and provide a dc voltage output to the meter that is equal to the rms value of the RF voltage at the probe tip. External current shunts provide for measurements of high values of current.

12.13.3   Counters

Basic Architecture and Modes
The backbone of a counter/timer is a string of decade counter units (DCUs); each of the DCUs is in turn made up of four flip-flops. Each DCU drives the following DCU with its ÷10 output, and also provides binary-coded-decimal (BCD) output lines to drive a display that will indicate the count stored in the DCU. A simple TOTALIZE counter need be nothing more than a string of DCUs, each with its own display, as shown in Fig. 12.182. At any given moment (assuming no prior overflow), this counter will display the total number of events (pulses) that have been applied to its input.

This totalize function is converted to a FREQUENCY counter by adding a precisely time-controlled gate ahead of its input (see Fig. 12.183). If the gate is held open precisely 1 s, the resulting display will be the average frequency (cycles per second or Hz) of the input signal during that time. Decade multiples or submultiples of 1 s can be used as gate times merely by properly positioning a decimal point in the display.

One more simple change in architecture produces a TIMER to measure PERIOD, WIDTH, or TIME INTERVAL. In these modes, the signal is used to control the gate and the time reference (clock) transitions are counted as in Fig. 12.184.

The timing resolution is that of the clock period. In PERIOD mode, the gate is opened for one full period of the signal. In WIDTH mode, a pulse leading edge opens the gate and the trailing edge closes it. For TIME INTERVAL between events, two inputs (START and STOP, or commonly A and B) are provided.

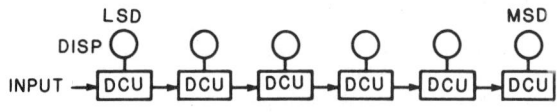

Fig. 12.182   Simple totalize counter.

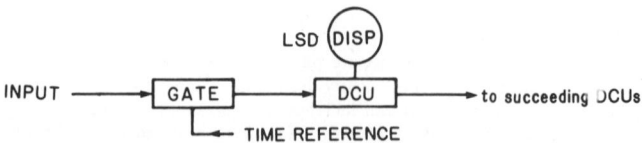

Fig. 12.183   Frequency counter.

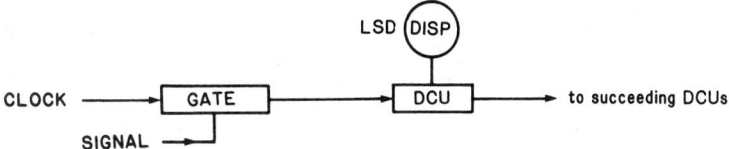

Fig. 12.184 Period, width, or interval timer.

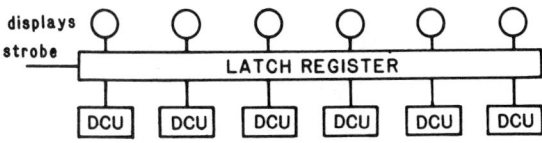

Fig. 12.185 Modified timer.

RATIO between two signals can also be measured by letting the lower-frequency signal control the gate while counting the cycles of the higher-frequency signal.

A degree of user convenience is added, as shown in Fig. 12.185, by providing a digital latch (register) between the BCD output of the DCUs and the numeric displays. The latch is strobed only at the completion of each measurement cycle. The displays thus indicate the value of the most recently completed measurement while the DCUs are engaged in "counting up" the next one. In totalize or manual stopwatch modes, the latch is disabled and the numeric displays follow the increasing count continuously.

Averaging features can be added to counters to improve accuracy and resolution. In PERIOD or RATIO mode, averaging is accomplished by adding a second counter chain of DCUs that monitor the input signal; this permits the gate to hold open for some decade multiple of signal periods rather than a single period. If the gate is held open for 1000 signal periods, then 1000 times as many clock pulses will be counted. The decimal point is moved three places left in the display and the resulting reading has 1000 times higher resolution.

In WIDTH and TIME INTERVAL modes, the gate cannot simply be held open while averaging since it is not desired to accumulate clock pulses between pulses. The gate must therefore still open for every pulse or pair of events, but the latch register line is not strobed until some decade multiple of pulses or events have been counted by the second, averaging counter. Again, the decimal point is moved left one place for each $10 \times$ increase in averaging and higher resolution results.

An instrument that only measures frequency and totalizes is called a FREQUENCY COUNTER. One that adds a second input to provide time interval, ratio, and other modes is called a UNIVERSAL COUNTER-TIMER.

Signal Conditioning

The DCUs in modern counters are typically monolithic integrated circuits of logic families such as ECL, TTL, and MOS. Signals to be measured may cover a wide range of amplitudes, and average dc values above or below ground. Furthermore, it may be necessary to measure between positive or negative slopes or selectively, to define the threshold for counting at various dc voltage levels, and to reject noise. All these needs lead to signal conditioning in the front end of practical counters and timers.

All counter signal-conditioning circuits include a Schmitt trigger or other shaper to provide a clean, fast, and constant transition waveform of the proper logic levels to the input gate and DCUs. The shaper also provides amplitude hysteresis to help reject noise; the signal must pass through both lower and upper hysteresis levels before the Schmitt switches to the opposite logic state.

In addition to the shaper, a full performance countertime will let the user select positive or negative transitions for counting, allows selection of the dc level on the input waveform at which the shaper switches, permits attenuation of high-level input signals, and allows ac coupling to be selected for simpler operation in less critical applications.

Time Bases

The accurate gate times in FREQUENCY mode and clock rates in PERIOD, WIDTH, or TIME INTERVAL modes are invariably derived in modern counters from quartz-crystal-controlled oscillators. The fundamental oscillator frequency is usually between 1 and 10 MHz. Decade dividers are used to produce the gate times in FREQUENCY mode. The crystal frequency is the clock rate directly in all but the highest-performance counters; in those units, the oscillator frequency is multiplied up to a higher frequency by a phase-locked loop to provide better timing resolution.

The least expensive oscillators are neither temperature controlled nor compensated. Better stability is provided by temperature-compensated crystal oscillators (TCXOs), and still better by oven-stabilized crystal oscillators. Uncompensated oscillators typically vary by 1×10^{-5} to 1×10^{-6} over the temperature change 0–50°C; TCXOs can hold 5×10^{-7} over the same range, and oven-controlled oscillators achieve stabilities of 7×10^{-9}. Long-term aging rates vary from 1×10^{-5} per year for inexpensive crystals to less than 2×10^{-7} per year for better units.

High-Frequency Techniques

While improvements in semiconductor technology are constantly increasing the frequency at which gates can switch and flip-flops can toggle, there are still many applications for frequency measurements beyond these limits. Three basic techniques of frequency coverage extension are available: prescaling, heterodyne, and transfer oscillators. Prescaling provides moderate extension by putting one or more very fast dividers (flip-flops) ahead of the counter gate. Division ratios of 8, 10, and 16 are common. The counter gate is then held open longer than usual by this same division factor, so no change in display results. The advantages are that input amplifier, gate, and DCU bandwidth of the basic counter need not be increased. The disadvantage is that longer measurement times are required, by the prescaling factor, for any given resolution. The upper limit for prescaling techniques is still limited by toggle rates of the fastest semiconductor technology.

Heterodyne techniques basically work like superheterodyne receivers. A high-frequency local oscillator (LO) frequency, derived from the counter time base so that accuracy is not impaired, is fed to a mixer along with the signal. The resulting difference frequency is counted. Resolution versus measurement time is not affected. Relatively complex hardware is required for an automatic heterodyne counter if the display is to be the absolute signal frequency, since the instrument must determine which harmonic of the oscillator is acting as the local oscillator and whether the LO is higher or lower than the signal frequency.

Transfer oscillator techniques function by phase locking a high-order harmonic of a voltage-controlled oscillator to the signal, counting the VCO frequency itself, and determining the harmonic number to compute the signal frequency. Transfer oscillator techniques are not advantageous if the signal has large amounts of frequency modulation.

Error Sources and Dealing with Them

In FREQUENCY mode, counters produce error only via the ± 1 count ambiguity plus errors in the time the gate is held open. The gate is directly derived from the time base, so frequency error is time base error ± 1 count. These errors can be reduced by use of better crystal time bases or external atomic frequency standards, by frequent recalibration of the time base to remove the effects of drift, and by selection of long gate times so that ± 1 count equates to a tolerably small frequency value.

In all timing modes, significant additional errors can occur because the input signal is being used to open and close the gate. The rise and fall times of the signal take a finite time to pass through the hysteresis levels of the shaper. Any noise combined with the signal during this time can make the gate open earlier or later than it should have, producing a direct timing error. Reducing these errors requires maximizing the signal-to-noise ratio of the external signal and selecting a fast rise and fall signal if a choice is available. Averaging is also extremely helpful, assuming that the signal is repetitive. In PERIOD mode, triggering errors reduce directly with increasing averaging factor since the errors can occur only at the initial gate opening and final gate closing. In TIME INTERVAL or WIDTH modes, averaging reduces errors only due to the statistical likelihood of random errors offsetting one another; therefore, triggering error is reduced by the square root of the averaging factor in these modes.

12.13.4   Logic Probes and Analyzers

The measurement instrument types discussed so far are used principally to measure the analog characteristics of a signal, whether the signal is part of an analog or a digital system. Other instrument types basically ignore the analog characteristics and help perform analysis in the digital domain.

Logic Probes

A simple, single-channel logic probe simply converts logic levels to visible signals (see Fig. 12.186). The reference voltage of a parameter is set to the decision threshold value for the logic type being used. If the voltage at the probe tip is above this value, the indicator lights. Below this value, the indicator is off.

Practical instruments add a number of sophistications to this basic scheme. Dual thresholds are desirable for many applications, with one being set to the "guaranteed zero" and the other to the

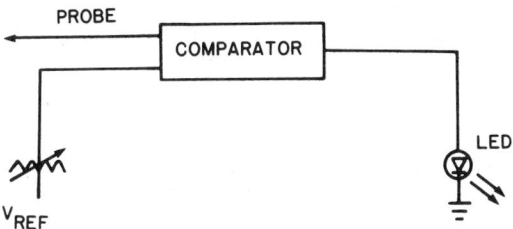

Fig. 12.186   Single-channel logic probe.

"guaranteed one" levels of the logic family. If the voltage being probed falls between those levels, an error signal can be generated, or a hysteresis-type comparator can be used that will not change state until the input has passed through both thresholds.

Under dynamic operating conditions at extremely low data rates, the simplified probe shown would blink off and on. At more practical rates, the indicator or human eye could not follow the data. Furthermore, narrow pulses would never be seen. Practical probes typically include a pulse stretcher to provide a visible signal, even with very narrow pulses. They also often cause a low-frequency toggling off and on of the indicator whenever the input data are changing logic states more than a few times per second, indicating to the user the presence of dynamic operation. Practical probes may include many channels of inputs and indicators; some are designed to clip onto the top of common logic IC packages. Consideration must be given to the speed and to resistive and capacitive loading of probe inputs for each application to make sure that they are adequate and will not degrade the performance of the logic circuits being tested.

Logic Analyzers

A logic analyzer adds the features of memory, triggering, and more sophisticated display and analysis to the simple logic probes just discussed. Typically, a logic analyzer consists of some number of logic probe channels driving the inputs of shift registers instead of simple indicators. A line from the clock of the unit under test controls when the data bits from the multiple channels are strobed into the register. Application of a trigger pulse disables the activity of writing into these registers; they then contain the last N pulses of the data stream. By also delaying the trigger pulse a selected number of system clock pulses before the write operation is stopped, the user can control whether the data stored precede the trigger, follow the trigger, or both. Then the stored data can be nondestructively clocked out to some display, on a repetitive basis, for convenient analysis at rates unrelated to the original clock rate. Common display techniques include the timing diagram, where each channel is represented on a CRT as a time-related line where up is true and down is false, or a binary state table with each channel represented by 1's and 0's. Three or four channels can be grouped and their combined states displayed as octal or hexadecimal states. Sophisticated analyzers can compare acquired data with previously stored "known correct" data and highlight discrepancies. Some analyzers have mapping modes that help analyze transitions from data word to word and ratio of time during which given words appear in a data stream.

12.13.5   Signal Sources

Extreme variety exists among test signal sources, spanning frequencies from millihertz to tens of gigahertz; amplitudes from submicrovolt to hundreds of volts; basic waveforms of sine, square or pulse, and ramps; frequency or amplitude modulation; specialized digital sources; and variations on the theme. Only the most commonly used general-purpose types will be discussed.

Pulse Generators

Architecture. The simplified block diagram shown in Fig. 12.187 describes the most general form of pulse generator. In the free-run mode, the period (or rate), delay after trigger, and width are all controllable independently via their separate generators. Rise and fall times are set by separate controls, and the output pulse amplitude and baseline dc offset may also be controlled. In the gated mode, the period generator is turned on and off by an external signal or control voltage. With SYNCHRONOUS GATING, any pulse already in progress will still be completed even if the gating signal stops during the pulse. Nonsynchronous gating will cause the last pulse in a burst to be shortened from the value set by the duration generator if the gating signal stops during a pulse. In the triggered mode, the period generator is disabled and one pulse is generated for each externally applied

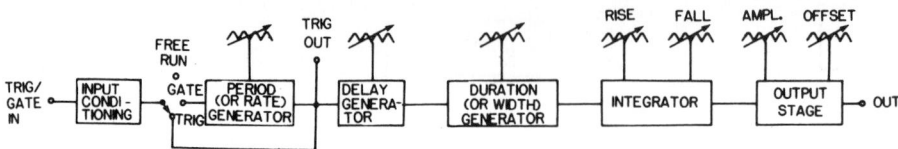

Fig. 12.187   General form of pulse generator.

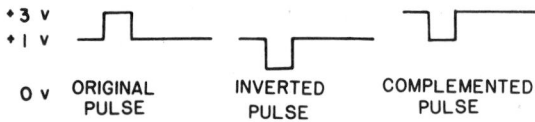

Fig. 12.188   Inverted pulse and complemented pulse outputs.

trigger pulse. A trigger output jack with a fixed-amplitude, fixed-duty-cycle (or fixed width) trigger pulse is usually provided.

Pulse generators with a delay capability also often have the ability to generate paired or double pulses, with the DELAY control determining the time between their leading edges and the DURATION control setting both their widths.

Output Stage. One common control approach for the user to set the pulse baseline and peak amplitude is with OFFSET and AMPLITUDE controls. OFFSET determines the baseline dc voltage and AMPLITUDE sets the pulse peak deviation from that baseline. An INVERT switch changes a positive-going pulse to negative-going without affecting the baseline or duty cycle (Fig. 12.188). A COMPLEMENT switch preserves the two original dc levels, but interchanges the high and low timing relationships to produce a duty cycle equal to 1-original duty cycle.

A second method for pulse amplitude control is via independent HIGH and LOW level controls. Within the amplitude restrictions of the unit, each control can be set independently to the desired dc level and the resulting pulse amplitude is the difference between the two levels.

Most output stages are either current sources (high impedance) or 50 Ω. The current source, assuming adequate current capability for a 50-Ω load, has the advantage of being able to deliver its full pulse amplitude into 50-Ω loads, open circuits, or anything in between. The disadvantage of the current source is that any energy reflected from an improper termination will be re-reflected by the generator and extremely distorted pulses will result. A 50-Ω source impedance absorbs any reflected energy due to poor terminations, producing maximum pulse "cleanliness." However, its maximum voltage across 50 Ω is only half its open-circuit value. Current source generators often also have a switchable internal 50-Ω termination, but significant reactance is usually also present due to the switch.

Word / Data Generators

Word or data generators are a special category or type of pulse generator designed for digital applications. Rather than being optimized for speed, pulse fidelity, and independent control of the pulse parameters, as is the case for pulse generators intended primarily for linear circuit or analog testing, word/data generators are optimized for digital testing. Their output pulses are treated as digital words or bit patterns.

Bit Patterns. The output pulses or bits can be coded in a variety of ways, as shown in Fig. 12.189. Each bit pattern has its own advantages and disadvantages in particular applications; for example, the NRZ coding schemes only have an average of half as many transitions compared to RZ and, therefore, would only require half the bandwidth in a tape recorder. Bipolar RZ, which has a transition either positive or negative each clock period, is self-clocking (does not require a separate clock when decoding).

Serial / Parallel. These bit patterns can be output in either a serial (sequentially) or parallel (simultaneously) format, but most generators offer only the serial format.

Synchronous / Asynchronous. The output format can also be either synchronous or asynchronous relative to the first bit or a separate clock, frame, or trigger input.

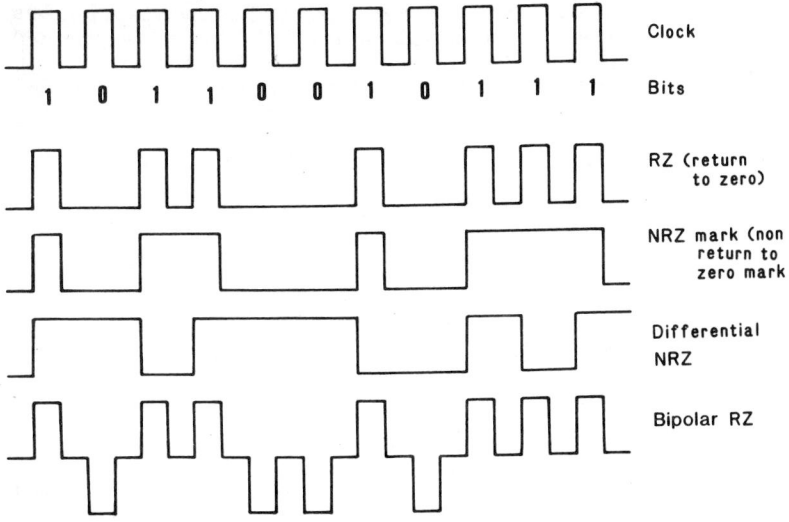

Fig. 12.189 Output pulse patterns.

Pseudo-Random Binary Sequence. Some generators offer pseudo-random binary sequences where the digital pattern or word is continuously changing or has noiselike properties.

Function Generators

This most versatile of all signal sources produces triangles, square waves, sine waves, ramps, and pulses and can be easily swept or frequency modulated.

Architecture. The triangle is the basic waveform of a function generator. It is produced by two current sources, a capacitor, and a switch, as shown in Fig. 12.190.

The constant current from the positive source causes a linearly increasing voltage ramp across the capacitor until a preset threshold is reached at the comparator. The comparator then causes the negative source to start charging the capacitor in the opposite direction until the lower comparator threshold is reached, and so on.

The square waveform is derived from the output of the comparator. Sine waves are produced from the triangle via diodes and resistors in a ladder configuration.

Ramps or pulses are obtained instead of triangles and square waves by upsetting the symmetry of the two current sources, producing longer transitions for one polarity than the other.

Frequency-range switching is obtained by switching in different values of timing capacitors, usually in decade steps. Within any given range, frequency is adjusted by simultaneously changing the

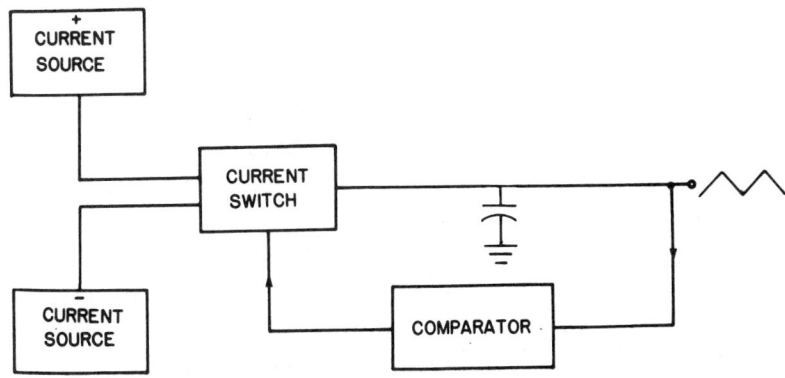

Fig. 12.190 Simplified function generator loop.

current values of both current sources. This is accomplished via a voltage control, either from the generator main dial potentiometer or a functionally parallel VCF (voltage-controlled frequency) input. The VCF input may be used to produce frequency modulation of the generator output. A linear ramp or sawtooth applied to the VCF input produces a swept output. For logarithmic sweep, desirable when sweeping over wide frequency ranges, an exponentially shaped ramp is used instead of a linear ramp. Some function generators incorporate a ramp or sawtooth generator internally in addition to the main generator in order to produce swept signals on a stand-alone basis. In such an instrument, the main dial controls the sweep start frequency (internal ramp at zero amplitude) and a second dial controls sweep stop frequency by adjusting the peak amplitude of the internal ramp.

Waveform Control. More flexible function generators offer a variable symmetry feature whereby a front-panel control increases the current from one current source while decreasing the other. The effect is to modify triangles into ramps and square waves into pulses, while producing minimal effects on frequency. Variable symmetry can typically modify the basic 50% duty cycle to a 5 or 95% duty cycle. Since most function generators have an output amplifier bandwidth and maximum loop frequency consistent with the 50% duty cycle case, it is common to reduce frequency by one decade when variable symmetry mode is selected since a 5% duty cycle at $f/10$ requires the same transition times as 50% duty cycle at frequency f.

The most sophisticated function generators provide a variable rise- and fall-time feature, similar to that of pulse generators, when in square-wave or pulse modes.

At low frequencies it is useful to provide a "hold" mode whereby the user can "freeze" the output waveform at any amplitude. Conceptually, this is accomplished by interrupting the current source. This feature is useful in checking the operating point of voltage- or current-sensitive comparators.

Triggering and Gating. As in pulse generators, a triggered mode generates one cycle of output waveform per input trigger pulse and a gated mode lets the generator run while the gating signal is in the "true" condition. These modes have added value in a function generator due to its waveform flexibility. Bursts of sine waves are used in acoustical testing and in remote control and signaling systems. Function generators with a triggered mode often have a phase control, permitting the triggered output waveform to begin at any angle through the cycle. With this feature it is possible to generate haver-sine and sine-squared pulses as testing impulses of known, limited bandwidth.

Output Stages. A function generator output stage provides the high-level signals needed for some applications, permits independent control of amplitude and dc offset, typically includes attenuators to furnish lower-level signals, and sometimes includes an amplitude modulator. Most function generator amplitude controls vary the peak-to-peak amplitude symmetrically about the dc offset value, in contrast to a pulse generator's amplitude control, which controls a peak unipolar excursion above (or below, but not both) the dc offset. Peak-to-peak voltage output of a function generator remains quite constant as the user switches between waveforms.

Sine-Wave Generators

A sine wave, the only waveform with all its energy at a single spectral line, plays a common part in testing applications. However, the techniques used to generate sine waves are radically different across the approximately 14 decades of frequency available from various generators.

RC Oscillators. In addition to function generators already discussed, the other common source of low-frequency sine waves is the resistance-capacitance (RC)-controlled oscillator. Such oscillators use an RC filter network such as a Wien bridge, twin tee, or bridged tee in the feedback loop of an amplifier. The architecture is such that the circuit oscillates only at the filter frequency. Harmonic distortion of these designs is lower than achievable by any other oscillator technique (without post-filtering) and amplitude flatness is good. Operation frequency is typically limited to the area of the spectrum from 10 Hz to 1 MHz.

High-Frequency Oscillators. A number of oscillator types are successfully used to generate sine waves from tens of kilohertz up into the gigahertz region. Oscillators with lumped inductance-capacitance (LC) frequency-determining elements are common to 100 MHz and above. In the UHF region, distributed-reactance, transmission line, and cavity approaches are typical. Quartz crystals are used whenever high orders of frequency stability are required. At lower frequencies, there is a relatively wide choice of active devices and reasonably independent choice of active-device type and frequency-determining elements, while at high frequencies the choices narrow and ultimately become highly integrated.

Synthesizer. A quartz crystal provides frequency stability many orders of magnitude better than any other practical, economical frequency-determining element. However, the operating frequency range of a crystal-controlled oscillator is almost infinitesimal and therefore unsuited as a general-pur-

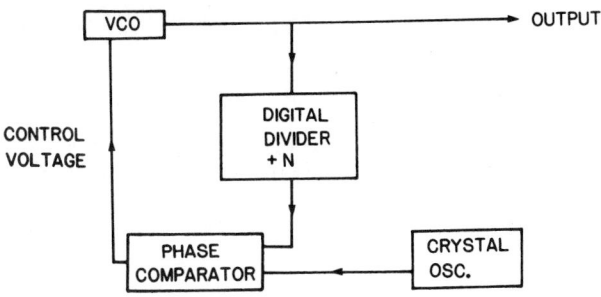

Fig. 12.191   Frequency synthesizer.

pose oscillator. The frequency synthesizer was developed as a means of providing users selectable sine-wave frequencies over a wide range, but with the inherent stability characteristics of a crystal.

In simplest form, as shown in Fig. 12.191, the synthesizer consists of some sort of voltage-controlled, variable-frequency oscillator (VCO), a digital divider chain, a phase comparator, and a crystal oscillator. The Nth subharmonic of the VCO frequency is compared to the crystal reference and a dc output generated by the comparator, which slews the VCO frequency and phase until phase equality is reached. At this point, the VCO output exhibits the same frequency stability as the crystal reference, but at a different frequency determined by the divider chain. Practical, flexible units are a good deal more complex with multiple divider chains and heterodyne frequency conversions. They can produce crystal-stable output frequencies in arbitrarily small increments from audio frequencies up into microwave.

Many synthesizers include internal modulators, calibrated attenuators, and excellent shielding to allow them to function as signal generators for demanding communications applications.

A disadvantage of frequency synthesizers (in addition to cost and complexity) is relatively poor spectral purity (spurious output signals) compared to simpler oscillator configurations.

REFERENCES

1. Motto, J. W., Jr., *Introduction to Solid State Power Electronics*, Youngwood, Pa., Westinghouse Electric Corp., Semiconductor Division, 1977.

2. *General Electric SCR Manual*, 5th ed., Syracuse, N.Y., General Electric Semiconductor Products Dept., 1972.

3. *General Electric Semiconductor Data Handbook*, published yearly, available from GE Semiconductor Products Dept., Electronics Park, Syracuse, N.Y.

4. Baascn, T. L., "Quantifying the Oscilloscopes," *Electron. Prod. Mag.*, Sept. 1970.

5. McQueen, J. G., "Monitoring of High Speed Waveform," *Electron. Eng.*, Oct. 1952.

6. Frye, G. S., and Nohman, N. S., "Random Sampling Oscillography," *IEEE Trans. Instrum. Meas.*, **IM-13**, Mar. 1964.

7. Prichard, R. L., Angell, J. B., Adler, R. B., Early, J. M., and Webster, W. M., "Transistor Internal Parameters for Small Signal Representation," *Proc. IRE*, **49**, 725 (1961) or Sze, S. M., *Physics of Semiconductor Devices*, New York, Wiley, 1969, p. 282.

8. Allison, R., "Silicon Bipolar Microwave Power Transistors," *IEEE Trans. Microwave Theory Tech.*, **MTT-27**, 415 (1979).

9. Sakai, T., "Step Electrode Transistor," *Dig. Tech. Paper ISSCC 1977*, p. 196.

10. Kirk, C. T., "A Theory of Transistor Cutoff Frequency (f_T) Fall-Off at High Current Density," *IEEE Trans. Electron. Devices*, **ED-9**, 164 (1962).

11. Poon, H. C., Gummel, H. K., and Schaffeter, D. L., "High Injection in Epitaxial Transistors," *IEEE Trans. Electron. Devices*, **ED-16**, 455 (May 1969).

12. Hsu, T. H., and Snapp, C. P., *IEEE Trans. Electron Devices*, **ED-25**, 723 (1978), also see Chen, J. T. C., and Snapp, C. P., "Bipolar Microwave Linear Power Transistor Design," *IEEE Trans. Microwave Theory Tech.*, **MTT-27**, 423 (1979).

13. Nielson, E. G., "Behavior of Noise Figure in Junction Transistors," *Proc. IRE*, **45**, 957 (1957).

14. *Hewlett-Packard Appl. Note 967*.

15. Schafft, H. A., "Second-Breakdown—A Comprehensive Review," *Proc. IEEE*, **55**, 1272 (1967).

16. Blumberg, R. J., and Brener, S., "A 1500 Gate, Random Logic, Large Scale Integrated (LSI) Master Slice," *IEEE J. Solid-State Circuits*, **SC-14**, 818 (1979).

17. Grebene, A. B., *Analog Integrated Circuit Design*, Huntington, N.Y., Krieger, 1978. Grebene, A. B., *Analog Integrated Circuits* (a collection of papers on analog circuit design), New York, IEEE, 1979.

18. Gordon, B. M., "Linear Electronic Analog/Digital Conversion Architectures, Their Origins, Parameters, Limitations, and Applications," *IEEE Trans. Circuits Syst.*, **CS-25**, 391 (July 1978).

19. Special Issue on Analog Circuits, *IEEE J. Solid-State Circuits*, **SC-14** (Dec. 1979).

20. Berger, H. H., and Wiedmann, S. K., "Schottky Transistor Logic," *Dig. Tech. Paper ISSCC 1975*, p. 172.

21. Berger, H. H., and Wiedmann, S. K., "Merged-Transistor Logic (MTL)—A Low-Cost Bipolar Logic Concept," *IEEE J. Solid-State Circuits*, **SC-7**, 340 (1970).

22. Tang, D. D., and Solomon, P. M., "Bipolar Transistor Design for Optimized Power-Delay Logic Circuit," *IEEE J. Solid-State Circuits*, **SC-14**, 679 (1979); also see Soloman, P. M., and Tang, D. D., "Bipolar Circuit Scaling," *Dig. Tech. Paper ISSCC 1979*, p. 86.

23. Lueke, G., Mize, J. P., and Carr, W. N., *Semiconductor Memory Design and Application*, McGraw-Hill, New York, 1973.

24. Kawarada, K., Suzuki, M., Mukai, H., Toyoda, K., and Kondo, Y., "A Fast 7.5 ns Access 1K-Bit RAM for Cache-Memory Systems," *IEEE J. Solid-State Circuits*, **SC-13**, 656 (1978).

25. Wiedmann, S. K., "Injection-Coupled Memory," *IEEE J. Solid-State Circuits*, **SC-8**, 332, (1973). Wiedmann, S. K., and Berger, H. H., "Small Size, Low Power Bipolar Memory Cell," *J. Solid-State Circuits*, **SC-6**, 283 (1971). Wiedmann, S. K., *Dig. Tech. Paper ISSCC 1980*, p. 223.

26. Sander, W. B., Shepherd, W. H., and Schinelle, R. D., "Dynamic I^2L Random-Access Memory Concepts with MOS Design," *Electronics*, Aug. 1976, p. 99.

27. Sze, S. M., *Physics of Semiconductor Devices*, New York, Wiley-Interscience, 1969.

28. Grove, A. S., *Physics and Technology of Semiconductor Devices*, New York, Wiley, 1967.

29. Deal, B. E., "The Current Understanding of Charges in Thermally Oxidized Silicon Structure," *J. Electrochem. Soc.*, **121**, 1986 (1974).

30. Deal, B. E., Snow, E. H., and Mead, C. A., "Barrier Energies in Metal-Silicon Dixode-Silicon Structures," *J. Phys. Chem. Solids*, **27**, 1873 (1966).

31. Deal, B. E., Sklar, M., Grove, A. S., and Snow, E. H., "Characteristics of the Surface State Charge (Q_{ss}) of Thermally Oxidized Silicon," *J. Electrochem. Soc.*, **114**, 226 (1967).

32. Aitken, A., MacArthur, A. T. P., Abbott, R., and Morris, J. D., "The Relative Performance and Merits of CMOS Technology," *IEEE Int. Electron. Devices Meet. Tech. Dig.*, Washington, D.C., 1976, p. 327.

33. Troutman, R. R., and Chakravarti, S. N., "Subthreshold Characteristics of Insulated Gate Field-Effect-Transistors," *IEEE Trans. Circuit Theory*, **CT-20** (6), 659 (1973).

34. Richman, P., *MOS-Field-Effect Transistors and Integrated Circuits*, New York, Wiley-Interscience, 1973.

35. Mai, C. C., Hswe, M., and Palmer, R. B., "Ion-Implantation Combined with Silicon-Gate Technology," *IEEE Trans. Electron. Devices*, **ED-19** (11), 1219 (1972).

36. Mead, C. A., and Conway, L. A., *Introduction to VLSI Systems*, Addison-Wesley, Reading, Mass., 1979.

37. Ganslen, F. H., "Geometry Effects of Small MOSFET Devices," *IEEE Int. Electron Devices Meet. Tech. Dig.*, Washington, D.C., 1977, p. 512.

38. Hoeneisen, B., and Mead, C. A., "Fundamental Limitations in Microelectronics," *Solid State Electron.*, **15**, 819, 981 (1972).

39. Pashley, R., Kokonnen, K., Boleky, E., Jecmen, R., Liu, S., and Owen, W., "HMOS Scales Traditional Devices to High Performance Levels," *Electronics*, Aug. 18, 1977, p. 95.

40. Crichlow, D. C., Dennard, R. H., and Schuster, S. E., "Design and Characteristics of n-Channel Insulated-Gate Field Effect Transistors," *IBM J. Res. Dev.*, **17**, 430 (1973).

41. Noble, W. P., and Cottrell, P. E., "Narrow Channel Effect in Insulated Gate Field Effect Transistors," *IEEE Int. Electron. Devices Meet. Tech. Dig.*, Washington, D.C., 1976, p. 582.

42. Troutman, R. R., and Fortino, A. G., "Simple Model for Threshold Volume in a Short-Channel IGFET," *IEEE Trans. Electron. Devices*, **ED-24** (10), 1266 (1977).

43. Altman, L., "Fine Technologies Squeezing More Performance from LSI Chips," *Electronics*, Aug. 18, 1977, p. 91.

44. Amelio, G. F., Tompsett, M. F., and Smith, G. E., "Experimental Verification of the Charge Coupled Device Concept," *Bell Syst. Tech. J.*, **49**, 593 (1970).

45. Weber, S., (Ed.), *Large and Medium Scale Integration*, New York, McGraw-Hill, 1974, p. 24.

46. Rideout, V. L., Gaensslen, F. H., and Le Blanc, A., "Device Design Considerations for Implanted n-Channel MOSFETs," *IBM J. Res. Dev.*, **19**, 50 (1975).

47. Geipel, H. J., and Tice, W. K., "Reduction of Leakage by Implantation Gettering in VLSI Circuits," *IBM J. Res. Dev.*, **24** (3), 310 (May 1980).

48. Stein, P. K., "The Engineering of Measuring Systems," *J. Met.*, **21**, 40–47 (Oct. 1969).

49. Stein, P. K., *Measurement Engineering*, **1**, Stein Engineering Services, Inc., Pheonix, Ariz., 1964.

50. Enke, C. G., "Data Domains—An Analysis of Digital and Analog Instrumentation Systems and Concepts," *Anal. Chem.*, **43** (1), 69A–80A (Jan. 1971).

BIBLIOGRAPHY FOR POWER ELECTRONICS I

Bedford, B. D., and Hoft, R. G., *Principles of Inverter Circuits*, New York, Wiley, 1964.

Humphrey, A. J., "Inverter Commutation Circuits," *IEEE Trans. Ind. Gen. Appl.*, **4** (1), 104–110, Jan.–Feb. 1968.

Kimbark, E. W., *Direct Current Transmission*, Vol. 1, New York, Wiley-Interscience, 1971.

Pollack, J. J., "Advanced Pulsewidth Modulated Inverter Techniques," *IEEE Trans. Ind. Appl.*, **8** (2), 145–155, Mar.–Apr. 1972.

BIBLIOGRAPHY FOR POWER ELECTRONICS II

Branson, L. K., *Introduction to Electronics*, Englewood Cliffs, N.J., Prentice-Hall, 1967.

Carson, R. S., *Principles of Applied Electronics*, New York, McGraw-Hill, 1961.

Fitzgerald, A. E., and Higgenbotham, D. E., *Electrical and Electronic Engineering Fundamentals*, New York, McGraw-Hill, 1964.

Gewartowski, J. W., and Watson, H. A., *Principles of Electron Tubes*, New York, Van Nostrand, 1965.

Ghausi, M., *Electronic Circuits*, New York, Van Nostrand Reinhold, 1971.

Gray, T. S., *Applied Electronics*, 2nd ed., New York, Wiley, 1954.

Gray, P. E., and Searle, C. L., *Electronic Principles*, New York, Wiley, 1969.

Millman, J., *Vacuum-Tube and Semiconductor Electronics*, New York, McGraw-Hill, 1958.

Van der Ziel, A., *Electronics*, Boston, Allyn and Bacon, 1966.

BIBLIOGRAPHY FOR OSCILLOSCOPES

"An Ultra Wideband Oscilloscope Based on Advanced Sampling Device," *Hewlett-Packard J.*, Oct. 1966.

"Basic Oscilloscope Operation," Technical Brief, Teletronix, 1978.

"Basic Oscilloscope Measurements: Set-Up and Analysis," Technical Brief, Teletronix, 1978.

"Basic Oscilloscope Measurements: Amplitude," Technical Brief, Teletronix, 1978.

"Basic Oscilloscope Measurements: Rise Time," Technical Brief, Teletronix, 1978.

"Basic Oscilloscope Measurements: Dual-Trace and X-Y Phase," Technical Brief, Teletronix, 1978.

"Basic Oscilloscope Measurements: Period and Frequency," Technical Brief, Teletronix, 1978.

"Coherent and Incoherent Sampling," *Hewlett-Packard J.*, July 1966.

Frequency divider extends automatic digital frequency measurements to 12.4 Ghz.

BIBLIOGRAPHY FOR MOS TECHNOLOGY

Penny, W. M., and Lau, L., (American Micro Systems, Inc.), *MOS Integrated Circuits*, New York, Van Nostrand Reinhold, 1972.

Ghandi, S. K., *The Theory and Practice of Microelectronics*, New York, Wiley, 1968.

Grove, A. S., *Physics and Technology of Semiconductor Devices*, New York, Wiley-Interscience, 1969.

Sze, S. M., *Physics of Semiconductor Devices*, New York, Wiley-Interscience, 1969.

Richman, P., *MOS Field Effect Transistors and Integrated Circuits*, New York, Wiley-Interscience, 1973.

Phillips, A. B., *Transistor Engineering*, New York, McGraw-Hill, 1962.

BIBLIOGRAPHY FOR SWITCHING CIRCUITS

Dietmeyer, D. L., *Logic Design of Digital Systems*, Boston, Allyn and Bacon, 1978.

Hill, F. J., and Peterson, G. R., *Introduction to Switching Theory and Logic Design*, New York, Wiley, 1974.

Krutz, R. L., *Microprocessors and Logic Design*, New York, Wiley, 1980.

Mead, C., and Conway, L., *Introduction to VLSI Systems*, Reading, Mass., Addison-Wesley, 1980.

Ross, C. H., Jr., *Fundamentals of Logic Design*, St. Paul, Minn., West Publishing, 1979.

BIBLIOGRAPHY FOR SENSORS AND TRANSDUCERS

Benedict, R. P., *Fundamentals of Temperature, Pressure and Flow Measurements*, New York, Wiley, 1977.

Fink, D. G., (Ed.-in-Chief), *Electronics Engineer's Handbook*, Sect. 10, "Transducer," by H. N. Norton, New York, McGraw-Hill, 1975.

Jones, B. E., *Instrumentation, Measurement and Feedback*, Maidenhead, England, McGraw-Hill, 1977.

Norton, H. N., *Sensor and Analyzer Handbook*, Englewood Cliffs, N.J., Prentice-Hall, 1982.

Standards and Practices for Instrumentation, 5th ed., Pittsburgh, Pa., Instrument Society of America, 1977.

BIBLIOGRAPHY FOR INSTRUMENTATION AMPLIFIERS

Analog Devices, Inc., *Data Acquisition Products Catalog*, Norwood, Mass., 1978.

Engineering Staff of Analog Devices, Inc., D. H. Sheingold, *Nonlinear Circuits Handbook*, Analog Devices, Norwood, Mass., 1976.

Fairchild Semiconductor, Inc., *The Voltage Regulator Applications Handbook*, Mountain View, Calif., 1974.

Graeme, J. G., *Applications of Operational Amplifiers, Third Generation Techniques*, New York, McGraw-Hill, 1973.

Graeme, J. G., Tobey, G. E., and Huelsman, L. P., *Operational Amplifiers, Design and Applications*, New York, McGraw-Hill, 1971.

Linear Applications Handbook, Vol. 1, National Semiconductor Corp., Santa Clara, Calif., 1973.

Linear Applications Handbook, Vol. 2, National Semiconductor Corp., Santa Clara, Calif., 1977.

Wait, J. V., Huelsman, L. P., and Korn, G. A., *Introduction to Operational Amplifier Theory and Applications*, New York, McGraw-Hill, 1975.

BIBLIOGRAPHY FOR INPUT / OUTPUT DEVICES: MICROPROCESSOR SYSTEM I / O

Bibbero, R. J., *Microprocessors in Instruments and Control*, New York, Wiley, 1977.

Eichenlaub, R. G., "Smart Instruments," *Proc. IEEE*, **66** (4), 423–428 (Apr. 1978).

Peatman, J. B., *Microcomputer Based Design*, New York, McGraw-Hill, 1977.

Randle, W. C., and Kerth, N., "Microprocessors in Instrumentation," *Proc. IEEE*, **66** (2), 172–181 (Feb. 1978).

Short, K. L., *Microprocessors and Programmed Logic*, Englewood Cliffs, N.J., Prentice-Hall, 1981.

CHAPTER 13
LIGHT, RADIATION, AND ACOUSTICS

M. Parker Givens

Institute of Optics
University of Rochester
Rochester, New York

Allan D. Pierce

School of Mechanical Engineering
Georgia Institute of Technology
Atlanta, Georgia

Yves H. Berthelot

School of Mechanical Engineering
Georgia Institute of Technology
Atlanta, Georgia

13.1 LIGHT AND RADIATION
M. Parker Givens

13.1.1 Introduction

Radiation is the transfer of energy through space without requiring any intervening medium; for example, the energy reaching the earth from the sun is classified as radiation.

The majority of this is in the form of electromagnetic waves, which have a wide range of frequencies. Of this, a relatively narrow frequency band between 4×10^{14} and 8×10^{14} Hz is capable of stimulating the visual system; this is light. Our attention will be directed primarily toward the visible part of the spectrum, but most of the principles are valid in other parts of the spectrum.

Electromagnetic radiation propagates through empty space with velocity c, which is one of the fundamental constants of nature; its approximate value is $2.998\ldots \times 10^8$ m/s. This velocity is

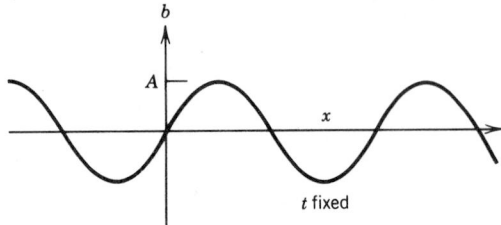

Fig. 13.1 Simple harmonic wave of the form $y = A \sin(2\pi/\lambda)(x - vt)$. Plot is for fixed $t = 0$. Over time, disturbance moves to right.

independent of frequency. In any material medium, the velocity of propagation v is less than c; the ratio $c/v \equiv n$ is called the **index of refraction** of the medium. The velocity v (and therefore n) depends upon the frequency of the radiation; this variation of velocity (or index) with frequency is known as dispersion.

The simplest wave to discuss is one in which some physical quantity varies sinusoidally with time at any point in space and this variation propagates with velocity v. Such a wave is represented by an equation of the form

$$b = A \sin \frac{2\pi}{\lambda}(x \pm vt) \tag{13.1}$$

where b represents the value of the quantity at position x and at time t and A is the maximum value of b and is called the *amplitude* of the wave. This equation represents a **plane wave**, that is, the quantity b is constant over a plane surface perpendicular to the x axis. The minus sign gives a wave propagating in the positive x direction; the plus sign gives a wave propagating in the negative x direction. The wavelength λ represents the smallest, nonzero distance for which $b(x + \lambda) = b(x)$ for all x; alternately, λ may be defined as the distance between adjacent crests of the wave. See Fig. 13.1, which is a plot of b as a function of x for some fixed t. Equation (13.1) is meaningful for all values of x and t; in this sense, it represents a wave of infinite extent in time and space. We can also define the period T of the wave as the time required to execute one cycle, or the smallest, nonzero, time for which $b(t + T) = b(t)$. Frequency ν is the reciprocal of the period; for period in seconds, the frequency unit is the hertz. Equation (13.1) is called a monochromatic wave since it contains only one frequency. These quantities are interrelated as follows:

$$v = \lambda \nu = \frac{\lambda}{T} \quad \text{or} \quad \lambda = vT$$

As the radiation passes from one medium to another, such as from glass to air, the frequency remains constant, but the velocity and wavelength change.

No real wave extends indefinitely in time but must begin and end; real waves cannot be monochromatic in the strictest interpretation of the term.

The methods of Fourier analysis enable us to construct a finite wave train as a sum of appropriately selected infinite wave trains of the proper phase. The amplitude as a function of time in the finite wave train and the amplitude as a function of frequency for the infinite components form a Fourier transform pair. Although the details in each case will depend upon the manner in which the wave builds up initially and dies away at the end, some "rule-of-thumb" statements are often helpful. These are:

$$\frac{\Delta \nu}{\bar{\nu}} \approx \frac{\bar{T}}{\Delta t} \approx \frac{\bar{\lambda}}{v \, \Delta t} \approx \frac{\Delta \lambda}{\bar{\lambda}}$$

where Δt is the duration of the wave train; $\Delta \nu$ is the frequency spread of the infinite, or monochromatic, components making up the wave train; $\bar{\nu}$, $\bar{T}$, and $\bar{\lambda}$ are the average frequency, period, and wavelength of the finite wave. These are rule-of-thumb or **order-of-magnitude** statements. A wave for which $\Delta \nu \ll \bar{\nu}$ is properly called **quasi-monochromatic** but is frequently called **monochromatic**. For many classical (i.e., prelaser) light sources, $\Delta \nu / \bar{\nu} \approx 10^{-5}$ and the wave is, for most practical purposes, monochromatic. Sunlight, on the other hand, has a very broad spectral range ($\Delta \nu$ comparable to $\bar{\nu}$) and may be described equally well as a series of randomly spaced short pulses or a

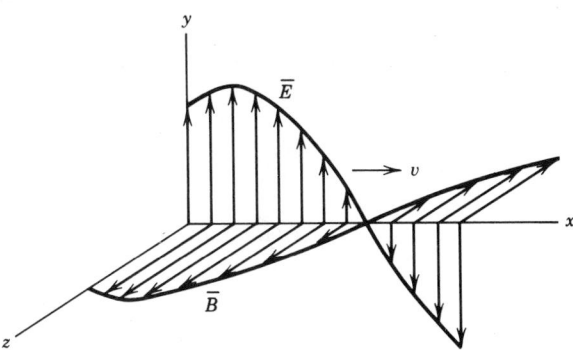

Fig. 13.2 Electromagnetic wave, where E, B, and v form right-handed orthogonal system as shown. Plot is for some fixed time; wave moves to right; E_y and B_z have maximum values for same value of x.

broad spectrum of randomly phased monochromatic waves. The two descriptions are equally valid and interchangeable.

As already mentioned, light is electromagnetic radiation. The quantities described by Eq. (13.1) are electric and magnetic fields E and B. These two fields each obey Eq. (13.1) and are in phase with each other. They are perpendicular to each other in space and each is perpendicular to the direction of propagation. This is illustrated in Fig. 13.2, which shows E in the y direction and B in the z direction for a wave propagated in the positive x direction. Here, E, B, and v form a right-handed orthogonal system so that in order to have a wave propagated in the negative x direction, either E or B must be reversed.

Within the constraint that it remain in a plane perpendicular to the direction of propagation, E may have any direction. Usually the direction of E changes in a random way, and the light is called unpolarized. If the direction of E remains constant, the light is called linearly polarized. Also, B is always perpendicular to E.

Theoretical considerations indicate that in vacuum $c = (\mu_0 \varepsilon_0)^{-1/2}$, which experiment confirms. Theory also predicts that in a medium $v = (\mu \varepsilon)^{-1/2}$ or $n = (\mu \varepsilon / \mu_0 \varepsilon_0)^{1/2}$. Here μ and ε are the permeability and permittivity of the medium; μ_0 and ε_0 are the corresponding quantities for vacuum. The prediction for the velocity in a real medium cannot be experimentally confirmed since μ and ε are frequency dependent and at optical frequencies the only available measurements are the measurements of n or v; there are no direct measurements of μ and ε. The velocity v in Eq. (13.1) is the phase velocity. For a finite wave train or pulse, the envelope of the pulse moves forward with the *group velocity* U. The value of U may be expressed in a variety of forms, including

$$U = \frac{c}{(d/dv)(nv)} = \frac{c}{n}\left(1 + \frac{\lambda}{n}\frac{dn}{d\lambda}\right) \tag{13.2}$$

For common transparent materials, $dn/d\lambda < 0$ and $U < v$. In nondispersive media (e.g., vacuum), $U = v$; λ is the wavelength in the medium.

One common aspect of wave propagation, as observed with water waves, is the tendency of the wave to spread into the shadow region behind barriers. This phenomenon is known as diffraction. Light also exhibits diffraction, but the effects are much smaller than for water waves because the wavelengths of light waves are so small ($\sim 5 \times 10^{-7}$ m). Diffraction effects are important if we attempt to pass light through openings only a few wavelengths wide or to focus the light into a very small spot. Otherwise, we may describe the light in terms of "rays" that represent the direction of energy flow and coincide with the direction of propagation. In a homogeneous isotropic medium, the rays are straight. That part of optics that may be treated by tracing rays is called geometrical optics.

13.1.2 Geometrical Optics

If a ray of light strikes a boundary separating two homogeneous isotropic media such as air and glass (see Fig. 13.3), a simple wave calculation will show and experiment will confirm the following statements:

1. The incident ray will be partially reflected at the boundary and partially transmitted (refracted) into the second medium.

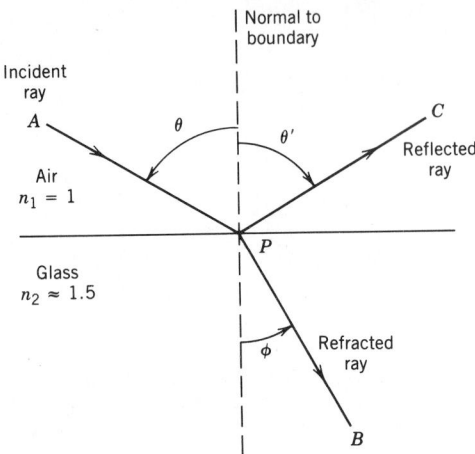

Fig. 13.3 Refraction at plane boundary separating two media with different indices of refraction. According to Snell's law, $n_1 \sin\theta = n_2 \sin\phi$. In the special case for which θ_B satisfies condition $\tan\theta_B = n_2/n_1$, reflected light is linearly polarized.

2. The incident ray, the reflected ray, the refracted ray, and the normal to the surface (erected at the point of incidence) are coplanar.
3. The angle of reflection θ' is equal to the angle of incidence θ; these angles are measured between the surface normal and the rays, as shown in Fig. 13.3.
4. The angle of refraction ϕ and the angle of incidence θ are related by the following equation, which is known as Snell's law:

$$\frac{\sin\theta}{\sin\phi} = \frac{v_1}{v_2} = \frac{n_2}{n_1} \tag{13.3}$$

where v_1 and v_2 are the velocities of propagation in medium 1 and medium 2 and n_1 and n_2 are the indices of refraction.

These statements can be proven without postulating that the light wave is electromagnetic. However, one must recognize the electromagnetic nature of the wave in order to calculate the fraction of the light that is reflected or transmitted. The path of light is reversible; that is, light will travel from B to A along the same path BPA.

It is a straightforward exercise in calculus to show that if A and B (of Fig. 13.3) are fixed points, one in each medium, and if P is an arbitrary point on the boundary, then the location of P that minimizes the propagation time from A to P to B is the same location of P for which Snell's law is satisfied. Also, for A and C as fixed points in the same medium, the choice of P that produces a minimum of $AP + PC$ is the same P for which $\theta = \theta'$. These are examples of Fermat's principle, which states that the path of an actual ray from one point to another is a path for which the transit time is stationary. By *stationary* we mean that the derivative of the transit time with respect to small changes in the path (such as small changes in the location of P) must be zero.

For a ray such as APB in Fig. 13.3, which passes through more than one medium, it is convenient to define the **optical path length** from A to B as $n_1 AP + n_2 PB$, or in case the ray passes through many media, the optical path length is $\Sigma_i l_i n_i$, where l_i is the path length in medium i and n_i is the index of medium i. The optical path length between two points A and B is the distance in vacuum that light could travel during the time required to propagate from A to B through the intervening media.

Snell's law and the law of reflection are sufficient to explain the image-forming properties of lenses and mirrors. We first turn our attention to lenses and to the case in which the same medium (air) is on both sides of the lens. The lens will have spherical surfaces and rotational symmetry about some line called the axis; distances are measured along and perpendicular to this axis. The following results are usually derived using small-angle approximations, $\sin\alpha = \alpha = \tan\alpha$, and are called **paraxial calculations**. Consider first a ray that is parallel to the axis at a distance h above the axis (see Fig. 13.4). Upon passing through the lens, the ray will be refracted according to Snell's law and cross the axis at

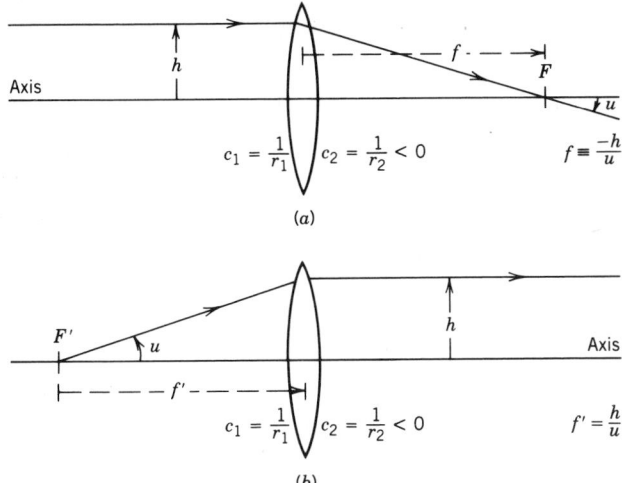

Fig. 13.4 Two focal points of **positive** lens. In (*a*) *F* is second (or back) focal point. In (*b*) *F'* is first (or front) focal point. For lens shown, r_1 (radius of curvature of front surface) is positive; r_2 (radius of curvature of second surface) is negative.

the point *F* in the figure. For a good lens, the point *F* is independent of *h*; it is called the focal point of the lens. As the ray passes through *F*, it has a slope of *u*; in the figure, *u* is negative. The ratio $h/(-u)$ is the focal length *f*. For a thin lens, *f* is the distance from the lens to *F*, where *F* is the second (or back) focal point. There is another point, *F'*, in front of the lens called the first (or front) focal point. Any ray that passes through *F'* (with slope *u*) and strikes the lens will be refracted to be parallel to the axis (at height *h*). The front focal length, $h/u \equiv f'$, will be equal to *f* provided there is the same medium on both sides of the lens. For a thin lens in air

$$\frac{1}{f'} = \frac{1}{f} = (n - 1)\left(\frac{1}{r_1} - \frac{1}{r_2}\right) \tag{13.4}$$

where *n* is the index of refraction of the lens material and r_1 and r_2 are the radii of curvature of the first and second surfaces of the lens; r_1 and r_2 are considered positive (negative) if the center of curvature of the surface is to the right or downstream (left or upstream) relative to the surface; and $1/f$ is the power of the lens. In Fig. 13.4, r_1 is positive and r_2 is negative; for this lens, the focal length *f* is positive. This lens is **convergent**.

Figure 13.5 shows the application of these definitions to a **negative**, or divergent, lens. In this case, r_1 is negative and r_2 is positive, making *f* negative. Notice that *F*, the **second** focal point, is to the left of the lens; the refracted ray does not pass through *F* but must be extended backward to intersect the axis (at *F*). In 13.5(*b*), the incident ray is headed toward *F'*, the **first** focal point, but is refracted by the lens to be parallel to the axis.

For both positive and negative lenses, a ray that crosses the axis at the center of the lens continues undeviated into the region beyond. This is called a chief ray.

It follows from Eq. (13.4) that positive, or convergent, lenses are thicker on axis than at the edge, whereas negative, or divergent, lenses are thinner on axis than at the edge. They may have a variety of shapes, as illustrated in Fig. 13.6.

Figure 13.7 shows a positive lens forming an image of point *A* at point *B*. All rays from *A* that pass through the lens converge to *B*, but only three are shown. It is assumed that the locations of *F* and *F'* are known; rays 1 and 3 are drawn to satisfy the definitions of these points. Ray 2 passes undeviated through the center of the lens. The image is inverted and the lateral magnification *m* is defined as y'/y or $B'B/A'A$; it is negative in the cases shown, indicating an inverted image. The image is real since the rays actually arrive at point *B*.

It is simple to calculate the image position by the equation

$$xx' = f^2 \tag{13.5}$$

where *x* is the distance from the object to the **first** focal point; it is taken as positive if (as shown in

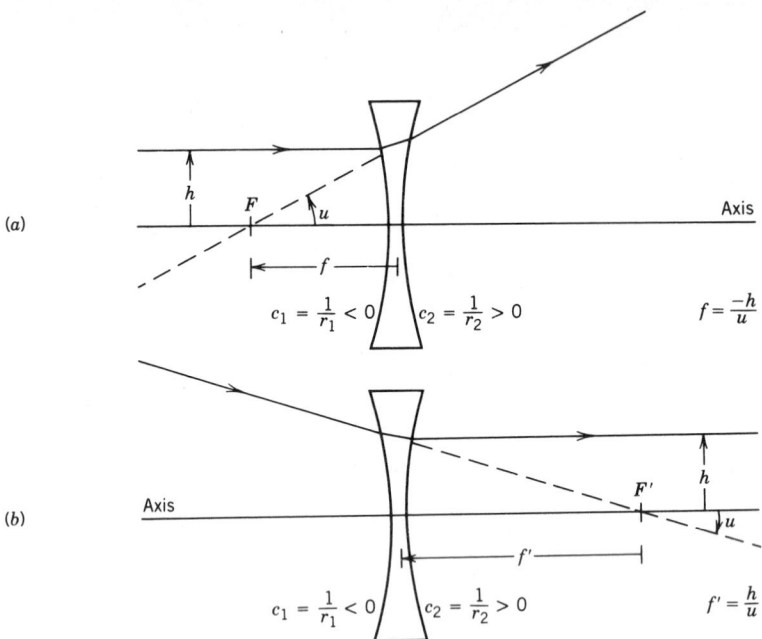

(a)

(b)

$$c_1 = \frac{1}{r_1} < 0 \qquad c_2 = \frac{1}{r_2} > 0 \qquad\qquad f = \frac{-h}{u}$$

$$c_1 = \frac{1}{r_1} < 0 \qquad c_2 = \frac{1}{r_2} > 0 \qquad\qquad f' = \frac{h}{u}$$

Fig. 13.5 Two focal points of **negative** lens. In (a) F, the second focal point, is in front of lens. In (b) F', first focal point, is behind lens. For lens shown, r_1 is negative and r_2 is positive.

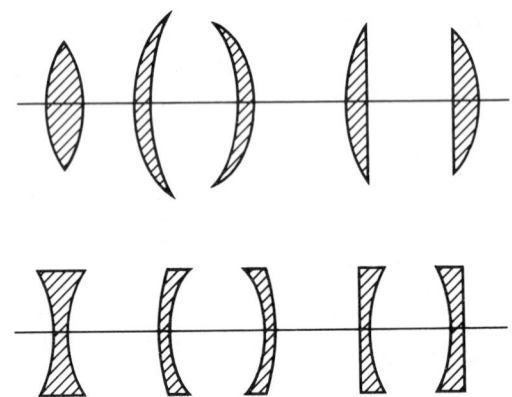

Fig. 13.6 Variety of positive lenses (upper group) and negative lenses (lower group).

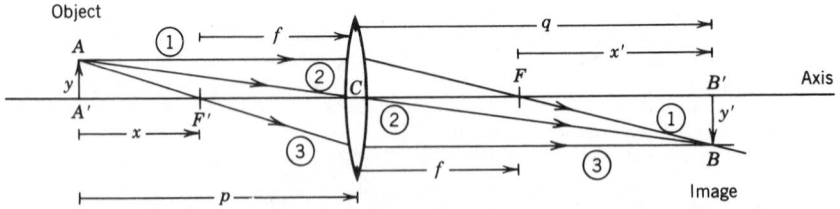

Fig. 13.7 Image formation by positive lens, illustrating quantities that appear in Eqs. (13.5)–(13.7). Also shown are three rays easily used in graphical ray tracing.

Fig. 13.7) the object-to-focal-point direction is the same as the direction of the light propagation. On the image side x' is the distance from the **second** focal point to the image; it is positive in Fig. 13.7. From Eq. (13.5), we see that x and x' always have the same sign. Since the product xx' is constant, moving the object to the right (toward F') moves the image to the right (away from F). The lateral magnification m is given as

$$m \equiv \frac{y'}{y} = -\frac{f}{x} = -\frac{x'}{f} \qquad (13.6)$$

Another pair of equations may be used

$$\frac{1}{p} + \frac{1}{q} = \frac{1}{f} \quad \text{and} \quad m = -\frac{q}{p} \qquad (13.7)$$

where p is the distance from the object to the lens and q is the distance from the lens to the image; they are considered positive if they are in the same direction as the light propagation. In Fig. 13.7, both p and q are positive. Equation (13.7) is very convenient for use with thin lenses. For thick lenses or lenses consisting of several elements, it is not obvious what point (or points) in the lens should be used for measuring p and q. By reversing the rays, A becomes the image of B, and A and B are said to be conjugate points.

Figure 13.8 shows the corresponding situation for a negative lens. Here the first focal point F' is to the right of the lens and the second focal point F is to the left of the lens. Ray 1 is parallel to the axis until it strikes the lens and is refracted along a line that appears to have come from F. Ray 3 is headed for F' but is refracted to be parallel to the axis. Ray 2 passes straight through the center of the lens. These rays do not intersect anywhere to the right of the lens but if extended backward appear to have intersected at B. Only ray 2 actually passes through B. Point B is a **virtual** image of A (in contrast to the real image formed in Fig. 13.7). Equations (13.5) and (13.6) or Eq. (13.7) work for this case, but notice the following: f is negative; x is measured from A' to F' as before, but F' is to the right of the lens; x' is measured from F to B', but F is to the left of the lens; p is positive; Eq. (13.7) gives a negative value for q, indicating that B' is to the left of the lens and therefore virtual; m is positive but less than 1, so the image is upright or erect and smaller than the object.

Equations (13.5)–(13.7) may be used to establish the information in Table 13.1.

The **focal length** of a thin positive lens may be calculated from Eq. (13.4) if the curvatures and the index of the glass are known. It may also be measured in the laboratory by setting up on an optical bench an experiment similar to Fig. 13.7, measuring the appropriate distances and calculating f. A small luminous source, such as the filament of an unfrosted light bulb, might serve as a suitable object. A ground glass screen is used to locate the image. Negative lenses cannot be measured in this way because the image in Fig. 13.8 is virtual and virtual images cannot be caught upon a screen. There are two ways around this problem.

Two thin lenses of focal lengths f_1 and f_2 when placed in contact are equivalent to a single lens of focal length f_c, given by the equation

$$\frac{1}{f_c} = \frac{1}{f_1} + \frac{1}{f_2} \qquad (13.8)$$

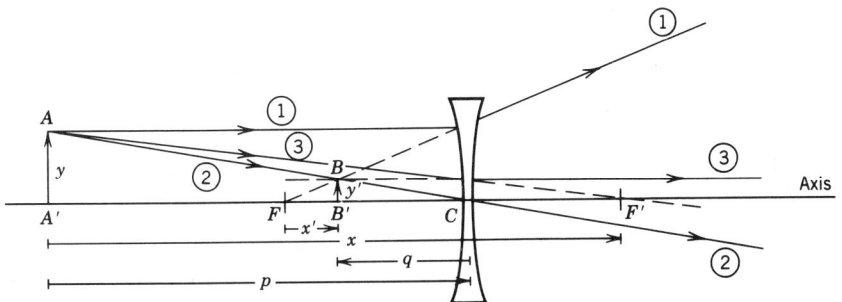

Fig. 13.8 Image formation by negative lens illustrating quantities from (13.5)–(13.7). Also shown are the rays easily used in graphical ray tracing.

TABLE 13.1   Images Formed by Thin Lenses

Lens	Object Position		Image Position		Nature of Image
Positive, or convex, $f > 0$	$x = \infty$	$p = \infty$	$x' = 0$	$q = f$	Real
	$\infty > x > f$	$\infty > p > 2f$	$0 < x' < f$	$f < q < 2f$	Real, inverted, $\|m\| < 1$
	$x = f$	$p = 2f$	$x' = f$	$q = 2f$	Real, inverted, $\|m\| = 1$
	$f > x > 0$	$2f > p > f$	$f < x' < \infty$	$2f < q < \infty$	Real, inverted, $\|m\| > 1$
	$0 > x > -f$	$f > p > 0$	$-\infty < x' < -f$	$-\infty < q < 0$	Virtual, erect, $\|m\| > 1$
Negative, or concave, $f < 0$	$x = \infty$	$p = \infty$	$x' = 0$	$q = f = -\|f\|$	Virtual
	$\infty > x > -f$	$\infty > p > 0$	$0 < x' < -f$	$f < q < 0$	Virtual, erect, $\|m\| < 1$

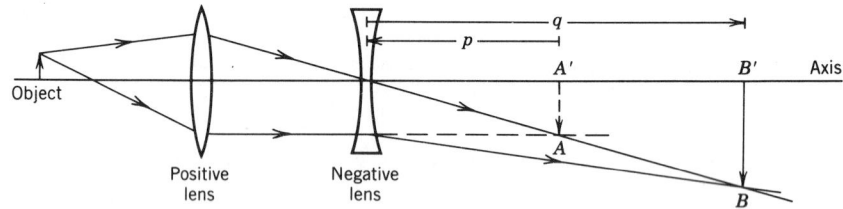

Fig. 13.9   Measuring focal length of negative lens using image AA' formed by positive lens as virtual object for negative lens (p is negative).

If f_1 is a negative lens under test, it may be combined with a positive lens of known focal length f_2. If $f_2 < |f_1|$, then the combined focal length f_c will be positive and can be measured by the experiment of Fig. 13.7; f_1 is then calculated from Eq. (13.8).

An alternate method is shown in Fig. 13.9. A positive lens is used to form a real image at $A'A$; its position is determined and recorded by observing the image on a screen. Being careful not to move the object or the positive lens, the negative lens to be measured is placed between the positive lens and the image $A'A$. The image now has a new location, $B'B$, which may be adjusted to some convenient position by moving the **negative** lens to the left or right. The initial image $A'A$ is the object for the negative lens; it is called a **virtual object** because the light is intercepted by the negative lens before it reaches point A. The distance from object to lens, p, is negative since it points upstream. The image distance q is the distance from the negative lens to the image $B'B$; in the figure it is positive (i.e., downstream). The focal length f of the negative lens may be calculated from Eq. (13.7) and the values of p and q just obtained.

A **spherical mirror**, concave toward the light source, forms images and behaves in many ways as a positive lens. There is, however, only one focal point F (see Fig. 13.10). A ray that is initially parallel to the axis is reflected to pass through F. A ray that passes through F before striking the mirror will be parallel to the axis after reflection. A ray that passes through the center of curvature C will strike

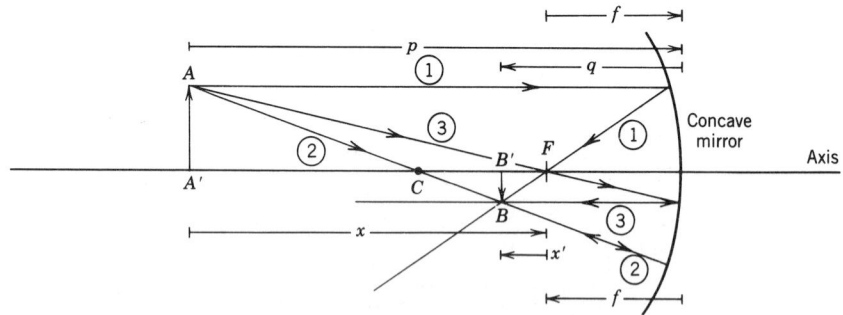

Fig. 13.10   Image formation by concave mirror. Focal point F is halfway between center of curvature C and mirror surface.

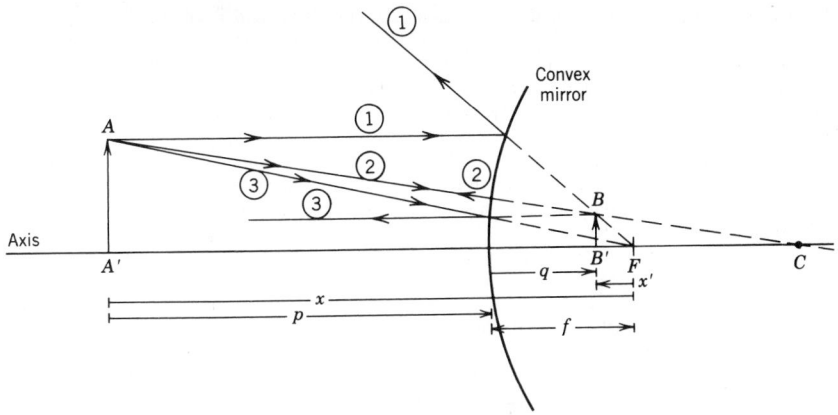

Fig. 13.11 Image formation by convex mirror.

the mirror normally and be reflected back on itself. These three rays are shown in the figure starting from A and intersecting at B, which is the image of A. The focal point F is midway between the center of curvature and the mirror surface.

Equations (13.5)–(13.7) may be used to calculate image locations and magnification. The focal length of a concave mirror is positive. The downstream direction is always positive, but this reverses when the rays reflect from the mirror. In Fig. 13.10, rays from the object that have not reached the mirror (called rays in object space) are downstream to the right, so this is the positive direction for p and x. Rays that have been reflected from the mirror (called rays in image space) are downstream to the left, so right to left is the positive direction for q and x', which locate the image. In the figure, x, x', p, and q are all positive.

A mirror that is **convex** toward the light source is similar to a negative lens. The focal point is behind the mirror surface and $|f| = \frac{1}{2}|r|$, where r is the radius of curvature. Image formation for this case is illustrated in Fig. 13.11. In the equations, f is negative; x and p are in object space and positive in the figure. Distances x' and q are part of image space where downstream is to the left, so x' is positive and q is negative in the figure.

Aberrations
In first-order optical calculations, it is found that all the rays from a point object that pass through the optical system converge upon the image point. If the object lies in a plane that is perpendicular to the axis of the system, the image will lie in a plane perpendicular to the axis of the system and be similar in shape to the object.

If we abandon first-order approximations ($\sin \theta \approx \theta$) and use more accurate calculations, we discover that the preceding statements are not true. The differences between the actual behavior of the optical system and the ideal (or first-order) behavior are called **aberrations**, which are generally classified as follows.

Chromatic Aberration
According to Eq. (13.4), the focal length of a lens depends upon the index of refraction of the glass from which it was made. Since the index is wavelength (or frequency) dependent, the focal length is also wavelength dependent. This is **chromatic aberration**. Usually the focal length is longer for red light than for blue light. By combining two lens elements (such as a positive lens of crown glass and a negative lens of flint glass), it is possible to create a "doublet" that has the same focal length for two specified wavelengths, such as a wavelength in the red and a wavelength in the blue. It will then have nearly the same focal length for the intervening wavelengths. Such a doublet is said to be **achromatic**. Objectives for small telescopes are usually of this design.

Spherical Aberration
See Fig. 13.4(a). If the point F at which the ray crosses the axis depends upon h, the lens has spherical aberration. For the simple lens shown, F moves slightly toward the lens as h increases. Likewise, in Figs. 13.7 and 13.10, all the rays from A' do not cross the axis at B'; rather, the crossing point depends to some extent upon height (or off-axis distance) at which the ray strikes the lens or mirror.

Spherical aberration and chromatic aberration exist for source and image points that are on axis or off axis. The following aberrations do not exist for object and image points on axis; they exist only for off-axis points and increase in magnitude as the object and image points are more off axis.

A lens is said to exhibit **field curvature** if the image of an off-axis point does not lie upon the paraxial image plane but departs from that plane by a distance proportional to the square of the angle off axis.

For **astigmatism**, the image of an off-axis point consists of two short line segments. These two line segments are at different distances from the lens. One line segment (called the sagittal image) is directed radially outward from the axis; the other (called the tangential image) is perpendicular to the radial line through its center. If a lens suffering from astigmatism is used to image an object consisting of radial lines and concentric circles, they will not be sharply imaged on the same plane. The radial lines will be imaged at the sagittal position; the concentric circles will be imaged at the tangential position. Midway between these two images is the position of "least confusion." The separation between the sagittal and tangential lines is a measure of astigmatism; it increases with the square of the angle off axis.

For a lens with **coma**, an image of a point source is formed at the paraxially predicted position by the rays that pass through the central portion of the lens (Fig. 13.12). Rays that pass through the lens at some fixed distance from the axis (i.e., through a circular zone) do not come to the paraxial image point. Instead, they intersect the image plane in a circle, the center of which is displaced from the paraxial image point by twice its radius. The image of a point produced by all the rays through the lens is a bright spot at the expected location with a 60° flare extending from it. It resembles a comet (hence the name *coma*). The size of the coma pattern increases linearly with angle off axis.

A lens with **distortion** produces a plane image of a plane object, and the rays that pass through the lens all reach the image point, but the image of an extended object (such as a square) is not the same shape as the object. Instead of having constant magnification m so that

$$y' = my$$

there is an additional term to give

$$y' = my + Cy^3$$

The second term on the right is small compared with the first term, but y'/y is no longer independent of y. In case $m < 0$ (real images) and positive C, we find that $|y'/y|$ decreases as $|y|$ increases; the image of a square has a barrel shape. Conversely, a lens for which C is negative images a square into a figure resembling a pin cushion.

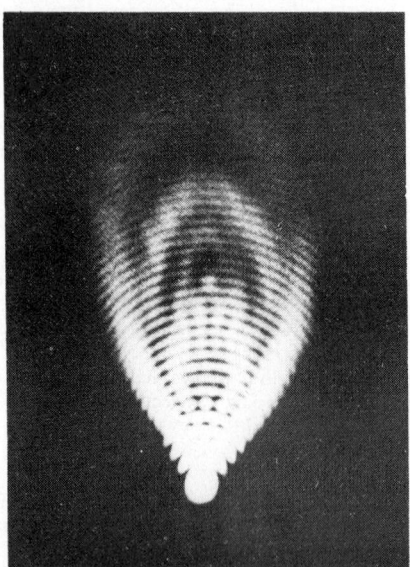

Fig. 13.12   Image of point source as formed by lens with coma (photograph by David Dutton).

The monochromatic aberrations (spherical, astigmatism, coma, field curvature, and distortion) can be calculated using third-order approximations, that is, approximations consistent with $\sin\theta \approx \theta - \frac{1}{6}\theta^3$. These are called Seidel, or third-order, aberrations. In third-order calculations there exist only the aberrations listed in the preceding (and linear combinations of them). We have also assumed the lens has rotational symmetry about its axis.

Most practical lenses consist of several elements or components, each of which is a simple lens. The lens designer reduces the aberrations by adjusting the following: (1) the distribution of the power among the elements; (2) the shapes, thicknesses, and spacings of the elements; (3) the position of the stop (or limiting aperture); and (4) the choice of glass in each element. Usually it is not possible to minimize all of the aberrations, and the designer must know the application for which the lens is intended and make some judgment about the relative importance of the various aberrations. A lens that is optimally corrected for one pair of conjugate points will not be optimally corrected for another pair of conjugate points; for this reason the tube length of a compound microscope is important (and has been standardized).

A few lenses have been produced in which one or more elements are made of inhomogeneous glass that contains a deliberately introduced gradient of the index of refraction. This is a recent method of controlling the aberrations.

The "ray" description of image formation by lenses and mirrors, which has occupied our attention in the previous sections, takes no account of the wave nature of light. We now recognize several situations in which the wave properties of light are important; these are part of a field known as **physical optics**.

13.1.3 Physical Optics

We set up the experiment diagrammed in Fig. 13.13. A source of light L is placed behind a slit S_0 on axis. Light from S_0 passes through two slits S_1 and S_2 in an opaque screen and continues on to a viewing screen. The slits S_1 and S_2 are symmetrically located about the axis at $\pm a/2$. The point P is an arbitrary point on the viewing screen at a distance y from the axis, R is the distance from S_0 to the slits S_1 and S_2, and D is the distance from these slits to the viewing screen. In a typical laboratory experiment, a would be in the range 0.5–1.0 mm and R and D in the range 50–100 cm. The three slits should be less than 0.1 mm wide.

The path difference Δ for the two possible paths is

$$\Delta \equiv S_0 S_2 P - S_0 S_1 P = S_2 P - S_1 P = \frac{ay}{D} \tag{13.9}$$

The light reaching P by way of S_2 is out of phase with the light reaching P by way of S_1. The phase difference δ is

$$\delta = \frac{2\pi}{\lambda}\Delta = \frac{2\pi}{\lambda}\frac{ay}{D} \tag{13.10}$$

Depending upon the value of δ, the two waves will interfere at P either constructively or destructively.

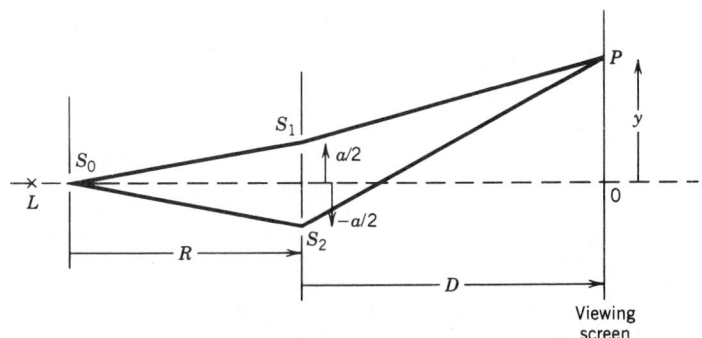

Fig. 13.13 Young's double-slit interference experiment. Light from the source slit S_0 passes through slits S_1 and S_2 and continues on to viewing screen. Interference pattern is formed on viewing screen. Intensity at point P depends upon path difference for two possible paths.

If the slits S_1 and S_2 are of equal width so that they contribute equally to the irradiance at P, then the irradiance at P is I_p:

$$I_p = 4I_0\cos^2\frac{\delta}{2} = 2I_0(1 + \cos\delta) \qquad (13.11)$$

where I_0 is the irradiance that would be produced at P by light from only one slit. From Eqs. (13.10) and (13.11), we see that I_p will have maxima at values of y given by

$$y = \frac{m\lambda D}{a} \qquad (13.12)$$

where $m = 0, \pm 1, \pm 2, \pm 3, \ldots$; m is the order number or order of interference. Halfway between each pair of consecutive maxima is a minimum at which $I_p = 0$.

This pattern of fluctuating irradiance is called a pattern of **interference fringes**, or a **fringe pattern**.

The minima in the fringe pattern are not always zero. In this case, we define the fringe contrast or visibility V as

$$V = \frac{I_{max} - I_{min}}{I_{max} + I_{min}} \qquad (13.13)$$

where I_{max} and I_{min} are the maximum and minimum values of I_p. For the fringe pattern described by Eq. (13.11), $V = 1$; this is its maximum possible value. The case $V = 0$ corresponds to uniform irradiance, that is, no fringe pattern, or $I_{min} = I_{max}$.

If the two slits do not contribute equally to the irradiance at P, the fringe visibility will be reduced. Let I_1 represent the irradiance at P due to slit S_1 and I_2 the irradiance due to S_2; then Eq. (13.11) becomes

$$I_p = I_1 + I_2 + 2\sqrt{I_1 I_2}\cos\delta \qquad (13.14)$$

giving a fringe pattern with visibility

$$V = \frac{2\sqrt{I_1 I_2}}{I_1 + I_2} \qquad (13.15)$$

If we keep S_1 and S_2 of equal width so that they contribute equally to the irradiance at P, the fringe visibility will be reduced if we increase the width of the source slit S_0. If S_0 is centered on axis and has width w, the fringe pattern is described by

$$I_p = 2I_0(1 + V\cos\delta) \qquad (13.16)$$

where the fringe visibility V is

$$V = \frac{\sin\beta}{\beta} \quad \text{where} \quad \beta = \frac{\pi wa}{\lambda R} \qquad (13.17)$$

The visibility V decreases from 1 (for very small values of w) to zero for $\beta = \pi$ or $w = \lambda R/a$. The visibility as a function of w is often identified with γ, the degree of coherence of the light emerging from slits S_1 and S_2. Here, γ is the normalized cross-correlation function of the waves emerging from the two slits. Loosely speaking, $|\gamma|$ represents the reliability with which the phase at S_2 could be predicted from a hypothetical determination of the phase at S_1. If β lies in the range $\pi < \beta < 2\pi$, the value of V (or γ) is negative. For the specific case $w = 1.4\lambda R/a$ or $\beta = 4.4$, we obtain from (13.17) that $V = -0.2$ or $\gamma = -0.2$. Fringes of contrast 0.2 are not very good but are easily recognizable. The negative value of γ indicates that there is a minimum rather than a maximum at the axial point ($y = 0$) of the viewing screen. The waves emerging from S_1 and S_2 are poorly phase correlated (only 0.2), and the most probable phase difference is 180° (minus sign). Since the separation of S_1 and S_2 is perpendicular to the direction of propagation, this is called spatial coherence.

Figure 13.14 illustrates the irradiance profile for fringe patterns with three different values of V.

Another interference experiment is illustrated in Figure 13.15. A broad source of light is used to illuminate two partially reflecting mirrors M_1 and M_2. The reflected light is viewed by the eye; in order to have both illumination and viewing directions nearly normal to the mirrors M_1 and M_2, a third mirror, M_0, is inserted as shown. Here, M_0 is partially reflecting and partially transmitting.

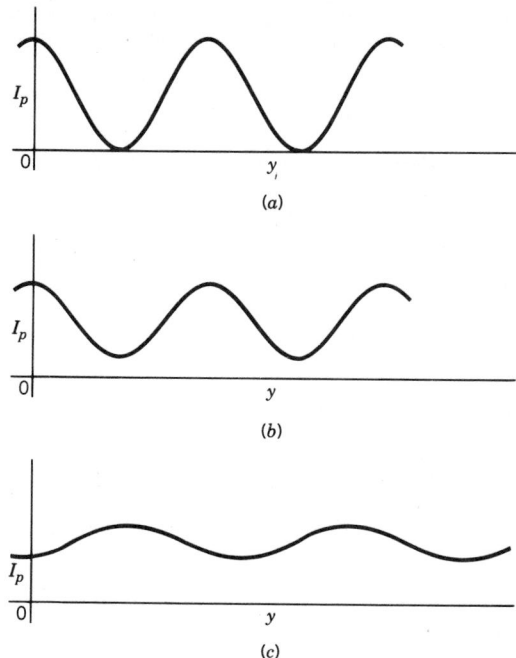

Fig. 13.14 Plots of I_p vs. y for fringes with different visibility V: (*a*) $V = 1.0$; (*b*) $V = 0.5$; (*c*) $V = -0.2$.

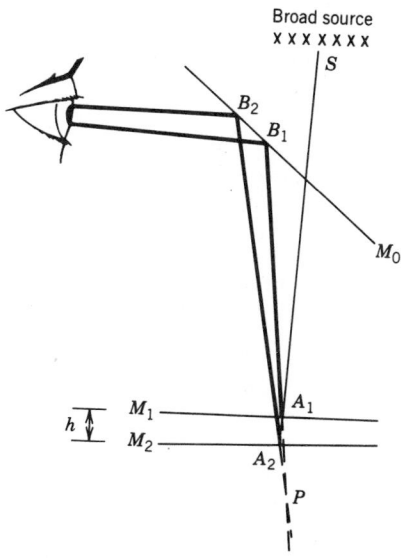

Fig. 13.15 Interference produced by light reflected from thin wedge. Effect may be observed in oil slick or in soap bubble.

In order to have interference effects, the two rays must be coherent. Since the source is broad, we have coherence only between rays reflected from M_1 at A_1 and from M_2 at A_2, where points S, A_1, and A_2 lie on a straight line, where S is an arbitrary point of source. The light reflected at A_1 will be reflected at B_1 into the eye by mirror M_0. Light reflected at A_2 reaches the eye by a similar path. In order for an interference pattern to be seen, both of these rays must enter the eye. Interference will take place where these rays intersect on the retina of the eye, but the eye (or the brain) will interpret this as an interference at P, the intersection point of lines $A_1 B_1$ and $A_2 B_2$. Interference such as this is observed in an oil film on wet pavement or sometimes in a soap bubble.

Both rays must enter the pupil of the eye; in order to obtain this, the mirrors M_1 and M_2 must be nearly parallel and **either** h (the separation of the mirrors) must be small **or** the ray SA_1 must be at near normal incidence. The mirror M_0 permits near normal illumination and viewing.

The path difference and therefore the phase difference between the two rays will be determined by the mirror separation h at the point A_1. If the space between the mirrors M_1 and M_2 is slightly wedged or if the mirrors are not exactly plane, then the locus of a maximum (or a minimum) in the fringe pattern represents the locus of points of constant h. A change from one bright (or dark) fringe to the next bright (or dark) fringe corresponds to a change of h by one-half wavelength.

Fringes of good contrast [corresponding to Eq. (13.11)] are obtained if the illumination is nearly monochromatic, if h is small, and if the two reflecting surfaces M_1 and M_2 have small but equal reflectances. The last condition ensures that the two rays $A_1 B_1$ and $A_2 B_2$ are of nearly equal irradiance. A clean glass surface in air reflects 4% and would serve very well as M_1 or M_2.

If the mirror M_2 is moved away from M_1 so as to increase h, this serves to introduce a time delay in the ray $A_2 B_2$ relative to $A_1 B_1$. If air separates the mirrors M_1 and M_2, the time delay τ is given by

$$\tau = \frac{2h}{c} \tag{13.18}$$

As h (and therefore τ) is increased, the fringe visibility gradually decreases, indicating a decrease in the coherence between the two waves or rays. In this case, the coherence decreases due to a time difference between the rays and is called temporal coherence. If a dispersive medium separates the two mirrors, the time delay should be calculated using the group velocity [U of Eq. (13.2)] instead of c. The largest value of τ for which fringes of reasonable visibility can be observed is called τ_c, the coherence time of the source; $c\tau_c$ is called the coherence length; τ_c is the same as Δt in the discussion following Eq. (13.1).

Aside from the interference fringes produced in Young's experiment (Fig. 13.13), there is another aspect of the experiment that is contrary to the ideas of ray optics. If light reaches P by two paths $S_0 S_1 P$ and $S_0 S_2 P$, then at least one (and usually both) of these paths must be bent. This ability of light to bend into regions that geometrical optics would call shadow regions is known as **diffraction**. Diffraction is a property of all **wave** propagation.

Practical diffraction problems can be solved by the Huygens–Fresnel method. The method is outlined in Fig. 13.16. Light from a small source S_0 spreads out to pass through an aperture A in an opaque screen. A wavefront (surface of constant phase) reaches the aperture and is transmitted through the open part but is blocked by the opaque portion of the screen. Each elemental area on the transmitted wavefront acts as a secondary source radiating spherical waves into the region to the right of A. To calculate the disturbance (i.e., electric field) at P, an arbitrary point on the viewing screen, we must sum (i.e., integrate) the contributions due to all the elemental areas of the aperture. The

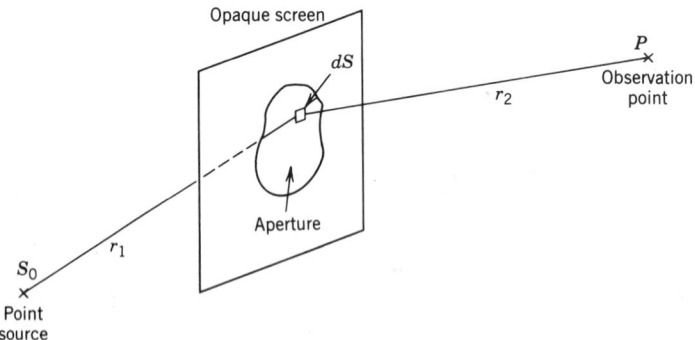

Fig. 13.16 Geometry of general diffraction problem. Here, S_0 is point source of light; disturbance at P calculated using Eq. (13.9) where r_1 and r_2 are distances illustrated.

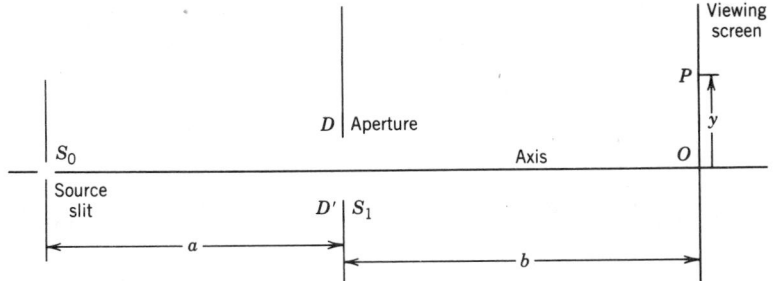

Fig. 13.17 Geometrical quantities used in calculating diffraction pattern produced by slit.

summation process must take into account the fact that the disturbances reaching P from the various elemental areas of the aperture are not in phase at P because contributions from the various elemental areas have traveled unequal distances from S_0 to P. Mathematically this is written

$$U(P) = -\frac{iA}{2\lambda} \int \int_A \frac{e^{ik(r_1+r_2)}}{r_1 r_2} dS \tag{13.19}$$

where $U(P)$ is the disturbance at P; dS is an element of area in the aperture; $k = 2\pi/\lambda$; and r_1 and r_2 are distances from S_0 to dS and from dS to P. The integration is over the open aperture. In many practical cases, the product $r_1 r_2$ in the denominator may be considered as a constant during the integration. The values of r_1 and r_2 in the exponent may not be considered as constant because the exponential is periodic and k is large; A is the open area of the aperture. (We have also assumed that r_1 and r_2 are nearly perpendicular to the plane of the opaque screen.) A complex quantity, $U(P)$ represents the amplitude and phase of the disturbance at P. The irradiance at P, $I(P) = |U(P)|^2 = U(P)U^*(P)$, where $U^*(P)$ is the complex conjugate of $U(P)$.

The integral in Eq. (13.19) is generally difficult to evaluate, but results are available for some simple cases.

If the aperture is a slit with length much larger than its width, we can consider this as a two-dimensional problem (as illustrated in Fig. 13.17), ignoring the dimension along the length of the slit. The point source may be replaced by S_0, a narrow slit (or line) source provided this is parallel to the diffracting slit S_1; use of the slit source rather than a point source will noticeably increase the irradiance on the viewing screen.

In discussing this problem, we take as axis the line from the source S_0 through the center of the diffracting slit S_1; it intersects the viewing screen at O. The slit S_1 is of width w and is symmetrically located about the axis so that the two edges D and D' of the slit are at $\pm \frac{1}{2}w$ from the axis. The distance $S_0 DO$ is larger than the axial distance $S_0 O$, and on the assumption that $w \ll a$ or $w \ll b$, the difference is

$$K \equiv S_0 DO - S_0 O = \frac{a+b}{2ab}\left(\frac{w}{2}\right)^2 \tag{13.20}$$

where a and b are the distances from source to diffracting slit and from diffracting slit to viewing screen, as shown in Fig. 13.17.

Some generality may be achieved by replacing w by the dimensionless parameter u:

$$u = w\sqrt{\frac{2(a+b)}{ab\lambda}} \tag{13.21}$$

and replacing y (the distance along the viewing screen from the axis to the point of observation P) by the dimensionless variable v:

$$v = y\sqrt{\frac{2a}{\lambda b(a+b)}} \tag{13.22}$$

With these changes $K = \frac{1}{16}u^2\lambda$.

Figure 13.18 shows the irradiance at P as a function of v for several values of the parameter u. In each case, the edge of the geometrical shadow is indicated by vertical lines.

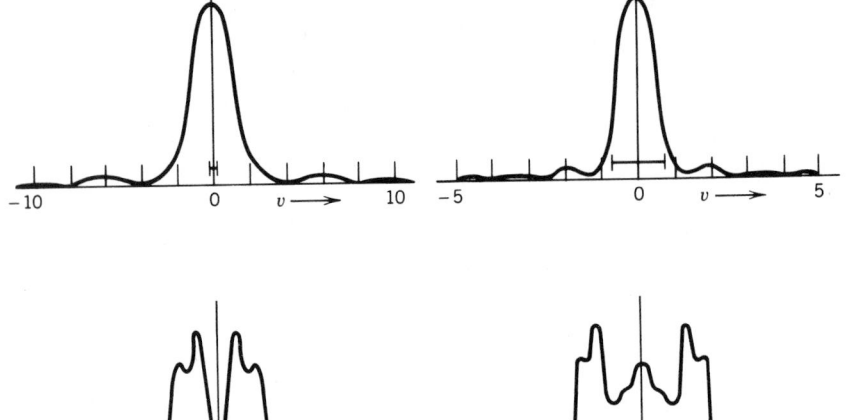

Fig. 13.18 Diffraction patterns of four slits of different widths. Each figure is plot of irradiance at P as function of v. (a) For narrow slit, $u = 0.5$. Other figures are for successively wider slits: (b) $u = 1.5$; (c) $u = 3.8$; (d) $u = 5.0$. In each case geometrically predicted width is indicated by horizontal line segment just above v axis.

Notice that for the larger values of u, the curves have several maxima and minima and that there is very little irradiance outside the geometrical shadow region. For the smaller values of u (e.g., $u = 1.5, 0.5$), the curves show less structure and for values of $u < 1$ spread out well beyond the geometrical shadow region.

The experiment of Fig. 13.17 may be repeated using a point source and a small circular aperture in place of the slit. The observed diffraction pattern on the viewing screen will have circular symmetry and the minima will be deeper than those observed with the slit. For $u > 1$, most of the light will be within the region of the geometrical shadow; and for $u < 1$, the light will spread out into the shadow region.

Diffraction problems are generally divided into two classes, known as **Fresnel diffraction** and **Fraunhofer diffraction.** Fresnel diffraction, or near-field diffraction, covers the cases for which u [of Eq. (13.21)] is larger than 1. Fraunhofer (or far-field) diffraction covers the cases for which $u < 1$. For a given aperture and wavelength, you can move from the Fresnel to the Fraunhofer region by increasing the distances a and b.

A lens may be placed on each side of the aperture as indicated in Fig. 13.19. Lens 1 forms an image of the source S_0 at S_0' and lens 2 images point O of the viewing screen to O'. In this situation,

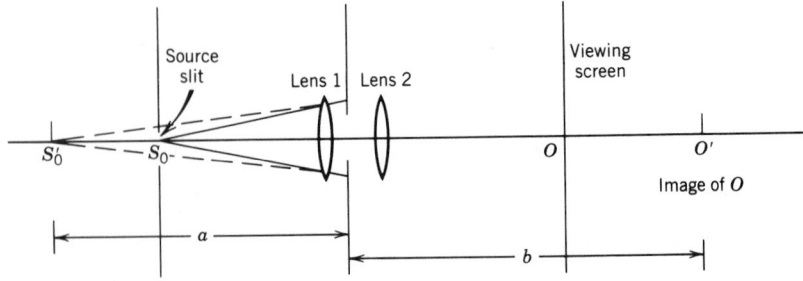

Fig. 13.19 Effect of using lenses in diffraction experiment. Source distance measured from S_0', image formed by lens 1 of real source slit. Screen distance is measured to O', image formed by lens 2 of actual viewing screen.

the distance a is measured from S_0' to the aperture and b is measured from the aperture to O'. If S_0 is at the front focal point of lens 1 and the viewing screen is at the second focal plane of lens 2, then a and b are infinite and $u = 0$ independent of the aperture size. Fraunhofer diffraction calculations assume that the preceding conditions are satisfied and some mathematical simplifications result therefrom. Even though the calculations are for the $u = 0$ case, the results are good approximations to any case for which $u \leq 1$. The irradiance distribution for the Fraunhofer diffraction pattern of a slit is

$$I(y) = I_0 \left(\frac{\sin \beta}{\beta} \right)^2 \tag{13.23}$$

In this equation, I_0 is the irradiance on axis; $I(y)$ is the irradiance at P that is at a distance y from the axis; $\beta = (\pi w y)/(\lambda f)$. Here w is the slit width and f is the focal length of lens 2. Figure 13.20(a) is a plot of Eq. (13.23); notice that it has a large central maximum with small secondary maxima on each side. The irradiance is zero at $\beta = \pi, 2\pi, \ldots$. Notice the similarity between this figure and the $u = 0.5$ case of Fig. 13.18.

For a circular aperture and a point source on axis, the Fraunhofer diffraction pattern will have circular symmetry. Calculations show that the irradiance $I(r)$ at a distance r from the axis in the rear

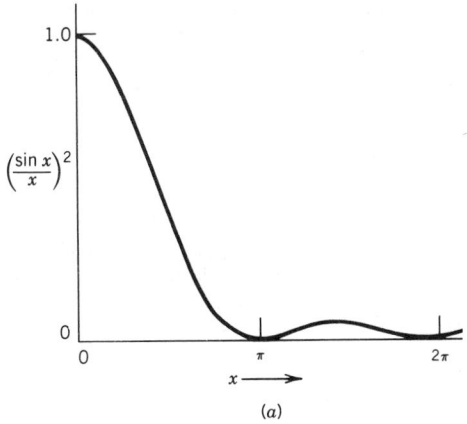

(a)

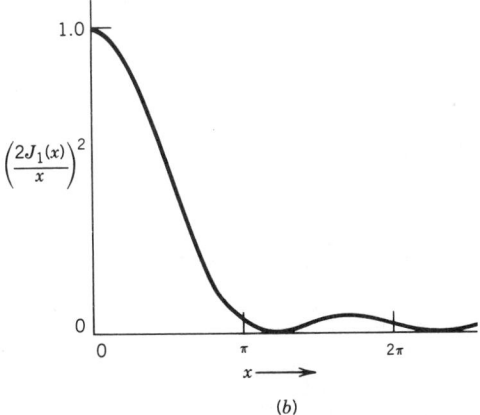

(b)

Fig. 13.20 (a) Function $(\sin x/x)^2$ as function of x. This is Fraunhofer diffraction pattern of single slit. (Pattern is symmetrical about $x = 0$.) (b) Function $(2J_1(x)/x)^2$ as function of x. This is Fraunhofer diffraction pattern of circular aperture. (Pattern has rotational symmetry about axis.)

focal plane of lens 2 is given by

$$I(r) = I_0 \left(\frac{2J_1(x)}{x} \right)^2 \tag{13.24}$$

where J_1 is the first-order Bessel function and $x = (\pi w r)/(\lambda f)$. In this case w represents the **diameter** of the circular aperture. Figure 13.20(b) is a plot of Eq. (13.24). The pattern has rotational symmetry about the $x = 0$ axis. For the circular aperture the zeros of irradiance are not quite equally spaced, occurring at $x = 1.220\pi, 2.23\pi, 3.328\pi, \dots$. The first dark ring is at $r = 1.22\lambda f/w$. Of all the optical power that passes through the aperture, 84% is focused inside this first dark ring, forming what is known as the Airy disk.

For many optical instruments (e.g., telescopes and cameras), the light enters through a circular aperture and the image of a distant point source such as a star is not a bright point but an Airy disk that has the size just discussed. It follows that two separate sources will not be recognized as separate (we say they will not be "resolved") unless their images are separated by a distance at least equal to the radius of the Airy disk. The angular separation θ of two just resolved stars is

$$\theta = 1.22 \frac{\lambda}{w} \tag{13.25}$$

As a rule of thumb the 1.22 is often ignored.

Photographers often express the size of the camera aperture in terms of the f-number, which we represent as $F^\#$. It is defined as $F^\# = f/w$. In terms of the f-number, the diameter d of the Airy disk is

$$d = 2\frac{1.22\lambda f}{w} = 2.44\lambda F^\# \tag{13.26}$$

Using a typical value of $\lambda = 0.55\ \mu$m, we obtain

$$d = 1.33 F^\# \quad (\mu\text{m})$$

The diameter of the Airy disk (in micrometers) is about equal to (or 33% larger than) the f-number used. This assumes the lens has no aberrations; the performance of such a lens is said to be **diffraction limited**.

Microscopists generally speak of the **numerical aperture** (NA) of the microscope:

$$\text{NA} \equiv n \sin i \tag{13.27}$$

where i is the half angle of the cone of light that enters the microscope from the object (see Fig. 13.21) and n is the index of refraction of the region between the object and the microscope objective. Usually $n = 1.0$, but in oil immersion microscopes, n is the index of the oil that fills the space between the object and the lens. Microscopes generally cover a small field of view and are diffraction limited. The separation s of two just resolved points is

$$s = \frac{\lambda}{2n \sin i} = \frac{\lambda}{2(\text{NA})} \tag{13.28}$$

where λ is the wavelength in a vacuum.

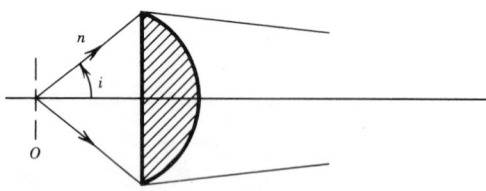

Fig. 13.21 Numerical aperture (NA) of microscope objective defined as NA $= n \sin i$. Here i is half-angle of cone of light entering objective from object O; n is index of refraction of medium between object and objective.

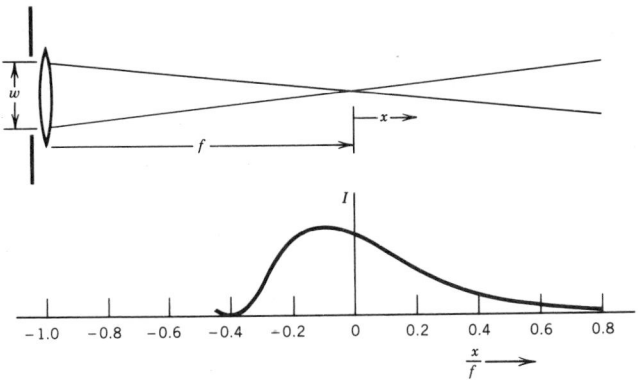

Fig. 13.22 Position of maximum irradiance shifted toward lens from geometrical focal point. Effect becomes significant only if w^2 is comparable to $4\lambda f$ or less. Lower curve of irradiance vs. x/f: $N = w^2/4\lambda f = 3$.

Equation (13.24) describes the irradiance as a function of r, the distance off axis, in a plane through the rear focal point and perpendicular to the axis. The **irradiance on axis** as a function of x, the distance to the right of the focal point, is given by

$$I(x) = I_0 \left(\frac{f}{f+x} \right)^2 \left(\frac{\sin(U_N/4)}{U_N/4} \right)^2 \qquad (13.29)$$

where I_0 is the irradiance at $x = 0$ and

$$U_N \equiv 2\pi N \frac{x}{f+x} \qquad (13.30)$$

and $N \equiv w^2/(4\lambda f)$. See Fig. 13.22 for a sketch of this experiment. An examination of Eq. (13.29) shows that the maximum value of $I(x)$ is not at $x = 0$ but is shifted toward the lens (i.e., to negative x) by the amount Δf given by

$$\frac{\Delta f}{f} = -\frac{1}{1 + N^2\pi^2/12} \qquad (13.31)$$

This shift is insignificant unless N is small ($N < 10$), which corresponds to a very large $F^\#$ or a very small cone angle in the converging light beam.

In spite of the shift just calculated, if one is given a fixed aperture size w with a distant **fixed** object on which it is required to produce maximum irradiance and the available variable is the focal length of the lens (or the curvature of the wavefront emerging from the lens), the optimum choice is to select the lens for which the distant object will be at the center of curvature of the emerging wavefront.

A beam with **Gaussian** irradiance profile, that is, a beam for which

$$I_p(r) = I_0 e^{-r^2/w^2} \qquad (13.32)$$

is unusual in the sense that as it propagates, the irradiance profile remains Gaussian. The curvature of the wavefront and the width w will vary as the beam propagates, but it will remain of the form given by Eq. (13.32), that is, it will remain Gaussian.

A **diffraction grating** may be used in spectroscopy to measure the wavelength of light or to study the distribution of optical power throughout the spectrum.

In its simplest form, the diffraction grating is an opaque screen in which a large number N of transparent slits have been made. The slits should be parallel, evenly spaced, and of the same width. We represent the width of each slit by w and the center-to-center spacing of adjacent slits by d. (Since gratings are routinely produced with d in the range of 3–5 μm and sometimes even smaller, their production requires great care and skill.)

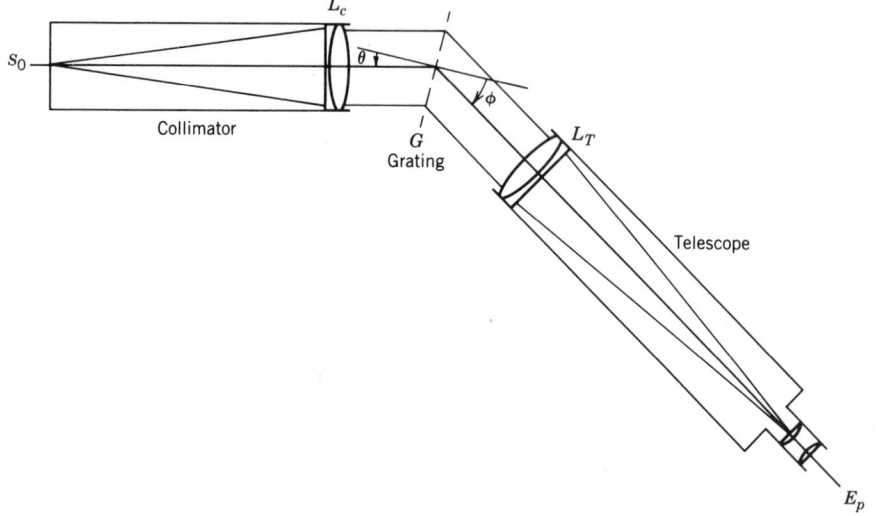

Fig. 13.23 Use of grating in spectroscopy. If grating space is known, measurement of angles θ and ϕ provides sufficient information to calculate λ, wavelength of light.

The grating is used in Fraunhofer diffraction, as shown in Fig. 13.23. This device is called a spectroscope. Light enters the collimator through slit S_0, which is at the front focal point of lens L_c. Parallel rays strike the grating G; the diffracted light is viewed through the telescope, which is focused for infinity. The telescope consists of an objective L_t and an eyepiece E_p. The light is incident upon the grating at angle θ and is viewed from the direction ϕ; both of these angles are measured from the normal to the plane of the grating and with a sign convention that makes both angles positive in Fig. 13.23. Since w and d are only a few times the wavelength of light, the diffraction angles will be so large that we cannot use small-angle approximations for $\sin \theta$ and $\sin \phi$.

A calculation applying Eq. (13.19) to this problem gives

$$I = I_0 \frac{\sin^2 \beta}{\beta^2} \frac{\sin^2 N\gamma}{\sin^2 \gamma} \tag{13.33}$$

where

$$\beta = \frac{\pi w}{\lambda} (\sin \theta + \sin \phi) \tag{13.34a}$$

and

$$\gamma = \frac{\pi d}{\lambda} (\sin \theta + \sin \phi) \tag{13.34b}$$

Equation (13.33) contains two factors, one of which depends upon β and therefore w; the other depends upon γ and therefore d. Here, I_0 is a normalizing factor and is equal to the value of I when $\theta = -\phi$ (i.e., the straight-through irradiance).

The factor involving γ has maxima (equal to N^2) when $\gamma = m\pi$, where $m = 0, \pm 1, +2, \ldots$. By way of Eq. (13.34b), this corresponds to maxima of I when

$$m\lambda = d(\sin \theta + \sin \phi) \tag{13.35}$$

The integer m is called the **order number**.

For large values of N, these maxima of I are very sharp (or narrow) functions of θ and ϕ. If γ changes from the value $m\pi$, which produces a maximum, by the amount $\pm \Delta \gamma = \pi/N$, the factor goes to zero. For good commercially available gratings, N may be in the range from 10^4 to 10^5 and $\Delta \gamma$ is correspondingly quite small.

If the source slit of the spectrometer (Fig. 13.23) is illuminated with monochromatic light and if d is known and θ and ϕ can be measured, then Eq. (13.35) enables us to calculate the wavelength λ. Conversely, if λ is known, we can calculate the grating space d. If the source slit of the spectrometer is sufficiently narrow, the half width $\Delta\phi$ of the image line may be calculated from Eq. (13.34b) using $\Delta\gamma = \pi/N$. Two spectral lines of wavelength λ and $\lambda + \Delta\lambda$ are said to be just resolved (Rayleigh criterion) when their angular separation has this value of $\Delta\phi$. It can be shown that

$$\frac{\lambda}{\Delta\lambda} = \frac{\gamma}{\Delta\gamma} = mN \qquad (13.36)$$

The ratio $\lambda/\Delta\lambda$ is known as the resolving power of the grating.

If the source of light contains several colors, the deviation $|\theta + \phi|$ of the light will, in any given order, be larger in the red than in the blue.

In Eq. (13.33), the factor involving β is the same factor found in Eq. (13.23); it describes the diffraction pattern of a single slit of width w. It does not alter the value of ϕ for any of the maxima, but it determines the relative irradiance of the various orders. For example, if we make $w = \frac{1}{2}d$, then the orders $m = \pm 2, \pm 4, \ldots$ will be missing because the zeros of this factor fall at the same values of ϕ as the even-numbered maxima given by Eq. (13.35).

The grating need not be composed of alternate opaque and transparent lines. It may be everywhere transparent but have a thickness variation that is periodic with period d in one dimension. One surface of such a grating would represent a small-scale copy of a sheet of corrugated steel. Alternately, the grating might be a transparent, uniformly thick sheet in which the index of refraction is a periodic function of one dimension with period d. In all these cases, Eqs. (13.34) and (13.35) are still valid, but the factor involving β has to be changed to represent the diffraction pattern of **one** period of the grating structure.

If the tool that cuts the grooves or lines on the grating is deliberately shaped to increase the irradiance in some order (say, order $m = 1$) at the expense of the irradiance in other orders (such as $m = 0, -1$), the resulting grating is said to be "blazed." In measuring weak spectral lines, this is desirable.

In Fraunhofer diffraction, the **amplitude distribution** in the observation plane is the Fourier transform of the **amplitude transmittance function** of the aperture (with a suitable scaling factor, which is wavelength dependent). Irradiance is proportional to the square of the amplitude. With this information, many Fraunhofer diffraction patterns are predictable.

Holography

Interference and diffraction form the basis of holography. Consider Fig. 13.24. A photographic plate P that is to become a hologram is exposed to two coherent beams of monochromatic light O and R that strike the plate from different directions. A wavefront for each beam is sketched in the figure. An interference pattern is formed on the plate. The fringe pattern formed will be irregular in shape and spacing unless **both** wavefronts have some simple form, such as plane, spherical, or cylindrical. The

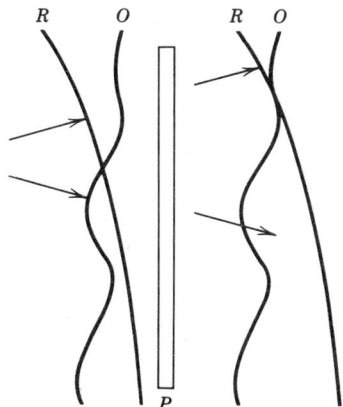

Fig. 13.24 Process of holography. Hologram is formed when photographic plate P records interference pattern produced by two overlapping wavefronts O and R. If plate is processed, returned to its original position, and illuminated by **one** wavefront (e.g., R), diffraction will reproduce copy of other wavefront (i.e., O).

photographic plate is exposed to this interference pattern for a suitable time and is then processed in the darkroom in the usual way. The photographic density as a function of position on the plate is a record of the irradiance in the interference pattern. The photographic plate is now a hologram. In some respects it resembles a grating, but the lines are unequally spaced and distorted as a means of carrying information about the shape of the two wavefronts.

If the hologram is now replaced in its original position (as in Fig. 13.24) and illuminated from the left by **either** of the beams (O or R), then to the right of P there will be observed both beams (O and R) as if they had been propagated through a window at P. This process is called wavefront reconstruction. The reconstructed wavefront will correspond in direction to the first order beam from a diffraction grating, but the irregular spacing and distortion of the lines will cause the wavefront to duplicate, in amplitude distribution and phase, the wavefront which was used to produce the hologram. There will usually be a third beam, called the conjugate beam, corresponding to order -1 of the grating. In some cases the form of the conjugate beam is easily predicted.

It is customary for one wave, R (called the reference wave), to be of some simple form, such as plane or spherical, that can be easily reproduced for use in the reconstruction process. The other wave, O, is from some object to be recorded; it carries information about the object and usually has rapid spatial variations of amplitude and phase. If such a complicated wave were used as the reference wave, then in the reconstruction process, the hologram would have to be repositioned very accurately. (Error from the exposure position must be small compared to the spatial scale of the amplitude and phase variations in the wavefront.)

The preceding statements may be derived from the interference and diffraction equations. Let us look at a simple example as illustrated in Fig. 13.25(a). A coherent, monochromatic beam of light such as an expanded laser beam falls upon an object (represented by a mug) and upon a polished steel ball. The light reflected from the ball provides a spherical wavefront at the photographic plate P; this is the reference wave. Light is also scattered by the object and reaches the plate P as a very irregular object wavefront carrying information about the object.

The plate is processed in the darkroom and becomes a hologram. This hologram is returned to its original position and is illuminated by light reflected by the ball, which is also in its original position. An eye looking through the hologram will see the object at its former position even though it has been removed. The hologram and the reference wave have reconstructed the object wavefront. This reconstructed object wavefront has entered the eye and produced therein the same image that would have been produced by the original object wavefront. The eye is using only that part of the hologram that is between the eye pupil and the reconstructed image. By moving the eye so as to look through a different part of the hologram, the observer sees the image from a different perspective and thereby observes that the image is three dimensional.

In this case, the conjugate wave gives rise to a conjugate image, as shown in Fig. 13.25(b). It may be seen by looking through the hologram. In this case, both the image and the conjugate image are virtual. Real images may be obtained at these same positions if the reconstructing reference wave is replaced by its "conjugate," a spherical wave that strikes the hologram from the left and is converging toward B.

If the reference wave is plane, then the image in the reconstruction is a duplicate of the object and is virtual. In this case, the conjugate image is real and located on the opposite side of the hologram from the object position. The conjugate reference wave is a wave of the same curvature as the reference wave but propagating in the reverse direction. In this case, it is plane and produces a real image with virtual conjugate image.

The hologram is a photographic record of an interference pattern. The average fringe space d is given by

$$d = \frac{\lambda}{\sin \theta} \tag{13.37}$$

where θ is the angle between the reference and object beams; see Fig. 13.25. It is necessary that the photographic emulsion (or whatever recording medium is used) be able to resolve lines of this spacing.

The recording system is very sensitive to vibration. The relative motion of one wavelength of light of the various parts during exposure will reduce the contrast of the interference pattern to nearly zero and make the "hologram" useless. It is usually necessary to isolate the hologram-recording system from building vibrations.

By using reasonably large values of θ ($\approx 30°$), it is possible to make the fringe spacing much less than the thickness of the photographic emulsion. In this case, the hologram has some of the properties of a blazed grating; that is, the reconstructed image is much brighter than the reconstructed conjugate image. It also makes the reconstructing angle of incidence more critical.

The main advantage of holography is its ability to record and reproduce the phase distribution in a wavefront. Other recording methods record only the irradiance distribution.

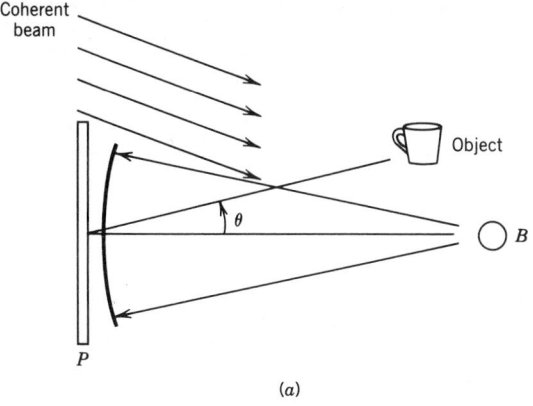

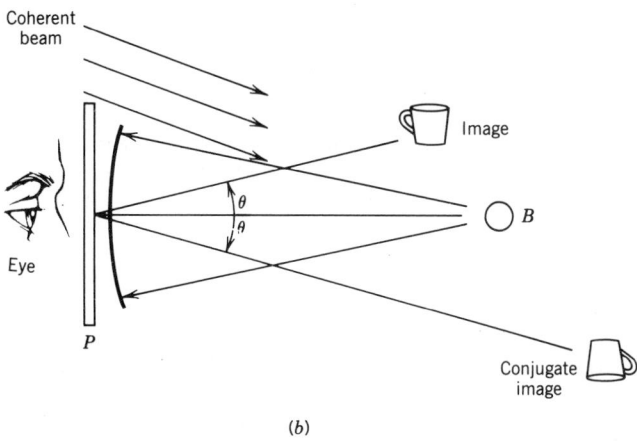

Fig. 13.25 Simple hologram experiment. Hologram is produced with arrangement of upper figure. Light reflected from polished steel ball *B* forms reference wave. Object wave is light scattered diffusely by object. In lower figure, object has been removed; reference wave falls on hologram. Eye, looking through hologram, sees image of object in its original position; conjugate image is also seen.

13.1.4 Light Sources

If we consider the light sources that we encounter in our daily lives, these would probably include (1) tungsten filament lamps and fluorescent lamps used to illuminate our homes and workplaces (2) mercury or sodium arcs frequently used for highway lighting, and (3) in recent years, lasers, which have become fairly common but must be used with caution.

The tungsten filament lamp is the most common example of light sources that are solids and produce light because they are hot (i.e., incandescent). The oil lamp of years past also belongs in this class, producing light because small specks of carbon (i.e., soot) are heated to incandescence by the flame.

Discussion of these sources usually begins with a consideration of "**blackbody**" **radiation**. This is because the radiation of light from a blackbody is understood from a theoretical view and provides a standard of comparison for other sources of radiation. The subject is treated in many textbooks of thermodynamics or modern physics. We shall be content to summarize the major results of the theory.

A blackbody is an object that absorbs all the radiation that falls upon it. In order to remain in thermal equilibrium with its surroundings, it must also emit radiation. A laboratory blackbody would consist of a cavity, the walls of which are maintained at some temperature *T*. There is thermal equilibrium between the walls and the radiation in the cavity except for a small hole in one wall that allows radiation to enter and leave the cavity. This hole is black in the sense that any radiation that enters the cavity through the hole has a negligible chance of finding its way back out. If the area of the hole is only a small part of the total wall area of the cavity, then the radiation escaping from the cavity through the hole is blackbody radiation at temperature T. The properties of this radiation are independent of the shape of the cavity and the material of which the walls are made. It is also assumed that the cavity is filled with a transparent medium of index of refraction $n = 1.0$.

Experimental investigations support the following theoretically derived equations that describe the radiation emitted from a blackbody:

$$\frac{d\Phi}{dA} \equiv M = \sigma T^4 \tag{13.38}$$

This is known as the Stefan–Boltzmann law, where M is the total radiant flux Φ per unit area (watts per square meters) emitted by the blackbody, T is the temperature (degrees Kelvin), σ is the Stefan–Boltzmann constant, which has the value 5.67×10^{-8} W/m^2 K^4. Here, M is called the **radiant exitance**.

It is also customary to define the radiance L, which is the radiant flux (or power) per unit solid angle per unit *projected* area. In Fig. 13.26, the solid angle $d\Omega$, in steradians, is defined as dA_2/r^2, where the area dA_2 is perpendicular to r and dA_1 is an element of area of the source (blackbody in our present discussion) but **projected** area means apparent area as seen from the direction θ; so projected area is $dA_1\cos\theta$:

$$L \equiv \frac{d^2\Phi}{d\Omega\, dA_1\cos\theta} \tag{13.39}$$

A source for which L is independent of θ is called a Lambertian source; it looks equally "bright" from all viewing directions. A blackbody is a Lambertian source; some other sources are approximately Lambertian. For a Lambertian source that radiates into a hemisphere,

$$M = \pi L \tag{13.40}$$

The spectral distribution of blackbody radiation is described by the Planck radiation law, which may be written in either of two forms. The first form is

$$L(\nu) = \frac{2h\nu^3}{c^2}\frac{1}{e^{h\nu/kT} - 1} \tag{13.41}$$

where $L(\nu)\,\Delta\nu$ is the radiance in the spectral region between frequencies ν and $\nu + \Delta\nu$; h is Planck's constant and k is the Boltzmann constant.

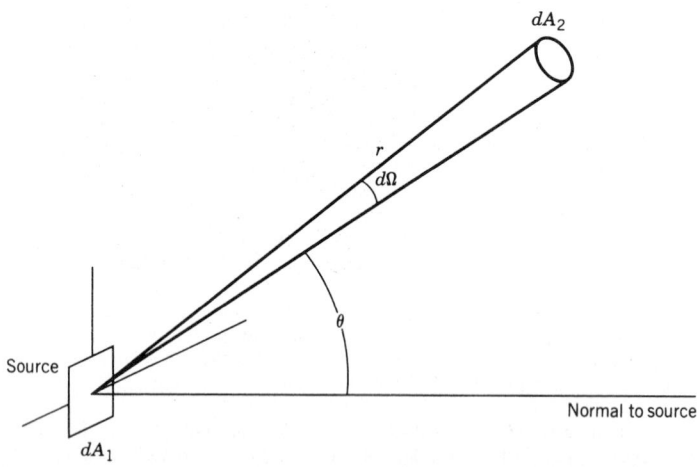

Fig. 13.26 Geometrical quantities used in defining radiance L.

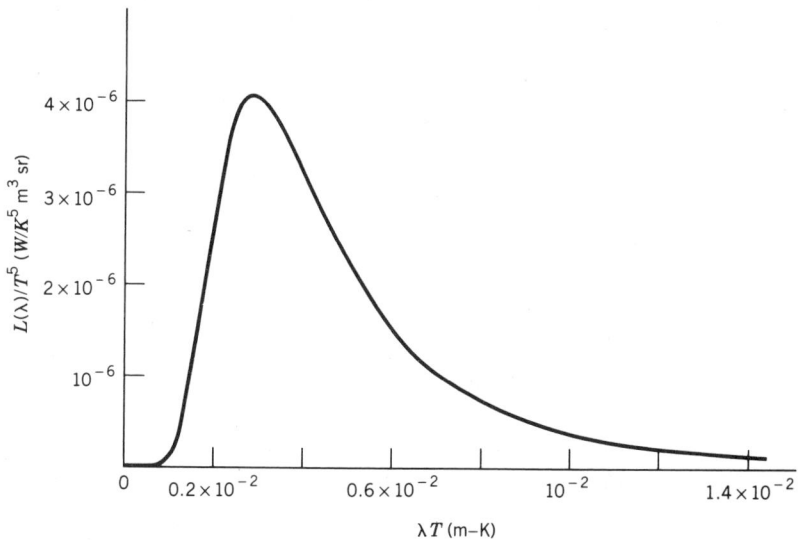

Fig. 13.27 Blackbody spectral distribution curve $L(\lambda)/T^5$ vs. λT. Power in watts; all lengths (including λ) in meters, and temperatures are degrees Kelvin. Curve has maximum value of 4.095×10^{-6} at $\lambda T = 2.898 \times 10^{-3}$.

If the spectral distribution is to be given in terms of wavelength, then Eq. (13.41) may be converted to

$$L(\lambda) = \frac{2hc^2}{\lambda^5} \frac{1}{e^{hc/\lambda kT} - 1} \tag{13.42}$$

Here $L(\lambda)\,\Delta\lambda$ is the radiance in the spectral region between wavelengths λ and $\lambda + \Delta\lambda$. The velocity of light in free space is represented by c.

If we divide both sides of the preceding equation by T^5, we see that $L(\lambda)/T^5$ depends upon λ and T only through the product λT. We can therefore plot a single graph of $L(\lambda)T^{-5}$ versus λT, which will serve for all wavelengths and temperatures. Figure 13.27 is such a plot.

The maximum value of $L(\lambda)T^{-5}$ is 4.095×10^{-6} if $L(\lambda)\,\Delta\lambda$ is in watts per square meter per steradian and $\Delta\lambda$ is measured in meters. The maximum occurs at $\lambda T = 2.898 \times 10^{-3}$ m K. The radiance $L = \int_0^\infty L(\lambda)\,d\lambda = \int_0^\infty L(\nu)\,d\nu$.

The surfaces of real solid objects are not black since they absorb only a fraction α of the radiation that falls upon them. The unabsorbed radiation will be reflected or scattered or, if the object is not too thick, transmitted. The fraction absorbed, α, will be a function of wavelength, temperature, and angle of incidence θ.

Heated solids radiate, but the spectral radiance $L(\lambda)$ is less than that of a blackbody at the same temperature. The emissivity of the material, $\varepsilon(\lambda)$, is the ratio of the radiance $L(\lambda)$ for the material at hand to the radiance $[L(\lambda)]_{BB}$ of a blackbody at the same temperature; $\varepsilon(\lambda)$ is not only a function of wavelength but depends also upon the temperature and the angle of viewing, θ. For given temperature, wavelength, and angle

$$\varepsilon(\lambda, \theta, T) = \alpha(\lambda, \theta, T) \tag{13.43}$$

Solids for which $\varepsilon(\lambda)$ varies slowly over the spectral range are known as "grey bodies." For example, the emissivity of tungsten at 2500 K varies in the visible spectrum from about 0.49 in the blue to about 0.40 in the red. It is considered a grey body. As a function of temperature, the value of $\varepsilon(\lambda)$ of tungsten at $\lambda = 800$ nm is 0.414 for $T = 2000$ K and 0.396 for $T = 3000$ K.

Radiation from the sun closely resembles, in spectral distribution, radiation from a blackbody at 6500 K. Selective absorption by the atmosphere removes the ultraviolet ($\lambda < 350$ nm) and various regions of the infrared so that the sunlight reaching the earth is different from blackbody radiation.

If light passes through an optical system, we can measure the radiance L_1 before it enters the system and also measure L_2 after it leaves the optical system. [Recall the definition of L as given by

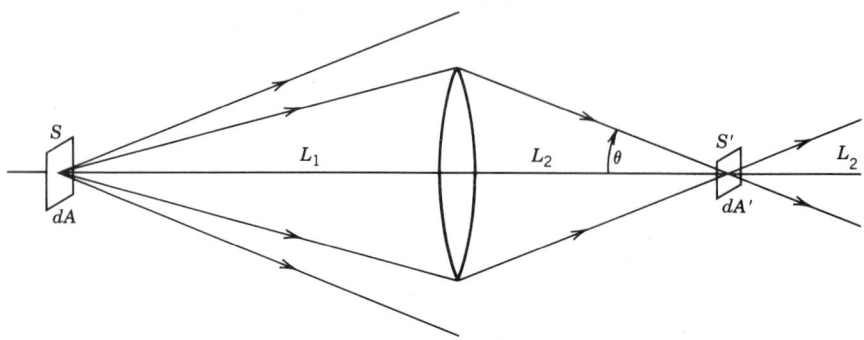

Fig. 13.28 Neglecting losses in lens, $L_2 = n_2^2 L_1/n_1^2$. Assuming there is no diffusing screen at image position S', L_2 remains the same to right of image; L_2 cannot be greater than L_1 if $n_2 = n_1$.

Eq. (13.39).] It can be shown that if there are no losses (e.g., by absorption or by reflection at surfaces), then

$$\frac{L_1}{n_1^2} = \frac{L_2}{n_2^2} \tag{13.44}$$

where n_1 and n_2 are the indices of refraction of the medium before and the medium after the optical system. In many cases, the same medium (e.g., air) exists on both sides of the optical system; in these cases $L_2 = L_1$. If there are losses in the system, then L_2 will be less than the value given by this calculation.

Consider the situation presented in Fig. 13.28. A source S of area dA is imaged by an optical system to S', which may serve as an apparent source for the region to the right of S'. The radiance of the beam to the left of the lens is L_1; neglecting losses and assuming air on both sides of the lens, the radiance to the right of the lens (both left and right of the image S') is $L_2 = L_1$. This is true only within the solid angle subtended by the lens at the image S'. Outside this solid angle L_2 is zero since no light reaches S' from the region beyond the circumference of the lens. As a secondary source, S' has radiance equal to L_1 out to this limiting angle and zero radiance for larger angles. No passive optical system can produce $L_2 > L_1$.

The irradiance (incident power per unit area) onto S' is E, given by

$$E = \frac{d\Phi}{dA} = \int L_2 \cos\theta \, d\Omega$$

where the integration is over the solid angle subtended at S' by the lens. This is the same irradiance that would be produced at S' if the source S filled the exit pupil of the lens. The value of E depends upon the solid angle subtended. In photography, the solid angle is usually represented through the f-number [see Eq. (13.26)] so that irradiance is proportional to $(F^\#)^{-2}$.

If one were to place a small luminous source, such as the filament of a small light bulb, at S and then place the pupil of his eye at S' while looking back at the lens, he would see the exit pupil of the lens entirely filled with light. Only if the pupil of the eye is located at S' will the exit pupil be entirely filled with light.

13.1.5 Lasers

In gases, the intermolecular spacing is generally so large that the molecules radiate independently of each other, and the spectrum produced is a line structure for monatomic molecules or a band structure of fine lines for polyatomic molecules. We comment briefly about the atomic (or monatomic) case. As an example, sodium vapor may be observed to emit light by introducing a little salt (NaCl) into a bunsen flame; alternatively, radiation from sodium vapor may be observed from the sodium arcs, which are widely used for highway lighting.

In the heat of the bunsen flame, the salt dissociates into sodium and chlorine atoms. Most of these atoms are in their lowest energy (or "ground") state. The atoms can, however, be put into more energetic (or "excited") states; the energy of the excited states is specific, depending upon the atom. The number of atoms per unit volume, N_1, in the excited state and the number of atoms per unit volume, N_0, in the ground state will be in the ratio

$$\frac{N_1}{N_0} = e^{-\Delta E/kT} \qquad (13.45)$$

where ΔE is the energy difference between an atom in the excited state and an atom in the ground state, T is the temperature (in Kelvin), and k is the Boltzmann constant. Clearly $N_1 < N_0$. The atom will have several excited states, each with its own ΔE, and there will be an equation similar to (13.45) for each state. The atoms in the more energetic states are said to be thermally excited. An excited atom may give up its excess energy in the form of light and return to the ground state. The frequency and wavelength of the emitted light are given by the equation

$$h\nu = \frac{hc}{\lambda} = \Delta E \qquad (13.46)$$

This bundle of light energy is called a **photon** or a **quantum**. Since each atom (e.g., sodium, potassium, mercury, and hydrogen) is characterized by its own set of excited states, it is also characterized by its own set of spectral lines.

In the case of the sodium arc lamp and other gaseous discharge lamps, the atomic excitations take place by inelastic collisions of the atoms with the moving electrons or ions that make up the electric current. Equation (13.45) is no longer an adequate description of the ratio N_1/N_0, but in most cases it remains true that $N_1 \ll N_0$; that is, only a small fraction of the atoms are excited.

If there are N_1 excited atoms per unit volume, there will be a rate of spontaneous return to the ground state. That rate will be

$$\frac{dN_1}{dt} = -A_{01}N_1 \qquad (13.47)$$

where A_{01} is a constant determined by the nature of the two states. Aside from the value of this constant, the rate of spontaneous return depends only upon N_1. Each atom that returns from the excited state to the ground state emits one quantum of light.

If light of the resonant frequency [i.e., the frequency given by Eq. (13.46)] passes through the gas, some of the atoms in the ground state will absorb a quantum and be excited into the more energetic state. This process is called absorption, and the rate at which it takes place is given by

$$\frac{dN_0}{dt} = -B_{10}N_0L \qquad (13.48)$$

where L is the radiance at the resonant frequency and B_{10} is a constant determined by the nature of the two states. The rate at which atoms are excited (or quanta absorbed) is proportional to L.

The atoms in the excited state are also affected by the light, causing some of them to emit a quantum and return to the ground state. This process is called stimulated emission. The rate at which it takes place is

$$\frac{dN_1}{dt} = -B_{01}N_1L \qquad (13.49)$$

This is in addition to the spontaneous emission. It can be shown that $B_{10} = B_{01}$ and $A_{01} = (h\nu^3/\pi c^2)B_{10}$. The photon emitted by stimulated emission is indistinguishable from the photon that stimulated it. There are now two photons instead of one; they have the same direction of propagation, the same frequency, the same phase, and the same polarization.

In most cases, $N_0 \gg N_1$, so that absorption predominates and stimulated emission is of little consequence. However, if it can be arranged so that $N_1 > N_0$, then the stimulated emission will exceed the absorption and the light can increase in L as it propagates. This might be called negative absorption. This condition must exist in order to produce a laser. The condition in which $N_1 > N_0$ is called a **population inversion**. It may be created in a variety of ways.

In the helium-neon laser, for example, the population inversion is produced as follows: The medium is a mixture of helium and neon atoms in the ratio of about $4:1$; the gas pressure of the mixture is about 1.0 torr to obtain a stable discharge. Electrical discharge in this gas mixture has little direct effect upon the neon but serves to excite some of the helium atoms into metastable states known as the 2^1S and the 2^3S states. These states are metastable in the sense that there are no allowed radiative transitions by which the atoms can return to the ground state; these helium atoms remain excited long enough to experience inelastic collisions with neon atoms. Fortunately, the excitation

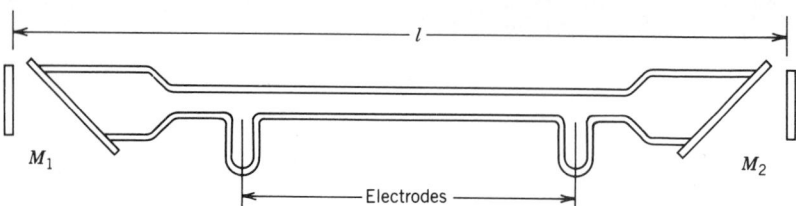

Fig. 13.29 Helium-neon laser with concave end mirrors and Brewster angle windows.

energy of the $3s_2$ state of a neon atom is the same as the 2^1S state of a helium atom; collision between an excited helium atom and an unexcited neon atom can result in energy transfer, producing a neon atom in the $3s_2$ state and a helium atom in the ground state. In the same way, a helium atom in the 2^3S state can excite a neon atom to the $2s_2$ state. In this way, a small but useful fraction of the neon atoms are excited into these two states even though the neon was not directly involved in the electrical discharge. At a lower energy than the two states we have been discussing is a state of the neon atom known as $2p_4$. There are essentially no neon atoms in this state, so there is a population inversion between states $3s_2$ and $2p_4$ and between states $2s_2$ and $2p_4$. These two population inversions can produce lasing at $\lambda = 632.8$ nm and $\lambda = 1152.3$ nm, respectively. For the process to run continuously, the neon atoms in the $2p_4$ state must return to the ground state. This involves a radiative transition to the $1s_5$ state and finally an inelastic collision of the neon atom with the walls of the tube.

Since the rate of stimulated emission is proportional to the spectral irradiance (or to the spectral energy density) and the rate of spontaneous emission is independent of the spectral irradiance, it follows that stimulated emission will become the dominant process when the spectral irradiance is large. To bring this about, the lasing medium [in our case the helium-neon (HeNe) gas mixture] is placed between two mirrors as indicated in Fig. 13.29. These mirrors should have high reflectivity; it is common for the reflectivities to exceed 99.5%. One of the mirrors should have a slight transmissivity (a few tenths of a percent) so that some of the light may escape from the space between the mirrors (called the cavity) into outside space.

If the mirror separation l and the wavelength λ_m are such that a round-trip path equals an integral number m of wavelengths, that is,

$$2l = m\lambda_m \tag{13.50}$$

there will be constructive interference for light reflected from the mirrors upon successive round trips. The **cavity** is said to be resonant at wavelength λ_m and at the corresponding frequency $\nu_m = c/\lambda_m$. The frequency spacing $\Delta\nu$ between ν_m and ν_{m+1} is $\Delta\nu = c/2l$. Cavity resonances are called **modes**. For the typical HeNe laser, l may be about 30 cm so that m is a very large number and $\Delta\nu$ is about 500 MHz.

If a resonance of the cavity exists at the same wavelength as the resonance of the neon states $3s_2$ and $2p_4$ and there is also a population inversion for these two states sufficiently large that the gain per pass exceeds the losses, then there will be lasing (oscillation at optical frequencies) within the cavity. The losses from the cavity include (1) the light that escapes through the partially transparent mirrors, (2) absorption by the mirrors, (3) diffraction and scattering from the beam inside the cavity, and (4) any reflection losses at the end windows of the discharge tube.

The spectral lines of neon have some width, primarily because of the Doppler broadening due to thermal motion. For an atom in motion, its Doppler-shifted resonant frequency must match the incident photon frequency in order to produce stimulated emission; that is, the Doppler-shifted frequency of the atom must match the cavity-resonant frequency for the system to lase. The spectral width of the neon resonance due to Doppler broadening is temperature dependent, but in a typical laser it is roughly 1.0 GHz. This width is sufficient to cover two, or sometimes three, cavity modes of a 30-cm laser cavity. These two or three modes will lase simultaneously.

In the first HeNe lasers, the mirrors M_1 and M_2 were plane mirrors. It was necessary that they be accurately parallel to each other; if they were not, a ray reflected back and forth between them would not remain in the discharge tube to be amplified by the active medium. It was later discovered that curved mirrors could be used, in which case the alignment is less critical. It can be shown by diffraction calculations or by ray tracing that a ray from M_1 to M_2 that is slightly off axis will remain trapped near the axis if the center of curvature of M_1 is to the right of M_2 and the center of curvature of M_2 is to the left of M_1. The axis is a line through the two centers of curvature; the axis should pass through the discharge tube.

A limiting case is to use one plane mirror, M_1, and let the center of curvature of the other (M_2) lie slightly to the left of M_1; this is called a hemispherical cavity. There are other mirror arrangements

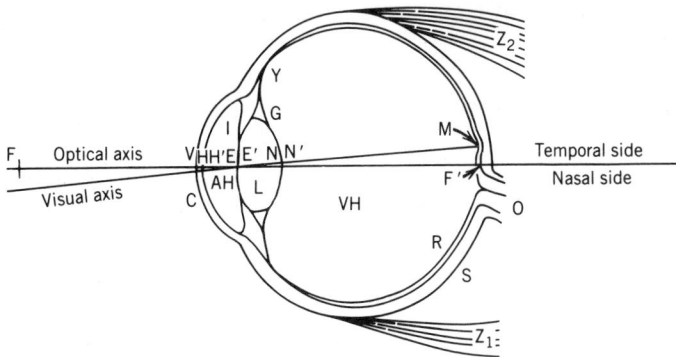

Fig. 13.30 Horizontal section of right human eye according to Helmholtz.

that form stable cavities (stable in the sense that rays are trapped near the axis), but the ones just mentioned are commonly used.

In contrast, there are unstable cavities in which a ray starting slightly off axis diverges from the axis and escapes from the cavity. As an example, we give a cavity for which the center of curvature of M_1 lies slightly to the left of M_2 and the center of curvature of M_2 lies slightly to the left of M_1. (One center is inside the cavity and the other outside the cavity.) The rays are not trapped in this cavity but escape after a few round trips; the losses in this cavity are very large. It does not lase.

In the earliest HeNe lasers, the reflection losses at the windows on the ends of the discharge tube were so great that the system could be made to lase only by eliminating these windows and attaching the cavity mirrors directly onto the discharge tube. Later it was realized that by attaching the windows at the Brewster angle (see Fig. 13.3), one polarization (the polarization with E parallel to the plane of incidence) would experience no reflection loss and the cavity mirrors could be mounted independently of the discharge tube. The light from such a laser is linearly polarized since the gain exceeds the losses only for the polarization, which experiences no loss at the Brewster angle windows.

The mirrors M_1 and M_2, which form the resonant cavity, are usually multilayer dielectric coatings since most metals do not have sufficiently high reflectivity. Also, with multilayer dielectric coatings the mirrors may be spectrally selective in their reflectivity. For example, the mirrors may be highly reflecting at $\lambda = 632.8$ nm with much lower reflectivity at $\lambda = 1152.3$ nm. In this case, the HeNe laser will lase at $\lambda = 632.8$ nm but not at 1152.3 nm. By exchanging the mirrors for a pair with high reflectivity at 1152.3 nm, we can cause the system to lase at that wavelength.

The modes considered a few paragraphs back are properly called longitudinal modes. A laser may also have several "transverse" modes, but the manufacturers of lasers usually suppress all the transverse modes except the TEM_{00} mode. In this mode the amplitude of the electric field at the output mirror as a function of distance off axis r is given by

$$E = E_0 e^{-r^2/w^2} \tag{13.51}$$

where E_0 is the amplitude on axis, E is the amplitude at a distance r from the axis, and w is a constant depending upon the geometry of the cavity. The surface of the output mirror is a surface of constant phase (i.e., a wavefront) for the emerging wave; the beam width w may be only a few millimeters.

We have given our attention to the HeNe laser because it is readily available and illustrates the principles involved. There are many other media in which population inversion can be produced and which can provide lasing if used in a suitable cavity.

13.1.6   The Eye and Vision

The eye is important because most of the information obtained in a lifetime is brought to the brain through the eye. For the student of optics, the eye is important because many optical instruments, for example, microscopes and telescopes, are used in conjunction with the eye so that the eye becomes a part of the optical system. The pupil of the eye may become the aperture stop of the system or in some cases the eye may limit the spatial frequency response or resolution of the optical system. A geometrical or physical description is inadequate because the eye is a living, functioning organ that should be considered in terms of physiology and neurology, but these fields are beyond the scope of this article and can be considered only superficially.

The Structure of the Eye

The human eye is an almost spherical organ about an inch in diameter. It is shown in cross section in Fig. 13.30. Six muscles, two of which are shown in the figure as Z_1 and Z_2, hold the eye in place and rotate it relative to the head. These muscles are attached to the sclera S, which is a tough white skin covering most of the eye. At the front of the eye the sclera is replaced by the cornea C, which is a transparent membrane through which light enters the eye. After entering through the cornea, light passes through the aqueous humor AH, the crystalline lens L, the vitreous humor VH, and finally reaches the retina R. The aqueous humor is a weak salt solution; the vitreous humor is a soft jelly consisting primarily of water. The fluids of the eye are slightly ($\sim$ 25 torr) above atmospheric pressure. This pressure helps to maintain the shape of the eyeball. The crystalline lens is a fibrous jelly contained in a thin membrane or sac; it is hard at the center and progressively softer toward the outside. The lens is held in place and attached to the ciliary muscle Y by the suspensory ligament G. When the ciliary muscle is relaxed, the second focal point is at the retina and distant objects are in focus. To view nearby objects, the ciliary muscle contracts, allowing the lens to become more nearly spherical. This is known as accommodation; with age, the lens becomes less elastic, and the ability to accommodate gradually decreases. The lens of the eye is not transparent to ultraviolet light.

The retina is the interior lining for a large part of the eyeball. It consists of rods and cones that are light-sensitive nerve endings, along with a delicate network of nerve fibers connecting the rods and cones to the optic nerve O and a network of capillary blood vessels that supply the necessary oxygen and nutrients. The yellow spot, or macula lutea M, which contains many cones and relatively few rods, is a slight depression in the retina; the central region, called the fovea centralis, contains cones exclusively, no rods. The macula lutea is about 2 mm in diameter, and the fovea centralis is about 0.25 mm in diameter. Cones in the fovea centralis are about 1.5 μm in diameter, increasing in size to about 5.5 μm in the outer portion of the macula lutea and several times this size in other portions of the retina. See Fig. 13.31. In the outer portion of the retina, the rods outnumber the cones by 10 : 1. Each human eye contains roughly 7 million cones and 120 million rods.

Vision in the fovea centralis is so much more acute than in the extra foveal region that the muscles surrounding the eye involuntarily rotate the eyeball until the object of interest is imaged upon the fovea centralis. The angle in **object space** covered by the fovea centralis is less than 1°; it is only a little more than sufficient to cover one letter of this printed page when the book is held at the usual reading distance of 25 cm. In reading or examining an extended object, the eye must move frequently. Extra foveal vision is not useful in observing details but enables one to be aware of objects around him. For a healthy eye, the total field is about 128°; an early sympton of glaucoma is the shrinking of the field of view.

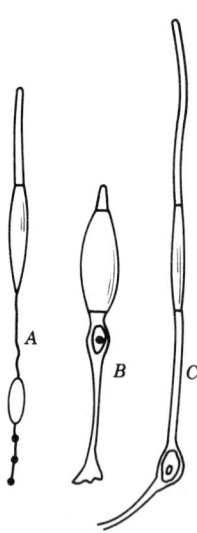

Fig. 13.31 Rods and cones of retina: *A*, rod; *B*, cone from extrafoveal region; *C*, cone from central fovea.

The mosaic structure of the cones in the fovea centralis limits the resolution of the eye. Considering the size of the cones, this varies from 0.3 to 1.0 min for the angular resolution in object space. These numbers should be slightly larger because the cones are separated by a small amount of inactive tissue. It is interesting to observe that this is comparable to the resolution limit set by diffraction at the pupil and also comparable to the limits produced by aberrations, primarily spherical aberration of the optical system. One minute is a good round number representing the overall resolution of the eye.

Part of the blood supply to the retina is provided by a network of blood vessels on the front of the retina. If one stares at a blue sky (or a white wall illuminated by blue or violet light), the red blood cells coursing through these blood vessels can be seen since they cast a shadow on the retina. Unlike the specks of dust that float upon the front of the eye or in the vitreous humor when one is tired, these shadows follow definite paths; that is, they are confined to the blood vessels. These shadows are called *muscae volitantes*, which means flying flies. Red blood cells are about 8 μm in diameter, so each one can cast a shadow over several cones of the fovea.

Adaptation of the Eye to Light

The iris diaphragm, I, is a ring-shaped involuntary muscle that controls the amount of light entering the eye. It is located just in front of the lens, and the diaphragm opening or pupil is the aperture stop of the eye. It varies in diameter from 2 to 8 mm. This is a factor of 4^2, or 16, in the area of the entrance pupil. The eye functions under illumination conditions that vary by a factor of $\sim 10^9$. Variation in pupil size is certainly not sufficient to account for this wide range; most of the adaptation to light and dark is accomplished by changing the sensitivity of the retina. The photosensitive chemicals (or pigments) in the rods and cones are bleached or altered by light and must be constantly reconstituted. Due to the lower rate at which the pigment is consumed in low illumination, the steady-state concentration of the pigments is higher and the retina more sensitive in low illumination than in high illumination.

Scotopic Vision

When the eye has been dark adapted (i.e., kept for half an hour or more in darkness comparable to outdoor illumination by a moonless night sky), the eye becomes sufficiently sensitive to see a small source of 2×10^{-8} cd at a distance of 3 m. Neglecting atmospheric absorption, this is equivalent to seeing a standard candle at a distance of 13 miles. Astronomers observe that except under unusually good conditions, stars of sixth magnitude represent the limit of vision of the unaided eye. This corresponds to seeing a standard candle at a distance of about 6.6 miles through the atmosphere. However one describes it, the dark-adapted eye is incredibly sensitive.

Vision by the dark-adapted eye is called **scotopic** vision and takes place in the rods of the eye, not in the cones. Since there are no rods in the fovea, the dark-adapted eye has no central vision, and in order to see an object in subdued light, one must look not at the object of interest but to the side so that the object of interest will be imaged on the outer part of the retina, which contains rods. There is no color in scotopic vision.

In the rods, the pigment that absorbs the light and somehow triggers the signal along the nerves to the brain is called rhodopsin. The chemical composition and structure is known to be a protein molecule combined with a molecule of retinal. Retinal is closely related to the compounds known as retinol (vitamin A) and carotene (the yellow pigment of carrots and many other yellow vegetables).

The spectral sensitivity of rod vision is shown in curve B of Fig. 13.32. The ordinate at each wavelength is inversely proportional to the minimum amount of energy that is just perceptable (i.e., to the threshold of vision). The curve is normalized to 1 at its peak. This closely matches the absorption curve of rhodopsin.

Photopic Vision

For conditions of ordinary illumination, the rhodopsin in the rods is almost completely bleached and vision is by the cones. This is called **photopic** vision, or cone vision. The spectral sensitivity for cone vision is shown by curve A of Fig. 13.32. Notice the shift of this curve toward the red relative to the scotopic curve B. Because of this shift, two nonluminous objects of different colors (e.g., yellow and blue-green) that appear "equally bright" in ordinary illumination will not appear equally bright in subdued illumination (e.g., twilight), the blue-green becoming much more conspicuous than the yellow. This shift in the spectral sensitivity and the resulting change in relative brightness of various colors is known as the Purkinje effect. It is a source of trouble in making visual comparisons of light sources of different colors.

The level of illumination at which the eye changes from photopic to scotopic vision (or vice versa) with the attendant change in spectral sensitivity, loss of color discrimination, and foveal vision is about the illumination level produced by the full moon on a clear night, or 0.16 lux.

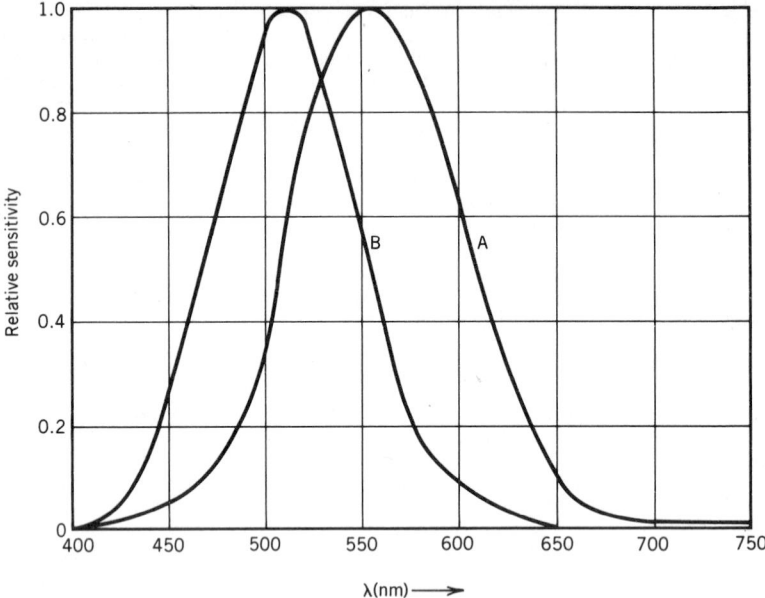

Fig. 13.32 Spectral sensitivity curves for normal human eye: A, light-adapted (photopic) eye; B, dark-adapted (scotopic) eye. Each curve is normalized to 1 at its maximum.

Color Vision

Color vision takes place in the cones. There are three different types of cones in the eye; the three differ in that they contain different photosensitive pigments and have distinct spectral response curves. There is no observable physical structure that enables one to distinguish between the three types; the photosensitive pigments are present in such low concentrations that it is difficult to distinguish even on this basis. The three pigments are probably three different protein molecules, each in combination with a molecule of retinal. Because of the chemical similarity of the three dyes to each other and to rhodopsin (which is much more abundant), they are difficult to isolate and identify. The spectral sensitivity of the three cone types is given in Fig. 13.33. The output of each cone is determined by the intensity reaching it, the wavelength of the light, and the spectral sensitivity of the cone for that wavelength; the same output signal could be obtained by use of a lower intensity at a wavelength closer to the peak of the sensitivity curve. Each cone is color blind (just as the rods are color blind); the sense of color is derived from the relative response of the three types of cones.

As shown in Fig. 13.33, the three cone types have peak sensitivities in the blue ($\sim$ 440 nm), green ($\sim$ 535 nm), and orange ($\sim$ 565 nm); they are labeled C, B, and A, respectively. For each spectral wavelength, the relative response of these three cone types is unique and determines the color sensation. If several wavelengths are present, each wavelength evokes a response in each cone type, and the relative size of the total response in each of the three cones determines the color sensation. Curve A peaks at 565 nm, which is in the orange; at this wavelength, the other curves, in particular B, still have nonzero values and the sensation of orange is produced by a signal from A and a weaker signal from B. At longer wavelengths, cones of type A respond less than they did at 565 nm, but the response from cones of type B decreases even more rapidly so that the signal from A makes up a larger fraction of the total output and the color sensation changes from orange to red.

Common forms of color blindness result from the absence of type A or type B cones. Protonopes are persons color blind due to lack of type A cones; duteranopes lack type B cones.

In ordinary vision, the output of type A cones is added to the output of type B cones (with perhaps a weak contribution from type C cones), and this sum is transmitted to the brain along the optic nerves. This sum encoded as nerve pulses per second is interpreted by the brain as luminosity (white) without color information. Color information is transmitted in two channels as the difference $(A - B)$ between the output of type A cones and type B cones and the difference $(A - C)$ between the output of type A cones and type C cones. The data is processed into these sum and differences in or near the eye and then transmitted along nerve fibers to the brain in the form of an increase or a decrease of the pulse frequency from the spontaneous value of the pulse frequency that exists when

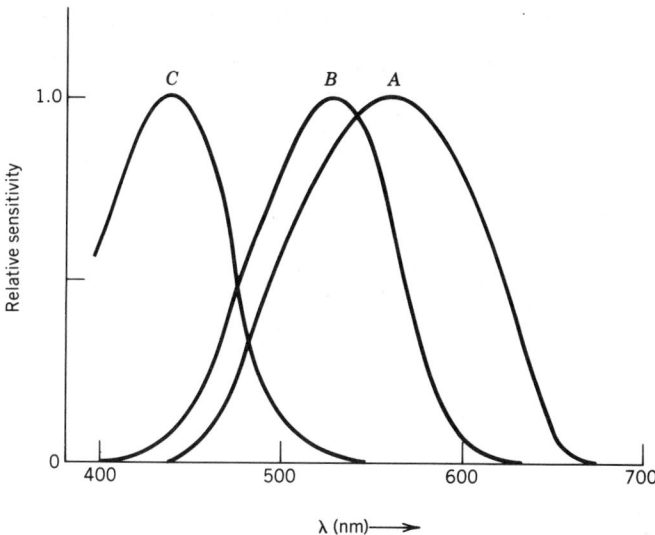

Fig. 13.33 Spectral sensitivity curves for three cone types of human eye. Each curve is normalized to 1 at its maximum.

TABLE 13.2

Color	Wavelength (nm)
Violet	< 450
Blue	450–500
Green	500–570
Yellow	570–590
Orange	590–610
Red	> 610

the eye is in the dark. Type C cones have little effect upon the sensation of brightness but are effective in producing color discrimination.

Although the system just described is believed by many to be the usual one, others are possible and sometimes effective because one can cover the right eye with a red filter and the left eye with a green filter (or vice versa) and obtain color vision. In this case, some of the data processing that usually takes place at the eye appears to be deferred to some later stage of the visual process, perhaps in the brain or perhaps at the optic chiasma, the point at which the two optic nerves (one from each eye) come together on their way to the brain.

For a person with normal vision, the colors associated with various portions of the spectrum are as shown in Table 13.2.

Colorimetry
The word *color* has several definitions. In one, it is associated with the properties of a dye; in another, it is a property of light; and in yet another, it is a physiological sensation produced in the brain by light entering the eye. In an earlier section, we have given a brief description of color vision. In this section, we use the word *color* as descriptive of the light entering the eye and present the methods used to give a quantitative description of the color. The branch of optics that deals with the quantitative specification of color is called colorimetry.

Color Mixing
In order to understand colorimetry, we must first establish the basic facts of color mixing, which are illustrated by the following experiment. We attempt to match all possible colors by mixing three "primaries." The selected primaries are monochromatic (or spectral) colors of wavelength 450 nm (blue), 550 nm (green), and 620 nm (red). We identify them as α, β, and γ, respectively. There is nothing unique about these particular wavelengths that entitle them to be primaries; we select them

because the experimental data using these three primaries was carefully determined in early color-mixing experiments. We now allow the eye to look at a white diffusing card. Two adjacent areas of the card are illuminated (1) by light of arbitrary or unknown color and (2) by a mixture of the three primaries. Area 2 is illuminated by all three primaries, and the amount of each primary is adjusted to obtain a match with the unknown. Most colors can be matched by this mixing process; a few cannot. In cases for which the unknown cannot be matched by the preceding process, a match can be obtained by moving one (or very rarely two) of the primaries from area 2 to area 1 and then adjusting the amount of each primary; this is equivalent to subtracting or using a negative amount of the moved primary in area 2. The use of three primaries widely spaced in the spectrum, as are the ones suggested here, reduces the number of cases in which a negative amount of any of the primaries is required. Neither the unknown nor the primaries need be monochromatic (spectrally pure) colors; a match can always be made. If the unknown is represented by U and the amount of each primary by A, B, and C, respectively, the experimental results may be represented by the equation

$$U = A + B + C \qquad (13.52)$$

which is interpreted to mean that the sensation of light and color produced by the unknown may be duplicated by the mixture of the three primaries. The values of A, B, and C are unique if U is given. The eye sees the overall effect of the mixture; it is not aware of the individual primaries that make up the mixture.

If we now restrict our unknown to spectrally pure (i.e., monochromatic) light and keep the power of the unknown constant but vary its wavelength, we can at each wavelength determine experimentally the power of each primary required to produce a match. The results of this experiment are given in Fig. 13.34, which gives the amount of each primary α, β, and γ required to match each spectral color. The curves have been normalized to $\beta = 100$ at 550 nm; α and γ are zero at this wavelength, which corresponds to the β primary. Each curve is normalized to 100 when the unknown wavelength is the same as that primary. For example, the curves indicate that a match is obtained for an

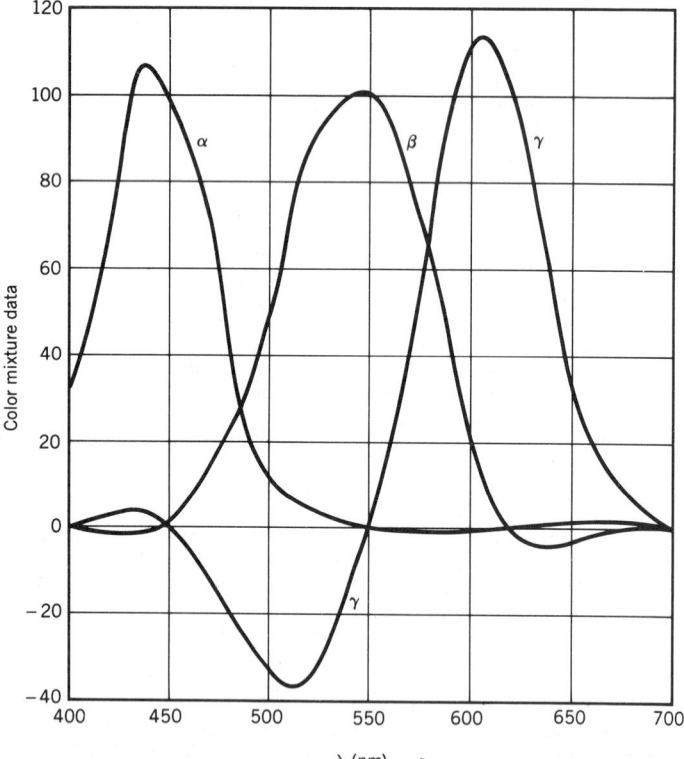

Fig. 13.34 Color mixture curves for matching spectrally pure colors by mixing primaries having wavelengths 450, 550, and 620 nm.

unknown at 500 nm by combining 47.5 units of α (light at 450 nm) with 125 units of β (light at 550 nm) and subtracting (i.e., adding to the unknown) 30.0 units of γ (light at 620 nm). Notice that the primaries do not add to 100; this is because the spectral sensitivity of the eye for each of the primaries differs from its sensitivity at the unknown wavelength. In this case, the most significant difference is a factor of about 3 between the sensitivity of the eye to the β primary and its sensitivity to the 500 nm unknown.

Similar curves for the mixing of other sets of monochromatic primaries could be determined experimentally, but it is unnecessary to do so because it is possible to deduce them from the curves already given. The process is straightforward but tedious, and we shall not describe it. It is also possible to specify a new set of primaries by giving the curves α', β', and γ', which give the mixing data required to match spectral colors using the new primaries. As long as the new curves α', β', and γ' (as functions of wavelength) are a linear combination of the experimental curves α, β, and γ (which were given in Fig. 13.34), the new system will give a satisfactory system of color specification. The requirement of algebraic linearity means that

$$\alpha' = K_{11}\alpha + K_{12}\beta + K_{13}\gamma$$
$$\beta' = K_{21}\alpha + K_{22}\beta + K_{23}\gamma$$
$$\gamma' = K_{31}\alpha + K_{32}\beta + K_{33}\gamma \tag{13.53}$$

where the K_{ij} are real and independent of wavelength but are otherwise subject to no restriction except that the determinant

$$\begin{vmatrix} K_{11} & K_{12} & K_{13} \\ K_{21} & K_{22} & K_{23} \\ K_{31} & K_{32} & K_{33} \end{vmatrix} \neq 0 \tag{13.54}$$

With such a wide choice of primaries and with the possibility of algebraic transformation from one set to another, color-mixing data could not provide any insight into the spectral sensitivity curves of Fig. 13.33, but this did not prevent the development of colorimetry in advance of a detailed understanding of color vision.

With so much freedom in the choice of the K_{ij} (or the curves α', β', and γ'), it s probably not surprising that some of the possible sets of primaries so described contain primaries that are not spectral colors (e.g., purples) or not spectrally pure (e.g., pinks). It is also true that many acceptable sets of primaries contain primary colors that are not real; by this we mean that they exist mathematically in terms of the mixing curves α', β', and γ', which produce real colors; in this sense, they are entirely satisfactory primaries, and yet they do not exist in the sense that the single primary alone cannot be seen as light and color.

Tristimulus Values and Trichromatic Coefficients

Since color specification is commercially important and since there is so much freedom in the choice of primaries, it was inevitable that there should develop some agreement on what set of primaries would be used. In 1931, the International Commission on Illumination (ICI) [also known by its French name, Commission Internationale de l'Eclairage (CIE)] agreed to express all color specifications in terms of three primaries defined by the color-mixing curves of Fig. 13.35. The letters $\bar{x}$, $\bar{y}$, and $\bar{z}$ have become standard, replacing the α, β, and γ used by earlier workers. The ordinates, called **tristimulus values**, are in arbitrary units and have been adjusted so that the areas under the three curves are equal. The shape of curve $\bar{y}$ was arbitrarily chosen to be the same as curve A of Fig. 13.32. Curves $\bar{x}$ and $\bar{z}$ have shapes selected so that the three primaries satisfy Eq. 13.53. For computational convenience it was also required that none of the curves is ever negative. None of the primaries defined by this set of mixing curves is real; they form a satisfactory base for the quantitative specification of color, but only real colors can be produced and mixed in the laboratory.

For monochromatic light with wavelength 500 nm (green), the tristimulus values are

$$\bar{x} = 0.0049 \qquad \bar{y} = 0.3230 \qquad \bar{z} = 0.2720$$

Define three new quantities x, y, and z such that

$$x \equiv \frac{\bar{x}}{\bar{x} + \bar{y} + \bar{z}} \qquad y \equiv \frac{\bar{y}}{\bar{x} + \bar{y} + \bar{z}} \qquad z \equiv \frac{\bar{z}}{\bar{x} + \bar{y} + \bar{z}}$$

These new quantities are called **trichromatic coefficients** and by definition have the property that $x + y + z = 1$; any two of the three quantities are sufficient to specify the color. For 500-nm light,

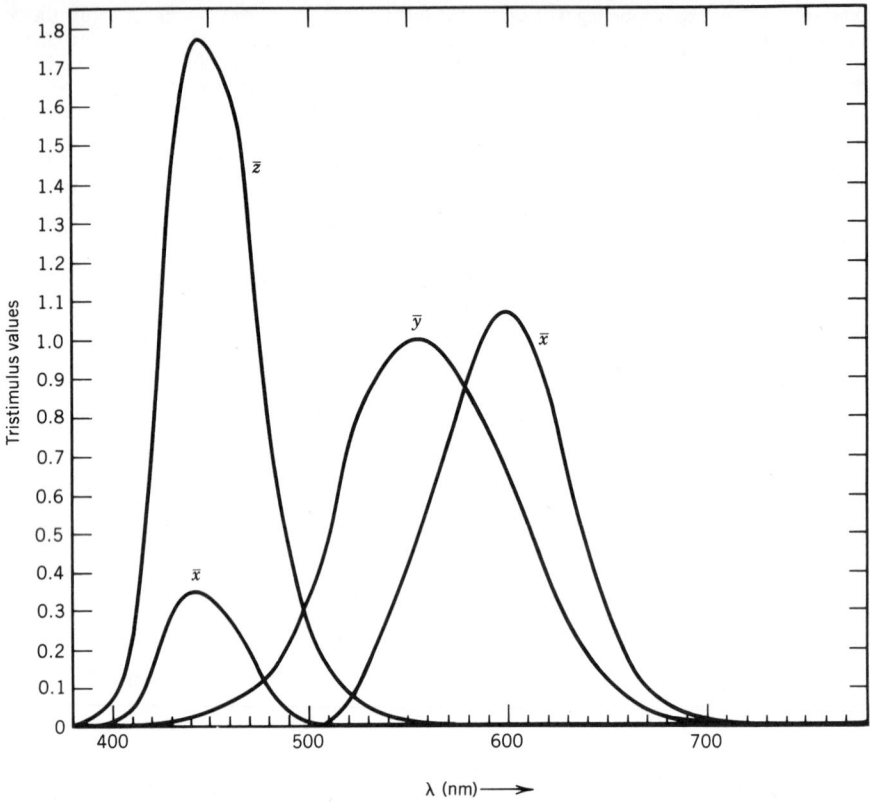

Fig. 13.35 Standard ICI (or CIE) tristimulus curves $\bar{x}$, $\bar{y}$, and $\bar{z}$ for unit power at indicated wavelength. Numerical values for these curves may be found in Refs. 1 and 2.

the values are

$$x = 0.0082 \qquad y = 0.5384 \qquad z = 0.4534$$

This system cannot contain any intensity information, only color information.

In this system, any spectral color may be specified by giving any two of the trichromatic coefficients; the values of x and y are usually given. If we plot on ordinary graph paper the values of x and y for the spectral colors, we obtain the curve of Fig. 13.36, where the wavelength (in nanometers) is shown at various places along the curve. A diagram such as this in which color information is plotted using the trichromatic coefficients is called a **chromaticity diagram**; the curve is known as the spectrum locus.

Trichromatic Coefficients for Non-monochromatic Light

In the previous section we defined the trichromatic coefficients of any monochromatic light using the ICI (or CIE) primaries. In very few cases is the light reaching the eye monochromatic; it is usually a mixture or distribution of spectral colors. If we represent the spectral distribution by the function $f(\lambda)$ defined so that $f(\lambda)\,d\lambda$ is the power (e.g., in watts) in the spectral interval between λ and $\lambda + d\lambda$, then we calculate the **tristimulus values** of the light by the equations

$$X \equiv \int_0^\infty \bar{x} f(\lambda)\,d\lambda \tag{13.55a}$$

$$Y \equiv \int_0^\infty \bar{y} f(\lambda)\,d\lambda \tag{13.55b}$$

$$Z \equiv \int_0^\infty \bar{z} f(\lambda)\,d\lambda \tag{13.55c}$$

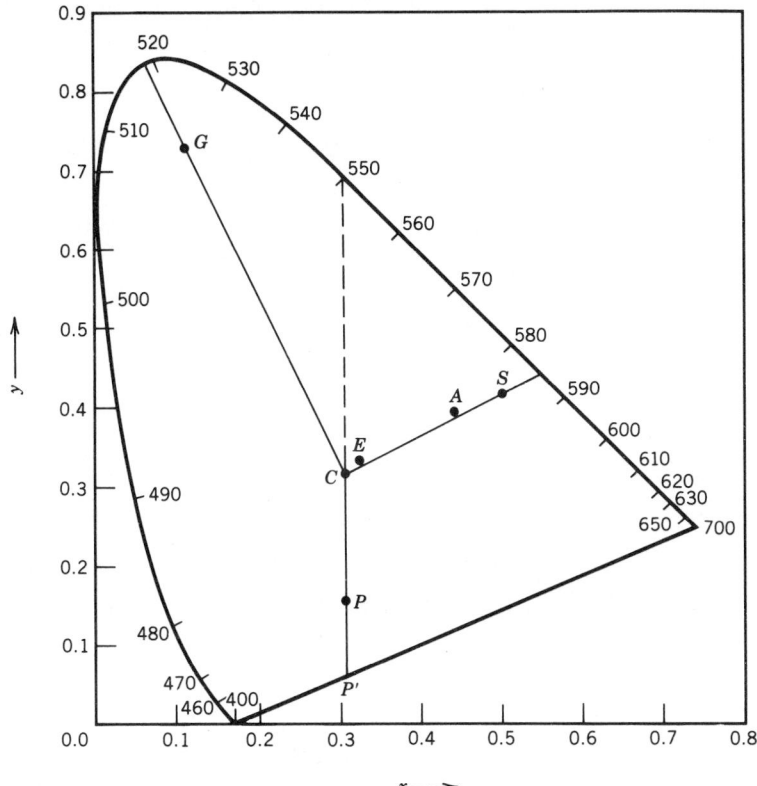

Fig. 13.36 Chromaticity diagram: Horseshoe curve, spectrum locus; E, source for which $f(\lambda)$ is constant; C, illuminant C, approximately daylight; A, illuminant A, illumination from tungsten filament lamp; S, light reflected from orange skin illuminated by illuminant C.

where $\bar{x}$, $\bar{y}$, and $\bar{z}$ are the functions represented in Fig. 13.35, the ICI color mixture curves. This process amounts to treating each spectral color in the light by the methods of the previous section and then adding (integrating) all of these effects together. By definition, the integrals are from zero to infinity; but since the functions $\bar{x}$, $\bar{y}$, and $\bar{z}$ are zero outside of the visible range, the integration is effectively limited to the visible-wavelength interval. The functions involved cannot be integrated by elementary methods; the integration is carried out numerically. The numerical data represented by the curves $\bar{x}$, $\bar{y}$, and $\bar{z}$ may be found in the original ICI report[1] or in any textbook on colorimetry.[2]

The three tristimulus values calculated by Eqs. (13.55) are converted to **trichromatic coefficients** by the equations

$$x = \frac{X}{X + Y + Z} \qquad y = \frac{Y}{X + Y + Z} \qquad z = \frac{Z}{X + Y + Z} \qquad (13.56)$$

We again have the property that

$$x + y + z = 1 \qquad (13.57)$$

and any two of these may be used to specify the color of the light; x and y are usually used. Information about the total intensity of the light has been lost, but all color information is retained.

As an example, consider light for which the spectral distribution $f(\lambda)$ is a constant. This is light for which, at any wavelength, a small wavelength interval $d\lambda$ contains the same power as an equal interval $d\lambda$ located at any other wavelength. Since the three curves x, y, and z of Fig. 13.35 have equal areas under them, the integrals of Eqs. (13.55) will, for this example, be equal, that is,

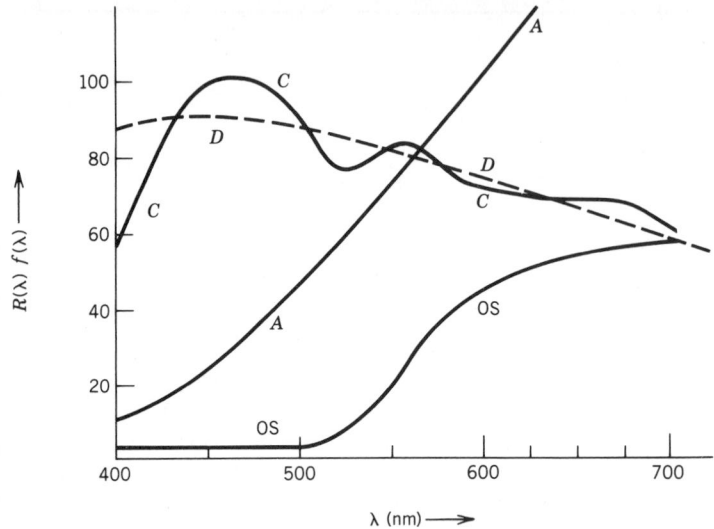

Fig. 13.37 Dashed curve D, curve A, and curve C are spectral distributions $f(\lambda)$ of average daylight, light from illuminant A, and light from illuminant C. (Vertical scale is arbitrary and not the same for the three curves.) Lower curve OS, spectral reflectance $R(\lambda)$ of orange skin; magnesium carbonate powder is taken as 100%.

$X = Y = Z$. When these are converted to the trichromatic coefficients x, y, and z, we obtain

$$x = y = z = 0.3333$$

On the chromaticity diagram (Fig. 13.36), this is represented by the point E at $(\frac{1}{3}, \frac{1}{3})$.

Another important example is light from a source known as illuminant C. Illuminant C is intended to have the same spectral distribution as average daylight, at least in the visible. It consists of a gas-filled tungsten lamp operated at the color temperature 2848 K combined with filters designed to alter the spectral distribution of the lamp to that of daylight. The spectral distribution of illuminant C is in Fig. 13.37. From this distribution, one can evaluate numerically the integrals of Eqs. (13.55) and then the trichromatic coefficients x, y, and z of Eqs. (13.56). The results of these calculations give point C at (0.3101, 0.3163) on the chromaticity diagram in Fig. 13.37. Light from illuminant C is generally considered to be "white," although the term *white light* has no universally accepted definition.

The Color of an Orange Skin
In Fig. 13.37, there is shown the reflectance of an orange skin as a function of wavelength, $R(\lambda)$. This curve may be obtained by illuminating the orange skin successively at several different wavelengths; at each wavelength, the reflected radiance is measured for the orange skin and for some white object. The reflectance $R(\lambda)$ at each wavelength is the ratio of these two measurements. Since the orange skin is a diffuse reflector, the white comparison object should be a diffuse reflector also. Freshly fallen snow is a good white diffuse reflector, but in the laboratory a powder of magnesium carbonate is more practical.

The observed color of the orange skin depends not only on its spectral reflectance but also on the spectral distribution of the illuminating light. Let us assume that illuminant C is used and we represent its spectral distribution by $C(\lambda)$. The spectral distribution of the light reflected from the orange skin is the product $C(\lambda)R(\lambda)$. We calculate the tristimulus values of this light from Eqs. (13.55):

$$X \equiv \int_0^\infty \bar{x} C(\lambda) R(\lambda)\, d\lambda = 341$$

$$Y \equiv \int_0^\infty \bar{y} C(\lambda) R(\lambda)\, d\lambda = 277$$

$$Z \equiv \int_0^\infty \bar{z} C(\lambda) R(\lambda)\, d\lambda = 50$$

and when these are normalized to the trichromatic coefficients, we have

$$x = 0.511 \qquad y = 0.414 \qquad z = 0.075$$

These locate a point (marked S) on the chromaticity diagram (see Fig. 13.36). This point is fairly close to the spectral locus for 586 nm. The light reflected from this orange skin is therefore close to the orange-yellow color of the sodium D lines.

If we used another illuminant instead of illuminant C, the location of point S representing the chromaticity of the light reflected by the orange skin would have to be recalculated and would probably have changed.

Two pieces of cloth that have the same spectral reflectance will always look alike, that is, have the same coordinates on the chromaticity diagram, as long as the same illuminant is used on each piece no matter what illuminant is used. It is possible, and sometimes happens, that two pieces of cloth that have different spectral reflectance curves may look alike when illuminant C is used but will be noticeably different when another illuminant, such as illuminant A, is used. Illuminant A is the gas-filled tungsten lamp operated at the color temperature 2848 K and used without filters; it is typical of the illumination produced by tungsten filament lamps. The chromaticity of illuminant A is represented by point A in Fig. 13.36. It more frequently happens that two pieces of cloth look alike under illuminant A but are noticeably different under illuminant C. Illuminant A is relatively weak in the short-wavelength region so that a match using this illuminant is relatively insensitive to the reflectance of the cloth for blue light. Illuminant C is slightly stronger at the shorter wavelengths than it is at the longer wavelength (see Fig. 13.37).

The Chromaticity Diagram as an Aid to Color Mixing

From the definitions of the trichromatic coefficients [Eqs. (55)], it follows that if we have two colors represented by points such as G and R (Fig. 13.38) of the chromaticity diagram, any additive mixture of these two colors will be represented by a point lying on the lie GR. If each component (G and R) is assigned a weight proportional to the sum of its tristimulus values ($X + Y + Z$), the point representing the chromaticity of the mixture will lie at the center of gravity of these weights. For example, if the mixture contains more of light G than of light R such that the sum $X + Y + Z$ for light G is twice the corresponding sum for light R, the mixture will have color represented by the point D on the line GR located so that the distance DR is twice the distance GD. Any color on the line from G to R may be obtained by additive mixing properly selected amounts of lights G and R. After obtaining light D in this way, light D may be mixed with some other light, such as that represented by B, to obtain any color along the line BD. It follows that by additive mixing properly selected amounts of the three lights represented by points G, R, and B, one can obtain any color within the triangle GRB. Colors outside this triangle cannot be produced by **additive** mixing of colors GRB.

Since all the real colors are mixtures of the spectral colors, they must lie in the area enclosed by the horseshoe-shaped spectrum locus curve and the straight line connecting the violet and red ends of the horseshoe.

If the triangle GRB is to enclose most of the real colors, the point G should lie close to the spectrum locus point for 520-nm (green) light and the points R and B should be near the red and violet ends of the spectrum locus curve. In this sense, red, green, and blue are desirable primaries.

Equations (13.55) and (13.56) give us a means of calculating the trichromatic coefficients (and therefore the location on the chromaticity diagram) for light with any given spectral distribution; the answer is unique. The reverse process is not unique. Given a light represented by a point such as G that has a specific set of trichromatic coefficients, this light may be matched by a mixture of two monochromatic colors with wavelengths 500 and 530 nm, by a pair with wavelengths 510 and 550 nm, or by several other pairs of monochromatic colors or a variety of continuously variable spectral distributions. These various matches are easily distinguished with the aid of a spectrometer, but to the unaided eye all look the same. There is no unique spectral distribution associated with a given point on the chromaticity diagram.

Color television is an example of additive color mixing. The screen consists of a mosaic (dots) of three different phosphors that can be excited independently; the three phosphors emit three different colors: red, green, and blue. At the customary viewing distance, the spacing of the dots is too small to be resolved by the eye and so the light from several dots is added together to give the sensation of color. The phosphors used give the colors represented by R', G', and B' in Fig. 13.38. The coordinates of these points are $R' = (0.670, 0.330)$, $G' = (0.210, 0.710)$, and $B' = (0.140, 0.080)$. By exciting these three phosphors in the proper ratio, any color within the triangle $R'G'B'$ can be produced.

In Fig. 13.38, we have added a curve representing the color of a blackbody at temperatures from 1000 to 4000 K. It is "cherry red' at 1000 K and progresses through orange toward white as the temperature rises. At very high temperatures (e.g., 15,000 K), the blackbody color is on the blue side of illuminant C.

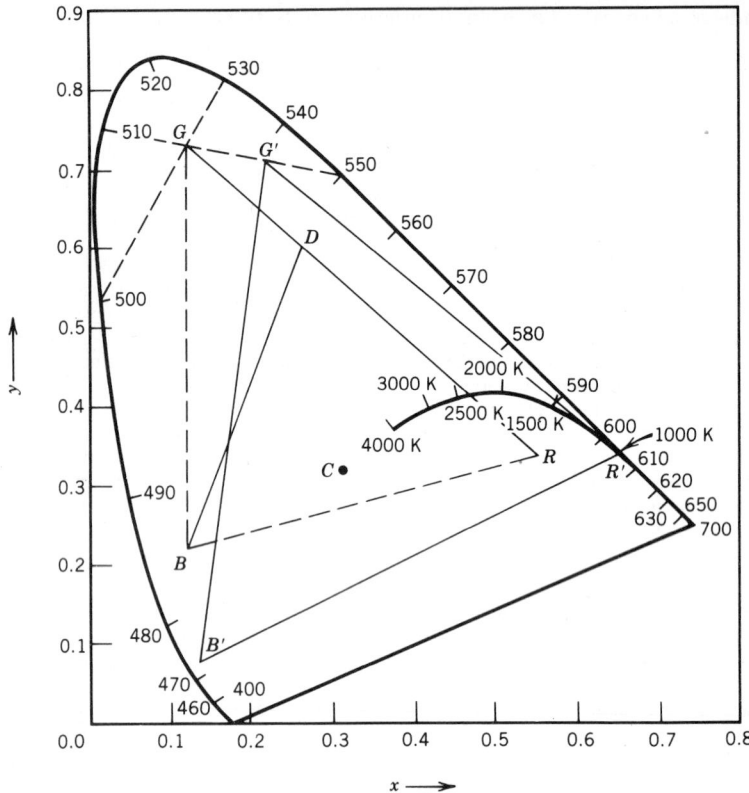

Fig. 13.38 Chromaticity diagram as aid to color mixing. Any color along line *GR* may be produced by adding colors *G* and *R*. Any color within triangle *GBR* may be produced by adding properly selected amounts of light of colors *G*, *B*, and *R*. Color television uses primaries *G'*, *B'*, and *R'*. Curve entering horseshoe near 610-nm locus gives color of blackbody radiation for several temperatures.

Dominant Wavelength and Purity

Dominant wavelength and purity are physical properties of light that evoke the physiological sensations called hue and saturation.

As an illustration, consider again the orange skin. The chromaticity of the light scattered from it is represented by the point *S* in Fig. 13.36. Continuing to use *C* as the white point, we see that a line *C* to *S* may be extended to intersect the spectrum locus at $\lambda = 587$ nm, which is labeled *D* and has coordinates (0.56, 0.44). It follows that the light from the orange skin may be color matched by a mixture of white light (illuminant C) and monochromatic light of wavelength 587 nm. The light from the orange skin is said to have a *dominant wavelength* of 587 nm. The distance from *C* to *S* divided by the distance from *C* through *S* to *D* is 0.83; light from the orange skin is said to have spectral purity *p* of 83%.

Similar procedures show that light represented by the point *G* has dominant wavelength of 519 nm and spectral purity of 80%. Specifying dominant wavelength and spectral purity is an alternate method of locating a point on the chromaticity diagram. For most people, these quantities are easier to interpret than the trichromatic coefficients *x* and *y*.

One runs into trouble for colors in the lower part of the diagram, for example, the color represented by the point *P*. A line from *C* to *P*, if extended, does not intersect the spectrum locus but intersects at *P'*, the straight line closing the bottom of the horseshoe. If the line is extended backward, it intersects the spectrum locus at *P''*, or 550 nm. The color represented by *P* is said to have dominant wavelength of −550 nm, or complementary, 550 nm. The spectral purity is the ratio of the distances $\overline{CP}/\overline{CP'}$, which is about 65% for this case.

Colors commonly called "pastel colors" are of low spectral purity. The color pink is a red of low spectral purity; but in every-day language, low purity is often indicated by some adjectives (e.g., "baby" blue and "apple" green).

Two colors are said to be complementary if they may be added to make white. In terms of the chromaticity diagram, two colors are complementary if the line joining the two points representing them passes through C. The negative, or complementary, greens are called magenta or purple; frequently they are incorrectly called red.

Average Reflectance

We have seen that the dominant wavelength and purity of light reflected by an object depends upon the spectral distribution of the illuminant. The average reflectance also depends upon the illuminant; the average reflectance will be large if the spectral distribution of the illuminant is large at the wavelengths for which the reflectance of the object is also large. The average reflectance depends in this same way upon the spectral sensitivity of the detector. For the light-adapted (photopic) eye, the spectral sensitivity is represented by curve A of Fig. 13.32, which is the same as curve $\bar{y}$ of Fig. 13.35. Since the eye has its maximum sensitivity in the wavelength interval near 550 nm, the averaging process must be weighted in favor of these wavelengths. The average reflectance r_a is calculated as

$$ r_a = \frac{\int_0^\infty r(\lambda)\bar{y}(\lambda)C(\lambda)\,d\lambda}{\int_0^\infty \bar{y}(\lambda)C(\lambda)\,d\lambda} \tag{13.58}$$

where $r(\lambda)$ is the spectral reflectance of the object; $r(\lambda)$ for an orange skin was given in Fig. 13.37. The spectral distribution of the illuminant is $C(\lambda)$, and $\bar{y}(\lambda)$ is the spectral sensitivity of the photopic eye. If some other detector were used, its spectral sensitivity would replace $\bar{y}(\lambda)$ in the equation.

For the orange skin, illuminant C, and the photopic eye we obtain an average reflectance

$$ r_a = 0.26 \quad \text{or } 26\% $$

Subtractive Color Mixing

If a white paper is used as a background for water colors, light must pass through the water color to get to the paper and after diffuse reflection from the paper again pass through the water color. Selective absorption by the dye in the water color gives the scattered light its color. Consider a dye that is absorbent and transmits only a little (say, 10%) for wavelengths shorter than 500 nm but is only slightly absorbent, transmitting 80 or 90%, for wavelengths longer than 500 nm. If this dye is painted on white paper, it will produce a yellow color. Another dye may transmit well for wavelengths shorter than 550 nm and absorb most of the light of longer wavelength; this dye will produce a blue color. If these two dyes do not react chemically and are used one on top of the other (or mixed together) so that light must pass through both of them, then most of the light between 500 and 550 nm will emerge but only a little of the light outside this wavelength interval will emerge. The resultant is a green color. This is a subtractive process in which yellow and blue give green; it must not be confused with the additive processes discussed earlier.

The principles in the water color experiment just described may be illustrated using a slide projector and two pieces of cellophane, one yellow and the other blue. Light projected onto a white wall through one piece of cellophane appears either yellow or blue, but when it passes through both pieces in series, it appears green.

In Fig. 13.39 are the transmission curves of a yellow filter (A), a blue filter (B), and the two in series (G). At each wavelength, the transmission represented by curve G is the product of the transmission for A and the transmission for B.

Color pictures and color slides (i.e., transparencies) use the subtractive method of producing colors. Three dyes are sufficient, and those that are most effective (i.e., produce the widest range of colors) are dyes that control the red, green, and blue. The dye that subtracts the red is blue in color and often is described as cyan. The dye that subtracts the green, leaving the red and blue, is magenta. The third dye subtracts the blue, leaving the green and longer wavelengths unaffected; it is yellow. By varying the concentration of each of these dyes, one can produce all real colors except the highly saturated ones. The available colors are sufficient for all ordinary use since spectrally pure colors are rare outside of the laboratory.

In color printing, it is often necessary to include a black and white image in addition to the three subtractive colors described. The use of black controls the average reflectance of a given area of the picture, not its color.

Good color reproduction in prints or in slides for projection necessitates prior knowledge of the illuminant used in viewing them. Slides are usually projected using a tungsten filament lamp so the dyes are adjusted on the assumption of illuminant A. In many cases, the color purity is increased deliberately and the blues emphasized because it is pleasing to have "bright colors" and "nice blue skys." Color pictures are more likely to be viewed in daylight and therefore are processed for use with illuminant C. If the pictures were produced photographically, the illuminant used for the initial exposure affects the final color.

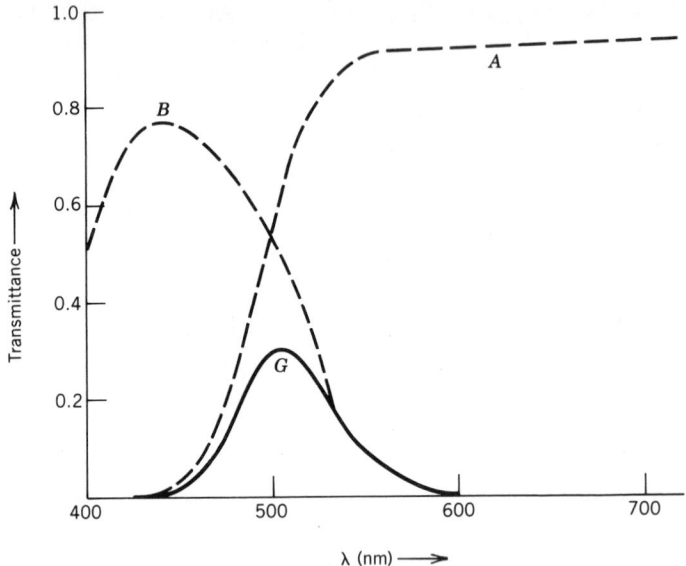

Fig. 13.39 Subtractive combination of yellow filter, *A*, and blue filter, *B*, to produce green, *G*. At each wavelength, *G* is product of *A* and *B* (*A* and *B* are Wratten filters 9 and 47A).

The Munsell System

In matching or specifying paints, the average reflectance is as important as the dominant wavelength and spectral purity. In the Munsell system, paint samples are assigned a "value" from zero to 10. Zero is black (i.e., nonreflecting) and 10 is white (100% diffuse reflecting); the intermediate shades of gray are equally spaced *subjectively*. At each value level is a plane polar arrangement in which the distance from the center indicates saturation or the sensation called **chroma** in subjectively equal steps, from zero or neutral at the center to 12 for a saturated color. Dominant waelength, or its subjective equivalent **hue**, is represented by the horizontal direction on the polar plot; five principal hues—red, yellow, green, blue, and purple—are recognized with the intermediate hues—yellow-red, green-yellow, and so on—making a total of 10 equally spaced hue segments. Each of these is divided into 10 numbered subdivisions (see Fig. 13.40). The Munsell quantities value, hue, and chroma are subjective, corresponding roughly to the physical quantities average reflectance, dominant wavelength, and purity. About a thousand distinguishable paints have been prepared and classified in this way. These samples preserved as an atlas are used to specify paints.

The average reflectance and ICI color specification (under illuminant C) for these samples have been measured, but equating the subjective and the physical quantities is difficult. The Munsell system predates the ICI system.

Photometric Units

In our earlier discussions of sources, such as the blackbody, we measured the radiated energy in physical units (i.e., in watts). The units used are called radiometric units. Long before it became possible to make such measurements, light sources and levels of illumination were compared and measured using the eye as the detector. The eye is quite good at judging the equality of illumination on two adjacent areas. A whole set of units evolved around this process and are still in use; these are known as photometric units, often distinguished from the corresponding radiometric units by including the word *luminous* in the name and the subscript v on the symbol.

The unit of luminous flux is the **lumen**. At the wavelength 555 nm, 1 W produces 683 lm; but since the sensitivity of the photopic eye follows curve *A* of Fig. 13.32 (or curve $\bar{y}$ of Fig. 13.35), the number of lumens per watt at other wavelengths is smaller, as indicated by this curve. For a nonmonochromatic source, the luminance (or luminous radiance) is

$$L_v = 683 \int_0^\infty L(\lambda)\,\bar{y}(\lambda)\,d\lambda \tag{13.59}$$

and has units of nits. A nit is one lumen per square meter per steradian.

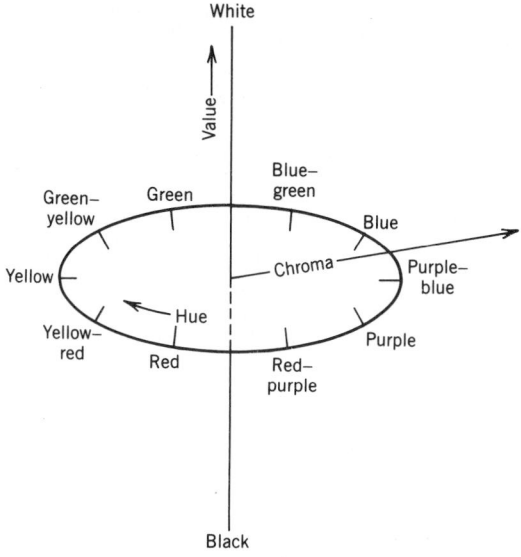

Fig. 13.40 Munsell representation of color data.

A small source that radiates 1 lm into each steradian is said to have a luminous intensity I_v of 1 candela. (It was formerly called a "standard candle".) Common units of illuminance are the foot-candle (1 lm/ft^2), the lux (1 lm/m^2), and the phot (1 lm/cm^2).

13.1.7 Detectors or Optical Transducers

Aside from the eye, there are numerous devices that are used to detect and measure radiant flux. These are usually separated into two groups: (1) thermal detectors and (2) quantum detectors. In thermal detectors, the radiation is absorbed and converted into heat, which raises the temperature of the detector. The temperature change causes a measurable change in some other physical property of the detector (e.g., its resistance). Thermal detectors are sensitive throughout the spectrum. For quantum detectors, the incident light (photons) affect the detector directly (i.e., without heating it); the best known of these is the photoelectric detector in which light causes electrons to be emitted from a surface.

Two thermal detectors are in common use. (1) In the **thermocouple**, two different materials (usually metals) are connected to form a closed circuit. One junction is exposed to the radiation and thereby heated slightly while the other junction is shielded from the radiation and remains at ambient temperature. The temperature difference between the two junctions produces an electromotive force (emf) in the circuit that may be measured. (This is known as the Peltier effect.) (2) The **bolometer** depends upon the change of electrical resistance with temperature. Two identical small detectors are arranged in a Wheatstone bridge circuit; again one detector is exposed to the radiation and the other shielded from the radiation. Small changes in the resistance of the exposed detector are taken as a measure of the incident flux.

If the bolometer element is small and thermally well insulated (except for the thin connecting wires) from its surroundings, then a very small optical or radiant power will produce a relatively large temperature rise; in this sense, the detector is very sensitive. However, it will cool slowly and be unable to respond to rapid fluctuations of the incident flux. For rapid response, the thermal isolation of the bolometer should be reduced. For any given application, one must find the optimum compromise between good sensitivity and rapid response.

Bolometer elements may be small flakes or ribbons of metal; nickel or platinum are commonly used. For metals, the resistance **increases** with increasing temperature, as represented by the equation

$$R = R_0[1 + \alpha(T - T_0)] \tag{13.60}$$

where R is the resistance at temperature T and R_0 is the resistance at ambient temperature T_0. The constant α depends upon the metal used, but values of 0.003–0.004 per degree Kelvin are typical.

Semiconductor bolometer elements (known as thermistors) are also available. For semiconductors, the resistance **decreases** with increasing temperature according to the equation

$$\frac{R(T)}{R(T_0)} = \frac{e^{\beta/T}}{e^{\beta/T_0}}$$ (13.61)

A typical value of β is 3600 K, which gives dR/dT equivalent to $\alpha = -0.04$ per degree Kelvin in Eq. (13.60). In this sense, the thermistor is about 10 times as sensitive as a metal bolometer. Since the resistance decreases with increasing temperature, it must be used with a suitably large series resistance to prevent self-burnout.

In detecting or measuring weak signals, the signal-to-noise ratio becomes important. It is important to realize that a small object (such as a bolometer element) that is in thermal equilibrium with its surroundings is not at a constant temperature but is constantly exchanging energy with its surroundings and fluctuating in temperature. It will experience a root-mean-square random fluctuation of temperature ΔT given by

$$\overline{\Delta T^2} = \frac{kT^2}{C}$$ (13.62)

where k is Boltzmann's constant, T is the absolute temperature of the surroundings, and C is the thermal capacity of the small object. Even if all the amplifying and/or measuring circuits could be noise free, the random temperature fluctuations given by this equation represent unavoidable noise. If the incoming radiation in this ideal case produces a temperature rise equal to ΔT, it is said to have **noise equivalent power** (NEP) and a signal-to-noise ratio of 1.0. In this respect, the thermistor has no advantage over the metal bolometer; it gives larger response to both signal and noise but does not improve the signal-to-noise ratio.

The simplest quantum detector is an evacuated glass tube containing two electrodes. The anode (or positive electrode) collects electrons emitted from the cathode (or negative electrode). Light striking the cathode, called the photocathode, causes the emission of electrons from the cathode; these are collected by the anode and are measured as current in an external circuit. Not every photon causes the emission of an electron, and the term *quantum efficiency* is used to represent the ratio of the number of electrons emitted to the number of photons incident on the cathode. Quantum efficiencies of 10–15% are typical of good photocathode surfaces.

Even in a light beam from a well-stabilized source, the photons do not arrive on the cathode at equally spaced times but at random time intervals. Also, the photons that produce electrons are randomly selected from those that arrive. The subject is usually treated by Poisson statistics, giving the result that if many measurements are made of n (the number of observed electrons in some constant time interval) and the average value of n is $\bar{n}$, then the departures of the individual measurements from the average, $\bar{n}$, will be given by

$$\overline{(n - \bar{n})^2} = \bar{n}$$ (13.63)

This statistical fluctuation of n about its average value $\bar{n}$ is known as photon noise; it arises from the same statistical considerations as "shot noise" in electric circuits. The signal-to-noise ratio is $\sqrt{\bar{n}}$, which may be increased by making $\bar{n}$ larger by (1) increasing the rate at which photons arrive or (2) increasing the observation time for each measurement. The time allotted to each measurement is often built into the associated amplifier, that is, the reciprocal of its bandwidth.

Quantum detectors are wavelength selective. The long-wavelength (low-frequency) limit is determined by the equation

$$\frac{hc}{\lambda} = h\nu \geq \phi e$$

where e is the electronic charge, ϕ is the "work function" of the photo cathode. Here, ϕe is the minimum energy required to remove an electron from the cathode into the vacuum, where ϕ is this energy expressed in electron volts and e is the electronic charge. The short-wavelength (high-frequency) limit is usually determined by the absorption of light by the glass walls of the vacuum tube. The sensitivity of photocathodes is wavelength dependent, and a variety of photocathodes are available having peak sensitivities at different regions of the spectrum.

At room temperature, there will be some emission from the cathode even in the dark. This is known as "dark current" and is due mainly to thermionic emission; it may be reduced by refrigerating the detector.

Some photodetectors realize an amplification of about 10 in the current by having a few torrs of gas in the tube. The electrons are accelerated toward the anode and gain enough energy to ionize some of the gas; the ions then contribute to the current. The recommended cathode-to-anode potential difference must be maintained. Too little will not provide the specified gain; too much will result in a glow discharge independent of light input (and damaging to the cathode). Recommended potential differences are usually 50–100 V.

The photomultiplier is a vacuum tube in which the photocathode is followed by several other electrodes called dynodes. Electrons emitted by the cathode are accelerated to the first dynode, which has a positive potential relative to the cathode on the order of 100 V. Each electron striking the dynode gives up its kinetic energy, thereby causing emission from the dynode of several (e.g., four) slow-moving electrons; this process is called secondary emission. These secondary electrons are accelerated to the next dynode where the process is repeated. It is repeated at each dynode until the electrons are finally collected on the anode and measured in some external circuit. If there is a gain of four electrons at each dynode and there are 10 dynodes, there will be ~ 10^6 electrons at the anode for each electron that left the cathode; under these conditions, one can observe individual photoemissive events and count their number. The photon noise is determined by the number of electrons leaving the cathode; the large gain makes the individual events easier to count; it does not improve the signal-to-noise ratio. The anode is usually at or near ground potential; to get 100 V for each of 10 dynodes requires the cathode to be −1000 V. The gain is sensitive to this voltage. The photomultiplier is useful primarily at low levels of illumination.

Advances in semiconductor science and technology have provided a number of solid quantum detectors that are more rugged and easier to use than the vacuum tube detectors of earlier years.

The electrical behavior of solids is usually described in terms of allowed energy bands for the electrons. For intrinsic semiconductors, the valence band contains all of the valence electrons of the solid and is filled by these electrons. There is no room for any net motion of these electrons. Above the valence band there is an energy region in which no electrons can exist. The width of this forbidden region is called the band gap ϕ (usually expressed in electron volts). Above the band gap is an energy band in which electrons are permitted and in which they are free to move; this is called the conduction band. Normally there are no electrons in the conduction band except for a negligible few that may be thermally excited there from the valence band. Light of wavelength shorter than that given by Eq. (13.63) can be absorbed by the semiconductor. A photon so absorbed can excite an electron from the valence band to the conduction band. An empty space, or hole, is left in the valence band; this hole acts as a small positive charge and can move through the solid in the valence band.

If the semiconductor just described is connected to a current meter and a source of small emf, the observed current is due to the motion of electrons and holes; it will depend upon the irradiance. This process is called photoconductivity, and a semiconductor used in this way is called a photoconductor. There are a number of photoconductors available, each having its characteristic band gap. A semiconductor made by mixing mercury telluride and cadmium telluride will have a band gap dependent upon the relative concentration of the components. In practice, one seeks a detector with band gap a little less than the quantum energy $h\nu$ of the radiation to be detected; in this way the detector becomes blind to undesired radiation at lower frequencies.

The electron has a mean lifetime before it recombines with a hole, after which it no longer contributes to the current. The lifetime is a random variable. This randomness contributes to the noise and is known as generation-recombination noise; it is larger than the photon noise of photoemissive detectors.

An extrinsic semiconductor is an intrinsic semiconductor (such as silicon) into which a small concentration of "impurity" has been introduced. Silicon, as carbon, has four valence electrons, and these are just sufficient to fill the valence band. The band gap for silicon is 1.14 eV. If a small concentration of an element with five valence electrons (e.g., phosphorus or arsenic) is included, each impurity atom will contribute four electrons to the valence band and the fifth electron will be loosely bound to its parent atom. A little energy, < 0.1 eV, will be sufficient to ionize this electron into the conduction band where it can move about and contribute to a measurable current. The ionization energy can be supplied by a photon. Impurities of this type are called donor impurities, and the semiconductor is known as n-type since negative charges (electrons) produce the current.

If the impurity is trivalent, such as boron or aluminum, then there is an electron missing (a hole). This hole is loosely bound to its parent atom. A little energy can move an electron from the valence band into this hole, leaving in the valence band a hole that is free to move. Impurities of this type are known as acceptor impurities, and the semiconductor is p-type since the hole behaves as a positive charge. Photoconductive detectors can be made of either p-type, n-type, or intrinsic semiconductors.

A short length of semiconductor that is p-type at one end but n-type at the other is called a **p-n junction diode**. This device is commonly used as a rectifier because it conducts electricity much better if a small potential difference (a few volts) is applied positive to the p-type and negative to the n-type than if the polarity is reversed. The first case is called forward biased (good conductivity); the second case is reverse biased (poor conductivity).

This device may be used as an optical detector. If the device is reverse biased, it will show a small dark current and, in addition, a current that is nearly linear with the irradiance at the junction between the p region and the n region. It is characterized by shot or photon noise, not generation-recombination noise. This is known as the photoconductive mode. These p-n junction photodiodes usually show faster response to changes of irradiance than photoconductors.

Junction photodiodes are also used in the photovoltaic mode. In this case, there is no external bias provided. A voltmeter (or sometimes an ammeter) is connected to the two ends of the diode. A potential difference (or current) is observed that is dependent on the irradiance at the junction. Usually the output is not a linear function of the irradiance.

The p-i-n photodiode is a p-n junction with an intrinsic region separating the p region from the n region. The radiation is absorbed in the intrinsic region. This device has a low internal capacitance, allowing it to respond quickly to changes of irradiance. It can respond in times shorter than 10^{-9} sec. It also functions over a very large dynamic range of irradiance.

This is certainly not an exhaustive list of the available photodetectors, but perhaps it gives a flavor of the subject. For details, one should consult the manufacturers' specification sheets.

The most commonly used figure of merit for detectors is the specific detectivity, or D^*. By definition,

$$D^* = \frac{\sqrt{A \, \Delta f}}{P_N}$$

where P_N is the incident radiant power that produces a signal just equal to the noise signal, A is the area of the detector, and Δf is the bandwidth of the detector and its amplifying circuits. (These are low-frequency circuits used to monitor fluctuations in radiant power.) Theoretical considerations show that if A and Δf are changed, P_N will change in such a way as to keep D^* constant. Therefore, D^* is characteristic of the process and material being used for detection and is not dependent upon detector geometry or bandwidth.

13.2   ACOUSTICS
by Allan D. Pierce and Yves H. Berthelot

13.2.1   Introduction

Acoustics, the science of sound, is concerned with waves of a mechanical nature. Such waves are found in gases, liquids, solids, granular media, and plasmas. The subject, consequently, has much overlap with many basic areas of physics and engineering science as well as with the environmental sciences of meteorology, geophysics, and oceanography. Because it is also associated with our natural ability to "hear" sound waves, acoustics is related to the art of music, to linguistics, to speech, to psychology, to other social sciences, and to medicine. Acoustics is a very old science, with origins dating back to antiquity. However, similar to many other areas of fundamental physics, quantitative understanding of acoustics was virtually nonexistent before the time of Mersenne (1588–1648), sometimes referred to as the father of acoustics, and of Galileo (1564–1642). Among the prominent past contributors to the understanding of acoustics are Newton, Euler, Lagrange, Laplace, Poisson, George Green, Stokes, Helmholtz, Kirchhoff, Rayleigh, and Sabine.

13.2.2   Frequency Spectrum

One of the principal terms used in describing sound is *frequency*. Although certainly not all acoustic disturbances are purely sinusoidal oscillations of a quantity (such as pressure or displacement) that oscillates with constant frequency about a mean value, it is usually possible to characterize (at least to an order of magnitude) any such disturbance by one or a limited number of representative frequencies, these being the reciprocals of the characteristic time scales. Such a description is highly useful because it gives one some insight into what detailed physical effects are relevant and of what instrumentation is applicable. The simplest sort of acoustic signal would be one where a quantity such as the fluctuating part p of the pressure is oscillating with time t as

$$p = A \sin (\omega t + \phi) \tag{13.64}$$

where the amplitude A and phase constant ϕ are independent of t (but possibly dependent on position). The quantity ω is called the **angular frequency** and has the units of radians divided by time (e.g., rad/sec or simply $\sec^{-1}$ when the unit of time is the second). The number f of repetitions per unit time is what one normally refers to as the **frequency** (without a modifier) such that $\omega = 2\pi f$. The value of f in hertz is the frequency in cycles (repetitions) per second.

The human ear responds almost exclusively to frequencies between roughly 20 Hz and 20 kHz. Consequently, sounds composed of frequencies below 20 Hz are said to be **infrasonic**; those composed of frequencies above 20 kHz are said to be **ultrasonic**. The scope of acoustics, given its general definition, is not limited to only audible frequencies.

The term *frequency spectrum* is often used in relation to sound. Often the use is somewhat loose, but there are circumstances for which the terminology can be made precise. If the fluctuating physical quantity p associated with the acoustic disturbance is a sum of sinusoidal disturbances, each of rms amplitude p_n and frequency f_n, no two frequencies being the same, then the set of numbers $\{p_n^2; n = 1, 2, \ldots\}$, given the set of frequencies to which they correspond, can be taken as a description of the spectrum of the signal. Also, if the sound is made up of many different frequencies, then one can use the sum of the p_n^2 that correspond to those frequencies within a given specified frequency band as a measure of the strength of the signal within that frequency band. This sum divided by the width of the frequency band often approaches a quasi limit as the bandwidth becomes small but with the number of included terms still being moderately large, with this quasi limit being a definite smooth function of the center frequency of the band. This quasi limit is called the **spectral density** p_f^2 of the signal. Although an idealized quantity, it can often be repetitively measured to relatively high accuracy; instrumentation for "measurement" of spectral densities or of integrals of spectral densities over specified (such as octave) frequency bands is widely available commercially.

The utility of the spectral density concept rests on a version of **Parseval's theorem**, which states that when the signal is a sum of discrete frequency components, if averages are taken over a sufficiently long time interval,

$$(p^2)_{av} = \sum p_n^2 \tag{13.65}$$

Consequently, in the quasi limit corresponding to the spectral density description, one has

$$(p^2)_{av} = \int_0^\infty p_f^2(f)\, df \tag{13.66}$$

The quantity $p_f^2(f)\Delta f$ is interpreted as the contribution to the mean squared acoustic pressure from a frequency band of width Δf centered at frequency f.

13.2.3 Basic Equations

Equations of Compressible Fluid Dynamics
For sound in fluids such as air and water, the primary equations governing sound are the partial differential equations governing unsteady compressible fluid flow. In many situations of interest, it is an excellent first approximation to neglect viscosity and thermal conductivity. Also, gravity has a minor influence on sound and can often be ignored in analytical studies. Thus, an idealized set of equations is typically used, these being

$$\frac{\partial \rho}{\partial t} + \nabla \cdot (\rho \mathbf{v}) = 0 \tag{13.67}$$

$$\rho \frac{D\mathbf{v}}{Dt} = -\nabla p \tag{13.68}$$

$$p = p(\rho, s) \tag{13.69}$$

$$\frac{Ds}{Dt} = 0 \tag{13.70}$$

with Stokes's abbreviation

$$\frac{D}{Dt} = \frac{\partial}{\partial t} + \mathbf{v} \cdot \nabla \tag{13.71}$$

for the total time derivative operator. Here ρ is the density of matter, mass per unit volume; $\mathbf{v}$ is the fluid velocity; and p is the total pressure. The conservation of mass is described by Eq. (13.67); the force balance on an infinitesimal fluid element is described by Eq. (13.68); the latter is referred to as

Euler's equation of fluid motion or as Newton's second law for fluids. Equation (13.69) is a thermodynamic equation of state; the total pressure p is given as a unique function of the density ρ and specific entropy s (entropy per unit mass of fluid).

Acoustic Field Equations

Sound results from a time-varying perturbation of pressure around a mean value p_0, the latter representing the ambient pressure in the medium. In air under standard conditions, the ambient pressure p_0 is the atmospheric pressure $p_{atm} \approx 1.01 \times 10^5$ Pa. Let p' represent the deviation between the total pressure p and the ambient pressure p_0; that is, let $p' = p - p_0$. Similarly, one introduces other variables: $\rho' = \rho - \rho_0$ and $s' = s - s_0$. Here, the primed variables are the acoustically induced perturbations. In the simplest idealization, the ambient variables (subscript 0) are considered as being independent of space and time. Also, the ambient fluid velocity v_0 is taken to be identically zero. (One regards p' as small if it is substantially less than $\rho_0 c^2$, and $|v'|$ as small if it is much less than c, where c is the sound speed defined further in what follows. It is not necessary that p' be much less than p_0 and certainly not necessary that $|v'|$ be less than $|v_0|$.)

Substitution of $p = p_0 + p'$, $s = s_0 + s'$, and so on, into the preceding set of equations leads to a set of equations in which the terms can be ordered as being of first, second, and so on, order in the primed variables. For example, the equation $p = p(\rho, s)$ leads to

$$p' = \left(\frac{\partial p}{\partial \rho}\right)_0 \rho' + \left(\frac{\partial p}{\partial s}\right)_0 s' + \frac{1}{2!}\left(\frac{\partial^2 p}{\partial \rho^2}\right)_0 (\rho')^2 + \cdots \tag{13.72}$$

In the indicated partial derivatives, p is regarded as a function of ρ and s, and the corresponding expression is subsequently evaluated with both arguments set to their ambient values. For acoustic disturbances, the usual assumption (**linear acoustics**) is that nonlinear (second-order or higher) terms can be neglected, so terms such as that involving $(\rho')^2$ can be neglected in Eq. (13.72). Since $Ds/Dt = 0$ leads to $\partial s'/\partial t = 0$ in the linear approximation and since s' is presumed to have been identically zero at some time in the past, one can set s' to zero, so one retains only the first term. The coefficient of ρ' in this term is denoted by c^2 since analysis leads to the identification of the square root c of this coefficient as being the **speed of sound**.

The linearization of the fluid dynamic equations yields the **acoustic field equations**

$$\frac{\partial \rho'}{\partial t} + \nabla \cdot (\rho_0 v) = 0 \tag{13.73a}$$

$$\rho_0 \frac{\partial v}{\partial t} + \nabla p' = 0 \tag{13.73b}$$

$$p' = c^2 \rho' \qquad c^2 = \left(\frac{\partial p}{\partial \rho}\right)_0 \tag{13.73c}$$

The derivation assumes that the fluid medium is homogeneous (ambient properties independent of spatial coordinates) and quiescent (ambient properties independent of time; and $v_0 = 0$). The standard form in which these can be rewritten is

$$\frac{\partial p'}{\partial t} + \rho_0 c^2 \nabla \cdot v = 0 \tag{13.74a}$$

$$\rho_0 \frac{\partial v}{\partial t} + \nabla p' = 0 \tag{13.74b}$$

The latter equations have the advantage that they involve only two dependent variables, p' and v; they also are what one would get even were one to allow for the possibility that ρ_0 and c vary with position. Furthermore, they remain good approximations for all but extremely low infrasonic frequencies when one incorporates gravity into the derivation, with account taken of the associated variation of density with height (as for the air in the atmosphere). Because the equations just given do not involve the ambient pressure p_0 or the density perturbation ρ', it is customary to delete the prime on p' and the subscript zero on ρ_0 in discussions of linear acoustics. Thus in the remainder of this chapter, p is the acoustic part of the total pressure and ρ is the ambient density unless stated otherwise.

The Wave Equation

The preceding field equations, upon elimination of **v**, lead to the partial differential equation

$$\rho c^2 \nabla \cdot \left(\frac{1}{\rho} \nabla p \right) - \frac{\partial^2 p}{\partial t^2} = 0 \qquad (13.75)$$

A very good approximation to this when the ambient density ρ is slowly varying with position (but not necessarily constant) is

$$c^2 \nabla^2 \left(\frac{p}{\sqrt{\rho}} \right) - \frac{\partial^2}{\partial t^2} \left(\frac{p}{\sqrt{\rho}} \right) = 0 \qquad (13.76)$$

When the ambient density ρ is constant, one obtains the **wave equation**

$$\nabla^2 p - \frac{1}{c^2} \frac{\partial^2 p}{\partial t^2} = 0 \qquad (13.77)$$

where the field variable is simply p.

The wave equation is sometimes rewritten in a more compact form

$$\Box^2 p = 0 \qquad (13.78)$$

where the operator

$$\Box^2 = \nabla^2 - c^{-2} \frac{\partial^2}{\partial t^2} \qquad (13.79)$$

is called the **d'Alembertian** because d'Alembert was the first (1747) to derive the one-dimensional version of Eq. (13.77) for the case of a vibrating string.

Energy Conservation Corollary

Another consequence of Eqs. (13.74) is the **acoustic energy conservation corollary**

$$\frac{\partial w}{\partial t} + \nabla \cdot \mathbf{I} = 0 \qquad (13.80)$$

with

$$w = \frac{1}{2} \rho v^2 + \frac{1}{2} \frac{1}{\rho c^2} p^2 \qquad (13.81a)$$

and

$$\mathbf{I} = p \mathbf{v} \qquad (13.81b)$$

identified as the **acoustic energy density** and **acoustic intensity** (energy flux vector), respectively.

Sound in Solids

Sound in solids is governed by analogous equations. The idealization of a linear elastic homogeneous solid is often used, and for such a case, the appropriate acoustic equations are

$$\rho \frac{\partial^2 \xi_i}{\partial t^2} = \sum_{j=1}^{3} \frac{\partial \sigma_{ij}}{\partial x_j} \qquad (13.82a)$$

$$\sigma_{ij} = 2\mu \varepsilon_{ij} + \lambda \delta_{ij} \sum_{k=1}^{3} \varepsilon_{kk} \qquad (13.82b)$$

$$\varepsilon_{ij} = \frac{1}{2} \left(\frac{\partial \xi_i}{\partial x_j} + \frac{\partial \xi_j}{\partial x_i} \right) \qquad (13.82c)$$

The first of the preceding relations is known as **Cauchy's equation of motion**. Here ρ is the ambient density; $\xi_i(\mathbf{x}, t)$ is the ith Cartesian component of the displacement of the particle nominally at $\mathbf{x}$ from its ambient position. The quantities σ_{ij} are the **stress tensor** components, while the quantities ε_{ij} are the **strain tensor components**. Equation (13.82b) gives the **stress-strain relations** for an isotropic solid. The symbol δ_{ij} is the **Kronecker delta**, which is 1 if $i = j$ and zero otherwise. The **Lamé constants** λ and μ (the latter being the same as the **shear modulus** G) are related to the **elastic modulus** E and **Poisson's ratio** ν by the relations

$$\lambda = \frac{\nu E}{(1 + \nu)(1 - 2\nu)} \tag{13.83a}$$

$$\mu = G = \frac{E}{2(1 + \nu)} \tag{13.83b}$$

Alternative quantities that are convenient to use are

$$c_1^2 = \frac{\lambda + 2\mu}{\rho} \tag{13.84a}$$

$$c_2^2 = \frac{\mu}{\rho} \tag{13.84b}$$

The quantities c_1 and c_2 are referred to as the **dilatational** and **shear wave** speeds, respectively.

When λ and μ are both constant, the substitution of the stress-strain relation into Cauchy's equation of motion leads to

$$\frac{\partial^2 \boldsymbol{\xi}}{\partial t^2} = \left(c_1^2 - c_2^2 \right) \nabla (\nabla \cdot \boldsymbol{\xi}) + c_2^2 \nabla^2 \boldsymbol{\xi} \tag{13.85}$$

This leads to the ordinary wave equation (13.77) for two special circumstances. If the displacement field is **irrotational** such that $\nabla \times \boldsymbol{\xi} = 0$ (as is so for sound in fluids), then one can set $\boldsymbol{\xi} = \nabla \Phi$, where the displacement potential Φ is constant outside the acoustically perturbed region. In this circumstance, one finds that Φ and the components ξ_i satisfy

$$\nabla^2 \Phi - \frac{1}{c_1^2} \frac{\partial^2 \Phi}{\partial t^2} = 0 \tag{13.86}$$

The other circumstance is when the displacement field is **solenoidal**, such that $\nabla \cdot \boldsymbol{\xi} = 0$. Then it is possible to set $\boldsymbol{\xi} = \nabla \times \boldsymbol{\Psi}$, where the components of the vector $\boldsymbol{\Psi}$ satisfy the wave equation

$$\nabla^2 \boldsymbol{\Psi} - \frac{1}{c_2^2} \frac{\partial^2 \boldsymbol{\Psi}}{\partial t^2} = 0 \tag{13.87}$$

For relatively general circumstances, one may decompose any displacement field as

$$\boldsymbol{\xi} = \nabla \Phi + \nabla \times \boldsymbol{\Psi} \tag{13.88}$$

where Φ satisfies Eq. (13.86) and the components of $\boldsymbol{\Psi}$ satisfy Eq. (13.87).

An **energy conservation corollary** of the form (13.80) also holds for sound in solids. The appropriate identifications for the energy density w and the components I_i of the intensity are

$$w = \frac{1}{2} \rho \sum_i \left(\frac{\partial \xi_i}{\partial t} \right)^2 + \frac{1}{2} \sum_{i,j} \varepsilon_{ij} \sigma_{ij} \tag{13.89a}$$

$$I_i = - \sum_j \sigma_{ij} \frac{\partial \xi_j}{\partial t} \tag{13.89b}$$

13.2.4   Plane, Spherical, and Cylindrical Waves

Plane Waves

A solution of the wave equation that plays a central role in many acoustical concepts is that of a **plane traveling wave**, which is such that all acoustic field quantities vary with time and with one Cartesian coordinate, taken here as x, but are independent of y and z. The Laplacian ∇^2 reduces thus to $\partial^2/\partial x^2$, and the d'Alembertian can be expressed as the product of two first-order operators, so that the wave equation takes the form

$$\left(\frac{\partial}{\partial x} - \frac{1}{c}\frac{\partial}{\partial t}\right)\left(\frac{\partial}{\partial x} + \frac{1}{c}\frac{\partial}{\partial t}\right)p = 0 \tag{13.90}$$

the solution of which is given by

$$p(x,t) = f(x - ct) + g(x + ct) \tag{13.91}$$

where f and g are two arbitrary functions. The quantity $f(x - ct)$ represents a plane wave traveling forward in the positive x direction at a velocity c, while $g(x + ct)$ represents a plane wave traveling backward in the minus x direction, also at a velocity c. For a traveling plane wave, not only the shape, but also the amplitude is conserved during propagation. A typical situation for which wave propagation can be adequately described with traveling plane waves is that of low-frequency sound propagation in a duct. Diverging or converging (focused) waves can often be approximately regarded as planar within regions of restricted extent.

The fluid velocity that corresponds to the preceding plane wave solution has $v_y = 0$, $v_z = 0$, while

$$v_x = \frac{1}{\rho c}f(x - ct) - \frac{1}{\rho c}g(x + ct) \tag{13.92}$$

The general rule that emerges from this is that for a plane wave propagating in the direction corresponding to unit vector $\mathbf{n}$, the acoustic part of the pressure is given by

$$p = f(\mathbf{n} \cdot \mathbf{x} - ct) \tag{13.93}$$

for some generic function $f(\xi)$, while the acoustically induced fluid velocity is

$$\mathbf{v} = \frac{\mathbf{n}}{\rho c}p \tag{13.94}$$

Because the fluid velocity is in the same direction as that of the wave propagation, such waves are said to be **longitudinal**. (Electromagnetic plane waves in free space, on the other hand, are transverse.)

Plane Waves in Solids
Plane acoustic waves in isotropic elastic solids have similar properties. **Dilatational plane waves** governed by Eq. (13.86) and that propagate in the x direction with speed c_1 are described by the relations

$$\xi_x = F(x - c_1 t) \qquad \xi_y = \xi_z = 0 \tag{13.95a}$$

$$\sigma_{xx} = \rho c_1^2 F'(x - c_1 t) \qquad \sigma_{yy} = \sigma_{zz} = \rho\left(c_1^2 - 2c_2^2\right)F'(x - c_1 t) \tag{13.95b}$$

$$\sigma_{xy} = \sigma_{yz} = \sigma_{zx} = 0 \tag{13.95c}$$

Similarly, a shear wave **polarized** in the y direction and propagating in the x direction with speed c_2 is described by

$$\xi_y = F(x - c_2 t) \qquad \xi_x = \xi_z = 0 \tag{13.96a}$$

$$\sigma_{yx} = \rho c_2^2 F'(x - c_2 t) = \sigma_{xy} \tag{13.96b}$$

with all of the other stress components being identically zero.

Dilatational elastic waves are longitudinal, while shear waves are transverse.

Spherical Waves
Another type of propagation of fundamental importance is that of a **spherically symmetric wave** spreading out radially from a source in an unbounded medium. The symmetry implies that the acoustic field variables are a function of only the radial coordinate r and time t. The Laplacian reduces then to

$$\nabla^2 p = \frac{\partial^2 p}{\partial r^2} + \frac{2}{r}\frac{\partial p}{\partial r} = \frac{1}{r}\frac{\partial^2(rp)}{\partial r^2} \tag{13.97}$$

so that the corresponding wave equation becomes

$$\frac{\partial^2(rp)}{\partial r^2} - \frac{1}{c^2}\frac{\partial^2(rp)}{\partial t^2} = 0 \tag{13.98}$$

the solution of which is given by

$$p(r,t) = \frac{f(r - ct)}{r} + \frac{g(r + ct)}{r} \tag{13.99}$$

Causality considerations (no sound before source is turned on) lead to the conclusion that the second term on the right side is not an appropriate solution of the wave equation when the source is concentrated near the origin. The expression $f(r - ct)/r$, which describes the acoustic pressure in an outgoing spherically symmetric wave, has the property that listeners at different radii will receive (with a time shift corresponding to the propagation time) waveforms of the same shape but of different amplitudes. The factor $1/r$ is characteristic of spherical spreading and implies that the peak waveform amplitudes in a spherical wave decrease with radial distance as $1/r$.

Cylindrical Waves

For **cylindrically symmetric waves**, there is no dependence on the azimuthal angle θ or on the axial coordinate z, so the Laplacian in cylindrical coordinates reduces to

$$\nabla^2 = \frac{1}{r}\frac{\partial}{\partial r}\left(r\frac{\partial}{\partial r}\right) \tag{13.100}$$

where r is here the radial distance from the z axis. Consequently, the wave equation takes the form

$$\frac{\partial^2(\sqrt{r}\,p)}{\partial r^2} - \frac{1}{c^2}\frac{\partial^2(\sqrt{r}\,p)}{\partial t^2} + \frac{\sqrt{r}\,p}{4r^2} = 0 \tag{13.101}$$

The waveform shapes of outward-propagating waves governed by this partial differential equation tend to distort with increasing propagation distance, especially at small r. However, at larger values of r, it is often a good approximation to neglect the last term in Eq. (13.101) so one obtains the approximate solution

$$p(r,t) = \frac{f(r - ct)}{\sqrt{r}} \tag{13.102}$$

which is similar to the expression for an outgoing spherical wave, only here the amplitude drops off with r as $1/\sqrt{r}$.

The fluid velocity induced by outgoing spherical or cylindrical waves is not as simply related to the corresponding acoustic pressure as that induced by a plane wave, although symmetry directs that the velocity must be in the appropriate radial direction when the propagation is cylindrically or spherically symmetric. Detailed expressions can be derived using Eqs. (13.74), but a simple approximate result emerges in the limit of large radial distance r:

$$v_r \approx \frac{p}{\rho c} \tag{13.103}$$

just as for a plane wave. (Here large r implies large compared to a characteristic wavelength or compared to c divided by a characteristic angular frequency.)

13.2.5 Speed of Sound

Speed of Sound in Gases

The speed of sound for waves in fluids can either be measured directly or inferred from thermodynamic information regarding the equation of state. For example, air is normally regarded as being an ideal gas, which obeys the relations

$$p = \frac{R_0}{M}\rho T = \rho R T \tag{13.104a}$$

$$p = K(s)\rho^\gamma \tag{13.104b}$$

TABLE 13.3   Acoustic Properties of Some Solids

Materials	ρ (kg/m³)	λ (N/m²), $\times 10^{10}$	μ (N/m²) $\times 10^{10}$,	c_1 (m/s)	c_2 (m/s)
Aluminum	2,700	6.1	2.5	6,410	3,040
Brass (70Cu–30Zn)	8,500	11.3	3.8	4,700	2,110
Copper	8,900	13.1	4.6	5,010	2,270
Cast iron	7,600	6.9	6.0	4,990	2,810
Lead	11,400	3.3	0.54	1,960	690
Nickel	8,900	16.4	8.0	6,030	3,000
Silver	10,400	8.5	2.7	3,660	1,610
Steel, 347 stainless	7,880	11.3	7.57	5,790	3,100
Fused silica	2,190	1.6	3.1	5,970	3,760
Pyrex glass	2,320	2.3	2.5	5,610	3,280
Lucite	1,180	0.562	0.143	2,680	1,100
Polystyrene	1,060	0.319	0.133	2,350	1,120

Adapted from W. P. Mason, *Acoustic Properties of Solids*, in *American Institute of Physics Handbook*, 3rd ed., McGraw-Hill, New York, 1972, pp. 3–104.

Here $R_0 = 8314$ J/(kg · K) is the universal gas constant, M is the average molecular weight (29.0 for air), R is an abbreviation for R_0/M, T is the absolute temperature, γ is the specific heat ratio (approximately 1.4 for air and for gases composed of diatomic molecules), and $K(s)$ is a function of entropy s per unit mass only. (The pressure appearing in Eqs. (13.104) is understood to be the total pressure.)

Equation (13.73), which yields the sound speed c given the relationship (13.104b), leads to

$$c^2 = \left(\frac{\partial p}{\partial \rho}\right)_0 = \frac{\partial\left(K\rho_0^\gamma\right)}{\partial \rho_0} = \gamma K \rho_0^{\gamma-1} = \frac{\gamma p_0}{\rho_0} \tag{13.105}$$

or equivalently, with the use of the ideal-gas equation (13.104a),

$$c = \sqrt{\gamma R T_0} \tag{13.106}$$

where $R = 287$ J/(kg · K) in air. The speed of sound is thus proportional to the square root of the absolute temperature. For dry air at 0°C, $c = 331$ m/s; and for dry air at 20°C, the sound speed is 343 m/s. An approximate rule of thumb when the temperature is close to room temperature is that c increases by 0.6 m/s for every degree centigrade of temperature increase.

Speed of Sound in Water

For liquids, the speed of sound can be regarded as a function of temperature and pressure as well as composition. The speed of sound in **seawater**, for example, differs slightly from that of pure water because of desolved salts (**salinity**). A simple approximate formula for the speed of sound (in meters per second) in seawater is

$$c = 1490 + (3.6)\,\Delta T + (1.6 \times 10^{-6})\,p_{abs} + (1.3)\,\Delta S \tag{13.107}$$

where p_{abs} is the absolute pressure in pascals, while ΔT and ΔS are the deviations of the temperature and salinity from 10°C and from 35 parts per thousand.

Table 13.3 gives the sound velocity in some media commonly encountered in acoustical applications.

13.2.6   Waves of Constant Frequency

Insofar as the governing equations are linear with coefficients independent of time, disturbances that vary sinusoidally with time can propagate without change of frequency. Such sinusoidally varying disturbances of constant frequency have the same repetition period (reciprocal of frequency) at every point, but the phase will in general vary from point to point.

For a plane wave traveling along the positive x axis at a velocity c, one can represent a harmonic acoustic pressure disturbance by

$$p = |P|\sin[k(ct - x) + \phi_0] \tag{13.108}$$

TABLE 13.4 Acoustic Properties of Some Liquids and Gases under Atmospheric Pressure

Medium	Temperature (°C)	Density (kg/m^3)	Sound Velocity (m/s)	Characteristic Impedance (mks rayls), $\times 10^6$
Water (fresh)	20	998	1,481	1.48
Water (sea)	13	1,026	1,500	1.54
Alcohol	20	790	1,150	0.91
Castor oil	20	950	1,540	1.45
Mercury	20	13,600	1,450	19.7
Turpentine	20	870	1,250	1.11
Glycerine	20	1,260	1,980	2.5
Air	0	1.293	331.6	428 Rayls
	20	1.21	343	415
Oxygen	0	1.43	3.17	453
Hydrogen	0	0.09	1269.5	114

From L. E. Kinsler and A. R. Frey, *Fundamentals of Acoustics*, 2nd ed., Wiley, New York, 1962, p. 503.

where $|P|$ is the amplitude of the disturbance, ϕ_0 is a phase constant, and k is a constant termed the **wave number**. The **wavelength** λ is the increment in propagation distance x required to change the argument of the sine by 2π radians, so we identify $k = 2\pi/\lambda$. Also, the increment in t required to change the argument by 2π is the **period** T, which is the reciprocal of the frequency f, so one has the simple rule that

$$\lambda = \frac{c}{f} \qquad (13.109)$$

relating wavelength, sound speed, and frequency. Alternately, since the angular frequency ω is the

TABLE 13.5 Typical Loss Factors (Flexural) at Audio Frequencies for Common Materials

Material	Loss factor η
Aluminum	10^{-4}
Brass, bronze	$< 10^{-3}$
Brick	$1 \times 10^{-2} - 2 \times 10^{-2}$
Concrete	
Light	1.5×10^{-2}
Porous	1.5×10^{-2}
Dense	$1 \times 10^{-2} - 5 \times 10^{-2}$
Copper	2×10^{-3}
Cork	$0.13 - 0.17$
Glass	$0.6 \times 10^{-3} - 2 \times 10^{-3}$
Gypsum board	$0.6 \times 10^{-3} - 3 \times 10^{-2}$
Lead	$0.5 \times 10^{-3} - 2 \times 10^{-3}$
Magnesium	10^{-4}
Masonry blocks	$5 \times 10^{-3} - 7 \times 10^{-3}$
Oak, fir	$0.8 \times 10^{-2} - 1 \times 10^{-2}$
Plaster	5×10^{-3}
Plexiglass, Lucite	$2 \times 10^{-2} - 4 \times 10^{-2}$
Plywood	$1 \times 10^{-2} - 1.3 \times 10^{-2}$
Sand, dry	$0.6 - 0.12$
Steel, iron	$1 \times 10^{-4} - 6 \times 10^{-4}$
Tin	2×10^{-3}
Wood fiberboard	$1 \times 10^{-2} - 3 \times 10^{-2}$
Zinc	3×10^{-4}

Source: E. E. Ungar, "Damping of Panels," in L. L. Beranek, ed., *Noise and Vibration Control*, McGraw-Hill, New York, 1971, p. 453.

From A. D. Pierce, *Acoustics: An Introduction to Its Physical Principles and Applications*, McGraw-Hill, New York, 1981, p. 147.

TABLE 13.6 Flow Resistivity of Porous Materials of Various Densities

Material	Density (kg/m^3)	Flow Resistivity $(10^3\ N\ s/m^4)$
Fiberglas AA	11.2	58
	7.4	34
Fiberglas H-33	41.6	29
Rock wool (Johns-Manville Stonefelt, type M)	54.1	28
	42.6	31
Kaowool Blanket B (Babcock and Wilcox)	50	65
Wood fiber	32.2	39
Ultralite No. 200 (Gustin Bacon Co.)	20.0	7
	100.0	90
Ultrafine No. 1001 (Certain-teed)	40	30
Acoustiform-Mat Ceiling Board (Celotex)	160	70
Thermafiber insulating blanket (U.S. Gypsum)	30	3.5

Source: L. L. Beranek, *J. Acoust. Soc. Am.*, **19**: 556–568 (1947); D. A. Bies, "Acoustical Properties of Porous Materials," in L. L. Beranek, ed., *Noise and Vibration Control*, McGraw-Hill, New York, 1971, pp. 250–251.

From A. D. Pierce, *Acoustics: An Introduction to Its Physical Principles and Applications*, McGraw-Hill, New York, 1981, p. 147.

TABLE 13.7 Representative Absorption Coefficients of Surfaces

Material	Absorption Coefficient α					
	125 Hz	250 Hz	500 Hz	1000 Hz	2000 Hz	4000 Hz
Brick, unglazed	0.03	0.03	0.03	0.04	0.05	0.07
Plaster, gypsum or lime						
On brick	0.01	0.02	0.02	0.03	0.04	0.05
On concrete block	0.12	0.09	0.07	0.05	0.05	0.04
Concrete block						
Coarse	0.36	0.44	0.31	0.29	0.39	0.25
Painted	0.10	0.05	0.06	0.07	0.09	0.08
Plywood, 1-cm-thick paneling	0.28	0.22	0.17	0.09	0.10	0.11
Cork, 2.5 cm thick with airspace behind	0.14	0.25	0.40	0.25	0.34	0.21
Glass, typical window	0.35	0.25	0.18	0.12	0.07	0.04
Drapery						
Lightweight, flat on wall	0.03	0.04	0.11	0.17	0.24	0.35
Heavyweight, draped to half area	0.14	0.35	0.55	0.72	0.70	0.65
Floor						
Concrete	0.01	0.01	0.02	0.02	0.02	0.02
Linoleum on	0.02	0.03	0.03	0.03	0.03	0.02
Heavy carpet on	0.02	0.06	0.14	0.37	0.66	0.65
Wood	0.15	0.11	0.10	0.07	0.06	0.07
Ceiling						
Gypsum board	0.29	0.10	0.05	0.04	0.07	0.09
Plastered	0.14	0.10	0.06	0.05	0.04	0.03
Plywood, 1 cm thick	0.28	0.22	0.17	0.09	0.10	0.11
Suspended acoustical tile, 2 cm thick	0.76	0.93	0.83	0.99	0.99	0.94
Gravel, loose and moist, 10 cm thick	0.25	0.60	0.65	0.70	0.75	0.80
Grass, 5 cm high	0.11	0.26	0.60	0.69	0.92	0.99
Rough soil	0.15	0.25	0.40	0.55	0.60	0.60
Water surface, as in a pool	0.01	0.01	0.01	0.02	0.02	0.03

Source: M. D. Egan, *Concepts in Architectural Acoustics*, McGraw-Hill, New York, 1972, pp. 32–34.

From A. D. Pierce, *Acoustics: An Introduction to Its Physical Principles and Applications*, McGraw-Hill, New York, 1981, p. 256.

rate at which the phase increases with time, one identifies

$$\omega = ck \qquad (13.110)$$

When the disturbance is not a plane traveling wave but is nevertheless of constant frequency, it is convenient to use a **complex-number representation** such that

$$p = \text{Re}\{\hat{p}e^{-i\omega t}\} \qquad (13.111)$$

where $\hat{p}$ is the **complex amplitude** of the acoustic pressure and in general varies with position. For a plane wave traveling in the positive x direction, one would have

$$\hat{p}(x) = Pe^{ikx} \qquad (13.112)$$

where P is a complex number independent of position.

For constant frequency disturbances, the substitution of the complex-number representation into the wave equation yields the **Helmholtz equation**

$$\nabla^2 \hat{p} + k^2 \hat{p} = 0 \qquad (13.113)$$

for the complex pressure amplitude. The solution of this, which corresponds to an outgoing spherically symmetric wave of constant frequency, is

$$p = A\frac{e^{ikr}}{r} \qquad (13.114)$$

while that which corresponds to a symmetrically outgoing wave in cylindrical coordinates is

$$p = AH_0^{(1)}(kr) \qquad (13.115)$$

in terms of the **Hankel function** of the first kind and of zero order. The correspondence with Eq. (13.102) is assured by the asymptotic limit for the Hankel function:

$$\lim_{kr \to \infty} H_0^{(1)}(kr) = \left(\frac{2}{\pi kr}\right)^{1/2} e^{-i\pi/4} e^{ikr} \qquad (13.116)$$

13.2.7   Acoustic Intensity and Power

Many sound fields can be idealized as being **steady**, such that long-term time averages are insensitive to the duration and the center time of the averaging interval. Constant-frequency sounds and continuous noises fall into this category. For such sounds, the time derivative of the acoustic energy density will average out to zero over a sufficiently long time period, so the acoustic energy corollary (13.80) yields the time average relation

$$\nabla \cdot \mathbf{I}_{av} = 0 \qquad (13.117)$$

so the time-averaged vector intensity field is solenoidal in regions that do not contain acoustic sources. This same relation holds for any frequency component of the acoustic field or for the net contribution to the field from any given frequency band. In what follows, the intensity $\mathbf{I}_{av}$ is understood to refer to such a time average for some specified frequency band.

An interpretation of Eq. (13.117) as a **conservation law** follows if we integrate it over an arbitrary fixed volume V within the fluid and reexpress the volume integral of $\nabla \cdot \mathbf{I}_{av}$ as a surface integral by means of **Gauss's theorem**. Doing this gives

$$\iint_S \mathbf{I}_{av} \cdot \mathbf{n} \, dS = 0 \qquad (13.118)$$

where $\mathbf{n}$ is the unit normal vector pointing out of the surface S enclosing V. This relation states that the net acoustic power flowing out of any region not containing sources must be zero on the time average and for any given frequency band.

For a closed surface that encloses one or more sources such that Eq. (13.117) does not apply at every point within the volume, the preceding reasoning allows one to define the time-averaged net

acoustic power of these sources as

$$\mathcal{P}_{av} = \int\int_{S} \mathbf{I}_{av} \cdot \mathbf{n} \, dS \tag{13.119}$$

It follows from Eq. (13.117) that the acoustic power of a source computed in such a manner will be the same for any two choices for the surface S provided that both surfaces enclose the same source and no other sources. The value of the integral is independent of the size and of the shape of S. This result is of fundamental importance for the measurement of source power. Instrumentation to measure the time-averaged intensity directly has become widely available in recent years and is often used in determining the principal noise sources in complicated environments.

13.2.8   Decibels

A large bulk of the literature in acoustics (especially so in noise control, underwater sound, and architectural acoustics) uses **decibel scales** to report amplitude and sound power measurements for steady and quasi-steady acoustic fields.

The **sound pressure level** in decibels is defined by the relation

$$L_p = 10 \log_{10} \frac{p_{rms}^2}{p_{ref}^2} \tag{13.120}$$

where p_{rms} is the root-mean-square acoustic pressure corresponding to the frequency band of interest. The **reference pressure** p_{ref} is 20×10^{-6} Pa for sound in air and 1 μPa for sound in water. For disturbances varying at a fixed frequency, the rms pressure is equal to the peak pressure amplitude divided by $\sqrt{2}$.

In noise control applications, the measured and reported levels often correspond to pressure signals for which different frequency components are weighted differently, rather than uniformly, as in Eq. (13.65). Sound level meters can do such weighting automatically in the process of determining a decibel reading. The three commonly available weightings are the A, B, and C weightings (shown in Fig. 13.41); the A weighting is by far the most frequently used, and the term *sound level* without other qualifications ordinarily implies that an A weighting has been applied.

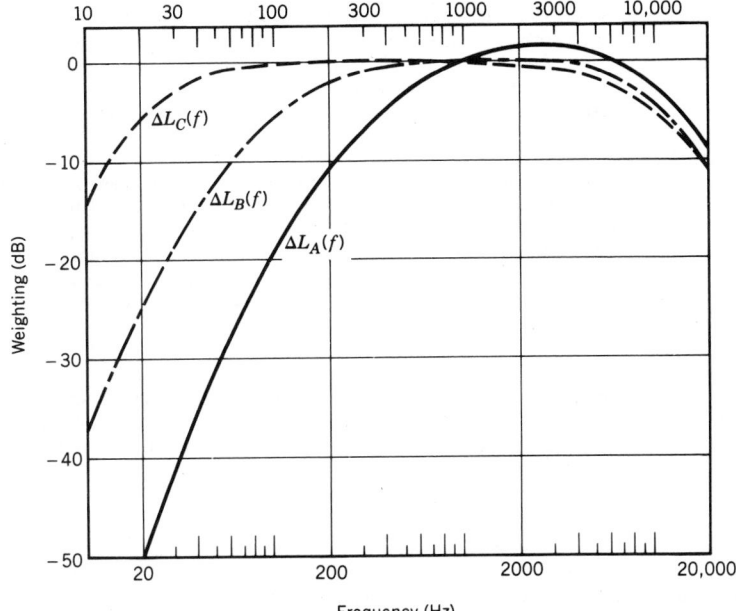

Fig. 13.41 Relative response functions for A, B, and C weightings. (From A. D. Pierce, *Acoustics: An Introduction to Its Physical Principles and Applications*, McGraw-Hill, New York, 1981, p. 67.)

The **sound power level** L_W in decibels is defined as

$$L_W = 10 \log_{10} \frac{\mathscr{P}_{av}}{\mathscr{P}_{ref}} \tag{13.121}$$

where $\mathscr{P}_{av}$ is the time-averaged power and the reference power $\mathscr{P}_{ref}$ is 10^{-12} W. The power used in computing the power level may correspond to a selected frequency band or to a weighted sum over frequencies.

13.2.9 Impedance

Specific Acoustic Impedance

The ratio of the complex amplitude of a sinusoidally varying force to the complex amplitude of the resulting velocity at a point on a vibrating object is called the **mechanical impedance** at that point. It is a complex number and usually a function of frequency. Other definitions of impedance are also in widespread use in acoustics. The **specific acoustic impedance** or **unit area acoustic impedance** $Z_S(\omega)$ for a surface is defined as

$$Z_S(\omega) = \frac{\hat{p}}{\hat{v}_{in}} \tag{13.122}$$

where $\hat{v}_{in}$ is the component of the fluid velocity directed into the surface under consideration. Typically, the specific acoustic impedance, often referred to briefly as impedance without any adjective, is used to describe the acoustic properties of materials. In many cases, surfaces of materials abutting fluids can be characterized as **locally reacting**, so that Z_S is independent of the detailed nature of the acoustic pressure field. In particular, the locally reacting hypothesis implies that the velocity of the material at the surface is unaffected by pressures other than in the immediate vicinity of the point of interest. At a nominally motionless and passively responding surface and when the hypothesis is valid, the appropriate boundary condition on the complex amplitude $\hat{p}$ that satisfies the Helmholtz equation (13.113) is

$$i\omega\rho\hat{p} = -Z_S \nabla\hat{p} \cdot \mathbf{n} \tag{13.123}$$

where $\mathbf{n}$ is the unit normal vector pointing out of the material into the fluid. A surface that is **perfectly rigid** has $|Z_S| = \infty$. The other extreme, where $Z_S = 0$, corresponds to the ideal case of a **pressure release surface**. This is, for example, what is normally assumed for the upper surface of the ocean in underwater sound. Since a **passive surface** absorbs energy from the sound field, the time-averaged intensity component into the surface should be positive or zero. This observation leads to the requirement that the real part (**specific acoustic resistance**) of the impedance should always be nonnegative. The imaginary part (**specific acoustic reactance**) may be either positive or negative.

Characteristic Impedance

For extended substances, a related definition is that of **characteristic impedance** Z_{char}, defined as the ratio of $\hat{p}$ to the complex amplitude $\hat{v}$ of the fluid velocity in the direction of propagation when a plane wave is propagating through the substance. As indicated by Eq. (13.94), this characteristic impedance is ρc regardless of frequency and position in the field. The MKS units (kg m^{-3} m/s) of specific acoustic impedance are referred to as **mks rayls** (after Rayleigh). The characteristic impedance of air under standard conditions is approximately 400 mks rayls and that of water is approximately 1.5×10^6 mks rayls.

Radiation Impedance

The **radiation impedance** Z_{rad} is defined as $\hat{p}/\hat{v}_n$, where $\hat{v}_n$ corresponds to the outward normal component of velocity at a vibrating surface. For outward propagating spherically symmetric waves, where $\hat{p} = Ar^{-1}e^{ikr}$, the surface can be taken as any sphere concentric with the origin, so the radiation impedance at radius r is

$$Z_{rad} = \frac{\hat{p}}{\hat{v}_r} = \rho c \frac{1}{1 + i/kr} \tag{13.124}$$

The time-averaged intensity in the radial direction is therefore

$$I_{av} = \tfrac{1}{2} \operatorname{Re}(p v_r^*) = \tfrac{1}{2}|\hat{p}|^2 \operatorname{Re}(Z_{rad}^{-1}) = \frac{1}{2\rho c}|\hat{p}|^2 \tag{13.125}$$

The latter is the same as results for a plane wave even though for a spherical wave the acoustic pressure is not in phase with the radial velocity except in the asymptotic limit of large radial distances.

Acoustic Impedance

The term *acoustic impedance* Z_A is reserved for the ratio of $\hat{p}$ to the **volume velocity** complex amplitude. Here volume velocity is the net volume of fluid flowing past a specified surface element per unit time in a specified directional sense. One may speak, for example, of the acoustic impedance of an orifice in a wall, of the acoustic impedance at the mouth of a Helmholtz resonator, and of the acoustic impedance at the end of a pipe.

13.2.10   Simple Sources of Sound

Point Sources

Sources in acoustics are incorporated into the governing equations through either **boundary conditions** or **source terms**. Sources that are some distance from bounding surfaces and that are small compared to a wavelength can frequently be described by source terms. The simplest such source would be one that causes a net amount of mass of fluid to flow out of or into a surface that encases it. This mass passing out per unit time divided by the ambient density ρ is a quantity $Q_S(t)$ termed the **source strength function** or the **source volume velocity**. If such a source is concentrated at a point $\mathbf{x}_0$, then the appropriate **inhomogeneous wave equation** that would replace Eq. (13.77) would be

$$\nabla^2 p - \frac{1}{c^2}\frac{\partial^2 p}{\partial t^2} = -\rho \dot{Q}_S(t)\delta(\mathbf{x} - \mathbf{x}_0) \tag{13.126}$$

where $\delta(\mathbf{x})$ is the **Dirac delta function**, which has a volume integral of unity and is concentrated at the point where its argument vanishes.

The solution of the preceding inhomogeneous wave equation for an isolated source in an unbounded region is

$$p = \frac{\rho}{4\pi r}\dot{Q}_S\left(t - \frac{r}{c}\right) = \frac{S[t - (r/c)]}{r} \tag{13.127}$$

where

$$S(t) = \frac{\rho}{4\pi}\frac{d}{dt}Q_S(t) \tag{13.128}$$

is called the **monopole strength**. The replacement of the argument by the **retarded time** $t - [r/c]$ accounts for the transit time lag r/c for the sound to propagate from the source to the listener.

The classic example of a **monopole source** is a pulsating small sphere. In such a case, $Q_S(t)$ can be taken as the time derivative of the instantaneous volume within the sphere. Thus the radiated acoustic pressure is proportional to the **volume acceleration**, the second derivative with respect to time of the sphere's volume, or

$$p = \frac{\rho}{4\pi r}\left(\frac{d^2 V}{dt^2}\right)_{t \to t - [r/c]} \tag{13.129}$$

This rule holds for bodies that are not necessarily spherically shaped provided the largest dimensions are small compared to a characteristic wavelength and provided the acoustic pressure is measured at a sufficiently large distance from the source.

Another example is that of a very small (relative to a wavelength) **piston** mounted in an infinite **rigid baffle**. If the piston has area A and outward normal velocity $v_n(t)$, then symmetry (or, equivalently, the inclusion of an **image source**) requires that the radiated sound be the same as from an isolated source with twice the volume velocity of the piston, so one would have

$$p = \frac{\rho}{2\pi r}A\dot{v}_n\left(t - \frac{r}{c}\right) \tag{13.130}$$

Multiple and Distributed Sources

For assemblages of sources, each concentrated at a point, the generalization of Eq. (13.126) is to replace the right side by a sum of individual source terms; the solution to this inhomogeneous wave equation is

$$p = \sum_n \frac{S[t - (R_n/c)]}{R_n} \qquad (13.131)$$

Here R_n is the distance of the listener from the nth source.

When the source is continuously distributed in space, the point source term is replaced by a smoothly varying function, so the inhomogeneous wave equation is

$$\nabla^2 p - \frac{1}{c^2}\frac{\partial^2 p}{\partial t^2} = -\rho \dot{q}_S(\mathbf{x}, t) = -4\pi s(\mathbf{x}, t) \qquad (13.132)$$

where here the function q_S is termed the **source strength density** (source strength per unit volume). For an unbounded medium, the solution of this latter equation is given by the definite integral

$$p(\mathbf{x}, t) = \int\!\!\int\!\!\int \frac{1}{R} s\left(\mathbf{x}_0, t - \frac{R}{c}\right) dV_0 \qquad (13.133)$$

where the volume integration ranges over source position $\mathbf{x}_0$, and where $R = |\mathbf{x} - \mathbf{x}_0|$ is the distance between listener and source positions.

Thermoacoustic Sources

The differential equation (13.132) arises when one suddenly adds **heat** to a fluid, as with a laser or by combusion, such that the entropy s per unit mass changes according to the thermodynamic relation

$$\rho T \frac{Ds}{Dt} = h \qquad (13.134)$$

and such that, to first order,

$$\frac{\partial p'}{\partial t} = c^2 \frac{\partial \rho'}{\partial t} + \left(\frac{\partial p}{\partial s}\right)_0 \frac{h}{\rho T}$$

$$= c^2 \frac{\partial \rho'}{\partial t} + \frac{c^2 \beta}{c_p} h \qquad (13.135)$$

where h is the heat added per unit time and unit volume, β (equal to $1/T_0$ for an ideal gas) is the volume expansion coefficient, and c_p (equal to $\gamma R/[\gamma - 1]$ for an ideal gas) is the specific heat at constant pressure. Equation (13.135) changes the basic linear acoustic equations (13.74) to

$$\frac{\partial p'}{\partial t} + \rho c^2 \nabla \cdot \mathbf{v} = \frac{c^2 \beta}{c_p} h \qquad (13.136a)$$

$$\rho_0 \frac{\partial \mathbf{v}}{\partial t} + \nabla p' = 0 \qquad (13.136b)$$

so the **energy conservation corollary** (13.80) becomes

$$\frac{\partial w}{\partial t} + \nabla \cdot \mathbf{I} = \frac{\beta}{\rho c_p} ph \qquad (13.137)$$

and the wave equation becomes

$$\nabla^2 p - \frac{1}{c^2}\frac{\partial^2 p}{\partial t^2} = -\frac{\beta}{c_p}\frac{\partial h}{\partial t} \qquad (13.138)$$

The appropriate identification for the monopole strength density function $s(\mathbf{x}, t)$ in the integral expression (13.133) is consequently

$$s(\mathbf{x}, t) = \frac{1}{4\pi}\frac{\beta}{c_p}\frac{\partial h}{\partial t} \qquad (13.139)$$

13.2.11 Dipoles, Quadrupoles, and Spherical Harmonics

Multipole Series
Radiation fields from sources of limited spatial extent in unbounded environments can be described
either in terms of **multipoles** or **spherical harmonics**. That such descriptions are feasible can be
demonstrated with the aid of the constant-frequency version of Eq. (13.133):

$$\hat{p}(\mathbf{x}) = \int\int\int \hat{s}(\mathbf{x}_0)\frac{e^{ikR}}{R}\,dV_0 \tag{13.140}$$

which is the solution of the inhomogeneous Helmholtz equation with a continuous distribution of
monopole sources taken into account.

   The **multipole series** results from the preceding when one expands $R^{-1}e^{ikR}$ in a power series in
the coordinates of the source position $\mathbf{x}_0$ and then integrates term by term. Up through second order
one obtains

$$\hat{p} = \hat{S}\frac{e^{ikr}}{r} - \sum_{\nu=1}^{3} \hat{D}_\nu \frac{\partial}{\partial x_\nu}\frac{e^{ikr}}{r} + \sum_{\mu,\nu=1}^{3} \hat{Q}_{\mu\nu}\frac{\partial^2}{\partial x_\mu\,\partial x_\nu}\frac{e^{ikr}}{r} \tag{13.141}$$

where

$$\hat{S} = \int\int\int \hat{s}(\mathbf{x})\,dV \tag{13.142a}$$

$$\hat{D}_\nu = -\int\int\int x_\nu\hat{s}(\mathbf{x})\,dV \tag{13.142b}$$

$$\hat{Q}_{\mu\nu} = \frac{1}{2!}\int\int\int x_\mu x_\nu\hat{s}(\mathbf{x})\,dV \tag{13.142c}$$

The three terms in Eq. (13.141) are said to be the **monopole, dipole,** and **quadrupole** terms,
respectively. The coefficients $\hat{S}$, $\hat{D}_\nu$, and $\hat{Q}_{\mu\nu}$ are similarly labeled. The D_ν are the components of a
dipole moment vector, while the $Q_{\mu\nu}$ are the components of a **quadrupole moment tensor**. The general
validity of such a description extends beyond the manner of derivation and is not restricted to sound
generated by a continuous-source distribution embedded in the fluid. It applies in particular to the
sound radiated by a vibrating body of arbitrary shape. The directional characteristics of these
elementary sources are shown in Fig. 13.42.

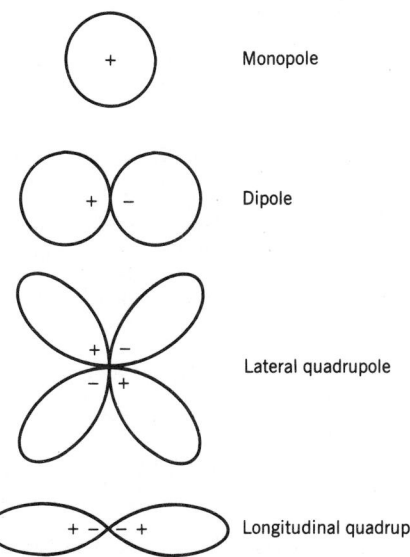

Fig. 13.42 Directional characteristics of elementary acoustic sources.

Acoustically Compact Sources

If the source is **acoustically compact** such that its largest dimension is much less than a wavelength, the multipole series converges rapidly, so one typically only need retain the first nonzero term. Sources exist whose net monopole strength is zero, and sources also exist whose dipole moment vector components all are zero as well. Consequently, compact sources are frequently classed as monopole, dipole, and quadrupole sources. The prototype of a monopole source is a body of oscillating volume. One for a dipole source is a rigid solid undergoing translational oscillations; another would be a vibrating plate or shell whose thickness changes negligibly. In the former case, the detailed theory shows that in the limit of sufficiently low frequency, the dipole moment vector is

$$\hat{D}_\nu = \hat{F}_\mu + m_d \hat{a}_{C,\nu} \tag{13.143}$$

where $\hat{F}_\nu$ is associated with the **force** that the moving body exerts on the surrounding fluid and where $\hat{a}_{C,\nu}$ is associated with the **acceleration** of the geometric center of the body. The quantity m_d is the mass of fluid displaced by the body.

The simplest example of a dipole source is that of a rigid sphere transversely oscillating along the z axis about the origin. If the radius of the sphere is a and if $ka \ll 1$, then the force and acceleration have only a z component, and

$$\hat{F}_z = \tfrac{1}{2} m_d \hat{a}_{C,z} \tag{13.144}$$

The dipole moment when the center velocity has amplitude $\hat{v}_C$ is consequently

$$\hat{D}_z = -\tfrac{3}{2} i\omega \left(\tfrac{4}{3} \rho \pi a^3 \right) \hat{v}_C \tag{13.145}$$

Taking into account that the derivative $\partial r / \partial z$ is $\cos\theta$, one finds the acoustic field from Eq. (13.141) to be

$$\hat{p} = -\hat{D}_z \cos\theta \frac{d}{dr} \frac{e^{ikr}}{r} \tag{13.146}$$

When $kr \gg 1$, this approaches the limiting form

$$\hat{p} \to -ikD_z \cos\theta \frac{e^{ikr}}{r} \tag{13.147}$$

The far-field intensity is $|\hat{p}|^2/(2\rho c)$ and is directed in the radial direction in this asymptotic limit. The drop of intensity as $1/r^2$ with increasing radial distance is the same as for spherical spreading, but the intensity varies with direction as $\cos^2\theta$.

Vibrating bodies that radiate as quadrupole sources (and that therefore have no dipole radiation) usually do so because of symmetry. Vibrating bells and tuning forks are typically quadrupole radiators.

As shown in Fig. 13.42, a dipole source can be represented by two similar monopole sources 180° out of phase with each other and very close together. Since they are radiating out of phase, there is no total mass flow input into the medium. Such a dipole source will have a net acoustic power output substantially lower than that of either of the component monopoles when radiating individually. Similarly, a quadrupole can be formed by two identical but oppositely directed dipoles brought very close together. If the two dipoles have a common axis, then a **longitudinal quadrupole** results; when they are side by side, a **lateral quadrupole** results. In either case, the quadrupole radiation is much weaker than would be that from either dipole when radiating separately.

Spherical Harmonics

The closely related general description of source radiation in terms of **spherical harmonics** results from Eq. (13.140) when one inserts the expansion

$$\frac{e^{ikR}}{R} = \sum_{l=0}^{\infty} (2l+1) j_l(kr_0) h_l^{(1)}(kr) P_l(\cos\Theta) \tag{13.148}$$

where j_l is the **spherical Bessel function** and $h_l^{(1)}$ is the **spherical Hankel function** of order l and of the first kind. (The expansion here assumes $r > r_0$; otherwise, one interchanges r and r_0.) The quantity $P_l(\cos\Theta)$ is the **Legendre function** of order l; the angle Θ is that angle between the directions of x and $\mathbf{x}_0$. Alternately, one has

$$P_l(\cos\Theta) = \sum_{m=-l}^{l} \frac{(l-|m|)!}{(l+|m|)!} Y_l^m(\theta,\phi) Y_l^{-m}(\theta_0,\phi_0) \qquad (13.149a)$$

$$Y_l^m(\theta,\phi) = e^{im\phi} P_l^{|m|}(\cos\theta) \qquad (13.149b)$$

where the functions $P_l^{|m|}(\cos\theta)$ are the **associated Legendre functions**. The quantities $Y_l^m(\theta,\phi)$ are here referred to as the **spherical harmonics**. [The value of $P_0(\cos\Theta)$ is identically 1.]

If such an expansion is inserted into Eq. (13.140) and if r is understood to be sufficiently large that there are no sources beyond that radius, one has

$$\hat{p}(r,\theta,\phi) = a_{00} h_0^{(1)}(kr) + \sum_{l=1}^{\infty} \sum_{m=-l}^{l} a_{lm} h_l^{(1)}(kr) Y_l^m(\theta,\phi) \qquad (13.150)$$

with

$$a_{lm} = ik(2l+1) \frac{(l-|m|)!}{(l+|m|)!} \int\int\int \hat{s}(r_0,\theta_0,\phi_0) j_l(kr_0) Y_l^{-m}(\theta_0,\phi_0) \, dV_0 \qquad (13.151)$$

The indicated volume integrations are to be carried out in spherical coordinates. The general result (13.150) holds for any source of limited extent; any such wave field in an unbounded medium must have such an expansion in terms of spherical Hankel functions and spherical harmonics.

The spherical Hankel functions have the asymptotic (large r) form

$$h_l^{(1)}(kr) \rightarrow (-i)^{(l+1)} \frac{e^{ikr}}{kr} \qquad (13.152)$$

so the acoustic radiation field must asymptotically approach

$$\hat{p} \rightarrow \hat{F}(\theta,\phi) \frac{e^{ikr}}{r} \qquad (13.153)$$

where the function $\hat{F}(\theta,\phi)$ is a function of θ and ϕ that has an expansion in terms of spherical harmonics. In this asymptotic limit the acoustic intensity is in the radial direction and given by

$$I_{r,av} = \frac{1}{2} \frac{|\hat{F}|^2}{\rho c r^2} \qquad (13.154)$$

For fixed θ and ϕ, the time-averaged intensity must asymptotically decrease as $1/r^2$. The coefficient of $1/r^2$ in the preceding describes the far-field **radiation pattern** of the source (in units of watts per steradian).

Although the two types of expansions, multipoles and spherical harmonics, are related, the relationship is not trivial. The quadrupole term in Eq. (13.141), for example, cannot be equated to the sum of the $l = 2$ terms in Eq. (13.150). It is possible to have spherically symmetric quadrupole radiation, so an $l = 0$ term would have to be included.

13.2.12 Helmholtz–Kirchhoff Integral Relations

The Helmholtz–Kirchhoff Corollary

The analysis of radiation of sound from a vibrating body of limited extent in an unbounded region is often aided by an **integral corollary of the Helmholtz equation** (13.113), which dates back to nineteenth-century works of Helmholtz and Kirchhoff. One considers a closed surface S where the outward normal component of the particle velocity has complex amplitude $\hat{v}_n(\mathbf{x}_S)$ and complex pressure amplitude $\hat{p}_S(\mathbf{x}_S)$ at a point $\mathbf{x}_S$ on the surface. For notational convenience, one introduces a quantity

$$\hat{f}_s = -i\omega\rho \hat{v}_n \qquad (13.155)$$

where $\hat{v}_n(\mathbf{x}_S)$ is the normal component $\mathbf{n}(\mathbf{x}_S) \cdot \hat{\mathbf{v}}(\mathbf{x}_S)$ of the complex fluid velocity vector amplitude $\hat{\mathbf{v}}(\mathbf{x})$ at the surface. One can regard $\hat{f}_s$ as a convenient grouping of symbols, either as a constant times the normal velocity or as a constant times the normal acceleration, as the normal component of the apparent body force per unit volume exerted on the fluid at the surface. Because of the latter identification, the use of the symbol f seems appropriate. The subscript S is used to denote values appropriate to the surface.

Then, given that there are no sources outside the surface, a mathematical derivation yields, for the complex pressure amplitude $\hat{p}$ at a point $\mathbf{x}$ *outside* the surface,

$$\hat{p}(\mathbf{x}) = \mathcal{M}\{\mathbf{x}, \hat{p}_S, \hat{f}_S\} \tag{13.156}$$

where

$$\mathcal{M}\{\mathbf{x}, \hat{p}_S, \hat{f}_S\} = \frac{1}{4\pi} \int\int \left[\hat{f}_S(\mathbf{x}'_s) G(\mathbf{x}|\mathbf{x}'_S) + \hat{p}_S(\mathbf{x}'_S)\mathbf{n}(\mathbf{x}'_S) \cdot \{\nabla' G(\mathbf{x}|\mathbf{x}')\}_{\mathbf{x}'=\mathbf{x}'_S} \right] dS' \tag{13.157}$$

The quantity

$$G(\mathbf{x}|\mathbf{x}') = \frac{e^{ikR}}{R} \tag{13.158}$$

is the so-called **free-space Green function** with

$$R = |\mathbf{x} - \mathbf{x}'| \tag{13.159}$$

denoting the distance between "source" and "receiver" points. In the integrand of Eq. (13.157), the point $\mathbf{x}'_S$ (after evaluation of any requisite normal derivatives) is understood to range over the surface S, with the point $\mathbf{x}$ held fixed during the integration. The unit outward normal vector $\mathbf{n}(\mathbf{x}'_S)$ points out of the enclosed volume V at the surface point $\mathbf{x}'_S$.

In Eq. (13.157) one should note that the integral $\mathcal{M}$ is a function of the point $\mathbf{x}$ but a **functional** (function of a function) of the function arguments $\hat{p}_S$ and $\hat{f}_S$.

Integral Relations between Fields at Surfaces

The functions $\hat{p}_S$ and $\hat{f}_S$ cannot be independently prescribed on the surface S. Specifying either one is a sufficient inner boundary condition on the Helmholtz equation. The corollary (13.156) applies only if both functions correspond to a **physically realizable** radiation field **outside** the surface S (see Fig. 13.43). If this is so and if the point $\mathbf{x}$ is formally set to a point **inside** the enclosed volume, then analogous mathematics involving the properties of the Green's function leads to the deduction.

$$\mathcal{M}\{\mathbf{x}, \hat{p}_S, \hat{f}_S\} = 0 \tag{13.160}$$

which is a general relation between the surface value functions $\hat{p}_S$ and $\hat{f}_S$.

Equation (13.156) allows one to derive two additional relations (distinguished by subscripts I and II) between the surface values of $\hat{p}_S$ and $\hat{f}_S$. One results when the off-surface point $\mathbf{x}$ is allowed to approach an arbitrary but fixed surface point $\mathbf{x}_S$. For one of the terms in the integral defining the quantity $\mathcal{M}$, the limit as $\mathbf{x} \to \mathbf{x}_S$ of the integral is not the same as the integral over the limit of the integrand as $\mathbf{x} \to \mathbf{x}_S$. With this subtlety taken into account, one obtains

$$\hat{p}_S(\mathbf{x}_S) - \mathcal{L}_I\{\mathbf{x}_S, \hat{p}_S\} = \mathcal{H}_I\{\mathbf{x}_S, \hat{f}_S\} \tag{13.161}$$

where

$$\mathcal{L}_I\{\mathbf{x}_S, \hat{p}_S\} = \frac{1}{2\pi} \int\int \hat{p}(\mathbf{x}'_S)\mathbf{n}(\mathbf{x}'_S) \cdot \{\nabla' G(\mathbf{x}_S|\mathbf{x}')\}_{\mathbf{x}'=\mathbf{x}'_S} dS' \tag{13.162}$$

$$\mathcal{H}_I\{\mathbf{x}_S, \hat{f}_S\} = \frac{1}{2\pi} \int\int \hat{f}(\mathbf{x}'_S) G(\mathbf{x}'_S|\mathbf{x}_S) dS' \tag{13.163}$$

such that the symbols $\mathcal{L}_I$ and $\mathcal{H}_I$ can be regarded as linear operators that operate on the surface values of $\hat{p}_S$ and $\hat{f}_S$, respectively, with the resultant in each case being a function of the position of the surface point $\mathbf{x}_S$.

The second type of surface relationship is obtained by taking the gradient of both sides of Eq. (13.156), subsequently setting $\mathbf{x}$ to $\mathbf{x}_S + \varepsilon\mathbf{n}(\mathbf{x}_S)$, where $\mathbf{x}_S$ is an arbitrary point on the surface, taking the dot product with $\mathbf{n}(\mathbf{x}_S)$, then taking the limit as $\varepsilon \to 0$. The order of the processes, doing the integration and taking the limit, cannot be blindly interchanged, and some mathematical manipulations making use of the properties of the Green's function are necessary before one can obtain a

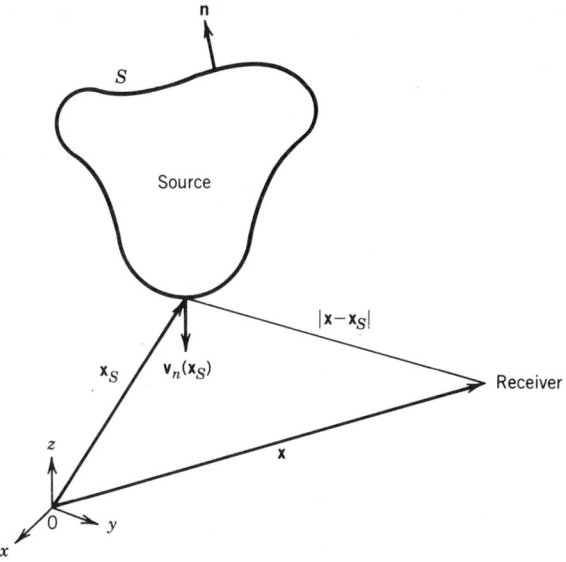

Fig. 13.43 Geometry for radiation problems.

relation in which all integrations are performed after all necessary limits are taken. The result is

$$-\mathscr{L}_{\mathrm{II}}\{\mathbf{x}_S, \hat{p}_S\} = \hat{f}_S(\mathbf{x}_S) + \mathscr{H}_{\mathrm{II}}\{\mathbf{x}_S, \hat{f}_S\}$$ (13.164)

where

$$\mathscr{L}_{\mathrm{II}}\{\mathbf{x}, \hat{p}_S\} = [\mathbf{n}(\mathbf{x}_S) \times \nabla] \cdot \frac{1}{2\pi} \int\int [\mathbf{n}(\mathbf{x}_S') \times \nabla' \hat{p}_S(\mathbf{x}_S')] G(\mathbf{x}_S|\mathbf{x}_S') \, dS'$$

$$+ \frac{k^2}{2\pi} \int\int \mathbf{n}(\mathbf{x}_S) \cdot \mathbf{n}(\mathbf{x}_S') \hat{p}_S(\mathbf{x}_S') G(\mathbf{x}_S|\mathbf{x}_S') \, dS'$$ (13.165)

and

$$\mathscr{H}_{\mathrm{II}}\{\mathbf{x}_S, \hat{f}_S\} = \frac{1}{2\pi} \int\int \hat{f}_S(\mathbf{x}_S')\mathbf{n}(\mathbf{x}_S) \cdot \{\nabla G(\mathbf{x}|\mathbf{x}_S')\}_{\mathbf{x}=\mathbf{x}_S} \, dS'$$ (13.166)

In regard to Eq. (13.165), one should note that the operator $\mathbf{n}(\mathbf{x}_S) \times \nabla$ involves only derivatives tangential to the surface, so that the integral on which it acts need only be evaluated at surface points $\mathbf{x}_S$.

A variety of numerical techniques have been used and discussed in the recent literature to solve either (13.160), (13.161), or (13.164) or some combination of these for the surface pressure $\hat{p}_S$ given the surface force function $\hat{f}_S$. Once this is done, the radiation field at any external point $\mathbf{x}$ is found by numerical integration of the corollary integral relation (13.156).

13.2.13   Attenuation of Sound

Attenuation Coefficient
Plane waves of constant frequency propagating through bulk materials have amplitudes that typically decrease exponentially with increasing propagation distance such that

$$|\hat{p}(x)| = |\hat{p}(0)|e^{-\alpha x}$$ (13.167)

The quantity α is the **plane wave attenuation coefficient** and has the units of nepers per meter; it is an intrinsic frequency-dependent property of the material. This exponential decrease of amplitude is called attenuation or absorption of sound and is associated with the transfer of acoustic energy to the internal energy of the material. If the linear equations of acoustics, (13.73) and (13.74), are extended and supplemented to include the mechanics of absorption processes, then the energy conservation law that results as a corollary will be of the form

$$\frac{\partial w}{\partial t} + \nabla \cdot \mathbf{I} = -\mathcal{D} \tag{13.168}$$

where $\mathcal{D}$ is a nonnegative expression that represents the energy dissipated per unit time and volume. For constant-frequency plane waves propagating in the x direction, the time average of $\partial w/\partial t$ is zero, while the time averages of $\mathbf{I}$ and $\mathcal{D}$ will both be quadratic in the wave amplitude $|\hat{p}|$. The identification of α is then such that

$$\mathcal{D}_{av} = -2\alpha I_{av} \tag{13.169}$$

and such that (13.168) yields

$$\frac{dI_{av}}{dx} = -2\alpha I_{av} \tag{13.170}$$

so the intensity has the exponential decrease

$$I_{av}(x) = I_{av}(0)e^{-2\alpha x} \tag{13.171}$$

The decrease of pressure level (dB) over a propagation distance of x is consequently

$$\Delta L_p = -20\log_{10}\frac{I_{av}(x)}{I_{av}(0)} = (40\log_{10}e)\alpha x \tag{13.172}$$

where the coefficient in the latter expression is approximately 17.37.

Classical and Relaxation Models for Attenuation

The actual value of α is rarely accounted for by shear viscosity and thermal conduction alone. Explanation of sound attenuation in pure water requires the inclusion of a bulk viscosity into the analysis. For seawater, the chemical relaxation effect of dissolved salts is a dominant contributor to the attenuation coefficient. In air, the relaxation of internal vibrations of diatomic molecules is a dominant contributor. The relatively small (and highly variable) number of water vapor molecules in the air has a significant effect on attenuation because collisions of diatomic molecules with H_2O molecules are much more likely to cause a transition between one internal vibrational quantum state and another.

The general theory of sound propagation in a fluid with viscosity, thermal conduction, and internal relaxation taken into account leads for cases of typical interest to the approximate expression

$$\alpha = \alpha'_{cl} + \sum_{\nu}\alpha_{\nu} \tag{13.173}$$

where

$$\alpha'_{cl} = \frac{\omega^2}{c^3}\delta'_{cl} \tag{13.174}$$

is the classical attenuation coefficient, which varies as the square of the frequency. (The prime implies that bulk viscosity has been included.) The quantities α_{ν} correspond to the various internal relaxation processes, where

$$\alpha_{\nu} = \frac{1}{\lambda}(\alpha_{\nu}\lambda)_{max}\frac{2\omega\tau_{\nu}}{1 + (\omega\tau_{\nu})^2} \tag{13.175}$$

The various quantities that appear in these expressions are discussed in what follows.

The quantity δ'_{cl} is the diffusion parameter (units of cubic meters per second) for classical absorption processes and given by

$$2\rho\delta'_{cl} = \tfrac{4}{3}\mu_S + \mu_B + (\gamma - 1)\frac{\kappa}{c_p} \tag{13.176}$$

with μ_S denoting the shear viscosity, μ_B denoting the bulk viscosity, κ denoting the thermal conductivity, and γ denoting the specific heat ratio.

In the expression (13.175) for α_ν, the quantity $(\alpha_\nu \lambda)_{max}$ is a frequency-independent constant that corresponds to the maximum absorption per wavelength associated with the ν-type relaxation process. The quantity τ_ν is the characteristic relaxation time for this process.

Attenuation in Air

Regarding the material constants needed to evaluate the preceding expressions, those for **air** are (approximately)

$$\mu_S = \mu_0 \left(\frac{T}{T_0} \right)^{3/2} \frac{T_0 + T_S}{T + T_S} \tag{13.177}$$

$$\mu_B = 0.6\mu_S \tag{13.178}$$

$$\kappa = \kappa_0 \left(\frac{T}{T_0} \right)^{3/2} \frac{T_0 + T_A e^{-T_B/T_0}}{T + T_A e^{-T_B/T}} \tag{13.179}$$

where $T_S = 110.4$ K, $T_A = 245.4$ K, $T_B = 27.6$ K, $T_0 = 300$ K, $\mu_0 = 1.846 \times 10^{-5}$ kg/m s, and $\kappa_0 = 2.624 \times 10^{-2}$ W/m K. Also one has

$$(\alpha_\nu \lambda)_{max} = \frac{\pi}{2} \frac{(\gamma - 1) c_{v\nu}}{c_p} \tag{13.180}$$

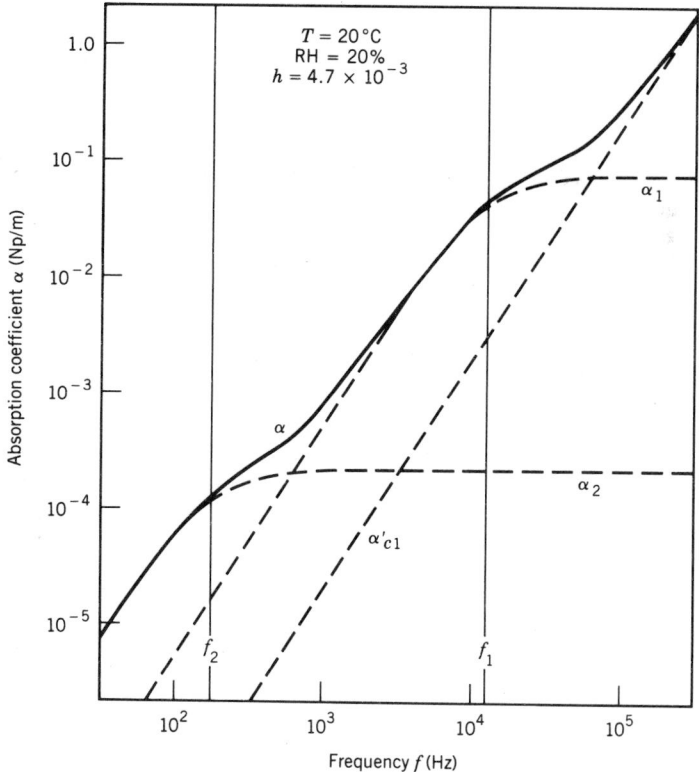

Fig. 13.44 Log-log plot of sound absorption coefficient vs. frequency for sound in air at 20°C at 1 atm pressure and with water vapor fraction h of 4.676×10^{-3} (RH = 20%). Two relaxation frequencies are 12,500 (O_2) and 173 Hz (N_2). (From A. D. Pierce, *Acoustics: An Introduction to Its Physical Principles and Applications*, McGraw-Hill, New York, 1981, p. 560.)

where

$$c_{v\nu} = \frac{n_\nu}{n} R \left(\frac{T_\nu^*}{T} \right)^2 e^{-T_\nu^*/T} \qquad (13.181)$$

and $R = 287$ J/kg K, $T_1^* = 2239$ K, $T_2^* = 3352$ K, $n_1/n = 0.21$, $n_2/n = 0.78$, and $\gamma = 1.4$. (The subscripts 1 and 2 correspond to oxygen and nitrogen, respectively.) The corresponding relaxation times are computed from the formulas

$$\frac{p_{\text{ref}}}{p_{\text{abs}}} \frac{1}{2\pi\tau_1} = 24 + 4.41 \times 10^6 h \frac{0.05 + 100h}{0.391 + 100h} \qquad (13.182a)$$

$$\frac{p_{\text{ref}}}{p_{\text{abs}}} \frac{1}{2\pi\tau_2} = \left(\frac{T_{\text{ref}}}{T} \right)^{1/2} (9 + 3.5 \times 10^4 h e^{-F}) \qquad (13.182b)$$

$$F = 6.142 \left[\left(\frac{T_{\text{ref}}}{T} \right)^{1/3} - 1 \right] \qquad (13.183)$$

$$h = 10^{-2} (\text{RH}) \frac{p_{\text{vp}}(T)}{p} \qquad (13.184)$$

where $p_{\text{ref}} = 1.013 \times 10^5$ Pa and $T_{\text{ref}} = 293.16$ K. Here h is the fraction of air molecules that are H_2O molecules, RH is the relative humidity expressed as a percentage, and p_{vp} is the vapor pressure of water at temperature T. A table of the vapor pressure of water may be found in various references; some representative values are 872, 1228, 1705, 2338, 4243, and 7376 Pa at temperatures of 5, 10, 15, 20, 30, and 40° C. Fig. 13.44 shows the variation of α with frequency.

Attenuation in Water
For pure water, the viscosity and thermal conductivity are given approximately by

$$\mu_S = 1.002 \times 10^{-3} e^{-0.0248\Delta T} \qquad (13.185)$$

$$\mu_B = 3\mu_S \qquad (13.186)$$

$$\kappa = 0.597 + 0.0017\Delta T - 7.5 \times 10^{-6} (\Delta T)^2 \qquad (13.187)$$

where ΔT is the temperature relative to 20°C. The thermal conduction, however, is unimportant for plane wave absorption because the specific heat ratio γ is very close to unity at ordinary temperatures for water. Also, the α_ν terms in Eq. (13.173) need not be included in the calculation of the absorption coefficient α.

For seawater, approximate formulas derived from a combination of experiment and theory are

$$\frac{\alpha'_{cl}}{f^2} = \left(55.9 - 2.37 T_C + 0.0477 T_C^2 - 0.000348 T_C^3 \right)$$
$$\times \left(1 - 3.84 \times 10^{-4} P_{\text{atm}} + 7.57 \times 10^{-8} P_{\text{atm}}^2 \right) \times 10^{-15} \qquad (13.188)$$

$$\frac{2}{c} (\alpha_1 \lambda)_{\text{max}} = \frac{S}{35} (1.03 + 0.0236 T_C - 0.000522 T_C^2) \times 10^{-8} \qquad (13.189a)$$

$$\frac{2}{c} (\alpha_2 \lambda)_{\text{max}} = \frac{S}{35} (5.62 + 0.0752 T_C)$$
$$\times \left(1 - 10.3 \times 10^{-4} P_{\text{atm}} + 3.7 \times 10^{-7} P_{\text{atm}}^2 \right) \times 10^{-8} \qquad (13.189b)$$

$$f_1 = \frac{1}{2\pi\tau_1} = 1320 T e^{-1700/T} \qquad (13.190a)$$

$$f_2 = \frac{1}{2\pi\tau_2} = 15.5 \times 10^6 T e^{-3052/T} \qquad (13.190b)$$

Here T_C is temperature in degrees centigrade, T is absolute temperature, while P_{atm} is the absolute temperature in atmospheres; S is the salinity in parts per thousand. The subscripts 1 and 2 refer to boric acid, $B(OH)_3$, and magnesium sulfate, $MgSO_4$. In the preceding equations, the units of α are in nepers per meter, those of c are meters per second, and those of the frequency f are hertz.

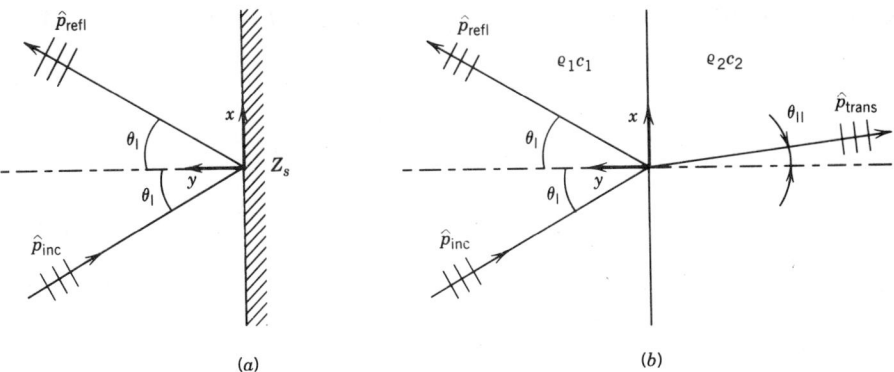

Fig. 13.45 Geometry associated with (a) reflection of plane waves off locally reacting surface of finite impedance Z_S and (b) reflection and transmission of plane waves between two fluid media of characteristic impedance $\rho_1 c_1$ and $\rho_2 c_2$.

13.2.14   Reflection and Transmission of Sound

Oblique Reflection at a Surface

When a plane wave reflects at a surface with finite specific acoustic impedance Z_S (see Fig. 13.45), a reflected wave is formed such that the angle of incidence θ_I equals the angle of reflection (**law of mirrors**). Here both angles are reckoned from the line normal to the surface and correspond to the directions of the two waves. If one takes the y axis as pointing out of the surface and the surface as coinciding with the $y = 0$ plane, then an incident plane wave propagating obliquely in the positive x direction will have a complex pressure amplitude given by

$$\hat{p}_{in} = \hat{f} e^{ik_x x} e^{-ik_y y} \tag{13.191}$$

where $\hat{f}$ is a constant. (For transient reflection, the quantity $\hat{f}$ can be taken as the Fourier transform of the incident pressure pulse at the origin.) The two indicated wave number components are $k_x = k \sin \theta_I$ and $k_y = k \cos \theta_I$. The reflected wave has a complex pressure amplitude given by

$$\hat{p}_{refl} = \mathscr{R}(\theta_I, \omega) \hat{f} e^{ik_x x} e^{ik_y y} \tag{13.192}$$

where the quantity $\mathscr{R}(\theta_I, \omega)$ is the **pressure amplitude reflection coefficient**.

Analysis that makes use of the boundary condition (13.123) leads to

$$\mathscr{R}(\theta_I, \omega) = \frac{\xi(\omega) \cos \theta_I - 1}{\xi(\omega) \cos \theta_I + 1} \tag{13.193}$$

where

$$\xi(\omega) = \frac{Z_S}{\rho c} \tag{13.194}$$

is the ratio of the specific acoustic impedance of the surface to the characteristic impedance of the medium.

Reflection at an Interface between Two Fluids

The preceding relations also apply, with an appropriate identification of the quantity Z_S, to **sound reflection at an interface between two fluids** with different sound speeds and densities. Translational symmetry requires that the disturbance in the second fluid have the same apparent phase velocity (ω/k_x) (**trace velocity**) along the x axis as does the disturbance in the first fluid. This requirement is known as the **trace velocity matching principle** and leads to the observation that k_x is the same in both fluids. One distinguishes two possibilities: The trace velocity is higher than the sound speed c_2 or lower than c_2.

For the first possibility, one has

$$c_2 < \frac{c_1}{\sin \theta_I} \tag{13.195}$$

and a propagating plane wave (transmitted wave) is excited in the second fluid, with complex pressure amplitude

$$\hat{p}_{\text{trans}} = \mathcal{T}(\omega, \theta_{\text{I}}) \hat{f} e^{ik_x x} e^{ik_2 y \cos \theta_{\text{II}}} \tag{13.196}$$

where $k_2 = \omega/c_2$ is the wave number in the second fluid and θ_{II} (angle of refraction) is the angle at which the transmitted wave is propagating. The trace velocity matching principle leads to Snell's law:

$$\frac{\sin \theta_{\text{I}}}{c_1} = \frac{\sin \theta_{\text{II}}}{c_2} \tag{13.197}$$

The change in propagation direction from θ_{I} to θ_{II} is the phenomenon of refraction. The requirement that the pressure be continuous across the interface yields the relation

$$1 + \mathcal{R} = \mathcal{T} \tag{13.198a}$$

while the continuity of the normal component of the fluid velocity yields

$$\frac{\cos \theta_{\text{I}}}{\rho_1 c_1} (1 - \mathcal{R}) = \frac{\cos \theta_{\text{II}}}{\rho_2 c_2} \mathcal{T} \tag{13.198b}$$

and from these one derives

$$\mathcal{R} = \frac{Z_{\text{II}} - Z_{\text{I}}}{Z_{\text{II}} + Z_{\text{I}}} \tag{13.199}$$

with

$$Z_{\text{I}} = \frac{\rho_1 c_1}{\cos \theta_{\text{I}}}, \qquad Z_{\text{II}} = \frac{\rho_2 c_2}{\cos \theta_{\text{II}}} \tag{13.200}$$

The other possibility, which is the opposite of that in Eq. (13.195), can only occur when $c_2 > c_1$ and, moreover, only if θ_{I} is greater than the critical angle θ_{cr}, where

$$\theta_{\text{cr}} = \arcsin \frac{c_1}{c_2} \tag{13.201}$$

In this circumstance, an inhomogeneous plane wave propagating in the x direction but dying out exponentially in the positive y direction is excited in the second medium. Instead of Eq. (13.196), one has

$$\hat{p}_{\text{trans}} = \mathcal{T}(\omega, \theta_{\text{I}}) \hat{f} e^{ik_x x} e^{-\beta k_2 y} \tag{13.202}$$

with

$$\beta = \left[\left(\frac{c_2}{c_1} \right)^2 \sin^2 \theta_{\text{I}} - 1 \right]^{1/2} \tag{13.203}$$

The previously stated equations governing the reflection and transmission coefficients are still applicable provided one replaces $\cos \theta_{\text{II}}$ by $i\beta$. This causes the magnitude of the reflection coefficient $\mathcal{R}$ to becomes unity so the time-averaged incident energy is totally reflected. Acoustic energy is present in the second fluid, but its time average over a wave period stays constant once the steady state is reached.

Theory of the Impedance Tube

Impedance tubes, shown in Fig. 13.46, are commonly used in the measurement of specific acoustic impedances; the underlying theory is based for the most part on Eqs. (13.191)–(13.194). The incident and the reflected waves propagate along the axis of a cylindrical tube with the sample surface at one end. A loudspeaker at the other end creates a sinusoidal pressure disturbance that propagates down the tube. Reflections from the end covered with the test material create an incomplete standing-wave pattern inside the tube. The wavelength of the sound emitted by the source can be adjusted, but it should be kept substantially larger than the pipe diameter so that the plane wave assumption holds.

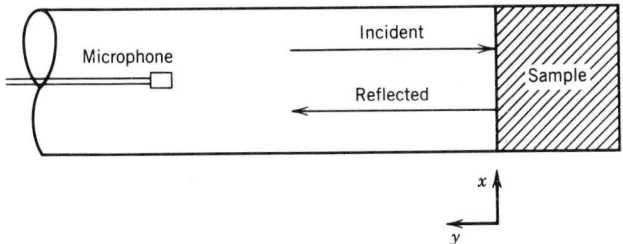

Fig. 13.46   Schematic diagram of impedance tube.

With k_x identified as being zero, the complex amplitude that corresponds to the sum of the incident and reflected waves has an absolute magnitude given by

$$|\hat{p}| = |\hat{f}| |1 + \mathcal{R}e^{2iky}| \qquad (13.204)$$

where y is now the distance in front of the sample. The second factor varies with y and repeats at intervals of a half-wavelength and varies from a minimum value of $1 - |\mathcal{R}|$ to a maximum value of $1 + |\mathcal{R}|$. Consequently, the ratio of the peak acoustic pressure amplitude $|\hat{p}|_{max}$ (which occurs at one y position) to the minimum acoustic pressure amplitude $|\hat{p}|_{min}$ (which occurs at a position a quarter-wavelength away) determines the magnitude of the reflection coefficient via the relation

$$\frac{|\hat{p}|_{min}}{|\hat{p}|_{max}} = \frac{1 - |\mathcal{R}|}{1 + |\mathcal{R}|} \qquad (13.205)$$

The phase δ of the reflection coefficient can be determined with use of the observation that the peak amplitudes occur at y values where $\delta + 2ky$ is an integer multiple of 2π, while the minimum amplitudes occur where it is π plus an integer multiple of 2π. Once the magnitude and phase of the reflection coefficient are determined, the specific acoustic impedance can be found from Eqs. (13.193) and (13.194).

Transmission through Walls and Slabs

The analysis of transmission of sound through a wall or a partition is often based on the idealization that the wall is of unlimited extent. If the incoming plane wave has an angle of incidence θ_I (see Fig. 13.47) and if the fluid on the opposite side of the wall has the same sound speed, then the trace velocity-matching principle requires that the transmitted wave be propagating in the same direction. A common assumption when the fluid is air is that the compression in the wall is negligible, so the wall is treated as a **slab** that has a uniform velocity v_{sl} throughout its thickness. The slab moves under the influence of the incident, reflected, and transmitted sound fields according to the relation (corresponding to Newton's second law)

$$m_{sl}\frac{\partial v_{sl}}{\partial t} = p_{front} - p_{back} + \text{bending term} \qquad (13.206)$$

where m_{sl} is the mass per unit surface area of the slab. The front side is here taken as the side facing the incident wave; the transmitted wave propagates away from the back side. The "bending term" (discussed further in what follows) accounts for any tendency of the slab to resist bending. If the slab is regarded as nonporous, then the normal component of the fluid velocity both at the front and the back is regarded the same as the slab velocity itself. If it is taken as **porous** then these continuity equations are replaced by the relations

$$v_{front} - v_{sl} = v_{back} - v_{sl} = \frac{1}{R_f}(p_{front} - p_{back}) \qquad (13.207)$$

where R_f is the **specific flow resistance**. The latter can be measured in steady flow for specific materials. For a homogeneous material, it is given by the product of the slab thickness h and the **flow resistivity**, the latter being a commonly tabulated property of porous materials.

In general, when one considers the reflection at and transmission through a slab, one can define a **slab specific impedance** Z_{sl} such that, with regard to complex amplitudes,

$$\hat{p}_{front} - \hat{p}_{back} = Z_{sl}\hat{v}_{front} = Z_{sl}\hat{v}_{back} \qquad (13.208)$$

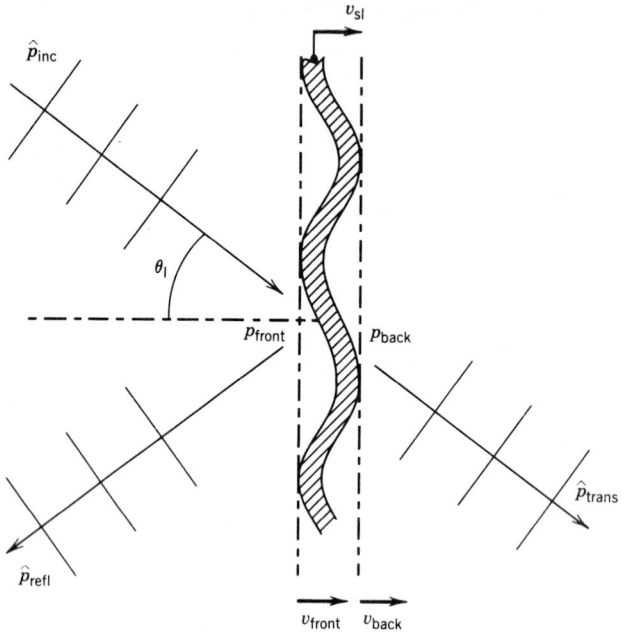

Fig. 13.47 Sound transmission through thin slab ($v_{\text{front}} = v_{\text{back}}$ = fluid velocity in vicinity of slab that moves at velocity v_{s1}).

where Z_{s1} depends on the angular frequency ω and the trace velocity $v_{\text{tr}} = c/\sin\theta_I$ of the incident wave over the surface of the slab. The value of the slab specific impedance can be derived using considerations such as correspond to Eqs. (13.207) and (13.208). In terms of the slab specific impedance, the transmission coefficient $\mathcal{T}$ is

$$\mathcal{T} = \left(1 + \frac{1}{2}\frac{Z_{s1}}{\rho c}\cos\theta_I\right)^{-1} \tag{13.209}$$

It is customary to describe transmission of sound through walls in terms of a **transmission loss** R_{TL}, reported in decibels and defined as

$$R_{\text{TL}} = 10\log\frac{1}{\tau} \tag{13.210}$$

where τ is the fraction of incident power that is transmitted. Here τ is $|\mathcal{T}|^2$, so one has

$$R_{\text{TL}} = 10\log\left(\left|1 + \frac{1}{2}\frac{Z_{s1}}{\rho c}\cos\theta_I\right|^2\right) \tag{13.211}$$

Transmission through Limp Plates

If the slab can be idealized as a **limp plate** (no resistance to bending) and not porous, the slab specific impedance is $Z_{s1} = -i\omega m_{s1}$, and one obtains

$$R_{\text{TL}} = 10\log\left[1 + \left(\frac{\omega m_{s1}}{2\rho c}\right)^2\cos^2\theta_I\right] \approx 10\log\left[\left(\frac{\omega m_{s1}}{2\rho c}\right)^2\cos^2\theta_I\right] \tag{13.212}$$

where the latter version, which typically holds at moderate to high audible frequencies, is known as the **mass law transmission loss** because it predicts a 6-dB increase in R_{TL} with a doubling of the slab mass m_{s1} per unit area.

Transmission through Porous Blankets
For a **porous blanket** that has a specific flow resistance, the specific slab impedance becomes

$$Z_{sl} = \left(\frac{1}{R_f} - \frac{1}{i\omega m_{sl}} \right)^{-1} \tag{13.213}$$

and the resulting transmission loss can be found with a substitution into Eq. (13.211).

Transmission through Elastic Plates
If the slab is idealized as a **Bernoulli–Euler plate** with elastic modulus E, Poisson's ratio v, and thickness h, the bending term in Eq. (13.206) has a complex amplitude $-B_{pl}k_x^4 \hat{v}_{sl}/(-i\omega)$, where

$$B_{pl} = \frac{1}{12} \frac{Eh^3}{1 - v^2} \tag{13.214}$$

is the **plate bending modulus**. The slab specific impedance is consequently

$$Z_{sl} = -i\omega m_{sl} \left[1 - \left(\frac{f}{f_c} \right)^2 \sin^4\theta_I \right] \tag{13.215}$$

where

$$f_c = \frac{c^2}{2\pi} \left(\frac{m_{sl}}{B_{pl}} \right)^{1/2} \tag{13.216}$$

is the **coincidence frequency**, the frequency at which the phase velocity of freely propagating bending waves in the plate equals the speed of sound in the fluid.

Although the simple result (13.215) predicts that the transmission loss is zero at a frequency of $f_c/\sin^2\theta_I$, the presence of damping processes in the plate causes R_{TL} to always have a nonzero value. A simple way of taking this into account makes use of a **loss factor** η (assumed much less than unity) for the plate, which corresponds to the fraction of stored elastic energy that is dissipated through damping processes during 1 rad (cycle period divided by 2π). The loss factor can be formally introduced into the mathematical model by the replacement of the real elastic modulus by the complex number $(1 - i\eta)E$. When this is done, one finds

$$Z_{sl} = \omega\eta m_{pl} \left(\frac{f}{f_c} \right)^2 \sin^4\theta_I - i\omega m_{sl} \left[1 - \left(\frac{f}{f_c} \right)^2 \sin^4\theta_I \right] \tag{13.217}$$

for the slab impedance that is to be inserted into Eq. (13.211). The extra term ordinarily has very little effect on the transmission loss except when the (normally dominant) imaginary term is close to zero. When the frequency f is $f_c/\sin^2\theta_I$, one finds

$$R_{TL} = 20 \log \left(1 + \frac{1}{2} \frac{\omega\eta m_{pl}}{\rho c} \cos\theta_I \right) \tag{13.218}$$

instead of a zero transmission loss.

13.2.15 Propagation in Pipes and Ducts

Guided Modes
Pipes or ducts act as guides of acoustic waves, and the net flow of energy, other than that associated with wall dissipation, is along the direction of the duct. The general theory of guided waves applies and leads to a representation in terms of **guided modes**.

If the duct axis is the x axis and the duct cross section is independent of x, the guided mode series has the form

$$\hat{p} = \sum_n X_n(x)\Psi_n(y, z) \tag{13.219}$$

where the $\Psi_n(y, z)$ are eigenfunctions of the equation

$$\left(\frac{\partial^2}{\partial y^2} + \frac{\partial^2}{\partial z^2} \right) \Psi_n + \alpha_n^2 \Psi_n = 0 \tag{13.220}$$

with the α_n^2 being the corresponding eigenvalues. The appropriate boundary condition, if the duct walls are idealized as being perfectly rigid, is that the normal component of the gradient of Ψ_n vanishes at the walls. Typically, the Ψ_n are required to conform to some normalization condition, such as

$$\int\int \Psi_n^2 \, dA = A \tag{13.221}$$

where A is the duct cross-sectional area.

The general theory leads to the conclusion that one can always find a **complete set** of Ψ_n, which with the rigid wall boundary condition imposed are such that the cross-sectional eigenfunctions are orthogonal so that

$$\int\int \Psi_n \Psi_m \, dA = 0 \tag{13.222}$$

if n and m correspond to different guided modes. The eigenvalues α_n^2, moreover, are all real and nonnegative. However, for cross sections that have some type of symmetry, it may be that more than one linearly independent eigenfunction Ψ_n (modes characterized by different values of the index n) correspond to the same numerical value of α_n^2. In such cases the eigenvalue is said to be **degenerate**.

The variation of guided mode amplitudes with source excitation is ordinarily incorporated into the **axial wave functions** $X_n(x)$, which satisfy the one-dimensional Helmholtz equation

$$\frac{d^2 X_n}{dx^2} + (k^2 - \alpha^2) X_n = 0 \tag{13.223}$$

Here $k = \omega/s$ is the free-space wave number. The form of the solution depends on whether α_n^2 is greater or less than k_2. If $\alpha_n^2 < k^2$, the mode is said to be a **propagating mode**, and the solution for X_n is

$$X_n = A_n e^{ik_n x} + B_n e^{-ik_n x} \tag{13.224}$$

where $k_n = (k^2 - \alpha_n^2)^{1/2}$ is the **modal wave number**. If, on the other hand, the value of α_n^2 is greater than k^2, the mode is **evanescent** (not propagating), and

$$X_n = A_n e^{-\beta_n x} + B_n e^{\beta_n x} \tag{13.225}$$

where $\beta_n = (\alpha_n^2 - k^2)^{1/2}$. Basic considerations ordinarily rule out any waves that grow exponentially with distance from the source, so only the term that corresponds to exponentially dying waves is kept in the description of sound fields in ducts.

Cylindrical Ducts

For the frequently encountered case of a duct with circular cross section and radius a, the index n is replaced by an index set (q, m, s), and the eigenfunctions Ψ_n become

$$\Psi_n = K_{qm} J_m\left(\frac{\eta_{qm} r}{a}\right) \begin{Bmatrix} \cos\phi \\ \sin\phi \end{Bmatrix} \tag{13.226}$$

where either the cosine ($s = 1$) or the sine ($s = -1$) corresponds to an eigenfunction. The quantities K_{qm} are normalization constants, and the J_m are Bessel functions of order m. The corresponding eigenvalues are

$$\alpha_n = \frac{\eta_{qm}}{a} \tag{13.227}$$

where the η_{qm} are the roots of $\eta J_m'(\eta) = 0$ arranged in ascending order with the index q ranging upward from 1. The smaller roots for the axisymmetric modes are $\eta_{1,0} = 0.00$, $\eta_{2,0} = 3.83171$, and $\eta_{3,0} = 7.01559$, while those corresponding to $m = 1$ are $\eta_{1,1} = 1.84118$, $\eta_{2,1} = 5.33144$, and $\eta_{3,1} = 8.53632$.

Low-Frequency Model for Duct Propagation

In many situations of interest, the frequency of the acoustic disturbance is so low that only one guided mode can propagate and all of the other modes are evanescent. Given that the walls can be idealized as rigid, there is always one mode that can propagate, this being the **plane wave mode** for which the eigenvalue α_0 is identically zero. The other modes will all be evanescent if the value of k is less than the corresponding α_n for each such mode. This would be so if the frequency is less than the lowest **cutoff frequency** for any of the nonplanar modes. For the circular duct case discussed previously, for example, this would require that

$$f < \frac{1.84118}{2\pi} \frac{c}{a} \tag{13.228}$$

When the single-guided-mode assumption is valid and even if the duct cross-sectional area should vary with distance along the duct, the acoustic field equations can be replaced to a good approximation by the **acoustic transmission line equations**

$$\frac{\partial p}{\partial t} + \frac{\rho c^2}{A} \frac{\partial U}{\partial x} = 0 \tag{13.229a}$$

$$\rho \frac{\partial U}{\partial t} = -A \frac{\partial p}{\partial x} \tag{13.229b}$$

where $U = A v_x$ is the **volume velocity**, the volume of fluid passing through the duct per unit time.

Abrupt Changes in Duct Cross-Sectional Area
One of the implications of the low-frequency model described by Eqs. (13.229) is that the volume velocity and the pressure are both continuous even when the duct has a sudden change in cross-sectional area. The pressure continuity assumption is not as universally applicable but becomes better with decreasing frequency. An improved model sets the difference of the complex amplitudes of the upstream and downstream pressures just ahead and after the junction to

$$\hat{p}_{\text{ahead}} - \hat{p}_{\text{after}} = Z_J \hat{U}_{\text{junction}} \tag{13.230}$$

where the **junction's acoustic impedance** Z_J is taken in the simplest approximation as $-i\omega M_{A,J}$, where $M_{A,J}$ is a real number independent of frequency that is called the **acoustic inertance** of the junction. Approximate expressions for this acoustic inertance can be found in the literature; a simple rule is that it is ordinarily less than $8\rho/3A_{\min}$, where $A_{\min}$ is the smaller of the two cross-sectional areas. One may note, moreover, that $Z_J \to 0$ when the frequency goes to zero.

When an incident wave is incident at a junction, reflected and transmitted waves are created in the two ducts. The pressure amplitude reflection and transmission coefficients are

$$\mathcal{R} = \frac{Z_J + \rho c/A_2 - \rho c/A_1}{Z_J + \rho c/A_2 + \rho c/A_1} \tag{13.231a}$$

$$\mathcal{T} = \frac{2\rho c/A_2}{Z_J + \rho c/A_2 + \rho c/A_1} \tag{13.231b}$$

Sound Attenuation in Ducts
The presence of the duct walls affects the attenuation of sound and causes the **attenuation coefficient** to ordinarily be much higher than for plane waves in open space. With viscosity and thermal conduction taken into account and with the assumption that the duct walls are much better heat conductors than the fluid itself, the approximate attenuation coefficient for the plane wave guided mode is

$$\alpha = \alpha_{\text{walls}} = \left(\frac{\omega}{8\rho c^2}\right)^{1/2} \left[\mu_s^{1/2} + (\gamma - 1)(\kappa/c_p)^{1/2}\right] \frac{L_P}{A} \tag{13.232}$$

where L_P is the length of the perimeter of the duct cross section.

A related effect is that sound travels slightly slower in ducts than in an open environment. The phase velocity is

$$v_{\text{ph}} = c - \frac{c^2 \alpha_{\text{walls}}}{\omega} \tag{13.233}$$

The group velocity is midway between the phase velocity and the open-space sound velocity c (which is still less than c).

13.2.16 Mufflers and Acoustic Filters

Two-Port Model

The analysis of mufflers is often based on the idealization that their acoustic transmission characteristics are independent of sound amplitudes, so they act as linear devices. The muffler is regarded as an insertion into a duct that reflects waves back toward the source and alters the transmission of sound beyond the muffler. The properties of the muffler vary with frequency, so the theory analyzes the muffler's effects on individual frequency components. The frequency is assumed to be sufficiently low that only the plane wave mode propagates in the inlet and outlet ducts. Because of the assumed linear behavior of the muffler and the single-mode assumption, the muffler conforms to the model of a **linear two-port model**, the ports being the inlet and outlet. The model leads to the prediction that one may characterize the muffler at any given frequency by a matrix such that

$$\begin{pmatrix} \hat{p}_{in} \\ \hat{p}_{out} \end{pmatrix} = \begin{pmatrix} K_{11} & K_{12} \\ K_{21} & K_{22} \end{pmatrix} \begin{pmatrix} \hat{U}_{in} \\ \hat{U}_{out} \end{pmatrix} \tag{13.234}$$

where the coefficients K_{ij} represent the acoustical properties of the muffler. Reciprocity requires that the determinant of the matrix be unity. Also, for a symmetric muffler, K_{12} and K_{21} must be identical.

The effectiveness of the sound reduction is measured in terms of the **insertion loss** of the muffler, that is, by the drop in sound pressure level caused by the insertion of the muffler into the duct.

It is ordinarily a good approximation that the waves reflected at the entrance of the muffler back to the source are significantly attenuated so that they have negligible amplitude when they return to the muffler. Similarly, the assumption is ordinarily valid that the waves transmitted beyond the muffler do not return to the muffler (anechoic termination). With these assumptions, the insertion loss is given by

$$IL = 10\log_{10}\left(\frac{1}{4}\left| K_{11} + K_{22} + \frac{\rho c}{A}K_{21} + \frac{A}{\rho c}K_{12} \right|^2 \right) \tag{13.235}$$

where A is the cross section of the duct ahead of and behind the muffler.

Acoustic mufflers can be divided into two broad categories: reactive mufflers and dissipative mufflers. In a reactive muffler, the basic property of the muffler is that it reflects a substantial fraction of the incident acoustic energy back toward the source. The dissipation of energy within the muffler itself plays a minor role; the reflection is caused primarily by the geometrical characteristics of the muffler. In a dissipative muffler, however, a low transmission of sound is achieved by internal dissipation of acoustic energy within the muffler. Absorbing material along the walls is ordinarily used to achieve this dissipation.

Nonreflecting Dissipative Muffler

When a segment of pipe of cross section A and length L is covered with a duct lining material that attenuates the amplitude of traveling plane waves by a factor of $e^{-\alpha L}$, the muffler's K matrix is

$$[K] = \begin{pmatrix} \cos(kl + i\alpha L) & -i\dfrac{\rho c}{A}\sin(kl + i\alpha L) \\[2mm] -i\dfrac{A}{\rho c}\sin(kl + i\alpha L) & \cos(kl + i\alpha L) \end{pmatrix} \tag{13.236}$$

so that the insertion loss reduces to

$$IL = 10\log_{10}e^{2\alpha L} = 8.68\alpha L \tag{13.237}$$

Expansion Chamber Muffler

The simplest reactive muffler is the expansion chamber, shown in Fig. 13.48, which consists of a duct of length L and cross section A_M connected at both ends to a pipe of smaller cross section A_P. The K matrix for such a muffler is found directly from Eq. (13.236) by setting α to zero (no dissipation in the chamber) but replacing A by A_M. The corresponding result for the insertion loss is

$$IL = 10\log_{10}\left[\cos^2 kL + \frac{1}{4}(m + m^{-1})^2\sin^2 kl \right] \tag{13.238}$$

where m is the expansion ratio A_M/A_P. The insertion loss is thus periodic with frequency. A maximum occurs when the length L is an odd multiple of quarter-wavelengths. The maximum

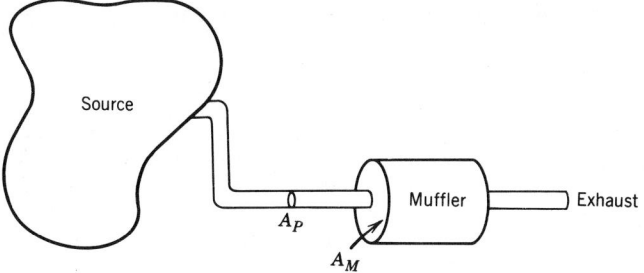

Fig. 13.48 Schematic diagram of expansion chamber of cross section A_M used as muffler on exhaust pipe of cross section A_P.

insertion loss is given by $10\log_{10}[\frac{1}{4}(m + m^{-1})^2]$. For example, a cylindrical expansion chamber of twice the diameter of the exhaust pipe to which it is connected yields a peak insertion loss of about 6.5 dB.

Helmholtz Resonator

A Helmholtz resonator, shown in Fig. 13.49, consists of a cavity of volume V with a neck of length l and cross section S. In the limit where the acoustic frequency is sufficiently low that the wavelength is much larger than any dimension of the resonator, the compressible fluid in the resonator acts as a spring with spring constant

$$k_{sp} = \frac{\rho c^2 S^2}{V} \tag{13.239}$$

and the fluid in the neck behaves as a lumped mass

$$m = \rho S l' \tag{13.240}$$

where l' is l plus **end corrections** for the two ends of the neck. If l is somewhat larger than the neck radius a and if both ends are terminated by a flange, then the two end corrections are each $0.82a$. The resonance frequency ω_r (in radians per second) of the resonator is

$$\omega_r = 2\pi f_r = \left(\frac{k_{sp}}{m}\right)^{1/2} = (M_A C_A)^{-1/2} \tag{13.241}$$

where

$$M_A = \frac{\rho l'}{S} \tag{13.242}$$

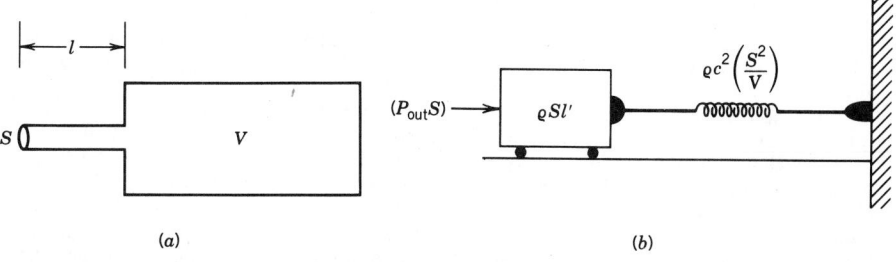

(a)   (b)

Fig. 13.49 (a) Helmholtz resonator; (b) its mechanical analog. (Adapted from A. D. Pierce, *Acoustics: An Introduction to Its Physical Principles and Applications*, McGraw-Hill, New York, 1981, p. 331.)

is the acoustic inertance of the neck and

$$C_A = \frac{V}{\rho c^2} \qquad (13.243)$$

is the acoustic compliance of the cavity. The ratio of the complex pressure amplitude just outside the mouth of the neck to the complex volume velocity amplitude of flow into the neck is the acoustic impedance Z_{HR} of the Helmholtz resonator and is given, with the neglect of damping, by

$$Z_{HR} = -i\omega M_A + \frac{1}{-i\omega C_A} \qquad (13.244)$$

and this vanishes at the resonance frequency.

If a Helmholtz resonator is inserted as a side branch into the wall of a duct, it acts as a reactive muffler that has a high insertion loss near the resonance frequency of the resonator. The analysis assumes that the acoustic pressures in the duct just before and just after the resonator are the same as the pressure at the mouth of the resonator. Also, the acoustical analog of Kirchhoff's circuit law for currents applies such that the volume velocity flowing in the duct ahead of the resonator equals the sum of the volume velocities flowing into the resonator and through the duct just after the resonator. These relations with Eq. (13.244) allow one to work out expressions for the amplitude reflection and transmission coefficients at the resonator, the latter being

$$\mathcal{T} = \frac{2Z_{HR}}{2Z_{HR} + \rho c/A_D} \qquad (13.245)$$

From this, the insertion loss is found to be

$$\text{IL} = 10 \log\left(1 + \frac{1}{4\beta^2 \left(f/f_r - f_r/f\right)^2}\right) \qquad (13.246)$$

where $\beta^2 = (M_A/C_A)(A_D/\rho c)$. The insertion loss is formally infinite at the resonance frequency, but if β is large compared with unity, the bandwidth over which appreciable insertion loss occurs is small compared with f_r.

13.2.17   Room Acoustics

Model of a Reverberant Field
A simple approximate theory due to Sabine, Franklin, and Jaeger is often used in the analysis of sound fields in rooms with highly reflecting walls. The volume integral of the acoustic energy conservation corollary, excluding sources and making use of Gauss's theorem with regard to the volume integral over $\nabla \cdot \mathbf{I}$, leads to the result

$$\frac{d}{dt} \iiint w \, dV = \mathcal{P}_{\text{source}} - \mathcal{P}_{\text{dis}} \qquad (13.247)$$

where w is the energy density in the room, $\mathcal{P}_{\text{source}}$ is the power supplied by sources, and $\mathcal{P}_{\text{dis}}$ is the energy absorbed per unit by the walls of the room. (The energy dissipated within the air itself is usually negligible at audible frequencies compared to that dissipated at the walls.) For continuous sound sources, the quantities that appear in the preceding equation can be regarded as short-period time averages such that the fluctuations over time scales on the order of any representative frequency are smoothed out. Also, these quantities can be regarded as pertaining to any given frequency band of interest.

A principal assumption that is made is that the sound fills the room in such a way that the average energy per unit volume in any region is nearly the same as in any other region. This implies that the local spatial average of the energy density w is independent of position. A related assumption is that the acoustic field within any local region can be regarded as a superposition of a large number of plane waves that are uniformly distributed with regard to propagation direction. This implies that the acoustic energy incident per unit time and per unit area on the walls is the same for all points on the walls. (Here the term *wall* includes floor and ceiling.) For a plane wave propagating in one direction the energy incident per unit area per unit time on a surface perpendicular to the propagation direction is cw because the energy travels with the sound speed c. However, if the surface is highly reflecting, half of the energy near the wall is traveling away from the wall so the energy incident per unit area and time is $\frac{1}{2}wc$. In a highly reverberant room, where all propagation directions are equally likely, the energy impinging on the wall per unit area and time can be argued from elementary ray acoustic considerations to be

$$\int_0^{2\pi}\int_0^{\pi/2}\frac{cw}{4\pi}\cos\theta\,d\Omega = \tfrac{1}{4}cw \qquad (13.248)$$

where the integration is over all solid angles that correspond to directions pointing into the wall, with θ denoting the angle of incidence. The integrand factor $w/4\pi$ is the energy per unit volume that is propagating in a given small range of directions per unit solid angle. The factor $\cos\theta$ takes into account the fact that any ray tube that has direction θ and intersects area dA on the surface must have itself a cross-sectional area $dA\cos\theta$.

The Absorption Coefficient
A fraction α of the energy incident on a wall segment will be absorbed by the wall, so the total power dissipated at the walls is

$$\mathcal{P}_{dis} = \tfrac{1}{4}cw\int\int\alpha\,dA \qquad (13.249)$$

where the integral extends over all wall area. The quantity α is called the **absorption coefficient** and is presumed to be dependent only on the local nature of the wall surface and frequency. Thus, in a room with various types of wall coverings, α will vary with position. Because the derivation of Eq. (13.249) presumes that the energy is equally distributed in propagation direction, the quantity α is also called the **random incidence absorption coefficient**. If the wall surface is locally reacting and has specific acoustic impedance Z_S, then the model just described, in conjunction with the theory of plane wave reflection at a surface of finite impedance, leads to

$$\alpha = 8\beta_R\int_0^{\pi/2}\frac{\cos^2\theta\,\sin\theta\,d\theta}{(\beta_R+\cos\theta)^2+\beta_I^2} \approx 8\beta_R \qquad (13.250)$$

where

$$\beta = \beta_R + i\beta_I = \frac{\rho c}{Z_S} \qquad (13.251)$$

The approximation $8\beta_R$ holds if the magnitude of the wall impedance is much higher than ρc.

The Reverberation Time
The insertion of the preceding described relation, between power dissipated at the walls and the absorption coefficient, into the integrated energy conservation relation yields the ordinary differential equation

$$V\frac{dw}{dt} = \mathcal{P}_{source} - \frac{cw}{4}\int\int\alpha\,dA \qquad (13.252)$$

for the energy density w. Here V is the total room volume. A simple implication of this equation is that whenever all sources of sound in the room are suddenly turned off, the acoustic energy density in the room must decrease exponentially according to the relation

$$w(t) = w(0)e^{-t/\tau} \qquad (13.253)$$

where the characteristic decay time τ is

$$\tau = \frac{4V}{cA\alpha_{av}} \qquad (13.254)$$

with α_{av} denoting the area-averaged absorption coefficient of the walls.
   The energy density in a reverberant field can be related to the mean-squared acoustic pressure

$$w = \frac{(p^2)_{av}}{\rho c} \qquad (13.255)$$

where the average need only be carried over a relatively small number of characteristic wave periods; thus Eq. (13.253) implies that the sound pressure level should decrease linearly with time as

$$\frac{dL_p}{dt} = -\frac{1}{\tau} 10 \log e = -\frac{60}{T_{60}} \tag{13.256}$$

The latter equality defines the **reverberation time** T_{60} as the time over which the sound pressure level would (ideally) decrease by 60 dB and over which the energy density would decrease by a factor of 10^{-6}. Comparison of Eq. (13.256) and Eq. (13.254) yields

$$e^{-t/\tau} = 10^{-6t/T_{60}} \tag{13.257a}$$

$$T_{60} = \frac{24V \ln 10}{cA\alpha_{av}} \tag{13.257b}$$

Measurement of the reverberation time consequently yields a measurement of the area-averaged absorption coefficient α_{av}.

Reverberation Room Measurement of Source Power
Among the many applications of the theory (and its extensions) outlined is the measurement of the total acoustic power output of a continuous source of sound. If the source is placed in a **reverberation chamber**, a room where the overall tenets of the reverberation field assumptions are fulfilled to a high degree of satisfaction, the time derivative of the energy density eventually becomes constant so the energy conservation relation, with the substitution (13.255), reduces to

$$\mathcal{P}_{source} = \frac{c}{4} \frac{(p^2)_{av}}{\rho c^2} A\alpha_{av} = \frac{(6 \ln 10)V}{T_{60}} \frac{(p^2)_{av}}{\rho c^2} \tag{13.258}$$

Radius of Reverbation
The basic assumption that the sound energy uniformly fills the room is not valid very close to a source, where the direct sound from the source may be a stronger contributor to the energy density. A simple modification to take this possibility into account assumes that the direct sound (that which has not yet been reflected from any walls) decays with a spherical spreading. In such a case, Eq. (13.258) is replaced by

$$(p^2)_{av} = \left(\frac{Q_\theta}{4\pi r^2} + \frac{4}{R_{rc}} \right) \rho c \mathcal{P}_{source} \tag{13.259}$$

where the directivity factor Q_θ for a nominally spherically symmetric radiator is 1 if the source is well removed from any surfaces and 2 if the source rests on the floor. The quantity R_{rc} is the **room constant** given by

$$R_{rc} = \frac{\alpha_{av} A}{1 - \alpha_{av}} \tag{13.260}$$

At the **radius of reverberation** r_0 the two terms in Eq. (13.259) are of equal contribution, so one has

$$r_0 = \left(\frac{R_{rc}Q_\theta}{16\pi} \right)^{1/2} \tag{13.261}$$

At a distance of $3r_0$ from the source, the error incurred by neglecting the direct wave contribution is only 0.5 dB.

13.2.18   Nonlinear Acoustics

Planar Nonlinear Waves
Even when the nonlinear terms in the governing fluid dynamic equations are small and negligible for short-term predictions, it is possible that they can have an appreciable **accumulative effect** over long periods of time or over long distances of propagation. The primary model for understanding such

accumulative effects is that of plane wave (one-dimensional) propagation governed by the equations for inviscid compressible flow. These are the same as Eqs. (13.67)–(13.71); for the present circumstances of interest there is only an x component of the fluid velocity and there is no dependence on y and z. These full nonlinear equations have an exact solution that corresponds to a traveling wave in the positive x direction. This solution, known in fluid dynamics as a **simple wave**, results when one assumes at the outset that the entropy s per unit mass is a constant and that both ρ and v are given functions of the pressure p. (Here p is the total pressure.) The fluid dynamic equations allow such a dependence if $\rho(p)$ corresponds to the fluid's equation of state and if the relation between v and p is such that

$$\frac{dv}{dp} = \frac{1}{\rho c} \tag{13.262}$$

where c is the abbreviation for

$$c = \left(\frac{d\rho}{dp}\right)^{-1/2} \tag{13.263}$$

Then, if $\rho(p)$ is given and if $v(p)$ is constructed according to Eq. (xx), the only remaining implication of the fluid dynamic equations is the nonlinear partial differential equation

$$\frac{\partial p}{\partial t} + (v + c)\frac{\partial p}{\partial x} = 0 \tag{13.264}$$

The only dependent variable here is the pressure p since v and ρ are known functions of p.
For an ideal gas, for example, $\rho(p)$ is $Kp^{1/\gamma}$ and

$$c(p) = \left(\frac{\gamma}{K}\right)^{1/2} p^{(\gamma+1)/2\gamma} \tag{13.265}$$

$$v(p) = \left(\frac{\gamma}{K}\right)^{1/2}\left(\frac{2}{\gamma-1}\right)\left[p^{(\gamma-1)/2\gamma} - p_0^{(\gamma-1)/2\gamma}\right] \tag{13.266}$$

where p_0 is the pressure at which $v(p) = 0$.
The solution of Eq. (13.264) can be written in an **implicit** form either as

$$x = \{v(p) + c(p)\}t + x_0(p) \tag{13.267a}$$

or as

$$p = F\big(x - [v(p) + c(p)]t\big) \tag{13.267b}$$

where the functions $x_0(p)$ and $F(x)$ are arbitrary but inverses of each other. The first version has the simple interpretation that a portion of a waveform with a given pressure amplitude, which is initially at x_0, moves with a constant speed $c(p) + v(p)$ such that the pressure amplitude at the corresponding point appears to stay constant. The lines of t versus x in the (t, x) plane that correspond to Eq. (13.267a) for fixed p are called **characteristics**. These are straight lines that cross the x axis at $x_0(p)$ and have slope $dt/dx = 1/(v + c)$. The alternate version, Eq. (13.267b), is also instructive because it resembles the general expression (13.93) for a traveling plane wave in linear acoustics theory. The distinction here is that the ambient sound speed for low-amplitude disturbances has been replaced by $c(p) + v(p)$. There are two modifications to the wave speed: The first is because the sound speed depends upon wave amplitude and the second is because the wave travels relative to the fluid itself, so if the wave causes the fluid to have a velocity v, then the wave speed relative to a fixed reference frame is increased by v. The latter is sometimes referred to as the **convective increment to the sound speed**.

A fundamental mathematical difficulty with the solution given in the preceding was a matter of considerable discussion during the nineteenth century. Even if the waveform of p versus x at $t = 0$ should have a smooth well-behaved form, the evolution predicted by either equation indicates that eventually there will be some time t_{sh} at which $\partial p/\partial x$ becomes infinite. For times thereafter there will be some range (depending on time) of x values where there are three values of p that correspond to the same x. The time t_{sh} is the time at which two adjacent characteristics intersect. The ambiguity arises because one can have three characteristics passing through a single point. The resolution of this ambiguity is that, given that the dissipative processes are sufficiently weak, **shock waves** must form in the fluid and the waveform becomes discontinuous although still single valued in x. The model corresponding to Eqs. (13.267) still applies for the waveform before and after the shock, but the shock location (the knowledge of which tells which characteristics to discard) has to be determined by considerations that lie outside the strict confines of the model based on the partial differential equations for inviscid compressible flow. The equations that allow one to determine the shock location are known as the **Rankine–Hugoniot equations** and are developed in most texts on compressible flow.

The Parameter of Nonlinearity

The standard approximation of nonlinear acoustics is to replace Eq. (13.264) by a second-order approximation whereby first-order terms in the pressure perturbation p' are kept such that

$$v \approx \frac{p'}{\rho_0 c_0}$$ (13.268a)

$$c(p) \approx c_0 + \left(\frac{dc}{dp}\right)_0 p'$$ (13.268b)

where the subscript zero indicates that the quantity is evaluated at the ambient pressure p_0. With this approximation, the pressure perturbation p' satisfies

$$\frac{\partial p'}{\partial t} + (c_0 + \beta u)\frac{\partial p'}{\partial x} = 0$$ (13.269)

where

$$u = \frac{p'}{\rho_0 c_0}$$ (13.270)

corresponds to the fluid velocity in the first approximation and

$$\beta = 1 + \rho_0 c_0 \left(\frac{dc}{dp}\right)_0$$ (13.271)

is the **parameter of nonlinearity**. Equations (13.268) are equivalently written in the simpler appearing form

$$\frac{\partial u}{\partial t} + (c + \beta u)\frac{\partial u}{\partial x} = 0$$ (13.272)

where here and in the remainder of this section the subscript 0 is omitted on the ambient sound speed (and also on the ambient density). The symbol p is used in what follows to denote the pressure perturbation since there is no further need in the mathematics to explicitly refer to the ambient pressure. The first-order partial differential equation (13.272) is sometimes referred to as the **inviscid Burgers equation** to distinguish it from the version that explicitly makes allowance for dissipative processes.

The parameter of nonlinearity $\beta = (\gamma + 1)/2$ for an ideal gas. For pure water, β ranges from 3.1 to 4.1 when the temperature ranges from 0 to 100°C. The values for seawater are slightly higher, with β equalling 3.6 at 20°C. It has been discovered in recent years that there are some substances for which β can be negative, but it is typically positive.

Since the derivation yielding Eq. (13.272) assumes $\beta u \ll c$, an equivalent approximation is

$$\frac{\partial u}{\partial x} + \left(\frac{1}{c} - \frac{\beta u}{c^2}\right)\frac{\partial u}{\partial t} = 0$$ (13.273)

Generation of Overtones

The dependence of the propagation speed on the pressure amplitude implies (for positive β) that peaks on a pressure waveform travel faster than troughs. The waveform distorts as it propagates such that portions where p increases with time at a given x tend to increase faster with time at larger values of x. This phenomena is referred to as **waveform steepening**. For a waveform that is sinusoidal in time at $x = 0$, the first appearance of a shock will be at a distance $x = \bar{x}$, where

$$\bar{x} = \frac{1}{\beta \varepsilon k}$$ (13.274)

where $\varepsilon = u_{pk}/c$ or, equivalently, $\varepsilon = p_{pk}/\rho c^2$ is a nondimensional measure of the amplitude of the pulse.

For $x > \bar{x}$, it becomes necessary to incorporate the presence of shocks in the model; the augmented theory is called the **weak shock theory** and is based on the observation that the Rankine–Hugoniot relations imply that shocks of low to moderate amplitudes propagate at a speed approximately given by

$$\left(\frac{dx}{dt}\right)_{shock} = c + \frac{\beta}{2}(u_a + u_b) \tag{13.275}$$

where u_a and u_b are the values of the particle velocity just before and after the shock, respectively.

In general, the periodic signal received at a distance x can be described in terms of its Fourier components by

$$u = u_{pk} \sum_{n=1}^{\infty} B_n \sin n\omega\tau \tag{13.276}$$

where $\tau = t - x/c$ is the retarded time and B_n is the amplitude of the nth harmonic. Let $\sigma = \beta\varepsilon k x = x/\bar{x}$ be the distance coordinate normalized by the shock formation distance. Then, provided that no shock has yet formed (so $\sigma < 1$), a classic solution due to Fubini of Eq. (13.273) yields

$$B_n = \frac{2J_n(n\sigma)}{n\sigma} \tag{13.277}$$

where J_n is the cylindrical Bessel function of order n.

Formation of Sawtooth Waveforms

At large values of σ, the weak shock model predicts that the initially sinusoidal waveform evolves into one resembling a **sawtooth** with a periodic succession of shocks, each shock lifting the acoustic pressure abruptly from $-P_{max}(x)$ to $+P_{max}(x)$, the acoustic pressure then decreasing linearly with time over an interval corresponding to the wave period from $+P_{max}(x)$ to $-P_{max}(x)$ (see Fig. 13.50). Here the peak amplitude is asymptotically given by

$$P_{max}(x) = \frac{\pi p_{pk,0}}{1 + x/\bar{x}} \tag{13.278}$$

where $p_{pk,0}$ is the peak amplitude of the sinusoidal waveform at $x = 0$.

If $x \gg \bar{x}$, then the 1 in the denominator of Eq. (13.278) can be neglected, and one obtains

$$P_{max}(x) = \frac{\pi \rho c^3}{\beta \omega x} \tag{13.279}$$

which is independent of the initial amplitude. This independence is associated with the **saturation effect**. At a given value of x the weak shock theory predicts that the peak amplitude cannot exceed the value given by the right side of Eq. (13.279) regardless of how high one makes the sinusoidal amplitude at $x = 0$.

The attenuation of sawtooth amplitude with x is caused by energy dissipation at a shock. Such dissipation is caused by the inherent dissipation mechanisms within the fluid but in the limit of small dissipation is independent of the nature of such mechanisms. The energy loss per unit time and per unit area of shock front is

$$-\left(\frac{dE}{dt}\right)_{dis} = \frac{\beta}{6\rho^2 c^3}(\Delta p)_{sh}^3 \tag{13.280}$$

where $(\Delta p)_{sh}$ is the total pressure jump at the shock.

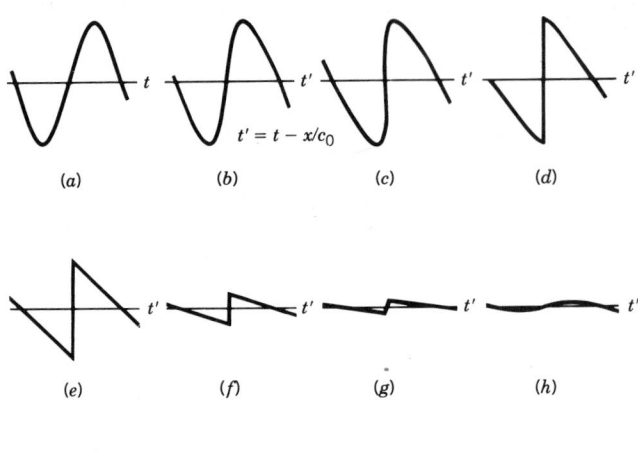

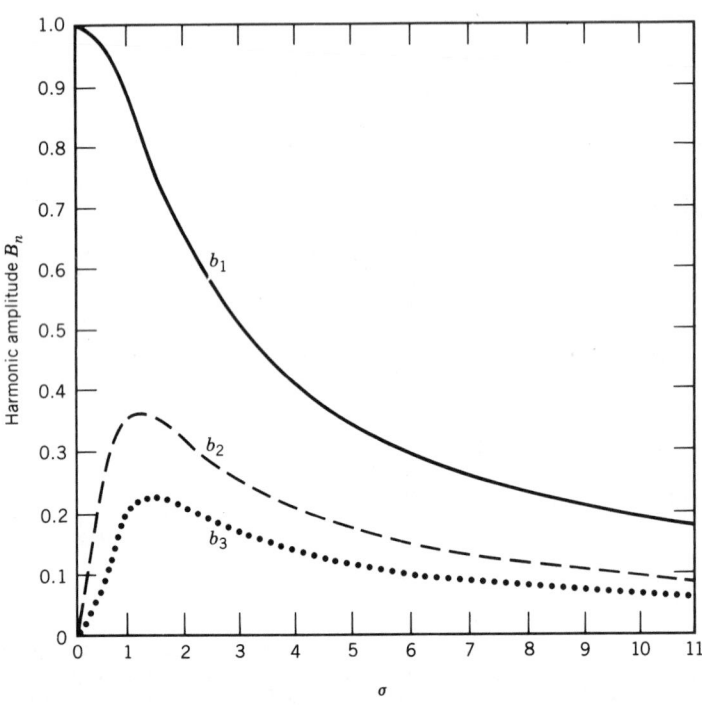

Fig. 13.50 (*a*) Case history of high-intensity sound wave. (*b*) Harmonic amplitudes of originally sinusoidal wave: (*a*) source waveform, $x = 0$; (*b*) distortion becoming noticeable; (*c*) shock formation, $x = \bar{x}$; (*d*) maximum shock amplitude; (*e*) full sawtooth shape, $x = l$; (*f*) decaying sawtooth; (*g*) shock beginning to disperse; (*h*) old age, $x = x_{max}$. *Source:* M. F. Hamilton, "Fundamentals and Applications of Nonlinear Acoustics," in *Nonlinear Wave Propagation in Mechanics*, Vol. 77, T. W. Wright, Ed., *American Society of Mechanical Engineers*, New York, 1986.)

Nonplanar Waves

The theory for the propagation of plane waves of finite amplitude has been extended by Blackstock to cylindrical and spherical waves. A general form of the wave equation, where cubic and higher order nonlinearities have been ignored, is

$$c^2 \nabla^2 \phi - \phi_{tt} = \left[\nabla\phi \cdot \nabla\phi + \frac{\beta - 1}{c^2} \phi_t^2 \right]_t \qquad (13.281)$$

where ϕ is the velocity potential (the gradient of which is the acoustically induced flow velocity) and the subscript t denotes a time derivative. This equation applies also to standing-wave fields. For outgoing cylindrically symmetric or radially symmetric waves, the **Blackstock equation** is

$$u_r + \frac{a}{r} u - \frac{\beta}{c_0^2} u u_\tau = 0 \qquad (13.282)$$

provided that the radial distance of observation r is large in comparison with a typical acoustic wavelength. Here τ abbreviates the retarded time $t - r/c$, and a is a parameter which is 0, $\frac{1}{2}$, or 1 for plane, cylindrical, or spherical waves.

Burgers Equation

When viscosity and thermal conductivity are explicitly taken into account, the nonlinear traveling-wave equation (13.272) is modified to the **Burgers equation**

$$\frac{\partial u}{\partial t} + (c + \beta u)\frac{\partial u}{\partial x} = \delta'_{cl}\frac{\partial^2 u}{\partial x^2} \qquad (13.283)$$

where δ'_{cl} is the same parameter that appears in the classical theory of plane wave attenuation given by Eq. (13.176). An equation that is approximately equivalent to this is what results when Eq. (13.273) is modified in a similar manner, the result being the **Mendousse equation**

$$\frac{\partial u}{\partial x} + \left(\frac{1}{c} - \frac{\beta u}{c^2} \right)\frac{\partial u}{\partial t} = \frac{\delta'_{cl}}{c^3}\frac{\partial^2 u}{\partial t^2} \qquad (13.284)$$

Extensive discussions and explicit solutions of both of these equations may be found in the literature.

REFERENCES

1. Judd, D. B., "The 1931 I.C.I. Standard Observer and Coordinate System for Colorimetry", *J. Opt. Soc. Am.* **23**, 359 (1933).
2. Billmeyer, F. W., and Saltzman, M., *Principles of Color Technology*, New York, Wiley, 2nd ed., 1981.

BIBLIOGRAPHY FOR LIGHT AND RADIATION

General Optics Textbooks

Born, M. and Wolf, E., *Principles of Optics*, Pergamon Press, 6th ed., 1980.
Hecht, E. and Zajac, A., *Optics*, Reading, Mass., Addision-Wesley, 2nd ed., 1979.
Jenkins, F. A., and White, H. E., *Fundamentals of Optics*, New York, McGraw-Hill, 4th ed., 1976.
Klein, M. V., *Optics*, New York, Wiley, 1970.

Lens Design and Testing

Malacara, D., *Optical Shop Testing*, New York, Wiley, 1978.
Smith, W. J., *Modern Optical Engineering: The Design of Optical Systems*, New York, McGraw-Hill, 1966.

Interference, Diffraction, and Holography

Collier, R. J., Burckhardt, C. B., and Lin, L. H., *Optical Holography*, New York, Academic Press, 1971.

Françon M., *Optical Interferometry*, New York, Academic Press, 1966.

Givens, M. P., "Focal Shifts in Diffracted Converging Spherical Waves," *Optics Communications* **41**, 145 (1982).

Givens, M. P., "Introduction to Holography," *Am. J. Phys.* **35**, 1056 (1967).

Givens, M. P., "Image Location and Magnification in Holography," *Am. J. Phys.* **40**, 1311 (1972).

Goodman, J. W., *Introduction to Fourier Optics*, New York, McGraw-Hill, 1968.

Li, Y., and Wolf, E., "Focal Shifts in Diffracted Converging Spherical Waves," *Optics Communications* **39**, 211 (1981).

Smith, H. M., *Principles of Holography*, New York, Wiley, 1969.

Solymar, L., and Cooke, D. J., *Volume Holography and Volume Gratings*, New York, Academic Press, 1981.

Steel, W. H., *Interferometry*, Cambridge University Press, 1967.

Lasers

Lengyel, B. A., *Lasers*, New York, Wiley, 2nd ed., 1971.

Siegman, A. E., *An Introduction to Lasers and Masers*, New York, McGraw-Hill, 1968.

Svelto, O., *Principles of Lasers*, Plenum Press, 2nd ed., 1982.

Yariv, A., *Introduction to Optical Electronics*, New York, Holt, Rinehart, and Winston, 3rd ed., 1985.

Color and Vision

Evans, R. M., *An Introduction to Color*, New York, Wiley, 1948.

Wright, W. D., *The Measurement of Color*, New York, Van Nostrand, 1969.

Light Sources and Detectors

Boyd, R. W., *Radiometry and the Detection of Optical Radiation*, New York, Wiley, 1983.

Kingston, R. H., *Detection of Optical and Infrared Radiation*, New York, Springer, 1978.

Kruse, P. W., McGlauchlin L. D., and McQuistan, R. B., *Elements of Infrared Technology*, New York, Wiley, 1962.

Smith, R. A., Jones, F. E., and Chasmar, R. P., *The Detection and Measurement of Infrared Radiation*, Oxford University Press, 1957.

BIBLIOGRAPHY FOR ACOUSTICS

Achenbach, J. D., *Wave Propagation in Elastic Solids*, North-Holland, Amsterdam, 1973.

Beranek, L. L., *Acoustics*, 1954; published by the American Institute of Physics for the Acoustical Society of America, New York, 1986.

Beranek, L. L., *Acoustic Measurements*, New York, Wiley, 1960.

Beranek, L. L., *Noise Reduction*, New York, McGraw-Hill, 1960.

Beyer, R. T., *Nonlinear Acoustics*, Washington D.C., Naval Sea Systems Command, 1974.

Blake, W. K., *Mechanics of Flow-Induced Sound and Vibration*, New York, Academic Press, 1986.

Boyles, C. A., *Acoustic Waveguides*, New York, Wiley, 1980.

Brekhovskikh, L. M., *Waves in Layered Media*, New York, Academic Press, 1960.

Brekhovskikh, L. M., and Lysanov, Yu., *Fundamentals of Ocean Acoustics*, Berlin, Springer-Verlag, 1982.

Bruneau, M., *Introduction aux Théories de l'Acoustique*, Université du Maine, Le Mans, France, 1983.

Clay, C. S., and Medwin, H. *Acoustical Oceanography: Principles and Applications*, New York, Wiley, 1977.

Cremer, L., Heckl, M., and Ungar, E. E., *Structure-borne Sound*, rev., New York, Springer-Verlag, 1973.

Crocker, M. J., and Price, A. J., *Noise and Noise Control*, Cleveland, CRC Press, 1975.

Dowling, A. P., and Ffowcs-Williams, J. E., *Sound and Sources of Sound*, Chichester, England, Ellis Horwood Publishers, 1983.

Goldstein, M. E., *Aeroacoustics*, New York, McGraw-Hill, 1976.

Harris, C. M., *Handbook of Noise Control*, 2nd ed., New York, McGraw-Hill, 1979.

Helmholtz, H., *On the Sensation of Tone*, 2nd ed., 1885; New York, Dover, 1954.

Hunt, F. V., 1954, *Electroacoustics*, New York, The American Institute of Physics for the Acoustical Society of America, 1982.

Junger, M. C., and Feit, D., *Sound, Structures, and Their Interaction*, 2nd ed., Cambridge, Mass., The MIT Press, 1986.

Kinsler, L. E., and Frey, A. R., *Fundamentals of Acoustics*, 2nd ed., New York, Wiley, 1962.

Knudsen, V. O., and Harris, C. M., 1950, *Acoustical Designing in Architecture*, New York, The American Institute of Physics for the Acoustical Society of America, 1978.

Kuttruff, H., *Room Acoustics*, London, Applied Science Publishers, 1973.

Lamb, H., *Hydrodynamics*, 6th ed., New York, Dover, 1945.

Lighthill, J., *Waves in Fluids*, Cambridge, England, Cambridge University Press, 1978.

Lindsay, R. B., *Acoustics: Historical and Philosophical Development*, Stroudsburg, Penn., Dowden, Hutchinson & Ross, 1972.

Lindsay, R. B., *Physical Acoustics*, Stroudsburg, Penn., Dowden, Hutchinson & Ross, 1974.

Mason W. P., and Thurston, R. N., Eds. *Physical Acoustics, Principles and Methods*, Vols. 1–17, New York, Academic Press, 1964–1984.

Meyer, E., and Neumann, E.-G., *Physical and Applied Acoustics*, New York, Academic Press, 1972.

Morse, P. M., *Vibration and Sound*, 1936, New York, The American Institute of Physics for the Acoustical Society of America, 1976.

Morse, P. M., and Feshbach, H., *Methods of Theoretical Physics*, Vols. 1 and 2, New York, McGraw-Hill, 1953.

Morse, P. M., and Ingard, K. U., *Theoretical Acoustics*, New York, McGraw-Hill, 1968.

Officer, C. B., *Introduction to the Theory of Sound Transmission*, New York, McGraw-Hill, 1958.

Pierce, A. D., *Acoustics: An Introduction to Its Physical Principles and Applications*, New York, McGraw-Hill, 1981.

Lord Rayleigh, *The Theory of Sound*, Vols. 1 and 2, 2nd ed., New York, Dover. 1945.

Rossing, T. D., *The Science of Sound*, Reading, Mass., Addison-Wesley, 1982.

Rudenko, O. V., and Soluyan, S. I., *Theoretical Foundations of Nonlinear Acoustics*, New York, Plenum, 1977.

Skudrzyk, E. J., *The Foundations of Acoustics: Basic Mathematics and Basic Acoustics*, New York, Springer-Verlag, 1971.

Sommerfeld, A., *Partial Differential Equations in Physics*, in *Lectures on Theoretical Physics*, Vol. 6, New York, Academic Press, 1949.

Stephens, R. W. B., and Bate, A. E., *Acoustics and Vibrational Physics*, 2nd ed., London, Edward Arnold, 1966.

Stump, F. B., *Analytical Acoustics*, Ann Arbor, Ann Arbor Science Publishers, 1980.

Temkin, S., *Elements of Acoustics*, New York, Wiley, 1981.

Tolstoy, I., *Wave Propagation*, New York, McGraw-Hill, 1973.

Urick, R. J., *Principles of Underwater Sound*, 3d ed., New York, McGraw-Hill, 1983.

Wood, A., *Acoustics*, New York, Dover, 1960.

Major Technical Journals in Acoustics

Journal of the Acoustical Society of America
Journal of Sound and Vibration
Acustica
Applied Acoustics
Ultrasonics
Soviet Physics. Acoustics
IEEE Journal of Oceanic Engineering
IEEE Transactions on Sonics and Ultrasonics
IEEE Transactions on Acoustics, Speech, and Signal Processing
American Institute of Aeronautics and Astronautics Journal
Journal of the Acoustical Society of Japan
Journal of Vibration, Acoustics, Stress, and Reliability in Design (ASME Transactions)
Noise Control Engineering

CHAPTER 14
CHEMISTRY

D. A. Kohl

The University of Texas at Austin
Austin, Texas

14.1 ATOMIC STRUCTURE AND THE PERIODIC TABLE

Atoms. Atoms contain a dense, positively charged nucleus surrounded by a cloud of negatively charged orbital electrons that occupy discrete energy levels and orbital configurations. The nucleus consists of Z positively charged protons and $A - Z$ uncharged neutrons. Atomic number Z determines the name of the element and the number of orbital electrons in a neutral atom. Mass number A equals the number of protons plus neutrons in the nucleus. Gain or loss of electrons results in ionized atoms with a net charge. Chemical reactions are concerned only with electronic structure changes, while nuclear reactions involve changes in the constitution of the nucleus. Isotopes are atoms with the same atomic number, the same number of electrons, and the same chemical reactions but differing in mass number because the nucleus contains a different $(A - Z)$ number of neutrons.

Atomic Weights of Elements. The mass of an element's naturally occurring distribution of isotopes is its atomic weight. This may be expressed in atomic mass units (amu) or in grams for an Avogadro number (6.023×10^{23}) of atoms. Atomic weights are relative to 12.000 for the $^{12}_{6}C$ isotope of carbon, an atom whose $Z = 6$ and $A = 12$. Examples: Naturally occurring chlorine contains 75.53 atom % of $^{35}_{17}Cl$ and 24.47% of $^{37}_{17}Cl$; it is assigned an atomic weight of 35.453. Naturally occurring tin consists of a presumably geographically invariant distribution of 10 stable isotopes ranging from mass number 112 to 124. These yield a weighted average atomic weight of 118.69.

Atomic Radii. The radii of atoms, though nebulous and difficult to estimate, range from 0.4 to 2.5 Å ($1\text{Å} = 10^{-8}$ cm). The radii may be deduced from quantum mechanical calculations or from experimentally determined crystal radii. Ionic radius differs from atomic radius due to a greater or lesser occupancy of the outer electron orbitals.

Periodicity of the Elements. The periodic chart is an arrangement of elements in rows and columns in order of increasing atomic number according to similarity of chemical behavior. Differences in electron energy levels, deduced from transitions that absorb or emit spectral energy, yield a similar and consistent pattern of electron buildup in numbers and energy levels.

Electronic Orbitals. The rationalization of spectroscopic data via quantum mechanics leads to four types of quantum numbers for describing the building up of shells of electron orbitals with discrete energy levels for each orbital. This buildup, with electrons, individually occupying the lowest vacant energy level, results in calculated spectra that closely match the experimental. These four types of quantum numbers are:

(a) **Principal quantum number n** represents a major grouping of energy levels, where n can be $1, 2, 3, \ldots$, with 1 representing the lowest energy level.

(b) **Azimuthal quantum number l** denotes different shapes of orbitals within the major grouping. These shapes are labeled s, p, d, and f for $l = 0, 1, 2,$ and 3 respectively, where l may range from 0 to $n - 1$.

(c) **Magnetic quantum number m_l** has $2l + 1$ possible values: $-l, -l + 1, -l + 2, \ldots, 0, \ldots, l - 1, l$.

(d) **Spin magnetic quantum number m_S** may have a value of $+\frac{1}{2}$ or $-\frac{1}{2}$ depending on the direction of electronic spin.

Azimuthal Characteristics of s, p, d, and f Electronic Orbitals. Directionality of the most probable electronic position about the nucleus is provided by sets of mathematically symmetrical and orthogonal orbitals shown in Fig. 14.1. Each orbital can contain two electrons of opposite spin. An electron has the highest probability of being located inside the surfaces of the revolution shown. The size of these surfaces (distance from the nucleus) increases with the principal quantum number.

Electronic Structure of Elements in the Ground State. As atomic number increases through the periodic chart, the buildup of electronic structure follows a regular pattern. One s, three p, five d, and seven f orbitals can contain 2, 6, 10, and 14 electrons (with different magnetic and spin quantum numbers), respectively. Electron energy levels associated with quantum number designations are sequentially filled, beginning with the lowest energy level available. Figure 14.2 shows the pattern of electron occupancy of orbitals for elements from hydrogen ($Z = 1$) to lawrencium ($Z = 103$) in their lowest, or ground, energy state.

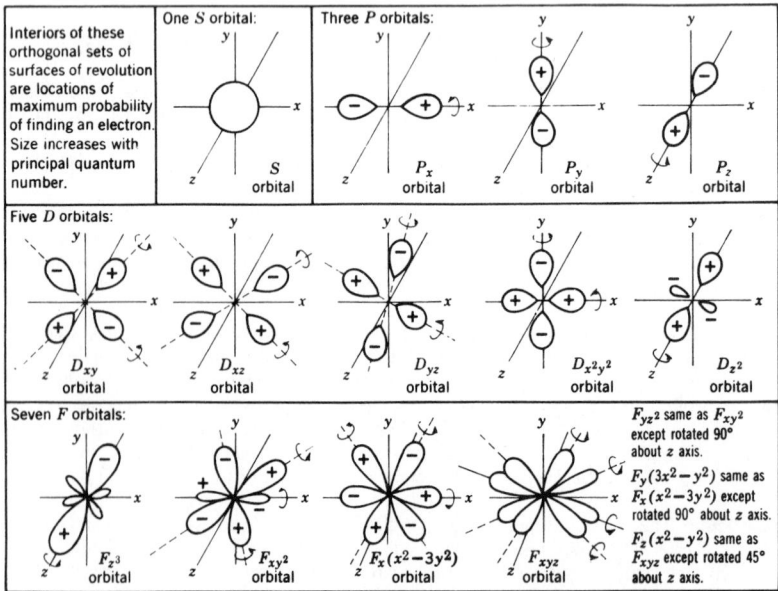

Fig. 14.1   Azimuthal characteristics of electronic orbitals.

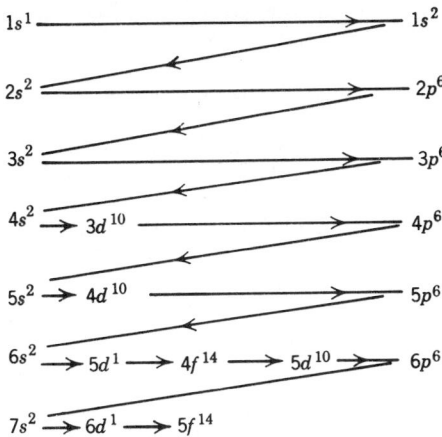

Fig. 14.2 Progressive filling of electron orbitals: $3d^{10}$ indicates 10-electron (complete) occupancy of the five d orbitals existent in principal quantum number 3 energy levels. These are filled after the $4s$ orbital is occupied by two electrons. Interruption of progressive buildup of the $5d$ sequence by 14-electron filling of $4f$ orbitals is partially explained by coulombic shielding of the positively charged nucleus by inner portions of the electron cloud.

Oxidation States of the Elements. The oxidation state of the elements is the atomic charge attainable by chemical methods. This is correlated with the systematic buildup of electrons in orbitals. The addition or loss of electrons is subject to energy barriers that become essentially insurmountable with filled shells exemplified by the noble-gas structures. In the discussion of groups of elements in each vertical column of the periodic chart, the known oxidation states are cited and related to electronic structure. This does not mean that all can exist as stable states in contact with oxygen, environmental moisture, or aqueous solution, for many of the oxidation states are particularly reactive. Electronic structure notation uses the structure of the prior noble gas plus the additional electronic orbitals that are filled for each element.

(a) **Inert gases He, Ne, Ar, Kr, Xe, and Rn** have filled p orbitals, and this maximum-stability, minimum-energy configuration exists as a neutral atom with nuclear charge equaling orbital electron charge. The energy barrier for gain or loss of electrons is very high and the oxidation state is zero although a small number of compounds with fluorine and oxygen have been prepared such as $XeOF_4$. With the exception of xenon (XeF_6) and radon compounds, where inner electrons shield the nuclear charge and reduce this potential barrier, stable compound formation has not been accomplished.

(b) **F, Cl, Br, I, and At,** lacking one electron of filled p orbitals, readily gain one electron; can lose one p orbital electron, reaching stability of a half-filled orbital; can lose five p orbital electrons, leaving a filled s orbital below; or may further lose both of those s electrons. Oxidation states of $-1, 0, +1, +3, +5,$ or $+7$ are possible.

(c) **O, S, Se, Te, and Po** lack two p electrons of the inert-gas configuration and easily reach the oxidation state of -2. Peroxides are a covalent combination of two oxygen atoms with -2 for the pair. Higher atomic number members of the group also can lose all four p electrons and may also lose the two s electrons lying beneath. Oxidation states of $-2, +4,$ and $+6$ result.

(d) **N, P, As, Sb, and Bi** with three p electrons can either gain or lose three and may also lose two s electrons beneath (exception: NO or NO_2 gases). Oxidation states of ± 3 and $+5$ exist for the group.

(e) **C, Si, Ge, Sn, and Pb** with two p electrons can lose these plus two s electrons beneath, resulting in $+2$ and $+4$ oxidation states for the group. Carbon forms numerous covalent bonds with an oxidation state of -4.

(f) **B, Al, Ga, In, and Tl** lose their single p electron plus the two s electrons below for an oxidation state of $+3$. Thallium has an additional stable $+1$ oxidation state.

(g) **Zn, Cd, and Hg,** with 10 electrons filling the d orbitals in stable configurations, lose two s electrons, forming $+2$. Exception: mercurous dimer Hg_2^{2+}

(h) **Cu, Ag,** and **Au,** with filled d orbitals, can lose their one s electron, forming $+1$ oxidation state. Exceptions for the column are Cu^{+2} and Au^{+3} due to removal of additional d electrons over low energy barriers.

(i) **Ni, Pd,** and **Pt,** with some overlap of d and s orbital energies, exhibit oxidation states of $+2$, $+4$, and $+6$.

(j) **Co, Rh, Ir,** and the **Fe, Ru,** and **Os** transition groups tend to have two electrons in the s orbital at the expense of vacant d orbitals. Oxidation states of $+2$, $+3$, $+4$, and $+6$ occur, with the $+8$ oxidation state reached in Ru and Os.

(k) **Mn, Tc,** and **Re,** with five-electron, half-filled d orbitals and filled s orbital, tend to lose the s orbital electrons for $+2$ and any in the d orbitals for a maximum oxidation state of $+7$. Exception: Re also forms a -1 ion; it is the only transition element to do so.

(l) **Cr, Mo,** and **W,** with four electrons in the d orbitals and two in the s orbital, can have oxidation states ranging from $+2$ to $+6$.

(m) The **V, Nb,** and **Ta** group, with three or four electrons in the d orbitals and one or two in the s orbital, can have oxidation states ranging from $+2$ to $+5$.

(n) The **Ti, Zr,** and **Hf** groups, with two d and two s electrons, tends to have an oxidation state of $+4$. Exception: Ti^{+3} also occurs.

(o) The **Sc, Y, La,** and **Ac** group, with one d and two s electrons, can lose these for an oxidation state of $+3$.

(p) The fourteen **rare earths (lanthanides),** following lanthanum and beginning with cerium, have the lanthanum $+3$ oxidation state and structure with approximately progressive filling of the seven $5f$ orbitals. Ce, Nd, Pr, and Nd form $+4$ ions while Sm, Eu, Tm, and Yb also form $+2$ ions.

(q) The fourteen **actinides,** beginning with thorium, paralleling cerium above, approximately repeat the structure of the rare earths with 14 electron filling of $5f$ orbitals. Some exceptions occur, forming the half-filled orbital situation, and the $+3$ oxidation state can be increased to $+6$ due to much electronic shielding of the nuclear positive charge.

(r) **Beryllium** and the **alkaline earth** group have two electrons filling the s orbital. Loss of these electrons results in a $+2$ oxidation state.

(s) **Hydrogen** and the **alkali metals** have a single electron in the s orbital. Its loss results in a $+1$ oxidation state. Exception: Gain of an electron by hydrogen to fill the s orbital results in the hydride H^{-1}.

Excited Electronic States and Ionization Energy. Excited electronic states are produced through energy absorption which raises an electron's energy level to that of an unfilled orbital. Loss of energy by photoemission may be very fast (10^{-8} sec) or very slow depending on the particular state. Energy absorption of magnitude greater than the highest available energy level is synonomous with having achieved an electronic escape velocity and ionization occurs. This ionization energy ranges from 3.9 eV (90 kcal/g atom) for cesium to 24.5 eV for helium.

Electron Affinity. The counterpart to ionization potential, electron affinity is energy released when an isolated monatomic gaseous atom gains an electron and becomes an anion. For halogens this energy is on the order of 3–4 eV; experimental problems have precluded measurement for many of the other elements that form negative ions.

Electronegativity. The electron-attracting power of the elements has been estimated from chemical bond energies and from average electron densities. The result is a consistent series of relative numbers based on zero for inertness of the noble gases and ranging from a maximum of 4.0 for fluorine (which readily forms the anion F^-) to 2.5 for carbon (which forms equally shared covalent bonds) to a minimum of 0.7 for cesium (which forms the cation Cs^+). Large differences in electronegativities of two chemically bonded atoms indicate an ionic bond with the more electronegative atom monopolizing the electron pair. Small differences in electronegativity indicate a covalent bond with approximately equal sharing of the electron pair.

14.2 MOLECULAR STRUCTURE AND CHEMICAL BONDING

Molecules. Molecules consist of groups of atoms held together in geometric arrangement by chemical bonds. Chemical bonds are based on electronic configuration of the individual atoms and electron occupancy of vacancies existent in the atomic orbitals. Mutual occupancy of orbitals of two atoms by an electron pair constitutes a chemical bond between atoms and results in joint configuration of greater stability, or lower energy, than possessed by the individual atoms. The total charge on a molecule is zero, with the number of orbital electrons matching the summation of positive nuclear charges.

Chemical Bonds. Mutual occupancy of orbitals, the shared electron pair, leads to two general classifications of chemical bonds. These are ionic and covalent. An ionic bond exists where the shared electron pair is monopolized by one atom, giving that portion of the molecule a net negative charge and leaving an electron deficiency, or net positive charge, on the other. A covalent bond results when both atoms have approximately the same electronegativity or electron affinity; the electron pair is shared without monopoly by either atom. One shared electron pair constitutes one chemical bond, commonly designated by a dash between symbols of the elements. The spatial geometry of molecules is related to orientation of the atomic orbitals involved in the bond.

Bond Energy. Energy liberated on bond formation, or energy required to rupture a bond, is the bond energy. This varies with the parent atoms and the particular molecular configuration involved (see Table 14.10 in Section 14.9). The summation of all bond energies is related to the heat of formation of a compound from its elements.

Chemical Compounds and Ions. Compounds can be classified into categories according to their ions in aqueous solution. Ionic compounds are those with ionic bonds between major atomic groupings, and these ionize in aqueous solution for form cations which are positive and anions which are negative. The naturally occurring combinations are used for inorganic chemical nomenclature. Metals, having low ionization potential, readily form positive cations, in agreement with their electronic configuration. Examples: Na^+, Ca^{2+}, Al^{3+}, Fe^{2+} and Fe^{3+}, Zn^{2+}, Ag^+, and so on. Halogens, having high electron affinity, form halide ions: F^-, Cl^-, Br^-, and I^-. Oxygen forms oxides and readily combines with other atoms to form MnO_4^-, $Cr_2O_7^{2-}$, SO_4^{2-}, SO_3^{2-}, NO_2^-, NO_3^-, PO_4^{-3}, CO_3^{-2}, SiO_3^{-2}, and similar ions.

Electrical Neutrality of Compounds. This results from summation of component oxidation states and permits assignment of an oxidation state to component atoms in the molecule.

Example 14.1. Determine oxidation state of Cr in $NaCrO_4$.
    Sodium's electronic structure, revealed by its position on the periodic chart, definitely assigns its ion as Na^+. Oxygen, except in peroxides, assumes -2 oxidation state. Chromium's usual oxidation states are $+2$, $+3$, and $+6$, as shown on the periodic chart. Therefore, the oxidation state of Cr is $+6$, as required for compound electrical neutrality.

Inorganic Chemical Nomenclature. Systematic and unambiguous nomenclature for inorganic and organic compounds and ions is necessitated by proliferation of known species. Traditional (often trivial) names of common compounds are paralleled by more cumbersome names following rules of the International Union of Pure and Applied Chemistry (IUPAC). Traditional nomenclature for inorganic compounds (Table 14.1) uses simple prefix and suffix terminology.

Acids and Bases. Three acid-base definitions are recognized:

(a) The traditional Arrhenius definition that acids yield H^+ and bases yield OH^- suffices for inorganic reactions in aqueous solution.
(b) Brönsted introduced the broader definition that acids are proton (H^+) donors and bases are proton acceptors. Since H^+ in aqueous solution exists hydrated as the hydronium ion H_3O^+, water is by definition a base because it can accept a proton. In nonaqueous media this definition recognizes hydrides, alcoholates, alkyllithium compounds, and ammonia as the powerful bases they are. Brönsted broadens the traditional definition because OH^- does accept a proton.
(c) Lewis defined acids as molecules or ions capable of accepting electron pairs and bases as molecules or ions capable of donating electron pairs. Thus $AlCl_3$, BF_3, $FeCl_3$, $ZnCl_2$, and $SnCl_4$ are Lewis acids because they lack pairs of electrons from filled orbitals. These strong acids are important in nonaqueous organic reactions. Lewis bases have an uncoordinated electron pair available; for example, on the nitrogen atom of ammonia, $:NH_3$.

The Brönsted definition of proton donors and acceptors, broadened by the Lewis definition of electron pair donors and acceptors, is necessary to explain reactions in nonaqueous media. The remaining sections on inorganic chemistry utilize the traditional simplicity of acid H^+ and base OH^- because these ions do result when any of the more broadly defined acids and bases react with water.

TABLE 14.1   Common Terminology

Cations (+ions)	Anions (−ions)
-ic. Higher usual oxidation state of cation. Examples: Fe^{+3} is ferric and Cr^{3+} is chromic.	*-ide.* Usual oxidation state of anion in binary compounds. Examples: oxide O^{2-}, chloride Cl^-, and hydride H^-.
-ous. Lower usual oxidation state of cation. Examples: Au^+ is aurous and Cr^{2+} is chromous.	*-ate.* Higher usual positive oxidation state of major element in oxygenated acid anions. Examples: ZnO_2^{-2} is zincate, SO_4^{-2} is sulfate, and ClO_3^- is chlorate.
-yl. Oxygenated cation containing the -ic or higher usual oxidation state of the major element. Examples UO_2^{2+} is uranyl, TiO^{2+} is titanyl.	*-ite.* Lower usual positive oxidation state of major element in oxygenated acid anions. Examples: NO_2^- is nitrite, SO_3^{-2} is sulfite, and ClO_2^- is chlorite.
No suffix is used for cations that form only one ion. Examples: Zn^{+2} is zinc and Na^+ is sodium.	*hypo -ite* or *-ous.* Lower oxidation state than -ite or -ous implies. Example: ClO^- is hypochlorite.
Instead of the prefix or suffix notation, oxidation states are also denoted by Roman numerals following the name, such as chromium(III), instead of Cr^{+3} or chromic.	*per- -ate* or *-ic.* Higher oxidation state than -ate or -ic implies. Example: ClO_4^- is perchlorate and MnO_4^- is permanganate.
	pyro-. Infers a dimeric structure (a di- -ate) of the oxygenated acid anion. Example: $P_2O_7^{4-}$ is pyrophosphate.
	thio-. Means single surfur substitution for oxygen in oxygenated acid anions. Example: $S_2O_2^{-2}$ is thiosulfate (SO_4^{-2} with S replacing one O).

Strong and Weak Acids and Bases. The strong or weak designation refers to extent of ionization; this does not refer to concentration in solution. Strong acids and bases are completely dissociated into their ions, in water, while weak ones are only slightly dissociated. A 1 M solution of a strong acid such as HCl yields a hydrogen ion concentration, H^+, of 1 mol/l, while a 1 M solution of a weak acid such as acetic, CH_3COOH, yields a H^+ of only 4×10^{-3} mol/l. A similar classification of bases is made based on the extent of ionization in the solvent employed.

Complex Ions and Coordination Compounds. Unfilled electron orbitals of transition metal elements, those occurring in the center of the periodic chart, permit forming complex ions and coordination compounds that appear inconsistent with the usual oxidation state of the metallic element. But complex ions are consistent with the metal's electronic structure; its unfilled orbitals are filled by the sharing of electron pairs to reach a stable configuration. Ligands (see Table 14.2) are groups attached to the central metal atom. Each ligand contributes an electron pair to be shared with and occupy vacant orbitals. Coordination number is the number of ligands surrounding the central metal atom. Solvation in water (hydration) or in liquid ammonia (ammoniation) and formation of hydrated and double salts and the numerous complex ions of inorganic and metal organic chemistry are examples of complex formation. The geometry of the complexes is spatially related to the hybridized orbitals involved. In the following examples, carbon monoxide, ammonia, and the cyanide ion are ligands:

$Ni(CO)_4$ — Nickel carbonyl

$Cr(NH_3)_6^{3+}$ — Hexamminechromium III ion

$Fe(CN)_6^{3-}$ — Ferricyanide, or hexacyanoiron III ion

In nickel carbonyl, nickel has an elemental, or zero, oxidation state with one $4s$ and three $4p$ orbitals unfilled before reaching the next noble-gas structure, Kr. An oxygen from each of the four CO molecules provides an electron pair which effectively occupies one of the four equal SP^3 hybrid orbitals to generate a stable compound.

The hexamminechromium III ion consists of Cr^{3+} symmetrically surrounded by six ammonia molecules, each providing an electron pair from its nitrogen atom. The resultant complex ion has gained stability by having all the $4d$ orbitals filled and the $4p$ orbitals half-filled. Further electron sharing with three negative ions will yield a hexamminechromium III salt; this effectively gives the chromium atom the stable electronic structure of Kr.

TABLE 14.2 Coordination Compound Ligands

F^-	NH_3, PH_3	$S_2O_3^{2-}$ thiosulfate
Cl^-	$NH_2-CO-NH_2$ (urea)	$C_2O_4^{2-}$ oxalate
Br^-		
CN^-	H_2O	S_x^{2-} polysulfide
OH^-	CO, NO	
SCN^-	Various unsaturated hydrocarbons, including acetylide ion $(HC\equiv C)^-$ and cyclopentadienyl ion $(C_5H_5)^-$	

The ferricyanide ion consists of Fe^{3+} surrounded by six CN^- ions, each N of which shares an electron pair to occupy the six d^2sp^3 hybrid orbitals. Three of these same cyanide ions further share an electron to fill the remaining three vacancies and give the iron atom an effective Kr electronic structure. Charge on the remaining three CN^- ions gives the very stable ferricyanide ion its 3 − charge.

A similar rationale can be applied to other coordination compounds composed of a central metal atom or ion and mixtures of ligands.

Chelate Compounds. Chelating (clawlike) ligands have a stereospecific configuration that permits multiple attachement to the metallic ion to form complexes. Examples are EDTA (ethylenediamine tetraacetate ion), the acetylacetonate ion, and ethylenediamine. Multiple attachment can consist of a chemical bond plus coordination with oxygen or nitrogen atoms of the ligands.

Dissociation of Complex Ions. The variable stability of complexes is represented by equilibrium constants for their dissociation. Many complexes are very stable, while others can exist only in specific environments.

14.3 CHEMICAL REACTIONS AND STOICHIOMETRY

Chemical Reactions. Chemical reactions fit into two broad categories: equilibrium-controlled conversions of reactants to products without changes in oxidation state and equilibrium control of oxidation-reduction reactions where changes in oxidation state occur.

Conversion of reactants to products can proceed only to an equilibrium condition because no driving force exists to go beyond equilibrium. Products can be maximized by continuously removing product from the reaction system so conversion can proceed without reaching equilibrium and stopping.

Le Chatelier's Principle. This principle is basic to the understanding of all chemical equilibria. It is universally applicable and states that systems adjust to reduce applied stress. For example, equilibrium of the reaction $aA + bB \leftrightarrows cC + dD$ can be shifted to the right when excess concentration of reactants A or B are present, or when products C or D are removed from the reaction mixture. Conversely, equilibria may be shifted to the left by adjustment of reactant or product concentrations.

Equilibrium-controlled Conversions. Many chemical operations use processes which consist of mixture of chemical species that proceed toward equilibrium and whose conversion to a desired product is governed by product removal from the reaction system. Product removal from the reaction system, as shown in Table 14.3, occurs through removal of ions by (a) forming slightly ionized products, (b) forming slightly soluble products, (c) complex ion formation, and (d) gas evolution.

Oxidation-Reduction (Redox) Reactions. Some chemical reactions change element oxidation states as molecular configurations and chemical bondings are rearranged. Redox terminology is given in Table 14.4. Electron transfer takes place from an oxidation source to a reduction sink. Electrons are not lost, only transferred. The number of electrons transferred determines the relative stoichiometry between their oxidation half-reaction source and their reduction half-reaction sink. Redox reactions are recognized by finding changes in oxidation state. Except in disproportionation, oxidation occurs for one atom and reduction occurs for another. Changes in oxidation state are determined by considering the following:

(a) Elemental state of the elements; that is, S, O_2, H_2, Fe, and so on; oxidation state is 0.

(b) Oxygen in other molecules: Oxidation state is -2 (exception, when covalently bonded in peroxide, -1).

TABLE 14.3   Equilibrium-controlled Conversions

Product Removal Method	Example	Net Reaction of Example	Equilibrium Governing Conversion Completeness
Slightly ionized product	Acid base neutralization, formation of weak acid or base	$H^+ + OH^- \rightleftarrows H_2O$ $H^+ + CH_3COO^- \rightleftarrows$ CH_3COOH	Dissociation of water Dissociation of weak acid or base
Slightly soluble products (least soluble product of several alternatives will prevail)	Precipitation reactions leaving uninvolved ions remaining in solution	$Ca^{2+} + SO_4^{2-} \rightleftarrows CaSO_4$	Solubility product of precipitate
Complete ion formation (most stable complex of several alternatives will prevail)	Removal of ions from solution (convert them to soluble species that do not participate in reaction)	$Ag^+ + 2CN^- \rightleftarrows Ag(CN)_2^-$	Dissociation of complex ion
Volatile product formation	Loss of CO_2 from aqueous solution	$H_2CO_3 \rightleftarrows H_2O + CO_2$	Solubility of gas at reaction temperature

TABLE 14.4   Redox Terminology

Oxidation	Process producing electrons; electron source; occurs at anode in electrochemistry; electron loss produces increase in oxidation state.
Oxidizing agent	Species that is reduced is electron sink and consumes electrons produced by oxidation of other species.
Oxidation half-reaction	Charge-balanced equation for electron source (reversing equation makes it reduction half-reaction).
Oxidation potential	Energetics of oxidation half-reaction, expressed thermally as free energy change or electrically as voltage, under standard conditions.
Disproportionation	Self-redox; reaction occurring when metastable oxidation state simultaneously oxidizes and reduces atom to more stable higher and lower oxidation states.
Reduction	Process consuming electrons; electron sink; occurs at cathode in electrochemistry; electron gain causes reduction in oxidation state.
Reducing agent	Species that is oxidized is electron source and produces electrons for reduction of other species.
Reduction half-reaction	Charge-balanced equation for electron sink.
Reduction potential	Energetics of reduction half-reaction. Positive sign means reaction can proceed spontaneously as written. Reversal of reduction half-reaction changes it to oxidation half-reaction and changes algebraic sign of potential.

(c) Hydrogen in other molecules: Oxidation state is $+1$ (exception in hydride, -1).

(d) Other elements according to position on periodic chart and list of usual oxidation states.

(e) Summation of oxidation states in any molecule is 0, and in any ion it equals the net charge on the ion.

(f) Consult a table of reduction potentials (Table 14.10) for typical half-reactions of the elements in acidic or neutral and in basic aqueous media.

Stoichiometric Equations. Stoichiometric equations can be written for reactions relating initial reactants and final products without regard to details such as equilibria, steps in the reaction mechanism, kinetics, percentage of conversion in a closed system, intermediate products, excess reactants required in the process, process variables, solvents, catalysts, and so. The balanced stoichiometric equation relates mass and moles of each product species to mass and moles of each of the reactants, assuming that the reaction goes to completion as written. The stoichiometric equation is the balanced chemical equation.

Balanced chemical equations involve three basic principles: (1) a **mass balance**, based on conservation of mass, which requires that total mass in equal total mass out; (2) an **atom balance**, based on constant mass of each atomic species involved in ordinary nonnuclear reactions, which requires that there is no change in the numbers of each atomic species; and (3) an **electron balance**, based on conservation of the number of electrons involved in oxidation-reduction reactions.

Example 14.2.

1. **Reaction not involving change in oxidation states** in a mixture of ions that may proceed to complete conversion because a product is removed from the reaction system. The stoichiometric equation $CaCl_2 + Na_2SO_4 \rightarrow CaSO_4 + 2NaCl$ can be written as a net ionic reaction: $Ca^{2+} + SO_4^{2-} \rightarrow CaSO_4$ for the formation of slightly soluble $CaSO_4$, whose solubility is quantified by its solubility product (see Section 14.6). Atom balance and mass balance apply; but electron balance, although valid, is not required since no oxidation-reduction occurs.

2. **Reaction involving redox** requires atom, mass, and electron balances. A gross reaction for dissolving copper in concentrated nitric acid (with liberation of brown N_2O_4 gas resulting from air oxidation of NO, the reaction product) can be written as two half-reactions:

(a) $NO_3^- + 4H^+ + 3e^- \rightarrow NO + 2H_2O$
(b) $Cu \rightarrow Cu^{2+} + 2e^-$

Charge balance requires electron balance so twice (a) provides $6e^-$ that match the requirement of three times (b) for the net result $2NO_3^- + 8H^+ + 3Cu \rightarrow 2NO + 4H_2O + 3Cu^{2+}$. It is apparent that $8H^+$ must have come from $8HNO_3$, and the product $3Cu^{2+}$ must be accompanied by $6NO_3^-$ for charge balance. The final stoichiometric equation is $3Cu + 8HNO_3 \rightarrow 2NO + 4H_2O + 3Cu(NO_3)_2$. Equations (a) and (b) can be found among the equations in the table of reduction potentials (Table 14.10).

3. **Balance an unbalanced redox equation** given $Ag_2S + CN^- + O_2 \rightarrow S + Ag(CN)_2^- + OH^-$. Atom, mass, and electron balances are required. Water can participate in the reaction, but H^+ cannot exist in alkaline solution. Determine which one element is oxidized and between which oxidation states. Do the same for the element reduced. (a) Reduction of O_2 from zero to -2 oxidation state in basic solution. (b) Oxidation of S from -2 to zero oxidation state in basic solution. Write net reactions for (a) and (b) or find those reactions in the table of reduction potentials.
(a) $O_2 + 2H_2O + 4e^- \rightarrow 4OH^-$
Note the need for H_2O inclusion to balance this half-reaction.
(b) $S^{2-} \rightarrow S + 2e^-$ or $Ag_2S \rightarrow S + 2Ag^+ + 2e^-$
Equation (a) supplies $4e^-$, meeting the electron needs of twice equation (b) for the net result:

$$2Ag_2S + O_2 + 2H_2O \rightarrow 4OH^- + 2S + 4Ag^+$$

This matches that given, except it needs the CN^- added to give the $Ag(CN)_2^-$ complex. The final stoichiometric equation is

$$2Ag_2S + O_2 + 2H_2O + 8CN^- \rightarrow 4OH^- + 2S + 4Ag(CN)_2^-$$

A redox equation cannot be correctly balanced unless the electron balance requirement is first met. Trial and error on the basis of an atom balance usually leads to erroneous results.

4. **Balance the disproportionation**: $ClO^- \rightarrow ClO_3^- + Cl^-$. The oxidation state of chlorine changes from $+1$ to $+5$ and -1. The simplified oxidation half-reaction is $Cl^+ \rightarrow Cl^{5+} + 4e^-$. The simplified reduction half-reaction is $Cl^+ + 2e^- \rightarrow Cl^-$. Electron balance equates the oxidation half-reaction with twice the reduction half-reaction. Addition produces a net skeleton reaction of $3Cl^+ \rightarrow Cl^{5+} + 2Cl^-$. The final redox equation is $3ClO^- \rightarrow ClO_3^- + 2Cl^-$ after oxygens are appropriately inserted to obtain the required mass and atom balance for the equation.

Molar Volume of Gases. At standard conditions of $0°C$ and 1 atm ($32°F$, 14.7 psia), the molar volume of ideal gases is approximately 22.4 l/g mol or 359 ft³/lb mol. Application of the ideal-gas law ($Pv = RT$) using ratios of absolute pressure and absolute temperature permits estimation of molar volumes at other temperatures and pressures. These molar volumes of gases can usefully elaborate the mass and molar quantities of gases directly available from the stoichiometric equations.

Material Balances. Material balances for processes can be written on many bases, for example, per ton mole of product, per hour of operation, and per batch. All are based on a stoichiometric equation for the process and on a definition of the system such that mass or molar balances may be made at the steady state using "out less in equals no accumulation" as the guiding principle. Selection of the element(s) used in the material balance depends upon the system; for example, the nitrogen of air in combustion processes passes through unchanged and quantities of other gases may be related to it via gas analyses.

Example 14.3.

1. Determine the amount of CaO required to neutralize a short ton of 37 wt % H_2SO_4 waste.

(a) Establish a balanced stoichiometric equation. Add the molecular weights of the species involved:

$$CaO + H_2SO_4 \rightarrow CaSO_4 + H_2O$$
$$56 \qquad 98 \qquad \quad 136 \qquad 18$$

Note that a material balance exists.

(b) Determine the weight of 100% H_2SO_4 that must be neutralized:

$$2000\,(0.37) = 740 \text{ lb}$$

(c) Ratio reactants and products according to the stoichiometric equation:

$$\frac{\text{wt CaO}}{56} = \frac{740}{98} = \frac{\text{wt CaSO}_4}{136} = \frac{\text{wt H}_2\text{O}}{18} \qquad \text{wt CaO} = 423 \text{ cb}$$

2. Per 1000 standard cubic feet (scf) of methane, calculate maximum carbon black available from partial combustion, scf dry air required for that combustion, and composition of wet flue gas.

(a) Establish balanced stoichiometric equation. Add atomic and molecular weights of all reactants and products:

$$CH_4 + O_2 \rightarrow C + 2H_2O$$
$$16 \quad\; 32 \quad\; 12 \quad 2(18)$$

Note that 1 lb mol of CH_4 can produce 1 lb atom of carbon black.

(b) Convert 1000 scf CH_4 to pound moles and pounds (359 scf gas = 1 lb mol; therefore,

$$\frac{1000}{359} = 2.79 \text{ lb mol, or } 44.7 \text{ lb, } CH_4$$

(c) Ratio reactants and products according to the stoichiometric equation:

$$\frac{44.7}{16} = \frac{\text{lb O}_2}{32} = \frac{\text{lb mol O}_2}{1} = \frac{\text{lb C}}{12} = \frac{\text{lb mol H}_2\text{O}}{2}$$

Obtain 2.79 lb mol O_2, 2.79 lb mol (or 33.5 lb) C, and 5.58 lb mol H_2O.

(d) Dry air composition is 21 vol % = 21 mol % oxygen and 79 mol % nitrogen and inerts:

$$\text{Dry air required} = 2.79\frac{100}{21} = 13.3 \text{ lb mol, or } 13.3\,(359), = 4770 \text{ scf}$$

(e) Assume complete electrostatic precipitation of carbon black and no water condensation. Then the flue gases consist of nitrogen, inerts, and H_2O vapor (nitrogen + inerts in = nitrogen + inerts out):

$$13.3(0.79) = 10.5 \text{ lb mol N + inerts}$$

$$\underline{5.58 \text{ lb mol H}_2\text{O}}$$

$$\text{total: } 16.08 \text{ lb mol}$$

$$N_2 = \frac{10.5}{16.08} = 65.3 \text{ vol \%}$$

$$H_2O = \frac{5.58}{16.08} = 34.7 \text{ vol \%}$$

(Ideal gas behavior has been assumed in the moles-to-volume conversion.)

3. Products of hydrocarbon combustion analyzed on a dry basis are 10.34 vol % CO_2, 0.80 vol % CO, 5.16 vol % O_2, and the remainder N_2. Water is undetermined. Calculate the carbon/hydrogen weight ratio of the hydrocarbon fuel, assuming all hydrogen burns to water before any carbon is converted to other products, and the percentage of excess air entering the combustion zone.

(a) Establish 100 lb mol dry product as the basis of calculation. On a molar basis, the products are

$$10.34 \text{ lb mol } CO_2$$
$$0.80 \text{ lb mol } CO$$
$$5.16 \text{ lb mol } O_2$$
$$83.70 \text{ lb mol } N_2$$

$$\text{total: } \overline{100.00 \text{ lb mol dry products}}$$

(b) Use a nitrogen balance to determine the O_2 entering the combustion zone:

$$83.70 \text{ lb mol } N_2 \text{ in } = 83.70 \text{ lb mol } N_2 \text{ out}$$

$$O_2 \text{ of air entering } = 83.70\frac{21}{79} = 22.25 \text{ lb mol}$$

(c) Determine the pound moles of oxygen accounted for in known products:

$$\text{in } CO_2: 10.34 \text{ lb mol}$$
$$\text{in } CO: \frac{0.80}{2} = 0.40 \text{ lb mol}$$
$$\text{as } O_2: 5.16 \text{ lb mol}$$

$$\overline{\text{total: } 15.90 \text{ lb mol}}$$

(d) Remainder of oxygen must exist as undetermined water:

$$22.25 - 15.90 = 6.35 \text{ lb mol } O_2 \text{ in water}$$

Therefore

$$2(6.35) = 12.70 \text{ lb mol } H_2O \text{ are in products}$$

(e) Hydrogen in products $= 12.70(1) = 12.70 \text{ lb mol} = 25.4 \text{ lb.}$
(f) Carbon in products $= 10.34(1) + 0.80(1) = 11.14 \text{ lb atom} = 133.5 \text{ lb; therefore}$

$$\text{C/H wt ratio } = \frac{133.5}{25.4} = 5.26$$

(g) Percentage of excess air:

$$100 \times \frac{(\text{total } O_2 \text{ entering}) - (O_2 \text{ required for complete combustion})}{O_2 \text{ required for complete combustion}}$$

(h) Alternate method for calculating percentage of excess air:

Free oxygen in combustion products $= 5.16 \text{ lb mol}$, oxygen required to burn CO to CO_2

$$= \frac{0.80}{2} = 0.40 \text{ lb mol}$$

Therefore excess O_2 beyond that required for complete combustion is 4.76 lb mol:

$$\text{Percentage of excess } O_2 = \text{percentage of excess air}$$

$$= 100 \times \frac{4.76}{17.49} = 27.2\%$$

Mixture and Dilution Calculations. Mixtures with properties linearly dependent on mass, volume, or mole fractions of components in the mixture can be handled by a simple summation:

$$P = \sum_i X_i P_i \tag{14.1}$$

where

$$1 = \sum_i X_i \tag{14.2}$$

X_i is the mass, volume, or mole fraction, as appropriate, of component i, and P_i is the property P of pure i. For example,

$$P \text{ of a binary mixture} = X_i P_i + (1 - X_i) P_i \tag{14.3}$$

Example 14.4.

1. A mixture of density 3.6 is desired from components having densities of 2.6 and 7.5:

$$1(3.6) = (x)2.6 + (1 - x)7.5 \qquad x = \frac{3.9}{4.9} = 0.796, \qquad 1 - x = 0.204$$

Since density is weight per volume, this is a weight summation when x is the volume fraction.

2. Dilute 37 wt % solution with a 3 wt % solution to make a 20 wt % solution:

$$1(20) = x(37) + (1 - x)3 \qquad x = 0.50, \qquad (1 - x) = 0.50$$

Since (wt) $\times$ (wt %) = wt, this is also a weight summation when x is mass or weight fraction. Densities and assumption of volume additivity of ideal solutions are not required unless volume fractions are sought.

14.4. CHEMICAL THERMODYNAMICS

Chemical thermodynamics deals with work-energy relationships and the driving forces of chemical reactions. It is based on a wider variety of energy terms than the expansion-compression work, thermal, flow, kinetic, and potential energy terms of engineering thermodynamics. Basic relationships presented in the engineering thermodynamics section are augmented with other forms of work and energy that are applicable to the systems considered.

Energy Terms of Thermodynamics. Energy terms are the product of an intensive property (independent of quantity) and an extensive property (varies with amount) as shown in Table 14.5. When any of these terms are involved in a chemical thermodynamic system, they are included in the basic equations of engineering thermodynamics.

Chemical Potential μ. For a multicomponent system with N_i moles of component i,

$$\mu_i = \left(\frac{\partial G}{\partial N_i}\right)_{T, P, N_{j \neq i}} = \left(\frac{\partial A}{\partial N_i}\right)_{T, V, N_{j \neq i}} \tag{14.4}$$

where G and A are the Gibbs free energy and the Helmholtz function, respectively.

Chemical Energy. As other forms of energy, all products of an intensive and an extensive property, chemical energy added to a system is $\mu_i \, dN_i$ for one component and $\sum \mu_i \, dN_i$ for all components of the system.

TABLE 14.5   Energy Terms of Thermodynamics

Kind of Energy	Intensive Property	Extensive Property	Energy Term
Expansion-contraction work	P, pressure	dV, volume	$P\,dV$
Thermal	T, temperature	dS, entropy	$T\,dS$
Mechanical	F, force	dX, displacement	$F\,dX$
Electrical	E, voltage	dQ, charge	$E\,dQ$ or $EF\,dn$
Magnetic	H, magnetic field intensity	dM, magnetization	$H\,dM$, work of magnetization
Surface	γ, surface tension	dA, area	$\gamma\,dA$, work of increasing area
Chemical	μ, chemical potential (partial molal free energy)	dN, number of moles	$\mu\,dN$, energy due to undergoing reaction or crossing phase boundaries

Chemical Thermodynamic Equations. Basic differential equations of engineering thermodynamics are expanded by adding the chemical energy and other appropriate energy terms. The thermal energy sign convention is used for chemical energy; positive terms are energy added and negative terms are energy removed from the system.

Relation of chemical potential to internal energy:

$$dU = T\,dS - P\,dV + \sum_i \mu_i\,dN_i \tag{14.5}$$

Relation to enthalpy:

$$dH = T\,dS + V\,dP + \sum_i \mu_i\,dN_i \tag{14.6}$$

Relation to Gibbs free energy:

$$dG = -S\,dT + V\,dP + \sum_i \mu_i\,dN_i \tag{14.7}$$

Relation to Helmholtz work function:

$$dA = -S\,dT - P\,dV + \sum_i \mu_i\,dN_i \tag{14.8}$$

For constant-temperature ($dT = 0$) and constant-pressure ($dP = 0$) reaction systems, Eq. (14.7) reduces to the most widely used relationship:

$$dG = \sum_i \mu_i\,dN_i \tag{14.9}$$

Standard Thermodynamic Data. Augmenting usual engineering sources of thermodynamic data (steam tables, gas tables, and special compilations for working fluids), useful tabulations of thermochemical data are available at 25°C (298 K).

$\Delta H_f{}^\circ$ = heat of formation from elements, kcal/g mole
($\Delta H_f{}^\circ$ of all elements in standard state arbitrarily taken as zero)

$\Delta G_f{}^\circ$ = Gibbs free energy of formation from elements, kcal/g mol
(some references use ΔF°)

S° = entropy, cal/g mol K

$C_p{}^\circ$ = heat capacity, cal/g mol K

New compilations may be expressed in joules rather than calories, where 1 cal = 4.184 J. The degree superscript designates standard state at 25°C and unit activity or 1 atm fugacity. For a pure compound with more than one allotropic form, this refers to the stable modification. Aqueous solution data are given at 1 molal activity. The standard state is independent of pressure or concentration variables because activity (as a thermodynamically corrected concentration) and fugacity (as a corrected pressure to give ideal-gas behavior) have been introduced as substitute variables.

Activity a. Activity is a thermodynamically effective concentration used in lieu of actual concentrations to compensate for deviations of gases from ideality, incomplete ionization of strong electrolytes in solution, and other discrepancies between calculated and experimental behavior.

Fugacity f. Fugacity deviations from the model behavior of a gas equals its activity. Fugacity is an effective partial pressure expressed in atmospheres. For ideal gases fugacity f exactly equals partial pressure. Since real gases in the standard state (in their usual phase at 1 atm pressure and 25°C) deviate slightly from ideality, the activity of real gases is defined as the ratio of fugacity f to fugacity $f°$ in a standard state where $f° = 1$ atm. For all practical purposes, $f° = 1$ atm pressure. Nonidealities of real gases at other temperatures and pressures are handled by an activity coefficient γ.

Activity Coefficient γ. Activity coefficient is the ratio of fugacity to partial pressure or the ratio of activity to molal concentration.

For ideal gases $\gamma = 1$.

For real gases γ varies with T and P and is the ratio f/p, where f is the fugacity of the gas and p is its partial pressure.

For ions in solution $\gamma = a/m$, where a is the activity of the ion and m is its molality in solution.

Activity of Gases in Mixtures. The activity of any gas in a mixture (Lewis and Randall rule) is calculated as

$$a_i = p_i \gamma_i \qquad (14.10)$$

where p_i is the partial pressure in atmospheres and γ_i is the fugacity coefficient of pure i at the temperature and total pressure of the mixture.

Calculation of $\Delta G°$ at Elevated Temperatures. The effect of temperature on $\Delta G°$ at constant pressure and at constant number of moles can be calculated from thermodynamic data at another temperature, which can be taken to be 298 K. At any temperature T,

$$\Delta G_T° = \Delta H_T° - T\Delta S_T° \qquad (14.11)$$

$$\Delta H_T° = \Delta H_{298}° + \int_{298}^{T} \Delta C_P \, dT' \qquad (14.12)$$

$$\Delta S_T° = \Delta S_{298}° + \int_{298}^{T} \Delta C_P \, dT'/T' \qquad (14.13)$$

where ΔC_P is the difference in heat capacity between products and reactants. If any phase changes occur between T and 298 K, there are additional contributions to Eqs. (14.12) and (14.12). In the absence of any phase changes, Eqs. (14.11)–(14.13) can be combined to obtain

$$\Delta G_T° = \Delta H_{298}° - T\Delta S_{298}° + \int_{298}^{T} \left(1 - \frac{T}{T'}\right) \Delta C_P \, dT' \qquad (14.14)$$

If ΔC_P is represented by a power series in T', the required integration in the last term is straightforward. (If ΔC_P is small and/or T is close to 298 K, the value of the integral is approximately zero.) An example is: Calculate $\Delta G_{298}°$ for the reaction: $CO + \frac{1}{2} O_2 \leftrightarrows CO_2$ at unit fugacity and activity of all products and reactants at (a) 298 K and (b) 800 K.

(a) At standard conditions 25°C (298 K) and 1 atm

	CO	O_2	CO_2
$\Delta H_f°$ (kcal/mol)	−26.42	0	−94.05
$\Delta S°$ (cal/K mol)	47.30	49.00	51.06

Therefore,

$$\Delta H^\circ_{298} = \Delta H_f^\circ \text{ (products)} - \Delta H_f^\circ \text{ (reactants)}$$

$$= -94.05 - (-26.42) - \tfrac{1}{2}(0) = -67.63 \text{ kcal/mol}$$

$$\Delta S^\circ_{298} = 51.06 - 47.30 - \tfrac{1}{2}(49.00) = -20.74 \text{ cal/K mol}$$

and

$$\Delta G^\circ_{298} = \Delta H^\circ_{298} - 298\,\Delta S^\circ_{298} = -67.63 - 298(-20.74/100)$$

$$= -61.45 \text{ kcal/mol}$$

(b) The heat capacity data are

For CO: $C_p = 6.3424 + 1.8363 \times 10^{-3}T - 0.2801 + 10^{-6}T^2$, cal/g mol K

For O_2: $G_p = 6.0954 + 3.2533 \times 10^{-3}T - 1.0171 \times 10^{-6}T^2$

For CO_2: $C_p = 6.393 + 10.100 \times 10^{-3}T - 3.405 \times 10^{-6}T^2$

Here, ΔC_p for the reaction equals the C_p for products $- C_p$ for the reactants:

$$\Delta C_p = 1\left(C_{p,CO_2}\right) - \tfrac{1}{2}\left(C_{p,O_2}\right) = -2.997 + 6.637 \times 10^{-3}T - 2.716 \times 10^{-6}T^2$$

$$= \alpha + \beta T + \gamma T^2$$

At $T = 800$ K, one finds ΔG°_{800} from Eq. (14.14) and

$$\Delta G^\circ_{800} = -67.63 - 800\left(-\frac{20.74}{1000}\right) + 0.19 = -50.85 \text{ kcal/mol}$$

Note that in this case the value of the integral, 0.19 kcal/mol, made a minor contribution to the net result.

Calculation of ΔG at Elevated Pressure* At constant temperature and constant number of moles, $dT = 0$ and $dN_i = 0$, Eq. (14.7) reduces to $dG = +V\,dP$. The calculation of ΔG for ideal gases, real gases, solids, and liquids at any pressure is dependent only on an expression for V as a function of P. For **ideal gases** $PV = RT$, and $dG = (RT/P)\,dP = RT\,d(\ln P)$. Integrating between limits gives

$$\Delta G = RT \ln\frac{P_2}{P_1} \qquad\qquad (14.15)$$

For **real gases** more complicated equations of state relating V and P can be substituted for the ideal-gas law and the integration performed. Deviations from ideality can also be incorporated by the use of the fugacity f_i and

$$\Delta G = RT \ln\frac{f_2}{f_1} \qquad\qquad (14.16)$$

For **incompressible liquids and solids** V is constant, so

$$\Delta G = V(P_2 - P_1) \qquad\qquad (14.17)$$

using appropriate units. The magnitude of this change for all condensed phases is very small.
For **compressible liquids and solids** V is dependent on isothermal compressibility β, $V = V_0(1 - \beta P)$:

$$dG = V_0(1 - \beta P)\,dP = V_0\,dP - \beta V_0 P\,dP \qquad\qquad (14.18)$$

---

*Note: ΔG is not the same as the pressure-independent ΔG° introduced in the preceding.

or

$$\Delta G = V_0(P_2 - P_1) - \frac{\beta V_0}{2}(P_2^2 - P_1^2)$$

Standard Gibbs Free-Energy Change and Reactivity. The ΔG_T° for a reaction is

$$\Delta G_T^\circ = \sum_i \nu_i G_{T_i}^\circ \qquad (14.19)$$

where ν_i is the stoichiometric coefficient of species i, taken as positive for products and negative for reactants. Here, ΔG_T° is the standard Gibbs free-energy change for a reaction at temperature T when all reactants and products are at unit activity (unit fugacity for gases). The ΔG_T° is independent of pressure; its magnitude as a driving force for predicting reactivity is very important:

If ΔG_T° is negative: Reaction can occur as written; a forward driving force or

                                                  chemical potential exists

If ΔG_T° is zero: No driving force exists; system is at equilibrium

If ΔG_T° is positive;   Reverse reaction can occur; a reverse driving force or chemical potential exists

$$(14.20)$$

Since the rate of reaction is dependent on temperature and activation energy, the value of ΔG° provides no information about the rate.

Example 14.5

(1) Reaction solid 1 CaO + gaseous 1 CO_2 = solid 1 $CaCO_3$

$\Delta G_f^\circ =$	$\Delta G_f^\circ =$	$\Delta G_f^\circ =$
-144.4	-94.26	-269.78
kcal/gmole	kcal/gmole	kcal/gmole

where ΔG° for the reaction $= (1)(-269.78) - (1)(-144.4) - (1)(-94.26) = -31.1$ kcal. For the reaction as written at 298 K, the process is spontaneous.
(2) Phase change:

$$C_{\text{graphite}} = C_{\text{diamond}}$$
$$\Delta G_f^\circ = 0 \quad \Delta G_f^\circ = +0.68$$

$\Delta G_{298}^\circ = 1(+0.68) - 1(0) = +0.68$ kcal for the reaction as written. The positive value of ΔG° indicates that the reaction as written cannot proceed at the standard condition, but the reverse reaction of diamond transition to graphite would be spontaneous if kinetic factors were favorable.

Thermodynamics of Equilibrium. At equilibrium the capability of any system for doing work is zero because there is no net energy available as the driving force to modify the status quo. In chemical systems at constant temperature and pressure, the change in the Gibbs free energy, ΔG, is zero at equilibrium between reactants and products; it is negative when a thermodynamic driving force for forward reaction is existent. (For equilibrium at constant temperature and constant volume, the change in the Helmholtz work function ΔA is zero.) Since most significant chemical equilibria are at constant temperature and pressure, this section contains only the relationships involving Gibbs free energy. The conditions for chemical equilibrium are

$$dG = 0 \quad \text{at constant } T, P \qquad dA = 0 \quad \text{at constant } T, V$$
$$dG = \sum_i \mu_i \, dN_i = 0 \qquad\qquad dA = \sum_i \mu_i \, dN_i = 0 \qquad (14.21)$$

Standard Gibbs Free-Energy Change and Equilibrium Constant. The magnitude of the driving force ΔG_T° at unit activity or unit fugacity of reactants and products, available to cause the reaction to proceed, was calculated for temperature T. The standard free energy ΔG_T° was considered pressure independent because unit activities or fugacities were involved.

Here, ΔG_T° is related to the equilibrium constant K, where K is expressed using activities (or fugacities of gases). This K is pressure independent because the product of any set of equilibrium fugacities, each to the power of its stoichiometric coefficient, is a constant:

$$\Delta G_T^\circ = -RT \ln K \qquad (14.22)$$

where K is expressed using activities (or fugacities) of gases.

Temperature Dependence of Equilibrium Constant. The temperature dependence of ΔG_T° directly affects K. The temperature dependence of K can be obtained by combining Eqs. (14.11) and (14.22) to yield

$$\ln K = -\frac{\Delta H_T^\circ}{RT} + \frac{\Delta S_T^\circ}{R} \qquad (14.23)$$

If T is close to 298 K and/or ΔC_P for the reaction is approximately zero,

$$\ln K = -\frac{\Delta H_{298}^\circ}{RT} + \frac{\Delta S_{298}^\circ}{R} \qquad (14.24)$$

Pressure Dependence of Equilibrium Constant. For an equilibrium constant expressed in activities and fugacity, the effect of pressure is introduced through change of activity and fugacity terms.

For **gases**: Fugacity equals the partial pressure of ideal gases and corrects for nonlinearities in the partial pressure of real gases.

For **liquids** or **solids**: Activities of solids and liquids are essentially invariant. The change in ΔG with pressure for these condensed phases, calculated from Eq. (14.17) $[\Delta G = V_0(P_2 - P_1)]$ is of small magnitude.

For **solutes**: The changes of activity coefficient and equilibrium constant with pressure are given by

$$\frac{d(\ln \gamma_i)}{dP} = \frac{\bar{v}_i}{RT} \qquad (14.25)$$

where $\bar{v}_i$ is the partial molal volume of component i. The magnitude of pressure dependence of the equilibrium constant is extremely small:

$$\frac{d(\ln K)}{dP} = \frac{\sum_i \nu_i \bar{v}_i}{RT} \qquad (14.26)$$

where $\sum \nu_i \bar{v}_i$ is the difference in the partial molal volumes of products and reactants.

Thermodynamics and Phase Equilibria. Phase equilibrium exists, as with other equilibria, when change in Gibbs free energy $\Delta G = 0$. Since the chemical potential, μ_i of component i is defined as $(\partial G/\partial N_i)_{T, P, N_{j \ne i}}$, μ is the partial molal free energy G_i. At constant temperature and pressure, $\Delta G = 0 = \sum_i \mu_i \, dN_i$. Phase equilibrium exists (with dN_i moles of component i transferred between phases) when the partial molal free energy $\bar{G}_i$ or chemical potential μ_i of each component is the same in all phases:

$$\mu_i \text{ (phase 1)} = \mu_i \text{ (phase 2)} \qquad (14.27)$$

Since $\bar{G}_i = \bar{G}_i^\circ + RT \ln a_i$, or $\mu_i = \mu_i^\circ + RT \ln a_i$, it follows that activities in each phase can be used with standard free energies to determine chemical potentials. For liquid phases, activities can be related to concentrations, and for vapor phases, activities are related to partial pressures. These relationships allow calculation for the effects of changes in temperature, pressure, and concentrations.

Partial Molal Quantities. Mixture nonidealities are often empirically treated by summing partial molal quantities contributed by each component. Partial molal quantities, such as molal volume or molal free energy, are of interest in phase equilibria:

$$dV = \sum_i \bar{v}_i \, dN_i \qquad dG = \sum_i \bar{G}_i \, dN_i \qquad (14.28)$$

where the overbar indicates partial molal quantities.

14.5 THERMOCHEMISTRY

$\Delta H°$ **Standard Heats of Reaction and Formation.** Energy is transferred in all reactions and phase changes. The enthalpy change during an endothermic reaction is positive, in agreement with the basic sign convention of thermodynamics. Conversely, the enthalpy change in an exothermic process is negative.

Data at a standard condition of 25°C (298 K) and 1 atm (or at unit fugacity of gases and unit activity of solutes) are designated by the ° superscript. Standard heats of formation of compounds from their elements are available, all based on a reference state of $\Delta H° = 0$ for elements in their stable phase at the standard condition. These data are given in kilocalories per gram mole.

Example 14.6

$$H_2(g) + \tfrac{1}{2}O_2(g) \rightarrow H_2O(g) \qquad \Delta H_f° = -57.80 \text{ kcal/g mol water (g)}$$

$$H_2(g) + \tfrac{1}{2}O_2(g) \rightarrow H_2O(l) \qquad \Delta H_f° = -68.32 \text{ kcal/g mol water (l)}$$

Parentheses enclose the phase of each species involved.

Additivity of Extensive State Functions. Extensive properties that are state functions, such as H, G, and S, are additive over successive reactions carried out at constant temperature and pressure according to the principle known as Hess's law:

$$\Delta X = \sum_i (\nu_i X_i)_1 + \sum_i (\nu_i X_i)_2 + \cdots \qquad (14.29)$$

where ν_i is the stoichiometric coefficient of species i. Thus ΔX for a reaction may be obtained from the summation of a series of simpler or better known reactions that give the same overall chemical balance. Thus the heat of a reaction may be obtained from the heats of combustion or formation of the compounds involved in the reaction. If X is any extensive property of the system and X_i is the corresponding molal property of pure i for the reaction,

$$\Delta X = \sum_i \nu_i X_i \qquad (14.30)$$

In simultaneous or successive reactions, where the numerical subscripts refer to the different reactions,

$$dN_i = (dN_i)_1 + (dN_i)_2 + \cdots \qquad (14.31)$$

and

$$N_i = N_i° + (\nu_i \Delta N)_1 + (\nu_i \Delta N)_2 + \cdots \qquad (14.32)$$

Estimating Heat of Formation and Reaction from Bond Energies. In the absence of heat-of-formation data for organic compounds, the standard heat of formation can be estimated by summing average nonpolar energies for each of the bonds of the compound. These energies are for gaseous molecules and are to be used with caution since bond strengths are influenced by molecular configuration, being reduced in resonance structures and increased in polar compounds. Heats of formation can be expected within $\pm 10\%$ of the experimentally determined values using the values of Table 14.6.

Example 14.7. Heat of formation of acetaldehyde $H\!-\!\overset{\displaystyle H}{\underset{\displaystyle H}{C}}\!-\!\overset{H}{C}\!\!=\!\!O$:

$$2C(s) + 2H_2 + \tfrac{1}{2}O_2 \rightarrow CH_3CHO$$

Energy evolved to make products bonds:

$$4\,C\!-\!H \text{ bonds at } 98.7 \qquad = \; -394.8$$

$$1\,C\!-\!C \text{ bond} \qquad\qquad = \; -82.6$$

$$1\,C\!\!=\!\!O \text{ aldehyde type bond} = \; -176$$

$$\overline{ -653.4 \text{ kcal/mole}}$$

TABLE 14.6 Bond Energies, kcal / g mol at 25°C^a

H—H	104.2	C—N	72.8
O=O	119.1	C=N	147
N≡N	225.8	C≡N	212.6
C=O (carbon monoxide type)	255.8	C—O	85.5
C—H	98.7	C=O (carbon dioxide type)	192
N—H	93.4	C=O (aldehyde type	176
O—H	110.6	C=O (ketone type)	179
S—H	83	H—F	134.6
P—H	76	H—Cl	103.2
N—N	39	H—B	87.5
N=N	100	H—I	71.4
O—O	35	C—F	116
S—S	54	C—Cl	81
N—O	53	C—Br	68
N=O	145	C—I	51
F—F	36.6	C—S	65
Cl—Cl	58.0	N—F	65
Br—Br	46.1	N—Cl	46
I—I	36.1	O—F	45
C—C	82.6	O—Cl	52
C=C	145.8	O—Br	48
C≡N	199.6		

^{a}These bond energies apply to energies of formation of gaseous molecules from gaseous atoms. For graphitic carbon to gaseous carbon atoms the heat of sublimation and atomization is 172 kcal/g mol. Heat of vaporization of water is 10.5 kcal/g mol.

Energy added to break reactant bonds:

2 H—H bonds at 104.2	=	+ 208.4
$\frac{1}{2}$ O=O bond at 119.1	=	+ 59.6
2 (heat of sublimation and atomization of graphite at 172)	=	+ 344

$$+ 612.0 \text{ kcal/mol}$$

$$\text{Net enthalpy-energy change for reaction} = -653.4 + 612.0$$
$$= -41.4 \text{ kcal/mol}$$

Compare with −39.8 actually evolved.

Estimating Heats of Reaction. Standard heats of reaction can be estimated by considering only the energy added to break specific bonds of reactants and energy released in forming the new product bonds. Where reactants and products are liquid or solid, enthalpies of vaporization or sublimation must be used to get reactants and products into and out of the vapor phase where the bond energies are valid.

ΔH for Changes of Phase. Thermal energy changes during phase transitions are significant factors in the overall energetics of chemical reaction systems. The numerical value of enthalpy of vaporization, sublimation, fusion, and other phase transitions is $f(T, P)$.

Trouton's Rule for Estimating ΔH_v. Based on an estimated average entropy change of 21 cal/K g mol occurring at the normal boiling point at constant pressure with no change in chemical potential between phases,

$$dH = T\,dS + V\,dP + \sum_i \mu_i\,dN_i \tag{14.33}$$

reduces to $\Delta H_v = 21$ (T K). This generality applies for a number of nonpolar materials but is grossly inaccurate for others where $\Delta S \neq 21$.

Clausius–Clapyron Equation for Phase Transitions

$$\frac{dP}{dT} = \frac{\Delta H}{T\Delta V} = \frac{\Delta S}{\Delta V} \tag{14.34}$$

where ΔH and ΔS are enthalpy and entropy changes of phase transitions at temperature T, and ΔV is the associated molal volume change. An integrated form, assuming ideal gas behavior of the vapor phase and negligible volume for condensed phases, is applicable to sublimation and vaporization:

$$\frac{d \ln P}{dT} = \frac{\Delta H}{RT^2} \quad \text{or} \quad \frac{d(\ln P)}{d(1/T)} = -\frac{\Delta H}{R} \tag{14.35}$$

Enthalpies of Solution, Dilution, Solvation, Adsorption, Crystallization, and Mixing. Very limited published data usually require direct experimental measurement in all but the most widely used industrial systems. Energies involved in solvation, association, and hydration are small relative to those normally associated with chemical bond rupture and formation.

Heat Capacity. The basic relationship that $\Delta H = \int C_p \, dT$ for temperature changes of a system plus enthalpy of phase transitions provides the basis for thermal calculations at other than the standard temperature. Further thermodynamic use of heat capacity data is in calculation of entropy changes via $\Delta S = \int (C_p/T) \, dT$ and calculation of Gibbs free-energy changes using the definitive relationship $G = H - TS$.

Heat Capacity Equations. Power functions of temperature best fit molar C_p data over a wide temperature range. These are of the form $C_p = \alpha + \beta T + \gamma T^2$. Since C_p is expressed as either cal/g mole K or Btu/lb mole °R, care must be used in either inserting T in Kelvins in the heat capacity equation or else using degrees Rankine divided by 1.8 to obtain the proper numerical value of C_p. Typical heat capacity equations are given in Table 14.7.

Mean Molar Heat Capacity. Integration of heat capacity equations between 25°C (298 K, 77°F, 537°R) and temperature T yields a mean molar heat capacity over the range 25°C to T that grossly simplifies thermal calculations. Graphs of these results for the usual gases are given by Smith and Van Ness.[1] Mean molar heat capacities between any pair of temperatures can be obtained by difference.

TABLE 14.7　Molar Heat Capacities of Gases in Ideal Gaseous State[a]

Compound	Formula	a	$b \times 10^3$	$c \times 10^6$	$d \times 10^9$	Range (K)	Accuracy (%)
Bromine	Br_2	8.4228	0.9379	−0.3555	—	300–1500	0.5
Chlorine	Cl_2	7.5755	2.4244	−0.9650	—	300–1500	0.5
Fluorine	F_2	6.115	5.864	−4.186	0.9797	273–2000	0.5
Hydrogen	H_2	6.9469	−0.1999	0.4808	—	300–1500	0.3
Iodine	I_2	8.504	1.3135	−1.0684	0.3125	273–1773	0.1
Nitrogen	N_2	6.4492	1.4125	−0.0807	—	300–1500	1
Oxygen	O_2	6.0954	3.2533	−1.0171	—	300–1500	0.5
Sulfur	S_2	6.499	5.298	−3.888	0.9520	273–1773	0.5
Air		6.557	1.477	−0.2148	—	273–3773	1
Ammonia	NH_3	6.189	7.887	−0 728	—	273–1000	0.5
Carbon dioxide	CO_2	6.393	10.100	−3.405	—	300–1500	1
Carbon monoxide	CO	6.3424	1.8363	−0.2801	—	300–1500	0.5
Cyanogen	$(CN)_2$	9.892	14.484	−6.207	—	291–1000	0.5
Hydrogen bromide	HBr	5.5776	0.9549	0.1581	—	300–1500	0.5
Hydrogen chloride	HCl	6.732	0.4325	0.3697	—	300–1500	0.5
Hydrogen cyanide	HCN	5.974	10.208	−4.317	—	300–1000	0.5
Hydrogen iodide	HI	6.702	0.4546	1.216	−0.4813	273–1873	0.5
Hydrogen sulfide	H_2S	6.385	5.704	−1.210	—	298–1500	1
Nitric oxide	NO	7.020	−0.370	2.546	−1.087	298–1500	0.5
Nitrous oxide	N_2O	6.529	10.515	−3.571	—	298–1500	1
Phosgene, carbonyl chloride	$COCl_2$	10.35	1.653	−8.408	—	273–973	0.5

TABLE 14.7 *(Continued)*

Compound	Formula	a	$b \times 10^3$	$c \times 10^6$	$d \times 10^9$	Range (K)	Accuracy (%)
Phosphine	PH_3	4.496	14.372	−4.072	—	298–1500	0.5
Phosphorus pentachloride	PCl_5	4.739	107.329	−119.2	—	298–500	2
Sulfur dioxide	SO_2	6.147	13.844	−9.103	2.057	273–1773	0.5
Sulfur trioxide	SO_3	6.077	23.537	−0.687	—	298–1200	1.5
Stannic chloride	$SnCl_4$	21.72	6.33	—	—	273–573	1
Water	H_2O	7.219	2.374	0.267	—	298–1500	0.5
Methane	CH_4	3.381	18.044	−4.300	—	298–1500	1
Ethane	C_2H_6	2.247	38.201	−11.049	—	298–1500	0.5
Propane	C_3H_8	2.410	57.195	−17.533	—	298–1500	1.5
n-Butane	C_4H_{10}	4.453	72.270	−22.214	—	298–1500	1.5
2-Methyl propane	C_{10}	3.332	75.214	−23.743	—	298–1500	1.5
n-Pentane	C_5H_{12}	5.910	88.449	−27.388	—	298–1500	1.5
n-Hexane	$C_6H_{13}4$	7.477	104.422	−32.471	—	298–1500	1.5
2,2-Dimethyl butane	C_6H_{14}	0.593	133.001	−52.878	—	298–1000	0.5
n-Heptane	C_7H_{16}	9.055	120.352	−37.528	—	298–1500	1.5
n-Octane	C_8H_{18}	10.626	136.398	−42.592	—	298–1500	1.5
2,2,4-Trimethyl pentane	C_8H_{18}	−3.2	152.5	—	—	400–500	2
Ethene	C_2H_4	2.830	28.601	−8.726	—	298–1500	1.3
Propene	C_3H_6	2.253	45.116	−13.740	—	298–1500	1
1-Butene	C_4H_8	5.132	61.760	−19.322	—	298–1500	1.5
cis-2-Butene	C_4H_8	1.625	64.836	−20.047	—	298–1500	1.3
trans-2-Butene	C_4H_8	4.967	59.961	−18.147	—	298–1500	0.8
Ethyne	C_2H_2	7.331	12.622	−3.889	—	298–1500	1.5
Propyne	C_3H_4	6.334	30.990	−9.457	—	298–1500	1
2-Butyne	C_4H_6	5.700	48.207	−14.479	—	298–1500	0.5
Benzene	C_6H_6	−0.409	77.621	−26.429	—	298–1500	3
Toluene	$C_6H_5CH_3$	0.576	93.493	−31.227	—	298–1500	2.5
Cyclopropane	C_3H_6	−6.481	82.06	−55.77	15.61	273–973	0.5
Cyclopentane	C_5H_{10}	−5.763	97.377	−31.328	—	298–1500	2.4
Cyclohexane	C_6H_{12}	−7.701	125.675	−41.584	—	298–1500	2
Methyl cyclohexane	$C_6H_{11}CH_3$	−4.624	140.87	−46.698	—	298–1500	2
Formaldehyde	$HCHO$	4.498	13.953	−3.730	—	291–1500	0.5
Acetaldehyde	CH_3CHO	7.422	29.029	−8.742	—	298–1500	1.5
Acetone	$(CH_3)_2CO$	5.371	49.227	−15.182	—	298–1500	1.5
Methanol	CH_3OH	4.398	24.274	−6.855	—	273–1000	2
Ethanol	C_2H_5OH	6.990	39.741	−11.926	—	298–1500	0.8
2-Propanol	$(CH_3)_2CHOH$	0.7936	85.02	−50.16	11.56	273–1473	0.2
n-Propanol	$C_2H_5CH_2OH$	−1.307	92.35	−58.00	14.14	273–1473	0.5
Ethyl ether	$(C_2H_5)_2O$	−24.83	338.7	−593	—	300–400	2
Ethylene oxide	$(CH_2)_2O$	−1.12	4.925	−23.89	3.149	273–973	0.3
Bromomethane	CH_3Br	4.184	22.445	−7.496	—	300–1200	1
Chloromethane	CH_3Cl	3.563	22.998	7.571	—	273–773	0.5
Fluoromethane	CH_3F	3.616	18.239	−2.035	—	298–600	0.5
Iodomethane	CH_3I	4.105	24.487	−9.733	—	300–600	0.5
Dichloromethane	CH_2Cl_2	4.309	31.67	−16.35	—	250–600	0.5
Fluorochloromethane	CH_2FCl	4.292	27.025	−10.605	—	250–600	0.5
Tribromomethane, bromoform	$CHBr_3$	9.356	32.319	−21.272	—	300–600	0.1
Trichloromethane, chloroform	$CHCl_3$	7.052	35.598	−21.686	—	273–773	0.5
Methyl cyanide	CH_3CN	5.018	27.935	−9.302	—	291–1200	0.5

Reprinted with permission from Wilson and Ries, *Principles of Chemical Engineering Thermodynamics*, New York, McGraw-Hill, 1956.

$^a C_p^\circ = a + bT + cT^2 + dT^3$, where T is in Kelvins.

TABLE 14.8   Contributory Atomic C_p° Values for Kopp's Rule[a]

C	1.8
H	2.3
B	2.7
Si	3.5
O	4.0
F	5.0
N	5.0
P	5.4
S	5.4
Mg	5.7
Al	5.8

[a]For all elements of higher molecular weight use an average value of 6.4 cal/g atom K (Ref. 2).
Example: $Ag_2CO_3 = 2(6.4) + 1.8 + 3(4.0) = 26.6$ cal/g mol K, (actual 26.8)

Heat Capacities of Ideal Gases. The C_p molar heat capacity at a constant pressure is independent of pressure. The C_v molar heat capacity at constant volume is independent of volume. $C_p - C_v = R$, where $R = 1.985$ cal/g mol K. The approximate values C_v of $\frac{3}{2}R$ for monatomic, $\frac{5}{2}R$ for diatomic, and $\frac{7}{2}R$ for triatomic ideal gases are inadequate for most engineering purposes.

Zero-Pressure Heat Capacity of Real Gases, C_p°. Zero-pressure heat capacities for an ideal-gas state can be calculated from spectroscopic data. Since C_p for ideal gases is independent of pressure and is a function of temperature only, it can be used for real gases under T, P conditions where the real gas does not grossly deviate from ideal-gas behavior. For conditions near critical and near the vapor-liquid phase change, recourse is made to more detailed compilations of specific thermal data for that system (steam tables, gas tables, etc.).

C_p of Liquids and Solids. The relative incompressibility of liquids and solids usually reduces the difference between C_p and C_v to less than the experimental error in their measurement. The specific heat of liquids and solids increases with temperature but to a lesser extent than for gases. Experimental data are reported on both molar and unit mass bases at one temperature, average over a temperature range, or as power functions of T.

Kopp's Rule. The molar C_p° of solid molecules at 25°C is approximated by summation of contributory atomic C_p° values for constituents. Average values are given in Table 14.8.

C_p of Solutions and Mixtures. In lieu of experimental data, the heat capacity of a solution or mixture can be estimated by

$$C_{p, \text{mixture}} = \sum_i y_i C_{pi} \tag{14.36}$$

where y_i is the mole fraction and C_{pi} is the molar heat capacity of component i.

Effect of Temperature on Heat of Reaction. Consider the reactants as one state and the products as another and that thermodynamically one can proceed from one state function to another regardless of the path. From reactants at temperature T, one can change temperature (using heat capacities and the enthalpies of any phase changes) to 25°C, then utilize the standard heat of reaction at 25°C and 1 atm, and then change products back to temperature T.

Example 14.8. Calculate the heat of the ammonia synthesis reaction, $N_2 + 3H_2 \rightarrow 2NH_3$, as a function of temperature T. Given is the standard heat of reaction $\Delta H^\circ = -22.08$ kcal for the equation as written at 25°C (298 K) and 1 atm.
The heat capacity equations for reactants and products are

$$H_2: \quad 6.62 + 0.0081 \ T \, (K) \text{ cal/g mol K or}$$

$$6.62 + \frac{0.00081}{1.8} T \, (°R) \frac{\text{Btu}}{\text{lb mol °R}}$$

$$N_2: \quad 6.50 + 0.00100 \ T \text{ K}$$

$$NH_3: \quad 6.70 + 0.00630 \ T \text{ K}$$

1. Cool reactants from T down to 298°K:

$$\Delta H_1 = \sum \int_T^{298} C_p \, dt = 1\left[(6.50)(298 - T) + 0.00100\frac{(298^2 - T^2)}{2}\right]$$

$$+ 3\left[(6.62)(298 - T) + 0.00081\frac{(298^2 - T^2)}{2}\right]$$

$$= 26.36(298 - T) + 0.00343\left(\frac{298^2 - T^2}{2}\right)$$

$$= -26.36(T - 298) - 0.00343\left(\frac{T^2 - 298^2}{2}\right)$$

2. Standard heat of reaction $\Delta H° = -22.08$ kcal/g mole quantities for $N_2 + 3H_2 = 2NH_3$ at the 298 K. Given is $\Delta H° = -11.04$ kcal for $\frac{1}{2}N_2 + \frac{3}{2}H_2 = NH_3$ in tables. Doubling the stoichiometric quantities doubles heat of reaction. Multiply by 1000 to convert the Btu/lb mole quantities.

3. Heat products from 298 K up to T K:

$$\Delta H_2 = \int_{298}^T C_p \, dt = 2\left[6.70(T - 298) + 0.00630\left(\frac{T^2 - 298^2}{2}\right)\right]$$

$$= 13.40(T - 298) + 0.01260\left(\frac{T^2 - 298^2}{2}\right)$$

4. Algebraically sum the three ΔH values to get ΔH_T for the reaction at temperature T K:

$$\Delta H_1 = -26.36(T - 298) - 0.00343\left(\frac{T^2 - 298^2}{2}\right)$$

$$\Delta H° = -22.08$$

$$\Delta H_2 = +13.40(T - 298) + 0.01260\left(\frac{T^2 - 298^2}{2}\right)$$

$$\overline{\Delta H_T = -22.08 - 12.96(T - 298) + 0.00917\left(\frac{T^2 - 298^2}{2}\right)}$$

kcal/g mol quantities for the equation as written.

5. Convert to Btu and lb mole units for T °R since

$$298 \text{ K} = 537 \text{ °R}$$

$$\Delta H_T = -22.08(1000) - \frac{12.96}{1.8}(T - 537) + \frac{0.00917}{1.8}\left(\frac{T^2 - 537^2}{2}\right)$$

Btu/lb mol quantities for the equation as written.

Combustion Calculations

(a) For purposes of combustion calculations **dry air** is assumed to have a molecular weight of 29 and the following composition:

Oxygen, 21 vol % = 21 mol %

Nitrogen, 79 vol % = 79 mol %

(The nitrogen includes 1% argon and traces of carbon dioxide.) This composition allows combustion calculation on the basis of 100 lb mol (2900 lb) dry air entering a reaction zone with 21 lb mol oxygen available for reaction and 79 lb mol nitrogen and inerts that pass through unreacted.

(b) The **moisture** in ambient air entering a combustion zone can often be ignored in approximate combustion calculations; however, it cannot be ignored in the exhaust gases. Gases saturated with water vapor contain a partial pressure of water vapor equal to the vapor pressure of water at that temperature. Saturation is 100% relative humidity; and the dew point, being the temperature of condensation initiation, is then the ambient temperature. The vapor pressure of water is available from steam tables.

Example 14.9. At a dew point of 77°F, the vapor pressure of water from steam tables (saturated steam, temperature table) is 0.459 psia. Applying Dalton's law of partial pressures, the mole fraction of water vapor in gases at 1 atm is 0.459 psia/14.7 psia = 0.031. At a dew point of 160°F, the vapor pressure is 4.739 psia and the mol fraction of water vapor in gases at 28 in. Hg pressure is

$$\frac{4.739 \text{ psia}}{\frac{28}{30}(14.7 \text{ psia})} = 0.347$$

(c) **Psychrometric charts**, useful for air conditioning and humidification calculations, express moisture content in grains of water per pound of dry air (7000 grains = 1 lb) and utilize wet-bulb (essentially dew point) and dry-bulb (ambient) temperatures as variables for determining percentage relative humidity.

(d) **Percentage excess air**. Stoichiometric air is the amount of air theoretically necessary to burn all hydrogen in fuels to water and to burn all carbon to carbon dioxide. With insufficient air, hydrogen is preferentially oxidized to water and carbon may be released as elemental carbon (soot), carbon monoxide, or mixtures of carbon monoxide and dioxide. Significant amounts of excess air and residence time in the combustion zone are required for complete combustion of fuel-contained carbon to carbon dioxide.

(e) **Basis for calculation**. All stoichiometric calculations should be preceded by a basis of calculation in order to readily permit scale-up or scale-down of the result. Examples of usual bases are per 100 lb of fuel, per 100 lb mol of a reactant, and per ton of a product.

(f) **Heat balances**. Thermal calculations for reaction systems are most easily made for steady-state conditions involving an accounting calculation which equates all forms of energy in plus heat of reaction to all forms of energy out plus losses. Forms of energy involved are thermal, chemical, electrical, and mechanical in the form of shaft work, kinetic, potential, and flow energies plus enthalpies of all phase transitions involved. Heat balances are necessarily predicated on valid material balances, and vice versa. Calculation can be based on mass flow of some component, a unit of operating time, or any other useful basis chosen to simplify calculation and utilize available thermal data.

14.6 CHEMICAL EQUILIBRIUM

Equilibrium Constant. The equilibrium condition for any reaction $a\text{A} + b\text{B} + \cdots = c\text{C} + d\text{D} + \cdots$ is defined by the expression

$$K = \frac{[\text{C}]^c [\text{D}]^d \cdots}{[\text{A}]^a [\text{B}]^b \cdots} \tag{14.37}$$

The exponents are stoichiometric coefficients of the equation as written, and bracketed terms are the activities of each species existent at equilibrium. The wide numerical range of K encountered has promoted usage of $pK = -\log_{10} K$ for tabulated data. The equilibrium constant is an expression relating activities of reactants and products such that the net chemical potential is zero at the temperature and pressure of reaction. The standard free energy of a reaction is based on the unit activity of all reactants and products. The K varies with temperature and has a pressure dependence introduced via fugacity of gaseous species, being derived from the relationship $\Delta G_f^\circ = -RT \ln K$ [see Eq. (14.22)].

For reactions where equilibrium constant data are unavailable, combining K's from other known reactions will yield the one desired. This is a consequence of the additivity of reactions and their thermodynamic state functions.

Example 14.10

$$CO_2 = CO + \tfrac{1}{2}O_2$$

$$H_2O = H_2 + \tfrac{1}{2}O_2 \qquad K = \frac{[H_2][O_2]^{1/2}}{[H_2O]} = 7.25 \times 10^{-1} \quad \text{at 1000 K}$$

$$CO_2 + H_2 = CO + H_2O \qquad K = \frac{[CO][H_2O]}{[CO_2][H_2]} = 8.70 \times 10^{-11} \quad \text{at 1000 K}$$

$$\therefore CO_2 = CO + \tfrac{1}{2}O_2 \qquad K = \frac{[CO][O_2]^{1/2}}{[CO_2]} = (7.25 \times 10^{-1})(8.70 \times 10^{-11})$$

$$= 6.3 \times 10^{-11} \quad \text{at 1000 K}$$

The activities of solids and solvents are substantially invariant, each in its own phase; they are assigned numerical values of 1. The activity of gases is their fugacity in atmospheres. Fugacity is related to partial pressure via a fugacity coefficient $\gamma = f/p$ that varies with pressure and temperature. The fugacity coefficient of ideal gases is 1, but the deviations of real gases necessitate its consideration. For real gases it is expendient to divide K, based on activities, into two parts:

$$K = K_\gamma K_p \qquad (14.38)$$

where K_γ is an activity coefficient product and K_p is the pressure product.

Equilibrium Composition of Reaction Mixtures. The effects of pressure and temperature on equilibrium composition, qualitatively predicted by Le Chatelier's principle, (Section 14.3), are increased conversion with pressure rise if moles of reactants exceeds moles of products and decreased conversion with temperature rise in an exothermic reaction. Quantitatively, equilibrium composition may be calculated from the equilibrium constant.

Sample Calculation of Equilibrium Composition for Gaseous Reactions

$$\tfrac{3}{2}H_2(g) + \tfrac{1}{2}N_2(g) = NH_3(g) \qquad K = \frac{[NH_3]^1}{[H_2]^{3/2}[N_2]^{1/2}} = 0.0431 \text{ atm}^{-1} \quad \text{at 600 K}$$

Note that the equilibrium constant is dimensional, and its units are useful in reconstructing the equilibrium constant expression and the reaction equation on which it is based.

Example 14.11

$$K = \frac{(f_{NH_3})^1}{(f_{H_2})^{3/2}(f_{N_2})^{1/2}} = 0.0431 \text{ atm}^{-1}$$

applies for NH_3 synthesis. For a stoichiometric ratio of hydrogen and nitrogen without inerts, calculate the mole fraction of ammonia in the equilibrium mixture at (a) 1 atm and (b) 100 atm.

At 1 atm total pressure, where the assumption of ideal-gas behavior for this system may be valid, the activity coefficient γ is assumed and confirmed to be 1 for all species. At 100 atm γ should be estimated for each species by entering the $\gamma = f/p$ chart (see Section 14.7) using reduced temperature and pressure based on the reaction temperature and total pressure. Significant deviation of K_γ from 1 signals the necessity of its inclusion in the calculation.

	P_C (atm)	T_C (K)	T_R	P_R at $P = 100$ atm	$\gamma = f/p$ at $P = 100$ atm	P_R at $P = 1$ atm	$\gamma = f/p$ at $P = 1$ atm
NH_3	111.3	405.5	0.90	1.48	ca ~ 0.92	0.009	~ 1
H_2	12.8	33.3	7.8	18.	ca ~ 1.1	0.078	~ 1
N_2	33.5	126.2	2.98	4.76	ca ~ 1.0	0.030	~ 1

$$K_\gamma = \frac{(0.92)^1}{(1.1)^{3/2}(1.0)^{1/2}} = \frac{0.92}{(1.16)(1.0)} = 0.79 \quad \text{at 100 atm}$$

$$= 1.02 \quad \text{at 1 atm}$$

Calculation of mole fraction NH_3:
At total $P = 1$ atm, $K_\gamma = 1$. Let

$$X = \text{partial pressure of } N_2 \text{ in the reaction mixture}$$
$$3X = \text{partial pressure of } H_2$$
$$1 - 4X = \text{partial pressure of } NH_3$$

$$K = K_\gamma K_p = 0.0431 \text{ atm}^{-1} = \frac{1 - 4X}{(3X)^{3/2}(X)^{1/2}}$$

$$1 - 4X = 0.0431(5.2\,X^2)$$

This is solved by the quadratic formula or successive approximation to give $X \cong 0.246$:

$$\text{Mole fraction } y_{NH_3} = \frac{p_{NH_3}}{p_{\text{total}}} = \frac{1 - 4X}{1} = 0.014$$

At total $P = 100$ atm, $K = 0.79$. Let

$$X = p_{N_2}$$
$$3X = p_{H_2}$$

$$K = 0.0431 = \frac{0.79(100 - 4X)}{(3X)^{3/2}(X)^{1/2}}$$

$$100 - 4X = p_{NH_3}$$

$$100 - 4X = \frac{0.0431}{0.79}(5.2\,X^2)$$

$$X = 13.0$$

$$\text{mole fraction } y_{NH_3} = \frac{p_{NH_3}}{p_{\text{total}}} = \frac{100 - 4X}{100} = 0.48$$

Applications to Solution Chemistry. Equilibrium constants have important quantitative applications in solution chemistry, and specific K's are given appropriate subscripts:

$$K_A = \text{ionization constant for weak acids}$$
$$K_B = \text{ionization constant for weak bases}$$
$$K_W = \text{dissociation constant for water}$$
$$K_D = \text{dissociation constant for complex ions}$$
$$K_{sp} = \text{solubility product for slightly soluble species}$$

Solutions and Solution Concentrations. Solutes are smaller amounts of any phase dissolved in larger amounts of liquid or solid solvents, resulting in single-phase solutions.
Solution concentrations are expressed in three ways:

(1) **Molar** concentration M is gram moles of solute per liter of solution. Small temperature dependence exists due to density changes.
(2) **Molal** concentration m is gram moles of solute per kilogram of solvent. Molal concentrations are not temperature dependent. Since volume of solution = (grams solute + grams solvent)/(solution density), molality m approximately equals molarity M only in dilute aqueous solutions.
(3) **Mole fraction** x is the ratio of moles of component i to total moles of all species present in that phase:

$$x_i \text{ or } y_i = \frac{N_i}{N_{\text{total}}} \tag{14.39}$$

where N is number of moles.

Activity-Concentration Relationship for Solutes. Activity coefficient γ is the ratio of activity a to molality m. At infinite dilution $\gamma = 1$ for all solutes, but γ varies with temperature, concentration, and the species involved (see Section 14.9). Activity is correctly used for equilibrium calculations, although molar concentrations are widely used as approximations.

Hydrogen Ion Concentration pH. The pH is an exponential notation using the operator p meaning $-\log_{10}$ of to express a wide range of $[H^+]$. At $10^{-7}\ M\ [H^+]$, resulting from dissociation of water at ambient temperature, pH = 7.

14.7 PHASE EQUILIBRIA

Temperature-Pressure Phase Diagrams. Pure chemical species, in the absence of thermal decomposition, have temperature-pressure phase diagrams of the same general shape exemplified by Fig. 14.3. In this figure, changes of state occur along transition lines, that is, sublimation along AB, fusion along BC, and vaporization along BD. Each phase transition is accompanied by volume, enthalpy, and entropy changes. Phase equilibrium exists at any T, P condition on a transition line. If multiple solid phases exist, transition occurs at T, P along lines such as EF.

Excepting material (H_2O, Bi) which expand on solidification, BC slopes only slightly to the right. For different species the coordinates of triple and critical points differ, but the diagrams retains the same general form. Logarithmic compression of scales is often helpful. The normal boiling point occurs along transition curve BD at $P = 1$ atm. The normal freezing point occurs approximately at the temperature of the triple point because of the almost vertical slope of line BC.

Equations of State. The PVT relations for gases are most simply expressed by the ideal-gas law $Pv = RT$, where v is molar volume, applicable with small error at pressures well below the critical pressure and at temperatures well above the condensation temperature. More precise equations of state are necessary for gases near their condensation conditions where atomic size becomes important and interaction must be considered. Several modifications are used, and all contain additional terms to improve precision near the liquid state. One of the several historical and more precise equations of state is the van der Waals equation

$$\left(P - \frac{a}{v^2}\right)(v - b) = RT \tag{14.40}$$

where a and b are dimensional and specific van der Waals constants applicable to individual gases. Constant a is a measure of the attractive force between molecules, and constant b is due to incompressibility and finite molecular volume.

The Redlich–Kwong equation is another example:

$$P = \frac{RT}{v - b} - \frac{a}{T^{1/2}v(v - b)} \tag{14.41}$$

where a and b are constants calculated from P_c and T_c available in Table 14.9.

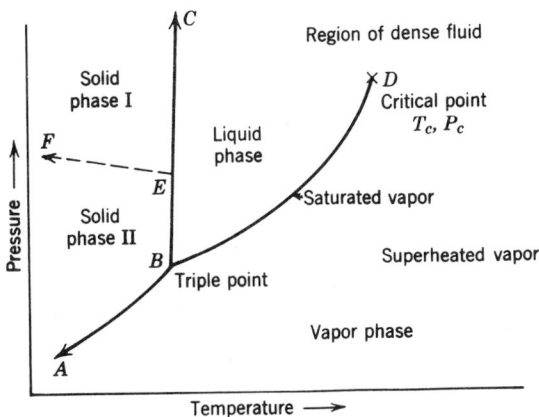

Fig. 14.3 Generalized temperature-pressure phase diagram.

TABLE 14.9 Critical Constants; Latent Heat of Vaporization at Normal Boiling Point

Compound	Formula	T_B (K)	P_c (atm)	ρ_c (g/cm³)	T_c (K)	ΔH_v at T_B (cal/g mol)
Bromine	Br_2	—	102	1.18	584	—
Chlorine	Cl_2	239.1	76.1	0.573	417	4878
Hydrogen	H_2	20.39	12.797	0.0310	33.24	216
Nitrogen	N_2	77.36	33.5	0.311	126.0	1333
Oxygen	O_2	90.19	50.14	0.430	154.78	1630
Sulfur	S_2	717.8	120	0.4	1313	2500
Air		—	37.2	0.35	132.5	—
Ammonia	NH_3	239.8	111.5	0.235	405.6	5581
Carbon dioxide	CO_2	194.7	72.9	0.459	304.1	6100
Carbon monoxide	CO	81.7	34.53	0.301	133.0	1444
Carbon disulfide	CS_2	319.4	76	0.4	546.2	6400
Hydrogen bromide	HBr	206.4	84		363	4210
Hydrogen chloride	HCl	188.1	81.6	0.421	324.6	3860
Hydrogen cyanide	HCN	298.86	50.0	0.20	456.7	6027
Hydrogen sulfide	H_2S	212.8	88.9	0.349	373.6	4463
Nitric oxide	NO	121.4	65	0.52	179	3292
Nitrous oxide	N_2O	184.7	71.7	0.45	309.7	3956
Phosgene	$COCl_2$	280.7	56	0.52	455.0	5832
Phosphine	PH_3	185.4	64.5	0.30	324.5	3490
Sulfur dioxide	SO_2	263.1	77.7	0.518	430.4	5950
Sulfur trioxide	SO_3	316.5	83.8	0.633	491.5	9990
Stannic chloride	$SnCl_4$	386	37	0.74	522	8300
Water	H_2O	373.2	218.4	0.323	647.3	9717
Methane	CH_4	111.67	45.8	0.162	191.0	1955
Ethane	C_2H_4	184.5	48.2	0.203	305.5	3517
Propane	C_3H_8	231.1	42.0	0.022	370.0	4487
n-Butane	C_4H_{10}	272.7	37.5	0.228	425.2	5350
2-Methyl propane	C_4H_{10}	261.4	36.0	0.221	408.15	5089
n-Pentane	C_5H_{12}	309.2	33.3	0.232	470.1	6160
n-Hexane	C_6H_{14}	341.9	29.9	0.234	507.9	6900
2,2-Dimethyl butane	C_6H_{14}	322.9	30.7	0.242	489.4	6290
n-Heptane	C_7H_{16}	371.6	27.0	0.235	540.2	7580
n-Octane	C_8H_{18}	398.8	24.6	0.233	569.4	8215
2,2,4-Trimethyl pentane	C_8H_{18}	372.4	25.4	0.243	544	7410
Ethene	C_2H_4	169.5	50.0	0.227	282.5	3237
Propene	C_3H_6	225.5	45.6	0.233	364.9	4405
1-Butene	C_4H_8	266.9	39.6	0.232	419.7	5240
cis-2-Butene	C_4H_8	276.9	40.8	0.239	428.2	5580
trans-2-Butene	C_4H_8	274.05	40.8	0.239	428.2	5440
Ethane	C_2H_2	184.7	61.6	0.231	308.7	4270
Propyne	C_2H_4	—	52.8	—	401	—
2-Butyne	C_4H_6	—	60	—	489	—
Benzene	C_6H_6	353.3	48.3	0.304	562.1	7350
Toluene	C_7H_8	383.8	41.6	0.291	593.8	8000
Cyclopropane	C_3H_6	240.2	54	—	398	—
Cyclopentane	C_5H_{10}	322.4	44.6	0.270	511.8	6525
Cyclohexane	C_6H_{12}	353.9	40	0.272	553.7	7190
Methyl cyclohexane	C_7H_{14}	374.1	34.32	0.285	572.3	7580
Acetaldehyde	CH_3CHO	293.3	44	0.26	461	6500
Acetone	$(CH_3)_2CO$	329.35	46.6	0.273	508.7	7100
Methanol	CH_3OH	337.9	78.5	0.272	513.2	8430
Ethanol	C_2H_5OH	351.7	63.0	0.2755	516	9220
i-Propanol	$(CH_3)_2CHOH$	355.36	53	0.27	509	9650
n-Propanol	C_3H_7OH	370.5	50.2	0.273	537	9890
Methyl ether	$(CH_3)_2O$	248.3	53	0.271	400.1	5141
Ethyl ether	$(C_2H_5)_2O$	307.8	35.6	0.263	467.0	6220
Ethylene oxide	$(CH_2)_2O$	283.7	71	0.31	469	6101
Chloromethane	CH_3Cl	248.9	65.9	0.35	416.3	5150
Fluoromethane	CH_3F	195.1	58.0	0.300	317.8	4230
Trichloromethane, chloroform	$CHCl_3$	334.4	55	0.516	536	7020
Methyl cyanide, acetonitrile	CH_3CN	354.7	47.7	0.240	547.9	7830

Source: Kobe and Lynn, *Chem. Rev.*, Vol. 52, p. 117, 1953.

Virial coefficients are used in more easily computed power series expansion used as equations of state:

$$Pv = RT + \frac{\beta}{v} + \frac{\gamma}{v^2} + \frac{\delta}{v^3} \quad \text{and} \quad Pv = RT + BP + CP^2 + DP^3 + \cdots \qquad (14.42)$$

where $\beta, \gamma, \delta,$ and B, C, D are second, third and fourth virial coefficients.

Reduced Equations of State. Coefficients of van der Waals equation of state, a and b, vary for individual gases and also vary to some extent with temperature. Coefficients a and b and the gas constant R can be generalized and expressed in terms of the critical constants for all gases using consistent units:

$$a = 3V_c^2 P_c = \left(\frac{3}{4}\right)^3 \frac{R^2 T_c^2}{P_c} \qquad (14.43)$$

$$b = \frac{V_c}{3} = \frac{RT_c}{8P_c} \qquad (14.44)$$

$$R = \frac{8}{3} \frac{P_c V_c}{T_c} \qquad (14.45)$$

Using reduced variables P_R, V_R, and T_R, the van der Waals equation of state yields a dimensionless reduced equation of state:

$$\left(P_R + \frac{3}{V_R^2}\right)(3V_R - 1) = 8T_R \qquad (14.46)$$

This infers that all gases have the same V_R at equal values of P_R and of T_R. This reduced equation of state is generally applicable at elevated pressures and near the critical point.

Corresponding States. The reduced equation of state, similar shapes of the basic T, P phase diagrams, maximum ΔH_v and ΔS_v at the triple point, and the disappearance of ΔH_v and ΔS_v at the critical point have led to the use of reduced pressure, volume, and temperature for correlation of other properties. These reduced variables P_R, V_R, and T_R are defined as ratios of the existent to the critical variable:

$$P_R = \frac{P}{P_c} \qquad V_R = \frac{V}{V_c} \qquad T_R = \frac{T}{T_c} \qquad (14.47)$$

Two gases are in corresponding states when they have the same values for two of the three reduced variables. Corresponding states permit property estimation by relating gases or liquids to comparable species where more complete PVT data are available.

Compressibility. The PVT relations of any gas may be expressed as

$$Pv = ZRT \qquad (14.48)$$

where Z is the compressibility factor or ratio of the volume of a real gas to that of an ideal gas. If the parameters P and T are replaced by their ratios to the corresponding critical values, called reduced properties, all gases are in corresponding states and can be approximately represented by a general chart (Fig. 14.4) of Z versus P_R on lines of constant T_R.

Most individual gases deviate from such a general graph by 2–7%. Hydrogen and helium, however, deviate so much that the reduced properties for them are computed by adding 8 atm to the critical pressure and 8 K to the critical temperature.

Generalized Fugacity Chart. A generalized fugacity chart can be constructed from the compressibility chart by integration, along each isotherm, of

$$\ln\frac{f}{p} = \int_0^{P_R}(Z - 1)\, d(\ln P_R) \qquad (14.49)$$

The result of these integrations is shown in Fig. 14.5.

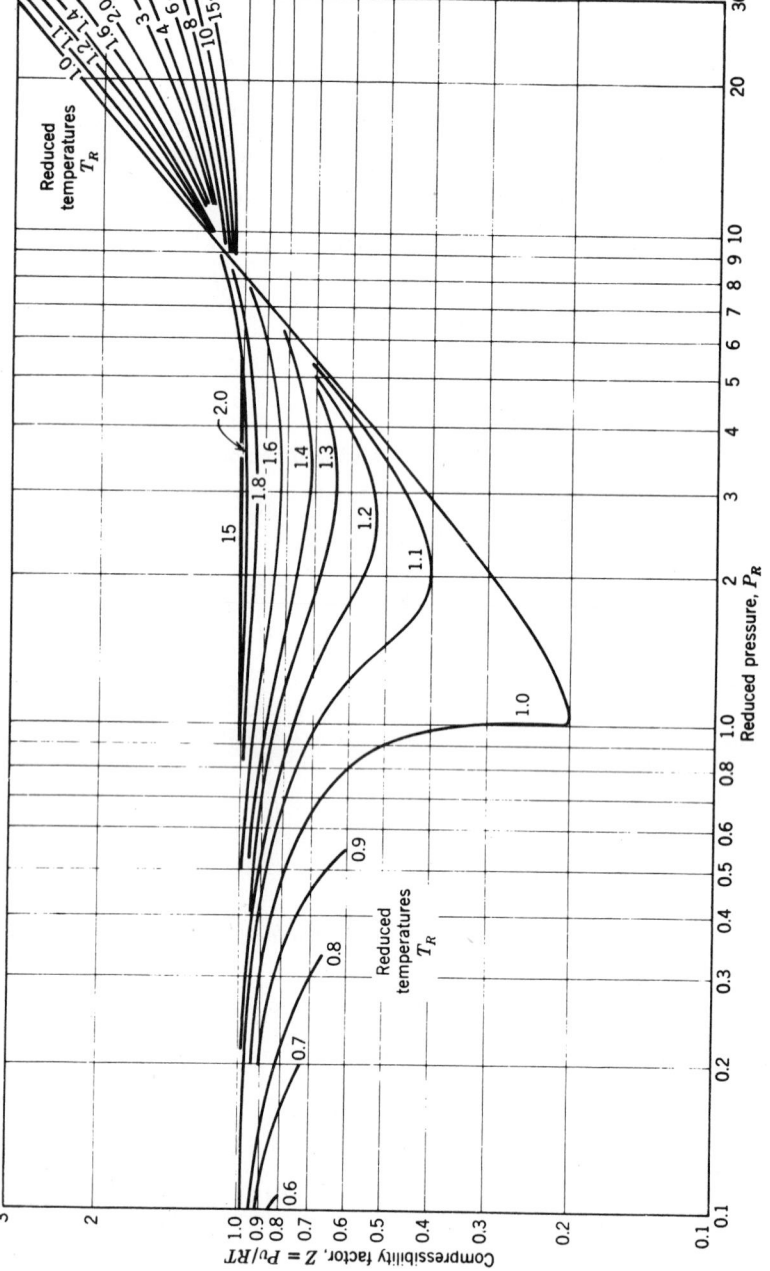

Fig. 14.4 Compressibility factor. (Reprinted by permission from M. Souders, *Engineers Companion*, Wiley, New York, 1966.)

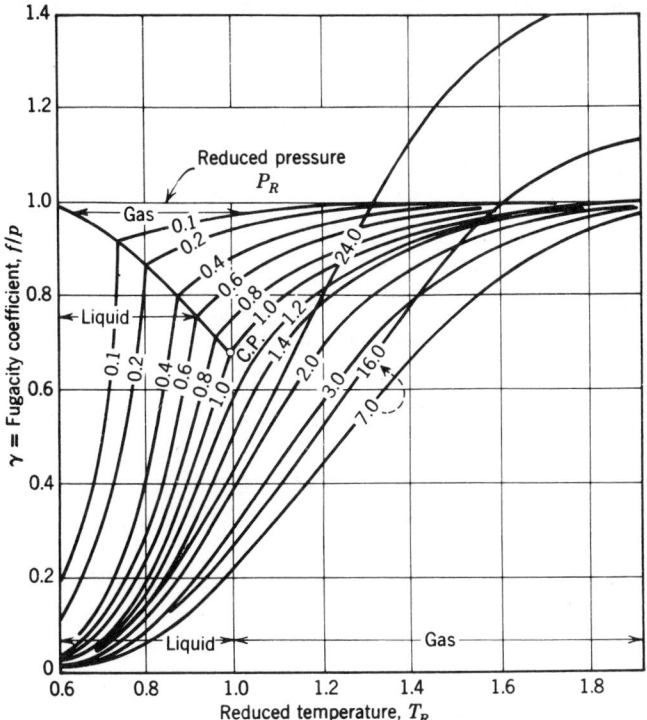

Fig. 14.5 Fugacity of gases and liquids. (Reprinted from O. A. Hougen, K. M. Watson, and R. A. Ragatz, *Chemical Process Principles*, Wiley, New York, 1959.)

Gas Mixtures. Mixtures of ideal gas have a compressibility $Z = 1$ at low pressures; however, mixtures of real gases deviate from ideality as pressure is increased. A compressibility factor for the mixture can be determined to simplify calculation. At pressures below 50 atm Dalton's law of additive partial pressures can be used for calculating mixture pressure, while above 300 atm Amagat's law of additive molal volumes allows better calculation of mixture volume. The pseudocritical point method may be used to estimate mixture compressibility between 50 and 300 atm.

Dalton's Law of Additive Partial Pressures. For real gases below 50 atm, where the actual volume of the molecules is small relative to the volume occupied and where molecular interaction does not have a large effect, Dalton's law is valid:

$$P = \sum_i p_i \qquad (14.50)$$

where $p_i = Py_i$ and P is total pressure, p_i is partial pressure of species i, and y_i is the mole fraction of species i.

Amagat's Law of Additive Molal Volumes. Amagat's law is valid at pressures above 300 atm:

$$V = \sum_i v_i \qquad (14.51)$$

where $v_i = Vy_i$ and v_i is the molal volume of species i.

Molal Average Compressibility

$$Z = \sum_i y_i Z_i \qquad (14.52)$$

where Z_i is determined using T_R and P_R for species i at the temperature of the mixture and at p_i for low total pressures (50 atm maximum) or at P for high total pressures (300 atm minimum).

Pseudocritical Point. In the absence of experimentally determined critical conditions for gas mixtures, pseudocritical temperatures and pressures can be estimated as molal average values derived from components:

$$T_{pc} = \sum_i y_i T_{ci} \quad \text{and} \quad P_{pc} = \sum_i y_i P_{ci} \tag{14.53}$$

where T_{ci} and P_{ci} are critical temperature and pressure of pure component i and the subscript pc refers to pseudocritical. These values may be used for determining a reduced temperature and pressure for the mixture, from which a compressibility may be estimated.

Examples of Gas Mixture Calculations. Calculate molal volume at 100°C (373 K) of a mixture containing 30 mol % CO_2 and 70% N_2 at (1) 20 atm, (2) 70 atm, and (3) 400 atm.

1. At 20 atm use Dalton's law and molal average compressibility:

For CO_2	For N_2
$y_i = 0.3$	$y_i = 0.7$
$p_i = 20(0.3) = 6$ atm^a	$p_j = 20(0.7) = 14$ atm
$T_c = 304$ K	$T_c = 126$ K
$P_c = 72.9$ atm	$P_c = 33.5$ atm
$T_R = \dfrac{373}{304} = 1.23$	$T_R = \dfrac{373}{126} = 2.96$
$P_R = \dfrac{6 \text{ atm}}{72.9} = 0.082$	$P_R = \dfrac{14 \text{ atm}}{33.5} = 0.42$
$Z_i = 0.98$	$Z_j = 0.99$

$$Z_{mixture} = (0.3)(98) + (0.7)(99)$$
$$= 0.987$$
$$v = \frac{ZRT}{P}$$
$$= \frac{(0.987)(0.0821)}{(20) \text{ atm}} 1 \text{ atm } (373) \text{ K/g mole K}$$
$$= 1.5 \text{ l/mol mixture}$$

^{a}This assumes that Dalton's law holds.

2. At 70 atm use pseudocritical point to determine mixture compressibility:

Mixture $T_{pc} = 0.3(304) + 0.7(126) = 179$ K

Mixture $P_{pc} = 0.3(72.9) + 0.7(33.5) = 45.3$ atm

$$T_R = \frac{373}{179} = 2.08$$

$$P_R = \frac{70}{45.3} = 1.54$$

$$Z_{mixture} = 0.980$$

$$v = \frac{ZRT}{P} = \frac{(0.980)(0.0821)(373)}{70} = 0.43 \text{ l/mol mixture}$$

3. At 400 atm use Amagat's law and molal average compressibility:

For CO_2	For N_2
$T_R = \dfrac{373}{304} = 1.23$	$T_R = \dfrac{373}{126} = 2.96$
$P_R = \dfrac{400}{72.9} = 5.49$	$P_R = \dfrac{400}{33.5} = 8.38$
$Z_i = 0.76$	$Z_i = 1.08$

$$Z_{mixture} = (0.3)(0.76) + (0.7)(1.08) = 0.984$$

$$v = \frac{ZRT}{P} = \frac{(0.984)(0.0821)(373)}{400} = 0.075 \text{ l/mol mixture}$$

Ideal Solutions. The activity of each constituent of ideal-liquid solutions is equal to its mole fraction under all conditions of temperature, pressure, and concentration. The solution volume exactly equals the summation of component volumes. The enthalpy of mixing of components is zero. The total vapor pressure is the summation of the contribution of individual components following Raoult's law. This also applies to solutions containing nonvolatile components. The freezing point of solvent in ideal solutions occurs at that temperature where the vapor pressure of the solution equals the vapor pressure of the solid solvent.

Real Solutions. Actual liquid solutions deviate from the preceding conditions of ideality. Most significant are positive or negative vapor pressure deviations from direct summation of component contributions; these affect behavior on distillation for separating the components. Deviations from ideality increase with solute concentration; that is, dilute solutions behave reasonably ideally.

Henry's Law. The solubility, and hence activity, of gaseous solute in liquid solvent is directly proportional to the partial pressure (fugacity) of the solute vapor phase in equilibrium with the solution. If the solution and vapor phases behave ideally, the model fraction of the gas solute in solution at low concentration equals solute activity ($X_i = a_i$). Henry's law is

$$Py_i = h_T x_i \qquad \qquad (14.54)$$

where y_i and x_i are mole fractions of i in the vapor and liquid phases, P is total pressure, and h_T is Henry's law of coefficient for i at temperature T. The solubility of gases in liquids is inversely proportional to temperature, and frequently chemical similarity between solute and solvent leads to a higher solubility. Henry's law applies to the same molecular species of solute in the solution and in the gas phases. For example, for NH_3, CO_2, SO_2, H_2S, Cl_2, HCl, and so on, in water it applies to the unhydrated, undissociated species in equilibrium.

For nonreactive, unhydrated, and undissociated gases such as O_2, N_2, H_2, CO, and CH_4 in water, h_T is on the order of 10^{-5} atm^{-1} between 0 and 80°C. Similar data of different magnitudes are existent for nonaqueous solvents, but a wider range of this type of data is presented in vapor-liquid equilibrium diagrams.

Raoult's Law. For ideal-vapor and ideal-liquid solutions at any temperature, the equilibrium partial pressure of any component of a liquid solution equals the product of the vapor pressure of that pure component and its mole fraction in solution:

$$Py_i = p_i^\circ x_i = p_i \qquad \qquad (14.55)$$

where y_i and x_i are the mole fractions of i in the vapor and liquid phase, P is total pressure, p_i° is the vapor pressure of pure i, and p_i is the actual partial pressure of i, all at temperature T. For ideal solutions Raoult's law holds for all mole fractions of a component, while for real solutions it can be assumed valid only for mole fractions near unity. Interaction between dissimilar species in the liquid phase leads to large deviations from ideal solution behavior and hence divergence from conformity with Raoult's law. Raoult's law can be considered a special case of Henry's law where Henry's law coefficient h_T is equal to the vapor pressure of pure component i, p_i° at that temperature.

Nernst Distribution Law. The equilibrium distribution of solute between two immiscible liquid phases at temperature T is constant regardless of concentration and is equal to the ratio of solute activities in the two phases:

$$K = \frac{a_1}{a_2} \qquad \qquad (14.56)$$

where a_1 is the activity in phase 1 and a_2 is the activity in phase 2. In dilute solutions molar concentrations can be used to replace activities.

Gibbs Phase Rule. The phase rule gives the relationship to define heterogeneous phase equilibria:

$$P + F = C + 2 \tag{14.57}$$

where P is the number of separate phases involved, C is the number of components (minimum number of chemical species necessary to define composition of all phases), and F is the number of degrees of freedom or number of variables, such as temperature, pressure, or concentration, of each component that can be varied independently without changing the number of phases.

Example 14.12

1. **Water at the triple point**. Solid, liquid, and vapor phases are in equilibrium. A pure compound is a single component, and its concentration in each phase is determined to be 100 mol %. Therefore $P = 3$, $C = 1$, and F is to be determined by substituting these values in Eq. (14.57):

$$3 + F = 1 + 2$$

   Hence $F = 0$, and the system is invariant; temperature and pressure are fixed, and a shift in either will reduce the number of phases in equilibrium.

2. **For a three-component system, defined by T, P but not composition**, determine the number of phases present; $F = 2$, $C = 3$, and P is to be determined:

$$P + 2 = 3 + 2$$

   Therefore $P = 3$ phases. Further defining the concentration of one of the three components raises F to 3 and reduces the number of phases to 2. Defining the concentration of two components also establishes the concentration of the third, raises F to 4, and reduces the number of phases to 1.

3. **Two-component system at liquid-vapor equilibrium (2 phases)**. The concentrations in the two phases are not independent; $P = 2$, $C = 2$, and F is to be determined:

$$2 + F = 2 + 2$$

   Two of the three independent variables (T and P, or concentration of only one component) are required to define the system; the third is fixed.

Colligative Properties of Solutions. Colligative properties depend on the number, rather than the nature, of particles in solution: nonelectrolytes, being undissociated, yield one particular per molecules unless they are dimerized. Strong electrolytes ionized to an extent indicated by their activity yield several ions. Colligative properties are (1) boiling point elevation as a direct consequence of vapor pressure reduction; (2) freezing point depression, also as a direct consequence of vapor pressure reduction; and (3) osmotic pressure. Since ideal solutions are expected at low concentrations, changes in these colligative properties are predictable.

Binary Solution Vapor-Liquid Equilibria. In the vapor phase

$$P_{\text{total}} = p_i + p_j$$
$$N_{\text{total}} = N_i + N_j$$
$$y_i = \frac{p_i}{P_{\text{total}}}$$
$$y_i + y_j = 1 \tag{14.58}$$

where p_i is the partial pressure, y_i is the mole fraction, and N_i is the number of moles of i.
   In the liquid phase

$$x_i = \frac{N_i}{N_{\text{total}}}$$
$$x_i + x_j = 1$$
$$N_{\text{total}} = N_i + N_j \tag{14.59}$$

where x_i is the mole fraction of i.

(a) *Ideal Solutions.* Each component of an ideal solution obeys Raoult's law, [Eq. (14.55)], relating concentrations in vapor and liquid phases.

(b) *Real Solutions.* Numerical distillation calculations often use a vapor-liquid equilibrium ratio K for each component:

$$K_i = \frac{y_i}{x_i} = \frac{f \text{ of pure } i \text{ at its vapor pressure at } T \text{ of system}}{f \text{ of pure } i \text{ at } T, P \text{ of system}} \tag{14.60}$$

The volatility ratio between components of binary solutions is

$$\alpha = \frac{k_i}{k_j} = \frac{y_i x_j}{x_i y_y} \tag{14.61}$$

where α is the volatility ratio. For ideal binary solutions, where α is constant, manipulation of Eqs. (14.60) and (14.61) relates the mole fraction in the vapor phase y_i to the volatility ratio α and the mole fraction in the liquid phase x_i:

$$y_i = \frac{\alpha x_i}{1 + x_i(\alpha - 1)} \tag{14.62}$$

Binary Solution Vapor Pressure–Composition Diagrams

1. Ideal solutions follow Raoult's law [Eq. (14.55)]. As shown in Fig. 14.6(*a*), the vapor pressure of each component is linear and proportional to the mole fraction, and the vapor pressure of the mixture is the simple sum of the component vapor pressures. The diagrams shown represent one fixed temperature.

2. Real solutions show significant deviations from linearity of individual vapor pressures with the mole fraction. At low mole fractions in the liquid phase. Henry's law [Eq. (14.54)] is followed, while at high mole fractions, Raoult's law tends to be followed. These deviations from linearity are shown in Figs. 14.6(*b*),(*c*).

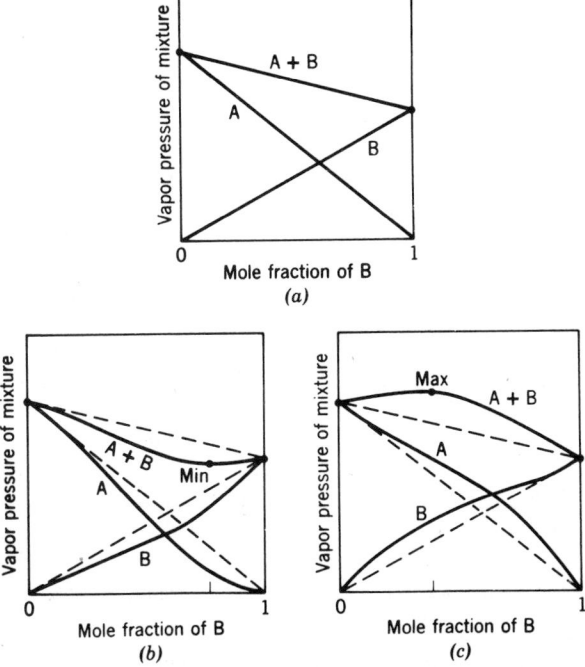

Fig. 14.6 Binary solution vapor pressure–composition diagrams.

Boiling Point–Composition Diagrams. Figure 14.7 shows three types of boiling point–composition diagrams. These are drawn for a single pressure. Diagram 14.7(*a*) is the usual case for ideal solutions; it corresponds with the vapor pressure diagram 14.6(*a*). At any given temperature, vapor composition *y* is in equilibrium with liquid composition *x*. This ideal type of boiling point diagram exists for many combinations of chemically similar materials.

Diagrams 14.7(*b*),(*c*) correspond to the vapor pressure diagrams 14.6(*b*),(*c*).

Azeotropes (constant-boiling mixtures) exist at 1 and 2. Their composition can be altered by changing pressure. These arise from nonideality of the solutions. For collected data on azeotropes, see *Handbook of Chemistry and Physics*.[3]

Freezing Point–Composition Diagrams. The freezing point is essentially independent of pressure, and the binary freezing point–composition diagrams of Fig. 14.8 apply over an extended pressure range. The usual case of two similar and mutually soluble components is to form a continuous series of solid and liquid solutions as shown in Fig. 14.8(*a*). At any one temperature the composition of the liquid phase differs from that of the solid phase in equilibrium with it. Nonidealities in the liquid and solid solutions can give rise to a composition of minimum freezing point typified by Fig. 14.8(*b*). Where components of the binary mixture have limited mutual solubility, a eutectic composition exists, and this is a mechanical mixture of the two components. A phase diagram idealized by Fig. 14.8(*c*) then occurs. Maximum-melting-point mixtures are rare, and the existence of a maximum is due to formation of a definite compound. Figure 14.8(*d*) shows the existence of an equimolar compound $A_1 B_1$. The resultant phase diagram then shows two eutectic diagrams placed side by side. One of these eutectics exists for component A and compound $A_1 B_1$; the other exists for compound $A_1 B_1$ and component B.

Numerous proliferations of these basic types of phase diagrams exist. All may be interpreted on the basis of the temperature and composition range for mutual solubility, the existence of nonideal solutions, and the formation of compounds stable over an often limited temperature range.

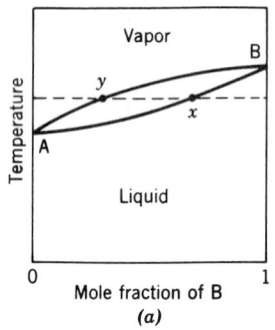

(a)

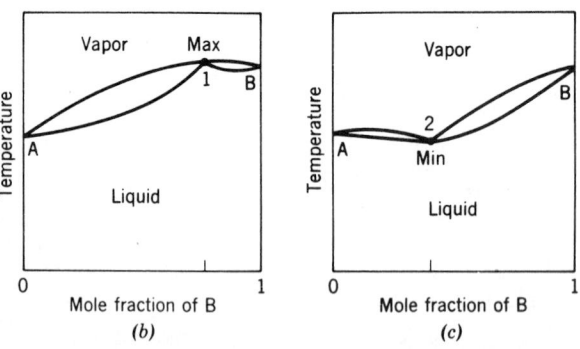

(b)   (c)

Fig. 14.7 Binary solution boiling point–composition diagrams.

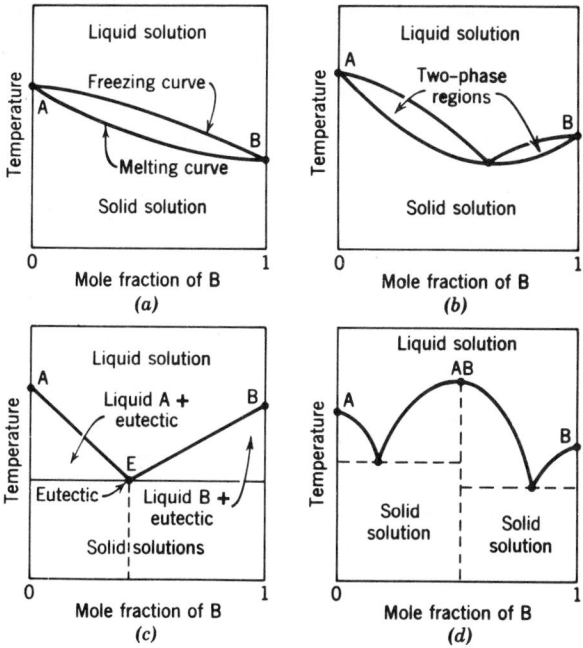

Fig. 14.8 Binary solution freezing point–composition diagrams.

14.8 CHEMICAL REACTION RATES

Process Kinetics and Conversion. The kinetics of chemical reactions vary widely, ranging from a rapid approach to equilibrium for ionic reactions in aqueous media (primarily limited by diffusion and mixing) to a much slower approach to equilibrium for organic reactions involving covalent bonds. The net reaction rate at equilibrium is zero, and further conversion of reactants to products depends upon the rate of product removal from the reaction system so the reaction can again proceed at a finite rate toward equilibrium. Only temperature can shift the equilibrium constant of the reaction system; however, the physical control of reactant and product concentrations can allow a reaction to proceed to complete conversion of reactants to product despite an equilibrium situation that may permit a maximum of a few percent of product in the system at any instant.

Kinetics of Reaction. This is concerned with the rate of approach to a temperature-dependent equilibrium condition. The rate varies with the following:

(a) *Displacement from Equilibrium.* This is dependent on the reactant and product concentrations.

(b) *Temperature.* In addition to changing the numerical value of the equilibrium constant and altering the diffusional mobility, increased temperature raises the fraction of molecules with enough energy to react.

(c) *Catalysis.* This modifies the reaction mechanism and may provide a lower energy path from reactants to products and vice versa. Catalysis has no effect on equilibrium.

(d) *Inhibition.* Inhibitors block an existent low-energy path between reactants and products, for example, by an essentially irreversible reaction with a catalytic surface.

Reaction Mechanism and Reaction Rate. Most reactions proceed via a sequence of several reversible steps. For example, the net reaction $a\mathrm{A} + b\mathrm{B} \rightleftarrows c\mathrm{C}$ might actually involve the following elementary steps:

$$\mathrm{A} \underset{K_{-1}}{\overset{k_1}{\rightleftarrows}} 2\mathrm{D}, \qquad \mathrm{D} + \mathrm{B} \underset{k_{-2}}{\overset{k_2}{\rightleftarrows}} \mathrm{E}, \qquad \mathrm{E} + \mathrm{A} \underset{k_{-3}}{\overset{k_3}{\rightleftarrows}} \mathrm{C}$$

where D and E are intermediates. The rate law for each of the species follow directly from the mechanism, for example,

$$d[E]/dt = k_2[B][D] - k_{-2}[E] - k_3[A][E] + k_{-3}[C]$$

where the brackets denote the concentration of the species. (There are analogous expressions for, e.g., $d[A]/dt$.) Determination of the overall rate of reaction requires the solution of the set of coupled differential equations for all of the species, and the resulting rate law will be a complex function of $[A]$, $[B]$, $[C]$, and the k's. While the expression for the equilibrium constant K is related to the stoichiometry of just the net reaction, the time dependence of the approach to equilibrium depends on the details of the mechanism.

Activation Energy. An enthalpy-versus-reaction diagram (Fig. 14.9) can be ideally drawn for any reaction. An activation energy barrier exists between reactants and products, and enthalpy of reaction ΔH separates the enthalpies of reactants and of products. Reactant molecules must receive a minimum activation energy E_a before they are energetic enough to react. For the reverse reaction, product molecules must receive $E_a + \Delta H$ before they can react in the reverse direction. Catalysts modify the intermediate condition and result in a reduced activation energy by providing a lower energy path between reactants and products; by so doing, they effectively speed *both* forward and reverse reactions without affecting the equilibrium condition.

Collision Theory. Energy of activation results from the statistical probability that some molecules, after collision with others, conserving momentum, have energy enough for reaction to occur. An elevated temperature increases the Maxwell–Boltzmann energy distribution of molecules and raises the number of molecules having the energy required for the reaction.

Rate Dependence on Temperature. The population of high-energy molecules is directly correlated with the temperature dependence on the reaction rate using the Arrhenius equation (which has a form similar to the dependence of the equilibrium constant on temperature)

$$\frac{d \ln k}{dT} = \frac{E_a}{RT^2} \qquad (14.63)$$

where E_a is the activation energy, R is the gas constant, and T is the absolute temperature. Integration of Eq. (14.63) yields $\ln k = -E_a/RT + \text{const}$. A plot of $\ln k$ versus $1/T$ is a straight line with slope equal to $-E_a/R$, from which the activation energy can be obtained. It shows that a rate at 25°C will double for a 10°C temperature rise when the activation energy is 12.5 kcal/g mol. A lower activation energy gives a smaller rate increase, and a higher energy gives a larger rate increase. *Caution*: Changes of mechanism, via catalysis or otherwise, will alter the E_a of a reaction.

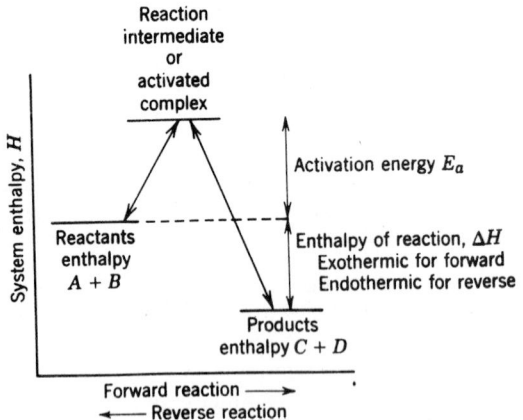

Fig. 14.9 Enthalpy-reaction diagram.

Reaction Mechanisms. The low statistical probability of trimolecular collisions essentially precludes the possibility that any reaction can proceed in one step according to the stoichiometric equation. More probable is a sequence of bimolecular reactions forming intermediate species, some of which may be too short-lived to isolate. Each of these stepwise reactions has its equilibrium constant and rates dependent on species concentrations. Usually one step is limiting and becomes the rate-determining factor for the entire reaction. The intermediate species may be an activated complex that rearranges to produce the reaction products.

Different reaction mechanisms have been postulated and proved for many reactions:

1. Simple bimolecular collision.
2. Chain reaction for polymerization, with initiation, propagation, and termination steps.
3. Free-radical, photoinitiated reactions.
4. Acid- or base-catalyzed intermediate steps.
5. Activated complex on a heterogeneous catalyst surface, with adsorption-desorption contributing to the overall rate.
6. Solvolysis participation in the reaction sequences.

Many processes have been studied in detail, with reaction sequences, proven or postulated, consistent with experimentally determined kinetics.

Initial Rate Equation. A reaction rate equation can be experimentally deduced and, from this, a reaction mechanism postulated. The initial rate equation has the generalized form

$$-\frac{dC_A}{dt} = kC_A^a C_B^b \ldots \tag{14.64}$$

where $-dC_A/dt$ is the rate of decrease of reactant A and is dependent upon a product of concentration terms. The order of a reaction is the sum of the exponents of concentration terms in the rate equation. It need not be a whole number; it may be 0, $\frac{1}{2}$, 1, $\frac{3}{2}$, or 2. There is no necessary relationship between the reaction order and the form of the overall stoichiometric relation. Where one of the reactants is a solvent, its concentration does not effectively change, and it will not appear in the rate equation.

Heterogeneous (Contact) Catalysis. Contact catalysis for gas phase reactions involves surface adsorption and reaction at active sites. Adsorption can be either physical, with a mono- or multimolecular gas layer, or chemical (chemisorption), with a monomolecular layer in some activated state. A large surface area is required. Steps involved in a catalyzed reaction are

1. Adsorption of reactants (favored at low temperature),
2. activation,
3. reaction, and
4. desorption of products (favored at high temperature).

For monolayer coverage, the amount of adsorption from the gas phase can be represented by the Langmuir adsorption isotherm

$$y = \frac{ap}{1 + bp} \tag{14.65}$$

where y is the fraction of area covered, p is pressure, and a and b are empirical constants applicable at one temperature only.

Adsorption from the liquid phase is represented by the Freundlich equation

$$y = aC^{1/b} \tag{14.66}$$

where C is the concentration of the solute and a and b are empirical and applicable at one temperature only.

Catalyst poisoning involves blocking active sites on the catalyst surface by adsorbed material that is not readily desorbed.

Photochemical Reactions. Many reactions proceed via UV energy absorption. Photons from mercury vapor at 2536 Å $\cong$ 113 kcal/g mol are absorbed, one maximum per molecule, and have energy sufficient to disrupt chemical bonds (usual range of 80–100 kcal) and form reactive free-radical species. Photons of IR wavelength 9000 Å $\cong$ 31.8 kcal/g mol do not have sufficient energy for bond disruption, although they do increase the overall thermal energy of a system.

An Einstein is an Avogadro number (6.02×10^{23}) of photons.

14.9 ELECTROCHEMISTRY

Chemical Energy of Redox Reactions. The chemical energy of redox reactions can be equated at constant T, P with Gibbs free energy using Eq. (14.7):

$$dG = -S\,dT + V\,dP + \sum_i \mu_i\,dN_i$$

Since dT and dP are both zero at constant T, P, then $dG = \sum_i \mu_i\,dN_i$. Gibbs free energy then can be manifest as electrical energy by an electron flow through an external conductor provided there is electrical isolation of the oxidation reaction (electron source) and reduction reaction (electron sink). The external conductor transfers electrons from source to sink, and a liquid junction permits internal migration of ions for preservation of charge neutrality of the system as a whole.

Chemical energy at constant T, P can be considered convertible to electrical and/or thermal energy, for it can be shown by conservation of energy that $\sum_i \mu_i\,dN_i = -E\,dq - T\,dS$. In the absence of $T\,dS$ thermal energy,

$$dG = -E\,dq \qquad (14.67)$$

where E is in volts, dq is in coulombs per gram mole, and dG is energy in joules per gram mole. Joules are convertible to calories by $J = 4.185$ J/cal.

Without source-sink electrical isolation, an internal short circuit results in energy conversion to thermal. Thus at constant T, P

$$dG = -T\,dS \qquad (14.68)$$

where T is in degrees Kelvin and dS is in calories per gram mole per degrees Kelvin.

Faraday F. This is the 96,500 C charge carried by 1 g equivalent or an Avogadro number of electrons. Hence

$$dq = F\,dn \qquad (14.69)$$

where dn is the number of gram equivalents.

Standard Potential, $E°$ Volts. The superscript refers to the standard conditions of 25°C (298 K), 1 atm fugacity (pressure), and 1 M activity of reactants and products, the same conditions as for $\Delta G°$, thus establishing dimensionally correct conversion of energy terms at the standard thermodynamic reaction condition:

$$dG° = -\frac{dn\,FE°}{J} \quad \text{or} \quad \Delta G° = -\frac{nFE°}{J} \qquad (14.70)$$

in integrated form. If the equilibrium constant for the reaction is known at 298 K, $\Delta G_T° = -RT \ln K$ can be used to calculate $\Delta G°$, which can then be related to $E°$ via Eq. (14.70). $\Delta G°$ indicates the magnitude of displacement from equilibrium that exists at 298 K when all reactants and products are at unit activity:

$$\Delta G° = RT \ln(1) - RT \ln K = -RT \ln K = -\frac{nFE°}{J} \qquad (14.71)$$

Potential $E_T°$ at Temperatures Other Than 298 K. The $\Delta G_T°$ calculated for reactions at other temperatures [via. Eq. (14.11)] can be directly related to $E_T°$. Similarly, experimental values of the equilibrium constant K can be related to $\Delta G_T°$ [via Eq. (14.23)] and then to $E_T°$ using

$$\Delta G_T° = -RT \ln K = -\frac{nFE_T°}{J} \qquad (14.72)$$

Potential E at Other Concentrations—Nernst Equation. No driving force exists at equilibrium where $\Delta G = 0$, and the driving force is directly related to displacement from equilibrium. Therefore, an E value can be corrected for reactant and product activities other than 1 at 298 K by using the general equation

$$G = RT \ln K' - RT \ln K = \frac{-nFE}{J} \tag{14.73}$$

where K' has the form and exponents of the reaction equilibrium constant but uses actual activities of products and reactants in the system rather than those existent at equilibrium.

Combining Eqs. (14.71) and (14.73) yields the Nernst equation

$$E = E_0 \frac{-JRT}{nF} \ln K' \tag{14.74}$$

At $T = 298$ K, this equation reduces to the usual equation

$$E = E° - \frac{0.0592}{n} \log_{10} K' \tag{14.75}$$

Both equations reduce E to zero at equilibrium concentrations. At temperatures other than 298 K, $E_T°$ instead of $E°$ may be used in the general equation to obtain E actual.

Standard Reduction Potential $E°$. Being related to both $\Delta G°$ and to equilibrium constant K at standard conditions, $E°$, is a measure of the chemical energy available from redox reactions. Tabulated potentials for a number of half-reactions in aqueous solution can be combined to give the standard potential of all possible combinations of these half-reactions. Table 14.10 gives the standard potentials for reduction half-reactions in 1 M acid and neutral media and in 1 M base media because the reactions and their potentials change in many cases. The table is reversible because a reduction half-reaction (with its $E°$), on reversal, becomes an oxidation half-reaction (with the same numerical $E°$ but of opposite sign).

Example 14.13

Reduction:	$Zn^{2+} + 2e^- \rightarrow Zn$	$E° = -0.76$ V in neutral or acid
Oxidation:	$Zn \rightarrow Zn^{2+} + 2e^-$	$E° = +0.76$ V in neutral or acid
Reduction:	$ZnO_2^{2-} + 2H_2O + 2e^- \rightarrow Zn + 4OH^-$	$E_B° = -1.22$ V in base
Oxidation:	$Zn + 4OH^- \rightarrow ZnO_2^{2-} + 2H_2O + 2e^-$	$E_B° = +1.22$ V in base

(a) *Sign convention.* Since $E°$ for $2H^+ + 2e^- = H_2$ is taken as a zero reference potential, all positive $E°$ values indicate the reaction can proceed as written, and all negative $E°$ values indicate that the reverse reaction (with a positive $E°$ value) can occur.

(b) *Oxidation Potential.* Reversal of any reduction half-reaction and change of sign of the reduction potential generates an oxidation half-reaction with its oxidation potential.

Reduction Potentials for Other Than Tabulated Half-Reactions. These may be obtained by adding or subtracting known half-reactions provided correction is made for the different numbers of electrons involved. This is done by summing voltage equivalents $nE°$ and dividing by n, the number of electrons involved in the final derived equation.

Example 14.14

$$\begin{array}{lll} Fe^{2+} + 2e^- \rightarrow Fe & E° = -0.44 \text{ V} & nE° = -0.88 \\ \underline{Fe^{3+} + e^- \rightarrow Fe^{2+}} & \underline{E° = +0.77 \text{ V}} & \underline{nE = +0.77} \\ Fe^{3+} + 3e^- \rightarrow Fe & & nE° = -0.11 \end{array}$$

Therefore, $E° = -0.11/3 = -0.04$ V.

Doubling stoichiometric coefficients of an equation does not change $E°$ since $E°$ is a molal quantity.

(a) Relationship of Tabulated Reduction Potentials to "Electromotive Series". A compilation of standard reduction potentials, in acid, of metal-to-ion reduction half-reactions arranged in order of decreasing $E°$ values comprises the usual abbreviated electromotive series tabulations.

TABLE 14.10 Reduction Potentials

Acid Solution	E^0	Basic Solution	E_B°
Aluminum			
$Al^{3+} + 3e^- \to Al$	-1.66	$H_2AlO_3^- + H_2O + 3e^- \to Al + 4OH^-$	-2.35
Antimony			
$Sb + 3H^+ + 3e^- \to SbH_3$	-0.51	$Sb + 3H_2O + 3e^- \to 3bH_3 + 3OH^-$	~ -1.3
$SbO^+ + 2H^+ + 3e^- \to Sb + 3H_2O$	$+0.21$	$SbO_2^- - 2H_2O + 3e^- \to Sb + 4OH^-$	-0.66
$Sb_2O_5 + 6H^+ + 4e^- \to 2SbO^+ + 3H_2O$	$+0.58$	$H_3SbO_6^{4-} + H_2O + 2e^- \to SbO_2^- + 5OH^-$	~ -0.4
Arsenic			
$As + 3H^+ + 3e^- \to AsH_3$	-0.60	$As + 3H_2O + 3e^- \to AsH_3 + 3OH^-$	-1.43
$HAsO_2 + 3H^+ + 3e^- \to As + 2H_2O$	$+0.25$	$AsO_2^- + 2H_2O + 3e^- \to As + 4OH^-$	-0.68
$H_3AsO_4 + 2H^+ + 2e^- \to HAsO_2 + 2H_2O$	$+0.56$	$AsO_4^{3-} + 2H_2O + 2e^- \to AsO_2^- + 4OH^-$	-0.67
Barium			
$Ba^{2+} + 2e^- \to Ba$	-2.90	$Ba(OH)_2 \cdot 8H_2O + 2e^- \to Ba + 2OH^- + 8H_2O$	-2.97
Beryllium			
$Be^{2+} + 2e^- \to Be$	-1.85	$Be_2O_3^{2-} + 3H_2O + 4e^- \to 2Be + 6OH^-$	-2.62
Bismuth			
$BiO^+ + 2H^+ + 3e^- \to Bi + H_2O$	$+0.32$	$Bi_2O_3 + 3H_2O + 6e^- \to 2Bi + 6OH^-$	-0.46
Boron			
$H_3BO_3 + 3H^+ + 3e^- \to B + 3H_2O$	-0.87	$H_2BO_3^- + 3e^- \to B + 4OH^-$	-1.79
Bromine			
$Br_2 + 2e^- \to 2Br^-$	$+1.07$	$Br_2 + 2e^- \to 2Br^-$	$+1.07$
$HBrO + H^+ + e^- \to \frac{1}{2}Br_2 + H_2O$	$+1.59$	$BrO^- + H_2O + 2e^- \to Br^- + 2OH^-$	$+0.76$
$BrO_3^- + 6H^+ + 5e^- \to \frac{1}{2}Br_2 + 3H_2O$	$+1.52$	$BrO_3^- + 3H_2O + 6e^- \to Br^- + 6OH^-$	$+0.61$
Cadmium			
$Cd^{2+} + 2e^- \to Cd$	-0.40	$Cd(OH)_2 + 2e^- \to Cd + 2OH^-$	-0.81
Calcium			
$Ca^{2+} + 2e^- \to Ca$	-2.87	$Ca(OH)_2 + 2e^- \to Ca + 2OH^-$	-3.03
Cerium			
$Ce^{4+} + e^- \to Ce^{3+}$	$+1.61$		
$Ce^{3+} + 3e^- \to Ce$	-2.48 or -2.34		
Cesium			
$Cs^+ + e^- \to Cs$	$+2.92$	$Cs^+ + e^- \to Cs$	-2.92

	Acidic solution	E°	Basic solution	E°
Chlorine	$Cl_2 + 2e^- \rightarrow 2Cl^-$	+1.36		
	$HClO + H^+ + e^- \rightarrow \tfrac{1}{2}Cl_2 + H_2O$	+1.63	$ClO^- + H_2O + e^- \rightarrow \tfrac{1}{2}Cl_2 + 2OH^-$	+0.40
	$ClO_3^- + 6H^+ + 5e^- \rightarrow \tfrac{1}{2}Cl_2 + 3H_2O$	+1.47	$ClO_3^- + 2H_2O + 4e^- \rightarrow ClO^- + 4OH^-$	+0.50
	$ClO_4^- + 2H^+ + 2e^- \rightarrow ClO_3^- + H_2O$	+1.20	$ClO_4^- + H_2O + 2e^- \rightarrow ClO_3^- + 2OH^-$	+0.36
Chromium	$Cr^{3+} + 3e^- \rightarrow Cr$	−0.74	$CrO_4^{2-} + 4H_2O + 3e^- \rightarrow Cr(OH)_3 + 5OH^-$	−0.13
	$Cr^{3+} + e^- \rightarrow Cr^{2+}$	−0.41	$CrO_2^- + 2H_2O + 3e^- \rightarrow Cr + 4OH^-$	−1.2
	$Cr_2O_7^{2-} + 14H^+ + 6e^- \rightarrow 2Cr^{3+} + 7H_2O$	+1.33	$Cr(OH)_3 + 3e^- \rightarrow Cr + 3OH^-$	−1.3
Cobalt	$Co^{2+} + 2e^- \rightarrow Co$	−0.28	$Co(OH)_2 + 2e^- \rightarrow Co + 2OH^-$	−0.73
	$Co^{3+} + e^- \rightarrow Co^{2+}$	+1.82	$Co(OH)_3 + e^- \rightarrow Co(OH)_2 + OH^-$	+0.14
Copper	$Cu^{2+} + 2e^- \rightarrow Cu$	+0.34	$Cu_2O + H_2O + 2e^- \rightarrow 2Cu + 2OH^-$	−0.36
	$Cu^{2+} + e^- \rightarrow Cu^+$	+0.15	$2Cu(OH)_2 + e^- \rightarrow Cu_2O + 4OH^-$	−0.08
Fluorine	$F_2 + 2e^- \rightarrow 2F^-$	+2.85		
Gallium	$Ga^{3+} + 3e^- \rightarrow Ga$	−0.53	$H_2GaO_3^- + H_2O + 3e^- \rightarrow Ga + 4OH^-$	−1.22
Germanium	$GeO_2 + 4H^+ + 4e^- \rightarrow Ge + 2H_2O$	−0.15	$HGeO_3^- + 2H_2O + 4e^- \rightarrow Ge + 5OH^-$	−1.0
Gold	$Au^{3+} + 3e^- \rightarrow Au$	+1.50	$H_2AuO_3^- + H_2O + 3e^- \rightarrow Au + 4OH^-$	+0.7
	$Au^{3+} + 2e^- \rightarrow Au^+$	+1.41		
Hafnium	$Hf^{4+} + 4e^- \rightarrow Hf$	−1.70	$HfO(OH)_2 + H_2O + 4e^- \rightarrow Hf + 4OH^-$	−2.50
Hydrogen	$H_2 + 2e^- \rightarrow 2H^-$	−2.23	$2H_2O + 2e^- \rightarrow H_2 + 2OH^-$	−0.83
	$2H^+ + 2e^- \rightarrow H_2$	0 reference	$H_2 + 2e^- \rightarrow 2H^-$	−2.23
			$2H^+ \text{ (at pH 7)} + 2e^- \rightarrow H_2$	−0.41
Indium	$In^{3+} + 3e^- \rightarrow In$	−0.34	$In(OH)_3 + 3e^- \rightarrow In + 3OH^-$	−1.0

TABLE 14.10 Reduction Potentials (Continued)

Acid Solution	E^0	Basic Solution	E_B°
Iodine			
$I_2 + 2e^- \rightarrow 2I^-$	+0.54	$I_2 + 2e^- \rightarrow 2I^-$	+0.54
$HIO + H^+ + e^- \rightarrow \frac{1}{2}I_2 + H_2O$	+1.45	$2IO^- + 2H_2O + 2e^- \rightarrow \frac{1}{2}I_2 + 4OH^-$	+0.45
$IO_3^- + 5H^+ + 4e^- \rightarrow HIO + 2H_2O$	+1.14	$IO_3^- + 2H_2O + 4e^- \rightarrow IO^- + 4OH^-$	+0.14
Iridium			
$Ir^{3+} + 3e^- \rightarrow Ir$	+1.15	$Ir_2O_3 + 3H_2O + 6e^- \rightarrow Ir + 6OH^-$	+0.1
$IrO_2 + 4H^+ + E^- \rightarrow Ir^{3+} + H_2O$	+0.7	$2IrO_2 + H_2O + 2e^- \rightarrow Ir_2O_3 + 2OH^-$	+0.1
Iron			
$Fe^{2+} + 2e^- \rightarrow Fe$	-0.44	$Fe(OH)_2 + 2e^- \rightarrow Fe + 2OH^-$	-0.88
$Fe^{3+} + e^- \rightarrow Fe^{2+}$	+0.78	$Fe(OH)_3 + e^- \rightarrow Fe(OH)_2 + OH^-$	-0.56
Lanthanum			
$La^{3+} + 3e^- \rightarrow La$	-2.52	$La(OH)_3 + 3e^- \rightarrow La + 3OH^-$	-2.90
Lead			
$Pb^{2+} + 2e^- \rightarrow Pb$	-0.13	$PbO_2 + H_2O + 2e^- \rightarrow PbO + 2OH^-$	+0.25
$PbO_2 + 4H^+ + 2e^- \rightarrow Pb^{2+} + 2H_2O$	+1.46	$HPbO_2^- + H_2O + 2e^- \rightarrow Pb + 3OH^-$	-0.54
Lithium			
$Li^+ + e^- \rightarrow Li$	-3.05	$Li^+ + e^- \rightarrow Li$	-3.05
Magnesium			
$Mg^{2+} + 2e^- \rightarrow Mg$	-2.37	$Mg(OH)_2 + 2e^- \rightarrow Mg + 2OH^-$	-2.69
Manganese			
$Mn^{2+} + 2e^- \rightarrow Mn$	-1.18	$Mn(OH)_2 + 2e^- \rightarrow Mn + 2OH^-$	-1.55
$MnO_2 + 4H^+ + 2e^- \rightarrow Mn^{2+} + 2H_2O$	+1.23	$Mn(OH)_3 + e^- \rightarrow Mn(OH)_2 + OH^-$	-0.4
$MnO_4^- + 4H^+ + 3e^- \rightarrow MnO_2 + 2H_2O$	+1.70	$MnO_2 + H_2O + 2e^- \rightarrow Mn(OH)_2 + 2OH^-$	-0.05
		$MnO_4^- + 2H_2O + 3e^- \rightarrow MnO_2 + 4OH^-$	+0.59
Mercury			
$Hg^{2+} + 2e^- \rightarrow Hg$	+0.85	$HgO + H_2O + 2e^- \rightarrow Hg + 2OH^-$	+0.10
$2Hg^{2+} + 2e^- \rightarrow Hg_2^{2+}$	+0.92		
Molybdenum			
$Mo^{3+} + 3e^- \rightarrow Mo$	-0.20	$MoO_2^- + 2H_2O + 4e^- \rightarrow Mo + 4OH^-$	-0.87
$MoO_2 + 4H^+ + 2e^- \rightarrow Mo^{3+} + 2H_2O$	0.0	$MoO_4^{2-} + 2H_2O + 2e^- \rightarrow MoO_2^- + 4OH^-$	-1.40
$H_2MoO_4 + 2H^+ + e^- \rightarrow MoO_2^+ + 2H_2O$	+0.4	$MoO_4^- + 4H_2O + 6e^- \rightarrow Mo + 8OH^-$	-1.05
Nickel			
$Ni^{2+} + 2e^- \rightarrow Ni$	-0.25	$Ni(OH_2) + 2e^- \rightarrow Ni + 2OH^-$	-0.72
$NiO_2 + 4H^+ + 2e^- \rightarrow Ni^{2+} + 2H_2O$	+1.78	$NiO_2 + 2H_2O + 2e^- \rightarrow Ni(OH)_2 + 2OH^-$	+0.49

Element / Reaction	E	Reaction	E
Niobium			
$Nb^{3+} + 3e^- \rightarrow Nb$	-1.10		
$Nb_2O_5 + 10H^+ + 10e^- \rightarrow 2Nb + 5H_2O$	-0.65		
Nitrogen			
$2NO_3^- + 4H^+ + 2e^- \rightarrow N_2O_4 + 2H_2O$	$+0.80$	$2NO_3^- + 2H_2O + 2e^- \rightarrow N_2O_4 + 4OH^-$	-0.86
$NO_3^- + 4H^+ + 3e^- \rightarrow NO + 3H_2O$	$+0.96$	$NO_3^- + H_2O + 2e^- \rightarrow NO_2^- + 2OH^-$	$+0.01$
$NO_3^- + 2H^+ + e^- \rightarrow HNO_2 + H_2O$	$+0.94$		
Osmium			
$OsO_4 + 8H^+ + 8e^- \rightarrow Os + 4H_2O$	$+0.85$	$HOsO_5^- + 4H_2O + 8e^- \rightarrow Os + 9OH^-$	$+0.02$
Oxygen			
$O_2 + 4H^+ + 4e^- \rightarrow 2H_2O$	$+1.23$	$O_3 + H_2O + 2e^- \rightarrow O_2 + 2OH^-$	$+1.24$
$O_3 + 2H^+ + 2e^- \rightarrow O_2 + H_2O$	$+2.07$	$O_2 + 2H_2O + 4e^- \rightarrow 4OH^-$	$+0.40$
		$O_2 + 4H^+ \text{ (at pH 7)} + 4e^- = 2H_2O$	$+0.81$
Palladium			
$Pd^{2+} + 2e^- \rightarrow Pd$	$+0.99$	$Pd(OH)_2 + 2e^- \rightarrow Pb + 2OH^-$	$+0.07$
$Pd^{4+} + 2e^- \rightarrow Pd^{2+}$	$+1.60$	$Pd(OH)_4 + 2e^- \rightarrow Pd(OH)_2 + 2OH^-$	$+0.73$
Phosphorus			
$P + 3H^+ + 3e^- \rightarrow PH_3$	-0.07	$P + 3H_2O + 3e^- \rightarrow PH_3 + 3OH^-$	-0.89
$H_3PO_2 + H^+ + e^- \rightarrow P + 2H_2O$	-0.51	$H_2PO_2^- + e^- \rightarrow P + 2OH^-$	-2.05
$H_3PO_3 + 2H^+ + 2e^- \rightarrow H_3PO_2 + H_2O$	-0.51	$HPO_3^{2-} + 2H_2O + 2e^- \rightarrow H_2PO_2^- + 3OH^-$	-1.57
$H_3PO_4 + 2H^+ + 2e^- \rightarrow H_3PO_3 + H_2O$	-0.28	$PO_4^{3-} + 2H_2O + 2e^- \rightarrow HPO_3^{2-} + 3OH^-$	-1.12
Platinum			
$Pt^{2+} + 2e^- \rightarrow Pt$	$+1.20$	$Pt(OH)_2 + 2e^- \rightarrow Pt + 2OH^-$	$+0.15$
$Pt(OH)_2 + 2H^+ + 2e^- \rightarrow Pt + 2H_2O$	$+0.98$	$Pt(OH)_6^{2-} + 2e^- + Pt(OH)_2 + 4OH^-$	$+0.20$
Potassium			
$K^+ + e^- \rightarrow K$	-2.93	$K^+ + e^- \rightarrow K^+$	-2.93
Rhenium			
$ReO_2 + 4H^+ + 4e^- \rightarrow Re + 2H_2O$	$+0.25$	$ReO_2 + H_2O + 4e^- \rightarrow Re + 4OH^-$	-0.58
$ReO_4^- + 4H^+ + 3e^- \rightarrow ReO_2 + 2H_2O$	$+0.51$	$ReO_4^- + 2H_2O + 3e^- \rightarrow ReO_2 + 4OH^-$	-0.59
Rhodium			
$Rh^{3+} + 3e^- \rightarrow Rh$	$+0.8$	$Rh_2O_3 + 3H_2O + 6e^- \rightarrow 2Rh + 6OH^-$	$+0.04$
Rubidium			
$Rb^+ + 3e^- \rightarrow Rb$	-2.93	$Rb^+ + e^- \rightarrow Rb$	-2.93
Ruthenium			
$RuO_2 + 4H^+ + 4e^- \rightarrow Ru + 2H_2O$	$+0.79$	$RuO_2 + 2H_2O + 4e^- \rightarrow Ru + 4OH^-$	-0.04
		$RuO_4 + H_2O + 4e^- \rightarrow RuO_2 + 4OH^-$	$+0.58$

TABLE 14.10 Reduction Potentials (Continued)

Acid Solution	E^0	Basic Solution	E°_B
Scandium			
$Sc^{3+} + 3e^- \rightarrow Sc$	-2.08	$Sc(OH)_3 + 3e^- \rightarrow Sc + 3OH^-$	~ -0.26
Selenium			
$Se + 2H^+ + 2e^- \rightarrow H_2Se$	-0.40	$Se + 2e^- \rightarrow Se^{2-}$	-0.92
$H_2SeO_3 + 4H^+ + 4e^- \rightarrow Se + 3H_2O$	$+0.74$	$SeO_3^{2-} + 3H_2O + 4e^- \rightarrow Se + 6OH^-$	-0.37
$SeO_4^{2-} + 4H^+ + 2e^- \rightarrow H_2SeO_3 + H_2O$	$+1.15$	$SeO_4^{2-} + H_2O + 2e^- \rightarrow SeO_3^{2-} + 2OH^-$	$+0.05$
Silicon			
$H_2SiO_3 + 4H^+ + 4e^- \rightarrow Si + 3H_2O$	-0.87	$SiO_3^{2-} + 3H_2O + 4e^- \rightarrow Si + 6OH^-$	-1.73
Silver			
$Ag^+ + e^- \rightarrow Ag$	$+0.80$	$Ag_2O + H_2O + 2e^- \rightarrow 2Ag + 2OH^-$	$+0.34$
Sodium			
$Na^+ + e^- \rightarrow Na$	-2.71	$Na^+ + e^- \rightarrow Na$	-2.71
Strontium			
$Sr^{2+} + 2e^- \rightarrow Sr$	-2.89	$Sr(OH)_2 \cdot 8H_2O + 2e^- \rightarrow Sr + 2OH^- + 8H_2O$	-2.99
Sulfur			
$S + 2H^+ + 2e^- \rightarrow H_2S$	$+0.14$	$S + 2e^- \rightarrow S^{2-}$	-0.51
$S_2O_3^{2-} + 6H^+ + 4e^- \rightarrow 2S + 3H_2O$	$+0.50$	$S_2O_3^{2-} + 3H_2O + 4e^- \rightarrow 2S + 6OH^-$	-0.74
$S_4O_6^{2-} + 2e^- \rightarrow 2S_2O_3^{2-}$	$+0.08$	$S_4O_6^{2-} + 2e^- \rightarrow 2S_2O_3^{2-}$	$+0.08$
$4H_2SO_3 + 4H^+ + 6e^- \rightarrow S_4O_6^{2-} + 6H_2O$	$+0.51$	$2SO_3^{2-} + 3H_2O + 4e^- \rightarrow S_2O_3^{2-} + 6OH^-$	-0.58
$SO_4^{2-} + 4H^+ + 2e^- \rightarrow H_2SO_3 + H_2O$	$+0.17$	$SO_4^{2-} + H_2O + 2e^- \rightarrow SO_3^{2-} + 2OH^-$	-0.93
Tantalum			
$Ta_2O_5 + 10H^+ + 10e^- \rightarrow 2Ta + 5H_2O$	-0.81		
Tellurium			
$Te + 2H^+ + 2e^- \rightarrow H_2Te$	-0.72	$Te + 2e^- \rightarrow Te^{2-}$	-1.14
$TeO_2 + 4H^+ + 4e^- \rightarrow Te + 2H_2O$	$+0.53$	$TeO_3^{2-} + 3H_2O + 4e^- \rightarrow Te + 6OH^-$	-0.57
$H_6TeO_6 + 2H^+ + 2e^- \rightarrow TeO_2 + 4H_2O$	$+1.02$	$TeO_4^{2-} + H_2O + 2e^- \rightarrow TeO_3^{2-} + 2OH^-$	$+0.4$
Tallium			
$Tl^+ + e^- \rightarrow Tl$	-0.34	$Tl(OH) + e^- \rightarrow Tl + OH^-$	-0.34
$Tl^{3+} + 2e^- \rightarrow Tl^+$	$+1.25$	$Tl(OH)_3 + 2e^- \rightarrow Tl(OH) + 2OH^-$	-0.05

Thorium			
$Th^{4+} + 4e^- \rightarrow Th$	-1.90	$Th(OH)_4 + 4e^- \rightarrow Th + 4OH^-$	-2.48
Tin			
$Sn^{2+} + 2e^- \rightarrow Sn$	-0.14	$HSnO_2^- + H_2O + 2e^- \rightarrow Sn + 3OH^-$	-0.93
$Sn^{4+} + 2e^- \rightarrow Sn^{2+}$	$+0.15$	$Sn(OH)_6^{2-} + 2e^- \rightarrow HSnO_2^- + H_2O + 3OH^-$	-0.90
Titanium			
$Ti^{2+} + 2e^- \rightarrow Ti$	-1.63	$TiO_2(xH_2O) + 4e^- \rightarrow Ti + 4OH^- +(x-2)H_2O$	-1.69
$Ti^{3+} + e^- \rightarrow Ti^{2+}$	-0.37		
$TiO^{2+} + 2H^+ + e^- \rightarrow Ti^{3+} + H_2O$	$+0.10$		
Tungsten			
$WO_2 + 4H^+ + 4e^- \rightarrow W + 2H_2O$	-0.12	$WO_4^{2-} + 4H_2O + 6e^- \rightarrow W + 8OH^-$	-1.05
$W_2O_6 + 2H^+ + 2e^- \rightarrow 2WO_2 + H_2O$	-0.04		
$2WO_3 + 2H^+ + 2e^- \rightarrow W_2O_5 + H_2O$	-0.03		
Uranium			
$U^{3+} + 3e^- \rightarrow U$	-1.80	$U(OH)_3 + 3e^- \rightarrow U + 3OH^-$	-2.17
$U^{4+} + e^- \rightarrow U^{3+}$	-0.61	$U(OH)_4 + e^- \rightarrow U(OH)_3 + OH^-$	-2.14
$UO_2^+ + 4H^+ + e^- \rightarrow U^{4+} + 2H_2O$	$+0.62$	$Na_2UO_4 + 4H_2O + 2e^- \rightarrow U(OH)_4$	-1.61
$UO_2^{2+} + e^- \rightarrow UO_2^+$	$+0.05$	$+2Na^+ + 4OH^-$	
Vanadium			
$V^{2+} + 2e^- \rightarrow V$	-1.18	$VO_3^- + 3H_2O + 5e^- \rightarrow V + 6OH^-$	-1.15
$V^{3+} + e^- \rightarrow V^{2+}$	-0.26		
$VO^{2+} + 2H^+ + e^- \rightarrow V^{3+} + H_2O$	$+0.36$		
$VO_2^+ + 2H^+ + e^- \rightarrow VO^{2+} + H_2O$	$+1.00$		
Yttrium			
$Y^{3+} + 3e^- \rightarrow Y$	-2.37	$Y(OH)_3 + 3e^- \rightarrow Y + 3OH^-$	-2.8
Zinc			
$Zn^+ + 2e^- \rightarrow Zn$	-0.76	$Zn(OH)_2 + 2e^- \rightarrow Zn + 2OH^-$	-1.24
		$ZnO_2^{2-} + 2H_2O + 2e^- \rightarrow Zn + 4OH^-$	-1.22
Zirconium			
$Zr^{4+} + 4e^- \rightarrow Zr$	-1.53	$H_2ZrO_3 + H_2O + 4e^- \rightarrow Zr + 4OH^-$	-2.36

TABLE 14.11   Summary of Air and Water Reduction Potentials

Acidic Solution, pH 0	Neutral, pH 7	Basic Solution, pH 14
$O_2 + 4H^+ + 4e^- \rightarrow 2H_2O$	$O_2 + 4H^+ + 4e^- \rightarrow 2H_2O$	$O_2 + 2H_2O + 4e^- \rightarrow 4OH^-$
$E^\circ = +1.23$ V	$E^\circ = +0.81$ V	$E_B^\circ = +0.40$ V
$2H^+ + 2e^- \rightarrow H_2$	$2H^+ + 2e^- \rightarrow H_2$	$2H_2O + 2e^- \rightarrow H_2 + 2OH^-$
$E^\circ = 0$ V	$E^\circ = -0.41$ V	$E_B^\circ = -0.83$ V

Table of Standard Reduction Potentials in Aqueous Solution. The potentials listed are derived from Latimer with additions from the recent literature. The E° values in volts for the reduction equations as written are based on a zero-reference voltage for the hydrogen couple at 25°C. Not all reactions are physically reversible or achievable as written under usual laboratory conditions. They are for 1 M activities for water-soluble species, for 1 atm fugacity for gases, and do not take into account overvoltages necessary for gas evolution at electrodes. Within the limitations imposed, these remain one of the most reliable sources of relative potential for reaction, although they infer nothing about reaction rate.

Significance of Reduction Half-Reactions and Potentials for Predicting Chemical Reactivity:

1. Reaction with acids, water, and bases. Stability of solutions to air oxidation.
2. Redox reactions. A source of balanced half-reactions that may be added to give balanced redox reactions.
3. A systematic guide to descriptive inorganic chemistry and usual oxidation states.
4. Compound disproportionation (self-oxidation and reduction).

Reactions with Acids, Bases, Water, or Air. Such reactions add complexity to all chemical reactions and offer competing reactions. If a potential exists to oxidize or reduce water (consider the pH of the system), that oxidation or reduction can proceed. Atmospheric oxygen is easily reduced (is a moderately strong oxidizing agent), and its reactions in systems exposed to air may be significant. These important potentials, listed under hydrogen and oxygen in the reduction potential table, are pH dependent and are summarized in Table 14.11.

Example 14.15

1. Air can oxidize Sn^{2+} to Sn^{4+} in neutral solution:

$$O_2 + 4H^+ + 4e^- \rightarrow 2H_2O \qquad\qquad E^\circ = +0.81 \text{ V}$$
$$\underline{2[Sn^{2+} \rightarrow Sn^{4+} + 2e^-]} \qquad\qquad\quad \underline{E^\circ = -0.15 \text{ V}}$$
$$2Sn^{2+} + O_2 + 4H^+ \rightarrow 2Sn^{4+} + 2H_2O \quad\; E^\circ = +0.66 \text{ V}$$

2. Gold is dissolved by dilute cyanide solution with air oxidation in alkaline solution:

$$O_2 + 2H_2O + 4e^- \rightarrow 4OH^- \qquad\qquad\qquad\quad E_B^\circ = +0.40 \text{ V}$$
$$\underline{4[2CN^- + Au \rightarrow Au(CN)_2^- + e^-]} \qquad\qquad\quad\; \underline{E_B^\circ = +0.60 \text{ V}}$$
$$4Au + 8CN^- + O_2 + 4H^+ \rightarrow 2H_2O + 4Au(CN)_2^- \quad E_B^\circ = +1.00 \text{ V}$$

3. Iron rusting in neutral solution:

$$O_2 + 4H^+ \text{(at pH 7)} + 4e^- \rightarrow 2H_2O \qquad E^\circ = +0.81 \qquad\qquad\qquad\qquad\text{(a)}$$
$$Fe + 2OH^- \rightarrow Fe(OH)_2 + 2e^- \qquad E^\circ = (+0.88) \text{ (this } E^\circ \text{ is reduced at}$$
$$\qquad\qquad\qquad\qquad\qquad\qquad\qquad\qquad\quad \text{pH 7 to } \sim +0.6 \text{ V)} \qquad\qquad\text{(b)}$$
$$Fe(OH)_2 + OH^- \rightarrow Fe(OH)_3 + e^- \qquad E_B^\circ = +0.56 \text{ V} \qquad\qquad\qquad\text{(c)}$$

The reduction of oxygen reaction (a) and the oxidation of iron reaction (b) are followed by further reactions (a) and (c). An alkaline solution reduces potential (a) to +0.40 and an acid solution increases it to +1.23. Many reaction schemes are postulated for rusting; this one is plausible. Overall corrosion rate is increased at low pH and is dependent on oxygen diffusion rate.

4. Aluminum dissolves in a base to liberate hydrogen:

$$2[Al + 4OH^- \rightarrow H_2AlO_3^- + H_2O + 3e^-] \qquad E_B^\circ = +2.35 \text{ V}$$
$$\underline{3[2H_2O + 2e^- \rightarrow H_2 + 2OH^-]} \qquad\qquad\qquad\;\; \underline{E_B^\circ = -0.83 \text{ V}}$$
$$2Al + 2OH^- + 4H_2O \rightarrow 2H_2AlO_3^- + 3H_2 \qquad\quad E_B^\circ = +1.52 \text{ V}$$

5. Copper will not dissolve in a nonoxidizing acid:

$$\begin{aligned} Cu &\rightarrow Cu^{2+} + 2e^- & E° &= -0.34 \text{ V} \\ \underline{2H^+ + 2e^- \rightarrow H_2} & & \underline{E° = 0 \text{ V}} \\ Cu + 2H^+ &\rightarrow Cu^{2+} + H_2 & E° &= -0.34 \text{ V and reaction cannot proceed spontaneously} \end{aligned}$$

6. Copper will dissolve in oxidizing acids (example: nitric):

$$\begin{aligned} 3[Cu &\rightarrow Cu^{2+} + 2e^-] & E° &= -0.34 \text{ V} \\ \underline{2[NO_3^- + 4H^+ + 3e^- \rightarrow 2H_2O + NO]} & & \underline{E° = +0.96 \text{ V}} \\ 3Cu + 2NO_3^- + 8H^+ &\rightarrow 3Cu^{2+} + 4H_2O + 2NO & E° &= +0.62 \text{ V} \end{aligned}$$

Electrolytic Cells. Electrolysis cells and the charging of batteries are comparable because an external applied voltage is used, depending on polarity, to augment the rate or to reverse the direction of spontaneous reaction. For example, charging a lead storage battery requires an opposing voltage greater than that of the spontaneous discharge reaction to reverse the direction of the oxidation and reduction reactions. Electrolysis cells can be operated using either molten salts or aqueous solutions, both of which are conductive.

(a) Molten Salt Cells. These cells are simple and straightforward in their operation.

Example 14.16. Electrolysis of fused $MgCl_2$ (with NaCl added to reduce melting point) liberates chlorine at the anode and magnesium metal at the cathode:

$$\begin{aligned} \text{Oxidation at anode: } 2Cl^- &\rightarrow Cl_2 + 3e^- & E° &= -1.36 \text{ V} \\ \text{Reduction at cathode: } Mg^{2+} + 2e^- &\rightarrow Mg & \underline{E° = -2.34 \text{ V}} \\ & & E° &= -3.70 \text{ V} \end{aligned}$$

Possible competitive reactions do not occur because the minimum applied potential of 3.70 V (plus IR losses) reduces the magnesium ion more easily than the sodium ion:

$$Na^+ + e^- \rightarrow Na \qquad E° = -2.71 \text{ V}$$

Although $E°$ is not directly applicable because of the nonaqueous media, elevated temperature, and high ionic concentrations, it can be helpful as a very crude approximation.

(b) Aqueous Solution Electrolysis Cells. These cells are used for electroplating, electrolytic copper purification, and many chemical reactions. Competing reactions such as electrolysis of water and overvoltages required for oxygen and hydrogen bubble formation and evolution are complications encountered.

Example 14.17. Electrolysis in Aqueous Solution of Electrolytic Copper. Purification of impure copper requires its separation from Fe, Ag, Ni, Sb, and As metals. An electrolytic cell with an impure copper anode, $CuSO_4$ solution as electrolyte, and a pure copper cathode is used.

Reactions at Anode		Reactions Possible at Cathode
		At low concentrations:
$Fe \rightarrow Fe^{2+} + 2e^-$	$E° = +0.44$ V	$Fe^2 + 2e^- \rightarrow Fe,\ E° = -0.44$ V
$Cu \rightarrow Cu^{2+} + 2e^-$	$E° = -0.34$ V	At high concentrations:
$Ag \rightarrow Ag^+ + e^-$	$E° = -0.80$ V	$Cu^{2+} + 2e^- \rightarrow Cu,\ E° = 0.34$ V

By application of a low voltage, Cu and the more easily oxidized metals Fe, Ni, Sb, and As are selectively dissolved at the anode, leaving Ag and metals that are more difficult to oxidize. Copper metal is deposited at the cathode and the more easily oxidized (less easily reduced) ions Fe^{2+}, Ni^{2+}, Sb^{3+}, and As^{3+} remain and accumulate in solution, replacing the Cu^{2+} of the electrolyte.

Overvoltage for Gas Evolution. Gaseous products can be formed and react with an anode at voltages consistent with their $E°$ values; however, bubble formation and evolution from inert anodes require an overvoltage. Overvoltage is experimentally dependent upon anode surface and current density. The approximate values in Table 14.12 show the large potentials required and the facts that hydrogen and halogen have very low overvoltages on platinum black and that oxygen, though high, is minimum on that particular surface. Reaction with noninert electrodes occurs whenever possible. The effect of overvoltage on the selection of alternate reactions that may occur is shown in the following

TABLE 14.12   Approximate Overvoltages for Gas Evolution

	At 0.01 A/cm^2			At 0.1 A/cm^2			At 1 A/cm^2		
	O_2	H_2	Cl_2	O_2	H_2	Cl_2	O_2	H_2	Cl_2
Graphite	0.80	0.70	—	1.09	0.89	0.25	1.24	1.17	0.50
Pt black	0.40	0.03	0.02	0.64	0.04	0.03	0.79	0.05	0.08
Pt smooth	0.72	0.07	0.03	1.28	0.29	0.05	1.38	0.68	0.24
Ni	0.35	0.74	—	0.73	1.05	—	0.87	1.24	—
Cu	0.42	0.58	—	0.66	0.80	—	0.84	1.25	—
Ag	0.58	0.76	—	0.98	0.98	—	1.14	1.10	—
Au	0.67	0.39	—	1.24	0.59	—	1.68	0.80	—
Fe	—	0.56	—	—	0.82	—	—	1.29	—

example: Electrolysis of NaCl (basic aqueous solution) forms H_2 and Cl_2 on graphite electrodes:

$$2[Na^+ + e^- \rightarrow Na] \qquad\qquad E° = -2.71 \text{ V} \qquad\qquad (a)$$
$$2Cl^- \rightarrow Cl_2 + 2e^- \qquad\qquad E° = -1.36 \text{ V} \qquad\qquad (b)$$
$$2[2H_2O + 2e^- \rightarrow H_2 + 2OH^-] \qquad E_B° = -0.83 \text{ V} \qquad\qquad (c)$$
$$4OH^- \rightarrow O_2 + 2H_2O + 4e^- \qquad E_B° = -0.40 \text{ V} \qquad\qquad (d)$$

Reduction reactions (a) or (c) can occur; (c), the easiest reaction, occurs and liberates hydrogen (despite the hydrogen evolution overvoltage). Oxidation reactions (b) or (d) can occur; they are dependent on the low concentration of OH^-, the high concentration of Cl^-, and the much lower overvoltage for Cl_2 than for O_2 evolution, and Cl_2 is preferentially evolved. The net result is that reactions (b) and (c) occur.

Definitions

Anode. Oxidation occurs at this electrode; site where reducing agent is oxidized; source of electrons to external circuit whether battery or electrolytic cell; electrode labeled negative terminal on batteries (physically labeled for condition of discharge); anode of a discharging battery becomes its cathode during charging (or acting as an electrolytic cell); name of electrode connected to positive terminal of external voltage source for charging a battery or for electrolytic cell; anions (negative ions) migrate toward it in an electrolytic cell.

Cathode. Reduction occurs at this electrode. Its characteristics are reversed from those of the anode.

Refer to redox terminology of Table 14.4.

Batteries. Batteries are redox cells that are physically arranged for external flow and internal ion mobility. When different electrolytes are used, gel structures or a porous membrane prevent mixing. The electrolyte(s) must permit ionization of the reactant species and be conductive. Mobility of ions in solution provides the mechanism for maintaining a charge balance and electrical neutrality within the electrolyte. By convention, batteries are labeled negative at the anode of the spontaneous discharge reaction.

(a) Voltage. Open-circuit terminal voltage is concentration and temperature dependent due to changes in chemical potential and is reduced by internal IR drop under current flow conditions. The current available depends on the reaction rate, which is a function of temperature and is not directly predictable from $E°$ values. For practical battery configurations $E°$ provides a reasonable voltage estimate.

(b) Polarization. Polarization is a localized accumulation of reaction products at an electrode until their concentration is reduced by diffusion, precipitation, formation of complexes, or further reaction to form new species not involved in the electrode reaction. Depolarization is the process of reducing the localized high concentration of reaction products; if depolarization is by a diffusional process, the cell may be reversible because the products retain their chemical identity and physical availability.

(c) Irreversibility. Although all redox reactions are ideally reversible near equilibrium, many cells operate far from equilibrium and with secondary reactions or physical configurations that render the cell partially, if not completely, irreversible for practical purposes. This situation results in having

many cells that can convert their chemical energy to electrical and a limited few that are reversible for use as practical and economic storage batteries.

(d) *Battery Reactions.* Battery reactions on discharge are listed in what follows. Only a few are designed for reversibility.

1. *Lead Storage Battery*, (H_2SO_4 Electrolyte) (Reversible). Oxidation reaction at lead plates, labeled electrically negative:

$$Pb + HSO_4^- \rightarrow PbSO_4 + 2e^- + H^+ \qquad E° = +0.36 \text{ V}$$

Reduction reaction at lead dioxide plates, labeled electrically positive:

$$PbO_2 + 3H^+ + HSO_4^- + 2e^- \rightarrow PbSO_4 + 2H_2O \qquad E° = +1.68 \text{ V}$$

2. *Mercury Cell* (KOH Electrolyte Saturated with ZnO). Oxidation (negative terminal):

$$Zn + 4OH^- \rightarrow ZnO_2^{2-} + 2H_2O + 2e^- \qquad E_B° = +1.22 \text{ V}$$

Reduction (positive terminal):

$$HgO + H_2O + 2e^- \rightarrow Hg + 2OH^- \qquad E_B° = +0.10 \text{ V}$$

3. *Zinc–Silver Peroxide Cell* (KOH Electrolyte Saturated with ZnO). Oxidation (negative terminal):

$$Zn + 4OH^- \rightarrow ZnO_2^{2-} + 2H_2O + 2e^- \qquad E_B° = +1.22 \text{ V}$$

Reduction (positive terminal):

$$Ag_2O + H_2O + 3e^- \rightarrow 2Ag + 2OH^- \qquad E_B° = +0.34 \text{ V}$$

4. *LeClanche Cell* (Flashlight Battery, NH_4Cl Electrolyte). Oxidation (negative terminal):

$$Zn \rightarrow Zn^{2+} + 2e^- \qquad E° = +0.76 \text{ V}$$

Reduction (positive terminal):

$$2MnO_2 + 2NH_4^+ + 2e^- \rightarrow Mn_2O_3 + H_2O + 2NH_3 \left(\text{for complexing } Zn^{2+} \right) \quad E° = +0.74 \text{ V}$$

5. *Alkaline Zinc Manganese Dioxide* "Alkaline Flashlight Battery" (KOH Electrolyte). Oxidation (negative terminal):

$$Zn + 4OH^- \rightarrow ZnO_2^{2-} + 2H_2O + 2e^- \qquad E_B° = +1.22 \text{ V}$$

Reduction (positive terminal):

$$MnO_2 + 2H_2O + e^- \rightarrow Mn(OH)_3 + OH^- \qquad E_B° = +0.35 \text{ V}$$

6. *Nickel Cadmium Storage Battery* (KOH Electrolyte) (Reversible). Oxidation (negative terminal):

$$Cd + 2OH^- \rightarrow Cd(OH)_2 + 2e^- \qquad E_B° = +0.81 \text{ V}$$

Reduction (positive terminal):

$$NiO_2 + 2H_2O + 2e^- \rightarrow Ni(OH)_2 \qquad E_B° = +0.49 \text{ V}$$

14.10 ORGANIC CHEMISTRY

Tetrahedral Carbon. The electronic orbitals of carbon are consistent with the formation of four covalent bonds. Symmetry leads to a tetrahedral spatial orientation. The ability of carbon to bond to carbon leads to a multiplicity of organic compounds based on linear, branched chain and ring sequences of carbon atoms.

Carbon-Carbon Bonds. Saturation refers to carbon-carbon single bonds, with the remaining three bonds to another carbon, to hydrogen, or to substituent groups. Double and triple bonds between carbon atoms are strained and quite reactive; compounds containing these are termed *unsaturated* because each double bond replaces two substituents.

14.10.1 Classes of Compounds Based on Structure of Carbon Chain

Aliphatic Compounds. Chains and branched chains of tetrahedral carbon atoms covalently bonded to each other give rise to a proliferation of aliphatic compounds. Bonds between carbon atoms forming the backbone chain may be single, double, or triple, and the remaining bonds available at each carbon atom are made to hydrogen or other substituent groups. Compounds have backbone and branch combinations ranging from C_1 to C_{40} or more. Since substituent groups other than hydrogen are possible at the carbon atoms and can be placed in different sequences along the backbone, the numbers of organic compounds become very large—before considering the effect of spatial configuration that often permits one part of the molecule to combine with another reactive part of the same or a kindred molecule.

Cyclic Compounds. Ring structures of aliphatic compounds exist subject to bond strain and the possible three-dimensional configurations of the carbon atom chain. Cyclohexane, C_6H_{12} [Fig. 14.10(a)], and cyclopentane, C_5H_{10} [Fig. 14.10(b)], most commonly exist as the backbone structures. These consist of tetrahedral carbon atoms connected by single bonds and with two hydrogen atoms existent at each apex. Double bonds between carbon atoms can occur: cyclohexene, C_6H_{10} [Fig. 14.10(c)] and cyclopentadiene, C_5H_5 [Fig. 14.10(d)], are examples.

Note that these can be variously oriented on paper just as the molecules can be in space, so there is no correct or preferred orientation. The presence of a double bond limits the number of hydrogen atoms or substituent groups to one at the double-bonded carbon atoms.

Aromatic Compounds. A particularly stable minimum-energy configuration of six carbon atoms [Fig. 14.10(e)] consists of equivalent carbon atoms in planar configuration with single bonds to a substituent at each apex of the hexagon. The absence of other substituent groups infers the presence of hydrogen, and the figure drawn signifies benzene, C_6H_6. Other groups may replace any or all hydrogen atoms subject to steric (space) considerations; two or three large groups or six small groups may each replace a hydrogen. Higher analogs of benzene [e.g., Figs. 14.10(f)–(h)] are condensed

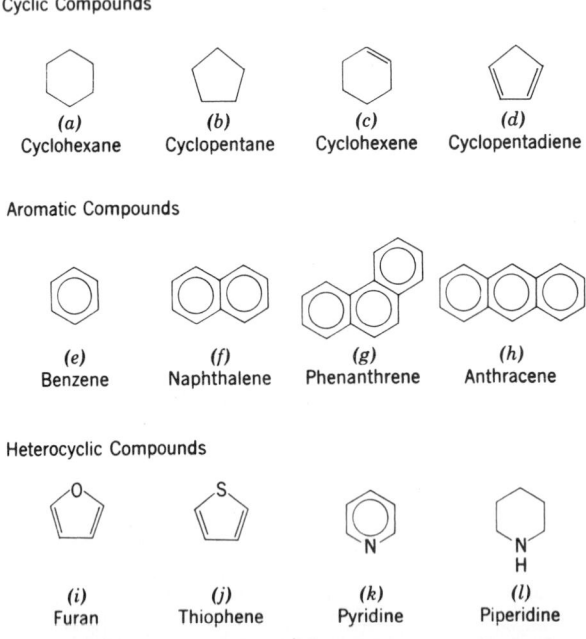

Fig. 14.10 Cyclic, aromatic, and heterocyclic compounds.

planar ring structures containing H atoms at the protruding corners that are subject to similar space-limited substitutions.

Heterocyclic Compounds. Heterocyclic compounds are ring structures with O, S, or N substituted for one or more of the carbon atoms. Bonds at these substituent atoms are those expected from their electronic structure. Examples are furan, C_4H_4O [Fig. 14.10(i)], thiophene, C_4H_4S [Fig. 14.10(j)], pyridine, C_5H_5N [Fig. 14.10(k)], a nitrogen analog of the aromatic hydrocarbon benzene, and piperidine, $C_5H_{10}NH$ [Fig. 14.10(l)], an alicyclic amine.

14.10.2   Abbreviated Organic Chemical Nomenclature

Organic nomenclature is systematic and dependent upon an extensive set of rules of the IUPAC that discourages use of older trivial names. Use of the structural formula is encouraged because it is unambiguous. A few of the simpler rules are helpful: Ring compounds are considered the basic unit and all appendages as secondary. If required, positions around ring compounds are numbered sequentially. Carbon atoms of linear and branched chain compounds are numbered along the longest string, starting at the end nearest a substituent group or branch. The positions of substituents and side chains are located by the number of the carbon atom where they are attached. Multiple bonds are located by the lowest numbered of the two numbered carbon atoms involved. Some of the prefix and suffix terms used in organic nomenclature are listed in Table 14.13.

TABLE 14.13   Organic Chemical Nomenclature

-ane	Saturated hydrocarbon compound
-ene	Unsaturated hydrocarbon compound with double bond
-yne	Unsaturated hydrocarbon compound with triple bond
-yl	Hydrocarbon group as substituent (see next section)
cis-	Refers to stereochemistry
trans-	Refers to stereochemistry
ortho-	1,2 positions on benzene ring (abbreviated *o-*)
meta-	1,3 positions on benzene ring (abbreviated *m-*)
para-	1,4 positions on benzene ring (abbreviated *p-*)
d-	Dextro-rotatory optical isomer
l-	Levo-rotatory optical isomer
r-	Racemic, mixture of *d* and *l* isomers
n-	Normal, or linear chain
iso-	Branched chain; usually with isopropyl group
t-	Tertiary, three carbons attached to one
sec-	Secondary, two carbons attached to one
neo-	Four carbons attached to one
di-	Two
tri-	Three
cyclo-	Ring structure

Three parts of a simple compound name are:
  1. A position number
  2. name of substituent group (see next section)
  3. name of parent structure
Examples:

3-Methyl-4-bromo-hex-1-ene

1,3-Di-bromo-benzene

t-Butyl-benzene

1,3-Cyclo-hexa-di-ene

Common usage of trivial names that are unrelated to structure will continue. The only recourse in a practical situation is to locate the compound and its physical constants in the *Merck Index*[4] or the *Handbook of Chemistry and Physics*.[3]

14.10.3   Classification of Compounds into Unreactive and Functional Groups

Although the number of known organic compounds is astronomic, the number of inexpensive and commercially available ones is drastically limited. Their chemistry ranges from complete oxidation (during combustion) to conversion into specific compounds with academic or commercial application.

Reactivity of organic compounds occurs only at selected sites on the molecule. Synthetic procedures attack substituent groups, more reactive-than-normal hydrogen atoms, and multiple bonds in the carbon chains; the remainder of the molecule remains unchanged in the process. This permits classification of compounds into a reactive portion of the molecule and one that is unaffected by the synthetic process. The reactive portion is a substituent or functional group. The unaffected portion is the remainder of the compound and is called an aliphatic (alkyl), cyclic aromatic (aryl), or a heterocyclic R group according to its structure; this R group is the backbone that carries the reactive substituent group (Tables 14.14 and 14.15).

14.10.4   Organic Reactions

Organic reactions make changes in the bonding to carbon atoms; their course is greatly influenced by reaction conditions that are experimentally established on the basis of reaction mechanism studies. Complications arise because several different reactions may occur simultaneously accompanied by cyclization, molecular rearrangements, and oxidation-reduction. All occur in the direction of minimum energy and increased stability. An encyclopedic literature is replete with tens of thousands of reactions, several hundred of which are generally useful in synthesis and are "name reactions" honoring their early investigators. Studies of rate, catalysis, stepwise mechanism, and stereochemistry have coalesced much of the accumulated information into a few generalized categories of related mechanisms.

TABLE 14.14   Alkyl and Aryl R Groups

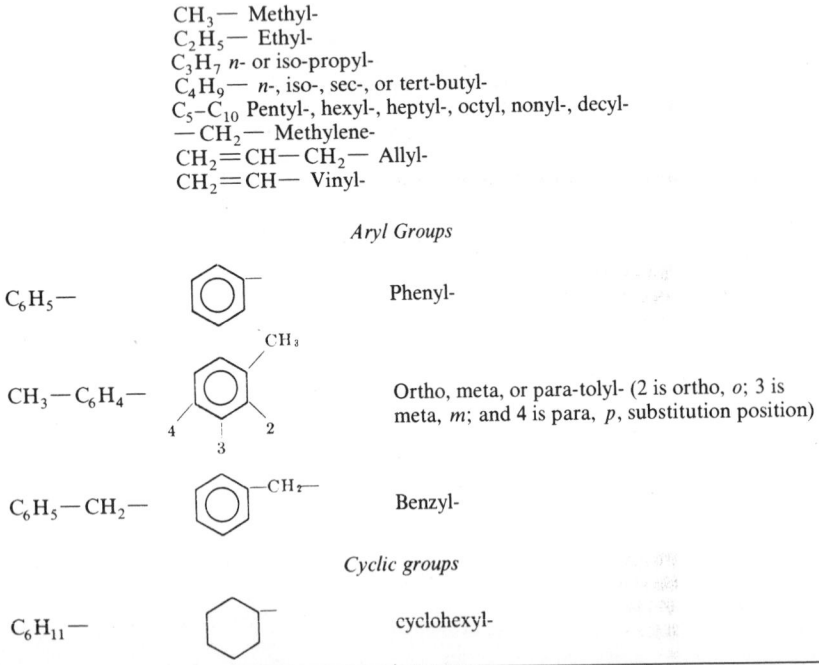

Alkyl Groups

CH_3— Methyl-
C_2H_5— Ethyl-
C_3H_7 *n*- or iso-propyl-
C_4H_9— *n*-, iso-, sec-, or tert-butyl-
C_5-C_{10} Pentyl-, hexyl-, heptyl-, octyl, nonyl-, decyl-
—CH_2— Methylene-
CH_2=CH—CH_2— Allyl-
CH_2=CH— Vinyl-

Aryl Groups

C_6H_5—   Phenyl-

CH_3—C_6H_4—   Ortho, meta, or para-tolyl- (2 is ortho, *o*; 3 is meta, *m*; and 4 is para, *p*, substitution position)

C_6H_5—CH_2—   Benzyl-

Cyclic groups

C_6H_{11}—   cyclohexyl-

TABLE 14.15 Functional Groups

—OH	alcohol when attached to alkyl R, phenol when attached to aryl R	—SH	thioalcohol, mercaptan
—O—	ether	—S—	sulfide
—O—O—	peroxide	—S—S—	disulfide
—O—OH	hydroperoxide	$-\overset{O}{\underset{\parallel}{S}}-$	sulfoxide
—O—N=O	nitrite	$-\overset{O}{\overset{\parallel}{S}}\underset{\diagdown O}{-}$	sulfone
$-O-\overset{\displaystyle O}{\overset{\diagup}{\underset{\diagdown}{N}}}_{\textstyle O}$	nitrate	$-\overset{\displaystyle O}{\overset{\parallel}{S}}{\underset{\diagdown OH}{}}$	sulfinic acid
$-\overset{\displaystyle O}{\overset{\diagup}{\underset{\diagdown}{N}}}_{\textstyle O}$	nitro compound	$-\overset{\displaystyle O}{\overset{\parallel}{S}}{\underset{\diagdown O}{}}$—OH	sulfonic acid
—NH₂	primary amine	$-\overset{\displaystyle O}{\overset{\parallel}{C}}$—OH	carboxylic acid
$-\overset{\textstyle H}{\underset{\diagup}{N}}-$	sec-amine	$-\overset{O}{\overset{\parallel}{C}}-O-\overset{O}{\overset{\parallel}{C}}-$	acid anhydride
$-\overset{\diagup}{\underset{\diagdown}{N}}$	tert-amine	$-\overset{\displaystyle O}{\overset{\parallel}{C}}-O-$	ester
=NH	imide	$-\overset{\displaystyle O}{\overset{\parallel}{C}}-\overset{\diagup}{\underset{\diagdown}{N}}$	amide
—C≡N	nitrile		
—X	halide, where X = F, Cl, Br, or I	$-\overset{\displaystyle O}{\overset{\parallel}{C}}{\underset{\diagdown X}{}}$	acid halide
—H	hydrogen		
$\overset{\diagdown}{\diagup}C=C\overset{\diagup}{\diagdown}$	double bond	$-\overset{O}{\overset{\parallel}{C}}-$	ketone
—C≡C—	triple bond	$-\overset{\displaystyle O}{\overset{\parallel}{C}}{\underset{\diagdown H}{}}$	aldehyde

Substitution. In substitution reactions a group attached to a carbon atom is removed and another enters in its place. Reactions are designated S_{N_1}, S_{N_2}, S_{E_1}, or S_{E_2} dependent on the nucleophilic or electrophilic nature of the reagent and the unimolecular or bimolecular dependence of the reaction rate. Nonpolar substitutions are dependent on a free-radical mechanism. Solvolysis reactions, such as hydrolysis of esters or amides, and organic acid-base neutralizations are substitution reactions. Other reactions occur at both saturated and unsaturated carbon atoms. Intra- and intermolecular substitutions are the mechanism for rearrangements and cyclization (ring formation).

Addition. Addition of functional groups occurs at unsaturated carbon atoms, increasing the number of groups attached at those positions. The reagent species may be nucleophilic, electrophilic, or free radical.

Elimination. Reducing the number of functional groups bound to one carbon atom necessitates the formation of an unsaturated carbon bond to the adjacent carbon. Elimination thus involves removal of two functional groups from contiguous carbon atoms and is the reverse of addition reactions. Elimination reactions often occur simultaneously and competitively with substitution reactions at saturated carbon atoms.

Rearrangement. Many reactions are encountered where functional groups migrate within the molecule and result in products other than anticipated. These migrations tend to occur at adjacent carbon atoms; they may involve five- and six-membered heterocyclic ring intermediates or neighboring group participation in the reaction. The driving force for any rearrangement, as for any reaction, is increased stability of the resultant product species.

Oxidation-Reduction. Redox reactions are selectively applied to carbon-hydrogen bonds and carbon-carbon multiple bonds, but few oxidative procedures allow selective breaking of a specific carbon-carbon single bond without disrupting the entire molecule. The removal of hydrogen to form multiple carbon-carbon bonds or to make new bonds between carbon and oxygen, sulfur, nitrogen, or halogens is called oxidation. Primary alcohols are oxidized to carboxylic acids, and acids are reduced to alcohols.

14.10.5   Reagents of Organic Reactions

Further classification is made according to classes of reagent, the species which attacks a substrate to produce products. Reagents are collected into three classes based on their electronic structure:

(a) *Nucleophilic.* Lewis bases can donate unshared electron pairs, reducing agents.

(b) *Electrophilic.* Lewis acids can accept unshared electron pairs, oxidizing agents.

(c) *Free Radical.* A reactive transient fragment that has one or more unpaired electrons. Examples: $Cl \cdot$, $\equiv C \cdot$, and $=C:$ are chlorine, carbon, and carbene radicals, respectively.

The more common reagents are summarized in Table 14.16.

14.10.6   Catalysis

Catalysis is absolutely dependent on the reaction mechanism; other than transition metal participation in intermediate species, acid and base catalyses are most important:

1. Acid catalysis facilitates production of electrophilic reagent species. Lewis acids provide acid catalysis.

2. Base catalysis facilitates production of nucleophilic reagent species. Lewis bases provide base catalysis.

3. Free-radical initiation: Heat, light, or peroxide can initiate free-radical formation, independent of both solvent and acid or base catalysis.

TABLE 14.16   Nucleophilic and Electrophilic Reagents

Nucleophilic Reagent, Bases and Reducing Agents, Donors of Unshared e^- Pair		Electrophilic Reagent, Acids and Oxidizing Agents, Acceptors of e^- pair	
I^-	iodide ion	H_3O^+	hydronium ion
OH^-	hydroxyl ion	$\begin{cases} R_2C\overset{+}{=}OH \\ R_2\overset{+}{C}\!-\!OH \end{cases}$	these occur after protonation of ketones
RO^-	alcoholate ion		
RS^-	mercaptide ion	BF_3	boron trifluoride
CN^-	cyanide ion	$AlCl_3$	aluminum chloride
$H_2\ddot{O}\!:$	water	Cl_2	(their cleavage forms X^+ and
		Br_2	X^-) X^+ is actual electro-
$R\ddot{O}H$	alcohol	I_2	philic reagent
$:NH_3$	ammonia	NO_2^+	nitronium ion (formed after
Br^-	bromide ion		protonation of nitric acid)
Cl^-	chloride ion		
$\overset{\diagdown}{\underset{\diagup}{-}}C^-\!:$	carbanion	$\overset{\diagdown}{\underset{\diagup}{-}}C^+$	carbonium ion

TABLE 14.17   Solvents for Organic Reactions

Polar	Nonpolar
Water	Hydrocarbons
Formic acid	Acetone—will dissolve some ionic species
Dimethylsulfoxide	Ethers
Dimethylformamide	Ethylene glycol dimethyl ether (diglyme)
Methanol	Tetrahydrofuran (THF)
Ethanol	Dioxane
Acetic acid	
Nitromethane	
Acetonitrile	
Liquid ammonia	

14.10.7   Solvents

The solvent for reactions may be one of the reactants or be introduced to facilitate mutual contact via solution. The choice is necessarily dependent on reaction conditions to avoid undesired solvent participation:

(a) Polar solvents are chosen to solubilize ionic species.

(b) Nonpolar solvents dissolve un-ionized molecules.

Table 14.17 summarizes the more widely used polar and nonpolar solvents.

14.11   NUCLEAR REACTIONS

The conventional notation for a nuclear reaction is

$$\left(_{z}C^{A}\right)_{1} + \left(_{z}C^{A}\right)_{2} \rightarrow \left(_{z}C^{A}\right)_{3} + \left(_{z}C^{A}\right)_{4} + Q \tag{14.76}$$

where z is the number of protons, A is the mass number, C is the chemical symbol for the atom, electron, or nucleon, and Q is the energy released.

The conservation equations are

$$\sum Z_i = 0 \quad \text{and} \quad \sum A_i = 0 \tag{14.77}$$

$$\text{Initial mass} - \text{final mass} = Q \tag{14.78}$$

An example is the **fission reaction**:

$$_{92}U^{235} + _{0}n^{1} \rightarrow _{92}U^{236} \rightarrow _{38}Sr^{94} + _{54}Xe^{140} + 2\,_{0}n^{1} + Q$$

The strontium and xenon products are highly radioactive and decay further to other products. The final result is a spectrum of products.

Table 14.18 gives frequently encountered constants.

TABLE 14.18   Some Nuclear Constants

	amu	grams $\times 10^{-24}$	Rest Energy (MeV)
Unit mass, m	1	1.6605655	931.5016
Electron, $_{-1}e^{0}$ or β^{-}	0.00054858026	0.0009109534	.5110041
Proton, p or $_{1}p^{1}$	1.007276470	1.6726485	938.2592
Neutron, n or $_{0}n^{1}$	1.008665012	1.6749543	939.5527
Hydrogen atom, $_{1}H^{1}$	1.007825	1.673559	938.770

Charge on electron, $1.6021892 \times 10^{-19}$ Coulomb
Radius of electron, $2.8179380 \times 10^{-13}$ cm

Source: Reproduced with permission from *CRC Handbook of Chemistry and Physics*, Robert E. Weast, (ed.), CRC Press, Boca Raton, Florida. Copyright © 1983 by CRC Press.

14.11.1   Modes of Radioactive Decay

Negative beta e^- (electron) emission:

$$_0n^1 \longrightarrow {}_{-1}e^0 + {}_1H^1 + \text{neutrino}$$

$$_{38}Sr^{94} \xrightarrow{\beta^-} {}_{39}Y^{94} \xrightarrow{\beta^-} {}_{40}Zr^{94}$$

$$_{54}Xe^{140} \longrightarrow 4\left({}_{-1}e^0\right) + {}_{58}Ce^{140}$$

Positive beta β^+ (positron) emission:

$$_7N^{13} \longrightarrow {}_{+1}e^0 + {}_6C^{13} + \text{neutrino}$$

$$_{+1}e^0 + {}_{-1}e^0 \longrightarrow 2 \text{ gammas of } 0.51 \text{ MeV each}$$

Alpha emission:

$$_{94}Pu^{239} \longrightarrow {}_2He^{2+} + {}_{92}U^{235}$$

Neutron emission:

$$_{53}I^{137} \xrightarrow{\beta^-} {}_{54}Xe^{137} \longrightarrow {}_0n^1 + {}_{54}Xe^{136}$$

Orbital electron (K) capture:

$$_{29}Cu^{64} + {}_{-1}e^0 \longrightarrow {}_{28}Ni^{64}$$

Gamma emission by (a) ejection of a gamm photon from an excited nucleus, (b) isomeric transition of a nucleus from one energy level to another, and (c) annihilation of an electron following positive beta emission.

Nuclei with an excess of neutrons are usually electron emitters. Among nuclei with a deficiency of neutrons, the heavy ones usually decay by alpha emission and the light ones by positron emission or orbital electron capture. Gamma emission often accompanies other types of decay.

14.11.2   Decay with Time

From statistical considerations the rate of decay (alpha emission) is proportional to the number N of radioactive nuclei present,

$$\frac{dN}{dt} = -\lambda N \tag{14.79}$$

where the proportionality constant λ is called the **distintegration constant** or the **radioactive decay constant**. Integration of Eq. (14.79) gives

$$N = N_0 e^{-\lambda t} \tag{14.80}$$

where N_0 is the initial number of radioactive nuclei. The **half-life** $t_{1/2}$, or the time required for one-half of the original atoms to decay, by substitution in Eq. (14.80) is

$$t_{1/2} = \frac{0.693}{\lambda} \tag{14.81}$$

When there is more than one radioisotope in the decay chain

$$P \xrightarrow{\lambda_1} Q \xrightarrow{\lambda_2} R \xrightarrow{\lambda_3}$$

for the second member or daughter Q,

$$\frac{N_Q}{N_{P_1}} = \frac{\lambda_1}{\lambda_2 - \lambda_1} e^{-\lambda_1 t} + \frac{\lambda_1}{\lambda_1 - \lambda_2} e^{-\lambda_2 t} \qquad (14.82)$$

If the half-life of the parent P is longer than the half-life of the daughter Q ($\lambda_2 > \lambda_1$), after a lapse of time $e^{-\lambda_2 t}$ becomes negligible and

$$\frac{N_Q}{N_P} = \frac{\lambda_1}{\lambda_2 - \lambda_1} e^{-\lambda_1 t} \qquad (14.83)$$

With native radioisotopes the half-life of the parent is very long compared to that of the daughter, so that Eq. (14.83) reduces to

$$\frac{N_Q}{N_P} = \frac{\lambda_1}{\lambda_2} e^{-\lambda_1 t} \qquad (14.84)$$

Half-life and emitted particle energy for selected radioisotopes are shown in Table 14.19.

TABLE 14.19   Selected Radioisotopes

Atomic No.	Element	Mass No.	β (MeV)	γ (MeV)	Half-life
13	Aluminum	29	2.5, 1.4	1.28, 2.43	6.6 months
51	Antimony	122	1.40, 1.97	0.69, 0.56	2.8 days
		124	0.61, 2.37	0.60	60 days
		125	0.30, 0.16	0.43, 0.60	2.4 yr
33	Arsenic	77	0.68	0.24	39 hr
4	Beryllium	10	0.56	None	2.7×10^6 yr
83	Bismuth	210	1.16	None	5.0 days
35	Bromine	82	0.44	0.77, 0.55	36 hr
48	Cadmium	115	0.59, 1.11	0.52, 0.49	53 hr
20	Calcium	45	0.25	None	165 days
		49	2.0, 0.95	3.10, 4.05	8.8 months
6	Carbon	11	$0.98 \, \beta^+$	0.51	20.4 months
		14	0.16	None	5.7×10^3 yr
58	Cerium	141	0.44, 0.58	0.14	33 days
17	Chlorine	36	0.17	None	3.2×10^5 yr
		38	4.81, 1.1, 2.77	2.15, 1.60	37 months
27	Cobalt	60	0.31	1.17, 1.33	5.3 yr
29	Copper	61	$1.22 \, \beta^+$	0.28, 0.66	3.3 hr
63	Europium	154	0.15, 1.85	0.12, 0.72, 1.28	16 yr
		155	0.15, 0.25	0.089	1.7 yr
9	Fluorine	18	$0.65 \, \beta^+$	0.51	1.87 hr
31	Gallium	72	0.64, 0.96	0.84, 2.20	14.1 hr
32	Germanium	71	e^- capture	None	11 days
		77	2.20, 1.38	0.22, 0.26	12 hr
79	Gold	198	0.96	0.41	2.7 days
		199	0.30, 0.25, 0.46	0.21, 0.16	3.2 days
72	Hafnium	181	0.41	0.13, 0.48	45 days
1	Hydrogen	3	0.018	None	12.3 yr
53	Iodine	131	0.61, 0.34	0.36, 0.080, 0.28, 0.64	8.0 days
77	Iridium	192	0.67, 0.53, 0.24	0.32, 0.47, 0.60	74 days
		194	2.24, 1.90	0.33, 0.65	19 hr
26	Iron	59	0.46, 0.27	1.10, 1.29	45 days
36	Krypton	85	0.67, 0.15	—	10.5 yr
57	Lanthanum	140	1.34, 1.10, 0.42	1.60, 0.49	40 hr
12	Magnesium	27	1.75, 1.59	0.84, 1.02	9.5 months
80	Mercury	203	0.21	0.28	46 days
42	Molybdenum	99	1.23, 0.45	0.04, 0.74, 0.78	67 hr

TABLE 14.19   *(Continued)*

Atomic No.	Element	Mass No.	β (MeV)	γ (MeV)	Half-life
60	Neodymium	147	0.83, 0.38	0.09, 0.53	11 days
41	Niobium	95	0.16	0.77	35 days
7	Nitrogen	13	$1.19\,\beta^+$	0.51	10 months
76	Osmium	191	0.14	0.04, 0.13	15.0 days
		193	1.13, 1.06	1.58, 0.07, 0.14, 0.56	32 hr
15	Phosophorus	32	1.71	None	14.3 days
78	Platinum	197	0.67, 0.48	0.08, 0.19	18 hr
19	Potassium	40	$1.32, e^-$ capture	1.42	1.3×10^9 yr
		42	3.54, 1.98	1.52	12.4 hr
59	Praeseodymium	142	2.17, 0.59	1.60	19.3 hr
		143	0.93	Nòne	13.7 days
61	Promethium	147	0.23	None	2.5 yr
75	Rhenium	186	1.07, 0.93	0.12, 0.14	91 hr
45	Rhodium	105	0.57, 0.25	0.33	36 hr
37	Rubidium	86	1.76, 0.68	1.08	18.7 days
44	Ruthenium	103	0.22, 0.14, 0.70	0.50	41 days
62	Samarium	153	0.70, 0.64, 0.81	0.103	47 hr
21	Scandium	46	0.36	0.89, 1.12	85 days
14	Silicon	31	1.48	None	2.6 hr
47	Silver	110	0.087, 0.53	0.66, 0.88	250 days
		111	1.05, 0.71	0.25, 0.34	7.5 days
11	Sodium	22	$0.54\,\beta^+$	1.28, 0.51	2.6 yr
		24	1.39	1.37, 2.75	15.0 hr
38	Strontium	89	1.46	None	50 days
		90	0.54	None	28 yr
16	Sulfur	35	0.17	None	87 days
73	Tantalum	182	0.18, 0.36, 0.44	0.10, 1.12, 1.23	115 days
43	Technetium	97	Isotope transition	0.09	> 91 days
					3.6×10^6 yr
		99	0.30	0.67, 0.75	1.5×10^6 yr
52	Tellurium	127	Isotope transition, 0.70	0.089	105 days, 9.3 hr
		129	IT, 1.45, 1.00	0.03, 0.10, 0.47	32 days, 72 months
		131	IT, 2.14, 1.69	0.15	30 hr, 25 months
81	Thallium	204	0.77	None	3.9 yr
22	Titanium	51	2.13, 1.52	0.32	5.8 months
74	Tungsten	185	0.43	None	75 days
		187	0.63, 1.32, 0.34	Many 0.13 to 0.87	24 hr
54	Xenon	133	0.35	—	5.3 days
39	Yttrium	90	2.27	None	64 hr

Source: W. H. Sullivan, *Trilinear Chart of Nuclides*, 2nd ed., U.S. Government Printing Office, Washington, DC, 1957, and subsequent revision sheets.

14.12   INSTRUMENTAL METHODS OF CHEMICAL ANALYSIS

Modern chemical methods rely heavily on large-scale instrumentation for analysis of samples. For pure samples, one not only needs to identify the material but also to elucidate the geometrical structure of the individual molecular units. For mixtures, identification of the various components and measurement of their relative concentrations is usually required. In a reactive system, the time dependence of the concentrations of all species, including transients, is the important variable. There is no single instrument that is suitable for all of these applications and in general several methods must be used. The following briefly describes the most important techniques, their common "names," and their principal applications.

   X-ray Diffraction. The diffraction pattern of X-rays from a pure, single crystal can yield a complete geometrical structure of the individual molecular units. Data collection in commercial units is totally automated while interpretation of the data requires extensive computations on large computers. The method is applicable to any size molecular unit (including macromolecules) although there is some loss of accuracy as the size of the molecular unit increases.

Infrared (IR) Spectroscopy. Absorption of energy from the IR region of the electromagnetic spectrum results in excitation of internal vibrations in molecules. Since the available spectral range is large (wavelengths of 2–1000 μm) and the absorption spectrum consists of a small number of relatively narrow bands, the spectrum provides a "fingerprint" for identification. Furthermore, absorption in certain wavelength regions is characteristic of particular molecular fragments. For example, the presence of molecules that contain the carbon-hydrogen group leads to absorption near 3000 cm^{-1}. Because of the fingerprint character, IR spectroscopy is suitable for analysis of mixtures. Infrared spectroscopy is applicable to samples in any phase and to surface studies.

Ultraviolet and Visible Spectroscopy (UV / VIS). Absorption of light in this region of spectrum excites electronic states of molecules and can lead to chemical reaction (photolysis), ionization, or emission of light. This type of spectroscopy can be used for chemical analysis by measuring the wavelength of the absorbed and/or emitted light. Both the absorption and emission spectra in condensed phases tend to be broad and unstructured compared to IR spectra. While most conventional instruments use standard light sources, spectrometers equipped with tunable, pulsed lasers are able to detect dynamical effects on ultrashort time scales. The UV/VIS methods are applicable to any phase and to surface studies.

Emission Spectroscopy. A solid or liquid sample excited by a flame or an electric arc reaches very high temperatures, and the result is the emission of light. Since the sample is broken down into its component atoms or their ions, the spectrum consists of sharp lines superimposed on the light from the excitation source. The method is very sensitive (parts per million) and requires only milligram quantities of sample.

Neutron Activation Analysis. Irradiation of sample with thermal neutrons from a reactor produces radioactive isotopes of many of the elements. The isotopes decay with the emission of gamma and/or beta rays with energies characteristic of each particular isotope. The sensitivity of neutron activation analysis varies from element to element, but in the most favorable cases, the detection limit is about 10^{-11} g.

Mass Spectrometry. This is a gas phase technique in which molecular ions (usually positive ions) formed with the sample are accelerated and collimated into a beam. When the ion beam is passed through suitable electric and/or magnetic fields, the ions are sorted and detected according to their mass-to-charge ratio. The method is fast, exceedingly sensitive, and universally applicable since ions can be formed from any type of sample. Only nanogram quantities of sample are required, and masses up to tens of thousand amu can be detected by the most sophisticated instruments. Low-resolution instruments easily resolve ions with one mass unit difference while high-resolution machines can distinguish two ions with the same nominal mass. For example, the masses of N_2 and CO are 28.006 and 27.995 amu, respectively. A common technique for analyzing mixtures is to connect a gas chromatograph to the inlet of a mass spectrometer, that is, a GCMS system. *Chromatography* refers to a variety of methods of separating the components of a mixture. In all cases there is a stationary phase (a solid or a liquid supported on a solid) and a mobile phase (a gas or a liquid) that contains the mixture. A physical separation occurs as the mobile phase passes over the stationary phase if the components of the mixture have sufficiently different adsorption affinities for the stationary phase. Among the many variants are PC (paper, TLC (thin layer), LC (liquid), GLC (gas-liquid), GSC (gas-solid), and HPLC (high-performance liquid).

Nuclear Magnetic Resonance (NMR). Owing to their magnetic properties, certain nuclei in the presence of a strong magnetic field (10,000–100,000 G) absorb energy in the radio-frequency region of the electromagnetic spectrum. Most applications involve the proton, ^{1}H; however, other nuclei such as ^{19}F, ^{13}C, and ^{31}P can also be used. Although an isolated nucleus absorbs energy at a particular frequency in a given magnetic field strength, the absorption frequency shifts and also may split into multiples depending on the intramolecular environment. The great utility of NMR is that the so-called chemical shift for a particular local environment is approximately transferable from molecule to molecule. Furthermore, detailed structural information can be obtained from the multiplet structure.

Electron Spin Resonance (ESR). Also called electron paramagnetic resonance (EPR), ESR is based on the absorption of microwave radiation by unpaired electrons in the presence of strong magnetic fields. Electron spin resonance is a very sensitive technique for the analysis of free radicals, transition metals, and molecular triplet states.

14.13 POLYMER CHEMISTRY

While the terms *dimer*, *trimer*, and so on, specify molecules with a well-defined number of repeating units, the term *polymer* (literally "many parts") is a loosely defined term for macromolecules with one or more repeating units. While there is no upper or lower limit, the number of units in a polymer can easily be in the thousands. The polymerization process for producing synthetic polymers is a chain reaction between individual molecules (monomers) that have two or more potentially reactive sites. Some polymerization reactions only require the presence of a catalyst for initiation and propagation, while some cases require a special intiator such as a free radical. Termination of the chain reaction occurs when all of the available monomeric material is consumed or when a species with only one reactive site reacts.

Addition Polymers. This type of polymer is formed from monomers that contain one or more double bonds, requires a free-radical iniator, and proceeds via a sequence of addition reactions. Consider a mixture of free radicals, X; and olefins,

Note that a large amount of iniator will enhance the initial rate of reaction, but it will also minimize the number of monomeric units in the chain. In addition to their trade names, names of addition polymers have the form poly(monomer name). Examples are polyethylene, polystyrene, and polyte-trafluoroethylene (Teflon).

The presence of a pair of conjugated double bonds in each monomer is a particularly important variant. In this case the propagation step is:

Natural rubber, an example of this type of polymer, contains only the cis form. The vulcanization process involves reaction of sulfur with the double bonds of adjacent polymer strands.

Condensation Polymers. A condensation reaction has the form

$$A-B + C-D \longrightarrow A-C + B-D$$

where B and D are minor fragments. The reactants in the formation of a condensation polymer must have two or more reactive functional groups suitable for a condensation reaction. Names for this type of polymer have the form poly(type of linkage). Common examples are polyester, polyamide, and polycarbonate. For example, the chain reaction that produces a polyester involves the reaction between an acid group on one monomer with an alcohol group on another monomer to produce an ester linkage and one water molecule.

14.14   BIOCHEMICAL STRUCTURES

Amino acids have the general form

$$R-\underset{\underset{H}{|}}{\overset{\overset{NH_2}{|}}{C}}-CO_2H$$

and the names of most amino acids have the suffix -ine. For R = H, the name is glycine. Twenty different amino acids occur in proteins.

   Proteins are condensation polymers of amino acids with the monomeric units linked together by amide (also called peptide) bonds. Because the R groups in the amino acids vary considerably, the structure and properties of proteins reflect the type and order of amino acids in the polymer chain.

   Carbohydrates (saccharides) are a class of molecules that includes simple sugars as well as polymeric materials. The most common simple sugar is glucose, $C_6H_{12}O_6$. Important disaccharides are sucrose (table sugar), lactose (milk sugar), and maltose. The common polysaccharides, starch and cellulose, are condensation polymers of glucose.

   Fats and oils are triesters of glycerol and fatty acids. If the fatty acids contain one (or more) carbon-carbon double bonds, the fat or oil is said to be unsaturated (polyunsaturated); otherwise it is saturated.

   Deoxyribonucleic acid (DNA) and ribonucleic acid (RNA) are the two types of nucleic acids. The DNA polymer has a backbone consisting of alternating sugar (ribose) and phosphate groups. Attached to each sugar is one of four nitrogen bases: adenine, cytosine, guanine, or thymine. The human body's unique genetic information is encoded in the DNA by the sequential order of the attached bases.

   Enzymes alone or in conjunction with their **coenzymes** are the biochemical catalysts, and their names bear the suffix -ase. Because most biochemical reactions have very specific shape and stereochemical requirements, a particular enzyme generally is only a catalyst for a very specific reaction.

List of Symbols

A	Helmholtz work function	V	volume
a	activity	v	molal volume
C	heat capacity, concentration	x	mole fraction in liquid phase
E	potential, V	y	mole fraction in vapor phase
E_a	activation energy	Z	compressibility factor
F	Faraday constant	β	isothermal compressibility
f	fugacity	γ	activity or fugacity coefficient
G	Gibbs free energy	Δ	change, in final to initial
H	enthalpy	μ	chemical potential
J	energy conversion, j/cal	ν	stoichiometric coefficient
K	equilibrium constant	ρ	density
k	reaction rate constant		
M	molarity		*Superscripts*
m	molality	$-$	partial molal quantity
N	number of moles	$\circ$	standard condition, defined
n	number of equivalents		
P	total pressure		*Subscripts*
p	partial pressure	c	critical
R	gas constant	i, j	refer to components
S	entropy	pc	pseudocritical
T	temperature	R	reduced variable
t	time		

REFERENCES

1. Smith, J. M. and Van Ness, H. C., *Introduction to Chemical Engineering Thermodynamics*, 3rd ed, New York, McGraw-Hill, 1975.
2. Dulong and Petit
3. R. C. Weast, ed., *Handbook of Chemistry and Physics*, 66th ed., Boca Raton, FL, 1985.
4. Windholz, M., *The Merck Index*, 10th ed., Rahway, NJ, 1983.

BIBLIOGRAPHY

Atkins, P. W., *Physical Chemistry*, 2nd ed., San Francisco, Freeman, 1982.

Bard, A. J., and Faulkner, L. R., *Electrochemical Methods*, New York, Wiley, 1980.

Davis, R. E., Gailey, K. D., and Whitten, K. W., *Principles of Chemistry*, New York, Saunders, 1984.

Dean, J. A., ed., *Lange's Handbook of Chemistry*, 13th ed., New York, McGraw-Hill, 1985.

Morrison, R. T., and Boyd, R. N., *Organic Chemistry*, 5th ed., Boston, Allyn and Bacon, 1987.

Perry, R. H., and Green, D. W., *Perry's Chemical Engineers Handbook*, 6th ed., New York, McGraw-Hill, 1984.

Selected Values of Chemical Thermodynamic Properties, NBS Circular 500, Washington, DC, U.S. Government Printing Office, 1952.

Stryer, L., *Biochemistry*, 2nd ed., San Francisco, Freeman, 1981.

Willard, H. H., Merritt, L. L., and Dean, J. A., *Instrumental Methods of Analysis*, 5th ed., New York, 1974.

CHAPTER 15

ENGINEERING ECONOMICS

John A. White

School of Industrial and Systems Engineering
Georgia Institute of Technology
Atlanta, Georgia

15.1 INTRODUCTION

One of the earliest engineering axioms is that it costs an engineer $1 to do what it costs others $10 to do. In essence, *engineering decisions are economic decisions*. The specifications developed by an engineer to produce a product or build a structure will impact the cost of the undertaking. Very few situations exist in which there are not alternative materials, processes, and physical designs available to the engineer; and each alternative has different costs associated with it.

Furthermore, the economics of engineering decisions must be viewed from a broad perspective. In particular, one should not define too narrowly the costs associated with engineering design alternatives. As an example, with computer-aided design it is possible to draw upon past experience and design quickly a product having similar characteristics to others designed in the past. Although such an approach might minimize the design cost, there is no guarantee that the **life-cycle cost** of the product will be minimized.

The life-cycle cost of a product is the total cost of design, production, maintenance, and subsequent replacements over the life of the product. For some products, such as airplanes, the life cycle might be 30 or more years; for others, such as computers, the life cycle might be 5 years or less.

The design of a product has an economic impact on all subsequent functions performed. As an example, the specification of screws or bolts, rather than snap fasteners, will affect the cost of assembly; the physical design of the product to allow all assembly to be performed along a single axis will reduce both the cost of assembly and the cost of maintaining the product. Likewise, the specification of the processes used to produce a product will have a major impact on its cost. The generation of a hole by blanking or drilling versus forming, for example, will affect the cost of manufacturing the product.

Despite the fact that the economic aspects of engineering have been recognized for many years, few practicing engineers seem to pay enough attention to the economic impact of their decisions. Too frequently, the engineer is concerned only with the technological viability of a design, rather than both its technological and economic viability. Furthermore, those who do consider the economic impact of their decisions tend to take too narrow a view of the total system.

An example of a narrow perspective is the engineer who designs products with the sole objective of automatic assembly. First, the economic impact of automatic assembly versus manual assembly should be examined. Second, the product should not be designed just on the basis of assembly costs but on the basis of all costs, including those associated with designing, manufacturing, assembling, inspecting, packaging, packing, storing, handling, identifying, distributing, installing, using, and maintaining the product.

Among the economic decisions confronting engineers, the following are typical of the type considered in this chapter: replacement of machinery and equipment, process modification, capacity expansion, new product or process introduction, lease versus buy, and make versus buy. To facilitate the comparison of such alternatives, a number of generally accepted methods will be presented. While the methods will be presented in the context of engineering alternatives, they may be applied in a variety of contexts, including personal finances.

In performing economic comparisons of engineering alternatives, it is recommended that a cash flow approach be taken. A cash flow occurs when money actually changes hands from one individual to another or from one organization to another. Thus, only those receipts and expenditures that occur as a direct result of the alternatives under consideration are involved in the economic comparison.

In the subsequent analysis of cash flows, an **end-of-period convention** will be used. Specifically, when the period is equal to a year, any receipts or disbursements occurring during a year will be shown as occurring at the end of the year. The present will be considered to be the end-of-year zero.

15.2   TIME VALUE OF MONEY OPERATIONS

Just as the spacing of forces along a beam is a primary consideration in engineering mechanics, the spacing of cash flows over time is a primary consideration in engineering economics. A fundamental concept underlying much of the material covered in this chapter is that *money has a time value.* Hence, the value of a given sum of money depends on when the money is to be received or spent. To understand the notion of the time value of money, suppose a particular computer costs $750,000, and the supplier offers to let you pay $750,000 when you install the computer or pay $750,000 when you no longer have use for the computer. Obviously, this is a ridiculous situation, but it serves to illustrate the fact that money has a time value. No one, regardless of the inflation rate, would prefer to pay $750,000 at the time of installation since one could invest the money in a relatively risk-free savings account and earn at least 5% interest on the money; so long as one earned something on the investment, it would be better to postpone payment on the computer.

Likewise, if you are on the receiving end, you would prefer to receive $750,000 today, rather than receive $750,000 at some time in the future. An exception to this is a situation in which income taxes would be adversely affected if the money were received today; in such a case, the income might be deferred to some later date more favorable from an income tax point of view. However, in this situation, when the after-tax amounts are compared, the choice will not be between the same amounts of money. Given the choice between receiving $750,000 *after taxes* today and receiving $750,000 *after taxes* at some time in the future, the early receipt would be preferred. So, the rule still applies: Money has a time value.

Because people favor current consumption to postponed consumption, it is necessary to compensate or reward postponed consumption in some way. The method that has emerged is to employ interest charges or a "return on investment." Without such compensation, consumption would not be postponed and capital formation would not be encouraged.

In assigning a value to the interest rate to be used, a number of considerations are involved. Among the more common factors influencing the selection of interest rates are the following:

1. The opportunity for investment elsewhere, sometimes termed the **opportunity cost**
2. The cost of capital to the firm
3. The magnitude of risk associated with the investment proposal
4. The duration of the investment opportunity
5. The amount of money involved in the investment
6. The kind of investment under consideration, for example, cost reduction versus a new venture
7. The amount of money available for investment
8. The degree of competition for investment funds

In the subsequent treatment of engineering economics, it is useful to consider the interest rate to represent a minimum attractive rate of return (MARR) on investment capital. If a project proves to be profitable after considering the MARR, then it will be recommended. Furthermore, if the money is not invested in the project under consideration, then it will be assumed that the money can be invested elsewhere and earn a return at least equal to the MARR.

The MARR is also referred to as a **hurdle rate** since it establishes the lower bound or benchmark against which the acceptability of a project is measured. The "height of the hurdle," or MARR value, will determine the number of competitors in an economic comparison. Also, it can have a significant impact on the profitability and viability of a company. If the hurdle is set too low, the company runs the risk of undertaking investments it cannot afford; however, if the hurdle is set too high, the firm runs the risk of rejecting projects that would contribute significantly to its profitability. Of the two risks, the latter is more prevalent, that is, establishing too high a value for the MARR.

Many firms, as an example, establish double-digit hurdle rates when there are few if any investment opportunities available elsewhere that will yield double-digit returns. Furthermore, as noted by Kaplan,[1] the real cost of capital is generally lower than many managers think; double-digit costs of capital, for instance, seldom occur.

In this chapter, all interest calculations are based on compound interest rather than simple interest. With compound interest, interest is charged on interest charges to date plus the initial principal amount. Additionally, all compounding is considered to occur at discrete and equally spaced intervals in time, that is, continuous compounding is not considered. Furthermore, only discretely valued cash flows are considered, that is, continuous flows are not considered. A more comprehensive treatment of interest calculations is provided in the references.

15.2.1 Single Sums of Money

In considering the time value of money, it is convenient to represent mathematically the relationship between the current or present value of a single sum of money and its future value. Letting time be measured in years, if a single sum of money P is invested for n years at an annual compound interest rate of $i\%$ per year, then its future value or future worth F will be given by

$$F = P(1 + i)^n \tag{15.1}$$

An underlying assumption of Eq. (15.1) is that interest is compounded annually and at a constant rate of interest. Thus, at the end of the first year P would have increased in value by an amount Pi; thus the value at the end of the first year would equal $P(1 + i)$. At the end of the second year, the $P(1 + i)$ would have increased in value by an amount $P(1 + i)i$ for a total of $P(1 + i) + P(1 + i)i$ or $P(1 + i)^2$. Similar results apply each year until at the end of n years P has increased in value to an amount equal to $P(1 + i)^n$.

To facilitate the determination of the future worth of a single sum of money, the following mnemonic convention is used

$$F = P(F|P\,i\%, n) \tag{15.2}$$

where $F|P\,i\%, n)$, reads "the F given P factor at $i\%$ interest per period for n interest periods." Alternately, the quantity $(1 + i)^n$ is referred to as the **single sum, compound amount factor** or the **single sum, future worth factor**. For convenience in computing future worth equivalents, values of the single sum, compound amount factor have been tabulated and are provided in the appendix.

Conversely, if a future value F is known and one wishes to compute its present value or present worth equivalent, then

$$P = F(1 + i)^{-n} \tag{15.3}$$

or

$$P = F(P|F\,i\%, n) \tag{15.4}$$

where the conversion factor is referred to as the **single sum, present worth factor**. The mnemonic designation for the single sum, present worth factor, $(P|F\,i\%, n)$, reads "the P given F factor at $i\%$ interest per period for n interest periods"; its values are also tabulated in the appendix.

In performing an economic comparison of engineering alternatives, it is recommended that cash flow diagrams be used to depict the timing and magnitudes of the cash flow amounts involved. Figure 15.1 depicts the relationship between P and F, where F occurs n periods after P.

15.2.2 Series of Cash Flows

Rather than a single sum of money being converted to either a future worth equivalent or a present worth equivalent, it is frequently the case that a series of cash flows is involved. Annual expenditures for maintenance, energy, and labor charges are examples of commonly occurring cash flow series. Of

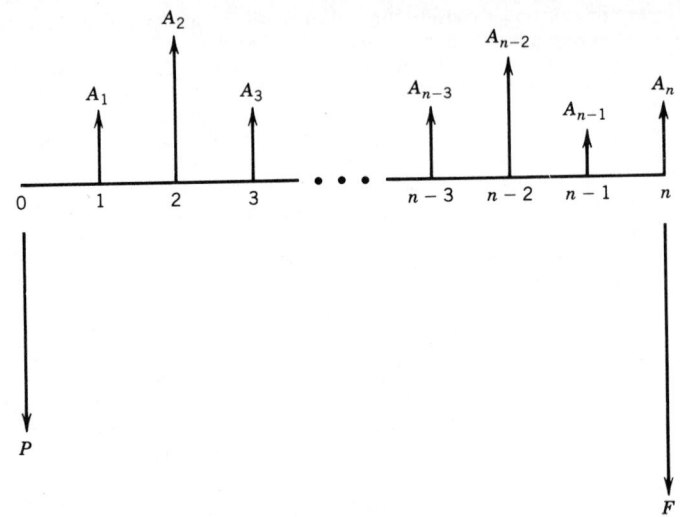

Fig. 15.1 Cash flow diagram of the time relationship between P and F. From Ref. 3.*

course, in such a situation, one can convert each cash flow to the desired equivalence and sum the results to obtain the overall single sum equivalent of the series.

Figure 15.2 depicts a cash flow diagram involving a single sum, present worth amount P, annual cash flows of $A_1, A_2, \ldots, A_n$, and a single sum, future worth equivalent F. Note that P occurs one period before the first A, F occurs at the same time as A_n, and P occurs n periods before F.

With interest at $i\%$ per period, the present worth equivalent of the cash flow series $\{A_1, A_2, \ldots, A_n\}$ can be obtained as follows:

$$P = A_1(P|F\,i\%,1) + A_2(P|F\,i\%,2) + \cdots + A_n(P|F\,i\%,n) \tag{15.5}$$

Similarly, the future worth equivalent of the cash flow series can be obtained by summing the future worth equivalents of the individual cash flows:

$$F = A_n + A_{n-1}(F|P\,i\%,1) + \cdots + A_1(F|P\,i\%, n-1) \tag{15.6}$$

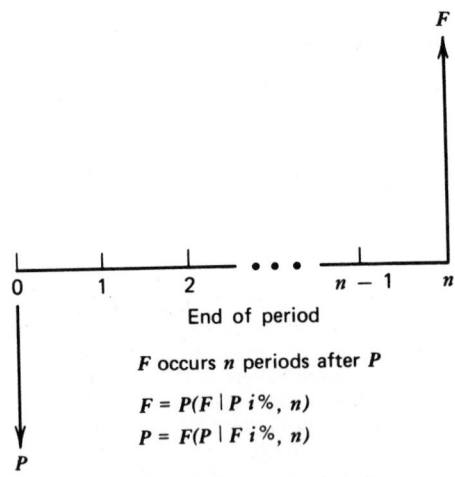

F occurs n periods after P

$$F = P(F\,|\,P\,i\%,\,n)$$
$$P = F(P\,|\,F\,i\%,\,n)$$

Fig. 15.2 Relationships among P, F, and A_k, $k = 1, \ldots, n$. From Ref. 3.

*Figures 15.1 to 15.5 are reprinted from J. A. White, M. H. Agee, and K. E. Case, *Principles of Engineering Economic Analysis*, 2nd edition, copyright © 1984 by John Wiley & Sons, Inc.

TABLE 15.1   Present Worth Equivalent Calculations for Example 15.1

End of Year	Cash Flow	Present Worth Factor	Present Worth Equivalent
0	− $875,000	1.0000	− $875,000.00
1	− 88,750	0.8333	− 73,955.38
2	− 88,750	0.6944	− 61,628.00
3	− 88,750	0.5787	− 51,359.63
4	− 88,750	0.4823	− 42,804.13
5	− 88,750	0.4019	− 35,668.63
6	− 88,750	0.3349	− 29,722.38
7	− 88,750	0.2791	− 24,770.13
8	− 88,750	0.2326	− 20,643.25
9	− 88,750	0.1938	− 17,199.75
10	− 48,750	0.1615	− 7,873.13
		Total	− $1,240,624.41

Example 15.1. To illustrate the conversion of a series of cash flows to a single sum equivalent at a point in time designated as either the present or the future, consider the investment of $875,000 in equipment to manufacture electronic components. The annual labor costs are estimated to be $60,000; the annual maintenance charges are estimated to be $20,000; and other annual costs are estimated to total $8,750. At the end of a 10-year planning period, the equipment is anticipated to have a resale (salvage) value of $40,000.

Using a minimum attractive rate of return of 20% before taxes, what would be the present worth equivalent for the investment? The cash flow profile and the calculations involved are given in Table 15.1. From Table 15.1, the present worth equivalent is a cost of $1,240,624.41.

Using the same interest rate, the future worth equivalent can be obtained directly from the present worth equivalent:

$$F = -\$1,240,624.41\,(F|P\ 20\%,10)$$
$$= -\$1,240,624.41\,(6.1917)$$
$$= -\$7,681,574.16$$

Alternatively, the future worth can be determined for each cash flow and the resulting future worths summed to obtain the total. Table 15.2 presents the calculations involved in computing the future worth term by term. The resulting future worth, − $7,681,572.13, differs from that just obtained by $2.03 due to roundoff error.

Uniform Series of Cash Flows

A common series of cash flows is the **uniform series**, where each cash flow amount in the series has the same value A. Specifically,

$$A_k = A \qquad k = 1, 2, \ldots, n \tag{15.7}$$

Figure 15.3 depicts the relationship between P, A, and F when a uniform series exists.

TABLE 15.2   Future Worth Equivalent Calculations for Example 15.1

End of Year	Cash Flow	Compound Amount Factor	Future Worth Equivalent
0	− $875,000	6.1917	− $5,417,737.50
1	− 88,750	5.1598	− 457,932.25
2	− 88,750	4.2998	− 381,607.25
3	− 88,750	3.5832	− 318,009.00
4	− 88,750	2.9860	− 265,007.50
5	− 88,750	2.4883	− 220,836.63
6	− 88,750	2.0736	− 184,032.00
7	− 88,750	1.7280	− 153,360.00
8	− 88,750	1.4400	− 127,800.00
9	− 88,750	1.2000	− 106,500.00
10	− 48,750	1.0000	− 48,750.00
		Total	− $7,681,572.13

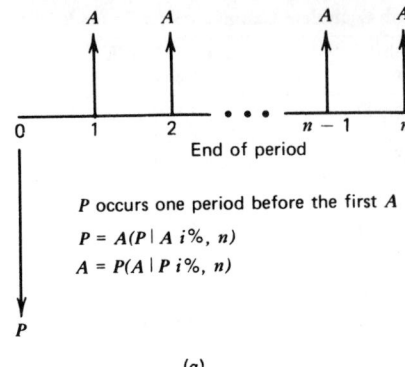

Fig. 15.3 (*a*) Cash flow diagram of the relationship between *P* and *A*. From Ref. 3.

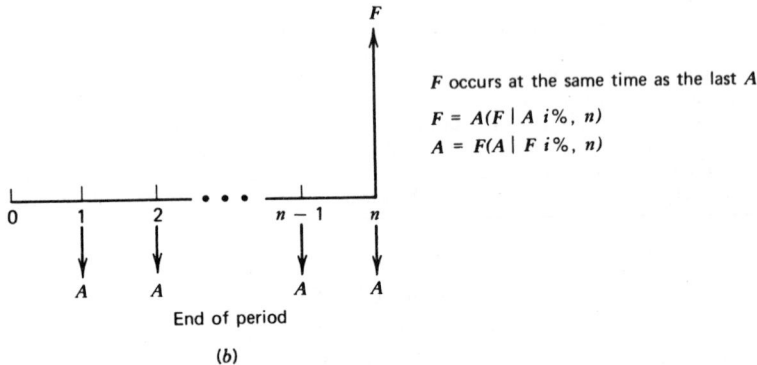

Fig. 15.3 (*b*) Cash flow diagram of the relationship between *A* and *F*. From Ref. 3.

In Example 5.1, the annual expenditures formed a uniform series of \$88,750 per year for 10 years. Because of the presence of the uniform series, an alternative approach can be used to determine the present worth equivalent for Example 15.1. Specifically, to obtain the overall present worth equivalent, sum the present worth equivalent of the initial investment (−\$875,000), the present worth equivalent of the uniform annual series (−\$88,750), and the present worth equivalent of the salvage value (\$40,000). To obtain the present worth equivalent of the uniform series, sum the present worth factors for years 1 through 10 and multiply the product by the uniform annual series amount of −\$88,750; to obtain the present worth equivalent for the salvage value, multiply the present worth factor for year 10 (0.1615) by the salvage value (\$40,000). Hence,

$$P = -\$875,000 - \$88,750(0.8333 + 0.6944 + 0.5787 + 0.4823 + 0.4019 + 0.3349$$
$$+ 0.2791 + 0.2326 + 0.1938 + 0.1615) + \$40,000(0.1615)$$
$$= -\$875,000 - \$88,750(4.1925) + \$40,000(0.1615)$$
$$= -\$1,240,624.38$$

(The difference of \$0.03 in the two approaches is due to roundoff errors in Table 15.1.)

In a similar fashion, the future worth equivalent can be obtained for a uniform series by summing the single sum, compound amount factors. For the previous example, the result would be as follows:

$$F = -\$875,000(6.1917) - \$88,750(1.0000 + 1.2000 + 1.4400 + 1.7280 + 2.0736$$
$$+ 2.4883 + 2.9860 + 3.5832 + 4.2998 + 5.1598) + \$40,000$$
$$= -\$875,000(6.1917) - \$88,750(25.9587) + \$40,000$$
$$= -\$7,681,572.13$$

Since uniform series of cash flows occur rather frequently, the sum of the individual single sum, present worth factors and the individual single sum, compound amount factors are tabulated and given in the interest tables in the appendix. Designated mnemonically as $(P|A\ i\%, n)$, the uniform series, present worth factor converts a uniform series of cash flows, each of magnitude \$A, to a single sum, present worth equivalent, of magnitude \$P. Designated mnemonically as $(F|A\ i\%, n)$, the uniform series, compound amount factor converts a uniform series of cash flows, each of magnitude \$A, to a single sum, future worth equivalent, of magnitude \$P. For $i = 20\%$ and $n = 10$, $(P|A\ 20\%, 10) = 4.1925$ and $(F|A\ 20\%, 10) = 25.9587$, which are identical to the values obtained by summing the individual present worth and compound amount factors.

In general, the present worth equivalent of a uniform series of cash flows can be obtained from the relationship

$$P = A \left[\frac{(1 + i)^n - 1}{i(1 + i)^n} \right] \tag{15.8}$$

or

$$P = A(P|A\ i\%, n) \tag{15.9}$$

where the factor is referred to as the **uniform series, present worth factor**.

Conversely, the uniform series equivalent of a present worth amount can be obtained from the reciprocal relationship

$$A = P \left[\frac{i(1 + i)^n}{(1 + i)^n - 1} \right] \tag{15.10}$$

or

$$A = P(A|P\ i\%, n) \tag{15.11}$$

where the conversion factor, in this case, is referred to as the **capital recovery factor**.

Likewise, the future worth equivalent of a uniform series of cash flows can be obtained from the relationship

$$F = A \left[\frac{(1 + i)^n - 1}{i} \right] \tag{15.12}$$

or

$$F = A(F|A\ i\%, n) \tag{15.13}$$

where the conversion factor is called the **uniform series, compound amount factor**.

Conversely, the reciprocal relationship between the uniform series and the future worth equivalent is given by

$$A = F \left[\frac{i}{(1 + i)^n - 1} \right] \tag{15.14}$$

or

$$A = F(A|F\ i\%, n) \tag{15.15}$$

where the conversion factor is referred to as the **sinking fund factor**.

Gradient Series of Cash Flows

Another common type of cash flow series is the **gradient series**, which increases in magnitude by a constant amount G from one period to the next. In this case,

$$A_k = (k - 1)G \qquad k = 1, 2, \ldots, n \tag{15.16}$$

Figure 15.4 depicts a combination of a uniform series (\$A) and a gradient series in which the gradient series increases by \$G per period, with the first \$G occurring at the end of the second period. The

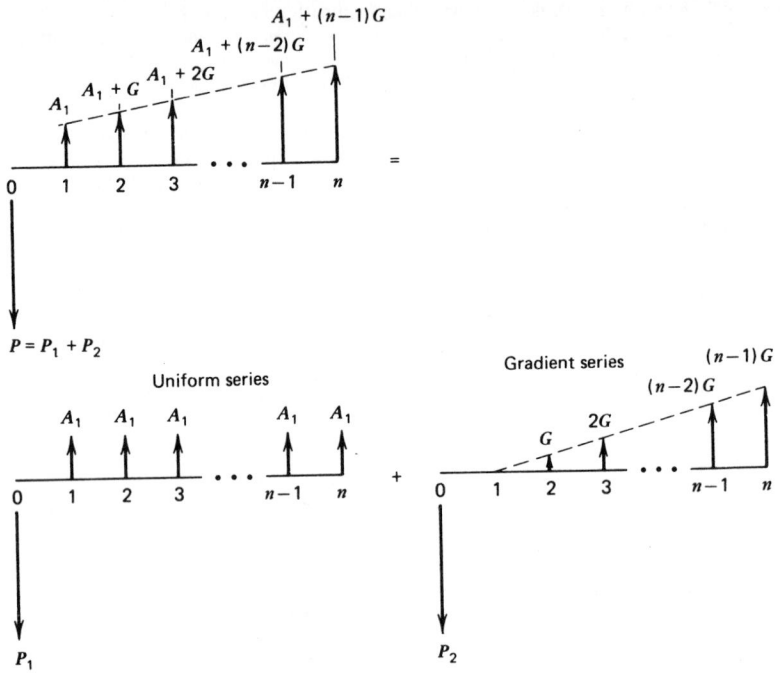

Fig. 15.4 Cash flow diagram of a combination of uniform and gradient series. From Ref. 3.

relationship between P and G is as follows:

$$P = G\left[\frac{1 - (1 + in)(1 + i)^{-n}}{i^2}\right] \tag{15.17}$$

or

$$P = G(P|G\ i\%, n) \tag{15.18}$$

where the conversion factor is referred to as the **gradient series, present worth factor** and is tabulated in the appendix.

The relationship between A and G is given by

$$A = G\left\{\frac{(1 + i)^n - (1 + in)}{i[(1 + i)^n - 1]}\right\} \tag{15.19}$$

or

$$A = G(A|G\ i\%, n) \tag{15.20}$$

where the factor is referred to as the **gradient-to-uniform series conversion factor**.

Geometric Series of Cash Flows

The final type series to be considered is the **geometric series of cash flows**. The geometric series increases by a constant *percentage* from one period to the next; whereas the gradient series increases by a constant *amount*. A cash flow diagram of the geometric series is given in Fig. 15.5. If j denotes the percentage change in the magnitude of a cash flow from one period to the next, the relationship between the size of the $(k - 1)$st cash flow and the kth cash flow is given by

$$A_k = A_{k-1}(1 + j) \qquad k = 2,\ldots, n \tag{15.21}$$

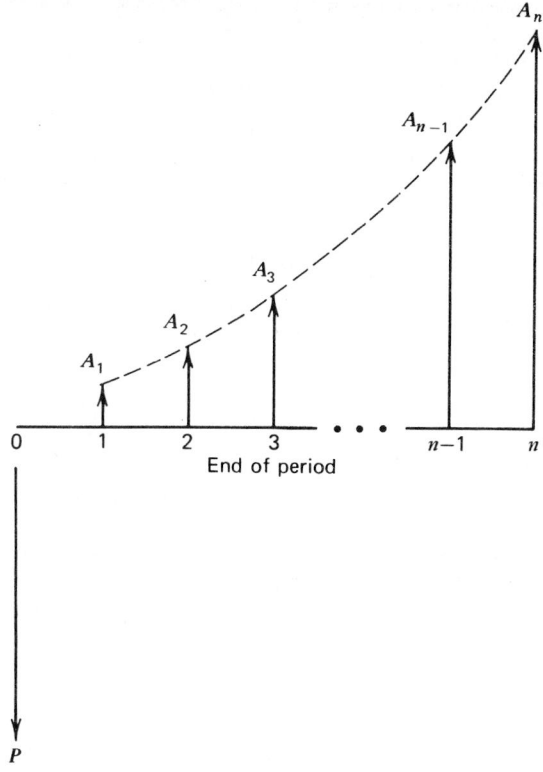

Fig. 15.5 Cash flow diagram of the geometric series. From Ref. 3.

Likewise, the relationship between the size of the kth cash flow and the first cash flow is given by

$$A_k = A_1(1+j)^{k-1} \qquad k = 2,\dots, n \tag{15.22}$$

The present worth equivalent of a geometric series of cash flows is obtained as follows:

$$P = \begin{cases} A_1\left[\dfrac{1-(1+j)^n(1+i)^{-n}}{i-j}\right] & \text{for } i \neq j \\[2ex] \dfrac{nA_1}{1+i} & \text{for } i = j \end{cases} \tag{15.23}$$

or

$$P = A_1(P|A_1\ i\%,\ j\%,\ n) \tag{15.24}$$

where the conversion factor is called the **geometric series, present worth factor**. For the case of $i \neq j$ in Eq. (15.21), P can be obtained by using a combination of other conversion factors. Specifically,

$$P = A_1\left[\frac{1-(F|P\,j\%,\,n)(P|F\,i\%,\,n)}{i-j}\right] \tag{15.25}$$

The future worth equivalent of a geometric series of cash flows can be obtained from the following relationship:

$$
F = \begin{cases} A_1 \left[\dfrac{(1+i)^{-n} - (1+j)^n}{i-j} \right] & i \neq j \\[3mm] nA_1(1+i)^{n-1} & i = j \end{cases} \qquad (15.26)
$$

or

$$
F = A_1(F|A_1 \; i\%, \; j\%, \; n) \qquad (15.27)
$$

where the conversion factor is denoted the **geometric series, future worth factor**. For the case of $i \neq j$, from Eq. (15.24) the future worth can be obtained as follows:

$$
F = A_1 \left[\frac{(F|P\,i\%,\,n) - (F|P\,j\%,\,n)}{i-j} \right] \qquad (15.28)
$$

Example 15.2. In order to illustrate the concepts presented to date, recall Example 15.1 involving the investment of $875,000 in production equipment. Suppose labor costs increase at a rate of 10% per year; maintenance costs increase by $2000 per year; and the other annual costs continue as a uniform series.

The present worth equivalent of labor costs can be obtained using Eq. (15.23) with $A_1 = \$60,000$, $i = 20\%$, $j = 10\%$, and $n = 10$. The resulting present worth is

$$
P = -\$60,000 \left[\frac{1 - (F|P\,10\%,10)(P|F\,20\%,10)}{0.20 - 0.10} \right]
$$

$$
= -\$60,000 \left[\frac{1 - (2.5937)(0.1615)}{0.10} \right]
$$

$$
= -\$348,670.47
$$

The present worth equivalent of maintenance costs can be obtained using a combination of Eq. (15.8) and (15.16), with $A = -\$20,000$, $G = -\$2,000$, $i = 20\%$, and $n = 10$. The resulting present worth is

$$
P = -\$20,000(P|A\,20\%,10) - \$2,000(P|G\,20\%,10)
$$

$$
= -\$20,000(4.1925) - \$2,000(12.8871)
$$

$$
= -\$109,624.20
$$

The present worth equivalent for the combination of the other annual costs and the salvage value is obtained as follows:

$$
P = -\$8,750(P|A\,20\%,10) + \$40,000(P|F\,20\%,10)
$$

$$
= -\$8,750(4.1925) + \$40,000(0.1615)
$$

$$
= -\$30,224.38
$$

Hence, the present worth equivalent for the investment is

$$
P = -\$875,000 - \$348,670.47 - \$109,624.20 - \$30,224.38
$$

$$
= -\$1,363,519.05
$$

Subsequently, we will find that one method of comparing engineering investment alternatives is to compare them on the basis of their present worth equivalents, annual worth equivalents, or future worth equivalents. It is relatively easy to compute the two remaining equivalencies, given any one of the three. For the example, the annual worth equivalent for the investment in processing equipment is

$$
A = P(A|P\,20\%,10)
$$

$$
= -\$1,363,519.05(0.2385)
$$

$$
= -\$325,199.28/yr
$$

and the future worth equivalent is

$$F = P(F|P\ 20\%, 10)$$
$$= -\$1,363,519.05(6.1917)$$
$$= -\$8,442,500.90$$

15.3   COMPARING ENGINEERING ALTERNATIVES

The comparison of engineering design alternatives involves the consideration of many factors, not the least of which are technological and monetary aspects of the design. Quality, safety, performance, delivery, risk, appearance, marketing, environmental, employment, and competitive pressures, for example, might dominate the final selection of the design. However, for now, the focus will be on comparing engineering alternatives on the basis of their economic impact on the organization.

To facilitate the comparison of engineering alternatives, the following systematic approach is recommended:

1. Define the set of feasible alternatives to be compared.
2. Define the planning horizon to be used in the comparison.
3. Develop the cash flow profiles for each alternative.
4. Specify the MARR.
5. Compare the alternatives using a specified method of comparison.
6. Perform supplementary analyses.
7. Select the preferred alternative.

15.3.1   Defining the Alternatives

The alternatives to be considered must be defined very carefully so as not to prematurely eliminate viable candidates. However, at the same time, care must be taken to only include those alternatives that appear to satisfy the noneconomic considerations that will influence the ultimate selection.

It is frequently the case that the "do nothing" alternative will be included. Such an alternative is intended to represent business as usual or maintaining the status quo; however, it is rarely the case that business conditions stand still. Doing nothing does not mean that nothing will be done; rather, it might mean that management has opted to pass up the opportunity to influence future events.

When the do nothing alternative is in consideration, extreme care must be taken not to underestimate the "cost of doing nothing." For many firms, business as usual is their most expensive alternative; "standing pat" or maintaining the status quo for too long can prove to be a disasterous course for many businesses.

Over a period of time, a series of decisions to "not make capital investments," or to do nothing, can result in a firm losing market share due to obsolete products or processes. Considered individually, each decision might seem to be the most economic at the time. However, collectively, the decisions might prove to be myopic, due to their being based on too narrow a perspective. In the defense of the decisions to do nothing, they are made in the absence of perfect knowledge of the future.

Ironically, it is because management cannot accurately predict the economic outcome of change that the absence of change occurs and doing nothing is preferred. Yet, there is also uncertainty concerning the economic impact of doing nothing—of failing to change. One must guard against taking too narrow a view of the economic impact of choosing to follow the path of doing nothing. In considering the long-term impact of doing nothing, managers must decide if they want to be "change masters" or to be mastered by change.

15.3.2   Defining the Planning Horizon

The selection of the planning horizon is also a very important decision since it establishes the width of the "window" through which the economic performance of each alternative will be viewed. In making the selection, consider the person who looks only a foot ahead when traveling along an unfamiliar path; that person might encounter dead ends and might be required to backtrack many times. Likewise, the person who looks a mile ahead and fails to see the short-term obstacles is very likely to stumble and fall along the journey. Adopting too short a planning horizon might result in the elimination of those alternatives that require radical change but will produce dramatic payoffs in the long run; however, adopting too long a planning horizon might result in the firm going out of business before realizing the promised benefits.

Experience suggests that a planning horizon of less than 5 years will support only small incremental changes. Thus, depending on management's objectives, a planning horizon of from 5 to 10 years is recommended.

It is important to distinguish between the length of the planning horizon, the working life of equipment, and the depreciable life of equipment. The working life is the actual period of time the equipment is capable of being used, for example, 20 years; the depreciable life of the equipment is the allowable period of time for depreciating the asset. The planning horizon might have no relationship to the other two periods of time; it is simply the time frame to be used in comparing the alternatives and should realistically represent the period of time over which reasonably accurate cash flow estimates can be provided.

15.3.3 Developing Cash Flow Profiles

In developing the cash flow profiles for each alternative, best estimates are needed of the annual costs and incomes anticipated over the planning horizon, including the salvage values of all assets at the end of the planning horizon. Again, care must be taken to assess the impact of each alternative on the economic performance of the firm, including the economic impact of the do nothing alternative, if it is a feasible candidate.

In determining the costs associated with each alternative, one should not blindly accept cost reduction estimates from a cost-accounting system. Such systems typically include a proration of overhead on the basis of direct costs. However, reductions in direct costs do not necessarily produce proportionate reductions in overhead costs. Again, in using a cash flow approach, the actual costs and savings to be realized should be included in the cash flows for each period. Furthermore, the cash flow estimates should reflect the impact of the alternative on the entire system, rather than just the impact on an individual department. Finally, it is important for consideration to be given to the impact of the investment decision on both revenues and costs.

15.3.4 Specifying the MARR

The comparison of alternatives on the basis of the time value of money requires an explicit determination of the time value to be used. We have previously defined the time value of money as a minimum attractive rate of return, a hurdle rate, an interest rate, and a discount rate. Regardless of its label, a specified interest rate is needed to convert the mix of cash flows occurring throughout the planning horizon to a meaningful economic measure that can be used rationally to compare engineering alternatives.

A number of common factors influencing the selection of the interest rate to be used were listed previously. Although some firms treat the MARR as a parameter whose value depends on the type investment in question, others use a fixed value for the MARR. Regardless of the basis used for determining the value of the MARR, its value should be greater than the cost of securing capital; likewise, it should represent the opportunity cost associated with investing in the candidate alternative as opposed to other available alternatives. Furthermore, care should be taken to ensure that arbitrarily specified values for the MARR do not counteract management's desire to maintain state-of-the-art competency.

Kaplan[1] observed that the use of discounted cash flow methods "most often goes wrong when companies set arbitrarily high hurdle rates for evaluating new investment projects." He noted that "the discounting function serves only to make cash flows received in the future equivalent to cash flows received now. For this narrow purpose—the only purpose, really, of discounting future cash flows—companies should use a discount rate based on the project's opportunity cost of capital (that is, the return available in the capital markets for investments of the same risk)" (pp. 87–88).

15.3.5 Comparing the Alternatives

Among the various measures of economic worth used to compare engineering investment alternatives, the following are the most popular:

1. Present worth (PW) method
2. Annual worth (AW) method
3. Internal rate of return (IRR) method
4. Payback period (PBP) method

Others that are used to varying degrees are the future worth method, the external rate of return method, the savings/investment ratio method, the benefit-cost ratio method, and the capitalized worth method. For a consideration of the latter, see Refs. 2 and 3.

Present Worth Method

For each alternative, the present worth method of comparing engineering investment alternatives converts all cash flows over the planning horizon to a single-sum equivalent at time zero using the MARR; the alternative having the greatest present worth equivalent is the economic choice.

Annual Worth Method

The annual worth method of comparing engineering investment alternatives converts, for each alternative, all cash flows occurring during the planning horizon to an equivalent uniform annual series of cash flows using the MARR; the alternative having the greatest annual worth equivalent is the economic choice. The relationship between the present worth for an alternative and its annual worth is simply

$$PW = AW(P|A \text{ MARR}, n)$$

where n is the length of the planning horizon.

Internal Rate of Return Method

The internal rate of return method compares the MARR with the rates of return resulting from the incremental investments among alternatives. Incremental investments are justified if they are greater than the MARR; otherwise, they are not justified.

Payback Period Method

The payback period method determines for each alternative, at a zero interest rate, how long it will take to recover the initial investment. The alternative that "pays back" the investment the quickest is recommended.

With the exception of the payback period method, the other three measures of economic worth are equivalent methods. Thus, the same recommendation results using present worth, annual worth, or internal rate of return.

Example 15.3. In order to illustrate the application of the four measures of economic worth, consider again the investment of $875,000 in production equipment with annual expenditures for labor, maintenance, and other charges totaling $88,750. At the end of the 10-year planning horizon the equipment is anticipated to have a salvage value of $40,000. Alternatively, the current equipment can continue to be used; it has annual costs of $350,000, and its salvage value in 10 years will be negligible. Yet another alternative is to invest in more sophisticated equipment that will cost $1,250,000, have annual costs of $25,000, and a salvage value of $50,000. A MARR of 20% is to be used to compare the alternatives.

Present Worth Analysis. The present worth of the first alternative has been determined previously to be $PW_1(20\%) = -\$1,240,624$. The present worth of the second alternative (continue with the present equipment) can be found as follows:

$$PW_2(20\%) = -\$350,000(P|A \ 20\%, 10)$$

$$= -\$350,000(4.1925)$$

$$= -\$1,467,375$$

The present worth for the third alternative can be determined as follows:

$$PW_3(20\%) = -\$1,250,000 - \$25,000(P|A \ 20\%, 10) + \$50,000(P|F \ 20\%, 10)$$

$$= -\$1,250,000 - \$25,000(4.1925) + \$50,000(0.1615)$$

$$= -\$1,346,738$$

Hence, on the basis of the present worth method, the economic choice is to invest the $875,000 in the production equipment. The next best alternative is to invest in the more sophisticated equipment. The least economic performance would result from continued use of the existing equipment.

Annual Worth Analysis. The annual worths of the three alternatives can be determined easily from the present worths already obtained. Specifically,

$$A_{W1}(20\%) = -\$1,240,624(A|P\ 20\%,10)$$
$$= -\$1,240,624(0.2385)$$
$$= -\$295,889/\text{yr}$$
$$AW_2(20\%) = -\$1,467,375(A|P\ 20\%,10)$$
$$= -\$1,467,375(0.2385)$$
$$= -\$349,969/\text{yr}$$
$$AW_3(20\%) = -\$1,346,738(A|P\ 20\%,10)$$
$$= -\$1,346,738(0.2385)$$
$$= -\$321,197/\text{yr}$$

and the same recommendations result.

Internal Rate of Return Analysis. The application of the internal rate of return approach requires additional steps. Specifically, the alternatives are first ranked in increasing order of initial investments. Next, the individual rates of return are determined for the alternatives if they exist. Beginning with the alternative having the smallest initial investment and an individual rate of return greater than the MARR, the rate of return is determined on successive increments of investment.

For the example, there are no positive cash flows other than the salvage values. Hence, none of the alternatives have a rate of return greater than the MARR. In this case, we compute the rate of return on the increment of investment required to change from the current equipment to the alternative with the smallest investment, that is, $875,000. Table 15.3 depicts the steps involved in determining the internal rate of return resulting from an investment of $875,000, which produced annual savings of $261,250 for 10 years, plus a single sum of $40,000 at the end of 10 years. The present worth for the incremental cash flow profile is positive using an interest rate of 25% and negative using an interest rate of 30%. Interpolating to determine the interest rate that will yield a zero present worth indicates the internal rate of return is approximately equal to 27.45%. Since the value obtained is greater than the MARR, alternative 1 (making the $875,000 investment) is preferred to alternative 2 (doing nothing).

Next, the rate of return is determined on the increment of investment required to step up to the more sophisticated equipment. To do so requires an investment of $375,000 in order to save $63,750 per year in annual expenditures, and to receive an additional $10,000 at the end of the tenth year. From Table 15.4, the present worth on the incremental investment is positive using a discount rate of 12% and negative using a discount rate of 15%. By interpolating, the internal rate of return on the incremental investment is approximately equal to 12.84%. Since the value obtained is less than the MARR, the additional investment should not be made. Thus, the economic choice is to invest the $875,000 in production equipment and obtain a rate of return of 27.45% over the 10-year period.

Payback Period Analysis. Since individual investments do not payback for this example, due to the absence of positive cash flows sufficient to offset the investments, the payback period method is applied by examining the savings produced by making the investments. The investment of $875,000 produces annual savings of $261,250 per year; therefore, the payback period would be

$$\text{PBP} = \frac{\$875,000}{\$261,250} = 3.35\ \text{yr}$$

TABLE 15.3 Internal Rate of Return for Alternative 1 in Example 15.3

End of Year	Cash Flows for Alternatives 1 and 2		Incremental Cash Flow Alt 1–Alt 2
	Alternative 2	Alternative 1	
0	—	− $875,000	− $875,000
1–10	− 350,000	− 88,750	261,250
10	—	40,000	40,000
$PW_{1\text{-}2}(25\%)$			$62,089
$PW_{1\text{-}2}(30\%)$			− $64,446
$\text{IRR} = 25\% + \dfrac{5\%(62,089)}{62,089 + 64,446}$			
$\text{IRR} = 27.45\%$			

TABLE 15.4   Internal Rate of Return for Alternative 3 in Example 15.3

End of Year	Cash Flows for Alternatives 1 and 3		Incremental Cash Flow Alt 3–Alt 1
	Alternative 3	Alternative 1	
0	− $1,250,000	− $875,000	− $375,000
1–10	− 25,000	− 88,750	63,750
10	50,000	40,000	10,000
$PW_{3-1}(12\%)$			$20,573
$PW_{3-1}(15\%)$			− $52,580
$IRR = 12\% + \dfrac{3\%(20,573)}{20,573 + 52,580}$			
$IRR = 12.84\%$			

The payback period for the more expensive investment would be

$$PBP = \frac{\$1,250,000}{\$325,000} = 3.85 \text{ yr}$$

Depending on the minimum allowable payback period, the firm might choose to reject both investment alternatives. Those firms that use the payback period method typically require a payback of less than 2 or 3 years. For them, neither of the alternatives would be satisfactory.

The payback period method ignores the timing of cash flows and the duration of the investment alternative. Despite these deficiencies, it continues to be one of the most popular methods of evaluating the economic performance of investment alternatives. The following reasons account for its popularity:

1. It does not require interest rate calculations.
2. It does not require a determination of the MARR.
3. It is easily explained and understood.
4. It tends to reflect management's attitude when capital is limited.
5. It serves as a hedge against uncertainty of future cash flows.
6. It provides a rough measure of the liquidity of the investment.

In spite of its popularity, we recommend that it be used as a supplementary method of evaluating the economic performance of engineering investment alternatives.

15.3.6   Performing Supplementary Analyses

The sixth step in performing economic comparisons of engineering investment alternatives is to perform supplementary analyses. Among the supplementary analyses that might be performed are breakeven analyses, sensitivity analyses, and risk analyses. A breakeven analysis is performed when the value of one or more parameters is not known with certainty, but a judgment can be made as to whether its value is greater than or less than some breakeven value.

A sensitivity analysis involves an explicit consideration of the economic impact of changes in the values of one or more parameters. The objective is to determine how sensitive the selection decision is to the values of the parameters used in the analysis. The parameters subject to change might include the planning horizon, the MARR, and any of the cash flows. The process of providing estimates of cash flows and determining the planning horizon and MARR is not a precise, errorless process. Sensitivity analyses are performed in an attempt to cope with the inexact aspects of economic justifications.

Risk analyses are performed by explicitly representing the possible values of parameters in the form of probability distributions. By treating the parameters as random variables, simulation models can be developed of the economic performance of the alternatives. Subsequent comparisons between alternatives can then be made on the basis of the probability distributions for the measure of merit being used, for example, probability distribution for present worth.

Explicit consideration of breakeven, sensitivity, and risk analysis is beyond the scope of this handbook. For additional material treating these subjects, see Refs. 2 and 3.

15.3.7 Selecting the Preferred Alternative

The final step in performing a comparison of engineering investment alternatives involves the selection of the preferred alternative. The final selection decision might be quite different from the one recommended as "the economic choice." Multiple criteria typically exist, rather than the single criterion of maximizing the economic worth. (However, it should be noted that any long-term decision process that consistently selects alternatives that do not earn a rate of return at least equal to the cost of capital will result in financial ruin for the firm. Hence, even when multiple criteria exist, the alternative ultimately selected should not be uneconomic.)

The presence of multiple criteria, coupled with the risks and uncertainties associated with estimating future outcomes, result in the selection process being quite complicated. To make the process easier, the engineer is encouraged to address as many as possible of management's concerns in comparing the investment alternatives. To the extent that the so-called intangibles are made tangible, the selection decision will agree with the engineer's recommendation.

15.4 INCOME TAX AND INFLATION CONSIDERATIONS

Before concluding the treatment of engineering economics, brief consideration will be given to the impact of *income taxes* and *inflation* on the economic performance of investment alternatives.

15.4.1 Impact of Income Taxes

Although the treatment of economic analysis, to this point, has been limited to a consideration of cash flows, in considering the impact of income taxes on the economic performance of engineering investment alternatives it will be necessary to consider a concept that, in fact, is not a cash flow. The concept to be considered is **depreciation**.

Simply stated, depreciation charges are an artifice developed to allow managers to reduce the net worth of assets that decline in value over time. Since depreciation charges are not really cash flows, why consider them?

For our purposes, the only reason for considering depreciation charges is that under current tax law they are an allowable corporate deduction in determining the income tax liability for a firm. Since income taxes are cash flows and depreciation charges affect the size of the income tax paid, the concept of depreciation is treated.

Income taxes are generally computed as a percentage of a firm's taxable income for the year. In recent years, income taxes have amounted to as much as 50% of taxable income, where the latter is defined as gross income or revenue subject to taxation, less certain allowable deductions. The deductions tend to be either *expenses* or *depreciation*. Deductible expenses include labor, material, administrative, sales, power, and interest costs, among others.

Investments in buildings and equipment are not deductible expenses. However, in order to encourage investments in such capital equipment, a portion of the initial investment can be deducted each year for a specified number of years depending on the type asset. The allowable portion that can be deducted is called depreciation.

Details concerning current depreciation laws are beyond the scope of the handbook. Furthermore, over the past decade there have been a number of changes in the treatment of depreciation and the determination of allowable depreciation charges. For these reasons, when an economic comparison is to be performed "after-taxes," it is advisable to gain an up-to-date understanding of current depreciation and tax laws before proceeding.

Prior to the early 1980s a number of different depreciation methods were used in the United States to determine taxable income. Among the more commonly used were straight-line depreciation, declining balance, double declining balance, and sum-of-the-year's-digits. After 1981, the accelerated cost recovery system (ACRS) was established. Detailed aspects of the ACR system were modified by the Tax Reform Act of 1986; the modified accelerated cost recovery system (MACRS) depreciation charges represent various combinations of declining balance, double declining balance, and straight-line depreciation charges.[4]

Under MACRS depreciation, an asset is classified as either 3-year, 5-year, 7-year, 10-year, 15-year, 20-year, or real property. The classification is based on the assets depreciation range (ADR), although there are numerous exceptions.

The 3-year class includes most property with an ADR midpoint of 4 or less years; examples include special tools and devices used in the manufacture of rubber products, special handling devices such as returnable pallets, and over-the-road tractors, among others.

The 5-year class includes property with an ADR midpoint of more than 4 but less than 10 years; examples include computers and peripherals, typewriters, calculators, automobiles, busses, general-purpose trucks, and production equipment for certain industry categories, for example, textiles, carpets, and electronic components.

The 7-year class includes property with an ADR midpoint of 10 or more but less than 16 years; examples include office furniture, railroad cars and locomotives, and most production equipment used to manufacture wood products, sporting goods, jewelry, musical instruments, films, and tapes, among others.

The 10-year class includes property with an ADR midpoint of 16 or more but less than 20 years; examples include assets used to manufacture tobacco and certain food products, as well as that used in petroleum refining.

The 15-year class includes property with an ADR midpoint of 20 or more but less than 25 years; examples are telephone distribution equipment, sewage treatment plants, and assets used to produce electricity for sale.

The 20-year class includes property with an ADR midpoint of 25 years or more, excluding real property with a class life of over 27.5 years; examples include sewer pipe, farm buildings, railroad steam or electric generating equipment, and telephone distribution plants.

The real property class is divided into residential and nonresidential; residential property is recovered over 27.5 years and nonresidential property is recovered over 31.5 years.

All ACR depreciation calculations use a half-year convention and all property is treated as if it were placed in service at midyear.[4] Likewise, if an asset is retired or disposed of during a year, a half-year allowance is provided. Thus, the depreciation charges are spread over one year more than the class designation, for example, 4 tax years for 3-year property.

The MACRS deduction is calculated by applying appropriate percentages to the cost basis for a specified recovery property class and year; the **cost basis** is usually the initial investment in the property. Depending on the recovery property class, the percentage used will depend on the month of the year in which the property was "placed in service." For our purposes, it is assumed that the property is placed in service at the beginning of the tax year and disposed of at the end of the tax year.

With any depreciation plan, the unrecovered investment at the end of year t, called the book value, is designated B_t and is given by

$$B_t = P - (D_1 + D_2 + \cdots + D_t) \qquad (15.29)$$

or

$$B_t = P[1 - (d_1 + d_2 + \cdots + d_t)] \qquad (15.30)$$

where n = allowable write-off or depreciation period
  P = cost basis for the property
  D_t = depreciation charge for year t, $t = 1, \ldots, n$,
    = $d_t P$
  d_t = depreciation percentage allowed for year t for the specific recovery property class, $t = 1, \ldots, n$, and given in Table 15.5.

Example 15.4. As an example of the calculation of annual depreciation charges and book values, suppose the earlier investment in production equipment is for 5-year recovery property. Thus, the MACRS percentages will be 20.00, 32.00, 19.20, 11.52, 11.52, and 5.76% for the 6-year period. The annual depreciation charges and book values will be as follows:

$D_1 = \$875,000(0.2000) = \$175,000$ $B_1 = \$875,000 - \$175,000 = \$700,000$
$D_2 = \$875,000(0.3200) = \$280,000$ $B_2 = \$700,000 - \$280,000 = \$420,000$
$D_3 = \$875,000(0.1920) = \$168,000$ $B_3 = \$420,000 - \$168,000 = \$252,000$
$D_4 = \$875,000(0.1152) = \$100,800$ $B_4 = \$252,000 - \$100,800 = \$151,200$
$D_5 = \$875,000(0.1152) = \$100,800$ $B_5 = \$151,200 - \$100,800 = \$50,400$
$D_6 = \$875,000(0.0576) = \$50,400$ $B_6 = \$50,400 - \$50,400 \quad = \$0$

Before demonstrating the impact of income taxes on the economic performance of an engineering investment alternative, it is necessary to consider the relationship between before-tax cash flows and after-tax cash flows. Specifically,

$$ATCF_t = BTCF_t - T_t \qquad (15.33)$$

TABLE 15.5   MACRS Percentages for 3-Year, 5-Year, 7-Year, 10-Year, 15-Year, and 20-Year Property[a]

Year	3-Year	5-Year	7-Year	10-Year	15-Year	20-Year
1	33.33%	20.00%	14.29%	10.00%	5.00%	3.75%
2	44.44	32.00	24.49	18.00	9.50	7.22
3	14.82	19.20	17.49	14.40	8.55	6.68
4	7.41	11.52	12.49	11.52	7.70	6.18
5		11.52	8.92	9.22	6.93	5.71
6		5.76	8.92	7.37	6.23	5.29
7			8.92	6.55	5.90	4.89
8			4.48	6.55	5.90	4.52
9				6.55	5.90	4.46
10				6.55	5.90	4.46
11				3.29	5.90	4.46
12					5.90	4.46
13					5.90	4.46
14					5.90	4.46
15					5.90	4.46
16					2.99	4.46
17						4.46
18						4.46
19						4.46
20						4.46
21						2.24

[a]The table reflects the official statements concerning the formulas to be used to compute the allowable depreciation deductions.

where $ATCF_t$ = after-tax cash flow for year t
$BTCF_t$ = before-tax cash flow for year t
T_t = income tax for year t
$= r_t TI_t$
r_t = income tax rate for year t
TI_t = taxable income for year t
$= BTCF_t - D_t$

Thus,

$$ATCF_t = BTCF_t - r_t(BTCF_t - D_t) \qquad (15.34)$$

or

$$ATCF_t = BTCF_t(1 - r_t) - r_t D_t \qquad (15.35)$$

Example 15.5. Recall Examples 15.3 and 15.4. For each alternative, the after-tax present worth is to be determined based on MACRS depreciation, a 34% income tax rate, an after-tax MARR of 10%, and a 10-year planning horizon. For the case of alternatives 1 and 3, 5-year recovery property is assumed. For alternative 2, it is assumed that there are no depreciation charges either because equipment has been fully depreciated or because the activity is wholly labor intensive; hence, the annual expenditures are "fully expensed," that is, written off in the year in which they occur.

From Tables 15.6–15.8, the after-tax present worths for the three alternatives are −$994,687, −$968,468, and −$1,010,035, respectively. Thus, alternative 2—do nothing is preferred to investing in either type of new equipment. If it had been necessary to choose one of the investment alternatives, then the $875,000 investment would be preferred to the $1,250,000 investment.

15.4.2   Impact of Inflation

For at least the last decade, explicit consideration of the impact of inflation on economic justifications has been recommended for those investments that require consideration of income tax effects, as well as those that are multinational. In this section, the impact of inflation will be described. For more extensive treatments of the subject, refer to recently published textbooks on engineering economic analysis.

TABLE 15.6 After-Tax Present Worth Calculations for Alternative 1 in Example 15.5

End of Year	Before-Tax Cash Flow	Depreciation Charge	Taxable Income	Income Tax	After-Tax Cash Flow
0	− $875,000				− $875,000
1	− 88,750	175,000	− 263,750	− 89,675[a]	925
2	− 88,750	280,000	− 368,750	− 125,375	36,625
3	− 88,750	168,000	− 256,750	− 87,295	− 1,455
4	− 88,750	100,800	− 189,550	− 64,447	− 24,303
5	− 88,750	100,800	− 189,550	− 64,447	− 24,303
6	− 88,750	50,400	− 139,150	− 47,311	− 41,439
7	− 88,750		− 88,750	− 30,175	− 58,575
8	− 88,750		− 88,750	− 30,175	− 58,575
9	− 88,750		− 88,750	− 30,175	− 58,575
10	− 88,750		− 88,750	− 30,175	− 58,575
10	40,000		40,000	13,600	26,400

$PW_1(10\%) = -\$994,697$

[a] A negative income tax means a tax savings will occur due to the investment in the alternative.

TABLE 15.7 After-Tax Present Worth Calculations for Alternative 2 in Example 15.5

End of Year	Before-Tax Cash Flow	Depreciation Charge	Taxable Income	Income Tax	After-Tax Cash Flow
0	—				
1–10	− 350,000	—	− 350,000	− 119,000	− 231,000
10	—				

$PW_2(10\%) = -\$968,468$

TABLE 15.8 After-Tax Present Worth Calculations for Alternatives 3 in Example 15.5

End of Year	Before-Tax Cash Flow	Depreciation Charge	Taxable Income	Income Tax	After-Tax Cash Flow
0	− $1,250,000				− $1,250,000
1	− 25,000	250,000	− 275,000	− 93,500	68,500
2	− 25,000	400,000	− 425,000	− 144,500	119,500
3	− 25,000	240,000	− 265,000	− 90,100	65,100
4	− 25,000	144,000	− 169,000	− 57,460	32,460
5	− 25,000	144,000	− 169,000	− 57,460	32,460
6	− 25,000	72,000	− 97,000 .	− 32,980	7,980
7	− 25,000		− 25,000	− 8,500	− 16,500
8	− 25,000		− 25,000	− 8,500	− 16,500
9	− 25,000		− 25,000	− 8,500	− 16,500
10	− 25,000		− 25,000	− 8,500	− 16,500
10	50,000		50,000	17,000	33,000

$PW_3(10\%) = -\$1,010,035$

The cash flows associated with an engineering investment alternative generally consist of a number of different components of cost. Among the more common components are labor costs, energy costs, land costs, material costs, cost of supplies, taxes, and investments in equipment and facilities, among others. Also, from the previous section, the after-tax cash flow will be affected by the allowable depreciation charges for equipment and facilities.

Of the cash flow components listed, it is not usual to find that costs change over the life of an investment. Such changes can result because of natural phenomena, for example, increased labor costs because people advance in pay with experience, increased maintenance costs because equipment wears out with age and requires more maintenance; decreased costs for computers because of improved yields, increased productivity, and technological breakthroughs in processing, and decreased material costs because of reduced scrap. However, in recent years, a major source of change for costs has been inflation.

Simply stated, inflation is a measure of the buying power of money. When sharp increases occur in the amount of currency in circulation, sudden decreases can occur in the purchasing power of money due to rising prices. Generally, the inflation rate for a country is determined using a "market basket" approach; in particular, it is the average rate of increase in prices for a wide range of goods and services. The Consumer Price Index (CPI) is one commonly used measure of inflation in the United States.

Engineers must be very careful to separate inflation effects from natural phenomena, where the latter includes changes that would occur in the absence of inflation, due to natural business conditions. As a result of the combination of changes due to inflation and natural causes, the costs associated with cash flow component k will increase at an overall rate of r_k percent, where r is made up of the inflation rate (j) and the rate of change in costs for cash flow component k due to natural growth (g_k), referred to as the growth rate. The relationship among the rates is given by

$$r_k = j + g_k + j(g_k) \tag{15.36}$$

To perform an economic analysis of an engineering investment under inflation, it is useful to distinguish between the **real discount rate** (i_r) and the **combined discount rate** (i_c). The former does not include an adjustment for inflation; instead, it represents the MARR in the absence of inflation; the latter does account for inflation. The relationship between the real discount rate and the combined discount rate is the same as that between the growth rate and inflation rate; namely,

$$i_c = j + i_r + j(i_r) \tag{15.37}$$

Having distinguished between the real and combined discount rates, before-tax economic analyses under inflation can be performed using either of the following two procedures:

1. Express all cash flows in terms of their "then-current" amounts and use the combined discount rate.

2. Express all cash flows in terms of their "constant worth" amounts and use the real discount rate.

However, as will be illustrated subsequently, when performing after-tax analyses, it is recommended that then-current cash flow estimates be used.

Example 15.6. In order to illustrate the relationship between the real and combined discount rates, consider the investment of $60,000 in a program to reduce material requirements in a production process. As a result of the study, the annual material requirement is reduced by 10,000 lb. The present unit cost of the material is $2/lb. Over a 5-year period, in the absence of inflation, the **real internal rate of return** will be slightly less than 20%.

Suppose the price of material increases at an annual rate of 8% due to inflation. The cost per pound will be $2.16 for the first year, $2.33 the second year, $2.52 the third year, $2.72 the fourth year, and $2.94 the fifth year. Performing the calculations to determine the **combined internal rate of return** on the investment yields a value of approximately 29.5%.

As expected, the two values obtained are approximately given by the relation

$$0.295 \approx 0.08 + 0.20 + 0.08(0.20)$$

with the discrepancy due to roundoff and interpolation errors.

Example 15.7. To illustrate the impact of inflation on the after-tax performance of investment alternatives, consider an investment of $100,000 in 5-year class equipment. Over the 6-year depreciation period, suppose the annual returns from the investment are a decreasing gradient series, with $40,000 the first year and decreases of $5000 per year, as depicted in Table 15.9. The after-tax internal rate of return based on constant worth cash flows can be shown to be approximately 13.75%.

Suppose inflation occurs at an annual rate of 6% and affects the annual returns accordingly. Table 15.10 provides the details of the after-tax analysis on the inflated returns for the investment using then-current cash flow estimates for the cash flows. The combined internal rate of return is approximately 19.25%. However, since the combined rate of return equals the sum of the inflation rate, the real return on investment, and their product, a calculation establishes that the real after-tax of return is 12.5%, rather than the 13.75% obtained using constant worth cash flow estimates. Thus, depending on the MARR used, the investment might be considered to be attractive using constant worth cash flow estimates, when it is actually unattractive financially due to the impact of inflation. (In this case, there was little difference in the rates of return obtained. However, in general, the differences in results increase with increasing inflation.)

After-tax analyses that result in favorable recommendations in the absence of inflation can be reversed when inflation effects are considered. Different recommendations can occur when using

TABLE 15.9 After-Tax Analysis Using Constant Worth Cash Flow Estimates for Example 15.7

End of Year	Before-Tax Cash Flow	Depreciation Charge	Taxable Income	Income Tax	After-Tax Cash Flow
0	− $100,000				− $100,000
1	40,000	20,000	20,000	6,800	33,200
2	35,000	32,000	3,000	1,020	33,980
3	30,000	19,200	10,800	3,672	26,328
4	25,000	11,520	13,480	4,583	20,417
5	20,000	11,520	8,480	2,883	17,117
6	15,000	5,760	9,240	3,142	11,858

TABLE 15.10 After-Tax Analysis Using Then-Current Cash Flow Estimates for Example 15.7

End of Year	Before-Tax Cash Flow	Depreciation Charge	Taxable Income	Income Tax	After-Tax Cash Flow
0	− $100,000				− $100,000
1	42,400	20,000	22,400	7,616	34,784
2	39,326	32,000	7,326	2,491	36,835
3	35,730	19,200	16,530	5,620	30,110
4	31,563	11,520	20,043	6,815	24,748
5	26,764	11,520	15,244	5,183	21,581
6	21,278	5,760	12,038	4,093	17,185

constant worth and then-current cash flow estimates *because depreciation charges do not increase with inflation*. For this reason, explicit consideration of inflation becomes important when after-tax analyses are performed. Under current U.S. tax law, allowable depreciation write-offs are not indexed to inflation. As a result, when inflation rates are high, capital investments become more difficult to justify and capital spending declines.

15.5 SUMMARY

In summary, the success of engineering design efforts is intimately related to the economic impact of the designs on the firm. As a result, the engineer must recommend design alternatives that yield positive economic benefits to the firm.

To facilitate a comparison of engineering investment alternatives, a number of techniques and philosophies have been presented in the chapter. The presentation was based on the fundamental concept that money has a time value, with the value determined by the opportunity for investment elsewhere. It was recommended that a systematic seven-step procedure be used in comparing engineering alternatives. Furthermore, each step must be considered carefully; the determination of the length of the planning horizon and the value of the MARR are just as important as the estimation of cash flows, generation of feasible investment alternatives, comparison of alternatives, and the supplementary analyses performed. The selection decision can be influenced significantly by the care with which the economic analysis is performed.

The coverage of engineering economic analysis concluded with a consideration of the impact of income taxes and inflation on the economic performance of investment alternatives. From the examples, it is clear that consideration of income taxes and inflation conditions can have an impact on the economic viability of engineering designs.

REFERENCES

1. Kaplan, R. S., "Must CIM be Justified by Faith Alone?", *Harvard Business Review*, Vol. 64, No. 2, March–April, 1986, pp. 87–95.
2. Canada, J. R. and White, J. A., *Capital Investment Decision Analysis for Management and Engineering*, Englewood Cliffs, N.J., Prentice-Hall, 1980.
3. White, J. A., Agee, M. H., and Case, K. E., *Principles of Engineering Economic Analysis*, 2nd edition, New York, Wiley, 1984.
4. Bradford, W. M., *The Prentice-Hall Business Tax Deduction Master Guide: 1987 Edition*, Englewood Cliffs, N.J., Prentice-Hall, 1987.

APPENDIX

TABLES OF DISCRETE COMPOUNDING

TABLE 15.A.1 Discrete Compounding: $i = 5\%$

	Single Payment		Uniform Series				Gradient Series	
	Compound Amount Factor	Present Worth Factor	Compound Amount Factor	Sinking Fund Factor	Present Worth Factor	Capital Recovery Factor	Uniform Series Factor	Present Worth Factor
n	To Find F Given P $F\|P\,i,n$	To Find P Given F $P\|F\,i,n$	To Find F Given A $F\|A\,i,n$	To Find A Given F $A\|F\,i,n$	To Find P Given A $P\|A\,i,n$	To Find A Given P $A\|P\,i,n$	To Find A Given G $A\|G\,i,n$	To Find P Given G $P\|G\,i,n$
1	1.0500	0.9524	1.0000	1.0000	0.9524	1.0500	0.0000	0.0000
2	1.1025	0.9070	2.0500	0.4878	1.8594	0.5378	0.4878	0.9070
3	1.1576	0.8638	3.1525	0.3172	2.7232	0.3672	0.9675	2.6347
4	1.2155	0.8227	4.3101	0.2320	3.5460	0.2820	1.4391	5.1028
5	1.2763	0.7835	5.5256	0.1810	4.3295	0.2310	1.9025	8.2369
6	1.3401	0.7462	6.8019	0.1470	5.0757	0.1970	2.3579	11.9680
7	1.4071	0.7107	8.1420	0.1228	5.7864	0.1728	2.8052	16.2321
8	1.4775	0.6768	9.5491	0.1047	6.4632	0.1547	3.2445	20.9700
9	1.5513	0.6446	11.0266	0.0907	7.1078	0.1407	3.6758	26.1268
10	1.6289	0.6139	12.5779	0.0795	7.7217	0.1295	4.0991	31.6520
11	1.7103	0.5847	14.2068	0.0704	8.3064	0.1204	4.5144	37.4988
12	1.7959	0.5568	15.9171	0.0628	8.8633	0.1128	4.9219	43.6241
13	1.8856	0.5303	17.7130	0.0565	9.3936	0.1065	5.3215	49.9879
14	1.9799	0.5051	19.5986	0.0510	9.8986	0.1010	5.7133	56.5538
15	2.0789	0.4810	21.5786	0.0463	10.3797	0.0963	6.0973	63.2880
16	2.1829	0.4581	23.6575	0.0423	10.8378	0.0923	6.4736	70.1597
17	2.2920	0.4363	25.8404	0.0387	11.2741	0.0887	6.8423	77.1405
18	2.4066	0.4155	28.1324	0.0355	11.6896	0.0855	7.2034	84.2043
19	2.5270	0.3957	30.5390	0.0327	12.0853	0.0827	7.5569	91.3275
20	2.6533	0.3769	33.0660	0.0302	12.4622	0.0802	7.9030	98.4884
21	2.7860	0.3589	35.7193	0.0280	12.8212	0.0780	8.2416	105.6672
22	2.9253	0.3418	38.5052	0.0260	13.1630	0.0760	8.5730	112.8461
23	3.0715	0.3256	41.4305	0.0241	13.4886	0.0741	8.8971	120.0086
24	3.2251	0.3101	44.5020	0.0225	13.7986	0.0725	9.2140	127.1402
25	3.3864	0.2953	47.7271	0.0210	14.0939	0.0710	9.5238	134.2275
26	3.5557	0.2812	51.1135	0.0196	14.3752	0.0696	9.8266	141.2585
27	3.7335	0.2678	54.6691	0.0183	14.6430	0.0683	10.1224	148.2225
28	3.9201	0.2551	58.4026	0.0171	14.8981	0.0671	10.4114	155.1101
29	4.1161	0.2429	62.3227	0.0160	15.1411	0.0660	10.6936	161.9126
30	4.3219	0.2314	66.4388	0.0151	15.3725	0.0651	10.9691	168.6225
31	4.5380	0.2204	70.7608	0.0141	15.5928	0.0641	11.2381	175.2333
32	4.7649	0.2099	75.2988	0.0133	15.8027	0.0633	11.5005	181.7391
33	5.0032	0.1999	80.0638	0.0125	16.0025	0.0625	11.7566	188.1351
34	5.2533	0.1904	85.0670	0.0118	16.1929	0.0618	12.0063	194.4168
35	5.5160	0.1813	90.3203	0.0111	16.3742	0.0611	12.2498	200.5806
40	7.0400	0.1420	120.7997	0.0083	17.1591	0.0583	13.3775	229.5451
45	8.9850	0.1113	159.7001	0.0063	17.7741	0.0563	14.3644	255.3145
50	11.4674	0.0872	209.3479	0.0048	18.2559	0.0548	15.2233	277.9148
55	14.6356	0.0683	272.7124	0.0037	18.6335	0.0537	15.9664	297.5103
60	18.6792	0.0535	353.5835	0.0028	18.9293	0.0528	16.6062	314.3430
65	23.8399	0.0419	456.7979	0.0022	19.1611	0.0522	17.1541	328.6909
70	30.4264	0.0329	588.5283	0.0017	19.3427	0.0517	17.6212	340.8408
75	38.8327	0.0258	756.6533	0.0013	19.4850	0.0513	18.0176	351.0720
80	49.5614	0.0202	971.2285	0.0010	19.5965	0.0510	18.3526	359.6460
85	63.2543	0.0158	1245.0867	0.0008	19.6838	0.0508	18.6346	366.8005
90	80.7306	0.0124	1594.6067	0.0006	19.7523	0.0506	18.8712	372.7488
95	103.0346	0.0097	2040.6926	0.0005	19.8059	0.0505	19.0689	377.6773
100	131.5012	0.0076	2610.0239	0.0004	19.8479	0.0504	19.2337	381.7490

TABLE 15.A.2   Discrete Compounding: $i = 6\%$

	Single Payment		Uniform Series				Gradient Series	
	Compound Amount Factor	Present Worth Factor	Compound Amount Factor	Sinking Fund Factor	Present Worth Factor	Capital Recovery Factor	Uniform Series Factor	Present Worth Factor
n	To Find F Given P $F\|P\,i, n$	To Find P Given F $P\|F\,i, n$	To Find F Given A $F\|A\,i, n$	To Find A Given F $A\|F\,i, n$	To Find P Given A $P\|A\,i, n$	To Find A Given P $A\|P\,i, n$	To Find A Given G $A\|G\,i, n$	To Find P Given G $P\|G\,i, n$
1	1.0600	0.9434	1.0000	1.0000	0.9434	1.0600	0.0000	0.0000
2	1.1236	0.8900	2.0600	0.4854	1.8334	0.5454	0.4854	0.8900
3	1.1910	0.8396	3.1836	0.3141	2.6730	0.3741	0.9612	2.5692
4	1.2625	0.7921	4.3746	0.2286	3.4651	0.2886	1.4272	4.9455
5	1.3382	0.7473	5.6371	0.1774	4.2124	0.2374	1.8836	7.9345
6	1.4185	0.7050	6.9753	0.1434	4.9173	0.2034	2.3304	11.4594
7	1.5036	0.6651	8.3938	0.1191	5.5824	0.1791	2.7676	15.4497
8	1.5938	0.6274	9.8975	0.1010	6.2098	0.1610	3.1952	19.8416
9	1.6895	0.5919	11.4913	0.0870	6.8017	0.1470	3.6133	24.5768
10	1.7908	0.5584	13.1808	0.0759	7.3601	0.1359	4.0220	29.6023
11	1.8983	0.5268	14.9716	0.0668	7.8869	0.1268	4.4213	34.8702
12	2.0122	0.4970	16.8699	0.0593	8.3838	0.1193	4.8113	40.3369
13	2.1329	0.4688	18.8821	0.0530	8.8527	0.1130	5.1920	45.9629
14	2.2609	0.4423	21.0151	0.0476	9.2950	0.1076	5.5635	51.7128
15	2.3966	0.4173	23.2760	0.0430	9.7122	0.1030	5.9260	57.5546
16	2.5404	0.3936	25.6725	0.0390	10.1059	0.0990	6.2794	63.4592
17	2.6928	0.3714	28.2129	0.0354	10.4773	0.0954	6.6240	69.4011
18	2.8543	0.3503	30.9057	0.0324	10.8276	0.0924	6.9597	75.3569
19	3.0256	0.3305	33.7600	0.0296	11.1581	0.0896	7.2867	81.3062
20	3.2071	0.3118	36.7856	0.0272	11.4699	0.0872	7.6051	87.2304
21	3.3996	0.2942	39.9927	0.0250	11.7641	0.0850	7.9151	93.1136
22	3.6035	0.2775	43.3923	0.0230	12.0416	0.0830	8.2166	98.9412
23	3.8197	0.2618	46.9958	0.0213	12.3034	0.0813	8.5099	104.7007
24	4.0489	0.2470	50.8156	0.0197	12.5504	0.0797	8.7951	110.3812
25	4.2919	0.2330	54.8645	0.0182	12.7834	0.0782	9.0722	115.9731
26	4.5494	0.2198	59.1564	0.0169	13.0032	0.0769	9.3414	121.4684
27	4.8223	0.2074	63.7058	0.0157	13.2105	0.0757	9.6029	126.8599
28	5.1117	0.1956	68.5281	0.0146	13.4062	0.0746	9.8568	132.1420
29	5.4184	0.1846	73.6398	0.0136	13.5907	0.0736	10.1032	137.3095
30	5.7435	0.1741	79.0582	0.0126	13.7648	0.0726	10.3422	142.3587
31	6.0881	0.1643	84.8017	0.0118	13.9291	0.0718	10.5740	147.2864
32	6.4534	0.1550	90.8898	0.0110	14.0840	0.0710	10.7988	152.0901
33	6.8406	0.1462	97.3432	0.0103	14.2302	0.0703	11.0166	156.7680
34	7.2510	0.1379	104.1838	0.0096	14.3681	0.0696	11.2276	161.3191
35	7.6861	0.1301	111.4347	0.0090	14.4982	0.0690	11.4319	165.7427
40	10.2857	0.0972	154.7619	0.0065	15.0463	0.0665	12.3590	185.9568
45	13.7646	0.0727	212.7435	0.0047	15.4558	0.0647	13.1413	203.1096
50	18.4202	0.0543	290.3357	0.0034	15.7619	0.0634	13.7964	217.4573
55	24.6503	0.0406	394.1719	0.0025	15.9905	0.0625	14.3411	229.3222
60	32.9877	0.0303	533.1279	0.0019	16.1614	0.0619	14.7909	239.0427
65	44.1450	0.0227	719.0828	0.0014	16.2891	0.0614	15.1601	246.9450
70	59.0759	0.0169	967.9319	0.0010	16.3845	0.0610	15.4613	253.3271
75	79.0569	0.0126	1300.9485	0.0008	16.4558	0.0608	15.7058	258.4526
80	105.7959	0.0095	1746.5984	0.0006	16.5091	0.0606	15.9033	262.5491
85	141.5788	0.0071	2342.9807	0.0004	16.5489	0.0604	16.0620	265.8093
90	189.4644	0.0053	3141.0735	0.0003	16.5787	0.0603	16.1891	268.3945
95	253.5462	0.0039	4209.1016	0.0002	16.6009	0.0602	16.2905	270.4373
100	339.3018	0.0029	5638.3633	0.0002	16.6175	0.0602	16.3711	272.0469

TABLE 15.A.3   Discrete Compounding: $i = 7\%$

	Single Payment		Uniform Series				Gradient Series	
	Compound Amount Factor	Present Worth Factor	Compound Amount Factor	Sinking Fund Factor	Present Worth Factor	Capital Recovery Factor	Uniform Series Factor	Present Worth Factor
	To Find F Given P	To Find P Given F	To Find F Given A	To Find A Given F	To Find P Given A	To Find A Given P	To Find A Given G	To Find P Given G
n	$F\|P\,i,n$	$P\|F\,i,n$	$F\|A\,i,n$	$A\|F\,i,n$	$P\|A\,i,n$	$A\|P\,i,n$	$A\|G\,i,n$	$P\|G\,i,n$
1	1.0700	0.9346	1.0000	1.0000	0.9346	1.0700	0.0000	0.0000
2	1.1449	0.8734	2.0700	0.4831	1.8080	0.5531	0.4831	0.8734
3	1.2250	0.8163	3.2149	0.3111	2.6243	0.3811	0.9549	2.5060
4	1.3108	0.7629	4.4399	0.2252	3.3872	0.2952	1.4155	4.7947
5	1.4026	0.7130	5.7507	0.1739	4.1002	0.2439	1.8650	7.6467
6	1.5007	0.6663	7.1533	0.1398	4.7665	0.2098	2.3032	10.9784
7	1.6058	0.6227	8.6540	0.1156	5.3893	0.1856	2.7304	14.7149
8	1.7182	0.5820	10.2598	0.0975	5.9713	0.1675	3.1465	18.7889
9	1.8385	0.5439	11.9780	0.0835	6.5152	0.1535	3.5517	23.1404
10	1.9672	0.5083	13.8164	0.0724	7.0236	0.1424	3.9461	27.7156
11	2.1049	0.4751	15.7836	0.0634	7.4987	0.1334	4.3296	32.4665
12	2.2522	0.4440	17.8885	0.0559	7.9427	0.1259	4.7025	37.3506
13	2.4098	0.4150	20.1406	0.0497	8.3577	0.1197	5.0648	42.3302
14	2.5785	0.3878	22.5505	0.0443	8.7455	0.1143	5.4167	47.3718
15	2.7590	0.3624	25.1290	0.0398	9.1079	0.1098	5.7583	52.4461
16	2.9522	0.3387	27.8881	0.0359	9.4466	0.1059	6.0897	57.5271
17	3.1588	0.3166	30.8402	0.0324	9.7632	0.1024	6.4110	62.5923
18	3.3799	0.2959	33.9990	0.0294	10.0591	0.0994	6.7225	67.6220
19	3.6165	0.2765	37.3790	0.0268	10.3356	0.0968	7.0242	72.5991
20	3.8697	0.2584	40.9955	0.0244	10.5940	0.0944	7.3163	77.5091
21	4.1406	0.2415	44.8652	0.0223	10.8355	0.0923	7.5990	82.3393
22	4.4304	0.2257	49.0057	0.0204	11.0612	0.0904	7.8725	87.0793
23	4.7405	0.2109	53.4361	0.0187	11.2722	0.0887	8.1369	91.7201
24	5.0724	0.1971	58.1767	0.0172	11.4693	0.0872	8.3923	96.2545
25	5.4274	0.1842	63.2490	0.0158	11.6536	0.0858	8.6391	100.6765
26	5.8074	0.1722	68.6765	0.0146	11.8258	0.0846	8.8773	104.9813
27	6.2139	0.1609	74.4838	0.0134	11.9867	0.0834	9.1072	109.1655
28	6.6488	0.1504	80.6977	0.0124	12.1371	0.0824	9.3289	113.2264
29	7.1143	0.1406	87.3465	0.0114	12.2777	0.0814	9.5427	117.1621
30	7.6123	0.1314	94.4608	0.0106	12.4090	0.0806	9.7487	120.9718
31	8.1451	0.1228	102.0730	0.0098	12.5318	0.0798	9.9471	124.6550
32	8.7153	0.1147	110.2181	0.0091	12.6466	0.0791	10.1381	128.2120
33	9.3253	0.1072	118.9334	0.0084	12.7538	0.0784	10.3219	131.6435
34	9.9781	0.1002	128.2587	0.0078	12.8540	0.0778	10.4987	134.9507
35	10.6766	0.0937	138.2368	0.0072	12.9477	0.0772	10.6687	138.1352
40	14.9745	0.0668	199.6350	0.0050	13.3317	0.0750	11.4233	152.2927
45	21.0024	0.0476	285.7490	0.0035	13.6055	0.0735	12.0360	163.7559
50	29.4570	0.0339	406.5286	0.0025	13.8007	0.0725	12.5287	172.9051
55	41.3150	0.0242	575.9282	0.0017	13.9399	0.0717	12.9215	180.1243
60	57.9464	0.0173	813.5200	0.0012	14.0392	0.0712	13.2321	185.7677
65	81.2728	0.0123	1146.7546	0.0009	14.1099	0.0709	13.4760	190.1452
70	113.9893	0.0088	1614.1333	0.0006	14.1604	0.0706	13.6662	193.5185
75	159.8759	0.0063	2269.6558	0.0004	14.1964	0.0704	13.8136	196.1035
80	224.2342	0.0045	3189.0608	0.0003	14.2220	0.0703	13.9273	198.0748
85	314.5000	0.0032	4478.5703	0.0002	14.2403	0.0702	14.0146	199.5717
90	441.1025	0.0023	6287.1797	0.0002	14.2533	0.0702	14.0812	200.7042
95	618.6692	0.0016	8823.8477	0.0001	14.2626	0.0701	14.1319	201.5581
100	867.7156	0.0012	12381.6524	0.0001	14.2693	0.0701	14.1703	202.2001

TABLE 15.A.4   Discrete Compounding: $i = 8\%$

	Single Payment		Uniform Series				Gradient Series	
	Compound Amount Factor	Present Worth Factor	Compound Amount Factor	Sinking Fund Factor	Present Worth Factor	Capital Recovery Factor	Uniform Series Factor	Present Worth Factor
	To Find F Given P	To Find P Given F	To Find F Given A	To Find A Given F	To Find P Given A	To Find A Given P	To Find A Given G	To Find P Given G
n	$F\|P\,i,n$	$P\|F\,i,n$	$F\|A\,i,n$	$A\|F\,i,n$	$P\|A\,i,n$	$A\|P\,i,n$	$A\|G\,i,n$	$P\|G\,i,n$
1	1.0800	0.9259	1.0000	1.0000	0.9259	1.0800	0.0000	0.0000
2	1.1664	0.8573	2.0800	0.4808	1.7833	0.5608	0.4808	0.8573
3	1.2597	0.7938	3.2464	0.3080	2.5771	0.3880	0.9487	2.4450
4	1.3605	0.7350	4.5061	0.2219	3.3121	0.3019	1.4040	4.6501
5	1.4693	0.6806	5.8666	0.1705	3.9927	0.2505	1.8465	7.3724
6	1.5869	0.6302	7.3359	0.1363	4.6229	0.2163	2.2763	10.5233
7	1.7138	0.5835	8.9228	0.1121	5.2064	0.1921	2.6937	14.0242
8	1.8509	0.5403	10.6366	0.0940	5.7466	0.1740	3.0985	17.8061
9	1.9990	0.5002	12.4876	0.0801	6.2469	0.1601	3.4910	21.8081
10	2.1589	0.4632	14.4866	0.0690	6.7101	0.1490	3.8713	25.9768
11	2.3316	0.4289	16.6455	0.0601	7.1390	0.1401	4.2395	30.2657
12	2.5182	0.3971	18.9771	0.0527	7.5361	0.1327	4.5957	34.6339
13	2.7196	0.3677	21.4953	0.0465	7.9038	0.1265	4.9402	39.0463
14	2.9372	0.3405	24.2149	0.0413	8.2442	0.1213	5.2731	43.4723
15	3.1722	0.3152	27.1521	0.0368	8.5595	0.1168	5.5945	47.8857
16	3.4259	0.2919	30.3243	0.0330	8.8514	0.1130	5.9046	52.2640
17	3.7000	0.2703	33.7502	0.0296	9.1216	0.1096	6.2037	56.5883
18	3.9960	0.2502	37.4502	0.0267	9.3719	0.1067	6.4920	60.8426
19	4.3157	0.2317	41.4463	0.0241	9.6036	0.1041	6.7697	65.0134
20	4.6610	0.2145	45.7620	0.0219	9.8181	0.1019	7.0369	69.0898
21	5.0338	0.1987	50.4229	0.0198	10.0168	0.0998	7.2940	73.0629
22	5.4365	0.1839	55.4567	0.0180	10.2007	0.0980	7.5412	76.9257
23	5.8715	0.1703	60.8933	0.0164	10.3711	0.0964	7.7786	80.6726
24	6.3412	0.1577	66.7647	0.0150	10.5288	0.0950	8.0066	84.2997
25	6.8485	0.1460	73.1059	0.0137	10.6748	0.0937	8.2254	87.8041
26	7.3964	0.1352	79.9544	0.0125	10.8100	0.0925	8.4352	91.1842
27	7.9881	0.1252	87.3507	0.0114	10.9352	0.0914	8.6363	94.4390
28	8.6271	0.1159	95.3388	0.0105	11.0511	0.0905	8.8289	97.5687
29	9.3173	0.1073	103.9659	0.0096	11.1584	0.0896	9.0133	100.5739
30	10.0627	0.0994	113.2831	0.0088	11.2578	0.0888	9.1897	103.4558
31	10.8677	0.0920	123.3458	0.0081	11.3498	0.0881	9.3584	106.2163
32	11.7371	0.0852	134.2135	0.0075	11.4350	0.0875	9.5197	108.8575
33	12.6760	0.0789	145.9505	0.0069	11.5139	0.0869	9.6737	111.3819
34	13.6901	0.0730	158.6266	0.0063	11.5869	0.0863	9.8208	113.7924
35	14.7853	0.0676	172.3167	0.0058	11.6546	0.0858	9.9611	116.0920
40	21.7245	0.0460	259.0564	0.0039	11.9246	0.0839	10.5699	126.0422
45	31.9204	0.0313	386.5054	0.0026	12.1084	0.0826	11.0447	133.7331
50	46.9016	0.0213	573.7698	0.0017	12.2335	0.0817	11.4107	139.5928
55	68.9138	0.0145	848.9224	0.0012	12.3186	0.0812	11.6902	144.0065
60	101.2570	0.0099	1253.2122	0.0008	12.3766	0.0808	11.9015	147.3000
65	148.7796	0.0067	1847.2463	0.0005	12.4160	0.0805	12.0602	149.7387
70	218.6061	0.0046	2720.0767	0.0004	12.4428	0.0804	12.1783	151.5326
75	321.2039	0.0031	4002.5518	0.0002	12.4611	0.0802	12.2658	152.8449
80	471.9541	0.0021	5886.9258	0.0002	12.4735	0.0802	12.3301	153.8001
85	693.4553	0.0014	8655.6953	0.0001	12.4820	0.0801	12.3772	154.4925
90	1018.9136	0.0010	12723.9219	0.0001	12.4877	0.0801	12.4116	154.9925
95	1497.1182	0.0007	18701.4805	0.0001	12.4917	0.0801	12.4365	155.3525
100	2199.7566	0.0005	27484.4649	0.0000	12.4943	0.0800	12.4545	155.6107

TABLE 15.A.5   Discrete Compounding: $i = 9\%$

	Single Payment		Uniform Series				Gradient Series	
	Compound Amount Factor	Present Worth Factor	Compound Amount Factor	Sinking Fund Factor	Present Worth Factor	Capital Recovery Factor	Uniform Series Factor	Present Worth Factor
	To Find F Given P	To Find P Given F	To Find F Given A	To Find A Given F	To Find P Given A	To Find A Given P	To Find A Given G	To Find P Given G
n	$F\|P\,i,n$	$P\|F\,i,n$	$F\|A\,i,n$	$A\|F\,i,n$	$P\|A\,i,n$	$A\|P\,i,n$	$A\|G\,i,n$	$P\|G\,i,n$
1	1.0900	0.9174	1.0000	1.0000	0.9174	1.0900	0.0000	0.0000
2	1.1881	0.8417	2.0900	0.4785	1.7591	0.5685	0.4785	0.8417
3	1.2950	0.7722	3.2781	0.3051	2.5313	0.3951	0.9426	2.3860
4	1.4116	0.7084	4.5731	0.2187	3.2397	0.3087	1.3925	4.5113
5	1.5386	0.6499	5.9847	0.1671	3.8897	0.2571	1.8282	7.1110
6	1.6771	0.5963	7.5233	0.1329	4.4859	0.2229	2.2498	10.0924
7	1.8280	0.5470	9.2004	0.1087	5.0330	0.1987	2.6574	13.3746
8	1.9926	0.5019	11.0285	0.0907	5.5348	0.1807	3.0512	16.8877
9	2.1719	0.4604	13.0210	0.0768	5.9952	0.1668	3.4312	20.5711
10	2.3674	0.4224	15.1929	0.0658	6.4177	0.1558	3.7978	24.3728
11	2.5804	0.3875	17.5603	0.0569	6.8052	0.1469	4.1510	28.2481
12	2.8127	0.3555	20.1407	0.0497	7.1607	0.1397	4.4910	32.1590
13	3.0658	0.3262	22.9534	0.0436	7.4869	0.1336	4.8182	36.0731
14	3.3417	0.2992	26.0192	0.0384	7.7862	0.1284	5.1326	39.9633
15	3.6425	0.2745	29.3609	0.0341	8.0607	0.1241	5.4346	43.8069
16	3.9703	0.2519	33.0034	0.0303	8.3126	0.1203	5.7245	47.5849
17	4.3276	0.2311	36.9737	0.0270	8.5436	0.1170	6.0024	51.2821
18	4.7171	0.2120	41.3013	0.0242	8.7556	0.1142	6.2687	54.8860
19	5.1417	0.1945	46.0184	0.0217	8.9501	0.1117	6.5236	58.3868
20	5.6044	0.1784	51.1601	0.0195	9.1285	0.1095	6.7674	61.7770
21	6.1088	0.1637	56.7645	0.0176	9.2922	0.1076	7.0006	65.0510
22	6.6586	0.1502	62.8733	0.0159	9.4424	0.1059	7.2232	68.2048
23	7.2579	0.1378	69.5319	0.0144	9.5802	0.1044	7.4357	71.2360
24	7.9111	0.1264	76.7898	0.0130	9.7066	0.1030	7.6384	74.1433
25	8.6231	0.1160	84.7009	0.0118	9.8226	0.1018	7.8316	76.9265
26	9.3992	0.1064	93.3239	0.0107	9.9290	0.1007	8.0156	79.5863
27	10.2451	0.0976	102.7231	0.0097	10.0266	0.0997	8.1906	82.1241
28	11.1671	0.0895	112.9682	0.0089	10.1161	0.0989	8.3571	84.5419
29	12.1722	0.0822	124.1354	0.0081	10.1983	0.0981	8.5154	86.8423
30	13.2677	0.0754	136.3074	0.0073	10.2737	0.0973	8.6657	89.0280
31	14.4618	0.0691	149.5751	0.0067	10.3428	0.0967	8.8083	91.1025
32	15.7633	0.0634	164.0368	0.0061	10.4062	0.0961	8.9436	93.0691
33	17.1820	0.0582	179.8002	0.0056	10.4644	0.0956	9.0718	94.9315
34	18.7284	0.0534	196.9822	0.0051	10.5178	0.0951	9.1933	96.6935
35	20.4140	0.0490	215.7106	0.0046	10.5668	0.0946	9.3083	98.3590
40	31.4094	0.0318	337.8821	0.0030	10.7574	0.0930	9.7957	105.3762
45	48.3272	0.0207	525.8582	0.0019	10.8812	0.0919	10.1603	110.5561
50	74.3574	0.0134	815.0825	0.0012	10.9617	0.0912	10.4295	114.3251
55	114.4081	0.0087	1260.0903	0.0008	11.0140	0.0908	10.6261	117.0362
60	176.0310	0.0057	1944.7886	0.0005	11.0480	0.0905	10.7683	118.9683
65	270.8455	0.0037	2998.2830	0.0003	11.0701	0.0903	10.8702	120.3344
70	416.7293	0.0024	4619.2149	0.0002	11.0845	0.0902	10.9427	121.2942
75	641.1897	0.0016	7113.2188	0.0001	11.0938	0.0901	10.9940	121.9646
80	986.5496	0.0010	10950.5547	0.0001	11.0999	0.0901	11.0299	122.4302
85	1517.9287	0.0007	16854.7695	0.0001	11.1038	0.0901	11.0551	122.7533
90	2335.5198	0.0004	25939.1328	0.0000	11.1064	0.0900	11.0726	122.9758
95	3593.4878	0.0003	39916.5430	0.0000	11.1080	0.0900	11.0847	123.1287
100	5529.0234	0.0002	61422.5391	0.0000	11.1091	0.0900	11.0930	123.2335

TABLE 15.A.6   Discrete Compounding: $i = 10\%$

	Single Payment		Uniform Series				Gradient Series	
	Compound Amount Factor	Present Worth Factor	Compound Amount Factor	Sinking Fund Factor	Present Worth Factor	Capital Recovery Factor	Uniform Series Factor	Present Worth Factor
	To Find F Given P	To Find P Given F	To Find F Given A	To Find A Given F	To Find P Given A	To Find A Given P	To Find A Given G	To Find P Given G
n	$F\|P\,i,n$	$P\|F\,i,n$	$F\|A\,i,n$	$A\|F\,i,n$	$P\|A\,i,n$	$A\|P\,i,n$	$A\|G\,i,n$	$P\|G\,i,n$
1	1.1000	0.9091	1.0000	1.0000	0.9091	1.1000	0.0000	0.0000
2	1.2100	0.8264	2.1000	0.4762	1.7355	0.5762	0.4762	0.8264
3	1.3310	0.7513	3.3100	0.3021	2.4869	0.4021	0.9366	2.3291
4	1.4641	0.6830	4.6410	0.2155	3.1699	0.3155	1.3812	4.3781
5	1.6105	0.6209	6.1051	0.1638	3.7908	0.2638	1.8101	6.8618
6	1.7716	0.5645	7.7156	0.1296	4.3553	0.2296	2.2236	9.6842
7	1.9487	0.5132	9.4872	0.1054	4.8684	0.2054	2.6216	12.7631
8	2.1436	0.4665	11.4359	0.0874	5.3349	0.1874	3.0045	16.0287
9	2.3579	0.4241	13.5795	0.0736	5.7590	0.1736	3.3724	19.4215
10	2.5937	0.3855	15.9374	0.0627	6.1446	0.1627	3.7255	22.8913
11	2.8531	0.3505	18.5312	0.0540	6.4951	0.1540	4.0641	26.3963
12	3.1384	0.3186	21.3843	0.0468	6.8137	0.1468	4.3884	29.9012
13	3.4523	0.2897	24.5227	0.0408	7.1034	0.1408	4.6988	33.3772
14	3.7975	0.2633	27.9750	0.0357	7.3667	0.1357	4.9955	36.8005
15	4.1772	0.2394	31.7725	0.0315	7.6061	0.1315	5.2789	40.1520
16	4.5950	0.2176	35.9497	0.0278	7.8237	0.1278	5.5493	43.4164
17	5.0545	0.1978	40.5447	0.0247	8.0216	0.1247	5.8071	46.5820
18	5.5599	0.1799	45.5992	0.0219	8.2014	0.1219	6.0526	49.6396
19	6.1159	0.1635	51.1591	0.0195	8.3649	0.1195	6.2861	52.5827
20	6.7275	0.1486	57.2750	0.0175	8.5136	0.1175	6.5081	55.4069
21	7.4002	0.1351	64.0025	0.0156	8.6487	0.1156	6.7189	58.1095
22	8.1403	0.1228	71.4027	0.0140	8.7715	0.1140	6.9189	60.6893
23	8.9543	0.1117	79.5430	0.0126	8.8832	0.1126	7.1085	63.1462
24	9.8497	0.1015	88.4973	0.0113	8.9847	0.1113	7.2881	65.4813
25	10.8347	0.0923	98.3470	0.0102	9.0770	0.1102	7.4580	67.6964
26	11.9182	0.0839	109.1817	0.0092	9.1609	0.1092	7.6186	69.7941
27	13.1100	0.0763	121.0998	0.0083	9.2372	0.1083	7.7704	71.7773
28	14.4210	0.0693	134.2098	0.0075	9.3066	0.1075	7.9137	73.6496
29	15.8631	0.0630	148.6308	0.0067	9.3696	0.1067	8.0489	75.4147
30	17.4494	0.0573	164.4939	0.0061	9.4269	0.1061	8.1762	77.0766
31	19.1943	0.0521	181.9432	0.0055	9.4790	0.1055	8.2962	78.6396
32	21.1138	0.0474	201.1376	0.0050	9.5264	0.1050	8.4091	80.1078
33	23.2251	0.0431	222.2513	0.0045	9.5694	0.1045	8.5152	81.4856
34	25.5476	0.0391	245.4765	0.0041	9.6086	0.1041	8.6149	82.7773
35	28.1024	0.0356	271.0239	0.0037	9.6442	0.1037	8.7086	83.9872
40	45.2592	0.0221	442.5921	0.0023	9.7791	0.1023	9.0962	88.9526
45	72.8904	0.0137	718.9038	0.0014	9.8628	0.1014	9.3740	92.4545
50	117.3906	0.0085	1163.9070	0.0009	9.9148	0.1009	9.5704	94.8889
55	189.0588	0.0053	1880.5872	0.0005	9.9471	0.1005	9.7075	96.5620
60	304.4810	0.0033	3034.8110	0.0003	9.9672	0.1003	9.8023	97.7011
65	490.3694	0.0020	4893.6953	0.0002	9.9796	0.1002	9.8672	98.4706
70	789.7449	0.0013	7887.4531	0.0001	9.9873	0.1001	9.9113	98.9871
75	1271.8921	0.0008	12708.9258	0.0001	9.9921	0.1001	9.9410	99.3318
80	2048.3936	0.0005	20473.9531	0.0000	9.9951	0.1000	9.9609	99.5607
85	3298.9583	0.0003	32979.6094	0.0000	9.9970	0.1000	9.9742	99.7121
90	5313.0039	0.0002	53120.0742	0.0000	9.9981	0.1000	9.9831	99.8119
95	8556.6484	0.0001	85556.5000	0.0000	9.9988	0.1000	9.9889	99.8774
100	13780.5625	0.0001	137795.6875	0.0000	9.9993	0.1000	9.9927	99.9202

TABLE 15.A.7   Discrete Compounding: $i = 12\%$

	Single Payment		Uniform Series				Gradient Series	
	Compound Amount Factor	Present Worth Factor	Compound Amount Factor	Sinking Fund Factor	Present Worth Factor	Capital Recovery Factor	Uniform Series Factor	Present Worth Factor
n	To Find F Given P $F\|P\,i, n$	To Find P Given F $P\|F\,i, n$	To Find F Given A $F\|A\,i, n$	To Find A Given F $A\|F\,i, n$	To Find P Given A $P\|A\,i, n$	To Find A Given P $A\|P\,i, n$	To Find A Given G $A\|G\,i, n$	To Find P Given G $P\|G\,i, n$
1	1.1200	0.8929	1.0000	1.0000	0.8929	1.1200	0.0000	0.0000
2	1.2544	0.7972	2.1200	0.4717	1.6901	0.5917	0.4717	0.7972
3	1.4049	0.7118	3.3744	0.2963	2.4018	0.4163	0.9246	2.2208
4	1.5735	0.6355	4.7793	0.2092	3.0373	0.3292	1.3589	4.1273
5	1.7623	0.5674	6.3528	0.1574	3.6048	0.2774	1.7746	6.3970
6	1.9738	0.5066	8.1152	0.1232	4.1114	0.2432	2.1720	8.9302
7	2.2107	0.4523	10.0890	0.0991	4.5638	0.2191	2.5515	11.6443
8	2.4760	0.4039	12.2997	0.0813	4.9676	0.2013	2.9131	14.4715
9	2.7731	0.3606	14.7757	0.0677	5.3283	0.1877	3.2574	17.3563
10	3.1058	0.3220	17.5487	0.0570	5.6502	0.1770	3.5847	20.2541
11	3.4785	0.2875	20.6546	0.0484	5.9377	0.1684	3.8953	23.1289
12	3.8960	0.2567	24.1331	0.0414	6.1944	0.1614	4.1897	25.9523
13	4.3635	0.2292	28.0291	0.0357	6.4236	0.1557	4.4683	28.7024
14	4.8871	0.2046	32.3926	0.0309	6.6282	0.1509	4.7317	31.3624
15	5.4736	0.1827	37.2797	0.0268	6.8109	0.1468	4.9803	33.9202
16	6.1304	0.1631	42.7533	0.0234	6.9740	0.1434	5.2147	36.3670
17	6.8660	0.1456	48.8836	0.0205	7.1196	0.1405	5.4353	38.6973
18	7.6900	0.1300	55.7497	0.0179	7.2497	0.1379	5.6427	40.9080
19	8.6128	0.1161	63.4396	0.0158	7.3658	0.1358	5.8375	42.9979
20	9.6463	0.1037	72.0524	0.0139	7.4694	0.1339	6.0202	44.9676
21	10.8038	0.0926	81.6987	0.0122	7.5620	0.1322	6.1913	46.8188
22	12.1003	0.0826	92.5025	0.0108	7.6446	0.1308	6.3514	48.5543
23	13.5523	0.0738	104.6028	0.0096	7.7184	0.1296	6.5010	50.1776
24	15.1786	0.0659	118.1551	0.0085	7.7843	0.1285	6.6406	51.6929
25	17.0000	0.0588	133.3337	0.0075	7.8431	0.1275	6.7708	53.1047
26	19.0400	0.0525	150.3338	0.0067	7.8957	0.1267	6.8921	54.4177
27	21.3249	0.0469	169.3738	0.0059	7.9426	0.1259	7.0049	55.6369
28	23.8838	0.0419	190.6987	0.0052	7.9844	0.1252	7.1098	56.7674
29	26.7499	0.0374	214.5825	0.0047	8.0218	0.1247	7.2071	57.8141
30	29.9599	0.0334	241.3324	0.0041	8.0552	0.1241	7.2974	58.7821
31	33.5551	0.0298	271.2922	0.0037	8.0850	0.1237	7.3811	59.6761
32	37.5817	0.0266	304.8472	0.0033	8.1116	0.1233	7.4586	60.5010
33	42.0915	0.0238	342.4290	0.0029	8.1354	0.1229	7.5302	61.2613
34	47.1424	0.0212	384.5203	0.0026	8.1566	0.1226	7.5965	61.9613
35	52.7995	0.0189	431.6629	0.0023	8.1755	0.1223	7.6577	62.6052
40	93.0508	0.0107	767.0901	0.0013	8.2438	0.1213	7.8988	65.1159
45	163.9872	0.0061	1358.2275	0.0007	8.2825	0.1207	8.0572	66.7343
50	289.0012	0.0035	2400.0127	0.0004	8.3045	0.1204	8.1597	67.7625

TABLE 15.A.8   Discrete Compounding: $i = 15\%$

	Single Payment		Uniform Series				Gradient Series	
	Compound Amount Factor	Present Worth Factor	Compound Amount Factor	Sinking Fund Factor	Present Worth Factor	Capital Recovery Factor	Uniform Series Factor	Present Worth Factor
	To Find F Given P	To Find P Given F	To Find F Given A	To Find A Given F	To Find P Given A	To Find A Given P	To Find A Given G	To Find P Given G
n	$F\|P\,i,n$	$P\|F\,i,n$	$F\|A\,i,n$	$A\|F\,i,n$	$P\|A\,i,n$	$A\|P\,i,n$	$A\|G\,i,n$	$P\|G\,i,n$
1	1.1500	0.8696	1.0000	1.0000	0.8696	1.1500	0.0000	0.0000
2	1.3225	0.7561	2.1500	0.4651	1.6257	0.6151	0.4651	0.7561
3	1.5209	0.6575	3.4725	0.2880	2.2832	0.4380	0.9071	2.0712
4	1.7490	0.5718	4.9934	0.2003	2.8550	0.3503	1.3263	3.7864
5	2.0114	0.4972	6.7424	0.1483	3.3522	0.2983	1.7228	5.7751
6	2.3131	0.4323	8.7537	0.1142	3.7845	0.2642	2.0972	7.9368
7	2.6600	0.3759	11.0668	0.0904	4.1604	0.2404	2.4498	10.1924
8	3.0590	0.3269	13.7268	0.0729	4.4873	0.2229	2.7813	12.4807
9	3.5179	0.2843	16.7858	0.0596	4.7716	0.2096	3.0922	14.7548
10	4.0456	0.2472	20.3037	0.0493	5.0188	0.1993	3.3832	16.9795
11	4.6524	0.2149	24.3493	0.0411	5.2337	0.1911	3.6549	19.1289
12	5.3502	0.1869	29.0017	0.0345	5.4206	0.1845	3.9082	21.1849
13	6.1528	0.1625	34.3519	0.0291	5.5831	0.1791	4.1438	23.1352
14	7.0757	0.1413	40.5047	0.0247	5.7245	0.1747	4.3624	24.9725
15	8.1371	0.1229	47.5804	0.0210	5.8474	0.1710	4.5650	26.6930
16	9.3576	0.1069	55.7175	0.0179	5.9542	0.1679	4.7522	28.2960
17	10.7613	0.0929	65.0751	0.0154	6.0472	0.1654	4.9251	29.7828
18	12.3754	0.0808	75.8363	0.0132	6.1280	0.1632	5.0843	31.1565
19	14.2318	0.0703	88.2118	0.0113	6.1982	0.1613	5.2307	32.4213
20	16.3665	0.0611	102.4436	0.0098	6.2593	0.1598	5.3651	33.5822
21	18.8215	0.0531	118.8100	0.0084	6.3125	0.1584	5.4883	34.6448
22	21.6447	0.0462	137.6315	0.0073	6.3587	0.1573	5.6010	35.6150
23	24.8914	0.0402	159.2763	0.0063	6.3988	0.1563	5.7040	36.4988
24	28.6252	0.0349	184.1677	0.0054	6.4338	0.1554	5.7979	37.3023
25	32.9189	0.0304	212.7929	0.0047	6.4642	0.1547	5.8834	38.0314
26	37.8568	0.0264	245.7118	0.0041	6.4906	0.1541	5.9612	38.6918
27	43.5353	0.0230	283.5684	0.0035	6.5135	0.1535	6.0319	39.2890
28	50.0656	0.0200	327.1038	0.0031	6.5335	0.1531	6.0960	39.8283
29	57.5754	0.0174	377.1694	0.0027	6.5509	0.1527	6.1541	40.3146
30	66.2117	0.0151	434.7449	0.0023	6.5660	0.1523	6.2066	40.7526
31	76.1435	0.0131	500.9566	0.0020	6.5791	0.1520	6.2541	41.1466
32	87.5650	0.0114	577.1001	0.0017	6.5905	0.1517	6.2970	41.5006
33	100.6998	0.0099	664.6650	0.0015	6.6005	0.1515	6.3357	41.8184
34	115.8047	0.0086	765.3648	0.0013	6.6091	0.1513	6.3705	42.1034
35	133.1754	0.0075	881.1694	0.0011	6.6166	0.1511	6.4019	42.3587
40	267.8633	0.0037	1779.0879	0.0006	6.6418	0.1506	6.5168	43.2830
45	538.7686	0.0019	3585.1248	0.0003	6.6543	0.1503	6.5830	43.8051
50	1083.6563	0.0009	7217.7070	0.0001	6.6605	0.1501	6.6205	44.0958

TABLE 15.A.9   Discrete Compounding: $i = 18\%$

	Single Payment		Uniform Series				Gradient Series	
	Compound Amount Factor	Present Worth Factor	Compound Amount Factor	Sinking Fund Factor	Present Worth Factor	Capital Recovery Factor	Uniform Series Factor	Present Worth Factor
n	To Find F Given P $F\|P\,i,n$	To Find P Given F $P\|F\,i,n$	To Find F Given A $F\|A\,i,n$	To Find A Given F $A\|F\,i,n$	To Find P Given A $P\|A\,i,n$	To Find A Given P $A\|P\,i,n$	To Find A Given G $A\|G\,i,n$	To Find P Given G $P\|G\,i,n$
1	1.1972	0.8353	1.0000	1.0000	0.8353	1.1972	0.0000	0.0000
2	1.4333	0.6977	2.1972	0.4551	1.5330	0.6523	0.4551	0.6977
3	1.7160	0.5827	3.6305	0.2754	2.1157	0.4727	0.8807	1.8632
4	2.0544	0.4868	5.3466	0.1870	2.6025	0.3843	1.2770	3.3234
5	2.4596	0.4066	7.4010	0.1351	3.0090	0.3323	1.6450	4.9497
6	2.9447	0.3386	9.8606	0.1014	3.3486	0.2986	1.9852	6.6477
7	3.5254	0.2837	12.8053	0.0781	3.6323	0.2753	2.2987	8.3496
8	4.2207	0.2369	16.3307	0.0612	3.8692	0.2585	2.5866	10.0081
9	5.0531	0.1979	20.5514	0.0487	4.0671	0.2459	2.8500	11.5913
10	6.0497	0.1653	25.6045	0.0391	4.2324	0.2363	3.0902	13.0790
11	7.2427	0.1381	31.6541	0.0316	4.3705	0.2288	3.3085	14.4596
12	8.6711	0.1153	38.8968	0.0257	4.4858	0.2229	3.5062	15.7282
13	10.3812	0.0963	47.5680	0.0210	4.5821	0.2182	3.6848	16.8842
14	12.4286	0.0805	57.9492	0.0173	4.6626	0.2145	0.8456	17.9301
15	14.8797	0.0672	70.3778	0.1421	4.7298	0.2114	3.9898	18.8710
16	17.8143	0.0561	85.2575	0.1173	4.7859	0.2089	4.1190	19.7130
17	21.3276	0.0469	103.0719	0.0970	4.8328	0.2069	4.2342	20.4632
18	25.5337	0.0392	124.3994	0.0804	4.8720	0.2053	4.3369	21.1290
19	30.5694	0.0327	149.9332	0.0667	4.9047	0.2039	4.4280	21.7178
20	36.5982	0.0273	180.5026	0.0554	4.9320	0.2028	4.5087	22.2370
21	43.8160	0.0228	217.1008	0.0461	4.9548	0.2018	4.5801	22.6934
22	52.4573	0.0191	260.9169	0.0383	4.9739	0.2011	4.6430	23.0938
23	62.8028	0.0159	313.3742	0.0319	4.9898	0.2004	4.6984	23.4441
24	75.1886	0.0133	376.1771	0.0266	5.0031	0.1999	4.7471	23.7500
25	90.0171	0.0111	451.3657	0.0222	5.0142	0.1994	4.7897	24.0166
26	107.7701	0.0093	541.3829	0.0185	5.0235	0.1991	4.8270	24.2486
27	129.0242	0.0078	649.1530	0.0154	5.0313	0.1988	4.8597	24.4501
28	154.4700	0.0065	778.1772	0.0129	5.0377	0.1985	4.8881	24.6249
29	184.9342	0.0054	932.6473	0.0107	5.0431	0.1983	4.9129	24.7763
30	2221.4064	0.0045	1117.5816	0.0089	5.0476	0.1981	4.9344	24.9072
31	265.0716	0.0038	1338.9880	0.0075	5.0514	0.1980	4.9532	25.0204
32	317.3483	0.0032	1694.0597	0.0062	5.0546	0.1978	4.9694	25.1181
33	379.9349	0.0026	1921.4082	0.0052	5.0572	0.1977	4.9835	25.2023
34	454.8647	0.0022	2306.4138	0.0043	5.0594	0.1977	4.9956	25.2749
35	544.5719	0.0018	2756.2080	0.0036	5.0612	0.1976	5.0062	25.3373
40	1339.4308	0.0007	6786.5789.	0.0015	5.0668	0.1974	5.0407	25.5398
45	3294.4681	0.0003	16699.6920	0.0006	5.0690	0.1973	5.0569	25.6333
50	8103.0839	0.0001	41082.0140	0.0002	5.0699	0.1972	5.0644	25.6760

TABLE 15.A.10   Discrete Compounding: $i = 20\%$

	Single Payment		Uniform Series				Gradient Series	
	Compound Amount Factor	Present Worth Factor	Compound Amount Factor	Sinking Fund Factor	Present Worth Factor	Capital Recovery Factor	Uniform Series Factor	Present Worth Factor
n	To Find F Given P $F\|P\,i,n$	To Find P Given F $P\|F\,i,n$	To Find F Given A $F\|A\,i,n$	To Find A Given F $A\|F\,i,n$	To Find P Given A $P\|A\,i,n$	To Find A Given P $A\|P\,i,n$	To Find A Given G $A\|G\,i,n$	To Find P Given G $P\|G\,i,n$
1	1.2000	0.8333	1.0000	1.0000	0.8333	1.2000	0.0000	0.0000
2	1.4400	0.6944	2.2000	0.4545	1.5278	0.6545	0.4545	0.6944
3	1.7280	0.5787	3.6400	0.2747	2.1065	0.4747	0.8791	1.8519
4	2.0736	0.4823	5.3680	0.1863	2.5887	0.3863	1.2742	3.2986
5	2.4883	0.4019	7.4416	0.1344	2.9906	0.3344	1.6405	4.9061
6	2.9860	0.3349	9.9299	0.1007	3.3255	0.3007	1.9788	6.5806
7	3.5832	0.2791	12.9159	0.0774	3.6046	0.2774	2.2902	8.2551
8	4.2998	0.2326	16.4991	0.0606	3.8372	0.2606	2.5756	9.8831
9	5.1598	0.1938	20.7989	0.0481	4.0310	0.2481	2.8364	11.4335
10	6.1917	0.1615	25.9587	0.0385	4.1925	0.2385	3.0739	12.8871
11	7.4301	0.1346	32.1504	0.0311	4.3271	0.2311	3.2893	14.2330
12	8.9161	0.1122	39.5805	0.0253	4.4392	0.2253	3.4841	15.4667
13	10.6993	0.0935	48.4966	0.0206	4.5327	0.2206	3.6597	16.5883
14	12.8392	0.0779	59.1959	0.0169	4.6106	0.2169	3.8175	17.6008
15	15.4070	0.0649	72.0351	0.0139	4.6755	0.2139	3.9588	18.5095
16	18.4884	0.0541	87.4421	0.0114	4.7296	0.2114	4.0851	19.3208
17	22.1861	0.0451	105.9306	0.0094	4.7746	0.2094	4.1976	20.0419
18	26.6233	0.0376	128.1166	0.0078	4.8122	0.2078	4.2975	20.6805
19	31.9480	0.0313	154.7399	0.0065	4.8435	0.2065	4.3861	21.2439
20	38.3376	0.0261	186.6879	0.0054	4.8696	0.2054	4.4643	21.7395
21	46.0051	0.0217	225.0255	0.0044	4.8913	0.2044	4.5334	22.1742
22	55.2061	0.0181	271.0305	0.0037	4.9094	0.2037	4.5941	22.5546
23	66.2474	0.0151	326.2366	0.0031	4.9245	0.2031	4.6475	22.8867
24	79.4968	0.0126	392.4839	0.0025	4.9371	0.2025	4.6943	23.1760
25	95.3962	0.0105	471.9807	0.0021	4.9476	0.2021	4.7352	23.4276
26	114.4754	0.0087	567.3770	0.0018	4.9563	0.2018	4.7709	23.6460
27	137.3705	0.0073	681.8525	0.0015	4.9636	0.2015	4.8020	23.8353
28	164.8446	0.0061	819.2229	0.0012	4.9697	0.2012	4.8291	23.9991
29	197.8135	0.0051	984.0676	0.0010	4.9747	0.2010	4.8527	24.1406
30	237.3762	0.0042	1181.8811	0.0008	4.9789	0.2008	4.8731	24.2628
31	284.8513	0.0035	1419.2573	0.0007	4.9824	0.2007	4.8908	24.3681
32	341.8215	0.0029	1704.1087	0.0006	4.9854	0.2006	4.9061	24.4588
33	410.1860	0.0024	2045.9295	0.0005	4.9878	0.2005	4.9194	24.5368
34	492.2232	0.0020	2456.1167	0.0004	4.9898	0.2004	4.9308	24.6038
35	590.6680	0.0017	2948.3391	0.0003	4.9915	0.2003	4.9406	24.6614
40	1469.7708	0.0007	7343.8516	0.0001	4.9966	0.2001	4.9728	24.8469
45	3657.2590	0.0003	18281.3008	0.0001	4.9986	0.2001	4.9877	24.9316
50	9100.4336	0.0001	45497.1641	0.0000	4.9995	0.2000	4.9945	24.9698

TABLE 15.A.11   Discrete Compounding: $i = 25\%$

	Single Payment		Uniform Series				Gradient Series	
	Compound Amount Factor	Present Worth Factor	Compound Amount Factor	Sinking Fund Factor	Present Worth Factor	Capital Recovery Factor	Uniform Series Factor	Present Worth Factor
n	To Find F Given P $F\|P\,i,n$	To Find P Given F $P\|F\,i,n$	To Find F Given A $F\|A\,i,n$	To Find A Given F $A\|F\,i,n$	To Find P Given A $P\|A\,i,n$	To Find A Given P $A\|P\,i,n$	To Find A Given G $A\|G\,i,n$	To Find P Given G $P\|G\,i,n$
1	1.2500	0.8000	1.0000	1.0000	0.8000	1.2500	0.0000	0.0000
2	1.5625	0.6400	2.2500	0.4444	1.4400	0.6944	0.4444	0.6400
3	1.9531	0.5120	3.8125	0.2623	1.9520	0.5123	0.8525	1.6640
4	2.4414	0.4096	5.7656	0.1734	2.3616	0.4234	1.2249	2.8928
5	3.0518	0.3277	8.2070	0.1218	2.6893	0.3718	1.5631	4.2035
6	3.8147	0.2621	11.2588	0.0888	2.9514	0.3388	1.8683	5.5142
7	4.7684	0.2097	15.0735	0.0663	3.1611	0.3163	2.1424	6.7725
8	5.9605	0.1678	19.8419	0.0504	3.3289	0.3004	2.3872	7.9469
9	7.4506	0.1342	25.8023	0.0388	3.4631	0.2888	2.6048	9.0207
10	9.3132	0.1074	33.2529	0.0301	3.5705	0.2801	2.7971	9.9870
11	11.6415	0.0859	42.5661	0.0235	3.6564	0.2735	2.9663	10.8460
12	14.5519	0.0687	54.2077	0.0184	3.7251	0.2684	3.1145	11.6020
13	18.1899	0.0550	68.7596	0.0145	3.7801	0.2645	3.2437	12.2617
14	22.7374	0.0440	86.9495	0.0115	3.8241	0.2615	3.3559	12.8334
15	28.4217	0.0352	109.6868	0.0091	3.8593	0.2591	3.4530	13.3260
16	35.5271	0.0281	138.1085	0.0072	3.8874	0.2572	3.5366	13.7482
17	44.4089	0.0225	173.6356	0.0058	3.9099	0.2558	3.6084	14.1085
18	55.5112	0.0180	218.0446	0.0046	3.9279	0.2546	3.6698	14.4147
19	69.3889	0.0144	273.5557	0.0037	3.9424	0.2537	3.7222	14.6741
20	86.7362	0.0115	342.9446	0.0029	3.9539	0.2529	3.7667	14.8932
21	108.4202	0.0092	429.6807	0.0023	3.9631	0.2523	3.8045	15.0777
22	135.5252	0.0074	538.1008	0.0019	3.9705	0.2519	3.8365	15.2326
23	169.4065	0.0059	673.6262	0.0015	3.9764	0.2515	3.8634	15.3625
24	211.7582	0.0047	843.0327	0.0012	3.9811	0.2512	3.8861	15.4711
25	264.6975	0.0038	1054.7910	0.0009	3.9849	0.2509	3.9052	15.5618
26	330.8721	0.0030	1319.4888	0.0008	3.9879	0.2508	3.9212	15.6373
27	413.5901	0.0024	1650.3611	0.0006	3.9903	0.2506	3.9346	15.7002
28	516.9878	0.0019	2063.9502	0.0005	3.9923	0.2505	3.9457	15.7524
29	646.2349	0.0015	2580.9390	0.0004	3.9938	0.2504	3.9551	15.7957
30	807.7935	0.0012	3227.1726	0.0003	3.9950	0.2503	3.9628	15.8316
31	1009.7417	0.0010	4034.9663	0.0002	3.9960	0.2502	3.9693	15.8614
32	1262.1773	0.0008	5044.7070	0.0002	3.9968	0.2502	3.9746	15.8859
33	1577.7217	0.0006	6306.8828	0.0002	3.9975	0.2502	3.9791	15.9062
34	1972.1519	0.0005	7884.6055	0.0001	3.9980	0.2501	3.9828	15.9229
35	2465.1887	0.0004	9856.7578	0.0001	3.9984	0.2501	3.9858	15.9367

TABLE 15.A.12   Discrete Compounding: $i = 30\%$

	Single Payment		Uniform Series				Gradient Series	
	Compound Amount Factor	Present Worth Factor	Compound Amount Factor	Sinking Fund Factor	Present Worth Factor	Capital Recovery Factor	Uniform Series Factor	Present Worth Factor
n	To Find F Given P $F\|P\,i,n$	To Find P Given F $P\|F\,i,n$	To Find F Given A $F\|A\,i,n$	To Find A Given F $A\|F\,i,n$	To Find P Given A $P\|A\,i,n$	To Find A Given P $A\|P\,i,n$	To Find A Given G $A\|G\,i,n$	To Find P Given G $P\|G\,i,n$
1	1.3000	0.7692	1.0000	1.0000	0.7692	1.3000	0.0000	0.0000
2	1.6900	0.5917	2.3000	0.4348	1.3609	0.7348	0.4348	0.5917
3	2.1970	0.4552	3.9900	0.2506	1.8161	0.5506	0.8271	1.5020
4	2.8561	0.3501	6.1870	0.1616	2.1662	0.4616	1.1783	2.5524
5	3.7129	0.2693	9.0431	0.1106	2.4356	0.4106	1.4903	3.6297
6	4.8268	0.2072	12.7560	0.0784	2.6427	0.3784	1.7654	4.6656
7	6.2749	0.1594	17.5828	0.0569	2.8021	0.3569	2.0063	5.6218
8	8.1573	0.1226	23.8577	0.0419	2.9247	0.3419	2.2156	6.4800
9	10.6045	0.0943	32.0150	0.0312	3.0190	0.3312	2.3963	7.2343
10	13.7858	0.0725	42.6195	0.0235	3.0915	0.3235	2.5512	7.8872
11	17.9216	0.0558	56.4053	0.0177	3.1473	0.3177	2.6833	8.4452
12	23.2981	0.0429	74.3269	0.0135	3.1903	0.3135	2.7952	8.9173
13	30.2875	0.0330	97.6250	0.0102	3.2233	0.3102	2.8895	9.3135
14	39.3737	0.0254	127.9124	0.0078	3.2487	0.3078	2.9685	9.6437
15	51.1859	0.0195	167.2862	0.0060	3.2682	0.3060	3.0344	9.9172
16	66.5416	0.0150	218.4721	0.0046	3.2832	0.3046	3.0892	10.1426
17	86.5041	0.0116	285.0137	0.0035	3.2948	0.3035	3.1345	10.3276
18	112.4553	0.0089	371.5176	0.0027	3.3037	0.3027	3.1718	10.4788
19	146.1919	0.0068	483.9729	0.0021	3.3105	0.3021	3.2025	10.6019
20	190.0494	0.0053	630.1648	0.0016	3.3158	0.3016	3.2275	10.7019
21	247.0643	0.0040	820.2144	0.0012	3.3198	0.3012	3.2480	10.7828
22	321.1836	0.0031	1067.2788	0.0009	3.3230	0.3009	3.2646	10.8482
23	417.5386	0.0024	1388.4624	0.0007	3.3254	0.3007	3.2781	10.9009
24	542.8001	0.0018	1806.0000	0.0006	3.3272	0.3006	3.2890	10.9433
25	705.6401	0.0014	2348.7998	0.0004	3.3286	0.3004	3.2979	10.9773
26	917.3323	0.0011	3054.4414	0.0003	3.3297	0.3003	3.3050	11.0045
27	1192.5320	0.0008	3971.7727	0.0003	3.3305	0.3003	3.3107	11.0263
28	1550.2915	0.0006	5164.3047	0.0002	3.3312	0.3002	3.3153	11.0437
29	2015.3775	0.0005	6714.5938	0.0001	3.3317	0.3001	3.3189	11.0576
30	2619.9920	0.0004	8729.9727	0.0001	3.3321	0.3001	3.3219	11.0687
31	3405.9902	0.0003	11349.9688	0.0001	3.3324	0.3001	3.3242	11.0775
32	4427.7852	0.0002	14755.9670	0.0001	3.3326	0.3001	3.3261	11.0845
33	5756.1211	0.0002	19183.7461	0.0001	3.3328	0.3001	3.3276	11.0901
34	7482.9570	0.0001	24939.8672	0.0000	3.3329	0.3000	3.3288	11.0945
35	9727.8438	0.0001	32422.8086	0.0000	3.3330	0.3000	3.3297	11.0980

CHAPTER 16

PROPERTIES OF MATERIALS

George E. Dieter

School of Engineering
University of Maryland
College Park, Maryland

Jack H. Westbrook

Sci-Tech Knowledge Systems
Scotia, New York

16.1 INTRODUCTION TO PROPERTIES, PROCESSING, AND SELECTION OF MATERIALS

16.1.1 Properties of Materials

The performance of a material is expressed in terms of physical, mechanical, chemical, electrical, nuclear, and thermal properties. Material properties are the link between the basic composition and structure of the material and the performance in service of a part that is made from the material.

The world of engineering has grown to encompass a large number and variety of engineering materials, Fig. 16.1. Metals and alloys, the most common engineering materials in terms of tonnage used, are divided into two classes: ferrous metals and alloys (those that contain a large percentage of iron) and nonferrous metals and alloys (those that contain no, or only a small amount of, iron). Polymeric materials consist of long chains or networks of organic (carbon-containing) molecules. These versatile materials comprise the category that has been the most rapidly growing in volume of use over the past 20 years. Ceramic materials are inorganic materials that consist of metallic and nonmetallic elements chemically bonded together. These materials are generally hard and mechanically brittle and possess outstanding high-temperature strength. Other materials used in large quantities in engineering construction are wood, concrete, and glass. Finally, there are two important new categories of engineering materials. Composite materials are mixtures of two or more materials in

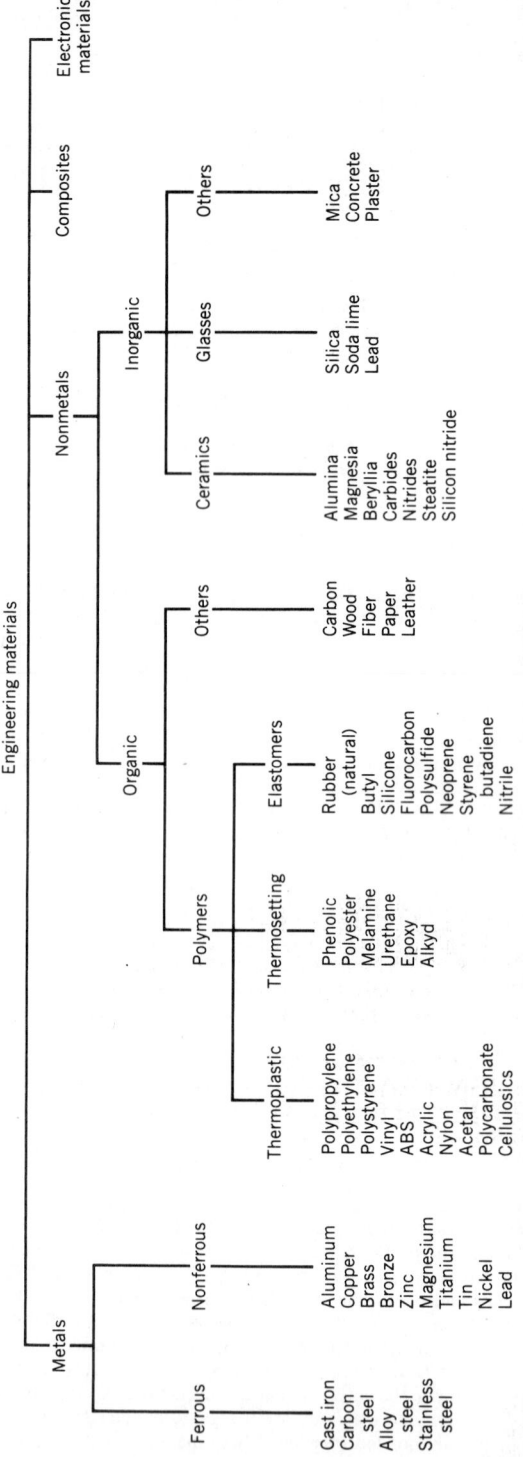

Fig. 16.1 Classification of engineering materials. (From Dieter, G. E., *Engineering Design*, © 1983, McGraw-Hill, New York. Reproduced with permission.)

which one material is usually present in the form of fibers. Two examples are fiberglass reinforcing materials in a polymeric matrix and SiC whiskers in an aluminum matrix. Electronic materials such as are incorporated in an integrated circuit are not a major material category in terms of volume of use, but they are extremely important materials in terms of their impact on technology.

The properties of materials are determined by the structure of material. At the most basic level there is atomic structure, as defined by forces within atoms and the arrangement of atoms in space. However, there are many aspects of structure at a level larger than the atomic level that control properties: crystal structure, microstructure, macrostructure, and so on. In metals the lattice defect known as a dislocation is very important, and the interaction of dislocations with many aspects of microstructure, such as grain boundaries or dispersed second-phase particles, determines mechanical properties. In polymers the length and morphology of a polymer chain determines properties, while in ceramics the properties of microcracks control mechanical properties.

Structure is determined most fundamentally by composition selection, but it can be altered significantly by thermal treatment and deformation processing. Thus, the listing of the properties of materials must consider the processing the material has received. Therefore, structure-sensitive properties are not single values that depend only on composition but they vary with such factors as heat treatment or degree of plastic deformation. A general background on the way structure determines the properties of engineering materials can be found in general texts on materials science.

16.1.2   Design Requirements

Table 16.1 provides a fairly complete listing of material properties used to describe the performance of a material in a particular engineering design. These properties have evolved with time because they are reasonably easy to measure and fairly reproducible and are associated with some well-defined material response. Individual properties relate to one another and to the basic information needs of the engineer in a kind of taxonomy of materials information, as shown in Fig. 16.2. Note that only selected items have been expanded for illustration; the reader may readily detail others appropriate to specific interests.

Figure 16.3 shows the relation between common mechanical failure modes of materials and frequently measured material properties. While in a few instances a single material property controls a mode of failure, it is much more common for two or three properties to interact in a complex way to control the material behavior. Moreover, the service conditions met by materials in many designs are more complex than the test conditions ordinarily used to measure material properties. For example, sometimes the stress level is not constant but varies with time in a random way. In another instance the material may be subjected to a complex superposition of environments such as an alternating stress (fatigue) at high temperature (creep) in a highly oxidizing environment (corrosion). Often specialized service simulation tests are developed to "screen materials" for complex service conditions. Finally, the best candidate materials that pass the screening tests are evaluated in field-trials or tests of design prototypes to evaluate their performance under actual service conditions.

The material properties required for design usually are formalized through specifications. The more generic manifestation of a specification is a standard. Design standards set the minimum performance characteristics of a material, such as the minimum yield strength. The subject of material standards and specifications has been discussed at some length in the Kirk–Othmer *Encyclopedia of Chemical Technology*.[1]

In addition to the technical properties that define performance, great attention in materials selection should be given to the properties that determine producibility. These characteristics of a material are often difficult to catalog and specify. The subject of selection of materials for ease of manufacture is beyond the scope of this chapter (see Chapters 29 and 31).

16.1.3   Selection of Materials

Circumstances initiating the materials selection process comprise the following

New product development

Product modification: changed operating conditions and cost reduction

Materials substitution arising from problems (e.g., service failures, supply problems, or changed legal requirements) and changes in manufacturing method

The usual materials selection problem is concerned with arriving at a decision among many candidate materials for which several properties are important but not necessarily of equal weight. A rational approach to materials selection separates the required properties into three categories: (1) go–no go parameters, (2) nondiscriminating parameters, and (3) discriminating parameters.

Go–no go parameters are property requirements that must meet a certain fixed minimum value. Corrosion resistance or weldability are examples of material properties in this category. For a go–no

TABLE 16.1   Material Performance Characteristics

Physical Properties	*Mechanical Properties*	*Thermal Properties*
Crystal structure	Hardness	Conductivity
Density	Modulus of elasticity	Specific heat
Melting point	Tension	Coefficient of expansion
Vapor pressure	Compression	Emissivity
Viscosity	Poisson's ratio	Absorptivity
Porosity	Stress-strain curve	Ablation rate
Permeability	Yield strength	Fire resistance
Reflectivity	Tension	Maximum/minimum operating
Transparency	Compression	temperature
Other optical properties	Shear	
Dimensional stability	Ultimate strength	*Chemical Properties*
	Tension	
Electrical/Magnetic Properties	Shear	Position in
	Bearing	electromotive series
Conductivity	Fatigue properties	Corrosion and degradation
Dielectric constant	Smooth	Atmospheric
Coercive force	Notched	Salt water
Hysteresis	Corrosion fatigue	Acids
Susceptibility	Rolling contact	Hot gases
Permeability	Fretting	Ultraviolet
Remanence	Charpy transition	Oxidation
	temperature	Thermal stability
	Fracture toughness (K_{Ic})	Biological stability
Nuclear Properties	High temperature	Stress corrosion
	Creep	Hydrogen embrittlement
	Stress rupture	Hydraulic permeability
Half-life	Damping properties	
Cross-section	Wear properties	*Fabrication Properties*
Stability	Galling	
	Abrasion	Castability
	Erosion	Heat treatability
	Cavitation	Hardenability
	Spalling	Formability
	Ballistic impact	Machinability
		Weldability

Source: G. E. Dieter, *Engineering Design*, © 1983, McGraw-Hill, New York. Reproduced with permission.

go parameter, there is no special advantage if the property exceeds the threshold, nor would a larger value of one parameter make up for a deficiency in another go–no go parameter.

Nondiscriminating parameters are requirements that must be met if a material is to be considered at all. Examples in this category are material availability or ease of manufacture. Similar to the go–no go criterion, this category represents a situation that does not allow stringent comparison or quantitative evaluation.

The preceding two categories can eliminate many candidate materials during the screening process. The final selection is based on discriminating parameters to which quantitative values can be assigned. Usually a weighted property index is used for making the decision between a number of candidate materials. Each competing material property is assigned a weighting factor W_i based on an estimate of the importance of the property. Usually the sum of the weighting factors equals unity. For a group of candidate materials a scaled property β is determined:

$$\beta_i = \frac{\text{numerical value of property} \times 100}{\text{largest value of materials under consideration}} \qquad (16.1)$$

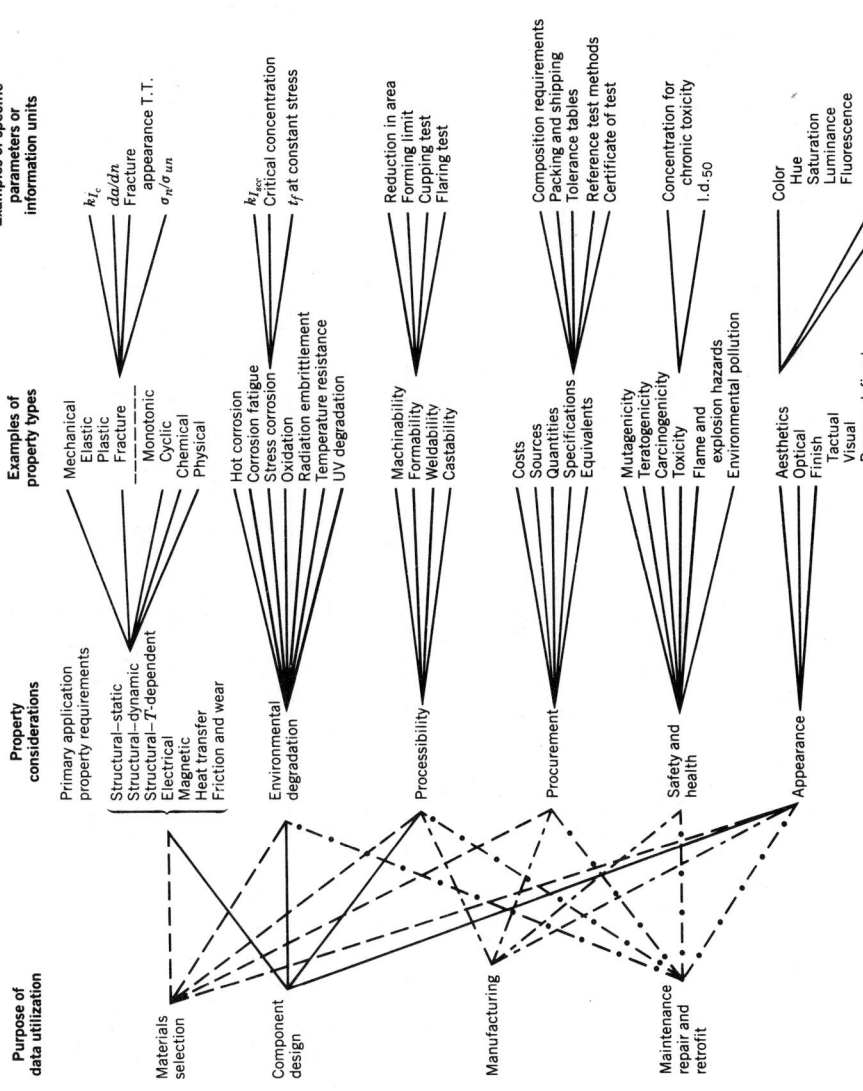

Fig. 16.2 Taxonomy of materials information (after Westbrook in *Factual Material Data Banks.* H. Kröckel, K. Reynard, & G. Steven, eds. EEC, Luxembourg, 1985).

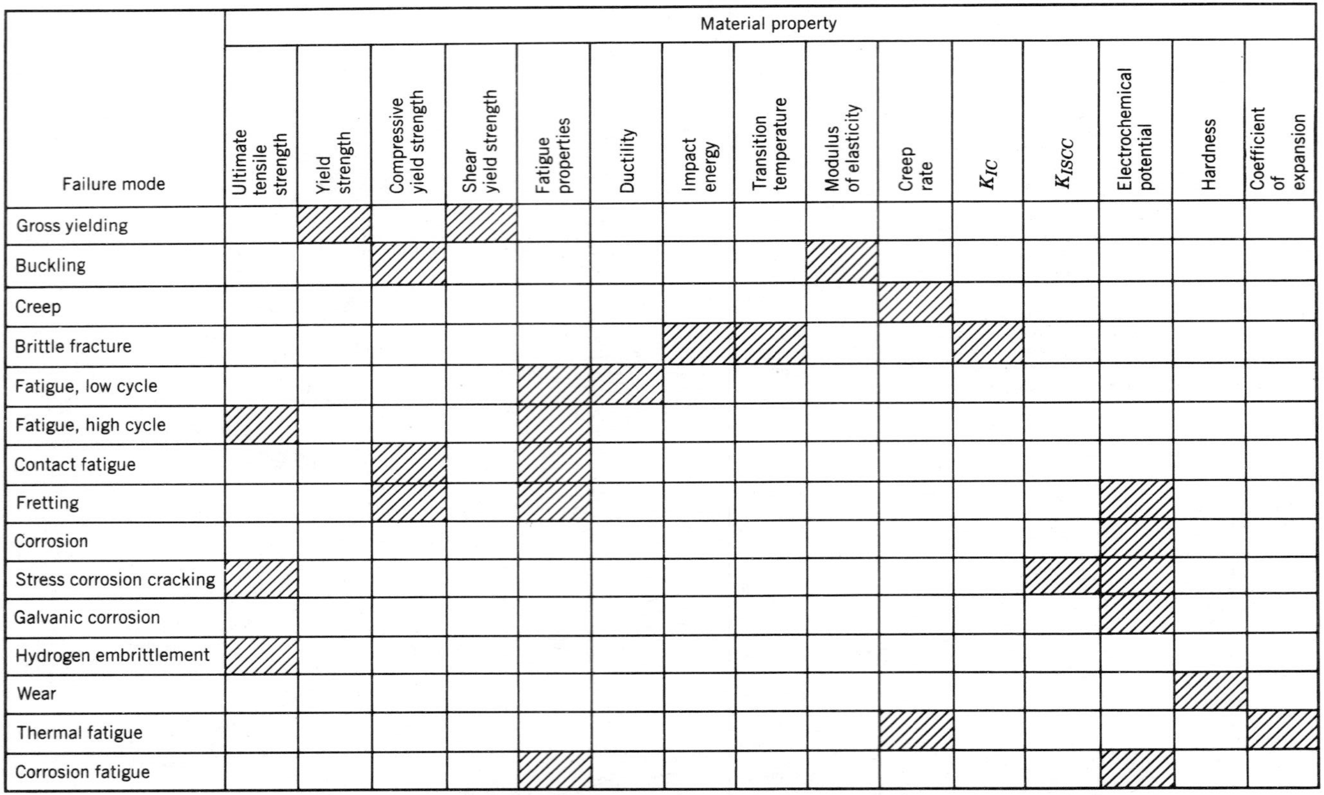

Material property

Failure mode	Ultimate tensile strength	Yield strength	Compressive yield strength	Shear yield strength	Fatigue properties	Ductility	Impact energy	Transition temperature	Modulus of elasticity	Creep rate	K_{Ic}	K_{Iscc}	Electrochemical potential	Hardness	Coefficient of expansion
Gross yielding		▨		▨											
Buckling			▨						▨						
Creep										▨					
Brittle fracture	▨					▨	▨	▨			▨				
Fatigue, low cycle					▨										
Fatigue, high cycle					▨										
Contact fatigue			▨		▨										
Fretting													▨		
Corrosion													▨		
Stress corrosion cracking	▨											▨	▨		
Galvanic corrosion													▨		
Hydrogen embrittlement	▨														
Wear														▨	
Thermal fatigue										▨					▨
Corrosion fatigue					▨								▨		

Fig. 16.3 Relations between failure modes and mechanical properties; shaded block at intersection of material property and failure mode indicates that particular material property is influential in controlling a particular failure mode. (From C. O. Smith and B. E. Boardman, *Metals Handbook*, 9th ed., Vol. I. American Society for Metals, Metals Park, OH, 1980. Copyright American Society for Metals, 1980.)

The weighted property index is

$$\gamma = \sum_{i=1}^{N} \frac{\beta_i W_i}{C_i} \tag{16.2}$$

where C_i is the cost per unit weight of the material. The use of a cost in this equation emphasizes the very great importance of material cost in making a selection of a material.

16.1.4   Costs of Materials

Once the performance requirements have been achieved, materials selection comes down to purchasing properties at the best available price. The basic cost of a material depends upon (1) scarcity, as influenced by either the concentration of the metal in the ore or the cost of the chemical feedstock; (2) the cost and amount of energy required to process the material; and (3) the basic supply and demand for the material. In general, large-volume usage materials such as crushed rock and cement have low prices while scarce materials such as industrial diamonds have high prices. Data for the price-volume relationship of materials can be plotted to reveal an exclusion curve (Fig. 16.4). This curve shows that

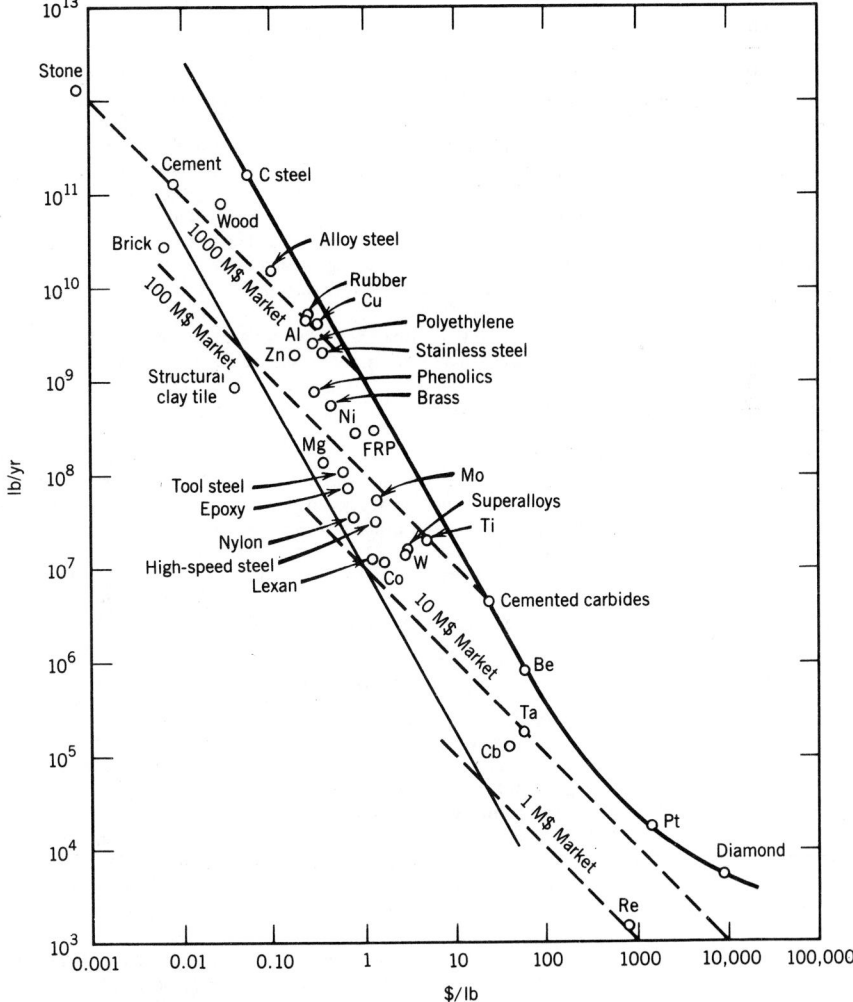

Fig. 16.4 Price-volume relationship for annual U. S. consumption of structural materials. (1962 data) (Westbrook, unpublished).

TABLE 16.2   Formulas for Cost per Unit Property^a

Cross Sections and Loading Conditions	Variable Dimension^b	Cost of Unit Strength	Cost of Unit Stiffness
Solid cylindrical bar in tension or compression	Diameter	$\rho m/\sigma$	$\rho m/E$
Solid cylindrical bar in bending	Diameter	$\rho m/\sigma^{2/3}$	$\rho m/E^{1/2}$
Solid cylindrical bar in torsion	Diameter	Same as for bending	$\rho m/G^{1/2}$
Solid cylindrical bar as slender column	Diameter	—	$\rho m/E^{1/2}$
Solid rectangle in bending	Depth	$\rho m/\sigma^{1/2}$	$\rho m/E^{1/3}$
Thin-walled cylindrical vessel	Wall thickness	$\rho m/\sigma$	—

Source: G. E. Dieter, *Engineering Design,* © 1983, McGraw-Hill, New York. Reproduced with permission.
^{a}Symbols: σ, yield strength; E, Young's modulus; G, shear modulus; ρ, density; m, material cost, $/lb.
^{b}Dimension varied to maintain equal strength or stiffness.

given a certain price for a material, its usage is limited in volume. Or conversely, if a certain volume of use is projected, the cost (price) must lie below a defined level. There are two further aspects of material costs other than the intrinsic scarcity/availability of the basic material. As more work is invested in the processing of a material, the cost increases, that is, value is added. Thus, hot-rolled steel billets cost about 40% more than pig iron, while the value of hot-rolled steel bar is nearly double that of pig iron. A cold-rolled steel sheet is about 2.5 times as costly as the slab from which it was rolled. The second factor is the degree of technological maturity of the material. When first introduced, a new material will be relatively costly and experience a low volume of use; its datum point in Fig. 16.4 will be below the curve at the lower right. With further technical development, its datum point moves up and to the left, perhaps at some time approaching the exclusion curve. Under competition from other materials, its datum point may subsequently move down, away from the limiting curve, indicative of lower usage, irrespective of cost.

Properties of metals are often improved by alloying. While alloys generally are more expensive than the pure base metal, the cost of an alloy may not simply be the weighted average of the cost of the elements. Often a large fraction of the cost of an alloy is the need to control one or more impurities to very low levels. This can require extra refining steps or the use of expensive high-purity raw materials.

The prices quoted for materials in trade magazines often do not include the many price extras that can be imposed. Price extras can result from requiring special chemistry, nonroutine testing or inspection, tighter than normal tolerances on dimensions, special packing, transportation charges, or purchase in less than standard quantities. These and other price extras need to be kept in mind by designers when selecting and purchasing materials.

A useful criterion in the selection of materials is the concept of cost per unit property of a material. These comparisons can be based on either the strength or the stiffness of the materials. Formulas for the cost per unit property are given in Table 16.2.

16.1.5   Sources of Information on Materials

Technical information sources are used in the broadest sense for a variety of purposes: tutorial, to learn something of a new field; recall, to retrieve for immediate use those facts or data not retained in memory; searching, to extract from the universe of knowns (in print or otherwise) those items pertinent to the case at hand; or data entry, to input necessary data to calculations, process control, and testing and analysis, each of which in turn yields new information.

In materials engineering, information is required for materials selection, component design, manufacturing, and maintenance, repair, and retrofit. Each of these functional uses of information requires a different mix of property considerations: property requirements for primary application, environmental degradation information, safety and health, processibility, appearance, and material procurement, as detailed in Fig. 16.2.

The information needs of the materials engineer are different and in many ways more demanding and difficult to satisfy than those of the materials scientist. The engineer must not only have accessible the best existing value for the property in question but must also know the limits of uncertainty for the value so that the reliability of a design may be estimated. Rather than working solely with pure chemical species, the engineer is frequently confronted with complex mixtures, alloys, and blends, whose properties and behavior must also be known or reliably estimated. Furthermore, many properties of commercial materials are not fixed values defined by pressure, temperature, and similar variables but are history and structure dependent. In addition, the engineer will frequently require data that cannot be expressed in terms of a single property or combination of properties but must be related to a performance test or service application.

Materials information sources may be categorized into several different groupings:

> Guides to technical information sources
> Encyclopedias
> Dictionaries, thesauri, and glossaries
> Data sources
> Auxiliary information services for data
> Primary literature
> Reviews
> Materials economics and business information
> Materials policy
> Sources on people
> Special information sources (specifications and standards, patents, symbols, safety, and health)

Only the starred group is discussed here. Reference is made to a more detailed article by Westbrook[2] for a fuller description and discussion of the other groups.

Data Sources
Five broad categories of data sources may be distinguished:

1. *Handbooks*. Printed compilations of numeric data usually in tabular form and often accompanied by tutorial text and graphical or pictorial illustrations.
2. *Graphical and Pictorial Reference Works*. Printed compilations of nonnumeric data, for example, phase diagrams, transformation or TTT curves, and Pourbaix charts, usually in a standardized format for ready comparison.
3. *Data Centers and Projects*. Activities of defined and limited technical scope staffed by specialists who are able to respond to phone or letter inquiries for specific data.
4. *Data Journals and Depositories*. A relatively new medium especially adapted to data archiving in print, tape, or microform.
5. *Machine-readable Databases*. Digitized information usually stored magnetically on tape or disk accessed either directly (on-line) or indirectly (off-line) with specially designed search and retrieval software.

Some 1250 data sources in these five categories have been listed and their materials/properties coverage analyzed in an extensive report by Wawrousek et al.[3]

REFERENCES

1. Westbrook, J. H., "Materials Standards and Specifications," Kirk–Othmer *Encyclopedia of Chemical Technology*, 3rd ed., New York, Wiley, 1981, p. 32.
2. Westbrook, J. H., "Materials Information Sources," in *Encyclopedia of Materials Science and Engineering*, M. B. Bever, ed., Pergamon, Oxford 1986, p. 527.
3. Wawrousek H., Westbrook J. H., and Grattidge, W., "Data Sources of Mechanical and Physical Properties of Engineering Materials," *Physik Daten* No. 30-1, 1989, Fachinformationszentrum-Energie, Physik, Mathematik, Gmbh, Karlsruhe, FRG.

BIBLIOGRAPHY FOR INTRODUCTION TO PROPERTIES, PROCESSING, AND SELECTION OF MATERIALS

Crane, F. F. A., and Charles, J. A., *Selection and Uses of Engineering Materials*, Butterworths, 1984.

Dieter, G. E., *Engineering Design*, New York, McGraw-Hill, 1983.

Kutz, M., *Mechanical Engineering Handbook*, New York, Wiley, 1986.

Smith, W. F., *Principles of Materials Science and Engineering*, New York, McGraw-Hill, 1986.

Westbrook, J. H., "Extraction and Compilation of Numerical and Factual Data," AGARD Lecture Series #130, "Development and Use of Numerical and Factual Data Bases," Gaithersburg, London and Lisbon, October 1983.

TABLE 16.3   Physical Properties of Elemental Metals

Name	Atomic Weight	Density g/cm³	Density Temperature,ᵃ °C	Melting Point, °C	Latent Heat of Fusion, g-cal/g	Boiling Point, °C	Latent Heat of Vaporization, g-cal/g	Specific Heat g-cal/g °C or Btu/lb °F	Specific Heat Temperature °C
Aluminum	26.97	2.70	20	657.1	93.0	2056	1950–2000	0.226	0–100
Antimony	121.76	6.618	20	630.5	38.26	1440	373	0.0493	20–100
Arsenic	74.91	5.73	14	814	—	615 Sublimes	74	0.0822	18
Barium	137.36	3.5	20	850	345.5	1637	628	0.068	−185 to +20
Beryllium	9.02	1.85	20	1285	318	2780	—	0.425	0–100
Bismuth	209.00	9.781	20	271.3	12.46	1450	221	0.029	20
Cadmium	112.41	8.648	20	320.9	13.17	766	227.5	0.0547	27.9
Calcium	40.08	1.55	20	851	78	1487	—	0.157	0–100
Cerium	140.13	6.90	20	775	—	1400	—	0.0511	20–100
Cesium	132.91	1.873	20	26.0	3.76	836.6	131.4	0.0521	0–26
Chromium	52.01	7.14	25	1550	31.75	2482	14.71	0.12	18–100
Cobalt	58.94	8.9	21	1480	58.38	2900	—	0.0989	20
Copper	63.57	8.94	20	1083	50.6	2595	1756	0.0918	18–100
Gold	197.2	19.3	20	1063	16.11	2966	446	0.0308	18
Iridium	193.1	22.42	17	2408.8	26.1	4900	340	0.0323	18–100
Iron, 99.97%	55.84	7.87	20	1535	65	2998	1110	0.1075	20
Lead	207.21	11.342	20	327.4	6.26	1744	323	0.0297	0
Lithium	6.94	0.534	20	186	32.81	1372	—	0.79	50
Magnesium	24.32	1.74	20	651	64.8	1107	1300–1500	0.249	0–100
Manganese	54.93	7.44	20	1242	64.8	2151	1044	0.107	20–100
Mercury	200.61	13.546	20	−38.87	2.66	356.9	71	0.0332	17
Molybdenum	95.95	10.2	—	2620	—	4803	176.8	0.0647	0
Nickel	58.69	8.85	20	1452	73.8	2900	1010	0.112	0
Osmium	190.82	22.48	20	2700	—	5493	350	0.0311	19.98
Palladium	106.7	12.0	20	1555	34.2	3980	610	0.0587	0
Platinum	195.23	21.45	20	1773.5	27.1	4389	637	0.0319	20–100
Potassium	39.10	0.87	20	63.5	14.5	773.9	513	0.177	3.4
Rhodium	102.91	12.44	20	1966	—	4500	620	0.0598	10–97
Silver	107.88	10.5	20	960.5	24.3	2001	551.6	0.0558	0
Sodium	22.997	0.9712	20	97.5	27.53	892	1170	0.295	0
Strontium	87.63	2.60	—	771	—	1383	1045	0.0735	15
Tantalum	180.88	16.6	—	2850	—	6093	—	0.0356	58
Tellurium	127.61	6.24	20	452.2	7.305	1087	159	0.0468	15–100
Thallium	204.39	11.85	20	303.3	7.185	1457	220	0.0368	20–100
Thorium	232.12	11.5	17	1845	—	5200	—	0.0276	0–100
Tin	118.70	7.30	—	231.89	14.4	2270	655	0.548	25
Titanium	47.90	4.5	18	1800	—	5100	1320	0.142	0–100
Tungsten	183.92	19.3	—	3370	44.0	5927	1183	0.034	100
Uranium	238.07	18.7	13	1690	—	3500	—	0.0276	0–98
Vanadium	50.95	5.68	—	1710	—	3000	—	0.1153	0–100
Zinc	65.38	7.14	—	419.45	24.09	907.2	426.8	0.0931	20–100
Zirconium	91.22	6.43	—	1700	—	5050	—	0.063	0–100

TABLE 16.3 *(Continued)*

Name	Thermal Coefficient of Linear Expansion		Thermal Conductivity		Electrical Resistivity		Temperature Coefficient of Resistivity	
	$\times 10^4/°C$	Temperature °C	g-cal per s · cm · °C	Temperature °C	$\mu\Omega$-cm	Temperature[a] °C	Temperature Coefficient	Temperature °C
Aluminum	0.2545	20–300	0.52	50	2.655	20	0.00445	0–100
Antimony	‖0.1129	20	0.0444	0	39.1	0	0.0036	20
	⊥ 0.080	20						
Arsenic	0.0386	20	—	—	35	0	0.0042	20
Barium	—	—	—	—	9.8	20	0.0033	20
Beryllium	0.123	20	—	—	18.5	20	—	—
Bismuth	‖0.1345	20	0.0200	18	115.0	18	0.004	20
	⊥ 0.103	20						
Cadmium	‖0.54	0	0.217	18	7.59	18	0.0042	0–100
	⊥ 0.20	0						
Calcium	0.25	0–21	—	—	4.59	20	0.00364	0–600
Cerium	—	—	—	—	78	20	—	—
Cesium	0.97	0–26	—	—	20	0	0.00478	−80 to +25
Chromium	0.081	20–100	0.165	—	13.1	0	—	—
Cobalt	0.1208	20	0.165	—	9.7	20	0.00658	0–100
Copper	0.1642	20	0.923	18	1.682	20	0.00382	20
Gold	0.144	17–100	0.707	17	2.42	20	0.0034	20
Iridium	0.0641	20	0.141	17	6.08	0	0.00411	0–100
Iron, 99.97%	0.119	0–100	0.19	0	9.8	20	0.0065	0–100
Lead	0.295	20–100	0.083	18	20.65	20	0.00422	18
Lithium	0.56	0–178	0.167	0	8.55	0	0.0047	0
Magnesium	0.257	20–300	0.376	0–100	4.4611	20	0.0040	20
Manganese	0.228	0–100	—	—	5.0	—	—	—
Mercury	[b]		0.020	0	95.783	20	0.00089	20
Molybdenum	0.0549	25–100	0.346	17	4.77	0	0.0034	0–100
Nickel	0.137	25–100	0.14	0–100	6.9	20	0.006	20
Osmium	0.057	40	—	—	9.0	20	—	—
Palladium	0.1160	20	0.161	18	10	20	0.0033	20
Platinum	0.088	20	0.1664	18	9.83	0	0.003	20
Potassium	0.83	0–50	0.237	0	7.0	0	0.0055	0
Rhodium	0.089	6–21	0.213	0	4.93	0	0.0043	0–100
Silver	0.189	0–100	0.974	18	1.629	18	0.0038	20
Sodium	0.71	−190 to −17	0.3225	0	4.6	0	0.0044	0
Strontium	—	—	—	—	22.76	20	—	—
Tantalum	0.0655	0–100	0.130	17	12.4	20	0.00383	0–100
Tellurium	0.168	20	0.0143	45	200,000	19.6	—	—
Thallium	0.280	40	0.09315	0	18.1	0	0.0040	0
Thorium	—	—	—	—	18	20	0.0021	20–1800
Tin	‖0.224[c]	20	0.157	0	11.5	20	0.0042	20
	⊥ 0.464[c]	20						
Titanium	0.0714	—	—	—	42	20	—	—
Tungsten	0.0444	27	0.476	17	5.48	20	0.0045	18
Uranium	—	—	—	—	60 ± 18	20	—	—
Vanadium	—	—	—	—	26.0	—	—	—
Zinc	[dd]	20–100 20–100	0.268	18	[e]	0	0.0037	20
Zirconium	—	—	—	—	40	0	0.0044	—

[a] Where temperature is not given, ordinary temperature is understood.

[b] $\dfrac{1}{v}\dfrac{dv}{dt} = 182.0 \times 10^{-6}$ at 20°C.

[c] Single crystal: 22.4 ‖ crystal axis; 46.4 ⊥ crystal axis.

[d] 32.5 pure, hot-rolled zinc with grain; 23.0 pure, hot-rolled zinc across rolling direction.

[e] Single crystal: 6.2 ‖ crystal axis; 5.8 ⊥ crystal axis.

TABLE 16.4 Chemical Composition of Some Common Metals

Item No.	Metal	C	Mn	Fe	Cr	Ni	Cu	Si	Mo	Other
					Iron Base Alloys					
1	Steel, low carbon	0.10–0.30	0.50–0.80	Balance	—	—	—	0.10 min	—	
2	Cast iron	2.75–3.50	0.50–0.80	Balance	—	—	—	1.90–2.25	—	
3	Ni Resist, Type 1	3.0[a]	1.0–1.5	Balance	1.75–2.50	13.5–17.5	7.0	1.0–2.5	—	
4a–h	Cr–Mo steels	0.15[a]	1.0[a]	Balance	1.0–9.0	—	—	1.0[a]	0.5	
5	12 Cr steel	0.15[a]	1.0[a]	Balance	11.50–13.0	—	—	0.5[a]	—	
6	Stainless, 304	0.08[a]	2.0[a]	Balance	18.0–20.0	8.0–11.0	—	1.0[a]	—	
7	Stainless, 316	0.10[a]	2.0[a]	Balance	16.0–18.0	10.0–14.0	—	1.0[a]	3.0	
8	Stainless, 317	0.10[a]	2.0[a]	Balance	18.0–20.0	11.0–14.0	—	1.0[a]	4.0	
9	Worthite	0.07[a]	0.6	Balance	20.0	24.0	1.75	3.25	3.0	
10	Durimet 20	0.07[a]	0.65–0.85	Balance	19.0–21.0	28.0–30.0	4.0	1.0	3.0	
11	25 Cr–12 Ni steel	0.20[a]	2.0[a]	Balance	22.0–24.0	12.0–15.0	—	1.0[a]	—	
12	25 Cr–20 Ni steel	0.25[a]	2.0[a]	Balance	24.0–26.0	19.0–22.0	—	1.5[a]	—	
13	Incoloy 800	0.10[a]	1.5[a]	Balance	19.0–22.0	32.0–36.0	0.50	1.0[a]	—	Al 0.38; Ti 0.38
14	Silicon iron	0.85[a]	0.50–0.65	Balance	—	—	—	14.5	—	
15	Durichlor	0.85[a]	0.65	Balance	—	—	—	14.5	3.0	
					Copper Base Alloys					
16	Copper	—	—	—	—	—	99.9	—	—	
17	Tin bronze	—	—	—	—	—	90.0	—	—	Sn 10.0
18	Aluminum bronze	—	—	—	—	—	95.0	—	—	Al 5.0
19	Ampco No. 18	—	—	3.5	—	—	Balance	—	—	Al 11.0
20	Red brass	—	—	—	—	—	85.0	—	—	Zn 15.0
21	Yellow brass	—	—	—	—	—	65.0	—	—	Zn 32.0; Sn 1.0; Pb 2.0
22	Muntz metal	—	—	—	—	—	60.0	—	—	Zn 40.0
23	Admiralty	—	—	0.06[a]	—	—	70.0	—	—	Sn 1.0; Zn balance
24	Silicon bronze	—	—	—	—	—	97.0	3.0	—	
25	70 Cu–30 Ni	—	—	—	—	30.0	70.0	—	—	

Nickel Base Alloys

26	Monel 400	0.15ᵃ	1.0	1.4	—	67.0	30.0	0.1ᵃ	—	
27	Monel K500	0.15ᵃ	0.75	0.9	—	66.0	29.0	0.5ᵃ	—	Al 2.75; Ti 0.75
28	Nickel 200	0.05ᵃ	0.20	0.15	—	99.4	0.10	0.05	—	
29	Inconel 600	0.08ᵃ	0.25	7.0	15.0	77.0	0.20	0.25	—	
30	Hastelloy B	0.12ᵃ	1.0ᵃ	4.0–7.0	1.0ᵃ	60.0–65.0	—	1.0	28.0	Co 2.5ᵃ
31	Hastelloy C	0.15ᵃ	1.0ᵃ	4.0–7.0	13.0–16.0	55.0–60.0	—	1.0ᵃ	17.0	W 3.5–5.5; Co 2.5ᵃ
32	Hastelloy D	0.12ᵃ	0.80–1.25	1.0ᵃ	1.0ᵃ	Balance	4.0	8.5–10.0	—	
33	Hastelloy F	0.05ᵃ	1.0–2.0ᵃ	Balance	21.0–23.0	44.0–47.0	—	1.0ᵃ	6.5	W 1.0; Cb 2.0; Co 2.5ᵃ
34	Hastelloy X	0.15ᵃ	0.75ᵃ	Balance	22.0	45.0	—	—	9.0	Co 1.5

Superalloys

35	Nimonic 80	0.04	0.56	Balance	21.0	74.0	—	0.47	—	Ti 2.5; Al 0.6
36	Inconel X 750	0.08	0.53	6.0	15.0	70.0	—	0.40	—	Cb 1.0; Ti 2.5; Al 0.6
37	M-252	0.15	1.00	Balance	19.0	48.5	—	0.70	10.0	W 7.5; Co 10.0; Ti 2.5
38	Refractaloy 26	0.05	0.70	17.0	18.0	37.0	—	0.80	3.0	Co 20; Ti 2.8; Al 0.2
39	A-286	0.05	1.35	1.0	15.5	26.0	—	0.95	1.25	Ti 1.9; Al 0.2; V 0.3
40	Discaloy	0.03	0.47	Balance	13.0	25.0	—	0.50	3.0	Ti 2.5; Al 0.2
41	N155 Alloy	0.10	1.50	Balance	20.0	20.0	—	0.50	3.0	W 2.0; Co 20.0; Cb 1.0
42	S-590	0.44	0.60	Balance	20.0	20.0	—	0.40	4.0	W 4.0; Co 20.0; Cb 4.0
43	S-816	0.38	1.30	2.95	20.0	20.0	—	0.59	4.0	W 4.0; Co 40.0; Cb 4.0
44	Haynes alloy 31	0.50	1.0ᵃ	2.0ᵃ	25.0	10.0	—	1.0ᵃ	—	W 7.5; Co balance
45	Haynes alloy 25	0.08	1.50	3.0ᵃ	20.0	10.0	—	1.0ᵃ	—	W 15.0; Co balance
46	Haynes alloy 21	0.24	1.0ᵃ	2.0ᵃ	29.0	2.5	—	1.0ᵃ	6.0	B 0.007; Co balance
47	16 Cr-25 Ni-6 Mo	0.15	1.14	Balance	16.0	25.0	—	0.57	6.0	N 0.13
48	19-9 DL	0.26	0.52	Balance	19.0	9.0	—	—	1.2	W 1.2; Cb 0.29; Ti 0.21
49	17-4 PH	0.05	—	—	16.5	4.0	4.0	—	—	Cb 0.25

Other Metals

50	Aluminum	—	—	—	—	—	—	—	—	Al 99.6
51	Lead	—	—	—	—	—	—	—	—	Pb 99.9
52	Titanium	0.20ᵃ	—	—	—	—	—	—	—	Ti 99.0 minimum
53	Zirconium	—	—	—	—	—	—	—	—	Zr 99.5
54	Tantalum	—	—	—	—	—	—	—	—	Ta 99.8 minimum

Courtesy of Shell Development Company.

ᵃMaximum.

TABLE 16.5 Physical Properties of Some Common Metals

Item No.	Metal	Melting Point °F	Melting Point °C	Electrical Resistance, μΩ-cm 70°F	Density g/cm³	Density lb/ft³	Density lb/in.³	Specific Heat, cal/g °C or Btu/lb °F	Brinell Hardness (Annealed)	Machinability Rating, SAE 1112 = 100
	Iron Base Alloys									
1	Steel, low carbon	2760	1516	10	7.86	491	0.284	0.11	130	70
2	Cast iron	2150	1176	66	7.20	449	0.260	0.13	180	50
3	Ni Resist, Type 1	2250	1231	140	7.30	456	0.264	0.11	170	30
4a–h	Cr–Mo. steels	2500	1371	10	7.86	491	0.284	0.11	180	70
5	12 Cr steel	2720	1495	9	7.75	484	0.280	0.11	150	70
6	Stainless 304	2590	1420	72	8.02	501	0.290	0.12	160	50
7	Stainless 316	2550	1398	74	8.02	501	0.290	0.12	165	50
8	Stainless 317	2550	1298	74	8.02	501	0.290	0.12	165	50
9	Worthite alloy 20	2650	1455	75	8.02	501	0.290	0.12	160	60
10	Durimet 20	2650	1454	75	8.02	501	0.290	0.12	160	60
11	25 Cr–12 Ni steel	2650	1454	78	8.02	501	0.290	0.12	160	50
12	25 Cr–20 Ni steel	2650	1454	78	8.02	501	0.290	0.12	165	50
13	Incoloy 800	2525	1385	93	8.05	503	0.291	0.12	184	45
14	Silicon iron	2300	1260	63	7.00	437	0.253	0.13	500	0
15	Durichlor	2300	1260	63	7.20	449	0.260	0.13	500	0
	Copper Base Alloys									
16	Copper	1980	1080	2	8.91	556	0.332	0.09	42	61
17	Tin bronze	1830	999	—	8.78	548	0.317	0.09	60	130
18	Aluminum bronze	1900	1038	19	7.78	486	0.281	0.09	190	200
19	Ampco No. 18	1900	1038	19	7.60	474	0.274	0.09	166	200
20	Red brass	1880	1025	5	8.75	546	0.316	0.09	50	100
21	Yellow brass	1710	931	7	8.47	529	0.307	0.09	55	200
22	Muntz metal	1660	904	7	8.39	524	0.303	0.09	80	200
23	Admiralty	1720	937	7	8.53	533	0.308	0.09	60	200
24	Silicon bronze	1870	1019	27	8.53	533	0.308	0.09	70	60
25	70 Cu–30 Ni	2240	1226	37	8.94	558	0.323	0.09	70	120

Nickel Base Alloys

No.	Material									
26	Monel 400	2460	1350	48	8.83	551	0.319	0.11	125	44
27	Monel K500	2460	1348	48	8.47	529	0.306	0.10	160	50
28	Nickel 200	2640	1449	7	8.91	556	0.322	0.11	100	20
29	Inconel 600	2600	1426	103	8.43	526	0.304	0.11	150	40
30	Hastelloy B	2460	1348	135	9.24	577	0.334	0.09	230	23
31	Hastelloy C	2380	1303	133	8.94	558	0.323	0.09	185	23
32	Hastelloy D	2050	1120	113	7.80	487	0.282	0.11	400	0
33	Hastelloy F	2350	1286	112	8.20	512	0.296	0.10	170	—
34	Hastelloy X	2350	1315	118	8.23	514	0.297	0.11	183	30

Superalloys

No.	Material									
35	Nimonic 80	2590	1420	124	8.25	515	0.298	0.10	185	—
36	Inconel X-750	2600	1425	122	8.25	515	0.298	0.11	176	20
37	M-252	2500	1370	92	8.25	515	0.298	0.10	—	—
38	Refractaloy 26	2450	1344	91	8.21	513	0.297	0.11	250	—
39	A 286	2550	1400	98	7.94	496	0.287	0.11	300	—
40	Discaloy	2520	1380	98	7.97	498	0.288	0.11	310	45
41	N-155	2500	1371	—	8.20	512	0.296	0.10	185	—
42	S-590	2400	1315	93	8.36	519	0.301	0.10	—	—
43	S-816	2400	1315	98	8.66	541	0.313	0.10	—	—
44	Haynes alloy 31	2500	1370	88	8.61	538	0.311	0.10	340	—
45	Haynes alloy 25	2570	1409	88	9.15	571	0.330	0.09	250	—
46	Haynes alloy 21	2470	1353	87	8.30	518	0.300	0.11	237	—
47	16 Cr-25 Ni-6 Mo	2550	1398	75	8.07	504	0.292	0.11	165	—
48	19-9 DL	2600	1427	78	7.94	496	0.287	0.10	200	—
49	17-4 PH	—	—	77	7.80	487	0.282	0.13	400	—

Other Metals

No.	Material									
50	Aluminum	1220	660	3	2.72	170	0.098	0.22	20	300–1500
51	Lead	621	327	21	11.35	709	0.410	0.03	4	—
52	Titanium	3135	1725	42	4.50	281	0.163	0.13	150	500
53	Zirconium	3355	1845	40	6.45	403	0.233	0.07	—	—
54	Tantalum	5425	2996	12	16.60	—	0.600	0.04	100	—

Courtesy of Shell Development Company.

TABLE 16.6 Tensile Strengths of Some Common Metals at Different Temperatures

psi, $\times 10^3$

Item No.	Metals	−325°F	−200°F	−100°F	70°F	200°F	400°F	600°F	800°F	1000°F	1200°F	1400°F	1600°F	1800°F	2000°F
	Iron Base Alloys														
1	Steel, low carbon	93	87	76	60	65	68	64	52	29	15	7	4	3	2
2	Cast iron	—	—	—	—	—	—	—	—	—	—	—	—	—	—
3	Ni Resist, Type 1	—	—	—	—	—	—	—	—	—	—	—	—	—	—
4a	½ Mo	—	—	—	63	67	70	68	60	46	28	12	—	—	—
4b	1 Cr-½ Mo	—	—	—	65	67	68	66	62	52	32	11	—	—	—
4c	1¼ Cr-½ Mo	—	—	—	67	70	74	75	70	55	31	12	—	—	—
4d	2 Cr-½ Mo	—	—	—	65	—	—	65	60	46	29	11	—	—	—
4e	2¼ Cr-1 Mo	—	—	—	70	70	70	69	67	53	33	14	—	—	—
4f	5 Cr-½ Mo	—	—	—	71	67	62	58	54	44	25	13	—	—	—
4g	7 Cr-½ Mo	—	—	—	80	78	75	69	59	53	29	14	—	—	—
4h	9 Cr-1 Mo	—	—	—	84	82	77	72	64	53	30	13	—	—	—
5	12 Cr Type 410	158	140	128	89	83	76	73	66	45	22	9	9	7	4
6	Stainless 304	232	200	168	85	79	74	73	69	58	45	30	16	9	6
7	Stainless 316	201	175	130	90	85	81	78	76	70	57	36	23	14	8
8	Stainless 317	—	—	—	—	—	—	—	—	—	—	—	—	—	—
9	Worthite alloy 20	—	—	—	—	—	—	—	—	—	—	—	—	—	—
10	Durimet 20	—	—	—	—	—	—	—	—	—	—	—	—	—	—
11	25 Cr-12 Ni steel	—	—	—	80	79	77	73	69	63	52	35	22	12	7
12	25 Cr-20 Ni steel	—	—	—	88	86	83	81	77	70	58	41	27	17	10
13	Incoloy 800	—	—	—	82	—	77	76	75	72	54	32	—	—	—
14	Silicon iron	—	—	—	—	—	—	—	—	—	—	—	—	—	—
15	Durichlor	—	—	—	—	—	—	—	—	—	—	—	—	—	—
	Copper Base Alloys														
16	Copper	50	44	37	32	31	23	20	17	9.0	4.5	—	—	—	—
17	Tin bronze	—	—	—	—	—	—	—	—	—	—	—	—	—	—
18	Aluminum bronze	102	93	85	68	63	57	38	20	11	—	—	—	—	—
19	Ampco No. 18	102	93	85	85	85	76	59	26	17	—	—	—	—	—
20	Red brass	—	—	—	42	42	32	22	14	11	—	—	—	—	—
21	Yellow brass	—	—	—	50	50	48	38	25	—	—	—	—	—	—
22	Muntz metal	—	—	—	55	—	—	—	—	—	—	—	—	—	—
23	Admiralty	73	63	58	53	50	—	33	—	—	—	—	—	—	—
24	Silicon bronze	92	182	172	58	55	50	40	27	17	—	—	—	—	—
25	70 Cu-30 Ni	—	—	—	55	—	47	—	—	—	—	—	—	—	—

Nickel Base Alloys

No.	Material														
26	Monel 400	115	107	95	81	79	78	75	64	46	27	18	9	5	—
27	Monel K500	171	182	172	160	150	149	146	124	95	80	45	21	6.0	3.0
28	Nickel 200	110	95	82	67	67	67	66	44	32	22	14	—	—	—
29	Inconel 600	117	102	92	85	81	78	79	83	79	71	47	23	15	11
30	Hastelloy B	172	144	142	131	124	119	118	116	108	74	68	53	32	18
31	Hastelloy C	160	143	134	121	116	111	106	102	99	97	79	56	—	—
32	Hastelloy D	—	—	—	102	96	92	89	84	80	72	53	29	15	6.0
33	Hastelloy F	—	—	—	114	106	103	100	100	94	83	63	37	23	13
34	Hastelloy X	—	—	—	—	—	—	—	—	—	—	—	—	—	—

Superalloys

No.	Material														
35	Nimonic 80	—	—	—	155	—	134	128	—	—	113	89	40	12	—
36	Inconel X750	213	200	188	178	173	167	158	152	142	125	92	34	9.0	—
37	M-252	—	—	—	180	180	180	180	180	180	160	135	80	—	—
38	Refractaloy 26	—	—	—	154	153	151	148	145	143	136	105	48	—	—
39	A-286	205	185	172	145	141	138	138	138	132	104	62	—	—	—
40	Discaloy	—	—	—	145	145	142	137	129	125	104	75	38	—	—
41	N155	—	—	—	118	117	114	110	103	93	80	60	—	—	—
42	S-590	171	159	150	140	142	145	144	140	127	98	66	60	25	—
43	S-816	—	—	—	140	132	126	124	125	121	112	90	—	—	—
44	Haynes alloy 31	—	—	—	172	163	149	135	121	105	91	74	—	—	—
45	Haynes alloy 25	—	—	—	150	150	148	143	134	120	98	68	—	—	—
46	Haynes alloy 21	—	—	—	154	153	150	143	133	120	100	78	—	—	—
47	16 Cr–25 Ni–6 Mo	—	—	—	107	105	101	98	93	86	72	50	—	—	—
48	19-9 DL	—	—	—	118	108	100	96	93	90	75	55	—	—	—
49	17-4 PH	260	240	220	200	185	175	160	150	95	40	—	—	—	—

Other Metals

No.	Material														
50	Aluminum 5052-H34	50	45	30	28	28	18	7.5	—	—	—	—	—	—	—
51	Lead	—	—	—	2.5	—	—	—	—	—	—	—	—	—	—
52	Titanium A55	140	110	100	85	67	50	38	33	22	—	—	—	—	—
53	Zirconium	—	—	—	49	42	25	18	14	—	—	—	—	—	—
54	Tantalum	160	—	82	54	54	54	55	57	50	37	—	23	21	15

Courtesy of Shell Development Company.

TABLE 16.7   Moduli of Elasticity of Some Common Metals at Different Temperatures

psi, $\times 10^6$

Item No.	Metals	−325°F	−200°F	−100°F	70°F	200°F	400°F	600°F	800°F	1000°F	1200°F	1400°F	1600°F	1800°F	2000°F
	Iron Base Alloys														
1	Steel, low carbon	30.0	29.5	29.0	29.0	28.7	25.7	27.0	23.4	15.4	—	—	—	—	—
2	Cast iron	—	—	—	13.4	13.2	12.6	11.7	10.2	—	—	—	—	—	—
3	Ni Resist, Type 1	—	—	—	—	—	—	—	—	—	—	—	—	—	—
4a	½ Mo	31.0	30.6	30.4	29.9	29.5	28.6	27.4	25.7	23.0	15.6	—	—	—	—
4b	1 Cr–½ Mo	31.0	30.6	30.4	29.9	29.5	28.6	27.4	25.7	23.0	15.6	—	—	—	—
4c	1¼ Cr–½ Mo	31.0	30.6	30.4	29.9	29.5	28.6	27.4	25.7	23.0	15.6	—	—	—	—
4d	2 Cr–½ Mo	31.0	30.6	30.4	29.9	29.5	29.6	27.4	25.7	23.0	15.6	—	—	—	—
4e	2¼ Cr–1 Mo	31.0	30.6	30.4	29.9	29.5	28.6	27.4	25.7	23.0	15.6	—	—	—	—
4f	5 Cr–½ Mo	29.4	28.5	28.1	27.4	27.0	26.4	25.4	24.2	22.8	20.8	18.1	—	—	—
4g	7 Cr–½ Mo	29.4	28.5	28.1	27.4	27.0	26.4	25.4	24.2	22.8	20.8	18.1	—	—	—
4h	9 Cr–1 Mo	29.4	28.5	28.1	27.4	27.0	26.4	25.4	24.2	22.8	20.8	18.1	—	—	—
5	12 Cr–steel	30.8	30.3	29.8	29.2	28.7	27.7	26.0	23.1	18.6	12.2	—	—	—	—
6	Stainless 304	29.4	28.5	28.1	27.4	27.1	26.4	25.4	24.2	22.5	21.1	19.4	—	—	—
7	Stainless 316	29.4	28.5	28.1	28.1	28.1	26.9	25.4	24.2	22.8	21.5	20.0	—	—	—
8	Stainless 317	—	—	—	—	—	—	—	—	—	—	—	—	—	—
9	Worthite alloy 20	—	—	—	—	—	—	—	—	—	—	—	—	—	—
10	Durimet 20	—	—	—	—	—	—	—	—	—	—	—	—	—	—
11	25 Cr–12 Ni steel	—	—	28.6	28.2	28.2	26.8	25.5	23.1	22.6	21.8	20.5	19.2	—	—
12	25 Cr–20 Ni steel	—	—	—	28.5	—	—	26.6	24.2	23.0	21.8	21.0	19.2	—	—
13	Incoloy 800	—	—	—	—	—	—	—	—	23.6	22.4	—	19.6	18.0	—
14	Silicon Iron	—	—	—	—	—	—	—	—	—	—	—	—	—	—
15	Durichlor	—	—	—	—	—	—	—	—	—	—	—	—	—	—
	Copper Base Alloys														
16	Copper	17.0	16.7	16.5	16.0	15.6	14.0	12.0	9.0	—	—	—	—	—	—
17	Tin bronze	14.2	13.8	13.5	13.0	12.7	12.0	11.3	—	—	—	—	—	—	—
18	Aluminum bronze	22.0	—	—	17.0	17.0	15.3	11.9	5.8	—	—	—	—	—	—
19	Ampco No. 18	22.0	—	—	17.0	17.0	15.3	11.9	7.5	6.0	—	—	—	—	—
20	Red brass	—	—	—	15.0	—	—	—	—	—	—	—	—	—	—
21	Yellow brass	15.0	14.7	14.5	14.0	13.7	13.0	12.2	—	—	—	—	—	—	—
22	Muntz metal	15.0	14.7	14.5	14.0	13.7	10.0	—	—	—	—	—	—	—	—
23	Admiralty	—	—	—	15.0	15.0	13.5	7.5	—	—	—	—	—	—	—
24	Silicon bronze	—	—	—	15.0	—	—	—	—	—	—	—	—	—	—
25	70 Cu–30 Ni	20.5	20.0	19.5	18.9	18.4	17.6	16.7	15.3	—	—	—	—	—	—

Nickel Base Alloys

#	Material														
26	Monel 400	26.8	26.6	26.4	26.0	26.8	25.6	25.6	24.8	23.7	22.6	21.3	18.3	—	—
27	Monel K500	—	—	—	26.0	—	—	—	—	—	—	—	—	—	—
28	Nickel 200	31.5	30.9	30.3	30.0	29.6	28.6	27.4	25.5	22.0	18.0	15.5	14.0	—	—
29	Inconel 600	32.6	31.9	31.5	31.0	31.0	31.0	29.5	28.0	25.0	20.0	17.0	14.0	—	—
30	Hastelloy B	—	—	—	31.0	—	—	—	—	—	—	—	—	—	—
31	Hastelloy C	—	—	—	29.8	29.8	29.8	29.8	29.8	24.8	24.5	22.7	19.5	15.0	10.5
32	Hastelloy D	—	—	—	—	—	—	—	—	—	—	—	—	—	—
33	Hastelloy F	T	—	—	29.0	29.0	29.0	29.0	27.7	24.8	20.8	19.5	18.9	—	—
34	Hastelloy X	—	—	—	28.6	28.6	23.8	25.5	21.4	24.3	22.5	20.1	18.7	9.1	6.1

Superalloys

#	Material														
35	Nimonic 80	—	—	—	27.0	—	—	—	—	27.0	—	—	14.0	6.0	—
36	Inconel X750	—	—	—	31.0	30.4	29.2	28.0	26.6	25.0	23.0	20.2	16.0	—	—
37	M-252	—	—	—	29.9	29.5	28.9	28.1	27.0	26.0	24.3	22.5	—	—	—
38	Refractaloy 26	—	—	—	30.6	—	—	—	—	26.3	25.0	23.0	—	—	—
39	A-286	29.6	—	28.4	29.1	28.4	27.2	26.2	24.8	23.5	22.2	20.6	19.5	—	—
40	Discaloy	—	—	—	28.0	27.0	25.0	24.0	23.0	22.0	21.0	—	18.9	—	—
41	N155 Alloy	—	—	—	29.3	28.8	27.6	26.3	25.2	24.2	—	—	—	—	—
42	S-590	—	—	—	31.1	—	—	—	—	—	—	—	—	—	—
43	S-816	—	—	—	35.2	34.5	33.3	32.2	31.0	29.9	24.6	27.1	25.1	—	—
44	Haynes alloy 31	—	—	—	28.0	—	—	—	—	33.5	28.4	—	19.0	—	—
45	Haynes alloy 25	—	—	—	32.6	32.1	31.0	29.6	28.5	27.1	25.0	23.5	22.0	21.0	—
46	Haynes alloy 21	—	—	—	36.0	—	—	—	—	32.6	—	—	15.4	—	—
47	16 Cr–25 Ni–6 Mo	—	—	—	32.5	—	—	—	—	—	17.9	—	—	—	—
48	19-9 DL	—	—	—	29.5	28.7	27.2	26.0	24.6	23.3	22.1	20.7	—	—	—
49	17-4 PH	29.8	29.0	28.7	28.5	27.5	26.0	24.8	23.3	—	—	—	—	—	—

Other Metals

#	Material														
50	Aluminum	11.3	11.1	10.9	10.6	10.4	9.5	—	—	—	—	—	—	—	—
51	Lead	—	—	—	2.0	—	—	—	—	—	—	—	—	—	—
52	Titanium A55	17.3	—	16.3	16.0	16.0	15.8	14.6	13.6	11.2	—	—	—	—	—
53	Zirconium	—	—	14.0	11.0	—	—	—	—	—	—	—	—	—	—
54	Tantalum	—	—	—	27.0	—	—	—	—	—	—	—	—	—	—

Courtesy of Shell Development Company.

TABLE 16.8 Yield Strengths of Some Common Metals at Different Temperatures

Yield strengths, psi, ×10³

Item No.	Metals	−325°F	−200°F	−100°F	70°F	200°F	400°F	600°F	800°F	1000°F	1200°F	1400°F	1600°F	1800°F	2000°F
	Iron Base Alloys														
1	Steel, low carbon	—	—	—	40	38	35	29	23	17	10	3	—	—	—
2	Cast iron	—	—	—	—	—	—	—	—	—	—	—	—	—	—
3	Ni Resist, Type 1	—	—	—	—	—	—	—	—	—	—	—	—	—	—
4a	½ Mo	—	—	—	40	38	36	33	29	23	15	5	—	—	—
4b	1 Cr-½ Mo	—	—	—	45	41	34	28	23	21	17	6	—	—	—
4c	1¼ Cr-½ Mo	—	—	—	48	48	46	42	34	26	17	7	—	—	—
4d	2 Cr-½ Mo	—	—	—	41	—	—	—	25	20	15	7	—	—	—
4e	2¼ Cr-1 Mo	—	—	—	43	42	39	37	33	28	21	8	—	—	—
4f	5 Cr-½ Mo	—	—	—	31	28	26	24	21	18	13	7	—	—	—
4g	7 Cr-½ Mo	—	—	—	56	42	38	37	35	37	20	6	—	—	—
4h	9 Cr-1 Mo	—	—	—	45	46	44	41	35	30	18	7.0	—	—	—
5	12 Cr Type 405	148	115	94	47	—	—	—	—	24	13	5	2	—	—
6	Stainless 304	50	45	42	35	28	23	19	16	14	12	11	—	—	—
7	Stainless 316	108	63	55	42	39	35	31	27	24	21	18	—	—	—
8	Stainless 317	—	—	—	—	—	—	—	—	—	—	—	—	—	—
9	Worthite alloy 20	—	—	—	—	—	—	—	—	—	—	—	—	—	—
10	Durimet 20	—	—	—	—	—	—	—	—	—	—	—	—	—	—
11	25 Cr-12 Ni steel	—	—	—	57	55	50	46	41	36	32	26	—	—	—
12	25 Cr-20 Ni steel	119	88	63	33	32	29	28	26	23	21	18	—	—	—
13	Incoloy 800	—	—	—	43	—	36	34	33	32	29	23	14	7	—
14	Silicon iron	—	—	—	—	—	—	—	—	—	—	—	—	—	—
15	Durichlor	—	—	—	—	—	—	—	—	—	—	—	—	—	—
	Copper Base Alloys														
16	Copper	12	—	10	10	6	6	—	—	—	—	—	—	—	—
17	Tin bronze	—	11	—	20	—	—	—	—	—	—	—	—	—	—
18	Aluminum bronze	—	—	—	40	—	—	—	—	—	—	—	—	—	—
19	Ampco No. 18	—	—	—	33	33	33	33	20	10	—	—	—	—	—
20	Red brass	—	—	—	15	—	—	—	—	—	—	—	—	—	—
21	Yellow brass	—	—	—	20	19	19	—	—	—	—	—	—	—	—
22	Muntz metal	—	—	—	20	24	23	—	—	—	—	—	—	—	—
23	Admiralty	30	29	28	28	40	—	20	—	—	—	—	—	—	—
24	Silicon bronze	50	47	44	42	—	—	—	—	—	—	—	—	—	—
25	70 Cu-30 Ni	—	—	—	22	—	—	—	—	—	—	—	—	—	—

Nickel Base Alloys

No.	Material														
26	Monel 400	50	45	40	32	29	26	22	21	20	15	11	6.5	2.5	—
27	Monel K500	153	143	134	111	108	103	105	105	92	80	30	—	—	—
28	Nickel 200	28	—	27	22	22	20	20	16	13	10	7	—	—	—
29	Inconel 600	—	61	40	36	32	28	27	28	22	22	19	—	—	—
30	Hastelloy B	83	79	59	56	52	46	42	42	42	42	40	39	18	9.7
31	Hastelloy C	96	—	65	58	55	51	47	45	44	43	41	38	—	—
32	Hastelloy D	—	—	—	—	—	—	—	—	—	—	—	—	—	—
33	Hastelloy F	—	—	—	52	—	—	—	—	—	—	—	—	—	—
34	Hastelloy X	—	—	—	52	51	49	43	44	42	40	38	26	16	8.0

Superalloys

No.	Material														
35	Nimonic 80	131	125	120	87	—	81	110	108	—	77	68	30	8	—
36	Inconel X-750	—	—	—	122	116	112	117	115	107	105	92	74	55	—
37	M-252	—	—	—	122	120	118	—	—	111	108	104	70	—	—
38	Refractaloy 26	122	115	110	91	—	—	—	—	—	—	—	47	—	—
39	A-286	—	—	—	100	95	94	93	92	85	89	85	—	—	—
40	Discaloy	—	—	—	106	105	104	100	94	91	85	52	—	—	—
41	N155	—	—	—	57	56	55	53	51	48	43	35	26	15	10
42	S-590	—	—	—	75	78	82	83	82	78	70	58	36	21	14
43	S-816	110	95	85	67	58	46	44	44	44	44	40	35	23	13
44	Haynes alloy 31	—	—	—	87	84	80	75	71	65	58	42	38	26	13
45	Haynes alloy 25	—	—	—	70	67	64	60	56	52	47	41	34	24	—
46	Haynes alloy 21	—	—	—	82	77	70	62	55	53	51	47	40	20	10
47	16 Cr–25 Ni–6 Mo	—	—	—	46	45	41	36	33	30	31	34	26	13	—
48	19-9 DL	—	—	—	69	65	56	50	45	40	37	33	—	—	—
49	17-4 PH	—	—	—	—	—	—	—	—	—	—	33	—	—	—

Other Metals

No.	Material														
50	Aluminum 5052-H34	30	27	25	25	25	15	5	—	—	—	—	—	—	—
51	Lead	—	—	—	1.3	—	—	—	—	—	—	—	—	—	—
52	Titanium A55	—	—	74	63	47	33	25	20	14	—	—	—	—	—
53	Zirconium	—	—	—	28	24	14	8	7	17	—	—	—	—	—
54	Tantalum	134	—	—	38	—	—	23	21	—	15	13	—	—	—

Courtesy of Shell Development Company.

TABLE 16.9 Thermal Conductivities of Some Common Metals at Different Temperatures

Btu/hr ft² °F ft

Item No.	Metals	−325°F	−200°F	−100°F	70°F	200°F	400°F	600°F	800°F	1000°F	1200°F	1400°F	1600°F	1800°F	2000°F
	Iron Base Alloys														
1	Steel, low carbon	—	25.8	—	30	27.6	26.8	25.5	24.5	23.2	22.2	21.1	—	—	—
2	Cast iron	—	—	—	—	—	—	—	—	—	—	—	—	—	—
3	Ni Resist, Type 1	—	—	—	—	—	—	—	—	—	—	—	—	—	—
4a	½ Mo	—	—	—	28	25.8	24.6	23.1	21.6	20.3	—	—	—	—	—
4b	1 Cr-½ Mo	—	—	—	19.2	19.1	18.7	18.5	18.2	18.0	17.7	17.5	—	—	—
4c	1¼ Cr-½ Mo	—	—	—	18.8	18.3	17.9	17.5	17.0	16.9	—	—	—	—	—
4d	2 Cr-½ Mo	—	—	—	17.1	17.0	16.9	16.9	16.9	16.9	16.7	16.7	—	—	—
4e	2¼ Cr-1 Mo	—	—	—	15.0	15.1	15.1	15.2	15.4	15.5	15.9	—	—	—	—
4f	5 Cr-½ Mo	—	—	—	15.5	15.6	15.8	16.0	16.2	16.3	16.4	16.6	—	—	—
4g	7 Cr-½ Mo	—	—	—	—	—	—	—	—	—	—	—	—	—	—
4h	9 Cr-1 Mo	—	—	—	14.8	15.0	15.2	15.4	15.6	15.8	—	—	—	—	—
5	12 Cr Type 410	—	—	—	13.0	14.4	14.7	15.4	15.9	—	—	—	—	—	—
6	Stainless 304	5.0	6.7	8.1	9.4	10.0	10.9	11.8	12.7	13.7	14.6	15.5	—	—	—
7	Stainless 316	5.0	6.7	8.1	9.4	—	—	—	13.0	13.0	—	—	—	—	—
8	Stainless 317	—	—	—	—	—	—	—	—	—	—	—	—	—	—
9	Worthite alloy 20	—	—	—	—	—	—	—	—	—	—	—	—	—	—
10	Durimet 20	—	—	—	—	—	—	—	—	—	—	—	—	—	—
11	25 Cr-12 Ni steel	—	—	—	8.0	8.50	9.50	10.7	11.7	12.9	14.1	15.3	—	—	—
12	25 Cr-20 Ni steel	—	—	—	8.0	8.00	11.0	—	13.0	13.0	—	—	—	—	—
13	Incoloy 800	—	—	—	8.0	—	—	—	—	—	—	—	—	—	—
14	Silicon iron	—	—	—	—	—	—	—	—	—	—	—	—	—	—
15	Durichlor	—	—	—	—	—	—	—	—	—	—	—	—	—	—
	Copper Base Alloys														
16	Copper	333	225	225	225	222	219	216	214	209	207	205	—	—	—
17	Tin bronze	—	—	—	—	—	—	—	—	—	—	—	—	—	—
18	Aluminum bronze	—	—	—	32.7	35.3	41.8	49.0	55.5	—	—	—	—	—	—
19	Ampco No. 18	—	—	—	—	—	—	—	—	—	—	—	—	—	—
20	Red brass	—	—	—	92	—	—	—	—	—	—	—	—	—	—
21	Yellow brass	35	47	56	69	—	—	—	—	—	—	—	—	—	—
22	Muntz metal	—	—	—	73	—	—	—	—	—	—	—	—	—	—
23	Admiralty	—	48	55	64	—	—	—	—	—	—	—	—	—	—
24	Silicon bronze	—	—	—	19	—	—	—	—	—	—	—	—	—	—
25	70 Cu-30 Ni	—	—	—	17	—	—	—	—	—	—	—	—	—	—

Nickel Base Alloys

#	Material														
26	Monel 400	9.4	10.8	11.6	12.6	13.8	16.0	18.0	20.0	22.0	24.6	25.8	27.6	30.0	—
27	Monel K500	—	—	—	10.1	11.3	13.0	14.8	16.6	18.3	20.1	21.8	—	—	—
28	Nickel 200	—	—	—	32.5	31.9	31.2	30.9	30.8	30.7	30.6	—	—	—	—
29	Inconel 600	—	—	—	8.6	9.1	10.1	11.1	12.1	13.7	14.3	15.5	16.7	—	—
30	Hastelloy B	—	—	—	6.5	—	7.1	7.6	8.1	9.0	9.7	—	—	—	—
31	Hastelloy C	—	—	—	7.3	5.60	6.55	7.45	8.40	9.30	10.2	—	—	—	—
32	Hastelloy D	—	—	—	—	—	—	—	—	—	—	—	—	—	—
33	Hastelloy F	—	—	—	9.4	—	—	—	—	—	—	—	—	—	—
34	Hastelloy X	—	—	—	5.25	6.33	7.30	8.45	9.80	11.2	12.4	13.7	15.0	16.3	—

Superalloys

#	Material														
35	Nimonic 80	—	5.83	6.16	7.0	7.0	8.0	9.0	10.0	—	—	—	—	14.0	—
36	Inconel X750	—	—	—	6.92	7.42	8.17	9.08	10.0	10.9	11.9	12.8	13.7	—	—
37	M-252	—	—	—	6.83	7.35	8.20	9.00	9.67	10.5	11.3	12.3	12.5	—	—
38	Refractaloy 26	—	—	—	—	7.80	7.9	8.0	8.0	8.2	—	—	—	—	—
39	A-286	—	—	—	7.75	—	7.3	—	—	—	14.0	14.2	15.0	—	—
40	Discaloy	—	—	—	8.8	8.20	9.20	9.95	10.9	12.0	13.1	—	—	—	—
41	N155	—	—	—	—	—	—	7.5	10.2	10.4	—	13.1	—	—	—
42	S-590	—	—	—	7.2	—	8.0	9.1	11.3	12.5	13.1	—	—	—	—
43	S-816	—	—	—	8.6	7.90	9.0	10.2	11.3	12.4	—	—	—	—	—
44	Haynes alloy 31	—	—	—	5.41	—	8.5	10.1	10.5	11.5	12.5	13.7	—	—	—
45	Haynes alloy 25	—	—	—	—	6.20	7.48	8.73	9.95	11.2	12.3	—	15.1	—	—
46	Haynes alloy 21	—	—	—	—	—	8.40	9.30	10.0	11.4	—	—	—	—	—
47	16 Cr–25 Ni–6 Mo	—	—	—	9.0	—	—	—	—	—	—	13.0	—	—	—
48	19-9 DL	—	—	—	8.2	8.30	9.10	9.90	10.7	11.5	12.3	—	—	—	—
49	17-4 PH	—	—	—	10.2	—	—	—	—	—	—	—	—	—	—

Other Metals

#	Material														
50	Aluminum	—	124	—	131	133	137	141	—	—	—	—	—	—	—
51	Lead	—	—	—	20	—	—	—	—	—	—	—	—	—	—
52	Titanium A55	—	—	11.8	11.5	10.9	10.4	10.5	10.7	11.3	12.1	—	—	—	—
53	Zirconium	—	—	—	14.0	—	—	—	—	—	—	—	—	—	—
54	Tantalum	—	—	—	31.8	—	—	—	—	—	—	—	—	—	—

Courtesy of Shell Development Company.

Ferrous Alloys

TABLE 16.10   Analyses of Pig Irons or Cast Irons

Trade Name	Total C (%)	Si (%)	Maximum S (%)	P (%)	Mn (%)
Low phosphorus (acid open-hearth)	—	0.50–3.00	0.035	0.035 maximum	0.75–1.25
Intermediate low phosphorus	4.00–4.50	1.00–3.00	0.05	0.036–0.075	0.75–1.25
Bessemer	3.50–4.00	1.00–3.00	0.05	0.076–0.100	1.00–1.25
Malleable	3.75–4.50	0.75–3.50	0.05	0.101–0.300	0.50–1.25
Basic, northern	3.50–4.00	1.00–1.50	0.05	0.400 maximum	1.01–2.00
Basic, southern	3.50–4.00	1.00–1.50	0.05	0.700–0.900	0.40–0.75
Foundry, northern, low phosphorus	4.00–4.50	1.75–3.50	0.05	0.300–0.500	0.50–1.25
Foundry, northern, high phosphorus	4.00–4.50	1.75–3.50	0.05	0.501–0.700	0.50–1.25
Foundry, southern	3.50–4.00	1.75–3.50	0.05	0.700–0.900	0.40–0.75
Charcoal iron, southern	—	0.50–3.00	0.035	0.035 maximum	0.40–1.00
Silvery pig iron	0.75–1.00	5.00–17.00	0.05	0.300 maximum	1.00–2.00

TABLE 16.11   AISI–SAE System of Designations

Numerals and Digits[a]	Type of Steel and/or Nominal Alloy Content
	Carbon Steels
10xx	Plain carbon (Mn 1.00% maximum)
11xx	Resulfurized
12xx	Resulfurized and rephosphorized
15xx	Plain carbon (maximum Mn range, 1.00–1.65%)
	Manganese Steels
13xx	Mn 1.75
	Nickel Steels
23xx	Ni 3.50
25xx	Ni 5.00
	Nickel-Chromium Steels
31xx	Ni 1.25; Cr 0.65 and 0.80
32xx	Ni 1.75; Cr 1.07
33xx	Ni 3.50; Cr. 1.50 and 1.57
34xx	Ni 3.00; Cr 0.77
	Molybdenum Steels
40xx	Mo 0.20 and 0.25
44xx	Mo 0.40 and 0.52
	Chromium-Molybdenum Steels
41xx	Cr 0.50, 0.80 and 0.95; Mo 0.12, 0.20, 0.25, and 0.30

TABLE 16.11 *(Continued)*

Numerals and Digits[a]	Type of Steel and/or Nominal Alloy Content
	Nickel-Chromium-Molybdenum Steels
43xx	Ni 1.82; Cr 0.50 and 0.80; Mo 0.25
43BVxx	Ni 1.82; Cr 0.50; Mo 0.12 and 0.25; V 0.03 min
47xx	Ni 1.05; Cr 0.45; Mo 0.20 and 0.35
81xx	Ni 0.30; Cr 0.40; Mo 0.12
86xx	Ni 0.55; Cr 0.50; Mo 0.20
87xx	Ni 0.55; Cr 0.50; Mo 0.25
88xx	Ni 0.55; Cr 0.50; Mo 0.35
93xx	Ni 3.25; Cr 1.20; Mo 0.12
94xx	Ni 0.45; Cr 0.40; Mo 0.12
97xx	Ni 0.55; Cr 0.20; Mo 0.20
98xx	Ni 1.00; Cr 0.80; Mo 0.25
	Nickel-Molybdenum Steels
46xx	Ni 0.85 and 1.82; Mo 0.20 and 0.25
48xx	Ni 3.50; Mo 0.25
	Chromium Steels
50xx	Cr 0.27, 0.40, 0.50, and 0.65
51xx	Cr 0.80, 0.87, 0.92, 0.95, 1.00, and 1.05
	Chromium Steels
50xxx	Cr 0.50 ⎫
51xxx	Cr 1.02 ⎬ C 1.00 minimum
52xxx	Cr 1.45 ⎭
	Chromium-Vanadium Steels
61xx	Cr 0.60, 0.80, and 0.95; V 0.10 and 0.15 minimum
	Tungsten-Chromium Steel
72xx	W 1.75; Cr 0.75
	Silicon-Manganese Steels
92xx	Si 1.40 and 2.00; Mn 0.65, 0.82, and 0.85; Cr 0.00 and 0.65
	High-Strength Low-Alloy Steels
9xx	Various SAE grades
	Boron Steels
xxBxx	B denotes boron steel
	Leaded Steels
xxLxx	L denotes leaded steel

Courtesy of ASM *Metals Handbook, Desk Edition.* H. E. Boyer and T. L. Gall, (eds.); ASM, Metals Park, Ohio, 1985.

[a]*xx* in last two (or three) digits of designations indicates that carbon content (in hundredths of a percent) is to be inserted.

TABLE 16.12 Influence of Alloying Elements upon Properties of Steel[a]

	Carbon	Manganese	Silicon	Aluminum	Nickel	Chromium	Molybdenum	Vanadium	Tungsten	Cobalt	Copper	Sulfur	Phosphorus	Titanium	Tantalum	Niobium
Yield strength	↗	↗	↗	—	↗	↗	↗	↗								
Tensile strength	⇗	⇗	⇗	—	↗	⇗	↗	↗	↗	↗	↗	—	↗	↗	↗	↗
Elongation	⇙	↙	↙	—	↙	↙	↙	—	↙	↙	⇙	—	↙			
Tensile strength at elevated temperature	↗	—	—	—	—	⇗	⇗	↗	↗	↗	—	—	↗			
Creep strength	↗	↗	↗	↙	↗	↗	⇗									
Fatigue strength	↗	—	—	—	—	↗	⇗	⇗	↗	↗						
Ac1 point	—	⇙	⇗	↗	⇙	⇗	↗	↗	↗							
Ac3 point	—	↙	↗	↗	↙	↗	↗	↗	↗	↙	↙	—	↗	↗	↗	↗
Austenite field	—	↗	↙	↙	↗	⇙	↙	↙	↙	↗	↗	—	↙	↙	↙	↙
Grain growth	↗	⇗	↗	⇙	↙	↗	—	⇙	—	↙	—	—	↗	↙		
Susceptibility to overheating	—	↗	↙	—	↙	↙	—	↙	↙	↙	—	—	↗			
Oxidation resistance	—	—	⇗	⇗	—	⇗	—	—	—	↗						
Red shortness	—	—	—	—	—	—	—	—	—	—	—	⇗				
Critical cooling rate	—	↙	↙	—	⇙	↙	↙	↙	↙	↗						
Hardenability	⇗	⇗	⇗	—	⇗	⇗	↗	↗	↗							
Hardness	⇗	↗	↗	—	↗	↗	↗	↗	↗	↗	↗	—	↗	↗	↗	↗
Tempering stability	↙	—	↗	—	↗	↗	⇗	⇗	↗	⇗	—	—	—	↗	↗	↗
Carbide formation	⇗	—	—	—	—	⇗	↗	⇗	↗	—	—	—	—	⇗	⇗	⇗

Elements taken individually in considering their influence.

[a] Influence of element is (↗) increased, (⇗) greatly increased, (↙) decreased, and (⇙) greatly decreased.

TABLE 16.13A AISI–SAE Standard Carbon Steels

AISI–SAE No.	UNS No.	Composition[a] (%)			
		C	Mn	Maximum P	Maximum S
		Nonresulfurized Grades			
MANGANESE 1.00% MAXIMUM					
1005[b]	G10050	0.06 max	0.35 max	0.040	0.050
1006[b]	G10060	0.08 max	0.25–0.40	0.040	0.050
1008	G10080	0.10 max	0.30–0.50	0.040	0.050
1010	G10100	0.08–0.13	0.30–0.60	0.040	0.050
1012	G10120	0.10–0.15	0.30–0.60	0.040	0.050
1015	G10150	0.13–0.18	0.30–0.60	0.040	0.050
1016	G10160	0.13–0.18	0.60–0.90	0.040	0.050
1017	G10170	0.15–0.20	0.30–0.60	0.040	0.050
1018	G10180	0.15–0.20	0.60–0.90	0.040	0.050
1019	G10190	0.15–0.20	0.70–1.00	0.040	0.050
1020	G10200	0.18–0.23	0.30–0.60	0.040	0.050
1021	G10210	0.18–0.23	0.60–0.90	0.040	0.050
1022	G10220	0.18–0.23	0.70–1.00	0.040	0.050
1023	G10230	0.20–0.25	0.30–0.60	0.040	0.050
1025	G10250	0.22–0.28	0.30–0.60	0.040	0.050
1026	G10260	0.22–0.28	0.60–0.90	0.040	0.050
1029	G10290	0.25–0.31	0.60–0.90	0.040	0.050
1030	G10300	0.28–0.34	0.60–0.90	0.040	0.050
1035	G10350	0.32–0.38	0.60–0.90	0.040	0.050
1037	G10370	0.32–0.38	0.70–1.00	0.040	0.050
1038	G10380	0.35–0.42	0.60–0.90	0.040	0.050
1039	G10390	0.37–0.44	0.70–1.00	0.040	0.050
1040	G10400	0.37–0.44	0.60–0.90	0.040	0.050
1042	G10420	0.40–0.47	0.60–0.90	0.040	0.050
1043	G10430	0.40–0.47	0.70–1.00	0.040	0.050
1044	G10440	0.43–0.50	0.30–0.60	0.040	0.050
1045	G10450	0.43–0.50	0.60–0.90	0.040	0.050

TABLE 16.13A *(Continued)*

AISI–SAE No.	UNS No.	Composition[a] (%)			
		C	Mn	Maximum P	Maximum S
1046	G10460	0.43–0.50	0.70–1.00	0.040	0.050
1049	G10490	0.46–0.53	0.60–0.90	0.040	0.050
1050	G10500	0.48–0.55	0.60–0.90	0.040	0.050
1053	G10530	0.48–0.55	0.70–1.00	0.040	0.050
1055	G10550	0.50–0.60	0.60–0.90	0.040	0.050
1059[b]	G10590	0.55–0.65	0.50–0.80	0.040	0.050
1060	G10600	0.55–0.65	0.60–0.90	0.040	0.050
1064[b]	G10640	0.60–0.70	0.50–0.80	0.040	0.050
1065[b]	G10650	0.60–0.70	0.60–0.90	0.040	0.050
1069[b]	G10690	0.65–0.75	0.40–0.70	0.040	0.050
1070	G10700	0.65–0.75	0.60–0.90	0.040	0.050
1078	G10780	0.72–0.85	0.30–0.60	0.040	0.050
1080	G10800	0.75–0.88	0.60–0.90	0.040	0.050
1084	G10840	0.80–0.93	0.60–0.90	0.040	0.050
1086[b]	G10860	0.80–0.93	0.30–0.50	0.040	0.050
1090	G10900	0.85–0.98	0.60–0.90	0.040	0.050
1095	G10950	0.90–1.03	0.30–0.50	0.040	0.050

MANGANESE MAXIMUM OVER 1.00%

AISI–SAE No.	UNS No.	C	Mn	Maximum P	Maximum S
1513	G15130	0.10–0.16	1.10–1.40	0.040	0.050
1522	G15220	0.18–0.24	1.10–1.40	0.040	0.050
1524	G15240	0.19–0.25	1.35–1.65	0.040	0.050
1526	G15260	0.22–0.29	1.10–1.40	0.040	0.050
1527	G15270	0.22–0.29	1.20–1.50	0.040	0.050
1541	G15410	0.36–0.44	1.35–1.65	0.040	0.050
1548	G15480	0.44–0.52	1.10–1.40	0.040	0.050
1551	G15510	0.45–0.56	0.85–1.15	0.040	0.050
1552	G15520	0.47–0.55	1.20–1.50	0.040	0.050
1561	G15610	0.55–0.65	0.75–1.05	0.040	0.050
1566	G15660	0.60–0.71	0.85–1.15	0.040	0.050

Free-Machining Grades

RESULFURIZED

AISI–SAE No.	UNS No.	C	Mn	Maximum P	Maximum S
1110	G11100	0.08–0.13	0.30–0.60	0.040	0.08–0.13
1117	G11170	0.14–0.20	1.00–1.30	0.040	0.08–0.13
1118	G11180	0.14–0.20	1.30–1.60	0.040	0.08–0.13
1137	G11370	0.32–0.39	1.35–1.65	0.040	0.08–0.13
1139	G11390	0.35–0.43	1.35–1.65	0.040	0.13–0.20
1140	G11400	0.37–0.44	0.70–1.00	0.040	0.08–0.13
1141	G11410	0.37–0.45	1.35–1.65	0.040	0.08–0.13
1144	G11440	0.40–0.48	1.35–1.65	0.040	0.24–0.33
1146	G11460	0.42–0.49	0.70–1.00	0.040	0.08–0.13
1151	G11510	0.48–0.55	0.70–1.00	0.040	0.08–0.13

RESULFURIZED AND REPHOSPHORIZED

AISI–SAE No.	UNS No.	C	Mn	Maximum P	Maximum S
1211	G12110	0.13 max	0.60–0.90	0.07–0.12	0.10–0.15
1212	G12120	0.13 max	0.70–1.00	0.07–0.12	0.16–0.23
1213	G12130	0.13 max	0.70–1.00	0.07–0.12	0.24–0.33
1215	G12150	0.09 max	0.75–1.05	0.04–0.09	0.26–0.35
12L14[c]	G12144	0.15 max	0.85–1.15	0.04–0.09	0.26–0.35

Source: Steel Products Manual, American Iron and Steel Institute, August 1977. (Reviewed April 1985)

[a] The following notes refer to boron, copper, lead, and silicon additions: Boron: Standard killed carbon steels, which are generally fine grain, may be produced with a boron treatment addition to improve hardenability. Such steels are produced to a range of 0.0005–0.003% B. These steels are identified by inserting the letter "B" between the second and third numerals of the AISI or SAE number, e.g., 10B46. Copper: When copper is required, 0.20% minimum is generally specified. Lead: Standard carbon steels can be produced with a lead range of 0.15–0.35% to improve machinability. Such steels are identified by inserting the letter "L" between the second and third numerals of the AISI or SAE number, e.g., 12L15 and 10L45. Silicon: It is not common practice to produce the 12XX series of resulfurized and rephosphorized steels to specified limits for silicon because of its adverse effect on machinability. When silicon ranges or limits are required for resulfurized or nonresulfurized steels, however, these values apply: a range of 0.08% Si for Si maximum up to 0.15% inclusive, a range of 0.10% Si for Si maximum over 0.15–0.20% inclusive, a range of 0.15% Si for Si maximum over 0.20–0.30% inclusive, and a range of 0.20% Si for Si maximum over 0.30–0.60% inclusive. Example: Si maximum is 0.25%, range is 0.10–0.25%.

[b] Standard grades for wire rods and wire only.

[c] 0.15–0.35% Pb.

TABLE 16.13B AISI–SAE Standard Alloy Steels

AISI–SAE No.	UNS No.	Composition[a] (%)							
		C	Mn	Maximum P	Maximum S	Si	Ni	Cr	Mo
1330	G13300	0.28–0.33	1.60–1.90	0.035	0.040	0.15–0.35	—	—	—
1335	G13350	0.33–0.38	1.60–1.90	0.035	0.040	0.15–0.35	—	—	—
1340	G13400	0.38–0.43	1.60–1.90	0.035	0.040	0.15–0.35	—	—	—
1345	G13450	0.43–0.48	1.60–1.90	0.035	0.040	0.15–0.35	—	—	—
4023	G40230	0.20–0.25	0.70–0.90	0.035	0.040	0.15–0.35	—	—	0.20–0.30
4024	G40240	0.20–0.25	0.70–0.90	0.035	0.035–0.050	0.15–0.35	—	—	0.20–0.30
4027	G40270	0.25–0.30	0.70–0.90	0.035	0.040	0.15–0.35	—	—	0.20–0.30
4028	G40280	0.25–0.30	0.70–0.90	0.035	0.035–0.050	0.15–0.35	—	—	0.20–0.30
4037	G40370	0.35–0.40	0.70–0.90	0.035	0.040	0.15–0.35	—	—	0.20–0.30
4047	G40470	0.45–0.50	0.70–0.90	0.035	0.040	0.15–0.35	—	—	0.20–0.30
4118	G41180	0.18–0.23	0.70–0.90	0.035	0.040	0.15–0.35	—	0.40–0.60	0.08–0.15
4130	G41300	0.28–0.33	0.40–0.60	0.035	0.040	0.15–0.35	—	0.80–1.10	0.15–0.25
4137	G41370	0.35–0.40	0.70–0.90	0.035	0.040	0.15–0.35	—	0.80–1.10	0.15–0.25
4140	G41400	0.38–0.43	0.75–1.00	0.035	0.040	0.15–0.35	—	0.80–1.10	0.15–0.25
4142	G41420	0.40–0.45	0.75–1.00	0.035	0.040	0.15–0.35	—	0.80–1.10	0.15–0.25
4145	G41450	0.43–0.48	0.75–1.00	0.035	0.040	0.15–0.35	—	0.80–1.10	0.15–0.25
4147	G41470	0.45–0.50	0.75–1.00	0.035	0.040	0.15–0.35	—	0.80–1.10	0.15–0.25
4150	G41500	0.48–0.53	0.75–1.00	0.035	0.040	0.15–0.35	—	0.80–1.10	0.15–0.25
4161	G41610	0.56–0.64	0.75–1.00	0.035	0.040	0.15–0.35	—	0.70–0.90	0.25–0.35
4320	G43200	0.17–0.22	0.45–0.65	0.035	0.040	0.15–0.35	1.65–2.00	0.40–0.60	0.20–0.30
4340	G43400	0.38–0.43	0.60–0.80	0.035	0.040	0.15–0.35	1.65–2.00	0.70–0.90	0.20–0.30
E4340[b]	G43406	0.38–0.43	0.65–0.85	0.025	0.025	0.15–0.35	1.65–2.00	0.70–0.90	0.20–0.30
4615	G46150	0.13–0.18	0.45–0.65	0.035	0.040	0.15–0.35	1.65–2.00	—	0.20–0.30
4620	G46200	0.17–0.22	0.45–0.65	0.035	0.040	0.15–0.35	1.65–2.00	—	0.20–0.30
4626	G46260	0.24–0.29	0.45–0.65	0.035	0.040	0.15–0.35	0.70–1.00	—	0.15–0.25
4720	G47200	0.17–0.22	0.50–0.70	0.035	0.040	0.15–0.35	0.90–1.20	0.35–0.55	0.15–0.25
4815	G48150	0.13–0.18	0.40–0.60	0.035	0.040	0.15–0.35	3.25–3.75	—	0.20–0.30
4817	G48170	0.15–0.20	0.40–0.60	0.035	0.040	0.15–0.35	3.25–3.75	—	0.20–0.30
4820	G48200	0.18–0.23	0.50–0.70	0.035	0.040	0.15–0.35	3.25–3.75	—	0.20–0.30
5117	G51170	0.15–0.20	0.70–0.90	0.035	0.040	0.15–0.35	—	0.70–0.90	—
5120	G51200	0.17–0.22	0.70–0.90	0.035	0.040	0.15–0.35	—	0.70–0.90	—
5130	G51300	0.28–0.33	0.70–0.90	0.035	0.040	0.15–0.35	—	0.80–1.10	—
5132	G51320	0.30–0.35	0.60–0.80	0.035	0.040	0.15–0.35	—	0.75–1.00	—
5135	G51350	0.33–0.38	0.60–0.80	0.035	0.040	0.15–0.35	—	0.80–1.05	—
5140	G51400	0.38–0.43	0.70–0.90	0.035	0.040	0.15–0.35	—	0.70–0.90	—
5150	G51500	0.48–0.53	0.70–0.90	0.035	0.040	0.15–0.35	—	0.70–0.90	—
5155	G51550	0.51–0.59	0.70–0.90	0.035	0.040	0.15–0.35	—	0.70–0.90	—
5160	G51600	0.56–0.64	0.75–1.00	0.035	0.040	0.15–0.35	—	0.70–0.90	—
E51100[b]	G51986	0.98–1.10	0.25–0.45	0.025	0.025	0.15–0.35	—	0.90–1.15	—
E52100[b]	G52986	0.98–1.10	0.25–0.45	0.025	0.025	0.15–0.35	—	1.30–1.60	—
6118	G61180	0.16–0.21	0.50–0.70	0.035	0.040	0.15–0.35	—	0.50–0.70	0.10–0.15
6150	G61500	0.48–0.53	0.70–0.90	0.035	0.040	0.15–0.35	—	0.80–1.10	0.15 V min

Grade	C	Mn	P	S	Si	Ni	Cr	Mo
8615	0.13–0.18	0.70–0.90	0.035	0.040	0.15–0.35	0.40–0.70	0.40–0.60	0.15–0.25
8617	0.15–0.20	0.70–0.90	0.035	0.040	0.15–0.35	0.40–0.70	0.40–0.60	0.15–0.25
8620	0.18–0.23	0.70–0.90	0.035	0.040	0.15–0.35	0.40–0.70	0.40–0.60	0.15–0.25
8622	0.20–0.25	0.70–0.90	0.035	0.040	0.15–0.35	0.40–0.70	0.40–0.60	0.15–0.25
8625	0.23–0.28	0.70–0.90	0.035	0.040	0.15–0.35	0.40–0.70	0.40–0.60	0.15–0.25
8627	0.25–0.30	0.70–0.90	0.035	0.040	0.15–0.35	0.40–0.70	0.40–0.60	0.15–0.25
8630	0.28–0.33	0.70–0.90	0.035	0.040	0.15–0.35	0.40–0.70	0.40–0.60	0.15–0.25
8637	0.35–0.40	0.75–1.00	0.035	0.040	0.15–0.35	0.40–0.70	0.40–0.60	0.15–0.25
8640	0.38–0.43	0.75–1.00	0.035	0.040	0.15–0.35	0.40–0.70	0.40–0.60	0.15–0.25
8642	0.40–0.45	0.75–1.00	0.035	0.040	0.15–0.35	0.40–0.70	0.40–0.60	0.15–0.25
8645	0.43–0.48	0.75–1.00	0.035	0.040	0.15–0.35	0.40–0.70	0.40–0.60	0.15–0.25
8655	0.51–0.59	0.75–1.00	0.035	0.040	0.15–0.35	0.40–0.70	0.40–0.60	0.15–0.25
8720	0.18–0.25	0.70–0.90	0.035	0.040	0.15–0.35	0.40–0.70	0.40–0.60	0.20–0.30
8740	0.38–0.43	0.75–1.00	0.035	0.040	0.15–0.35	0.40–0.70	0.40–0.60	0.20–0.30
8822	0.20–0.25	0.75–1.00	0.035	0.040	0.15–0.35	0.40–0.70	0.40–0.60	0.30–0.40
9260	0.56–0.64	0.75–1.00	0.035	0.040	1.80–2.20	—	—	—
Standard Boron Grades (0.005–0.003% B)								
50B44	0.43–0.48	0.75–1.00	0.035	0.040	0.15–0.35	—	0.40–0.60	—
50B46	0.44–0.49	0.75–1.00	0.035	0.040	0.15–0.35	—	0.20–0.35	—
50B50	0.48–0.53	0.75–1.00	0.035	0.040	0.15–0.35	—	0.40–0.60	—
50B60	0.56–0.64	0.75–1.00	0.035	0.040	0.15–0.35	—	0.40–0.60	—
51B60	0.56–0.64	0.75–1.00	0.035	0.040	0.15–0.35	—	0.70–0.90	—
81B45	0.43–0.48	0.75–1.00	0.035	0.040	0.15–0.35	0.20–0.40	0.35–0.55	0.08–0.15
94B17	0.15–0.20	0.75–1.00	0.035	0.040	0.15–0.35	0.30–0.60	0.30–0.50	0.08–0.15
94B30	0.28–0.33	0.75–1.00	0.035	0.040	0.15–0.35	0.30–0.60	0.30–0.50	0.08–0.15

Source: Steel Products Manual, American Iron and Steel Institute, August 1977. (Reviewed April 1985)

[a] Small quantities of certain elements are present that are not specified or required. These incidental elements may be present to the following maximum amounts: Cu, 0.35%; Ni, 0.25%; Cr, 0.20%; and Mo, 0.60%. Standard alloy steels can also be produced with a lead range of 0.15–0.35%. Such steels are identified by inserting the letter "L" between the second and third numerals of the AISI or SAE number, e.g., 41L40.

[b] Electric furnace practice steel.

Table 16.14 (Part I and Part II) is offered as a guide to show the potential user what to expect of a given grade of steel in the indicated condition. Data were obtained from specimens 0.505 in. in diameter machined from 1-in. rounds; gage lengths were 2 in. Average properties of hot-rolled, normalized, and annealed material are listed (Part I), while properties of quenched and tempered grades are for single heats (Part II). Sources of the data are Bethlehem Steel Corporation and Republic Steel Corporation.

Because of the many variables that affect a steel's properties, however, these listed properties should not be considered either as average or typical. Both strengths and ductilities may range up and down from the values given, depending on the compositions of individual heats of the same grade, section sizes, and internal structures. Properties of carbon steels and many alloy steels are also affected by residual elements (particularly nickel, chromium, and molybdenum) even though their amounts are limited to maximums by AISI and SAE specifications.

Fine-grained steels normally have better impact strength than coarse-grained types, a factor that should be considered when reviewing the results of Izod tests. Hardness values are not always related to corresponding tensile strengths. In particular, this effect occurs with carbon steels because they are shallow hardening. Hardness tests were made on surfaces, and these hardnesses will not reflect the tensile strengths obtained with specimens representing bar centers. (Center hardnesses are usually lower than surface hardnesses.)

Hot-rolled properties for alloy steels are not given because these grades are customarily heat treated. Because the samples were small enough to assure full quenching, values indicate strengths and ductilities which may be obtained with hardened fine-grained steels of a similar section size at room temperature.

TABLE 16.14 Properties of Selected Carbon and Alloy Steels

Part I. Hot Rolled, Normalized, and Annealed

AISI No.[a]	Treatment	Yield Strength (psi)	Tensile Strength (psi)	Elongation (%)	Reduction in Area (%)	Hardness (Bhn)	Impact Strength Izod (ft-lb)
1015	As rolled	45,500	61,000	39.0	61.0	126	81.5
	Normalized (1700°F)	47,000	61,500	37.0	69.6	121	85.2
	Annealed (1600°F)	41,250	56,000	37.0	69.7	111	84.8
1020	As rolled	48,000	65,000	36.0	59.0	143	64.0
	Normalized (1600°F)	50,250	64,000	35.8	67.9	131	86.8
	Annealed (1600°F)	42,750	57,250	36.5	66.0	111	91.0
1022	As rolled	52,000	73,000	35.0	67.0	149	60.0
	Normalized (1700°F)	52,000	70,000	34.0	67.5	143	86.5
	Annealed (1600°F)	46,000	65,250	35.0	63.6	137	89.0
1030	As rolled	50,000	80,000	32.0	57.0	179	55.0
	Normalized (1700°F)	50,000	75,500	32.0	60.8	149	69.0
	Annealed (1550°F)	49,500	67,250	31.2	57.9	126	51.2
1040	As rolled	60,000	90,000	25.0	50.0	201	36.0
	Normalized (1650°F)	54,250	85,500	28.0	54.9	170	48.0
	Annealed (1450°F)	51,250	75,250	30.2	57.2	149	32.7
1050	As rolled	60,000	105,000	20.0	40.0	229	23.0
	Normalized (1650°F)	62,000	108,500	20.0	39.4	217	20.0
	Annealed (1450°F)	53,000	92,250	23.7	39.9	187	12.5
1060	As rolled	70,000	118,000	17.0	34.0	241	13.0
	Normalized (1650°F)	61,000	112,500	18.0	37.2	229	9.7
	Annealed (1450°F)	54,000	90,750	22.5	38.2	179	8.3
1080	As rolled	85,000	140,000	12.0	17.0	293	5.0
	Normalized (1650°F)	76,000	146,500	11.0	20.6	293	5.0
	Annealed (1450°F)	54,500	89,250	24.7	45.0	174	4.5
1095	As rolled	83,000	140,000	9.0	18.0	293	3.0
	Normalized (1650°F)	72,500	147,000	9.5	13.5	293	4.0
	Annealed (1450°F)	55,000	95,250	13.0	20.6	192	2.0
1117	As rolled	44,300	70,600	33.0	63.0	143	60.0
	Normalized (1650°F)	44,000	67,750	33.5	63.8	137	62.8
	Annealed (1575°F)	40,500	62,250	32.8	58.0	121	69.0
1118	As rolled	45,900	75,600	32.0	70.0	149	80.0
	Normalized (1700°F)	46,250	69,250	33.5	65.9	143	76.3
	Annealed (1450°F)	41,250	65,250	34.5	66.8	131	78.5
1137	As rolled	55,000	91,000	28.0	61.0	192	61.0
	Normalized (1650°F)	57,500	97,000	22.5	48.5	197	47.0
	Annealed (1450°F)	50,000	84,750	26.8	53.9	174	36.8
1141	As rolled	52,000	98,000	22.0	38.0	192	8.2
	Normalized (1650°F)	58,750	102,500	22.7	55.5	201	38.8
	Annealed (1500°F)	51,200	86,800	25.5	49.3	163	25.3
1144	As rolled	61,000	102,000	21.0	41.0	212	39.0
	Normalized (1650°F)	58,000	96,750	21.0	40.4	197	32.0
	Annealed (1450°F)	50,250	84,750	24.8	41.3	167	48.0
1340	Normalized (1600°F)	81,000	121,250	22.0	62.9	248	68.2
	Annealed (1475°F)	63,250	102,000	25.5	57.3	207	52.0
3140	Normalized (1600°F)	87,000	129,250	19.7	57.3	262	39.5
	Annealed (1500°F)	61,250	100,000	24.5	50.8	197	34.2
4130	Normalized (1600°F)	63,250	97,000	25.5	59.5	197	63.7
	Annealed (1585°F)	52,250	81,250	28.2	55.6	156	45.5
4140	Normalized (1600°F)	95,000	148,000	17.7	46.8	302	16.7
	Annealed (1500°F)	60,500	95,000	25.7	56.9	197	40.2
4150	Normalized (1600°F)	106,500	167,500	11.7	30.8	321	8.5
	Annealed (1500°F)	55,000	105,750	20.2	40.2	197	18.2
4320	Normalized (1640°F)	67,250	115,000	20.8	50.7	235	53.8
	Annealed (1560°F)	61.625	84,000	29.0	58.4	163	81.0
4340	Normalized (1600°F)	125,000	185,500	12.2	36.3	363	11.7
	Annealed (1490°F)	68,500	108,000	22.0	49.9	217	37.7
4620	Normalized (1650°F)	53.125	83,250	29.0	66.7	174	98.0
	Annealed (1575°F)	54,000	74,250	31.3	60.3	149	69.0

TABLE 16.14 *(Continued)*

Part I. Hot Rolled, Normalized, and Annealed

AISI No.[a]	Treatment	Yield Strength (psi)	Tensile Strength (psi)	Elongation (%)	Reduction in Area (%)	Hardness (Bhn)	Impact Strength Izod (ft-lb)
4820	Normalized (1580°F)	70,250	109,500	24.0	59.2	229	81.0
	Annealed (1500°F)	67,250	98,750	22.3	58.8	197	68.5
5140	Normalized (1600°F)	68,500	115,000	22.7	59.2	229	28.0
	Annealed (1525°F)	42,500	83,000	28.6	57.3	167	30.0
5150	Normalized (1600°F)	76,750	126,250	20.7	58.7	255	23.2
	Annealed (1520°F)	51,750	98,000	22.0	43.7	197	18.5
5160	Normalized (1575°F)	77,000	138,750	17.5	44.8	269	8.0
	Annealed (1495°F)	40,000	104,750	17.2	30.6	197	7.4
6150	Normalized (1600°F)	89,250	136,250	21.8	61.0	269	26.2
	Annealed (1500°F)	59,750	96,750	23.0	48.4	197	20.2
8620	Normalized (1675°F)	51,750	91,750	26.3	59.7	183	73.5
	Annealed (1600°F)	55,875	77,750	31.3	62.1	149	82.8
8630	Normalized (1600°F)	62,250	94,250	23.5	53.5	187	69.8
	Annealed (1550°F)	54,000	81,750	29.0	58.9	156	70.2
8650	Normalized (1600°F)	99,750	148,500	14.0	40.4	302	10.0
	Annealed (1465°F)	56,000	103,750	22.5	46.4	212	21.7
8740	Normalized (1600°F)	88,000	134,750	16.0	47.9	269	13.0
	Annealed (1500°F)	60,250	100,750	22.2	46.4	201	29.5
9255	Normalized (1650°F)	84,000	135,250	19.7	43.4	269	10.0
	Annealed (1550°F)	70,500	112,250	21.7	41.1	229	6.5
9310	Normalized (1630°F)	82,750	131,500	18.8	58.1	269	88.0
	Annealed (1550°F)	63,750	119,000	17.3	42.1	241	58.0

Part II. Quenched and Tempered

AISI No.[a]	Tempering Temperature °F	Tensile Strength (psi)	Yield Strength (psi)	Elongation (%)	Reduction in Area (%)	Hardness (Bhn)
1030[b]	400	123,000	94,000	17	47	495
	600	116,000	90,000	19	53	401
	800	106,000	84,000	23	60	302
	1,000	97,000	75,000	28	65	255
	1,200	85,000	64,000	32	70	207
1040[b]	400	130,000	96,000	16	45	514
	600	129,000	94,000	18	52	444
	800	122,000	92,000	21	57	352
	1,000	113,000	86,000	23	61	269
	1,200	97,000	72,000	28	68	201
1040	400	113,000	86,000	19	48	262
	600	113,000	86,000	20	53	255
	800	110,000	80,000	21	54	241
	1,000	104,000	71,000	26	57	212
	1,200	92,000	63,000	29	65	192
1050[b]	400	163,000	117,000	9	27	514
	600	158,000	115,000	13	36	444
	800	145,000	110,000	19	48	375
	1,000	125,000	95,000	23	58	293
	1,200	104,000	78,000	28	65	235
1050	400	—	—	—	—	—
	600	142,000	105,000	14	47	321
	800	136,000	95,000	20	50	277
	1,000	127,000	84,000	23	53	262
	1,200	107,000	68,000	29	60	223
1060	400	160,000	113,000	13	40	321
	600	160,000	113,000	13	40	321
	800	156,000	111,000	14	41	311
	1,000	140,000	97,000	17	45	277
	1,200	116,000	76,000	23	54	229

TABLE 16.14 *(Continued)*

Part II. Quenched and Tempered

AISI No.[a]	Tempering Temperature °F	Tensile Strength (psi)	Yield Strength (psi)	Elongation (%)	Reduction in Area (%)	Hardness (Bhn)
1080	400	190,000	142,000	12	35	388
	600	189,000	142,000	12	35	388
	800	187,000	138,000	13	36	375
	1,000	164,000	117,000	16	40	321
	1,200	129,000	87,000	21	50	255
1095[b]	400	216,000	152,000	10	31	601
	600	212,000	150,000	11	33	534
	800	199,000	139,000	13	35	388
	1,000	165,000	110,000	15	40	293
	1,200	122,000	85,000	20	47	235
1095	400	187,000	120,000	10	30	401
	600	183,000	118,000	10	30	375
	800	176,000	112,000	12	32	363
	1,000	158,000	98,000	15	37	321
	1,200	130,000	80,000	21	47	269
1137	400	157,000	136,000	5	22	352
	600	143,000	122,000	10	33	285
	800	127,000	106,000	15	48	262
	1,000	110,000	88,000	24	62	229
	1,200	95,000	70,000	28	69	197
1137[b]	400	217,000	169,000	5	17	415
	600	199,000	163,000	9	25	375
	800	160,000	143,000	14	40	311
	1,000	120,000	105,000	19	60	262
	1,200	94,000	77,000	25	69	187
1141	400	237,000	176,000	6	17	461
	600	212,000	186,000	9	32	415
	800	169,000	150,000	12	47	331
	1,000	130,000	111,000	18	57	262
	1,200	103,000	86,000	23	62	217
1144	400	127,000	91,000	17	36	277
	600	126,000	90,000	17	40	262
	800	123,000	88,000	18	42	248
	1,000	117,000	83,000	20	46	235
	1,200	105,000	73,000	23	55	217
1330[b]	400	232,000	211,000	9	39	459
	600	207,000	186,000	9	44	402
	800	168,000	150,000	15	53	335
	1,000	127,000	112,000	18	60	263
	1,200	106,000	83,000	23	63	216
1340	400	262,000	231,000	11	35	505
	600	230,000	206,000	12	43	453
	800	183,000	167,000	14	51	375
	1,000	140,000	120,000	17	58	295
	1,200	116,000	90,000	22	66	252
4037	400	149,000	110,000	6	38	310
	600	138,000	111,000	14	53	295
	800	127,000	106,000	20	60	270
	1,000	115,000	95,000	23	63	247
	1,200	101,000	61,000	29	60	220
4042	400	261,000	241,000	12	37	516
	600	234,000	211,000	13	42	455
	800	187,000	170,000	15	51	380
	1,000	143,000	128,000	20	59	300
	1,200	115,000	100,000	28	66	238
4130[b]	400	236,000	212,000	10	41	467
	600	217,000	200,000	11	43	435

TABLE 16.14 *(Continued)*

Part II. Quenched and Tempered

AISI No.[a]	Tempering Temperature °F	Tensile Strength (psi)	Yield Strength (psi)	Elongation (%)	Reduction in Area (%)	Hardness (Bhn)
	800	186,000	173,000	13	49	380
	1,000	150,000	132,000	17	57	315
	1,200	118,000	102,000	22	64	245
4140	400	257,000	238,000	8	38	510
	600	225,000	208,000	9	43	445
	800	181,000	165,000	13	49	370
	1,000	138,000	121,000	18	58	285
	1,200	110,000	95,000	22	63	230
4150	400	280,000	250,000	10	39	530
	600	256,000	231,000	10	40	495
	800	220,000	200,000	12	45	440
	1,000	175,000	160,000	15	52	370
	1,200	139,000	122,000	19	60	290
4340	400	272,000	243,000	10	38	520
	600	250,000	230,000	10	40	486
	800	213,000	198,000	10	44	430
	1,000	170,000	156,000	13	51	360
	1,200	140,000	124,000	19	60	280
5046	400	253,000	204,000	9	25	482
	600	205,000	168,000	10	37	401
	800	165,000	135,000	13	50	336
	1,000	136,000	111,000	18	61	282
	1,200	114,000	95,000	24	66	235
50B46	400	—	—	—	—	560
	600	258,000	235,000	10	37	505
	800	202,000	181,000	13	47	405
	1,000	157,000	142,000	17	51	322
	1,200	128,000	115,000	22	60	273
50B60	400	—	—	—	—	600
	600	273,000	257,000	8	32	525
	800	219,000	201,000	11	34	435
	1,000	163,000	145,000	15	38	350
	1,200	130,000	113,000	19	50	290
5130	400	234,000	220,000	10	40	475
	600	217,000	204,000	10	46	440
	800	185,000	175,000	12	51	379
	1,000	150,000	136,000	15	56	305
	1,200	115,000	100,000	20	63	245
5140	400	260,000	238,000	9	38	490
	600	229,000	210,000	10	43	450
	800	190,000	170,000	13	50	365
	1,000	145,000	125,000	17	58	280
	1,200	110,000	96,000	25	66	235
5150	400	282,000	251,000	5	37	525
	600	252,000	230,000	6	40	475
	800	210,000	190,000	9	47	410
	1,000	163,000	150,000	15	54	340
	1,200	117,000	118,000	20	60	270
5160	400	322,000	260,000	4	10	627
	600	290,000	257,000	9	30	555
	800	233,000	212,000	10	37	461
	1,000	169,000	151,000	12	47	341
	1,200	130,000	116,000	20	56	269
51B60	400	—	—	—	—	600
	600	—	—	—	—	540
	800	237,000	216,000	11	36	460
	1,000	175,000	160,000	15	44	355
	1,200	140,000	126,000	20	47	290

TABLE 16.14 *(Continued)*

Part II. Quenched and Tempered

AISI No.[a]	Tempering Temperature °F	Tensile Strength (psi)	Yield Strength (psi)	Elongation (%)	Reduction in Area (%)	Hardness (Bhn)
6150	400	280,000	245,000	8	38	538
	600	250,000	228,000	8	39	483
	800	208,000	193,000	10	43	420
	1,000	168,000	155,000	13	50	345
	1,200	137,000	122,000	17	58	282
81B45	400	295,000	250,000	10	33	550
	600	256,000	228,000	8	42	475
	800	204,000	190,000	11	48	405
	1,000	160,000	149,000	16	53	338
	1,200	130,000	115,000	20	55	280
8630	400	238,000	218,000	9	38	465
	600	215,000	202,000	10	42	430
	800	185,000	170,000	13	47	375
	1,000	150,000	130,000	17	54	310
	1,200	112,000	100,000	23	63	240
8640	400	270,000	242,000	10	40	505
	600	240,000	220,000	10	41	460
	800	200,000	188,000	12	45	400
	1,000	160,000	150,000	16	54	340
	1,200	130,000	116,000	20	62	280
86B45	400	287,000	238,000	9	31	525
	600	246,000	225,000	9	40	475
	800	200,000	191,000	11	41	395
	1,000	160,000	150,000	15	49	335
	1,200	131,000	127,000	19	58	280
8650	400	281,000	243,000	10	38	525
	600	250,000	225,000	10	40	490
	800	210,000	192,000	12	45	420
	1,000	170,000	153,000	15	51	340
	1,200	140,000	120,000	20	58	280
8660	400	—	—	—	—	580
	600	—	—	—	—	535
	800	237,000	225,000	13	37	460
	1,000	190,000	176,000	17	46	370
	1,200	155,000	138,000	20	53	315
8740	400	290,000	240,000	10	41	578
	600	249,000	225,000	11	45	495
	800	208,000	197,000	13	50	415
	1,000	175,000	165,000	15	55	363
	1,200	143,000	131,000	20	60	302
9255	400	305,000	297,000	1	3	601
	600	281,000	260,000	4	10	578
	800	233,000	216,000	8	22	477
	1,000	182,000	160,000	15	32	352
	1,200	144,000	118,000	20	42	285
9260	400	—	—	—	—	600
	600	—	—	—	—	540
	800	255,000	218,000	8	24	470
	1,000	192,000	164,000	12	30	390
	1,200	142,000	118,000	20	43	295
94B30	400	250,000	225,000	12	46	475
	600	232,000	206,000	12	49	445
	800	195,000	175,000	13	57	382
	1,000	145,000	135,000	16	65	307
	1,200	120,000	105,000	21	69	250

Source: Materials and Processing Databook, Metal Progress, June 1985, pp. 16–18.

[a]All grades are fine grained except for those in the 1100 series, which are coarse grained. Normalizing and annealing temperatures are given in parentheses.

[b]Water quenched.

TABLE 16.15   ASTM Specifications that Incorporate AISE–SAE Designations

A29	Carbon and alloy steel bars, hot rolled and cold finished, generic
A108	Standard-quality cold-finished carbon steel bars
A295	High carbon-chromium ball- and roller-bearing steel
A304	Alloy steel bars having hardenability requirements
A322	Hot-rolled alloy steel bars
A331	Cold-finished alloy steel bars
A434	Hot-rolled or cold-finished quenched and tempered alloy steel bars
A505	Hot-rolled and cold-rolled alloy steel sheet and strip, generic
A506	Regular-quality hot-rolled and cold-rolled alloy steel sheet and strip
A507	Drawing quality hot-rolled and cold-rolled alloy steel sheet and strip
A510	Carbon steel wire rods and coarse round wire, generic
A534	Carburizing steels for antifriction bearings
A535	Special-quality ball- and roller-bearing steel
A544	Scrapless nut quality carbon steel wire
A545	Cold heading quality carbon steel wire for machine screws
A546	Cold heading quality medium-high carbon steel wire for hexagon-head bolts
A547	Cold heading quality alloy steel wire for hexagon head bolts
A548	Cold heading quality carbon steel wire for tapping or sheet metal screws
A549	Cold heading quality carbon steel wire for wood screws
A575	Merchant-quality hot-rolled carbon steel bars
A576	Special-quality hot-rolled carbon steel bars
A634	Aircraft-quality hot-rolled and cold-rolled alloy steel sheet and strip
A646	Premium-quality alloy steel blooms and billets for aircraft and aerospace forgings
A659	Commercial-quality hot-rolled carbon steel sheet and strip
A680	Untempered spring quality cold-rolled hard carbon steel strip
A682	Cold-rolled spring quality carbon steel strip, generic
A684	Untempered spring quality cold-rolled soft carbon steel strip
A689	Carbon and alloy steel bars for springs
A711	Carbon and alloy steel blooms, billets and slabs for forging
A713	High-carbon spring steel wire for heat-treated components

TABLE 16.16   Generic ASTM Specifications

A6	Rolled steel structural plate, shapes, sheet piling and bars, generic
A20	Steel plate for pressure vessels, generic
A29	Carbon and alloy steel bars, hot rolled and cold finished, generic
A505	Alloy steel sheet and strip, hot rolled and cold rolled, generic
A510	Carbon steel wire rod and coarse round wire, generic
A568	Carbon and HSLA, hot-rolled and cold-rolled steel sheet and hot-rolled strip, generic
A646	Premium-quality alloy steel blooms and billets for aircraft and aerospace forgings
A711	Carbon and alloy steel blooms, billets and slabs for forging

Source: Metals Handbook, (9th edition), ASM, Metals Park, Ohio, 1978.

TABLE 16.17   Composition Ranges and Limits for Sheet and Strip, Plain Carbon and HSLA Grades

ASTM Specification	Description[a]	Maximum C	Maximum Mn	Maximum P	Maximum S	Other
A611	CRSQ					
	Grades A, B, C	0.20	0.60	0.04	0.04	[b]
	Grade E	0.20	0.90	0.04	0.04	[b]
A366	CRCQ	0.15	0.60	0.035	0.04	[b]
A109	CR strip					
	Tempers 1, 2, 3	0.25	0.60	0.035	0.04	[b]
	Tempers 4, 5	0.15	0.60	0.035	0.04	[b]
A619	CRDQ	0.10	0.50	0.025	0.035	[b]
A620	CR DQSK	0.10	0.50	0.025	0.035	[c]
A570	HR SQ					
	Grades A, B, C	0.25	0.25–0.60	0.04	0.04	[b]
	Grades D, E	0.25	0.60–0.90	0.04	0.04	[b]
A569	HR CQ	0.15	0.60	0.035	0.04	[b]
A621	HR DQ	0.10	0.50	0.025	0.035	—
A622	HR DQSK	0.10	0.50	0.025	0.035	[c]
A414	Pressure vessel					
	Grade A	0.15	0.90	0.035	0.04	[b]
	Grade B	0.22	0.90	0.035	0.04	[b]
	Grade C	0.25	0.90	0.035	0.04	[b]
	Grade D	0.25	1.20	0.035	0.04	[b]
	Grade E	0.27	1.20	0.035	0.04	[b]
	Grade F	0.31	1.20	0.035	0.04	[b]
	Grade G	0.31	1.35	0.035	0.04	[b]
A606	HSLA	0.22	1.25	—	0.05	[d]
A607	Grade 45	0.22	1.35	0.04	0.05	[e]
	Grade 50	0.23	1.35	0.04	0.05	[e]
	Grade 55	0.25	1.35	0.04	0.05	[e]
	Grade 60	0.26	1.50	0.04	0.05	[e]
	Grade 65	0.26	1.50	0.04	0.05	[e]
	Grade 70	0.26	1.65	0.04	0.05	0.012 max N[e]
A715	Basic composition	0.15	1.65	0.025	0.035	0.012 max N

Type 1: 0.05 min Ti, 0.10 max Si[f]
Type 2: 0.02 min V, 0.60 max Si,[g] 0.005 min N[f, g]
Type 3: 0.005 min Nb, 0.08 max V,[g] 0.60 max Si,[g] 0.020 max N[f, g]
Type 4: 0.05 min Zr, 0.90 max Si, 0.80 max Cr,[g] 0.10 max Ti,[g]
    0.0025 max B,[g] 0.005–0.06 Nb[f, h]
Type 5: 0.03 min Nb,[j] 0.20 min Mo,[j] 0.30 max Si[f]
Type 6: 0.005–0.10 Nb, 0.90 max Si[f]
Type 7: 0.005 min Nb or V, or both, 0.60 max Si, 0.020 max N[f]
Type 8: 0.005–0.15 Nb, 0.05 min Zr

[a] CR, cold rolled; SQ, structural quality; DQ, drawing quality; DQSK, drawing quality special killed; CQ, commerical quality.
[b] Cu when specified as Cu-bearing steel: 0.20% min.
[c] Aluminum as deoxidizer usually exceeds 0.010% in product.
[d] Other elements may be added if necessary to meet mechanical and corrosion requirements.
[e] 0.005 minimum Nb or 0.01 minimum V for all grades.
[f] These elements are added to basic composition.
[g] Not added to grades 50 and 60.
[h] Might not be added to grade 50.
[j] Available as grade 80 only.

TABLE 16.18 Composition Ranges and Limits for Carbon Steel Structural Shapes and Plate

ASTM Specification	Form, Type, or Grade	UNS Designation	Heat Composition Ranges and Limits[a] (%)			
			Maximum C	Mn	Si	Cu[b]
A36	Plate	—	0.29	0.80–1.20	—	0.20
	Shapes	K02600	0.26	[c]	[d]	0.20
	Bars	—	0.29	0.60–0.90	—	0.20
A283	Plate	—	—	—	—	0.20
A284	Grade A	K01804	0.24	0.90 max	0.10–0.30	—
	Grade B	K02001	0.24	0.90 max	0.15–0.30	—
	Grade C	K02401	0.36	0.90 max	0.15–0.30	—
	Grade D	K02702	0.35	0.90 max	0.15–0.30	—
A529	Plate, bars, and shapes	K02703	0.27	1.20 max	—	0.20
A573	Grade 58	K02301	0.23	0.60–0.90	0.10–0.35	—
	Grade 65	K02404	0.26	0.85–1.20	0.15–0.30	—
	Grade 70	K02701	0.28	0.85–1.20	0.15–0.30	—
A678	Grade A	K01600	0.16	0.90–1.50	0.15–0.50	0.20
	Grade B	K02002	0.20	0.70–1.60	0.15–0.50	0.20
	Grade C	K02204	0.22	1.00–1.60	0.20–0.50	0.20

Source: Metals Handbook, (9th edition), ASM, Metals Park, Ohio, 1978.
[a] Typical limits on phosphorus and sulfur content are 0.040% maximum phosphorus and 0.050% maximum sulfur.
[b] Minimum copper content applicable only if copper-bearing steel is specified.
[c] 0.85–1.35% manganese required for shapes heavier than 634 kg/m (426 lb/ft).
[d] 0.15–0.30% silicon required for shapes heavier than 634 kg/m (426 lb/ft).

TABLE 16.19 Properties and Applications of Carbon and Low-Alloy[a] Steel Castings

| | Mechanical Properties, Minimum Specification Values | | | | | Charpy Impact, ft-lb[b] | | | |
| | | | | | | 70°F | | −40°F | |
Typical Specifications	Tensile Strength, 1000 psi	Yield Strength, 1000 psi	Elong in 2 in. (%)	Red in Area (%)	Hardness Bhn[b]	Key-hole	V-Notch	Key-hole	V-Notch
Carbon Steels									
ASTM A27-80, Grade 60-30	60	30	24	35	131	30	12	8	5
ASTM A27-80, Grade 65-35	65	35	24	35	131	30[e]	35[e]	15[e]	12[e]
MIL-S-15083B, Class 70-36	70	36	22	30	143	30[e]	30[e]	13[e]	12[e]
MIL-S-15083B, Class 80-40	80	40	17	25	163	25[e]	35[e]	12[e]	10[e]
SAE J435c, Grade 0050A	85	45	16	24	170–229	20	26	10	10
SAE J435c, Grade 0050B	100	70	10	15	207–255	30	40	15	12
Low-Alloy Steels[a]									
ASTM A352-79, Grade LC1	65	35	24	35	137	50	60	18	20
ASTM A217-80, Grade WC4	70	40	20	35	143	48	55	25	22
ASTM A148-80, Grade 80-50	80	50	22	35	170	45[e]	48[e]	25[e]	18[e]
ASTM A148-80, Grade 90-60	90	60	20	40	192	40[e]	40[e]	20[e]	16[e]
ASTM A148-80, Grade 105-85	105	85	17	35	217	50	58	40	40
ASTM A148-80, Grade 120-95	120	95	14	30	262	43	45	35	31
ASTM A148-80, Grade 150-125	150	125	9	22	311	28	30	20	17
ASTM A148-80, Grade 175-145	175	145	6	12	352	22	24	15	12

TABLE 16.19 (Continued)

Endurance Limit, 1000 psi[b]		Machinability Speed Index[b]		Heat Treatment[d]	Other Current Specifications	Application and Outstanding Characteristics
Unnotched	Notched	HSS[c]	Carbide Tool			
30	19	160	400	A, N, NT, QT	ASTM A216, WCA; AAR M201-53, Grade AU, Grade AA; MIL-S-15083B, Class B; ABS Class 1, hull	Excellent weldability; can be case hardened or carburized, low electrical resistivity, desirable magnetic properties
30	19	135	230	A, N, NT, QT	SAE Grade 0030; Federal QQ-S-681d, Class 65-35; Lloyds Class A; ASTM A352 LCB; MIL-S-15083B, Grade 65-35	Excellent machinability and weldability; combine moderate strength, good ductility, and machinability
35	22	135	230	A, NT, QT	ASTM A27, Grade 70-36; ASTM A216, WCB; AAR M201-53, Grade B; ABS Class 2	
37	26	135	400	A, NT, QT	SAE Grade 080; Federal QQ-S-681d, Class 80-40	High-strength, good machinability, toughness and excellent fatigue resistance; readily weldable
39	28	120	325	N, NT		
45	31	80	310	QT	Federal QQ-S-681d, Class 0050	High hardness, wear resistance
32	20	130	400	NT, QT	ASTM A352, LC2, LC3; ASTM A217, Grade WC1; MIL-S-870B	Suitable for high- and low-temperature service; excellent weldability
35	23	120	230	NT	ASTM A217, WC5, WC6, WC9; MIL-S-15464B, Class 1, 2, and 3	Excellent weldability; combine moderate strength, high toughness and machinability; suitable for high-temperature service
39	25	110	240	A, N, NT, QT	ASTM A148, Grade 80-40; SAE Grade 080; Federal QQ-S-681d, 80-50; MIL-S-15083B, 80-50	
42	31	95	290	A, N, NT, QT	ASTM A217, C5; SAE Grade 090; QQ-S-681d, Class 90-60; AAR M201-47, Class C; MIL-S-15083B, 90-60	Excellent combination of strength and toughness. Certain steels in this group are deep-hardening grades are suitable for high- and low-temperature service. High resistance to impact. Readily weldable.
53	34	90	310	N, NT, QT	QQ-S-681d, 105-85; MIL-S-15083B, 105-85; SAE Grade 0105	
62	37	75	180	N, NT, QT	SAE Grade 0120; QQ-S-681d, 120-54; MIL-S-15083B, 120-95	
74	44	45	200	N, NT, QT	SAE Grade 0150; QQ-S-681d, 150-125; MIL-S-15083B, 150-125	Deep hardening, high strength, resistance to wear and fatigue
84	48	35	180	N, NT, QT	SAE Grade 0175; QQ-S-681d, 175-145	High strength and hardness, resistance to wear and fatigue

Source: Steel Founders' Society of America, Cleveland (reviewed February 1985).
[a] Below 8% total alloy content.
[b] Typical properties; should not be used as design or specification limit.
[c] Machinability speed index for standard 18-4-1 high-speed steel based on cutting speed which gives a tool life of 1 hr. For carbide tools, cutting speed for tool life of 1 hr based on 0.015-in. wearland.
[d] Abbreviations: A, annealed; N, normalized; NT, normalized and tempered; QT, liquid quenched and tempered.
[e] Normalized and tempered or quenched and tempered.

TABLE 16.20 Specifications and Properties of Gray and White Iron Castings

Specification No.[a]	Grade or Class	Minimum Tensile Strength (psi)	Brinell Hardness	Other Requirements	Microstructure	Typical Applications
ASTM A126-84	A	21,000		Composition: P_{max}, 0.75; S_{max}, 0.15; 1.2 in. diameter × 12 in. transverse test is optional.		Stock valves, flanges, and pipe fittings and castings not requiring critical tensile test evaluation
	B	31,000				
	C	41,000				
ASTM A48-83.	20[b]	20,000[c]		Test bar size shall be related in cooling rate to the critical section of the casting and so specified; at least two test bars shall be cast and prepared for each casting lot, the lot size being designated; test bars shall be cast in dry silica sand molds similar to that in which the castings are poured; tension test shall be under true axial loading;		Small or thin-sectioned castings requiring good appearance, good machinability, and close dimensions
	25[b]	25,000[c]				
	30[b]	30,000[c]				General machinery, municipal and water works, light compressors
	35[b]	35,000[c]				
ANSI G25.1	40[b]	40,000[c]		hardness, chemical composition, microstructure, pressure tightness; radiographic soundness, dimension, surface finish etc., can be established as requirements upon written agreement between manufacturer and purchaser.		Machine tools, medium gear blanks, heavy compressors
	45[b]	45,000[c]				
	50[b]	50,000[c]				Dies, crankshafts, high-pressure cylinders, heavy-duty machine tool parts, large gears, press frames
	55[b]	55,000[c]				
	60[b]	60,000[c]				
ASTM A159-83.				Total Carbon (%)	Microstructure	
	G1800	—	187 maximum[e]	—	Ferritic-pearlitic	For machinability where higher strength not necessary
	G2500	—	170–229[e]	—	Ferritic-pearlitic	Small cylinder blocks and heads, pistons, clutch plates, pump bodies, gear boxes, housings, light-duty brake drums
	G2500a	—	170–229[e]	3.40 minimum mandatory	"A" graphite size 2–4 15% maximum ferrite	Brake drums and clutch plates to minimize heat checking
	G3000	—	187–241[e]	—	Pearlitic	Cylinder blocks, heads, liners, flywheels, pistons, medium-duty brake drums, clutch plates
SAE J431c[d]	G3500	—	207–255[e]	—	Pearlitic	Truck cylinder blocks and heads, heavy flywheels and transmission cases, differential carriers
	G3500b	—	207–255[e]	3.40 minimum mandatory	"A" graphite size 3–5, 5% maximum ferrite or carbide	Brake drums and clutch plates for heavy service requiring heat resistance and higher strength

Specification	Grade/Class	Tensile strength (psi)	Hardness, Brinell	Microstructure / Composition	Applications
G3500c		—	207–255[e]	"A" graphite size 3–5, 5% maximum ferrite or carbide; 3.50 minimum mandatory	Extra-heavy-duty service brake drums
G4000		—	217–269[e]	Pearlitic; —	Diesel engine castings, liners, cylinders, pistons, heavy parts in general
G4000d		—	241–321	—; 0.85–1.25% Cr, 0.40–0.60% Mo; minimum $\frac{1}{8}$ in. of carbides at cam nose	Alloyed automobile engine camshafts
ASTM A278-84, ASME SA278	40[f]	40,000[h]	—	Castings and test bars must be stress relieved by prescribed methods. Carbon equivalent maximum = 3.8; $P_{max} = 0.25$; $S_{max} = 0.12$; [CE = %C + 0.3(%Si + %P)]	Pressure containing parts for use to 650°F; valve bodies, papermill drier rolls, chemical process equipment, pressure vessel castings
	50	50,000[h]	—		
	60	60,000[h]	—		
	70[g]	70,000[h]	—		
	80[g]	80,000[h]	—		
ASTM A319-71 (1985)	I	Low strength[h]	Maximum hardness at casting locations to be machined shall be agreed on by manufacturer and purchaser	(see Type composition table below)	Nonpressure-containing parts at elevated temperatures; stoker and fire box parts, grate bars, process furnace parts, ingot molds, glass molds, caustic pots, metal melting pots. Class I: superior thermal shock resistance, low strength; Class II: average thermal shock resistance; Class III: high strength at temperature
	II	Above 30,000 may be expected[i]			
	III	As high as 40,000 may be expected[h]			
ASTM A823-84	A-SA	30,000	163–207	Statically cast, permanent mold gray iron castings; grade letters designate whether castings are annealed (A) or normalized (N); solid (S) or cored (C); test bar as-cast diameter is 0.88 in. (A), 1.20 in. (B), 2.00 in. (C), or as agreed (S).	Automobile, truck, appliance, and machinery castings in quantity
	A-SB	25,000	163–207		
	A-SC	20,000	163–207		
	A-SS	18,000	143–207		
	A-CA	30,000	143–207		
	A-CB	25,000	143–207		
	A-CC	20,000	143–207		
	N-SA	30,000	170–229		
	N-SB	25,000	170–229		
	N-SC	20,000	170–229		
	N-SS	18,000	149–229		
	N-CA	30,000	170–229		
	N-CB	25,000	170–229		
	N-CC	20,000	170–229		

ASTM A319-71 Type composition:

Carbon Equivalent	Carbon Minimum	Type[j]	Cr
3.81–4.40	3.50	A	0.20–0.40
3.51–4.10	3.20	B	0.41–0.65
3.20–3.80	2.80	C	0.66–0.95
		D	0.96–1.20

TABLE 16.20 (Continued)

Specification No.[a]	Grade or Class		Minimum Tensile Strength (psi)	Brinell Hardness	Composition (%)								Typical Applications
					TC	Si	Mn	Ni	Cr	Cu	S	Other	
ASTM A436-84[k]	1	Minimum	25,000[l]	131–183	—	1.00	0.50	13.50	1.50	5.50	—	—	Valve guides, insecticide pumps, flood gates, piston ring bands
		Maximum			3.00	2.80	1.50	17.50	2.50	7.50	0.12	—	
	1b	Minimum	30,000[l]	149–212	—	1.00	0.50	13.50	2.50	5.50	—	—	Seawater valve and pump bodies, pump section belt
		Maximum			3.00	2.80	1.50	17.50	3.50	7.50	0.12	—	
	2	Minimum	25,000[l]	118–174	—	1.00	0.50	18.00	1.50	—	—	—	Fertilizer applicator parts, pump impellers, pump casings, plug valves
		Maximum			3.00	2.80	1.50	22.00	2.50	0.50	0.12	—	
	2b	Minimum	30,000[l]	171–248	—	1.00	0.50	18.00	3.00[m]	—	—	—	Caustic pump casings, valves, pump impellers
		Maximum			3.00	2.80	1.50	22.00	6.00	0.50	0.12	—	
	3	Minimum	25,000[l]	118–159	—	1.00	0.50	28.00	2.50	—	—	—	Turbocharger housings, pumps and liners, stove tops, steam piston valve rings, pumps and valves
		Maximum			2.60	2.00	1.50	32.00	3.50	0.50	0.12	—	
	4	Minimum	25,000[l]	149–212	—	5.00	0.50	29.00	4.50	—	—	—	Range tops
		Maximum			2.60	6.00	1.50	32.00	5.50	0.50	0.12	—	
	5	Minimum	20,000[l]	99–124	—	1.00	0.50	34.0	—	—	—	—	Glass rolls and molds, machine tools, gauges, optical parts requiring minimum expansion and good damping qualities, solder rails and pots
		Maximum			2.40	2.00	1.50	36.00	0.10	0.50	0.12	—	
	6	Minimum	25,000[l]	124–174	—	1.50	0.50	18.00	1.00	3.50	—	1.00 Mo	Valves
		Maximum			3.00	2.50	1.50	22.00	2.00	5.50	0.12		

Specification No.	Grade or Class		Minimum Tensile Strength (psi)	Brinell Hardness	Composition (%)								Typical Applications
					TC	Si	Mn	Ni	Cr	Mo	S	Other	
ASTM A532-82[n]	I Type A	Minimum	—	550 sand cast, 600 chill cast	3.0	—	—	3.3	1.4	—	—	—	—
		Maximum			3.6	0.8	1.3	5.0	4.0	1.0	0.15	0.30 P	
	I Type B	Minimum	—	550 sand cast, 600 chill cast	2.5	—	—	3.3	1.4	—	—	—	—
		Maximum			3.0	0.8	1.3	5.0	4.0	1.0	0.15	0.30 P	
	I Type C	Minimum	—	550 sand cast, 600 chill cast	2.9	—	—	2.7	1.1	—	—	—	—
		Maximum			3.7	0.8	1.3	4.0	1.5	1.0	0.15	0.30 P	
	I Type D	Minimum	—	550 sand cast, 600 chill cast	2.5	1.0	—	5.0	7.0	—	—	—	Can be hardened to BHN 600 minimum or annealed to BHN 400 maximum
		Maximum			3.6	2.2	1.3	7.0	11.0	1.0	0.15	0.10 P	
	II Type A	Minimum	—	550 sand cast, 600 hardened	2.4	—	0.5	—	11.0	0.5	—	—	Can be annealed to BHN 400 maximum
		Maximum			2.8	1.0	1.5	0.5	14.0	1.0	0.06	0.10 P, 1.2 Cu	

II Type B	450 sand cast, 600 hardened	Minimum Maximum	2.4 2.8	— 1.0	0.5 1.5	— —	14.0 18.0	1.0 3.0	— 0.06	— 0.10 P, 1.2 Cu	—	Can be annealed to BHN 400 maximum
II Type C	550 sand cast, 600 hardened	Minimum Maximum	2.8 3.6	— 1.0	0.5 1.5	— —	14.0 18.0	2.3 3.5	— 0.06	— 0.10 P, 1.2 Cu	—	Can be annealed to BHN 400 maximum
II Type D	450 sand cast, 600 hardened	Minimum Maximum	2.0 2.6	— 1.0	0.5 1.5	— —	18.0 23.0	1.5 —	— 0.06	— 0.10 P, 1.2 Cu	—	Can be annealed to BHN 400 maximum
II Type E	450 sand cast, 600 hardened	Minimum Maximum	2.6 3.2	— 1.0	0.5 1.5	— —	18.0 23.0	1.0 2.0	— 0.06	— 0.10 P, 1.2 Cu	—	Can be annealed to BHN 400 maximum
III Type A	450 sand cast, 600 hardened	Minimum Maximum	2.3 3.0	— 1.0	0.5 1.5	— —	23.0 28.0	1.5 1.5	— 0.06	— 0.10 P, 1.2 Cu	—	Can be annealed to BHN 400 maximum
ASTM A518-80	—	Minimum Maximum	0.7 1.1	14.20 14.75	1.5		0.5	0.5	—	0.5 Cu	^{p} ^{o}	Pumps and piping for corrosive liquids

ASTM Tensile Bar Dimensions

Controlling Section of Castings (in.)	Test Bar	Minimum Cast Bar Length (in.)	Maximum Cast Bar Length (in.)	Cast Bar Average Diameter (in.)	Machined Bar Diameter (in.)
0.25–0.50	A	5.0	6.0	0.88	0.50
0.51–1.00	B	6.0	9.0	1.20	0.75
1.00–2.00	C	7.0	10.0	2.00	1.25
Under 0.25, over 2.0	S	Intended for use when standard bars are not satisfactory; all dimensions shall be agreed upon by manufacturer and purchaser			

Source: Charles F. Walton, P. E.; revised February 1985. You are invited to contact Mr. Walton directly for more information on iron castings. He may be reached at 3626 Stoneleigh Rd., Cleveland Heights, Ohio 44121.

^{a} Abbreviations: ANSI, American National Standards Institute; ASME, American Society of Mechanical Engineers; ASTM, American Society for Testing and Materials.
^{b} Each class number is followed by a letter, A, B, C, or S, indicating test bar size required for class. All test bars must be separately cast and machined, and tension test result is required for casting qualification. At least two test bars are required for each lot of castings intended to conform to this specification.
^{c} Test bar size shall be determined by controlling casting section if test bar is not specified. Recommended dimensions are listed under "ASTM Tensile Bar Dimensions" at end of table.
^{d} Automotive castings cast in sand molds.
^{e} Hardness of casting in properly prepared area or areas established by agreement and shown on drawings.
^{f} Classes 20, 25, 30, and 35 are covered but limited to use below 450°F.
^{g} Not in ASME SA278.
^{h} Required test bar size is as listed under "ASTM Tensile Bar Dimensions" at end of table.
^{i} Low strength is desired for thermal shock resistance. Strength may be specified where essential (up to strength prescribed for Class 40, ASTM A48-74).
^{j} Subdivision of class when chromium present as alloying element. Other alloys to increase strength and to improve and stabilize structure for elevated temperature service may be used in all classes.
^{k} Austenitic gray iron castings with heat, wear, and corrosion resistance. Austenitic matrix contains uniformly distributed graphite flakes plus some carbides.
^{l} Test bars machined from 1 in. keel block, or a "Y" block in ½-, 1-, or 3-in. size by option of purchaser.
^{m} When some machining is required, 3.00–4.00% Cr range is recommended.
^{n} Abrasion resisting white irons for mining, milling, and earth-handling uses.
^{o} Transverse strength shall be 930 lb minimum with 0.026 in. minimum deflection on special test bar.
^{p} Heat treatment required.

TABLE 16.21 Specifications and Properties of Ductile (Nodular) Iron Castings

Specification No.	Class or Grade	Minimum Tensile Strength^a (psi)	Minimum Yield Strength^b (psi)	Elongation in 2 in. (%)
ASTM A536-84	60-40-18	60,000	40,000	18
	65-45-12	65,000	45,000	12
	80-55-06	80,000	55,000	6
	100-70-03	100,000	70,000	3
	120-90-02	120,000	90,000	2
	60-42-10	60,000	42,000	10
	70-50-05	70,000	50,000	5
	80-60-03	80,000	60,000	—
SAE J434c	D-4018	—	—	—
	D-4512	—	—	—
	D-5506	—	—	—
	D-7003	—	—	—
	DQ & T	—	—	—
ASTM A395-80, ASME SA395	60-40-18	60,000^e	40,000	18
ASTM A476-82 (metric version available)	80-60-03^g	80,000	60,000	3
MIL-I-24137 (ships) amended	Class A	60,000	45,000	15

TABLE 16.21 *(Continued)*

Heat Treatment	Other Requirements	Typical Applications
May be annealed — — Usually normalized Quenched and tempered — —	Chemical composition is subordinate to mechanical properties; content of any chemical element may be specified by mutual agreement	Pressure castings such as valve and pump bodies; machinery castings subject to shock and fatigue loading; crankshafts, gears, and rollers; high-strength gears, automotive and machine components; pinions, gears, rollers, slides; for special applications such as pipe and fittings

Heat Treatment	Hardness,[c] Bhn	Microstructure[d]	Typical Applications
May be annealed	170 maximum	Ferritic	Steering knuckles
—	156–217	Ferritic-pearlitic	Disc brake calipers
—	187–255	Ferritic-pearlitic	Crankshafts
May be normalized	241–302	Pearlitic	Gears
Quenched and tempered	Range specified	Martensitic	Rocker arms

Heat Treatment	Composition, %						Typical Applications
		TC	Si	P	Other	CE[f]	Bhn	
Ferritized by annealing, Bhn 143–187	Min Max	3.0 —	— 2.50	— 0.08	— —	— —	— —	Valves and fittings for steam and chemical plant equipment
To be used in as-cast condition; hardness shall be minimum Bhn 201	Min Max	3.0 —	— 3.0	— 0.08	— 0.05 S	3.8 4.5	— —	Paper mill dryer rolls; used up to 450°F
Shall be ferritized by annealing to Bhn 190 max[h]	Min Max	3.0 —	[i] 2.50	[i] 0.08	— —	— 4.3	— 190	Shipboard electric equipment, engine blocks, pumps, compressors gears, valves, clamps, and hydraulic equipment

TABLE 16.21 Specifications and Properties of Ductile (Nodular) Iron Castings *(Continued)*

Specification No.	Class or Grade	Minimum Tensile Strength[a] (psi)	Minimum Yield Strength[b] (psi)	Elongation in 2 in. (%)	Heat Treatment
		High-Alloy Ductile (Nodular) Iron Castings			
	D-2[j]	58,000	30,000	8	—
	D-2B	58,000	30,000	7	—
ASTM A439-83	D-2C	58,000	28,000	20	—
	D-3[j]	55,000	30,000	6	—
	D-3A	55,000	30,000	10	—
	D-4	60,000	—	—	—
	D-5	55,000	30,000	20	—
	D-5B	55,000	30,000	6	—
ASTM A571-84 metric version available, ASME SA571[l]	Class 1 Class 2	65,000 60,000	30,000 25,000	30 ⎫ 25 ⎭	Annealed
MIL-I-24137 (ships) amended	Class B Class C	55,000 50,000	30,000 25,000	7 ⎫ 20 ⎭	Stress relief, 1200°F (carbide solution at 1750°F if necessary)

TABLE 16.21 *(Continued)*

	TC	Si	Mn	P	Ni	Cr	Bhn	Typical Applications
Min	—	1.50	0.70	—	18.00	1.75	139 ⎫	Valve stem bushings, valve and
Max	3.00	3.00	1.25	0.08	22.00	2.75	202 ⎭	pump bodies in petroleum, salt water, and caustic service; manifolds; turbocharger housings; air compressor parts
Min	—	1.50	0.70	—	18.00	2.75	148 ⎫	Turbocharger housings, rolls
Max	3.00	3.00	1.25	0.08	22.00	4.00	211 ⎭	
Min	—	1.00	1.80	—	21.00	—	121 ⎫	Electrode guide rings, steam
Max	2.90	3.00	2.40	0.08	24.00	0.50[b]	171 ⎭	turbine dubbing rings
Min	—	1.00	—	—	28.00	2.50	139 ⎫	Turbocharger nozzles and housings,
Max	2.60	2.80	1.00[k]	0.08	32.00	3.50	202 ⎭	steam turbine diaphragms, gas compressor diffusers
Min	—	1.00	—	—	28.00	1.00	131 ⎫	High-temperature bearing rings
Max	2.60	2.80	1.00[k]	0.08	32.00	1.50	193 ⎭	requiring gall resistance
Min	—	5.00	—	—	28.00	4.50	202 ⎫	Diesel engine manifolds,
Max	2.60	6.00	1.00[k]	0.08	32.00	5.50	273 ⎭	manifold joints
Min	—	1.00	—	—	34.00	—	131 ⎫	Guidance system housings, gas
Max	2.40	2.80	1.00[k]	0.08	36.00	0.10	185 ⎭	turbine shroud rings, glass rolls
Min	—	1.00	—	—	34.00	2.00	139 ⎫	Optical system mirrors and parts
Max	2.40	2.80	1.00[k]	0.08	36.00	3.00	193 ⎭	for dimensional stability, compressor stators
Min	2.20	1.50	3.75	—	21.00	—	121 ⎫	Compressors, expanders, pumps
Max	2.70	2.50	4.50	0.08	24.00	0.20	111– ⎬	and other pressure-containing
							171 ⎭	parts requiring stable austenitic matrix at −423°F
Min	2.40	1.80	0.80	—	18.00	1.70	⎫	
Max	3.00	3.20	1.50	0.20	22.00	2.40	190 ⎬	Resistance to heat, corrosion,
Min	2.70	2.00	1.90	—	20.00	—	⎪	shock; nonmagnetic; shipboard
							⎬	use and propellers
Max	3.10	3.00	2.50	0.15	23.00	0.50	175 ⎭	

Other Requirements / Composition, %

Source: Charles F. Walton, P.E.; revised February 1985. You are invited to contact Mr. Walton directly for more information on iron castings. He may be reached at 3626 Stoneleigh Rd., Cleveland Heights, Ohio 44121.

[a]Test bars machined from 1-in. keel block or Y block in $\frac{1}{2}$-, 1-, or 3-in. size by option of purchaser.
[b]As determined by "offset method" at 0.2%.
[c]Hardness in properly prepared area is established by agreement and shown on drawings.
[d]Graphite shall be at least 80% spheroidal conforming to Types I and II in ASTM A247.
[e]Test specimen shall be machined from test coupon based on size of controlling section of casting. Recommended dimensions are: 1 in. and under controlling section of casting, 1 in. coupon (may be Y or keel block); 1–3 in. 3 in. coupon (must be Y block); over 3 in., larger coupons may be used by agreement.
[f]Percentage of carbon equivalent = % TC + 0.3(% Si + % P).
[g]Tensile strength, yield strength, and elongation properties obtained from 1-in.-thick test coupon. For 3-in.-thick coupon, tensile and yield strength specifications are identical; elongation is 1%.
[h]One metallographic test shall be made for each lot after annealing. Microstructure at 50 × shall show a matrix of 90% ferrite minimum with no primary carbides; all graphite spheroidal.
[i]For castings with $\frac{1}{2}$-in. sections and smaller, 2.75 Si maximum and 0.08 P are allowed or 3.00 Si maximum with 0.05 P maximum.
[j]Additions of 0.7–1.0 Mo will increase the mechanical properties above 800°F.
[k]Not intentionally added.
[l]Notched impact of 15 ft-lb at −320°F.

TABLE 16.22 Properties and Applications of Malleable Iron Castings

Specification No.	Type	Class or Grade	Minimum Tensile Strength (psi)	Minimum Yield Strength (psi)	Elongation (%)
ASTM A 47-84,[a]	Ferritic	32510	50,000	32,000	10
ANSI G48.1-1969, ASME SA47		35018	53,000	35,000	18
ASTM A 197-79, ANSI G49.1-1948 (R1972), ASME SA197	Cupola		40,000	30,000	5
ASTM A 220-84, ANSI G48.2-1972	Pearlitic	40010	60,000	40,000	10
		45008	65,000	45,000	8
		45006	65,000	45,000	6
		50005	70,000	50,000	5
		60004	80,000	60,000	4
		70003	85,000	70,000	3
		80002	95,000	80,000	2
		90001	105,000	90,000	1

ASTM A 338-84[b]

Specification No.	Type	Class or Grade	Hardness (Bhn)	Heat Treatment
		M3210	156 maximum	Annealed
		M4504	163–217	Air quenched and tempered
ASTM A 602-70 (1982)	Automotive	M5003	187–241	Air quenched and tempered
SAE J158a		M5503	187–241	Liquid quenched and tempered
		M7002	229–269	Liquid quenched and tempered
		M8501	269–302	Liquid quenched and tempered

TABLE 16.22 *(Continued)*

Microstructure	Typical Applications
Temper carbon and ferrite	General engineering service at normal and elevated temperatures for good machinability and excellent shock resistance
Free of primary graphite	Pipe fittings and valve parts for pressure service
Temper carbon necessary in matrix without primary cementite or graphite	General engineering service at normal and elevated temperatures; dimensional tolerance range for castings stipulated
	Flanges, pipe fittings, and valve parts for railroad, marine, and other heavy-duty service up to 650°F
Ferritic	Good machinability; steering gear housings, carriers, and mounting brackets
Ferrite and tempered pearlite[c]	Compressor crankshafts and hubs
Ferrite and tempered pearlite[c]	For selective hardening, planet carriers, transmission gears, differential cases
Tempered martensite	For machinability and improved response to induction hardening
Tempered martensite	Strength; connecting rods and universal joint yokes
Tempered martensite	High strength and wear resistance; gears

Source: Charles F. Walton, P. E.; revised February 1985. You are invited to contact Mr. Walton directly for more information on iron castings. He may be reached at 3626 Stoneleigh Rd., Cleveland Heights, Ohio 44121.
[a]Metric version of A47 available.
[b]Property requirements as specified for each application.
[c]May be all tempered martensite for some applications.

TABLE 16.23 Classification and Selection of Tool Steels

AISI Type	Composition (%)						Typical Applications
	C	W	Mo	Cr	V	Other	
Water Hardening							
W1	0.60, 1.40^a	—	—	—	—	—	Low carbon: blacksmith tools, blanking tools, caulking tools, cold chisels, forging dies, rammers, rivet sets, shear blades, punches, sledges. Medium carbon: arbors, beading tools, blanking dies, reamers, bushings, cold heading dies, chisels, coining dies, countersinks, drills, forming dies, jeweler dies, mandrels, punches, shear blades, woodworking tools. High carbon: glass cutters, jeweler dies, lathe tools, reamers, taps and dies, twist drills, woodworking tools. Vanadium content of W2 imparts finer grain, greater toughness and shallow hardenability.
W2	0.60, 1.40^a	—	—	—	0.25	—	
W5	1.10	—	—	0.50	—	—	Heavy stamping and drawing dies, tube drawing mandrels, large punches, reamers, razor blades, coldforming rools and dies, wear plates
Shock Resisting							
S1	0.50	2.50	—	1.50	—	—	Bolt header dies, chipping and caulking chisels, pipe cutters, concrete drills, expander rolls, forging dies, forming dies, grippers, mandrels, punches, pneumatic tools, scarfing tools, swaging dies, shear blades, track tools, master hobs
S2	0.50	—	0.50	—	—	1.00 Si	Hand and pneumatic chisels, drift pins, forming tools, knockout pins, mandrels, nail sets, pipe cutters, rivet sets cutters, rivet sets, screwdriver bits, shear blades, spindles, stamps, tool shanks, track
S4	0.55	—	2.00	—	—	0.80 Si	
S5	0.55	—	0.40	—	—	0.80 Mn, 2.00 Si	Hand and pneumatic chisels, drift pins, forming tools, knockout pins, mandrels, nail sets, pipe cutters, rivet sets and busters, screwdriver bits, shear blades, spindles, stamps, tool shanks, track tools, tools lathe and screw machine collets, bending dies, punches, rotary shears
S6	0.45	—	0.40	1.50	—	1.40 Mn, 2.25 Si	Shear blades, aluminum impact extrusion dies and punches, rivet sets, cold coining dies, cold header punches, knockout punches
S7	0.50	—	1.40	3.25	—	—	Shear blades, punches, slitters, chisels, forming dies, hot header dies, blanking dies, rivet sets, gripper dies, engraving dies, plastic molds, die-casting dies, master hobs, beading tools, caulking tools, chuck jaws, clutches, pipe cutters, swaging dies
Oil Hardening							
O1	0.90	0.50	—	0.50	—	1.00 Mn	Blanking dies, plastic mold dies, drawing dies, trim dies, paper knives, shear blades, taps, reamers, tools, gages, bending and forming dies, bushings, punches
O2	0.90	—	—	—	—	1.60 Mn	Blanking, stamping, trimming, cold forming dies and punches, cold forming rolls, threading dies and taps, reamers, gages, plugs and master tools, circular cutters and saws, thread roller dies, bushings, plastic molding dies

Type	C	W	Mo	Cr	V	Other
O6[c]	1.45	—	0.25	—	—	0.80 Mn, 1.00 Si
O7	1.20	1.75	—	0.75	—	—

Applications:

O6[c]: Blanking dies, forming dies, mandrels, punches, cams, brake dies, deep-drawing dies, cold forming rollers, bushings, gages, blanking and forming punches, piercing and perforating dies, taps, paper-cutting dies, wear plates, tool shanks, jigs, machine spindles, arbors, guides in grinders and straighteners

O7: Mandrels, slitters, skiving knives, taps, reamers, drills, blanking and forming dies, gages, chasers, brass finishing tools, dental burrs, paper knives, roll-turning tools, burnishing dies, pipe-threading dies, rubber-cutting knives, woodworking tools, hand reamers, scrapers, spinning tools, broaches, blanking and cold forming punches

Cold Work (Medium Alloy, Air Hardening)

Type	C	W	Mo	Cr	V	Other
A2	1.00	—	1.00	5.00	—	—
A4	1.00	—	1.00	1.00	—	2.00 Mn
A6	0.70	—	1.25	1.00	—	2.00 Mn
A7	2.25	1.00[b]	1.00	5.25	4.75	—
A8	0.55	1.25	1.25	5.00	—	—
A9	0.50	—	1.40	5.00	1.00	1.50 Ni
A10[c]	1.35	—	1.50	—	—	1.80 Mn, 1.25 Si, 1.80 Ni
A11	2.45	—	1.30	5.25	9.75	0.50 Mn, 0.90 Si

Applications:

A2: Thread-rolling dies, extrusion dies, trimming dies, blanking dies, coining dies, mandrels, shear blades, slitters, spinning rolls, forming rolls, gages, beading dies, burnishing tools, ceramic tools, embossing dies, plastic molds, stamping dies, bushings, punches, liners for brick molds

A4: Blanking dies, forming dies, trimming dies, punches, shear blades, mandrels, bending dies, forming rolls, broaches, knurling tools, gages, arbors, bushings, slitting cutters, cold threading rollers, drill bushing, master hobs, cloth-cutting knives, pilot pins, punches, engraver rolls

A6: Blanking dies, forming dies, coining dies, trimming dies, punches, shear blades, spindles, master hobs, retaining rings, mandrel, plastic dies

A7: Brick mold liners, drawing dies, briquetting dies, liners for shot-blasting equipment and sand slingers, burnishing tools, gages, forming dies

A8: Cold slitters, shear blades, hot pressing dies, blanking dies, beading tools, cold forming dies, punches, coining dies, trimming dies, master hobs, rolls, forging die inserts, compression molds, notching dies, slitter knives

A9: Solid cold heading dies, die inserts, heading hammers, coining dies, forming dies and rolls, die casings, gripper dies. Hot work applications: punches, piercing tools, mandrels, extrusion tooling, forging dies, gripper dies, die casings, heading dies, hammers, coining and forming dies

A10[c]: Blanking dies, forming dies, gages, trimming shears, punches, forming rolls, wear plates, spindle arbors, master cams and shafts, stripper plates, retaining rings

A11: Punches and dies for blanking, piercing, and forming; dies for cold drawing; rolls

TABLE 16.23 (*Continued*)

AISI Type	Composition (%)						Typical Applications
	C	W	Mo	Cr	V	Other	
Cold Work (High Carbon, High Chromium)							
D2	1.50	—	1.00	12.00	1.00	—	Blanking dies, cold forming dies, drawing dies, lamination dies, thread-rolling dies, shear blades, slitter knives, forming rolls, burnishing tools, punches, gages, knurling tools, lathe centers, broaches, cold extrusion dies, mandrels, swaging dies, cutlery
D3	2.25	—	—	12.00	—	—	Blanking dies, cold forming dies, drawing dies, lamination dies, thread rolling dies, shear blades, slitter knives, forming rolls, seaming rolls, burnishing tools, punches, gages, crimping dies, swaging dies
D4	2.25	—	1.00	12.00	—	—	Blanking dies, brick molds, burnishing tools, thread-rolling dies, hot swaging dies, wire-drawing dies, forming tools and rolls, gages, punches, trimmer dies, dies for deep drawing
D5	1.50	—	1.00	12.00	—	3.00 Co	Cold forming dies, thread-rolling dies, blanking dies, coining dies, trimming dies, draw dies, shear blades, punches, quality cutlery, rolls
D7	2.35	—	1.00	12.00	4.00	—	Brick mold liners and die plates, briquetting dies, grinding wheel molds, dies for deep drawing, flattening rolls, shot and sandblasting liners, slitter knives, wear plates, wire-drawing dies, Sendzimir mill rolls, ceramic tools and dies, lamination dies
Hot Work (Chromium)							
H10	0.40	—	2.50	3.25	0.40	—	Mandrels, extrusion and forging dies, die holders, bolsters and dummy blocks, punches, die inserts, gripper and header dies, hot shears, aluminum die-casting dies, inserts for forging dies and upsetters, shell-piercing tools
H11	0.35	—	1.50	5.00	0.40	—	Die-casting dies, punches, piercing tools, mandrels, extrusion tooling, forging dies, high-strength structural components
H12	0.35	1.50	1.50	5.00	0.40	—	Extrusion dies, dummy blocks, holders, gripper and header dies, forging die inserts, punches, mandrels, sleeves for cold heading dies
H13	0.35	—	1.50	5.00	1.00	—	Die-casting dies and inserts, dummy blocks, cores, ejector pins, plungers, sleeves, slides, extrusion dies, forging dies and inserts
H14	0.40	5.00	—	5.00	—	—	Backer blocks, die holders, aluminum and brass extrusion dies, press liners, dummy blocks, forging dies and inserts, gripper dies, shell-forging points and mandrels, hot punches, pushout rings, dies and inserts for brass forging
H19	0.40	4.25	—	4.25	2.00	4.25 Co	Extrusion and die inserts, dummy blocks, punches, forging dies and die inserts, mandrels, hot punch tools

Hot Work (Tungsten)

Type							Application
H21	0.35	9.00	—	3.50	—	—	Mandrels, hot blanking dies, hot punches, blades for flying shear, hot trimming dies, extrusion and die-casting dies for brass, dummy blocks, piercer points, gripper dies, hot nut tools (crowners, cutoffs, side dies, piercers), hot headers
H22	0.35	11.00	—	2.00	—	—	Mandrels, hot blanking dies, hot punches, blades for flying shear, hot trim dies, extrusion dies, dummy blocks, piercer points, gripper dies
H23	0.30	12.00	—	12.00	—	—	Extrusion and die-casting dies for brass, brass and bronze permanent molds
H24	0.45	15.00	—	3.00	—	—	Punches and shear blades for brass, hot blanking and drawing dies, trimming dies, dummy blocks, hot press dies, hot punches, gripper dies, hot forming rolls, hot shear blades, swaging dies, hot heading dies, extrusion dies
H26	0.50	18.00	—	4.00	1.00	—	Mandrels, hot blanking dies, hot punches, blades for flying shear, hot trimming dies, extrusion dies, dummy blocks, piercer points, gripper dies, pipe-threading dies, nut chisels, forging press inserts, extrusion dies for brass and copper

Hot Work (Molybdenum)

Type							Application
H42	0.60	6.00	5.00	4.00	2.00	—	Cold trimming dies, hot upsetting dies, dummy blocks, header dies, hot extrusion dies, cold header and extrusion dies and die inserts, hot forming and swaging dies, nut piercers, hot punches, mandrels, chipping chisels

High Speed (Tungsten)

Type							Application
T1	0.75[a]	18.00	—	4.00	1.00	—	Drills, taps, reamers, hobs, lathe and planer tools, broaches, crowners, burnishing dies, cold extrusion dies, cold heading die inserts, lamination dies, chasers, cutters, taps, end mills, milling cutters
T4	0.75	18.00	—	4.00	1.00	5.00 Co	Lathe and planer tools, drills, boring tools, broaches, roll-turning tools, milling cutters, shaper tools, form tools, hobs, single-point cutting tools
T5	0.80	18.00	—	4.00	2.00	8.00 Co	Lathe and planer tools, form tools, cutoff tools, heavy-duty tools requiring high red hardness
T6	0.80	20.00	—	4.50	1.50	12.00 Co	Heavy-duty lathe and planer tools, drills, checking tools, cutoff tools, milling cutters, hobs
T8	0.75	14.00	—	4.00	2.00	5.00 Co	Boring tools, lathe tools, heavy-duty planer tools, tool bits, single-point cutting tools for stainless steel
T15	1.50	12.00	—	5.00	4.00	5.00 Co	Form tools, lathe and planer tools, broaches, milling cutters, blanking dies, punches, heavy-duty tools requiring good wear resistance

TABLE 16.23 (Continued)

High Speed (Molybdenum)

AISI Type	Composition (%)						Typical Applications
	C	W	Mo	Cr	V	Other	
M1	0.85"	1.50	8.50	4.00	1.00	—	Drills, taps, end mills, reamers, milling cutters, hobs, punches, lathe and planer tools, form tools, saws, chasers, broaches, routers, woodworking tools
M2	0.85, 1.00"	6.00	5.00	4.00	2.00	—	Drills, taps, end mills, reamers, milling cutters, hobs, form tools, saws, lathe and planer tools, chasers, broaches and boring tools
M3-1	1.05	6.00	5.00	4.00	2.40	—	Drills, taps, end mills, reamers and counterbores, broaches, hobs, form tools, lathe and planer tools, cheeking tools, milling cutters, slitting saws, punches, drawing dies, routers, woodworking tools
M3-2	1.20	6.00	5.00	4.00	3.00	—	Drills, taps, end mills, reamers and counterbores, broaches, hobs, form tools, lathe and planer tools, cheeking tools, slitting saws, punches, drawing dies, woodworking tools
M4	1.30	5.50	4.50	4.00	4.00	—	Broaches, reamers, milling cutters, chasers, form tools, lathe and planer tools, cheeking tools, blanking dies and punches for abrasive materials, swaging dies
M6	0.80	4.00	5.00	4.00	1.50	12.00 Co	Lathe tools, boring tools, planer tools, form tools, milling cutters
M7	1.00	1.75	8.75	4.00	2.00	—	Drills, taps, end mills, reamers, routers, saws, milling cutters, lathe and planer tools, chasers, borers, woodworking tools, hobs, form tools, punches
M10	0.85, 1.00"	—	8.00	4.00	2.00	—	Drills, taps, reamers, chasers, end mills, lathe and planer tools, woodworking tools, routers, saws, milling cutters, hobs, form tools, punches, broaches
M33	0.90	1.50	9.50	4.00	1.15	8.00 Co	Drills, taps, end mills, lathe tools, milling cutters, form tools, chasers
M34	0.90	2.00	8.00	4.00	2.00	8.00 Co	Drills, taps, end mills, lathe tools, milling cutters, form tools, chasers
M36	0.80	6.00	5.00	4.00	2.00	8.00 Co	Heavy-duty lathe and planer tools, boring tools, milling cutters, drills, cutoff tools, tool holder bits
M41	1.10	6.75	3.75	4.25	2.00	5.00 Co	Drills, end mills, reamers, form cutters, lathe tools, hobs, broaches, milling cutters, twist drills, end mills; Hardenable to Rockwell C 67–70
M42	1.10	1.50	9.50	3.75	1.15	8.00 Co	
M46	1.25	2.00	8.25	4.00	3.20	8.25 Co	Milling cutters, lathe tools, form tools, reamers, end mills
M48	1.50	10.00	5.25	3.75	3.10	9.00 Co	
M61	1.80	12.50	6.50	4.00	5.00	—	Honing tools, broaches, hobs, punches, taps, drills, cutters, chasers
M62	1.30	6.25	10.50	3.75	2.00	—	Honing tools, broaches, hobs, punches, taps, drills, cutters, chasers

Intermediate High Speed

						Applications
M50	0.85	—	4.00	4.00	1.00	—	Bearings, chasers, drills, hydraulic pump components, punches, router bits, taps, woodworking tools
M52	0.90	1.25	4.00	4.00	2.00	—	Bearings, chasers, drills, hydraulic pump components, punches, router bits, taps, woodworking tools

Special Purpose (Low Alloy)

						Applications
L2	0.50:1.10ᵃ	—	—	1.00	0.20	—	Automotive gears and forgings, arbors, crank pins, chuck jaws and liners, chain feed sprockets, dogs, drift pins, die rings, friction feed disks, gun barrels, gun hoops, jack screws, lead and feed screws, machine shafts, pinions, rivet sets, shear blades, spindles, wrenches, die insert holders
L6	0.70ᵃ	—	0.25ᵇ	0.75	—	1.50 Ni	Arbors, blanking dies, forming dies, disk saws, drift pins, brake dies, hand stamps, hobs, lead and feed screws, machine parts, punches, pawls, pinions, shear blades, spindles, spring collets, swages, tool shanks, metal slitters, wood-cutting saws

Mold

						Applications
P6	0.10	—	—	1.50	—	3.50 Ni	Heavy-duty gears, shafts, bearings, plastic molding dies (hobbed and carburized)
P20	0.35	—	0.40	1.70	—	—	Molds for zinc and plastic articles, holding blocks for die-casting dies
P21	0.20	—	—	—	—	4.00 Ni, 1.20 Al	Thermoplastic injection molds, zinc die-casting dies, holding blocks for plastic and die-casting dies

Source: AISI Tool Steel Products Manual and producers of tool steels (reviewed April 1985).
ᵃOther carbon contents may be available.
ᵇOptional.
ᶜContains free graphite in microstructure to improve machinability. Some of these grades can be made with added sulfur to improve machinability.

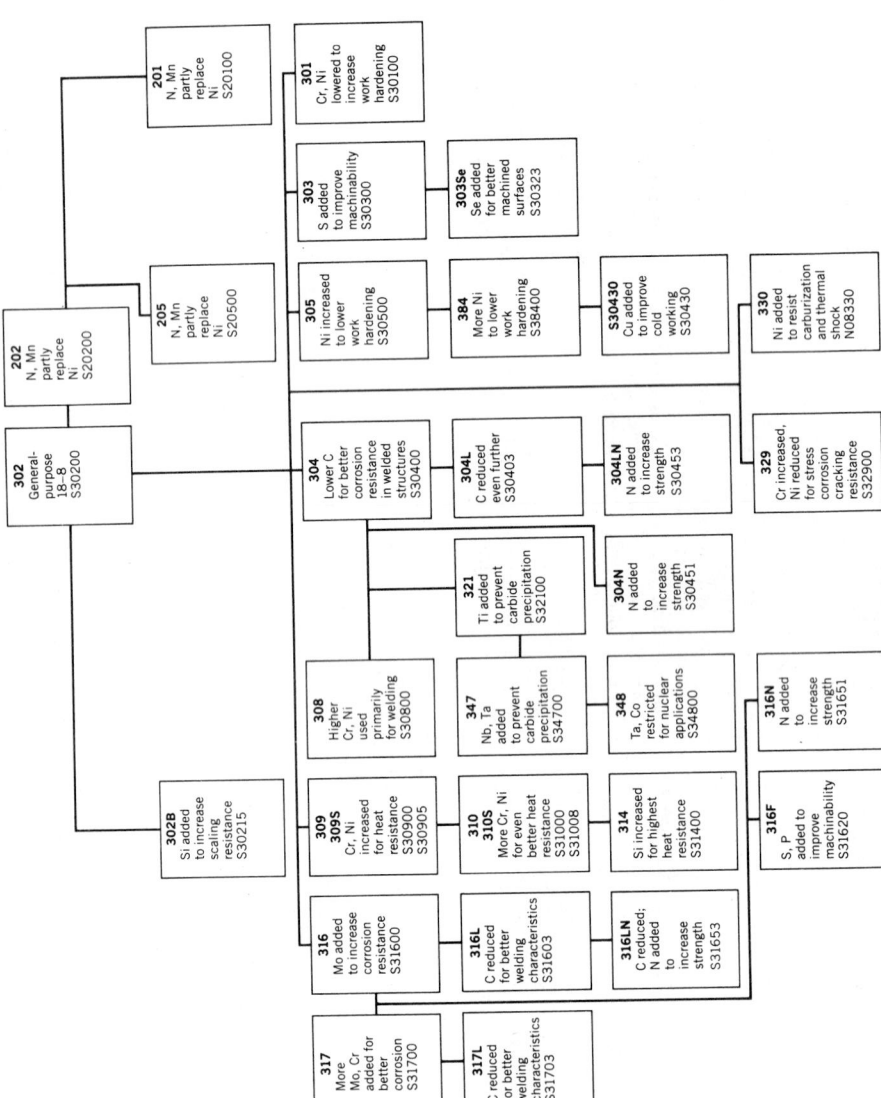

Fig. 16.5 Family relationships for standard austenitic stainless steels (see Table 16.24, Part A, pages 16 · 60 to 16 · 64). (From *Metals Handbook*, 9th ed., Vol. I, American Society for Metals; Metals Park, Ohio, 1980. Copyright American Society for Metals, 1980.)

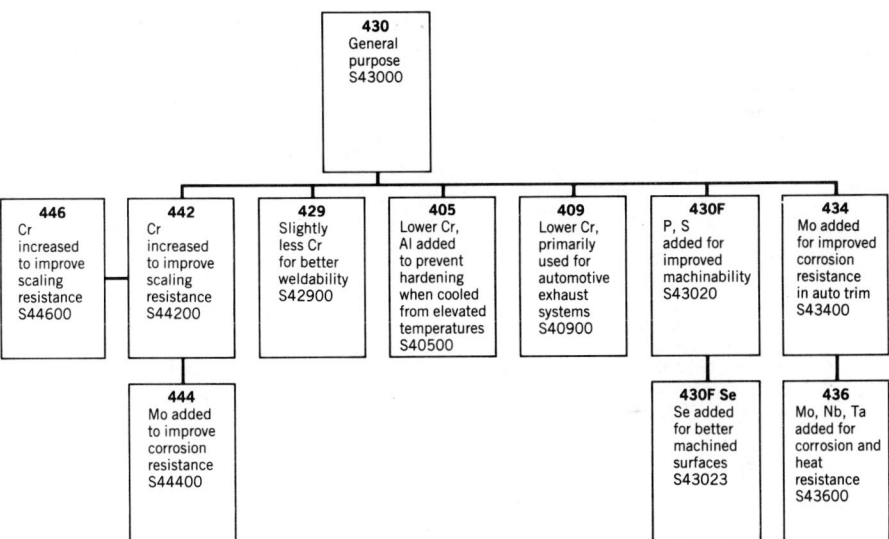

Fig. 16.6 Family relationships for standard ferritic stainless steels (see Table 16.24, Part A, pages 16 · 64 to 16 · 67). (From *Metals Handbook*, 9th ed., Vol. I, American Society for Metals, Metals Park, Ohio, 1980. Copyright American Society for Metals, 1980.)

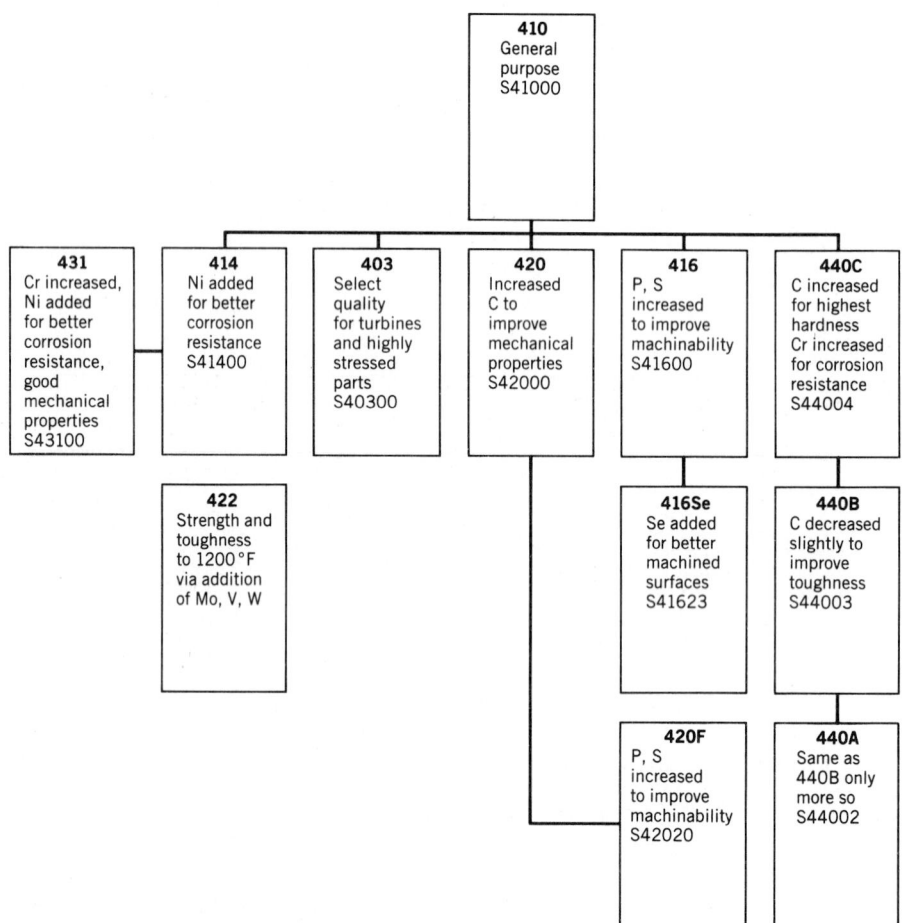

Fig. 16.7 Family relationships for standard martensitic stainless steels (see Table 16.24, Part A, pages 16 · 68 to 16 · 69). (From *Metals Handbook*, 9th ed., Vol. I, American Society for Metals, Metals Park, Ohio, 1980. Copyright American Society for Metals, 1980.)

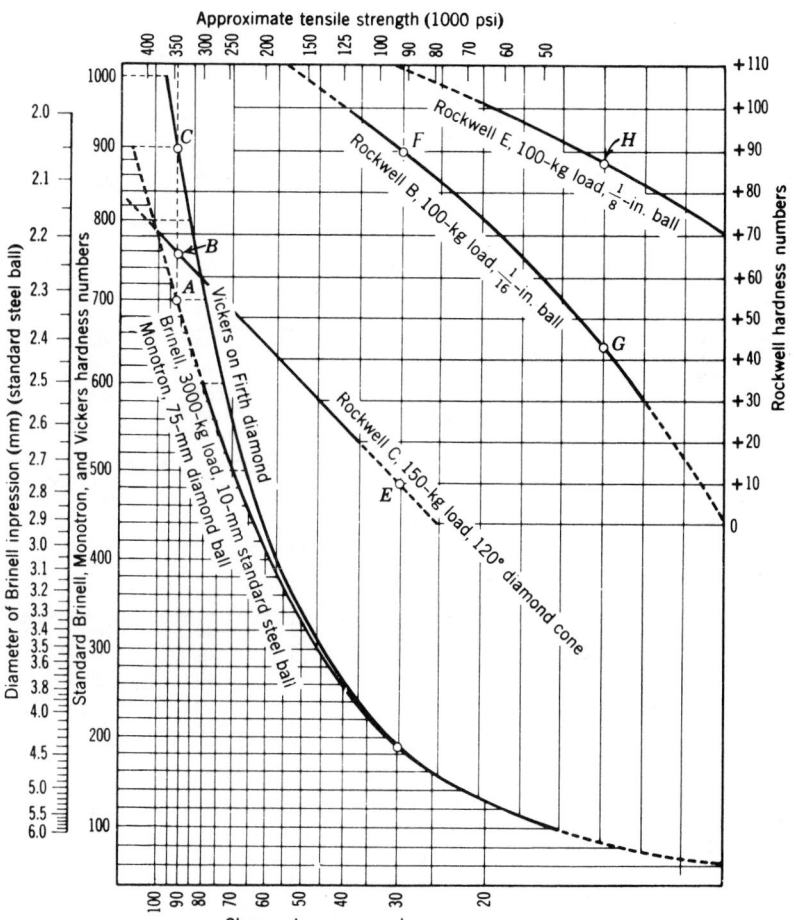

Fig. 16.8 Approximate relationship between hardnesses determined by various testing systems. Verticals represent equivalent hardness. For instance: Brinell or Monotron 700 (point *A*) equals Rockwell C-65 (point *B*) equals Vickers 900 (point *C*) equals Scleroscope 89 (bottom scale) and has about 340,000 psi tensile (top scale). Again: Rockwell B-91 (point *F*) equals Rockwell C-10 (point *E*) equals Brinell, Monotron, or Vickers 187 and Scleroscope 29 and has about 92,000 psi tensile strength. Finally, Rockwell B-43 (point *G*) equals Rockwell E-87 (point *H*). Chart based on data from Westinghouse Research Laboratories, Bureau of Standards, and the ASM *Metals Handbook*. (Courtesy of Shell Development Company.)

TABLE 16.24   Standard Stainless and Heat-Resisting Steels: Part A

AISI Type (UNS)	Typical Composition[a] (%)	Form[b]	Tensile Strength (10^3 psi)	Yield Strength (10^3 psi)	Elongation (%)	Hardness
AUSTENITIC[d]						
201 (S20100)	16–18 Cr, 3.5–5.5 Ni, 0.15 C, 5.5–7.5 Mn, 1.0 Si, 0.060 P, 0.030 S, 0.25 N	Sheets Strips Tubing	95 95 95	38 38 38	40 40 40	R_B 90 R_B 90 R_B 90
202 (S20200)	17–19 Cr, 4–6 Ni, 0.15 C, 7.5–10.0 Mn, 1.0 Si, 0.060 P, 0.030 S, 0.25 N	Sheets Strips Tubing	90 90 90	38 38 38	40 40 40	R_B 90 R_B 90 R_B 90
205 (S20500)	16.5–18 Cr, 1–1.75 Ni, 0.12–0.25 C, 14–15.5 Mn, 1.0 Si, 0.060 P, 0.030 S, 0.32–0.40 N	Plates	110	65	30	R_B 98
301 (S30100)	16–18 Cr, 6–8 Ni, 0.15 C, 2.0 Mn, 1.0 Si, 0.045 P, 0.030 S	Plates Sheets Strips Tubing	75 75 75 75	30 30 30 30	40 40 40 35	R_B 92 R_B 92 R_B 92 R_B 90
302 (S30200)	17–19 Cr, 8–10 Ni, 0.15 C, 2.0 Mn, 1.0 Si, 0.045 P, 0.030 S, 0.10 N	Bars Plates Sheets Strips Tubing Wire	75 75 75 75 75 75	30 30 30 30 30 30	50 50 50 50 50 50	Bhn 150 R_B 80 R_B 85 R_B 85 R_B 85 R_B 83
302B (S30215)	17–19 Cr, 8–10 Ni, 0.15 C, 2.0 Mn, 2.0–3.0 Si, 0.045 P, 0.030 S	Bars Plates Sheets Strips Wire	75 75 75 75 75	40 40 40 40 35	50 50 55 55 50	R_B 95 R_B 95 R_B 95 R_B 95 R_B 95
303 (S30300)	17–19 Cr, 8–10 Ni, 0.15 C, 2.0 Mn, 1.0 Si, 0.20 P, 0.15 S minimum, 0.60 Mo (optional)	Bars Tubing Wire	85 85 85	35 38 35	50 53 50	Bhn 160, R_B 76
303Se (S30323)	17–19 Cr, 8–10 Ni, 0.15 C, 2.0 Mn, 1.0 Si, 0.20 P, 0.060 S, 0.15 Se minimum					
304 (S30400)	18–20 Cr, 8–10.50 Ni, 0.08 C, 2.0 mn, 1.0 Si, 0.045 P, 0.030 S, 0.10 N	Bars Plates Sheets	75 75 75	30 30 30	30 40 40	Bhn 149 Bhn 149 R_B 80
304H (S30409)	Same as 304 but 0.04–0.10 C and no nitrogen	Strips Tubing Wire	75 75 75	30 30 30	40 40 35	R_B 80 R_B 80 R_B 83
304L (S30403)	18–20 Cr, 8–12 Ni, 0.03 C, 2.0 Mn, 0.75 Si, 0.045 P, 0.030 S, 0.10 N	Plate, sheet, and strip Bars Wire Pipe, tubing Forgings	70 70 70 70 65	25 25 25 25 25	40 40 35 35 40	R_B 88 — — R_B 90 —
304LN (S30453)	18–20 Cr, 8–12 Ni, 0.03 C, 2.0 Mn, 0.75 Si, 0.045 P, 0.030 S, 0.10–0.16 N	Plate, sheet, and strip Bars Pipe, tubing	75 75 75	40 30 35	40 30 35	R_B 92 — —

TABLE 16.24 PART A *(Continued)*

Nominal Properties of Annealed Material at Low Temperature

Temperature (°F)	Tensile Strength (10^3 psi)	Yield Strength (10^3 psi)	Elongation (%)	Reduction in Area (%)	Izod Impact Energy (ft-lb)
+70	—	—	—	—	110–120,
−300					38–70
+70	100	55	55	—	110–120
−100	145	95	38	—	—
−300	200	150	15	—	42–120
−423	220	170	5	—	—
—	—	—	—	—	200^e
+70	105	40	60	70	100
+32	155	43	53	64	110
−40	180	48	42	63	110
−80	195	50	40	62	110
−320	275	75	30	57	110
+70	94	37	68	78	110
+32	122	40	65	76	110
−40	145	48	60	73	110
−80	161	50	57	70	110
−320	219	68	46	70	110
−423	250	125	41	55	—
+70	f	f	f	f	90
+70	100	40	67	67	85
+32	114	40	61	65	90
−40	145	40	45	62	100
−80	162	40	40	60	106
−320	235	37	35	52	125
−452	267	—	30	37	—
+70	95	35	65	71	110
+32	130	34	55	68	110
−40	155	34	47	64	110
−80	170	34	39	63	110
−320	221	39	40	55	110
−423	243	50	40	50	110

TABLE 16.24   PART A   *(Continued)*

AISI Type (UNS)	Typical Composition[a] (%)	Form[b]	Tensile Strength (10^3 psi)	Yield Strength (10^3 psi)	Elongation (%)	Hardness
(S30430)	17–19 Cr, 8–10 Ni, 0.08 C, 2.0 Mn, 1.0 Si, 0.045 P, 0.030 S, 3–4 Cu	Bars Wire	75 80	30 30	70	R_B 70
304N (S30451)	18–20 Cr, 8–10.5 Ni, 0.08 C, 2.0 Mn, 0.75 Si, 0.045 P, 0.030 S, 0.10–0.16 N	Bars Sheets	80 80	35 35	30 30	Bhn 217 R_B 95
305 (S30500)	17–19 Cr, 10.50–13 Ni, 0.12 C, 2.0 Mn, 1.0 Si, 0.045 P, 0.030 S	Bars Plates Sheets Strips Tubing Wire	75 70 70 70 75 75	30 25 25 25 30 30	— 55 50 50 50 60	— — R_B 80 R_B 80 R_B 80 R_B 77
308 (S30800)	19–21 Cr, 10–12 Ni, 0.08 C, 2.0 Mn, 1.0 Si, 0.045 P, 0.030 S	Bars Plates Sheets Strips Tubing Wire	75 75 75 75 75 75[f]	30 30 30 30 30 30[f]	55 55 50 50 50 50[a]	R_B 80 Bhn 150 R_B 80 R_B 80 R_B 80
309 (S30900)	22–24 Cr, 12–15 Ni, 0.20 C, 2.0 Mn, 1.0 Si, 0.045 P, 0.030 S	Bars Plates Sheets	75 75 75	40 40 40	40 40 40	R_B 83 Bhn 170 R_B 85
309S (S30908)	22–24 Cr, 12–15 Ni, 0.08 C, 2.0 Mn, 1.0 Si, 0.045 P, 0.030 S	Strips Tubing Wire	75 75 90	40 40 45	40 35 35	R_B 85 R_B 85 R_B 98
310 (S31000)	24–26 Cr, 19–22 Ni, 0.25 C, 2.0 Mn, 1.5 Si, 0.045 P, 0.030 S	Bars Plates Sheets	75 75 75	30 30 30	40 40 40	R_B 89 Bhn 170 R_B 85
310S (S31008)	24–26 Cr, 19–22 Ni, 0.08C, 2.0 Mn, 1.5 Si, 0.045 P, 0.030 S	Stripes Tubing[h] Wire	75 55 90	30 30 45	40 35 35	R_B 85 R_B 85 R_B 98
316 (S31600)	16–18 CR, 10–14 Ni, 0.08 C, 2.0 Mn, 1.0 Si, 0.045 P, 0.030 S, 2.0–3.0 Mo, 0.10 N	Bars Plates Sheets Strips Tubing Wire	75 75 75 75 75 75	30 30 30 30 30 30	40 40 40 40 35 35	R_B 78 Bhn 149 R_B 79 R_B 79 R_B 79 R_B 78
316H (S31609)	Same as 316 but 0.04–0.10 C and no nitrogen					
316L (S31603)	16–18 Cr, 10–14 Ni, 2–3 Mo, 0.03 C, 2.0 Mn, 0.75 Si, 0.045 P, 0.030 S, 0.10 N	Plate, sheet, and strip Bar Wire Pipe, tubing Forgings	70 75 70 70 65	25 30 25 25 25	40 40 35 35 40	R_B 95 — — R_B 90 —
316LN (S31653)	16–18 Cr, 10–14 Ni, 2–3 Mo, 0.03 C, 2.0 Mn, 0.75 Si, 0.045 P, 0.030 S, 0.10–0.16 N	Plate, sheet, and strip Pipe, tubing	75 75	30 30	40 35	R_B 95 —
316F (S31620)	16–18 Cr, 10–14 Ni, 0.08 C, 2.0 Mn, 1.0 Si, 0.20 P, 0.10 S min, 1.75–2.50 Mo	Bars Sheets	85 85	35 38	40 40	Bhn 143 R_B 85
316N (S31651)	16–18 Cr, 10–14 Ni, 0.08 C, 2.0 Mn, 1.0 Si, 0.045 P, 0.030 S, 2–3 Mo, 0.10–0.16 N	Bars Sheets	80 80	35 35	30 30	Bhn 180 R_B 85

Mechanical Properties of Annealed Material at Room Temperature

TABLE 16.24 PART A *(Continued)*

Nominal Properties of Annealed Material at Low Temperature

Temperature (°F)	Tensile Strength (10^3 psi)	Yield Strength (10^3 psi)	Elongation (%)	Reduction in Area (%)	Izod Impact Energy (ft-lb)
—	—	—	—	—	240[e]
—	—	—	—	—	—
+70	—	—	—	—	110
+70	—	—	—	—	110
+70	—	—	—	—	110
+70	86	37	55	70	110
+32	85	32	64	75	110
−40	95	39	57	75	110
−80	100	40	55	75	110
−320	152	74	54	64	85
−423	176	108	56	61	—
+70	85	37	50	76	110
+32	90	39	50	75	110
−40	104	41	50	75	110
−80	118	44	50	73	110
−320	185	75	50	76	—
−423	210	84	50	60	—
—	—	—	—	—	—
—	—	—	—	—	—

TABLE 16.24 PART A *(Continued)*

AISI Type (UNS)	Typical Composition[a] (%)	Form[b]	Tensile Strength (10^3 psi)	Yield Strength (10^3 psi)	Elongation (%)	Hardness
			Mechanical Properties of Annealed Material at Room Temperature			
317 (S31700)	18–20 Cr, 11–15 Ni, 0.08 C, 2.0 Mn, 1.0 Si, 0.045 P, 0.030 S, 3.0–4.0 Mo, 0.10 N	Bars	75	30	40	Bhn 160
		Plates	75	30	35	Bhn 160
		Sheets	75	30	35	R_B 85
		Strips	75	30	35	R_B 85
		Tubing	75	30	35	R_B 85
317L (S31703)	18–20 Cr, 11–15 Ni, 0.03 C, 2.0 Mn, 1.0 Si. 0.045 P, 0.030 S, 3.0–4.0 Mo, 0.10 N	Plates	75	30	55	R_B 80
		Sheets	75	30	55	R_B 85
		Tubing	85	35	55	
		Bars	85	35	45	
321 (S32100)	17–19 Cr, 9–12 Ni, 0.08 C, 2.0 Mn, 1.0 Si, 0.045 P, 0.030 S, Ti = 5(C + N) min. 0.10 N	Bars	75	30	40	Bhn 150
		Plates	75	30	40	Bhn 160
		Sheets	75	30	40	R_B 80
		Strips	75	30	40	R_B 80
321H (S32109)	Same as 321 but 0.04–0.10 C	Tubing	75	30	35	R_B 80
		Wire	75	30	40	R_B 89
329 (S32900)	23–28 Cr, 2.5–5 Ni, 0.08 C, 2.0 Mn, 0.75 Si, 0.040 P, 0.030 S, 1–2 Mo	Bars	105	80	25	Bhn 230
		Strips	90	70	25	Bhn 230
		Plate	90	70	15	
330 (N08330)	17–20 Cr, 34–37 Ni, 0.08 C, 2.0 Mn, 0.75–1.50 Si, 0.040 P, 0.030 S	Bars	70	30	45	R_B 80
		Plates	70	30	45	R_B 80
		Sheets	70	30	40	
		Strips	70	30	40	
347 (S34700)	17–19 Cr, 9–13 Ni, 0.08 C, 2.0 Mn, 1.0 Si, 0.045 P, 0.030 S, (Cb + Ta) = 10(C) minimum to 1.10 maximum					
347H (S34709)	Same as 347 but 0.04–0.10 C and 1.0 (Cb + Ta) maximum	Bars	75	30	40	Bhn 160
		Plates	75	30	40	Bhn 160
		Sheets	75	30	40	R_B 85
348 (S34800)	17–19 Cr, 9–13 Ni, 0.08 C, 2.0 Mn, 1.0 Si, 0.045 P, 0.030 S, (Cb + Ta) = 10(C) minimum to 1.10 maximum, 0.10 Ta, 0.20 Co	Strips	75	30	35	R_B 85
		Tubing	75	30	35	R_B 85
		Wire	100[g]	70[g]	40[g]	R_B 95[g]
348H (S34809)	Same as 348 but 0.04–0.10 C and 1.0 (Cb + Ta) maximum					
384 (S38400)	15–17 Cr, 17–19 Ni, 0.08 C, 2.0 Mn, 1.0 Si, 0.045 P, 0.030 S	Wire	80	30	38	R_B 70
		Bar	75	30	—	—
FERRITIC[d]						
405 (S40500)	11.5–14.5 Cr, 0.08 C, 1.0 Mn, 1.0 Si, 0.040 P, 0.030 S, 0.1–0.3 Al	Bars	70	40	20	Bhn 150
		Plates	70	30	25	Bhn 150
		Sheets	70	30	25	R_B 75
		Tubing	60	30	25	R_B 80
		Wire	70[g]	45[g]	20[g]	

TABLE 16.24 PART A *(Continued)*

Nominal Properties of Annealed Material at Low Temperature

Temperature (°F)	Tensile Strength (10^3 psi)	Yield Strength (10^3 psi)	Elongation (%)	Reduction in Area (%)	Izod Impact Energy (ft-lb)
Same as type 316	Same as type 316	Same as type 316	Same as type 316	Same as type 316	Same as type 316
—	—	—	—	—	—
+70	89	37	62	50	110
+32	99	38	58	50	110
−40	117	44	58	50	115
−80	130	45	57	50	117
−320	208	64	44	50	110
−423	238	92	35	—	—
—	—	—	—	—	40^e
—	—	—	—	—	240^e
+70	93	38	55	69	110
+32	105	42	62	72	110
−40	117	44	63	71	117
−80	130	45	57	70	110
−320	200	47	43	65	95
−423	228	65	39	53	60
—	—	—	45	55	—
—	—	—	35	—	—
+70	Approximately same as type 410 in annealed condition				

TABLE 16.24 Part A *(Continued)*

AISI Type (UNS)	Typical Composition[a] (%)	Form[b]	Mechanical Properties of Annealed Material at Room Temperature			
			Tensile Strength (10^3 psi)	Yield Strength (10^3 psi)	Elongation (%)	Hardness
409 (S40900)	10.5–11.75 Cr, 0.08 C, 1.0 Mn, 1.0 Si, 0.045 P, 0.045 S, Ti = 6(C) minimum to 0.75 maximum	Bars	60	30	20	R_B 75
		Plates	60	30	20	R_B 75
		Sheets	60	30	20	R_B 75
		Sheets	60	30	20	R_B 75
		Strips	60	30	20	R_B 75
		Tubing	60	30	20	—
429 (S42900)	14–16 Cr, 0.12 C, 1.0 Mn, 1.0 Si, 0.040 P, 0.030 S	Bars	70	45	30	Bhn 156
		Plates	65	40	30	Bhn 163
430 (S43000)	16–18 Cr, 0.12 C, 1.0 Mn, 1.0 Si, 0.040 P, 0.030 S	Bars	70	45	40	Bhn 155
		Plates	65	40	30	Bhn 160
		Sheets	65	50	30	R_B 85
		Strips	65	50	30	R_B 85
		Tubing	60	40	35	R_B 80
		Wire	70	40	40	R_B 82
430F (S43020)	16–18 Cr, 0.12 C, 1.25 Mn, 1.0 Si, 0.060 P, 0.15 S minimum 0.60 Mo (optional)	Bars	80	55	25	Bhn 170
		Wire	95[g]	85[g]	10[g]	R_B 92[g]
434 (S43400)	16–18 Cr, 0.12 C, 1.0 Mn, 1.0 Si, 0.040 P, 0.030 S, 0.75–1.25 Mo	Sheets	77	53	23	R_B 83
		Strips	77	53	23	R_B 83
		Wire	79	60	33	R_B 90
436 (S43600)	16–18 Cr, 0.12 C, 1.0 Mn, 1.0 Si, 0.040 P, 0.030 S, 0.75–1.25 Mo, 5(Cb + Ta) × C minimum to 0.70 maximum	Sheets	77	53	23	R_B 83
		Strips	77	53	23	R_B 83
439 (S43035 or S43900)	17–19 Cr, 0.50 Ni, 0.07 C, 1.0 Mn, 1.0 Si, 0.040 P, 0.030 S, 0.15 Al, 0.04 N, Ti = 0.20 + 4(C + N) minimum to 1.10 maximum	Bars	70	40	20	—
		Plates	65	30	22	R_B 88
		Sheets	65	30	22	R_B 88
		Strips	65	30	22	R_B 88
		Tubing	60	30	20	R_B 90
442 (S44200)	18–23 Cr, 0.20 C, 1.0 Mn, 1.0 Si, 0.040 P, 0.030 S 17.5–19.5 Cr, 1.0 Ni,	Bars	80	45	20	R_B 90
444 (S44400)	0.025 C, 1.0 Mn, 1.0 Si, 0.040 P, 0.030 S, 1.75–2.5 Mo, 0.035 N, (Ti + Cb) = 0.20 + 4(C + N) minimum to 0.80 maximum	Plates	60	40	20	R_B 95
		Sheets	60	40	20	R_B 95
		Strips	60	40	20	R_B 95
		Tubing	60	45	20	R_B 95
446 (S44600)	23–27 Cr, 0.20 C, 1.5 Mn, 1.0 Si, 0.040 P, 0.030 S, 0.25 N	Bars	70	40	20	R_B 86
		Plates	75	40	20	R_B 84
		Sheets	75	40	20	R_B 83
		Strips	75	40	20	R_B 83
		Tubing	70	40	18	R_B 84
		Wire	70[g]	40[g]	20[g]	R_B 92[g]

TABLE 16.24   PART A   *(Continued)*

Nominal Properties of Annealed Material at Low Temperature

Temperature (°F)	Tensile Strength (10^3 psi)	Yield Strength (10^3 psi)	Elongation (%)	Reduction in Area (%)	Izod Impact Energy (ft-lb)
—	—	—	—	45	—
				45	
				45	
				45	
				45	
—	—	—	—	—	—
+70	65	38	37	73	35
+32	69	40	37	72	20
−40	76	41	36	72	10
−80	81	44	36	70	8
−320	90	87	2	4	2
+70					5.5
−100	—	—	—	—	4
−300					1
—	—	—	—	—	—
—	—	—	—	—	—
—	—	—	—	—	—
+70	—	—	—	—	5
—	—	—	—	—	—
+70	—	—	—	—	2

TABLE 16.24   **Part A** *(Continued)*

AISI Type (UNS)	Typical Composition[a] (%)	Form[b]	Tensile Strength (10³ psi)	Yield Strength (10³ psi)	Elongation (%)	Hardness
MARTENSITIC[d]						
403 (S40300)	11.5–13.0 Cr, 0.15 C, 1.0 Mn, 0.5 Si, 0.040 P, 0.030 S	Bars	75	40	35	R_B 82
		Sheets	70	45	25	R_B 80
		Strips	70	45	25	R_B 80
		Tubing	75	40	35	R_B 80
		Wire	95[g]	80[g]	15[g]	R_B 92[g]
410 (S41000)	11.5–13.5 Cr, 0.15 C, 1.0 Mn, 1.0 Si, 0.040 P, 0.030 S	Bars	70	35	35	R_B 82
		Plates	60	30	30	Bhn 150
		Sheets	60	30	25	R_B 80
410S (S41008)	Same as 410 but 0.08 C	Strips	60	30	25	R_B 80
		Tubing	60	30	30	R_B 82
		Wire	70	40	30	R_B 82
414 (S41400)	11.5–13.5 Cr, 1.25–2.50 Ni, 0.15 C, 1.0 Mn, 1.0 Si, 0.040 P, 0.030 S	Bars	115	90	20	Bhn 235
		Plates	115	90	20	Bhn 235
		Sheets	115	105	15	R_B 98
		Strips	115	105	15	R_B 98
		Wire	135[g]	115[g]	10[g]	R_C 29[g]
416 (S41600)	12–14 Cr, 0.15 C, 1.25 Mn, 1.0 Si, 0.060 P, 0.15 S minimum, 0.60 Mo (optional)	Bars	75	40	30	R_B 82
		Tubing	75	40	30	R_B 82
		Wire	75	40	20	R_B 82
420 (S42000)	12–14 Cr, 0.15 C minimum, 1.0 Mn, 1.0 Si, 0.040 P, 0.030 S	Bars	95	50	25	R_B 92
		Wire	95	50	20	R_B 92
420F (S42020)	12–14 Cr, over 0.15 C, 1.25 Mn, 1.0 Si, 0.060 P, 0.15 S minimum, 0.60 Mo maximum (optional)	Bars	95	55	22	Bhn 220
		Wire	100[g]	80[g]	15[g]	R_B 99[g]
422 (S42200)	11.0–12.5 Cr, 0.5–1.0 Ni, 0.20–0.25 C, 0.5–1.0 Mn, 0.5 Si, 0.025 P, 0.025 S, 0.9–1.25 Mo, 0.2–0.3 V, 0.9–1.25 W	Bars	140	110	13	Bhn 320
		Sheet	180	140	5	
431 (S43100)	15–17 Cr, 1.25–2.50 Ni, 0.20 C, 1.0 Mn, 1.0 Si, 0.040 P, 0.030 S	Bars	125	95	20	Bhn 260
		Wire	135[g]	115[g]	10[g]	R_C 29[g]
440A (S44002)	16–18 Cr, 0.60–0.75 C, 1.0 Mn, 1.0 Si, 0.040 P, 0.030 S, 0.75 Mo	Bars	105	60	20	R_B 95
		Wire	140	60	18	R_B 95
		Strip	100	60		
440B (S44003)	16–18 Cr, 0.75–0.95 C, 1.0 Mn, 1.0 Si, 0.040 P, 0.030 S, 0.75 Mo	Bars	107	62	18	R_B 96
		Wire	140	62	16	R_B 96
440C (S44004)	16–18 Cr, 0.95–1.20 C, 1.0 Mn, 1.0 Si, 0.040 P, 0.030 S, 0.75 Mo	Bars	110	65	14	R_B 97
		Wire	110	65	13	R_B 97

TABLE 16.24　PART A　*(Continued)*

Nominal Properties of Annealed Material at Low Temperature

Temperature (°F)	Tensile Strength (10^3 psi)	Yield Strength (10^3 psi)	Elongation (%)	Reduction in Area (%)	Izod Impact Energy (ft-lb)
Same as type 410	Same as type 410	Same as type 410	Same as type 410	Same as type 410	Same as type 410
+70	110	87	21	68	85
+32	115	89	24	69	40
−40	122	90	23	64	25
−80	128	94	22	60	25
−320	158	148	10	11	5
70	—	—	—	—	40–80
+70					20–64
−100	—	—	—	—	50
−300					3
+70					10
+32	—	—	—	—	10
−40					8
−80					7
—	—	—	—	—	—
—	—	—	—	—	—
+70					50
+32	—	—	—	—	50
−40					30
−80					17
—	—	—	—	—	—
—	—	—	—	—	—
—	—	—	—	—	—

TABLE 16.24   Part A *(Continued)*

AISI Type (UNS)	Typical Composition[a] (%)	Form[b]	Mechanical Properties of Annealed Material at Room Temperature			
			Tensile Strength (10^3 psi)	Yield Strength (10^3 psi)	Elongation (%)	Hardness
HEAT RESISTING						
501 (S50100)	4–6 Cr, 0.10 C minimum, 1.0 Mn, 1.0 Si, 0.040 P, 0.030 S, 0.40–0.65 Mo	Bars	70	30	28	Bhn 160 ⎫
		Plates	70	30	28	Bhn 160 ⎬
502 (S50200)	1.0 Si, 0.040 P, 0.030 S, 0.40–0.65 Mo	Bars	65	25	30	Bhn 150 ⎫
		Plates	65	25	30	Bhn 150 ⎪
		Sheets	70	—	30	R_B 75 ⎬
		Strips	70	—	30	R_B 75 ⎪
		Wire	75	30	30	R_B 72 ⎭
503 (S50300)	6–8 Cr, 0.15 C, 1.0 Mn, 1.0 Si, 0.040 P, 0.040 S, 0.45–0.65 Mo	Plates	60–85	30	18	— ⎫
504 (S50400)	Same as 503 but 8–10 Cr and 0.90–1.10 Mo	Forgings	60	30	20	Bhn 201 ⎬
PRECIPITATION HARDENING[p]						
(S13800)	12.25–13.25 Cr, 7.5–8.5 Ni, 0.05 C, 0.10 Mn, 0.10 Si, 0.010 P, 0.008 S, 0.90–1.35 Al, 2.0–2.5 Mo, 0.010 N	Bars	160	120	17	R_C 33 ⎫
		Plates	160	120	17	R_C 33 ⎬
(S15500)	14–15.5 Cr, 3.5–5.5 Ni, 0.07 C, 1.0 Mn, 1.0 Si, 0.04 P, 0.03 S, 2.5–4.5 Cu, 0.15–0.45 (Cb + Ta)	Bars	160	145	15	R_C 35 ⎫
		Plates	160	145	15	R_C 35 ⎪
		Sheets	160	145	15	R_C 35 ⎬
		Strips	160	145	15	R_C 35 ⎭
(S17400)	15.5–17.5 Cr, 3–5 Ni, 0.07 C, 1.0 Mn, 1.0 Si, 0.040 P, 0.030 S, 3–5 Cu, 0.15–0.45 (Cb + Ta)	Bars	160	145	15	R_C 35 ⎫
		Plates	160	145	15	R_C 35 ⎬
		Sheets	160	145	5	R_C 35 ⎭
(S17700)	16–18 Cr, 6.5–7.75 Ni, 0.09 C, 1.0 Mn, 1.0 Si, 0.040 P, 0.040 S, 0.75–1.50 Al	Bars	130	40	10	R_B 90 ⎫
		Plates	130	40	10	R_B 90 ⎬
		Sheets	130	40	35	R_B 85 ⎭

Source: Committee of Stainless Steel Producers. American Iron and Steel Institute (revised April 1985). Based on the "Stainless and Heat Resisting Steels" Section of the AISI *Steel Products Manual*, latest edition.

[a] Single values are maximums, except as noted.

[b] Forms listed are only those for which mechanical properties are given. Most types are available in many forms.

[c] Followed by rapid cooling. *H* is hardening temperature; *T* is tempering.

[d] Austenitic: hardenable by cold working; not hardenable by heat treatment. Ferritic: not hardenable by heat treatment or cold working. Martensitic: hardenable by heat treatment.

[e] Charpy V-notch.

[f] Not applicable. Silicon added to type 302 for oxidation resistance.

[g] Soft temper.

TABLE 16.24 PART A *(Continued)*

Nominal Properties of Annealed Material at Low Temperature

Temperature (°F)	Tensile Strength (10^3 psi)	Yield Strength (10^3 psi)	Elongation (%)	Reduction in Area (%)	Izod Impact Energy (ft-lb)
—	—	—	—	—	—
—	—	—	—	—	—
—	—	—	—	—	—
—	—	—	—	—	60^e
					60^e
—	—	—	—	—	30^e
					30^e
					30^e
					30^e
—	—	—	—	—	30^e
					30^e
—	—	—	—	—	—

^{h}Composition for type 310 tubing varies slightly from AISI values; for standard compositions, refer to ASTM A213.

^{i}Stabilizing temperature, 1550–1650°F.

^{j}Hardening and tempering temperature, °F.

^{k}Full anneal, followed by slow cooling.

^{l}Low anneal.

^{m}Tempering within the range of 800–1100°F is not recommended because of resulting low and erratic impact properties and reduced corrosion resistance. Time at temperature and temperatures may vary depending on part size.

^{n}Retarded cool.

^{o}Retarded cool and anneal.

^{p}Mechanical properties are for a solution-treated condition.

TABLE 16.24 Standard Stainless and Heat-Resisting Steels: Part B

AISI Type (UNS)	Mechanical Properties at Elevated Temperatures					Scaling Temperature		Initial Forging Temperature (°F)	Thermal Treatment			Characteristics and Typical Applications
	Creep Strength (Load for 1% Elongation in 10 000 hr, 10^3 psi)					Maximum Continuous Service in Air (°F)	Maximum Intermittent Service in Air (°F)		Annealing Temperature (°F)^c	Stress-Relief Annealing Temperature (°F)	Melting Range (°F)	
	1000 F°	1100 F°	1200 F°	1300 F°	1500 F°							
AUSTENITIC												
201 (S20100)	—	—	—	—	—	1550	1450	2100–2250	1850–2050	—	—	High work-hardening rate; low nickel equivalent of type 301. Flatware; automobile wheel covers, trim.
202 (S20200)	—	—	—	—	—	1550	1450	2100–2250	1850–2050	—	—	General-purpose low nickel equivalent of type 302. Kitchen equipment; hub caps; milk handling.
205 (S20500)	—	—	—	—	—	—	—	2250	1950	—	—	Lower work-hardening rate than type 202; used for spinning and special drawing operations. Nonmagnetic and cryogenic parts.
301 (S30100)	19	12.5	8	4.5	1.8	1650	1500	2100–2300	1850–2050	400–750	2550–2590	High work-hardening rate; used for structural applications where high strength plus high ductility is required. Railroad cars; trailer bodies; aircraft structurals; fasteners; automobile wheel covers, trim; pole line hardware.
302 (S30200)	20	12.5	7.5	4.3	1.5	1650	1500	2100–2300	1850–2050	400–750	2550–2590	General-purpose austenitic stainless steel. Trim; food-handling equipment; aircraft cowlings; antennas; springs; cookware; building exteriors; tanks; hospital, household appliances; jewelry; oil-refining equipment; signs.
302B (S30215)	—	—	7	4.5	1	1750	1600	2050–2250	1850–2050	—	2500–2550	More resistant to scale than type 302. Furnace parts; still liners; heating elements; annealing covers; burner sections.
303 (S30300)	16.5	11.5	6.5	3.5	0.7	1650	1400	2100–2350	1850–2050	400–750	2550–2590	Free-machining modification of type 302, for heavier cuts. Screw machine products; shafts; valves; bolts; bushings; nuts.
303 Se (S30323)	16.5	11.5	6.5	3.5	0.7	1650	1400	2100–2350	1850–2050	400–750	2550–2590	Free-machining modification of type 302, for lighter cuts; used where hot working or cold heading may be involved. Aircraft fittings; bolts; nuts; rivets; screws; studs.

Type (UNS)											Characteristics and typical uses
304 (S30400), 304H (S30409)	20	12	7.5	4	1.5	1650	2100–2300	1850–2050	400–750	2550–2650	Low-carbon modification of type 302 for restriction of carbide precipitation during welding. Chemical and food-processing equipment; brewing equipment; cryogenic vessels; gutters; downspouts; flashings. 304H: Improved high-temperature strength.
304L (S30403)	20	12	7.5	4	1.5	1650	2100–2300	1850–2050	400–750	2550–2650	Extra-low-carbon modification of type 304 for further restriction of carbide precipitation during welding. Coal hopper linings; tanks for liquid fertilizer and tomato paste.
304LN (S30453)	20	12	7.5	4	1.5	1650	2100–2300	1850–2050	400–750	2250–2650	Used in welded fabrication where its mechanical strength exceeds that of type 304L.
(S30430)	—	—	—	—	—	—	2100–2300	1850–2050	—	2550–2650	Lower work-hardening rate than type 305. Severe cold-heading applications.
304N (S30451)	—	—	—	—	—	—	2100–2300	1850–2050	—	2550–2650	Higher nitrogen than type 304 to increase strength with minimum effect on ductility and corrosion resistance, more resistant to increased magnetic permeability. Type 304 applications requiring higher strength.
305 (S30500)	19	12.5	8	4.5	2	1650	2100–2300	1850–2050	—	2550–2650	Low work-hardening rate; used for spin forming, severe drawing, cold heading, and forming. Coffee urn tops; mixing bowls; reflectors.
308 (S30800)	—	—	—	—	—	1700	2100–2300	1850–2050	1550	2550–2590	Higher alloy steel having high corrosion and heat resistance. Welding filler metals to compensate for alloy loss in welding; industrial furnaces.
309 (S30900)	16.5	12.5	10	6	3	1950	2050–2250	1900–2050	—	2550–2650	High-temperature strength and scale resistance. Aircraft heaters; heat-treating equipment; annealing covers; furnace parts; heat exchangers; heat-treating trays; oven linings; pump parts.
309S (S30908)	16.5	12.5	10	6	3	1950	2050–2250	1900–2050	—	2250–2650	Low-carbon modification of type 309. Welded constructions; assemblies subject to moist corrosion conditions.
310 (S31000)	33	23	15	10	3	2050	2000–2250	1900–2100	400–750	2550–2650	Higher elevated temperature strength and scale resistance than type 309. Heat exchangers; furnace parts; combustion chambers; welding filler metals; gas turbine parts; incinerators; recuperators; rolls for roller hearth furnaces.
310S (S31008)	33	23	15	10	3	2050	2000–2250	1900–2100	400–750	2550–2650	Low-carbon modification of type 310. Welded constructions; jet engine rings.

TABLE 16.24 PART B *(Continued)*

AISI Type (UNS)	Mechanical Properties at Elevated Temperatures							Initial Forging Temperature (°F)	Thermal Treatment		Melting Range (°F)	Characteristics and Typical Applications
	Creep Strength (Load for 1% Elongation in 10 000 hr, 10^3 psi)					Scaling Temperature			Annealing Temperature (°F)c	Stress-Relief Annealing Temperature (°F)		
	1000 F°	1100 F°	1200 F°	1300 F°	1500 F°	Maximum Continuous Service in Air (°F)	Maximum Intermittent Service in Air (°F)					
316 (S31600), and 316H (S31609)	25	17.4	11.6	7.5	2.4	1650	1550	2100–2300	1850–2050	400–750	2500–2550	Higher corrosion resistance than types 302 and 304; high creep strength. Chemical and pulp-handling equipment; photographic equipment, brandy vats; fertilizer parts; ketchup cooking kettles; yeast tubs.
316L (S31603), and 316LN (S31653)	25	17.4	11.6	7.5	2.4	1650	1550	2100–2300	1850–2050	400–750	2500–2550	Extra-low-carbon modification of type 316. Welded construction where intergranular carbide precipitation must be avoided. Type 316 applications requiring extensive welding.
316F (S31620)	—	—	—	—	—	—	—	2200	2000	—	2500–2550	Higher phosphorus and sulfur than type 316 to improve machining and nonseizing characteristics. Automatic screw machine parts.
316N (S31651)	—	—	—	—	—	—	—	2100–2300	1850–2050	—	2500–2550	Higher nitrogen than type 316 to increase strength with minimum effect on ductility and corrosion resistance. Type 316 applications requiring extra strength.
317 (S31700)	23	16.8	11.2	6.9	2.0	1700	1600	2100–2300	1850–2050	—	2500–2550	Higher corrosion and creep resistance than type 316. Dyeing and ink-manufacturing equipment.
317L (S31703)	—	—	—	—	—	—	—	2250	1900–2000	—	2500–2550	Extra-low-carbon modification of type 317 for restriction of carbide precipitation during welding. Welded assemblies.

Type												Applications
321 (S32100) and 321H (S32109)	18	17	9	5	1.5	1650	1550	2100–2300	1750–2050	400–750[i]	2550–2600	Stabilized for weldments subject to severe corrosive conditions and for service from 800 to 1600°F. Aircraft exhaust manifolds; boiler shells; process equipment; expansion joints; cabin heaters; fire walls; flexible couplings; pressure vessels.
329 (S32900)	—	—	—	—	—	—	—	2000	1750–1800	H1350[c]	—	Austenitic-ferritic type with general corrosion resistance similar to type 316 but with better resistance to stress-corrosion cracking; capable of age hardening. Valves; valve fittings; piping; pump parts.
330 (N08330)	—	—	—	—	—	—	—	2100–2150	1950–2150	—	2550–2600	Good resistance to carburization and to heat and thermal shock. Heat treating fixtures.
347 (S34700), and 347H (S34709)	32	23	16	10	2	1650	1550	2100–2300	1850–2050	400–750[i]	2550–2600	Similar to type 321 with higher creep strength. Airplane exhaust stacks; welded tank cars for chemicals; jet engine parts.
348 (S34800), and 348H (S34809)	32	23	16	10	2	1650	1550	2100–2300	1850–2050	400–750[i]	2500–2600	Similar to type 321; low retentivity. Tubes and pipes for radioactive systems; nuclear energy uses.
384 (S38400)	—	—	—	—	—	—	—	2100–2250	1900–2100	—	2550–2650	Suitable for severe cold heading or cold forming; lower cold-work hardening rate than type 304. Bolts; rivets; screws; instrument parts.
FERRITIC[d]												
405 (S40500)	8.4	—	—	—	—	1400	1450	1950–2050	Low anneal, 1350–1500	—	2700–2790	Nonhardenable grade for assemblies where air-hardening types such as 410 or 403 are objectionable. Annealing boxes; quenching racks; oxidation-resistant partitions.
409 (S40900)	—	—	—	—	—	—	—	—	1625	—	2600–2750	General-purpose construction stainless. Automotive exhaust systems; transformer and capacitor cases; dry fertilizer spreaders; tanks for agricultural sprays.

TABLE 16.24   PART B   (Continued)

Mechanical Properties at Elevated Temperatures

AISI Type (UNS)	Creep Strength (Load for 1% Elongation in 10 000 hr, 10³ psi)					Scaling Temperature		Thermal Treatment			Melting Range (°F)	Characteristics and Typical Applications
	1000 F°	1100 F°	1200 F°	1300 F°	1500 F°	Maximum Continuous Service in Air (°F)	Maximum Intermittent Service in Air (°F)	Initial Forging Temperature (°F)	Annealing Temperature (°F)c	Stress-Relief Annealing Temperature (°F)		
429 (S42900)	—	—	—	—	—	—	—	1900–2050	1450–1550	—	2650–2750	Improved weldability as compared to type 430. Nitric acid and nitrogen fixation equipment.
430 (S43000)	8.5	4.7	2.6	1.4	—	1550	1650	1900–2050	Low anneal, 1400–1500	—	2600–2750	General-purpose nonhardenable chromium type. Decorative trim, nitric acid tanks; annealing baskets; combustion chambers; dishwashers; heaters; mufflers; range hoods; recuperators; restaurant equipment.
430F (S430200)	8.5	4.6	1.9	1.3	—	1500	1600	1950–2100	Low anneal, 1250–1400	—	2600–2750	Free-machining modification of type 430 for heavier cuts. Screw machine parts.
434 (S43400)	—	—	—	—	—	—	—	1900–2050	1450–1550	—	2600–2750	Modification of type 430 designed to resist atmospheric corrosion in presence of winter road-conditioning and dust-laying compounds. Automotive trim and fasteners.
436 (S43600)	—	—	—	—	—	—	—	1900–2050	1450–1550	—	2600–2750	Similar to types 430 and 434. Used where low "roping" or "ridging" required. General corrosion and heat-resistant applications such as automobile trim.
439 (S43035) or (S43900)	—	—	—	—	—	—	—	—	—	—	—	—
442 (S44200)	8.5	5	1.6	1	0.6	1800	1900	1600–2100	1300	—	2600–2750	High-chromium steel, principally for parts which must resist high service temperatures without scaling. Furnace parts; nozzles; combustion chambers.

Type (UNS)											Hardening and Tempering Temperature (F°)		Uses
444 (S44400)	—	—	—	—	—	—	—	—	—	—	—	—	
446 (S44600)	6.4	2.9	1.4	0.6	0.4	—	1950	2050	1950-2050	1450-1600	—	2600-2750	High resistance to corrosion and scaling at high temperatures especially for intermittent service; often used in sulfur-bearing atmosphere. Annealing boxes; combustion chambers; glass molds; heaters; pyrometer tubes; recuperators; stirring rods; valves.
MARTENSITIC[d]													
403 (S40300)	11	4.5	2	1.4	—	—	1300	1450	2000-2200[i]	1500-1650[k] / 1200-1400[l]	H1700-1850,[c] T400-1400[j,m]	2700-2790	"Turbine quality" grade. Steam turbine blading and other highly stressed parts including jet engine rings.
410 (S41000)	11.5	4.3	2	1.5	—	—	1300	1450	2000-2200[n]	1500-1650[k] / 1200-1400[l]	H1700-1850,[c] T400-1400[m]	2700-2790	General-purpose heat-treatable type. Machine parts; pump shafts; bolts; bushings; coal chutes; cutlery; finishing tackle; hardware; jet engine parts; mining machinery; rifle barrels; screws; valves.
410S (S41008)	11.5	4.3	2	1.5	—	—	1300	1450	2000-2200[n]	1500-1650[k] / 1200-1400[l]	H1700-1850,[c] T400-1400[m]	2700-2790	
414 (S41400)	—	—	—	—	—	—	1300	1450	2100-2200	— / 1200-1300[l]	H1800-1900,[c] T400-1300[m]	—	High-hardenability steel. Springs; tempered rules, machine parts, bolts; mining machinery; scissors; ships' bells; spindles; valve seats.
416 (S41600)	11	4.6	2	1.2	—	—	1250	1400	2100-2300[n]	1500-1650[k] / 1200-1400[l]	H1700-1850,[c] T400-1400[m]	2700-2790	Free-machining modification of type 410; for heavier cuts. Aircraft fittings; bolts; nuts; fire extinguisher inserts; rivets; screws.
420 (S42000)	9.2	4.2	2	1	—	—	1200	1400	2000-2200[o]	1550-1650[k] / 1350-1450[l]	H1800-1900,[c] T300-700	2650-2750	Higher carbon modification of type 410. Cutlery; surgical instruments; valves; wear-resisting parts; glass molds; hand tools; vegetable choppers.
420F (S42020)	—	—	—	—	—	—	—	—	2050-2250	1550-1650[n]	H1800-1900,[c] T300-700	2650-2750	Free-machining modification of type 420. Applications similar to those for type 420 requiring better machinability.
422 (S42200)	—	—	—	—	—	—	—	—	2100	1350-1450	H1900[c]	2675-2700	High strength and toughness at service temperatures up to 1200°F. Steam turbine blades; fasteners.

TABLE 16.24 Standard Stainless and Heat-Resisting Steels: Part B *(Continued)*

AISI Type (UNS)	Mechanical Properties at Elevated Temperatures								Thermal Treatment			Characteristics and Typical Applications
	Creep Strength (Load for 1% Elongation in 10 000 hr, 10³ psi)					Scaling Temperature		Initial Forging Temperature (°F)	Annealing Temperature (°F)^c	Hardening and Tempering Temperature (°F)	Melting Range (°F)	
	1000 F°	1100 F°	1200 F°	1300 F°	1500 F°	Maximum Continuous Service in Air (°F)	Maximum Intermittent Service in Air (°F)					
431 (S43100)	6.8	3.5	—	—	—	1500	1600	2100–2250°	— 1150–1225^l	H1800–1900^c T 400–1200^m	—	Special-purpose hardenable steel used where particularly high mechanical properties are required. Aircraft fittings; beater bars; paper machinery; bolts.
440A (S44002)	—	—	—	—	—	1400	1500	1900–2200°	1550–1650,^{k} 1350–1450^l	H1850–1950,^{c} T 300–800	2500–2750	Hardenable to higher hardness than type 420 with good corrosion resistance. Cutlery; bearings; surgical tools.
440B (S44003)	—	—	—	—	—	1400	1500	1900–2150°	1550–1650,^{k} 1350–1450^l	H1850–1950,^{c} T 300–800	2500–2750	Cutlery grade. Cutlery, valve parts; instrument bearings.
440C (S44004)	—	—	—	—	—	1400	1500	1900–2100°	1550–1650,^{k} 1350–1450^l	H1850–1950,^{c} T 300–800	2500–2750	Yields highest hardnesses of hardenable stainless steels. Balls; bearings; races; nozzles; balls and seats for oil well pumps; valve parts.
HEAT RESISTING												
501 (S50100)	—	—	—	—	—	—	—	2100–2200°	1525–1600,^{k} 1325–1375^l	H1600–1700m^c T 400–1400	2700–2800	Heat resistance; good mechanical properties at moderately elevated temperatures. Heat exchangers; petroleum refining equipment.
502 (S50200)	—	—	—	—	—	—	—	2100–2200	1525–1600m^k 1325–1375^l	—	2700–2800	More ductility and less strength than type 501. Heat exchangers; petroleum refining equipment; gaskets.
503 (S50300) 504 (S50400)	—	—	—	—	—	—	—	—	—	—	—	—
(S13800)	—	—	—	—	—	—	—	2150	—	H950–1150^c	2560–2625	Martensitic (maraging) stainless that can be hardened by a single low-temperature heat treatment. Aircraft parts, forged.

(S15500)	—	—	—	—	—	—	2150	—	—	—	—	H900–1150[c]	—	2560–2625	Martensitic (maraging) stainless with high strength, hardness, and corrosion resistance. Gears; cams; cutlery; shafting; aircraft parts.
(S17400)	—	—	—	—	—	—	2150	—	—	—	—	H900–1150[c]	—	2560–2625	Similar to S15500 but with slightly higher chromium content. Gears; springs; cutlery; fasteners; aircraft and turbine parts.
(S17700)	—	—	—	—	—	—	2150	—	—	—	—	H900–1050[c]	—	2560–2625	Semiaustenitic stainless. Can be cold drawn and then hardened by a low-temperature heat treatment. Springs; knives; pressure vessels.

Source: Committee of Stainless Steel Producers. American Iron and Steel Institute (revised April 1985). Based on the "Stainless and Heat Resisting Steels" Section of the AISI *Steel Products Manual,* latest edition.

[a]Single values are maximums, except as noted.

[b]Forms listed are only those for which mechanical properties are given. Most types are available in many forms.

[c]Followed by rapid cooling. *H* is hardening temperature; *T* is tempering.

[d]Austenitic: hardenable by cold working; not hardenable by heat treatment. Ferritic: not hardenable by heat treatment or cold working. Martensitic: hardenable by heat treatment.

[e]Charpy V-notch.

[f]Not applicable. Silicon added to type 302 for oxidation resistance.

[g]Soft temper.

[h]Composition for type 310 tubing varies slightly from AISI values; for standard compositions, refer to ASTM A213.

[i]Stabilizing temperature, 1550–1650°F.

[j]Hardening and tempering temperature, °F.

[k]Full anneal, followed by slow cooling.

[l]Low anneal.

[m]Tempering within the range of 800–1100°F is not recommended because of resulting low and erratic impact properties and reduced corrosion resistance. Time at temperature and temperatures may vary depending on part size.

[n]Retarded cool.

[o]Retarded cool and anneal.

[p]Mechanical properties are for a solution-treated condition.

TABLE 16.25 Properties of Aluminum

		Sample Purity, Percent
Atomic weight,	26.97	—
Boiling point,	2060°F	99.996
Crystal form,	Face-centered cubic	—
Mean coefficient of expansion		
20–300°C,	0.0000257	99.95 (cast)
20–500°C,	0.0000277	99.95 (cast)
Density at 20°C (68°F),	2.70 g/cm³	99.971
		Wrought annealed
Density at melting point (solid),	2.55 g/cm³	—
Density at melting point (liquid),	2.38 g/cm³	—
Electric resistivity at 20°C,	2.688 $\mu\Omega$-cm	99.968 (hard drawn)
	17.01 Ω (mil, ft)	Commercial
Temperature coefficient at 20°C[a]	0.00403	Commercial
Electric conductivity at 20°C		
Mass, International Annealed Copper Standard,	212.9%	—
Volume, International Annealed Copper Standard,	64.9%	99.996
Freezing point,	660.2°C	99.996
Heat of vaporization,	1950–2000 g-cal/g	—
Latent heat of fusion	94.6 g-cal/g	99.996
	170.27 Btu/lb	—
Mechanical properties:		
Tensile strength,	9000 psi	Annealed
Yield strength,[b]	3000 psi	Sheet
Elongation in 2 in.,	60%	99.95
Brinell hardness, 10-mm ball, 500-kg load,	17	99.996 (annealed)
Modulus of elasticity,	10,000,000 psi	—
Modulus of rigidity (torsion),	3,870,000 psi	Commercial
Poisson's ratio,	0.33	—
Total reflectivity, for white light	87%	—
Mean specific heat, 0–100°C	0.226 g-cal/g °C	—
Thermal conductivity		
cgs units	0.52	99.66
Watts/cm² °C cm	2.17	99.66
Btu/hr ft² °F in.	1509	—

[a]AIEE Standard.
[b]Set = 0.2%.

TABLE 16.26 Aluminum Alloy Designation Systems

Alloy Type[a]	Four-Digit Designation
Wrought alloys	
99.00% (minimum) aluminum	1XXX
Copper	2XXX
Manganese	3XXX
Silicon	4XXX
Magnesium	5XXX
Magnesium and silicon	6XXX
Zinc	7XXX
Others	8XXX
Casting alloys	
99.00% (minimum) aluminum	1XX.X
Copper	2XX.X
Silicon with added copper and/or magnesium	3XX.X
Silicon	4XX.X
Magnesium	5XX.X
Zinc	7XX.X
Tin	8XX.X
Others	9XX.X

Source: *Encyclopedia of Materials Science and Engineering*, M. B. Bever, (ed.), Pergamon Press, Oxford, 1986.

[a] Designations are based on aluminum content or main alloying elements.

TABLE 16.27 Key to Temper Designations for Aluminum Alloys

F = as fabricated (no treatment subsequent to forming)
O = annealed (wrought products only)
H = strain hardened (cold worked)

	Relative Tensile Strength (O = 1.0)
H1 = strain hardened only	
H12 = cold-worked, one-quarter hard temper (20% for 1100)[a]	1.18
H14 = cold-worked, one-half hard temper (40% for 1100)	1.36
H16 = cold-worked, three-quarters hard temper (60% for 1100)	1.54
H18 = cold-worked hard temper (80% for 1100)	1.72
H19 = cold-worked, extra hard temper (90% for 1100)	1.81

H2 = strain hardened and then partly annealed
    H22 = one-quarter hard plus a light anneal
    H24, H26, H28 similarly
H3 = strain hardened and then stabilized
    H32 = one-quarter hard plus a stabilizing heat treatment
    H34, H36, H38, H39 similarly
W = solution heat treated and quenched (unstable temper)
T = heat treated to produce tempers other than F, O, or H[b]
    T2 = annealed (cast products only)
    T3 = solution heat treated and then cold worked
    T4 = solution heat treated and naturally aged to a stable condition
    T5 = artificially aged only (no solution heat treatment—as in certain permanent-mold alloys and extrusions)
    T6 = solution heat treated and then artificially aged (aged at an elevated temperature)
    T7 = solution heat treated and then stabilized (somewhat overaged to provide dimensional stability)
    T8 = solution heat treated, cold worked, and then artificially aged
    T9 = solution heat treated, artificially aged, and then cold worked
    T10 = artificially aged and then cold worked

From *Handbook of Engineering Materials*, Miner and Seastone, New York, Wiley, 1955.
[a] Percentage of reduction in cross-sectional area.
[b] For casting alloys the T designations may be followed by additional numbers designating a specific heat treatment; see *Metals Handbook*, under the specific casting alloy.

TABLE 16.28   Nominal Composition of Wrought Aluminum Alloys

Item No.	Alloy	Percentage of Alloying Elements[a]								
		Silicon	Copper	Manganese	Magnesium	Chromium	Nickel	Zinc	Lead	Bismuth
1	1350[b]	99.50% minimum aluminum[c]								
3	1100	99.00% minimum aluminum								
5	2011	—	5.5	—	—	—		—	0.40	0.40
6	2014	0.8	4.4	0.8	0.40					
9	2024	—	4.5	0.6	1.5					
10	2124	—	4.4	0.6	1.5					
13	2219[d]	—	6.3	0.30	—					
14	3003	—	—	1.2	—					
15	3004	—	—	1.2	1.0					
17	4032	12.2	0.9	—	1.1	—	0.9			
18	5005	—	—	—	0.8					
20	5052	—	—	—	2.5	0.25				
22	5083	—	—	0.8	4.45	0.10				
23	5086	—	—	0.45	4.0	0.10				
29	5456	—	—	0.8	5.25	0.10				
32	5652	—	—	—	2.5	0.25				
34	6061	0.6	0.25	—	1.0	0.25				
36	6063	0.40	—	—	0.7					
37	6066	1.3	0.9	0.9	1.1					
38	6151	1.0	—	—	0.6	0.25				
39	7049	—	1.6	—	2.4	0.16	—	7.7		
40	7050[d]	—	2.3	—	3.2		—	6.2		
44	7075	—	1.6	—	2.5	0.30	—	5.6		
46	7178	—	2.0	—	2.7	0.30	—	6.8		

Source: From *Alcoa Aluminum Handbook*, 1984 edition.

[a] Aluminum and normal impurities constitute remainder.
[b] Formerly EC.
[c] Boron 0.02.
[d] Titanium 0.06; vanadium 0.10; zirconium 0.18.

TABLE 16.29 Typical Physical Properties of Wrought Aluminum Alloys

Item No.	Alloy and Temper	Specific Gravity	Weight (lb/in.³)	Melting Range Approximate (°F)	Electrical Conductivity at 20°C (68°F), Percentage of International Annealed Copper Standard	Thermal Conductivity at 25°C (77°F), cgs units[a]
1, 1a	1350-O, 1350-H19	2.70	0.098	1195–1215	63, 62.5	0.57, 0.56
3, 3a	1100-O, 1100-H18	2.71	0.098	1190–1215	59[b]	0.53
5a, 5b	2011-T3, 2011-T8	2.82	0.102	995–1190	39, 45	0.36, 0.41
6, 6a, 6b	2014-O, 2014-T4, 2014-T6	2.80	0.101	950–1180	50, 34, 40	0.46, 0.32, 0.37
9, 9a–9f	2024-O; 2024-T3, -T36, -T4, 2024-T6, -T81, -T86	2.77	0.100	935–1180	50; 30, 38	0.46; 0.29, 0.36
10, 10a	2124-O, 2124-T851	2.77	0.100	935–1180	50, 39	0.26, 0.37
13, 13a–13e	2219-O; 2219-T31, -T37; 2219-T62, -T81, -T87	2.83	0.103	1010–1190	44; 28, 32	0.41; 0.27, 0.30
14, 14a	3003-O, 3003-H18	2.73	0.099	1190–1210	46[b]	0.42
15, 15a	3004-O, 3004-H38	2.72	0.098	1165–1210	42[b]	0.39
17, 17a	4032-O, 4032-T6	2.69	0.097	990–1060	40, 35	0.37, 0.33
18, 18a	5005-O, 5005-H38	2.69	0.097	1170–1210	54[b,c]	0.49[c]
20, 20a	5052-O, 5052-H38	2.68	0.097	1125–1200	35[b]	0.33
22, 22a	5083-O, 5083-H321	2.66	0.096	1075–1185	29[b]	0.28
23, 23a	5086-O, 5086-H34	2.66	0.096	1085–1185	32[b]	0.30
29, 29a	5456-O, 5456-H321	2.66	0.096	1055–1180	29[b]	0.28
32, 32a	5652-O, 5652-H38	2.68	0.097	1125–1200	35[b]	0.33
34, 34a, 34b	6061-O, 6061-T4, 6061-T6	2.70	0.098	1100–1205	47, 40, 43	0.43, 0.37, 0.40
36, 36c, 36d–36e	6063-O, 6063-T5, 6063-T6, -T83	2.70	0.098	1140–1210	58, 55, 53	0.52, 0.50, 0.48
37, 37a	6066-O, 6066-T6	2.72	0.098	1045–1195	40, 37	0.37, 0.35
38, 38a, 38b	6151-O, 6151-T4, 6151-T6	2.70	0.098	1025–1200	54, 42, 45	0.49, 0.39, 0.41
39b	7049-T73,	2.84	0.103	890–1175,	40,	0.37
40a	7050-T74	2.83	0.102	910–1165	41	0.37
44a	7075-T6	2.80	0.101	890–1175	33	0.31
46a	7178-T6	2.81	0.102	890–1165	31	0.30

Source: From *Alcoa Aluminum Handbook*, 1984 edition.

[a]cgs units = cal/cm cm² °C s.
[b]Average for range.
[c]When fabricated for use as electrical conductors, average values for electrical conductivity of the -O and -H19 tempers are 56 and 55%, respectively, and the corresponding values for thermal conductivities are 0.51 and 0.50.

TABLE 16.30 Typical Mechanical Properties of Wrought Aluminum Alloys[a]

Item No.	Alloy and Temper	Tensile Strength (psi)		Elongation in 2 in. (%)		Hardness (Brinell Number, 500-kg load, 10-mm ball)	Shearing Strength (psi)	Fatigue, Endurance Limit[b] (psi)	Modulus of Elasticity[c] (psi)
		Ultimate	Yield	1/16-in. Thick Specimen	1/2-in. Diameter Specimen				
1	1350-O[d]	10,000	4,000	—	9[f]	—	8,000	—	10.0×10^6
1a	1350-H19	27,000	24,000	—	—	—	15,000	7,000	10.0×10^6
3	1100-O	13,000	5,000	35	40	23	9,000	5,000	10.0×10^6
3a	1100-H18	24,000	22,000	5	15	44	13,000	9,000	10.0×10^6
5a	2011-T3	55,000[g]	43,000[g]	—	15	95	32,000	18,000	10.2×10^6
5b	2011-T8	59,000	45,000	—	12	100	35,000	18,000	10.2×10^6
6	2014-O	27,000	14,000	—	18	45	18,000	13,000	10.6×10^6
6a	2014-T4	62,000	42,000[h]	—	20	105	38,000	20,000	10.6×10^6
6b	2014-T6	70,000[i]	60,000[i]	—	13	135	42,000	18,000	10.6×10^6
C 6	Alclad 2014-O	25,000	10,000	21	—	—	18,000	—	10.5×10^6
C 6a	Alclad 2014-T3	63,000[j]	40,000[j]	20	—	—	37,000	—	10.5×10^6
C 6b	Alclad 2014-T4	61,000[j]	37,000[j]	22	—	—	37,000	—	10.5×10^6
C 6c	Alclad 2014-T6	68,000[j]	60,000[j]	10	—	—	41,000	—	10.5×10^6
9	2024-O	27,000	11,000	20	22	47	18,000	13,000	10.6×10^6
9a	2024-T3	70,000	50,000	18	—	120	41,000	20,000	10.6×10^6
9b	2024-T36	72,000	57,000	13	—	130	42,000	18,000	10.6×10^6
9c	2024-T4	68,000[i]	47,000[i]	20	19	120	41,000	20,000	10.6×10^6
9d	2024-T6	69,000	57,000	—	10	125	43,000	18,000	10.6×10^6
9e	2024-T81	70,000	65,000	6	—	128	43,000	18,000	10.6×10^6
9f	2024-T86	75,000	71,000	6	—	135	45,000	18,000	10.6×10^6
C 9	Alclad 2024-O	26,000	11,000	20	—	—	18,000	—	10.6×10^{6d}
C 9a	Alclad 2024-T3	65,000[k]	45,000[k]	18	—	—	40,000	—	10.6×10^{6d}
C 9b	Alclad 2024-T36	67,000[k]	53,000[k]	11	—	—	41,000	—	10.6×10^{6d}
C 9c	Alclad 2024-T4	64,000[k]	42,000[k]	19	—	—	40,000	—	10.6×10^{6d}
C 9e	Alclad 2024-T81	65,000[k]	60,000[k]	6	—	—	40,000	—	10.6×10^{6d}
C 9f	Alclad 2024-T86	70,000[k]	66,000[k]	6	—	—	42,000	—	10.6×10^{6d}
13	2219-O	25,000	10,000	20	—	—	—	—	10.6×10^6
13a	2219-T31	54,000	37,000	17	—	100	33,000	—	10.6×10^6
13b	2219-T37	60,000	49,000	11	—	117	37,000	—	10.6×10^6
13c	2219-T62	61,000	42,000	11	—	115	37,000	15,000	10.6×10^6

	Alloy and temper	Tensile strength	Yield strength	Elongation	Elongation	Brinell hardness	Shear strength	Fatigue limit	Modulus of elasticity
13d	2219-T81	70,000	53,000	11	—	130	41,000	15,000	10.6×10^6
13e	2219-T87	70,000	58,000	10	—	130	41,000	15,000	10.6×10^6
14	3003-O	16,000	6,000	30	40	28	11,000	7,000	10.0×10^6
14a	3003-H18	29,000	27,000	4	10	55	16,000	10,000	10.0×10^6
C 14	Alclad 3003-O	16,000	6,000	30	40	—	11,000	—	10.0×10^6
C 14b	Alclad 3003-H12	19,000	18,000	10	20	—	12,000	—	10.0×10^6
C 14c	Alclad 3003-H14	22,000	21,000	8	16	—	14,000	—	10.0×10^6
C 14d	Alclad 3003-H16	26,000	25,000	5	14	—	15,000	—	10.0×10^6
C 14a	Alclad 3003-H18	29,000	27,000	4	10	—	16,000	—	10.0×10^6
15	3004-O	26,000	10,000	20	25	45	16,000	14,000	10.0×10^6
15a	3004-H38	41,000	36,000	5	6	77	21,000	18,000	10.0×10^6
C 15	Alclad 3004-O0	26,000	10,000	20	25	—	16,000	—	10.0×10^6
C 15a	Alclad 3004-H32	31,000	25,000	10	17	—	17,000	—	10.0×10^6
C 15b	Alclad 3004-H34	35,000	29,000	9	12	—	18,000	—	10.0×10^6
C 15c	Alclad 3004-H36	38,000	33,000	5	9	—	20,000	—	10.0×10^6
C 15d	Alclad 3004-H32	41,000	36,000	5	6	—	21,000	—	10.0×10^6
17a	4032-T6	55,000	46,000	—	9	120	38,000	16,000	11.4×10^6
18	5005-O	18,000	6,000	30	—	30	11,000	—	10.0×10^6
18a	5005-H38	29,000	27,000	5	—	51	16,000	—	10.0×10^6
20	5052-O	28,000	13,000	25	30	47	18,000	16,000	10.2×10^6
20a	5052-H38	42,000	37,000	7	14	77	24,000	20,000	10.2×10^6
22	5083-O	42,000	21,000	22	25	67	25,000	22,000	10.3×10^6
22a	5083-H321	46,000	33,000	—	16	—	—	23,000	10.3×10^6
23	5086-O	38,000	17,000	22	30	60	23,000	21,000	10.3×10^6
23a	5086-H34	47,000	37,000	10	14	82	28,000	23,000	10.3×10^6
29	5456-O	45,000	23,000	24	20	70	27,000	22,000	10.3×10^6
29a	5456-H321	51,000	37,000	16	16	90	30,000	23,000	10.3×10^6
32	5652-O	28,000	13,000	25	30	47	18,000	16,000	10.2×10^6
32a	5652-H38	42,000	37,000	7	8	77	24,000	20,000	10.2×10^6
34	6061-O	18,000	8,000	25	30	30	12,000	9,000	10.0×10^6
34a	6061-T4	35,000	21,000	22	25	65	24,000	13,000	10.0×10^6
34b	6061-T6	45,000^m	40,000^m	12	17	95	30,000	14,000	10.0×10^6
C 34	Alclad 6061-O	17,000	7,000	25	—	—	11,000	—	$10.0 \times 10^6{}^f$
C 34a	Alclad 6061-T4	33,000	19,000	22	—	—	22,000	—	$10.0 \times 10^6{}^f$
C 34b	Alclad 6061-T6	42,000	37,000	12	—	—	27,000	—	$10.0 \times 10^6{}^f$
36	6063-O	13,000	7,000	—	—	25	10,000	8,000	10.0×10^6
36a	6063-T4	25,000	13,000	22	—	—	16,000	—	10.0×10^6
36c	6063-T5	27,000	21,000	12	22	60	17,000	10,000	10.0×10^6

TABLE 16.30   (Continued)

Item No.	Alloy and Temper	Tensile Strength (psi)		Elongation in 2 in. (%)		Hardness (Brinell Number, 500-kg load, 10-mm ball)	Shearing Strength (psi)	Fatigue, Endurance Limit[b] (psi)	Modulus of Elasticity[c] (psi)
		Ultimate	Yield	$\frac{1}{16}$-in. Thick Specimen	$\frac{1}{2}$-in. Diameter Specimen				
36d	6063-T6	35,000	31,000	12	18	74	22,000	10,000	10.0×10^6
36e	6063-T83	37,000	35,000	9	—	80	22,000	—	10.0×10^6
37	6066-O	22,000	12,000	—	18	43	14,000	—	10.0×10^6
37a	6066-T6	57,000	52,000	—	12	120	34,000	16,000	10.0×10^6
38b	6151-T6	48,000	43,000	—	17	100	32,000	12,000	10.2×10^6
39b	7049-T73	75,000	65,000	—	12	135	44,000	—	10.4×10^6
40b	7050-T73510	72,000	63,000	—	12	—	—	—	10.4×10^6
40c	7050-T7451	76,000	68,000	—	11	—	44,000	—	10.4×10^6
40d	7050-T7651	80,000	71,000	—	11	—	47,000	—	10.4×10^6
44a	7075-T6	83,000[m, o]	73,000[m, o]	11	11	150	48,000	22,000	10.4×10^{6f}
C 44	Alclad 7075-O	32,000	14,000	17	—	—	22,000	—	10.4×10^{6f}
C 44a	Alclad 7075-T6	76,000	67,000	11	—	—	46,000	—	10.4×10^{6f}
46a	7178-T6	88,000[n]	78,000[n]	10	11	160	52,000	22,000	10.4×10^6
C 46	Alclad 7178-O	32,000	14,000	16	—	—	22,000	—	10.4×10^{6f}
C 46a	Alclad 7178-T6	81,000	71,000	10	—	—	49,000	—	10.4×10^{6f}

Source: From *Alcoa Aluminum Handbook,* 1984 edition.

[a]These typical properties are average for various forms, sizes, and methods of manufacture and may not exactly describe any one particular product or size.

[b]Based on 500,000,000 cycles of completely reversed stress using the R. R. Moore type of machine and specimen.

[c]Average of tension and compression moduli. Compression modulus is about 2% greater than tension modulus.

[d]Electrical conductor grade, 99.60% minimum aluminum.

[e]EC-O wire will have an elongation of approximately 23% in 10 in.

[f]EC-H19 wire will have an elongation of approximately $1\frac{1}{2}$% in 10 in.

[g]Sizes greater than $1\frac{1}{2}$ inches will have strengths slightly lower than these value.

[h]Die forgings will have a yield strength approximately 20% lower than this value.

[i]Extruded products more than $\frac{3}{4}$ in. thick will have strengths 15–20% higher than these values.

[j]Sheet less than 0.040 in. thick will have strengths slightly lower than these values.

[k]Sheet more than 0.062 in. thick will have strengths slightly higher than these values.

[l]Value shown is primary modulus. Secondary modulus is from 3 to 10% lower depending on thickness of cladding.

[m]Die forgings will have strengths approximately 5% higher than this value.

[n]Extruded products will have strengths approximately 10% higher than these values.

[o]Die forgings have strengths approximately 4% lower than these values.

TABLE 16.31 Typical Tensile Properties of Wrought Aluminum Alloys At Various Temperatures[a]

Item No.	Alloy and Temper	Temperature (°F)	Tensile Stength, psi		Elongation in 2 in. (%)
			Ultimate	Yield[b]	
3	1100-O	−320	24,000	6,000	55
		−112	15,000	5,500	48
		−18	14,000	5,000	46
		75	13,000	5,000	45
		212	11,000	5,000	45
		300	8,500	4,500	55
		400	6,000	3,500	65
		500	4,000	2,000	75
		600	2,500	1,500	80
		700	2,000	1,000	85
3a	1100-H18	−320	35,000	26,000	30
		−112	26,000	23,000	16
		−18	24,000	22,000	15
		75	24,000	22,000	15
		212	22,000	19,000	15
		300	18,000	14,000	20
		400	6,000	3,500	65
		500	4,000	2,000	75
		600	2,500	1,500	80
		700	2,000	1,000	85
5a	2011-T3	75	55,000	43,000	15
		212	47,000	34,000	16
		300	28,000	19,000	25
		400	16,000	11,000	35
		500	6,500	4,000	45
		600	3,500	2,000	90
		700	2,500	1,500	125
6b	2014-T6	−320	84,000	69,000	14
		−112	74,000	63,000	14
		−18	72,000	61,000	13
		75	70,000	60,000	13
		212	63,000	56,000	14
		300	40,000	35,000	15
		400	16,000	13,000	35
		500	9,500	7,500	45
		600	6,500	5,000	65
		700	4,500	3,500	70
9a	2024-T3	−320	85,000	62,000	18
		−112	75,000	52,000	17
		−18	72,000	51,000	17
		75	70,000	50,000	17
		212	66,000	48,000	16
		300	55,000	50,000	11
		400	29,000	22,000	23
		500	12,000	9,000	55
		600	8,000	6,000	75
		700	5,500	4,000	100
9c	2024-T4	−320	81,000	57,000	19
		−112	71,000	49,000	19
		−18	70,000	48,000	19
		75	68,000	47,000	19
		212	64,000	45,000	19
		300	45,000	36,000	17
		400	27,000	20,000	27
		500	12,000	9,000	55
		600	8,000	6,000	75
		700	5,500	4,000	100
9d	2024-T6	−320	83,000	68,000	11
		−112	72,000	60,000	10
		−18	71,000	58,000	10

TABLE 16.31   *(Continued)*

Item No.	Alloy and Temper	Temperature (°F)	Tensile Stength, psi		Elongation in 2 in. (%)
			Ultimate	Yield[b]	
		75	69,000	57,000	10
		212	65,000	55,000	10
		300	45,000	36,000	17
		400	27,000	20,000	27
		500	12,000	9,000	55
		600	8,000	6,000	75
		700	5,500	4,000	100
9e	2024-T81	−320	85,000	77,000	8
		−112	75,000	68,000	7
		−18	72,000	67,000	7
		75	70,000	65,000	7
		212	66,000	61,000	8
		300	55,000	50,000	11
		400	29,000	22,000	23
		500	12,000	9,000	55
		600	8,000	6,000	75
		700	5,500	4,000	100
9f	2024-T86	−320	91,000	86,000	5
		−112	80,000	77,000	5
		−18	78,000	74,000	5
		75	75,000	71,000	5
		212	70,000	67,000	6
		300	55,000	51,000	11
		400	20,000	16,000	28
		500	12,000	9,000	55
		600	8,000	6,000	75
		700	5,500	4,000	100
13c	2219-T62	−320	78,000	52,000	14
		−112	65,000	44,000	12
		−18	63,000	44,000	12
		75	61,000	42,000	11
		212	53,000	39,000	14
		300	44,000	30,000	15
		400	33,000	24,000	18
		500	27,000	20,000	18
		600	9,000	7,500	35
		700	4,500	3,500	100
13d	2219-T81	−320	86,000	63,000	13
		−112	73,000	56,000	12
		−85	72,000	54,000	11
		75	70,000	53,000	11
		212	63,000	50,000	13
		300	48,000	38,000	15
		400	34,000	29,000	16
		500	28,000	23,000	16
		600	7,000	6,500	40
		700	4,500	3,500	100
14	3003-O	−320	33,000	8,500	46
		−112	20,000	7,000	42
		−18	17,000	6,500	41
		75	16,000	6,000	40
		212	13,000	5,500	43
		300	11,000	5,000	47
		400	8,500	4,500	60
		500	6,000	3,500	65
		600	4,000	2,500	70
		700	3,000	2,000	70
14a	3003-H18	−320	41,000	33,000	23
		−112	32,000	29,000	11

TABLE 16.31 *(Continued)*

Item No.	Alloy and Temper	Temperature (°F)	Tensile Stength, psi		Elongation in 2 in. (%)
			Ultimate	Yield[b]	
		−18	30,000	28,000	10
		75	29,000	27,000	10
		212	26,000	21,000	10
		300	23,000	16,000	11
		400	14,000	9,000	18
		500	7,500	4,000	60
		600	4,000	2,500	70
		700	3,000	2,000	70
15	3004-O	−320	42,000	13,000	38
		−112	28,000	11,000	30
		−18	26,000	10,000	26
		75	26,000	10,000	25
		212	26,000	10,000	25
		300	22,000	10,000	35
		400	14,000	9,500	55
		500	10,000	7,500	70
		600	7,500	5,000	80
		700	5,000	3,000	90
15a	3004-H38	−320	58,000	43,000	20
		−112	44,000	38,000	10
		−18	42,000	36,000	7
		75	41,000	36,000	6
		212	40,000	36,000	7
		300	31,000	27,000	15
		400	22,000	15,000	30
		500	12,000	7,500	50
		600	7,500	5,000	80
		700	5,000	3,000	90
17a	4032-T6	−320	66,000	48,000	11
		−112	58,000	46,000	10
		−18	56,000	46,000	9
		75	55,000	46,000	9
		212	50,000	44,000	9
		300	37,000	33,000	9
		400	13,000	9,000	30
		500	8,000	5,500	50
		600	5,000	3,000	70
		700	3,500	2,000	90
20	5052-O	−320	44,000	16,000	45
		−112	30,000	13,000	36
		−18	28,000	13,000	32
		75	28,000	13,000	30
		212	28,000	13,000	37
		300	24,000	13,000	50
		400	18,000	11,000	60
		500	12,000	8,000	85
		600	7,500	5,000	110
		700	5,000	3,000	130
20a	5052-H38	−320	59,000	44,000	24
		−112	45,000	39,000	20
		−18	43,000	37,000	16
		75	42,000	37,000	14
		212	41,000	37,000	17
		300	34,000	29,000	25
		400	23,000	15,000	50
		500	12,000	8,000	75
		600	7,500	5,000	110
		700	5,000	3,000	130
22	5083-O	−320	60,000	24,000	36
		−112	43,000	21,000	30

TABLE 16.31 *(Continued)*

Item No.	Alloy and Temper	Temperature (°F)	Tensile Stength, psi		Elongation in 2 in. (%)
			Ultimate	Yield[b]	
		−18	42,000	21,000	27
		75	42,000	21,000	25
		212	41,000	21,000	37
		300	30,000	19,000	50
		400	22,000	17,000	60
		500	17,000	11,000	85
		600	11,000	7,500	110
		700	6,000	4,500	130
23	5086-O	−320	56,000	20,000	45
		−112	39,000	17,000	36
		−18	38,000	17,000	32
		75	38,000	17,000	30
		212	37,000	17,000	37
		300	29,000	17,000	60
		400	22,000	16,000	60
		500	17,000	11,000	85
		600	11,000	7,500	110
		700	6,000	4,500	130
29	5456-O	−320	63,000	27,000	30
		−112	46,000	23,000	26
		−18	45,000	23,000	22
		75	45,000	23,000	20
		212	43,000	22,000	30
		300	30,000	20,000	50
		400	23,000	17,000	60
		500	17,000	11,000	85
		600	11,000	7,500	110
		700	6,000	4,500	130
32	5652-O	−320	44,000	16,000	46
		−112	29,000	13,000	35
		−18	28,000	13,000	32
		75	28,000	13,000	30
		212	28,000	13,000	30
		300	23,000	13,000	50
		400	17,000	11,000	60
		500	12,000	7,500	80
		600	7,500	5,500	110
		700	5,000	3,100	130
32a	5652-H34	−320	55,000	36,000	28
		−112	40,000	32,000	21
		−18	38,000	31,000	18
		75	38,000	31,000	16
		212	38,000	31,000	18
		300	30,000	27,000	18
		400	24,000	15,000	45
		500	12,000	7,500	80
		600	7,500	5,500	110
		700	5,000	3,100	130
32b	5652-H38	−320	60,000	44,000	25
		−112	44,000	38,000	18
		−18	42,000	37,000	15
		75	42,000	37,000	14
		212	40,000	36,000	16
		300	34,000	28,000	24
		400	25,000	15,000	45
		500	12,000	7,500	80
		600	7,500	5,500	110
		700	5,000	3,100	130
34b	6061-T6	−320	60,000	47,000	22

TABLE 16.31 *(Continued)*

Item No.	Alloy and Temper	Temperature (°F)	Tensile Stength, psi Ultimate	Yield[b]	Elongation in 2 in. (%)
		−112	49,000	42,000	18
		−18	47,000	41,000	17
		75	45,000	40,000	17
		212	42,000	38,000	18
		300	34,000	31,000	20
		400	19,000	15,000	28
		500	7,500	5,000	60
		600	4,500	2,500	85
		700	3,000	2,000	95
36b	6063-T42	−320	34,000	16,000	44
		−112	26,000	15,000	36
		−18	24,000	14,000	34
		75	22,000	13,000	33
		212	22,000	14,000	18
		300	21,000	15,000	20
		400	9,000	6,500	40
		500	4,500	3,500	75
		600	3,000	2,500	80
		700	2,500	2,000	105
36c	6063-T5	−320	37,000	24,000	28
		−112	29,000	22,000	24
		−18	28,000	22,000	23
		75	27,000	21,000	22
		212	24,000	20,000	18
		300	20,000	18,000	20
		400	9,000	6,500	40
		500	4,500	3,500	75
		600	3,000	2,500	80
		700	2,500	2,000	105
36d	6063-T6	−320	47,000	36,000	24
		−112	38,000	33,000	20
		−18	36,000	32,000	19
		75	35,000	31,000	18
		212	31,000	28,000	15
		300	21,000	20,000	20
		400	9,000	6,500	40
		500	4,500	3,500	75
		600	3,000	2,500	80
		700	2,500	2,000	105
38b	6151-T6	75	48,000	43,000	17
		212	42,000	39,000	19
		300	27,000	25,000	22
		400	12,000	9,500	40
		500	6,500	5,500	50
		600	5,000	4,500	50
		700	4,000	3,500	50
39c	7049-T76	75	83,000	75,000	—
		212	73,000	67,000	—
		300	63,000	58,000	—
		400	49,000	46,000	—
		500	24,000	22,000	—
39b	7049-T73	75	23,000	65,000	11
		212	67,000	63,000	14
		300	57,000	55,000	17
		400	40,000	40,000	22
		500	18,000	18,000	30
40c	7050-T7451	75	83,000	73,000	14
		212	71,000	68,000	11
		300	60,000	60,000	16

TABLE 16.31   *(Continued)*

Item No.	Alloy and Temper	Temperature (°F)	Tensile Stength, psi		Elongation in 2 in. (%)
			Ultimate	Yield[b]	
		400	43,000	43,000	20
		500	20,000	20,000	25
44a	7075-T6	−320	102,000	92,000	9
		−112	90,000	79,000	11
		−18	86,000	75,000	11
		75	83,000	73,000	11
		212	70,000	65,000	14
		300	31,000	27,000	30
		400	16,000	13,000	55
		500	11,000	9,000	65
		600	8,000	6,500	70
		700	6,000	4,500	70
46a	7178-T6, T651	−320	106,000	94,000	5
		−112	94,000	84,000	8
		−18	91,000	81,000	9
		75	88,000	78,000	11
		212	73,000	68,000	14
		300	31,000	27,000	40
		400	15,000	12,000	70
		500	11,000	9,000	76
		600	8,500	7,000	80
		700	6,500	5,500	80
46b, 46c	7178-T76, T7651	−320	106,000	89,000	10
		−112	91,000	78,000	10
		−18	88,000	76,000	10
		75	83,000	73,000	11
		212	69,000	64,000	17
		300	31,000	27,000	40
		400	15,000	12,000	70
		500	11,000	9,000	76
		600	8,500	7,000	80
		700	6,500	5,500	80

Source: From *Alcoa Aluminum Handbook*, 1984 edition.

[a]Lowest strengths during 10,000 h of heating at testing temperature under no load; stress applied at 5000 psi/min to yield strength and then at strain rate of 0.05 in./in. min to failure. Under some conditions of temperature and time the application of heat will adversely affect certain other properties of some alloys. For specific information concerning the suitability of the various alloys for use at elevated temperatures, the nearest sales office of Aluminum Company of America should be consulted.

[b]Offset equals 0.2%.

TABLE 16.32   Thermal Expansion of Wrought Aluminum Alloys

Equations of linear thermal expansion for aluminum alloys:[a]

1. $L_{t(0 \text{ to } -320°F)} = L_0[1 + C(11.74t - 0.00125t^2 - 0.0000248t^3)10^{-6}]$
2. $L_{t(0-1000°F)} = L_0[1 + C(12.19t + 0.003115t^2)10^{-6}]$
     L_0 = length at 0°F
     L_t = length at temperature t°F within range indicated
     C = alloy constant[b]

Alloy	C	Alloy	C
1350	1.000	5086	1.010
1060	1.000	5154	1.015
1100	1.000	5254	1.015
1345	1.000	5357	1.005
2011	0.980	5454	1.005
2014	0.955	5456	1.015
2017	0.970	5457	1.005
2018	0.950	5557	1.000
2024	0.970	5652	1.010
2025	0.965	6053	0.980
2117	0.990	6061	0.990
2218	0.950	6062	0.990
2219	0.955	6063	0.995
2618	0.945	6151	0.985
3003	0.985	6262	0.990
3004	0.985	6463	0.995
3105	0.995	6563	1.000
4032	0.825	6951	0.990
4043	0.940	7072	1.000
5005	1.005	7075	0.990
5050	1.005	7076	0.980
5052	1.010	7079	1.005
5056	1.025	7178	0.995
5083	1.010		

Source: From *Alcoa Aluminum Handbook*, 1984 edition.

[a]Empirical equations derived from expansion determinations on annealed high-purity aluminum within temperature range −320 to 1000°F, C = 1.000.
[b]Constant C established from determinations made on Alcoa alloys in annealed tempers, i.e., in states approaching dimensional stability through wide temperature ranges. With heat-treatable alloys, application of equation 2 is restricted to temperatures below 600°F. With wrought alloys 7075, 7076, 7079, and 7178, application is restricted to temperatures below 400°F. Values of constant when applied to alloys in their heat-treated tempers are approximately 0.015 greater than given. With these tempers, application is further restricted to temperatures which do not appreciably exceed those used in the final aging treatments.

TABLE 16.33   Relative Typical Characteristics of Wrought Aluminum Alloys[a]

Item No.	Alloy and Temper	Resistance to Corrosion	Work ability (Cold)	Machinability	Brazeability	Weldability Gas	Arc	Resistance Spot and Seam	Forgeability
1	1350-O	A	A	E	A	A	A	B	—
1a	1350-H18	A	B	D	A	A	A	A	—
3	1100-O	A	A	E	A	A	A	B	A
3a	1100-H18	A	C	D	A	A	A	A	A
5a	2011-T3	D	C	A	D	D	D	D	—
5b	2011-T8	D	D	A	D	D	D	D	—
6a	2014-T4	D	C	B	D	D	B	B	C
6b	2014-T6	D	D	B	D	D	B	B	C
9a	2024-T3	D	C	B	D	D	B	B	—
9c	2024-T4	D	C	B	D	D	B	B	—
9b	2024-T36	D	D	B	D	D	C	B	—
9e	2024-T81	D	D	B	D	D	C	B	—
13a	2219-T31	D	C	B	D	A	A	A	C
13b	2219-T37	D	D	B	D	A	A	A	C
13d	2219-T81	D	D	B	D	A	A	A	C
13e	2219-T87	D	D	B	D	A	A	A	C
14	3003-O	A	A	E	A	A	A	A	A
14a	3003-H18	A	C	D	A	A	A	A	A
15	3004-O	A	A	D	B	B	A	B	—
15a	3004-H38	A	C	C	B	B	A	A	—
17a	4032-T6	C	—	B	D	D	B	C	—
18	5005-O	A	A	E	B	A	A	B	—
18a	5005-H38	A	C	C	B	A	A	A	—
20	5052-O	A	A	D	C	A	A	B	—
20a	5052-H38	A	C	C	C	A	A	A	—
22	5083-O	A	B	D	D	C	A	B	—
22a	5083-H321	A	C	D	D	C	A	A	—
23	5086-O	A	A	D	D	C	A	B	—
23a	5086-H34	A	B	C	D	C	A	A	—
29	5456-O	A	B	D	D	C	A	B	—
29a	5456-H321	A	C	D	D	C	A	A	—
32	5652-O	A	A	D	C	A	A	B	—
32b	5652-H38	A	C	C	C	A	A	A	—
34	6061-O	B	A	D	A	A	A	B	—
34a	6061-T4	B	B	C	A	A	A	A	—
34b	6061-T6	B	C	C	A	A	A	A	—
36	6063-O	A	A	D	A	A	A	B	—
36c	6063-T5	A	B	C	A	A	A	A	—
36d	6063-T6	A	C	C	A	A	A	A	—
36e	6063-T83	A	C	C	A	A	A	A	—
37	6066-O	C	B	D	D	D	B	B	A
37a	6066-T6	C	C	B	D	D	B	B	A
39a	7049-T3	C	D	B	D	D	C	B	A
40	7050-T(all)	C	D	B	D	D	C	B	—
44a	7075-T6	C	D	B	D	D	C	B	D
46a	7178-T6	C	D	B	D	D	C	B	D

Source: From *Alcoa Aluminum Handbook*, 1984 edition.

[a]Resistance to corrosion, workability (cold), machinability, and foregeability ranges A, B, C, and D are relative ratings in decreasing order of merit. Weldability and brazeability ratings, A, B, C, and D are relative ratings defined as follows: A, generally weldable by all commercial procedures and methods; B, weldable with special technique or on specific applications which justify preliminary trials or testing to develop welding procedure and weld performance; C, limited weldability because of crack sensitivity or loss in resistance to corrosion and all mechanical properties; D, no commonly used welding methods have so far been developed.

TABLE 16.34 Nominal Composition of Aluminum Casting Alloys

Alloy No.		Process	Nominal Percentage of Alloying Elements[a]							
UNS	AA	Type	Si	Cu	Mn	Mg	Ni	Zn	Sn	Other
A02010	201.0	Sand	—	4.6	0.35	0.35	—	—	—	Ag 0.7, Ti 0.25
A02420	242.0	Sand, Prm. mold	—	4.0	—	1.5	2.0			
A02950	295.0	Sand	0.8	4.5						
A03550	355.0	Sand, Prm. mold	5.0	1.2	—	0.5				
A03560	356.0	Sand, Prm. mold	7.0	—	—	0.32				
A03600	360.0	Die	9.5	—	—	0.5				
A03800	380.0	Die	8.5	3.5						
A03900	390.0	Die	17	4.5	—	0.6				
A04130	413.0	Die	12							
A05130	513.0	Prm. mold, Die	—	—	—	4.0	—	1.8		
A05180	518.0	Die	—	—	—	8.0				
A05200	520.0	Sand	—	—	—	10.0				
A07120	712.0	Sand, Prm. mold	—	—	—	0.6	—	5.8	—	Cr 0.5, Ti 0.2
A07130	713.0	Sand, Prm. mold	—	0.7	—	0.35	—	7.5		
A08500	850.0	Sand, Prm. mold	—	1.0	—	—	1.0	—	6.2	
A13560	A356.0	Sand, Prm. mold	7	—	—	0.35				
A13570	A357.0	Prm. mold	7	—	—	0.5	—	—	—	Be 0.05, Ti 0.15
A13600	A360.0	Die	9.5	—	—	0.5				
A24430	B443.0	Sand, Prm. mold	5.2	—	—					
A33550	C355.0	Sand, Prm. mold	5.0	1.2	—	0.5				

[a]Aluminum and impurities constitute remainder.

TABLE 16.35 Typical Physical Properties of Aluminum Casting Alloys

Alloy No.		Density (g/cm³)	Approximate Melting Range (°C)	Electrical Conductivity (% IACS)	Thermal Conductivity (cgs units)	Thermal Expansion, 20–100°C (10^{-6}/K)	Modulus of Elasticity (GPa)
UNS	AA						
A02010	201.0	2.77	535–650	30	0.41	19.3	71.0
A02420	242.0	2.82	530–635	38	0.35	22.5	71.0
A02950	295.0	2.82	520–645	37	0.33	23.0	70.0
A03550	355.0	2.71	545–620	39	0.35	22.4	70.3
A03560	356.0	2.69	555–615	39	0.36	21.5	72.4
A03600	360.0	2.63	555–595	28	0.27	21.0	75.0
A03800	380.0	2.74	540–595	23	0.23	21.0	76.1
A03900	390.0	2.73	505–650	27	0.32	18.0	81.2
A04130	413.0	2.66	575–582	31	0.29	20.4	—
A05130	513.0	2.65	580–640	—	—	24.0	—
A05180	518.0	2.57	535–620	25	0.23	24.1	—
A05200	520.0	2.57	450–605	21	0.21	25.0	66.0
A07120	712.0	2.81	570–615	35	0.33	24.7	71.0
A07130	713.0	2.81	595–640	35	0.33	24.0	—
A08500	850.0	2.88	225–650	47	0.43	23.1	71.0
A13560	A356.0	2.69	555–615	39	0.36	21.5	72.4
A13570	A357.0	2.68	555–615	—	0.36	21.6	71.7
A13600	A360.0	2.63	555–595	30	0.29	21.0	71.0
A24430	B443.0	2.69	575–630	37	0.34	22.0	71.0
A33550	C355.0	2.71	545–620	39	0.35	22.4	70.3

TABLE 16.36   Typical Tensile Properties for Separately Cast Test Bars of Common Aluminum Casting Alloys

Alloy	Product[a]	Temper	Tensile Strength MPa	ksi	Yield Strength[b] MPa	ksi	Elongation[c] (%)
201.0	S	T4	365	53	215	31	20
	S	T6	485	70	435	63	7
	S	T7	460	67	415	60	4.5
242.0	S	T21	185	27	125	18	1.0
	S	T571	220	32	205	30	0.5
	S	T77	205	30	160	23	2.0
	P	T571	275	40	235	34	1.0
	P	T61	325	47	290	42	0.5
295.0	S	T4	220	32	110	16	8.5
	S	T6	250	36	165	24	5.0
	S	T62	285	41	220	32	2.0
355.0	S	T51	195	28	160	23	1.5
	S	T6	240	35	175	25	3.0
	S	T61	270	39	240	35	1.0
	S	T7	265	38	250	36	0.5
	S	T71	175	35	200	29	1.5
	P	T51	210	30	165	24	2.0
	P	T6	290	42	190	27	4.0
	P	T62	310	45	280	40	1.5
	P	T7	280	40	210	30	2.0
	P	T71	250	36	215	31	3.0
356.0	S	T51	175	25	140	20	2.0
	S	T6	230	33	165	24	3.5
	S	T7	235	34	210	30	2.0
	S	T71	195	28	145	21	3.5
	P	T6	265	38	185	27	5.0
	P	T7	220	32	165	24	6.0
357.0, A357.0	S	T62	360	52	290	42	8.0
360.0	D	F	325	47	170	25	3.0
A360.0	D	F	320	46	165	24	5.0
380.0	D	F	330	48	165	24	3.0
390.0	D	F	280	41	240	35	1.0
	D	T5	300	43	260	38	1.0
413.0	D	F	300	43	140	21	2.5
B443.0	P	F	159	23	62	9	10.0
518.0	D	F	310	45	190	28	5.0–8.0
520.0	S	T4	330	48	180	26	16
712.0	S	F	240	35	170	25	5.0
713.0	S	T5	210	30	150	22	3.0
	P	T5	220	32	150	22	4.0
850.0	P	T5	160	23	75	11	10.0

Source: Metals Handbook, (9th edition), ASM, Metals Park, Ohio, 1978.

[a]S, sand casting; P, permanent mold casting; D, die casting.

[b]0.2% offset.

[c]With 12.7-mm ($\frac{1}{2}$-in.) diameter specimen.

TABLE 16.37 Typical Tensile Properties of Cast Aluminum Alloys at Elevated Temperatures[a]

Alloy and Temper Properties	Test Temperature, °C					
	RT	150°	205°	260°	315°	370°
201-T7						
TS	67	55	47	28	20	
YS	60	52	45	27	19	
El	4.5	6–8.5	9	14	12	
242 sand cast						
TS	32	30	26	13	8	
YS	30	28	21	8	4	
El		0.5	1.0	8.0	20	
242 permanent mold cast						
TS	40	37	28	13	8	
YS	34	33	22	8	4	
El	1.0	1.0	2.0	15.0	35	
295-T4						
TS	32	28	15	9	4	
YS	16	15	9	6	3	
El	8.5	5.0	15	25	75	
295-T6						
TS	36	28	15	9	4	
YS	24	20	9	6	3	
El	5.0	5.0	15	25	75	
355-T6, sand cast						
TS	35	33	17	9.5	6	
YS	25	25	13	5.0	3	
El	3.0	1.5	8.0	16	36	
355-T6, permanent mold cast						
TS	42	32	19	9.5	6	
YS	27	25	13	5.0	3	
El	4	10	20	40	50	
356.0-T6						
TS	33	23	12	7.5	4.0	
YS	24	20	8.5	5.0	3.0	
El	3.5	6.0	18	35	60	
360.0-F						
TS	47	35	22	12	7	4.5
YS	25	24	14	7.5	4.5	3.0
El	3.0	4.0	8.0	20	35	40
A360.0-F						
TS	46	34	21	11	6.5	4.0
YS	24	23	13	6.5	4.0	2.5
El	5	5	14	30	45	45
520.0-F						
TS	46	35	22	15	10.5	
YS	25	19	11.5	7.5	3.5	
El	14	16	40	55	70	

[a]RT, room temperature; TS, tensile strength (ksi); YS, yield strength (ksi); El, elongation (%).

TABLE 16.38 Characteristics of Common Aluminum Alloys Used in Sand and Permanent Mold[a] Casting (a) (b)

Alloy	Type of Mold[b]	Fluidity	Resistance to Hot Cracking	Pressure Tightness	Heat Treatment	Strength at Elevated Temperatures	General Corrosion Resistance	Machining	Polishing	Anodizing Appearance	Weldability	Typical Applications
242.0	S or P	3	4	4	Yes	1	4	2	2	3	4	Heavy-duty pistons and air-cooled cylinder heads
295.0	S	3	4	4	Yes	3	4	2	2	2	3	Machinery and aircraft structural members
355.0	S or P	1	1	1	Yes	2	3	3	3	4	1	Timing gears, impellers, compressors, aircraft components requiring high strength
C355.0	S or P	1	1	1	Yes	2	3	3	3	4	1	Same as 355.0
356.0	S or P	1	1	1	Yes	3	2	3	4	4	1	Machine tool parts, aircraft wheels, pump parts, marine hardware, valve bodies
A356.0	S or P	1	1	1	Yes	3	2	3	4	4	1	Same as 356.0
A357.0	S or P	1	1	1	Yes	2	2	3	4	4	1	Same as 355.0
B443.0	S or P	1	1	1	No	4	2	5	4	4	1	Carburetor bodies, waffle irons
513.0	P	4	4	4	No	3	1	1	1	1	3	Ornamental hardware and architectural fittings
520.0	S	4	4	5	Yes	5	1	1	1	1	4	Aircraft fittings, truck and bus frame components, levers, brackets
713.0	S or P	3	4	4	No	4	3	1	1	1	4	General-purpose casting alloy for applications requiring strength without heat treatment
850.0	S or P	4	5	5	Yes	5	4	1	3	...	5	Bearings, bushings

Source: From Standards for Aluminum Sand and Permanent Mold Castings, The Aluminum Association, 1977.

[a]Characteristics are comparatively related from 1 to 5; 1 is the highest or best possible rating.

[b]S, sand; P, permanent.

TABLE 16.39 Characteristics of Aluminum Die Casting Alloys[a]

Alloy	Approximate Melting Temperature (°C)	Resistance to			Die Filling Capacity	Machining	Polishing	Electroplating	Anodized Surface		Elevated Temperature Strength	Pressure Tightness	Typical Applications
		Hot Cracking	Die Soldering	Corrosion					Appearance	Protection			
360.0	557–596	1	2	2	3	3	3	2	3	3	1	2	Frying skillets, cover plates; instrument cases, parts requiring corrosion resistance
A360.0	557–596	1	2	2	3	3	3	2	3	3	1	2	Same as 360.0
380.0	538–593	2	1	4	2	3	3	1	3	4	3	2	Lawn mower housings, gear cases, cylinder heads for air-cooled engines
A380.0	583–593	2	1	4	2	3	3	1	3	4	3	2	Street lamp housings, typewriter frames, dental equipment
413.0	574–582	1	1	2	1	4	5	3	5	3	3	1	Pistons, connecting rods and housings for outboard motors
518.0	535–621	5	5	1	5	1	1	5	1	1	4	5	Escalator parts, conveyor components, aircraft and marine hardware and fittings

Source: From ASTM B85.

[a] Relative rating of die casting alloys from 1 to 5; 1 is highest or best possible rating. Rating of 5 in one or more categories does not rule an alloy out of commercial use if other attributes are favorable; however, ratings of 5 may present manufacturing difficulties.

Copper Alloys

TABLE 16.40 Classification of Coppers

Designations	Type of Copper	UNS No.	Form in Which Copper is Available[a]							
			From Refiners				From Fabricators			
			Wire Bars	Billets	Cakes	Ingots and Ingot Bars	Flat Products	Pipe and Tube	Rod and Wire	Shapes
CATH	Electrolytic cathode		Cathodes only							
		Tough-Pitch Coppers								
ETP	Electrolytic tough pitch	C11000	x	x	x		x	x	x	x
RHC	Remelted, high-conductivity tough pitch	C11010	x	x	x	x	x	x	x	x
ETP	Electrolytic tough pitch (anneal resist)	C11100	x	x	x		x	x	x	x
CRTP	Chemically refined tough pitch	C11030	x	x	x		x	x	x	x
FRHC	Fire-refined, high-conductivity tough pitch	C11020	x	x	x	x	x	x	x	x
ETP[b]	Silver bearing, tough pitch	C11300, C11400, C11500, C11600	x	x	x	x	x	x	x	x
FRTP	Fire refined, tough pitch	C12500		x	x	x	x	x	x	
FRSTP	Fire refined tough pitch with silver	C12700, C12800, C12900, C1300		x	x	x	x	x	x	x
		Oxygen-Free Coppers (Without use of Deoxidants)								
OFE	Oxygen free, electronic	C10100	x	x	x		x	x	x	x
OF	Oxygen free	C10200	x	x	x		x	x	x	x
OFS	Oxygen free, silver bearing	C10400, C10500, C10700	x	x	x		x	x	x	x
OFXLP	Oxygen free, extra low phosphorus	C10300	x	x	x		x	x	x	x
OFLP	Oxygen free, low phosphorus	C10800	x	x	x		x	x	x	x

Deoxidized Coppers

Designation	Material	UNS No.							
DLP	Phosphorized, low-residual phosphorus	C12000	×	×	×	×	×	×	
DLPS[c]	Phosphorized, low-residual phosphorus, silver bearing	C12100	×	×	×	×	×	×	×
DHP[d]	Phosphorized, high-residual phosphorus	C12200	×	×	×	×	×	×	×
DHPS[c]	Phosphorized, high-residual phosphorus, silver bearing	C12300	×		×	×	×	×	
DPA	Phosphorized, arsenic bearing	C14200	×	×	×	×	×		×
DPTE[e]	Phosphorized, tellurium bearing	C14500	×			×	×		

Other Coppers

Designation	Material	UNS No.							
	Sulfur bearing	C14700	×		×	×		×	
	Zirconium bearing	C15000	×	×	×	×		×	×

Source: ASTM B224-80.

[a] The "×" indicates commercial availability.
[b] This includes Types ETP, CRTP, and FRHC coppers to which silver has been added in amounts agreed upon.
[c] This includes oxygen-free copper to which phosphorus and silver have been added in amounts agreed upon.
[d] This includes oxygen-free copper to which phosphorus has been added.
[e] This includes oxygen-free tellurium-bearing copper to which phosphorus has been added in amounts agreed upon.

TABLE 16.41   Mechanical and Physical Properties of Electrolytic Copper (Tough Pitch)[a]

	Hard[b]	Soft[c]	Forgings Hot	Forgings Cold[d]	Forgings Cold[e]
Mechanical Properties					
Tensile strength, psi ($\times 10^3$)	55	33	33–36	35–50	55
Apparent elastic limit, psi ($\times 10^3$)	40	4	4–8	8–35	44
Yield strength, 0.5% extension, psi ($\times 10^3$)	48	8	8	22–46	50
Yield strength, 0.2% offset, psi ($\times 10^3$)	49	7	7	21–45	45
Yield strength, 0.1% offset, psi ($\times 10^3$)	43	6	6	10–40	44
Elongation, % in 2 in.	10	50	50–45	40–10	5
Reduction of area, %	45	65	60–40	60–50	40
Endurance limit, psi ($\times 10^3$)	15	10	10	12–15	15
Rockwell hardness F, $\frac{1}{16}$-in. ball, 60-kg load	90	25	25–65	65–85	90
Rockwell hardness B, $\frac{1}{16}$-in. ball, 100-kg load	55	—	—	15–50	55
Brinell hardness, 10-mm ball, 500-kg load	194	40	40–60	60–83	39
Modulus of elasticity, psi			16,000,000		

Physical Data

Melting point, solidus	1949°F
Coefficient of expansion, from 68 to 572°F	0.0000098/°F
Electrical conductivity,[d] IACS, 68°F (vol and wt basis)	101%
Thermal conductivity,[d] at 68°F	227 Btu/ft^2 ft hr °F
Density	0.322 lb/in.^{3}
Electrical resistivity, at 68°F (annealed)	10.3 Ω/mil ft
Specific heat, at 68°F	0.092 Btu/lb °F
Specific gravity	8.89–8.94
Forging range	1250–1450°F
Forging quality	Good
Type of structure	Single phase, alpha

[a] Copper, 99.94%; oxygen, 0.04%; phosphorus, none; silver, none.
[b] Refers to rod previously hard drawn 50%; rod under 1 in. in diameter; ready to finish; grain size, 0.030 mm.
[c] Refers to 1100°F anneal for 1 hr.
[d] Material cold forged from soft rod (5–40% reduction).
[e] Material cold forged from cold-worked condition (40%).

TABLE 16.42 Chemical and Physical Properties of Various Wrought Copper Base Alloys[a]

Item No.	Material	Form	Copper	Zinc	Lead	Tin	Nickel	Other	Tensile Strength (psi) Hard[b]	Soft	Elongation in 2 in. (%) Hard[b]	Soft	Yield Point (psi) Hard[b]	Soft	Johnson's Elastic Limit (psi) Hard[b]	Soft	Modulus of Elasticity ($\times 10^{-6}$ psi) Hard[b]
1	Commercial bronze 95% (gilding metal)	S	95.00	5.00	—	—	—	—	60,000	35,000	4	45	55,000	11,000	—	—	15.0
2	Commercial bronze 90%	S	90.00	10.00	—	—	—	—	67,000	37,000	3	40	53,000	11,000	—	—	15.0
3	Red brass 80%	S	80.00	20.00	—	—	—	—	85,000	43,000	4	50	—	—	—	—	15.0
4	Cartridge brass	S	70.00	30.00	—	—	—	—	90,000	45,000	4	50	—	—	—	—	14.0
5	Yellow brass	S	65.00	35.00	—	—	—	—	90,000	45,000	5	60	—	—	55,000	7,500	14.0
6	Brass wire	R	63.00	37.00	—	—	—	—	70,000	50,000	12	50	—	—	—	—	—
6a	Brass wire	S	63.00	37.00	—	—	—	—	84,000	48,000	4	50	—	—	—	—	—
6b	Brass wire	W	63.00	37.00	—	—	—	—	125,000	50,000	2[c]	50[c]	—	—	—	—	—
7	Muntz metal	S	60.00	40.00	—	—	—	—	80,000	52,000	9.5	48	—	20,000	—	—	12.8
8	Leaded commercial bronze	R	88.50	10.00	1.50	—	—	—	60,000	35,000	3	30	—	—	—	—	15.0
9	Leaded brass	R	64.00	35.00	1.00	—	—	—	90,000	45,000	5	60	70,000	12,000	65,000	9,000	15.0
10	Free-cutting brass	R	62.00	35.00	3.00	—	—	—	70,000	50,000	10	50	55,000	17,000	39,000	12,000	15.0
11	Admiralty	S	70.00	29.00	—	1.00	—	—	95,000	45,000	3	65	75,000	12,000	65,000	11,000	—
11a	Admiralty	W	70.00	29.00	—	1.00	—	—	125,000	57,000	2[c]	42[c]	—	—	—	—	—
11b	Admiralty	T	70.00	29.00	—	1.00	—	—	100,000	55,000	4	65	—	—	—	—	—
12	Naval brass (Tobin bronze)	R	60.00	39.25	—	0.75	—	—	85,000	59,000	15	50	60,000	25,000	—	—	15.0
12a	Naval brass (Tobin bronze)	S	60.00	39.25	—	0.75	—	—	100,000	58,000	4	40	80,000	20,000	72,500	16,000	—
13	Phosphor bronze A	S	95.00	—	—	5.00	—	—	100,000	50,000	3	50	87,000	20,000	67,000	17,000	15.0
14	Phosphor bronze B	S	92.00	—	—	8.00	—	—	110,000	55,000	5	60	90,000	24,000	75,000	20,000	14.0
15	Cupro-nickel	R	70.00	—	—	—	30.00	—	85,000	55,000	15	45	78,000	20,000	69,000	16,000	—
16	18% nickel silver A	S	65.00	17.00	—	—	18.00	—	90,000	60,000	3	34	90,000	26,000	72,000	26,000	—
17	18% nickel silver B	S	56.00	26.00	—	—	18.00	—	110,000	60,000	2	45	100,000	20,000	90,000	17,000	18.0*
18	5% aluminum bronze	S	95.00	—	—	—	—	Al 5.00	105,000	52,000	5	70	—	—	—	—	—
18a	8% aluminum bronze	R	92.00	—	—	—	—	Al 8.00	100,000	60,000	4	60	—	—	—	—	—
19	10% aluminum bronze	R	90.00	—	—	—	—	Al 10.00	125,000[d]	78,000	5[d]	36	67,000	41,000	—	—	—
20	Manganese bronze	R	59.00	39.00	Iron 1.00	1.00	Mn 1.00	Mn 0.30	90,000	65,000	10	35	65,000	25,000	65,000	22,000	—
21	High-silicon bronze	R	96.00	—	—	—	Mn 1.00	Si 3.00	113,000	65,000	5	70	—	—	94,000	21,000	15.0
22	Low-silicon bronze	R	98.25	—	—	—	Mn 0.25	Si 1.50	100,000	40,000	6	60	65,000	12,000	55,000	7,000	—
23	Extruded architectural bronze shapes	—	57.00	40.00	Lead 3.60	0.34	—	Iron 0.16	88,000	60,000	15	25	60,000	35,000	51,000	25,000	—
24	Beryllium copper	S	97.40	—	Be 2.25	—	—	Ni 0.35	118,000	70,000	4.3	45	105,000	31,000	79,000	18,000	17.2
24a	Beryllium copper	S	97.40	—	Be 2.25	—	—	Ni 0.35	192,000	175,000[e]	2[d]	6.3[e]	138,000[d]	114,000[e]	130,000[d]	87,000[f]	18.4[d]

TABLE 16.42 *(Continued)*

Item No.	Material	Form	Shearing Strength (psi) Hard	Soft	Brinell Hardness No. 10-mm Ball, 500-kg Load Hard	Soft	Rockwell B Hardness No. 1/16-in. Ball, 100 kg Hard	Soft	Melting Point (°C)	Specific Gravity	Density (lb/in.³)	Coefficient of Expansion (×10⁷)	Electrical Properties at 20°C Resistivity (Ω/mil ft)	Conductivity (% IACS)	Thermal Conductivity	Uses and Remarks
1	Commercial bronze 95% (gilding metal)	S	—	—	110	43	68	4	1065^l	8.866^k	0.320	181	18.98	54.6	0.576	For jewelry trade and manufacturing where soft, pliable metal is required.
2	Commercial bronze 90%	S	—	25,000	115	50	75	1	1045^l	8.804^k	0.318	182	25.36	40.90	0.446	Window screen wire and automobile radiators on account of resistance to corrosion and atmospheric action.
3	Red brass 80%	S	43,000	27,000	150	53	86	11	1000^l	8.667^k	0.313	191	31.95^l	32.5^l	0.335	Color and resistance to corrosion and atmospheric action.
4	Cartridge brass	S	—	—	157	53	87	—	955^l	8.528^k	0.308	199	37.61	27.58	0.290	Primers, shot shells, cartridges, seamless tubes, etc.
5	Yellow brass	S	—	—	153	52	85	30	930^l	8.460^k	0.306	202	38.68	26.8	0.285	Large variety of articles, lamp fixtures, automobile radiators, and ornamental purposes. Not good for exposure to weather.
6	Brass wire	R	—	—	—	—	—	—	920^l	8.437^k	0.305	205	39.97	25.95	0.285	Rivets, pins, screws, and other heading operations.
7	Muntz metal	S	—	—	155	80	87	42	905^l	8.396	0.303	208	36.25	28.60	0.300	Bolts, nuts, sheathing, pin wire, etc.
8	Leaded commercial bronze	R	35,000	—	104	—	58	—	—	8.830^k	0.319	183	25.61^m	40.50^m	0.432	Free cutting.

No.	Material	Form														Uses
9	Leaded brass	R	—	—	—	—	—	—	—	8.562[k]	0.309	200	37.65	27.55	—	Special shapes where high lead is detrimental to bending or working of stock.
10	Free-cutting brass	R	36,000	28,000	120	54	77	16	885[″]	8.489[k]	0.307	204	41.46	25.0	0.258	Automatic machine work. Drills and turns easily.
11	Admiralty	S	—	—	—	—	—	—	935[″]	8.535[″]	0.308	202	42.07	24.65	0.263	Condenser tubes. Resists action of seawater.
12	Naval brass (Tobin bronze)	R	45,000	33,000	100	89	85	55	885[″]	8.404[″]	0.304	211	41.60	24.93	0.279	Piston rods, propeller shafts, nuts, bolts, plates, etc. Welding rod.
12a	Naval brass (Tobin bronze)	S	—	33,000	165	90	93	55	—	—	—	—	—	—	—	Same as 12.
13	Phosphor bronze A	S	33,000	—	190	60	96	30	1050[o]	8.87[″]	0.320	178	56.46	18.37	0.195	Springs, electric switches, window weight chain. Bronze chain in general.
14	Phosphor bronze B	S	60,000	—	200	70	104	47	1025[o]	8.815[″]	0.318	182	79.8	13.00	0.150	Electric switches, contact fingers, diaphragms, radio parts, etc.
15	Cupro-nickel	R	—	—	—	—	81	37	1225[p]	8.950	0.323	162[q]	218.4	4.75	0.069	Condenser tubes.
16	18% nickel silver A	S	—	—	170	70	91	50	1110[f]	8.752[f]	0.316	—	175.0	5.91	0.080	Silver-plated forks, spoons, knives, hollow ware, etc.
17	18% nickel silver B	S	65,000	—	190	70	90	50	1055[f]	8.68[″]	0.314	—	186.5[f]	5.56[f]	0.071	Similar to 30% nickel silver but of lower resistance.
18	5% aluminum bronze	S	—	—	176	67	93	20	1060[f]	8.176[b]	0.295	—	58.61	17.69	0.198	Diaphragms to withstand pressure; also for its color.
19	10% aluminum bronze	R	—	—	190	100	100	65	1040[b]	7.57[b]	0.273	—	76.80	13.5	0.157	Strength and resistance to ordinary corrosion and wear.
20	Manganese bronze	R	—	—	—	—	90	—	—	8.370[b]	0.302	—	42.15	24.6	0.241	Structural work due to strength and resistance to corrosion.
21	High-silicon bronze	R	60,000	33,000	200	70	95	40	1019[″]	8.539[b]	0.308	180	155.0	6.7	0.078	Strength and resistance to corrosion. Has strength of mild steel and corrosion resistance of copper. Welding rod. Subject to stress corrosion.

TABLE 16.42 (Continued)

Item No.	Material	Form	Shearing Strength (psi) Hard[b]	Soft	Brinell Hardness No., 10-mm Ball, 500-kg Load Hard[b]	Soft	Rockwell B Hardness No., $\frac{1}{16}$-in. Ball, 100 kg Hard[b]	Soft	Melting Point (°C)	Specific Gravity[g]	Density (lb/in.³)	Coefficient of Expansion ($\times 10^7$)	Electrical Properties at 20°C Resistivity (Ω/mil ft)	Conductivity (% IACS)	Thermal Conductivity[i]	Uses and Remarks
22	Low-silicon bronze	R	—	—	—	—	90	3	1055[s]	8.740[n]	0.316	—	86.4	12.0	0.129	Strength and resistance to corrosion, bolt stock, and sheet metal requiring high ductility.
23	Extruded architectural bronze shapes	—	—	—	—	—	84	63	884[n]	8.432[n]	0.305	—	—	—	—	Architectural shapes.
24	Beryllium copper	S	—	—	—	—	102	65–73	995[n]	—	0.297 ± 0.01	170	—	17 ±[e]	—	Springs, diaphragms, low-duty bushings and bearings, Bourdon tubes.
24a		S	—	—	—	—	114[d]	112.5[e]	—	—	0.297 ± 0.01	—	—	18-25[d]	0.25[e], 0.20[d]	High resistance to fatigue.

[a]Variations must be expected in practice. Abbreviations: R, rod; S, sheet; T, tube; W, wire.
[b]For some alloys the figures given are for a temper slightly different from that commonly known as "hard."
[c]Elongation in 10 in., percent
[d]Cold-worked and heat-treated.
[e]Annealed, quenched, and heat-treated.
[f]Tafel constitution diagram.
[g]Compared to water at 4°C.
[h]Average linear coefficient per °C from 25 to 300°C. Tests on rod. Scientific Paper 410, National Bureau of Standards.
[i]G-cal, per sec per sq cm per °C per cm at 20°C.
[j]Bauer and Hansen constitution diagram.
[k]From Scientific Paper 410, National Bureau of Standards.
[l]Soft.
[m]Hard at 25°C.
[n]Determination.
[o]Heycock-Neville constitution diagram.
[p]Guertler-Tammann constitution diagram.
[q]Corning Glass Works.
[r]Stockdale constitution diagram.
[s]Smith constitution diagram.

TABLE 16.43 Sand-Cast Copper-Base Alloys

Item No.	Name	Range of Composition (%)						Tension 1000 psi		Elongation in 2 in. (%)	Reduction in Area (%)	Brinell Hardness	Impact Strength Izod (ft-lb)	Remarks and Uses
		Cu	Al	Zn	Si	Mn	Fe	Yield Strength	Ultimate Strength					
		Cast Copper												
1, 1a	Pure copper	99.6, 99.9	—	—	—	—	—	8–10	20–25	20–25	75	40	—	Deoxidized with Si, B_2C_3, or carbon-free Mn. Electrical uses if pure.
		High-Tensile Bronzes and Brasses												
2	Aluminum bronze	80–90	7–12	—	—	0–3	1–5	25–45	65–90	5–30	—	110–180	30–36	Aluminum bronzes. These alloys should not be slowly cooled from a high temperature if they are to be stressed at ordinary temperatures. Resistant to cold dilute H_2SO_4, cold weak HCl, seawater, etc. Pipe fittings, valves, and other equipment for chemical industry.
3	Silicon bronze	94–98	—	—	2–4	0–1	—	18–30	40–60	20–50	15–22	80–100	—	Resistant to H_2SO_4, HCl (in absence of air), seawater, caustic soda, and phenol.
4	Manganese bronze	56–65	0–6	30–44	—	0.25–5.0	0–5	25–80	60–110	8–30	5–40	80–200	7–40	Manganese-bronze. Propellers, engine frames, parts requiring strength and toughness. Resistant to seawater.
5	Nickel silver	60–75	3–5	—	1–10	20–30	0–1	17–40	40–65	15–25	—	75–150	15–30	Nickel-bronze. Valve facing for severe conditions. Abrasion resistant. For hardness.

TABLE 16.43 (Continued)

Nominal Composition: Bronze and Red Brass for General Engineering Work

Item No.	Name	Range of Composition (%)						Tension 1000 psi				Brinell Hardness	Impact Strength Izod (ft-lb)	Remarks and Uses
		Cu	Al	Zn	Si	Mn	Fe	Yield Strength	Ultimate Strength	Elongation in 2 in. (%)	Reduction in Area (%)			
6	Bronze	88	8	—	4	—	—	16–21	35–45	25–40	16–33	60–72	11–16	Bronze for water-tight castings, under-water fittings, machine parts. Non-corrosive.
7	Gear bronze	90	10	—	—	—	—	18–20	30–35	5–15	6–8	65–85	—	Gear bronze. Very resistant to abrasion. For heavy-duty gears and worm wheels.
8	Leaded red brass	85	5	5	5	—	—	17–24	33–46	15–35	12–32	55–65	—	Red brass for pump bodies, valves, steam fittings, bearing backs, and metal patterns.
9	Leaded semi-red brass	83	4	6	7	—	—	12–17	30–38	15–27	12–25	50–60	—	Semi-red brass. Very resistant to atmospheric corrosion. Overhead electrical fittings and for oil and water pumps.
10	Hard bronze	80–85	15–20	—	—	—	—	18–32	25–35	0–1	0–1	70–160	—	Hardness. Difficult to machine.
Yellow Brass														
11	Yellow brass	60–72	0.5–1.5	0.5–4.0	25–40	—	—	10–20	27–45	15–40	15–40	40–75	—	Common castings where low cost and good machining properties are main considerations.

Nominal Composition: High-Conductivity Alloys, Precipitation Hardening

Item No.	Name	Cu	Cr	Si	Be	Co	Yield Strength	Ultimate Strength	Elongation in 2 in. (%)	Reduction in Area (%)	Brinell Hardness	Percentage of Conductivity	Remarks and Uses
12	Cr Copper	Remainder	0.8	—	—	—	35	52	18	—	100	85	High electrical conductivity combined with good mechanical properties.
12a	Be Copper	Remainder	—	—	0.45	2.5	70	85	7	—	160	45	Same as 12.

TABLE 16.44 Chemical Requirements For Brass Alloy Die Castings[a]

Element[c]	Composition[b] (%)					
	C85800[d]	C86500[d]	C86800[d]	C87800[d]	C87900[d]	C99750[d]
Copper	57.0 minimum	55.0–60.0	54.0–57.0	80.0–83.0	63.0–67.0	55.0–61.0
Tin	1.5	1.5	1.5	0.25	0.25	0.35
Lead	1.5	0.30	0.20	0.15	0.25	2.5
Zinc	31.0–41.0	36.0–42.0	31.0–39.0	12.0–16.0	30.0–36.0	17.0–23.0
Iron	0.55	2.5	2.5	0.25	0.50	1.5[e]
Nickel	0.50	0.8	0.8	0.20	0.50	5.0
Aluminum	0.50	1.5	2.0	0.15	0.15	3.0
Manganese	0.25	1.5	2.5–4.0	0.15	0.15	17.0–23.0
Silicon	0.25	—	—	3.8–4.2	0.75–1.2	—
Magnesium	—	—	—	0.01	—	—
Total named elements, minimum	99.5	99.5	99.5	99.8	99.5	99.5

Source: ASTM B176-79.

[a] For purposes of acceptance and rejection, the observed value or calculated value obtained from analysis should be rounded to the nearest unit in the last right-hand place of figures, used in expressing the specified limit, in accordance with the rounding procedure prescribed in Section 3 of Recommended Practice E 29.
[b] Maximum, unless shown as range or minimum.
[c] Amount of arsenic, antimony, and sulfur shall not exceed 0.05% for each, and phosphorus shall not exceed 0.01%.
[d] ASTM alloy designations established in accordance with Recommended Practice B 275. UNS designations established in accordance with Recommended Practice E 527. Prior to 1952 alloys C85800, C87800, and C87900 were designated A, C, and B, respectively.
[e] Iron content shall not exceed nickel content.

TABLE 16.45 Typical Mechanical Properties of Test Specimens of Brass Die Casting Alloys[a]

Copper Alloy UNS No.	Tensile Strength, ksi (MPa)	Yield Strength, (0.2% Offset), ksi (MPa)	Elongation in 2 in. (50 mm), %	Impact Strength, Charpy, ft lbf (J)	Hardness, Rockwell B	Modulus of Elasticity, ×10³ ksi (GPa)
C85800	55 (379)	30 (207)	15	40 (54)	55–60	15,000 (103)
C87800	85 (586)	50 (345)	25	70 (95)	85–90	20,000 (138)
C87900	70 (483)	35 (241)	25	50 (68)	68–72	15,000 (103)

Source: ASTM B176-79.

[a] Typical die cast properties of alloys C86500, C86800, and C99750 are under development; ksi = 1000 psi.

Magnesium Alloys

TABLE 16.46   Typical Mechanical Properties of Magnesium Forgings

Alloy	Condition	Tension			Compression	Hardness	
		Typical Tensile Strength (psi)	Typical[a] Yield Strength (psi)	Typical Elongation in 2 in. (%)	Yield Strength (psi)	Brinell, 500-kg Load, 10-mm Ball	Rockwell E
AZ61A-F	As forged	45,000	33,000	16	19,000	60	72
M1A	As forged	36,000	23,000	7	—	47	54
AZ80A-F	As forged	46,000	31,000	8	25,000	69	80
AZ80A-T8	Heat treated and aged	52,000	34,000	5	27,000	72	82
HM21A	T8, solution treated, cw, and aged	35,000	26,000	12	21,000		
HM31A	T5 heat treated and aged	44,000	38,000	8	27,000		
ZK60A	F as forged	49,000	37,000	14	27,000		
ZK60A	T5 heat treated and aged	52,000	43,000	12	31,000		

[a]Yield strength defined as stress at which stress-strain curve deviates 0.2% from the modulus line.

TABLE 16.47   Typical Mechanical Properties of Magnesium Plate, Sheet, and Strip

Alloy	Condition	Tension			Compression	Shear	Hardness		
		Typical Tensile Strength (psi)	Typical Yield Strength (psi)	Typical Elongation in 2 in. (%)	Yield Strength (psi)	Ultimate Stress (psi)	Brinell, 500-kg Load, 10-mm Ball	Rockwell E	Maximum Service Temp (°C)
AZ31B	Annealed	37,000	22,000	21	16,000	21,000	56	67	100
	Hard rolled	43,000	33,000	11	26,000	23,000	73	83	—
M1A	Annealed	33,000	18,000	16	12,000	18,000	48	55	—
	Hard rolled	37,000	28,000	7	20,000	17,000	56	67	—
HK31A	Annealed	33,000	18,000	20	13,000	—	—	55	—
	H24 temper	38,000	29,000	12	23,000	—	—	68	315
HM21A	T8 temper	35,000	26,000	12	21,000	18,000	—	—	345

[a]Yield strength defined as stress at which stress-strain curve deviates 0.2% from modulus line.

TABLE 16.48 Physical Constants of Magnesium Alloys in Common Use[a]

UNS No.	ASTM Designation	Typical Composition (%)							Specific Gravity	Density (lb/in.^{3})	Melting Point (°F)	Thermal Conductivity (68°F)	Electric Resistivity ($\mu\Omega$/cm), 20°C (68°F)	Uses
		Aluminum	Zinc	Manganese[b]	Other[c]	Th	Zr	Magnesium						
M10100	AM100A	10.0	0.2[c]	0.13	0.55	—	—	Balance	1.81	0.066	1100	41	14.5	Sheet and plate
M11311	AZ31B	3.0	1.0	0.2	0.77	—	—	Balance	1.77	0.064	1160	56	10.0	Extrusions, sheets, and forgings
M11610	AZ61A	6.5	0.95	0.15	0.46	—	—	Balance	1.80	0.065	1145	46	12.5	Extrusions and forgings
M11630	AZ63A	6.0	3.0	0.15	0.71	—	—	Balance	1.83	0.067	1135	44	12.8	Sand and permanent mold castings
M11800	AZ80A	8.5	0.5	0.15	0.66	—	—	Balance	1.80	0.065	1130	44	14.5	Extrusions and forgings
M11912	AZ91B	9.0	0.7	0.13	1.11	—	—	Balance	1.81	0.066	875–1105	41	17.0	Die castings
M11920	AZ92A	9.0	2.0	0.10	0.41	—	—	Balance	1.82	0.066	830–1100	39	16.0	Sand and permanent mold castings
M15100	M1A	—	—	1.2	—	—	—	Balance	1.76	0.064	1200	73	5.0	Extrusions, forgings, sheets, and special sand castings
M13310	HK31A	—	—	—	—	3.3	0.7	Balance	1.79	0.065	1090	66	6.1	Sheet and plate, high-temperature service
M13210	HM21A	—	—	0.6	—	2.0	—	Balance	1.78	0.064	1120	77	5.2	Sheet, plate, and forgings
M13312	HM31A	—	—	1.2	—	3.0	—	Balance	1.81	0.065	1120	60	6.6	Weldable, high-temperature service
M16210	ZK21A	—	2.3	—	—	—	0.45[b]	Balance	1.78	0.064	—	—	5.4	Extruded bars and shapes
M16510	ZK51A	—	4.6	—	0.41	—	0.7	Balance	1.81	0.066	1040	63	6.2	Sand castings
M16610	ZK61A	—	6.0	—	0.41	—	0.7	Balance	1.83	0.066	985	—	—	Sand castings
M16600	ZK60A	—	5.5	—	—	—	0.45[b]	Balance	1.83	0.066	1175	—	6.0	Extruded bars and shapes

[a] Modulus of elasticity, 6.5×10^6 psi. Modulus of rigidity, 2.4×10^6 psi. Coefficient of thermal expansion, 16×10^{-6} in./in. °F (68–750°F). Poisson's ratio, 0.35.
[b] Minimum.
[c] Maximum.

TABLE 16.49 Typical Mechanical Properties of Magnesium Sand, Die, and Permanent Mold-Casting Alloys[a]

Alloy	Condition	Tension — Typical Tensile Strength (psi)	Tension — Typical Yield Strength (psi)	Tension — Typical Elongation in 2 in. (%)	Compression — Yield Strength (psi)	Hardness — Brinell, 500-kg Load, 10-mm Ball	Hardness — Rockwell F	Shear — Shear Ultimate Stress (psi)	Impact Izod (ft-lbs)
AZ92A-F	As cast	24,000	14,000	2	14,000	65	77	19,000	1
AZ92A-T4	Heat treated	40,000	14,000	10	14,000	63	75	20,000	4
AZ92A-T5	Heat treated and aged	40,000	23,000	2	23,000	84	90	21,000	1
AM100A-F	As cast	22,000	12,000	2	12,000	54	65	17,000	2
AM100A-T4	Heat treated	40,000	13,000	6	13,000	52	62	19,000	4
AM100A-T6	Heat treated and aged	40,000	16,000	4	16,000	69	80	21,000	2
AZ83A-F	As cast	29,000	14,000	6	14,000	50	59	18,000	3
AZ83A-T4	Heat treated	40,000	14,000	12	14,000	55	66	19,000	5
AZ83A-T6	Heat treated and aged	40,000	19,000	5	19,000	73	83	21,000	2
M1A-F	As cast	14,000	4,500	5	4,500	33	3	11,000	9
AZ91B	Die cast	34,000	23,000	3	22,000	67	75	20,000	—
ZK51A-T5	Heat treated and aged	40,000	26,000	8	20,000	62	72	22,000	—
ZK61A-T6	Heat treated and aged	45,000	28,000	10	—	—	—	—	—

[a]Yield strength defined as stress at which stress-strain curve deviates 0.2% from modulus line.

Superalloys

TABLE 16.50 Nominal Compositions of Wrought Superalloys

Alloy	UNS No.	Cr	Ni	Co	Mo	W	Nb	Ti	Al	Fe	C	Other
Iron Base Solid-Solution Alloys												
Incoloy 800	N08800	21.0	32.5	—	—	—	—	0.38	0.38	45.7	0.05	—
N-155	R30155	21.0	20.0	20.0	3.00	2.5	1.0	—	—	32.2	0.15	0.15 N; 0.02Zr La; 0.02 Zr
Cobalt Base Solid-Solution Alloys												
Haynes 25 (L-605)	R30605	20.0	10.0	50.0	—	15.0	—	—	—	3.0	0.10	1.5 Mo
Haynes 188	R30188	22.0	22.0	37.0	—	14.5	—	—	—	3.0ᵃ	0.10	0.90 La
Nickel Base Solid-Solution Alloys												
Hastelloy X	N06002	22.0	49.0	1.5ᵃ	9.0	0.6	—	—	2.0	15.8	0.15	
Inconel 600	N06600	15.5	76.0	—	—	—	—	—	—	8.0	0.08	0.25ᵃ Cu
Inconel 625	N06625	21.5	61.0	—	9.0	—	3.6	0.2	0.2	2.5	0.05	
Iron Base Precipitation-Hardening Alloys												
A-286	K66286	15.0	26.0	—	1.25	—	—	2.0	0.2	55.2	0.04	0.005 B, 0.3 V
Nickel Base Precipitation-Hardening Alloys												
Inconel 706	N09706	16.0	41.5	—	—	—	—	1.75	0.2	37.5	0.03	2.9 (Nb + Ta), 0.15ᵃ Cu
Inconel 718	N07718	19.0	52.5	—	3.0	—	5.1	0.9	0.5	18.5	0.08ᵃ	0.15ᵃ Cu
Inconel 751	N07751	15.5	72.5	—	—	—	1.0	2.3	1.2	7.0	0.05	0.25ᵃ Cu
Inconel X750	N07750	15.5	73.0	—	—	—	1.0	2.5	0.7	7.0	0.04	0.25ᵃ Cu
M252	N07252	19.0	56.5	10.0	10.0	—	—	2.6	1.0	< 0.75	0.15	0.005 B
René 41	N07041	19.0	55.0	11.0	10.0	—	—	3.1	1.5	< 0.3	0.09	0.01 B
Udimet 500	N07500	19.0	48.0	19.0	4.0	—	—	3.0	3.0	4.0ᵃ	0.08	0.005 B
Waspaloy	N07001	19.5	57.0	13.5	4.3	—	—	3.0	1.4	2.0ᵃ	0.07	0.006 B, 0.09 Zr

Source: *Metals Handbook*, (9th edition), ASM, Metals Park, Ohio, 1978.
ᵃMaximum.

TABLE 16.51 Nominal Compositions of Cast Superalloys (%)

Alloy	UNS No.	Cr	Ni	Co	Mo	W	Nb	Ti	Al	Fe	C	Other
Cobalt Base												
FSX 414	—	29	10	ᵃ	—	7	—	—	—	1.0	0.25	0.01 B
Mar M 509	—	24	10	ᵃ	—	7	—	0.2	—	—	0.60	3.5 Ta, 0.5 Zr
X-40	R30031	25	10	ᵃ	—	7.5	—	—	—	—	0.50	0.5 Mn, 0.5 Si
WI-52	—	21	—	ᵃ	—	11	2	—	—	1.7	0.45	
Nickel Base												
B-1900	—	8	ᵃ	10	6	—	—	1.0	6	—	0.10	4 Ta, 0.015 B, 0.1 Zr
Mar M 200 (DS)	N13009	9	ᵃ	10	—	12	—	2.0	5	—	0.13	1 Nb, 0.015 B, 0.05 Zr
Mar M 246	N13246	9	ᵃ	10	2.5	10	—	1.5	5.5	—	0.15	1.5 Ta, 0.015 B, 0.05 Zr

ᵃBalance of percent composition.

TABLE 16.52   Typical Mechanical Properties of Wrought Cobalt Base and Nickel Base Superalloys

Temperature		Tensile Strength		Yield Strength		Elongation
°C	°F	MPa	ksi	MPa	ksi	(%)

Cobalt Base Alloys

HAYNES 25 (L-605) SHEET

°C	°F	MPa	ksi	MPa	ksi	(%)
21	70	1010	146	460	67	64
540	1000	800	116	250	36	59
650	1200	710	103	240	35	35
760	1400	455	66	260	38	12
870	1600	325	47	240	35	30

HAYNES 188, SHEET

°C	°F	MPa	ksi	MPa	ksi	(%)
21	70	960	139	485	70	56
540	1000	740	107	305	44	70
650	1200	710	103	305	44	61
760	1400	635	92	290	42	43
870	1600	420	61	260	38	73

Nickel Base Alloys

HASTELLOY X, SHEET

°C	°F	MPa	ksi	MPa	ksi	(%)
21	70	785	114	360	52	43
540	1000	650	94	290	42	45
650	1200	570	83	275	40	37
760	1400	435	63	260	38	37
870	1600	255	37	180	26	50

INCONEL 600, BAR

°C	°F	MPa	ksi	MPa	ksi	(%)
21	70	620	90	250	36	47
540	1000	580	84	195	28	47
650	1200	450	65	180	26	39
760	1400	185	27	115	17	46
870	1600	105	15	62	9	80

INCONEL 625, BAR

°C	°F	MPa	ksi	MPa	ksi	(%)
21	70	855	124	490	71	50
540	1000	745	108	405	59	50
650	1200	710	103	420	61	35
760	1400	505	73	420	61	42
870	1600	285	41	475	40	125

INCONEL 706, BAR

°C	°F	MPa	ksi	MPa	ksi	(%)
21	70	1300	188	980	142	19
540	1000	1120	163	895	130	19
650	1200	1010	147	825	120	21
760	1400	690	100	675	98	32

INCONEL 718, BAR

°C	°F	MPa	ksi	MPa	ksi	(%)
21	70	1430	208	1190	172	21
540	1000	1280	185	1060	154	18
650	1200	1230	178	1020	148	19
760	1400	950	138	740	107	25
870	1600	340	49	330	48	88

INCONEL 718, SHEET

°C	°F	MPa	ksi	MPa	ksi	(%)
21	70	1280	185	1050	153	22
540	1000	1140	166	945	137	26
650	1200	1030	150	870	126	15
760	1400	675	98	625	91	8

TABLE 16.52   *(Continued)*

Temperature		Tensile Strength		Yield Strength		Elongation
°C	°F	MPa	ksi	MPa	ksi	(%)
INCONEL X 750, BAR						
21	70	1120	162	635	92	24
540	1000	965	140	580	84	22
650	1200	825	120	565	82	9
760	1400	485	70	455	66	9
870	1600	235	34	165	24	47
M-252, BAR						
21	70	1240	180	840	122	16
540	1000	1230	178	765	111	15
650	1200	1160	168	745	108	11
760	1400	945	137	715	104	10
870	1600	510	74	485	70	18
RENÉ 41, BAR						
21	70	1420	206	1060	154	14
540	1000	1400	203	1010	147	14
650	1200	1340	194	1000	145	14
760	1400	1100	160	940	136	11
870	1600	620	90	550	80	19
UDIMET 500, BAR						
21	70	1310	190	840	122	32
540	1000	1240	180	795	115	28
650	1200	1210	176	760	110	28
760	1400	1040	151	730	106	39
870	1600	640	93	495	72	20
WASPALOY, BAR						
21	70	1280	185	795	115	25
540	1000	1170	170	725	105	23
650	1200	1120	162	690	100	34
760	1400	795	115	675	98	28
870	1600	525	76	515	75	35

Source: *Metals Handbook*, (9th edition), ASM, Metals Park, Ohio, 1978.

TABLE 16.53   Typical Mechanical Properties of Cast Cobalt Base and Nickel Base Superalloys

| Temperature | | Tensile Strength | | Yield Strength | | Elongation |
°C	°F	MPa	ksi	MPa	ksi	(%)
			B1900			
21	70	974	141	828	120	8
540	1000	1007	146	870	126	7
650	1200	1015	147	925	134	6
760	1400	953	138	808	117	4
870	1600	794	115	696	101	4
			Mar M-200 (DS)			
21	70	1000	145	862	125	10
540	1000	1015	147	876	127	10
650	1200	1022	148	890	129	9
760	1400	1049	152	925	134	4.5
870	1600	918	133	781	113	4.5
			Mar M-246			
21	70	966	140	862	125	5
540	1000	1006	145	862	125	5
650	1200	1035	150	862	125	5
760	1400	1035	150	862	125	5
870	1600	862	125	690	100	5
			FSX-414			
21	70	739	107	441	64	11
540	1000	538	78	242	35	15
650	1200	483	70	214	31	15
760	1400	400	58	193	28	18
870	1600	310	45	166	24	23
			Mar M-509			
21	70	786	114	573	83	4
540	1000	650	83	400	58	6
650	1200	559	81	372	54	7
760	1400	600	83	366	53	10
870	1600	352	51	290	42	20
			WI-52			
21	70	752	109	586	85	5
540	1000	745	108	441	64	7
650	1200	739	107	400	58	8
760	1400	607	88	345	50	9
870	1600	414	60	276	40	11
			X-40			
21	70	745	108	524	76	9
540	1000	552	80	276	40	17
650	1200	517	75	252	38	12
760	1400	483	70	—	—	—
870	1600	324	47	—	—	—

TABLE 16.54 Typical Rupture Strengths of Selected Superalloys

Temperature		Stress Rupture at 100 h		Stress Rupture at 1000 h	
°C	°F	MPa	ksi	MPa	ksi
Iron Base					
INCOLOY 800					
650	1200	220	32	145	21
760	1400	115	17	69	10
870	1600	45	6.5	33	4.8
N-155, SHEET					
980	1800	39	5.6	20	2.9
N-155, BAR					
650	1200	360	52	295	43
730	1350	195	28	150	22
870	1600	97	14	66	9.5
Nickel Base					
INCONEL 625					
650	1200	440	64	370	54
815	1500	130	19	93	13.5
870	1600	72	10.5	48	7
INCONEL 718					
540	1000	—	—	951	138
595	1100	860	125	760	110
650	1200	690	100	585	85
INCONEL X750					
540	1000	—	—	827	120
870	1600	83	12	45	6.5
925	1700	58	8.4	21	3.1
B-1900					
815	1500	505	73	380	55
870	1600	385	56	255	37
1090	2000	62	9	34	5
MAR M-200 (DS)					
815	1500	580	84	—	—
980	1800	200	29	140	20
1040	1900	140	20	97	14
Cobalt Base					
L605					
790	1350	255	37	—	—
815	1500	150	22	117	17
870	1600	—	—	86	12.6
FSX-414					
815	1500	160	23	115	16.5
870	1600	110	16	83	12
980	1800	55	8	34	5
1090	2000	24	3.5	—	—

Titanium Alloys

TABLE 16.55 Chemical Composition (%) of Titanium Strip, Sheet, and Plate

Element	R50250, Grade 1	R50400, Grade 2	R50550, Grade 3	R50700, Grade 4	R56400, Grade 5	R54520, Grade 6	R52400, Grade 7	R58030, Grade 10	R52250, Grade 11	R53400, Grade 12
Nitrogen, maximum	0.03	0.03	0.05	0.05	0.05	0.05	0.03	0.05	0.03	0.03
Carbon, maximum	0.10	0.10	0.10	0.10	0.10	0.10	0.10	0.10	0.10	0.08
Hydrogen, maximum[a]	0.015	0.015[b]	0.015	0.015	0.015	0.020	0.015	0.020	0.015	0.015
Iron, maximum	0.20	0.30	0.30	0.50	0.40	0.50	0.30	0.35	0.20	0.30
Oxygen, maximum	0.18	0.25	0.35	0.40	0.20	0.20	0.25	0.18	0.18	0.25
Aluminum	—	—	—	—	5.5–6.75	4.0–6.0	—	—	—	—
Vanadium	—	—	—	—	3.5–4.5	—	—	—	—	—
Tin	—	—	—	—	—	2.0–3.0	—	3.75–5.25	—	—
Palladium	—	—	—	—	—	—	0.12–0.25	—	0.12–0.25	—
Molybdenum	—	—	—	—	—	—	—	10.0–13.0	—	0.2–0.4
Zirconium	—	—	—	—	—	—	—	4.50–7.50	—	—
Nickel	—	—	—	—	—	—	—	—	—	0.6–0.9
Residuals[c,d] (each), maximum	0.1	0.1	0.1	0.1	0.1	0.1	0.1	0.1	0.1	0.1
Residuals[c,d] (total), maximum	0.4	0.4	0.4	0.4	0.4	0.4	0.4	0.4	0.4	0.4
Titanium[e]	Remainder	Remainder	Remainder	Remainder	Remainder	Remainder	Remainder	Remainder	Remainder	Remainder

Source: ASTM B265-79.

[a]Lower hydrogen may be obtained by negotiation with manufacturer.
[b]Editorially corrected.
[c]Need not be reported.
[d]Residual is element present in metal or alloy in small quantities inherent to manufacturing process but not added intentionally.
[e]Percentage of titanium determined by difference.

TABLE 16.56　Comparison of Typical Room Temperature Properties of Wrought, Cast, and P/M Titanium Products

Condition	Tensile Strength		Yield Strength		Elongation (%)	Reduction in area (%)	Impact Strength[a]	
	MPa	ksi	MPa	ksi			J	ft·lb
Unalloyed Ti (UNS R50550)								
Wrought bar, annealed	550	80	480	70	18	33	35	26
Cast bar, as cast	635	92	510	74	20	31	26	19
P/M compact, annealed[b]	480	70	370	54	18	22	—	—
Ti–5Al–2.5Sn–ELI (UNS R54523)								
Wrought bar, annealed	815	118	710	103	19	34	—	—
Cast bar, as cast	795	115	725	105	10	17	—	—
P/M compact, annealed and forged[c]	795	115	715	104	16	27	—	—
Ti–6Al–4V (UNS R56400)								
Wrought bar, annealed	1000	145	925	134	16	34	22	16
Cast bar								
As cast	1025	149	880	128	12	19	19	14
Annealed	1015	147	890	129	10	16	—	—
Solution treated and aged[d]	1180	171	1085	157	6	11	—	—
P/M compact								
Annealed[b]	825–855	120–124	740–785	107–114	5–8	8–14	—	—
Annealed and forged[c]	925	134	840	122	12	27	—	—
Solution treated and aged[d]	965	140	895	130	4	6	—	—
Ti–6Al–6V–2Sn (UNS R56620)								
Wrought bar, annealed	1125	163	1055	153	16	38	20	15
Cast bar, as cast	1105	160	965	140	6	11	14	10
P/M compact, annealed[b]	965	140	840	122	5	5	—	—

Source: Metals Handbook, Desk Edition, H. E. Boyer and T. L. Gall, (eds.), ASM, Metal Park, Ohio, 1985.

[a] Charpy, at −40°C (−40°F).
[b] About 94% dense.
[c] Almost 100% dense.
[d] Aging treatment not specified.

Zinc Alloys

TABLE 16.57 Commercial Grades of Zinc

| Grade | UNS No. | Composition (%) | | | |
		Maximum Lead	Maximum Iron	Maximum Cadmium	Minimum Zinc by Difference[a]
Special High Grade[b]	Z13001	0.003	0.003	0.003	99.990
High Grade	Z15001	0.07	0.02	0.03	99.90
Intermediate[c]	Z16001[c]	0.20	0.03	0.40	99.5
Brass Special[c]	Z17001[c]	0.6	0.03	0.50	99.0
Prime Western	Z19001	1.6	0.05	0.50	98.0

Source: ASTM B6-77.

[a]Analysis need not regularly be made for copper, tin, and aluminum in any grade. Nevertheless, it is understood that the minimum percentage of zinc (by difference) takes into account the copper, tin, and aluminum contents, if any, in addition to the impurities listed.
[b]Analysis need not regularly be made for tin in Special High Grade but, if found, shall not exceed 0.001%.
[c]No longer active, reference only.

TABLE 16.58 Composition and Mechanical Properties of Zinc Die Castings

	UNS Z33521 (Alloy AG40A)[a]	UNS Z35530 (Alloy AC41A)[a]	UNS Z35630 (Zn-11A1)	UNS Z35840 (ZA27)
Composition,[b] wt %				
Copper	0.25 max[c]	0.75–1.25	0.5–1.25	2.0–2.5
Aluminum	3.5–4.3	3.5–4.3	10.5–11.5	25–28
Magnesium	0.020–0.05	0.03–0.08	0.015–0.030	0.01–0.02
Iron, maximum	0.100	0.100	0.075	0.10
Lead, maximum	0.005	0.005	0.004	0.004
Cadmium, maximum	0.004	0.004	0.003	0.003
Tin, maximum	0.003	0.003	0.002	0.002
Zinc	Remainder	Remainder	Remainder	Remainder
Mechanical Properties				
Charpy impact strength, ft-lb, $\frac{1}{4} \times \frac{1}{4}$ in. bar, as-cast	20	20	—	—
Charpy impact strength, ft-lb after 8 yr aging indoors	2	2	—	—
Tensile strength, psi, as-cast	40,300	45,400	42,500	61,000
Tensile strength, 8 yr indoor age	34,400	37,200	—	—
Elongation in 2 in., as-cast, %	5	3	1–3	3–6
Elongation in 2 in., 8 yr indoor age, %	8	5	—	—
Expansion, in./in. after 8 yr	0.0001	0.0001	—	—
Other Properties and Constants, As-Cast				
Brinell hardness	74	79	115	105
Compression strength, lb/in.^{2}	60,500	87,300	—	—
Electric conductivity, mhos/cm at 20°C	157,000	153,000	—	—
Melting point, °C	380.9	380.6	379	378
Modulus of rupture, lb/in.^{2}	95,000	105,000	—	—
Shearing strength, lb/in.^{2}	30,900	38,400	37,000	42,000

TABLE 16.58 *(Continued)*

	UNS Z3352 (Alloy AG40A)[a]	UNS Z35530 (Alloy AC41A)[a]	UNS Z35630 (Zn-11Al)	UNS Z35840 (ZA27)
Solidification point, °C	380.6	380.4	377	375
Solidification shrinkage, in./ft	0.14	0.14	0.15	0.15
Specific gravity	6.6	6.7	6.03	5.01
Specific heat, cal/g °C	0.10	0.10	—	—
Thermal conductivity, cal/sec cm^3 °C	0.27	0.26	—	—
Thermal expansion, per °C	27.4×10^{-6}	27.4×10^{-6}	27.9×10^{-6}	26×10^{-6}
Transverse deflection, in.	0.27	0.16	—	—
Wt, lb/in.^{3}	0.24	0.24	0.22	0.18

Sources: ASTM B86-83, B669-84 and *Die Casting for Engineers*, The New Jersey Zinc Company.

[a]Designations in parentheses are former alloy designations.
[b]Zinc alloy die castings may contain nickel, chromium, silicon, and manganese in amounts of 0.02, 0.035, and 0.5%, respectively. No harmful effects have ever been noted due to presence of these elements in these concentrations and, therefore, analyses are not required for these elements.
[c]For majority of commercial applications, copper content in range of 0.25–0.75% will not adversely affect serviceability of die castings and should not serve as basis for rejection.

16.2.2 Polymers*

The use of plastics has increased almost 10-fold in the last 20 years. Plastics have come on the scene as the result of a continual search for man-made substances that can perform better or can be produced at a lower cost than natural materials such as wood, glass, and metals, which require mining, refining, processing, milling, and machining. Plastics can also increase productivity by producing finished parts and by consolidating parts. These increases in productivity have led to fantastic growth. For example, the total volumes of raw steel and plastics production in the United States in 1978 were 5.67×10^8 and 5.77×10^8 ft^3, respectively. Thus, we are producing a larger volume of plastics than of steel.

Plastics can be classified several ways. From a performance standpoint they can be either commodity or engineering plastics. During fabrication a polymer will behave as a thermoplastic or a thermoset. Thermoplastics can be repeatedly softened by heat and shaped. By contrast, a thermoset can be shaped and cured by heat only once. Other classification schemes include crystalline and amorphous or synthetic and natural. These classifications will be useful in differentiating plastics.

Commodity Thermoplastics
The commodity thermoplastics include polyolefins and side-chain-substituted vinyl polymers.

Polyethylene. Polyethylenes (PEs) have the largest volume use of any plastic. They are prepared by the catalytic polymerization of ethylene. Depending on the mode of polymerization, one can obtain a high-density (HDPE) or a low-density polyethylene (LDPE) polymer. LDPE is prepared under more vigorous conditions, which result in short-chain branching. Linear low-density polyethylene (LLDPE) is prepared by introducing short branching via copolymerization of PE with a small amount of long-chain olefin.

Polyethylenes are crystalline thermoplastics that exhibit toughness, near-zero moisture absorption, excellent chemical resistance, excellent electrical insulating properties, low coefficient of friction, and ease of processing. Their heat deflection temperatures are reasonable but not high. HDPE exhibits greater stiffness, rigidity, improved heat resistance, and increased resistance to permeability than LDPE. Some typical properties of PEs are listed in Table 16.59.

*Reprinted from E. N. Peters, Chapter 17 in *Mechanical Engineers Handbook*, M. Kutz, ed. John Wiley & Sons, 1986.

TABLE 16.59   Typical Property Values of Polyethylenes

	HDPE	LDPE
Density, Mg/m^3	0.96–0.97	0.91–0.93
Tensile modulus, GPa	0.76–1.0	—
Tensile strength, MPa	25–32	4–20
Elongation at break, %	500–700	275–600
Flexural modulus, GPa	0.8–1.0	0.21
Vicat soft point, °C	120–129	80–96
Brittle temperature, °C	-100 to -70	-85 to -35
Hardness, Shore	D60–D69	D45–D52
Dielectric constant, 10^6 Hz	—	2.3
Dielectric strength, MV/m	—	9–21
Dissipation factor, 10^6 Hz	—	0.0002
Linear mold shrinkage, in./in.	0.007–0.009	0.015–0.035

Uses. HDPE's major use is in blow-molded bottles, automotive gas tanks, drums, and carboys; injection-molded material-handling pallets, trash and garbage cans, and household and automotive parts; and extruded pipe.

LDPEs find major applications in film form for food packaging, as a vapor barrier film; for extruded wire and cable insulation; and for bottles, closures, and toys.

Polypropylene. Polypropylene (PP) is prepared by the catalyzed polymerization of propylene. PP is a highly crystalline thermoplastic that exhibits low density, rigidity, excellent chemical resistance, negligible water absorption, and excellent electrical properties. Its properties appear in Table 16.60.

Uses. End uses for PP are in blow-molded bottles, closures, automotive parts, appliances, housewares, and toys. PP can be extruded into fibers and filaments for use in carpets, rugs, and cordage.

Polystyrene. Catalytic polymerization of styrene yields polystyrene (PS), a clear, amorphous polymer with moderately high heat deflection temperature. PS has excellent electrical insulating properties; however, it is brittle under impact and exhibits very poor resistance to surfactants and solvents. Its properties appear in Table 16.61.

TABLE 16.60   Typical Property Values of Polypropylene

Density, Mg/m^3	0.90–0.91
Tensile modulus, GPa	1.8
Tensile strength, MPa	37
Elongation at break, %	10–50
Heat deflection temperature at 0.45 MPa, °C	100–105
Heat deflection temperature at 1.81 MPa, °C	60–65
Vicat soft point, °C	130–148
Linear thermal expansion, mm/mm K	3.8×10^{-5}
Hardness, Shore	D76
Volume resistivity, Ω cm	1.0×10^{17}
Linear mold shrinkage, in./in.	0.01–0.02

TABLE 16.61 Typical Properties of Styrene Thermoplastics

Property	PS	SAN	IPS/HIPS	ABS
Density, Mg/m^3	1.050	1.080	1.020–1.040	1.050–1.070
Tensile modulus, GPa	2.76–3.10	3.4–3.9	2.0–2.4	2.5–2.7
Tensile strength, MPa	41.4–51.7	65–76	26–40	36–39
Yield elongation, %	1.5–2.5	—	—	2.2–2.6
Heat deflection temperature at 264 psi, °C	82–93	101–103	81	80–95
Vicat soft point, °C	98–107	110	88–101	90–99
Notched Izod, kJ/m	0.02	0.02	0.10–0.32	0.09–0.48
Linear thermal expansion, 10^{-5} mm/mm K	5–7	6.4–6.7	7.0–7.5	7.5–9.5
Hardness, Rockwell	M60–M75	M80–M83	M45, L55	R69–R115
Linear mold shrinkage, in./in.	0.007	0.003–0.004	0.007	0.0055

Uses. Ease of processing, rigidity, clarity, and low cost combine to support applications in toys, displays, and housewares. PS foams can readily be prepared and are characterized by excellent low thermal conductivity, high strength-to-weight ratio, low water absorption, and excellent energy absorption. These attributes have made PS foam of special interest as insulation boards for construction, protective packaging materials, insulated drinking cups, and flotation devices.

SAN (Styrene / AcryloNitrile Copolymer). Copolymerization of styrene with a moderate amount of acrylonitrile provides a clear, amorphous polymer (SAN) with increased heat deflection temperature and chemical resistance compared to polystyrene. However, the impact resistance is still poor.

Uses. SAN is utilized in typical PS-type applications where a slight increase in the heat deflection temperature and/or chemical resistance are needed, such as housewares and appliances.

Impact Polystyrene. Copolymerization of styrene with butadiene can reduce the brittleness of PS but only at the expense of rigidity and heat deflection temperature. These opaque materials (IPS and HIPS) generally exhibit poor weathering characteristics.

Uses. IPS and HIPS are used for small appliance, radio, and TV housings.

ABS. ABS is a terpolymer prepared from the combination of acrylonitrile, butadiene, and styrene monomers. Compared to PS, ABS exhibits good impact strength, improved chemical resistance, and similar heat deflection temperature and rigidity. ABS is also opaque.

Uses. The previously mentioned properties of ABS are suitable for tough consumer products; automotive parts; business machine housings; telephones; luggage; and pipe, fittings, and conduit.

Polyvinyl Chloride. The catalytic polymerization of vinyl chloride yields polyvinyl chloride. It is commonly referred to as PVC or vinyl and is second only to polyethylene in volume use. Normally, PVC has a low degree of crystallinity and good transparency. The high chlorine content of the polymer produces advantages in flame resistance, fair heat deflection temperature, good electrical resistance, and chemical resistance. However, PVC is difficult to process. The chlorine atoms also have a tendency to split out under the influence of heat during processing and heat and light during end use in finished products, producing discoloration and embrittlement. Therefore, special stabilizer systems are often used with PVC to retard degradation.

There are two major subclassifications of PVC, rigid and flexible. Properties appear in Table 16.62.

TABLE 16.62   Typical Property Values for Polyvinyl Chloride Materials

Property	General Purpose	Rigid	Rigid Foam	Plasticized	Copolymer
Linear mold shrinkage, in./in.	0.003	—	—	—	—
Density, Mg/m^3	1.400	1.340–1.390	0.750	1.290–1.340	1.370
Tensile modulus, GPa	3.45	2.41–2.45	—	—	3.15
Tensile strength, MPa	8.7	37.2–42.4	> 13.8	14–26	52–55
Elongation at break, %	113	—	> 40	250–400	—
Notched Izod, kJ/m	0.53	0.74–1.12	> 0.06	—	0.02
Heat deflection temperature at 1.81 MPa, °C	77	73–77	65	—	65
Brittle temperature, °C	—	—	—	−60 to −30	—
Hardness	D85 (Shore)	R107–R112 (Rockwell)	D55 (Shore)	A71–A96 (Shore)	—
Linear thermal expansion, 10^{-5} mm/mm K	7.00	5.94	5.58	—	—

Rigid PVC. PVC alone is a fairly good rigid polymer but is difficult to process and has low impact strength. Both of these properties are improved by the addition of elastomers or impact-modified graft copolymers such as ABS and impact acrylic polymers. These improve the melt flow during processing and improve the impact strength without seriously lowering the rigidity or the heat deflection temperature. With this improved balance of properties, rigid PVCs are used in such applications as door and window frames; pipe, fittings, and conduit; building panels and siding; rainwater gutters and downspouts; credit cards; and flooring.

Plasticized PVC. Flexible PVC is a plasticized material. Thus PVC is softened by the addition of compatible, nonvolatile, liquid plasticizers. The plasticizers, which are usually used in > 20 parts per hundred (pph) resin level, lower the crystallinity in PVC and act as internal lubricants to give a clear, flexible plastic. Plasticized PVC is used for wire and cable insulation, outdoor apparel, rainwear, flooring, interior wall coverings, upholstery, automotive seat coverings, garden hose, toys, clear tubing, shoes, tablecloths and shower curtains.

PVC is also available in liquid formulations known as plastisols or organosols. These materials are used in coating fabrics, paper, and metal and are rotationally cast into dolls, balls, and so on.

Foamed PVC. Rigid PVC can be foamed to a low-density cellular material that is used for decorative moldings and trim.

Foamed plastisols add greatly to the softness and energy absorption already inherent in plasticized PVC, giving richness and warmth to leatherlike upholstery, clothing, shoe fabrics, handbags, luggage, and auto door panels and energy absorption for quiet and comfort in flooring, carpet backing, auto headliners, and so on.

PVC Copolymer. Copolymerization of vinyl chloride with 10–15% vinyl acetate gives a vinyl copolymer with improved flexibility and less crystallinity than PVC, making such copolymers easier to process without detracting seriously from the rigidity and heat deflection temperature. These copolymers find primary applications in phonograph records, flooring, and solution coatings.

Poly(vinylidene Chloride). Poly(vinylidene chloride) is prepared by the catalytic polymerization of 1,1-dichloroethylene. This crystalline polymer exhibits high strength, abrasion resistance, high melting point, better than ordinary heat resistance (100°C maximum service temperature), and outstanding impermeability to oil, grease, water vapor, oxygen, and carbon dioxide. It is used for packaging films, coatings, and monofilaments.

When the polymer is extruded into film, quenched, and oriented, the crystallinity is fine enough to produce high clarity and flexibility. These properties contribute to widespread use in packaging film, especially for food products that require impermeable barrier protection.

Poly(vinylidene chloride) and/or copolymers with vinyl chloride, alkyl acrylate, or acrylonitrile are used in coating paper, paperboard, or other films to provide more economical, impermeable materials.

A small amount of poly(vinylidene chloride) is extruded into monofilament and tape that is used in outdoor furniture upholstery.

Poly(methyl Methacrylate). The catalytic polymerization of methylmethacrylate yields poly(methyl methacrylate) (PMMA), a strong, rigid, clear, amorphous polymer. PMMA has excellent resistance to weathering, low water absorption, and good electrical resistivity. PMMA properties appear in Table 16.63.

TABLE 16.63   Typical Properties of Poly(methyl Methacrylate)

Density, Mg/m^3	1.180–1.190
Tensile modulus, GPa	3.10
Tensile strength, MPa	72
Elongation at break, %	5
Notched Izod, kJ/m	0.4
Heat deflection temperature at 1.81 MPa, °C	96
Hardness, Rockwell	M90–M100
Linear thermal expansion, 10^{-5} mm/mm K	6.3
Continuous-service temperature, °C	88
Linear mold shrinkage, in./in.	0.002–0.008

Uses. PMMA is used for glazing, lighting diffusers, skylights, outdoor signs, and automobile tail lights.

Poly(ethylene Terephthalate). Poly(ethylene terephthalate) (PET) is prepared from the condensation polymerization of dimethyl terephthalate and ethylene glycol. It is a crystalline polymer that exhibits high modulus, high strength, high melting point, good electrical properties, and moisture and solvent resistance.

Uses. Primary applications of PET include fibers for wash and wear, wrinkle-resistant fabrics, and films that are used in food packaging, electrical applications (e.g., capacitors), magnetic recording tape, and graphic arts.

Engineering Thermoplastics

Engineering polymers comprise a special high-performance segment of synthetic plastic materials that offer premium properties. When properly formulated, they may be shaped into mechanically functional, semiprecision parts or structural components. Mechanically functional implies that the parts may be subject to mechanical stress, impact, flexure, vibration, sliding friction, temperature extremes, hostile environments, and so on, and continue to function.

As substitutes for metal in the construction of mechanical apparatus, engineering plastics offer advantages such as transparency, light weight, self-lubrication, and economy in fabricating and decorating. Replacement of metals by plastic is favored as the physical properties and operating temperature ranges of plastics improve and the cost of metals and their fabrication increases.

Polyesters (Thermoplastic). Poly(butylene terephthalate) (PBT) is prepared from the condensation polymerization of butanediol with dimethyl terephthalate. PBT is a crystalline polymer. It seems to have a unique and favorable balance of properties between nylons and acetal resins. PBT has lower moisture absorption, extremely good self-lubrication, fatigue resistance, solvent resistance, and good maintenance of mechanical properties at elevated temperatures. Properties appear in Table 16.64.

TABLE 16.64   Typical Properties of Poly(butylene Terephthalate)

Property	PBT	PBT + 40% Glass Fiber
Density, Mg/m^3	1.300	1.600
Flexural modulus, GPa	2.4	9.0
Flexural strength, MPa	88	207
Elongation at break, %	300	3
Notched Izod, kJ/m	0.06	0.12
Heat deflection temperature at 1.81 MPa, °C	54	232
Heat deflection temperature at 0.45 MPa, °C	154	232
Hardness, Rockwell	R117	M86
Linear thermal expansion, 10^{-5} mm/mm K	9.54	1.89
Linear mold shrinkage, in./in.	0.020	< 0.007

Uses. Applications of PBT include gears, rollers, bearings, housings for pumps and appliances, impellers, pulleys, switch parts, automotive components, and electrical/electronic components.

Polyamides. The two major types of polyamides are nylon 6 and nylon 66. Nylon 6, or polycaprolactam, is prepared by the polymerization of caprolactam. Poly(hexamethylene adipate), or nylon 66, is derived from the condensation of hexamethylene diamine with adipic acid. These polyamides are crystalline polymers. Their key features include a high degree of solvent resistance, toughness, and fatigue resistance. Nylons do exhibit a tendency to creep under applied load. Their properties appear in Table 16.65.

Uses. The largest application of nylons is in fibers. Molded applications include automotive components, related machine parts (gears, cams, pulleys, rollers, boat propellers, etc.), appliance parts, and electrical insulation.

Polyacetals. Polyacetals are prepared via the polymerization of formaldehyde or its copolymerization with ethylene oxide. Polyacetals are crystalline polymers that exhibit rigidity, high strength, solvent resistance, fatigue resistance, toughness, self-lubricity, and cold-flow resistance. They also exhibit a tendency to thermally unzip and hence are difficult to flame retard. Properties appear in Table 16.66.

TABLE 16.65   Typical Properties of Nylons

Property	Nylon 6	Nylon 6 +40% Glass Fiber	Nylon 66	Nylon 66 +40% Glass Fiber
Density, Mg/m^3	1.130	1.460	1.140	1.440
Flexural modulus, GPa	2.8	10.3	2.8	9.3
Flexural strength, MPa	113	248	—	219
Elongation at break, %	150	3	60	4
Notched Izod, kJ/m	0.06	0.16	0.05	0.14
Heat deflection temperature at 1.81 MPa, °C	64	216	90	250
Heat deflection temperature at 0.45 MPa, °C	170	218	235	260
Hardness, Rockwell	R119	M92	R121	R119
Linear thermal expansion, 10^{-5} mm/mm K	8.28	2.16	8.10	3.42
Linear mold shrinkage, in./in.	0.013	0.003	0.0150	0.0025

TABLE 16.66   Typical Properties of Polyacetals

Property	Polyacetal	Polyacetal +40% Glass Fiber
Density, Mg/m^3	1.420	1.740
Flexural modulus, GPa	2.7	11.0
Flexural strength, MPa	107	117
Elongation at break, %	75	1.5
Notched Izod, kJ/m	0.12	0.05
Heat deflection temperature at 1.81 MPa, °C	124	164
Heat deflection temperature at 0.45 MPa, °C	170	167
Hardness, Rockwell	M94	R118
Linear thermal expansion, 10^{-5} mm/mm K	10.4	3.2
Linear mold shrinkage, in./in.	0.02	0.003

Uses. Applications of polyacetals include moving parts in appliances and machines (gears, bearings, bushings, etc.), in automobiles (e.g., door handles), and in plumbing (valves, pumps, faucets, etc.).

Polyphenylene Sulfide. The condensation polymerization of dichlorobenzene and sodium sulfide yields a crystalline polymer, polyphenylene sulfide. It is characterized by high heat resistance, rigidity, excellent chemical resistance, low friction coefficient, good abrasion resistance, and electrical properties. Polyphenylene sulfides are somewhat difficult to process due to their very high melting temperature, relatively poor flow characteristics, and some tendency for slight cross-linking during processing. Properties appear in Table 16.67.

Uses. The unreinforced resin is used only in coatings. The reinforced materials are used in aerospace applications, pump components, electrical/electronic components, appliance parts, and automotive applications.

Polycarbonates. Most commercial polycarbonates are derived from bisphenol A and phosgene. Polycarbonates are transparent, amorphous polymers. They are among the stronger, tougher, and more rigid thermoplastics. Polycarbonates also show resistance to creep and excellent electrical-insulating characteristics. Polycarbonate properties appear in Table 16.68.

TABLE 16.67   Typical Properties of Polyphenylene Sulfide

Property	Polyphenylene Sulfide +40% Glass Fiber
Density, Mg/m^3	1.640
Tensile modulus, GPa	7.7
Tensile strength, MPa	135
Flexural modulus, GPa	11.7
Flexural strength, MPa	200
Elongation at break, %	1.3
Notched Izod, kJ/m	0.08
Heat deflection temperature at 1.81 MPa, °C	> 260
Hardness, Rockwell	R123
Linear thermal expansion, 10^{-5} mm/mm K	4.0
Linear mold shrinkage, in./in.	0.004
Constant-service temperature, °C	232

TABLE 16.68   Typical Properties of Polycarbonates

Property	Polycarbonate	Polycarbonate +40% Glass Fiber
Density, Mg/m^3	1.200	1.520
Tensile modulus, GPa	2.4	11.6
Tensile strength, MPa	65	158
Flexural modulus, GPa	2.3	9.7
Flexural strength, MPa	93	186
Elongation at break, %	110	4
Notched Izod, kJ/m	0.86	0.13
Heat deflection temperature at 1.81 MPa, °C	132	146
Heat deflection temperature at 0.45 MPa, °C	138	154
Hardness, Rockwell	M70	M93
Linear thermal expansion, 10^{-5} mm/mm K	6.75	1.67
Linear mold shrinkage, in./in.	0.006	0.0015
Constant-service temperature, °C	121	135

Uses. Applications of polycarbonates include safety glazing, safety shields, nonbreakable windows, automobile tail lights, electrical relay covers, various appliance parts and housings, power tool housings, automotive fender extensions, and blow-molded bottles.

Polysulfone. Polysulfone is prepared from the condensation polymerization of bisphenol A and dichlorodiphenyl sulfone. This transparent, amorphous resin is characterized by excellent thermooxidative resistance, hydrolytic stability, and creep resistance. Polysulfone properties appear in Table 16.69.

Uses. Typical applications of polysulfones include microwave cookware, medical equipment where sterilization by steam is required, coffee makers, and electrical-electronic components.

Modified Polyphenylene Ether. Blends of poly(2,6-dimethyl phenylene ether) with styrenics (HIPS, ABS, etc.) form a family of modified polyphenylene-ether-based resins. Depending on the blend, these materials have a broad temperature use range. They are characterized by outstanding dimensional stability at elevated temperatures, outstanding hydrolytic stability, long-term stability under load, and excellent dielectric properties over a wide range of frequencies and temperatures. Their properties appear in Table 16.70.

Uses. Modified polyphenylene ether applications include automotive applications (dashboards, trim, etc.), TV cabinets, electrical connectors, pumps, plumbing fixtures, and small appliance and business machine housings.

TABLE 16.69   Typical Properties of Polysulfone

Density, Mg/m^3	1.240
Tensile modulus, GPa	2.48
Tensile strength, MPa	70
Flexural modulus, GPa	2.69
Flexural strength, MPa	106
Elongation at break, %	75
Notched Izod, kJ/m	0.07
Heat deflection temperature at 1.81 MPa, °C	174
Hardness, Rockwell	M69
Linear thermal expansion, 10^{-5} mm/mm K	5.6
Linear mold shrinkage, in./in.	0.007
Constant-service temperature, °C	150

TABLE 16.70 Typical Properties of Modified Polyphenylene Ethers

Property	190 Grade	225 Grade	300 Grade
Density, Mg/m³	1.080	1.090	1.060
Tensile modulus, GPa	2.5	2.4	—
Tensile strength, MPa	48	55	76
Flexural modulus, GPa	2.2	2.4	2.4
Flexural strength, MPa	56.5	76	104
Elongation at break, %	35	—	—
Notched Izod, kJ/m	0.37	0.32	0.53
Heat deflection temperature at 1.81 MPa, °C	88	107	149
Heat deflection temperature at 0.45 MPa, °C	96	118	157
Hardness, Rockwell	R115	R116	R119
Linear thermal expansion, 10^{-5} mm/mm K	—	—	5.9
Linear mold shrinkage, in./in.	0.006	0.006	0.006
Constant-service temperature, °C	—	95	—

Polyimides. Polyimides are a class of polymers prepared from the condensation reaction of a carboxylic acid dianhydride with a diamine. Thermoplastic and thermoset grades of polyimides are available. The thermoset polyimides are among the most heat resistant polymers; for example, they can withstand temperatures up to 250°C. Thermoplastic polyimides, which can be processed by standard techniques, fall into two main categories: polyetherimides and polyamideimides.

In general, polyimides have very good electrical properties, very good wear resistance, superior dimensional stability, outstanding flame resistance and high strength and rigidity. Polyimide properties appear in Table 16.71.

Uses. Polyimide applications include gears, bushings, bearings, seals, insulators, electrical/electronic components (printed wiring boards, connectors, etc.), microwave oven components, and structural components.

Fluorinated Thermoplastics

In general, fluoropolymers, fluoroplastics, or fluorocarbons are a family of fluorine-containing thermoplastics that exhibit some unusual properties. These properties include inertness to most chemicals, resistance to high temperature, extremely low coefficient of friction, and excellent dielectric properties. Properties appear in Table 16.72.

Mechanical properties are normally low but can be improved when reinforced with glass or carbon fibers or molybdenum disulfide fillers.

TABLE 16.71 Typical Properties of Polyimides

Property	Polyimide	Polyetherimide Unfilled	Polyetherimide 30% Glass Reinforced	Polyamideimide Unfilled	Polyamideimide 30% Glass Reinforced
Density, Mg/m³	—	1.27	1.51	1.38	1.57
Tensile modulus, GPa	2.65	0.30	0.90	—	1.15
Tensile strength, MPa	196.2	104.8	168.9	117.2	195.2
Elongation at break, %	90	60	3	10	5
Notched Izod, kJ/m	—	0.6	0.11	0.13	0.11
Heat deflection temperature at 1.81 MPa, °C	—	392	410	260	274
Heat deflection temperature at 0.45 MPa, °C	—	410	414	—	—
Hardness, Rockwell	—	R109	M125	E78	E94
Linear thermal expansion, 10^{-5} mm/mm K	—	5.6	2.0	3.60	1.80
Linear mold shrinkage, in./in.	—	0.5	0.2	—	0.25

TABLE 16.72   Typical Properties of Fluoropolymers

Property	PTFE	CTFE	FEP	ETFE	ECTFE
Density, Mg/m^3	2.160	2.100	2.150	1.700	1.680
Tensile modulus, GPa	—	14.3	—	—	—
Tensile strength, MPa	27.6	39.4	20.7	44.8	48.3
Elongation at break, %	~ 275	~ 150	~ 300	100–300	200
Notched Izod, kJ/m	—	0.27	0.15	—	—
Heat deflection temperature at 1.81 MPa, °C	—	75	—	71	77
Heat deflection temperature at 0.45 MPa, °C	—	126	—	104	116
Hardness	D55–65 (Shore)	D75–80 (Shore)	D55 (Shore)	D75 (Shore)	R93 (Rockwell)
Linear thermal expansion, 10^{-5} mm/mm K	9.9	4.8	9.3	13.68	—
Dielectric strength, MV/m	23.6	19.7	82.7	7.9	19.3
Dielectric constant at 10^2 Hz	2.1	3.0	2.1	2.6	2.5
Dielectric constant at 10^3 Hz	2.1	2.7	—	2.6	2.5
Constant-service temperature, °C	260	199	204	—	150–170
Linear mold shrinkage, in./in.	0.033–0.053	0.008	—	—	< 0.025

The difficulty of fluorochemical synthesis makes the price of fluoropolymers relatively high, and their uses are largely restricted to critical specialty applications.

Poly(tetrafluoroethylene). Prepared from tetrafluoroethylene, poly(tetrafluoroethylene) (PTFE) is a crystalline, very heat-resistant (up to 500°F), and outstanding chemical-resistant polymer, and it has the lowest coefficient of friction of any polymer. PTFE does not soften like other thermoplastics and has to be processed by nonconventional techniques (PTFE powder is compacted to the desired shape and sintered).

Uses. PFTE applications include nonstick coatings on cookware; nonlubricated bearings; chemical-resistant pipe fittings, valves, and pump parts; high-temperature electrical parts; and gaskets, seals, and packings.

Poly(chlorotrifluoroethylene). Poly(chlorotrifluoroethylene) (CTFE) is less crystalline and exhibits higher rigidity and strength than PTFE; it is chemical resistant and has heat resistance up to 390°F. Unlike PTFE, it can be molded and extruded by conventional processing techniques.

Uses. CTFE applications include electrical insulation, cable jacketing, electrical and electronic coil forms, pipe and pump parts, valve diaphragms, and coatings for corrosive process industries and other industrial parts.

Fluorinated Ethylene–Propylene. (FEP) Copolymerization of tetrafluoroethylene with some hexafluoropropylene produces a polymer with less crystallinity, lower melting point, and improved impact strength than PTFE. This copolymer can be molded by thermoplastic molding techniques.

Uses. Fluorinated ethylene–propylene applications include wire insulation and jacketing, high-frequency connectors, coils, gaskets, and tube sockets.

Polyvinylidene Fluoride. Polyvinylidene fluoride has high tensile strength and better ability to be processed but less thermal and chemical resistance than the previous fluoropolymers.

Uses. Polyvinylidene fluoride applications include insulation, seals and gaskets, diaphragms, and piping.

Poly(ethylene Trifluoroethylene). Copolymerization of ethylene with trifluoroethylene produces poly(ethylene trifluoroethylene) (ETFE) with good high-temperature and chemical resistance; ETFE can be processed by conventional techniques.

Uses. ETFE applications include molded labware, valve liners, electrical connectors, and coil bobbins.

Poly(ethylene Chlorotrifluoroethylene). The copolymer of ethylene and chlorotrifluoroethylene, poly(ethylene chlorotrifluoroethylene) (ECTFE), is strong and chemical and impact resistant. It can be processed by conventional techniques.

Uses. ECTFE applications include wire and cable coatings, chemical-resistant coatings and linings, molded labware, and medical packaging.

Poly(vinyl Fluoride). Poly(vinyl fluoride) films exhibit excellent outdoor durability.

Uses. Poly(vinyl fluoride) uses include glazing, lighting, and coatings on presurfaced exterior building panels.

Thermosets
Thermosetting polymers are used in molded and laminated plastics. They are first polymerized to a low-molecular-weight polymer, which is still soluble, fusible, and highly reactive during final processing. Thermosets are generally catalyzed and/or heated to finish the polymerization reaction, cross-linking them to almost infinite molecular weight. This step is often referred to as cure. Such cured polymers cannot be reprocessed or reshaped.

Phenolic. Phenolic resins combine the high reactivity of phenol and formaldehyde to form prepolymers and oligomers called resoles and novalaks. These materials are combined with fibrous fillers to give a phenolic resin, which when heated provides rapid complete cross-linking to highly cured structures. The high cross-linked aromatic structure has high hardness, rigidity, strength, heat resistance, chemical resistance, and good electrical properties.

Uses. Phenolic applications include automotive uses, distributor caps, rotors, and brake linings; appliance parts, pot handles, knobs, and bases; electrical/electronic components, connectors, circuit breakers, and switches; and adhesives in laminates (e.g., plywood).

Epoxy Resins. Epoxy resins are low-molecular-weight materials that are liquid either at room temperature or on warming. Each polymer chain usually contains two or more epoxide groups. The high reactivity of the epoxide groups with amines, anhydrides, and other curing agents provides facile conversion into highly cross-linked materials. Cured epoxy resins exhibit hardness, strength, heat resistance, electrical resistance, and broad chemical resistance.

Uses. Epoxy applications include glass-reinforced, high-strength composites used in aerospace, pipes, tanks, pressure vessels; encapsulation or casting of various electrical and electronic components; adhesives; protective coatings in appliances, flooring, and industrial equipment; and sealants.

Unsaturated Polyesters. Properly formulated glass-reinforced unsaturated polyesters are commonly referred to as sheet molding compound (SMC) or reinforced plastics. Unsaturated polyesters are thermosets and are quite distinct from thermoplastic polyesters. In general, unsaturated polyesters are dissolved in styrene monomer to produce any viscosity material desired for impregnation and lamination of glass fibers. The low-molecular-weight polyesters have fumarate ester units that provide easy reactivity with styrene monomer.

In combination with reinforcing materials such as glass fibers, cured resins offer outstanding strength, high rigidity, high strength-to-weight ratio, impact resistance, and chemical resistance.

Uses. The prime use of unsaturated polyesters is in combination with glass fibers in high-strength composites; these include transportation markets (body parts and components for automobiles, trucks, trailers, buses, and aircraft), marine uses (small- to medium-size boat hulls and associated marine equipment), building panels, housings, bathroom components (bathtubs and shower stalls), appliances, and electronic/electrical components.

Alkyd Resins. Alkyd resins are based on branched prepolymers from glycerol, phthalic anhydride, and glyceryl esters of fatty acids. Alkyds have excellent heat resistance; are dimensionally stable at high temperatures; and have excellent dielectric strength (> 14 MV/m), high resistance to electrical leakage, and excellent arc resistance.

Uses. Alkyd resin applications include drying oils in enamel paints; lacquers for automobiles and appliances; and molding compounds when formulated with reinforcements for electrical applications (circuit breaker insulation, encapsulation of capacitors and resistors, and coil forms).

Diallyl Phthalate. Diallyl phthalate (DAP) is the most widely used compound in the allylic family. These low-molecular-weight prepolymers can be reinforced and compression molded into highly cross-linked, completely cured products.

The most outstanding properties of DAP are excellent dimension stability and high insulation resistance. In addition, DAP has high dielectric strength, excellent arc resistance, and chemical resistance.

Amino Resins. The two main members of the amino family of thermosets are the melamine and urea resins. They are prepared from the reaction of melamine and urea with formaldehyde. In general, these materials exhibit extreme hardness, scratch resistance, electrical resistance, and chemical resistance.

Uses. DAP applications include electronic parts, electrical connectors, bases, and housings. DAP is also used as a coating and impregnating material.

Uses. Melamine resins find use in colorful, rugged dinnerware; decorative laminates (countertops, tabletops, and furniture surfacing); electrical applications (switchboard panels, circuit breaker parts, arc barriers, and armature and slot wedges); and adhesives and coatings.

Urea resins are used in particle board binders, decorative housings, closures, electrical parts, coatings, and paper and textile treatment.

General-Purpose Elastomers

Elastomers are materials that can be stretched substantially beyond their original length and can retract rapidly and forcibly to essentially their original dimensions (on release of the force).

The optimum properties and/or economics of many rubbers are obtained through formulating with reinforcing agents, fillers, extending oils, vulcanizing agents, antioxidants, pigments, and so on. End-use markets for formulated rubbers include automobile tire products (including tubes, retread applications, valve stems, and innerliners), adhesives, cements, caulks, sealants, latex foam products, hose (automotive, industrial, and consumer applications), belting (V-conveyor and trimming), footwear (heels, soles, slab stock, boots, and canvas), and molded, extruded, and calendered products (athletic goods, flooring, gaskets, household products, O-rings, blown sponge, thread, and rubber sundries). Properties appear in Table 16.73.

Specialty Elastomers

Specialty rubbers are more costly and hence are produced in smaller volume than the general-purpose rubbers. They fill the need for high-performance materials. Properties and uses are summarized in Table 16.74.

TABLE 16.73 Properties of General-Purpose Elastomers

Rubber	ASTM Nomenclature	Temperature Use Range (°C)	Outstanding Property	Property Deficiency
Butadiene rubber	BR	−100 to 90	Very flexible; resistance to abrasive wear	Sensitive to oxidation; poor resistance to fuels and oil
Natural rubber	NR	−50 to 180	Similar to BR but less resilient	Similar to BR
Isoprene rubber	IR	−50 to 80	Similar to BR but less resilient	Similar to BR
Isobutylene-isoprene rubber (butyl rubber)	IIR	−45 to 150	High flexibility; low permeability to air	
Chloroprene	CR	−40 to 115	Flame resistant; fair fuel and oil resistance; increased resistance toward oxygen, ozone, heat, light	Poor low-temperature flexibility
Nitrile-butadiene rubber	NBR	−50 to 80	Good resistance to fuels, oils, solvents; improved abrasion resistance	Lower resilience; higher hysteresis; poor electrical properties; poorer low-temperature flexibility
Styrene-butadiene rubber	SBR	−50 to 80	Relatively low cost	Less resilience; higher hysteresis; limited low-temperature flexibility
Ethylene-propylene copolymer	EPM	−50 to < 175	Resistance to ozone and weathering	Poor hydrocarbon and oil resistance
Ethylene-propylene-diene terpolymer	EPDM	−50 to < 175	Resistance to ozone and weathering	Poor hydrocarbon and oil resistance
Polysulfide	T	−45 to 120	Chemical resistance; resistance to ozone and weathering	Creep; low resilience

TABLE 16.74 Properties of Specialty Rubbers

Rubber	ASTM Nomenclature	Temperature Use Range (°C)	Outstanding Property	Typical Applications
Silicones (polydimethylsiloxane)	MQ	−100 to 300	Wide temperature range; resistance to aging, ozone, sunlight; very high gas permeability	Seals specialty molded and extruded goods, adhesives, biomedical
Fluoroelastomers	CFM	−40 to 200	Resistance to heat, oils, chemicals	Seals such as O-rings, corrosion-resistant coatings
Acrylic	AR	−40 to 200	Oil, oxygen, ozone, and sunlight resistance	Seals, hose
Epichlorohydrin	ECO	−18 to 150	Resistance to oils, fuels; some flame resistance; low gas permeability	Hose, tubing, coated fabrics, vibration isolators
Chlorosulfonated	CSM	−40 to 150	Resistance to oils, ozone weathering, oxidizing chemicals	Automotive hose, wire and cable, linings for reservoirs
Chlorinated polyethylene	CM	−40 to 150	Resistance to oils, ozone, chemicals	Impact modifier, automotive applications
Ethylene-acrylic	—	−40 to 175	Resistant to ozone, weathering	Seals, insulation vibration damping
Propylene oxide	—	−60 to 150	Low-temperature properties	Motor mounts

TABLE 16.75   Physical Properties of Textile Fibers

Fiber Type	Diameter (μm)	Density (g/cm³)	Tenacity (gf/tex)	Breaking Extension (%)	Initial Modulus (gf/tex)	Work of Rupture (g/tex)	Moisture Regain[a] (%)	Melting Point (°C)
Natural								
Cotton	11–22	1.52	35	7	500	1.3	7	Decomposes
Flax	5–40	1.52	55	3	1840	0.8	7	Decomposes
Jute	8–30	1.52	50	2	1750	0.5	12	Decomposes
Sisal	8–40	1.52	40	2	2500	0.5	8	Decomposes
Wool	18–44	1.31	12	40	250	3	14	Decomposes
Silk	10–15	1.34	40	23	750	6	10	Decomposes
Glass	≥ 5	2.54	76	2–5	3000	1	0	800
Asbestos	0.01–0.30	2.5	—	—	1300	—	1	1500
Regenerated								
Viscose rayon	≥ 12	1.45–1.54	20	20	500	3	13	Decomposes
Acetate	≥ 15	1.32	13	24	350	2	6	230
Triacetate	≥ 15	1.32	12	30	300	2	4	230
Synthetic								
Nylon 6	≥ 14	1.14	32–65	30–55	250	6–7	2.8–5	225
Nylon 66	≥ 14	1.14	32–65	16–66	250	6–7	2.8–5	250
Qiana	≥ 10	1.03	25	26–36	—	—	2.5	274
Nomex	≥ 12	1.38	36–50	22–32	—	7.5	6.5	Decomposes > 380
Kevlar	≥ 12	1.44	200	2–4	4500–8000	—	4	Decomposes > 500
Polyester	≥ 12	1.34–1.38	25–54	12–55	1000	2–9	0.4	250
Acrylic	≥ 12	1.16–1.17	18–30	20–50	650	5	1.5	Sticks at 235
Polypropylene		0.91	60	20	800	8	0.1	165
Polyethylene		0.95	30–60	10–45	—	3	0	115
Spandex		1.21	6–8	444–555	—	18	1.3	230

After Goswani et al., *Textile Yarns: Technology, Structure and Applications*, Wiley-Interscience, NY, 1977.
[a]At 65% relative humidity.

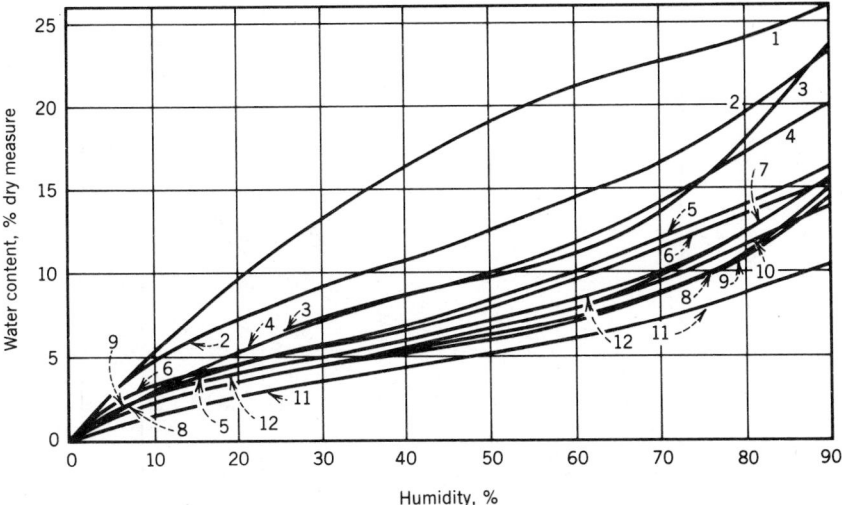

Fig. 16.9 Hygroscopic moisture (25°C) of natural fiber textile materials: 1, absorbent cotton; 2, wool, worsted; 3, silk, new yellow; 4, jute; 5, Manila hemp; 6, sisal hemp; 7, Indian cotton; 8, cotton cloth; 9, Egyptian cotton; 10, American cotton; 11, linen; 12, flax. (From *International Critical Tables*, Vol. 2, p. 323.)

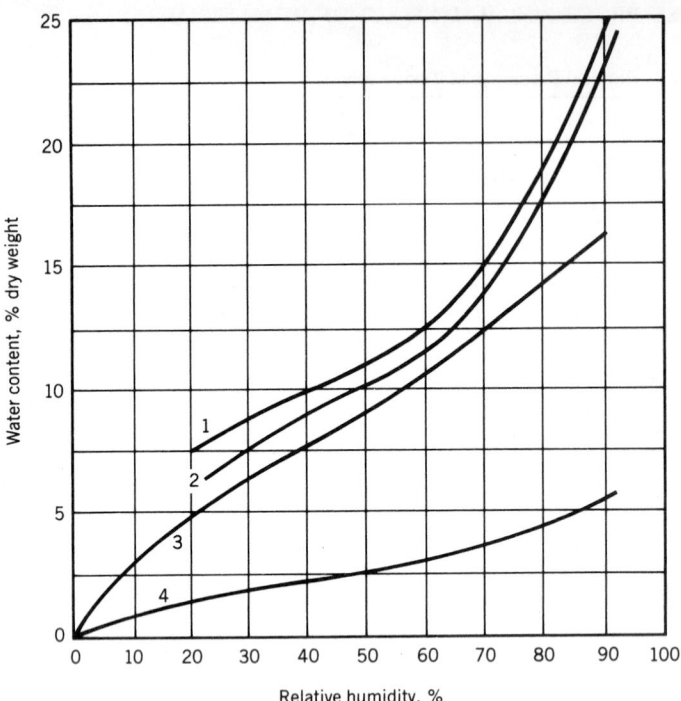

Fig. 16.10 Hygroscopic moisture (25°C) of artificial textile fibers compared with crude constituents and natural silk: 1, viscose rayon (artificial silk); 2, natural silk, new yellow; 3, nitrocellulose; 4, cellulose acetate. (From *International Critical Tables*, Vol. 2, p. 323.)

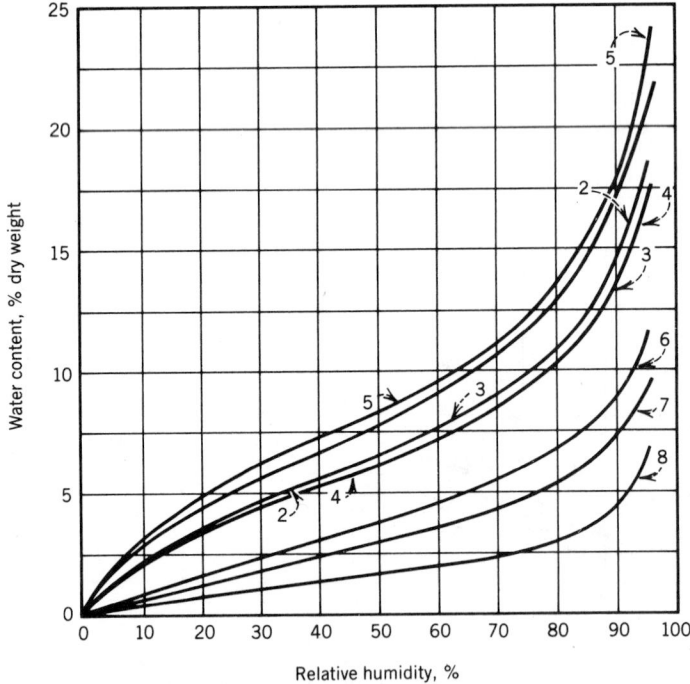

Fig. 16.11 Hygroscopic moisture of various fibrous materials prepared for electrical insulation: 1, Manila paper; 2, red rope paper; 3, pressboard; 4, leatheroid paper; 5, silk; 6, red rope paper (varnished); 7, empire cloth; 8, asbestos paper. (From *International Critical Tables*, Vol. 2, p. 323.)

TABLE 16.76   Properties of Reinforcement Fibers

	Density (g/cm^3)	Modulus $(10^6 psi)$	Tensile Strength[a] $(10^3 psi)$	Failure Strain[a] (%)	Thermal Expansion Axial Coefficient $(10^{-6}/°C)$
PAN "graphite"					
High strength	1.8	34	360	1.1	−0.4
High modulus	1.9	53	260	0.5	−0.5
Ultrahigh modulus	2.0	75	150	0.2	−1.1
High strain	1.8	33–37	440–770	1.3	—
Intermediate modulus	1.8	42	750	1.8	—
Pitch-based "graphite"					
P55	2.02	55	200	0.36	−0.9
P75	2.00	75	200	0.27	−0.7
P100	2.15	100	250	0.25	−0.9
Aramid					
Kevlar 49	1.44	18	330	1.8	−2.0
Kevlar 29	1.44	12	330	2.8	—
Polyethylene	0.97	17	375	3.8	−10.0
Boron (W core)	2.5	58	400	0.7	4.9
Silicon carbide					
Monofilament, carbon core	3.1	62	300	0.6	—
Multifilament yarn "Nicalon"	2.6	26–29	350	1.2	3.1
Glass					
Al_2O_3–SiO_2–B_2O_3 (Nextel 312)	2.7	22	250	1.1	2.6
S-2 glass	2.6	10.5	250	2.4	5.0
E-Glass	2.5	12.6	360	2.9	5.6
Alumina	3.95	57	290	—	8

[a]In a composite.

TABLE 16.77 General Properties Chart of Epoxy-based Advanced Composite Materials, Unidirectional Laminates[a]

Mechanical Property[b]	Boron-Epoxy		LHS Graphite-Epoxy		UHM Graphite-Epoxy		Unidirectional Fiberglass (7743)-Epoxy		Kevlar 49-Epoxy	
	RT	350°F	RT	350°F	RT	350°F	RT	350°F	RT	350°F
Longitudinal tensile strength, ksi	200	150	190	178	85	80	108	80	200	150
Longitudinal tensile modulus, 10^6 psi	30.0	26.0	18.5	18.0	41.0	41.0	6.1	5.1	11.0	8.0
Longitudinal flexural strength, ksi	245	180	210	157	105	96	150	69	90	60
Longitudinal flexural modulus, 10^6 psi	27.0	20.0	17.0	16.0	35.0	31.0	6.2	3.7	10.0	8.0
Longitudinal compression strength, ksi	375	140	126	116	65	60	79	62.5	40	30
Longitudinal compression modulus, 10^6 psi	33.0	31.0	18.0	18.0	40.0	40.0	4.9	4.9	8.0	6.0
Longitudinal shear strength, horizontal, ksi	13.0	5.0	7.0	5.0	5.4	4.5	4.3	2.0	7.0	5.0
Transverse tensile strength, ksi	7.0	3.0	8.0	4.6	3.0	3.0	8.6	8.0	4.0	2.0
Transverse tensile modulus, 10^6 psi	3.0	1.0	2.0	1.0	0.9	0.9	2.2	0.75	1.0	0.75
Transverse flexural strength, ksi	11.0	8.0	9.0	4.6	3.0	1.5	—	—	—	—
Transverse flexural modulus, 10^6 psi	2.0	0.8	2.0	1.0	—	—	—	—	—	—
Density, lb/in.^{3}	0.075	—	0.055	—	0.061	—	0.065	—	0.050	—
Layer thickness, in.	0.00510	—	0.00525	—	0.00525	—	0.0092	—	0.0050	—

Source: Materials and Processes, (3rd edition), J. F. Young and R. S. Shane, (eds.), Marcel Dekker, New York, 1985.
[a]Polymer matrix composites; significant strength < 200°C. Abbreviations: LHS, low cost, high strength; UHM, ultrahigh modulus; RT, room temperature.
[b]Minimum average value.

TABLE 16.78 Effects of Fiber Type and Length on Properties of Various Composites[a]

	PPS with Short E Glass	PPS with Short PAN Carbon	Polycarbonate Short E-glass	Polycarbonate Long E-glass	Nylon 66 Short E-glass	Nylon 66 Long E-glass	PEEK with Short PAN Carbon	PEEK with Long PAN Carbon
Fiber content, wt %	50	45	40	40	40	40	40	40
Specific gravity	1.75	1.52	1.50	1.50	1.45	1.45	1.44	1.46
Ultimate tensile strength, psi	21,000	24,000	21,000	22,000	26,000	32,000	38,500	42,500
Tensile modulus, 10^6 psi	2.9	5.9	1.7	2.6	1.9	2.6	5.1	5.6
Flexural strength, $\frac{1}{8}$ in., psi	33,000	38,000	26,000	37,000	40,000	49,000	50,300	65,700
Flexural modulus, $\frac{1}{8}$ in., 10^3 psi	2,400	4,600	1,400	1,950	1,700	1,700	3,200	4,200
Compressive strength, psi	29,200	29,500	22,000	32,000	23,000	37,000	37,100	42,400
Shear strength, psi	10,400	9,200	—	—	—	—	15,200	17,100
Izod impact, ft-lb/in.								
Notched, $\frac{1}{4}$ in.	1.7	1.0	2.2	8.9	2.6	11.1	1.1	2.2
Unnotched, $\frac{1}{4}$ in.	8.0	6.0	12	23	15	31	8.0	13.0

Source: Advanced Materials and Processes, ASM, Metal Park, Ohio, Oct. 1985.

[a]Abbreviations: PPS, polyphenylene sulfide; PEEK, polyetheretherketone. Short, 50 length-to-diameter aspect ratio; long, 300 aspect ratio.

TABLE 16.79 General Design Practice for Laminated Composites

Practice	Reason
Filamentary controlled laminates: minimum of three-layer orientations	To prevent matrix and stiffness degradation
0/90/ ± 45 laminate with minimum of one layer in each direction	0° layers for longitudinal load, 90° layers for transverse load, ±45° layers for shear load
+45° and −45° ply in contact with each other	To minimize interlaminar shear
45° layers added in pairs (±)	In-plane shear carried by tension and compression in 45° layers
Adding plies try to maintain symmetry	To minimize warping, interlaminar shear
Minimize stress concentrations	Composites essentially elastic to failure
±45° plies, at least one pair on extremes of laminate; however, specific design requirements (applied moments) 0° or 90° plies may be more advantageous in direction of moments	Increases buckling for thin laminates; better damage tolerance
Maintain homogeneous stacking sequence, banding several plies of the same orientation together	Increased strength

Source: Materials and Processes, (3rd edition), J. F. Young and R. S. Shane, (eds.), Marcel Dekker, New York, 1985, p. 799.

TABLE 16.80 General Design Practice for Joints

Practice	Reason
Three-diameters edge distance and six-diameters pitch for bolted joints	Bearing strength
Design bonded step joints rather than scarf joints	More consistent results, design flexibility
Bonded joints: no 90° plies in contact with metals	Reduction in lap shear strength
Bonded joints ±45° plies on last step	To reduce peak loading (E for $\pm45°$, Boron-Expoxy = 3.5×10^6 cf. $E_{Ti} = 17 \times 10^6$)
When adding plies use 0.3-in. overlap in major load direction using wedge-type pattern	Requires approximately 0.3 in. to develop strength

Source: Materials and Processes, (3rd edition), J. F. Young and R. S. Shane, (eds.), Marcel Dekker, New York, 1985.

TABLE 16.81 Three-Dimensional Composites

Composite	Density (g/cm³)	Linear Density (tex)	Fiber Volume Fraction (%)	Sample Dimensions[a] (mm)	Breaking Load (kg)
Laminated woven	1.97	74.525	80	25.8 × 1.46	1550
Stitched woven	1.82	33.443	76	25.7 × 1.55	1355
Triaxial woven	2.04	113.949	83	25.6 × 2.32	2836
Three-dimensional braid	1.88	83.633	76	16.0 × 2.45	—

Composite	Tensile Properties				
	Rupture Strength (MPa)	Modulus (GPa)	Breaking Elongation (%)	Short-Beam Shear Strength MPa	Short-Beam Shear Strength ksi
Laminated woven	402	10.308	3.9	13.54	1.96
Stiched woven	334	10.121	3.3	13.32	1.93
Triaxial woven	468	9.55	4.9	20.58	2.98
Three-dimensional braid	740	13.96	5.3	33.32	4.83

Source: Advanced Materials and Processes, ASM, Metals Park, Ohio, Sept. 1985.
[a] Width times thickness.

TABLE 16.82 Fatigue Endurance of Reinforced Engineering Thermoplastics

Material	Glass Fibers (%)	Carbon Fibers (%)	Cyclic Failure Stress (psi)	
			10^4 cycles	10^7 cycles
Acetal copolymer	30	—	9,000	7,000
ETFE copolymer	30	—	—	3,500
ETFE copolymer	—	30	—	6,100
Nylon 6[a]	30	—	7,000	5,750
Nylon 6/6	—	—	6,500	5,200
Nylon 6/6[a]	—	—	3,400	3,100
Nylon 6/6[a]	30	—	8,000	5,900
Nylon 6/6[a]	40	—	9,000	7,000
Nylon 6/6	40	—	10,500	9,100
Nylon 6/6[a]	—	30	13,000	8,000
Nylon 6/6[a]	—	40	15,000	8,500
Nylon 6/10[a]	30	—	6,800	5,500
Nylon 6/10[a]	40	—	8,000	7,000
Polycarbonate	20	—	9,000	5,000
Polycarbonate	40	—	14,500	6,000
Polyester, PBT	30	—	11,000	5,100
Polyester, PBT	—	30	13,000	6,500
Polyetheretherketone	—	30	18,000	17,500
Polyethersulfone	30	—	16,000	5,000
Polyethersulfone	40	—	19,000	6,200
Polyethersulfone	—	30	22,000	6,700
Mod. polyphenylene oxide	30	—	7,200	4,750
Polyphenylene sulfide	—	30	13,000	9,500
Polypropylene	30	—	—	4,500
Polysulfone	30	—	14,000	4,500
Polysulfone	40	—	16,000	5,500
SAN	30	—	—	6,500
Styrene	30	—	—	6,000

Sources: *Guide to Engineered Materials*, p. 41, 1987; and *Advanced Materials and Processes*, Feb. 1986, ASM, Metals Park, Ohio. LNP Corporation data, 39th SPI RP/C Conference. Tests by ASTM D 671 at 1800 cycles/min.

[a] Moisture conditioned, 50% RH.

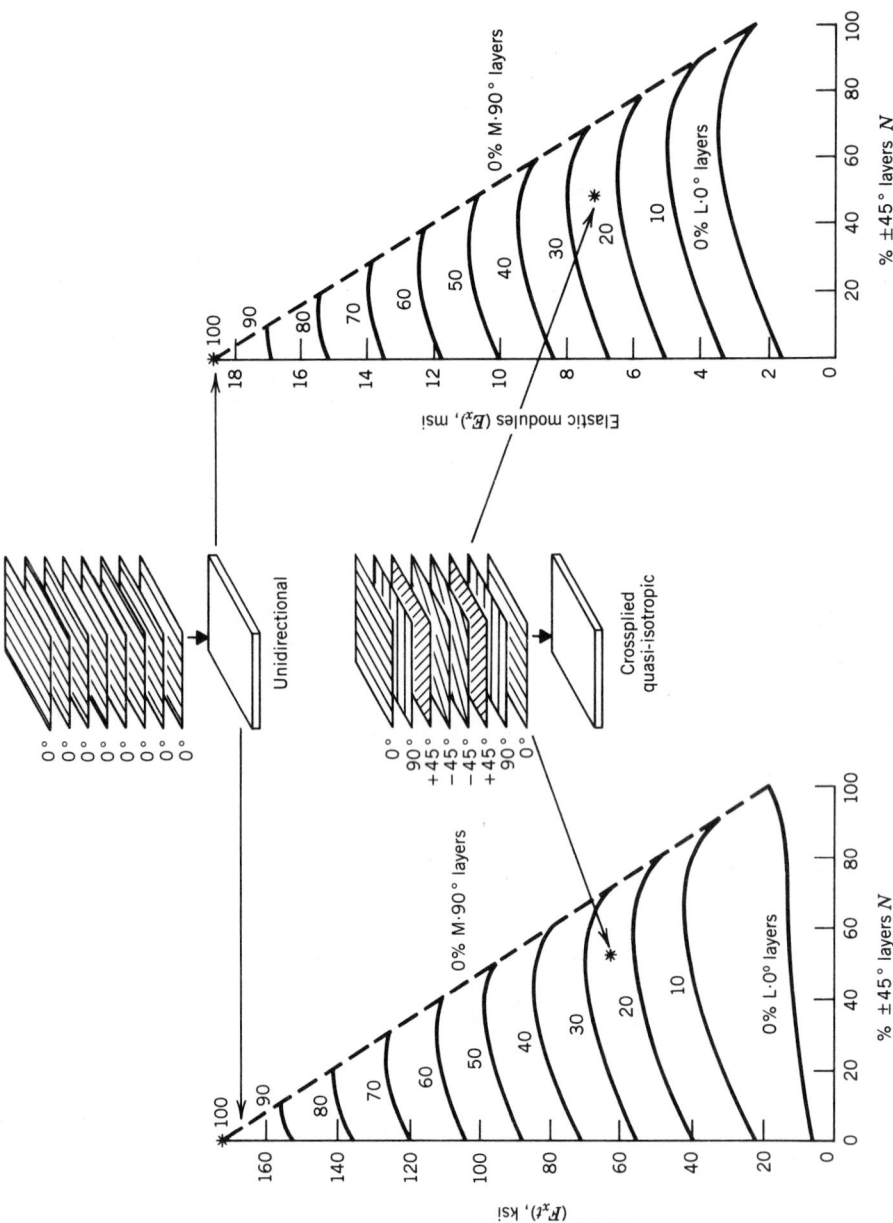

Fig. 16.12 Allowable tension stress and elastic moduli of graphite-epoxy laminates at room temperature. *Source: Materials and Processes,* (3rd edition), J. F. Young and R. S. Shane, (eds.), Marcel Dekker, New York, 1985.

TABLE 16.83 Wear Properties of Composites

Lubricant Additive		Reinforcement			Properties		
PTFE (wt %)	Silicone (wt %)	Glass Fiber (%)	Carbon Fiber (%)	Aramid Fiber (%)	Wear Factor[a]	Coefficient of Friction[b]	Limiting PV at 100 fpm (psi × fpm)
Acetal							
—	—	—	—	—	65	0.21	3,500
20	—	—	—	—	13	0.12	16,000
—	2	—	—	—	27	0.12	9,000
—	—	—	20	—	40	0.14	15,000
Nylon 6/6							
—	—	—	—	—	200	0.28	2,500
18	2	—	—	—	6	0.08	30,000
—	—	30	—	—	75	0.31	10,000
13	2	—	30	—	6	0.11	43,000
—	—	—	—	20	62	0.25	—
10	—	—	—	15	23	0.25	—
Polycarbonate							
—	—	—	—	—	2,500	0.38	500
20	—	—	—	—	70	0.14	22,000
13	2	30	—	—	27	0.19	30,000
—	—	—	30	—	85	0.17	5,500
Polyester, PBT							
—	—	—	—	—	210	0.25	—
20	—	—	—	—	15	0.17	15,500
13	2	30	—	—	12	0.12	24,000
—	—	—	30	—	24	0.15	22,000
—	—	—	—	20	95	0.21	—
Polyetheretherketone							
—	—	30	—	—	60	0.13	—
20	—	—	—	—	130	0.23	—
15	—	—	15	—	60	0.20	40,000
Polyetherimide							
—	—	—	—	—	4,000	0.17	—
—	—	—	30	—	75	0.23	—
15	—	30	—	—	35	0.20	—
Polyethersulfone							
15	—	30	—	—	60	0.20	30,000
15	—	—	30	—	40	0.17	33,000
Polyimide							
—	—	—	—	—	100	0.29	300,000
10	—	—	15	—	28	0.12	1,000,000
Modified polyphenylene oxide							
—	—	—	—	—	3,000	0.39	500
15	—	—	—	—	100	0.16	—
15	—	30	—	—	45	0.22	22,000

TABLE 16.83   *(Continued)*

Lubricant Additive		Reinforcement			Properties		
PTFE (wt %)	Silicone (wt %)	Glass Fiber (%)	Carbon Fiber (%)	Aramid Fiber (%)	Wear Factor[a]	Coefficient of Friction[b]	Limiting PV at 100 fpm (psi × fpm)
Polyphenylene sulfide							
—	—	—	—	—	540	0.24	3,000
20	—	—	—	—	55	0.10	—
15	—	30	—	—	110	0.17	35,000
—	—	—	30	—	160	0.20	20,000
13	2	30	—	—	50	0.22	30,000
Polypropylene							
20	—	—	—	—	33	0.11	5,000
15	—	30	—	—	36	0.09	12,000
Polyurethane							
—	—	—	—	—	340	0.37	1,500
15	—	30	—	—	35	0.25	10,000
Polysulfone							
—	—	—	—	—	1,500	0.37	5,000
—	—	30	—	—	160	0.22	—
15	—	30	—	—	55	0.19	35,000

Source: *Guide to Engineered Materials*, ASM, Metals Park, Ohio, 1987, p. 90. Data on thermoplastics against steel from LNP Corp.

[a] Wear factor units: 10^{-10} in.^{3}-min/ft-lb-hr.
[b] Dynamic, at 40 psi, 50 fpm.

TABLE 16.84 Typical Mechanical Properties of Metal-Matrix Composites[a]

Fiber	Matrix	Reinforcement (vol %)	Density[b] (g/cm³)	Longitudinal Tensile Strength[c] (MPa)	Longitudinal Modulus[c] (GPa)	Transverse Tensile Strength[c] (MPa)	Transverse Modulus[c] (GPa)
G-T 50	201 Al	30	2.380	620	170	50	30
G-T 50	201 Al	49	—	1120	160	—	—
G-GY 70	201 Al	34	2.380	660	210	30	30
G-GY 70	201 Al	30	2.436	550	160	70	40
G-HM pitch	6061 Al	41	2.436	620	320	—	—
G-HM pitch	AZ31 Mg	38	1.827	510	300	—	—
B on W, 142-μm fiber	6061 Al	50	2.491	1380	230	140	160
Borsic	Ti	45	3.681	1270	220	460	190
G-T 75	Pb	41	7.474	720	200	—	—
G-T 75	Cu	39	6.090	290	240	—	—
FP alumina	201 Al	50	3.598	1170	210	(140)	140
SiC	6061 Al	50	2.934	1480	230	(140)	140
SiC	Ti	35	3.931	1210	260	520	210
SiC whisker	Al	20	2.796	340	100	340	100
B$_4$C on B	Ti	38	3.737	1480	230	> 340	> 140
G-T 75	Mg	42	1.799	450	190	—	—
G-HM	Pb	35	7.750	500	120	—	—
G-T 75	Al–7% Zn	38	2.408	870	190	—	—
G-T 75	Zinc	35	5.287	770	120	—	—
G-T 50	Nickel	50	5.295	790	240	—	—
G-T 75	Nickel	50	5.342	828	310	30	40
G (81.3 μm)	2024 Al	50	2.436	760	140	—	—
G (142 μm)	2024 Al	60	2.436	1100	180	—	—
Superhybrid[d]	Graphitic, RE[e]	60	2.048	860	120	220	60
Superhybrid[d]	S-glass, RE[e]	60	2.159	740	80	190	30
Superhybrid[d]	Kevlar, RE[e]	60	1.799	700	80	190	10

Source: Encyclopedia of Composite Materials and Components, M. Grayson, (ed.), Wiley, New York, 1983.
[a]Significant strength > 200°C. G, graphite.
[b]To convert g/cm³ to lb/in.³ divide by 27.68.
[c]To convert MPa to psi, multiply by 145; to convert GPa to psi, multiply by 145,000.
[d]Metal foil clad, multi-ply laminates.
[e]RE, reinforced epoxy.

TABLE 16.85 Ceramic-Matrix Composites[a]

Matrix	Reinforcement	Room Temperature Flexural Strength (MPa)	Fracture Toughness, K_{Ic}, (MPa $\sqrt{m}$)	Elastic Modulus (GPa)
Al$_2$O$_3$	SiC whiskers	800	8.7	400
Al$_2$O$_3$	SiC-coated C fibers	—	10.5	—
ZrO$_2$	BN-coated SiC fibers	450	22.0	—
Al$_2$O$_3$–SiO$_2$ (mullite)	SiC fibers	386	3.5	—
Reaction-bonded Si$_3$N$_4$	SiC whiskers	900	20	—
Hot-pressed Si$_3$N$_4$	SiC whiskers	800	56	—
MoSi$_2$	SiC whiskers	310	8.2	—
SiC	SiC whiskers	< 630	—	240
Carbon	Carbon	2100	—	340

[a]Significant strength > 1000°C.

Paper and Paperboard

TABLE 16.86   Representative Data for Several Paper and Paperboard Grades

Paper Type	Apparent Density (kg/m^3)	Breaking Length (km)			Percentage of Stretch		Tensile Energy Absorption (kJ/m^2)		Compressive Strength (MPa)		Tear Resistance (mN)	
		MD^a	CD^a	ZD^a	MD	CD	MD	CD	MD	CD	MD	CD
Linerboard 42 lb	721	7.22	3.71	—	—	—	—	—	8.28	4.83	—	—
Laboratory BKSW^b												
Handsheet I	721	9.11	5.28	0.066	3.2	3.7	—	—	23.1	15.1	—	—
Handsheet II	673	14.3	3.15	0.056	—	—	—	—	27.9	9.33	—	—
Kraft envelope	695	8.42	4.32	—	1.2	2.6	11.2	11.0	—	—	863	912
Rag bond	687	4.9	3.47	—	1.8	4.7	6.29	13.2	—	—	637	376
Newsprint	596	3.65	1.84	—	1.1	1.4	1.78	1.29	—	—	118	226

Source: *Encyclopedia of Materials Science and Engineering*, M. B. Bever, (ed.), Pergamon Press, Oxford, 1986.
^{a}Orientation: MD, machine direction; CD, cross direction; ZD, thickness direction.
^{b}BKSW, bleached kraft softwood.

TABLE 16.87   Strength and Mechanical Properties of Laminated Paperboard^a

Property	Value
Density, g/cm^3	0.51–0.53
Modulus of elasticity (compression), MPa^b	
Along length of panel^c	2.1–2.7
Across length of panel^c	0.69–0.97
Modulus of rupture, MPa^b	
Span parallel to length of panel^c	9.7–13.1
Span perpendicular to length of panel^c	6.2–7.6
Tensile strength parallel to surface, MPa^b	
Along length of panel^c	11.7–14.5
Across length of panel^c	4.1–5.5
Compressive strength parallel to surface, MPa^b	
Along length of panel^c	4.8–6.2
Across length of panel^c	3.4–5.5
24-hr water absorption, wt %	10–170
Linear expansion of 50–90% rh,^{d} %	
Along length of panel^c	0.2–0.3
Across length of panel^c	1.1–1.3
Thermal conductivity (at 24°C), mW/m K^e	73

Source: *Encyclopedia of Composite Materials and Components*, M. Grayson, (ed.), Wiley, New York, 1983.

^{a}Data are general round-figure values, accumulated from numerous sources; for more exact figures on specific product, individual manufacturers should be consulted or actual tests made. Values are for general laboratory conditions of temperature and humidity.
^{b}To convert MPa to psi, multiply by 145.
^{c}Because of directional properties, values are presented for two principal directions, along usual length of panel (machine direction) and across it.
^{d}Measurements made on material at equilibrium at each condition at room temperature; rh, relative humidity.
^{e}To convert mW/m K to Btu in./h ft^2 °F, divide by 144.1.

TABLE 16.88 Mechanical Properties of Single-Wall Corrugated Board[a]

Flute	Nominal Component Basis Weight (g/m^2) Facings[b]	Medium	Minimum Bursting Strength (kPa)	Thickness (mm)	Cross Direction Edgewise Compressive Strength (kN/m)	Flexural Stiffness (kN/m) Machine Direction	Cross Direction
A	185	127	1206	4.7–5.0	—	26–39	11–15
A	205	127	1379	4.8–5.2	7.0–8.8	27–49	15–17
A	337	127	1896	5.1–5.5	8.8–10.5	52–70	18–30
A	439	127	2413	5.2–6.0	10.5–13.1	—	—
C	127	127	862	3.6–3.9	4.9–5.8	14–23	6–8
C	185	127	1206	3.7–4.2	7.0–8.8	18–26	7–10
C	205	127	1379	3.8–4.2	7.0–9.6	20–30	8–14
C	337	127	1896	4.3–4.7	9.6–13.1	28–40	15–20
C	439	127	2413	4.4–4.8	12.3–15.8	—	—
B	205	127	1379	3.0–3.2	7.0–8.8	12–15	4–6
B	337	127	1896	3.3–3.6	10.5–13.1	18–25	9–12
B	439	127	2413	3.4–3.8	—	—	—

Source: *Encyclopedia of Materials Science and Engineering*, M. B. Bever, (ed.), Pergamon Press, Oxford, 1986.
[a] Data from various sources and believed to encompass normally encountered ranges of strength.
[b] Total facing weight is twice indicated figure.

TABLE 16.89 Linerboard and Corrugating Medium, Mechanical Properties[a]

Property	Orientation	Test Ranges Linerboard	Medium
Nominal basis weight g/m^2	—	205	127
Thickness, mm	—	0.30–0.35	0.23–0.30
Bursting strength, kPa	—	690–830	—
Edgewise compressive strength,[b] kN/m	MD	2.7–3.5	—
	CD	1.9–2.6	0.9–1.4
Tensile strength, kN/m	MD	14–17	4.5–9.5
	CD	5–6.5	2.0–3.3
Stretch, %	MD	1.3–2.0	1.0–1.7
	CD	3.0–4.4	1.3–4.2
Elastic stiffness,[a] kN/m	MD	1500–1700	660–1100
	CD	500–1000	235–390
Tearing strength, N	MD	2400–3500	—
	CD	2900–4000	—
Concora flat crush, N	—	—	245–290
Water drop, sec/100 ml	—	—	15–600 +
Porosity (Bendtsen), ml/min	—	—	300–1400
Transverse shear stiffness,[c] kN/m	MD	—	~ 25–30
Kinetic friction coefficient vs. steel	MD	—	0.29–0.62

Source: *Encyclopedia of Materials Science and Engineering*, M. B. Bever, (ed.), Pergamon Press, Oxford, 1986.
[a] At 50% relative humidity and 22.5°C, determined by TAPPI test procedures where available.
[b] By ring crush test (TAPPI T818).
[c] Elastic modulus multiplied by thickness.

TABLE 16.90 Approximate Average Mechanical Properties of Softwood Linerboard as Function of Equilibrium Moisture Content for Machine (MD) and Cross-Machine (CD) Directions[a]

Equilibrium Moisture Content (%)	Tensile Strength (kg/cm²)		Modulus of Elasticity (10³ kg/cm²)		Strain to Failure (%)		TEA (kg cm cm³)		PL (kg/cm²)	
	MD	CD	MD	CD	MD	CD	MD	CD	MD	CD
4	600	300	70	30	2.5	1.2	4.7	6.0	235	85
8	475	250	55	25	3.5	1.5	5.0	6.5	160	60
12	350	150	45	20	4.7	1.8	4.3	5.7	100	35
16	300	125	35	15	5.5	1.9	3.3	4.5	72	22

After Benson, R. E., TAPPI, Vol. 54, p. 699, 1971.

[a]Abbreviations: TEA, tensile energy absorption; PL, value of tensile stress at proportional limit.

TABLE 16.91 Ranges of Numerical Values for Optical Properties of Eight Types of Papers[a]

Attributes and TAPPI Test Method(s)	Fine Papers (Office and Letter)	Printing and Book Papers		Newsprint	White Boxboard	Kraft and Shipping	Tissue	Speciality
		Uncoated	Coated					
Brightness: T452 or T525	65–90	65–90	65–90	45–70	60–90	< 50	50–80	65–90
Opacity: T425	80–99	80–99	80–99	80–99	> 98	80–100	50–90	70–100
Color: T524 for near-whites L	80–95	80–95	80–95	50–80	80–95	< 70	70–95	70–95
a	−5 to +5	−5 to +5	−5 to +5	−5 to +5	−5 to +5	0 to +20	−5 to +5	−5 to +5
b	−5 to +15	−5 to +15	−5 to +15	0 to +15	−5 to +15	0 to +25	−5 to +15	−5 to +15
Gloss: T480 (75°)	5–20	5–25	20–60	< 20	5–60	< 10	< 10	5–60

Source: TAPPI 1969 Optical Measurements Terminology, Technical Information Sheet 017-2. Technical Association of the Pulp and Paper Industry, Atlanta, Georgia.

[a]Whites and near whites only.

TABLE 16.92 Specifications of Optical Properties of Paper Used by One Publisher of Magazines[a]

Basis weight, I	36.0 ± 1.1	38.0 ± 1.1	40.0 ± 1.2	43.0 ± 1.3	45.0 ± 1.3	50.0 ± 1.5
Opacity	89.0–1.0	89.5–1.0	90.0–1.0	91.0–1.0	92.0–1.0	94.0–1.0
Brightness	70.5–1.0	70.5–1.0	70.5–1.0	70.5–1.0	70.5–1.0	70.5–1.0
Color (Hunter Lab) L	86.0	86.0	86.0	86.0	86.0	86.0
a	+0.4 ± 0.2	+0.4 ± 0.2	+0.4 ± 0.2	+0.4 ± 0.2	+0.4 ± 0.2	+4.4 ± 0.2
b	+4.4 ± 0.2	+4.4 ± 0.2	+4.4 ± 0.2	+4.4 ± 0.2	+4.4 ± 0.2	+4.4 ± 0.2
Gloss (75°) wire	48.0 ± 6.0	48.0 ± 6.0	48.0 ± 6.0	48.0 ± 6.0	50.0 ± 6.0	50.0 ± 6.0
felt	48.0 ± 6.0	48.0 ± 6.0	48.0 ± 6.0	48.0 ± 6.0	50.0 ± 6.0	50.0 ± 6.0

Source: Encyclopedia of Materials Science and Engineering, M. B. Bever, (ed.), Pergamon Press, Oxford, 1986.

[a]Coated letterpress magazine publication body and insert.

TABLE 16.93 Permeability Properties

Grade	Basic Weight (g/m^2)	Permeability (cm^2/min)
Wrapping tissue	12	130
Cigarette paper	25	930
Art printing	115	1
Blotting paper	130	4600
Card	247	20

After Corte, in Rance, H. E. (ed.), *1982 Handbook of Paper Science*, Vol. 2, Elsevier, Oxford, pp. 1–70.

Wood

TABLE 16.94 Allowable Stresses, Properties, and Uses for Standard Commercial Grades of Lumber

Species and Commercial Grade*	Allowable Unit Stress (lb/in.^{2})				Modulus of Elasticity, E	Density (lb/ft^3), 12% Moisture	Distinctive Uses
	Extreme Fiber in Bending f and Tension Parallel to Grain, t	Horizontal Shear, H	Compression Perpendicular to Grain, $c\perp$	Compression Parallel to Grain, $c\parallel$			
Ash, white[a]					1,500,000	41	Handles
2150 f grade	2150	145	600	1700			
1900 f grade	1900	145		1500			
1700 f grade	1700	145		1325			
1450 f grade	1450	120		1150			
1300 f grade	1300	120		1050			
1450 c grade	—	—		1450			
1200 c grade	—	—		1200			
1075 c grade	—	—		1075			
Beech[a]					1,600,000	45	Woodenware, flooring, furniture
2150 f grade	2150	145	600	1750			
1900 f grade	1900	145		1525			
1700 f grade	1700	145		1350			
1450 f grade	1450	120		1150			
1550 c grade	—	—		1550			
1450 c grade	—	—		1450			
1200 c grade	—	—		1200			
Birch[a]					1,600,000	44	Millwork, furniture
2150 f grade	2150	145	600	1750			
1900 f grade	1900	145		1525			
1700 f grade	1700	145		1350			
1450 f grade	1450	120		1150			
1550 c grade	—	—		1550			
1450 c grade	—	—		1450			
1200 c grade	—	—		1200			

Species and grade							Uses
Cypress, southern[a]							
1700 f grade	1700	145		360	1425	1,200,000	32	Greenhouses, tanks, construction
1300 f grade	1300	120			1125			
1450 c grade	—	—			1450			
1200 c grade	—	—			1200			
Douglas fir, coast region[b]							
2150 f dense select structural	2150	145	455	1550	1,600,000	34	Construction, plywood veneer
1900 f select structural	1900	120	415	1450			
1700 f dense no. 1	1700	145	455	1325			
1450 f no. 1	1450	120	390	1200			
1100 f no. 2	1100	110	390	1075			
1550 c dense select structural	—	—	455	1550			
1450 c select structural	—	—	415	1450			
1400 c dense no. 1	—	—	455	1400			
1200 c no. 1	—	—	390	1200			
Douglas fir, inland empire[c]							
Select structural	2150	145	455	1750	1,600,000	31	Construction
Structural	1900	100	400	1400	1,500,000		
Common structural	1450	95	380	1250	1,100,000	28	Construction
Hemlock, eastern[d]							
Select structural	1300	85	360	850			
Prime structural	1200	60		775			
Common structural	1100	60		650			
Utility structural	950	60		600			
Hemlock, west coast[b]							
1600 f select structural	1600	100	360	1100	1,400,000	29	Construction, boxes, crates, flooring
1450 f no. 1	1450	100		1075			
1100 f no. 2	1100	90		850			
1075 c no. 1	—			1075			

TABLE 16.94 (*Continued*)

Species and Commercial Grade	Allowable Unit Stress (lb/in.²)				Modulus of Elasticity E	Density (lb/ft³), 12% Moisture	Distinctive Uses
	Extreme Fiber in Bending f and Tension Parallel to Grain, t	Horizontal Shear, H	Compression Perpendicular to Grain c⊥	Compression Parallel to Grain, c‖			
Hickory^a							
2150 f grade	2150	145	720	1725	1,800,000	51	Tools, bats
1900 f grade	1900	145		1550			
1700 f grade	1700	145		1350			
1550 c grade	—	—		1550			
1450 c grade	—	—		1450			
1325 c grade	—	—		1325			
Maple, hard^a							
2150 f grade	2150	145	600	1750	1,600,000	44	Flooring, furniture
1900 f grade	1900	145		1525			
1700 f grade	1700	145		1425			
1450 f grade	1450	120		1150			
1550 c grade	—	—		1550			
1450 c grade	—	—		1450			
1200 c grade	—	—		1200			
Oak, red and white^a							
2150 f grade	2150	145	600	1550	1,500,000	47	Cooperage, flooring, furniture
1900 f grade	1900	145		1375			
1700 f grade	1700	145		1200			
1450 f grade	1450	120		1050			
1300 f grade	1300	120		950			
1325 c grade	—	—		1325			
1200 c grade	—	—		1200			
1075 c grade	—	—		1075			

Pine, southern longleaf[c]							
Select structural	2400	120	455	1750	1,600,000	41	Construction flooring
Prime structural	2000	120		1400			
Merchantable structural	1800	120		1300			
Structural, S.E. & S.							
No. 1 structural	1800	120		1300			
No. 1 dimension	1600	120		1150			
No. 1 dimension	1700	150		1400			
No. 2 stress dimension	1250	150		1025			
Pine, southern shortleaf[c]							
Dense select structural	2400	120	455	1750	1,600,000	36	Construction, boxes, crates
Dense structural	2000	120	455	1400			
Dense structural S.E. & S.	1800	120	455	1300			
Dense No. 1 structural	1600	120	455	1150			
No. 1 dense dimension	1700	150	455	1400			
No. 1 dimension	1450	125	390	1200			
No. 2 dense stress dimension	1250	100	455	1025			
No. 2 medium-grain stress dimension	1100	85	390	875			
Redwood[f]							
Dense structural	1700	110	320	1450	1,200,000	28	Tanks, planing, mill products
Heart structural	1300	95		1100			
Spruce, eastern[g]							
1450 f structural grade	1450	110	300	1050	1,200,000	28	Musical instruments
1300 f structural grade	1300	95		975			
1200 f structural grade	1200	95		900			
Tupelo[a]							
1700 f grade	1700	120	360	1225	1,200,000	35	Factory flooring, boxes, crates
1450 f grade	1450	120		1050			
1200 f grade	1200	120		875			
1075 c grade	—	—		1075			

*Rules under which graded; abbreviations: f = fine, c = common.
[a]National Hardwood Lumber Association; [b]West Coast Bureau of Lumber Grades and Inspection; [c]Western Pine Association; [d]Northern Hemlock and Hardwood Manufacturers Association; [e]Southern Pine Inspection Bureau of the Southern Pine Assocation; [f]California Redwood Association; [g]Northeastern Lumber Manufacturers Association, Inc.

TABLE 16.95 Strength Properties of Some Commercially Important Woods Grown in United States[a]

Commercial and Botanical Name of Specie	Moisture Content (%)	Specific Gravity[b]	Static Bending — Fiber Stress at Proportional Limit (lb/in.²)	Static Bending — Rupture Modulus (lb/in.²)	Static Bending — Elasticity Modulus (1000 lb/in.²)	Static Bending — Proportional Limit (in.-lb/in.³)	Static Bending — Maximum Load (in.-lb/in.³)	Impact Bending — Fiber Stress at Proportional Limit (lb/in.²)	Impact Bending — Height of Drop Causing Complete Failure (50-lb Hammer) (in.)	Compression Parallel to Grain — Fiber Stress at Proportional Limit (lb/in.²)	Compression Parallel to Grain — Maximum Crushing Strength (lb/in.²)	Compression Perpendicular to Grain — Fiber Stress at Proportional Limit (lb/in.²)	Shear Parallel to Grain — Maximum Shearing Strength (lb/in.²)	Hardness — End (lb)	Hardness — Side (lb)
Alder, red (*Alnus rubra*)	98	0.37	3,800	6,500	1170	0.70	8.0	8,000	22	2620	2960	310	770	550	440
	12	0.41	6,900	9,800	1380	1.85	8.4	11,600	20	4530	5820	540	1080	980	590
Ash, black (*Fraxinus nigra*)	85	0.45	2,600	6,000	1040	0.41	12.1	—	33	1690	2300	430	860	590	—
	12	0.49	7,200	12,600	1600	1.57	14.9	—	35	4520	5970	940	1570	1150	—
Ash, commercial white[c] (*Fraxinus* sp.)	43	0.54	5,300	9,500	1400	1.14	14.7	12,800	37	3360	4060	860	1350	1010	940
	12	0.58	8,900	14,600	1680	2.68	15.6	17,000	40	5580	7280	1510	1920	1680	1260
Ash, Oregon (*F. oregona*)	48	0.50	4,200	7,600	1130	0.92	12.2	8,900	39	2760	3510	650	1190	850	790
	12	0.55	7,000	12,700	1360	2.08	14.4	13,300	33	4100	6040	1540	1790	1430	1160
Aspen (*Populus tremuloides*)	94	0.35	3,200	5,100	860	0.69	6.4	7,000	22	1670	2140	220	660	280	300
	12	0.38	5,600	8,400	1180	1.53	7.6	9,000	21	3040	4250	460	850	510	350
Basswood (*Tilia glabra*)	105	0.32	2,700	5,000	1040	0.40	5.3	6,300	16	1690	2220	210	600	290	250
	12	0.37	5,900	8,700	1460	1.37	7.2	9,800	16	3800	4730	450	990	520	410
Beech (*Fagus grandifolia*)	54	0.56	4,300	8,600	1380	0.85	11.9	11,500	43	2550	3550	670	1290	970	850
	12	0.64	8,700	14,900	1720	2.63	15.1	16,000	41	4880	7300	1250	2010	1590	1300
Birch[d] (*Betula* sp.)	62	0.57	4,400	8,700	1560	0.79	15.9	11,100	48	2640	3510	550	1160	910	850
	12	0.63	10,100	16,700	2070	2.83	19.8	20,000	52	6200	8310	1250	2020	1660	1340
Birch, paper (*B. papyrifera*)	65	0.48	3,000	6,400	1170	0.45	16.2	8,000	49	1640	2360	340	840	470	560
	12	0.55	6,900	12,300	1590	1.80	16.0	12,400	34	3610	5690	740	1210	890	910
Butternut (*Juglans cinerea*)	104	0.36	2,900	5,400	970	0.52	8.2	7,300	24	2020	2420	270	760	410	390
	12	0.38	5,700	8,100	1180	1.59	8.2	11,200	24	4200	5110	570	1170	570	490

Species															
Cedar, Alaska (*Chamaecyparis nootkatensis*)	38	0.42	3,800	6,400	1140	0.77	9,100	9.2	27	2500	3050	430	840	540	440
	12	0.44	7,100	11,100	1420	2.06	12,200	10.4	29	5210	6310	770	1130	790	580
Cedar, eastern red (*Juniperus virginiana*)	35	0.44	3,400	7,000	650	1.08	7,000	15.0	35	2540	3570	860	1010	760	650
	12	0.47	3,800	8,800	880	1.01	8,500	8.3	22	—	6020	1140	—	900	—
Cedar, incense (*Libocedrus decurrens*)	108	0.35	3,900	6,200	840	0.94	7,300	6.4	17	2940	3150	460	830	570	390
	12	—	5,900	8,000	1040	1.67	9,600	5.4	17	4760	5200	730	880	830	470
Cedar, northern white (*Thuja occidentalis*)	55	0.29	2,600	4,200	640	0.60	5,300	5.7	15	1490	1990	290	620	320	230
	12	0.31	4,900	6,500	800	1.72	7,100	4.8	12	2630	3960	380	850	450	320
Cedar, Port Orford (*Chamaecyparis lawsoniana*)	43	0.40	4,000	6,200	1420	0.65	9,200	7.4	22	2770	3130	350	830	460	400
	12	0.42	7,700	11,300	1730	1.97	13,500	9.1	28	5890	6470	760	1080	730	560
Cedar, southern white (*C. thyoides*)	35	0.31	2,500	4,700	750	0.51	6,000	5.9	18	1660	2390	300	690	400	290
	12	0.32	4,800	6,800	930	1.46	7,600	4.1	13	2740	4700	500	800	520	350
Cedar, western red (*T. plicata*)	37	0.31	3,200	5,100	920	0.63	6,900	5.0	17	2470	2750	340	710	430	270
	12	0.33	5,300	7,700	1120	1.44	8,600	5.8	17	4360	5020	610	860	660	350
Cherry, black (*Prunus serotina*)	55	0.47	4,200	8,000	1310	0.80	10,200	12.8	33	2940	3540	440	1130	750	660
	12	0.50	9,000	12,300	1490	3.11	13,600	11.4	29	5960	7110	850	1700	1470	950
Chestnut (*Castanea dentata*)	122	0.40	3,100	5,600	930	0.59	7,900	7.0	24	2080	2470	380	800	530	420
	12	0.43	6,100	8,600	1230	1.78	10,700	6.5	19	3780	5320	760	1080	720	540
Cottonwood, eastern (*Populus deltoides*)	111	0.37	2,900	5,300	1010	0.49	7,200	7.3	21	1740	2280	240	680	380	340
	12	0.40	5,700	8,500	1370	1.39	7,300	7.4	20	3490	4910	470	930	580	430
Cottonwood, northern black (*P. trichocarpa hastata*)	132	0.32	2,900	4,800	1070	0.44	6,800	5.0	20	1760	2160	200	600	280	250
	12	0.35	5,300	8,300	1260	1.25	9,800	6.7	22	3270	4420	370	1020	540	350
Cypress, southern (*Taxodium distichum*)	91	0.42	4,200	6,600	1180	0.91	8,800	6.6	25	3100	3580	500	810	440	390
	12	0.46	7,200	10,600	1440	2.15	10,400	8.2	24	4740	6360	900	1000	660	510
Douglas fir (coast region) (*Pseudotsuga taxifolia*)	36	0.45	4,800	7,600	1550	0.85	9,800	6.8	24	3410	3890	510	930	510	480
	12	0.48	8,100	11,700	1920	1.96	12,700	8.6	30	6450	7420	910	1140	760	670

TABLE 16.95 *(Continued)*

Commercial and Botanical Name of Species	Moisture Content	Specific Gravity[b]	Static Bending					Impact Bending		Compression Parallel to Grain		Compression Perpendicular to Grain	Shear Parallel to Grain	Hardness (Load Required to Embed 0.444-in. Ball to Half Its Diameter)	
			Fiber Stress at Proportional Limit	Rupture Modulus	Elasticity Modulus	Proportional Limit	Maximum Load	Fiber Stress at Proportional Limit	Height of Drop Causing Complete Failure (50-lb Hammer)	Fiber Stress at Proportional Limit	Maximum Crushing Strength	Fiber Stress at Proportional Limit	Maximum Shearing Strength	End	Side
	(%)		(lb/in.²)	(lb/in.²)	(1000 lb/in.²)	(in.-lb/in.³)	(in.-lb/in.³)	(lb/in.²)	(in.)	(lb/in.²)	(lb/in.²)	(lb/in.²)	(lb/in.²)	(lb)	(lb)
Douglas fir ("Inland Empire" region) (*P. taxifolia*)	42	0.41	3,600	6,800	1340	0.55	6.9	8,700	22	2460	3240	500	870	530	470
	12	0.44	7,400	11,300	1610	1.91	8.6	11,800	27	5520	6700	950	1190	720	630
Douglas fir (Rocky Mountain region) (*P. taxifolia*)	38	0.40	3,600	6,400	1180	0.65	6.8	9,100	20	2540	3000	450	880	450	400
	12	0.43	6,300	9,600	1400	1.60	6.4	12,100	26	4660	6060	820	1070	740	630
Elm, American (*Ulmus americana*)	89	0.46	3,900	7,200	1110	0.81	11.8	—	38	1920	2910	440	1000	680	—
	12	0.50	7,600	11,800	1340	2.53	13.0	—	39	4030	5520	850	1510	1110	—
Elm, rock (*U. racemosa*)	48	0.57	4,600	9,500	1190	1.05	19.8	—	54	2970	3780	750	1270	980	—
	12	0.63	8,000	14,800	1540	2.45	19.2	—	56	4700	7050	1520	1920	1510	—
Elm, slippery (*U. fulva*)	85	0.48	4,000	8,000	1230	0.82	15.4	9,200	47	2790	3320	510	1110	750	660
	12	0.53	7,700	13,000	1490	2.35	16.9	15,300	45	4760	6360	1010	1630	1120	860
Fir, balsam (*Abies balsamea*)	117	0.34	3,000	4,900	960	0.52	4.7	6,900	16	2080	2400	210	610	290	290
	12	0.36	5,200	7,600	1230	1.23	5.1	7,800	20	3970	4530	380	710	510	400
Fir, commercial white[c] (*Abies sp.*)	108	0.36	3,800	5,800	1120	0.75	5.3	8,300	22	2470	2810	360	750	390	340
	12	0.38	6,300	9,300	1470	1.55	7.0	11,200	20	3870	5380	610	930	710	460
Gum, black (*Nyssa sylvatica*)	55	0.46	4,000	7,000	1030	0.91	8.0	9,800	30	2490	3040	600	1100	790	640
	12	0.50	7,300	9,600	1200	2.54	6.2	14,500	22	3470	5520	1150	1340	1240	810
Gum, red (*Liquidambar styraciflua*)	81	0.44	3,700	6,800	1150	0.81	9.4	10,000	33	2230	2840	460	1070	630	520
	12	0.49	8,100	11,900	1490	2.57	11.3	16,800	32	4700	5800	860	1610	950	690

Species																
Gum, tupelo (*N. aquatica*)	97	0.46	4,200	7,300	1050	0.98	8.3	9,000	30	2690	3370	590	1190	800	710
	12	0.50	7,200	9,600	1260	2.41	6.9	12,500	23	4280	5920	1070	1590	1200	880
Hackberry (*Celtis occidentalis*)	65	0.49	2,900	6,500	950	0.58	14.5	7,900	48	2070	2650	490	1070	760	700
	12	0.53	5,900	11,000	1190	1.72	12.8	13,700	43	3710	5440	1100	1590	1110	880
Hemlock, eastern (*Tsuga canadensis*)	111	0.38	3,800	6,400	1070	0.76	6.7	7,900	21	2600	3080	440	850	500	400
	12	0.40	6,100	8,900	1200	1.79	6.8	10,700	21	4020	5410	800	1060	810	500
Hemlock, western (*T. heterophylla*)	74	0.38	3,400	6,100	1220	0.57	6.8	8,100	22	2480	2990	390	810	520	430
	12	0.42	6,800	10,100	1490	1.82	7.5	12,400	26	5340	6210	680	1170	940	580
Hickory, pecan[f] (*Hicoria* sp.)	68	0.59	5,300	9,900	1380	1.18	19.3	14,200	60	3810	4320	980	1260	1274	1308
	12	0.65	9,100	16,300	1780	2.61	18.8	20,900	57	6360	8280	2040	1770	1930	1820
Hickory, true[g] (*Hicoria* sp.)	57	0.65	6,100	11,300	1570	1.34	28.9	15,700	88	3650	4570	1080	1360	—	—
	12	0.73	10,900	19,700	2180	3.07	27.2	22,900	75	—	8970	2310	2140	—	—
Honey locust (*Gleditsia triacanthos*)	63	0.60	5,600	10,200	1290	1.40	12.6	11,800	47	3320	4420	1420	1660	1440	1390
	12	—	8,800	14,700	1630	2.74	13.3	15,400	47	5250	7500	2280	2250	1860	1580
Larch, western (*Larix occidentalis*)	58	0.48	4,600	7,500	1350	1.01	7.1	9,400	24	3250	3800	560	920	470	450
	12	0.52	7,900	11,900	1710	2.46	8.0	15,100	32	5950	7490	1080	1360	1110	760
Locust, black (*Robinia pseudoacacia*)	40	0.66	8,800	13,800	1850	2.36	15.4	18,300	44	6120	6800	1430	1760	1640	1570
	12	0.69	12,800	19,400	2050	4.62	18.4	21,100	57	6800	10,180	2260	2480	1580	1700
Magnolia, cucumber (*Magnolia acuminata*)	80	0.44	4,200	7,400	1560	0.66	10.0	9,300	30	2810	3140	410	990	600	520
	12	0.48	8,000	12,300	1820	1.98	12.2	14,700	35	4840	6310	710	1340	950	700
Magnolia, evergreen (*M. grandiflora*)	105	0.46	3,600	6,800	1110	0.67	15.4	8,800	54	2160	2700	570	1040	780	740
	12	0.50	6,800	11,200	1400	1.90	12.8	13,600	29	3420	5460	1060	1530	1280	1020
Maple, bigleaf (*Acer macrophyllum*)	72	0.44	4,400	7,400	1100	1.02	8.7	8,500	23	2510	3240	550	1110	760	620
	12	0.48	6,600	10,700	1450	1.66	7.8	—	28	4790	5950	930	1730	1330	850
Maple, black (*A. nigrum*)	65	0.52	4,100	7,900	1330	0.70	12.8	10,200	48	2800	3270	740	1130	940	840
	12	0.57	8,300	13,300	1620	2.39	12.5	13,500	40	4600	6680	1250	1820	1700	1180
Maple, red (*A. rubrum*)	63	0.49	3,800	7,700	1390	0.71	11.4	—	32	2360	3280	500	1150	780	—
	12	0.54	8,700	13,400	1640	2.84	12.5	—	32	4650	6540	1240	1850	1430	—
Maple, silver (*A. saccharinum*)	66	0.44	3,100	5,800	940	0.61	11.0	6,800	29	1930	2490	460	1050	670	590
	12	0.47	6,200	8,900	1140	1.90	8.3	12,400	25	4360	5220	910	1480	1140	700
Maple, sugar (*A. saccharum*)	58	0.56	5,100	9,400	1550	1.03	13.3	12,200	40	2850	4020	800	1460	1070	970
	12	0.63	9,500	15,800	1830	2.76	16.5	20,600	39	5390	7830	1810	2330	1840	1450

TABLE 16.95 (Continued)

Commercial and Botanical Name of Species	Moisture Content	Specific Gravity[b]	Static Bending					Impact Bending		Compression Parallel to Grain		Compression Perpendicular to Grain	Shear Parallel to Grain	Hardness (Load Required to Embed 0.444-in. Ball to Half Its Diameter)	
			Fiber Stress at Proportional Limit	Rupture Modulus	Elasticity Modulus	Proportional Limit	Maximum Load	Fiber Stress at Proportional Limit	Height of Drop Causing Complete Failure (50-lb Hammer)	Fiber Stress at Proportional Limit	Maximum Crushing Strength	Fiber Stress at Proportional Limit	Maximum Shearing Strength	End	Side
Oak, red[h] (Quercus sp.)	80	0.57	4,400	8,500	1360	0.85	12.6	10,800	43	2590	3520	800	1220	1050	1030
	12	0.63	8,400	14,400	1810	2.30	15.0	17,000	43	4610	6920	1260	1830	1490	1300
Oak, white[i] (Quercus sp.)	70	0.59	4,700	8,100	1200	1.08	11.3	10,900	42	2940	3520	850	1270	1110	1070
	12	0.67	7,900	13,900	1620	2.31	13.3	17,400	39	4350	7040	1410	1890	1420	1330
Pine, lodgepole (Pinus contorta)	65	0.38	3,000	5,500	1080	0.49	5.6	7,200	20	2110	2610	310	680	320	330
	12	0.41	6,700	9,400	1340	1.97	6.8	9,600	20	4310	5370	750	880	530	480
Pine, northern white (P. strobus)	68	0.34	3,100	5,000	1020	0.54	5.2	6,700	17	2060	2490	290	660	310	310
	12	0.36	6,000	8,800	1280	1.59	6.7	9,500	19	3680	4840	550	860	500	400
Pine, Norway (P. resinosa)	54	0.44	3,700	6,400	1380	0.59	5.8	7,500	28	2410	3080	360	780	360	340
	12	0.48	9,400	12,500	1800	2.78	10.0	15,900	25	5330	7340	830	1230	670	580
Pine, ponderosa (P. ponderosa)	91	0.38	3,100	5,000	970	0.59	5.1	6,800	20	2070	2400	360	680	300	310
	12	0.40	6,300	9,200	1260	1.85	6.6	9,800	17	4060	5270	740	1160	550	450
Pines, southern yellow: Loblolly (P. taeda)	81	0.47	4,100	7,300	1410	0.68	8.2	8,900	30	2550	3490	480	850	420	450
	12	0.51	7,800	12,800	1800	1.92	10.4	12,100	30	4820	7080	980	1370	750	690
Longleaf (P. palustris)	63	0.54	5,200	8,700	1600	0.95	8.9	10,100	35	3430	4300	590	1040	550	590
	12	0.58	9,300	14,700	1990	2.44	11.8	15,400	34	6150	8440	1190	1500	920	870
Shortleaf (P. echinata)	81	0.46	3,900	7,300	1390	0.63	8.2	8,600	30	2500	3430	440	850	410	440
	12	0.51	7,700	12,800	1760	1.93	11.0	13,600	33	5090	7070	1000	1310	750	690

Species														
Pine, sugar (*P. lambertiana*) 137	0.35	3,400	5,100	940	0.70	5.4	7,400	17	2330	2530	350	680	320	310
12	0.36	5,700	8,000	1200	1.53	5.5	10,700	18	4140	4770	590	1050	530	380
Pine, western white														
(*P. monticola*) 54	0.36	3,400	5,200	1170	0.56	5.0	7,600	19	2430	2650	290	640	310	310
12	0.38	6,200	9,500	1510	1.47	8.8	11,900	23	4480	5620	540	—	440	370
Poplar, yellow														
(*Liriodendron tulipifera*) 64	0.38	3,400	5,400	1090	0.62	5.4	8,600	18	1930	2420	330	740	390	340
12	0.40	6,100	9,200	1500	1.43	6.8	13,500	20	3550	5290	580	1100	560	450
Redwood (virgin)														
(*Sequoia sempervirens*) 112	0.38	4,800	7,500	1180	1.18	7.4	8,900	21	3700	4200	520	800	570	410
12	0.40	6,900	10,000	1340	2.04	6.9	10,200	19	4560	6150	860	940	790	480
Spruce, eastern[f] (*Picea* sp.) 46	0.38	3,300	5,600	1110	0.57	6.5	7,000	21	2120	2600	290	710	390	340
12	0.40	6,500	10,100	1440	1.68	8.4	11,400	22	4160	5590	590	1070	630	490
Spruce, Engelmann														
(*P. engelmannii*) 100	0.31	2,500	4,200	830	0.43	4.9	5,800	14	1680	1980	290	590	250	240
12	0.33	6,000	8,500	1160	1.64	5.6	9,000	15	3580	4580	640	1010	450	310
Spruce, Sitka (*P. sitchensis*) 42	0.37	3,300	5,700	1230	0.53	6.3	8,400	24	2240	2670	340	760	430	350
12	0.40	6,700	10,200	1570	1.62	9.4	11,400	25	4780	5610	710	1150	760	510
Sugarberry (*Celtis laevigata*) 62	0.47	3,200	6,600	810	0.78	12.0	8,200	33	1990	2800	580	1050	840	740
12	0.51	6,200	9,900	1140	2.18	11.2	11,600	36	3970	5620	1240	1280	1280	960
Sycamore														
(*Platanus occidentalis*) 83	0.46	3,300	6,500	1060	0.60	7.5	8,800	26	2400	2920	450	1000	700	610
12	0.49	6,400	10,000	1420	1.66	8.5	10,500	26	3710	5380	860	1470	920	770
Tamarack (*Larix laricina*) 52	0.49	4,200	7,200	1240	0.84	7.2	7,800	28	2930	3480	480	860	400	380
12	0.53	8,000	11,600	1640	2.19	7.1	12,500	23	4780	7160	990	1280	670	590
Walnut, black (*Juglans nigra*) 81	0.51	5,400	9,500	1420	1.16	14.6	11,900	37	3520	4300	600	1220	960	900
12	0.55	10,500	14,600	1680	3.70	10.7	16,400	34	5780	7580	1250	1370	1050	1010

Source: *Wood Handbook*, U.S. Department of Agriculture, prepared by Forest Products Laboratory, Madison, Wis.

[a]Results of tests on small clear specimens in green and air-dry condition: Test specimens 2 × 2-in. section; bending specimens 30 in. long, others shorter depending on kind of test. Values in first line for each species are from tests of green material; those in second line are from tests of seasoned material adjusted to average air-dry condition of 12% moisture.

[b]Based on weight when oven dry and volume when green or at 12% moisture content.

[c]Average of Biltmore white ash (*Fraxinus biltmoreana*), blue ash (*F. quadrangulata*), green ash (*F. pennsylvanica lanceolata*), and white ash (*F. americana*).

[d]Average of sweet birch (*Betula lenta*) and yellow birch (*B. lutea*).

[e]Average of lowland white fir (*Abies grandis*) and white fir (*A. concolor*).

[f]Average of bitternut hickory (*Hicoria cordiformis*), nutmeg hickory (*H. myristicaeformis*), water hickory (*H. aquatica*), and pecan (*H. pecan*).

[g]Average of bigleaf shagbark hickory (*Hicoria laciniosa*), mockernut hickory (*H. alba*), pignut hickory (*H. glabra*), and shagbark hickory (*H. ovata*).

[h]Average of black oak (*Quercus velutina*), laurel oak (*Q. laurifolia*), pin oak (*Q. palustris*), red oak (*Q. borealis*), scarlet oak (*Q. coccinea*), southern red oak (*Q. rubra*), swamp red oak (*Q. rubra pagodaefolia*), water oak (*Q. nigra*), and willow oak (*Q. phellos*).

[i]Average of bur oak (*Quercus macrocarpa*), chestnut oak (*Q. montana*), post oak (*Q. stellata*), swamp chestnut oak (*Q. prinus*), swamp white oak (*Q. bicolor*), and white oak (*Q. alba*).

[j]Average of black spruce (*Picea mariana*), red spruce (*P. rubra*), and white spruce (*P. glauca*).

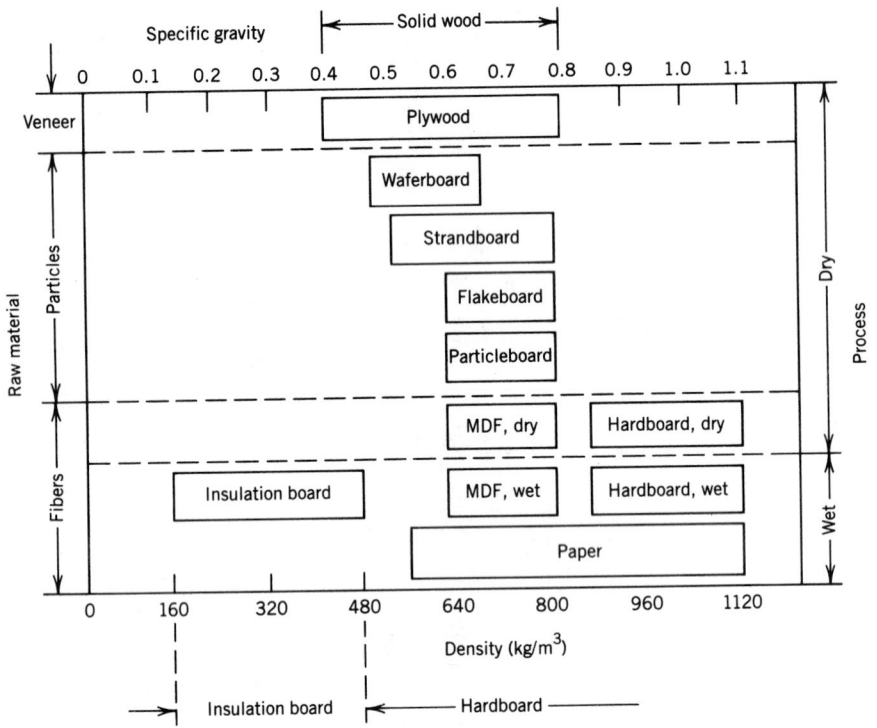

Fig. 16.13   Classification of wood composition boards. *Source: Encyclopedia of Materials Science and Engineering*, M. B. Bever, (ed.), Pergamon Press, Oxford, 1986.

TABLE 16.96 Strength and Mechanical Properties of Wood-Based Building Fiberboards[a]

Property	Structural Insulating Board	Medium-Density Hardboard	High-Density Hardboard	Tempered Hardboard	Special Densified Hardboard
Density, g/cm³	0.16–0.42	0.53–0.80	0.80–1.28	0.93–1.28	1.36–1.44
Modulus of elasticity, GPa[b]	0.17–0.86	2.24–4.83	2.76–5.52	4.48–7.59	8.62
Modulus of rupture, MPa[b]	1.4–5.5	13.1–41.4	20.7–48.3	38.6–69.0	69.0–86.2
Tensile strength parallel to surface, MPa[b]	1.4–3.4	6.9–27.6	20.7–41.4	24.8–53.8	53.8
Tensile strength perpendicular to surface, MPa[b]	0.069–0.17	0.28–1.4	0.52–2.8	1.1–3.1	3.4
Compressive strength parallel to surface, MPa[b]	—	6.9–24.1	12.4–41.4	25.5–41.4	183
Shear strength (in plane of board), MPa[b]	—	0.69–3.3	2.1–4.1	3.0–5.9	—
Shear strength (across plane of board), MPa[b]	—	4.1–17.2	13.8–20.7	19.3–23.4	—
24-hr water absorption, vol %	1–10	—	—	—	—
24-hr water absorption, wt %	—	5–20	3–30	3–20	0.3–1.2
Thickness swelling (24-hr soaking), %	—	2–10	10–25	8–15	—
Linear expansion (50–90% relative humidity), %[c]	0.2–0.5	0.2–0.4	0.15–0.45	0.15–0.45	—
Thermal conductivity (at 24°C), mW/m K[d]	39–65	65–108	108–202	108–216	267

Source: Encyclopedia of Composite Materials and Components, M. Grayson, (ed.), Wiley, New York, 1983.

[a]Data presented are general, round-figure values accumulated from numerous sources. For more exact figures on a specific product, individual manufacturers should be consulted or tests conducted. Values are for general laboratory conditions of temperature and humidity.
[b]To convert MPa to psi, multiply by 145; to convert GPa to psi, multiply by 145,000.
[c]Measurements made on material at equilibrium at each condition at room temperature.
[d]To convert mW/m K to Btu in./hr ft² °F, divide by 144.1.

TABLE 16.97 Example of Allowable Stresses for Plywood Conforming to U.S. Product Standard PS 1 for Softwood Plywood, Construction and Industrial, Dry-Location, Normal-Load Basis

Type of Stress	Species Group[e]	Stresses, MPa[b]		
		Exterior A-A, A-C, and C-C; Comparable Grades of Overlaid Plywood, Structural I A-A, A-C[a]	Exterior A-B, B-B, B-C C-C (Plugged), Plyform Class I, Plyform Class II; Comparable Grades of Overlaid Plywood Structural I C-D,[a] Structural II C-D,[d] Standard Sheathing,[c] All Interior Grades,[c]	All Other Grades of Interior Including Standard Sheathing
Extreme fiber in bending,	1	13.8	11.4	11.4
tension face grain	2, 3	9.7	8.3	8.3
parallel or	4	8.3	6.9	6.9
perpendicular to span				
(at 45° to face grain, use				
$\frac{1}{6}$ value)				
Compression parallel or	1	11.4	10.7	10.7
perpendicular to face	2, 3	8.3	7.6	7.6
grain (at 45° to face	4	6.9	6.6	6.6
grain, use $\frac{1}{3}$ value)				
Bearing (on face)	1	2.3	2.3	2.3
	2, 3	1.5	1.5	1.5
	4	1.1	1.1	1.1
Shear in plane	1	1.7	1.7	1.6
perpendicular to plies	2, 3	1.3	1.3	1.2
parallel or	4	1.2	1.2	1.1
perpendicular to face				
grain (at 45°, increase				
value by 100%)				
Shear, rolling, in plane of	All	0.4	0.4	0.3
plies parallel or				
perpendicular to face				
grain (at 45°, increase				
value by $\frac{1}{3}$)				
Modulus of elasticity in	1	—	12,400	—
bending face grain	2	—	10,300	—
parallel or	3	—	8,300	—
perpendicular to span	4	—	6,200	—

Source: *Plywood Design Specifications*, AIA File 19-F, American Plywood Assoc.
[a]Use group 1 stresses.
[b]To convert MPa to psi, multiply by 145.
[c]Exterior glue.
[d]Use group 3 stresses.
[e]NBS Voluntary Product Standard PS 1-74.

TABLE 16.98 Strength and Mechanical Properties of Mat-Formed (Platen-Pressed) Wood Particleboard[a]

Property	Low-Density Particleboard	Medium-Density Particleboard	High-Density Particleboard
Density, g/cm^3	0.40–0.59[b]	0.59–0.80	0.80–1.12
Modulus of elasticity (bending), MPa[c]	1.0–1.7[b]	1.7–4.8	2.4–6.9
Modulus of rupture, MPa	5.5–9.7[d]	11.0–55.2	16.6–51.7
Tensile strength			
Parallel to surface, MPa	—	3.4–27.6	6.9–34.5
Perpendicular to surface, MPa	0.14–0.21[d]	0.28–1.4	0.86–3.1
Compressive strength			
Parallel to surface, MPa	—	9.7–20.7	24.1–35.9
Shear strength			
In plane of board, MPa	—	0.60–3.1	1.4–5.5
Across plane of the board, MPa	—	1.4–12.4	—
24 hr water absorption, wt %	—	10–50	15–40
Thickness swelling from 24 hr soaking, %	—	5–50	15–40
Linear expansion[e] (50–90% relative humidity), %	0.30[f]	0.2–0.6	0.2–0.85
Thermal conductivity (at 24°C), mW/m K[g]	79–108	108–144	144–180

Source: *Encyclopedia of Composite Materials and Components*, M. Grayson, (ed.), Wiley, New York, 1983.

[a] Data presented are general round-figure values accumulated from numerous sources. For more exact figures on a specific product, individual manufacturers should be consulted or actual tests made. Values are for general laboratory conditions of temperature and humidity.

[b] Lower limit is for boards as generally manufactured; still lower density products with lower properties may be made.

[c] To convert MPa to psi, multiply by 145.

[d] Only limited production of low-density particleboard so values presented are specification limits.

[e] Measurements made on material at equilibrium at each condition at room temperature.

[f] Maximum permitted by specification.

[g] To convert mW/m K to Btu in./hr ft^2 °F, divide by 144.1.

16.2.3 Inorganic Materials

Cement and Concrete

TABLE 16.99   Types of Portland Cement

Type	Use
	Non-air-entraining[a]
I	General concrete construction when special properties specified for other types not required; when no type is indicated, Type I assumed
II	General concrete construction exposed to moderate sulfate action or where moderate heat of hydration required
III	Construction when high early strength required
IV	Construction when low heat of hydration required; this type not generally carried in stock
V	Construction when high sulfate resistance required; not generally carried in stock
	Air-entraining[a]
IA, IIA, IIIA	Same as corresponding Types I, II, and III of non-air-entraining cement but imparting to concrete properties of greatly improved resistance to severe weathering and to deleterious effect of applications of sodium or calcium salts to pavement surfaces for snow and ice removal

[a]From ASTM Designation C 150-85a

TABLE 16.100 Chemical Requirements for Portland Cements^a (%)

	Type I, Type IA	Type II, Type IIA	Type III, Type IIIA	Type IV, Type IVA	Type V, Type VA
Minimum silicon dioxide (SiO$_2$)	—	21.0	—	—	—
Maximum aluminum oxide (Al$_2$O$_3$)	—	6.0	—	—	—
Maximum ferric oxide (Fe$_2$O$_3$)	—	6.0	—	6.5	—
Maximum magnesium oxide (MgO)	5.0	5.0	5.0	5.0	5.0
Maximum sulfur trioxide (SO$_3$)					
When 3CaO · Al$_2$O$_3$ is 8% or less	2.5	2.5	3.0	2.3	2.3
When 3CaO · Al$_2$O$_3$ is more than 8%	3.0	—	4.0	—	—
Maximum loss on ignition	3.0	3.0	3.0	2.5	3.0
Maximum insoluble residue	0.75	0.75	0.75	0.75	0.75
Maximum tricalcium silicate (3CaO · SiO$_2$)^b	—	—	—	35	—
Minimum dicalcium silicate (2CaO · SiO$_2$)^b	—	—	—	40	—
Maximum tricalcium aluminate (3CaO · Al$_2$O$_3$)^b	—	8	15^c	7	5
Maximum sum of tricalcium silicate and tricalcium aluminate	—	58^d	—	—	—
Maximum tetracalcium aluminoferrite plus twice tricalcium aluminate^b (4CaO · Al$_2$O$_3$ · Fe$_2$O$_3$ + 2(3CaO · Al$_2$O$_3$), or solid solution (4CaO · Al$_2$O$_3$ · Fe$_2$O$_3$ + 2CaO · Fe$_2$O$_3$), as applicable	—	—	—	—	20.0

Source: ASTM Designation C 150-85a (Portland Cement).

^{a}Types I–V, non-air-entraining; types IA–VA, air-entraining.
^{b}Expressing chemical limitations by means of calculated assumed compounds does not necessarily mean that oxides are actually or entirely present as such compounds. When ratio of percentages of aluminum oxide to ferric oxide is 0.64 or more, percentages of tricalcium silicate, dicalcium silicate, tricalcium aluminate, and tetracalcium aluminoferrite shall be calculated from chemical analysis as follows:

$$\text{Tricalcium silicate} = (4.071 \times \text{percent CaO}) - (7.600 \times \text{percent SiO}_2)$$
$$- (6.718 \times \text{percent Al}_2\text{O}_3) - (1.430 \times \text{percent Fe}_2\text{O}_3)$$
$$- (2.852 \times \text{percent SO}_3)$$
$$\text{Dicalcium silicate} = (2.867 \times \text{percent SiO}_2) - (0.7544 \times \text{percent C}_3\text{S})$$
$$\text{Tricalcium aluminate} = (2.650 \times \text{percent Al}_2\text{O}_3) - (1.692 \times \text{percent Fe}_2\text{O}_3)$$
$$\text{Tetracalcium aluminoferrite} = 3.043 \times \text{percent Fe}_2\text{O}_3$$

When alumina-ferric oxide ratio is less than 0.64, calcium aluminoferrite solid solution (expressed as ss(C$_4$AF + C$_2$F) is formed. Contents of this solid solution (ss) and of tricalcium silicate shall be calculated by

$$\text{ss}(\text{C}_4\text{AF} + \text{C}_2\text{F}) = (2.100 \times \text{percent Al}_2\text{O}_3) + (1.702 \times \text{percent Fe}_2\text{O}_3)$$
$$\text{Tricalcium silicate} = (4.071 \times \text{percent CaO}) - (7.600 \times \text{percent SiO}_2) - (4.479 \times \text{percent Al}_2\text{O}_3)$$
$$- (2.859 \times \text{percent Fe}_2\text{O}_3) - (2.852 \times \text{percent SO}_3)$$

No tricalcium aluminate will be present in cements of this composition. Dicalcium silicate shall be calculated as previously shown. In calculations of C$_3$A, values of Al$_2$O$_3$ and Fe$_2$O$_3$ determined to nearest 0.01% shall be used. In calculation of other compounds oxides determined to nearest 0.1% shall be used. Values for C$_3$A and for sum of C$_4$AF + 2C$_3$A shall be reported to nearest 0.1%. Values for other compounds shall be reported to nearest 1%.
^{c}When moderate sulfate resistance is required for Type III or Type IIIA cement, tricalcium aluminate shall be limited to 8%. When high sulfate resistance is required, tricalcium aluminate shall be limited to 5%.
^{d}This limit applies when moderate heat of hydration is required and tests for heat of hydration are not requested.

TABLE 16.101 Approximate Relative Strengths of Concrete as Affected by Type of Cement

Type of Portland Cement	Compressive Strength, Percentage of Strength of Normal Portland Cement Concrete[a]		
	3 days	28 days	3 months
1. Normal	100	100	100
2. Modified	80	85	100
3. High early strength	190	130	115
4. Low heat	50	65	90
5. Sulfate resistant	65	65	85

Source: *Design and Control of Concrete Mixtures*, Portland Cement Association Bulletin T12-10, 10th ed., 1952.

[a] Values based on concrete moist-cured until tested.

TABLE 16.102 Maximum Sizes of Aggregate Recommended for Various Types of Construction

Minimum Dimension of Section (in.)	Maximum Size of Aggregate[a] (in.)			
	Reinforced Walls, Beams, and Columns	Unreinforced Walls	Heavily Reinforced Slabs	Lightly Reinforced or Unreinforced Slabs
$2\frac{1}{2}$–5	$\frac{1}{2}-\frac{3}{4}$	$\frac{3}{4}$	$\frac{3}{4}$–1	$\frac{3}{4}$–$1\frac{1}{2}$
6–11	$\frac{3}{4}$–$1\frac{1}{2}$	$1\frac{1}{2}$	$1\frac{1}{2}$	$1\frac{1}{2}$–3
12–29	$1\frac{1}{2}$–3	3	$1\frac{1}{2}$–3	3
30 or more	$1\frac{1}{2}$–3	6	$1\frac{1}{2}$–3	3–6

Source: *Recommended Practice for Selecting Proportions for Concrete*, American Concrete Institute Standard 613-54.

[a] Based on sieves with square openings.

TABLE 16.103 Recommended Slumps for Various Types of Construction

Types of Construction	Slump[a] (in.)	
	Maximum	Minimum
Reinforced foundation walls and footings	5	2
Plain footings, caissons, and substructure walls	4	1
Slabs, beams, and reinforced walls	6	3
Building columns	6	3
Pavements	3	2
Heavy-mass construction	3	1

Source: *Recommended Practice for Selecting Proportions for Concrete*, American Concrete Institute Standard 613-54.

[a] When high-frequency vibrators are used, the values given should be reduced about one-third.

TABLE 16.104 Compressive Strength of Concrete for Various Water-Cement Ratios

Water-Cement Ratio (gal/bag of cement)[b]	Probable Compressive Strength at 28 days[a] (psi)	
	Non-Air-Entrained Concrete	Air-Entrained Concrete
4	6000	4800
5	5000	4000
6	4000	3200
7	3200	2600
8	2500	2000
9	2000	1600

Source: *Recommended Practice for Selecting Proportions for Concrete*, American Concrete Institute Standard 613-54.

[a]Average strengths are for concretes containing not more than percentages of entrained and/or entrapped air shown in Table 16.105. For constant water-cement ratio, strength of concrete is reduced as air content is increased. For air contents higher than those listed in Table 16.105, strengths will be proportionally less than those listed here. Strengths are based on 6 × 12-in. cylinders moist cured under standard conditions for 28 days. See Method of Making and Curing Concrete Compression and Flexure Test Specimens in the Field (ASTM Designation C 31).

[b]1 bag = 94 lb, 1 ft³.

TABLE 16.105 Approximate Mixing Water Requirements for Different Slumps and Maximum Sizes of Aggregates

	Water[a] (gal/yd³ of Concrete for Indicated Maximum Sizes of Aggregate)							
	$\frac{3}{8}$ in.	$\frac{1}{2}$ in.	$\frac{3}{4}$ in.	1 in.	$1\frac{1}{2}$ in.	2 in.	3 in.	6 in.
Non-Air-Entrained Concrete								
Slump (in.)								
1–2	42	40	37	36	33	31	29	25
3–4	46	44	41	39	36	34	32	28
6–7	49	46	43	41	38	36	34	30
Approximate amount of entrapped air in non-air-entrained concrete, %	3	2.5	2	1.5	1	0.5	0.3	0.2
Air-Entrained Concrete								
Slump, in.								
1–2	37	36	33	31	29	27	25	22
3–4	41	39	36	34	32	30	28	24
6–7	43	41	38	36	34	32	30	26
Recommended average total air content, %	8	7	6	5	4.5	4	3.5	3

Source: *Recommended Practice for Selecting Proportions for Concrete*, American Concrete Institute Standard 613-54.

[a]These quantities of mixing water are for use in computing cement factors for trial batches. They are maxima for reasonably well-shaped angular coarse aggregates graded within limits of accepted specifications. If *more* water is required than shown, cement factor, estimated from these quantities, *should* be increased to main desired water-cement ratio, except as otherwise indicated by laboratory tests for strength. If *less* water is required than shown, cement factor, estimated from these quantities, *should not* be decreased except as indicated by laboratory tests for strength.

TABLE 16.106 Maximum Permissible Water-Cement Ratios (gal / bag) for Different Types of Structures and Degrees of Exposure

Type of structures	Severe wide range in temperature or frequent alternations of freezing and thawing (air-entrained concrete only)[a]			Mild temperature rarely below freezing, or rainy, or acid		
	In air	At the water line or within the range of fluctuating water level or spray		In air	At the water line or within the range of fluctuating water level or spray	
		In fresh water	In seawater or in contact with sulfates[b]		In fresh water	In seawater or in contact with sulfates[b]
A. Thin sections such as reinforced piles and pipe	5.5	5.0	4.5	6	5.5	4.5
B. Thin sections such as railings, curbs, sills, ledges, ornamental or architectural concrete, and all sections with less than 1-in. concrete cover over reinforcement	5.5	—	—	6	5.5	—
C. Moderate sections, such as retaining walls, abutments, piers, girders, beams	6.0	5.5	5.0	c	6.0	5.0
D. Exterior portions of heavy (mass) sections	6.5	5.5	5.0	c	6.0	5.0
E. Concrete deposited by tremie under water	—	5.0	5.0	—	5.0	5.0
F. Concrete slabs laid on ground	6.0	—	—	c	—	—
G. Concrete protected from weather, interiors of buildings, concrete below ground	c	—	—	c	—	—
H. Concrete which will later be protected by enclosure or backfill but which may be exposed to freezing and thawing for several years before such protection is offered	6.0	—	—	c	—	—

Source: Design of Concrete Mixtures, Portland Cement Association Bulletin ST 100 (1963); adapted from Recommended Practice for Selecting Proportions for Concrete, American Concrete Institute Standard 613-54.

[a]Air-entrained concrete should be used under all conditions involving severe exposure and may be used under mild exposure conditions to improve workability of mixture.

[b]Soil or groundwater containing sulfate concentrations of more than 0.2%. For moderate sulfate resistance, the tricalcium aluminate content of cement should be limited to 8% and for high sulfate resistance to 5%.

[c]Water-cement ratio should be selected on basis of strength and workability requirements but should not exceed 9 gal/bag.

TABLE 16.107 Approximate Compositions of Commercial Glasses[a]

No.	Designation	SiO_2	Na_2O	K_2O	CaO	Percentage of MgO	BaO	PbO	B_2O_3	Al_2O_3
1	Silica glass (fused silica)	99.5 +	—	—	—	—	—	—	—	—
2	96% silica glass	96.3	< 0.2	< 0.2	—	—	—	—	2.9	0.4
3	Soda lime, window sheet	71–73	13–15	—	8–10	1.5–3.5	—	—	—	0.5–1.5
4	Soda lime, plate glass	71–73	12–14	—	10–12	1–4	—	—	—	0.5–1.5
5	Soda lime, containers	70–73	13–16		10–13		—	—	—	1.5–2.5
6	Soda lime, electric lamp bulbs	73.6	16	0.6	5.2	3.6	—	—	—	1
7	Lead-alkali silicate, electrical	63	7.6	6	0.3	0.2	—	21	0.2	0.6
8	Lead-alkali silicate, high lead	35	—	7.2	—	—	—	58	—	—
9	Alumino-borosilicate (apparatus)	74.7	6.4	0.5	0.9	—	2.2	—	9.6	5.6
10	Borosilicate, low expansion	80.5	3.8	0.4	—	—	—	Li_2O	12.9	2.2
11	Borosilicate, low electrical loss	70.0	—	0.5	—	—	—	1.2	28.0	1.1
12	Borosilicate, tungsten sealing	67.3	4.6	1.0	—	0.2	—	—	24.6	1.7
13	Alumino-silicate	57	1.0	—	5.5	12	—	—	4	20.5

Source: Glass Engineering Handbook, Shand, E. B. (ed.), Corning Glass Works, Corning, New York, 1955.

[a] In commercial glasses iron may be in form of Fe_2O_3 to extent of 0.02–0.1% or more. In infrared absorbing glasses it is in the form of FeO in amounts from 0.5 to 1%.

TABLE 16.108   Typical Values for the Properties of Technical Glasses[a]

Glass Code[a]	Type	Color[b]	Principal Use	Forms Usually Available[c]	Corrosion Resistance[d] Class	Weathering	Water	Acid
0010	Potash Soda Lead	Cl	Lamp tubing	T	I	2	2	2
0080	Soda Lime	Cl	Lamp bulbs	BMT	I	3	2	2
0120	Potash Soda Lead	Cl	Lamp tubing	TM	I	2	2	2
0330	Glass-Ceramic	Gr	Bench tops	RS	I	—	1	3
1720	Aluminosilicate	Cl	Ignition tube	BT	I	1	1	3
1723	Aluminosilicate	Cl	Electron tube	BT	I	1	1	3
1990	Potash Soda Lead	Cl	Iron sealing	—	II	3	3	4
2405	Borosilicate	R	General	BPU	I	—	—	—
2473	Soda Zinc	R	Lamp bulbs	B	I	2	2	2
3320	Borosilicate	Ca	Tungsten sealing	—	I	(1)	(1)	(2)
6720	Soda Zinc	Op	General	P	I	—	1	2
6750	Soda Barium	Op	Lighting ware	BPR	I	—	2	2
7040	Borosilicate	Cl	Kovar sealing	BT	II	(3)	(3)	(4)
7050	Borosilicate	Cl	Series sealing	T	II	(3)	(3)	(4)
7052	Borosilicate	Cl	Kovar sealing	BMPT	II	(2)	(2)	(4)
7056	Borosilicate	Cl	Kovar sealing	BTP	II	2	2	4
7070	Borosilicate	Cl	Low loss electr.	BMPT	I	(2)	(2)	(2)
7251	Borosilicate	Cl	Sealed beam lamps	P	I	(1)	(2)	(2)
7570	High Lead	Cl	Solder sealing	—	II	1	1	4
7720	Borosilicate	Cl	Tungsten sealing	BPT	I	(2)	(2)	(2)
7740	Borosilicate	Cl	General	BPSTU	I	(1)	(1)	(1)
7760	Borosilicate	Cl	General	BP	I	2	2	2
7800	Soda Ba Borosilicate	Cl	Pharmaceutical	T		1	1	1
7900	96% Silica	Cl	High temp.	BPTUM	I	1	1	1
7913	96% Silica	Cl	High temp.	BPRST	I	1	1	1
7940	Fused Silica	Cl	Optical	U	I	1	1	1
7971	Titanium Silicate	Cl	Optical	U	—	1	1	1
8160	Potash Soda Lead	Cl	Electron tubes	PT	II	2	2	3
8161	Potash Lead	Cl	Electron tubes	PT	I	2	1	4
9606	Glass-Ceramic	Wh	Missile nose cones	C	II	—	1	4
9608	Glass-Ceramic	Wh	Cooking ware	BP	I	—	1	2
9741	Borosilicate	Cl	Ultraviolet transmission	BUT	II	(3)	(3)	(4)

Source: Courtesy of Corning Glass Works.

[a]Glasses 7905, 7910, 7911, 7912, 7913 and 7917 for special ultraviolet and infrared applications. Glass 1720 is available with improved ultraviolet transmittance (designated glass 9730). Glass 7760 also available with special transmission suitable for sun lamps.

[b]Abbreviations for color: Cl, clear; Gr, grey; R, red; Ca, canary yellow; Op, opal; Wh, white.

[c]Abbreviations for available forms: B, blown ware; M, multiform; U, panels; P, pressed ware; R, rolled sheet; C, castings; S, plate glass; T, tubing and rod.

[d]Since weathering is determined primarily by clouding which changes transmission, a rating for the opal glasses is omitted. Values in parentheses are for borosilicate glasses which may rate differently if subjected to excessive heat treatment. All data subject to normal manufacturing variations.

TABLE 16.108 *(Continued)*

Thermal Expansion[e] (Multiply By 10^7 cm/cm · °C)		Upper Working Temperatures (Mechanical Considerations Only)[g]				Thermal Shock Resistance Plate 15 × 15 cm[h]		
		Annealed		Tempered		Annealed		
0 to 300°C	25°C to Setting Point	Normal Service °C	Extreme Service °C	Normal Service °C	Extreme Service °C	3.2 mm Thick °C	6.4 mm Thick °C	12.7 mm Thick °C
93.5	101	110	380	—	—	65	50	35
93.5	105	110	460	220	250	65	50	35
89.5	97	110	380	—	—	65	50	35
9.7	—	538	—	—	—	—	—	—
42	52	200	650	400	450	135	115	75
46	54	200	650	400	450	125	100	70
124	136	100	310	—	—	45	35	25
43	53	200	480	—	—	135	115	75
91	—	110	460	—	—	65	50	35
40	43	200	480	—	—	145	110	80
78.5	90	110	480	220	275	70	60	40
88	—	110	420	220	220	65	50	35
47.5	54	200	430	—	—	—	—	—
46	51	200	440	235	235	125	100	70
46	53	200	420	210	210	125	100	70
51.5	56	200	460	—	—	—	—	—
32	39	230	430	230	230	180	150	100
36.7	38.1	230	460	260	260	160	130	90
84	92	100	300	—	—	—	—	—
36	43	230	460	260	260	160	130	90
32.5	35	230	490	260	290	160	130	90
34	37	230	450	250	250	160	130	90
50	53	200	460	—	—	—	—	—
8	5[f]	800	1100	—	—	—	—	—
7.5	5.5[f]	900	1200	—	—	—	—	—
5.5	3.5[f]	900	1100	—	—	—	—	—
0.5	−2	800	1100	—	—	—	—	—
91	100	100	380	—	—	65	50	35
90	99	100	390	—	—	—	—	—
57	—	700	—	—	—	200	170	130
4-20	—	700	800	—	—	—	—	—
39.5	50	200	390	—	—	150	120	80

[e]Code 9608 may be produced in a range of expansion values depending upon intended application.
[f]Extrapolated values.
[g]Normal service: No breakage from excessive thermal shock is assumed. Extreme limits: Glass will be very vulnerable to thermal shock. Recommendations in this range are based on mechanical stability considerations only. Tests should be made before adopting final designs. These data approximate only.
[h]These data approximate only. Based on plunging sample into cold water after oven heating. Resistance of 100°C (212°F) means no breakage if heated to 110°C (230°F) and plunged into water at 10°C (50°F). Tempered samples have over twice the resistance of annealed glass.

TABLE 16.108 *(Continued)*

| Glass Code | Type | Thermal Stress Resistance °C[i] | Viscosity Data[j] | | | | Knoop Hardness KHN$_{100}$[k] | Density g/cm^3 |
			Strain Point °C	Anneal- ing Point °C	Soften- ing Point °C	Working Point °C		
0010	Potash Soda Lead	19	392	432	626	983	363	2.86
0080	Soda Lime	16	473	514	696	1005	465	2.47
0120	Potash Soda Lead	20	395	435	630	985	382	3.05
0330	Glass-Ceramic	178	—	—	—	—	522	2.54
1720	Aluminosilicate	28	667	712	915	1202	513	2.52
1723	Aluminosilicate	26	665	710	908	1168	514	2.64
1990	Potash Soda Lead	14	340	370	500	756	—	3.50
2405	Borosilicate	37	501	537	765	1083	—	2.48
2473	Soda Zinc	19	466	509	697	—	—	2.65
3320	Borosilicate	43	493	540	780	1171	—	2.27
6720	Soda Zinc	20	505	540	780	1023	—	2.58
6750	Soda Barium	18	447	485	676	1040	—	2.59
7040	Borosilicate	37	449	490	702	1080	—	2.24
7050	Borosilicate	39	461	501	703	1027	—	2.24
7052	Borosilicate	41	436	480	712	1128	375	2.27
7056	Borosilicate	33	472	512	718	1058	—	2.29
7070	Borosilicate	66	456	496	—	1068	—	2.13
7251	Borosilicate	48	500	544	780	1167	—	2.25
7570	High Lead	21	342	363	440	558	—	5.42
7720	Borosilicate	49	484	523	755	1146	—	2.35
7740	Borosilicate	54	510	560	821	1252	418	2.23
7760	Borosilicate	52	478	523	780	1198	442	2.24
7800	Soda Barium Borosilicate	33	533	576	795	1189	—	2.36
7900	96% Silica	207	820	910	1500	—	463	2.18
7913	96% Silica	220	890	1020	1530	—	487	2.18
7940	Fused Silica	286	956	1084	1580	—	489	2.20
7971	Titanium Silicate	3370	—	1000	1500	—	—	2.21
8160	Potash Soda Lead	18	397	438	632	973	—	2.98
8161	Potash Lead	22	400	435	600	862	—	3.99
9606	Glass-Ceramic	16	—	—	—	—	657	2.6
9608	Glass-Ceramic	—	—	—	—	—	593	2.5
9741	Borosilicate	54	408	450	705	1161	—	2.16

[i] Resistance in °C (°F) is the temeprature differential between the two surfaces of a tube or a constrained plate that will cause a tensile stress of 0.7 kg/mm^2 (1000 psi) on the cooler surface.
[j] These data subject to normal manufacturing variations.
[k] Determined by ASTM standard C730-85.

TABLE 16.108 *(Continued)*

Young's Modulus Multiply By 10^3 kg/mm^2	Poisson's Ratio	Log$_{10}$ Volume Resistivity ohm · cm			Dielectric Properties at 1 MHz, 20°C^l			Refractive Index^m
		25°C	250°C	350°C	Power Factor %	Dielectric Constant	Loss Factor %	
6.3	.21	17.+	8.9	7.0	.16	6.7	1.	1.539
7.1	.22	12.4	6.4	5.1	.9	7.2	6.5	1.512
6.0	.22	17.+	10.1	8.0	.12	6.7	.8	1.560
8.8	.26	—	—	—	—	—	—	—
8.9	.24	17.+	11.4	9.5	.38	7.2	2.7	1.530
8.8	.24	17.+	13.5	11.3	.16	6.3	1.0	1.547
5.9	.25	17.+	10.1	7.7	.04	8.3	.33	—
6.9	.21	—	—	—	—	—	—	1.507
6.7	.22	—	—	—	—	—	—	1.52
6.6	.19	—	8.6	7.1	.30	4.9	1.5	1.481
7.1	.21	—	—	—	—	—	—	1.507
—	—	—	—	—	—	—	—	1.513
6.0	.23	—	9.6	7.8	.20	4.8	1.0	1.480
6.1	.22	16.	8.8	7.2	.33	4.9	1.6	1.479
5.8	.22	17.	9.2	7.4	.26	4.9	1.3	1.484
6.5	.21	—	10.2	8.3	.27	5.7	1.5	1.487
5.2	.22	17.+	11.2	9.1	.06	4.1	.25	1.469
6.5	.19	18.	8.1	6.6	.45	4.85	2.18	1.476
5.6	.28	17.+	10.6	8.7	.22	15.	3.3	1.86
6.4	.20	16.	8.8	7.2	.27	4.7	1.3	1.487
6.4	.20	15.	8.1	6.6	.50	4.6	2.6	1.474
6.3	.20	17.	9.4	7.7	.18	4.5	.79	1.473
—	—	—	7.0	5.7				1.491
6.9	.19	17.	9.7	8.1	.05	3.8	.19	1.458
6.9	.19	17.+	9.7	8.1	.04	3.8	.15	1.458
7.4	.16	17.+	11.8	10.2	.001	3.8	.0038	1.459
6.9	.17	20.3	12.2	10.1	(< .002)	(4.0)	(< .008)	1.484
—	—	17.+	10.6	8.4	.09	7.0	.63	1.553
5.5	.24	17.+	12.0	9.9	.06	8.3	.50	1.659
12	.24	16.7	10.0	8.7	.30	5.6	1.7	—
8.8	.25	13.4	8.1	6.8	.34	6.9	2.3	—
5.0	.23	17.+	9.4	7.6	.32	4.7	1.5	1.468

^{l}Values in parentheses are calculated at 10 kHz.
^{m}Refractive Index may be at either the sodium yellow line (589.3 nm) or the helium yellow line (587.6 nm). Values at these wavelengths do not vary in the first three places beyond the decimal point.

TABLE 16.109   Properties of Four Commercial Glass Ceramics[a]

Property	CER VIT C-101	Corning 9608	Corning 9606	Corning 9658
Thermal				
Coefficient of expansion, $10^{-7}/°C$				
from 0 to 38°C	0	—	—	—
from 25 to 300°C	—	13	57	—
from 25 to 400°C	—	—	—	94
Thermal conductivity at 25°C				
W/m K	1.675	2.008	mean 3.398 from 20 to 800°C	1.675
Mechanical				
Young's modulus at 20°C, GPa[b]	92.4	86.2	120	64.1
Shear modulus at 20°C, GPa	36.5	—	47.6	25.5
Modulus of rupture at 20°C, MPa[c]	55	83	241	103
Density at 20°C, g/cm³	2.50	2.50	2.60	2.52
Hardness, Knoop, kgf/mm² [d]				
100-g loading	—	593	657	—
200-g loading	540	—	—	—
500-g loading	—	—	—	250
Chemical				
Corrosion rate, mg/cm³ after 24 hr				
5% HCl at 95°C	0.3	0.12	—	87
5% NaOH at 85°C	3.5	—	—	8.5[e]
H₂O at 100°C	0.3	—	—	—
Electrical				
Dielectric constant				
100 Hz	11.7 (25°C)	—	—	—
100 kHz	—	—	—	5.92 (25°C)
1 MHz	9.4 (25°C)	6.9 (20°C)	5.6 (20°C)	—
1 MHz	14.2 (250°C)	—	—	—
Dissipation factors				
100 Hz	0.076 (25°C)	—	—	—
10 kHz	—	—	—	0.003 (25°C)
1 MHz	0.023 (25°C)	0.023 (20°C)	0.017 (20°C)	—
1 MHz	0.093 (250°C)	—	—	—
$\log_{10}$ of volume resistivity, Ω-cm				
25°C	14	13.4	16.7	—
250°C	8	8.1	10	10.8

Source: Property summaries provided by Owens-Illinois, Inc., and Corning Glass Works.
[a] Basic chemical system: CER VIT C-101, Li_2O-Al_2O_3-SiO_2; Corning 9608, Li_2O-Al_2O_3-SiO_2; Corning 9606, MgO-Al_2O_3-SiO_2; Corning 9658, K_2O-MgO-Al_2O_3-B_2O_3-SiO_2-F.
[b] To convert GPa to psi, multiply by 145,000.
[c] To convert MPa to psi, multiply by 145.
[d] 1 kgf/mm² = 9.8 MPa.
[e] After 6 hr.

TABLE 16.110 Properties of Ceramics

Material	Melting Point or Range (°C)	Density (g/cm^3)	Thermal Expansion Coefficient[a] (10^{-6} in./in. °C)	Thermal Conductivity (cal/s cm^2 °C cm)	Elastic Modulus (10^6 psi)	Electrical Resistivity (Ω-cm)	Dielectric Constant (K')	Compression Strength (ksi)	Tensile Strength (ksi)	Flexural Strength (ksi)	Room Temperature Fracture Toughness, K_{Ic} (MN/m$^{3/2}$)
Al$_2$O$_3$	2045	3.65	8	0.09	55	10^{16}	8.6–10.6	90%, 350; 99%, 375	90%, 20; 99%, 30	90%, 46; 99%, 50	4.2–5.8
SiO$_2$ (fused)	—	2.20	0.9	0.004	10	10^{15}	3.8	—	—	—	—
SiO$_2$ (quartz)	1728	2.65	9^a 14^b	—	—	10^{14}	4.6	—	—	—	—
MgO	2852	3.58	9	0.13	30	10^{19}	9.6	200	15	25	1.2
BeO	2570	3.01	10	1	45	10^{20}	12	114	20	40	—
ZrO$_2$ (stabilized)	2700	6.27	10	0.0045	30	10^7	96	300	20	25	—
MgAl$_2$O$_4$ (spinel)	2135	3.58	9	0.036	34	—	—	270	20	—	—
Al$_6$Si$_2$O$_{13}$ (mullite)	1850	3.16	5	0.015	21	—	—	40	2.5	9	2.2
Si$_3$N$_4$ (hot pressed)	1900	3.18	3.2	0.10	44	10^{14}	9.4	500		125	4.0–6.0
SiC	2500^c	3.21	4.5	0.25	68	10		sintered, 560; cast, 20	sintered, 25; cast, 3.5	sintered, 80; cast, 10	4.0
TiC	3100	4.95	7.2	0.04	67	1 × 10^{-4}				124	—
HfC	3890	12.6	6.6	—	51	0.4 × 10^{-4}				—	—
B$_4$C (hot pressed)	2450	2.51	4.5	0.07	65	0.2 to 7		414		71	6
TiB$_2$	~ 2600	4.51	8.1	0.6	75–83	4 × 10^{-5}				102–145	6–8
Sialon (low poly type)	—	3.23	3.0	0.05	44	—		> 500	60	137	7.7

[a]Parallel to c axis.
[b]Perpendicular to c axis.
[c]Decomposes or sublimes.
[d]25–1000°C.

*Refractories**

TABLE 16.111 Approximate Chemical Analyses of Refractories (%)

Type of Brick	SiO_2	$Al_2O_3 + TiO_2$[a]	Fe_2O_3	FeO	CaO	MgO	$Cr_2O_3 + Al_2O_3$	Alkalies
Fireclay								
Superduty	51.0	45.0	2.2	—	0.3	0.4	—	1.0
High heat duty (aluminous)	55.0	41.0	2.2	—	0.3	0.4	—	1.2
Intermediate heat duty	61.0	34.0	3.0	—	0.3	0.4	—	1.5
High heat duty (siliceous)	75.0	23.0	1.4	—	0.2	0.2	—	0.2
High alumina								
50% alumina class	43.0	53.0	1.8	—	0.2	0.5	—	1.4
70% alumina class	23.0	74.0	1.7	—	0.1	0.3	—	0.9
Silica	95.5	1.2	1.0	—	2.1	0.2	—	0.2
Magnesite (fired)	3	1	7	16	3	85	—	—
Chrome (fired)	5	—	—	—	—	15.5	63[b]	—

[a]TiO_2 content averages ~ 5% of combined $Al_2O_3 + TiO_2$. Thus for brick with 53% $Al_2O_3 + TiO_2$, TiO_2 content is ~ 2.7% and Al_2O_3 is ~ 50.3%.
[b]In brick made from different raw materials, Al_2O_3 content will vary from 12 to 30% and Cr_2O_3 will vary from 33 to 44%.
*See also Tables 16.149 to 16.151 and 16.153

TABLE 16.112 Physical Properties of Some Typical Fired Refractory Brick

	Magnesia (95% MgO)	Chrome (30% Cr_2O_3)	90% Alumina	70% Alumina	Zircon	Fireclay (Missouri Superduty)	Silicon Carbide	Silica (Superduty)
Bulk density kg/m³	2805–2950	3060–3140	2900–2965	2530–2600	3605–3720	2310–2370	2565–2660	1780–1875
Porosity, %	15–19	16–20	14–18	17.5–21.5	19–23	11–14	11–15	20–24
Cold crushing strength, MPa	48–70	35–55	62–95	27–18	48–76	12–21	69–83	27–41
Modulus of rupture, MPa	17–24	14–21	17–21	7.6–11	15–23	4.8–6.9	21–24	4.1–6.9
Reheat test, permanent linear change (%) after heating to								
1600°C	—	—	—	+3.5–6.0	—	0.0–0.9	—	—
1650°C	—	—	—	—	0.0	—	—	—
1725°C	−0.2–1.0	—	+0.1–1.0	—	—	—	−0.1–0.1	—
Load test, 170 kPa: temperature to which withstands load, °C	1620	1400	1760	1450	1600	1450	1650	1680

TABLE 16.113 Fireclay and High-Alumina Refractory Brick Classified According to Classes and Subdivided into Types

Class	Type	Pyrometric Cone Equivalent, min	Panel Spalling Loss, max, percent	Hot Load Subsidence, max, percent	Reheat Shrinkage, max, percent	Modulus of Rupture, min, psi (MPa)	Other Test Requirements
Fireclay brick							
Super-duty	Regular	33	8 at 3000 F (1650 C)	—	1.0 at 2910 F (1600 C)	600 (4.14)	—
	Spall-resistant	33	4 at 3000 F (1650 C)	—	1.0 at 2910 F (1600 C)	600 (4.14)	—
	Slag-resistant	33	—	—	—	1000 (6.89)	Bulk density, min, 140 lb/ft³ (2243 kg/m³)
High-duty	Regular	31 1/2					—
	Spall-resistant	31 1/2	10 at 2910 F (1600 C)	—	—	500 (3.45)	—
	Slag-resistant	31 1/2	—	—	—	1200 (8.27)	Bulk density, min, 137 lb/ft³ (2194 kg/m³) or max porosity 15 percent
Semi-silica		—	—	1.5 at 2460 F (1350 C)	—	300 (2.07)	Silica content, min, 72 percent
Medium-duty		29	—	—	—	500 (3.45)	—
Low-duty		15	—	—	—	600 (4.14)	—
High-alumina brick							
50 percent alumina		34	—	—	—	—	Alumina content, 50 ± 2.5 percent
60 percent alumina		35	—	—	—	—	Alumina content, 60 ± 2.5 percent
70 percent alumina		36	—	—	—	—	Alumina content, 70 ± 2.5 percent
80 percent alumina		37	—	—	—	—	Alumina content, 80 ± 2.5 percent
85 percent alumina		—	—	—	—	—	Alumina content, 85 ± 2.0 percent
90 percent alumina		—	—	—	—	—	Alumina content, 90 ± 2.0 percent
99 percent alumina		—	—	—	—	—	Alumina content, min, 97 percent

Source: ASTM C27-1984.

TABLE 16.114 Pyrometric Cone Equivalents for Fire Clay Mortars[a]

Class	Pyrometric Cone Equivalent not Lower Than
Superduty	No. 31
High heat duty	No. 28
Intermediate heat duty	No. 26
Low heat duty	No. 16

Source: ASTM C105-47 (1958), reapproved 1981.

TABLE 16.115 Temperature Equivalents for Pyrometric Cones Used in Refractory Testing

Cone No.	End Point[a] °F	End Point[a] °C	Cone No.	End Point[a] °F	End Point[a] °C
12	2439	1337	31	3061	1683
13	2460	1349	31 1/2	3090	1699
14	2548	1398	32	3123	1717
15	2606	1430	32 1/2	3135	1724
16	2716	1491	33	3169	1743
17	2754	1512	34	3205	1763
18	2772	1522	35	3245	1785
19	2806	1541	36	3279	1804
20	2847	1564	37	3308	1820
23	2921	1605	38	3335	1835
26	2950	1621	39	3389	1865
27	2984	1640	40	3425	1885
28	2995	1646	41	3578	1970
29	3018	1659	42	3659	2015
30	3029	1665			

Source: ASTM C24-1984.
[a]These temperatures, which were determined for a heating rate of 150°C/hr for cones 12–37, 100°C/hr for cone 38, and 600°C/hr for cones 39–42, other conditions being the same as specified, apply satisfactorily for all the conditions of this test method but do not apply to conditions of the commercial firing and use of refractory materials.

TABLE 16.116 Approximate Temperatures (°C) of Initial Reactions between Refractories

Composition	Magnesia	Magnesia-chrome	Chrome	90% Alumina	70% Alumina
Magnesia	—	≥ 1700	1700	≥ 1700	1600
Magnesia-chrome	≥ 1700	—	≥ 1700	1650	1650
Chrome	1700	≥ 1700	—	1650	1600
90% alumina	≥ 1700	1600	1650	—	≥ 1700
70% alumina	1650	1550	1600	≥ 1700	—
Fireclay	1400	1600	1600	1700	1700
Silicon carbide	1500	1500	1500	1650	1650
Silica, conventional	1500	1500	1650	1650	1500
Silica, superduty	1500	1550	1650	1650	1650

Source: Harbison-Walker Refractories Co. (1961).

TABLE 16.117 Mean Specific Heats (kJ / kg K) of Refractory Materials between 0°C and Indicated Temperature

Temperature (°C)	Fireclay Brick	Silica Brick	Magnesia Brick	Chrome Brick	99% Alumina Brick
0	0.808	0.708	0.871	0.712	0.716
200	0.863	0.883	0.971	0.762	0.896
400	0.913	0.988	1.043	0.808	0.976
600	0.976	1.051	1.097	0.850	1.034
800	1.022	1.080	1.130	0.879	1.063
1000	1.063	1.110	1.168	0.909	1.093
1200	1.097	1.139	1.206	0.930	1.118
1400	1.122	1.164	1.239	0.942	1.139
1500	1.143	1.177	1.256	0.950	1.164

Source: Harbison-Walker Refractories Co. (1961).

TABLE 16.118 Thermal Expansion of Refractories

Material	Firing Temperature (°C)	Mean Coefficient of Expansion from 0°F to Shrinkage Temperature (in./in./°C × 10⁻⁶)	Maximum Coefficient of Expansion between 300 and 700°C (in./in./°C × 10⁻⁶)	Maximum Irreversible Contraction or Expansion[a] up to 1700°C (%)	Temperature at which Shrinkage or Expansion Begins (°C)
Silica	—	8.3	100.0	1	1550
Kaolin	1300	4.7	7.9	1	1050
	1430	6.8	8.7	1	1380
	1500	5.3	7.0	1	1580
	1620	4.3	6.7	1	1610
Fireclay					
Missouri	—	5.4	8.0	Large	1300
Pennsylvania	—	5.1	6.4	−5	1250
Colorado	—	5.4	17.4	Large	1220
Maryland	—	4.5	8.0	−3	1100
Silicon carbide	—	4.3	4.8	0	1700 +
Zircon					
White	1650	6.4	9.2	Large	1510
Brown	1590	4.2	4.8	2	1550
Zirconia	1675	5.9	8.7	1	1600
Mullite	1785	5.3	8.2	0	1700 +
Magnesite					
Pure	1680	14.2	15.1	0	1700 +
Commercial	—	14.7	21.0	2	1440
Chrome (commercial)	—	10.4	12.4	2	1540
Spinel	1690	7.6	11.0	1	1600
Lime	1740	13.8	14.5	0	1700 +
Alumina	1650	7.7	8.2	1	1580
Insulating refractory	—	7.4	48.0	Large	1050
Zircon (electrically fused)	—	4.2	14.0	—	—
Magnesite (electrically fused)	10.2	12.0	—	—	—
Topaz	1675	4.3	15.1	—	—
Zircon (purified)	—	3.8	15.1	—	—

Source: F. H. Norton, *Refractories*, 3rd ed., New York, McGraw-Hill, 1949.

[a] Minus sign indicates expansion.

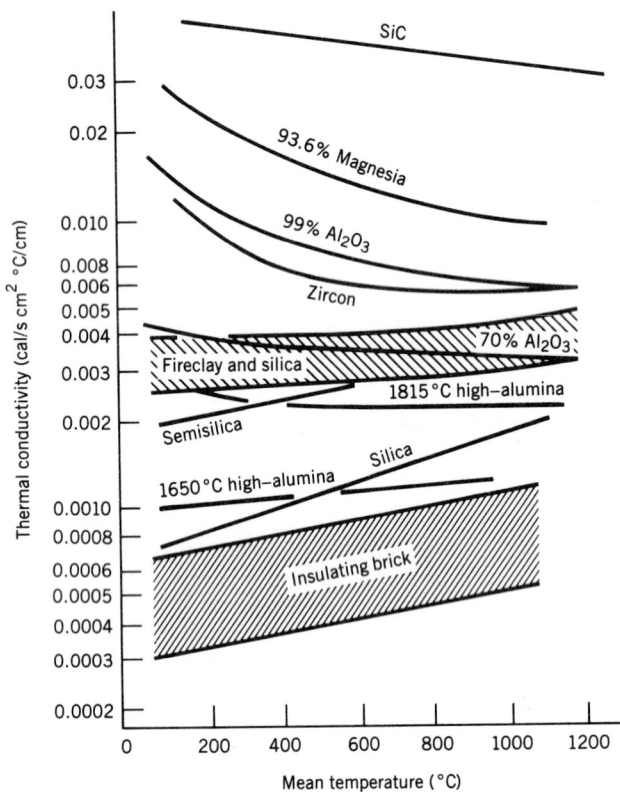

Fig. 16.14 Thermal conductivity of various types of refractory brick (after Ruh, E. and McDowell, J. S., "Thermal conductivity of refractory brick," *J. Am. Ceram. Soc.* **45**, 189–195, 1962).

16.2.4 Miscellaneous Properties of Materials

TABLE 16.119 Density of Miscellaneous Nonmetallic Solids

Name	Density[a] (g/cm³)	(lb/ft³)	Name	Density[a] (g/cm³)	(lb/ft³)
Agate	2.5–2.7	156–168	Glass		
Amber	1.06–1.11	66–69	Common	2.5–2.75	156–172
Asbestos	2.0–2.8	125–175	Crystal	2.90–3.00	181–187
Asphalt	1.1–1.5	69–94	Flint	3.2–4.7	200–294
Basalt	2.4–3.1	150–193	Plate	2.45–2.72	153–170
Bauxite	2.55	159	Glue	1.27	80
Beeswax	0.96–0.97	60–61	Gneiss	2.4–2.7	150–169
Biotite	2.7–3.1	168–193	Granite	2.65–2.7	165–169
Borax	1.7–1.8	106–112	Gravel, dry		
Brick			Loose	1.4–1.7	87–106
Soft	1.6	100	Packed	1.6–1.9	100–119
Common	1.79	112	Wet	1.9	119
Hard	2.0	125	Gypsum	2.31–2.33	144–145
Pressed	2.16	135	Hematite	4.9–5.3	306–330
Fire	2.24–2.4	140–150	Hornblende	3.0	187
Sand lime	2.18	136	Iodine	4.94	308
Brickwork			Ivory	1.83–1.92	114–120
Mortar	1.6	100	Lava		
Cement	1.79	112	Basaltic	2.8–3.0	175–187
Carbon			Trachytic	2.0–2.7	125–168
Diamond	3.52	220	Leather	0.86–1.02	54–64
Graphite	2.25	140	Lime		
Cement			Mortar	1.65–1.78	103–111
Natural	2.8–3.2	175–200	Quick (in bulk)	0.8–0.96	50–60
Portland	3.05–3.15	190–197	Slaked	1.3–1.4	81–87
Loose	1.44	90	Limestone	2.00–2.9	125–181
Barreled	1.84	115	Litharge		
Slag	1.9–2.3	119–144	Artificial	9.3–9.4	580–587
Chalk	1.9–2.8	119–175	Natural	7.8–8.0	487–499
Charcoal			Magnesia, carbonate	2.4	150
Oak	0.57	35	Magnetite	4.9–5.2	306–324
Pine	0.28–0.44	17–27	Marble	2.6–2.84	162–177
Clay	1.8–2.6	112–162	Masonry,		
Coal			Dry rubble	2.24–2.56	140–160
Anthracite	1.4–1.8	87–112	Dressed	2.24–2.88	140–180
Bituminous	1.2–1.6	75–100	Mica	2.6–3.2	162–200
Charcoal	0.27–0.58	17–36	Muscovite	2.76–3.00	172–187
Lignite	1.1–1.4	69–87	Oligoclase	2.65–2.67	165–167
Coke	1.0–1.7	62–106	Orthoclase	2.58–2.61	161–163
Concrete		144	Paper	0.7–1.15	44–72
Corundum	3.9–4.0	244–250	Pitch	1.07	67
Dolomite	2.84	177	Plaster-of-Paris	1.5–1.8	94–112
Earth			Porcelain	2.3–2.5	143–156
Dry, loose	1.2	75	Porphyry	2.6–2.9	162–181
packed	1.5	94	Pumice	0.37–0.90	23–56
Moist, loose	1.3	81	Pyrite	4.95–5.1	309–318
packed	1.6	100	Quartz	2.65	165
Mud, flowing	1.7	106	Quartzite	2.73	170
packed	1.8	112	Resin	1.07	67
Emery	4.0	250	Riprap		
Feldspar	2.55–2.75	159–172	Limestone	1.3–1.4	81–87
Flint	2.63	164	Sandstone	1.4	87
Garnet	3.15–4.3	197–268	Shale	1.7	106
Gas carbon	1.88	117	Rock salt	2.18	136
Gelatin	1.27	80			

TABLE 16.119   *(Continued)*

Name	Density[a] (g/cm³)	(lb/ft³)	Name	Density[a] (g/cm³)	(lb/ft³)
Rubber			Slate	2.6–3.3	162–205
Caoutchouc	0.92–0.96	57–60	Soapstone	2.6–2.3	162–175
Manufactured	1.0–2.0	62–125	Starch	1.53	95
Salt	0.78–1.25	49–78	Stone		
Sand			Various	2.16–3.4	135–212
Dry	1.44–1.76	90–110	Crushed	1.6	100
Wet	1.89–2.07	118–129	Sugar	1.61	100
Selenium	4.82	301	Sulfur	2.0–2.1	125–131
Serpentine	2.50–2.65	156–165	Talc	2.7–2.8	168–175
Shale	2.6–2.9	162–181	Tar, bituminous	1.02	66
Silicon	2.42	151	Terracotta	1.9	119
Slag			Tile	1.76–1.92	110–120
Bank	1.1–1.2	69–75	Tourmaline	3.0–3.2	187–200
Bank screenings	1.5–1.9	94–119	Traprock	2.72–3.4	170–212
Furnace	2.0–3.9	125–244	Wood	See Table 16.95	See Table 16.95
Machine	1.5	94			
Sand	0.8–0.9	50–56			

[a] Ordinary temperatures understood.

TABLE 16.120   **Specific Heat of Miscellaneous Nonmetallic Solids**

Name	Temperature (°C)	Specific Heat (g-cal/g, Btu/lb)	Name	Temperature (°C)	Specific Heat (g-cal/g, Btu/lb)
Asbestos	0	0.25	Granite	20–100	0.20
Bakelite	—	0.3–0.4	Ice	−20	0.48
Basalt	—	0.20		0	0.505
Carbon	−76–0	0.126	India rubber, Para	?–100	0.481
Graphite	26–76	0.165	Limestone	—	0.217
Diamond	0	0.147	Marble	18	0.21
Calcspar	0–100	0.2005	Mica	20–98	0.2061
Cellulose	—	0.32	Paraffin	0–20	0.694
Chalk	20–99	0.214	Porcelain	15–950	0.26
Clay	0	0.224	Quartz	0	0.17
Coal	—	0.26–0.37	Rock salt	13–45	0.219
Coke	21–400	0.265	Sand	—	0.191
Concrete	70–312	0.156	Selenium	20.5	0.077
Ebonite	20–100	0.40	Silicon	18.2–99.1	0.181
Glass			Sulfur		
Normal thermometer			Rhombic	15–96	0.176
Jena 16III	19–100	0.1988	Monoclinic	0–52	0.181
Crown	0	0.16–0.20	Woods, general	—	0.45–0.65
Flint	10–50	0.117			

TABLE 16.121 Thermal Coefficient of Linear Expansion of Miscellaneous Nonmetallic Solids[a]

Name	Temperature (°C)	Thermal Coefficient of Expansion, per °C (×10⁴)
Amber	0–30	0.50
Bakelite, bleached	20–60	0.22
Caoutchouc	16.7–25.3	0.770
Carbon		
Diamond	50	0.012
Graphite	50	0.0786
Celluloid	20–80	1.09
Ebonite	25.3–35.4	0.842
Fluorspar, CaF_2	0–100	0.1950
Glass		
Tube	0–100	0.0833
Plate	0–100	0.0891
Crown (mean)	0–100	0.0897
Flint	50–60	0.0788
Jena thermometer, 16[III]		
Normal	0–100	0.081
Jena thermometer, 56[III]	0–100	0.058
Jena thermometer, 59[III]	−191 to +16	0.424
Gutta percha	20	1.983
Ice	−20 to −1	0.51
Iceland spar		
Parallel to axis	0–80	0.2631
Perpendicular to axis	0–80	0.0544
Limestone	25–100	0.09
Marble	15–100	0.117
Paraffin	0–16	1.0662
Paraffin	16–38	1.3030
Porcelain	20–790	0.0413
Porcelain, Bayeux	1000–1400	0.0553
Quartz:		
Parallel to axis	0–80	0.0797
Parallel to axis	−190 to +16	0.0521
Perpendicular to axis	0–80	0.1337

Name	Temperature (°C)	Thermal Coefficient of Expansion, per °C (×10⁴)
Quartz glass	−190 to +16	−0.0026
	16–500	0.0057
	16–1000	0.0058
Rock salt	40	0.4040
Rubber, hard	0	0.691
Selenium	−160	0.300
Silicon	0–100	0.660
Fluorspar, CaF_2 (Tourmaline:)	−3 to +18	0.0249
Parallel to longitudinal axis	0–100	0.0937
Parallel to horizontal axis	0–10	0.0773
Vulcanite	0–18	0.6360
Wedgwood ware	0–100	0.0890
Wood, parallel to fiber		
Ash	0–100	0.0951
Beech	2.34	0.0257
Chestnut	2.34	0.0649
Elm	2.34	0.0565
Mahogany	2.34	0.0361
Maple	2.34	0.0638
Oak	2.34	0.0492
Pine	2.34	0.0541
Walnut	2.34	0.0658
Wood, perpendicular to fiber		
Beech	2.34	0.614
Chestnut	2.34	0.325
Elm	2.34	0.443
Mahogany	2.34	0.404
Maple	2.34	0.484
Oak	2.34	0.544
Pine	2.34	0.341
Walnut	2.34	0.484
Wax, white	10–26	2.300
	26–31	3.120

[a]Coefficient of cubical expansion may be taken as 3 times the linear coefficient.

16.3 PROPERTIES OF MATERIALS BY APPLICATION

16.3.1 Bearing Materials

TABLE 16.122 Nominal Composition (%) of Tin-Base Bearing Alloys

Designation	Sn[a]	Sb	Pb (Maximum)	Cu	Fe (Maximum)	As (Maximum)	Bi (Maximum)	Zn (Maximum)	Al (Maximum)	Others (Total Maximum)
ASTM B23 Alloys										
Alloy 1	91.0	4.5	0.35	4.5	0.08	0.10	0.08	0.005	0.005	—
Alloy 2	89.0	7.5	0.35	3.5	0.08	0.10	0.08	0.005	0.005	—
Alloy 3	84.0	8.0	0.35	8.0	0.08	0.10	0.08	0.005	0.005	—
Alloy 11	87.5	6.8	0.50	5.8	0.08	0.10	0.08	0.005	0.005	—
SAE Alloys										
SAE 11	86.0	6.0–7.5	0.50	5.0–6.5	0.08	0.10	0.08	0.005	0.005	0.20
SAE 12	88.0	7.0–8.0	0.50	3.0–4.0	0.08	0.10	0.08	0.005	0.005	0.20

Source: Metals Handbook. (9th edition), ASM, Metals Park, Ohio, 1985.

[a] Desired in ASTM alloys: specified minimum in SAE alloys.

TABLE 16.123 Properties of Selected ASTM B23 Tin-Base Bearing Alloys

Designation	Specific Gravity	Compressive Yield Strength[a,b] 20°C (68°F) MPa	ksi	100°C (212°F) MPa	ksi	Compressive Ultimate Strength[a,c] 20°C (68°F) MPa	ksi	100°C (212°F) MPa	ksi	Brinell Hardness[d] 20°C	100°C	Solid Temperature °C	°F	Liquid Temperature °C	°F	Pouring Temperature °C	°F
Alloy 1	7.34	30.3	4.40	18.3	2.65	88.6	12.85	47.9	6.95	17.0	8.0	223	433	371	700	440	825
Alloy 2	7.39	42.1	6.10	20.7	3.00	102.7	14.90	60.0	8.70	24.5	12.0	241	466	354	669	425	795
Alloy 3	7.46	45.5	6.60	21.7	3.15	121.3	17.60	68.3	9.90	27.0	14.5	240	464	422	792	490	915

Source: Metals Handbook. (9th edition), ASM, Metals Park, Ohio, 1985.

[a] Compression test specimens were cylinders $1\frac{1}{2}$ in. long and $\frac{1}{2}$ in. in diameter machined from chill castings 2 in. long and $\frac{3}{4}$ in. in diameter.

[b] Values for yield point from stress-strain curves at a deformation of 0.125% reduction of gage length.

[c] Values for ultimate strength taken as unit load necessary to produce deformation of 25% of length of specimen.

[d] Tests made on bottom face of parallel machined specimens cast at room temperature in steel mold 2 in. in diameter by $\frac{5}{8}$ in. deep. Brinell hardness values are averages of three impressions on each alloy using a 10-mm ball and applying 500-kg load for 30 s.

TABLE 16.124 Nominal Composition (%) of Lead-Base Bearing Alloys

Designation	Pb	Sb	Sn	Cu (Maximum)	Fe (Maximum)	As	Bi (Maximum)	Zn (Maximum)	Al (Maximum)	Cd (Maximum)	Other
ASTM B23 Alloys											
Alloy 7[a]	Remainder	15.0	10.0	0.50	0.1	0.45	0.10	0.005	0.005	0.05	—
Alloy 8	Remainder	15.0	5.0	0.50	0.1	0.45	0.10	0.005	0.005	0.05	—
Alloy 13[b]	Remainder	10.0	6.0	0.50	0.1	0.25[a]	0.10	0.005	0.005	0.05	—
Alloy 15[c]	Remainder	16.0	1.0	0.50	0.1	1.10	0.10	0.005	0.005	0.05	—
Other Alloys											
SAE 16	Remainder	3.5	4.5	0.10	—	0.05[a]	0.10	0.005	0.005	0.005	—
AAR M501[d]	Remainder	8.75	3.5	0.50	—	0.20[a]	—	—	—	—	—
SAE 19	Remainder	—	10.0	—	—	—	—	—	—	—	—
SAE 190	Remainder	—	7.0	3.0	—	—	—	—	—	—	—
Proprietary Alloys											
A	95.65	—	3.35	0.08	—	—	—	—	—	—	0.67 Ca
B	83.30	12.54	0.84	0.10	—	3.05	—	—	—	—	—
C	Remainder	10.0	3.0	0.20	—	—	—	—	—	—	2.0 Ag

Source: Metals Handbook. (9th edition), ASM, Metals Park, Ohio, 1985.

[a]Also SAE 14.
[b]Also SAE 13.
[c]Also SAE 5.
[d]Association of American Railroads, Specification M501; also ASTM B67.

TABLE 16.125 Properties of Selected ASTM B23 Lead-Base Bearing Alloys

Designation	Specific Gravity	Compressive Yield Strength[a,b]				Compressive Ultimate Strength[a,c]				Hardness[d]		Solid Temperature		Liquid Temperature		Pouring Temperature	
		20°C		100°C		20°C		100°C		20°C	100°C	°C	°F	°C	°F	°C	°F
		MPa	ksi	MPa	ksi	MPa	ksi	MPa	ksi								
Alloy 7	9.73	24.5	3.55	11.0	1.60	107.9	15.65	42.4	6.15	22.5	10.5	240	464	268	514	338	640
Alloy 8	10.04	23.4	3.40	12.1	1.75	107.6	15.60	42.4	6.15	20.0	9.5	237	459	272	522	340	645
Alloy 15	10.05	—	—	—	—	—	—	—	—	21.0	13.0	248	479	281	538	350	662

Source: Metals Handbook. (9th edition), ASM, Metals Park, Ohio, 1985.

[a]Compression test specimens cylinders 1.5 in. long, 0.5 in. in diameter, machined from chill castings 2 in. long, 0.75 in. in diameter.

[b]Values taken from stress-strain curves at deformation of 0.125% reduction of gage length.

[c]Values taken as unit load necessary to produce deformation of 25% of length of specimen.

[d]Tests made on bottom face of parallel-machined specimens cast at room temperature in steel mold 2 in. in diameter ×0.625 in. deep. Values listed are averages of three impressions on each alloy using 10-mm ball and applying 500-kg load for 30 s.

TABLE 16.126 Designations and Nominal Compositions of Copper-Base Bearing Alloys

UNS No.	SAE	Other	Former SAE	Cu	Sn	Pb	Zn	Other	Form	Use
	Designations			Nominal composition (%)						
Commercial Bronze										
C22000	795	—	—	90	0.5	—	9.5	—	Wrought strip	Solid bronze bushings and washers
Unleaded Tin Bronzes										
C90300	C90300	—	620	88	8	0	4	—	Cast tubes	Solid bronze bearings
C90500	C90500	—	62	88	10	0	2	—	Cast tubes	Solid bronze bearings
C91100	—	—	—	84	16	0	0	—	Cast tubes	Solid bronze bearings
C91300	—	—	—	81	19	0	0	—	Cast tubes	Solid bronze bearings
Low-Lead Tin Bronzes										
C92200	C92200	—	622	88.5	6	1.5	4	—	Cast tubes	Solid bronze bearings
C92300	C92300	—	621	87.0	8.5	0.5	4	—	Cast tubes	Solid bronze bearings
C92700	C92700	—	63	87.5	10	2	0.5	—	Cast tubes	Solid bronze bearings
Medium-Lead Tin Bronzes										
C54400	791	—	—	88	4	4	4	—	Wrought strip	Solid bronze bushings and washers
—	—	F32/62	—	87	4	4	3	2 Fe	Cast on steel back	Bimetal bushings and washers, trimetal intermediate layer
C83600	C83600	—	40	85	5	5	5	—	Cast tubes	Solid bronze bearings and bronze bearing backs
C93200	C93200	—	660	83	7	7	3	—	Cast tubes	Solid bronze bearings
C93600	793	—	—	85	4	8	3	—	Cast on steel back	Bimetal surface layer
—	798	—	—	88	4	8	—	—	Sintered on steel back	Bimetal surface layer
C93700	C93700	—	64	80	10	10	—	—	Cast tubes	Solid bronze bearings and bronze bearing backs
—	792	—	—	80	10	10	—	—	Cast on steel back	Bimetal surface layer and trimetal intermediate layer
—	797	—	—	80	10	10	—	—	Sintered on steel back	Bimetal surface layer

TABLE 16.126 *(Continued)*

UNS No.	SAE	Other	Former SAE	Cu	Sn	Pb	Zn	Other	Form	Use
	Designations			Nominal composition (%)						
High-Lead Tin Bronze										
C93800	C93800	—	67	78	6	16	—	—	Cast tubes	Solid bronze bearings and bronze bearing backs
—	—	AMS 4825	—	74	10	16	—	—	Cast on steel back	Bimetal surface layer
—	794	—	—	71.5	3.5	23	2	—	Cast on steel back	Bimetal surface layer
—	799	—	—	74	3	23	—	—	Sintered on steel back	Bimetal surface layer
—	—	AMS 4824	—	75	1	24	—	—	Cast on steel back	Trimetal intermediate layer
—	—	F780	—	74	2.5	23.5	—	—	Sintered on steel back	Trimetal intermediate layer

Source: Metals Handbook, (9th edition), ASM, Metals Park, Ohio, 1985.

TABLE 16.127　Designations and Nominal Composition of Aluminum-Base Bearing Alloys

No.	SAE	AA	Other	Al	Si	Cu	Ni	Mg	Sn	Other	Form	Typical Applications
	Designation			Nominal Composition (%)								
High-Tin Aluminum Alloy												
1	783	8081	A98081	79	—	1	—	—	20	—	Wrought strip, O temper, bonded to steel back	Bimetal surface layer
High-Lead Aluminum Alloys												
2	—	—	F-66	85	4	1	—	—	1.5	8.5 Pb	Powder rolled and sintered strip, O temper, bonded to steel back	Bimetal surface layer
3	—	—	AL-6	88	4	0.5	—	0.5	1	6 Pb	Wrought strip, O temper, bonded to steel back	Bimetal surface layer
Low-Tin Aluminum Alloys												
4	770	850.0	A08500	91.5	0.7	1	1	—	6.5	—	Cast tubes T101 temper[a]	Solid aluminum bearings; aluminum bearing backs

Source: Metals Handbook, (9th edition), ASM, Metals Park, Ohio, 1985.
[a]Artificially aged and cold pressed.

TABLE 16.128 Operating Limits for Selected Nonmetallic Bearing Materials

Bearing Material	Load Capacity (psi)	Maximum Temperature (°F)	Speed (fpm)	PV Limit[a]
Phenolics	6,000	200	2,500	15,000
Nylon	1,000	200	1,000	3,000
TFE	500	500	50	1,000
Reinforced TFE	2,500	500	1,000	10,000
TFE fabric	60,000	500	50	25,000
Polycarbonate	1,000	220	1,000	3,000
Acetal	1,000	180	1,000	3,000
Carbon-graphite	600	750	2,500	15,000
Rubber	50	150	4,000	—
Wood	2,000	100	2,000	12,000

Source: Materials and Processes, (3rd edition), Young, J. F., and Shane, R. S., (eds.), Marcel Dekker, New York, 1985.

[a]Symbols: P, psi load; V, surface speed (fpm).

16.3.2 Conducting Materials

TABLE 16.129 Selected Properties of Aluminum and Copper Conductors

	Aluminum			Copper		
	EC-Al (1350)[a]	Al–Mg (5005)[b]	Al–MgSi (6201)[c]	OFHC[d]	Phosphor bronze[e] (95-5)	Cartridge brass[f] (70-30)
Density, g/cm³	2.7	2.7	2.69	8.94	8.86	8.53
Melting point, °C	660	652	654	1083	1060	955
Coefficient of linear thermal expansion, 10^{-6} K^{-1}	23.6	23.7	23.4	17.0	17.8	20.3
Thermal conductivity, W/cm K	2.34	2.05	2.05	3.91	0.84	1.2
Resistivity, $\mu\Omega$-cm	2.8	3.32	3.2	1.7	8.7	6.4
Temperature coefficient of resistivity, 10^{-3} K^{-1}	4.46	4.03	4.03	3.93	4.0	1.0
Elastic modulus, GPa	69	69.6	69.6	115	110	110
Yield strength, MPa	28	193	310	69	140	110
Tensile strength, MPa	83	200	330	220	345	330
Specific heat capacity, J/g K	0.9	0.9	0.9	0.385	0.38	0.38
Current-carrying capacity, %	80	—	—	100	—	—

Source: Metals Handbook, (9th edition), ASM, Metals Park, Ohio, 1978.

[a]Annealed.
[b]0.8% Mg, fully cold worked (H19).
[c]0.7% Si–0.8% Mg, solution treated, cold worked, aged (temper T81).
[d]Annealed, grain size 0.050 mm (OS050).
[e]94.8% Cu–5% Sn–0.2% P, annealed, grain size 0.035 mm (OS035).
[f]70% Cu–30% Zn, annealed (OS050).

TABLE 16.130 Comparison of Copper and Aluminum Wires For Equal Resistance per Unit Length

Property	Copper	Aluminum
Cross section	1	1.61
Diameter	1	1.27
Weight	1	0.488
Breaking strength	1	0.64

TABLE 16.131   International Annealed Copper Standard

The International Electrotechnical Commission and The American Institute of Electrical Engineers have adopted the following normal values for standard annealed copper:

1. Volume resistance, at 20°C, of wire 1 m long and of uniform cross section of 1 mm^2 is $\frac{1}{58}$ Ω = 0.017241 Ω.
2. Density, at 20°C, is 8.89 g/cm^3.
3. Constant mass temperature coefficient of resistance, at 20°C, measured between two potential points rigidly fixed to wire, is 0.00393 = 1/254.45 °C^{-1}.
4. Hence mass resistance, at 20°C, of wire of uniform section 1 m in length and weighing 1 g is $\frac{1}{58}$ × 8.89 = 0.15328 Ω.

Resistivities in various units (at 20°C) are as follows:

Volume Resistivity	Mass Resistivity
0.017241 Ω (m, mm^2)	0.015328 Ω (m, g)
1.7241 $\mu\Omega$-cm	875.20 Ω (mile, lb)
0.67879 $\mu\Omega$-in.	
10.371 Ω (mil, ft)	

Source: Miner and Seastone, *Handbook of Engineering Materials*, New York, Wiley, 1955.

TABLE 16.132   Equivalent Resistivity Values

Volume Conductivity (% IACS at 68°F)	Equivalent Resistivity at 68°F	
	Ω-circular mil/ft	$\mu\Omega$-in.
52.5	19.754	1.2929
53.5	19.385	1.2687
53.8	19.277	1.2617
53.9	19.241	1.2593
54.0	19.206	1.2570
54.3	19.099	1.2501
55.0	18.856	1.2341
56.0	18.520	1.2121
56.5	18.356	1.2014
57.0	18.195	1.1908
59.0	17.578	1.1505
59.5	17.430	1.1408
61.0	17.002	1.1128
61.2	16.946	1.1091
61.3	16.918	1.1073
61.4	16.891	1.1055
61.5	16.863	1.1037
61.8	16.782	1.0983
62.0	16.727	1.0948
62.1	16.700	1.0931
62.2	16.674	1.0913
62.3	16.647	1.0896
62.4	16.620	1.0878

Source: Aluminum Standards and Data, Aluminum Association, Washington, D.C., 1984.

TABLE 16.133   Properties and Prices of Liquid Metal Conductors

Material	Melting point (°C)	Density (g/cm³ at 20°C)	Viscosity (cP at 100°C)	Resistivity (μΩ-cm at 20°C)	Price ($/kg)	Price ($/cm³)
Cs	28.5	1.878	0.47	36.6 (30°C)	495	0.95
Na	97.8	0.968	0.680	4.69	1.52	0.0015
K	63.7	0.855	0.46	13.2 (64°C)	—	—
Na–78 wt% K[a]	−12.6	0.867	0.505	38.0	—	—
Ga	29.9	6.095 (29.9°C)	1.73	25.9 (29.75°C)	630	3.8
Ga–16.5 at % In[a]	15.7	6.31	—	24.0	—	—
Ga–18.9 at % In, 10.1 at % Sn[a]	10.7	—	—	—	—	—
Hg	−38.9	13.55	1.21	95.8	10.6	0.14
Hg–34.7 at % In[a]	−37.2	—	—	47	—	—
Hg–50.0 at % In[b]	−19.3	—	—	40.6	—	—
Hg–61.5 at % In[a]	−31.0	—	—	37	—	—
In	114.8	7.30	—	29.1 (154°C)	73	0.53

Source: Encyclopedia of Materials Science and Engineering, M. B. Bever (ed.), Pergamon Press, Oxford, 1986.

[a] Eutectic composition.
[b] Intermetallic compound composition.

TABLE 16.134 Wire Table, Standard Annealed Copper[a]

Gage No. AWG	Diameter (mils), 20°C	Cross Section, 20°C		Ω/1000 ft,[b] 20°C (68°F)	lb/1000 ft	ft/lb	ft/Ω,[c] 20°C (68°F)	Ω/lb, 20°C (68°F)	lb/Ω, 20°C (68°F)
		Circular mils	in.²						
0000	460.0	211,600	0.1662	0.04901	640.5	1.561	20,400	0.00007652	13,070
000	409.6	167,800	0.1318	0.06180	507.9	1.968	16,180	0.0001217	8,219
00	364.8	133,100	0.1045	0.07793	402.8	2.482	12,830	0.0001935	5,169
0	324.9	105,500	0.08289	0.09827	319.5	3.130	10,180	0.0003076	3251
1	289.3	83,690	0.06573	0.1239	253.3	3.947	8,070	0.0004891	2,044
2	257.6	66,370	0.05213	0.1563	200.9	4.977	6,400	0.0007778	1,286
3	229.4	52,640	0.04134	0.1970	159.3	6.276	5,075	0.001237	808.6
4	204.3	41,740	0.03278	0.2485	126.4	7.914	4,025	0.001966	508.5
5	181.9	33,100	0.02600	0.3133	100.2	9.980	3,192	0.003127	319.8
6	162.0	26,250	0.02062	0.3951	79.46	12.58	2,531	0.004972	201.1
7	144.3	20,820	0.01635	0.4982	63.02	15.87	2,007	0.007905	126.5
8	128.5	16,510	0.01297	0.6282	49.98	20.01	1,592	0.01257	79.55
9	114.4	13,090	0.01028	0.7921	39.63	25.23	1,262	0.01999	50.03
10	101.9	10,380	0.008155	0.9989	31.43	31.82	1,001	0.03178	31.47
11	90.74	8,234	0.006467	1.260	24.92	40.12	794.0	0.05053	19.79
12	80.81	6,530	0.005129	1.588	19.77	50.59	629.6	0.08035	12.45
13	71.96	5,178	0.004067	2.003	15.68	63.80	499.3	0.1278	7.827
14	64.08	4,107	0.003225	2.525	12.43	80.44	396.0	0.2032	4.922
15	57.07	3,257	0.002558	3.184	9.858	101.4	314.0	0.3230	3.096
16	50.82	2,583	0.002028	4.016	7.818	127.9	249.0	0.5136	1.947
17	45.26	2,048	0.001609	5.064	6.200	161.3	197.5	0.8167	1.224
18	40.30	1,624	0.001276	6.385	4.917	203.4	156.6	1.299	0.7700
19	35.89	1,288	0.001012	8.051	3.899	256.5	124.2	2.065	0.4843
20	31.96	1,022	0.0008023	10.15	3.092	323.4	98.50	3.283	0.3046

Gage No. AWG	Diameter (mils), 20°C	Cross Section, 20°C		Ω/1000 ft,[b] 20°C (68°F)	lb/1000 ft	ft/lb	ft/Ω,[c] 20°C (68°F)	Ω/lb, 20°C (68°F)	lb/Ω, 20°C (68°F)
		Circular mils	in.²						
21	28.46	810.1	0.0006363	12.80	2.452	407.8	78.11	5.221	0.1915
22	25.35	642.4	0.0005046	16.14	1.945	514.2	61.95	8.301	0.1205
23	22.57	509.5	0.0004002	20.36	1.542	648.4	49.13	13.20	0.07576
24	20.10	404.0	0.0003173	25.67	1.223	817.7	38.96	20.99	0.04765
25	17.90	320.4	0.0002517	32.37	0.9699	1,031	30.90	33.37	0.02997
26	15.94	254.1	0.0001996	40.81	0.7692	1,300	24.50	53.06	0.01885
27	14.20	201.5	0.0001583	51.47	0.6100	1,639	19.43	84.37	0.01185
28	12.64	159.8	0.0001255	64.90	0.4837	2,067	15.41	134.2	0.007454
29	11.26	126.7	0.00009953	81.83	0.3836	2,607	12.22	213.3	0.004688
30	10.03	100.5	0.00007894	103.2	0.3042	3,287	9.691	339.2	0.002948
31	8.928	79.70	0.00006260	130.1	0.2413	4,145	7.685	539.3	0.001854
32	7.950	63.21	0.00004964	164.1	0.1913	5,227	6.095	857.6	0.001166
33	7.080	50.13	0.00003937	206.9	0.1517	6,591	4.833	1,364	0.0007333
34	6.305	39.75	0.00003122	260.9	0.1203	8,310	3.833	2,168	0.0004612
35	5.615	31.52	0.00002476	329.0	0.09542	10,480	3.040	3,448	0.0002901
36	5.000	25.000	0.00001964	414.8	0.07568	13,210	2.411	5,482	0.0001824
37	4.453	19.83	0.00001557	523.1	0.06001	16,660	1.912	8,717	0.0001147
38	3.965	15.72	0.00001235	659.6	0.04759	21,010	1.516	13,860	0.00007215
39	3.531	12.47	0.000009793	831.8	0.03774	26,500	1.202	22,040	0.00004538
40	3.145	9.888	0.000007766	1,049	0.02993	33,410	0.9534	35,040	0.00002854

[a] American Wire Gage (B.& S.). English units.
[b] Resistance at stated temperatures of wire whose length is 1000 ft at 20°C.
[c] Length at 20°C of wire whose resistance is 1 Ω at stated temperatures.

TABLE 16.135 Specifications for Copper Wire

Diameter[c] (in.)	Area at 20°C (68°F)		Hard Drawn[a]		Medium Hard Drawn[b]			Soft[c]	Tinned Soft[a]
	(cir mils)	(sq in.)	Tensile Strength (psi)	Elongation in 60 in. (%)	Minimum Tensile Strength (psi)	Maximum Tensile Strength (psi)	Elongation, min, per cent in 60 in.	Elongation in 10 in. (%)	
0.4600	211,600	0.1662	49,000	3.75[f]	42,000	49,000	3.75[f]	35	30
0.4096	167,800	0.1318	51,000	3.25[f]	43,000	50,000	3.60[f]	35	30
0.3648	133,100	0.1045	52,000	2.80[f]	44,000	51,000	3.25[f]	35	30
0.3249	105,600	0.08291	54,500	2.40[f]	45,000	52,000	3.00[f]	35	30
0.2893	83,690	0.06573	56,100	2.17[f]	46,000	53,000	2.75[f]	30	25
0.2576	66,360	0.05212	57,600	1.98[f]	47,000	54,000	2.50[f]	30	25
0.2294	52,620	0.04133	59,000	1.79[f]	48,000	55,000	2.25[f]	30	25
0.2043	41,740	0.03278	60,100	1.24	48,330	55,330	1.25	30	25
0.1819	33,090	0.02599	61,200	1.18	48,660	55,660	1.20	30	25
*0.1650	27,220	0.02138	62,000	1.14	—	—	—	—	—
0.1620	26,240	0.02061	62,100	1.14	49,000	56,000	1.15	30	25
0.1443	20,820	0.01635	63,000	1.09	49,330	56,330	1.11	30	25
*0.1340	17,960	0.01410	63,400	1.07	—	—	—	—	—
0.1285	16,510	0.01297	63,700	1.06	49,600	56,660	1.08	30	25
0.1144	13,090	0.01028	64,300	1.02	50,000	57,000	1.06	30	25
*0.1040	10,820	0.008495	64,800	1.00	—	—	—	—	—
0.1019	10,380	0.008155	64,900	1.00	50,330	57,330	1.04	25	20
*0.0920	8,460	0.00665	65,400	0.97	—	—	—	—	—
0.0907	8,230	0.00646	65,400	0.97	50,660	57,660	1.02	25	20
0.0808	6,530	0.00513	65,700	0.95	51,000	58,000	1.00	25	20
*0.0800	6,400	0.00503	65,700	0.94	—	—	—	—	—
0.0720	5,180	0.00407	65,900	0.92	51,330	58,330	0.98	25	20
*0.0650	4,220	0.00332	66,200	0.91	—	—	—	—	—
0.0641	4,110	0.00323	66,200	0.90	51,660	58,660	0.96	25	20
0.0571	3,260	0.00256	66,400	0.89	52,000	59,000	0.94	25	20

0.0508	2,580	0.00203	66,600	0.87	52,330	59,330	0.92	25	20
0.0453	2,050	0.00161	66,800	0.86	52,660	59,660	0.90	25	20
0.0403	1,620	0.00128	67,000	0.85	53,000	60,000	0.88	25	20
0.0359	1,290	0.00101	—	—	—	—	—	25	20
0.0320	1,020	0.000804	—	—	—	—	—	25	20
0.0285	812	0.000638	—	—	—	—	—	25	20
0.0253	640	0.000503	—	—	—	—	—	25	20
0.0226	511	0.000401	—	—	—	—	—	25	20
0.0201	404	0.000317	—	—	—	—	—	20	15
0.0179	320	0.000252	—	—	—	—	—	20	15
0.0159	253	0.000199	—	—	—	—	—	20	15
0.0142	202	0.000158	—	—	—	—	—	20	15
0.0126	159	0.000125	—	—	—	—	—	20	15
0.0113	128	0.000100	—	—	—	—	—	20	15
0.0100	100	0.0000785	—	—	—	—	—	15	10
0.0089	79.2	0.0000622	—	—	—	—	—	15	10
0.0080	64.0	0.0000503	—	—	—	—	—	15	10
0.0071	50.4	0.0000396	—	—	—	—	—	15	10
0.0063	39.7	0.0000312	—	—	—	—	—	15	10
0.0056	31.4	0.0000246	—	—	—	—	—	15	10
0.0050	25.0	0.0000196	—	—	—	—	—	15	10
0.0045	20.2	0.0000159	—	—	—	—	—	15	10
0.0040	16.0	0.0000126	—	—	—	—	—	15	10
0.0035	12.2	0.00000962	—	—	—	—	—	15	10
0.0031	9.61	0.00000755	—	—	—	—	—	15	10

[a] From ASTM B 1-56.
[b] From ASTM B 2-52.
[c] From ASTM B 3-63. No requirement for tensile strength specified.
[d] From ASTM B 33-63. No requirement for tensile strength specified.
[e] Diameters marked by asterisks are often specified for communication lines but are not in the American Wire Gage (B & S Wire Gage) series, as are other diameters listed.
[f] Elongation in 10 in.

TABLE 16.136   Temperature Coefficients of Copper for Different Initial Temperatures (°C) and Different Conductivities

Percentage of Conductivity[b]	Ohms (m, g) at 20°C	Factor a_{t_1}[a]					
		a_0	a_{15}	a_{20}	a_{25}	a_{30}	a_{50}
96	0.16134	0.00403	0.00380	0.00373	0.00367	0.00360	0.00336
96	0.15966	0.00408	0.00385	0.00377	0.00370	0.00364	0.00339
97	0.15802	0.00413	0.00389	0.00381	0.00374	0.00367	0.00342
97.3	0.15753	0.00414	0.00390	0.00382	0.00375	0.00368	0.00343
98	0.15640	0.00417	0.00393	0.00385	0.00378	0.00371	0.00345
99	0.15482	0.00422	0.00397	0.00389	0.00382	0.00374	0.00348
100	0.15328	0.00427	0.00401	0.00393	0.00385	0.00378	0.00352
101	0.15176	0.00431	0.00405	0.00397	0.00389	0.00382	0.00355

[a]For use in the formula:

$$R_t = R_{t_1}\left[1 + a_{t_1}(t - t_1)\right]$$

where R_t is resistance at any temperature t (°C) and R_{t_1} is resistance at initial temperature t_1.
[b]For any conductivity other than shown:

$$a_{t_1} = \left(\frac{1}{n(0.00393)} + (t_1 - 20)\right)^{-1}$$

where n is percentage of conductivity of copper (within commercial ranges) relative to that of International Annealed Copper Standard at 20°C, expressed decimally (e.g., if percentage conductivity is 90, $n = 0.90$), and t_1 is any temperature.

TABLE 16.137 Thermal Spectrum for Electrical Materials

Class of Materials	70–90°C (160–195°F)	90–120°C (195–250°F)	120–170°C (250–340°F)	170–250°C (340–480°F)	250–400°C (480–750°F)	400–650°C (750–1200°F)	650–1000°C (1200–1830°F)	1000–1500°C (1830–2730°F)
Electrical insulation	Untreated cotton, paper, silk	Oil-filled or varnished cotton, paper, silk, enamels such as Formvar, nylon	Varnished glass and mica, Mylar and other polyester films, enamels such as polyurethane, polyester, epoxy, and combinations	Polyimide, silicone, TFE-fluorocarbon, silicone-varnished glass fibers, mica, etc., resins plus ceramic, polyimide plus glass fiber	Ceramic-coated wires, glass-bonded fiberglass, glass enamel, glass-bonded mica and asbestos	Glass-bonded fibers, glass plus ceramic, glass-bonded synthetic mica, glass enamel	Glass plus refractories, crystallized glass, quartz, ceramic fibers plus glass	Pure refractory oxides, sapphire, beryllia
Conductors	Copper, aluminum	Copper, aluminum	Copper, aluminum	Nickel-plated copper, aluminum copper at 180°C	Nickel-plated copper, nickel clad copper, aluminum	Nickel-clad copper, stainless-steel-clad copper, Cufenic (nickel-iron-clad copper), nickel-clad silver	Inconel plus barrier over dispersion-strengthened copper, Inconel-clad silver	Platinum
Magnetic materials	Iron	Iron	Iron	Iron	Iron	Iron (to 500°C,) cobalt alloys	Cobalt alloys, cobalt	None available

Source: W. W. Pendleton, "Advanced Magnet Wire Systems," *Electro-Technology,* October 1963.

TABLE 16.138   Allowable Ampacities of Insulated
Copper Conductors[a]

Conductor Size, AWG or MCM	T, TW, 140°F (60°C)	RHW,[b] THW 167°F (75°C)	THHN 185°F (85–90°C)
14	15	15	25[c]
12	20	20	30[c]
10	30	30	40[c]
8	40	45	50
6	55	65	70
4	70	85	90
3	80	100	105
2	95	115	120
1	110	130	140
0	125	150	155
00	145	175	185
000	165	200	210
0000	195	230	235
250	215	255	270
300	240	285	300
350	260	310	325
400	280	335	360
500	320	380	405
600	355	420	455
700	385	460	490
750	400	475	500
800	410	490	515
900	435	520	555
1000	455	545	585
1250	495	590	645
1500	520	625	700
1750	545	650	735
2000	560	665	775

Correction Factors For Room Temperatures Over 86°F (30°C)

104°F, 40°C	0.82	0.88	0.90
113°F, 45°C	0.71	0.82	0.85
122°F, 50°C	0.58	0.75	0.80
131°F, 55°C	0.41	0.67	0.74
140°F, 60°C	—	0.58	0.67
158°F, 70°C	—	0.35	0.52
167°F, 75°C	—	—	0.43
176°F, 80°C	—	—	0.30
194°F, 90°C	—	—	—
212°F, 100°C	—	—	—
248°F, 120°C	—	—	—
284°F, 140°C	—	—	—

Source: Adapted from *National Electrical Code*, 1968 edition. Copyright National Fire Protection Association, Boston, Mass.

[a] Not more than three conductors in raceway or cable or direct burial [based on room temperature of 86°F (30°C)]. Description of insulated conductors most commonly used in general service: T, thermoplastic (permissible location, dry), TW, moisture-resistant thermoplastic (dry or wet); RHW, moisture- and heat-resistant rubber (dry or wet); THW, moisture- and heat-resistant thermoplastic (dry or wet); THHN, heat-resistant thermoplastic (dry). For other types of insulated conductors, see *National Electrical Code.*

[b] For over 2000-V service, insulation shall be ozone-resistant.

[c] Ampacities for type THHN conductors for sizes AWG 14, 12, and 10 shall be the same as designated for 167°F (75°C) conductors in this table.

TABLE 16.139   Wire Table, Aluminum[a]

Gage No.	Diameter (mils)	Cross Section Circular mils	Cross Section in.^{2}	Ω/1000 ft	lb/1000 ft	lb/Ω	ft/Ω
0000	460	212,000	0.166	0.0804	195	2420	12,400
000	410	168,000	0.132	0.101	154	1520	9,860
00	365	133,000	0.105	0.128	122	957	7,820
0	325	106,000	0.0829	0.161	97.0	602	6,200
1	289	83,700	0.0657	0.203	76.9	379	4,920
2	258	66,400	0.0521	0.256	61.0	238	3,900
3	229	52,600	0.0413	0.323	48.4	150	3,090
4	204	41,700	0.0328	0.408	38.4	94.2	2,450
5	182	33,100	0.0260	0.514	30.4	59.2	1,950
6	162	26,300	0.0206	0.648	24.1	37.2	1,540
7	144	20,800	0.0164	0.817	19.1	23.4	1,220
8	128	16,500	0.0130	1.03	15.2	14.7	970
9	114	13,100	0.0103	1.30	12.0	9.26	770
10	102	10,400	0.00815	1.64	9.55	5.83	610
11	91	8,230	0.00647	2.07	7.57	3.66	484
12	81	6,530	0.00513	2.61	6.00	2.30	384
13	72	5,180	0.00407	3.29	4.76	1.45	304
14	64	4,110	0.00323	4.14	3.78	0.911	241
15	57	3,260	0.00256	5.22	2.99	0.573	191
16	51	2,580	0.00203	6.59	2.37	0.360	152
17	45	2,050	0.00161	8.31	1.88	0.227	120
18	40	1,620	0.00128	10.5	1.49	0.143	95.5
19	36	1,290	0.00101	13.2	1.18	0.0897	75.7
20	32	1,020	0.000802	16.7	0.939	0.0564	60.0
21	28.5	810	0.000636	21.0	0.745	0.0355	47.6
22	25.3	642	0.000505	26.5	0.591	0.0223	37.8
23	22.6	509	0.000400	33.4	0.468	0.0140	29.9
24	20.1	404	0.000317	42.1	0.371	0.00882	23.7
25	17.9	320	0.000252	53.1	0.295	0.00555	18.8
26	15.9	254	0.000200	67.0	0.234	0.00349	14.9
27	14.2	202	0.000158	84.4	0.185	0.00219	11.8
28	12.6	160	0.000126	106	0.147	0.00138	9.39
29	11.3	127	0.0000995	134	0.117	0.000868	7.45
30	10.0	101	0.0000789	169	0.0924	0.000546	5.91
31	8.9	79.7	0.0000626	213	0.0733	0.000343	4.68
32	8.0	63.2	0.0000496	269	0.0581	0.000216	3.72
33	7.1	50.1	0.0000394	339	0.0461	0.000136	2.95
34	6.3	39.8	0.0000312	428	0.0365	0.0000854	2.34
35	5.6	31.5	0.0000248	540	0.0290	0.0000537	1.85
36	5.0	25.0	0.0000196	681	0.0230	0.0000338	1.47
37	4.5	19.8	0.0000156	858	0.0182	0.0000212	1.17
38	4.0	15.7	0.0000123	1,080	0.0145	0.0000134	0.924
39	3.5	12.5	0.00000979	1,360	0.0115	0.00000840	0.733
40	3.1	9.9	0.00000777	1,720	0.0091	0.00000528	0.581

[a] Hard-drawn aluminum wire at 20°C (68°F), American Wire Gage (B.& S.). English units.

TABLE 16.140 Allowable Ampacities of Insulated Aluminum Conductors[a]

Size of Conductor AWG or MCM	T, TW, 140°F (60°C)	RHW, THW, 167°F (75°C)	THHN, 185°F (85–90°C)
12	15	15	25[b]
10	25	25	30[b]
8	30	40	40[b]
6	40	50	55
4	55	65	70
3	65	75	80
2	75	90[c]	95
1	85	100[c]	110
0	100	120[c]	125
00	115	135[c]	145
000	130	155[c]	165
0000	155	180[c]	185
250	170	205	215
300	190	230	240
350	210	250	260
400	225	270	290
500	260	310	330
600	285	340	370
700	310	375	395
750	320	385	405
800	330	395	415
900	355	425	455
1000	375	445	480
1250	405	485	530
1500	435	520	580
1750	455	545	615
2000	470	560	650

Correction Factors For Room Temperatures Over 86° F (30° C)

104°F,	40°C	0.82	0.88	0.90
113°F,	45°C	0.71	0.82	0.85
122°F,	50°C	0.58	0.75	0.80
131°F,	55°C	0.41	0.67	0.74
140°F,	60°C	—	0.58	0.67
158°F,	70°C	—	0.35	0.52
167°F,	75°C	—	—	0.43
176°F,	80°C	—	—	0.30
194°F,	90°C	—	—	—
212°F,	100°C	—	—	—
248°F,	120°C	—	—	—
284°F,	140°C	—	—	—

Source: Adapted from *National Electrical Code*, 1968 edition. Copyright National Fire Protection Association. Boston, Mass.

[a] Not more than three conductors in raceway or cable or direct burial (based on room temperature of 86°F, 30°C). Description of insulated conductors most commonly used in general service: T, thermoplastic (permissible location, dry); TW, moisture-resistant thermoplastic (dry or wet); RHW, moisture- and heat-resistant rubber (dry or wet); THW, moisture- and heat-resistant thermoplastic (dry or wet); THHN, heat-resistant thermoplastic (dry). For other types of insulated conductors, see *National Electrical Code*.

[b] Ampacities of Type THHN conductors for sizes AWG 12, 10, and 8 shall be the same as designated for 167°F (75°C) conductors in this table.

[c] For three-wire single-phase service and subservice circuits, allowable ampacity of RHW aluminum conductors; No. 2, 100 A; No. 1, 110 A; No. 1/0, 125 A; No. 2/0, 150 A; No. 3/0, 170 A, and No. 4/0, 200 A.

TABLE 16.141   Specifications for Carbon Brushes[a]

	Grade A	Grade B	Grade D	Grade E
Specific resistance				
at 30°C, Ω-in.	0.00176–0.0021	0.0015–0.0019	0.0006–0.0012	0.000005–0.000025
at 250°C, Ω-in.	0.0013–0.0017	0.0011–0.0015	0.0004–0.0010	—
Contact drop, V, at	1.5 minimum,	1.8 minimum,	1.6 minimum,	Low
rated current	3.2 maximum	3.0 maximum	3.0 maximum	
Coefficient of friction				
(no current)	0.50 maximum	0.55 maximum	0.60 maximum	0.40 maximum
(carrying current)	0.35 maximum	0.40 maximum	0.45 maximum	—
Transverse strength, psi	2100 minimum	3000 minimum	1000 minimum	2500 minimum
Hardness, scleroscope	43–58	50–65	8–16	8–20
Ash, %	0.25 maximum	0.25 maximum	2.3 maximum	—
Density, lb/in.^{3}	0.054–0.058	0.058–0.065	0.042–0.050	—
Graphite, %	—	—	—	20–30

[a]Designations from U.S. Navy.

TABLE 16.142   Approximate Current-carrying Capacities of Carbon Electrodes

Nominal Diameter or Side (in.)	Area (in.^{2})	Current-carrying Capacity (A)	Current Density (A/in.^{2})
Round Electrodes			
8	50	2000–3000	40–60
10	79	3000–4800	40–60
12	113	4500–6800	40–60
14	154	5400–8500	35–55
17	227	7900–12,500	35–55
20	314	11,000–17,300	35–55
24	452	15,800–24,800	35–55
30	707	24,700–35,300	35–50
35	962	28,800–38,400	30–40
40	1257	37,700–50,200	30–40
Square or Rectangular Electrodes			
8	64	2500–3800	40–60
10	100	4000–6000	40–60
12	144	5700–8600	40–60
14	196	6800–10,800	35–55
16	256	9000–14,000	35–55
20	400	14,000–22,000	35–55
24	576	20,200–25,900	35–45
24 × 30	720	21,600–28,800	30–40

Source: Miner and Seastone, *Handbook of Engineering Materials*, New York, Wiley, 1955.

TABLE 16.143   Approximate Current-carrying Capacities of Graphite Electrodes

Nominal Diameter or Side (in.)	Area (in.²)	Current-carrying Capacity (A)	Current Density (A/in.²)
Round Electrodes			
2	3.1416	600–1000	200–320
$2\frac{1}{2}$	4.9087	800–1500	160–310
3	7.0686	1200–2100	170–300
4	12.566	1800–3000	140–240
$5\frac{1}{8}$	20.629	2300–4100	110–200
6	28.274	3100–5400	110–190
7	38.485	4200–6900	110–180
8	50.265	5500–9000	110–180
9	63.617	6400–10,800	100–170
10	78.540	7800–12,500	100–160
12	113.10	11,300–17,000	100–150
14	153.94	15,400–21,500	100–140
16	201.06	20,100–26,100	100–130
17	226.98	22,700–28,400	100–125
18	254.47	25,500–30,500	100–120
20	314.16	28,300–34,600	90–110
24	452	38,000–41,000	85–90
30	707	56,000–60,000	80–85
Square Electrodes			
2	4	650–1000	160–250
4	16	1900–3000	120–180
6	36	3600–4900	100–140
8	64	6400–8300	100–130
10	100	9000–13,000	90–130
12	144	13,000–17,300	90–120
16.35	267.32	21,400–26,700	80–100

Source: Miner and Seastone, *Handbook of Engineering Materials*, New York, Wiley, 1955.

TABLE 16.144   Characteristics of High-Resistivity Conductors

Composition (wt %)	Room Temperature Resistivity ($\mu\Omega$-cm)	Temperature Coefficient of Resistance (ppm/°C)	Thermoelectric Potential versus Cu (μV/°C)
Radio alloys			
98 Cu–2 Ni	5.0	1350 (25–105°C)	−13 (25–105°C)
78 Cu–22 Ni	30.0	160 (25–105°C)	−36 (0–75°C)
Manganins			
87 Cu–13 Mn	48.0	±15 (15–35°C)	1 (0–50°C)
83 Cu–13 Mn–4 Ni	48.0	±15 (15–35°C)	−1 (0–50°C)
Constantans			
57 Cu–13 Ni	49.0	±20 (25–105°C)	−43 (25–105°C)
55 Cu–15 Ni	50.0	±40 (20–1000°C)	−42 (0–75°C)
Nickel-chromium type (Nichromes)			
72 Ni–20 Cr–3 Al–5 Mn	135.5	±20 (−55–105°C)	−0.1 (25–105°C)
75 Ni–20 Cr–3 Al–2 (Cu, Fe, or Mn)	133.0	±20 (−55–105°C)	−0.1 (25–105°C)
Nickel-based			
76 Ni–17 Cr–4 Si–3 Mn	133.0	±20 (−55–150°C)	−1 (20–100°C)
71 Ni–29 Fe	20.0	4500 (25–105°C)	−40 (25–105°C)
94 Ni–3 Mn–2 Al–1 Si	31.5	2400 (20–100°C)	
Fe–Cr–Al			
81 Fe–15 Cr–4 Al (Alloy 750)	125.0	±50	−1.2 (0–100°C)
72.5 Fe–22 Cr–5.5 Al (Alloy 875)	145.0	±50	−2.6 (0–100°C)
Elemental			
99.8 Ni	8.0	6000 (20–35°C)	−22 (0–75°C)
99.99 Fe	9.7	6100 (20°C)	12.2 (0–100°C)
99.99 Cu	1.7	4270 (0–50°C)	0
Cr (not specified)	13.0	2900 (20–150°C)	10 (25–200°C)
99.99 Pt	10.6	3920 (0–100°C)	7.6 (0–100°C)

Source: Metals Handbook, 9th ed., Vol. 3, 1980. American Society for Metals, Metals Park, Ohio.

16.3.3   Superconducting Materials

TABLE 16.145   Properties of Candidate Materials for Conductor Development

Superconductor	T_c (K)	B_{c_2} at 4.2 K (T)	J_c at 4T and 4.2 K (GA/m^{-2})	Fabrication
Ductile Nb-based body-centered-cubic alloys (wt %)				
Nb–(25–33) Zr	11–11.5	7–8	1 ⎫	Multifilamentary wire by
Nb–(35–55) Ti	9–9.5	12	1.5 ⎭	conventional processing
A15 compounds (brittle)				
V_3Si	17.1	23	—	
V_3Ga	15.3	23	25 ⎫	Tape by diffusion;
Nb_3Sn	18.3	26	10 ⎭	multifilamentary wire
Nb_3Al	18.9	30	—	by bronze process
Nb_3Ga	20.3	33	—	
Nb_3Ge	22.5	37	—	Tape by CVD
$Nb_3(Al, Ge)$	21.0	41	—	
B1 compounds				
NbN	17.3	15	—	
Nb(CN)	17.8	12	1	Chemical deposition on carbon yarn
Laves phases, V_2Hf and alloys	9–10.5	20–26	1–5	Ductile at high temperatures
Chevrel phases, $PbMo_6S_4$	14.5	50	0.05	Difficult
RE-Ba-Cu oxides, $YBa_2Cu_3O_{7-x}$	95	—	< 0.01	Difficult, sensitive to stoichiometry, not yet commercialized

Source: Encyclopedia of Materials Science and Engineering, M. B. Bever (ed.), Pergamon Press, Oxford, 1986.

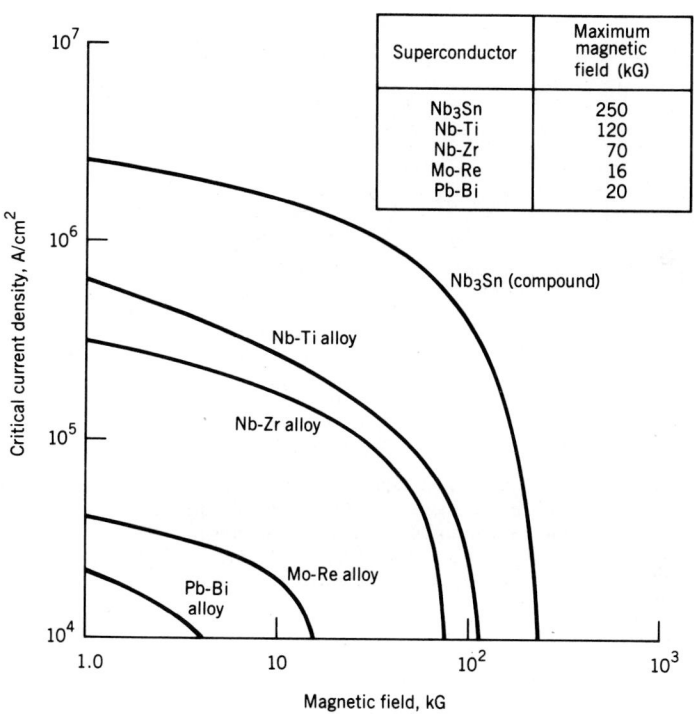

Fig. 16.15 Critical current density related to critical magnetic field for untreated superconductors in electromagnets (heat treatment increases current-carrying capacity by as much as 4 to 1). *Source:* Javitz, A. E. (ed.), *Materials Science and Technology for Design Engineers*, Hayden Book Co., New York, 1972.

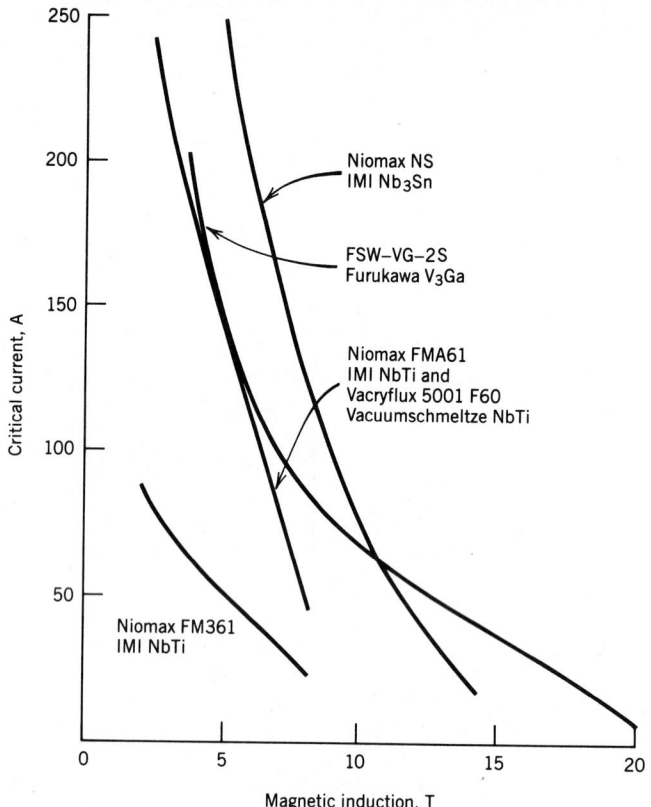

Fig. 16.16 Critical current versus magnet induction for some commercial conductors normalized to 0.4 mm diameter at 4.2 K. The 61-filament NbTi carries a higher current than the 361-filament material, but the latter is more stable and therefore can be used with a higher filling factor. Nb_3Sn is superior in performance to NbTi at all fields: V_3Ga is superior to Nb_3Sn above ~ 11 T (1 T = 10 kG). *Source: Encyclopedia of Materials Science and Engineering*, M. B. Bever (ed.), Pergamon Press, Oxford, 1986.

TABLE 16.146 Critical Currents in Amperes for Some Superconducting Wires

Transverse Magnetic Field (kG)	Niobium-Tin, Diameter, (in.)			Niobium-Titanium, Diameter, (in.)			Niobium-Zirconium, Diameter, (in.)			
	0.005	0.0075	0.010	0.013	0.018	0.024	0.005	0.0075	0.010	0.030[a]
20	140	320	560	144	260	430	13	32	54	377
60	76	174	305	58	105	166	5	11.5	20	142
110	42	96	168	6	11	16	—	—	—	—
160	15	34	59	—	—	—	—	—	—	—
200	4.5	10	18	—	—	—	—	—	—	—
230	0.6	1.4	2.5	—	—	—	—	—	—	—

Source: Javitz, A. E. (ed.), *Materials Science and Technology for Design Engineers*, Hayden Book Co., New York, 1972. Ni-Ti data from H. L. Saums, "Magnet Wire, Strip Hollow Conductors, and Superconductors," *Insulation Directory/Encyclopedia*, June/July, 1967.

[a] Seven strands of 0.010-in. round wire.

TABLE 16.147   Available Superconductive Wire and Cable[a]

Composition	Construction	Insulation
Niobium-zirconium (Nb-Zr) alloys (25 or 33% Zr)	1. Round, solid wire, copper plated	Formvar, epoxy, fused nylon
	2. Seven-stranded wire, copper-plated (interconductor spaces may be filled with indium)	Polyester film
	3. Nb-Zr strands embedded in copper matrix (interstrand space completely filled with copper; round and rectangular wires available)	Formvar (solid construction with smooth exterior permits very thin coatings needed in cryogenic applications)
Niobium-tin compound (Nb$_3$Sn)	1. Triple composite with inner core of Nb$_3$Sn and niobium sheath, plus Monel outer sheath	—
	2. Composite ribbon type, comprising vacuum-deposited, pure, single-phase crystalline Nb$_3$Sn on flexible stainless-steel ribbon substrate electroplated with silver	—
	3. Another ribbon construction with pure niobium ribbon substrate (copper plating optional)	—
Niobium-titanium (Ni–Ti) alloys (usually 48% Ti)	1. Copper-clad solid wire, metallurgically bonded (may be cabled alone or with copper strands for higher currents; indium or tin fillers optional for improved contact betweenstrands)	Formvar may be used (only one-third or one-half of cable surface may be covered to permit better access to liquid helium)
	2. Ni–48% Ti filaments embedded in matrix of OFHC copper (oxygen-free high-conductivity grade)	"Skip" insulation of nylon or polyester (to permit better access of cryogenic liquid to conductor)

Source: Javitz, A. E. (ed.), *Materials Science and Technology for Design Engineers*, Hayden Book Co., New York, 1972.

[a] Cladding or plating provides the following functions: (1) improved performance with respect to higher values of critical current and critical magnetic field; (2) establishment of contiguous heat sink of high thermal conductivity, thus providing stabilizing effect against local instabilities in conductor; (3) establishment of good surface for soft solder joints to coil leads; (4) shunt circuit capability to produce current decay when quenching occurs. Superconductor wires, whether solid, stranded, or flat-ribbon types, are available in range of sizes, depending on composition and construction. Detailed information is available from producers.

Thermal

16.3.4 Insulating Materials

TABLE 16.148 Conductivities, Conductances, and Resistances of Building and Insulating Materials[a]

Material	Description	Density (lb/ft³)	Mean Temperature (°F)	Conductivity k	Conductance C	Resistance[b] R Per inch thickness (1/k)	For thickness listed (1/C)
Building board[c]: boards, panels, flooring, sheathing, etc.	Asbestos-cement board	120	75	4.0	—	0.25	—
	Asbestos-cement board, $\frac{1}{8}$ in.	120	75	—	33.00	—	0.033
	Gypsum or plaster board, $\frac{3}{8}$ in.	50	75	—	3.10	—	0.32
	Gypsum or plaster board, $\frac{1}{2}$ in.	50	75	—	2.25	—	0.45
	Plywood	34	75	0.80	—	1.25	—
	Sheathing, wood fiber (impreg. or coated)	20	75	0.38	—	2.63	—
		22	75	0.41	—	2.44	—
		25	75	0.44	—	2.27	—
	Wood fiber board, laminated or homogeneous	26	75	0.42	—	2.38	—
		33	75	0.55	—	1.82	—
	Wood fiber, hardboard type	65	75	1.40	—	0.72	—
	Wood fiber, hardboard type, $\frac{1}{4}$ in.	65	75	—	5.60	—	0.18
	Wood subfloor, $\frac{25}{32}$ in.	—	—	—	1.02	—	0.98
	Wood, hardwood finish, $\frac{3}{4}$ in.	—	—	—	1.47	—	0.68
Building paper	Vapor, permeable felt	—	75	—	16.70	—	0.06
	Vapor, seal, two layers of mopped 15-lb felt	—	75	—	8.35	—	0.12
	Vapor, seal, plastic film	—	75	—	—	—	Negl.
Finish flooring materials	Carpet and fibrous pad	—	75	—	0.48	—	2.08
	Carpet and rubber pad	—	75	—	0.81	—	1.23
	Cork tile, $\frac{1}{8}$ in.	—	75	—	3.60	—	0.28
	Terrazzo, 1 in.	—	75	—	12.50	—	0.08
	Tile: asphalt, linoleum, vinyl, rubber	—	75	—	20.00	—	0.05
Insulating materials Blanket and batt	Cotton fiber	0.8–2.0[d]	75	0.26	—	3.85	—
	Mineral wool, fibrous form processed from rock, slag, or glass	0.5[d,e]	75	0.32	—	3.12	—
		1.5–4.0[d]	75	0.27	—	3.70	—
	Wood fiber	3.2–3.6[d]	75	0.25	—	4.00	—

	Material	Density	Temp				
Board and slabs	Cellular glass	9	90		0.41	—	2.44	—
			60		0.39	—	2.56	—
			30		0.37	—	2.70	—
			0		0.35	—	2.86	—
			−30		0.33	—	3.00	—
			−60		0.32	—	3.12	—
	Corkboard	6.5–8.0	90		0.28	—	3.57	—
			60		0.27	—	3.70	—
			30		0.26	—	3.85	—
			0		0.25	—	4.00	—
			−30		0.24	—	4.17	—
			−60		0.23	—	4.35	—
		12	90		0.31	—	3.22	—
			60		0.30	—	3.33	—
			30		0.29	—	3.45	—
			0		0.28	—	3.57	—
			−30		0.27	—	3.70	—
			−60		0.26	—	3.85	—
	Glass fiber	4–9	90		0.26	—	3.85	—
			60		0.24	—	4.17	—
			30		0.22	—	4.55	—
			0		0.21	—	4.76	—
			−30		0.19	—	5.26	—
			−60		0.18	—	5.56	—
	Expanded rubber, rigid	4.5	75		0.22	—	4.55	—
	Expanded polyurethane/ (R-11 blown), thickness 1 in. and greater	1.5–2.5	100		0.18	—	5.56	—
			75		0.17	—	5.88	—
			50		0.16	—	6.25	—
			25		0.17	—	5.88	—
			0		0.17	—	5.88	—
			−25		0.17	—	5.88	—
			−50		0.16	—	6.25	—
			−75		0.15	—	6.67	—
			−100		0.14	—	7.14	—
	Expanded polystyrene, extruded	1.9	75		0.26	—	3.85	—
			60		0.25	—	4.00	—
			30		0.24	—	4.17	—
			0		0.22	—	4.55	—
			−60		0.19	—	5.26	—

TABLE 16.148 *(Continued)*

Material	Description	Density (lb/ft³)	Mean Temperature (°F)	Conductivity k	Conductance C	Resistance,[b] R Per inch thickness (1/k)	For thickness listed (1/C)
	Expanded polystyrene, molded beads	1.0	75	0.28	—	3.57	—
			30	0.26	—	3.85	—
			0	0.24	—	4.17	—
			−60	0.21	—	4.76	—
	Mineral wool with resin binder	15	90	0.29	—	3.45	—
			60	0.28	—	3.57	—
			30	0.27	—	3.70	—
			0	0.25	—	4.00	—
			−30	0.24	—	4.17	—
			−60	0.23	—	4.35	—
	Mineral fiberboard, wet felted						
	Core or roof insulation	16–17	75	0.34	—	2.94	—
	Acoustical tile	18	75	0.35	—	2.86	—
	Acoustical tile	21	75	0.37	—	2.73	—
	Mineral fiberboard, wet molded						
	Acoustical tile[g]	23	75	0.42	—	2.38	—
	Wood or cane fiberboard						
	Acoustical tile,[g] ½ in.	—	75	—	0.84	—	1.19
	Acoustical tile,[g] ¾ in.	—	75	—	0.56	—	1.78
	Interior finish (plank, tile)	15	75	0.35	—	2.86	—
	Insulating roof deck						
	Approximately 1½ in.	—	75	—	0.24	—	4.17
	Approximately 2 in.	—	75	—	0.18	—	5.56
	Approximately 3 in.	—	75	—	0.12	—	8.33
	Wood shredded (cemented in preformed slabs)	22	75	0.60	—	1.67	—
Loose fill	Macerated paper or pulp products	2.5–3.5	75	0.28	—	3.57	—
	Mineral wool (glass, slag, or rock)	2.0–5.0	90	0.30	—	3.33	—
			60	0.27	—	3.70	—
			30	0.25	—	4.00	—
			0	0.23	—	4.35	—
			−30	0.20	—	5.00	—
			−60	0.18	—	5.56	—

Material	Density	Mean temp, °F	k		R	
Perlite (expanded)	5.0–8.0	90	0.38	—	2.63	—
		60	0.36	—	2.78	—
		30	0.34	—	2.94	—
		0	0.32	—	3.12	—
		−30	0.30	—	3.33	—
		−60	0.28	—	3.57	—
		−90	0.26	—	3.85	—
		−120	0.24	—	4.17	—
	10.0	−115	0.30	—	3.33	—
	7.5	−115	0.26	—	3.84	—
	5.0	−115	0.23	—	4.35	—
	3.0	−115	0.20	—	5.00	—
Sawdust or shavings	0.8–15	75	0.45	—	2.22	—
Silica aerogel	7.6	90	0.17	—	5.88	—
		60	0.16	—	6.25	—
		30	0.15	—	6.67	—
		0	0.15	—	6.67	—
		−30	0.14	—	7.14	—
		−60	0.13	—	7.69	—
Vermiculite (expanded)	7.0–8.2	90	0.48	—	2.08	—
		60	0.46	—	2.18	—
		30	0.44	—	2.27	—
		0	0.42	—	2.38	—
		−30	0.40	—	2.50	—
		−60	0.38	—	2.63	—
	4.0–6.0	90	0.45	—	2.22	—
		60	0.43	—	2.33	—
		30	0.40	—	2.50	—
		0	0.38	—	2.63	—
		−30	0.36	—	2.78	—
		−60	0.33	—	3.00	—
Wood fiber: redwood, hemlock, or fir	2.0–3.5	75	0.30	—	3.33	—
Wood fiber: redwood bark	3	90	0.31	—	3.22	—
	4	60	0.28	—	3.57	—
	4.5	30	0.26	—	3.85	—
		0	0.25	—	4.00	—
		−30	0.23	—	4.35	—
		−60	0.21	—	4.76	—
			0.20	—	5.00	—

TABLE 16.148 (Continued)

Material	Description	Density (lb/ft³)	Mean Temperature (°F)	Conductivity k	Conductance C	Resistance[b] R Per inch thickness (1/k)	Resistance[b] R For thickness listed (1/C)
Roof insulation[h]	Preformed, for use above deck						
	Approximately ½ in.	—	75	—	0.72	—	1.39
	Approximately 1 in.	—	75	—	0.36	—	2.78
	Approximately 1½ in.	—	75	—	0.24	—	4.17
	Approximately 2 in.	—	75	—	0.19	—	5.26
	Approximately 2½ in.	—	75	—	0.15	—	6.67
	Approximately 3 in.	—	75	—	0.12	—	8.33
	Cellular glass	—	75	0.39	—	2.56	—
Masonry-materials: concretes	Cement mortar	116	—	5.0	—	0.20	—
	Gypsum-fiber concrete, 87.5% gypsum, 12.5% wood chips	51	—	1.66	—	0.60	—
	Lightweight aggregates including expanded shale,	120	—	5.2	—	0.19	—
	clay or slate; expanded slags; cinders; pumice;	100	—	3.6	—	0.28	—
	perlite; vermiculite; also cellular concretes	80	—	2.5	—	0.40	—
		60	—	1.7	—	0.59	—
		40	—	1.15	—	0.86	—
		30	—	0.90	—	1.11	—
		20	—	0.70	—	1.43	—
	Sand and gravel or stone aggregate (oven dried)	140	—	9.0	—	0.11	—
	Sand and gravel or stone aggregate (not dried)	140	—	12.0	—	0.08	—
	Stucco	116	—	5.0	—	0.20	—
Masonry units	Brick, common[i]	120	75	5.0	—	0.20	—
	Brick, face[i]	130	75	9.0	—	0.11	—
	Clay tile, hollow:						
	1 cell deep, 3 in.	—	75	—	1.25	—	0.80
	1 cell deep, 4 in.	—	75	—	0.90	—	1.11
	2 cells deep, 6 in.	—	75	—	0.66	—	1.52
	2 cells deep, 8 in.	—	75	—	0.54	—	1.85
	2 cells deep, 10 in.	—	75	—	0.45	—	2.22
	3 cells deep, 12 in.	—	75	—	0.40	—	2.50

Concrete blocks, three oval core						
Sand and gravel aggregate						
4 in.	—	75	—	1.40	—	0.71
8 in.	—	75	—	0.90	—	1.11
12 in.	—	75	—	0.78	—	1.28
Cinder aggregate						
3 in.	—	75	—	1.16	—	0.86
4 in.	—	75	—	0.90	—	1.11
8 in.	—	75	—	0.58	—	1.72
12 in.	—	75	—	0.53	—	1.89
Lightweight aggregate (expanded shale, clay, slate or slag: pumice)						
3 in.	—	75	—	0.79	—	1.27
4 in.	—	75	—	0.67	—	1.50
8 in.	—	75	—	0.50	—	2.00
12 in.	—	75	—	0.44	—	2.27
Concrete blocks, rectangular core[j]						
Sand and gravel aggregate						
2 core, 8 in. 36 lb[k]	—	45	—	0.96	—	1.04
Same with filled cores[j]	—	45	—	0.52	—	1.93
Lightweight aggregate (expanded shale, clay, slate or slag, pumice)						
3 core, 6 in., 19 lb[k]	—	45	—	0.61	—	1.65
Same with filled cores[j]	—	45	—	0.33	—	2.99
2 core, 8 in., 24 lb[k]	—	45	—	0.46	—	2.18
Same with filled cores[j]	—	45	—	0.20	—	5.03
3 core, 12 in., 38 lb[k]	—	45	—	0.40	—	2.48
Same with filled cores[j]	—	45	—	0.17	—	5.82
Stone, lime or sand	—	75	12.50	—	0.08	—
Gypsum partition tile						
3 × 12 × 30 in. solid	—	75	—	0.79	—	1.26
3 × 12 × 30 in. 4-cell	—	75	—	0.74	—	1.35
4 × 12 × 30 in. 3-cell	—	75	—	0.60	—	1.67
Plastering materials						
Cement plaster, sand aggregate	116	—	5.0	—	0.20	—
Sand aggregate, $\frac{1}{2}$ in.	—	75	—	10.00	—	0.10
Sand aggregate, $\frac{3}{4}$ in.	—	75	—	6.66	—	0.15

TABLE 16.148 *(Continued)*

Material	Description	Density (lb/ft³)	Mean Temperature (°F)	Conductivity k	Conductance C	Resistance.[b] R Per inch thickness ($1/k$)	For thickness listed ($1/C$)
	Gypsum plaster						
	Lightweight aggregate, ½ in.	45	75	—	3.12	—	0.32
	Lightweight aggregate, ⅝ in.	45	75	—	2.67	—	0.39
	Lightweight aggregate on metal lath, ¾ in.	—	75	—	2.13	0.67	0.47
	Perlite aggregate	45	75	1.5	—	0.18	—
	Sand aggregate	105	75	5.6	—	—	0.09
	Sand aggregate, ½ in.	105	75	—	11.10	—	0.09
	Sand aggregate, ⅝ in.	105	75	—	9.10	—	0.11
	Sand aggregate, ¾ in.	105	75	—	7.70	—	0.1
	Sand aggregate on wood lath	—	75	—	2.50	—	0.40
	Vermiculite aggregate	45	75	1.7	—	0.59	—
Roofing	Asbestos-cement shingles	120	75	—	4.76	—	0.21
	Asphalt roll roofing	70	75	—	6.50	—	0.15
	Asphalt shingles	70	75	—	2.27	—	0.44
	Built-up roofing, ⅜ in.	70	75	—	3.00	—	0.33
	Slate ½ in.	—	75	—	20.00	—	0.05
	Wood shingles	—	75	—	1.06	—	0.94
Siding materials (on flat surface)	Shingles						
	Asbestos-cement	120	75	—	4.76	—	0.21
	Wood, 16 in., 7½-in. exposure	—	75	—	1.15	—	0.87
	Wood, double, 16 in., 12-in. exposure	—	75	—	0.84	—	1.19
	Wood, plus insulation backer board, 5/16 in.	—	75	—	0.71	—	1.40
	Siding						
	Asbestos-cement, ¼ in., lapped	—	75	—	4.76	—	0.21
	Asphalt roll siding	—	75	—	6.50	—	0.15
	Asphalt insulating siding (½ in. board)	—	75	—	0.69	—	1.46
	Wood, drop, 1 × 8 in.	—	75	—	1.27	—	0.79
	Wood, bevel, ½ × 8 in., lapped	—	75	—	1.23	—	0.81
	Wood, bevel, ¾ × 10 in., lapped	—	75	—	0.95	—	1.05
	Wood, plywood, ⅜ in., lapped	—	75	—	1.59	—	0.59
	Architectural glass	—	75	—	10.00	—	0.10

Woods						
Maple, oak, and similar hardwoods	45	75	1.10	—	0.91	—
Fir, pine, and similar softwoods	32	75	0.80	—	1.25	—
Fir, pine, and similar softwoods						
$\frac{25}{32}$ in.	32	75	—	1.02	—	0.98
$1\frac{5}{8}$ in.	32	75	—	0.49	—	2.03
$2\frac{5}{8}$ in.	32	75	—	0.30	—	3.28
$3\frac{5}{8}$ in.	32	75	—	0.22	—	4.55

Source: Reprinted by permission from *ASHRAE Handbook of Fundamentals,* 1967. Copyright ASHRAE.

[a] These constants are expressed in Btu per (hour) (square foot) (Fahrenheit degree temperature difference). Conductivities *k* are per inch thickness and conductances *C* are for thickness or construction stated, not per inch thickness. Representative values for dry materials selected by ASHRAE Technical Coimmittee 2.4 on insulation. They are intended as design (not specification) values for materials of building construction in normal use. For conductivity of particular product, user may obtain value supplied by manufacturer or secure results of unbiased tests.

[b] Resistance values are reciprocals of *C* before rounding off *C* to two decimal places.

[c] See also Insulating Materials, Board.

[d] Includes paper backing and facing if any. In cases where insulation forms boundary (highly reflective or otherwise) of air space, refer to Tables and to obtain insulating value of air space for appropriate effective emissivity and temperature conditions of space.

[e] Conductivity varies also with fiber diameter.

[f] These are values for aged board stock.

[g] Insulating values of acoustical tile vary depending on density of board and on type, size, and depth of perforations. Average conductivity *k* value is 0.42.

[h] U.S. Department of Commerce, *Simplified Practice Recommendation for Thermal Conductance Factors for Preferred Above Deck Roof Insulation,* No. R 257-55, recognizes specification of roof insulation on basis of *C* values shown. Roof insulation is made in thicknesses to meet these values. Therefore, thickness supplied by different manufacturers may vary depending on conductivity *k* of particular material.

[i] Face brick and common brick do not always have these specific densities. When density is different from that shown, there will be change in thermal conductivity.

[j] Data on rectangular core concrete blocks differs from data on oval core blocks due to core configuration, different mean temperatures, and possibly differences in unit weights. Weight data on oval core blocks tested are not available.

[k] Weights of units approximately $7\frac{1}{2}$ in. high and $15\frac{5}{8}$ in. long. These weights are given as means of describing blocks tested, but conductance values are all for 1 ft² of area.

[l] Vermiculite, perlite, or mineral wood insulation. Where insulation is used, vapor barriers and/or other precautions must be considered to keep insulation dry.

TABLE 16.149 Properties of High-Temperature Furnace Insulations[a]

Material	Density, lb/ft^3	Average Temperature, °F	Average Temperature, °C	Conductivity, k_B (Btu-ft/hr (ft^2) °F)	Conductivity, k_W (W-in./in^2 °C)	Resistivity, $1/k_W$ (in.^{2} °C/W in.)
Thermal insulations, fibrous						
Fiberfrax[b]	6	75	23.9	0.0189	0.000833	1205
		200	93.4	0.0250	0.00110	909
		801	427	0.0640	0.00281	356
		999	537	0.0808	0.00655	282
		1200	649	0.103	0.0045	222
		1350	732	0.121	0.00531	188
Fiberglas batts	3	68	20	0.0207	0.000912	1099
(basic form)		300	149	0.0342	0.00150	667
(TW-F)		399	204	0.0453	0.00212	472
	6	68	20	0.0191	0.000843	1190
		300	149	0.10271	0.00119	840
		399	204	0.0351	0.00154	649
	9	68	20	0.184	0.000806	1234
		300	149	0.0234	0.00103	971
		399	204	0.0287	0.00126	794
Glass wool	1.4	104	46	0.0239	0.00105	952
	2.1	100	38	0.0234	0.00103	971
		203	95	0.0310	0.00136	735
		302	150	0.0460	0.00202	495
	3.0	203	95	0.0260	0.00114	877
		401	205	0.0300	0.00132	758
		797	425	0.0476	0.00209	478
	4.8	100	38	0.0225	0.000989	1011
		203	95	0.0300	0.00132	758
		302	150	0.0392	0.00172	581
Kaolin wool	10.8	797	452	0.0592	0.00260	385
Refrasil[c]	3	199	93	0.0250	0.00110	909
		999	537	0.104	0.00455	220
		158	870	0.200	0.00880	114
	6	199	93	0.0216	0.000952	1053
		9099	537	0.0715	0.00314	318
		1598	870	0.150	0.00660	152
	9	199	93	0.0198	0.000870	1149
		999	537	0.0610	0.002868	373
		1598	870	0.111	0.00487	205
Rock wool (mineral wool)	7.0	117	47	0.0241	0.00106	943
	8–12	203	95	0.0316	0.00139	719
		401	205	0.0435	0.00191	524
		797	425	0.0651	0.00286	350

Source: Heat Transfer Databook, Genium Publishing Co., Schenectady, New York, 1987.

See also: Ceramics. Table 16.110 and Refractories, Fig. 16-14.

[a]All values shown are typical; actual values vary widely according to density, porosity, composition, and surface condition.

[b](White, fibrous, 50% SiO_2, 50% Al_2O_3, M.P. 1820C, max. temp. for use = 1370C. Fiber diam. 1 to 8 microns (5 av.), fiber length up to 3 inch, avg. length 0.5 inch.)

[c](Fibrous quartz bolts 100% SiO_2, max. temp. for use = 980C.)

TABLE 16.150 Typical Thermal Conductivity of Graphite Felt (Btu/hr ft² in. °F⁻¹)

Hot Face Temperature (°F)	Atmosphere	
	Vacuum	Inert
1000	0.5	1.0
2000	0.8	1.8
3000	1.5	3.5
4000	2.5	7.3
5000	4.5	11.5

Source: Courtesy of Ultra Carbon Corporation, Bay City, Michigan.

TABLE 16.151 Nominal Weight for Carbon and Graphite Felt

Nominal Thickness (in.)	Graphite Felt (lb/ft²)	Carbon Felt (lb/ft²)
$\frac{1}{8}$	0.048	0.052
$\frac{1}{4}$	0.10	0.11
$\frac{3}{8}$	0.17	0.18
$\frac{1}{2}$	0.20	0.21
$\frac{3}{4}$	0.31	0.32
1	0.48	0.51

Source: Courtesy of Ultra Carbon Corporation, Bay City, Michigan.

TABLE 16.152 Properties of Cryogenic Thermal Insulation

Class[a]	Descriptive name	Approximate Density (lbm/ft³)	(kg/m³)	Approximate Specific Heat (Btu/lbm °F)	(kJ/kg K)	Range of Mean Conductivities (Btu/hr ft °F)	(mW/m K)	Interspace Pressure (mm Hg)
2	Multilayer	5	80	0.22	0.92	0.000023–0.00012	0.04–0.2	10^{-4}
3	Opacified powder	7	110	0.23	0.96	0.00015–0.0004	0.26–0.7	10^{-4}
4	Evacuated powder	6	100	0.25	1.05	0.00057–0.00115	1.0–2.0	10^{-4}
5	Vacuum flask	—	—	—	—	0.0029	5.0	10^{-6}
6	Gas-filled powder	6	100	0.25	1.05	0.001–0.004	1.7–7.0	760
7	Expanded foam	2	30	0.4	1.67	0.0029–0.020	5.0–35	760
8	Fiber blanket	8	130	0.5	2.09	0.02–0.026	35–45	760

Source: Handbook of Materials Science, Vol. 1, C. T. Lynch, (ed.), CRC Press, Boca Raton, Florida, 1974.

[a] 1. Liquid and vapor shields: Very low temperature, valuable, or dangerous liquids such as helium or fluorine are often shielded by intermediate cryogenic liquid or vapor container that must in turn be insulated by one of the methods described below. 2. Multilayer reflecting shields: Foil or aluminized plastic alternated with paper-thin or plastic fiber sheets; lowest conductivity, low density, and heat storage; good stability; minimum support structure. 3. Opacified evacuated powders: Contain metallic flakes to reduce radiation; conform to irregular shapes. 4. Evacuated dielectric powders: Very fine powders of low-conductivity adsorbent; moderate vacuum requirement; minimum fire hazard in oxygen. 5. Vacuum flasks (Dewar): Tight shield-space with highly reflecting walls and high vacuum; minimum heat capacity; rugged; small thickness. 6. Gas-filled powders: Same powders as Class 4, but with air or inert gas; low cost; easy application; no vacuum requirement. 7. Expanded foams: Very light foamed plastic; inexpensive; minimum weight but bulky; self-supporting. 8. Porous fiber blankets: Blanket material of fine fibers, usually glass; minimum cost and easy installation but not an adequate insulation for most cryogenic applications.

TABLE 16.153   Examples of Insulating Materials with Maximum or Minimum Service Temperatures

Insulating Material	Service Temperature (°C)
High-temperature applications	
Alumina-silica fibers	1265 (max)
Potassium titanate fibers	1200 (max)
Diatomaceous silica	1040 (max)
Mineral fiber	1000 (max)
Perlite	875 (max)
Intermediate-temperature applications	
Hydrous calcium silicate	650 (max)
Glass fibers	540 (max)
Cellular glass	425 (max)
Cellulose fibers	95 (max)
Low-temperature applications	
Plastic foams	−140 (min)
Polystyrene	−130 (min)
Poly(vinyl chloride)	−40 (min)
Cellular rubber	−40 (min)
Cellular glass	−245 (min)
Mineral fibers	−270 (min)
Evacuated foil and fiber mats	−270 (min)

Source: Encyclopedia of Materials Science and Engineering, M. B. Bever (ed.), Pergamon Press, Oxford, 1986.

Acoustic

TABLE 16.154   Sound Absorption Coefficients[a]

Materials	125 cps	250 cps	500 cps	1000 cps	2000 cps	4000 cps
Brick, unglazed	0.03	0.03	0.03	0.04	0.05	0.07
Brick, unglazed, painted	0.01	0.01	0.02	0.02	0.02	0.03
Carpet, heavy, on concrete	0.02	0.06	0.14	0.37	0.60	0.65
Carpet, heavy, on 40 oz hairfelt or foam rubber	0.08	0.24	0.57	0.69	0.71	0.73
Carpet, heavy, with impermeable latex backing on 40 oz hairfelt or foam rubber	0.08	0.27	0.39	0.34	0.48	0.63
Concrete block, coarse	0.36	0.44	0.31	0.29	0.39	0.25
Concrete block, painted	0.10	0.05	0.06	0.07	0.09	0.08
Fabrics						
Light velour, 10 oz/yd^2, hung straight, in contact with wall	0.03	0.04	0.11	0.17	0.24	0.35
Medium velour, 14 oz/yd^2, draped to half area	0.07	0.31	0.49	0.75	0.70	0.60
Heavy velour, 18 oz/yd^2, draped to half area	0.14	0.35	0.55	0.72	0.70	0.65
Floors						
Concrete or terrazzo	0.01	0.01	0.015	0.02	0.02	0.02
Linoleum, asphalt, rubber, or cork tile on concrete	0.02	0.03	0.03	0.03	0.03	0.02
Wood	0.15	0.11	0.10	0.07	0.06	0.07
Wood parquet in asphalt on concrete	0.04	0.04	0.07	0.06	0.06	0.07
Glass						
Large panes of heavy plate glass	0.18	0.06	0.04	0.03	0.02	0.02
Ordinary window glass	0.35	0.25	0.18	0.12	0.07	0.04
Gypsum board, $\frac{1}{2}$ in. nailed to 2 × 4's 16 in. on center	0.29	0.10	0.05	0.04	0.07	0.09
Marble or glazed tile	0.01	0.01	0.01	0.01	0.02	0.02
Openings						
Stage, depending on furnishings	0.25–0.75	0.25–0.75	0.25–0.75	0.25–0.75	0.25–0.75	0.25–0.75
Deep balcony, upholstered seats	0.50–1.00	0.50–1.00	0.50–1.00	0.50–1.00	0.50–1.00	0.50–1.00
Grills, ventilating	0.15–0.50	0.15–0.50	0.15–0.50	0.15–0.50	0.15–0.50	0.15–0.50
Plaster, gypsum or lime, smooth finish on tile or brick	0.013	0.015	0.02	0.03	0.04	0.05
Plaster, gypsum or lime, rough finish on lath	0.02	0.03	0.04	0.05	0.04	0.03
Plaster, gypsum or lime, with smooth finish	0.02	0.02	0.03	0.04	0.04	0.03
Plywood paneling, $\frac{3}{8}$ in. thick	0.28	0.22	0.17	0.09	0.10	0.11
Water surface, as in a swimming pool	0.008	0.008	0.013	0.015	0.020	0.025
Air, sabins/1000 ft^3					2.3	7.2

Source: Handbook of Materials Science, Vol. 1, C. T. Lynch, (ed.), CRC Press, Boca Raton, Florida, 1974.
[a] The reverberation time, Tr, is defined as the time required for the reverberant sound intensity to decrease to one millionth of its initial intensity, or 60 dB. The time is given in seconds by the relation $T_r = 0.049 \, V/\alpha$, where V is the volume of the room in cubic feet and α is the absorption. The absorption α in sabins is computed by multiplying the area in square feet of each surface by its absorption coefficient and taking the sum of the products. The absorption of seats and audience is also computed on an area basis using the coefficients given below.

TABLE 16.155   Absorption of Seats and Audience[a]

Materials	125 cps	250 cps	500 cps	1000 cps	2000 cps	4000 cps
Audience, seated in upholstered seats[b]	0.60	0.74	0.88	0.96	0.93	0.85
Unoccupied cloth-covered upholstered seats[b]	0.49	0.66	0.80	0.88	0.82	0.70
Unoccupied leather-covered upholstered seats[b]	0.44	0.54	0.60	0.62	0.58	0.50
Wooden pews, occupied[b]	0.57	0.61	0.75	0.86	0.91	0.86
Chairs, metal or wood seats, each, unoccupied	0.15	0.19	0.22	0.39	0.38	0.30

Source: Sound Absorption Coefficients of Archietectural Acoustical Materials. Bulletin 22, Acoustical Materials Association, New York, 1962, p. 68. With permission.

[a] Values given are in sabins per per square foot of seating area or per unit.

[b] Per square foot of floor area.

Electrical

TABLE 16.156   Dielectric Constants for Ceramic Materials

Material	K'
NaCl	5.9
MgO	9.6
Al_2O_3	8.6–10.6
TiO_2	96
Porcelain	5.0
Fused SiO_2	3.8
High-lead glass	19.0
$BaTiO_3$	1600

Source: Materials and Processes, (3rd edition), J. F. Young and R. S. Shane, (eds.), Marcel Dekker, New York, 1985.

TABLE 16.157   Mica Properties and Data

Properties	Natural Muscovite Ruby Mica	Natural Phlogopite Amber Mica	Synthetic Fluorophlogopite
Specific gravity	2.8 avg	2.8 avg	2.88
Weight factor (1 lb)	10 in.^{3}	10 in.^{3}	—
Specific heat	0.207	0.207	0.194
Mohs hardness	2.8–3.2	2.5–3.0	3–3.51
Shore hardness	80–150	70–100	—
Volume resistance, Ω/cm	2×10^{13} to 1×10^{17}	Somewhat less than Muscovite	10^{12}–10^{16}
Dielectric strength (1–3 mils thick in air), volts per mil thickness[a]	Clear ruby, 3000–6000	3000–4200	3000–6000
Dielectric constant	6.5–8.7	5–6	6.5
Power factor (stained or better qualities)	0.0001–0.0004	0.004 to 0.07	0.0002
Modulus of elasticity, psi	~ 25×10^6	~ 25×10^6	~ 25×10^6
Compressive strength, psi	32,000	32,000	—
Thermal conductivity, $W/in.^2$ °C in.	0.009	—	—

Source: Insulation/Circuits Desk Manual, Lake Publishing Company, Libertyville, Illinois, 1981.

[a] The greater the thickness, the lower the dielectric strength permitted.

TABLE 16.158   Polymer Insulation Shown by Temperature Category

Temperature (°C)	Polymer Insulation
90	Cellulose acetate, cellulose acetate–butyrate, polyvinyl chloride
105	Acrylonitrile-butadiene-styrene, polystyrene, poly(vinyl formaldehyde), glycerol phthalate, methyl methacrylate, urea-formaldehyde, phenol-formaldehyde, chlorinated polyether, melamine-formaldehyde, ionomer polymer, methyl pentene olefin, polyallomer, polyamide, poly(phenylene oxide), polyarylether, polycarbonate, poly(ethylene maleate), polyethylene, polypropylene, polyurethane
130	Epoxy, fluorinated ethylene propylene, polyethylene-polytetrafluoroethylene, polyethylene-polychlorotrifluoroethylene, phenoxy, cross-linked polyethylene
155	Poly(ethylene terephthalate), poly(phenylene sulfide), poly(aryl sulfone), polysulfone
180	Poly(vinyl fluoride), poly(vinylidene fluoride)
200	Polyesterimide, polyesteramideimide, poly(methyl siloxane)
220	Polyamideimide, aromatic polyamide, diallyl phthalate, diallyl isophthalate, polychlorotrifluoroethylene
250	Polytetrafluoroethylene, polyimide

Source: Encyclopedia of Materials Science and Engineering, M. B. Bever (ed.), Pergamon Press, Oxford, 1986.

TABLE 16.159 Properties of Low-Voltage Ceramic Electrical Insulation

	Porcelain	Steatite	Forsterite	Zircon	Cordierite	Crushable Ceramics
Water absorption, %	0.2–2.5	0.08–0.00	0.00–10.00	5.0–0.00	0.0–18	15–22
Density, g/cm^3	2.2–2.4	2.5–2.9	2.8	2.8–3.8	1.7–2.4	2.0–2.4
Safe continuous operating temperature, °C	800	1000	1000–1300	1000–1200	1000–1250	1400–2000
Linear thermal expansion coefficient,[a] 10^{-6} K^{-1}	5.5	7.7–8.8	9.9–11.2	3.3–5.5	1.0–3.8	8.0–12.9
Tensile strength, MPa	10–31	58–71	70	34–104	10–55	—
Flexural strength, MPa	20–59	120–177	140	68–240	20–110	3–15
Compressive strength, MPa	207–448	500–1035	585	550–1240	80–340	13–50
Modulus of elasticity, GPa	—	100–110	145	—	60–117	—
Shear modulus, GPa	—	40–42	—	—	26–48	—
Poisson ratio	—	0.20–0.23	0.23	—	0.17–0.21	—
Thermal conductivity, W/m K						
25°C	0.9–1.8	2.8–6.0	3.0–7.9	6.5	1.6–3.2	—
500°C	—	2.4–4.0	2.2–4.2	—	1.6–2.1	—
Dielectric strength, kV/mm	3.5–8.0	8.0–9.2	8.0–9.5	0.6–8.6	2.2–8.0	1.8–2.2
Resistivity, Ω-cm						
25°C	1.8×10^{10}–10×10^{10}	$> 10^{14}$	$> 10^{14}$	1×10^9–1×10^{12}	10^{14} to $> 10^{14}$	$> 10^{14}$
100°C		2×10^{11}–1×10^{14}	5×10^{13}	1×10^5–1×10^9	3×10^6–2×10^{10}	$> 10^{14}$
300°C		2×10^4–5×10^6	1×10^6	—	7×10^5–3×10^6	1×10^6–1×10^9
Dielectric constant						
60 MHz		5.7–6.0	6.3			
1 MHz	5.0–7.5	5.4–6.3	6.4–6.2	8.0–7.5	4.0–5.4	4.3–5.8
100 MHz		5.4–5.8	6.1			
Dissipation factor						
60 Hz		0.010–0.003	0.0014	0.03		
1 MHz	0.009–0.012	0.004–0.0005	0.0010–0.0004	0.02–0.0005	0.010–0.004	0.015–0.006
100 MHz		0.003–0.0005	0.0003			
Loss factor						
60 Hz		0.090–0.020	0.009			
1 MHz	0.036–0.090	0.035–0.003	0.007–0.002	0.09–0.007	0.060–0.020	0.09–0.006
100 MHz		0.035–0.003	0.002			

Source: Encyclopedia of Materials Science and Engineering, M. B. Bever (ed.), Pergamon Press, Oxford, 1986.
[a]Room temperature to 700°C.

TABLE 16.160   Typical Physical Properties of Wet-Process, High-Voltage Ceramic Insulators

Modulus of rupture, 1.3 cm diameter bar, MPa	
Unglazed	70
Glazed	110
Flexural strength, MPa	80
Tensile strength, MPa	50
Compressive strength, MPa	390
Modulus of elasticity, GPa	80
Dielectric constant	5.75–6.00
Dielectric strength, 5 mm, kV/mm	28
Coefficient of linear expansion, 20–800°C, K^{-1}	6×10^{-6}
Resistivity, Ω-cm	
20°C	10^{14}
200°C	10^{8}
Power factor at 1 MHz	0.008
Hardness, Mohs' scale	7
Porosity, fuchsine dye penetration	0
Moisture absorption	0

Source: ASTM D-116-63.

TABLE 16.161 Typical Properties of Polymeric Films

Property[a]	Cellulose Acetate	Polystyrene (Oriented)	Polyethylene (Low Density)	Polycarbonate (Cast Film)	Polypropylene Unoriented	Polypropylene Biaxially Oriented	Fluorinated Ethylene Propylene	Ethylene Chlorotrifluoro Ethylene	Polyester	Polyether Sulfone	Polyvinylidene Fluoride	Polyimide	Polytetrafluoroethylene
Density, g/cm^3	1.27–1.31	1.05	0.91–0.925	1.20	0.89	0.89	2.15	1.68	1.39	1.37	1.76	1.42	2.2
Tensile strength, psi	7000–12,000	> 5000	> 1500	7500–9500	—	—	3000	8000	20,000–30,000	10,000–12,000	6000–6500	25,000	3300
MD	—	—	—	—	7000	25,000	—	—	—	—	—	—	—
TD	—	—	—	—	4000	23,000	—	—	—	—	—	—	2000
Break elongation, %	15–60	1.5–5.0	200–800	95–110	825	90	300	200	95–110	150	150–500	70	300 MD, 200 TD
Tear strength, g/mil	4–10	5	50–300	20–25	—	3–10	125	450	12–27	7–16	40–60	8	10–100
MD	—	—	—	—	20	—	—	—	—	—	—	—	—
TD	—	—	—	—	450	—	—	—	—	—	—	—	—
Service temperature, °C													
Maximum	90	105	100	130	130	130	130	130	150	180	180	250	250
Minimum	—	—	-45	-100	—	—	—	—	-60	—	—	—	—
Water absorption, In 24 hr, %	3.6–6.8	0.1	< 0.01	0.35	< 0.005	< 0.005	< 0.01	< 0.02	0.5–0.6	2.1	0.04	1.3	None
Volume resistivity, Ω-cm	10^{15}	1.6×10^{15}	10^{17}	4.7×10^{16}	3×10^{15}	5×10^{14}	$> 10^{18}$	10^{16}	10^{17}–10^{19}	10^{17}–10^{18}	2×10^{14}	10^{18}	$> 10^{12}$
Dielectric Constant													
60 Hz	—	—	2.2	—	—	2.2	2.0	2.4	3.3	3.5	8.4	—	2.1
1 kHz	3.6–5.0	2.5	2.2	2.99	2.2	2.2	2.0	—	2.95–3.29	3.5	8.0	3.5	2.1
1 MHz	3.2–5.0	2.5	2.2	2.93	2.1	—	2.0	—	—	3.5	6.6	3.4	2.1
Dielectric strength, V/mil	2800–3600	5000	1800	2250	—	5000	2000	5500	7000–7500	2400	1280	3600	1500
Dissipation factor													
60 Hz	—	< 0.0003	0.0003 max	—	0.0007	—	0.0002	< 0.005	0.002	—	0.049	—	0.0005
1 kHz	0.013–0.029	< 0.0003	—	0.13	—	0.0005	—	—	0.005–0.006	0.0035	0.018	0.0025	0.0003
1 MHz	0.023–0.048	< 0.0003	—	1.10	0.0002	0.0003	0.0015	0.013	—	0.006	0.170	0.010	0.0002

[a]Abbreviations: MD, machine direction; TD, transverse direction.

16.3.5   Magnetic Materials

Soft Magnetic Materials

TABLE 16.162   Silicon Contents, Densities, and Applications of Electrical Steel Sheet and Strip

AISI Type	Nominal Si + Al Content (%)	Assumed Density (Mg/m³)	Characteristics and Applications
			Lamination steel
—	0	7.85	High magnetic saturation; magnetic properties may not be guaranteed; intermittent-duty small motors
			Nonoriented Electrical Steels
M47	1.05	7.80	Ductile, good stamping properties, good permeability at high inductions; small motors, ballasts, relays
M45	1.85	7.75	Good stamping properties, good permeability at moderate
M43	2.35	7.70	and high inductions, good core loss; small generators, high-efficiency continuous-duty rotating machines, ac and dc
M36	2.65	7.70	Good permeability at low and moderate inductions, low
M27	2.80	7.70	core loss; high-reactance cores, generators, stators of high-efficiency rotating machines
M22	3.20	7.65	Excellent permeability at low inductions, lowest core loss;
M19	3.30	7.65	small-power transformers, high-efficiency rotating machines
M15	3.50	7.65	
			Oriented Electrical Steels
M6	3.15	7.65	Grain-oriented steel has highly directional magnetic
M5	3.15	7.65	properties with lowest core loss and highest permeability
M4	3.15	7.65	when flux path is parallel to rolling direction; heavier
M3	3.15	7.65	thicknesses used in power transformers, thinner thicknesses generally used in distribution transformers. Energy savings improve with lower core loss
			High-Permeability Oriented Steel
—	2.9–3.15	7.65	Low core loss at high operating inductions

Source: Metals Handbook, (9th edition), ASM, Metals Park, Ohio, 1978.

TABLE 16.163　Magnetic and Thermal Properties of Electrical Steel Sheet and Strip

AISI Type	Thickness (mm)[a]	ASTM Designation	Maximum Core Loss at 60 Hz[b]				Saturation Induction (T)	Thermal Conductivity (W/m K)
			Induction, 1.5 T		Induction, 1.7 T			
			W/lb	W/kg	W/lb	W/kg		
Fully Processed								
—	0.64	64F610	6.10	13.45	—	—	—	—
—	0.47	47F475	4.75	10.45	—	—	—	—
M-47	0.64	64F490	4.90	10.80	—	—	2.11	37.7
	0.47	47F400	4.00	10.14	—	—	2.11	37.7
M-45	0.64	64F360	3.60	7.94	—	—	2.07	25.1
	0.47	47F305	3.05	6.72	—	—	2.07	25.1
M-43	0.64	64F270	2.70	5.95	—	—	2.04	20.9
	0.47	47F230	2.30	5.07	—	—	2.04	20.9
M-36	0.64	64F240	2.40	5.29	—	—	2.02	18.8
	0.47	47F205	2.05	4.52	—	—	2.02	18.8
	0.36	36F190	1.90	4.19	—	—	2.02	18.8
M-27	0.64	64F225	2.25	4.96	—	—	2.02	18.8
	0.47	47F190	1.90	4.19	—	—	2.02	18.8
	0.36	36F180	1.80	3.97	—	—	2.02	18.8
M-22	0.64	64F218	2.18	4.81	—	—	2.00	18.8
	0.47	47F185	1.85	4.08	—	—	2.00	18.8
	0.36	36F168	1.68	3.70	—	—	2.00	18.8
M-19	0.64	64F208	2.08	4.59	—	—	1.99	16.7
	0.47	47F174	1.74	3.84	—	—	1.99	16.7
	0.36	36F158	1.58	3.48	—	—	1.99	16.7
M-15	0.47	47F168	1.68	3.70	—	—	1.98	16.7
	0.36	36F145	1.45	3.20	—	—	1.98	16.7
M-6	0.35	35G066	0.66	1.46	—	—	2.00	16.7
	0.35	35H094	—	—	0.94	2.07	2.00	16.7
M-5	0.30	356058	0.58	1.28	—	—	2.00	16.7
	0.30	30H083	—	—	0.83	1.83	2.00	16.7
M-4	0.27	27G053	0.53	1.17	—	—	2.00	16.7
	0.27	27H076	—	—	0.76	1.68	2.00	16.7
M-3	0.27	27H071[c]	—	—	0.76	1.57	2.00	16.7
—	0.35	35P076[d]	—	—	0.76	1.68	2.01	16.7
—	0.30	30P079[d]	—	—	0.70	1.54	2.01	16.7
—	0.27	27P066[d]	—	—	0.66	1.46	2.01	16.7
Semiprocessed								
M47	0.64	64S350	3.50	7.72	—	—	—	—
	0.47	47S300	3.00	6.61	—	—	—	—
M45	0.64	64S280	2.80	6.17	—	—	—	—
	0.47	47S250	2.50	5.51	—	—	—	—
M43	0.64	64S230	2.30	5.07	—	—	—	—
	0.47	47S200	2.00	4.41	—	—	—	—
M36	0.64	64S213	2.13	4.70	—	—	—	—
	0.47	47S188	1.88	4.14	—	—	—	—
M27	0.64	64S194	1.94	4.28	—	—	—	—
	0.47	47S178	1.78	3.92	—	—	—	—

Source: Metals Handbook, (9th edition), ASM, Metals Park, Ohio, 1978.

[a]0.64 mm is equivalent to 24-gage sheet; 0.47 mm to 26-gage sheet; and 0.36 mm to 29-gage sheet.

[b]Standard tests on all nonoriented sheets (M-47 to M-19 inclusive) are made on samples cut half with and half across the rolling direction; grain-oriented grades (ASTM designations G0 and H0) are tested in rolling direction only.

[c]Unofficial designation.

[d]Unofficial designations of tentative grades of high-permeability type.

TABLE 16.164 Magnetic and Physical Properties of Alloys with Moderately High Permeability at Low Field Strength and High Electrical Resistance

Alloy	Permeability Initial	Permeability Maximum	B Value at Maximum Permeability	Hysteresis Loss (erg/cm³ cycle)	Residual Induction (G)	Coercive Force (Oe)	Saturation (G)	Resistivity (μΩ-cm)	Specific Gravity
45 Permalloy	2,500	25,000	—	400	—	0.25	16,000	45	8.17
Sinimax	2,200	50,000	5,400	400	5,500	0.06	11,000	90	7.70
Thermenol	6,000	60,000	1,500	—	2,070	0.018	6,100	162	6.58
Monimax	3,000	60,000	6,200	800	8,000	0.06	14,500	80	8.27
High Permalloy 49, A-L 4750, Armco 48, Hipernik	5,000	70,000	4,500	300	10,000	0.50	16,000	48	8.25
4-79 Moly Permalloy, Hymu 80	20,000 min	90,000 min	4,000	200	4,000–5,500	0.03	7,000–7,800	58	8.74
Mumetal	20,000 min	100,000	2,000	—	2,300	0.05	6,500	60	8.58
1040 alloy	20,000 min	100,000	2,000	200	2,400	0.20	6,000	56	8.76
Supermalloy	55,000 min	300,000 min	4,000	20	4,000–5,500	0.006	6,800–7,800	65	8.77
Metglas	—	—	—	—	—	0.075	16,100	140	7.05

Source: Metals Handbook, (9th edition), ASM, Metals Park, Ohio, 1978.

TABLE 16.165 Typical Magnetic Properties of Alloys with High Permeability at Higher Field Strength Annealed for Optimum Magnetic Properties

Material	Permeability Initial	Permeability Maximum	Induction at Maximum Permeability (G)	Hysteresis Loss (erg/cm³ cycle)	Maximum Induction[a] (G)	Coercive Force (Oe)	Residual Induction (G)	Saturation Value (G)	Resistivity (μΩ-cm)
0.5% Si steel	280	3,000	8,000	2,300	10,000	0.90	—	20,500	28
Ingot iron	150	5,000	8,000	2,700	10,000	1.00	7,700	21,400	10
1.75% Si steel	280	5,000	6,000	2,100	10,000	0.80	—	20,000	37
2V Permendur, 49% Co, 2% V	800	5,000	13,000	—	15,000	1.00	10,000	23,500	40
3.0% Si steel	290	8,000	8,000	1,600	10,000	0.70	—	20,100	47
Grain-oriented 3.0% Si steel	1,400	50,000	10,000	400	15,000	0.09	12,000	20,100	50
Supermendur, 49% Co, 2% V	800	70,000	20,000	1,500	21,000	0.23	21,400	24,000	40
50% Ni iron	3,500	100,000	5,500	250	10,000	0.05	9,000	15,500	50
Grain-oriented 50% Ni iron	500	200,000	8,000	450	15,000	0.02	9,000	16,000	50

Source: Metals Handbook, (9th edition), ASM, Metals Park, Ohio, 1978.

[a]Maximum induction for hysteresis loss, coercive force, and residual induction measurements.

TABLE 16.166 Typical Magnetic Properties for Various Fe-Ni and Fe-Co Alloys

Material	Nominal Composition[a]	Typical Anneal[b]	Permeability At $B = 20$	Permeability Maximum	Saturation Induction (G)	Coercivity, H_c (Oe)	Retentivity, B_r (G)
45 Permalloy	45 Ni	1920°F	2,500	30,000	16,000	0.20	8,000
4750 alloy	47–50 Ni	H₂, 2150°F	4,000	50,000	16,000	0.06	8,000
Carpenter 49 alloy	47–50 Ni	H₂, 2050°F	4,000	50,000	16,000	0.07	8,000
Conpernik	50 Ni	H₂, 2050°F	4,000	50,000	16,000	0.07	8,000
Orthonol	50 Ni[c]	—	1,500	2,000	15,600	0.20	14,500
78 Permalloy	78 Ni	H₂, 1825°F	8,000	60,000	10,700	0.05	6,000
4-79 Moly Permalloy	79 Ni, 4 Mo	1920°F	20,000	100,000	8,700	0.03	5,000
Hymu 80	79 Ni, 4 Mo	2000°F, Q	20,000	100,000	8,700	0.05	5,000
Supermalloy	79 Ni, 5 Mo	2000°F, Q	75,000	800,000	8,000	0.006	5,000
Mumetal	77 Ni, 5 Cu, 2.75 Cr	H₂, 2375°F, Q	20,000	100,000	6,500	0.05	3,000
Permendur	50 Co	2050°F	800	5,000	24,500	2.00	14,000
2V Permendur	49 Co, 2 V	1470°F	800	8,000	24,000	1.2	14,000
Hiperco 2.7	37 Co, 0.6 Cr	1470°F	650	10,000	24,200	1.00	13,000
Supermendur	49 Co, 2 V	—	—	60,000	24,000	0.20	21,500
2-81 Moly Permalloy powder	81 Ni, 2 Mo	1200°F	125	130	—	—	—
Carbonyl iron powder	—	—	60	150	—	—	—

Source: Metals Handbook, (9th edition), ASM, Metals Park, Ohio, 1978.
[a] Remainder iron plus deoxidizer.
[b] H₂, annealed in hydrogen; Q, quenched or controlled cooled.
[c] Grain oriented.

TABLE 16.167 Direct-Current Magnetic Properties of Relay Steels and Alloys after Annealing

Metal	Magnetizing Force, H_{max} from B = 10,000	Coercive Force[a] H_c (Oe)	B_r (kG) from B = 10,000	Permeability, μ Initial	Permeability, μ Maximum	Flux Density at Maximum Permeability (kG)	Saturation Induction (kG)	Resistivity ($\mu\Omega$-cm)
Low-Carbon Iron and Steel								
Low-carbon iron	2.0–5.8	0.80–1.70	7.0–8.5	200[b]	2,200–5,500	6.4–7.5	21.5	10
1010 steel	3.0	1.00–2.00	8.4	500–1000[c]	3,800	7.5	21.0	12
Silicon Steels								
1% Si	1.7–3.2	0.40–0.80	2.8–8.5	400,[b] 650[c]	1,700–6,000	5.0–9.0	20.6	23
2.5% Si	0.8–1.90	0.13–0.70	2.3–6.7	900[b]	1,800–11,000	4.0–8.0	20.0	41
3% Si	1.30–2.00	0.23–0.65	3.9–7.6	550[b]	7,500–10,000	4.7–8.0	18.4–20.0	48
3% Si, grain oriented	0.22	0.05–0.10	6.6–8.6	2,500	55,000–60,000	8.0–10.0	19.7	48
Stainless Steels								
Type 430[d]	—	3.00–4.00	8.0–9.5	230[c]	1,100–1,600	6.0–8.2	14.7	60
Type 416[e]	—	4.00–6.00	9.0–11.0	200[c]	800–1,000	9.0–10.5	15.0	57
Type 410[e]	—	4.50–7.50	8.0–12.0	110–180[c]	800–1,000	6.0–7.0	16.0	57
Type 443[f]	—	6.50–7.50	3.2–3.8	60[c]	450–550	2.5–3.5	12.0	68
Type 446[g]	—	3.00–4.00	4.5–5.0	100[c]	800–900	5.7–5.9	12.0	61
Nickel Irons								
50% Ni[h]	0.20–0.50	0.06–0.14	5.5–8.0	2,500–3500[b]	30,000–120,000	5.0–7.0	16.0	48
78% Ni	0.20	0.05	8.0	8,000[b]	100,000	5.0	10.7	16
77% Ni (Cu, Cr)[i]	0.10	—	4.0	20,000	150,000	2.0	7.5	60
79% Ni (Mo)[j]	—	—	3.2–3.5	—	70,000–75,000	3.2–4.0	8.0	58

Source: Metals Handbook, (9th edition), ASM, Metals Park, Ohio, 1978.

[a] From B = 10,000.
[b] At B = 20.
[c] At B = 200.
[d] Coercive force from saturation is 3.50 to 4.50 oersteds.
[e] Coercive force from saturation is 5.00 to 7.00 oersteds.
[f] Coercive force from saturation is 7.30 oersteds.
[g] Coercive force from saturation is 3.50 to 5.00 oersteds.
[h] Coercive force from saturation is 0.15 oersteds.
[i] Coercive force from B = 5000 is 0.02 oersteds.
[j] For 79% Ni iron with molybdenum, coercive force from B = 5000 is 0.010 to 0.12 oersteds.

TABLE 16.168 Recommended Materials for Motors and Generators[a]

Type of Motor	Material Thickness		Material
	mm	in.	
Starting Motors			
Automotive	0.35	0.014	1008
	0.46	0.0185	1008
	0.63	0.025	1008
Medium, 1–100 hp	0.35	0.014	M-36
	0.46	0.0185	M-43
	0.63	0.025	1008
Large, 100 hp minimum	0.35	0.014	M-19
	0.46	0.0185	M-36, M-27
	0.63	0.025	M-43, M-36
Motors and Generators for Intermittent Operation[b]			
Miniature	0.46	0.0185	M-50, M-43
	0.63	0.025	1008, M-50
Gyros	0.35	0.014	M-15
	0.46	0.0185	M-15
Selsyns	0.35	0.014	M-15, 45–50% Ni iron
Fractional, $\frac{1}{4}$ hp	0.63	0.025	1008
Fractional, $\frac{1}{2}$ hp	0.63	0.025	M-43
Fractional, $\frac{3}{4}$ hp	0.63	0.025	M-36
Medium and large	0.46	0.0185	M-43, M-36, M-27
	0.63	0.025	M-43, M-36, M-27
Motors and Generators for Continuous Operation			
Fractional, $\frac{1}{4}$ hp	0.63	0.025	1008
Fractional, $\frac{1}{2}$ hp	0.63	0.025	M-43
Fractional, $\frac{3}{4}$ hp	0.63	0.025	M-36
Medium, 1–100 hp	0.35	0.014	M-22, M-19
	0.46	0.0185	M-36, M-27
Large, 100–5000 hp	0.35	0.014	M-19, M-15
	0.46	0.0185	M-27, M-19, M-15
	0.63	0.025	M-27, M-19
Large, > 5000 hp	0.35	0.014	M-15, M-6
	0.46	0.0185	M-19, M-15

Frequency	Thickness		Material
	μm	mils	
High-Frequency Motors			
To 400 cycles	180	7	3% Si steel[c]
	380	15	M-19, M-15
800–1200 cycles	125	5	3% Si steel[c]
	180	7	3% Si steel[c]
Servomotors	125	5	3% Si steel[c]
Synchronous motors	100–355	4 to 14	45–50% Ni iron

Source: Metals Handbook, (9th edition), ASM, Metals Park, Ohio, 1978.
[a]Where more than one grade of silicon sheet is shown, they are listed in order of increasing cost and efficiency as result of electrical properties.
[b]1008 steel is used in 0.76 mm (0.030 in.) for all applications in this category.
[c]Cold rolled, nonoriented.

TABLE 16.169   Recommended Materials for Transformers

Type	Material thickness		Material
	mm	in.	
	Continuous Duty[a]		
Distribution	0.27	0.011	M-3, M-4
	0.30	0.012	M-5
	0.35	0.014	M-6
Power	0.30	0.012	M-5
	0.35	0.014	M-6
Voltage regulator	0.30	0.012	M-5
	0.35	0.014	M-15
	0.63	0.025	M-22
Welding transformer	0.30	0.012	M-5
	0.35	0.014	M-6
	0.63	0.025	M-43, M-36, M-27

Application	Standard Electrical Steels	Other Alloys
	Special Application Transformers	
Instrument	M-15, M-6, 0.30–0.63 mm (0.012–0.025 in.)	Vanadium Permendur, 70–80% Ni iron, 45–50% Ni iron
Radio, power	M-27, M-22, M-19, 0.30–0.35 mm (0.012–0.014 in.)	Vanadium Permendur, 70–80% Ni iron
Radio, audio	M-19, M-17, M-15, M-6, M-5, 0.35–0.46 mm (0.014–0.0185 in.)	45–50% Ni iron
Radar pulse transformers	25–100 μm (1–4 mils): oriented 3% Si steel; 125–180 μm (5–7 mils): nonoriented 3% Si steel	Oriented 45–50% Ni iron, 4-79 Moly Permalloy, Supermalloy, Supermendur, Monimax, Sinimax, 13–100 μm (0.5–4 mils)
Chokes, power	M-22, M-19, M-15, M-6	—
Chokes, radio	—	Carbonyl irons, ferrites
Ballasts	M-27, M-22, 0.46–0.63 mm (0.0185–0.025 in.)	—
Miscellaneous bell-ringing and toy	1008	—

Source: Metals Handbook, (9th edition), ASM, Metals Park, Ohio, 1974.

[a]For core laminations for welding transformers, M-27 and M-22 are recommended in 0.46-mm (0.0185-in.) sheet.

Permanent Magnetic Materials

TABLE 16.170   Nominal Compositions, Curie Temperatures, and Magnetic Orientations of Selected Permanent Magnet Materials

Designation	Nominal Composition	Approximate Curie Temperature °C	Approximate Curie Temperature °F	Magnetic Orientation[a]
3.5% Cr steel	Fe–3.5 Cr–1 C	745	1370	No
6% W steel	Fe–6 W–0.5 Cr–0.7 C	760	1400	No
17% Co steel	Fe–17 Co–8.25 W–2.5 Cr–0.7 C	—	—	No
36% Co steel	Fe–36 Co–3.75 W–5.75 Cr–0.8 C	890	1630	No
Cast Alnico 1	Fe–12 Al–21 Ni–5 Co–3 Cu	780	1440	No
Cast Alnico 2	Fe–10 Al–19 Ni–13 Co–3 Cu	810	1490	No
Cast Alnico 3	Fe–12 Al–25 Ni–3 Cu	760	1400	No
Cast Alnico 4	Fe–12 Al–27 Ni–5 Co	800	1475	No
Cast Alnico 5	Fe–8.5 Al–14.5 Ni–24 Co–3 Cu	900	1650	Y, H
Cast Alnico 5DG	Fe–8.5 Al–14.5 Ni–24 Co–3 Cu	900	1650	Y, H, C
Cast Alnico 5–7	Fe–8.5 Al–14.5–Ni–24Co–3Cu	900	1650	Y, H, C
Cast Alnico 6	Fe–8 Al–16 Ni–24 Co–3 Cu–2 Ti	860	1580	Y, H
Cast Alnico 7	Fe–8 Al–18 Ni–24 Co–4 Cu–5 Ti	840	1540	Y, H
Cast Alnico 8	Fe–7 Al–15 Ni–35 Co–4 Cu–5 Ti	860	1580	Y, H
Cast Alnico 9	Fe–7 Al–15 Ni–35 Co–4 Cu–5 Ti	—	—	Y, H, C
Cast Alnico 12	Fe–6 Al–18 Ni–35 Co–8 Ti	—	—	No
Sintered Alnico 2	Fe–10 Al–17 Ni–12.5 Co–6 Cu	810	1490	No
Sintered Alnico 4	Fe–12 Al–28 Ni–5 Co	800	1475	No
Sintered Alnico 5	Fe–8.5Al–14.5 Ni–24 Co–3 Cu	900	1650	Y, H
Sintered Alnico 6	Fe–8 Al–16 Ni–24 Co–3 Cu–2 Ti	860	1580	Y, H
Sintered Alnico 8	Fe–7 Al–15 Ni–35 Co–4 Cu–5 Ti	860	1580	Y, H
Bonded ferrite A	$BaO–6\ Fe_2O_3$ + organics	450	840	No, P
Bonded ferrite B	$BaO–6\ Fe_2O_3$ + organics	450	840	No
Sintered ferrite 1	$BaO–6\ Fe_2O_3$	450	840	No, P
Sintered ferrite 2	$BaO–6\ Fe_2O_3$	450	840	Y, A
Sintered ferrite 3	$BaO–6\ Fe_2O_3$	450	840	Y, A
Sintered ferrite 4	$SrO–6\ Fe_2O_3$	460	860	Yes
Sintered ferrite 5	$SrO–6\ Fe_2O_3$	460	860	Yes
Lodex 30	9.9 Fe–5.5 Co–77.0 Pb–8.6 Sb	980	1800	Y, A
Lodex 31	16.0 Fe–9.0 Co–67.5 Pb–7.5 Sb	980	1800	Y, A
Lodex 32	19.2 Fe–10.8 Co–63.0 Pb–7.0 Sb	980	1800	Y, A
Lodex 33	21.9 Fe–12.3 Co–59.2 Pb–6.6 Sb	980	1800	Y, A
Lodex 36	9.9 Fe–5.5 Co–77 Pb–8.6 Sb	980	1800	No, E
Lodex 37	16 Fe–9 Co–67.5 Pb–7.5 Sb	980	1800	No, E
Lodex 38	19.2 Fe–10.8 Co–63 Pb–7.0 Sb	980	1800	No, E
Lodex 40	9.9 Fe–5.5 Co–77 Pb–8.6 Sb	980	1800	No, P
Lodex 41	16 Fe–9 Co–67.5 Pb–7.5 Sb	980	1800	No, P
Lodex 42	19.2 Fe–10.8 Co–63.0 Pb–7.0 Sb	980	1800	No, P
Lodex 43	21.9 Fe–12.3 Co–59.2 Pb–6.6 Sb	980	1800	No, P
P-6 alloy	45 Fe–45 Co–6 Ni–4 V	—	—	No
Cunife	20 Fe–20 Ni–60 Cu	410	770	Y, R
Cunico	29 Co–21 Ni–50 Cu	860	1580	No
Vicalloy I	39 Fe–51 Co–10 V	855	1570	No
Vicalloy II	35 Fe–52 Co–13 V	855	1570	Y, R
Remalloy 1	17 Mo–12 Co–71 Fe	900	1650	No
Remalloy 2	20 Mo–12 Co–68 Fe	900	1650	No

TABLE 16.170   (*Continued*)

Designation	Nominal Composition	Approximate Curie Temperature °C	°F	Magnetic Orientation[a]
Platinum cobalt	76.7 Pt–23.3 Co	480	900	No
Cobalt–rare earth 1	Co_5Sm	725	1340	Y, A
Cobalt–rare earth 2	Co_5Sm	725	1340	Y, A
Cobalt–rare earth 3	Co_5Sm	725	1340	Y, A
Cobalt–rare earth 4	$(Co, Cu, Fe)_7Sm_2$	—	—	Y, A

Source: Metals Handbook, (9th edition), ASM, Metals Park, Ohio, 1978.

[a]Abbreviations: Y, yes; H, orientation developed during heat treatment; C, columnar crystal structure developed; P or E, some orientation developed during pressing or extrusion; R, orientation developed by rolling or other mechanical working; A, orientation developed predominantly by magnetic alignment of powder prior to compacting but alignment influenced by pressing forces also.

TABLE 16.171 Nominal Magnetic Properties of Selected Permanent Magnetic Materials[a]

Designation	H_c (Oe)	H_{cf} (Oe)	B_r (G)	B_{is} (G)	$(BH)_{max}$ (Mg·Oe)	B_d (G)	H_d (Oe)	Required Magnetizing field (Oe)	Permeance Coefficient at $(BH)_{max}$	Average Recoil Permeability
3.5% Cr steel	66	—	9,500	—	0.29	—	—	—	—	—
6% W steel	74	—	9,500	—	0.33	—	—	—	—	—
17% Co steel	170	—	9,500	—	0.65	—	—	—	—	—
36% Co steel	240	—	9,750	—	0.93	—	—	—	—	—
Cast Alnico 1	440	455	7,100	10,500	1.4	4,500	305	2,000	14	6.8
Cast Alnico 2	550	580	7,250	10,900	1.6	4,500	350	2,500	12	6.4
Cast Alnico 3	470	485	7,000	10,000	1.4	4,300	320	2,500	13	6.5
Cast Alnico 4	730	770	5,350	8,600	1.3	3,000	420	3,000	8.0	4.1
Cast Alnico 5	620	625	12,500	13,500	5.25	10,200	525	3,000	18	4.3
Cast Alnico 5DG	650	655	12,900	14,000	6.1	10,500	580	3,500	17	4.0
Cast Alnico 5–7	730	735	13,200	14,000	7.4	11,500	640	3,500	17	3.8
Cast Alnico 6	750	—	10,500	13,000	3.7	7,100	525	4,000	13	5.3
Cast Alnico 7	1,050	—	8,570	9,450	3.7	—	—	5,000	8.2	—
Cast Alnico 8	1,600	1,720	8,300	10,500	5.0	5,060	950	8,000	5.0	3.0
Cast Alnico 9	1,450	—	10,500	—	8.5	—	—	7,000	7.0	—
Cast Alnico 12	950	—	6,000	—	1.7	3,150	540	5,000	5.6	—
Sintered Alnico 2	525	545	6,700	11,000	1.5	4,300	345	2,500	12	6.4
Sintered Alnico 4	700	760	5,200	—	1.2	3,000	400	3,500	—	7.5
Sintered Alnico 5	600	605	10,400	12,050	3.60	7,850	465	3,000	18	4.0
Sintered Alnico 6	760	790	8,800	11,500	2.75	5,500	500	4,000	12	4.5
Sintered Alnico 8	1,550	1,675	7,600	9,400	4.5	4,600	1,000	8,000	5.0	2.1
Bonded ferrite A	1,940	—	2,140	—	1.0	1,160	—	12,000	1.3	1.1
Bonded ferrite B	1,150	—	1,400	—	0.4	—	—	8,000	1.2	1.1
Sintered ferrite 1	1,800	3,450	2,200	—	1.0	1,100	900	10,000	1.2	1.2
Sintered ferrite 2	2,200	2,300	3,800	—	3.4	1,850	1,650	10,000	1.1	1.1
Sintered ferrite 3	3,000	3,650	3,200	—	2.5	1,600	1,600	10,000	1.1	1.1
Sintered ferrite 4	2,200	2,300	4,000	—	3.7	2,150	1,700	12,000	1.2	1.05
Sintered ferrite 5	3,150	3,590	3,550	—	3.0	1,730	1,730	15,000	1.0	1.05

Lodex 30	1,250	1,470	4,000	4,400	1.6	2,200	750	6,000	3.4	1.5
Lodex 31	1,140	1,180	6,300	7,000	3.4	4,400	770	6,000	5.3	1.9
Lodex 32	940	960	7,350	8,300	3.5	5,400	650	5,000	8.2	2.6
Lodex 33	865	875	8,000	9,200	3.2	5,850	545	5,000	10.5	3.0
Lodex 36	1,210	1,380	3,500	4,400	1.5	1,850	800	5,000	2.0	2.0
Lodex 37	1,000	1,080	5,450	7,000	2.1	3,150	670	5,000	5.8	3.0
Lodex 38	850	890	6,200	8,300	2.2	3,700	600	5,000	7.0	3.5
Lodex 40	1,100	1,400	2,700	4,400	0.8	1,400	600	5,000	2.0	2.5
Lodex 41	990	1,100	4,350	7,000	1.4	2,400	600	5,000	3.8	3.2
Lodex 42	845	920	5,300	8,300	1.4	2,750	510	5,000	7.6	3.5
Lodex 43	710	750	6,000	9,200	1.3	3,300	400	5,000	10	3.8
P-6 alloy	58	—	14,000	19,000	0.5	10,500	48	300	220	23
Cunife	550	555	5,400	5,900	1.5	4,000	325	2,500	12	3.7
Cunico	680	750	3,400	4,500	0.8	1,950	390	3,000	5.0	3.2
Vicalloy I	240	242	8,400	12,900	0.9	5,600	160	1,000	—	—
Vicalloy II	415	420	9,050	—	2.3	7,000	325	2,000	—	—
Remalloy I	250	—	9,700	14,200	1.0	6,100	155	1,000	40	13
Remalloy II	340	345	8,550	—	1.2	5,400	220	2,000	—	—
Platinum cobalt	4,450	5,400	6,450	—	9.2	3,500	2,700	20,000	1.2	1.2
Cobalt–rare earth 1	9,000	20,000	9,200	9,800	21	—	—	30,000	—	—
Cobalt–rare earth 2	8,000	>25,000	8,600	—	18	4,400	4,100	30,000	—	1.05
Cobalt–rare earth 3	6,700	>15,000	8,000	—	15	4,000	3,700	30,000	—	1.1
Cobalt–rare earth 4	5,700	6,500	9,400	—	21	4,600	4,600	>15,000	—	—

Source: Metals Handbook, (9th edition), ASM, Metals Park, Ohio, 1978.

[a]For nominal compositions see Table 16.170.

Soldering and Brazing

16.3.6 Joining Materials

TABLE 16.172 Properties of Lead-Tin Solder Alloys[a]

ASTM Alloy Grade	Nominal Composition (%)			Specific Gravity	Melting Range				Uses
	Tin	Lead	Antimony		Solidus		Liquidus		
					°C	°F	°C	°F	
Tin-Lead Alloys									
70A	70	30	—	8.32	183	361	192	378	For coating metals.
63A	63	37	—	8.40	183	361	183	361	As lowest melting (eutectic) solder for both by dip and by hand soldering methods
60A	60	40	—	8.65	183	361	190	374	"Fine solder"; for general purposes, but particularly where temperature requirements critical.
50A	50	50	—	8.85	183	361	216	421	For general purposes: Most popular of all
45A	45	55	—	8.97	183	361	227	441	For automobile radiator cores and roofing seams
40A	40	60	—	9.30	183	361	238	460	Wiping solder for joining lead pipes and cable sheaths; for automobile radiator cores and heating units
35A	35	65	—	9.50	183	361	247	477	General purpose and wiping solder
30A	30	70	—	9.70	183	361	255	491	For machine and torch soldering
25A	25	75	—	10.00	183	361	266	511	For machine and torch soldering
20B	20	80	—	10.20	183	361	277	531	For coating and joining metals; for filling dents or seams in automobile bodies
15B	15	85	—	10.50	227[f]	440[b]	288	550	For coating and joining metals
10B	10	90	—	10.80	268[f]	514[b]	299	570	For coating and joining metals
5A	5	95	—	11.30	270	518	312	594	For coating and joining metals

	Tin	Lead	Silver						Uses
Tin-Lead-Antimony Alloys									
40C	40	58	2	9.23	185	365	231	448	Same uses as (50–50) tin-lead but not recommended for use on galvanized iron
35C	35	62.3	1.8	9.44	185	365	243	470	For wiping and all uses except on galvanized iron
30C	30	68.4	1.6	9.65	185	364	250	482	For torch soldering or machine soldering, except on galvanized iron
25C	25	73.7	1.3	9.96	184	364	263	504	For torch and machine soldering, except on galvanized iron
20C	20	79	1	10.17	184	363	270	517	For machine soldering, and coating of metals, tipping, and like uses but not recommended for use on galvanized iron.
Tin-Antimony Alloys									
95TA	95	—	5	7.25	234	452	240	464	For joints on copper: electrical plumbing and heating
Silver-Lead Alloys									
2.5S	0	95.7	2.5	11.35	304	579	304	579	For use on copper, brass, and similar metals with torch heating; not recommended in humid environments due to its known susceptibility to corrosion
1.5S	1	97.5	1.5	11.28	309	588	309	588	For use on copper, brass, and similar metals with torch heating.

Source: From ASTM B 32-66T.

^{a}Federal specifications similar to, and also those for, ASTM alloy grades 70B, 63B, 60B, 50B, and 40B contain antimony in the range 0.20–0.50% in addition to a permissible limit of 0.25% bismuth. Such formulation is intended to provide for reliability of soldered joints below 32°F since there is possibility of failure of joints resulting from phase change of tin constituent of solder. Change from beta tin to alpha tin is accompanied by volume increase of 26%, which may be manifested in a powdery structural disintegration. It is recommended that the grade of solder metal be selected which contains least amount of tin required to give suitable flowing and adhesive qualities for work in hand. To provide for reliability of soldered connections closely adjacent to other electrical conductors in compact equipment, the following precaution is suggested in order to reduce troubles developing in the nature of leaky or short circuits: Possibility of growth of "tin whiskers" will be reduced when tin-lead solders containing 50% tin or less are used. When soldering silver film surfaces (preferably 0.0005 in. minimum thickness) grade A solders containing 60–70% tin with addition of 1–3% silver are suggested for use. Object is to avoid detachment of silver film by its solution in or migration into solder. Valuable information on use of solders and fluxes may be found in ASTM Special Technical Publication No. 189, "Symposium on Solder," June 1956. Alloys are completely solid below designated "solidus," and completely liquid only above designated "liquidus." In range of temperature between these two points alloys are partly solid and partly liquid. In 60• tin, 40% lead alloy amount of solid portion is so small in range given that it is practically unnoticeable. In 40% tin, 60% lead alloy proportion of solid and fluid metal in range given makes this alloy suitable for use as wiping solder.

^{b}For some engineering design purposes it is well to consider these alloys as having practically no mechanical strength at 183°C (361°F).

TABLE 16.173   Relative Solderability of Various Types of Copper Metals

Type of Copper Metal	Solderability and Remarks
Coppers[a]	Excellent; need only rosin or other noncorrosive flux
Copper-tin alloys	Good; easily soldered with activated rosin and intermediate fluxes
Copper-zinc alloys	Good; easily soldered with activated rosin and intermediate fluxes
Copper-nickel alloys	Good; easily soldered with intermediate and corrosive-type fluxes
Chromium copper and beryllium copper	Good; require intermediate and corrosive-type fluxes
Copper-silicon alloys	Fair; silicon produces refractory oxides that require use of corrosive fluxes
Copper-aluminum alloys	Difficult; may be soldered with help of very corrosive fluxes
High-strength manganese bronze	Not recommended; should be plated to ensure consistent solderability

Source: Metals Handbook, (9th edition), ASM, Metals Park, Ohio, 1978.
[a]Includes tough-pitch, oxygen-free, phosphorized, arsenical, silver-bearing, leaded, tellurium, and selenium coppers.

Welding

TABLE 16.174   Nominal Values of Physical Properties Important in Welding

Property	Aluminum Alloys			Low-Carbon Steel	Copper C11000
	1100-O	5454-H32	6061-T6		
Thermal conductivity					
W/m K	223	134	167	52	388
Btu/ft hr °F	129	77.4	96.5	30	224
Specific heat (volumetric)					
MJ/m^3 K	2.45	2.41	2.60	3.99	3.38
Btu/in.^{3} °F	0.0212	0.0209	0.0225	0.034	0.0295
Heat of fusion (volumetric)					
GJ/m^3	1.05	—	—	2.14	1.82
Btu/in.^{3}	16.4	—	—	33.1	10.6
Coefficient of linear thermal expansion[a]					
μm/m K	23.6	23.6	23.6	12.6	17.0
μin./in. °F	13.1	13.1	13.1	7.0	9.4
Melting temperature					
°C	645–655	600–645	580–650	1450–1520	1065–1083
°F	1190–1215	1115–1195	1080–1205	2640–2770	1949–1981
Electrical conductivity, % IACS	59	34	43	10	100
Density					
Mg/m^3	2.71	2.68	2.70	7.85	8.89
lb/in.^{3}	0.098	0.097	0.098	0.283	0.321

Source: Metals Handbook, (9th edition), ASM, Metals Park, Ohio, 1978.
[a]At 20–100°C (68–392°F).

TABLE 16.175　Weldability of Aluminum Alloys by Gas Metal–Arc and Gas Tungsten–Arc Processes

Readily Weldable

Wrought alloys
　Unalloyed aluminum, 1060, 1100, 1350 2219
　3003, 3004, 3105
　5005, 5050, 5052, 5056, 5083, 5086,
　5154, 5252, 5254, 5454, 5456, 5457, 5652, 5657
　6061, 6063, 6070, 6101, 6201, 6262, 6463, 7005
Casting alloys
　328.0, 355.0, C355.0, 356.0, A356.0, 357.0, A357.0, 359.0
　443.0, A443.0, B443.0

Weldable in Most Applications[a]

Wrought alloys: 2014, 4032, 6066
Casting alloys
　208.0, 308.0, 319.0, 332.0
　413.0, 712.0

Limited Weldability[b]

Wrought alloys: 2024, 2218, 2618
Casting alloys
　213.0, 222.0, 295.0, 296.0
　333.0, 336.0, 354.0
　512.0, 513.0, 514.0
　Die Casting alloys

Welding Not Recommended

Wrought alloys: 2011, 7075, 7178
Casting alloys
　242.0, 520.0, 535.0
　705.0, 707.0, 710.0, 711.0, 713.0, 771.0

Source: *Metals Handbook*, (9th edition), ASM, Metals Park, Ohio, 1978.
[a] May require special techniques for some applications.
[b] Require special techniques.

TABLE 16.176 Recommended Electrodes for Shielded Metal Arc Welding of Carbon and Low-Alloy ASTM Steels

ASTM Specification	Description	Grades	Recommended Electrodes[a]
Steel Plates, Sheets, Forgings, Shapes, and Castings			
A36-74	Structural, 36-000 psi YS minimum	—	[a]Note (2)
A113-70a	Railway rolling stock	All	[a]Note (2)
A131-74	Ship structurals	A, B, C, CS, D, E	[a]Note (2)
		AH, DH, EH	E7018
A148-73	Steel castings for structural use	80-40, 50	E8018-C3
		90-60	E9018-G
		105-85, 120-95	E11018-M
A184	Bar mats for concrete reinforcement	See A615, A616, A617	
A202-74a	Boiler and pressure vessel	A, B	E9018-G
A203-74a	Pressure vessel	A, B	E8018-C1
		D, E	E8018-C2
A204-74a	Boiler and pressure vessel	A, B	E7010-A1 or E7018-A1
		C	E8018-B2 [a]Note (3)
A205-74a	Boiler and pressure vessel	A, B	E8018-C3
A216	Carbon steel castings, high temperature	WCA	E7018, E8018-C3
		WCB, WCC	E7018, E8010-C3, E8018-B2
A225-74a	Boiler and pressure vessel	A, B	E8018-B2
A236-74	Railway forgings	A, B	E7018 or E7028
		C, D, E	E8018-C3
		F, G	E9018-G
		H	E11018-M
A238-71	Railway forgings	A	E8018-C3
		B	E9018-G
		C, D, E	E11018-M
A242-74	High-strength structurals	All	E7018 or E7028 [a]Note (4)
A266-69	Drum forgings	1	[a]Note (2)
		2	E7018
		3	E8018-C3
A283-74	Structural plates	All	[a]Note (2)
A284-70a	Carbon-silicon plates	All	[a]Note (2)
A285-74a	Flange and firebox plates	All	[a]Note (2)
A299-74a	Boiler plates	All	E8018-C3 and B-2
A302-74a	Boiler and pressure vessel	All	E8018-C3 and B-2
A328-70	Steel piling	All	E7018 or E7028
A336-70a	Alloy forgings	F1	E7018-A1
		F12	E8018-B2 [a]Note (3)
A352-74a	Low-temperature castings	LCA, LCB, LCC	E7018
		LC1	E7018-A1
		LC2	E8018-C1
		LC3	E8018-C2
A356-74	Steam-turbine castings	5	E8018-B1
		6	E8018-B2 [a]Note (3)
		8, 10	E9018-B3
A361-71	Galvanized sheets	—	[a]Notes (2), (5)
A366-72	Carbon steel sheets	—	[a]Note (2)
A372-74	Pressure vessel forgings	Class I	E7018 or E7028
		Class II	E8018-C3
		Class III	E9018-G
		Class IV	E11018-M
A387-74a	Cr-Mo boiler plates	A, B, C	E8018-B2 [a]Note (3)
		D	E9018-B3
A389-74a	High-temperature castings	C23	E8018-B2 [a]Note (3)
		C24	E9018-B3

TABLE 16.176 *(Continued)*

ASTM Specification	Description	Grades	Recommended Electrodes[a]
A410-72	Pressure-vessel plates		E8018-C2
A414-72	Flange and firebox sheets	A, B, C, D	[a]Note (2)
		E, F	E7018 or E7028
		G	E8018-C3
A424-73	Porcelain enameling sheets	—	E7018
A441-74	High-strength structurals	All	E7018 or E7028 [a]Note (4)
A442-74	Fine-grain plates	All	E7018 or E7028
A444-71	Galvanized sheets	A, B, C	[a]Notes (2), (5)
A446-72	Galvanized sheets	D, F	E7010-A1
A455-74c	C-Mn pressure vessel plates	All	E8018-C3 [a]Note (6)
A486-74	Highway bridge castings	70	E7018 or E7028
		90	E9018-G
A487-71a	Pressure vessel castings	8N, 9N	E8018-B3
		A, AN, AQ, B, N, C, CN	E8018-B3
		BQ, CQ	E8018-C3
A508	Pressure vessel forgings, quenched and tempered	1, 1a	E7018, E8018-C3, E8018-B2
		2, 3	E8018-C3, E8018-B2, E11018-M
		2a	E8018-B2, E11018-M
		4, 5	E11018-M
A514-74a	Quenched and tempered plates	All	E11018-M [a]Note (7)
A515-74b	High-temperature boiler plates	All	E7018 or E7028
A516-74a	Low-temperature pressure vessel plates	55, 60	E7018 or E7028
		65, 70	E7018 or E8018-C3
A517-74a	Quenched and tempered plates	All	E11018-M [a]Note (7)
A521	Closed die forgings	AA, AB, CE, CF, CF1	E8018-C3
		AC, AD, CG	E8018-B2
		CA, CC, CC1	E7018
		AF	E11018-M
A526-71 A528-71	Galvanized sheets	—	[a]Notes (2), (5)
A529-72	Structural, 42,000-psi YS min	—	[a]Note (2) or E8018-C3
A533-74	Quenched and tempered plates	Class 1	E8018-C3
		Classes 2, 3	E11018-M
A537-74	Pressure vessels and structures	Class 1	E7018 or E7028
		Class 2	E8018-C3
A538	High-strength structural	All	E7018
A541-73	Pressure-vessel forgings	Class 1	E7018 or E7028
		Classes 2, 3, 4	E8018-C3
		Class 5	E8018-B2 [a]Note (3)
		Class 6	E9018-B3
A543-74	Quenched and tempered plates	1, 2, 3	E11018-M [a]Note (7)
A570-72	Structural sheets and strip	All	[a]Note (2)
A572-74b	Structural plates	42, 45	[a]Note (2)
		50, 55	E7018 or E7028
		60, 65	E8018-C3
E573-74	Structural plates	65, 70	E7018 or E7028
E588-74a	High-strength structurals	All	E7018 or E7028 [a]Note (4)
A592	Quenched and tempered fittings for pressure vessels	All	E11018-M
A595	Structural tubes	A, B	[a]Note (2)
		C	E7018 or E8018-C3
E606-71	High-strength sheets	All	[a]Note (2)
E607-70	High-strength low-alloy sheets	45, 50, 55	[a]Note (2)
		60, 65	E8018-C3
		70	E9018-G

TABLE 16.176   *(Continued)*

ASTM Specification	Description	Grades	Recommended Electrodes[a]
E611-72	Cold-rolled sheets	A, B, C, D	[a] Note (2)
A612	Pressure vessel, low temperature	All	E8018-C3, E8018-C1, E11018-M
A615-74a	Reinforcement bars	40	[a] Note (2)
		60	E9018-G
		75	E11018-M
A616-72	Reinforcement bars	50	E8018-C3
		60	E9018-G
A617-74	Reinforcement bars	40	[a] Note (2)
A633	Normalized high-strength low-alloy structural	A, B	E7018, E8018-C3
		C, D	E7018, E8018-C3, E8018-B2
		E	E8018-C3, E8018-B2, E11018-M
A643	Carbon and alloy steel casting	A	E7018
		B	E8018-C3, E8018-B2, E11018-M
A656	High-strength structural	All	E8018-B2, E11018-M
A662	Pressure vessel, low temperature	A, B	E7018, E8018-C1 or C3
A668	Carbon and alloy forgings	A, B	[a] Note (2)
		C, D	E7018
		E, F	E8018-C3, E8018-B2
		G	E8018-C3
		H, J	E8018-B2
		K, L	R11018-M
A678	Quenched and tempered plate	A	E7018, E8018-C3, E8018-B2
		B	E8018-C3, E8018-B2, . . . , E11018-M
		C	E11018-M
A690	H-piles and sheet piling	—	E7018
A699	Plates, shapes, bars	All	E8018-B2, E11018-M
A706-74	Reinforcement bars	60	E9018-G
		60	E9018-G

Steel Pipes, Tubes, and Fittings

ASTM Specification	Description	Grades	Recommended Electrodes[a]
A53-73 A106-74 A120-73 A135-73d A139-74 A179-73 A192-73 A211-73 A214-71 A226-73 A252-74 A523-73 A587-73 A589-73	Mild-steel pipes	All	[a] Notes (2), (8)
A105-73	High-temperature fittings	I, II	E7018
A106-74	high-temperature pipes	A, B, C	E7018

TABLE 16.176 *(Continued)*

ASTM Specification	Description	Grades	Recommended Electrodes[a]
A155-74	High-temperature pipes	C45, C50, C55	[a]Note (2)
		KC- and KCF-55, 60	E7018 or E7028
		KC and KCF-65	E7018 or E7028
		CM65, 70	E7010-A1 or E7018-A1
		CM75	E8018-B2 [a]Note (3)
		CMS75, CMSH70	E8018-C3
		$\frac{1}{2}$, 1, $1\frac{1}{4}$ Cr	E8018-B2 [a]Note (3)
		$2\frac{1}{4}$ Cr	E9018-B3
E161-72	Still tubes	Low carbon	[a]Note (2)
		T1	E7010-A1 or E7018-A1
E178-73,			
E179-73	Boiler condenser tubes	All	[a]Note (2)
A181-68	General service fittings	I, II	E7018 or E7010-A1
A182-74	High-temperature fittings	F1	E7010-A1, E7018-A1
		F2, F11, F12	E8018-B2 [a]Note (3)
A199-73	Heat exchanger and condenser tubes	T11	E8018-B2 [a]Note (3)
A200-72	Refinery still tubes	T11	E8018-B2 [a]Note (3)
A209-73	C-Mo boiler tubes	T1, T1a, T1b	E7010-A1, E7018-A1
A210-73	Carbon-steel boiler tubes	A1	E7010-A1 [a]Note (3)
		C	E7010-A1
A213-74b	Boiler tubes	T2, T11, T12, T17	E8018-B2 [a]Note (3)
A214-74b	Condenser tubes	All	[a]Note (2)
A216-74b	High-temperature cast fittings	WCA, WCB, WCC	E7018 or E7018-A1
A217-74c	High-temperature cast fittings	WC1	E7010-A1
		WC4	E8018-C3
		WC6	E8018-B2 [a]Note (3)
A234-74	Wrought welding fittings	WPA, WPB, WPC	[a]Note (2)
		WP1	E7010-A1 [a]Note (8)
		WP11	E8018-B2 [a]Note (3)
A250-73	C-Mo boiler tubes	T1, T1a, T1b	E7010-A1 [a]Note (8)
A333-74,			
A334-74	Low-temperature pipes	1, 6	E7018 or E8018-C3
		3	E8018-C2
		7	E8018-C1
A335-74a	High-temperature pipes	P1	E7010-A1 [a]Note (8)
		P2, P11, P12	E8018-B2 [a]Note (3)
A350-74	Low-temperature fittings	LF1, LF2	E8018-C3
		LF3	E8018-C2
		LF5	E8018-C3
E369-73a	High-temperature pipes	See A335, A182	
A381-73	High-pressure pipes	Y35, Y42, Y46	[a]Notes (2), (8)
		Y52, Y56	[a]Note (10)
		Y60, Y65	E8018-C3 [a]Note (11)
A405-70	High-temperature pipes	P24	E8018-B2 [a]Note (3)
A420-73	Low-temperature pipes	See A203, A333, A334, A350	
A423-73	Low-alloy tubes	1, 2	E8018-C3 or E7018
A426-74	High-temperature cast pipes	See A335	
A498-73	Condenser tubes	See A199, A179, A213, A214, A334	
A500-74a	Structural tubing	A, B, C	E7018 [a]Note (9)
A501-74	Structural tubing		E7018 [a]Note (9)
A524-72a	Process piping	1, 2	E7010-A1 or E7018

TABLE 16.176 *(Continued)*

ASTM Specification	Description	Grades	Recommended Electrodes[a]
A556-73,	Feedwater heater tubes	A2, B2	E7018 [a] Note (9)
A557-73		C2	E7018
A587	Low-carbon steel pipe, chemical industry	All	E6010, E6011, E7018
A595	Structural tubing	A, B	E7018
A618-74	Structural tubing	I, II, III	E7018
A660	Cast high-temperature pipe	WCA	E6010, E6011, E7018
		WCB, WCC	E7018
A692	Seamless low-alloy steel tubes	—	E7018 or E8018-C3
A694	Carbon and alloy steel forgings	All	E7018 or E8018-C3
A696	Carbon steel bars	—	E7018
A699	Low-carbon alloy steel plates, shapes and bars	—	E8018-B2, E11018-M
A707	Carbon, alloy steel flanges	L1, L2, L3	E7018, E11018-C3
		L4, L5, L6	E8018-C1
A714	Low alloy pipe	I, II, III	E7018
		IV, V, VI	E8018-C3, E8018-C1
A727	Notch-tough carbon steel forgings	—	E7018, E8018-C3

Source: Lincoln Electric Co. (revised February 1985).

[a] Notes: 1. These electrode recommendations are based primarily on matching strength of base metal; yield and tensile. Other properties such as impact energy may well require greater selectivity of electrodes as well as very carefully controlled welding procedures. Since it is impossible to foresee all conditions of every application, electrodes other than those recommended here may also be satisfactory and should be tested before weldment is started. 2. Unless restricted by specifications, use any E60XX or E70XX electrode for steel grades with 60,000 psi or lower tensile strength; for steel grades with 60,000–70,000 psi tensile strength, use E70XX electrodes. When specific alloys are required in weld deposit letter-number suffix given for electrode indicates alloy present and amount thereof. 3. Do not use E8018-B2 for low-temperature applications. 4. Use E8018-C3 or E8010-B2 for best color match on unpainted steels with enhanced atmospheric corrosion resistance. Consult steel supplier. 5. Usually E6010 is most satisfactory electrode for galvanized sheet. 6. E7018 or E7028 for fillets or E8018-C3 for general-purpose welding can be used on these steels. If weldment is to be precipitation hardened or high weld strength is required, use E8018-B2. 7. E7018 or E8018-C3 are frequently used for fillet welds. However, redrying of these electrodes prior to use is recommended. 8. Use E7010-G, an electrode designed for field welding of pipe. 9. Unless restricted by specifications, any E60XX or E70XX electrode may also be used for grades with 60,000 psi or lower tensile strength; any E70XX electrode may be used for grades with 60,000–70,000-psi tensile strength. 10. Use special electrode designed for field welding of 5LX pipe, Grades X42–X65. 11. Special electrode designed for field welding of 5LX pipe, Grades X42–X65 may also be used.

TABLE 16.177   Electrodes for Mild and Low-Alloy Steels: AWS Classification System

a. Prefix: E designates electrode
b. First two or three digits: mechanical properties

Classification	Minimum Tensile Strength, psi (MPa)	Minimum Yield Strength, psi (MPa)	Minimum Elongation, %
E60XX	62,000 (427)	50,000 (345)	17
E70XX	70,000 (483)	57,000 (393)	22
E80XX	80,000 (552)	67,000 (462)	19
E90XX	90,000 (621)	77,000 (531)	17
E100XX	100,000 (690)	87,000 (600)	16
E110XX[a]	110,000 (758)	97,000 (669)	15
E120XX[a]	120,000 (827)	107,000 (738)	14

c. Third (or fourth) digit: applicable welding positions
  EXX1X: flat, horizontal, vertical, and overhead
  EXX2X: flat and horizontal fillet
  EXX4X: flat, horizontal, overhead, and vertical down

d. Last digit: electrode usability

Classification	Current[b]	Arc	Penetration	Covering-Slag	Iron Powder[c] %
EXX10	dcep	Digging	Deep	Cellulose-sodium	0–10
EXXX1	ac, dcep	Digging	Deep	Cellulose-potassium	0
EXXX2	ac, dcen	Medium	Medium	Rutile-sodium	0–10
EXXX3	ac, dcen, dcep	Soft	Light	Rutile-potassium	0–10
EXXX4	ac, dcen, dcep	Soft	Light	Rutile-iron powder	25–50
EXXX5	dcep	Medium	Medium	Low hydrogen-sodium	0
EXXX6	ac, dcep	Medium	Medium	Low hydrogen-potassium	0
EXXX7	ac, dcen	Medium	Medium	Iron oxide-iron powder	40–50
EXXX8	ac, dcep	Medium	Medium	Low hydrogen-iron	25–50
EXX20 and				powder	
EXX22 (single pass)	ac, dcen, dcep	Medium	Medium	Iron oxide-sodium	0

e. Suffix: chemical composition of deposited weld metal: suffix not applied to E60XX classification

Suffix	C	Mn	Si	Ni	Cr	Mo	V
			Weld Metal Composition[d], %				
A1	0.12	0.60 or 1.00[e]	0.40, 0.80[e]	—	—	0.40–0.65	—
B1	0.12	0.90	0.60, 0.80[e]	—	0.40–0.65	0.40–0.65	—
B2L	0.05	0.90	0.80–1.00	—	1.00–1.50	0.40–0.65	—
B2	0.12	0.90	0.60, 0.80[e]	—	1.00–1.50	0.40–0.65	—
B3L	0.05	0.90	0.80–1.00	—	2.00–2.50	0.90–1.20	—
B3	0.12	0.90	0.60, 0.80[e]	—	2.00–2.50	0.90–1.20	—
B4L	0.05	0.90	1.00	—	1.75–2.25	0.40–0.65	—
B5	0.07–0.15	0.40–0.70	0.30–0.60	—	0.40–0.60	1.00–1.25	0.05
C1	0.12	1.20	0.60, 0.80[e]	2.00–2.75	—	—	—
C2	0.12	1.20	0.60, 0.80[e]	3.00–3.75	—	—	—
C3	0.12	0.40–1.25	0.80	0.80–1.10	0.15	0.35	0.05
D1	0.12	1.25–1.75	0.60, 0.80[e]	—	—	0.25–0.45	—
D2	0.15	1.65–2.00	0.60, 0.80[e]	—	—	0.25–0.45	—
G[f]	—	1.00 min	0.80 min	0.50 min	0.30 min	0.20 min	0.10 min
M[g]	0.10	0.60–2.25[e]	0.60, 0.80[e]	1.25–2.50[e]	0.15–1.50[e]	0.25–0.55[e]	0.05

Source: Hobart Bros. Co. (reviewed February 1985).

[a] Low-hydrogen type coating only.
[b] Abbreviations: dcep, electrode positive-reverse polarity; dcen, electrode negative-straight polarity.
[c] Iron powder percentage based on weight of covering.
[d] Compositions are maximum unless otherwise indicated.
[e] Amount depends on electrode classification.
[f] Electrodes of G classification require only one of the alloying elements listed.
[g] The M suffix is used to cover military electrode classifications.

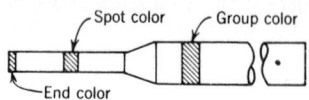

TABLE 16.178   Standard Color Markings for Electrode Identification

AWS–ASTM Classification	End Color	Spot Color	Group Color
Mild- and Low-Alloy Steel Electrodes			
E6010	None	None	None
E6011	None	Blue	None
E6012	None	White	None
E6013	None	Brown	None
E6020	None	Green	None
E7010-A1	Blue	White	None
E7011-A1	Blue	Yellow	None
E7020-A1	Blue	Yellow	Silver
E10013	Green	Brown	Silver
Iron Powder Electrodes			
E6010	None	None	None
E6024	None	Yellow	None
E6014	None	None	None
E7016	Blue	Orange	Green
E8016-B2	White	Gray	Green
E9016-B3	Brown	Blue	Green
Low-Hydrogen Electrodes			
E6016	None	Orange	None
E7016	Blue	Orange	Green
E8016	White	Orange	Green
E8016-B1	White	Black	Green
E8016-C1	White	Blue	Green
E8016-B2	White	Gray	Green
E9016-B3	Brown	Blue	Green
E9016	Brown	Orange	Green
E10015	Green	Red	Green
E10016	Green	Orange	Green
E12016	Yellow	Orange	Green
E12015	Yellow	Red	Green
E15016	None	None	None
Electrodes for Welding Cast Iron			
EST	Orange	None	None
ENI	Orange	Blue	White
Stainless Steel Electrodes			
E308-15	Yellow	None	Black
E308-16	Yellow	None	Yellow
E308-ELC-15	Brown	None	Black
E308-ELC-16	Brown	None	Yellow
E347-15	Yellow	Blue	Black
E347-16	Yellow	Blue	Yellow
E316-15	Yellow	White	Black
E316-16	Yellow	White	Yellow
E316-ELC-15	Brown	White	Black
E316-ELC-16	Brown	White	Yellow
E317-15	Yellow	Brown	Black

TABLE 16.178 *(Continued)*

AWS–ASTM Classification	End Color	Spot Color	Group Color
E317-16	Yellow	Brown	Yellow
E309-15	Black	None	Black
E309-16	Black	None	Yellow
E310-15	Red	None	Black
E310-16	Red	None	Yellow
E307-15	None	Black	Black
E307-16	None	Black	Yellow
E308-MO-15	None	None	Black
E308-MO-16	None	None	Yellow
E502-16	Gray	Blue	Yellow

Source: *Canadian Metalworking/Machine Production*, May 1960, by permission.

Adhesives

TABLE 16.179 Adhesives Commonly Used for Joining Various Materials[a]

Material	Adhesive	Table 16.180 Reference
ABS	Polyester;	a
	epoxy;	e
	alpha-cyanoacrylate;	c
	nitrile-phenolic	b
Aluminum and its alloys	Epoxy;	e
	epoxy-phenolic;	d
	nylon-epoxies;	f
	polyurethane rubber;	g
	polyesters;	a
	alpha-cyanoacrylate;	c
	polyamides;	h
	polyvinyl-phenolic;	i
	neoprene-phenolic	b
Brick	Epoxy;	e
	epoxy-phenolic;	d
	polyesters	a
Ceramics	Epoxy;	e
	cellulose esters;	j
	vinyl chloride-vinyl acetate;	k
	polyvinyl butyral	l
Chromium	Epoxy	e
Concrete	Polyester,	a
	epoxy	e
Copper and its alloys	Polyesters;	a
	epoxy;	e
	alpha-cyanoacrylate;	c
	polyamide;	h
	polyvinyl-phenolic;	i
	polyhydroxyether	m
Fluorocarbons	Epoxy;	e
	nitrile-phenolic;	b
	silicone	t
Glass	Epoxy;	e
	epoxy-phenolic;	d
	alpha-cyanoacrylate;	c
	cellulose esters;	j
	vinyl chloride-vinyl acetate;	k
	polyvinyl butyral	l

TABLE 16.179   *(Continued)*

Material	Adhesive	Table 16.180 Reference
Lead	Epoxy;	e
	vinyl chloride-vinyl acetate;	k
	polyesters	a
Leather	Vinyl chloride-vinyl acetate;	k
	polyvinyl butyral;	l
	polyhydroxyether;	m
	polyvinyl acetate;	n
	flexible adhesives	g
Magnesium	Polyesters;	a
	epoxy;	e
	polyamide;	h
	polyvinyl-phenolic;	i
	neoprene-phenolic;	b
	nylon-epoxy	f
Nickel	Epoxy;	e
	neoprene;	g
	polyhydroxyether	m
Paper	Animal glue;	o
	starch glue;	p
	urea-, melamine-, resorcinal- and phenol-formaldehyde;	q
	epoxy;	e
	polyesters;	a
	cellulose esters;	j
	vinyl chloride-vinyl acetate;	k
	polyvinyl butyral;	l
	polyvinyl acetate;	n
	polyamide;	h
	flexible adhesives	g
Phenolic and melamine	Epoxy;	e
	alpha-cyanoacrylate;	c
	flexible adhesives	g
Polyamide	Epoxy;	e
	flexible adhesives;	g
	phenol- and resorcinol-formaldehyde;	q
	polyester	a
Polycarbonate	Polyesters;	a
	epoxy;	e
	alpha-cyanoacrylate;	c
	polyurethane rubber	g
Polyester, glass reinforced	Polyester;	a
	epoxy;	e
	polyacrylates;	r
	nitrile-phenolic	b
Polyethylene	Polyester, isocyanate modified;	a
	butadiene-acrylonitrile;	g
	nitrile-phenolic	b
Polyformaldehyde	Polyester-isocyanate modified;	a
	butadiene-acrylonitrile;	g
	nitrile-phenolic	b
Polymethylmethacrylate	Epoxy;	e
	alpha-cyanoacrylate;	c
	polyester;	a
	nitrile-phenolic	b
Polypropylene	Polyester, isocyanate modified;	a
	nitrile phenolic;	b
	butadiene-acrylonitrile	g
Polystyrene	Vinyl chloride-vinyl acetate;	k
	polyesters	a
Polyvinyl chloride, flexible	Butadiene-acrylonitrile;	g
	polyurethane rubber	g

TABLE 16.179   *(Continued)*

Material	Adhesive	Table 16.180 Reference
Polyvinyl chloride, rigid	Polyesters;	a
	epoxy;	e
	polyurethane	g
Rubber, butadiene-styrene	Epoxy;	e
	butadiene-acrylonitrile;	g
	urethane rubber	g
Rubber, natural	Epoxy;	e
	flexible adhesives	h
Rubber, neoprene	Epoxy;	e
	flexible adhesives	h
Rubber, silicone	Silicone	t
Rubber, urethane	Flexible adhesives;	h
	silicone;	t
	alpha-cyanoacrylate	c
Silver	Epoxy;	e
	neoprene;	g
	polyhydroxyether	m
Steel	Epoxy;	e
	polyesters;	a
	polyvinyl butyral;	l
	alpha-cyanoacrylate;	c
	polyamides;	h
	polyvinyl-phenolic;	i
	nitrile-phenolic;	b
	neoprene-phenolic;	b
	nylon-epoxy	d
Stone	See Brick	
Tin	Epoxy	e
Wood	Animal glue;	o
	polyvinyl acetate;	n
	ethylene-vinyl acetate;	u
	urea-, melamine-, resorcinol-, and	
	phenol-formaldehyde	q

Source: Compiled from *Adhesive Age Magazine*, October 1973; G. L. Schneberger, private communication; and other sources.
[a]Adhesive suppliers should be consulted for additional information.

TABLE 16.180　Adhesive Commentary

Adhesive Type	Table Reference	Comments	Typical Cure Conditions
Polyesters and their variations	a	Used primarily for repairing fiberglass-reinforced polyester resins, ABS, and concrete. Generally unsaturated esters are polymerized with catalyst such as methyl ethyl ketone (MEK) peroxide and accelerator such as cobalt naphthenate. Coreactant solvent such as styrene may be present. Bonds are strong. Sometimes combined with polyisocyanates to control shrinkage stresses and reduce brittleness. Unreacted monomer, if present, keeps viscosity low for application, provides good wetting, enhances cross-linking. Occasionally used on metals.	Minutes to hours at room temperature
Nitrile phenolic, Neoprene phenolic	b	Blend of flexible nitrile or neoprene rubber with phenolic novolac resin. They combine impact resistance of rubber with strength of cross-linked phenolic. Inexpensive: produce strong durable bonds which resist water, salt spray, and other corrosive media well. Work-horses of adhesive tape industry although require high pressure, relatively long high temperature cures. Used for metals and some plastics including ABS, polyethylene, and polypropylene. Airframe components and automotive brakes typical examples.	up to 12 hr at 250–300°F (120–150°C)
Alpha-cyanoacrylate	c	Low-viscosity liquids polymerize or "cure" rapidly in presence of moisture or many metal oxides. Most surfaces can be bonded. Bonds are fairly strong but somewhat brittle. Used widely for assembly of jewelry and electronic components.	0.5–5 minutes at room temperature
Epoxy phenolic	d	Combination of epoxy resin with resol phenolic. Noted for strength retention at 300–500°F (150–250°C), strong bonds, and good moisture resistance. Normally stored refrigerated. Used for some metals, glass, and phenolic resins.	1 hr, 350°F (175°X)
Epoxy, amine amide, and anhydride cured	e	Epoxies noted for high tensile and low peel strengths. Cross-linked and in general have good high-temperature strength, resistance to moisture, and little tendency to react with acids, bases, salts, or solvents. There are important exceptions to these generalizations, however, which are often result of curing agent used. Primary amines give faster setting adhesives which are less flexible and less moisture resistant than is the case when polyamide curing agents are used. Anhydride cured epoxies generally have good high temperature strength but are subject to hydrolysis especially in the presence of acids or bases. Other important features of epoxies are their low shrinkage upon cure, their compatibility with a variety of fillers, their	Two part, room temperature, curable, 2 min to 24 hr; two part, heat cured, $\frac{1}{2}$–5 hr, 148–224°C; one part, heat cured, 5–60 min, 150–175°C

Material		Description	Processing/cure
Nylon epoxy	f	long life when properly applied and their easy modification with other resins. Cross link density is easily varied with epoxies; thus some control over brittleness, vapor permeation, and heat deflection is possible. These resins are widely used to bond metal, ceramics and rigid plastics (not polyolefins). Tensile shear strengths above 6,000 psi (41.4 MPa) and peel strength above 100 lb/in. (18 kg/cm) are possible when epoxy resins are modified with special low-melting nylons. These gains, however, are accompanied by loss of strength upon exposure to moist air, a tendency to creep under load and poor low temperature impact behavior. A phenolic primer may increase bond life and moisture resistance. Used primarily for aluminum, magnesium and steel.	1 hr, 300–350°F (150–175°C)
Flexible adhesives, natural rubber, butadiene, acrylonitrile, proprene, polyurethane, polyacrylates, silicones	g	These adhesives are flexible. Thus their load bearing ability is limited. They have excellent impact and moisture resistance. They are easily tackified and are used as pressure sensitive tapes or as contact cements. Urethane and silicone adhesives are lightly cross linked which gives them reasonable hot strength. They are also compatible with many surfaces but are somewhat costly and must be protected against moisture before use. They have good low temperature tensile shear and impact strength. The urethanes are two part products which require mixing before use. Silicones cure in the presence of atmospheric moisture. (See entries f and t).	Pressure-sensitive tape or solvent cements. Low temperature bake for urethane. Ambient cure for silicones.
Polyamides	h	These adhesives, which are chemically similar to nylon resins, have good strength at ambient temperatures and are fairly tough. They are available in a variety of molecular weights, softening ranges and melt viscosities. Often applied as hot melts, they have good adhesion to a variety of surfaces. The higher molecular weight varieties often have the best tensile properties. Lower molecular weight polyamides may be applied in solution.	Hot melt, cure by cooling
Polyvinyl-phenolic	i	These resins, which combine a resol phenolic resin with polyvinyl formal or polyvinyl butyral, were the first important synthetic structural adhesives. A considerable range of compositions are available with hot strength and tensile proprties increasing at the expense of impact and peel strength as the phenolic content rises. The durability of vinyl-phenolics is generally excellent. They are often selected for low cost applications where heat and pressure curing can be used.	300°F (150°C), 1 hr
Cellulose esters	j	Cellulose ester adhesives are usually high viscosity, inexpensive, rigid materials. They do not have high strength and are sensitive to heat and many solvents. ormally used for holding small parts or repairing wood, cardboard or plastic items. Model airplane cement is a common example.	Air dry

TABLE 16.180   (Continued)

Adhesive Type	Table Reference	Comments	Typical Cure Conditions
Vinyl chloride vinyl acetate	k	This is a combination of two resins which are sometimes used alone. They may be used as hot melts or as solution adhesives. Since thin films of vinyl chloride-vinyl acetate are somewhat flexible, they are often used for bonding metal foil, paper and leather. A range of compositions is available with a corresponding variety of properties.	Cooling (hot melt) or solvent loss
Polyvinylbutyral	l	A tough transparent resin which is used as a hot melt or heat cured solution adhesive. It has good adhesion to glass, wood, metal and textiles. It is flexible and can be modified with other resins or additives to give a range of properties. Not generally used as a structural adhesive, although structural phenolics sometimes incorporate polyvinylbutyral to give better impact resistance.	Cooling (hot melt), heating under pressure
Polyhydroxyether	m	These are resins based on hydroxylated polyethylene oxide polymers. Generally used as hot melts, they have only moderate strength, but are flexible and have fairly good adhesion.	Hot melt, cure by cooling
Polyvinyl acetate	n	This adhesive is generally supplied as a water emulsion (white glue) or used as a hot melt. It dries quickly and forms a strong bond. It is flexible and has low resistance to heat and moisture. Porous substrates are required when the resin is used as an emulsion.	Hot melt, cure by cooling, emulsion, air dry
Animal glue	o	Chemically, animal glues are proteins; they are polar, water soluble polymers with high affinity for paper, wood and leather surfaces. They easily form strong bonds but have poor resistance to moisture. They are being replaced in many areas by synthetic resin adhesives but their low cost is often an important advantage. They are usually applied as highly viscous liquids.	Air dry under pressure
Starch glue	p	These products, based on corn starch, have high affinity for paper but are used for little else. They are moisture sensitive and are applied as water dispersions.	Low temperature, dry
Urea-formaldehyde, melamine-formaldehyde, resorcinol-formaldehyde, phenol-formaldehyde	q	These thermosetting resins are widely used for wood bonding. Urea-formaldehyde is inexpensive but has low moisture resistance. It can be cured at room temperature if a catalyst is used. Melamine-formaldehyde resins have better moisture resistance but must be heat cured. Phenol-formaldehyde adhesives form strong, waterproof wood-to-wood bonds. The resorcinol-formaldehyde resin will cure at room temperature while phenol-formaldehyde requires heating. These resins are often combined resulting in an adhesive with intermediate processing or performance characteristics.	Up to 300°F (149°C) and 200 psi (1.38 MPa)

Polyacrylate esters	r	These resins are n-alkyl esters of acrylic acid. They have good flexibility and find frequent use for high quality pressure sensitive tapes and foams. They are not suitable for structural application is because of their poor heat resistance and their cold flow behavior. Frequently used on flexible substrates.	Pressure sensitive
Polysulfides	s	These resins have good moisture resistance and can range from thermoplastic to thermosetting depending on the degree of cross linking which is developed during cure. They are two or three part systems, the third part being a catalyst. Ventilation is generally required. They make excellent adhesive sealants for wood, metal, concrete and glass. Polysulfide resins may be combined with epoxies to flexibilize the latter.	Low pressures, moderate temperature
Silicones[a]	t	These expensive adhesives have high peel strength and excellent property retention at high and low temperatures. They resist all except the most corrosive environments and will adhere to nearly everything. They are usually formulated to react with atmospheric moisture and form lightly cross linked films.	Low pressure, room temperature
Ethylene-vinyl acetate	u	This copolymer is widely used as a hot melt adhesive because it is inexpensive, adheres to most surfaces and is available in a range of melting points. It is widely used for bookbinding and packaging.	Hot melt, cures by cooling
Urethanes, rigid	v	Rigid urethanes are highly cross linked. While somewhat expensive, they adhere well to most materials, especially plastics, and have good impact strength. Structural urethanes are two part systems and have good low temperature strength retention.	Low pressures up to 300°F

Source: G. L. Schneberger, private communication.
[a]See also Flexible adhesives.

16.3.7   Nuclear Materials

TABLE 16.181   Requirements of Materials for Nuclear Reactor Components

Component	Neutron Absorption Cross Section	Effect in Slowing Neutrons	Strength	Resistance to Radiation Damage	Thermal Conductivity	Corrosion Resistance	Cost	Other	Typical Material
Moderator and reflector	Low	High	Adequate	—	—	High	Low	Low atomic weight	H_2O, Be
Fuel	Low	—	Adequate	High	High	High	Low	—	U, Th
Control rod	High	—	Adequate	Adequate	High	—	—	—	Cd, B_4C
Shield	High	High	High	—	—	—	—	High γ radiation absorption	Concrete
Cladding	Low	—	Adequate	—	High	High	Low	—	Al
Structural	Low	—	High	Adequate	—	High	—	Low corrosion rate	Zr, stainless steel
Coolant	Low	—	—	—	High	—	—	high heat capacity	H_2O, Na, NaK

Source: T. Baumeister et al., *Marks' Standard Handbook for Mechanical Engineers*, 6th ed., © 1958, McGraw-Hill, New York. Reproduced with permission.

TABLE 16.182 Moderating Properties of Materials

Moderator	Slowing-Down Power (cm^{-1})	Moderating Ratio
H$_2$O	1.53	70
D$_2$O	0.177	21,000
He	1.6 × 10^{-5}	83
Be	0.16	150
BeO	0.11	180
C (graphite)	0.063	170

Source: T. Baumeister et al., *Marks' Standard Handbook for Mechanical Engineers*, 6th ed., © 1958, McGraw-Hill, New York. Reproduced with permission.

TABLE 16.183 Slow-Neutron Absorption Cross Sections

Low		Intermediate		High	
Element	Cross Section (barns)	Element	Cross Section (barns)	Element	Cross Section (barns)
Oxygen	0.0016	Zinc	1.0	Manganese	12
Carbon	0.0045	Columbium	1.2	Tungsten	18
Beryllium	0.009	Barium	1.2	Tantalum	21
Fluorine	0.01	Strontium	1.3	Chlorine	32
Bismuth	0.015	Nitrogen	1.7	Cobalt	35
Magnesium	0.07	Potassium	2.0	Silver	60
Silicon	0.1	Germanium	2.3	Lithium	67
Phosphorus	0.15	Iron	2.4	Gold	95
Zirconium	0.18	Molybdenum	2.4	Hafnium	100
Lead	0.18	Gallium	2.8	Mercury	340
Aluminum	0.22	Chromium	2.9	Iridium	470
Hydrogen	0.32	Thallium	3.3	Boron	715
Calcium	0.42	Copper	3.6	Cadmium	3,000
Sodium	0.48	Nickel	4.5	Samarium	8,000
Sulfur	0.49	Tellurium	4.5	Gadolinium	36,000
Tin	0.6	Vanadium	4.8		
		Antimony	5.3		
		Titanium	5.8		

Source: T. Baumeister et al., *Marks' Standard Handbook for Mechanical Engineers*, 6th ed., © 1958, McGraw-Hill, New York. Reproduced with permission.

TABLE 16.184 Resistance of Materials to Liquid Sodium and NaK[a]

Temperature (°F)	Good	Limited	Poor
< 1000	Carbon steels, low-alloy steels, alloy steels, stainless steels, nickel alloys, cobalt alloys, refractory metals, beryllium, aluminum oxide, magnesium oxide, aluminum bronze	Gray cast iron, copper, aluminum alloys, magnesium alloys, glasses	Sb, Bi, Cd, Ca, Au, Pb, Se, Ag, S, Sn, Teflon
1000–1600	Armco iron, stainless steels, nickel alloys,[b] cobalt alloys, refractory metals[c]	Carbon steels, alloy steels, Monel, titanium, zirconium, beryllium, aluminum oxide, magnesium oxide	Gray cast iron, copper alloys, Teflon, Sb, Bi, Cd, Ca, Au, Pb, Se, Ag, S, Sn, Pt, Si, magnesium alloys

Source: T. Baumeister et al., *Marks' Standard Handbook for Mechanical Engineers*, 6th ed., © 1958, McGraw-Hill, New York. Reproduced with permission.

[a]For more complete details, see Ukanwa, A. O., *Handbook of Liquid Metals* (1976). Marshall Space Flight Center, Alabama.
[b]Except Monel.
[c]Except titanium and zirconium.

TABLE 16.185 Mechanical Properties of Special Metals for Nuclear Reactors

Material	Longitudinal Yield Strength (kpsi)	Longitudinal Ultimate Strength (kpsi)	Longitudinal Elongation[b] (%)	Transverse Ultimate Strength (kpsi)	Transverse Elongation[b] (%)	Charpy Impact (ft-lb)	Elastic Modulus (10^6 psi)	°F	Ultimate Strength (kpsi)	Elongation (%)
Beryllium										
Cast, extruded, and annealed	—	40	1.82	16.6	0.18	—	44	392	62	23.5
Flake, extruded, and annealed	—	63.7	5.0	25.5	0.30	4.1	—	752	43	29
Powder, hot extruded	39.5	81.8	15.8	45.2	2.3		—	1112	23	8.5
Powder, vacuum hot pressed	32.1	45.2	2.3	45.2	2.3	0.8	—	1472	5.2	10.5
Zirconium										
Kroll, 50% CW	—	82.6	—	—	—	14.8	14	250	32	—
								500	23	—
								700	17	—
								900	12	—
								1500	3	—
Kroll, annealed	—	49.0	—	—	—	—	—			
Iodide	15.9	35.9	31	—	—	2.5–6.0	—			
Zircaloy-2[c]	45	70	14	67	15	3–5	14	600	31	24
Uranium	25	53	< 10	—	—	15	—	302	27	—
								1112	12	—
Thorium	27	37.5	40	—	—	—	—	570	22	—
								930	17.5	—

Source: T. Baumeister et al., *Marks' Standard Handbook for Mechanical Engineers*, 6th ed. © 1958. McGraw-Hill, New York. Reproduced with permission.

[a] All elevated-temperature data on beryllium for hot-extruded powder; on zirconium for iodide material.

[b] Beryllium is extremely notch sensitive. Tabulated data have been obtained under very carefully controlled conditions, but ductility values in practice will be found in general to be much lower and essentially zero in transverse direction.

[c] Zr + 1.5 Sn + Fe, Cr, and Ni; hot-rolled strip.

PROPERTIES OF MATERIALS

TABLE 16.186 Chemical Composition (wt %) Specifications for Representative LWR Pressure Vessel and Piping Steels

ASME Material Specification	C	Si	Mn	P	S	Ni	Cr	Mo
Reactor Pressure Vessel Steels[a]								
SA-533 grade B, class 1 plate	0.25 max	0.15–0.30	1.15–1.50	0.035 max	0.040 max	0.40–0.70	—	0.45–0.60
SA-508 class 2 forging[b]	0.27 max	0.15–0.40	0.50–1.00	0.025 max	0.025 max	0.50–1.00	0.25–0.45	0.55–0.70
SA-508 class 3 forging	0.25 max	0.15–0.40	1.20–1.50	0.025 max	0.025 max	0.40–1.00	0.25 max	0.45–0.60
Large-Diameter Piping Steels								
SA-106 grade B	0.30 max	0.10 min	0.29–1.06	0.048 max	0.058 max	—	—	—
SA-516 grade 70 plate	0.28 max	0.15–0.30	0.85–1.20	0.035 max	0.040 max	—	—	—
SA-358 type 304L plate[c]	0.03 max	1.00 max	2.00 max	0.045 max	0.030 max	8.0–12.0	18.0–20.0	—
SA-376 grade TP 204	0.08 max	0.75 max	2.00 max	0.040 max	0.030 max	8.0–11.0	18.0–20.0	—
SA-376 grade TP 316	0.08 max	0.75 max	2.00 max	0.040 max	0.030 max	11.0–14.0	16.0–18.0	2.00–3.00

Source: Encyclopedia of Materials Science and Engineering, M. B. Bever (ed.), Pergamon Press, Oxford, 1986.

[a]Supplementary requirements are usually imposed by manufacturers to limit sulfur (0.015 maximum), phosphorus (0.012 maximum), and copper (0.10 maximum).
[b]Limited to 0.05 maximum vanadium.
[c]Nitrogen content restricted to 0.10 maximum.

TABLE 16.187   Tensile requirements[a] for LWR Pressure Vessel and Piping Steels

ASME Material Specification	Strength (MPa)		Total Elongation[b] (%)
	Yield	Ultimate	
Reactor Pressure Vessel Steels			
SA-533 grade B, class 1 plate	345	550–670	18
SA-508 class 2 forging[c]	345	550–725	18
SA-508 class 3 forging[c]	345	550–725	18
Large-diameter Piping Steels			
SA-106 grade B	240	415	22
SA-516 grade 70 plate	260	485–620	21
SA-358 type 304L plate	170	485	40
SA-376 grade TP 304	210	520	35
SA-376 grade TP 316	210	520	35

Source: Encyclopedia of Materials Science and Engineering, M. B. Bever (ed.), Pergamon Press, Oxford, 1986.
[a] Single values are minimum requirements.
[b] Total elongation measured in 50 mm.
[c] Minimum reduction of area specified as 38%.

16.3.8   Electronic Materials

TABLE 16.188   Materials for Electronic Components

Device Class	Materials	Production[a] (billion $)
Active Components		
(a) Semiconductor devices,	Silicon, III–V compounds	9.0
discrete		3.1
integrated		5.9
(b) Magnetic memories, core and bubble	Ferrites, garnets	NA
(c) Optically pumped lasers	Ruby, garnets, other oxides	NA
(d) Piezoelectric transducers, filters and oscillators	Quarts, $AlPO_4$, lead zirconate titanate	NA
(e) Nonlinear optical and electrooptic devices	Niobates, germanates	NA
(f) Pyro- and ferroelectric devices	Titanates	NA
Passive Components		
(g) Capacitors	Tantalum oxide, aluminum oxide, other oxides	1.1
(h) Resistors	Carbon, metal film	0.7
(i) Coils, transformers	Copper	1.0
(j) Convectors	Base metal, noble metal plated or inlayed	2.0
(k) Optical fibers	Silica-based doped glass	NA
(l) Printed memory boards	Polymer, copper, solder, noble metal	NA
Total sum of NA's [items (b)–(d), (k) and (l) predominate]		~ 10

Source: U.S. Industrial Outlook, Chapter 26, U.S. Government Printing Office, 1982, and Hodgson, M. (ed.), *Yearbook and Directory*, Semiconductor Industry Association, Cupertino, CA, (1981).
[a] 1981 Values.

Substrates

TABLE 16.189 Laminate Selector Chart

Laminate	Cost	Electrical Properties	Chemical Resistance	Thermal Stability	Arc Resistance	Flame Resistance	Humidity Resistance	Dimensional Stability	Mechanical Properties
Phenolic	X							X	X
Melamine		X			X	X		X	X
Epoxy			X		X		X		
Polyester	X		X		X	X		X	
Phenylsilane				X					X
Silicone				X	X	X			
Diallyl phthalate		X	X				X	X	
Teflon^a			X			X	X		
Polyimide		X	X	X		X			
Polybenzimidazole				X					X

Source: C. A. Harper, (ed.), *Handbook of Materials and Processes for Electronics*, © 1970, McGraw-Hill, New York. Reproduced with permission.

^aTrademark of E. I. du Pont de Nemours & Co., Wilmington, Del.

TABLE 16.190 NEMA Laminate Selector Chart

Grade	X	XP	XPC	XX	XXP	XXXP	XXXPC	FR-2	FR-3	C	CE	L	LE	CF	A	AA	G-2	G-3	G-5	G-7	G-9	G-10	G-11	FR-4	FR-5	GPO-1	GPO-2	N-1
Mechanical application	×	×								×		×			×	×												
Electrical application				×	×	×	×	×	×		×		×	×			×	×	×	×	×	×	×	×	×	×	×	×
Punching qualities	×	×	×	×	×	×	×	×	×					×														
Tubes	×	×	×	×						×	×	×	×		×	×	×	×	×	×		×	×	×	×			
Rods	×	×	×	×						×	×	×	×		×	×	×	×	×	×		×	×	×	×			
Flame resistance								×	×						×	×	×	×	×	×						×	×	×
Impact resistance										×	×		×	×	×	×	×	×	×	×				×	×	×	×	×
Flexural strength										×	×		×	×	×	×	×	×	×	×				×	×	×	×	×
Thermal stability															×	×	×	×			×	×	×	×	×	×	×	
Cold flow		×																			×	×	×	×	×	×	×	×
Humidity resistance				×	×	×	×	×	×						×	×			×	×	×	×	×	×	×	×	×	×
Electronic application				×	×	×	×	×	×											×	×	×	×	×	×			
Machinability			×							×	×	×	×		×	×	×	×	×	×	×	×	×	×	×	×	×	×
Arc resistance																											×	×

Source: C. A. Harper, (ed.), *Handbook of Materials and Processes for Electronics,* © 1970, McGraw-Hill, New York. Reproduced with permission.

TABLE 16.191   Properties of Laminated and Reinforced Plastics[a]

NEMA Minimum or Maximum Average Values, Sheets[b]

NEMA Grade	Minimum Flexural Strength, 1/16" Thick (psi) LW	CW	Minimum Izod Impact Strength, notch, edge (ft-lb/in) LW	CW	Minimum Bond Strength (lb)	Water Absorption, Maximum, 1/16" (%)	Minimum Dielectric Breakdown,[d] (kV)	Maximum Dielectric Constant, 1 MHz 1/32" or more	Maximum Dissipation Factor 1 MHz 1/32" or more	Minimum Arc Resistance (sec)
X	25,000	22,000	0.55	0.50	700	6.00				
XP	13,000	11,000				3.60	40			
XPC	10,000	8,000				5.50				
XX	15,000	14,000	0.40	0.35	80	2.00	40	5.5	0.045	
XXP	14,000	12,000			0	1.80	60	5.0	0.040	
XXX	13,500	11,800	0.40	0.35	950	1.40	50	5.3	0.038	
XXXP	12,000	10,500				1.00	60	4.6	0.038	
XXXPC	12,000	10,500				0.75	60	4.6	0.038	
C	17,000	16,000	1.90	1.70	1800	4.40	15			
CE	16,500	14,000	1.60	1.40	1800	2.20	35			
L	15,000	14,000	1.35	1.10	1600	2.50	15			
LE	15,000	13,500	1.25	1.00	1600	1.95	40	5.8	0.055	
A	13,000	11,000	0.60	0.60	700	1.50	5			
AA	16,000	14,000	3.60	3.00	1800	3.00				
G-3	50,000	40,000	6.5	5.5	850	2.70				
G-5	50,000	40,000	7.0 (to 1/2")	5.5	1570	2.70	23	7.8	0.020	180
G-7	20,000	18,000	6.5	5.5	650	0.55	32	4.2	0.003	180
G-9	60,000	40,000	7.0 (to 1/2")	5.5	1700	0.80	60	7.2	0.017	180
G-10	60,000	50,000	7.0	5.5	2000	0.25	45	5.2	0.025	
G-11	60,000	50,000	7.0	5.5	1600	0.25	45	5.2	0.025	
N-1	10,000	9,500	3.0	2.0	1000	0.60	60	3.9	0.038	
FR-1	13,000					3.60	40			
FR-2	12,000	10,500				0.75	60	4.6	0.038	
FR-3	20,000	16,000				0.65	60	4.6	0.035	
FR-4	60,000	50,000	7.0	5.5	2000	0.25	45	5.2	0.025	
FR-5	60,000	50,000	7.0	5.5	1600	0.25	45	5.2	0.025	
GPO-1	18,000	18,000	8.0	8.0	850	1.00	40	4.3[c]	0.03[c]	100
GPO-2	18,000	18,000	8.0	8.0	850	0.90	40			100
CEM-1	35,000	28,000	1.8	1.2		0.30	45	5.0	0.035	
CEM-3	40,000	32,000				0.25	45	5.2	0.025	

Source: Insulation / Circuits Desk Manual, Lake Publishing Corp., Libertyville, Il.
[a]All tests conducted in accordance with applicable NEMA and/or ASTM standards.
[b]See NEMA Pub. No. LI 1-1971 (R 1976). Standards Publication for Industrial Laminated Thermosetting Products, regarding test methods, conditions, etc.
[c]These are only typical values obtained from a number of sources and should not be used in establishing specifications or standards—consult manufacturers.
[d]Parallel to lamination, Step-by-Step, 1/32" to 1" thickness.

	Typical Values, Sheets[c]										
	Strength					Dielectric Strength Perp. to Lam., (vpm)		Thickness Range (in.)			
	Tension (psi)		Compression (psi)		Rock-well Hard-ness							
NEMA Grade	LW	CW	Flat	Edge	M Scale	Sp. Gr.	Short time	Step by Step	Minimum	Maximum	Base Material	Resin
X	20,000	16,000	36,000	19,000	110	1.36	700	500	0.010	2	Paper	Phenolic
XP	12,000	9,000	25,000		95	1.33	650	450	0.010	1/4	Paper	Phenolic
XPC	10,500	8,500	22,000		75		600	425	1/32	1/4	Paper	Phenolic
XX	16,000	13,000	34,000	23,000	105	1.34	700	500	0.010	2	Paper	Phenolic
XXP	11,000	8,500	25,000		100	1.32	700	500	0.015	1/4	Paper	Phenolic
XXX	15,000	12,000	32,000	25,500	110	1.32	650	450	0.015	2	Paper	Phenolic
XXXP	12,400	9,500	25,000		105	1.30	650	450	0.015	1/4	Paper	Phenolic
XXXPC	12,400	9,500	25,000		105	1.31	650	450	1/32	1/4	Paper	Phenolic
C	10,000	8,000	37,000	23,500	103	1.36	150		1/32	10	Cotton	Phenolic
CE	9,000	7,000	39,000	24,500	105	1.33	500	300	1/32	2	Cotton	Phenolic
L	13,000	9,000	35,000	23,500	105	1.35	150		0.010	2	Cotton	Phenolic
LE	12,000	8,500	37,00	25,000	105	1.33	500	300	0.015	2	Cotton	Phenolic
A	10,000	8,000	40,000	17,000	111	1.72	225	135	0.025	2	Asbestos paper	Phenolic
AA	12,000	10,000	38,000	21,000	103	1.70			1/16	2	Asbestos fabric	Phenolic
G-3	23,000	20,000	50,000	17,500	100	1.65	700	500	0.010	2	Cont. Gl.	Phenolic
G-5	37,000	30,000	70,000	25,000	120	1.90	350	220	0.010	31/2	Cont.l Gl.	Melamine
G-7	23,000	18,500	45,000	14,000	100	1.68	400	350	0.010	2	Cont. Gl.	Silicone
G-9	40,000	25,000	65,000			1.90	400	350	0.010	31/2	Cont. Gl.	Melamine
G-10	35,000	30,000	70,000	30,000	110	1.75	700	500	0.010	1	Cont. Gl.	Epoxy
G-11	35,000	30,000	70,000	30,000	110	1.75	700	500	0.010	1	Cont. Gl.	Epoxy
N-1	8,500	8,000	28,000		105	1.15	600	450	0.010	1	Nylon	Phenolic
FR-1									1/32	1/4	Paper	Phenolic
FR-2	12,400	9,500	25,000		105	1.30	650	450	1/32	1/4	Paper	Phenolic
FR-3	12,000	9,000	28,000		105	1.45	600	500	1/32	1/4	Paper	Epoxy
FR-4	35,000	30,000	70,000	30,000	110	1.75	700	500	0.010	1	Cont. Gl.	Epoxy
FR-5	35,000	30,000	70,000	30,000	110	1.75	700	500	0.010	1	Cont. Gl.	Epoxy
GPO-1	12,000	10,000	30,000	20,000	100	1.5–1.9	400		1/16	2	Gl. Mat	Polyester
GPO-2	10,000	9,000	30,000	20,000	100	1.5–1.9			1/16	2	Gl. Mat	Polyester
CEM-1									1/32	3/32	Cont. Gl- Paper core	Epoxy
CEM-3									1/32	3/32	Cont. CL glass mat	Epoxy

TABLE 16.192   Some Typical Properties of Polycrystalline Ceramic Substrate Materials

	96% Al_2O_3, Thick-Film Substrate	99.5% Al_2O_3 Thin-Film Substrate	99.5% BeO, Thin- or Thick-Film Substrate
Surface finish as fired, μm CLA	0.5	0.13	0.2
Density, g/cm³	3.70	3.85	2.88
Compressive strength, MPa	2600	2700	1500
Flexural strength, MPa	320	470	225
Modulus of elasticity, GPa	325	350	325
Thermal expansion, linear coefficient, K^{-1}			
25–300°C	6.4×10^{-6}	6.5×10^{-6}	7.5×10^{-6}
25–700°C	7.5×10^{-6}	7.3×10^{-6}	8.4×10^{-6}
25–900°C	7.9×10^{-6}	7.7×10^{-6}	8.7×10^{-6}
Thermal conductivity W/m K			
25°C	35.1	36.7	250
300°C	17.1	18.7	121
500°C	10.8	11.7	75
Dielectric strength,[a] kV/mm	31.5	30.5	—
Volume resistivity, Ω-cm			
25°C	$> 10^{14}$	$> 10^{14}$	$> 10^{14}$
300°C	1×10^6	9×10^{12}	8×10^{12}
700°C	3.5×10^6	1.6×10^4	3×10^9
Dielectric constant at 25°C, 1 MHz	9.3	10.0	6.0
Dissipation factor at 25°C, 1 MHz	0.0003	0.0009	0.0003

Source: *Encyclopedia of Materials Science and Engineering*, M. B. Bever (ed.), Pergamon Press, Oxford, 1986.
[a]At 60 Hz, samples 0.635 mm thick.

TABLE 16.193 Properties of Glass Substrate Materials

Glass Type	Soda Lime	Alkali Zinc Boro-silicate	Lime Aluminosilicate, Alkali Free	Lime Aluminosilicate, Alkali Free	Barium Alumino-Silicate, Alkali Free	Alkali Boro-silicate	96% Silica	Fused Silica
Code number[a]	0080	0211	1715	1723	7059	7740	7900	7940
Annealing point, °C	512	542	866	710	650	565	910	1050
Softening point. °C	696	720	1060	910	872	820	1500	1580
Thermal expansion coefficient, $\times 10^6$, in./in. °C	9.2	7.2	3.5	4.6	4.5	5.25	0.8	0.36
Thermal conductivity at 25°C, cal/cm s °C	0.0023	—	—	0.0032	—	0.0027	0.0038	0.0034
Density	2.47	2.57	2.48	2.63	2.76	2.23	2.18	2.20
Dielectric constant at 25°C, MHz	6.9	6.6	5.9	6.4	5.8	4.6	3.9	3.0
Loss tangent at 23°C, MHz	0.01	0.0047	0.0024	0.0013	0.0011	0.0062	0.0006	0.00002
Log volume resistivity at 250°C, Ω-cm	6.4	8.3	13.6	14.1	13.5	8.1	9.7	11.8
Dielectric strength at 25°C kV (rms)	0.35	2	>10	>10	>10	2	7	>10
Weatherability g/cm²	>3.0	0.05–0.25	<0.01	<0.01	<0.01	0.05–0.25	<0.01	<0.01
Chemical durability, mg/cm²								
In 5% HCl for 24 hr	0.02	0.03	0.10	0.4	5.5	0.005	0.001	0.001
In 5% NaOH for 6 hr	0.5	2	1.2	0.3	3.7	1.1	1.1	0.7
In 0.02 N Na_2CO_2 for 6 hr	0.1	0.1	0.15	0.1	0.3	0.1	0.03	0.03

Source: C. W. Harper, (ed.), *Handbook of Materials and Processes for Electronics*, ©1970, McGraw-Hill, New York. Reproduced with permission.

[a]Code numbers of Corning Glass Works, Corning, NY.

Dielectric Films

TABLE 16.194 Organic Film Selector Chart

Film	Cost	Thermal Stability	Dielectric Constant	Dissipation Factor	Strength	Electric Strength	Water Absorption	Folding Endurance
Cellulose	Low	Low	Medium	Medium	High	Medium	High	Low
FEP fluorocarbon	High	High	Low	Low	Low	High	Very low	Medium
Polyamide	Medium	Medium	Medium	Medium	High	Low	High	Very high
PTFE polytetrafluoroethylene	High	High	Low	Low	Low	Low	Very low	Medium
Acrylic	Medium	Low	Medium	Medium	Medium	Low	Medium	Medium
Polyethylene	Low	Low	Low	Low	Low	Low	Low	High
Polypropylene	Low	Medium	Low	Low	Low	Medium	Low	High
Polyvinyl fluoride	High	High	High	High	High	Medium	Low	High
Polyester	Medium	Medium	Medium	Low	High	High	Low	Very high
Polytrifluorochloroethylene	High	High	Low	Low	Medium	Medium	Very low	Medium
Polycarbonate	Medium	Medium	Medium	Medium	Medium	Low	Medium	Low
Polyimide	Very high	High	Medium	Low	High	High	High	Medium

Source: C. W. Harper, (ed.), *Handbook of Materials and Processes for Electronics.* © 1970, McGraw-Hill, New York. Reproduced with permission.

TABLE 16.195 Properties of Various Ceramic Dielectric Films

Material	Method of Deposition	Dielectric Constant, ε	tan δ	Breakdown Stress (Mv/cm)	Capacitance ($\mu F/cm^2$)
Silicon dioxide (SiO_2)	Reactive sputtering	4	0.001	3.0	0.015
Magnesium fluoride (MgF_2)	Evaporation	5	0.016	1.0	0.01
Silicon monoxide ($SiOx$)	Evaporation	5–7	0.010	1.2	0.01
Aluminum oxide (Al_2O_3)	Plasma oxidation	8	0.005	2.0	0.10
Aluminum oxide (Al_2O_3)	Anodic oxidation	8	0.005	4.0	0.20
Tantalum oxide (Ta_2O_5)	Reactive sputtering	20	0.003	1.0	0.10
Tantalum oxide (Ta_2O_5)	Anodic oxidation	27	0.005	3.0	0.15
Titanium oxide ($TiOx$)	Anodic oxidation	30–40	0.030	1.0	0.30
Lead titanate ($PbTiO_3$)	Reactive sputtering	80	0.040	0.6	0.20

Source: C. W. Harper, (ed.), *Handbook of Materials and Processes for Electronics*, © 1970, McGraw-Hill, New York. Reproduced with permission.

Semiconductors

TABLE 16.196 Properties of Silicon and Germanium Semiconductors

Property	Germanium	Silicon
Atomic number	32	14
Atomic weight	72.60	28.08
Lattice constant, A	5.657	5.431
Density, g/cm^3	5.323	2.328
Dielectric constant	16.0	12.0
Magnetic susceptibility, cgs units	-0.12×10^{-6}	-0.13×10^{-6}
Debye temperature, K	290	—
C_{11}, elastic constant, dyn/cm^2	12.98×10^{11}	16.740×10^{11}
C_{22}, elastic constant, dyn/cm^2	4.88×10^{11}	6.523×10^{11}
Volume compressibility, cm^2/dyn	1.3×10^{-12}	0.98×10^{-12}
Linear expansion coefficient, ppm/°C	6.1 (0–300°C)	4.2 (10–50°C)
Thermal conductivity, cal (sec cm °C)$^{-1}$	0.14 (25°C)	0.20 (20°C)
Specific heat, cal (g °C)$^{-1}$	0.074 (0–100°C)	0.181 (18–100°C)
Latent heat of fusion, cal/mol	8300	9450
Melting point	936°C	1420°C
Boiling point	2700°C	2600°C
Intrinsic resistivity at 300 K	47 Ω-cm	63,600 Ω-cm
Intrinsic density of carriers at 300 K	2.5×10^{13} cm^{-3}	6.8×10^{10} cm^{-3}
Mobilities at 300 K, cm^2 (V sec)$^{-1}$		
Electron	3600	1200
Hole	1700	250
Ionization Energy, eV		
Valence band	$0.75-0.0001T$	$1.12-0.0003T$
B, Al, Ga, In (low concentrations)	0.01 –	0.08
P, As, Sb (low concentrations)	0.01 +	0.05

Source: M. J. Sinnott, *The Solid State for Engineers*, Wiley, New York, 1958.

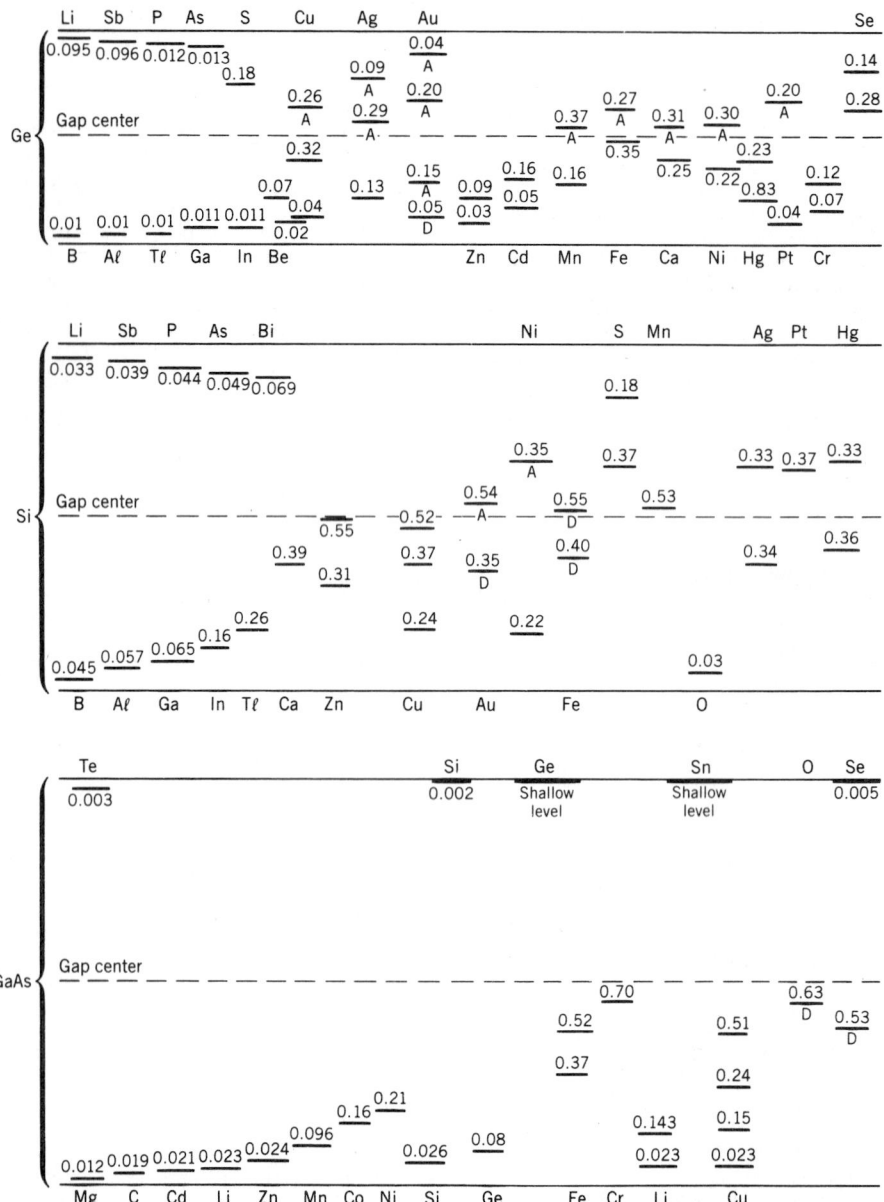

Fig. 16.17 Impurity energy levels for various elements in Ge, Si and GaAs at room temperature. (From C. A. Harper, (ed.), *Handbook of Materials and Processes for Electronics*, © 1970, McGraw-Hill, New York. Reproduced with permission.)

TABLE 16.197 Some Properties of II–VI and IV–VI Compounds

Compound	Melting Point (°C)	Crystal Structure[a]	Lattice Constant (Å)	Density (g/cm^3)	Energy gap[b] (eV)
II–VI Compounds					
ZnO	~ 1975	W	$a = 3.250$, $c = 5.207$	5.68	3.435
ZnS	1830	Z → W	5.409	4.10	3.839 (Z) 3.912 (W)
ZnSe	1515	Z	5.669	5.26	2.818
ZnTe	1290	Z	6.103	5.64	2.391
CdS	1397	W	$a = 4.137$, $c = 6.716$	4.82	2.583
CdSe	1259	W	$a = 4.298$, $c = 7.016$	5.68	1.841
CdTe	1092	Z	6.481	5.85	1.606
HgS	~ 1750	C → Z	$a = 4.146$, $c = 9.497$	8.10	2.090
HgSe	799	Z	6.085	8.24	Semimetal
HgTe	670	Z	6.460	8.09	Semimetal
IV–VI compounds					
GeS	674	O	$a = 4.30$, $b = 3.65$, $c = 10.44$	4.24	1.8
GeSe	670	O	$a = 4.38$, $b = 3.38$, $c = 10.80$	5.56	1.53
GeTe	724	R → NaCl	5.987, $\alpha = 88.3°$	6.19	0.1
SnS	881	O	$a = 4.33$, $b = 3.98$, $c = 11.19$	5.20	1.08
SnSe	860	O	$a = 4.44$, $b = 4.15$, $c = 11.50$	6.19	0.90
SnTe	806	NaCl	6.314	6.45	0.18
PbS	1111	NaCl	5.936	7.60	0.41
PbSe	1081	NaCl	6.126	8.27	0.27
PbTe	924	NaCl	6.462	8.24	0.31

Source: *Encyclopedia of Materials Science and Engineering*, M. B. Bever (ed.), Pergamon Press, Oxford, 1986.
[a] W, wurtzite; Z, zinc blende; C, cinnabar; O, orthorhombic; R, rhombohedral.
[b] At ~ 2 K for II–VI compounds; at 300 K for IV–VI compounds.

TABLE 16.198 Selected Properties of III–V Compounds

Compound	E_g at 300 K (eV)	Melting Point[a] (°C)	Crystal Structure[b]	Conductor type	Mobilities at 300 K μ_e	Mobilities at 300 K μ_h	Conduction Band Minimum[c]	Effective Masses $m_e^*(m_0)$	Effective Masses $m_h^*(m_0)$	a_0 (Å)	Dielectric Constant[d] ε_∞	Dielectric Constant[d] ε_0	Thermal Conductivity (W/cm K)
BN	10.0	(> 2000)	S / W	n, p	—	—	—	—	—	3.615; 2.504 (a); 6.661 (c)	4.5	7.1	—
BP	2.1	(> 1500)	S	n, p	80	70	I	—	—	4.54	—	—	—
B₆P	3.3	(> 1500)	R	n[a], p	—	150	I	—	—	5.984 (a_0); 11.85 (c_0)	—	—	—
BAs	1.46	(> 1500)	S	p	—	100	I	—	—	4.777	—	—	—
AlN	5.9	(> 1800)	W	—	—	—	I	—	—	3.111 (a_0); 4.978 (c_0)	—	—	—
AlP	2.45	2000	S	n	80	—	X	0.35	—	5.4625	—	—	~ 0.9
AlAs	2.16	1700	S	n, p	280	—	X	0.35	0.11	5.6622	8.16	10.06	~ 0.8
AlSb	1.62	1050	S	n	200	550	X	0.39	0.5	6.1355	10.24	14.4	0.54
GaN	3.39	~ 1500	W	n, p[a]	150	—	Γ	0.19	0.6 / 0.14	3.189 (a); 5.186 (c)	5.80	9.87	—
GaP	2.26	1470	S	n, p	190	120	X	0.35	0.86	5.4512	9.04	11.1	0.97
GaAs	1.43	1238	S	n, p	9000	435	Γ	0.068	0.5	5.6532	10.9	13.2	0.54
GaSb	0.72	706	S	n, p	6000	1420	Γ	0.047 (Γ) / 0.036 (L)	0.33	6.095	14.4	15.7	0.35
InN	1.95	—	W	—	—	—	Γ	—	—	3.533 (a); 5.693 (c)	—	—	—
InP	1.35	1062	S	n, p	4600	150	Γ	0.072	~ 0.8	5.8688	9.52	12.35	0.68
InAs	0.356	942	S	n, p	33000	460	Γ	0.026	0.41	6.0584	11.8	14.55	0.26
InSb	0.18	530	S	n, p	78000	1700	Γ	0.015	0.4	6.4794	15.7	17.72	0.18

[a] Values in parentheses are uncertain.
[b] S, sphalerite (zinc blende); W, wurtzite; R, rhombohedral.
[c] I, indirect (minima position not reported).
[d] ε_∞ and ε_0 are the high-frequency and dc dielectric constants, respectively.

Conductive Films

TABLE 16.199   Electrical Properties of Conductive Thin Films

Films	Substrate	Resistivity Ω/square[a]	Resistance Range (Ω)	Maximum Temperature Coefficient (ppm/°C)	Maximum Operating Temperature (°C)	Stability, Loaded (%, hr, °C)
			Vacuum Deposited			
Nickel-chromium (80/20)	Glass	100–300	$10\text{–}15 \times 10^3$	100	100	0.2, 2000, 70
	Ceramic	100–1,000	$10\text{–}1 \times 10^6$	100	150	1.0, 7000, 70
	Ceramic	$1\text{–}10 \times 10^3$	—	100	—	0.4, 2000, 70
Tantalum	Glass	50–600	$10\text{–}1.5 \times 10^6$	±150	—	1.0, 1000, 100
	Glass	4000	—	−150	—	—
Tantalum-gold	Glass	50	—	±50	—	0.25, 1700, 150
Tantalum nitride	Glass	10–100	$10\text{–}300 \times 10^3$	−60	—	1.0, 1000, 150
Chromium	Glass	200–1000	—	±100	300	—
Chromium-nickel (95/5)	Glass	5–50	—	±20	—	—
Titanium	Glass	50	—	Near zero	—	0.5, 1000, 65
Stainless steel	Glass	100	—	200	100	0.2, 1000, 100
Cermet (Cr-SiO)	Silicon monoxide	250	—	−50	300	0.2, 1000, 200 2.0, 1000, 300
	Ceramic	$2\text{–}20 \times 10^3$	—	±100	—	0.2, 2000, 125
Silicon	Quartz	200	—	−200	600	—
Chromium-silicon	Glass	$100\text{–}10 \times 10^3$	—	±500	250	1.0, 1000, 250
Germanium	Glass	5000	—	2000	300	1.0, yr, 50
Tungsten	Glass	200–600	—	Near zero	125	10 ppm, hr, 125
	Glass	2000	—	−200	—	—
Platinum	Quartz	2	—	200	—	—
Ruthenium		30	—	600	—	—
Osmium		50	—	250	—	—
Niobium	Quartz	50	—	150–300	—	—
Vanadium		150–300	—	−450 to −780	—	—
Zirconium		300	—	1000	—	—
Molybdenum	Quartz	100	—	350	—	—
Rhenium		100–800	—	50–150	—	—
Tungsten		100–300	—	50–200	—	—
			Chemically Deposited			
Metal oxide (SnO-SbO)	Glass	10–400	$10\text{–}2 \times 10^6$	±250	150	0.5, 2000, 70
	Ceramic	—	$10\text{–}500 \times 10^3$	±250	150	1.0, 2000, 70
Conductive glaze (Pd-PdO-Ag)	Ceramic	$10\text{–}100 \times 10^3$	$50\text{–}500 \times 10^3$	±250	500	5, 1000, 70
Vishalloy (Cr-Ni)	Glass	0.6	$15\text{–}500 \times 10^3$	±0.5	125	0.02, yr, 70
	Plastic	—	—	—	—	—
Nickel-cobalt	Ceramic	$1\text{–}10 \times 10^3$	$40\text{–}360 \times 10^3$	±30	70	0.2, 2000, 70
Gold-platinum	Glass	1–100	$10\text{–}3 \times 10^6$	350	200	0.1, long term
Cracked carbon	Ceramic	$10\text{–}10 \times 10^3$	$10\text{–}22 \times 10^6$	−500	100	0.2, 1000, 70

Source: C. W. Harper, (ed.), *Handbook of Materials and Processes for Electronics*, © 1970, McGraw-Hill, New York. Reproduced with permission.

[a] Film resistivity is measured in ohms per square and is the dc resistance measured between opposite faces of a square. The advantage in using ohms per square to specify resistance is that the measurement is independent of the size of the square. A uniform film of given thickness will have the same resistance per square centimeter as per square inch. Ohms per square resistivity is controlled by the film thickness.

TABLE 16.200   Properties of Metal Film Conductors for Interconnects

Metal	Melting Point (°C)	Volume Conductivity at 20°C	Specific Heat (cal/g °C)	Adherence Quality to SiO_2	Etching Quality	TC Bonds with Au Wire	Other Pertinent Information
Silver	961	108.5	−2.59	Poor	Good	Feasible	Forms solid solutions with gold
Copper	1083	100	−34.9	Fair	Good	Feasible	Forms solid solutions with gold
Gold	1063	77.7	+39	Poor	Excellent	Feasible	Eutectic with Si at 377°C
Aluminum	660	61.2	−376 (Al_2O_3)	Excellent	Excellent	Feasible	Brittle $AuAl_2$ with Au
Magnesium	650	38.7	−136.1	Very good	Good	Feasible	Chemically reactive
Rhodium	1906	38.3	—	Poor	Good	Feasible	High cost
Tungsten	3410	32.5	−182.5 (WO_3)	Very good	Good	Difficult	Alternative for Mo
Molybdenum	2625	30.9	−162 (MoO_2)	Very good	Very good	Difficult	Slight mutual solubility with Au
Cobalt	1495	27.6	−51 (CoO)	Good	Good	Difficult	Difficult to deposit
Nickel	1455	25.2	−51.7	Good	Good	Difficult	Forms solid solution with Au
Platinum	3224	16.3	—	Poor	Good	Feasible	Expensive
Palladium	1554	20.0	−52.2	Poor	Good	Feasible	Expensive
Chromium	1890	13.8	−250 (Cr_2O_3)	Excellent	Good	Difficult	Au solubility ~ 20 wt % at 900°C
Tantalum	2850	13.2	−471 (Ta_2O_3)	Excellent	Excellent	Difficult	High resistance
Lead	621	11.1	−45.3	Poor	Fair	Difficult	—
Vanadium	1860	6.5	−271 (V_2O_3)	Excellent	Fair	Difficult	Solution in Au $1\frac{1}{2}$ wt % at 500°C
Zirconium	1750	4.2	−244 (ZrO_2)	Excellent	Good	Difficult	—
Titanium	1820	3.2	−204 (TiO_2)	Excellent	Excellent	Difficult	Au solution 7.8 wt % at 700°C

Source: C. W. Harper, (ed.), *Handbook of Materials and Processes for Electronics*, © 1970, McGraw-Hill, New York. Reproduced with permission.

TABLE 16.201 Comparison of Thick and Thin Resistive Films

Criterion	Thick Film	Thin Film
Initial resistor tolerances, %	±10–20 as fired	±5 as deposited
Trimming tolerance, %	±0.5	±0.1
Power-handling capabilities, W/in.^{2}	~ 50	~ 15
Resistor temperature coefficients, ppm/°C	±100	0
Resistance, Ω/square[a]	1–50,000	0.1–1000
Line width capabilities, mils	5 ± 1	0.2 ± 0.02
Capacitance limitations	10,000 pF practical	0.01 μF practical
10,000-hr drift, %	±1	±0.1
Cost each per 100 circuits	$30 each, plus $1000 tooling	$55 each, plus $2500 tooling

Source: C. W. Harper, (ed.) *Handbook of Materials and Processes for Electronics*, © 1970, McGraw-Hill, New York. Reproduced with permission.
[a]See footnote *a* of Table 16.199.

TABLE 16.202 Resistance (Ω) of Films One Unit in Length for Various Resistivities

Line width (units)	Resistivity Ω/square[a]								
	25	50	100	150	200	500	1,000	1,400	2,000
0.006	4,166	8,333	16,666	25,000	33,332	83,333	166,666	233,324	333,333
0.007	3,560	7,120	14,240	21,360	28,480	71,200	142,400	200,360	284,800
0.008	3,140	6,280	12,560	18,840	25,120	62,800	125,600	175,840	251,200
0.009	2,790	5,580	11,160	16,740	22,320	55,800	111,600	156,240	223,200
0.010	2,500	5,000	10,000	15,000	20,000	50,000	100,000	140,000	200,000
0.011	2,275	4,550	9,100	13,650	18,200	45,500	91,000	127,400	182,000
0.012	2,080	4,160	8,320	12,480	16,640	41,600	83,200	116,480	166,400
0.013	1,920	3,840	7,680	11,420	15,360	38,400	76,800	107,520	153,600
0.014	1,780	3,560	7,120	10,680	14,240	33,600	67,200	99,680	134,400
0.015	1,665	3,330	6,660	10,000	13,320	33,300	66,600	93,240	133,200
0.016	1,565	3,130	6,260	9,390	12,520	31,300	62,600	87,640	125,200
0.017	1,470	2,940	5,880	8,820	11,760	29,400	58,800	82,320	117,600
0.018	1,385	2,770	5,540	8,310	11,080	27,700	55,400	77,560	110,800
0.019	1,315	2,630	5,260	7,890	10,520	26,300	52,600	73,640	105,200
0.020	1,250	2,500	5,000	7,500	10,000	25,000	50,000	70,000	100,000
0.021	1,170	2,340	4,680	7,020	9,360	23,400	46,800	65,520	93,600
0.022	1,135	2,270	4,540	6,810	9,080	22,700	45,400	63,560	90,800
0.023	1,085	2,170	4,340	6,510	8,680	21,700	43,400	60,760	86,800
0.024	1,040	2,080	4,160	6,240	8,320	20,800	41,600	58,240	83,200
0.025	1,000	2,000	4,000	6,000	8,000	20,000	40,000	56,000	80,000
0.050	500	1,000	2,000	3,000	4,000	10,000	20,000	28,000	40,000
0.100	250	500	1,000	1,500	2,000	5,000	10,000	14,000	20,000

Source: C. W. Harper, (ed.), *Handbook of Materials and Processes for Electronics*, © 1970, McGraw-Hill, New York. Reproduced with permission.
[a]See footnote *a* of Table 16.199.

Photoresists

TABLE 16.203 Properties of Photoresists[a]

Supplier	Name	Remarks
Negative Resists		
Du Pont	Riston	Supplied as film; for plating and etching
Dynachem	DCR 3140	General-purpose resist
	DCR 3154	Improved adhesion to aluminum and improved resistance to alkaline etches
	DCR 3118, 3118H	Roller-coating formulation
	DCR 3116	Provides heavy resist layer
	DCR 3170	Microelectronic formulation
Philip A. Hunt	Waycoat No. 10	General-purpose etching and plating resist
	Waycoat No. 20	Especially useful as plating resist
Kodak	KPR	Used on copper and copper-based alloys; used on clear and light-colored anodized aluminum
	KPR2	Used on copper and copper-based alloys; electroplating resist
	KPR3	Formulated for dip-coating systems
	KPR4	Formulated for roller-coating systems; good for plated-through holes
	KOR	Similar to KPR 2; possesses greater spectral sensitivity than other products
	KMER	Used on all surfaces except copper and copper alloys and clear and light-colored anodized aluminum
	KTFR	Microelectronic formulation
	KPL	Used to increase viscosity of KPR; rarely used alone
Norland	Photoresist 30	One-part water-based resist for stainless, regular steel, nickel, copper, brass
	Photoresist 22	Two-part water-based system for Kovar-type metals
Positive Resists		
GAF	—	Positive resist in field testing
Shipley		
	AZ-111	General-purpose photoresist
	AZ-119	As above but formulated for roller coating
	AZ-340	Used for circuit boards and plated-through holes
	AZ-345	Higher solids and viscosity than AZ-340 to provide lands around plated-through holes
	AZ-1350	Used for microelectronic applications
	AZ-1350H	As above, but formulated for roller coating

Source: C. W. Harper, (ed.), *Handbook of Materials and Processes for Electronics*, © 1970, McGraw-Hill, New York. Reproduced with permission.

[a] Thicknesses, 5000–10,000 Å. Negative resists develop patterns corresponding to transparent (exposed) areas of mask; typical resolution, 2.5 μm; typical sensitivity, 8–20 mJ/cm^2. Positive resists develop patterns corresponding to opaque (unexposed) areas of mask; typical resolution, 0.5 μm; typical sensitivity, 8–20 mJ/cm^2.

TABLE 16.204 Polymer Materials Evaluated as Electron Resists

Polymer	Type[a]	Sensitivity at 10 kV μC cm^{-2}	Etch Resistance[b]	Lithographic Performance	Resolution[c] (μm)
KMER	N	6	A, B, I	poor	2.0
Polystyrene	N	200	A, B, I	excellent	< 0.5
Poly(glycydyl methacrylate-co-3-chlorostyrene) (GMC)	N	< 1	A, B, I	excellent	< 1.0
Poly(glycidyl methacrylate-co-ethyl acrylate) (COP)	N	0.4	A, B, I	good	1.5
Poly(diallyl orthophthalate)	N	1	A, B, I	fair	1.0
Polysioxanes	N	100	I[d]	good	0.5
Polymethyl methacrylate	P	60	A, B, I	excellent	0.2
Shipley AZ (1350)	P	15	A, I	fair	1.0
Poly(butene-l-sulfone) (PSB)	P	0.8	A, B	excellent	0.25
Poly(2-methylpentene-1-sulfone)	P	0.2	A, B	fair	< 0.5
Poly(methyl isopropenyl ketone)	P	6	A	—	0.5
Cross-linked poly(methyl methacrylate)	P	10	A, B, I	excellent	< 0.5
Novolac PMPS (NPR)	P	2[e]	A, B, I	excellent	< 0.5

Source: *Encyclopedia of Materials Science and Engineering*, M. B. Bever (ed.), Pergamon Press, Oxford, 1986.
[a]N, negative; P, positive.
[b]A = acid resistant, B = base, I = ion and plasma.
[c]For equal lines and spaces.
[d]Converted to SiO and are not good etch masks.
[e]PMPS vapor develops at this dose.

16.3.9 Optical Materials

TABLE 16.205 Optical Properties of Common Transmitting Materials[a]

Material	Solar Transmittance (%)	Infrared Transmittance (%)	Photodegradation Resistance[b]
Sheets			
Float-glass sheet 3.2 mm thick	84	2	Excellent
Low-iron glass sheet, 3.2 mm thick	90	2	Excellent
Acryclic sheet, 3.2 mm thick	90	2	Excellent
Polycarbonate sheet, 3.2 mm thick	84	2	Fair
Fiberglass reinforced, polyester sheet, 0.6 mm thick	87	8	Fair
Thin Films			
Polyethylene film, 0.1 mm	92	80	Poor
Polyester film, 0.13 mm	87	20	Poor
Poly(vinyl fluoride) film, 0.1 mm	92	20	Good to excellent
Polytetrafluoroethylene, 0.05 mm	96	26	Excellent
Poly(vinylidene fluoride), 0.1 mm	93	23	Excellent

Source: *Encyclopedia of Materials Science and Engineering*, M. B. Bever (ed.), Pergamon Press, Oxford, 1986.
[a]Approximate property values. Actual properties of materials may vary from manufacturer to manufacturer and lot to lot.
[b]Degradation is subjective consensus based on similar uses in glazing applications.

TABLE 16.206   Properties of Common Absorber Coatings[a]

Material	Solar Absorptance (%)	Thermal Emittance at 25°C (%)	Stability Maximum (°C)
Black paints (low-temperature organic binders)	92–96	90–92	100
Black paints (high-temperature inorganic binders)	95–96	84–85	600–800
Black chrome (electrodeposited)	94–96	5–10	300
Black cobalt (electrodeposited)	93–96	14–18	700
Tabor black (electrodeposited NiS/ZnS)	91	14	100
Platinum aluminum-oxide cermets	95	8–15	500

Source: Encyclopedia of Materials Science and Engineering, M. B. Bever (ed.), Pergamon Press, Oxford, 1986.
[a]Approximate property values. Actual properties of materials may vary from manufacturer to manufacturer and lot to lot.

TABLE 16.207   Characteristics of Fiber-optic Materials

Wavelength Region ("Window")	Major Use	Typical Fibers	Typical Attenuation (dB/km)	Bandwidth (MHz/km)
~ 630 nm	Short-haul data transmission	Plastic	1200	—
~ 850 nm[a]	General purpose	Plastic-clad silica,	15	40
		Graded-index glass,	—	—
		Step index glass	4.5	20
~ 1300 nm[b]	Long-haul trunk lines	Graded-index silica,	1.5	400
		Single-mode silica	1.0	—
~ 1550 nm[c]	Long-haul trunk lines	High-grade silica (single mode)	< 0.2	—

[a]First window.
[b]Second window.
[c]Third window.

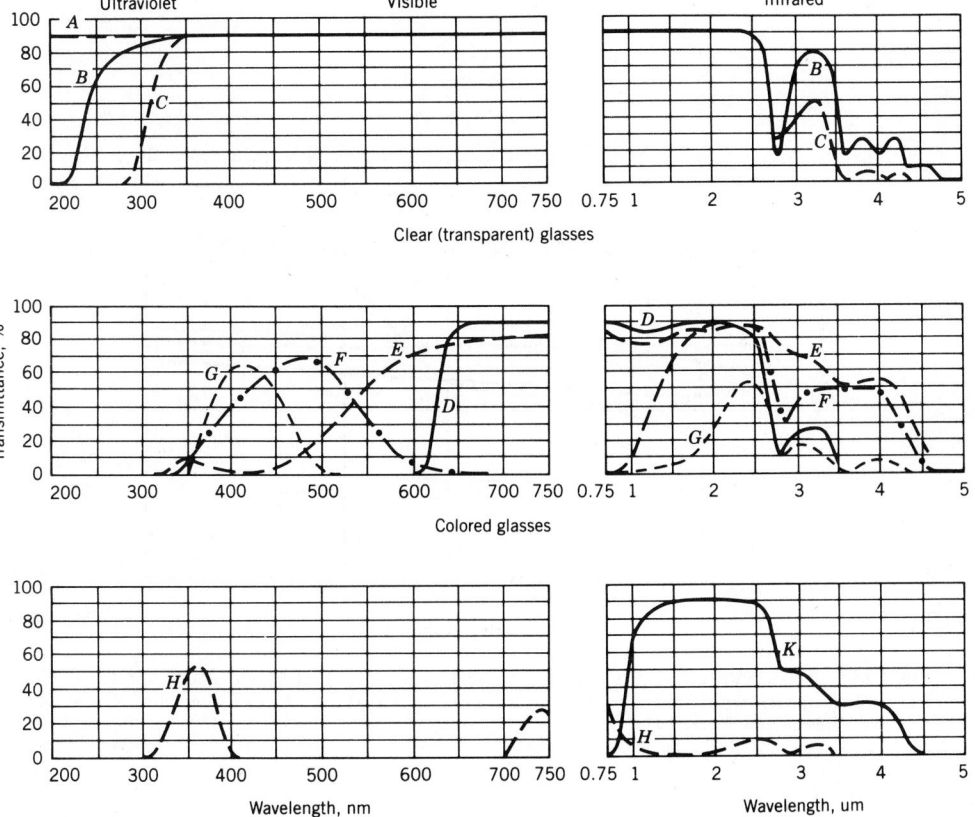

Fig. 16.18 Curve of transmittance versus wave-length for selected glasses.

Curve	Glass Type	Corning Glass No.	Corning Filter Specification	Approximate Thickness (mm)
A	Clear silica glass (very pure)	7940	—	5
B	Clear 96% silica glass	7910	9–54	2
C	Clear borosilicate glass	7740	0–53	2
D	Red	2408	2–60	3
E	Amber	3307	3–77	3
F	Green	4445	4–74	2.5
G	Blue	5543	5–60	5
H	Ultraviolet transmitting	5874	7–39	5
K	Infrared transmitting	2540	7–56	2.5

(From *Glass Engineering Handbook*, E. B. Shand, (ed.), Corning Glass Works, Corning, New York, 1955.)

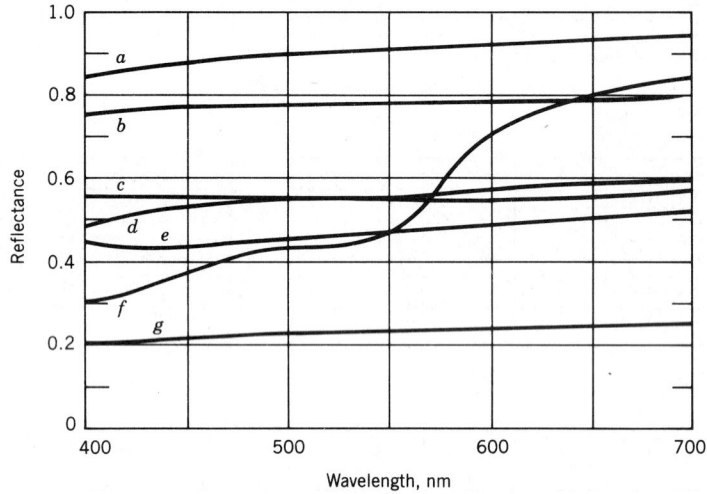

Fig. 16.19 Reflectance curves from various metallic surfaces: (*a*) silver, (*b*) aluminum, (*c*) steel, (*d*) chromium, (*e*) molybdenum, (*f*) copper, and (*g*) graphite. (From R. M. Evans, *An Introduction to Color*, Wiley, New York, 1948.)

TABLE 16.208   Characteristics of Typical Solid-State Laser Crystals

Crystal	Active Center	Concentration of Active Centers (cm^{-3})	Spontaneous Fluorescence Lifetime	Wavelength
Ruby	Cr^{3+}	1.6×10^{19}	3 ms	694.3 nm
YAG-Nd	Nd^{3+}	1.4×10^{20}	230 μs	1.061 μm, 1.064 μm, 1.839 μm (77 K), 0.946 μm (77 K), 1.318 μm
NaF-(F_2^+)	(F_2^+)* color center	2×10^{17}	40 ns	Tunable, 0.99–1.22 μm
$Nd_{0.5}La_{0.5}P_5O_4$	Nd^{3+}	2×10^{21}	150 μs	1.05 μm

Source: *Encyclopedia of Materials Science and Engineering*, M. B. Bever (ed.), Pergamon Press, Oxford, 1986.

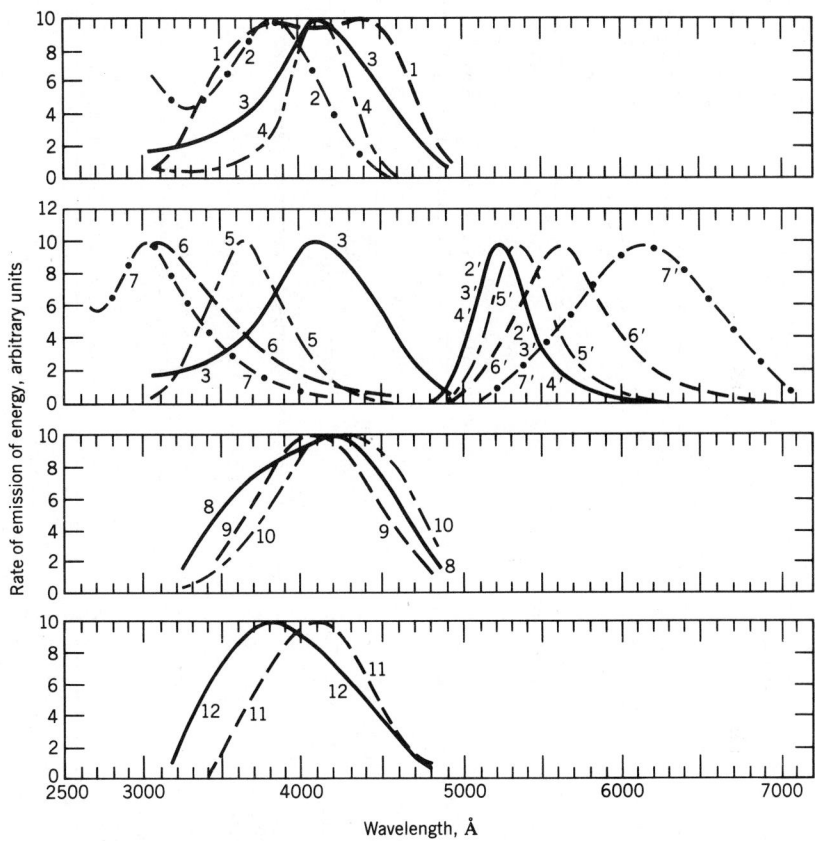

Fig. 16.20 Spectral distribution curves for complex oxide phosphors: 1, SiO_2, 1300°C, 1 hr; 2, rbhdl.-ZnO · $2SiO_2$, 1250°C, 1 hr; 3, rbhdl.-2ZnO · SiO_2, 1250°C, 1 hr; 4, rbhdl.-3ZnO · SiO_2, 1250°C, 1 hr; 5, rbhdl.-2ZnO · GeO_2, 1200°C, 1 hr; 6, β-2ZnO · SiO_2, 1600°C, 10 minutes, quenched; 7, rbhdl.-2BeO · SiO_2, 1400°C, 1 hr; 3′, rbhdl.-2ZnO · SiO_2 · $0.012MnSiO_3$, 1250°C, 1 hr; 5′, rbhdl.-2ZnO · GeO_2 · $0.012MnSiO_3$, 1100°C, 1 hr; 6′, β-2ZnO · SiO_2 · $0.012MnSiO_3$, 1600°C, 10 minutes, quenched; 7′, rbhdl.-3ZnO · BeO · $2.2SiO_2$ · $0.384MnSiO_3$, 1200°C, 1 hr; 8, CaO · SiO_2 · 0.005Re, 1200°C, 7 hr, quenched; 9, SrO · SiO_2 · 0.006Re, 1550°C, 1 hr, quenched; 10, BaO · SiO_2 · 0.006Re, 1500°C, 4 hr, quenched; 11, rhomb.-2MgO · SiO_2, 1600°C, 20 minutes; 12, tetr.-ZrO_2 · SiO_2 · 0.005V, 1600°C, 40 minutes. (From Leverenz, *An Introduction to the Luminescence of Solids*, Wiley, New York, 1950.)

16.3.10 Abrasive Materials

TABLE 16.209 Natural Abrasives

Abrasive	Mineral Species	Mohs Hardness	Important Grinding and Polishing Applications
Diamond	Diamond	10	Carbides, stone, gems
Corundum	Al_2O_3	9	Optical glass
Aluminum oxide	Calcined bauxite	8.5–9	Glass, ceramics, metals
Emery	Al_2O_3 plus hematite or magnetite, quartz, and spinel	7.5–8.5	Emery paper and cloth for metal finishing
Garnet	$Fe_3Al_2(SiO_4)_3$	7.5–8	"Garnet" paper for finishing wood
Crushed quartz	SiO_2	7	wood finishing
Sand	SiO_2	7	Pressure blasting, buffing, polishing cpds
Tripoli	Microcrystalline, friable quartz	7	Hand soaps, dentrifrices, scouring, buffing, polishing cpds
Amorphous silica	Same as above	7	Same as above
Flint and chert	Microcrystalline quartz	7	"Flint" paper
Pumice	Volcanic glass, SiO_2–Al_2O_3–K_2O–Na_2O	6–7	Cleaning and scouring cpds
Olivine	Mg_2SiO_4–Fe_2SiO_4	6–7	Blasting medium
Kaolin	Calcined Al_2O_3–$2SiO_2$–$2H_2O$	6–7	Automotive polish, silver polish, tooth paste
"Mild polish"	SnO_2	6–7	Soft metals
Feldspar	Aluminum silicate minerals, e.g., orthoclase, microcline, anorthite	6	Scouring powder for glass and enamel
Nepheline syenite	Na–K–Al silicate rock	5.5–6	Same as feldspar
Green rouge	Cr_2O_3	5.5	Soft metals
Rouge	Hydrated iron oxide	5–5.5	Glass polishing
Diatomite	Hydrated SiO_2	5	Silver polish, automobile buffing, polishing cpds
Tricalcium phosphate	Hydrated $Ca_3(PO_4)_2$	4.5–5	Dentrifrice
Chalk	$CaCO_3$ or $CaMg(CO_3)_2$	3.5–4.0	Polishing wood, plastics, semiprecious stones
Talc	$4SiO_2 \cdot 3MgO \cdot H_2O$	1–1.5	Polishing cereal grains, leather, synthetic fibers

TABLE 16.210   Artificial Abrasives

Type	Knoop Hardness	Instability Temperature (°C)	Thermal Conductivity (cal/sec cm °C)	Particular Applications
Diamond	9000–10,000	Oxidation, 600–800; graphitization, 900–1000	5	Sawing and drilling stone, grinding optical glass and carbide tools, oil well drilling, grinding and broaching Al-Si alloys
CBN[a]	4500–5000	Oxidation, 1300 graphitization, 1600	3	Grinding hard steels and cast irons
B$_4$C	2800	Oxidation, 900	0.07	Lapping paste, dressing tools for grinding wheels
SiC	2500	Oxidation, 1600	0.21	Grinding cast iron and ceramics

[a] Cubic boron nitride.

16.3.11 Corrosion-resistant Materials

TABLE 16.211   Electromotive Force Series

Anodic end		Anodic end	
1	Lithium	49	25% Cr–20% Ni steel, type 310 (active)
2	Rubidium	50	18% Cr–12% Ni–3% Mo steel, type 316 (active)
3	Potassium		
4	Strontium	51	Hastelloy C (59% Ni, 17% Mo, 5% Fe, 14% Cr, 5% W, 0.1% C)
5	Barium		
6	Calcium	52	Lead
7	Sodium	53	Tin
8	Magnesium and its alloys	54	Iron (Fe^{3+})
9	Aluminum	55	Hydrogen
10	Beryllium	56	Antimony
11	Uranium	57	Bismuth
12	Manganese	58	Arsenic
13	Tellurium		
14	Zinc	59	Muntz metal (60% Cu, 40% Zn)
15	Chromium	60	Manganese bronze (66.5% Cu, 19% Zn, 6% Al, 4% Mn)
16	Sulfur		
17	Gallium	61	Naval brass (addition of $\frac{3}{4}$% Sn to Muntz metal)
18	Iron (Fe^{2+})		
19	Galvanized steel	62	Nickel (active)
20	Galvanized wrought iron	63	60% Ni–15% Cr (active)
21	Al-Zn-Mg alloys (e.g., Al-75S)	64	Inconel (78% Ni, 13.5% Cr, 6% Fe) (active)
22	Al–high Mg alloys (e.g., Al-220)		
23	A^b Al–low Mg alloys (e.g., Al-4S)	65	80% Ni–20% Cr (active)
24	Aluminum plus very low percentage of alloying constituents (e.g., Al-53S)	66	Hastelloy A (60% Ni, 20% Ni, 20% Mo, 20% Fe, 0.1%C)
25	Alclads	67	Hastelloy B (65% Ni, 30% Mo, 5% Fe, 0.1% C)
26	Cadmium	68	Yellow brass (58–70% Cu, 0.50–1.5% Sn, 0.75–3.5% Pb, balance Zn)
27	Al-Si-Mg alloys^c (e.g., Al-356)		
28	Al-Cu alloys^c (with or without small additions of Mg) (e.g., Al-24S)	B^c	
29	Al-Cu alloys^c (with or without small additions of Zn) (e.g., Al-113)	69	Admiralty brass (71% Cu, 28% Zn, 1% Sn)
30	Al-Cu-Si alloys^c (e.g., Al-108)	70	Aluminum bronze (addition of 2–2.5% Al to 75% Cu–25% Zn alloy)
31	Mild steel		
32	Copper steel		
33	SAE 4140	71	Red brass (85% Cu, 15% Zn)
34	SAE 3140		
35	Wrought iron	72	Copper (-ic)
36	Cast iron	73	Oxygen
37	4–6% chromium steel, type 501 or 502 (active)	74	Polonium
		75	Copper (-ous)
38	12–14% chromium steel, types 403, 410, 416 (active)	76	Iodine
		77	Tellurium
39	16–18% chromium steel, type 440 (active)	78	Silicon bronze (1.0–3.0%Si)^e
40	23–30% chromium steel, type 446 (active)	79	Nickel-silver
		80	Ambrac (5.0% Zn, 20% Ni, balance Cu)
41	Indium^d		
42	Thallium^d	81	70% Cu–30% Ni
43	Cobalt^d	82	Comp. G-bronze (88% Cu, 2% Zn, 10% Sn)
44	Ni-resist cast iron	C^f	
45	50–50 lead tin solder	83	Comp. M-bronze (88% Cu, 3% Zn, 6.5% Sn, 1.5% Pb)
46	17% Cr–7% Ni steel, type 301 (active)		
47	18% Cr–8% Ni steel, types 302, 303, 304, 321, 347 (active)	84	Silver solder
		85	Nickel (passive)
48	23% Cr–14% Ni steel, type 309 (active)	86	60% Ni–15% Cr (passive)

TABLE 16.211   *(Continued)*

87		Inconel (passive)	99	Mercury
88		80% Ni–20% Cr (passive)	100	Silver
89		Titanium	101	Lead (Pb^{4+})
90		Monel (70% Ni, 30% Cu)	102	Palladium
91		12–14% Cr steel, types 403, 410, 416 (passive)	103	Bromine
			104	Chlorine
92		16–18% Cr steel, type 440 (passive)	105	Graphite[g]
			106	Gold (-ic)
93		17% Cr–7% Ni steel, type 301 (passive)	107	Gold (-ous)
			108	Platinum
94	C^f	18% Cr–8% Ni steel, types 302, 303, 304, 321, 347 (passive)	109	Fluorine Cathodic End
95		23% Cr–14% Ni steel, type 309 (passive)		
96		23–30% chromium steel, type 446 (passive)		
97		25% Cr–20% Ni steel, type 310 (passive)		
98		18% Cr–12% Ni–3% Mo steel, type 316 (passive)		

Source: A. E. Durkin and Carroll Seversike, Thomson Laboratory, General Electric Company, Lynn, Massachusetts.

[a] Select contacting combinations as close together as possible in this series. Keep dissimilar metals as far apart as possible and particularly avoid threaded connections between widely dissimilar metals. Avoid combinations where area of less noble metal is small; i.e., use a more noble fastening for less corrosion-resistant part, like Monel rivets on steel tank instead of steel rivets on Monel tank. Insulate joints where possible. Paint both noble and less noble if paint insulation is used. When brazing, choose brazing alloy more noble than at least one of the metals to be jointed.

[b] Members of group A are listed relative to each other, but position of group with respect to group between chromium and ferrous iron is uncertain. The group, however, is listed correctly between zinc and cadmium.

[c] Heat treatment of aluminum and its alloys has effect upon their potentials. They fall in aluminum group, however, regardless of heat treatment.

[d] Indium, thallium, and cobalt are listed relative to each other, but their position relative to stainless steels is uncertain. They are properly placed between aluminum and lead.

[e] Members of group B are listed relative to each other, but position of group relative to ferric iron to arsenic is uncertain. Group is, however, properly placed between tin and copper.

[f] Members of group C are listed relative to each other, but position of this group withy respect to iodine tellurium and mercury is uncertain. However, group is properly placed between copper and silver.

[g] Graphite lies between lead and gold, but its position relative to palladium, bromine, and chlorine is uncertain.

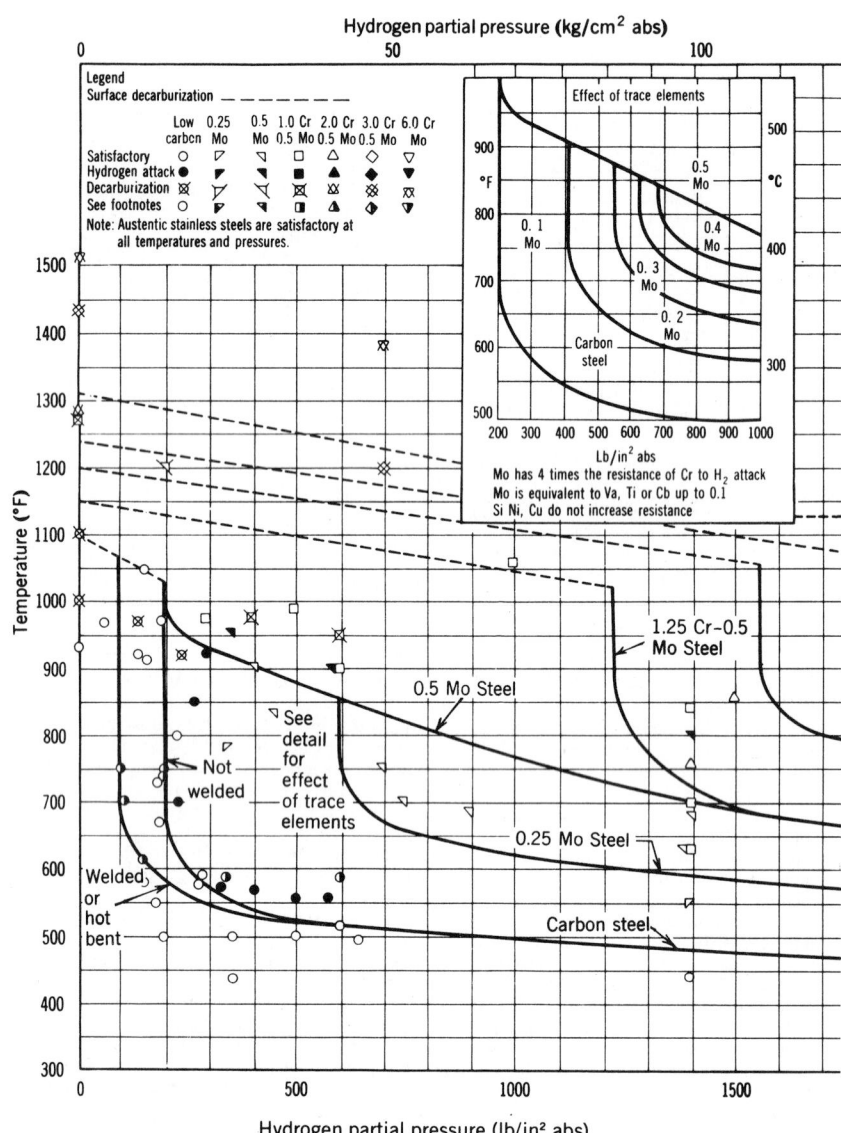

Fig. 16.21 Operating limits for steels in hydrogen service. (Courtesy of Shell Development Company.)

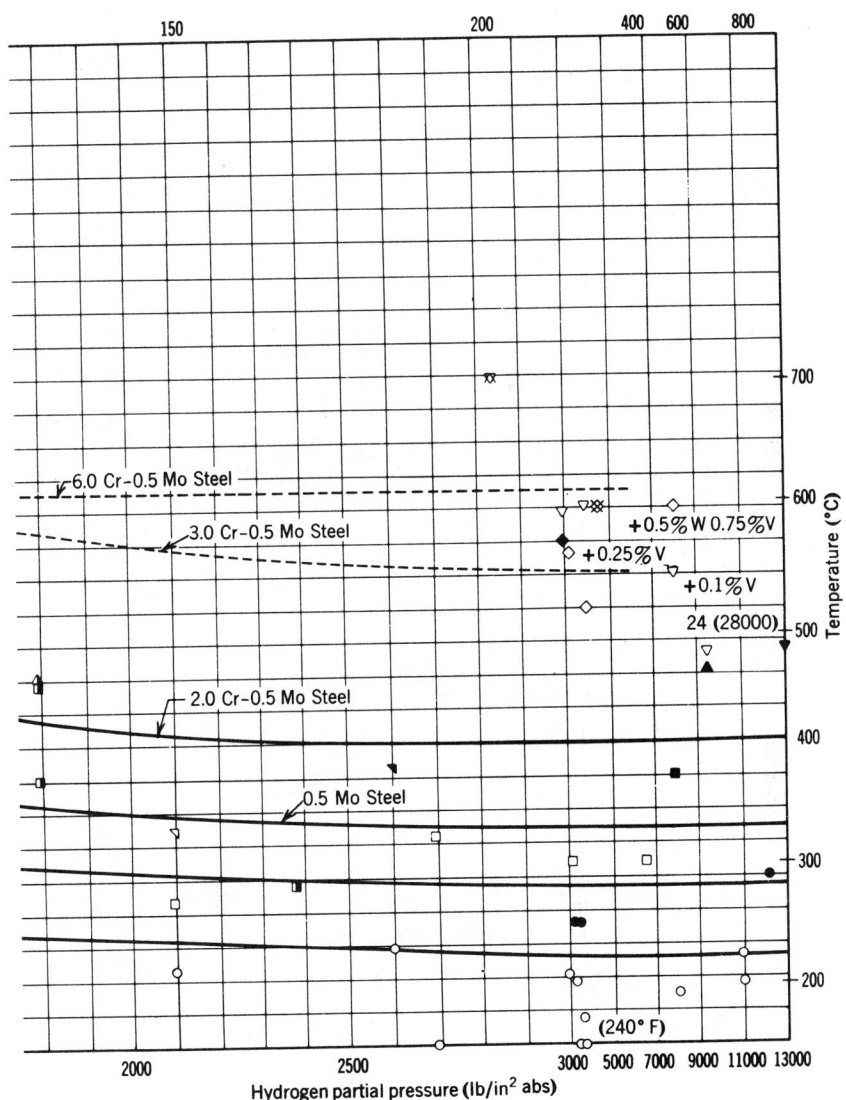

Fig. 16.21 *(Continued)*

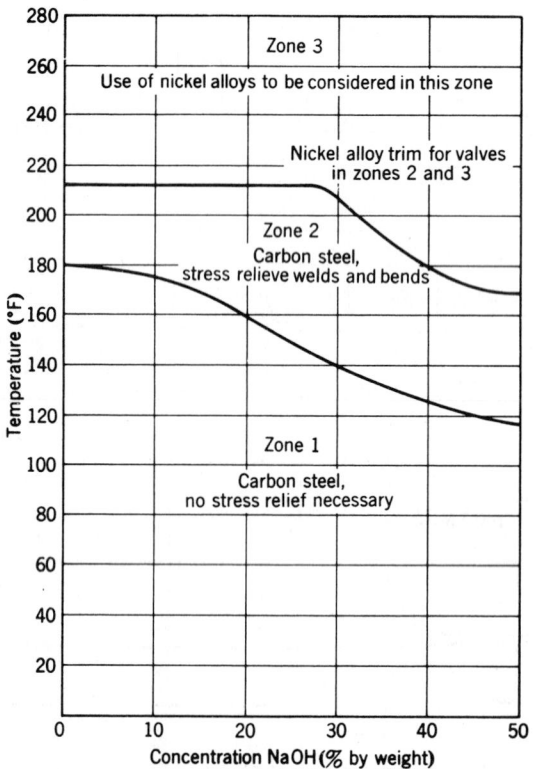

Fig. 16.22 Suitability of steels in caustic soda service.

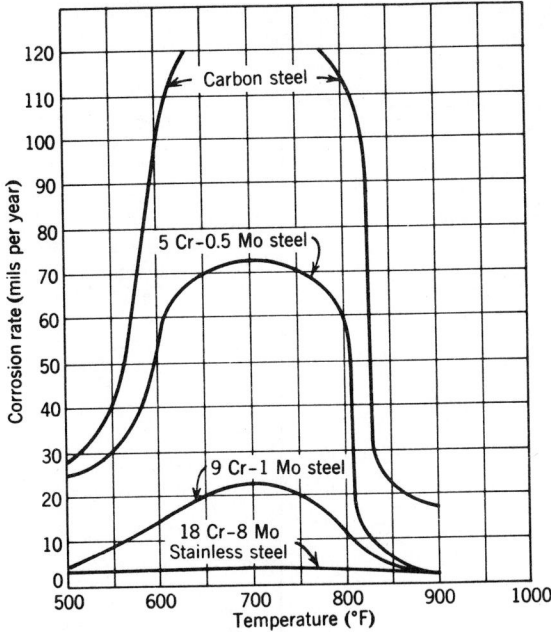

Fig. 16.23   Corrosion rates for steels in crude oil containing 1.5% sulfur.

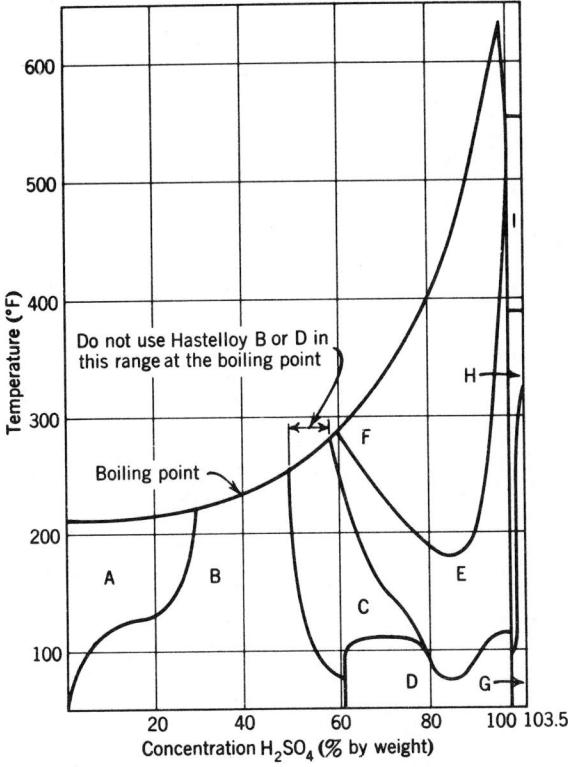

Fig. 16.24   Corrosion resistance of various materials to sulfuric acid.

TABLE 16.212   Key For Sulfuric Acid Chart[a]

Area A

10% aluminum bronze
   (air free)
Illium G
Glass
Hastelloy B and D
Durimet 20
Worthite
Lead
Copper (air free)
Monel (air free)
Haveg 43
Rubber (up to 170°F)
Impervious graphite
Tantalum
Gold
Platinum
Silver
Zirconium
Nionel
Tungsten
Molybdenum
Type 316 stainless
   (up to 10% aerated)

Area B

Glass
Silicon iron
Hastelloy B and D
Durimet 20 (up to 150°F)
Worthite (up to 150°F)
Lead
Copper (air free)
Monel (air free)
Haveg 43
Rubber (up to 170%F)
10% aluminum bronze
   (air free)
Ni Resist (up to 20% at 75%F)
Impervious graphite
Tantalum
Gold
Platinum
Silver
Zirconium
Nionel
Tungsten
Molybdenum
Type 316 stainless (up to 25%
   at 75°F) aerated

Area C

Glass
Silicon iron
Hastelloy B and D
Durimet 20 (up to 150°F)
Worthite (up to 150°F)
Lead
Monel (air free)
Impervious graphite
Tantalum
Gold
Platinum
Zirconium
Molybdenum

Area D

Steel
Glass
Silicon iron
Hastelloy B and D
Lead (up to 96% H_2SO_4)
Durimet 20
Worthite
Ni Resist
Type 316 stainless
   (above 80%)
Impervious graphite
   (up to 96% H_2SO_4)
Tantalum
Gold
Platinum
Zirconium

Area E

Glass
Silicon iron
Hastelloy B and D
Durimet 20 (up to 150°F)
Worthite (up to 150°F)
Lead (up to 175%F
   and 96% H_2SO_4)
Impervious graphite
   (up to 175°F and 96% H_2SO_4)
Tantalum
Gold
Platinum

TABLE 16.212 *(Continued)*

Area F	*Area H*
Glass	Glass
Silicon iron	18 Cr–8 Ni
Tantalum	Durimet 20
Gold	Worthite
Platinum	Gold
	Platinum
Area G	
Glass	
Steel	
18 Cr–8 Ni	
Durimet 20	*Area I*
Worthite	
Hastelloy C	Glass
Gold	Gold
Platinum	Platinum

*Materials having reported corrosion rate less than 0.020 in./yr.

TABLE 16.213   Key For Hydrofluoric Acid Chart[a]

Area A	*Area B*
Monel (air free)	Monel (air free)
Copper (air free)	70 Cu–30 Ni (air free)
70 Cu–30 Ni (air free)	Copper (air free)
Lead (air free)	Lead (air free)
Nickel (air free)	Nickel (air free)
Alloy 20	Alloy 20
Ni Resist	Hastelloy C
Hastelloy C	Platinum
Platinum	Silver
Silver	Gold
Gold	Impervious graphite
Impervious graphite	Rubber
Haveg 43	Haveg 43
Rubber	
25 Cr–20 Ni steel	

TABLE 16.213   *(Continued)*

Area C	*Area E*
Monel (air free)	Monel (air free)
70 Cu–30 Ni (air free)	70 Cu–30 Ni (air free)
Copper (air free)	Lead (air free)
Lead (air free)	Hastelloy C
Alloy 20	Platinum
Hastelloy C	Silver
Platinum	Gold
Silver	Impervious graphite
Gold	Haveg 43
Impervious graphite	
Haveg 43	*Area F*
Rubber	Monel (air free)
	Hastelloy C
	Platinum
	Silver
	Gold
Area D	Haveg 43
Monel (air free)	
70 Cu–30 Ni (air free)	*Area G*
Copper (air free)	Carbon steel
Lead (air free)	Monel (air free)
Hastelloy C	Hastelloy C
Platinum	Platinum
Silver	Silver
Gold	Gold
Impervious graphite	Haveg 43
Haveg 43	

[a]Materials having reported corrosion rate less than 0.020 in./yr.

TABLE 16.214 Key for Hydrochloric Acid Chart[a]

Area A	*Area C*
Chlorimet 2	Chlorimet 2
Glass	Glass
Silver	Silver
Platinum	Platinum
Tantalum	Tantalum
Hastelloy B	Hastelloy B (chlorine free)
Durichlor (FeCl$_3$ free)	Durichlor (FeCl$_3$ free)
Haveg	Haveg
Saran	Saran
Rubber	Rubber
Silicon bronze (air free)	Molybdenum
Copper (air free)	Zirconium
Nickel (air free)	Impervious graphite
Monel (air free)	
Zirconium	*Area D*
Tungsten	Chlorimet 2
Titanium, up to 10% HCl at room temperature	Glass
	Silver
Worthite, up to 2% HCl at room temperature	Platinum
	Tantalum
	Hastelloy B (chlorine free)
Area B	Durichlor (FeCl$_3$ free)
Chlorimet 2	Monel (air free) (up to 0.5% HCl)
Glass	Zirconium
Silver	Impervious graphite
Platinum	Tungsten
Tantalum	
Hastelloy B	*Area E*
Durichlor (FeCl$_3$ free)	Chlorimet 2
Haveg	Glass
Saran	Silver
Rubber	Platinum
Silicon bronze (air free)	Tantalum
Zirconium	Hastelloy B (chlorine free)
Molybdenum	Zirconium
Impervious graphite	Impervious graphite

[a] Materials having reported corrosion rate less than 0.020 in./yr.

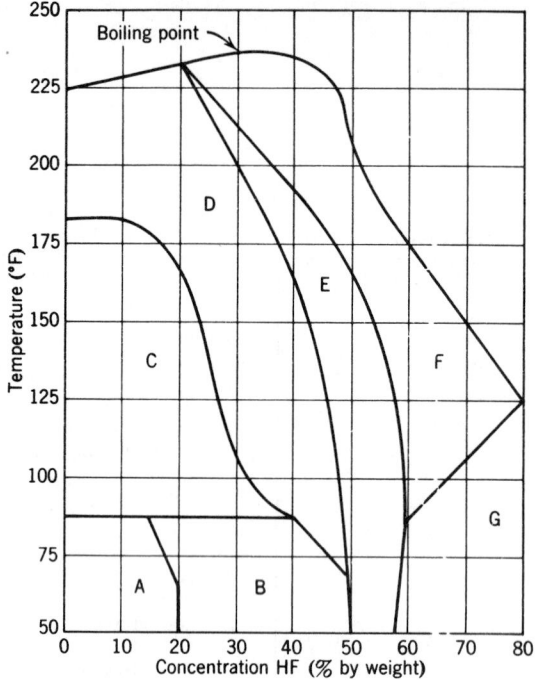

Fig. 16.25 Corrosion resistance of various materials to hydrofluoric acid.

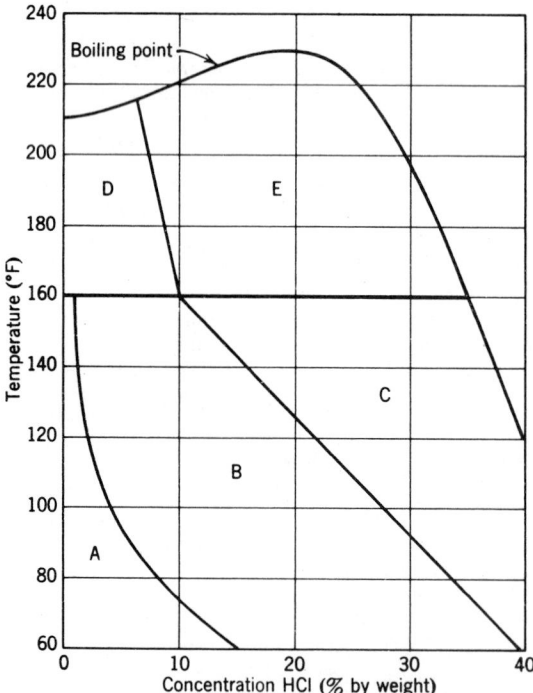

Fig. 16.26 Corrosion resistance of various materials to hydrochloric acid.

TABLE 16.215　Corrosion Ratings of Commercial Alloys of Iron Nickel, and Chromium[a]
Data courtesy of International Nickel Company.

Material	American Iron and Steel Institute Type Number	Outdoor Air			Specific Industrial Atmospheres (Wet)					Scaling Temperature,[b] (°F)
		Rural	Urban	Marine	Ammonia	Hydrogen Sulfide	Hydrogen Chloride	Sulfur Dioxide	Chlorine	
Low-carbon steel	—	D+	D	D	B	D	D	D	D	1050
Copper-bearing steel	—	D+	D+	D	B	D	D	D	D	1000
4–6% Cr steel	502	D+	D+	D+	B	D+	D	D	D	1200
12–14% Cr steel	420	B+	B	C−	A	C	D	D	D	1250
12–14% Cr, 0.60% Mo	416	B+	B	C−	A	C	D	D	D	1250
16–18% Cr	430	A	B+	C	A	B	D	D	D	1600
23–30% Cr	446	A	A	C+	A	A	D	B	D	2100
7% Ni–17% Cr	301	A	A	B+	A	A	D	C	C	Not used for oxidation resistance
8% Ni–18% Cr	302 303 304	A	A	A−	A	A	D	B	C	1650
8% Ni–18% Cr, 1% Cb	347	A	A	A−	A	A	D	B	C	1650
8% Ni–18% Cr, 0.5% Ti	321	A	A	A−	A	A	D	B	C	1650
14% Ni–23% Cr	309	A	A	A	A	A	D	B	C	2000
12% Ni–18% Cr, 3% Mo	316	A	A	A+	A	A	C	A	B	1650

TABLE 16.215 (*Continued*)

Material	American Iron and Steel Institute Type Number	Specific Industrial Atmospheres (Wet)								Scaling Temperature,[b] (°F)
		Outdoor Air			Ammonia	Hydrogen Sulfide	Hydrogen Chloride	Sulfur Dioxide	Chlorine	
		Rural	Urban	Marine						
20% Ni–25% Cr Nickel	310	A	A$^+$	A$^+$	A	A	D	B	C	2100
	—	A	A$^-$	A$^+$	C–E[f]	B	C–B	C	B	1900,[c] 1000,[d] 700[e]
30% Ni–70% Cu Monel	—	A	A$^-$	A	C–E[f]	C$^-$	C$^+$	B	C$^+$	—
	—	A	A	A	C–E[f]	B	B	C	B	1000,[c] 1000,[d] 650[e]
80% Ni–20% Cr	—	A	A	A$^+$	A	A	B	B	B	2100
Inconel–80% Ni, 7% Fe–3% Cr	—	A	A	A$^+$	A	A	B	B	B	2000,[c] 1500,[d] 1000[e]

[a] A. Practically complete resistance, or alloy is best of materials within its class. B. Good resistance, as proved by being in common use, may replace materials given A rating to secure some other advantage. C. Adequate resistance under favorable conditions which should be investigated beforehand. D. Sufficient resistance if adequate precautions are taken to reduce effect of corrosive conditions, as by coatings, cathodic protection, redesign, etc., or where appearance is not important and appreciable corrosion may be provided for or tolerated. E. Poor resistance; use only if no better material is available. Plus and minus signs are used to permit better differentiation between corrosion-resistant qualities.

[b] Values assume substantially constant-temperature operations; should be lowered for cyclic heating and cooling to extent dependent upon frequency and range of temperature fluctuations.

[c] Scaling temperature in low-sulfur atmosphere.

[d] Scaling temperature in high-sulfur oxidizing atmosphere.

[e] Scaling temperature in high-sulfur reducing atmosphere.

[f] Not recommended primarily to resist ammonia attack but may be used where resistance to ammonia in low concentrations is incidental requirement.

INDEX

Periodic Table of the Elements

Atomic number / Symbol / Atomic weight

Example: H 1 1.0079

Period	IA	IIA	IIIB	IVB	VB	VIB	VIIB	VIII	VIII	VIII	IB	IIB	IIIA	IVA	VA	VIA	VIIA	0 (Noble gases)
1	1 H 1.0079																	2 He 4.00260
2	3 Li 6.941	4 Be 9.01218											5 B 10.81	6 C 12.011	7 N 14.0067	8 O 15.9994	9 F 18.998403	10 Ne 20.179
3	11 Na 22.98977	12 Mg 24.305											13 Al 26.98154	14 Si 28.0855	15 P 30.97376	16 S 32.06	17 Cl 35.453	18 Ar 39.948
4	19 K 39.0983	20 Ca 40.08	21 Sc 44.9559	22 Ti 47.90	23 V 50.9415	24 Cr 51.996	25 Mn 54.9380	26 Fe 55.847	27 Co 58.9332	28 Ni 58.70	29 Cu 63.546	30 Zn 65.38	31 Ga 69.72	32 Ge 72.59	33 As 74.9216	34 Se 78.96	35 Br 79.904	36 Kr 83.80
5	37 Rb 85.4678	38 Sr 87.62	39 Y 88.9059	40 Zr 91.22	41 Nb 92.9064	42 Mo 95.94	43 Tc (98)	44 Ru 101.07	45 Rh 102.9055	46 Pd 106.4	47 Ag 107.868	48 Cd 112.41	49 In 114.82	50 Sn 118.69	51 Sb 121.75	52 Te 127.60	53 I 126.9045	54 Xe 131.30
6	55 Cs 132.9054	56 Ba 137.33	57 La* 138.9055	72 Hf 178.49	73 Ta 180.9479	74 W 183.85	75 Re 186.207	76 Os 190.2	77 Ir 192.22	78 Pt 195.09	79 Au 196.9665	80 Hg 200.59	81 Tl 204.37	82 Pb 207.2	83 Bi 208.9804	84 Po (209)	85 At (210)	86 Rn (222)
7	87 Fr (223)	88 Ra 226.0254	89 Ac† 227.0278	104 Unq (261)	105 Unp (262)	106 Unh (263)												

*Lanthanides:

58 Ce 140.12	59 Pr 140.9077	60 Nd 144.24	61 Pm (145)	62 Sm 150.4	63 Eu 151.96	64 Gd 157.25	65 Tb 158.9254	66 Dy 162.50	67 Ho 164.9304	68 Er 167.26	69 Tm 168.9342	70 Yb 173.04	71 Lu 174.967

†Actinides:

90 Th 232.0381	91 Pa 231.0359	92 U 238.029	93 Np 237.0482	94 Pu (244)	95 Am (243)	96 Cm (247)	97 Bk (247)	98 Cf (251)	99 Es (254)	100 Fm (257)	101 Md (258)	102 No (259)	103 Lr (260)